REFERENCE DATA FOR RADIO ENGINEERS

REFERENCE DATA FOR RADIO ENGINEERS

HOWARD W. SAMS & CO., INC.
INDIANAPOLIS/KANSAS CITY/NEW YORK
A SUBSIDIARY OF
INTERNATIONAL TELEPHONE AND TELEGRAPH CORPORATION

ITT

SIXTH EDITION

SECOND PRINTING—1977

International Standard Book Number: 0-672-21218-8
Library of Congress Catalog Card Number: 75-28960

FOREWORD

This sixth edition of *Reference Data for Radio Engineers* grew from a 60-page brochure of that title originally compiled by W. L. McPherson of ITT's Standard Telephones and Cables Limited, which company published it in England in 1942. Its immediate acceptance prompted the parent company, International Telephone and Telegraph Corporation, to arrange for a United States version that in five editions dated 1943, 1946, 1949, 1956, and 1968 had a total sale of 450,000 copies. The book has become a first place for the busy radio engineer to look for all kinds of data. It has also been widely used in colleges by both faculties and students.

To maintain a high standard in a book of such wide scope, it has been necessary to go outside of the ITT System for objective reviews of material to guide revisions, for critical reviews of manuscripts, and for the preparation of some new material. Such contributions are gratefully acknowledged from H. R. Romig; D. J. LeVine; N. Marchand of Marchand Laboratories; L. J. Lidofsky of Columbia University; Colin Cherry of London University; R. C. Barker of Yale University; J. D. Kraus of Ohio State University; J. A. Pierce of Harvard University; G. A. Deschamps of University of Illinois; J. R. Ragazzini of New York University; E. A. Guillemin of Massachusetts Institute of Technology; and from N. Marrcuvitz, W. K. Kahn, and T. Tamir of Brooklyn Polytechnic Institute. The following persons from National Bureau of Standards and Environmental Science Services Administration also generously contributed to the book: J. W. Herbstreit, R. T. Disney, W. Q. Crichlow, P. L. Rice, D. D. Crombie, A. F. Barghausen, G. W. Haydon, R. S. Lawrence, and M. S. Cord.

Special acknowledgments are given to additional contributors to the sixth edition: D. Davis of Synergetic Audio Concepts, L. S. Golding of Digital Communications Corporation and J. E. D. Ball of Public Broadcasting Service, J. L. Hilburn and D. E. Johnson of Louisiana State University, F. M. Mims III, R. E. Taylor of the National Aeronautics and Space Administration, and J. L. Walters and R. J. Adams of the Naval Research Laboratory.

The difficulty in identifying the authors of material carried over during the more than 30-year history of the book makes it impossible for us to list their names despite the debt we owe them. However, special acknowledgment is made of the valuable contributions of A. G. Kandoian as Chairman of the Editorial Boards for the third through fifth editions and of H. P. Westman who edited the fourth and fifth editions. The following currently active International System personnel are acknowledged as contributors:

E. Baguley	H. G. Busignies	S. H. Dodington
E. E. Benham	A. Casabona	J. G. Dunn
R. A. Bones	R. Clayton	E. Eberhardt
T. Brown	D. K. Coles	J. A. Fingerett
J. H. Brundage	C. R. Cook	M. T. Fujita
F. X. Bucher	A. E. Cookson	D. S. Girling
J. A. Budek	M. Dishal	F. F. Hall

v

I. W. Hammer

D. E. Herrington

J. L. Jatlow

P. King

J. Kylander

P. Lighty

W. Litchman

J. G. Litterick

C. W. Moody

J. M. Moore

H. G. Nordlin

J. E. Obst

J. A. O'Connell

C. P. Oliphant

J. Polyzou

M. C. Poylo

L. G. Rado

D. S. Ridler

L. Rosenberg

J. E. Schlaikjer

H. H. Smith

R. Smith

T. L. Squires

J. L. Storr-Best

J. G. Tatum

L. F. Turner

R. Vachss

J. M. Valentine

R. Weber

A. K. Wing, Jr.

J. Youlios

TABLE OF CONTENTS

WAVELENGTH-FREQUENCY CONVERSION

Figure 1 permits conversion between frequency and wavelength; by use of multiplying factors shown, this graph will cover any portion of the electromagnetic-wave spectrum.

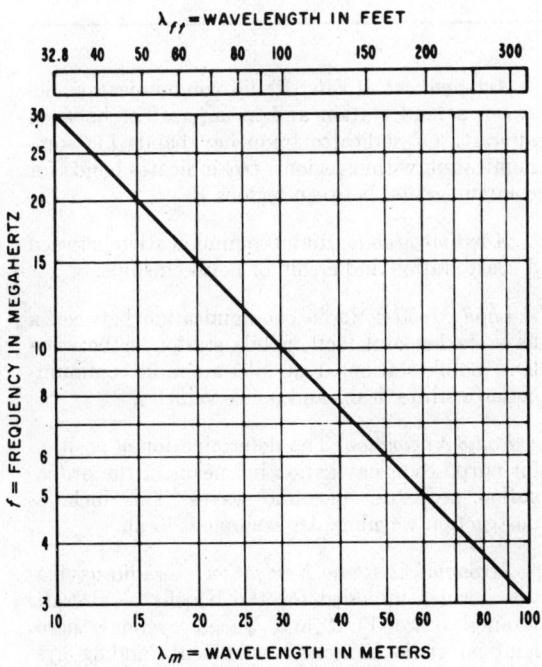

Fig. 1—Wavelength-frequency conversion.

For Frequencies in Megahertz from		Multiply f by	Multiply λ by
0.03–	0.3	0.01	100
0.3 –	3.0	0.1	10
3.0 –	30	1.0	1.0
30 –	300	10	0.1
300 –	3 000	100	0.01
3 000 –	30 000	1 000	0.001
30 000 –	300 000	10 000	0.0001

Conversion Equations

Propagation velocity

$$c \approx 3 \times 10^8 \text{ meters/second}$$

Wavelength in meters

$$\lambda_m = \frac{300\ 000}{f \text{ in kilohertz}} = \frac{300}{f \text{ in megahertz}}$$

Wavelength in centimeters

$$\lambda_{cm} = \frac{30}{f \text{ in gigahertz}}$$

Wavelength in feet

$$\lambda_{ft} = \frac{984\ 000}{f \text{ in kilohertz}} = \frac{984}{f \text{ in megahertz}}$$

Wavelength in inches

$$\lambda_{in} = \frac{11.8}{f \text{ in gigahertz}}$$

$$\begin{aligned}
1 \text{ angstrom unit Å} &= 3.937 \times 10^{-9} \text{ inch} \\
&= 1 \times 10^{-10} \text{ meter} \\
&= 1 \times 10^{-4} \text{ micrometer}
\end{aligned}$$

$$\begin{aligned}
1 \text{ micrometer } \mu m &= 3.937 \times 10^{-5} \text{ inch} \\
&= 1 \times 10^{-6} \text{ meter} \\
&= 1 \times 10^4 \text{ angstrom units.}
\end{aligned}$$

(Note that the term "micrometer" has superceded the term "micron.")

Nomenclature of Frequency Bands

Table 1 is adapted from the Radio Regulations of the International Telecommunication Union, Article 2, Section 11, Geneva; 1959.

Letter Designations for Frequency Bands

Letter designations commonly used for microwave bands (particularly in references to radar equipment) are shown in Table 2. These designations have no official international standing, and various engineers have used limits for the bands and subbands other than those listed in the table. Subband code letters should be used as sub-

TABLE 1—NOMENCLATURE OF FREQUENCY BANDS.

Band Number*	Frequency Range	Metric Subdivision	Adjectival Designation	
2	30 to 300 hertz	Megametric waves	ELF	Extremely low frequency
3	300 to 3000 hertz	———	VF	Voice frequency
4	3 to 30 kilohertz	Myriametric waves	VLF	Very-low frequency
5	30 to 300 kilohertz	Kilometric waves	LF	Low frequency
6	300 to 3000 kilohertz	Hectometric waves	MF	Medium frequency
7	3 to 30 megahertz	Decametric waves	HF	High frequency
8	30 to 300 megahertz	Metric waves	VHF	Very-high frequency
9	300 to 3000 megahertz	Decimetric waves	UHF	Ultra-high frequency
10	3 to 30 gigahertz	Centimetric waves	SHF	Super-high frequency
11	30 to 300 gigahertz	Millimetric waves	EHF	Extremely high frequency
12	300 to 3000 gigahertz or 3 terahertz	Decimillimetric waves	—	

* "Band Number N" extends from 0.3×10^N to 3×10^N hertz. The upper limit is included in each band; the lower limit is excluded.

scripts in designating particular frequency ranges; for example, L_x indicates the band between 0.950 and 1.150 gigahertz.

FREQUENCY ALLOCATIONS BY INTERNATIONAL TREATY

The following information is adapted from the Radio Regulations of the ITU, Geneva, 1959, corrected to July 1965. Some 400 footnotes describing special conditions pertaining to allocations within particular frequency bands and much other detailed information are not reproduced here. Copies of the Radio Regulations may be obtained from the Secretary General, International Telecommunication Union, Place des Nations, CH-1211, Geneva 20, Switzerland.*

For purposes of frequency allocations, the world has been divided into regions shown in Fig. 2.

See Article 5, Section 1 of the ITU Radio Regulations for definitions of the regions and of lines A, B, and C.

Frequency bands are allocated to services defined as follows:

Fixed: Radio communication between specified fixed points. Examples are point-to-point high-frequency circuits and microwave links.

Mobile: Radio communication between stations intended to be used while in motion or during halts at unspecified points or between such stations and fixed stations.

* In the official documents of ITU and FCC the following terms are combined as single words: radiobeacon, radiocommunication, radiodetermination, radiolocation, radionavigation, radiorange, radiosonde, radiotelegraphy, and radiotelephony.

Aeronautical Mobile: Radio communication between a land station and an aircraft or between aircraft. (*R* indicates frequency bands for communication within regions. *OR* indicates bands for communication between regions.)

Maritime Mobile: Radio communication between a coast station and a ship or between ships.

Land Mobile: Radio communication between a base station and land mobile station or between land mobile stations. Examples are radio communication with taxicabs and police vehicles.

Radio Navigation: The determination of position for purposes of navigation by means of the propagation properties of radio waves. This includes obstruction warning. An example is loran.

Aeronautical Radio Navigation: A radio navigation service intended for the benefit of aircraft. Examples are VOR and Tacan systems, aeronautical radio beacons, instrument landing systems, radio altimeters, and airborne obstruction-indicating radar.

Maritime Radio Navigation: A radio navigation service intended for the benefit of ships. Examples are coastal radio beacons, direction-finding stations, and shipboard radar.

Radio Location: The determination of position for purposes other than those of navigation by means of the propagation properties of radio waves. Examples are land radars, coastal radars, and tracking systems.

Broadcasting: Radio communication intended for direct reception by the general public. Examples are amplitude-modulation broadcasting on me-

TABLE 2—LETTER DESIGNATIONS FOR MICROWAVE BANDS.

Subband	Frequency in Gigahertz	Wavelength in Centimeters
	P Band	
	0.225	133.3
	0.390	76.9
	L Band	
	0.390	76.9
p	0.465	64.5
c	0.510	58.8
l	0.725	41.4
y	0.780	38.4
t	0.900	33.3
s	0.950	31.6
x	1.150	26.1
k	1.350	22.2
f	1.450	20.7
z	1.550	19.3
	S Band	
e	1.55	19.3
f	1.65	18.3
t	1.85	16.2
c	2.00	15.0
q	2.40	12.5
y	2.60	11.5
g	2.70	11.1
s	2.90	10.3
a	3.10	9.67
w	3.40	8.32
h	3.70	8.10
z*	3.90	7.69
d	4.20	7.14
	5.20	5.77
	X Band	
a	5.20	5.77
q	5.50	5.45
y*	5.75	5.22
d	6.20	4.84
b	6.25	4.80
r	6.90	4.35
c	7.00	4.29
	8.50	3.53

Subband	Frequency in Gigahertz	Wavelength in Centimeters
	X Band—Continued	
l	9.00	3.33
s	9.60	3.13
x	10.00	3.00
f	10.25	2.93
k	10.90	2.75
	K Band	
	10.90	2.75
p	12.25	2.45
s	13.25	2.26
e	14.25	2.10
c	15.35	1.95
u†	17.25	1.74
t	20.50	1.46
q†	24.50	1.22
r	26.50	1.13
m	28.50	1.05
n	30.70	0.977
l	33.00	0.909
a	36.00	0.834
	Q Band	
	36.0	0.834
a	38.0	0.790
b	40.0	0.750
c	42.0	0.715
d	44.0	0.682
e	46.0	0.652
	V Band	
	46.0	0.652
a	48.0	0.625
b	50.0	0.600
c	52.0	0.577
d	54.0	0.556
e	56.0	0.536
	W Band	
	56.0	0.536
	100.0	0.300

* C Band includes S_z through X_y (3.90–6.20 gigahertz).
† K_1 Band includes K_u through K_q (15.35–24.50 gigahertz).

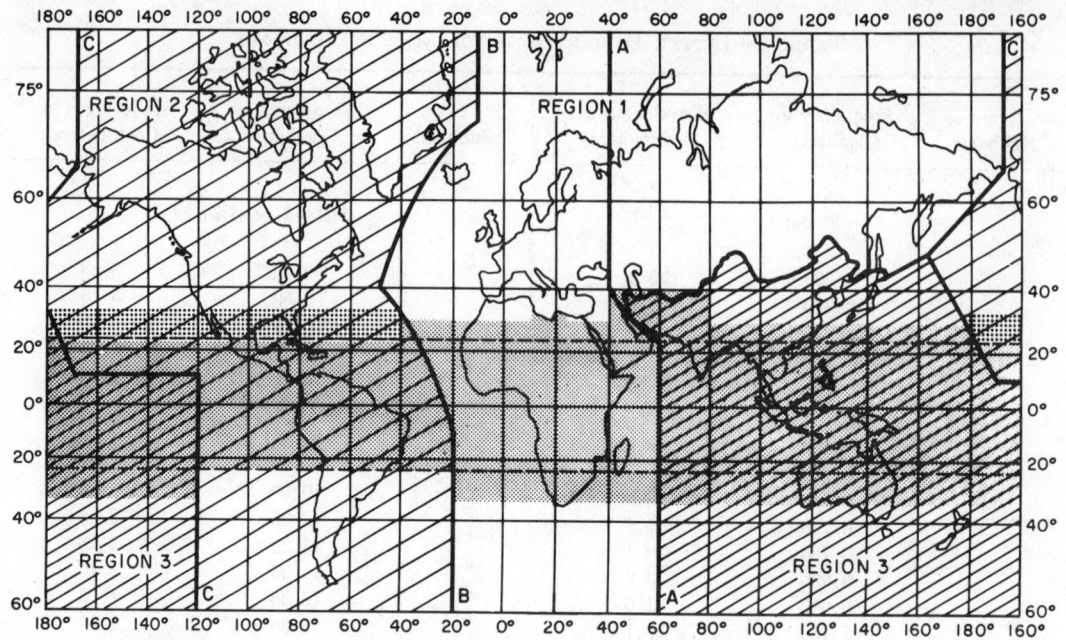

Fig. 2—Regions defined in table of frequency allocations. Shaded area represents tropical zone.

dium and high frequencies, frequency-modulation broadcasting, and television.

Amateur: Radio communication carried on by persons interested in the radio technique solely with a personal aim and without pecuniary interest.

Space: Radio communication between space stations.

Earth-Space: Radio communication between earth stations and space stations. An example is between the earth and a satellite.

Radio Astronomy: Astronomy based on the reception of radio waves of cosmic origin.

Standard Frequency: Radio transmission of specified frequencies of stated high precision, intended for general reception for scientific, technical, and other purposes.

The following allocations pertain to Region 2 (the western hemisphere). An asterisk (*) indicates that the allocation also pertains on a worldwide basis. Frequency assignments in the United States comply with the following table but they have been further divided (especially above 30 megahertz) among particular types of service as shown in the listings later in this chapter.

Services printed in small capitals (example: FIXED) are primary services.

Services printed in italics (example: *Radio location*) are permitted services and have equal rights with primary services except that the primary services have prior choice of frequencies.

Services printed in lower-case type (example: Mobile) are secondary services which shall not cause harmful interference to, or claim protection from, stations of a primary or permitted service.

The order of listing does not indicate relative priority within each category.

Kilohertz	Service
Below 10.00	(not allocated)*
10.00–14.00	RADIO NAVIGATION*
	*Radio location**
14.00–19.95	FIXED*
	MARITIME MOBILE*
19.95–20.05	STANDARD FREQUENCY*
20.05–70.00	FIXED*
	MARITIME MOBILE*
70.00–90.00	FIXED
	MARITIME MOBILE
	MARITIME RADIO NAVIGATION
	Radio location
90.00–110.0	RADIO NAVIGATION
	FIXED
	Maritime mobile
110.0–130.0	FIXED
	MARITIME MOBILE
	MARITIME RADIO NAVIGATION
	Radio location
130.0–160.0	FIXED
	MARITIME MOBILE
160.0–200.0	FIXED
200.0–285.0	AERONAUTICAL RADIO NAVIGATION
	Aeronautical mobile
285.0–325.0	MARITIME RADIO NAVIGATION (radio beacons)
	Aeronautical radio navigation

Kilohertz	Service
325.0–405.0	AERONAUTICAL RADIO NAVIGATION*
	Aeronautical mobile*
405.0–415.0	MARITIME RADIO NAVIGATION
	(radio direction-finding)
	Aeronautical radio navigation
	Aeronautical mobile
415.0–490.0	MARITIME MOBILE*
	(radiotelegraphy only)
490.0–510.0	MOBILE*
	(distress and calling)
510.0–525.0	MOBILE
	Aeronautical radio navigation
525.0–535.0	MOBILE
	Broadcasting
	Aeronautical radio navigation
535.0–1605	BROADCASTING*
1605–1800	FIXED
	MOBILE
	AERONAUTICAL RADIO NAVIGATION
	Radio location
1800–2000	AMATEUR
	FIXED
	MOBILE EXCEPT AERONAUTICAL
	RADIO NAVIGATION
2000–2065	FIXED
	MOBILE
2065–2107	MARITIME MOBILE
2107–2170	FIXED
	MOBILE
2170–2194	MOBILE*
	(distress and calling)
2194–2300	FIXED
	MOBILE
2300–2495	FIXED
	MOBILE
	BROADCASTING
2495–2505	STANDARD FREQUENCY
2505–2850	FIXED
	MOBILE
2850–3025	AERONAUTICAL MOBILE (R)*

Megahertz	Service
3.025–3.155	AERONAUTICAL MOBILE (OR)*
3.155–3.200	FIXED*
	MOBILE EXCEPT AERONAUTICAL (R)*
3.200–3.400	FIXED*
	MOBILE EXCEPT AERONAUTICAL*
	BROADCASTING*
3.400–3.500	AERONAUTICAL MOBILE (R)*
3.500–4.000	AMATEUR
	FIXED
	MOBILE EXCEPT AERONAUTICAL (R)
4.000–4.063	FIXED*
4.063–4.438	MARITIME MOBILE*
4.438–4.650	FIXED
	MOBILE EXCEPT AERONAUTICAL (R)
4.650–4.700	AERONAUTICAL MOBILE (R)*
4.700–4.750	AERONAUTICAL MOBILE (OR)*

Megahertz	Service
4.750–4.850	FIXED
	BROADCASTING
4.850–4.995	FIXED*
	LAND MOBILE*
	BROADCASTING*
4.995–5.005	STANDARD FREQUENCY*
5.005–5.060	FIXED*
	BROADCASTING*
5.060–5.250	FIXED*
5.250–5.450	FIXED
	LAND MOBILE
5.450–5.680	AERONAUTICAL MOBILE (R)
5.680–5.730	AERONAUTICAL MOBILE (OR)*
5.730–5.950	FIXED*
5.950–6.200	BROADCASTING*
6.200–6.525	MARITIME MOBILE*
6.525–6.685	AERONAUTICAL MOBILE (R)*
6.685–6.765	AERONAUTICAL MOBILE (OR)*
6.765–7.000	FIXED*
7.000–7.300	AMATEUR
7.300–8.195	FIXED*
8.195–8.815	MARITIME MOBILE*
8.815–8.965	AERONAUTICAL MOBILE (R)*
8.965–9.040	AERONAUTICAL MOBILE (OR)*
9.040–9.500	FIXED*
9.500–9.775	BROADCASTING*
9.775–9.995	FIXED*
9.995–10.005	STANDARD FREQUENCY*
10.005–10.10	AERONAUTICAL MOBILE (R)*
10.10–11.175	FIXED*
11.175–11.275	AERONAUTICAL MOBILE (OR)*
11.275–11.40	AERONAUTICAL MOBILE (R)*
11.40–11.70	FIXED*
11.70–11.975	BROADCASTING*
11.975–12.33	FIXED*
12.33–13.20	MARITIME MOBILE*
13.20–13.26	AERONAUTICAL MOBILE (OR)*
13.26–13.36	AERONAUTICAL MOBILE (R)*
13.36–14.00	FIXED*
14.00–14.35	AMATEUR*
14.35–14.99	FIXED*
14.99–15.01	STANDARD FREQUENCY*
15.01–15.10	AERONAUTICAL MOBILE (OR)*
15.10–15.45	BROADCASTING*
15.45–16.46	FIXED*
16.46–17.36	MARITIME MOBILE*
17.36–17.70	FIXED*
17.70–17.90	BROADCASTING*
17.90–17.97	AERONAUTICAL MOBILE (R)*
17.97–18.03	AERONAUTICAL MOBILE (OR)*
18.03–18.052	FIXED*
18.052–18.068	FIXED*
	Space research*
18.068–19.99	FIXED*
19.99–20.01	STANDARD FREQUENCY*
20.01–21.00	FIXED*
21.00–21.45	AMATEUR*
21.45–21.75	BROADCASTING*
21.75–21.85	FIXED*

Megahertz	Service
21.85–21.87	RADIO ASTRONOMY*
21.87–22.00	AERONAUTICAL FIXED*
	AERONAUTICAL MOBILE (R)*
22.00–22.72	MARITIME MOBILE*
22.72–23.20	FIXED*
23.20–23.35	AERONAUTICAL FIXED*
	AERONAUTICAL MOBILE (OR)*
23.35–24.99	FIXED*
	LAND MOBILE*
24.99–25.01	STANDARD FREQUENCY*
25.01–25.07	FIXED*
	MOBILE EXCEPT AERONAUTICAL*
25.07–25.11	MARITIME MOBILE*
25.11–25.60	FIXED*
	MOBILE EXCEPT AERONAUTICAL*
25.60–26.10	BROADCASTING*
26.10–27.50	FIXED*
	MOBILE EXCEPT AERONAUTICAL*
27.50–28.00	METEOROLOGICAL AIDS
	FIXED
	MOBILE
28.00–29.70	AMATEUR*
29.70–30.005	FIXED*
	MOBILE*
30.005–30.01	FIXED*
	MOBILE*
	SPACE OPERATION*
	(satellite identification)
	SPACE RESEARCH*
30.01–37.75	FIXED*
	MOBILE*
37.75–38.25	FIXED*
	MOBILE*
	Radio astronomy*
38.25–50.00	FIXED
	MOBILE
50.00–54.00	AMATEUR
54.00–73.00	FIXED
	MOBILE
	BROADCASTING
73.00–74.60	RADIO ASTRONOMY
74.60–75.40	AERONAUTICAL RADIO NAVIGATION
75.40–88.00	FIXED
	MOBILE
	BROADCASTING
88.00–108.0	BROADCASTING
108.0–117.975	AERONAUTICAL RADIO NAVIGATION*
117.975–132.0	AERONAUTICAL MOBILE (R)*
132.0–136.0	AERONAUTICAL MOBILE*
136.0–137.0	SPACE RESEARCH
	(space to earth)
137.0–138.0	METEOROLOGICAL SATELLITE*
	SPACE OPERATION*
	(telemetering and tracking)
	SPACE RESEARCH*
	(space to earth)

Megahertz	Service
138.0–143.6	FIXED
	MOBILE
	Radio location
	Space research
	(space to earth)
143.6–143.65	FIXED
	MOBILE
	SPACE RESEARCH
	(space to earth)
	Radio location
143.65–144.0	FIXED
	MOBILE
	Radio location
	Space research
	(space to earth)
144.0–148.0	AMATEUR
148.0–149.9	FIXED
	MOBILE
149.9–150.05	RADIO NAVIGATION-SATELLITE*
150.05–174.0	FIXED
	MOBILE
174.0–216.0	FIXED
	MOBILE
	BROADCASTING
216.0–220.0	FIXED
	MOBILE
	RADIO LOCATION
220.0–225.0	AMATEUR
	RADIO LOCATION
225.0–267.0	FIXED
	MOBILE
267.0–272.0	FIXED*
	MOBILE*
	Space operation*
	(telemetering)
272.0–273.0	FIXED*
	MOBILE*
	SPACE OPERATION*
	(telemetering)
273.0–328.6	FIXED*
	MOBILE*
328.6–335.4	AERONAUTICAL RADIO NAVIGATION*
	(glide-path systems)
335.4–399.9	FIXED*
	MOBILE*
399.9–400.05	RADIO NAVIGATION-SATELLITE*
400.05–400.15	STANDARD FREQUENCY SATELLITE*
400.15–401.0	METEOROLOGICAL AIDS*
	METEOROLOGICAL-SATELLITE*
	(maintenance telemetering)
	SPACE RESEARCH*
	(telemetering and tracking)
401.0–402.0	METEOROLOGICAL AIDS*
	SPACE OPERATION*
	(telemetering)
	Fixed*
	Meteorological satellite*
	(earth to space)
	Mobile except aeronautical*

Megahertz	Service	Megahertz	Service
402.0–403.0	METEOROLOGICAL AIDS* Fixed* Meteorological Satellite* (earth to space) Mobile except aeronautical*	1670–1690	METEOROLOGICAL AIDS* METEOROLOGICAL SATELLITE* (space to earth) FIXED* MOBILE EXCEPT AERONAUTICAL*
403.0–406.0	METEOROLOGICAL AIDS* Fixed* Mobile except aeronautical*	1690–1700	METEOROLOGICAL AIDS METEOROLOGICAL-SATELLITE (space to earth)
406.0–406.1	MOBILE SATELLITE* (earth to space)	1700–1710	FIXED MOBILE SPACE RESEARCH (space to earth)
406.1–410.0	FIXED* MOBILE EXCEPT AERONAUTICAL* RADIO ASTRONOMY*	1710–1770	FIXED MOBILE
410.0–420.0	FIXED* MOBILE EXCEPT AERONAUTICAL*	1770–1790	FIXED MOBILE Meteorological-satellite
420.0–450.0	RADIO LOCATION Amateur	1790–2290	FIXED MOBILE
450.0–460.0	FIXED* MOBILE*	2290–2300	SPACE RESEARCH (space to earth) FIXED MOBILE
460.0–470.0	FIXED* MOBILE* Meteorological satellite*	2300–2450	RADIO LOCATION Amateur Fixed Mobile
470.0–890.0	BROADCASTING	2450–2500	FIXED MOBILE RADIO LOCATION
890.0–942.0	FIXED RADIO LOCATION	2500–2535	BROADCASTING SATELLITE FIXED FIXED SATELLITE (space to earth) MOBILE EXCEPT AERONAUTICAL
942.0–960	FIXED	2535–2655	BROADCASTING SATELLITE FIXED MOBILE EXCEPT AERONAUTICAL
960.0–1215	AERONAUTICAL RADIO NAVIGATION*	2655–2690	BROADCASTING SATELLITE FIXED FIXED-SATELLITE (earth to space) MOBILE EXCEPT AERONAUTICAL
1215–1300	RADIO LOCATION* Amateur*	2690–2700	RADIO ASTRONOMY*
1300–1350	AERONAUTICAL RADIO NAVIGATION* Radio location*	2700–2900	AERONAUTICAL RADIO NAVIGATION* Radio location*
1350–1400	RADIO LOCATION	2900–3100	RADIO NAVIGATION* (ground-based radars) Radio location*
1400–1427	RADIO ASTRONOMY*		
1427–1429	FIXED* MOBILE EXCEPT AERONAUTICAL* SPACE OPERATION* (telecommand)	**Gigahertz**	**Service**
1429–1435	FIXED MOBILE	3.100–3.300	RADIO LOCATION*
1435–1525	MOBILE Fixed	3.300–3.400	RADIO LOCATION Amateur
1525–1535	SPACE OPERATION (telemetering) Fixed Mobile Earth Exploration Satellite	3.400–3.500	RADIO LOCATION FIXED SATELLITE (space to earth) Amateur
1535–1542.5	MARITIME MOBILE SATELLITE*		
1542.5–1543.5	AERONAUTICAL MOBILE SATELLITE (R)* MARITIME MOBILE SATELLITE*		
1543.5–1558.5	AERONAUTICAL MOBILE SATELLITE (R)*		
1558.5–1636.5	AERONAUTICAL RADIO NAVIGATION*		
1636.5–1644	MARITIME MOBILE SATELLITE*		
1644–1645	AERONAUTICAL MOBILE SATELLITE (R)* MARITIME MOBILE SATELLITE*		
1645–1660	AERONAUTICAL MOBILE SATELLITE (R)*		
1660–1670	METEOROLOGICAL AIDS* RADIO ASTRONOMY*		

Gigahertz	Service
3.500–3.700	FIXED
	MOBILE
	RADIO LOCATION
	FIXED SATELLITE
	(space to earth)
3.700–4.200	FIXED
	MOBILE
	FIXED SATELLITE
	(space to earth)
4.200–4.400	AERONAUTICAL RADIO NAVIGATION*
4.400–4.700	FIXED*
	MOBILE*
	FIXED SATELLITE*
	(earth to space)
4.700–4.990	FIXED*
	MOBILE*
4.990–5.000	RADIO ASTRONOMY
5.000–5.250	AERONAUTICAL RADIO NAVIGATION*
5.250–5.255	RADIO LOCATION*
	Space research*
5.255–5.350	RADIO LOCATION*
5.350–5.460	AERONAUTICAL RADIO NAVIGATION*
	Radio location*
5.460–5.470	RADIO NAVIGATION*
	Radio location*
5.470–5.650	MARITIME RADIO NAVIGATION*
	Radio location*
5.650–5.670	RADIO LOCATION*
	Amateur*
5.670–5.725	RADIO LOCATION*
	Amateur*
	Space research*
	(deep space)
5.725–5.925	RADIO LOCATION
	Amateur
5.925–6.425	FIXED*
	MOBILE*
	FIXED SATELLITE*
	(earth to space)
6.425–7.250	FIXED*
	MOBILE*
7.250–7.300	FIXED SATELLITE
	(space to earth)
7.300–7.450	FIXED*
	FIXED SATELLITE*
	(space to earth)
	MOBILE*
7.450–7.550	FIXED*
	FIXED SATELLITE*
	(space to earth)
	METEOROLOGICAL SATELLITE*
	(space to earth)
	MOBILE*
7.550–7.750	FIXED*
	FIXED SATELLITE*
	(space to earth)
	MOBILE*
7.750–7.900	FIXED*
	MOBILE*

Gigahertz	Service
7.900–7.975	FIXED*
	MOBILE*
	FIXED SATELLITE
	(earth to space)
7.975–8.025	FIXED SATELLITE
	(earth to space)
8.025–8.175	EARTH EXPLORATION SATELLITE
	(space to earth)
	FIXED
	FIXED SATELLITE
	(earth to space)
	MOBILE
8.175–8.215	EARTH EXPLORATION SATELLITE
	(space to earth)
	FIXED
	FIXED SATELLITE
	(earth to space)
	METEOROLOGICAL SATELLITE
	(earth to space)
	MOBILE
8.215–8.400	EARTH EXPLORATION SATELLITE
	(space to earth)
	FIXED
	FIXED SATELLITE
	(earth to space)
	MOBILE
8.400–8.500	FIXED*
	MOBILE*
	SPACE RESEARCH*
	(space to earth)
8.500–8.750	RADIO LOCATION*
8.750–8.850	RADIO LOCATION*
	AERONAUTICAL RADIO NAVIGATION*
	(airborne doppler aids)
8.850–9.000	RADIO LOCATION*
9.000–9.200	AERONAUTICAL RADIO NAVIGATION*
	(ground-based radars)
	Radio location*
9.200–9.300	RADIO LOCATION*
9.300–9.500	RADIO NAVIGATION*
	Radio location*
9.500–9.800	RADIO LOCATION*
9.800–10.00	RADIO LOCATION*
	Fixed*
10.00–10.50	RADIO LOCATION*
	Amateur*
10.50–10.55	RADIO LOCATION
	(continuous-wave systems only)
10.55–10.60	FIXED*
	MOBILE*
	Radio location*
10.60–10.68	FIXED*
	MOBILE*
	RADIO ASTRONOMY*
	Radio location*
10.68–10.70	RADIO ASTRONOMY*
10.70–10.95	FIXED*
	MOBILE*

Gigahertz	Service
10.95–11.20	FIXED
	FIXED SATELLITE
	(space to earth)
	MOBILE
11.20–11.45	FIXED*
	MOBILE*
11.45–11.70	FIXED*
	FIXED SATELLITE*
	(space to earth)
	MOBILE*
11.70–12.20	BROADCASTING
	BROADCASTING SATELLITE
	FIXED
	FIXED SATELLITE
	(space to earth)
	MOBILE EXCEPT AERONAUTICAL
12.20–12.50	BROADCASTING
	FIXED
	MOBILE EXCEPT AERONAUTICAL
12.50–12.75	FIXED
	FIXED SATELLITE
	(earth to space)
	MOBILE EXCEPT AERONAUTICAL
12.75–13.25	FIXED*
	MOBILE*
13.25–13.40	AERONAUTICAL RADIO NAVIGATION*
13.40–14.00	RADIO LOCATION*
14.00–14.30	RADIO NAVIGATION*
	FIXED SATELLITE
	(earth to space)
14.30–14.40	RADIO NAVIGATION SATELLITE*
	FIXED SATELLITE
	(earth to space)
14.40–14.50	FIXED*
	FIXED SATELLITE*
	(earth to space)
	MOBILE*
14.50–15.35	FIXED*
	MOBILE*
15.35–15.40	RADIO ASTRONOMY*
15.40–15.70	AERONAUTICAL RADIO NAVIGATION*
15.70–17.70	RADIO LOCATION*
17.70–19.70	FIXED*
	FIXED SATELLITE*
	(space to earth)
	MOBILE*
19.70–21.20	FIXED SATELLITE*
	(space to earth)
21.20–22.00	EARTH EXPLORATION SATELLITE*
	(space to earth)
	FIXED*
	MOBILE*
22.00–23.60	FIXED
	MOBILE
23.60–24.00	RADIO ASTRONOMY*
24.00–24.05	AMATEUR*
24.05–24.25	RADIO LOCATION*
	Amateur*

Gigahertz	Service
24.25–25.25	RADIO NAVIGATION*
25.25–27.50	FIXED*
	MOBILE*
27.50–29.50	FIXED*
	FIXED SATELLITE*
	(earth to space)
	MOBILE*
29.50–31.00	FIXED SATELLITE*
	(earth to space)
31.00–31.30	FIXED*
	MOBILE*
	Space research*
31.30–31.50	RADIO ASTRONOMY*
31.50–31.80	SPACE RESEARCH
31.80–32.30	RADIO NAVIGATION*
	Space research*
32.30–33.40	RADIO NAVIGATION
33.40–34.20	RADIO LOCATION*
34.20–35.20	RADIO LOCATION*
	Space research*
35.20–36.00	RADIO LOCATION*
36.00–40.00	FIXED*
	MOBILE*
40.00–41.00	FIXED SATELLITE
	(space to earth)
41.00–43.00	BROADCASTING SATELLITE
43.00–48.00	AERONAUTICAL MOBILE SATELLITE
	AERONAUTICAL RADIO NAVIGATION SATELLITE
	MARITIME MOBILE SATELLITE
	MARITIME RADIO NAVIGATION SATELLITE
48.00–50.00	Not allocated
50.00–51.00	FIXED SATELLITE
	(earth to space)
51.00–52.00	EARTH EXPLORATION SATELLITE
	SPACE RESEARCH
52.00–54.25	SPACE RESEARCH (Passive)
54.25–58.20	INTERSATELLITE
58.20–59.00	SPACE RESEARCH (Passive)
59.00–64.00	INTERSATELLITE
64.00–65.00	SPACE RESEARCH (Passive)
65.00–66.00	EARTH EXPLORATION SATELLITE
	SPACE RESEARCH
66.00–71.00	AERONAUTICAL MOBILE SATELLITE
	AERONAUTICAL RADIO NAVIGATION SATELLITE
	MARITIME MOBILE SATELLITE
	MARITIME RADIO NAVIGATION SATELLITE
71.00–84.00	Not allocated
84.00–86.00	BROADCASTING SATELLITE
86.00–92.00	RADIO ASTRONOMY
	SPACE RESEARCH (Passive)
92.00–95.00	FIXED SATELLITE
	(earth to space)
95.00–101.0	AERONAUTICAL MOBILE SATELLITE
	AERONAUTICAL RADIO NAVIGATION SATELLITE
	MARITIME MOBILE SATELLITE
	MARITIME RADIO NAVIGATION SATELLITE

Gigahertz	Service
101.0–102.0	SPACE RESEARCH (Passive)
102.0–105.0	FIXED SATELLITE (space to earth)
105.0–130.0	INTERSATELLITE
130.0–140.0	RADIO ASTRONOMY SPACE RESEARCH (Passive)
140.0–142.0	FIXED SATELLITE (earth to space)
142.0–150.0	AERONAUTICAL MOBILE SATELLITE AERONAUTICAL RADIO NAVIGATION SATELLITE MARITIME MOBILE SATELLITE MARITIME RADIO NAVIGATION SATELLITE
150.0–152.0	FIXED SATELLITE (space to earth)
152.0–170.0	Not allocated
170.0–182.0	INTERSATELLITE
182.0–185.0	SPACE RESEARCH (Passive)
185.0–190.0	INTERSATELLITE
190.0–200.0	AERONAUTICAL MOBILE SATELLITE AERONAUTICAL RADIO NAVIGATION SATELLITE MARITIME MOBILE SATELLITE MARITIME RADIO NAVIGATION SATELLITE
200.0–220.0	Not allocated
220.0–230.0	FIXED SATELLITE
230.0–240.0	RADIO ASTRONOMY SPACE RESEARCH (Passive)
240.0–250.0	Not allocated
250.0–265.0	AERONAUTICAL MOBILE SATELLITE AERONAUTICAL RADIO NAVIGATION SATELLITE MARITIME MOBILE SATELLITE MARITIME RADIO NAVIGATION SATELLITE
265.0–275.0	FIXED SATELLITE
Above 275.0	Not allocated

FREQUENCY ALLOCATIONS IN UNITED STATES

The following listings are abstracted from Part 2 of the Rules and Regulations of the Federal Communications Commission as revised to September 1974. Several hundred footnotes in the Regulations, describing special conditions for use and for assignment of many frequency bands or individual frequencies, have been omitted.

Since changes are frequently made in the Rules and Regulations, the latest issue always should be consulted. It may be obtained from the Superintendent of Documents, Government Printing Office, Washington, D.C. Additional guidance may be obtained from the Federal Communications Commission, Washington, D.C.[1]

[1] In the official documents of ITU and FCC the following terms are combined as single words: radiobeacon, radiocommunication, radiodetermination, radiolocation, radionavigation, radiorange, radiosonde, radiotelegraphy, and radiotelephony.

Aeronautical Mobile (Ground-air-ground and air-air communication.)

General

200.0	– 285.0	kHz
325.0	– 415.0	
2850	–3155	
3.400	– 3.500	MHz
4.650	– 4.750	
5.450	– 5.730	
6.525	– 6.765	
8.815	– 9.040	
10.005	– 10.10	
11.175	– 11.40	
13.20	– 13.36	
15.01	– 15.10	
17.90	– 18.03	
21.87	– 22.00	
23.20	– 23.35	
123.5875	– 136.0	

Calling and Distress
500 kHz, telegraph calling frequency
2182 kHz, telephone calling frequency
156.8 MHz, telephone calling frequency
243.0 MHz, survival craft and equipment

Airdrome Control
118.0 –121.4 MHz

Aero Search and Rescue
123.1 MHz

Aero Utility
121.6 –121.925 MHz

Private Aircraft
121.975–123.075 MHz

Flight Test
123.125–123.275 MHz
123.325–123.475
123.525–123.575

Aviation Instructional
123.300 MHz
123.500

Civil Air Patrol
26.62 MHz
143.90
148.15

Telemetering
1435   –1535

Aeronautical Radio Navigation (Radio beacons, radio ranges, landing systems, airborne radar, etc.)

General

200.0	– 285.0	kHz
325.0	– 405.0	
1605	–1715	
960.0	–1215	MHz

1300   –1350 MHz
1558.5 –1636.5
   5.000–   5.250 GHz
   5.350–   5.460
   9.000–   9.200
  15.40 – 15.70

Direction Finding
   405.0–415.0 kHz
   (410 kHz is direction-finding frequency)

Marker–Beacon
   74.60–75.40 MHz
   (75.0 MHz is marker frequency)

VOR (omnidirectional range and localizer)
   108.0–117.975 MHz

Glide Path
   328.6 –335.4 MHz

Altimeter
   4.200– 4.400 GHz

Airborne Doppler Radar
   8.800 GHz (government)
  13.25 – 13.40 GHz

Amateur

1800   –2000 kHz
  3.500–   4.000 MHz
  7.000–   7.300
 14.00 – 14.35
 21.00 – 21.45
 28.00 – 29.70
 50.00 – 54.00
144.0 – 148.0
220.0 – 225.0
420.0 – 450.0
1215   –1300
2300   –2450
  3.300–   3.500 GHz
  5.650–   5.925
 10.00 – 10.50
 24.00 – 24.25
 48.00 – 50.00
 71.00 – 84.00
152.0 – 170.0
200.0 – 220.0
240.0 – 250.0
Above 275.0

Broadcasting

Standard Amplitude-Modulation Broadcasting
  535.0 –1605 kHz

Frequency-Modulation Broadcasting
  88.00 – 108.0 MHz

Television Broadcasting
  54.00 – 72.00 MHz
  76.00 – 88.00
174.0 – 216.0
470.0 – 806.0

International Amplitude-Modulation Broadcasting
  5.950–   6.200 MHz
  9.500–   9.775
 11.70 – 11.975
 15.10 – 15.45
 17.70 – 17.90
 21.45 – 21.75
 25.60 – 26.10

Citizens Radio (Personal radio services)

  26.96 – 27.23   MHz
462.5375–462.7375
467.5375–467.7375

Fixed (Point-to-point radio services in which neither terminal is mobile)

International Fixed Public and Aeronautical Fixed
  14.00 – 19.95   kHz (IFP only)
  20.05 – 59.00   (IFP only)
  61.00 – 90.00   (not Aero)
 110.0 – 200.0   (not Aero)
1605   –1750
2107   –2170
2194   –2495
2505   –2850
  3.155–   3.400 MHz
  4.000–   4.063
  4.438–   4.650
  4.750–   4.995
  5.005–   5.450
  5.730–   5.950
  6.765–   7.000
  7.300–   8.195
  9.040–   9.500
  9.775–   9.995
 10.10 – 11.175
 11.40 – 11.70
 11.975– 12.33
 13.36 – 14.00
 14.35 – 14.99
 15.45 – 16.46
 17.36 – 17.70
 18.03 – 19.99
 20.01 – 21.00
 21.75 – 21.85
 21.87 – 22.00 (Aero only)
 22.72 – 23.20
 23.20 – 23.35 (Aero only)
 23.35 – 24.99
 26.95 – 26.96 (IFP only)
 29.80 – 29.89
 29.91 – 30.00

Fixed in Alaska, Hawaii, and United States Possessions
 110.0 – 200.0   kHz (Alaska)
1605   –1750   (Alaska)
2107   –2170   (Alaska)

2194	−2495 kHz	(Alaska)
2505	−2850	(Alaska)
3.155–	3.200 MHz	(Alaska and Puerto Rico)
3.200–	3.400	(Alaska)
4.000–	4.063	(Alaska)
4.438–	4.650	(Alaska)
4.750–	4.995	(Alaska)
5.005–	5.450	(Alaska)
5.730–	5.950	(Alaska)
6.765–	7.000	(Alaska)
7.300–	8.195	(Alaska)
9.040–	9.500	(Alaska)
947.0 –	952.0	(Alaska, Hawaii, and U.S. possessions)
952.0 –	960.0	(Puerto Rico and Virgin Islands)

Disaster

    1750   −1800   kHz

Zone and Interzone Police

        5.005–   5.450 MHz
        7.300–   8.195

Omnidirectional (Point-to-Point)

    2150   −2180 MHz

Domestic Fixed Public (Point-to-point services by common carriers within United States)

        2110   −2130   MHz
        2160   −2180
            3.700–   4.200 GHz
            5.925–   6.425
            10.70 –   11.70
            17.70 –   19.70
            21.2 –   23.6

Operational Fixed and International Control (Point-to-point services not for public use, as well as links between control centers and stations for international service)

    72.00 – 73.00 MHz (Operational Fixed only)
    75.40 – 76.00   (Operational Fixed only)
    952.0 – 960.0
    1850   −1990
    2130   −2150
    2180   −2200
    2500   −2690   (Not International Control)
    6.525–   6.575 GHz (Operational Fixed only)
    6.575–   6.875
    12.20 – 12.70

*Aural Broadcast (Studio-transmitter link)**
    947.0 – 952.0   MHz

Instructional Television
    2500   −2690   MHz

* See Broadcast Remote Pickup under Land Mobile.

Television Pickup (Intercity relay and studio-transmitter link)

    1990   −2110   MHz
    6.875–   7.125 GHz
    12.70 – 13.20

Cable Television Relay
    12.70 – 12.95 GHz

Industrial, Scientific, and Medical Equipment
    13.56   MHz
    27.12
    40.68   (government)
    915.0   (government)
    5.800   GHz
    24.125

Government (Armed Forces and other departments of the national government)

General

510.0	–	535.0	kHz
25.33	–	25.60	MHz
27.54	–	28.00	
29.89	–	29.91	
30.00	–	30.56	
32.00	–	33.00	
34.00	–	35.00	
36.00	–	37.00	
38.00	–	39.00	
40.00	–	42.00	
46.60	–	47.00	
49.60	–	50.00	
138.0	–	144.0	
148.0	–	149.9	
150.05	–	150.8	
157.0375–		157.1875	
173.4	–	174.0	
225.0	–	328.6	
335.4	–	399.9	
410.0	–	420.0	
902.0	–	928.0	
1350	−	1400	
1429	−	1435	
1710	−	1850	
2200	−	2290	
2700	−	2900	
4.400	–	4.990	GHz
7.125	–	8.400	
14.5	–	15.35	
20.2	–	21.2	
25.25	–	27.5	
30.0	–	31.0	
36.0	–	38.6	
54.25	–	58.2	
59.0	–	64.0	
92.0	–	93.0	
102.0	–	103.0	
105.0	–	110.0	
117.5	–	122.5	
140.0	–	141.0	
150.0	–	151.0	
170.0	–	175.0	
189.0	–	190.0	

Land Mobile (Communication on land between base stations and mobile stations or between mobile stations)

Public Safety (Police, fire, highway, forestry, and emergency services)

1605	–1750	kHz
2107	–2170	
2194	–2495	
2505	–2850	
3.155 –	3.400	MHz
30.56 –	32.00	
33.01 –	33.11	
33.41 –	34.00	
35.19 –	35.69	
37.01 –	37.43	
37.89 –	38.00	
39.00 –	40.00	
42.00 –	42.95	
43.19 –	43.69	
44.61 –	46.60	
47.00 –	47.69	
150.98 –	151.4825	
153.7325–	154.46	
154.6375–	156.25	
158.715 –	159.48	
162.0125–	173.2	
451.0 –	454.0	
456.0 –	459.0	
460.0 –	462.5375	
462.7375–	467.5375	
467.7375–	512.0	
1427	–1435	

Zone and Interzone Police

2804	kHz
2808	
2812	
5005–5450	
7300–8195	

Disaster

1750–1800 kHz

Industrial (Power, petroleum, pipeline, forest products, factories, builders, ranchers, motion picture, press relay, etc.)

1605	–1750	kHz
2107	–2170	
2194	–2495	
2505	–2850	
3.155 –	3.400	MHz
4.438 –	4.650	
25.01 –	25.33	
27.28 –	27.54	
29.70 –	29.80	
30.56 –	32.00	
33.11 –	33.41	
35.00 –	35.19	
35.69 –	36.00	
37.00 –	37.01	
37.43 –	37.89	
42.95 –	43.19	

47.43 –	49.60	MHz
151.4975–	152.0	
152.465 –	152.495	
152.855 –	153.7325	
154.46 –	154.6375	
157.725 –	157.755	
158.115 –	158.475	
173.2 –	173.4	
216.0 –	220.0	
451.0 –	454.0	
456.0 –	459.0	
460.0 –	462.5375	
462.7375–	465.5375	
467.7375–	512.0	

Land Transportation (Taxis, trucks, buses, railroads)

30.56 –	32.00	MHz
33.00 –	34.00	
43.68 –	44.61	
150.8 –	150.98	
152.255 –	152.465	
157.45 –	157.725	
159.48 –	161.575	
451.0 –	454.0	
456.0 –	459.0	
460.0 –	462.5375	
462.7375–	467.5375	
467.7375–	512.0	
1427	–1435	

Operational Land and Operational Mobile

6.525 –	6.575	GHz
10.55 –	10.68	

Domestic Public

35.19 –	35.69	MHz
43.19 –	43.69	
152.0 –	152.255	
152.495 –	152.855	
157.755 –	158.115	
158.475 –	158.715	
454.0 –	455.0	
459.0 –	460.0	
470.0 –	512	

Broadcast Remote Pickup

1605	–1715	kHz
26.10 –	26.48	MHz
161.625 –	161.775	
166.25		
170.15		
450.0 –	451.0	
455.0 –	456.0	

Television Pickup

1990	–2110	MHz
6.875 –	7.125	GHz
12.70 –	13.20	

Common Carrier

6.425 –	6.525	GHz
11.70 –	12.20	
27.5 –	29.5	

Maritime Mobile (Communication between coast stations and ships, or between ships)

General

110.0	– 160.0 kHz
415.0	– 490.0 (telegraphy)
1605	–1750
2000	–2035
2092.5	–2170
2190.5	–2495
2505	–2850
3.155 –	3.400 MHz
4.1395–	4.166
6.2104–	6.2165
6.2445–	6.248
8.2812–	8.288
8.328 –	8.3315
12.421 –	12.4315
12.4795–	12.483
16.565 –	16.576
16.6365–	16.640
22.0945–	22.112
22.1605–	22.164

Calling, Safety, and Distress

500 kHz, telegraph calling frequency
2182 kHz, telephone calling frequency
156.8 MHz, telephone calling frequency
243.0 MHz, survival craft and equipment

Coast Stations (Telegraphy and facsimile)

2035	–2065 kHz (Telegraphy only)
4.231 –	4.361 MHz
6.3455–	6.514
8.4595–	8.7285
12.689 –	13.1075
16.9175–	17.255
22.374 –	22.6245

Ship Stations (Telegraphy)

4.166 –	4.178 MHz
4.187 –	4.231
6.248 –	6.267
6.2805 –	6.3455
8.34175–	8.356
8.374 –	8.4595
12.483 –	12.534
12.561 –	12.689
16.640 –	16.712
16.748 –	16.9175
22.164 –	22.2225
22.2675 –	22.374

Ship Stations (Wide-band telegraphy, facsimile, and special)

2068.5	–2078.5 kHz
6.2165–	6.2445 MHz
8.288 –	8.328
12.4315–	12.4795
16.576 –	16.6365
22.112 –	22.1605

Coast Stations (Telephony)

2065	–2068.5 kHz
2078.5	–2089.5
2173.5	–2190.5
4.1395–	4.1425
4.361 –	4.438
6.2104–	6.2165
6.514 –	6.525
8.2812–	8.288
8.7285–	8.815
12.421 –	12.4315
13.1075–	13.20
16.565 –	16.576
17.255 –	17.360
22.0945–	22.112
22.6245–	22.720
156.250 –	157.0375
161.575 –	161.625
161.775 –	162.0125

Ship Stations (Telephony)

2065	–2068.5 kHz
2078.5	–2089.5
2170	–2190.5
4.063 –	4.1395 MHz
6.200 –	6.2104
8.195 –	8.2812
12.330 –	12.421
16.460 –	16.565
22.000 –	22.0945
156.250 –	157.0375
157.1875–	157.450

Ship Calling (Telegraphy)

2089.5	–2092.5 kHz
4.178 –	4.187 MHz
6.267 –	6.2805
8.356 –	8.374
12.534 –	12.561
16.712 –	16.748
22.2225–	22.2675

Intership (Telephony)

2638 kHz
2738

Meteorological Aids

Radiosondes

400.15 –	406.0 MHz
1660	–1700

Ground-Based Radars

5.600–	5.650 GHz
9.300–	9.500

Meteorological Satellite

137.0–138.0 MHz

Radio Astronomy

```
        21.850–  21.870 MHz
        73.00 –  74.60
       406.1  – 410
      1400     –1427
      1660     –1670
      2690     –2700
         4.990–   5.000 GHz
        10.68 –  10.70
        15.35 –  15.40
        23.6  –  24.0
        31.20 –  31.50
        86.0  –  92.0
       130.0  – 140.0
       230.0  – 240.0
```

Standard Frequencies

```
        19.95 –  20.05 kHz
        59.00 –  61.00
      2495     –2505
         4.995–   5.005 MHz
         9.995–  10.005
        14.99 –  15.01
        19.99 –  20.01
        24.99 –  25.01
```

Radio Location (Coastal radar, tracking systems, etc.)

```
        70.00 –  90.00  kHz
       110.0  – 130.0
      1605     –1800
      2450     –2500 MHz
      2900     –3700
         5.250–   5.650 GHz
         8.5  –  10.55
        13.4  –  14.0
        15.7  –  17.7
        33.4  –  36.0
```

Radio Navigation (Radio beacons, shipboard radar, navigational systems, direction finding, etc.)*

General

```
        10.00 –  14.00  kHz
        90.00 – 110.0
         5.460–   5.470 GHz
         9.300–   9.500
        14.00 –  14.30
        24.25 –  25.25
        31.80 –  33.40
```

Maritime Radio Navigation

```
       285.0  – 325.0   kHz
      2900     –3100    MHz
         5.470–   5.650 GHz
```

* See listings under Aeronautical Radio Navigation.

Maritime Direction Finding

```
       405.0  – 415.0   kHz (410 kHz is
                            the direction-
                            finding
                            frequency)
```

Loran

```
        90.0  – 110.0   kHz (100 kHz is
                            Loran C fre-
                            quency)
      1800     –2000        (Loran A)
```

Land Radio Navigation
```
      1638 kHz
      1708
```

Satellites

Aeronautical Mobile
```
      1542.5–1558.5 MHz
      1644.0–1660.0
        43.0–  48.0 GHz
        66.0–  71.0
        95.0– 101.0
       142.0– 150.0
       190.0– 200.0
       250.0– 265.0
```

Aeronautical Radionavigation
```
        43.0– 48.0 GHz
        66.0– 71.0
        95.0–101.0
       142.0–150.0
       190.0–200.0
       250.0–265.0
```

Broadcasting
```
      2500  –2690   MHz
        41.0–  43.0 GHz
        84.0–  86.0
```

Earth Exploration
```
        21.2–22.0 GHz
        51.0–52.0
        65.0–66.0
```

Fixed
```
      2500    –2535   MHz
      2655    –2690
      3700    –4200
      5925    –6425
      6625    –6875
        10.95–  11.2  GHz
        11.45–  12.2
        12.5–  12.75
        14.0 –  14.5
        17.7 –  20.2
        27.5 –  30.0
        40.0 –  41.0
        50.0 –  51.0
        93.0 –  95.0
       103.0 – 105.0
       141.0 – 142.0
       151.0 – 152.0
       220.0 – 230.0
       265.0 – 275.0
```

Intersatellite

    110.0–117.5 GHz
    122.5–130.0
    175.0–182.0
    185.0–189.0

Maritime Mobile

    1535 –1543.5 MHz
    1636.5–1645
    43.0– 48.0 GHz
    66.0– 71.0
    95.0– 101.0
    142.0– 150.0
    190.0– 200.0
    250.0– 265.0

Maritime Radionavigation

    43.0– 48.0 GHz
    66.0– 71.0
    95.0–101.0
    142.0–150.0
    190.0–200.0
    250.0–265.0

Meteorological

    401– 403 MHz
    1670–1710

Radionavigation

    149.9–150.05 MHz
    399.9–400.05

Standard Frequency

    400.05–400.15 MHz

Space

Operation

    137.0– 138.0 MHz
    401.0– 402.0
    1427 –1429

Research

    136.0 – 138.0 MHz
    400.15– 401
    1700 –1710
    2290 –2300
    8400 –8500
    13.4 – 14.0 GHz
    14.4 – 14.5
    31.5 – 31.8
    51.0 – 52.0
    52.0 – 54.25 (Passive)
    58.2 – 59.0 (Passive)
    64.0 – 65.0 (Passive)
    65.0 – 66.0
    86.0 – 92.0 (Passive)
    101.0 – 102.0 (Passive)
    130.0 – 140.0 (Passive)
    182.0 – 185.0 (Passive)
    230.0 – 240.0 (Passive)

INTERNATIONAL CALL-SIGN PREFIXES

AAA–ALZ	United States
AMA–AOZ	Spain
APA–ASZ	Pakistan
ATA–AWZ	India
AXA–AXZ	Australia
AYA–AZZ	Argentina
A2A–A2Z	Botswana
A3A–A3Z	Tonga
A5A–A5Z	Bhutan
BAA–BZZ	China
CAA–CEZ	Chile
CFA–CKZ	Canada
CLA–CMZ	Cuba
CNA–CNZ	Morocco
COA–COZ	Cuba
CPA–CPZ	Bolivia
CQA–CRZ	Portuguese Territories
CSA–CUZ	Portugal
CVA–CXZ	Uruguay
CYA–CZZ	Canada
C2A–C2Z	Nauru
C3A–C3Z	Andorra
DAA–DTZ	Germany
DUA–DZZ	Philippines
EAA–EHZ	Spain
EIA–EJZ	Ireland
EKA–EKZ	USSR
ELA–ELZ	Liberia
EMA–EOZ	USSR
EPA–EQZ	Iran
ERA–ERZ	USSR
ESA–ESZ	Estonia (USSR)
ETA–ETZ	Ethiopia
EUA–EWZ	Belorussia (USSR)
EXA–EZZ	USSR
FAA–FZZ	France and Territories
GAA–GZZ	United Kingdom
HAA–HAZ	Hungary
HBA–HBZ	Switzerland
HCA–HDZ	Ecuador
HEA–HEZ	Switzerland
HFA–HFZ	Poland
HGA–HGZ	Hungary
HHA–HHZ	Haiti
HIA–HIZ	Dominican Republic
HJA–HKZ	Colombia
HLA–HMZ	Korea
HNA–HNZ	Iraq
HOA–HPZ	Panama
HQA–HRZ	Honduras
HSA–HSZ	Thailand
HTA–HTZ	Nicaragua
HUA–HUZ	El Salvador
HVA–HVZ	Vatican State
HWA–HYZ	France and Territories
HZA–HZZ	Saudi Arabia
IAA–IZZ	Italy and Territories
JAA–JSZ	Japan
JTA–JVZ	Mongolia

JWA–JXZ	Norway	VZA–VZZ	Australia
JYA–JYZ	Jordan	WAA–WZZ	United States
JZA–JZZ	West Irian	XAA–XIZ	Mexico
KAA–KZZ	United States	XJA–XOZ	Canada
LAA–LNZ	Norway	XPA–XPZ	Denmark
LOA–LWZ	Argentina	XQA–XRZ	Chile
LXA–LXZ	Luxembourg	XSA–XSZ	China
LYA–LYZ	Lithuania (USSR)	XTA–XTZ	Upper Volta
LZA–LZZ	Bulgaria	XUA–XUZ	Khmer Republic
L2A–L9Z	Argentina	XVA–XVZ	Vietnam
MAA–MZZ	United Kingdom	XWA–XWZ	Laos
NAA–NZZ	United States	XXA–XXZ	Portuguese Territories
OAA–OCZ	Peru	XYA–XZZ	Burma
ODA–ODZ	Lebanon	YAA–YAZ	Afghanistan
OEA–OEZ	Austria	YBA–YHZ	Indonesia
OFA–OJZ	Finland	YIA–YIZ	Iraq
OKA–OMZ	Czechoslovakia	YJA–YJZ	New Hebrides
ONA–OTZ	Belgium	YKA–YKZ	Syria
OUA–OZZ	Denmark	YLA–YLZ	Latvia (USSR)
PAA–PIZ	Netherlands	YMA–YMZ	Turkey
PJA–PJZ	Netherlands West Indies	YNA–YNZ	Nicaragua
PKA–POZ	Indonesia	YOA–YRZ	Romania
PPA–PYZ	Brazil	YSA–YSZ	El Salvador
PZA–PZZ	Surinam	YTA–YUZ	Yugoslavia
QAA–QZZ	(Service abbreviations)	YVA–YYZ	Venezuela
RAA–RZZ	(USSR)	YZA–YZZ	Yugoslavia
SAA–SMZ	Sweden	ZAA–ZAZ	Albania
SNA–SRZ	Poland	ZBA–ZJZ	British Territories
SSA–SSM	Egypt	ZKA–ZMZ	New Zealand
SSN–STZ	Sudan	ZNA–ZOZ	British Territories
SUA–SUZ	Egypt	ZPA–ZPZ	Paraguay
SVA–SZZ	Greece	ZQA–ZQZ	British Territories
TAA–TCZ	Turkey	ZRA–ZUZ	South Africa
TDA–TDZ	Guatemala	ZVA–ZZZ	Brazil
TEA–TEZ	Costa Rica	2AA–2ZZ	United Kingdom
TFA–TFZ	Iceland	3AA–3AZ	Monaco
TGA–TGZ	Guatemala	3BA–3BZ	Mauritius
THA–THZ	France and Territories	3CA–3CZ	Equatorial Guinea
TIA–TIZ	Costa Rica	3DA–3DM	Swaziland
TJA–TJZ	Cameroon	3DN–3DZ	Fiji
TKA–TKZ	France and Territories	3EA–3FZ	Panama
TLA–TLZ	Central African Republic	3GA–3GZ	Chile
TMA–TMZ	France and Territories	3HA–3UZ	China
TNA–TNZ	Congo	3VA–3VZ	Tunisia
TOA–TQZ	France and Territories	3WA–3WZ	Vietnam
TRA–TRZ	Gabon	3XA–3XZ	Guinea
TSA–TSZ	Tunisia	3YA–3YZ	Norway
TTA–TTZ	Chad	3ZA–3ZZ	Poland
TUA–TUZ	Ivory Coast	4AA–4CZ	Mexico
TVA–TXZ	France and Territories	4DA–4IZ	Philippines
TYA–TYZ	Dahomey	4JA–4LZ	USSR
TZA–TZZ	Mali	4MA–4MZ	Venezuela
UAA–UQZ	USSR	4NA–4OZ	Yugoslavia
URA–UTZ	Ukraine (USSR)	4PA–4SZ	Sri Lanka
UUA–UZZ	USSR	4TA–4TZ	Peru
VAA–VGZ	Canada	4UA–4UZ	United Nations
VHA–VNZ	Australia	4VA–4VZ	Haiti
VOA–VOZ	Canada	4WA–4WZ	Yemen (YAR)
VPA–VSZ	British Territories	4XA–4XZ	Israel
VTA–VWZ	India	4YA–4YZ	International Civil Aviation Organization
VXA–VYZ	Canada		

4ZA–4ZZ	Israel
5AA–5AZ	Libya
5BA–5BZ	Cyprus
5CA–5GZ	Morocco
5HA–5IZ	Tanzania
5JA–5KZ	Colombia
5LA–5MZ	Liberia
5NA–5OZ	Nigeria
5PA–5QZ	Denmark
5RA–5SZ	Malagasy Republic
5TA–5TZ	Mauretania
5UA–5UZ	Niger
5VA–5VZ	Togo
5WA–5WZ	Western Samoa
5XA–5XZ	Uganda
5YA–5ZZ	Kenya
6AA–6BZ	Egypt
6CA–6CZ	Syria
6DA–6JZ	Mexico
6KA–6NZ	Korea
6OA–6OZ	Somali Republic
6PA–6SZ	Pakistan
6TA–6UZ	Sudan
6VA–6WZ	Senegal
6XA–6XZ	Malagasy Republic
6YA–6YZ	Jamaica
6ZA–6ZZ	Liberia
7AA–7IZ	Indonesia
7JA–7NZ	Japan
7OA–7OZ	Yemen (PDRY)
7PA–7PZ	Lesotho
7QA–7QZ	Malawi
7RA–7RZ	Algeria
7SA–7SZ	Sweden
7TA–7YZ	Algeria
7ZA–7ZZ	Saudi Arabia
8AA–8IZ	Indonesia
8JA–8NZ	Japan
8OA–8OZ	Botswana
8PA–8PZ	Barbados
8QA–8QZ	Maldive Islands
8RA–8RZ	Guyana
8SA–8SZ	Sweden
8TA–8YZ	India
8ZA–8ZZ	Saudi Arabia
9AA–9AZ	San Marino
9BA–9DZ	Iran
9EA–9FZ	Ethiopia
9GA–9GZ	Ghana
9HA–9HZ	Malta
9IA–9JZ	Zambia
9KA–9KZ	Kuwait
9LA–9LZ	Sierra Leone
9MA–9MZ	Malaysia
9NA–9NZ	Nepal
9OA–9TZ	Zaire
9UA–9UZ	Burundi
9VA–9VZ	Singapore
9WA–9WZ	Malaysia
9XA–9XZ	Rwanda
9YA–9ZZ	Trinidad and Tobago

DESIGNATION OF EMISSIONS

The full designation of an emission consists of the symbol for that emission, as given in Table 3, preceded by a number indicating the necessary bandwidth in kilohertz. Bandwidths are generally expressed to a maximum of 3 significant figures. Necessary bandwidth and examples of designations are shown in Table 4.

Emissions are classified and symbolized according to the type of modulation, type of transmission, and supplementary characteristics.

NECESSARY BANDWIDTHS *

The necessary bandwidth is the minimum value of bandwidth sufficient to ensure the transmission of information at the rate and with the quality required for the system employed. Emissions needed for satisfactory functioning of the receiving equipment such as the carrier in reduced-carrier systems, or a vestigial sideband, are included in the necessary bandwidth.

For the determination of necessary bandwidth, Table 4 may be considered a guide. In formulating the table, the following terms have been employed:

B_n = Necessary bandwidth in hertz

B = Telegraph speed in bauds

N = Maximum possible number of black plus white elements to be transmitted per second, in facsimile and television

M = Maximum modulation frequency in hertz

C = Subcarrier frequency in hertz

D = Half the difference between the maximum and minimum values of the instantaneous frequency: Instantaneous frequency is the rate of change of phase

t = Pulse duration in seconds

K = An overall numerical factor which varies according to the emission and which depends on the allowable signal distortion

N_c = Number of baseband telephone channels in radio systems employing multichannel multiplex telephony

P = Continuity pilot subcarrier frequency in hertz (may exceed value of M)

R = Maximum transmission speed in binary digits (bits) per second

S = Number of equivalent nonredundant signaling states

FREQUENCY TOLERANCES

The following information is abstracted from the Radio Regulations of the International Telecommunication Union, Geneva, 1959, Appendix 3.

* From FCC Rules and Regulations, Vol. 2, 1972.

TABLE 3—DESIGNATION OF EMISSIONS.

Type of Modulation of Main Carrier	Type of Transmission	Supplementary Characteristics	Symbol
Amplitude modulation	With no modulation	—	A0
	Telegraphy without the use of a modulating audio frequency (by on-off keying)	—	A1
	Telegraphy by the on-off keying of an amplitude-modulating audio frequency or audio frequencies, or by the on-off keying of the modulated emission (special case: an unkeyed emission amplitude modulated)	—	A2
	Telephony	Double sideband, full carrier	A3
		Single sideband, reduced carrier	A3A
		Single sideband, suppressed carrier	A3J
		Two independent sidebands, reduced carrier	A3B
	Facsimile (with modulation of main carrier either directly or by a frequency-modulated subcarrier)	—	A4
		Single sideband, reduced carrier	A4A
	Television	Vestigial sideband	A5C
	Multichannel voice-frequency telegraphy	Single sideband, reduced carrier	A7A
	Cases not covered by the above, such as a combination of telephony and telegraphy	Two independent sidebands	A9B
Frequency (or phase) modulation	With no modulation	—	F0
	Telegraphy by frequency-shift keying without the use of a modulating audio frequency: one of two frequencies being emitted at any instant	—	F1
	Telegraphy by the on-off keying of a frequency-modulating audio frequency or by the on-off keying of a frequency-modulated emission (special case: an unkeyed emission, frequency modulated)	—	F2
	Telephony	—	F3
	Facsimile by direct frequency modulation of the carrier	—	F4
	Television	—	F5
	Four-frequency diplex telegraphy	—	F6
	Cases not covered by the above, in which the main carrier is frequency modulated	—	F9
Pulse modulation	A pulsed carrier without any modulation intended to carry information, for example, radar	—	P0

TABLE 3—*Continued*

Type of Modulation of Main Carrier	Type of Transmission	Supplementary Characteristics	Symbol
	Telegraphy by the on-off keying of a pulsed carrier without the use of a modulating audio frequency	—	P1D
	Telegraphy by the on-off keying of a modulating audio frequency or audio frequencies, or by the on-off keying of a modulated pulsed carrier (special case: an unkeyed modulated pulsed carrier)	Audio frequency or audio frequencies modulating the amplitude of the pulses	P2D
		Audio frequency or audio frequencies modulating the width (or duration) of the pulses	P2E
		Audio frequency or audio frequencies modulating the phase (or position) of the pulses	P2F
	Telephony	Amplitude modulated pulses	P3D
		Width (or duration) modulated pulses	P3E
		Phase (or position) modulated pulses	P3F
		Code modulated pulses (after sampling and quantization)	P3G
	Cases not covered by the above in which the main carrier is pulse modulated	—	P9

Applicable dates and certain exceptions which appear in the Regulations have been omitted.

Frequency tolerance is defined as the maximum permissible departure by the center frequency of the frequency band occupied by an emission from the assigned frequency or, by the characteristic frequency of an emission from the reference frequency. The frequency tolerance is expressed in parts in 10^6 or in some cases, in hertz.

The power shown for the various categories of stations is the mean power defined as power supplied to the antenna transmission line by a transmitter during normal operation, averaged over a time sufficiently long compared with the period of the lowest frequency encountered in the modulation. A time of 0.1 second during which the mean power is greatest will be selected normally.

Radio determination stations include radio navigation stations such as radio beacons, marker beacons, instrument landing systems, navigational radio, loran, decca, et cetera, and it also includes radio location stations such as radar used for purposes other than radio navigation. Where specific frequencies are not assigned to radar stations, the bandwidth occupied by the emission shall be maintained wholly within the band allocated to the service and the indicated tolerance does not apply.

Frequency Bands and Station Categories	Tolerance (Parts in 10^6)
Band: 10–535 kHz	
Fixed stations:	
10–50 kilohertz	1000
50–535 kilohertz	200
Land stations:	
Coast stations	
Power $\leq$200 watts	500
Power >200 watts	200
Aeronautical stations	100
Mobile stations:	
Ship stations	1000
Aircraft stations	500
Emergency ship transmitters	5000
Survival-craft stations	5000
Radio determination stations	100
Broadcasting stations	10 Hz
Band: 535–1605 kHz	
Broadcasting stations	10 Hz
Stations covered by the North American Regional Broadcasting Agreement	20 Hz
Band: 1605–4000 kHz	
Fixed stations:	
Power $\leq$200 watts	100
Power >200 watts	50

Frequency Bands and Station Categories	Tolerance (Parts in 10^6)
Land stations:	
Power ≤200 watts	100
Power >200 watts	50
Mobile stations:	
Ship stations	200
Aircraft stations	100
Survival-craft stations	300
Land mobile stations	200
Radio determination stations:	
Power ≤200 watts	100
Power >200 watts	50
Broadcasting stations	20

Band: 4–29.7 MHz

Fixed stations:	
Power ≤500 watts	50
Power >500 watts	15
Land stations:	
Coast stations:	
Power ≤500 watts	50
Power >500 <5000 watts	30
Power >5000 watts	15
Aeronautical stations:	
Power ≤500 watts	100
Power >500 watts	50
Base stations:	
Power ≤500 watts	100
Power >500 watts	50
Mobile stations:	
Ship stations:	
Class A1 emission	200
Emission other than class A1	50
Aircraft stations	100
Survival-craft stations	200
Land mobile stations	200
Broadcasting stations	15

Band: 29.7–100 MHz

Fixed stations:	
Power ≤200 watts	50
Power >200 watts	30
Land stations:	
Power ≤15 watts	50
Power >15 watts	20
Mobile stations:	
Power ≤5 watts	100
Power >5 watts	50
Radio determination stations	200
Broadcasting stations (other than television):	
Power ≤50 watts	50
Power >50 watts	20
Television broadcasting stations:	
Power ≤50 watts	100
Power >50 watts	1000 Hz

Band: 100 – 470 MHz

Fixed stations:	
Power ≤50 watts	50
Power >50 watts	20
Land stations:	
Coast stations	20
Aeronautical stations	50
Base stations:	
Power ≤5 watts	50
Power >5 watts	20
Mobile stations:	
Ship and survival-craft stations:	
In the band 156–174 megahertz	20
Outside this band	50
Aircraft stations	50
Land mobile stations:	
Power ≤5 watts	50
Power >5 watts	20
Radio determination stations	50
Broadcasting stations (other than television)	20
Television broadcasting stations:	
Power ≤100 watts	100
Power >100 watts	1000 Hz

Band: 470–2450 MHz

Fixed stations:	
Power ≤100 watts	300
Power >100 watts	100
Land stations	300
Mobile stations	300
Radio determination stations	500
Broadcasting stations (other than television)	100
Television broadcasting stations:	
Power ≤100 watts	100
Power >100 watts	1000 Hz

Band: 2450–10 500 MHz

Fixed stations:	
Power ≤100 watts	300
Power >100 watts	100
Land stations	300
Mobile stations	300
Radio determination stations	2000

Band: 10.5–40 GHz

Fixed stations	500
Radio determination stations	7500

NOTE: Requirements in the USA with respect to frequency tolerances are in all cases at least as restrictive (and for some services more restrictive) than the tolerances specified by the International Convention. For details, consult the Rules and Regulations of the Federal Communications Commission.

Spurious-Emission Tolerances (ITU, Geneva, 1959)

Spurious emission occurs on a frequency or frequencies outside the necessary band, and the level of this spurious emission may be reduced without affecting transmission of information.

TABLE 4—DETERMINATION OF NECESSARY BANDWIDTH.

Description and Class of Emission	Necessary Bandwidth in Hertz	Examples — Details	Designation of Emission
	Amplitude Modulation		
Continuous-wave telegraphy, A1	$B_n = BK$ where $K = 5$ for fading circuits $K = 3$ for nonfading circuits	Morse code at 25 words per minute, $B = 20$, $K = 5$ Bandwidth: 100 hertz	0.1A1
		Four-channel time-division multiplex, 7-unit code, 42.5 bauds per channel, $B = 170$, $K = 5$ Bandwidth: 850 hertz	0.85A1
Telegraphy, modulated by an audio frequency, A2	$B_n = BK + 2M$ where $K = 5$ for fading circuits $K = 3$ for nonfading circuits	Morse code at 25 words per minute, $B = 20$, $M = 1000$, $K = 5$ Bandwidth: 2100 hertz	2.1A2
Telephony, A3	$B_n = M$ for single sideband $B_n = 2M$ for double sideband	Double-sideband telephony, $M = 3000$ Bandwidth: 6000 hertz	6A3
		Single-sideband telephony, reduced carrier, $M = 3000$ Bandwidth: 3000 hertz	3A3A
		Telephony, two independent sidebands, $M = 3000$ Bandwidth: 6000 hertz	6A3B
Sound broadcasting, A3	$B_n = 2M$ M may vary between 4000 and 10 000 depending on quality desired.	Speech and music, $M = 4000$ Bandwidth: 8000 hertz	8A3
Facsimile, carrier modulated by tone and by keying, A4	$B_n = KN + 2M$ where $K = 1.5$	The total number of picture elements (black plus white) transmitted per second is equal to the circumference of the cylinder multiplied by the number of lines per unit length and by the speed of rotation of the cylinder in revolutions per second. Diameter of cylinder = 70 millimeters Number of lines per millimeter = 5 Speed of rotation = 1 rotation per second $N = 1100$ $M = 1900$ Bandwidth: 5450 hertz	5.45A4

TABLE 4—Continued

Description and Class of Emission	Necessary Bandwidth in Hertz	Examples	
		Details	Designation of Emission
Television (visual and aural), A5 and F3	Refer to relevant CCIR documents for the bandwidths of the commonly used television systems.	Number of lines=525 Number of lines per second=15 750 Video bandwidth: 4.2 megahertz Total visual bandwidth: 5.75 megahertz FM aural bandwidth including guard bands: 250 000 hertz Total bandwidth: 6 megahertz	5750A5C 250F3
Composite transmission, A9	$B_n=2M$ (double sideband)	Television relay, video limited to 4 megahertz, audio on 6.5 megahertz FM subcarrier, subcarrier deviation=50 kilohertz. M=subcarrier frequency plus its maximum deviation = 6.55×10^6. Bandwidth: 13.1×10^6 hertz	13 100A9
Composite transmission, A9	$B_n=2M$ (double sideband)	Microwave relay system providing 10 telephone channels occupying baseband between 4 and 164 kilohertz $M=164\ 000$ Bandwidth: 328 000 hertz	328A9
Composite transmission A9Y, digital modulation using DSB–AM	$B_n=\dfrac{2RK}{\log_2 S}$	Microwave radio relay specifications: digital modulation used to send 5 megabits per second by use of amplitude modulation of the main carrier with four signaling states. $R=5\times10^6$ bits per second $K=1$ $S=4$ $B_n=5$ MHz	5000A9Y
		Frequency Modulation	
Frequency-shift telegraphy, F1	$B_n=2.6D+0.55B$ for $1.5\leq 2D/B\leq5.5$ $B_n=2.1D+1.9B$ for $5.5\leq 2D/B\leq20$	Four-channel time-division multiplex with 7-unit code, 42.5 bauds per channel, $B=170$, $D=200$; $2D/B=2.35$; therefore the first equation in column 2 applies. Bandwidth: 613 hertz	0.6F1

TABLE 4—Continued

Description and Class of Emission	Necessary Bandwidth in Hertz	Examples	
		Details	Designation of Emission
Commercial telephony, F3	$B_n = 2M + 2DK$ K is normally 1 but under certain conditions a higher value may be necessary.	For an average case of commercial telephony, $D = 15\,000$, $M = 3000$. Bandwidth: 36 000 hertz	36F3
Sound broadcasting, F3	$B_n = 2M + 2DK$	$D = 75\,000$, $M = 15\,000$ and assuming $K = 1$ Bandwidth: 180 000 hertz	180F3
Facsimile, F4	$B_n = KN + 2M + 2D$ where $K = 1.5$	(See facsimile, amplitude modulation) Diameter of cylinder = 70 millimeters Number of lines per millimeter = 5 Speed of rotation = 1 rotation per second $N = 1100$. $M = 1900$ $D = 10\,000$ Bandwidth: 25 450 hertz	25.5F4
Four-frequency diplex telegraphy, F6	If the channels are not synchronized, $B_n = 2.6D + 2.75B$, where B is the speed of the higher-speed channel. If the channels are synchronized the bandwidth is as for F1, B being the speed of either channel.	Four-frequency diplex system with 400-hertz spacing between frequencies, channels not synchronized, 170 bauds keying in each channel, $D = 600$, $B = 170$ Bandwidth: 2027 hertz	2.05F6
Composite transmission, F9	$B_n = 2P + 2DK$ where $K = 1$	Microwave radio relay system specifications: 60 telephone channels occupying baseband between 60 and 300 kHz; rms per-channel deviation 200 kHz; continuity pilot at 331 kHz produces 100 kHz rms deviation of main carrier. $D = (200 \times 10^6 \times 3.76 \times 2.02)$ Hz $= 1.52 \times 10^6$ Hz; $P = 0.331 \times 10^6$ Hz. Bandwidth: 3.702×10^6 Hz.	3700F9

TABLE 4—*Continued*

Designation and Class of Emission	Necessary Bandwidth in Hertz	Examples	
		Details	Designation of Emission
Composite transmission, F9	$B_n = 2M + 2DK$ where $K=1$	Microwave radio relay system specifications: 960 telephone channels occupying baseband between 60 and 4028 kHz; rms per-channel deviation 200 kHz; continuity pilot at 4715 kHz produces 140 kHz rms deviation of main carrier. $D = (200 \times 10^3 \times 3.76 \times 5.5)$ Hz $= 4.13 \times 10^6$ Hz; $M = 4.028 \times 10^6$ Hz; $P = 4.715 \times 10^6$ Hz; $(2M + 2DK) > 2P$. Bandwidth: 16.32×10^6 Hz.	16 300F9
Composite transmission, F9	$B_n = 2P$	Microwave radio relay system specifications: 600 telephone channels occupying baseband between 60 and 2540 kHz; rms per-channel deviation 200 kHz; continuity pilot at 8500 kHz produces 140 kHz rms deviation of main carrier. $D = (200 \times 10^3 \times 3.76 \times 4.36)$ Hz $= 3.28 \times 10^6$ Hz; $M = 2.54 \times 10^6$ Hz; $K = 1$; $P = 8.5 \times 10^6$ Hz; $(2M + 2DK) < 2P$. Bandwidth: 17×10^6 Hz.	17 000F9
Composite transmission, F9	$B_n = 2M + 2DK$ where $K = 1$	TV microwave relay system specifications: Aural program on 7.5 MHz, aural subcarrier deviation ±150 kHz; continuity pilot at 8.5 MHz produces 140 kHz rms deviation of main carrier; $D = 3.7 \times 10^6$ Hz (visual) plus 0.3×10^6 Hz (aural). $M = (7.5 + 0.15) \times 10^6$ Hz; $P = 8.5 \times 10^6$ Hz; $D = (3.7 + 0.3) \times 10^6$ Hz; $(2M + 2DK) > 2P$. Bandwidth: 23.3×10^6 Hz.	23 300F9
Composite transmission, F9	$B_n = 2P$	TV microwave relay system specifications: Aural program on 6.0 MHz subcarrier; aural subcarrier deviation ±150 kHz; continuity pilot at 8.5 MHz produces 50 kHz rms deviation of main carrier; $D = 2 \times 10^6$ Hz (visual) plus 0.2×10^6 Hz (aural). $D = (2.0 + 0.2) \times 10^6$ Hz; $M = 6.15 \times 10^6$ Hz; $K = 1$; $P = 8.5 \times 10^6$ Hz; $(2M + 2DK) < 2P$. Bandwidth: 17×10^6 Hz.	17 000F9

TABLE 4—*Continued*

Description and Class of Emission	Necessary Bandwidth in Hertz	Examples	
		Details	Designation of Emission
Composite transmission, F9	$B_n = 2M + 2DK$ where $K = 1$	Stereophonic FM broadcasting (United States system) with multiplexed subsidiary communications subcarrier, $M = 75\,000$, $D = 75\,000$ Bandwidth: 300 000 hertz	300F9
Composite transmission, F9Y, digital modulation using PSK	$B_n = \dfrac{2RK}{\log_2 S}$	Microwave radio relay system specifications: digital modulation used to send 10 megabits per second by use of phase-shift keying with four signaling states. $R = 10 \times 10^6$ bits per second $K = 1$ $S = 4$ $B_n = 10$ MHz	10 000F9Y
Composite transmission, F9Y, digital modulation using FSK	$B_n = \dfrac{R}{\log_2 S} + 2DK$	Microwave radio relay system specifications: digital modulation used to send 10 megabits per second by use of frequency-shift keying with four signaling states and 2-MHz peak deviation of the main carrier. $R = 10 \times 10^6$ bits per second $D = 2$ MHz $K = 1$ $S = 4$ $B_n = 9$ MHz	9000F9Y
	Pulse Modulation		
Unmodulated pulse, P0	$B_n = 2K/t$ K depends on the ratio of pulse duration to pulse rise time. Its value usually falls between 1 and 10 and in many cases it does not need to exceed 6.	$t = 3 \times 10^{-6}$, $K = 6$ Bandwidth: 4×10^6 hertz	4000P0
Modulated pulse, P2 or P3	The bandwidth depends on the particular types of modulation used, many of these being still in the development stage.		
Composite transmission, P9	$B_n = 2K/t$ where $K = 1.6$	Microwave relay, pulse-position modulated by 36-channel baseband; pulse width at half amplitude = 0.4 microsecond Bandwidth: 8×10^6 hertz	8000P9

Spurious emissions include harmonics, parasitic emissions, and intermodulation products.

The mean power of any spurious emission supplied to the antenna transmission line shall not exceed the values specified below. Spurious radiation from any other part of the installation shall not have an effect greater than would occur if the antenna system were supplied with the maximum permissible spurious power.

Fundamental Frequency	Below Mean Power of Fundamental (dB)	Maximum Power of Spurious Emissions
Below 30 MHz:		
Transmitter power ≤ 50 kW	40	50 mW
Transmitter power > 50 kW	60	50 mW
30 MHz to 235 MHz:		
Transmitter power ≤ 25 W	40	10 μW
Transmitter power > 25 W	60	1 mW
Above 235 MHz		As low as practicable

STANDARD FREQUENCY SOURCES

There are the following general types of stable oscillators:

(**A**) Free-running crystal oscillators.

(**B**) Crystal oscillators locked to an atomic transition frequency.

(**C**) Atomic beam devices (using cesium or thallium).

(**D**) Gas-cell devices (using mostly rubidium).

Some of these types, such as the atomic beam devices and the hydrogen maser, have excellent reproducibility because of the small interactions between the atoms and the environment. They are suitable for primary standards. In other devices, mostly gas cells, the frequency of atomic transition is affected by interaction between the atoms and the buffer gas or the walls of the cell, and the frequency of transition must be calibrated since it cannot be reproduced with sufficient accuracy. These devices, however, can be very useful as secondary, or working, standards if they have good stability and good reliability.

The performances of the types of oscillators most commonly used as sources for standard frequencies are given in Tables 5 and 6.

TIME SCALES

Ephemeris Time

In 1956 the International Committee of Weights and Measures defined the fundamental unit of time, the second, as 1/31 556 925.9747 of the tropical year for 12^h ephemeris time January 0, 1900 (January 0, 1900= December 31, 1899). The tropical year is the interval between two consecutive returns of the sun to the vernal equinox. It consisted of 365 days, 5 hours, 48 minutes, and 46 seconds at that time and has been decreasing by about 5.3 milliseconds per year.

A constant second has thus been defined in relation to the tropical year at a particular time. Although the ephemeris second has been defined very exactly, it cannot be measured by astronomical observations with anything like the precision implied by so many digits, but actually only to parts in 10^9.

In practice, ephemeris time is determined by observations of the moon and subsequent reference to lunar ephemeris tables.

Atomic Time

Atomic resonance can be used to provide time scales which are presumably uniform. The agreement between groups of measurements of certain atomic transition frequencies by different instruments is of the order of 5 parts in 10^{12}. The frequency standard maintained by the National Bureau of Standards consists of atomic standards which are stable to 1 part in 10^{12}.

The zero-field $(4,0)-(3,0)$ transition for cesium has been measured in terms of the ephemeris second to be 9 192 631 770$\pm$20 hertz. The uncertainty of 2 parts in 10^9 is avoided by temporarily defining the standard to be used in determining the second to be exactly equal to the above number of vibrations with no uncertainty. This standard was adopted at the 12th General Conference of Weights and Measures in 1964. Other standards may be adopted in future years if more-precise measurements can be made using them. The frequency of the hydrogen maser has been measured as 1 420 405 751.73$\pm$0.03 hertz on this time scale.

Sidereal Time (θ)

This time scale is based on the mean time of rotation of the earth about its axis in relation to the vernal-equinox point in the sky. It is determined by observing the meridian transits of stars. The mean sidereal day is 23 hours, 56 minutes, 4.09 seconds. Because of variations in the rotational speed of the earth as outlined below, sidereal time is not perfectly uniform.

Universal-Time Scales (UT)

A universal-time scale, also known as Greenwich Mean Time or Greenwich Civil Time, is based on the mean angle of rotation of the earth about its axis in relation to the sun. It is referred to the

TABLE 5—CHARACTERISTICS OF ATOMIC FREQUENCY STANDARDS.

Characteristic	Hydrogen Maser	Cesium Beam (24-Inch) Controlled Oscillator	Rubidium Gas Cell Controlled Oscillator
Intrinsic reproducibility of present devices	No general published results Measured: $\pm 2 \times 10^{-12}$ and less	$\pm 3 \times 10^{-12}$	Does not apply
Stability (rms deviation from the mean):			
One second	5×10^{-13}	5×10^{-11}	1×10^{-11}
One minute	6×10^{-11}	6×10^{-12}	2×10^{-12}
One hour	3×10^{-14}	8×10^{-13}	1×10^{-12}
One day	2×10^{-14}	3×10^{-13}	5×10^{-12}
Systematic drift	None detectable with resolution of 1×10^{-12} per year	$< 3 \times 10^{-12}$ per year	$< 3 \times 10^{-11}$ per month
Volume (with power supply)	16.4 ft³	1.6 ft³	0.6 ft³
Weight (with power supply)	800 lb	64 lb	20 lb
Power demand (115 volts ac)	200 watts	60 watts	40 watts
Relative cost	5.5	1.5	1.0

TABLE 6—PHYSICAL DATA OF AVAILABLE ATOMIC FREQUENCY STANDARDS.

Characteristic	Hydrogen Maser	Cesium Beam (24-Inch) Controlled Oscillator	Rubidium Gas Cell Controlled Oscillator
Nominal resonance frequency	1420.405 751 MHz	9192.631 770 MHz	6834.682 608 MHz
Resonance width	1 Hz	250 Hz	200 Hz (typical)
Atomic interaction time	0.5 second	2.5×10^{-3} second. Interaction length, $L-25$ (typical)	2×10^{-3} second (typical)
Atomic resonance events per second	10^{12}	10^{6}	10^{12}
Principal frequency offsets:			
Magnetic	$f - f_o = 2750B^2$ (gauss) 5×10^{-13} (typical)	$f - f_o = 427B^2$ (gauss) 1×10^{-10} (typical)	$f - f_o = 574B^2$ (gauss) 1×10^{-9} (typical)
2nd-order doppler	4×10^{-11} ($\Delta f/°T = 1.4 \times 10^{-13}/K$)	3×10^{-13}	8×10^{-13}
Collisions	2×10^{-11}	None	3×10^{-7} (typical)
State selection method	Atomic beam deflection in hexapole magnets	Atomic beam reflection in dipole or multi-pole magnets	Optical pumping
Resonance detection method	Atomic microwave radiation (active maser oscillation)	Surface ionization of deflected atoms	Optical absorption
Temperature of resonating atoms	300 K	360 K	330 K

TABLE 7—SERVICES PROVIDED BY RADIO STATIONS OF THE NATIONAL BUREAU OF STANDARDS.

Services	WWV	WWVH	WWVB
Standard radio frequencies:			
60 kilohertz			×
2.5 megahertz	×	×	
5 megahertz	×	×	
10 megahertz	×	×	
15 megahertz	×	×	
20 megahertz	×	×	
25 megahertz	×		
Standard audio frequencies:			
440 hertz	×	×	
500 hertz	×	×	
600 hertz	×	×	
Standard time intervals	×	×	×
Time signals	×	×	×
Time code	×	×	×
$UT1$ corrections	×	×	×
Radio propagation forecasts	×		
Geophysical alerts	×	×	

The station locations and radiated powers are as follows:

WWV—Fort Collins, Colorado (10 kilowatts for 5, 10, and 15 MHz; 2.5 kilowatts for 2.5, 20, and 25 MHz)
    40°40′49″N, 105°02′27″W.

$WWVH$—Kekaha, Kauai, Hawaii (10 kilowatts for 5, 10, and 15 MHz; 5 kilowatts for 2.5 MHz; 2.5 kilowatts for 20 MHz)
    21°59′26″N, 159°46′00″W.

$WWVB$—Fort Collins, Colorado (13 kilowatts)
    40°40′28.3″N, 105°02′39.5″W.

$WWVL$—Fort Collins, Colorado
    40°40′51.3″N, 105°03′00.0″W.

prime meridian that passes through Greenwich, England.

Since actual solar days vary throughout the year, a mean solar day of 24 hours is used to denote one revolution. Determinations of the earth's rotation relative to the sun are made by observing mean sidereal rotation of the earth and converting it to mean solar rotation by ephemeris tables based on the accumulated data of many astronomical observatories.

Mean solar rotation derived from uncorrected astronomical observations is denoted $UT0$.

Annual variations occur in the speed of rotation of the earth and are probably due to seasonal changes in the wind patterns of the Northern and Southern Hemispheres. There is also a semiannual variation due chiefly to tidal action of the sun, which distorts the shape of the earth slightly. The cumulative effect of these variations is that the earth is late about 30 milliseconds or 0.45 arc second near June 1 and is ahead about 30 milliseconds or 0.45 arc second near October 1 each year. $UT0$ corrected for these periodic variations is denoted $UT1$.

Irregular variations in the speed of rotation of the earth also occur. They may be due to turbulent motions in the core of the earth. In addition, friction of the ocean tides causes a decrease in speed of about one millisecond per century. Observations of these effects throughout the world are reported to the Bureau International de l'Heure at Paris, which issues corrections for $UT1$ to establish $UT2$.

For navigators and space scientists, time signals must be correlated with the earth's rotation, and for this reason most radio time signals include $UT2$ or earth rotation information.

"Standard Times"

"Standard times" are based on UT or on $UT2$. Although the term "standard time" is a misnomer, as implied by the foregoing, common usage dictates its retention in this section.

The world is divided into 24 zones, each 15° of longitude, or 1 hour angle, apart. The meridian of Greenwich, England, is the center of the zero zone which extends to 7.5° east and west. Proceeding

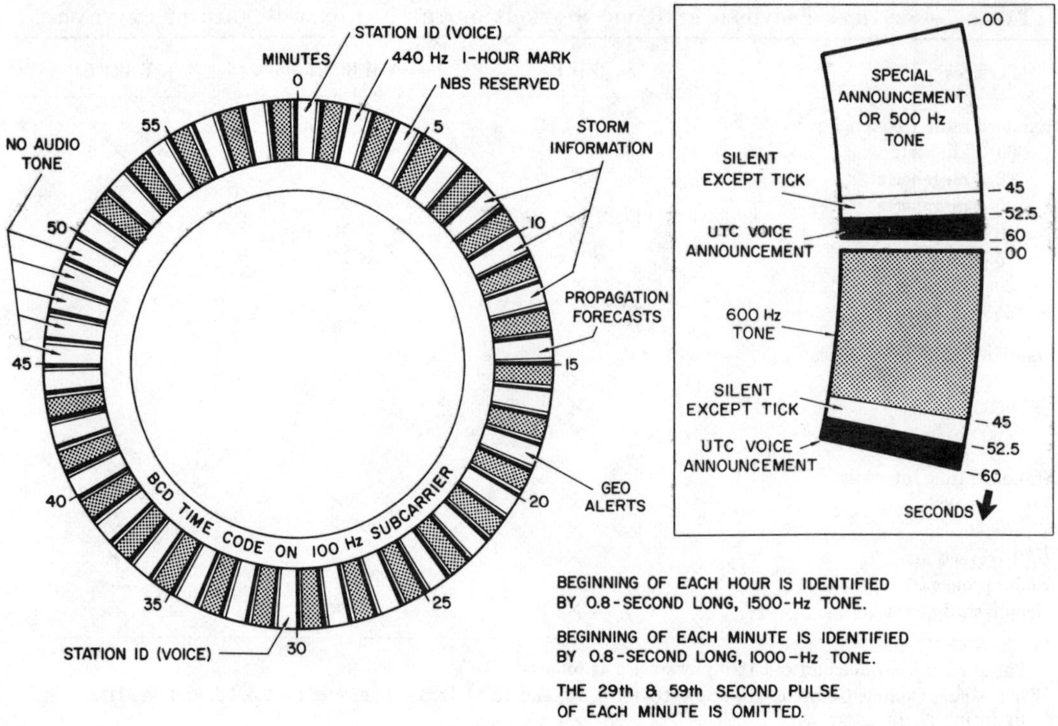

STATION ID (VOICE)
MINUTES
440 Hz 1-HOUR MARK
NBS RESERVED
STORM INFORMATION
NO AUDIO TONE
PROPAGATION FORECASTS
GEO ALERTS
BCD TIME CODE ON 100 Hz SUBCARRIER
STATION ID (VOICE)
SPECIAL ANNOUNCEMENT OR 500 Hz TONE
SILENT EXCEPT TICK
UTC VOICE ANNOUNCEMENT
600 Hz TONE
SILENT EXCEPT TICK
UTC VOICE ANNOUNCEMENT
SECONDS
BEGINNING OF EACH HOUR IS IDENTIFIED BY 0.8-SECOND LONG, 1500-Hz TONE.
BEGINNING OF EACH MINUTE IS IDENTIFIED BY 0.8-SECOND LONG, 1000-Hz TONE.
THE 29th & 59th SECOND PULSE OF EACH MINUTE IS OMITTED.

(A) *WWV.*

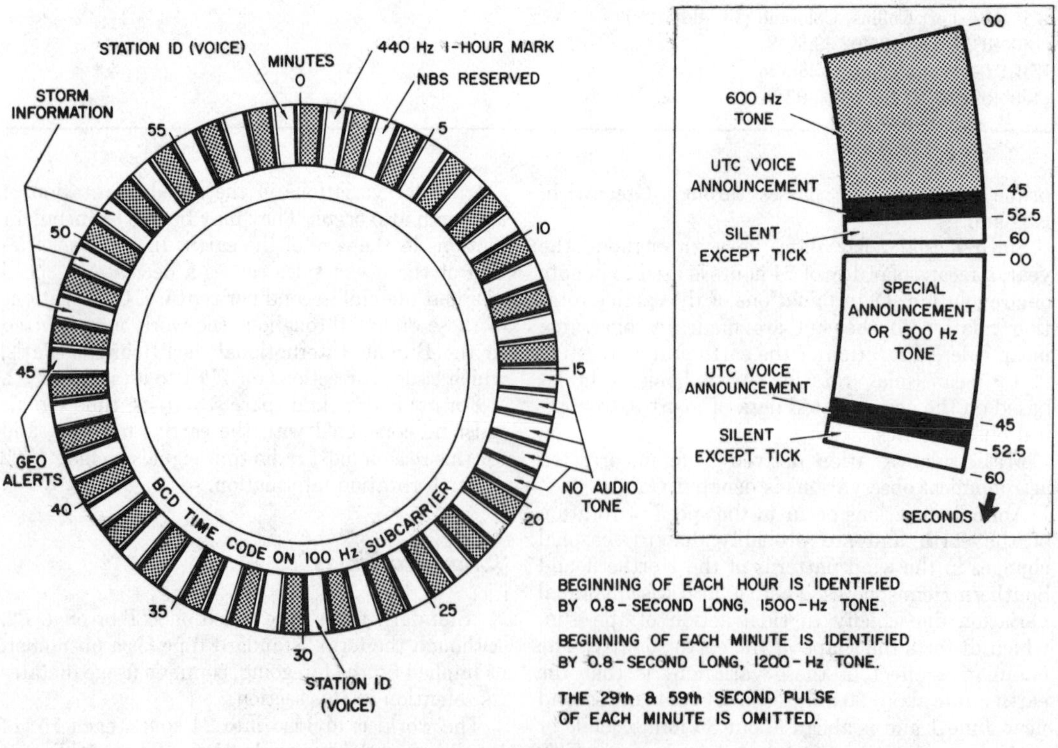

STATION ID (VOICE)
MINUTES
440 Hz 1-HOUR MARK
NBS RESERVED
STORM INFORMATION
GEO ALERTS
NO AUDIO TONE
BCD TIME CODE ON 100 Hz SUBCARRIER
STATION ID (VOICE)
600 Hz TONE
UTC VOICE ANNOUNCEMENT
SILENT EXCEPT TICK
SPECIAL ANNOUNCEMENT OR 500 Hz TONE
UTC VOICE ANNOUNCEMENT
SILENT EXCEPT TICK
SECONDS
BEGINNING OF EACH HOUR IS IDENTIFIED BY 0.8-SECOND LONG, 1500-Hz TONE.
BEGINNING OF EACH MINUTE IS IDENTIFIED BY 0.8-SECOND LONG, 1200-Hz TONE.
THE 29th & 59th SECOND PULSE OF EACH MINUTE IS OMITTED.

(B) *WWVH.*

Fig. 3—Hourly broadcast schedules of *WWV* and *WWVH*.

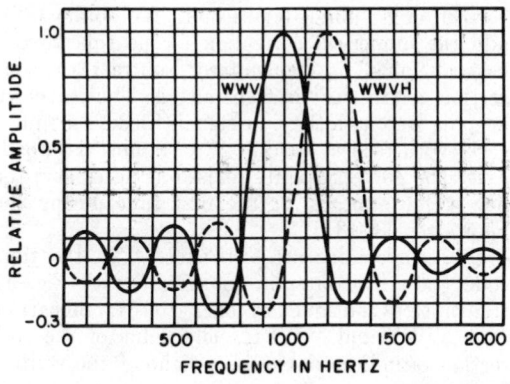

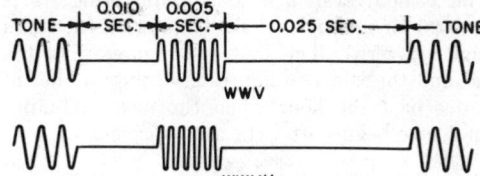

WWV AND WWVH SECONDS PULSES

THE SPECTRA ARE COMPOSED OF DISCRETE FREQUENCY COMPONENTS AT INTERVALS OF 1.0 HERTZ. THE COMPONENTS AT THE SPECTRAL MAXIMA HAVE AMPLITUDES OF 0.005 VOLT FOR A PULSE AMPLITUDE OF 1.0 VOLT. THE WWV PULSE CONSISTS OF FIVE CYCLES OF 1000 HERTZ. THE WWVH PULSE CONSISTS OF SIX CYCLES OF 1200 HERTZ.

Fig. 4—Waveforms and pulse durations of *WWV* and *WWVH*.

eastward from Greenwich the zones are numbered 1 to 12 with the prefix "plus" to indicate the hour angle to be added to universal time to obtain local "standard time." Proceeding westward, the zones are numbered 1 to 12 with the prefix "minus" to indicate the hour angle to be subtracted from universal time to obtain local "standard time." For example, Washington, D.C., at longitude 77° West, is in time zone -5.

The actual boundaries of time zones are defined by law or custom in the various countries and generally do not coincide with the theoretical zone, even in some places at sea. In many areas local legal "standard time" differs by 60 or 30 minutes from theoretical standard time. A chart in Chapter 48 shows "standard time" in principal cities, and most atlases contain time-zone maps.

STANDARD FREQUENCIES AND TIME SIGNALS

The national Bureau of Standards operates standard frequency and time stations WWV, WWVB, and WWVL at Fort Collins, Colorado, and WWVH at Kekaha, Kauai, Hawaii.* The services provided by WWV, WWVH, and WWVB are listed in Table 7. WWVL is a vlf station broadcasting experimental programs, usually involving multiple frequencies, on an intermittent basis only.

WWV and WWVH

The program from WWV and WWVH is continuous, 24 hours a day. The content is summarized in Fig. 3. In addition to off-the-air reception, the broadcasts of WWV may be heard via telephone by dialing (303) 499-7111, Boulder, Colorado. The telephone user will hear the broad-

* Information on these services may be obtained from the Frequency-Time Broadcast Services Section, Time and Frequency Division, NBS, Boulder, CO 80302.

casts as transmitted from the station. Because of the instabilities and variable delays of propagation by telephone, the accuracy of the telephone time signals should not be expected to be better than 30 milliseconds. This service is automatically limited to 3 minutes per call. Similar time-of-day broadcasts from WWVH can be heard by dialing (808) 355-4363 on the Island of Kauai. (Neither of the listed telephone numbers is a toll-free number.)

All carrier and modulation frequencies at WWV and WWVH are derived from cesium-controlled oscillators and are accurate to within ± 1 part in 10^{11}. Deviations are normally less than 1 part in 10^{12} from day to day. Changes in the propagation medium (causing Doppler shifts, diurnal shifts, etc.) result in fluctuations in the received carrier frequencies which may be much greater than the uncertainties of the frequencies as transmitted.

In conformity with the UTC scale, the carrier and modulation frequencies of WWV and WWVH are no longer offset significantly from nominal values. The frequency offset of UTC was made permanently zero relative to International Atomic Time (TAI) effective 0000 hours UTC January 1, 1972. Previously, the fractional frequency offset was -150 parts in 10^{10} during 1960 and 1961; -130 parts in 10^{10} during 1962 and 1963; -150 parts in 10^{10} during 1964 and 1965; and -300 parts in 10^{10} from 1966 through 1971.

Standard Time Intervals: Seconds pulses at precise intervals are derived from the same frequency standard that controls the radio carrier frequencies; for example, they commence at intervals of 5,000,000 cycles of the 5-MHz carrier. The 1-second markers are transmitted throughout all programs of WWV and WWVH except that the 29th and 59th markers of each minute are omitted. Details of these pulses are shown in Fig. 4. (Compositions of the hour and minute markers are listed in Fig. 3.)

Both WWV and WWVH broadcast voice announcements of Coordinated Universal Time

(UTC) each minute. The reference time scale is the Coordinated Time Scale maintained by the National Bureau of Standards, UTC(NBS). The UTC(NBS) scale includes small frequency offsets relative to the NBS primary frequency standard for coordination purposes.

The 24-hour system is used. Numbering starts with 0000 for midnight at the Greenwich Meridian (longitude zero). The first two figures give the hour, and the last two figures give the number of minutes past the hour when the next 800-millisecond tone begins after the announcement.

Time Corrections: Prior to January 1, 1972, time signals broadcast from WWV and WWVH were kept in close agreement with UT2 (astronomical time) by making step adjustments of 100 milliseconds as necessary. On December 31, 1971 at 23 hours 59 minutes 60.107600 seconds UTC, the UTC(NBS) scale was retarded 0.107600 second to give it an initial difference of exactly 10 seconds late with respect to the International Atomic Time (TAI) scale as maintained by the Bureau International d l'Heure (BIH).

Since the new UTC rate (effective January 1, 1972) is no longer adjusted periodically to agree with the earth's rotation rate, UTC departs more rapidly than before from earth rotation time (known as UT1), gaining about 1 second per year. Corrections to UTC are now made in step adjustments of exactly 1 second (called a leap second) as directed by BIH. The leap-second adjustments ensure that UTC signals as broadcast never differ from UT1 by more than about $+0.7$ second. (The corrections no longer relate to UT2.)

The leap-second adjustments are made as necessary at the end of the UTC month, preferably on December 31 or June 30. Thus, when required, a leap second is inserted between the end of the 60th second of the last minute of the last day of a month and the beginning of the next minute. The minute during which the correction is made contains either 59 or 61 seconds, depending on whether the correction is negative or positive.

For those applications in which it is necessary to date events on the UT1 time scale to better than ± 0.7 second, coded corrections are provided in the broadcast formats to give UT1-UTC values to a resolution of 0.1 second. The method involves the use of a system of double second pulses. Doubling of the first through the seventh seconds pulses indicates a "plus" correction; doubling the ninth through the fifteenth pulses indicates a "minus" correction. The eighth seconds pulse is not used. The amount of correction in units of 0.1 second is determined by counting the number of seconds pulses that are doubled. For example, if the first, second, and third seconds pulses are doubled, the UT1 correction is "plus 0.3 second." Or if the ninth, tenth, eleventh, twelfth, thirteenth, and fourteenth seconds pulses are doubled, the UT1

correction is "minus 0.6 second." To obtain UT1, add the numerical correction to the time broadcast if "plus" is transmitted; subtract the correction if "minus" is transmitted. Thus, a clock keeping step with the time signals broadcast will be early with respect to UT1 if a "minus" is broadcast. The corrections are revised as necessary; the new value appears for the first time during the hour after 0000 UTC.

UT1 corrections are also encoded in the time code (described later in connection with Fig. 5) transmitted continuously on a 100-Hz subcarrier from WWV and WWVH. The value of the correction is indicated by the weight of the control functions that occur at the end of the code frame. The "plus and minus" indication is encoded in the first control function. If the first control function is a binary one, the correction is positive; if it is a binary zero, the correction is negative. The correction is expressed to the nearest 0.1 second.

Propagation Forecasts: A forecast of radio propagation conditions is broadcast from WWV in voice at 14 minutes after each hour. The announcements are short-term forecasts for propagation along paths in the North Atlantic area. The announcement consists of the statement, "The radio propagation quality forecast at ... (0100, 0700, 1300, or 1900 UTC) is ... (excellent, very good, good, fair-to-good, fair, poor-to-fair, poor, very poor, or useless). Current geomagnetic activity is ... (quiet, unsettled, or disturbed)." The propagation forecast announcements are repeated in synoptic form, consisting of a phonetic and a numeral. The phonetic identifies the radio quality at the time the forecast is made. The numeral indicates the radio propagation quality expected during the six-hour period after the forecast is issued. The meanings of the phonetics and numerals are:

Phonetic	Meaning
Whiskey	Disturbed
Uniform	Unsettled
November	Normal
Numeral	
One	Useless
Two	Very poor
Three	Poor
Four	Poor-to-fair
Five	Fair
Six	Fair-to-good
Seven	Good
Eight	Very good
Nine	Excellent

For example, if propagation conditions are normal and expected to be good during the next six hours, the coded forecast announcement would be "November Seven."

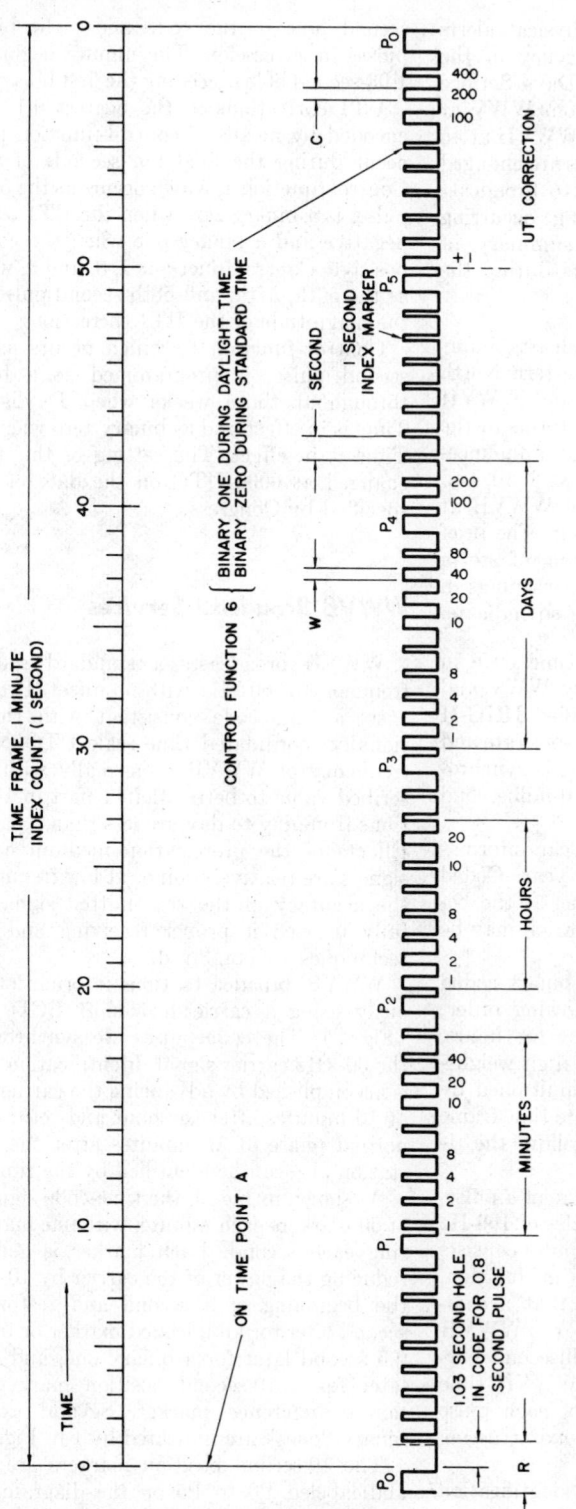

Fig. 5—Time code format of *WWV* and *WWVH*.

TIME FRAME 1 MINUTE
INDEX COUNT (1 SECOND)

CONTROL FUNCTION 6 { BINARY ONE DURING DAYLIGHT TIME
BINARY ZERO DURING STANDARD TIME

1 SECOND
INDEX MARKER P₅

1 SECOND

UTI CORRECTION

ON TIME POINT A

1.03 SECOND HOLE
IN CODE FOR 0.8
SECOND PULSE

R   FRAME REFERENCE MARKER (P₀ AND 1.03 SECOND "HOLE")

P₀-P₅ POSITION IDENTIFIERS (0.8 SECOND DURATION)

W   WEIGHTED CODE DIGIT (0.5 SECOND DURATION)

C   WEIGHTED CONTROL ELEMENT (0.5 SECOND DURATION)

DURATION OF INDEX MARKERS, UNWEIGHTED CODE AND CONTROL ELEMENTS = 0.2 SECOND

IN THIS EXAMPLE:

UTC AT POINT A = 173 DAYS, 21 HOURS, 10 MINUTES

UTI AT POINT A = 173 DAYS, 21 HOURS, 10 MINUTES, 0.3 SECOND

Geophysical Alerts: Current geophysical alerts declared by the World Warning Agency of the International Ursigram and World Days Service (IUWDS) are broadcast in voice from WWV at 18 minutes after each hour and from WWVH at 45 minutes after each hour. The messages are changed daily at 0400 UTC with provisions to broadcast real-time data alerts of outstanding occurring events. These are followed by a summary of selected solar and geophysical events during the previous 24 hours.

Weather Information: WWV broadcasts information about major storms in the western North Atlantic and eastern North Pacific, and WWVH broadcasts information about major storms in the eastern and central North Pacific. These announcements are given in voice from WWV at 8, 10, and 12 minutes after each hour and from WWVH at 17, 19, and 51 minutes after each hour. The brief messages are designed to tell mariners of storm threats in their areas. If there are no warnings in the designated areas, the broadcasts so indicate.

WWV/WWVH Time Code: The time code in Fig. 5 is transmitted continuously by WWV and WWVH. The code format is a modified IRIG-H time code. The code is produced at a 1-pps rate and is carried on a 100-Hz subcarrier that is synchronous with the code pulses so that 10-millisecond resolution is readily obtained.

The code contains UTC time-of-year information in minutes, hours, and day of year. Coded time-of-day information refers to time at the beginning of the frame. Seconds information may be obtained by counting the pulses.

Each minute contains seven binary-coded decimal (BCD) groups, in the following order: two groups for minutes, two groups for hours, and three groups for day of year. The digit weighting is 1-2-4-8 for each BCD group, multiplied by 1, 10, or 100 as appropriate. A complete time frame is 1 minute. The binary groups follow the 1-minute reference marker.

"On-time" occurs at the leading edge of a pulse. A binary zero pulse consists of 20 cycles of 100-Hz amplitude modulation; a binary one pulse consists of 50 cycles of 100-Hz amplitude modulation. Because of the 10-millisecond hole that accompanies each seconds marker in the WWV/WWVH format, however, the leading 30-millisecond portion of each marker in the WWV/WWVH time code is deleted. The leading edge of each pulse coincides with a positive-going zero-axis crossing of the 100-Hz modulating frequency.

The code contains six position identification markers per minute and a minute reference marker. Each 6-per-minute position identification marker consists of an 0.8-second pulse preceding a code group. The 1-per-minute reference marker consists of a 0.8-second pulse followed by a 1.03-

second hole in the code and eight binary zero pulses in succession. The minute begins with the 1.03-second hole preceding the first binary zero.

UT1 corrections to the nearest 0.1 second are encoded by means of control-function pulses that occur during the final ten seconds of the frame. Control function 1, which occurs as the 50th second pulse, is a binary zero when the UT1 correction is negative and a binary one when the correction is positive. Control functions 7, 8, and 9, which occur as the 56th, 57th, and 58th second pulses, identify the magnitude of the UT1 correction.

Control function 6, which occurs as the 55th second pulse, is programmed as a binary one throughout those weeks when Daylight Saving Time is in effect and as binary zero when Standard Time is in effect. The setting of this function is changed at 0000 UTC on the date of change as specified by Congress.

WWVB Broadcast Services

WWVB broadcasts a standard radio carrier frequency of 60 kHz with no offset. It also broadcasts a time code consistent with the internationally coordinated time scale UTC(NBS). The frequency of WWVB is normally within its prescribed value to better than 1 part in 10^{11}. Deviations from day to day are less than 5 parts in 10^{12}. Effects of the propagation medium on received signals are relatively minor at low frequencies, and the accuracy of the transmitted signals may be fully utilized if proper receiving and averaging techniques are employed.

WWVB broadcasts time information continuously using a carrier-level-shift BCD time code (Fig. 6). The code pulses are synchronized with the 60-kHz carrier signal. Identification of WWVB is accomplished by advancing the carrier phase 45° at 10 minutes after the hour and returning to the normal phase at 15 minutes after the hour. The station also can be identified by the time code.

As shown in Fig. 6, the time-code signal consists of 60 markers each minute, with one marker occurring each second. Each marker is generated by reducing the power of the carrier by 10 decibels at the beginning of a second and restoring it 0.2 second later for an uncoded marker or binary zero, 0.5 second later for a binary one, and 0.8 second later for a 10-second position marker or for a minute reference marker. Several examples of binary "ones" are indicated by I in Fig. 6.

The 10-second position markers are blackened and labeled P0 to P5 on the diagram. Uncoded markers occur periodically in the 5th, 15th, 25th, 35th, 45th, and 55th seconds of each minute; they are shaded in the figure. Other uncoded markers occur in the 11th, 12th, 21st, 22nd, 36th, 56th, 57th, 58th, and 59th seconds as indicated by *U* in

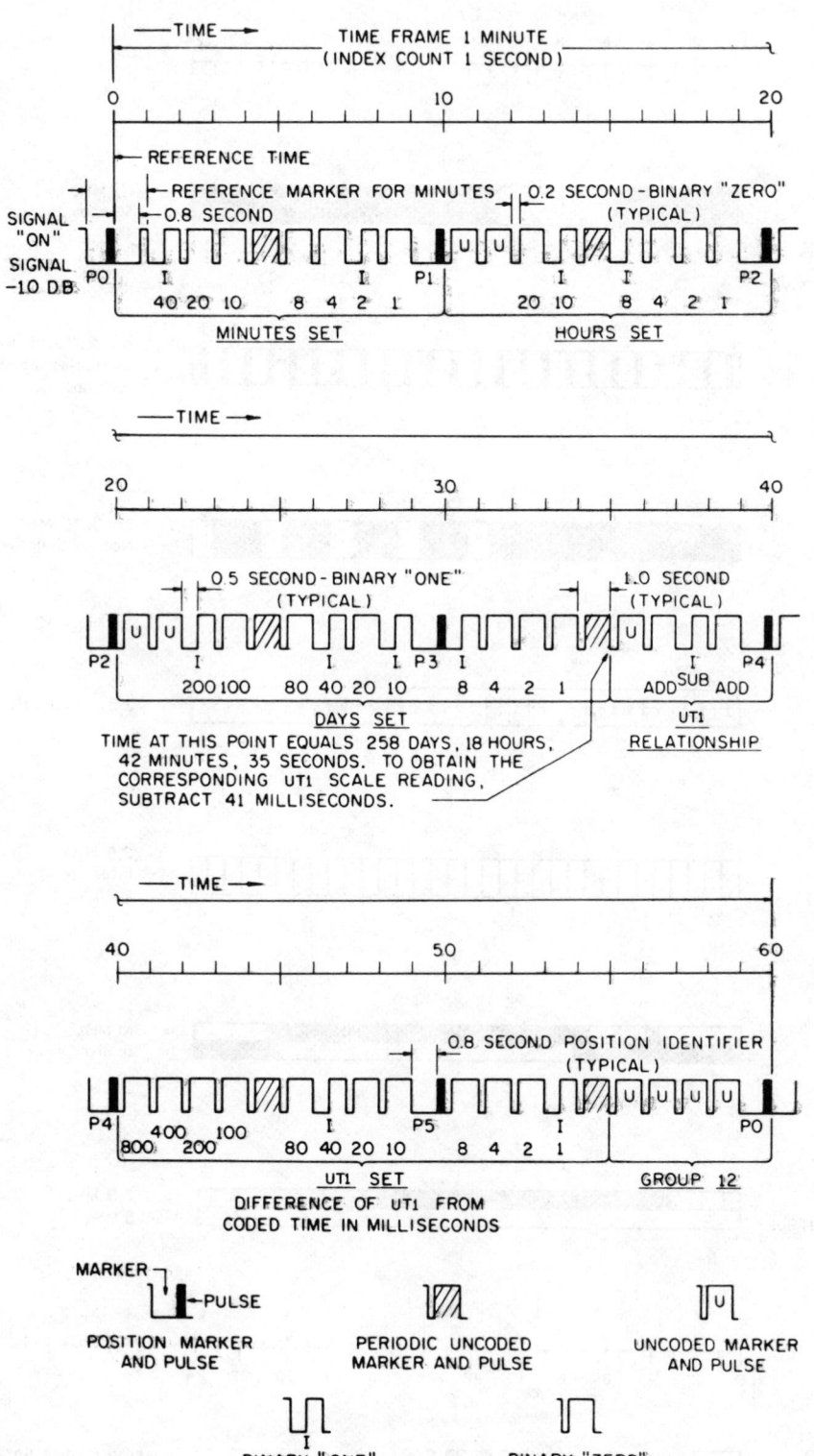

Fig. 6—Time code format of *WWVB*.

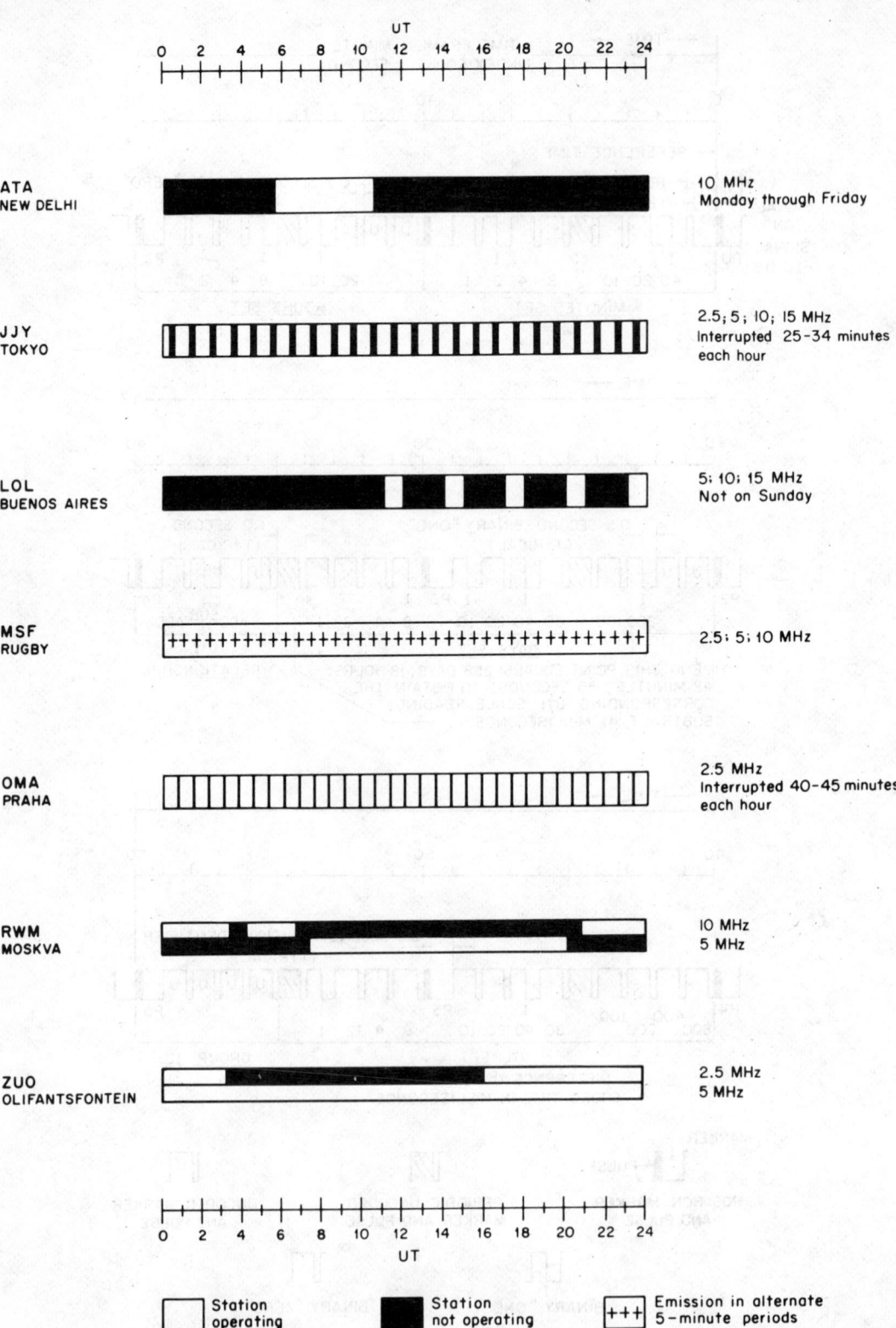

Fig. 7—Daily emission schedules of standard frequency and time stations (other than National Bureau of Standards) operating on high frequency.

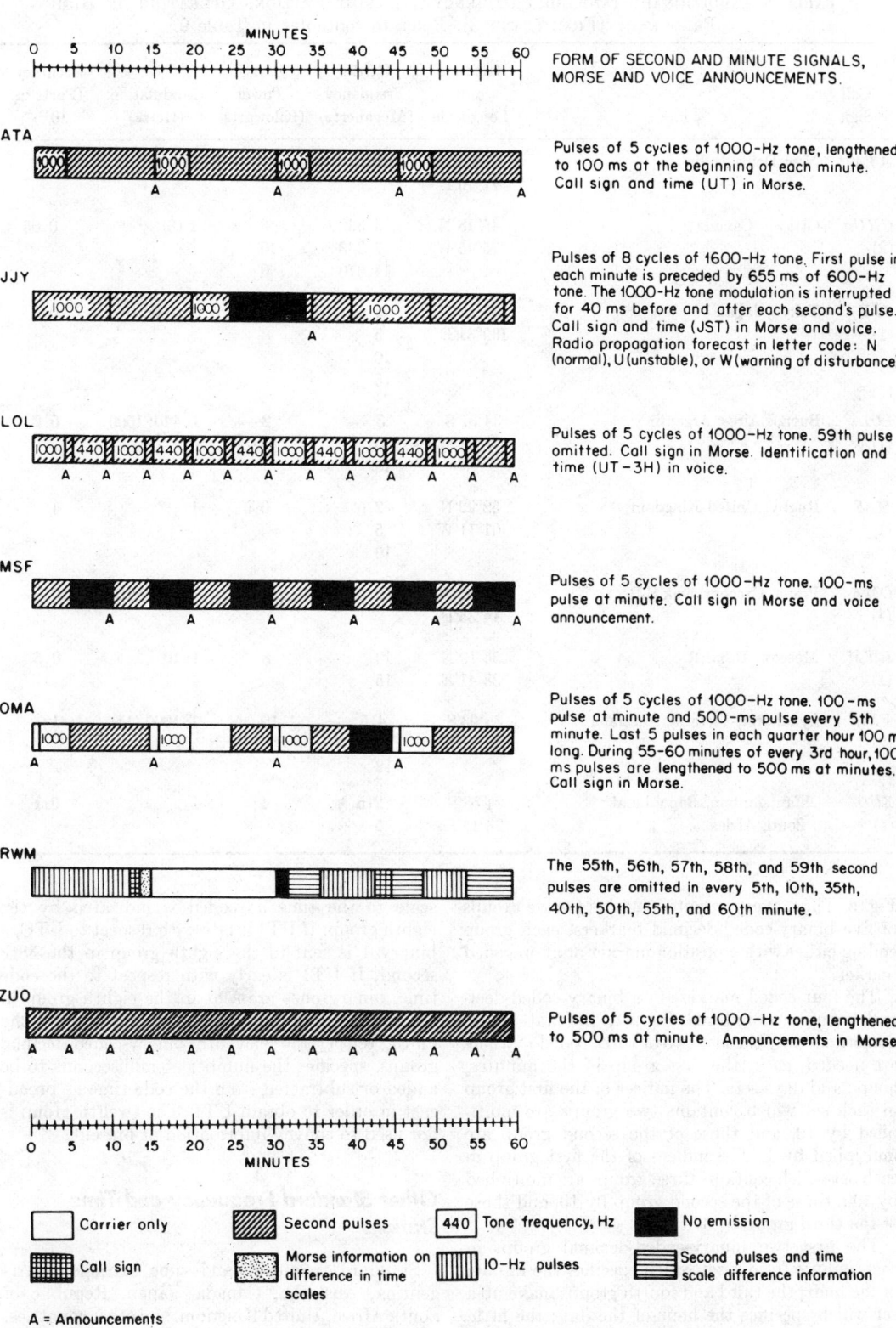

Fig. 8—Hourly modulation schedules of stations in Fig. 7.

TABLE 8—ADDITIONAL STANDARD FREQUENCY AND TIME STATIONS OPERATING ON HIGH FREQUENCY (FIGS. 7 AND 8). Refer to footnotes in Table 9.

Call Sign	Place	Latitude and Longitude	Carrier Frequency (Megahertz)	Carrier Power (Kilowatts)	Modulation (Hertz)	Accuracy (Parts in 10^{10})
ATA	New Delhi, India	28°34′N 77°19′E	10	2	1; 1000	100
CHU (2)	Ottawa, Canada	45°18′N 75°45′W	3.330 7.335 14.670	3 10 3	1 (8)	0.05
JJY (1)	Tokyo, Japan	35°42′N 139°31′E	2.5 5 10 15	2	1; 1000 (9)	0.5
LOL (1)	Buenos Aires, Argentina	34°37′S 58°21′W	5 10 15	2	1; 440; 1000	0.2
MSF (1)	Rugby, United Kingdom	52°22′N 01°11′W	2.5 5 10	0.5	1	1
OMA (1)	Praha, Czechoslovak S.R.	50°07′N 14°35′E	2.5	1	1; 1000 (10)	10
RWM (1)	Moscow, U.S.S.R.	55°19′N 38°41′E	10 15	8	1; 10	0.5
VNG (2)	Lyndhurst, Victoria, Australia	38°03′S 145°16′E	4.5 7.5 12	10	1; 1000 (11)	1
ZUO (1)	Olifantsfontein, Republic of South Africa	24°58′S 28°14′E	2.5 5	4	1	0.1

Fig. 6. Thus, every minute contains twelve groups of five binary-coded-decimal markers, each group ending either with a position marker or an uncoded marker.

The four coded markers in a binary-coded-decimal group are indexed 8-4-2-1, in that order. Sometimes only the last two or three coded markers are needed, as in the first group in the minutes, hours, and days sets. The indices of the first group in each set which contains two groups are multiplied by 10, and those of the second group are multiplied by 1. The indices of the first group in each set which contains three groups are multiplied by 100, those of the second group by 10, and those of the third group by 1.

The first two binary-coded-decimal groups in each minute form a set which specifies the minute of the hour; the third and fourth groups make up a set which specifies the hour of the day; the fifth, sixth, and seventh groups form a set which specifies the day of the year. The relationship of the UT1 scale to the time as coded is indicated by the eighth group. If UT1 is late with respect to UTC, a binary 1 is sent in the eighth group in the 38th second. If UT1 is early with respect to the code time, binary ones are sent in the eighth group in the 37th and 39th seconds. A set, made up of the ninth, tenth, and eleventh binary-coded-decimal groups, specifies the number of milliseconds to be added or subtracted from the code time as broadcast in order to obtain UT1. The twelfth group is not used to convey information at present.

Other Standard Frequency and Time Stations

Standard frequency and time stations in Argentina, Australia, Canada, Japan, Republic of South Africa, United Kingdom, and other countries, as well as the United States Navy, are listed in Tables 8 and 9.

TABLE 9—ADDITIONAL PRINCIPAL STANDARD FREQUENCY AND TIME STATIONS OPERATING ON LOW FREQUENCY AND VERY-LOW FREQUENCY.

Call Sign	Place	Latitude and Longitude	Carrier Frequency (kHz)	Power to Antenna (kW)	Estimated Radiated Power (kW)	Modulation (Hz)	Time Signal	Accuracy (Parts in 10^{10})
DCF 77 (2)	Mainflingen, West Germany	50°01′N 09°00′E	77.5	38	—	1	Continuous (14)	0.1
GBR (2)(3)	Rugby, United Kingdom	52°22′N 01°11′W	15.95 16	750	60 (6)	1 (12)	4×5 min per day (15)	0.2
Loran-C	Carolina Beach, North Carolina	34°04′N 77°55′W	100	—	1000 (7)	(19)	Continuous	0.05
MSF	Rugby, United Kingdom	52°22′N 01°11′W	60	50	—	1 (13)	Continuous	0.1
NAA (2)(4)(5)	Cutler, Maine	44°38′N 67°16′W	17.8	2000	1000 (6)	nil	nil	0.1
NBA (2)(4)(5)	Balboa, Panama Canal Zone	09°03′N 79°38′W	24	300	150 (6)	nil	(16)	0.1
NLK (2)(5)	Jim Creek, Washington	48°12′N 121°55′W	18.6	1200	250 (6)	nil	nil	0.1
NPM (2)(4)(5)	Lualualei, Hawaii	21°25′N 158°09′W	23.4	1000	140 (6)	nil	nil	0.1
NSS (2) (5)	Annapolis, Maryland	38°59′N 76°27′W	21.4	1000	85 (2)	nil	(17)	0.1
OMA	Podebrady, Czechoslovak S.R.	50°08′N 15°08′E	50	5	—	1 (12)	23 hours per day (18)	10

Notes Associated with Tables of Standard Frequency and Time Stations

(1) Station follows UTC system as specified in ITU Recommendation 460-1. Time signals remain within about 0.8 second of UT1 by means of occasional 1-second steps as directed by BIH.

(2) Station follows one of the systems referred to in ITU Recommendation 460-1.

(3) FSK is used alternately with CW; both carriers are frequency controlled.

(4) FSK is used. Phase stable on assigned frequency.

(5) Station primarily for communication purposes; data subject to change.

(6) Estimated radiated power.

(7) Peak radiated power.

(8) Seconds pulses consist of 300 cycles of 1000-Hz tone. The first pulse in each second is prolonged.

(9) Seconds pulses consist of 8 cycles of 1600-Hz tone; first pulse of each minute is preceded by 655 ms of 600-Hz tone. There is 1000-Hz modulation between minutes 0–10, 20–25, 34–35, 40–50, and 59–60 except 40 ms before and after each seconds pulse.

(10) Audio-frequency modulation is replaced by time signals during the period 1800–0600 hours UT.

(11) Seconds pulses consist of 50 cycles of 1000-Hz tone, shortened to 5 cycles from the 55th to the 58th second (59th pulse is omitted). At the 5th, 10th, 15th, etc., minutes, pulses from the 50th to the 58th second are shortened to 5 cycles. Voice identification is given between the 20th and 50th pulses in the 15th, 30th, 45th, and 60th minutes.

(12) A1 telegraphy signals.

(13) The carrier is interrupted for 100 ms at each second and 500 ms at each minute. From 1430 to 1530 hours UT, A2 pulses are transmitted in the same form as for MSF high-frequency transmission.

(14) 100-ms carrier interruptions at beginning of each second except 59th second of each minute. Minute, hour, day, month, year, and day of week are transmitted each minute in BCD code beginning with 20th second and ending with 58th second; binary zero is 0.1-s wide second marker, and binary one is 0.2-s wide marker.

(15) 0255–0300, 0855–0900, 1455–1500, and 2055–2100 hours UT.

(16) Time signal on FSK 5 minutes before each even hour except 2355 to 2400 hours UTC.

(17) Temporarily suspended; FSK time signals planned.

(18) From 1000 to 1100 hours UT, transmissions without keying except for call sign at beginning of each quarter hour.

(19) Pulse repetition, 99 300 microseconds. Time pulses appear in groups of 9.

CHAPTER 2

INTERNATIONAL TELECOMMUNICATION RECOMMENDATIONS

INTERNATIONAL STANDARDS

Administrations and operating companies throughout the world carry on studies of technical and other problems related to the interworking of their respective national telecommunication systems to provide a worldwide telecommunications network. Two international committees exist for this purpose: The International Telegraph and Telephone Consultative Committee (CCITT), and The International Radio Consultative Committee (CCIR). They operate under the auspices of the International Telecommunication Union (ITU). They promulgate their decisions in the form of Recommendations, which are published by the ITU. Generally, these Recommendations cover features of international circuits, but where essential, they deal with relevant characteristics of the national systems which may form part of international connections. This compendium collects, in condensed form, major Recommendations dealing with telephone, telegraph, and data-transmission circuits and equipment.

Recommendations of the CCITT

The CCITT develops new Recommendations, and updates existing ones, through the activities of Study Groups, whose reports are acted on at Plenary Assemblies, which meet at intervals of 3 or 4 years. The resulting Recommendations of the Second Plenary Assembly, New Delhi, 1960, were published by the ITU in a number of volumes, called collectively the Red Book. The subsequent study periods culminated in the Third Plenary Assembly, Geneva, 1964 (Blue Book); the Fourth Plenary Assembly, Mar del Plata, 1968 (White Book); and the Fifth Plenary Assembly, Geneva, 1972 (Green Book). This compendium refers to Green Book Recommendations, designated thus: (G.101), (H.31), (V.2), etc.

Recommendations of the CCIR

The CCIR also functions with Study Groups and Plenary Assemblies. The Eleventh Plenary Assembly was held at Oslo in 1966, the Twelfth Plenary Assembly at New Delhi in 1970, and the Thirteenth Plenary Assembly at Geneva in 1974. After each Plenary Assembly, the ITU publishes volumes which contain the currently accepted Recommendations, including such Recommendations of the Plenary Assemblies at London (1953), Warsaw (1956), Los Angeles (1959), and Geneva (1963) which are still in effect. No color coding is used. This compendium deals with those Recommendations which treat point-to-point radio relay systems. A purpose of those Recommendations is to make the performance of such systems compatible with metallic line systems which follow the CCITT Recommendations. References to the CCIR Recommendations are made thus: (CCIR, 391).

ZERO-RELATIVE-LEVEL POINTS AND RELATIVE LEVELS

Many CCITT and CCIR Recommendations specify signal or noise levels at "a point of zero relative level," or in dBm0 or pWp0, etc., where "0" (zero) stands for "measured at or referred to a point of zero relative level."

(A) In 2-wire switching systems, the sending end terminals of a long-distance circuit have long been considered to be at a point of zero relative level. The relative levels of all other points are calculated from this reference point, as the algebraic sum of all transmission losses and gains from it to the point in question. Any point in a circuit with the same relative level as the sending terminals is a point of zero relative level, which may be written 0 dBr (dB relative level). The American term for relative level is transmission level. Thus: "Zero-transmission-level point" (0TLP).

For convenience in comparing circuit noise performance, it is customary to convert absolute noise measurements made at the receiving ends of circuits having various relative levels, to absolute power levels at a zero-relative-level point. For example, −50 dBmp of noise at a −7–dBr point would be reported as −43 dBm0p. Signaling-tone levels are

TABLE 1—RELATIVE LEVELS, PLUS LEVELS OF ABSOLUTE AND REFERRED POWER, FOR A 2-WIRE CIRCUIT.

	Sending Point	Circuit Loss: 5 dB Direction→	Receiving Point
	O——————————————————————O		
		2-Wire Circuit	
Relative levels, or transmission levels	0 dBr		−5 dBr
"Milliwatt" test tone	0 dBm		−5 dBm absolute power
At any point	0 dBm0		0 dBm0 referred power
Signaling tone	−10 dBm		−15 dBm absolute power
At any point	−10 dBm0		−10 dBm0 referred power
Circuit noise, picked up along circuit	—		−65 dBmp absolute power
Circuit noise, referred to 0-dBr point	—		−60 dBm0p referred power

similarly expressed. For example, a tone introduced at a −3.5–dBr point with an absolute power level of −18.5 dBm may be referred to as a −15–dBm0 signal. The latter designation would apply to such a tone no matter where it appeared; the "0" denotes that its level is referred to a point of zero relative level. (Refer to Table 1.)

Statistics of speech power, requirements for linearity and limiting, system loading factors, crosstalk, and noise have become well known in terms of their values at points of zero relative level. The proper performance of voice repeaters, carrier terminal and line equipment, radio relay systems, etc., depends on adherence to the relative levels for which they were designed. Many relative levels associated with such equipment have been standardized.

(**B**) In 4-wire switching systems, it is often considered desirable to handle speech and signaling at lower values of absolute power through the switching equipment than is customary in 2-wire systems. In 1964, the CCITT adopted a relative level of −3.5 dBr for the sending end of a 4-wire circuit, at the "virtual" switching points. These are theoretical points; their exact location depends on national practice, and the CCITT considers it unnecessary to define them. (In the American commercial system, −2 dBr is widely used.) (G.101, Section B)

Therefore, to ensure that carrier and other transmission equipment will be subjected to the same absolute speech and signaling power levels as in 2-wire systems, determination of relative levels in 4-wire circuits must take into account the relative level of the virtual switching points. In a 4-wire circuit, there may be no actual point of zero relative level. Nevertheless, standards will continue to refer many requirements to a zero relative point. (Refer to Table 2.)

Currently, many transmission measurements are made with a standard 800- or 1000-hertz test tone, with an absolute power of 1 milliwatt at a zero-

TABLE 2—RELATIVE LEVELS, PLUS LEVELS OF ABSOLUTE AND REFERRED POWER, FOR ONE DIRECTION OF A 4-WIRE CIRCUIT.

	Sending Point	Circuit Loss: 5 dB Direction→	Receiving Point
	O——————————————————————O		
		One Side of 4-Wire Circuit	
Relative levels, or transmission levels	−3.5 dBr		−8.5 dBr
"Milliwatt" test tone	−3.5 dBm		−8.5 dBm absolute power
At any point	0 dBm0		0 dBm0 referred power
Signaling tone	−13.5 dBm		−18.5 dBm absolute power
At any point	−10 dBm0		−10 dBm0 referred power
Circuit noise, picked up along circuit	—		−68.5 dBmp absolute power
Circuit noise, referred to 0-dBr point	—		−60 dBm0p referred power

relative-level point: a power of 0 dBm0. The actual level applied is adjusted to the relative level of the sending point. The test-tone level in dBm will be numerically equal to the relative level in dBr at any point in the circuit, but it is not proper to express *relative* levels in dBm, since dBm represents absolute power levels. If the standard-test-tone power is ever changed to another value, such as −10 dBm0, as has been tentatively proposed, the distinction between relative levels and test-tone levels will be more apparent.

PSOPHOMETRIC NOISE AND POWER

Psophometric Noise Weighting

The CCITT calls a noise measuring set a "psophometer." A psophometer includes a device for measuring power through a weighting network. For measurements on commercial telephone circuits, a weighting characteristic is used which results in the objective instrument measurements approximately parallelling the results of subjective tests with human observers using modern telephone sets. The CCITT weighting characteristic for commercial circuits is nominally identical with the American F1A line weighting. Psophometric noise power may be expressed in dBm0 "psophometrically weighted," or dBm0p. The conventional conversion equation used between dBm0p and dBa0 (F1A) is

$$dBm0p = dBa0 - 84.$$

Psophometric Weighting for Commercial Telephone Circuits:

Frequency (hertz)	Level (dB)	Frequency (hertz)	Level (dB)
100	−41.0	1350	−0.65
150	−29.0	1500	−1.30
200	−21.0	1750	−2.22
250	−15.0	2000	−3.00
300	−10.6	2250	−3.60
400	−6.3	2500	−4.20
500	−3.6	2750	−4.87
600	−2.0	3000	−5.60
700	−0.9	3500	−8.5
800	0.0	4000	−15.0
900	+0.6	4200	−18.7
1000	+1.0	4500	−25.0
1100	+0.6	4700	−29.4
1200	0.0	5000	−36.0

(Extracted from G.223, which gives weighting for 84 frequencies.)

Psophometric Weighting Factor: If uniform-spectrum random noise is measured in a 3.1-kilohertz band with a flat attenuation/frequency characteristic, the noise level must be reduced by 2.5

decibels to obtain the psophometric power level. For another bandwidth B, the weighting factor will be equal to

$$2.5 + 10 \log_{10}(B/3.1) \text{ decibels.}$$

When $B = 4$ kilohertz, for example, this gives a weighting factor of 3.6 decibels. (G.223)

Psophometric Power: Where power addition of noise can be assumed, it has been found convenient for calculations and design of international circuits to use the concept of "psophometric power."

psophometric power

$= $ (psophometric voltage)^2/600

$= $ (psophometric emf)^2/(4×600).

A convenient unit is the picowatt (pW) $= 10^{-12}$ watt, so that

psophometric power in pW

$= $ (psophometric voltage in mV)^2/0.0024. (G.212)

CONVENTIONAL TELEPHONE SIGNAL

For the calculation or measurement of crosstalk noise between adjacent channels, of the balance return loss for echo, and generally speaking, when it is desired to simulate the speech currents transmitted by a telephone channel, the CCITT recommends the use of a conventional telephone signal. This signal may be produced by passing the output of a generator of a uniform-spectrum random noise signal ("white noise") through a weighting network with a characteristic as shown in Fig. 1. The amount of this signal that appears in another circuit because of crosstalk, etc., is measured with a psophometer or weighted-noise measuring set, with standard psophometric weighting for commercial telephone circuits. (G.227)

TELEPHONE CIRCUIT LOADING

Nominal Mean Power During Busy Hour

To simplify calculations when designing carrier systems on cables or radio links, the CCITT has adopted a *conventional* value to represent the *mean absolute power level*, at a point of zero relative level, of the speech-plus-signaling currents, etc., transmitted over a telephone channel in one direction of transmission during the busy hour, which is −15 dBm (−1.73 nepers) (mean power = 31.6 microwatts); this is the mean with time and the mean for a large batch of circuits. This total mean power of about 32 microwatts is conventionally distributed as follows (nominal mean power): 10 microwatts, all signaling and tones; 22 microwatts,

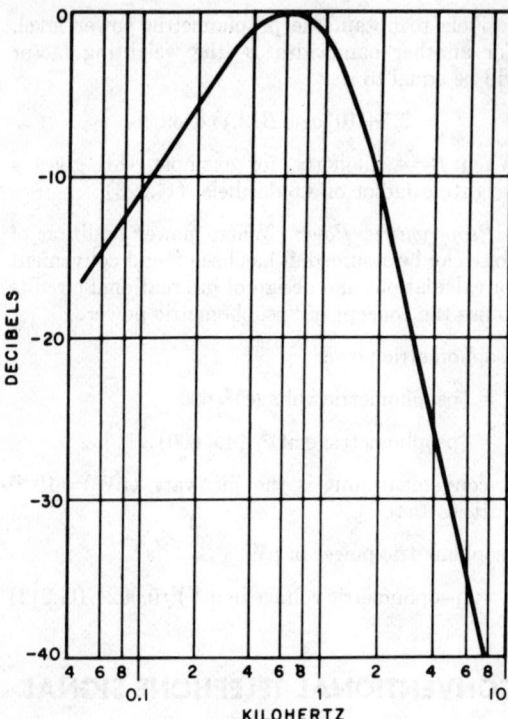

Fig. 1—Relative response curve for the weighting network of the conventional telephone signal generator. (G. 227)

to include speech currents (including echoes), carrier leak, and telegraph signals, based on a speech activity factor of 0.25 for one direction of a telephone channel. No account is taken of pilot signals, which are assumed to be an integral part of the carrier system, not affecting telephone channel power. (G.223, Section 1)

Conventional Load

It will be assumed for the calculation of inter-modulation noise below the overload point that the multiplex signal during the busy hour can be represented by a uniform-spectrum random noise signal, the mean absolute power level of which, at a zero-relative-level point, $n(\bar{P})$, is given by

$$n(\bar{P}) = -15 + 10 \log_{10} N \text{ dB}$$

for $N = 240$ or more

and

$$n(\bar{P}) = -1 + 4 \log_{10} N \text{ dB}$$

for values of N between 12 and 240

where N is the total number of telephone channels in the system.

Typical values so calculated are as follows

N	$n(\bar{P})$, dB
12	3.3
24	4.5
36	5.2
48	5.7
60	6.1
120	7.3
⋮	⋮
240	8.8
300	9.8
600	12.8
960	14.8
1800	17.5
2700	19.3

Assumed: No pre-emphasis, and use of independent amplifiers for each direction. (G.223, Section 2)

POWER LEVELS

Maximum Power Level for Signaling Pulses

For crosstalk reasons, each component of a short-duration signal should not exceed the following absolute power levels, at a zero-relative-level point.

Signaling Frequency (hertz)	Absolute Power Level at Zero-Relative-Level Point (dBm0)
800	−1
1200	−3
1600	−4
2000	−5
2400	−6
2800, 3200	−8

(G.224)

Private Telegraph Transmission on a Rented International Circuit, with Alternative Private Telephone Service

The frequency of 1500 hertz is recommended for private telegraph transmission between subscribers permanently connected via a rented international circuit. The permissible power for a continuously transmitted telegraph marking signal is 0.3 milliwatt at a zero-relative-level point (−5 dBm0). (H.31)

Simultaneous Communication by Telephony and Telegraphy on a Telephone Circuit

A continuously transmitted telegraph signal should not exceed a level of -13 dBm0. There should not be more than 3 telephone circuits of this type per group, nor more than the number of supergroups in a wide-band system. (H.32)

Phototelegraphy Transmissions Over Telephone Circuits which are Entirely 4-Wire Between Phototelegraph Stations

The sent voltage for the phototelegraph signal corresponding to maximum amplitude should be so adjusted that the absolute power level of the signal, at a zero-relative-level point, is 0 dBm0 for amplitude-modulation facsimile, and -10 dBm0 for frequency-modulation facsimile. In the former, the "black" level is about 30 decibels lower than the "white" level. (H.41)

Power Levels for Data Transmission Over Telephone Circuits

Private Wires on Carrier Systems:

(A) Maximum power output of subscriber's apparatus into line: 1 milliwatt.

(B) Continuous-tone systems (for example, frequency modulation): Maximum power level at zero-relative-level point: -10 dBm0, to be reduced to or below -20 dBm0 when data transmission is discontinued for any appreciable time.

(C) Noncontinuous-tone systems (for example, amplitude modulation): Maximum power level at zero-relative-level point: -6 dBm0, provided that busy-hour mean power in both directions of transmission added does not exceed 64 microwatts (-15 dBm0 mean level in each direction simultaneously). Also, the level of tones above 2400 hertz should conform to recommendations for signaling tones in G.224. (H.51)

Switched Telephone Network:

(A) Maximum power output of subscriber's apparatus into line: 1 milliwatt.

(B) Continuous-tone systems (for example frequency- or phase-modulation systems): The power level at the subscriber's equipment should be adjusted not to exceed -10 dBm0 (simplex operation) or -13 dBm0 (duplex operation).

(C) Noncontinuous-tone systems (for example amplitude-modulation or multifrequency systems): Higher levels may be used, if mean power at international-circuit input is limited to 64 microwatts in any hour (both directions), i.e., -15 dBm0 in each direction simultaneously. (H.51)

HYPOTHETICAL REFERENCE CIRCUITS

General Definitions (G.212)

Hypothetical Reference Circuit: A hypothetical circuit of defined length and with a specified number of terminal and intermediate equipments, this number being sufficient but not excessive. It forms the basis for the study of certain characteristics of long-distance circuits, for example, noise.

Hypothetical Reference Circuit for Telephony: This is a complete telephone circuit, between audio-frequency terminals, established on a hypothetical international telephone carrier system and having a specified length and a specified number of modulations and demodulations of the groups, supergroups, and mastergroups, these numbers being reasonably great but not having their maximum possible values.

Various hypothetical reference circuits for telephony have been defined by the CCITT and CCIR to allow the coordination of the different specifications concerning the constituent parts of the multi-channel carrier telephone systems, so that the complete telephone circuits set up on these systems can meet CCITT standards.

Homogeneous Section: A section without diversion or modulation of any one of the mastergroups, supergroups, groups, or channels established on the system being considered, except for those modulations or demodulations defined at the ends of the section. All the hypothetical reference circuits defined below consist of homogeneous sections of equal length (6 or 9 sections). It is assumed that at the end of each homogeneous section, the channels, groups, supergroups, and mastergroups, as appropriate, are connected through at random.

Important Hypothetical Reference Circuits Defined by CCITT and CCIR

Hypothetical Reference Circuit on Symmetric Cable Pairs: This circuit is 2500 kilometers long and is set up on a symmetric-cable-pair carrier system. For each direction of transmission, it has a total of 3 pairs of channel modulators and demodulators, 6 pairs of group modulators and demodulators, and 6 pairs of supergroup modulators and demodulators.

Figure 2A shows that there are 15 modulations and 15 demodulations for each direction of transmission, assuming that single-stage translations are used. There are 6 homogeneous sections of equal length. (G.322)

Hypothetical Reference Circuit for 4-Megahertz Systems on Coaxial Cable: This circuit is 2500 kilometers long and is set up on a 4-megahertz

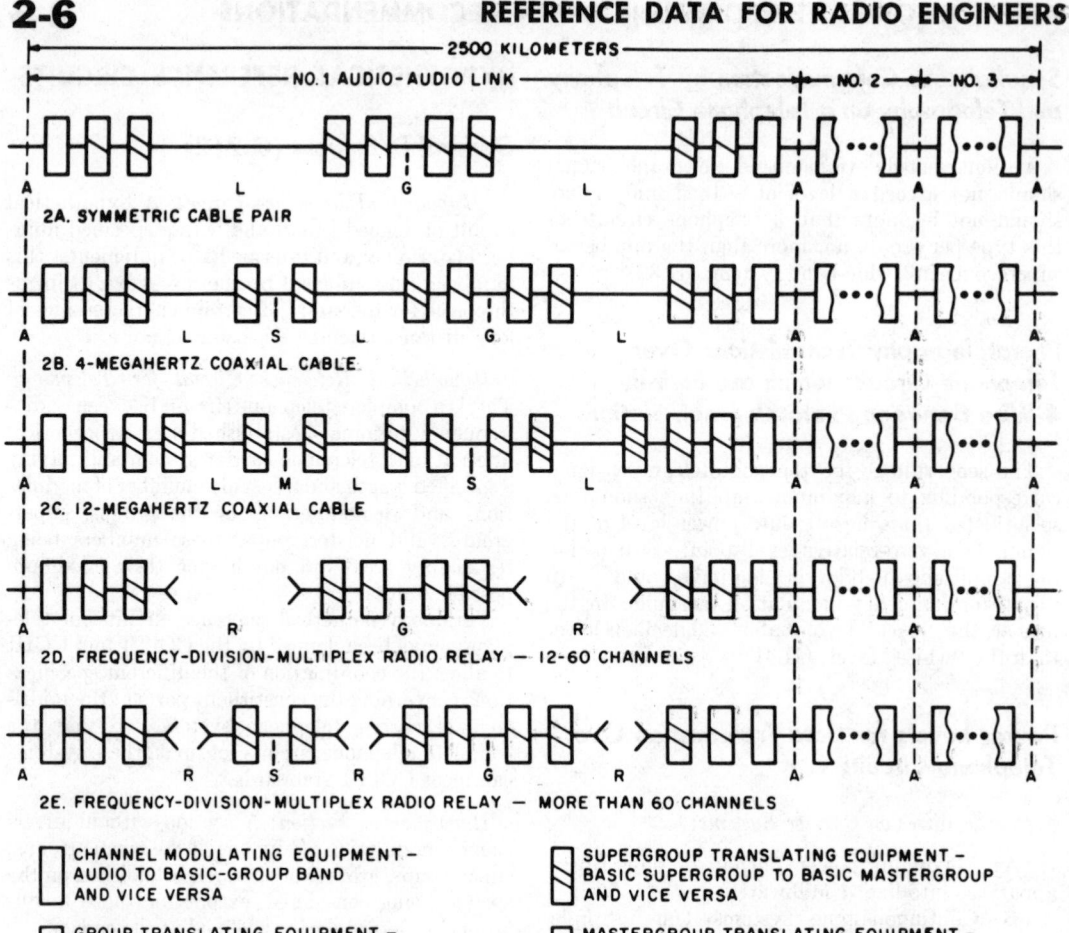

Fig. 2—Typical hypothetical reference circuits for systems using frequency-division multiplex.

carrier system on 0.104/0.375-inch coaxial cable pairs. For each direction of transmission, it has a total of 3 pairs of channel modulators and demodulators, 6 pairs of group modulators and demodulators, and 9 pairs of supergroup modulators and demodulators.

Figure 2*B* shows that there are 18 modulations and 18 demodulations for each direction of transmission, assuming that single-stage translations are used. There are 9 homogeneous sections of equal length. (G.338)

Hypothetical Reference Circuit for 12-Megahertz Systems on Coaxial Cable: This circuit is 2500 kilometers long and is set up on a 12-megahertz carrier system on 0.104/0.375-inch coaxial cable pairs. For each direction of transmission, it has a total of 3 pairs of channel modulators and demodulators, 3 pairs of group modulators and demodulators, 6 pairs of supergroup modulators and demodulators, and 9 pairs of mastergroup modulators and demodulators.

Figure 2*C* shows that there are 21 modulations and 21 demodulations for each direction of transmission, assuming that single-stage translations are used. There are 9 homogeneous sections of equal length. (G.332)

Hypothetical Reference Circuit over Radio Relay Systems with Frequency-Division Multiplex, Providing 12 to 60 Channels: This circuit is 2500 kilometers long and is set up on a carrier system providing 12 to 60 channels per radio channel over line-of-sight and near-line-of-sight radio relay systems. For each direction of transmission it has a total of 3 sets of channel modulators and demodulators, 6 sets of group modulators and demodulators, 6 sets of supergroup modulators and demodulators, and 6 sets of radio modulators and demodulators. The circuit is divided into 6 homogeneous sections of equal length. (See Fig. 2*D*.) (CCIR, 391)

Hypothetical Reference Circuit over Radio Relay

Systems with Frequency-Division Multiplex, Providing More than 60 Channels: This circuit is 2500 kilometers long, and is set up on a carrier system providing more than 60 channels per radio channel over line-of-sight or near-line-of-sight radio relay systems. For each direction of transmission it has a total of 3 sets of channel modulators and demodulators, 6 sets of group modulators and demodulators, 9 sets of supergroup modulators and demodulators, and 9 sets of radio modulators and demodulators. The circuit is divided into 9 homogeneous sections of equal length. (See Fig. 2E.) (CCIR, 392)

Hypothetical Reference Circuit over Transhorizon Radio Relay Systems: This circuit is 2500 kilometers long and is set up over a transhorizon radio relay system for telephony using frequency-division multiplex. For each direction of transmission it has a total of 3 sets of channel modulators and demodulators, 6 sets of group modulators and demodulators, and 9 sets of supergroup modulators and demodulators.

It is recommended that the hypothetical reference circuit not be divided into homogeneous sections of fixed length, because these systems, as distinct from line-of-sight systems, are usually composed of long radio sections, the lengths of which depend on local conditions and may vary considerably (between 100 and 400 kilometers). If a radio section under study is L kilometers long, the hypothetical reference circuit should be composed of $(2500/L)$ sections of this type in tandem, the value $(2500/L)$ being taken to the nearest whole number. (CCIR, 396–1)

Hypothetical Reference Circuit over Active Intercontinental Communication-Satellite Systems. This circuit has no fixed length. For intercontinental connections, satellite links should be capable of spanning 7500 kilometers. For great-circle distances up to 25 000 kilometers, it will be necessary to connect two or three satellite links in tandem. The basic hypothetical reference circuit shall consist of one earth–satellite–earth link, as shown in Fig. 3. It shall contain one pair of modulation and demodulation equipments for translation from the baseband to the radio-frequency carrier and back. (CCIR, 352-2)

TELEPHONE CIRCUIT CHARACTERISTICS

Impedance of International and National Trunks

All circuits (whether international circuits or national 2-wire or 4-wire trunk circuits) terminating at the same trunk exchange should have the same nominal value of impedance as seen from the switchboard or selectors. The preferred value is 600 ohms. (G.232, Section L)

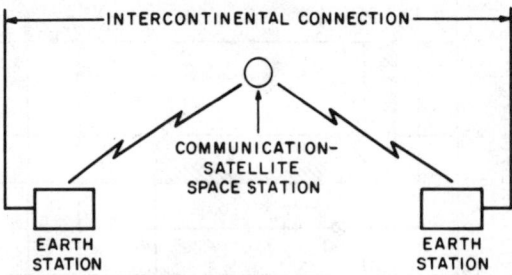

Fig. 3—Basic hypothetical reference circuit, active intercontinental communication-satellite system (CCIR, 352-2).

One-Way Propagation Time

The 1-way propagation time of connections when echo sources exist and echo suppressors are used is considered to be: acceptable without reservation, 0–150 milliseconds; provisionally acceptable, 150–400 milliseconds; and provisionally unacceptable, 400 milliseconds or higher (should not be used except under extraordinary circumstances).

On National Extension Circuits: Probable propagation time, most distant subscriber to international center: 12 plus (0.0064×distance in statute miles) milliseconds or 12 plus (0.004×distance in kilometers) milliseconds.

International Circuits, Terrestrial (Including Submarine Cable): 100 statute miles (160 kilometers) per millisecond, including effects of terminal or intermediate multiplex equipment.

International Circuits, Communication-Satellite Systems: Moving satellite, altitude 8700 miles (14 000 kilometers): 110 milliseconds. Geostationary satellite, altitude 22 500 miles (36 000 kilometers): 260 milliseconds. (G.114)

Group Delay Distortion

The permissible differences, for a worldwide chain of 12 circuits, between the minimum group delay throughout the frequency band transmitted, and the group delay at the upper and lower limits of this band in milliseconds are as follows:

Facility	At the Limits of the Frequency Band	
	Lower Limit	Upper Limit
International chain	30	15
Each of the national 4-wire extensions	15	7.5
Entire 4-wire chain	60	30

(G.113, P.13)

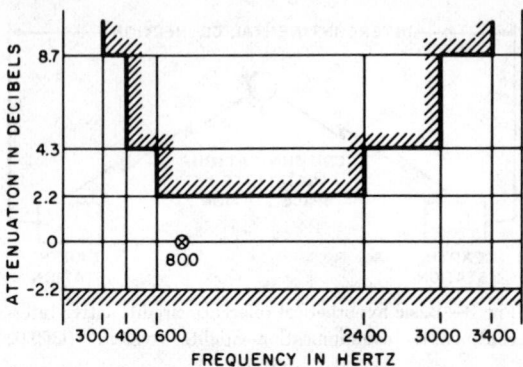

Fig. 4—Attenuation vs. frequency objective for worldwide chain of 12 circuits in terminal service (G. 132).

The CCITT does not recommend limiting values for the group delay at different frequencies, but offers information on typical values reported by four Administrations. Values are in milliseconds measured on a pair of equipments.

Administration	Frequency (hertz)				
	300	400	2000	3000	3400
Belgium (b)	4.0	2.7	1.0	1.3	2.6
France (b)	4.2	2.9	1.0	1.4	2.8
Federal Republic of Germany (b)	3.9	2.7	1.2	1.6	3.0
United Kingdom (a)	2.6	2.2	1	1.4	2.6
(b)	4.2	2.7	1.2	1.8	3.4
For 12 pairs (c)	50	35	14	22	41

(a) With in-band signaling.
(b) With out-band signaling.
(c) Typical maximum values in milliseconds for 12 pairs of equipments.

(G.232, Section C)

Attenuation Distortion in Worldwide Chain

The objectives for the variation with frequency of transmission loss in the terminal condition of a worldwide chain of 12 circuits (international plus national extensions), each one routed over a single group link, are shown in Fig. 4, which assumes that no use is made of high-frequency radio circuits or 3-kilohertz channel equipment. (G.132)

Variation of Transmission Loss with Time

(A) The standard deviation for the variation in transmission loss of a circuit should not exceed 1 decibel. This can be met for circuits on single-group links with automatic regulation, and it should be met on any national circuit whether regulated or not. For other international circuits, the standard deviation should not exceed 1.5 decibels. Standard deviation is sometimes called "distribution grade."

(B) The difference between the mean value of the transmission loss and the nominal value (that is, the bias) should not exceed 0.5 decibel. (G.151, Section C)

Linear Crosstalk

(A) Between circuits, the near-end or far-end crosstalk ratio (intelligible only), measured at audio frequency at a national or international center between two complete circuits in terminal service, should not be less than 58 decibels for 90 percent of all 2-circuit combinations, and 52 decibels for all combinations. 58 dB is the planning objective for modern systems for all combinations.

(B) Between the go and return channels of a 4-wire circuit, the near-end crosstalk ratio must be at least 43 decibels.

Note: The above refers to telephone circuits which are not equipped with (or used in conjunction with) modern echo suppressors designed for long propagation times. Circuits which can form part of switched connections with a long propagation time and lie between terminal half-echo suppressors of modern design should conform to high standards. (G.151, Section D)

Frequency Accuracy of Virtual Carrier Frequencies on an International Circuit

As the channel of any international telephone circuit should be suitable for voice-frequency telegraphy, the accuracy of the virtual carrier frequencies should be such that the difference between an audio frequency applied to one end of the circuit and the frequency received at the other end should not exceed 2 hertz, even when there are intermediate modulating and demodulating processes.

Considering either the hypothetical reference circuit proper to each system or the worldwide hypothetical reference connection, the accuracy of restitution of frequency which is recommended above should be assured if the channel and group carrier frequencies of the various stages have the following accuracies.

Virtual channel carrier frequencies in a group	$\pm 10^{-6}$
Group and supergroup carrier frequencies	$\pm 10^{-7}$
Mastergroup and supermastergroup carrier frequencies	
for 12-MHz system	$\pm 5 \times 10^{-8}$
for 60-MHz system	$\pm 10^{-8}$

(G.225, Section a)

For 12-channel open wire carrier systems, channel carriers (G.311) $\pm 5 \times 10^{-6}$

CIRCUIT NOISE OBJECTIVES

Noise Characteristics for 2500-Kilometer Hypothetical Reference Circuits

The following design objectives should apply to any telephone channel provided by multichannel carrier systems on cable and radio relay links, having the same composition as the appropriate hypothetical reference circuit of 2500 kilometers. The noise is to be measured at or referred to a point of zero relative level. The objectives are intended to ensure adequate speech and signaling performance. They include an allowance of 2500 pWp0 due to frequency-division-multiplex equipment. (See Fig. 5.)

(**A**) The mean psophometric power during any hour shall not exceed 10 000 pWp0.

(**B**) The mean noise power over 1 minute shall not exceed 10 000 pWp0 for more than 20 percent of any month.

(**C**) The mean noise power over 1 minute shall not exceed 50 000 pWp0 for more than 0.1 percent of any month.

(**D**) The unweighted noise power, measured or calculated with an integrating time of 5 milliseconds, shall not exceed 1 000 000 pW0 (10^6 pW0) for more than 0.01 percent (10^{-4}) of any month. However, this is reduced to 0.001 percent (10^{-5}) of any month, or 0.1 percent of any hour, if the telephone channel is to be used for 50-baud amplitude-modulated voice-frequency telegraph, if recommended quality of telegraph operation is to be obtained. This stiffer requirement need not apply to frequency-modulated voice-frequency telegraph systems operating at 50 bauds.

In a part of a hypothetical reference circuit consisting of one or more homogeneous sections, the values of noise power in (**A**) and (**B**) above, and the small percentages of time in (**C**) and (**D**) above, shall be considered to be proportional to the number of homogeneous sections involved.

These are design objectives, "and it is not intended that they be quoted in specifications for equipment or used for acceptance tests." (G.222, and CCIR, 393-2)

Noise Characteristics for Long Circuits Not More than 2500 Kilometers in Length

Circuits on Land, in Submarine-Cable Systems, or on Radio Relay Systems: The mean psophometric power during any hour should be of the order of 4 pWp0 per kilometer, including noise due to frequency-division-multiplex equipment, except for very short circuits or those with a very complicated composition. (G.151, Section A)

Circuits on Transhorizon Relay Systems: The CCIR divides transhorizon systems into two classes, from the viewpoint of performance.

Transhorizon systems of the first class are those intended to operate between two points which might, without excessive difficulty, be connected by line-of-sight radio relay or cable systems. The hypothetical reference circuit described for transhorizon systems applies to this class. The noise power at the end of this circuit will be calculated by statistical combination of the noise power in each of its radio sections. The statistical noise distribution curve should meet requirements (**A**) through (**D**) above.

Transhorizon systems of the second class are those used between points for which other transmission systems cannot be used without excessive difficulty. If the design objectives given above cannot be met without excessive difficulty, the calculated noise-power distribution should meet the following objectives: The mean psophometric power during 1 minute must not exceed 25 000 pWp0 for more than 20 percent of any month, nor exceed 63 000 pWp0 for more than 0.5 percent of any month. Also, the unweighted noise power (integrated over 5 milliseconds) stated in (**D**) above, should not be exceeded for more than 0.05 percent of the most unfavorable month.

The objectives for both classes are provisional. (CCIR, 397–2)

Noise Characteristics for International Circuits More than 2500 Kilometers in Length

Circuits in Cable or Radio Relay Systems, with no Long Submarine-Cable Section: Such circuits, be-

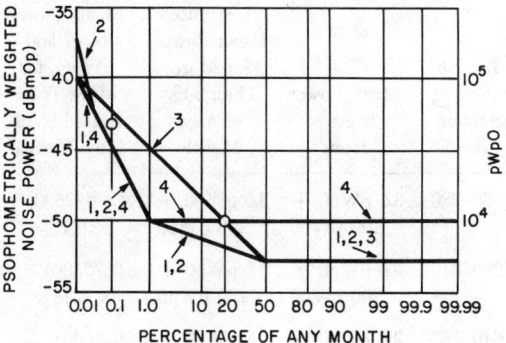

THE NOISE FIGURES INCLUDE 2500 pW FOR TERMINAL EQUIPMENT

O = DESIGN OBJECTIVES, INCLUDING TERMINAL NOISE

THE NUMBERS 1 TO 4 ARE USED TO DISTINGUISH THE CURVES

Fig. 5—Examples of distribution curves for the psophometrically weighted 1-minute mean noise power at the end of the hypothetical reference circuit (CCIR, 393–2).

tween 2500 and 25 000 kilometers in length, are generally carried in land cable or radio relay systems designed to 2500-kilometer objectives, and the number of channel demodulations seldom exceeds that in the corresponding part of the worldwide hypothetical reference chain. Automatic regulation should be used on each group link. The noise objectives are:

(**A**) The mean psophometric noise power during any hour, due to the line, should not exceed 3 pWp0 per kilometer; preferably 2 pWp0 per kilometer or even less. It is noted that in some countries, long overland systems (5000 kilometers or more) have the same objectives as the submarine-cable system (1 pWp0/km).

(**B**) For systems up to about 7500 kilometers long, the 1-minute mean power should not exceed 50 000 pWp0 for more than 0.3 percent of any month.

(**C**) For systems up to about 7500 kilometers long, the unweighted noise power, measured or calculated with an integrating time of 5 milliseconds, should not exceed 1 000 000 pW0 for more than 0.03 percent of any month.

Note: (**B**) and (**C**) are prorated from 2500-kilometer objectives, and proportional values should be used for lengths between 2500 and 7500 kilometers. The CCITT does not yet recommend short-term objectives for systems longer than 7500 kilometers. (G.153, Section A)

Circuits with a Long Submarine-Cable Section: The circuit noise attributable to the submarine section, without compandors, should not exceed a mean psophometric noise value of 3 pWp0 per kilometer in the worst hour on the worst channel. The mean noise power for each direction of transmission, over all channels used for the longest circuits, should not exceed 1 pWp0 per kilometer. No objectives are given for circuits equipped with compandors. The other parts of the circuit should conform to normal 2500-kilometer objectives. (G.153, Section B)

Circuits on Communication-Satellite Systems: The psophometrically weighted noise power at a point of zero relative level in any telephone circuit in the basic hypothetical reference circuit (as defined in Recommendation 352) should not exceed the provisional values of 10 000 pWp0 mean in any hour, 10 000 pWp0 one-minute mean for more than 20 percent of any month, 50 000 pWp0 one-minute mean for more than 0.3 percent of any month, and 1 000 000 pW0 (5-millisecond integrating time) for more than 0.03 percent of any month. The multiplex noise is not included in these figures. (CCIR, 353-2)

Circuits on Open Wire Carrier Systems: In systems of about 10 000 kilometers in length, with lines of strict regularity of construction, accurately operating automatic line regulators, adjustment of line levels to take account of special climatic conditions, and carefully chosen repeater spacings, the following objectives may apply: The mean psophometric power, during any hour, at the end of a circuit of about 10 000 kilometers, taking into account all noise that exists, with the exception of noise due to radio transmitters, should not exceed 50 000 pWp0. This is for a reasonable distribution of wet weather in the territory crossed by the circuit. (G.153, Section D)

Noise in Actual Circuits

Radio Relay Circuits, 280 to 2500 Kilometers in Length: In a telephone channel of an actual radio relay system using frequency-division multiplex, whose composition does not differ appreciably from the hypothetical reference circuit and whose length L is between 280 and 2500 kilometers, the psophometrically weighted noise power, excluding multiplex noise, should not exceed (A) $3L$ pWp0 mean power in any hour, (B) $3L$ pWp0 one-minute mean power for more than 20 percent of any month, and (C) 47 500 pWp0 one-minute mean power for more than $(L/2500) \times 0.1$ percent of any month, as a planning objective. (CCIR, 359-3)

Radio Relay Circuits, Over Real Links, 50 to 2500 Kilometers in Length: If, for planning reasons, the composition of a real link differs substantially from the hypothetical reference circuit, the noise power of a circuit of length L, carried in one or more baseband sections of frequency-division-multiplex radio links, should not exceed the following values.

Length of System (km)	Mean Power in Any Hour	1-Minute Mean Power for More Than 20% of Any Month	1-Minute Mean Power of 47 500 pWpO, for More Than Stated % of Any Month
50–280	$3L$ pWp0 + 200 pWp0	$3L$ pWp0 + 200 pWp0	(280/2500) × 0.1%
280–840	$3L$ pWp0 + 200 pWp0	$3L$ pWp0 + 200 pWp0	(L/2500) × 0.1%
840–1670	$3L$ pWp0 + 400 pWp0	$3L$ pWp0 + 400 pWp0	(L/2500) ×0.1%
1670–2500	$3L$ pWp0 + 600 pWp0	$3L$ pWp0 + 600 pWp0	(L/2500) × 0.1%

Note 1: Frequency-division-multiplex noise is not included.

Note 2: The hourly mean-noise-power objective and its subdivision are under study.

National Circuits on Carrier Systems over Very Short Distances: Assuming that circuits in an international connection, making use of frequency-division-multiplex carrier systems over very short distances, can be limited in number to 4, the mean psophometric power should not exceed 2000 pWp0 per circuit during any hour, including crosstalk.

For CCITT Recommendations for pulse-code-modulation systems see G.123.

Design Objectives for Noise Produced by Modulating Equipments

The mean psophometric power, which corresponds to the noise produced by all modulating equipment mentioned in the definition of the hypothetical reference circuit in question, should not exceed 2500 pWp0. This value includes noise due to various causes, such as thermal noise, intermodulation, crosstalk, power supplies, etc. Its allocation between the various equipments can be left somewhat to the discretion of designers, but the following values are given as a guide to the target design values.

	pWp0
One pair of channel modulators	200–400
One pair of group modulators	60–100
One pair of supergroup modulators	60–100
One pair of mastergroup modulators	40–60

(G.222, Section d)

CCITT AND TELEGRAPHY

The CCITT Blue Book contains the Recommendations adopted by the Third Plenary Assembly in Geneva, in 1964. The Recommendations on Telegraph Technique are included in Volume VII, and those on Data Transmission are included in Volume VIII. The Recommendations on Telegraph Operations and Tariffs are contained in Volume II of the Red Book and in Documents AP III–64, –67, and –74. The latter (AP III–74) has been importantly revised by CCITT Circular No. 15 dated 12 November 1964 entitled "List of Destination Indicators."

Numbering

There is a worldwide system of Destination Indicators for the telegraph-message retransmission network. These indicators consist of two letters signifying the country and its telegraph network (if more than one) followed by two letters signifying the town on that network. Examples: Vienna AUWI, Panama City (Tropical Radio) PAPA, Balboa (ITTCACR) PZBA, Stockholm SWSM, San Francisco (ITT Worldcom) UISF.

The CCITT has approved a worldwide numbering system for telex services. The telex destination code consists of 2 or 3 numerical digits signifying the country or network within the country. The destination code is followed by the telex subscriber's national number, also consisting of numerical digits.

The telex system provides also for *designation* codes, for identifying the country and network of

TABLE 3—CCITT SIGNALING SYSTEMS.

No.	Systems
1	500/20-hertz system used in the international manual service (ringdown).
2	600/750-hertz 2-frequency system. Never used in international service.

International Automatic and Semiautomatic Systems

No.	Systems
3	For unidirectional operation of circuits. Uses 1 in-band frequency (2280 hertz) for the transmission of both line and interregister signals; used for terminal traffic; in general not to be used for new installations.
4	For unidirectional operation of circuits (circuits seized from one end only). Uses 2 in-band frequencies (2040 and 2400 hertz) for the end-to-end transmission of both line and register signals; used for international intracontinental traffic; suitable for terminal and transit traffic; in the latter case 2 or 3 circuits equipped with System No. 4 may be switched in tandem. Suitable for submarine- or land-cable circuits and microwave radio circuits; not applicable to TASI-equipped systems. Capable of interworking with System No. 5.
5	For both-way operation of circuits. Uses 2 in-band signaling frequencies (2400 and 2600 hertz) for the link-by-link transmission of line signals, and 6 in-band frequencies (700, 900, 1100, 1300, 1500, and 1700 hertz) in a 2-out-of-6 code (numerical information transmitted *en bloc*) for the link-by-link transmission of register signals; used for intercontinental traffic. Suitable for submarine- or land-cable circuits and microwave links, whether or not TASI is used; suitable for terminal or transit traffic—in the latter case, 2 or more circuits equipped with System No. 5 may be switched in tandem but are subject to possible undesirable delays if all are TASI-equipped. Capable of interworking with System No. 4.
6	A proposed system to be free from some limitations of Systems No. 3, 4, and 5; expected to use voice channel for interregister signaling, plus a separate channel for line signaling and "management" signaling (changing of routing, et cetera); not expected to be in use before 1970.

TABLE 4—LINE SIGNALS IN CCITT SYSTEMS.

Signal	Direction	CCITT No. 3 (1VF)	CCITT No. 4 (2VF)	CCITT No. 5 (2VF)
Seize, terminal	↑	X	PX	X
Seize, transit	↑		PY	
Start pulsing, terminal	↓	X	X	Y
Start pulsing, transit	↓		Y	
End of pulsing (ST)	↑	250 ms*	xSxSx*	1500+1700 hertz*
Busy	↓	XX	PX	Y
Acknowledge	↑			X
Answer	↓	XSX	PY	X
Acknowledge	↑			X
Clear back (on-hook)	↓	XX	PX	Y
Acknowledge	↑			X
Ring forward	↑	XSX	PYY	Y(850±200 ms)
Clear forward (disconnect)	↑	XXSXX	PXX	X+Y
Release guard (disconnect acknowledge)	↓	XXSXX	PYY	X+Y

X: 2280±6 hertz, 150±30 ms
XX: 2280±6 hertz, 600±120 ms
S: 100±20 ms silence

X: 2040±6 hertz, 100±20 ms
Y: 2400±6 hertz, 100±20 ms
XX: 350±70 ms
YY: 350±70 ms
S: 35±7 ms; x: 2040 hertz, 35 ms
P: (2040 hertz, 2400 hertz), 150+30 ms

X: 2400±6 hertz
Y: 2600±6 hertz

* Combination No. 15 of address code.

the originator of a communication. The designation code consists of two letters, the same two letters that compose the first half of the message-retransmission-system *destination* indicator.

Examples of destination codes are:

North and Central America: 200 Cuba, 205 Puerto Rico (RCA), 206 Puerto Rico (ITTWC), 207 Puerto Rico (C & W), 21 Canada (except TWX), 22 Mexico, 25 USA (TWX), 271 Guatemala, 275 British Honduras, 290 Bermuda, 292 Virgin Islands.

South America: 304 Surinam, 305 Paraguay, 31 Venezuela, 36 Peru, 381 Brazil (Radio Brazil), 383 Brazil (PTT), 387 Argentina (ITTCM), 390 Netherlands Antilles, 391 Trinidad.

Europe: 400 Canary Islands, 403 Spain, 409 Algeria, 41 Germany, 46 Belgium, 492 Syria, 496 Kuwait, 501 Iceland, 51 United Kingdom, 57 Finland.

Eastern Europe: 601 Greece, 606 Israel, 61 Hungary, 64 USSR, 65 Romania.

Pacific: 702 Guam, 704 Hawaii (RCA), 705 Hawaii (ITTWC), 71 Australia, 72 Japan, 75 Philippines.

Asia: 801 Korea, 802 Hong Kong, 81 India, 85 China, 88 Iran.

Africa: 901 Libya, 907 Southern Rhodesia, 91 United Arab Republic, 94 Ghana, 95 South Africa, 972 Dahomey, 975 Niger, 981 Congo (Brazzaville), 982 Congo (Leopoldville), 991 Angola, 992 Mozambique.

CCITT AND TELEPHONY

International Country Codes

The addressing signals of worldwide automatic telephony consist of the national telephone number, as used for long-distance dialing within a country, prefixed by a country code. Country codes are grouped by continental regions; for example, the country codes of all South American countries begin with "5." Where the national numbering system includes more than one country, the country code may also include the countries included in the national system. Thus the country code for the United States—"1"—includes Canada and some other countries. The following are examples of some country codes, grouped by world numbering regions or zones as assigned by the Third Plenary Assembly of the CCITT in Geneva in 1964.

Zone 1—Code 1: USA, Canada, Mexico and Central America, Bahamas, Bermuda, Jamaica, French Antilles, Netherlands Antilles.

Zone 2—Africa: 51 countries, 48 country codes (Algeria, Morocco, Tunisia, Libya in one group—the Maghreb— code 21). United Arab Republic 20, South Africa 27, 45 three-digit codes.

Zones 3 and 4—Europe, Iceland, Malta, Cyprus: 17 two-digit and 13 three-digit country codes. Examples: France 33, Spain 34, Italy 39, United Kingdom 44, Germany 49, Iceland 354, Finland 401, Hungary 402.

Zone 5—South America and Cuba: 6 two-digit and 8 three-digit country codes. Examples: Cuba 53, Argentina 54, Brazil 55, Chile 56, Colombia 57, Venezuela 58, Peru 596.

Zone 6—Southwestern Pacific: 6 two-digit and 14 three-digit country codes. Examples: Malaysia 60, Australia 61, Indonesia 62, Philippines 63, New Zealand 64, Thailand 66, Guam 682.

Zone 7—Country code 7: Soviet Union.

Zone 8—Northwestern Pacific: 4 two-digit and 6 three-digit country codes. Examples: Japan 81, Korea 82, Vietnam 84, China (Formosa) 85, Hong Kong 852, Mongolia 854, Laos 856.

Zone 9—East: 5 two-digit and 15 three-digit country codes. Examples: India 91, Burma 95, Iran 98, Lebanon 961, Saudi Arabia 966, Israel 972, Nepal 977.

TELEPHONE SIGNALING

CCITT signaling systems have been standardized for international use. General descriptions are

TABLE 5—MULTIFREQUENCY NUMERICAL CODE USED BY CCITT (2-OUT-OF-6).

Digit	Frequencies	Weighting	
1	700+ 900	0+ 1	
2	700+1100	0+ 2	
3	900+1100	1+ 2	
4	700+1300	0+ 4	
5	900+1300	1+ 4	
6	1100+1300	2+ 4	
7	700+1500	0+ 7	
8	900+1500	1+ 7	
9	1100+1500	2+ 7	
0	1300+1500	4+ 7	
Code 11	700+1700	0+11	for inward
Code 12	900+1700	1+11	operators
KP	1100+1700	2+11	start of pulsing
KP2	1300+1700	4+11	transit traffic
ST	1500+1700	7+11	end of pulsing

TABLE 6—NUMERICAL 4-BY-4 MULTIFREQUENCY CODE.

Touch-Tone or Touch Calling

Low group (hertz)				
697	1	2	3	
770	4	5	6	
852	7	8	9	
941	spare	0	spare	
	1209	1336	1477	(1633) High group (hertz)

US Air Force 412L

Low group (hertz)				
1020	1	2	3	
1140	4	5	6	
1260	7	8	9	
1380		0		
	1620	1740	1860	(1980) High group (hertz)

Note: Each digit is composed of one frequency from the low group and one frequency from the high group. The frequencies have been chosen to minimize voice simulation.

given in Table 3, and some of the signaling characteristics are given in Table 4.

Signals in communications are used for passing information, for identifying the called subscriber or addressee (with resulting internal system signals concerned with the establishment of a connection), and for supervising and controlling the connection once it has been established.

Information Signals may be analog (voice, telemetry, or facsimile) or digital (teleprinter or data).

Addressing Signals may be dial pulse, multifrequency, or binary. They are not needed once a communication has been established.

TABLE 7—US ARMY TA–341/PT NUMERICAL CODE.

Digit	Frequencies
1	2100+2300
2	2300+2500
3	1900+2700
4	1900+2100
5	2500+2700
6	2300+2700
7	2100+2500
8	1900+2300
9	2100+2700
0	1900+2500

(A) *Dial Pulse* signals consist of a series of from 1 to 10 pulses representing the corresponding numerical digits 1 to 9 and 0. The pulses are breaks in a continuous direct current on the line, usually lasting from 58 to 67 percent of the time interval between the starts of successive pulses. These breaks in direct current may have to be converted into pulses in a tone, or to frequency

TABLE 8—NUMERICAL CODE, 2-VOICE-FREQUENCY SIGNALING SYSTEM, CCITT No. 4.

	Successive Elements			
Digit	1	2	3	4
1	y	y	y	x
2	y	y	x	y
3	y	y	x	x
4	y	x	y	y
5	y	x	y	x
6	y	x	x	y
7	y	x	x	x
8	x	y	y	y
9	x	y	y	x
0	x	y	x	y

Note: The 2 frequencies are sent one at a time, with a silent space between pulses. The duration of both frequency and silent periods is 35 ± 7 milliseconds. Frequencies: $x=2040\pm6$ hertz; $y=2400\pm6$ hertz. Power level: -9 decibels.

shifts between tones, in order to pass through some media or multiplexing systems. Dial pulse speeds are usually 10 pulses per second, although high-speed dials of nearly twice that speed are sometimes used.

(**B**) *Multifrequency* signals represent numerical digits by 1 pulse (or sometimes 4 pulses) of a specific frequency combination. The 2-out-of-6 and 4-by-4 multifrequency codes are shown in Tables 5 and 6. Inherent in both is the constant-ratio error-control principle (the simultaneous receipt of 3 or more, or 1 only, frequencies indicates an error). Table 7 shows the US Army 2-out-of-5

numerical code. The CCITT 2-voice-frequency code, consisting of 2 frequencies sent one at a time in 4 pulses, is given in Table 8.

(**C**) *Binary* signals for addressing are usually in a numeric or alphanumeric code used also for information signals, such as in telegraph addresses or headings. Multifrequency signals are sometimes directly converted to binary signals by changing the 2-frequencies-out-of-6 code to a corresponding 2-time-slots-out-of-6 synchronous 6-element binary code. Another example of a binary code used only for addressing is the CCITT 1-voice-frequency code, a 4-element start-stop code given in Table 9. (The CCITT 2-voice-frequency code is not truly binary since it uses a third condition—no tone—in addition to the 2 tones.)

Supervisory or Line Signals

Supervisory or line signals cannot be generated exclusively by registers because they are required during the entire use of the connection, after the registers that established the connection have been disconnected. Supervisory signals are an extension of the original basic signals of ringing and of closing a line loop to allow direct current to flow. Supervisory signals may be classified as spurt (discontinuous) and continuous.

Continuous Signals are based on conditions of on-hook and off-hook, representing the condition of blocked or flowing direct current on the subscriber's line, and their extension to trunk signaling is given in Table 10. Either condition is continuous and may be detected at any time. On the other hand, discontinuous signals must be recorded, and the condition represented is presumed to continue until a new signal is sent. Supervisory signaling in

TABLE 9—NUMERICAL CODE, 1-VOICE-FREQUENCY SIGNALING SYSTEM, CCITT NO. 3.

Digit	Start	1	2	3	4	Stop
			Time Elements			
1	1					1
2	1			1		
3	1			1		1
4	1		1			
5	1		1			1
6	1		1	1		
7	1		1	1		1
8	1	1				
9	1	1				1
0	1	1			1	

Note: "1" signifies frequency present. Length of each time element is 50 milliseconds±1 percent. Frequency: 2280±6 hertz. Power level: −6 decibels.

TABLE 10—ON-HOOK AND OFF-HOOK SIGNALS.

Direct Current Telephone Line	Trunk
On-hook signifies loop is open to direct current supplied from other end.	If idle, signals on-hook to other end. Seizure at calling end signals off-hook to called end. While calling end awaits answer, called end signals on-hook to calling end.
Off-hook signifies loop is closed, allowing relay at other end to operate. Signaling in reverse direction is ringdown.	Answer results in signaling off-hook from called end.
	If called end is not ready to receive address signals when seized, it signals off-hook to calling end until ready.

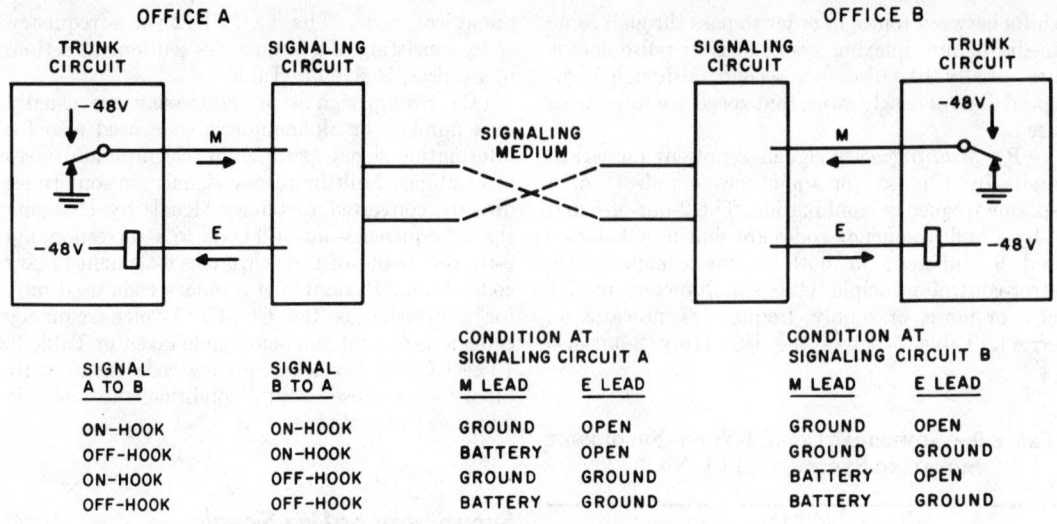

SIGNAL A TO B	SIGNAL B TO A	CONDITION AT SIGNALING CIRCUIT A		CONDITION AT SIGNALING CIRCUIT B	
		M LEAD	E LEAD	M LEAD	E LEAD
ON-HOOK	ON-HOOK	GROUND	OPEN	GROUND	OPEN
OFF-HOOK	ON-HOOK	BATTERY	OPEN	GROUND	GROUND
ON-HOOK	OFF-HOOK	GROUND	GROUND	BATTERY	OPEN
OFF-HOOK	OFF-HOOK	BATTERY	GROUND	BATTERY	GROUND

Fig. 6—E and M signaling. The E lead receives open or ground signals from the signaling circuit. The M lead sends ground or battery signals to the signaling circuit.

a backward direction (toward the calling end) is also needed in automatic working and is also described as on-hook or off-hook, although on 2-wire metallic circuits the signaling condition is usually a reversal of flow of the direct current rather than an interruption.

Continuous supervisory signaling over longer distances is effected by use of signaling paths distinct from the voice path. These signaling paths may be telegraph legs of a composite telegraph

SIGNAL	TONE	OPERATION	LEAD	CONDITION
ON-HOOK	ON	SENDING	M	GROUND
		RECEIVING	E	OPEN
OFF-HOOK	OFF	SENDING	M	BATTERY
		RECEIVING	E	GROUND

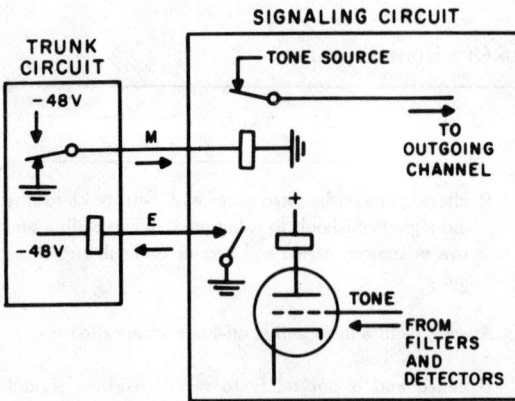

Fig. 7—Tone signaling.

system, simplexing of the voice pair, or special tones inside or outside of the voice channel. Whatever the paths, they are extensions of separate direct-current leads from the trunk circuits, known as E and M leads. The relation between the conditions of these leads and the on-hook or off-hook signaling conditions they represent is shown in Fig. 6. The usual method of extending the E and M leads over tone channels is shown in Fig. 7. The frequency of the tone used is preferably higher than 2000 hertz and is usually 2600 hertz on 4-wire circuits. A 3825-hertz signaling system is known as "out-of-band," and must be built into the carrier system using it.

Spurt Signaling avoids the necessity of using distinct signaling paths. For manual operation, ring-down signaling is usually satisfactory. For semiautomatic or automatic operation, however, more-elaborate systems are required. Voice-frequency signals are used and they are distinguished from voice transmissions by filters and timing. Single-frequency systems use 1 or 2 long or short pulses of the specific signaling frequency, with the pulses carefully timed to within minimum and maximum limits. Two-frequency systems use a gate-opening prefix pulse of the 2 frequencies together, followed, without a silent period, by a long or short pulse of either of the 2 frequencies. The submarine-cable signaling system (CCITT No. 5) uses 2 frequencies in the "compel" mode, wherein the signal frequencies are sent until acknowledged, and the acknowledging signal is sent until the original signal is stopped. No gate-opening prefix is used.

TABLE 11—TABLE OF PROPOSED STANDARD OF AUDIBLE TONES IN NORTH AMERICA (FROM CCITT DOCUMENT AP III-84).

Use	Frequencies§ (hertz) 350	440	480	620	Power per Frequency at Exchange Where Tone is Applied (dBm0)	Cadence
Dial tone	x	x			−13	Continuous
Busy tone			x	x	−24	0.5 second on 0.5 second off
Reorder tone*			x	x	−24	0.2 second on 0.3 second off or 0.3 second on 0.2 second off
Audible ringing tone		x	x		−16	2 seconds on 4 seconds off
High tone†			x		−16	Varies according to use
Pre-emption tone‡		x		x	−18	Single 200/500-ms pulse
Call-waiting tone		x			−13	Single 500-ms pulse

Notes:
  * A possible alternative is the use of a call-failure tone, which would identify the office and type of condition that prevented the successful completion of the call.
  † High tone is used in many ways. For example:
  (A) Spurts of tone to indicate specific orders to operators in the manual service (order tones).
  (B) To inform operators of lines that are temporarily out of service (permanent signal tone).
  (C) To alert customers that their services are in a permanent off-hook condition.
  ‡ Pre-emption tones are used in certain private switched networks which may interconnect with national networks.
  § Frequency limits are ±0.5 percent of nominal.

Ringdown Signals

Ringdown signals are spurts of ringing current (16 to 25 hertz) applied usually through the ringing key of an operator and intended to operate a bell, ringer, or drop at the called end. The current may be generated by a manually operated magneto or by a ringing machine with or without automatically inserted silent periods. Ringing to telephone subscribers in automatic central offices is stopped or "tripped" automatically by relay action resulting from the subscriber's off-hook condition. Ringing signals may be converted to 500 or 1000 hertz, usually interrupted at a 20-hertz rate, to pass through voice channels of carrier equipment. A ringing signal to a manual switchboard usually lights a switchboard lamp, which can be darkened again only by local action and not by stopping or repeating the ringing signal. This characteristic makes ringdown operation unsuitable for fully automatic operation. Ringdown signaling over carrier circuits has the advantages of simplicity and of not requiring the distinct signaling channels of E and M systems.

Tones

A special case of signaling is that of information in the form of tones to the subscriber of a telephone system. The basic tones are dial tone, busy tone, and ring-back tone (representing the ringing of the called subscriber's line). The dial tone is generated at the subscriber's local switching center, but the busy tone and ring-back tone, plus special tones such as no-such-number and line-out-of-order, are generated at the called subscriber's switching center and should be standardized for universal intelligibility. A proposal by the American Telephone and Telegraph Company for standard tones is described in Table 11. On some international calls, the busy and ring-back tones are generated locally in accordance with spurt supervisory signals from the called end.

Alternative Routing

Switching systems in which the complete called number is recorded in the first center to which the subscriber is connected permit a translation of digits from those identifying the called subscriber to those most conveniently used by the switching mechanism to establish the desired connection. This translation of digits permits the controlling mechanism to use a variety of routes in such a way that if the first-choice route is occupied or disabled a second-choice route may be tried, etc. Safeguards are required to prevent doubling back of routes and dead-end choices. In the near future many sophisticated plans may be expected that depend on rapid analysis of network possibilities by data-transmission means. The signals used for this purpose are interregister signals and are classed with addressing signals in that they are used in establishing a connection to the desired addressee, not being needed after a connection has been established.

CHAPTER 3
UNITS, CONSTANTS, AND CONVERSION FACTORS

SYMBOLS FOR UNITS

Unit symbols are letters, combinations of letters, or other characters that may be used in place of the names of the units. The following list covers symbols for *units* only. It does not cover letter symbols for physical *quantities*.

> Example: In the expression $I = 150$ mA, I is the symbol for a physical quantity (current) and mA is the symbol for a unit (milliamperes) of that quantity.

The symbols in the list at the end of this section are consistent in nearly all respects with the recommendations of the International Organization for Standardization (ISO). Most of the list and much of the text material in this section have been taken from ANSI Y10.19-1969.*

The basic symbols are short; thus multiplication and division may be indicated in ways that resemble those used with algebraic quantities. When an unfamiliar unit symbol is first used in text, it should be followed by its name in parentheses; only the symbol need be used thereafter.

Symbols for units are written in lower-case letters, except for the first letter if the name of the unit is derived from a proper name, and except for a very few that are not formed from letters. Every effort should be made to follow the distinction between upper- and lower-case letters, even if the symbols appear in applications where the other lettering is in upper-case style.

Symbols for units are printed in roman (upright) type. Their form is the same for both singular and plural, and they are not followed by a period. When there is risk of confusion in using the standard symbols "in" for inch and "l" for liter, the name of the unit should be spelled out.

When a compound unit is formed by multiplication of two or more other units, its symbol consists of the symbols for the separate units joined by a raised dot (for example, N·m for newton meter). The dot may be omitted in the case of familiar compounds (such as watthour, whose symbol is Wh), if no confusion would result.* Hyphens should not be used in symbols for compound units.

Positive and negative exponents may be used with unit symbols, but care must be taken in text to avoid confusion with superscripts that indicate footnotes or references.

When a compound unit is formed by division of one unit by another, its symbol consists of the symbols for the separate units either separated by a solidus (for example, m/s for meter per second) or multiplied using negative powers (for example, $m \cdot s^{-1}$ for meter per second). In simple cases use of the solidus is recommended, but in no case should more than one solidus on the same line, or a solidus followed by a product, be included in such a combination unless parentheses are inserted to avoid ambiguity. In complicated cases negative powers should be used.

The following prefixes are used to indicate multiples or submultiples of units:

Multiple	Prefix	Symbol
10^{18}	exa	E
10^{15}	peta	P
10^{12}	tera	T
10^{9}	giga	G
10^{6}	mega	M
10^{3}	kilo	k
10^{2}	hecto	h
10	deka	da
10^{-1}	deci	d
10^{-2}	centi	c
10^{-3}	milli	m
10^{-6}	micro	μ†
10^{-9}	nano	n
10^{-12}	pico	p
10^{-15}	femto	f
10^{-18}	atto	a

* Extracted from American National Standard *Letter Symbols for Units Used in Science and Technology*, ANSI Y10.19-1969, with permission of the publisher, The American Society of Mechanical Engineers, United Engineering Center, 345 East 47th Street, New York, New York 10017.

* When a unit symbol prefix is identical to a unit symbol, special care must be taken. m·N indicates the product of the units *meter* and *newton*, while mN is the symbol for *millinewton*.

† Lower-case u is frequently used in typing.

Symbols for prefixes are printed in roman type, without space between the prefix and the symbol for the unit. The distinctions between upper- and lower-case letters must be observed.

Compound prefixes should not be used:

Use		Do not use	
tera	T	megamega	MM
giga	G	kilomega	kM
nano	n	millimicro	mμ
pico	p	micromicro	$\mu\mu$

When a symbol representing a unit that has a prefix carries an exponent, this indicates that the multiple (or submultiple) unit is raised to the power expressed by the exponent. For example:

$$2 \text{ cm}^3 = 2(\text{cm})^3 = 2(10^{-2}\text{m})^3 = 2 \cdot 10^{-6}\text{m}^3$$

$$1 \text{ ms}^{-1} = 1(\text{ms})^{-1} = 1(10^{-3}\text{s})^{-1} = 10^3\text{s}^{-1}$$

Numerical Values

To facilitate the reading of numbers, the digits may be separated into groups of three, counting from the decimal sign toward the left and the right. The groups should be separated by a small space, but not by a comma or a point. In numbers of four digits, the space is usually not necessary. For example:

2.141 596   73 772   7372   0.133 47

In English-language documents the recommended decimal sign is a period. (The British recommended decimal sign is a raised period ($\cdot$).) For all other languages the recommended decimal sign is a comma.

If the magnitude of a number is less than unity, the decimal sign should be preceded by a zero.

The sign of multiplication of numbers is a cross ($\times$) or a raised dot ($\cdot$).

SI Units

In the following list some units are identified as SI units. These units belong to the International System of Units (Système International d'Unités), which is the name given in 1960 by the Conférence Générale des Poids et Mesures to the coherent system of units based on the following basic units and quantities:

Unit	Quantity
meter	length
kilogram	mass
second	time
ampere	electric current
kelvin	temperature
candela	luminous intensity
mole	amount of substance

The SI units include as subsystems the MKS system of units, which covers mechanics, and the MKSA (*Meter Kilogram Second Ampere*) or Georgi system, which covers mechanics, electricity, and magnetism.

Letter Symbols Alphabetically by Name of Unit

Unit	Symbol	Notes
ampere	A	SI unit of electric current
ampere (turn)	At	SI unit of magnetomotive force
ampere-hour	Ah	
ampere per meter	A/m	SI unit of magnetic field strength
angstrom	Å	1 Å $= 10^{-10}$ m
apostilb	asb	1 asb $= (1/\pi)$ cd/m^2 A unit of luminance. The SI unit, candela per square meter, is preferred.
atmosphere:		
standard atmosphere	atm	1 atm $= 101\ 325$ N/m^2
technical atmosphere	at	1 at $= 1$ kgf/cm^2
atomic mass unit (unified)	u	The (unified) atomic mass unit is defined as one-twelfth of the mass of an atom of the ^{12}C nuclide. Use of the old atomic mass unit (amu), defined by reference to oxygen, is deprecated.
bar	bar	1 bar $= 100\ 000$ N/m^2
barn	b	1 b $= 10^{-28}$ m^2
baud	Bd	Unit of signaling speed equal to one element per second.
becquerel	Bq	1 Bq $= 1$ s^{-1} SI unit of radioactivity.
bel	B	
billion electronvolts		The name *billion electronvolts* is deprecated; see *gigaelectronvolt* (GeV).
bit	b	

Unit	Symbol	Notes
British thermal unit	Btu	
calorie (International Table calorie)	cal_{IT}	1 cal° = 4.1868 J The 9th Conférence Générale des Poids et Mesures adopted the joule as the unit of heat, avoiding the use of the calorie as far as possible.
calorie (thermochemical calorie)	cal	1 cal = 4.1840 J (See note for International Table calorie.)
candela	cd	SI unit of luminous intensity.
candela per square inch	cd/in^2	Use of the SI unit, candela per square meter, is preferred.
candela per square meter	cd/m^2	SI unit of luminance. The name nit has been used.
candle		The unit of luminous intensity has been given the name *candela*; use of the word *candle* for this purpose is deprecated.
centimeter	cm	
circular mil	cmil	$1\ cmil = (\pi/4) \cdot 10^{-6}\ in^2$
coulomb	C	SI unit of electric charge.
cubic centimeter	cm^3	
cubic foot	ft^3	
cubic foot per minute	ft^3/min	
cubic foot per second	ft^3/s	
cubic inch	in^3	
cubic meter	m^3	
cubic meter per second	m^3/s	
cubic yard	yd^3	
curie	Ci	Unit of activity in the field of radiation dosimetry.
cycle	c	
cycle per second	c/s	Deprecated. Use hertz
decibel	dB	
degree (plane angle)	°	
degree (temperature):		Note that there is no space between the symbol ° and the letter. The use of the word *centigrade* for the Celsius temperature scale was abandoned by the Conférence Générale des Poids et Mesures in 1948.
degree Celsius	°C	
degree Fahrenheit	°F	
degree Kelvin		See Kelvin.
degree Rankine	°R	
dyne	dyn	
electronvolt	eV	
erg	erg	
erlang	E	Unit of telephone traffic.
farad	F	SI unit of capacitance.
foot	ft	
footcandle	fc	The name *lumen per square foot* (lm/ft^2) is recommended for this unit. Use of the SI unit of illuminance, the lux (lumen per square meter), is preferred.
footlambert	fL	If luminance is to be measured in English units, the candela per square inch (cd/in^2) is recommended. Use of the SI unit, the candela per square meter, is preferred.
foot per minute	ft/min	
foot per second	ft/s	
foot per second squared	ft/s^2	
foot pound-force	$ft \cdot lb_f$	
gal	Gal	$1\ Gal = 1\ cm/s^2$
gallon	gal	The gallon, quart, and pint differ in the US and the UK, and their use is deprecated.

Unit	Symbol	Notes
gauss	G	The gauss is the electromagnetic CGS (Centimeter Gram Second) unit of magnetic flux density. The SI unit, tesla, is preferred.
gigaelectronvolt	GeV	
gigahertz	GHz	
gilbert	Gb	The gilbert is the electromagnetic CGS (Centimeter Gram Second) unit of magnetomotive force. Use of the SI unit, the ampere (or ampere-turn), is preferred.
grain	gr	
gram	g	
gray	Gy	1 Gy= 1 J/kg SI unit of absorbed dose.
henry	H	
hertz	Hz	SI unit of frequency.
horsepower	hp	Use of the SI unit, the watt, is preferred.
hour	h	Time may be designated as in the following example: $9^h46^m30^s$.
inch	in	
inch per second	in/s	
joule	J	SI unit of energy.
joule per Kelvin	J/K	SI unit of heat capacity and entropy.
Kelvin	K	SI unit of temperature (formerly called *degree Kelvin*). The symbol K is now used without the symbol °.
kiloelectronvolt	keV	
kilogauss	kG	
kilogram	kg	SI unit of mass.
kilogram-force	kg$_f$	In some countries the name *kilopond* (kp) has been adopted for this unit.
kilohertz	kHz	
kilojoule	kJ	
kilohm	kΩ	
kilometer	km	
kilometer per hour	km/h	
kilopond	kp	See kilogram-force.
kilovar	kvar	
kilovolt	kV	
kilovoltampere	kVA	
kilowatt	kW	
kilowatthour	kWh	
knot	kn	1 kn= 1 nmi/h
lambert	L	The lambert is the CGS (Centimeter Gram Second) unit of luminance. The SI unit, candela per square meter, is preferred.
liter	l	
liter per second	l/s	
lumen	lm	SI unit of luminous flux.
lumen per square foot	lm/ft^2	Use of the SI unit, the lumen per square meter, is preferred.
lumen per square meter	lm/m^2	SI unit of luminous exitance.
lumen per watt	lm/W	SI unit of luminous efficacy.
lumen second	lm·s	SI unit of quantity of light.
lux	lx	1 lx= 1 lm/m^2. SI unit of illuminance.
maxwell	Mx	The maxwell is the electromagnetic CGS (Centimeter Gram Second) unit of magnetic flux. Use of the SI unit, the weber, is preferred.
megaelectronvolt	MeV	
megahertz	MHz	

Unit	Symbol	Notes
megavolt	MV	
megawatt	MW	
megohm	MΩ	
meter	m	SI unit of length.
mho	mho	1 mho$=1\ \Omega^{-1}=1$ S
microampere	μA	
microbar	μbar	
microfarad	μF	
microgram	μg	
microhenry	μH	
micrometer	μm	
micron		The name *micrometer* (μm) is preferred.
microsecond	μs	
microwatt	μW	
mil	mil	1 mil$=0.001$ in.
mile		
nautical	nmi	
statute	mi	
mile per hour	mi/h	
milliampere	mA	
millibar	mbar	mb may be used.
milligal	mGal	
milligram	mg	
millihenry	mH	
milliliter	ml	
millimeter	mm	
conventional millimeter of mercury	mm Hg	1 mm Hg$=133.322$ N/m^2.
millimicron		The name *nanometer* (nm) is preferred.
millisecond	ms	
millivolt	mV	
milliwatt	mW	
minute (plane angle)	...′	
minute (time)	min	Time may be designated as in the following example: $9^h46^m30^s$.
mole	mol	SI unit of amount of substance.
nanoampere	nA	
nanofarad	nF	
nanometer	nm	
nanosecond	ns	
nanowatt	nW	
nautical mile	nmi	
neper	Np	
newton	N	SI unit of force.
newton meter	N·m	See pascal.
newton per square meter	N/m^2	
nit	nt	1 nt$=1$ cd/m^2. See candela per square meter.
oersted	Oe	The oersted is the electromagnetic CGS (Centimeter Gram Second) unit of magnetic field strength. Use of the SI unit, the ampere per meter, is preferred.
ohm	Ω	SI unit of electric resistance.
ounce (avoirdupois)	oz	
pascal	Pa	SI unit of pressure or stress. 1 Pa$=1$ N/m^2
picoampere	pA	
picofarad	pF	
picosecond	ps	
picowatt	pW	
pint	pt	The gallon, quart, and pint differ in the US and the UK, and their use is deprecated.

Unit	Symbol	Notes
pound	lb	
poundal	pdl	
pound-force	lb$_f$	
pound-force foot	lb$_f$·ft	
pound-force per square inch	lb$_f$/in^2	
pound per square inch		Although use of the abbreviation psi is common, it is not recommended. See pound-force per square inch.
quart	qt	The gallon, quart, and pint differ in the US and the UK, and their use is deprecated.
rad	rd	Unit of absorbed dose in the field of radiation dosimetry.
radian	rad	SI unit of plane angle.
rem	rem	Unit of dose equivalent in the field of radiation dosimetry.
revolution per minute	r/min	Although use of the abbreviation rpm is common, it is not recommended.
revolution per second	r/s	
roentgen	R	Unit of exposure in the field of radiation dosimetry.
second (plane angle)	...''	
second (time)	s	SI unit of time. Time may be designated as in the following example: 9^{h}46^{m}30^s.
siemens	S	SI unit of conductance. 1 S=1 Ω^{-1}
square foot	ft^2	
square inch	in^2	
square meter	m^2	
square yard	yd^2	
steradian	sr	SI unit of solid angle.
stilb	sb	1 sb=1 cd/cm^2 A CGS unit of luminance. Use of the SI unit, the candela per square meter, is preferred.
tesla	T	SI unit of magnetic flux density. 1 T=1 Wb/m^2.
tonne	t	1 t=1000 kg.
(unified) atomic mass unit	u	See atomic mass unit (unified).
var	var	Unit of reactive power.
volt	V	SI unit of electromotive force.
voltampere	VA	SI unit of apparent power.
watt	W	SI unit of power.
watthour	Wh	
watt per steradian	W/sr	SI unit of radiant intensity.
watt per steradian square meter	W/(sr·m^2)	SI unit of radiance.
weber	Wb	SI unit of magnetic flux. 1 Wb=1 V·s.
yard	yd	

CONVERSION FACTORS

To Convert	Into	Multiply by	Conversely, Multiply by
acres	square feet	4.356×10^4	2.296×10^{-5}
acres	square meters	4047	2.471×10^{-4}
acres	square yards	4.84×10^3	2.066×10^{-4}
acres	hectares	0.4047	2.471
ampere-hours	coulombs	3600	2.778×10^{-4}
amperes per sq cm	amperes per sq inch	6.452	0.1550
ampere-turns	gilberts	1.257	0.7958
ampere-turns per cm	ampere-turns per inch	2.540	0.3937
angstroms	nanometers	10^{-1}	10
ares	square meters	10^2	10^{-2}

To Convert	Into	Multiply by	Conversely, Multiply by
atmospheres	bars	1.0133	0.9869
atmospheres	mm of mercury at 0°C	760	1.316×10^{-3}
atmospheres	feet of water at 4°C	33.90	2.950×10^{-2}
atmospheres	inches of mercury at 0°C	29.92	3.342×10^{-2}
atmospheres	kg per sq meter	1.033×10^4	9.678×10^{-5}
atmospheres	newtons per sq meter	1.0133×10^5	0.9869×10^{-5}
atmospheres	pounds per sq inch	14.70	6.804×10^{-2}
barns	square meters	10^{-28}	10^{28}
bars	newtons per square meter	10^5	10^{-5}
bars	hectopiezes	1	1
bars	baryes (dyne per sq cm)	10^6	10^{-6}
bars	pascals (newtons per sq meter)	10^5	10^{-5}
baryes	newtons per sq meter	10^{-1}	10
Btu	foot-pounds	778.3	1.285×10^{-3}
Btu	joules	1054.8	9.480×10^{-4}
Btu	kilogram-calories	0.2520	3.969
Btu	horsepower-hours	3.929×10^{-4}	2545
bushels	cubic feet	1.2445	0.8036
calories (I.T.)	joules	4.1868	0.238
calories (thermochem)	joules	4.184	0.239
carats (metric)	grams	0.2	5
Celsius (centigrade)	Fahrenheit	$°C \times 9/5 = °F - 32$ $(°C + 40) \times 9/5 = (°F + 40)$	
chains (surveyor's)	feet	66	1.515×10^{-2}
circular mils	square centimeters	5.067×10^{-6}	1.973×10^5
circular mils	square mils	0.7854	1.273
cords	cubic meters	3.625	0.2758
cubic feet	cords	7.8125×10^{-3}	128
cubic feet	gallons (liq US)	7.481	0.1337
cubic feet	liters	28.32	3.531×10^{-2}
cubic inches	cubic centimeters	16.39	6.102×10^{-2}
cubic inches	cubic feet	5.787×10^{-4}	1728
cubic inches	cubic meters	1.639×10^{-5}	6.102×10^4
cubic inches	gallons (liq US)	4.329×10^{-3}	231
cubic meters	cubic feet	35.31	2.832×10^{-2}
cubic meters	cubic yards	1.308	0.7646
degrees (angle)	radians	1.745×10^{-2}	57.30
dynes	pounds	2.248×10^{-6}	4.448×10^5
dynes	newtons	10^{-5}	10^5
electron volts	joules	1.602×10^{-19}	0.624×10^{19}
ergs	foot-pounds	7.376×10^{-8}	1.356×10^7
ergs	joules	10^{-7}	10^7
fathoms	feet	6	0.16667
fathoms	meters	1.8288	0.5467
feet	centimeters	30.48	3.281×10^{-2}
feet	varas	0.3594	2.782
feet of water at 4°C	inches of mercury at 0°C	0.8826	1.133
feet of water at 4°C	kg per sq meter	304.8	3.281×10^{-3}
feet of water at 4°C	pounds per sq foot	62.43	1.602×10^{-2}
fermis	meters	10^{-15}	10^{15}
footcandles	lumens per sq meter	10.764	0.0929
footlamberts	candelas per sq meter	3.4263	2.919×10^{-1}
foot-pounds	horsepower-hours	5.050×10^{-7}	1.98×10^6
foot-pounds	kilogram-meters	0.1383	7.233
foot-pounds	kilowatt-hours	3.766×10^{-7}	2.655×10^6
gallons (liq US)	cubic meters	3.785×10^{-3}	264.2
gallons (liq US)	gallons (liq Br Imp) (Canada)	0.8327	1.201
gammas	teslas	10^{-9}	10^9
gausses	lines per sq inch	6.452	0.1550

To Convert	Into	Multiply by	Conversely, Multiply by
gausses	teslas	10^{-4}	10^4
gilberts	amperes	7.9577×10^{-1}	1.257
grains (for humidity calculations)	pounds (avoirdupois)	1.429×10^{-4}	7000
grams	dynes	980.7	1.020×10^{-3}
grams	grains	15.43	6.481×10^{-2}
grams	ounces (avoirdupois)	3.527×10^{-2}	28.35
grams	poundals	7.093×10^{-2}	14.10
grams per cm	pounds per inch	5.600×10^{-3}	178.6
grams per cu cm	pounds per cu inch	3.613×10^{-2}	27.68
grams per sq cm	pounds per sq foot	2.0481	0.4883
hectares	square meters	10^4	10^{-4}
hectares	acres	2.471	0.4047
horsepower (boiler)	Btu per hour	3.347×10^4	2.986×10^{-5}
horsepower (metric) (542.5 ft-lb per second)	Btu per minute	41.83	2.390×10^{-2}
horsepower (metric) (542.5 ft-lb per second)	foot-lb per minute	3.255×10^4	3.072×10^{-5}
horsepower (metric) (542.5 ft-lb per second)	kg-calories per minute	10.54	9.485×10^{-2}
horsepower (550 ft-lb per second)	Btu per minute	42.41	2.357×10^{-2}
horsepower (550 ft-lb per second)	foot-lb per minute	3.3×10^4	3.030×10^{-5}
horsepower (550 ft-lb per second)	kilowatts	0.745	1.342
horsepower (metric) (542.5 ft-lb per second)	horsepower (550 ft-lb per second)	0.9863	1.014
horsepower (550 ft-lb per second)	kg-calories per minute	10.69	9.355×10^{-2}
inches	centimeters	2.540	0.3937
inches	feet	8.333×10^{-2}	12
inches	miles	1.578×10^{-5}	6.336×10^4
inches	mils	1000	0.001
inches	yards	2.778×10^{-2}	36
inches of mercury at 0°C	lbs per sq inch	0.4912	2.036
inches of water at 4°C	kg per sq meter	25.40	3.937×10^{-2}
inches of water at 4°C	ounces per sq inch	0.5782	1.729
inches of water at 4°C	pounds per sq foot	5.202	0.1922
inches of water at 4°C	in of mercury	7.355×10^{-2}	13.60
inches per ounce	meters per newton (compliance)	9.136×10^{-2}	10.95
joules	foot-pounds	0.7376	1.356
joules	ergs	10^7	10^{-7}
kilogram-calories	kilogram-meters	426.9	2.343×10^{-3}
kilogram-calories	kilojoules	4.186	0.2389
kilogram-meters	joules	0.102	9.81
kilogram force	newtons	9.81	0.102
kilograms	tons, long (avdp 2240 lb)	9.842×10^{-4}	1016
kilograms	tons, short (avdp 2000 lb)	1.102×10^{-3}	907.2
kilograms	pounds (avoirdupois)	2.205	0.4536
kilograms per kilometer	pounds (avdp) per mile (stat)	3.548	0.2818
kg per sq meter	pounds per sq foot	0.2048	4.882
kilometers	feet	3281	3.048×10^{-4}
kilopond force	newtons	9.81	0.102
kilowatt-hours	Btu	3413	2.930×10^{-4}
kilowatt-hours	foot-pounds	2.655×10^6	3.766×10^{-7}
kilowatt-hours	joules	3.6×10^6	2.778×10^{-7}
kilowatt-hours	kilogram-calories	860	1.163×10^{-3}

To Convert	Into	Multiply by	Conversely, Multiply by
kilowatt-hours	kilogram-meters	3.671×10^5	2.724×10^{-6}
kilowatt-hours	pounds carbon oxidized	0.235	4.26
kilowatt-hours	pounds water evaporated from and at 212°F	3.53	0.283
kilowatt-hours	pounds water raised from 62° to 212°F	22.75	4.395×10^{-2}
kips	newtons	4.448×10^3	2.248×10^{-4}
knots* (naut mi per hour)	feet per second	1.688	0.5925
knots	meters per minute	30.87	0.03240
knots	miles (stat) per hour	1.1508	0.8690
lamberts	candelas per sq cm	0.3183	3.142
lamberts	candelas per sq inch	2.054	0.4869
lamberts	candelas per sq meter	3.183×10^3	3.142×10^{-4}
leagues	miles (approximately)	3	0.33
links (surveyor's)	chains	0.01	100
links	inches	7.92	0.1263
liters	bushels (dry US)	2.838×10^{-2}	35.24
liters	cubic centimeters	1000	0.001
liters	cubic meters	0.001	1000
liters	cubic inches	61.02	1.639×10^{-2}
liters	gallons (liq US)	0.2642	3.785
liters	pints (liq US)	2.113	0.4732
$\log_e$ or ln	$\log_{10}$	0.4343	2.303
lumens per sq foot	foot-candles	1	1
lux	lumens per sq foot	0.0929	10.764
maxwells	webers	10^{-8}	10^8
meters	yards	1.094	0.9144
meters	varas	1.179	0.848
meters per min	feet per minute	3.281	0.3048
meters per min	kilometers per hour	0.06	16.67
microhms per cu cm	microhms per inch cube	0.3937	2.540
microhms per cu cm	ohms per mil foot	6.015	0.1662
microns	meters	10^{-6}	10^6
miles (nautical)*	feet	6076.1	1.646×10^{-4}
miles (nautical)	meters	1852	5.400×10^{-4}
miles (nautical)	miles (statute)	1.1508	0.8690
miles (statute)	feet	5280	1.894×10^{-4}
miles (statute)	kilometers	1.609	0.6214
miles per hour	kilometers per minute	2.682×10^{-2}	37.28
miles per hour	feet per minute	88	1.136×10^{-2}
miles per hour	kilometers per hour	1.609	0.6214
millibars	inches of mercury (32°F)	0.02953	33.86
millibars (10^3 dynes per sq cm)	pounds per sq foot	2.089	0.4788
mils	meters	2.54×10^{-5}	3.94×10^4
nepers	decibels	8.686	0.1151
newtons	dynes	10^5	10^{-5}
newtons	kilograms	0.1020	9.807
newtons	poundals	7.233	0.1383
newtons	pounds (avoirdupois)	0.2248	4.448
oersteds	amperes per meter	7.9577×10	1.257×10^{-2}
ounce-inches	newton-meters	7.062×10^{-3}	1.416×10^2
ounces (fluid)	quarts	3.125×10^{-2}	32
ounces (avoirdupois)	pounds	6.25×10^{-2}	16
pascals	newtons per sq meter	1	1
pascals	pounds per sq inch	1.45×10^{-4}	6.895×10^3
piezes	newtons per sq meter	10^3	10^{-3}
piezes	sthenes per sq meter	1	1

To Convert	Into	Multiply by	Conversely, Multiply by
pints	quarts (liq US)	0.50	2
poises	newton-seconds per sq meter	10^{-1}	10
pounds of water (dist)	cubic feet	1.603×10^{-2}	62.38
pounds of water (dist)	gallons	0.1198	8.347
pounds per inch	kg per meter	17.86	0.05600
pounds per foot	kg per meter	1.488	0.6720
pounds per mile (statute)	kg per kilometer	0.2818	3.548
pounds per cu foot	kg per cu meter	16.02	6.243×10^{-2}
pounds per cu inch	pounds per cu foot	1728	5.787×10^{-4}
pounds per sq foot	pounds per sq inch	6.944×10^{-3}	144
pounds per sq foot	kg per sq meter	4.882	0.2048
pounds per sq inch	kg per sq meter	703.1	1.422×10^{-3}
poundals	dynes	1.383×10^{4}	7.233×10^{-5}
poundals	pounds (avoirdupois)	3.108×10^{-2}	32.17
quarts	gallons (liq US)	0.25	4
rods	feet	16.5	6.061×10^{-2}
slugs (mass)	pounds (avoirdupois)	32.174	3.108×10^{-2}
sq inches	circular mils	1.273×10^{6}	7.854×10^{-7}
sq inches	sq centimeters	6.452	0.1550
sq feet	sq meters	9.290×10^{-2}	10.76
sq miles	sq yards	3.098×10^{6}	3.228×10^{-7}
sq miles	acres	640	1.562×10^{-3}
sq miles	sq kilometers	2.590	0.3861
sq millimeters	circular mils	1973	5.067×10^{-4}
steres	cubic meters	1	1
stokes	sq meters per second	10^{-4}	10^{4}
(temp rise, °C) $\times$ (U.S. gal water)/minute	watts	264	3.79×10^{-3}
tonnes	kilograms	10^{3}	10^{-3}
tons, short (avoir 2000 lb)	tonnes (1000 kg)	0.9072	1.102
tons, long (avoir 2240 lb)	tonnes (1000 kg)	1.016	0.9842
tons, long (avoir 2240 lb)	tons, short (avoir 2000 lb)	1.120	0.8929
tons (US shipping)	cubic feet	40	0.025
torrs	newtons per sq meter	133.32	7.5×10^{-3}
watts	Btu per minute	5.689×10^{-2}	17.58
watts	ergs per second	10^{7}	10^{-7}
watts	foot-lb per minute	44.26	2.260×10^{-2}
watts	horsepower (550 ft-lb per second)	1.341×10^{-3}	745.7
watts	horsepower (metric) (542.5 ft-lb per second)	1.360×10^{-3}	735.5
watts	kg-calories per minute	1.433×10^{-2}	69.77
watt-seconds (joules)	gram-calories (mean)	0.2389	4.186
webers per sq meter	gausses	10^{4}	10^{-4}
yards	feet	3	0.3333

* Conversion factors for the nautical mile and, hence, for the knot, are based on the International Nautical Mile. which was adopted by the U.S. Department of Defense and the U.S. Department of Commerce, effective 1 July 1954, See, "Adoption of International Nautical Mile," *National Bureau of Standards Technical News Bulletin*, vol. 38, p. 122; August 1954. The International Nautical Mile has been in use by many countries for various lengths of time.

Note: Pounds are avoirdupois in every entry except where otherwise indicated.

Examples

(**A**) Required, the conversion factor for pounds (avoirdupois) to grams. Duplication of entries in the table has been reduced to the minimum. An entry will be found for kilograms to pounds, from which the required factor is obviously 453.6.

(**B**) Convert inches per pound to meters per kilogram. A number of conversions have been collected under the name, pounds. The desired factor appears under pounds per inch. Since the reciprocal is tabulated, the factors must be interchanged, so the desired one is 0.05600.

FUNDAMENTAL PHYSICAL CONSTANTS†

Quantity	Symbol	Value	Error (ppm)	SI	cgs
				Units	
Velocity of light	c	2.9979250(10)	0.33	10^8 m sec^{-1}	10^{10} cm sec^{-1}
Fine structure constant, $[\mu_0 c^2/4\pi](e^2/\hbar c)$	α α^{-1}	7.297351(11) 137.03602(21)	1.5 1.5	10^{-3}	10^{-3}
Electron charge	e	1.6021917(70) 4.803250(21)	4.4 4.4	10^{-19} C	10^{-20} emu 10^{-10} esu
Planck's constant	h $\hbar = h/2\pi$	6.626196(50) 1.0545919(80)	7.6 7.6	10^{-34} J·sec 10^{-34} J·sec	10^{-27} erg·sec 10^{-27} erg·sec
Avogadro's number	N	6.022169(40)	6.6	10^{26} kmole^{-1}	10^{23} mole^{-1}
Atomic mass unit	amu	1.660531(11)	6.6	10^{-27} kg	10^{-24} g
Electron rest mass	m_e $m_e{}^*$	9.109558(54) 5.485930(34)	6.0 6.2	10^{-31} kg 10^{-4} amu	10^{-28} g 10^{-4} amu
Proton rest mass	M_p $M_p{}^*$	1.672614(11) 1.00727661(8)	6.6 0.08	10^{-27} kg amu	10^{-24} g amu
Neutron rest mass	M_n $M_n{}^*$	1.674920(11) 1.00866520(10)	6.6 0.10	10^{-27} kg amu	10^{-24} g amu
Ratio of proton mass to electron mass	M_p/m_e	1836.109(11)	6.2		
Electron charge to mass ratio	e/m_e	1.7588028(54) 5.272759(16)	3.1 3.1	10^{11} C kg^{-1}	10^7 emu g^{-1} 10^7 esu g^{-1}
Magnetic flux quantum, $[c]^{-1}(hc/2e)$	Φ_0 h/e	2.0678538(69) 4.135708(14) 1.3795234(46)	3.3 3.3 3.3	10^{-15} T·m^2 10^{-15} J·sec C^{-1}	10^{-7} G·cm^2 10^{-7} erg·sec emu^{-1} 10^{-17} erg·sec esu^{-1}
Quantum of circulation	$h/2m_e$ h/m_e	3.636947(11) 7.273894(22)	3.1 3.1	10^{-4} J·sec kg^{-1} 10^{-4} J·sec kg^{-1}	erg·sec g^{-1} erg·sec g^{-1}
Faraday constant, Ne	F	9.648670(54) 2.892599(16)	5.5 5.5	10^7 C kmole^{-1}	10^3 emu mole^{-1} 10^{14} esu mole^{-1}

Quantity	Symbol	Value	Error (ppm)	Units SI	Units cgs
Rydberg constant, $[\mu_0 c^2/4\pi]^2(m_e e^4/4\pi\hbar^3 c)$	R_∞	1.09737312(11)	0.10	10^7 m^{-1}	10^5 cm^{-1}
Bohr radius, $[\mu_0 c^2/4\pi]^{-1}(\hbar^2/m_e e^2) = \alpha/4\pi R_\infty$	α_0	5.2917715(81)	1.5	10^{-11} m	10^{-9} cm
Classical electron radius $[\mu_0 c^2/4\pi](e^2/m_e c^2) = \alpha^3/4\pi R_\infty$	r_0	2.817939(13)	4.6	10^{-15} m	10^{-13} cm
Electron magnetic moment in Bohr magnetons	μ_e/μ_B	1.0011596389(31)	0.0031		
Bohr magneton, $[c](e\hbar/2m_e c)$	μ_B	9.274096(65)	7.0	10^{-24} J T^{-1}	10^{-21} erg G^{-1}
Electron magnetic moment	μ_e	9.284851(65)	7.0	10^{-24} J T^{-1}	10^{-21} erg G^{-1}
Gyromagnetic ratio of protons in H_2O	γ'_p $\gamma'_p/2\pi$	2.6751270(82) 4.257597(13)	3.1 3.1	10^8 rad sec$^{-1}\cdot$T^{-1} 10^7 Hz T^{-1}	10^4 rad sec$^{-1}\cdot$G^{-1} 10^3 Hz G^{-1}
γ'_p corrected for diamagnetism of H_2O	γ_p $\gamma_p/2\pi$	2.6751965(82) 4.257707(13)	3.1 3.1	10^8 rad sec$^{-1}\cdot$T^{-1} 10^7 Hz T^{-1}	10^4 rad sec$^{-1}\cdot$G^{-1} 10^3 Hz G^{-1}
Magnetic moment of protons in H_2O in Bohr magnetons	μ'_p/μ_B	1.52099312(10)	0.066	10^{-3}	10^{-3}
Proton magnetic moment in Bohr magnetons	μ_p/μ_B	1.52103264(46)	0.30	10^{-3}	10^{-3}
Proton magnetic moment	μ_p	1.4106203(99)	7.0	10^{-26} J T^{-1}	10^{-23} erg G^{-1}
Magnetic moment of protons in H_2O in nuclear magnetons	μ'_p/μ_n	2.792709(17)	6.2		
μ'_p/μ_n corrected for diamagnetism of H_2O	μ_p/μ_n	2.792782(17)	6.2		
Nuclear magneton, $[c](e\hbar/2M_p c)$	μ_n	5.050951(50)	10	10^{-27} J T^{-1}	10^{-24} erg G^{-1}

Compton wavelength of the electron, $h/m_e c$	λ_C	2.426096(74)	3.1	10^{-12} m	3.1	10^{-10} cm
	$\lambda_C/2\pi$	3.861592(12)	3.1	10^{-13} m	3.1	10^{-11} cm
Compton wavelength of the proton, $h/M_p c$	$\lambda_{C,p}$	1.3214409(90)	6.8	10^{-15} m	6.8	10^{-13} cm
	$\lambda_{C,p}/2\pi$	2.103139(14)	6.8	10^{-16} m	6.8	10^{-14} cm
Compton wavelength of the neutron, $h/M_n c$	$\lambda_{C,n}$	1.3196217(90)	6.8	10^{-15} m	6.8	10^{-13} cm
	$\lambda_{C,n}/2\pi$	2.100243(14)	6.8	10^{-16} m	6.8	10^{-14} cm
Gas constant, R_0	R_0	8.31434(35)	42	10^3 J kmole^{-1}·K^{-1}	42	10^7 erg mole^{-1}·K^{-1}
Boltzman's constant R_0/N	k	1.380622(59)	43	10^{-23} J K^{-1}	43	10^{-16} erg K^{-1}
Stefan–Boltzman constant, $\pi^2 k^4/60\hbar^3 c^2$	σ	5.66961(96)	170	10^{-8} W m^{-2} K^4	170	10^{-5} erg sec^{-1}·cm^{-2}·K^{-4}
First radiation constant, $8\pi hc$	c_1	4.992579(38)	7.6	10^{-24} J·m	7.6	10^{-15} erg·cm
Second radiation constant, hc/k	c_2	1.438833(61)	43	10^{-2} m·K	43	cm·K
Gravitational constant	G	6.6732(31)	460	10^{-11} N·m^2 kg^{-2}	460	10^{-8} dyn·cm^2 g^{-2}
kx-unit-to-angstrom conversion factor, $\Lambda = \lambda(\text{Å})/\lambda(\text{kxu}); \lambda(\text{CuK}\alpha_1) \equiv 1.537400$ kxu	Λ	1.0020764(53)	5.3			
Å-to-angstrom conversion factor, $\Lambda^* = \lambda(\text{Å})/\lambda(\text{Å}^*); \lambda(\text{WK}\alpha_1) \equiv 0.2090100$ Å*	Λ^*	1.0000197(56)	5.6			

The first factor, in brackets, is to be included only if all quantities are expressed in SI units. With the exception of the auxiliary constants which have been taken to be exact, the uncertainties of these constants are correlated, and therefore the general law of error propagation must be used in calculating additional quantities requiring two or more of these constants. The numbers in parentheses are the standard-deviation uncertainties in the last digits of the quoted value, computed on the basis of internal consistency.

† Compiled by B. N. Taylor, W. H. Parker, and D. N. Langenberg. Reprinted from *Reviews of Modern Physics*, vol. 41, p. 375, 1969.

Note that the unified atomic mass scale ^{12}C≡12 has been used throughout, that amu=atomic mass unit, C=coulomb, G=gauss, Hz=hertz=cycles/sec, J=joule, K=kelvin (degrees kelvin), T=tesla (10^4 G), V=volt, and W=watt. In cases where formulas for constants are given (e.g., R_∞), the relations are written as the product of two factors. The second factor, in parentheses, is the expression to be used when all quantities are expressed in cgs units, with the electron charge in electrostatic units.

ENERGY CONVERSION FACTORS†

Quantity	Value	Unit	Error (ppm)
1 kg	5.609538(24)	10^{29} MeV	4.4
1 amu	931.4812(52)	MeV	5.5
Electron mass	0.5110041(16)	MeV	3.1
Proton mass	938.2592(52)	MeV	5.5
Neutron mass	939.5527(52)	MeV	5.5
1 electron volt	1.6021917(70)	10^{-19} J	4.4
		10^{-12} erg	
	2.4179659(81)	10^{14} Hz	3.3
	8.065465(27)	10^{5} m^{-1}	3.3
		10^{3} cm^{-1}	
	1.160485(49)	10^{4} K	42
Energy-wavelength conversion	1.2398541(41)	10^{-6} eV•m	3.3
		10^{-4} eV•cm	
Rydberg constant, R_∞	2.179914(17)	10^{-18} J	7.6
		10^{-11} erg	
	13.605826(45)	eV	3.3
	3.2898423(11)	10^{15} Hz	0.35
	1.578936(67)	10^{5} K	43
Bohr magneton, μ_B	5.788381(18)	10^{-5} eV T^{-1}	3.1
	1.3996108(43)	10^{10} Hz T^{-1}	3.1
	46.68598(14)	m^{-1}•T^{-1}	3.1
		10^{-2} cm^{-1}•T^{-1}	
	0.671733(29)	K T^{-1}	43
Nuclear magneton, μ_n	3.152526(21)	10^{-8} eV T^{-1}	6.8
	7.622700(42)	10^{6} Hz T^{-1}	5.5
	2.542669(14)	10^{-2} m^{-1}•T^{-1}	5.5
		10^{-4} cm^{-1}•T^{-1}	
	3.65846(16)	10^{-4} K T^{-1}	44
Gas constant, R_0	8.20562(35)	10^{-2} m³•atm kmole^{-1}•K^{-1}	42
Standard volume of ideal gas, V_0	22.4136	m³ kmole^{-1}	

† Compiled by B. N. Taylor, W. H. Parker, and D. N. Langenberg. Reprinted from Reviews of Modern Physics, vol. 41, 1969. The numbers in parentheses are the standard-deviation uncertainties in the last digits of the quoted value, computed on the basis of internal consistency.

DECIBELS AND POWER, VOLTAGE, AND CURRENT RATIOS

The decibel, abbreviated dB, is a unit used to express the ratio between two amounts of power, P_1 and P_2, existing at two points. By definition number of $dB = 10 \log_{10}(P_1/P_2)$. It is also used to express voltage and current ratios: number of $dB = 20 \log_{10}(V_1/V_2) = 20 \log_{10}(I_1/I_2)$.

Strictly, it can be used to express voltage and current ratios only when the voltages or currents in question are measured at places having identical impedances.

Power Ratio	Voltage and Current Ratio	Decibels	Nepers	Power Ratio	Voltage and Current Ratio	Decibels	Nepers
1.0233	1.0116	0.1	0.01	19.953	4.4668	13.0	1.50
1.0471	1.0233	0.2	0.02	25.119	5.0119	14.0	1.61
1.0715	1.0351	0.3	0.03	31.623	5.6234	15.0	1.73
1.0965	1.0471	0.4	0.05	39.811	6.3096	16.0	1.84
1.1220	1.0593	0.5	0.06	50.119	7.0795	17.0	1.96
1.1482	1.0715	0.6	0.07	63.096	7.9433	18.0	2.07
1.1749	1.0839	0.7	0.08	79.433	8.9125	19.0	2.19
1.2023	1.0965	0.8	0.09	100.00	10.0000	20.0	2.30
1.2303	1.1092	0.9	0.10	158.49	12.589	22.0	2.53
1.2589	1.1220	1.0	0.12	251.19	15.849	24.0	2.76
1.3183	1.1482	1.2	0.14	398.11	19.953	26.0	2.99
1.3804	1.1749	1.4	0.16	630.96	25.119	28.0	3.22
1.4454	1.2023	1.6	0.18	1000.0	31.623	30.0	3.45
1.5136	1.2303	1.8	0.21	1584.9	39.811	32.0	3.68
1.5849	1.2589	2.0	0.23	2511.9	50.119	34.0	3.91
1.6595	1.2882	2.2	0.25	3981.1	63.096	36.0	4.14
1.7378	1.3183	2.4	0.28	6309.6	79.433	38.0	4.37
1.8197	1.3490	2.6	0.30	10^4	100.000	40.0	4.60
1.9055	1.3804	2.8	0.32	$10^4 \times 1.5849$	125.89	42.0	4.83
1.9953	1.4125	3.0	0.35	$10^4 \times 2.5119$	158.49	44.0	5.06
2.2387	1.4962	3.5	0.40	$10^4 \times 3.9811$	199.53	46.0	5.29
2.5119	1.5849	4.0	0.46	$10^4 \times 6.3096$	251.19	48.0	5.52
2.8184	1.6788	4.5	0.52	10^5	316.23	50.0	5.76
3.1623	1.7783	5.0	0.58	$10^5 \times 1.5849$	398.11	52.0	5.99
3.5481	1.8836	5.5	0.63	$10^5 \times 2.5119$	501.19	54.0	6.22
3.9811	1.9953	6.0	0.69	$10^5 \times 3.9811$	630.96	56.0	6.45
5.0119	2.2387	7.0	0.81	$10^5 \times 6.3096$	794.33	58.0	6.68
6.3096	2.5119	8.0	0.92	10^6	1 000.00	60.0	6.91
7.9433	2.8184	9.0	1.04	10^7	3 162.3	70.0	8.06
10.0000	3.1623	10.0	1.15	10^8	10 000.0	80.0	9.21
12.589	3.5481	11.0	1.27	10^9	31 623	90.0	10.36
15.849	3.9811	12.0	1.38	10^{10}	100 000	100.0	11.51

To convert:

Decibels to nepers, multiply by 0.1151

Decibels per statute mile to nepers per kilometer, multiply by 7.154×10^{-2}

Decibels per nautical mile to nepers per kilometer, multiply by 6.215×10^{-2}

Nepers to decibels, multiply by 8.686

Nepers per kilometer to decibels per statute mile, multiply by 13.978

Nepers per kilometer to decibels per nautical mile, multiply by 16.074.

Where the power ratio is less than unity, it is usual to invert the fraction and express the answer as a decibel loss.

UNIT CONVERSION TABLE*

Quantity	Symbol	Equation in MKS(R) (Rationalized) Units	MKS(R) (Rationalized) Unit	MKS (NR) Units	Pract Units	esu	emu	MKS(NR) (Nonrationalized) Unit
					Equivalent Number of			
length	l		meter (m)	1	10^2	10^2	10^2	meter (m)
mass	m		kilogram	1	10^3	10^3	10^3	kilogram
time	t		second	1	1	1	1	second
force	$\mathbf{F}$	$F=ma$	newton	1	10^5	10^5	10^5	newton
work, energy	W	$W=\mathbf{F}l$	joule	1	1	10^7	10^7	joule
power	P	$P=W/t$	watt	1	1	10^7	10^7	watt
electric charge	q		coulomb	1	1	3×10^9	10^{-1}	coulomb
volume charge density	ρ	$\rho=q/v$	coulomb/m³	1	10^{-6}	3×10^3	10^{-7}	coulomb/m³
surface charge density	σ	$\sigma=q/A$	coulomb/m²	1	10^{-4}	3×10^5	10^{-5}	coulomb/m²
electric dipole moment	$\mathbf{p}$	$p=ql$	coulomb-meter	1	10^2	3×10^{11}	10	coulomb-meter
polarization	$\mathbf{P}$	$\mathbf{P}=\mathbf{p}/v$	coulomb/m²	1	10^{-4}	3×10^5	10^{-5}	coulomb/m²
electric field strength	$\mathbf{E}$	$\mathbf{E}=\mathbf{F}/q$	volt/m	1	10^{-2}	$10^{-4}/3$	10^6	volt/m
permittivity	ϵ	$F=q^2/4\pi\epsilon l^2$	farad/m	4π	$4\pi\times10^{-9}$	$36\pi\times10^9$	$4\pi\times10^{-11}$	
displacement	$\mathbf{D}$	$\mathbf{D}=\epsilon\mathbf{E}$	coulomb/m²	4π	$4\pi\times10^{-4}$	$12\pi\times10^5$	$4\pi\times10^{-5}$	
displacement flux	Ψ	$\Psi=DA$	coulomb	4π	4π	$12\pi\times10^9$	$4\pi\times10^{-1}$	
emf, electric potential	V	$V=El$	volt	1	1	$10^{-2}/3$	10^8	volt
current	I	$I=q/t$	ampere	1	1	3×10^9	10^{-1}	ampere
volume current density	$\mathbf{J}$	$\mathbf{J}=I/A$	ampere/m²	1	10^{-4}	3×10^5	10^{-5}	ampere/m²
surface current density	$\mathbf{K}$	$\mathbf{K}=I/l$	ampere/m	1	10^{-2}	3×10^7	10^{-3}	ampere/m
resistance	R	$R=V/I$	ohm	1	1	$10^{-11}/9$	10^9	ohm
conductance	G	$G=1/R$	mho	1	1	9×10^{11}	10^{-9}	mho
resistivity	ρ	$\rho=RA/l$	ohm-meter	1	10^2	$10^{-9}/9$	10^{11}	ohm-meter
conductivity	γ	$\gamma=1/\rho$	mho/meter	1	10^{-2}	9×10^9	10^{-11}	mho/meter
capacitance	C	$C=q/V$	farad	1	1	9×10^{11}	10^{-9}	farad
elastance	S	$S=1/C$	daraf	1	1	$10^{-11}/9$	10^9	daraf
magnetic charge	m		weber	$1/4\pi$	$10^8/4\pi$	$10^{-2}/12\pi$	$10^8/4\pi$	
magnetic dipole moment	$\mathbf{m}$	$m=ml$	weber-meter	$1/4\pi$	$10^{10}/4\pi$	$1/12\pi$	$10^{10}/4\pi$	
magnetization	$\mathbf{M}$	$\mathbf{M}=\mathbf{m}/v$	weber/m²	$1/4\pi$	$10^4/4\pi$	$10^{-6}/12\pi$	$10^4/4\pi$	
magnetic field strength	$\mathbf{H}$	$\mathbf{H}=nI/l$	ampere-turn/m	4π	$4\pi\times10^{-3}$	$12\pi\times10^7$	$4\pi\times10^{-3}$	
permeability	μ	$F=m^2/4\pi\mu l^2$	henry/m	$1/4\pi$	$10^7/4\pi$	$10^{-13}/36\pi$	$10^7/4\pi$	
induction	$\mathbf{B}$	$\mathbf{B}=\mu\mathbf{H}$	weber/m²	1	10^4	$10^{-6}/3$	10^4	weber/m²
induction flux	Φ	$\Phi=BA$	weber	1	10^8	$10^{-2}/3$	10^8	weber
mmf, magnetic potential	M	$M=Hl$	ampere-turn	4π	$4\pi\times10^{-1}$	$12\pi\times10^9$	$4\pi\times10^{-1}$	
reluctance	$\mathcal{R}$	$\mathcal{R}=M/\Phi$	amp-turn/weber	4π	$4\pi\times10^{-9}$	$36\pi\times10^{11}$	$4\pi\times10^{-9}$	
permeance	$\mathcal{P}$	$\mathcal{P}=1/\mathcal{R}$	weber/amp-turn	$1/4\pi$	$10^9/4\pi$	$10^{-11}/36\pi$	$10^9/4\pi$	
inductance	L	$L=\Phi/I$	henry	1	1	$10^{-11}/9$	10^9	henry

* Compiled by J. R. Ragazzini and L. A. Zadeh, Columbia University, New York.
  (G) = Gaussian unit.

Equivalent Number of			Practical (CGS) Unit	Equivalent Number of			Equivalent Number of	
Pract Units	esu	emu		esu	emu	esu	emu	emu
10^2	10^2	10^2	centimeter (cm)	1	1	centimeter (cm) (G)	1	centimeter (cm)
10^3	10^3	10^3	gram	1	1	gram (G)	1	gram
1	1	1	second	1	1	second (G)	1	second
10^5	10^5	10^5	dyne	1	1	dyne (G)	1	dyne
1	10^7	10^7	joule	10^7	10^7	erg (G)	1	erg
1	10^7	10^7	watt	10^7	10^7	erg/second (G)	1	erg/second
1	3×10^9	10^{-1}	coulomb	3×10^9	10^{-1}	statcoulomb (G)	$10^{-10}/3$	abcoulomb
10^{-6}	3×10^3	10^{-7}	coulomb/cm³	3×10^9	10^{-1}	statcoulomb/cm³(G)	$10^{-10}/3$	abcoulomb/cm³
10^{-4}	3×10^5	10^{-5}	coulomb/cm²	3×10^9	10^{-1}	statcoulomb/cm²(G)	$10^{-10}/3$	abcoulomb/cm²
10^2	3×10^{11}	10	coulomb-cm	3×10^9	10^{-1}	statcoulomb-cm (G)	$10^{-10}/3$	abcoulomb-cm
10^{-4}	3×10^5	10^{-5}	coulomb/cm²	3×10^9	10^{-1}	statcoulomb/cm²(G)	$10^{-10}/3$	abcoulomb/cm²
10^{-2}	$10^{-4}/3$	10^6	volt/cm	$10^{-2}/3$	10^8	statvolt/cm (G)	3×10^{10}	abvolt/cm
10^{-9}	9×10^9	10^{-11}		9×10^{18}	10^{-2}	(G)	$10^{-20}/9$	
10^{-4}	3×10^5	10^{-5}		3×10^9	10^{-1}	(G)	$10^{-10}/3$	
1	3×10^9	10^{-1}		3×10^9	10^{-1}	(G)	$10^{-10}/3$	
1	$10^{-2}/3$	10^8	volt	$10^{-2}/3$	10^8	statvolt (G)	3×10^{10}	abvolt
1	3×10^9	10^{-1}	ampere	3×10^9	10^{-1}	statampere (G)	$10^{-10}/3$	abampere
10^{-4}	3×10^5	10^{-5}	ampere/cm²	3×10^9	10^{-1}	statampere/cm² (G)	$10^{-10}/3$	abampere/cm²
10^{-2}	3×10^7	10^{-3}	ampere/cm	3×10^9	10^{-1}	statampere/cm (G)	$10^{-10}/3$	abampere/cm
1	$10^{-11}/9$	10^9	ohm	$10^{-11}/9$	10^9	statohm (G)	9×10^{20}	abohm
1	9×10^{11}	10^{-9}	mho	9×10^{11}	10^{-9}	statmho (G)	$10^{-20}/9$	abmho
10^2	$10^{-9}/9$	10^{11}	ohm-cm	$10^{-11}/9$	10^9	statohm-cm (G)	9×10^{20}	abohm-cm
10^{-2}	9×10^9	10^{-11}	mho/cm	9×10^{11}	10^{-9}	statmho/cm (G)	$10^{-20}/9$	abmho/cm
1	9×10^{11}	10^{-9}	farad	9×10^{11}	10^{-9}	statfarad (cm) (G)	$10^{-20}/9$	abfarad
1	$10^{-11}/9$	10^9	daraf	$10^{-11}/9$	10^9	statdaraf (G)	9×10^{20}	abdaraf
10^8	$10^{-2}/3$	10^8		$10^{-10}/3$	1		3×10^{10}	unit pole (G)
10^{10}	$1/3$	10^{10}		$10^{-10}/3$	1		3×10^{10}	pole-cm (G)
10^4	$10^{-6}/3$	10^4		$10^{-10}/3$	1		3×10^{10}	pole/cm² (G)
10^{-3}	3×10^7	10^{-3}	oersted	3×10^{10}	1		$10^{-10}/3$	oersted (G)
10^7	$10^{-13}/9$	10^7	gauss/oersted	$10^{-20}/9$	1		9×10^{20}	gauss/oersted (G)
10^4	$10^{-6}/3$	10^4	gauss	$10^{-10}/3$	1		3×10^{10}	gauss (G)
10^8	$10^{-2}/3$	10^8	maxwell (line)	$10^{-10}/3$	1		3×10^{10}	maxwell (line) (G)
10^{-1}	3×10^9	10^{-1}	gilbert	3×10^{10}	1		$10^{-10}/3$	gilbert (G)
10^{-9}	9×10^{11}	10^{-9}	gilbert/maxwell	9×10^{20}	1		$10^{-20}/9$	gilbert/maxwell (G)
10^9	$10^{-11}/9$	10^9	maxwell/gilbert	$10^{-20}/9$	1		9×10^{20}	maxwell/gilbert (G)
1	$10^{-11}/9$	10^9	henry	$10^{-11}/9$	10^9	stathenry (G)	9×10^{20}	abhenry (cm) (G)

The velocity of light was taken as 3×10^{10} centimeters/second in computing the conversion factors. Equations in the third column are for dimensional purposes only.

GREEK ALPHABET

Name	Capital	Small	Commonly Used to Designate
Alpha	A	α	Angles, coefficients, attenuation constant, absorption factor, area
Beta	B	β	Angles, coefficients, phase constant
Gamma	Γ	γ	Complex propagation constant (cap), specific gravity, angles, electrical conductivity, propagation constant
Delta	Δ	δ	Increment or decrement (cap or small), determinant (cap), permittivity (cap), density, angles
Epsilon	E	ϵ	Dielectric constant, permittivity, base of natural logarithms, electric intensity
Zeta	Z	ζ	Coordinates, coefficients
Eta	H	η	Intrinsic impedance, efficiency, surface charge density, hysteresis, coordinates
Theta	Θ	$\vartheta\ \theta$	Angular phase displacement, time constant, reluctance, angles
Iota	I	ι	Unit vector
Kappa	K	κ	Susceptibility, coupling coefficient
Lambda	Λ	λ	Permeance (cap), wavelength, attenuation constant
Mu	M	μ	Permeability, amplification factor, prefix micro
Nu	N	ν	Reluctivity, frequency
Xi	Ξ	ξ	Coordinates
Omicron	O	o	
Pi	Π	π	3.1416
Rho	P	ρ	Resistivity, volume charge density, coordinates
Sigma	Σ	σ	Summation (cap), surface charge density, complex propagation constant, electrical conductivity, leakage coefficient, deviation
Tau	T	τ	Time constant, volume resistivity, time-phase displacement, transmission factor, density
Upsilon	Υ	υ	
Phi	Φ	$\phi\ \varphi$	Scalar potential (cap), magnetic flux, angles
Chi	X	χ	Electric susceptibility, angles
Psi	Ψ	ψ	Dielectric flux, phase difference, coordinates, angles
Omega	Ω	ω	Resistance in ohms (cap), solid angle (cap), angular velocity

Note: Small letter is used except where capital (Cap) is indicated.

USEFUL NUMERICAL DATA

1 cubic foot of water at 4°C (weight)	62.43 lb
1 foot of water at 4°C (pressure)	0.4335 lb/in.^{2}
Velocity of light in vacuum, c	186 280 mi/second $= 2.998 \times 10^{10}$ cm/second
Velocity of sound in dry air at 20°C, 760 mm Hg	1127 ft/second
Degree of longitude at equator	68.703 statute miles, 59.661 nautical miles
Acceleration due to gravity at sea level, 40° latitude, g	32.1578 ft/second^2
$(2g)^{1/2}$	8.020
1 inch of mercury at 4°C	1.132 ft water $= 0.4908$ lb/in.^{2}
Base of natural logs ϵ	2.718
1 radian	$180° \div \pi = 57.3°$
360 degrees	2π radians
π	3.1416
Sin 1′	0.00029089
Arc 1°	0.01745 radian
Side of square	0.707 $\times$ (diagonal of square)

FRACTIONS OF AN INCH WITH METRIC EQUIVALENTS

Fractions of an inch	Decimals of an inch	Millimeters	Fractions of an inch	Decimals of an inch	Millimeters
1/64	0.0156	0.397	33/64	0.5156	13.097
1/32	0.0313	0.794	17/32	0.5313	13.494
3/64	0.0469	1.191	35/64	0.5469	13.891
1/16	0.0625	1.588	9/16	0.5625	14.288
5/64	0.0781	1.984	37/64	0.5781	14.684
3/32	0.0938	2.381	19/32	0.5938	15.081
7/64	0.1094	2.778	39/64	0.6094	15.478
1/8	0.1250	3.175	5/8	0.6250	15.875
9/64	0.1406	3.572	41/64	0.6406	16.272
5/32	0.1563	3.969	21/32	0.6563	16.669
11/64	0.1719	4.366	43/64	0.6719	17.066
3/16	0.1875	4.763	11/16	0.6875	17.463
13/64	0.2031	5.159	45/64	0.7031	17.859
7/32	0.2188	5.556	23/32	0.7188	18.256
15/64	0.2344	5.953	47/64	0.7344	18.653
1/4	0.2500	6.350	3/4	0.7500	19.050
17/64	0.2656	6.747	49/64	0.7656	19.447
9/32	0.2813	7.144	25/32	0.7813	19.844
19/64	0.2969	7.541	51/64	0.7969	20.241
5/16	0.3125	7.938	13/16	0.8125	20.638
21/64	0.3281	8.334	53/64	0.8281	21.034
11/32	0.3438	8.731	27/32	0.8438	21.431
23/64	0.3594	9.128	55/64	0.8594	21.828
3/8	0.3750	9.525	7/8	0.8750	22.225
25/64	0.3906	9.922	57/64	0.8906	22.622
13/32	0.4063	10.319	29/32	0.9063	23.019
27/64	0.4219	10.716	59/64	0.9219	23.416
7/16	0.4375	11.113	15/16	0.9375	23.813
29/64	0.4531	11.509	61/64	0.9531	24.209
15/32	0.4688	11.906	31/32	0.9688	24.606
31/64	0.4844	12.303	63/64	0.9844	25.003
1/2	0.5000	12.700	—	1.0000	25.400

TEMPERATURE CONVERSION TABLE

To use the table, find the temperature reading you have in the middle column. If the reading you have is in degrees Celsius (Centigrade), read the Fahrenheit equivalent in the right-hand column. If the reading you have is in degrees Fahrenheit, read the Celsius equivalent in the left-hand column.

−459.4 to 0

C	F	F
−273	−459.4	
−268	−450	
−262	−440	
−257	−430	
−251	−420	
−246	−410	
−240	−400	
−234	−390	
−229	−380	
−223	−370	
−218	−360	
−212	−350	
−207	−340	
−201	−330	
−196	−320	
−190	−310	
−184	−300	
−179	−290	
−173	−280	
−169	−273	−459.4
−168	−270	−454
−162	−260	−436
−157	−250	−418
−151	−240	−400
−146	−230	−382
−140	−220	−364
−134	−210	−346
−129	−200	−328
−123	−190	−310
−118	−180	−292
−112	−170	−274
−107	−160	−256
−101	−150	−238
− 96	−140	−220
− 90	−130	−202
− 84	−120	−184
− 79	−110	−166
− 73	−100	−148
− 68	− 90	−130
− 62	− 80	−112
− 57	− 70	− 94
− 51	− 60	− 76
− 46	− 50	− 58
− 40	− 40	− 40
− 34	− 30	− 22
− 29	− 20	− 4
− 23	− 10	+ 14
− 17.8	− 0	+ 32

0 to 100

C	F	F	C	F	F
−17.8	0	32	10.0	50	122.0
−17.2	1	33.8	10.6	51	123.8
−16.7	2	35.6	11.1	52	125.6
−16.1	3	37.4	11.7	53	127.4
−15.6	4	39.2	12.2	54	129.2
−15.0	5	41.0	12.8	55	131.0
−14.4	6	42.8	13.3	56	132.8
−13.9	7	44.6	13.9	57	134.6
−13.3	8	46.4	14.4	58	136.4
−12.8	9	48.2	15.0	59	138.2
−12.2	10	50.0	15.6	60	140.0
−11.7	11	51.8	16.1	61	141.8
−11.1	12	53.6	16.7	62	143.6
−10.6	13	55.4	17.2	63	145.4
−10.0	14	57.2	17.8	64	147.2
− 9.4	15	59.0	18.3	65	149.0
− 8.9	16	60.8	18.9	66	150.8
− 8.3	17	62.6	19.4	67	152.6
− 7.8	18	64.4	20.0	68	154.4
− 7.2	19	66.2	20.6	69	156.2
− 6.7	20	68.0	21.1	70	158.0
− 6.1	21	69.8	21.7	71	159.8
− 5.6	22	71.6	22.2	72	161.6
− 5.0	23	73.4	22.8	73	163.4
− 4.4	24	75.2	23.3	74	165.2
− 3.9	25	77.0	23.9	75	167.0
− 3.3	26	78.8	24.4	76	168.8
− 2.8	27	80.6	25.0	77	170.6
− 2.2	28	82.4	25.6	78	172.4
− 1.7	29	84.2	26.1	79	174.2
− 1.1	30	86.0	26.7	80	176.0
− 0.6	31	87.8	27.2	81	177.8
0.0	32	89.6	27.8	82	179.6
0.6	33	91.4	28.3	83	181.4
1.1	34	93.2	28.9	84	183.2
1.7	35	95.0	29.4	85	185.0
2.2	36	96.8	30.0	86	186.8
2.8	37	98.6	30.6	87	188.6
3.3	38	100.4	31.1	88	190.4
3.9	39	102.2	31.7	89	192.2
4.4	40	104.0	32.2	90	194.0
5.0	41	105.8	32.8	91	195.8
5.6	42	107.6	33.3	92	197.6
6.1	43	109.4	33.9	93	199.4
6.7	44	111.2	34.4	94	201.2
7.2	45	113.0	35.0	95	203.0
7.8	46	114.8	35.6	96	204.8
8.3	47	116.6	36.1	97	206.6
8.9	48	118.4	36.7	98	208.4
9.4	49	120.2	37.2	99	210.2
			37.8	100	212.0

Notes:
1. Kelvins (Celsius absolute) = °C + 273.18.

100 to 1000						1000 to 2000				
C	F	C		F		C	F	C		F
38	100	212	260	500	932	538	1000	1832	816	1500	2732
43	110	230	266	510	950	543	1010	1850	821	1510	2750
49	120	248	271	520	968	549	1020	1868	827	1520	2768
54	130	266	277	530	986	554	1030	1886	832	1530	2786
60	140	284	282	540	1004	560	1040	1904	838	1540	2804
66	150	302	288	550	1022	566	1050	1922	843	1550	2822
71	160	320	293	560	1040	571	1060	1940	849	1560	2840
77	170	338	299	570	1058	577	1070	1958	854	1570	2858
82	180	356	304	580	1076	582	1080	1976	860	1580	2876
88	190	374	310	590	1094	588	1090	1994	866	1590	2894
93	200	392	316	600	1112	593	1100	2012	871	1600	2912
99	210	410	321	610	1130	599	1110	2030	877	1610	2930
100	212	413.6	327	620	1148	604	1120	2048	882	1620	2948
104	220	428	332	630	1166	610	1130	2066	888	1630	2966
110	230	446	338	640	1184	616	1140	2084	893	1640	2984
116	240	464	343	650	1202	621	1150	2102	899	1650	3002
121	250	482	349	660	1220	627	1160	2120	904	1660	3020
127	260	500	354	670	1238	632	1170	2138	910	1670	3038
132	270	518	360	680	1256	638	1180	2156	916	1680	3056
138	280	536	366	690	1274	643	1190	2174	921	1690	3074
143	290	554	371	700	1292	649	1200	2192	927	1700	3092
149	300	572	377	710	1310	654	1210	2210	932	1710	3110
154	310	590	382	720	1328	660	1220	2228	938	1720	3128
160	320	608	388	730	1346	666	1230	2246	943	1730	3146
166	330	626	393	740	1364	671	1240	2264	949	1740	3164
171	340	644	399	750	1382	677	1250	2282	954	1750	3182
177	350	662	404	760	1400	682	1260	2300	960	1760	3200
182	360	680	410	770	1418	688	1270	2318	966	1770	3218
188	370	698	416	780	1436	693	1280	2336	971	1780	3236
193	380	716	421	790	1454	699	1290	2354	977	1790	3254
199	390	734	427	800	1472	704	1300	2372	982	1800	3272
204	400	752	432	810	1490	710	1310	2390	988	1810	3290
210	410	770	438	820	1508	716	1320	2408	993	1820	3308
216	420	788	443	830	1526	721	1330	2426	999	1830	3326
221	430	806	449	840	1544	727	1340	2444	1004	1840	3344
227	440	824	454	850	1562	732	1350	2462	1010	1850	3362
232	450	842	460	860	1580	738	1360	2480	1016	1860	3380
238	460	860	466	870	1598	743	1370	2498	1021	1870	3398
243	470	878	471	880	1616	749	1380	2516	1027	1880	3416
249	480	896	477	890	1634	754	1390	2534	1032	1890	3434
254	490	914	482	900	1652	760	1400	2552	1038	1900	3452
			488	910	1670	766	1410	2570	1043	1910	3470
			493	920	1688	771	1420	2588	1049	1920	3488
			499	930	1706	777	1430	2606	1054	1930	3506
			504	940	1724	782	1440	2624	1060	1940	3524
			510	950	1742	788	1450	2642	1066	1950	3542
			516	960	1760	793	1460	2660	1071	1960	3560
			521	970	1778	799	1470	2678	1077	1970	3578
			527	980	1796	804	1480	2696	1082	1980	3596
			532	990	1814	810	1490	2714	1088	1990	3614
			538	1000	1832				1093	2000	3632

2. Degrees Rankine (Fahrenheit absolute) = °F + 459.72.

TIME CONVERSION NOMOGRAPH*

The duration of component and equipment tests and requirements for equipment reliability are usually stated in hours. Such figures become more meaningful when converted to days, months, or years. For example, Fig. 1 shows that a "1000-hour test" requires approximately 42 days for completion.

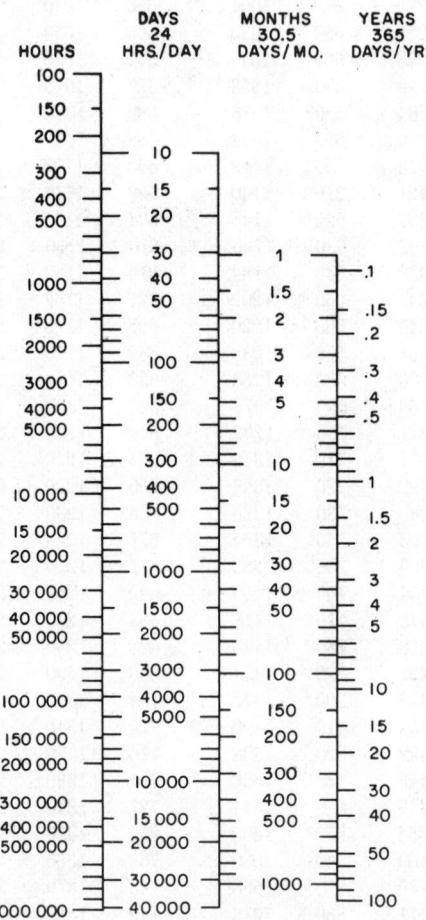

Fig. 1—Time conversion nomograph. *From R. F. Graf, " 'Reliability' in Terms of Time," Electronic Industries, p. 95; April 1958. © 1958, Chilton Company, Philadelphia, Pa.*

* Extracted from: R. F. Graf, " 'Reliability' in Terms of Time," *Electronic Industries*, p. 95; April 1958. © *1958, Chilton Company, Philadelphia, Pa.*

CHAPTER 4
PROPERTIES OF MATERIALS

GENERAL PROPERTIES OF THE ELEMENTS (TABLE 1)*

Atomic Number Z represents the number of protons per atom.

Mass Number $Z+N$ is equal to the number of protons Z plus the number of neutrons N present in the nuclei. Mass numbers of the most abundant isotopes are given in order of decreasing abundance. For example, Cadmium Cd-48 mass numbers 114–112 means that cadmium atoms of greater abundance (28.86%) have a mass number of 114; that is, that the nucleus of Cd^{114} has $114-48=66$ neutrons while isotope Cd^{112} of lower abundance (24.07%) has $112-48=64$ neutrons.

Atomic Radii. The values listed provide a comparison of sizes (deduced from interatomic spacing of bound atoms).

Gram Atomic Volume in cubic centimeters gives the volume occupied in the solid state by an atom at its melting point. The gram atomic volume contains the Avogadro number of atoms (6.0225×10^{23}).

Electronegativity represents the relative tendency of an atom to attract shared electron pairs. The highest electronegativity is assigned to fluorine with the value 3.90.

* Tables 1 through 6 of this chapter are partly based on data from the following sources: "Textbook of Chemistry", McGraw-Hill Book Company; 1961. "Handbook of Chemistry and Physics," 55th ed., CRC Press, Inc.; 1974. "Fundamentals of Chemistry," John Wiley & Sons. "The Encyclopedia of Electrochemistry," Reinhold Publishing Corp.; 1964. "American Institute of Physics Handbook," 3rd ed., McGraw-Hill Book Co.; 1972. "Lange's Handbook of Chemistry," 11th ed., McGraw-Hill Book Co.; 1973.

First Ionization Potential is the work in electronvolts required to pull 1 electron off an isolated neutral atom.

$$1 \text{ electron-volt} = 3.8 \times 10^{-20} \text{ calorie}$$

$$= 1.602 \times 10^{-12} \text{ erg}$$

$$= 1.602 \times 10^{-19} \text{ joules.}$$

Electron Work Function, expressed in electronvolts, represents the energy that must be supplied to an electron to cross over the surface barrier of a metal. That energy may be supplied by heat (thermionic work function), by light (photoelectric work function), or by contact with a dissimilar metal (contact potential).

Electrochemical Equivalents are expressed in ampere-hours per gram liberated at the electrode.

PERIODIC CLASSIFICATION OF THE ELEMENTS (TABLE 2)

Oxidation Number is defined as the charge which an atom appears to have in a compound when electrons are counted according to certain rules:

(**A**) In the free elements each atom has an oxidation number of 0.

(**B**) Electrons shared between two unlike atoms are counted with the more electronegative atom.

(**C**) Electrons shared between two like atoms are divided equally between the sharing atoms.

When an atom loses an electron its oxidation number increases by one.

Groups 1A and 2A of the periodic table having respectively oxidation numbers 1 and 2 (in all compounds) form only +1 ions (group 1A) and +2 ions (group 2A).

TABLE 1—PROPERTIES

	Symbol	Atomic Number Z	Mass Number $Z+N$	Atomic Weight	Atomic Radii (angstrom units)	Gram Atomic Volume (cm³)
Actinium	Ac	89	227	227		
Aluminum	Al	13	27	26.98	1.25	10
Americium	Am	95	243	243		
Antimony	Sb	51	121–123	121.75	1.41	18
Argon	Ar or A	18	40	39.948	1.74	24
Arsenic	As	33	75	74.92	1.21	16
Astatine	At	85	210	210		
Barium	Ba	56	138	137.34	1.98	38
Berkelium	Bk	97	247	247		
Beryllium	Be	4	9	9.012	0.89	5
Bismuth	Bi	83	209	208.98	1.52	21
Boron	B	5	11	10.81	0.88	5
Bromine	Br	35	79–81	79.904	1.14	23
Cadmium	Cd	48	114–112	112.40	1.41	13
Calcium	Ca	20	40	40.08	1.74	26
Californium	Cf	98	251	251		
Carbon	C	6	12	12.011	0.77	5
Cerium	Ce	58	140	140.12	1.65	21
Cesium	Cs	55	133	132.905	2.35	71
Chlorine	Cl	17	35	35.453	0.99	19
Chromium	Cr	24	52	51.996	1.17	7
Cobalt	Co	27	59	58.933	1.16	7
Copper	Cu	29	63	63.546	1.17	7
Curium	Cm	96	247	247		
Dysprosium	Dy	66	164–162–163	162.50	1.59	19
Einsteinium	Es or E	99	254	254		
Erbium	Er	68	166–168–167	167.26	1.57	18
Europium	Eu	63	153–151	151.96	1.85	29
Fermium	Fm	100	257	257		
Fluorine	F	9	19	18.998	0.64	15
Francium	Fr	87	223	223		
Gadolinium	Gd	64	158–160–156	157.25	1.61	20
Gallium	Ga	31	69–71	69.72	1.25	12
Germanium	Ge	32	74–72–70	72.59	1.22	13
Gold	Au	79	197	196.967	1.34	10
Hafnium	Hf	72	180–178–177	178.49	1.44	13
Helium	He	2	4	4.003		32
Holmium	Ho	67	165	164.93	1.58	19
Hydrogen	H	1	1	1.008	0.37	13
Indium	In	49	115	114.82	1.50	16
Iodine	I	53	127	126.904	1.33	26
Iridium	Ir	77	193–191	192.22	1.26	9
Iron	Fe	26	56	55.847	1.17	7
Krypton	Kr	36	84–86	83.80	1.89	33
Lanthanum	La	57	139	138.905	1.69	22
Lawrencium	Lw	103	257	257		
Lead	Pb	82	208–206–207	207.2	1.54	18
Lithium	Li	3	7	6.940	1.23	13
Lutetium	Lu	71	175	174.97	1.56	18
Magnesium	Mg	12	24	24.305	1.36	14
Manganese	Mn	25	55	54.938	1.17	7

OF THE ELEMENTS.

Electro-negativity, Relative Scale	First Ionization Potential (electron volts)	Electron Work Function			Electrochemical Equivalent	
		Thermionic	Photoelectric	Contact	Valence* Involved	Amp-Hours per Gram
1.1	6.9				3	0.35
1.5	5.98		4.08	3.38	3	2.98
	6.05					
2.05	8.64		4.01	4.14	5	1.1
0	15.76				n	0.67
2.0	9.81		5.11		5	1.79
2.2						
0.9	5.21	2.11	2.48	1.73	2	0.39
1.5	9.32		3.92	3.10	2	5.94
1.9	7.29		4.25	4.17	5	0.64
2.0	8.3		4.5		3	7.43
2.85	11.81				1	0.335
1.7	8.99		4.07	4.0	2	0.477
1.0	6.11	2.24	2.706	3.33	2	1.337
2.6	11.26	4.34	4.81		4	8.93
1.1	5.6	2.6	2.84		3	0.574
0.7	3.89	1.81	1.92	4.46	1	0.2
3.15	12.97				1	0.756
1.6	6.76	4.60	4.37	4.38	3	1.546
1.8	7.86	4.40	4.20	4.21	2	0.91
1.9	7.72	4.26	4.18	4.46	2	0.84
1.2	5.93				3	0.495
1.2	6.10				3	0.48
1.1	5.67				3	0.53
3.9	17.42				1	1.41
0.65						
1.1	6.16				3	0.513
1.6	5.99	4.12		3.80	3	1.15
1.9	7.89		4.5	4.5	4	1.48
2.4	9.22	4.32	4.82	4.46	3	0.41
1.3	7.0	3.53			4	0.600
0	24.59				n	6.698
1.2	6.02				3	0.488
2.2	13.59				1	26.59
1.7	5.78				3	0.700
2.65	10.45		6.8		1	0.211
2.2	9.1	5.3		4.57	4	0.555
1.8	7.87	4.25	4.33	4.40	3	1.440
0	13.99				n	0.32
1.1	5.61	3.3			3	0.579
1.8	7.42		4.05	3.94	4	0.517
1.0	5.39		2.35	2.49	1	3.862
1.2	6.15				3	0.46
1.2	7.64		3.68	3.63	2	2.204
1.5	7.43	3.83	3.76	4.14	4	1.952

TABLE 1—

	Symbol	Atomic Number Z	Mass Number $Z+N$	Atomic Weight	Atomic Radii (angstrom units)	Gram Atomic Volume (cm³)
Mendelevium	Md or Mv	101	256	256		
Mercury	Hg	80	202–200–199	200.59	1.44	14
Molybdenum	Mo	42	98–96–92–95	95.94	1.29	9
Neodymium	Nd	60	142–144–146	144.24	1.64	21
Neon	Ne	10	20	20.179	1.31	17
Neptunium	Np	93	237	237.048		
Nickel	Ni	28	58	58.71	1.15	6
Niobium	Nb	41	93	92.906	1.34	11
Nitrogen	N	7	14	14.007	0.70	14
Nobelium	No	102	254	254		
Osmium	Os	76	192–190–189	190.2	1.26	9
Oxygen	O	8	16	15.999	0.66	11
Palladium	Pd	46	108–106–105	106.4	1.28	9
Phosphorus	P	15	31	30.974	1.10	17
Platinum	Pt	78	195–194–196	195.09	1.29	9
Plutonium	Pu	94	242	242		
Polonium	Po	84	209	210	1.53	
Potassium	K	19	39	39.098	2.03	46
Praseodymium	Pr	59	141	140.907	1.65	21
Promethium	Pm	61	145	145		
Protactinium	Pa	91	231	231.036		
Radium	Ra	88	226	226.025		45
Radon	Rn	86	222	222	2.14	50
Rhenium	Re	75	187–185	186.2	1.28	9
Rhodium	Rh	45	103	102.905	1.25	8
Rubidium	Rb	37	85–87	85.468	2.16	56
Ruthenium	Ru	44	102–104–101	101.07	1.24	8
Samarium	Sm	62	152–154–147	150.35	1.66	20
Scandium	Sc	21	45	44.956	1.44	15
Selenium	Se	34	80–78	78.96	1.17	16
Silicon	Si	14	28	28.086	1.17	12
Silver	Ag	47	107–109	107.868	1.34	10
Sodium	Na	11	23	22.99	1.57	24
Strontium	Sr	38	88	87.62	1.92	34
Sulfur	S	16	32	32.064	1.04	16
Tantalum	Ta	73	181	180.948	1.34	11
Technetium	Tc	43	99	98.906		
Tellurium	Te	52	130–128–126	127.60	1.37	21
Terbium	Tb	65	159	158.925	1.59	19
Thallium	Tl	81	205–203	204.37	1.55	17
Thorium	Th	90	232	232.038	1.65	20
Thulium	Tm	69	169	168.934	1.56	18
Tin	Sn	50	120–118	118.69	1.40	16
Titanium	Ti	22	48	47.90	1.32	11
Tungsten	W	74	184–186–182	183.85	1.30	10
Uranium	U	92	238	238.029	1.42	13
Vanadium	V	23	51	50.94	1.22	8
Xenon	Xe	54	132–129–131	131.30	2.09	43
Ytterbium	Yb	70	174–172–173	173.04	1.70	25
Yttrium	Y	39	89	88.906	1.62	21
Zinc	Zn	30	64–66–68	65.38	1.25	9
Zirconium	Zr	40	90–94–92	91.22	1.45	14

CONTINUED.

Electro-negativity, Relative Scale	First Ionization Potential (electron volts)	Electron Work Function			Electrochemical Equivalent	
		Thermionic	Photoelectric	Contact	Valence* Involved	Amp-Hours per Gram
1.9	10.43		4.53	4.50	2	0.267
1.8	7.10	4.20	4.25	4.28	6	1.67
1.1	5.49	3.3			3	0.557
0	21.56				n	1.33
1.3	5.8					
1.8	7.63	5.03	5.01	4.96	2	0.913
1.6	6.88	4.01	4.5			
3.05	14.53				5	9.57
2.2	8.7			4.55	4	0.56
3.5	13.62				2	3.35
2.2	8.33	4.99	4.97	4.49	4	1.005
2.15	10.48				5	4.33
2.2	9.0	5.32	5.22	5.36	4	0.549
	5.8					
2.0	8.43				6	0.766
0.8	4.34		2.24	1.60	1	0.685
1.1	5.42	2.7			3	0.571
1.1	5.55					
1.5					5	0.580
0.9	5.28				2	0.237
0	10.75				n	0.121
1.9	7.87	5.1	5.0		7	1.007
2.2	7.46	4.80	4.57	4.52	4	1.042
0.8	4.18		2.09		1	0.314
2.2	7.37			4.52	4	1.054
1.1	5.63	3.2			3	0.535
1.3	6.54				3	1.783
2.45	9.75		4.8	4.42	6	2.037
1.9	8.15	3.59	4.52	4.2	4	3.821
1.9	7.57	3.56	4.73	4.44	1	0.248
0.9	5.14		2.28	1.9	1	1.166
1.0	5.69		2.74		2	0.612
2.6	10.36				6	5.01
1.3	7.88	4.19	4.14	4.1	5	0.741
1.9	7.28					
2.3	9.01		4.76	4.70	6	1.260
1.2	5.98				3	0.505
1.8	6.11		3.68	3.84	3	0.393
1.3	6.95	3.35	3.47	3.46	4	0.462
1.2	6.18				3	0.475
1.8	7.34		4.38	4.09	4	0.903
1.5	6.82	3.95	4.06	4.14	4	2.238
1.7	7.98	4.52	4.49	4.38	6	0.874
1.7	6.08	3.27	3.63	4.32	6	0.676
1.6	6.74	4.12	3.77	4.44	5	2.63
0	12.13				n	0.204
1.2	6.25				3	0.465
1.3	6.38				3	0.904
1.6	9.39		3.73	3.78	2	0.820
1.6	6.84	4.21	3.82	3.60	4	1.175

* n=nonvalent.

TABLE 2—PERIODIC CLASSIFICATION OF THE ELEMENTS.

PERIOD	1A	2A	3B	4B	5B	6B	7B	8	8	8	1B	2B	3A	4A	5A	6A	7A	0
1	$+1\ -1$ H 1																	0 He 2
2	$+1$ Li 3	$+2$ Be 4											$+3$ B 5	$+2\ -4\ +4$ C 6	$+1\ +2\ +3\ +4\ +5\ -2\ -3$ N 7	-2 O 8	-1 F 9	0 Ne 10
3	$+1$ Na 11	$+2$ Mg 12											$+3$ Al 13	$+2\ -4\ +4$ Si 14	$+3\ +5\ -3$ P 15	$+4\ -2\ +6$ S 16	$+1\ -1\ +5\ +7$ Cl 17	0 Ar 18
4	$+1$ K 19	$+2$ Ca 20	$+3$ Sc 21	$+2\ +3\ +4$ Ti 22	$+2\ +3\ +5$ V 23	$+2\ +3\ +6$ Cr 24	$+2\ +3\ +7$ Mn 25	$+2\ +3$ Fe 26	$+2\ +3$ Co 27	$+2\ +3$ Ni 28	$+1\ +2$ Cu 29	$+2$ Zn 30	$+3$ Ga 31	$+2\ +4$ Ge 32	$+3\ -3\ +5$ As 33	$+4\ -2\ +6$ Se 34	$+1\ -1\ +5$ Br 35	0 Kr 36
5	$+1$ Rb 37	$+2$ Sr 38	$+3$ Y 39	$+4$ Zr 40	$+3\ +5$ Nb 41	$+6$ Mo 42	$+4\ +6\ +7$ Tc 43	$+3$ Ru 44	$+3\ +4$ Rh 45	$+2\ +4$ Pd 46	$+1$ Ag 47	$+2$ Cd 48	$+3$ In 49	$+2\ +4$ Sn 50	$+3\ -3\ +5$ Sb 51	$+4\ -2\ +6$ Te 52	$+1\ -1\ +5\ +7$ I 53	0 Xe 54
6	$+1$ Cs 55	$+2$ Ba 56	◆ 57–71	$+4$ Hf 72	$+5$ Ta 73	$+6$ W 74	$+4\ +6\ +7$ Re 75	$+3\ +4$ Os 76	$+3\ +4$ Ir 77	$+2\ +4$ Pt 78	$+1\ +3$ Au 79	$+1\ +2$ Hg 80	$+1\ +3$ Tl 81	$+2\ +4$ Pb 82	$+3\ +5$ Bi 83	$+2\ +4$ Po 84	At 85	0 Rn 86
7	$+1$ Fr 87	$+2$ Ra 88	★ 89–103	— 104	— 105													

LIGHT METALS — HEAVY METALS (BRITTLE, DUCTILE, LOW-MELTING) — NON METALS — INERT GAS

GROUP 1 (H), He

TRANSITION ELEMENTS { BETWEEN GROUPS 2A AND 3A.

◆ LANTHANIDES (RARE EARTHS)	$+3$ La 57	$+3\ +4$ Ce 58	$+3$ Pr 59	$+3$ Nd 60	$+3$ Pm 61	$+2\ +3$ Sm 62	$+2\ +3$ Eu 63	$+3$ Gd 64	$+3$ Tb 65	$+3$ Dy 66	$+3$ Ho 67	$+3$ Er 68	$+3$ Tm 69	$+2\ +3$ Yb 70	$+3$ Lu 71
★ ACTINIDES	$+3$ Ac 89	$+4$ Th 90	$+4\ +5$ Pa 91	$+3\ +4\ +5\ +6$ U 92	$+3\ +4\ +5\ +6$ Np 93	$+3\ +4\ +5\ +6$ Pu 94	$+3\ +4\ +5\ +6$ Am 95	$+3$ Cm 96	$+3\ +4$ Bk 97	Cf 98	Es 99	Fm 100	Md 101	No 102	Lw 103

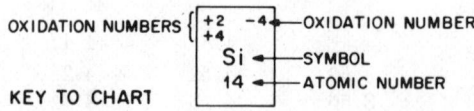

OXIDATION NUMBERS { $+2\ -4$ / $+4$ } — OXIDATION NUMBER
Si — SYMBOL
14 — ATOMIC NUMBER

KEY TO CHART

ELECTRONIC CONFIGURATION OF THE ELEMENTS (TABLE 3)

$n=$ principal quantum number
Maximum electron population of any energy level$= 2n^2$.

OXIDATION POTENTIALS OF THE ELEMENTS

In all galvanic cells the observed voltage E_c arises from two independent voltages: oxidation potential E_{ox} and reduction potential E_{red}.

$$E_c = E_{ox} + E_{red}.$$

To determine independently the oxidation potential E_{ox} of a single electrode, the potential of the standard hydrogen electrode at 25°C (H_2 pressure 1 atmosphere and $1mH^+$ concentration) is given arbitrarily the value 0. Then in a cell with the standard hydrogen electrode, the measured voltage is attributed to the half reaction at the other electrode.

The standard hydrogen electrode can play the role of an anode (oxidation process) or a cathode (reduction process) for electrode elements having respectively a negative or positive oxidation number.

Lithium (Li) is the best reducing agent since it has the highest tendency ($+3.045$ volts) to give up electrons (oxidation process). The fluoride ion (F^-) is the worst reducing agent.

The reducing agents arranged in decreasing order as in Table 4 are also called electromotive series.

TABLE 3.

Period	Atomic Number Z	Elements	1 K s	2 L s	2 L p	3 M s	3 M p	3 M d	4 N s	4 N p	4 N d	4 N f	5 O s	5 O p	5 O d	5 O f	6 P s	6 P p	6 P d	6 P f	7 Q s
I	1	H	1																		
	2	He	2																		
II	3	Li	2	1																	
	4	Be	2	2																	
	5	B	2	2	1																
	6	C	2	2	2																
	7	N	2	2	3																
	8	O	2	2	4																
	9	F	2	2	5																
	10	Ne	2	2	6																
III	11	Na	2	2	6	1															
	12	Mg	2	2	6	2															
	13	Al	2	2	6	2	1														
	14	Si	2	2	6	2	2														
	15	P	2	2	6	2	3														
	16	S	2	2	6	2	4														
	17	Cl	2	2	6	2	5														
	18	Ar	2	2	6	2	6														
IV	19	K	2	2	6	2	6		1												
	20	Ca	2	2	6	2	6		2												
	21	Sc	2	2	6	2	6	1	2												
	22	Ti	2	2	6	2	6	2	2												
	23	V	2	2	6	2	6	3	2												
	24	Cr	2	2	6	2	6	5	1												
	25	Mn	2	2	6	2	6	5	2												
	26	Fe	2	2	6	2	6	6	2												
	27	Co	2	2	6	2	6	7	2												
	28	Ni	2	2	6	2	6	8	2												
	29	Cu	2	2	6	2	6	10	1												
	30	Zn	2	2	6	2	6	10	2												
	31	Ga	2	2	6	2	6	10	2	1											
	32	Ge	2	2	6	2	6	10	2	2											
	33	As	2	2	6	2	6	10	2	3											
	34	Se	2	2	6	2	6	10	2	4											
	35	Br	2	2	6	2	6	10	2	5											
	36	Kr	2	2	6	2	6	10	2	6											
V	37	Rb	2	2	6	2	6	10	2	6			1								
	38	Sr	2	2	6	2	6	10	2	6			2								
	39	Y	2	2	6	2	6	10	2	6	1		2								
	40	Zr	2	2	6	2	6	10	2	6	2		2								
	41	Nb	2	2	6	2	6	10	2	6	4		1								
	42	Mo	2	2	6	2	6	10	2	6	5		1								
	43	Tc	2	2	6	2	6	10	2	6	6		1								
	44	Ru	2	2	6	2	6	10	2	6	7		1								
	45	Rh	2	2	6	2	6	10	2	6	8		1								
	46	Pd	2	2	6	2	6	10	2	6	10										
	47	Ag	2	2	6	2	6	10	2	6	10		1								
	48	Cd	2	2	6	2	6	10	2	6	10		2								
	49	In	2	2	6	2	6	10	2	6	10		2	1							
	50	Sn	2	2	6	2	6	10	2	6	10		2	2							
	51	Sb	2	2	6	2	6	10	2	6	10		2	3							

TABLE 3—CONTINUED.

Period	Z	Elements	1 K s	2 L s	2 L p	3 M s	3 M p	3 M d	4 N s	4 N p	4 N d	4 N f	5 O s	5 O p	5 O d	5 O f	6 P s	6 P p	6 P d	6 P f	7 Q s
	52	Te	2	2	6	2	6	10	2	6	10		2	4							
	53	I	2	2	6	2	6	10	2	6	10		2	5							
	54	Xe	2	2	6	2	6	10	2	6	10		2	6							
VI	55	Cs	2	2	6	2	6	10	2	6	10		2	6			1				
	56	Ba	2	2	6	2	6	10	2	6	10		2	6			2				
	57	La	2	2	6	2	6	10	2	6	10		2	6	1		2				
	58	Ce	2	2	6	2	6	10	2	6	10	2	2	6			2				
	59	Pr	2	2	6	2	6	10	2	6	10	3	2	6			2				
	60	Nd	2	2	6	2	6	10	2	6	10	4	2	6			2				
	61	Pm	2	2	6	2	6	10	2	6	10	5	2	6			2				
	62	Sm	2	2	6	2	6	10	2	6	10	6	2	6			2				
	63	Eu	2	2	6	2	6	10	2	6	10	7	2	6			2				
	64	Gd	2	2	6	2	6	10	2	6	10	7	2	6	1		2				
	65	Tb	2	2	6	2	6	10	2	6	10	9	2	6			2				
	66	Dy	2	2	6	2	6	10	2	6	10	10	2	6			2				
	67	Ho	2	2	6	2	6	10	2	6	10	11	2	6			2				
	68	Er	2	2	6	2	6	10	2	6	10	12	2	6			2				
	69	Tm	2	2	6	2	6	10	2	6	10	13	2	6			2				
	70	Yb	2	2	6	2	6	10	2	6	10	14	2	6			2				
	71	Lu	2	2	6	2	6	10	2	6	10	14	2	6	1		2				
	72	Hf	2	2	6	2	6	10	2	6	10	14	2	6	2		2				
	73	Ta	2	2	6	2	6	10	2	6	10	14	2	6	3		2				
	74	W	2	2	6	2	6	10	2	6	10	14	2	6	4		2				
	75	Re	2	2	6	2	6	10	2	6	10	14	2	6	5		2				
	76	Os	2	2	6	2	6	10	2	6	10	14	2	6	6		2				
	77	Ir	2	2	6	2	6	10	2	6	10	14	2	6	7		2				
	78	Pt	2	2	6	2	6	10	2	6	10	14	2	6	9		1				
	79	Au	2	2	6	2	6	10	2	6	10	14	2	6	10		1				
	80	Hg	2	2	6	2	6	10	2	6	10	14	2	6	10		2				
	81	Tl	2	2	6	2	6	10	2	6	10	14	2	6	10		2	1			
	82	Pb	2	2	6	2	6	10	2	6	10	14	2	6	10		2	2			
	83	Bi	2	2	6	2	6	10	2	6	10	14	2	6	10		2	3			
	84	Po	2	2	6	2	6	10	2	6	10	14	2	6	10		2	4			
	85	At	2	2	6	2	6	10	2	6	10	14	2	6	10		2	5			
	86	Rn	2	2	6	2	6	10	2	6	10	14	2	6	10		2	6			
VII	87	Fr	2	2	6	2	6	10	2	6	10	14	2	6	10		2	6			1
	88	Ra	2	2	6	2	6	10	2	6	10	14	2	6	10		2	6			2
	89	Ac	2	2	6	2	6	10	2	6	10	14	2	6	10		2	6	1		2
	90	Th	2	2	6	2	6	10	2	6	10	14	2	6	10		2	6	2		2
	91	Pa	2	2	6	2	6	10	2	6	10	14	2	6	10	2	2	6	1		2
	92	U	2	2	6	2	6	10	2	6	10	14	2	6	10	3	2	6	1		2
	93	Np	2	2	6	2	6	10	2	6	10	14	2	6	10	4	2	6	1		2
	94	Pu	2	2	6	2	6	10	2	6	10	14	2	6	10	6	2	6			2
	95	Am	2	2	6	2	6	10	2	6	10	14	2	6	10	7	2	6			2
	96	Cm	2	2	6	2	6	10	2	6	10	14	2	6	10	7	2	6	1		2
	97	Bk	2	2	6	2	6	10	2	6	10	14	2	6	10	9	2	6			2
	98	Cf	2	2	6	2	6	10	2	6	10	14	2	6	10	10	2	6			2
	99	Es	2	2	6	2	6	10	2	6	10	14	2	6	10	11	2	6			2
	100	Fm	2	2	6	2	6	10	2	6	10	14	2	6	10	12	2	6			2
	101	Md	2	2	6	2	6	10	2	6	10	14	2	6	10	13	2	6			2
	102	No	2	2	6	2	6	10	2	6	10	14	2	6	10	14	2	6			2
	103	Lw	2	2	6	2	6	10	2	6	10	14	2	6	10	14	2	6	1		2

TABLE 4—OXIDATION POTENTIALS OF ELEMENTS.

Element or Reducing Agent	Reaction	E_{ox} (volts)	Element or Reducing Agent	Reaction	E_{ox} (volts)
Lithium	$Li \rightarrow Li^+ + 1e$	+3.045	Manganese	$Mn \rightarrow Mn^{++} + 2e$	+1.18
Potassium	$K \rightarrow K^+ + 1e$	+2.925	Vanadium	$V \rightarrow V^{++} + 2e$	+1.18
Rubidium	$Rb \rightarrow Rb^+ + 1e$	+2.925	Niobium	$Nb \rightarrow Nb^{+++} + 3e$	+1.1
Cesium	$Cs \rightarrow Cs^+ + 1e$	+2.923	Tellurium	$Te^{--} \rightarrow Te + 2e$	+0.92
Radium	$Ra \rightarrow Ra^{++} + 2e$	+2.92	Selenium	$Se^{--} \rightarrow Se + 2e$	+0.78
Barium	$Ba \rightarrow Ba^{++} + 2e$	+2.90	Zinc	$Zn \rightarrow Zn^{++} + 2e$	+0.763
Strontium	$Sr \rightarrow Sr^{++} + 2e$	+2.89	Chromium	$Cr \rightarrow Cr^{++} + 2e$	+0.56
Calcium	$Ca \rightarrow Ca^{++} + 2e$	+2.87	Gallium	$Ga \rightarrow Ga^{+++} + 3e$	+0.56
Sodium	$Na \rightarrow Na^+ + 1e$	+2.714	Sulfur	$S^{--} \rightarrow S + 2e$	+0.51
Lanthanum	$La \rightarrow La^{+++} + 3e$	+2.52	Iron	$Fe \rightarrow Fe^{++} + 2e$	+0.44
Cerium	$Ce \rightarrow Ce^{+++} + 3e$	+2.48	Cadmium	$Cd \rightarrow Cd^{++} + 2e$	+0.403
Neodymium	$Nd \rightarrow Nd^{+++} + 3e$	+2.44	Indium	$In \rightarrow In^{+++} + 3e$	+0.345
Samarium	$Sm \rightarrow Sm^{+++} + 3e$	+2.41	Thallium	$Tl \rightarrow Tl^+ + 1e$	+0.336
Gadolinium	$Gd \rightarrow Gd^{+++} + 3e$	+2.40	Cobalt	$Co \rightarrow Co^{++} + 2e$	+0.277
Magnesium	$Mg \rightarrow Mg^{++} + 2e$	+2.37	Nickel	$Ni \rightarrow Ni^{++} + 2e$	+0.246
Yttrium	$Y \rightarrow Y^{+++} + 3e$	+2.37	Molybdenum	$Mo \rightarrow Mo^{+++} + 3e$	+0.2
Dysprosium	$Dy \rightarrow Dy^{+++} + 3e$	+2.35	Tin	$Sn \rightarrow Sn^{++} + 2e$	+0.136
Ytterbium	$Yb \rightarrow Yb^{+++} + 3e$	+2.27	Lead	$Pb \rightarrow Pb^{++} + 2e$	+0.126
Lutetium	$Lu \rightarrow Lu^{+++} + 3e$	+2.25	Hydrogen	$H_2 \rightarrow 2H^+ + 2e$	0.000
Hydrogen	$H^- \rightarrow \frac{1}{2}H_2 + 1e$	+2.25	Copper	$Cu \rightarrow Cu^{++} + 2e$	−0.337
Scandium	$Sc \rightarrow Sc^{+++} + 3e$	+2.08	Iodine	$2I^- \rightarrow I_2 + 2e$	−0.536
Plutonium	$Pu \rightarrow Pu^{+++} + 3e$	+2.02	Rhodium	$Rh \rightarrow Rh^+ + 1e$	−0.6
Thorium	$Th \rightarrow Th^{++++} + 4e$	+1.90	Silver	$Ag \rightarrow Ag^+ + 1e$	−0.7995
Neptunium	$Np \rightarrow Np^{+++} + 3e$	+1.86	Mercury	$Hg \rightarrow Hg^{++} + 2e$	−0.854
Beryllium	$Be \rightarrow Be^{++} + 2e$	+1.85	Palladium	$Pd \rightarrow Pd^{++} + 2e$	−0.987
Uranium	$U \rightarrow U^{+++} + 3e$	+1.80	Bromine	$2Br^- \rightarrow Br_2(liq) + 2e$	−1.065
Hafnium	$Hf \rightarrow Hf^{++++} + 4e$	+1.70	Platinum	$Pt \rightarrow Pt^{++} + 2e$	−1.2
Aluminum	$Al \rightarrow Al^{+++} + 3e$	+1.66	Chlorine	$2Cl^- \rightarrow Cl_2 + 2e$	−1.36
Titanium	$Ti \rightarrow Ti^{++} + 2e$	+1.63	Gold	$Au \rightarrow Au^+ + 1e$	−1.68
Zirconium	$Zr \rightarrow Zr^{++++} + 4e$	+1.53	Fluorine	$2F^- \rightarrow F_2 + 2e$	−2.87

TABLE 5—PHYSICAL PROPERTIES

	Symbol	Atomic Number	Density at 20°C (gr/cm³)	Relative Hardness	Melting Point (°C)	Boiling Point (°C)
Actinium	Ac	89			1 050	3 200
Aluminum	Al	13	2.70	2.9	660	2 467
Americium	Am	95	13.67		994	2 600
Antimony	Sb	51	6.62	3	630.5	1 750
Argon	Ar or A	18	1.78*		−189.2	−185.7
Arsenic (gray)	As	33	5.73	3.5	820‡	615¶
Astatine	At	85			302	337
Barium	Ba	56	3.5		725	1 640
Berkelium	Bk	97				
Beryllium	Be	4	1.82	3	1 278	2 970
Bismuth	Bi	83	9.80	2.5	271.3	1 560
Boron	B	5	2.46	9.5	2 300	2 550¶
Bromine	Br	35	3.12		−7.2	58.8
Cadmium	Cd	48	8.65	2.0	320.9	765
Calcium	Ca	20	1.54		842	1 487
Californium	Cf	98				
Carbon	C	6	2.22	10†	>3 500	4 827
Cerium	Ce	58	6.9	2.5	795	3 468
Cesium	Cs	55	1.87	0.2	28.5	690
Chlorine	Cl	17	3.21*		−100.98	−34.7
Chromium	Cr	24	7.14	9	1 890	2 672
Cobalt	Co	27	8.9	5	1 495	2 900
Copper	Cu	29	8.96	3	1 083	2 595
Curium	Cm	96	13.51		1 340	
Dysprosium	Dy	66	8.54		1 407	2 600
Einsteinium	Es or E	99				
Erbium	Er	68	9.05		1 497	2 900
Europium	Eu	63	5.26		826	1 439
Fermium	Fm	100				
Fluorine	F	9	1.69* at 15°C		−220	−188
Francium	Fr	87			27	677
Gadolinium	Gd	64	7.89		1 312	3 000
Gallium	Ga	31	5.91	1.5	29.78	2 403
Germanium	Ge	32	5.36	6.2	937.4	2 830
Gold	Au	79	19.3	2.5	1 063	2 966
Hafnium	Hf	72	13.31		2 220	4 602
Helium	He	2	0.1664*		<−272§	−268.94
Holmium	Ho	67	8.803		1 461	2 600
Hydrogen	H	1	0.08375*		−259.14	−252.8
Indium	In	49	7.31	1.2	156	2 050
Iodine	I	53	4.93		113.5	184.3
Iridium	Ir	77	22.4	6.15	2 410	4 527
Iron	Fe	26	7.87	4	1 535	3 000
Krypton	Kr	36	3.448*		−156.6	−152.3
Lanthanum	La	57	6.15		920	3 469
Lawrencium	Lw	103				
Lead	Pb	82	11.34	1.5	327.4	1 744
Lithium	Li	3	0.53	0.6	179	1 336
Lutetium	Lu	71	9.84		1 652	3 327
Magnesium	Mg	12	1.74	2	651	1 100
Manganese	Mn	25	7.44	5.0	1 244	2 097
Mendelevium	Md or Mv	101				
Mercury	Hg	80	13.55	1.5	−38.87	356.9

OF THE ELEMENTS.

Latent Heat of Fusion (cal/gr)	Specific Heat at 20°C (cal/gr °C)	Thermal Conductivity at 20°C (watts/cm °C)	Linear Thermal Expansion per °C at 20°C ($\times 10^{-6}$)	Elasticity Modulus (kg/mm²)	Tensile Strength (kg/mm²)
93	0.226	2.18	22.9	7 250	6.3
38.3	0.049	0.19	8.5–10.8	7 900	1.05
6.7	0.125	1.7×10^{-4}			
	0.082		4.7		
324	0.425	1.64	12	30 000	12.0
12.5	0.0294	0.084	13.3	3 200	
	0.307		2		
16.2	0.107				
13.2	0.055	0.91	29.8	5 500	7.2
	0.145		25	2 100	5.7
	0.165	0.24	0.6–4.3	500	
	0.042				9.05
3.8	0.052		97		
23	0.226	0.072×10^{-4}			
75.6	0.11	0.69	6.2		
58.4	0.1001	0.69	12.3	21 000	24.4
50.6	0.0921	3.94	16.5	11 000	22.5
10.1					
19.2	0.079		18		
	0.073				
16.1	0.031	2.96	14.2	7 300	11.5
	1.25	13.9×10^{-4}			
15	3.415	17×10^{-4}			
	0.057	0.24	33		0.3
15.8	0.052	43.5×10^{-4}	93		
33	0.032	1.4	6.5	52 500	
65	0.108	0.79	11.7	20 000	20.5
		0.89×10^{-4}			
	0.045				
6.3	0.030	0.35	28.7	1 800	1.33
159	0.79	0.71	56		
88	0.249	1.55	25.2	4 600	9.15
64.8	0.107		23	16 000	39.0
2.7	0.033	0.084			

TABLE 5—

	Symbol	Atomic Number	Density at 20°C (gr/cm³)	Relative Hardness	Melting Point (°C)	Boiling Point (°C)
Molybdenum	Mo	42	10.2	6	2 610	4 800
Neodymium	Nd	60	7.05		1 024	3 027
Neon	Ne	10	0.8387*		−248.7	−245.9
Neptunium	Np	93	20.45		640	3 902
Nickel	Ni	28	8.9	5	1 453	2 732
Niobium	Nb	41	8.57		2 468	5 127
Nitrogen	N	7	1.1649*		−209.9	−195.8
Nobelium	No	102				
Osmium	Os	76	22.48	7.0	3 000	5 000
Oxygen	O	8	1.3318*		−218.4	−183
Palladium	Pd	46	12	4.8	1 552	2 927
Phosphorus	P	15	1.82		44.1	280
Platinum	Pt	78	21.45	4.3	1 769	3 827
Plutonium	Pu	94	19.82		639.5	3 235
Polonium	Po	84	9.2		254	962
Potassium	K	19	0.86	0.5	63.65	774
Praseodymium	Pr	59	6.63		935	3 127
Promethium	Pm	61			1 027	2 027
Protactinium	Pa	91	15.37		1 227	4 027
Radium	Ra	88	5		700	1 525
Radon	Rn	86	4.40*		−71	−61.8
Rhenium	Re	75	20		3 180	5 627
Rhodium	Rh	45	12.44	6	1 966	3 727
Rubidium	Rb	37	1.53	0.3	38.5	688
Ruthenium	Ru	44	12.2	6.5	2 250	3 900
Samarium	Sm	62	7.7		1 072	1 900
Scandium	Sc	21	2.5		1 539	2 727
Selenium	Se	34	4.81	2	217	685
Silicon	Si	14	2.4	7	1 410	2 355
Silver	Ag	47	10.49	2.7	960.8	2 212
Sodium	Na	11	0.97	0.4	97	892
Strontium	Sr	38	2.6	1.8	769	1 384
Sulfur	S	16	2.07	2.0	116	444.6
Tantalum	Ta	73	16.6	7	2 996	5 425
Technetium	Tc	43	11.49		2 200	4 700
Tellurium	Te	52	6.24	2.3	449.5	990
Terbium	Tb	65	8.27		1 356	2 800
Thallium	Tl	81	11.85	1.2	303.5	1 457
Thorium	Th	90	11.5		1 800	4 200
Thulium	Tm	69	9.33		1 545	1 727
Tin	Sn	50	7.3	1.8	231.89	2 270
Titanium	Ti	22	4.54	4	1 675	3 260
Tungsten	W	74	19.3	7	3 410	5 660
Uranium	U	92	18.7		1 133	3 818
Vanadium	V	23	5.68		1 890	3 400
Xenon	Xe	54	5.495*		−111	−107
Ytterbium	Yb	70	6.98		824	1 427
Yttrium	Y	39	5.51		1 495	2 927
Zinc	Zn	30	7.14	2.5	419.4	907
Zirconium	Zr	40	6.4	4.7	1 852	4 377

* per liter.   † diamond   ‡ 36 atm   § 26 atm   ¶ sublimes

CONTINUED.

Latent Heat of Fusion (cal/gr)	Specific Heat at 20°C (cal/gr °C)	Thermal Conductivity at 20°C (watts/cm °C)	Linear Expansion per °C at 20°C ($\times 10^6$)	Elasticity Modulus (kg/mm²)	Tensile Strength (kg/mm²)
69.0	0.065	1.46	4.9	35 000	120.0
	0.045				
		4.57×10^{-4}			
73.8	0.112	0.9	13.3	21 000	32.3
68.0	0.064	0.52	7.1		
6.2	0.247				
34.0	0.031	0.61	5		
3.3	0.218				
34.2	0.059	0.70	11.8	12 000	14.0
5.0	0.177		125		
27.1	0.032	0.69	8.9	15 000	16
3.0	0.032	0.08	54		
14.5	0.177	0.99	83		
	0.458				
	0.035				
50.0	0.060	1.5	8.1	3 000	
6.1	0.080		90		
	0.061		9.1		
16.0	0.077	0.005	37		
430.0	0.176	0.84	2.8–7.3	11 000	
24.3	0.056	4.08	18.9	7 200	15.1
27.5	0.295	1.35	71		
25					
9.3	0.175	26.4×10^{-4}	6.4		
41.0	0.036	0.54	6.6	19 000	50
25.3	0.047	0.060	16.8	2 100	1.12
7.2	0.031	0.39	28		
17.0	0.028	0.41	11.1		56
14.4	0.054	0.64	23	41 100	1.4
100.0	0.142	0.2	8.5	8 500	
44.0	0.034	1.99	4.3	35 000	270
12.0	0.028	0.25	13.4		
98.0	0.115	0.60	8		
		5.9×10^{-4}			
24.1	0.09	1.1	17–39	8 400	10.5
	0.066		5.6	7 500	30

In air all the metals from lithium down to copper rust with comparative ease. The metals below copper do not rust.

All metals in the series above hydrogen (H_2) are found in nature in the combined state as sulfides, carbonates, etc. Metals below hydrogen (H_2) are frequently found in the free state.

PHYSICAL PROPERTIES OF THE ELEMENTS (TABLE 5)

GALVANIC SERIES IN SEA WATER

Two dissimilar metals connected by a conductor form in sea water a galvanic cell. If the two metals are in different groups of Table 6 (separated by spaces), the metal coming first in the series—starting from corroded end to protected end—will be anodic, (i.e., corroded by the metal contained in the group farther from the corroded end). If the two metals are in the same group, no appreciable corrosive action will take place.

TEMPERATURE—EMF CHARACTERISTICS OF THERMOCOUPLES* (FIG. 1)

Electromotive Force and Other Properties (Tables 7–8)

* R. L. Weber, "Temperature Measurement and Control," Blakiston Co., Philadelphia, Pennsylvania; 1941: pp. 68–71.

TABLE 6—GALVANIC SERIES IN SEA WATER.

Corroded end (anodic)

Magnesium
Magnesium alloys

Zinc
Galvanized steel
Galvanized wrought iron

Aluminum:
　52SH, 4S, 3S, 2S, 53ST
　Aluminum clad

Cadmium

Aluminum:
　A17ST, 17ST, 24ST

Mild steel
Wrought iron
Cast iron

Ni-resist

13% chromium stainless steel
　(type 410–active)

50–50 lead–tin solder

18–8 stainless steel type 304
　(active)
18–8–3 stainless steel type 316
　(active)

Lead
Tin

Muntz metal
Manganese bronze
Naval brass

Nickel (active)
Inconel (active)

Yellow brass
Admiralty brass
Aluminum bronze
Red brass
Copper
Silicon bronze
Ambrac
70–30 copper–nickel
Comp. G, bronze
Comp. M, bronze

Nickel (passive)
Inconel (passive)

Monel

18–8 stainless steel type 304
　(passive)
18–8–3 stainless steel type 316
　(passive)

Protected end (cathodic or most noble)

TABLE 7—THERMOCOUPLES AND THEIR CHARACTERISTICS.

Type	Copper/Constantan	Iron/Constantan	Chromel/Constantan	Chromel/Alumel	Platinum/Platinum Rhodium (10)	Platinum/Platinum Rhodium (13)	Carbon/Silicon Carbide
Composition, percent	100 Cu/60 Cu 40 Ni	100 Fe/60 Cu 40 Ni	90 Ni 10 Cr/55 Cu 45 Ni	90 Ni 10 Cr/94 Ni 2 Al 3 Mn 1 Si	Pt/90 Pt 10 Rh	Pt/87 Pt 13 Rh	C/SiC
*Range of application, °C	−200 to +300	−200 to +1100	0 to +1100	−200 to +1200	0 to +1450	0 to +1450	0 to +2000
Resistivity, microhm-cm	1.75	10	70	29.4	10	21	
Temperature coefficient of resistivity, per °C	0.0039	0.005	0.00035	0.000125	0.0030	0.0018	
Melting temperature, °C	1085	1535	1400	1430	1755	1700	3000 2700
emf in millivolts; reference junction at 0°C	100°C 4.24mV 200 9.06 300 14.42	100°C 5.28mV 200 10.78 400 21.82 600 33.16 800 45.48 1000 58.16	100°C 6.3mV 200 13.3 400 28.5 600 44.3	100°C 4.1mV 200 8.13 400 16.39 600 24.90 800 33.31 1000 41.31 1200 48.85 1400 55.81	100°C 0.643mV 200 1.436 400 3.251 600 5.222 800 7.330 1000 9.569 1200 11.924 1400 14.312 1600 16.674	100°C 0.646mV 200 1.464 400 3.398 600 5.561 800 7.927 1000 10.470 1200 13.181 1400 15.940 1600 18.680	1210°C 353.6mV 1300 385.2 1360 403.2 1450 424.9
Influence of temperature and gas atmosphere	Subject to oxidation and alteration above 400°C due Cu, above 600°C due constantan wire. Ni-plating of Cu tube gives protection in acid-containing gas. Contamination of Cu affects calibration greatly. Resistance to oxid. atm. good. Resistance to reducing atmosphere good. Requires protection from acid fumes.	Oxidizing and reducing atmosphere have little effect on accuracy. Best used in dry atmosphere. Resistance to oxidation good to 400°C. Resistance to reducing atmosphere good. Protect from oxygen, moisture, sulphur.	Chromel attacked by sulphurous atmosphere. Resistance to oxidation good. Resistance to reducing atmosphere poor.	Resistance to oxidizing atmosphere very good. Resistance to reducing atmosphere poor. Affected by sulphur, reducing or sulphurous gas, SO_2 and H_2S.	Resistance to oxidizing atmosphere very good. Resistance to reducing atmosphere poor. Susceptible to chemical alteration by As, Si, P vapor in reducing gas (CO_2, H_2, H_2S, SO_2). Pt corrodes easily above 1000°. Used in gas-tight protecting tube.		Used as tube element. Carbon sheath chemically inert.
Particular applications	Low temperature, industrial. Internal-combustion engine. Used as a tube element for measurements in steam line.	Low temperature, industrial. Steel annealing, boiler flues, tube stills. Used in reducing or neutral atmosphere.		Used in oxidizing atmosphere. Industrial. Ceramic kilns, tube stills, electric furnaces.	International Standard 630 to 1065°C.	Similar to Pt/PtRh(10) but has higher emf.	Steel furnace and ladle temperatures. Laboratory measurements.

* For prolonged use; can be used at higher temperature for short periods.

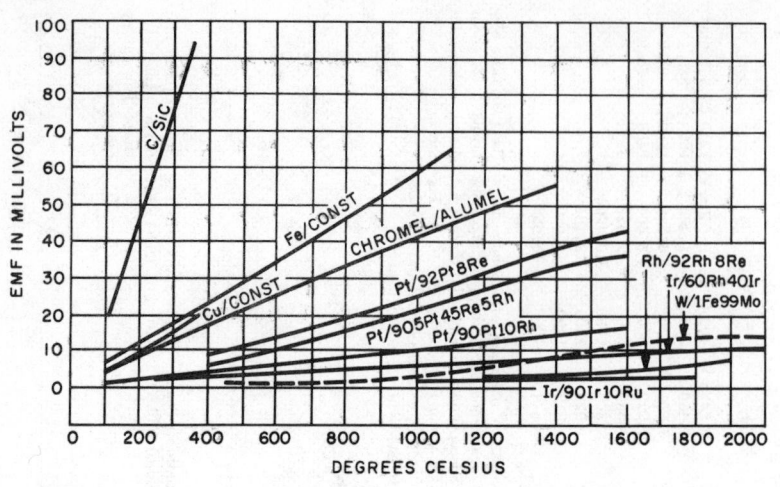
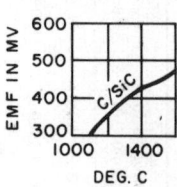

Fig. 1.

TABLE 8—THERMAL ELECTROMOTIVE FORCE OF PLATINUM–RHODIUM ALLOYS VERSUS PLATINUM. *From "Smithsonian Physical Tables," 9th revised edition, vol. 120, Smithsonian Institution, Washington, D. C.; 1969.*

	Electromotive Force (millivolts)							
	Percent Rhodium							
Temp., °C	0.5	1.0	5.0	10.0	20.0	40.0	80.0	100.0
0	0.00	0.00	0.00	0.00	0.00	0.00	0.00	0.00
100	+0.10	+0.18	+0.54	+0.64	+0.63	+0.65	+0.62	+0.70
200	0.20	0.37	1.16	1.43	1.44	1.52	1.49	1.61
300	0.29	0.57	1.82	2.32	2.40	2.55	2.55	2.68
400	0.39	0.76	2.49	3.25	3.47	3.70	3.77	3.91
500	0.48	0.94	3.17	4.22	4.63	4.97	5.12	5.28
600	0.58	1.12	3.86	5.22	5.87	6.36	6.60	6.77
700	0.67	1.30	4.55	6.26	7.20	7.85	8.20	8.40
800	0.76	1.48	5.25	7.33	8.59	9.45	9.92	10.16
900	0.85	1.66	5.96	8.43	10.06	11.16	11.76	12.04
1000	0.94	1.84	6.68	9.57	11.58	12.98	13.73	14.05
1100	1.03	2.02	7.42	10.74	13.17	14.90	15.81	16.18
1200	1.13	2.20	8.16	11.93	14.84	16.91	17.99	18.42

Table 9—Melting Points of Mixtures of Metals. *From "Smithsonian Physical Tables," 9th revised edition, vol. 120, Smithsonian Institution, Washington, D. C.; 1969.*

| | | | | | Melting Points, °C | | | | | | |
| | | | | Percentage of second metal in metals column | | | | | | | |
Metals	0	10	20	30	40	50	60	70	80	90	100
Pb Sn	327	295	276	262	240	220	190	185	200	216	232
Bi	327	290	—	—	179	145	126	168	205	—	271
Te	327	710	790	880	917	760	600	480	410	425	452
Ag	327	460	545	590	620	650	705	775	840	905	961
Na	327	360	420	400	370	330	290	250	200	130	97.5
Cu	327	870	920	925	945	950	955	985	1005	1020	1083
Sb	327	250	275	330	395	440	490	525	560	600	631
Al Sb	660	750	840	925	945	950	970	1000	1040	1010	631
Cu	660	630	600	560	540	580	610	755	930	1055	1083
Au	660	675	740	800	855	915	970	1025	1055	675	1063
Ag	660	625	615	600	590	580	575	570	650	750	961
Zn	660	640	620	600	580	560	530	510	475	425	419
Fe	660	860	1015	1110	1145	1145	1220	1315	1425	1500	1533
Sn	660	645	635	625	620	605	590	570	560	540	232
Sb Bi	631	610	590	575	555	540	520	470	405	330	271
Ag	631	595	570	545	520	500	505	545	680	850	961
Sn	631	600	570	525	480	430	395	350	310	255	232
Zn	631	555	510	540	570	565	540	525	510	470	419
Ni Sn	1453	1380	1290	1200	1235	1290	1305	1230	1060	800	232
Na Bi	97.5	425	520	590	645	690	720	730	715	570	271
Cd	97.5	125	185	245	285	325	330	340	360	390	321
Cd Ag	321	420	520	610	700	760	805	850	895	940	961
Tl	321	300	285	270	262	258	245	230	210	235	303
Zn	321	280	270	295	313	327	340	355	370	390	419
Au Cu	1063	910	890	895	905	925	975	1000	1025	1060	1083
Ag	1063	1062	1061	1058	1054	1049	1039	1025	1006	982	961
Pt	1063	1125	1190	1250	1320	1380	1455	1530	1610	1685	1769
K Na	63	17.5	−10	−3.5	5	11	26	41	58	77	97.5
Hg	63	—	—	—	—	90	110	135	162	265	—
Tl	63	133	165	188	205	215	220	240	280	305	303
Cu Ni	1083	1180	1240	1290	1320	1335	1380	1410	1430	1440	1453
Ag	1083	1035	990	945	910	870	830	788	814	875	961
Sn	1083	1005	890	755	725	680	630	580	530	440	232
Zn	1083	1040	995	930	900	880	820	780	700	580	419
Ag Zn	961	850	755	705	690	660	630	610	570	505	419
Sn	961	870	750	630	550	495	450	420	375	300	232
Na Hg	97.5	90	80	70	60	45	22	55	95	215	—

TABLE 10—MELTING POINTS IN °C OF LOW-MELTING-POINT ALLOYS. *From "Smithsonian Physical Tables," 9th revised edition, vol. 120, Smithsonian Institution, Washington, D. C.; 1969.*

	Percent						
Cadmium	10.8	10.2	14.8	13.1	6.2	7.1	6.7
Tin	14.2	14.3	7.0	13.8	9.4	—	—
Lead	24.9	25.1	26.0	24.3	34.4	39.7	43.4
Bismuth	50.1	50.4	52.2	48.8	50.0	53.2	49.9
Solidification at	65.5°	67.5°	68.5°	68.5°	76.5°	89.5°	95°

	Percent								
Lead	32.0	25.8	25.0	43.0	33.3	10.7	50.0	35.8	20.0	70.9
Tin	15.5	19.8	15.0	14.0	33.3	23.1	33.0	52.1	60.0	9.1
Bismuth	52.5	54.4	60.0	43.0	33.3	66.2	17.0	12.1	20.0	20.0
Solidification at	96°	101°	125°	128°	145°	148°	161°	181°	182°	234°

TEMPERATURE CHARTS OF METALS (TABLES 9–10)

SURFACE TENSIONS OF MATERIALS

To extend the surface of a liquid, some molecules from the interior must be brought to the surface and some work must be done against the inward attractive forces.

In extending the surface of a liquid, the work per unit area extended is by definition the surface tension of the liquid. The work per unit area expressed in ergs/cm² may also be expressed in dynes/cm. Then the concept of work per unit area can be replaced by a hypothetical tension γ, which is the normal tensile force per unit of length traced on the surface of the liquid.

Surface tensions of liquids and particularly solids at melting point play an important role in zone-leveling and zone-refining processes, in which the material in rod form is often in a vertical position and the molten zone is held in place by its surface tension. The maximum height of the stable molten zone is given by

$$h_z(\text{cm}) = 2.8\,(\gamma/\rho g)^{1/2}$$

from Heywang, ρ being the density in gr/cm³, and g = gravity.

As the molten zone is moving along the rod, the impurity will travel in the same direction as the molten zone if the impurity depresses the freezing point and in the opposite direction if the impurity raises the freezing point.

Some values of γ, ρ, and h_z are given in Table 11.

Surface tensions of some liquids at room temperature are given below (in dynes/cm).

Water	71.9
Alcohol	21.8
Mercury	513
Mercury under water	392
Turpentine	28.8
Glycerine	63

Rise in capillarity tube, diameter = $2r$ (Fig. 2)

$$h = 2\gamma/\rho g r.$$

Angle of contact α

$$\cos\alpha = r\rho g h/2\gamma.$$

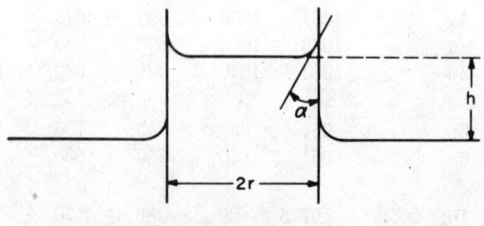

Fig. 2.

TABLE 11—SURFACE TENSIONS OF ELEMENTS AND OTHER CONSTANTS. Taylor's (1954) estimated values are given in parentheses ( ) where no experimental values were available. Room temperature values for density and italicized values for surface tension were used in calculating the surface tension/density ratios and the Heywang zone lengths. Various tables of reference (Honig, 1962; Taylor, 1954) were used to obtain many values for the other physical constants.

Element	Atomic Number	Density, $\rho(\text{g/cm}^3)$ Room Temp.	Melting Point	Surface Tension γ at Melting Point (dynes/cm)	Ratio: Surface Tension/Density	Stable Zone Height h_z(cm) from Heywang's Equation
Ag	47	10.49	9.3	800	76	0.78
Al	13	2.70	2.37	*840*, 915	311	1.58
As	33	5.72	—	—	—	—
Au	79	19.32	—	*580*, 1000	30	0.49
B	5	2.34	2.08	1060	453	1.91
Ba	56	3.5	—	224	64	0.72
Be	4	1.85	—	(1620)	876	2.64
Bi	83	9.80	10.06	388	40	0.56
C	6	2.25	—	—	—	—
Ca	20	1.55	—	*337*	218	1.32
Cd	48	8.65	8.02	*630*, 550	73	0.76
Ce	58	6.77	—	(610)	90	0.84
Co	27	8.85	—	*1880*	212	1.30
Cr	24	7.19	—	(1420)	198	1.25
Cs	55	1.90	1.84	(55)	29	0.48
Cu	29	8.96	—	*1265*, 1103	141	1.06
Dy	66	8.55	—	—	—	—
Er	68	9.15	—	—	—	—
Eu	63	5.24	—	—	—	—
Fe	26	7.87	—	*1754*, 1510	223	1.34
Ga	31	5.91	—	*358*, 735	61	0.70
Gd	64	7.86	—	—	—	—
Ge	32	5.32	5.575	600	113	0.95
Hf	72	13.09	—	1460	112	0.95
Ho	67	8.80	—	—	—	—
I	53	4.94	—	—	—	—
In	49	7.31	7.03	*513*, 599	70	0.75
Ir	77	22.5	—	(2310)	103	0.90
K	19	0.86	—	86	100	0.89
La	57	6.19	—	—	—	—
Li	3	0.534	0.51	(430)	805	2.54
Lu	71	9.85	—	—	—	—
Mg	12	1.74	1.57	*556*, 987	320	1.60
Mn	25	7.43	6.54	(1050)	141	1.06
Mo	42	10.22	—	2080	204	1.28

TABLE 11—CONTINUED.

Element	Atomic Number	Density, $\rho(g/cm^3)$ Room Temp.	Density, $\rho(g/cm^3)$ Melting Point	Surface Tension γ at Melting Point (dynes/cm)	Ratio: Surface Tension/Density	Stable Zone Height h_z (cm) from Heywang's Equation
Na	11	0.97	0.93	192.2	198	1.26
Nb	41	8.57	—	(2030)	237	1.38
Nd	60	7.00	—	688	98	0.89
Ni	28	8.9	—	*1735*, 1660, 1670	195	1.25
Os	76	22.57	—	(2450)	108	0.93
P	15	1.82	—	—	—	—
Pb	82	11.4	10.6	440	39	0.56
Pd	46	12.02	—	1470	122	0.99
Pr	59	6.77	—	—	—	—
Pt	78	21.45	—	1819 (at 2000°C)	85	0.83
Ra	88	5.00	—	—	—	—
Rb	37	1.53	—	(75)	49	0.63
Re	75	21.04	—	(2480)	118	0.97
Rh	45	12.44	10.65	1940	156	1.12
Ru	44	12.2	—	—	—	—
S	16	2.07	—	—	—	—
Sb	51	6.62	6.50	383	58	0.68
Sc	21	2.99	—	—	—	—
Se	34	4.79	—	*720*, 712, 732	150	1.57
Si	14	2.33	—	*72*	30.9	0.35
Sm	62	7.49	—	—	—	—
Sn	50	7.30	—	*566*, 625	78	0.79
Sr	38	2.60	—	288	111	0.94
Ta	73	16.6	—	*1910*, 2300	115	0.96
Tb	65	8.25	—	—	—	—
Te	52	6.24	—	(300)	48	0.62
Th	90	11.66	—	(1075)	92	0.86
Ti	22	4.51	—	*1390*, 1588, 1460	309	1.57
Tl	81	11.85	11.29	*465*, 401	39	0.56
Tm	69	9.31	—	—	—	—
U	92	19.07	—	(1020)	53	0.65
V	23	6.1	—	(1710)	280	1.50
W	74	19.3	—	2300	119	0.98
Y	39	4.47	—	—	—	—
Yb	70	6.96	—	—	—	—
Zn	30	7.13	6.7	824	116	0.96
Zr	40	6.49	—	*1400*, 1390	216	1.32

CONDUCTING MATERIALS (TABLES 12–14)

Conducting materials (Table 12) can be classified as follows:

Conductors: Resistivities from 10^{-6} to 10^{-4} ohm-cm (1 to 100 microhm-cm). Conductivities from 10^4 to 10^6 mho-cm^{-1}.

Semiconductors: Resistivities from 10^{-4} to 10^9 ohm-cm. Conductivities from 10^{-9} to 10^4 mho-cm^{-1}.

Insulators: Resistivities from 10^9 to 10^{25} ohm-cm. Conductivities from 10^{-25} to 10^{-9} mho-cm^{-1}.

TABLE 12—RESISTIVITIES OF METALS AND ALLOYS.

Material	Form	Resistivity (ohm-cm$\times 10^{-6}$)	Temperature (°C)	Temperature Coefficient
Alumel	solid	33.3	0	0.0012
Aluminum	liquid	20.5	670	
	solid	2.62	20	0.0039
Antimony	liquid	123	800	
	solid	39.2	20	0.0036
Arsenic	solid	35	0	0.0042
Beryllium		4.57	20	
Bismuth	liquid	128.9	300	
	solid	115	20	0.004
Boron		1.8×10^{12}	0	
Brass (66 Cu 34 Zn)		3.9	20	0.002
Cadmium	liquid	34	400	
	solid	7.5	20	0.0038
Carbon	diamond	5×10^{20}	15	
	graphite	1400	20	−0.0005
Cerium		78	20	
Cesium	liquid	36.6	30	
	solid	20	20	
		18.83	0	
Chromax (15 Cr, 35 Ni, balance Fe)		100	20	0.00031
Chromel	solid	70–110	0	0.00011–0.000054
Chromium		2.6	0	
Cobalt		9.7	20	0.0033
Constantan (55 Cu, 45 Ni)		44.2	20	+0.0002
Copper (commercial annealed)	liquid	21.3	1083	
	solid	1.7241	20	0.0039
Gallium	liquid	27	30	
	solid	53	0	
Germanium		45	20	
German silver (18% Ni)		33	20	0.0004
Gold	liquid	30.8	1063	
	solid	2.44	20	0.0034
		2.19	0	
Hafnium		32.1	20	
Indium	liquid	29	157	
	solid	9	20	0.00498
Iridium		5.3	20	0.0039
Iron		9.71	20	0.0052–0.0062
Kovar A (29 Ni, 17 Co, 0.3 Mn, balance Fe)		45–85	20	
Lead	liquid	98	400	
	solid	21.9	20	0.004
PbO$_2$		92		

TABLE 12—CONTINUED.

Material	Form	Resistivity (ohm-cm×10⁻⁶)	Temperature (°C)	Temperature Coefficient
Lithium	liquid	45	230	0.003
	solid	9.3	20	0.005
Magnesium		4.46	20	0.004
Manganese		5	20	
Manganin (84 Cu, 12 Mn, 4 Ni)		44	20	±0.0002
Mercury	liquid	95.8	20	0.00089
	solid	21.3	−50	
Molybdenum		5.17	0	
		4.77	20	0.0033
MnO₂		6 000 000	20	
Monel metal (67 Ni, 30 Cu, 1.4 Fe, 1 Mn)	solid	42	20	0.002
Neodymium	solid	79	18	
Nichrome (65 Ni, 12 Cr, 23 Fe)	solid	100	20	0.00017
Nickel	solid	6.9	20	0.0047
Nickel–silver (64 Cu, 18 Zn, 18 Ni)	solid	28	20	0.00026
Niobium		12.4	20	
Osmium		9	20	0.0042
Palladium		10.8	20	0.0033
Phosphor bronze (4 Sn, 0.5 P, balance Cu)		9.4	20	0.003
Platinum		10.5	20	0.003
Plutonium		150	20	
Potassium	liquid	13	62	
	solid	7	20	0.006
Praseodymium		68	25	
Rhenium		19.8	20	
Rhodium		5.1	20	0.0046
Rubidium		12.5	20	
Ruthenium		10	20	
Selenium	solid	1.2	20	
Silicon		85×10³	20	
Silver		1.62	20	0.0038
Sodium	liquid	9.7	100	
	solid	4.6	20	
Steel (0.4–0.5 C, balance Fe)		13–22	20	0.003
Steel, manganese (13 Mn, 1 C, 86 Fe)		70	20	0.001
Steel, stainless (0.1 C, 18 Cr, 8 Ni, balance Fe)		90	20	
Strontium		23	20	
Sulfur		2×10²³	20	
Tantalum		13.1	20	0.003
Thallium		18.1	20	0.004
Thorium		18	20	0.0021
Tin		11.4	20	0.0042
Titanium		47.8	25	
Tungsten		5.48	20	0.0045
Tophet A (80 Ni, 20 Cr)		108	20	0.00014
Uranium		29	0	0.0021
W₂O₅		450	20	
WO₃		2×10¹¹	20	
Zinc	liquid	35.3	420	
	solid	6	20	0.0037
Zirconium		40	20	0.0044

SEMICONDUCTING MATERIALS (TABLES 15–18)

INSULATING MATERIALS

The permittivity ϵ of an insulating material is defined by

$$\epsilon = \epsilon_0 + (P/E)$$

where ϵ_0 = permittivity of free space, P = flux density from dipoles within the dielectric medium, E = electric-field intensity, and P/E = electric susceptibility.

(*Continued on p. 4-27*)

TABLE 13—ELECTRICAL RESISTIVITY OF ROCKS AND SOILS. *From "Smithsonian Physical Tables," 9th revised edition, vol. 120, Smithsonian Institution, Washington, D. C.; 1969.*

Igneous Rocks	Resistivity (ohm-cm)
Granite	10^7–10^9
Lava flow (basic)	10^6–10^7
Lava, fresh	3×10^5–10^6
Quartz vein, massive	$> 10^6$

Metamorphic Rocks	
Marble, white	10^{10}
Marble	4×10^8
Marble, yellow	10^{10}
Schist, mica	10^7
Shale, Nonesuch	10^4
Shale, bed	10^5

Sedimentary Rocks	
Limestone	10^4
Limestone, Cambrian	10^4–10^5
Sandstone, eastern	3×10^3–10^4
Sandstone	10^5

Unconsolidated Materials	
Clay, blue	2×10^4
Clayey earth	10^4–4×10^4
Clay, fire	2×10^5
Gravel	10^5
Sand, dry	10^5–10^6
Sand, moist	10^5–10^6

TABLE 14—SUPERCONDUCTIVITY OF SOME METALS, ALLOYS, AND COMPOUNDS.

Material	Critical Temperature (K)
NbC	10.1
Niobium	9.22
TaC	9.2
Pb–As–Bi	9.0
Pb–Bi–Sb	8.9
Pb–Sn–Bi	8.5
Pb–As	8.4
MoC	7.7
Lead	7.2
N_2Pb_5	7.2
Bi_6Tl_3	6.5
Sb_2Tl_7	5.5
Lanthanum	5.2
Tantalum	4.4
Vanadium	4.3
TaSi	4.2
Mercury	4.15
PbS	4.1
Hg_5Tl_7	3.8
Tin	3.71
Indium	3.38
ZrB	2.82
WC	2.8
Rhenium	2.57
Mo_2C	2.4
Thallium	2.4
W_2C	2.05
Au_2Bi	1.84
CuS	1.6
TiN	1.4
Thorium	1.32
VN	1.3
Aluminum	1.15
Gallium	1.12
TiC	1.1
Zinc	0.95
Uranium	0.75
Osmium	0.71
Zirconium	0.54
Cadmium	0.54
Titanium	0.53
Ruthenium	0.47
Hafnium	0.35

TABLE 15—PROPERTIES OF ELEMENTAL SEMICONDUCTORS. *From Clauser et al, "Encyclopedia of Engineering Materials and Processes," Reinhold Publishing Corporation, New York; 1963.*

Element	Crystal Structure	Density (gm/cm³)	Melting Point (°C)	Linear Coefficient of Expansion (10⁻⁶/°C)	Energy Band Gap at 300 K, (eV)	Electron Mobility (cm²/volt sec)	Hole Mobility (cm²/volt sec)	
							Light Mass	Heavy Mass
B	—	2.34	2 075	—	1.4	1	—	2
C (diamond)	Cub. (f.c.), O_h^7	3.51	3 800	1.18	5.3	1 800	—	1 600
Si	Cub. (f.c.), O_h^7	2.33	1 417	4.2	1.09	1 500	1 500	480
Ge	Cub. (f.c.), O_h^7	5.32	937	6.1	0.66	3 900	14 000	1 860
α-Sn	Cub. (f.c.), O_h^7	5.75	231.9	—	0.08	144 000*	—	1 600*
As	Hex. (rhomb.), D_{3d}^5	5.73	814	3.86	1.2	—	—	—
Sb	Hex. (rhomb.), D_{3d}^5	6.68	630.5	10.88	0.11	—	—	—
α-S	Rhomb. (f.c.), V_h^{24}	2.07	112.8	64.1	2.6	—	—	—
Se	Hex., D_3^4	4.79	217	36.8	1.8	—	—	1
Se	Amorphous	4.82	—	—	2.3	0.005	—	0.15
Te	Hex., D_3^4	6.25	452	16.8	0.38	1 100	10 000	700

* Values for carrier mobilities are experimental values obtained at 300 K, except in the case of gray tin where the 77 K values are given.

TABLE 16—III-V COMPOUNDS AND THEIR PROPERTIES. *From Clauser et al, "Encyclopedia of Engineering Materials and Processes," Reinhold Publishing Corporation, New York; 1963.*

Compound	Crystal Structure	Density (gm/cm³)	Melting Point (°C)	Linear Coefficient of Expansion (10⁻⁶/°C)	Energy Band Gap at 300 K (eV)	Electron Mobility* (cm²/volt sec) Light Mass	Heavy Mass	Hole Mobility* (cm²/volt sec) Light Mass	Heavy Mass
BN	Hex. (graphite), D^4_{6h}	2.2	3 000	—	—	—	—	—	—
BN	Cub. (ZnS), T^2_d	—	—	—	4.6	—	—	—	—
BP	Cub. (ZnS), T^2_d	—	>3 000	—	—	—	—	—	500–1000
BAs	Cub. (ZnS), T^2_d	—	—	—	—	—	—	—	—
AlN	Hex., C^4_{6v}	3.26	>2 700	—	—	—	—	—	—
AlP	Cub. (ZnS), T^2_d	—	>2 100	—	2.42	—	—	—	—
AlAs	Cub. (ZnS), T^2_d	—	1 600	—	2.16	—	—	—	—
AlSb	Cub. (ZnS), T^2_d	4.28	1 065	—	1.6	—	180–230	—	420–500
GaN	Hex., C^4_{6v}	—	1 500	—	3.25	—	—	—	150–250
GaP	Cub. (ZnS), T^2_d	4.13	1 450	5.3	2.25	—	120–300	—	70–150
GaAs	Cub. (ZnS), T^2_d	5.32	1 238	5.7	1.43	8 600–11 000	1 000	3 000	426–500
GaSb	Cub. (ZnS), T^2_d	5.62	706	6.9	0.70	5 000–40 000	1 000	7 000	700–1 200
InN	Hex., C^4_{6v}	—	1 200	—	—	—	—	—	—
InP	Cub. (ZnS), T^2_d	4.79	1 062	4.5	1.27	4 800–6 800	—	—	150–200
InAs	Cub. (ZnS), T^2_d	5.67	942	5.3	0.33	33 000–40 000	—	8 000	450–500
InSb	Cub. (ZnS), T^2_d	5.78	530	5.5	0.17	78 000	—	12 000	750

* Since InSb is the only III-V compound which has been prepared with an impurity concentration low enough that the characteristic lattice carrier mobility can be determined, the best experimental value is given followed by the theoretical estimate for higher-purity material. All mobility values are for 300 K.

TABLE 17—SEMICONDUCTOR APPLICATIONS AND MATERIALS USED (INCLUDING SOME INSULATORS).
*From Clauser et al, "Encyclopedia of Engineering Materials and Processes," Reinhold Publishing Corporation,
New York; 1963.*

1. TRANSISTORS, DIODES, RECTIFIERS,
   and RELATED DEVICES

Transistors	Ge, Si, GaAs
Diodes	
Switching Diodes	Ge, Si, GaAs
Varactor Diodes	Ge, Si, GaAs
Tunnel Diodes	Ge, Si, GaAs, GaSb
Photodiodes	Ge, Si, GaAs
Zener Diodes	Ge, Si, GaAs
Microwave Diodes	Ge, Si, GaAs
Magnetodiodes	InSb
Power Rectifiers	Cu_2O, Se, Si, Ge
Varistors	SiC, Cu_2O

2. LUMINESCENT DEVICES

Electroluminescence	ZnS, ZnO
Phosphors	ZnS, CdS, ZnO, $ZnSiO_3$, $MgWO_4$, $SrWO_4$
Lasers	Al_2O_3, CaF_2, $CaWO_4$, BaF_2, $SrMoO_4$, SrF_2, LaF_2, LaF_3, As_2S_3, $CaMoO_4$, $KMgF_3$, $(BaMg)_2P_2O_7$

3. FERRITES

Soft	ZnO, MnO, NiO, Fe_2O_3
Permanent	$BaFe_{12}O_{19}$

4. SPECIAL RESISTORS

Thermistors	B, U_3O_8, Si, $(NiMn)O_2$
Photoconductors	Ge, Se, CdS, CdSe, GaAs, InSb, PbSe, PbTe
Particle Detectors	CdS, Diamond, Si, GaAs
Magnetoresistors	InSb, InAs
Piezoresistors	Si, PbTe, GaSb
Cryosars	Ge
Bokotrons	Ge
Helicons	Fe_2S
Oscillistors	Ge, Si, InSb
Chargistors	Ge, Si

5. PHOTOVOLTAIC AND HALL EFFECT

Photovoltaic Cells	Se, Si, GaAs
Photoelectromagnetic Cells	InSb, InAs
Hall Effect Devices	InSb, InAs, GaAs

6. OPTICAL MATERIALS

Infrared Lenses and Domes	Ge, Si, Se, Al_2O_3, As_2S_3, MgO, TiO_2, $SrTiO_3$

7. THERMOELECTRIC DEVICES

Generators	PbTe, Bi_2Te_3, ZnSb, GeTe, MnTe, CeS
Refrigerators	Bi_2Te_3

8. PIEZOELECTRIC DEVICES　　　　　$BaTiO_3$, $PbTiO_3$, $(PbZr)TiO_3$, $PbNbO_2$, CdS, GaAs

9. XEROGRAPHY　　　　　　　　　　　Se, ZnO

10. ELECTRON EMISSION　　　　　　　BaO, SrO

TABLE 18—COMPOUND SEMICONDUCTORS. *From Clauser et al, "Encyclopedia of Engineering Materials and Processes," Reinhold Publishing Corporation, New York; 1963*

Formula No.	Typical Compounds
I-V	KSb, K$_3$Sb, CsSb, Cs$_3$Sb, Cs$_3$Bi
I-VI	Ag$_2$S, Ag$_2$Se, Cu$_2$S, Cu$_2$Te
II-IV	Mg$_2$Si, Mg$_2$Ge, Mg$_2$Sn, Ca$_2$Si, Ca$_2$Sn, Ca$_2$Pb, MnSi$_2$, CrSi$_2$
II-V	ZnSb, CdSb, Mg$_3$Sb$_2$, Zn$_3$As$_2$, Cd$_3$P$_2$, Cd$_3$As$_2$
II-VI	CdS, CdSe, CdTe, ZnS, ZnSe, ZnTe, HgS, HgSe, HgTe, MoTe$_2$, RuTe$_2$, MnTe$_2$, BeS, MgS, CaS
III-VI	Al$_2$S$_3$, Ga$_2$S$_3$, Ga$_2$Se$_3$, Ga$_2$Te$_3$, In$_2$S$_3$, In$_2$Se$_3$, In$_2$Te$_3$, GaS, GaSe, GaTe, InS, InSe, InTe
IV-IV	SiC
IV-VI	PbS, PbSe, PbTe, TiS$_2$, GeTe
V-VI	Sb$_2$S$_3$, Sb$_2$Se$_3$, Sb$_2$Te$_3$, As$_2$Se$_3$, As$_2$Te$_3$, Bi$_2$S$_3$, Bi$_2$Se$_3$, Bi$_2$Te$_3$, Ce$_2$S$_3$, Gd$_2$Se$_3$
A^{I}B^{II}C$_2^{VI}$	CuFeS$_2$
A^{I}B^{III}C$_2^{VI}$	C$_4$AlS$_2$, CuInS$_2$, CuInSe$_2$, CuInTe$_2$, AgInSe$_2$, AgInTe$_2$, CuGaTe$_2$
A^{I}B^{V}C$_2^{VI}$	AgSbSe$_2$, AgSbTe$_2$, AgBiS$_2$, AgBiSe$_2$, AgBiTe$_2$, AuSbTe$_2$, Au(SbBi)Te$_2$
A$_3^I$B^{V}C$_3^{VI}$	Cu$_3$SbS$_3$, Cu$_3$AsS$_3$
A$_3^I$B^{V}C$_4^{VI}$	Cu$_3$AsSe$_4$
A^{II}B^{IV}C$_2^V$	ZnSnAs$_2$
Oxides	SrO, BaO, MnO, NiO, Fe$_2$O$_3$, BaFe$_{12}$O$_{19}$, Al$_2$O$_3$, In$_2$O$_3$, TiO$_2$, BeO, MgO, CaO, CdO, ZnO, SiO$_2$, GeO$_2$, ZnSiO$_3$, MgWO$_3$, CuO

To deal with a dimensionless coefficient, it is customary to use the relative dielectric constant ϵ_r defined by

$$\epsilon_r = \epsilon/\epsilon_0.$$

The relative dielectric constant ϵ_r is a function of temperature and frequency. From Table 19, which gives the values of ϵ_r as a function of frequencies at room temperature, it is easy to get

$$\epsilon = \epsilon_r \epsilon_0.$$

In the MKS rationalized system of units the permittivity of vacuum is equal to

$$\epsilon_0 = 10^{-9}/36\pi = 8.854 \times 10^{-12} \text{ farad/meter}$$

and we have

Coulomb Law

$$F = (1/4\pi\epsilon_0\epsilon_r)(q_1q_2/R^2)$$

Gauss Law

$$\Phi = (\epsilon_0\epsilon_r)^{-1} \sum q_i.$$

The dissipation factor of an insulating material (Table 19) is defined as the ratio of the energy dissipated to the energy stored in the dielectric per hertz, or as the tangent of the loss angle. For dissipation factors less than 0.1, the dissipation factor may be considered equal to the power factor of the dielectric, which is the cosine of the phase angle by which the current leads the voltage.

Many of the materials listed are characterized by a peak dissipation factor occurring somewhere in the frequency range, this peak being accompanied by a rapid change in the dielectric constant. These effects are the result of a resonance phenomenon occurring in polar materials. The position of the dissipation-factor peak in the frequency spectrum is very sensitive to temperature. An increase in the temperature increases the frequency at which the peak occurs, as illustrated qualitatively in Fig. 3. Nonpolar materials have very low losses without a noticeable peak; the dielectric constant remains essentially unchanged over the frequency range.

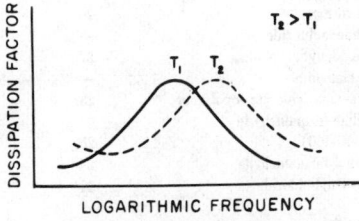

Fig. 3.

TABLE 19—CHARACTERISTICS

Material Composition	$T°C$	Dielectric Constant at (Frequency in Hertz)					
		60	10^3	10^6	10^8	$3×10^9$	$2.5×10^{10}$
Ceramics:							
Aluminum oxide	25	—	8.83	8.80	8.80	8.79	—
Barium titanate†	26	1250	1200	1143	—	600	100
Calcium titanate	25	168	167.7	167.7	167.7	165	—
Magnesium oxide	25	—	9.65	9.65	9.65	—	—
Magnesium silicate	25	6.00	5.98	5.97	5.96	5.90	—
Magnesium titanate	25	—	13.9	13.9	13.9	13.8	13.7
Oxides of aluminum, silicon, magnesium, calcium, barium	24	—	6.04	6.04	—	5.90	—
Porcelain (dry process)	25	5.5	5.36	5.08	5.04	—	—
Steatite 410	25	5.77	5.77	5.77	5.77	5.7	—
Strontium titanate	25	—	233	232	232	—	—
Titanium dioxide (rutile)	26	—	100	100	100	—	—
Glasses:							
Iron-sealing glass	24	8.41	8.38	8.30	8.20	7.99	7.84
Soda-borosilicate	25	—	4.97	4.84	4.84	4.82	4.65
100% silicon dioxide (fused quartz)	25	3.78	3.78	3.78	3.78	3.78	3.78
Plastics:							
Alkyd resin	25	—	5.10	4.76	4.55	4.50	—
Cellulose acetate-butyrate, plasticized	26	3.60	3.48	3.30	3.08	2.91	—
Cresylic acid–formaldehyde, 50% α-cellulose	25	5.45	4.95	4.51	3.85	3.43	3.21
Cross-linked polystyrene	25	2.59	2.59	2.58	2.58	2.58	—
Epoxy resin (Araldite CN–501)	25	—	3.67	3.62	3.35	3.09	—
Epoxy resin (Epon resin RN–48)	25	—	3.63	3.52	3.32	3.04	—
Foamed polystyrene, 0.25% filler	25	1.03	1.03	1.03	—	1.03	1.03
Melamine—formaldehyde, α-cellulose	24	—	7.57	7.00	6.0	4.93	—
Melamine—formaldehyde, 55% filler	26	—	6.00	5.75	5.5	—	—
Phenol—formaldehyde (Bakelite BM 120)	25	4.90	4.74	4.36	3.95	3.70	3.55
Phenol—formaldehyde, 50% paper laminate	26	5.25	5.15	4.60	4.04	3.57	—
Phenol—formaldehyde, 65% mica, 4% lubricants	24	5.1	5.03	4.78	4.72	4.71	—
Polycarbonate	—	3.17	3.02	2.96	—	—	—
Polychlorotrifluoroethylene	25	2.72	2.63	2.42	2.32	2.29	2.28
Polyethylene	25	2.26	2.26	2.26	2.26	2.26	2.26
Polyethylene-terephthalate	—	3.16	3.12	2.98	—	—	—
Polyethylmethacrylate	22	—	2.75	2.55	2.52	2.51	2.5
Polyhexamethylene-adipamide (nylon)	25	3.7	3.50	3.14	3.0	2.84	2.73
Polyimide	—	—	3.5	3.4	—	—	—
Polyisobutylene	25	2.23	2.23	2.23	2.23	2.23	—
Polymer of 95% vinyl-chloride, 5% vinyl-acetate	20	—	3.15	2.90	2.8	2.74	—
Polymethyl methacrylate	27	3.45	3.12	2.76	—	2.60	—
Polyphenylene oxide	—	2.55	2.55	2.55	—	2.55	—
Polypropylene	—	2.25	2.25	2.55	—	—	—
Polystyrene	25	2.56	2.56	2.56	2.55	2.55	2.54
Polytetrafluoroethylene (teflon)	22	2.1	2.1	2.1	2.1	2.1	2.08
Polyvinylcyclohexane	24	—	2.25	2.25	2.25	2.25	—
Polyvinyl formal	26	3.20	3.12	2.92	2.80	2.76	2.7
Polyvinylidene fluoride	—	8.4	8.0	6.6	—	—	—
Urea-formaldehyde, cellulose	27	6.6	6.2	5.65	5.1	4.57	—
Urethane elastomer	—	6.7–7.5	6.7–7.5	6.5–7.1	—	—	—
Vinylidene-vinyl chloride copolymer	23	5.0	4.65	3.18	2.82	2.71	—
100% aniline-formaldehyde (Dilectene-100)	25	3.70	3.68	3.58	3.50	3.44	—
100% phenol-formaldehyde	24	8.6	7.15	5.4	4.4	3.64	—
100% polyvinyl-chloride	20	3.20	3.10	2.88	2.85	2.84	—
Organic Liquids:							
Aviation gasoline (100 octane)	25	—	—	1.94	1.94	1.92	—
Benzene (pure, dried)	25	2.28	2.28	2.28	2.28	2.28	2.28

of Insulating Materials.*

| Dissipation Factor at | | | | | | Dielectric Strength in Volts/Mil at 25°C | DC Volume Resistivity in Ohm-cm at 25°C | Thermal Expansion (Linear) in Parts/°C | Softening Point in °C | Moisture Absorption in Percent |
| (Frequency in Hertz) | | | | | | | | | | |
60	10^3	10^6	10^8	3×10^9	2.5×10^{10}					
—	0.00057	0.00033	0.00030	0.0010	—	—	—	—	1400–1430	0.1
0.056	0.0130	0.0105	—	0.30	0.60	75	10^{12}—10^{13}	—	1510	<0.1
0.006	0.00044	0.0002	—	0.0023	—	100	10^{12}—10^{14}	—	1510	<0.1
—	<0.0003	<0.0003	<0.0003	—	—	—	—	—	—	—
0.012	0.0034	0.0005	0.0004	0.0012	—	—	>10^{14}	9.2×10^{-6}	1350	0.1–1
—	0.0011	0.0004	0.0005	0.0017	0.0065	—	—	—	—	—
—	0.0019	0.0011	—	0.0024	—	—	—	7.7×10^{-6}	1325	—
0.03	0.0140	0.0075	0.0078	—	—	—	—	—	—	—
—	0.0030	0.0007	0.0006	0.00089	—	—	—	—	—	—
—	0.0011	0.0002	0.0001	—	—	100	10^{12}—10^{14}	—	1510	0.1
—	0.0015	0.0003	0.00025	—	—	—	—	—	—	—
—	0.0004	0.0005	0.0009	0.00199	0.0112	—	10^{10} at 250°	132×10^{-7}	484	Poor
—	0.0055	0.0036	0.0030	0.0054	0.0090	—	7×10^7 at 250°	50×10^{-7}	693	—
0.0009	0.00075	0.0001	0.0002	0.00006	0.00025	410 (0.25″)	>10^{19}	5.7×10^{-7}	1667	—
—	0.0236	0.0149	0.0138	0.0108	—	—	—	—	—	—
0.0045	0.0097	0.018	0.017	0.028	—	250–400 (0.125″)	—	$11-17\times10^{-5}$	60–121	2.3
0.098	0.033	0.036	0.055	0.051	0.038	1020 (0.033″)	3×10^{13}	3×10^{-5}	>125	1.2
0.0004	0.0005	0.0016	0.0020	0.0019	—	—	—	—	—	—
—	0.0024	0.019	0.034	0.027	—	405 (0.125″)	>3.8×10^7	4.77×10^{-5}	109 (distortion)	0.14
—	0.0038	0.0142	0.0264	0.021	—	—	—	—	85	low
<0.0002	<0.0001	<0.0002	—	0.0001	—	—	—	—	99 (stable)	0.4–0.6
—	0.0122	0.041	0.085	0.103	—	300–400	—	—	—	0.6
—	0.0119	0.0115	0.020	—	—	—	—	1.7×10^{-5}	—	—
0.08	0.0220	0.0280	0.0380	0.0438	0.0390	300 (0.125″)	10^{11}	$30-40\times10^{-6}$	<135 (distortion)	<0.6
0.025	0.0165	0.034	0.057	0.060	—	—	—	—	—	—
0.015	0.0104	0.0082	0.0115	0.0126	—	—	—	—	—	—
0.009	0.0021	0.010	—	—	—	364 (0.125″)	2×10^{16}	7×10^{-5}	135 (deflection)	—
0.015	0.0270	0.0082	—	0.0028	0.0053	—	10^{18}	—	—	—
<0.0002	<0.0002	<0.0002	0.0002	0.00331	0.0006	1200 (0.033″)	10^{17}	19×10^{-5} (varies)	95–105 (distortion)	0.03
0.0021	0.0047	0.016	—	—	—	4000 (0.002″)	—	—	60 (distortion)	low
—	0.0294	0.0090	—	0.0075	0.0083	—	—	—	—	—
0.018	0.0186	0.0218	0.0200	0.0117	0.0105	400 (0.125″)	8×10^{14}	10.3×10^{-5}	65 (distortion)	1.5
—	0.002	0.003	—	—	—	570	—	—	—	—
0.0004	0.0001	0.0001	0.0003	0.00047	—	600 (0.010″)	—	—	25 (distortion)	low
—	0.0165	0.0150	0.0080	0.0059	—	—	—	—	—	—
0.064	0.0465	0.0140	—	0.0057	—	990 (0.030″)	>5×10^{16}	$8-9\times10^{-5}$	70–75 (distortion)	0.3–0.6
0.0004	0.0003	0.0007	—	0.0011	—	500 (0.125″)	10^{17}	5.3×10^{-5}	195 (deflection)	—
<0.0005	<0.0005	<0.0005	—	—	—	650 (0.125″)	6×10^{16}	$6-8.5\times10^{-5}$	99–116 (deflection)	—
<0.00005	<0.00005	0.00007	<0.0001	0.00033	0.0012	500–700 (0.125″)	10^{18}	$6-8\times10^{-5}$	82 (distortion)	0.05
<0.0005	<0.0003	<0.0002	<0.0002	0.00015	0.0006	1000–2000 (0.005″–0.012″)	10^{17}	9.0×10^{-5}	66 (distortion) (stable to 300)	0.00
—	0.0002	<0.0002	<0.0002	0.00018	—	—	—	—	—	—
0.003	0.0100	0.019	0.013	0.0113	0.0115	860 (0.034″)	>5×10^{16}	7.7×10^{-5}	190	1.3
0 049	0.018	0.17	—	—	—	260 (0.125″)	2×10^{14}	12×10^{-5}	148 (deflection)	—
0.032	0.024	0.027	0.050	0.0555	—	375 (0.085″)	—	2.6×10^{-5}	152 (distortion)	2
0.016	0.055	—	—	—	—	450–500 (0.125″)	2×10^{11}	$10-20\times10^{-5}$	—	—
0.042	0.063	0.057	0.0180	0.0072	—	300 (0.125″)	10^{14}—10^{16}	15.8×10^{-5}	150	<0.1
0.0033	0.0032	0.0061	0.0033	0.0026	—	810 (0.068″)	10^{16}	5.4×10^{-5}	125	0.06–0.08
0.15	0.082	0.060	0.077	0.052	—	277 (0.125″)	—	$8.3-13\times10^{-5}$	50 (distortion)	0.42
0.0115	0.0185	0.0160	0.0081	0.0055	—	400 (0.125″)	10^{14}	6.9×10^{-5}	54 (distortion)	0.05–0.15
—	—	—	0.0001	0.0014	—	—	—	—	—	—
<0.0001	<0.0001	<0.0001	<0.0001	<0.0001	<0.0001	—	—	—	—	—

| Material Composition | $T°C$ | \multicolumn{6}{c}{Dielectric Constant at} |
| | | \multicolumn{6}{c}{(Frequency in Hertz)} |
		60	10^3	10^6	10^8	3×10^9	2.5×10^{10}
Organic Liquids:—Continued							
Carbon tetrachloride	25	2.17	2.17	2.17	2.17	2.17	—
Ethyl alcohol (absolute)	25	—	—	24.5	23.7	6.5	—
Ethylene glycol	25	—	—	41	41	12	—
Jet fuel (JP-3)	25	—	—	2.08	2.08	2.04	—
Methyl alcohol (absolute analytical grade)	25	—	—	31	31.0	23.9	—
Methyl or ethyl siloxane polymer (1000 cs)	22	2.78	2.78	2.78	—	2.74	—
Monomeric styrene	22	2.40	2.40	2.40	2.40	2.40	—
Transil oil	26	2.22	2.22	2.22	2.20	2.18	—
Vaseline	25	2.16	2.16	2.16	2.16	2.16	—
Waxes:							
Beeswax, yellow	23	2.76	2.66	2.53	2.45	2.39	—
Dichloronaphthalenes	23	3.14	3.04	2.98	2.93	2.89	—
Polybutene	25	2.34	2.34	2.34	2.30	2.27	—
Vegetable and mineral waxes	25	2.3	2.3	2.3	2.3	2.25	—
Rubbers:							
Butyl rubber	25	2.39	2.38	2.35	2.35	2.35	—
GR-S rubber	25	2.96	2.96	2.90	2.82	2.75	—
Gutta-percha	25	2.61	2.60	2.53	2.47	2.40	—
Hevea rubber (pale crepe)	25	2.4	2.4	2.4	2.4	2.15	—
Hevea rubber, vulcanized (100 pts pale crepe, 6 pts sulfur)	27	2.94	2.94	2.74	2.42	2.36	—
Neoprene rubber	24	6.7	6.60	6.26	4.5	4.00	4.0
Organic polysulfide, fillers	23	—	2260	110	30	16	13.6
Silicone-rubber compound	25	—	3.35	3.20	3.16	3.13	—
Woods:‡							
Balsawood	26	1.4	1.4	1.37	1.30	1.22	—
Douglas fir	25	2.05	2.00	1.93	1.88	1.82	1.78
Douglas fir, plywood	25	2.1	2.1	1.90	—	—	1.6
Mahogany	25	2.42	2.40	2.25	2.07	1.88	1.6
Yellow birch	25	2.9	2.88	2.70	2.47	2.13	1.87
Yellow poplar	25	1.85	1.79	1.75	—	1.50	1.4
Miscellaneous:							
Amber (fossil resin)	25	2.7	2.7	2.65	—	2.6	—
DeKhotinsky cement	23	3.95	3.75	3.23	—	2.96	—
Gilsonite (99.9% natural bitumen)	26	2.69	2.66	2.58	2.56	—	—
Shellac (natural XL)	28	3.87	3.81	3.47	3.10	2.86	—
Mica, glass-bonded	25	—	7.45	7.39	—	—	—
Mica, glass, titanium dioxide	24	—	9.3	9.0	—	—	—
Ruby mica	26	5.4	5.4	5.4	5.4	5.4	—
Paper, royalgrey	25	3.30	3.29	2.99	2.77	2.70	—
Selenium (amorphous)	25	—	6.00	6.00	6.00	6.00	6.00
Asbestos fiber–chrysotile paper	25	—	4.80	3.1	—	—	—
Sodium chloride (fresh crystals)	25	—	5.90	5.90	—	—	5.90
Soil, sandy dry	25	—	2.91	2.59	2.55	2.55	—
Soil, loamy dry	25	—	2.83	2.53	2.48	2.44	—
Ice (from pure distilled water)	−12	—	—	4.15	3.45	3.20	—
Freshly fallen snow	−20	—	3.33	1.20	1.20	1.20	—
Hard-packed snow followed by light rain	−6	—	—	1.55	—	1.5	—
Water (distilled)	25	—	—	78.2	78	76.7	34

* Mostly taken from "Tables of Dielectric Materials," vols. I–IV, prepared by the Laboratory for Insulation Research of the Massachusetts Institute of Technology, Cambridge, Massachusetts, January 1953; from "Dielectric Materials and Applications," A. R. von Hippel, editor, John Wiley & Sons, New York, N. Y., 1954; and from "Modern Plastics Encyclopedia," Joel Frados, editor, 1301 Avenue of the Americas, New York, N. Y., 1962. Materials listed are typical of a class. Further data should be sought for a particular material of interest.

Continued.

	Dissipation Factor at					Dielectric Strength in Volts/Mil at 25°C	DC Volume Resistivity in Ohm-cm at 25°C	Thermal Expansion (Linear) in Parts/°C	Softening Point in °C	Moisture Absorption in Percent
		(Frequency in Hertz)								
60	10^3	10^6	10^8	3×10^9	2.5×10^{10}					
0.007	0.0008	<0.00004	<0.0002	0.0004	—	—	—	—	—	—
—	—	0.090	0.062	0.250	—	—	—	—	—	—
—	—	0.030	0.045	1.00	—	—	—	—	—	—
—	—	0.0001	—	0.0055	—	—	—	—	—	—
—	—	0.20	0.038	0.64	—	—	—	—	—	—
0.0001	0.00008	<0.0003	—	0.0096	—	—	—	—	—	—
0.01	0.005	<0.0003	—	0.0020	—	300 (0.100″)	3×10^{12}	—	—	0.06
0.001	<0.00001	<0.0005	0.0048	0.0028	—	300 (0.100″)	—	—	−40 (pour point)	—
0.0004	0.0002	<0.0001	<0.0004	0.00066	—	—	—	—	—	—
—	0.0140	0.0092	0.0090	0.0075	—	—	—	—	45–64 (melts)	—
0.10	0.0110	0.0003	0.0017	0.0037	—	—	—	—	35–63 (melts)	nil
0.0002	0.0003	0.00133	0.00133	0.0009	—	—	—	—	—	—
0.0009	0.0006	0.0004	0.0004	0.00046	—	—	—	—	57	—
0.0034	0.0035	0.0010	0.0010	0.0009	—	—	—	—	—	—
0.0008	0.0024	0.0120	0.0080	0.0057	—	870 (0.040″)	2×10^{15}	—	—	—
0.0005	0.0004	0.0042	0.0120	0.0060	—	—	10^{15}	—	—	—
0.0030	0.0018	0.0018	0.0050	0.0030	—	—	—	—	—	—
0.005	0.0024	0.0446	0.0180	0.0047	—	—	—	—	—	—
0.018	0.011	0.038	0.090	0.034	0.025	300 (0.125″)	8×10^{12}	—	—	nil
—	1.29	0.39	0.28	0.22	0.10	—	—	—	—	—
—	0.0067	0.0030	0.0032	0.0097	—	—	—	—	—	—
0.058	0.0040	0.0120	0.0135	0.100	—	—	—	—	—	—
0.004	0.0080	0.026	0.033	0.027	0.032	—	—	—	—	—
0.012	0.0105	0.0230	—	—	0.0220	—	—	—	—	—
0.008	0.0120	0.025	0.032	0.025	0.020	—	—	—	—	—
0.007	0.0090	0.029	0.040	0.033	0.026	—	—	—	—	—
0.004	0.0054	0.019	—	0.015	0.017	—	—	—	—	—
0.001	0.0018	0.0056	—	0.0090	—	2300 (0.125″)	Very high	—	200	—
0.049	0.0335	0.024	—	0.021	—	—	—	9.8×10^{-5}	80–85	—
0.006	0.0035	0.0016	0.0011	—	—	—	—	—	155 (melts)	—
0.006	0.0074	0.031	0.030	0.0254	—	—	10^{16}	—	80	low after baking
—	0.0019	0.0013	—	—	—	—	—	—	—	—
—	0.0125	0.0026	—	0.0040	—	—	—	—	400	<0.5
0.005	0.0006	0.0003	0.0002	0.0003	—	3800–5600 (0.040″)	5×10^{13}	—	—	—
0.010	0.0077	0.038	0.066	0.056	—	202 (0.125″)	—	—	—	—
—	0.0004	<0.0003	<0.0002	0.00018	0.0013	—	—	—	—	—
—	0.15	0.025	—	—	—	—	—	—	—	—
—	<0.0001	<0.0002	—	—	<0.0005	—	—	—	—	—
—	0.08	0.017	—	0.0062	—	—	—	—	—	—
—	0.05	0.018	—	0.0011	—	—	—	—	—	—
—	—	0.12	0.035	0.0009	—	—	—	—	—	—
—	0.492	0.0215	—	0.00029	—	—	—	—	—	—
—	—	0.29	—	0.0009	—	—	—	—	—	—
—	—	0.040	0.005	0.157	0.2650	—	10^6	—	—	—

† Dielectric constant and dissipation factor depend on electrical field strength.

‡ Field perpendicular to grain.

Another effect that contributes to dielectric losses is that of ionic or electronic conduction. This loss, if present, is important usually at the lower end of the frequency range only, and is distinguished by the fact that the dissipation factor varies inversely with frequency. Increase in temperature increases the loss due to ionic conduction because of increased ionic mobility.

The data given on dielectric strength are accompanied by the thickness of the specimen tested because the dielectric strength, expressed in volts/mil, varies inversely with the square root of thickness, approximately.

The direct-current volume resistivity of many materials is influenced by changes in temperature or humidity. The values given in the table may be reduced several decades by raising the temperature toward the higher end of the working range of the material, or by raising the relative humidity of the air surrounding the material to above 90 percent.

MAGNETIC MATERIALS (TABLES 20–22)

The permeability μ of a magnetic material is defined by

$$\mu = \mu_0 + (M/H)$$

where μ_0 = permeability of free space, M = flux density or magnetic polarization, H = magnetic-field strength, and M/H = magnetic susceptibility.

To deal with a dimensionless coefficient, it is customary to use the relative permeability μ_r defined by

$$\mu_r = \mu/\mu_0 = 1 + M/\mu_0 H.$$

The relative permeability μ_r of magnetic materials is given in Table 20. We have $\mu = \mu_0\mu_r$, with $\mu_0 = 1.257 \times 10^{-6}$ henry per meter in the MKS rationalized system of units.

For ferromagnetic materials μ_r is larger than 1, for paramagnetic materials μ_r is slightly larger than 1, and for diamagnetic materials μ_r is less than 1.

Applications in Rationalized Systems

Coulomb Law

$$F = (1/4\pi\mu_r\mu_0)(m_1 m_2/r^2).$$

Field Strength

$$H = (1/4\pi\mu_r\mu_0)(m/r^2).$$

F is in newtons if m_1 and m_2 are in webers and r in meters. H is in ampere-turns per meter if m is in webers and r in meters.

In ferromagnetic materials the permeability is a function of temperature, pressure, frequency, and magnetic strength.

In paramagnetic and diamagnetic materials the permeability is independent of magnetic strength but depends also on temperature, pressure, and frequency.

FERRITES

Ferrite is the common term applied to a wide range of different ceramic ferromagnetic materials.

(*Continued on p. 4-36*)

TABLE 20—RELATIVE PERMEABILITY μ_r OF VARIOUS MATERIALS.

	μ_r Maximum	μ_r at Small Magnetization
Ferromagnetic:		
Cobalt	60	60
Nickel	50	50
Cast iron	90	60
Silicon iron	7 000	3 500
Transformer iron	5 500	3 000
Very pure iron	8 000	4 000
Machine steel	450	300
Paramagnetic:		
Aluminum	1.000 000 65	
Beryllium	1.000 000 79	
MnSO$_4$	1.000 100	
NiCl	1.000 040	
Diamagnetic:		
Bismuth	0.999 998 600	
Paraffin	0.999 999 42	
Silver	0.999 999 81	
Wood	0.999 999 50	

Table 21—Properties of High-Permeability Materials. These properties are expressed in the CGS electromagnetic system (typical values).*

Name	Composition, %	Permeability		Coercivity, H_c (Oe)	Retentivity, B_r (G)	B (max) (G)	Resistivity ($\mu\Omega$-cm)
		Initial	Maximum				
Iron, pure (laboratory conditions)	Annealed	25 000	350 000	0.05	12 000	14 000	9.7
Iron, Swedish		250	5 500	1.0	13 000	21 000	10
Iron, cast		100	600	4.5	5 300	20 000	30
Iron, silicon	4 Si, bal. Fe (hot rolled)	500	7 000	0.3	7 000	20 000	50
Rhometal	36 Ni, bal. Fe	1 000	5 000	0.5	3 600	10 000	90
Permalloy 45	45 Ni, bal. Fe	2 500	25 000	0.3		16 000	45
Mumetal	71–78 Ni, 4.3–6 Cu, 0–2 Cr, bal. Fe	20 000	100 000	0.05	6 000	7 200	25–50
Supermalloy	79 Ni, 5 Mo, bal. Fe	100 000	1 000 000	0.002		8 000	60
HyMu80	80 Ni, bal. Fe	20 000	100 000	0.05		8 700	57
Alfenol	16 Al, bal. Fe	3 450	116 000	0.025	3 800	7 825	150
Permendure	50 Co, 1–2 V, bal Fe	800	4 500	2.0	14 000	24 000	26
Sendust	10 Si, 5 Al, bal. Fe (cast)	30 000	120 000	0.05	5 000	10 000	60–80
Ferroxcube 3	Mn-Zn-Ferrite	1 000	1 500	0.1	1 000	3 000	$>10^6$
Ferroxcube 101	Ni-Zn-Ferrite	1 100		0.18	1 100	2 300	$>10^5$

* The above values are approximate only and vary with heat treatment and mechanical working of the material. Data sources include: P. R. Bardell, "Magnetic Materials in the Electrical Industry," courtesy Philosophical Library, New York; 1955. R. M. Bozorth, "Ferromagnetism," D. Van Nostrand Company, Inc., Princeton, N.J.; 1951. Technical catalog data by Allegheny Ludlum Steel Corporation, Pittsburgh; and Ferroxcube Corporation, Saugerties, N.Y.

TABLE 22—TYPICAL PROPERTIES OF HIGH-REMANENCE MAGNETIC MATERIALS
(CGS electromagnetic system).*

Name and Composition, %	H(max) (Oe)	B(max) (G)	Retentivity B_r(G)	Coercivity H_c(Oe)	Ext. energy $B_d H_d$ (max)	Form†	Resistivity ($\mu\Omega$–cm)
Carbon steel 1 C, 0.5 Mn, bal. Fe	300	14 800	8 600	48	180 000	Bar F-P-M	
Tungsten steel 5 W, 1 C, bal. Fe	300	14 500	10 300	70	320 000	Bar H-F-M	
Chromium steel 3.5 Cr, 1 C, bal. Fe	300	13 500	9 000	63	290 000	Bar H-F-M	
Cobalt steel 36 Co, 35 Cr, 3 W, 0.85 C, bal. Fe	1 000	15 500	9 000	210	936 000	Bar F-P-M	
Alnico 1 12 Al, 20 Ni, 5 Co, bal. Fe	2 000	12 350	7 100	400	1 300 000	Cast H-B	75
Alnico 2 (sintered) 10 Al, 17 Ni, 12.5 Co, 6 Cu, bal. Fe	2 000	12 000	6 900	520	1 430 000	Sintered H	
Alnico 2 (cast) 10 Al, 17 Ni, 12.5 Co, 6 Cu, bal. Fe	2 000	12 600	7 200	540	1 600 000	Cast H-B	65
Alnico 3 (section over ⅝″×⅜″) 12 Al, 25 Ni, bal. Fe	2 000	12 000	6 700	450	1 380 000	Cast H-B	60
Alnico 4 (cast and sintered) 12 Al, 28 Ni, 5 Co, bal. Fe	3 000	11 850	5 200	700	1 200 000	C or S H-B	75
Alnico 5 8 Al, 14 Ni, 24 Co, 3 Cu, bal. Fe	2 000	15 700	12 000	575	4 500 000	Cast H-B	47
Alnico 6 8 Al, 15 Ni, 24 Co, 3 Cu, 1.25 Ti, bal. Fe	3 000	14 300	10 000	750	3 500 000	Cast H-B	
Alnico 12 6 Al, 18 Ni, 35 Co, 8 Ti, bal. Fe	3 000	12 800	5 800	950	1 500 000	Cast H-B	
Cunife I 60 Cu, 20 Ni, 20 Fe	2 400	8 400	5 800	600	1 960 000	Wire-Tape D-M	45

TABLE 22—CONTINUED.

Name and Composition, %	H(max) (Oe)	B(max) (G)	Retentivity B_r(G)	Coercivity H_e(Oe)	Ext. energy B_dH_d (max)	Form†	Resistivity ($\mu\Omega$–cm)
Cunife II 50 Cu, 20 Ni, 2.5 Co, bal. Fe	2 400	8 000	7 300	260	780 000	D-M	
Cunico 50 Cu, 20 Ni, 29 Co	3 200	8 000	3 400	660	800 000	Strip D-M	
Vectolite 30 Fe$_2$O$_3$, 44 Fe$_3$O$_4$, 26 Co$_2$O$_3$	3 000	4 800	1 600	1 000	600 000	Sintered B	
Sintered oxide 30 Fe$_2$O$_3$, 40 Fe$_3$O$_4$, 26 Co$_2$O$_3$	3 000	4 800	900	1 000	500 000	Powder B-weak	10^{12}
Silmanal 86.75 Ag, 8.8 Mn, 4.45 Al	20 000	20 830 $B_i=830$	550	6 000	75 600	Rod W	26
77% platinum alloy 77 Pt, 23 Co	15 000	22 900	4 500	2 000	3 800 000		50
Vicalloy 52 Co, 11 V, 37 Fe			14 100 9 400	200 400	1 350 000 2 400 000	Wire strip	
Remalloy or Comol 17 Mo, 12 Co, bal. Fe	1 000	17 000	10 500	250	1 100 000	Rolled F-M-P	
17% Cobalt 17 Co, 0.75 C, 3 Na, 2.5 Cr, 8 W	1 000	15 000	9 500	150	650 000	Rolled H-A-F	
Iron oxide powder 8 Fe$_2$O$_3$, Fe$_3$O$_4$	Density 2.81 g/cm^3		2 265	550			
Iron oxide powder 8 Fe$_2$O$_3$, Fe$_3$O$_4$	Density 4.96 g/cm^3		7 480	395			
Magnetite Plated coating 80 Co, 20 Ni	Density 2.62 g/cm^3		1 600 10 000	190 250			

* The values are approximate only. Data sources include R. M. Bozorth, "Ferromagnetism," D. Van Nostrand Company, Inc., Princeton, N. J.; 1951.

† A—Annealed for machining; B—Brittle; D—Ductile; F—Hot Forged; H—Hard; M—Machinable; P—Punch; W—Workable.

TABLE 23—FERRITE CHARACTERISTICS.

Ferrite	Saturation Moment in Gauss	Curie Temperature in °C	Saturation Moment in Bohr Magnetons n_B	X-ray Density	Lattice Constant	First-Order Anisotropy Constant K_1	Saturation Magneto-striction $\lambda_s \times 10^6$
$NiFe_2O_4$	3400	585	2.3	5.38	8.34	−0.06	−22
$Ni_{0.8}Zn_{0.2}Fe_2O_4$	4600	460	3.5	—	—	—	−18.5
$Ni_{0.6}Zn_{0.4}Fe_2O_4$	5800	360	4.8	—	—	—	−15.0
$Ni_{0.5}Zn_{0.4}Fe_2O_4$	5500	290	5.0	—	—	—	−8.3
$Ni_{0.3}Zn_{0.5}Fe_2O_4$	2600	85	4.0	—	—	−0.004	−1.0
$MnFe_2O_4$	5200	300	0	5.00	8.50	−0.04	−14
$Mn_{0.5}Zn_{0.5}Fe_2O_4$	—	100	6.0	—	—	−0.004	—
$FeFe_2O_4$	6000	585	4.1	5.24	8.39	−0.135	+41
$CoFe_2O_4$	5000	520	3.8	5.29	8.38	+2000	−250
$CuFe_2O_4$	1700	455	1.3	5.35	$\begin{cases} a=8.24 \\ c=8.68 \end{cases}$	—	—
$Li_{0.5}Fe_{2.5}O_4$	3900	670	2.6	4.75	8.33	—	—
$MgFe_2O_4$	1400	440	1.1	4.52	8.36	−0.05	—
$MgAlFeO_4$	—	—	0.3	—	—	—	—
$NiAl_{0.25}Fe_{1.75}O_4$	1300	506	1.30	—	8.31	—	—
$NiAl_{0.45}Fe_{1.55}O_4$	900	465	0.61	—	8.28	—	—
$NiAl_{0.62}Fe_{1.38}O_4$	0	360	0	—	8.25	—	—
$NiAlFeO_4$	900	198	0.64	5.00	8.20	—	—

Specifically, the term applies to those materials with the spinel crystal structures having the general formula XFe_2O_4, where X is any divalent metallic ion having the proper ionic radius to fit in the spinel structure. Several ceramic ferromagnetic materials have been prepared that do not have the basic formula XFe_2O_4 but common usage has included them in the family of ferrite materials.

The behavior of the conductivity and dielectric constant of ferrites is not well understood. They behave as if they consisted of large regions of fairly low-resistance material separated by thin layers of a relatively poor conductor. Therefore, the dielectric constant and conductivity show a relaxation as a function of frequency with the relaxation frequency varying from 1000 to several million hertz. Most ferrites appear to have relatively high resistivities ($\approx 10^6$ ohm-centimeters) if they are prepared carefully so as to avoid the presence of any divalent iron in the material. However, if the ferrite is prepared with an appreciable amount of divalent iron, then both the conductivity and dielectric constant are very high. Relative dielectric constants as high as 100 000 and resistivities less than 1 ohm-centimeter have been measured in several ferrites having a small amount of divalent iron in their composition.

Tables 23, 24, and 25 list some of the pertinent information with respect to the more-important ferrites. Properties such as electrical conductivity and dielectric constant, which are extremely structure-sensitive, are not listed since slight changes in method of preparation can cause these properties to change by several orders of magnitude. Also not included is the initial permeability of ferrite materials since this is also a structure-sensitive property. The initial permeability of most ferrites lies between 100 and 2000. In general, the ferrites listed in the tables have the following properties in common.

Thermal conductivity

$$= 1.5 \times 10^{-2} \text{ calorie/second/centimeter}^2\text{/degree C}$$

Specific heat

$$= 0.2 \text{ calorie/gram/degree C}$$

Young's modulus

$$= 1.5 \times 10^{12} \text{ dynes/centimeter}^2.$$

MAGNETOSTRICTION (FIG. 4)

The static strain $\Delta l/l$ produced by a direct-current polarizing flux density B_0 is given by

$$\Delta l/l = cB_0^2$$

c being a material constant expressed in $\text{m}^4/\text{weber}^2$.

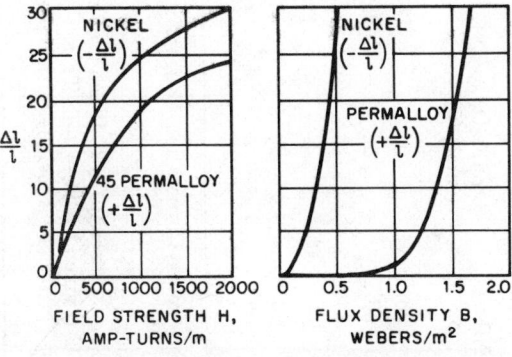

Fig. 4—Strain versus field strength (left) and versus flux density (right). *From "Sonics," p. 173, editors T. F. Hueter and R. H. Bolt, John Wiley & Sons, Inc., New York; 1955.*

If a small alternating-current driving field is superimposed on a large constant polarizing field B_0, we have

$$d(\Delta l/l) = 2cB_0B = \beta B.$$

The magnetostriction stress constant Λ in newtons/weber is

$$\Lambda = \beta Y_0 = 2cB_0Y_0$$

Y_0 being the Young's modulus for a free bar.

Nickel contracts with increasing B so Λ is negative. Permalloy and Alfer expand and their Λ is positive.

Table 26 gives values for the three important transducer materials in both MKS and CGS units: annealed nickel, 45 *Permalloy* (45-percent nickel, 55-percent iron) and *Alfer* (an alloy of 13-percent aluminum and 87-percent iron). For nickel the values for two different polarizing conditions are given: 160 ampere-turns and 1200 ampere-turns, the latter appearing in parentheses. Table 27 compares properties of six magnetostriction materials.

PIEZOELECTRICITY (TABLES 28–29)

Table 28 lists piezoelectric strain coefficients[*] d_{nm}, which are ratios of piezoelectric polarization components to components of applied stress at constant electric field (direct piezoelectric effect) and also ratios between piezoelectric strain components to applied electric field components at constant mechanical stress (converse effect). The subscripts $n=1$ to 3 indicate electric field components, $m=1$ to 6 mechanical stress or strain components. These components are referred to orthogonal coordinate axes. For correlation of these to crystallographic axes, we follow Standards on Piezoelectric Crystals.

In the monoclinic system, indices 2 and 5 refer to the symmetry (b) axis, in distinction from the older convention relating indices 3 and 6 to the symmetry axis. Crystal classes are designated by international (Hermann–Mauguin) symbols. A dash in place of a coefficient indicates that it is equal by symmetry from another listed coefficient; a blank space indicates that the coefficient is zero by sym-

(*Continued on p. 4-54*)

[*] The coefficient d_{14} of Rochelle salt is extremely dependent on temperature and on amplitude. The ratio of d_{14} to dielectric constant K is, however, nearly constant; $4\pi d_{14}/K = g_{14} = 6.4 \times 10^{-7}$ statvolt cm/dyne.

TABLE 24—FERROXCUBE TYPE FERRITES.
(A) General Applications for Commercial Materials*

Designation	MnFe₂O₄ / NiFe₂O₄, mole %	ZnFe₂O₄, mole %	μ_0	B, gauss ($H=10$ Oe) 20°C	100°C	B_r, G	H_c, Oe	Loss factor† tanδ/μ×10⁶ 100 kHz	1 MHz	ρ, Ω-cm‡	TF ×10⁻⁶§	T_c/°C§	Density, g/cm³
3B¶	57.5	42.5	900	3450	2300	1500	0.5	—	—	20	3	150	4.9
3B7	—	—	2300	3400	—	900	0.2	5	120	100	0	170	4.8
3B9	—	—	1800	—	—	1200	0.3	6	120	300	1.8	145	—
3B11	—	—	1500	—	—	—	—	15	—	—	1.4	125	—
3C5	—	33	800	4300	3500	2200	0.2	20	350	60	—	200	4.9
3D3	67	—	750	—	—	—	—	8	30	150	2	150	4.9
3E1	54	46	2700	3500	—	1600	0.15	15	—	30	4	125	4.9
3E2A	—	—	5000	4200	2800	700	0.08	15	—	10	—	170	4.8
3H1	61	39	2300	3500	—	—	—	5	—	100	1.5	170	4.9
4A	36	64	600	2900	2700	1800	0.4	22	100	10⁵	6	125	4.6
4B	50	50	250	3300	2700	1950	1.9	35	70	10⁵	8	250	4.2
4C	64	36	125	2800	2500	2000	3	105	120	10⁵	12	350	4.3
4C4	—	—	125	3000	—	2350	3	40	40	10⁵	−5	350	4.1
4D	80	20	50	2500	2300	1500	5	160	160	10⁵	15	400	4.0
4E	100	0	15	1900	1800	1100	12	300	300	10⁵	15	500	—

(For rows 4A–4E the second column is NiFe₂O₄, mole %.)

* Courtesy of Ferroxcube Corporation, Saugerties, N.Y.
† tanδ/μ=1/μQ.
‡ Minimum value.
§ TF= temperature factor=$\Delta\mu/\mu_0^2 T$. When used with magnetic circuit with air gap and effective permeability μ_{eff}, then the temperature coefficient of the magnetic circuit TC is given by TC= $\mu_{eff}\times$TF.
¶ All varieties of Ferroxcube 3 include in the compositions controlled amounts of ferrous ferrite.

TABLE 24—Continued.

(B) Special Applications for Microwave Devices*

[Applications based on (1) Faraday rotation (F.R.), (2) ferromagnetic resonance (Res.), (3) field displacement (F.D.)]

Designation	Basic composition†	T_c, °C	ρ, MΩ-cm	$4\pi M_s$, G	Proposed use (F.R., Res., F.D.)	Freq., MHz
5B1	MnO·XAl$_2$O$_3$·$(1-x)$Fe$_2$O$_3$‡	140	17	1 000	F.R. and F.D.	2 000–3 500
5A1	MnO·XAl$_2$O$_3$·$(1-x)$Fe$_2$O$_3$‡	160	0.12	1 200	Res.	2 000–5 000
5A2	MnO·XAl$_2$O$_3$·$(1-x)$Fe$_2$O$_3$‡	180	3.4	1 450	F.R. and F.D.	3 500–4 200
5D3	NiFe$_2$O$_4$ 80%, CuFe$_2$O$_4$ 20%§	500	21	2 900	F.R. and F.D.	4 200–6 000
					Res.	>5 000
					F.R. and F.D.	>9 000
5D1	NiFe$_2$O$_4$ 80%, ZnFe$_2$O$_4$ 20%	460	0.1	3 900	Res. (low power)	>8 000
5E1	NiFe$_2$O$_4$ 57%, ZnFe$_2$O$_4$ 43%	280	5 000	4 900	Res. (high power)	>8 000
					F.R. and F.D.	>14 000

* From "American Institute of Physics Handbook," 2nd edition, McGraw-Hill Book Co.; 1963.
† May have minor amounts of other components.
‡ The fraction x of Al$_2$O$_3$ adjusted to produce the desired properties.
§ Percentages are mole percentages.

(C) Special Applications for Magnetostrictive Transducers*

Designation	Basic Composition†	T_c, °C	Optimum Magnetostrictive Coupling Coef K_{opt}	Mechanical Quality Factor Q_{mech}	Biasing Field for Max K (oersteds)
7A1	Ni$_{(1-x)}$Cu$_x$Fe$_2$O$_4$	530	0.25–0.32	>2 000	12–19
7A2	Ni$_{(1-x)}$Cu$_x$Fe$_2$O$_4$	530	0.20–0.26	>2 500	9–15
7B	NiFe$_2$O$_4$	590	0.19–0.22	>4 000	25–50

* From "American Institute of Physics Handbook," 2nd edition, McGraw-Hill Book Co.; 1963.
† Contains controlled amounts of cobalt ferrite (see C. M. v. d. Burgt, *Philips Tech. Rev.*, vol. 18, pp. 285–298; 1957).

TABLE 25—NOMINAL FERRITE VALUES.

Material Code	Q-2	Q-3	Q-4	MF-6784	2285	2285A	C-10	C-20	K-12	U-17	U-60	SF	J	W
Manufacturer	1	1	1	1	2	2	3	3	4	4	4	5	5	5
Recommended frequency range in megahertz	10–80	50–225	200–500	10–75	30–80	30–300	30–80	30–80	3–40	10–220		10–250	40–250	40–500
Initial permeability, μ'	40	16	4.5	35	13.3	7.5	15	10	24	10	9			
Magnetic loss tangent, $\tan\delta_m$	0.006	0.02	0.010	0.0125			0.005	0.006	0.09	0.009	0.06			
Dielectric constant, ϵ'	10				10.5	10.5	11.0	11.0	11	11	11			
Dielectric loss tangent, $\tan\delta_d$	0.005						0.002	0.001						
Quality factor, μQ	5900	2380			2300	2500								
Temperature coefficient, $\%\Delta\mu/°C$	0.10	0.10		0.05	0.36	0.36			0.0144			0.003	0.017	0.013
Saturation magnetization, B_s in gauss	2400	2600			2100		2600	1900	1450		1100			
Residual magnetization, B_r in gauss	750	1470			1600									
Maximum permeability, μ_{max}	115	42			38					15				
Coercivity, H_c in oersteds	4.7	21			21				12	10				

Manufacturer code

1. Indiana General, Electronics Division
2. Stackpole Carbon Co., Electronic Components Division
3. Orbit Industries, Division of Countis Industries
4. Siemens America Inc., Components Division
5. Arnold Engineering Co.

Table 26—Magnetomechanical Coefficients of Three Important Magnetostrictive Materials at Internal Polarizing Field H_0*. From "Sonics," p. 175, editors T. F. Hueter and R. H. Bolt, John Wiley & Sons, Inc., New York; 1955.

Coefficient	Annealed Nickel†		45 Permalloy		Alfer (13% Al, 87% Fe)	
	MKS	CGS	MKS	CGS	MKS	CGS
H_0						
Amp/m	160 (1200)		600		800	
Oersted		2 (15)		7.5		10
B_0						
Volt·sec/m²	0.25 (0.51)		1.43		1.15	
Gauss		2500 (5100)		14 300		11 500
μ	1250 (340)		1900		1150	
μ_i	137 (41)		230		190	
$\Delta l/l$ at H_0	-8×10^{-6} (-26×10^{-6})		14×10^{-6}		26×10^{-6}	
c						
m⁴/weber²	-1×10^{-4}		6.9×10^{-6}		19.5×10^{-6}	
Gauss⁻²		-1×10^{-12}		6.9×10^{-14}		19.5×10^{-14}
Λ						
Newton/weber	-4.8×10^6 (-20×10^6)		2.7×10^6		6.7×10^6	
Dynes/gauss·cm³		-4.8×10^3 (-20×10^3)		2.7×10^3		6.7×10^3
Y_0						
Newtons/m²	20×10^{10}		13.8×10^{10}		15×10^{10}	
Dynes/m²		20×10^{11}		13.8×10^{11}		15×10^{11}
ρ						
kg/m³	8.7×10^3		8.25×10^3		6.7×10^3	
g/cm³		8.7		8.25		6.7
$k_c\%$ (electromechanical coupling factor)	14 (31)		12.4		27	
ρ_c						
Ohm·m	7×10^{-8}		7×10^{-7}		9×10^{-7}	
Ohm·cm		7×10^{-6}		7×10^{-5}		9×10^{-5}

* The number of external ampere-turns per meter required to produce H_0 depends on the shape of the core. For closed magnetic loops H_{ext} is equal to H_0. For rod-shaped cores external fields larger than H_0 are necessary to compensate for the demagnetizing effect of the poles at the free ends.

† The values for two different polarizing conditions are given: 160 and 1200 ampere-turns/meter, the latter in parentheses.

TABLE 27—OTHER CONSTANTS FOR SOME MAGNETOSTRICTION MATERIALS.

Composition	$\Delta l/l$ at Saturation of B	Λ (newtons/weber)	μ_i (henries/meter)	Young's modulus (newtons/m²)	k_c (%)	Curie Temperature (°C)
99.9% Ni annealed	-33×10^{-6}	-20×10^6	4.3×10^{-6}	2.0×10^{11}	31	358
2V 49 Co 49 Fe (2V Permadur)	$+70 \times 10^{-6}$	—	—	1.7×10^{11}	20–37	980
45 Ni 55 Fe (45 Permalloy)	$+27 \times 10^{-6}$	2.7×10^6	2.9×10^{-4}	1.4×10^{11}	12	440
13 Al 87 Fe (13 Alfer)	$+40$	6.7×10^6	2.4×10^{-4}	1.5×10^{11}	27	500
Fe_3O_4	$+40$	-90×10^6	1.9×10^{-2}	1.8×10^{11}	3	190
Ferrite 7 Al	-28	-28 to -44×10^6	$4-5 \times 10^{-5}$	1.68 to 1.75×10^{11}	25–30	640

TABLE 28—PIEZOELECTRIC STRAIN COEFFICIENTS FOR VARIOUS MATERIALS. *From "Smithsonian Physical Tables," 9th revised edition, vol. 120, p. 432, Smithsonian Institution, Washington, D. C.; 1969.*

(**A**) Cubic and Tetragonal Crystals	Composition	Class	d_{14}	d_{36}
Sphalerite	ZnS	43	9.7	—
Sodium chlorate	$NaClO_3$	23	5.2	—
Sodium bromate	$NaBrO_3$	23	7.3	—
"ADP"	NH_4H_2PO	$\overline{4}2$	-1.5	$+48.0$
"KDP"	KH_2PO_4	$\overline{4}2$	$+1.3$	$+21$
"ADA"	$NH_4H_2AsO_4$	$\overline{4}2$	$+41$	$+31$
"KDA"	KH_2AsO_4	$\overline{4}2$	$+23.5$	$+22$

(**B**) Trigonal Crystals	Class	d_{11}	d_{14}	d_{15}	d_{22}	d_{31}	d_{33}
Quartz	32	$+6.9$	-2.0				
Tourmaline	3			$+11.0$	-0.94	$+0.96$	$+5.4$

(**C**) Orthorhombic Crystals	Class	d_{14}	d_{25}	d_{36}
Epsomite	222	-6.2	-8.2	-11.5
Iodic acid	222	57	46	70
Rochelle salt (30°C)	222	$+1500$	-160	$+35$
$NaNH_4$ tartrate	222	$+56$	-150	$+28$
LiK tartrate	222	9.6	33.6	22.8
$LiNH_4$ tartrate	222	13.2	19.6	14.8
$(NH_4)_2$ oxalate	222	50	11	25

	d_{15}	d_{24}	d_{31}	d_{32}	d_{33}
K pentaborate	9.5	1.7	-5.4	0	$+5.6$

TABLE 28—CONTINUED.

(D) Monoclinic Crystals (Class 2)	d_{14}	d_{16}	d_{21}	d_{22}	d_{23}	d_{25}	d_{34}	d_{36}
Lithium sulfate	+14.0	−12.5	+11.6	−45.0	−5.5	+16.5	−26.4	+10.0
Tartaric acid	+24.0	+15.8	−2.3	−6.5	−6.3	+1.1	−32.4	+35.0
K₂ tartrate (DKT)	−25	+6.5	−2.2	+8.5	−10.4	−22.5	+29.4	−66.0
(NH₄)₂ tartrate	+9.3	−8.5	+17.6	−26.2	+1.8	−5.9	−14.0	+5.6
EDT (ethylene diamine tartrate)	−31.1	−36.5	+30.6	+6.6	−33.8	−54.3	−51	−56.9
Cane sugar	−3.7	−7.2	+4.4	−10	+2.2	−2.6	−1.3	+1.3

(E) Polarized Polycrystalline Substance	d_{15}	d_{31}	d_{33}
Barium titanate ceramic K=1700	750	−235	+570

Note: If the sign of a coefficient is not given, it is unknown (not necessarily positive).

TABLE 29—CONSTANTS OF SOME PIEZOELECTRIC MATERIALS.* *From J. R. Frederick, "Ultrasonic Engineering," p. 66, John Wiley & Sons, Inc., New York; 1965.*

Physical Property	Quartz 0° X-cut	Lithium Sulfate 0° Y-cut	Barium Titanate Type B	Lead Zirconate-Titanate		Lead Meta-Niobate	Units
				PZT-4	PZT-5		
Density ρ	2.65	2.06	5.6	7.6	7.7	5.8	10^3 kg/m³
Acoustic impedance ρc	15.2	11.2	24	30.0	28.0	16	10^6 kg/m²s
Frequency thickness constant ft	2870	2730	2740	2000	1800	1400	kHz/mm
Maximum operating temperature	550	75	70–90	250	290	500	°C
Dielectric constant	4.5	10.3	1700	1300	1700	225	—
Electromechanical coupling factor for thickness mode k_{33}	0.1	0.35	0.48	0.64	0.675	0.42	—
Electromechanical coupling factor for radial mode k_p	0.1	—	0.33	0.58	0.60	0.07	—
Elastic quality factor Q	10^6	—	400	500	75	11	—
Piezoelectric modulus for thickness mode d_{33}	2.3	16	149	285	374	85	10^{-12} m/V
Piezoelectric pressure constant g_{33}	58	175	14.0	26.1	24.8	42.5	10^{-3}(V/m)/(N/m²)
Volume resistivity at 25°C	>10^{12}	—	>10^{11}	>10^{12}	>10^{13}	10^9	—
Curie temperature	575	—	115	320	365	550	°C
Young's modulus E	8.0	—	11.8	8.15	6.75	2.9	10^{10} N/m²
Rated dynamic tensile strength	—	—	—	3500	4000	—	psi

* The properties of the ceramic materials can vary with slight changes in composition and processing, and hence the values that are shown should not be taken as exact.

TABLE 30—Simplified Equations for Sound Intensity. *From "Sonics", p. 105, editors T. F. Hueter and R. H. Bolt, John Wiley & Sons, Inc. New York; 1955.*

Material	Mode of Vibration	Crystal Cut	Effective* Piezo Modulus H (coulombs/m²)	Sound Velocity c (m/sec)	Density ρ (kg/m³)	Sound Intensity in Water for Airbacked Transducer† $\mathcal{G}$ (watts/m²)	Units (rms values for V and E)
Quartz	Thickness	X	$H=e_{11}$ $=5.2\times10^4$ esu $=0.17$ coulomb/m²	5.72×10^3	2.65×10^3	$f_0^2V^2\times10^4$	V in kV f in MHz
	Longitudinal	X	$H=d_{11}/s'$ $=5.4\times10^4$ esu $=0.18$ coulomb/m²	5.44×10^3	2.65×10^3	$0.087\times E^2\times10^4$	E in kV/cm
Ammonium di-hydrogen phosphate (ADP)	Longitudinal	45°Z	$H=d_{36}/2s'$ $=1.4\times10^4$ esu $=0.042$ coulomb/m²	3.28×10^3	1.8×10^3	$0.6E^2\times10^4$	E in kV/cm
Rochelle salt (0°C)	Longitudinal	45°X	$H=d_{14}/2s'$ $=1.7\times10^6$ esu $=5.67$ coulombs/m²	3.4×10^3	1.77×10^3	$85E^2\times10^4$	E in kV/cm
Barium titanate (40°C)	Thickness	Polarized normal to thickness	$H=e_{33}$ $=3$ to 5×10^6 esu $=10$ to 17 coulombs/m²	5×10^3	5.5×10^3	$0.005f_0^2V^2\times10^4$ to $0.014f_0^2V^2\times10^4$	V in volts f in MHz

* The quantity s' in the relationship $H=d_{th}/s'$ is the effective compliance in the direction of longitudinal vibration.
† Without the factor 10^4, this column gives the sound intensity in watts/cm².

TABLE 31—VELOCITIES, DENSITIES, AND CHARACTERISTIC IMPEDANCES OF VARIOUS METALS. *From J. R. Frederick, "Ultrasonic Engineering," p. 363, John Wiley & Sons., Inc., New York; 1965.*

	Velocities							Charac-teristic impedance ρc
	Longitudinal				Shear		Density ρ	
	Bulk		Bar					Bulk
Metals	m/s	in./s	m/s	in./s	m/s	in./s	(kg/m³)	(kg/m²s)
	$\times 10^3$	$\times 10^5$	$\times 10^3$	$\times 10^5$	$\times 10^3$	$\times 10^5$	$\times 10^3$	$\times 10^6$
Aluminum	6.40	2.50	5.15	2.03	3.13	1.24	2.7	17.3
Beryllium	12.89	17.	...	...	8.88	3.51	1.8	23.2
Brass, 70-30	4.37	1.73	3.40	1.34	2.10	0.83	8.5	37.0
Cast iron	3.50-5.60	1.38-2.22	3.0-4.7	1.19-1.86	2.2-3.2	0.87-1.31	7.2	25.0-40.0
Copper	4.80	1.90	3.65	1.45	2.33	0.92	8.9	42.5
Gold	3.24	1.28	2.03	...	1.20	0.47	19.3	63.0
Iron	5.96	2.35	5.18	2.05	3.22	1.28	7.9	46.8
Lead	2.40	0.95	1.25	0.49	0.79	0.31	11.3	27.2
Magnesium	5.74	2.27	4.90	1.94	3.08	1.22	1.7	9.9
Mercury	1.45	0.57	...	...	...	...	13.6	19.6
Molybdenum	6.25	2.47	...	...	3.35	1.33	10.2	63.7
Nickel	5.48	2.17	4.70	1.86	2.99	1.19	8.9	48.5
Platinum	3.96	1.57	2.80	1.11	1.67	0.66	21.4	85.0
Steel, mild	6.10	2.41	5.05	2.00	3.24	1.29	7.9	46.7
Silver	3.70	1.46	2.67	1.06	1.70	0.67	10.5	36.9
Tin	3.38	1.34	2.74	1.09	1.61	0.66	7.3	24.7
Titanium	5.99	2.37	...	...	3.12	1.24	4.50	27.0
Tungsten	5.17	2.04	...	...	2.88	1.14	19.3	100.0
Uranium	3.37	1.33	...	...	2.02	0.80	18.7	63.0
Zinc	4.17	1.65	3.81	1.51	2.48	0.98	7.1	29.6
Zirconium	4.65	1.84	...	...	2.30	0.91	6.4	29.8
Other Solid Materials:								
Crown glass	5.66	2.24	5.30	2.10	3.42	1.35	2.5	14.0
Granite	...	...	3.95	1.56	...	...	2.75	...
Ice	3.98	1.58	...	...	1.99	0.79	0.9	3.6
Nylon	1.8-2.2	0.71-0.87	...	...	...	...	1.1-1.2	2.0-2.7
Paraffin, hard	2.2	0.87	...	...	...	...	0.83	1.8
Plexiglas or Lucite	2.68	1.06	1.8	0.71	1.32	0.52	1.20	3.2
Polystyrene	2.67	1.06	...	...	...	...	1.06	2.8
Quartz, fused	5.57	2.21	5.37	2.12	3.52	1.39	2.6	14.5
Teflon	1.35	0.53	...	...	...	...	2.2	3.0
Tungsten carbide	6.66	2.64	...	...	3.98	1.57	10.0-15.0	66.5-98.5
Wood, oak	...	...	4.1	1.62	...	...	0.8	...
Fluids:								
Benzene	1.32	0.52	...	...	...	...	0.88	1.16
Castor oil	1.54	0.61	...	...	...	...	0.95	1.45
Glycerine	1.92	0.76	...	...	...	...	1.26	2.5
Methyl iodide	0.98	0.39	...	...	...	...	3.23	3.2
Oil, SAE 20	1.74	0.69	...	...	...	...	0.87	1.5
Water, fresh	1.48	0.59	...	...	...	...	1.00	1.48

Note: The values that are shown should not be taken as exact values because of the effects of variations in composition and processing. They are adequate for most practical purposes, however.

TABLE 32—ADHESIVES CLASSIFIED BY CHEMICAL COMPOSITION. *From Clauser et al, "Encyclopedia of Engineering Materials and Processes," Reinhold Publishing Corporation, New York; 1963.*

Group→	Natural	Thermoplastic	Thermosetting	Elastomeric	Alloys*
Types within group	Casein, blood albumin, hide, bone, fish, starch (plain and modified); rosin, shellac, asphalt; inorganic (sodium silicate, litharge-glycerin)	Polyvinyl acetate, polyvinyl alcohol, acrylic, cellulose nitrate, asphalt, oleoresin	Phenolic, resorcinol, phenol-resorcinol, epoxy, epoxy-phenolic, urea, melamine, alkyd	Natural rubber, reclaim rubber, butadiene-styrene (GR-S), neoprene, acrylonitrile-butadiene (Buna-N), silicone	Phenolic-polyvinyl butyral, phenolic-polyvinyl formal, phenolic-neoprene rubber, phenolic-nitrile rubber, modified epoxy
Most used form	Liquid, powder	Liquid, some dry film	Liquid, but all forms common	Liquid, some film	Liquid, paste, film
Common further classifications	By vehicle (water emulsion is most common but many types are solvent dispersions)	By vehicle (most are solvent dispersions or water emulsions)	By cure requirements (heat and/or pressure most common but some are catalyst types)	By cure requirements (all are common); also by vehicle (most are solvent dispersions or water emulsions)	By cure requirements (usually heat and pressure except some epoxy types); by vehicle (most are solvent dispersions or 100% solids); and by type of adherends or end-service conditions

Bond characteristics	Wide range, but generally low strength; good resistance to heat, chemicals; generally poor moisture resistance	Good to 200–500 F; poor creep strength; fair peel strength	Good to 200–500 F; good creep strength; fair peel strength	Good to 150–400 F; never melt completely; low strength; high flexibility	Balanced combination of properties of other chemical groups depending on formulation; generally higher strength over wider temperature range
Major type of use†	Household, general purpose, quick set, long shelf life	Unstressed joints; designs with caps, overlaps, stiffeners	Stressed joints at slightly elevated temperature	Unstressed joints on lightweight materials; joints in flexure	Where highest and strictest end-service conditions must be met; sometimes regardless of cost, as military uses
Materials most commonly bonded	Wood (furniture), paper, cork, liners, textiles, some metals and plastics. Industrial uses giving way to other groups	Formulation range covers all materials, but emphasis on nonmetallics—especially wood, leather, cork, paper, etc.	Epoxy-phenolics for structural uses of most materials; others mainly for wood; alkyds for laminations; most epoxies are modified (alloys)	Few used "straight" for rubber, fabric, foil, paper, leather, plastics, films; also as tapes. Most modified with synthetic resins	Metals, ceramics, glass, thermosetting plastics; nature of adherends often not as vital as design or end-service conditions (i.e., high strength, temperature)

* "Alloy," as used here, refers to formulations containing resins from two or more *different* chemical groups. There are also formulations which benefit from compounding two resin types from the same chemical group (e.g., epoxy-phenolic).

† Although some uses of the "nonalloyed" adhesives absorb a large percentage of the quantity of adhesives sold, the uses are narrow in scope; from the standpoint of diversified applications, by far the most important use of any group is the forming of adhesive alloys.

TABLE 33—SOLID COPPER—COMPARISON OF GAUGES.

American (B & S) Wire Gauge	Birmingham (Stubs') Iron Wire Gauge	British Standard (NBS) Wire Gauge	Diameter		Area			Weight	
			Mils	Milli-meters	Circular Mils	Square Milli-meters	Square Inches	per 1000 Feet in Pounds	per Kilometer in Kilograms
—	0	—	340.0	8.636	115 600	58.58	0.09079	350	521
0	—	—	324.9	8.251	105 500	53.48	0.08289	319	475
—	—	0	324.0	8.230	105 000	53.19	0.08245	318	472
—	1	1	300.0	7.620	90 000	45.60	0.07069	273	405
1	—	—	289.3	7.348	83 690	42.41	0.06573	253	377
—	2	—	284.0	7.214	80 660	40.87	0.06335	244	363
—	—	—	283.0	7.188	80 090	40.58	0.06290	242	361
—	—	2	276.0	7.010	76 180	38.60	0.05963	231	343
—	3	—	259.0	6.579	67 080	33.99	0.05269	203	302
2	—	—	257.6	6.544	66 370	33.63	0.05213	201	299
—	—	3	252.0	6.401	63 500	32.18	0.04988	193	286
—	4	—	238.0	6.045	56 640	28.70	0.04449	173	255
—	—	4	232.0	5.893	53 820	27.27	0.04227	163	242
3	—	—	229.4	5.827	52 630	26.67	0.04134	159	237
—	5	—	220.0	5.588	48 400	24.52	0.03801	147	217
—	—	5	212.0	5.385	44 940	22.77	0.03530	136	202
4	—	—	204.3	5.189	41 740	21.18	0.03278	126	188
—	6	—	203.0	5.156	41 210	20.88	0.03237	125	186
—	—	6	192.0	4.877	36 860	18.68	0.02895	112	166
5	—	—	181.9	4.621	33 100	16.77	0.02600	100	149
—	7	—	180.0	4.572	32 400	16.42	0.02545	98.0	146
—	—	7	176.0	4.470	30 980	15.70	0.02433	93.6	139
—	8	—	165.0	4.191	27 220	13.86	0.02138	86.2	123
6	—	—	162.0	4.116	26 250	13.30	0.02062	79.5	118
—	—	8	160.0	4.064	25 600	12.97	0.02011	77.5	115
—	9	—	148.0	3.759	21 900	11.10	0.01720	66.3	98.6
7	—	—	144.3	3.665	20 820	10.55	0.01635	63.0	93.7
—	—	9	144.0	3.658	20 740	10.51	0.01629	62.8	93.4
—	10	—	134.0	3.404	17 960	9.098	0.01410	54.3	80.8
8	—	—	128.8	3.264	16 510	8.366	0.01297	50.0	74.4
—	—	10	128.0	3.251	16 380	8.302	0.01267	49.6	73.8
—	11	—	120.0	3.048	14 400	7.297	0.01131	43.6	64.8

TABLE 33—CONTINUED.

American (B & S) Wire Gauge	Birming-ham (Stubs') Iron Wire Gauge	British Standard (NBS) Wire Gauge	Diameter		Area			Weight	
			Mils	Milli-meters	Circular Mils	Square Milli-meters	Square Inches	per 1000 Feet in Pounds	per Kilometer in Kilograms
—	—	11	116.0	2.946	13 460	6.818	0.01057	40.8	60.5
9	—	—	114.4	2.906	13 090	6.634	0.01028	39.6	58.9
—	12	—	109.0	2.769	11 880	6.020	0.009331	35.9	53.5
—	—	12	104.0	2.642	10 820	5.481	0.008495	32.7	48.7
10	—	—	101.9	2.588	10 380	5.261	0.008155	31.4	46.8
—	13	—	95.00	2.413	9 025	4.573	0.007088	27.3	40.6
—	—	13	92.00	2.337	8 464	4.289	0.006648	25.6	38.1
11	—	—	90.74	2.305	8 234	4.172	0.006467	24.9	37.1
—	14	—	83.00	2.108	6 889	3.491	0.005411	20.8	31.0
12	—	—	80.81	2.053	6 530	3.309	0.005129	19.8	29.4
—	—	14	80.00	2.032	6 400	3.243	0.005027	19.4	28.8
—	15	15	72.00	1.829	5 184	2.627	0.004072	16.1	23.4
13	—	—	71.96	1.828	5 178	2.624	0.004067	15.7	23.3
—	16	—	65.00	1.651	4 225	2.141	0.003318	12.8	19.0
14	—	—	64.08	1.628	4 107	2.081	0.003225	12.4	18.5
—	—	16	64.00	1.626	4 096	2.075	0.003217	12.3	18.4
—	17	—	58.00	1.473	3 364	1.705	0.002642	10.2	15.1
15	—	—	57.07	1.450	3 257	1.650	0.002558	9.86	14.7
—	—	17	56.00	1.422	3 136	1.589	0.002463	9.52	14.1
16	—	—	50.82	1.291	2 583	1.309	0.002028	7.82	11.6
—	18	—	49.00	1.245	2 401	1.217	0.001886	7.27	10.8
—	—	18	48.00	1.219	2 304	1.167	0.001810	6.98	10.4
17	—	—	45.26	1.150	2 048	1.038	0.001609	6.20	9.23
—	19	—	42.00	1.067	1 764	0.8938	0.001385	5.34	7.94
18	—	—	40.30	1.024	1 624	0.8231	0.001276	4.92	7.32
—	—	19	40.00	1.016	1 600	0.8107	0.001257	4.84	7.21
—	—	20	36.00	0.9144	1 296	0.6567	0.001018	3.93	5.84
19	—	—	35.89	0.9116	1 288	0.6527	0.001012	3.90	5.80
—	20	—	35.00	0.8890	1 225	0.6207	0.0009621	3.71	5.52
—	21	21	32.00	0.8128	1 024	0.5189	0.0008042	3.11	4.62
20	—	—	31.96	0.8118	1 022	0.5176	0.0008023	3.09	4.60

TABLE 34—ANNEALED COPPER.

| AWG B & S Gauge | Diameter in Mils | Cross Section | | Ohms per 1000 Ft at 20° C (68° F) | Lb per 1000 Ft | Ft per Lb | Ft per ohm at 20° C (68° F) | Ohms per Lb at 20° C (68° F) |
		Circular mils	Square Inches					
0000	460.0	211 600	0.1662	0.04901	640.5	1.561	20 400	0.00007652
000	409.6	167 800	0.1318	0.06180	507.9	1.968	16 180	0.0001217
00	364.8	133 100	0.1045	0.07793	402.8	2.482	12 830	0.0001935
0	324.9	105 500	0.08289	0.09827	319.5	3.130	10 180	0.0003076
1	289.3	83 690	0.06573	0.1239	253.3	3.947	8 070	0.0004891
2	257.6	66 370	0.05213	0.1563	200.9	4.977	6 400	0.0007778
3	229.4	52 640	0.04134	0.1970	159.3	6.276	5 075	0.001237
4	204.3	41 740	0.03278	0.2485	126.4	7.914	4 025	0.001966
5	181.9	33 100	0.02600	0.3133	100.2	9.980	3 192	0.003127
6	162.0	26 250	0.02062	0.3951	79.46	12.58	2 531	0.004972
7	144.3	20 820	0.01635	0.4982	63.02	15.87	2 007	0.007905
8	128.5	16 510	0.01297	0.6282	49.98	20.01	1 592	0.01257
9	114.4	13 090	0.01028	0.7921	39.63	25.23	1 262	0.01999
10	101.9	10 380	0.008155	0.9989	31.43	31.82	1 001	0.03178
11	90.74	8 234	0.006467	1.260	24.92	40.12	794	0.05053
12	80.81	6 530	0.005129	1.588	19.77	50.59	629.6	0.08035
13	71.96	5 178	0.004067	2.003	15.68	63.80	499.3	0.1278
14	64.08	4 107	0.003225	2.525	12.43	80.44	396.0	0.2032
15	57.07	3 257	0.002558	3.184	9.858	101.4	314.0	0.3230
16	50.82	2 583	0.002028	4.016	7.818	127.9	249.0	0.5136
17	45.26	2 048	0.001609	5.064	6.200	161.3	197.5	0.8167
18	40.30	1 624	0.001276	6.385	4.917	203.4	156.6	1.299
19	35.89	1 288	0.001012	8.051	3.899	256.5	124.2	2.065
20	31.96	1 022	0.0008023	10.15	3.092	323.4	98.50	3.283
21	28.46	810.1	0.0006363	12.80	2.452	407.8	78.11	5.221
22	25.35	642.4	0.0005046	16.14	1.945	514.2	61.95	8.301
23	22.57	509.5	0.0004002	20.36	1.542	648.4	49.13	13.20
24	20.10	404.0	0.0003173	25.67	1.223	817.7	38.96	20.99
25	17.90	320.4	0.0002517	32.37	0.9699	1 031.0	30.90	33.37
26	15.94	254.1	0.0001996	40.81	0.7692	1 300	24.50	53.06
27	14.20	201.5	0.0001583	51.47	0.6100	1 639	19.43	84.37
28	12.64	159.8	0.0001255	64.90	0.4837	2 067	15.41	134.2
29	11.26	126.7	0.00009953	81.83	0.3836	2 607	12.22	213.3
30	10.03	100.5	0.00007894	103.2	0.3042	3 287	9.691	339.2
31	8.928	79.70	0.00006260	130.1	0.2413	4 145	7.685	539.3
32	7.950	63.21	0.00004964	164.1	0.1913	5 227	6.095	857.6
33	7.080	50.13	0.00003937	206.9	0.1517	6 591	4.833	1 364
34	6.305	39.75	0.00003122	260.9	0.1203	8 310	3.833	2 168
35	5.615	31.52	0.00002476	329.0	0.09542	10 480	3.040	3 448
36	5.000	25.00	0.00001964	414.8	0.07568	13 210	2.411	5 482
37	4.453	19.83	0.00001557	523.1	0.06001	16 660	1.912	8 717
38	3.965	15.72	0.00001235	659.6	0.04759	21 010	1.516	13 860
39	3.531	12.47	0.000009793	831.8	0.03774	26 500	1.202	22 040
40	3.145	9.888	0.000007766	1049.0	0.02993	33 410	0.9534	35 040

TABLE 35—HARD-DRAWN COPPER.

AWG B & S Gauge	Wire Diameter in Inches	Breaking Load in Pounds	Tensile Strength in lb/in.²	Weight		Maximum Resistance (ohms per 1000 feet at 68° F)	Cross-Sectional Area	
				Pounds per 1000 Feet	Pounds per Mile		Circular Mils	Square Inches
4/0	0.4600	8 143	49 000	640.5	3 382	0.05045	211 600	0.1662
3/0	0.4096	6 722	51 000	507.9	2 682	0.06361	167 800	0.1318
2/0	0.3648	5 519	52 800	402.8	2 127	0.08021	133 100	0.1045
1/0	0.3249	4 517	54 500	319.5	1 687	0.1011	105 500	0.08289
1	0.2893	3 688	56 100	253.3	1 338	0.1287	83 690	0.06573
2	0.2576	3 003	57 600	200.9	1 061	0.1625	66 370	0.05213
3	0.2294	2 439	59 000	159.3	841.2	0.2049	52 630	0.04134
4	0.2043	1 970	60 100	126.4	667.1	0.2584	41 740	0.03278
5	0.1819	1 591	61 200	100.2	529.1	0.3258	33 100	0.02600
—	0.1650	1 326	62 000	82.41	435.1	0.3961	27 225	0.02138
6	0.1620	1 280	62 100	79.46	419.6	0.4108	26 250	0.02062
7	0.1443	1 030	63 000	63.02	332.7	0.5181	20 820	0.01635
—	0.1340	894.0	63 400	54.35	287.0	0.6006	17 956	0.01410
8	0.1285	826.0	63 700	49.97	263.9	0.6533	16 510	0.01297
9	0.1144	661.2	64 300	39.63	209.3	0.8238	13 090	0.01028
—	0.1040	550.4	64 800	32.74	172.9	0.9971	10 816	0.008495
10	0.1019	529.2	64 900	31.43	165.9	1.039	10 380	0.008155
11	0.09074	422.9	65 400	24.92	131.6	1.310	8 234	0.006467
12	0.08081	337.0	65 700	19.77	104.4	1.652	6 530	0.005129
13	0.07196	268.0	65 900	15.68	82.77	2.083	5 178	0.004067
14	0.06408	213.5	66 200	12.43	65.64	2.626	4 107	0.003225
15	0.05707	169.8	66 400	9.858	52.05	3.312	3 257	0.002558
16	0.05082	135.1	66 600	7.818	41.28	4.176	2 583	0.002028
17	0.04526	107.5	66 800	6.200	32.74	5.266	2 048	0.001609
18	0.04030	85.47	67 000	4.917	25.96	6.640	1 624	0.001276

TABLE 36—TENSILE STRENGTH OF COPPER WIRE.

AWG B & S Gauge	Wire Diameter in Inches	Hard Drawn		Medium-Hard Drawn		Soft or Annealed	
		Minimum Tensile Strength lb/in.²	Breaking Load in Pounds	Minimum Tensile Strength lb/in.²	Breaking Load in Pounds	Maximum Tensile Strength lb/in.²	Breaking Load in Pounds
1	0.2893	56 100	3 688	46 000	3 024	37 000	2 432
2	0.2576	57 600	3 003	47 000	2 450	37 000	1 929
3	0.2294	59 000	2 439	48 000	1 984	37 000	1 530
4	0.2043	60 100	1 970	48 330	1 584	37 000	1 213
5	0.1819	61 200	1 591	48 660	1 265	37 000	961.9
—	0.1650	62 000	1 326	—	—	—	—
6	0.1620	62 100	1 280	49 000	1 010	37 000	762.9
7	0.1443	63 000	1 030	49 330	806.6	37 000	605.0
—	0.1340	63 400	894.0	—	—	—	—
8	0.1285	63 700	826.0	49 660	643.9	37 000	479.8
9	0.1144	64 300	661.2	50 000	514.2	37 000	380.5
—	0.1040	64 800	550.4	—	—	—	—
10	0.1019	64 900	529.2	50 330	410.4	38 500	314.0
11	0.09074	65 400	422.9	50 660	327.6	38 500	249.0
12	0.08081	65 700	337.0	51 000	261.6	38 500	197.5

TABLE 37—COPPER-CLAD STEEL

AWG B & S Gauge	Cross-Sectional Area			Weight			Resistance Ohms/1000 ft at 68°F		Breaking Load Pounds		Attenuation in Decibels/Mile*				Characteristic Impedance*	
	Diam Inch	Circular Mils	Square Inch	Pounds per 1000 Feet	Pounds per Mile	Feet per Pound	40% Conduct	30% Conduct	40% Conduct	30% Conduct	40% Cond Dry	40% Cond Wet	30% Cond Dry	30% Cond Wet	40% Cond	30% Cond
4	0.2043	41 740	0.03278	115.8	611.6	8.63	0.6337	0.8447	3 541	3 934	—	—	—	—	—	—
5	0.1819	33 100	0.02600	91.86	485.0	10.89	0.7990	1.065	2 938	3 250	—	—	—	—	—	—
6	0.1620	26 250	0.02062	72.85	384.6	13.73	1.008	1.343	2 433	2 680	0.078	0.086	0.103	0.109	650	686
7	0.1443	20 820	0.01635	57.77	305.0	17.31	1.270	1.694	2 011	2 207	0.093	0.100	0.122	0.127	685	732
8	0.1285	16 510	0.01297	45.81	241.9	21.83	1.602	2.136	1 660	1 815	0.111	0.118	0.144	0.149	727	787
9	0.1144	13 090	0.01028	36.33	191.8	27.52	2.020	2.693	1 368	1 491	0.132	0.138	0.169	0.174	776	852
10	0.1019	10 380	0.008155	28.81	152.1	34.70	2.547	3.396	1 130	1 231	0.156	0.161	0.196	0.200	834	920
11	0.0907	8 234	0.006467	22.85	120.6	43.76	3.212	4.28	896	975	0.183	0.188	0.228	0.233	910	1 013
12	0.0808	6 530	0.005129	18.12	95.68	55.19	4.05	5.40	711	770	0.216	0.220	0.262	0.266	1 000	1 120
13	0.0720	5 178	0.004067	14.37	75.88	69.59	5.11	6.81	490	530						
14	0.0641	4 107	0.003225	11.40	60.17	87.75	6.44	8.59	400	440						
15	0.0571	3 257	0.002558	9.038	47.72	110.6	8.12	10.83	300	330						
16	0.0508	2 583	0.002028	7.167	37.84	139.5	10.24	13.65	250	270						
17	0.0453	2 048	0.001609	5.684	30.01	175.9	12.91	17.22	185	205						
18	0.0403	1 624	0.001276	4.507	23.80	221.9	16.28	21.71	153	170						
19	0.0359	1 288	0.001012	3.575	18.87	279.8	20.53	27.37	122	135						
20	0.0320	1 022	0.0008023	2.835	14.97	352.8	25.89	34.52	100	110						
21	0.0285	810.1	0.0006363	2.248	11.87	444.8	32.65	43.52	73.2	81.1						
22	0.0253	642.5	0.0005046	1.783	9.413	560.9	41.17	54.88	58.0	64.3						
23	0.0226	509.5	0.0004002	1.414	7.465	707.3	51.92	69.21	46.0	51.0						
24	0.0201	404.0	0.0003173	1.121	5.920	891.9	65.46	87.27	36.5	40.4						
25	0.0179	320.4	0.0002517	0.889	4.695	1 125	82.55	110.0	28.9	32.1						
26	0.0159	254.1	0.0001996	0.705	3.723	1 418	104.1	138.8	23.0	25.4						
27	0.0142	201.5	0.0001583	0.559	2.953	1 788	131.3	175.0	18.2	20.1						
28	0.0126	159.8	0.0001255	0.443	2.342	2 255	165.5	220.6	14.4	15.9						
29	0.0113	126.7	0.0000995	0.352	1.857	2 843	208.7	278.2	11.4	12.6						
30	0.0100	100.5	0.0000789	0.279	1.473	3 586	263.2	350.8	9.08	10.0						
31	0.0089	79.70	0.0000626	0.221	1.168	4 521	331.9	442.4	7.20	7.95						
32	0.0080	63.21	0.0000496	0.175	0.926	5 701	418.5	557.8	5.71	6.30						
33	0.0071	50.13	0.0000394	0.139	0.734	7 189	527.7	703.4	4.53	5.00						
34	0.0063	39.75	0.0000312	0.110	0.582	9 065	665.4	887.0	3.59	3.97						
35	0.0056	31.52	0.0000248	0.087	0.462	11 430	839.0	1 119	2.85	3.14						
36	0.0050	25.00	0.0000196	0.069	0.366	14 410	1 058	1 410	2.26	2.49						
37	0.0045	19.83	0.0000156	0.055	0.290	18 180	1 334	1 778	1.79	1.98						
38	0.0040	15.72	0.0000123	0.044	0.230	22 920	1 682	2 243	1.42	1.57						
39	0.0035	12.47	0.00000979	0.035	0.183	28 900	2 121	2 828	1.13	1.24						
40	0.0031	9.89	0.00000777	0.027	0.145	36 440	2 675	3 566	0.893	0.986						

* DP insulators, 12-inch wire spacing at 1000 hertz.

TABLE 38—CONDUCTOR SIZE FOR 2-PERCENT VOLTAGE DROP.

Current in Amperes	Single-phase—110 volts — Distance in Feet									Single-phase—220 volts — Distance in Feet								
	25	50	75	100	150	200	300	400	500	25	50	75	100	150	200	300	400	500
1	—	—	—	—	—	—	14	12	10	—	—	—	—	—	—	—	—	14
1.5	—	—	—	—	14	14	12	10	10	—	—	—	—	—	—	14	14	12
2	—	—	—	—	14	12	10	10	8	—	—	—	—	—	—	14	12	12
3	—	—	14	14	12	10	8	8	6	—	—	—	—	14	14	12	10	10
4	—	—	14	12	10	10	8	6	6	—	—	—	—	14	12	10	10	8
5	—	14	12	12	10	8	6	6	4	—	—	—	14	12	12	10	8	8
6	—	14	12	10	8	8	6	4	4	—	—	14	14	12	10	8	8	6
7	—	14	12	10	8	8	6	4	2	—	—	14	14	12	10	8	8	6
8	—	12	10	10	8	6	4	2	2	—	—	14	12	10	10	8	6	6
9	—	12	10	8	8	6	4	2	2	—	14	14	12	10	8	8	6	4
10	14	12	10	8	6	6	4	2	2	—	14	12	12	10	8	6	6	4
12	14	10	8	8	6	4	2	1	1	—	14	12	10	8	8	6	4	4
14	14	10	8	8	6	4	2	0	0	—	14	12	10	8	8	6	4	2
16	12	10	8	6	4	4	2	0	00	—	12	10	10	8	6	4	4	2
18	12	8	8	6	4	2	1	00	00	14	12	10	8	8	6	4	2	2
20	12	8	6	6	4	2	1	00	000	14	12	10	8	6	6	4	2	2
25	10	8	6	4	2	2	0	000	0000	14	10	8	8	6	4	2	2	1
30	10	6	4	4	2	1	00	—	—	12	10	8	6	4	4	2	1	0
35	10	6	4	2	2	0	000	—	—	12	10	8	6	4	2	2	0	00
40	8	6	4	2	1	00	0000	—	—	12	8	6	6	4	2	1	00	0000
45	8	4	4	2	0	00	—	—	—	10	8	6	4	4	2	0	00	0000
50	8	4	2	2	0	000	—	—	—	10	8	6	4	2	2	0	000	0000
60	6	4	2	1	00	0000	—	—	—	10	6	4	4	2	1	00	0000	—
70	6	2	2	0	000	—	—	—	—	10	6	4	2	2	0	000	—	—
80	6	2	1	00	0000	—	—	—	—	8	6	4	2	1	00	0000	—	—
90	4	2	0	00	—	—	—	—	—	8	4	4	2	0	00	—	—	—
100	4	2	0	000	—	—	—	—	—	8	4	2	2	0	000	—	—	—
120	4	1	00	0000	—	—	—	—	—	6	4	2	1	00	0000	—	—	—

Current in Amperes	Three-phase—220 volts — Distance in Feet									Three-phase—440 volts — Distance in Feet								
	25	50	75	100	150	200	300	400	500	25	50	75	100	150	200	300	400	500
1	—	—	—	—	—	—	—	—	—	—	—	—	—	—	—	—	—	—
1.5	—	—	—	—	—	—	—	14	14	—	—	—	—	—	—	—	—	—
2	—	—	—	—	—	—	14	14	12	—	—	—	—	—	—	—	—	—
3	—	—	—	—	—	14	12	12	10	—	—	—	—	—	—	—	14	14
4	—	—	—	—	14	14	12	10	10	—	—	—	—	—	—	14	14	12
5	—	—	—	—	14	12	10	10	8	—	—	—	—	—	—	14	12	12
6	—	—	—	14	12	12	10	8	8	—	—	—	—	—	14	12	12	10
7	—	—	14	14	12	10	8	8	6	—	—	—	—	14	14	12	10	10
8	—	—	14	14	12	10	8	6	6	—	—	—	—	14	14	12	10	10
9	—	—	14	12	10	10	8	6	6	—	—	—	—	14	12	10	10	8
10	—	—	14	12	10	10	8	6	6	—	—	—	—	14	12	10	10	8
12	—	14	12	12	10	8	6	6	4	—	—	—	14	12	12	10	8	8
14	—	14	12	10	8	8	6	4	4	—	—	14	14	12	10	8	8	6
16	—	14	12	10	8	8	6	4	2	—	—	14	14	12	10	8	8	6
18	—	12	10	10	8	6	4	4	2	—	—	14	12	10	10	8	6	6
20	—	12	10	10	8	6	4	2	2	—	—	14	12	10	10	8	6	6
25	14	12	10	8	6	6	4	2	1	—	14	12	12	10	8	6	6	4
30	14	10	8	8	6	4	2	2	0	—	14	12	10	8	8	6	6	4
35	12	10	8	6	4	4	2	1	0	—	12	10	10	8	6	4	4	4
40	12	10	8	6	4	2	2	0	00	—	12	10	10	8	6	4	2	2
45	12	8	6	6	4	2	1	0	000	14	12	10	8	6	6	4	2	2
50	12	8	6	4	4	2	0	00	000	14	12	10	8	6	6	4	2	1
60	10	8	6	4	2	2	0	000	—	14	10	8	8	6	4	2	2	0
70	10	6	4	4	2	1	00	0000	—	12	10	8	6	4	4	2	1	0
80	10	6	4	2	2	0	000	—	—	12	10	8	6	4	2	2	0	00
90	8	6	4	2	1	0	0000	—	—	12	8	6	6	4	2	1	0	000
100	8	6	4	2	0	00	—	—	—	12	8	6	6	4	2	0	00	000
120	8	4	2	2	0	0000	—	—	—	10	8	6	4	2	2	0	000	0000

metry. If the sign of a coefficient is not given it is unknown, not necessarily positive.

Unit for $d_{nm} = 10^{-8}$ statcoulomb/dyne

$$= \tfrac{1}{3} \times 10^{-12} \text{ coulomb/newton}$$

$$= 10^{-8} \text{ cm/statvolt}$$

$$= \tfrac{1}{3} \times 10^{-12} \text{ meter/volt.}$$

Coupling factor k is defined practically by

$k^2 =$ (mechanical energy converted into electric charge)/(mechanical energy put into the crystal)

The converse effect is also true. The same type of relationship holds and the coupling coefficient

TABLE 39—FUSING CURRENTS IN AMPERES. *Courtesy of Automatic Electric Company; Chicago, Ill.*

AWG B & S Gauge	Diam d in Inches	Copper ($K=10\,244$)	Aluminum ($K=7585$)	German Silver ($K=5230$)	Iron ($K=3148$)	Tin ($K=1642$)
40	0.0031	1.77	1.31	0.90	0.54	0.28
38	0.0039	2.50	1.85	1.27	0.77	0.40
36	0.0050	3.62	2.68	1.85	1.11	0.58
34	0.0063	5.12	3.79	2.61	1.57	0.82
32	0.0079	7.19	5.32	3.67	2.21	1.15
30	0.0100	10.2	7.58	5.23	3.15	1.64
28	0.0126	14.4	10.7	7.39	4.45	2.32
26	0.0159	20.5	15.2	10.5	6.31	3.29
24	0.0201	29.2	21.6	14.9	8.97	4.68
22	0.0253	41.2	30.5	21.0	12.7	6.61
20	0.0319	58.4	43.2	29.8	17.9	9.36
19	0.0359	69.7	51.6	35.5	21.4	11.2
18	0.0403	82.9	61.4	42.3	25.5	13.3
17	0.0452	98.4	72.9	50.2	30.2	15.8
16	0.0508	117	86.8	59.9	36.0	18.8
15	0.0571	140	103	71.4	43.0	22.4
14	0.0641	166	123	84.9	51.1	26.6
13	0.0719	197	146	101	60.7	31.7
12	0.0808	235	174	120	72.3	37.7
11	0.0907	280	207	143	86.0	44.9
10	0.1019	333	247	170	102	53.4
9	0.1144	396	293	202	122	63.5
8	0.1285	472	349	241	145	75.6
7	0.1443	561	416	287	173	90.0
6	0.1620	668	495	341	205	107

is numerically identical to what it was before, namely

$k^2 =$ (electrical energy converted into mechanical energy)/(electrical energy put into the crystal)

d is the measure of the deflection caused by an applied voltage or the amount of charge produced by a given force (units= meters per volt or coulomb per newton)

g denotes a field produced in a piezoelectric crystal by an applied stress unit:

$$\frac{\text{volts/meter}}{\text{newtons/square meter}}.$$

Relations between g and d are

$$g = d/\epsilon_r \epsilon_0$$

where $\epsilon_r =$ relative permittivity of the dielectric, $\epsilon_0 =$ permittivity of free space $= 8.85 \times 10^{-12}$ farad/meter, $E =$ Young's modulus, and $k^2 = gdE$.

ACOUSTIC PROPERTIES OF SOME MATERIALS (TABLES 30–31)

ADHESIVES (TABLE 32)

SHOP DATA

Wire Tables

Temperature coefficient of resistance: The resistance of a conductor at temperature T in degrees Celsius is given by

$$R = R_{20}[1 + \alpha_{20}(T - 20)]$$

where R_{20} is the resistance at 20 degrees Celsius and α_{20} is the temperature coefficient of resistance at 20 degrees Celsius. For copper, $\alpha_{20} = 0.00393$. That

is, the resistance of a copper conductor increases approximately $\frac{4}{10}$ of 1 percent per degree Celsius rise in temperature.

Modulus of elasticity is 17 000 000 lb/inch². Coefficient of linear expansion is 0.0000094/degree Fahrenheit.

Weights are based on a density of 8.89 grams/cm³ at 20 degrees Celsius (equivalent to 0.00302699 lb/circular mil/1000 feet).

The resistances are maximum values for hard-drawn copper and are based on a resistivity of 10.674 ohms/circular-mil foot at 20 degrees Celsius (97.16 percent conductivity) for sizes 0.325 inch and larger, and 10.785 ohms/circular-mil foot at 20 degrees Celsius (96.16 percent conductivity) for sizes 0.324 inch and smaller. (Refer to Tables 33–37.)

Voltage Drop in Long Circuits

Table 38 shows the conductor size (AWG or B&S gauge) necessary to limit the voltage drop to 2 percent maximum for various loads and distances. The calculations are for alternating-current circuits in conduit.

Fusing Currents of Wires

The current I in amperes at which a wire will melt can be calculated from

$$I = Kd^{3/2}$$

where d is the wire diameter in inches and K is a constant that depends on the metal concerned. Table 39 gives the fusing currents in amperes for 5 commonly used types of wire. Owing to the wide variety of factors that can influence the rate of heat loss, these figures must be considered as only approximations.

TABLE 40—PHYSICAL PROPERTIES OF VARIOUS WIRES. *Reprinted by permission from*

Property	Copper Annealed	Copper Hard-Drawn	Aluminum 99 Percent Pure
Conductivity, Matthiessen's standard in percent	99 to 102	96 to 99	61 to 63
Ohms/mil-foot at 68° F=20° C	10.36	10.57	16.7
Circular-mil-ohms/mile at 68° F=20° C	54 600	55 700	88 200
Pounds/mile-ohm at 68° F=20° C	875	896	424
Mean temp coefficient of resistivity/°F	0.00233	0.00233	0.0022
Mean temp coefficient of resistivity/°C	0.0042	0.0042	0.0040
Mean specific gravity	8.89	8.94	2.68
Pounds/1000 feet/circular mil	0.003027	0.003049	0.000909
Weight in pounds/inch³	0.320	0.322	0.0967
Mean specific heat	0.093	0.093	0.214
Mean melting point in °F	2 012	2 012	1 157
Mean melting point in °C	1 100	1 100	625
Mean coefficient of linear expansion/°F	0.00000950	0.00000950	0.00001285
Mean coefficient of linear expansion/°C	0.0000171	0.0000171	0.0000231
Solid wire Ultimate tensile strength	30 000 to 42 000	45 000 to 68 000	20 000 to 35 000
Average tensile strength	32 000	60 000	24 000
(Values in Elastic limit	6 000 to 16 000	25 000 to 45 000	14 000
pounds/in.²) Average elastic limit	15 000	30 000	14 000
Modulus of elasticity	7 000 000 to 17 000 000	13 000 000 to 18 000 000	8 500 000 to 11 500 000
Average modulus of elasticity	12 000 000	16 000 000	9 000 000
Concentric strand Tensile strength	29 000 to 37 000	43 000 to 65 000	25 800
Average tensile strength	35 000	54 000	—
Elastic limit	5 800 to 14 800	23 000 to 42 000	13 800
(Values in Average elastic limit	—	27 000	—
pounds/in.²) Modulus of elasticity	5 000 000 to 12 000 000	12 000 000	Approx 10 000 000

"Transmission Towers," American Bridge Company, Pittsburgh, Pa.; 1925: p. 169.

Iron (Ex BB)	Steel (Siemens-Martin)	Crucible Steel, High Strength	Plow Steel, Extra-High Strength	Copper-Clad 30% Cond	Copper-Clad 40% Cond
16.8	8.7	—	—	29.4	39.0
62.9	119.7	122.5	125.0	35.5	26.6
332 000	632 000	647 000	660 000	187 000	140 000
4 700	8 900	9 100	9 300	2.775	2.075
0.0028	0.00278	0.00278	0.00278	0.0024	—
0.0050	0.00501	0.00501	0.00501	0.0044	0.0041
7.77	7.85	7.85	7.85	8.17	8.25
0.002652	0.002671	—	—	0.00281	0.00281
0.282	0.283	0.283	0.283	0.298	0.298
0.113	0.117	—	—	—	—
2 975	2 480	—	—	—	—
1 635	1 360	—	—	—	—
0.00000673	0.00000662	—	—	0.0000072	0.0000072
0.0000120	0.0000118	—	—	0.0000129	0.0000129
50 000 to 55 000	70 000 to 80 000	—	—	—	—
55 000	75 000	125 000	187 000	60 000	100 000
25 000 to 30 000	35 000 to 50 000	—	—	—	—
30 000	38 000	69 000	130 000	30 000	50 000
22 000 000 to 27 000 000	22 000 000 to 29 000 000	—	—	—	—
26 000 000	29 000 000	30 000 000	30 000 000	19 000 000	21 000 000
—	74 000 to 98 000	85 000 to 165 000	140 000 to 245 000	70 000 to 97 000	—
—	80 000	125 000	180 000	80 000	—
—	37 000 to 49 000	—	—	—	—
—	40 000	70 000	110 000	—	—
—	12 000 000	15 000 000	15 000 000	—	—

Physical Properties (Tables 40–41)

TABLE 41—PHYSICAL PROPERTIES OF STRANDED COPPER (AWG)*.

Circular Mils	AWG B & S Gauge	Number of Wires	Individual Wire Diam in Inches	Cable Diam in Inches	Area in Square Inches	Weight in Lb per 1000 Ft	Weight in Lb per Mile	*Maximum Resistance in Ohms/1000 Ft at 20° C
211 600	4/0	19	0.1055	0.528	0.1662	653.3	3 450	0.05093
167 800	3/0	19	0.0940	0.470	0.1318	518.1	2 736	0.06422
133 100	2/0	19	0.0837	0.419	0.1045	410.9	2 170	0.08097
105 500	1/0	19	0.0745	0.373	0.08286	325.7	1 720	0.1022
83 690	1	19	0.0664	0.332	0.06573	258.4	1 364	0.1288
66 370	2	7	0.0974	0.292	0.05213	204.9	1 082	0.1624
52 640	3	7	0.0867	0.260	0.04134	162.5	858.0	0.2048
41 740	4	7	0.0772	0.232	0.03278	128.9	680.5	0.2582
33 100	5	7	0.0688	0.206	0.02600	102.2	539.6	0.3256
26 250	6	7	0.0612	0.184	0.02062	81.05	427.9	0.4105
20 820	7	7	0.0545	0.164	0.01635	64.28	339.4	0.5176
16 510	8	7	0.0486	0.146	0.01297	50.98	269.1	0.6528
13 090	9	7	0.0432	0.130	0.01028	40.42	213.4	0.8233
10 380	10	7	0.0385	0.116	0.008152	32.05	169.2	1.038
6 530	12	7	0.0305	0.0915	0.005129	20.16	106.5	1.650
4 107	14	7	0.0242	0.0726	0.003226	12.68	66.95	2.624
2 583	16	7	0.0192	0.0576	0.002029	7.975	42.11	4.172
1 624	18	7	0.0152	0.0456	0.001275	5.014	26.47	6.636
1 022	20	7	0.0121	0.0363	0.0008027	3.155	16.66	10.54

* The resistance values in this table are trade maxima for soft or annealed copper wire and are higher than the average values for commercial cable. The following values for the conductivity and resistivity of copper at 20 degrees Celsius were used:

Conductivity in terms of International Annealed Copper Standard: 98.16 percent

Resistivity in pounds per mile-ohm: 891.58

The resistance of hard-drawn copper is slightly greater than the values given, being about 2 percent to 3 percent greater for sizes from 4/0 to 20 AWG.

TABLE 42—MACHINE-SCREW DIMENSIONS AND OTHER DATA.

Screw		Threads/in.		Clearance Drill*		Tap Drill† — Diameter			Head — Round		Head — Flat	Head — Fillister		Head — Hex Nut			Washer		
No.	Diam	Coarse	Fine	No.	Diam	No.	Inches	mm	Max OD	Max Height	Max OD	Max OD	Max Height	Across Flat	Across Corner	Thickness	OD	ID	Thickness
0	0.060	—	80	52	0.064	56	0.047	1.2	0.113	0.053	0.119	0.096	0.059	0.156	0.171	0.046	—	—	—
1	0.073	64	72	47	0.079	53	0.060	1.5	0.138	0.061	0.146	0.118	0.070	0.156	0.171	0.046	—	—	—
2	0.086	56	64	42	0.094	50	0.070	1.8	0.162	0.070	0.172	0.140	0.083	0.187	0.205	0.062	¼	0.093	0.032
3	0.099	48	56	37	0.104	47 / 45	0.079 / 0.082	2.0 / 2.1	0.187	0.078	0.199	0.161	0.095	0.187	0.205	0.062	¼	0.105	0.020
4	0.112	40	48	31	0.120	43 / 42	0.089 / 0.094	2.3 / 2.4	0.211	0.086	0.225	0.183	0.107	0.250	0.275	0.093	5⁄16	0.125	0.032
5	0.125	40	44	29	0.136	38 / 37	0.102 / 0.104	2.6 / 2.6	0.236	0.095	0.252	0.205	0.120	0.312	0.344	0.109	3⁄8	0.140	0.032
6	0.138	32	40	27	0.144	36 / 33	0.107 / 0.113	2.7 / 2.9	0.260	0.103	0.279	0.226	0.132	0.312	0.344	0.109	5⁄16 / 3⁄8	0.156	0.026 / 0.046
8	0.164	32	36	18	0.170	29 / 29	0.136 / 0.136	3.5 / 3.5	0.309	0.119	0.332	0.270	0.156	0.344	0.373	0.125	3⁄8 / 7⁄16	0.186	0.032 / 0.046
10	0.190	24	32	9	0.196	25 / 21	0.150 / 0.159	3.8 / 4.0	0.359	0.136	0.385	0.313	0.180	0.375	0.413	0.125	7⁄16 / 1⁄2	0.218	0.036 / 0.063
12	0.216	24	28	2	0.221	16 / 14	0.177 / 0.182	4.5 / 4.6	0.408	0.152	0.438	0.357	0.205	0.437	0.488	0.156	1⁄2 / 9⁄16	0.250	0.063
¼	0.250	20	28	—	17⁄64	7 / 3	0.201 / 0.213	5.1 / 5.5	0.472	0.174	0.507	0.414	0.237	0.437 / 0.500	0.488 / 0.577	0.203 / 0.250	9⁄16 / 5⁄8	0.281	0.040 / 0.063

All dimensions in inches except where noted.

* Clearance-drill sizes are practical values for use of the engineer or technician doing his own shop work.

† Tap-drill sizes are for use in hand tapping material such as brass or soft steel. For copper, aluminum, Norway iron, cast iron, bakelite, or for very thin material, the drill should be a size or two larger in diameter than shown.

TABLE 43—DRILL SIZES. *From New Departure Handbook.*

Drill	Inches	Drill	Inches	Drill	Inches	Drill	Inches
0.10 mm	0.003937	no 59	0.041000	no 41	0.096000	4.30 mm	0.169291
0.15 mm	0.005905	1.05 mm	0.041338	2.45 mm	0.096456	no 18	0.169500
0.20 mm	0.007874	no 58	0.042000	no 40	0.098000	$^{11}\!/_{64}$ in.	0.171875
0.25 mm	0.009842	no 57	0.043000	2.50 mm	0.098425	no 17	0.173000
0.30 mm	0.011811	1.10 mm	0.043307	no 39	0.099500	4.40 mm	0.173228
no 80	0.013000	1.15 mm	0.045275	no 38	0.101500	no 16	0.177000
no 79½	0.013500	no 56	0.046500	2.60 mm	0.102362	4.50 mm	0.177165
0.35 mm	0.013779	$^{3}\!/_{64}$ in.	0.046875	no 37	0.104000	no 15	0.180000
no 79	0.014000	1.20 mm	0.047244	2.70 mm	0.106299	4.60 mm	0.181102
no 78½	0.014500	1.25 mm	0.049212	no 36	0.106500	no 14	0.182000
no 78	0.015000	1.30 mm	0.051181	2.75 mm	0.108267	no 13	0.185000
$^{1}\!/_{64}$ in.	0.015625	no 55	0.052000	$^{7}\!/_{64}$ in.	0.109375	4.70 mm	0.185039
0.40 mm	0.015748	1.35 mm	0.053149	no 35	0.110000	4.75 mm	0.187007
no 77	0.016000	no 54	0.055000	2.80 mm	0.110236	$^{3}\!/_{16}$ in.	0.187500
0.45 mm	0.017716	1.40 mm	0.055118	no 34	0.111000	4.80 mm	0.188976
no 76	0.018000	1.45 mm	0.057086	no 33	0.113000	no 12	0.189000
0.50 mm	0.019685	1.50 mm	0.059055	2.90 mm	0.114173	no 11	0.191000
no 75	0.020000	no 53	0.059500	no 32	0.116000	4.90 mm	0.192913
no 74½	0.021000	1.55 mm	0.061023	3.00 mm	0.118110	no 10	0.193500
0.55 mm	0.021653	$^{1}\!/_{16}$ in.	0.062500	no 31	0.120000	no 9	0.196000
no 74	0.022000	1.60 mm	0.062992	3.10 mm	0.122047	5.00 mm	0.196850
no 73½	0.022500	no 52	0.063500	$^{1}\!/_{8}$ in.	0.125000	no 8	0.199000
no 73	0.023000	1.65 mm	0.064960	3.20 mm	0.125984	5.10 mm	0.200787
0.60 mm	0.023622	1.70 mm	0.066929	3.25 mm	0.127952	no 7	0.201000
no 72	0.024000	no 51	0.067000	no 30	0.128500	$^{13}\!/_{64}$ in.	0.203125
no 71½	0.025000	1.75 mm	0.068897	3.30 mm	0.129921	no 6	0.204000
0.65 mm	0.025590	no 50	0.070000	3.40 mm	0.133858	5.20 mm	0.204724
no 71	0.026000	1.80 mm	0.070866	no 29	0.136000	no 5	0.205500
no 70	0.027000	1.85 mm	0.072834	3.50 mm	0.137795	5.25 mm	0.206692
0.70 mm	0.027559	no 49	0.073000	no 28	0.140500	5.30 mm	0.208661
no 69½	0.028000	1.90 mm	0.074803	$^{9}\!/_{64}$ in.	0.140625	no 4	0.209000
no 69	0.029000	no 48	0.076000	3.60 mm	0.141732	5.40 mm	0.212598
no 68½	0.029250	1.95 mm	0.076771	no 27	0.144000	no 3	0.213000
0.75 mm	0.029527	$^{5}\!/_{64}$ in.	0.078125	3.70 mm	0.145669	5.50 mm	0.216535
no 68	0.030000	no 47	0.078500	no 26	0.147000	$^{7}\!/_{32}$ in.	0.218750
no 67	0.031000	2.00 mm	0.078740	3.75 mm	0.147637	5.60 mm	0.220472
$^{1}\!/_{32}$ in.	0.031250	2.05 mm	0.080708	no 25	0.149500	no 2	0.221000
0.80 mm	0.031496	no 46	0.081000	3.80 mm	0.149606	5.70 mm	0.224409
no 66	0.032000	no 45	0.082000	no 24	0.152000	5.75 mm	0.226377
no 65	0.033000	2.10 mm	0.082677	3.90 mm	0.153543	no 1	0.228000
0.85 mm	0.033464	2.15 mm	0.084645	no 23	0.154000	5.80 mm	0.228346
no 64	0.035000	no 44	0.086000	$^{5}\!/_{32}$ in.	0.156250	5.90 mm	0.232283
0.90 mm	0.035433	2.20 mm	0.086614	no 22	0.157000	ltr A	0.234000
no 63	0.036000	2.25 mm	0.088582	4.00 mm	0.157480	$^{15}\!/_{64}$ in.	0.234375
no 62	0.037000	no 43	0.089000	no 21	0.159000	6.00 mm	0.236220
0.95 mm	0.037401	2.30 mm	0.090551	no 20	0.161000	ltr B	0.238000
no 61	0.038000	2.35 mm	0.092519	4.10 mm	0.161417	6.10 mm	0.240157
no 60½	0.039000	no 42	0.093500	4.20 mm	0.165354	ltr C	0.242000
1.00 mm	0.039370	$^{3}\!/_{32}$ in.	0.093750	no 19	0.166000	6.20 mm	0.244094
no 60	0.040000	2.40 mm	0.094488	4.25 mm	0.167322	ltr D	0.246000

TABLE 43—CONTINUED.

Drill	Inches	Drill	Inches	Drill	Inches	Drill	Inches
6.25 mm	0.246062	ltr O	0.316000	10.00 mm	0.393700	17.50 mm	0.688975
6.30 mm	0.248031	8.10 mm	0.318897	ltr X	0.397000	$\frac{45}{64}$ in.	0.703125
ltr E }		8.20 mm	0.322834	ltr Y	0.404000	18.00 mm	0.708660
¼ in. }	0.250000	ltr P	0.323000	$\frac{13}{32}$ in.	0.406250	$\frac{23}{32}$ in.	0.718750
6.40 mm	0.251968	8.25 mm	0.324802	ltr Z	0.413000	18.50 mm	0.728345
6.50 mm	0.255905	8.30 mm	0.326771	10.50 mm	0.413385	$\frac{47}{64}$ in.	0.734375
ltr F	0.257000	$\frac{21}{64}$ in.	0.328125	$\frac{27}{64}$ in.	0.421875	19.00 mm	0.748030
6.60 mm	0.259842	8.40 mm	0.330708	11.00 mm	0.433070	¾ in.	0.750000
ltr G	0.261000	ltr Q	0.332000	$\frac{7}{16}$ in.	0.437500	$\frac{49}{64}$ in.	0.765625
6.70 mm	0.263779	8.50 mm	0.334645	11.50 mm	0.452755	19.50 mm	0.767715
$\frac{17}{64}$ in.	0.265625	8.60 mm	0.338582	$\frac{29}{64}$ in.	0.453125	$\frac{25}{32}$ in.	0.781250
6.75 mm	0.265747	ltr R	0.339000	$\frac{15}{32}$ in.	0.468750	20.00 mm	0.787400
ltr H	0.266000	8.70 mm	0.342519	12.00 mm	0.472440	$\frac{51}{64}$ in.	0.796875
6.80 mm	0.267716	$\frac{11}{32}$ in.	0.343750	$\frac{31}{64}$ in.	0.484375	20.50 mm	0.807085
6.90 mm	0.271653	8.75 mm	0.344487	12.50 mm	0.492125	$\frac{13}{16}$ in.	0.812500
ltr I	0.272000	8.80 mm	0.346456	½ in.	0.500000	21.00 mm	0.826770
7.00 mm	0.275590	ltr S	0.348000	13.00 mm	0.511810	$\frac{53}{64}$ in.	0.828125
ltr J	0.277000	8.90 mm	0.350393	$\frac{33}{64}$ in.	0.515625	$\frac{27}{32}$ in.	0.843750
7.10 mm	0.279527	9.00 mm	0.354330	$\frac{17}{32}$ in.	0.531250	21.50 mm	0.846455
ltr K	0.281000	ltr T	0.358000	13.50 mm	0.531495	$\frac{55}{64}$ in.	0.859375
$\frac{9}{32}$ in.	0.281250	9.10 mm	0.358267	$\frac{35}{64}$ in.	0.546875	22.00 mm	0.866140
7.20 mm	0.283464	$\frac{23}{64}$ in.	0.359375	14.00 mm	0.551180	⅞ in.	0.875000
7.25 mm	0.285432	9.20 mm	0.362204	$\frac{9}{16}$ in.	0.562500	22.50 mm	0.885825
7.30 mm	0.287401	9.25 mm	0.364172	14.50 mm	0.570865	$\frac{57}{64}$ in.	0.890625
ltr L	0.290000	9.30 mm	0.366141	$\frac{37}{64}$ in.	0.578125	23.00 mm	0.905510
7.40 mm	0.291338	ltr U	0.368000	15.00 mm	0.590550	$\frac{29}{32}$ in.	0.906250
ltr M	0.295000	9.40 mm	0.370078	$\frac{19}{32}$ in.	0.593750	$\frac{59}{64}$ in.	0.921875
7.50 mm	0.295275	9.50 mm	0.374015	$\frac{39}{64}$ in.	0.609375	23.50 mm	0.925195
$\frac{19}{64}$ in.	0.296875	⅜ in.	0.375000	15.50 mm	0.610235	$\frac{15}{16}$ in.	0.937500
7.60 mm	0.299212	ltr V	0.377000	⅝ in.	0.625000	24.00 mm	0.944880
ltr N	0.302000	9.60 mm	0.377952	16.00 mm	0.629920	$\frac{61}{64}$ in.	0.953125
7.70 mm	0.303149	9.70 mm	0.381889	$\frac{41}{64}$ in.	0.640625	24.50 mm	0.964565
7.75 mm	0.305117	9.75 mm	0.383857	16.50 mm	0.649605	$\frac{31}{32}$ in.	0.968750
7.80 mm	0.307086	9.80 mm	0.385826	$\frac{21}{32}$ in.	0.656250	25.00 mm	0.984250
7.90 mm	0.311023	ltr W	0.386000	17.00 mm	0.669290	$\frac{63}{64}$ in.	0.984375
$\frac{5}{16}$ in.	0.312500	9.90 mm	0.389763	$\frac{43}{64}$ in.	0.671875	1 in.	1.000000
8.00 mm	0.314960	$\frac{25}{64}$ in.	0.390625	$\frac{11}{16}$ in.	0.687500		

Machine Screws and Drill Sizes (Tables 42–43)

Head Styles—Method of Length Measurement (Fig. 5)

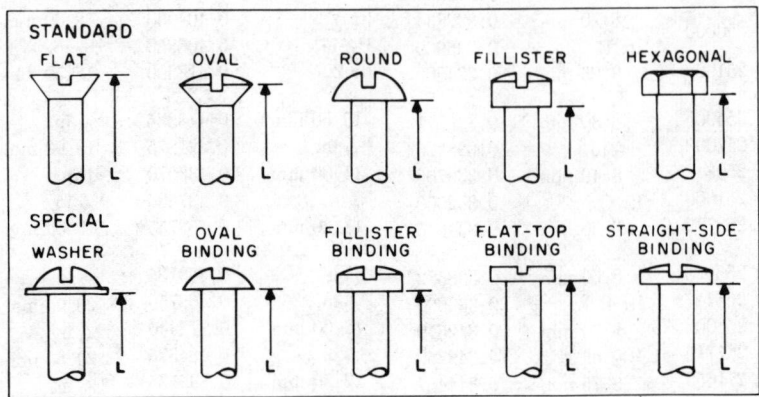

Fig. 5

Sheet-Metal Gauges

Systems in Use: Materials are customarily made to certain gauge systems. While materials can usually be had specially in any system, some usual practices are shown in Tables 44 and 45.

TABLE 44—COMMON GAUGE PRACTICES.

Material	Sheet	Wire
Aluminum	B&S	AWG (B&S)
Brass, bronze, sheet	B&S	—
Copper	B&S	AWG (B&S)
Iron, steel, band, and hoop	BWG	—
Iron, steel, telephone, and telegraph wire	—	BWG
Steel wire, except telephone and telegraph	—	W&M
Steel sheet	US	—
Tank steel	BWG	—
Zinc sheet	"Zinc gauge" proprietary	—

TABLE 45—COMPARISON OF GAUGES. *Courtesy of Whitehead Metal Products Co., Inc.*

Gauge	AWG B&S	Birmingham or Stubs BWG	Wash. & Moen W&M	British Standard NBS SWG	London or Old English	United States Standard US	American Standard Preferred Thickness*
0000000	—	—	0.490	0.500	—	0.50000	—
000000	0.5800	—	0.460	0.464	—	0.46875	—
00000	0.5165	—	0.430	0.432	—	0.43750	—
0000	0.4600	0.454	0.3938	0.400	0.454	0.40625	—
000	0.4096	0.425	0.3625	0.372	0.425	0.37500	—
00	0.3648	0.380	0.3310	0.348	0.380	0.34375	—
0	0.3249	0.340	0.3065	0.324	0.340	0.31250	—
1	0.2893	0.300	0.2830	0.300	0.300	0.28125	—
2	0.2576	0.284	0.2625	0.276	0.284	0.265625	—
3	0.2294	0.259	0.2437	0.252	0.259	0.250000	0.224
4	0.2043	0.238	0.2253	0.232	0.238	0.234375	0.200
5	0.1819	0.220	0.2070	0.212	0.220	0.218750	0.180
6	0.1620	0.203	0.1920	0.192	0.203	0.203125	0.160
7	0.1443	0.180	0.1770	0.176	0.180	0.187500	0.140
8	0.1285	0.165	0.1620	0.160	0.165	0.171875	0.125
9	0.1144	0.148	0.1483	0.144	0.148	0.156250	0.112
10	0.1019	0.134	0.1350	0.128	0.134	0.140625	0.100
11	0.09074	0.120	0.1205	0.116	0.120	0.125000	0.090
12	0.08081	0.109	0.1055	0.104	0.109	0.109375	0.080
13	0.07196	0.095	0.0915	0.092	0.095	0.093750	0.071
14	0.06408	0.083	0.0800	0.080	0.083	0.078125	0.063
15	0.05707	0.072	0.0720	0.072	0.072	0.0703125	0.056
16	0.05082	0.065	0.0625	0.064	0.065	0.0625000	0.050
17	0.04526	0.058	0.0540	0.056	0.058	0.0562500	0.045
18	0.04030	0.049	0.0475	0.048	0.049	0.0500000	0.040
19	0.03589	0.042	0.0410	0.040	0.040	0.0437500	0.036
20	0.03196	0.035	0.0348	0.036	0.035	0.0375000	0.032
21	0.02846	0.032	0.03175	0.032	0.0315	0.0343750	0.028
22	0.02535	0.028	0.02860	0.028	0.0295	0.0312500	0.025
23	0.02257	0.025	0.02580	0.024	0.0270	0.0281250	0.022
24	0.02010	0.022	0.02300	0.022	0.0250	0.0250000	0.020
25	0.01790	0.020	0.02040	0.020	0.0230	0.0218750	0.018
26	0.01594	0.018	0.01810	0.018	0.0205	0.0187500	0.016
27	0.01420	0.016	0.01730	0.0164	0.0187	0.0171875	0.014
28	0.01264	0.014	0.01620	0.0148	0.0165	0.0156250	0.012
29	0.01126	0.013	0.01500	0.0136	0.0155	0.0140625	0.011
30	0.01003	0.012	0.01400	0.0124	0.01372	0.0125000	0.010
31	0.008928	0.010	0.01320	0.0116	0.01220	0.01093750	0.009
32	0.007950	0.009	0.01280	0.0108	0.01120	0.01015625	0.008
33	0.007080	0.008	0.01180	0.0100	0.01020	0.00937500	0.007
34	0.006305	0.007	0.01040	0.0092	0.00950	0.00859375	0.006
35	0.005615	0.005	0.00950	0.0084	0.00900	0.00781250	—
36	0.005000	0.004	0.00900	0.0076	0.00750	0.007031250	—
37	0.004453	—	0.00850	0.0068	0.00650	0.006640625	—
38	0.003965	—	0.00800	0.0060	0.00570	0.006250000	—
39	0.003531	—	0.00750	0.0052	0.00500	—	—
40	0.003145	—	0.00700	0.0048	0.00450	—	—

* These thicknesses are intended to express the desired thickness in decimal fractions of an inch. They have no relation to gauge numbers; they are approximately related to the AWG sizes 3–34.

Antifreeze Solutions (Table 46)

TABLE 46—COMMERCIAL ANTIFREEZE SOLUTIONS.

Percent by Volume	Percent by Volume in Water with Freezing Points and Specific Gravities				
	10	20	30	40	50
Typical commercial methanol antifreeze	−5.2°C	−12.0°C	−21.1°C	−32.2°C	−45.0°C
Sp. gr. at 15°C/15°C*	0.986	0.975	0.963	0.950	0.935
Typical commercial ethanol antifreeze	−3.3°C	−7.7°C	−14.2°C	−22.0°C	−30.6°C
Sp. gr. at 15°C/15°C*	0.988	0.977	0.967	0.955	0.938
Commercial glycerine† antifreeze	−1.6°C	−4.7°C	−9.5°C	−15.4°C	−23.0°C
Sp. gr. at 15°C/15°C*	1.023	1.048	1.074	1.101	1.128
Typical commercial ethylene glycol† antifreeze	−3.8°C	−8.8°C	−15.5°C	−24.3°C	−36.5°C
Sp. gr. at 15°C/15°C*	1.015	1.030	1.045	1.060	1.074

* Specific gravity is measured for mixture at 15°C referred to water at 15°C.

† Glycerine and ethylene glycol are practically nonvolatile. All types must be suitably inhibited to prevent cooling-system corrosion. Commercial antifreeze solutions based on ethylene glycol and on glycerine are in use at the present time.

GENERAL STANDARDS

Standardization of electronic components or parts is handled by several cooperating agencies.

In the US, the Electronic Industries Association (EIA)*, and the American National Standards Institute (ANSI)† are active in the commercial field. Electron-tube and semiconductor-device standards are handled by the Joint Electron Device Engineering Council (JEDEC), a cooperative effort of EIA and the National Electrical Manufacturers Association (NEMA)‡. Military (MIL) standards are issued by the US Department of Defense or one of its agencies such as the Defense Electronics Supply Center (DESC).

International standardization in the electronics field is carried out by the various Technical Committees of the International Electrotechnical Commission (IEC)§. A list of the available IEC Recommendations is included in the ANSI Index (outside the US, consult the national standardization agency or the IEC). IEC documents may be used directly or their recommendations may be incorporated in whole or in part in national standards issued by the EIA or ANSI. A few broad areas may be covered by standards issued by the International Standards Organization (ISO).

These organizations establish standards for electronic components or parts (and in some cases, for equipments) to provide interchangeability among different products regarding size, performance, and identification; minimum number of sizes and designs; and uniform testing of products for acceptance. This chapter presents a brief outline of the requirements, characteristics, and designations for the major types of component parts used in electronic equipment. Such standardization offers economic advantages to both the parts user and the parts manufacturer, but is not intended to prevent the manufacture and use of other parts under special conditions.

There is a trend away from circuits assembled of separate components and wiring and toward integrated circuits that have the various resistors, capacitors, inductors, and semiconductor devices deposited on a common substrate. Such integrated circuits offer advantages of minimum size, good stability (with proper protection), and economical manufacture in large quantities. However, field repair and revisions of the circuit values after manufacture are not possible—failure of any portion of an integrated circuit necessitates replacement of the entire unit. As of this printing (1975) integrated-circuit standards are still being developed; however, standard outlines, systems of nomenclature, methods of reporting technical characteristics, and some standards for testing procedures have been established.

Color Coding

The color code of Table 1 is used for marking electronic parts.

Tolerance

The maximum deviation allowed from the specified nominal value is known as the tolerance. It is usually given as a percentage of the nominal value, though for very small capacitors the tolerance may be specified in picofarads (pF). For critical applications it is important to specify the permissible tolerance; where no tolerance is specified, components are likely to vary by ±20 percent from the nominal value.

Do not assume that a given lot of components will have values distributed throughout the acceptable range of values. A lot ordered with a

---

* EIA Engineering Dept., Washington, D. C. Index of standards is available. EIA was formerly Radio-Electronics-Television Manufacturers' Association (RETMA).

† ANSI, New York, New York. Index of standards is available. ANSI was formerly the USA Standards Institute (USASI).

‡ NEMA, New York, New York. Index of standards is available.

§ IEC, Central Office; Geneva, Switzerland. The US National Committee for the IEC operates within the ANSI.

TABLE 1—STANDARD COLOR CODE OF ELECTRONICS INDUSTRY.

Color	Significant Figure	Decimal Multiplier	Tolerance in Percent*	Voltage Rating	Characteristic
Black	0	1	±20 (M)	—	A
Brown	1	10	±1 (F)	100	B
Red	2	100	±2 (G)	200	C
Orange	3	1 000	±3	300	D
Yellow	4	10 000	GMV‡	400	E
Green	5	100 000	±5(J)†, (0.50(D))§	500	F
Blue	6	1 000 000	±6, (0.25(C))§	600	G
Violet	7	10 000 000	±12.5, (0.10(B))§	700	—
Gray	8	0.01†	±30, (0.05(N))§	800	I
White	9	0.1†	±10†	900	J
Gold	—	0.1	±5 (J), (0.50(E))‖	1 000	—
Silver	—	0.01	±10 (K)	2 000	—
No Color	—	—	±20	500	—

* Tolerance letter symbol as used in type designations has tolerance meaning as shown. ±3, ±6, ±12.5, and ±30 percent are tolerances for USA Std 40-, 20-, 10-, and 5-step series, respectively.
† Optional coding where metallic pigments are undesirable.
‡ GMV is −0 to +100-percent tolerance or Guaranteed Minimum Value.
§ For some film and other resistors only.
‖ For some capacitors only.

±20% tolerance may include *no* parts having values within 5% of the desired nominal value; these may have been sorted out before shipment. The manufacturing process for a given lot may produce parts in a narrow range of values only, not necessarily centered in the acceptable tolerance range.

Preferred Values

To maintain an orderly progression of sizes, preferred numbers are frequently used for the nominal values. A further advantage is that all parts manufactured are salable as one or another of the preferred values. Each preferred value differs from its predecessor by a constant multiplier, and the final result is conveniently rounded to two significant figures.

ANSI Standard Z17.1-1973 covers a series of preferred numbers based on $(10)^{1/5}$ and $(10)^{1/10}$ as listed in Table 2. This series has been widely used for fixed wire-wound power-type resistors and for time-delay fuses.

Because of the established practice of using ±20-, ±10-, and ±5-percent tolerances, a series of values based on $(10)^{1/6}$, $(10)^{1/12}$, and $(10)^{1/24}$ has been adopted by the EIA, and is now an ANSI Standard (C83.2-1971) (EIA RS-385). It is widely used for such small electronic components as fixed composition resistors and fixed ceramic, mica, and molded paper capacitors. These values are listed in Table 2. (For series with smaller steps, consult the ANSI or EIA Standard.)

Voltage Rating

Distinction must be made between the breakdown-voltage rating (test volts) and the working-voltage rating. The maximum voltage that may be applied (usually continuously) over a long period of time without causing the part to fail determines the working-voltage rating. Application of the test voltage for more than a very few minutes, or even repeated applications of short duration, may result in permanent damage or failure of the part.

Characteristic

This term is frequently used to include various qualities of a part such as temperature coefficient of capacitance or resistance, Q value, maximum permissible operating temperature, stability when subjected to repeated cycles of high and low temperature, and deterioration when it is subjected to moisture either as humidity or water immersion.

TABLE 2—PREFERRED VALUES*.

Name of Series	USA Standard Z17.1-1973†		USA Standard C83.2-1971‡		
	"5"	"10"	±20%(E6)	±10%(E12)	±5%(E24)
Percent step size	60	25	≈40	20	10
Step multiplier	$(10)^{1/5}=1.58$	$(10)^{1/10}=1.26$	$(10)^{1/6}=1.46$	$(10)^{1/12}=1.21$	$(10)^{1/24}=1.10$
Values in the series (Use decimal multipliers for smaller or larger values)	10	10	10	10	10
	—	12.5 } (12)	—	—	11
	—		—	12	12
	—	—	—	—	13
	—	—	15	15	15
	16	16	—	—	16
	—	—	—	18	18
	—	20	—	—	20
	—	—	22	22	22
	—	—	—	—	24
	25	25	—	—	—
	—	—	—	27	27
	—	31.5 } (32)	—	—	30
	—		—	—	—
	—	—	33	33	33
	—	—	—	—	36
	—	—	—	39	39
	40	40	—	—	—
	—	—	—	—	43
	—	—	47	47	47
	—	50	—	—	—
	—	—	—	—	51
	—	—	—	56	56
	—	—	—	—	62
	63	63	—	—	—
	—	—	68	68	68
	—	—	—	—	75
	—	80	—	—	—
	—	—	—	82	82
	—	—	—	—	91
	100	100	100	100	100

* ANSI Standard C83.2-1971 applies to most electronics components. It is the same as EIA Standard RS-385 (formerly GEN-102) and agrees with IEC Publication 63. ANSI Standard Z17.1-1973 covers preferred numbers and agrees with ISO 3 and ISO 497.

† "20" series with 12-percent steps ($(10)^{1/20}=1.22$ multiplier) and a "40" series with 6-percent steps ($(10)^{1/40}=1.059$ multiplier) are also standard.

‡ Associate the tolerance ±20%, ±10%, or ±5% only with the values listed in the corresponding column. Thus, 1200 ohms may be either ±10 or ±5, but not ±20 percent; 750 ohms may be ±5, but neither ±20 nor ±10 percent.

One or two letters are assigned in EIA or MIL type designations, and the characteristic may be indicated by color coding on the part. An explanation of the characteristics applicable to a component or part will be found in the following sections covering that part.

ENVIRONMENTAL TEST METHODS

Since many component parts and equipments have the same environmental exposure, environmental test methods are becoming standardized. The principal standards follow.

EIA Standard RS-186-D (*ANSI C83.58-1972*): Standard Test Methods for Passive Electronic Component Parts.

IEC Publication 68: Basic Environmental Testing Procedures for Electronic Components and Electronic Equipment (published in several parts).

MIL-STD-202E: Military Standard Test Methods for Electronic and Electrical Component Parts.

MIL-STD-810C: Military Standard Environmental Test Methods.

ASTM* Standard Test Methods—Primarily applicable to the materials used in electronic component parts.

Wherever the test methods in these standards are reasonably applicable, they should be specified in preference to other methods. This simplifies testing of a wide variety of parts, testing in widely separated locations, and comparison of data.

When selecting destructive environmental tests to determine the probable life of a part, distinguish between the environment prevailing during normal equipment operation and the environment used to accelerate deterioration. During exposure to the latter environment, the item may be out of tolerance with respect to its parameters in its normal operating-environment range. Accelerated tests are most meaningful if some relation between the degree of acceleration and component life is known. Such acceleration factors are known for many insulation systems.

STANDARD AMBIENT CONDITIONS FOR MEASUREMENT

	Standard	Temperature (°C)	Relative Humidity (%)	Barometric Pressure mm Hg	mbar
Normal range	RS-186-D	15–35	45–75	650–800	860–1060
	IEC-68	15–35	45–75	(645–795)	860–1060
	MIL-STD-202E	15–35	45–75	650–800	(866–1066)
	MIL-STD-810C	13–33	20–80	650–775	(866–1033)
Closely controlled range	IEC-68	20±1	65±2	(645–795)	860–1060
	IEC-68	23±1	50±2	(645–795)	860–1060
	MIL-STD-202E	23±1	50±2	650–800	(866–1066)
	MIL-STD-810C	23±1.4	50±5	650–775	(866–1033)
	IEC-68	27±1	65±2	(645–795)	860–1060
	RS-186-D	25±2	50±2	650–800	860–1060

Notes:

1. Use the closely controlled range only if the properties are sensitive to temperature or humidity variations, or for referee conditions in case of a dispute. The three temperatures 20°, 23°, and 27°C correspond to normal laboratory conditions in various parts of the world.

2. Rounded derived values are shown in parentheses ( ).

3. 25±2°C, 20 to 50% relative humidity (RH) has been widely used as a closely controlled ambient for testing electronics components.

OTHER STANDARD ENVIRONMENTAL TEST CONDITIONS

Ambient Temperature

Dry heat, °C: +30, +40, (+49), +55, (+68), +70, (+71), +85, +100, +125, +155, +200 (values in parentheses not universally used).
Cold, °C: −10, −25, −40, −55, −65.

* ASTM = American Society for Testing and Materials; Philadelphia, Pa. Index of standards is available.

Constant-Humidity Tests

40°C, 90 to 95% RH; 4, 10, 21, or 56 days.
66°C, ≈100% RH: 48, 96, or 240 hours (primarily for small items).

Cycling Humidity Tests

Figure 1 shows a number of cycling humidity tests. (See applicable chart in standard for full details.) Preconditioning is customary before starting cycle series. RH = relative humidity.

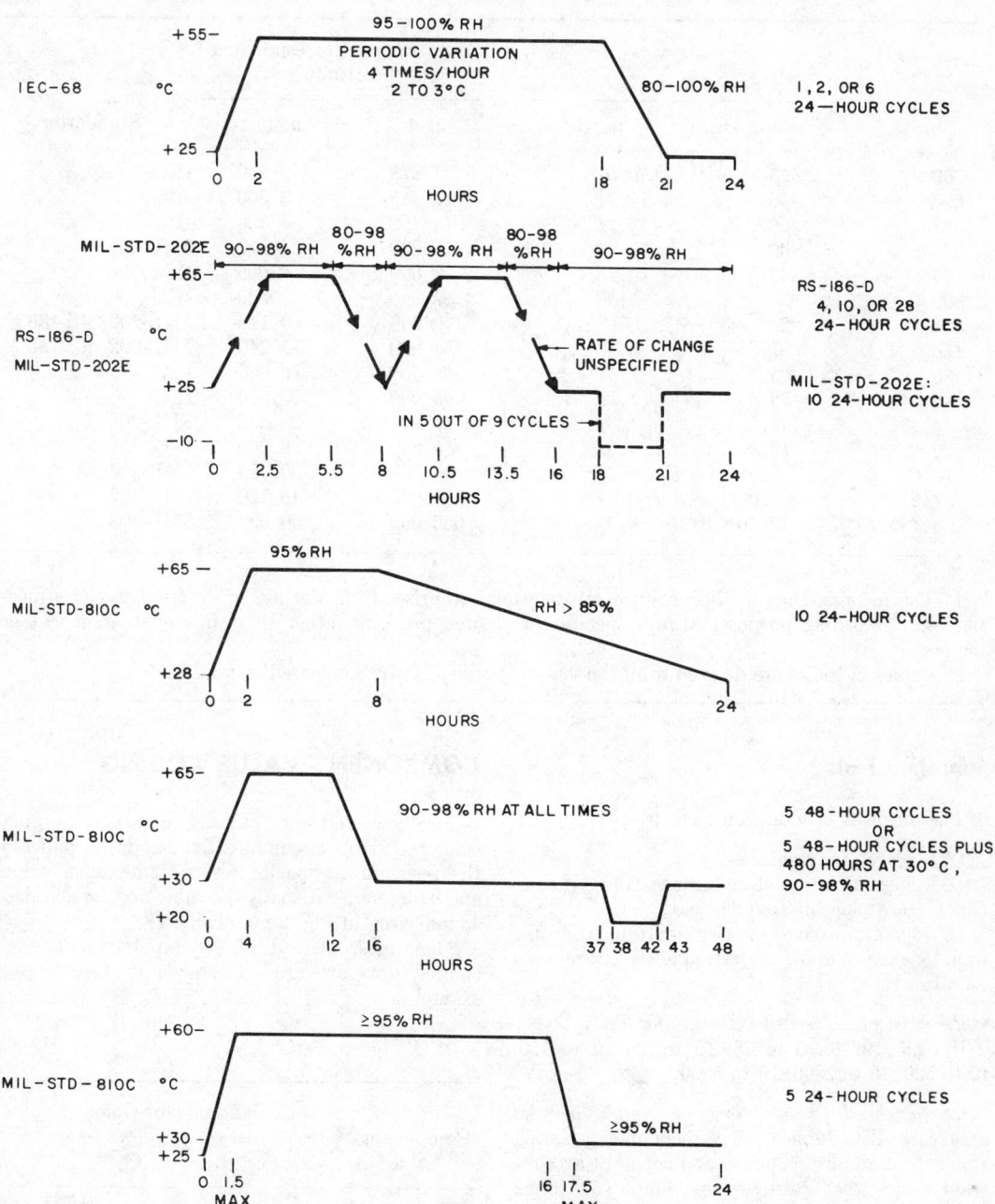

Fig. 1—Cycling humidity tests. Relative humidity for RS–186–D is 90–95% but may be uncontrolled during temperature changes.

High Altitude Tests

Pressure			Approximate Corresponding Altitude		
mbar	mm Hg	in. Hg	feet	meters	Standard
700	*525*	*20.67*	*7 218*	2 200	IEC
600	*450*	*17.72*	*11 483*	3 500	IEC
533	*400*	*15.74*	*14 108*	4 300	IEC
586	439	17.3	15 000	4 572	MIL-202
466	*349*	13.75	20 000	*6 096*	RS-186
300	*225*	*8.86*	*27 900*	8 500	IEC
300	226	8.88	30 000	9 144	MIL-202, RS-186
116	87.0	3.44	50 000	15 240	MIL-202, RS-186
85	*63.8*	*2.51*	*52 500*	16 000	IEC
44	*33.0*	*1.30*	*65 600*	20 000	IEC
44.4	33.0	1.31	70 000	21 336	MIL-202
20	*17.2*	*0.677*	*85 300*	26 000	IEC
10.6	8.00	0.315	100 000	30 480	MIL-202
1.28	1.09	0.043	150 000	45 720	MIL-202
3.18×10^{-6}	2.40×10^{-6}	9.44×10^{-8}	656 000	200 000	MIL-202

Notes:

1. The inconsistency in the pressure–altitude relation arises from the use of different model atmospheres. For testing purposes always specify the desired pressure rather than an elevation in feet or meters.

2. Values in italics are derived from the values specified in the associated standard.

Vibration Tests

The purposes of vibration tests are:

(**A**) Search for resonance.
(**B**) Determination of endurance (life) at resonance (or at specific frequencies).
(**C**) Determination of deterioration resulting from long exposure to swept frequency (or random vibration).

Recommended Frequency Ranges for Tests:
Hertz: 1 to 10, 5 to 35, 10 to 55, 10 to 150, 10 to 500, 10 to 2000, 10 to 5000.

Recommended Combinations of Amplitude and Frequency: IEC Publication 68 recommends testing at constant amplitude below and constant acceleration above the crossover frequency (57 to 62 hertz). MIL-STD-202 and MIL-STD-810 also follow this principle but use different crossover points and low-frequency severities. The choice of frequency range and vibration amplitude or acceleration should bear some relation to the actual service environment. Successful completion of 10^7 vibration cycles indicates a high probability of no failures in a similar service environment. Resonances may make the equipment output unusable, although the mechanical life may be adequate.

COMPONENT VALUE CODING

Axial-lead and some other components are often color coded by circumferential bands to indicate the resistance, capacitance, or inductance value and its tolerance. Usually the value may be decoded as indicated in Fig. 2 and Table 1.

Sometimes, instead of circumferential bands, colored dots are used as shown in Fig. 3 and examples.

Examples:

Component Value	Band or Dot Color			
	A	B	C	D
3300±20%	Orange	Orange	Red	Black or omitted
5.1±10%	Green	Brown	Gold or white	Silver
1.8 megohms ±5% (as applied to a resistor)	Brown	Gray	Green	Gold

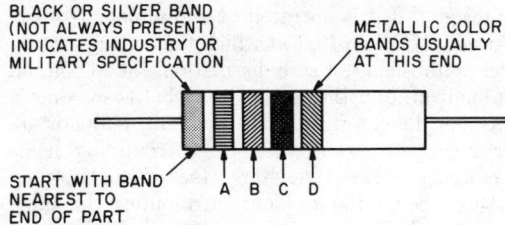

BLACK OR SILVER BAND
(NOT ALWAYS PRESENT)
INDICATES INDUSTRY OR
MILITARY SPECIFICATION

METALLIC COLOR
BANDS USUALLY
AT THIS END

START WITH BAND
NEAREST TO
END OF PART

A B C D

Fig. 2—Component value coding. The code of Table 1 determines values. Band A color = First significant figure of value in ohms, picofarads, or microhenries. Band B color = Second significant figure of value. Band C color = Decimal multiplier for significant figures. Band D color = Tolerance in % (if omitted, the broadest tolerance series of the part applies).

Semiconductor-Diode Type Number Coding

The sequential number portion (following the "1N" of the assigned industry type number) may be indicated by color bands* as shown in Fig. 4.

Colors have the numerical significance given in Table 1.

Bands J, K, L, M represent the digits in the sequential number (for 2-digit numbers, band J is black).

Band N is used to designate the suffix letter as follows.

Color	Suffix Letter	Number
Black	—	0
Brown	A	1
Red	B	2
Orange	C	3
Yellow	D	4
Green	E	5
Blue	F	6
Violet	G	7
Gray	H	8
White	J	9

Band N may be omitted in 2- or 3-digit number coding if not required, but will always be present

* EIA Standard RS-236-B.

on 4-digit number coding (black if no suffix letter is required).

Example:

Band	Band Color		
J	Red	Red	Orange
K	Green	Green	Blue
L	Yellow	Yellow	Violet
M	—	—	Red
N	—	Red	Black
	1N254	1N254B	1N3672

A single band indicates the cathode end of a diode or rectifier.

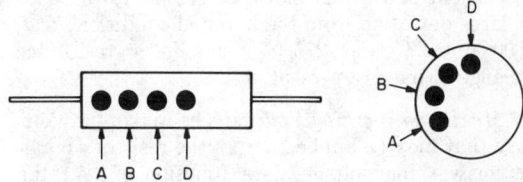

A B C D

Fig. 3—Alternative methods of component value coding.

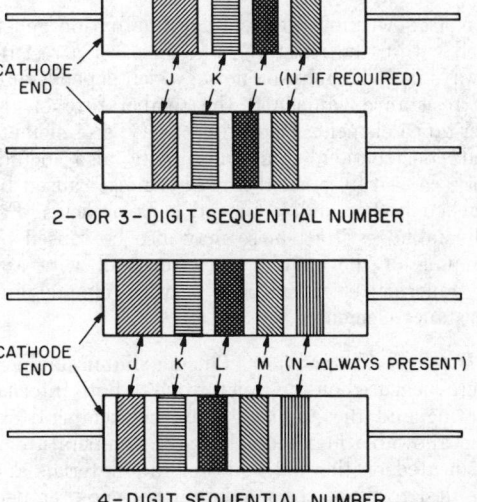

CATHODE
END

J K L (N–IF REQUIRED)

2- OR 3- DIGIT SEQUENTIAL NUMBER

CATHODE
END

J K L M (N–ALWAYS PRESENT)

4–DIGIT SEQUENTIAL NUMBER

Fig. 4—Semiconductor-diode value coding.

RESISTORS

Definitions

Wattage Rating: The maximum power that the resistor can dissipate, assuming (A) a specific life, (B) a standard ambient temperature, and (C) a stated long-term drift from its no-load value. Increasing the ambient temperature or reducing the allowable deviation from the initial value (more-stable resistance value) requires derating the allowable dissipation. With few exceptions, resistors are derated linearly from full wattage at rated temperature to zero wattage at the maximum temperature.

Temperature Coefficient (Resistance–Temperature Characteristic): The magnitude of change in resistance due to temperature, usually expressed in percent per degree Celsius or parts per million per degree Celsius (ppm/°C). If the changes are linear over the operating temperature range, the parameter is known as "temperature coefficient"; if nonlinear, the parameter is known as "resistance–temperature characteristic". A large temperature coefficient and a high hot-spot temperature cause a large deviation from the nominal condition; e.g., 500 ppm/°C and 275°C result in a resistance change of over 12 percent.

Maximum Working Voltage: The maximum voltage that may be applied across the resistor (maximum working voltage) is a function of (A) the materials used, (B) the allowable resistance deviation from the low-voltage value, and (C) the physical configuration of the resistor. Carbon composition resistors are more voltage-sensitive than other types.

Noise: An unwanted voltage fluctuation generated within the resistor. Total noise of a resistor always includes Johnson noise, which depends only on resistance value and the temperature of the resistance element. Depending on type of element and construction, total noise may also include noise caused by current flow and noise caused by cracked bodies and loose end caps or leads. For adjustable resistors, noise may also be caused by jumping of the contact over turns of wire and by imperfect electrical path between contact and resistance element.

Hot-Spot Temperature: The maximum temperature measured on the resistor due to both internal heating and the ambient operating temperature. The allowable maximum hot-spot temperature is predicated on thermal limits of the materials and the design. Since the maximum hot-spot temperature may not be exceeded under normal operating conditions, the wattage rating of the resistor must be lowered if it is operated at an ambient temperature higher than that at which the wattage rating was established. At zero dissipation, the maximum ambient around the resistor may be its maximum hot-spot temperature. The ambient temperature for a resistor is affected by surrounding heat-producing devices; resistors stacked together do not experience the ambient surrounding the stack except under forced cooling.

Critical Resistance Value: A resistor of specified power and voltage ratings has a critical resistance value above which the allowable voltage limits the permissible power dissipation. Below the critical resistance value, the maximum permitted voltage across the resistor is never reached at rated power.

Inductance and Other Frequency Effects: For other than wire-wound resistors, the best high-frequency performance is secured if (A) the ratio of resistor length to cross section is a maximum, and (B) dielectric losses are kept low in the base material and a minimum of dielectric binder is used in composition types.

Carbon composition types exhibit little change in effective dc resistance up to frequencies of about 100 kHz. Resistance values above 0.3 megohm start to decrease in resistance at approximately 100 kHz. Above 1 MHz all resistance values decrease.

Wire-wound types have inductive and capacitive effects and are unsuited for use above 50 kHz, even when specially wound to reduce the inductance and capacitance. Wire-wound resistors usually exhibit an increase in resistance at high frequencies because of "skin" effect.

Film types have the best high-frequency performance. The effective dc resistance for most resistance values remains fairly constant up to 100 MHz and decreases at higher frequencies. In general, the higher the resistance value the greater the effect of frequency.

Established-Reliability Resistors: Some resistor styles can be purchased with maximum-failure-rate guarantees. Standard-failure-rate levels are:

%/1000 hours—1.0; 0.1; 0.01; 0.001.

Resistance Value and Tolerance Choice: A calculated circuit-resistance nominal value should be checked to determine the allowable deviation in that value under the most unfavorable circuit, ambient, and life conditions. A resistor type, resistance value, and tolerance should be selected considering (A) standard resistance values (specials are uneconomical in most cases), (B) purchase tolerance, (C) resistance value changes caused by temperature, humidity, voltage, etc., and (D) long-term drift.

RESISTORS—FIXED COMPOSITION

Color Code

EIA-standard and MIL-specification requirements for color coding of fixed composition resistors are identical (see Fig. 2). The exterior body color of insulated axial-lead composition resistors is usually tan, but other colors (except black) are permitted. Noninsulated axial-lead composition resistors have a black body color.

If 3 significant figures are required, Fig. 5 shows the resistor markings (EIA Std RS-279).

Another form of resistor color coding (MIL-STD-1285A) is shown in Fig. 6. Colors have the following significance.

	Brown	Red	Orange	Yellow	Green	White
Failure Rate Level:						
Letter	M	P	R	S	—	—
Rate (%/1000 hrs)	1.0	0.1	0.01	0.001	—	—
Terminal	—	—	—	—	—	Solderable
Special	—	—	—	—	Fig. 6	—

Tolerance

Standard resistors are furnished in ±20-, ±10-, and ±5-percent tolerances, and in the preferred-value series of Table 2. "Even" values, such as 50 000 ohms, may be found in old equipment, but they are seldom used in new designs.

Temperature and Voltage Coefficients

Resistors are rated for maximum wattage at an ambient temperature of 70 degrees Celsius; above these temperatures up to the maximum allowable hot-spot temperature of 130 or 150 degrees Celsius, it is necessary to operate at reduced wattage ratings. Resistance values are a function of voltage as well as temperature; present specifications allow a maximum voltage coefficient of resistance as given in Table 3 and permit a resistance–temperature characteristic as in Table 4.

A 1000-hour rated-load life test should not cause a change in resistance greater than 12% for $\frac{1}{8}$-watt resistors and 10% for all other ratings. A severe cycling humidity test may cause resistance changes of 10% average and 15% maximum; 250 hours at 40°C and 95% relative humidity may cause up to 10% change. Five temperature-change cycles, −55°C to +85°C, should not change the resistance value by more than 4% from the 25°C value. Soldering the resistor in place may cause a resistance change of 3%. Always allow $\frac{1}{4}$-inch-minimum lead length; use heat-dissipating clamps when soldering confined assemblies. The preceding summary indicates that close tolerances cannot be maintained over a wide range of load and ambient conditions.

Noise

Composition resistors above 1 megohm have high Johnson noise levels, precluding their use in critical applications.

RF Effects

The end-to-end shunted capacitance effect may be noticeable because of the short resistor bodies

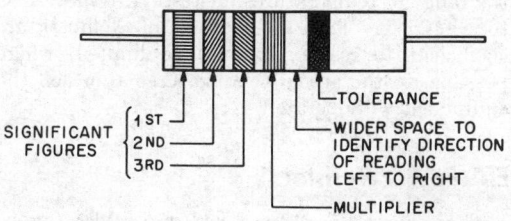

Fig. 5—Resistor value color code for 3 significant figures. Colors of Table 1 determine values.

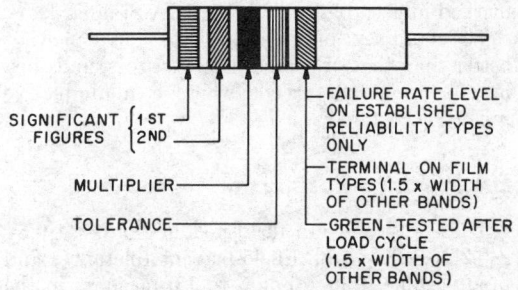

Fig. 6—Resistor color code per MIL-STD-1285A.

TABLE 3—STANDARD RATINGS FOR COMPOSITION RESISTORS.

Watts	Working Volts (maximum)	Hot-Spot Temperature (°C) (maximum)	Critical Resistance (megohms)	Voltage Coefficient* (%/Volt) (maximum)
$\frac{1}{8}$	150	150	0.22	0.05
$\frac{1}{4}$	250	130	0.25	0.035
$\frac{1}{2}$	350	130	0.25	0.035
1	500	130	0.25	0.02
2	500	130	0.12	0.02

* Applicable only to resistors of 1000 ohms and over.

and small internal distance between the ends. Operation at VHF or higher frequencies reduces the effective resistance because of dielectric losses (Boella effect).

Good Design Practice

Operate at $\frac{1}{2}$ allowable wattage dissipation for expected ambient temperature. Provide adequate heat sink. Mount no other heat-dissipating parts within 1 diameter. Use only in applications where 15% change from installed value is permissible or where environment is controlled to reduce resistance value change.

RESISTORS—FIXED WIRE-WOUND

Fixed wire-wound resistors are available as low-power insulated types, precision types, and power types.

EIA Low-Power Insulated Resistors †

These resistors are furnished with power ratings from 1 watt through 15 watts, in tolerances of ±5 and ±10 percent, and in resistance values from 0.1 ohm to 30 000 ohms in the preferred-value series of Table 2. They may be color coded as described in Fig. 2, but band A will be twice the width of the other bands. They may also be typographically marked in accordance with the EIA Standard.

The stability of these resistors is somewhat better than that of composition resistors, and they may be preferred except where a noninductive resistor is required.

EIA Precision Resistors ‡

These resistors are furnished in ±1.0-, ±0.5-, ±0.25-, ±0.1-, and ±0.05-percent tolerances and in any value from 1.0 ohm to 1.0 megohm in the

† EIA Standard RS-344.
‡ EIA Standard RS-229-A.

preferred-value series of Table 2. Power ratings range from 0.1 watt to 0.5 watt. The maximum ambient temperature for full-wattage rating is 125°C. If the resistor is mounted in a confined area or may be required to operate in higher ambient temperatures (145°C maximum), the allowable dissipation must be reduced in accordance with the EIA Standard.

These resistors have an inherently low noise level, approaching the thermal agitation level, and their stability is excellent—the typical change in resistance for the lifetime of the resistor will not exceed 50 percent of the initial resistance tolerance when used within the specified design limits of the EIA Standard.

The temperature coefficient of resistance over the range −55°C to +145°C, referred to 25°C, may have maximums as follows:

Value	EIA Standard
Above 10 ohms	±0.002%/°C
5 ohms to 10 ohms	±0.006%/°C
Below 5 ohms	±0.010%/°C

Where required, temperature coefficients of less than ±20 ppm/°C can be obtained by special selection of the resistance wire. Temperature coefficients of ±10 ppm/°C may be obtained by limiting the range of temperatures for testing from −40°C to +105°C. The application of temperature coefficient to resistors should be limited, where possible, to the actual temperatures at which the equipment will operate.

EIA Power Resistors*

These resistors are furnished in 3 styles (strip; tubular, open end; and axial lead) and 24 power ratings ranging from 1 watt to 210 watts in toler-

* EIA Standard RS-155-B.

ances of ±1.0 percent and ±5 percent. Resistance values range from 1.0 ohm to 182 kilohms in the preferred-value series of Table 2.

Axial-lead types are available in 2 general inductance classifications—inductive winding and noninductive winding. The noninductive styles have a maximum resistance value of 1/2 the maximum resistance of inductive styles because of the special manner in which they are wound. The inductance of noninductive styles must not exceed 0.5 microhenry when measured at a test frequency of 1.0 megahertz ±5%. However, these resistors should not be used in very-high-frequency circuits where the inductance may affect circuit operation.

The maximum ambient temperature for full wattage rating for these resistors is 25°C. When the resistors are operated at ambient temperatures above 25°C, the wattage dissipation must be reduced in accordance with the EIA Standard.

RESISTORS—FIXED FILM

Film-type resistors use a thin layer of resistive material deposited on an insulating core. The low-power types are more stable than the usual composition resistors. Except for very high-precision requirements, film-type resistors are a good alternative to accurate wire-wound resistors, being both smaller and less expensive and having excellent noise characteristics.

The power types are similar in size and performance to conventional wire-wound power resistors. While their 200°C maximum operating temperature limits the power rating, the maximum resistance value available for a given physical size is much higher than that of the corresponding wire-wound resistor.

Construction

For low-resistance values, a continuous film is applied to the core, a range of values being obtained by varying the film thickness. Higher resistances are achieved by the use of a spiral pattern, a coarse spiral for intermediate values and a fine spiral for high resistance. Thus, the inductance is greater in high values, but it is likely to be far less than in wire-wound resistors. Special high-frequency units having greatly reduced inductance are available.

Resistive Films

Resistive-material films presently used are micro-crystalline carbon, boron–carbon, and various metallic oxides or precious metals.

Deposited-carbon resistors have a negative temperature coefficient of 0.01 to 0.05 percent/°C for low resistance values and somewhat larger for higher values. Cumulative permanent resistance changes of 1 to 5 percent may result from soldering, overload, low-temperature exposure, and aging. Additional changes up to 5 percent are possible from moisture penetration and temperature cycling.

The introduction of a small percentage of boron in the deposited-carbon film results in a more stable unit. A negative temperature coefficient of 0.005 to 0.02 percent/°C is typical. Similarly, a metallic dispersion in the carbon film provides a negative coefficient of 0.015 to 0.03 percent/°C. In other respects, these materials are similar to standard deposited carbon. Carbon and boron–carbon resistive elements have the highest random noise of the film-type resistors.

Metallic-oxide and precious-metal-alloy films permit higher operating temperatures. Their noise characteristics are excellent. Temperature coefficients are predominantly positive, varying from 0.03 to as little as 0.0025 percent/°C.

Applications

Power ratings of film resistors are based on continuous direct-current or on root-mean-square operation. Power derating is necessary for operation at ambient temperatures above the rated temperature. In pulse applications, the power dissipated during each pulse and the pulse duration are more significant than average power conditions. Short high-power pulses may cause instantaneous local heating sufficient to alter or destroy the film. Excessive peak voltages may result in flashover between turns of the film element. Derating under these conditions must be determined experimentally.

Film resistors are fairly stable up to about 10 megahertz. Because of the extremely thin resistive film, skin effect is small. At frequencies above 10 megahertz, it is advisable to use only unspiraled units if inductive effects are to be minimized (these are available in low resistance values only).

Under extreme exposure, deposited-carbon resistors deteriorate rapidly unless the element is protected. Encapsulated or hermetically sealed units are preferred for such applications. Open-circuiting in storage as the result of corrosion under the end caps has been reported in all types of film resistors. Silver-plated caps and core ends effectively overcome this problem.

Technical Characteristics

Stable equivalents for composition resistors; axial leads; data for MIL "RL" series.

Watts	$\frac{1}{4}$	$\frac{1}{2}$	1	2
Voltage rating	250	350	500	500
Critical resistance (megohms)	0.25	0.25	0.25	0.12

Maximum temp for full load—70°C; for 0 load—150°C

Resistance–temperature characteristic: ±200 ppm/°C maximum

Life-test resistance change: ±2% maximum

Moisture resistance test: ±1.5% maximum change

Resistance values: E24 series, same as composition resistors; tolerances 2% or 5%.

The MIL "RN" series of film resistors is more stable than the "RL" series and is available in a wider range of ratings. Commercial equivalents are also offered. Where stability and reliability are desired, the "RN" series is economically very competitive with the "RC" or "RL" series.

High-stability film resistors; axial leads; data for MIL "RD" series. Uninsulated commercial versions have lower temperature limits and greater resistance change. Color coding is the same as previously indicated for composition resistors.

	Watts							
	$\frac{1}{20}$	$\frac{1}{10}$	$\frac{1}{8}$	$\frac{1}{4}$	$\frac{1}{2}$	$\frac{3}{4}$	1	2
Characteristic	Voltage Rating							
B	—	—	—	—	—	—	500	750
D	—	—	200	300	350	500	—	—
C, E	200	200	250	300	350	—	500	—

	Characteristic			
	B	C	D	E
Maximum temp (°C):				
Full load	70	125	70	125
0 load	150	175	165	175
Life-test resistance change (max)	±1%	±0.5%	±1%	±0.5%
Resistance-temperature characteristic (max ppm/°C)	±500	±50	+200 −500	±25
Moisture resistance test (max change)	±1.5%	±0.5%	±1.5%	±0.5%

Resistance Values: E96 series (E48 preferred): 1% tolerance; E192 series (E96 preferred): 0.5%, 0.25%, 0.1% tolerances.

Power-type film resistors, uninsulated.

	Axial-Lead and MIL "RD" Series			Commercial Tab-Terminal Styles				
Watts	2	4	8	7	23	25	55	115
Voltage rating	350	500	750	525	1380	2275	3675	7875
Critical resistance (kilohms)	61	62	70	39	83	208	245	540
Maximum temperature (°C):								
Full load	25	25	25	25	25	25	25	25
0 load	235 275 (MIL)	235 275 (MIL)	235 275 (MIL)	235	235	235	235	235

Life-test resistance change: ±5% maximum
Resistance–temperature characteristic: ±500 ppm/°C maximum
Moisture resistance test: ±3% maximum change
Resistance values: E12 series (approx. for MIL)
Tolerances: Axial lead: $\frac{1}{2}$%, 1%, 2% (MIL "RD" Series), 5%, 10%.
Tab lead: 1%, 2%, 5%, 10%, 20%.

TABLE 4—TEMPERATURE COEFFICIENT OF RESISTANCE FOR COMPOSITION RESISTORS.

	Charac-teristic*	Percent Maximum Allowable Change from Resistance at 25 Degrees Celsius†					
At −55°C ambient	F	±6.5	±10	±13	±15	±20	±25
At +105°C ambient	F	±5	±6	±7.5	±10	±15	±15
Nominal resistance in ohms		0 to 1 000	>1 000 to 10 000	>10 000 to 0.1 meg	>0.1 meg to 1.0 meg	>1 meg to 10 meg	>10 meg

* Resistance—temperature.

† Up to 1 megohm, data also apply to MIL Established Reliability characteristic G (= former GF).

RESISTORS, ADJUSTABLE

General

Adjustable resistors are available with several types of resistance elements: carbon composition, wire-wound, and metallic film.

The wattage rating of an adjustable resistor is based on using the full resistance element; if only a portion of the element is used, the allowable wattage is reduced approximately in the same proportion as the resistance (for a linear-resistance-vs-rotation characteristic).

At the extremes of movable-contact travel there is a residual (hop-off) resistance between the end contact and the movable contact having a maximum value (for linear tapers) approximately as follows:

Element Type	Residual Resistance (ohms)
Composition, linear taper:	
100 000 ohms	0.05 % total R
25 000, 50 000 ohms	35*
5000, 10 000 ohms	25*
Wire-wound, low power, 1 turn	1 ohm or 0.5%* of total R†
Wire-wound, low power, 1 turn, reliable	1 ohm or 3% of total R†
Wire-wound, lead-screw actuated	1 ohm or 2% of total R†
Wire-wound, lead-screw actuated, reliable	1 ohm or 0.25% of total R†
Wire-wound, precision	5 ohms or 1% of total R†
Wire-wound, power, 1 turn	0.2 ohm or 0.2% of total R†
Nonwire-wound, lead-screw actuated	20 ohms or 2% of total R†

*EIA standard.
†Whichever is greater.

The usual form is a movable arm attached to a rotating shaft with an angular travel (electrical rotation) of about 300°. Multi-turn lead-screw-actuated movable-contact resistors are available with linear or circular resistance elements; locking is less essential in high shock or vibration environments for this construction. Multi-turn movable-arm styles are available for high-resistance or high-resolution applications.

The common low-power resistors may be fitted with a power switch that is actuated in the first 20° of mechanical motion preceding the electrical rotation.

Composition Resistors

Carbon composition elements may be formed on an insulating base such as paper-base phenolic laminate, or they may be hot molded integral with a molded base; the latter is more durable. Thin-layer composition elements may be noisy and tend to wear away with frequent use. The resistance change is continuous.

In addition to a linear-rotation-vs-resistance characteristic, composition elements can have a wide range of nonlinear curves (tapers). Standard tapers are shown in Fig. 7. Commercial units are available with 2 taps in addition to the usual 3 terminals, permitting frequency-sensitive circuit adjustments such as are required for tone controls.

Usual ratings are as follows:

	Single-Turn Shaft				Lead-Screw Actuated
Watts at 70°C*	$\frac{1}{2}$	$\frac{3}{4}$	2, 3	5	$\frac{1}{4}$
Max operating temp	120°C	120°C	120°C	120°C	125°C
Case size in inches (nominal, typical)	0.50 dia	0.72 dia	1.15 dia	1.62×2.15	0.28×0.33×1.25
Resistance range (min ohms to max megohms)	100 to 5	500 to 2.5	50 to 5	50 to 2.5	100 to 2.5
Life expectancy–shaft rotations:					
Hot-molded	25 000	—	>25 000	100 000	25 000
EIA RS-303	10 000	—	10 000	—	—

* Derate 50% for other than linear taper, plus 50% if not mounted on metal panel (e.g., printed-wiring board).

Hot-molded elements may change up to 5% in resistance value after 100 hours at 40°C, 95% relative humidity.

Film- and Ceramic-Element Resistors

Adjustable resistors using a metallic-film or conductive-ceramic element are stepless and have less inherent noise than units using a composition element.

They are available as linear and circular lead-screw-actuated resistors and as single-turn resistors. They may be operated from −65°C to +150°C, but operation above 150°C is limited primarily by the lubricant requirements for the moving parts, and by the housing material. Life expectancies are of the order of 10^7 shaft revolutions for precision single-turn designs.

Usual ratings for nonwire-wound lead-screw-actuated adjustable resistors are as follows:

	Linear Type–Rectangular Case		Circular Type–Square Case		
Watts	$\frac{3}{4}$	$\frac{1}{4}$	$\frac{1}{2}$	$\frac{1}{2}$	0.2
Temp for rated watts	85°C	25°C	85°C	85°C	85°C
Max operating temp	150°C	85°C	150°C	150°C	150°C
Case size in inches (nominal)	0.30×0.37×1.25 0.20×0.33×1.25	0.28×0.36×1.00 0.16×0.31×0.75	0.50 sq	0.38 sq	0.25 sq
Resistance range (min ohms to max megohms)	10 to 1	10 to 1	10 to 1	10 to 1	10 to 1
Life expectancy–end-to-end travel of wiper on resistance element (cycles)	200	200	200	200	200

Wire-Wound Resistors

These are available as general-purpose low-power resistors, precision resistors, and power resistors.

Wire-wound elements are usually not suitable for frequency-sensitive rf circuits because of inductive and capacitive effects. High-resistance wire-wound units are more subject to winding damage because of the fine wire required; they are impractical above a resistance value determined by the winding space available and the smallest resistance wire that can be space-wound.

Resistors wound with wire smaller than 0.001 inch in diameter are not as rugged as those using larger-diameter wire; composition or film elements may be more suitable for high resistance values.

The usual rotation-vs-resistance characteristic is linear, but by tapering or otherwise shaping the winding card, a limited amount of taper, log characteristic, sine wave, or similar function can be generated. By changing the resistance wire size or material, different linear tapers may be obtained on a uniform-width winding card.

Wire-wound elements, with rare exceptions, change value in steps. There may be 20 steps in a small low-resistance unit or 2000 steps (turns) in a single-turn high-resistance unit; in some cases this abrupt step characteristic may provide an unwanted signal in the circuit. Noise arises from motion of the wiper from turn to turn. Resolution may be improved by using multi-turn adjustable resistors. (If the desired resistance value is low enough, resolution may also be improved by using a resistor in which the wiper is in continuous contact with a single-turn wire.)

Usual ratings follow (linear tapers).

Wire-Wound Low-Power Adjustable Resistors:

	Bushing-Mounted Single-Turn Shaft		
Watts	1*	2*	4*
Temp for rated watts	40°C	40°C	40°C
Max operating temp	105°C	105°C	105°C
Case size in inches (nominal)	0.75 dia	1.30 dia	1.72 dia
Resistance range (min ohms to max megohms)	15 to 0.0025	3 to 0.015	3 to 0.025
Wire diameter in inches	0.00156	0.00156	0.00156
Rotational life–shaft turns	25 000	25 000	25 000

	Lead-Screw Actuated			
	Linear Type—Rectangular Case		Circular Type—Square Case	
Watts	$\frac{3}{4}$	$\frac{3}{4}$	$\frac{3}{4}$	$\frac{3}{4}$
Temp for rated watts	85°C	85°C	85°C	85°C
Max operating temp	105°C	150°C	150°C	150°C
Case size in inches (nominal)	0.19×0.32×1.00	0.20×0.32×1.25	0.20×0.50 sq	0.15×0.38 sq
Resistance range (min ohms to max megohms)	100 to 0.005	10 to 0.01	10 to 0.01	10 to 0.005
Wire diameter (inches)	0.001	0.001	0.001	0.001
Rotational life–shaft turns	2 500	4 000	4 000	3 000

	Multi-Turn Shaft		
	10t	10t	40t
Watts	1.5	5	20
Temp for rated watts	40°C	40°C	40°C
Max operating temp	85°C	85°C	85°C
Case size in inches (nominal)	0.88 dia	1.82 dia	3.32 dia
Resistance range (min ohms to max megohms)	10 to 0.125	10 to 0.648	125 to 5.3
Rotational life–shaft turns	2×10^6	2×10^6	2×10^6

* At 40°C, derate 50% for other than linear tapers, plus additional for other than metal panel mounting.

Wire-Wound High-Power Adjustable Resistors:

Watts at 25°C:					
MIL	6	25	50	75	100
EIA	12.5	25	50	75	100
Max operating temp	340°C	340°C	340°C	340°C	340°C
Body diameter in inches	0.90	1.68	2.41	2.81	3.19
Resistance range* (min	1 to	2 to	1 to	2 to	2 to
ohms to max megohms)	0.0035	0.005	0.01	0.01	0.01

* Based on 0.0024-inch-diameter wire.

CAPACITORS

Definitions

Dielectric: A dielectric is a material that can withstand high electric stress without appreciable conduction. When such stress is applied, energy in the form of an electric charge is held by the dielectric. Most of this stored energy is recovered when the stress is removed. The only perfect dielectric in which no conduction occurs and from which the whole of the stored energy may be recovered is a perfect vacuum.

Relative Capacitivity: The relative capacitivity or relative permittivity or dielectric constant is the ratio by which the capacitance is increased when another dielectric replaces a vacuum between two electrodes.

Dielectric Absorption: Dielectric absorption is the absorption of charge by a dielectric when subjected to an electric field by other than normal polarization. This charge is not recovered instantaneously when the capacitor is short-circuited, and a decay current will continue for many minutes. If the capacitor is short-circuited momentarily a new voltage will build up across the terminals afterward. This is the source of some danger with high-voltage dc capacitors or with ac capacitors not fitted with a discharge resistor. The phenomenon may be used as a measure of dielectric absorption.

Tangent of Loss Angle: This is a measure of the energy loss in the capacitor. It is expressed as $\tan\delta$ and is the power loss of the capacitor divided by its reactive power at a sinusoidal voltage of specified frequency. (This term also includes power factor, loss factor, and dielectric loss. The true power factor is $\cos(90-\delta)$.)

Insulation Resistance: This is a measure of the conduction in the dielectric. Because this conduction takes a very long time to reach a stable value, it is usually measured after 1 minute of electrification for nonelectrolytic types and 3 minutes for electrolytics. It is measured preferably at the rated working voltage or at a standardized voltage.

The insulation resistance is usually multiplied by the capacitance to give the ohm-farad value, which is the apparent discharge time constant (seconds). This is a figure of merit for the dielectric, although for small capacitances a maximum value of insulation resistance is usually also specified.

In electrolytics the conduction is expressed as leakage current at rated working voltage. It is calculated as $\mu A/\mu FV$, which is the reciprocal of the ohm-farad value. In this case a maximum value of leakage current is specified for small capacitances.

Leakage Current: The current flowing between two or more electrodes by any path other than the interelectrode space is termed the leakage current, and the ratio of this to the test voltage is the insulation resistance.

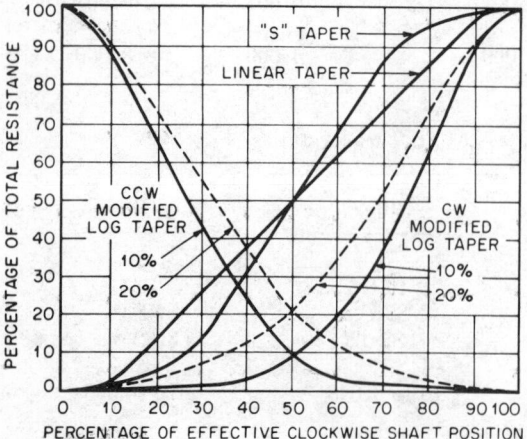

Fig. 7—Nominal resistance-vs-rotation characteristics for standard tapers.

Impedance: Impedance is the ratio of voltage to current at a specified frequency. At high frequencies the inductance of leads becomes a limiting factor, in which case a transfer impedance method may be employed. This then measures the impedance of the shunt path only.

DC or AC Capacitor: A dc capacitor is designed to operate on direct current only. It is normally not suitable for use above 200 volts ac because of the occurrence of discharges in internal gas bubbles. An ac capacitor is designed to have freedom from internal discharges and low tangent of loss angle to minimize internal heating. It is normally not suitable for use on dc.

Rated Voltage and Temperature: The rated voltage is the direct operating voltage that may be applied continuously to a capacitor at the rated temperature.

Category Voltage and Temperature: The category voltage is the voltage which may be applied to the capacitor at the maximum category temperature. It differs from the rated voltage by a derating factor.

Ripple Voltage: If alternating voltages are present in addition to direct voltage, the working voltage of the capacitor is taken as the sum of the direct voltage and the peak alternating voltages. This sum must not exceed the value of the rated voltage.

In electrolytics the permissible ripple may be expressed as a rated ripple current.

Surge Voltage: This is a voltage above the rated voltage which the capacitor will withstand for a short time.

Voltage Proof Test: This is the highest possible voltage that may be applied without breakdown to a capacitor during qualification approval testing to prove the dielectric. The repeated application of this voltage may cause failure.

Forming Voltage (*Electrolytics*): The voltage at which the anodic oxide has been formed. The thickness of the oxide layer is proportional to this voltage.

Burnout Voltage (Metallized Types): The voltage at which metallized types burn out during manufacture.

Self-Healing Failure (Metallized Types): A momentary partial discharge of a capacitor resulting from a localized failure of the dielectric. Burning away the metallized electrode isolates the fault and effectively restores the properties of the capacitor.

Volt-Ampere Rating (VA): This is the reactive power in a capacitor when an ac voltage is applied. VA $\cos\theta$ gives the amount of heat generated in the capacitor. Since the amount of heat that can be dissipated is limited, the VA must also be limited and in some cases a VA rating is quoted. (Note that $\cos\theta = \cos(90-\delta) \approx \tan\delta$, when δ is small.)

Scintillation: Minute and rapid fluctuations of capacitance formerly exhibited by silvered mica or silvered ceramic types but overcome by modern manufacturing techniques.

Internal Discharge: Partial discharge of a capacitor due to ionization of the gas in a bubble in the dielectric. On ac this may occur in unsuitable dielectrics above 200 volts and is a major cause of failure. On dc such discharges are very infrequent and normally are not a cause of failure.

CLASSES OF CAPACITORS

Modern electronics circuits require the smallest possible capacitors, which are usually made with the thinnest possible dielectric material since they are for operation at low voltages. There are three broad classes of capacitors.

(A) Low-loss capacitors with good capacitance stability. These are usually of mica, glass, ceramic, or a low-loss plastic such as polystyrene.

(B) Capacitors of medium loss and medium stability, usually required to operate over a fairly wide range of ac and dc voltages. This need is met by paper, plastic film, or high-*K* ceramic types. The first two of these may have electrodes of metal foil or electrodes of evaporated metal which have a self-healing characteristic.

(C) Capacitors of the highest possible capacitance per unit volume. These are the electrolytics, which are normally made either of aluminum or tantalum. Both of these metals form extremely thin anodic oxide layers of high dielectric constant and good electrical characteristics. Contact with this oxide layer is normally by means of a liquid electrolyte that has a marked influence on the characteristics of the capacitor. In solid tantalum the function of the electrolyte is performed by a manganese-dioxide semiconductor.

PLASTIC FILM CAPACITORS

Advances in organic chemistry have made it possible to produce materials of high molecular weight. These are formed by joining together a number of basic elements (monomers) to produce a polymer. Some of these have excellent dielectric characteristics.

Physically they can be classified as thermoplastic or thermosetting. In the former case the molecule consists of long chains with little or no branching, while in the latter the molecules are crosslinked. Thermosetting materials have no clearly defined melting point and are usually hard and brittle, making them unsuitable for the manufacture of plastic films. A cast film is usually amorphous but by extrusion, stretching, and heat treatment, oriented crystalline films are produced with good flexibility and dielectric characteristics.

The electrical properties of these plastics depend on the structure of the molecule. If the molecule is not symmetrical it will have a dipole moment giving increased dielectric constant. On the other hand, the dielectric constant and tan δ are then dependent on frequency. Generally speaking, non-polar materials have electrical characteristics that are independent of frequency, while polar materials exhibit a decrease in capacitance with increasing frequency, and tan δ may pass through a maximum in the frequency range.

Figures 8, 9, and 10 show some characteristics of several types of capacitors. At the present time two classes of plastic film capacitors are recognized.

(**A**) *Polystyrene Capacitors.* Polystyrene is a nonpolar plastic and has excellent electrical characteristics which are independent of frequency.

(**B**) *Polyester Films.* Strictly speaking these are the polyethylene terephthalates (Mylar, Melinex, Hostaphan) but the polycarbonates are now included in this group because they have similar electrical characteristics.

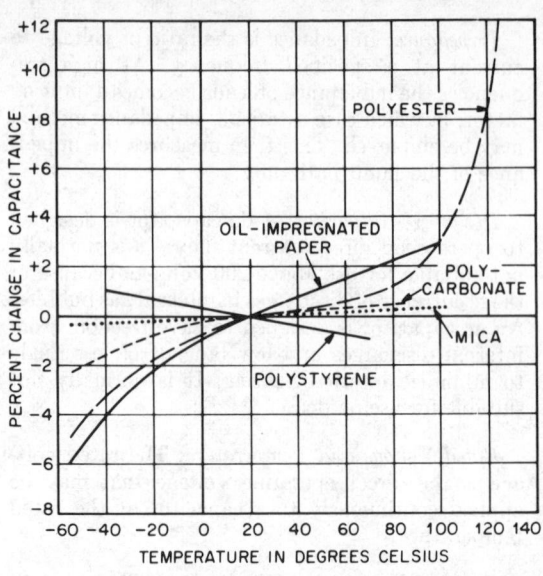

Fig. 9—Variation of tan δ with temperature for various plastic films compared with oil-impregnated paper.

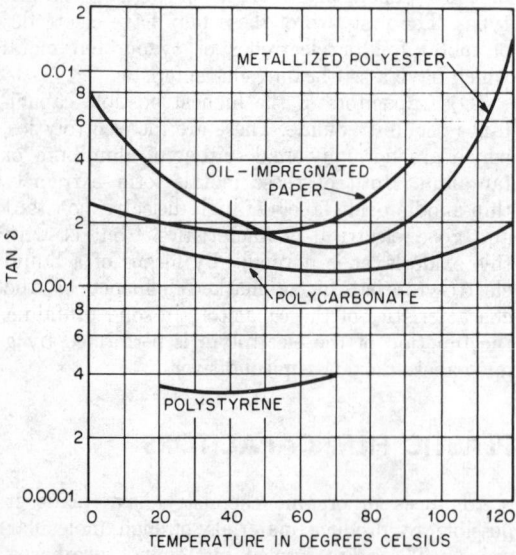

Fig. 8—Typical capacitance characteristics of various capacitors as a function of temperature. Measured on 0.1-microfarad capacitors at 1000 hertz.

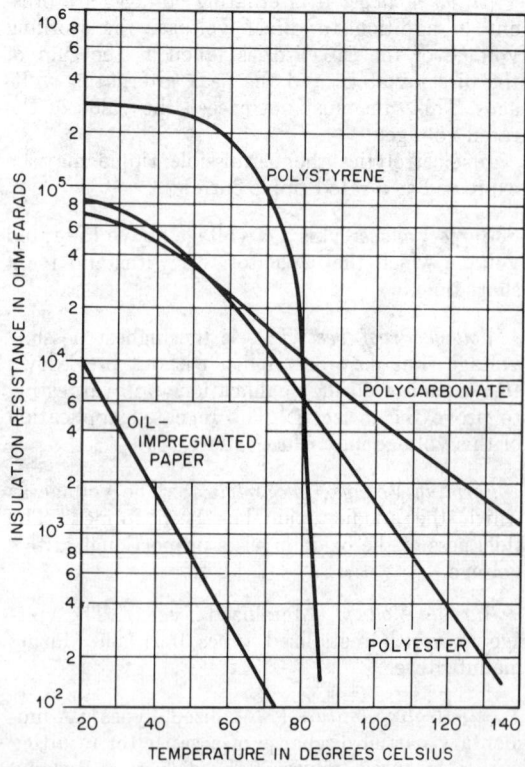

Fig. 10—Variation of insulation resistance with temperature for various plastic films compared with oil-impregnated paper.

Plastic films for capacitor manufacture are usually of the oriented crystalline type because of their good combinations of characteristics. One important feature of some of these films is that they tend to shrink back to their original shape after being heated. This fact is sometimes exploited in manufacturing the capacitor.

Moisture usually has little effect on the dielectric properties of plastic films, and capacitors made from them require less protection than paper or mica types. This, together with simple processing, has permitted them to be mass-produced at relatively low cost.

The electrical characteristics of capacitors made with these materials depend on the construction employed. The resistance of metallized electrodes and the shape and method of connection to the unit are particularly important.

Many plastic materials are being used in the manufacture of capacitors for which there are no internationally agreed specifications. In this case it is necessary to obtain the relevant data from the manufacturer. There is no doubt that in the future the range of plastic film capacitors will be considerably extended.

Polystyrene Film Capacitors with Foil Electrodes

Polystyrene has excellent electrical characteristics. The film employed is of the oriented crystalline type, which makes it flexible and suitable for forming into thin films. On heat treatment the film shrinks considerably and this is used in the manufacturing process to obtain capacitance stability. The film is affected by greases and solvents, and care must be taken both in manufacture and in use to ensure that capacitors do not come into contact with these materials.

The power factor of polystyrene is low over the whole frequency range, but the resistance of the electrodes may result in an increase of power factor at high frequencies in the larger values, as shown in Table 5.

Polyester Film Capacitors with Foil Electrodes

This generic title is usually used to apply to polyethylene terephthalate. It is a slightly polar plastic film suitable for operation up to temperatures of 125°C. Capacitors are available with foil electrodes.

TABLE 5—POWER FACTOR OF POLYSTYRENE AT VARIOUS FREQUENCIES.

Frequency (hertz)	Nominal Capacitance (pF)			
	Up to 1 000	1 000 to 10 000	10 000 to 100 000	Above 100 000
800	0.0003	0.0003	0.0003	0.0003
10 000	0.0003	0.0003	0.0003	0.001
100 000	0.0003	0.0005	0.001– 0.003	—
1 000 000	0.001	0.002	0.005– 0.02	—

Plastic Film Capacitors with Metallized Electrodes

These are tending to supersede metallized paper capacitors on dc applications because of superior electrical characteristics, less tendency for self-healing to occur during service, higher and more stable insulation resistance, and approximately the same space factors. Two types of film are generally used, polyethylene terephthalate and polycarbonate. For some purposes these are comparable, but polycarbonate has a lower loss angle and smaller change of capacitance with temperature. Polycarbonate is also available in thinner films, giving an advantage of space factor.

ELECTROLYTIC CAPACITORS

Electrolytic capacitors (Fig. 11) employ for at least one of their electrodes a "valve metal". This metal, when operated in an electrolytic cell as the

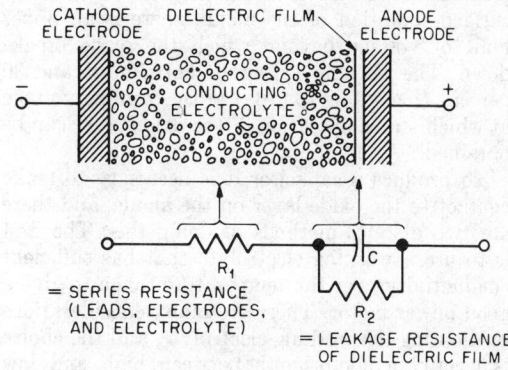

Fig. 11—Basic cell and simplified equivalent circuit for polar electrolytic capacitor.

anode, forms a layer of dielectric oxide. The most commonly used metals are aluminum or tantalum. The valve-metal behavior of these metals was known about 1850. Tantalum electrolytic capacitors were introduced in the 1950's because of the need for highly reliable miniature capacitors in transistor circuits over a wide temperature range. These capacitors were made possible by improved refining and powder metallurgy techniques.

The term "electrolytic capacitor" is applied to any capacitor in which the dielectric layer is formed by an electrolytic method. The capacitor does not necessarily contain an electrolyte.

The oxide layer is formed by placing the metal in a bath containing a suitable forming electrolyte, and applying voltage between the metal as anode and another electrode as cathode. The oxide grows at a rate determined by the current flowing, but this rate of growth decreases until the oxide has reached a limiting thickness determined by the voltage. For most practical purposes it may be assumed that the thickness of the oxide is proportional to the forming voltage.

Properties of aluminum and tantalum and their oxides follow.

Metal	Density	Principal Oxide	Dielectric Constant	Thickness (Å/V)
Aluminum	2.7	Al_2O_3	8	13.5
Tantalum	16.6	Ta_2O_5	27.6	17

The structure of these oxide layers plays an important part in determining their performance. Ideally they are amorphous but aluminum tends to form two distinct layers, the outer one being porous. Tantalum normally forms an amorphous oxide which, under conditions of a high field strength of the oxide layer, may become crystalline. Depending on the forming electrolyte and the surface condition of the metal, there is an upper limit of voltage beyond which the oxide breaks down. The working voltage is between 25 and 90 percent (according to type) of the forming voltage at which stable operation of the oxide layer can be obtained.

To produce a capacitor it is necessary to make contact to the oxide layer on the anode, and there are two distinct methods of doing this. The first is to use a working electrolyte that has sufficient conductivity over the temperature range to give a good power factor. There are many considerations in choosing the working electrolyte, and the choice is usually a compromise between high and low temperature performance. The working electrolyte also provides a rehealing feature in that any faults in the oxide layer will be repaired by further anodization.

In aluminum electrolytic capacitors the working electrolyte must be restricted to those materials in which aluminum and its oxide are inert. Corrosion can be minimized by using the highest possible purity of aluminum. This also reduces the tendency of the oxide layer to dissolve in the electrolyte, giving a better shelf life.

Tantalum, on the other hand, is very inert and therefore allows a wider choice of electrolyte. Since there is no gas evolution, better methods of sealing can be employed. The characteristics of aluminum and tantalum capacitors are shown in Figs. 12 and 13.

A major problem with all electrolytic capacitors is to ensure that the electrolyte is retained within the case under all operating conditions. In the aluminum capacitor, allowance must be made for gas evolution on forming. Even the tantalum

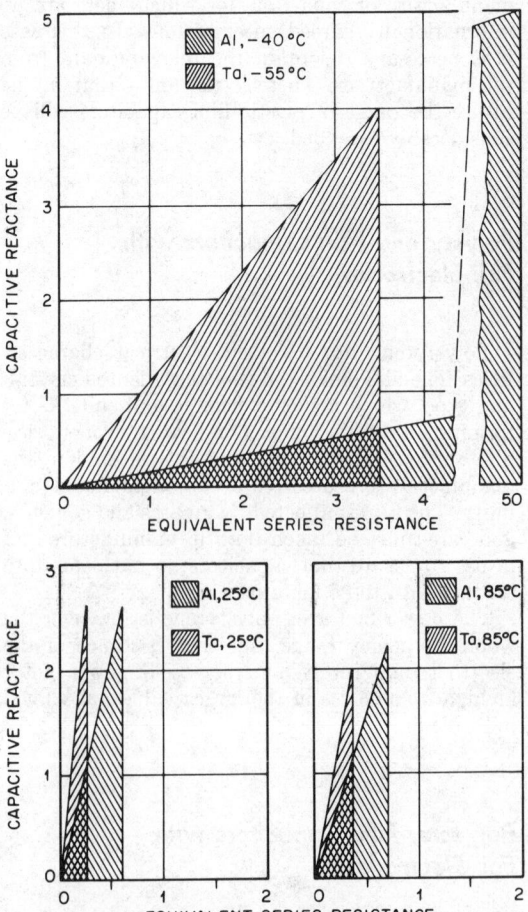

Fig. 12—Typical 120-hertz impedance diagrams for aluminum (Al) and tantalum (Ta) plain-foil polar electrolytic capacitors of 150-volt rating at low, high, and room temperatures. Resistance and reactance are drawn to the same arbitrary scale for all charts.

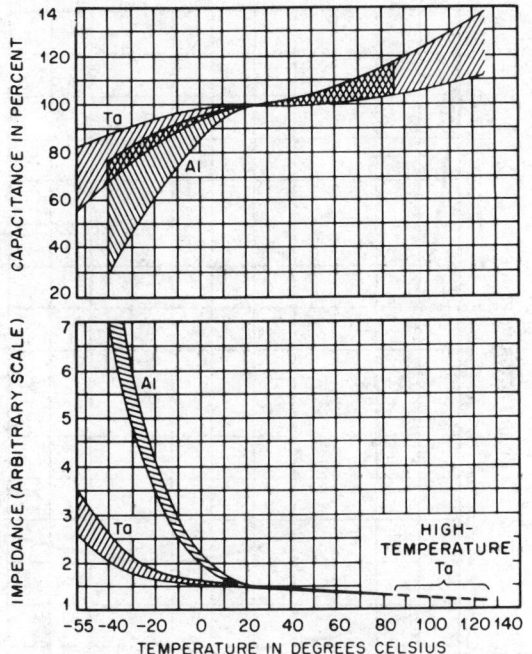

Fig. 13—Capacitance (top) and 120-hertz impedance (below) as a function of temperature for aluminum (Al) and tantalum (Ta) electrolytic capacitors.

be noted that, even when the case is not connected to one terminal, a low-resistance path exists between it and the electrodes. The case must be insulated from the chassis, particularly if the chassis and the negative terminal are not at the same potential.

Aluminum Electrolytics

This is the most widely known electrolytic capacitor and is used extensively in radio and television equipment. It has a space factor about 6 times better than the equivalent paper capacitor. Types of improved reliability are now available using high-purity (better than 99.99%) aluminum.

Conventional aluminum electrolytic capacitors which have gone 6 months or more without voltage applied may need to be reformed. Rated voltage is applied from a dc source with an internal resistance of 1500 ohms for capacitors with a rated voltage exceeding 100 volts, or 150 ohms for capacitors with a rated voltage equal to or less than 100 volts. The voltage must be applied for one hour after reaching rated value with a tolerance of ±3 percent. The capacitor is then discharged through a resistor of 1 ohm/volt.

Tantalum Foil-Type Electrolytics

This type of capacitor was introduced around 1950 to provide a more reliable type of electrolytic capacitor without shelf-life limitation. It was made possible by the availability of thin high-purity annealed-tantalum foils and wires. Plain-foil types were introduced first, followed by etched types. The purity, and particularly the surface purity, of these materials plays a major part in determining the leakage current and their ability to operate at the higher working voltages.

These capacitors are smaller than their aluminum counterparts and will operate at temperatures up to about 125°C (Figs. 14–16). The plain-foil types usually exhibit less variation of capacitance with temperature or frequency.

Tantalum Electrolytics with Porous Anode and Liquid Electrolyte

This is the first type of tantalum electrolytic capacitor to be introduced and still has the best space factor. Types using sulphuric-acid electrolyte have excellent electrical characteristics up to about 70 working volts. Other types contain neutral electrolytes.

Basically it consists of a sintered porous anode of tantalum powder housed in a silver or silver-

capacitor usually must employ only organic materials for sealing, and these do not provide completely hermetic sealing. All organic materials have finite moisture transmission properties and, therefore, at the maximum category temperature the high vapor pressure of the electrolyte results in some diffusion.

An elegant solution to this problem was found by Bell Telephone Laboratories, using a semiconductor instead of an electrolyte. The semiconductor is manganese dioxide in a polycrystalline form and has a higher conductivity than conventional electrolyte systems. This material also provides a limited self-healing feature at a fault, resulting in oxidation of the tantalum and reduction of the manganese dioxide to a nonconducting form.

Electrolytic capacitors take many forms and the anode may be of foil, wire, or a porous sintered body. The foil may be either plain or etched. The porous body may be made with fine or with coarse particles and the body itself may be short and fat or long and thin. The aluminum-foil capacitor has a space factor about 6 times better than that of the equivalent paper capacitor, while tantalum capacitors are even smaller and enjoy a space factor up to 20 times better.

To operate electrolytic capacitors in series–parallel, stabilizing resistors should be used to equalize the voltage distribution. It should also

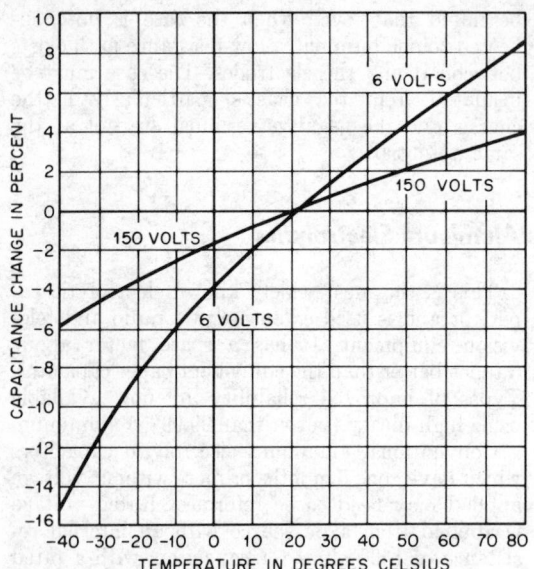

Fig. 14—Variation of capacitance with temperature for plain-tantalum-foil electrolytic capacitors.

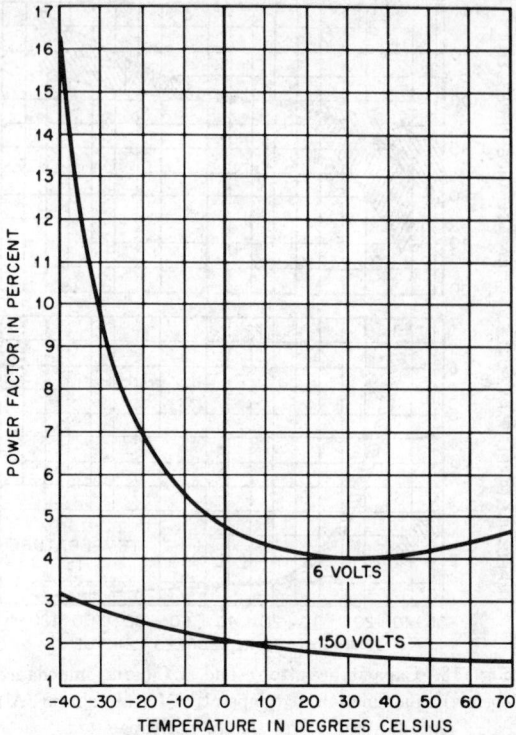

Fig. 15—Variation of power factor with temperature for plain-tantalum-foil electrolytic capacitors.

plated container. The porous anode is made by pressing a tantalum high-purity powder into a cylindrical body and sintering in vacuum at about 2000°C.

Tantalum Electrolytics with Porous Anode and Solid Electrolyte

This is the so-called solid tantalum capacitor originally developed by Bell Telephone Laboratories. It developed from the porous-anode type with liquid electrolyte by replacing the liquid with a semiconductor. This overcame the problem of sealing common to all other types of electrolytic capacitors. Since there is no liquid electrolyte it is possible to use a conventional hermetic seal.

CERAMIC CAPACITORS

Electrical ceramics have a wide range of electrical characteristics, which makes them the most versatile of capacitor dielectric materials. The low-K materials have virtually linear characteristics and their properties are independent of frequency over the normal range. The materials are usually magnesium titanate (which has a positive temperature coefficient of capacitance) and calcium titanate (which has a negative one). By combining these, a range of controlled temperature coefficients can be obtained. These materials, together with other additions, produce a range of ceramics with di-

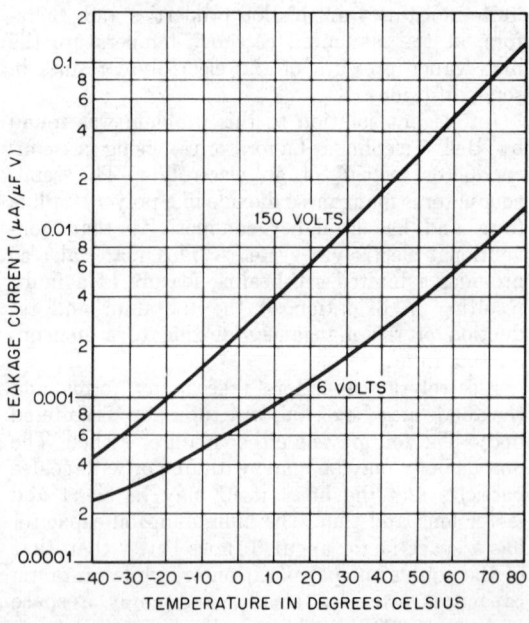

Fig. 16—Variation of leakage current with temperature for plain-tantalum-foil electrolytic capacitors.

electric constant of between 5 and 110 and temperature coefficients from $+150$ to -4700 ppm/°C with tolerances as small as ±15 ppm/°C. Their properties are largely independent of voltage. For these reasons they are sometimes known as the temperature-compensating ceramics. Glass and porcelain capacitors also have similar characteristics although their temperature coefficients lie in the range 0 to 140 only.

The high-K materials are the ferroelectrics. Because of their crystal structure, they sometimes have very high values of internal polarization, giving very high effective dielectric constants. In this way these materials are comparable with ferromagnetic materials. Above the Curie temperature a change of domain structure occurs which results in a change of electrical characteristics. This region is known as the para-electric region. In common with the ferromagnetic materials, a hysteresis effect is apparent and this makes the capacitance voltage-dependent.

The ferroelectrics are based on barium titanate, which has a peak dielectric constant of 6000 at the Curie point of 120°C. Additions of barium stanate, barium zirconate, or magnesium titanate reduce this dielectric constant but make it more uniform over the temperature range. Thus a family of materials can be obtained with a Curie point at about room temperature and with the dielectric constant falling off on either side. The magnitude of this change increases with increasing dielectric constant. These materials exhibit a decrease of capacitance with time and, as a result of the hysteresis effect, with increasing voltage.

Low-K ceramics are suitable for resonant-circuit or filter applications, particularly where temperature compensation is a requirement. Disc and tubular types are the best forms for this purpose. Stability of capacitance is good, being next to that of mica and polystyrene capacitors.

High-K ceramics are suitable for coupling and decoupling applications where low tangent of loss angle and stable capacitance are not requirements. Inductance in the leads and element causes parallel resonance in the megahertz region. Care is necessary in their application above about 50 megahertz for tubular styles and about 500 megahertz for disc types.

Color Code

The significance of the various colored dots for EIA Standard RS-198-B (ANSI C83.4-1972) fixed ceramic dielectric capacitors is explained by Figs. 17 and 18, and may be interpreted from Table 6.

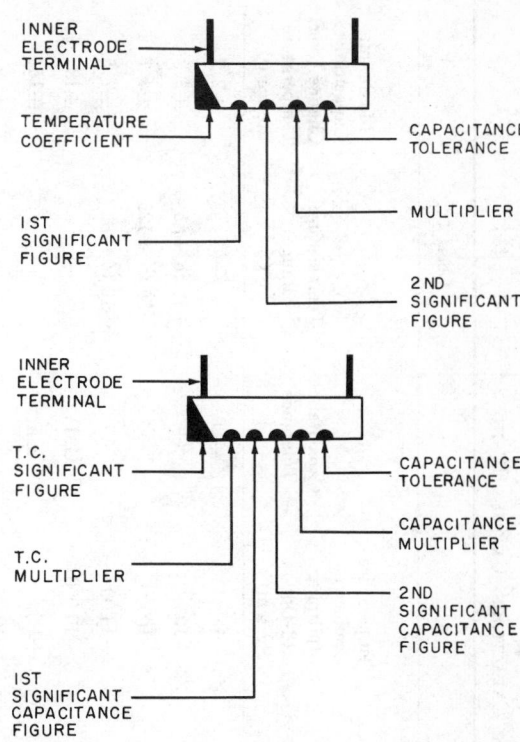

Fig. 17—Color coding of EIA Class-1 ceramic dielectric capacitors. See Table 6 for color code. Tubular style shown to illustrate identification of inner electrode. For disc or plate styles, color code will read from left to right as observed with lead wires downward. Five-dot system at top; six-dot system at bottom.

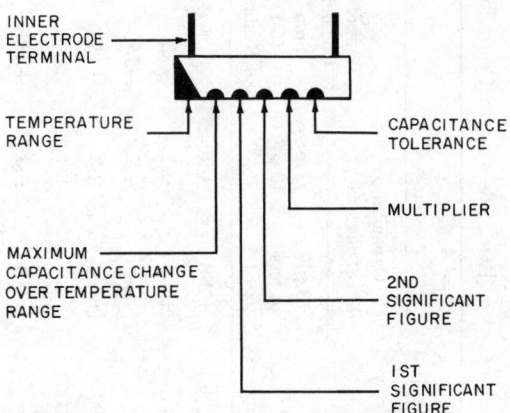

Fig. 18—Color coding of EIA Class-2 ceramic dielectric capacitors. See Table 6 for color code. Tubular style shown to illustrate identification of inner electrode. For disc or plate styles, color code will read from left to right as observed with lead wires downward.

TABLE 6—COLOR CODE FOR CERAMIC DIELECTRIC CAPACITORS, CLASSES 1 AND 2.*

Color	Digit	Multiplier	Capacitance Tolerance — 10 pF or less (pF)	Capacitance Tolerance — Over 10 pF (%)	Temperature Coefficient ppm/°C (5-Dot System)	Temperature Coefficient Significant Figure (6-Dot System)	Temperature Coefficient Multiplier (6-Dot System)	Capacitance Tolerance (%)	Temperature Range (°C)	Maximum Capacitance Change Over Temperature Range (%)
Black	0	1	±2.0	±20	0	0.0	−1	±20	—	±2.2
Brown	1	10	±0.1	±1	−33	—	−10	—	+10 to +85	±3.3
Red	2	100	—	±2	−75	1.0	−100	—	−55 to +125	±4.7
Orange	3	1 000	—	±3	−150	1.5	−1 000	—	+10 to +65	±7.5
Yellow	4	10 000	—	—	−220	2.2	−10 000	GMV	—	±10
Green	5	—	±0.5	±5	−330	3.3	+1	±5	—	±15
Blue	6	—	—	—	−470	4.7	+10	—	—	±22
Violet	7	—	—	—	−750	7.5	+100	—	—	+22, −33
Gray	8	0.01	±0.25	—	+150 to −1500	(−1000 to −5200 ppm/°C, With Black Multiplier)	+1 000	+80, −20	—	+22, −56
White	9	0.1	±1.0	±10	+100 to −750	—	+10 000	±10	—	+22, −82
Silver	—	—	—	—	—	—	—	—	−30 to +85	±1.5
Gold	—	—	—	—	—	—	—	—	−55 to +85	±1

* EIA Standard RS-198-B (ANSI C83.4-1972). This standard classifies ceramic dielectric, fixed capacitors as follows:

Class 1—Temperature compensating ceramics suited for resonant circuit or other applications where high Q and stability of capacitance characteristics are required.

Class 2—Ceramics suited for bypass and coupling applications, or for frequency discriminating circuits where high Q and stability of capacitance characteristics are not of major importance.

Class 3—Low-voltage ceramics specifically suited for transistorized or other electronic circuits for bypass, coupling, or frequency determination where dielectric losses, high insulation resistance, and capacitance stability are not of major importance.

Note: Where size permits, EIA Class-3 ceramics are typographically marked as follows:
(1) Capacitance value in microfarads. (2) Rated voltage. (3) Manufacturer's mark or EIA source code.
(4) Capacitance value tolerance or appropriate code letter; either ±20% (Code M) or +80, −20% (Code Z).
(5) Temperature stability code (see the EIA or ANSI Standard).

Temperature Coefficient

Standard temperature coefficients of capacitance expressed in parts per million per degree Celsius are: $+150$, $+100$, $+33$, 0, -33, -75, -150, -220, -330, -470, -750, -1500, -2200, -3300, and -4700.

PAPER FOIL-TYPE CAPACITORS

General

Paper consists of a network of cellulose fibers, usually produced today from kraft (wood pulp). Very careful control in manufacture is necessary to produce extremely fine tissues with the requisite chemical purity and freedom from unreduced fibers and conducting particles. Paper normally has a moisture content of up to about 10 percent by weight, and this is removed before the capacitor is impregnated with an oil or wax.

Although paper capacitors have been largely replaced by plastic film types in electronic circuits, they are nevertheless still unsurpassed for high-voltage dc and ac power applications. They fall within the category of medium loss and medium stability, and provide an economical solution in this field.

The proper application of paper capacitors is a complex problem requiring consideration of the equipment, duty cycle, desired capacitor life, ambient temperature, applied voltage and waveform, and the capacitor impregnant characteristics.

Construction

The paper capacitor element is manufactured by winding together two aluminum foils interleaved with an appropriate number of layers of capacitor tissue. At least two layers normally are used to avoid the effects of conducting particles in the layer. For higher voltages a large number of layers is preferred.

Contact to the aluminum foils is usually by means of tinned copper tapes, which are preferably welded to the aluminum foil. Circular elements are wound on small-diameter mandrels but, if a large mandrel is used, the element may be flattened to fit conveniently in a rectangular case.

A wide variety of materials is used as impregnants for paper capacitors and these are classified generally as follows.

(A) Waxes. These are normally suitable for dc applications only. Chlorinated naphthalene has a high dielectric constant and therefore gives a cost and space saving. Chlorinated impregnants, however, must be stabilized to avoid electrochemical deterioration. Waxes are not suitable for ac applications because of the formation of gas voids due to shrinkage on cooling.

(B) Mineral oil, polyisobutylene, and silicone oils. These are relatively nonpolar impregnants suitable for either dc or ac applications. Oils have the advantage over waxes of not forming voids, minimizing internal discharges which would otherwise give rise to early failure on ac.

(C) Askarels. The most important of these are the chlorinated diphenyls with trade names such as Aroclor, Diconal, Inertene, and Pyranol. These are polar impregnants having a dielectric constant of about 6, but whose characteristics vary markedly with temperature and frequency. They are used mostly for ac power applications in which the higher dielectric constant gives a cost and space saving as well as an improvement in internal discharge characteristics. Dc operation is satisfactory only if the impregnant is stabilized.

Life Derating

The energy content of a capacitor may be found from

$$W = CE^2/2 \text{ watt-second}$$

where $C =$ capacitance in farads, and $E =$ applied voltage in volts.

In multiple-section capacitors, the sum of the watt-second ratings should be used to determine the proper derating of the unit.

Experiment has shown that the life of paper-dielectric capacitors having the usual oil or wax impregnants is approximately inversely proportional to the 5th power of the applied voltage.

Desired life in years (at ambient $\approx 45°C$)	1	2	5	10	20
Applied voltage in percent of rated voltage	100	85	70	60	53

The above life derating is to be applied together with the ambient-temperature derating to determine the adjusted-voltage rating of the paper capacitor for a specific application.

Waveform

Normal filter capacitors are rated for use with direct current. Where alternating voltages are present, the adjusted voltage rating of the capacitor should be calculated as the sum of the direct voltage and the peak value of the alternating voltage. The alternating component must not exceed 20 percent of the rating at 60 hertz, 15 percent at 120 hertz, 6 percent at 1000 hertz, or 1 percent at 10 000 hertz.

Where ac rather than dc conditions govern, this

fact must be included in the capacitor specification, and capacitors specially designed for ac service should be procured.

Where heavy transient or pulse currents are present, standard capacitors may not give satisfactory service unless an allowance is made for the unusual conditions.

Applications

Paper capacitors cover a wide range of applications as follows.

(**A**) Low-voltage dc. Tubular capacitors for coupling and decoupling in electronics circuits.

(**B**) High-voltage dc. Capacitors in smoothing filters, power-separating filters, energy-storage capacitors, etc.

(**C**) Low-voltage ac. Motor start, fluorescent lighting, interference suppression, and power-factor correction.

(**D**) High-voltage ac. Power-factor correction, power-line coupling, distribution capacitors for high-voltage switchgear, voltage-dividing capacitors for ac voltage measurement, etc.

METALLIZED PAPER CAPACITORS

General

Metallized capacitors use an evaporated metal film as an electrode instead of the conventional aluminum foil, as shown in Fig. 19. The metal film is sufficiently thin so that in the event of breakdown self-healing can occur. This consists of burning away the film in the area of the fault to isolate it. Higher working stresses can thus be tolerated even with single layers of tissue, giving an advantage of space factor in the lower voltage ratings. At 200 volts they are one quarter of the size of conventional paper types.

These capacitors are classified into type 1, which does not normally self-heal in service, and type 2, in which self-healing may occur under working conditions. Healing requires a low-impedance circuit and does not occur unless adequate volt-amperes are supplied for a very short time.

Construction

The capacitors are normally made by winding together two sheets of tissue that have been metallized with an offset electrode pattern. The metallizing in paper capacitors is usually zinc, although aluminum is regarded as superior particularly for plastic film types.

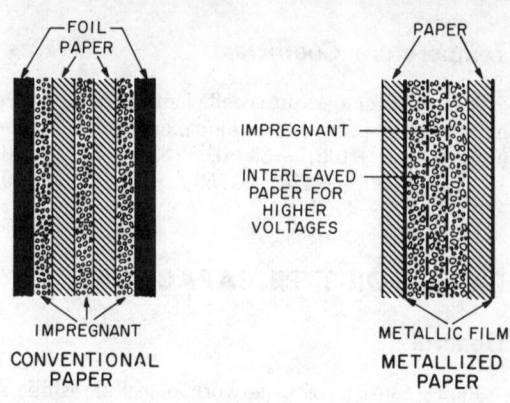

Fig. 19—Construction of conventional and metallized-type paper capacitors.

Connection to the film electrode is by spraying the end of the unit with metal. This is a critical operation which can result in poor contact with the electrodes if not done right. Types of metal vary widely and solder is satisfactory for making connection with zinc. With aluminum it is usually preferable to apply pure copper or zinc and to follow this by solder.

Applications

Direct-Current: Metallized paper capacitors are often used for coupling and decoupling applications in which small size is particularly important. In digital or other circuits where noise is important, it is essential to use a type-1 capacitor. If a type-2 capacitor is operated at 75 percent of rated voltage, self-healing is usually negligible but space advantage is less significant. The lower insulation-resistance performance of type-2 capacitors should be noted.

Alternating-Current: Special designs have been developed for use on ac, particularly for motor-starting and fluorescent-lighting applications.

MICA CAPACITORS

General

Mica capacitors fall within the classification of low loss and good capacitance stability. Mica is one of the earliest dielectric materials used and has an unrivalled combination of physical and electrical characteristics. It is of mineral origin and, because of its monoclinic structure, can be readily slit into thin plates. It has a dielectric constant of about 6 (largely independent of frequency) together with a very low loss.

Construction

Types of mica capacitors follow.

(**A**) Clamp type with tin-foil electrodes. This form of construction was used in the manufacture of the earliest filter-type capacitors and is still used today in the construction of standard capacitors. It has been largely superseded for other applications.

(**B**) Eyelet construction. Silvered mica plates are eyeletted together to form stacks. Because of the possibility of relative movement and bowing of the plates, this construction has poor stability of both capacitance and temperature coefficient.

(**C**) Bonded silvered mica construction. Mica plates are silvered on both sides with appropriate electrode areas, stacked, and bonded together by firing. The result is dimensionally stable stacks of good stability of capacitance and with characteristics representative of the mica itself.

(**D**) Button styles. These are circular capacitors with a metal band round the outer periphery forming one terminal, and an eyeletted connection at the center forming the other. They are particularly suitable for high-frequency operation if mounted in a truly coaxial arrangement. These are discussed in detail later in this chapter.

The environmental protection necessary in the mica capacitor depends on the application. For some applications unprotected capacitors can be used, but the adsorption of water on the surface gives rise to relatively poor electrical characteristics. The two main methods of protection are as follows.

(**A**) *Epoxy resin molding.* Impregnated and wired stacks are molded in an appropriate resin having good moisture-protection properties. The advantage of this construction is that the molding has regular faces with good dimensional control.

(**B**) *Dipping.* A coating of resin is applied to the capacitor either by dipped or by fluidized-bed methods.

Applications

Because of their low temperature coefficient of capacitance and good capacitance stability both with temperature and frequency, mica capacitors are invaluable for filter applications.

Type Designation

A comprehensive numbering system, the type designation, is used to identify mica capacitors. Type designations are of the form shown in Fig. 20.

MIL specifications now require type designation marking. Color coding is now used only for EIA standard capacitors.

Component Designation: Fixed mica-dielectric capacitors are identified by the symbol CM. For EIA, a prefix letter R is always included, and dipped types are identified by the symbol DM.

Case Designation: The case designation is a two-digit symbol that identifies a particular case size and shape.

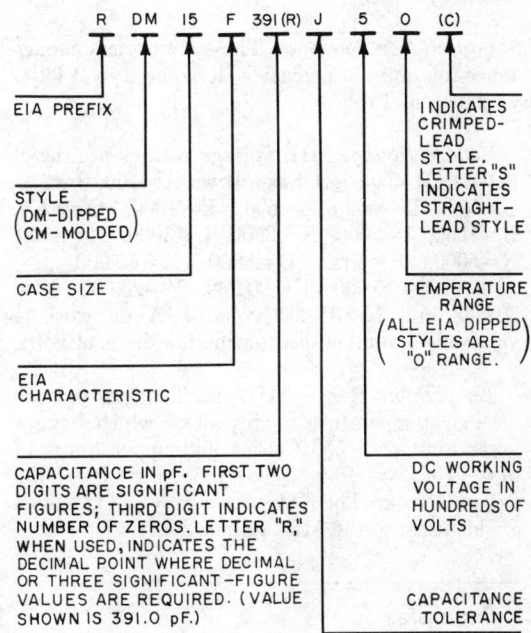

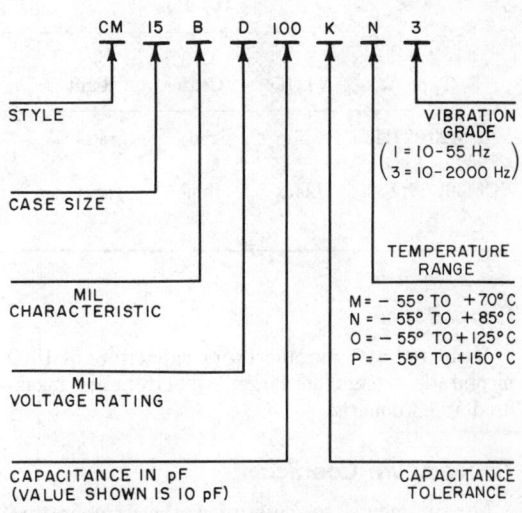

Fig. 20—Type designation for mica-dielectric capacitors. EIA at top; MIL at bottom.

Characteristic: The EIA or MIL characteristic is indicated by a single letter in accordance with Table 7.

Capacitance Value: The nominal capacitance value in picofarads is indicated by a 3-digit number. The first two digits are the first two digits of the capacitance value in picofarads. The final digit specifies the number of zeros that follow the first two digits. For EIA, if more than two significant figures are required, an additional digit is used, and the letter "R" is inserted to designate the decimal position.

Capacitance Tolerance: The symmetrical capacitance tolerance in percent is designated by a letter as shown in Table 1.

Voltage Rating: MIL voltage ratings are designated by a single letter as follows. A=100, B=250, C=300, D=500, E=600, F=1000, G=1200, H=1500, J=2000, K=2500, L=3000, M=4000, N=5000, P=6000, Q=8000, R=10 000, S=12 000, T=15 000, U=20 000, V=25 000, W=30 000, and X=35 000 volts. EIA dc working voltage is a number designating hundreds of volts.

Temperature Range: MIL specifications provide for four temperature ranges all of which have a lower limit of $-55°$ Celsius and upper limits of M=+70, N=+85, O=+125, and P=+150 degrees Celsius. The EIA uses only N and O, which are identical to the MIL standard.

TABLE 7—FIXED-MICA-CAPACITOR REQUIREMENTS BY EIA AND MIL CHARACTERISTIC.

EIA Standard and MIL-Specification Requirements		
EIA or MIL Charac- teristic	Maximum Capacitance Drift	Maximum Range of Temperature Coefficient (ppm/°C)
B	Not Specified	Not Specified
C	$\pm(0.5\%+0.1 \text{ pF})$	±200
D	$\pm(0.3\%+0.1 \text{ pF})$	±100
E	$\pm(0.1\%+0.1 \text{ pF})$	-20 to $+100$
F	$\pm(0.05\%+0.1 \text{ pF})$	0 to $+70$

Vibration Grade: The MIL vibration grade is a number, 1 corresponding to vibration from 10 to 55 hertz at 10g for 4.5 hours and 3 corresponding to 10 to 2000 hertz at 20g for 12 hours.

Color Coding

The significance of the various colored dots for EIA-standard and former MIL-specification mica capacitors is explained by Fig. 21. The meaning of each color may be interpreted from Table 1.

Examples

	Top Row			Bottom Row			
					Tolerance	Multiplier	
Type	Left	Center	Right	Left	Center	Right	Description
RCM20B221K	white	red	red	brown	silver	brown	220 pF ±10%, EIA characteristic B.
CM30C681J	black	blue	gray	red	green	brown	680 pF ±5%, MIL characteristic C.

Capacitance

Measured at 1 megahertz for capacitors of 1000 picofarads or smaller; larger capacitors are measured at 1 kilohertz.

Temperature Coefficient

Measurements to determine the temperature coefficient of capacitance and the capacitance drift are based on one cycle over the following temperature values (all in degrees Celsius).

$$+25, -55, -40, -10, +25, +45, +65, +70, +85, +125, +150, +25.$$

Measurements at +85, +125, and +150 are not made if these values are not within the applicable temperature range of the capacitor.

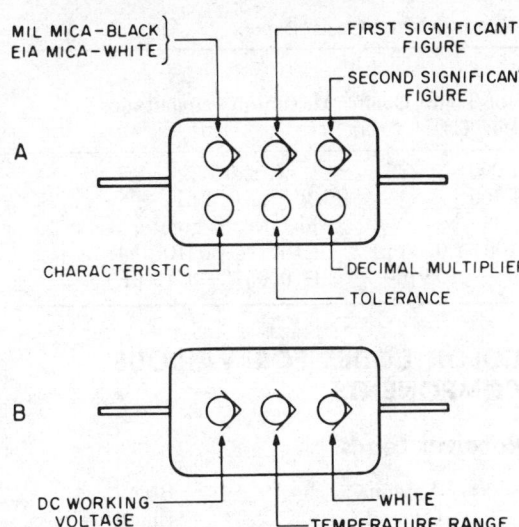

Fig. 21—Standard code for fixed mica capacitors. See color code in Table 1. *A* is the basic 6-dot form. The 9-dot form with *B* on the other side of the capacitor is used if the additional data are required.

Dissipation Factor

EIA and MIL specifications require that for molded and dipped capacitors the dissipation factor not exceed the values shown in Fig. 22. For potted and cast epoxy capacitors, the dissipation factor shall not exceed 0.35 percent from 1 to 1000 picofarads and 0.15 percent above 1000 picofarads.

High-Potential or Withstanding-Voltage Test

Molded or dipped mica capacitors are subjected to a test potential of twice their direct-current voltage rating.

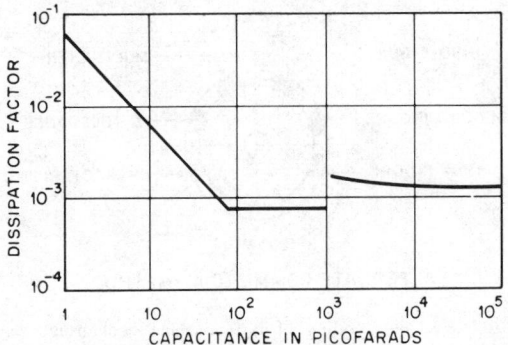

Fig. 22—EIA and MIL maximum dissipation factor at 1 megahertz for capacitance of 1000 picofarads or less and at 1 kilohertz for capacitance greater than 1000 picofarads.

Humidity and Thermal-Shock Tests

EIA Standard RS-153-B capacitors must withstand 5 cycles of −55, +25, +85 or +125 (as applicable), and +25 degrees Celsius thermal shock followed by a humidity test of 10 cycles (each of 24 hours) given for EIA Standard RS-186-D in Fig. 1. Units must pass withstanding-voltage test. Capacitance may not change by more than 1.0 percent or 1.0 picofarad, whichever is greater. Insulation resistance must meet or exceed 30 percent of the initial requirements at 25°C (50 000 megohms for capacitances of 20 000 picofarads or less; 1000 ohm-farads for larger capacitances).

MIL Specification MIL-C-5D capacitors must withstand 5 cycles of −55; +25; +85, +125, or +150 (as applicable); and +25 degrees Celcius thermal shock followed by a humidity test of 10 cycles (each of 24 hours) given for MIL-STD-202E in Fig. 1. Units must pass withstanding-voltage test. Capacitance may not change by more than ±(0.2 percent+0.5 picofarad). Insulation resistance must meet or exceed 30 percent of the initial requirements at 25°C (100 000 megohms for 10 000 picofarads or less; 1000 megohm-microfarads for larger capacitances).

Life

Capacitors are given accelerated life tests at 85 degrees Celsius with 150 percent of rated voltage applied for 2000 hours for MIL specification or 250 hours for EIA standard. If capacitors are rated above +85°C, the test will be at their maximum rated temperature.

BUTTON MICA CAPACITORS

Color Code

Button mica capacitors are color coded in several different ways, of which the two most widely used methods are shown in Fig. 23.

Characteristic

The characteristics for button mica capacitors are given in Table 8. Typical initial Q values are approximately 500 for capacitors 5 to 50 picofarads, 700 for capacitors 51 to 100 picofarads, and 1000 for capacitors 101 to 5000 picofarads. Initial insulation resistance shall exceed 7500 megohms for resin-sealed units and 50 gigaohms for hermetically sealed units. Withstanding-voltage tests are made at twice rated voltage.

Table 8—Requirements for Button Mica Capacitors.

Characteristic		Max Range of Temp Coeff (ppm/°C)	Maximum Capacitance Drift
MIL	Commercial		
—	C	±200	±0.5%
D	—	±100	±0.3% or 0.3 pF, whichever is greater
—	E	(−20 to +100)+0.05 pF	±(0.1%+0.10 pF)
—	F	(0 to +70)+0.05 pF	±(0.05%+0.10 pF)

Thermal-Shock and Humidity Tests

After 10 cycles of 24 hours each of the cycling for MIL-STD-202E given in Fig. 1, capacitors should have the following properties.

	Resin Seal	Hermetic Seal
Insulation Resistance in megohms (minimum)	500	4000
Q factor, % of initial requirement	50	75
Capacitance change, maximum % of initial capacitance	3	2

Thermal shock and immersion tests are also applied; they have less effect than the cycling humidity test above.

I.F. TRANSFORMER FREQUENCIES

Recognized standard frequencies* for receiver intermediate-frequency transformers are:

Standard broadcast (540 to 1600 kilohertz)	455 and 260 kilohertz
Standard broadcast (vehicular)	262.5 kilohertz
Very-high-frequency broadcast	10.7 megahertz
Very-, ultra-, and super-high-frequency equipment	30, 60, 100 megahertz (common practice)

Television:

Sound carrier	41.25 megahertz
Picture carrier	45.75 megahertz

* EIA Standard REC-109-C.

COLOR CODES FOR VARIOUS COMPONENTS

Receiver Leads*

Antenna	Blue
Ground	Black

Intermediate-Frequency Transformers

Primary:

Plate	Blue
B+	Red

Secondary:

Grid or diode	Green
Grid return	White

For full-wave transformer:

Second diode	Violet

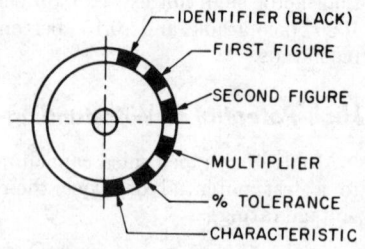

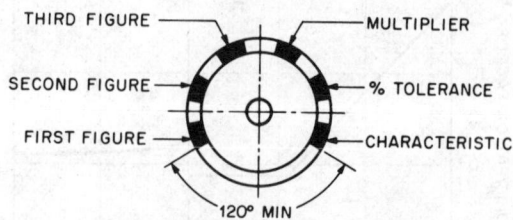

ALTERNATE COMMERCIAL METHOD

Fig. 23—Color coding of button mica capacitors. See Table 1 for color code. Commercial color code for characteristic not standardized; varies with manufacturer.

* EIA Standard RS-336.

Traveling-Wave Tubes and Klystrons*

Body (Tracer)	TWT Element		Klystron
Black	Grounds or grounded elements		
Brown†	Heater or filament off ground		
Yellow†	Cathode or common heater–cathode		
Red	Collector		Collector, if isolated
Orange	Helix 1	1st	
Orange (green)	Helix 2	2nd	Reflector, phase modulation
Orange (blue)	Helix 3	3rd	element, electrostatic
Orange (gray)	Helix 4	4th	focussing element, etc.
Orange (black)	—	5th	
Orange (white)	—	6th	
Green	Grid 1		Grid 1
Blue	Grid 2		Grid 2
Gray	Grid 3		Grid 3
White	Grid 4		Grid 4
Green (black)	Grid 5		Grid 5
Blue (black)	Grid 6		Grid 6
Gray (black)	Grid 7		Grid 7
White (black)	Grid 8		Grid 8

† If individual leads are provided for elements having internal connections, the body color will identify the major element and a tracer will identify the internally connected element. For example: Brown (yellow) indicates heater lead internally connected to the cathode.

Note:

No attempt is made to separate anode, control grid, modulating anode, etc. Elements are numbered sequentially according to their relative position starting from the cathode. If two elements are equidistant from the cathode, the lower number indicates the lower-voltage element. Grids or helices not available externally are omitted from the number sequence.

Crossed-Field Devices*—Magnetrons, BWO's, etc.

Color	Magnetron	Voltage-Tunable Magnetron	Backward-Wave Oscillator (M-type)
Black	Body or other grounded elements		
Brown	Heater or filament off ground		
Yellow	Cathode or common heater–cathode		
Red	Anode	Anode	Delay line
Orange	—	—	Sole
Green	—	Injector	Grid
Blue	—	—	Accelerator
White	—	Cold cathode	—
Gray	Turnoff electrode in crossed-field amplifiers		

Electromagnet Leads (When Electromagnet is Integral Part of Microwave-Tube Package)

Orange tracer on black, red, white, or other color body.

* EIA Standard RS-235-C.

Hall Generator Devices—Wiring*

Black	Control Current–Negative
Red	Control Current–Positive
Yellow	Output–Negative
Blue	Output–Positive

Stereo Pickup Leads†

No. of Leads	Right High	Right Low	Left High	Left Low	Return or Ground
3	red	—	white	—	black
4	red	green	white	blue	—
5	red	green	white	blue	black

* EIA Standard RS-336.
† EIA Standard RS-243.

OTHER MARKING CODES

Single Band or Mark at One End

Semiconductor diode	cathode end
Capacitor, wound-foil type	outside-electrode end
Capacitor, ceramic-tube type	inner electrode

Symbols

Electrolytic capacitor sections—terminal markings.

$\bigcirc$	$\square$	$\triangle$	No mark
Highest	Intermediate	Lower	Lowest

The sequence applies first to voltage rating of section. If voltage ratings are the same, the sequence applies to capacitance value. For fewer than 4 terminals any mark may be omitted, but interpretation of marks used follows preceding rules.

Connections

Stereo Headphone 3-Contact Plugs:

Sleeve (barrel)	common
Ring	left phone
Tip	right phone
Polarization	with + on tip or ring, diaphragm moves toward listener's ear (− on barrel).

Phasing of Microphones‡

Terminal or Lead:

In phase	red (or 1)
Out of phase	black (or 2)
Ground	G

‡ EIA Standard RS-221.

PRINTED CIRCUITS

A printed circuit is a conductive circuit pattern applied to one or both sides of an insulating substrate. The conductive pattern can be formed by any of several techniques after which component lead holes are drilled or punched in the substrate and components are installed and soldered in place. Printed-circuit construction is ideal for assembly of circuits which employ miniature solid-state components. Its advantages over conventional chassis and point-to-point wiring include:

(A) Considerable space savings over conventional construction methods is usually a result.

(B) A complex circuit may be modularized by using several small printed circuits instead of a single larger one. Modularization simplifies troubleshooting, circuit modification, and mechanical assembly in an enclosure.

(C) Soldering of component leads may be accomplished in an orderly sequence by hand or all the leads may be soldered simultaneously by dip or wave soldering.

(D) A more uniform product is produced because wiring errors are eliminated and because distributed capacitances are constant from one production unit to another.

(E) The printed-circuit method of construction lends itself to automatic assembly and testing machinery.

(F) The printed circuit consists of printed wiring but may also include printed components such as capacitors and inductors. Capacitors can be produced by printing conducting areas on opposite sides of the wiring board, using the board material as the dielectric. Spiral-type inductors can also be printed. Both types of components are illustrated in Fig. 24.

(G) Using appropriate base materials, flexible cables or flexible circuits can be built.

(H) By using several layers of circuits (in proper registry) in a sandwich construction, with the conductors separated by insulating layers, relatively complex wiring can be provided.

Printed-Circuit Base Materials

Rigid printed-circuit base materials are available in thicknesses varying from $\frac{1}{64}$ to $\frac{1}{2}$ inch. The important properties of the usual materials are given in Table 9. For special applications, other rigid or flexible materials are available as follows.

(A) Glass-cloth Teflon (polytetrafluoroethylene, PTFE) laminate.

(B) Kel-F (polymonochlorotrifluoroethylene) laminate.

(C) Silicone rubber (flexible).

(D) Glass-mat–polyester-resin laminate.

(E) Teflon film.

(F) Ceramic.

The most widely used base material is NEMA-XXXP paper-base phenolic.

Conductor Materials

Copper is used almost exclusively as the conductor material, although silver, brass, and aluminum also have been used. The common thicknesses of foil are 0.0014 inch (1 oz/sq ft) and 0.0028 inch (2 oz/sq ft). The current-carrying capacity of a copper conductor may be determined from Fig. 25.

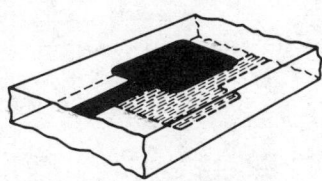

PRINTED-CIRCUIT CAPACITOR

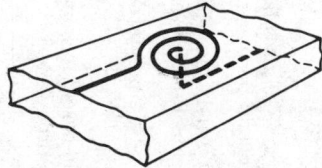

PRINTED-CIRCUIT INDUCTOR

Fig. 24—Formation of reactive elements by printed-circuit methods.

Manufacturing Processes

The most widely used production methods are:

(A) Etching process, wherein the desired circuit is printed on the metal-clad laminate by photographic, silk-screen, photo-offset, or other means, using an ink or lacquer resistant to the etching bath. The board is then placed in an etching bath that removes all of the unprotected metal (ferric chloride is a commonly used mordant for copper-clad laminates). After the etching is completed, the ink or lacquer is removed to leave the conducting pattern exposed.

(B) Plating process, wherein the designed circuit pattern is printed on the unclad base material using an electrically conductive ink and, by electroplating, the conductor is built up to the desired thickness. This method lends itself to plating through punched holes in the board for making connections from one side of the board to the other.

(C) Other processes, including metal spraying and die stamping.

Circuit-Board Finishes

Conductor protective finishes are required on the circuit pattern to improve shelf-storage life of the circuit boards and to facilitate soldering. Some of the most widely used finishes are:

(A) Hot-solder coating (done by dip-soldering in a solder bath) is a low-cost method and gives good results where coating thickness is not critical.

(B) Silver plating is used as a soldering aid but is subject to tarnishing and has a limited shelf life.

(C) Hot-rolled or plated solder coat gives good solderability and uniform coating thickness.

(D) Other finishes for special purposes are gold plate for corrosion resistance and solderability, and electroplated rhodium over nickel for wear resistance. Insulating coatings such as acrylic, polystyrene, epoxy, or silicone resin are sometimes applied to circuit boards to improve circuit performance under high humidity or to improve the anchorage of parts to the board. Conformal coatings are relatively thick and tend to smooth the irregular contour of the mounted items; they add less mass than encapsulation. A protective organic coating (unless excessively thick) will not improve the electrical properties of an insulating base material during long exposure to high humidity. On two-sided circuit boards, where the possibility of components shorting out the circuit patterns exists, a thin sheet of insulating material is sometimes laminated over the circuit before the parts are inserted.

TABLE 9—PROPERTIES OF TYPICAL PRINTED-CIRCUIT DIELECTRIC BASE MATERIALS.

Material	Comparable MIL Type	Punchability	Mechanical Strength	Moisture Resistance	Insulation	Arc Resistance	Abrasive Action on Tools	Max Temperature (°C)*
NEMA type XXXP paper-base phenolic	—	Good	Good	Good	Good	Poor	No	105
NEMA type XXXPC paper-base phenolic	—	Very good	Good	Very good	Good	Poor	No	105
NEMA type FR-2 paper-base phenolic, flame resistant	—	Very good	Good	Very good	Good	Poor	No	105
NEMA type FR-3 paper-base epoxy, flame resistant	PX	Very good	Very good	Very good	Very good	Good	No	105
NEMA type FR-4 glass-fabric-base epoxy, general purpose, flame resistant	GF	Fair	Excellent	Excellent	Excellent	Very good	Yes	130 (125)
NEMA type FR-5 glass-fabric-base epoxy, temperature and flame resistant	GH	Fair	Excellent	Excellent	Excellent	Very good	Yes	155 (150)
NEMA type G-10 glass-fabric-base epoxy, general purpose	GE	Fair	Excellent	Excellent	Excellent	Very good	Yes	130 (125)
NEMA type G-11 glass-fabric-base epoxy, temperature resistant	GB	Poor	Excellent	Excellent	Excellent	Very good	Yes	155 (150)
Glass-fabric-base polytetrafluoroethylene	GT	—	Good	Excellent	Excellent	Excellent	—	(150)
Glass-fabric-base fluorinated ethylene propylene	FEP	—	Good	Excellent	Excellent	Excellent	—	(150)

* MIL-STD-275C rating shown in parentheses if different from industry rating.

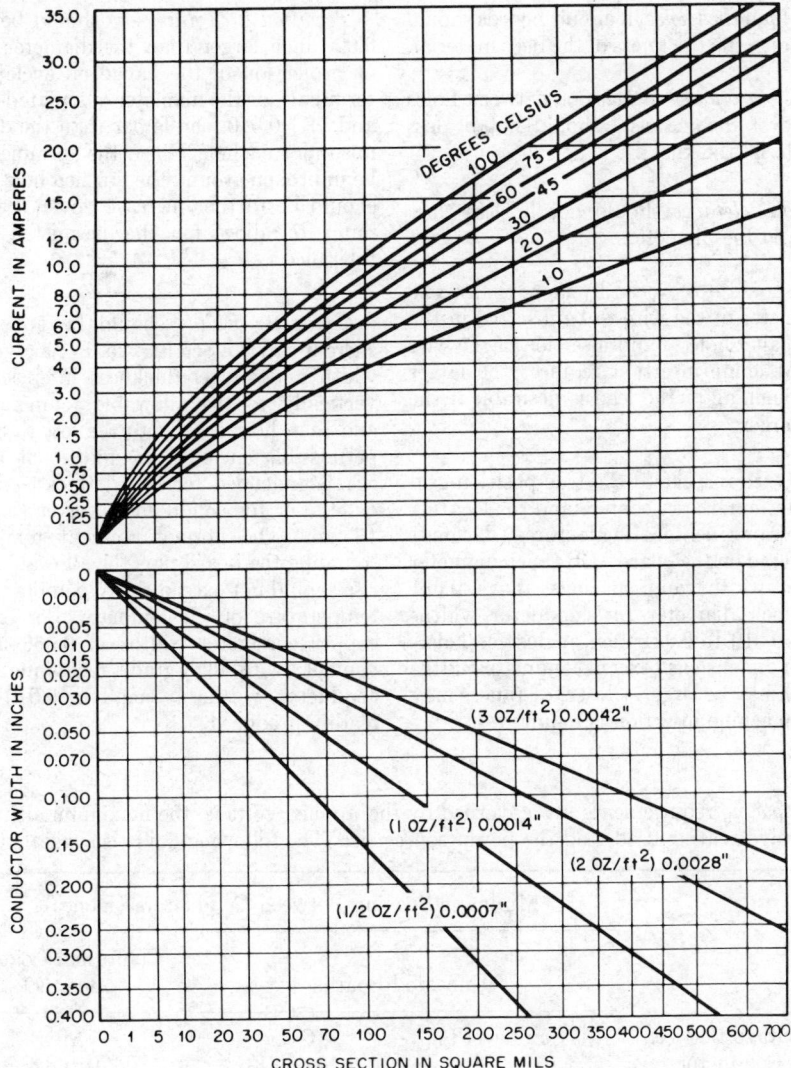

Fig. 25—Current-carrying capacity and sizes of etched copper conductors for various temperature rises above ambient. *From MIL–STD–275C, 9 January 1970.*

Design Considerations

Before a printed-circuit layout is made, the circuit must be breadboarded and tested under the anticipated final operating conditions. This procedure will permit operating deficiencies and quirks to be detected and corrected before the time-consuming process of producing the circuit board is begun. It is important to note that certain circuits may operate differently on a printed-circuit board than on a breadboard and appropriate corrective steps may be necessary. For example, inductive coupling between foil patterns may cause unwanted oscillation in high-frequency or amplifier circuits.

Modular Layout: All features (terminal areas, contacts, board boundaries, holes, etc.) should be arranged to be centered at the intersections of a 0.100-, 0.050-, or 0.025-inch rectangular grid, with preference in the order stated. Many components are available with leads spaced to match the standard grids. Devices with circular lead configurations and a few other multilead devices are exceptions that require special attention and dimensioning. Following this grid-layout principle simplifies drafting and subsequent machine operations in board manufacture and assembly.

Drilled holes must be employed if the stated requirements for punched holes cannot be met, or if the material is not of a punching grade. Drilling is less detrimental to the laminate surrounding the hole; punching may cause crazing or separation of the laminate layers.

Diameter of punched holes in circuit boards should not be less than $\frac{2}{3}$ the thickness of the base material.

Distance between punched holes or between holes and the edge of the material should not be less than the material thickness.

Punched-hole tolerance should not be less than ±0.005 inch on the diameters.

Hole sizes should not exceed by more than 0.020 inch the diameter of the wire to be inserted in the hole. With smaller holes, hand insertion of the wire is difficult. Machine insertion requires the larger allowance. Clinching of the lead is desirable if the clearance is larger.

Tolerances with respect to the true-position-grid location for terminal area centers and for locating edges of boards or other locating features (datums) should not exceed on the board: 0.014 inch diameter for conductor widths and spacings above 0.031 inch; 0.010 inch diameter for conductor widths and spacings 0.010 to 0.031 inch, inclusive. Tolerances on other dimensions (except conductor widths and spacings) may be larger. Closer tolerances may be needed if machine insertion is required.

Terminal area diameters should be at least (A) 0.020 inch larger than the diameter of the flange or projection of the flange on eyelets or standoff terminals, or the diameter of a plated-through hole, and (B) 0.040 inch larger than the diameter of an unsupported hole. Since the terminal area should be unbroken around the finished hole, the diameter should be further increased over the above minimum to allow for the permitted hole-position tolerance.

Conductor widths should be adequate for the current carried. See Fig. 25. For a given conductor-width and copper-thickness intersection, proceed vertically to the allowable temperature-rise line and then horizontally to the left to determine the permissible current. An additional 15% derating is recommended for board thicknesses of $\frac{1}{32}$ inch or less, or for conductors thicker than 0.004 inch (3 oz). The normal ambient temperature surrounding the board plus the allowable temperature rise should not exceed the maximum safe operating temperature of the laminate. For ordinary work copper conductor widths of 0.060 inch are convenient; with high-grade technique (extra cost) conductor widths as small as 0.010 inch can be readily produced.

Conductor spacing requirements are governed by the applied voltage, the maximum altitude, the conductor protective coating used, and the power-source size. The following guide is suggested.*

	Minimum Spacing Between Conductors (inches)		
	Uncoated Boards		Conformal Coated Boards
Voltage Between Conductors	Sea Level to 10 000 Ft	Over 10 000 Ft	All Altitudes
0–30	0.025	0.025	0.010
0–50	0.025	0.025	—
0–150	0.025	—	—
31–50	0.025	0.025	0.015
51–100	0.025	0.060	0.020
101–170	—	0.125	0.030
101–300	—	—	0.030
151–300	0.050	—	0.030
171–250	0.050	0.250	0.030
251–500	—	0.500	—
301–500	0.100	0.500	0.060
Above 500	0.0002 per volt	0.001 per volt	0.00012 per volt

* From MIL-STD-275C.

Preparation of Artwork

Workmanship: In preparing the master artwork for printed circuits, careful workmanship and accuracy are important. When circuits are reproduced by photographic means, much retouching time is saved if care is taken with the original artwork.

Materials: Artwork should be prepared on a dimensionally stable material. Tracing paper and bristol board are now outmoded, and specially treated (toothed) polyethylene terephthalate (Mylar, Cronar) base drafting films are used for most printed-circuit layouts. The layout pattern may be produced using one of the following methods:

1. Hand application of opaque, permanent black ink,
2. Pressure-sensitive tape,
3. Hand-cut stencil made from self-adhesive opaque film, or
4. Preformed self-adhesive layout patterns.

Scale: Artwork should be prepared to a scale that is two to five times oversize. Photographic reduction to final negative size should be possible, however, in one step.

Bends: Avoid the use of sharp corners when laying out the circuit. See Fig. 26.

Holes: The centers of holes to be manually drilled or punched in the circuit board should be indicated by a circle of $\frac{1}{32}$-inch diameter (final size after reduction). See Fig. 27. This feature is not needed on each board if templates or numerically controlled machine tools are used for hole preparation; however, it is still a convenience for checking drawings, master artwork, and photographic negatives.

Registration of Reverse Side: When drawing the second side of a printed-circuit board, corresponding centers should be taken directly from the back of the drawing of the first side.

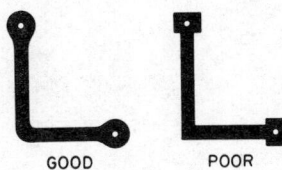

GOOD   POOR

Fig. 26—Proper design of bends for printed-circuit conductors.

Fig. 27—Indication for hole.

Reference Marks: In addition to the illustration of the circuit pattern, the trim line, registration marks, and two scale dimensions at right angles should be shown. Nomenclature, reference designations, operating instructions, and other information may also be added.

Assembly

All components should be inserted on one side of the board if practicable. In the case of boards with the circuit on one side only, the parts should be inserted on the side opposite the circuit. This allows all connections to be soldered simultaneously by dip-soldering.

Dip-soldering consists of applying a flux, usually a rosin–alcohol mixture, to the circuit pattern and then placing the board in contact with molten solder. Slight agitation of the board will insure good fillets around the wire leads. In good present technique, the circuit board with its components assembled (on one side only) has its conductor pattern passed through the crest of a "wave" of molten solder; all junctions are soldered as the board progresses through the wave. The flux, board temperature, solder temperature, and immersion time are interrelated and must be adjusted for best results. Long exposure to hot solder is detrimental to the insulating material and to the adhesive that joins the copper foil to the insulation. For hand dipping, a 5-second dip in a 60/40 tin–lead solder bath maintained at a temperature of 450 degrees Fahrenheit will give satisfactory results.

After solder-dipping, the residual flux should be removed by a suitable solvent. Be sure the solvent is compatible with the materials used in the component parts mounted on the board; solvents frequently dissolve cements or plastics and marking inks, or cause severe stress cracking of plastics.

To secure the advantages of machine assembly:

(**A**) Components should be of similar size and shape, or separate inserting heads will be required for each different shape of item.

(**B**) Components of the same size and shape must be mounted using the same terminal lead spacing at all points.

(**C**) Different values of a part, or even different parts of similar shape and size (if axial-lead style) may be specially sequenced in a lead-taped package for insertion by one programmed insertion head.

(**D**) A few oddly sized or shaped components may be economically inserted by hand after the machine insertion work is completed.

Reference

An excellent reference on microelectronic printed-circuit techniques is J. A. Scarlett, "Printed Circuit Boards for Microelectronics," Van Nostrand Reinhold Co., New York, 1970.

CHAPTER 6

FUNDAMENTALS OF NETWORKS

INDUCTANCE OF SINGLE-LAYER SOLENOIDS*

The approximate value of the low-frequency inductance of a single-layer solenoid is†

$$L = Fn^2d \text{ microhenries}$$

where F=form factor, a function of the ratio d/l (value of F may be read from Fig. 1), n= number of turns, d=diameter of coil (inches) between centers of conductors, and l=length of coil (inches) = n times the distance between centers of adjacent turns.

The equation is based on the assumption of a uniform current sheet, but the correction due to the use of spaced round wires is usually negligible for practical purposes. For higher frequencies, skin effect alters the inductance slightly. This effect is not readily calculated, but is often negligibly small. However, it must be borne in mind that the equation gives approximately the true value of inductance. In contrast, the apparent value is affected by the shunting effect of the distributed capacitance of the coil.

Example: Required, a coil of 100 microhenries inductance, wound on a form 2 inches in diameter by 2 inches winding length. Then $d/l = 1.00$, and $F = 0.0173$ in Fig. 1.

$$n = (L/Fd)^{1/2}$$
$$= [100/(0.0173 \times 2)]^{1/2}$$
$$= 54 \text{ turns.}$$

Reference to Table 1 will assist in choosing a desirable size of wire, allowing for a suitable spacing between turns according to the application of the coil. A slight correction may then be made for the increased diameter (diameter of form, plus two

* Calculation of copper losses in single-layer solenoids is treated in F. E. Terman, *"Radio Engineers Handbook,"* 1st edition, McGraw-Hill Book Company, Inc., New York, N.Y.; 1943: pp. 77–80.

† Equations and Fig. 1 are derived from equations and tables in Bureau of Standards Circular No. C74.

times radius of wire), if this small correction seems justified.

Approximate Equation

For single-layer solenoids of the proportions normally used in radio work, the inductance is

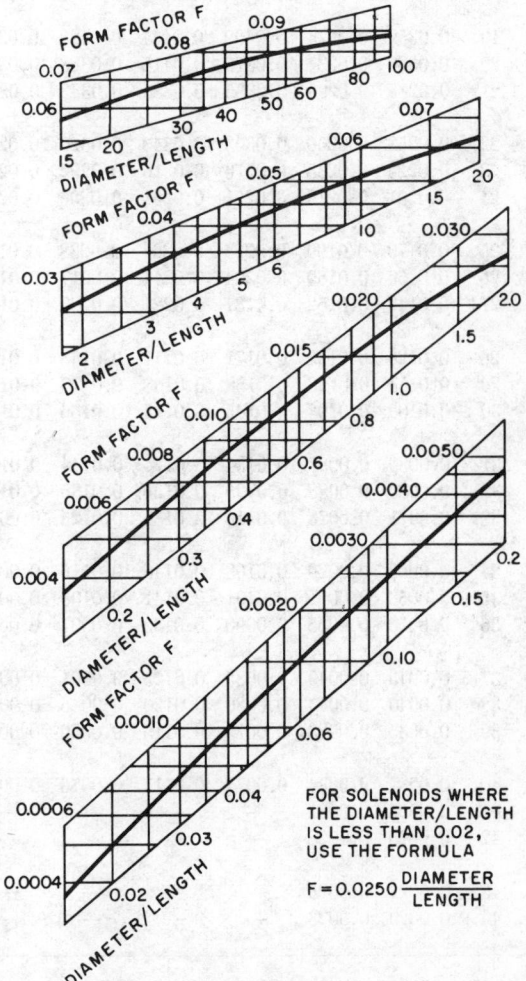

Fig. 1—Inductance of a single-layer solenoid, form factor = F.

6-1

TABLE 1—MAGNET-WIRE DATA.

AWG B & S gauge	Bare nom diam (in.)	Enam nom diam (in.)	SCC* diam (in.)	DCC* diam (in.)	SCE* diam (in.)	SSC* diam (in.)	DSC* diam (in.)	SSE* diam (in.)	Bare min diam (in.)	Bare max diam (in.)	Enameled min diam (in.)	Enameled diam* (in.)
10	0.1019	0.1039	0.1079	0.1129	0.1104	—	—	—	0.1009	0.1029	0.1024	0.1044
11	0.0907	0.0927	0.0957	0.1002	0.0982	—	—	—	0.0898	0.0917	0.0913	0.0932
12	0.0808	0.0827	0.0858	0.0903	0.0882	—	—	—	0.0800	0.0816	0.0814	0.0832
13	0.0720	0.0738	0.0770	0.0815	0.0793	—	—	—	0.0712	0.0727	0.0726	0.0743
14	0.0641	0.0659	0.0691	0.0736	0.0714	—	—	—	0.0634	0.0647	0.0648	0.0664
15	0.0571	0.0588	0.0621	0.0666	0.0643	0.0591	0.0611	0.0613	0.0565	0.0576	0.0578	0.0593
16	0.0508	0.0524	0.0558	0.0603	0.0579	0.0528	0.0548	0.0549	0.0503	0.0513	0.0515	0.0529
17	0.0453	0.0469	0.0503	0.0548	0.0523	0.0473	0.0493	0.0493	0.0448	0.0457	0.0460	0.0473
18	0.0403	0.0418	0.0453	0.0498	0.0472	0.0423	0.0443	0.0442	0.0399	0.0407	0.0410	0.0422
19	0.0359	0.0374	0.0409	0.0454	0.0428	0.0379	0.0399	0.0398	0.0355	0.0363	0.0366	0.0378
20	0.0320	0.0334	0.0370	0.0415	0.0388	0.0340	0.0360	0.0358	0.0316	0.0323	0.0326	0.0338
21	0.0285	0.0299	0.0335	0.0380	0.0353	0.0305	0.0325	0.0323	0.0282	0.0287	0.0292	0.0303
22	0.0253	0.0266	0.0303	0.0343	0.0320	0.0273	0.0293	0.0290	0.0251	0.0256	0.0261	0.0270
23	0.0226	0.0238	0.0276	0.0316	0.0292	0.0246	0.0266	0.0262	0.0223	0.0228	0.0232	0.0242
24	0.0201	0.0213	0.0251	0.0291	0.0266	0.0221	0.0241	0.0236	0.0199	0.0203	0.0208	0.0216
25	0.0179	0.0190	0.0224	0.0264	0.0238	0.0199	0.0219	0.0213	0.0177	0.0181	0.0186	0.0193
26	0.0159	0.0169	0.0204	0.0244	0.0217	0.0179	0.0199	0.0192	0.0158	0.0161	0.0166	0.0172
27	0.0142	0.0152	0.0187	0.0227	0.0200	0.0162	0.0182	0.0175	0.0141	0.0144	0.0149	0.0155
28	0.0126	0.0135	0.0171	0.0211	0.0183	0.0146	0.0166	0.0158	0.0125	0.0128	0.0132	0.0138
29	0.0113	0.0122	0.0158	0.0198	0.0170	0.0133	0.0153	0.0145	0.0112	0.0114	0.0119	0.0125
30	0.0100	0.0108	0.0145	0.0185	0.0156	0.0120	0.0140	0.0131	0.0099	0.0101	0.0105	0.0111
31	0.0089	0.0097	0.0134	0.0174	0.0144	0.0109	0.0129	0.0119	0.0088	0.0090	0.0094	0.0099
32	0.0080	0.0088	0.0125	0.0165	0.0135	0.0100	0.0120	0.0110	0.0079	0.0081	0.0085	0.0090
33	0.0071	0.0078	0.0116	0.0156	0.0125	0.0091	0.0111	0.0100	0.0070	0.0072	0.0075	0.0080
34	0.0063	0.0069	0.0108	0.0148	0.0116	0.0083	0.0103	0.0091	0.0062	0.0064	0.0067	0.0071
35	0.0056	0.0061	0.0101	0.0141	0.0108	0.0076	0.0096	0.0083	0.0055	0.0057	0.0059	0.0063
36	0.0050	0.0055	0.0090	0.0130	0.0097	0.0070	0.0090	0.0077	0.0049	0.0051	0.0053	0.0057
37	0.0045	0.0049	0.0085	0.0125	0.0091	0.0065	0.0085	0.0071	0.0044	0.0046	0.0047	0.0051
38	0.0040	0.0044	0.0080	0.0120	0.0086	0.0060	0.0080	0.0066	0.0039	0.0041	0.0042	0.0046
39	0.0035	0.0038	0.0075	0.0115	0.0080	0.0055	0.0075	0.0060	0.0034	0.0036	0.0036	0.0040
40	0.0031	0.0034	0.0071	0.0111	0.0076	0.0051	0.0071	0.0056	0.0030	0.0032	0.0032	0.0036
41	0.0028	0.0031	—	—	—	—	—	—	0.0027	0.0029	0.0029	0.0032
42	0.0025	0.0028	—	—	—	—	—	—	0.0024	0.0026	0.0026	0.0029
43	0.0022	0.0025	—	—	—	—	—	—	0.0021	0.0023	0.0023	0.0026
44	0.0020	0.0023	—	—	—	—	—	—	0.0019	0.0021	0.0021	0.0024

* Nominal bare diameter plus maximum additions.

For additional data on copper wire, see Chapters 4 and 13.

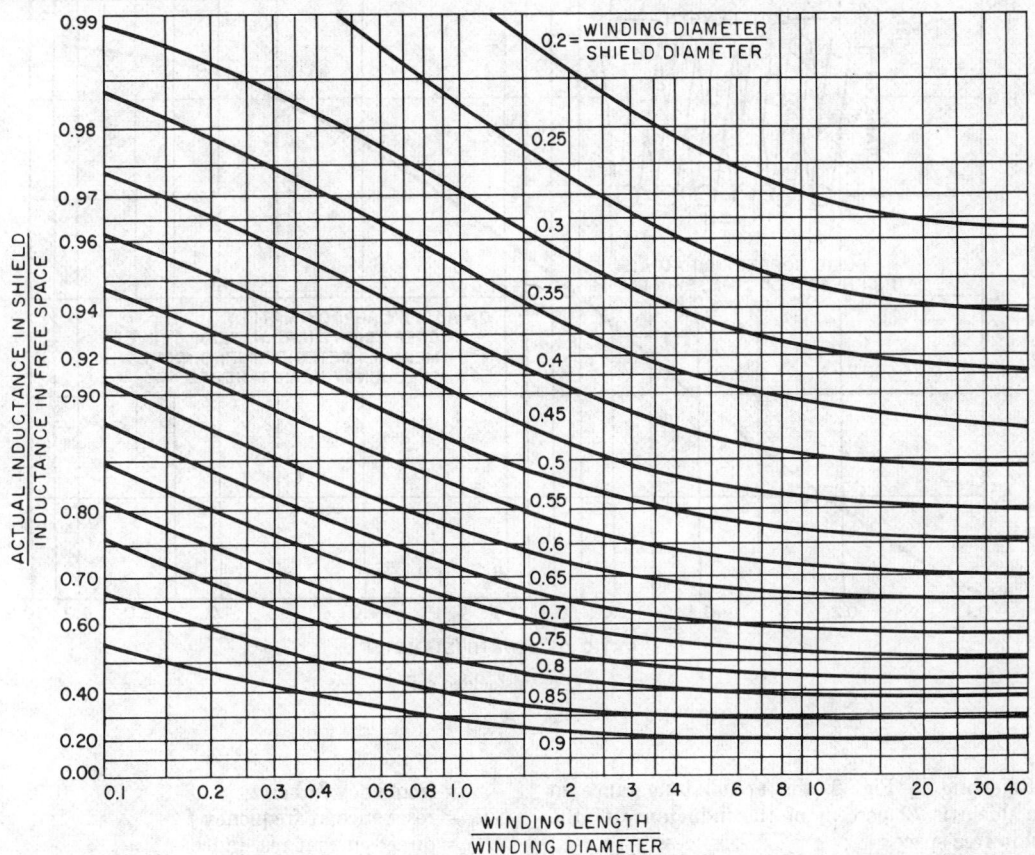

Fig. 2—Curves for determination of inductance decrease when a solenoid is shielded. *By permission of RCA, copyright proprietor.*

given to an accuracy of about 1 percent by

$$L = n^2[r^2/(9r+10l)] \text{ microhenries}$$

where $r = d/2$.

General Remarks

In the use of various charts, tables, and calculators for designing inductors, the following relationships are useful in extending the range of the devices. They apply to coils of any type or design.

(A) If all dimensions are held constant, inductance is proportional to n^2.

(B) If the proportions of the coil remain unchanged, then for a given number of turns the inductance is proportional to the dimensions of the coil. A coil with all dimensions m times those of a given coil (having the same number of turns) has m times the inductance of the given coil. That is, inductance has the dimensions of length.

Decrease of Solenoid Inductance by Shielding*

When a solenoid is enclosed in a cylindrical shield, the inductance is reduced by a factor given in Fig. 2. This effect has been evaluated by considering the shield to be a short-circuited single-turn secondary. The curves in Fig. 2 are reasonably accurate provided the clearance between each end of the coil winding and the corresponding end of the shield is at least equal to the radius of the coil. For square shield cans, take the equivalent shield diameter (for Fig. 2) as being 1.2 times the width of one side of the square.

Example: Let the coil winding length be 1.5 inches and its diameter 0.75 inch, while the shield diameter is 1.25 inches. What is the reduction of inductance due to the shield? The proportions are

(winding length)/(winding diameter) = 2.0

(winding diameter)/(shield diameter) = 0.6.

* RCA Application Note No. 48; 12 June 1935.

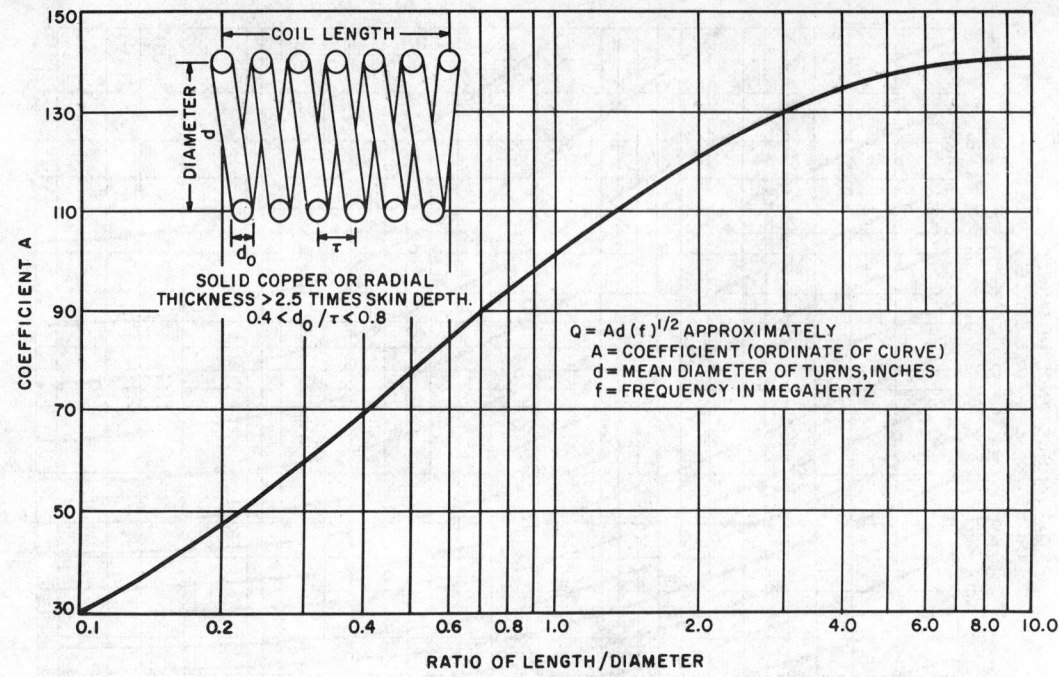

Fig. 3— Q of unshielded coil.

Referring to Fig. 2, the actual inductance in the shield is 72 percent of the inductance of the coil in free space.

Q of Unshielded Solenoid

Figure 3 can be used to obtain the unloaded Q of an unshielded solenoid.

REACTANCE CHARTS

Figures 4, 5, and 6 give the relationships of capacitance, inductance, reactance, and frequency. Any one value may be determined in terms of two others by use of a straightedge laid across the correct chart for the frequency under consideration.

Example: Given a capacitance of 0.001 microfarad, find the reactance at 50 kilohertz and inductance required to resonate. Place a straightedge through these values and read the intersections on the other scales, giving approximately 3200 ohms and 11 millihenries. See Fig. 5.

SKIN EFFECT

Symbols

A = correction coefficient
D = diameter of conductor in inches

f = frequency in hertz
R_{ac} = resistance at frequency f
R_{dc} = direct-current resistance
R_{sq} = resistance per square
T = thickness of tubular conductor in inches
T_1 = depth of penetration of current
δ = skin depth
λ = free-space wavelength in meters
μ_r = relative permeability of conductor material ($\mu_r = 1$ for copper and other nonmagnetic materials)
ρ = resistivity of conductor material at any temperature
ρ_c = resistivity of copper at 20°C = 1.724 microhms-centimeter.

Skin Depth

The skin depth is that distance below the surface of a conductor where the current density has diminished to $1/e$ of its value at the surface. The thickness of the conductor is assumed to be several (perhaps at least three) times the skin depth. Imagine the conductor replaced by a cylindrical shell of the same surface shape but of thickness equal to the skin depth, with uniform current density equal to that which exists at the surface of the actual conductor. Then the total current in

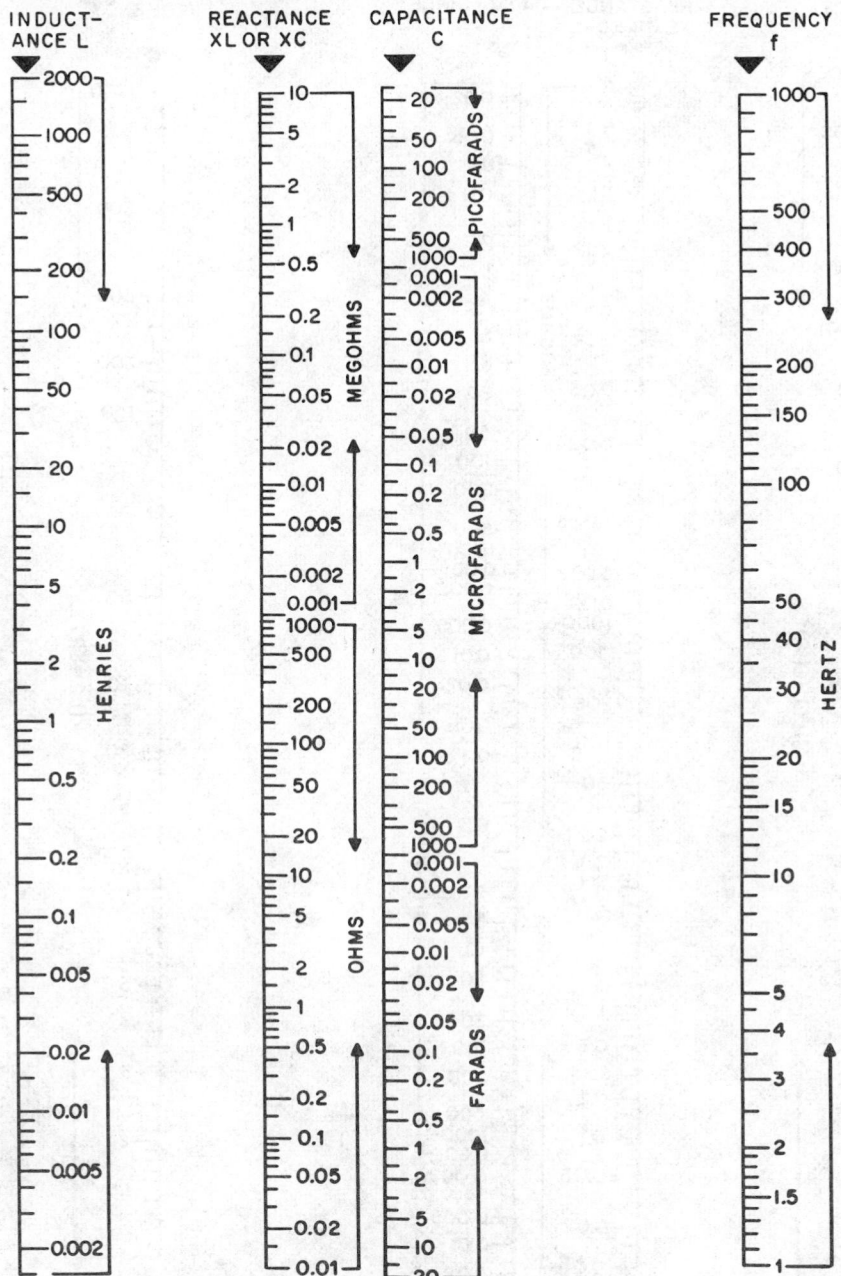

Fig. 4—Chart covering 1 hertz to 1000 hertz.

the shell and its resistance are equal to the corresponding values in the actual conductor.

The skin depth and the resistance per square (of any size), in meter–kilogram–second (rationalized) units, are

$$\delta = (\lambda/\pi\sigma\mu c)^{1/2} \text{ meter}$$

$$R_{sq} = 1/\delta\sigma \text{ ohm}$$

where

c = velocity of light *in vacuo*

$= 2.998 \times 10^8$ meter/second

$\mu = 4\pi \times 10^{-7} \mu_r$ henry/meter

$1/\sigma = 1.724 \times 10^{-8} \rho/\rho_c$ ohm-meter.

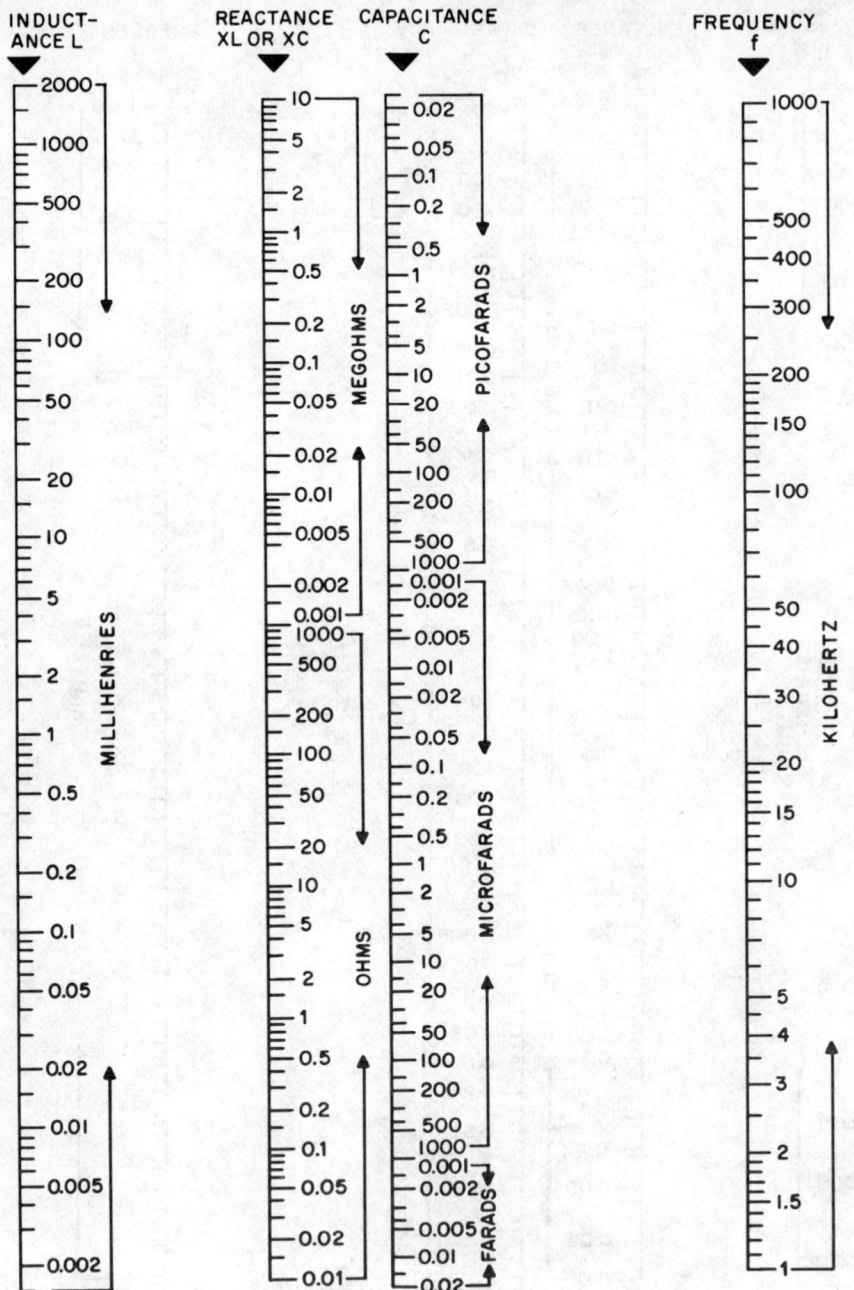

Fig. 5—Chart covering 1 kilohertz to 1000 kilohertz.

For numerical computations

$$\delta = (3.82 \times 10^{-4} \lambda^{1/2}) k_1$$

$$= (6.61/f^{1/2}) k_1 \text{ centimeter}$$

$$\delta = (1.50 \times 10^{-4} \lambda^{1/2}) k_1$$

$$= (2.60/f^{1/2}) k_1 \text{ inch}$$

$$\delta_m = (2.60/f_{mc}^{1/2}) k_1 \text{ mil}$$

$$R_{sq} = (4.52 \times 10^{-3}/\lambda^{1/2}) k_2$$

$$= (2.61 \times 10^{-7} f^{1/2}) k_2 \text{ ohm}$$

where

$$k_1 = [(1/\mu_r) \rho/\rho_c]^{1/2}$$

$$k_2 = (\mu_r \rho/\rho_c)^{1/2}$$

$k_1, k_2 =$ unity for copper.

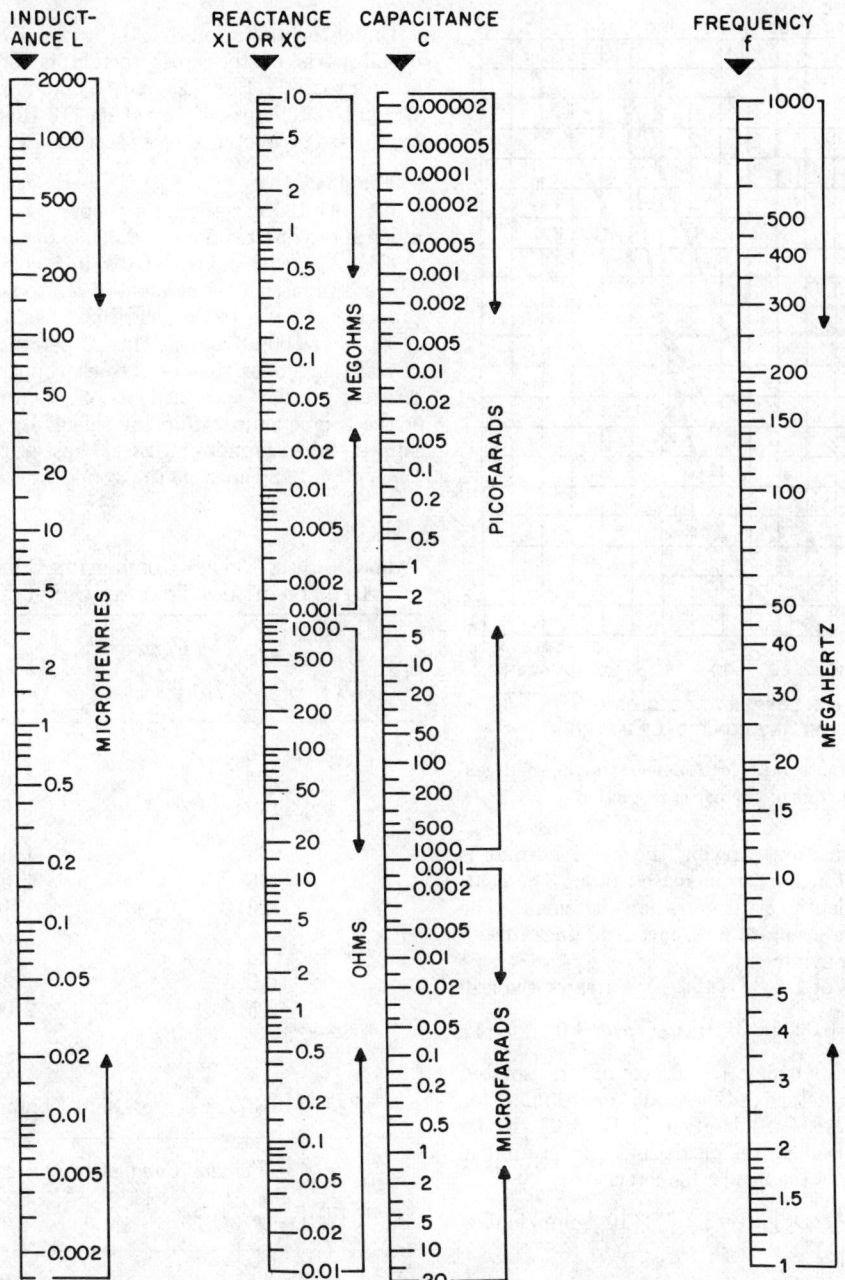

Fig. 6—Chart covering 1 megahertz to 1000 megahertz.

Example: What is the resistance/foot of a cylindrical copper conductor of diameter D inches?

$$R = (12/\pi D) R_{sq}$$

$$= (12/\pi D) \times 2.61 \times 10^{-7} (f^{1/2})$$

$$= 0.996 \times 10^{-6} (f^{1/2})/D \text{ ohm/foot.}$$

If $D = 1.00$ inch and $f = 100 \times 10^6$ hertz, then $R = 0.996 \times 10^{-6} \times 10^4 \approx 1 \times 10^{-2}$ ohm/foot.

General Considerations

Figure 7 shows the relationship of R_{ac}/R_{dc} versus $D(f^{1/2})$ for copper, or versus $D(f^{1/2})(\mu_r \rho_c/\rho)^{1/2}$ for any conductor material, for an isolated straight solid conductor of circular cross section. Negligible error in the equations for R_{ac} results when the conductor is spaced at least $10D$ from adjacent conductors. When the spacing between axes of

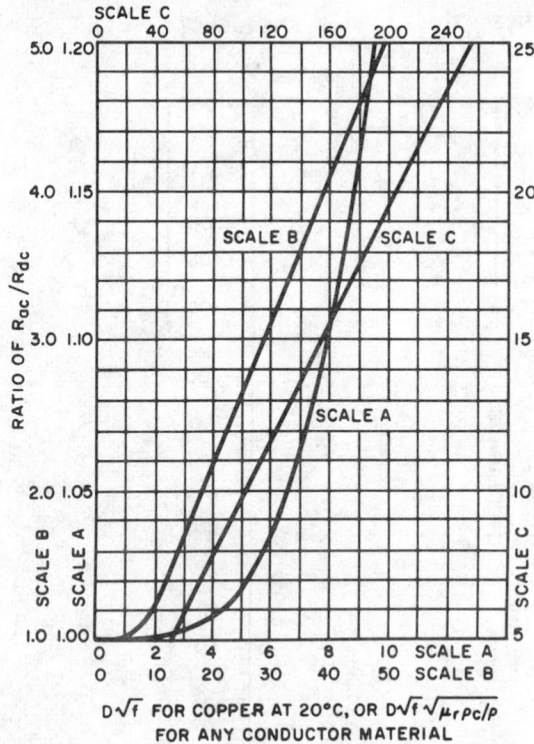

Fig. 7—Resistance ratio for isolated straight solid conductors of circular cross section.

parallel conductors carrying the same current is $4D$, the resistance R_{ac} is increased about 3 percent, when the depth of penetration is small. The equations are accurate for concentric lines due to their circular symmetry.

For values of $D(f^{1/2})(\mu_r\rho_c/\rho)^{1/2}$ greater than 40

$$R_{ac}/R_{dc}=0.0960D(f^{1/2})(\mu_r\rho_c/\rho)^{1/2}+0.26. \quad (1)$$

The high-frequency resistance of an isolated straight conductor (either solid or tubular for $T<D/8$ or $T_1<D/8$) is given in Eq. (2). If the current flow is along the inside surface of a tubular conductor, D is the inside diameter.

$$R_{ac}=A[(f^{1/2})/D][\mu_r(\rho/\rho_c)]^{1/2}\times10^{-6}\ \text{ohm/foot.}$$
$$(2)$$

The values of the correction coefficient A for solid conductors and for tubular conductors are given in Table 2.

The value of $T(f^{1/2})(\mu_r\rho_c/\rho)^{1/2}$ that just makes $A=1$ indicates the penetration of the currents below the surface of the conductor. Thus, approximately

$$T_1=[3.5/(f^{1/2})](\rho/\mu_r\rho_c)^{1/2}\ \text{inch.} \quad (3)$$

When $T_1<D/8$ the value of R_{ac} as given by Eq. (2) (but not the value of R_{ac}/R_{dc} in Table 2, "Tubular Conductors") is correct for any value $T\ge T_1$.

Under the limitation that the radius of curvature of all parts of the cross section is appreciably greater than T_1, Eqs. (2) and (3) hold for isolated straight conductors of any shape. In this case the term $D=$ (perimeter of cross section)$/\pi$.

Examples:
(**A**) At 100 megahertz, a copper conductor has a depth of penetration $T_1=0.00035$ inch.

(**B**) A steel shield with 0.005-inch copper plate, which is practically equivalent in R_{ac} to an isolated copper conductor 0.005-inch thick, has a value of $A=1.23$ at 200 kilohertz. This 23-percent increase in resistance over that of a thick copper sheet is satisfactorily low as regards its effect on the losses of the components within the shield. By comparison, a thick aluminum sheet has a resistance $(\rho/\rho_c)^{1/2}=1.28$ times that of copper.

TABLE 2—SKIN-EFFECT CORRECTION COEFFICIENT *A* FOR SOLID AND TUBULAR CONDUCTORS.

Solid Conductors	
$D(f^{1/2})[\mu_r(\rho_c/\rho)]^{1/2}$	A
>370	1.000
220	1.005
160	1.010
98	1.02
48	1.05
26	1.10
13	1.20
9.6	1.30
5.3	2.00
<3.0	$R_{ac}\approx R_{dc}$

$$R_{dc}=(10.37/D^2)(\rho/\rho_c)\times10^{-6}\ \text{ohm/foot}$$

Tubular Conductors		
$T(f^{1/2})[\mu_r(\rho_c/\rho)]^{1/2}$	A	R_{ac}/R_{dc}
$=B$ where $B>3.5$	1.00	$0.384B$
3.5	1.00	1.35
3.15	1.01	1.23
2.85	1.05	1.15
2.60	1.10	1.10
2.29	1.20	1.06
2.08	1.30	1.04
1.77	1.50	1.02
1.31	2.00	1.00
$=B$ where $B<1.3$	$2.60/B$	1.00

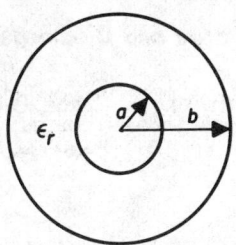

Fig. 8—Coaxial cylindrical capacitor.

EQUATIONS FOR SIMPLE R, L, AND C NETWORKS*

Self-Inductance of Circular Ring of Round Wire at Radio Frequencies, for Nonmagnetic Materials

$L = (a/100)$

$$\times [7.353 \log_{10}(16a/d) - 6.386] \text{ microhenrys}$$

where $a =$ mean radius of ring in inches, $d =$ diameter of wire in inches, and $a/d > 2.5$.

Capacitance

For Parallel-Plate Capacitor:

$C = 0.0885 \epsilon_r [(N-1) A]/t$

$\quad = 0.225 \epsilon_r [(N-1) A''/t''] \text{ picofarads}$

where

$\quad A =$ area of one side of one plate in square centimeters
$\quad A'' =$ area in square inches
$\quad N =$ number of plates

$t =$ thickness of dielectric in centimeters
$t'' =$ thickness in inches
$\epsilon_r =$ dielectric constant relative to air.

This equation neglects "fringing" at the edges of the plates.

For Coaxial Cylindrical Capacitor (Fig. 8): Per unit axial length

$C = 2\pi \epsilon_r \epsilon_v / [\log_e (b/a)]$

$\quad = \{ (5 \times 10^6 \epsilon_r) / [c^2 \log_e (b/a)] \} \text{ farads/meter}$

where

$\quad c =$ velocity of light in vacuo, meters per second
$\quad\quad = 2.998 \times 10^8$
$\quad \epsilon_r =$ dielectric constant relative to air
$\quad \epsilon_v =$ permittivity of free space in farads/meter
$\quad\quad = 8.85 \times 10^{-12}$

$C = 0.2416 \epsilon_r / [\log_{10} (b/a)] \text{ picofarads/centimeter}$
$\quad = 0.614 \epsilon_r / [\log_{10} (b/a)] \text{ picofarads/inch}$
$\quad = 7.36 \epsilon_r / [\log_{10} (b/a)] \text{ picofarads/foot.}$

When $1.0 < (b/a) < 1.4$, then with accuracy of 1 percent or better

$$C = 8.50 \epsilon_r \frac{(b/a) + 1}{(b/a) - 1} \text{ picofarads/foot.}$$

T–π or Y–Δ Transformation

The two networks (Fig. 9) are equivalent, as far as conditions at the terminals are concerned, provided the listed equations are satisfied (either the impedance equations or the admittance equations may be used)

$$Y_1 = 1/Z_1 \quad Y_c = 1/Z_c, \text{ etc.}$$

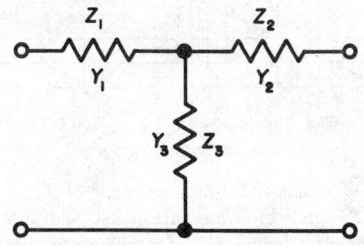

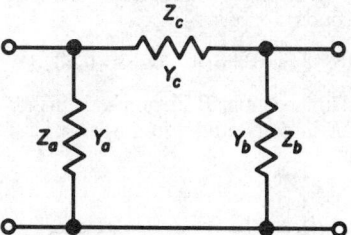

Impedance Equations:

$Z_c = (Z_1 Z_2 + Z_1 Z_3 + Z_2 Z_3)/Z_3$
$Z_a = (Z_1 Z_2 + Z_1 Z_3 + Z_2 Z_3)/Z_2$
$Z_b = (Z_1 Z_2 + Z_1 Z_3 + Z_2 Z_3)/Z_1$

$Z_1 = Z_a Z_c/(Z_a + Z_b + Z_c)$
$Z_2 = Z_b Z_c/(Z_a + Z_b + Z_c)$
$Z_3 = Z_a Z_b/(Z_a + Z_b + Z_c)$

Admittance Equations:

$Y_c = Y_1 Y_2/(Y_1 + Y_2 + Y_3)$
$Y_a = Y_1 Y_3/(Y_1 + Y_2 + Y_3)$
$Y_b = Y_2 Y_3/(Y_1 + Y_2 + Y_3)$

$Y_1 = (Y_a Y_b + Y_a Y_c + Y_b Y_c)/Y_b$
$Y_2 = (Y_a Y_b + Y_a Y_c + Y_b Y_c)/Y_a$
$Y_3 = (Y_a Y_b + Y_a Y_c + Y_b Y_c)/Y_c$

Fig. 9—T or Y network (left) and π or Δ network.

* Many equations for computing capacitance, inductance, and mutual inductance will be found in Bureau of Standards Circular No. C74, obtainable from the Superintendent of Documents, Government Printing Office, Washington, D.C. 20402.

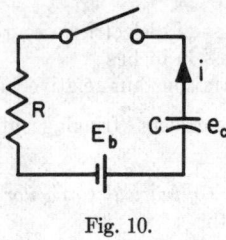

Fig. 10.

TRANSIENTS—ELEMENTARY CASES

The complete transient in a linear network is, by the principle of superposition, the sum of the individual transients due to the store of energy in each inductor and capacitor and to each external source of energy connected to the network. To this is added the steady-state condition due to each external source of energy. The transient may be computed as starting from any arbitrary time $t=0$ when the initial conditions of the energy of the network are known.

Time Constant (Designated T)

The time constant of the discharge of a capacitor through a resistor is the time $t_2 - t_1$ required for the voltage or current to decay to $1/\epsilon$ of its value at time t_1. For the charge of a capacitor the same definition applies, the voltage "decaying" toward its steady-state value. The time constant of discharge or charge of the current in an inductor through a resistor follows an analogous definition.

Energy stored in a capacitor $= \frac{1}{2} C E^2$ joules (watt-seconds)

Energy stored in an inductor $= \frac{1}{2} L I^2$ joules (watt-seconds)

$\epsilon = 2.718 \quad 1/\epsilon = 0.3679 \quad \log_{10}\epsilon = 0.4343$

T and t in seconds, R in ohms, L in henries, C in farads, E in volts, and I in amperes.

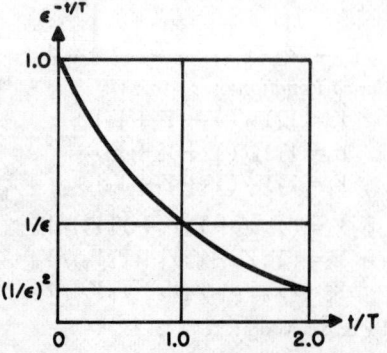

Fig. 11—Capacitor discharge.

Capacitor Charge and Discharge

Closing of switch (Fig. 10) occurs at time $t=0$. Initial conditions (at $t=0$): Battery $= E_b$; $e_c = E_0$. Steady state (at $t=\infty$): $i=0$; $e_c = E_b$.

Transient:

$$i = [(E_b - E_0)/R] \exp(-t/RC)$$

$$= I_0 \exp(-t/RC)$$

$$\log_{10}(i/I_0) = -(0.4343/RC)t$$

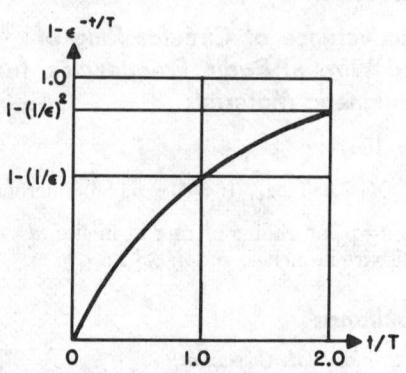

Fig. 12—Capacitor charge.

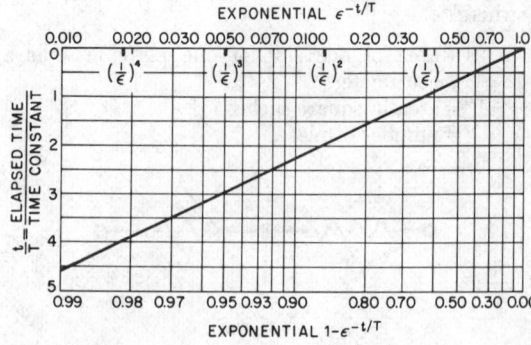

Fig. 13—Exponential functions $\exp(-t/T)$ and $1 - \exp(-t/T)$ applied to transients in R–C and L–R circuits. Use exponential $\exp(-t/T)$ for charge or discharge of capacitor or discharge of inductor:

(current at time t)/(initial current).

Discharge of capacitor:

(voltage at time t)/(initial voltage).

Use exponential $1 - \exp(-t/T)$ for charge of capacitor:

(voltage at time t)/(battery or final voltage).

Charge of inductor:

(current at time t)/(final current).

$$e_c = E_0 + C^{-1} \int_0^t i\, dt$$

$$= E_0 \exp(-t/RC) + E_b[1 - \exp(-t/RC)].$$

Time constant:

$$T = RC.$$

Figure 11 shows current:

$$i/I_0 = \exp(-t/T).$$

Figure 11 shows discharge (for $E_b = 0$):

$$e_c/E_0 = \exp(-t/T).$$

Figure 12 shows charge (for $E_0 = 0$):

$$e_c/E_b = 1 - \exp(-t/T).$$

These curves are plotted for a wider range in Fig. 13.

Two Capacitors

Closing of switch (Fig. 14) occurs at time $t = 0$.
Initial conditions (at $t = 0$):

$$e_1 = E_1; \quad e_2 = E_2.$$

Steady state (at $t = \infty$):

$$e_1 = E_f; \quad e_2 = -E_f; \quad i = 0$$
$$E_f = (E_1 C_1 - E_2 C_2)/(C_1 + C_2)$$
$$C' = C_1 C_2/(C_1 + C_2).$$

Transient:

$$i = [(E_1 + E_2)/R] \exp(-t/RC')$$
$$e_1 = E_f + (E_1 - E_f) \exp(-t/RC')$$
$$= E_1 - (E_1 + E_2)(C'/C_1)[1 - \exp(-t/RC')]$$
$$e_2 = -E_f + (E_2 + E_f) \exp(-t/RC')$$
$$= E_2 - (E_1 + E_2)(C'/C_2)[1 - \exp(-t/RC')].$$

Original energy $= \frac{1}{2}(C_1 E_1^2 + C_2 E_2^2)$ joule

Final energy $= \frac{1}{2}(C_1 + C_2) E_f^2$ joule

Loss of energy $= \int_0^\infty i^2 R\, dt$

$$= \frac{1}{2} C'(E_1 + E_2)^2 \text{ joule}.$$

(Loss is independent of the value of R.)

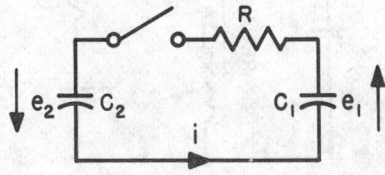

Fig. 14.

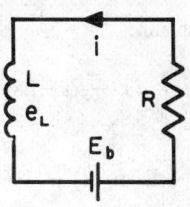

Fig. 15.

Inductor Charge and Discharge (Fig. 15)

Initial conditions (at $t = 0$):

$$\text{Battery} = E_b; \quad i = I_0.$$

Steady state (at $t = \infty$):

$$i = I_f = E_b/R.$$

Transient, plus steady state:

$$i = I_f[1 - \exp(-Rt/L)] + I_0 \exp(-Rt/L)$$
$$e_L = -L\, di/dt$$
$$= -(E_b - RI_0) \exp(-Rt/L).$$

Time constant:

$$T = L/R.$$

Figure 11 shows discharge (for $E_b = 0$):

$$i/I_0 = \exp(-t/T).$$

Figure 12 shows charge (for $I_0 = 0$):

$$i/I_f = [1 - \exp(-t/T)].$$

These curves are plotted for a wider range in Fig. 13.

Series R-L-C Circuit Charge and Discharge (Fig. 16)

Initial conditions (at $t = 0$):

$$\text{Battery} = E_b; \quad e_c = E_0; \quad i = I_0.$$

Steady state (at $t = \infty$):

$$i = 0; \quad e_c = E_b.$$

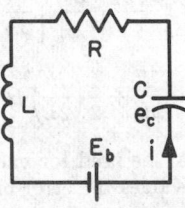

Fig. 16.

Differential equation:

$$E_b - E_0 - C^{-1} \int_0^t i \, dt - Ri - L(di/dt) = 0$$

when

$$L(d^2i/dt^2) + R(di/dt) + (i/C) = 0.$$

Solution of equation:

$$i = \exp(-Rt/2L)\left[\frac{2(E_b - E_0) - RI_0}{R(D^{1/2})} \sinh(Rt/2L)\right.$$

$$\left. \times (D^{1/2}) + I_0 \cosh(Rt/2L)(D^{1/2})\right]$$

where $D = 1 - (4L/R^2C)$.

Case 1: When L/R^2C is small

$$i = (1 - 2A - 2A^2)^{-1}\left\{\left[\frac{E_b - E_0}{R} - I_0(A + A^2)\right]\right.$$

$$\times \exp\left(-\frac{t}{RC}(1 + A + 2A^2)\right)$$

$$+ \left[I_0(1 - A - A^2) - \frac{E_b - E_0}{R}\right]$$

$$\left. \times \exp\left(-\frac{Rt}{L}(1 - A - A^2)\right)\right\}$$

where $A = L/R^2C$.

For practical purposes, the terms A^2 can be neglected when $A < 0.1$. The terms A may be neglected when $A < 0.01$.

Case 2: When $4L/R^2C < 1$ for which $D^{1/2}$ is real

$$i = \frac{\exp(-Rt/2L)}{D^{1/2}}\left\{\left[\frac{E_b - E_0}{R} - \tfrac{1}{2}I_0(1 - D^{1/2})\right]\right.$$

$$\times \exp\left(\frac{Rt}{2L}D^{1/2}\right)$$

$$\left. + \left[\tfrac{1}{2}I_0(1 + D^{1/2}) - \frac{E_b - E_0}{R}\right]\exp\left(-\frac{Rt}{2L}D^{1/2}\right)\right\}.$$

Case 3: When D is a small positive or negative quantity

$$i = \exp(-Rt/2L)\left\{\frac{2(E_b - E_0)}{R}\left[\frac{Rt}{2L} + \frac{1}{6}\left(\frac{Rt}{2L}\right)^3 D\right]\right.$$

$$\left. + I_0\left[1 - \frac{Rt}{2L} + \frac{1}{2}\left(\frac{Rt}{2L}\right)^2 D - \frac{1}{6}\left(\frac{Rt}{2L}\right)^3 D\right]\right\}.$$

This equation may be used for values of D up to ± 0.25, at which values the error in the computed current i is approximately 1 percent of I_0 or of $(E_b - E_0)/R$.

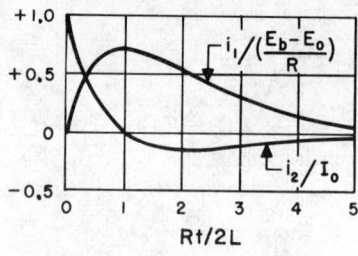

Fig. 17—Transients for $4L/R^2C = 1$.

Case 3A: When $4L/R^2C = 1$ for which $D = 0$, the equation reduces to

$$i = \exp(-Rt/2L)\left[\frac{E_b - E_0}{R}\frac{Rt}{L} + I_0\left(1 - \frac{Rt}{2L}\right)\right]$$

or $i = i_1 + i_2$, plotted in Fig. 17. For practical purposes, this equation may be used when $4L/R^2C = 1 \pm 0.05$ with errors of 1 percent or less.

Case 4: When $4L/R^2C > 1$ for which $D^{1/2}$ is imaginary

$$i = \exp(-Rt/2L)\left[\left(\frac{E_b - E_0}{\omega_0 L} - \frac{RI_0}{2\omega_0 L}\right)\right.$$

$$\left. \times \sin\omega_0 t + I_0 \cos\omega_0 t\right]$$

$$= I_m \exp(-Rt/2L)\sin(\omega_0 t + \psi)$$

where

$$\omega_0 = \left[(LC)^{-1} - (R^2/4L^2)\right]^{1/2}$$

$$I_m = (\omega_0 L)^{-1}\{[E_b - E_0 - \tfrac{1}{2}(RI_0)]^2 + \omega_0^2 L^2 I_0^2\}^{1/2}$$

$$\psi = \tan^{-1}\{\omega_0 L I_0 / [E_b - E_0 - \tfrac{1}{2}(RI_0)]\}.$$

The envelope of the voltage wave across the inductor is

$$\pm \exp(-Rt/2L)[\omega_0(LC)^{1/2}]^{-1}$$

$$\times \{[E_b - E_0 - \tfrac{1}{2}(RI_0)]^2 + \omega_0^2 L^2 I_0^2\}^{1/2}.$$

Example: Relay with transient-suppressing capacitor (Fig. 18). The switch is closed until time $t = 0$, then opened.

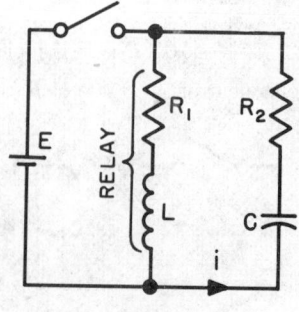

Fig. 18.

Let $L=0.10$ henry, $R_1=100$ ohms, and $E=10$ volts.

Suppose we choose $C=10^{-6}$ farad, and $R_2=100$ ohms.

Then $R=200$ ohms, $I_0=0.10$ ampere, $E_0=10$ volts, $\omega_0=3\times10^3$, and $f_0=480$ hertz.

Maximum peak voltage across L (envelope at $t=0$) is approximately 30 volts. Time constant of decay of envelope is 0.001 second.

Nonoscillating Condition: It is preferable that the circuit be just nonoscillating (Case 3A) and that it present a pure resistance at the switch terminals for any frequency.

$$R_2=R_1=R/2=100 \text{ ohms}.$$

$$4L/R^2C=1$$

$$C=10^{-5} \text{ farad}=10 \text{ microfarads}.$$

At the instant of opening the switch, the voltage across the parallel circuit is $E_0-R_2I_0=0$.

Series R-L-C Circuit with Sinusoidal Applied Voltage

By the principle of superposition, the transient and steady-state conditions are the same for the actual circuit and the equivalent circuit shown in Fig. 19, the closing of the switch occurring at time $t=0$. In the equivalent circuit, the steady state is due to the source e acting continuously from time $t=-\infty$, while the transient is due to short-circuiting the source $-e$ at time $t=0$.

Source:

$$e=E\sin(\omega t+\alpha).$$

Steady state:

$$i=(E/Z)\sin(\omega t+\alpha-\phi)$$

where

$$Z=\{R^2+[\omega L-(1/\omega C)]^2\}^{1/2}$$

$$\tan\phi=(\omega^2LC-1)/\omega CR.$$

The transient is found by determining current $i=I_0$ and capacitor voltage $e_c=E_0$ at time $t=0$, due to the source $-e$. These values of I_0 and E_0 are then substituted in the equations of Case 1, 2, 3, or 4, above, according to the values of R, L, and C.

At time $t=0$, due to the source $-e$

$$i=I_0=-(E/Z)\sin(\alpha-\phi)$$

$$e_c=E_0=(E/\omega CZ)\cos(\alpha-\phi).$$

This form of analysis may be used for any periodic applied voltage e. The steady-state current and the capacitor voltage for an applied voltage

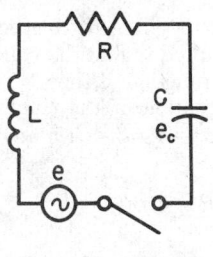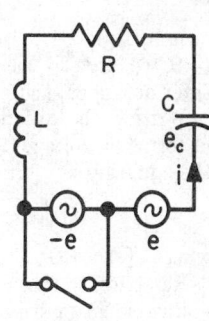

Fig. 19—Actual circuit (left) and equivalent circuit.

$-e$ are determined, the periodic voltage being resolved into its harmonic components for this purpose, if necessary. Then the instantaneous values $i=I_0$ and $e_c=E_0$ at the time of closing the switch are easily found, from which the transient is determined. It is evident, from this method of analysis, that the waveform of the transient need bear no relationship to that of the applied voltage, depending only on the constants of the circuit and the hypothetical initial conditions I_0 and E_0.

TRANSIENTS—OPERATIONAL CALCULUS AND LAPLACE TRANSFORMS

Among the various methods of operational calculus used to solve transient problems, one of the most efficient makes use of the Laplace transform.

If we have a function $v=f(t)$, then by definition the Laplace transform is $\mathcal{L}[f(t)]=F(p)$, where

$$F(p)=\int_0^\infty \exp(-pt)f(t)\ dt. \qquad (4)$$

The inverse transform of $F(p)$ is $f(t)$. Most of the mathematical functions encountered in practical work fall in the class for which Laplace transforms exist. Transforms of functions are given in Chapter 46.

In the following, an abbreviated symbol such as $\mathcal{L}[i]$ is used instead of $\mathcal{L}[i(t)]$ to indicate the Laplace transform of the function $i(t)$.

The electrical (or other) system for which a solution of the differential equation is required, is considered only in the time domain $t\geq0$. Any currents or voltages existing at $t=0$, before the driving force is applied, constitute initial conditions. Driving force is assumed to be 0 when $t<0$.

Example

Take the circuit of Fig. 20, in which the switch is closed at time $t=0$. Before the closing of the switch, suppose the capacitor is charged; then at $t=0$, we have $v=V_0$. It is required to find the voltage v across capacitor C as a function of time.

Writing the differential equation of the circuit in terms of voltage, and since $i=dq/dt=C(dv/dt)$, the equation is

$$e(t)=v+Ri=v+RC(dv/dt) \qquad (5)$$

where $e(t)=E_b$.

Referring to the table of transforms, the applied voltage is E_b multiplied by unit step, or $E_b S_{-1}(t)$; the transform for this is E_b/p. The transform of v is $\mathcal{L}[v]$. That of $RC(dv/dt)$ is $RC[p\mathcal{L}[v]-v(0)]$, where $v(0)=V_0=$ value of v at $t=0$. Then the transform of (5) is

$$E_b/p=\mathcal{L}[v]+RC[p\mathcal{L}[v]-V_0].$$

Rearranging, and resolving into partial fractions

$$\mathcal{L}[v]=\frac{E_b}{p(1+RCp)}+\frac{RCV_0}{1+RCp}$$

$$=E_b[p^{-1}-(p+1/RC)^{-1}]+\frac{V_0}{p+1/RC}. \qquad (6)$$

Now we must determine the equation that would transform into (6). The inverse transform of $\mathcal{L}[v]$ is v, and those of the terms on the right-hand side are found in the table of transforms. Then, in the time domain $t \geq 0$

$$v=E_b[1-\exp(-t/RC)]+V_0\exp(-t/RC). \qquad (7)$$

This solution is also well known by classical methods. However, the advantages of the Laplace transform method become more and more apparent in reducing the labor of solution as the equations become more involved.

Circuit Response Related to Unit Impulse

Unit impulse (see Laplace transforms) has the dimensions of time^{-1}. For example, suppose a capacitor of 1 microfarad is suddenly connected to a battery of 100 volts, with the circuit inductance and resistance negligibly small. Then the current is 10^{-4} coulomb multiplied by unit impulse.

The general transformed equation of a circuit or system may be written

$$\mathcal{L}[i]=\phi(p)\mathcal{L}[e]+\psi(p). \qquad (8)$$

Here $\mathcal{L}[i]$ is the transform of the required current (or other quantity) and $\mathcal{L}[e]$ is the transform of the applied voltage or driving force $e(t)$. The transform of the initial conditions, at $t=0$, is included in $\psi(p)$.

First considering the case when the system is initially at rest, $\psi(p)=0$. Writing i_a for the current in this case

$$\mathcal{L}[i_a]=\phi(p)\mathcal{L}[e]. \qquad (9)$$

Now apply unit impulse $S_0(t)$ (multiplied by 1 volt-second), and designate the circuit current in this case by $B(t)$ and its transform by $\mathcal{L}[B]$. The Laplace transform of $S_0(t)$ is 1, so

$$\mathcal{L}[B]=\phi(p). \qquad (10)$$

Equation (9) becomes, for any driving force

$$\mathcal{L}[i_a]=\mathcal{L}[B]\mathcal{L}[e]. \qquad (11)$$

Applying the convolution function (Laplace transform)

$$i_a=\int_0^t B(t-\lambda)e(\lambda) \, d\lambda$$

$$=\int_0^t B(\lambda)e(t-\lambda) \, d\lambda. \qquad (12)$$

To this there must be added the current i_0 due to any initial conditions that exist. From (8)

$$\mathcal{L}[i_0]=\psi(p). \qquad (13)$$

Then i_0 is the inverse transform of $\psi(p)$.

Circuit Response Related to Unit Step

Unit step is defined and designated $S_{-1}(t)=0$ for $t<0$ and equals unity for $t>0$. It has no dimensions. Its Laplace transform is $1/p$. Let the circuit current be designated $A(t)$ when the applied voltage is $e=S_{-1}(t)\times(1 \text{ volt})$. Then, the current i_a for the case when the system is initially at rest, and for any applied voltage $e(t)$, is given by any of

$$i_a=A(t)e(0)+\int_0^t A(t-\lambda)e'(\lambda) \, d\lambda$$

$$=A(t)e(0)+\int_0^t A(\lambda)e'(t-\lambda) \, d\lambda$$

$$=A(0)e(t)+\int_0^t A'(t-\lambda)e(\lambda) \, d\lambda$$

$$=A(0)e(t)+\int_0^t A'(\lambda)e(t-\lambda) \, d\lambda \qquad (14)$$

where A' is the first derivative of A and similarly for e' of e.

As an example, consider the problem of Fig. 20 and Eqs. (5) to (7) above. Suppose $V_0=0$, and that the battery is replaced by a linear source

$$e(t)=Et/T_1$$

where T_1 is the duration of the voltage rise in seconds. By Eq. (7), setting $E_b=1$

$$A(t)=1-\exp(-t/RC).$$

Then using the first equation in (14) and noting that $e(0)=0$, and $e'(t)=E/T_1$ when $0\leq t\leq T_1$, the solution is

$$v=(Et/T_1)-(ERC/T_1)[1-\exp(-t/RC)].$$

This result can, of course, be found readily by direct application of the Laplace transform to Eq. (5) with $e(t)=Et/T_1$.

Heaviside Expansion Theorem

When the system is initially at rest, the transformed equation is given by Eq. (9) and may be written

$$\mathcal{L}[i_a]=[M(p)/G(p)]\mathcal{L}[e]. \quad (15)$$

$M(p)$ and $G(p)$ are rational functions of p. In the following, $M(p)$ must be of lower degree than $G(p)$, as is usually the case. The roots of $G(p)=0$ are p_r, where $r=1, 2, \cdots, n$, and there must be no repeated roots. The response may be found by application of the Heaviside expansion theorem.

For a force $e=E_{max}\exp(j\omega t)$ applied at time $t=0$

$$\frac{i_a(t)}{E_{max}}=\frac{M(j\omega)}{G(j\omega)}\exp(j\omega t)+\sum_{r=1}^{n}\frac{M(p_r)\exp(p_r t)}{(p_r-j\omega)G'(p_r)}$$

$$(16A)$$

$$=\frac{\exp(j\omega t)}{Z(j\omega)}+\sum_{r=1}^{n}\frac{\exp(p_r t)}{(p_r-j\omega)Z'(p_r)}. \quad (16B)$$

The first term on the right-hand side of either form of (16) gives the steady-state response, and the second term gives the transient. When $e=E_{max}\cos\omega t$, take the real part of (16), and similarly for $\sin\omega t$ take the imaginary part. $Z(p)$ is defined in Eq. (19). If the applied force is the unit step, set $\omega=0$ in Eq. (16).

Application to Linear Networks

The equation for a single mesh is of the form

$$A_n(d^n i/dt^n)+\cdots+A_1(di/dt)$$

$$+A_0 i+B\int i\,dt=e(t). \quad (17)$$

System Initially at Rest: Then, Eq. (17) trans-

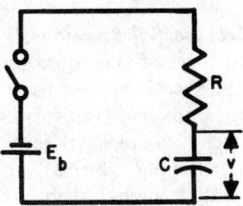

Fig. 20.

forms into

$$(A_n p^n+\cdots+A_1 p+A_0+Bp^{-1})\mathcal{L}[i]=\mathcal{L}[e] \quad (18)$$

where the expression in parentheses is the operational impedance, equal to the alternating-current impedance when we set $p=j\omega$.

If there are m meshes in the system, we get m simultaneous equations like (17) with m unknowns $i_1, i_2, \cdots, i_m$. The m algebraic equations like (18) are solved for $\mathcal{L}[i_1]$, etc., by means of determinants, yielding an equation of the form of (15) for each unknown, with a term on the right-hand side for each mesh in which there is a driving force. Each such driving force may of course be treated separately and the responses added.

Designating any two meshes by the letters h and k, the driving force $e(t)$ being in either mesh and the mesh current $i(t)$ in the other, then the fraction $M(p)/G(p)$ in (15) becomes

$$M_{hk}(p)/G(p)=1/Z_{hk}(p)=Y_{hk}(p) \quad (19)$$

where $Y_{hk}(p)$ is the operational transfer admittance between the two meshes. The determinant of the system is $G(p)$, and $M_{hk}(p)$ is the cofactor of the row and column that represent $e(t)$ and $i(t)$.

System Not Initially at Rest: The transient due to the initial conditions is solved separately and added to the above solution. The driving force is set equal to zero in (17), $e(t)=0$, and each term is transformed according to

$$\mathcal{L}[d^n i/dt^n]=p^n\mathcal{L}[i]-\sum_{r=1}^{n}p^{n-r}[d^{r-1}i/dt^{r-1}]_{t=0}$$

$$(20A)$$

$$\mathcal{L}\left[\int_0^t i\,dt\right]=p^{-1}\mathcal{L}[i]+p^{-1}\left[\int i\,dt\right]_{t=0} \quad (20B)$$

where the last term in each equation represents the initial conditions. For example, in (20B) the last term would represent, in an electrical circuit, the quantity of electricity existing on a capacitor at time $t=0$, the instant when the driving force $e(t)$ begins to act.

Resolution into Partial Fractions: The solution of the operational form of the equations of a system involves rational fractions that must be simplified before finding the inverse transform. Let the fraction be $h(p)/g(p)$ where $h(p)$ is of lower degree than $g(p)$, for example $(3p+2)/(p^2+5p+8)$. If $h(p)$ is of equal or higher degree than $g(p)$, it can be reduced by division.

The reduced fraction can be expanded into partial fractions. Let the factors of the denominator be $(p-p_r)$ for the n nonrepeated roots p_r of the equation $g(p)=0$, and $(p-p_a)$ for a root p_a repeated m times.

$$\frac{h(p)}{g(p)} = \sum_{r=1}^{n} \frac{A_r}{p-p_r} + \sum_{r=1}^{m} \frac{B_r}{(p-p_a)^{m-r+1}}. \quad (21A)$$

There is a summation term for each root that is repeated. The constant coefficients A_r and B_r can be evaluated by reforming the fraction with a common denominator. Then the coefficients of each power of p in $h(p)$ and the reformed numerator are equated and the resulting equations solved for the constants. More formally, they may be evaluated by

$$A_r = \frac{h(p_r)}{g'(p_r)} = \left[\frac{h(p)}{g(p)/(p-p_r)}\right]_{p=p_r} \quad (21B)$$

$$B_r = [1/(r-1)!]f^{(r-1)}(p_a) \quad (21C)$$

where

$$f(p) = (p-p_a)^m[h(p)/g(p)]$$

and $f^{(r-1)}(p_a)$ indicates that the $(r-1)$th derivative of $f(p)$ is to be found, after which we set $p=p_a$.

Fractions of the form $(A_1p+A_2)/(p^2+\omega^2)$ or, more generally

$$\frac{A_1p+A_2}{p^2+2ap+b} = \frac{A(p+a)+B\omega}{(p+a)^2+\omega^2} \quad (22A)$$

where $b>a^2$ and $\omega^2=b-a^2$ need not be reduced further. From the Laplace transforms the inverse transform of (22A) is

$$\exp(-at)(A\cos\omega t+B\sin\omega t) \quad (22B)$$

where

$$A = \frac{h(-a+j\omega)}{g'(-a+j\omega)} + \frac{h(-a-j\omega)}{g'(-a-j\omega)} \quad (22C)$$

$$B = j\left[\frac{h(-a+j\omega)}{g'(-a+j\omega)} - \frac{h(-a-j\omega)}{g'(-a-j\omega)}\right]. \quad (22D)$$

Similarly, the inverse transform of the fraction

$$[A(p+a)+B\alpha]/[(p+a)^2-\alpha^2]$$

is $\exp(-at)(A\cosh\alpha t+B\sinh\alpha t)$, where A and B are found by (22C) and (22D), except that $j\omega$ is replaced by α and the coefficient j is omitted in the expression for B.

CHAPTER 7
FILTERS, IMAGE-PARAMETER DESIGN

GENERAL

The basic filter half section and the full sections derived from it are shown in Fig. 1. The fundamental filter equations follow, with filter characteristics and design equations next. Also given is the method of building up a composite filter and the effect of the design parameter m on the image-impedance characteristic. An example of the design of a low-pass filter completes the chapter. Note that while the impedance characteristics and design equations are given for the half sections as shown, the attenuation and phase characteristics are for full sections, either T or π.

FUNDAMENTAL FILTER EQUATIONS

Image Impedances Z_T and Z_π

The element-value design equations to be given are derived by assuming that the network is terminated with impedances that change with frequency in accordance with the following image-impedance equations. Unfortunately, this assumption can be only approximately satisfied.

Z_T = mid-series image impedance

= impedance looking into 1–2 (Fig. 1A) with Z_π connected across 3–4.

Z_π = mid-shunt image impedance

= impedance looking into 3–4 (Fig. 1A) with Z_T connected across 1–2.

Equations for the above are

$$Z_T = (Z_1Z_2 + Z_1^2/4)^{1/2}$$

$$= (Z_1Z_2)^{1/2}(1 + Z_1/4Z_2)^{1/2} \text{ ohm}$$

$$Z_\pi = \frac{Z_1Z_2}{(Z_1Z_2 + Z_1^2/4)^{1/2}}$$

$$= \frac{(Z_1Z_2)^{1/2}}{(1 + Z_1/4Z_2)^{1/2}} \text{ ohm}$$

$$Z_TZ_\pi = Z_1Z_2.$$

Image Transfer Constant

The transfer constant $\theta = \alpha + j\beta$ of a network is defined as half the natural logarithm of the complex ratio of the steady-state volt-amperes entering and leaving the network when the latter is terminated in its image impedance. The real part α of the transfer constant is called the image attenuation constant, and the imaginary part β is called the image phase constant.

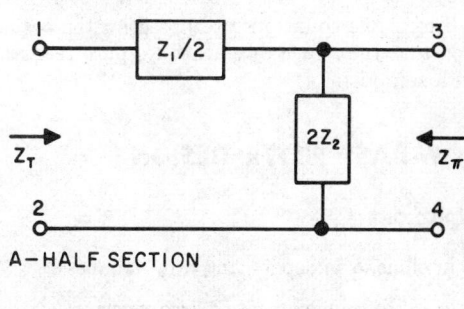

A—HALF SECTION

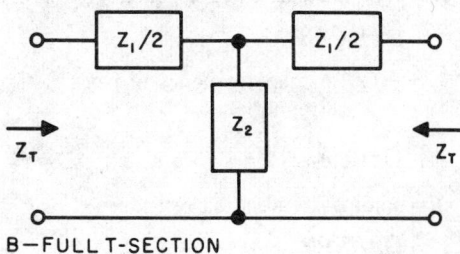

B—FULL T-SECTION

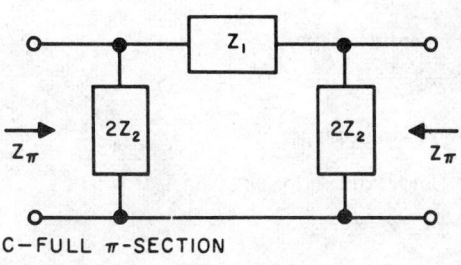

C—FULL π-SECTION

Fig. 1—Basic filter sections.

7-1

Equations in terms of full sections are

$$\cosh\theta = 1 + Z_1/2Z_2.$$

Pass Band:

$\alpha = 0$, for frequencies making $-1 \leq Z_1/4Z_2 \leq 0$

$\beta = \cos^{-1}(1 + Z_1/2Z_2)$

$\quad = \pm 2 \sin^{-1}(-Z_1/4Z_2)^{1/2}$ radian

Image impedance = pure resistance.

Stop Band:

$\begin{cases} \alpha = \cosh^{-1} \mid 1 + Z_1/2Z_2 \mid \\ \quad = 2 \sinh^{-1}(Z_1/4Z_2)^{1/2} \text{ neper} \quad \text{for } Z_1/4Z_2 > 0 \\ \beta = 0 \text{ radian} \end{cases}$

$\begin{cases} \alpha = \cosh^{-1} \mid 1 + Z_1/2Z_2 \mid \\ \quad = 2 \cosh^{-1}(-Z_1/4Z_2)^{1/2} \text{ neper} \quad \text{for } Z_1/4Z_2 < -1 \\ \beta = \pm\pi \text{ radian} \end{cases}$

Image impedance = pure reactance.

The above equations are based on the assumption that the impedance arms are pure reactances with zero loss.

LOW-PASS FILTER DESIGN

Notations

Z in ohms, α in nepers, and β in radians

$\omega_c = 2\pi f_c =$ angular cutoff frequency

$\quad = 1/(L_k C_k)^{1/2}$

$\omega_\infty = 2\pi f_\infty =$ angular frequency of peak attenuation

$m = (1 - \omega_c^2/\omega_\infty^2)^{1/2}$

$R =$ nominal terminating resistance

$\quad = (L_k/C_k)^{1/2}$

$\quad = (Z_{Tk}Z_{\pi k})^{1/2}$

For constant-k type:

$R^2 = Z_{1k}Z_{2k} = k^2$

For m-derived type:

Curves drawn for $m \approx 0.6$

$R^2 = Z_{T2}Z_{\pi 1}$

$\quad = Z_{1(\text{series}-m)}Z_{2(\text{shunt}-m)}$

$\quad = Z_{1(\text{shunt}-m)}Z_{2(\text{series}-m)}$

Constant-k

Half Section

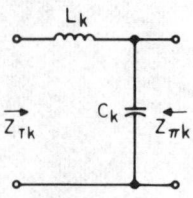

Impedance Characteristics

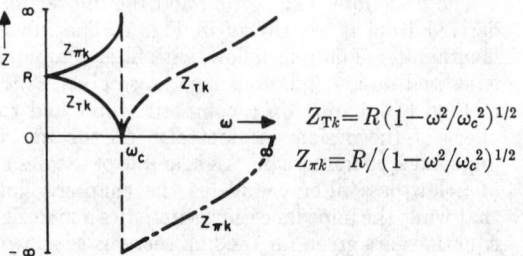

$Z_{Tk} = R(1 - \omega^2/\omega_c^2)^{1/2}$

$Z_{\pi k} = R/(1 - \omega^2/\omega_c^2)^{1/2}$

Full-Section Attenuation α and Phase β Characteristics

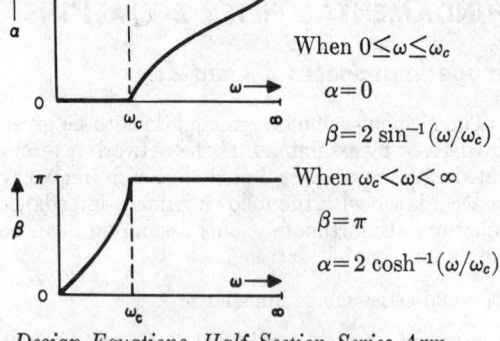

When $0 \leq \omega \leq \omega_c$

$\alpha = 0$

$\beta = 2 \sin^{-1}(\omega/\omega_c)$

When $\omega_c < \omega < \infty$

$\beta = \pi$

$\alpha = 2 \cosh^{-1}(\omega/\omega_c)$

Design Equations, Half-Section Series Arm

$$L_k = R/\omega_c$$

Design Equations, Half-Section Shunt Arm

$$C_k = 1/\omega_c R$$

Series m-Derived

Half Section

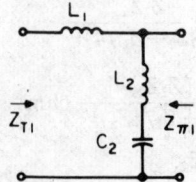

Impedance Characteristics

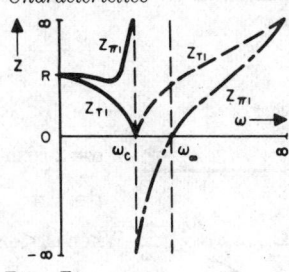

$$Z_{T1} = Z_{Tk}$$

$$Z_{\pi 1} = \frac{R(1-\omega^2/\omega_\infty^2)}{(1-\omega^2/\omega_c^2)^{1/2}}$$

$$= \frac{R[1-(\omega^2/\omega_c^2)(1-m^2)]}{(1-\omega^2/\omega_c^2)^{1/2}}$$

Full-Section Attenuation α and Phase β Characteristics

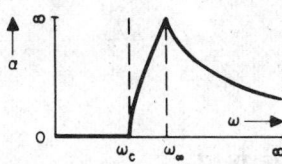

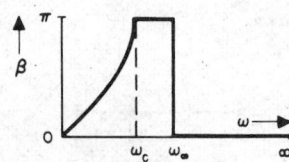

When $\omega_c < \omega < \omega_\infty$, $\beta = \pi$ and

$$\alpha = \cosh^{-1}\left[2\frac{1/\omega_\infty^2 - 1/\omega_c^2}{1/\omega_\infty^2 - 1/\omega^2} - 1\right]$$

$$= \cosh^{-1}\left[2\frac{m^2}{\omega_c^2/\omega^2 - (1-m^2)} - 1\right]$$

When $0 \leq \omega \leq \omega_c$, $\alpha = 0$ and

$$\beta = \cos^{-1}\left[1 - 2\frac{1/\omega_\infty^2 - 1/\omega_c^2}{1/\omega_\infty^2 - 1/\omega^2}\right]$$

$$= \cos^{-1}\left[1 - 2\frac{m^2}{\omega_c^2/\omega^2 - (1-m^2)}\right]$$

When $\omega_\infty < \omega < \infty$, $\beta = 0$ and

$$\alpha = \cosh^{-1}\left[1 - 2\frac{1/\omega_\infty^2 - 1/\omega_c^2}{1/\omega_\infty^2 - 1/\omega^2}\right]$$

$$= \cosh^{-1}\left[1 - 2\frac{m^2}{\omega_c^2/\omega^2 - (1-m^2)}\right]$$

Design Equations, Half-Section Series Arm

$$L_1 = mL_k$$

Design Equations, Half-Section Shunt Arm

$$L_2 = [(1-m^2)/m]L_k$$

$$C_2 = mC_k$$

Shunt m-Derived

Half Section

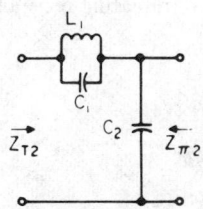

Impedance Characteristics

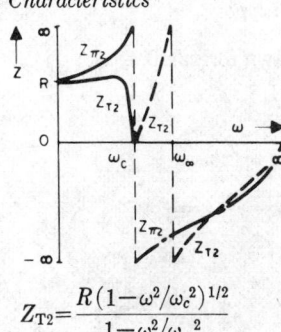

$$Z_{T2} = \frac{R(1-\omega^2/\omega_c^2)^{1/2}}{1-\omega^2/\omega_\infty^2}$$

$$= \frac{R(1-\omega^2/\omega_c^2)^{1/2}}{1-(\omega^2/\omega_c^2)(1-m^2)}$$

$$= R^2/Z_{\pi 1}$$

$$Z_{\pi 2} = Z_{\pi k}$$

Full-Section Attenuation α and Phase β Characteristics

Same as for series m-derived.

Design Equations, Half-Section Series Arm

$$L_1 = mL_k$$

$$C_1 = [(1-m^2)/m]C_k$$

Design Equations, Half-Section Shunt Arm

$$C_2 = mC_k$$

HIGH-PASS FILTER DESIGN

Notations

Z in ohms, α in nepers, and β in radians

$\omega_c = 2\pi f_c =$ angular cutoff frequency

$$= 1/(L_k C_k)^{1/2}$$

$\omega_\infty = 2\pi f_\infty =$ angular frequency of peak attenuation

$$m = (1 - \omega_\infty^2/\omega_c^2)^{1/2}$$

$R =$ nominal terminating resistance

$$= (L_k/C_k)^{1/2}$$

$$= (Z_{Tk} Z_{\pi k})^{1/2}$$

For constant-k type:

$$R^2 = Z_{1k} Z_{2k} = k^2$$

For m-derived type:

Curves drawn for $m \approx 0.6$

$$R^2 = Z_{T2} Z_{\pi 1}$$

$$= Z_{1(\text{series}-m)} Z_{2(\text{shunt}-m)}$$

$$= Z_{1(\text{shunt}-m)} Z_{2(\text{series}-m)}$$

Constant-k

Half Section

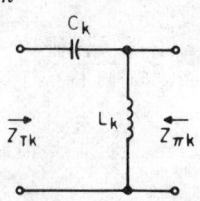

Impedance Characteristics

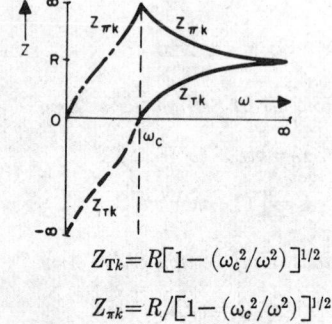

$$Z_{Tk} = R[1 - (\omega_c^2/\omega^2)]^{1/2}$$

$$Z_{\pi k} = R/[1 - (\omega_c^2/\omega^2)]^{1/2}$$

Full-Section Attenuation α and Phase β Characteristics

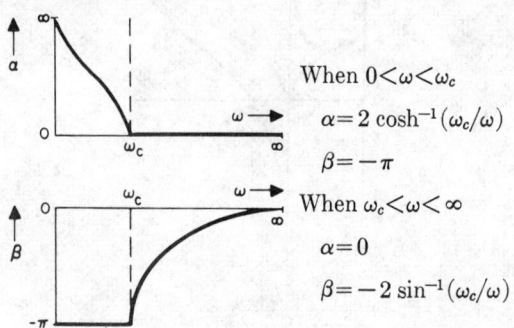

When $0 < \omega < \omega_c$

$$\alpha = 2 \cosh^{-1}(\omega_c/\omega)$$

$$\beta = -\pi$$

When $\omega_c < \omega < \infty$

$$\alpha = 0$$

$$\beta = -2 \sin^{-1}(\omega_c/\omega)$$

Design Equations, Half-Section Series Arm

$$C_k = 1/\omega_c R$$

Design Equations, Half-Section Shunt Arm

$$L_k = R/\omega_c$$

Series m-Derived

Half Section

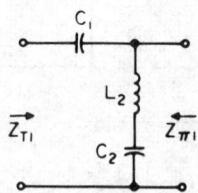

Impedance Characteristics

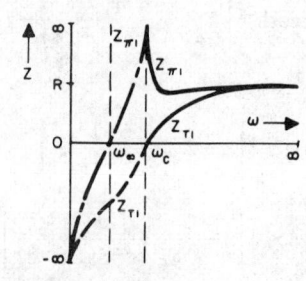

$$Z_{T1} = Z_{Tk}$$

$$Z_{\pi 1} = \frac{R[1 - (\omega_\infty^2/\omega^2)]}{(1 - \omega_c^2/\omega^2)^{1/2}}$$

$$= \frac{R[1 - (\omega_c^2/\omega^2)(1 - m^2)]}{(1 - \omega_c^2/\omega^2)^{1/2}}$$

Full-Section Attenuation α and Phase β Characteristics

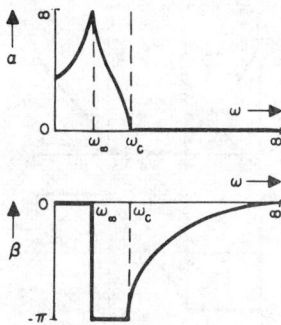

When $\omega_\infty < \omega < \omega_c$, $\beta = -\pi$ and

$$\alpha = \cosh^{-1}\left[2\frac{\omega_c^2 - \omega_\infty^2}{\omega^2 - \omega_\infty^2} - 1\right]$$

$$= \cosh^{-1}\left[2\frac{m^2}{(\omega^2/\omega_c^2) - (1-m^2)} - 1\right]$$

When $0 < \omega < \omega_\infty$, $\beta = 0$ and

$$\alpha = \cosh^{-1}\left[1 - 2\frac{\omega_\infty^2 - \omega_c^2}{\omega_\infty^2 - \omega^2}\right]$$

$$= \cosh^{-1}\left[1 + 2\frac{m^2}{(1-m^2) - (\omega^2/\omega_c^2)}\right]$$

When $\omega_c < \omega < \infty$, $\alpha = 0$ and

$$\beta = \cos^{-1}\left[1 - 2\frac{\omega_c^2 - \omega_c^2}{\omega_\infty^2 - \omega^2}\right]$$

$$= \cos^{-1}\left[1 + 2\frac{m^2}{(1-m^2) - (\omega^2/\omega_c^2)}\right]$$

Design Equations, Half-Section Series Arm

$$C_1 = C_k/m$$

Design Equations, Half-Section Shunt Arm

$$L_2 = L_k/m$$

$$C_2 = [m/(1-m^2)]C_k$$

Shunt m-Derived

Half Section

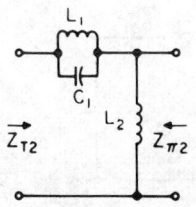

Impedance Characteristics

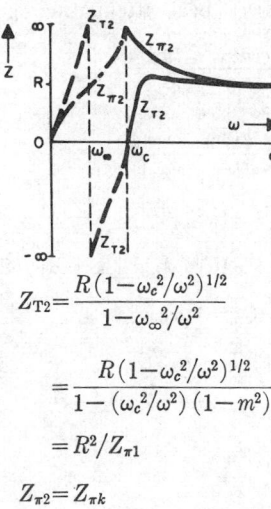

$$Z_{T2} = \frac{R(1 - \omega_c^2/\omega^2)^{1/2}}{1 - \omega_c^2/\omega^2}$$

$$= \frac{R(1 - \omega_c^2/\omega^2)^{1/2}}{1 - (\omega_c^2/\omega^2)(1-m^2)}$$

$$= R^2/Z_{\pi1}$$

$$Z_{\pi2} = Z_{\pi k}$$

Full-Section Attenuation α and Phase β Characteristics

Same as for series m-derived.

Design Equations, Half-Section Series Arm

$$L_1 = [m/(1-m^2)]L_k$$

$$C_1 = C_k/m$$

Design Equations, Half-Section Shunt Arm

$$L_2 = L_k/m$$

BAND-PASS FILTER DESIGN

Notations

The notations apply to the charts on band-pass filter design that appear on the following pages.

Z in ohms, α in nepers, and β in radians

$\omega_1 = 2\pi f_1 =$ lower cutoff angular frequency

$\omega_2 = 2\pi f_2 =$ upper cutoff angular frequency

$\omega_0 = (\omega_1\omega_2)^{1/2} =$ midband angular frequency

$\omega_2 - \omega_1 =$ width of pass band

$R =$ nominal terminating resistance

$\omega_{1\infty} = 2\pi f_{1\infty} =$ lower angular frequency of peak attenuation

$\omega_{2\infty} = 2\pi f_{2\infty} =$ upper angular frequency of peak attenuation

$$m_1 = \frac{(\omega_1\omega_2/\omega^2_{2\infty})g + h}{1 - (\omega^2_{1\infty}/\omega^2_{2\infty})}$$

$$m_2 = \frac{g + h(\omega^2_{1\infty}/\omega_1\omega_2)}{1 - (\omega^2_{1\infty}/\omega^2_{2\infty})}$$

$$g = \{[1 - (\omega^2_{1\infty}/\omega_1^2)][1 - (\omega^2_{1\infty}/\omega_2^2)]\}^{1/2}$$

$$h = \{[1 - (\omega_1^2/\omega^2_{2\infty})][1 - (\omega_2^2/\omega^2_{2\infty})]\}^{1/2}$$

$$L_{1k}C_{1k} = L_{2k}C_{2k} = 1/\omega_1\omega_2 = 1/\omega_0^2$$

$$R^2 = L_{1k}/C_{2k} = L_{2k}/C_{1k}$$

$$= Z_{1k}Z_{2k} = k^2$$

$$= Z_{Tk}Z_{\pi k}$$

$\left. \begin{array}{l} = Z_{1(\text{series}-m)}Z_{2(\text{shunt}-m)} \\[4pt] = Z_{2(\text{series}-m)}Z_{1(\text{shunt}-m)} \\[4pt] = Z_{T(\text{shunt}-m)}Z_{\pi(\text{series}-m)} \end{array} \right\}$ for any one pair of m-derived half sections.

$$Z_{T(\text{series}-m)} = Z_{Tk}$$

$$Z_{\pi(\text{shunt}-m)} = Z_{\pi k}$$

Constant-k

Half Section

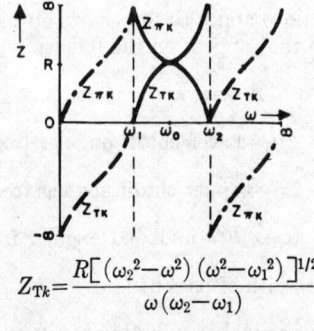

Impedance Characteristics

$$Z_{Tk} = \frac{R[(\omega_2^2 - \omega^2)(\omega^2 - \omega_1^2)]^{1/2}}{\omega(\omega_2 - \omega_1)}$$

$$Z_{\pi k} = \frac{R\omega(\omega_2 - \omega_1)}{[(\omega_2^2 - \omega^2)(\omega^2 - \omega_1^2)]^{1/2}}$$

Full-Section Attenuation α and Phase β Characteristics

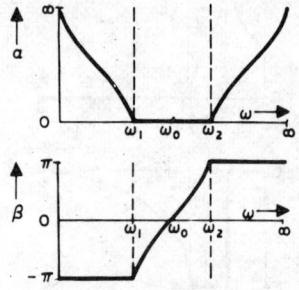

When $\omega_2 < \omega < \infty$, $\beta = \pi$ and

$$\alpha = 2\cosh^{-1}\left[\frac{\omega^2 - \omega_0^2}{\omega(\omega_2 - \omega_1)}\right]$$

When $0 < \omega < \omega_1$, $\beta = -\pi$ and

$$\alpha = 2\cosh^{-1}\left[\frac{\omega_0^2 - \omega^2}{\omega(\omega_2 - \omega_1)}\right]$$

When $\omega_1 < \omega < \omega_2$, $\alpha = 0$ and

$$\beta = 2\sin^{-1}\left[\frac{\omega^2 - \omega_0^2}{\omega(\omega_2 - \omega_1)}\right]$$

Frequencies of Peak α

$$\omega_{1\infty} = 0$$

$$\omega_{2\infty} = \infty$$

Design Equations, Half-Section Series Arm

$$L_{1k} = R/(\omega_2 - \omega_1)$$

$$C_{1k} = (\omega_2 - \omega_1)/R\omega_0^2$$

Design Equations, Half-Section Shunt Arm

$$L_{2k} = R(\omega_2 - \omega_1)/\omega_0^2$$

$$C_{2k} = 1/R(\omega_2 - \omega_1)$$

3-Element Series I

Half Section

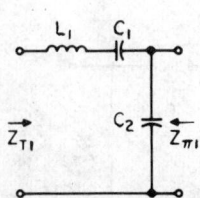

Impedance Characteristics

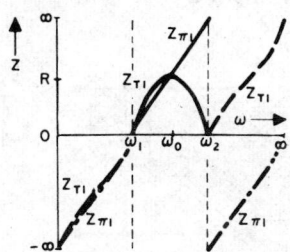

$$Z_{\pi1} = \frac{R(\omega_2+\omega_1)}{\omega}\left[\frac{\omega^2-\omega_1^2}{\omega_2^2-\omega^2}\right]^{1/2}$$

Full-Section Attenuation α and Phase β Characteristics

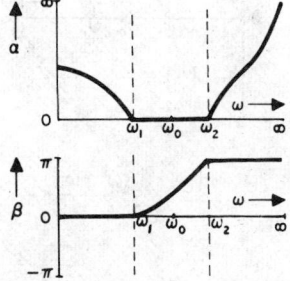

When $0<\omega<\omega_1$, $\beta=0$ and

$$\alpha = \cosh^{-1}\left[1-2\frac{\omega^2-\omega_1^2}{\omega_2^2-\omega_1^2}\right]$$

When $\omega_1<\omega<\omega_2$, $\alpha=0$ and

$$\beta = \cos^{-1}\left[1-2\frac{\omega^2-\omega_1^2}{\omega_2^2-\omega_1^2}\right]$$

When $\omega_2<\omega<\infty$, $\beta=\pi$ and

$$\alpha = \cosh^{-1}\left[2\frac{\omega^2-\omega_1^2}{\omega_2^2-\omega_1^2}-1\right]$$

Conditions

$$m_1 = 1$$
$$m_2 = \omega_1/\omega_2$$

Frequencies of Peak α

$$\omega_{2\infty} = \infty$$

Design Equations, Half-Section Series Arm

$$L_1 = L_{1k}$$
$$C_1 = C_{1k}/m_2$$

Design Equations, Half-Section Shunt Arm

$$C_2 = \left[(1-m_2)/(1+m_2)\right]C_{2k}$$

3-Element Shunt I

Half Section

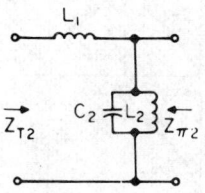

Impedance Characteristics

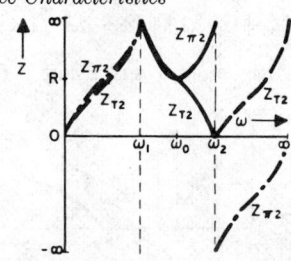

$$Z_{T2} = \left[R\omega/(\omega_2+\omega_1)\right]\left[(\omega_2^2-\omega^2)/(\omega^2-\omega_1^2)\right]^{1/2}$$
$$= R^2/Z_{\pi1}$$
$$Z_{\pi2} = Z_{\pi k}$$

Full-Section Attenuation α and Phase β Characteristics

Same as for 3-element series I.

Conditions

Same as for 3-element series I.

Frequencies of Peak α

Same as for 3-element series I.

Design Equations, Half-Section Series Arm

$$L_1 = \left[(1-m_2)/(1+m_2)\right]L_{1k}$$

Design Equations, Half-Section Shunt Arm

$$L_2 = L_{2k}/m_2$$
$$C_2 = C_{2k}$$

3-Element Series II

Half Section

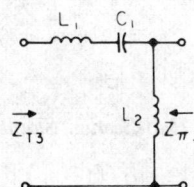

Impedance Characteristics

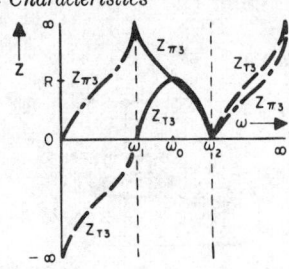

$$Z_{T3} = Z_{Tk}$$

$$Z_{\pi 3} = \left[R\omega(\omega_2+\omega_1)/\omega_2^2 \right]\left[(\omega_2^2-\omega^2)/(\omega^2-\omega_1^2) \right]^{1/2}$$

Full-Section Attenuation α and Phase β Characteristics

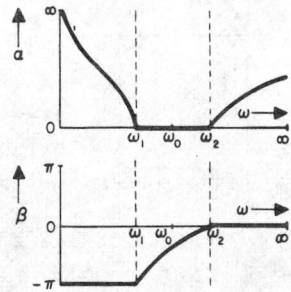

When $0<\omega<\omega_1$, $\beta=-\pi$ and

$$\alpha = \cosh^{-1}\left[2\,\frac{\omega_1^2(\omega_2^2-\omega^2)}{\omega^2(\omega_2^2-\omega_1^2)} - 1 \right]$$

When $\omega_1<\omega<\omega_2$, $\alpha=0$ and

$$\beta = \cos^{-1}\left[1-2\,\frac{\omega_1^2(\omega_2^2-\omega^2)}{\omega^2(\omega_2^2-\omega_1^2)} \right]$$

When $\omega_2<\omega<\infty$, $\beta=0$ and

$$\alpha = \cosh^{-1}\left[1-2\,\frac{\omega_1^2(\omega_2^2-\omega^2)}{\omega^2(\omega_2^2-\omega_1^2)} \right]$$

Conditions

$$m_1=\omega_1/\omega_2$$

$$m_2=1$$

Frequencies of Peak α

$$\omega_{1\infty}=0$$

Design Equations, Half-Section Series Arm

$$L_1=m_1 L_{1k}$$

$$C_1=C_{1k}$$

Design Equations, Half-Section Shunt Arm

$$L_2=\left[(1+m_1)/(1-m_1)\right]L_{2k}$$

3-Element Shunt II

Half Section

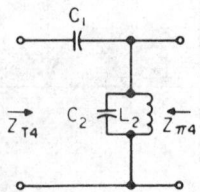

Impedance Characteristics

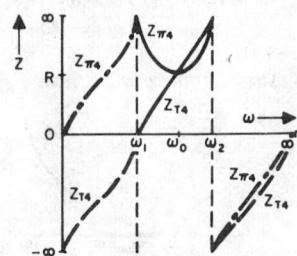

$$Z_{T4} = \left[R\omega_2^2/\omega(\omega_2+\omega_1) \right]\left[(\omega^2-\omega_1^2)/(\omega_2^2-\omega^2) \right]^{1/2}$$

$$= R^2/Z_{\pi 3}$$

$$Z_{\pi 4}=Z_{\pi k}$$

Full-Section Attenuation α and Phase β Characteristics

Same as for 3-element series II.

Conditions

Same as for 3-element series II.

Frequencies of Peak α

Same as for 3-element series II.

Design Equations, Half-Section Series Arm

$$C_1=\left[(1+m_1)/(1-m_1)\right]C_{1k}$$

Design Equations, Half-Section Shunt Arm

$$L_2=L_{2k}$$

$$C_2=m_1 C_{2k}$$

4-Element Series I

Half Section

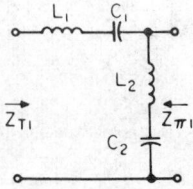

Impedance Characteristics

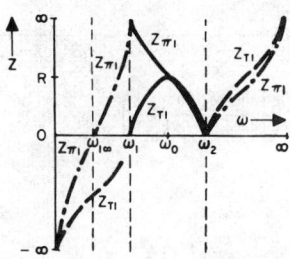

$Z_{T1} = Z_{Tk}$

$$Z_{\pi 1} = \left[R/\omega(\omega_2-\omega_1)\right]\left[\frac{\omega_2{}^2-\omega^2}{\omega^2-\omega_1{}^2}\right]^{1/2}$$

$$\times\left[(\omega^2-\omega_1{}^2)+m_1{}^2(\omega_2{}^2-\omega^2)\right]$$

Full-Section Attenuation α and Phase β Characteristics

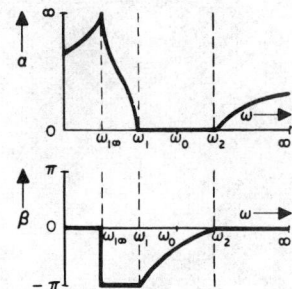

When $\omega_1<\omega<\omega_2$, $\alpha=0$ and

$$\beta = \cos^{-1}A$$

When $0<\omega<\omega_{1\infty}$, $\beta=0$ and

$$\alpha = \cosh^{-1}A$$

When $\omega_{1\infty}<\omega<\omega_1$, $\beta=-\pi$ and

$$\alpha = \cosh^{-1}(-A)$$

When $\omega_2<\omega<\infty$, $\beta=0$ and

$$\alpha = \cosh^{-1}A$$

Conditions

$$A = 1 - \frac{2}{1+\left[(\omega^2-\omega_1{}^2)/m_1{}^2(\omega_2{}^2-\omega^2)\right]}$$

$$m_2 = \left[\frac{1-(\omega^2{}_{1\infty}/\omega_1{}^2)}{1-(\omega^2{}_{1\infty}/\omega_2{}^2)}\right]^{1/2}$$

$$m_1/m_2 = \omega_1/\omega_2$$

Frequencies of Peak α

$$\omega_{1\infty} = \left[\frac{\omega_1{}^2-\omega_2{}^2m_1{}^2}{1-m_1{}^2}\right]^{1/2}$$

Design Equations, Half-Section Series Arm

$$L_1 = m_1 L_{1k}$$

$$C_1 = C_{1k}/m_2$$

Design Equations, Half-Section Shunt Arm

$$L_2 = \left[(1-m_1{}^2)/m_1\right]L_{1k}$$

$$C_2 = \left[m_2/(1-m_2{}^2)\right]C_{1k}$$

4-Element Shunt I

Half Section

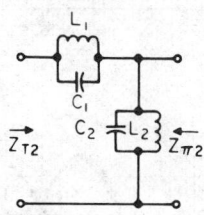

Impedance Characteristics

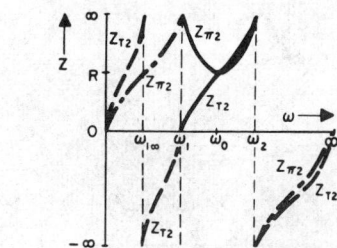

$$Z_{T2} = \frac{R\omega(\omega_2-\omega_1)}{(\omega^2-\omega_1{}^2)+m_1{}^2(\omega_2{}^2-\omega^2)}\left(\frac{\omega^2-\omega_1{}^2}{\omega_2{}^2-\omega^2}\right)^{1/2}$$

$$= R^2/Z_{\pi 1}$$

$$Z_{\pi 2} = Z_{\pi k}$$

Full-Section Attenuation α and Phase β Characteristics

Same as for 4-element series I.

Conditions

Same as for 4-element series I.

Frequencies of Peak α

Same as for 4-element series I.

Design Equations, Half-Section Series Arm

$$L_1 = \left[m_2/(1-m_2{}^2)\right]L_{2k}$$

$$C_1 = \left[(1-m_1{}^2)/m_1\right]C_{2k}$$

Design Equations, Half-Section Shunt Arm

$$L_2 = L_{2k}/m_2$$

$$C_2 = m_1 C_{2k}$$

4-Element Series II

Half Section

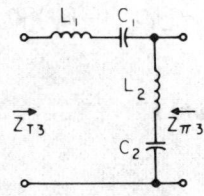

Impedance Characteristics

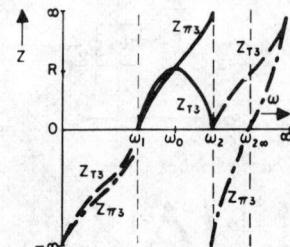

$$Z_{T3} = Z_{Tk}$$

$$Z_{\pi3} = [R/\omega(\omega_2 - \omega_1)]\left(\frac{\omega^2 - \omega_1^2}{\omega_2^2 - \omega^2}\right)^{1/2}$$

$$\times [(\omega_2^2 - \omega^2) + m_1^2(\omega^2 - \omega_1^2)]$$

Full-Section Attenuation α and Phase β Characteristics

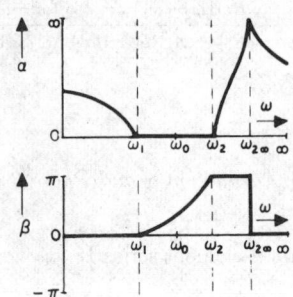

When $\omega_2 < \omega < \omega_{2\infty}$, $\beta = \pi$ and

$$\alpha = \cosh^{-1}(-B)$$

When $0 < \omega < \omega_1$, $\beta = 0$ and

$$\alpha = \cosh^{-1}B$$

When $\omega_1 < \omega < \omega_2$, $\alpha = 0$ and

$$\beta = \cos^{-1}B$$

When $\omega_{2\infty} < \omega < \infty$, $\beta = 0$ and

$$\alpha = \cosh^{-1}B$$

Conditions

$$B = 1 - \frac{2}{1 + [(\omega_2^2 - \omega^2)/m_1^2(\omega^2 - \omega_1^2)]}$$

$$m_1 = \left(\frac{1 - (\omega_2^2/\omega_{2\infty}^2)}{1 - (\omega_1^2/\omega_{2\infty}^2)}\right)^{1/2}$$

$$m_1/m_2 = \omega_2/\omega_1$$

Frequencies of Peak α

$$\omega_{2\infty} = \left(\frac{m_1^2\omega_1^2 - \omega_2^2}{m_1^2 - 1}\right)^{1/2}$$

Design Equations, Half-Section Series Arm

$$L_1 = m_1 L_{1k}$$

$$C_1 = C_{1k}/m_2$$

Design Equations, Half-Section Shunt Arm

$$L_2 = [(1 - m_1^2)/m_1]L_{1k}$$

$$C_2 = [m_2/(1 - m_2^2)]C_{1k}$$

4-Element Shunt II

Half Section

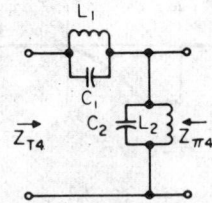

Impedance Characteristics

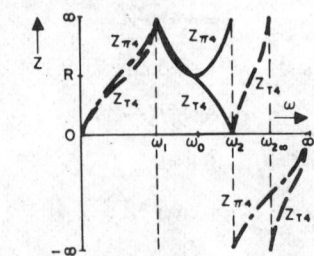

$$Z_{T4} = \frac{R\omega(\omega_2 - \omega_1)}{(\omega_2^2 - \omega^2) + m_1^2(\omega^2 - \omega_1^2)}\left(\frac{\omega_2^2 - \omega^2}{\omega^2 - \omega_1^2}\right)^{1/2}$$

$$= R^2/Z_{\pi3}$$

$$Z_{\pi4} = Z_{\pi k}$$

Full-Section Attenuation α and Phase β Characteristics

Same as for 4-element series II.

Conditions

Same as for 4-element series II.

Frequencies of Peak α

Same as for 4-element series II.

Design Equations, Half-Section Series Arm

$$L_1 = [m_2/(1-m_2^2)]L_{2k}$$

$$C_1 = [(1-m_1^2)/m_1]C_{2k}$$

Design Equations, Half-Section Shunt Arm

$$L_2 = L_{2k}/m_2$$

$$C_2 = m_1 C_{2k}$$

5-Element Series I

Half Section

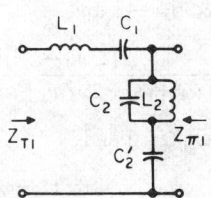

Impedance Characteristics

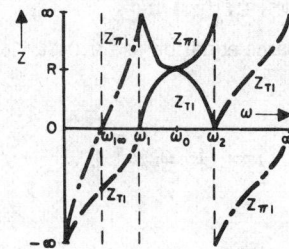

$$Z_{T1} = Z_{Tk}$$

$$Z_{\pi 1} = R\left[\frac{\omega^2(\omega_2^2+\omega_1^2-2\omega_0^2 m_2)+\omega_0^4(m_2^2-1)}{\omega(\omega_2-\omega_1)[(\omega_2^2-\omega^2)(\omega^2-\omega_1^2)]^{1/2}}\right]$$

Full-Section Attenuation α and Phase β Characteristics

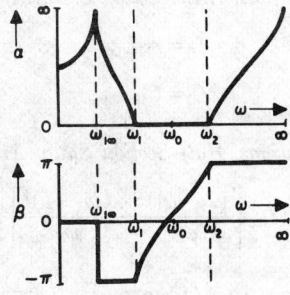

When $\omega_1 < \omega < \omega_2$, $\alpha = 0$ and

$$\beta = \cos^{-1}\left[1 - \frac{2(\omega^2-\omega_0^2 m_2)^2}{\omega^2(\omega_2^2+\omega_1^2-2\omega_0^2 m_2)+\omega_0^4(m_2^2-1)}\right]$$

When $0 < \omega < \omega_{1\infty}$, $\beta = 0$ and

$$\alpha = \cosh^{-1}\left[1 - \frac{2(\omega^2-\omega_0^2 m_2)^2}{\omega^2(\omega_2^2+\omega_1^2-2\omega_0^2 m_2)+\omega_0^4(m_2^2-1)}\right]$$

When $\omega_{1\infty} < \omega < \omega_1$, $\beta = -\pi$ and

$$\alpha = \cosh^{-1}\left[\frac{2(\omega^2-\omega_0^2 m_2)^2}{\omega^2(\omega_2^2+\omega_1^2-2\omega_0^2 m_2)+\omega_0^4(m_2^2-1)} - 1\right]$$

When $\omega_2 < \omega < \infty$, $\beta = \pi$ and

$$\alpha = \text{same equation as for } 0 < \omega < \omega_{1\infty}$$

Conditions

$$m_1 = 1$$

$$m_2 = (\omega_{1\infty}^2/\omega_0^2) + [(1-\omega_{1\infty}^2/\omega_1^2)(1-\omega_{1\infty}^2/\omega_2^2)]^{1/2}$$

Frequencies of Peak α

$$\omega_{1\infty} = \omega_0^2\left[\frac{1-m_2^2}{\omega_2^2+\omega_1^2-2\omega_0^2 m_2}\right]^{1/2}$$

$$\omega_{2\infty} = \infty$$

Design Equations, Half-Section Series Arm

$$L_1 = L_{1k}$$

$$C_1 = C_{1k}/m_2$$

Design Equations, Half-Section Shunt Arm

$$L_2 = \frac{L_{1k}}{m_2}\left[\frac{(\omega_2-\omega_1)^2}{\omega_0^2} - \frac{(1-m_2)^2}{m_2}\right]$$

$$C_2 = C_{1k} \Big/ \left[\frac{(\omega_2-\omega_1)^2}{\omega_0^2} - \frac{(1-m_2)^2}{m_2}\right]$$

$$C_2' = [m_2/(1-m_2^2)]C_{1k}$$

5-Element Shunt I

Half Section

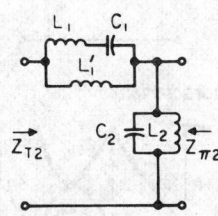

Impedance Characteristics

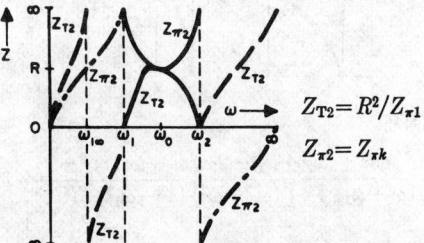

$$Z_{T2} = R^2/Z_{\pi 1}$$

$$Z_{\pi 2} = Z_{\pi k}$$

Full-Section Attenuation α and Phase β Characteristics

Same as for 5-element series I.

Conditions

Same as for 5-element series I.

Frequencies of Peak α

Same as for 5-element series I.

Design Equations, Half-Section Series Arm

$$L_1 = L_{2k} \bigg/ \left[\frac{(\omega_2 - \omega_1)^2}{\omega_0^2} - \frac{(1 - m_2)^2}{m_2} \right]$$

$$C_1 = \frac{C_{2k}}{m_2} \left[\frac{(\omega_2 - \omega_1)^2}{\omega_0^2} - \frac{(1 - m_2)^2}{m_2} \right]$$

$$L_1' = [m_2/(1 - m_2^2)] L_{2k}$$

Design Equations, Half-Section Shunt Arm

$$L_2 = L_{2k}/m_2$$

$$C_2 = C_{2k}$$

5-Element Series II

Half Section

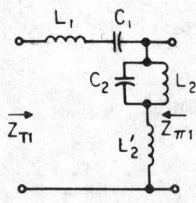

Impedance Characteristics

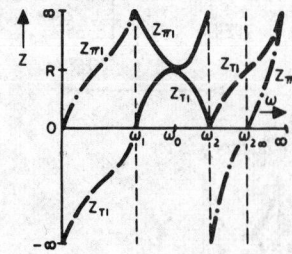

$$Z_{T1} = Z_{Tk}$$

$$Z_{\pi 1} = \frac{\omega R}{(\omega_2 - \omega_1)} \frac{\omega_2^2 + \omega_1^2 - 2\omega_0^2 m_1 + \omega^2 (m_1^2 - 1)}{[(\omega_2^2 - \omega^2)(\omega^2 - \omega_1^2)]^{1/2}}$$

Full-Section Attenuation α and Phase β Characteristics

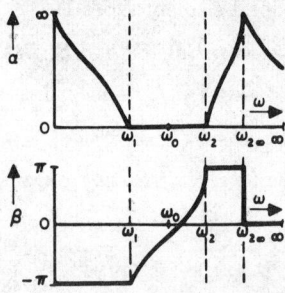

When $\omega_2 < \omega < \omega_{2\infty}$, $\beta = \pi$ and

$$\alpha = \cosh^{-1} \left[1 - \frac{2(\omega^2 m_1 - \omega_0^2)^2}{\omega^2 [\omega_2^2 + \omega_1^2 - 2\omega_0^2 m_1 + \omega^2 (m_1^2 - 1)]} \right]$$

When $0 < \omega < \omega_1$, $\beta = -\pi$ and

$$\alpha = \cosh^{-1} \left[\frac{2(\omega^2 m_1 - \omega_0^2)^2}{\omega^2 [\omega_2^2 + \omega_1^2 - 2\omega_0^2 m_1 + \omega^2 (m_1^2 - 1)]} - 1 \right]$$

When $\omega_1 < \omega < \omega_2$, $\alpha = 0$ and

$$\beta = \cos^{-1} \left[1 - \frac{2(\omega^2 m_1 - \omega_0^2)^2}{\omega^2 [\omega_2^2 + \omega_1^2 - 2\omega_0^2 m_1 + \omega^2 (m_1^2 - 1)]} \right]$$

When $\omega_{2\infty} < \omega < \infty$, $\beta = 0$ and

$$\alpha = \text{same equation as for } 0 < \omega < \omega_1$$

Conditions

$$m_1 = (\omega_0^2/\omega_{2\infty}^2) + [(1 - \omega_1^2/\omega_{2\infty}^2)(1 - \omega_2^2/\omega_{2\infty}^2)]^{1/2}$$

$$m_2 = 1$$

Frequencies of Peak α

$$\omega_{1\infty} = 0$$

$$\omega_{2\infty} = \left(\frac{\omega_2^2 + \omega_1^2 - 2\omega_0^2 m_1}{1 - m_1^2} \right)^{1/2}$$

Design Equations, Half-Section Series Arm

$$L_1 = m_1 L_{1k}$$

$$C_1 = C_{1k}$$

Design Equations, Half-Section Shunt Arm

$$L_2 = L_{1k} \left[\frac{(\omega_2 - \omega_1)^2}{\omega_0^2} - \frac{(m_1 - 1)^2}{m_1} \right]$$

$$C_2 = m_1 C_{1k} \bigg/ \left[\frac{(\omega_2 - \omega_1)^2}{\omega_0^2} - \frac{(m_1 - 1)^2}{m_1} \right]$$

$$L_2' = [(1 - m_1^2)/m_1] L_{1k}$$

5-Element Shunt II

Half Section

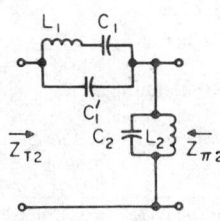

Impedance Characteristics

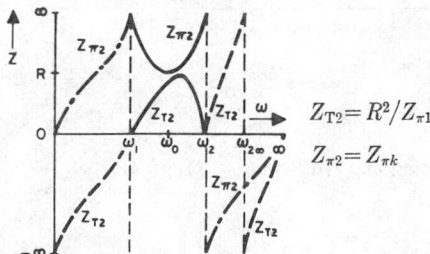

$$Z_{T2} = R^2/Z_{\pi 1}$$

$$Z_{\pi 2} = Z_{\pi k}$$

Full-Section Attenuation α and Phase β Characteristics

Same as for 5-element series II.

Conditions

Same as for 5-element series II.

Frequencies of Peak α

Same as for 5-element series II.

Design Equations, Half-Section Series Arm

$$L_1 = m_1 L_{2k} \Big/ \left[\frac{(\omega_2-\omega_1)^2}{\omega_0^2} - \frac{(m_1-1)^2}{m_1} \right]$$

$$C_1 = C_{2k} \left[\frac{(\omega_2-\omega_1)^2}{\omega_0^2} - \frac{(m_1-1)^2}{m_1} \right]$$

$$C_1' = \left[(1-m_1^2)/m_1 \right] C_{2k}$$

Design Equations, Half-Section Shunt Arm

$$L_2 = L_{2k}$$

$$C_2 = m_1 C_{2k}$$

6-Element Series

Half Section

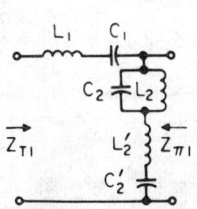

Impedance Characteristics

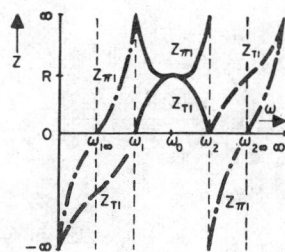

$$Z_{T1} = Z_{Tk}$$

$$Z_{\pi 1} = \frac{R}{\omega(\omega_2-\omega_1)}$$

$$\times \frac{(\omega_2^2-\omega^2)(\omega^2-\omega_1^2) + (\omega_0^2 m_2 - \omega^2 m_1)^2}{\left[(\omega_2^2-\omega^2)(\omega^2-\omega_1^2) \right]^{1/2}}$$

Full-Section Attenuation α and Phase β Characteristics

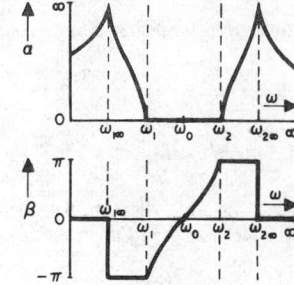

When $\omega_1 < \omega < \omega_2$, $\alpha = 0$ and

$$\beta = \cos^{-1}\left[1 - \frac{2(\omega^2 m_1 - \omega_0^2 m_2)^2}{(\omega^2 m_1 - \omega_0^2 m_2)^2 + (\omega_2^2-\omega^2)(\omega^2-\omega_1^2)} \right]$$

When $\omega_2 < \omega < \omega_{2\infty}$, $\beta = \pi$ and

$$\alpha = \cosh^{-1}$$

$$\times \left[\frac{2(\omega^2 m_1 - \omega_0^2 m_2)^2}{(\omega^2 m_1 - \omega_0^2 m_2)^2 + (\omega_2^2-\omega^2)(\omega^2-\omega_1^2)} + 1 \right]$$

When $0 < \omega < \omega_{1\infty}$, $\beta = 0$ and

$$\alpha = \cosh^{-1}$$

$$\times \left[1 - \frac{2(\omega^2 m_1 - \omega_0^2 m_2)^2}{(\omega^2 m_1 - \omega_0^2 m_2)^2 + (\omega_2^2-\omega^2)(\omega^2-\omega_1^2)} \right]$$

When $\omega_{1\infty} < \omega < \omega_1$, $\beta = -\pi$ and

$$\alpha = \cosh^{-1}$$

$$\times \left[\frac{2(\omega^2 m_1 - \omega_0^2 m_2)^2}{(\omega^2 m_1 - \omega_0^2 m_2)^2 + (\omega_2^2-\omega^2)(\omega^2-\omega_1^2)} - 1 \right]$$

When $\omega_{2\infty} < \omega < \infty$, $\beta = 0$ and

α = same equation as for $0 < \omega < \omega_{1\infty}$

Conditions

$$m_1 = \frac{g(\omega_0^2/\omega_{2\infty}^2)+h}{1-(\omega_{1\infty}^2/\omega_{2\infty}^2)}$$

$$m_2 = \frac{g+h(\omega_{1\infty}^2/\omega_0^2)}{1-(\omega_{1\infty}^2/\omega_{2\infty}^2)}$$

Frequencies of Peak α

$$\omega_{1\infty}^2 + \omega_{2\infty}^2 = \frac{\omega_2^2 + \omega_1^2 - 2\omega_0^2 m_1 m_2}{1-m_1^2}$$

$$\omega_{1\infty}^2 \times \omega_{2\infty}^2 = \omega_0^4 \left(\frac{1-m_2^2}{1-m_1^2}\right)$$

Design Equations, Half-Section Series Arm

$$L_1 = m_1 L_{1k}$$

$$C_1 = C_{1k}/m_2$$

Design Equations, Half-Section Shunt Arm

$$L_2 = \frac{L_{1k}}{m_2}\left[\frac{(\omega_2-\omega_1)^2}{\omega_0^2} - \frac{(m_1-m_2)^2}{m_1 m_2}\right]$$

$$L_2' = \left[(1-m_1^2)/m_1\right]L_{1k}$$

$$C_2 = \frac{m_1 C_{1k}}{\left[(\omega_2-\omega_1)^2/\omega_0^2\right]-\left[(m_1-m_2)^2/m_1 m_2\right]}$$

$$C_2' = \left[m_2/(1-m_2^2)\right]C_{1k}$$

6-Element Shunt

Half Section

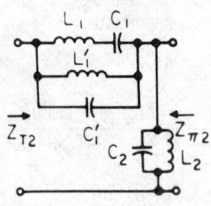

Impedance Characteristics

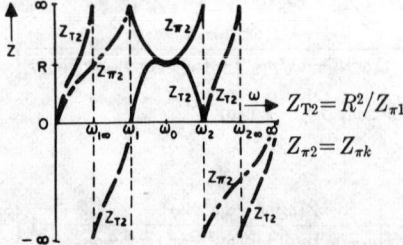

$$Z_{T2} = R^2/Z_{\pi 1}$$

$$Z_{\pi 2} = Z_{\pi k}$$

Full-Section Attenuation α and Phase β Characteristics

Same as for 6-element series.

Conditions

Same as for 6-element series.

Frequencies of Peak α

Same as for 6-element series.

Design Equations, Half-Section Series Arm

$$L_1 = \frac{m_1 L_{2k}}{\left[(\omega_2-\omega_1)^2/\omega_0^2\right]-\left[(m_1-m_2)^2/m_1 m_2\right]}$$

$$C_1 = \frac{C_{2k}}{m_2}\left[\frac{(\omega_2-\omega_1)^2}{\omega_0^2} - \frac{(m_1-m_2)^2}{m_1 m_2}\right]$$

$$L_1' = \left[m_2/(1-m_2^2)\right]L_{2k}$$

$$C_1' = \left[(1-m_1^2)/m_1\right]C_{2k}$$

Design Equations, Half-Section Shunt Arm

$$L_2 = L_{2k}/m_2$$

$$C_2 = m_1 C_{2k}$$

BAND-STOP FILTER DESIGN

Notations

Z in ohms, α in nepers, and β in radians

$\omega_1 = $ lower cutoff angular frequency

$\omega_2 = $ upper cutoff angular frequency

$\omega_0 = (\omega_1\omega_2)^{1/2} = 1/(L_{1k}C_{1k})^{1/2}$

$\quad = 1/(L_{2k}C_{2k})^{1/2}$

$\omega_2 - \omega_1 = $ width of stop band

$\omega_{1\infty} = $ lower angular frequency of peak attenuation

$\omega_{2\infty} = $ upper angular frequency of peak attenuation

$R = $ nominal terminating resistance

$R^2 = L_{1k}/C_{2k} = L_{2k}/C_{1k}$

$\quad = Z_{1k}Z_{2k} = Z_{Tk}Z_{\pi k} = k^2$

$\quad = Z_{1(\text{series}-m)}Z_{2(\text{shunt}-m)}$

$\quad = Z_{2(\text{series}-m)}Z_{1(\text{shunt}-m)}$

$\quad = Z_{T2}Z_{\pi 1}$

Constant-k

Half Section

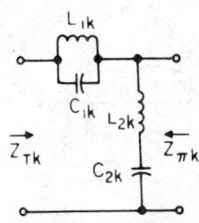

Impedance Characteristics

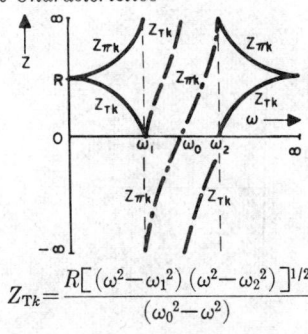

$$Z_{Tk} = \frac{R\left[(\omega^2 - \omega_1^2)(\omega^2 - \omega_2^2)\right]^{1/2}}{(\omega_0^2 - \omega^2)}$$

$$Z_{\pi k} = \frac{R(\omega_0^2 - \omega^2)}{\left[(\omega^2 - \omega_1^2)(\omega^2 - \omega_2^2)\right]^{1/2}}$$

For the pass bands, use $|\omega_0^2 - \omega^2|$ in the above equations.

Full-Section Attenuation α and Phase β Characteristics

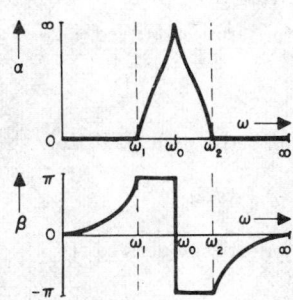

When $\omega = \omega_0$

$$\alpha = \infty$$

When $\omega_0 < \omega < \omega_2$

$$\alpha = 2 \cosh^{-1} \frac{\omega(\omega_2 - \omega_1)}{\omega^2 - \omega_0^2}$$

$$\beta = -\pi$$

When $\omega_2 < \omega < \infty$

$$\alpha = 0$$

$$\beta = 2 \sin^{-1} \frac{\omega(\omega_2 - \omega_1)}{\omega_0^2 - \omega^2}$$

When $\omega_1 < \omega < \omega_0$

$$\alpha = 2 \cosh^{-1} \frac{\omega(\omega_2 - \omega_1)}{\omega_0^2 - \omega^2}$$

$$\beta = \pi$$

When $0 < \omega < \omega_1$

$$\alpha = 0$$

$$\beta = 2 \sin^{-1} \frac{\omega(\omega_2 - \omega_1)}{\omega_0^2 - \omega^2}$$

Frequencies of Peak α

$$\omega_\infty = \omega_0$$

Design Equations, Half-Section Series Arm

$$L_{1k} = R(\omega_2 - \omega_1)/\omega_1\omega_2$$

$$C_{1k} = 1/R(\omega_2 - \omega_1)$$

Design Equations, Half-Section Shunt Arm

$$L_{2k} = R/(\omega_2 - \omega_1)$$

$$C_{2k} = (\omega_2 - \omega_1)/\omega_1\omega_2 R$$

Series m-Derived

Half Section

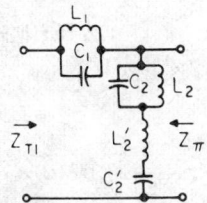

Impedance Characteristics

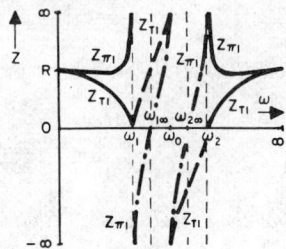

(Curves drawn for $m = 0.6$)

$$Z_{T1} = Z_{Tk}$$

$$Z_{\pi 1} = R \left\{ \frac{1 - (1 - m^2)\left[\omega(\omega_2 - \omega_1)/(\omega_0^2 - \omega^2)\right]^2}{\{1 - \left[\omega(\omega_2 - \omega_1)/(\omega_0^2 - \omega^2)\right]^2\}^{1/2}} \right\}$$

Full-Section Attenuation α and Phase β Characteristics

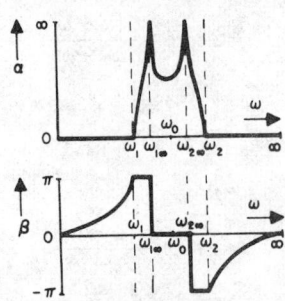

(Curves drawn for $m = 0.6$)

When $\omega_2 < \omega < \infty$, $\alpha = 0$ and

$\beta =$ same equation as for $0 < \omega < \omega_1$

When $\omega_{2\infty} < \omega < \omega_2$, $\beta = -\pi$ and

$\alpha =$ same equation as for $\omega_1 < \omega < \omega_{1\infty}$

When $0 < \omega < \omega_1$, $\alpha = 0$ and

$$\beta = \cos^{-1}\left[1 - \frac{2\omega^2 m^2 (\omega_2 - \omega_1)^2}{(\omega^2 - \omega_1^2)(\omega^2 - \omega_2^2) + \omega^2 m^2 (\omega_2 - \omega_1)^2}\right]$$

When $\omega_1 < \omega < \omega_{1\infty}$, $\beta = \pi$ and

$$\alpha = \cosh^{-1}\left[\frac{2\omega^2 m^2 (\omega_2 - \omega_1)^2}{(\omega^2 - \omega_1^2)(\omega^2 - \omega_2^2) + \omega^2 m^2 (\omega_2 - \omega_1)^2} - 1\right]$$

When $\omega_{1\infty} < \omega < \omega_{2\infty}$, $\beta = 0$ and

$$\alpha = \cosh^{-1}\left[1 - \frac{2\omega^2 m^2 (\omega_2 - \omega_1)^2}{(\omega^2 - \omega_1^2)(\omega^2 - \omega_2^2) + \omega^2 m^2 (\omega_2 - \omega_1)^2}\right]$$

Conditions

$$m = \left\{1 - \left[(\omega_{2\infty} - \omega_{1\infty})^2 / (\omega_2 - \omega_1)^2\right]\right\}^{1/2}$$

Frequencies of Peak α

$$\omega_{1\infty}\omega_{2\infty} = \omega_0^2$$

Design Equations, Half-Section Series Arm

$$L_1 = mL_{1k}$$

$$C_1 = C_{1k}/m$$

Design Equations, Half-Section Shunt Arm

$$L_2 = \left[(1 - m^2)/m\right]L_{1k}$$

$$C_2 = \left[m/(1 - m^2)\right]C_{1k}$$

$$L_2' = L_{2k}/m$$

$$C_2' = mC_{2k}$$

Shunt m-Derived

Half Section

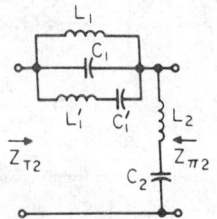

Impedance Characteristics

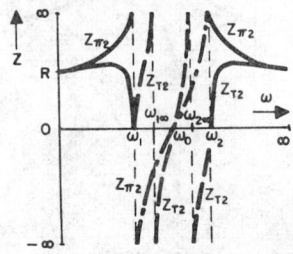

(Curves drawn for $m = 0.6$)

$$Z_{T2} = R^2/Z_{\pi 1}$$

$$Z_{\pi 2} = Z_{\pi k}$$

Full-Section Attenuation α and Phase β Characteristics

Same as for series m-derived.

Conditions

Same as for series m-derived.

Frequencies of Peak α

Same as for series m-derived.

Design Equations, Half-Section Series Arm

$$L_1 = mL_{1k}$$

$$C_1 = C_{1k}/m$$

$$L_1' = \left[m/(1 - m^2)\right]L_{2k}$$

$$C_1' = \left[(1 - m^2)/m\right]C_{2k}$$

Design Equations, Half-Section Shunt Arm

$$L_2 = L_{2k}/m$$

$$C_2 = mC_{2k}$$

BUILDING UP A COMPOSITE FILTER

The intermediate sections (Fig. 2) are matched on an image-impedance basis, but the attenuation characteristics of the sections may be varied by suitably choosing the infinite attenuation frequencies of each section. Thus, the frequencies attenuated only slightly by one section may be strongly attenuated by other sections. However, the image impedance will be far from constant in the pass band and therefore the use of true resistors for terminations will change the attenuation shape.

Some improvement in the uniformity of the image impedance is obtained by using suitably designed terminating half sections. For these terminating sections, a value of $m \approx 0.6$ is usually used (Fig. 3).

EXAMPLE OF LOW-PASS IMAGE-PARAMETER DESIGN

To cut off at 15 kilohertz; to give peak attenuation at 30 kilohertz; with a load resistance of 600 ohms; and using a constant-k midsection and an m-derived midsection. Full T-sections will be used.

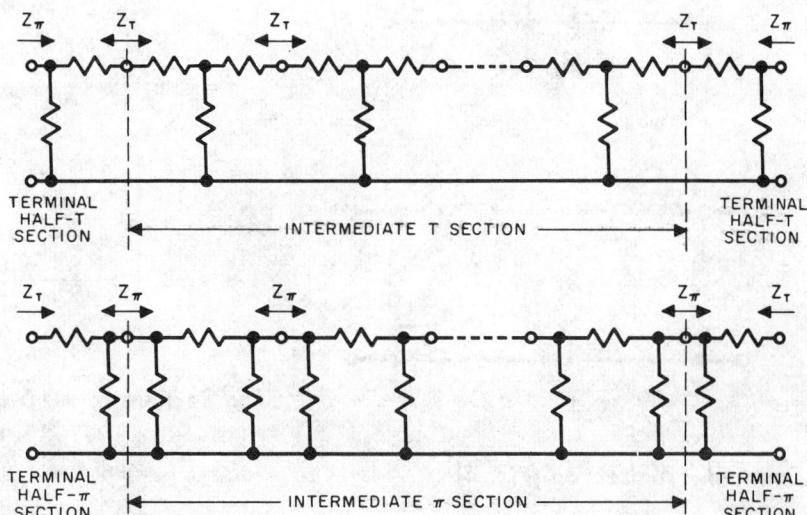

Fig. 2—Method of building up a composite filter.

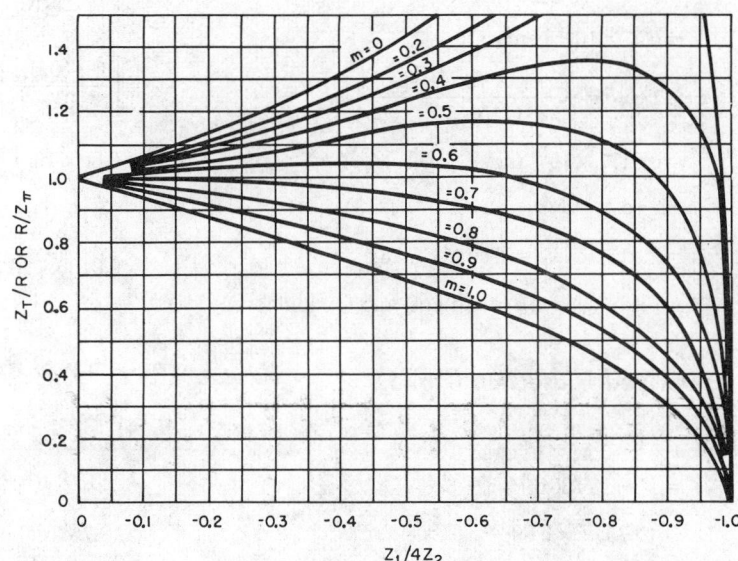

Fig. 3—Effect of design parameter m on the image-impedance characteristics in the pass band.

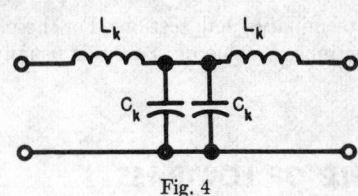

Fig. 4

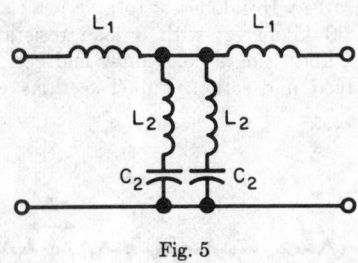

Fig. 5

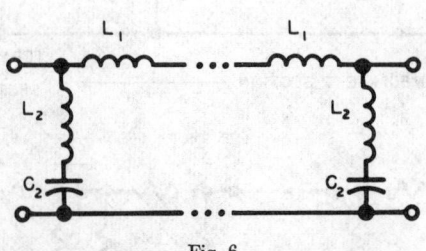

Fig. 6

$$L_1 = mL_k = 0.866(6.37 \times 10^{-3})$$

$$= 5.52 \times 10^{-3} \text{ henry}$$

$$L_2 = [(1-m^2)/m]L_k$$

$$= \left[\frac{1-(0.866)^2}{0.866}\right](6.37 \times 10^{-3})$$

$$= 1.84 \times 10^{-3} \text{ henry}$$

$$C_2 = mC_k = 0.866(0.0177 \times 10^{-6})$$

$$= 0.0153 \times 10^{-6} \text{ farad}$$

$$\alpha = \cosh^{-1}\left[1 - \frac{2m^2}{(\omega_c^2/\omega^2) - (1-m^2)}\right]$$

$$= \cosh^{-1}\left[1 - \frac{1.5}{(225/f^2) - 0.25}\right]$$

$$\beta = \cos^{-1}\left[1 - \frac{2m^2}{(\omega_c^2/\omega^2) - (1-m^2)}\right]$$

$$= \cos^{-1}\left[1 - \frac{1.5}{(225/f^2) - 0.25}\right]$$

End Sections m = 0.6 (Fig. 6)

$$L_1 = mL_k = 0.6(6.37 \times 10^{-3})$$

$$= 3.82 \times 10^{-3} \text{ henry}$$

$$L_2 = [(1-m^2)/m]L_k$$

$$= \left[\frac{1-(0.6)^2}{0.6}\right](6.37 \times 10^{-3})$$

$$= 6.80 \times 10^{-3} \text{ henry}$$

$$C_2 = mC_k = 0.6(0.0177 \times 10^{-6})$$

$$= 0.0106 \times 10^{-6} \text{ farad.}$$

Constant-k Midsection (Fig. 4)

$$L_k = R/\omega_c = \frac{600}{(6.28)(15 \times 10^3)}$$

$$= 6.37 \times 10^{-3} \text{ henry}$$

$$C_k = 1/\omega_c R = \frac{1}{(6.28)(15 \times 10^3)(600)}$$

$$= 0.0177 \times 10^{-6} \text{ farad}$$

$$\alpha = 2\cosh^{-1}(\omega/\omega_c) = 2\cosh^{-1}(f/15)$$

$$\beta = 2\sin^{-1}(\omega/\omega_c) = 2\sin^{-1}(f/15)$$

where α is in nepers, β in radians, and f in kilohertz.

m-Derived Midsection (Fig. 5)

$$m = (1 - \omega_c^2/\omega_\infty^2)^{1/2}$$

$$= (1 - 15^2/30^2)^{1/2}$$

$$= (0.75)^{1/2} = 0.866$$

Frequency of Peak Attenuation f_∞

$$f_\infty = [f_c^2/(1-m^2)]^{1/2}$$

$$= \{(15 \times 10^3)^2/[1-(0.6)^2]\}^{1/2}$$

$$= 18.75 \text{ kilohertz.}$$

Filter Showing Individual Sections (Fig. 7)

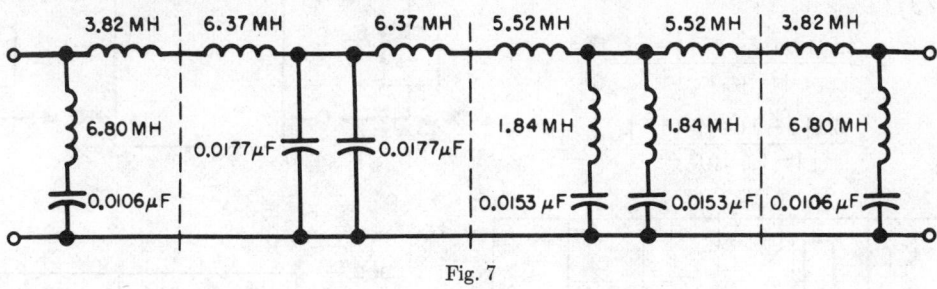

Fig. 7

Filter After Combining Elements (Fig. 8)

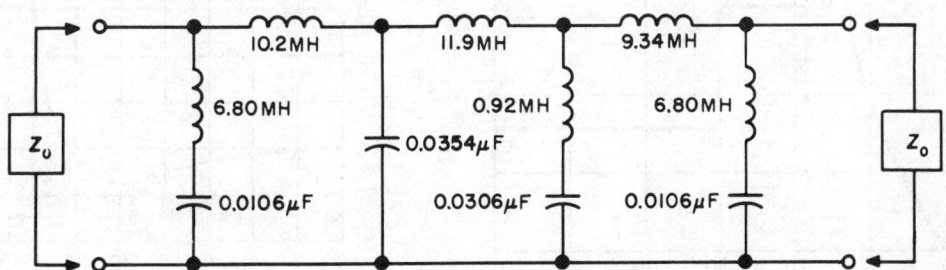

Fig. 8

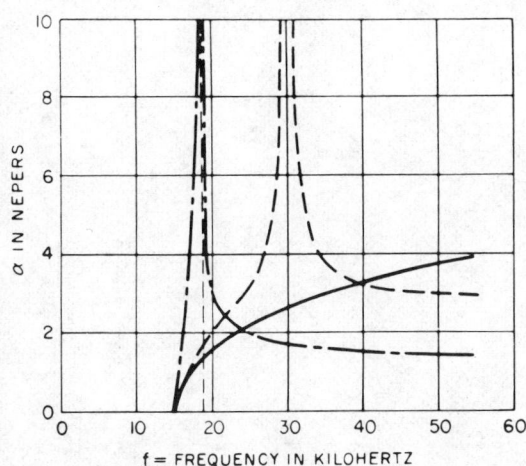

Fig. 9—Image-terminated attenuation of each section. Solid line, constant-k midsection. Dashed, m-derived midsection. Dash–dot, m-derived ends.

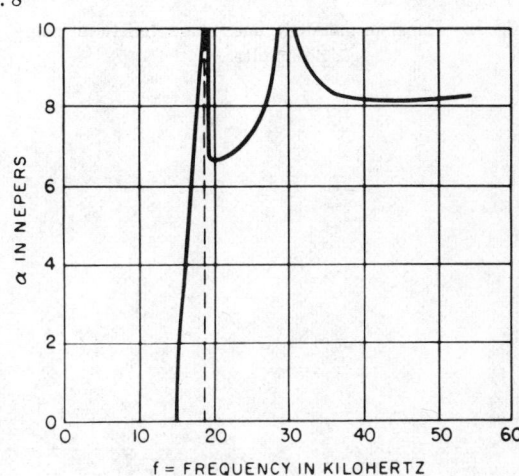

Fig. 10—Image-terminated attenuation of composite filter.

Image Attenuation and Phase Characteristics

Figures 9 through 12 are the image-terminated attenuation and phase characteristics. These shapes are not obtainable when 600-ohm resistors are used in place of the terminating Z_0.

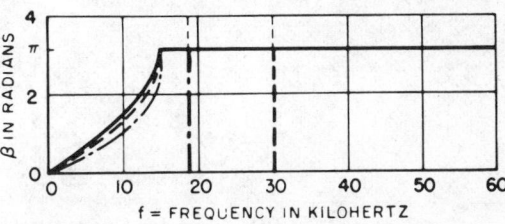

Fig. 11—Image-terminated phase characteristic of each section. Solid line, constant-k midsection. Dashed, m-derived midsection. Dash–dot, m-derived ends.

Impedance required for Proper Termination (Fig. 13)

$$Z_0 = \frac{R[1-(\omega^2/\omega_c^2)(1-m^2)]}{(1-\omega^2/\omega_c^2)^{1/2}}$$

$$= \frac{600[1-0.64(f/15)^2]}{[1-(f/15)^2]^{1/2}}$$

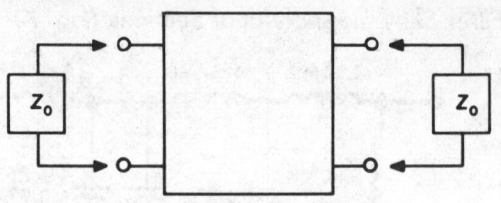

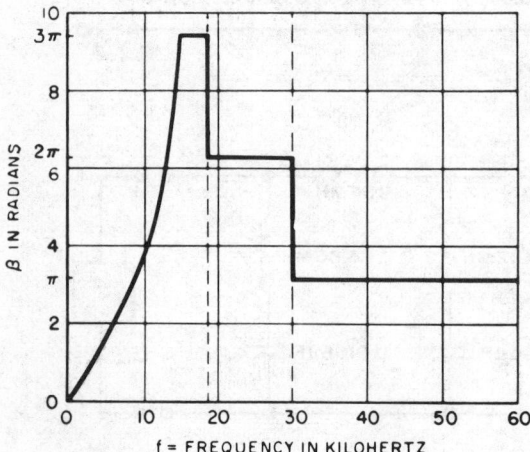

Fig. 12—Image-terminated phase characteristic of composite filter.

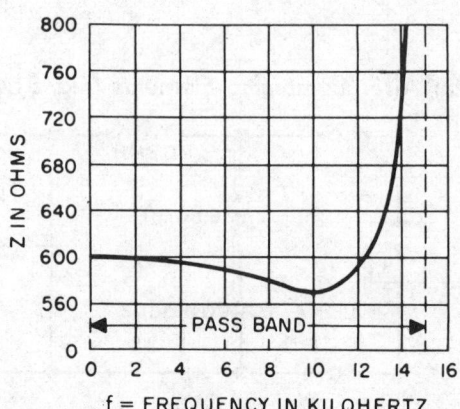

Fig. 13

CHAPTER 8
FILTERS, MODERN-NETWORK-THEORY DESIGN

The design information in this chapter results from the application of modern network theory to electric wave filters. Only design results are supplied, and a careful study of the cited references is required for an understanding of the synthesis procedures that underlie these results.

LIMITATIONS OF IMAGE-PARAMETER THEORY

Consider the simple low-pass ladder network of Fig. 1A. Two simultaneous design equations, (1) and (2), are provided by classical image-parameter theory (refer to Chapter 7).

$$(Z_1/4Z_2)_{f=fc} = -1 \text{ and } 0 \qquad (1)$$

$$Z_{0T} = (Z_1 Z_2)^{1/2}[1+(Z_1/4Z_2)]^{1/2}. \qquad (2)$$

Z_1 and Z_2, the full series- and shunt-arm impedances, respectively, must be suitably related to make (1) true at the desired cutoff frequencies, and the generator and load impedance must satisfy (2). Under the image-parameter theory, the resulting attenuation for the low-pass case is

$$V_p/V = 1.0, \qquad (\omega/\omega_c) < 1$$

$$= \exp[(n-1)\cosh^{-1}(\omega/\omega_c)], \quad (\omega/\omega_c) > 1$$

$$(3)$$

where n is the number of arms in the network of Fig. 1 and V_p/V and ω are as in Fig. 3. It is this attenuation shape that is plotted in the tabulations of Chapter 7.

Equation (1) offers no problems. The application of (2) to Fig. 1 demands *terminating impedances that are physically impossible with a finite number of elements.* The generator and load impedances for Fig. 1A must be pure resistances of $(L/C)^{1/2}$ ohms at zero frequency. As frequency increases, the value of resistance must decrease to a short-circuit at the cutoff frequency, and with further increase in frequency must behave like a pure inductance starting at zero value at the cutoff

frequency and increasing to $L/2$ at infinite frequency.

The physical impracticability of devising such terminating impedances is why element values obtained by (1) cannot simultaneously satisfy (2). The relative attenuation indicated by (3) is similarly incorrect and cannot be realized in practice.

Lattice-configuration filters also require impractical terminating impedances when designed by image-parameter theory. (Constant-resistance lattices are an exception but are seldom used for filtering.) The practical use of resistive terminations automatically makes element values computed on the basis of ideal impedance terminations incorrect.

For more than four decades, filters have been designed according to the image-parameter theory. Their commercial acceptance is due in no small part to the highly approximate requirements for most filters. Where more-exact characteristics are required, shifting of element values in the actual filter has usually resulted in an acceptable design. For precise amplitude and phase response in the pass band, the simple and approximate solutions obtained through image-parameter theory must give way to equations based on modern network theory.

MODERN-NETWORK-THEORY DESIGN

Relative Attenuation

A typical low-pass filter with resistive generator and load is shown in Fig. 1B. It is composed of lumped inductors, capacitors, and the resistive elements unavoidably associated therewith. The circuit equations for the complete network can be written by applying Kirchhoff's laws. Modern network theory does just this and then solves the equations to find the network parameters that will produce optimum performance in some desired respect.

A block diagram of a generalized filter is illustrated in Fig. 2. This may be of low-pass, high-pass, band-pass, band-rejection, phase-compen-

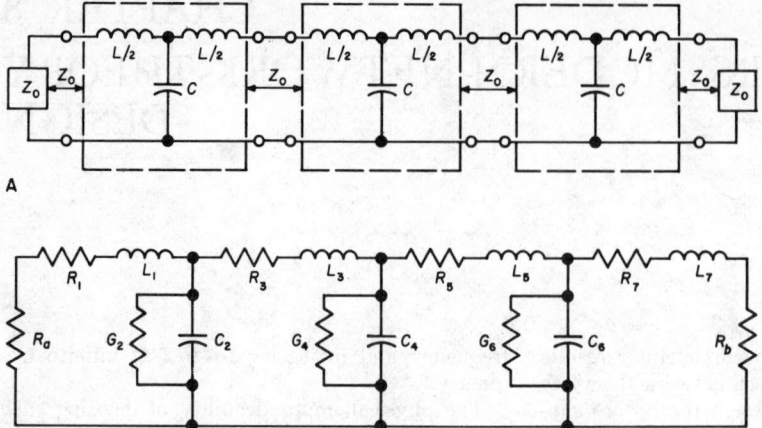

A

B

Fig. 1—A 7-element low pass filter considered on the basis of image-parameter theory at A and of modern network theory at B.

sating, or other type. The elements of the filter include resistors, capacitors, self- and mutual-inductors, and possibly coupling elements such as electron tubes or transistors, all according to the design. The terminations shown are a constant-voltage generator (the same voltage at all frequencies) with a series resistor at the input and a resistive load. (Frequently it is preferable to stipulate a constant-current generator with a shunt conductance.) The generator and load resistors need not be equal and they can be assigned any value between zero and infinity. Characteristic impedance plays no part in the modern network theory of filters.

Either or both the generator or load can be reactive, in which case the reactances are absorbed inside the block of Fig. 2 as specified parts of the filter. Either, but not both, R_a or R_b can be zero or infinite.

The term *bandwidth* as used herein has two different meanings, according to the type of filter. For low- or high-pass filters, it is synonymous with the actual frequency of the point in question, or equivalent to the number of hertz in a band terminated on one side by zero frequency and on the other by the actual frequency. The actual frequency can be anywhere in the pass or the reject region. For symmetrical band-pass (Fig. 4) and band-reject filters, it is the difference in hertz between two particular frequencies (anywhere in the pass or reject regions) with the requirement that their geometric mean be equal to the geometric midfrequency f_0 of the pass or reject band.

Typical filter characteristics are plotted in Fig. 3 for a low-pass filter. In Fig. 3A, the magnitude

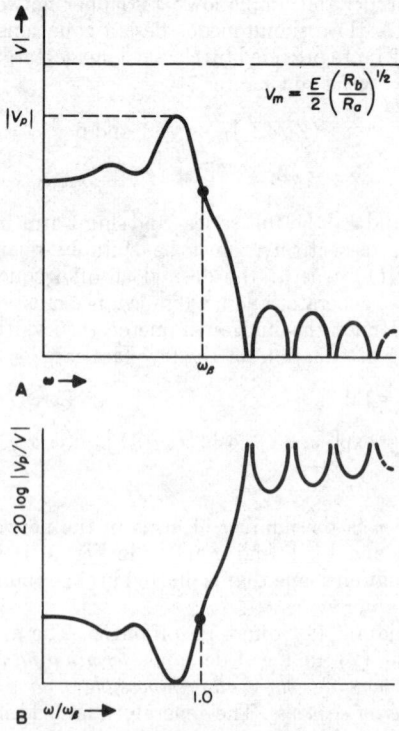

Fig. 3—Low-pass-filter output voltage versus frequency at A; attenuation versus normalized frequency at B. A is the actual voltage across the load as a function of frequency and is for the low-pass case. B uses the information in A to produce a plot of *relative* attenuation against *relative* bandwidth.

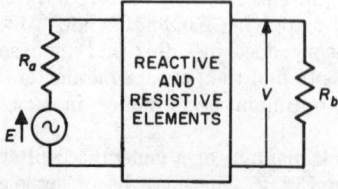

Fig. 2—Block diagram of a filter. The generator and load must be considered part of the filter.

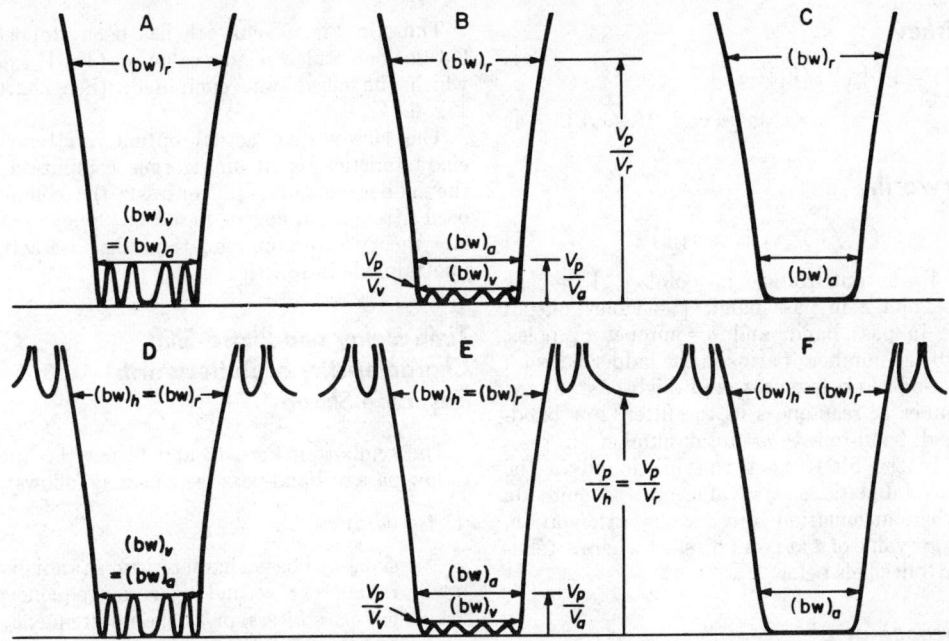

Fig. 4—A, B, C, are the optimum relative attenuation shapes of (4) and (5) that can be produced by networks supplying only transfer-function poles. D, E, F, are the optimum relative attenuation shapes of (8), (12), (13), (16) that can be produced by networks supplying both transfer-function poles and zeros.

of the output voltage V is plotted against radian bandwidth ω. Several specific points are indicated on the diagram. V_p is the peak voltage output, while V_m is the maximum voltage that could be developed across the load were it matched to the generator through an ideal network. Symbol ω_β designates a specified frequency or bandwidth where some particular characteristic is exhibited by the filter, such as the point where the response is 3 decibels down from the peak, for example.

The characteristic of major interest to the filter engineer is the plot, shown in Fig. 3B, of relative attenuation versus relative bandwidth. Relative attenuation is defined as the ratio of the peak output voltage V_p to the voltage output V at the frequency being considered. Relative bandwidth is defined as the ratio of the bandwidth being considered to a clearly specified reference bandwidth (e.g., the 3-decibel-down bandwidth).

It should be noted that the elements of a filter are not uniquely fixed if only a certain relative attenuation shape is specified; in general it is possible also to demand that at one frequency the absolute magnitude of some transfer function be optimized.

The complex relative attenuation of a complete filter (including generator and load) composed of lumped linear passive elements is always equal to a constant multiplied by the ratio of two polynomials in ($j\omega$). The complex roots of the numerator polynomial are commonly called attenuation poles or transfer-function zeros; the complex roots of the denominator polynomial are commonly called attenuation zeros or transfer-function poles. Modern filter theory has derived various expressions for optimum relative attenuation shapes that can be physically realized from these complex expressions. The shapes are optimum in that they give the maximum possible rate of cutoff between the accept and reject bands for a given number of filter components, with a specified allowable equal ripple in the accept band, and a specified required equal ripple in the reject band. See Fig. 4 for typical shapes of attenuation characteristic for band-pass filters.

CHEBISHEV AND BUTTERWORTH PERFORMANCE WITH FILTERS SUPPLYING ONLY TRANSFER-FUNCTION POLES

The attenuation-curve shapes illustrated in Figs. 4A and 4B are termed Chebishev and that in Fig. 4C is termed Butterworth. The equations for these shapes are (4) and (5), respectively. The Butterworth shape is the same as the limiting case of the Chebishev shape when we set $V_p/V_v = 1.0$.

Chebishev

$$(V_p/V)^2 = 1 + [(V_p/V_v)^2 - 1]$$
$$\times \cosh^2[n \cosh^{-1}(x/x_v)]. \quad (4)$$

Butterworth

$$(V_p/V)^2 = 1 + (x/x_{3\mathrm{dB}})^{2n} \quad (5)$$

where V = output voltage at point x, V_p = peak output voltage in pass band, V_v = valley output voltage in pass band, and n = number of poles, equal to the number of arms in the ladder network being used. For low-pass and high-pass filters, n = number of reactances in the filter. For band-pass and band-reject, n = total number of resonators in the filter, x = a variable found in the following tabulations, x_v = value of x at point on skirt where attenuation equals valley attenuation, and $x_{3\mathrm{dB}}$ = value of x at point on skirt where attenuation is 3 decibels below V_p.

Significance of x

Low-Pass Filters:

$$x = \omega = 2\pi f.$$

High-Pass Filters:

$$x = -1/\omega = -1/2\pi f.$$

Symmetrical Band-Pass Filters:

$$x = (\omega/\omega_0 - \omega_0/\omega) = (f_2 - f_1)/f_0 = (\mathrm{bw})/f_0.$$

Symmetrical Band-Reject Filters:

$$x = -1/(\omega/\omega_0 - \omega_0/\omega) = -f_0/(\mathrm{bw})$$

where $f_0 = (f_1 f_2)^{1/2}$ = midfrequency of the pass or reject band and f_1, f_2 = two frequencies where the characteristic exhibits the same attenuation.

Working charts for these filters, derived from (4) and (5), are presented in Figs. 5 through 12 for values of n from 1 through 8, respectively. These curves give

$$(V_p/V)_{\mathrm{dB}} = 20 \log_{10}(V_p/V)$$

versus $x/x_{3\mathrm{dB}}$.

For low-pass and band-pass filters

$$x/x_{3\mathrm{dB}} = (\mathrm{bw})/(\mathrm{bw})_{3\mathrm{dB}}.$$

For high-pass and band-reject filters, the scale of the abscissa gives $(\mathrm{bw})_{3\mathrm{dB}}/(\mathrm{bw})$.

In Figs. 5 through 12, the family of curves toward the right side gives the attenuation shape for points where it is less than 3 decibels, while those toward the left are for the reject band (greater than 3 decibels). Each curve of the former family has been stopped where the attenuation is equal to that of the peak-to-valley ratio.

Thus, in Fig. 6, curve 3 has been stopped at 0.3 decibel, which is the value of $(V_p/V_v)_{\mathrm{dB}}$ for which the curve was computed. (See chart on Fig. 6.)

The curves give actual optimum attenuation characteristics based on rigorous computation of the ladder network. In contrast, the commonly used attenuation curves based on "image-parameter theory" are approximations that are actually unattainable in practice.

Time-Delay and Phase-Shift Characteristics of Butterworth Response Shape

The symbols of Figs. 13 and 14 may be applied to low-pass or band-pass responses as follows:

$t_0 = d\theta/dw$

     = slope of phase characteristic in radians per radian per second at zero frequency for low-pass filters, or at the midfrequency for band-pass filters

t = slope of phase characteristic at a frequency Δf removed from zero frequency for low-pass filters, or from the midfrequency for band-pass filters

$\Delta f_{3\mathrm{dB}}$ = 3-decibel-down bandwidth of the low-pass filter or half the total 3-decibel-down bandwidth of the band-pass filter, in hertz.

Low- and Band-Pass Filters—Required Unloaded Q

Filters supplying only transfer-function poles can be constructed that will actually give the attenuation shapes predicted by modern network theory. To attain this result, it is required that the unloaded Q of each element be greater than a certain minimum value[*]. The $q_{\min}$ column on the charts in Figs. 5 through 12 is used in the following manner to obtain this minimum allowable

---

[*] S. Darlington, "Synthesis of Reactance 4-Poles," *Journal of Mathematics and Physics*, vol. *18*, pp. 257–353; September 1939. Also, M. Dishal, "Design of Dissipative Band-Pass Filters Producing Desired Exact Amplitude-Frequency Characteristics," *Proceedings of the IRE*, vol. *37*, pp. 1050–1069; September 1949: also, *Electrical Communication*, vol. *27*, pp. 56–81; March 1950. Also, M. Dishal, "Concerning the Minimum Number of Resonators and the Minimum Unloaded Q Needed in a Filter," *Transactions of the IRE Professional Group on Vehicular Communication*, vol. PGVC–*3*, pp. 85–117; June 1953: also, *Electrical Communication*, vol. *31*, pp. 257–277; December 1954.

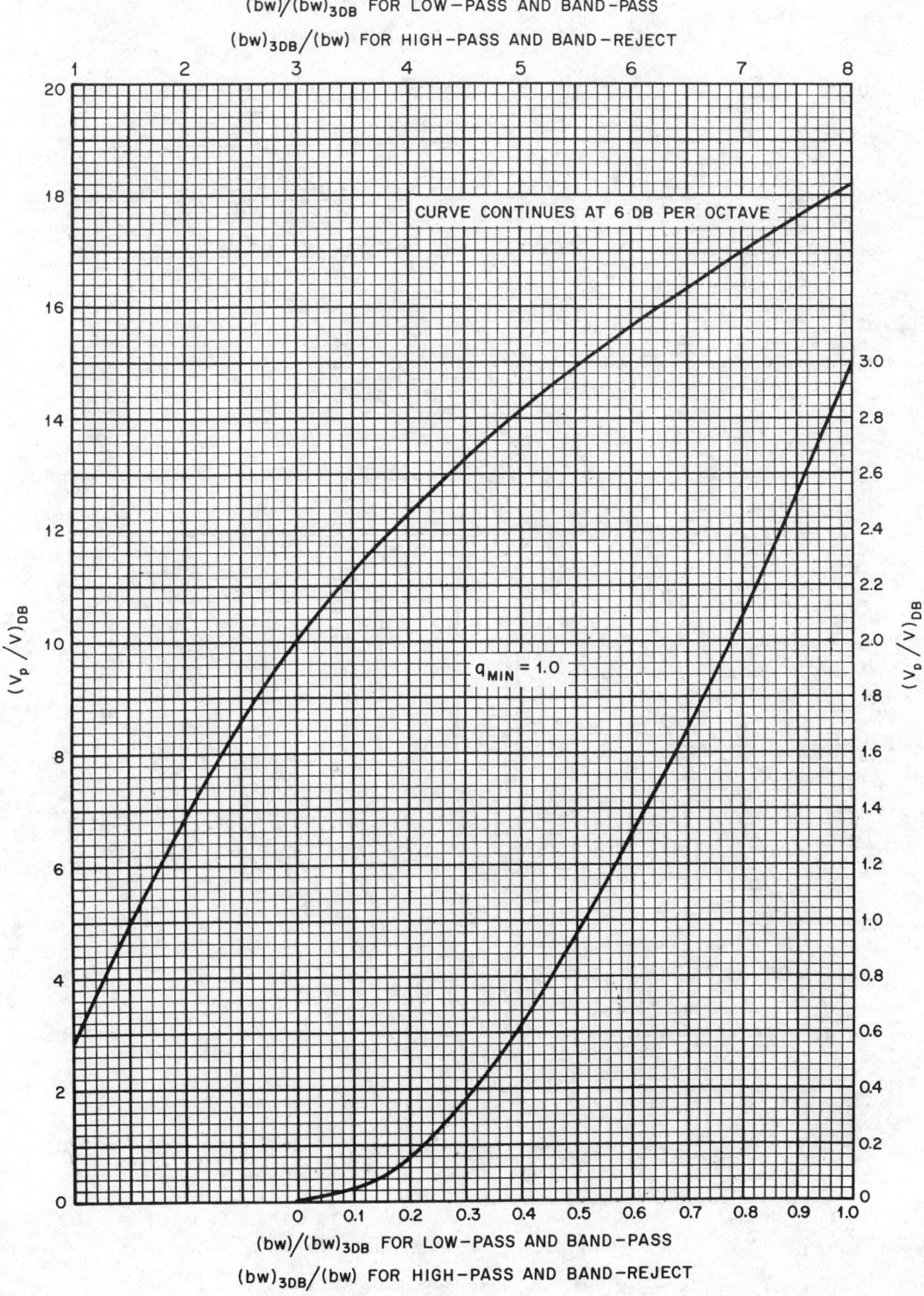

Fig. 5—Relative attenuation for a 1-pole network.

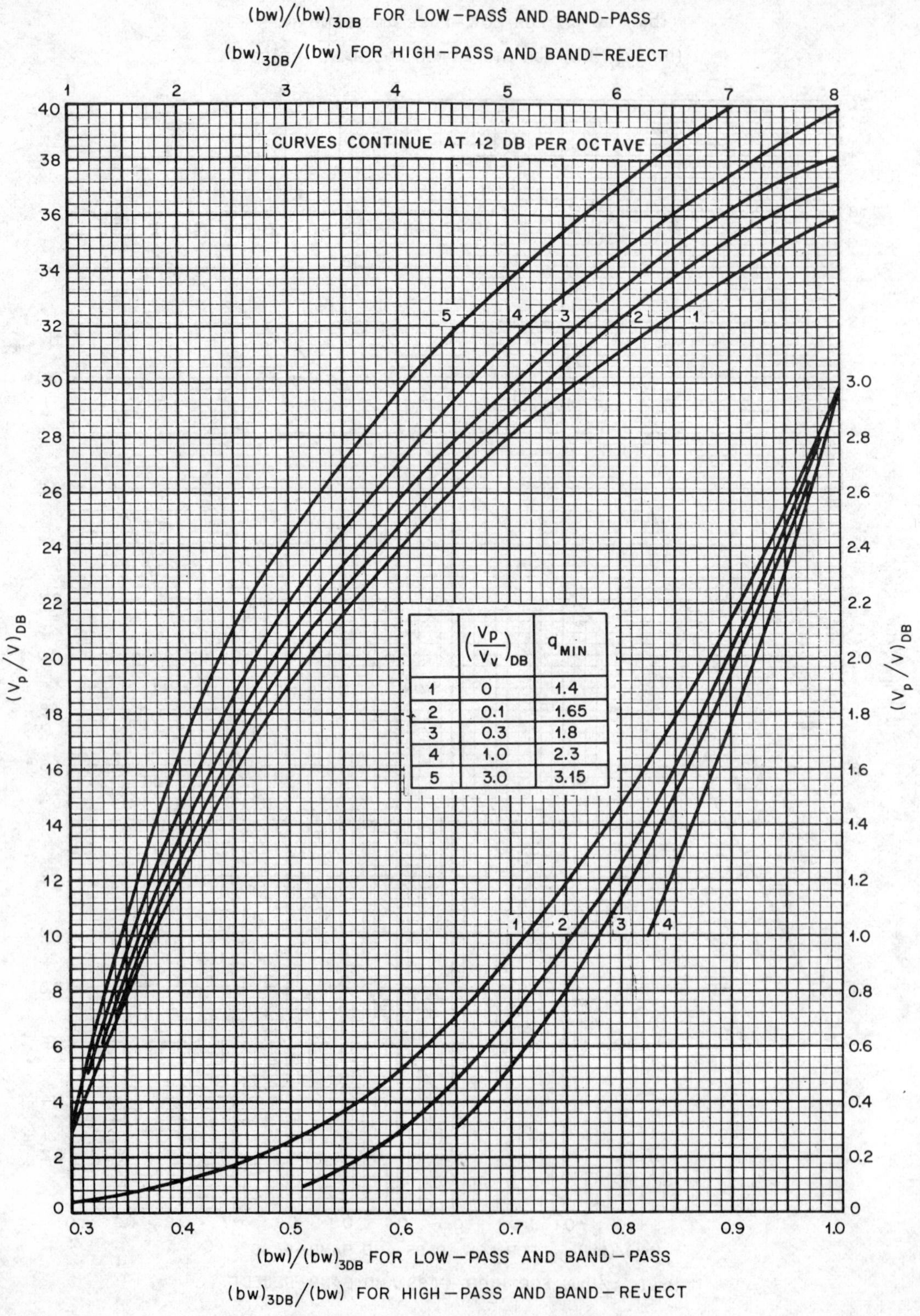
(bw)/(bw)3DB FOR LOW-PASS AND BAND-PASS
(bw)3DB/(bw) FOR HIGH-PASS AND BAND-REJECT
1 2 3 4 5 6 7 8
CURVES CONTINUE AT 12 DB PER OCTAVE
0 2 4 6 8 10 12 14 16 18 20 22 24 26 28 30 32 34 36 38 40
(Vp/V)DB
0 0.2 0.4 0.6 0.8 1.0 1.2 1.4 1.6 1.8 2.0 2.2 2.4 2.6 2.8 3.0
(Vp/V)DB
5 4 3 2 1
1 2 3 4
0.3 0.4 0.5 0.6 0.7 0.8 0.9 1.0
(bw)/(bw)3DB FOR LOW-PASS AND BAND-PASS
(bw)3DB/(bw) FOR HIGH-PASS AND BAND-REJECT

	$\left(\frac{V_p}{V_v}\right)_{DB}$	q_{MIN}
1	0	1.4
2	0.1	1.65
3	0.3	1.8
4	1.0	2.3
5	3.0	3.15

Fig. 6—Relative attenuation for a 2-pole network.

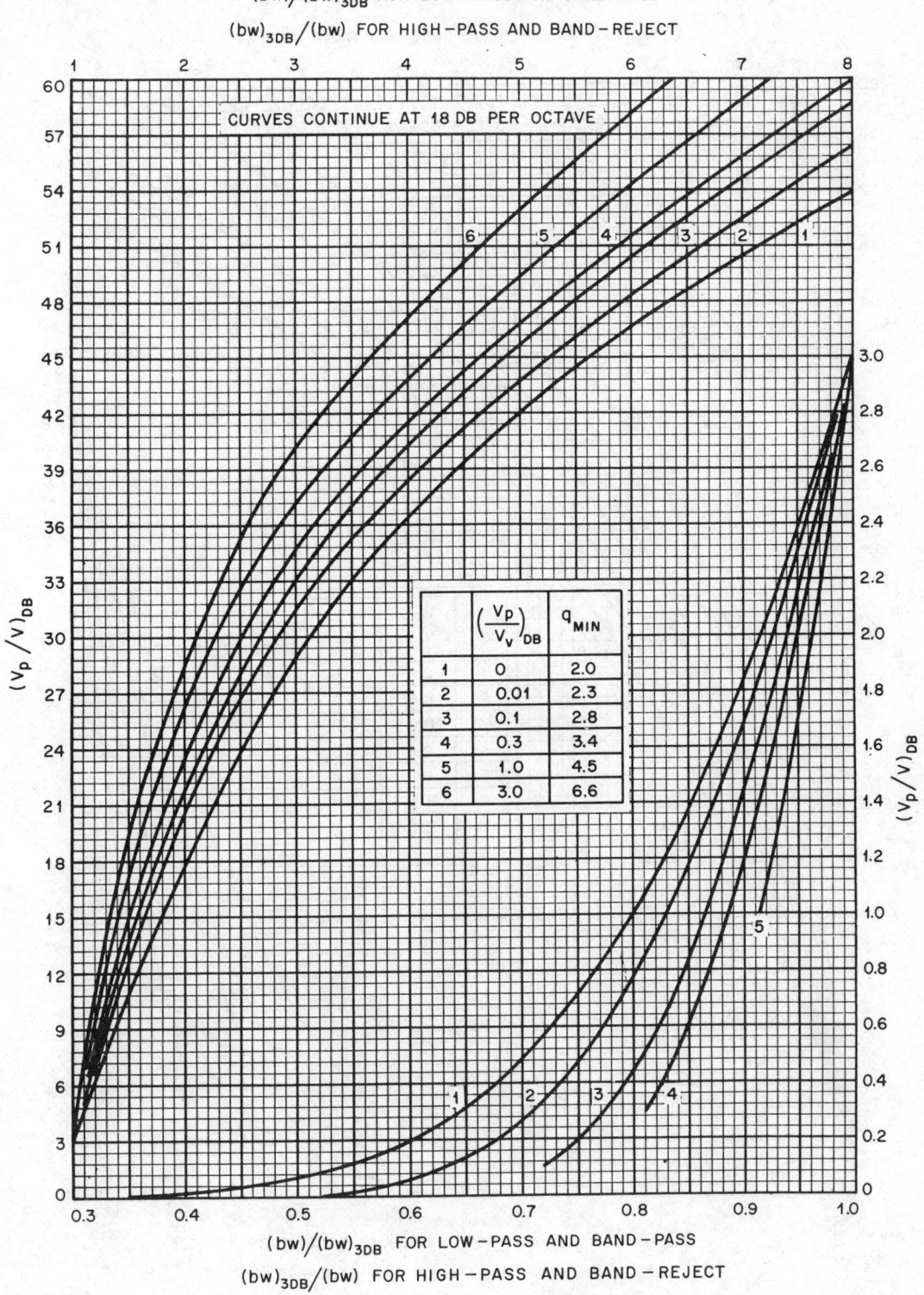

Fig. 7—Relative attenuation for a 3-pole network.

$(bw)/(bw)_{3DB}$ FOR LOW-PASS AND BAND-PASS

$(bw)_{3DB}/(bw)$ FOR HIGH-PASS AND BAND-REJECT

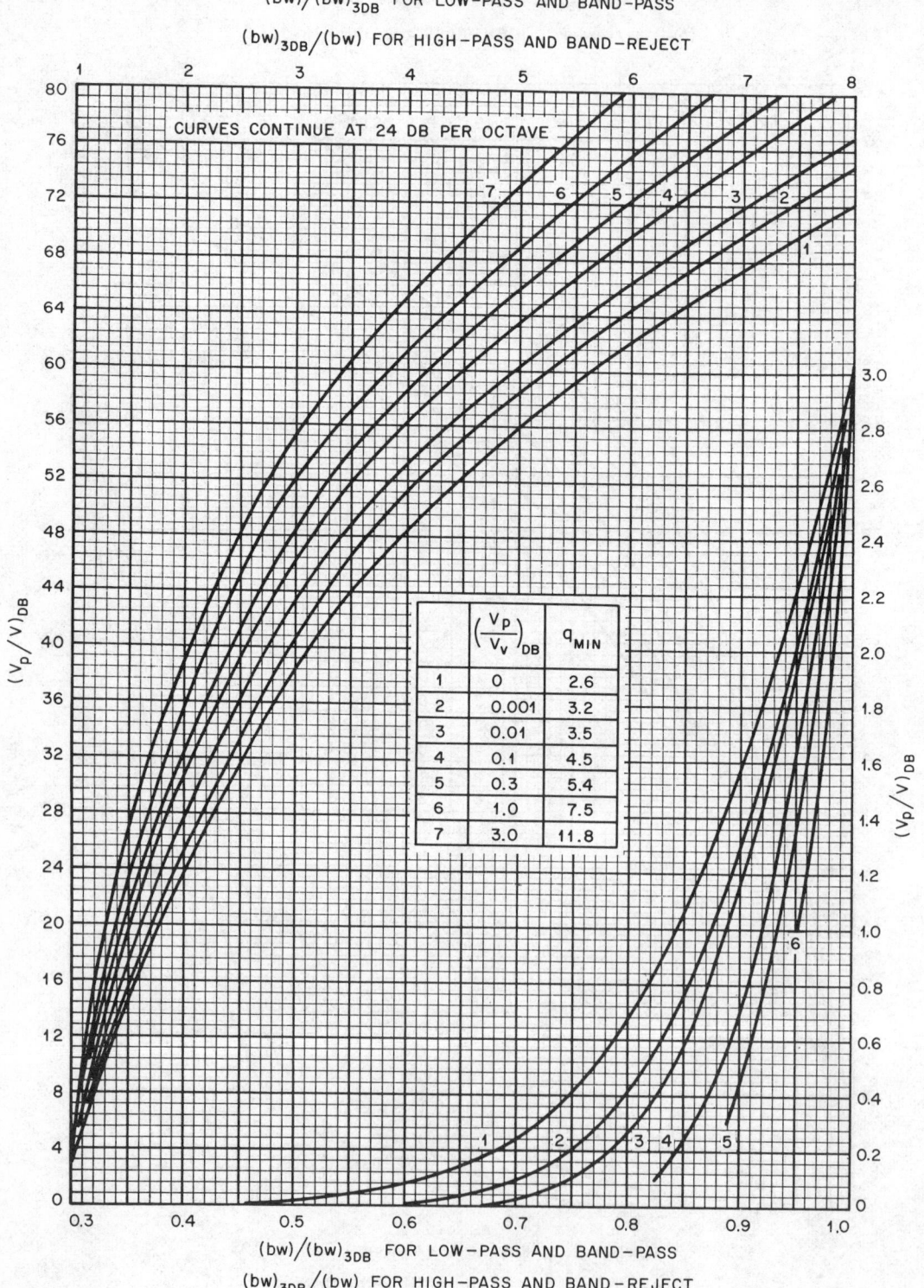

	$\left(\dfrac{V_p}{V_v}\right)_{DB}$	q_{MIN}
1	0	2.6
2	0.001	3.2
3	0.01	3.5
4	0.1	4.5
5	0.3	5.4
6	1.0	7.5
7	3.0	11.8

CURVES CONTINUE AT 24 DB PER OCTAVE

$(bw)/(bw)_{3DB}$ FOR LOW-PASS AND BAND-PASS

$(bw)_{3DB}/(bw)$ FOR HIGH-PASS AND BAND-REJECT

Fig. 8—Relative attenuation for a 4-pole network.

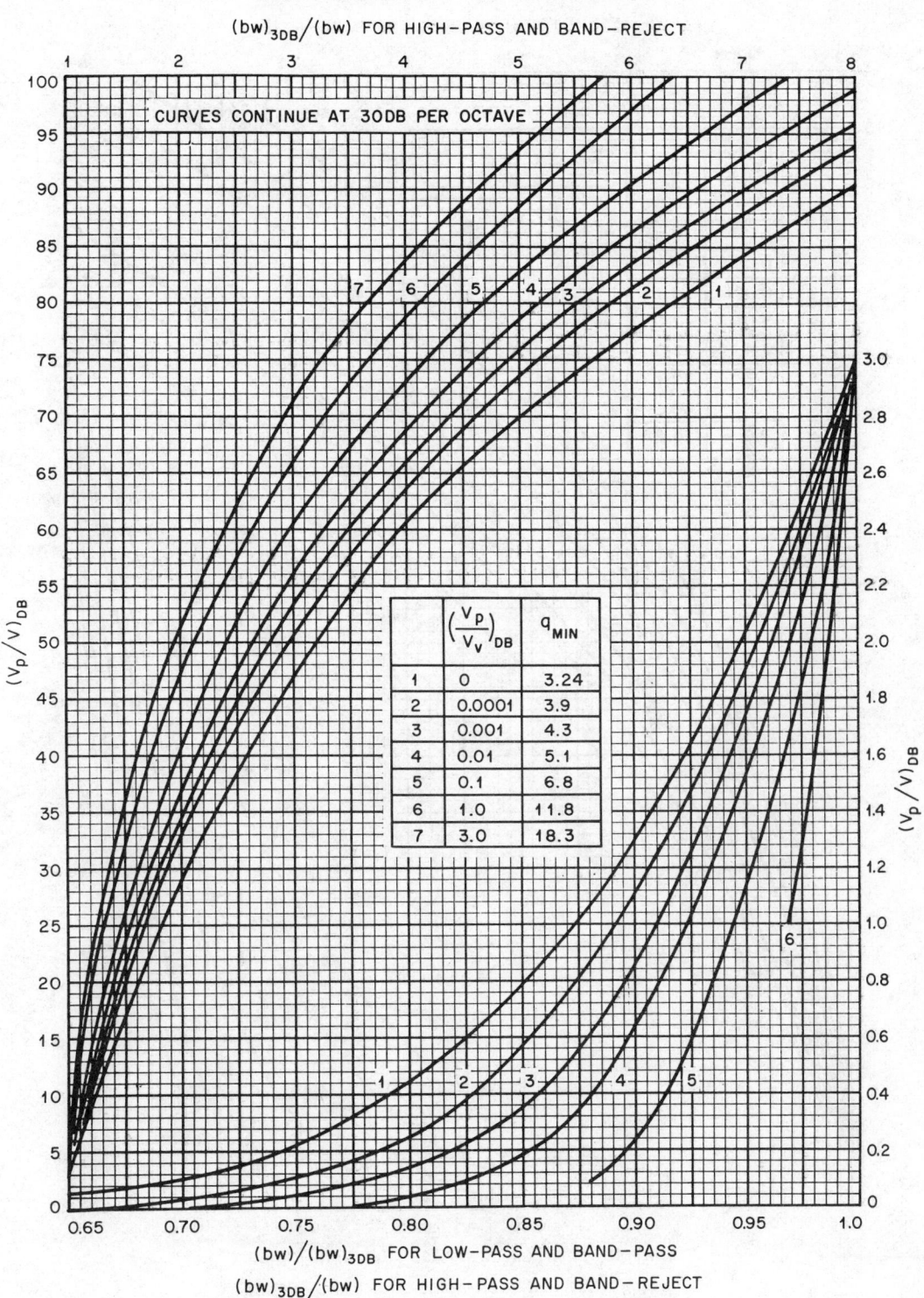

Fig. 9—Relative attenuation for a 5-pole network.

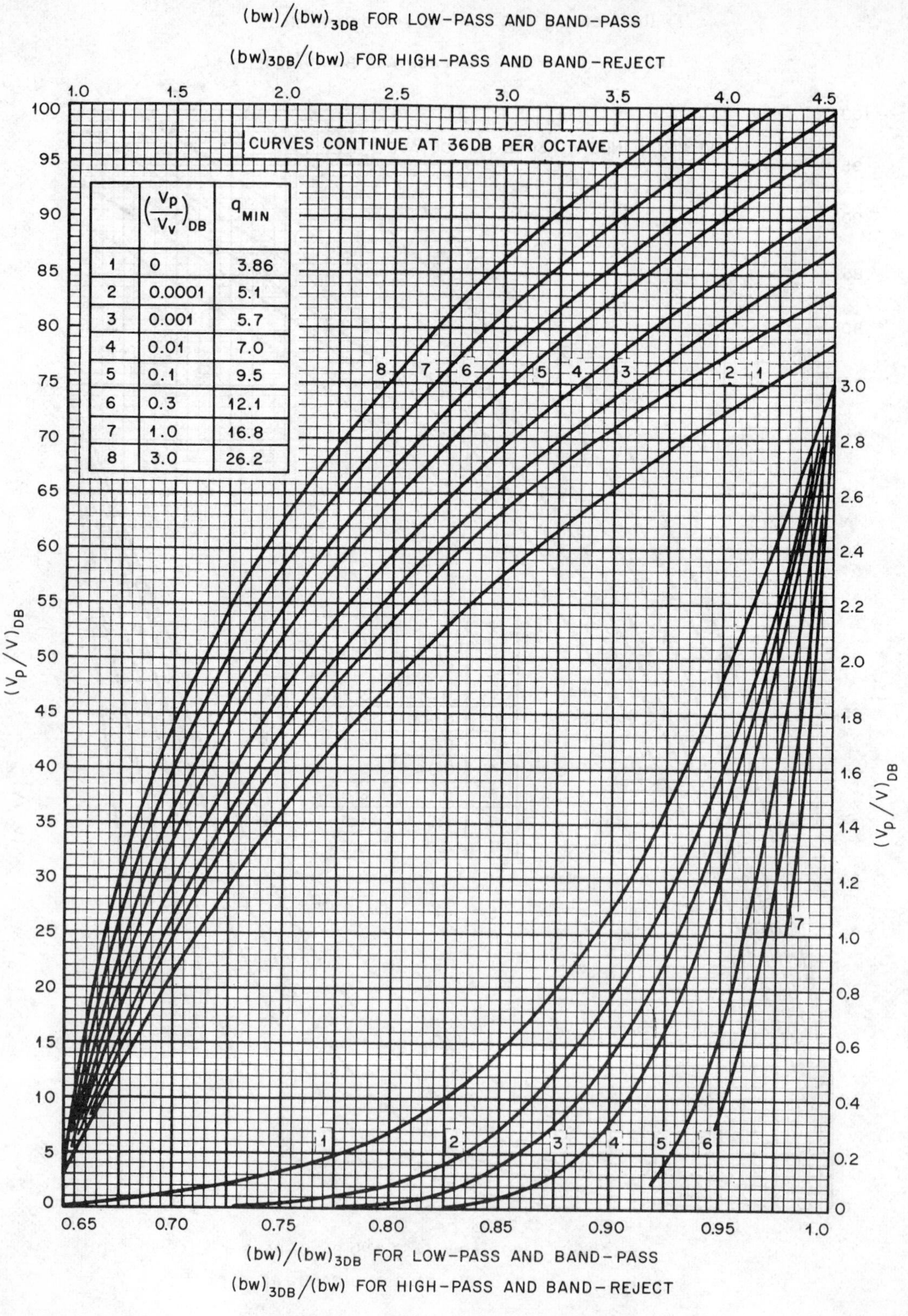

(bw)/(bw)$_{3DB}$ FOR LOW-PASS AND BAND-PASS

(bw)$_{3DB}$/(bw) FOR HIGH-PASS AND BAND-REJECT

Fig. 10—Relative attenuation for a 6-pole network.

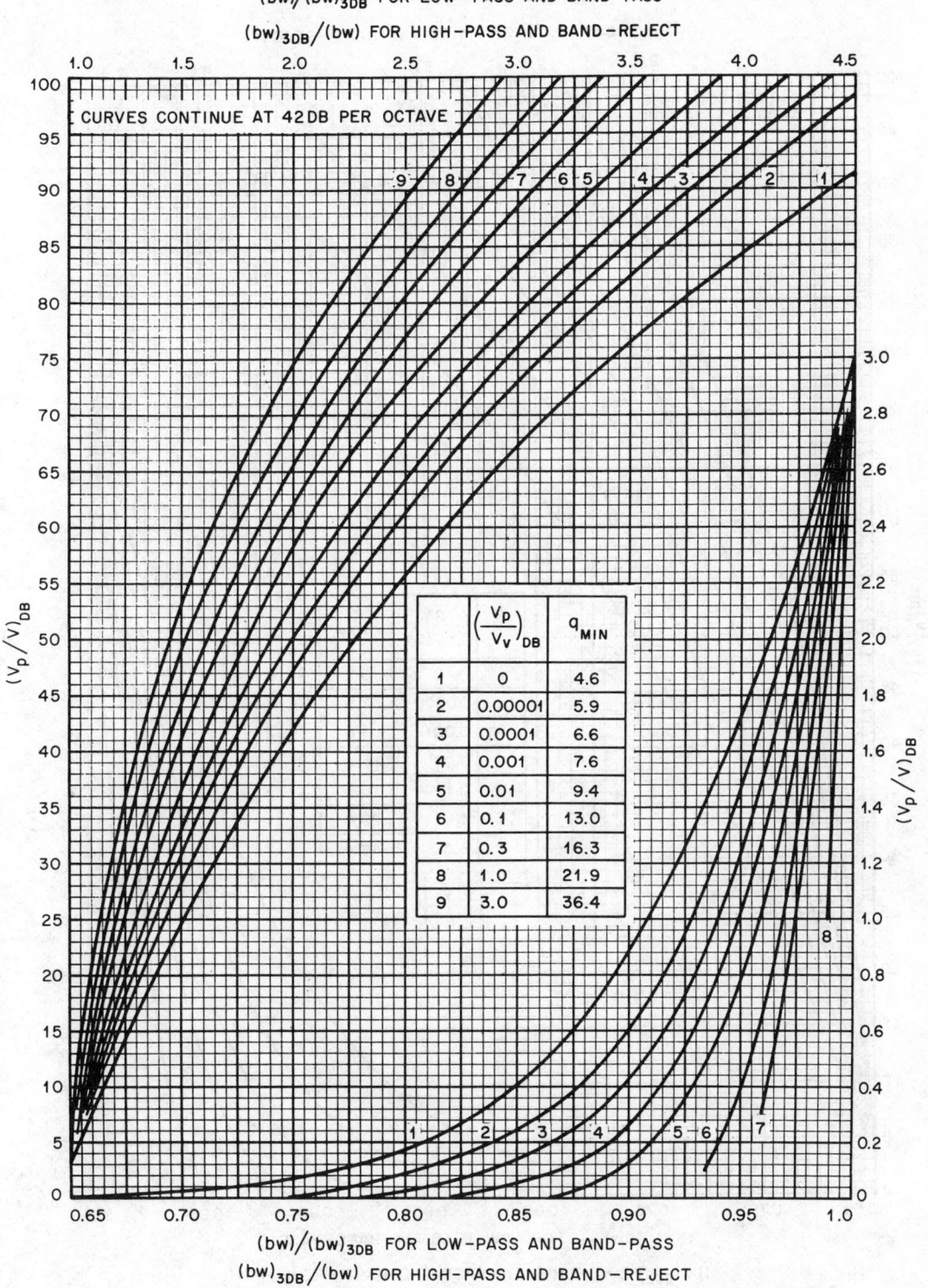

Fig. 11—Relative attenuation for a 7-pole network.

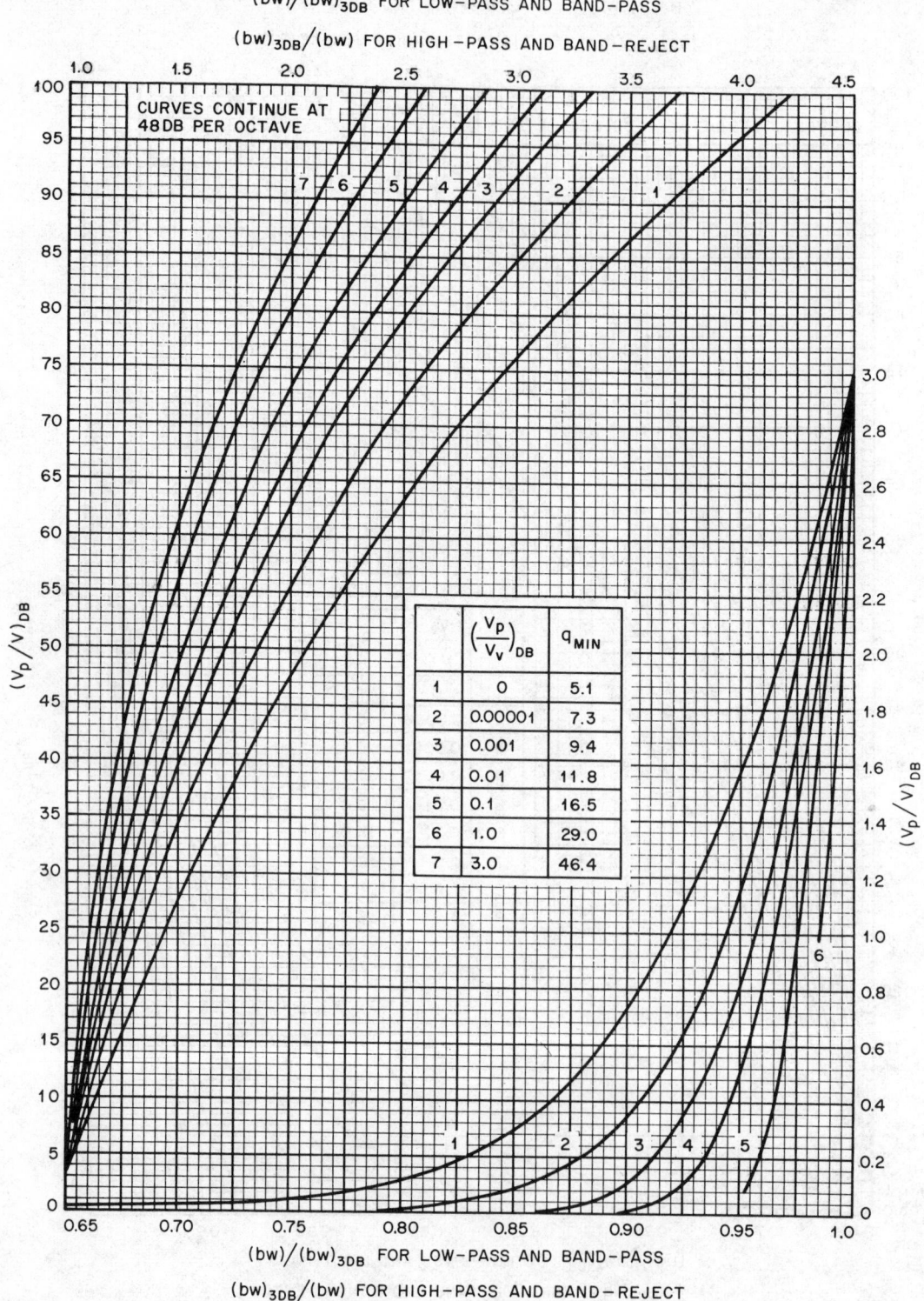

Fig. 12—Relative attenuation for an 8-pole network.

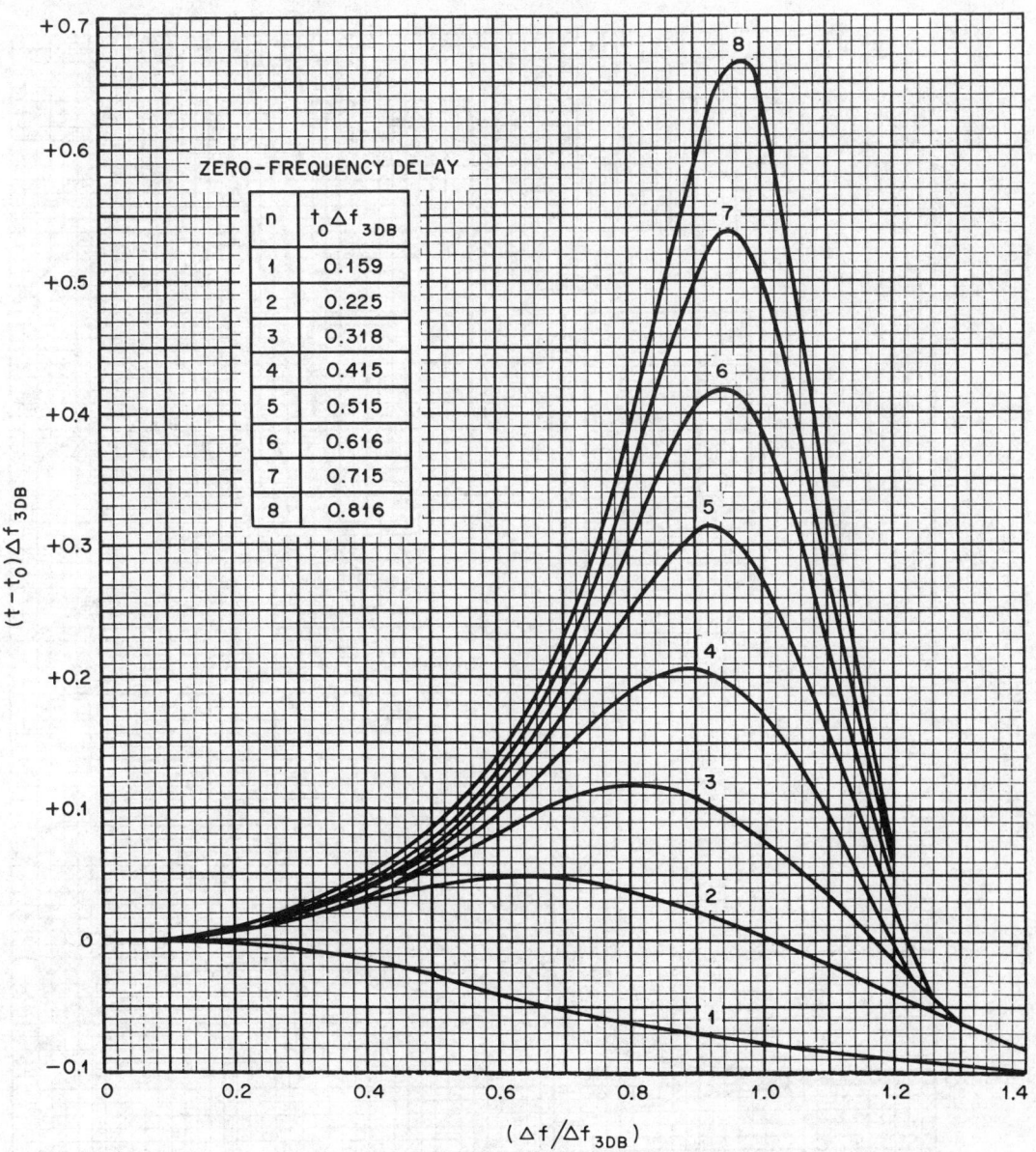

ZERO-FREQUENCY DELAY

n	$t_0 \Delta f_{3DB}$
1	0.159
2	0.225
3	0.318
4	0.415
5	0.515
6	0.616
7	0.715
8	0.816

Fig. 13—Delay distortion of Butterworth response shape.

value: For the internal reactances of low-pass circuits

$$Q_{min} = q_{min}.$$

For the internal resonators of band-pass circuits

$$Q_{min} = q_{min} [f_0 / (bw)_{3dB}].$$

Examples

(A) In a low-pass filter without any peaks of infinite attenuation at a finite frequency, how few elements are required to satisfy the following specifications, and what minimum Q must they have? Response to be 1 decibel down at 30 kilohertz and 50 decibels down at not more than 75 kilohertz, compared with the peak response.

The allowable ripple is 1 decibel in the pass band. Then

$$(bw)_{50dB} / (bw)_{1dB} < 75/30 = 2.5$$

$$(V_p / V_v)_{dB} \leq 1.0 \text{ decibel.}$$

Since $(bw)_{1dB}$ will be slightly less than $(bw)_{3dB}$,

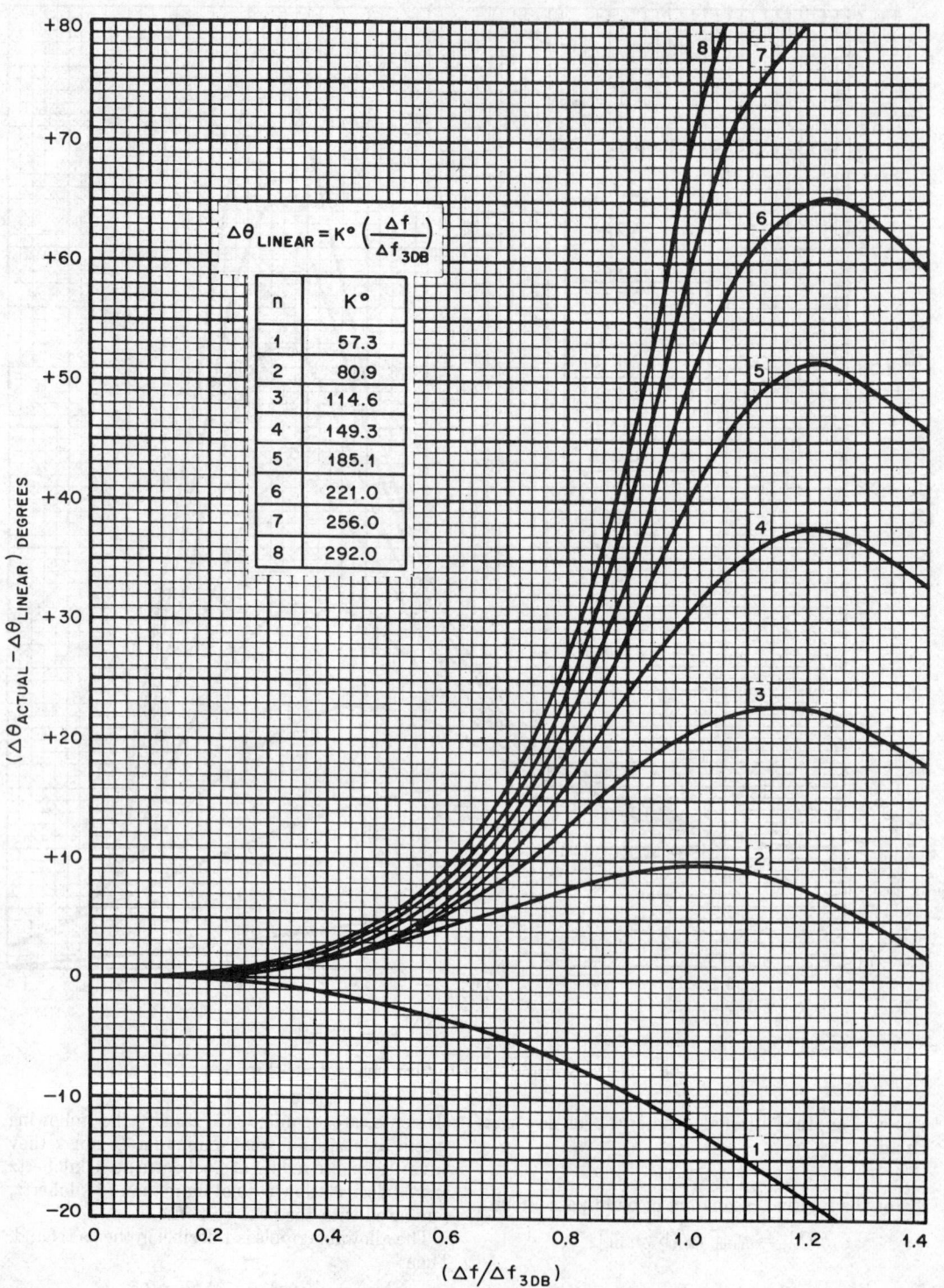

Fig. 14—Phase-shift distortion of Butterworth response shape.

we must have $(bw)_{50dB}/(bw)_{3dB}$ a little less than 2.5 when $(V_p/V)_{dB}=50$ decibels. Consulting Figs. 5 through 12 and examining curves for $(V_p/V_v)_{dB}=1.0$, it is found that a 5-pole network (Fig. 9) is the least that will meet the requirements. Here, curve 6 gives

$$(bw)_{50dB}/(bw)_{3dB}=2.14$$

while

$$(bw)_{1dB}/(bw)_{3dB}=0.97.$$

Then

$$(bw)_{50dB}/(bw)_{1dB}=2.14/0.97=2.20.$$

The 3-decibel frequency will be

$$30(bw)_{3dB}/(bw)_{1dB}=30/0.97=31 \text{ kilohertz}.$$

At this frequency, the Q of each capacitor and inductor must be at least equal to $q_{min}=11.8$ as shown in the table on Fig. 9.

(B) Consider a band-pass filter with requirements similar to the above: bandwidth 1 decibel down to be 30 kilohertz, 50 decibels down at 75-kilohertz bandwidth, and 1-decibel allowable ripple. Further, let the midfrequency be $f_0=500$ kilohertz. The solution at first is the same as above, and a 5-pole network is required.

The 3-decibel bandwidth is 31 kilohertz and the Q of each resonator must be at least

$$11.8 f_0/(bw)_{3dB}=11.8\times500/31=190$$

where 11.8 is q_{min} as read from the table on Fig. 9. If a Q of 190 is not practical to attain, a greater number of resonators can be used. Suppose 7 resonators or poles are tried, per Fig. 11. Then curve 2 gives

$$(bw)_{50dB}/(bw)_{1dB}=2.10/0.93=2.26.$$

The table shows the peak-to-valley ratio of 10^{-5} decibel and $q_{min}=5.9$. The 3-decibel bandwidth is $30/0.93=32.2$ kilohertz. Then, the minimum Q of each resonator can be $5.9\times500/32.2=92$, which is less than half that required if 5 resonators are used.

(C) In the band-pass filter, suppose the filter is subdivided into N identical stages in cascade, isolated by active devices or decoupling capacitors or resistors. For each stage the response requirements are the original number of decibels divided by N. For $N=2$ stages

$$(bw)_{25dB}/(bw)_{0.5dB}<2.5$$

$$(V_p/V_v)_{dB}\leq0.5 \text{ decibel}.$$

Proceeding as before, it is found that a 3-pole network (Fig. 7) for each stage will just suffice, curve 4 giving

$$(V_p/V_v)_{dB}=0.3$$

and

$$(bw)_{25dB}/(bw)_{0.5dB}=2.1/0.84=2.5.$$

To find the required minimum Q of each of the 6 resonators, the 3-decibel bandwidth of each stage is

$$30/0.84=35.8 \text{ kilohertz}.$$

For curve 4, $q_{min}=3.4$, so the minimum allowable Q for each resonator is

$$3.4\times500/35.8=47.5.$$

Maximally Linear Phase Response

In the design of filters where the linearity of the phase characteristic inside the pass band is important, certain changes in design are necessary compared with the previously considered cases. For filters supplying only transfer-function poles, rate of change of phase with frequency becomes more and more linear as the number of arms is increased, provided the design produces a complex relative attenuation characteristic given by the polynomial of (6)*.

$$\frac{V_p}{V}=\frac{n!}{(2n)!}\sum_{r=0}^{n}\frac{2^r(2n-r)!}{r!(n-r)!}\left[j(x/x_\beta)^r\right] \quad (6)$$

where r is a series of integers. The magnitude of (6) is plotted in Figs. 15 and 16 for several values of n.

The former is for the relative attenuation inside the 3-decibel points and the latter for the response outside these points. The curves for $n=\infty$ are plotted from (7), which is the Gaussian shape that the attenuation characteristic approaches as n approaches infinity.

$$10 \log(V_p/V)^2=3(x/x_{3dB})^2. \quad (7)$$

With a filter supplying only transfer-function poles, a maximally linear phase response can be produced only at the limitation of a rounded attenuation shape in the pass band as illustrated in Figs. 15 and 16.

The column labeled q_{min} on Fig. 15 gives the minimum allowable Q, measured at the 3-decibel-down frequency, of the inductors and capacitors of a low-pass filter. For band-pass filters, the minimum allowable unloaded Q at the midfrequency f_0 is $q_{min}f_0/(bw)_{3dB}$. For the phase response figures on Fig. 15, the symbols are as follows.

* W. E. Thomson, "Networks with Maximally Flat Delay," *Wireless Engineer*, vol. 29, pp. 256–263; October 1952.

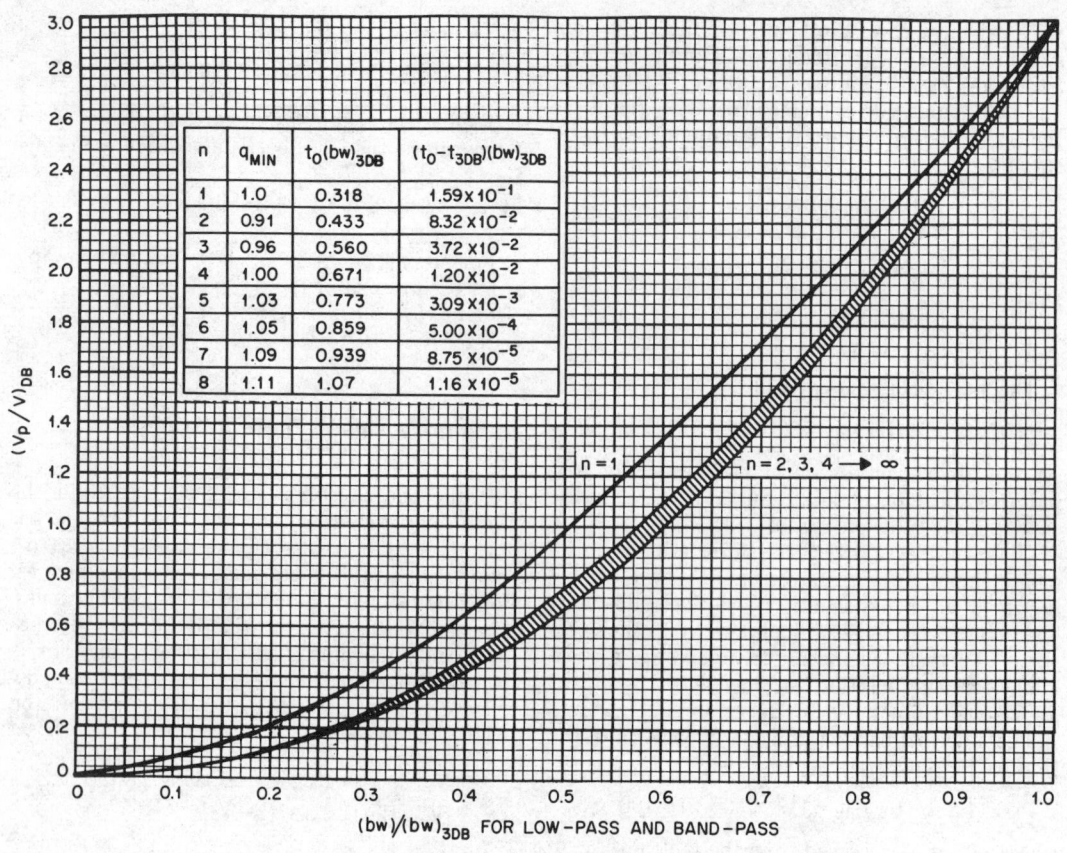

n	q_{MIN}	$t_0(bw)_{3DB}$	$(t_0-t_{3DB})(bw)_{3DB}$
1	1.0	0.318	1.59×10^{-1}
2	0.91	0.433	8.32×10^{-2}
3	0.96	0.560	3.72×10^{-2}
4	1.00	0.671	1.20×10^{-2}
5	1.03	0.773	3.09×10^{-3}
6	1.05	0.859	5.00×10^{-4}
7	1.09	0.939	8.75×10^{-5}
8	1.11	1.07	1.16×10^{-5}

(bw)/(bw)$_{3DB}$ FOR LOW–PASS AND BAND–PASS

(bw)$_{3DB}$/(bw) FOR HIGH–PASS AND BAND–REJECT

Fig. 15—Attenuation shape within the 3-decibel-down pass band for n-pole maximally flat time-delay filters.

Low-Pass Filter:

$$t_0 = d\theta/d\omega$$

= slope of phase characteristic at zero frequency in radians per radian per second

t_{3dB} = slope at f_{3dB}

f_{3dB} = frequency of 3-decibel-down response

$(bw)_{3dB} = 2f_{3dB}$.

Band-Pass Filter:

t_0 = slope at midfrequency

t_{3dB} = slope at 3-decibel-down bandwidth

$(bw)_{3dB}$ = total 3-decibel bandwidth.

The column $(t_0-t_{3dB})(bw)_{3dB}$ shows the *group-delay* distortion over the pass band. It shows numerically that the phase slope becomes much more constant as the number of elements is increased, in a filter designed for this purpose.

FILTERS SUPPLYING BOTH TRANSFER-FUNCTION POLES AND ZEROS

Typical attenuation curves for these filters are shown in Figs. 4D, E, and F. The modern network theory of these filters has been treated by Norton and by Darlington.* The attenuation shapes produced may be called elliptic and inverse-hyperbolic and are optimum in the sense that the rate of cutoff between the accept and reject bands is a maximum. Equation (8) gives the elliptic-function shape.

$$(V_p/V)^2 = 1 + [(V_p/V_v)^2 - 1]$$
$$\times cd_v^2[n(K_v/K_f)cd_f^{-1}(x/x_v)] \quad (8)$$

where $cd = (cn/dn)$, the ratio of the two elliptic

* S. Darlington, "Synthesis of Reactance 4-Poles," *Journal of Mathematics and Physics*, vol. *18*, pp. 257–353, September 1939.

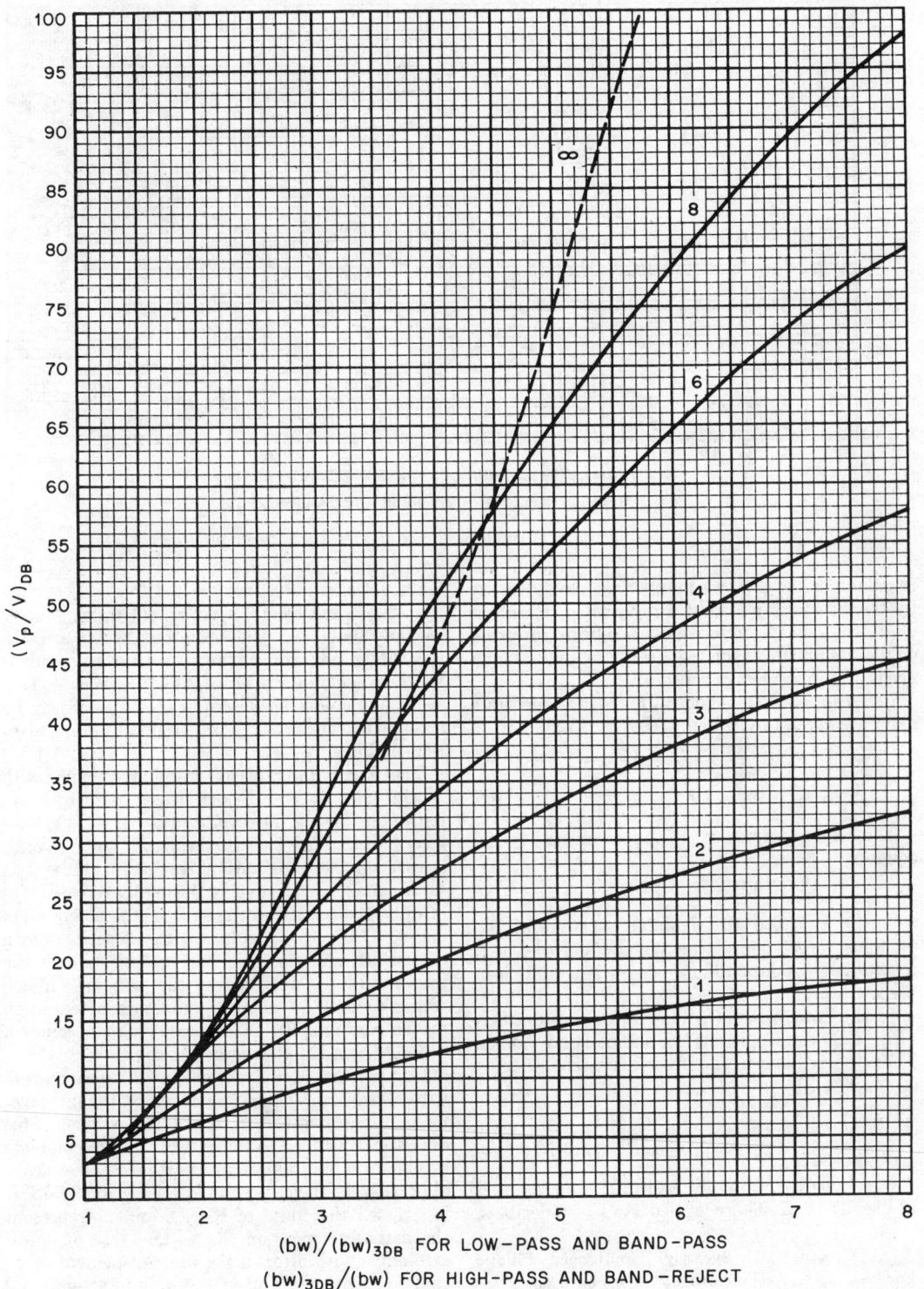

Fig. 16—Attenuation shape beyond 3-decibel-down pass band for *n*-pole maximally flat time-delay filters.

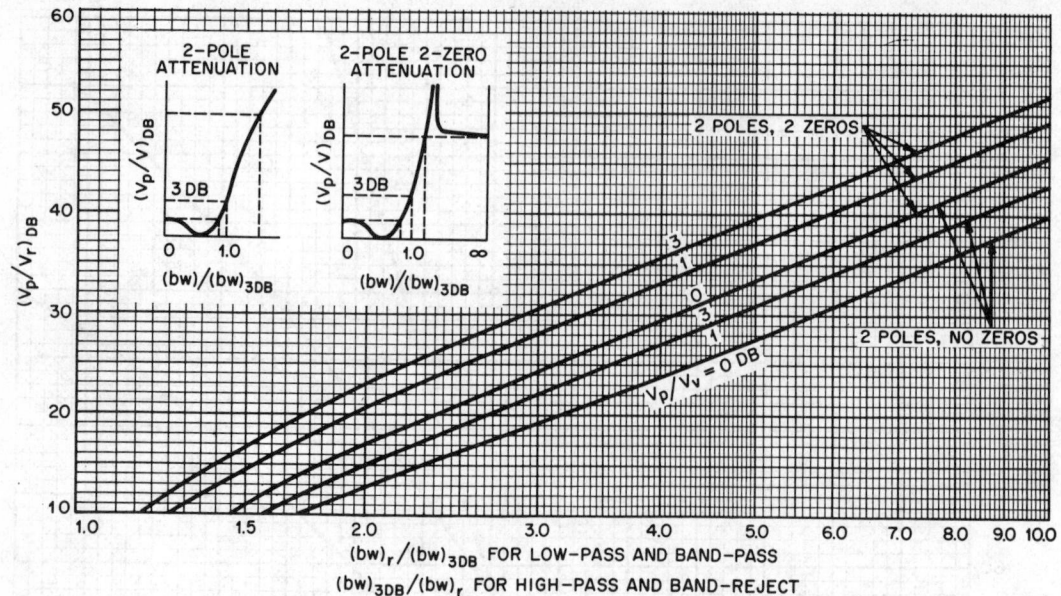

Fig. 17—Maximum rate of cutoff for 2-pole and for 2-pole 2-zero filters.

functions cn and dn*, $n=$ number of poles supplied by the filter, $x=$ a bandwidth variable described under (5), and K_v, $K_f=$ complete elliptic integrals of the first kind, evaluated for the modulus value given by the respective subscript.

Referring to the symbols on Fig. 4, the moduli v and f are given in (9) and (10).

$$v=\{[(V_p/V_v)^2-1]/[(V_p/V_h)^2-1]\}^{1/2} \quad (9)$$

$$f=x_v/x_n=(bw)_v/(bw)_h. \quad (10)$$

These are not independent, but must satisfy the equation

$$\log q_v = n \log q_f \quad (11)$$

where q_k is called the modular constant of the modulus value k, the latter being equal to v or f, respectively. A tabulation of $\log q$ is available in the literature.†

In the limit, when $V_p/V_v=1.0$ or zero decibels (Fig. 4F), the ripples in the accept band vanish. Then (8) reduces to the inverse hyperbolic shape of (12).

$$(V_p/V)^2=1+\frac{(V_p/V_h)^2-1}{\cosh^2[n\cosh^{-1}(x_h/x)]}. \quad (12)$$

Curves plotted from (8) and (12) are presented in Figs. 17 to 22. Those labeled $V_p/V_v=0$ decibels,

---

* G. W. and R. M. Spencely, "Smithsonian Elliptic Function Tables," (Publication 3863), Smithsonian Institution, Washington, D. C.; 1947.

† E. Jahnke and F. Emde, "Table of Functions with Formulas and Curves," 4th Edition, Dover Publications, New York, 1945: pp. 49–51.

for n poles, m zeros, are plotted from (12) while the others are from (8). For all these shapes, $n=$ the number of poles= the number of arms in the ladder network. When n is an even number, the number of zeros $m=n$. When n is odd, $m=n-1$. The following description of Fig. 17 can be extended to cover the entire group of figures mentioned above.

The maximum rates of cutoff obtainable with 2-pole no-zero and 2-pole 2-zero networks are plotted in Fig. 17 for several ratios of V_p/V_v. Two insert sketches drawn in the figure show typical shapes of the attenuation curves for these two cases. The main curves give the relative coordinates of only two points on the skirt of the attenuation curve. These two points are the 3-decibel-down bandwidth and the "hill bandwidth" (where the response first equals that of the "response hills", where occurs the uniform minimum attenuation in the reject band). Thus each point specifies a different relative attenuation shape.

Comparison of the curves for 2 poles no-zero with those for 2 poles 2 zeros shows the improvement in cutoff rate that is obtainable when zeros are correctly added to the network. More-complete attenuation information on the 2-pole no-zero configuration has been presented on Fig. 6. Again, it is stressed that data of Figs. 6 and 17 represent the actual attenuation shapes and rate of cutoff attainable with filters using finite-Q elements (except for a rounding off of the infinite attenuation peaks). In contrast, the rates of cutoff and the attenuation shapes predicted by the simple "image" theory are unobtainable in physically realizable networks.

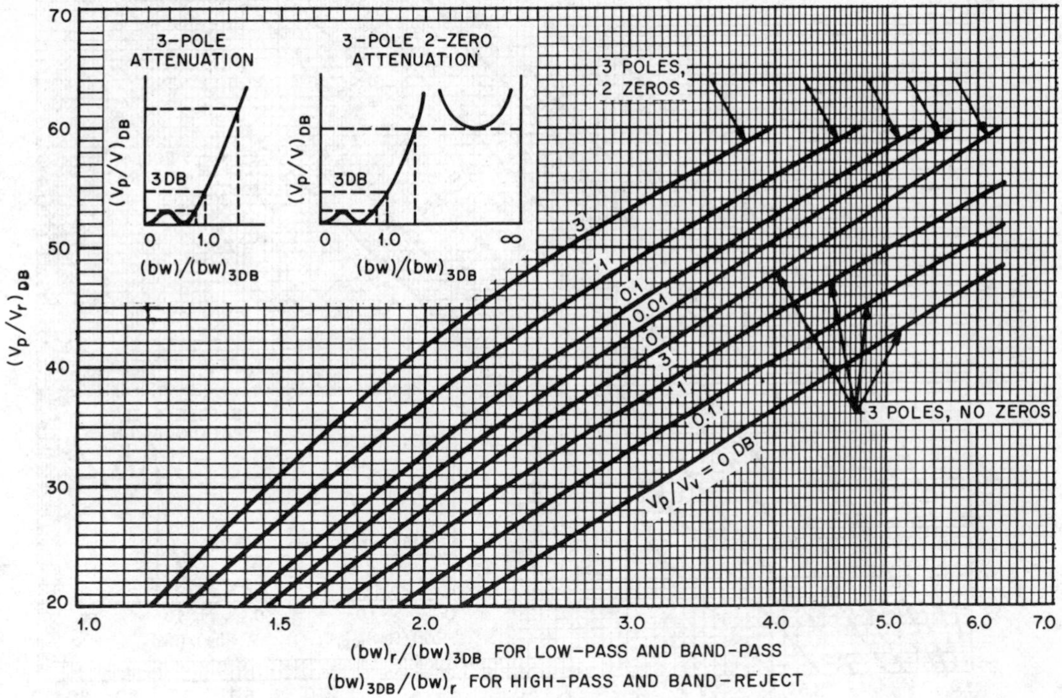

Fig. 18—Maximum rate of cutoff for 3-pole and for 3-pole 2-zero filters.

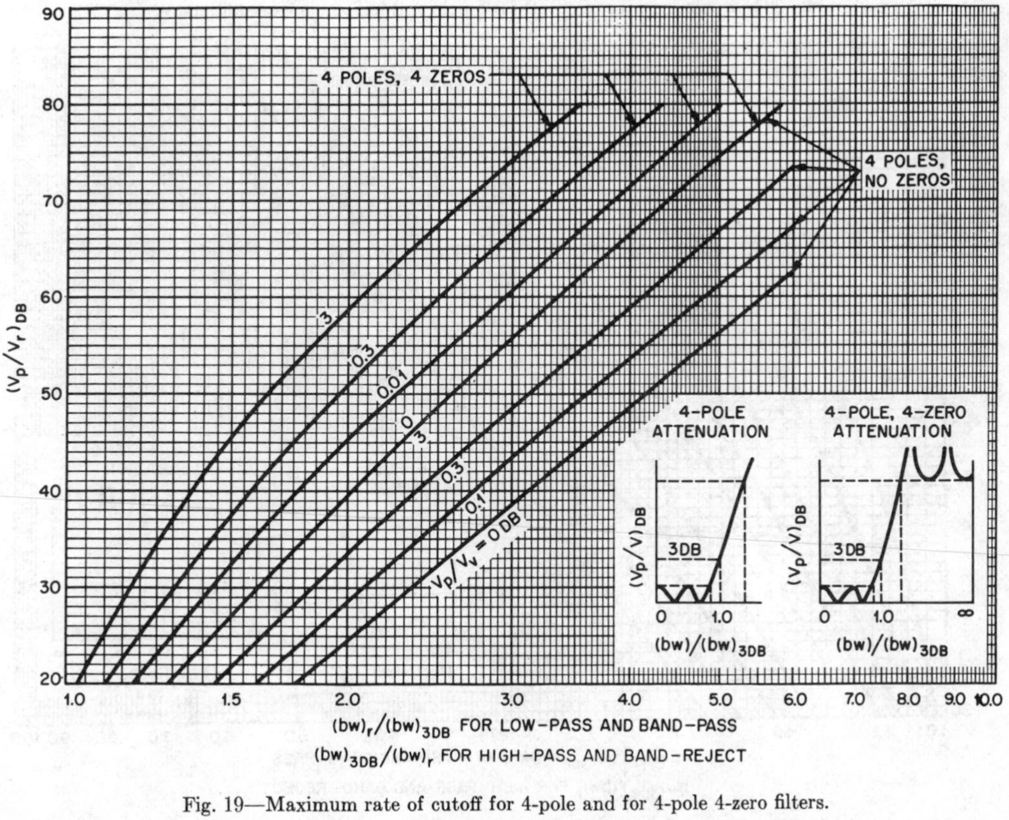

Fig. 19—Maximum rate of cutoff for 4-pole and for 4-pole 4-zero filters.

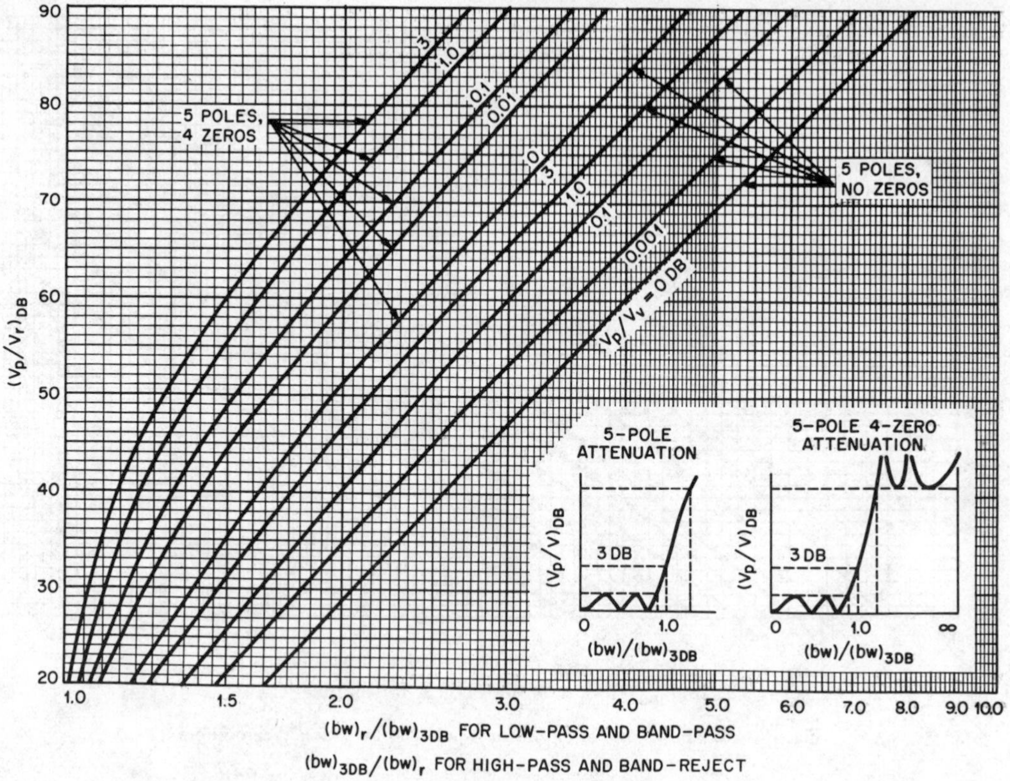

Fig. 20—Maximum rate of cutoff for 5-pole and for 5-pole 4-zero filters.

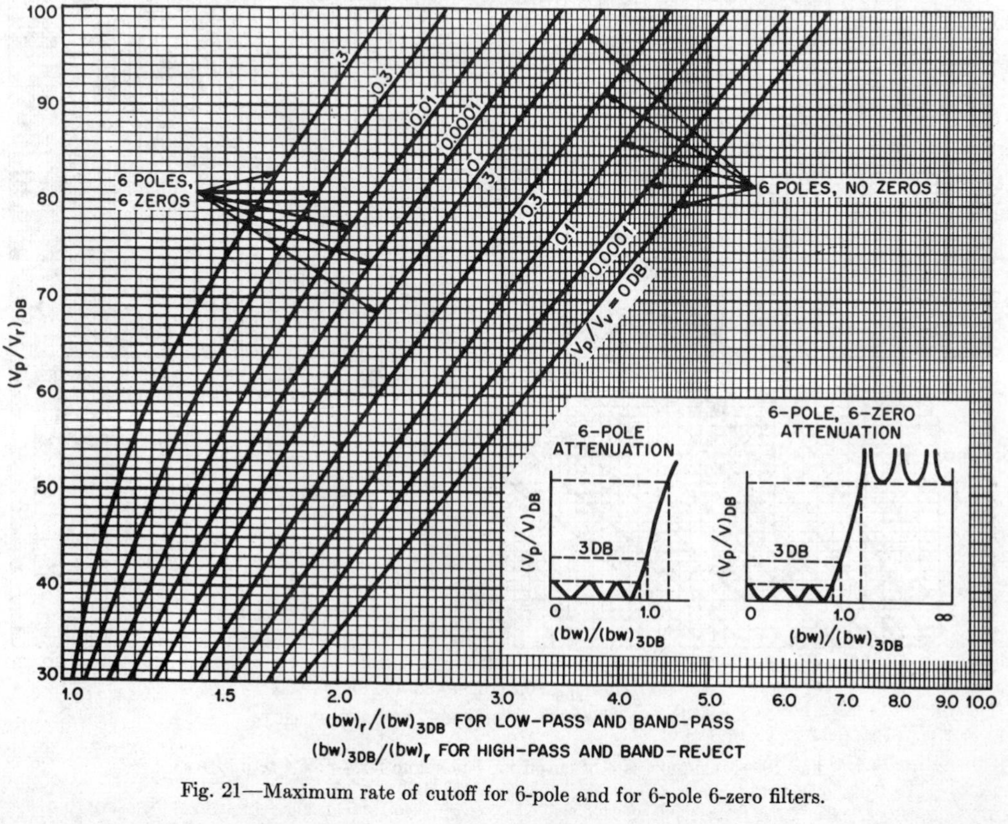

Fig. 21—Maximum rate of cutoff for 6-pole and for 6-pole 6-zero filters.

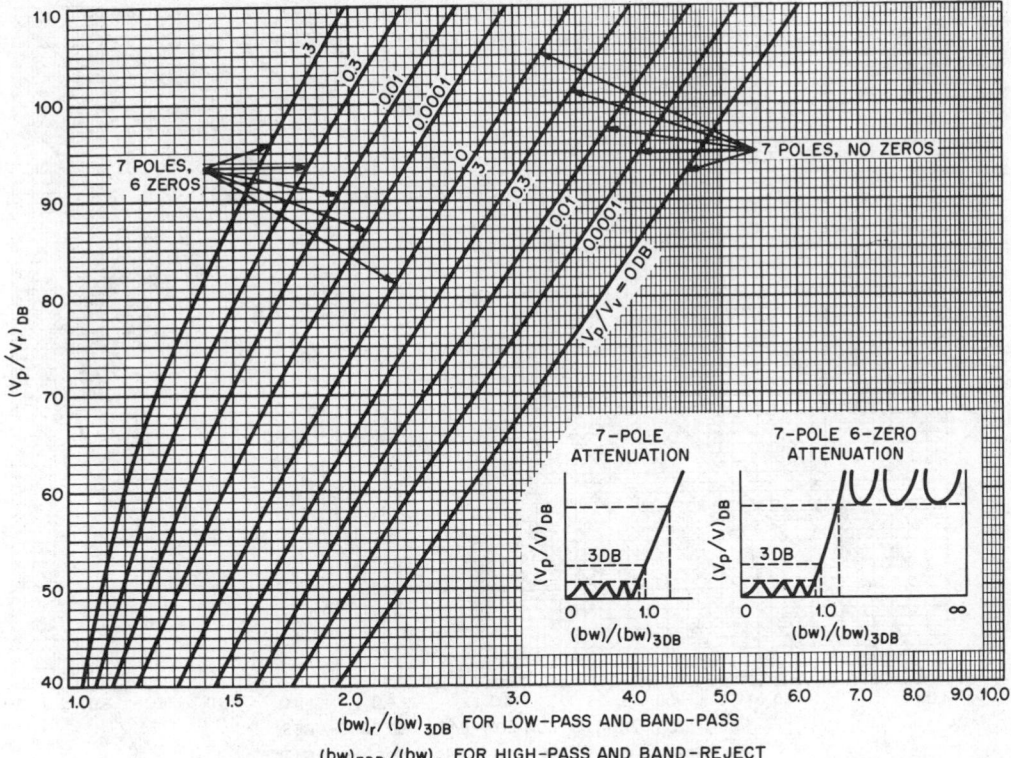

Fig. 22—Maximum rate of cutoff for 7-pole and for 7-pole 6-zero filters.

The rates of cutoff shown are the best that are possible of attainment with the specified number of poles and zeros, and with equal-ripple-type behavior.

Resistive Terminations and n Even

It is evident from the attenuation shapes of Figs. 17, 19, and 21 that for n even, the optimum shape given by (8) produces a finite attenuation at an infinite frequency. This requires a completely reactive termination at one end of the network. If resistive terminations must be used, then the optimum shape that is practically realizable with an even number of arms is given by

$$(V_p/V)^2 = 1 + [(V_p/V_v)^2 - 1] cd_v^2 [n(K_v u/K_f)] \tag{13}$$

where

$$u = sc_f^{-1}\{[(x_v/x)^2 - 1]^{1/2}[dn_f(K_f/n)]/f'\}. \tag{14}$$

The modulus v is given by (9) and the modulus f by (10). Solving (13) then gives the ratio of hill-to-valley bandwidth as

$$x_h/x_v = [fcd_f(K_f/n)]^{-1}. \tag{15}$$

This optimum attenuation shape (13) produces two fewer points of infinite rejection, or response zeros, than response poles. In contrast, (8) requires an equal number of zeros and poles.

If the ripples in the pass band approach zero decibels ($V_p/V_v = 1$) then, as a limit, (13) becomes

$$(V_p/V)^2 = 1 + \frac{(V_p/V_h)^2 - 1}{\cosh^2(n \cosh^{-1} y)} \tag{16}$$

where

$$y = \left[\left(\frac{x_h}{x} \cos \frac{90}{n}\right)^2 + \sin^2 \frac{90}{n}\right]^{1/2}.$$

Based on (13) and (16), the rates of cutoff have been plotted in Figs. 23 and 24 for 4-pole 2-zero and for 6-pole 4-zero filters. Figure 6 already has presented the data for a 2-pole no-zero network, the simplest case. An increase in rate of cutoff results when $n-2$ response zeros are suitably added to n response poles as shown by the curves in Figs. 23 and 24.

CIRCUIT-ELEMENT VALUES

This section concerns the values of the circuit elements required to produce the optimum relative-attenuation shapes of ladder-network filters supplying only transfer-function poles. There are two

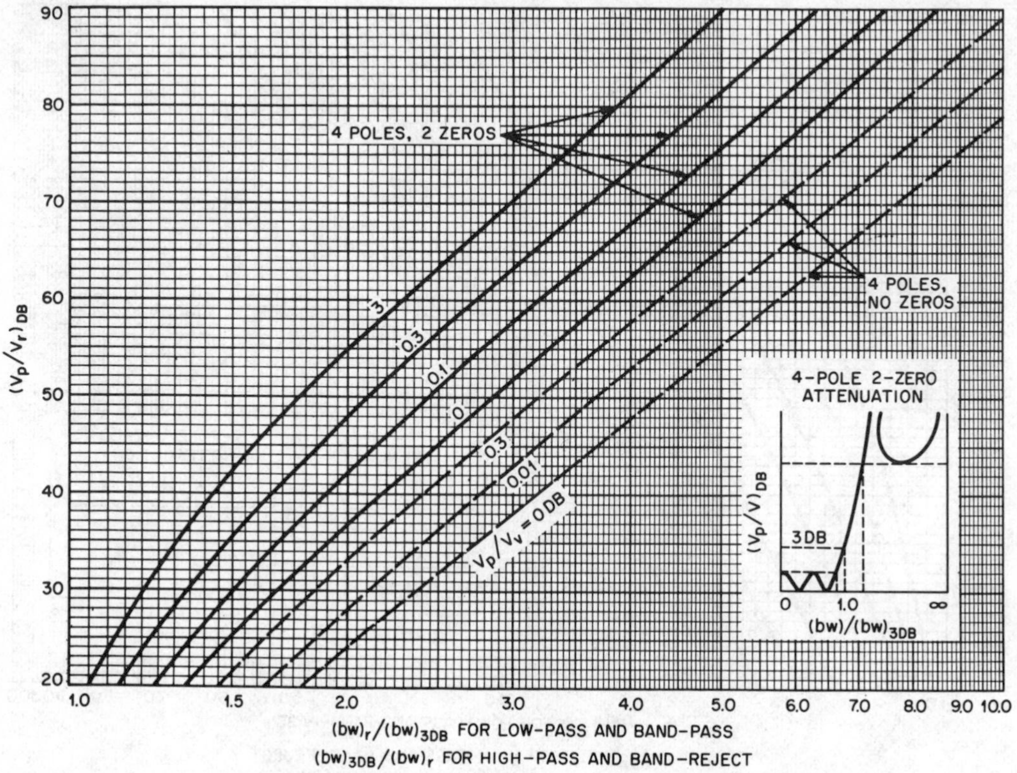

Fig. 23—Maximum rate of cutoff for 4-pole and for 4-pole 2-zero filters.

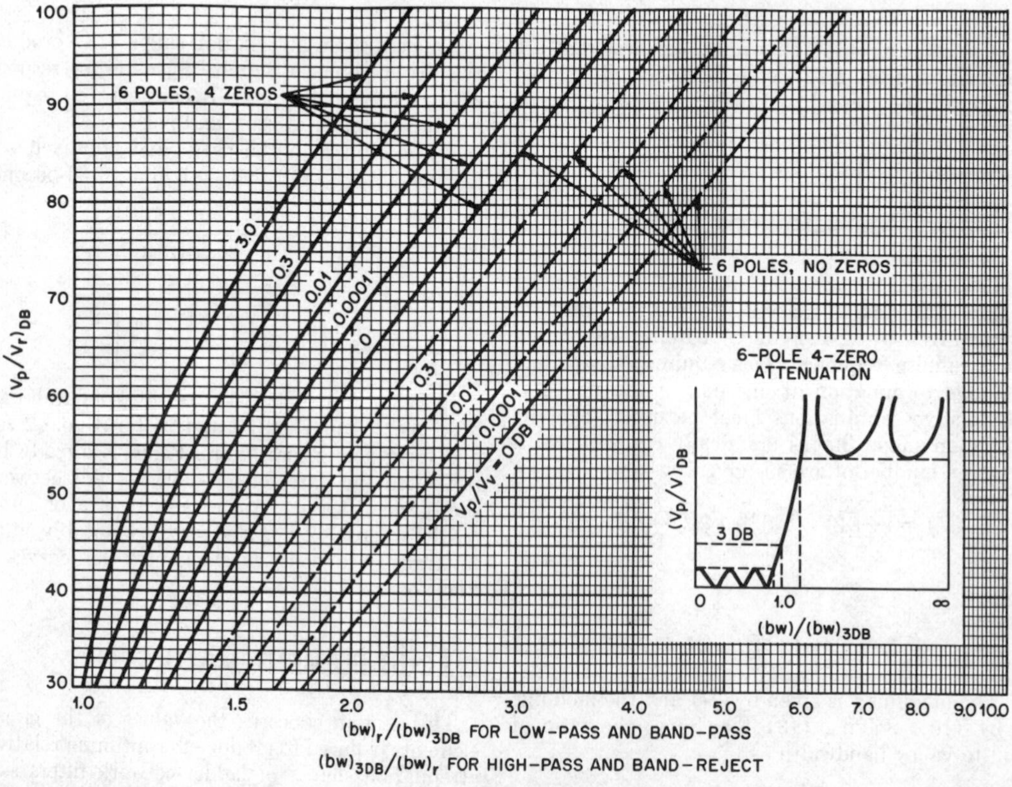

Fig. 24—Maximum rate of cutoff for 6-pole and for 6-pole 4-zero filters.

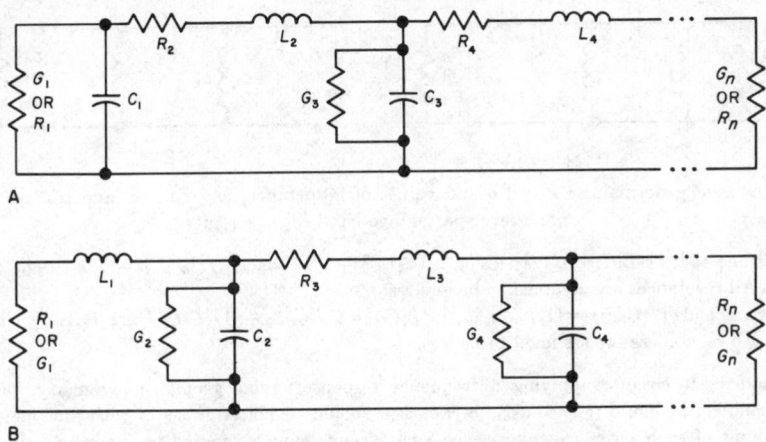

A

B

Fig. 25—Relations among normalized k and q and values of inductance, capacitance, and resistance for low-pass and large-percentage-band-pass circuits.

A—Shunt arm at one end. $1/(C_1L_2)^{\frac{1}{2}} = k_{12}\omega_{3\,dB}$, $1/(L_2C_3)^{\frac{1}{2}} = k_{23}\omega_{3\,dB}$, $1/(C_3L_4)^{\frac{1}{2}} = k_{34}\omega_{3\,dB}$, etc.
$G_1/C_1 = (1/q_1)\omega_{3\,dB}$, $q_2 = (\omega_{3\,dB}\ L_2)/R_2$, $q_3 = (\omega_{3\,dB}\ C_3)/G_3$, $q_4 = (\omega_{3\,dB}\ L_4)/R_4$, etc.
B—Series arm at one end. $1/(L_1C_2)^{\frac{1}{2}} = k_{12}\omega_{3\,dB}$, $1/(C_2L_3)^{\frac{1}{2}} = k_{23}\omega_{3\,dB}$, $1/(L_3C_4)^{\frac{1}{2}} = k_{34}\omega_{3\,dB}$, etc.
$R_1/L_1 = (1/q_1)\omega_{3\,dB}$, $q_2 = (\omega_{3\,dB}\ C_2)/G_2$, $q_3 = (\omega_{3\,dB}\ L_3)/R_3$, $q_4 = (\omega_{3\,dB}\ C_4)/G_4$, etc.

To design a band-pass circuit that supplies a frequency response having geometric symmetry, the total required 3-decibel-down bandwidth should replace $\omega_{3\,dB}$, an inductor should be connected across each shunt capacitor, and a capacitor put in series with each series inductor, each such circuit being resonated to the geometric mean frequency

$$f_0 = (f_1 f_2)^{\frac{1}{2}}.$$

convenient ways of expressing the element values for these ladder networks.

(**A**) The reactive and resistive components of each element may be related to one of the terminating resistances (or to a completely arbitrary normalizing resistance R_0) and also to a definite bandwidth, usually the 3-decibel-down value. The numerical results are called ladder-network coefficients or singly loaded Q's.

(**B**) The reactive component of each element may be related to the reactive part of the immediately preceding element, and to a definite bandwidth such as the 3-decibel-down value. These numerical results are called the normalized coefficients of coupling. The resistive component of each element is related to its reactive part and the numerical values are called normalized decrements or, when inverted, normalized Q's.

The latter form of normalized coefficients of coupling k and normalized Q's ($=q$) will be used

because the numerical values may be applied directly to the adjustment and checking of actual filters.

Figures 25 through 29 relate the normalized k and q to the inductance, capacitance, and resistance values for various types of filters.

For low-pass filters, Fig. 25 shows that k gives the ratio of resonant frequency of two immediately adjacent elements to the overall 3-decibel-down frequency. The resonant frequency of C_1 and L_2 in this example must be k_{12} times the required overall 3-decibel-down bandwidth.

Figure 25 also gives, as the inverse of q, the ratio of the 3-decibel-down bandwidth of a single element resulting from the resistive load and losses associated with it, to the required 3-decibel-down bandwidth of the overall filter. Thus, $1/R_1C_1$ is the 3-decibel-down radian bandwidth of C_1 and the conductance G_1 that must be shunted across it. If C_1 and G_1 are properly chosen, the measured

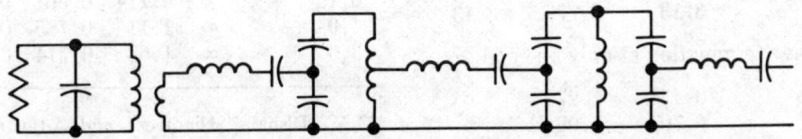

Fig. 26—Band-pass circuit supplying a frequency response having geometric symmetry. Parallel and series circuits must alternate, and Figs. 1 and 2 of Chapter 9 give the relationship between element values and resulting actual coefficient of coupling $K\{ = k[(\text{bw})_{3\,dB}/f_0]\}$. Any adjacent pair of resonators may be coupled by any of the methods shown.

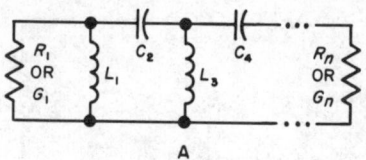
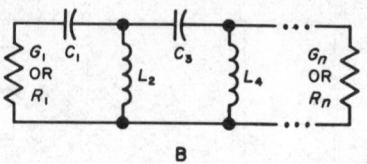

A　　　　　　　　　　　　　　B

Fig. 27—Relations among normalized k and q and values of inductance, capacitance, and resistance for high-pass and large-percentage-band-reject circuits.

A—Shunt arm at one end. $1/(L_1 C_2)^{\frac{1}{2}} = (1/k_{12})\omega_{3dB}$, $1/(C_2 L_3)^{\frac{1}{2}} = (1/k_{23})\omega_{3dB}$, $1/(L_3 C_4)^{\frac{1}{2}} = (1/k_{34})\omega_{3dB}$, etc. $(R_1/L_1) = q_1 \omega_{3dB}$. All reactances are assumed to be lossless.
B—Series arm at one end. $1/(C_1 L_2)^{\frac{1}{2}} = (1/k_{12})\omega_{3dB}$, $1/(L_2 C_3)^{\frac{1}{2}} = (1/k_{23})\omega_{3dB}$, $1/(C_3 L_4)^{\frac{1}{2}} = (1/k_{34})\omega_{3dB}$, etc. $(G_1/C_1) = q_1 \omega_{3dB}$. All reactances are assumed to be lossless.

To design a band-reject circuit supplying a frequency response having geometric symmetry, the total required 3-decibel-down bandwidth should replace ω_{3dB}, a capacitor should be placed in series with each shunt inductor, and an inductor in shunt of each series capacitor, each such circuit being resonated to the geometric mean frequency

$$f_0 = (f_1 f_2)^{\frac{1}{2}}.$$

bandwidth of these elements at their 3-decibel-down point will be $1/q_1$ times the required overall 3-decibel-down bandwidth of the filter.

The legend of Fig. 25 shows how it is applicable also to large-percentage band-pass filters.

When the procedure of Fig. 25 is used to design band-pass circuits of medium- or small-percentage bandwidths, the resulting element value ratios become impractical to obtain. The circuit configuration shown in Fig. 26 gives practical element value ratios for these two cases, while still providing true geometric symmetry.

Figure 27 gives the required information for high-pass and large-percentage band-reject filters.

Similar data are given in Fig. 28 for small-percentage band-pass filters. It should be noted that the required actual coefficient of coupling between resonant circuits, $M_{ab}/(L_a L_b)^{1/2}$ for example, is obtained by multiplying the required overall fractional 3-decibel-down bandwidth by

the normalized coefficient of coupling. The required actual resonant-circuit Q results from multiplying the fractional midfrequency by q. An experimental procedure for checking k and q values is available.[*] Fractional midfrequency $f_0/(\text{bw})_{3dB} =$ reciprocal of fractional 3-decibel-down bandwidth.

Figure 29 supplies the data for small-percentage band-reject filters.

Butterworth, Chebishev, and Maximally Linear Phase Designs

Elegant closed-form equations for k and q values producing optimum Chebishev and Butterworth

TABLE 1—TWO-POLE NO-ZERO FILTER 3-DECIBEL-DOWN k AND q VALUES.

$(V_p/V_v)_{dB}$	q_1	k_{12}	q_2
Maximum-power-transfer terminations			
Linear phase	0.576	0.899	2.15
0	1.414	0.707	1.414
0.3	1.82	0.717	1.82
1.0	2.21	0.739	2.21
3.0	3.13	0.779	3.13
Resistive termination at only one end			
Linear phase	0.455	1.27	∞
0	0.707	1.00	∞
0.3	0.910	0.904	∞
1.0	1.11	0.866	∞
3.0	1.56	0.840	∞

TABLE 2—THREE-POLE NO-ZERO FILTER 3-DECIBEL-DOWN k AND q VALUES.

$(V_p/V_v)_{dB}$	q_2	q_1	k_{12}	k_{23}	q_3
Maximum-power-transfer terminations					
Linear phase	∞	0.338	1.74	0.682	2.21
0	∞	1.00	0.707	0.707	1.00
0.1	∞	1.43	0.665	0.665	1.43
1.0	∞	2.21	0.645	0.645	2.21
3.0	∞	3.36	0.647	0.647	3.36
Resistive termination at only one end					
Linear phase	∞	0.293	2.01	0.899	∞
0	∞	0.500	1.22	0.707	∞
0.1	∞	0.714	0.961	0.661	∞
1.0	∞	1.11	0.785	0.645	∞
3.0	∞	1.68	0.714	0.649	∞

[*] M. Dishal, "Alignment and Adjustment of Synchronously Tuned Multiple-Resonant-Circuit Filters," *Proceedings of the IRE*, vol. 39, pp. 1448–1455; November 1951: Also, *Electrical Communication*, vol. 29, pp. 154–164; June 1952.

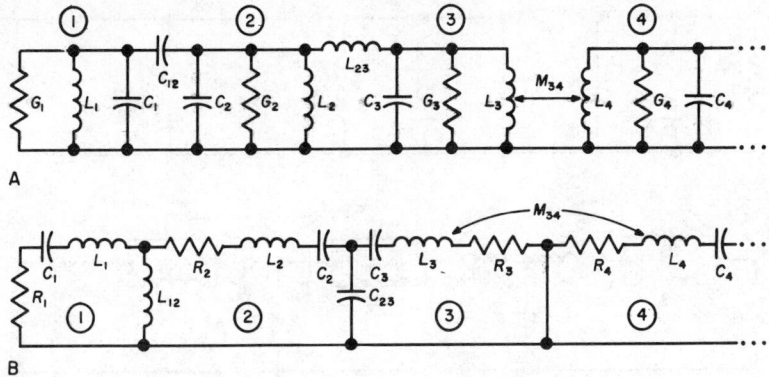

Fig. 28—Relations among normalized k and q and values of inductance, capacitance, and resistance for small-percentage-band-pass circuits.

A—Parallel-resonant circuits. $C_{12}/(C_1 C_2)^{\frac{1}{2}} \doteq k_{12}[(\text{bw})_{3\text{dB}}/f_0]$, $(L_2 L_3)^{\frac{1}{2}}/L_{23} \doteq k_{23}[(\text{bw})_{3\text{dB}}/f_0]$, $M_{34}/(L_3 L_4)^{\frac{1}{2}} \doteq k_{34}[(\text{bw})_{3\text{dB}}/f_0]$, etc. $Q_1 = q_1[f_0/(\text{bw})_{3\text{dB}}]$, $q_2 = Q_2/[f_0/(\text{bw})_{3\text{dB}}]$, $q_3 = Q_3/[f_0/(\text{bw})_{3\text{dB}}]$, $q_4 = Q_4/[f_0/(\text{bw})_{3\text{dB}}]$, etc. Any adjacent pair of resonators may be coupled by any of the three methods shown. Each node must resonate at f_0 with all other nodes short-circuited.

B—Series-resonant circuits. $L_{12}/(L_1 L_2)^{\frac{1}{2}} \doteq k_{12}[(\text{bw})_{3\text{dB}}/f_0]$, $(C_2 C_3)^{\frac{1}{2}}/C_{23} \doteq k_{23}[(\text{bw})_{3\text{dB}}/f_0]$, $M_{34}/(L_3 L_4)^{\frac{1}{2}} \doteq k_{34}[(\text{bw})_{3\text{dB}}/f_0]$, etc. $Q_1 = q_1[f_0/(\text{bw})_{3\text{dB}}]$, $q_2 = Q_2/[f_0/(\text{bw})_{3\text{dB}}]$, $q_3 = Q_3/[f_0/(\text{bw})_{3\text{dB}}]$, $q_4 = Q_4/[f_0/(\text{bw})_{3\text{dB}}]$. Any adjacent pair of resonators may be coupled by any of the three methods shown. Each mesh must resonate at f_0 with all other meshes open-circuited.

response shapes for filters having any number of total arms may be obtained if lossless reactances are used.* The design data in Tables 1 through 7 are based on such equations. The k and q values for the maximally linear phase shape result from the Darlington synthesis procedure applied to (6). The tables provide data for two limiting cases of terminations; maximum-power-transfer loading at the two ends of the filter and resistive loading at only one end.

For Tables 1 through 7, the $(V_p/V_v)_{\text{dB}}$ column gives the ripple in decibels in the pass band, and the corresponding curves on Figs. 5 through 12 give the complete attenuation shape.

For low-pass circuits, $q_{2,3,4,\cdots}$, is the required unloaded Q, measured at the required 3-decibel-down frequency, of the internal inductors and capacitors to be used. For band-pass circuits, the unloaded resonator Q required in the internal resonators is obtained by multiplying the required 3-decibel fractional midfrequency $[f_0/(\text{bw})_{3\text{dB}}]$ by $q_{2,3,4,\cdots}$.

For the detailed way in which the q and k

* V. Belevitch, "Tchebyshev Filters and Amplifier Networks," *Wireless Engineer*, vol. 29, pp. 106–110; April 1952. H. J. Orchard, "Formulas for Ladder Filters," *Wireless Engineer*, vol. 30, pp. 3–5; January 1953. E. Green, "Exact Amplitude–Frequency Characteristics of Ladder Networks," *Marconi Review*, vol. 16, no. 108, pp. 25–68; 1953. M. Dishal, "Two New Equations for the Design of Filters," *Electrical Communication*, vol. 30, pp. 324–337; December 1952.

columns fix the required element values, see Figs. 25 through 29 and related discussion.

The first column of each table gives the peak-to-valley ratio within the pass band.

To be exactly correct, the design values given in the tables require that (as shown in the second column of each table except Table 1) infinite unloaded Q's be available for the internal elements of the filter. It should be realized that designs can be made using elements having finite unloaded Q's; these designs are given in Figs. 30 through 56.

Proceeding across each table, figuratively from the left end of the filter, the next column gives q_1 from which, with the aid of Figs. 25 through 29, the relation between the terminating resistance R_1 and the first reactance element is obtained. The next column for k_{12} (with Figs. 25 through 29) provides for the relation between the first and second reactances. Continuing across each table, all relations between adjacent elements are obtained including that of the right-hand terminating resistance.

Example: Reverting to the previous example, a filter is required having $(\text{bw})_{50\text{dB}}/(\text{bw})_{1\text{dB}} = 2.5$, and $V_p/V_v < 1$ decibel. The 5-pole no-zero response with a pass-band peak-to-valley ratio of 1 decibel in Fig. 9 satisfied the requirement.

Table 4 is for 5-pole networks and, if the terminations are to be maximum-power-transfer loads, the upper part of the table should be used. If a shunt capacitance is to appear at one end of the low-pass filter, Fig. 25A will apply.

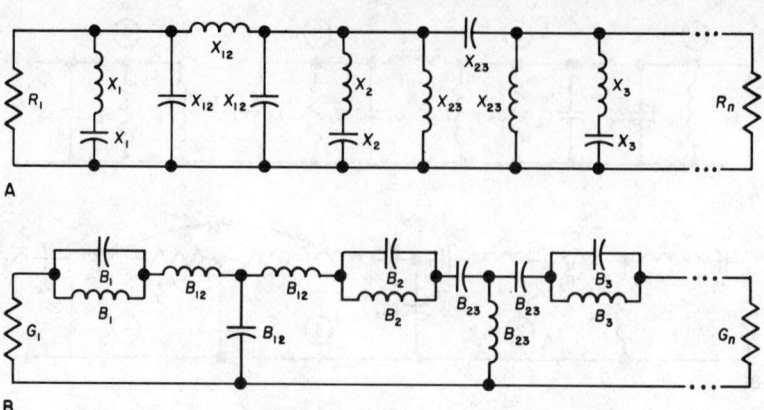

Fig. 29—Relations among normalized k and q and values of inductance, capacitance, and resistance for small-per-centage-band-reject circuits.

A—Series-resonant circuits. $X_{12}/(X_1 X_2)^{\frac{1}{2}} = (1/k_{12})[(\text{bw})_{3\text{dB}}/f_0]$, $X_{23}/(X_2 X_3)^{\frac{1}{2}} = (1/k_{23})[(\text{bw})_{3\text{dB}}/f_0]$, etc. $X_1/R_1 = (1/q_1)$ $[f_0(\text{bw})_{3\text{dB}}]$, $X_n/R_n = (1/q_n)[f_0/(\text{bw})_{3\text{dB}}]$. All resonant circuits are assumed to be lossless. Any adjacent pair of resonators may be coupled by either of the two π (or their dual T) couplings shown. The reactances X are measured at the midfrequency of the reject band.

B—Parallel-resonant circuits. $B_{12}/(B_1 B_2)^{\frac{1}{2}} = (1/k_{12})[(\text{bw})_{3\text{dB}}/f_0]$, $B_{23}/(B_2 B_3)^{\frac{1}{2}} = (1/k_{23})[(\text{bw})_{3\text{dB}}/f_0]$, etc. $B_1/G_1 = (1/q_1)$ $[f_0/(\text{bw})_{3\text{dB}}]$, $B_n/G_n = (1/q_n)[f_0/(\text{bw})_{3\text{dB}}]$.

All resonant circuits are assumed to be lossless. Any adjacent pair of resonators may be coupled by either of the two T (or their dual π) couplings shown. The susceptances B are measured at the midfrequency of the reject band.

TABLE 3—FOUR-POLE NO-ZERO FILTER 3-DECIBEL-DOWN k AND q VALUES.

$(V_p/V_v)_{\text{dB}}$	$q_{2,3}$	q_1	k_{12}	k_{23}	k_{34}	q_4
Maximum-power-transfer terminations						
Linear phase	∞	2.24	0.644	1.175	2.53	0.233
0	∞	0.766	0.840	0.542	0.840	0.766
0.01	∞	1.05	0.737	0.541	0.737	1.05
0.1	∞	1.34	0.690	0.542	0.690	1.34
1.0	∞	2.21	0.638	0.546	0.638	2.21
3.0	∞	3.45	0.624	0.555	0.624	3.45
Resistive termination at only one end						
Linear phase	∞	0.211	2.78	1.29	0.828	∞
0	∞	0.383	1.56	0.765	0.644	∞
0.01	∞	0.524	1.20	0.666	0.621	∞
0.1	∞	0.667	1.01	0.626	0.618	∞
1.0	∞	1.10	0.781	0.578	0.614	∞
3.0	∞	1.72	0.692	0.567	0.609	∞

TABLE 4—FIVE-POLE NO-ZERO FILTER 3-DECIBEL-DOWN k AND q VALUES.

$(V_p/V_v)_{\text{dB}}$	$q_{2,3,4}$	q_1	k_{12}	k_{23}	k_{34}	k_{45}	q_5
Maximum-power-transfer terminations							
Linear phase	∞	0.175	3.36	1.56	1.06	0.631	2.26
0	∞	0.618	1.0	0.556	0.556	1.0	0.618
0.001	∞	0.822	0.845	0.545	0.545	0.845	0.822
0.1	∞	1.29	0.703	0.535	0.535	0.703	1.29
1.0	∞	2.21	0.633	0.538	0.538	0.633	2.21
3.0	∞	3.47	0.614	0.538	0.538	0.614	3.47
Resistive termination at only one end							
Linear phase	∞	0.162	3.62	1.68	1.14	0.804	∞
0	∞	0.309	1.90	0.900	0.655	0.619	∞
0.001	∞	0.412	1.48	0.760	0.603	0.606	∞
0.1	∞	0.649	1.044	0.634	0.560	0.595	∞
1.0	∞	1.105	0.779	0.570	0.544	0.595	∞
3.0	∞	1.74	0.679	0.554	0.542	0.597	∞

TABLE 5—SIX-POLE NO-ZERO FILTER 3-DECIBEL-DOWN k AND q VALUES.

$(V_p/V_v)_{\text{dB}}$	$q_{2,3,4,5}$	q_1	k_{12}	k_{23}	k_{34}	k_{45}	k_{56}	q_6
Maximum-power-transfer terminations								
0	∞	0.518	1.17	0.606	0.518	0.606	1.17	0.518
0.0001	∞	0.679	0.967	0.573	0.518	0.573	0.967	0.679
0.01	∞	0.936	0.810	0.550	0.518	0.550	0.810	0.936
0.1	∞	1.27	0.716	0.539	0.518	0.539	0.716	1.27
1.0	∞	2.21	0.631	0.531	0.520	0.531	0.631	2.21
3.0	∞	3.51	0.610	0.532	0.524	0.532	0.610	3.51
Resistive termination at only one end								
Linear phase	∞	0.129	4.55	2.09	1.42	1.09	0.803	∞
0	∞	0.259	2.26	1.05	0.732	0.606	0.606	∞
0.0001	∞	0.340	1.76	0.869	0.650	0.573	0.596	∞
0.01	∞	0.468	1.34	0.725	0.591	0.550	0.591	∞
0.1	∞	0.637	1.06	0.642	0.560	0.539	0.589	∞
1.0	∞	1.12	0.771	0.566	0.533	0.531	0.589	∞
3.0	∞	1.75	0.673	0.546	0.529	0.531	0.591	∞

TABLE 6—SEVEN-POLE NO-ZERO FILTER 3-DECIBEL-DOWN k AND q VALUES.

$(V_p/V_v)_{dB}$	$q_{2,3,4,5,6}$	q_1	k_{12}	k_{23}	k_{34}	k_{45}	k_{56}	k_{67}	q_7
			Maximum-power-transfer terminations						
0	∞	0.445	1.34	0.669	0.528	0.528	0.669	1.34	0.445
0.00001	∞	0.580	1.10	0.611	0.521	0.521	0.611	1.10	0.580
0.001	∞	0.741	0.930	0.579	0.519	0.519	0.579	0.930	0.741
0.01	∞	0.912	0.830	0.560	0.519	0.519	0.560	0.830	0.912
0.1	∞	1.26	0.723	0.541	0.517	0.517	0.541	0.723	1.26
1.0	∞	2.25	0.631	0.530	0.517	0.517	0.530	0.631	2.25
3.0	∞	3.52	0.607	0.529	0.519	0.519	0.529	0.607	3.52
			Resistive termination at only one end						
Linear phase	∞	0.105	5.53	2.53	1.72	1.33	1.08	0.804	∞
0	∞	0.223	2.62	1.20	0.824	0.659	0.579	0.598	∞
0.00001	∞	0.290	2.05	0.981	0.710	0.601	0.552	0.589	∞
0.001	∞	0.370	1.64	0.830	0.642	0.570	0.541	0.588	∞
0.01	∞	0.456	1.38	0.744	0.602	0.551	0.538	0.588	∞
0.1	∞	0.629	1.08	0.648	0.560	0.531	0.530	0.587	∞
1.0	∞	1.12	0.770	0.564	0.530	0.521	0.527	0.587	∞
3.0	∞	1.76	0.669	0.542	0.523	0.520	0.528	0.588	∞

TABLE 7—EIGHT-POLE NO-ZERO FILTER 3-DECIBEL-DOWN k AND q VALUES.

$(V_p/V_v)_{dB}$	$q_{2,3,4,5,6,7}$	q_1	k_{12}	k_{23}	k_{34}	k_{45}	k_{56}	k_{67}	k_{78}	q_8
			Maximum-power-transfer terminations							
0	∞	0.391	1.52	0.734	0.551	0.510	0.551	0.734	1.52	0.391
0.00001	∞	0.545	1.16	0.640	0.534	0.510	0.534	0.640	1.16	0.545
0.001	∞	0.717	0.960	0.592	0.524	0.510	0.524	0.592	0.960	0.717
0.01	∞	0.896	0.843	0.567	0.520	0.510	0.520	0.567	0.843	0.896
0.1	∞	1.25	0.727	0.545	0.516	0.510	0.516	0.545	0.727	1.25
1.0	∞	2.20	0.633	0.530	0.514	0.511	0.514	0.530	0.633	2.20
3.0	∞	3.53	0.605	0.527	0.515	0.513	0.515	0.527	0.605	3.53
			Resistive termination at only one end							
0	∞	0.199	2.98	1.36	0.920	0.721	0.615	0.562	0.591	∞
0.00001	∞	0.272	2.17	1.04	0.749	0.627	0.567	0.543	0.587	∞
0.001	∞	0.358	1.69	0.859	0.660	0.580	0.544	0.535	0.584	∞
0.01	∞	0.448	1.40	0.755	0.610	0.556	0.533	0.529	0.583	∞
0.1	∞	0.627	1.08	0.651	0.564	0.534	0.523	0.525	0.582	∞
1.0	∞	1.10	0.779	0.567	0.531	0.519	0.516	0.523	0.583	∞
3.0	∞	1.76	0.668	0.541	0.522	0.516	0.516	0.524	0.585	∞

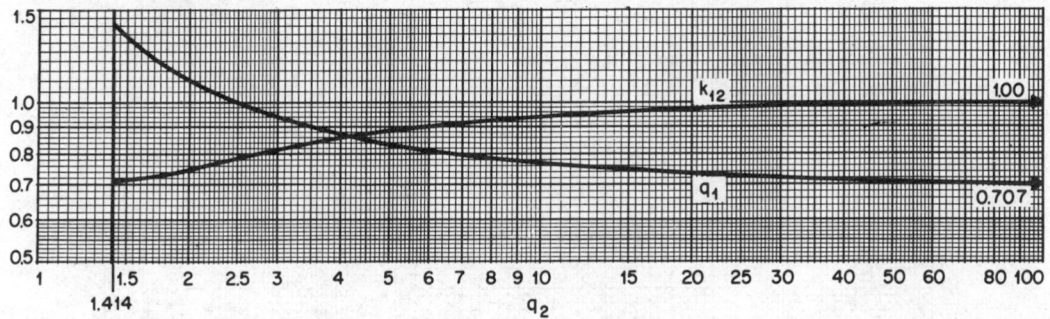

Fig. 30—(Loaded on one side only) 2-pole filter of finite-Q elements producing a maximally flat amplitude shape. See curve 1 of Fig. 6.

Reading along the row for $(V_p/V_v)_{dB}=1$, the second column gives normalized unloaded Q's of infinity at the overall 3-decibel-down frequency, which for this example is 31 kilohertz. Realize that much-lower unloaded-Q designs can be accomplished.

The required value of $q_1=2.21$ is found in the third column. From Fig. 25A, $1/R_1 C_1 = \omega_{3dB}/2.21 =$ $0.451\omega_{3dB}$, from which R_1 or C_1 may be obtained. Experimentally, the 3-decibel-down bandwidth of $R_1 C_1$ must measure 0.451 times the required 3-decibel-down bandwidth or $31\times0.451=14$ kilohertz.

From the table a value of 0.633 is obtained for k_{12} and from Fig. 25A it is found that $1/(C_1 L_2)^{1/2}=$ $0.633\omega_{3dB}$. This means that a resonant circuit made

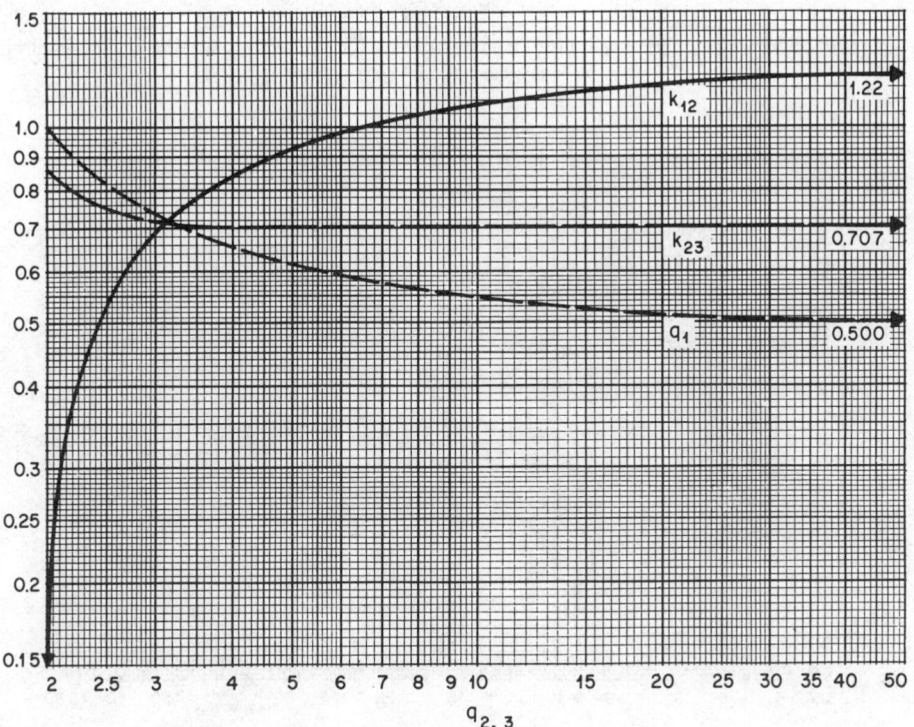

Fig. 31—(Loaded on one side only) 3-pole filter of finite-Q elements producing a maximally flat amplitude shape. See curve 1 of Fig. 7.

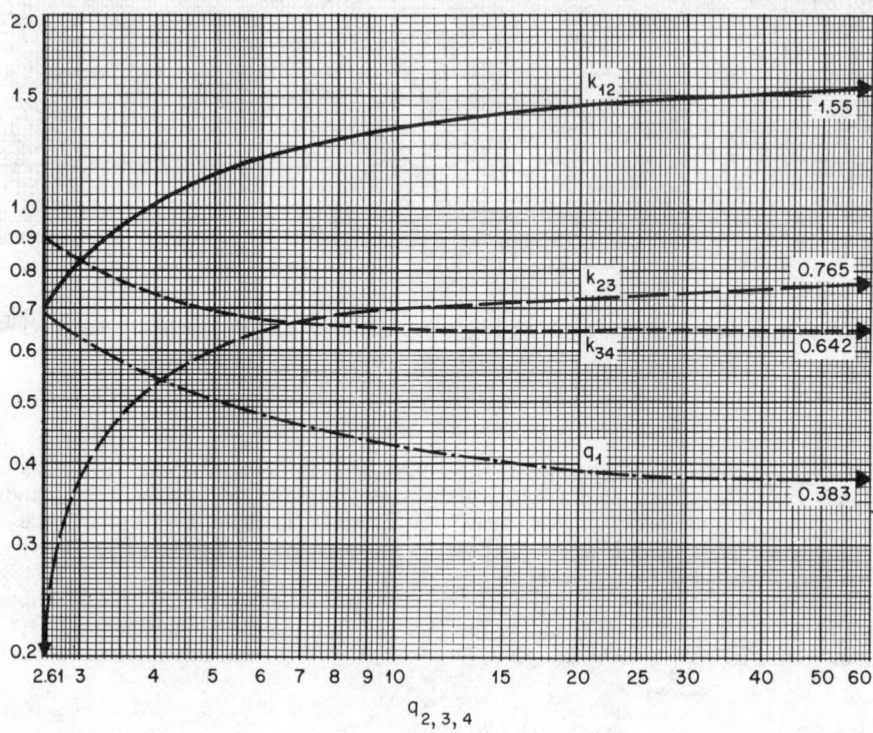

Fig. 32—(Loaded on one side only) 4-pole filter of finite-Q elements producing a maximally flat amplitude shape
See curve 1 of Fig. 8.

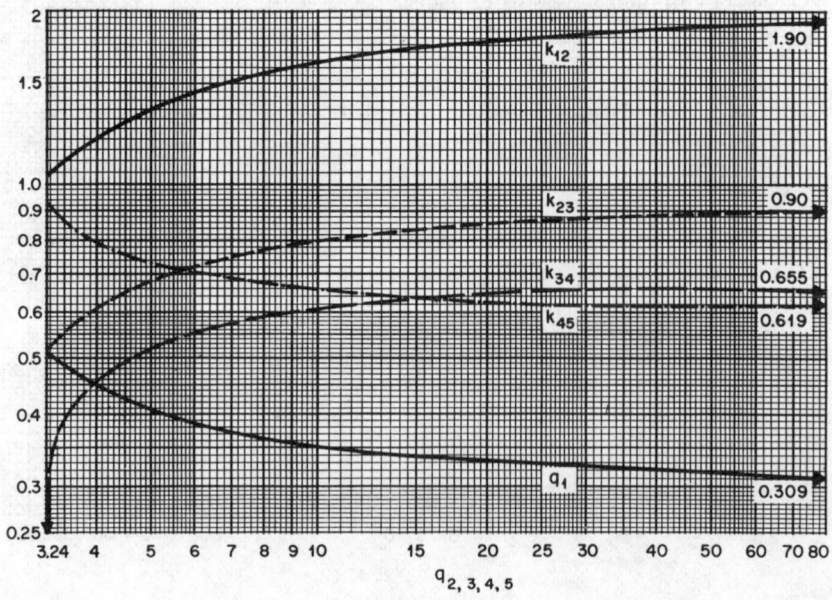

Fig. 33—(Loaded on one side only) 5-pole filter of finite-Q elements producing a maximally flat amplitude shape.
See curve 1 of Fig. 9.

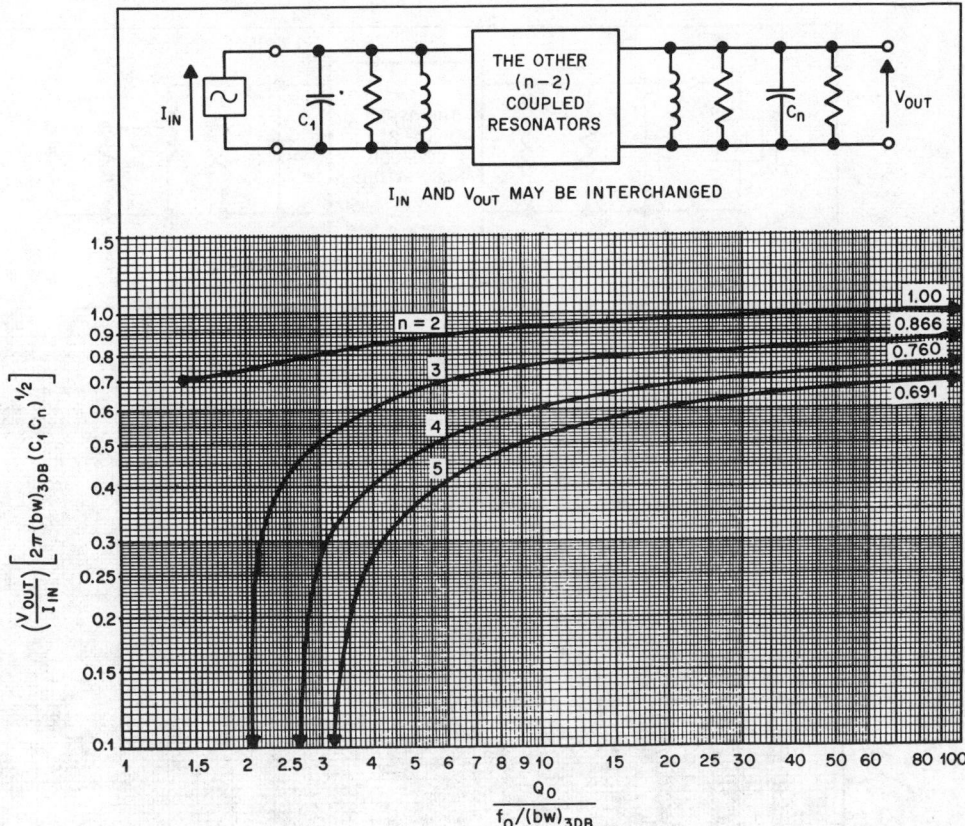

Fig. 34—Midfrequency transfer-impedance magnitude for designs of Figs. 30 through 33 with direct resistive loading on one side only.

up of C_1 and L_2 must tune to 0.633 times the required 3-decibel-down bandwidth or $31 \times 0.633 = 19.7$ kilohertz.

In this fashion, all the remaining elements are determined. Any one of them may be set arbitrarily (for instance, the input load resistance R_1), but once it has been set, all other values are rigidly determined by the k and q factors.

Elements of Lower Q

Designs may be accomplished for elements having finite unloaded Q's. These designs are necessary for small-percentage band-pass filters. As is evident from Fig. 28, the Q of the internal resonators measured at the midfrequency must be the normalized q multiplied by the fractional midfrequency $f_0/(\mathrm{bw})_{3\mathrm{dB}}$. If the bandwidth percentage is small, the fractional midfrequency and therefore the actually required Q will be large.

Practical values of end q's and all k's will result

if the internal elements have finite q's above the minimum values given in Figs. 5–12. For a required response shape, the resulting data can be expressed as in Figs. 30 through 56. These curves are for zero-decibel ripple (Butterworth) and for the maximally linear phase shape.

There are three possible generator and load conditions.

(**A**) Resistive generator and resistive load. It is usually desirable to maximize the ratio of the power delivered to the load to that available from the generator. The generator resistance and the load resistance must be transformed onto their associated resonators to obtain the required q_1 and q_n.

(**B**) Resistive generator and reactive load or vice versa. The function to be considered here is the transfer impedance or admittance. Again, the resistive impedance must be transformed onto the associated resonator.

(**C**) Reactive generator and load. The transfer impedance or admittance is the significant factor,

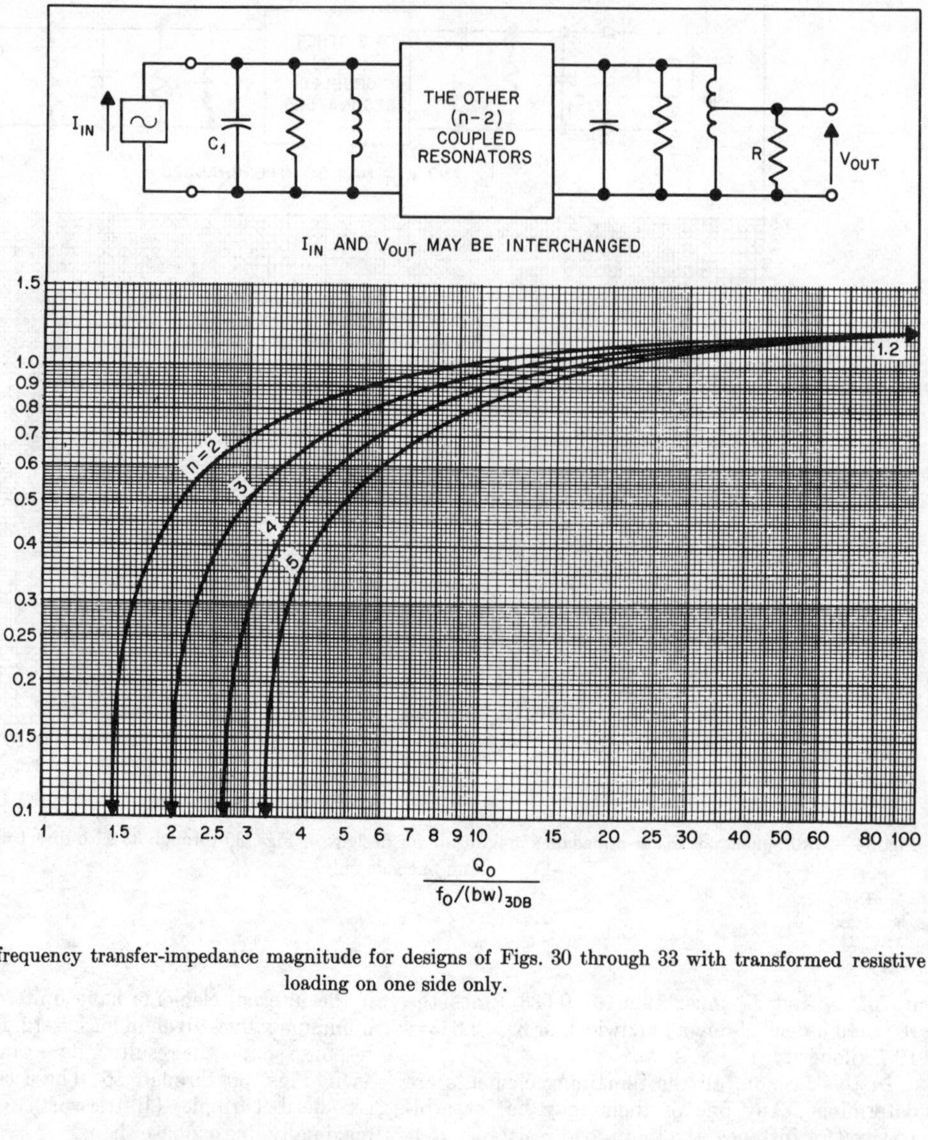

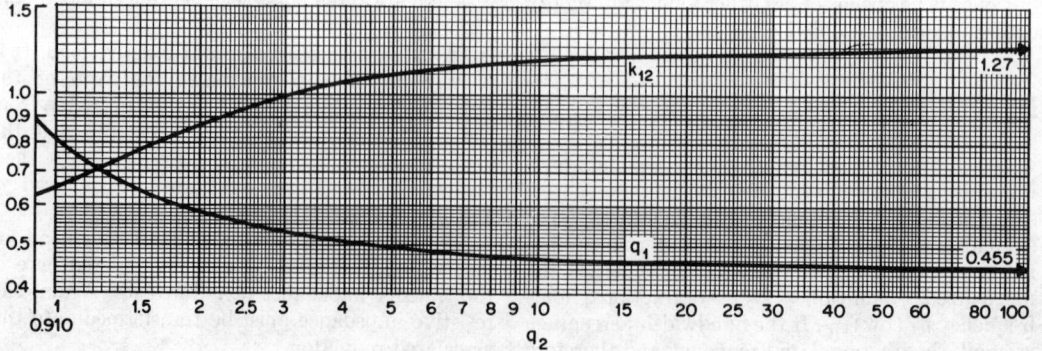

Fig. 35—Midfrequency transfer-impedance magnitude for designs of Figs. 30 through 33 with transformed resistive loading on one side only.

Fig. 36—(Loaded on one side only) 2-pole filter of finite-Q elements producing a maximally linear phase shape. See Figs. 15 and 16.

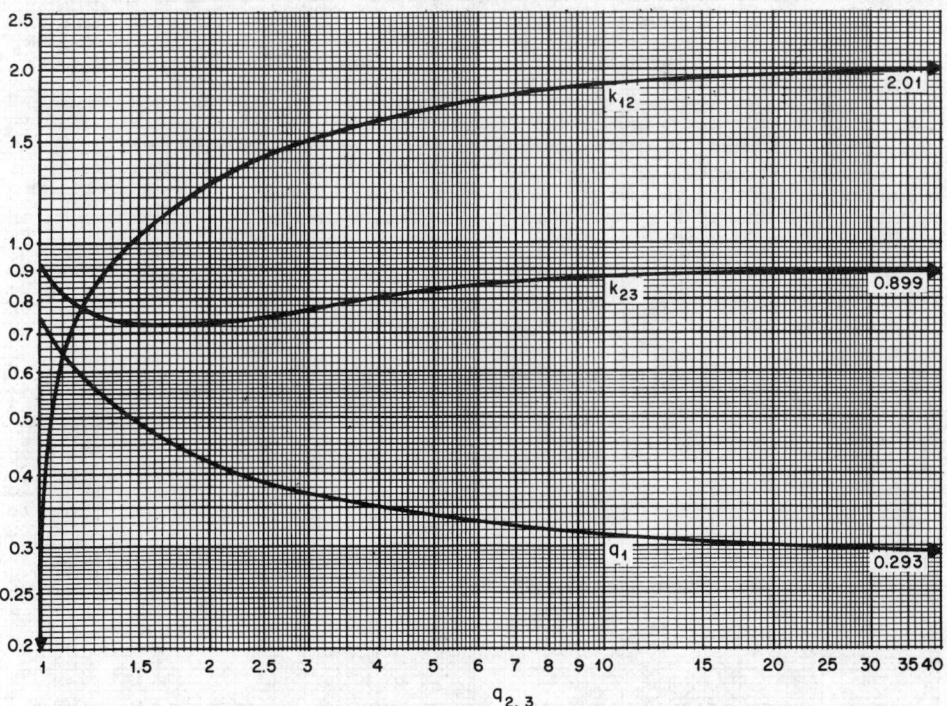

Fig. 37—(Loaded on one side only) 3-pole filter of finite-Q elements producing a maximally linear phase shape. See Figs. 15 and 16.

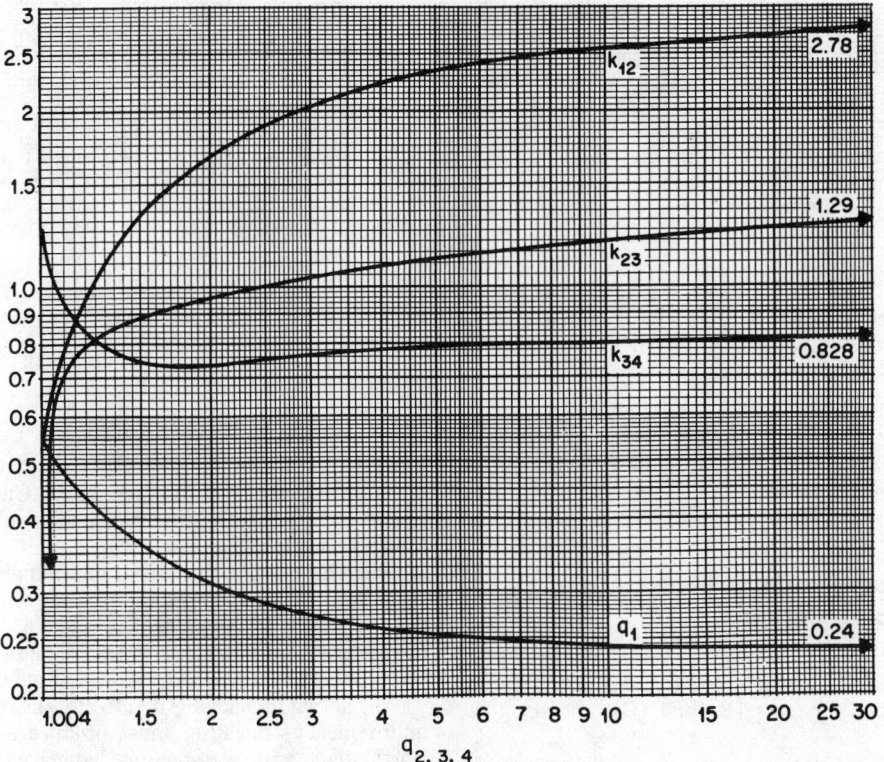

Fig. 38—(Loaded on one side only) 4-pole filter of finite-Q elements producing a maximally linear phase shape. See Figs. 15 and 16.

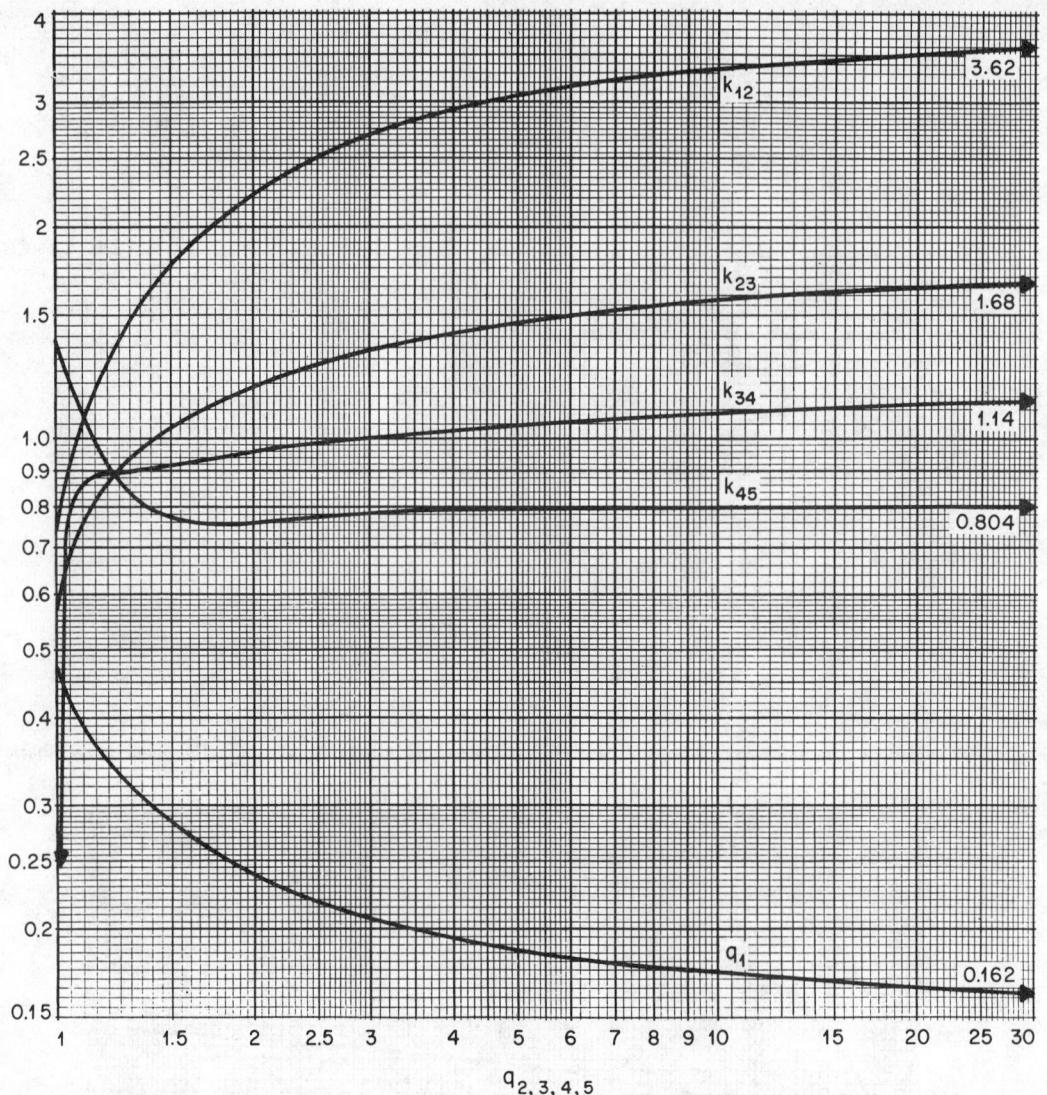

Fig. 39—(Loaded on one side only) 5-pole filter of finite-Q elements producing a maximally linear phase shape. See Figs. 15 and 16.

and a loading resistance must be added to either or both end resonators.

Maximum-Transfer Impedance or Admittance Designs

Figures 30 through 41 give optimum design information for cases (**B**) and (**C**), above.

Example:

(**A**) The filter to be designed must have a rela-

tive attenuation of $(bw)_{70dB}/(bw)_{3dB}=5$ and there must be no ripple in the pass band. Curve 1 of Fig. 9 satisfies these conditions and calls for a 5-pole network.

(**B**) The specified fractional midfrequency is 20 (pass band = 5 percent of the midfrequency), the Q_{min} from Fig. 9 becomes $3.24 \times 20 = 65$. Assume further that resonators with unloaded midfrequency Q's of 100 are available. As the normalized unloaded q is the actual unloaded Q divided by the fractional midfrequency, the filter must produce a Butterworth shape with 5 resonators having normalized unloaded q's of $100/20 = 5$.

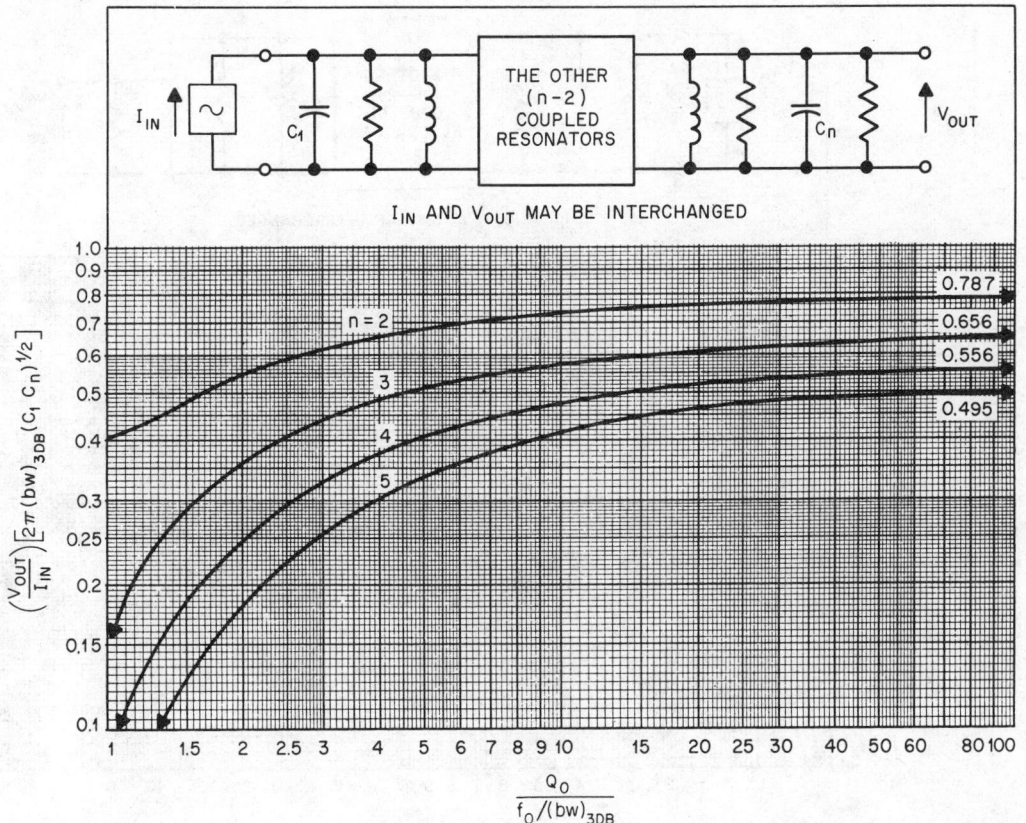

Fig. 40—Midfrequency transfer-impedance magnitude for designs of Figs. 36 through 39 with direct resistive loading on one side only.

Assuming a high-impedance filter to be required, the network of Fig. 42 might well be used. High-side capacitance coupling will be employed and the element values will be obtained from Fig. 33.

(**A**) The q_1 curve of Fig. 33 intersects the abscissa value of 5 at 0.405. By tapping a resistive generator or load onto it, or placing a resistor across it, the resonator C_1L_1 must be loaded to produce an actual Q of $0.405 f_0/(bw)_{3dB} = 8.1$ (see Fig. 28A).

(**B**) As a convenience, the same size of inductor may be used for resonating each node, say 4 millihenries. For a required midfrequency of 80 kilohertz for this example, each node total capacitance will be 1000 picofarads.

(**C**) Again from Fig. 33, we get k_{12} of 1.35 for an abscissa value of 5. From Fig. 28

$$C_{12} = 1.35[(bw)_{3dB}/f_0](C_1 C_2)^{1/2}$$

$$= 1.35 \times 0.05 \times 1000$$

$$= 67.5 \text{ picofarads.}$$

At the midfrequency of 80 kilohertz, node 1 must be resonant when all other nodes are short-circuited. To produce the required capacitance in shunt of L_1, C_a must be $1000 - 67.5 = 932.5$ picofarads.

(**D**) From Fig. 33, a value of 0.67 is obtained for k_{23}, and $C_{23} = 0.67 \times 0.05 \times 1000 = 33.5$ picofarads. To resonate node 2 at the midfrequency with all other nodes short-circuited, $C_b = 1000 - 33.5 - 67.5 = 899$ picofarads.

(**E**) Additional computations give values for C_{34} of $0.53 \times 0.05 \times 1000 = 26.5$ picofarads

$$C_c = 1000 - 33.5 - 26.5 = 940$$

$$C_{45} = 0.73 \times 0.05 \times 1000 = 36.5$$

$$C_d = 1000 - 36.5 - 26.5 = 937$$

and

$$C_e = 1000 - 36.5 = 963.5 \text{ picofarads.}$$

All inductances will be identical and of 4 milli-

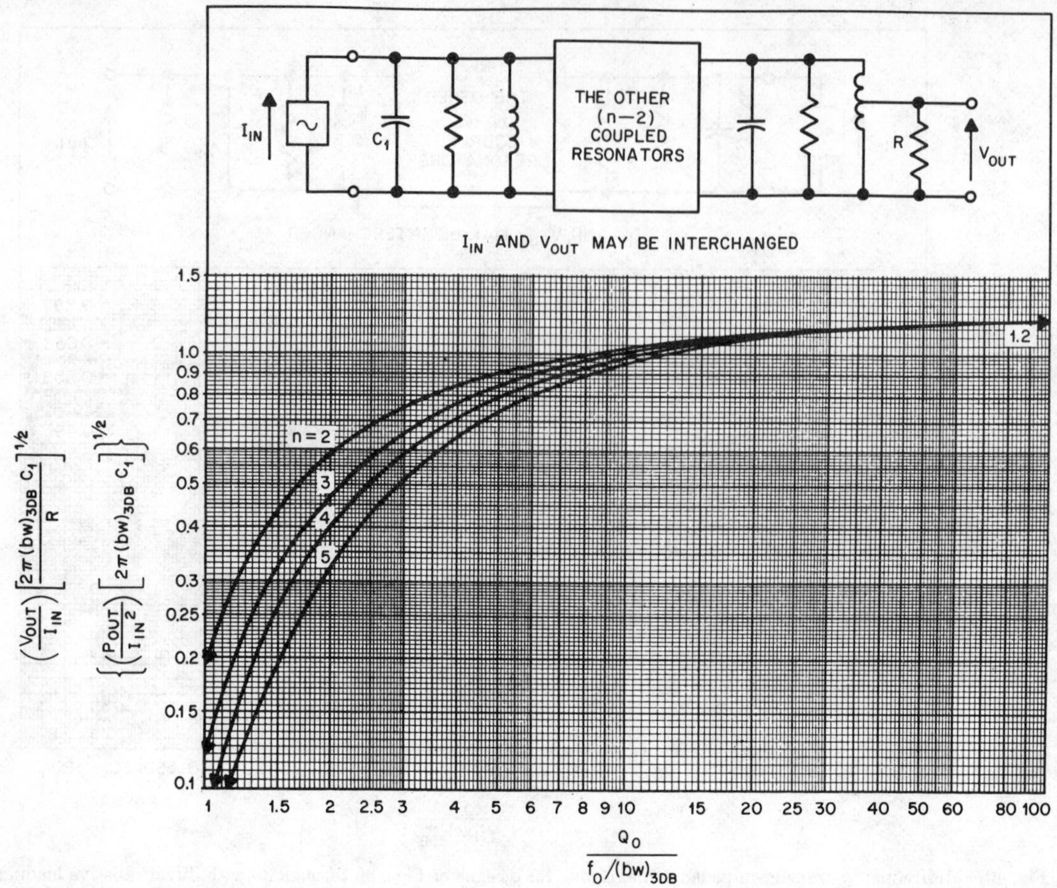

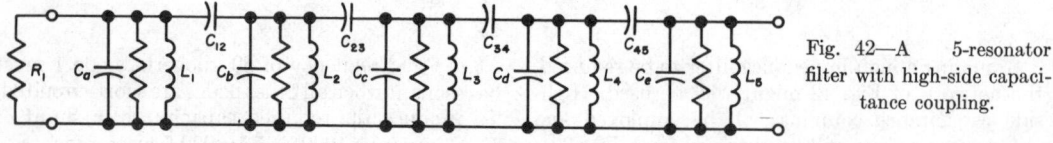

Fig. 41—Midfrequency transfer-impedance magnitude for designs of Figs. 36 through 39 with transformed resistive loading on one side only.

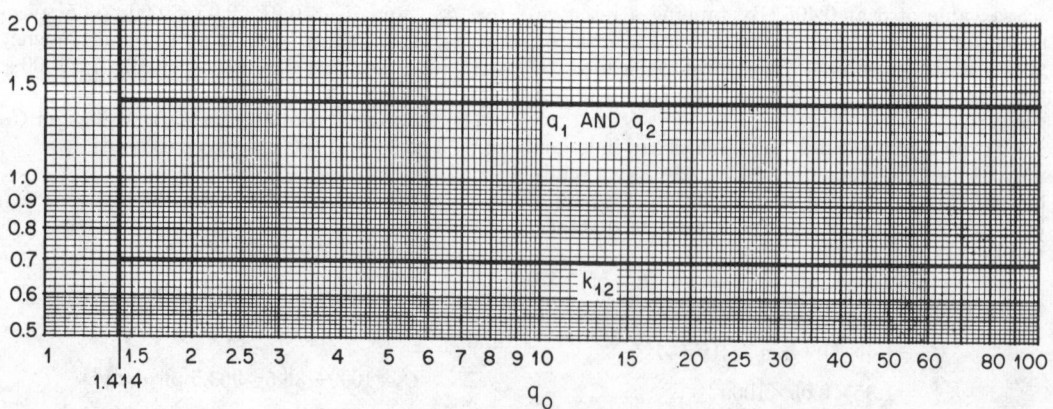

Fig. 42—A 5-resonator filter with high-side capacitance coupling.

Fig. 43—Resistive generator—resistive load 2-pole filter of finite-Q elements producing a maximally flat amplitude shape. See curve 1 of Fig. 6.

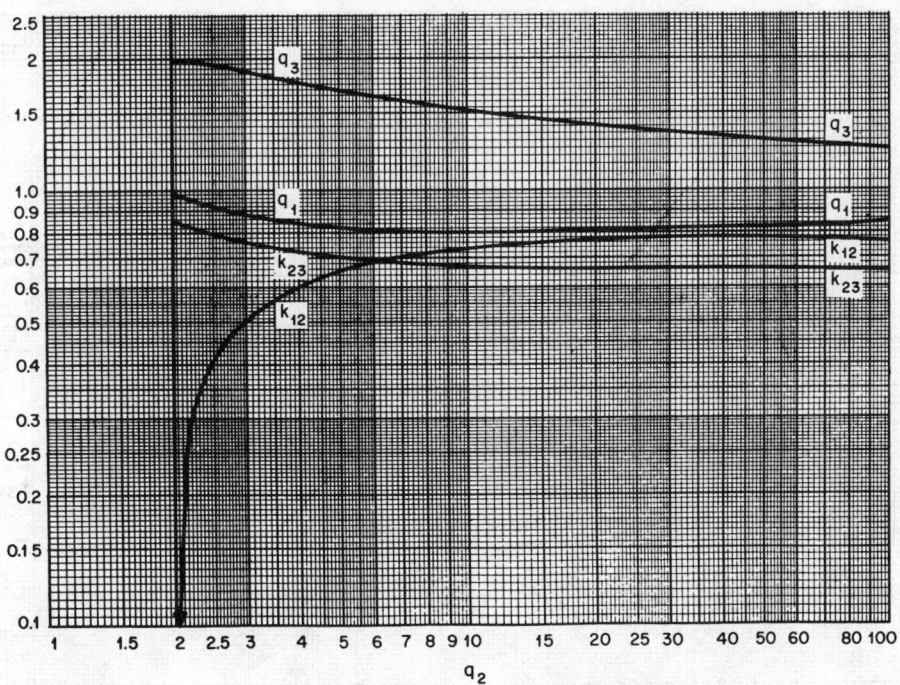

Fig. 44—Resistive generator—resistive load 3-pole filter of finite-Q elements producing a maximally flat amplitude shape. See curve 1 of Fig. 7.

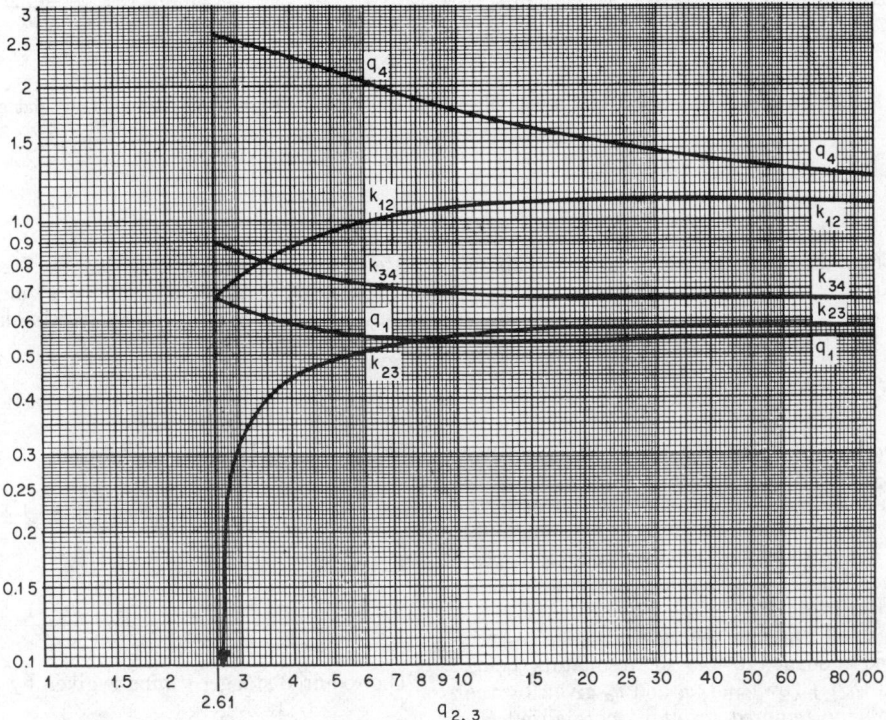

Fig. 45—Resistive generator—resistive load 4-pole filter of finite-Q elements producing a maximally flat amplitude shape. See curve 1 of Fig. 8.

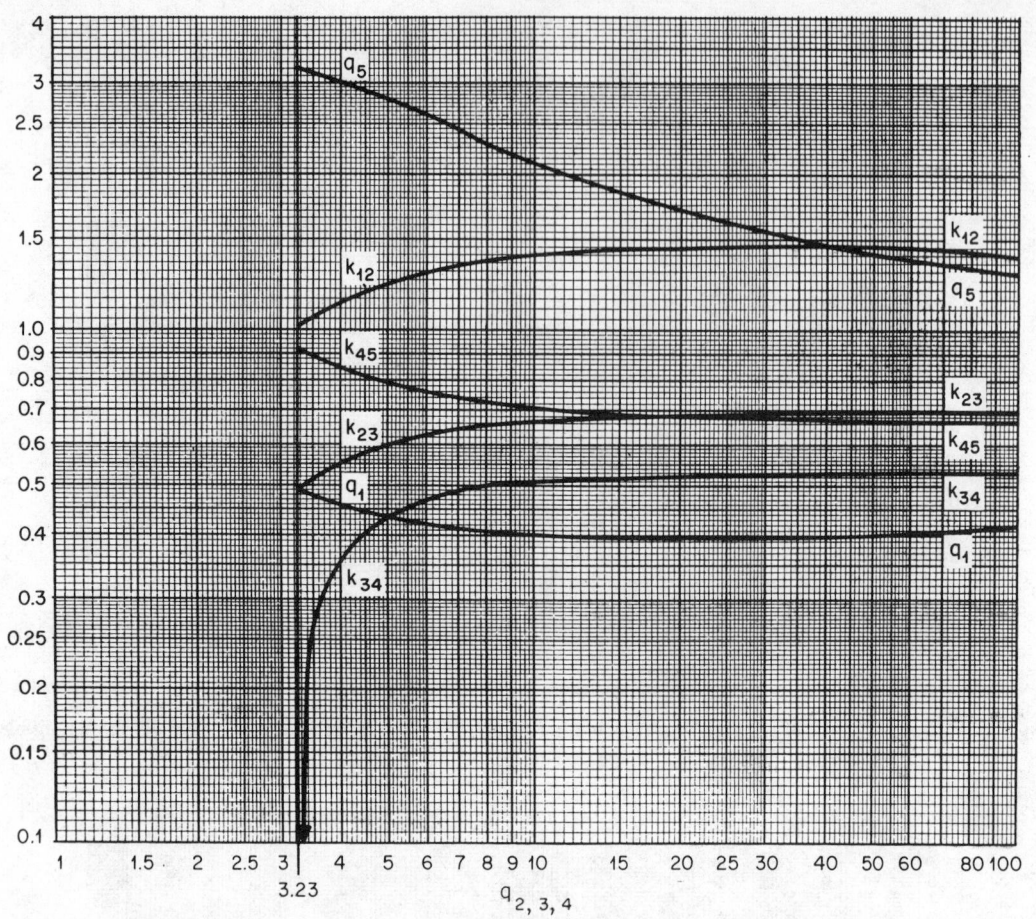

Fig. 46—Resistive generator—resistive load 5-pole filter of finite-Q elements producing a maximally flat amplitude shape. See curve 1 of Fig. 9.

henries, and there will be no inductive coupling among them.

Maximum-Power-Transfer Designs

Maximally Flat Amplitude Shape: When a filter of finite Q is to be placed between a resistive generator and a resistive load, and the Butterworth response shape is desired, Figs. 43 through 49 give the design data that produce maximum possible power transfer between the resistive generator and resistive load.

In Figs. 43 through 49, the abscissa is the normalized unloaded Q of the resonators being used, i.e., $Q_0/[f_0(\text{bw})_{3\text{dB}}]$. q_1 and q_n given by the graphs are the required resultant normalized Q's of the end resonators due to their unloaded Q's, plus the coupled loading from the resistive generator and resistive load.

Maximally Linear Phase Shape: Figures 50 through 56 give (loaded on both sides) maximum-power-transfer designs when the maximally linear phase shape is desired.

STAGGER TUNING OF SINGLE-TUNED INTERSTAGES

Butterworth Response (Figs. 4 and 57)

The required Q's are given by

$$Q_m^{-1} = \frac{(\text{bw})_\beta/f_0}{[(V_p/V_\beta)^2 - 1]^{1/2n}} \sin\left(\frac{2m-1}{n} 90°\right).$$

The required stagger tuning is given by

$$(f_a - f_b)_m = \frac{(\text{bw})_\beta}{[(V_p/V_\beta)^2 - 1]^{1/2n}} \cos\left(\frac{2m-1}{n} 90°\right)$$

$$(f_a + f_b)_m = 2f_0.$$

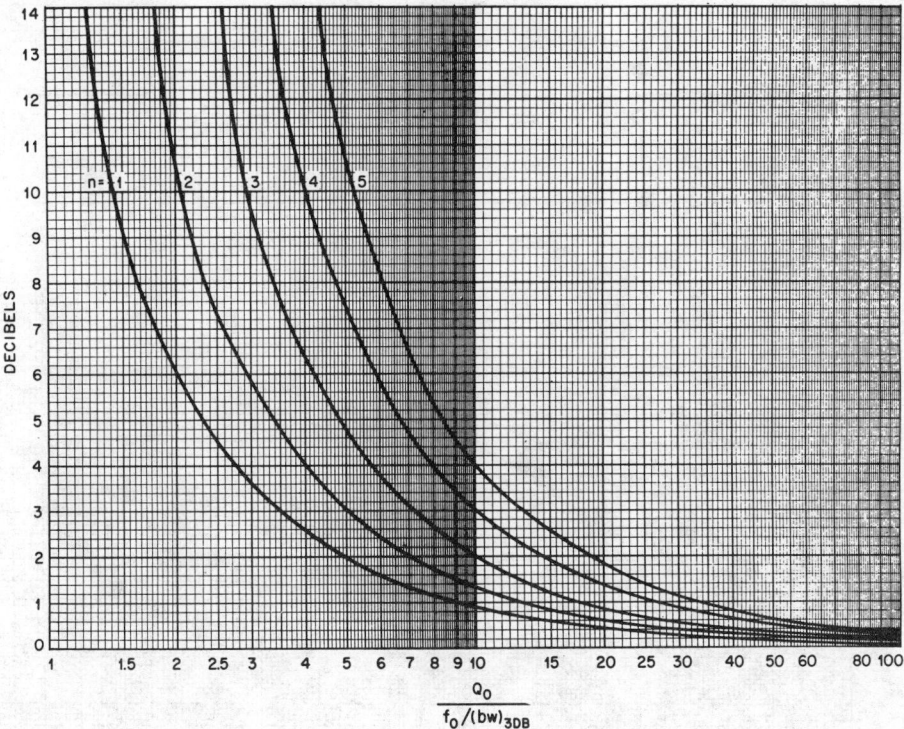

Fig. 47—Ratio of power available from generator to power delivered to load for designs of Figs. 43 through 46.

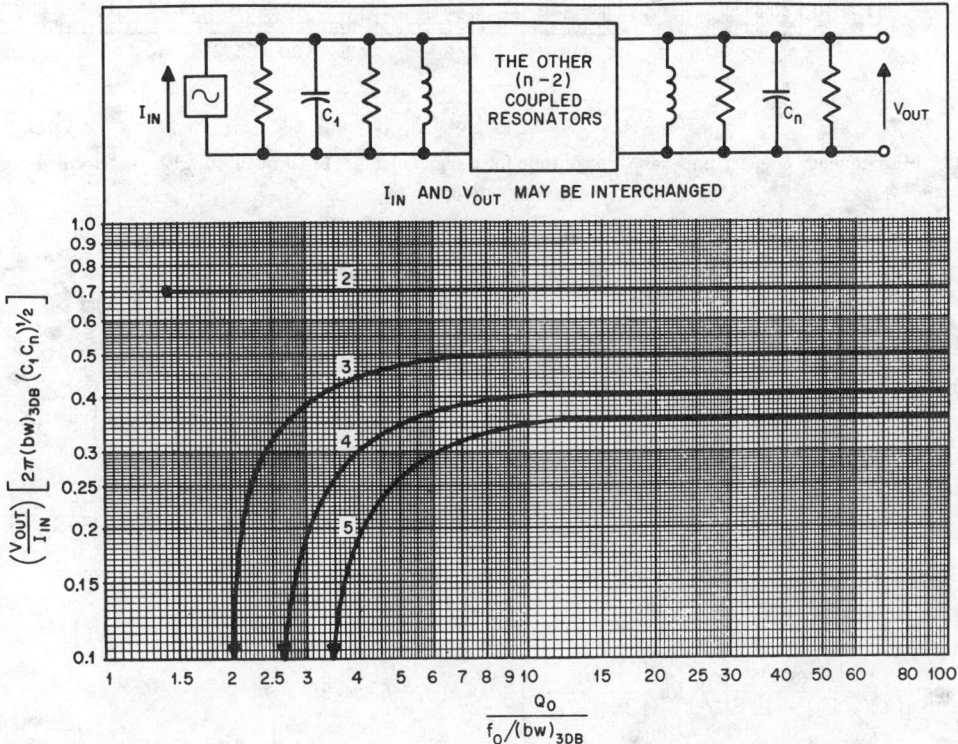

Fig. 48—Midfrequency transfer-impedance magnitude for designs of Figs. 43 through 46 with direct resistive loading on both sides.

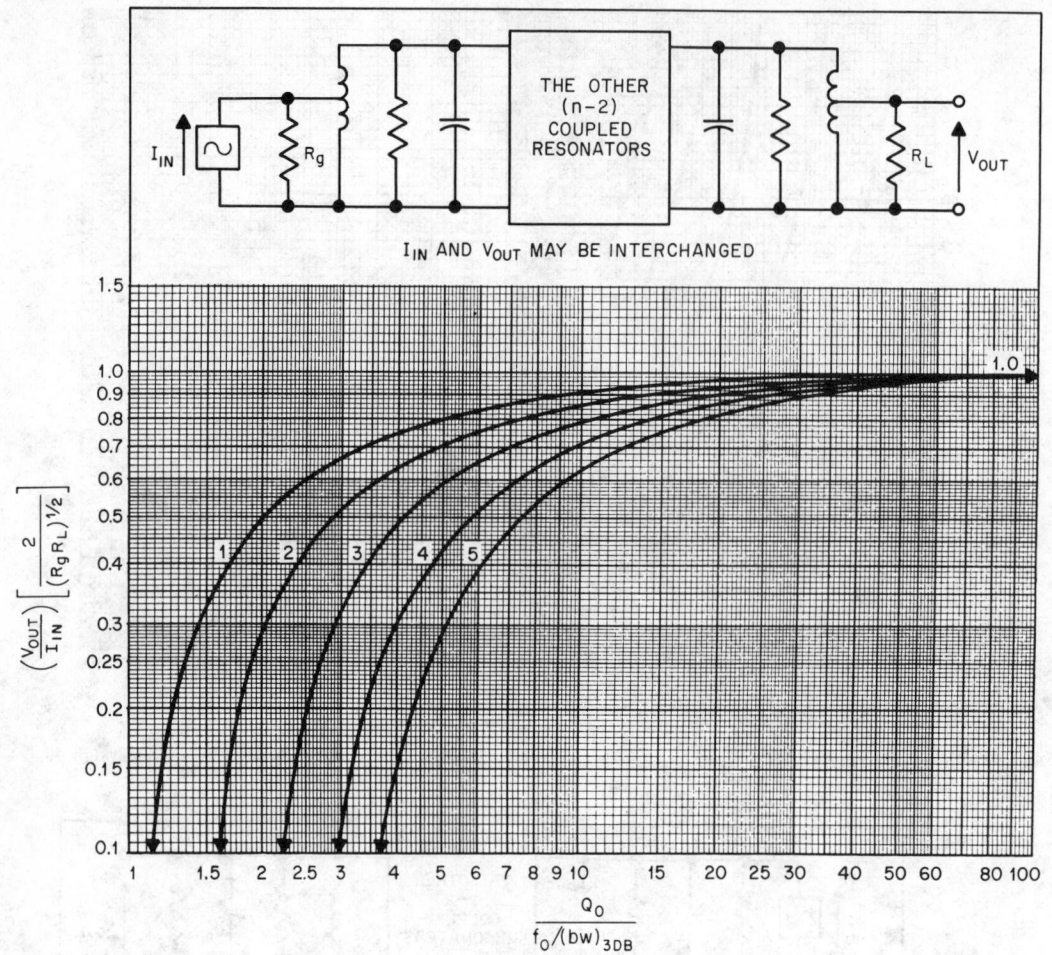

Fig. 49—Midfrequency transfer-impedance magnitude for designs of Figs. 43 through 46 with transformed resistive loading on both sides.

The amplitude response is given by

$$V_p/V = \{1+[(V_p/V_\beta)^2-1][(bw)/(bw)_\beta]^{2n}\}^{1/2}$$

$$\frac{(bw)}{(bw)_\beta} = \left[\frac{(V_p/V)^2-1}{(V_p/V_\beta)^2-1}\right]^{1/2n}$$

$$n = \log\left[\frac{(V_p/V)^2-1}{(V_p/V_\beta)^2-1}\right]\Big/\left\{2\log\left[\frac{(bw)}{(bw)_\beta}\right]\right\}.$$

$$\text{Stage gain} = \frac{g_m}{2\pi(bw)_\beta C}[(V_p/V_\beta)^2-1]^{1/2n}$$

or

$$n = \log\left\{\frac{(\text{total gain})}{[(V_p/V_\beta)^2-1]^{1/2}}\right\}\Big/\log\left[\frac{g_m}{2\pi(bw)_\beta C}\right]$$

where g_m = geometric-mean transconductance of n active devices, and C = geometric-mean capacitance.

Chebishev Response (Figs. 4 and 58)

The required Q's are given by

$$Q_m^{-1} = \frac{(bw)_\beta}{f_0} S_n \sin\left(\frac{2m-1}{n}90°\right)$$

$$S_n = \sinh\left\{n^{-1}\sinh^{-1}\frac{1}{[(V_p/V_\beta)^2-1]^{1/2}}\right\}.$$

The required stagger tuning is given by

$$(f_a-f_b)_m = (bw)_\beta C_n \cos\left(\frac{2m-1}{n}90°\right)$$

$$(f_a+f_b)_m = 2f_0$$

$$C_n = \cosh\left\{n^{-1}\sinh^{-1}\frac{1}{[(V_p/V_\beta)^2-1]^{1/2}}\right\}.$$

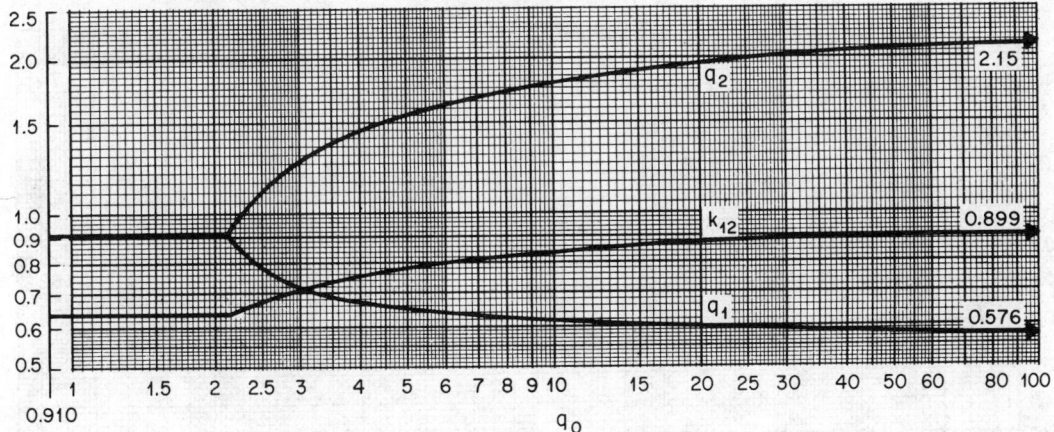

Fig. 50—Resistive generator—resistive load 2-pole filter of finite-Q elements producing a maximally linear phase shape. See Figs. 15 and 16.

Shape outside pass band is

$$\frac{V_p}{V}=\left\{1+\left[\left(\frac{V_p}{V_\beta}\right)^2-1\right]\right.$$

$$\left.\times\left\{\cosh^2\left[n\cosh^{-1}\frac{(\text{bw})}{(\text{bw})_\beta}\right]\right\}\right\}^{1/2}$$

$$\frac{(\text{bw})}{(\text{bw})_\beta}=\cosh\left\{n^{-1}\cosh^{-1}\left[\frac{(V_p/V)^2-1}{(V_p/V_\beta)^2-1}\right]^{1/2}\right\}$$

$$n=\cosh^{-1}\left[\frac{(V_p/V)^2-1}{(V_p/V_\beta)^2-1}\right]^{1/2}\Big/\cosh^{-1}\left[\frac{(\text{bw})}{(\text{bw})_\beta}\right].$$

Shape inside pass band is

$$\frac{V_p}{V}=\left\{1+\left[\left(\frac{V_p}{V_\beta}\right)^2-1\right]\left\{\cos^2\left[n\cos^{-1}\frac{(\text{bw})}{(\text{bw})_\beta}\right]\right\}\right\}^{1/2}$$

$$\frac{(\text{bw})_{\text{crest}}}{(\text{bw})_\beta}=\cos\left(\frac{2m-1}{n}90°\right)$$

$$\frac{(\text{bw})_{\text{trough}}}{(\text{bw})_\beta}=\cos\left(\frac{2m}{n}90°\right).$$

$$\text{Stage gain}=\frac{g_m}{2^{1/n}\pi(\text{bw})_\beta C}\left[(V_p/V_\beta)^2-1\right]^{1/2n}$$

$$n=\log\left[\frac{(\text{total gain})}{\frac{1}{2}[(V_p/V_\beta)^2-1]^{1/2}}\right]\Big/\log\left[\frac{g_m}{\pi(\text{bw})_\beta C}\right]$$

where g_m= geometric-mean transconductance of n active devices, and C= geometric-mean capacitance.

QUARTZ-CRYSTAL BAND-PASS FILTERS FOR $f_0/(bw)_{3dB}$ BETWEEN 10 AND 100

When a filter requires simultaneously a small-percentage bandwidth and a high rate of cutoff, it is usually not practical to obtain sufficiently high unloaded Q in ordinary L-C resonators. When the ultimate attenuation required is no greater than approximately 30 dB, and the percentage bandwidth required is between 10% and 1%, then a single half-lattice filter configuration involving two L-C resonators and $(n-2)$ quartz crystal resonators can be successfully used. It should be realized that piezoelectric crystal resonators often exhibit spurious responses, and the presence of these (usually on the high-frequency side of the response) must be considered. The following sections first give the design procedure for the full-lattice configuration and then present the half-lattice equivalent.

High-Impedance Lattice

An "open-circuited" lattice is shown in Fig. 59. The arrangements of the impedance arms Z_A and Z_B are shown in Fig. 60. In each arm there is an L-C parallel-resonant circuit shunted by $(n/2)-1$ quartz crystals. The number of complex poles in the transfer function is equal to the n. The L-C circuit is loaded by R_p to give the required $Q_p=\omega_0 C_p R_p$. Its capacitance includes those of the crystal holders, and it is resonant to $(f_0+\Delta f_p)$ as shown

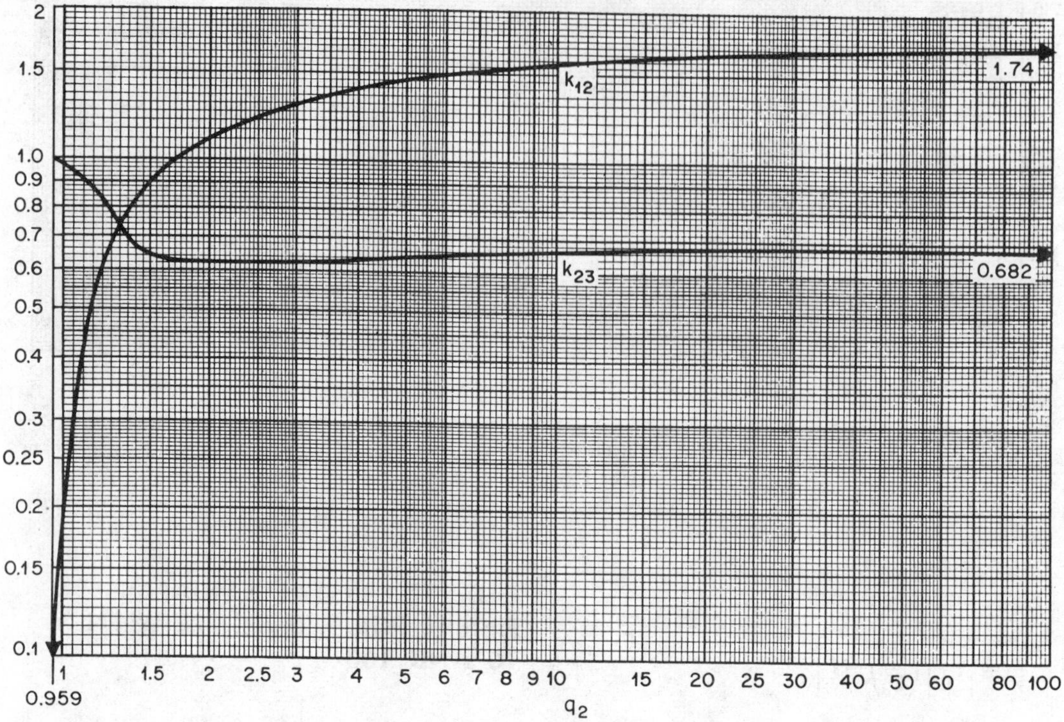

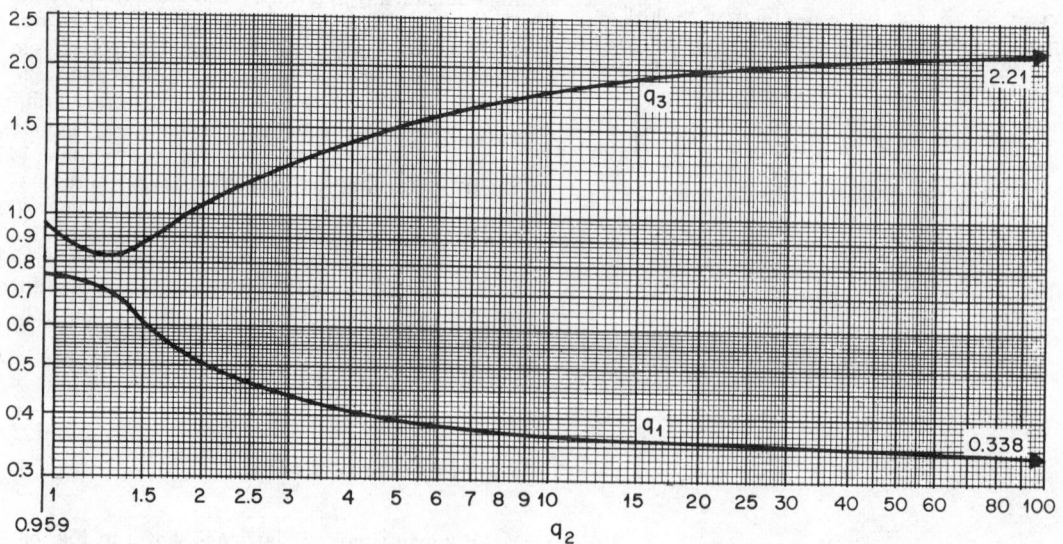

Fig. 51—Resistive generator—resistive load 3-pole filter of finite-Q elements producing a maximally linear phase shape. See Figs. 15 and 16.

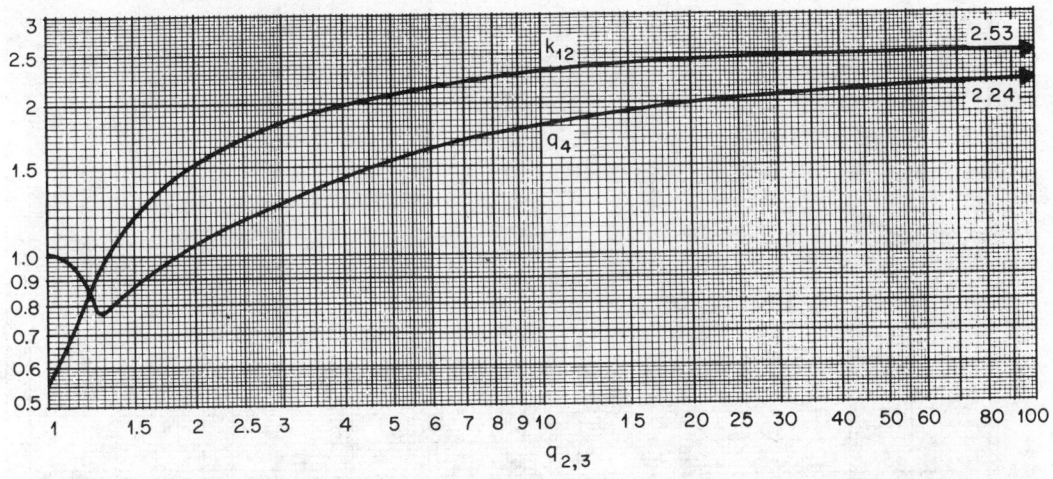

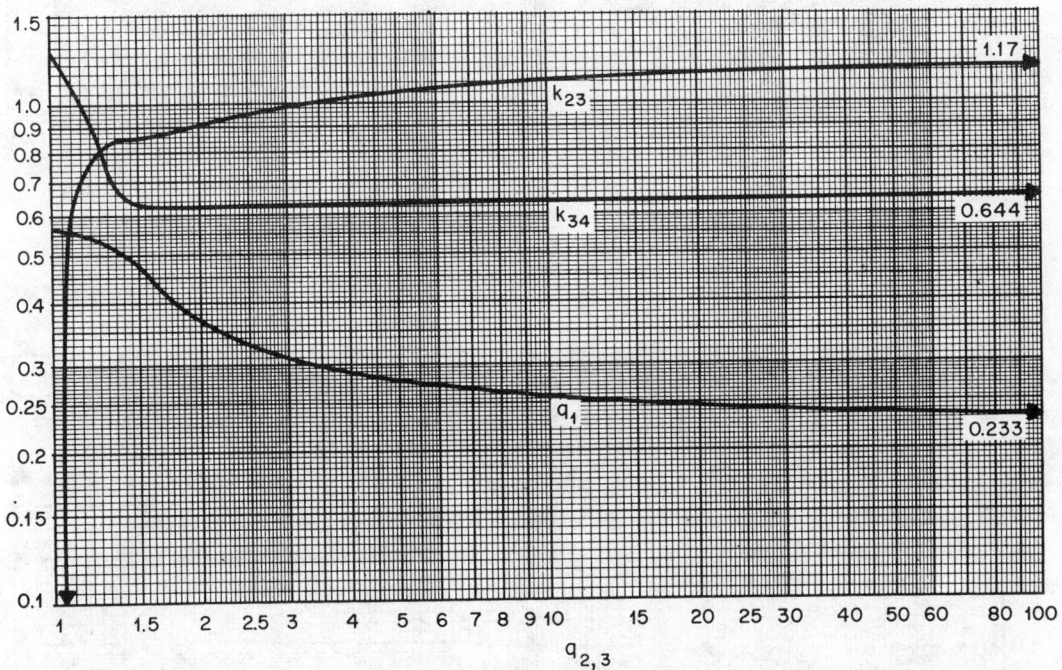

Fig. 52—Resistive generator—resistive load 4-pole filter of finite-Q elements producing a maximally linear phase shape. See Figs. 15 and 16.

in the diagrams. The motional capacitance C_1, C_2, C_3, etc., must have a particular value, and each crystal must be resonant to a particular frequency, $(f_0 \pm \Delta f_1)$, $(f_0 \pm \Delta f_2)$, etc.

Frequently, divided-electrode crystals are used so that one crystal can be used for the identical resonators in the two series arms, and likewise in the lattice arms.

The structure can be modified by converting the lattice to its equivalent in accordance with Fig. 61. The elements Z that are lifted out of the

arms and shunted across the terminals consist of L_p, R_p, and most of C_p.

Design Information

The data of Fig. 62 are for the Chebishev and Butterworth response shapes of 4-pole no-zero networks for which the relative attenuation is plotted in Fig. 8. Similarly, Figs. 63 through 65 are for 6-pole, 8-pole, and 10-pole no-zero networks, the

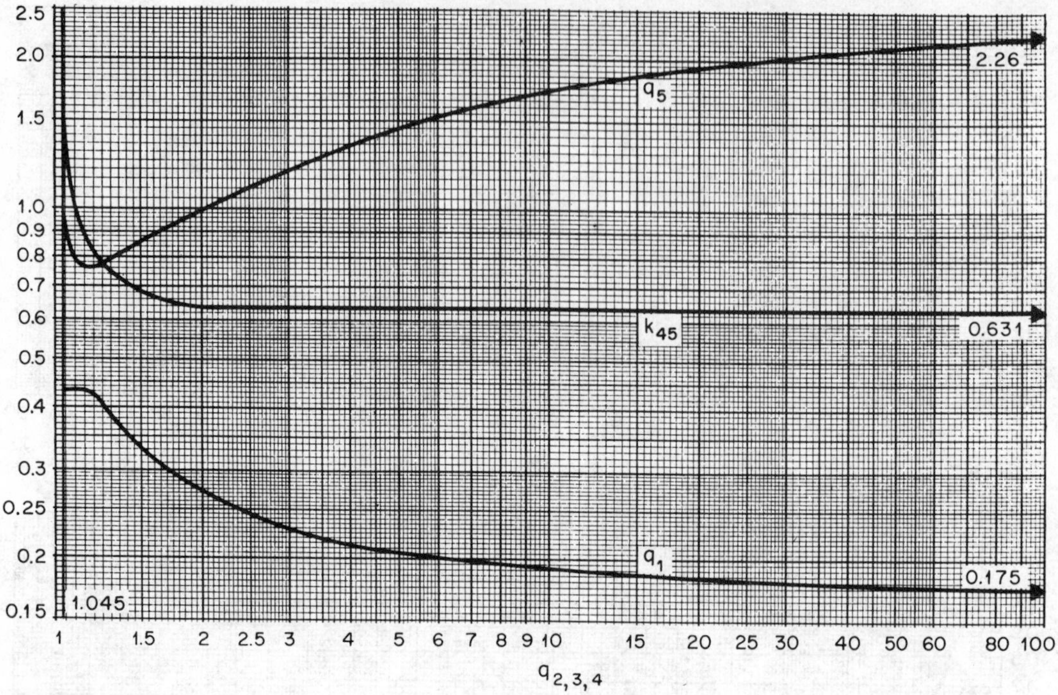

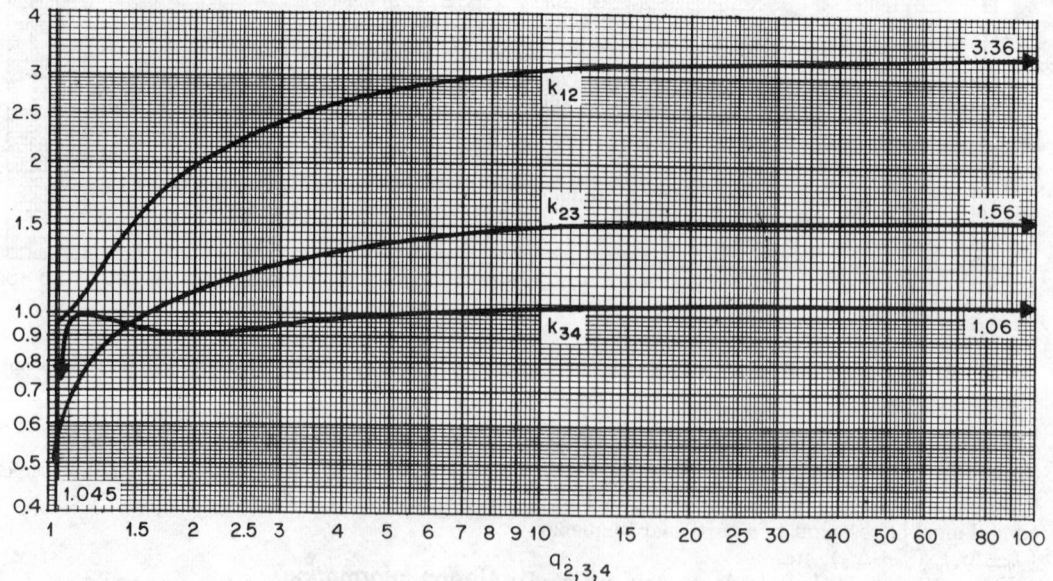

Fig. 53—Resistive generator—resistive load 5-pole filter of finite-Q elements producing a maximally linear phase shape. See Figs. 15 and 16.

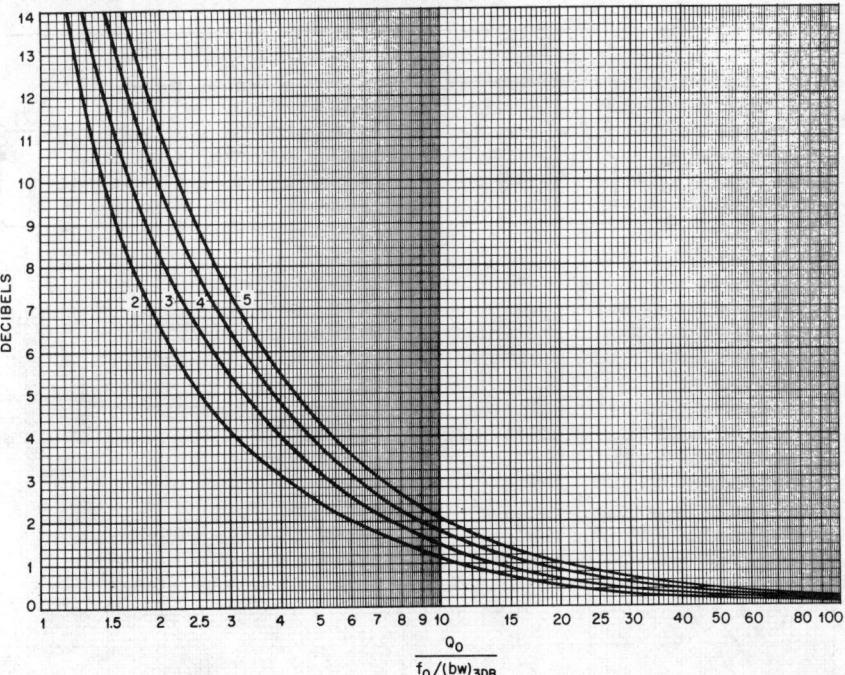

Fig. 54—Ratio of power available from generator to power delivered to load for designs of Figs. 50 through 53.

THE OTHER
(n − 2)
COUPLED
RESONATORS

I_{IN} AND V_{OUT} MAY BE INTERCHANGED

$\left(\dfrac{V_{OUT}}{I_{IN}}\right)\left[2\pi(bw)_{3DB}(C_1 C_n)^{1/2}\right]$

$\dfrac{Q_0}{f_0/(bw)_{3DB}}$

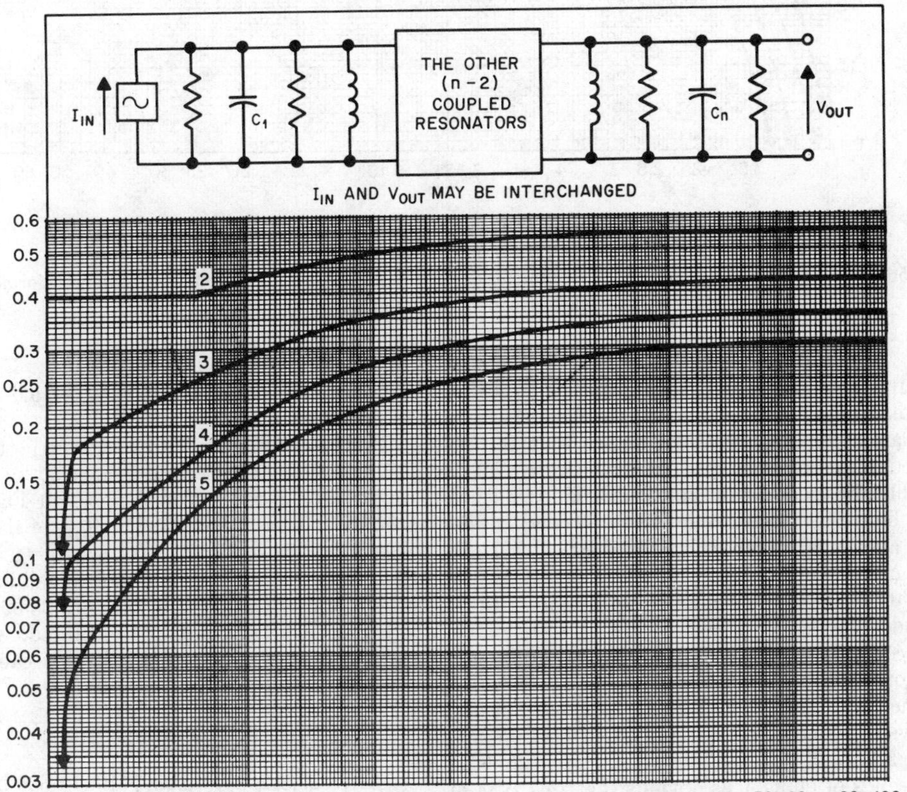

Fig. 55—Midfrequency transfer-impedance magnitude for designs of Figs. 50 through 53 with direct resistive loading on both sides.

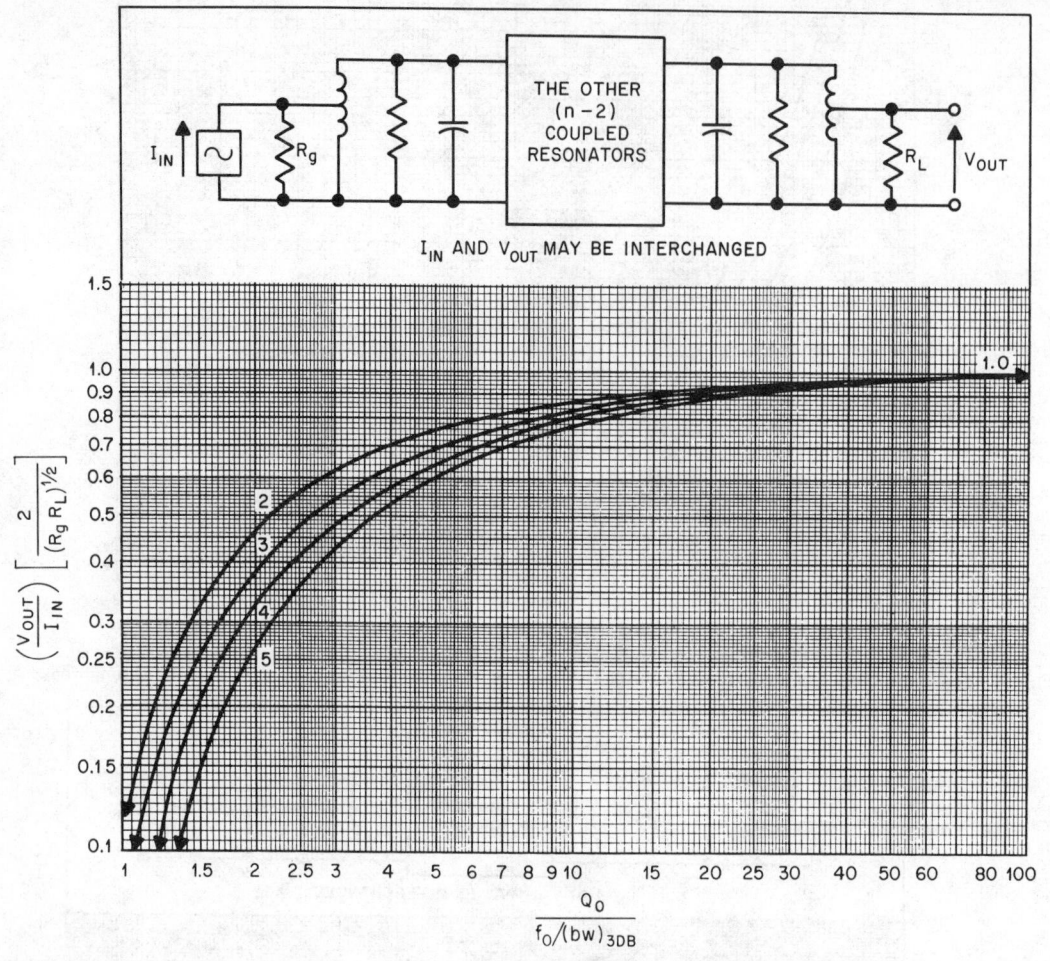

Fig. 56—Midfrequency transfer-impedance magnitude for designs of Figs. 50 through 53 with transformed resistive loading on both sides.

relative attenuation of the former two being plotted in Figs. 10 and 12, respectively.

Examination of the tables shows that the required Q_p of the L-C parallel-resonant circuit is roughly the same as the fractional midfrequency. This limits the practical design to $f_0/(\text{bw})_{3\text{dB}}$ less than about 250. A lower limit to the $f_0/(\text{bw})_{3\text{dB}}$ is of the order of 10 due to the fact that C_p/C_1 is roughly equal to the square of $f_0/(\text{bw})_{3\text{dB}}$, and C_p includes the capacitances of the crystal holders and coil and stray distributed capacitances, so it cannot be reduced indefinitely.

The impedance Z in Fig. 61 must include the equivalent-generator and equivalent-load impedances. Since R_p often comes to some hundreds of thousands of ohms, it is obvious that this type of filter requires a very-high-impedance equivalent generator and load.

Example: Required, a filter for $f_0 = 175$ kilohertz,

$(\text{bw})_{3\text{dB}} = 2.0$ kilohertz, $(\text{bw})_{60\text{dB}} < 5.0$ kilohertz, $(V_p/V_v)_{\text{dB}} < 0.3$.

Then, $f_0/(\text{bw})_{3\text{dB}} = 87.5$ and $(\text{bw})_{60\text{dB}}/(\text{bw})_{3\text{dB}} < 2.5$. The latter requirement is satisfied by the curve for $(V_p/V_v)_{\text{dB}} = 0.1$-decibel ripple on Fig. 10 with a 6-pole, no-zero network. The $Q_{\min}$ of the internal resonators must be $q_{\min}f_0/(\text{bw})_{3\text{dB}} = 9.5 \times 87.5 = 831$. This is far beyond L-C possibilities, but crystal unloaded Q usually exceeds 25 000.

In Fig. 63, let $C_1 = 0.020$ picofarad, which can be obtained. Lower values for C_2 can also be realized.

$$C_2 = C_1/2.53 = 0.00800 \text{ picofarad.}$$

$$\Delta f_1 = 0.339\Delta f_{3\text{dB}} = 0.339 \times 1000 = 339 \text{ hertz.}$$

Then the first crystal in arm A is series-resonant at 175 kilohertz minus 339 hertz (in arm B plus 339 hertz).

Similarly, $\Delta f_2 = 0.859 \times 1000 = 859$ hertz. In the parallel-resonant circuits

$$C_p = 2.73 C_1 [f_0/(\text{bw})_{3\text{dB}}]^2$$

$$= 2.73 \times 0.020 \times (87.5)^2$$

$$= 422 \text{ picofarads.}$$

Since $F_p = 0$, they are parallel-resonant at 175 kilohertz. The loaded $Q_p = 1.28 \times 87.5 = 112$. The equivalent

$$R_p = Q_p/2\pi f_0 C_p$$

$$= 112/2\pi \times 175 \times 422 \times 10^{-9}$$

$$= 240\ 000 \text{ ohms.}$$

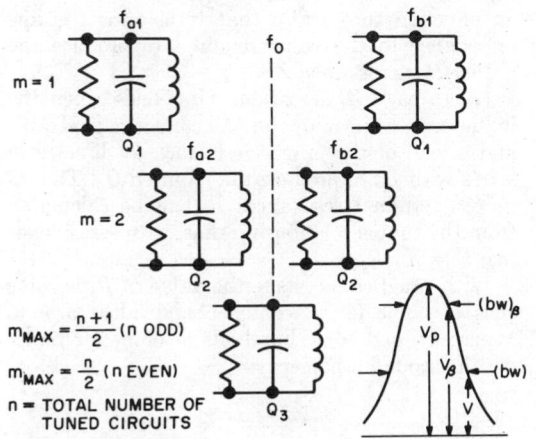

Fig. 57—Stagger-tuned interstages for Butterworth response. Each circuit is coupled to the next only by an isolating active element.

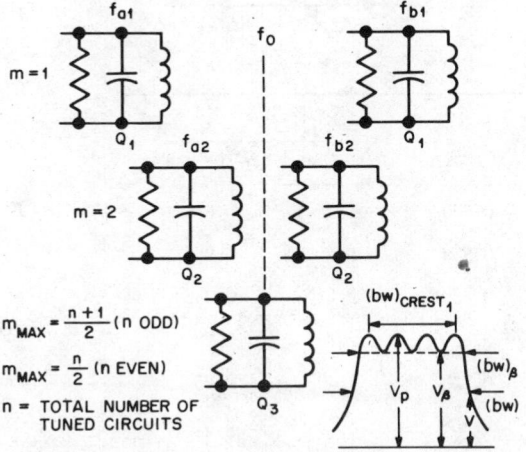

Fig. 58—Stagger-tuned interstages for Chebishev response. Each circuit is coupled to the next only by an isolating active element.

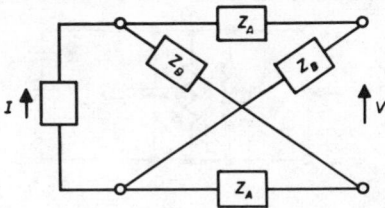

Fig. 59—High-impedance lattice section.

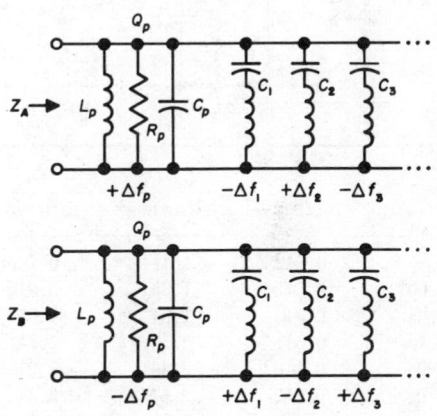

Fig. 60—Detailed structure of the lattice arms indicated in Fig. 59.

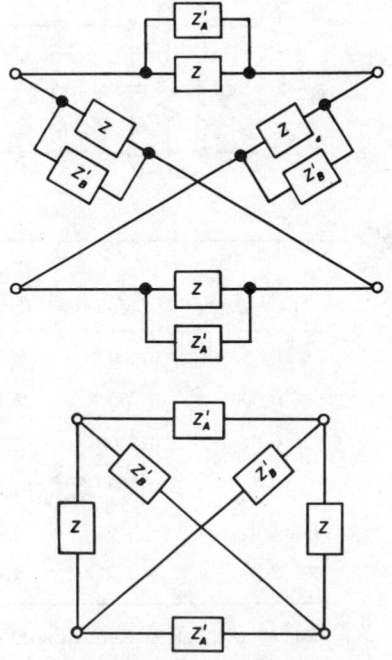

Fig. 61—Equivalent lattices.

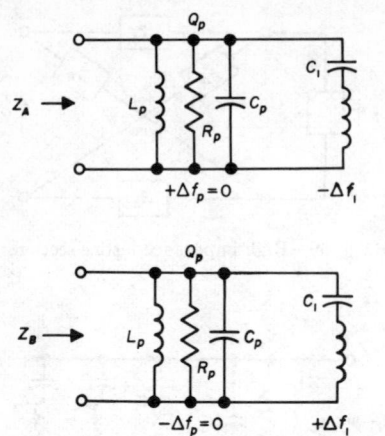

$(V_p/V_t)_{dB}$	$\Delta f_1/\Delta f_{3dB}$	$\dfrac{C_p/C_1}{[f_0/(bw)_{3dB}]^2}$	$\dfrac{Q_p}{f_0/(bw)_{3dB}}$
0	0.542	1.414	0.766
0.001	0.541	1.66	0.912
0.01	0.540	1.84	1.05
0.1	0.541	2.10	1.34
1.0	0.546	2.46	2.21
3.0	0.552	2.57	3.44

Fig. 62—Four-pole no-zero lattice-filter design for Chebishev response. Note that Δf_{3dB} is one-half the total 3-decibel bandwidth, or, $2\Delta f_{3dB} = (bw)_{3dB}$.

If the unloaded Q of the inductor L_p is 200, the added loading due to generator or load must be in excess of one-half megohm.

Low-Impedance Generator and Load

A low-impedance generator and/or load may be used with the above filter design by the following procedure: After the arms of Fig. 60 have been designed, convert the resulting lattice of Fig. 59 to the configuration of Fig. 61 so that the Z across each end of the filter consists of L_p, R_p, and most of C_p. Then use either of the following two steps.

(**A**) Couple the generator to one L_p and the load to the other L_p via mutual inductance, with an effective turns ratio that transforms the low impedance to the value required to produce the proper R_p across each Z.

(**B**) In each Z, across the filter ends, open the inductor L_p at its midpoint and connect a generator and a load of the proper resistance R_s directly in series with L_p to produce the required Q_p. The required terminal resistances R_s can be calculated from the simple relationship that, with series loading, $Q_p = X_p/R_s$.

With practical crystals, the value of R_s is some tens of ohms for percentage bandwidths around 1 percent, and some hundreds of ohms for bandwidths around 5 percent.

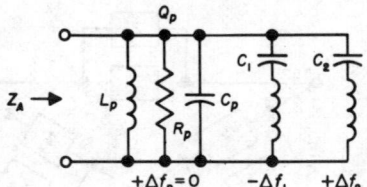

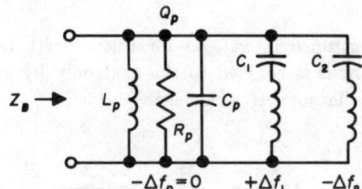

$(V_p/V_v)_{dB}$	$\Delta f_1/\Delta f_{3dB}$	C_1/C_2	$\Delta f_2/\Delta f_{3dB}$	$\dfrac{C_p/C_1}{[f_0/(bw)_{3dB}]^2}$	$\dfrac{Q_p}{f_0/(bw)_{3dB}}$
0	0.400	2.30	0.920	1.05	0.518
0.0001	0.370	2.40	0.889	1.51	0.680
0.01	0.350	2.47	0.869	2.14	0.936
0.1	0.339	2.53	0.859	2.73	1.28
1.0	0.330	2.57	0.850	3.49	2.25
3.0	0.332	2.58	0.858	3.72	3.51

Fig. 63—Six-pole no-zero lattice-filter design for Chebishev response. Note that Δf_{3dB} is one-half the total 3-decibel bandwidth, or that $2\Delta f_{3dB} = (bw)_{3dB}$.

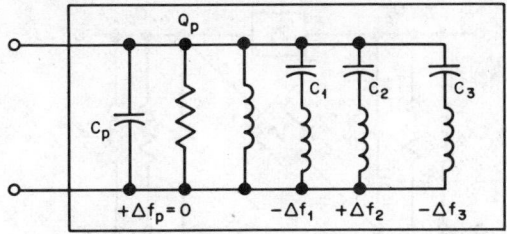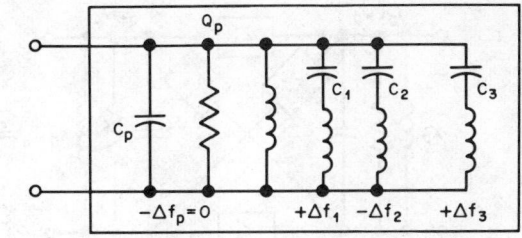

$(V_p/V_v)_{dB}$	$\Delta f_1/\Delta f_{3dB}$	C_1/C_2	$\Delta f_2/\Delta f_{3dB}$	C_1/C_3	$\Delta f_3/\Delta f_{3dB}$	$\dfrac{C_p/C_1}{[f_0/(bw)_{3dB}]^2}$	$\dfrac{Q_p}{f_0/(bw)_{3dB}}$
0	0.303	1.08	0.856	2.16	1.063	1.04	0.390
0.00001	0.275	1.24	0.762	2.95	0.996	1.60	0.548
0.001	0.259	1.34	0.715	3.51	0.966	2.20	0.720
0.01	0.249	1.40	0.691	3.93	0.951	2.77	0.900
0.1	0.241	1.46	0.699	4.35	0.938	3.62	1.29
1.0	0.235	1.50	0.655	4.62	0.931	4.70	2.21
3.0	0.234	1.52	0.653	4.75	0.932	5.10	3.53

Fig. 64—Eight-pole no-zero lattice-filter design for Chebishev response.

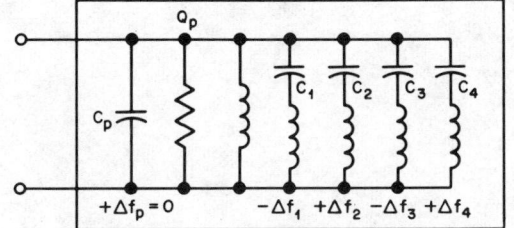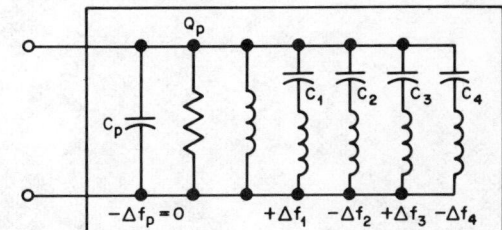

$(V_p/V_v)_{dB}$	$\Delta f_1/\Delta f_{3dB}$	C_1/C_2	$\Delta f_2/\Delta f_{3dB}$	C_1/C_3	$\Delta f_3/\Delta f_{3dB}$	C_1/C_4	$\Delta f_4/\Delta f_{3dB}$	$\dfrac{C_p/C_1}{[f_0/(bw)_{3dB}]^2}$	$\dfrac{Q_p}{f_0/(bw)_{3dB}}$
0	0.244	1.27	0.687	0.948	1.118	1.33	1.181	1.021	0.314
0.00001	0.213	1.17	0.608	1.38	0.924	2.85	1.035	1.90	0.509
0.001	0.200	1.19	0.571	1.65	0.862	4.15	0.997	2.72	0.690
0.01	0.193	1.21	0.552	1.83	0.834	5.10	0.981	3.48	0.879
0.1	0.186	1.23	0.534	2.00	0.810	6.18	0.969	4.60	1.24
1.0	0.181	1.26	0.516	2.16	0.793	7.10	0.961	6.07	2.29
3.0	0.180	1.26	0.516	2.22	0.787	7.50	0.958	6.90	6.30

Fig. 65—Ten-pole no-zero lattice-filter design for Chebishev response.

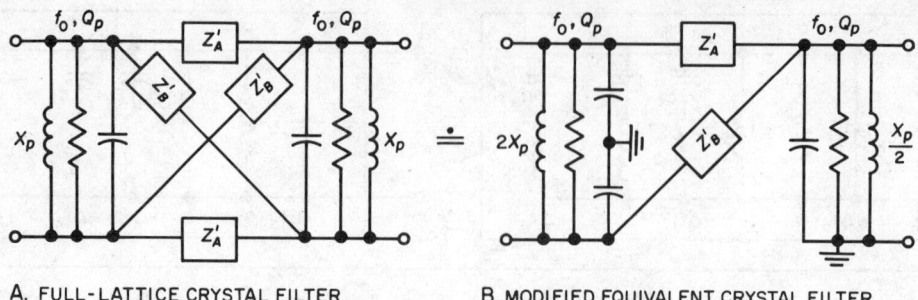

A. FULL-LATTICE CRYSTAL FILTER   B. MODIFIED EQUIVALENT CRYSTAL FILTER

Fig. 66—Modification of L–C resonators to halve the required number of crystals.

Lattice Equivalent*

An important lattice equivalent (Fig. 66) halves the number of crystals required for the full-lattice filter. After the full-lattice design is completed, it is merely necessary to double the reactances of one L-C resonator and to center-tap it; halve the reactances of the second L-C resonator and ground

its bottom side; and then, as shown in Fig. 66B, two arms of the full lattice may be omitted. This equivalence is valid when dealing with small-percentage bandwidths and with high L-C-resonator loaded Q's (Q_p).

For large-percentage bandwidths and/or low loaded Q's, it is necessary to use an inductive center tap with a coupling coefficient between the two sides of the coil (L_p) approaching unity. The use of a capacitive center tap greatly simplifies the problem of "trimming-in" the tap point, which is always necessary in practice.

---

* M. Dishal, "Practical Modern Network Theory Design Data for Crystal Filters," *IRE 1957 National Convention Record*, Part 8.

CHAPTER 9

FILTERS, SIMPLE BAND-PASS DESIGN

COEFFICIENT OF COUPLING

Several types of coupled circuits are shown in Table 1 and Figs. 1 and 2, together with equations for the coefficient of coupling. Also shown is the dependence of bandwidth on resonance frequency. This dependence is only a rough approximation to show the trend and may be altered radically if L_m, M, or C_m is adjusted as the circuits are tuned to various frequencies.

$K_{12}=$ coefficient of coupling between resonant circuits

$X_{10}=$ reactance of inductor (or capacitor) of first circuit at f_0

$X_{20}=$ reactance of similar element of second circuit at f_0

$(\text{bw})_C=$ bandwidth with capacitive tuning

$(\text{bw})_L=$ bandwidth with inductive tuning.

GAIN AT RESONANCE

Single Circuit

In Table 1A

$$E_0/E_g = -g_m |X_{10}| Q$$

where $E_0=$ output volts at resonance frequency f_0, $E_g=$ input volts to active device, and $g_m=$ transconductance of active device.

Pair of Coupled Circuits (Figs. 3 and 4)

In B through F in Table 1

$$E_0/E_g = jg_m (X_{10}X_{20})^{1/2}Q[K_{12}Q/(1+K_{12}^2Q^2)].$$

This is maximum at critical coupling, where $K_{12}Q=1$.

$Q=(Q_1Q_2)^{1/2}=$ geometric-mean Q for the two circuits, as loaded with active device input and output impedances.

For circuits with critical coupling and over-coupling, the approximate gain is

$$|E_0/E_g| \approx 0.1 g_m/(C_1C_2)^{1/2}(\text{bw})$$

where (bw) is the useful pass band in megahertz, g_m is in micromhos, and C is in picofarads.

SELECTIVITY FAR FROM RESONANCE

The selectivity curves of Fig. 5 are based on the presence of only a single type of coupling between the circuits. The curves are useful beyond the peak region treated on pp. 9–5 through 9–9.

In the equations for selectivity in Table 1

$E=$ output volts at signal frequency f for same value of E_g as that producing E_0.

For Inductive Coupling

$$A = \frac{Q^2}{1+K_{12}^2Q^2}\left[\left(\frac{f}{f_0}-\frac{f_0}{f}\right)^2 - K_{12}^2\left(\frac{f}{f_0}\right)^2\right]$$

$$\approx \frac{Q^2}{1+K_{12}^2Q^2}\left(\frac{f}{f_0}-\frac{f_0}{f}\right)^2.$$

For Capacitive Coupling

A is defined by a similar equation, except that the neglected term is $-K_{12}^2(f_0/f)^2$. The 180-degree phase shift far from resonance is indicated by the minus sign in the expression for E_0/E.

Example. The use of the curves in Figs. 5, 6, and 7, is indicated by the following example. Given the circuit of Table 1C with input to PB across capacitor C_1. Let $Q=50$, $K_{12}Q=1.50$, and $f_0=16.0$ megahertz. Required is the response of $f=8.0$ megahertz.

Here, $f/f_0=0.50$ and curve C, Fig. 5, gives -75 decibels. Then applying the corrections from Figs. 6 and 7 for Q and $K_{12}Q$, we find

$$\text{response} = -75+12+4 = -59 \text{ decibels.}$$

TABLE 1—SEVERAL TYPES OF COUPLED CIRCUITS, SHOWING COEFFICIENT OF COUPLING AND SELECTIVITY EQUATIONS.

Diagram	Coefficient of Coupling	Approximate Bandwidth Variation with Frequency	Selectivity Far From Resonance — Equation*	Curve in Fig. 5
A			Input to PB or to $P'B'$: $E_0/E = jQ[(f/f_0) - (f_0/f)]$	A
B	$K_{12} = L_m/[(L_1+L_m)(L_2+L_m)]^{1/2}$ $= \omega_0^2 L_m (C_1 C_2)^{1/2}$ $\approx L_m/(L_1 L_2)^{1/2}$	$(\mathrm{bw})_C \propto f_0$ $(\mathrm{bw})_L \propto f_0^3$	Input to PB: $E_0/E = -A(f/f_0)$	C
			Input to $P'B'$: $E_0/E = -A(f_0/f)$	D
C	$K_{12} = M/(L_1 L_2)^{1/2}$ $= \omega_0^2 M (C_1 C_2)^{1/2}$ M may be positive or negative	$(\mathrm{bw})_C \propto f_0$ $(\mathrm{bw})_L \propto f_0^3$	Input to PB: $E_0/E = -A(f/f_0)$	C
			Input to $P'B'$: $E_0/E = -A(f_0/f)$	D

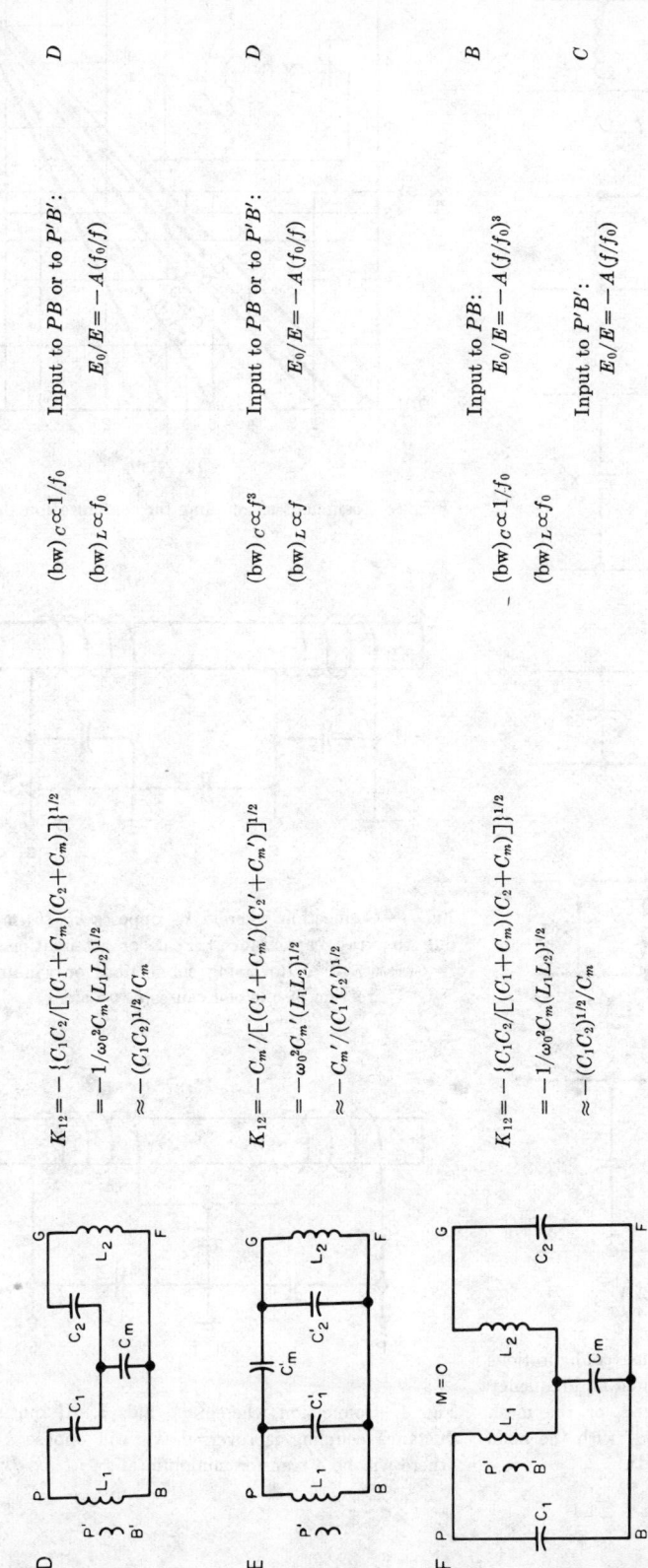

D

$$K_{12} = -\{C_1C_2/[(C_1+C_m)(C_2+C_m)]\}^{1/2}$$
$$= -1/\omega_0^2 C_m(L_1L_2)^{1/2}$$
$$\approx -(C_1C_2)^{1/2}/C_m$$

$(bw)_C \propto 1/f_0$
$(bw)_L \propto f_0$

Input to PB or to $P'B'$:
$$E_0/E = -A(f_0/f)$$

D

$$K_{12} = -C_m'/[(C_1'+C_m)(C_2'+C_m')]^{1/2}$$
$$= -\omega_0^2 C_m'(L_1L_2)^{1/2}$$
$$\approx -C_m'/(C_1'C_2')^{1/2}$$

$(bw)_C \propto f^3$
$(bw)_L \propto f$

Input to PB or to $P'B'$:
$$E_0/E = -A(f_0/f)$$

B

$$K_{12} = -\{C_1C_2/[(C_1+C_m)(C_2+C_m)]\}^{1/2}$$
$$= -1/\omega_0^2 C_m(L_1L_2)^{1/2}$$
$$\approx -(C_1C_2)^{1/2}/C_m$$

$(bw)_C \propto 1/f_0$
$(bw)_L \propto f_0$

Input to PB:
$$E_0/E = -A(f/f_0)^3$$

C

Input to $P'B'$:
$$E_0/E = -A(f/f_0)$$

*Where $A = [Q^2/(1+K_{12}^2Q^2)][(f/f_0)-(f_0/f)]^2$.

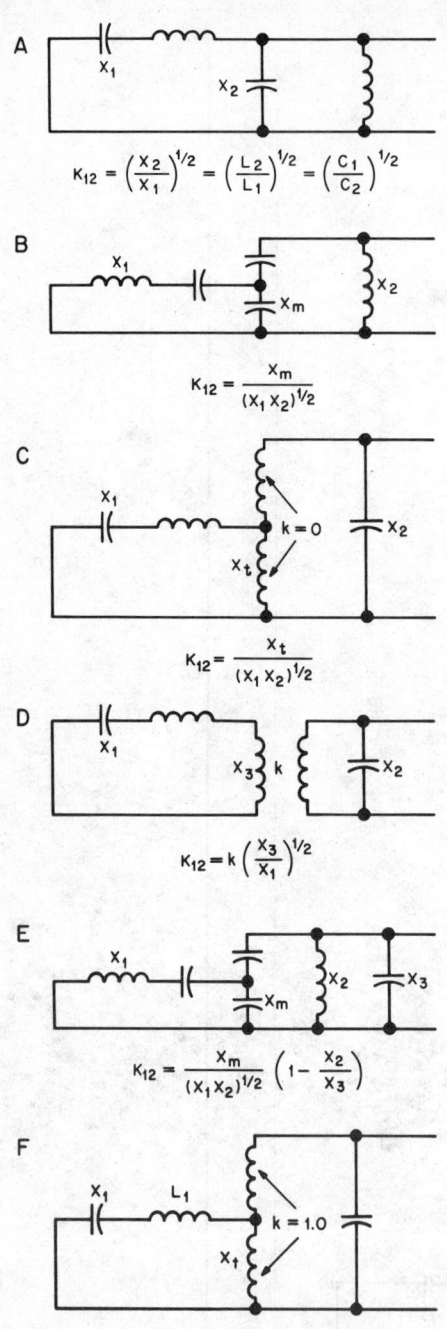

$$K_{12} = \left(\frac{x_2}{x_1}\right)^{1/2} = \left(\frac{L_2}{L_1}\right)^{1/2} = \left(\frac{C_1}{C_2}\right)^{1/2}$$

$$K_{12} = \frac{x_m}{(x_1 x_2)^{1/2}}$$

$$K_{12} = \frac{x_t}{(x_1 x_2)^{1/2}}$$

$$K_{12} = k\left(\frac{x_3}{x_1}\right)^{1/2}$$

$$K_{12} = \frac{x_m}{(x_1 x_2)^{1/2}}\left(1 - \frac{x_2}{x_3}\right)$$

$$K_{12} = \left(\frac{x_t}{x_1}\right)^{1/2} = \left(\frac{L_t}{L_1}\right)^{1/2}$$

Fig. 1—Additional coefficient-of-coupling configurations (the node resonator is tuned to the desired midfrequency with the mesh resonator open-circuited, or the mesh resonator is tuned to this midfrequency with the node resonator short-circuited).

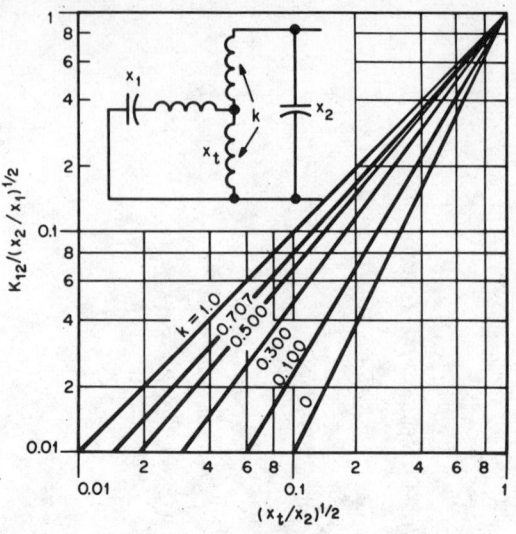

Fig. 2—Coefficient of coupling for configuration shown.

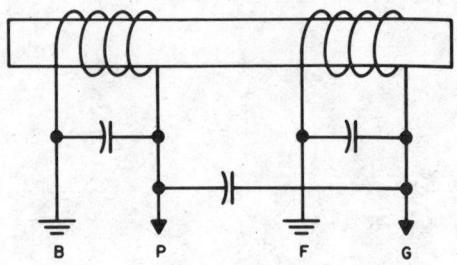

Fig. 3—Connection wherein k_m opposes k_c. (k_c may be due to stray capacitance.) Peak of attenuation is at $f = f_0(-k_m/k_c)^{1/2}$. Reversing connections or winding direction of one coil causes k_m to aid k_c.

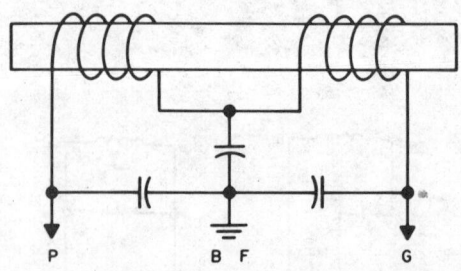

Fig. 4—Connection wherein k_m aids k_c. If mutual-inductance coupling is reversed, k_m will oppose k_c and there will be a transfer minimum at $f = f_0(-k_m/k_c)^{1/2}$.

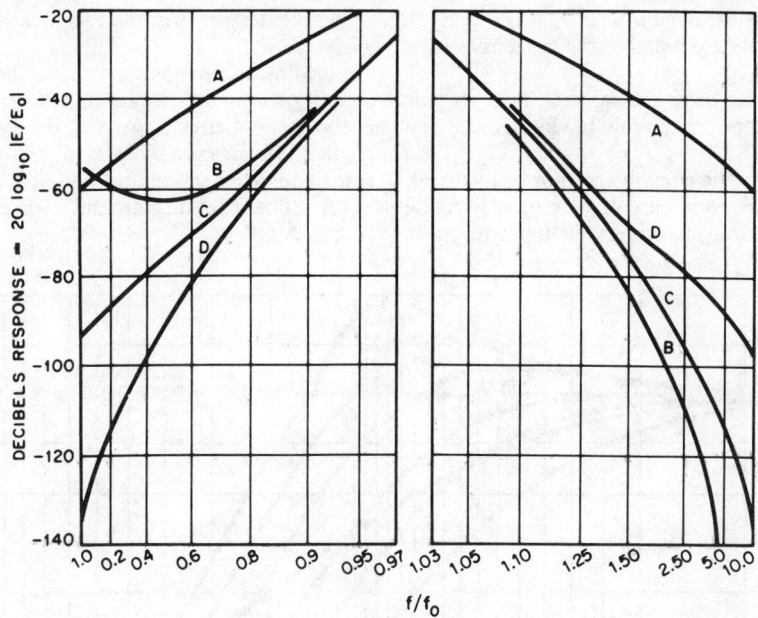

Fig. 5—Selectivity for frequencies far from resonance. $Q = 100$ and $|K_{12}|Q = 1.0$.

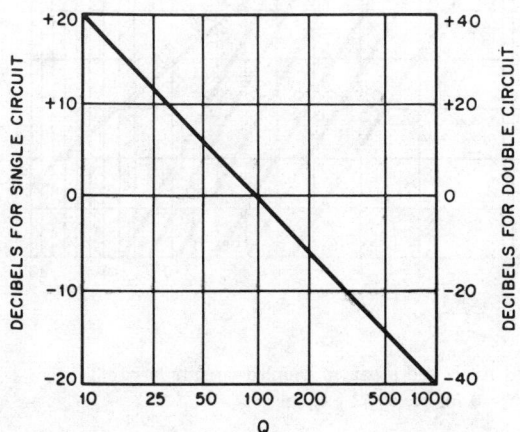

Fig. 6—Correction for $Q \neq 100$.

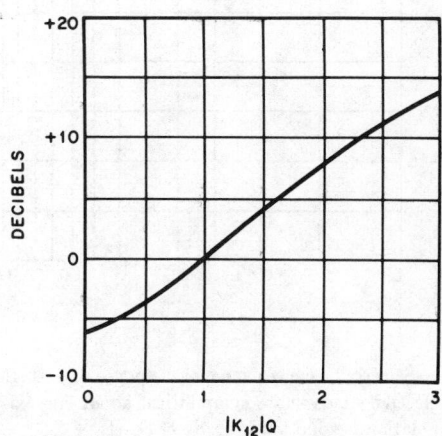

Fig. 7—Correction for $|K_{12}|Q \neq 1.0$.

SELECTIVITY OF SINGLE- AND DOUBLE-TUNED CIRCUITS NEAR RESONANCE

Equations and curves are presented for the selectivity and phase shift:

Of n single-tuned circuits
Of m pairs of coupled tuned circuits.

Fig. 8—Single-tuned circuit.

The conditions assumed are

(**A**) All circuits are tuned to the same frequency f_0.

(**B**) All circuits have the same Q, or each pair of circuits includes one circuit having Q_1 and the other having Q_2.

(**C**) Otherwise the circuits need not be identical.

(**D**) Each successive circuit or pair of circuits is isolated from the preceding and following ones by active devices, with no regeneration around the system.

Certain approximations have been made to simplify the equations. In most actual applications of the types of circuits treated, the error involved is negligible from a practical standpoint. Over the narrow frequency band in question, it is assumed that

(**A**) The reactance around each circuit is equal to $2X_0 \, \Delta f/f_0$.

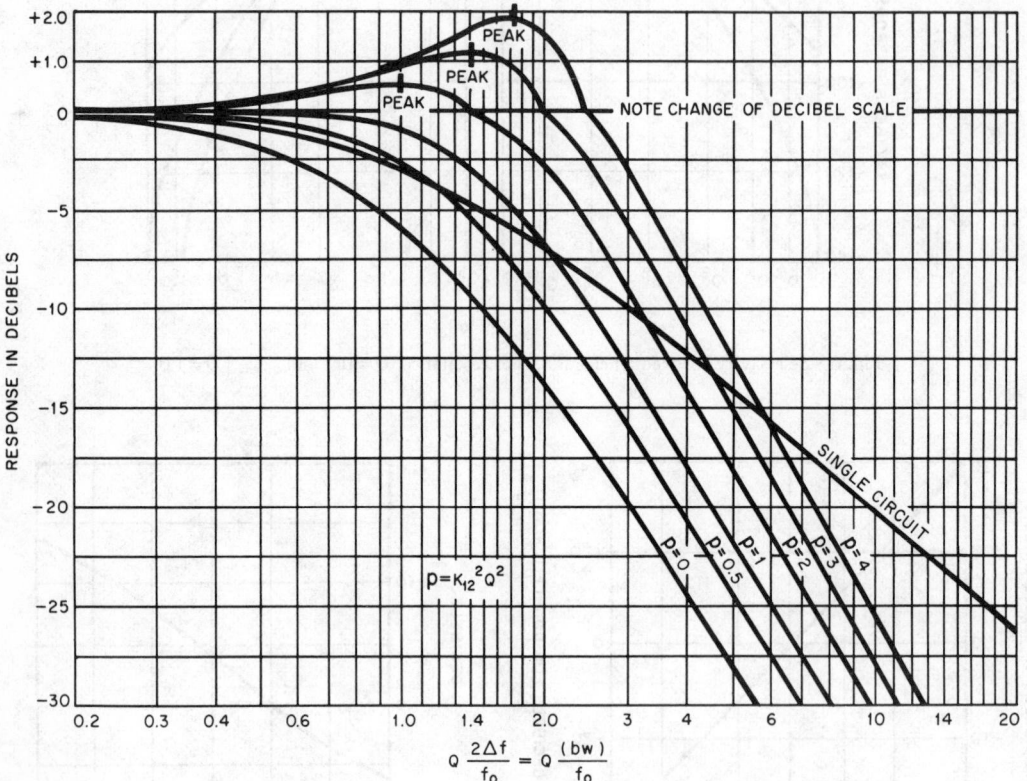

Fig. 9—Selectivity curves showing response of a single circuit $n=1$, and a pair of coupled circuits $m=1$. The selectivity curves are symmetrical about the axis $Q \, \Delta f/f_0 = 0$ for practical purposes. Extrapolation beyond lower limits of chart:

		Useful Limit	
Δ Response for Doubling Δf	Circuit	at $(bw)/f_0$	Error Becomes
-6 dB	←single→	0.6	1 to 2 dB
-12 dB	←pair→	0.4	3 to 4 dB

Example of the use of Figs. 9 and 10. Suppose there are three single-tuned circuits ($n=3$). Each circuit has a $Q=200$ and is tuned to 1000 kilohertz. The results are shown in the following table:

Abscissa $Q \, (bw)/f_0$	Bandwidth (kilohertz)	Ordinate dB Response for $n=1$	Decibels Response for $n=3$	ϕ^* for $n=1$	ϕ^* for $n=3$
1.0	5.0	-3.0	-9	$\mp 45°$	$\mp 135°$
3.0	15	-10.0	-30	$\mp 71\frac{1}{2}°$	$\mp 215°$
10.0	50	-20.2	-61	$\mp 84°$	$\mp 252°$

* ϕ is negative for $f > f_0$, and vice versa.

(**B**) The resistance of each circuit is constant and equal to X_0/Q.

(**C**) The coupling between two circuits of a pair is reactive and constant. (When an untuned link is used to couple the two circuits, this condition frequently is far from satisfied, resulting in a lopsided selectivity curve.)

(**D**) The equivalent input voltage, taken as being in series with the tuned circuit (or the first of a pair), is assumed to bear a constant proportionality to the input voltage of the active device or other driving source, at all frequencies in the band.

(**E**) Likewise, the output voltage across the circuit (or the final circuit of a pair) is assumed to be proportional only to the current in the circuit.

The following symbols are used in the equations in addition to those defined previously.

$\Delta f/f_0 = (f-f_0)/f_0$ = (deviation from resonance frequency)/(resonance frequency)

(bw) = bandwidth = $2\Delta f$

X_0 = reactance at f_0 of inductor in tuned circuit

n = number of single-tuned circuits

m = number of pairs of coupled circuits

ϕ = phase shift of signal at f relative to shift at f_0 as signal passes through cascade of circuits

$p = K_{12}^2 Q^2$ or $p = K_{12}^2 Q_1 Q_2$, a parameter determining the form of the selectivity curve of coupled circuits

$B = p - \frac{1}{2}[(Q_1/Q_2) + (Q_2/Q_1)]$.

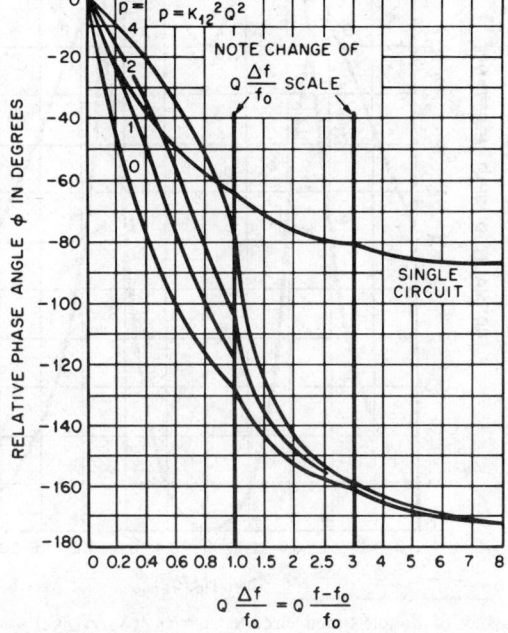

Fig. 10—Phase-shift curves for a single circuit $n=1$ and a pair of coupled circuits $m=1$. For $f>f_0$, ϕ is negative, while for $f<f_0$, ϕ is positive. The numerical value is identical in either case for the same $|f-f_0|$.

Selectivity and Phase Shift of Single-Tuned Circuits (Fig. 8)

$$E/E_0 = \{[1 + (2Q\,\Delta f/f_0)^2]^{-1/2}\}^n$$

$$\Delta f/f_0 = \pm (2Q)^{-1}[(E_0/E)^{2/n} - 1]^{1/2}$$

Decibel response = $20 \log_{10}(E/E_0)$

(dB response of n circuits) = $n \times$ (dB response of single circuit)

$\phi = n \tan^{-1}(-2Q\,\Delta f/f_0)$.

These equations are plotted in Figs. 9 and 10.

Q Determination by 3-Decibel Points

For a single-tuned circuit, when

$E/E_0 = 0.707$ (3 decibels down)

$Q = f_0/2\Delta f$

= (resonance frequency)/(bandwidth)$_{\text{3dB}}$.

Selectivity and Phase Shift of Pairs of Coupled Tuned Circuits (Fig. 11)

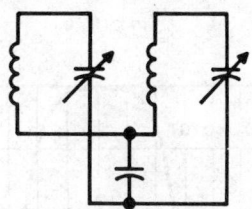

Fig. 11—One of several types of coupling.

Case 1: When $Q_1 = Q_2 = Q$:

These equations can be used with reasonable accuracy when Q_1 and Q_2 differ by ratios up to 1.5 or even 2 to 1. In such cases use the value $Q = (Q_1 Q_2)^{1/2}$.

$$E/E_0 = \left[\frac{p+1}{\{[(2Q\,\Delta f/f_0)^2 - (p-1)]^2 + 4p\}^{1/2}}\right]^m$$

$$\Delta f/f_0 = \pm (2Q)^{-1}$$
$$\times \{(p-1) \pm [(p+1)^2(E_0/E)^{2/m} - 4p]^{1/2}\}^{1/2}.$$

For very small values of E/E_0 the equations reduce to

$$E/E_0 = [(p+1)/(2Q\,\Delta f/f_0)^2]^m$$

Decibel response = $20 \log_{10}(E/E_0)$

(dB response of m pairs of circuits) = $m \times$ (dB response of one pair)

$$\phi = m \tan^{-1}\left[\frac{-4Q\,\Delta f/f_0}{(p+1) - (2Q\,\Delta f/f_0)^2}\right].$$

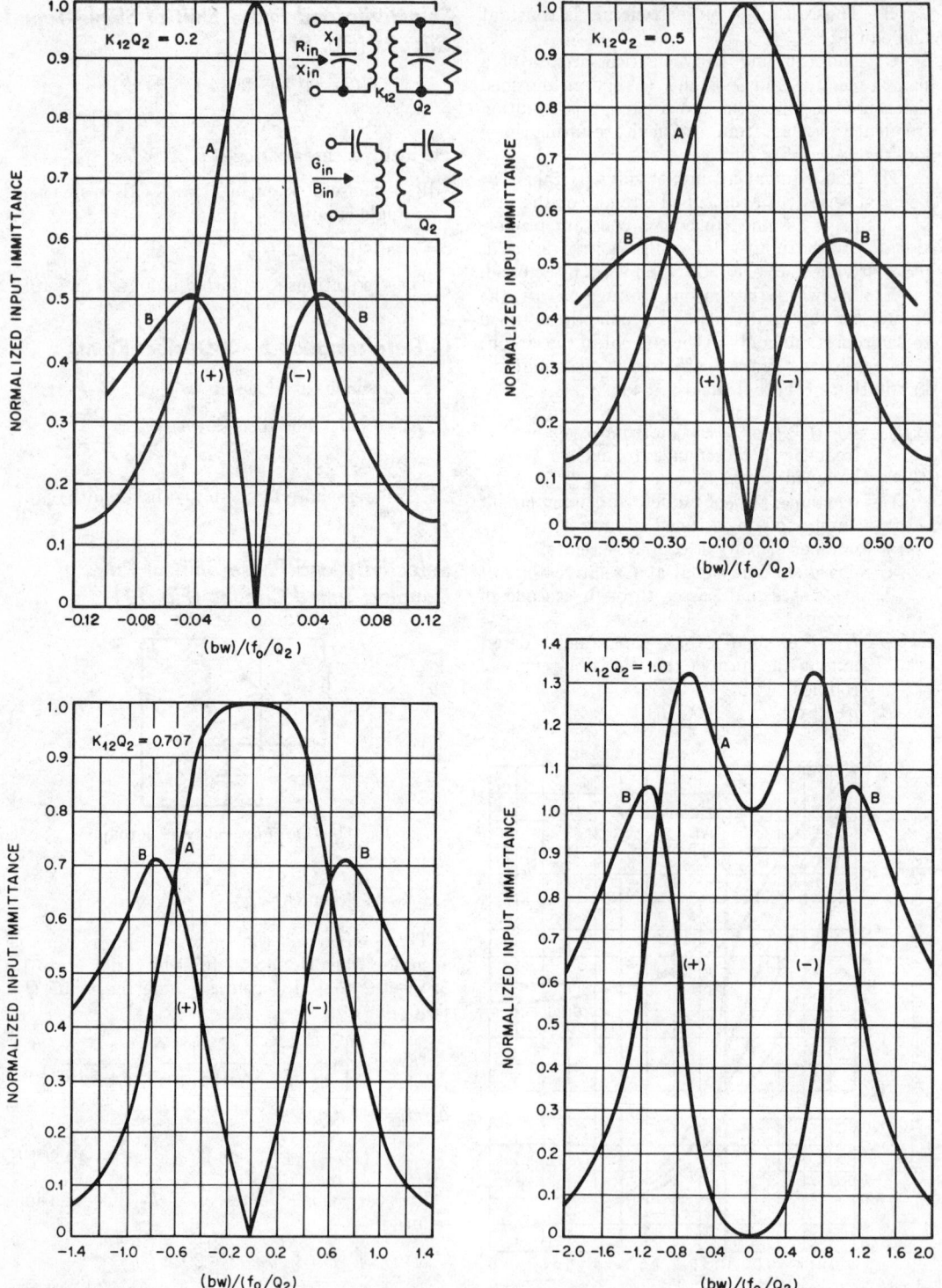

Fig. 12—Normalized input immittance vs. normalized frequency of double-tuned circuits. $A = R_{in}/(X_1/K_{12}^2Q_2)$ or $G_{in}/(B_1/K_{12}^2Q_2)$. $B = X_{in}/(X_1/K_{12}^2Q_2)$ or $B_{in}/(B_1/K_{12}^2Q_2)$.

As p approaches zero, the selectivity and phase shift approach the values for n single circuits, where $n = 2m$ (gain also approaches zero).

The above equations are plotted in Figs. 9 and 10.

For Overcoupled Circuits $(p > 1)$:

Location of peaks:

$$(f_{peak} - f_0)/f_0 = \pm (2Q)^{-1}(p-1)^{1/2}.$$

Amplitude of peaks:

$$E_{peak}/E_0 = [(p+1)/2(p^{1/2})]^m.$$

Phase shift at peaks:

$$\phi_{peak} = m \tan^{-1}[\mp (p-1)^{1/2}].$$

Approximate pass band (where $E/E_0 = 1$) is

$$(f_{unity} - f_0)/f_0 = \sqrt{2}[(f_{peak} - f_0)/f_0]$$
$$= \pm Q^{-1}[\tfrac{1}{2}(p-1)]^{1/2}.$$

Case 2: General Equation for Any Q_1 and Q_2:

$$E/E_0 = \left[\frac{p+1}{\{[(2Q\,\Delta f_0/f_0)^2 - B]^2 + (p+1)^2 - B^2\}^{1/2}}\right]^m$$

$$B = p - \tfrac{1}{2}[(Q_1/Q_2) + (Q_2/Q_1)]$$

$$\Delta f/f_0 = \pm (2Q)^{-1}$$
$$\times \{B \pm [(p+1)^2(E_0/E)^{2/m} - (p+1)^2 + B^2]^{1/2}\}^{1/2}$$

$$\phi = m \tan^{-1}$$
$$\times \left(-\frac{2Q\,\Delta f/f_0[(Q_1/Q_2)^{1/2} + (Q_2/Q_1)^{1/2}]}{(p+1) - (2Q\,\Delta f/f_0)^2}\right).$$

For Overcoupled Circuits:

Location of peaks:

$$(f_{peak} - f_0)/f_0 = \pm B^{1/2}/2Q$$
$$= \pm \tfrac{1}{2}[K_{12}^2 - \tfrac{1}{2}(1/Q_1^2 + 1/Q_2^2)]^{1/2}.$$

Amplitude of peaks:

$$E_{peak}/E_0 = \{(p+1)/[(p+1)^2 - B^2]^{1/2}\}^m.$$

Case 3: Peaks Just Converged to a Single Peak:

Here $B = 0$ or $K_{12}^2 = \tfrac{1}{2}(1/Q_1^2 + 1/Q_2^2)$

$$E/E_0 = \{2/[(2Q'\,\Delta f/f_0)^4 + 4]^{1/2}\}^m$$

where

$$Q' = 2Q_1Q_2/(Q_1 + Q_2)$$

$$\Delta f/f_0 = \pm (\sqrt{2}/4)(1/Q_1 + 1/Q_2)[(E_0/E)^{2/m} - 1]^{1/4}$$

$$\phi = m \tan^{-1}\left[-\frac{4Q'\,\Delta f/f_0}{2 - (2Q'\,\Delta f/f_0)^2}\right].$$

The curves of Figs. 9 and 10 may be applied to this case, using the value $p = 1$ and substituting Q' for Q.

NODE INPUT IMPEDANCE OR MESH INPUT ADMITTANCE OF A DOUBLE-TUNED CIRCUIT

Figure 12 gives the normalized input immittance versus the normalized frequency of double-tuned circuits.

CHAPTER 10
ACTIVE FILTER DESIGN

The information presented in this chapter may be used by a filter designer to obtain complete designs of a wide variety of types of active filters using standard circuit element values. Only minor computations, such as obtaining a scale factor SF and multiplying the resistances of a given table by SF, are required. For the interested reader, brief descriptions of active filters of various types are given, but to use the results of the chapter it is not necessary to read or understand the textual material, since a self-contained summary follows the discussion of each filter type. Detailed examples of complete designs are provided in the various sections; these examples may be used as guidelines, if desired.

INTRODUCTION

An electric filter is a network which transforms an input signal in some specified manner to produce a desired output signal. In most cases, filters are frequency-selective, phase-shifting, or time-delay devices. These concepts may be more specifically defined by considering the general filter representation of Fig. 1. The transfer function is

$$H(s) = \frac{V_2(s)}{V_1(s)} \qquad (1)$$

where V_1 is the input voltage and V_2 is the output voltage.

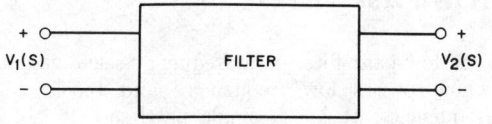

Fig. 1—A general filter representation.

In the steady-state case, $s = j\omega$, where $\omega = 2\pi f$ rad/s and f is the frequency in hertz. The transfer function is then complex, and is given by

$$H(j\omega) = |H(j\omega)| \exp j\phi(\omega) \qquad (2)$$

where $|H(j\omega)|$ is the amplitude and $\phi(\omega)$ is the phase. The plots of $|H(j\omega)|$ and $\phi(\omega)$ versus ω are the amplitude and phase responses, respectively.

A frequency-selective filter is one which passes signals of certain frequencies and blocks those of other frequencies. The range, or band, of frequencies which pass is the passband in which the amplitude is relatively large, and ideally a constant. The frequencies which are blocked constitute the stopband in which the amplitude is relatively small, and ideally zero. Of course, a filter may be designed to have more than one passband and/or stopband. Examples of frequency-selective filters are low-pass, high-pass, bandpass, and band-reject filters, which will be described later.

An all-pass filter is one which passes all frequencies equally well. Its amplitude response is constant for all frequencies, and its phase response is a function of frequency. It is thus a phase-shifting filter.

The time delay, $T(\omega)$, of a filter is defined in terms of its phase response by

$$T(\omega) = -\frac{d}{d\omega}\phi(\omega) \qquad (3)$$

Ideally, the phase response is linear, say $\phi(\omega) = -\omega T_0$, and thus the time delay is ideally a constant, $T(\omega) = T_0$. Time-delay filters are those designed to have a time delay which is approximately constant over a specified frequency range.

Ideal behavior is impossible to achieve, except in the all-pass case. Therefore, filter design consists of obtaining a transfer function which approximates the ideal response to some prescribed degree and constructing a network which realizes the transfer function. The transfer function is a ratio of two polynomials, and the order of the filter is defined as the degree of the denominator polynomial. In general, the higher the order of the filter, the better the approximation to the ideal case; however, higher order is accompanied by more complexity and higher cost.

Passive filters are those whose components are resistors, inductors, and capacitors and are quite useful for operation in certain frequency ranges. However, for low-frequency operation, say 1 Hz to approximately 0.5 MHz, inductors are undesirable because of their size and considerable departure from ideal behavior. Also, inductors, unlike resistors and capacitors, are not readily adaptable to integrated-circuit techniques, which have become extremely important in recent years.

Active filters are constructed with resistors, capacitors, and one or more active devices, such as transistors, controlled sources, etc. They are extremely useful for operation at the lower frequencies, where they are much to be preferred over passive filters. The active element which is most often used*, and the one exclusively considered in this chapter, is the integrated circuit (IC) operational amplifier.

CIRCUIT ELEMENTS

Operational Amplifiers

The operational amplifier (or "op amp") is a multiterminal device which is often represented in the ideal case by the three-terminal device of Fig. 2. The terminals shown are the inverting input (1), the noninverting input (2), and the output (3) terminals. Ideally, the voltage between the input terminals is zero, and the current into the input terminals is zero. This is a consequence of the ideal op amp's properties of infinite input resistance, zero output resistance, and infinite gain.

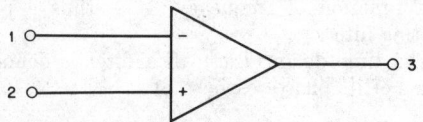

Fig. 2—Symbol for ideal operational amplifier.

Practical op amps approximate the ideal only over a limited frequency range. Typical bipolar op amps, such as the popular 741 type, have open-loop gains of 200 000 at dc, decreasing monotonically to about 1000 at 1 kHz. Input resistances of bipolar devices range from approximately 200 kΩ to 2 MΩ. Typical FET op amps, such as the 536 type, have similar gain characteristics, with input resistances as high as 10^{14} Ω.

Other terminals not shown in Fig. 2 normally include power-supply terminals, compensation terminals, and offset-null terminals. Most op amps have two power-supply terminals which operate from ±5 to ±22 volts. Some (e.g., the 741) are internally compensated and have no compensation terminals. Others (e.g., the 709) require an external compensating network of resistances and capacitances to obtain the desired operating characteristics. Externally compensated op amps generally have performance characteristics at higher frequencies superior to those of internally compensated op amps. Offset null terminals are provided on some op amps, such as the 741, to set the dc output voltage to zero when no signal is applied to the input terminals. The manner in

which these additional terminals are connected is generally specified by the manufacturer.

Another important consideration in the selection of an op amp is that of slew rate, which is the maximum rate of change in the output voltage in volts per microsecond. Therefore, in applications requiring large output voltage swings, op amps with high slew rates may be required. Op amps are available with slew rates as high as 320 V/μs.

Resistors

Three commonly used types of resistors are carbon composition, metal film, and wirewound*. Carbon composition resistors are the most common and least expensive. They are generally satisfactory in simpler filter designs if temperature considerations are not critical. Metal film and wirewound resistors are widely used in filter design and have better temperature characteristics. The wirewound resistor generally has the best performance characteristics of the three, but it is also the most expensive.

Capacitors

Capacitors are the most critical elements in active filter design†. Those types commonly used include ceramic disk, Mylar, polystyrene, and teflon. The ceramic disk capacitor is the least expensive and is of the poorest quality. It should be used only in the most noncritical applications. Mylar capacitors are widely available, relatively inexpensive, and suitable in most simpler designs. Polystyrene and teflon capacitors are of the highest quality and yield best results in complex designs.

LOW-PASS FILTERS

A low-pass filter is a frequency-selective filter which passes low frequencies and blocks high frequencies. It has a single passband, $0 < \omega < \omega_c$, and a single stopband, $\omega > \omega_c$. The frequency ω_c (or in hertz, $f_c = \omega_c/2\pi$) which separates the two bands is the cutoff frequency.

As an example, the amplitude response of an ideal low-pass filter is represented by the broken line of Fig. 3. The ideal is impossible to realize, since the transfer function of an actual filter is

* J. L. Hilburn and D. E. Johnson, *Manual of Active Filter Design*, McGraw-Hill Book Co., New York, 1973.

* J. G. Graeme, G. E. Tobey, and L. P. Huelsman (eds), *Operational Amplifiers: Design and Applications*, McGraw-Hill Book Co., New York, 1971.
† *Loc. cit.*

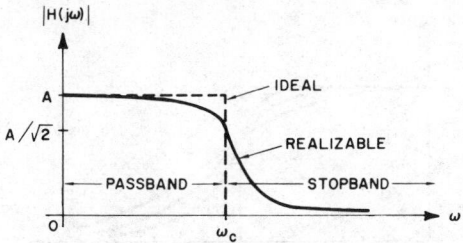

Fig. 3—Ideal and realizable low-pass responses.

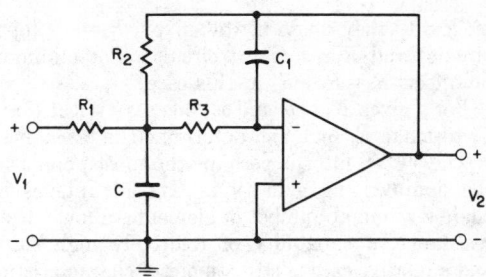

Fig. 4—A second-order MFB low-pass filter.

a ratio of polynomials and its amplitude cannot possess the sharp corners of the ideal response. A realizable approximation to the ideal is represented by the solid line of Fig. 3. In the realizable case, the division between the passband and stopband is not clearly distinguishable, and the cutoff frequency is defined as the point at which $|H(j\omega)|$ is $1/\sqrt{2} = 0.707$ times its maximum value, shown as A in Fig. 3. In decibels, the cutoff frequency is the point at which the amplitude is 3 dB below its maximum value.

The gain of a low-pass filter is the value of its amplitude at zero frequency, or equivalently, gain $= H(0)$, where $H(s)$ is the transfer function.

The simplest transfer function which approximates the ideal low-pass characteristic is the ratio of a constant to a polynomial. It is called an all-pole-function because all its poles, and none of its zeros, are finite.

Second-Order Designs

A second-order all-pole approximation to the ideal low-pass filter with cutoff frequency ω_c is achieved by the transfer function

$$\frac{V_2}{V_1} = \frac{Kb\omega_c^2}{s^2 + a\omega_c s + b\omega_c^2} \tag{4}$$

where K, a, and b are appropriately chosen constants. The gain is K, and the coefficients a and b are determined by the type of low-pass filter under consideration, such as Butterworth or Chebishev, to be considered later.

One of the simplest active filters which realizes the second-order low-pass filter is the multiple-feedback (MFB) network* of Fig. 4. This circuit realizes (4) with an inverting gain $-K(K>0)$ for

$$R_2 = \frac{2(K+1)}{\{aC + [a^2C^2 - 4bCC_1(K+1)]^{1/2}\}\omega_c}$$

$$R_1 = R_2/K$$

$$R_3 = 1/bCC_1\omega_c^2 R_2 \tag{5}$$

*J. G. Graeme, G. E. Tobey, and L. P. Huelsman (eds), *ibid.*

Thus for a given K, a, b, and ω_c, one may select C and C_1 (with C_1 sufficiently small to make the denominator of R_2 real) and determine the resistances.

The MFB filter is one of the most commonly used filters with inverting gain, because of its minimal number of circuit elements. Other advantages are its low output resistance, which makes it convenient for cascading with other stages, and its good stability characteristics. A disadvantage, however, is that relatively high gains are not possible.

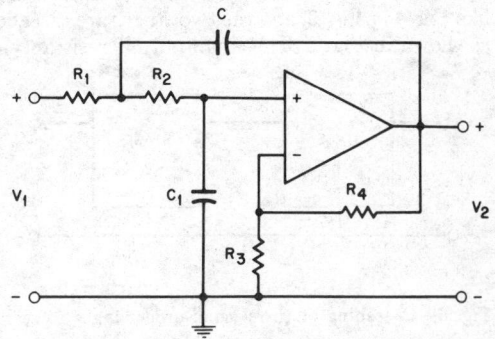

Fig. 5—A second-order VCVS low-pass filter.

Another commonly used circuit which realizes the second-order low-pass filter is the voltage-controlled voltage source (VCVS) network* of Fig. 5. This circuit yields a noninverting gain $K>0$, realizing (4) with

$$R_1 = \frac{2}{(aC + \{[a^2 + 4b(K-1)]C^2 - 4bCC_1\}^{1/2})\omega_c}$$

$$R_2 = 1/bCC_1 R_1 \omega_c^2$$

$$R_3 = K(R_1 + R_2)/(K-1), \qquad K \neq 1$$

$$R_4 = K(R_1 + R_2) \tag{6}$$

The latter two equations give $K = 1 + R_4/R_3$ in a way which minimizes the dc offset of the op amp.

* R. P. Sallen and E. L. Key, "A Practical Method of Designing RC Active Filters," *IRE Trans. on Circuit Theory*, CT-2, pp. 74–85, March 1955.

If $K=1$, they may be taken as $R_3=\infty$ (open circuit) and $R_4=0$ (short circuit). For minimum dc offset, $R_4=R_1+R_2$ in this case.

For a given K, a, b and ω_c, one may select C and C_1 to make R_1 real, and determine the resistances.

The VCVS filter is perhaps the most popular of the noninverting gain types. Its advantages include a minimal number of elements, a low output resistance, a capability of relatively high gains, and a relative ease of adjustment of characteristics. For example, the gain may be set precisely by adjusting R_3 and R_4. This circuit, however, is not as stable as the MFB filter.

Fourth-Order Designs

A fourth-order all-pole low-pass filter transfer function is a ratio of a constant to a fourth-degree polynomial. One method of realizing the function is to factor it into two second-order factors like (4), given by

$$\frac{V_2}{V_1}=\frac{K_1 b_1 \omega_c^2}{s^2+a_1\omega_c s+b_1\omega_c^2}\cdot\frac{K_2 b_2 \omega_c^2}{s^2+a_2\omega_c s+b_2\omega_c^2} \quad (7)$$

Each factor is then realized by a network, or stage, like Fig. 4 or Fig. 5, and the two stages are cascaded as shown in Fig. 6. The small output resistance of

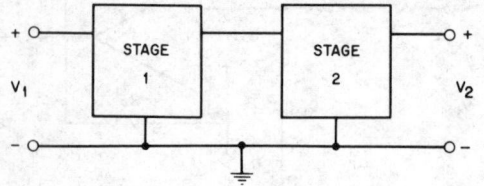

Fig. 6—Cascading of two second-order stages to realize a fourth-order filter.

practical op amps prevents one stage from loading down the other.

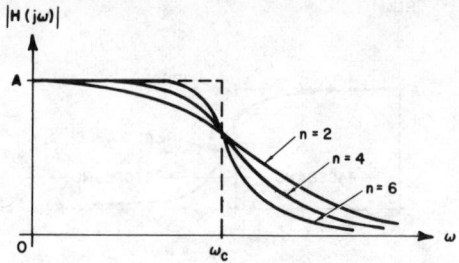

Fig. 7—Butterworth low-pass amplitude responses.

The gains of the stages are K_1 and K_2, respectively, and the overall gain is $K=K_1 K_2$.

Butterworth Filters

One method of approximating the ideal low-pass response of Fig. 3 is with the amplitude function

$$|H(j\omega)|=\frac{A}{[1+(\omega/\omega_c)^{2n}]^{1/2}}; \qquad n=1,2,3,\cdots \quad (8)$$

This is the amplitude response of the Butterworth low-pass filter of order n, which is exceedingly flat (so-called maximally flat) in the passband $0<\omega<\omega_c$. As n increases, (8) better approximates the ideal response, as may be seen in Fig. 7 for $n=2$, 4, and 6.

For low-pass Butterworth designs, the coefficients of (4) for the second-order case ($n=2$) and (7) for the fourth-order case ($n=4$) are given in Table 1.

To obtain a practical design of a second-order Butterworth filter, one may select standard values of C and C_1, and substitute the appropriate values of a and b given by Table 1 and the desired values of ω_c and K into (5) for the MFB realization of Fig. 4, or into (6) for the VCVS realization of Fig. 5.

TABLE 1—LOW-PASS COEFFICIENTS FOR SECOND- AND FOURTH-ORDER DESIGNS.

		Butterworth	Chebishev		
n			0.1 dB	0.5 dB	1 dB
2	a	1.41421	2.37209	1.42562	1.09773
	b	1.00000	3.31329	1.51620	1.10251
4	a_1	0.76537	0.52827	0.35071	0.27907
	b_1	1.00000	1.32981	1.06352	0.98650
	a_2	1.84776	1.27536	0.84668	0.67374
	b_2	1.00000	0.62282	0.35641	0.27940

TABLE 2—SECOND-ORDER MFB LOW-PASS BUTTERWORTH DESIGNS.

Gain	C_1	R_1	R_2	R_3
2	$0.150C$	1.612	3.223	2.068
10	$0.033C$	1.021	10.211	2.968

TABLE 3—SECOND-ORDER VCVS LOW-PASS BUTTERWORTH DESIGNS.

Gain	C_1	R_1	R_2	R_3	R_4
2	$1C$	0.707	1.414	4.242	4.242
10	$2C$	0.290	1.723	2.237	20.131

To facilitate the design procedure and eliminate most of the calculations, one may use Table 2 for the MFB filter and Table 3 for the VCVS filter. The procedure is as follows: Select a standard value of the capacitance C. Determine a scale factor SF from

$$SF = \frac{500}{\pi f_c C'} \qquad (9)$$

where f_c is the desired cutoff frequency in hertz and C' is C in microfarads. For example, if $C = 0.01 \ \mu F$, then $C' = 0.01$. For a gain of 2 or 10, multiply the resistances in Table 2 or Table 3 by SF. The results are the filter elements in kilohms. Capacitance C_1 is given by the table in terms of C. The filter should be constructed in accordance with Fig. 4 or Fig. 5, using the indicated capacitances and standard values of resistances as close as possible to the calculated values.

For fourth-order designs, the Butterworth coefficients a_1, b_1, a_2, and b_2 of Table 1 should be used along with the gains K_1 and K_2 of each stage. (The filter gain, it will be recalled, is $K = K_1 K_2$.) Each stage should then be determined as in the second-order case and the two stages cascaded as in

Fig. 6. The process may be considerably shortened for a gain of 4 (2 per stage) by the use of Tables 4 and 5 for the MFB and VCVS filters, respectively. The procedure for each stage is identical to that of the second-order design, described earlier. That is, for each stage, select C, determine SF from (9), and obtain C_1 and the unscaled resistances for the stage from Table 4 or 5, as appropriate. The unscaled resistances multiplied by SF are the filter resistances in kilohms.

A detailed example, illustrating both the direct calculation method and the use of the tables, is given in a later section.

Chebishev Filters

Another method of approximating the ideal low-pass response of Fig. 3 is by means of the Chebishev low-pass filter. Its amplitude function is given by

$$|H(j\omega)| = \frac{A}{[1 + \epsilon^2 C_n^2(\omega/\omega_c)]^{1/2}}; \qquad n = 1,2,3,\cdots$$

$$(10)$$

TABLE 4—FOURTH-ORDER MFB LOW-PASS BUTTERWORTH DESIGNS.

Stage	Gain	C_1	R_1	R_2	R_3
1	2	$0.047C$	3.286	6.572	3.238
2	2	$0.220C$	1.100	2.200	2.066

TABLE 5—FOURTH-ORDER VCVS LOW-PASS BUTTERWORTH DESIGNS.

Stage	Gain	C_1	R_1	R_2	R_3	R_4
1	2	$1C$	1.307	0.765	4.144	4.144
2	2	$1C$	0.541	1.848	4.778	4.778

where A and ϵ are constants and C_n is the Chebishev polynomial of degree n given by

$$C_n(x) = \cos(n \cos^{-1}x) \qquad (11)$$

The amplitude response is characterized by ripples (which are equal in magnitude) in the channel $0 < \omega < \omega_c$, and a monotonically decreasing response in the band $\omega > \omega_c$. This is shown in Fig. 8 for $n = 2$, 3, and 4. The Chebishev filter has a much sharper cutoff than the Butterworth filter for a given allowable deviation in the band $0 < \omega < \omega_c$. In this respect, it is the best of the all-pole filters.*

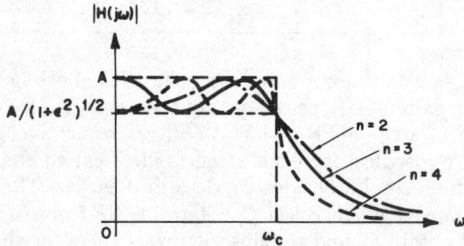

Fig. 8—Chebishev low-pass amplitude responses.

TABLE 6—RATIO OF CONVENTIONAL CUTOFF f_3 db TO RIPPLE CHANNEL TERMINAL f_c FOR LOW-PASS CHEBISHEV FILTERS.

dB	f_3 db$/f_c$	
	$n = 2$	$n = 4$
0.1	1.943	1.213
0.5	1.390	1.093
1	1.218	1.053

As is clear from Fig. 8, if $\epsilon = 1$, ω_c is the cutoff frequency in the conventional sense, because the amplitude is $A/\sqrt{2}$ at ω_c. However, if $0 < \epsilon < 1$, the cutoff frequency is greater than ω_c. In any case, ω_c is the terminal frequency of the ripple channel and will be used in lieu of the cutoff frequency. Table 6 gives the conventional cutoff frequency in terms of $f_c = \omega_c/2\pi$ for various ripple widths so that a designer may find from the desired cutoff frequency the appropriate ω_c to be used in (10).

The bottom of the ripple channel is $10 \log_{10}(1 + \epsilon^2)$ decibels below the top, since

$$dB = 20 \log_{10}A - 20 \log_{10}[A/(1+\epsilon^2)^{1/2}]$$

or

$$dB = 10 \log_{10}(1+\epsilon^2) \qquad (12)$$

The ripple width may be used to characterize the Chebishev filter, as was done in Table 6. For example, a 1-dB filter is one for which dB = 1 in (12), or $\epsilon = 0.50885$.

The coefficients in (4) and (7) for second- and fourth-order Chebishev functions are given, for the 0.1-, 0.5-, and 1-dB cases, in Table 1.

To obtain second- or fourth-order Chebishev low-pass filter designs, one may use the same procedure as in the Butterworth case. The only difference is in the values of the coefficients in Table 1. Also, as in the Butterworth case, the procedure is greatly simplified by the use of tables. Tables 7 and 8 apply for the second-order MFB and VCVS Chebishev filters, respectively, and Tables 9 and 10 apply for the fourth-order cases. Capacitor C is selected for the second-order case or for a stage of the fourth-order case, SF is calculated from (9), and the unscaled resistances in the appropriate table are multiplied by SF to obtain the filter resistances. Capacitance C_1 is determined from the table.

TABLE 7—SECOND-ORDER MFB LOW-PASS CHEBISHEV DESIGNS.

Gain	C_1	R_1	R_2	R_3	dB
2	$0.100C$	0.820	1.641	1.840	
10	$0.033C$	0.672	6.717	1.362	0.1
2	$0.100C$	1.590	3.180	2.074	
10	$0.022C$	1.011	10.105	2.967	0.5
2	$0.050C$	1.635	3.270	5.548	
10	$0.020C$	1.390	13.904	3.262	1

* P. I. Richards, "Universal Optimum-Response Curve for Arbitrarily Selected Coupled Resonators," *Proc. IRE*, vol. 34, pp. 624–629, September 1946.

TABLE 8—SECOND-ORDER VCVS LOW-PASS CHEBISHEV DESIGNS.

Gain	C_1	R_1	R_2	R_3	R_4	dB
2	$1C$	0.422	0.716	2.275	2.275	
10	$2C$	0.163	0.927	1.211	10.902	0.1
2	$1C$	0.701	0.940	3.283	3.283	
10	$2C$	0.247	1.335	1.758	15.819	0.5
2	$1C$	0.911	0.996	3.813	3.813	
10	$2C$	0.296	1.533	2.032	18.289	1

TABLE 9—FOURTH-ORDER MFB LOW-PASS CHEBISHEV DESIGNS.

Stage	Gain	C_1	R_1	R_2	R_3	dB
1	2	$0.015C$	4.124	8.247	6.079	
2	2	$0.200C$	1.831	3.662	2.192	0.1
1	2	$0.0082C$	6.171	12.342	9.291	
2	2	$0.100C$	2.167	4.334	6.474	0.5
1	2	$0.005C$	7.215	14.431	14.049	
2	2	$0.100C$	2.946	5.893	6.074	1

TABLE 10—FOURTH-ORDER VCVS LOW-PASS CHEBISHEV DESIGNS.

Stage	Gain	C_1	R_1	R_2	R_3	R_4	dB
1	2	$1C$	1.893	0.397	4.580	4.580	
2	2	$1C$	0.784	2.048	5.664	5.664	0.1
1	2	$1C$	2.851	0.330	6.362	6.362	
2	2	$1C$	1.181	2.376	7.113	7.113	0.5
1	2	$1C$	3.583	0.283	7.732	7.732	
2	2	$1C$	1.484	2.411	7.791	7.791	1

Practical Design Example

As an example, suppose it is desired to construct a low-pass VCVS Butterworth filter of second order having a gain of $K = 2$ and a cutoff frequency of 1000 Hz. Assume capacitance values of $C = C_1 = 0.01 \, \mu F = 10^{-8}$ F. From Table 1, $a = 1.41421 = \sqrt{2}$ and $b = 1$, so that by (6),

$R_1 = 2/(\sqrt{2}\,(10^{-8}) + \{[(\sqrt{2})^2 + 4(1)(2-1)](10^{-16})$
$$- 4(1)(10^{-16})\}^{1/2})(2\pi)(10^3)\,\Omega$$

$$= 11.254 \text{ k}\Omega$$

$R_2 = 1/[(1)(10^{-16})(11.254)(10^3)(2\pi)^2(10^6)]\,\Omega$

$$= 22.508 \text{ k}\Omega$$

$$R_3 = \frac{2(11.254 + 22.508)}{2-1} = 67.524 \text{ k}\Omega$$

$$R_4 = 2(11.254 + 22.508) = 67.524 \text{ k}\Omega$$

To use the tables to obtain the design, note that $C' = 0.01$ so that (9) yields

$$SF = \frac{500}{\pi(1000)(0.01)} = 15.915$$

By Table 3 for a gain of 2, $C_1 = C = 0.01\ \mu\text{F}$, and the resistances are

$$R_1 = 0.707\,SF = 11.252 \text{ k}\Omega$$

$$R_2 = 1.414\,SF = 22.504 \text{ k}\Omega$$

$$R_3 = R_4 = 4.242\,SF = 67.511 \text{ k}\Omega$$

Standard values of resistance as close as possible to the calculated values should be used in constructing the circuit in accordance with Fig. 5.

Design Summary of Low-Pass Filters

For the desired filter type (Butterworth or Chebishev) with gain K and cutoff frequency f_c(hertz), select the circuit (Fig. 4 for the MFB filter or Fig. 5 for the VCVS filter) and the capacitance C to be used. In the Chebishev case, the design formulas and tables are given with f_c as the terminal frequency of the ripple channel. If the conventional cutoff frequency $f_{3\text{ db}}$ is specified, then the proper value of f_c to be used is found from Table 6. A good choice of C is a standard value approximately equal to $10/f_c\ \mu\text{F}$.

Method 1:

(**A**) Obtain coefficients a and b in the second-order case, or a_1, a_2, b_1, and b_2 in the fourth-order case, from Table 1.

(**B**) Select a standard value of C_1 so that

$$C_1 < \frac{a^2 C}{4b(K+1)} \qquad \text{(MFB)}$$

$$C_1 < \frac{[a^2 + 4b(K-1)]C}{4b} \qquad \text{(VCVS)}$$

for the second-order case, or for each stage in the fourth-order case.

(**C**) Obtain the resistances from (5) for the MFB filter and from (6) for the VCVS filter.

Method 2:

(**A**) Find SF from (9).

(**B**) From the appropriate one of Tables 2 through 5 and 7 through 10, find C_1. Determine the resistances in kilohms by multiplying the table values by SF.

In either method, the filter should be constructed from standard values of resistances as close as possible to the calculated values. Generally, Mylar capacitors and 5% carbon composition resistors yield suitable designs. In all designs, a dc return to ground is required for each input terminal of the op amp. In the selection of an op amp, the open-loop gain should be at least 50 times the gain of the filter, and the peak-to-peak output voltage should not exceed $10^6/\pi f_a$ times the slew rate of the op amp, where f_a is the highest frequency of interest in the passband. A specific example of a low-pass design, given earlier, may be used as a guideline.

HIGH-PASS FILTERS

A high-pass filter passes high frequencies and blocks low frequencies. Thus it has $0 < \omega < \omega_c$ as the stopband and $\omega > \omega_c$ as the passband, where, as in the low-pass case, ω_c is the cutoff frequency. An ideal response, along with a realizable approximation, is shown in Fig. 9.

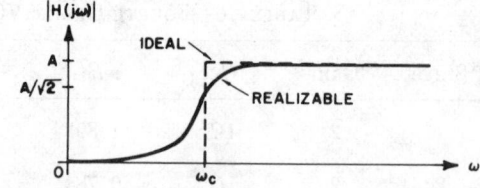

Fig. 9—Ideal and realizable high-pass responses.

The transfer function of a high-pass filter may be obtained from that of a low-pass filter by replacing s by $1/s$. This has the effect of interchanging the low and high frequencies, and thus the stopband and passband. In the second-order case, effecting the transformation $s \to 1/s$ in the low-pass function (4) results in the second-order high-pass function,

$$\frac{V_2}{V_1} = \frac{Ks^2}{s^2 + \dfrac{a}{b\omega_c}s + \dfrac{1}{b\omega_c^2}} \qquad (13)$$

In the fourth-order case the transformation is applied to (7), resulting in

$$\frac{V_2}{V_1}=\frac{K_1 s^2}{s^2+\dfrac{a_1}{b_1\omega_c}s+\dfrac{1}{b_1\omega_c^2}}\cdot\frac{K_2 s^2}{s^2+\dfrac{a_2}{b_2\omega_c}s+\dfrac{1}{b_2\omega_c^2}} \quad (14)$$

In (13) and (14), the a's and b's are the low-pass coefficients of (4) and (7).

For a Butterworth or Chebishev high-pass filter, (13) and (14) are obtained in the second- and fourth-order cases by using the appropriate coefficients from Table 1. The Butterworth response is maximally flat, and the Chebishev response has ripples in the passband ($\omega>\omega_c$). In the Chebishev case, ω_c is the beginning of the ripple channel rather than the conventional 3-dB cutoff frequency. To relate the two, one may use the reciprocals of the values in Table 6, since $\omega_{3\ db}<\omega_c$ in the high-pass case.

The gain of a high-pass filter is defined as the value of its transfer function as $s\rightarrow\infty$. From (13) and (14) it may be seen that the gain of the second-order filter is K and that of the fourth-order filter is $K=K_1 K_2$, which is precisely that of its low-pass counterpart.

Second-Order Designs

Like its low-pass counterpart, the second-order high-pass filter may be realized by an MFB circuit or a VCVS circuit. The MFB circuit* is shown in Fig. 10 and realizes (13) if

$$R_1=\frac{aK\omega_c}{C(2K+1)}$$

$$R_2=\frac{b(2K+1)\omega_c}{aC}$$

$$C_1=C/K \quad (15)$$

where the gain is $-K$ (inverting).

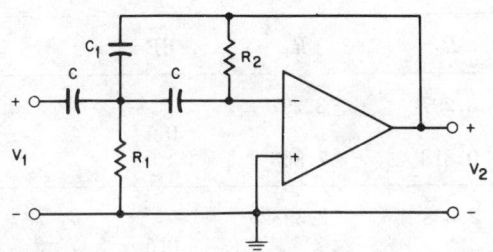

Fig. 10—A second-order MFB high-pass filter.

* J. G. Graeme, G. E. Tobey, and L. P. Huelsman (eds), *op cit.*

Since $K=C/C_1$, C may be selected arbitrarily, and for given values of a, b, and ω_c, values of C_1, R_1, and R_2 may be calculated. It should be noted that the gain K is restricted to a ratio of standard capacitances.

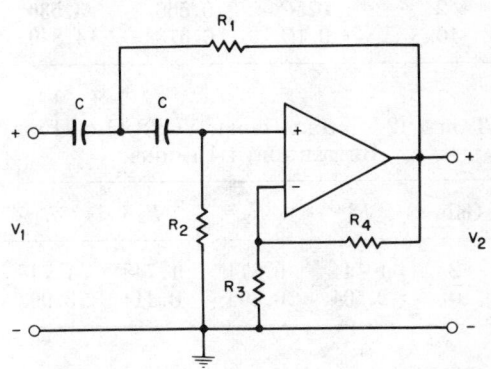

Fig. 11—A VCVS second-order high-pass filter.

A VCVS realization* of (13) is shown in Fig. 11, where

$$R_2=\frac{4b\omega_c}{C\{a+[a^2+8b(K-1)]^{1/2}\}}$$

$$R_1=b\omega_c^2/C^2 R_2$$

$$R_3=\frac{KR_2}{K-1}, \qquad K\neq1$$

$$R_4=KR_2 \quad (16)$$

The gain is given by

$$K=1+R_4/R_3$$

and R_3 and R_4 are selected to minimize the dc offset of the op amp. If $K=1$, then R_3 is infinite (open circuit) and $R_4=0$. For minimum dc offset, let $R_4=R_2$.

Second-order high-pass Butterworth and Chebishev filters may be constructed in exactly the same manner as their low-pass counterparts described in the previous section. Eq. (15) or (16) may be used directly for chosen practical values of C and/or C_1. A shortcut procedure, using tables, is also provided for certain gains. The procedure and the tables are given later in this section.

Fourth-Order Designs

Fourth-order MFB and VCVS high-pass filters may be constructed by realizing each factor of the transfer function (14) by a section such as Fig. 10

* R. P. Sallen and E. L. Key, *op cit.*

TABLE 11—SECOND-ORDER MFB HIGH-PASS BUTTERWORTH DESIGNS.

Gain	C_1	R_1	R_2
2	$0.5C$	0.566	3.536
10	$0.1C$	0.673	14.849

TABLE 12—SECOND-ORDER VCVS HIGH-PASS BUTTERWORTH DESIGNS.

Gain	R_1	R_2	R_3	R_4
2	1.144	0.874	1.748	1.748
10	2.504	0.399	0.444	3.993

TABLE 13—FOURTH-ORDER MFB HIGH-PASS BUTTERWORTH DESIGNS.

Stage	Gain	C_1	R_1	R_2
1	2	$0.5C$	0.306	6.533
2	2	$0.5C$	0.739	2.706

or 11, as described previously. Cascading the two sections then results in the desired fourth-order filter. A shortcut procedure, using Tables 13, 14, 17, and 18, is described at the end of this section, for

TABLE 15—SECOND-ORDER MFB HIGH-PASS CHEBISHEV DESIGNS.

Gain	C_1	R_1	R_2	dB
2	$0.5C$	0.949	6.984	
10	$0.1C$	1.130	29.332	0.1
2	$0.5C$	0.570	5.318	
10	$0.1C$	0.679	22.334	0.5
2	$0.5C$	0.439	5.022	
10	$0.1C$	0.523	21.091	1

certain gains in the case of Butterworth and Chebishev filters.

Design Summary of High-Pass Filters

For the desired filter type (Butterworth or Chebishev) with gain K and cutoff frequency f_c (hertz), select the circuit (Fig. 10 for the MFB filter or Fig. 11 for the VCVS filter) and the capacitance C to be used. In the Chebishev case, the design formulas and tables are given with f_c as the beginning frequency of the ripple channel. If the conventional cutoff $f_{3\,db}$ is specified, then the

TABLE 14—FOURTH-ORDER VCVS HIGH-PASS BUTTERWORTH DESIGNS.

Stage	Gain	R_1	R_2	R_3	R_4
1	2	0.924	1.082	2.165	2.165
2	2	1.307	0.765	1.531	1.531

TABLE 16—SECOND-ORDER VCVS HIGH-PASS CHEBISHEV DESIGNS.

Gain	R_1	R_2	R_3	R_4	dB
2	2.010	1.648	3.297	3.297	
10	4.500	0.736	0.818	7.363	0.1
2	1.297	1.169	2.338	2.338	
10	2.993	0.507	0.563	5.066	0.5
2	1.066	1.034	2.069	2.069	
10	2.519	0.438	0.486	4.377	1

TABLE 17—FOURTH-ORDER MFB HIGH-PASS CHEBISHEV DESIGNS.

Stage	Gain	C_1	R_1	R_2	dB
1	2	$0.5C$	0.211	12.586	
2	2	$0.5C$	0.510	2.442	0.1
1	2	$0.5C$	0.140	15.162	
2	2	$0.5C$	0.339	2.105	0.5
1	2	$0.5C$	0.112	17.675	
2	2	$0.5C$	0.269	2.074	1

TABLE 18—FOURTH-ORDER VCVS HIGH-PASS CHEBISHEV DESIGNS.

Stage	Gain	R_1	R_2	R_3	R_4	dB
1	2	0.958	1.388	2.776	2.776	
2	2	0.962	0.648	1.295	1.295	0.1
1	2	0.822	1.294	2.587	2.587	
2	2	0.684	0.521	1.042	1.042	0.5
1	2	0.776	1.272	2.544	2.544	
2	2	0.578	0.483	0.966	0.966	1

proper value of f_c to be used is found by using the *reciprocal* of the figures in Table 6. A good choice of C is a standard value approximately equal to $10/f_c \mu$F.

Method 1:

(**A**) Obtain coefficients a and b in the second-order case, or a_1, a_2, b_1, and b_2 in the fourth-order case, from Table 1.

(**B**) Obtain the resistances from (15) for the MFB filter and from (16) for the VCVS filter. Note: For the MFB filter, the gain K and the selection of C must be such that $C_1 = C/K$ is a standard value of capacitance.

Method 2:

(**A**) Find SF from (9).

(**B**) From the appropriate one of Tables 11 through 18, find the resistances in kilohms by multiplying the table resistance values by SF. In the MFB case, C_1 is also found in the tables.

In either method, the filter should be constructed with standard values of resistances as close as possible to the calculated values. The suggestions in the low-pass design summary apply except that the dc return to ground is already satisfied by R_2 in both circuits.

BANDPASS FILTERS

A bandpass filter is one with a single passband, $\omega_1 < \omega < \omega_2$, and two stopbands, $0 < \omega < \omega_1$ and $\omega > \omega_2$. An ideal and a realizable bandpass response are shown in Fig. 12. The frequencies ω_1 and ω_2 are the cutoff frequencies, and in the nonideal case

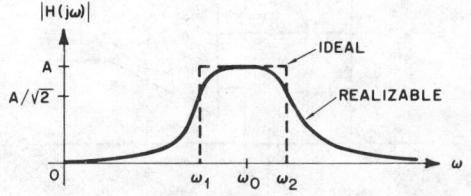

Fig. 12—Ideal and realizable bandpass responses.

TABLE 19—SECOND-ORDER MFB BANDPASS DESIGNS.

Q	Gain	C_1	R_1	R_2	R_3
2	2	$1C$	1.000	0.333	4.000
2	10	$2C$	0.200	1.000	3.000
4	2	$1C$	2.000	0.133	8.000
4	10	$2C$	0.400	0.105	6.000
6	2	$1C$	3.000	0.086	12.000
6	10	$2C$	0.600	0.061	9.000
8	2	$1C$	4.000	0.063	16.000
8	10	$2C$	0.800	0.044	12.000
10	2	$1C$	5.000	0.051	20.000
10	10	$2C$	1.000	0.034	15.000

they are the points at which the amplitude equals $1/\sqrt{2}$ times its maximum value.

The frequency ω_0, about which the passband is approximately centered, is the center frequency, and the bandwidth B is defined by $B=\omega_2-\omega_1$. The quality factor Q is defined by $Q=\omega_0/B$, so that large Q implies small bandwidth and vice-versa. The gain of a bandpass filter is the value of its amplitude at the center frequency. That is, $K=|H(j\omega_0)|$.

Bandpass transfer functions may be obtained from low-pass functions by replacing s in the low-pass function for $\omega_c=1$ by

$$S=\frac{s^2+\omega_0{}^2}{Bs} \tag{17}$$

Thus applying (17) to the first-order low-pass function,

$$\frac{V_2}{V_1}=\frac{K}{s+1}$$

results in the second-order bandpass function,

$$\frac{V_2}{V_1}=\frac{KBs}{s^2+Bs+\omega_0{}^2}=\frac{\dfrac{K\omega_0}{Q}s}{s^2+\dfrac{\omega_0}{Q}s+\omega_0{}^2} \tag{18}$$

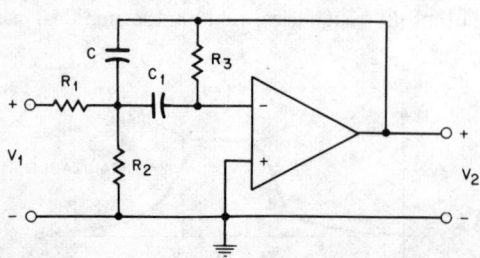

Fig. 13—A second-order MFB bandpass filter.

with center frequency $\omega_0=(\omega_1\omega_2)^{1/2}$, quality factor Q (or bandwidth $B=\omega_0/Q$), and gain K. A fourth-order bandpass function results when (17) is applied to the second-order function (4) with $\omega_c=1$, as given later.

Second-Order Designs

An MFB second-order bandpass filter* which realizes (18) is shown in Fig. 13. For a given ω_0, Q, and K, the resistances are

$$R_1=\frac{Q}{C\omega_0 K}$$

$$R_2=\frac{Q}{\omega_0[C(Q^2-K)+C_1Q^2]}$$

$$R_3=\frac{Q}{\omega_0}\left(\frac{1}{C}+\frac{1}{C_1}\right) \tag{19}$$

Capacitances C and C_1 are arbitrary and may be given standard values so that R_2 is positive. The filter yields an inverting gain of $-K$ ($K>0$).

For certain values of Q and K, Table 19 is provided for rapid designs. A scale factor SF is calculated from

$$SF=\frac{500}{\pi f_0 C'} \tag{20}$$

where $f_0=\omega_0/2\pi$ and C' is the value of C in microfarads. Capacitance C_1 is then as shown in the table, and the circuit resistances are those of the table multiplied by SF. A complete design example is given later in the section.

The MFB bandpass filter has a minimal number of elements, has an inverting gain, and is capable of values of Q up to about 10 for moderate gains.

* L. P. Huelsman, *Theory and Design of Active RC Circuits*, McGraw-Hill Book Co., New York, 1968.

Another popular second-order bandpass filter is the VCVS circuit* of Fig. 14. It realizes (18) for

$$R_1 = \frac{2Q}{C\omega_0 K}$$

$$R_2 = \frac{2Q}{C\omega_0\{-1+[(K-1)^2+8Q^2]^{1/2}\}}$$

$$R_3 = \frac{1}{C^2\omega_0^2}\left(\frac{1}{R_1}+\frac{1}{R_2}\right)$$

$$R_4 = 2R_3 \qquad\qquad\qquad (21)$$

For given values of K, ω_0, and Q and a selected value of C, one may calculate the required resistances. The resistance of R_4 is chosen to minimize the dc offset of the op amp.

* W. J. Kerwin and L. P. Huelsman, "The Design of High Performance Active RC Band-Pass Filter," *IEEE International Convention Record*, vol. 14, pt. 10, pp. 74–80, 1960.

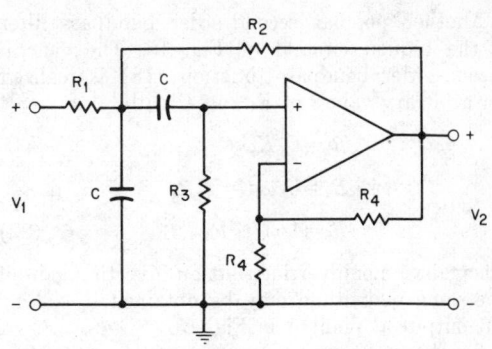

Fig. 14—A second-order VCVS bandpass filter.

Advantages of the VCVS bandpass filter are in general the same as those of the VCVS low-pass filter, discussed previously. It yields a noninverting gain, and for moderate gains, it can easily achieve a value of Q up to 10. Table 20 may be used with (20) to obtain a shortcut practical design for certain values of Q and gain.

TABLE 20—SECOND-ORDER VCVS BANDPASS DESIGNS.

Q	Gain	R_1	R_2	R_3	R_4
2	2	2.000	0.843	1.686	3.372
2	10	0.400	0.415	4.908	9.815
4	2	4.000	0.772	1.545	3.089
4	10	0.800	0.594	2.932	5.864
6	2	6.000	0.750	1.500	3.000
6	10	1.200	0.659	2.351	4.702
8	2	8.000	0.739	1.478	2.956
8	10	1.600	0.685	2.084	4.169
10	2	10.000	0.733	1.465	2.930
10	10	2.000	0.697	1.934	3.868

Fig. 15—A second-order biquad bandpass filter.

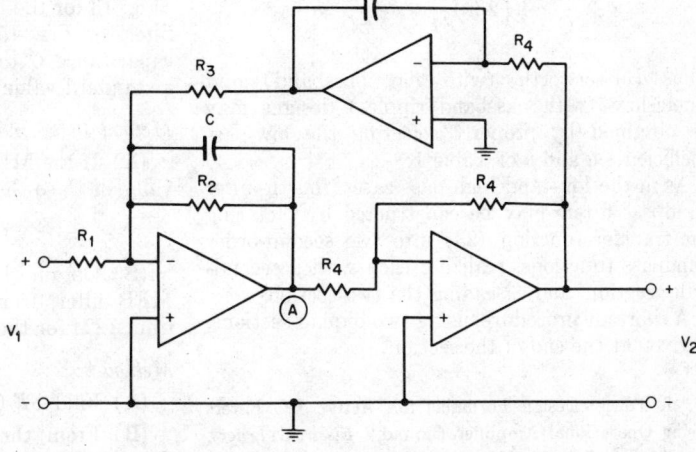

Another popular second-order bandpass filter is the biquad circuit* of Fig. 15. The general second-order bandpass function (18) is realized for arbitrary values of R_4 and C with

$$R_1 = Q/K\omega_0 C$$

$$R_2 = Q/\omega_0 C = KR_1$$

$$R_3 = 1/\omega_0^2 C^2 R_4 \qquad (22)$$

The gain is noninverting, but an inverting gain of the same magnitude may be obtained by taking the output at point A in Fig. 15.

The biquad circuit requires more elements than the MFB or VCVS circuits, but it is capable of achieving values of Q up to 100. Also, it has very low sensitivity to element changes and is very easy to tune. The center frequency can be adjusted by varying R_3, Q can be adjusted by varying R_2, and the gain can be adjusted by varying R_1.

A shortcut design procedure is provided by means of Table 21 and (20).

TABLE 21—SECOND-ORDER BIQUAD BANDPASS DESIGNS†

R_1	R_2	R_3, R_4
Q/K	Q	1

† K = gain, Q = quality factor

Fourth-Order Designs

Applying transformation (17) to the second-order low-pass function (4) with $\omega_c = 1$ results in the fourth-order bandpass function,

$$\frac{V_2}{V_1} = \frac{\omega_0^2 b K s^2/Q^2}{s^4 + \dfrac{a\omega_0}{Q}s^3 + \left(2 + \dfrac{b}{Q^2}\right)\omega_0^2 s^2 + \dfrac{a\omega_0^3}{Q}s + \omega_0^4} \qquad (23)$$

Thus Butterworth (with flat passband) and Chebishev (with passband ripples) designs may be obtained by properly selecting the low-pass coefficients a and b of Table 1.

As in the low- and high-pass cases, fourth-order bandpass filters may be constructed by factoring the transfer function (23) into two second-order bandpass functions, realizing each with a second-order section, and cascading the two sections.

A shortcut procedure, using two biquad sections, is given at the end of the section.

---

* J. Tow, "Design Formulas for Active RC Filters Using Operational Amplifier Biquad," *Electron. Letters*, pp. 339–341, July 24, 1969.

Practical Design Example

Suppose it is desired to obtain a second-order bandpass VCVS filter with a center frequency of $f_0 = 1000$ Hz, a gain of $K = 10$, and $Q = 10$, with $C = 0.1\,\mu\text{F} = 10^{-7}$ F. By (21) the resistances are

$$R_1 = \frac{2(10)}{(10^{-7})(2\pi)(10^3)(10)}\Omega = 3.183\ \text{k}\Omega$$

$$R_2 = \frac{2(10)}{(10^{-7})(2\pi)(10^3)\{-1+[(10-1)^2+8(10)^2]^{1/2}\}}\Omega$$

$$= 1.110\ \text{k}\Omega$$

$$R_3 = \frac{1}{(10^{-14})(2\pi)^2(10^6)}\left[\frac{1}{3.183}+\frac{1}{1.110}\right](10^{-3})\Omega$$

$$= 3.078\ \text{k}\Omega$$

$$R_4 = 2R_3 = 6.156\ \text{k}\Omega$$

Alternately, to use Table 20, it is necessary to find SF from (20):

$$SF = \frac{500}{\pi(10^3)(0.1)} = 1.5915$$

Multiplying the appropriate resistances from Table 20 by SF results in the network resistances, in kilohms.

$$R_1 = 2.000\,SF = 3.183$$

$$R_2 = 0.697\,SF = 1.109$$

$$R_3 = 1.934\,SF = 3.078$$

$$R_4 = 3.868\,SF = 6.156$$

Design Summary of Bandpass Filters

For the desired gain K, quality factor Q, and center frequency f_0 (hertz), select the circuit (Fig. 13 for the MFB filter, Fig. 14 for the VCVS filter, or Fig. 15 for the biquad filter) and the capacitance C to be used. A good choice of C is a standard value approximately equal to $10/f_0\,\mu\text{F}$.

Method 1 (second-order cases only):

(A) If the MFB filter is used, select a standard value of C_1 so that

$$C_1 > C(K-Q^2)/Q^2$$

(B) Obtain the resistances from (19) for the MFB filter, from (21) for the VCVS filter, or from (22) for the biquad filter.

Method 2:

(A) Find SF from (20).

(B) From the appropriate one of Tables 19 through 25, determine the resistances in kilohms

TABLE 22—FOURTH-ORDER BIQUAD BANDPASS BUTTERWORTH DESIGNS.*

Q	Stage	R_1	R_2	R_3, R_4
2	1	2.000/K	2.404	0.700
	2	2.000/K	3.436	1.429
4	1	4.000/K	5.197	0.838
	2	4.000/K	6.205	1.194
6	1	6.000/K	8.013	0.888
	2	6.000/K	9.016	1.125
10	1	10.000/K	13.659	0.932
	2	10.000/K	14.661	1.073
20	1	20.000/K	27.793	0.965
	2	20.000/K	28.793	1.036
50	1	50.000/K	70.215	0.986
	2	50.000/K	71.214	1.014

* $K =$ gain per stage

TABLE 23—FOURTH-ORDER BIQUAD BANDPASS CHEBISHEV DESIGNS (0.1 dB).*

Q	Stage	R_1	R_2	R_3, R_4
2	1	1.099/K	1.260	0.494
	2	1.099/K	2.548	2.023
4	1	2.197/K	2.878	0.707
	2	2.197/K	4.073	1.415
6	1	3.296/K	4.538	0.794
	2	3.296/K	5.715	1.260
10	1	5.494/K	7.887	0.871
	2	5.494/K	9.056	1.148
20	1	10.987/K	16.300	0.933
	2	10.987/K	17.465	1.071
50	1	27.469/K	41.583	0.973
	2	27.469/K	42.747	1.028

* $K =$ gain per stage.

TABLE 24—FOURTH-ORDER BIQUAD BANDPASS CHEBISHEV DESIGNS (0.5 dB).*

Q	Stage	R_1	R_2	R_3, R_4
2	1	1.624/K	2.250	0.604
	2	1.624/K	3.726	1.656
4	1	3.248/K	4.988	0.778
	2	3.248/K	6.413	1.286
6	1	4.873/K	7.769	0.846
	2	4.873/K	9.185	1.182
10	1	8.121/K	13.359	0.904
	2	8.121/K	14.770	1.106
20	1	16.243/K	27.371	0.951
	2	16.243/K	28.780	1.051
50	1	40.606/K	69.448	0.980
	2	40.606/K	70.855	1.020

* K = gain per stage

by multiplying the table values by SF. In the MFB case, C_1 is found in the table also. Tables 19 through 21 are second-order designs, and Tables 22 through 25 are cascaded fourth-order designs.

In either method, the filter should be constructed with standard values of resistances as close as possible to the calculated values. The suggestions for the low-pass design summary apply except that the dc return to ground is already satisfied in every case.

The example of a practical design, given earlier, may be used as a guideline.

BAND-REJECT FILTERS

A band-reject, or notch, filter is one with a single stopband, $\omega_1 < \omega < \omega_2$, and two passbands, $0 < \omega < \omega_1$ and $\omega > \omega_2$. An ideal and a physically realizable band-reject response are shown in Fig. 16. The cutoff frequencies are ω_1 and ω_2, defined in the nonideal case as the points at which the amplitude equals $1/\sqrt{2}$ times its maximum value.

The frequency ω_0, about which the stopband is approximately centered, is the center frequency.

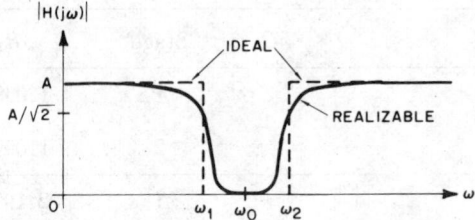

Fig. 16—Ideal and realizable band-reject responses.

The bandwidth is $B = \omega_2 - \omega_1$, and the quality factor is $Q = \omega_0/B$. The gain of the band-reject filter is the value of its amplitude at zero frequency or at infinite frequency, shown in Fig. 16 as A.

Band-reject transfer functions may be obtained from low-pass functions by replacing s in the low-pass function for $\omega_0 = 1$ by

$$S = \frac{Bs}{s^2 + \omega_0^2} \qquad (24)$$

Applying (24) to the first-order function

$$\frac{V_2}{V_1} = \frac{K}{s+1}$$

TABLE 25—FOURTH-ORDER BIQUAD BANDPASS CHEBISHEV DESIGNS (1 dB).*

Q	Stage	R_1	R_2	R_3, R_4
2	1	$1.905/K$	2.986	0.639
	2	$1.905/K$	4.673	1.565
4	1	$3.809/K$	6.557	0.799
	2	$3.809/K$	8.202	1.251
6	1	$5.715/K$	10.174	0.861
	2	$5.715/K$	11.811	1.161
10	1	$9.524/K$	17.440	0.914
	2	$9.524/K$	19.073	1.094
20	1	$19.047/K$	35.641	0.956
	2	$19.047/K$	37.272	1.046
50	1	$47.619/K$	90.289	0.982
	2	$47.619/K$	91.920	1.018

* $K =$ gain per stage

results in the second-order band-reject function

$$\frac{V_2}{V_1} = \frac{K(s^2 + \omega_0^2)}{s^2 + Bs + \omega_0^2} = \frac{K(s^2 + \omega_0^2)}{s^2 + (\omega_0/Q)s + \omega_0^2} \quad (25)$$

having center frequency ω_0, quality factor Q (or bandwidth $B = \omega_0/Q$), and gain K.

Second-Order Designs

A VCVS second-order band-reject filter* which realizes (25) is shown in Fig. 17. The gain is $K = 1$, and for specified ω_0 and Q and an arbitrary value of C, the resistances are given by

$$R_1 = 1/(2\omega_0 Q C)$$

$$R_2 = 2Q/(\omega_0 C)$$

$$R_3 = 2Q/[\omega_0 C(4Q^2 + 1)] \quad (26)$$

Values of Q up to 10 may be readily obtained.

Equations (26) may be used, for a specified value of C, to obtain a band-reject filter design.

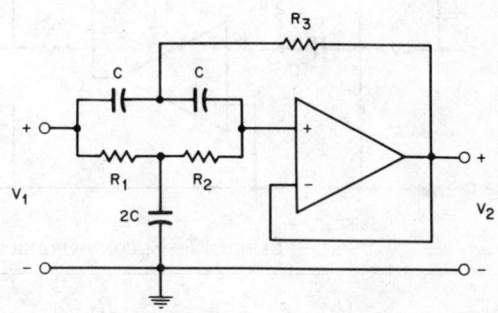

Fig. 17—A second-order VCVS band-reject filter.

For certain values of Q, Table 26 may be used in a shortcut procedure. A value of SF is found from (20), where C' is the value of C in microfarads, and the resistances are those of the table multiplied by SF.

An MFB circuit which also realizes (25) is shown in Fig. 18.* For given K, ω_0, and Q, there

---

* R. M. Inigo, "Active Filter Realization Using Finite-Gain Voltage Amplifiers," *IEEE Transactions on Circuit Theory,* vol. CT-17, pp. 445–448, Aug. 1970.

* L. P. Huelsman, *Active Filters: Lumped, Distributed, Integrated, Digital, and Parametric,* McGraw-Hill Book Co., New York, 1970.

are many ways to solve for the circuit elements. One way is to choose C and R_6 arbitrarily, with the remaining resistances given by

$$R_1 = Q/(2\omega_0 C)$$

$$R_2 = Q/[2\omega_0 C(Q^2 - 1)] = R_1/(Q^2 - 1), \qquad Q > 1$$

$$R_3 = 2Q/(\omega_0 C) = 4R_1$$

$$R_4 = K R_6$$

$$R_5 = 2R_6 \qquad\qquad (27)$$

The gain for the circuit is $-K$ (inverting).

The MFB filter requires more elements than the VCVS band-reject filter, but it has the advantage that the gain can be specified. For moderate gains, the filter may easily achieve values of Q up to 25. Instead of using the equations, a shortcut design procedure, using (20) and Table 27 as described for the VCVS filter, is available.

Practical Design Example

Suppose it is desired to construct a second-order band-reject MFB filter with $f_0 = \omega_0/2\pi = 1000$ Hz,

TABLE 26—SECOND-ORDER VCVS BAND-REJECT DESIGNS.*

Q	R_1	R_2	R_3
1	0.500	2.000	0.400
2	0.250	4.000	0.235
3	0.167	6.000	0.162
4	0.125	8.000	0.123
5	0.100	10.000	0.099
6	0.083	12.000	0.083
7	0.071	14.000	0.071
8	0.063	16.000	0.062
9	0.056	18.000	0.055
10	0.050	20.000	0.050

* Gain = 1 in all cases.

$Q = 6$, and an inverting gain of $K = 10$, using 0.01-μF capacitances. By (27), the resistances are

$$R_1 = \frac{6}{2(2\pi)(10^3)(10^{-8})}\Omega = 47.747 \text{ k}\Omega$$

$$R_2 = 47.747/35 = 1.364 \text{ k}\Omega$$

$$R_3 = 4(47.747) = 190.988 \text{ k}\Omega$$

$$R_4 = 10R_6, \qquad R_5 = 2R_6$$

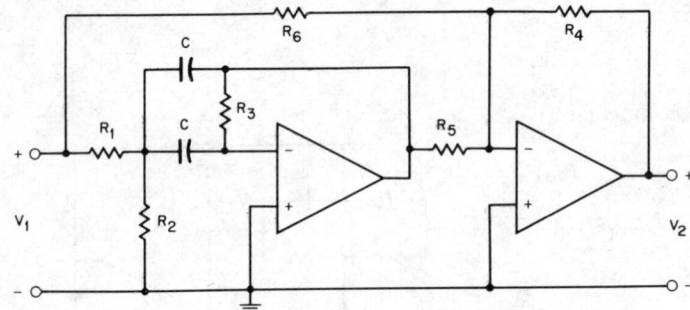

Fig. 18—A second-order MFB band-reject filter

TABLE 27—SECOND-ORDER MFB BAND-REJECT DESIGNS.*

Q	Gain	R_1	R_2	R_3	R_4
2	2	1.000	0.333	4.000	2.000
2	10	1.000	0.333	4.000	10.000
4	2	2.000	0.133	8.000	2.000
4	10	2.000	0.133	8.000	10.000
6	2	3.000	0.086	12.000	2.000
6	10	3.000	0.086	12.000	10.000
8	2	4.000	0.063	16.000	2.000
8	10	4.000	0.063	16.000	10.000
10	2	5.000	0.051	20.000	2.000
10	10	5.000	0.051	20.000	10.000
12	2	6.000	0.042	24.000	2.000
12	10	6.000	0.042	24.000	10.000

* In all cases, $R_5 = 2$ and $R_6 = 1$.

Selecting $R_6 = 10$ kΩ yields $R_4 = 100$ kΩ and $R_5 = 20$ kΩ.

For the shortcut method using Table 27, it is necessary first to find SF, given by (20) as

$$SF = 500/[\pi(10^3)(0.01)] = 15.915$$

Multiplying the appropriate resistances of the table by SF yields the network resistances in kilohms:

$$R_1 = 3SF = 47.745$$

$$R_2 = 0.086SF = 1.369$$

$$R_3 = 12SF = 190.98$$

$$R_4 = 10SF = 159.15$$

$$R_5 = 2SF = 31.83$$

$$R_6 = 1SF = 15.92$$

If R_6 had been selected as 15.92 kΩ instead of 10 kΩ in the first method, the values of R_4, R_5, and R_6 obtained by the two methods would agree.

Design Summary of Band-Reject Filters

For the desired gain K, quality factor Q, and center frequency f_0 (hertz), select the circuit (Fig. 17 for the VCVS filter or Fig. 18 for the MFB filter) and the capacitance C to be used. A good choice of C is a standard value approximately equal to $10/f_0$ μF.

Method 1:

Obtain the resistances from (26) for the VCVS filter or from (27) for the MFB filter. Note: The gain of the VCVS filter is restricted to 1 and that of the MFB filter is inverting.

Method 2:

(A) Find SF from (20).

(B) From the appropriate one of Tables 26 and 27, determine the resistances in kilohms by multiplying the table values by SF.

In either method, the filter should be constructed with standard values of resistances as close as possible to the calculated values. The suggestions for the low-pass design summary apply except that the dc return to ground is already satisfied in the MFB circuit.

The example of a practical design, given earlier, may be used as a guideline.

ALL-PASS FILTERS

An all-pass, or phase-shifting, filter has a constant amplitude response and a phase response that varies with frequency. A second-order all-pass filter has a transfer function

$$\frac{V_2}{V_1} = \frac{K(s^2 - a\omega_0 s + b\omega_0^2)}{s^2 + a\omega_0 s + b\omega_0^2} \tag{28}$$

which has a constant amplitude or gain K, and a phase given by

$$\phi(\omega) = -2 \tan^{-1}[a\omega_0\omega/(b\omega_0^2 - \omega^2)] \tag{29}$$

The numbers a and b are constants and are the transfer-function coefficients when $\omega_0 = 1$.

From (29), the phase shift at $\omega_0 = 2\pi f_0$ is given by

$$\phi_0 = \phi(\omega_0) = -2 \tan^{-1}[a/(b-1)] \tag{30}$$

Thus for a given ϕ_0, parameters a and b may be determined. Since a may be expressed in terms of b, an additional constraint, such as gain, dc offset minimization, or damping factor, may be satisfied.

Second-Order Design

An MFB filter which realizes (28) is shown in Fig. 19.* For a specified gain K $(0 < K < 1)$, phase

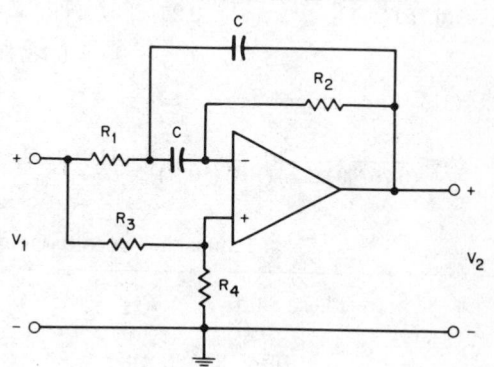

Fig. 19—A second-order MFB all-pass filter.

shift $\phi_0 = \phi(\omega_0)$ at a given frequency ω_0, and arbitrary value of C, the resistances are given by

$$R_2 = \frac{2}{a\omega_0 C}$$

$$R_1 = \frac{(1-K)}{4K} R_2$$

$$R_3 = R_2/K$$

$$R_4 = R_2/(1-K) \tag{31}$$

where for $0 < \phi_0 < 180°$,

* T. Deliyannis, "RC Active Allpass Sections," *Electron. Letters*, vol. 5, pp. 59–60, February 1969.

$$a=\frac{1-K}{2K\tan(\phi_0/2)}\left[-1+\left(1+\frac{4K}{1-K}\cdot\tan^2(\phi_0/2)\right)^{1/2}\right], \tag{32}$$

and for $-180°<\phi_0<0$,

$$a=\frac{1-K}{2K\tan(\phi_0/2)}\left[-1-\left(1+\frac{4K}{1-K}\cdot\tan^2(\phi_0/2)\right)^{1/2}\right] \tag{33}$$

The resistances have been calculated to minimize the dc offset of the op amp.

A shortcut design using equation (20) and Table 28 is available for certain cases of phase shift and gain. The procedure is identical to that of the band-reject filter case considered in the previous section.

Practical Design Example

Suppose an all-pass filter is desired with a phase shift $\phi_0=-90°$ at $f_0=1000$ Hz, a gain of 3/4, and capacitances of 0.01 μF. By (33), it is found that

$$a=\frac{0.25}{2(0.75)(-1)}\left[-1-\left(1+\frac{4(0.75)}{0.25}(1)\right)^{1/2}\right]$$
$$=0.76759$$

By (31) the resistances are

$$R_2=\frac{2}{(0.76759)(2\pi)(10^3)(10^{-8})}\Omega=41.469\text{ k}\Omega$$

$$R_1=\frac{0.25}{4(0.75)}(41.469)=3.456\text{ k}\Omega$$

$$R_3=\frac{41.469}{0.75}=55.292\text{ k}\Omega$$

$$R_4=\frac{41.469}{0.25}=165.875\text{ k}\Omega$$

To use Table 28, SF is given from (20) to be

$$SF=\frac{500}{\pi(10^3)(0.01)}=15.915$$

and the resistances, in kilohms, are

$$R_1=0.217\,SF=3.454$$

$$R_2=2.606\,SF=41.474$$

$$R_3=3.474\,SF=55.290$$

$$R_4=10.422\,SF=165.866$$

Design Summary of All-Pass Filters

For the gain K and phase shift ϕ_0 desired at the frequency f_0 (hertz), select the capacitance C to be used in the MFB filter of Fig. 19. A good choice of C is a standard value approximately equal to $10/f_0\ \mu$F.

Method 1:

(A) Obtain a from (32) or (33).

(B) Obtain the resistances from (31).

TABLE 28—SECOND-ORDER MFB ALL-PASS DESIGNS.*

Phase Shift	R_1	R_2	R_3	R_4
10°	1.948	23.374	31.165	93.495
20	1.026	12.317	16.423	49.268
30	0.735	8.824	11.765	35.296
40	0.597	7.169	9.558	28.675
50	0 518	6.219	8.292	24.875
60	0.467	5.605	7.473	22.420
70	0.431	5.175	6.900	20.700
80	0.405	4.855	6.474	19.421
90	0.384	4.606	6.141	18.422
−10	0.043	0.513	0.685	2.054
−20	0.081	0.974	1.299	3.897
−30	0.113	1.360	1.813	5.440
−40	0.139	1.674	2.232	6.696
−50	0.161	1.930	2.573	7.719
−60	0.178	2.141	2.855	8.564
−70	0.193	2.319	3.092	9.275
−80	0.206	2.472	3.295	9.886
−90	0.217	2.606	3.474	10.422

* Gain=0.75 in all cases.

Method 2:

(A) Find SF from (20).

(B) Determine the resistances in kilohms by multiplying the appropriate values in Table 28 by SF. Note: The gain using the table is restricted to 0.75.

In either method, the filter should be constructed with standard values of resistances as close as possible to the calculated values. The suggestions for the low-pass design summary apply except that the dc return to ground is already satisfied by R_2 and R_4.

The practical design example, given earlier, may be used as a guideline.

CONSTANT-TIME-DELAY FILTERS

A filter which achieves an approximately constant time delay over the range $0<\omega<\omega_0$ is the Bessel filter*, with a second-order transfer function given by

$$\frac{V_2}{V_1}=\frac{3K\omega_0^2}{s^2+3\omega_0 s+3\omega_0^2} \quad (34)$$

The gain is K, and the time delay at $\omega_0=2\pi f_0$ is given by

$$T_0= T(\omega_0)= 12/(13\omega_0) \text{ seconds} \quad (35)$$

Equation (34) is a low-pass function with the same form as (4) where $a=b=3$ and ω_c is replaced by ω_0. Therefore the MFB low-pass circuit of Fig. 4, using (5), or the VCVS low-pass circuit of Fig. 5, using (6), may be used to obtain a realization. The frequency ω_0 may be found from a given time delay T_0 by using (35). Conversely, if ω_0 is specified, the constant delay which results is then T_0.

TABLE 29—SECOND-ORDER MFB CONSTANT-
TIME-DELAY DESIGNS.

Gain	C_1	R_1	R_2	R_3
2	0.200C	0.691	1.382	1.206
10	0.047C	0.471	4.709	1.506

Tables 29 and 30 are available for shortcut designs. For a given T_0, $\omega_0=2\pi f_0$ is found from (35), and the result is then used to find SF from (20). The resistances are then those of the table multiplied by SF.

Practical Design Example

Suppose it is desired to construct an MFB constant-time-delay filter which has $T_0=100$ μs and a gain of 2; the capacitances are to be $C=0.01$ μF and $C_1=0.002$ μF. From (35), ω_0 is

$$\omega_0=\frac{12}{13(10^{-4})}=9231=2\pi f_0 \quad (36)$$

The pertinent design equations are (5) with $\omega_c=\omega_0$ and $a=b=3$, resulting in

$$R_2= 2(3)/\{3(10^{-8})+[9(10^{-16})$$
$$-4(3)(2)(10^{-17})(3)]^{1/2}\}(9231)\Omega$$
$$= 14.971 \text{ k}\Omega$$

$$R_1= 14.971/2= 7.486 \text{ k}\Omega$$

$$R_3=\frac{1}{3(2)(10^{-17})(9231)^2(14.971)(10^3)}\Omega= 13.065 \text{ k}\Omega$$

In the shortcut procedure, SF is given by (20) and (36) to be

$$SF=\frac{500}{(9231/2)(0.01)}= 10.833$$

The resistances (in kilohms) are the appropriate ones of Table 29 multiplied by SF, or

$$R_1= 0.691 SF= 7.486$$

$$R_2= 1.382 SF= 14.971$$

$$R_3= 1.206 SF= 13.065$$

Design Summary of Constant-Time-Delay Filters

For the desired gain K and time delay T_0, select the circuit (Fig. 4 for the MFB filter or

TABLE 30—SECOND-ORDER VCVS CONSTANT-TIME-DELAY DESIGNS.

Gain	C_1	R_1	R_2	R_3	R_4
2	1C	0.333	1.000	2.667	2.667
10	2C	0.158	1.054	1.346	12.118

* L. Storch, "Synthesis of Constant-Time-Delay Ladder Networks Using Bessel Polynomials," *Proc. of the IRE*, vol. 42, no. 11, pp. 1666–1675. November 1954.

Fig. 5 for the VCVS filter) and the capacitance C to be used. A good choice of C is a standard value approximately equal to $10/f_0$ μF. Calculate $f_0 = \omega_0/2\pi$ Hz from (35). (Alternately, ω_0 may be specified and T_0 calculated.)

Method 1:

This method is identical to that given earlier in the low-pass design summary, except that only second-order cases are considered, and $a = b = 3$.

Method 2:

(**A**) Find SF from (20).

(**B**) From the appropriate one of Tables 29 or 30, find C_1. Determine the resistances in kilohms by multiplying the table values by SF.

In either method, the filter should be constructed with standard values of resistances as close as possible to the calculated values. The suggestions for the low-pass design summary apply directly.

The practical time-delay design example, given earlier, may be used as a guideline.

DEFINITIONS

An attenuator is a network designed to introduce a known loss when working between resistive impedances Z_1 and Z_2 to which the input and output impedances of the attenuator are matched. Either Z_1 or Z_2 may be the source and the other the load. The attenuation of such networks expressed as a power ratio is the same regardless of the direction of working.

Three forms of resistance network that may be conveniently used to realize these conditions are shown on page 11-4. These are the T section, the π section, and the bridged-T section. Equivalent balanced sections also are shown. Methods are given for the computation of attenuator networks, the hyperbolic expressions giving rapid solutions with the aid of tables of hyperbolic functions in Chapter 47. Tables of the various types of attenuators are given on pages 11-6 through 11-9.

LADDER ATTENUATOR

Ladder attenuator, Fig. 1, input switch points P_0, P_1, P_2, P_3 at shunt arms. Also intermediate point P_m tapped on series arm. May be either unbalanced, as shown, or balanced.

Ladder, for design purposes, Fig. 2, is resolved into a cascade of π sections by imagining each shunt arm split into two resistors. Last section matches Z_2 to $2Z_1$. All other sections are symmetrical, matching impedances $2Z_1$, with a terminating resistor $2Z_1$ on the first section. Each section is designed for the loss required between the switch points at the ends of that section.

Input to P_0:

Loss in decibels $= 10 \log_{10}[(2Z_1+Z_2)^2/4Z_1Z_2]$

Input impedance $= Z_2/2$

Output impedance $= Z_1Z_2/(Z_1+Z_2)$.

Input to P_1, P_2, or P_3: Loss in decibels $= 3+$ (sum of losses of π sections between input and output). Input impedance $= Z_1$.

Input to P_m (on a symmetrical π section):

$$e_0/e_m = \frac{1}{2}\frac{m(1-m)(K-1)^2+2K}{K-m(K-1)}$$

where $e_0 =$ output voltage when $m=0$ (switch on P_1), $e_m =$ output voltage with switch on P_m, and $K =$ current ratio of the section (from P_1 to P_2) $(K>1)$.

Input impedance

$$Z_1' = Z_1\{m(1-m)[(K-1)^2/K]+1\}$$

maximum $Z_1' = Z_1\{[(K-1)^2/4K]+1\}$ for $m=0.5$.

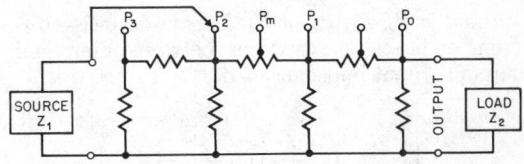

Fig. 1—Ladder attenuator.

The unsymmetrical last section may be treated as a system of voltage-dividing resistors. Solve for the resistance R from P_0 to the tap, for each value of

$$\left(\frac{\text{output voltage with input on } P_0}{\text{output voltage with input on tap}}\right).$$

A Useful Case

When $Z_1 = Z_2 = 500$ ohms.
Then loss on P_0 is 3.52 decibels.
Let the last section be designed for loss of 12.51 decibels. Then from Attenuator Network Design table, page 11-4.

$$R_{13} = 2444 \text{ ohms}$$

$$R_{23} = 654 \text{ ohms}$$

$$R_{12} = 1409 \text{ ohms}.$$

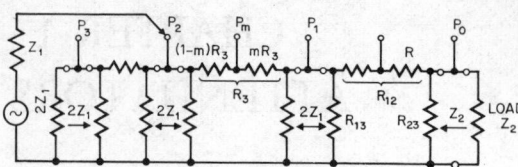

Fig. 2—Ladder attenuator resolved into a cascade of π sections.

The table shows the location of the tap and the input and output impedances for several values of loss, relative to the loss on P_0.

Relative Loss (dB)	Tap R (ohms)	Input Impedance (ohms)	Output Impedance (ohms)
0	0	250	250
2	170	368	304
4	375	478	353
6	615	562	394
8	882	600	428
10	1157	577	454
12	1409	500	473

Input to P_0: Output impedance $= 0.6Z$. (See Fig. 3.)

Input to P_0, P_1, P_2, or P_3: Loss in decibels $= 6 +$ (sum of losses of π sections between input and output). Input impedance $= Z$.

Input to P_m:

$$\frac{e_0}{e_m} = \frac{1}{4} \frac{m(1-m)(K-1)^2 + 4K}{K - m(K-1)}$$

Input impedance:

$$Z' = Z\left[\frac{m(1-m)(K-1)^2}{2K} + 1\right]$$

$$\text{maximum } Z' = Z\left[\frac{(K-1)^2}{8K} + 1\right] \text{ for } m = 0.5.$$

LOAD IMPEDANCE

Effect of Incorrect Load Impedance on Operation of an Attenuator

In the applications of attenuators, the question frequently arises as to the effect on the input impedance and the attenuation by the use of a load impedance different from that for which the network was designed. The following results apply to all resistive networks that, when operated between resistive impedances Z_1 and Z_2, present matching terminal impedances Z_1 and Z_2, respectively. The results may be derived in the general case by the application of the network theorems

and may be readily confirmed mathematically for simple specific cases such as the T section.

For the designed use of the network, let

$Z_1 =$ input impedance of properly terminated network

$Z_2 =$ load impedance that properly terminates the network

$N =$ power ratio from input to output

$K =$ current ratio from input to output

$K = i_1/i_2 = (NZ_2/Z_1)^{1/2}$ (different in the two directions except when $Z_2 = Z_1$).

For the actual conditions of operation, let

$$(Z_2 + \Delta Z_2) = Z_2(1 + \Delta Z_2/Z_2)$$
$$= \text{actual load impedance}$$
$$(Z_1 + \Delta Z_1) = Z_1(1 + \Delta Z_1/Z_1)$$
$$= \text{resulting input impedance}$$
$$(K + \Delta K) = K(1 + \Delta K/K)$$
$$= \text{resulting current ratio.}$$

While Z_1, Z_2, and K are restricted to real quantities by the assumed nature of the network, ΔZ_2 is not so restricted, for example,

$$\Delta Z_2 = \Delta R_2 + j\Delta X_2.$$

As a consequence, ΔZ_1 and ΔK can become imaginary or complex. Furthermore, ΔZ_2 is not restricted to small values.

The results for the actual conditions are

$$\frac{\Delta Z_1}{Z_1} = \frac{2\Delta Z_2/Z_2}{2N + (N-1)(\Delta Z_2/Z_2)}$$

and

$$\frac{\Delta K}{K} = \left(\frac{N-1}{2N}\right)\frac{\Delta Z_2}{Z_2}.$$

Certain Special Cases May Be Cited

Case 1: For small $\Delta Z_2/Z_2$

$$\Delta Z_1/Z_1 = (1/N)(\Delta Z_2/Z_2) \quad \text{or} \quad \Delta Z_1 = (1/K^2)\Delta Z_2$$
$$\Delta i_2/i_2 = -\tfrac{1}{2}(\Delta Z_2/Z_2)$$

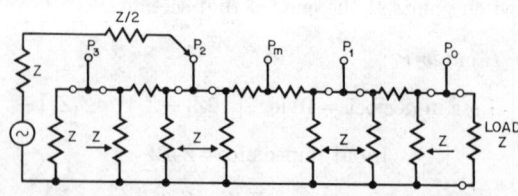

Fig. 3—A variation of the ladder attenuator, useful when $Z_1 = Z_2 = Z$. Simpler in design, with improved impedance characteristics, but having minimum insertion loss 2.5 decibels higher than attenuator of Fig. 2. All π sections are symmetrical.

but the error in insertion power loss of the attenuator is negligibly small.

Case 2: Short-circuited output

$$\Delta Z_1/Z_1 = -2/(N+1)$$

or

$$\text{input impedance} = \left(\frac{N-1}{N+1}\right) Z_1 = Z_1 \tanh\theta$$

where θ is the designed attenuation in nepers.

Case 3: Open-circuited output

$$\Delta Z_1/Z_1 = 2/(N-1)$$

or

$$\text{input impedance} = \left(\frac{N+1}{N-1}\right) Z_1 = Z_1 \coth\theta.$$

Case 4: For $N=1$ (possible only when $Z_1 = Z_2$ and directly connected)

$$\Delta Z_1/Z_1 = \Delta Z_2/Z_2$$

$$\Delta K/K = 0.$$

Case 5: For large N

$$\Delta K/K = \tfrac{1}{2}(\Delta Z_2/Z_2).$$

ATTENUATOR NETWORK DESIGN
(See table on pages 11-4—11-5)

Symbols

Z_1 and Z_2 are the terminal impedances (resistive) to which the attenuator is matched.

N is the ratio of the power absorbed by the attenuator from the source to the power delivered to the load.

K is the ratio of the attenuator input current to the output current into the load. When $Z_1 = Z_2$, $K = N^{1/2}$. Otherwise K is different in the two directions.

Attenuation in decibels $= 10 \log_{10}N$.
Attenuation in nepers $= \theta = \tfrac{1}{2}\log_e N$.

For a table of decibels versus power and voltage or current ratio, see Chapter 3. Factors for converting decibels to nepers, and nepers to decibels, are given at the foot of that table.

Notes on Error Equations

The equations and figures for errors, given in Tables 1 through 5, are based on the assumption that the attenuator is terminated approximately by its proper terminal impedances Z_1 and Z_2. They hold for deviations of the attenuator arms and load impedances up to ±20 percent or somewhat more.

The error due to each element is proportional to the deviation of the element, and the total error of the attenuator is the sum of the errors due to each of the several elements.

When any element or arm R has a reactive component ΔX in addition to a resistive error ΔR, the errors in input impedance and output current are

$$\Delta Z = A(\Delta R + j\Delta X)$$

$$\Delta i/i = B[(\Delta R + j\Delta X)/R]$$

where A and B are constants of proportionality for the elements in question. These constants can be determined in each case from the figures given for errors due to a resistive deviation ΔR.

The reactive component ΔX produces a quadrature component in the output current, resulting in a phase shift. However, for small values of ΔX, the error in insertion loss is negligibly small.

For the errors produced by mismatched terminal load impedance, refer to Case 1, page 11-2.

SYMMETRICAL *T* OR *H* ATTENUATORS

Interpolation of Symmetrical T or H Attenuators (Table 1)

Column R_1 may be interpolated linearly. Do not interpolate R_3 column. For 0 to 6 decibels interpolate the $1000/R_3$ column. Above 6 decibels, interpolate the column $\log_{10}R_3$ and determine R_3 from the result.

Errors in Symmetrical T or H Attenuators (Fig. 4)

Series Arms R_1 and R_2 in Error: Error in input impedances:

$$\Delta Z_1 = \Delta R_1 + (1/K^2)\Delta R_2$$

and

$$\Delta Z_2 = \Delta R_2 + (1/K^2)\Delta R_1.$$

Error in insertion loss, in decibels:

$$\text{decibels} = 4[(\Delta R_1/Z_1) + (\Delta R_2/Z_2)]\text{approximately.}$$

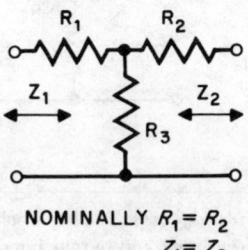

NOMINALLY $R_1 = R_2$
$Z_1 = Z_2$

Fig. 4.

ATTENUATOR NETWORK DESIGN *(See page 11-3 for symbols)*

Description	Unbalanced Configuration	Balanced Configuration	Hyperbolic Equations*
Unbalanced T and balanced H (refer to Table 5)			$R_3 = (Z_1 Z_2)^{1/2} / \sinh\theta$ $R_1 = (Z_1 / \tanh\theta) - R_3$ $R_2 = (Z_2 / \tanh\theta) - R_3$
Symmetrical T and H ($Z_1 = Z_2 = Z$) (refer to Table 1)			$R_3 = Z / \sinh\theta$ $R_1 = Z \tanh(\theta/2)$
Minimum-loss pad matching Z_1 and Z_2 ($Z_1 > Z_2$) (refer to Table 4)			$\cosh\theta = (Z_1/Z_2)^{1/2}$ $\cosh2\theta = 2(Z_1/Z_2) - 1$
Unbalanced π and balanced O			$R_3 = (Z_1 Z_2)^{1/2} \sinh\theta$ $1/R_1 = (1/Z_1 \tanh\theta) - (1/R_3)$ $1/R_2 = (1/Z_2 \tanh\theta) - (1/R_3)$
Symmetrical π and O ($Z_1 = Z_2 = Z$) (refer to Table 2)			$R_3 = Z \sinh\theta$ $R_1 = Z / \tanh(\theta/2)$
Bridged T and bridged H (refer to Table 3)			

* *Four-terminal networks*: The hyperbolic equations above are valid for passive linear 4-terminal networks in general, working between input and output impedances matching the respective image impedances. In this case: Z_1 and Z_2

Arithmetic Equations	Checking Equations

$R_3 = 2(NZ_1Z_2)^{1/2}/(N-1)$

$R_1 = Z_1[(N+1)/(N-1)] - R_3$

$R_2 = Z_2[(N+1)/(N-1)] - R_3$

$R_3 = \dfrac{2Z(N)^{1/2}}{N-1} = \dfrac{2ZK}{K^2-1}$

$\qquad = \dfrac{2Z}{K-1/K}$

$R_1 = Z[(N^{1/2}-1)/(N^{1/2}+1)] = Z[(K-1)/(K+1)]$

$\qquad = Z[1-2/(K+1)]$

$R_1R_3 = \dfrac{Z^2}{1+\cosh\theta} = Z^2\dfrac{2K}{(K+1)^2}$

$R_1/R_3 = \cosh\theta - 1 = 2\sinh^2(\theta/2)$

$\qquad = (K-1)^2/2K$

$\qquad Z = R_1[1+2(R_3/R_1)]^{1/2}$

$R_1 = Z_1[1-(Z_2/Z_1)]^{1/2}$

$R_3 = Z_2/[1-(Z_2/Z_1)]^{1/2}$

$R_1R_3 = Z_1Z_2$

$R_1/R_3 = (Z_1/Z_2) - 1$

$N = \{(Z_1/Z_2)^{1/2} + [(Z_1/Z_2)-1]^{1/2}\}^2$

$R_3 = \frac{1}{2}(N-1)(Z_1Z_2/N)^{1/2}$

$1/R_1 = (1/Z_1)[(N+1)/(N-1)] - (1/R_3)$

$1/R_2 = (1/Z_2)[(N+1)/(N-1)] - (1/R_3)$

$R_3 = Z[(N-1)/2(N)^{1/2}] = Z[(K^2-1)/2K]$

$\qquad = Z(K-1/K)/2$

$R_1 = Z[(N^{1/2}+1)/(N^{1/2}-1)] = Z[(K+1)/(K-1)]$

$\qquad = Z[1+2/(K-1)]$

$R_1R_3 = Z^2(1+\cosh\theta) = Z^2[(K+1)^2/2K]$

$R_3/R_1 = \cosh\theta - 1 = (K-1)^2/2K$

$\qquad Z = R_1/[1+2(R_1/R_3)]^{1/2}$

$R_1 = R_2 = Z$

$R_4 = Z(K-1)$

$R_3 = Z/(K-1)$

$R_3R_4 = Z^2$

$R_4/R_3 = (K-1)^2$

are the image impedances; R_1, R_2, and R_3 become complex impedances; and θ is the image transfer constant. $\theta = \alpha + j\beta$, where α is the image attenuation constant and β is the image phase constant.

TABLE 1—SYMMETRICAL T AND H ATTENUATOR VALUES. $Z = 500$ OHMS RESISTIVE (DIAGRAM ON PAGE 11-4).

Attenu-ation (dB)	Series Arm R_1 (ohms)	Shunt Arm R_3 (ohms)	$1000/R_3$	$\log_{10} R_3$
0.0	0.0	∞	0.0000	—
0.2	5.8	21 700	0.0461	—
0.4	11.5	10 850	0.0921	—
0.6	17.3	7 230	0.1383	—
0.8	23.0	5 420	0.1845	—
1.0	28.8	4 330	0.2308	—
2.0	57.3	2 152	0.465	—
3.0	85.5	1 419	0.705	—
4.0	113.1	1 048	0.954	—
5.0	140.1	822	1.216	—
6.0	166.1	669	1.494	2.826
7.0	191.2	558	—	2.747
8.0	215.3	473.1	—	2.675
9.0	238.1	405.9	—	2.608
10.0	259.7	351.4	—	2.546
12.0	299.2	268.1	—	2.428
14.0	333.7	207.8	—	2.318
16.0	363.2	162.6	—	2.211
18.0	388.2	127.9	—	2.107
20.0	409.1	101.0	—	2.004
22.0	426.4	79.94	—	1.903
24.0	440.7	63.35	—	1.802
26.0	452.3	50.24	—	1.701
28.0	461.8	39.87	—	1.601
30.0	469.3	31.65	—	1.500
35.0	482.5	17.79	—	1.250
40.0	490.1	10.00	—	1.000
50.0	496.8	3.162	—	0.500
60.0	499.0	1.000	—	0.000
80.0	499.9	0.1000	—	−1.000
100.0	500.0	0.01000	—	−2.000

Shunt Arm R_3 in Error (10 *Percent High*):

Designed Loss (dB)	Error in Insertion Loss (dB)	Error in Input Impedance $100(\Delta Z/Z)$ Percent
0.2	−0.01	0.2
1	−0.05	1.0
6	−0.3	3.3
12	−0.5	3.0
20	−0.7	1.6
40	−0.8	0.2
100	−0.8	0.0

Error in input impedance:

$$\frac{\Delta Z}{Z} = 2 \frac{K-1}{K(K+1)} \frac{\Delta R_3}{R_3}.$$

Error in output current:

$$\frac{\Delta i}{i} = \frac{K-1}{K+1} \frac{\Delta R_3}{R_3}.$$

See notes on page 11-3.

SYMMETRICAL π AND O ATTENUATORS

Interpolation of Symmetrical π and O Attenuators (Table 2)

Column R_1 may be interpolated linearly above 16 decibels, and R_3 up to 20 decibels. Otherwise interpolate the $1000/R_1$ and $\log_{10} R_3$ columns, respectively.

TABLE 2—SYMMETRICAL π AND O ATTENUATORS. $Z = 500$ OHMS RESISTIVE (DIAGRAM ON PAGE 11-4).

Attenu-ation (dB)	Shunt Arm R_1 (ohms)	$1000/R_1$	Series Arm R_3 (ohms)	$\log_{10} R_3$
0.0	∞	0.000	0.0	—
0.2	43 400	0.023	11.5	—
0.4	21 700	0.046	23.0	—
0.6	14 500	0.069	34.6	—
0.8	10 870	0.092	46.1	—
1.0	8 700	0.115	57.7	—
2.0	4 362	0.229	116.1	—
3.0	2 924	0.342	176.1	—
4.0	2 210	0.453	238.5	—
5.0	1 785	0.560	304.0	—
6.0	1 505	0.665	373.5	—
7.0	1 307	0.765	448.0	—
8.0	1 161.4	0.861	528.4	—
9.0	1 049.9	0.952	615.9	—
10.0	962.5	1.039	711.5	—
12.0	835.4	1.197	932.5	—
14.0	749.3	1.335	1 203.1	—
16.0	688.3	1.453	1 538	—
18.0	644.0	—	1 954	—
20.0	611.1	—	2 475	3.394
22.0	586.3	—	3 127	3.495
24.0	567.3	—	3 946	3.596
26.0	552.8	—	4 976	3.697
28.0	541.5	—	6 270	3.797
30.0	532.7	—	7 900	3.898
35.0	518.1	—	14 050	4.148
40.0	510.1	—	25 000	4.398
50.0	503.2	—	79 100	4.898
60.0	501.0	—	2.50×10^5	5.398
80.0	500.1	—	2.50×10^6	6.398
100.0	500.0	—	2.50×10^7	7.398

Errors in Symmetrical π and O Attenuators (Fig. 5)

Error in input impedance:

$$\frac{\Delta Z'}{Z'} = \frac{K-1}{K+1}\left(\frac{\Delta R_1}{R_1} + \frac{1}{K^2}\frac{\Delta R_2}{R_2} + \frac{2}{K}\frac{\Delta R_3}{R_3}\right).$$

Error in insertion loss, in decibels:
decibels $= -8(\Delta i_2/i_2)$ (approximately)

$$= 4\frac{K-1}{K+1}\left(-\frac{\Delta R_1}{R_1} - \frac{\Delta R_2}{R_2} + 2\frac{\Delta R_3}{R_3}\right).$$

See notes on page 11-3.

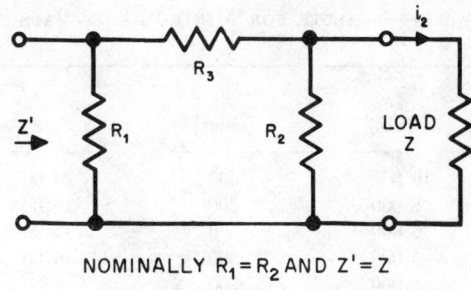

NOMINALLY $R_1 = R_2$ AND $Z' = Z$

Fig. 5.

BRIDGED T OR H ATTENUATORS

Interpolation of Bridged T or H Attenuators (Table 3)

Bridge Arm R_4: Use the formula $\log_{10}(R_4 + 500) = 2.699 + \text{decibels}/20$ for $Z = 500$ ohms. However, if preferred, the tabular values of R_4 may be interpolated linearly, between 0 and 10 decibels only.

Shunt Arm R_3: Do not interpolate R_3 column. Compute R_3 by $R_3 = 10^6/4R_4$, for $Z = 500$ ohms.

Note: For attenuators of 60 decibels and over, the bridge arm R_4 may be omitted provided a shunt arm is used having twice the resistance tabulated in the R_3 column. (This makes the input impedance 0.1 of 1 percent high at 60 decibels.)

Table 3—Values for Bridged T or H Attenuators. $Z = 500$ Ohms Resistive, $R_1 = R_2 = 500$ Ohms (Diagram on Page 11-4).

Attenuation (decibels)	Bridge Arm R_4 (ohms)	Shunt Arm R_3 (ohms)
0.0	0.0	∞
0.2	11.6	21 500
0.4	23.6	10 610
0.6	35.8	6 990
0.8	48.2	5 180
1.0	61.0	4 100
2.0	129.5	1 931
3.0	206.3	1 212
4.0	292.4	855
5.0	389.1	642
6.0	498	502
7.0	619	404
8.0	756	331
9.0	909	275.0
10.0	1 081	231.2
12.0	1 491	167.7
14.0	2 006	124.6
16.0	2 655	94.2
18.0	3 472	72.0
20.0	4 500	55.6
25.0	8 390	29.8
30.0	15 310	16.33
40.0	49 500	5.05
50.0	157 600	1.586
60.0	499 500	0.501
80.0	5.00×10^6	0.0500
100.0	50.0×10^6	0.00500

Errors in Bridged T or H Attenuators

Resistance of any one arm 10 percent higher than correct value:

Designed Loss (decibels)	A (decibels)	B (percent)	C (percent)
0.2	0.01	0.005	0.2
1	0.05	0.1	1.0
6	0.2	2.5	2.5
12	0.3	5.6	1.9
20	0.4	8.1	0.9
40	0.4	10	0.1
100	0.4	10	0.0

Element in Error (10 percent high)	Error in Loss	Error in Terminal Impedance
Series arm R_1 (analogous for arm R_2)	Zero	B, for adjacent terminals*
Shunt arm R_3	$-A$†	C
Bridge arm R_4	A‡	C

Notes:
* Error in impedance at opposite terminals is zero.
† Loss is lower than designed loss.
‡ Loss is higher than designed loss.

Error in input impedance:

$$\frac{\Delta Z_1}{Z_1} = \left(\frac{K-1}{K}\right)^2\frac{\Delta R_1}{R_1} + \frac{K-1}{K^2}\left(\frac{\Delta R_3}{R_3} + \frac{\Delta R_4}{R_4}\right).$$

For $\Delta Z_2/Z_2$ use subscript 2 in equation in place of subscript 1.

TABLE 4—VALUES FOR MINIMUM-LOSS PADS MATCHING Z_1 AND Z_2, BOTH RESISTIVE (DIAGRAM ON PAGE 11-4).

Z_1 (ohms)	Z_2 (ohms)	Z_1/Z_2	Loss (decibels)	Series Arm R_1 (ohms)	Shunt Arm R_3 (ohms)
10 000	500	20.00	18.92	9 747	513.0
8 000	500	16.00	17.92	7 746	516.4
6 000	500	12.00	16.63	5 745	522.2
5 000	500	10.00	15.79	4 743	527.0
4 000	500	8.00	14.77	3 742	534.5
3 000	500	6.00	13.42	2 739	547.7
2 500	500	5.00	12.54	2 236	559.0
2 000	500	4.00	11.44	1 732	577.4
1 500	500	3.00	9.96	1 224.7	612.4
1 200	500	2.40	8.73	916.5	654.7
1 000	500	2.00	7.66	707.1	707.1
800	500	1.60	6.19	489.9	816.5
600	500	1.20	3.77	244.9	1224.7
500	400	1.25	4.18	223.6	894.4
500	300	1.667	6.48	316.2	474.3
500	250	2.00	7.66	353.6	353.6
500	200	2.50	8.96	387.3	258.2
500	160	3.125	10.17	412.3	194.0
500	125	4.00	11.44	433.0	144.3
500	100	5.00	12.54	447.2	111.80
500	80	6.25	13.61	458.3	87.29
500	65	7.692	14.58	466.4	69.69
500	50	10.00	15.79	474.3	52.70
500	40	12.50	16.81	479.6	41.70
500	30	16.67	18.11	484.8	30.94
500	25	20.00	18.92	487.3	25.65

Error in output current:

$$\frac{\Delta i}{i} = \frac{K-1}{2K}\left(\frac{\Delta R_3}{R_3} - \frac{\Delta R_4}{R_4}\right).$$

See notes on page 11-3.

MINIMUM-LOSS PADS

Interpolation of Minimum-Loss Pads (Table 4)

Table 4 may be interpolated linearly with respect to Z_1, Z_2, or Z_1/Z_2 except when Z_1/Z_2 is between 2.0 and 1.2. The accuracy of the interpolated value becomes poorer as Z_1/Z_2 passes below 2.0 toward 1.2, especially for R_3.

For Other Terminations

If the terminating resistances are to be Z_A and Z_B instead of Z_1 and Z_2, respectively, the procedure is as follows. Enter the table at $Z_1/Z_2 = Z_A/Z_B$ and read the loss and the tabular values of R_1 and R_3. Then the series and shunt arms are, respectively, MR_1 and MR_3, where $M = Z_A/Z_1 = Z_B/Z_2$.

Errors in Minimum-Loss Pads

Impedance Ratio Z_1/Z_2	D Decibels*	E Percent*	F Percent*
1.2	0.2	+4.1	+1.7
2.0	0.3	7.1	1.2
4.0	0.35	8.6	0.6
10.0	0.4	9.5	0.25
20.0	0.4	9.7	0.12

* *Notes*:

Series arm R_1 10 percent high: Loss is increased by D decibels from above table. Input impedance Z_1 is increased by E percent. Input impedance Z_2 is increased by F percent.

Shunt arm R_3 10 percent high: Loss is decreased by D decibels from above table. Input impedance Z_2 is increased by E percent. Input impedance Z_1 is increased by F percent.

TABLE 5—VALUES FOR MISCELLANEOUS T AND H PADS (DIAGRAM ON PAGE 11-4).

Resistive Terminations			Attenuator Arms		
Z_1 (ohms)	Z_2 (ohms)	Loss (decibels)	Series R_1 (ohms)	Series R_2 (ohms)	Shunt R_3 (ohms)
5000	2000	10	3889	222	2222
5000	2000	15	4165	969	1161
5000	2000	20	4462	1402	639
5000	500	20	4782	190.7	319.4
2000	500	15	1763	165.4	367.3
2000	500	20	1838	308.1	202.0
2000	200	20	1913	76.3	127.8
500	200	10	388.9	22.2	222.2
500	200	15	416.5	96.9	116.1
500	200	20	446.2	140.2	63.9
500	50	20	478.2	19.07	31.94
200	50	15	176.3	16.54	36.73
200	50	20	183.8	30.81	20.20

Errors in input impedance:

$$\Delta Z_1/Z_1 = (1 - Z_2/Z_1)^{1/2}(\Delta R_1/R_1 + N^{-1}\,\Delta R_3/R_3)$$

$$\Delta Z_2/Z_2 = (1 - Z_2/Z_1)^{1/2}(\Delta R_3/R_3 + N^{-1}\,\Delta R_1/R_1).$$

Error in output current, working either direction:

$$\Delta i/i = \tfrac{1}{2}(1 - Z_2/Z_1)^{1/2}(\Delta R_3/R_3 - \Delta R_1/R_1).$$

See notes on page 11-3.

MISCELLANEOUS *T* AND *H* PADS (Table 5)

Errors in T and H Pads

Series Arms R_1 and R_2 in Error: Errors in input impedances are

$$\Delta Z_1 = \Delta R_1 + N^{-1}(Z_1/Z_2)\Delta R_2$$

and

$$\Delta Z_2 = \Delta R_2 + N^{-1}(Z_2/Z_1)\Delta R_1.$$

Error in insertion loss, in decibels

$$= 4(\Delta R_1/Z_1 + \Delta R_2/Z_2) \text{ approximately.}$$

Shunt Arm R_3 in Error (10 *Percent High*):

Z_1/Z_2	Designed Loss (decibels)	Error in Insertion Loss (decibels)	Error in Input Impedance $100(\Delta Z_1/Z_1)$	$100(\Delta Z_2/Z_2)$
2.5	10	−0.4	1.1%	7.1%
2.5	15	−0.6	1.2	4.6
2.5	20	−0.7	0.9	2.8
4.0	15	−0.5	0.8	6.0
4.0	20	−0.65	0.6	3.6
10	20	−0.6	0.3	6.1

$$\frac{\Delta Z_1}{Z_1} = \frac{2}{N-1}\left[\left(\frac{NZ_2}{Z_1}\right)^{1/2} + \left(\frac{Z_1}{NZ_2}\right)^{1/2} - 2\right]\frac{\Delta R_3}{R_3}$$

(for $\Delta Z_2/Z_2$ interchange subscripts 1 and 2)

$$\frac{\Delta i}{i} = \frac{N+1 - (N^{1/2})\left[(Z_1/Z_2)^{1/2} + (Z_2/Z_1)^{1/2}\right]}{N-1}\frac{\Delta R_3}{R_3}$$

where i is the output current.

CHAPTER 12

BRIDGES AND IMPEDANCE MEASUREMENTS

In the diagrams of bridges below, the source is shown as a generator, and the detector as a pair of headphones. The positions of these two elements may be interchanged as dictated by detailed requirements in any individual case, such as location of grounds. For all but the lowest frequencies, a shielded transformer is required at either the input or output (but not usually at both) terminals of the bridge. This is shown in some of the diagrams. The detector is chosen according to the frequency of the source. When insensitivity of the ear makes direct use of headphones impractical, a simple radio receiver or its equivalent is essential. Some selectivity is desirable to discriminate against harmonics, for the bridge is often frequency sensitive. The source may be modulated to obtain an audible signal, but greater sensitivity and discrimination against interference are obtained by the use of a continuous-wave source and a heterodyne detector. An amplifier and oscilloscope or an output meter are sometimes preferred for observing nulls. In this case it is convenient to have an audible output signal available for the preliminary setup and for locating trouble, since much can be deduced from the quality of the audible signal that would not be apparent from observation of amplitude only.

FUNDAMENTAL ALTERNATING-CURRENT OR WHEATSTONE BRIDGE

Balance condition is $Z_x = Z_s Z_a / Z_b$. Maximum sensitivity exists when Z_d is the conjugate of the bridge output impedance and Z_g the conjugate of its input impedance. Greatest sensitivity exists when bridge arms are equal, for example, for resistive arms

$$Z_d = Z_a = Z_b = Z_x = Z_s = Z_g.$$

BRIDGE WITH DOUBLE-SHIELDED TRANSFORMER

Shield on secondary may be floating, connected to either end, or to center of secondary winding. It may be in two equal parts and connected to opposite ends of the winding. In any case, its capacitance to ground must be kept to a minimum.

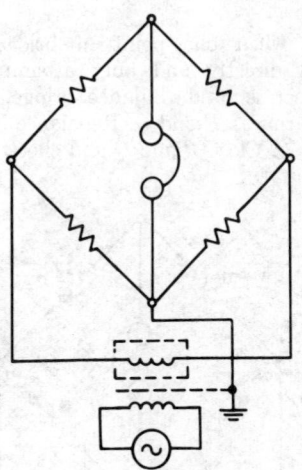

WAGNER EARTH CONNECTION

None of the bridge elements are grounded directly. First balance bridge with switch to B. Throw switch to G and rebalance by means of R and C. Recheck bridge balance and repeat as required. The capacitor balance C is necessary only when the frequency is above the audio range. The transformer may have only a single shield as

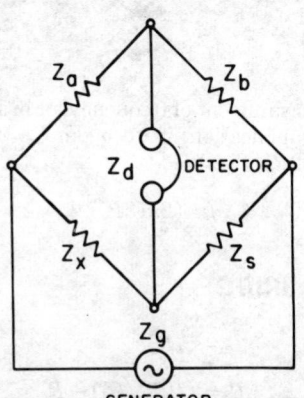

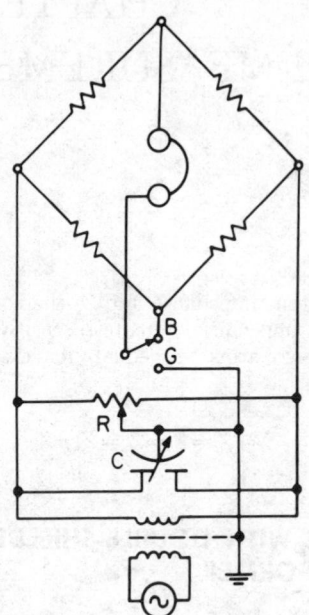

shown, with the capacitance of the secondary to the shield kept to a minimum.

CAPACITOR BALANCE

Useful when one point of bridge must be grounded directly and only a simple shielded transformer is used. Balance bridge, then open the two arms at P and Q. Rebalance by auxiliary capacitor C. Close P and Q and check balance.

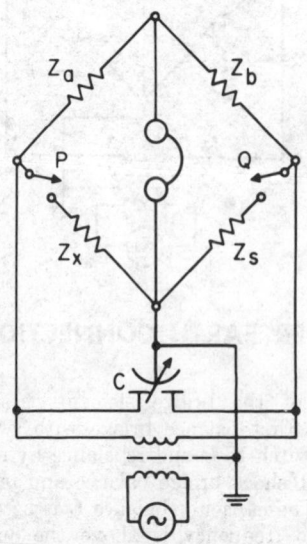

SERIES-RESISTANCE-CAPACITANCE BRIDGE

$$C_x = C_s R_b / R_a$$

$$R_x = R_s R_a / R_b.$$

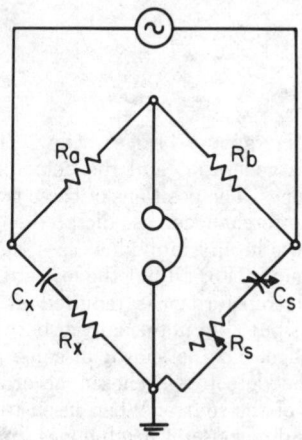

WIEN BRIDGE

$$C_x / C_s = (R_b / R_a) - (R_s / R_x)$$

$$C_s / C_x = 1 / \omega^2 R_s R_x.$$

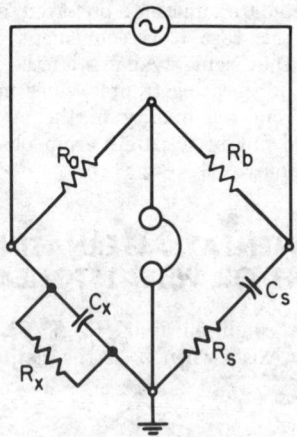

For measurement of frequency, or in a frequency-selective application, if we make $C_x = C_s$, $R_x = R_s$, and $R_b = 2R_a$, then

$$f = (2\pi C_s R_s)^{-1}.$$

OWEN BRIDGE

$$L_x = C_b R_a R_d$$

$$R_x = (C_b R_a / C_d) - R_c.$$

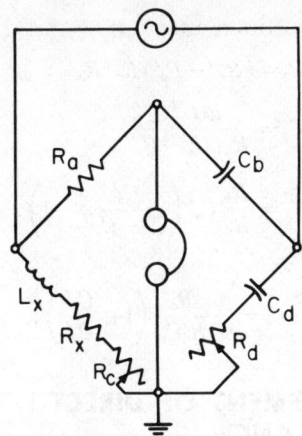

RESONANCE BRIDGE

$$\omega^2 LC = 1$$
$$R_x = R_s R_a / R_b.$$

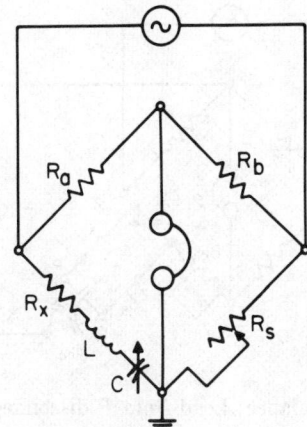

MAXWELL BRIDGE

$$L_x = R_a R_b C_s$$
$$R_x = R_a R_b / R_s$$
$$Q_x = \omega (L_x / R_x) = \omega C_s R_s.$$

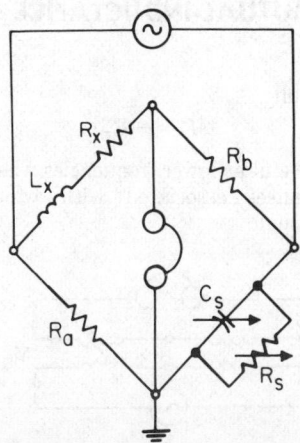

HAY BRIDGE

For measurement of large inductance.

$$L_x = R_a R_b C_s / (1 + \omega^2 C_s^2 R_s^2)$$
$$Q_x = \omega L_x / R_x = (\omega C_s R_s)^{-1}.$$

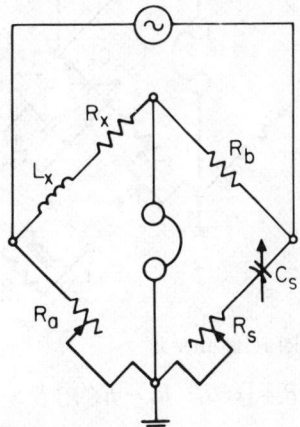

SCHERING BRIDGE

$$C_x = C_s R_b / R_a$$
$$1/Q_x = \omega C_x R_x = \omega C_b R_b.$$

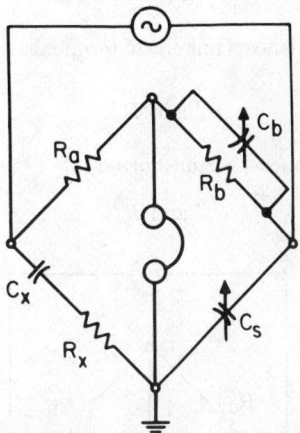

SUBSTITUTION METHOD FOR HIGH IMPEDANCES

Initial balance (unknown terminals x–x open):

$$C_s' \text{ and } R_s'.$$

Final balance (unknown connected to x–x):

$$C_s'' \text{ and } R_s''.$$

Then when $R_x > 10/\omega C_s'$, there results, with error < 1 percent

$$C_x = C_s' - C_s''.$$

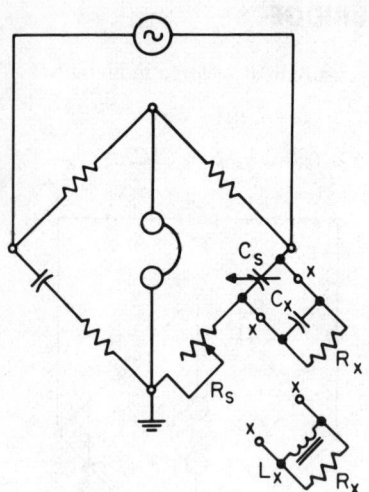

The parallel resistance is

$$R_x = [\omega^2 C_s'^2 (R_s' - R_s'')]^{-1}.$$

If unknown is an inductor

$$L_x = -(\omega^2 C_x)^{-1} = [\omega^2 (C_s'' - C_s')]^{-1}.$$

MEASUREMENT WITH CAPACITOR IN SERIES WITH UNKNOWN

Initial balance (unknown terminals x–x short-circuited):

$$C_s' \text{ and } R_s'.$$

Final balance (x–x unshorted):

$$C_s'' \text{ and } R_s''.$$

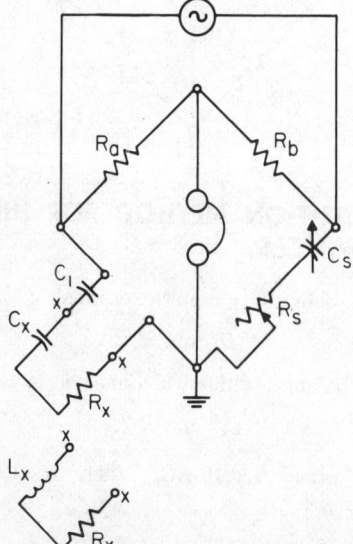

Then the series resistance is

$$R_x = (R_s'' - R_s') R_a / R_b$$

$$C_x = \frac{R_b C_s' C_s''}{R_a (C_s' - C_s'')}$$

$$= \frac{R_b}{R_a} C_s' \left(\frac{C_s'}{C_s' - C_s''} - 1 \right).$$

When $C_s'' > C_s'$

$$L_x = \frac{1}{\omega^2} \frac{R_a}{R_b C_s'} \left(1 - \frac{C_s'}{C_s''} \right).$$

MEASUREMENT OF DIRECT CAPACITANCE

Connection of N to N' places C_{nq} across phones, and C_{np} across R_b which requires only a small readjustment of R_s.

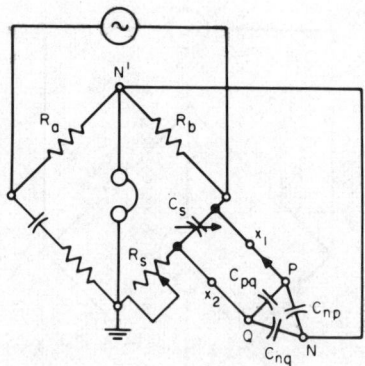

Initial balance: Lead from P disconnected from X_1 but lying as close to the connected position as practical.
Final balance: Lead connected to X_1.
By the substitution method above

$$C_{pq} = C_s' - C_s''.$$

FELICI MUTUAL-INDUCTANCE BALANCE

At the null

$$M_x = -M_s.$$

This is useful at lower frequencies where capacitive reactances associated with windings are negligibly small.

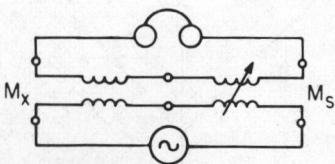

MUTUAL-INDUCTANCE CAPACITANCE BALANCE

Using low-loss capacitor, at the null

$$M_x = 1/\omega^2 C_s.$$

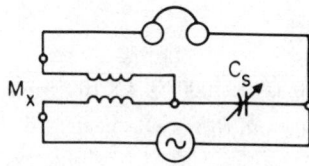

HYBRID-COIL METHOD

At the null

$$Z_1 = Z_2.$$

The transformer secondaries must be accurately matched and balanced to ground. This is useful at audio and carrier frequencies.

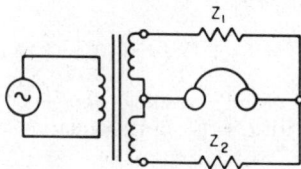

Q OF RESONANT CIRCUIT BY BANDWIDTH

For 3-decibel or half-power points. Source loosely coupled to circuit. Adjust frequency to each side of resonance, noting bandwidth when

$$v = 0.71 \times (v \text{ at resonance})$$
$$Q = (\text{resonance frequency})/(\text{bandwidth}).$$

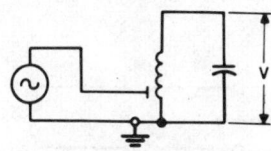

Q-METER (Boonton Radio Type 160A)

$R_1 = 0.04$ ohm
$R_2 = 100$ megohms
$V = $ vacuum-tube voltmeter
$I = $ thermal milliammeter
$L_x R_x C_0 = $ unknown coil plugged into COIL terminals for measurement.

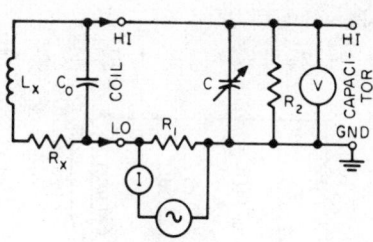

Correction of Q Reading

For distributed capacitance C_0 of coil

$$Q_{true} = Q[(C + C_0)/C]$$

where $Q = $ reading of Q-meter (corrected for internal resistors R_1 and R_2 if necessary), and $C = $ capacitance reading of Q-meter.

Measurement of C_0 and True L_x

C plotted against $1/f^2$ is a straight line.

$$L_x = \text{true inductance}$$
$$= \frac{1/f_2^2 - 1/f_1^2}{4\pi^2(C_2 - C_1)}$$

$C_0 = $ negative intercept
$f_0 = $ natural frequency of coil.

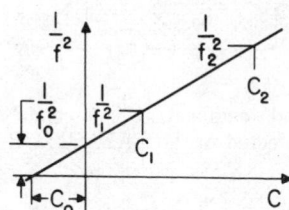

When only two readings are taken and $f_1/f_2 = 2.00$

$$C_0 = (C_2 - 4C_1)/3.$$

Using microhenries, megahertz, and picofarads

$$L_x = 19\,000/f_2^2(C_2 - C_1).$$

Measurement of Admittance

Initial reading $C'Q'$ (LR_p is any suitable coil):

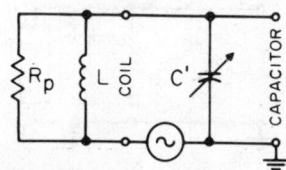

Final reading $C''Q''$:

$$1/Z = Y = G + jB = 1/R_p + j\omega C.$$

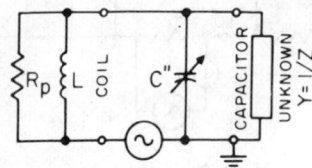

Then

$$C = C' - C''$$

$$1/Q = G/\omega C$$

$$= C'/C (1000/Q'' - 1000/Q') \times 10^{-3}.$$

If Z is inductive

$$C'' > C'.$$

Measurement of Impedances Lower Than Those Directly Measurable

For the initial reading $C'Q'$, CAPACITOR terminals are open.

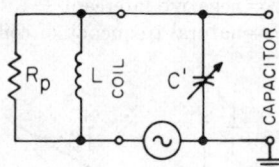

On second reading $C''Q''$, a capacitive divider $C_a C_b$ is connected to the CAPACITOR terminals.

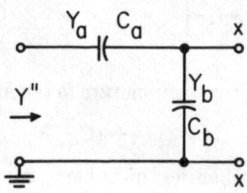

Final reading $C'''Q'''$, unknown connected to x-x.

$$Y_a = G_a + j\omega C_a \qquad Y_b = G_b + j\omega C_b$$

with G_a and G_b not shown in diagrams.

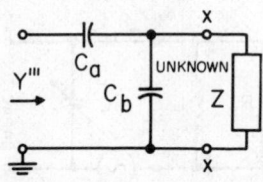

Then the unknown impedance is

$$Z = [Y_a/(Y_a + Y_b)]^2 (Y''' - Y'')^{-1}$$
$$- (Y_a + Y_b)^{-1} \text{ ohms}$$

where, with capacitance in picofarads and $\omega = 2\pi \times$ frequency in megahertz

$$(Y''' - Y'')^{-1}$$
$$= \frac{10^6/\omega}{C'(1000/Q''' - 1000/Q'') \times 10^{-3} + j(C'' - C''')}.$$

Usually G_a and G_b may be neglected, then there results

$$Z = \left(\frac{1}{1 + C_b/C_a}\right)^2 (Y''' - Y'')^{-1}$$
$$+ j\frac{10^6}{\omega(C_a + C_b)} \text{ ohms.}$$

For many measurements, C_a may be 100 picofarads. $C_b = 0$ for very low values of Z and for highly reactive values of Z. For unknowns that are principally resistive and of low or medium value, C_b may take sizes up to 300 to 500 picofarads. When $C_b = 0$

$$Z = (Y''' - Y'')^{-1} + j(10^6/\omega C_a) \text{ ohms}$$

and the "second" reading above becomes the "initial," with $C' = C''$ in the equations.

Measurement of Coupling Coefficient of Loosely Coupled Coils

The coefficient of coupling

$$k = M/(L_1 L_2)^{1/2}$$

between two high-Q coils can be obtained by measuring the inductance L with S_1 closed and again with S_1 open. From these two measurements

$$k = (1 - L_{closed}/L_{open})^{1/2}.$$

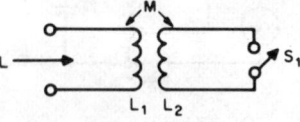

When the coil self-inductances are known, a measurement of L_a and L_b yields

$$k = (L_a - L_b)/4(L_1 L_2)^{1/2}.$$

If $L_1 = L_2$

$$k = (L_a - L_b)/(L_a + L_b).$$

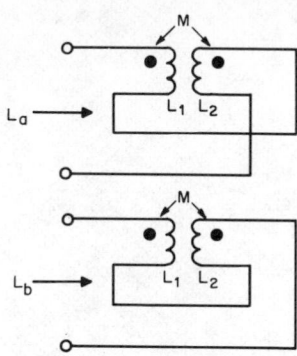

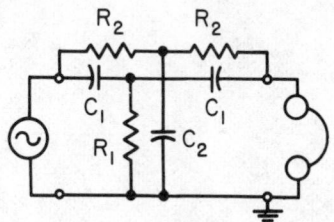

When used as a frequency-selective network, if we make $R_2 = 2R_1$ and $C_2 = 2C_1$, then

$$f = 1/2\pi C_1 R_2 = 1/2\pi C_2 R_1.$$

For additional information, see G. E. Valley, Jr. and H. Wallman, "Vacuum Tube Amplifiers," McGraw-Hill Book Company, Inc., New York, N.Y.; 1948; pp. 387–389.

Neither of the above methods provides adequate precision when two high-Q coils are only about critically coupled. In that case, the Q of each coil is measured with the other coil open-circuited. Then the coupled coils (L_1R_1 and L_2R_2) and a low-loss adjustable capacitor (C_2) are connected to the Q-meter as shown, C_2 is disconnected, and C is adjusted to maximize the Q-meter reading; C_2 is then connected and adjusted to minimize the reading. If the final reading is Q_0

$$K = [1/Q_2(1/Q_0 - 1/Q_1)]^{1/2}.$$

If the final reading is too small to be read accurately, Q_2 may be reduced by inserting a small resistance in series with L_2R_2.

TWIN-T ADMITTANCE-MEASURING CIRCUIT (General Radio Type 821-A)

This circuit may be used for measuring admittances in the range somewhat exceeding 400 kilohertz to 40 megahertz. It is applicable to the special measuring techniques described above for the Q-meter.

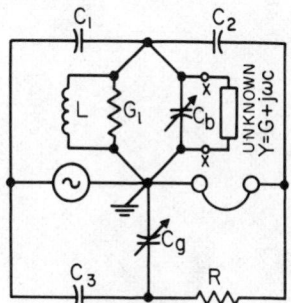

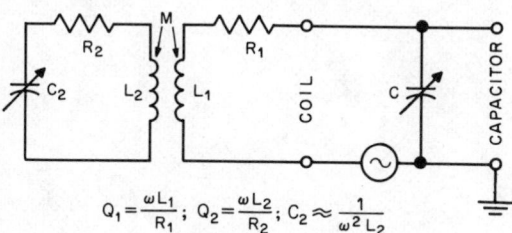

$$Q_1 = \frac{\omega L_1}{R_1}; \quad Q_2 = \frac{\omega L_2}{R_2}; \quad C_2 \approx \frac{1}{\omega^2 L_2}$$

PARALLEL-T (SYMMETRICAL)

Conditions for zero transfer are

$$\omega^2 C_1 C_2 = 2/R_2^2$$

$$\omega^2 C_1^2 = 1/2R_1 R_2$$

$$C_2 R_2 = 4C_1 R_1.$$

Conditions for null in output

$$G + G_l = R\omega^2 C_1 C_2 (1 + C_g/C_3)$$

$$C + C_b = 1/\omega^2 L - C_1 C_2 (1/C_1 + 1/C_2 + 1/C_3).$$

With the unknown disconnected, call the initial balance C_b' and C_g'. With unknown connected, final balance is C_b'' and C_g''. Then the components of the unknown $Y = G + j\omega C$ are

$$C = C_b' - C_b''$$

$$G = (R\omega^2 C_1 C_2/C_3)(C_g'' - C_g').$$

CHAPTER 13

MAGNETIC-CORE TRANSFORMERS AND REACTORS

MAGNETIC-CORE TRANSFORMERS

Magnetic-core transformers, with few exceptions, are closely coupled circuits for transmitting alternating-current energy and for matching impedances between the generator and the load. The major types of transformers follow.

Power transformers (25, 50, 60, 400 Hz): plate, filament, isolation, vibrator, auto.

Audio-frequency transformers (20–20 000 Hz): input, output, interstage, matching, driver, modulation

High-frequency transformers (2–1000 kHz): input, output, matching, modulation

Pulse transformers (repetition rates up to 4 MHz): isolation, step-up, step-down, blocking oscillator

The equivalent circuit of a generalized transformer is shown in Fig. 1.

POWER TRANSFORMERS

Power transformers operate from a nearly zero source impedance at a single low frequency, primarily to transfer power at convenient voltages.

The magnetic cores used for these types of transformers are usually Type E and I laminations of 4% silicon steel. Grain-oriented silicon-steel Type C cores are also used. They are wound in tape form on a rectangular mandrel, impregnated, and cut into halves. These halves are then banded around the wound transformer coil. Table 1 gives typical operating conditions for these types of cores.

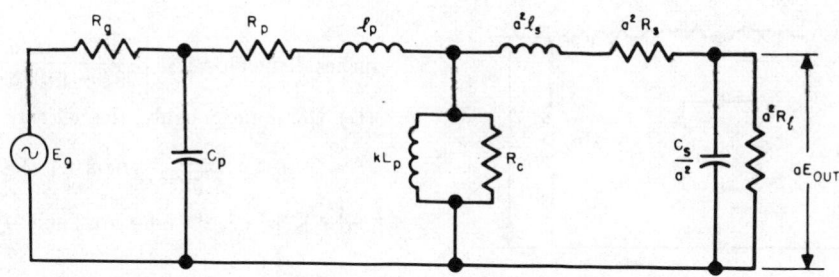

Fig. 1—Equivalent network of a transformer.

a = turns ratio = N_p/N_s

C_p = primary equivalent shunt capacitance

C_s = secondary equivalent shunt capacitance

E_g = root-mean-square generator voltage

E_{out} = root-mean-square output voltage

k = coefficient of coupling

L_p = primary inductance

l_p = primary leakage inductance

l_s = secondary leakage inductance

R_c = core-loss equivalent shunt resistance

R_g = generator impedance

R_l = load impedance

R_p = primary-winding resistance

R_s = secondary-winding resistance

TABLE 1—TYPICAL OPERATING CONDITIONS FOR CORE MATERIALS AT VARIOUS FREQUENCIES.

Frequency in Hertz	Lamination Thickness in Inches	Core Material	Core Flux Density B_m in Gauss	Approximate Core Loss in Watts/Lb	Approximate Exciting (VA)/Lb
25	0.025	2.5-percent silicon	14 000	0.65	4.0
60	0.014	4-percent silicon	12 000	1.0	6.0
60	0.014	Grain-oriented silicon	15 000	1.0	6.0
400	0.004	Grain-oriented silicon	10 000	4.5	10.0
800	0.004	Grain-oriented silicon	6 000	4.5	10.0

DESIGN OF POWER TRANSFORMERS FOR RECTIFIERS USING E AND I LAMINATIONS

The equivalent circuit of a power transformer for a vacuum-tube circuit is shown in Fig. 2.

(A) Determine total output volt-amperes and compute the primary and secondary currents from

$$E_p I_p \times 0.9 = (1/\eta) \left[(E_s I_{dc})_{pl} K + (EI)_{fil} \right]$$

$$I_s = K' I_{dc}$$

where the numeric 0.9 is the power factor, and the efficiency η and the K, K' factors are listed in Tables 2 and 3. $E_p I_p$ is the input volt-amperes, I_{dc} refers to the total direct-current component drawn by the supply, and the subscripts pl and fil refer to the volt-amperes drawn from the plate-supply and filament-supply (if present) windings,

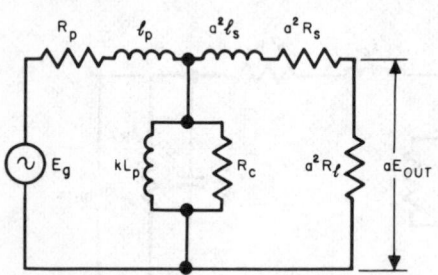

Fig. 2—Equivalent network of a power transformer. l_p and l_s may be neglected when there are no strict requirements on voltage regulation.

TABLE 2—FACTORS K AND K' FOR SINGLE-PHASE-RECTIFIER SUPPLIES.

Filter	K	K'
Full-wave:		
Capacitor input	0.707	1.06
Reactor input	0.5	0.707
Half-wave:		
Capacitor input	1.4	2.2
Reactor input	1.06	1.4

respectively. E_s is the total voltage across the secondary of the transformer.

$$E_s = 2.35 E_{dc}$$

for a single-phase full-wave rectifier. E_{dc} is the direct-current output voltage of the rectifier. Factor 2.35 is twice the ratio of root-mean-square to average values plus an allowance for 5-percent regulation.

(B) Compute the size of wire of each winding, on the basis of current densities given by the following.

For 60-hertz sealed units

$$\text{amperes/inch}^2 = 2470 - 585 \log W_{out}$$

or

$$\text{inches diameter} \approx 1.13 \left[\frac{I \text{ (in amperes)}}{2470 - 585 \log W_{out}} \right]^{1/2}.$$

For 60-hertz open units (uncased)

$$\text{amperes/inch}^2 = 2920 - 610 \log W_{out}$$

or

$$\text{inches diameter} \approx 1.13 \left[\frac{I \text{ (in amperes)}}{2920 - 610 \log W_{out}} \right]^{1/2}.$$

(C) Compute roughly, the net core area

$$A_c = \frac{(W_{out})^{1/2}}{5.58} (60/f)^{1/2} \text{ inches}^2$$

where f is in hertz (refer to Table 4). Select a

TABLE 3—EFFICIENCY OF VARIOUS SIZES OF POWER SUPPLIES.*

Output in Watts	Approximate Efficiency in Percent
20	70
30	75
40	80
80	85
100	86
200	90

* From "Radio Components Handbook," Technical Advertising Associates, Cheltenham, Pa.; May 1948: p. 92.

TABLE 4—Equivalent *EI* Ratings of Power Transformers. B_m = flux density in gauss. *EI* rating = volt-ampere capability at 60 and 400 hertz for various core sizes. Ratings are based on a rise above ambient of 50° C. These values should be reduced to obtain lower temperature rise when more than two windings are required or when extra-high-voltage operation is required. *Adapted from "Radio Components Handbook," Technical Advertising Associates, Cheltenham, Pa.; May 1948: p. 92.*

At 60 Hertz		At 400 Hertz		Current Density (amps/sq m)	EI-Type Lamination†	Tongue Width of Lamination g (in.)	Stack Height p (in.)	Average Copper Mean Length per Turn (MLT) (in.)	Core Weight (lb)
EI	B_m*	EI	B_m*						
3.9	14 000	9.5	5 000	3 200	EI-21	0.5	0.5	3.12	0.199
5.8	14 000	15.0	4 900	2 700	EI-625	0.625	0.625	3.62	0.361
13.0	14 000	30.0	4 700	2 560	EI-75	0.75	0.75	4.33	0.609
17.0	14 000	38.0	4 600	2 560	EI-75	0.75	1.00	4.83	0.812
24.0	13 500	50.0	4 500	2 330	EI-11	0.875	0.875	5.04	0.966
37.0	13 000	80.0	4 200	2 130	EI-12	1.00	1.00	5.71	1.43
54.0	13 000	110	4 000	2 030	EI-12	1.00	1.50	6.71	2.14
82.0	12 500	180	3 900	1 800	EI-125	1.25	1.25	7.21	2.83
110	12 000	230	3 900	1 770	EI-125	1.25	1.75	8.21	3.97
145	12 000	325	3 700	1 600	EI-13	1.50	1.50	8.63	4.92
195	11 000	420	3 500	1 500	EI-13	1.50	2.00	9.63	6.56
525	10 500	1 100	3 200	1 220	EI-19	1.75	1.75	12.8	9.75

* Refers to silicon steel 0.014 inch thick.

† Lamination designation and constants per Allegheny Ludlum Corp., Pittsburgh, Pa.

Table 5—Data on Metallic Core Materials.

Metal or Alloy	Material or Trade Name	Composition in Percent (remainder is iron)	Characteristic Property or Application	Permeability		Direct-Current Saturation in Kilogauss	Residual Induction in Kilogauss	Coercive Force in Oersteds	Resistivity in $\mu\Omega$-Centimeters	Curie Temperature in Degrees Celsius
				Initial	Maximum					
Silicon–iron*	Silicon–Iron	4 Si	Transformer	400	7 000	20	12	0.5	60	690
	Hypersil Trancor 3X Silectron	3.5 Si	Grain oriented	1 500	35 000	20	13.7	0.1 to 0.3	50	750
	Sendust	9.5 Si, 5.5 Al	High frequency, powder	30 000	120 000	10	5	0.05	80	—
Cobalt–iron*	Hyperco Permendur 2V	35 Co, 0.5 Cr 49 Co, 2 V	High saturation	650 800	10 000 4 500	24 24	>13 14	>1 2	28 25	970 980
Nickel–iron*	Perminvar 45–25 Perminvar 7–70 Conpernik	45 Ni, 25 Co 70 Ni, 7 Co 50 Ni	"Constant" permeability	400 850 1 500	2 000 4 000 2 000	15.5 12.5 16	3.3 2.4 —	1.2 0.6 —	20 15 45	715 650 —
	Isoperm 36 Isoperm 50	36 Ni, 9 Cu 50 Ni	High frequency	60 90	65 100	— 16	— —	— —	70 40	300 500
	Permalloy 45 Allegheny 4750 Armco 48 Nicaloi High Perm 49 Hipernik	45 Ni 47 to 50 Ni 48 Ni 49 Ni 49 Ni 50 Ni, Si, Mn	Combine good permeability and flux density	2 700 9 000 — — 5 000 4 000	23 000 50 000 — — 50 000 100 000	16.5 16 16 16 16 16	8 6.2† — — 6.5 8†	0.3 0.08† — — 0.03 0.03†	45 52 — — 43 45	440 430 — — 475 500
	Monimax Sinimax	47 Ni, 3 Mo 42 Ni, 3 Si	High resistivity	2 000 3 500	38 000 30 000	15 11	— —	0.06 0.1	80 90	390 290

Material	Composition	Remarks						
Permenorm 5000Z Permenite Deltamax Hypernik V Orthonik Orthonol	45 to 50 Ni		400 to 1 700	40 000 to 100 000	15.5 to 16	13 to 15	0.2 to 0.4	40 to 50	450 to 500
Permalloy 65	65 to 68 Ni	Rectangular hysteresis loop	1 500	250 000 to 600 000	13	13	0.03	20	600
Alloy 1040	72 Ni, 14 Cu, 3 Mo		40 000	100 000	6	2.5	0.02	55	290
Mumetal	77 Ni, 5 Cu, 2 Cr		20 000	100 000	8	6	0.05	60	400
Permalloy 78	78 Ni, 0.6 Mn		9 000	100 000	10.7	6	0.05	16	580
Mo-Permalloy 4-79	79 Ni, 4 Mo		20 000	75 000	8	5.5	0.05	55	—
Supermalloy	79 Ni, 5 Mo	Highest permeability, low saturation	55 000 to 150 000	500 000 to 1 000 000	6.8 to 7.8	—	0.002 to 0.05	65	400
Hymu 80	80 Ni		10 000	100 000	8	—	0.06	58	460
Ferrites‡								
3C3	MnZi	High-frequency transformers	2 200±20%		4.6	3.5	0.1	60×10⁶	150
3B7 and 3B9	MnZi	High-Q coils	2 300±20%		4.6	3.0	0.2	100×10⁶	170
3D3	NiZi	High-frequency	750±20%		4.7	3.0	0.5	150×10⁶	150
4C4	NiZi	High-Q coils	125±20%		4.1	2.0	4.5	10×10⁹	300

* Reprinted by permission from S. R. Hoh, "Evaluation of High-Performance Core Materials (Part 1)," *Tele-Tech and Electronic Industries*, vol. 12, pp. 86–89, 154–156; October 1953.

† $B_m = 10\,000$ gauss.

‡ Data furnished by Ferroxcube Corporation of America, Saugerties, N.Y.

Note 1—The table shows characteristics as listed by the manufacturers. The parameters of different lots of material may vary considerably from the above values. In the cases of residual induction and coercive force, the difference may amount to 50 percent.

Note 2—For information on ferrite materials, see Table 23 in Chapter 4, Properties of Materials.

lamination and core size from the manufacturer's data book that will nearly meet the space requirements, and provide core area for a flux density B_m not to exceed the values shown in Table 1. Further information on available core materials is given in Table 5.

(**D**) Compute the primary turns N_p from the transformer equation

$$N_p = \frac{3.49 E_p \times 10^6}{f A_c B_m}$$

with A_c in square inches and B_m in gauss. Then the secondary turns

$$N_s = 1.05 (E_s/E_p) N_p$$

(this allows 5 percent for IR drop of windings).

(**E**) Calculate the number of turns per layer that can be placed in the lamination window space, deducting from the latter the margin space given in Table 6 (see also Fig. 3).

(**F**) From (**D**) and (**E**) compute the number of layers n_l for each winding. Use interlayer insu-

TABLE 6—WIRE TABLE FOR TRANSFORMER DESIGN. The resistance R_T at any temperature T is given by $R_T = (234.5 + T)/(234.5 + t) \times r$, where $t =$ reference temperature of winding, and $r =$ resistance of winding at temperature t, all in degrees Celsius. Additional data on wire will be found in Chapter 4, Properties of Materials, and Chapter 6, Fundamentals of Networks.

AWG B & S Gauge	Diameter in Inches			Turns per Inch (Formvar)	Space Factor	Ohms per 1000 Ft†	Pounds per 1000 Ft	Margin m in Inches	Interlayer Insulation Thickness t(in.)‡
	Bare	Single Formvar*	Double Formvar						
10	0.1019	0.1039	0.1055	8	90	0.9989	31.43	0.25	0.010K
11	0.0907	0.0927	0.0942	9	90	1.260	24.92	0.25	0.010K
12	0.0808	0.0827	0.0842	10	90	1.588	19.77	0.25	0.010K
13	0.0719	0.0738	0.0753	12	90	2.003	15.68	0.25	0.010K
14	0.0641	0.0659	0.0673	13	90	2.525	12.43	0.25	0.010K
15	0.0571	0.0588	0.0602	15	90	3.184	9.858	0.25	0.010K
16	0.0508	0.0524	0.0538	17	90	4.016	7.818	0.1875	0.010K
17	0.0453	0.0469	0.0482	19	90	5.064	6.200	0.1875	0.007K
18	0.0403	0.0418	0.0431	21	90	6.385	4.917	0.1875	0.007K
19	0.0359	0.0374	0.0386	23	90	8.051	3.899	0.1562	0.007K
20	0.0320	0.0334	0.0346	26	90	10.15	3.092	0.1562	0.005K
21	0.0285	0.0299	0.0310	30	90	12.80	2.452	0.1562	0.005K
22	0.0253	0.0266	0.0277	33	90	16.14	1.945	0.125	0.003K
23	0.0226	0.0239	0.0249	37	90	20.36	1.542	0.125	0.003K
24	0.0201	0.0213	0.0223	42	90	25.67	1.223	0.125	0.002G
25	0.0179	0.0190	0.0200	47	90	32.37	0.9699	0.125	0.002G
26	0.0159	0.0169	0.0179	52	89	40.81	0.7692	0.125	0.002G
27	0.0142	0.0152	0.0161	57	89	51.47	0.6100	0.125	0.002G
28	0.0126	0.0135	0.0145	64	89	64.90	0.4837	0.125	0.0015G
29	0.0113	0.0122	0.0131	71	89	81.83	0.3836	0.125	0.0015G
30	0.0100	0.0109	0.0116	80	89	103.2	0.3042	0.125	0.0015G
31	0.0089	0.0097	0.0104	88	88	130.1	0.2413	0.125	0.0015G
32	0.0080	0.0088	0.0094	98	88	164.1	0.1913	0.0937	0.0013G
33	0.0071	0.0079	0.0084	110	88	206.9	0.1517	0.0937	0.0013G
34	0.0063	0.0070	0.0075	124	88	260.9	0.1203	0.0937	0.001G
35	0.0056	0.0062	0.0067	140	88	329.0	0.0954	0.0937	0.001G
36	0.0050	0.0056	0.0060	155	87	414.8	0.0757	0.0937	0.001G
37	0.0045	0.0050	0.0054	170	87	523.1	0.0600	0.0937	0.001G
38	0.0040	0.0045	0.0048	193	87	659.6	0.0476	0.0625	0.001G
39	0.0035	0.0040	0.0042	215	86	831.8	0.0377	0.0625	0.0007G
40	0.0031	0.0036	0.0038	239	86	1049	0.0299	0.0625	0.0007G

* Dimensions very nearly the same as for enamelled wire.

† Values are at 20 degrees Celsius.

‡ $K =$ kraft paper, $G =$ glassine.

lation of thickness t as given in Table 6, except that the voltage stress should be limited to 40 volts/mil.

(**G**) Calculate the coil-build a

$$a = 1.1[n_l(D+t) - t + t_c]$$

for each winding from (**B**) and (**F**), where $D=$ diameter of insulated wire, and $t_c=$ thickness of insulation under and over the winding; the numeric 1.1 allows for a 10-percent bulge factor. The total coil-build should not exceed 85–90 percent of the window width. (Note: Insulation over the core may vary from 0.025 to 0.050 inch for core-builds of $\frac{1}{2}$ to 2 inches.)

(**H**) Compute the mean length per turn (MLT), of each winding, from the geometry of core and windings as shown in Fig. 4. Compute length of each winding N(MLT).

$$(\text{MLT})_1 = 2(r+J) + 2(s+J) + \pi a_1$$

$$(\text{MLT})_2 = 2(r+J) + 2(s+J) + \pi(2a_1+a_2)$$

where $a_1=$ build of first winding, $a_2=$ build of second

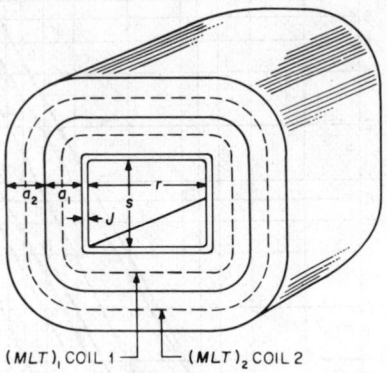

(MLT)₁ COIL 1   (MLT)₂ COIL 2

Fig. 4—Dimensions relating to coil mean length of turn (MLT).

winding, $J=$ thickness of winding form, and $r,s=$ winding-form dimensions.

(**I**) Calculate the resistance of each winding from (**H**) and Table 6 and determine IR drop and I^2R loss for each winding.

(**J**) Make corrections, if required, in the number of turns of the windings to allow for the IR drops, so as to have the required E_s.

$$E_s = (E_p - I_p R_p) N_s/N_p - I_s R_s.$$

(**K**) Compute core losses from weight of core and Table 1 or Fig. 5.

(**L**) Determine the percent efficiency η and voltage regulation (vr) from

$$\eta = \frac{W_{\text{out}} \times 100}{W_{\text{out}} + (\text{core loss}) + (\text{copper loss})}$$

$$(\text{vr}) = \frac{I_s[R_s + (N_s/N_p)^2 R_p]}{E_s}.$$

(**M**) For a more accurate evaluation of voltage regulation, determine leakage-reactance drop$=$ $I_{\text{dc}}\omega l_{\text{sc}}/2\pi$, and add to the above (vr) the value of $(I_{\text{dc}}\omega l_{\text{sc}})/2\pi E_{\text{dc}}$. Here, $l_{\text{sc}}=$ leakage inductance viewed from the secondary; see "Methods of Winding Transformers" (in this chapter) to evaluate l_{sc}.

(**N**) Bring out all terminal leads using the wire of the coil, insulated with suitable sleevings, for all sizes of wire heavier than 21; and by using 7-30 stranded and insulated wire for smaller sizes.

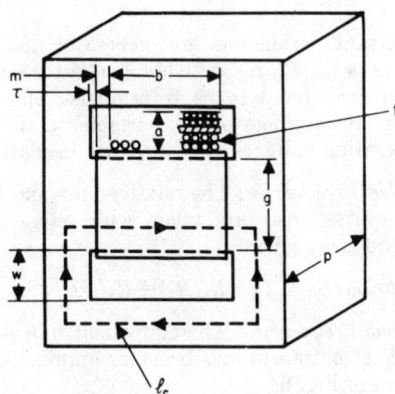

Fig. 3—Dimensions relating to the design of a transformer coil-build and core.

$A_c=$ core area $= (gp)k$

$a=$ height of coil

$\quad=$ coil-build

$b=$ coil width

$g=$ width of lamination tongue

$l_c=$ average length of magnetic-flux path

$k=$ stacking factor

$\quad \approx 0.90$ for 14-mil lamination

$\quad \approx 0.80$ for 2-mil lamination or ribbon-wound core

$m=$ marginal space given in Table 6

$p=$ height of lamination stack

$t=$ thickness of interlayer insulation

$w=$ width of core window

$\tau=$ window length tolerance

$\quad = 1/16$ inch, total.

Effect of Power Frequency on Design

Design procedure is similar to that described above for 60-hertz transformers except for the flux density at which the core is operated. Operation at lower frequencies requires a larger core (see equation in paragraph (**C**) above) although reduction of core loss partly compensates the size

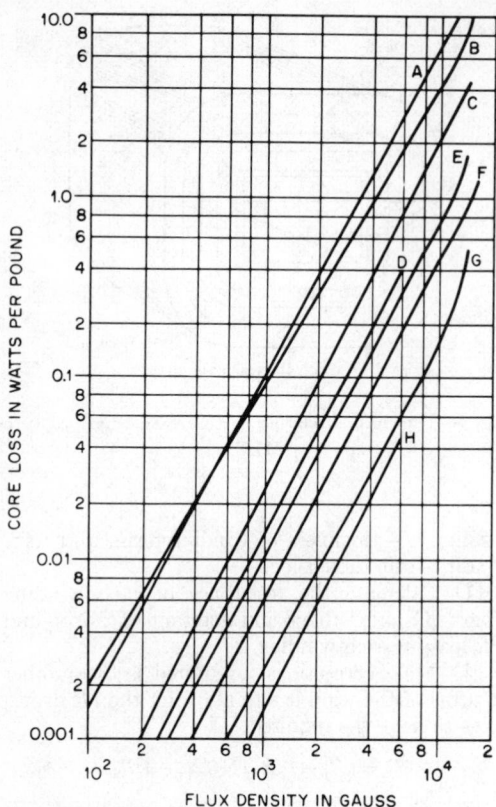

Fig. 5—Core loss characteristics.

A–400-hertz 0.014-inch grain-oriented silicon.

B–400-hertz 0.004-inch grain-oriented silicon.

C–400-hertz 0.006-inch low nickel.

D–400-hertz 0.006-inch high nickel.

E–60-hertz 0.014-inch 4% silicon.

F–60-hertz 0.012-inch grain-oriented silicon.

G–60-hertz 0.014-inch low nickel.

H–60-hertz 0.014-inch high nickel.

increase. As an example, a 25-hertz transformer is approximately twice as large as its 60-hertz equivalent.

Effect of Duty Cycle on Design

If a transformer is operated at different loads according to a regular duty cycle, the equivalent volt-ampere (VA) rating is

$$(VA)_{eq}$$

$$= \left[\frac{(VA)_1^2 t_1 + (VA)_2^2 t_2 + (VA)_3^2 t_3 + \cdots (VA)_n^2 t_n}{t_1 + t_2 + t_3 + \cdots t_n} \right]^{1/2}$$

where $(VA)_1 =$ output during time (t_1), etc.

Example: 5 kilovolt-ampere output, 1 minute on, 1 minute off.

$$(VA)_{eq} = \left[\frac{(5000)^2(1) + (0)^2(1)}{1+1} \right]^{1/2} = \left[\frac{(5000)^2}{2} \right]^{1/2}$$

$$= 5000/(2)^{1/2} = 3535 \text{ volt-amperes.}$$

AUDIO-FREQUENCY TRANSFORMERS

Audio-frequency transformers are used mainly for matching impedances and transmitting audio frequencies. They also provide isolation from direct currents and present balanced impedances to lines or circuits.

The magnetic core for this type of transformer is usually an E–I type using either audio-grade silicon steel or nickel-alloy steel (refer to Table 5). High-permeability nickel-alloy tape cores in toroidal form are used for extreme bandwidths.

DESIGN OF AUDIO-FREQUENCY TRANSFORMERS

Important parameters are: generator and load impedances R_g, R_l, respectively, generator voltage E_g, frequency band to be transmitted, efficiency (output transformers only), harmonic distortion, and operating voltages (for adequate insulation).

At Mid-Frequencies: The relative low- and high-frequency responses are taken with reference to mid-frequencies where

$$aE_{out}/E_g = \left[(1 + R_s/R_l) + R_1/a^2 R_l \right]^{-1}.$$

At Low Frequencies: The equivalent unity-ratio network of a transformer becomes approximately as shown in Fig. 6.

$$\text{Amplitude} = 1/\left[1 + (R'_{par}/X_m)^2 \right]^{1/2}$$

$$\text{Phase angle} = \tan^{-1}(R'_{par}/X_m)$$

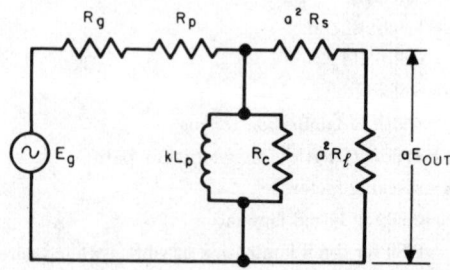

Fig. 6—Equivalent network of an audio-frequency transformer at low frequencies. $R_1 = R_g + R_p$, and $R_2 = R_s + R_l$. In a good output transformer, R_p, R_s, and R_c may be neglected. In input or interstage transformers, R_c may be omitted.

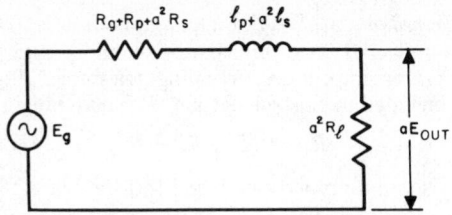

Fig. 7—Equivalent network of an audio-frequency transformer at high frequencies, neglecting the effect of the winding shunt capacitances.

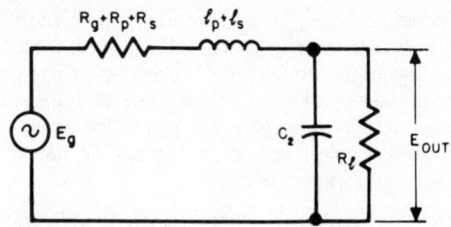

Fig. 9—Equivalent network of a 1:1-turns-ratio audio-frequency transformer at high frequencies when effect of winding shunt capacitances is appreciable. In a step-up transformer, C_2 = equivalent shunt capacitances of both windings. In a step-down transformer, C_2 shunts both leakage inductances and R_2.

where

$$R'_{par} = (R_1 R_2 a^2)/(R_1 + R_2 a^2), \quad R_1 = R_g + R_p,$$

$$R_2 = R_l + R_s, \quad X_m = 2\pi f L_p.$$

At High Frequencies: Neglecting the effect of winding and other capacitances (as in low-impedance-level output transformers), the equivalent unity-ratio network becomes approximately as in Fig. 7.

$$\text{Amplitude} = [1 + (X_l/R'_{se})^2]^{-1/2}$$

$$\text{Phase angle} = \tan^{-1}(X_l/R'_{se})$$

where

$$R'_{se} = R_1 + R_2 a^2, \quad X_l = 2\pi f l_{scp},$$

and l_{scp} = inductance measured across primary with secondary short-circuited = $l_p + a^2 l_s$.

These low- and high-frequency responses are shown on the curves of Fig. 8.

If at high frequencies the effect of winding and other capacitances is appreciable, the equivalent network on a 1:1-turns-ratio basis becomes as shown in Fig. 9. The relative high-frequency response of this network is given by

$$\frac{(R_1 + R_2)/R_2}{[(R_1/X_c + X_l/R_l)^2 + (X_l/X_c - R_g/R_l - 1)^2]^{1/2}}.$$

This high-frequency response is plotted in Figs. 10 and 11 for $R_1 = R_2$ (matched impedances), and $R_2 = \infty$ (input and interstage transformers) based on simplified equivalent networks as indicated.

Harmonic distortion requirements may constitute a deciding factor in the design of transformers. Such distortion is caused by either variations in load impedance or nonlinearity of magnetizing current. The percent harmonic voltage appearing in the output of a loaded transformer is given by

(percent harmonics)

$$= E_h/E_f = I_h/I_f (R'_{par}/X_m)[1 - (R'_{par}/4X_m)]$$

where $100 I_h/I_f$ = percent of harmonic current measured with zero-impedance source (values in Table 7 are for 4-percent silicon-steel core).*

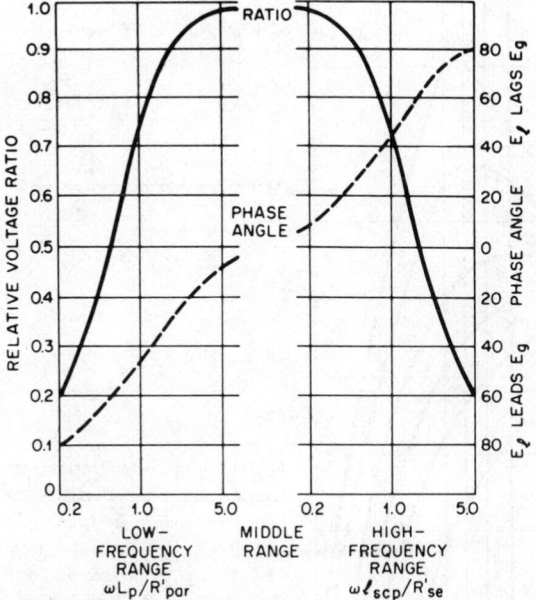

Fig. 8—Universal frequency and phase response of output transformers. *Courtesy McGraw-Hill Publishing Co.*

TABLE 7—HARMONICS PRODUCED BY VARIOUS FLUX DENSITIES B_m IN A 4-PERCENT SILICON-STEEL-CORE AUDIO TRANSFORMER.

B_m	Percent 3rd Harmonic	Percent 5th Harmonic
100	4	1.0
500	7	1.5
1 000	9	2.0
3 000	15	2.5
5 000	20	3.0
10 000	30	5.0

* N. Partridge, "Harmonic Distortion in Audio-Frequency Transformers," *Wireless Engineer*, vol. 19; September, October, and November, 1942.

Insertion Loss: Loss introduced in circuit by addition of transformer. At midband, loss is caused by winding resistance and core loss. Frequency discrimination adds to this at low and high frequencies. Insertion loss is input divided by output expressed in decibels or (in terms of measured voltages and impedance)

$$(\text{dB insertion loss}) = 10 \log\left[(E_g{}^2 R_l) / (4 E_g{}^2 R_g) \right].$$

Impedance Match: For maximum power transfer, the reflected load impedance should equal the generator impedance. Winding resistance should be included in this calculation: For matching

$$R_g = a^2 (R_l + R_s) + R_p.$$

Also, in a properly matched transformer

$$R_g = a^2 R_l = (Z_{oc} \times Z_{sc})^{1/2}$$

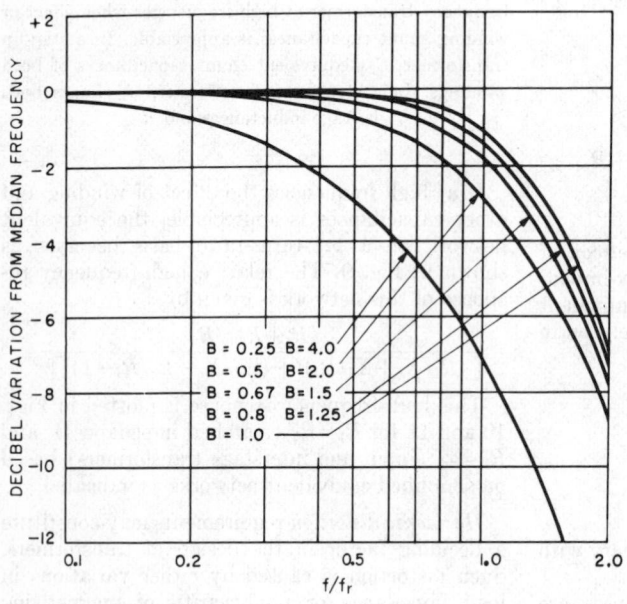

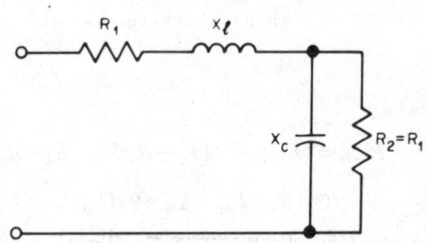

Fig. 10—Transformer characteristics at high frequencies for matched impedances. At frequency f_r, $X_l = X_c$ and $B = X_c / R_1$. Reprinted from "*Electronic Transformers and Circuits*," by R. Lee, 2nd ed., p. 151, 1955; by permission, John Wiley & Sons, New York.

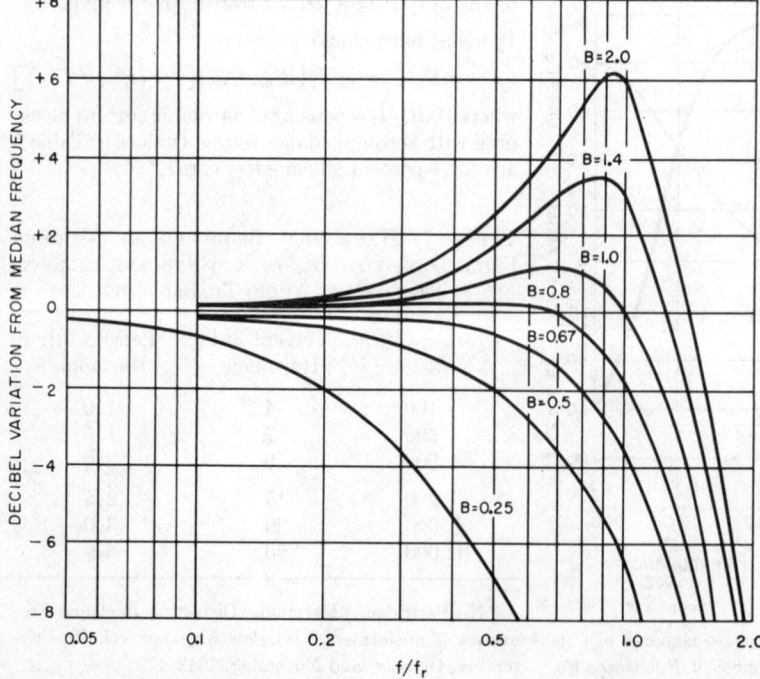

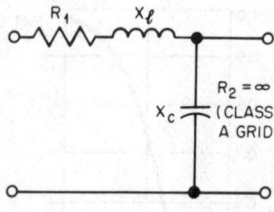

Fig. 11—Input- or interstage-transformer characteristics at high frequencies. At f_r, $X_l = X_c$ and $B = X_c / R_1$. Reprinted from "*Electronic Transformers and Circuits*," by R. Lee, 2nd ed., p. 153, 1955; by permission, John Wiley & Sons, New York.

where Z_{oc} = transformer primary open-circuit impedance, and Z_{sc} = transformer primary impedance with secondary winding short-circuited.

If more than one secondary is used, the turns ratio to match impedances properly depends on the power delivered from each winding.

$$N_s/N_p = (R_n/R_g \times W_n/W_p)^{1/2}.$$

Example: Using Fig. 12

$$N_2/N_p = (10/600 \times 10/16)^{1/2} = 0.102$$

$$N_3/N_p = (50/600 \times 5/16)^{1/2} = 0.161$$

$$N_4/N_p = (100/600 \times 1/16)^{1/2} = 0.102.$$

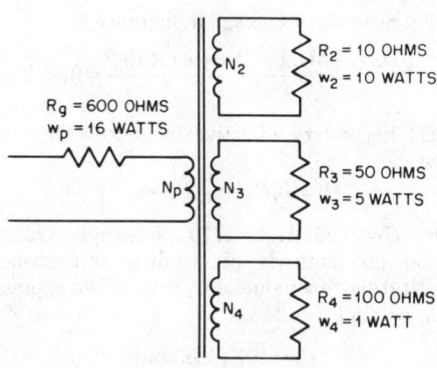

Fig. 12—Multisecondary audio transformer.

Example of Audio-Output-Transformer Design

This transformer is to operate from a 4000-ohm impedance; to deliver 5 watts to a matched load of 10 ohms; to transmit frequencies of 60 to 15 000 hertz with a V_{out}/V_{in} ratio of 71 percent of that at mid-frequencies (400 hertz); and the harmonic distortion is to be less than 2 percent. (See Figs. 6 and 7.)

(**A**) We have

$$E_s = (W_{out}R_l)^{1/2} = 7.1 \text{ volts}$$

$$I_s = W_{out}/E_s = 0.7 \text{ ampere}$$

$$a = (R_g/R_l)^{1/2} = 20.$$

Then

$I_p \approx 1.1 I_s/a = 0.039$ ampere, and $E_p \approx 1.1 a E_s = 156$.

(**B**) To evaluate the required primary inductance to transmit the lowest frequency of 60 hertz, determine

$$R'_{se} = R_1 + a^2 R_2 \text{ and } R'_{par} = (R_1 R_2 a^2)/(R_1 + R_2 a^2)$$

where $R_1 = R_g + R_p$, and $R_2 = R_l + R_s$. We choose winding resistances $R_s = R_p/a^2 \approx 0.05 R_1 = 0.5$ (for a copper efficiency = $R_l a^2 \times 100/[(R_1+R_s)a^2 + R_p] = 91$ percent). Then $R'_{se} = 2R_1 = 8400$ ohms, and $R'_{par} = R_1/2 = 2100$ ohms.

(**C**) To meet the frequency-response requirements, we must have according to Fig. 8

$$\omega_{low}L_p/R'_{par} = 1 = \omega_{high}l_{scp}/R'_{se}$$

which yield $L_p \approx 5.6$ henries and $l_{scp} = 0.089$ henry.

(**D**) Harmonic distortion is usually a more important factor in determining the minimum inductance of output transformers than is the attenuation requirement at low frequencies. Compute now the number of turns and inductance for an assumed $B_m = 5000$ for 4-percent silicon-steel core with type EI-12 punchings in square stack. From manufacturer's catalog, A_c(net) = 0.90 inch², and $l_c = 6.0$ inches. From Fig. 13, $\mu_{ac} \approx 5000$.

$$N_p = \frac{3.49 E_p \times 10^6}{f A_c B_m} = 2020$$

$$N_s = 1.1 N_p/a = 111$$

$$L_p \approx \frac{3.19 N_p^2 \mu_{ac} A_c}{l_c} \times 10^{-8} = 97 \text{ henries.}$$

At 60 hertz, $X_m = \omega L_p = 36\ 600$, and $R'_{par}/X_m \approx 0.06$. From values of I_h/I_f for 4-percent silicon steel (Table 7)

$$E_h/E_f = (I_h/I_f)(R'_{par}/X_m)[1 - (R'_{par}/4X_m)]$$

$$\approx 0.012 \text{ or } 1.2 \text{ percent.}$$

(**E**) Now see if core window is large enough to fit windings. Assuming a simple method of winding (secondary over the primary), compute from

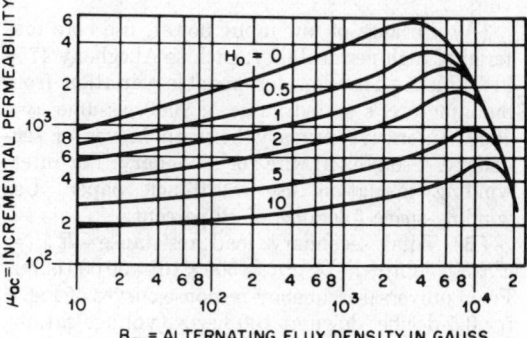

Fig. 13—Incremental permeability μ_{ac} characteristics of Allegheny audio-transformer "A" sheet steel at 60 hertz. No. 29 US gauge, L–7 standard laminations stacked 100 percent, interleaved. This is 4-percent silicon-steel core material. H_0 = magnetizing field in oersteds. *Courtesy of Allegheny-Ludlum Corp., Pittsburgh, Pa.*

geometry of core the approximate (MLT) for each winding (Fig. 4).

For the primary, $(\text{MLT})\approx0.42$ foot and

$$N_p(\text{MLT})\approx850 \text{ feet.}$$

For the secondary, $(\text{MLT})\approx0.58$ foot and

$$N_s(\text{MLT})\approx65 \text{ feet.}$$

For the primary, then, the size of wire is obtained from $R_p/N_p(\text{MLT})=0.236$ ohm/foot; and from Table 6, use No. 33. For the secondary, $R_s/N_s(\text{MLT})\approx0.008$, and size of wire is No. 18.

(F) Compute the turns/layer, number of layers, and total coil-build, as for power transformers. For an efficient design, $(\text{total coil-build})\approx(0.85 \text{ to } 0.90)\times(\text{window width})$.

(G) To determine if leakage inductance is within the required limit of (C) above, evaluate

$$l_{scp}=\frac{10.6N_p^2(\text{MLT})(2nc+a)}{n^2b\times10^9}=0.036 \text{ henry}$$

which is less than the limit 0.089 henry of (C). The symbols of this equation are defined in Fig. 17. If leakage inductance is high, interleave windings as indicated in the section on methods of winding transformers.

Example of Audio-Input-Transformer Design

This transformer must couple a 500-ohm line to the grids of 2 tubes in class-A push-pull. Attenuation to be flat to 0.5 decibel over 100 to 15 000 hertz; step-up$=1:10$; and input to primary is 2 volts.

(A) Because of low input power, use core material of high permeability, such as Allegheny 4750 in Table 5. To allow for possible variation from manufacturer's stated value of 9000, assume $\mu_0=4000$. Interleave primary between halves of secondary. Use No. 40 wire for secondary. For interwinding insulation use 0.010-inch paper. Use winding-space tolerance of 10 percent.

(B) Total secondary load resistance$=R'_{par}=a^2R_1R_2/(a^2R_1+R_2)\approx a^2R_1=500\times10^2=50\,000$ ohms. From universal-frequency-response curves of Fig. 8 for 0.5-decibel down at 100 hertz (voltage ratio$=0.95$)

$$(\omega_{low}L_s)/R'_{par}=3, \text{ or } L_s\approx240 \text{ henries.}$$

(C) Try Allegheny type EI-68 punchings, square stack. From manufacturer's catalog, $A_c=0.473$ inch^2, $l_c=4.13$ inches, and window dimensions$=\frac{11}{32}\times1\frac{1}{32}$ inches, interleaved singly; $l_g=0.0002$

inch. From

$$L=\frac{3.19N^2A_c}{l_g+l_c/\mu_0}\times10^{-8}$$

and above constants, compute

$$N_s=4400$$

$$N_p=N_s/a=440.$$

(D) Choose size of wire for primary winding, so that $R_p\approx0.1R_g=50$ ohms. From geometry of core, $(\text{MLT})\approx0.29$ foot; also, $R_p/N_p(\text{MLT})=0.392$, or No. 35 wire $(D=0.0062$ for No. 35F).

(E) Turns per layer of primary$=0.9b/d=110$; number of layers $n_p=N_p/110=4$; turns per layer of secondary $0.9b/d=200$; number of layers $n_s=N_s/200=22$.

(F) Secondary leakage inductance

$$l_{scs}=\frac{10.6N_s^2(\text{MLT})(2nc+a)\times10^{-9}}{n^2b}=0.35 \text{ henry.}$$

(G) Secondary effective layer-to-layer capacitance

$$C_e=(4C_l/3n_l)(1-n_l^{-1})$$

where $C_l=0.225A\epsilon/t=1770$ picofarads (refer to section on methods of winding transformers). Substituting this value of C_l into above expression of C_e, we find

$$C_e=107 \text{ picofarads.}$$

(H) Winding-to-core capacitance$=0.225A\epsilon/t\approx63$ picofarads (using 0.030-inch insulation between winding and core). Assuming tube and stray capacitances total 30 picofarads, total secondary capacitance $C_s\approx200$ picofarads.

(I) Series-resonance frequency of l_{sc} and C_s is

$$f_r=1/2\pi(l_{sc}C_s)^{1/2}=19\,200 \text{ hertz.}$$

At f_r, $B=X_c/R_1=1/2\pi f_rC_sR_1=0.83$; at 15 000 hertz, $f/f_r=0.78$. From Fig. 10, decibel variation from median frequency is seen to be less than 0.5.

If it is required to extend the frequency range, use Mumetal core material for its higher μ_0 (20 000). This will reduce the primary turns, the leakage inductance, and the winding shunt capacitance.

Considerations in Audio-Frequency-Transformer Design

Using modifications of the design procedures just described, it is possible to design the following types of transformers:

Modulation
Driver
Class-A output
Class-B output
Push–pull output

For further information consult R. Lee, "Electronic Transformers and Circuits," 2nd edition, John Wiley & Sons, New York; 1955.

HIGH-FREQUENCY TRANSFORMERS

High-frequency transformers are categorized by continuous-wave excitation over a broad band of frequencies to differentiate them from tuned transformers and pulse transformers. Their application is principally in the telephone carrier range limited to 1 megahertz. Their function is primarily the same as for audio-frequency transformers.

The principles discussed in the section on audio-frequency transformers can be used for designing high-frequency transformers, but the core materials, leakage inductance, and winding capacitance become more critical. Generally the smaller a transformer can be made and still meet the low-frequency requirements, the better the transformer will operate at high frequency.

Because of the high resistivity and permeability of ferrites, E-I or pot cores of this material (Table 5) are generally used.

For more-detailed information consult R. Lee, "Electronic Transformers and Circuits," 2nd edition, John Wiley & Sons, New York; 1955.

PULSE TRANSFORMERS

Pulse transformers are designed to transmit square waves or trains of pulses while maintaining as closely as possible the original shape (refer to the section on pulse modulation in Chapter 23). Fourier analysis shows that such pulse waveforms consist of a wide range of frequency components. Thus the transformer must have suitable bandwidth to maintain fidelity.

Pulse transformers can be analyzed by considering the leading edge, top, and trailing edge of the pulse separately. Figure 14 portrays a typical transformer output pulse compared with input pulse. Figure 15 shows the fundamental circuit, and Fig. 16 illustrates equivalent circuits for the various transient conditions.

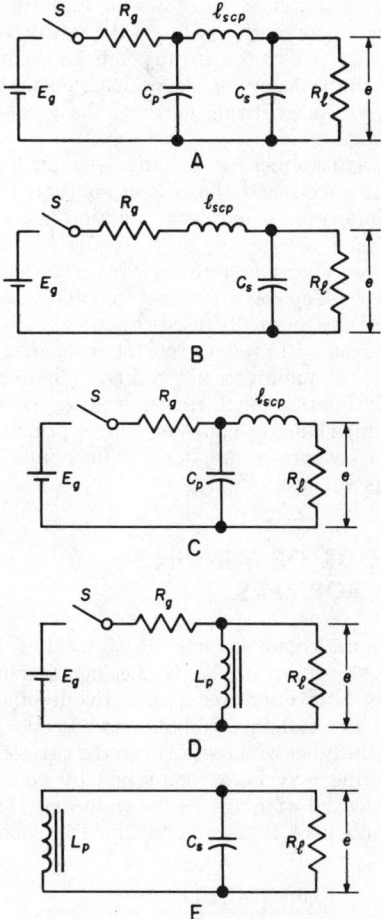

Fig. 16—Pulse-transformer equivalent circuits. A—Leading-edge equivalent circuit. B—Leading-edge equivalent circuit for step-up-ratio transformer. C—Leading-edge equivalent circuit for step-down-ratio transformer. D—Top-of-pulse equivalent circuit. E—Trailing-edge equivalent circuit.

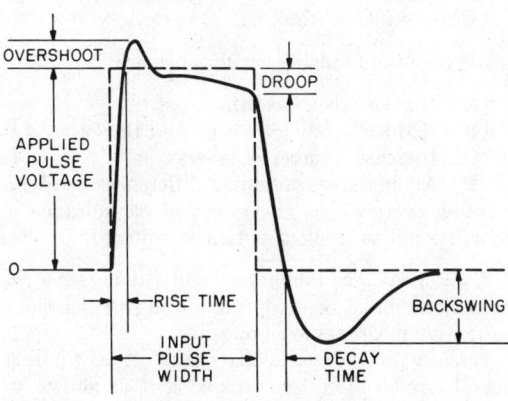

Fig. 14—Output pulse shape. In the strictest sense, pulse rise and decay times are measured between the 10- and 90-percent values; width is measured between the 50-percent values.

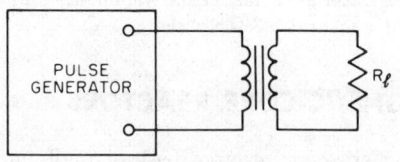

Fig. 15—Pulse-transformer circuit.

Leading-edge reproduction requires transmission of a wide band of frequencies and is controlled by leakage inductance l_{scp} and winding capacitances C_p and C_s as indicated in Fig. 16A, B, and C. Analysis for step-up and step-down transformers varies slightly as shown. Leakage inductance and winding capacitance must be minimized to achieve a sharp rise; however, output voltage may overshoot input voltage and oscillation may be encountered where very abrupt rise times are involved.

Pulse-top response depends on the magnitude of the open-circuit inductance of the transformer, as indicated in Fig. 16D. The greater the inductance L_p, the smaller the droop from input voltage level.

Control of the trailing edge of the pulse depends on the open-circuit inductance and secondary winding capacitance, as shown in Fig. 16E. The lower the capacitance, the faster the rate of voltage decay. Negative backswing depends on the magnitude of the transformer magnetizing current. The greater the magnetizing current, the greater the backswing.

Pulse-transformer design involves analysis of transient effects and thus direct solution is complex. Empirical or graphical solution* is usually used.

Low-loss core materials such as grain-oriented silicon-steel loop cores or nickel-iron alloys in 2-mil thickness are normally used. Small air gaps are commonly used to reduce remanent magnetism in core due to unidirectional pulses. Windings are normally interleaved to reduce leakage reactance. If load impedance is high, single-layer primary and secondary windings are best; if low, interleaved windings are best.

METHODS OF WINDING TRANSFORMERS

The most common methods of winding transformers are shown in Fig. 17. Leakage inductance is reduced by interleaving, i.e., by dividing the primary or secondary coil into two sections and placing the other winding between the two sections. Interleaving may be accomplished by concentric and by coaxial windings, as shown in Fig. 17B and C; reduction of leakage inductance is computed from

$$l_{sc} = \frac{10.6N^2(\text{MLT})(2nc+a)}{n^2 b \times 10^9} \text{ henries}$$

with dimensions in inches to be the same for Fig. 17B and C.

* R. Lee, "Electronic Transformers and Circuits," 2nd edition, John Wiley & Sons, New York; 1955: chapter 10, p. 292.

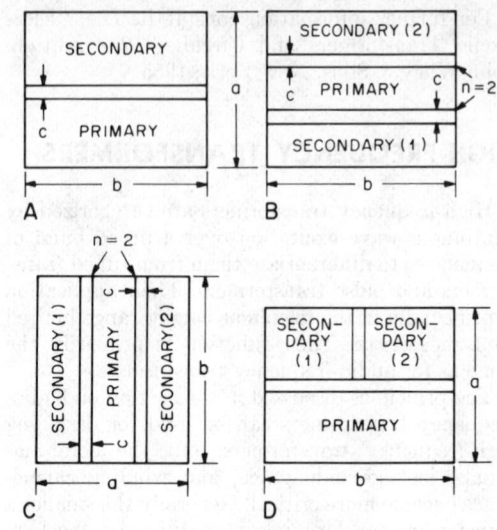

Fig. 17—Methods of winding transformers.

Means of reducing leakage inductance are:

(**A**) Minimize turns by using high-permeability core.
(**B**) Reduce build of coil.
(**C**) Increase winding width.
(**D**) Minimize spacing between windings.
(**E**) Use bifilar windings.

Means of minimizing capacitance are:

(**A**) Increase dielectric thickness t.
(**B**) Reduce winding width b and thus area A.
(**C**) Increase number of layers.
(**D**) Avoid large potential differences between winding sections, as the effect of capacitance is proportional to applied potential squared.

Note: Leakage inductance and capacitance requirements must be compromised in practice since corrective measures are opposite.

Effective interlayer capacitance of a winding may be reduced by sectionalizing it as shown in Fig. 17D. This can be seen from

$$C_e = (4C_l/3n_l)(1-1/n_l) \text{ picofarads}$$

where $n_l =$ number of layers, $C_l =$ capacitance of one layer to another $= 0.225A\epsilon/t$ picofarads, $A =$ area of winding layer $= (\text{MLT})b$ inches², $t =$ thickness of interlayer insulation in inches and $\epsilon =$ dielectric constant ≈ 3 for paper.

MAGNETIC-CORE REACTORS

Magnetic-core reactors basically fall into two categories: rectifier filter reactors and audio filter reactors. The rectifier filter reactor is used mainly in direct-current power supplies to smooth the output ripple voltage. (Refer to Chapter 14.) It carries all the direct current of the rectifier filter and must be designed not to saturate with direct current in the reactor winding. Audio filter reactors (sometimes called coils) are used principally in audio filters to provide the necessary inductance with relatively high quality factor Q. (Refer to Chapter 8.) The term "audio" filter reactor is used to differentiate it from "rectifier" filter reactor; its applications are not restricted to audio frequencies but extend into the megahertz range.

Rectifier Filter Reactors

This type of reactor must be made relatively large, depending on the inductance and direct cur-

rent. Consequently the core is usually either E-I laminations using 4% silicon steel or type "C" tape core loops using grain-oriented silicon steel.

Design of Rectifier Filter Reactors

These reactors carry direct current and are provided with suitable air gaps. Optimum design data may be obtained from Hanna curves, Fig. 18. These curves relate direct-current energy stored in the core per unit volume, LI_{dc}^2/V to magnetizing field NI_{dc}/l_c (where $l_c =$ average length of flux path in core), for an appropriate air gap. Heating is seldom a factor, but direct-current-resistance requirements affect the design; however, the transformer equivalent volt–ampere ratings of chokes given in Table 8 should be useful in determining

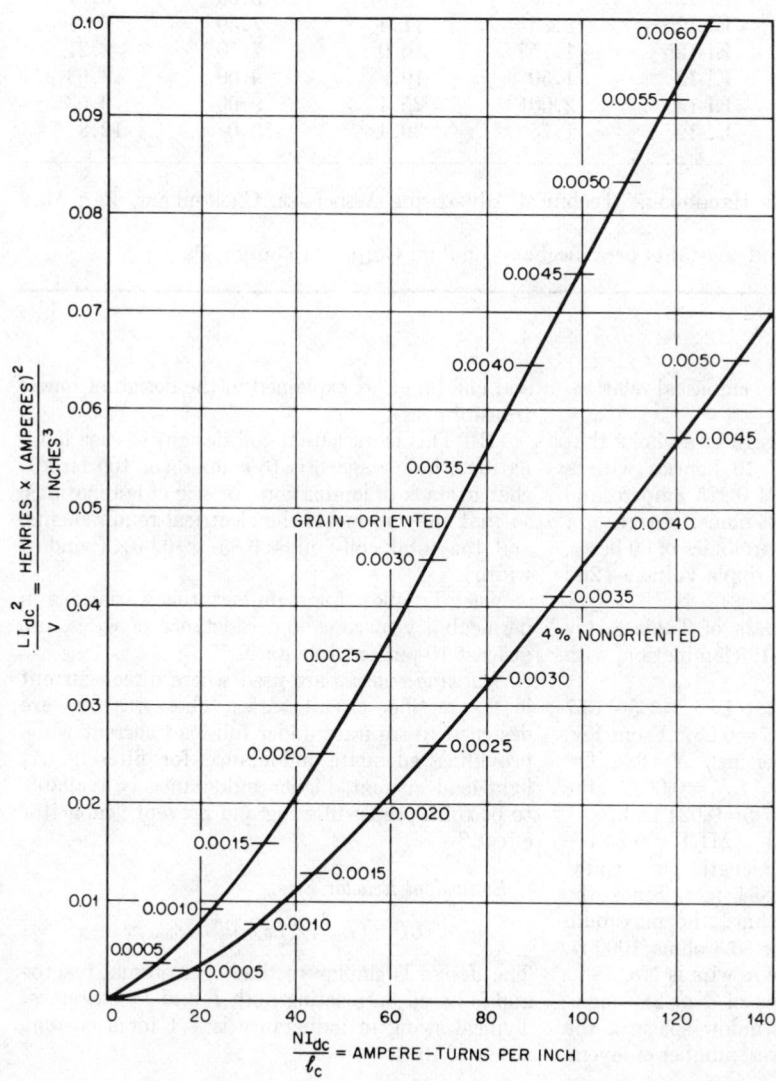

Fig. 18—Hanna curves for silicon steel. The numbers on the curves represent length of air gap l_g/length of flux path l_c, in inches.

TABLE 8—EQUIVALENT LI^2 RATING OF FILTER REACTOR FOR RECTIFIERS. $L=$ inductance in henries, $I=$ direct current in amperes. The rating is based on power supply frequencies up to 400 hertz and 50° C temperature rise above ambient. The LI^2 values should be reduced for lower temperature rises and high-voltage operation.

LI^{2*}	Current Density* (amps/sq in.)	EI-Type Lamination†	Stack Height p (in.)	Core Volume V (cu in.)	Magnetic Path Length l_c (in.)	Average Copper Mean Length per Turn (MLT) (in.)
0.0195	3200	EI-21	0.5	0.80	3.25	3.12
0.0288	2700	EI-625	0.625	1.45	3.75	3.62
0.067	2560	EI-75	0.75	2.51	4.50	4.33
0.088	2560	EI-75	1.00	3.35	4.50	4.83
0.111	2330	EI-11	0.875	3.88	5.25	5.04
0.200	2130	EI-12	1.00	5.74	6.00	5.71
0.300	2030	EI-12	1.50	8.61	6.00	6.71
0.480	1800	EI-125	1.25	11.4	7.50	7.21
0.675	1770	EI-125	1.75	16.0	7.50	8.21
0.850	1600	EI-13	1.50	19.8	9.00	8.63
1.37	1500	EI-13	2.00	26.4	9.00	9.63
3.70	1200	EI-19	1.75	39.4	13.0	12.8

* From "Radio Components Handbook," Technical Advertising Associates, Cheltenham, Pa.; May 1948: page 92.

† Lamination designation and constants per Allegheny Ludlum Corp., Pittsburgh, Pa.

their sizes. This is based on the empirical relationship $(VA)_{eq} = 188 LI_{dc}^2$.

As an example, take the design of a choke that is to have an inductance of 10 henries with a superimposed direct current of 0.225 ampere and a direct-current resistance ≤ 125 ohms. This reactor shall be used for suppressing harmonics of 60 hertz, where the alternating-current ripple voltage (2nd harmonic) is about 35 volts.

(**A**) $LI^2 = 0.51$. Based on data of Table 8, try 4% silicon-steel core, type EI-125 lamination, with a core buildup of 1.5 inches.

(**B**) From Table 8, $V = (11.4/1.25) \times 1.5 = 13.7$ in.^{3}, $l_c = 7.50$, $LI^2/V = 0.51/13.7 = 0.037$. From Fig. 18, $NI/l_c = 88$ ampere-turns per inch, $N = 88l_c/I = (88 \times 7.5)/0.225 = 2930$ turns. $l_g/l_c = 0.0032$, the length of air gap $l_g = 0.0032 \times 7.5 = 0.024$ inch.

(**C**) From Table 8, coil (MLT) $= (7.21 + 0.5)/12 = 0.643$ foot, and length of coil $= N \times (MLT) = 2930 \times 0.643 = 1884$ feet. Since the maximum resistance is 125 ohms, the maximum ohms/ft $= 125/1884 = 0.0663$ or 66.3 ohms/1000 ft. From Table 6, the nearest size of wire is No. 28.

(**D**) Now see if 2930 turns of No. 28 single-Formex wire will fit in the window space of the core. (Determine turns per layer, number of layers,

and coil-build, as explained in the design of power transformers.)

(**E**) This is an actual coil design; in case lamination window space is too small (or too large), change stack of laminations, or size of laminations, so that the coil meets the electrical requirements, and the total coil-build ≈ 0.85 to $0.90 \times$ (window width).

Note: To allow for manufacturing variations in permeability of cores and resistance of wires, use at least 10-percent tolerance.

Swinging reactors are used where direct current in the rectifier circuit varies. These reactors are designed to saturate under full-load current while providing adequate inductance for filtering. At light-load current, higher inductance is available to perform proper filtering and prevent "capacitor effect."

Equivalent Reactor Size:

$$LI^2 = (L_{max} \times L_{min})^{1/2} I_{dc(max)}^2.$$

The design is similar to that of a normal reactor and is based on meeting both L and I_{dc} extremes. Typical swing in inductance is 4:1 for a current swing of 10:1.

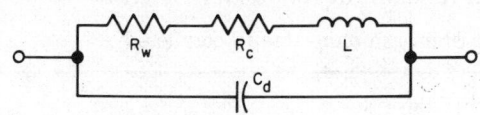

Fig. 19—Equivalent network of an audio filter coil.

L = calculated inductance based on turns and magnetic-core constant

C_d = distributed capacitance of winding

R_w = copper losses in winding

R_c = magnetic-core losses reflected in series with winding

Q = quality factor = $\omega L/(R_w+R_c)$, $\omega=2\pi f$.

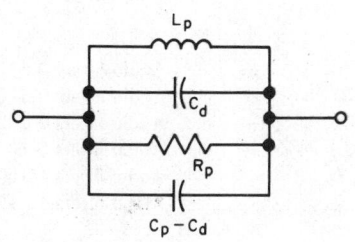

Fig. 20—Parallel tuned mesh.

L_p = equivalent parallel inductance

R_p = equivalent parallel resistance

C_d = distributed capacitance of winding

C_p = parallel tuning capacitance

Q_p = effective parallel Q

$L_p = L$, for $Q>10$

$R_p = (R_w+R_c)(Q^2+1)$, for $Q_p=Q$.

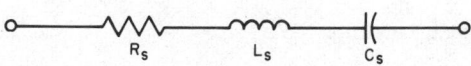

Fig. 21—Series tuned mesh.

L_s = equivalent series inductance

R_s = equivalent series resistance

C_s = series tuning capacitance

Q_s = effective series Q.

For $Q>10$

$L_s = L/(1-\omega^2 LC_d)$

$R_s = (R_w+R_c)/(1-\omega^2 LC_d)^2$

$Q_s = Q(1-\omega^2 LC_d)$

$L = L_s/(1+\omega^2 L_s C_d)$.

Audio Filter Reactors (Coils)

Audio filter reactors, hereafter referred to as coils, are inductances that have the following qualities:

(**A**) High quality factor Q.

(**B**) Stable inductance with variations in frequency, operating level, and temperature.

To design coils having these qualities, it is necessary to understand the equivalent circuit of a coil (Fig. 19) and its application to the parallel tuned mesh (Fig. 20) and series tuned mesh (Fig. 21). The distributed capacitance C_d is very important in series tuned meshes, as it affects both the inductance and the Q. If possible, it is desirable to keep the distributed capacitance C_d to one-tenth of the tuning capacitance C_s. Means of minimizing C_d are discussed in succeeding sections.

R_w represents the copper loss in the coil winding. It is made up of the dc resistance of the wire and the eddy-current losses generated in the winding by stray flux from the core cutting the winding. This eddy-current loss can be minimized at high frequencies by dividing the wire into many small wires. This type of wire is called Litz wire.

R_c represents the loss due to the magnetic core reflected in series with the coil winding. The calculation of this factor is discussed in succeeding sections.

There are three principal types of audio filter coils. Their use is usually dictated by the frequency range of interest and the desired Q. At frequencies below 300 hertz, laminated punchings of nickel–steel alloy usually are used. For frequencies between 300 and 5000 hertz, toroidal permalloy-dust cores are usually used. For frequencies above 5000 hertz, ferrite pot cores are usually used. (Refer to Table 9.)

Audio Filter Coils with Punching Laminated Cores

Externally this coil resembles a transformer. The punchings are usually "F" in shape to create an air gap in the center leg of the core. For high Q, nickel–steel laminations either 14 or 6 mils thick are used (Table 9). An air gap is placed in the core to reduce the effective permeability μ_e, which reduces the effective core losses and stabilizes the inductance.

The winding is usually layer wound as described in the section on methods of winding transformers. The techniques for reducing capacitance described

TABLE 9—CHARACTERISTICS OF SOME CORE MATERIALS FOR AUDIO FILTER COILS.

$R_c/(\mu_0 Lt) = aB_m + c + ef$, where R_c = series resistance in ohms due to core loss.*

Material or Alloy	Initial Permeability (μ_0)	Resistivity (ohm-cm)	Hysteresis Coefficient $(a \times 10^6)$	Residual Coefficient $(c \times 10^6)$	Eddy-Current Coefficient $(e \times 10^9)$	Gauge (mils)	Application and Frequency Range (kilohertz)
4% silicon steel	400	60×10^{-6}	120	75	870	14	Rectifier filters
Low nickel	3 500 to 10 000	44×10^{-6}	0.4	14	1 550	14	Audio filters up to 0.2
					284	6	Audio filters up to 10
High nickel	10 000 to 20 000	57×10^{-6}	0.05	0.05	950	14	Audio filters up to 0.2
					175	6	Audio filters up to 10
Molybdenum permalloy dust	200†	1.0	1.3	40	40	—	Audio filters 0.1–7
	160†	1.0	1.5	38	25	—	Audio filters 0.1–10
	125†	1.0	1.6	30	19	—	Audio filters 0.2–20
	60*	1.0	3.2	50	10	—	Audio filters 5–50
	26*	1.0	6.9	96	7.7	—	Audio filters 15–60
	14*	1.0	11.4	143	7.1	—	Audio filters 40–150
Carbonyl types:							
C	55	—	9	80	7	—	High-frequency filters
P	26	—	3.4	220	27	—	High-frequency filters
Th	16	—	2.5	80	8	—	High-frequency filters
Ferrites‡:							
3B7	2 300	100	§	§	§	—	Audio filters 0.2–300
3B9	1 800	100	§	§	§	—	Audio filters 0.2–300
3D3	750	10^5	§	§	§	—	HF filters 200–2 500
4C4	125	10^5	§	§	§	—	HF filters 1 000–20 000

* Data and coefficients a, c, and e are from V. E. Legg and F. J. Given, "Compressed Powdered Molybdenum Permalloy for High Quality Inductance Coils," *Bell System Technical Journal*, vol. 19, no. 3, pp. 385–406; July 1940.
† Data from Catalog PC-303, Magnetics, Inc., Butler, Pa.
‡ Data from Bulletin 220-C, Ferroxcube Corporation of America, Saugerties, N.Y.
§ See Fig. 27.

in that section can be used to reduce the distributed capacitance of the winding.

A detailed method for designing this type of coil can be found in "High Q Reactors for Low Frequencies," Bulletin A10, Magnetic Metals Co.. Camden, N.J.

Toroidal Audio Filter Coils

Toroidal coils are doughnut shaped, with the winding covering the entire core. The core is usually made of pressed molybdenum-permalloy dust, al-though some cores are made of carbonyl powdered iron for very-high-frequency applications (Table 9).

Permalloy-dust toroidal cores are made in various sizes with several material compositions to change the effective permeability μ_e for different frequency ranges. Figure 22 gives the Q-vs-frequency characteristics of four sizes of toroids with four different effective permeabilities. In addition, the inductance in millihenries per 1000 turns, A_L, is given for each size of core.

Figure 23 gives the number of turns of various sizes of heavy Formvar wire which can be placed

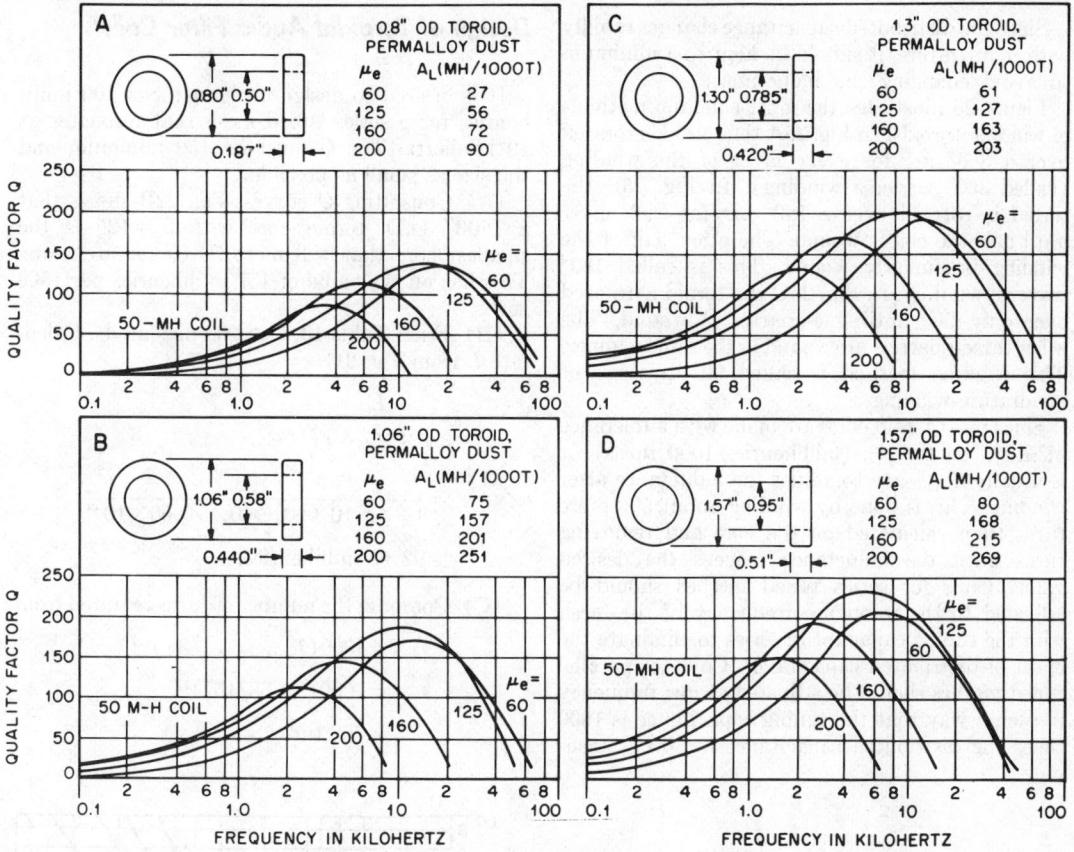

Fig. 22—Quality factor Q as a function of frequency for several sizes of permalloy-dust toroids. The data have been replotted from "Permalloy Powder Cores," Catalog PC–303, Magnetics, Inc., Butler, Pa.

on the toroid as well as the mean length of turn (MLT) for the respective toroidal cores of Fig. 22. These curves are based on a single winding wound with 180° traverse using commercially available toroidal winding machines.

Figure 24 gives the core loss factor in ohms per millihenry for various effective permeabilities and frequencies. This factor is multiplied by the in-ductance in millihenries to determine R_c. The curves are based on a low magnetic induction of 20 gauss. For higher levels, R_c can be calculated from the Legg coefficients given in Table 9.

Table 10 gives an approximate value of dis-tributed capacitance C_d that can be expected for the different sizes of cores and winding methods.

When maximum temperature stability of in-ductance is required, most manufacturers of perm-alloy toroidal cores can provide various types of stabilized cores. The stabilizations most used fol-low.

TABLE 10—ESTIMATION OF DISTRIBUTED CAPACI-TANCE FOR CORE SIZE AND WINDING METHOD.

Toroid Core Size	Distributed Capacitance C_d (picofarads)		
	360° Traverse	180° Traverse	90° Traverse
0.8″ O.D.	120	60	30
1.06″ O.D.	200	100	50
1.30″ O.D.	300	150	75
1.57″ O.D.	360	180	90

Identification	Inductance Temperature Stability (%)	Temperature Range (°C)
B	±0.1	+13 to +35
D	±0.1	0 to +55
W	±0.25	−55 to +85

Since the distributed capacitance changes rapidly with temperature, it should be kept to a minimum to avoid changing the inductance.

Figure 25 illustrates the most common methods of winding toroids. In Fig. 25A the toroid is rotated over a 360° arc for every layer of the winding (called 360° traverse winding). In Fig. 25B the toroid is rotated over a 180° arc for each layer until half the coil is wound. The other half of the winding is similarly wound. This is called 180° traverse winding. In Fig. 25C the toroid is rotated over only 90° until one-quarter is wound. The other three-quarters are wound in the same manner. This winding method is called 90° traverse or quadrature winding.

Since most toroid cores are made with a tolerance of ±8% on the A_L (millihenries/1000 turns), it is usually necessary to adjust the inductance after winding. This is done by winding about 5% more turns than calculated on the core and removing turns until the inductance reaches the desired value. Coils for series tuned meshes should be adjusted to the resonance frequency of the mesh with the tuning capacitor in series to eliminate the effect of distributed capacitance. Coils for parallel tuned meshes should be adjusted at low frequency in such a way that the tuning capacitance is 1000 times the distributed capacitance for 0.1% accuracy.

Design of Toroidal Audio Filter Coils*

It is desired to design an inductor of 100 millihenries for a series tuned mesh that resonates at 10 kilohertz. The Q must be 150 minimum and the size as small as possible.

(**A**) Consulting Q curves, Fig. 22B shows that a 1.06″ O.D. toroid core with $\mu_e = 125$ is the smallest core that will meet the Q requirements. This has an A_L value of 157 millihenries per 1000 turns.

(**B**) From Table 10, $C_d \approx 200$ picofarads. Calculate L from Fig. 21.

$$L = \frac{L_s}{1 + \omega^2 L_s C_d}$$

$$= \frac{100}{1 + (2\pi 10\ 000)^2 \times 0.1 \times 200 \times 10^{-12}}$$

$$= 92.68 \text{ millihenries.}$$

(**C**) Compute the number of turns required from

$$N = 1000 (L_{\text{millihenries}}/A_L)^{1/2}$$

$$= 1000 (92.68/157)^{1/2}$$

$$= 768 \text{ turns.}$$

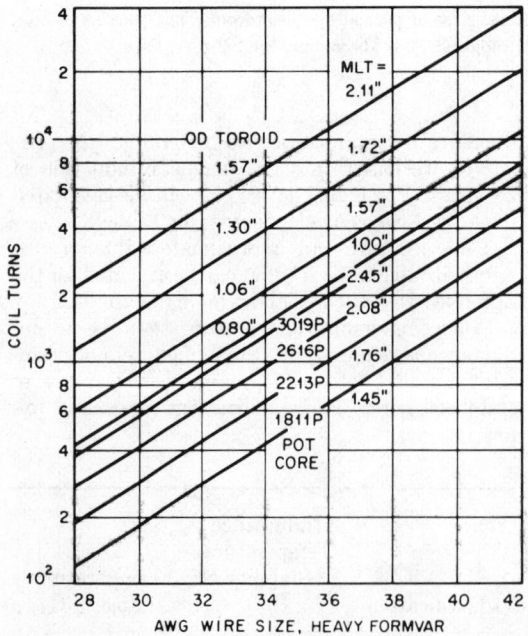

Fig. 23—Coil turns as a function of wire gauge for various core sizes and types.

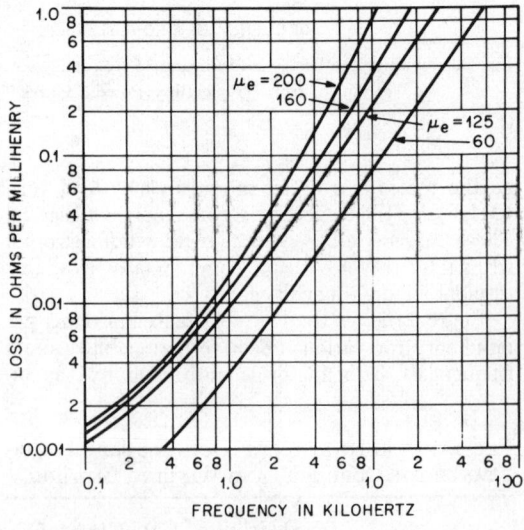

Fig. 24—Molybdenum permalloy-dust-core loss characteristics. $B_m = 20$ gauss. The data have been replotted from "Permalloy Powder Cores," Catalog PC–303, Magnetics, Inc., Butler, Pa.

* For more-detailed data on designing toroid coils, consult "Permalloy Powder Cores," Catalog PC–303, Magnetics, Inc., Butler, Pa.

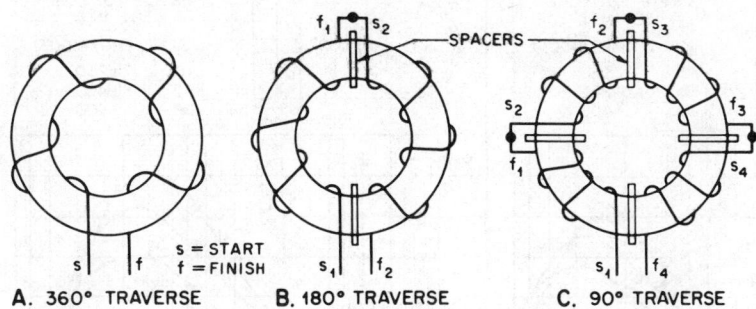

A. 360° TRAVERSE   B. 180° TRAVERSE   C. 90° TRAVERSE

s = START
f = FINISH

Fig. 25—Methods of winding toroid cores.

(**D**) Figure 23 gives the maximum size of wire and mean length of turn. Use No. 30 heavy Formvar for 768 turns. On a 1.06″ O.D. core the (MLT) = 1.57″.

(**E**) Calculate R_w from (MLT), N, and Table 6.

$$R_w = \frac{(\text{MLT}) \times N \times \text{ohms}/1000 \text{ ft}}{12\ 000}$$

$$= \frac{1.57 \times 768 \times 103.2}{12\ 000}$$

$$= 10.36 \text{ ohms.}$$

(**F**) Calculate R_c from R/L values of Fig. 24 for $\mu_e = 125$ and $f = 10$ kilohertz. $R/L = 0.23$ ohm/millihenry. $R_c = (R/L)L = 0.23 \times 92.68 = 21.32$ ohms.

(**G**) Calculate Q per Fig. 19.

$$Q = \frac{\omega L}{R_w + R_c} = \frac{2\pi 10 \times 92.68}{10.36 + 21.32} = 183.8.$$

(**H**) Calculate Q_s per Fig. 21.

$$Q_s = Q(1 - \omega^2 L C_d)$$

$$= 183.8[1 - (2\pi 10\ 000)^2 \times 0.09268 \times 200 \times 10^{-12}]$$

$$= 170.4.$$

This should approximate the measured Q.

(**I**) The coil is wound using 360° traverse winding and, since no special temperature stability is required, a standard unstabilized core is used.

Ferrite-Pot-Core Audio Filter Coils

A ferrite-pot-core coil resembles a thin cylinder with the core entirely surrounding the winding except where the leads protrude. The core is pressed

in two halves of green ferrite material and is fired like a ceramic. Each half has a recessed ring where the coil is placed after being wound. (See Fig. 26.) The cores are made of different compositions of ferrite to obtain different permeabilities, loss characteristics, and temperature coefficients. (See Table 9 and Fig. 27.)

To reduce the effective permeability μ_e, an air gap is machined into the center leg of the core. The larger the air gap the smaller the μ_e. Lowering the μ_e reduces the core losses, A_L (millihenries per 1000 turns), and temperature coefficient. Figure 28 gives Q-vs-frequency characteristics of four sizes of pot cores, each with three different effective

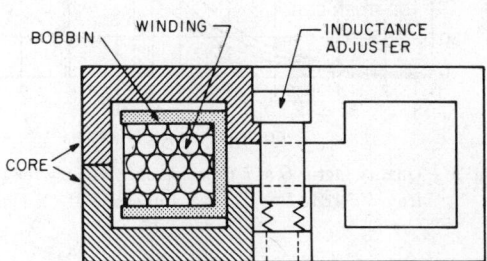

BOBBIN   WINDING   INDUCTANCE ADJUSTER
CORE

Fig. 26—Typical pot-core coil.

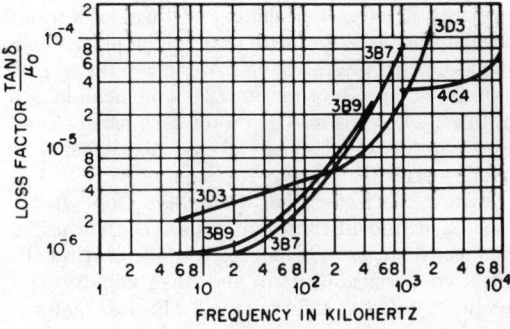

LOSS FACTOR $\frac{\text{TAN}\delta}{\mu_0}$

FREQUENCY IN KILOHERTZ

Fig. 27—Loss factor as a function of frequency. From "Ferrite Pot Cores," Bulletin 220-C, Ferroxcube Corporation of America, Saugerties, N.Y.

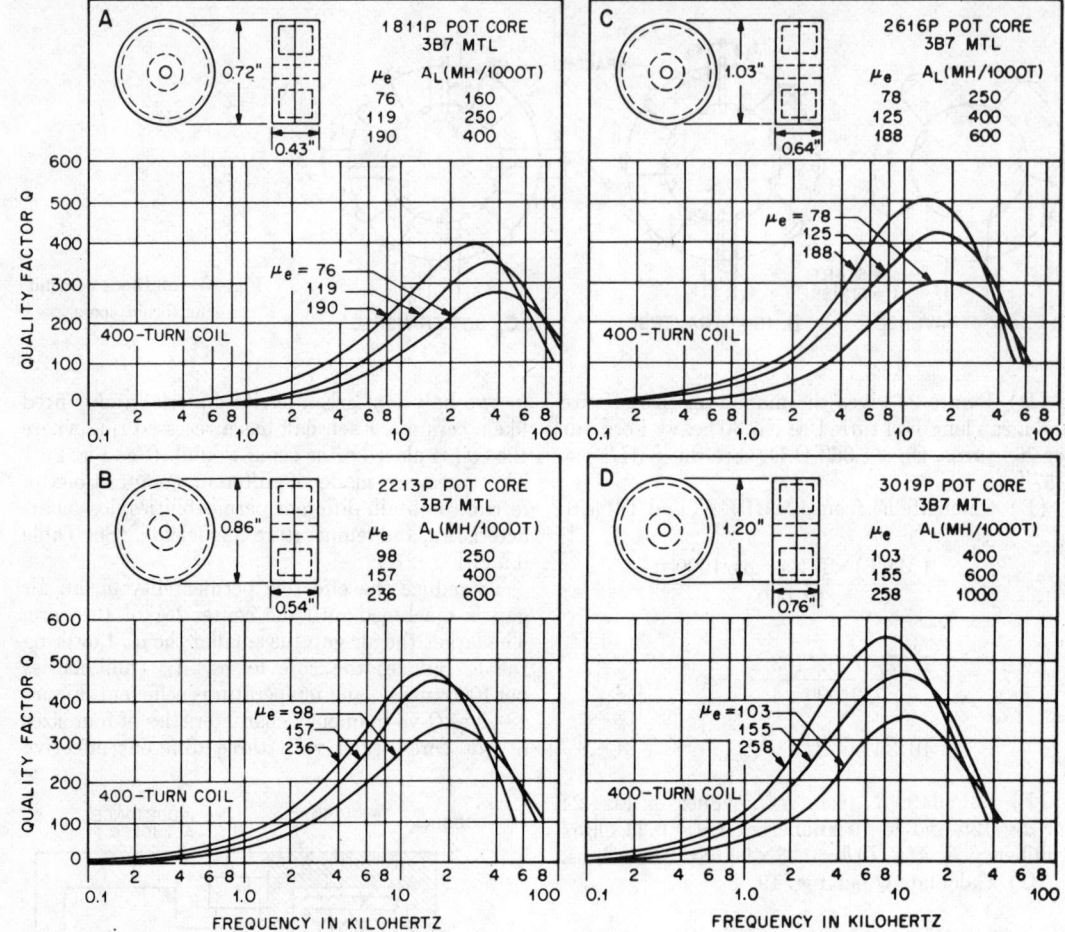

Fig. 28—Quality factor Q as a function of frequency for several types of ferrite pot cores. The data have been replotted from "Ferrite Pot Cores," Bulletin 220–C, Ferroxcube Corporation of America, Saugerties, N. Y.

permeabilities and A_L values. Q values are based on a single section winding using 400 turns of solid wire.

Figure 23 gives the number of turns of various sizes of double-coated wire that can be placed on a single-section plastic bobbin designed for a particular pot core shown in Fig. 28. The mean length of turn (MLT) is also given for each size of core. Litz wire should be used above 20 kilohertz to obtain better Q's.

Figure 27 gives the normalized loss factor $\tan\delta/\mu_0$ for the different materials over their useful frequency ranges. This factor is multiplied by the effective permeability μ_e and the inductive reactance $(2\pi fL)$ to determine R_c. This loss factor is based on low-level excitation, and losses will be greater for high-excitation applications.

Table 11 gives approximate values of distributed

TABLE 11—ESTIMATION OF DISTRIBUTED CAPACITANCE FOR CORE SIZE AND TYPE OF BOBBIN USED.

Ferrite Pot Core No.	Distributed Capacitance C_d (picofarads)		
	1-Section Bobbin	2-Section Bobbin	3-Section Bobbin
3019P	75	31	21
2616P	64	29	19
2213P	48	20	14
1811P	35	16	11

capacitance C_d that can be expected for different core sizes and numbers of sections in a bobbin. Since the core is external to the winding, grounding the core will increase these values.

The temperature stability is roughly proportional to the effective permeability μ_e. Ferroxcube 3B7 material has a flat temperature coefficient. Its temperature factor $(\pm 0.6 \times 10^{-6})$ multiplied by μ_e gives the temperature-coefficient limits of inductance from $+20$ to $+70°C$. For a positive temperature coefficient, 3B9 material is used. This material has a temperature factor of $+1.4$ to $+2.2 \times 10^{-6}$ from -30 to $+70°C$. Care must be taken to clamp and process these ferrite coils properly to realize these temperature coefficients. Distributed capacitance should also be minimized because of its effect on the inductance (Fig. 21).

Generally it is possible to obtain from the core manufacturer winding bobbins that have 1, 2, or 3 sections. Table 11 shows that the use of more than 1 section reduces the distributed capacitance but requires a smaller-diameter wire. Figure 29 illustrates the three bobbins mentioned. The sections are wound one at a time from start to finish without breaking the wire.

Because the ferrite pot core comes in two halves, it is necessary to clamp or cement them together in assembly. Clamping is preferred, as cement on the mating surfaces will decrease the A_L value and affect the temperature coefficient. Manufacturers of ferrite pot cores can furnish clamping hardware.

It is possible to vary the inductance of a pot core by threading a ferrite slug into the center hole of the core. As this slug bridges the air gap in the center leg of the core (Fig. 26), μ_e is increased and hence the inductance increases. In most applications the adjustment range is $+13\%$ with a sensitivity of better than 0.1%.

Most ferrite pot cores for this application have preadjusted air gaps to guarantee an A_L value within limits of $\pm 3\%$. If an inductance adjuster is used (Fig. 26), it is possible to adjust the coil to within 0.1%. Also, if a $\pm 2\%$ mesh tuning capacitor is used, the mesh can be tuned using

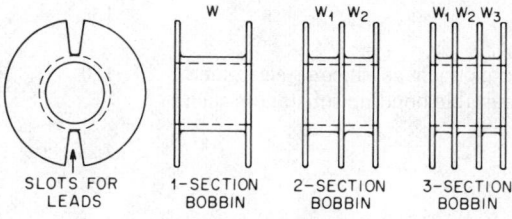

SLOTS FOR LEADS   1-SECTION BOBBIN   2-SECTION BOBBIN   3-SECTION BOBBIN

Fig. 29—Ferrite-pot-core bobbins available.

the inductance adjuster alone, thus eliminating the need for padding capacitors.

Design of Ferrite-Pot-Core Audio Filter Coils*

It is desired to design an inductor of 1000 millihenries for a series tuned mesh that resonates at 5 kilohertz. The Q should be near 200, the temperature coefficient less than ± 100 ppm/°C, and the coil adjustable to tune the mesh, with a fixed capacitor of $\pm 2\%$ tolerance.

(**A**) To obtain a flat temperature coefficient, 3B7 material should be used; for ± 100 ppm/°C, the effective permeability μ_e should be less than $100/0.6$, or 167.

(**B**) Consulting the Q curves, Fig. 28B shows that a 2213P pot core with $\mu_e = 157$ and $A_L = 400$ should meet the Q requirements.

(**C**) From Table 11, C_d is 48 picofarads for a 1-section coil. From Fig. 21

$$L = \frac{L_s}{1 + \omega^2 L_s C_d}$$

$$= \frac{1000}{1 + (2\pi 5000)^2 \times 1.0 \times 48 \times 10^{-12}}$$

$$= 954.8 \text{ millihenries.}$$

(**D**) Since we must use an inductance adjuster that increases the inductance by 13% maximum, the coil without adjuster should have an inductance of $954.8/1.06 = 900.7$ millihenries. This allows an inductance adjustment of $\pm 6\%$.

(**E**) Compute the number of turns required from

$$N = 1000 (L_{\text{millihenries}}/A_L)^{1/2}$$

$$= 1000 (900.7/400)^{1/2}$$

$$= 1500 \text{ turns.}$$

(**F**) Consult Fig. 23 to determine maximum wire size and mean length of turn: Use No. 38 heavy Formvar for 1500 turns. On a 2213P pot core, the (MLT) $= 1.76''$.

(**G**) Calculate R_w from (MLT), N, and Table 6.

$$R_w = \frac{1.76 \times 1500 \times 659.6}{12\,000} = 145.1 \text{ ohms.}$$

(**H**) Effective permeability μ_e with adjuster $= 157 \times 1.06 = 166.4$. From Fig. 27, estimate loss

---

* For more details about designing ferrite-pot-core coils, consult Bulletin 220–C, Ferroxcube Corporation of America, Saugerties, N. Y.

TABLE 12—CLASSIFICATION OF ELECTRICAL INSULATING MATERIALS. *Abridged from "General Principles Upon Which Temperature Limits are Based in the Rating of Electrical Equipment," AIEE Standard No. 1; Dec. 1962.*

Class	Insulating Material	Limiting Insulation Temperature (Hottest Spot) in °C
O	Materials or combinations of materials such as cotton, silk, and paper without impregnation*	90
A	Materials or combinations of materials such as cotton, silk, and paper when suitably impregnated or coated or when immersed in a dielectric liquid*	105
B	Materials or combinations of materials such as mica, glass fiber, asbestos, etc., with suitable bonding substances*	130
F	Same as for Class B	155
H	Materials or combinations of materials such as silicone elastomer, mica, glass fiber, asbestos, etc., with suitable bonding substances such as appropriate silicone resins*	180
C	*	220
Over C	Materials consisting entirely of mica, porcelain, glass, quartz, and similar inorganic materials*	Over 220

* (Other) materials or combinations of materials may be included in this class if by experience or accepted tests they can be shown to have comparable thermal life at the temperature given in the right-hand column.

These temperatures are, and have been in most cases over a long period of time, benchmarks descriptive of the various classes of insulating materials, and various accepted test procedures have been or are being developed for use in their identification. They should not be confused with the actual temperatures at which these same classes of insulating materials may be used in the various specific types of equipment nor with the temperatures on which specified temperature rise in equipment standards are based.

In the above definitions the words "accepted tests" are intended to refer to recognized test procedures established for the thermal evaluation of materials by themselves or in simple combinations. Experience or test data, used in classifying insulating materials, are distinct from the experience or test data derived for the use of materials in complete insulation systems. The thermal endurance of complete systems may be determined by suitable test procedures.

A material that is classified as suitable for a given temperature may be found suitable for a different temperature, either higher or lower, by an insulation system test procedure. For example, it has been found that some materials suitable for operation at one temperature in air may be suitable for a higher temperature when used in a system operated in an inert gas atmosphere. Likewise some insulating materials when operated in dielectric liquids will have lower or higher thermal endurance than in air.

It is important to recognize that other characteristics (in addition to thermal endurance) such as mechanical strength, moisture resistance, and corona endurance, are required in varying degrees in different applications for the successful use of insulating materials.

factor $\tan\delta/\mu_0$ for 5 kilohertz $= 0.5\times10^{-6}$. (This requires extrapolation of the curve for 3B7 material.) Calculate R_c from

$$R_c = 2\pi f L\mu_e(\tan\delta/\mu_0)$$

$$= 2\pi 5000\times0.9548\times166.4\times0.5\times10^{-6}$$

$$= 2.50 \text{ ohms.}$$

(**J**) Calculate Q per Fig. 19.

$$Q = \frac{2\pi f L}{R_w + R_c} = \frac{2\pi 5000\times0.9548}{145.1+2.5} = 203.2.$$

(**K**) Calculate Q_s per Fig. 21.

$$Q_s = Q(1-\omega^2 L C_d)$$

$$= 203.2[1-(2\pi 5000)^2\times0.9548\times48\times10^{-12}]$$

$$= 194.0.$$

(**L**) Temperature coefficient

$$= \mu_e\times\pm0.6\times10^{-6} = 166.4\times0.6\times10^{-6}$$

or ±99.8 ppm/°C from $+20$ to $+70$°C.

(**M**) The coil is wound on a 1-section bobbin for a 2213P core and is assembled with a 2213P $A_L = 400$ core and proper clamping assembly. The adjuster is screwed into the core and the inductance is checked to be sure there is a $\pm3\%$ tuning range of about 1000 millihenries at 5 kilohertz. Some adjustment of turns may be necessary to allow for the winding method, processing, and estimated C_d.

TEMPERATURE AND HUMIDITY

Table 12 lists the standard classes of insulating materials and their limiting operating temperatures. Table 13 compares the properties of five high-temperature wire-insulating coatings.

Open-type constructions generally permit greater

TABLE 13—COMPARISON OF FIVE HIGH-TEMPERATURE WIRE-INSULATING MATERIALS. *From J. Holland, "Choosing Wire Insulation For High Temperatures," Electronic Design, vol. 2, p. 14; July 1954.*

Characteristic	Modified Teflon	Teflon	Silicone Enamel DC1360	Formvar (vinyl acetal)	Plain Enamel
Upper temp limit	+250°C	+250°C	+180°C	+105°C	+80°C
Lower temp limit	−100°C	−100°C	−40°C	−40°C	−40°C
Dielectric strength	Excellent	Very good	Very good	Good	Good
Dielectric constant (60 Hz–30 000 MHz)	2.0–2.05*	2.0–2.05*	Inferior	Inferior	Inferior
Power factor (60 Hz–10 000 MHz)	0.0002*	0.0002*	Inferior, about 0.006–0.007	Inferior	Inferior
Space factor	Excellent	Excellent	Excellent	Excellent	Excellent
Solvent resistance	Excellent	Excellent	Fair	Fair	Poor
Abrasion resistance	Good	Fair	Very good	Excellent	Good
Thermoplastic flow	Good	Fair	Excellent	Excellent	Good
Crazing resistance	Excellent	Very good	Fair	Fair	Fair
Flame resistance	Excellent	Excellent	Fair	Poor	Poor
Fungus resistance	Excellent	Excellent	Good	Good	Poor
Moisture resistance	Excellent	Excellent	Good	Good	Good
Continuity of insulation	Excellent	Excellent	Good	Good	Good
Arc resistance	Excellent	Excellent	Good	Good	Good
Flexibility	Excellent	Very good	Good	Good	Good

* Stable at temperatures up to 250°C.

cooling than enclosed types, thus allowing smaller sizes for the same power ratings. Moderate humidity protection may be obtained by impregnating and dip-coating or molding transformers in polyester or epoxy resins; these units provide good heat dissipation but are not as good in this respect as completely open transformers.

Protection against the detrimental effects of humidity is commonly obtained by enclosing transformers in hermetically sealed metal cases. This is particularly important if very fine wire, high output voltage, or direct-current potentials are involved. Heat conductivity to the case exterior may be improved by the use of asphalt or thermosetting resins as filling materials. Best conductivity is obtained with high-melting-point silica-filled asphalts or resins of the polyester or epoxy types. Coils impregnated with these resins dissipate heat best since voids in the heat path may be eliminated.

Immersion in oil is an excellent means of removing heat from transformers. An air space or bellows must be provided to accommodate expansion of oil when heated.

DIELECTRIC INSULATION AND CORONA

For class A (Table 12), a maximum dielectric strength of 40 volts/mil is considered safe for small thicknesses of insulation. At high operating voltages, due regard must be paid to corona that occurs before dielectric breakdown and will in time deteriorate insulation and cause dielectric failure. Best practice is to operate insulation at least 25 percent below the corona starting voltage. Approximate 60-hertz root-mean-square corona voltage V is

$$\log \frac{V \text{ (in volts)}}{800} = \tfrac{2}{3} \log(100t)$$

where $t=$ total insulation thickness in inches. This may be used as a guide in determining the thickness of insulation. With the use of varnishes that require no solvents, but solidify by polymerization, the bubbles present in the usual varnishes are eliminated, and much higher operating voltages and, hence, reduction in the size of high-voltage units may be obtained. Epoxy resins and some polyesters belong in this group. In the design of high-voltage transformers, the creepage distance required between wire and core may necessitate the use of insulating channels covering the high-voltage coil, or taping of the latter. For units operating at 10 kilovolts or higher, oil insulation will greatly reduce creepage and, hence, the size of the transformer.

CHAPTER 14

RECTIFIERS AND FILTERS

BASIC RECTIFIER CHARACTERISTICS

All rectifiers exhibit relatively low resistance (forward resistance) when the anode is positive with respect to its cathode, and high resistance (reverse resistance—typically at least 10^3 times the forward resistance) when the anode is negative with respect to the cathode. Since semiconductors have replaced electron tubes in most rectifier applications, the discussion of electron-tube rectifiers is brief and most of the material deals with semiconductor rectifiers.

COMPARISON OF RECTIFIER TYPES

Electron-tube rectifiers have one advantage over semiconductors. The reverse resistance of a tube rectifier is essentially infinite (neglecting insulation leakage, etc.) while that of a semiconductor is finite (although considerably higher than the forward resistance). Semiconductors have the advantages of requiring no filament power and, for a given power rating, are much smaller in size. For many high-power applications, semiconductor rectifiers must be mounted on heat sinks. However, the combination of rectifier and heat sink generally occupies less space than an equivalent electron-tube device.

Table 1 compares the characteristics of silicon, germanium, selenium, and copper-oxide semiconductor rectifiers.

TYPICAL RECTIFIER CIRCUITS

Table 2 shows seven of the most commonly used power-rectifier circuits and general design information for each type. Their advantages, disadvantages, and common applications follow.

Single-Phase Half-Wave Rectifier: Since only half of the input wave is used, the efficiency is low and the regulation is relatively poor. Capacitors are commonly used in half-wave circuits to increase the output voltage and decrease the ripple. The output voltage and degree of filtering are determined by the value of capacity used in relation to

the load current. Transformer design is complicated, and the unidirectional secondary current flow causes core saturation and poor regulation. Most half-wave circuits operate either directly from ac lines, or at a high voltage with a relatively low current.

Single-Phase Full-Wave Center-Tap Rectifier: The efficiency is good but the transformer ac voltage is approximately 2.2 times the dc output voltage. The circuit requires a larger transformer than an equivalent bridge rectifier, with the added complication of a center tap. Each arm of the center-tap circuit must block the full terminal voltage of the transformer. Because of this, center-tap connections are economical only in voltage ranges where not more than one rectifier per arm is required. If series units must be used to obtain the required output voltage, a bridge circuit is preferable.

Single-Phase Full-Wave Bridge Rectifier: If single-phase full-wave output is required, a bridge circuit is commonly used. Efficiency is good and transformer design is easy. Filtering is simplified because the ripple frequency is twice the input frequency.

Three-Phase Wye (or Star) Half-Wave Rectifier: Commonly used if dc output-voltage requirements are relatively low and current requirements are moderately large. The dc output voltage is approximately equal to the phase voltage. However, each of the three arms must block the line-to-line voltage, which is approximately 2.5 times the phase voltage. For this reason, it is desirable to use a 3-phase half-wave connection only where one series unit per arm will provide the required dc output. The transformer design and utilization are somewhat complicated because there is a tendency to saturate the core with unidirectional current flow in each winding.

Three-Phase Full-Wave Bridge Rectifier: Commonly used if high dc power is required and if efficiency must be considered. The ripple component in the load is 4.2% at a frequency 6 times the input frequency, so additional filtering is required in most applications. The dc output voltage is approximately 25% higher than the phase voltage, and each arm must block only the phase voltage. Transformer utilization is good.

TABLE 1—COMPARISON OF SEMICONDUCTOR RECTIFIERS.

	Silicon	Germanium	Selenium	Copper Oxide
Size	Very small	Very small	Large	Large
Weight	Very light	Very light	Light	Heavy
Cooling	Natural or forced	Natural or forced	Natural or forced	Natural or forced
Life	Very long	Very long	60 000 to 100 000 hrs. normal; can vary according to rating and cooling means	Long
Aging (forward resistance)	None	None	Increases	Stabilizes
Forward loss at same current density	Good	Excellent (lowest)	Fair (highest)	Good
Approx. dc forward voltage drop per cell	0.9 volt	0.65 volt	1.0 volt	0.5 volt
Leakage (reverse) current	Excellent	Excellent	Good	Fair
Ability to recover from voltage transients	None	None	Excellent	Good
Unforming (loss of rectifier characteristic)	None	None	Some	None
Series operation of cells	Good	Good	Excellent	Excellent
Parallel operation of cells	Good	Good	Excellent	Excellent
Present cell operating temperature limit	200°C	105°C	130°C	75°C
Thermal capacity	Poor	Poor	Fair	Best
Efficiency at low voltage	Good	Excellent	Fair	Good
Humidity effects	Hermetically sealed	Hermetically sealed	Negligible	Negligible
Frequency response	Good	Good	Poor	Fair

Three-Phase Diametric Half-Wave Rectifier: The characteristics of this circuit approximate those of the 3-phase double-wye circuit without an interphase transformer. Popular applications include requirements for very high dc load currents in low- to medium-voltage ranges (approximately 6 to 125 volts dc).

Three-Phase Double-Wye Half-Wave Rectifier: A 3-phase double-wye connection is recommended if a very high direct current is required at a relatively low dc voltage. Each arm is required to block the full phase voltage of the secondary windings. The dc output current rating is double that of a three-phase bridge or half-wave connection. However, the output voltage is only 75% of the phase voltage. The transformer design is complicated by additional connections and extra insulation, and an interphase transformer (or balance coil) is required.

CIRCUIT DESIGN FOR SEMICONDUCTOR POWER RECTIFIERS

Tables 3 and 4 show the theoretical values of direct and alternating voltages, current, and power for the basic rectifier and transformer elements of single-phase and polyphase conversion circuits, based on perfect rectifiers and transformers. The equations and the values of the constants K and I_{ac} are approximate, but they are sufficiently accurate for practical design of small rectifier circuits.

Symbols for Tables 3 and 4

I_{ac} = transformer secondary current in root-mean-square amperes

I_{dc} = average load direct current in amperes

K = circuit form factor

n = number of cells in series in each arm of rectifier

V_{ac} = alternating root-mean-square input voltage per secondary winding (see diagrams)

$V_{ac\Delta}$ = phase-to-phase alternating input voltage for 3-phase full-wave bridge

V_{dc} = average value of direct-current output voltage

V_p = reverse root-mean-square voltage per cell (rating of rectifier cell)

ΔV = root-mean-square voltage drop per cell at I_{dc}.

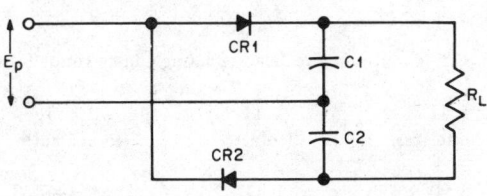

BASIC CIRCUIT

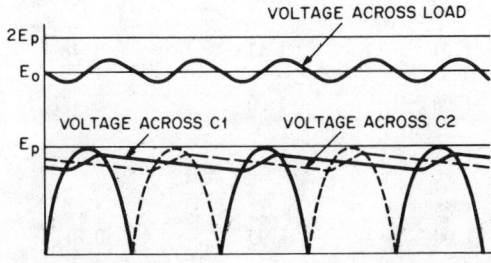

WAVEFORM

Fig. 1—Conventional voltage doubler.

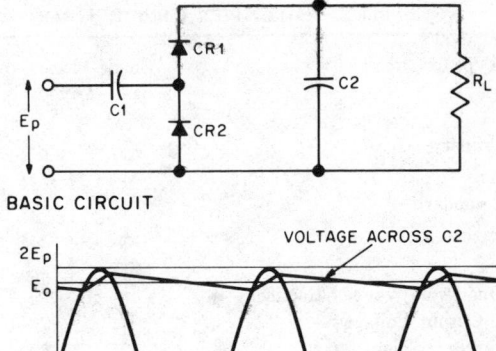

BASIC CIRCUIT

WAVEFORM

Fig. 2—Cascade voltage doubler.

More-rigorous equations, to be found in textbooks and technical papers, should be used in designing rectifiers with outputs in excess of about 10 kW and for accurate computations of regulation, efficiency, power factor, and overload characteristics of even small rectifiers.

VOLTAGE-MULTIPLIER RECTIFIER CIRCUITS

These circuits use the principle of charging capacitors in parallel from the ac input and adding the voltages across them in series to obtain dc voltages higher than the source voltage. Filtering must be of the capacitor-input type.

Conventional and Cascade Voltage Doublers: In the conventional circuit (Fig. 1), capacitors C1 and C2 are each charged, during alternate half-cycles, to the peak value of the alternating input voltage. The capacitors are discharged in series into load R_L, thus producing an output across the load of approximately twice the ac peak voltage.

In the cascade circuit (Fig. 2), C1 is charged to the peak value of the ac input voltage through rectifier CR2 during one half-cycle, and during the other half-cycle it discharges in series with the ac source through CR1 to charge C2 to twice the ac peak voltage.

The "conventional" circuit has slightly better regulation and, since the ripple frequency is twice the supply frequency, the output is easier to filter, the percentage ripple being approximately the same in both circuits. In addition, both capacitors are rated at the ac peak voltage, whereas C2 in the cascade circuit must be rated at twice this value. With both circuits, the peak inverse voltage across each rectifier is twice the ac peak. The cascade

TABLE 2—RECTIFIER CIRCUIT CHART. THE DATA ASSUME ZERO FORWARD DROP AND ZERO

Type of Circuit→		Single-Phase Half Wave	Single-Phase Center Tap	Single-Phase Bridge	Three-Phase Star (Wye)
Primary→					
Secondary→					
One Cycle Wave of Rectifier Output Voltage (No Overlap)					
Number of rectifier elements	=	1	2	4	3
RMS dc volts output	=	1.57	1.11	1.11	1.02
Peak dc volts output	=	3.14	1.57	1.57	1.21
Peak reverse volts per rectifier element	=	3.14	3.14	1.57	2.09
	=	1.41	2.82	1.41	2.45
	=	1.41	1.41	1.41	1.41
Average dc output current	=	1.00	1.00	1.00	1.00
Average dc output current per rectifier element	=	1.00	0.500	0.500	0.333
RMS current per rectifier element:					
Resistive load	=	1.57	0.785	0.785	0.587
Inductive load	=	—	0.707	0.707	0.578
Peak current per rectifier element:					
Resistive load	=	3.14	1.57	1.57	1.21
Inductive load	=	—	1.00	1.00	1.00
Ratio of peak to average current per element:					
Resistive load	=	3.14	3.14	3.14	3.63
Inductive load	=	—	2.00	2.00	3.00
% Ripple (rms of ripple/ average output voltage)	=	121%	48%	48%	18.3%
Ripple frequency	=	1	2	2	3
		Resistive Load		Inductive Load or Large Choke Input Filter	
Transformer secondary rms volts per leg	=	2.22	1.11 (to center-tap)	1.11 (total)	0.855 (to neutral)
Transformer secondary rms volts line-to-line	=	2.22	2.22	1.11	1.48
Secondary line current	=	1.57	0.707	1.00	0.578
Transformer secondary volt-amperes	=	3.49	1.57	1.11	1.48
Transformer primary rms amperes per leg	=	1.57	1.00	1.00	0.471
Transformer primary volt-amperes	=	3.49	1.11	1.11	1.21
Average of primary and secondary volt-amperes	=	3.49	1.34	1.11	1.35
Primary line current	=	1.57	1.00	1.00	0.817
Line power factor	=	—	0.900	0.900	0.826

Reverse Current in Rectifiers and No Alternating-Current Line or Source Reactance.

Three-Phase Bridge	Six-Phase Star (Three-Phase Diametric)	Three-Phase Double Wye with Interphase Transformer	Note: Assumes perfect rectifiers and zero reactance of ac line and source
			To Determine Actual Value of Parameter in Any Column, Multiply Factor Shown By Value of:
6	6	6	
1.00	1.00	1.00	× Average dc voltage output
1.05	1.05	1.05	× Average dc voltage output
1.05	2.09	2.42	× Average dc voltage output
2.45	2.83	2.83	× RMS secondary volts per transformer leg
1.41	1.41	1.41 (diametric)	× RMS secondary volts line-to-line
1.00	1.00	1.00	× Average dc output current
0.333	0.167	0.167	× Average dc output current
0.579	0.409	0.293	× Average dc output current
0.578	0.408	0.289	× Average dc output current
1.05	1.05	0.525	× Average dc output current
1.00	1.00	0.500	× Average dc output current
3.15	6.30	3.15	
3.00	6.00	3.00	
4.2%	4.2%	4.2%	
6	6	6	× Line frequency f

Inductive Load or Large Choke Input Filter

0.428 (to neutral)	0.740 (to neutral)	0.855 (to neutral)	× Average dc voltage output
0.740	1.48 (max)	1.71 (max-no load)	× Average dc voltage output
0.816	0.408	0.289	× Average dc output current
1.05	1.81	1.48	× DC watts output
0.816	0.577	0.408	× Average dc output current
1.05	1.28	1.05	× DC watts output
1.05	1.55	1.26	× DC watts output
1.41	0.817	0.707	× (Avg. load current × sec. leg voltage)/ primary line voltage
0.955	0.955	0.955	

TABLE 3—SINGLE-PHASE-RECTIFIER CIRCUITS, EQUATIONS, AND DESIGN CONSTANTS.

Constant	Half-Wave	Full-Wave Center Tap	Full-Wave Bridge
Circuit			
V_{ac}	$KV_{dc} + n\Delta V$	$KV_{dc} + n\Delta V$	$KV_{dc} + 2n\Delta V$
Resistive and inductive loads:			
n	$KV_{dc}/(V_p - \Delta V)$	$2KV_{dc}/(V_p - 2\Delta V)$	$KV_{dc}/(V_p - 2\Delta V)$
V_p	V_{ac}/n	$2V_{ac}/n$	V_{ac}/n
K	2.26	1.13	1.13
$I_{ac,rms}$	$1.57 I_{dc,Av}$	$0.785 I_{dc,Av}$	$1.11 I_{dc,Av}$
Battery and capacitive loads:			
n	$2KV_{dc}/(V_p - 2\Delta V)$	$2KV_{dc}/(V_p - 2\Delta V)$	$KV_{dc}/(V_p - 2\Delta V)$
V_p	$2V_{ac}/n$	$2V_{ac}/n$	V_{ac}/n
K	1.0	0.85	0.85
$I_{ac,rms}$	$2.3 I_{dc,Av}$	$1.15 I_{dc,Av}$	$1.65 I_{dc,Av}$

circuit, however, has the advantage of a common input and output terminal and, therefore, permits the combination of units to give higher-order voltage multiplications. The regulation of both circuits is poor, so that only small load currents can be drawn.

Bridge Voltage Doubler: This circuit (Fig. 3) is a combination of the conventional voltage doubler and the bridge rectifier circuit. If CR3 and CR4 are removed, the circuit becomes a "conventional" voltage doubler. If the two capacitors and the connection from their midpoint to the junction of CR3 and CR4 are removed, the circuit becomes a standard bridge rectifier.

Further Voltage Multiplication: The cascade voltage doubler shown in Fig. 2 can be combined several times to obtain higher dc voltages, as shown in Fig. 4. The voltage ratings of all the capacitors and rectifiers are twice the ac peak

voltage, but the capacitors must have the values shown. The value of C will be the same as that for the cascade voltage doubler (Fig. 2), which is the basic unit for the circuit in Fig. 4. The load current must be small. The increasing size of capacitors and the deterioration in regulation limit the voltages that can be obtained from this type of circuit.

SILICON RECTIFIERS

Ratings

Silicon-rectifier ratings* are generally expressed in terms of reverse-voltage ratings and of mean-

* For a complete list of silicon-rectifier ratings, refer to the EIA—JEDEC Recommendations for Letter Symbols, Abbreviations, Terms and Definitions for Semiconductor Device Data Sheets and Specifications, published by Electronics Industries Association, 2001 Eye Street, N.W., Washington, D.C. 20006.

TABLE 4—THREE-PHASE-RECTIFIER CIRCUITS, EQUATIONS, AND DESIGN CONSTANTS. USED FOR ALL LOADS.

Constant	Half-Wave	Full-Wave-Bridge
Circuit		
Input	$V_{ac} = KV_{dc} + n\Delta V$	$V_{ac\Delta} = KV_{dc} + 2n\Delta V$
n	$1.73 KV_{dc}/(V_p - 1.73\Delta V)$	$KV_{dc}/(V_p - 2\Delta V)$
V_p	$1.73 V_{ac}/n$	$V_{ac\Delta}/n$
K	0.855	0.74
$I_{ac,rms}$	$0.577 I_{dc,Av}$	$0.816 I_{dc,Av}$

forward-current ratings in a half-wave circuit operating from a 60-hertz sinusoidal supply and into a purely resistive load.

There are three reverse-voltage ratings of importance.

Peak transient reverse voltage $\quad V_{RM}$
Maximum repetitive reverse voltage $\quad V_{RM(rep)}$
Working peak reverse voltage $\quad V_{RM(wkg)}$

V_{RM} is the rated maximum value of any nonrecurrent surge voltage, and this value must not be exceeded under any circumstances, even for a microsecond.

$V_{RM(rep)}$ is the maximum value of reverse voltage that may be applied recurrently, e.g., in every cycle, and will include any circuit oscillatory voltage that may appear on the sinusoidal supply voltage.

$V_{RM(wkg)}$ is the crest value of the sinusoidal voltage of the supply at its maximum limit. The manufacturer generally recommends a $V_{RM(wkg)}$ that has an appreciable safety margin in relation to the V_{RM} to allow for the commonly experienced transient overvoltages on power mains.

Three forward-current ratings are similarly of importance.

Nonrecurrent surge current $\quad I_{FM(surge)}$
Repetitive peak forward current $\quad I_{FM(rep)}$
Average forward current $\quad I_{F(av)}$

Silicon diodes have comparatively small thermal mass, and care must be taken to ensure that short-term overload currents are limited. The nonrecurrent surge current is sometimes given as a single value which must not at any time be exceeded, but it is more generally given in the form of a graph of permissible surge current versus time.

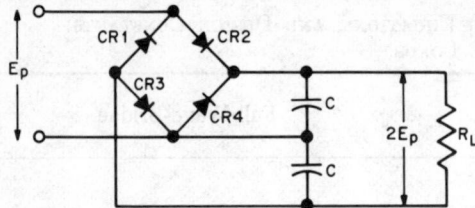

Fig. 3—Bridge voltage doubler.

It is important to observe whether the surge current scale is marked in peak, rms, or average value in order that the data be correctly interpreted.

The repetitive peak forward current is the peak value of the forward current reached in every cycle and excludes random peaks due to transients. Its relation to the average forward current depends on the circuit used and on the load that is applied. For example, the repetitive peak is about three times the average for a half-wave or bridge circuit working into a resistive load; it may be many times greater when the same circuits work into capacitive loads.

Forward Characteristics

The manufacturer generally supplies curves of instantaneous forward voltage versus instantaneous forward current at one or more operating temperatures; a typical characteristic is shown by the solid curve in Fig. 5.

Such curves are not exact for all rectifiers of a given type but are subject to normal production spreads. They are of particular importance in determining the power dissipated by the rectifier under given working conditions.

Calculation of power dissipation from the voltage/current curves need not be done in every instance since the manufacturer gives curves of power dissipation versus forward current for a limited number of commonly used circuits. However, cases do arise for which the particular form of circuit or load is not covered, and it is then necessary to calculate the dissipation for these particular conditions. The calculation can be greatly simplified, with little loss of accuracy in most cases, by approximating the actual V/I characteristic curve to a straight line, as shown by the broken line in Fig. 5. The approximate characteristic corresponds to that of a fixed voltage (the *threshold* voltage) plus a fixed resistance (the *slope* resistance). For any shape of current waveform, the power dissipated at constant voltage is the product of the average current and this fixed voltage, while the power dissipated at constant resistance is the product of the square of the rms current and this fixed resistance. Thus, the following simple equation can be used.

$$P = I_{F(av)} \times V_T + I_{F(rms)}^2 \times R_s$$

where P is the forward power dissipation, $I_{F(av)}$ is

the average forward current through the rectifier, averaged over one complete cycle, $I_{F(rms)}$ is the rms value of the forward current through the rectifier, V_T is the threshold voltage, and R_s is the slope resistance.

For the best accuracy, the straight-line approximation should be drawn through points on the current curve corresponding to 50% and 150% of the peak current at which the rectifier is to be used. Thus, in Fig. 5 the dotted line would correspond to a peak working current of 200 amperes.

Carrier Storage

On switching from forward conduction to reverse blocking, a silicon diode cannot immediately revert to its blocking state because of the presence of stored carriers at the junction. These have the effect of allowing current to flow in reverse, as through a forward-biased junction, when reverse voltage is applied. The current is limited only by the external voltage and circuit. However, the carriers are rapidly removed from the junction, both by internal recombination and by the sweeping effect of the reverse current, and when this has happened the diode reverts to its blocking condition in which only a low leakage current flows. This sudden cessation of a large reverse current can cause objectionable voltage transients if there is appreciable circuit inductance and surge suppression has not been included. The reverse current due to carrier storage is not excessive in normal operation of power rectifier circuits and

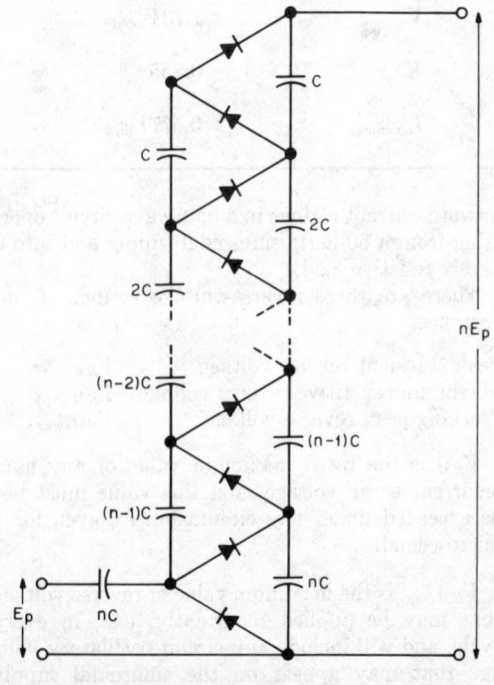

Fig. 4—Circuit for high-order voltage multiplication.

does not in itself constitute a hazard; however, its effect can sometimes lead to complications in switching arrangements. For example, in an inductively loaded circuit, the current will "freewheel" through the diodes after the supply has been removed until the inductive energy has been discharged. Should the supply be reapplied while this process is going on, some of the diodes will be required to conduct in a forward direction but others will be required to block; while the latter are recovering from the carrier storage injected by the free-wheeling current, the short-circuit across the supply can cause a damaging surge current to flow.

Parallel Operation

Silicon diodes are frequently used in parallel assemblies to provide large outputs of rectified current. However, diodes of any one type do not necessarily have exactly the same forward characteristic, and this leads to problems of current sharing.

When the manufacturer divides the production output of a particular type of diode into several grades according to their measured forward characteristics, it is important to use diodes all of one grade in a parallel assembly. It may still be necessary to apply a current derating factor, as recommended by the manufacturer, to determine the number of diodes that will be required to carry a given current.

In addition, care should be taken that the electrical connections to the individual diodes are in a reasonably symmetrical arrangement so that no extra unbalance is introduced by unequal external resistance. It is an advantage to mount as many of the diodes as possible on a common heat sink to facilitate heat transfer between them.

Individual resistances or reactances may be used in the connections to each diode to assist in forced sharing of current. However, these can be difficult to mount in association with normal air-cooled assemblies, and their expense may outweigh the cost of the additional diodes that would otherwise be necessary for safe rating.

Series Operation

When higher voltage outputs are required than are normally possible with existing ratings of single diodes, series assemblies of diodes are used. The problem here is to assure equal sharing of reverse voltage so that the voltage across any one diode does not exceed its rated value. One factor which tends to prevent equal voltage sharing is the dissimilarity between the reverse-leakage-current characteristics of several diodes of the same nomi-

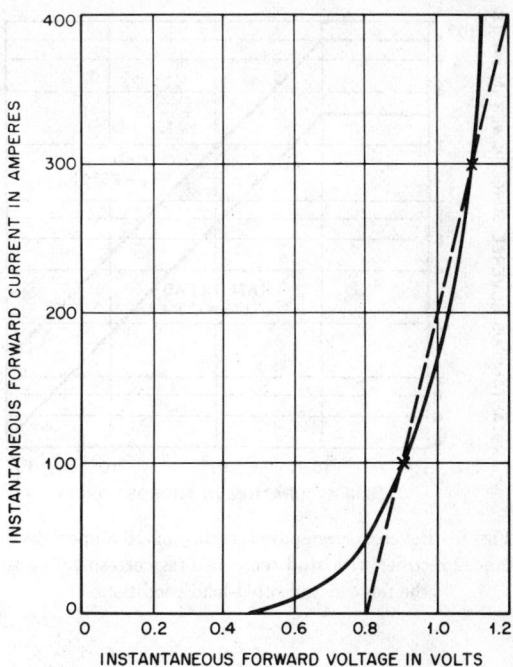

Fig. 5—Instantaneous forward voltage–current characteristic for a typical 100-ampere diode operated at 100°C junction temperature. The ideal threshold forward voltage V_T is the value where the broken line intercepts zero current.

nal type. This problem is generally overcome by applying shunt resistance across each rectifier in the chain such that the current they take when the diodes are reverse-biased is several times greater than the leakage current of the diodes.

Another factor is the difference between the extent of carrier storage present in the diodes after they have been carrying the same forward current. If no precautions are taken, the effect of the different recovery time of each diode (when reverse voltage is applied) will be to apply the full voltage across the first diode in the chain to recover. To overcome this problem, capacitors are connected across each diode in the chain to equalize the transient reverse voltage during the recovery period.

If very long chains of diodes are used to achieve very high voltage rectification, or if a high-frequency supply is used, the effect of stray capacitance to earth must be taken into account. The stray capacitance from each diode interconnection to earth, coupled with the diode junction capacitances, effectively form a capacitive ladder attenuator the effect of which is to apply an unfair share of the total reverse voltage across the early diodes in the chain (at the ac input end). The solution is to keep the ratio of shunt-to-stray-capacitance large, either by minimizing stray capacitance in the layout or by adding effective quantities of capacitance across each diode.

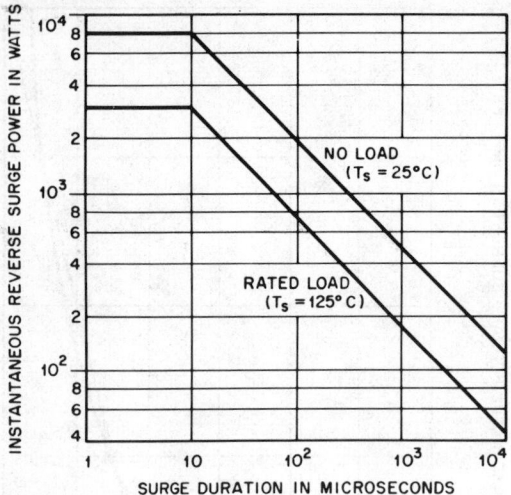

Fig. 6—Reverse surge-power rating of 10-ampere avalanche rectifier with stud temperatures corresponding to the no-load and rated-load conditions.

Heat Sinks

Heat-sink data are generally provided in the diode data sheets to enable the user to select a suitable size of heat sink in relation to the operating current, ambient temperature, and cooling medium (air, convection, or forced air). If the data are applicable to a number of different semiconductor devices, the information is generally set out in the form of graphs of temperature drop in the heat sink versus power dissipation in the device. It is then necessary to note from the silicon-diode data the maximum *case* (or *stud*) temperature that is permissible for a given flow of current. This temperature, less the temperature drop in the sink, is the maximum permissible ambient temperature for the particular working conditions.

Silicon junctions have a very low thermal mass, generally much lower than that of their associated heat sinks, and therefore the heat-sink calculations are only relevant to steady-state operating data. The heat sink will play little part in preventing a sudden burnout due to a heavy overcurrent transient, for example. The junction mass is so small, in fact, that its temperature will actually follow the cyclic variations of current at the supply frequency. This factor is taken into account by the manufacturer in arriving at a rated (mean) operating junction temperature at normal supply frequencies. At lower frequencies, the junction will more nearly reach its steady-state temperature corresponding to the peak of the alternating current; therefore care should be taken to derate appropriately if working at frequencies much lower than that of the normal supply.

SILICON AVALANCHE RECTIFIERS

The avalanche rectifier can withstand high reverse power dissipation without damage. As an example, an avalanche diode with normal forward rating of 10 amperes can dissipate a reverse transient power of 8 kilowatts for 10 microseconds without damage. Even for 10 milliseconds it can withstand a reverse power of 125 watts. Figure 6 illustrates the reverse power rating of such a diode when operated at no load (stud temperature= 25°C) and at rated load (stud temperature= 125°C).

The advantages of the avalanche diode follow.

(**A**) The elimination of surge absorption devices and networks required for conventional diodes.

(**B**) The elimination of voltage dividing networks from series-operated high-voltage diode chains.

(**C**) Fewer diodes are needed in series for a given voltage since there is no need to underrate them on reverse voltage to allow for transient peaks.

In very long chains of diodes designed for extra-high-voltage operation, or where a high supply frequency is used, it may still be advisable to use shunt dividing capacitors (but *not* resistors) even with avalanche rectifiers; this is because of the effects of carrier storage and stray capacitance.

ZENER DIODES

Zener is the name given to a class of silicon diodes having a sharp turnover characteristic at a particular reverse voltage, as shown in Fig. 7. If such a diode is operated in this part of its characteristic, no breakdown (in the sense of dielectric

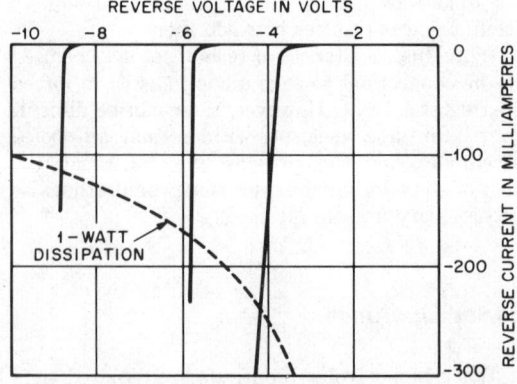

Fig. 7—Typical reverse characteristics for three low-voltage zener diodes of 1-watt rating. *From J. M. Waddell and D. R. Coleman, "Zener Diodes—Their Properties and Applications," Wireless World, vol. 66, no. 1, p. 18, fig. 2; January 1960.* ©1959, Iliffe Electrical Publications, Ltd., London, England.

breakdown) occurs and the process is reversible without damage. The steepness of the reverse part of the current–voltage characteristic in the turn-over region makes these diodes excellent elements for voltage reference and voltage regulation.

A factor of importance in circuit design is the slope resistance, which is the change of reverse voltage per unit change of current in the breakdown region. It is usual to distinguish between ac and dc slope resistance. The former is the dynamic resistance as measured at constant junction temperature. Under dc conditions, however, the junction temperature will stabilize at a new value for each different value of reverse current. The dc slope resistance will therefore be higher than the ac when the diode temperature coefficient is positive and lower when the coefficient is negative. Figure 8 shows typical ac slope resistances for diodes of different breakdown voltages, when run at various mean reverse currents. The temperature coefficient for a typical range of zener diodes is shown in Fig. 9. It will be seen that the coefficient changes from negative to positive in the region of 5 volts. Use is sometimes made of this phenomenon to match diodes of opposite coefficient to produce a series pair having a low effective temperature coefficient in combination.

The capacitance of a zener diode in its break-

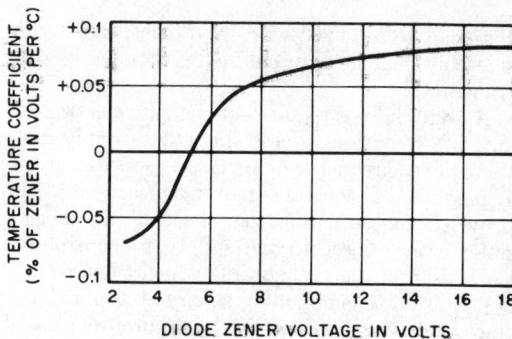

Fig. 9—Temperature coefficient for a typical range of low-voltage zener diodes. *From J. M. Waddell and D. R. Coleman, "Zener Diodes—Their Properties and Applications," Wireless World, vol. 66, no. 1, p. 18, fig. 4; January 1960. ©1959, Iliffe Electrical Publications, Ltd., London, England.*

down region is usually of no practical significance, since it is effectively shunted by the slope resistance of the diode in this region. At reverse voltages below breakdown, capacitance changes with voltage in a manner similar to that of a conventional diode, i.e., capacitance decreases with increase of voltage. Figure 10 illustrates typical values of capacitance for 1-watt zener diodes operated below their breakdown voltage.

Zener diodes have found a variety of low-voltage applications. In addition to voltage references and regulators, they are used as waveform clippers, voltage quantizers, and as coupling elements in dc logic circuits. Figures 11, 12, and 13 illustrate three simple applications of the zener diode.

THYRISTORS (SILICON CONTROLLED RECTIFIERS)

The thyristor* is much like a normal rectifier which has been modified to "block" in the forward direction until a small signal is applied to the control (gate) electrode. After the signal is applied, the device conducts in the forward direction with a forward characteristic very similar to that of a normal silicon rectifier and continues to conduct even after the control signal has been removed. The characteristics are similar to those of a gas thyratron except that the forward drop is about one tenth of that of a thyratron and the deionization time is shorter by several orders of magnitude. The maximum voltage that the thyristor will block is lower, in the present state of the art,

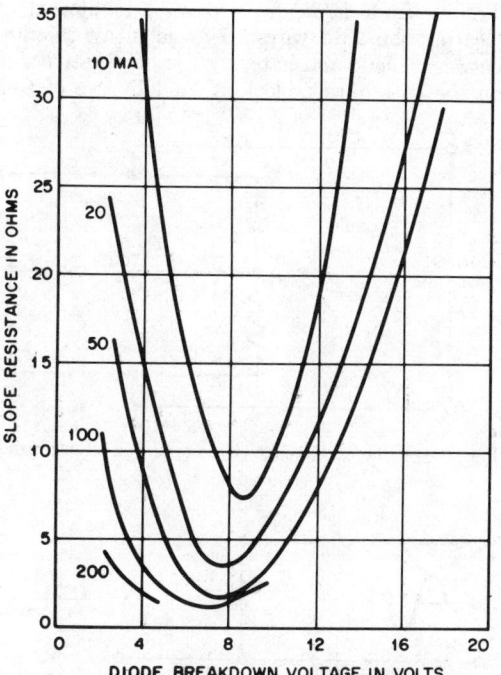

Fig. 8—Slope resistance for zener diodes of various breakdown voltages, at various operating currents. *From J. M. Waddell and D. R. Coleman, "Zener Diodes—Their Properties and Applications," Wireless World, vol. 66, no. 1, p. 18, fig. 3; January 1960. ©1959, Iliffe Electrical Publications, Ltd., London, England.*

* This term, which indicates a general class of solid-state controlled rectifiers, is used throughout this section instead of the term "silicon controlled rectifier" (a 4-layer *pnpn* device that is the most common member of the class).

than can be achieved with thyratrons, but is considerably higher than normally obtainable with transistors.

A small pulse is required at the gate electrode to switch a thyristor on, and the anode supply must be removed, reduced, or reversed to switch it off. If proportional control is required, means must be provided for adjusting the phase of the trigger pulse with respect to the supply to control the proportion of the cycle during which the thyristor is permitted to conduct. In ac circuit applications, the switchoff is obtained by the natural reversal of

the supply voltage every half-cycle; in dc circuits it is usual to charge a "commutating" capacitor during the "on" period and to apply this charge in negative polarity between the anode and cathode when it is desired to switch off.

Ratings

Voltage and current ratings are generally expressed in similar terms to those for silicon rectifiers, as discussed in a previous section. It is necessary, however, to add the following ratings.

Peak Forward Blocking Voltage is the maximum safe value that may be applied, under recurrent or nonrecurrent transient conditions, while the thyristor is in the blocking state. The thyristor will break over into a conducting state regardless of gate drive if either (A) too high a positive voltage is applied between anode and cathode or (B) a positive anode–cathode voltage is applied too quickly (dv/dt firing). Even small voltage pulses, if their leading edges are sufficiently steep at the anode, can turn the thyristor on. Firing by condition (A) is avoided by making the peak forward blocking voltage lower than the breakover voltage of any thyristor of a particular type and seeing that this voltage is not exceeded in practice. Trouble from dv/dt effects may be minimized by locating the gate wires to avoid stray coupling between anode and gate, by use of negative bias on the gate during blocking, and by use of C–R

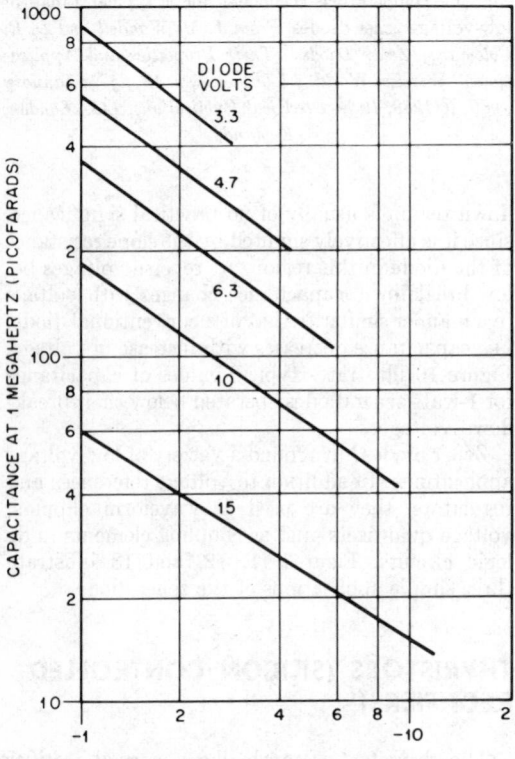

Fig. 10—Typical values of capacitance for 1-watt zener diodes of various voltages, operated below breakdown. *From J. M. Waddell and D. R. Coleman, "Zener Diodes— Their Properties and Applications," Wireless World, vol. 66, no. 1, p. 21, fig. 15; January 1960. ©1959, Iliffe Electrical Publications, Ltd., London, England.*

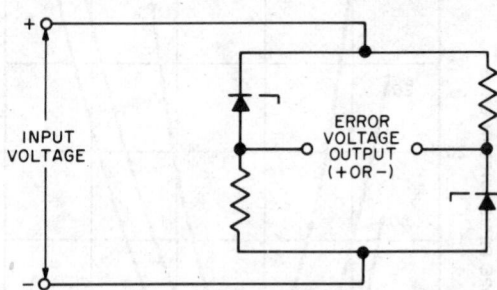

Fig. 12—Voltage reference or comparator circuit using zener diodes.

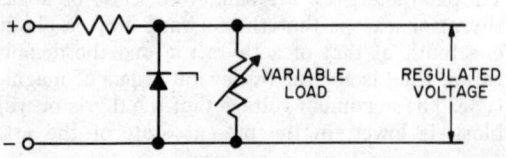

Fig. 11—Simple voltage regulator using a zener diode. *From J. M. Waddell and D. R. Coleman, "Zener Diodes— Their Properties and Applications," Wireless World, vol. 66, no. 1, p. 20, fig. 8; January 1960. ©1959, Iliffe Electrical Publications, Ltd., London, England.*

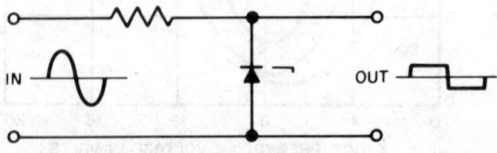

Fig. 13—Waveform clipper or surge limiter using a zener diode. *From J. M. Waddell and D. R. Coleman, "Zener Diodes—Their Properties and Applications," Wireless World, vol. 66, no. 1, p. 20, fig. 9; January 1960. ©1959, Iliffe Electrical Publications, Ltd., London, England.*

damping circuits between anode and cathode to slow down the rate of change of applied voltage.

Continuous Forward Blocking Voltage covers operation under dc conditions.

Peak Forward Gate Voltage is quoted for anode positive with respect to cathode and for anode negative with respect to cathode. The voltage rating is quite low in the latter case (typically 0.25 V) since reverse voltage rating is reduced by forward gate current.

Peak Reverse Gate Voltage is generally the same whether the anode is positive or negative with respect to the cathode.

Peak Forward Gate Current. Forward gate impedance is a finite value subject to quite large variations between samples and over a temperature range. It is usually necessary to plot a load line on gate current–voltage characteristics to determine the gate current that may flow due to a given external gate voltage and source resistance. Care must be taken that the rating is not exceeded with all known spreads of gate–cathode characteristic and temperature.

Gate Dissipation is generally given in terms both of average rating and of peak rating.

Characteristics

Characteristics of the thyristor important for circuit design are as follows.

Leakage Currents are specified for both forward and reverse blocking, at maximum applied voltage and at maximum rated temperature. Although these are low in comparison with forward conducting currents and can be neglected in assessing power losses, they must be taken into account in certain circumstances. An example would be a circuit in which a capacitor is slowly charged from

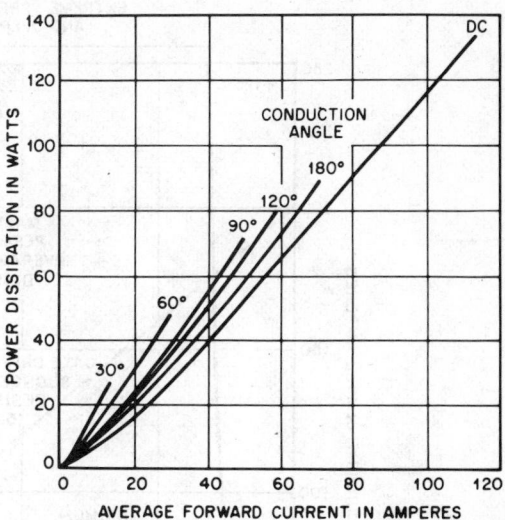

Fig. 15—Power dissipation vs average thyristor forward current in single-phase half-wave or bridge circuit feeding a resistive load. (70-ampere average rated thyristor.)

an external source and then suddenly discharged through a thyristor into a second circuit (as in pulse modulators); the capacitor charging operation may be affected by the amount of forward leakage current conducted by the thyristor in its blocking state.

Holding Current is the minimum anode–cathode current that will keep the thyristor conducting after it has been switched on. In some applications a thyristor with a high holding current is wanted so that it can be turned off easily without the need of reducing the anode current to a very low level. In other applications, where a low load current is normal, it might be desirable to have a low holding current to ensure that the thyristor latches on reliably with light loading.

Forward Voltage Drop is important in assessing power loss. The same methods of assessing power loss in terms of the forward current–voltage characteristic apply as in the case of silicon rectifiers. It is impracticable to measure the junction temperature under working conditions, and therefore the manufacturers list maximum values of stud or case temperature related to the forward current. This is expressed in the form of a graph, as shown in Fig. 14. Note that since the ratio of rms to average forward current varies with the angle of conduction, the power dissipation for any average current also varies with this angle. Figure 14 is drawn for fractional sine waves, as would apply to the cases of single-phase half-wave or bridge rectifier circuits working into a resistive load.

Figure 15 shows a typical relationship between power dissipation and average current for a 70-

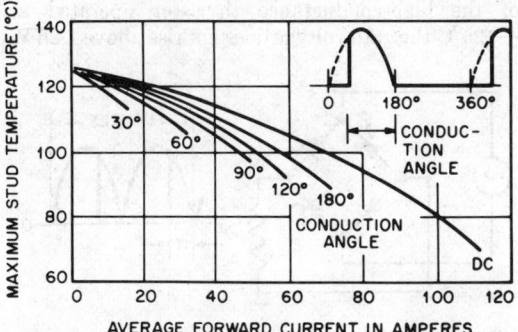

Fig. 14—Maximum permitted stud temperature vs average thyristor forward current in single-phase half-wave or bridge circuit feeding a resistive load. (70-ampere average rated thyristor.)

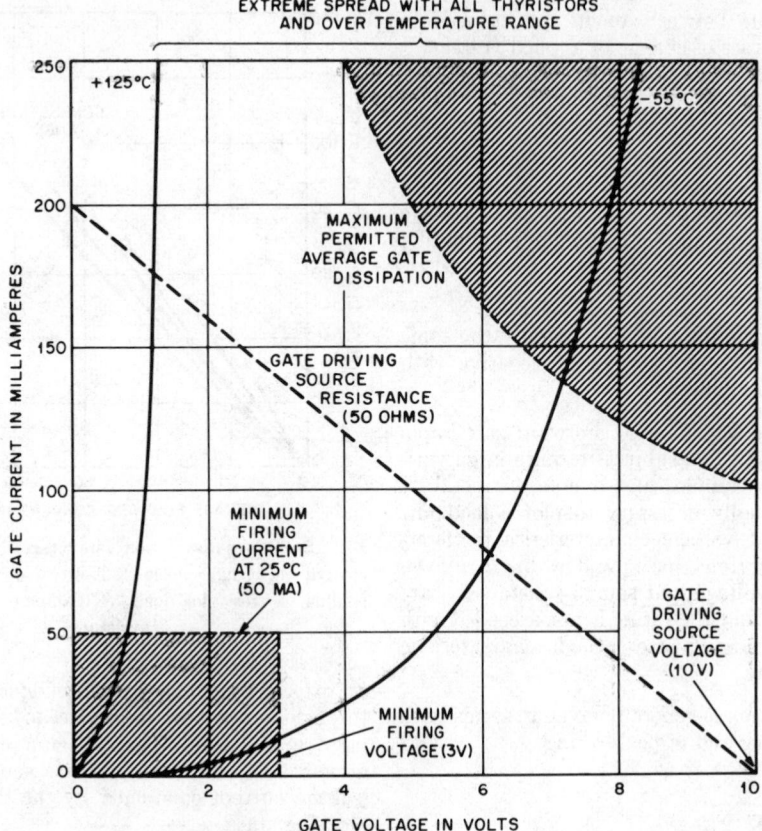

Fig. 16—Gate current–voltage characteristics for all samples of one type of thyristor (70-ampere rating) and for an extreme range of temperature. The maximum rated average gate dissipation is shown and a load line is drawn for typical values of gate driving source voltage and resistance.

ampere thyristor under the same circuit conditions. The two types of graphs illustrated by Figs. 14 and 15 together enable one to calculate the thermal resistance of heat sink required to keep the thyristor below its maximum temperature ratings under given working conditions of current and ambient temperature.

Gate Trigger Sensitivity is specified in terms of a minimum voltage and/or current that must be applied to ensure that all samples of a particular type of thyristor will be triggered into conduction. The minimum voltage is not temperature sensitive but the minimum trigger current varies considerably with temperature, more current being required to turn on at low temperature than at high. The basic requirements of a gate drive circuit are therefore that the driving voltage and source resistance must be such that either the minimum voltage or the minimum current (or both) are exceeded but that the rated gate dissipation is not exceeded. These points are illustrated in Fig. 16, which shows a portion of the gate current–voltage characteristics for a 70-ampere range of thyristors.

In Fig. 16 the two thyristor gate curves represent extremes of characteristics for all thyristors of this type operated over a wide temperature range. The shaded areas must be avoided because of the danger of excess gate dissipation or the danger of inadequate triggering level applied. Any driver load line which avoids these areas, such as the line illustrated, will be satisfactory. Note that in the case of the high-conductance thyristor operated at +125°C the gate voltage does not rise above 1.25 V.

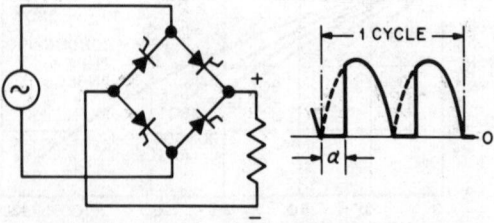

Fig. 17—Full-controlled single-phase thyristor bridge. The broken lines indicate the path of the normal waveform at full conduction, i.e., where $\alpha = 0°$ and the circuit behaves like a diode rectifier. α is the firing angle delay. Delay range required for this circuit is 0° to 180°.

Although this is less than the minimum 3 V specified, the thyristor will fire satisfactorily because the gate current is 175 mA, which is considerably greater than the 50 mA minimum specified. In Fig. 16 the power-dissipation curve corresponds to the average rated value. In practice one very rarely applies steady dc to the gate and, in the cases of pulses or half-sine waves, it is perfectly in order for the load line to cut into the area shown *provided that the peak gate dissipation is not exceeded* (a separate curve may be plotted for this) and so long as the average dissipation (averaged over a complete cycle) is within the limit given.

Switching Times of importance are the turnon and turnoff times, the latter generally being at least one order greater than the former. When a gate signal is applied to the thyristor, there is a finite delay time during which the anode current remains at its normal blocking level; this is followed by a "rise time" during which the anode current increases from its blocking level to a value determined by the external load circuit. Turnon time is the sum of these two times. For a given thyristor the turnon time is influenced by the magnitude of gate drive, the load current to be achieved, and, to a lesser extent, the applied anode supply voltage. The time is reduced by high gate drive, low load current, and high anode supply voltage. Turnoff time is similarly composed of two individual periods; the first is a storage time, analogous to that obtained with a saturated transistor, and the second is a recovery time. Forward voltage may not be reapplied before the completion of both phases of the turnoff process or the thy-

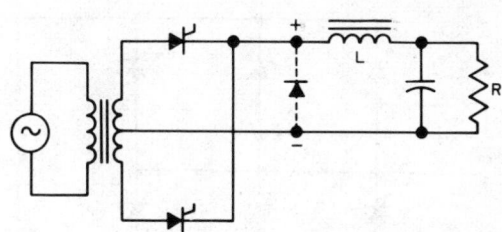

Fig. 19—Half-controlled single-phase thyristor bridge working into inductive load. Common-anode thyristor connection. Waveform and delay range as for Fig. 17.

ristor will conduct again. After this period, however, forward voltage may be applied and the thyristor will remain in its blocking state so long as the rate of rise of anode voltage is not allowed to exceed the specified maximum dv/dt, as already discussed.

THYRISTOR CONTROLLED RECTIFIER CIRCUITS

Figures 17 to 22 are basic circuits of thyristors used as controlled rectifiers. In many applications it is not necessary for all rectifier elements to be controllable, and it is common to find bridges composed of thyristors and ordinary diodes in equal numbers; such circuits are called *half-controlled*, whereas those containing only thyristors are termed *full-controlled*.

Figures 17 and 18 show a full-controlled single-phase thyristor bridge, the thyristor circuit being the same whether a resistive or inductive load is being driven. The voltage waveforms are different, however, in the two cases. The principal difference of practical importance is that the range of firing-pulse phase control is required to be different in the two cases. For the resistive load, phase control over the range 0° to 180° is necessary to obtain

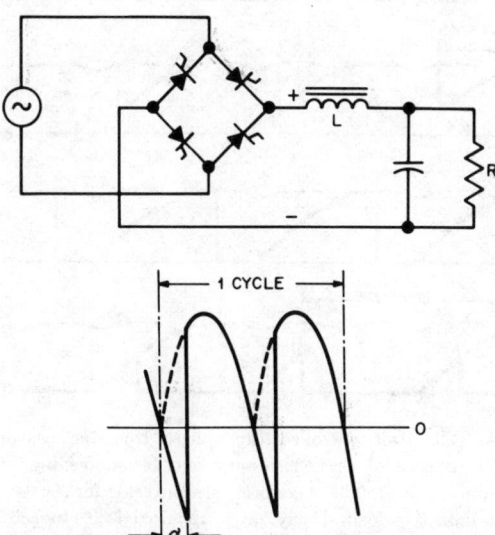

Fig. 18—Full-controlled single-phase thyristor bridge working into an inductive load where $\omega L \gg R$. Delay range required is 0° to 90°.

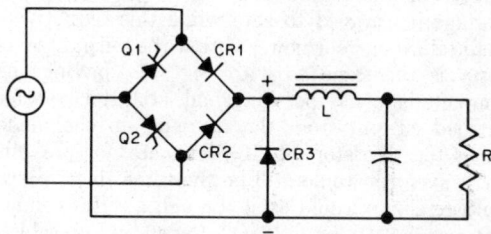

Fig. 20—Full-controlled single-phase push-pull thyristor rectifier working into inductive load. With diode connected, waveform and delay range are as for Fig. 17. Without diode, waveform and delay range are as for Fig. 18 (if $\omega L \gg R$). Addition of diode reduces critical inductance of L for continuous current.

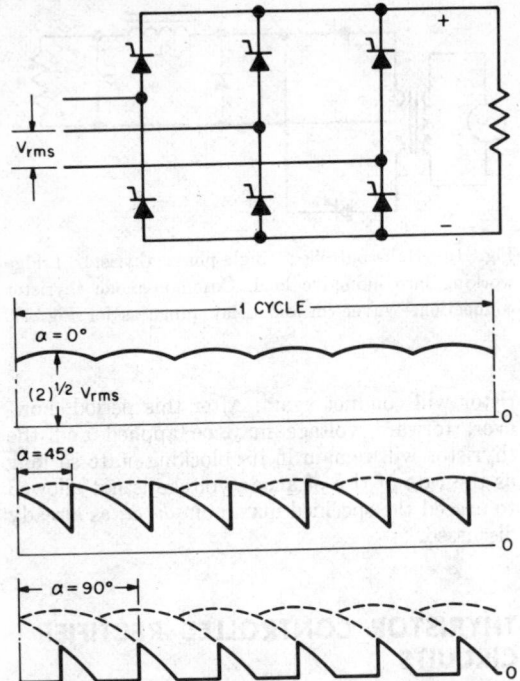

Fig. 21—Full-controlled three-phase thyristor bridge working into resistive load; waveforms of voltage output are shown for three values of α. Waveform develops characteristic flat portion when $\alpha \geq 60°$. Delay range required is 0° to 120°.

full control from maximum output voltage down to zero; for an almost pure inductive load (i.e., a very high $\omega L/R$ ratio) full-phase control is obtained with a range of only 0° to 90°.

Figure 19 shows a half-controlled single-phase rectifier driving an inductive load. For such loads a bypass diode*, CR3, must be added at the output; at the end of a voltage half-cycle, current still flows in the choke, but in this circuit the current is transferred at the end of the voltage half-cycle to the bypass diode. The two important effects of this diode are (A) that the output voltage is clamped to zero while this inductively maintained current flows, so that the output waveform is the same as with Fig. 17 (having the characteristic flat portion) and (B) the transfer of load current from the thyristor to the diode turns the thyristor off. If CR3 were not present, the waveform would still be the same since a zero voltage clamp would exist through a series combination CR1–Q1 or CR2–Q2, depending on which thyristor was conducting the previous half-cycle. However, this flow of uncontrolled current through the thyristor, bypassing the supply, is undesirable; for example, should it be required that the output

* Sometimes called a "commutating," "freewheel," or "flywheel" diode.

voltage be turned off by removal of gate pulses, this action may prove to be impossible. The thyristor can be held on continuously through the negative half-cycle by inductive circulating current and is then ready to conduct the next positive half-cycle. Thus the gate has lost control and a continuous half-wave output is produced. The bypass diode overcomes these difficulties by ensuring thyristor turnoff at the end of each voltage half-cycle.

Figure 20 is a push–pull controlled rectifier circuit. Figure 21 is a three-phase full-controlled rectifier circuit. Figure 22 shows the circuit of a three-phase half-controlled rectifier, the bypass diode being necessary only for inductive loads.

In all the cases illustrated in which a bypass diode is used, this diode must be rated continuously for a maximum average current equal to the load current if the full load current is to be drawn when

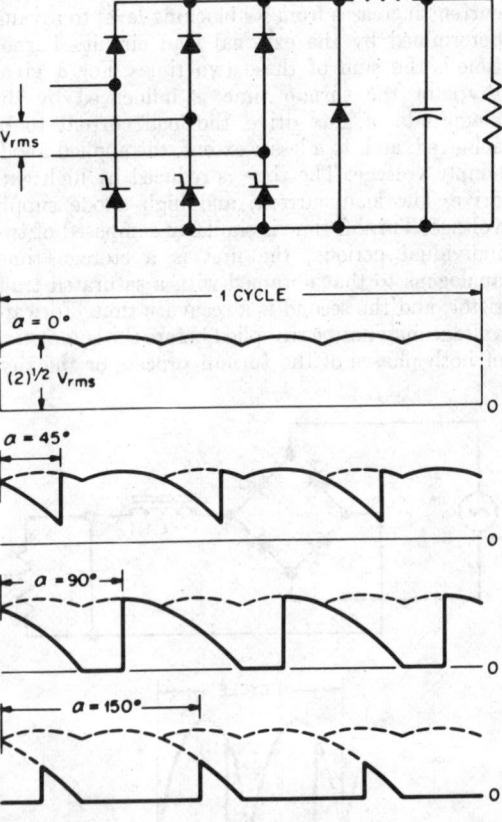

Fig. 22—Half-controlled three-phase thyristor bridge. The bypass diode is necessary only when feeding an inductive load. The waveforms are the same for resistive or inductive load. Delay range required is 0° to 180°. Note that the three thyristors could have been put in the positive bridge arms; it is more usual to put them in the negative arms because a common anode connection permits the use of a common heat sink when thyristors with anode studs are used.

TABLE 5—MEAN DC OUTPUT VOLTAGE FOR THYRISTOR CONTROLLED RECTIFIERS.

Circuit	V_{do}	$V_{d\alpha}$
Fig. 17	$2E_p/\pi$	$V_{do}\frac{1}{2}(1+\cos\alpha)$
Fig. 18	$2E_p/\pi$	$V_{do}\cos\alpha$
Fig. 19	$2E_p/\pi$	$V_{do}\frac{1}{2}(1+\cos\alpha)$
Fig. 20, no diode and inductive load	$2E_p/\pi$	$V_{do}\cos\alpha$
Fig. 20, all other cases	$2E_p/\pi$	$V_{do}\frac{1}{2}(1+\cos\alpha)$
Fig. 21	$3E_p/\pi$	$V_{do}\cos\alpha, \quad$ for $\alpha=0°$ to $60°$ $V_{do}[1+\cos(\alpha+60°)], \quad$ for $\alpha=60°$ to $120°$
Fig. 22	$3E_p/\pi$	$V_{do}\frac{1}{2}(1+\cos\alpha)$

the average output voltage is reduced almost to zero. In practice, a larger diode is often used in this position than in the bridge arms, or several diodes may be used in parallel.

For each of the circuits illustrated in Figs. 17 through 22, Table 5 gives equations for the average dc output voltage, $V_{d\alpha}$, at any angle α in terms of the maximum average dc output voltage, V_{do}, obtained at $\alpha=0°$. The table also shows the value of V_{do} for each circuit in terms of the peak sinusoidal input voltage, E_p. For the single-phase push–pull circuit, the peak input voltage is E_p-0-E_p, and for all the three-phase circuits E_p means the peak value of the line-to-line voltage.

THYRISTOR AC POWER CONTROL CIRCUITS

Figures 23, 24, and 25 show three basic circuits for single-phase ac control, and Figs. 26, 27, and 28 show the load voltage and current waveforms applicable to all three when driving purely resistive and inductive loads.

The load rms current I_{rms} at any phase delay angle α is given in terms of the normal full-load rms current at $\alpha=0°$, I_{rms-0}, by

$$I_{rms}=I_{rms-0}[1-(\alpha/\pi)+(2\pi)^{-1}\sin2\alpha]^{1/2}.$$

The load rms voltage bears the same relation to the full-load rms load voltage. This equation shows that, although the theoretical delay range for complete control is $0°$ to $180°$, a practical range of $20°$ to $160°$ gives a power control of approximately 99% to 1% of maximum.

Figure 23 requires a trigger pulse source having a pair of isolated outputs, insulated from each other by at least the peak value of the supply voltage; the two pulse trains must be phased $180°$ with respect to each other and must shift together, with respect to the supply voltage phase, when the power throughput is adjusted. Provided that two separate pulse trains are used, it is permissible for the two trains to be identical, operating at twice the supply frequency; the thyristor whose anode–cathode is forward-biased will fire when pulses are applied to both. Normally it is not considered safe to drive a thyristor gate positive while its anode is reverse-biased; however, in this circuit it is permissible since the thyristor that is fired immediately removes the reverse voltage from the other.

Figure 24 is a variation of the basic circuit and removes the necessity for isolation of the two pulse

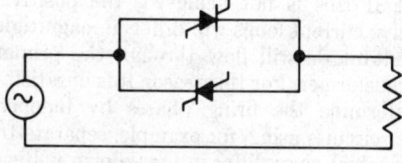

Fig. 23—Inverse–parallel ac control circuit.

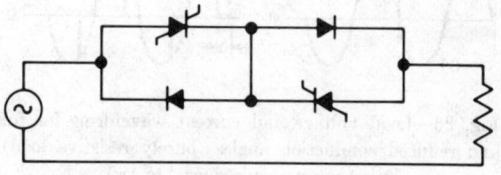

Fig. 24—Inverse–parallel bridge ac control circuit.

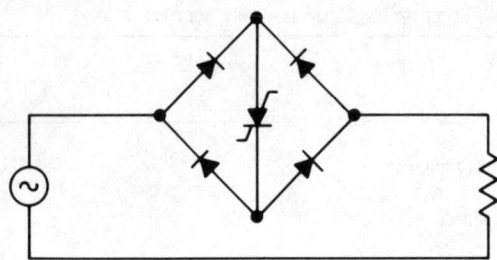

Fig. 25—Single thyristor ac control circuit.

trains from each other; the common thyristor cathode connection permits the use of a simple three-terminal push–pull pulse source. The advantage is gained, however, at the expense of two extra diodes, each of which must be rated for the same current and voltage as the thyristors; there will be twice the heat dissipation compared with the simpler circuit and it is therefore generally used only in low-power circuits (up to about 5 amperes rms load current). Since no reverse voltage, or at least not more than 1 or 2 volts, can be applied to the thyristors in the circuit of Fig. 24, this circuit may be preferred if transient spike voltages are likely to be present on the supply. In Fig. 24 the only resulting breakdown in the thyristors must be in the forward direction, which is not as dangerous as reverse breakdown because the thyristor is automatically turned on by the overvoltage and the voltage across it rapidly drops to zero.

Figure 25 is another variation that is used in low-power control systems. Its main advantage is the use of a single thyristor, with the consequent simplification of trigger circuits. Its main disadvantage is that the conducting path in each direction contains three devices, two diodes and a thyristor in series, with relatively high power dissipation.

With inductive loads, as shown in Fig. 28, there is a sharp transient change of load voltage at the end of each current loop. This may not be important to the load but it does stress the thyristors. Thus at the end of the first positive loop of current, as shown by the dotted waveform in the upper diagram of Fig. 28, the conducting thyristor turns off, thereby disconnecting the load from the supply;

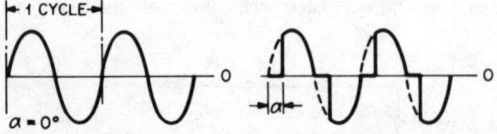

Fig. 26—Load voltage and current waveforms for full and reduced conduction angles (purely resistive load). Delay range required is 0° to 180°.

the voltage at the load therefore rapidly returns to ground. The effect of this transient voltage in the circuits of Figs. 23 and 24 is to apply a sharply rising positive anode voltage to the thyristor opposing the one which has been conducting. If the rated dv/dt is exceeded, this thyristor will then turn on and conduction will take place independent of the gate drive. The rate of rise of voltage may, however, be controlled by adding a resistor in shunt with the load; the rate of rise of voltage will then be determined by the time constant L/R where L is the load inductance and R is the series sum of the load resistance and the added resistor. This resistor has the further advantage of providing a direct resistive path for thyristor current so that

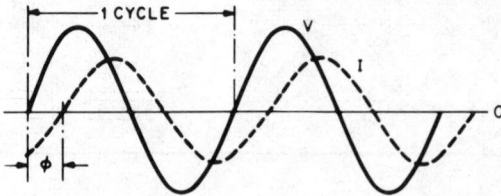

Fig. 27—Voltage and current with full conduction into inductive load.

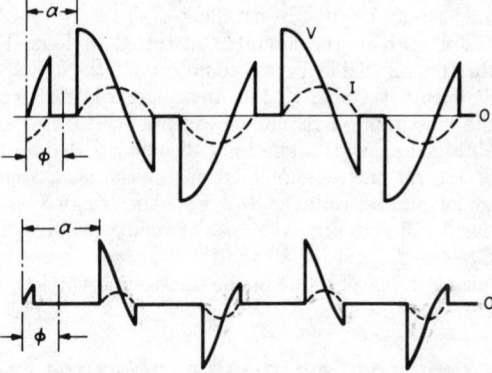

Fig. 28—Voltage and current with reduced conduction, $\alpha > \phi$.

the holding current is established quickly after turnon instead of at a time determined by the rate of rise of load current, which is fixed by the applied voltage and load inductance.

When an intervening transformer couples the load to the thyristor control circuit, it is particularly important to ensure that the two thyristors are fired exactly at 180° relative to each other. If this is not achieved, the positive and negative current loops will differ in magnitude and a resultant dc will flow through the primary of the transformer. For this reason it is unsatisfactory to determine the firing phases by independent trigger circuits using, for example, separate trigger diodes which may differ in breakdown voltage.

The above ac control circuits may be used, via transformers, to drive ordinary diode rectifier circuits; this is done where it is inconvenient to use thyristors in the rectifier circuit itself, because either the rectifier current or the voltage level is too high for existing thyristors.

There are similarly several basic methods of three-phase ac power control, as illustrated by Figs. 29 and 30. Figure 30 is used more frequently for its economy in parts. When driving a transformer–rectifier load, care must be taken to ensure that the phase of thyristor gate pulses bears the correct relationship with the commutation phases of the diode rectifier.

In Figs. 29 and 30 the required delay range of pulses is 0° to 210° with respect to the supply phase-to-neutral voltages. When driving a transformer–rectifier, the necessary delay range is reduced to 120° for Fig. 29 and to 180° for Fig. 30. For star–star or delta–delta transformer connection the zero firing phase corresponds to 30° lag behind the supply phase-to-neutral zero phase. For a delta–star connection the zero firing phase corresponds to the supply phase-to-neutral phase.

THYRISTOR SERVO SYSTEMS

Closed-loop feedback servo systems may be constructed using thyristors as the control element. Figure 31 illustrates three typical systems. Figure 31A represents a voltage-regulated power supply system, for either single- or three-phase inputs. A line transformer feeds a thyristor controlled rectifier of one of the types described, and the output passes through a smoothing filter to the dc load. A portion of the load voltage is compared with a zener diode (or similar) voltage reference and the difference is used to control the phase of trigger pulses applied to the thyristor gates. The usual feedback stability criteria apply. Assuming the use of a trigger phase-shift circuit having a reasonably linear phase-vs-input-voltage characteristic, the low-frequency gain of the system is greatest at the

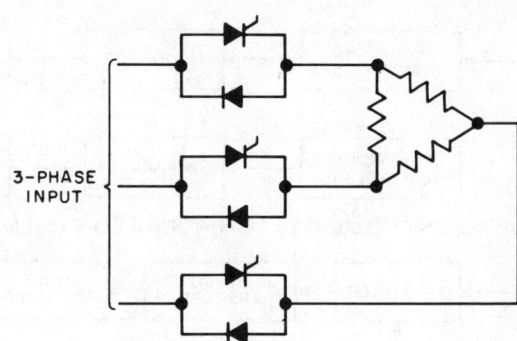

Fig. 30—Half-thyristor three-phase ac control.

steepest part of the $V_{d\alpha}$–α curve, as defined for various circuits by Table 5. The greatest value of the slope $dV_{d\alpha}/d\alpha$ occurs at $\alpha = 90°$ for all the equations shown in Table 5 except those for Fig. 21; here the maximum slope occurs at $\alpha = 60°$. The choke critical inductance and the input ripple to the smoothing filter are both much greater for controlled rectifier circuits than for conventional diode rectifiers. The normal single-stage choke-input filter will have a 12-dB-per-octave response slope at quite a low frequency. It is therefore necessary for stability to ensure that the feedback control circuit has a low cutoff frequency sufficient to reduce the loop gain well below unity at the smoothing-filter cutoff frequency.

Figure 31B illustrates thyristor ac control on the primary side of the rectifier transformer, with ordinary diode rectification on the secondary. With this arrangement it is important to protect the thyristors from the rapid dv/dt transient that can occur on actuating the circuit on the supply side. The solution is to apply capacitance on the mains side of the thyristor stack, between line and neutral or from line to line, which in conjunction with normal mains reactance will slow the rate of rise of thyristor anode voltage sufficiently.

Figure 31C involves no rectification. This is a simple oven temperature control in which a thermistor is used for temperature sensing.

THYRISTOR INVERTERS AND CHOPPERS

Application in pure dc circuits and in dc-to-ac conversion is more complex than in ac circuits because the means of turning the thyristors off cyclically is not automatically present in supply reversals. Inverters generally use a separate oscillator to provide timing for the gate circuits. Industrial sequence switching applications often charge a capacitor during the on period of a thyristor and discharge it in reverse through the

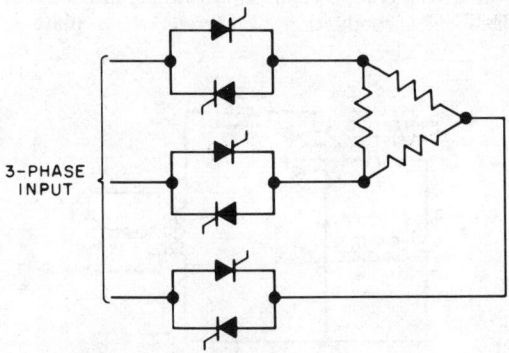

Fig. 29—Full-thyristor three-phase ac control.

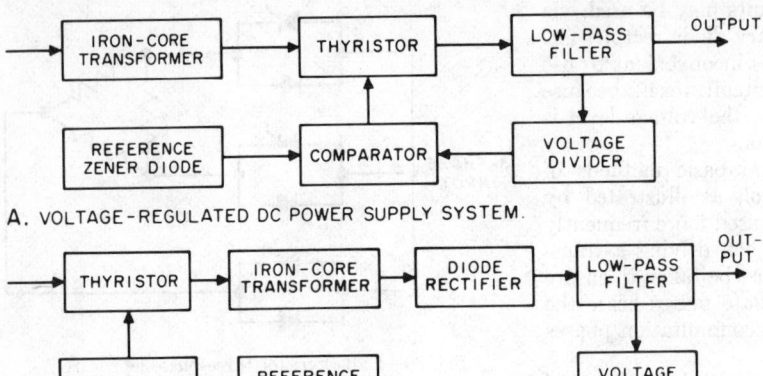

A. VOLTAGE-REGULATED DC POWER SUPPLY SYSTEM.

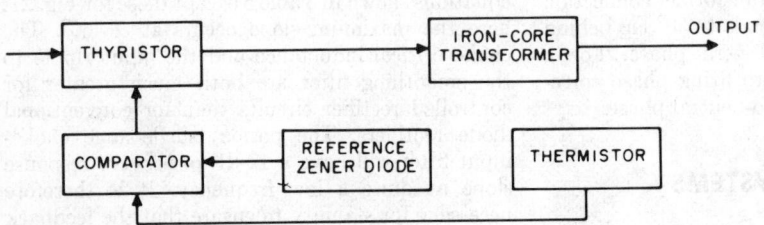

B. VOLTAGE-REGULATED DC POWER SUPPLY SYSTEM, USING THYRISTOR
   AC CONTROL ON THE TRANSFORMER PRIMARY.

C. TEMPERATURE-CONTROLLED AC POWER SUPPLY SYSTEM,
   USING THERMISTOR SENSING.

Fig. 31—Three examples of
closed-loop thyristor control
systems.

thyristor when it is to be turned off. Monostable switching is also achievable by using *L–C* circuits in which reverse oscillatory current is used to turn off the thyristor at a predetermined time after turnon.

High-power inverters use thyristors designed as gate-controlled switches or for turnoff type of operation. Conventional thyristors may be operated successfully from dc in ring counters and other low-power applications.

Direct-current chopper circuits for power control or for small servos have an advantage over conventional dc power circuits by being able to operate at higher frequencies. This improves the response time of control systems and enables smaller iron-core parts to be used. The upper frequency limit of operation is determined by thyristor turnoff time and the drop in efficiency created by switching

times that are an appreciable fraction of the overall cycle.

GRID-CONTROLLED GAS RECTIFIERS

Grid-controlled rectifiers are used to obtain closely controlled voltages and currents. They are commonly used in the power supplies of high-power radio transmitters. For low voltages, gas-filled tubes such as argon (those that are unaffected by temperature changes) are used. For higher voltages, mercury-vapor tubes are used to avoid flashback (conduction of current when plate is

Fig. 32—Critical grid voltage versus plate voltage.

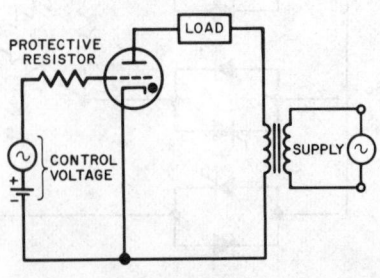

Fig. 33—Basic thyratron circuit. The grid voltage has
dc and ac components.

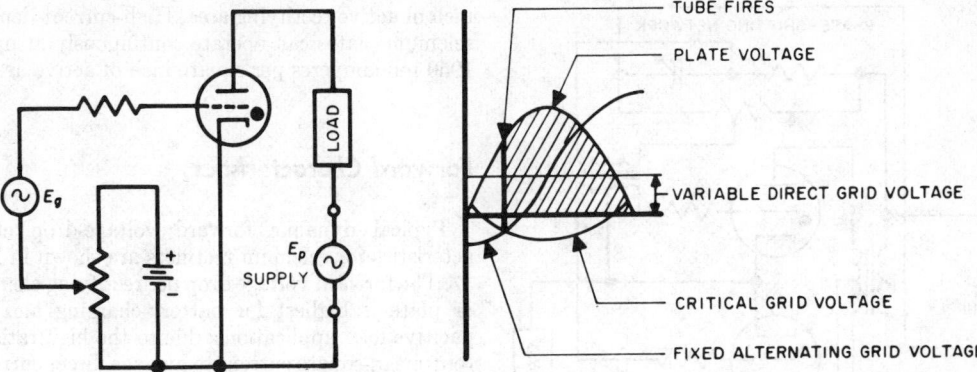

Fig. 34—Control of plate-current conduction period by means of adjustable direct grid voltage. E_g lags E_p by 90°.

negative). These circuits permit large power to be handled, with smooth and stable control of voltage, and permit the control of short-circuit currents through the load by automatic interruption of the rectifier output long enough to permit short-circuit arcs to clear, followed by immediate reapplication of voltage.

In a thyratron, the grid has a one-way control of conduction, and serves to fire the tube at the instant that it acquires a critical voltage. Relationship of the critical voltage to the plate voltage is shown in Fig. 32. Once the tube is fired, current flow is generally determined by the external circuit conditions; the grid then has no control, and plate current can be stopped only when the plate voltage drops to zero.

Basic Circuit

The basic circuit of a thyratron with ac plate and grid excitation is shown in Fig. 33. The average plate current may be controlled by maintaining the following.

(**A**) An adjustable direct grid voltage plus a fixed alternating grid voltage that lags the plate voltage by 90 degrees (Fig. 34).

(**B**) A fixed direct grid voltage plus an alternating grid voltage of adjustable phase (Fig. 35).

Phase Shifting

The phase of the grid voltage may be shifted with respect to the plate voltage as follows.

(**A**) Adjusting the indicated resistor in Fig. 35.

(**B**) Adjusting the inductance of the saturable reactor in Fig. 35.

(**C**) Adjusting the capacitor in Fig. 36.

On multiphase circuits, a phase-shifting transformer can be used.

For a stable output with good voltage regulation, it is necessary to use an inductor-input filter in the load circuit. The value of the inductance is critical, increasing with the firing angle. The design of the plate-supply transformer of a full-wave circuit (Fig. 36) is the same as that of an ordinary full-wave rectifier, to which the circuit of Fig. 36 is closely similar. Grid-controlled rectifiers yield larger harmonic output than ordinary rectifier circuits.

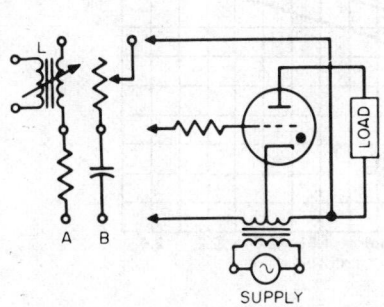

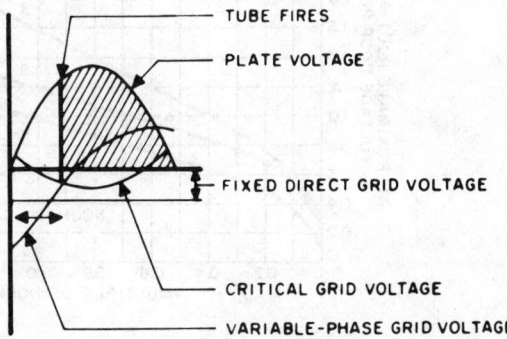

Fig. 35—Control of plate-current conduction period by fixed direct grid voltage (not indicated in schematic) and alternating grid voltage of adjustable phase. Either inductance–resistance (A) or capacitance–resistance (B) phase-shift networks may be used. *L* may be an adjustable inductor of the saturable-reactor type.

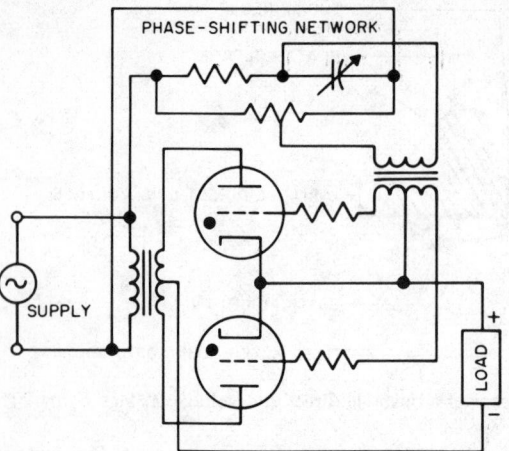

Fig. 36—Full-wave thyratron rectifier. The capacitor is the adjustable element in the phase-shift network and hence gives control of output voltage.

SELENIUM RECTIFIERS

Ratings

It is common practice to rate a selenium rectifier cell on the basis of the root-mean-square sinusoidal voltage that it can withstand in the reverse direction and on the average forward current that it will pass at a certain current density. Typical ratings at 35°C ambient temperature are: 26 volts rms per cell, and 600 milliamperes dc per square

inch of active rectifying area. High-current-density selenium plates can operate continuously at up to 1950 milliamperes per square inch of active area.

Forward Characteristics

Typical dynamic forward voltage-drop characteristics for selenium rectifiers are shown in Fig. 37. The forward voltage drop per rectifying element or plate is highest for battery-charging and capacitive load applications, due to the high ratio of root-mean-square current to average direct current.

Rating of a Selenium Rectifier Stack

Stacks are operated at a temperature that is a safe value with allowance for aging. Catalog rating is in most cases based on an ambient temperature of 35 degrees Celsius. Ratings for higher temperatures (Fig. 38) are based on reduction in forward current to reduce forward-current losses, reduction in reverse voltage to reduce reverse-current losses, or a combination of both forward-current and reverse-voltage reductions to obtain the desired operating temperature with good electrical efficiency. The forward voltage drop and consequent heating depend to a small degree on the temperature of the rectifier cell, as does also the reverse current.

The 35°C rating of a rectifier is based typically on a current density for a cell of 600 milliamperes

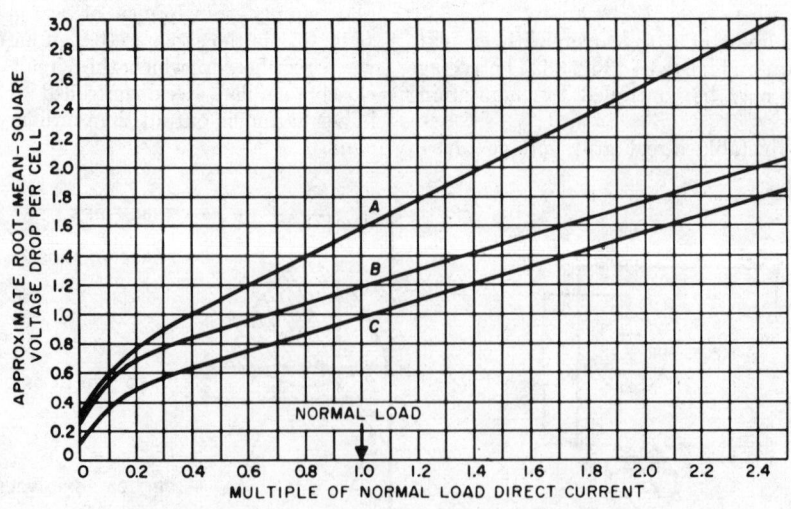

Fig. 37—Typical dynamic forward-voltage-drop curves for selenium-rectifier cells, at 65°C cell temperature. A—Battery or capacitive loads: single-phase half-wave, bridge, or center-tap. B—Resistive or inductive loads: single-phase half-wave, bridge, or center-tap; and 3-phase half-wave. C—All types of loads: 3-phase bridge or center-tap.

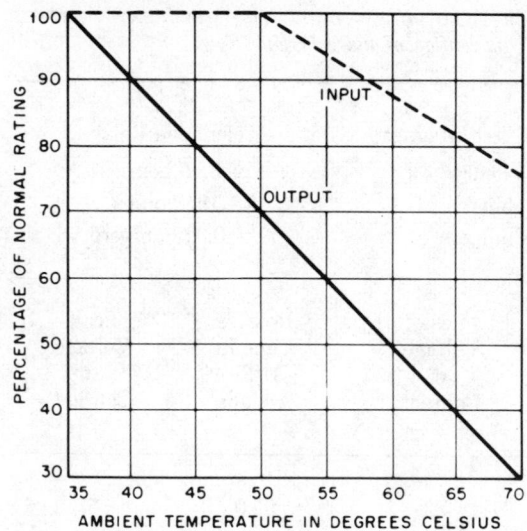

Fig. 38—Selenium-rectifier temperature derating curves (approximate), for root-mean-square alternating input voltage and average direct output current based on 35-degree-Celsius ambient.

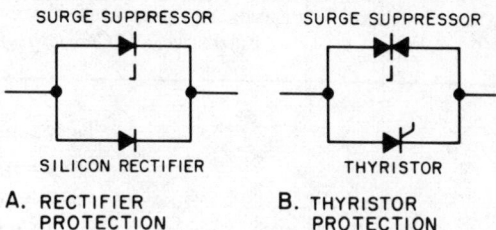

Fig. 41—Basic surge-suppressor configurations.

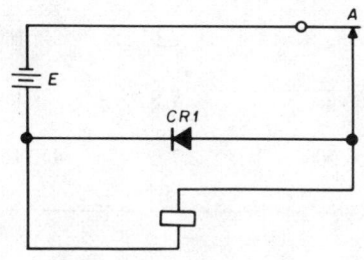

Fig. 39—Conventional method of using the selenium rectifier as a spark suppressor.

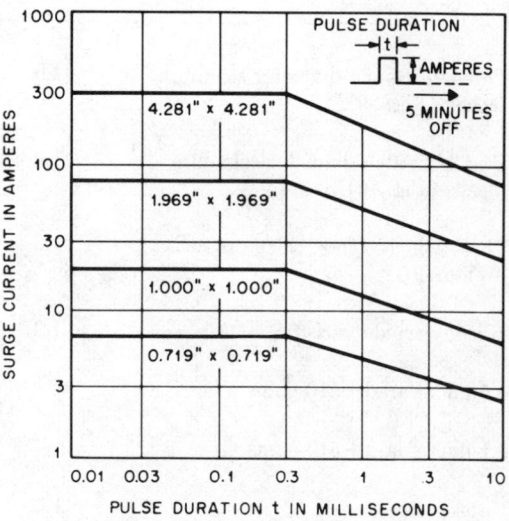

Fig. 42—Surge current vs duration for several sizes of selenium surge-suppressor plates.

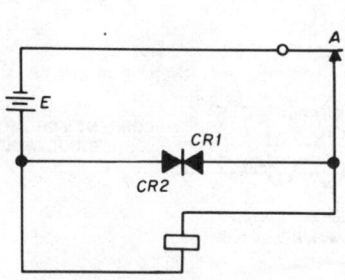

Fig. 40—Method of improving the release time by adding a second rectifier.

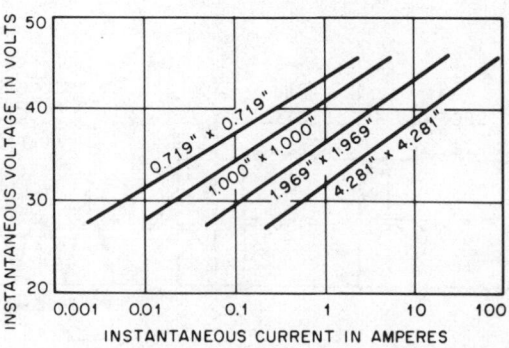

Fig. 43—Instantaneous reverse voltage vs instantaneous reverse current for several sizes of surge-suppressor plates.

TABLE 6—PEAK VOLTAGES AND RELEASE TIMES FOR ELECTROMAGNETS WITH DIFFERENT CONTACT PROTECTIONS. *Courtesy of Transactions of the AIEE.*

Contact Protection	Telephone Clutch Magnet $L=0.485$ henry $R=164$ ohms $I=0.293$ ampere		Telephone Relay $L=3.45$ henries $R=1650$ ohms $I=0.029$ ampere	
	Release Time in Milliseconds	Peak Voltage at Contact	Release Time in Milliseconds	Peak Voltage at Contact
Three 9/32-inch-diameter selenium cells (Fig. 39)	4.0	83	55.0	57
Two 9/32-inch-diameter selenium cells (Fig. 40)*	1.3	180	12.0	150
Three 1-inch square selenium cells (Fig. 40)*	1.3	192	10.9	169
Silicon-carbide varistor	1.3	210	12.8	140
0.5 microfarad+510 ohms	—	arcing	10.9	160
0.1 microfarad+510 ohms	—	arcing	7.9	259
Unprotected	1.0	400 to 900	7.6	450 to 750

* For each rectifier, CR1 and CR2.

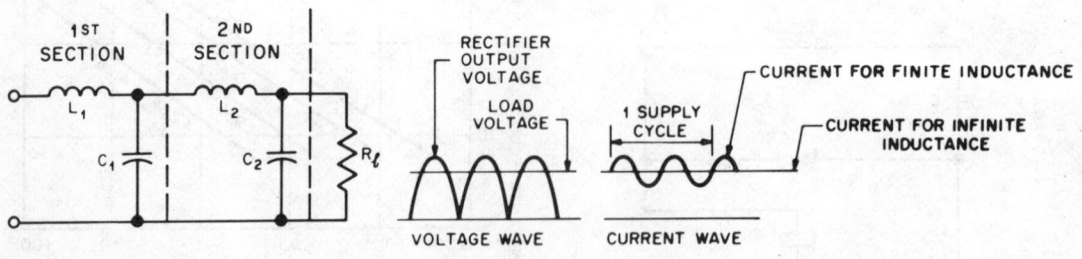

Fig. 44—Inductor-input filter.

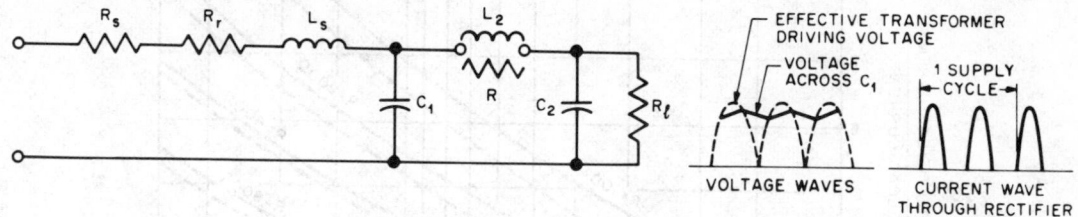

Fig. 45—Capacitor-input filter. C_1 is the input capacitor. $R_s = \frac{1}{2} \times$ (secondary-winding resistance). $L_s =$ leakage inductance viewed from $\frac{1}{2}$ secondary winding. $R_r =$ equivalent resistance of IR drop in rectifier element.

per square inch of active rectifying area. While each cell has this basic rating, it is common practice to increase the current density for the same temperature rise by increasing the space between cells or by using forced-air or oil cooling. The increase in spacing allows for current-density increases from 20 to 50 percent; the higher percentage applies to smaller cells. This causes some reduction in efficiency due to higher voltage drop.

The cells at each end of a stack have the lowest temperature due to greatest cooling there. Cell temperatures rise successively from each end toward the center of the stack. In a long stack, the temperatures of a number of the central cells are practically identical. As a consequence, some manufacturers raise the rating of stacks of 1 to 8 cells as much as 50 percent, and of stacks of 9 to 16 cells as much as 25 percent. These increases apply only to the normal-spaced convection-cooled ratings and not to the wide-spaced or forced-air-cooled or oil-cooled ratings.

For forced-air-cooled or oil-cooled rectifiers, experience shows that up to twice the normal rating is a good design figure to use when long life, good efficiency, and good voltage regulation are emphasized.

RECTIFIERS FOR MAGNETIC AMPLIFIERS

Rectifiers used in conjunction with magnetic amplifiers must have low reverse leakage currents to obtain as high a gain as possible with a given set of components. Rectifier leakage current behaves like negative feedback, thus reducing amplification. Changes in the rectifier operating temperature which result in changes in the reverse leakage current may also result in objectionable unbalances between associated amplifiers. For best amplifier performance, the reverse leakage of rectifiers for magnetic-amplifier applications should be held to

approximately 0.2 percent of the required forward current.

RECTIFIERS FOR SPARK QUENCHING

If the current in an inductive circuit is suddenly interrupted, the resulting surge can have several undesirable effects.

(**A**) Contact arcing, producing deterioration that eventually results in circuit failure due to mechanical locking or snagging, or to high contact resistance.

(**B**) High-voltage transients resulting in insulation breakdown.

(**C**) Wide-band electrical interference.

One method of suppressing surges is to shunt a selenium rectifier across the inductor as shown in Figs. 39 and 40.

The rectifier in Fig. 39 appreciably lengthens the release time (as when the electromagnet is a relay coil). By connecting the rectifier across contact A instead of across the coil, a release time only slightly lengthened is secured. This, however, is usually a less desirable connection, especially when there are several contacts controlling the same coil. Also, when contact A is open a small reverse current flows, of the order of 0.5 milliampere. The system of Fig. 39 is applicable to dc circuits only.

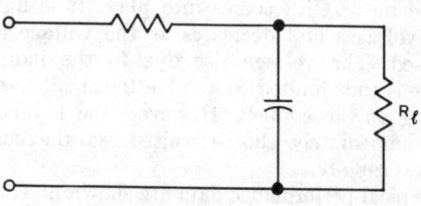

Fig. 46—Resistor-input filter.

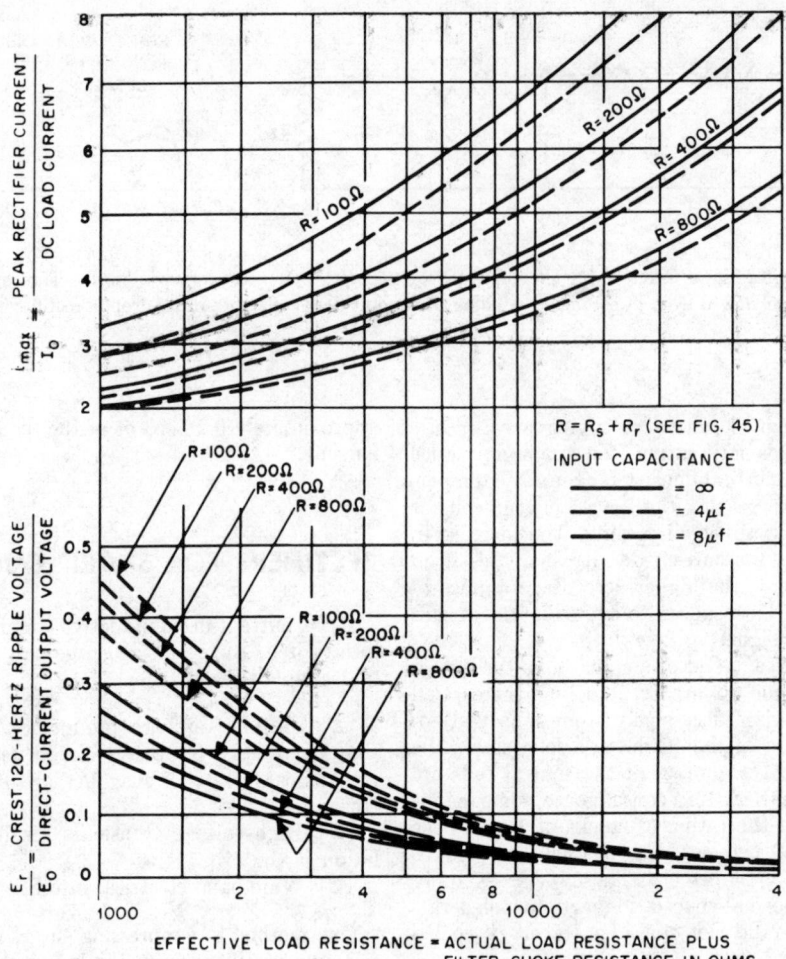

Fig. 47—Performance of capacitor-input filter for 60-hertz full-wave rectifier, assuming negligible leakage–inductance effect. *Adapted from "Radio Engineers Handbook," by F. E. Terman, 1st ed., p. 603; 1943. By permission McGraw-Hill Book Co., New York, N. Y.*

The system of Fig. 40 gives good protection with only a small lengthening of the release time over that when no protection is used. It is applicable to both ac and dc circuits. When contact A is closed, CR1 blocks current flow from the battery. On opening contact A, the reverse-resistance characteristic of CR2 comes into play. It is high at low voltages and decreases as the voltage is increased. The voltage rise due to the inductive surge is thus limited to a value too small to cause arcing at the contact. However, the inductor is not immediately short-circuited, so the current decays rapidly.

Typical performance data are shown in Table 6. For comparison, data are included for cases where a capacitor with series resistor is shunted across the coil; also for a silicon–carbide varistor in place of the rectifier shown in Fig. 39.

RECTIFIERS FOR SURGE SUPPRESSION

The arrangement shown in Fig. 40 that uses the reverse-resistance characteristic of a selenium rectifier for surge suppression has led to specially designed selenium stacks. Their application extends to the protection of silicon rectifiers and thyristors, which are more susceptible than selenium to the effects of overvoltage. The surge suppressor has a well-defined knee in its reverse voltage–current characteristic and a steep slope beyond the knee. The transient energy is dissipated as a reverse current in the selenium stack, and the transient

In Fig. 41A the suppressor is built with a sufficient number of plates to ensure that the power dissipated by reverse leakage current is low when the normal rms voltage is applied. Negligible forward current is taken by the suppressor, since a

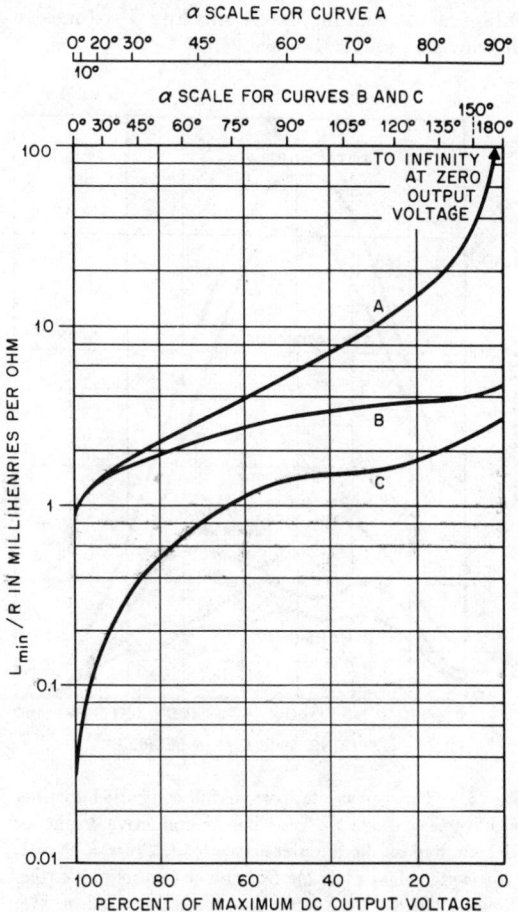

Fig. 48—L_{min}/R as a function of the percentage of maximum dc output voltage for thyristor rectifiers. The values of L_{min}/R given by the curves apply to a supply frequency of 60 hertz. Curve A is for full-controlled single-phase rectifiers such as those in Fig. 18 and in Fig. 20 with the bypass diode omitted. Curve B is for half-controlled single-phase rectifiers such as those in Fig. 19 and in Fig. 20 with the bypass diode included. Curve C is for a half-controlled three-phase rectifier such as that in Fig. 22. It could also apply to Fig. 30, feeding a transformer and three-phase diode bridge rectifier.

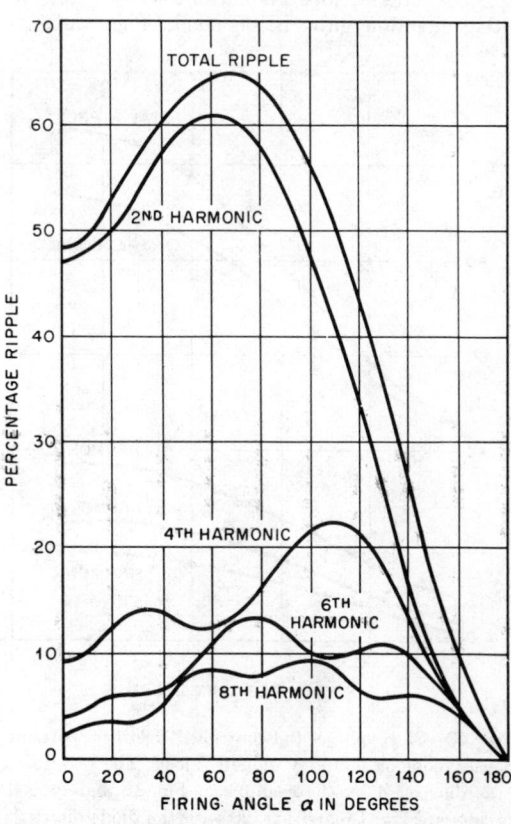

Fig. 49—Single-phase full-controlled rectifiers with bypass diode on resistive or inductive loads, or without bypass diode on resistive load; also applies to half-controlled rectifiers with or without bypass diode on resistive or inductive loads. The waveform is as illustrated by the example of Fig. 17 and typical circuits are Figs. 17, 19, and 20 with the diode inserted. *From R. Smith, "Harmonic Voltages in the Outputs of Controlled Rectifier Circuits," Electronic Engineering, vol. 36, no. 442, p. 833, fig. 1; © December 1964, Morgan Brothers (Publishers) Ltd., London, England.*

voltage is limited to a value determined by the turnover of the reverse characteristic of the stack.

The basic configurations for rectifier and thyristor protection are shown in Fig. 41, and it is possible to build combinations of these on a single spindle for composite duty.

single silicon rectifier junction has a much lower forward voltage drop per unit of current than the corresponding series assembly of selenium plates that is necessary to produce a similar reverse-voltage rating. A figure of between 2 and 2.5 is normally obtainable for the ratio of peak limited transient reverse voltage to normal rms working voltage, and lower ratios can be obtained by selection of selenium plates of particular voltage grades.

In Fig. 41B two suppressor sections are connected in opposition so that a voltage-limiting characteristic is obtained for both polarities of applied voltage; forward current cannot flow in either direction through this stack. This configur-

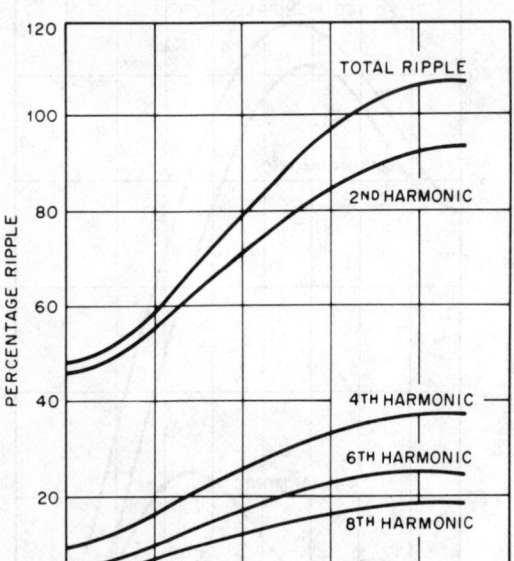

Fig. 50—Single-phase full-controlled rectifier without bypass diode on infinitely inductive load. The waveform is as illustrated by the example of Fig. 18 and typical circuits are Fig. 18, and Fig. 20 with the diode omitted. *From R. Smith, "Harmonic Voltages in the Outputs of Controlled Rectifier Circuits," Electronic Engineering, vol. 36, no. 442, p. 833, fig. 2; © December 1964, Morgan Brothers (Publishers) Ltd., London, England.*

ation is necessary for thyristor protection to ensure that both the reverse and the forward blocking voltages are limited.

Figure 42 shows typical surge-current ratings versus surge duration for several sizes of plates. Figure 43 shows the instantaneous reverse voltage developed at various reverse currents for the same range of plate sizes.

FILTERS FOR RECTIFIER CIRCUITS

Rectifier filters may be classified into three types:

Inductor Input (Fig. 44): Have good voltage regulation, high transformer utilization factor, and

low rectifier peak currents, but also give relatively low output voltage.

Capacitor Input (Fig. 45): Have high output voltage, but poor regulation, poor transformer utilization factor, and high peak currents. Used mostly in television and radio receivers.

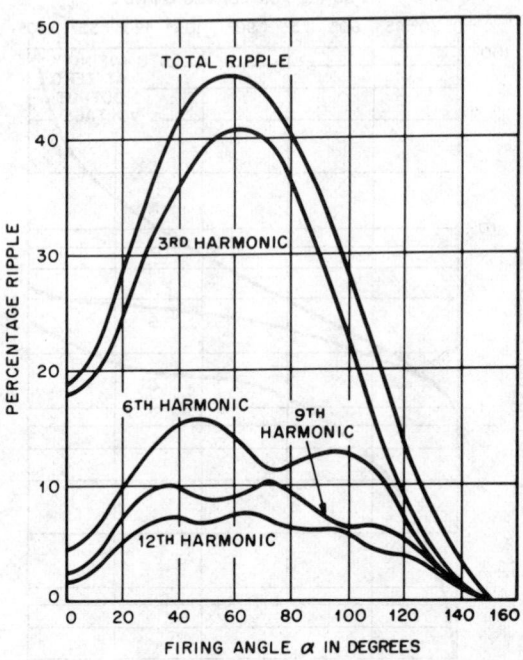

Fig. 51—Three-phase half-wave full-controlled rectifier with bypass diode on resistive or inductive loads, or without bypass diode on resistive load. *From R. Smith, "Harmonic Voltages in the Outputs of Controlled Rectifier Circuits," Electronic Engineering, vol. 36, no. 442, p. 833, fig. 3; © December 1964, Morgan Brothers (Publishers) Ltd., London, England.*

Resistor Input (Fig. 46): Used for low-current applications.

DESIGN OF INDUCTOR-INPUT FILTERS

The constants of Fig. 44 are determined from the following considerations:

(A) There must be sufficient inductance to ensure continuous operation of rectifiers and good voltage regulation. Increasing this critical value of inductance by a 25-percent safety factor, the minimum value becomes

$$L_{min} = (K/f_s) R_l \text{ henry} \tag{1}$$

where

f_s = frequency of source in hertz
R_l = maximum value of total load resistance in ohms
K = 0.060 for full-wave single-phase circuits
= 0.0057 for full-wave two-phase circuits
= 0.0017 for full-wave three-phase circuits.

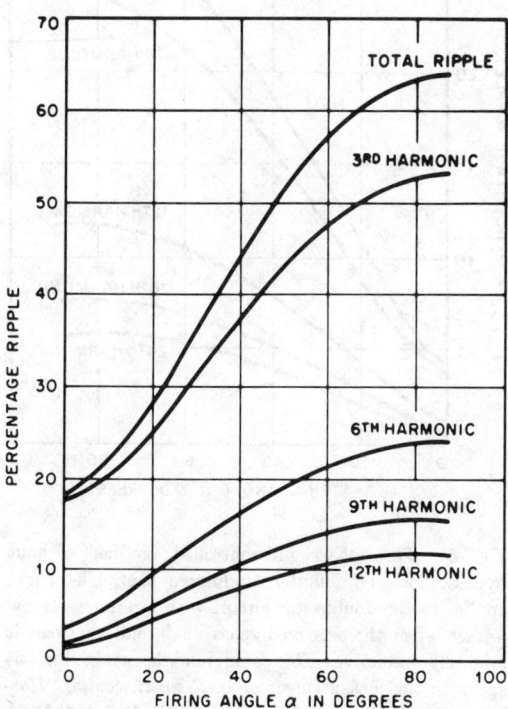

Fig. 52—Three-phase half-wave full-controlled rectifier without bypass diode on infinitely inductive load. *From R. Smith, "Harmonic Voltages in the Outputs of Controlled Rectifier Circuits," Electronic Engineering, vol. 36, no. 442, p. 833, fig. 4; © December 1964, Morgan Brothers (Publishers) Ltd., London, England.*

At 60 hertz, single-phase full-wave

$$L_{min} = R_l/1000 \text{ henry.} \tag{1A}$$

(B) The LC product must exceed a certain minimum, to ensure a required ripple factor

$$r = E_r/E_{dc}$$

$$= [\sqrt{2}/(p^2-1)][10^6/(2\pi f_s p)^2 L_1 C_1]$$

$$= K'/L_1 C_1 \tag{2}$$

where, except for single-phase half-wave, p = effective number of phases of rectifier, E_r = root-mean-square ripple voltage appearing across C_1, E_{dc} = direct-current voltage on C_1, and L_1 is in henries and C_1 in microfarads.

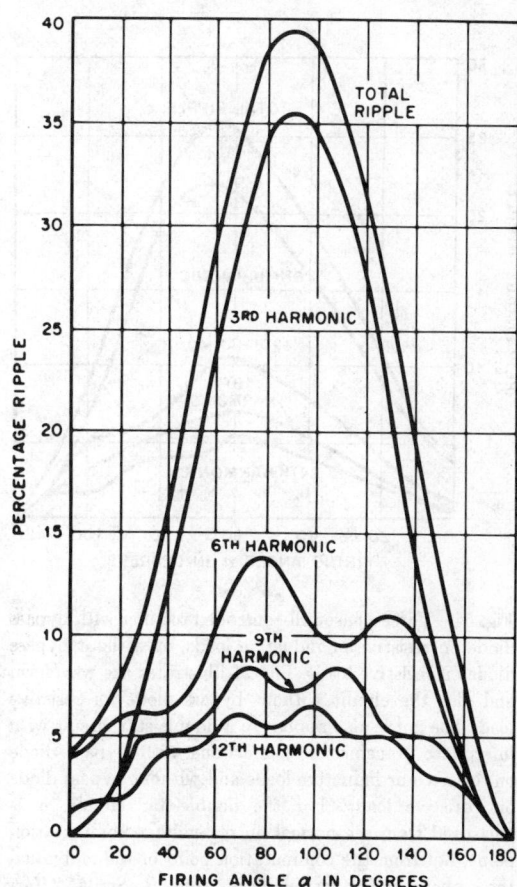

Fig. 53—Three-phase half-controlled rectifier with or without bypass diode on resistive or inductive loads. The waveform and the circuit are as shown by the example of Fig. 22. *From R. Smith, "Harmonic Voltages in the Outputs of Controlled Rectifier Circuits," Electronic Engineering, vol. 36, no. 442, p. 834, fig. 5; © December 1964, Morgan Brothers (Publishers) Ltd., London, England.*

For single-phase full-wave, $p = 2$ and

$$r = (0.83/L_1 C_1)(60/f_s)^2. \tag{2A}$$

For three-phase full-wave, $p = 6$ and

$$r = (0.0079/L_1 C_1)(60/f_s)^2. \tag{2B}$$

Equations (1) and (2) define the constants L_1 and C_1 of the filter, in terms of the load resistor R_l and allowable ripple factor r.

Swinging Chokes: Swinging chokes have inductances that vary with the load current. When the load resistance varies through a wide range, a swinging choke, with a bleeder resistor R_b (10 000 to 20 000 ohms) connected across the filter output, is used to guarantee efficient operation; i.e., $L_{min}=$

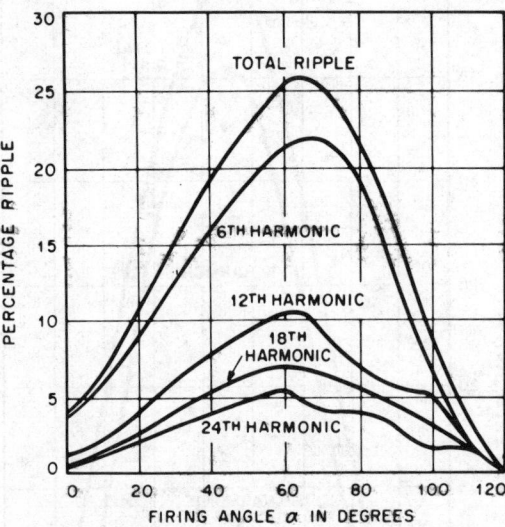

Fig. 54—Three-phase full-controlled rectifier with bypass diode on resistive or inductive loads, or without bypass diode on resistive loads. Fig. 21 illustrates the waveform and also the circuit without bypass diode for resistive load. The curve also applies to a double-star circuit with interphase transformer connections with bypass diode on resistive or inductive loads and without bypass diode on resistive loads. For the double-star circuit, α is measured from the normal output-voltage commutation point, not from the commutation point of the individual three-phase half-wave rectifiers. *From R. Smith, "Harmonic Voltages in the Outputs of Controlled Rectifier Circuits," Electronic Engineering, vol. 36, no. 442, p. 834, fig. 6;* © *December 1964, Morgan Brothers (Publishers) Ltd., London, England.*

$R_l'/1000$ for all loads, where $R_l' = (R_l R_b)/(R_l+R_b)$. Swinging chokes are economical because of their smaller relative size and result in adequate filtering in many cases.

Second Section: For further reduction of ripple voltage E_{r1}, a smoothing section (Fig. 44) may be added and will result in output ripple voltage E_{r2}.

$$E_{r2}/E_{r1} \approx 1/(2\pi f_r)^2 L_2 C_2 \qquad (3)$$

where $f_r =$ ripple frequency.

DESIGN OF CAPACITOR-INPUT FILTERS

The constants of the input capacitor (Fig. 45) are determined from the following.

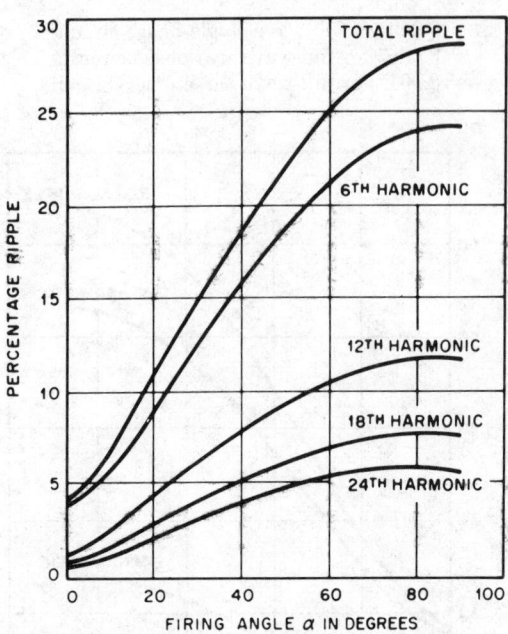

Fig. 55—Three-phase full-controlled rectifier without bypass diode on infinitely inductive load. The curve applies to the double-star circuit with interphase transformer when there is no bypass diode and the load is infinitely inductive. The same remarks as in Fig. 54 apply to the measurement of α. *From R. Smith, "Harmonic Voltages in the Outputs of Controlled Rectifier Circuits," Electronic Engineering, vol. 36, no. 442, p. 834, fig. 7;* © *December 1964, Morgan Brothers (Publishers) Ltd., London, England.*

(A) Degree of filtering required.

$$r = E_r/E_{dc}$$

$$= \sqrt{2}/2\pi f_r C_1 R_l$$

$$= (0.00188/C_1 R_l)(120/f_r) \qquad (4)$$

where $C_1 R_l$ is in microfarads×megohms, or farads×ohms.

(B) A maximum allowable C_1 (so as not to exceed the maximum allowable peak-current rating of the rectifier).

Unlike the inductor-input filter, the source impedance (transformer and rectifier) affects output dc and ripple voltages and the peak currents. The equivalent network is shown in Fig. 45.

Neglecting leakage inductance, the peak output ripple voltage E_{r1} (across the capacitor) and the

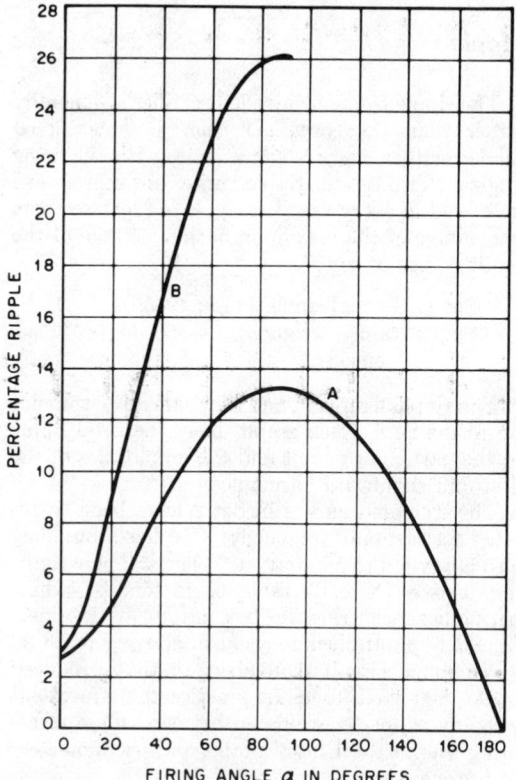

Fig. 56—Curve A is a weighted ripple curve, including all harmonics up to the 24th, weighted with respect to 800 hertz. This curve applies to the same circuits as for Fig. 49. Curve B is produced on the same basis but refers to the circuits as for Fig. 50. *From R. Smith, "Harmonic Voltages in the Outputs of Controlled Rectifier Circuits," Electronic Engineering, vol. 36, no. 442, p. 834, fig. 8; © December 1964, Morgan Brothers (Publishers) Ltd., London, England.*

peak rectifier current for varying effective load resistance are given in Fig. 47. If the load current is small, there may be no need to add the L-section consisting of an inductor and a second capacitor. Otherwise, with the completion of an L_2C_2 or RC_2 section (Fig. 45), greater filtering is obtained, the peak output-ripple voltage E_{r2} being given by (3) or by

$$E_{r2}/E_{r1} = 1/2\pi f_r RC_2. \qquad (5)$$

FILTERS FOR CONTROLLED RECTIFIERS

The same general principles apply to filters for controlled rectifier circuits as to those for ordinary rectifier circuits. Capacitive-input filters, however, are rarely used since a capacitive load restricts the range of conduction angle control that it is possible to obtain with thyristors.

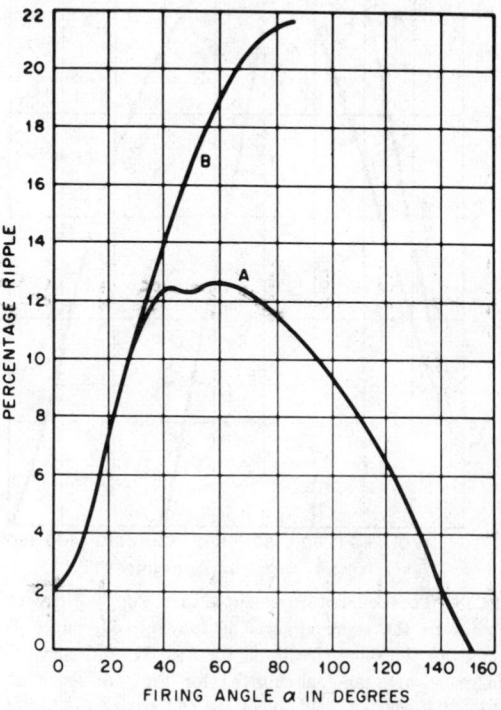

Fig. 57—These curves are weighted as in Fig. 56. Curve A applies to the same circuits as for Fig. 51 and curve B applies to the same circuits as for Fig. 52. *From R. Smith, "Harmonic Voltages in the Outputs of Controlled Rectifier Circuits," Electronic Engineering, vol. 36, no. 442, p. 835, fig. 9; © December 1964, Morgan Brothers (Publishers) Ltd., London, England.*

The two main differences to consider in filter design for controlled rectifiers, compared with that for ordinary rectifiers, are (A) the greatly increased values of critical inductance for the input choke and (B) the larger input ripples.

Empirical equations for the critical inductance L_{min} were given that apply to various configurations of diode rectifiers. In controlled rectifiers, L_{min} rises as the conduction angle is decreased by trigger control, i.e., as the firing angle delay α is increased. Figure 48 shows the ratio L_{min}/R as

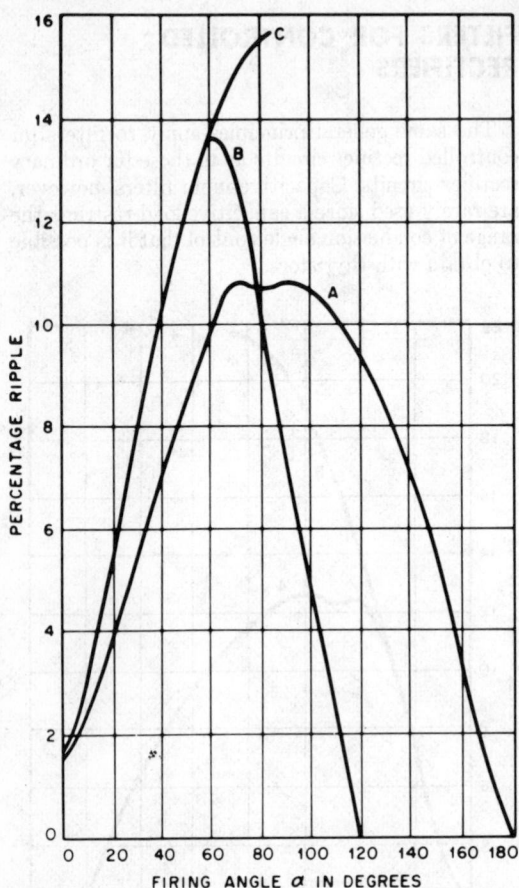

Fig. 58—These curves are weighted as in Fig. 56. Curve A applies to the same circuits as for Fig. 53, curve B applies to the same circuits as for Fig. 54, and curve C applies to the same circuits as for Fig. 55. *From R. Smith, "Harmonic Voltages in the Outputs of Controlled Rectifier Circuits," Electronic Engineering, vol. 36, no. 442, p. 835, fig. 10; © December 1964, Morgan Brothers (Publishers) Ltd., London, England.*

a function of the percentage of maximum output voltage obtained from three-phase half-controlled and from single-phase half- and full-controlled rectifier types discussed in the section on thyristors. Scales are included to show the corresponding values of α for various percentage outputs.

Ripple

The ripple from a controlled rectifier is generally larger than that obtained from a conventional diode rectifier, and its value varies with the firing angle α. Two sets of ripple curves are reproduced here, and in every case the ripple is expressed as a percentage of the maximum output voltage of the rectifier, i.e., at $\alpha = 0°$.

Set 1: Normal ripple (Figs. 49–55).
Set 2: Ripple weighted to 800 hertz (Figs. 56–58).

The ripple figures in Set 1 include all harmonics up to the 24th. Each graph shows the total ripple in the particular circuit and the amplitudes of the first four significant harmonics.

The weighted curves in Set 2 have been calculated for a supply frequency of 50 hertz but they also apply approximately at 60 hertz. The weighting uses a CCITT* table of factors by which harmonics occurring in the output of rectifiers should be multiplied to yield interfering noise on a telephone circuit equivalent to that produced by an 800-hertz tone. In practice the equivalent weighted ripple is evaluated by using noise measuring sets in the United States and psophometers in Europe.

* Refer to Directives of the Comité Consultatif International Téléphonique et Télégraphique (CCIF and CCIT), Geneva; 1952.

INTRODUCTION

The magnetic amplifier enjoyed its maximum prominence in power control and low-frequency signal-processing electronics from about 1947 to 1957. By 1957 junction transistors were readily available. Development rapidly shifted away from magnetic amplifiers toward transistor and semiconductor switch equivalents, combined magnetic/transistor amplifiers, and the host of new devices made possible by the joint use of square-loop cores and transistors. Complex magnetic-amplifier designs developed in this period must be looked on as an interim technology. Many magnetic amplifiers in present use and manufacture are the result of early design commitments which have not yet been replaced.

There are a few areas, however, in which magnetic amplifiers continue to excel. In power control, they tolerate extreme environmental and overload conditions that would be fatal to semiconductors. They may also generate less noise because of the slower switching saturable reactors. Perhaps most important, they permit the summing of a number of input signals that must remain electrically isolated. In instrumentation amplifiers, magnetic-amplifier circuits still offer high, drift-free gain with this summing feature. The development of magnetic-core transistor oscillators makes it possible to supply ac power of practically any desired frequency for these amplifiers, making them much smaller than they would be with power-frequency excitation. Similar circuits have come into increasing use in magnetometry where the unique direct transducing capability of the magnetic amplifier puts it in a class by itself.

The magnetic-core transistor oscillator, which is capable of inverting dc to ac up to about 100 kilohertz and by rectification can convert a single primary dc power source at high efficiency to several independent conductively isolated dc voltage supplies, today has all of the engineering prominence that magnetic amplifiers once enjoyed. Many of these newer circuits are, in fact, regulated and timed by magnetic-amplifier principles.

The history and present state of the art in magnetic amplifiers is documented in the proceedings of the Conference on Nonlinear Magnetics and Magnetic Amplifiers [1], and the more recent *IEEE Transactions on Magnetics* [2]. Several books ranging between texts and advanced treatments of design principles are also available [3-7]. The development of soft magnetic materials used in magnetic-amplifier circuits can be traced in the proceedings of the Annual Conferences on Magnetism and Magnetic Materials [8]. Most of the material to be found in these references deals with basic principles and the physical properties of materials. For the circuit engineer who wishes to capitalize quickly and effectively on the design rules that have emerged from this effort, the design manuals available from the manufacturers of magnetic-amplifier core materials tabulate the available material types and related design information such as wire holding capacity, insulation, and temperature characteristics. Having made the basic selection of the core windings and circuit configuration, the designer should expect to spend some time tailoring the circuit to meet design specifications.

PRINCIPLE OF THE MAGNETIC AMPLIFIER

The elementary principle of magnetic amplification can be conveniently represented in terms of a flux-actuated switch in series with a load. The magnetic material is characterized by the nearly rectangular hysteresis loop of Fig. 1. In this figure, the narrow hysteresis loop corresponds to the loop measured at dc and the wider loop is that measured at the power supply frequency. It should be noted that this dynamic loop widens with increasing frequency. Figure 2 shows the dc hysteresis loops corresponding to three of the materials listed in Table 1.

Figure 3 shows a winding on a core in series with a resistor representing the load. At the beginning of a positive half-cycle of the supply voltage, the

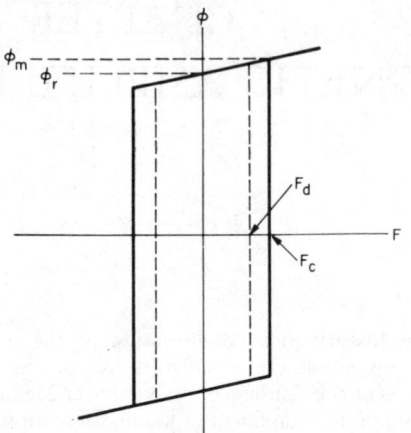

Fig. 1—Schematic representation of dynamic and dc (dotted lines) hysteresis loops.

core is in some initial flux state ϕ_0. Essentially all of the supply voltage is impressed on the core winding until the flux in the core reaches saturation. In saturation, the core becomes a very low impedance and practically all of the supply voltage appears across the resistor. This situation is diagrammed in Fig. 4A for a sinusoidal supply voltage and in Fig. 4B for a square-wave supply voltage. The lower portions of the diagrams show that the switching of the supply voltage from the core to the resistor becomes perfect when the width of the hysteresis loop is zero. In practical designs, the choice of power supply voltage and frequency, core size, winding, and load impedance is subject to the constraints of the problem. To approximate the above-mentioned ideal conditions is often the main object of the design.

Figure 5 shows integrals of portions of the supply-voltage integral in analytical form. It is clear from this figure that the average voltage applied to the load is a function of the switching angle α, which in turn depends on the initial flux ϕ_0. The half-cycle average of the load voltage is expressed in terms of ϕ_0 for the sine-wave and square-wave cases as follows.

Sine Wave	Square Wave
$\bar{v}_r = (2/T)(E_s/\omega)$ $\times (1+\cos\alpha)$	$\bar{v}_r = E_s[1-(\alpha/\pi)]$
$(E_s/\omega)(1-\cos\alpha)$ $= (E_s/\omega) - N\phi_0$	$(T/2)E_s(\alpha/\pi)$ $= N(\phi_m - \phi_0)$
$\cos\alpha = \phi_0/\phi_m$	$\alpha/\pi = \tfrac{1}{2}[1-(\phi_0/\phi_m)]$
$\bar{v}_r = (2/T)(E_s/\omega)$ $\times [1+(\phi_0/\phi_m)]$	$\bar{v}_r = E_s/2[1+(\phi_0/\phi_m)]$

It is further obvious that in order for there to be no output, and no excess flux capacity in the core (normal excitation), the flux linkage capacity of the core must be set equal to the volt-second capacity of the power supply. This results in the simple equations

$$E_s = B_m A\omega N \qquad \text{(sine wave)}$$

$$E_s = (2/\pi)B_m A\omega N \qquad \text{(square wave)}$$

relating the peak value of the supply voltage, the maximum flux density of the core (in webers/meter²), the material cross-section (in meters²), the angular frequency, and the number of turns.

Correspondingly, the exciting current for a given coercive force F_c in ampere-turns is $i_x = F_c/N = H_c l/N$, where H_c is in ampere-turns/meter and l is the mean length of the magnetic path in the core in meters. These equations in CGS units become

$$E_s = B_m A\omega N \times 10^{-8} \qquad \text{(sine wave)}$$

$$E_s = (2/\pi)B_m A\omega N \times 10^{-8} \qquad \text{(square wave)}$$

where $B_m = $ gauss, $A = $ cm², $E_s = $ volts (peak), and $N = $ turns.

$$I_m = 0.794 H_c l/N$$

where $H_c = $ oersteds, $l = $ cm, and $I = $ amperes.

Although the above discussion contains most of the ideas basic to magnetic amplifiers, no mention has been made of how ϕ_0 is related to the control

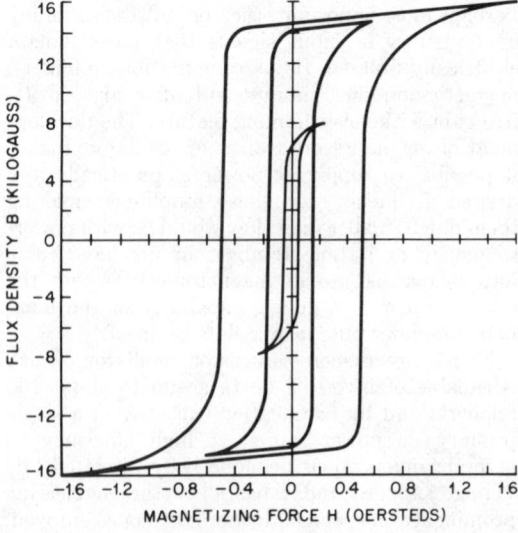

Fig. 2—Comparison of dc loops of three materials listed in Table 1.

TABLE 1—MATERIALS AND APPLICATIONS.

Material Letter Code	Trade Name	Cost/Core* Relative to Material A	Principal Use
A	Square Orthonol, Hipernik V, Orthonik, 49 Square Mu, Deltamax	1.00	High-gain amplifiers, oscillators, integrators, timers, memory devices.
N	Round Orthonol	1.00	Amplifier applications with slightly less gain, but less liability to triggering instability, than material A.
H	48 Alloy, Carpenter 49	1.00	Material A and N applications with lower sensitivity, lower losses, and less triggering instability. High-quality transformers.
D	Square Permalloy 80, Square Mu 79, Super Square Mu 79, Hy Ra 80, 4-79 Permalloy, Square Permalloy	1.14	High-gain amplifiers at low signal levels and low losses, low-power-consumption inverters and converters.
R	Round Permalloy 80, Hy Mu 80	1.14	High-quality low-loss inductors and transformers.
F	Supermalloy	1.63	Material D and R applications in which the lowest possible exciting currents and losses are required.
K	Magnesil, Selectron, Microsil, Hypersil, Supersil	0.70	Power amplifiers requiring lower gain and lower cost than material A applications. High-quality power transformers.
S	Supermendur	3.4	Material A and K applications where minimum size and weight and maximum operating temperatures are required.

* Based on 2-mil tape-wound core of about 3-inch diameter.

signal. Further, note that at the end of the half-cycle, ϕ is at saturation. Thus the core must be reset to ϕ_0 in the second half-cycle. If a similar voltage is to be applied to the load in the second half-cycle, a second core must be included in the circuit. Such resetting and output problems are responsible for the variety of amplifier circuit configurations that have been used.

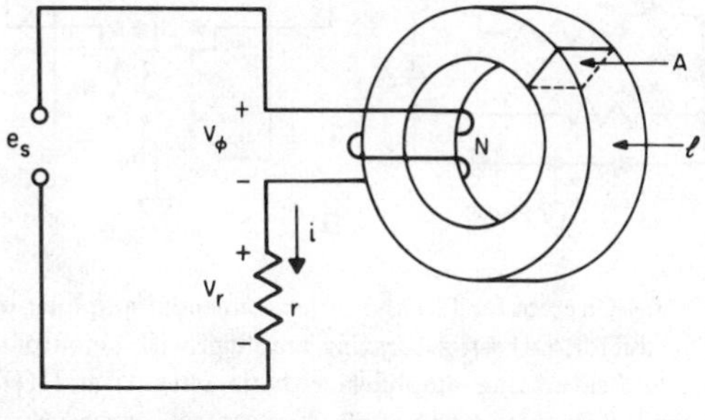

AMPLIFIER CONFIGURATIONS

The amplifier configuration is arranged with two considerations in mind. One is the method of control, and the other is the type of output desired. As seen from the discussion above, a core that is brought to saturation and is gating power to a load on a positive half-cycle must be reset to its initial state if it is to repeat this function on the next positive half-cycle. On the other hand it is usually desirable to deliver power to the load on both half-cycles. Thus a second core will be gating power to the load during the half-cycle in which

Fig. 3—Saturable reactor in series with a resistor.

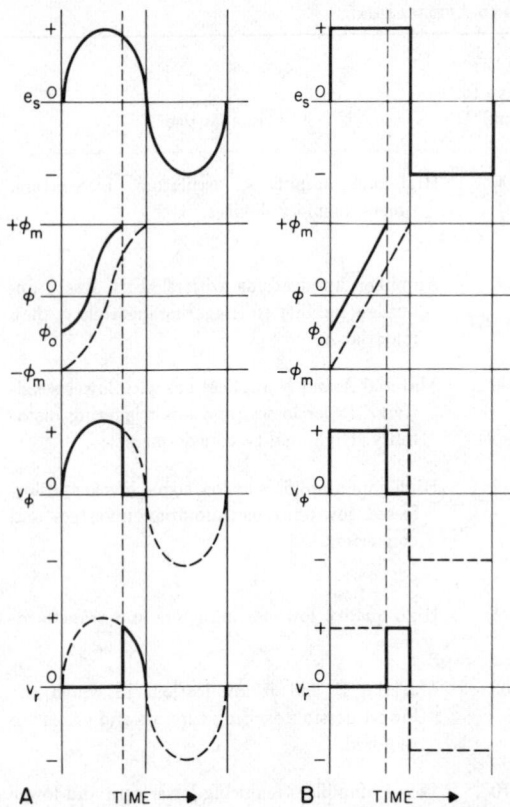

Fig. 4—Voltage and flux waveforms for the circuit of Fig. 3 for (A) sine-wave and (B) square-wave excitation.

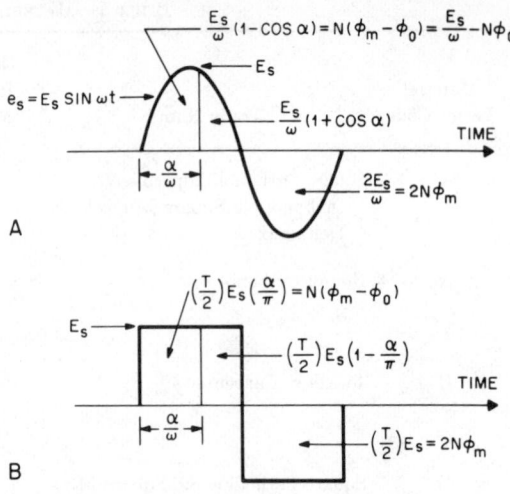

A

B

Fig. 5—Components of integrals of (A) sine-wave and (B) square-wave half-cycles.

the first core is being reset. Since the increment in flux linkage in the two cores is equal in the steady state, the core driven from the power supply can be used to reset the second core by transformer action through the control circuit. In single-phase circuits the roles of the cores interchange during alternate half-cycles. The use of one core to reset the other is fundamental to most amplifier configurations.

Several of the most common configurations are shown in Fig. 6. Figure 6A is the series-connected amplifier, sometimes called the transductor. It has been extensively analyzed [3, 9, 10] and is commonly used to measure large direct currents in electrochemical and power applications [11]. The details of the circuit operation are complicated but, in essence, at most one core is saturated at a time, gating power to the load. During this interval, the second (unsaturated) core, which has a very small coercive force, must therefore maintain the condition $N_L i_L - N_c i_c = F_c$, the dynamic coercive force of the unsaturated core. The control current i_c has the same waveform as the load current i_L.

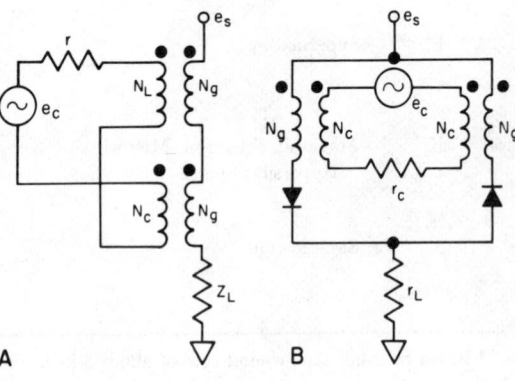

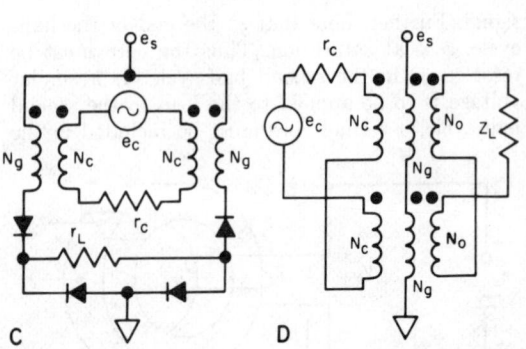

Fig. 6—Circuits for (A) the series-connected amplifier or transductor, (B) self-saturating amplifier with ac output, (C) self-saturating amplifier with dc output, and (D) second-harmonic modulator. N_c = control turns, N_g = gating turns, and N_o = output turns.

but has the same polarity on each half-cycle. Thus, the rectified average values of the load current and the control current, I_L and I_c, are related by the turns ratio.

$$I_L = (N_c/N_L)I_c + (1/N_L)F_c$$

a linear function with a constant offset as shown in Fig. 7A. In practice, the linearity of this function can be kept within about 0.1 percent, which makes it very useful for instrumentation.

The circuit in Fig. 6B is characterized by parallel-connected saturable reactors, so that the load current does not have to flow through an unsaturated core. Thus, since the resetting core is primarily transformer driven through the control circuit by the power-gating core, only the exciting current for the resetting core must be carried in the control circuit. In the steady state, the flux linkages coming from the winding on the gating core must equal the flux linkages delivered to the resetting core. With zero control voltage, the two flux linkages will differ by the integral of $i_c r_c$ over the half-cycle. The function of the control voltage is to offset this voltage drop to make the two flux linkages equal at the desired output level. The diodes decouple the cores from the power supply during their resetting half-cycles.

When the amplifier is delivering full output, there is essentially no flux excursion in the gating core. It therefore does not drive the resetting core. Since the resetting core must not be reset under these conditions, the control current must be just below the coercive direct current for the core. At zero output, the gating core drives the resetting core at normal power voltage, resulting in a control current equal to the normal power-frequency coercive current. Full control of the amplifier is obtained over a control-current range equal to the widening of the hysteresis loop from dc to the power-frequency loop, divided of course by the control-winding turns. The resulting control characteristic is shown in Fig. 7B, with reference to Fig. 1. Again, there is a minimum output corresponding to the exciting current for the gating core. Also, the control current and voltage are automatically rectified because of the half-cycle symmetry of the circuit as seen from the control terminals. The modification shown in Fig. 6C delivers dc to the load.

A fourth configuration (Fig. 6D) used for very-small-signal amplifiers and magnetometers [12, 13] takes advantage of transformer coupling of the output to eliminate the residual exciting-current component found in the other circuits. The fundamental component of the induced voltage is canceled out and, at input currents other than zero, there is a second-harmonic component in the output proportional to the input current. The

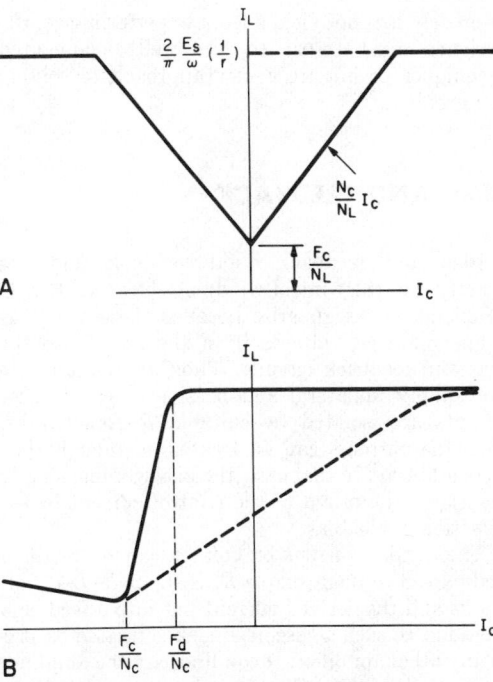

Fig. 7—Control characteristics for (A) the amplifier of Fig. 6A; and (B) for the amplifier of Fig. 6B and 6C. The dotted characteristic corresponds to nonsquare hysteresis loop, resistively shunted rectifiers, or negative feedback.

phase of this output reverses with the input-current polarity. Other examples of high-sensitivity amplifiers can be found in the literature [14]. For many power applications, 3-phase amplifiers are preferred for the usual reasons. The output is much easier to filter when dc is required [15].

Examples of parametric amplifiers and microwave magnetic amplifiers can be found in the literature.

MULTISTAGE AMPLIFIERS

Because of the bilateral characteristics of magnetic amplifiers, multistage amplifiers are difficult to analyze. In addition, their properties as measured experimentally are technically unattractive and they are difficult to design even by empirical techniques except when the coupling-circuit impedance isolates the circuits. In this case, gain and frequency response are usually sacrificed. With the advent of a highly developed transistor technology, it is rarely necessary to design multistage amplifiers. Where it is necessary, the problem is treated as though the first stage were driving any other R–L or R–L–C isolated dc load. If this approach does not yield adequate performance, the designer must resort to less well documented techniques or initiate a new approach to solving the problem.

BIAS AND FEEDBACK

Bias and feedback windings look and act exactly like the control windings shown in Fig. 6. Problems in design arise because these windings couple induced voltages from the cores into the bias and feedback circuits. Thus, the circuits are not simply unilateral and passive [16]. If these circuits are isolated by suitable $R-L$ networks, then the currents can be treated as though they were additive. In that case, the bias winding simply translates the origin of the control current in the direction of the bias.

The output current or voltage can be rectified and passed through simple R, $R-L$, or $R-L-C$ networks and the derived current put into a feedback winding. In such a case, feedback is treated as it is in any other amplifier. It can linearize the amplifier successfully if the range of output is lowered. If the amplifier is used as a power device and its range of output is fixed by performance specifications, feedback is of no help unless compensating non-linearity can be inserted in other amplifying stages. Feedback in such cases is useful primarily to lower output or impedance of the amplifier.

FREQUENCY RESPONSE

Analysis has shown that the voltage-gain bandwidth product [17] is

$$G_v \times (\text{bw}) = 4f/N \quad \text{(series-connected amplifier, Fig. 6A)}$$

$$G_v \times (\text{bw}) = 2f/N \quad \text{(self-saturating amplifier, Fig. 6B)}$$

where f is the frequency and N is the turns ratio N_c/N_g. These relations show again the advantage of high-frequency power supplies. They also suggest that if a design is adjusted to increase gain, the bandwidth (frequency response) will be reduced unless the turns ratios are adjusted to raise the gain–bandwidth product. High-gain amplifiers can be expected to have a bandwidth of about one-tenth the carrier frequency, which is frequently sufficient in instrumentation applications. In signal-processing applications, it is preferable to use several stages of low-gain wide-band amplification since the gains multiply and the bandwidths

go down more or less additively. It is in this area that magnetic amplifiers have been largely replaced by semiconductor circuits.

CHOICE OF CORE MATERIALS

A variety of core materials can be chosen for magnetic amplifiers. They can be obtained in the form of tape-wound cores, laminations, and encapsulated tape cores cut into two mating C-shaped pieces. The latter configurations permit the use of simple inexpensive lathe-wound windings which can be assembled on the core. The cut-C core configuration maintains the rolling direction of the tape along the primary magnetic path. The laminations are stamped out of continuous strip such that part of the magnetic path is along the direction of rolling and part is perpendicular to it. Since most good-quality strip is anisotropic, the resulting characteristics of the cores are not as good as they would be in the tape-wound configuration, which has the best magnetic qualities. As a result, only the lower-cost lower-quality magnetic materials are widely used in other than the tape-wound configuration. In addition to differences in processing and in the cost of raw materials, the above manufacturing considerations significantly contribute to the economic basis for choosing core materials. Cost is the dominant consideration in most amplifier designs. There are, however, extreme cases where only the highest-quality material can meet the technical specifications.

In general, the large, heavy, power-control applications make use of the least expensive materials in the least expensive configuration. In one important sense, they are sometimes technically superior as well. First, because of low remanence of the core, using these materials in a self-saturating configuration (Fig. 6B, 6C) results in a control characteristic which crosses the control-current axis as shown dotted in Fig. 7B. This automatically biases the amplifier near the desired operating point. In addition, the lack of squareness also causes fairly slow switching at the firing time of the power-gating core. The result is much less noise than found in the more objectionable gas tubes, semiconductors, and square-loop core circuits.

Table 1 lists the core materials available in tape-wound cores from most of the core manufacturers, plus a guide to their principal applications. An approximate cost ratio is given for 2-mil tape in a core of about 3 inches in diameter. This indicates the economic advantage of using materials no better than necessary. In lamination form, the cost per pound of the material is lower by about a factor of 5.

TABLE 2—TECHNICAL PROPERTIES OF MATERIALS.

Material Letter Code	Flux Density (kilogauss)	Squareness (B_r/B_m) (400 Hz) ccfr*	Coercive Force (oersted) (dc)	(400 Hz) ccfr*	Gain (kG/oe) (400 Hz) ccfr*	Curie Temp (°C)	Core Loss (mW/lb) (at +1 kG, 400 Hz)
A	14.2–15.8	0.94 up	0.1–0.2	0.45–0.65	310–715	500	56
N	14.0–15.6	0.85–0.95	0.07–0.17	0.10–0.20	260–500	500	42
H	11.5–14.0	0.80–0.92	0.05–0.15	0.08–0.15	280–550	500	19
D	6.6–8.2	0.80 up	0.02–0.04	0.022–0.044	550–1650	460	6.5
R	6.6–8.2	0.45–0.75	0.008–0.02	0.008–0.026	250–715	460	4.5
F	6.5–8.2	0.40–0.70	0.003–0.008	0.004–0.015	250–715	460	3.7
K	15.0–18.0	0.85 up	0.40–0.60	0.45–0.65	130–220	750	42
S	19.0–22.0	0.90 up	0.15–0.35	0.50–0.70	85–135	940	230

* The values are typical of cores with ID/OD ratio of about 0.80 and tape thickness of 0.002 in. Tests made are in accordance with AIEE Standards paper II-432. (ccfr = constant-current-flux reset.)

Table 2 summarizes the technical properties of these materials. Note that many of their properties are given in terms of the IEEE standard referenced in AIEE Standards paper II-432 [18]. The use of these standards in circuit design and component specification is highly recommended.

The control-current range for self-saturating amplifiers, as indicated in Fig. 7B, can be estimated by referring to the difference between the dc and 400-hertz coercive-force columns in Table 2. This value must be multiplied by about 0.8 times the mean magnetic path length of the core in centimeters to obtain control ampere-turns. The values are for 400 hertz and must be corrected experimentally for other frequencies. Studies of the properties of materials and their influence on circuits covering a wide range of frequencies, temperatures, and materials can be found in the published literature [19, 20].

For more-specific and detailed design information, the designer should use the referenced literature. Also, several core-materials manufacturers have prepared excellent booklets containing all the essential tables and nomograms for designing magnetic-core circuits.

OTHER DESIGN CONSIDERATIONS

The basic design calculations, as discussed above, pick the core size and number of turns to fit the frequency and voltage. For a given magnetic material, a larger core requires fewer turns to support a given voltage at a given frequency. The number of turns varies as the inverse of the cross-section. From this fact alone, exciting current rises linearly with cross-section. There is also a linear relation between the exciting current and the mean magnetic path length for a fixed H. Thus, the exciting current is proportional to the volume of the core, as indicated by the energy dissipated in the material.

As the frequency rises, it is possible to use a smaller core for a fixed voltage. Comparing a 400-hertz design with a 60-hertz design, for example, the cores in the 400-hertz unit would be smaller by about a factor of 7. Since this is true for transformers and inductors as well, high-frequency power supplies are commonly found on aircraft where space and weight are important. The higher supply frequency also puts the carrier farther above the modulation-signal frequency spectrum, making it easier to recover the signal.

Since in many small-signal applications it is not necessary to have a large supply voltage, it is common to change available dc signals to square-wave ac voltages in the range from 5 to 25 kHz and higher. This means very small cores and very compact, sensitive amplifiers, a combination that often yields better performance in low-noise low-signal applications than semiconductor circuits.

REFERENCES

1. Institute of Electrical and Electronics Engineers, 345 East 47th Street, New York, New York 10017, before 1965. Many papers published in *Communication and Electronics* as well.
2. *IEEE Transactions on Magnetics*, Vol. MAG-1, No. 1; March 1965.
3. H. F. Storm, "Magnetic Amplifiers," John Wiley & Sons, Inc., New York; 1955.

4. G. E. Lynn, T. J. Pula, J. F. Ringelman, and F. G. Timmel, "Self-Saturating Magnetic Amplifiers," McGraw-Hill Book Co., Inc., New York; 1960.

5. D. L. Lafuze, "Magnetic Amplifier Analysis," John Wiley & Sons, Inc., New York; 1962.

6. W. A. Geyger, "Magnetic Amplifier Circuits," McGraw-Hill Book Co., Inc., New York; 1957.

7. R. C. Barker, "Nonlinear Magnetics," *Electro-Technology*, Science and Engineering Series 51; March 1963.

8. Each proceedings is published as a special issue of the Journal of Applied Physics in the spring of each year.

9. R. C. Barker, "The Series Magnetic Amplifier, Parts I and II," *Communication and Electronics*, pp. 819–831; January 1957.

10. A. G. Milnes, "Transductors and Magnetic Amplifiers," Macmillan Co. Ltd., London; 1957.

11. A. B. Rosenstein, "160 000-Ampere High-Speed Magnetic-Amplifier Design," *AIEE Transactions*, Vol. 74, Part I, pp. 90–97; 1955.

12. D. I. Gordon, R. H. Lundsten, and R. A. Chiarodo, "Factors Affecting the Sensitivity of Gamma-Level Ring-Core Magnetometers," *IEEE Transactions on Magnetics*, Vol. MAG-1, No. 4, pp. 330–337; December 1965.

13. R. C. Barker, "On the Analysis of Second-Harmonic Modulators," *IEEE Transactions on Magnetics*, Vol. MAG-1, No. 4, pp. 337–341; December 1965.

See also S. Ohteru and H. Kobayashi, "A New Type Magnetic Modulator," *IEEE Transactions on Magnetics*, Vol. MAG-1, No. 1, pp. 56–62; March 1965.

14. H. E. Darling, "New Magnetic Amplifier Improves EMF to Current Converter," *IEEE Transactions on Magnetics*, Vol. MAG-3, No. 3, pp. 365–369; September 1967.

15. H. C. Bourne, Jr., and T. Kusuda, "A Three-Phase Magnetic Amplifier: Part II—Experimental Results," *IEEE Transactions on Magnetics*, Vol. MAG-3, No. 1, pp. 17–22; March 1967.

16. L. A. Finzi and J. J. Suozzi, "On the Feedback in Magnetic Amplifiers: Part II—Combined Magnetic and Electric Feedbacks," *AIEE Transactions*, Vol. 78, Part I, pp. 136–141; 1959.

17. R. C. Barker and G. M. Northrop, "Some Frequency Response Measurements on Magnetic Amplifiers," *Proceedings of the National Electronics Conference*, Vol. 12, pp. 444–453; 1956.

18. AIEE Standards Paper No. 432, obtainable from IEEE Headquarters [1].

19. M. Pasnak and R. Lundsten, "Effects of Ultrahigh Temperature on Magnetic Properties of Core Materials," *AIEE Transactions*, Vol. 78, Part I, pp. 1033–1039; 1959.

20. C. E. Ward and M. F. Littman, "Relation of D-C Magnetic Properties of Oriented 48-Per-Cent Nickel-Iron to Magnetic-Amplifier Performance," *AIEE Transactions*, Vol. 74, Part I, pp. 422–427; 1955.

CHAPTER 16

FEEDBACK CONTROL SYSTEMS

INTRODUCTION

A feedback control system (Fig. 1) is one in which the difference between a reference input and some function of the controlled variable is used to supply an actuating error to the control elements in the control system. The amplified actuating error signal is fed back to the system, thus tending to reduce the difference to zero. Supplemental power for signal amplification is available in such systems.

The two most common types of feedback control systems are regulators and servomechanisms. Fundamentally, the systems are similar but the choice of systems depends on the nature of reference inputs, the disturbance to which the control is subjected, and the number of integrating elements in the control. Thus, regulators are designed primarily to maintain the controlled variable or system output very nearly equal to a desired value in the presence of output disturbances. Generally, a regulator does not contain any integrating elements.

A servomechanism is a feedback control system in which the controlled variable is a position (or velocity). Ordinarily in a servomechanism the reference input is the input signal of primary importance; load disturbances, while they may be present, are of secondary importance. Generally, one or more integrating elements are contained in the forward transfer function of the servomechanism.

Control system problems may generally be classified into two types, linear and nonlinear. All physical systems are nonlinear and have time-varying parameters to some degree. Where the effect of the nonlinearity is very small or the time-varying parameters are slow with respect to time, linear fixed-parameter analysis of the system is adequate for engineering purposes.

The most important property of a linear system is that the principle of superposition applies. Thus, in a linear system the shape of the time response is independent of the size of the input or initial condition. A linear system, described by a linear constant coefficient differential equation, can be analyzed in a uniform fashion using a general technique.

In a control system where the control elements possess such properties as saturation, limiting, backlash, or hysteresis, the system exhibits nonlinear characteristics and the principle of superposition does not hold. The response of the system becomes dependent on the size of the input and initial condition. Exact analytical solutions to a nonlinear system (nonlinear differential equation) are very rare, and the analysis of the system must be by other means such as numerical or graphical methods.

In the following sections both the linear (Part 1) and nonlinear (Part 2) systems are discussed.

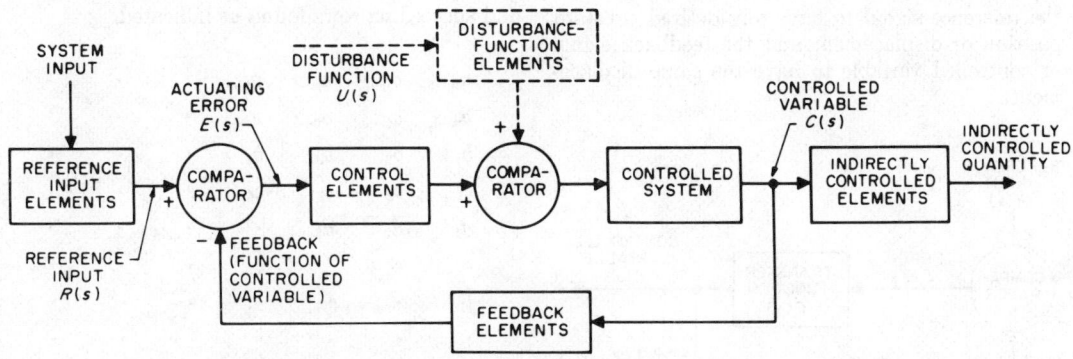

Fig. 1—Block diagram of feedback control system.

PART 1—LINEAR SYSTEMS

TYPES OF LINEAR SYSTEMS

The various types of feedback control systems can be described most effectively in terms of the simple closed-loop direct feedback system. Figure 2 shows such a system. $R(s)$, $C(s)$, and $E(s)$ are the Laplace transforms of the reference input, controlled variable, and error signal, respectively.

Note: The complex variable s instead of p will be employed in this chapter to conform with the general practice in the literature on feedback control systems.

For a typical linear system, $G(s)$ might appear as

$$C(s)/E(s) = G(s) = \frac{K(T_1 s + 1)(T_3 s + 1)}{s^n(T_2 s + 1)(T_4 s + 1)}.$$

The value of the exponent n, an integer, designates the type of the system. This in turn reveals the nature of the steady-state performance of the system as outlined below.

Type-0 System: A constant value of the controlled variable requires a constant error signal under steady-state conditions. A feedback control system of this type is generally referred to as a regulator system.

Type-1 System: A constant rate of change of the controlled variable requires a constant error signal under steady-state conditions. A type-1 feedback control system is generally referred to as a servomechanism system. For reference inputs that change with time at a constant rate, a constant error is required to produce the same steady-state rate of the controlled variable. When applied to position control, type-1 systems may also be referred to as "zero-displacement-error" systems. Under steady-state conditions, it is possible for the reference signal to have any desired constant position or displacement and the feedback signal or controlled variable to have the same displacement.

Type-2 System: A constant acceleration of the controlled variable requires a constant error under steady-state conditions for a type-2 system. Since these systems can maintain a constant value of controlled variable and a constant controlled variable speed with no actuating error, they are sometimes referred to as "zero-velocity-error" systems.

STABILITY OF LINEAR SYSTEMS

A linear control system is unstable when its response to any aperiodic bounded signal increases without bound. Mathematically, instability may be investigated by analysis of the closed-loop response of the system shown in Fig. 2.

$$(C/R)(s) = G(s)/[1 + G(s)]$$

$$s = \sigma + j\omega.$$

The stability of the system depends upon the location of the poles of $C(s)/R(s)$ or the zeros of $[1 + G(s)]$ in the complex s plane. Several methods of stability determination can be employed.

Routh's Criterion

A method due to Routh is constructed as follows. Let $D =$ numerator polynomial of $1 + G(s)$. Then form

$$D = \sum_{i=0}^{i=n} a_i s^i$$

where $a_n > 0$.

(**A**) Construct the table shown below, with the first two rows formed directly from the coefficients and succeeding rows found as indicated.

a_n	a_{n-2}	a_{n-4}	a_{n-6}	•	•	•
a_{n-1}	a_{n-3}	a_{n-5}	a_{n-7}	•	•	•
b_1	b_2	b_3	b_4	•	•	•
c_1	c_2	c_3	c_4	•	•	•
d_1	d_2	d_3	•	•	•	•
e_1	e_2	•	•	•	•	•
f_1	•	•	•	•	•	•
•	•	•	•	•	•	•
•	•	•	•	•	•	•
•	•	•	•	•	•	•

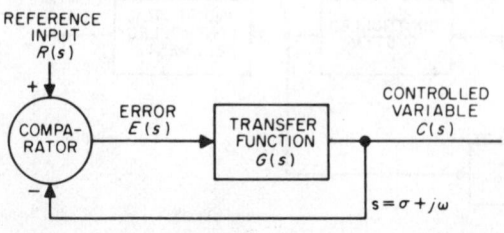

REFERENCE
INPUT
$R(s)$

Fig. 2—Single-loop system.

where

$$b_1 = (a_{n-1}a_{n-2} - a_{n-3}a_n)/a_{n-1}$$

$$b_2 = (a_{n-1}a_{n-4} - a_{n-5}a_n)/a_{n-1}$$

$$b_3 = (a_{n-1}a_{n-6} - a_{n-7}a_n)/a_{n-1}$$

$$c_1 = (b_1a_{n-3} - b_2a_{n-1})/b_1$$

$$c_2 = (b_1a_{n-5} - b_3a_{n-1})/b_1$$

$$c_3 = (b_1a_{n-7} - b_4a_{n-1})/b_1$$

$$d_1 = (c_1b_2 - b_1c_2)/c_1$$

$$d_2 = (c_1b_3 - b_1c_3)/c_1$$

$$d_3 = (c_1b_4 - b_1c_4)/c_1$$

$$\vdots$$

The table will consist of n rows.

(**B**) The system is stable; i.e., the polynomial has no right-half-plane zeros if every entry in the first column of the table is positive. If any com-

plete row is zero, the rest of the table cannot be formed. In such a case the polynomial always has zeros in the right-half plane or on the imaginary axis.

Nyquist Stability Criterion

A second method for determining stability is known as Nyquist stability criterion. This method consists in obtaining the locus of the transfer function $G(s)$ in the complex G plane for values of $s = j\omega$ for ω from $-\infty$ to $+\infty$. For single-loop systems, if the locus thus described encloses the point $-1 + j0$, the system is unstable; otherwise it is stable. Since the locus is always symmetrical about the real axis, it is sufficient to draw the locus for positive values of ω only. Figure 3 shows loci for several simple systems. Curves A and C represent stable systems and are typical of the

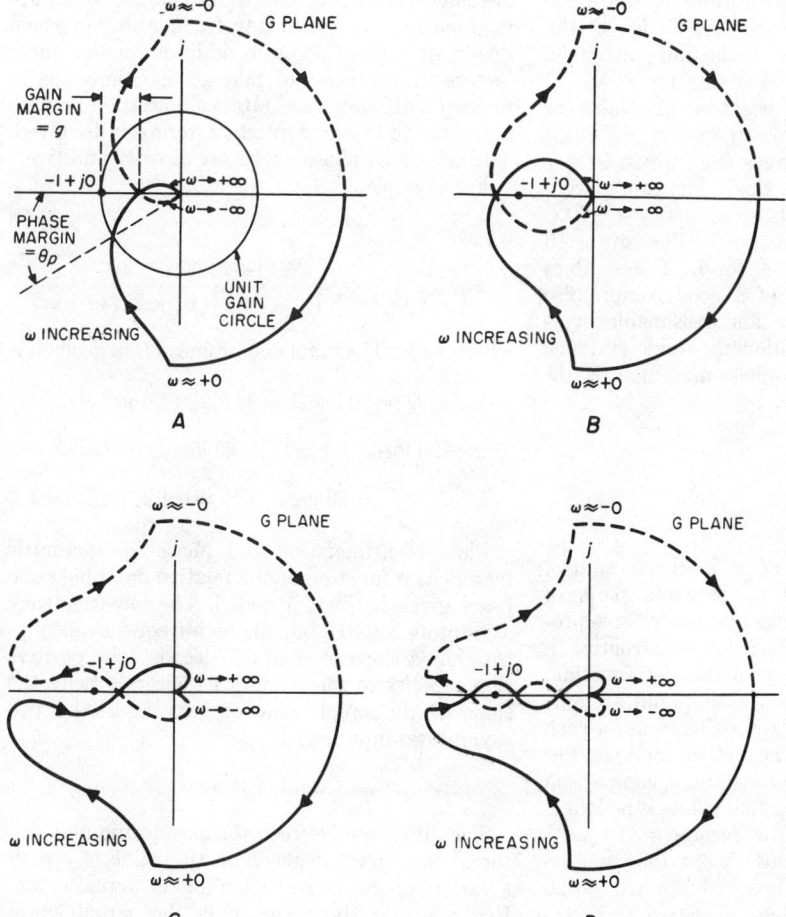

Fig. 3—Typical Nyquist loci.

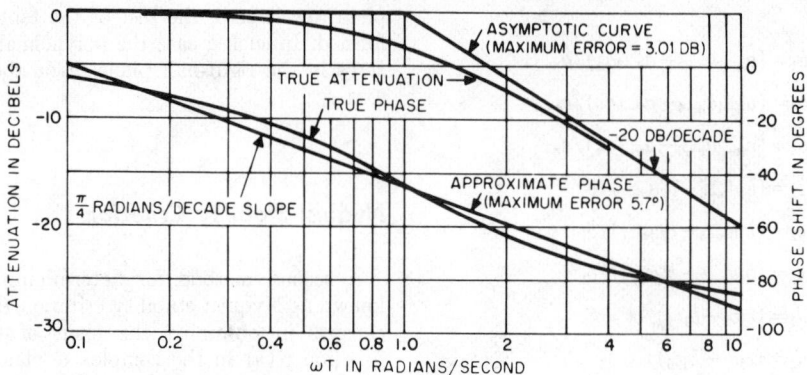

Fig. 4—Transfer-function plot. $G(j\omega) = 1/(1+j\omega T)$.

type-1 system; curve B is an unstable system. Curve D is conditionally stable; that is, for a particular range of values of gain K it is unstable. The system is stable both for larger and smaller values of gain. *Note*: It is unstable as shown.

Phase margin θ_p and gain margin g are also illustrated in Fig. 3A. The former is the angle between the negative real axis and $G(j\omega)$ at the point where the locus intersects the unit-gain circle. It is positive when measured as shown.

Gain margin g is the negative dB value of $G(j\omega)$ corresponding to the frequency at which the phase angle is 180 degrees (i.e., where $G(j\omega)$ intersects the negative real axis). The gain margin is often expressed in decibels, so that $g = -20 \times \log_{10} G(j\omega)$. Typical satisfactory values are -10 dB for g and an angle of 30° for θ_p. These values are selected on the basis of a good compromise between speed of response and reasonable over-shoot. Note that for conditionally stable systems, the terms gain margin and phase margin are without their usual significance.

Logarithmic Plots

The transfer function of a feedback control system can be described by separate plots of attenuation and phase versus frequency. This provides a very simple method for constructing a Nyquist diagram from a given transfer function. Use of logarithmic frequency scale permits simple straight-line (asymptotic) approximations for each curve. Figure 4 illustrates the method for a transfer function with a single time constant. A comparison between approximate and actual values is included.

Transfer functions of the form $G = (1 + j\omega T)$ have similar approximations except that the attenuation curve slope is inverted upward $(+20$ dB/decade$)$ and the values of phase shift are positive.

The transfer function of feedback control systems can often be expressed as a fraction with the numerator and denominator each composed of linear factors of the form $(Ts + 1)$. Certain types of control systems, such as hydraulic motors where compressibility of the oil in the pipes is appreciable or some steering problems where the viscous damping is small, give rise to transfer functions in which quadratic factors occur in addition to the linear factors. The process of taking logarithms (as in making a dB plot) facilitates computation because only the addition of product terms is involved. The associated phase angles are directly additive.

For example,

$$G(j\omega) = \frac{K(1+j\omega T_2)}{[T^2(j\omega)^2 + 2\zeta T(j\omega) + 1](1+j\omega T_1)(1+j\omega T_3)}$$

where $s = j\omega$. The exact magnitude of G in decibels is

$$20 \log_{10}|G| = 20 \log_{10}K + 20 \log_{10}|1+j\omega T_2|$$
$$- 20 \log_{10}|1+j\omega T_1| - 20 \log_{10}|1+j\omega T_3|$$
$$- 20 \log_{10}|T^2(j\omega)^2 + 2\zeta T(j\omega) + 1|.$$

Plots of attenuation and phase for quadratic factors as a function of the relative damping ratio ζ are given in Figs. 5 and 6. The low-frequency asymptote is 0 dB, but the high-frequency asymptote has a slope of ± 40 dB/decade (the positive slope applies to zero quadratic factors), twice the slope of the simple pole or zero case. The two asymptotes intersect at

$$\omega = 1/T.$$

The difference between the asymptotic plot and the actual curves depends on the value of ζ with a variety of shapes realizable for the actual curve. Regardless of the value of ζ, the actual curve approaches the asymptotes at both low and high

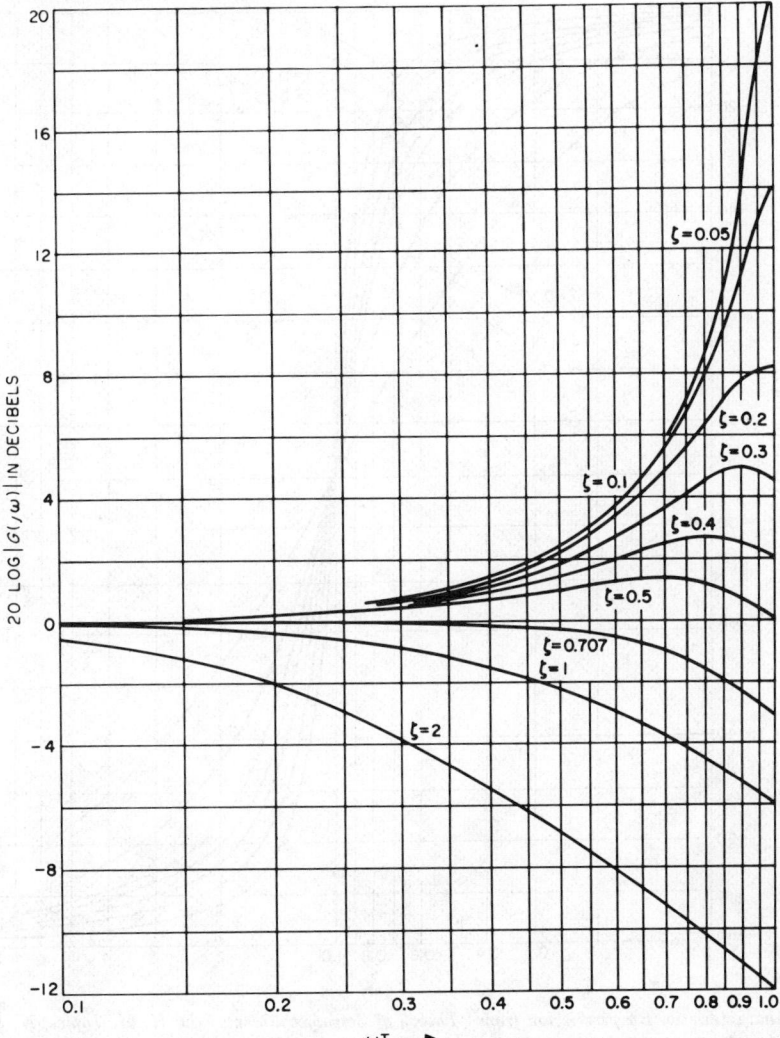

Fig. 5—Attenuation curve for quadratic factor. $G(j\omega) = 1/[T^2(j\omega^2) + 2\zeta T(j\omega) + 1]$. *By permission from "Automatic Feedback Control System Synthesis," by J. G. Truxal. Copyright 1955, McGraw-Hill Book Company, Inc.*

frequencies. In addition, the error between the asymptotic plot and the actual curve is geometrically symmetrical about the break frequency $\omega = 1/T$. As a result of this symmetry, the curves of Fig. 5 are plotted only for $\omega T \leq 1$. The error for $\omega = \alpha/T$ is identical with the error at $\omega = 1/\alpha T$.

Log Plots Applied to Transfer Functions

Nyquist's method, although yielding satisfactory results, has undesirable limitations when applied to system synthesis because the quantitative effect of parameter changes is not readily apparent. The use of attenuation–phase plots yields a more direct approach to the problem. The method* is based on the relation between phase and the rate of change of gain with frequency of networks. As a first approximation, which is valid for simple systems, a gain rate of change of 20 dB/decade corresponds to a phase shift of 90°. Since the stability of a system can be determined from its phase margin at unity gain (0 dB), simple criteria for the slope of the attenuation curve can be established. Thus it is obvious that to avoid insta-

* A theorem due to Bode shows that the phase angle of a network at any desired frequency is dependent on the rate of change of gain with frequency, where the rate of change of gain at the desired frequency has the major influence on the value of the phase angle at that frequency.

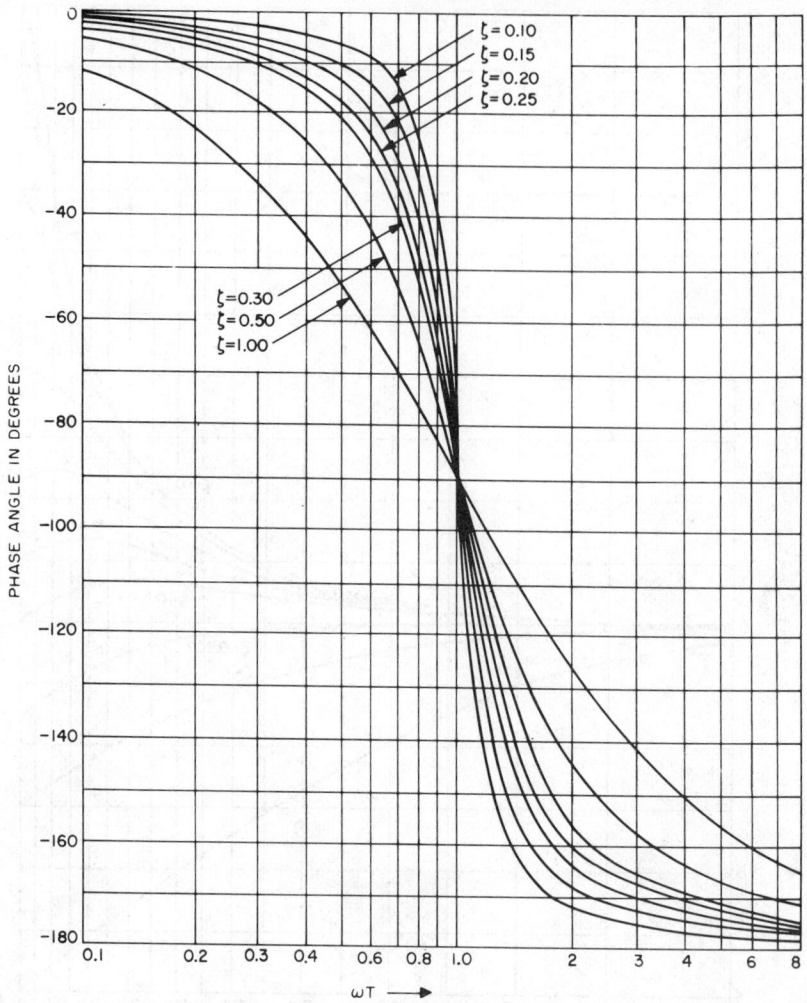

Fig. 6—Phase characteristic. *By permission from "Theory of Servomechanisms," by H. M. James, N. B. Nichols, and R. S. Phillips. Copyright 1947, McGraw-Hill Book Company, Inc.*

bility, the slope of the attenuation curve at unity gain must be appreciably less than −40 dB/decade (commonly about −33 dB/decade).

The design procedure is to construct asymptotic attenuation–phase curves as a first approximation. From this it can be determined whether the stability requirements are met. Refinements can be made by using the actual instead of asymptotic values for the curve as outlined in Fig. 4.

Figures 7 and 8 are examples of transfer functions plotted in this manner. In Fig. 7 a positive phase margin exists and the system is stable. Associated with the first-order pole at the origin is a uniform (low-frequency) slope of −20 dB/decade and −90° phase shift. This may be considered characteristic of the integrating action of a type-1 control system. Figure 8 is an unstable system. It has a negative phase margin (as a result of the steep slope of the attenuation curve). The former is stable, the latter is unstable.

Root-Locus Method

Root locus is a method of design due to Evans, based on the relation between the poles and zeros of the closed-loop system function and those of the open-loop transfer function. The rapidity and ease with which the loci can be constructed form the basis for the success of root-locus design methods, in much the same way that the simplicity of the gain and phase plots (Bode diagrams) makes design in the frequency domain so attractive. The root-locus plots can be used to adjust system gain, guide the design of compensation networks, or study the effects of changes in system parameters.

In the usual feedback control system, $G(s)$ is a rational algebraic function, the ratio of two polynomials in s; thus

$$G(s) = m(s)/n(s).$$

From Fig. 2

$$(C/R)(s) = G(s)/[1+G(s)]$$

$$= \frac{m(s)/n(s)}{1+[m(s)/n(s)]}$$

$$= m(s)/[m(s)+n(s)].$$

The zeros of the closed-loop system are identical with those of the open-loop system function.

The closed-loop poles are the values of s at which $m(s)/n(s) = -1$. The root-locus method is a graphic technique for determination of the zeros of $m(s)+n(s)$ from the zeros of $m(s)$ and $n(s)$. Root loci are plots in the complex s plane of the variations of the poles of the closed-loop system

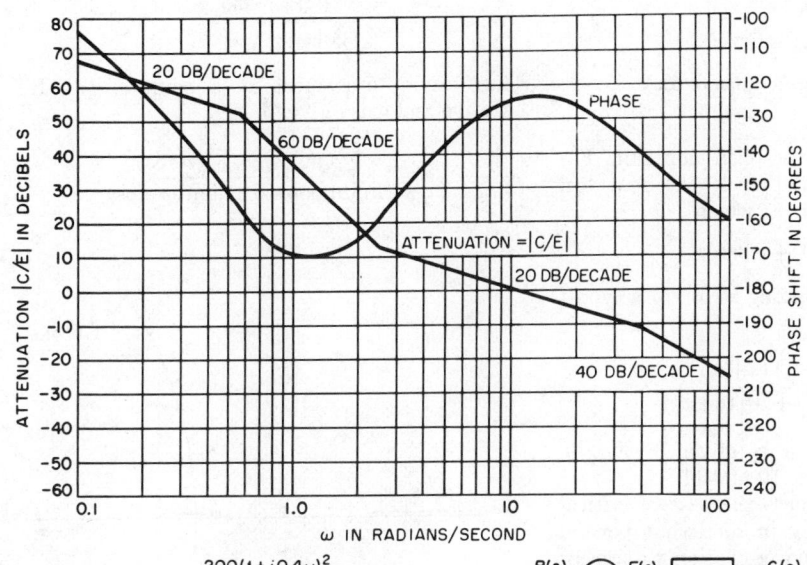

$$G = \frac{C}{E} = \frac{200(1+j\,0.4\omega)^2}{j\omega(1+j1.789\omega)^2(1+j\,0.25\omega)}$$

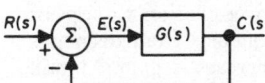

Fig. 7—Attenuation and phase shift for a stable system.

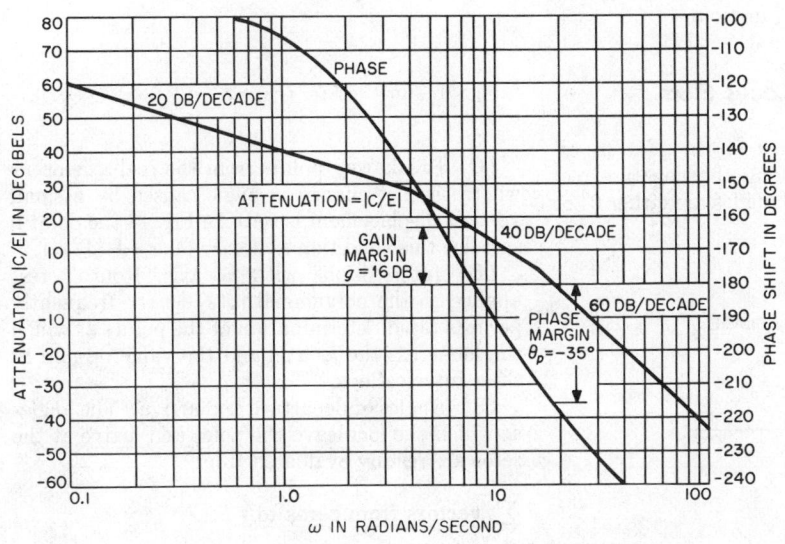

$$G = \frac{C}{E} = \frac{100}{j\omega(1+j\,0.25\omega)(1+j\,0.0625\omega)}$$

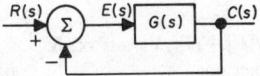

Fig. 8—Attenuation and phase shift for an unstable system.

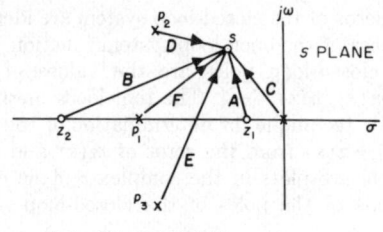

$$G(s) = K(AB/CDEF)$$

$$= \underline{/A} + \underline{/B} - \underline{/C} - \underline{/D} - \underline{/E} - \underline{/F}$$

Fig. 9—Graphic interpretation of $G(s)$.

function with changes in the open-loop gain. For the single-loop system of Fig. 2, the root loci constitute all s-plane points at which

$$\underline{/G(s)} = 180° + n\,360°$$

where n is any integer including zero. For a type-1 feedback control system

$$G(s) = \frac{K(s+z_1)(s+z_2)}{s(s+p_1)(s+p_2)(s+p_3)}.$$

A graphic interpretation is given in Fig. 9. Examples are given in Figs. 10 and 11.

Gain K_1, Fig. 11, produces the case of critical damping. An increase in gain somewhat beyond this value causes a damped oscillation to appear. The latter increases in frequency (and decreases in damping) with further increase in gain. At gain K_3 a sustained oscillation will result. Instability exists for gain greater than K_3, as at K_4. This corresponds to poles in the right half of the s plane for the closed-loop transfer function.

Aids in Sketching Root-Locus Plots

(A) The simplest portions of the plot to establish are the intervals along the negative real $(-\sigma)$ axis, because then all angles are either 0° or 180°.

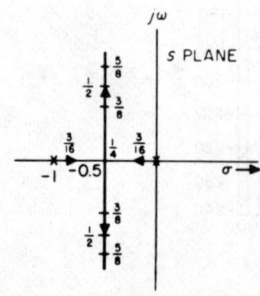

Fig. 10—Root loci for $G(s) = K/[s(s+1)]$. Values of K as indicated by fractions.

Complex pairs of zeros or poles contribute no net angle for points along the real axis.

Along the real axis, the locus will exist for intervals that have an *odd* number of zeros and poles to the right of the interval (Fig. 12).

(B) For very large values of s, all angles are essentially equal. The locus will thus finally approach asymptotes at the angles (Fig. 13) given by

$$\frac{180 + n\,360°}{(\text{poles}) - (\text{zeros})}.$$

These asymptotes meet at a point s_1 (on the negative real axis) given by

$$s_1 = \frac{\sum(\text{poles}) - \sum(\text{zeros})}{(\text{finite poles}) - (\text{finite zeros})}.$$

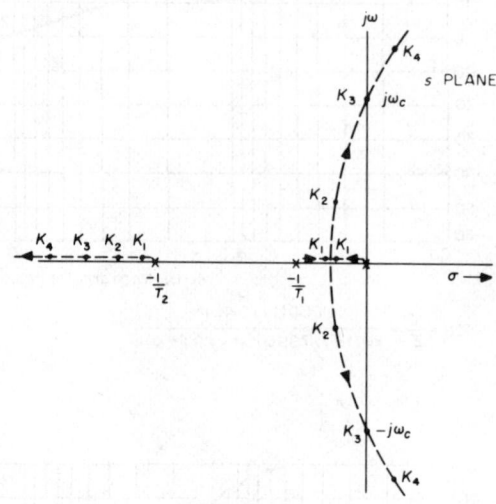

Fig. 11—Root loci for $G(s) = K/[s(T_1 s + 1)(T_2 s + 1)]$.

(C) Breakaway points from the real axis occur where the net change in angle caused by a small vertical displacement is zero. In Fig. 14 the point p satisfies this condition at $1/x_0 = (1/x_1) + (1/x_2)$.

(D) Intersections with $j\omega$ axis. Routh's test applied to the polynomial $m(s) + n(s)$ frequently permits rapid determination of the points at which the loci cross the $j\omega$ axis and the value of gain at these intersections.

(E) Angles of departure and arrival. The angles at which the loci leave the poles and arrive at the zeros are readily evaluated from

$$\sum\underline{/\text{vectors from zeros to } s}$$

$$-\sum\underline{/\text{vectors from poles to } s} = 180° + n\,360°.$$

For example, consider Fig. 15. The angle of departure of the locus from the pole at $(-1+j1)$ is

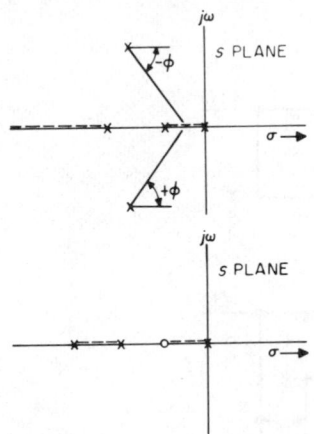

Fig. 12—Root-locus intervals along the real axis.

desired. If a test point is assumed only slightly displaced from the pole, the angles contributed by all critical frequencies (except the pole in question) are determined approximately by the vectors from these poles and zeros to $(-1+j1)$. The angle contributed by the pole at $(-1+j1)$ is then just sufficient to make the total angle 180°. In the example shown in the figure the departure angle is found from the relation

$$+45° - \underbrace{(135°}_{s+2} + \underbrace{90°}_{s} + \underbrace{26.6°}_{s+1+j1} + \underbrace{\theta}_{s+3} \underbrace{}_{s+1-j1}) = 180° + n\,360°.$$

Hence, $\theta = -26.6°$, the angle at which the locus leaves $(-1+j1)$.

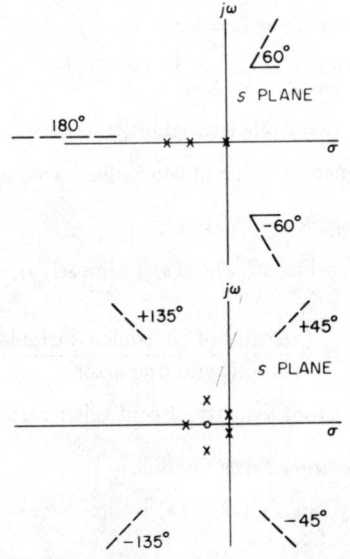

Fig. 13—Final asymptotes for root loci. Top, 60° asymptotes for system having 3 poles. Bottom, 45° asymptotes for system having an excess of 4 poles over zeros.

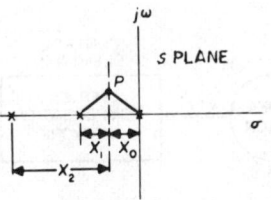

Fig. 14—Breakaway point.

METHODS OF STABILIZATION

Methods of stabilization for improving feedback-control-system response fall into the following basic categories:

(**A**) Series (cascade) compensation.
(**B**) Feedback (parallel) compensation.
(**C**) Load compensation.

In many cases any one of the above methods may be used to advantage and it is largely a question of practical considerations as to which is selected. Figure 16 illustrates the three methods.

Networks for Series Stabilization

Common networks for stabilization are shown in Fig. 17 with the transfer functions. The bridged-T network can be used for stabilization of ac systems although it has the disadvantage of requiring close control of the carrier frequency. Asymptotic attenuation and phase curves for the first three networks are shown in Figs. 18 and 19. The positive values of phase angle are to be associated with the phase-lead network whereas the negative values are to be applied to the phase-lag network. Figure 20 is a plot of the maximum phase shift for lag and

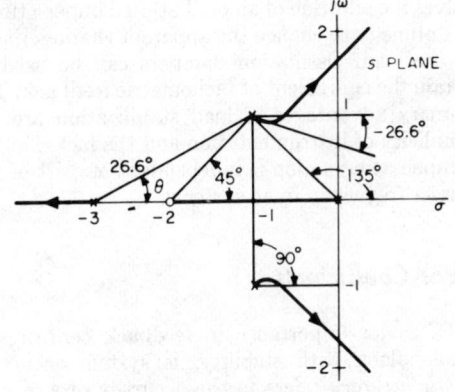

Fig. 15—Loci for

$$G(s) = K(s+2)/[s(s+3)(s^2+2s+2)].$$

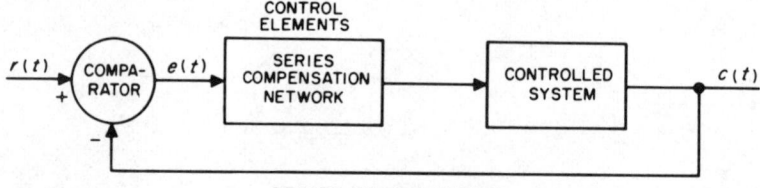

SERIES COMPENSATION

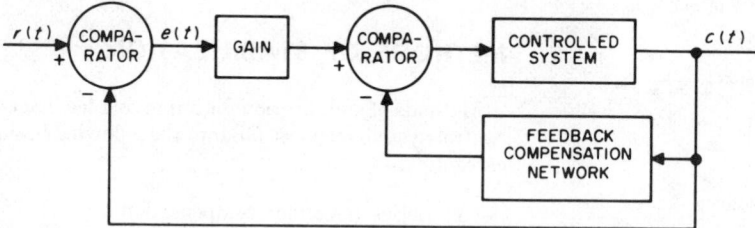

FEEDBACK COMPENSATION

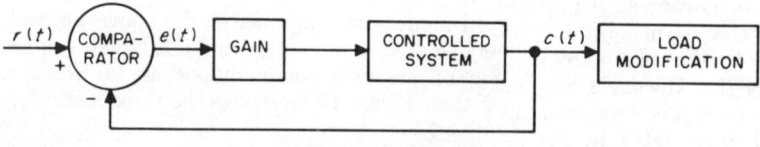

LOAD COMPENSATION

Fig. 16—Simple schemes for compensation.

lead networks as a function of the time-constant ratio.

Instead of direct feedback, the feedback connection may contain frequency-sensitive elements. Typical of such frequency-sensitive elements are tachometers or other rate- or acceleration-sensitive devices that may be fed back directly or through suitable stabilizing means.

Load Stabilization

The commonest form of load stabilization involves the addition of an oscillation damper (tuned or untuned) to change the apparent characteristics of the load. Oscillation dampers can be used to obtain the equivalent of tachometric feedback. The primary advantages of load stabilization are the simplicity of instrumentation and the fact that the compensating action is independent of drift of the carrier frequency in ac systems.

Error Coefficients

Of major importance in feedback control systems, along with stability, is system accuracy. *Static accuracy* refers to the accuracy of a system after the steady state is reached and is ordinarily measured with the system input constant or slowly varying. *Dynamic accuracy* refers to the ability of

the system to follow rapid changes of the input. The following refers to a system such as Fig. 2.

Static-Error Coefficients

Position Error Constant:

$$K_p = \lim_{s \to 0} [C(s)/E(s)] = \lim_{s \to 0} G(s)$$

$$= (\text{controlled variable})/(\text{actuating error})$$

for a constant value of controlled variable.

Velocity Error Constant:

$$K_v = \lim_{s \to 0} [sC(s)/E(s)] = \lim_{s \to 0} sG(s)$$

$$= \frac{(\text{velocity of controlled variable})}{(\text{actuating error})}$$

for a constant velocity of controlled variable.

Acceleration Error Constant:

$$K_a = \lim_{s \to 0} [s^2C(s)/E(s)] = \lim_{s \to 0} s^2G(s)$$

$$= \frac{(\text{acceleration of controlled variable})}{(\text{actuating error})}$$

for constant acceleration of the controlled variable.

Phase-Lag Network:

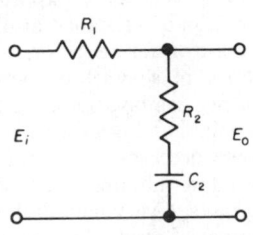

$$E_o/E_i = (T_2s+1)/(T_1s+1)$$

where

$$T_2 = R_2C_2$$

$$T_1 = (R_1+R_2)C_2.$$

Lead-Lag Network:

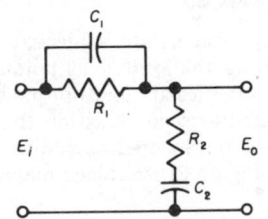

$$E_o/E_i = \frac{(T_1s+1)(T_2s+1)}{T_1T_2s^2+(T_1+T_2+T_{12})s+1}$$

where

$$T_1 = R_1C_1$$

$$T_2 = R_2C_2$$

$$T_{12} = R_1C_2$$

$$G_1 = (T_1+T_2)/(T_1+T_2+T_{12}).$$

Phase-Lead Network:

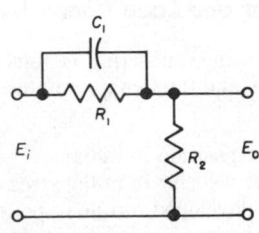

$$E_o/E_i = (T_2/T_1)[(T_1s+1)/(T_2s+1)]$$

where

$$T_1 = R_1C_1$$

$$T_2 = R_2R_1C_1/(R_1+R_2).$$

Bridged-T Network:

$$E_o/E_i = \frac{T_1T_3s^2+2T_1s+1}{T_1T_3s^2+(2T_1+T_3)s+1}$$

where

$$T_1 = R_1C$$

$$T_3 = R_3C.$$

Fig. 17—Networks for series stabilization.

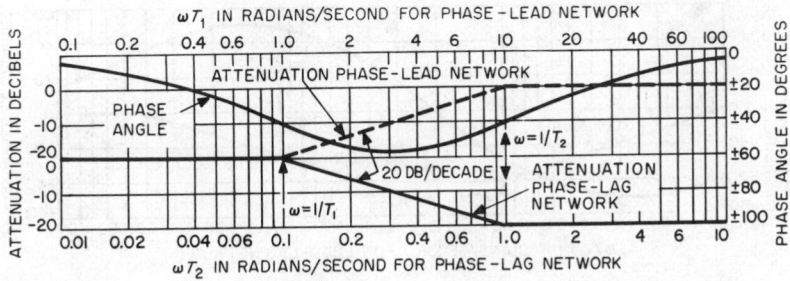

Fig. 18—Phase and attenuation for phase-lead and phase-lag networks. $T_1=10T_2$.

MULTIPLE INPUTS AND LOAD DISTURBANCES

Frequently systems are subjected to unwanted signals entering the system at points other than the input. Examples are load-torque disturbances, noise generated at a point within the system, etc. These may be represented as additional inputs to the system. Figure 21 is a block diagram of such a condition.

For linear operation

(A) $C/R = G_1 G_2 / (1 + H G_1 G_2)$

(B) $C/U = G_2 / (1 + H G_1 G_2)$.

Combining (A) and (B)

$$C/U = (1/G_1)(C/R).$$

If it is desired' that the sum of R and U be reproduced in the output (controlled variable), then G_1 should be equal to unity. If U is a disturbance to be minimized, then G_1 should be as large as possible. An example of such a disturbance is the torque produced on a radar antenna by wind forces.

Practical Application

An example of a common application is the positioning-type servomechanism shown in Fig. 22. Such a system ordinarily includes the following components: a comparator to measure the error, an amplifier, a second comparator or mixer to measure $(E_1 - B)$, a motor, and a tachometer.

For this system

$$C(s)/E(s) = G_1(s)G_2(s)/[1 + H(s)G_2(s)]$$

$$C(s)/R(s) = \frac{G_1(s)G_2(s)}{[1 + H(s)G_2(s) + G_1(s)G_2(s)]}.$$

CONTROL-SYSTEM COMPONENTS

Error-Measuring Systems: Potentiometers, Synchros

Commonly used error-measuring systems or comparators are shown in Fig. 23.

For synchros whose primary excitation is 115 volts, the error sensitivity is approximately 1 volt/degree for a load resistance of 10 000 ohms across the control-transformer rotor.

The static error of a synchro transmitter and control transformer combination is of the order of 18 minutes maximum and is a function of the rotor position. In some precision units, this error may be reduced to a few minutes of arc. In synchro-control transformers, a very undesirable characteristic is the presence of residual voltages at the null position. In well-designed units this voltage will be less than 30 millivolts.

Synchro errors can be materially reduced by the use of double-speed systems. Such systems consist of a dual set of synchro units whose shafts are geared in such a manner as to provide a "fine" and a "coarse" control. The synchro error can be effectively reduced by the factor of the gear ratio employed. Synchronizing networks are employed to provide for proper switching between the two sets of synchros.

Linear Motor and Load Characteristics

In the following, subscript m refers to motor, l refers to load, and 0 refers to combined motor and load.

$\theta =$ angular position in radians

$r =$ angular velocity in radians/sec $= d\theta/dt$

$T_m =$ motor-developed torque in pound-feet

$J_m =$ motor moment of inertia in slug-feet^2

$E_m =$ impressed volts

$k_t =$ motor stalled-torque constant in pound-feet/volt
$= [\Delta T_m / \Delta E_m]_{r_m}$

$f_m =$ motor internal-damping characteristic in pound-feet-seconds/radian
$= -[\Delta T_m / \Delta r_m]_{E_m}$

$r_m =$ motor torque-inertia constant in 1/second
$= T_m / J_m$

$J_l =$ load inertia in slug-feet^2

$f_l =$ load viscous-friction coefficient in pound-feet-seconds/radian

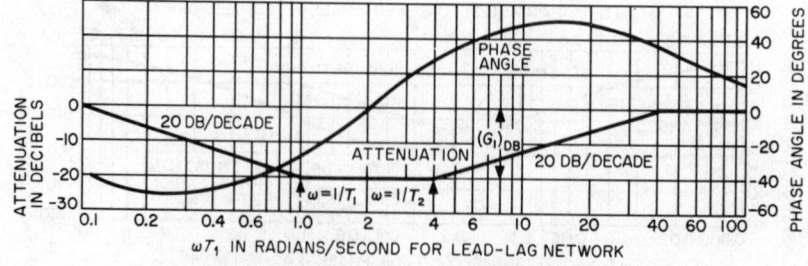

Fig. 19—Phase and attenuation for lead-lag network.

$G_1 = (T_1 + T_2)/(T_1 + T_2 + T_{12})$. $T_2 = T_1/4$ and $T_{12} = 11.25 T_1$.

F_l = load coulomb friction in pound-feet
N = motor-to-load gear ratio
  = θ_m/θ_l
f_0 = overall viscous-friction coefficient referred to load shaft
  = $f_l + N^2 f_m$
J_0 = overall inertia referred to load shaft
  = $J_l + N^2 J_m$
T_0 = overall time constant in seconds
  = J_0/f_0.

The ideal motor characteristics of Fig. 24 are quite representative of dc shunt motors. For alternating-current two-phase servomotors, one phase of which is excited from a constant-voltage source (the reference winding), the curves are approximately valid up to about 40 percent of synchronous speed.

The speed and load-transfer characteristics are given by

$$\theta_0(s) = [k_t N E_m(s) - F_l(s)] / (J_0 s^2 + f_0 s).$$

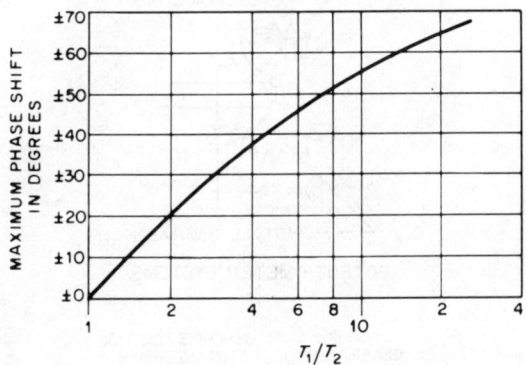

Fig. 20—Maximum phase shift for phase-lead (use positive angles) and phase-lag (negative angles) networks.

When the coulomb friction F_l can be neglected

$$G(s) = \theta_0(s) / E_m(s)$$
$$= k_t N / [f_0 s (T_0 s + 1)].$$

Rate Generators

A rate generator (or tachometer generator) is a precision electromechanical component resembling a small motor and having an output voltage proportional to its shaft rotational speed. Rate generators have extensive applications both as computing instruments and as stabilizing components of feedback control systems. An example of the latter is illustrated in Fig. 22. The use of the rate generator produces an effective viscous damping and also tends to linearize the servomechanism by inserting damping of a linear nature and of such magnitude that it swamps out the rather large nonlinear damping of the motor. To eliminate the backlash between rate generators and servomotors, they are often constructed as integral units having a common shaft. These units are available for dc or ac (either 400- or 60-hertz) operation.

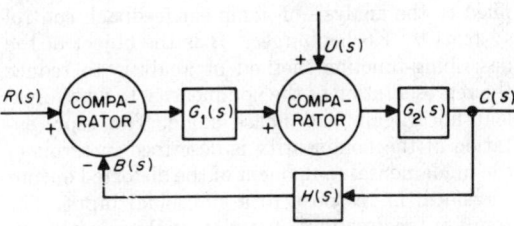

Fig. 21—Multiple-input control system.

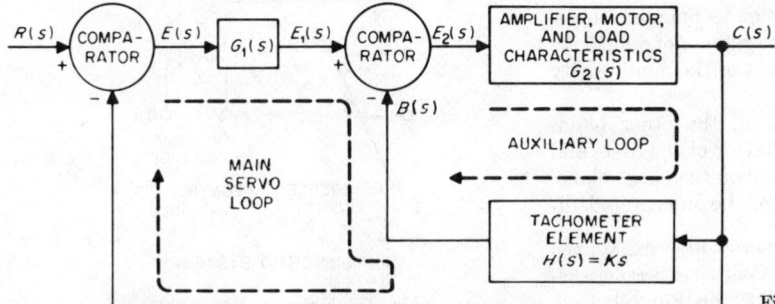

Fig. 22—Positioning-type servo.

PART 2—NONLINEAR SYSTEMS

SCOPE

All physical systems have nonlinearities and time-varying parameters to some degree. This is justified by the fact that any element may physically break down or exhibit deterioration as a result of time.

Linear systems, as described in the preceding sections, have a linear relationship between the variables described by a linear differential equation; the theory of superposition also applies. Analysis and synthesis techniques applicable to any linear system have been thoroughly investigated and developed.

Nonlinear systems have no general methods of analysis and synthesis. Therefore, for reasons of simplicity, they are often treated with linear approximations and, in many cases of small nonlinear effects, satisfactory results have been obtained. In general, however, linear methods become restrictive in their application and quite often unrealistic.

There are many different ways of solving nonlinear systems that may be applicable to a certain type of system but not to all. Among these methods, two techniques have proved useful in the study of nonlinear systems.

The describing-function technique was first applied to the analysis of nonlinear feedback control systems by Kochenburger.* It is the object of the describing-function method of analysis to reduce the representation of the nonlinearity to an equivalent linear gain and phase angle. The representation of the nonlinearity is described in terms of the fundamental component of the distorted output waveform in response to a sinusoidal input. The result of the describing-function analysis is a representation of the system in the frequency domain; however, the correlation between the time and frequency domains in comparison with linear system analysis is much less precise in nonlinear system analysis. The synthesis of nonlinear systems can be carried out with the describing-function technique in much the same way as is done with linear systems. The describing-function technique is therefore most useful in complex systems of relatively high order and where the effect of the nonlinearity is small but significant.

The phase-plane method, on the other hand, is a representation of the behavior of the first- and second-order systems portrayed on the phase plane. The phase-plane diagrams can be interpreted di-

rectly in terms of time domain; they thus are a useful tool in the study of transient response to any initial condition and in some cases to step and ramp inputs. The phase-plane method does not readily indicate the steps required to correct system performance, and synthesis is carried out by trial-and-error procedures.

This method can be extended to higher-order systems in the so-called phase space. However, an nth-order system in the nth-dimensional phase space compared with the phase plane is difficult to envision and interpret and may be less effective

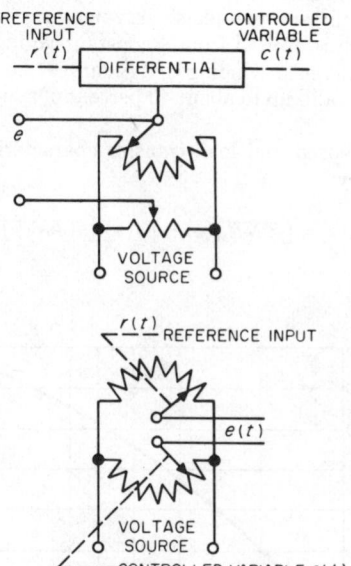

POTENTIOMETER SYSTEMS

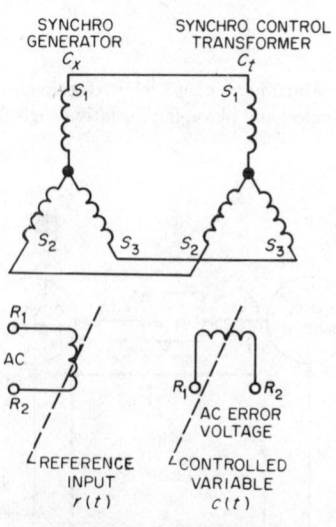

SYNCHRO SYSTEM

Fig. 23—Error-measuring systems.

---

* R. J. Kochenburger, "A Frequency Response Method For Analyzing and Synthesizing Contractor Servomechanisms," *AIEE Transactions*, vol. 69, pp. 270–284; 1950.

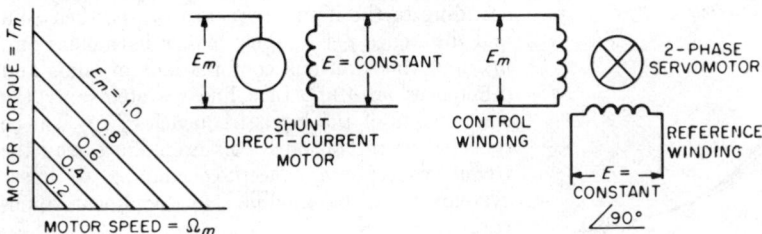

Fig. 24—Ideal motor curves.

in its usefulness. In summary, this method is useful for large nonlinearities of the second-order systems.

The two methods are in a sense complementary and constitute the principal tools for the study of nonlinear control problems.

CHARACTERISTICS OF NONLINEAR SYSTEMS

A comparison of linear and nonlinear systems yields the following representative characteristics.

Principle of Superposition Does Not Apply

In nonlinear systems, the response to a combination of individual signals at the input will not be the same as the response to the sum of those same signals. The impulse response (weighting function) describing the frequency response is therefore not applicable. The familiar Laplace transform operation $(s=\sigma+j\omega)$ and the transfer-function concept used extensively in linear systems cannot be directly applied without some modification. In fact, transfer functions are noncommutative in nonlinear systems. That is, in a linear system transfer, functions $G_1(s)$ and $G_2(s)$ can be cascaded as $G_1(s) \cdot G_2(s)$ or $G_2(s) \cdot G_1(s)$, whereas in nonlinear systems $N_1 \cdot N_2$ is not the same as $N_2 \cdot N_1$.

Nonlinear Response Is Dependent On Input Signal

In linear systems, the system response is strictly a function of the system parameters. A system proved to possess a good stable response to one type of input will behave similarly to any other type of input. Nonlinear systems, on the other hand, are dependent on the input-signal size and initial conditions as well as the system parameter. A stable response for one input signal may be unstable for another.

In nonlinear systems, phenomena which are ordinarily unexplainable by linear analytical methods (generation of new frequencies, jump resonance, limit cycles, etc.) exist. The following are examples of such unique nonlinear phenomena.

Jump Resonance

The phenomenon called jump resonance is observed in certain closed-loop systems with saturation, where the input–output amplitude ratio and phase angle as a function of frequency exhibit sudden discontinuities. The typical closed-loop gain characteristics of a saturating system with jump resonance are shown in Fig. 25. $|\theta_o/\theta_i|$ is plotted as a function of frequency for a fixed amplitude θ_i. As the frequency is increased from zero, the frequency response follows the curve along points A, B, and C. At point C a sudden discontinuous jump to D is observed with an incremental increase in frequency. Further increase in frequency leads to point E along the curve. If the frequency is reversed, the response retraces the path E, D and continues to point F at which a sudden jump to point G occurs and continues on through B and A of the gain curve. The phase-angle response behaves similarly. The overall response curve exhibits a hysteresis-type property or jump resonance. The system response dependent on the input-signal size is illustrated in Fig. 26. $|\theta_o/\theta_i|$ is plotted for different values of θ_i. For small-signal values of θ_i,

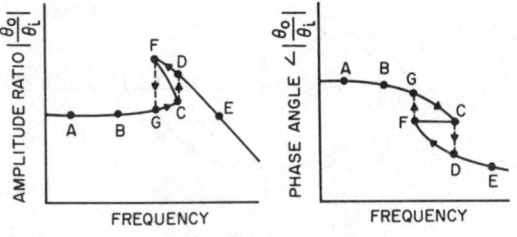

Fig. 25—Jump resonance.

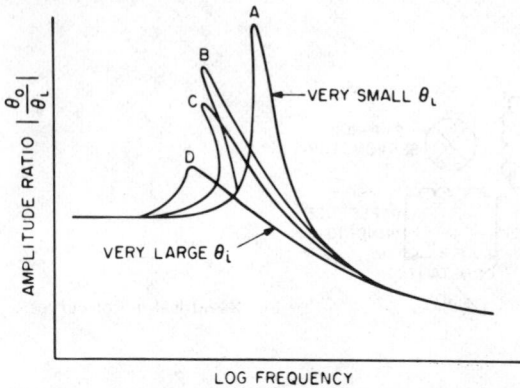

Fig. 26—Closed-loop frequency response of a saturating amplifier. *D. Graham and D. McRuer, "Analysis of Nonlinear Control Systems," fig. 1–18, ©1961, John Wiley & Sons, Inc., New York.*

the system behaves linearly, corresponding to curve A. As the input-signal value is increased, jump resonance is observed as shown in curves B and C.

Limit Cycle

Limit cycle is a phenomenon of oscillation peculiar to nonlinear systems. The oscillatory behavior, unexplainable in terms of linear theory, is characterized by a constant amplitude and frequency determined by the nonlinear properties of the system. Limit cycles are distinguishable from linear oscillation in that their amplitude of oscillation is independent of initial conditions. For instance, if a system has a stable limit cycle, the system will tend to fall into the limit cycle, with the output approaching the amplitude of that limit cycle regardless of the initial condition and forcing function. A limit cycle is easily recognized in the phase plane as an isolated closed path as shown in Fig. 27.

Often the system falls into a limit cycle in the presence of very small excitation or disturbances. This behavior is termed *soft self-excitation*. Conversely, for systems requiring forced excitation above a certain minimum amplitude or appropriate initial condition before entering a limit cycle, the term *hard self-excitation* is used. In any event, the presence of limit cycles in control systems is undesirable and makes isolation of these limit cycles important to the analysis of nonlinear systems. The phase-plane method described on p. 15-28 provides techniques for investigating limit cycles.

Generation of New Frequencies

In a nonlinear system the output of the non-linear device contains harmonic and subharmonic frequencies of the input signal. For instance, application of two sine waves of different frequencies f_1 and f_2 to the input will produce components corresponding to the input frequencies f_1, f_2, their sum and difference $f_1 \pm f_2$, their higher harmonics mf_1, nf_2, and their various combinations of sums and differences $mf_1 \pm nf_2$. In a linear system only the components of the input frequencies f_1, f_2 will be reflected in the output. This exemplifies why the frequency-response concept pertaining to linear systems must be modified for nonlinear applications.

Hysteresis

Multivalued functions exist when two or more function values correspond to the value of the variable. Multivalued functions are intrinsically nonlinear. Some examples of this are the hysteresis curves of magnetic materials and the backlash of a gear train.

TYPES OF NONLINEAR ELEMENTS

Because of the infinite variety of nonlinearities, specific classification becomes impossible. Two generally accepted classifications do exist; they are the *incidental* and *intentional* nonlinearities.

Incidental nonlinearities are extraneous to system design and are usually undesirable. Saturation, dead zone, and backlash are examples of non-linearities that may lead to inaccurate and poor response or even instability. There are cases, however, where incidental nonlinearities aid and improve the system response.

Intentional nonlinearities are functions purposely introduced into the system to compensate and improve performance. The relay servo system is typical of this type. Other classifications are the *single-valued* function and *multivalued* function. These functions are transfer characteristics describing the input-to-output relationship of the

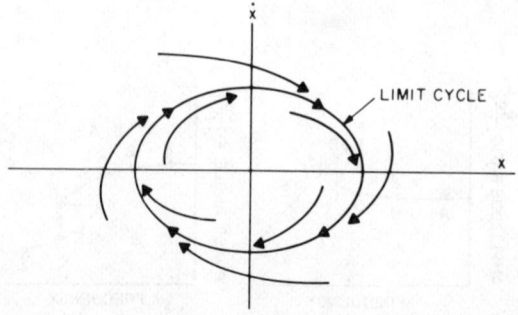

Fig. 27—Limit cycle.

nonlinear element. Single-valued functions have only one output value corresponding to the input value, while the multivalued functions have two or more output values. Several typical nonlinear characteristics commonly encountered in control systems are presented in Fig. 28.

In addition to the classifications described above, other functions to be found may include large and small values of continuous and discontinuous types of nonlinearities. More than one classification may apply to any nonlinear function.

ANALYTICAL METHODS FOR SOLVING NONLINEAR SYSTEMS

The analytical methods for solving nonlinear systems described in this section will be concentrated mainly on graphical methods for describing-function techniques and phase-plane methods. Other methods of analysis are:

(**A**) Direct Solution to the Nonlinear Differential Equation. There are certain nonlinear differential equations of lower order that are analytically solvable or integrable; however, they are very rare.

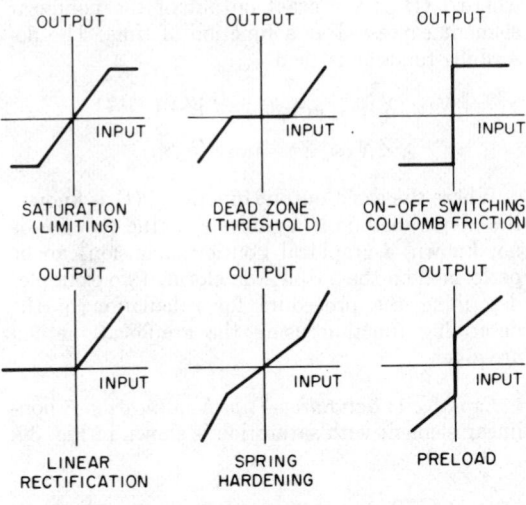

SINGLE-VALUED FUNCTIONS

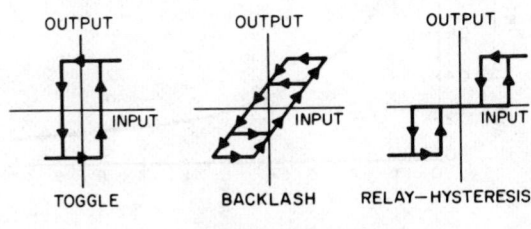

MULTIVALUED FUNCTIONS

Fig. 28—Types of nonlinear elements.

(**B**) Numerical Method. The numerical method is a step-by-step process obtaining the solution to the differential equation as a table of corresponding values of independent and dependent variables. In theory any equation can be solved numerically, although the process may be quite complex.

Neither (**A**) nor (**B**) is discussed here but further information may be found in references 5–8 of the bibliography.

Describing-Function Technique

This technique is valuable in the analysis and design of an important class of nonlinear feedback control systems, in which the output of the nonlinear element is filtered by a linear element having low-pass frequency characteristics as it travels around the control loop. The object of the describing-function technique is to represent the actual nonlinearity of the system in terms of an equivalent linear system by considering only the fundamental component of the output waveform of the nonlinear element subject to a sinusoidal input.

Describing Function: The describing-function analysis is made of the following basic assumptions.

(**A**) The input to the nonlinear element n is a sinusoidal signal, and only the fundamental component of the output of n contributes to the input. The output response of a nonlinear element to a periodic signal consists of the fundamental frequency component of the input signal and its harmonics. Generally, the harmonic components are smaller in amplitude compared with the fundamental component. Further, in most control systems the system behaves as a low-pass filter and the higher harmonics are attenuated. If the higher harmonics are sufficiently small they can be neglected and the equivalent linear approximation may be justified.

(**B**) There is only one nonlinear element in the system. All nonlinearities in the system are lumped into one single nonlinear element n. Figure 29 shows a block diagram of a closed-loop system containing a nonlinear element n.

(**C**) The output of the lumped nonlinear element is a function only of the present value and past history of the input, i.e., n is not a function of time.

The describing function of a nonlinear element

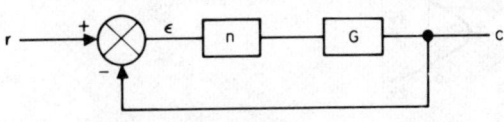

Fig. 29—Block diagram of nonlinear closed-loop system.

is defined as the ratio of the fundamental frequency component of the output as a complex quantity (amplitude and phase angle) to the amplitude of the sinusoidal input signal. If the input signal as applied to the nonlinear element n is described by

$$e_{in}(t) = X \sin\omega t$$

the output response $e_o(t)$ may generally take the form of

$$e_o(t) = (a_0 X/2) + a_1 X \sin\omega t + b_1 X \cos\omega t$$

$$+ \sum_{n=2}^{\infty} a_n X \sin n\omega t + \sum_{n=2}^{\infty} b_n X \cos n\omega t + \text{subharmonics}.$$

The $(a_0/2)$ term is the dc component; a_n and b_n are the harmonic components.

The fundamental frequency component of the output may be expressed in terms of amplitude and phase angle as

$$e_{o1} = A(\omega, X) X \sin[\omega t + \phi(\omega, X)].$$

$A(\omega, X) X$ is the amplitude and $\phi(\omega, X)$ is the phase angle of the fundamental component. Both amplitude and phase angle are a function of the frequency and amplitude of the input signal. The describing function $N(\omega, X)$ by definition is

$$N(\omega, X) = \{A(\omega, X) X \exp[j\phi(\omega, X)]/X\}$$

$$= A(\omega, X) \exp[j\phi(\omega, X)]$$

$$= A(\omega, X) \cos\phi(\omega, X)$$

$$+ jA(\omega, X) \sin\phi(\omega, X). \quad (1)$$

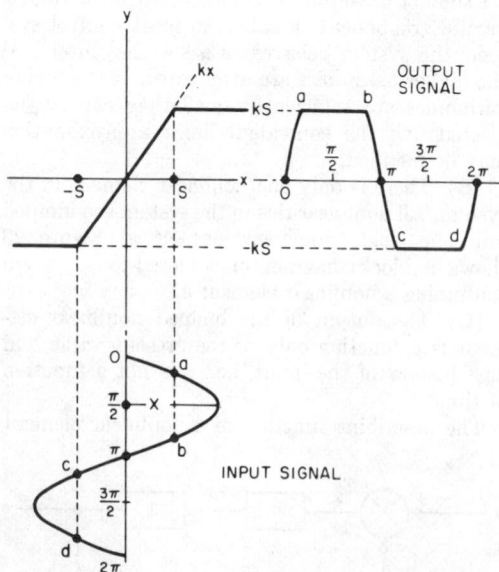

Fig. 30—Graphic representation of saturation.

The describing function $N(\omega, X)$ may be purely real or contain a phase angle depending on the type of nonlinearity. For single-valued nonlinear functions N is real, whereas for multivalued functions phase shift exists, generally lagging.

Calculation of a Describing Function: Calculation of the describing function involves performing a conventional Fourier analysis on the output waveform to obtain the fundamental component. The Fourier series expansion of the output waveform to an input sinusoidal $X \sin\omega t$ may be expressed as

$$e_o(t) = (a_0 X/2) + a_1 X \sin\omega t + b_1 X \cos\omega t$$

$$+ a_2 X \sin 2\omega t + b_2 X \cos 2\omega t + \cdots.$$

For the describing function, only the coefficients of the fundamental frequency component are required. The coefficients may be obtained from the integrals

$$a_1 = (\pi X)^{-1} \int_0^{2\pi} f_o(t) \sin\omega t \cdot d(\omega t) \quad (2)$$

$$b_1 = (\pi X)^{-1} \int_0^{2\pi} f_o(t) \cos\omega t \cdot d(\omega t) \quad (3)$$

where $f_o(t)$ is the exact output of the nonlinear element expressed as a function of time. The describing function is then

$$|N(\omega, X)| = |a_1 + jb_1| = (a_1^2 + b_1^2)^{1/2}$$

$$\angle N(\omega, X) = \tan^{-1}(b_1/a_1).$$

Where the exact output function $f_o(t)$ is known, the above method is applicable. If the function is not known, a graphical Fourier expansion can be performed on the output waveform. Two examples describing the procedure for calculation of the describing function using the graphical method are given.

Example 1: Saturation-Type Nonlinearity. A nonlinear element with saturation is shown in Fig. 30.

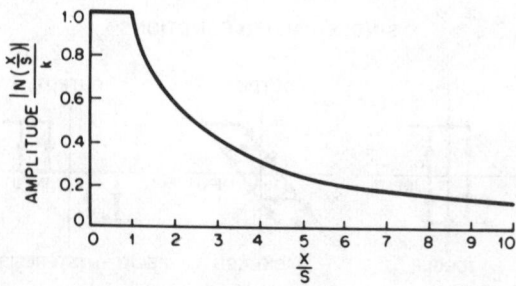

Fig. 31—Describing function for saturation (normalized amplitude).

The output y is held constant for input values greater than S. This region is called saturation or limiting. For input values less than S the output behaves linearly with the input. The input–output relationship can be expressed by

(A) $y = kx$ for $-S < x < S$

(B) $y = kS$ for $x > S$

(C) $y = -kS$ for $x < -S$.

The output is an odd function and thus only the sine term of the fundamental equation need be calculated. Furthermore, because of symmetry only the first quarter of the integration need be evaluated as follows:

$$a_1 = (4/\pi X) \int_0^{\pi/2} f_o(t) \sin\omega t \cdot d(\omega t).$$

If the input is $x = X \sin\omega t$, then the output $f_o(t) = y$ is expressed by

$$f_o(t) = kX \sin\omega t \qquad X < S$$

and for X greater than S

$$f_o(t) = kX \sin\omega t \qquad 0 < \omega t < \sin^{-1}(S/X)$$

$$= kS \qquad \sin^{-1}(S/X) < \omega t < \pi/2.$$

Therefore the coefficients a_1 are

$$a_1 = k \qquad \text{for } X < S$$

and for X greater than S

$$a_1 = (4/\pi X) \int_0^{\sin^{-1}(S/X)} kX \sin^2(\omega t) \cdot d(\omega t)$$

$$+ (4/\pi X) \int_{\sin^{-1}(S/X)}^{\pi/2} kS \sin\omega t \cdot d(\omega t)$$

$$= (2k/\pi)[\phi + \tfrac{1}{2}(\sin 2\phi)]$$

where $\phi = \sin^{-1}(S/X)$.

The describing function N is given by

$$N = k \qquad X < S$$

$$N = k(2/\pi)[\phi + \tfrac{1}{2}(\sin 2\phi)] \qquad X > S.$$

The variation of amplitude of N with respect to X/S is plotted in Fig. 31. The phase angle is zero over the entire range.

Example 2: Backlash-Type Nonlinearity. For the second example a simple backlash-type nonlinearity of Fig. 32 is evaluated. The backlash is a multivalued nonlinearity where the input–output relationship follows a different path dependent on

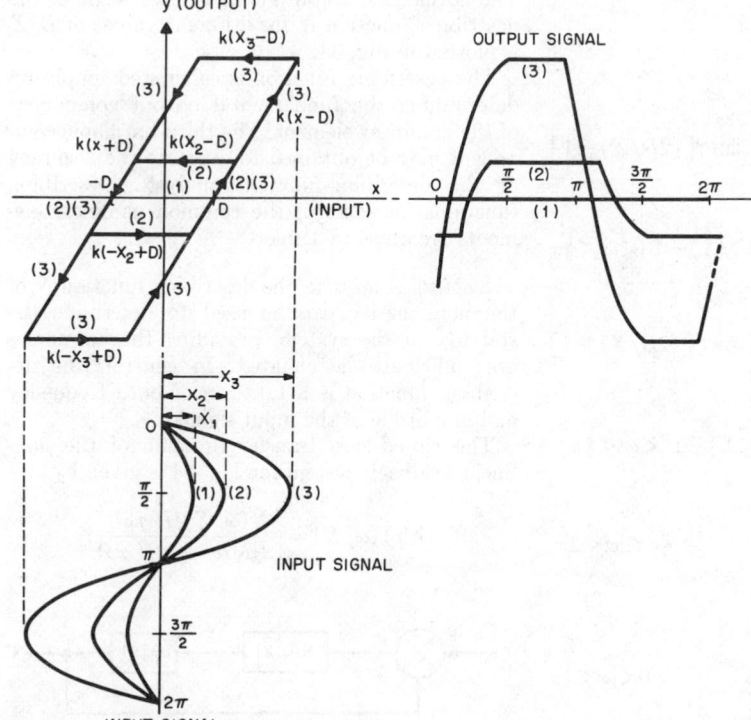

Fig. 32—Graphic representation of backlash nonlinearity. E. Levinson, "*Nonlinear Feedback Control Systems,*" *Electro-Technology*, p. 139, fig. 36; September 1962.

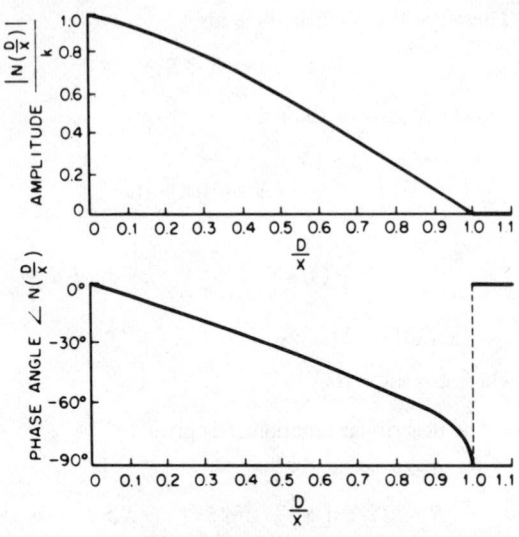

Fig. 33—Describing function for backlash (normalized amplitude and phase angle).

the input-signal amplitude (curves (1), (2), and (3)). After the steady state is established, the output $f_o(t)$ corresponding to different values of X of an input signal $x = X \sin\omega t$ are:

(A) For $X < D$ (curve 1):

$$f_o(t) = 0.$$

(B) For $D < X < 2D$ (curve 2):

$$f_o(t) = -k(X - D)$$

$$0 < \omega t < \sin^{-1}[(2D/X) - 1]$$

$$f_o(t) = k(X \sin\omega t - D)$$

$$\sin^{-1}[(2D/X) - 1] < \omega t < \tfrac{1}{2}\pi$$

$$f_o(t) = k(X - D)$$

$$\tfrac{1}{2}\pi < \omega t < \pi + \sin^{-1}[(2D/X) - 1]$$

$$f_o(t) = k(X \sin\omega t + D)$$

$$\pi + \sin^{-1}[(2D/X) - 1] < \omega t < \tfrac{3}{2}\pi$$

$$f_o(t) = -k(X - D)$$

$$\tfrac{3}{2}\pi < \omega t < 2\pi.$$

(C) For $X > 2D$ (curve 3):

$$f_o(t) = k(X \sin\omega t - D)$$

$$0 < \omega t < \tfrac{1}{2}\pi$$

$$f_o(t) = k(X - D)$$

$$\tfrac{1}{2}\pi < \omega t < \pi - \sin^{-1}[1 - (2D/X)]$$

$$f_o(t) = k(X \sin\omega t + D)$$

$$\pi - \sin^{-1}[1 - (2D/X)] < \omega t < \tfrac{3}{2}\pi$$

$$f_o(t) = -k(X - D)$$

$$\tfrac{3}{2}\pi < \omega t < 2\pi - \sin^{-1}[1 - (2D/X)]$$

$$f_o(t) = k(X \sin\omega t - D)$$

$$2\pi - \sin^{-1}[1 - (2D/X)] < \omega t < 2\pi.$$

Solving for a_1 and b_1 of equations (2) and (3), for the three conditions above, yields the describing-function terms

$$a_1 = 0 \qquad\qquad X < D$$

$$a_1 = (k/\pi)[\tfrac{1}{2}\pi + \theta + \tfrac{1}{2}(\sin 2\theta)] \qquad X > D$$

$$b_1 = 0 \qquad\qquad X < D$$

$$b_1 = (-k/\pi) \cos^2\theta \qquad\qquad X > D$$

where

$$\theta = \sin^{-1}[1 - (2D/X)]$$

or in terms of amplitude and phase angle

$$|N| = (k/\pi)\{[\tfrac{1}{2}\pi + \theta + \tfrac{1}{2}(\sin 2\theta)]^2 + \cos^4\theta\}^{1/2}$$

$$\angle N = \tan^{-1}\left[-\left(\frac{\cos^2\theta}{\tfrac{1}{2}\pi + \theta + \tfrac{1}{2}(\sin 2\theta)}\right)\right].$$

The normalized amplitude and phase angle of the describing function N for different values of D/X is plotted in Fig. 33.

The describing function is calculated simply by determining the fundamental output component of the nonlinear element. The third-harmonic component may be obtained to estimate the accuracy of the describing-function analysis. Describing functions for some of the common nonlinear elements are given in Table 1.

Stability Analysis: The describing function N of the nonlinearity can be used to determine the stability of the system, providing the harmonics are sufficiently attenuated. In general, the describing function is a function of both frequency and amplitude of the input signal.

The closed-loop transfer function of the nonlinear feedback system for Fig. 34 is given by

$$(c/r)(\omega, X) = \frac{N(\omega, X) G(j\omega)}{1 + N(\omega, X) G(j\omega)}.$$

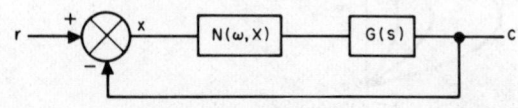

Fig. 34—Single-loop nonlinear system.

TABLE 1—DESCRIBING FUNCTIONS FOR COMMON NONLINEAR ELEMENTS.

DESCRIBING FUNCTION: $N(X) = a_1 + jb_1$ or $|N(X)| = (a_1^2 + b_1^2)^{1/2}$, $\angle N(X) = \tan^{-1}(b_1/a_1)$.

Characteristic	Describing-Function Coefficients	Characteristic	Describing-Function Coefficients
A	$a_1 = 4S/\pi X$ $b_1 = 0$	H	$a_1 = \tfrac{1}{2}k_1$ $b_1 = 0$
B	$a_1 = (4S/\pi X)\cos\theta$ $b_1 = 0$ $\theta = \sin^{-1}(D/X)$	I (SQUARE LAW)	$a_1 = 4X/3\pi$ $b_1 = 0$
C	$a_1 = (2k_1/\pi)[\theta + \tfrac{1}{2}(\sin 2\theta)]$ $b_1 = 0$ $\theta = \sin^{-1}(S/X)$	J	$a_1 = (k_1/\pi)[\tfrac{1}{2}\pi + \theta + \tfrac{1}{2}(\sin 2\theta)]$ $b_1 = -(k_1/\pi)\cos^2\theta$ $\theta = \sin^{-1}[1 - (2D/X)]$
D	$a_1 = (2k_1/\pi)[\tfrac{1}{2}\pi - \theta - \tfrac{1}{2}(\sin 2\theta)]$ $b_1 = 0$ $\theta = \sin^{-1}(D/X)$	K	$a_1 = (4L/\pi X)\cos\theta$ $b_1 = -(4L/\pi X)\sin\theta$ $\theta = \sin^{-1}(D/X)$
E	$a_1 = (2k_1/\pi)[\psi - \theta + \tfrac{1}{2}(\sin 2\psi) - \tfrac{1}{2}(\sin 2\theta)]$ $b_1 = 0$ $\psi = \sin^{-1}(S/X)$ $\theta = \sin^{-1}(D/X)$	L	$a_1 = (2L/\pi X)(\cos\theta + \cos\psi)$ $b_1 = (2L/\pi X)(\sin\psi - \sin\theta)$ $\psi = \sin^{-1}(P/X)$ $\theta = \sin^{-1}(Q/X)$
F	$a_1 = k_1 + (4A/\pi X)$ $b_1 = 0$		
G	$a_1 = k_2 - [(k_2 - k_1)/\pi](2\theta + \sin 2\theta)$ $b_1 = 0$ $\theta = \sin^{-1}(P/X)$		

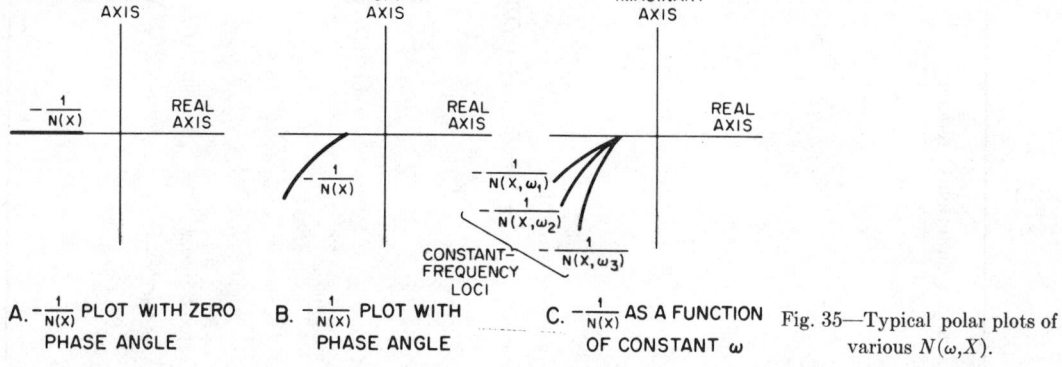

A. $-\frac{1}{N(x)}$ PLOT WITH ZERO PHASE ANGLE   B. $-\frac{1}{N(x)}$ PLOT WITH PHASE ANGLE   C. $-\frac{1}{N(x)}$ AS A FUNCTION OF CONSTANT ω

Fig. 35—Typical polar plots of various $N(\omega, X)$.

The characteristic equation of the system is

$$1 + N(\omega, X) G(j\omega) = 0 \qquad (4)$$

or

$$G(j\omega) = -[1/N(\omega, X)].$$

The condition of (4) must be satisfied for sustained oscillation of the output with zero input. Since $N(\omega, X)$ is a function of both frequency and amplitude, various combinations of ω and X can be found for oscillation. If there are no possible combinations satisfying the oscillatory condition, the system is stable. In the case of sustained oscillation, the oscillatory mode may be either stable or unstable. If a slight disturbance in amplitude or frequency occurs and the oscillation returns to its original value, the oscillation is stable (stable limit cycle). If the oscillation amplitude increases or decreases from the original value, the oscillation is unstable (unstable limit cycle). The stability of the closed-loop system may be evaluated analytically by directly solving the characteristic equation by any one of the modified linear graphical methods.

Polar Plot (Nyquist Diagram): The conventional Nyquist diagram must be modified to apply the Nyquist stability criteria to the frequency-response plot. In a linear system the critical point on the Nyquist diagram is -1. For nonlinear systems the $-[1/N(\omega, X)]$ locus corresponds to the critical point -1. To evaluate the stability of the system, both $-[1/N(\omega, X)]$ and the $G(j\omega)$ function are plotted on the polar plane. The describing function $N(\omega, X)$ generally is a function of both ω and X. If N is only a function of X, there will be one locus $-[1/N(x)]$ plotted as a function of X. If N is also a function of ω, a family of constant-frequency loci are plotted for different values of ω (see Fig. 35).

The stability of the system is determined by the following relationship between the $-[1/N(\omega, X)]$ locus and the $G(j\omega)$ plot (Fig. 36). If the $-[1/N(\omega, X)]$ lies to the left of the $G(j\omega)$ plot or is not enclosed, the system is *stable*. Conversely, if the $-[1/N(\omega, X)]$ lies to the right of the $G(j\omega)$ plot or is enclosed, the system is *unstable*. If the $-[1/N(\omega, X)]$ locus intersects with the $G(j\omega)$ plot, the system may have a *sustained oscillation*. In the case where N is a function of ω, the condition for sustained oscillation is satisfied if the ω of the $G(j\omega)$ plot at the intersecting point

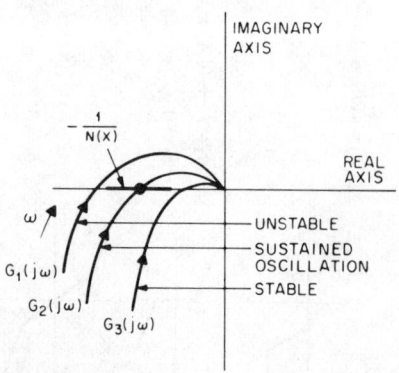

Fig. 36—Polar-plot stability criteria.

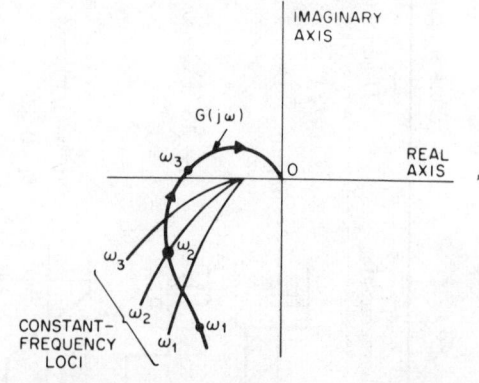

Fig. 37—Typical polar plot of $G(j\omega)$ and $-[1/N(\omega, X)]$ as a function of ω.

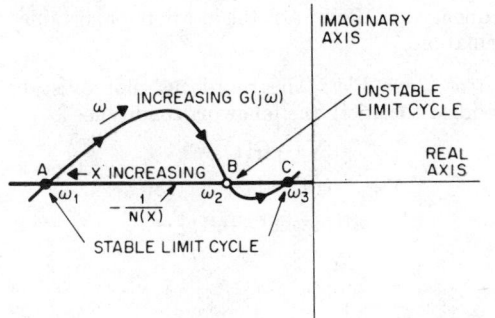

Fig. 38—Polar plot of stable and unstable limit cycles.

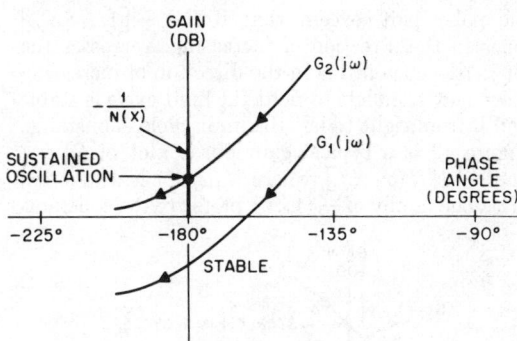

Fig. 40—Gain–phase plot stability criteria.

is the same ω of the $-[1/N(\omega, X)]$ locus (see Fig. 37).

The oscillation may be either stable or unstable. If the $G(j\omega)$ intersects with the $-[1/N(\omega, X)]$ locus at one point only, the oscillation is stable

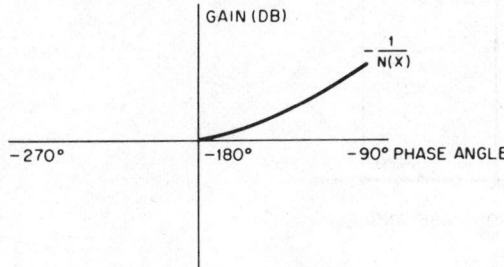

A. $-\frac{1}{N(X)}$ PLOT WITH ZERO PHASE ANGLE

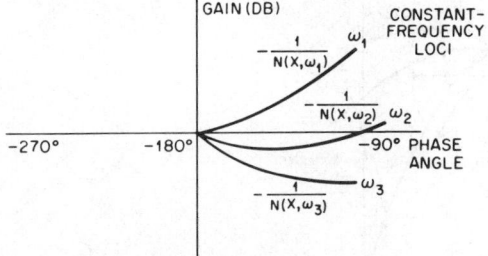

B. $-\frac{1}{N(X)}$ PLOT WITH PHASE ANGLE

C. $-\frac{1}{N(X)}$ PLOT AS A FUNCTION OF CONSTANT ω

Fig. 39—Typical gain–phase plots of various $N(\omega, X)$.

(stable limit cycle). If more points of intersection exist, the limit cycle may be either stable or unstable. The stability of the limit cycle is determined by the direction of the two loci at the crossover point.

By establishing the $G(j\omega)$ locus pointing in the direction of increasing frequency as a reference, if the $-[1/N(X)]$ locus pointing in the direction of increasing amplitude X crosses the $G(j\omega)$ locus from right to left, the limit cycle is stable. If the crossover occurs from left to right the limit cycle is unstable. A polar plot with both stable and unstable limit cycles is shown in Fig. 38.

Gain–Phase Plot: The gain–phase plot is the direct transfer of the polar plot from the polar coordinate to the rectangular coordinate. The ordinate is the gain in decibels and the abscissa is the phase angle in degrees.

The gain and phase angle of the two functions $G(j\omega)$ and $-[1/N(\omega, X)]$ are for $G(j\omega)$:

Gain $20 \log |G(j\omega)|$

Phase angle $\angle G(j\omega)$

and for $N(X, \omega)$:

Gain $-20 \log_{10} |N(\omega, X)|$

Phase angle $-180° - \angle N(\omega, X)$.

A typical gain–phase plot for various types of $N(\omega, X)$ is given in Fig. 39.

The system is stable if the $-(1/N)$ locus does not intersect with the $G(j\omega)$ plot. If the $-(1/N)$ locus intersects with the $G(j\omega)$ plot, the system has a sustained oscillation (Fig. 40).

In the case of sustained oscillation, there may be more than one point of intersection, as shown in Fig. 41. Points A and C are stable points (stable limit cycle) and point B is an unstable point (unstable limit cycle). The stability of the limit cycle is determined in a manner similar to that of the polar plot, except that if the $-[1/N(X)]$ locus in the direction of increasing X crosses the $G(j\omega)$ locus pointing in the direction of increasing frequency from left to right the limit cycle is stable and if from right to left the limit cycle is unstable. Figure 42 is a typical gain–phase plot of $G(j\omega)$ and $-[1/N(\omega, X)]$, where $N(\omega, X)$ is a function of ω. The family of $-(1/N)$ plots are the constant-frequency loci. Point A is the location for sustained oscillation.

Root-Locus Plot: The root-locus plot must be modified to satisfy the following condition.

$$|N|\cdot|G(s)|=1$$

$$\angle N+\angle G(s) = (1+2m)\,180°$$

$$m\,(\text{integer})=0, 1, \cdots.$$

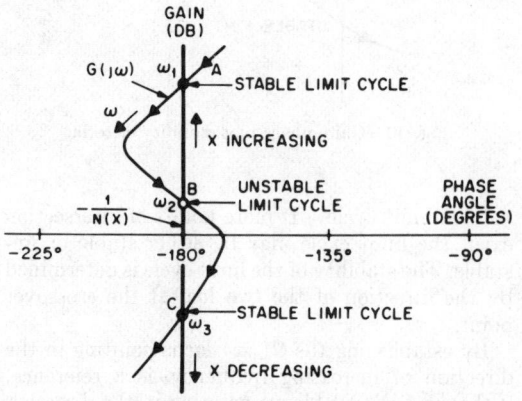

Fig. 41—Gain–phase plot of stable and unstable limit cycles.

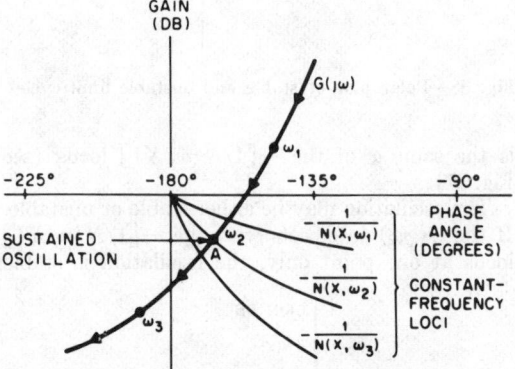

Fig. 42—Typical gain–phase plot of $G(j\omega)$ and $-[1/N(\omega, X)]$ as a function of ω.

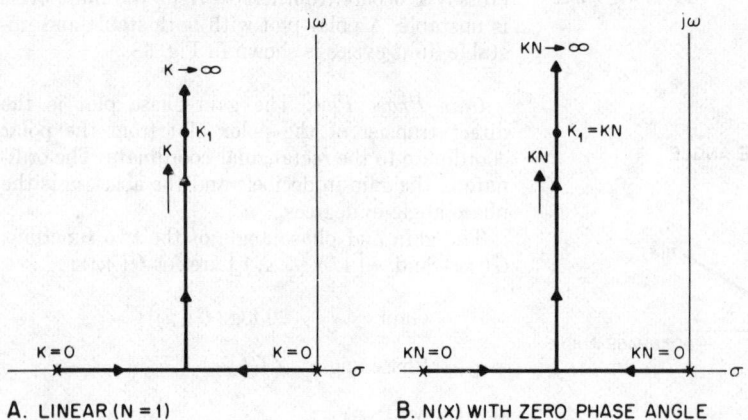

A. LINEAR (N = 1)

B. N(X) WITH ZERO PHASE ANGLE

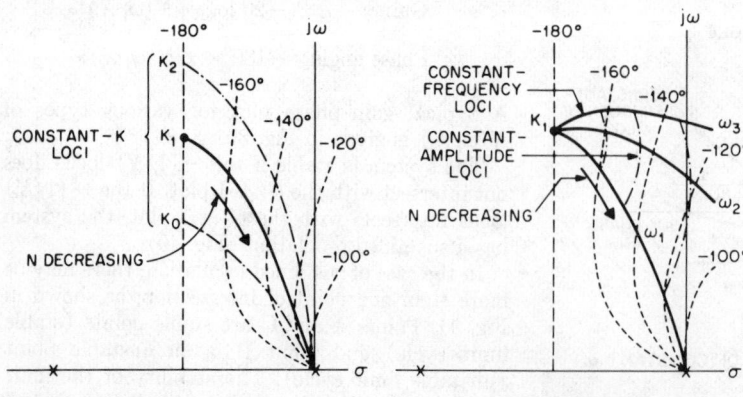

C. N(X) WITH PHASE ANGLE

D. N(ω, X) AS A FUNCTION OF ω

Fig. 43—Typical root-locus plots of various $N(\omega, X)$.

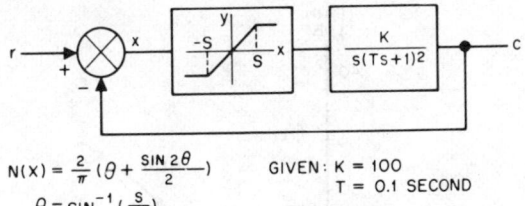

$$N(x) = \frac{2}{\pi}(\theta + \frac{SIN\,2\theta}{2})$$

$$\theta = SIN^{-1}(\frac{S}{X})$$

GIVEN: K = 100
T = 0.1 SECOND

Fig. 44—Nonlinear system with saturation.

For cases of single-valued nonlinear functions where the phase angle is zero, the root-locus plot is identical to the linear case, except for the gain constant K. The gain constant K of a linear system (Fig. 43A) is modified to KN (Fig. 43B). If the root locus remains in the left half of the s plane for all values of KN, the system is stable. Conversely if the root locus extends into the right half of the s plane, the system is unstable. The value of KN corresponding to the crossover point of the root locus at the $j\omega$ axis yields sustained oscillation.

If the describing function N involves a phase angle, which is the case of multivalued functions, the root locus must take into consideration the phase angle of N. The root locus is then plotted for various fixed values of K, resulting with a family of constant-K loci (Fig. 43C). The stability criteria of the linear root-locus case still apply. When the describing function is a function of ω as well as X, a root locus must be plotted for each ω

with the K value fixed, generating a family of constant-frequency loci (Fig. 43D). It is noted that for sustained oscillation the ω of the constant-frequency loci must intersect with the same ω of the $j\omega$ axis. When the root locus intersects the $j\omega$ axis more than once, stable and unstable limit cycles exist.

As an illustrated example, the stability of a closed-loop system with saturation shown in Fig. 44, and with backlash in Fig. 48, are evaluated using the analytical method, polar plot, gain–phase plot, and root-locus plot.

Example 3: *Closed-Loop System With Saturation.* The stability of the system to be evaluated is depicted in Fig. 44.

Analytical Method: The closed-loop transfer function is

$$(C/r)(\omega, R) = \frac{N(\omega, X)G(j\omega)}{1 + N(\omega, X)G(j\omega)}$$

where $G(j\omega)$ is given as

$$G(j\omega) = K/j\omega(j\omega T + 1)^2.$$

The characteristic equation of the system is

$$N(\omega, X)[K/j\omega(j\omega T + 1)^2] = -1.$$

The condition for oscillation is

$$|N(X)|[K/\omega(\omega^2 T^2 + 1)] = 1 \qquad (5)$$

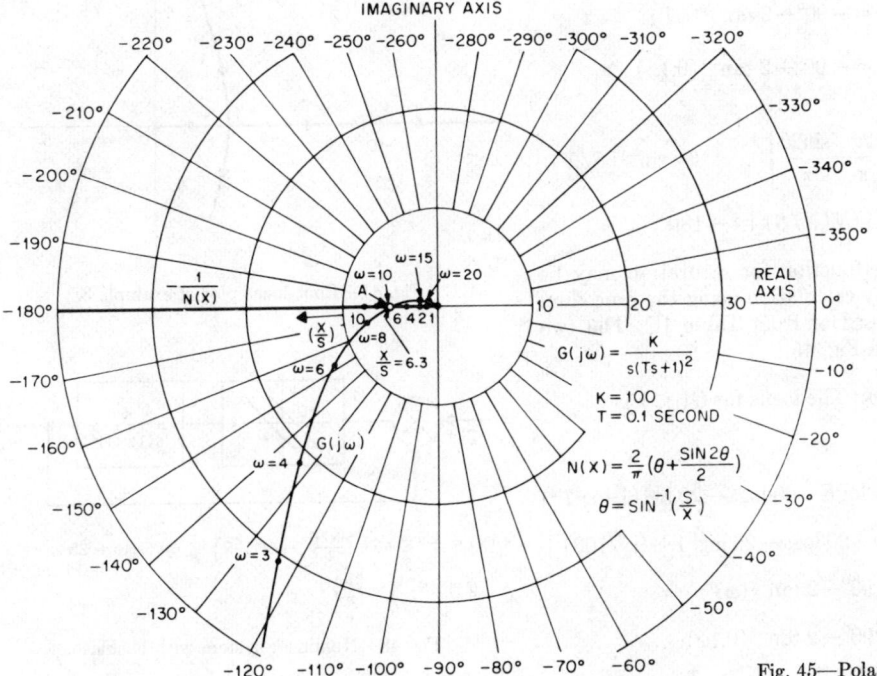

Fig. 45—Polar plot of example 3.

and

$$\angle N(X) - 90° - 2\tan^{-1}(\omega T) = -180°. \quad (6)$$

Solving (6) with phase angle $\angle N = 0$ yields

$$\omega T = 1 \quad\text{or}\quad \omega = 1/T$$

which satisfies the phase relation. Substituting this into (5) yields

$$|N(X)| = 2/KT.$$

For the example where $K = 100$, $T = 0.1$ second

$$N(S/X) = 0.2 \quad\text{and}\quad \omega = 10.$$

The expression for the describing function for saturation is given in Table 1C as

$$N(S/X) = (2k_1/\pi)[\theta + \tfrac{1}{2}(\sin 2\theta)], \quad \theta = \sin^{-1}(S/X)$$

and equating $N(S/X) = 0.2$, for $k = 1$ (unit slope) yields

$$S/X = 0.158 \quad\text{or}\quad X/S = 6.3.$$

The system has a sustained oscillation at a frequency of $f = 10/2\pi$ and the threshold-to-input-amplitude ratio S/X of 0.158, $(X/S = 6.3)$.

Polar Plot: The two loci, $G(j\omega)$ and $-[1/N(X/S)]$, are plotted on the polar plane for

$$G(\omega) = \frac{K}{\omega(1+\omega^2 T^2)} = \frac{100}{\omega[1+(\omega^2/100)]}$$

$$\angle G(\omega) = -90° - 2\tan^{-1}(\omega T)$$

$$= -90° - 2\tan^{-1}(0.1\omega)$$

and

$$|N(X/S)|^{-1} = \left[\frac{2\theta}{\pi} + \frac{\sin 2\theta}{\pi}\right]^{-1}, \quad \theta = \sin^{-1}(S/X)$$

$$\angle -[1/N(X/S)] = -180°.$$

The describing function for saturation may be obtained either by calculation using the procedure previously discussed or from Table 1C. The two loci are plotted in Fig. 45.

Gain–Phase Plot: The locus for $G(j\omega)$ is

$$20\log|G(j\omega)|$$

$$= 20\log K - 20\log\omega - 20\log(1+\omega^2 T^2)$$

$$= 40 - 20\log\omega - 20\log[1+(\omega^2/100)]$$

$$\angle G(j\omega) = -90° - 2\tan^{-1}(\omega T)$$

$$= -90° - 2\tan^{-1}(0.1\omega).$$

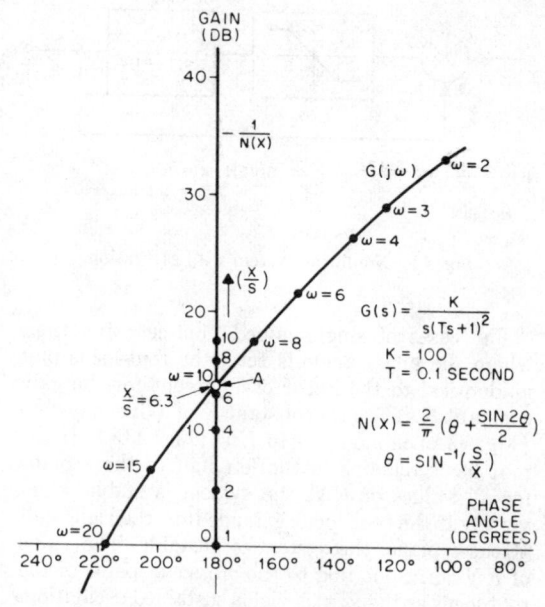

Fig. 46—Gain–phase plot of example 3.

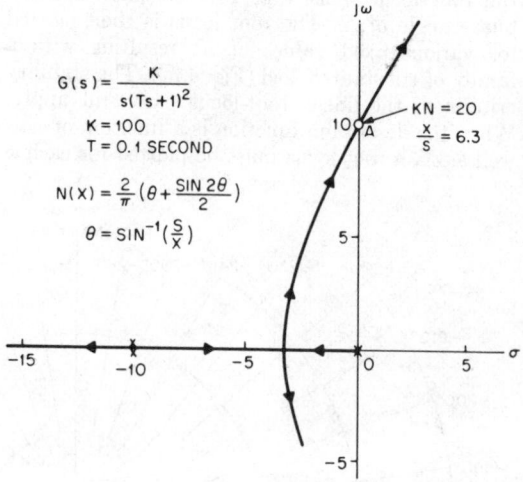

Fig. 47—Root-locus plot of example 3.

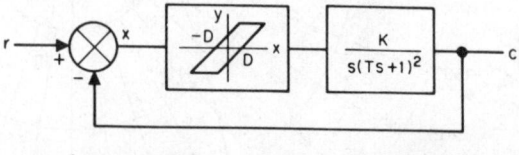

$$N(X) = \frac{1}{\pi}\left(\frac{\pi}{2} + \theta + \frac{\sin 2\theta}{2} - j\cos^2\theta\right) \qquad \text{GIVEN: } K = 25$$

$$\theta = \sin^{-1}\left(1 - \frac{2D}{X}\right) \qquad\qquad\qquad T = 0.1 \text{ SECOND}$$

Fig. 48—Nonlinear system with backlash.

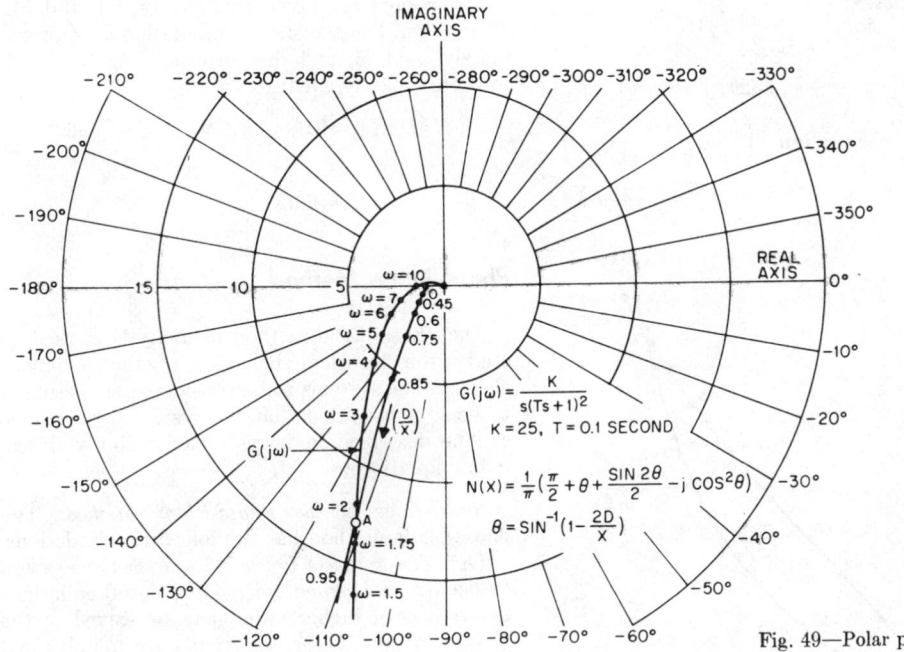

Fig. 49—Polar plot of example 4

The locus for $-[1/N(X)]$ is

$$20 \log |-(1/N)(X/S)|$$

$$= -20 \log[(2/\pi)\theta + \tfrac{1}{2}(\sin 2\theta)], \quad \theta = \sin^{-1}(S/X)$$

$$\angle -[1/N(X/S)] = -180°.$$

The results are plotted in Fig. 46.

Root-Locus Plot: The root locus is plotted for

$$s(sT+1) = |N(X/S)| K$$

$$= |(2/\pi)[\theta + \tfrac{1}{2}(\sin 2\theta)]| \cdot 100, \quad \theta = \sin^{-1}(S/X)$$

and

$$90 + 2 \tan^{-1}(\omega T) = +180° + \angle N(X/S)$$

$$= 180°.$$

The root locus for the saturation case is plotted in Fig. 47. All three graphical methods yield the same results, that is, sustained oscillation occurs at the crossover point A of the two loci, at $\omega = 10$ and $X/S = 6.3$.

Example 4: Closed-Loop System With Simple Backlash. A system block diagram of a simple backlash nonlinear system is given in Fig. 48. The stability of the system for $K = 25$, $T = 0.1$ second is evaluated using the polar plot, gain-phase plot, and root-locus plot. The loci to be plotted follow.

Polar Plot:

$$|G(\omega)| = \frac{25}{\omega[(\omega^2/100) + 1]}$$

$$\angle G(\omega) = -90° - 2 \tan^{-1}(0.1\omega)$$

and

$$1/|N(D/X)| = \pi/\{[\tfrac{1}{2}\pi + \theta + \tfrac{1}{2}(\sin 2\theta)]^2 + \cos^4\theta\}^{1/2}$$

$$\theta = \sin^{-1}[1 - (2D/X)]$$

$$\angle -[1/N(D/X)]$$

$$= -180° - \tan^{-1}\left[\frac{-\cos^2\theta}{\tfrac{1}{2}\pi + \theta + \tfrac{1}{2}(\sin 2\theta)}\right].$$

Gain–Phase Plot:

$$20 \log |G(\omega)| = 20 \log 25 - 20 \log \omega$$

$$- 20 \log[(\omega^2/100) + 1]$$

$$\angle G(\omega) = -90° - 2 \tan^{-1}(0.1\omega)$$

$$-20 \log |N(D/X)|$$

$$= -20 \log \left(\pi^{-1}\{[\tfrac{1}{2}\pi + \theta + \tfrac{1}{2}(\sin 2\theta)]^2 + \cos^4\theta\}^{1/2}\right)$$

$$\angle -[1/N(D/X)]$$

$$= -180° - \tan^{-1}\left(\frac{-\cos^2\theta}{\tfrac{1}{2}\pi + \theta + \tfrac{1}{2}(\sin 2\theta)}\right)$$

$$\theta = \sin^{-1}[1 - (2D/X)].$$

Root-Locus Plot:

$$s(0.1s+1)$$
$$= 25\left(\pi^{-1}\left\{\left[\tfrac{1}{2}\pi+\theta+\tfrac{1}{2}\left(\sin 2\theta\right)\right]^2+\cos^4\theta\right\}^{1/2}\right)$$
$$\angle s(0.1s+1)=180°+\tan^{-1}\left(\frac{-\cos^2\theta}{\tfrac{1}{2}\pi+\theta+\tfrac{1}{2}\left(\sin 2\theta\right)}\right)$$
$$\theta=\sin^{-1}\left[1-(2D/X)\right].$$

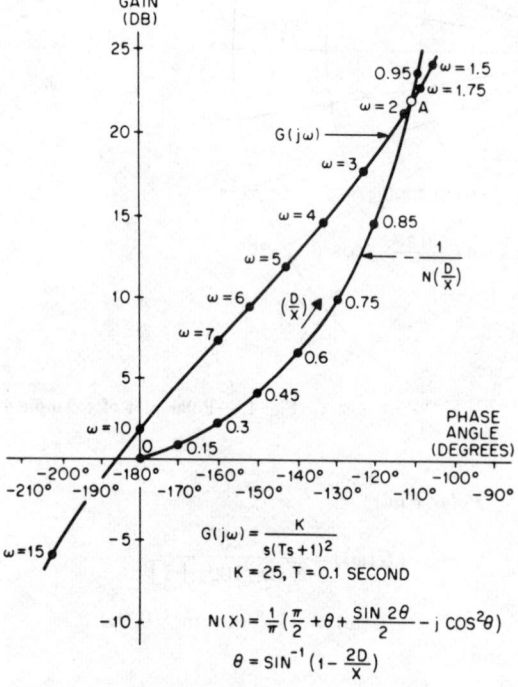

Fig. 50—Gain–phase plot of example 4.

The results are given in Figs. 49, 50, and 51. The system has a sustained oscillation at approximately $\omega = 1.85$, and the corresponding $N(D/X)$ is approximately equal to

$$|N(D/X)|=0.08 \qquad \angle N(D/X)=-69°$$

or

$$D/X = 0.925.$$

Phase-Plane Method

The phase-plane method of analysis is used to study the transient behavior of the nonlinear system. The systems to be considered are assumed to be so constituted that the system performance can be described in terms of an ordinary differential equation.

Restrictions of the Phase-Plane Method: The phase-plane method has the following restrictions.

(A) *The analysis is limited to systems described by the first and second order.* Differential equations to systems of higher order may be solved in the phase space; however, the results are complex and unwieldy.

(B) *The analysis can be used only for study of the transient response.* The forcing function of the differential equation is zero and, consequently, only the response to the initial condition is obtained. Simple forcing functions such as step and ramp functions with which, by appropriate substitutions, the characteristic equation may be made equal to zero can also be solved. It is extremely difficult to extend the forcing function to sinusoidal and complex functions.

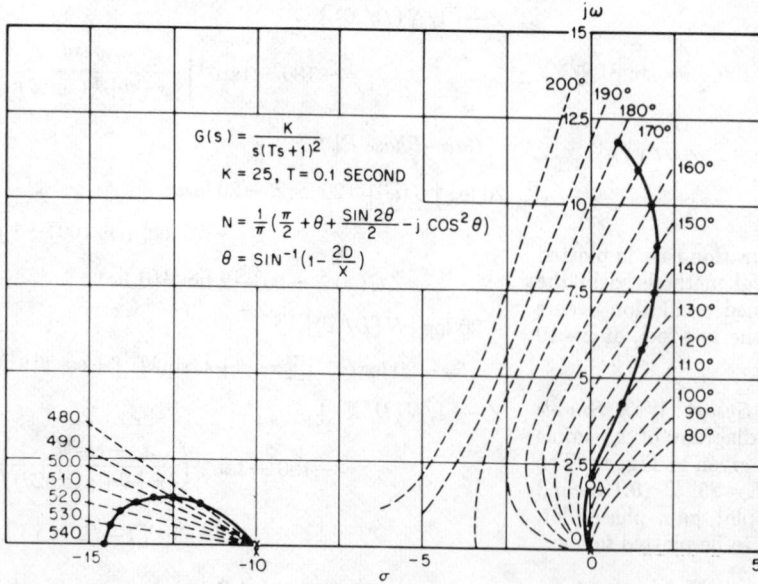

Fig. 51—Root-locus plot of example 4.

(C) *The analysis is limited to autonomous functions.* That is, the coefficients of the derivatives must be functions of x and $\dot{x}$ and not of time explicitly.

Phase Plane: The differential equation describing a second-order system may be expressed by

$$f_1(x, dx/dt, t)\,(d^2x/dt^2) + f_2(x, dx/dt, t)\,(dx/dt)$$
$$+ f_3(x, dx/dt, t)\,x = g(t).$$

The type of equations that can be evaluated in the phase plane is of the form

$$f_1(x, dx/dt)\,(d^2x/dt^2) + f_2(x, dx/dt)\,(dx/dt)$$
$$+ f_3(x, dx/dt)\,x = 0. \quad (7)$$

The equation in which t does not appear explicitly is called "autonomous." By defining

$$\dot{x} = dx/dt$$

the equation may be rewritten as

$$f_1(x, \dot{x})\,(d\dot{x}/dt) + f_2(x, \dot{x})\dot{x} + f_3(x, \dot{x})x = 0$$

or

$$dx/dt = P(x, \dot{x}) = \dot{x} \quad (8)$$

$$d\dot{x}/dt = Q(x, \dot{x})$$
$$= -[f_2(x, \dot{x})\dot{x} + f_3(x, \dot{x})x]/f_1(x, \dot{x}) \quad (9)$$

and further into the form of

$$\frac{(d\dot{x}/dt)}{(dx/dt)} = d\dot{x}/dx$$

$$= Q(x, \dot{x})/P(x, \dot{x})$$

$$= -[f_2(x, \dot{x})\dot{x} + f_3(x, \dot{x})x]/\dot{x}f_1(x, \dot{x}).$$
$$(10)$$

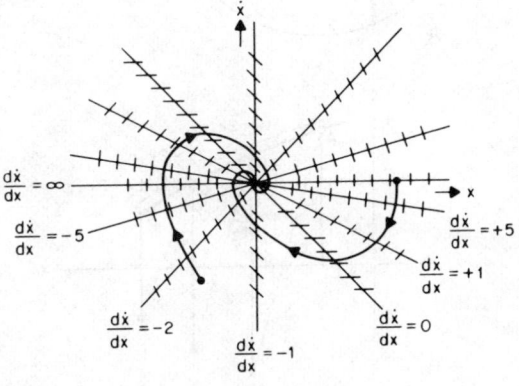

Fig. 52—Isocline method.

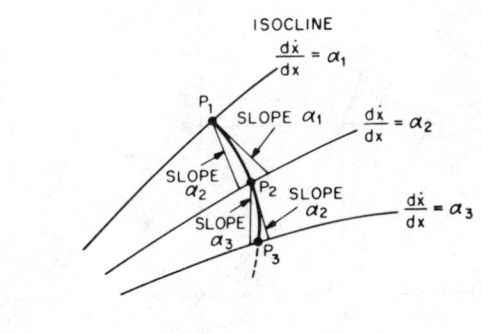

Fig. 53—Step-by-step slope averaging.

The second-order differential equation with respect to time is reduced to a first-order equation of x and $\dot{x}$.

The phase-plane diagram has the $\dot{x}$ as its ordinate and x as its abscissa. The plot of $\dot{x}$ as a function of x on the phase-plane diagram is termed *phase trajectory*. A family of phase trajectories is called the *phase portrait*.

The phase trajectory originates at a point corresponding to the initial condition $(x_0, \dot{x}_0)$ and moves to a new location at each increment of time. Generally, the increments of time are not portrayed on the trajectory and must be obtained by other means described in a later section. If the value of time at each point on the trajectory is obtained, the time response of $\dot{x}(t)$ and $x(t)$ can be plotted. The phase trajectory has a definite direction associated with time. When $\dot{x}$ is positive the trajectory moves from left to right, and for negative values of $\dot{x}$ all paths move from right to left. If the trajectory approaches the origin or some finite point on the phase plane as time goes to ∞, the system is stable. If the trajectory goes to ∞ with time, the system is unstable. If the trajectory approaches an enclosed path in the phase plane, the system has sustained oscillation. The enclosed path is called the limit cycle.

Construction of the Phase Portrait.

Method of Isoclines: The slope $d\dot{x}/dx$ of equation (10) is simply the slope of the trajectory in the phase plane. The locus of constant $d\dot{x}/dx$ is termed an *isocline* corresponding to the slope α, that is

$$\alpha = d\dot{x}/dx$$

$$= -[f_2(x, \dot{x})/f_1(x, \dot{x})]$$

$$\qquad - [f_3(x, \dot{x})/f_1(x, \dot{x})](x/\dot{x})$$

$$= -g(x, \dot{x}) - h(x, \dot{x})\,(x/\dot{x}).$$

The phase portrait is constructed by plotting a

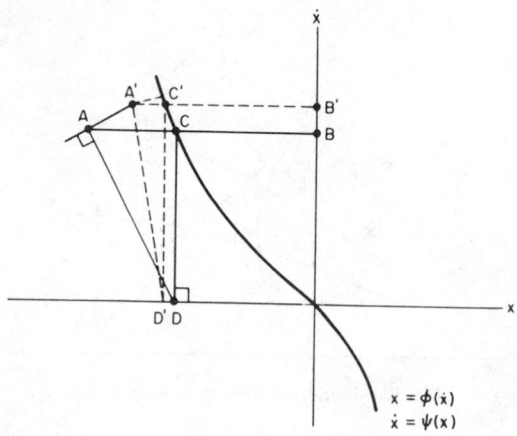

$$x = \phi(\dot{x})$$
$$\dot{x} = \psi(x)$$

Fig. 54—Lienard's method.

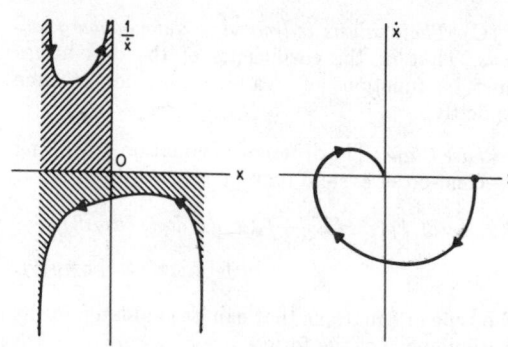

Fig. 56—Reciprocal plot of $\dot{x}$ for determining time.

large number of isoclines corresponding to the various slopes of the trajectory on the phase plane. All points located on the same isocline have the same slope α. Beginning at the location of the initial condition $(x_0, \dot{x}_0)$, the trajectory traverses in the clockwise direction, crossing each isocline at an angle corresponding to that slope α. Figure 52 shows the isocline for a damped, linear, second-order system. Isoclines for first- and second-order linear differential equations are straight lines.

For a more accurate method of sketching the trajectory, the method* of Fig. 53 is helpful. The

method is a step-by-step construction using the average slope between the two adjacent isoclines. At point P_1 the slopes α_1 and α_2 are projected on the isocline of α_2. The halfway point of the two projected points is defined as P_2. The same procedure is repeated, determining points P_3, P_4, $\cdots$. The trajectory is obtained by connecting the various points P.

Lienard's Method: Lienard's graphic construction determines the slope of the trajectory at any point without the use of isoclines. With the slope at one point known, a short line segment with that slope

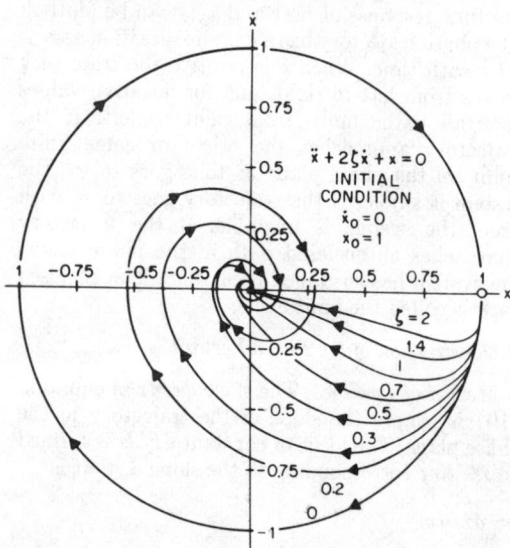

Fig. 55—Phase portrait of second-order system for various damping ratios ζ.

* A. A. Andronow and C. E. Chaikin, "Theory of Oscillations," Princeton University Press, Princeton, New Jersey; 1949: p. 248.

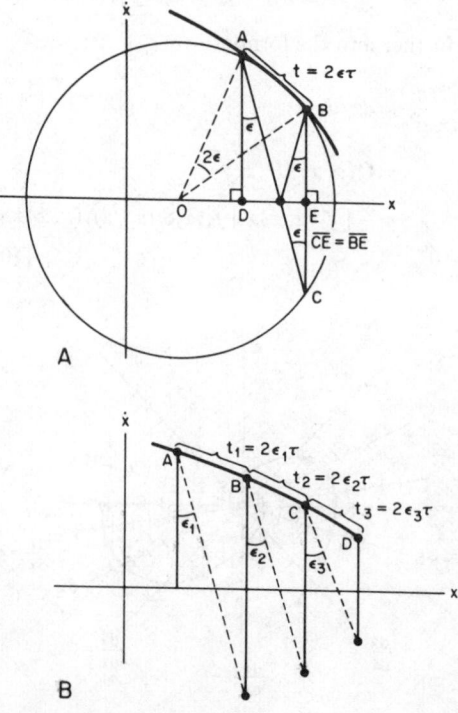

Fig. 57—Graphic construction for determining time.

can be projected to a new point and the procedure repeated. The construction procedure is as follows.

Referring to (10)

$$d\dot{x}/dx = -(1/\dot{x})\{[f_2(x,\dot{x})\dot{x} + f_3(x,\dot{x})x]/f_1(x,\dot{x})\}$$

$$= -(1/\dot{x})[f(x,\dot{x})].$$

The first step is to set

$$f(x,\dot{x}) = 0$$

and solve for either x or $\dot{x}$

$$x = \phi(\dot{x}) \quad \text{or} \quad \dot{x} = \psi(x).$$

The function $x = \phi(\dot{x})$ or $\dot{x} = \psi(x)$ is then plotted on the phase plane (Fig. 54).

The slope of the trajectory at location A is determined by the following procedure. Draw a horizontal line from A to B intersecting the curve $x = \phi(\dot{x})$ at C. Draw a line perpendicular to $\overline{AB}$ from C intersecting the x axis at D. Connect a line from A to D. The slope of the trajectory at point A is normal to the joining line $\overline{AD}$. Repeating the same procedure at a new point A' the trajectory is carried forward. The trajectory may be constructed to any desired accuracy by selecting a small enough segment $\overline{AA'}$. A phase portrait of a second-order system for different values of damping ratio is shown in Fig. 55.

Determination of Time on the Phase Plane.

Time From Reciprocal Plot: This method is based on the relationship of time and the reciprocal plot of $\dot{x}$. Since

$$\dot{x} = dx/dt$$

dt may be expressed as

$$dt = dx/\dot{x}.$$

Integrating both sides yields

$$t = \int_{x_0}^{x_1} (1/\dot{x})\,dx.$$

Since $\dot{x}$ as a function of x is known from the phase plane, the reciprocal $1/\dot{x}$ may be plotted as a function of x, and the integral under the curve between any two points is the time required for the trajectory to change from one point to the other.

A typical reciprocal plot of $\dot{x}$ to determine time is shown in Fig. 56. The integration is in the direction of increasing time from right to left in the lower half-plane and vice versa in the upper half-plane.

In the vicinity of the x axis, the function $1/\dot{x}$ approaches infinity. Even with the integral un-

bounded the integral is finite; however, it is not readily evaluated graphically.

As an alternative method, the value of time may be evaluated from integration involving $\dot{x}$ instead of x. From the original differential equation (9)

$$d\dot{x}/dt = Q(x,\dot{x})$$

dt is written as

$$dt = [1/Q(x,\dot{x})]d\dot{x}$$

and therefore

$$t = \int_{\dot{x}_1}^{\dot{x}_2} [1/Q(x,\dot{x})]d\dot{x}.$$

The expression $1/Q(x,\dot{x})$ may be plotted as a function of $\dot{x}$ and graphic integration may be performed.

The two methods may be used alternately if the integral goes to infinity.

Graphic Construction: This method is based on the approximation of the phase trajectory by a series of circular arcs centered on the x axis (Fig. 57A). Consider the section of the path $\overline{AB}$ to be a segment of an arc centered on the x-axis 0. The angle of the arc $\overline{AB}$ is $\angle A0B = 2\epsilon$. The time from A to B is given by $2\epsilon\tau$, where ϵ is measured in radians and τ is the ratio of the x and $\dot{x}$ scale factor.

$$\tau = \frac{\text{value of } x/\text{unit scale}}{\text{value of } \dot{x}/\text{unit scale}}.$$

On this basis, time from A to B of the path may be evaluated as follows.

Drop a line from A perpendicular to the x axis at point D. Drop a line from B perpendicular to the x axis, and scale an equal distance $\overline{BE}$ opposite the x-axis $\overline{CE}$. Draw a line $\overline{AC}$ connecting points A and C. Measure the angle $\angle CAD$ in radians. The time from point A to B is then

$$t = 2\angle CAD \cdot \tau.$$

Repeat the procedure as the points move along the trajectory. A typical example is shown in Fig. 57B.

Singular Points: In a second-order system, the differential equation of the system may be described by two variables x and $\dot{x}$ in the following form.

$$dx/dt = P(x,\dot{x})$$

$$d\dot{x}/dt = Q(x,\dot{x}).$$

The points where dx/dt and $d\dot{x}/dt$ vanish are called *singular points*. At a singular point the system is in a state of equilibrium.

The importance of a singular point in the phase plane is how the trajectories of the phase portrait behave in the vicinity of the singular point. When the trajectory converges to the singular point the system is stable, whereas if it diverges the system is unstable. Typical types of singular points are described below.

Types of Singular Points: Besides stable and unstable equilibrium, the singular points may be classified into node, focus, center, and saddle points.

Consider, for example, a singular point at $x = a$ and $\dot{x} = b$ of equations (8) and (9). At a singular point the derivatives dx/dt and $d\dot{x}/dt$ are both zero, and the location may be solved in the phase plane by setting (8) and (9) equal to zero. A singular point exists at $x = a$, $\dot{x} = b$ and the functions P and Q can be expressed in terms of the Taylor series about those points; then

$$dx/dt = c_1(x-a) + c_2(\dot{x}-b) + c_3(x-a)^2$$

$$+ c_4(x-a)(\dot{x}-b) + c_5(\dot{x}-b)^2 + \cdots$$

$$d\dot{x}/dt = d_1(x-a) + d_2(\dot{x}-b) + d_3(x-a)^2$$

$$+ d_4(x-a)(\dot{x}-b) + d_5(\dot{x}-b)^2 + \cdots.$$

Taking a sufficiently small region around the singular point, the derivatives are dominated by the linear terms and hence quantities c_1, c_2, d_1, and d_2. By changing the variables, the singular point may be moved to the origin. Then the system equation may be rewritten as

$$dx/dt = p_1 x + p_2 \dot{x} = 0$$

$$d\dot{x}/dt = q_1 x + q_2 \dot{x} = 0$$

and the characteristic equation is

$$\lambda^2 - (p_1 + q_2)\lambda + (p_1 q_2 - p_2 q_1) = 0.$$

The roots of the characteristic equation determine the nature of the critical points. The roots are

$$\lambda = \tfrac{1}{2}\{(p_1 + q_2) \pm [(p_1 + q_2)^2 - 4(p_1 q_2 - p_2 q_1)]^{1/2}\}.$$

There are six possible cases for the six types of singular points.

(**A**) The roots are real and are both negative.

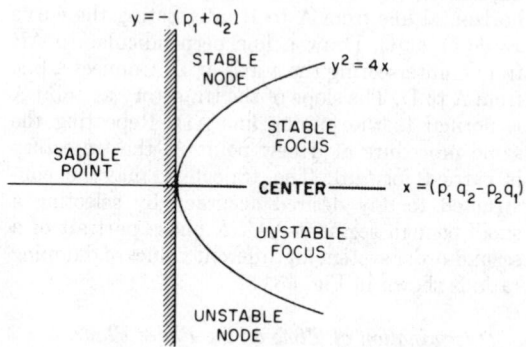

Fig. 59—Regions of various singular points.

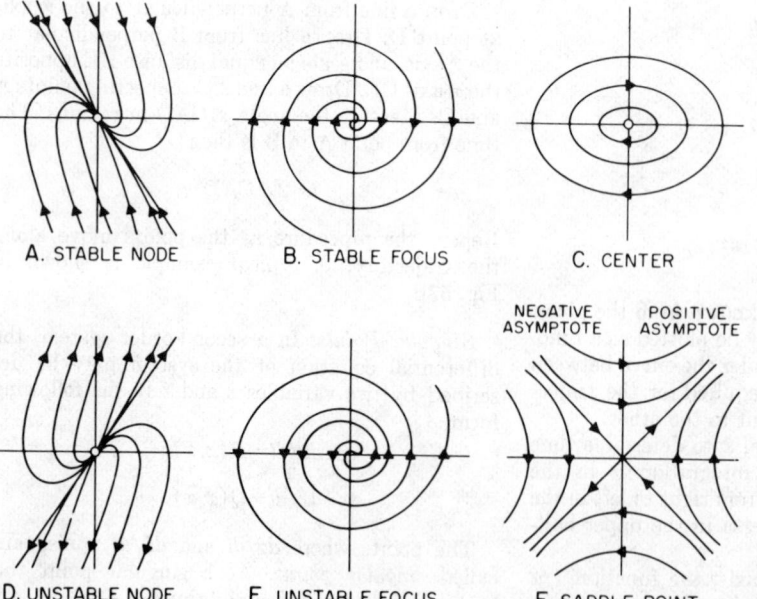

A. STABLE NODE   B. STABLE FOCUS   C. CENTER

D. UNSTABLE NODE   E. UNSTABLE FOCUS   F. SADDLE POINT

Fig. 58—Types of singular points.

If the initial condition is $x_0 \neq 0$, $\dot{x}_0 = 0$, the trajectory approaches the singular point without an overshoot or oscillation and is called a *stable node* (Fig. 58A).

(**B**) The roots are complex conjugate with negative real parts. The trajectory displays a spiraling response as it converges to the singular point as shown in Fig. 58B. This is called the *stable-focus* type singularity.

(**C**) The roots are conjugate and pure imaginary. The response exhibits a sustained oscillatory motion with the amplitude dependent on the initial condition. The trajectory displays a family of ellipses about the singular point, and is termed *center* (Fig. 58C).

(**D**) The roots are both positive real. The response in the time domain increases exponentially and is unstable. The portrait is the same as the stable node except the trajectory diverges from the singular point. This is termed *unstable node* (Fig. 58D).

(**E**) The roots are complex conjugate with positive real parts. The phase portrait is the same as the stable focus except the trajectory diverges from the singular points. This is termed the *unstable focus* (Fig. 58E).

(**F**) The roots are real with one negative and the other positive. The phase portrait consists of a family of curves of the hyperbolic type having $k_1 = (\lambda_1 - p_1)/p_2$ and $k_2 = (\lambda_2 - p_1)/p_2$ for its asymptotes. The direction of the paths is toward the singular point on the negative asymptote and away from the singular point on the positive asymptote. Singular points of this type are called *saddle points* and are unstable (Fig. 58F).

The six types of singularities correspond to the six regions of Fig. 59.

Limit Cycle and Existence Theorem.

Singular points alone do not provide a complete picture of the phase portrait required to determine the stability of the system.

A limit cycle differs from a center (singular point) in that a limit cycle results after a buildup or decay of a signal that eventually falls into an isolated closed path in the phase plane. A limit cycle may be either stable or unstable depending on whether all the neighboring paths spiral toward or spiral away from the limit cycle. The limit cycle may result from either soft or hard self-excitation. Figure 60 is a phase portrait of soft and hard self-excitation.

There is no concise way to determine the location of the limit cycle or whether a limit cycle

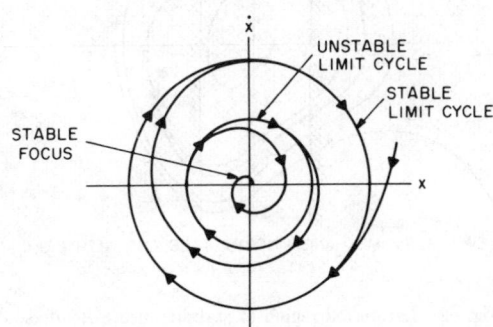

A. PORTRAIT OF SOFT SELF-EXCITATION

B. PORTRAIT OF HARD SELF-EXCITATION

Fig. 60—Phase portrait of a limit cycle.

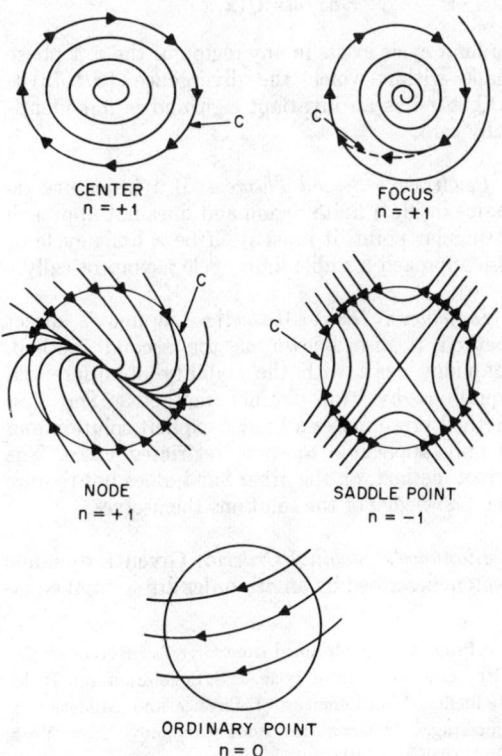

Fig. 61—Poincare's index. *From A. A. Andronow and C. E. Chaiken, "Theory of Oscillations," Princeton University Press, 1949: fig. 197, p. 211. Reprinted by permission.*

exists. Several theorems presented below may be useful.

Poincare's Index: The index of a closed curve C is determined by the number of revolutions of the vector $P(x, \dot{x})$, $Q(x, \dot{x})$ of the point $(x, \dot{x})$ on the curve C as it describes one complete cycle in the clockwise direction (Fig. 61).

(A) The index n of a closed curve C enclosing only ordinary points is zero.

(B) The index n of a closed curve C, enclosing a focus node or center is $+1$.

(C) The index n of a closed curve C enclosing a saddle point is -1.

(D) The index n of any closed curve C is the sum of the indices of all the singular points enclosed by C.

(E) The index n of every closed phase trajectory is $+1$. Therefore, a limit cycle must enclose at least one singular point other than a saddle point.

Bendixson's First Theorem*: For the system described by the equations

$$dx/dt = P(x, \dot{x})$$

$$d\dot{x}/dt = Q(x, \dot{x})$$

no limit cycle exists in any region of the x, $\dot{x}$ phase plane within which the divergence $(\partial P/\partial x) + (\partial Q/\partial \dot{x})$ has an invariant sign and is not identically zero.

Bendixson's Second Theorem: If a trajectory remains inside a finite region and does not approach a singular point, it must itself be a limit cycle or else approach a stable limit cycle asymptotically.

Liapunov's Direct Method†: Liapunov's direct method is also known as the second method. Liapunov dealt with the stability of differential equations by two distinct methods. The first method presupposes a known explicit solution and is only applicable to some restricted cases. The direct method, on the other hand, does not require the knowledge of the solutions themselves.

Liapunov's Stability Criteria: Given a dynamic system described by an nth-order differential equation

$$\dot{x}_1 = f_1(x_1, x_2, \cdots, x_n)$$

$$\dot{x}_2 = f_2(x_1, x_2, \cdots, x_n)$$

$$\dot{x}_n = f_n(x_1, x_2, \cdots, x_n).$$

If a positive (negative) definite Liapunov function V can be chosen, such that the derivative with respect to time $\dot{V} = g(x_1, x_2, \cdots, x_n)$ is negative (positive) definite $(\dot{V} < 0)$, then the system is asymptotically stable.

Asymptotic stability implies that the state of the system converges to a stable equilibrium point as time approaches infinity. The condition $\dot{V} \leq 0$ permits the system to have a limit cycle. This is termed stable in the mathematical sense. If $\dot{V} > 0$ (same sign as V) the state of the system diverges and is unstable. The function $V(x)$ has the following properties.

(A) $V(x)$ is continuous together with its first derivative in a certain finite region Ω about the equilibrium point.

(B) $V(x) = 0$ at the equilibrium point.

(C) $V(x)$ is always positive outside the equilibrium point in the region Ω.

(D) $V(x)$ is called a Liapunov function if in addition $\dot{V} \leq 0$ in Ω.

There are no general rules for finding a Liapunov function. Usually, because of mathematical convenience, the function is chosen to have a positive definite quadratic form. Figure 62 illustrates an estimated region of stability derived from Bendixson's first theorem and Liapunov's direct method. The dynamic system under evaluation is the Lewis

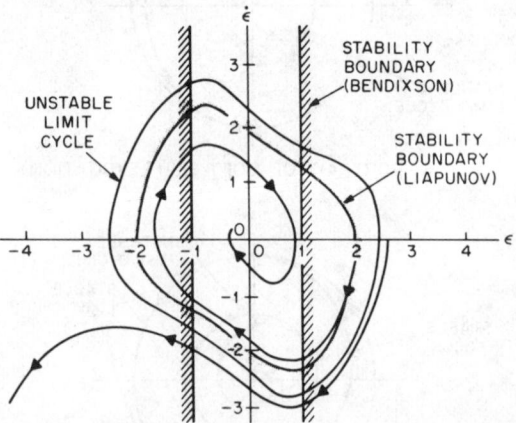

LEWIS SERVOMECHANISM NORMALIZED AND WITH $2\zeta = a = 1.0$
$$\ddot{\epsilon} + 2\zeta(1 - a|\epsilon|)\dot{\epsilon} + \epsilon = 0$$

Fig. 62—Estimated region of stability using Bendixson's and Liapunov's methods. *From D. Graham and D. McRuer, "Analysis of Nonlinear Control Systems," fig. 8-22, ©1961, John Wiley & Sons, New York.*

* Proof may be obtained from Green's theorem setting $Pd\dot{x} - Qdx = 0$ for limit cycle, I. S. Sokolnikof and R. M. Redheffer, "Mathematics of Physics and Modern Engineering," McGraw-Hill Book Company, New York, N. Y.; 1958: p. 391.

† J. LaSalle and S. Lefschetz, "Stability By Liapunov's Direct Method With Applications," Academic Press, New York; 1961.

servomechanism. It is noted that the region of stability obtained by Liapunov's method is much more accurate than Bendixson's results.

Application of the Phase-Plane Method to Non-linear Systems: An illustrated example is presented using the phase-plane method to evaluate a hardening-type nonlinear feedback system.

Example 5: Piecewise Linear Analysis of a Non-linear System. The block diagram of the nonlinear system to be evaluated is illustrated in Fig. 63. The system equation is given by

$$r - c = \epsilon$$

$$c = [K(\epsilon)/(Ts+1)s] \cdot \epsilon$$

or

$$T\ddot{\epsilon} + \dot{\epsilon} + K(\epsilon) = T\ddot{r} + \dot{r} = 0, \qquad r = \text{constant}.$$

Since

$$\ddot{\epsilon} = \dot{\epsilon}(d\dot{\epsilon}/d\epsilon)$$

the system equation may be rewritten as

$$d\dot{\epsilon}/d\epsilon = -T^{-1}\{[\dot{\epsilon}+K(\epsilon)]/\dot{\epsilon}\}$$

$$= Q(\epsilon, \dot{\epsilon})/P(\epsilon, \dot{\epsilon})$$

where

$$Q = -[\dot{\epsilon}+K(\epsilon)]/T$$

$$P = \dot{\epsilon}.$$

The singular point is determined by setting P and Q equal to zero.

For the slope K_1 the singular point is located at $\epsilon = 0$ and $\dot{\epsilon} = 0 [\therefore K(0) = 0]$ and for the slope K_2 the singular points are $\dot{\epsilon} = 0$, $K(\epsilon_1) = 0$; $\dot{\epsilon} = 0$,

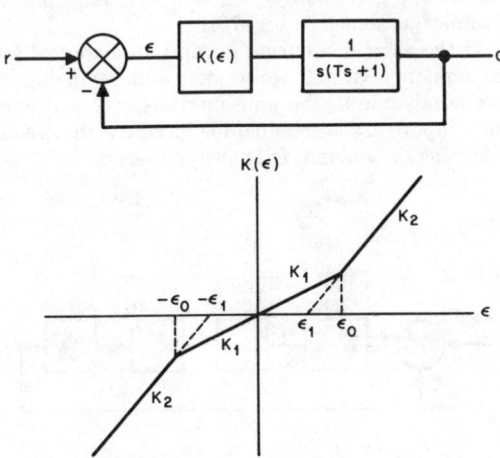

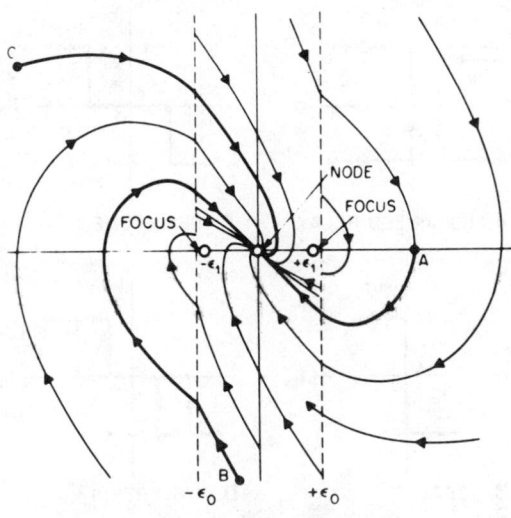

Fig. 64—Overall phase portrait.

$K(-\epsilon_1) = 0$. The singular point on the phase plane is illustrated at the origin, at $\epsilon = +\epsilon_1$, and at $\epsilon = -\epsilon_1$. The type of singularities at the singular point is determined from the roots of the characteristic equation (assuming here that $T = 0.5$, $K_1 = 0.5$, and $K_2 = 4.5$).

The characteristic equation of the feedback system is given by

$$\lambda^2 + (\lambda/T) + [K(\epsilon)/T] = 0.$$

For $K_1 = 0.5$:

$$\lambda^2 + 2\lambda + 1 = 0$$

stable node—roots are both negative real.

For $K_2 = 4.5$:

$$\lambda^2 + 2\lambda + 9 = 0$$

stable focus—roots are conjugate complex with negative real parts.

Since the $K(\epsilon)$ function goes through a transition in slope at $\epsilon = \pm\epsilon_0$, the transition line is drawn vertical to the ϵ axis at $+\epsilon_0$ and $-\epsilon_0$.

In the first region, $+\epsilon_0 < \epsilon < \infty$, the phase portrait is that of a stable focus at $\epsilon = +\epsilon_1$ (damping ratio $\zeta = 1/3$).

In the second region, $-\epsilon_0 < \epsilon < +\epsilon_0$, the phase portrait is that of a stable mode at the origin (damping ratio $\zeta = 1$).

In the third region, $-\infty < \epsilon < -\epsilon_0$, the phase portrait is that of a stable focus at $\epsilon = -\epsilon_1$ (damping ratio $\zeta = 1/3$).

Figure 64 is the overall phase portrait of the feedback system.

Fig. 63—Nonlinear system with spring "hardening".

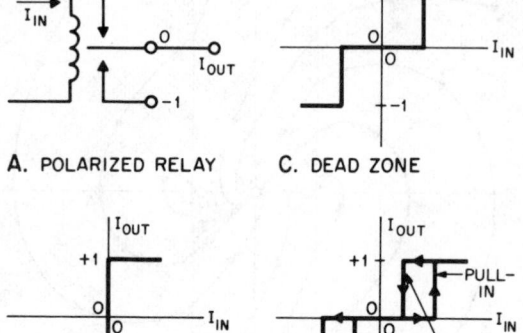

A. POLARIZED RELAY   C. DEAD ZONE

B. IDEAL   D. HYSTERESIS

Fig. 65—Static characteristic of a relay.

RELAY CONTROL SYSTEM AND OPTIMUM SWITCHING

The system performance is termed "optimum" if the system is controlled to zero error in the minimum period of time. Relays are often used to achieve optimum performance in regulators and servo systems. The advantage of a relay is that it is economical and provides full power or torque capabilities of the control element (motor) to the load.

Characteristics of a Relay

The static characteristics of a relay may be idealized as a three-valued nonlinear device. The relay characteristics are shown in Fig. 65. In general, the relay has a threshold value (dead zone), which must be exceeded to close the contacts (Fig. 65C). A hysteresis loop is also present, that is, the pull-in and drop-out currents of a relay are of different values (Fig. 65D). Besides the static characteristics, a time delay that corresponds to the pull-in time after excitation must be considered for a dynamic representation. A simple block diagram of a relay servo system is shown in Fig. 66. A phase portrait of various combinations of relay characteristics are also illustrated in Fig. 67. It is noted that in the case where the time constant T of the lead equalizer is very large (Fig. 67F), relay chattering occurs as shown in Fig. 68. The chattering mode slows down the response, wears out the relay contacts, and must be avoided.

It is evident from the phase portraits that switching in the second and fourth quadrants of the phase plane is preferable to the first and third quadrants. Also, systems with lead equalizers show the best response, while the worst response occurs with pure time delay.

Optimum Switching

To achieve optimum performance of a relay servo system, the relay must be controlled and switched to the proper forcing function at the proper time. As an example, consider the ideal relay servo system of Fig. 69. The phase trajectory approaches the origin only if the initial point is located on the zero trajectories. The initial point at any other location will result in a continuous oscillation. The zero trajectories in this case are the optimum switching line. The initial point P_1 will be accelerated at full torque to a new location on the zero trajectory P_2, at which switching takes place and the deceleration torque moves the system to the origin. Finally, the deceleration torque is removed and the system is in an equilibrium state. The initial point P_1 was moved to the origin in a minimum period of time.

The optimum response is achieved by a controller detecting the difference between the state of the system and the optimum switching lines, and commanding switching when the two states coincide.

In a second-order system expressed by

$$\ddot{\epsilon} + 2\zeta\dot{\epsilon} + \epsilon = \pm 1$$

for damping ratios of $\zeta \geq 1$, the switching lines are simple zero trajectories as shown in Fig. 70A, and for damping ratios $0 \leq \zeta < 1$ the switching line is complex as shown in Fig. 70B.

If the switching curves for optimum control are to remain invariant, only step and ramp inputs are admissible to the pure inertia servo, and only step inputs are admissible for inertia with viscous damping or coulomb friction-type servo.

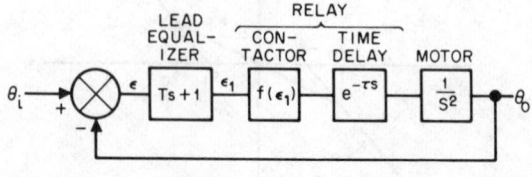

Fig. 66—Simple relay servo system. *From D. Graham and D. McRuer, "Analysis of Nonlinear Control Systems," fig. 9-4b, ©1961, John Wiley & Sons, New York.*

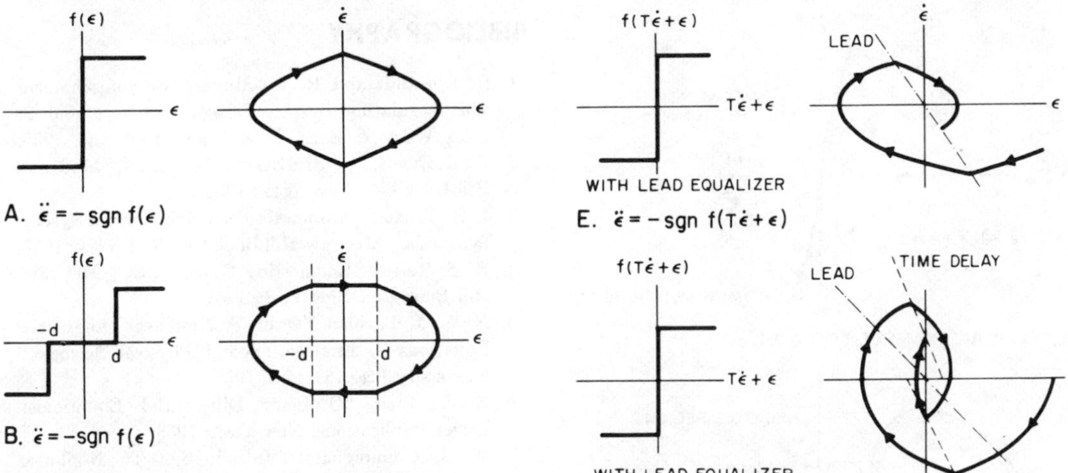

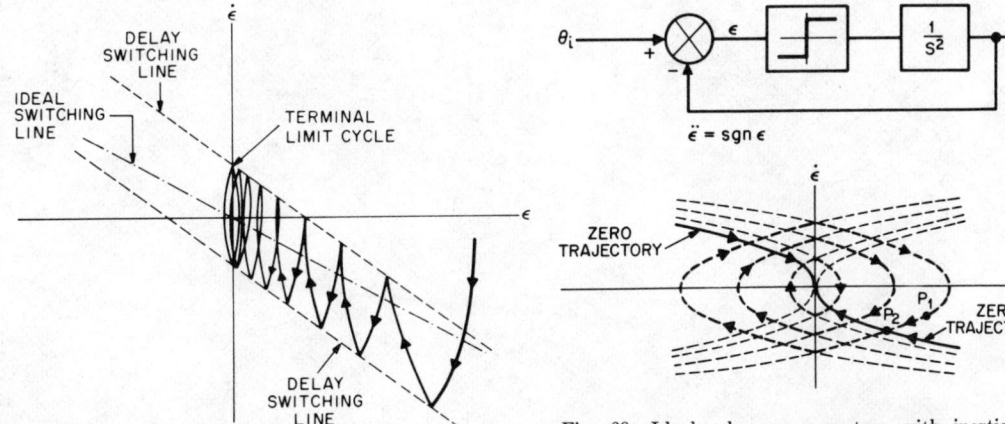

A. $\ddot{\epsilon} = -\operatorname{sgn} f(\epsilon)$

B. $\ddot{\epsilon} = -\operatorname{sgn} f(\epsilon)$

C. $\ddot{\epsilon} = -\operatorname{sgn} f(\epsilon)$

WITH TIME DELAY

D. $\ddot{\epsilon} = -\left[\operatorname{sgn} f(\epsilon)\right]e^{-\tau s}$

WITH LEAD EQUALIZER

E. $\ddot{\epsilon} = -\operatorname{sgn} f(T\dot{\epsilon}+\epsilon)$

WITH LEAD EQUALIZER, TIME DELAY

F. $\ddot{\epsilon} = -\left[\operatorname{sgn} f(T\dot{\epsilon}+\epsilon)\right]e^{-\tau s}$

WITH LEAD EQUALIZER, TIME DELAY

G. $\ddot{\epsilon} = -\left[\operatorname{sgn} f(T\dot{\epsilon}+\epsilon)\right]e^{-\tau s}$

Fig. 67—Various phase portraits of the relay-controlled inertial servo system of Fig. 66. *From D. Graham and D. McRuer, "Analysis of Nonlinear Control Systems," parts of Table 9–1,* © *1961, John Wiley & Sons, New York.*

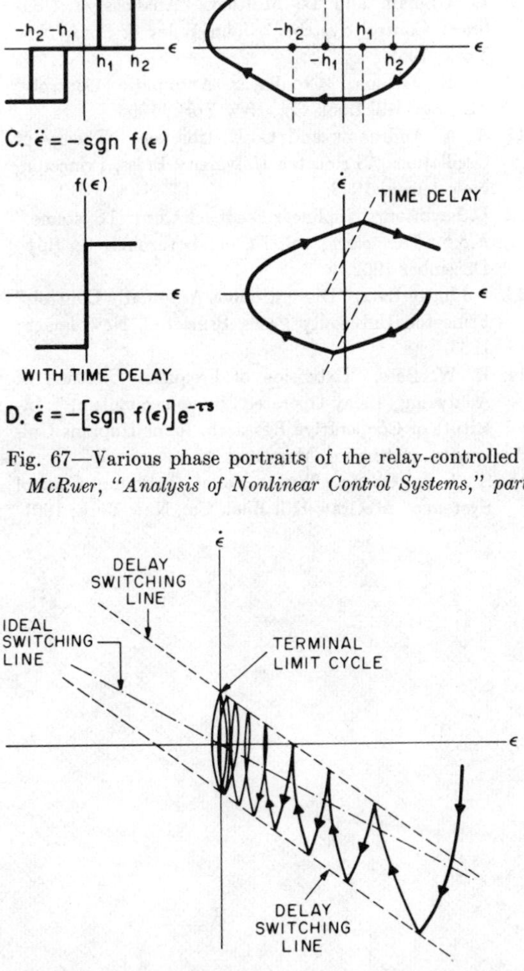

Fig. 68—Relay chattering. *From D. Graham and D. McRuer. "Analysis of Nonlinear Control Systems," fig. 9–8b,* ©*1961, John Wiley & Sons, New York.*

$\ddot{\epsilon} = \operatorname{sgn} \epsilon$

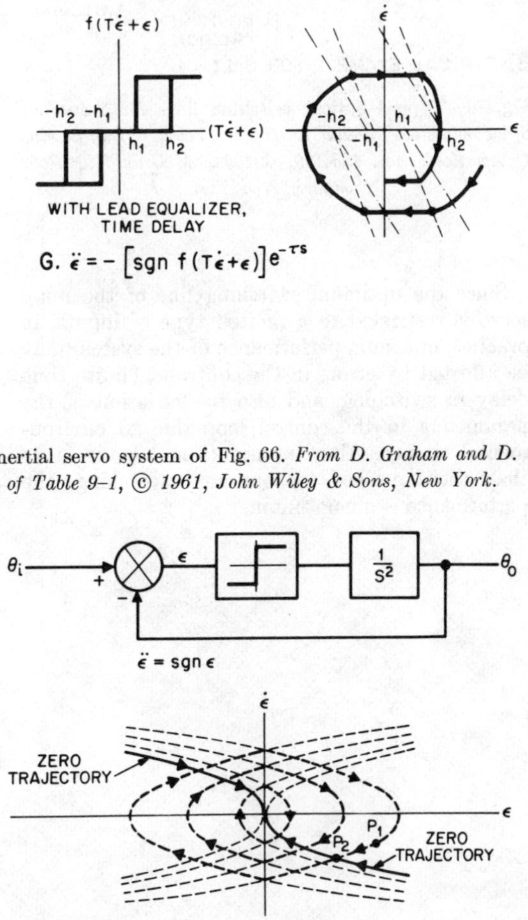

Fig. 69—Ideal relay servo system with inertia load. *From D. Graham and D. McRuer, "Analysis of Nonlinear Control Systems," fig. 9–9 (modified),* ©*1961, John Wiley & Sons, New York.*

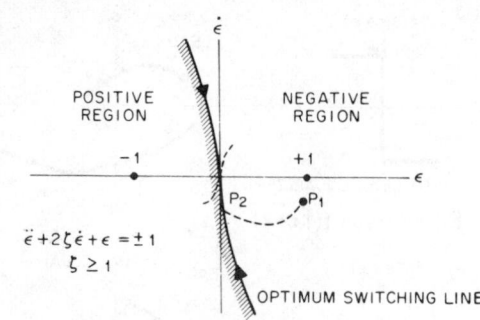

A. TYPICAL EXAMPLE FOR $\zeta \geq 1$

$$\ddot{\epsilon} + 2\zeta\dot{\epsilon} + \epsilon = \pm 1$$
$$\zeta \geq 1$$

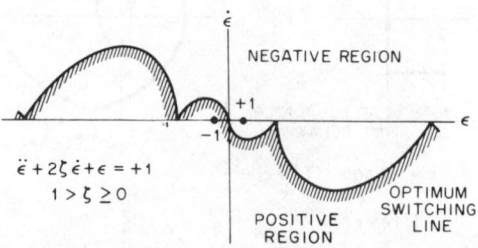

B. TYPICAL EXAMPLE FOR $0 \leq \zeta < 1$

$$\ddot{\epsilon} + 2\zeta\dot{\epsilon} + \epsilon = +1$$
$$1 > \zeta \geq 0$$

Fig. 70—Typical optical switching lines for a second-order system. *Similar to H. S. Tsien, "Engineering Cybernetics," pp. 136-139, ©1954, McGraw-Hill Book Company, New York.*

Since the optimum switching line of the relay servo is restricted to a limited type of inputs, in practice, optimum performance of the system may be affected by errors in the controller, finite time delay in switching, and also by variation of the parameters in the control loop due to environmental change. Also, for inputs and loads other than those derived for optimum switching, the performance is nonoptimum.

BIBLIOGRAPHY

1. H. Chestnut and R. W. Mayer, "Servomechanism and Regulating System Design," vols. 1 and 2, John Wiley & Sons, New York; 1951 and 1955.
2. W. R. Evans, "Control System Dynamics," McGraw-Hill Book Co., New York; 1954.
3. J. G. Truxal, "Automatic Feedback Control System Synthesis," McGraw-Hill Book Co., New York; 1955.
4. H. S. Tsien, "Engineering Cybernetics," McGraw-Hill Book Co., New York; 1954.
5. N. W. McLachlan, "Ordinary Nonlinear Differential Equations in Engineering and Physical Sciences," University Press, Oxford; 1955.
6. E. L. Ince, "Ordinary Differential Equations," Dover Publications, New York; 1953.
7. W. J. Cunningham, "Introduction To Nonlinear Analysis," McGraw-Hill Book Co., New York; 1958.
8. F. B. Hildebrand, "Introduction to Numerical Analysis," McGraw-Hill Book Co., New York; 1956.
9. D. Graham and D. McRuer, "Analysis of Nonlinear Control Systems," John Wiley & Sons, New York; 1961.
10. J. E. Gibson, "Nonlinear Automatic Control," McGraw-Hill Book Co., New York; 1963.
11. A. A. Andronow and C. E. Chaiken, "Theory of Oscillations," Princeton University Press, Princeton, New Jersey; 1949.
12. E. Levinson, "Nonlinear Feedback Control Systems," *Electro-Technology*, vol. 70, no. 1 through 6; July-December 1962.
13. I. Flugge-Lotz, "Discontinuous Automatic Control," Princeton University Press, Princeton, New Jersey; 1953.
14. R. W. Bass, "Extension of Frequency Method of Analyzing Relay-Operated Servomechanism," Institute of Cooperative Research, Johns Hopkins University, Baltimore, Maryland.
15. S. S. L. Chang, "Synthesis of Optimum Control Systems," McGraw-Hill Book Co., New York; 1961.

CHAPTER 17
ELECTRON TUBES

ELECTRON EMISSION

All electron tubes* depend for their operation on the flow of electrons within the tube, either through high vacuum or an ionized gas. The electrons are emitted from a cathode surface as a result of one of four processes that are distinguished on the basis of the mechanism by which the electrons are enabled to leave the surface. These processes are elevated temperature (thermionic or primary emission); bombardment by other particles, generally electrons (secondary emission); the action of a high electric field (field emission); or the incidence of photons (photoemission).

Thermionic Emission

Thermionic emission occurs when the electrons in the cathode material have enough thermal energy to overcome the forces at the surface and escape.

The thermal emission of electrons from metals obeys the Richardson–Dushman equation

$$J_0 = AT^2 \exp(-11600\phi_0/T)$$

where J_0 is emission density in amperes/cm², A is a constant [amperes/cm²(K)²], ϕ_0 is the work function (electronvolts), and T is temperature in kelvins. A and ϕ_0 are characteristic of the specific material.

The current density given by this equation is usually referred to as the saturation emission current density. Typical constants are given in Table 1 for several commonly used cathode materials.

The maximum current of which a cathode is capable at the operating temperature is known as the saturation current and is normally taken as the value at which the current first fails to increase as the three-halves power of the voltage causing the current to flow. Thoriated-tungsten filaments for continuous-wave operation are usually assigned an available emission of approximately half the

* J. W. Gewartowski and H. A. Watson, "Principles of Electron Tubes," Van Nostrand, Princeton, New Jersey; 1965. J. Millman and S. Seely, "Electronics," 1st ed., McGraw-Hill Book Company, New York; 1941. K. R. Spangenberg, "Vacuum Tubes," 1st ed., McGraw-Hill Book Company, New York; 1948. A. H. W. Beck, "Thermionic Valves, Their Theory and Design," Cambridge University Press, London, England; 1953. "Standards on Electron Tubes: Definitions of Terms, 1950," Institute of Radio Engineers, New York.

saturation value. Oxide-coated emitters do not have a well-defined saturation point and are designed empirically. The available emission from the cathode must be at least equal to the sum of the peak currents drawn by all the electrodes.

Figure 1 gives a plot of saturation current as a

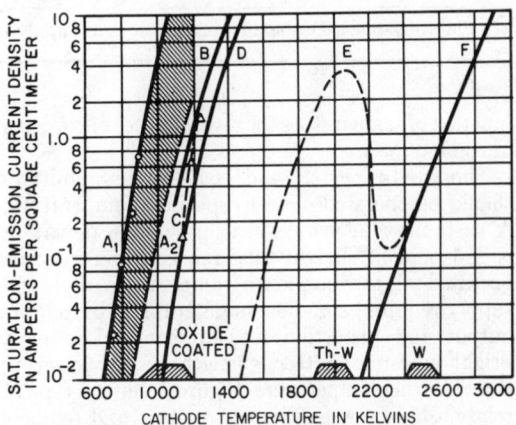

Fig. 1—Emission current density vs. cathode temperature for several types of thermionic emitters. The shaded blocks at the bottom of the figure show the normal operating range for three of the cathodes. Curves are given for (A) *The oxide-coated cathode.* Curve A₁ gives the saturation emission current density under pulsed conditions. Curve A₂ gives the direct-current saturation emission density. The position of this curve may vary substantially with environmental conditions. Direct-current densities much in excess of 0.5 amp/cm² lead to relatively short cathode life. (B) *The pressed nickel cathode.* Curve B shows the direct-current saturation emission current density obtained from a pressed nickel cathode. (C) *The impregnated nickel cathode.* Curve C shows the saturation emission current obtained from the impregnated nickel cathode. The measurements were taken with 40-microsecond pulses and a repetition rate of 60 pulses per second. (D) *Pressed and impregnated tungsten cathodes.* Curve D shows the saturation emission density obtained from pressed and impregnated tungsten cathodes based on $A = 2.5$ amps/cm² (K)² and $\phi_0 = 1.67$ electronvolts. (E) *The thoriated-tungsten cathode.* Curve E shows the measured saturation emission current density of an uncarburized thoriated-tungsten filament. (F) *Tungsten filaments.* Curve F shows the saturation emission current density of a tungsten filament based on $A = 70$ amps/cm² (K)² and $\phi_0 = 4.5$ electronvolts. *J. W. Gewartowski and H. A. Watson, "Principles of Electron Tubes," 1965: p. 42. Courtesy of D. Van Nostrand Company, Inc.*

TABLE 1—COMMONLY USED CATHODE MATERIALS.

Type	A	ϕ_0	Efficiency (milliamperes/ watt)	Specific Emission (amp/cm²)	Emissivity (watts/cm²)	Operating Temperature (K)	Resistance Ratio (hot/cold)
Bright tungsten (W)	70	4.50	5–10	0.25–0.7	70–84	2500–2600	14/1
Thoriated tungsten (Th–W)	4	2.65	40–100	0.5–3.0	26–28	1950–2000	10/1
Tantalum (Ta)	37	4.12	10–20	0.5–1.2	48–60	2380–2480	6/1
Oxide coated (Ba–Ca–Sr)	*	1.0–1.3	50–150	0.5–2.5	3–5	1000–1150	2.5 to 5.5/1
Impregnated	2.4	1.65		1.8–5.4	2.6–3.8	1300–1400	

* The Richardson-Dushman equation does not apply to a composite surface of this type.

function of temperature for several types of emitters in common use.

Thoriated-tungsten and oxide-coated emitters should be operated close to specified temperature. A customary allowable heating-voltage deviation is ±5 percent. Bright-tungsten emitters may be operated at the minimum temperature that will supply required emission as determined by power-output and distortion measurements. Life of a bright-tungsten emitter is lengthened by lowering the operating temperature. Figure 2 shows a typical relationship between filament voltage and temperature, life, and emission.

Mechanical stresses in filaments due to the magnetic field of the heating current are proportional to I_f^2. Current flow through a cold filament should

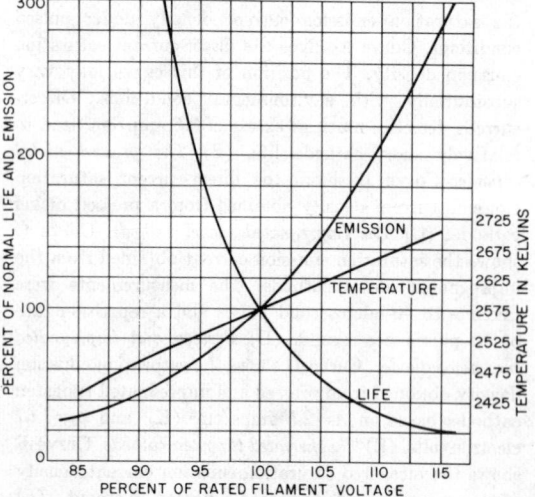

Fig. 2—Effect of change in filament voltage on the temperature, life, and emission of a bright-tungsten filament (based on 2575-kelvin normal temperature of filament).

be limited to 150 percent of the normal operating value for large tubes, and 250 percent for medium types. Excessive starting current may easily warp or break a filament.

Secondary Emission

When the surface of a solid is bombarded by charged particles having appreciable velocity, electrons are emitted from the solid. This is the process of secondary emission,* the most important case being when the bombarding particles are also electrons. One then differentiates between incident and emitted electrons by the terms primary and secondary, respectively. The latter term commonly describes all electrons collected from a secondary emitter; these electrons comprise three groups: (A) true secondaries, (B) inelastically reflected primaries, and (C) elastically reflected primaries. True secondaries are considered to be those of the solid which have been excited above the energy level required for escape across the surface barrier. The three groups are separable to a degree on the basis of energy as indicated in the energy distribution curve of Fig. 3. True secondaries constitute the bulk of emitted electrons at moderate primary energies and have a mode energy of at most a few electronvolts. Their distribution is almost independent of primary energy. Electrons in the rela-

* H. Bruining, "Physics and Applications of Secondary Electron Emission," McGraw-Hill Book Company, New York; 1954. O. Hackenberg and W. Brauer, "Secondary Electron Emission from Solids," Advances in Electronics and Electron Physics, Vol. XI, Academic Press; 1959. A. J. Dekker, "Secondary Electron Emission," Solid State Physics, Vol. 6, Academic Press, New York; 1958. R. Kollath, "Sekundärelektronen-Emission fester Körper bei Bestrahlung mit Elektronen," Handbuch der Physik, Band XXI, Springer-Verlag, Berlin; 1956.

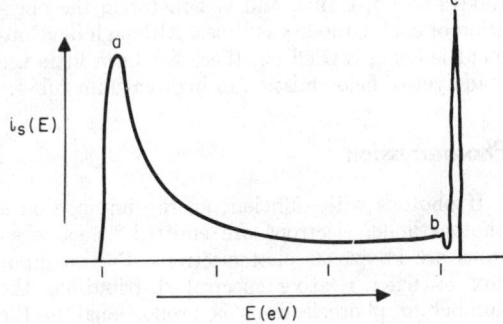

Fig. 3—Total energy distribution of secondary electrons.

tively flat interval within a, b constitute a mixture of true secondaries and inelastically reflected primaries. It has become customary to arbitrarily designate those emitted electrons having energies less than 50 electronvolts as true secondaries.

Total secondary yield δ, defined as the ratio of secondary to primary electron current, is independent of primary current but strongly dependent on primary energy as indicated in Fig. 4. The shape of the yield curve follows from generating and escape mechanisms; the former leading to an initial rise in yield with primary energy, and the latter causing an eventual reduction owing to increased penetration of primaries and a greater mean depth of escape of secondaries. Significant points of the yield curve are first and second crossover at which yield becomes unity, the maximum yield δ_m, and the primary energy eV_m at which the maximum occurs. For most insulators, first crossover occurs between 15 and 25 eV primary energy. Insulators generally exhibit higher yields than conductors, a property attributed to the absence of conduction electrons which tend to reduce the mean energy and the escape probability

of secondaries through collision losses within the solid. The yield of insulators decreases noticeably as temperature is increased, owing to increasing electron–phonon interaction.

Secondary yield increases with angle of primary incidence, the effect being most pronounced at high primary energies. Yield is also a function of surface structure and may be minimized by employing physical trapping such as provided by a porous surface. Lowest yields are obtained for porous carbon deposits and highest yields for single crystal insulators having low electron affinity. Secondary yield may also be influenced by internal electric fields which tend to assist or retard the escape of secondaries. If such fields are strongly dependent on charge transport within the bombarded material, yield may become dependent on primary current which can in turn give rise to anomalous time-dependent effects. Barring such effects, it appears that the interaction time for the secondary-emission process is of the order of 10^{-12} second.

When the rate of bombardment by primary electrons becomes very low, as in single-electron counting, the statistical nature of the secondary-emission process becomes evident. The probability of obtaining $0, 1, 2, \cdots, n$ secondaries per incident primary is given by the Poisson distribution.

Commonly used secondary-emission materials are silver-magnesium or copper-beryllium alloy processed to provide a high-yield partly conductive surface film. Typically such surfaces exhibit yields of 2.5 to 4 at 100 eV primary energy.

Secondary emission is employed advantageously in the operation of many electron devices, such as camera tubes, storage tubes, and image intensifiers. A most important application lies in secondary-electron multiplication, which provides a means for amplifying very weak electron currents as in photomultiplier tubes. A conventional electron multiplier consists of a number of secondary-emitting dynodes operated at progressively higher potentials and terminated by an electron-collecting electrode. Electrons incident on the first dynode are multiplied, the resultant secondaries are accelerated to the second dynode where the process is repeated, and so on throughout the multiplier structure.

If the dynodes exhibit uniform yield characteristics, the overall amplification G obtained from a multiplier having n dynodes is

$$G = g^n$$

where g is the gain per dynode. The actual gain g may be slightly less than the secondary yield δ because of multiplier geometry. In the absence of appreciable space-charge effects, maximum gain is realized when the available potential is uniformly distributed across the dynode chain. An empirical relation for g may be obtained by approximating

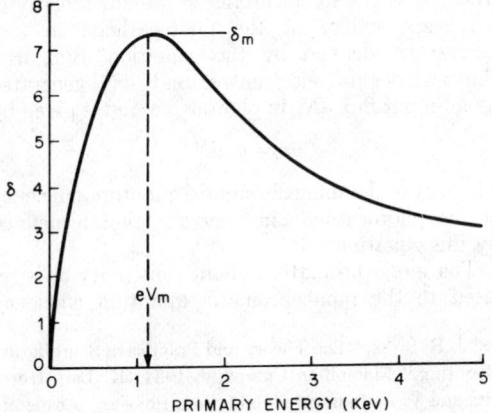

Fig. 4—Secondary-emission yield curve.

the initial portion of the yield curve, that is

$$g \cong \delta \cong A \Delta V^m$$

$\Delta V =$ interdynode potential (primary energy), and A and m are empirical constants. Using this approximation, one obtains

$$g_{opt} = \epsilon^m$$

$$n_{opt} = A^{1/m}(V/\epsilon)$$

($\epsilon =$ base of natural logarithms), where g_{opt} and n_{opt} are the optimum values of gain/dynode and number of dynodes, respectively, for maximizing overall amplification, given a total potential V available for distribution across the dynodes. The number of dynodes n is taken as the closest integer to the calculated value. In some cases, deviation from a uniform potential distribution and optimum gain conditions is advantageous, for example to reduce space-charge effects and improve time-delay and time-dispersion properties. Increasing the energy of electrons incident on the first dynode improves signal-to-noise ratio and single-electron-counting capability.

In addition to the conventional discrete dynode multiplier, a number of novel arrangements have been devised including crossed-field strip multipliers and tubular multipliers which commonly employ a continuous semiconductive dynode surface for potential distribution and multiplication. Tubular multipliers, when formed into a parallel array of small-diameter elements, may be employed for electron image intensification. Another form of multiplier commonly used for this purpose is the transmission secondary-emission multiplier, wherein secondaries exit from the side opposite primary incidence. The structure normally takes the form of a thin-film or porous supported layer, having the side of primary incidence made electrically conductive.

Field Emission

If an electric field of sufficient magnitude is offered to the surface of a metal, the potential barrier at the surface will be lowered, allowing the escape of electrons, and field emission* will result. The current has been found to vary with the applied field in accordance with

$$J = CE^2 \exp(-D/E) \text{ amperes/centimeter}^2$$

where J is the current density, E the electric field at the surface, and C and D are approximately constant coefficients with D determined mainly by the work function. Field emission must be taken into account in the design of very-high-voltage

---

* R. H. Fowler and L. Nordheim, *Royal Society Proceedings*, Vol. *119*, p. 173; 1928.

tubes and apparatus, and is a factor in the operation of cold-cathode gas tubes. Although development is being carried on, there has been little use made yet of field emission in high-vacuum tubes.

Photoemission

If photons with sufficient energy impinge on a photocathode, electrons are emitted.* Such electrons are known as photoelectrons. For an input flux of fixed relative spectral distribution, the number of photoelectrons is proportional to the intensity of the input flux while the energy of the photoelectrons is independent of this intensity. The maximum energy of emitted electrons expressed in volts V depends on the wavelength λ and temperature. At absolute zero, according to Einstein's law

$$e(V+\phi) = hc/\lambda$$

where $e =$ electron charge $= 1.6 \times 10^{-19}$ coulomb, $\phi =$ work function in volts, $h =$ Planck's constant $= 6.6 \times 10^{-34}$ joule-second, $c =$ velocity of light $= 3 \times 10^8$ meters/second, and $\lambda =$ wavelength in meters.

If a threshold wavelength λ_0 is defined by

$$e\phi = hc/\lambda_0$$

V is seen to be zero (except for thermal velocities) at the wavelength λ_0; for $\lambda > \lambda_0$, there is no photoelectric emission at absolute zero. At temperatures above absolute zero there is always a finite probability of some photoemission at all wavelengths due to the thermalization of the electron distribution.

Photocathode Response to Monochromatic Radiation: The output current dI_λ in amperes, generated by a photocathode subjected to a monochromatic input flux dW_λ in watts, is given by

$$dI_\lambda = s_\lambda \, dW_\lambda$$

where s_λ is the monochromatic radiant sensitivity (or responsivity) of the photocathode in amperes/watt defined by this equation. Similarly, the number of electrons/second dn_λ generated by an input flux dN_λ in photons/second is given by

$$dn_\lambda = q_\lambda \, dN_\lambda$$

where q_λ is the monochromatic quantum efficiency of the photocathode in electrons/photon defined by this equation.

The monochromatic radiant sensitivity s_λ is related to the monochromatic quantum efficiency

---

* J. B. Birks, "The Theory and Practice of Scintillation Counting," Macmillan Company; 1964. K. Lark-Horovitz and V. A. Johnson, "Solid State Physics," Academic Press, New York; 1959.

q_λ by

$$s_\lambda = e\lambda q_\lambda/hc = 8.08\times10^5\lambda q_\lambda.$$

Typical values of the monochromatic radiant sensitivity s_λ and corresponding monochromatic quantum efficiency q_λ as a function of wavelength λ are shown in Fig. 5 for some commonly used photocathodes, designated by their JEDEC registered "S numbers." Table 2 gives typical peak sensitivities for the various surfaces, while Table 3 indicates the general composition and other properties of the common surfaces.

Photocathode Response to Spectrally Distributed Sources: The total photocurrent I in amperes emitted by a photocathode subjected to a spectrally distributed input flux is given by

$$I = w_{\lambda\,max}s_{\lambda\,max}\int_0^\infty w_\lambda\sigma_\lambda\,d\lambda$$

$$= W_{\lambda_1\lambda_2}s_{\lambda\,max}\left(\int_0^\infty w_\lambda\sigma_\lambda\,d\lambda\bigg/\int_{\lambda_1}^{\lambda_2} w_\lambda\,d\lambda\right)$$

TABLE 2—TYPICAL PEAK PHOTOCATHODE SENSITIVITIES.

	Radiant Sensitivity		Quantum Efficiency	
S Number	$s_{\lambda\,max}$ (amp watt^{-1})	λ_{max} (nano-meters)	$q_{\lambda\,max}$ (electron photon^{-1})	λ_{max} (nano-meters)
S1	0.0025	800*	0.004	770*
S3	0.0019	420	0.0058	400
S4	0.042	420	0.13	380
S5	0.052	330	0.21	320
S8	0.0024	360	0.0082	350
S9	0.023	490	0.056	480
S10	0.021	440	0.061	420
S11	0.048	450	0.14	400
S13	0.048	450	0.20	<200
S17	0.083	500	0.26	350
S20	0.066	420	0.20	400
S21	0.024	460	0.07	370

* Neglecting short wavelength peak

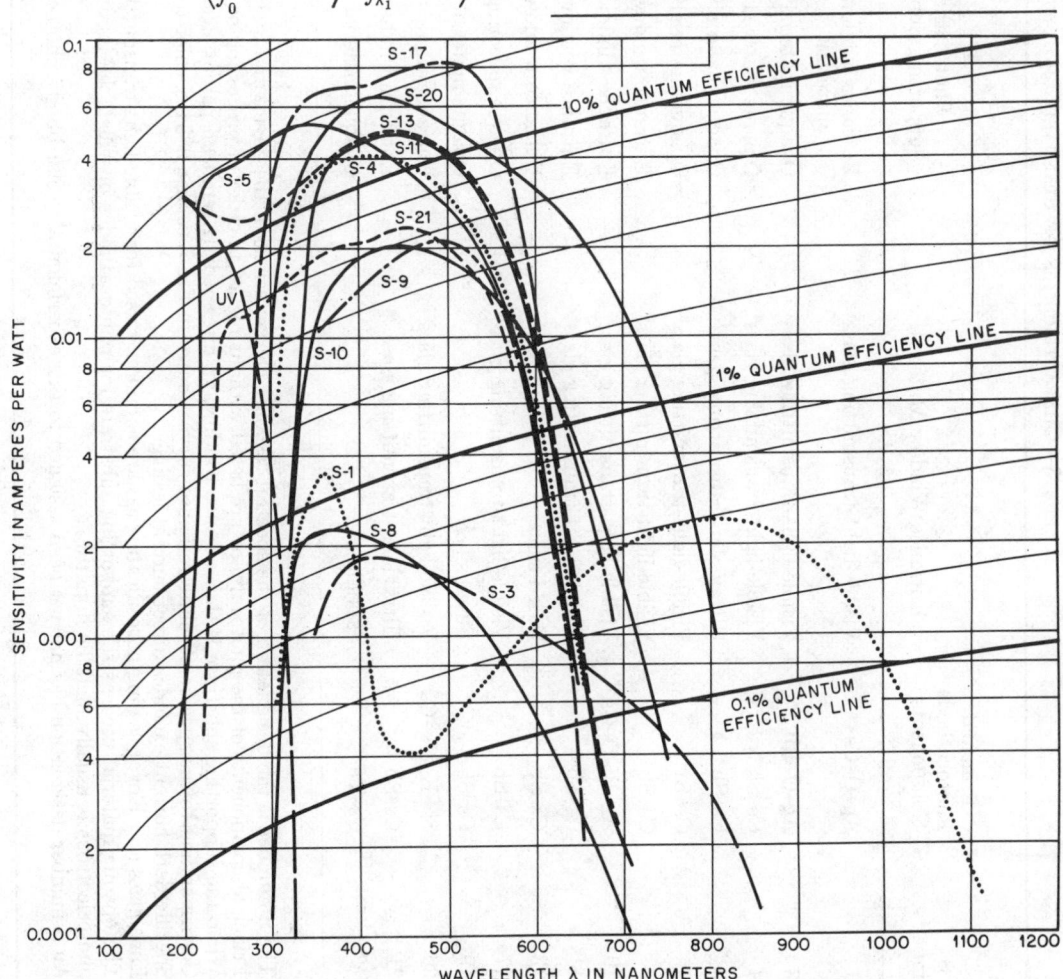

Fig. 5—Typical absolute spectral response characteristics of photoemissive devices.

TABLE 3—CHARACTERISTICS OF STANDARD PHOTOSURFACES.

S Number[1]	Principal Photocathode Components[2]	Entrance Window Material	Photocathode Supporting Substrate[3]	Typical Luminous Sensitivity[4] (μA/lumen)	Typical Photocathode Dark Current[5] at 25°C (A/cm²)
S1	Ag-O-Cs	Visible-light-transmitting glass[6]	Entrance window or opaque material[7]	25	10^{-11}–10^{-13}
S3	Ag-O-Rb	Visible-light-transmitting glass[6]	Opaque material[7]	6.5	10^{-12}
S4	Cs-Sb	Visible-light-transmitting glass[6]	Opaque material[7]	40	10^{-14}
S5	Cs-Sb	Ultraviolet-transmitting glass	Opaque material[7]	40	10^{-14}
S8	Cs-Bi	Visible-light-transmitting glass[6]	Opaque material[7]	3	10^{-14}–10^{-15}
S9	Cs-Sb	Visible-light-transmitting glass[6]	Entrance window	30	10^{-14}
S10	Ag-Bi-O-Cs	Visible-light-transmitting glass[6]	Entrance window	40	10^{-13}–10^{-14}
S11	Cs-Sb	Visible-light-transmitting glass[6]	Entrance window	60	10^{-14}–10^{-15}
S13	Cs-Sb	Fused silica	Entrance window	60	10^{-14}–10^{-15}
S17	Cs-Sb	Visible-light-transmitting glass[6]	Opaque reflecting material[7]	125	10^{-14}–10^{-15}
S19	Cs-Sb	Fused silica	Opaque material[7]	40	10^{-14}
S20	Sb-K-Na-Cs	Visible-light-transmitting glass[6]	Entrance window	150	10^{-15}–10^{-16}
S21	Cs-Sb	Ultraviolet-transmitting glass	Entrance window	30	10^{-14}
UV[8]	Cs-Te	Sapphire	Opaque material[7]	0	—

Notes:

1. The S number is the designation of the spectral response characteristic of the device and includes the transmission of the device window material.
2. Principal components of the photocathode are listed without regard to order of processing or relative proportions.
3. When the supporting substrate is the entrance window, an intermediate semitransparent electrically conductive layer may be used.
4. Corresponding to the specific absolute response curves shown in Fig. 5 using a 2854 K color-temperature tungsten-lamp test source.
5. Specific dark current excludes direct-current leakage.
6. Lime glass and Kovar sealing borosilicate glass are commonly used for visible-light-transmitting glass.
7. The opaque material used as the supporting substrate for photocathodes in which the input radiation is incident on the same side as the emitted photoelectrons is usually metallic in nature.
8. An S number designation has not yet been assigned to this experimental "solar blind" photoemissive surface.

where $w_{\lambda\,max}=$ peak input flux spectral density (watt/m), $s_{\lambda\,max}=$ peak monochromatic radiant sensitivity of the photocathode (amp/watt), $w_\lambda=$ relative spectral distribution of the input flux, $\sigma_\lambda=$ relative spectral distribution of the radiant sensitivity of the photocathode, $\lambda=$ wavelength, and $W_{\lambda_1\lambda_2}=$ flux in watts within the wavelength interval $\lambda_1\leq\lambda\leq\lambda_2$.

Typical values of the dimensionless "spectral matching factor" ratio

$$\int_0^\infty w_\lambda\sigma_\lambda\,d\lambda \Big/ \int_{\lambda_1}^{\lambda_2} w_\lambda\,d\lambda$$

appearing in the above relationship and describing the comparative spectral match between various photocathode and input-flux spectral distributions are shown in Table 4 evaluated for $\lambda_1=0$ and $\lambda_2=1.2\mu m=1.2\times10^{-6}$ meter.

If the input flux is measured in lumens L instead of watts, the resultant ratio I/L of emitted current I to flux input is designated as the photocathode luminous sensitivity S, and is given by

$$S=I/L=s_{\lambda\,max}\int_0^\infty w_\lambda\sigma_\lambda\,d\lambda \Big/ 680\int_0^\infty w_\lambda E_\lambda\,d\lambda$$

where $680=$ luminous equivalent of 555 Å radiation (lumens/watt), $E_\lambda=$ relative photopic human eye response, normalized to unity maximum, and

$$\int_0^\infty w_\lambda\sigma_\lambda\,d\lambda \Big/ \int_0^\infty w_\lambda E_\lambda\,d\lambda$$

is a dimensionless ratio that can be computed from the spectral-matching-factor data in Table 4 for various photocathode and input-flux spectral

distributions by dividing the spectral-matching-factor data for

$$\int_0^\infty w_\lambda\sigma_\lambda\,d\lambda \Big/ \int_{\lambda_1}^{\lambda_2} w_\lambda\,d\lambda$$

by the spectral-matching-factor data for

$$\int_0^\infty w_\lambda E_\lambda\,d\lambda \Big/ \int_{\lambda_1}^{\lambda_2} w_\lambda\,d\lambda.$$

In the special case where the input spectral distribution w_λ, designated as $w_\lambda(2854)$, corresponds to a standard 2854 K color-temperature tungsten lamp, the luminous sensitivity S, designated as $S(2854)$, and often used as a specification on photocathode sensitivity, is given by

$$S(2854)=s_{\lambda\,max}$$

$$\times\int_0^\infty w_\lambda(2854)\sigma_\lambda\,d\lambda \Big/ 681\int_0^\infty w_\lambda(2854)E_\lambda\,d\lambda.$$

This equation relates the peak cathode monochromatic radiant sensitivity $s_{\lambda\,max}$ (also commonly used to specify cathode sensitivity) to the standard luminous sensitivity $S(2854)$. Since the wavelength at which the cathode quantum efficiency has its peak value does not correspond, in general, to the wavelength at which the radiant sensitivity has its peak value (Fig. 5 and Table 2) no general relationship exists between the peak radiant sensitivity $s_{\lambda\,max}$ and the peak quantum efficiency $q_{\lambda\,max}$.

Photocathode Response with Optical Filter: The ratio of (A) the emitted photocurrent, $I(filter)$, with a filter inserted between a given flux source and the photocathode to (B) the current without

TABLE 4—SPECTRAL MATCHING FACTORS.

$$\int_0^\infty w_\lambda\sigma_\lambda\,d\lambda \Big/ \int_0^{1.2\mu} w_\lambda\,d\lambda \quad \text{and} \quad \int_0^\infty w_\lambda E_\lambda\,d\lambda \Big/ \int_{\lambda_1}^{\lambda_2} w_\lambda\,d\lambda$$

Source	λ_1	λ_2	Photocathode Type			Photopic Eye
			S1	S11	S20	
2854 K lamp	0	1200	0.52	0.060	0.112	0.071
5000 K blackbody	0	1200	0.53	0.26	0.34	0.140
Mean solar flux	0	1200	0.54	0.32	0.36	0.197
P1 phosphor	0	∞	0.28	0.28	0.69	0.768
P4 phosphor	0	∞	0.31	0.67	0.73	0.402
P11 phosphor	0	∞	0.22	0.91	0.88	0.201
P20 phosphor	0	∞	0.39	0.42	0.58	0.707
NaI(Th)	0	∞	0.53	0.88	0.90	0.046

Note: λ_1 and λ_2 are in nanometers.

TABLE 5—TYPICAL FILTER FACTORS.

| Filter | | | | | Filter Factor $\displaystyle\int_0^\infty t_\lambda w_\lambda \sigma_\lambda \, d\lambda \Big/ \int_0^\infty w_\lambda \sigma_\lambda \, d\lambda$ $w_\lambda = w_\lambda(2854)$ | | | |
| | | | | | Photocathode Type | | | |
Manufacturer	Glass Number	Thickness	Color Series	Description	S1	S4	S11	S20
Corning	2540	stock	CS 7–56	Infrared	0.108	0.000	0.000	0.000
Corning	5113	$\frac{1}{2}$ stock	CS 5–58	Blue	0.004	0.126	0.103	0.055
Corning	2403	stock	CS 2–58	Deep red	0.788	0.114	0.112	0.257

the filter, I(no filter), is called the filter factor $T(t_\lambda, w_\lambda, \sigma_\lambda)$ and is given by

$$T(t_\lambda, w_\lambda, \sigma_\lambda) = I(\text{filter})/I(\text{no filter})$$

$$= \int_0^\infty t_\lambda w_\lambda \sigma_\lambda \, d\lambda \Big/ \int_0^\infty w_\lambda \sigma_\lambda \, d\lambda$$

where t_λ is the transmission of the filter at a given wavelength λ, and the notation $T(t_\lambda, w_\lambda, \sigma_\lambda)$ indicates that the filter factor is a function not only of the filter transmission t_λ but also of the detector response σ_λ and the source distribution w_λ. Typical filter factors are given in Table 5.

The ratio of emitted photocurrent with the filter, I(filter), to the flux in lumens $L(2854)$, incident on the filter (not on the cathode) from a 2854 K source is designated as S(photocathode+filter) and is given by

$$S(\text{photocathode+filter}) = I(\text{filter})/L(2854)$$

$$- S(2854) \, T(t_\lambda, 2854, \sigma_\lambda).$$

The magnitude of the luminous sensitivity, S(photocathode+filter), in amperes per lumen is used to specify cathode sensitivity, or more precisely, cathode-plus-filter sensitivity, over a selected spectral region, where the filter is chosen to restrict the flux incident on the photocathode to the desired region. The sensitivity, S(photocathode+filter), is then designated as the "infrared" sensitivity, or "red" sensitivity, or "blue" sensitivity, et cetera, depending on the predominant spectral region passed by the filter.

ELECTRODE DISSIPATION

After the electron stream has given up the useful component of its energy, the remainder is dissipated as heat in some suitable part of the tube. Five processes are commonly used to remove this heat. The amount which can be removed depends on the area available, the temperature differential, and, in the cases of forced cooling, the coolant flow.

TABLE 6—TYPICAL OPERATING DATA FOR COMMON TYPES OF COOLING.

Type	Average Cooling Surface Temperature (°C)	Specific Dissipation of Cooling Surface (watts/cm²)	Cooling-Medium Supply
Radiation	400–1000	4–10	
Water	30–150	30–110	0.25–0.5 gallon/minute/kilowatt
Forced air	150–200	0.5–1	50–150 feet³/minute/kilowatt
Evaporative	100–120	80–125	Water-, air-, or convection-cooled condenser. A water-cooled condenser would require 0.07–0.1 gallon/minute/kilowatt
Conduction	100–250	5–30	Heat sink operating at 50–100°C

In computing cooling-medium flow, a minimum velocity sufficient to assure turbulent flow at the dissipating surface must be maintained. The figures for specific dissipation (Table 6) apply to clean cooling surfaces and may be reduced to a small fraction of the values shown by heat-insulating coatings such as scale or dust.

Radiation Cooling

In a radiation-cooled system, that portion of the tube on which the heat is dissipated is allowed to reach a temperature such that the heat is radiated to the surroundings. The amount of heat which

TABLE 7—TOTAL THERMAL EMISSIVITY ϵ_t OF ELECTRON-TUBE MATERIALS.

Material	Temperature (K)	Thermal Emissivity
Aluminum	450	0.1
Anode graphite	1000	0.9
Copper	300	0.07
Molybdenum	1300	0.13
Molybdenum, quartz-blasted	1300	0.5
Nickel	600	0.09
Tantalum	1400	0.18
Tungsten	2600	0.30

can be removed in this manner is given by the equation

$$P = \epsilon_t \sigma (T^4 - T_0^4)$$

where P = radiated power in watts/centimeter^2, ϵ_t = total thermal emissivity of the surface, σ = Stefan-Boltzmann constant = 5.67×10^{-12} watt-centimeters$^{-2} \times$ kelvins^{-4}, T = temperature of radiating surface in kelvins, and T_0 = temperature of surroundings in kelvins. Total thermal emissivity varies with the degree of roughness of the surface of the material and the temperature. Values for typical surfaces are in Table 7.

Water Cooling

For water cooling the water is circulated through a suitably designed structure. The amount of heat which can be removed by this process is given by

$$P = 264 Q_W (T_2 - T_1)$$

where P = power in watts, Q_W = flow in gallons per minute, and T_2, T_1 = outlet and inlet water temperatures, respectively, in kelvins.

This same relationship is given in the nomogram of Fig. 6 with the temperature rise in degrees Fahrenheit or Celsius and the power in kilowatts.

Forced Air Cooling

With forced air cooling a stream of air is forced past a suitable radiator. The heat which can be removed by this process is given by

$$P = 169 Q_A [(T_2/T_1) - 1]$$

where Q_A = air flow in feet^3/minute, other quantities as above.

Evaporative Cooling

A typical evaporative-cooled system consists of a tube with a specially designed anode immersed in a boiler containing distilled water. When power is dissipated on the anode, the water boils and the steam is conducted upward through an insulating pipe to a condenser. The condensate is then gravity fed back to the boiler, thus eliminating the pump required in a circulating water system.

For some transmitter applications the steam is directed downward to leave the space above the tube available for other components. Such a system requires a pump to return the condensate to the boiler, but even then the pump has to handle only about 0.05 of the amount of water required for a water-cooled system because of the exploitation of the latent heat of steam.

The size of the heat-exchanger equipment for an evaporative-cooled system is less than one-third of that required for a water-cooled system because of the greater mean temperature differential between the cooled liquid and the secondary coolant. Typical temperature differentials for the two systems are 75°C and 30°C, respectively.

The anode dissipation should not exceed 135 watts per square centimeter of external anode surface because at this point, often referred to as the "Leidenfrost" or "calefaction" point, the surface becomes completely covered with a sheath of vapor and the thermal conductivity between the anode and the cooling liquid drops to 30 watts per square centimeter, with resultant overheating of the anode. Special designs of the external anode surface (such as the "pineapple") allow up to 500 watts to be dissipated per square centimeter of internal anode surface.

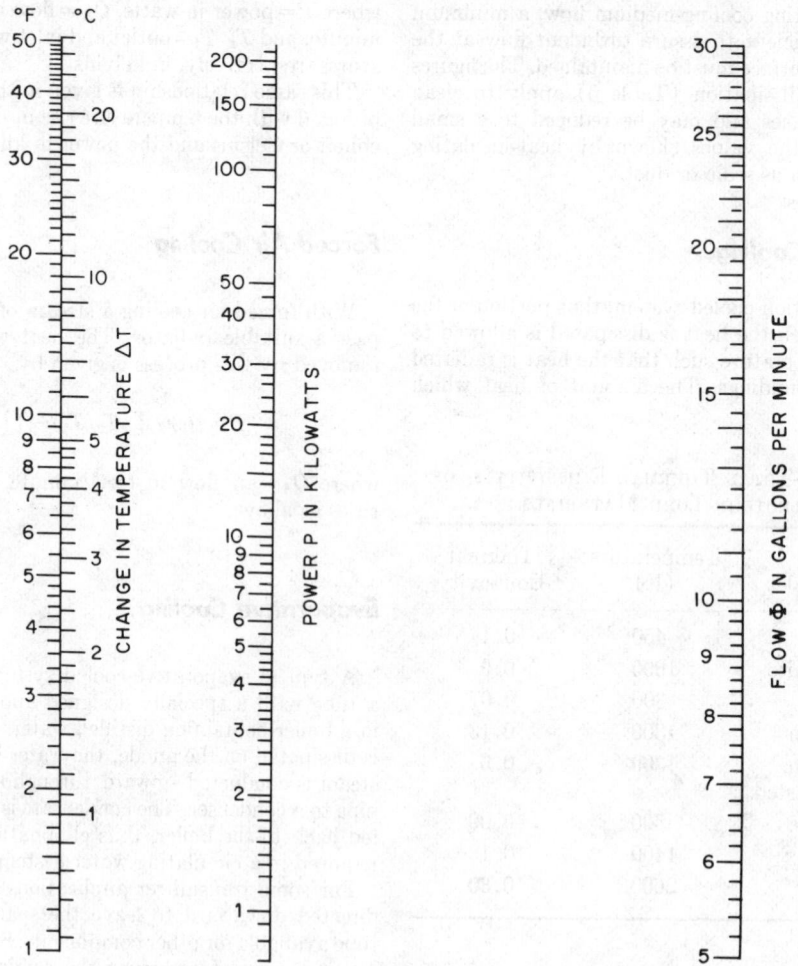

Fig. 6—Heat transfer in cooling water. $P=0.1466\Phi\Delta T$, for T in Fahrenheit. *Courtesy of Clyde G. Haehnle.*

Conduction Cooling

When an external heat sink is available, heat may be removed from the tube by conduction. Since the electrode where the heat appears is usually at an elevated potential, it is often necessary to conduct the heat through an electrical insulator.

Because of its relatively high thermal conductivity, beryllia ceramic can be used as a common insulator and thermal conductor between the anode of a tube and a heat sink.

Properties of beryllia:

Breakdown strength= 10 kV/mm
Dielectric constant= 6–8
Thermal conductivity= 2.62 watts/cm/°C at 20°C, 1.75 watts/cm/°C at 200°C
Dielectric loss factor= 4×10^{-5}

Tensile strength= 18 000 lbs/square inch
Compressive strength= 150 000 lbs/square inch.

The temperature drop in degrees across Celsius the beryllia ceramic is given by

$$t_1-t_2=dW_a/KA \quad \text{(for a parallel configuration)}$$

where $t_1=$ temperature of tube anode (typical maximum 250°C), $t_2=$ temperature of heat sink (typically 100°C), $d=$ thickness of beryllia in cm, $A=$ cross-sectional area of beryllia perpendicular to direction of heat flow, $K=$ thermal conductivity of beryllia in watts/cm/°C, and $W_a=$ power dissipated on anode in watts.

To the temperature drop across the beryllia ceramic must be added the temperature drop across the interfaces between the ceramic and the anode and heat sink, typically 20°C for clamped surfaces at a loading of 25 watts/cm².

Because of its toxic nature, care must be taken in handling and disposal of beryllia ceramic.

Grid Temperature

Operation of grids at excessive temperatures will result in one or more harmful effects: liberation of gas, high primary (thermal) emission, contamination of the other electrodes by deposition of grid material, and melting of the grid. Grid-current ratings should not be exceeded, even for short periods.

NOISE IN TUBES

There are several sources of noise in electron tubes*, some associated with the nature of electron emission and some caused by other effects in the tube.

Shot Effect

The electric current emitted from a cathode consists of a large number of electrons and consequently exhibits fluctuations which produce tube noise and set a limit to the minimum signal that can be amplified. The root-mean-square value of the fluctuating (noise) component of the plate current I_n is given in amperes by

$$I_n{}^2 = 2eI\Gamma^2\Delta f$$

where I = plate direct current in amperes, e = electron charge = 1.6×10^{-19} coulomb, Δf = bandwidth in hertz, and Γ^2 = space-charge reduction or smoothing factor. For temperature-limited cases, $\Gamma^2 = 1$. For space-charge-controlled regions

$$\Gamma^2 = 2kT_c g\theta/\sigma eI$$

where k = Boltzmann's constant = 1.380×10^{-23} joule/kelvin, T_c = cathode temperature in kelvins,

* S. Goldman, "Frequency Analysis, Modulation and Noise," McGraw-Hill Book Company, New York; 1948. A. van der Ziel, "Noise," Prentice-Hall, Englewood Cliffs, New Jersey; 1954. "Noise in Electron Devices," edited by L. D. Smullin and H. A. Haus, The Technology Press of Massachusetts Institute of Technology and John Wiley & Sons, New York; 1959. D. H. Bell, "Electron Noise," D. Van Nostrand Company, London; 1960. W. R. Bennet, "Electrical Noise," McGraw-Hill Book Company, New York; 1960. D. K. C. MacDonald, "Noise and Fluctuations: An Introduction," John Wiley & Sons, New York; 1962.

g = conductance or transconductance in mhos, which relates the output signal current to the input signal voltage*, θ = a factor which in most practical cases is nearly equal to its asymptotic value of $3[1 - (\pi/4)] = 0.644$, and σ = a tube parameter, related to the amplification factor and electrode spacings, which has a value of unity for diodes and varies between 0.5 and 1.0 for negative-grid tubes.

Partition Noise

Excess noise appears in multicollector tubes because of fluctuations in the division of the current between the different electrodes. In a grid-controlled tube, these fluctuations in current division reduce the effectiveness of the space-charge smoothing of the shot noise in the plate current. For a screen-grid tube, the root-mean-square noise currents in the cathode lead, the screen-grid lead, and the plate lead (I_{nk}, I_{nc2}, and I_n, respectively) are given by

$$I_{nk}{}^2 = 2eI_k\Gamma^2\Delta f$$

$$I_{nc2}{}^2 = 2eI_{c2}[(\Gamma^2 I_{c2} + I)/I_k]\Delta f$$

$$I_n{}^2 = 2eI[(\Gamma^2 I + I_{c2})/I_k]\Delta f$$

where I_k and I_{c2} are the cathode and screen-grid currents, respectively.

Flicker Effect

The mechanism is not completely understood but appears to depend on the field distribution in the surface layer of the cathode due to its porous structure. Because this same field distribution also will influence the cathode activity and temperature, flicker noise will depend on cathode activity and temperature in a complicated manner.

The flicker noise spectrum is usually of the form $f^{-\alpha}$ with α close to unity and thus is important only at low frequencies. The sensitivity of audio, subaudio, and direct-current amplifiers is limited by the flicker noise generated in the first tube.

Collision Ionization

Free gas ions can be generated by collisions with the electron stream. The electrons thus liberated and collected by the anode will appear as noise in the anode circuit. The ions that travel to the cathode will travel slowly through the po-

* For diodes, g is the conductance; for triode and pentode amplifiers, g is the transconductance g_m; and for triode or pentode mixers and converters, g is the conversion conductance g_c.

tential minimum and reduce the space charge, which in turn will reduce the space-charge smoothing effect. This also will increase the noise in the anode circuit.

Induced Noise

At high frequencies it is not necessary for electrons to reach an electrode for induced current to flow in the electrode leads. This noise is an important consideration in miniature tubes above 15 megahertz and becomes the principal limiting factor in low-noise amplifier design above about 100 megahertz. For microwave tubes, this is the dominant method by which beam noise is coupled to the output circuit.

Miscellaneous Noise

Other noise may be present due to microphonics, hum, leakage, charges on insulators, poor contacts, and secondary emission.

Evaluation of Tube Performance

There are two common ways of evaluating tube performance: equivalent noise input resistance value, and noise figure (or factor).

Equivalent Noise Input Resistance: A resistor generates an amount of thermal noise (also called Johnson noise or Brownian motion noise) given by

$$E_n{}^2 = 4kTR\Delta f$$

where $E_n =$ the open-circuit root-mean-square fluctuating voltage measured across the resistor terminal in volts, $T =$ resistor temperature in kelvins, and $R =$ resistor resistance in ohms. The equivalent noise input resistance in ohms R_{eq} is defined as that value of resistance which, when connected to the input of the tube and held at room temperature, will double the output noise power. This can be expressed as

$$4kT_0R_{eq}\Delta f g^2 = 2eI\Gamma_{eff}{}^2\Delta f$$

or

$$R_{eq} = eI\Gamma_{eff}{}^2/2kT_0 g^2$$

where $T_0 = 293$ K, $\Gamma_{eff}{}^2 =$ the effective space-charge reduction factor to include partition noise effects, and $g =$ the appropriate transconductance or conversion conductance as before. Practical approximations to R_{eq} are given in Table 8 for several tube functions.

Noise Figure: The noise figure of a tube is defined as the ratio of the available signal-to-noise ratio at the input to the signal-to-noise ratio at the

TABLE 8—APPROXIMATE EQUIVALENT NOISE RESISTANCES.

Function	Type	R_{eq}
Amplifying	Triode	$2.5/g_m$
	Pentode	$I/(I+I_{c2})$ $\times[(2.5/g_m)+(20I_{c2}/g_m{}^2)]$
Mixing	Triode	$4/g_c$
	Pentode	$I/(I+I_{c2})(4/g_c+20I_{c2}/g_c{}^2)$
Converting and mixing	Multigrid	$19I(I_k-I)/g_c{}^2I_k$

output. It is usually given the symbol F and is always greater than unity. For a more-detailed discussion of noise figure refer to the chapter on "Radio Noise and Interference."

Microwave Tubes

The noise appearing in the output circuit of a microwave tube is due in part to induced noise from the beam. Also, some of the electrons may be intercepted by the radio-frequency structure (microwave cavity, slow-wave circuit, et cetera) giving rise to partition noise. In well-designed low-noise tubes, however, this latter effect is kept negligibly small.

For lossless linear beam tubes (traveling-wave amplifiers, klystron amplifiers, backward-wave amplifiers), the minimum obtainable noise figure F_{min} for one-dimensional single-velocity small-signal theory and high gain has been found to be given by

$$F_{min} = 1 + (2\pi/kT_0)(S-\pi)$$

where $S-\pi$ is the basic noise parameter and is established in the region of the potential minimum of the beam. If certain assumptions concerning the potential minimum are made, such as full shot noise and uncorrelated current and velocity fluctuations,[*] then values for S and π can be obtained. They are given as

$$\pi = 0$$

$$S = [1-(\pi/4)]^{1/2}(kT_c/\pi)$$

therefore

$$F_{min} = 1 + (4-\pi)^{1/2}(T_c/T_0).$$

For $T_c/T_0 = 4$, $F_{min} \approx 4$. The assumptions made are not entirely valid, as shown by the fact that noise

---

[*] J. R. Pierce, "A Theorem Concerning Noise in Electron Streams," *Journal of Applied Physics*, Vol. 25, p. 931; 1954.

figures of less than 4 have been obtained experimentally. At the present time values of S and π/S are obtained by measurement.

LOW- AND MEDIUM-FREQUENCY TUBES

This section applies particularly to triodes and multigrid tubes operated at frequencies where electron-inertia effects are negligible. Traditionally the vacuum envelope of such tubes has been of glass with metal, usually copper, for the anode in larger sizes. In recent years the trend has been toward ceramic in place of glass for the external insulating portions of such tubes. Figure 7 shows a typical construction of a medium-power transmitting tube.

Ceramic-envelope tubes have the following advantages over glass tubes.

(**A**) The radio-frequency loss P_{rf} in the seals of a tube is given by

$$P_{rf} = Kf^{5/2}R^{1/2}\mu^{1/2}$$

where $K=$ constant, $f=$ frequency, $R=$ resistivity

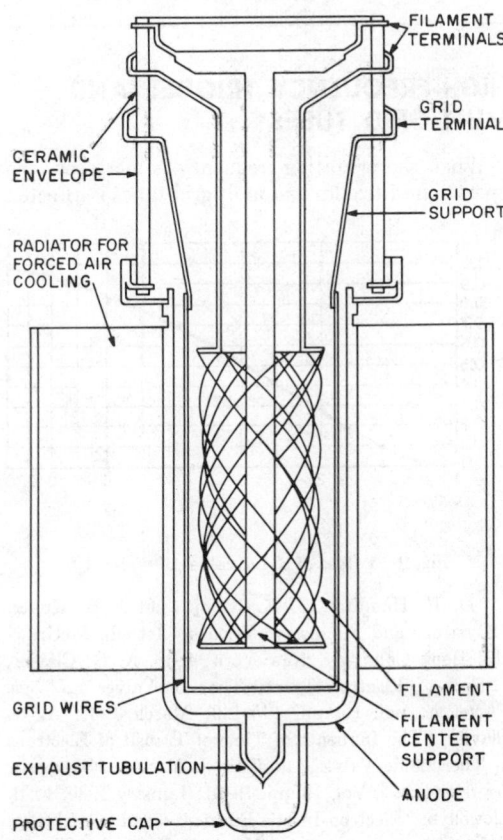

FILAMENT TERMINALS

GRID TERMINAL

CERAMIC ENVELOPE

GRID SUPPORT

RADIATOR FOR FORCED AIR COOLING

GRID WIRES

EXHAUST TUBULATION

PROTECTIVE CAP

FILAMENT

FILAMENT CENTER SUPPORT

ANODE

Fig. 7—Electrode arrangement of medium-power external-anode transmitting tube.

of the conducting material, and $\mu=$ permeability of the conducting material. In glass-to-metal seals, the metal is normally of a magnetic material such as Kovar. As Kovar has high resistivity and permeability, the radio-frequency losses at the seals are therefore high, and at high frequencies cracking and/or glass suck-in near the seals can result. With ceramic-to-metal seals this problem is minimized because the radio-frequency circulating currents at the seals flow through the metallizing and plating on the ceramic. The resistivity is low, and the permeability is unity.

(**B**) Ceramics have a lower dielectric loss than glass. Furthermore the loss factor of glass rapidly rises with temperature. This leads to a "runaway" condition, glass suck-in, and hence severe limitation of maximum frequency of operation of glass tubes.

(**C**) The safe operating temperature of a ceramic-to-metal seal may be between 220 and 250 degrees Celsius as against 180 degrees Celsius for Kovar glass seals.

(**D**) High bakeout temperature of ceramic-envelope tubes during evacuation increases reliability and life.

(**E**) Ceramic tubes withstand higher thermal and mechanical shocks than those with glass envelopes. They can also be manufactured to closer dimensional tolerances.

Coefficients

Amplification factor μ: Ratio of incremental plate voltage to control-electrode voltage change at a fixed plate current with constant voltage on other electrodes

$$\mu = \left[\frac{\delta e_b}{\delta e_{c1}}\right] \quad \begin{subarray}{l} I_b, E_{c2}, \cdots E_{cn} \text{ constant.} \\ r_l = 0 \end{subarray}$$

Transconductance s_m: Ratio of incremental plate current to control-electrode voltage change at constant voltage on other electrodes

$$s_m = \left[\frac{\delta i_b}{\delta e_{c1}}\right] \quad \begin{subarray}{l} E_b, E_{c2}, \cdots, E_{cn} \text{ constant.} \\ r_l = 0 \end{subarray}$$

When electrodes are plate and control grid, the ratio is the mutual conductance g_m

$$g_m = \mu/r_p.$$

Variational (AC) Plate Resistance r_p: Ratio of incremental plate voltage to current change at constant voltage on other electrodes

$$r_p = \left[\frac{\delta e_b}{\delta i_b}\right] \quad \begin{subarray}{l} E_{c1}, \cdots, E_{cn} \text{ constant.} \\ r_l = 0 \end{subarray}$$

Total (DC) Plate Resistance R_p: Ratio of total plate voltage to current for constant voltage on

TABLE 9—TUBE CHARACTERISTICS FOR UNIPOTENTIAL CATHODE AND NEGLIGIBLE SATURATION OF CATHODE.

Function	Parallel-Plane Cathode and Anode	Cylindrical Cathode and Anode
Diode anode current (amperes)	$G_1 e_b^{3/2}$	$G_1 e_b^{3/2}$
Triode anode current (amperes)	$G_2[(e_b+\mu e_c)/(1+\mu)]^{3/2}$	$G_2[(e_b+\mu e_c)/(1+\mu)]^{3/2}$
Diode perveance G_1	$2.3\times10^{-6}(A_b/d_b^2)$	$2.3\times10^{-6}(A_b/\beta^2 r_b^2)$
Triode perveance G_2	$2.3\times10^{-6}(A_b/d_b d_c)$	$2.3\times10^{-6}(A_b/\beta^2 r_b r_c)$
Amplification factor μ	$2.7d_c[(d_b/d_c)-1]/[\rho \log(\rho/2\pi r_g)]$	$(2\pi d_c/\rho)[\log(d_b/d_c)/\log(\rho/2\pi r_g)]$
Mutual conductance g_m	$1.5G_2[\mu/(\mu+1)](E'_g)^{1/2}$	$1.5G_2[\mu/(\mu+1)](E'_g)^{1/2}$
	$E'_g=(E_b+\mu E_c)/(1+\mu)$	$E'_g=(E_b+\mu E_c)/(1+\mu)$

where A_b=effective anode area in square centimeters, d_b=anode-cathode distance in centimeters, d_c=grid-cathode distance in centimeters, β=geometric constant, a function of ratio of anode-to-cathode radius; $\beta^2\approx1$ for $r_b/r_k>10$ (Fig. 9), ρ=pitch of grid wires in centimeters, r_g=grid-wire radius in centimeters, r_b=anode radius in centimeters, r_k=cathode radius in centimeters, and r_c=grid radius in centimeters.

Note:

These equations are based on theoretical considerations and do not provide accurate results for practical structures; however, they give a fair idea of the relationship between the tube geometry and the constants of the tube.

---

other electrodes

$$R_p=\left[\frac{E_b}{I_b}\right] \quad \begin{matrix} E_{c1},\cdots,E_{cn}\text{ constant}. \\ r_l=0 \end{matrix}$$

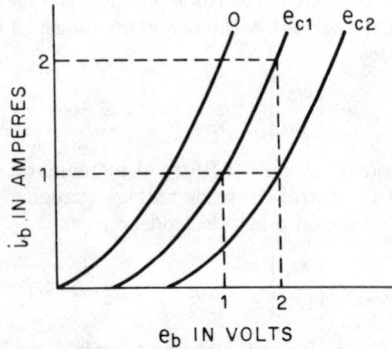

Amplification factor $\mu= (e_{b2}-e_{b1})/(e_{c2}-e_{c1})$
Mutual conductance $g_m= (i_{b2}-i_{b1})/(e_{c2}-e_{c1})$
Total plate resistance $R_p=e_{b2}/i_{b2}$
Variational plate resistance $r_p= (e_{b2}-e_{b1})/(i_{b2}-i_{b1})$.

Fig. 8—Graphic method of determining coefficients.

A useful approximation of these coefficients may be obtained from a family of anode characteristics, Fig. 8. Relationships between the actual geometry of a tube and its coefficients are given roughly in Table 9.

HIGH-FREQUENCY TRIODES AND MULTIGRID TUBES*

When the operating frequency is increased, the operation of triodes and multigrid tubes is affected

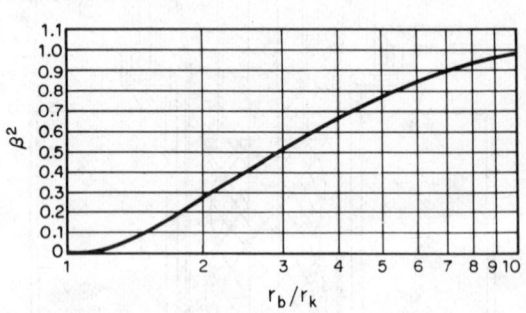

Fig. 9—Values of β^2 for values of $r_b/r_k<10$.

* D. R. Hamilton, J. K. Knipp, and J. B. Kuper, "Klystrons and Microwave Triodes," 1st ed., McGraw-Hill Book Company, New York; 1948. A. G. Clavier, "Effect of Electron Transit-Time in Valves," *L'Onde Electrique*, Vol. 16, pp. 145–149; March 1937. A. G. Clavier, "The Influence of Time of Transit of Electrons in Thermionic Valves," *Bulletin de la Société Française des Électriciens*, Vol. 19, pp. 79–91; January 1939. F. B. Llewellyn, "Electron-Inertia Effects," 1st ed., Cambridge University Press, London; 1941. A. H. W. Beck, "Thermionic Valves," Cambridge University Press, London; 1953.

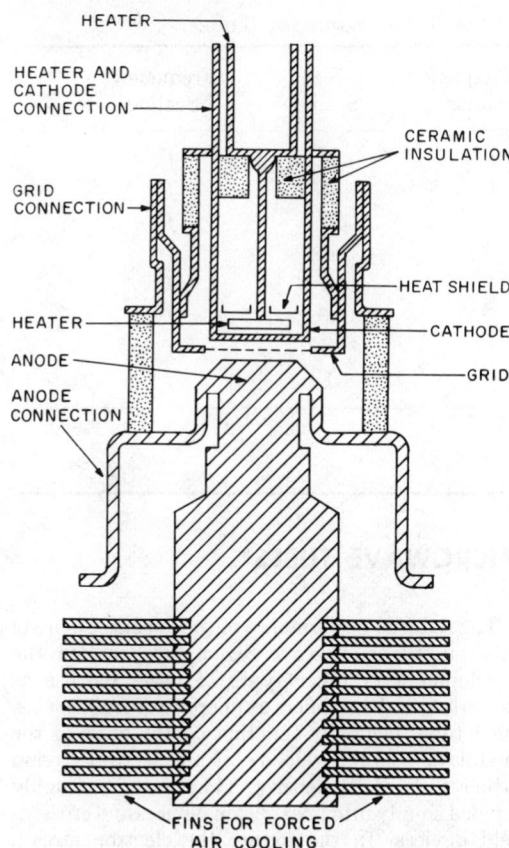

Fig. 10—Electrode arrangement of a small high-frequency
external-anode triode.

by electron-inertia effects. The design features that
distinguish the high-frequency tube shown in Fig.
10 from the lower-frequency tube (Fig. 7) are:
reduced cathode-to-grid and grid-to-anode spacings, high emission density, high power density,
small active and inactive capacitances, heavy
terminals, short support leads, and adaptability
to a cavity circuit.

Factors Affecting Ultra-High-Frequency Operation

Electron Inertia: The theory of electron-inertia
effects in small-signal tubes has been formulated;
no comparable complete theory is now available
for large-signal tubes.

When the transit time of the electrons from
cathode to anode is an appreciable fraction of one
radio-frequency cycle:

(**A**) Input conductance due to reaction of electrons with the varying field from the grid becomes
appreciable. This conductance, which increases as
the square of the frequency, results in lowered
gain, an increase in driving-power requirement,
and loading of the input circuit.

(**B**) Grid-anode transit time introduces a phase
lag between grid voltage and anode current. In
oscillators, the problem of compensating for the
phase lag by design and adjustment of a feedback
circuit becomes difficult. Efficiency is reduced in
both oscillators and amplifiers.

(**C**) Distortion of the current pulse in the grid-
anode space increases the anode-current conduction
angle and lowers the efficiency.

Electrode Admittances: In amplifiers, the effect
of cathode-lead inductance is to introduce a conductance component in the grid circuit. This effect
is serious in small-signal amplifiers because the
loading of the input circuit by the conductance
current limits the gain of the stage. Cathode-grid
and grid-anode capacitive reactances are of small
magnitude at ultra-high frequencies. Heavy currents flow as a result of these reactances and
tubes must be designed to carry the currents
without serious loss. Coaxial cavities are often
used in the circuits to resonate with the tube
reactances and to minimize resistive and radiation
losses. Two circuit difficulties arise as operating
frequencies increase:

(**A**) The cavities become physically impossible
as they tend to take the dimensions of the tube
itself.

(**B**) Cavity Q varies inversely as the square
root of the frequency, which makes the attainment
of an optimum Q a limiting factor.

Scaling Factors: For a family of similar tubes,
the dimensionless magnitudes such as efficiency
are constant when the parameter

$$\phi = fd/V^{1/2}$$

is constant, where f = frequency in megahertz,
d = cathode-to-anode distance in centimeters, and
V = anode voltage in volts.

Based on this relationship and similar considerations, it is possible to derive a series of factors
that determine how operating conditions will vary
as the operating frequency or the physical dimensions are varied (Table 10). If the tube is to be
scaled exactly, all dimensions will be reduced inversely as the frequency is increased, and operating
conditions will be as given in the "Size-Frequency
Scaling" column. If the dimensions of the tube are
to be changed but the operating frequency maintained, operation will be as in the "Size Scaling"
column. If the dimensions are to be maintained
but the operating frequency changed, operating
conditions will be as in the "Frequency Scaling"
column. These factors apply in general to all types
of tubes.

Table 10—Scaling Factors For Ultra-High-Frequency Tubes.

Quantity	Ratio	Size-Frequency Scaling	Size Scaling	Frequency Scaling
Voltage	V_2/V_1	1	d^2	f^2
Field	E_2/E_1	f	d	f^2
Current	I_2/I_1	1	d^3	f^3
Current density	J_2/J_1	f^2	d	f^3
Power	P_2/P_1	1	d^5	f^5
Power density	h_2/h_1	f^2	d^3	f^5
Conductance	G_2/G_1	1	d	f
Magnetic-flux density	B_2/B_1	f	1	f

d = ratio of scaled to original dimensions
f = ratio of original to scaled frequency

With present knowledge and techniques, it has been possible to reach certain values of power with conventional tubes in the ultra- and super-high-frequency regions. The approximate maximum values that have been obtained are plotted in Fig. 11.

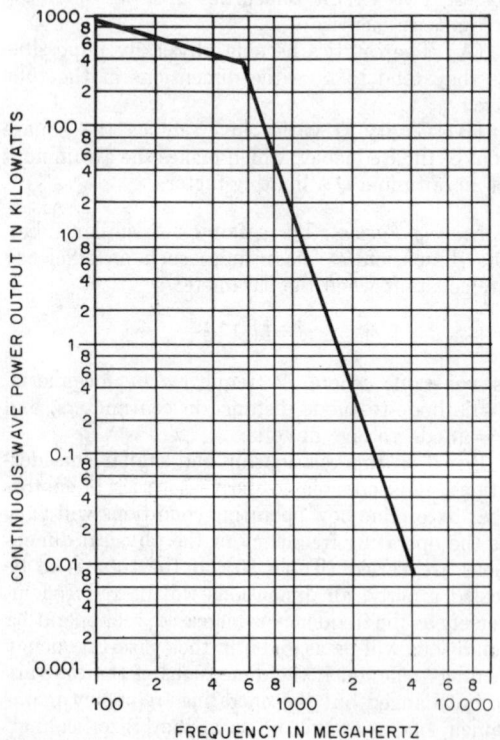

Fig. 11—Approximate maximum ultra-high-frequency continuous-wave power obtainable from a single triode or tetrode. These data are based on 1965 knowledge and techniques.

MICROWAVE TUBES*

The reduced performance of space-charge control tubes in the microwave region has fostered the development of other types of tubes for use as oscillators and amplifiers at microwave frequencies. Such tubes generally function on the basis of the modulation of the velocity of an electron stream rather than of its density. They may be roughly divided simply into linear beam devices and crossed-field devices. In the former the electron stream flows essentially linearly, often with a collimating magnetic field to counteract space-charge spreading; in the latter the electron stream follows a curved path under the action of orthogonal electric and magnetic fields. The linear beam devices are often referred to as O-type, while the crossed-field devices are referred to as M-type.

Terminology

Bunching: Any process that introduces a radio-frequency conduction-current component into a velocity-modulated electron stream as a direct result of the variation in electron transit time that the velocity modulation produces.

* A. H. W. Beck, "Space-Charge Waves and Slow Electromagnetic Waves," Pergamon Press, New York; 1958. R. G. E. Hutter, "Beam and Wave Electronics in Microwave Tubes," D. Van Nostrand Company, Princeton, N.J.; 1960. W. J. Kleen, "Electronics of Microwave Tubes," Academic Press, New York; 1958. J. C. Slater, "Microwave Electronics," D. Van Nostrand Company, Princeton, N.J.; 1950. J. F. Hull, "Microwave Tubes of the Mid-Sixties," 1965 *IEEE International Convention Record*, IEEE, New York.

Cavity Resonator: Any region bounded by conducting walls within which resonant electromagnetic fields may be excited.

Circuit Efficiency: The ratio of (A) the power of the desired frequency delivered to the output terminals of the circuit of an oscillator or amplifier to (B) the power of the desired frequency delivered by the electron stream to the circuit.

Coherent-Pulse Operation: Method of pulse operation in which the phase of the radio-frequency wave is maintained through successive pulses.

Drift Space: In an electron tube, a region substantially free of externally applied alternating fields in which a relative repositioning of the electrons is determined by their velocity distributions and the space-charge forces.

Duty Cycle: The product of the pulse duration and the pulse repetition rate. It is also the ratio of the average power output to the peak power output.

External Q: The reciprocal of the difference between the reciprocals of the loaded and unloaded Q's.

Frequency Pulling of an oscillator is the change in the generated frequency caused by a change of the load impedance.

Frequency Pushing of an oscillator is the change in frequency due to change in anode current (or in anode voltage).

Loaded Q of a specific mode of resonance of a system is the Q when there is external coupling to that mode. Note: When the system is connected to the load by means of a transmission line, the loaded Q is customarily determined when the line is terminated in its characteristic impedance.

Mode: One of the components of a general configuration of a vibrating system. A mode is characterized by a particular geometric pattern of the electromagnetic field and a resonant frequency (or propagation constant).

Noise Figure: The ratio in decibels of the total available output noise from an amplifier to the available noise which would be present at the output if the amplifier itself were noiseless, assuming a source temperature of 290 K.

Pulling Figure of an oscillator is the difference in megahertz between the maximum and minimum frequencies of oscillation obtained when the phase angle of the load-impedance reflection coefficient varies through 360 degrees, while the absolute value of this coefficient is constant and is normally equal to 0.20.

Pulse: Momentary flow of energy of such short time duration that it may be considered as an isolated phenomenon.

Pushing Figure of an oscillator is the rate of frequency pushing in megahertz per ampere or megahertz per volt.

Q: The Q of a specific mode of resonance of a system is 2π times the ratio of the stored electromagnetic energy to the energy dissipated per cycle when the system is excited in this mode.

Reflector: Electrode whose primary function is to reverse the direction of an electron stream. It is also called a *repeller*.

Reflex Bunching: Type of bunching that occurs when the velocity-modulated electron stream is made to reverse its direction by means of an opposing direct-current field.

Slow-Wave Structure: A microwave circuit, as used in beam-type microwave tubes, capable of propagating radio-frequency waves with phase velocities appreciably less than the velocity of light.

LINEAR BEAM TUBES

The principal types of linear beam tubes are the klystron, the traveling-wave amplifier, and the backward-wave oscillator.

Klystrons

A klystron* is an electron tube in which the following processes may be distinguished:
(A) Periodic variations of the longitudinal velocities of the electrons forming the beam in a region confining a radio-frequency field.
(B) Conversion of the velocity variation into conduction-current modulation by motion in a region free from radio-frequency fields.
(C) Extraction of the radio-frequency energy from the beam in another confined radio-frequency field.
The transit angles in the confined fields are made short $(\delta \doteq \pi/2)$ so that there is no appreciable conduction-current variation while traversing them.

* D. R. Hamilton, J. K. Knipp, and J. B. H. Kuper, "Klystrons and Microwave Triodes," McGraw-Hill Book Company, New York; 1948. A. H. W. Beck, "Velocity-Modulated Thermionic Valves," Cambridge University Press, London, England; 1948. A. H. W. Beck, "Thermionic Valves, Their Theory and Design," Cambridge University Press, London; 1953. A. H. W. Beck, "Space-Charge Waves and Slow Electromagnetic Waves," Pergamon Press, New York; 1958.

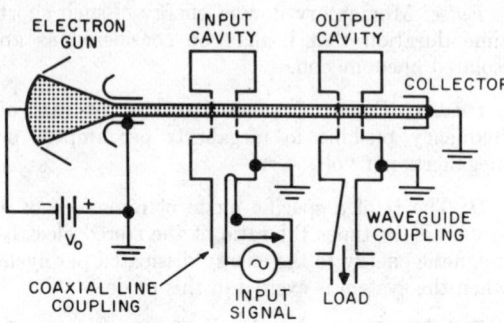

Fig. 12—Two-cavity klystron amplifier. *J. W. Gewartowski and H. A. Watson, "Principles of Electron Tubes," 1965: p. 296. Courtesy of D. Van Nostrand Company, Inc.*

Several variations of the basic klystron exist. Of these, the simplest is the 2-cavity amplifier or oscillator. The most important is the reflex klystron, used as a low-power oscillator. The multicavity high-power amplifier is also important.

Two-Cavity Klystron Amplifiers: An electron beam is formed in an electron gun and passed through the gaps associated with the two cavities (Fig. 12). After emerging from the second gap, the electrons pass to a collector designed to dissipate the remaining beam power without the production of secondary electrons. In the first gap, the electron beam is alternately accelerated and decelerated in succeeding half-periods of the radio-frequency cycle, the magnitude of the change in speed depending on the magnitude of the alternating voltage impressed on the cavity. The electrons then move in a drift space where there are no radio-frequency fields. Here, the electrons that were accelerated in the input gap during one half-cycle catch up with those that were decelerated in the preceding half-cycle, and a local increase of current density occurs in the beam. Analysis shows that the maximum of the current-density wave occurs at the position, in time and space, of those electrons that passed the center of the input gap as the field changed from negative to positive. There is therefore a phase difference of $\pi/2$ between the current wave and the voltage wave that produced it. Thus at the end of the drift space, the initially uniform electron beam has been altered into a beam showing periodic density variations. This beam now traverses the output gap and the variations in density induce an amplified voltage wave in the output circuit, phased so that the negative maximum corresponds with the phase of the bunch center. The increased radio-frequency energy has been gained by conversion from the direct-current beam energy.

The 2-cavity amplifier can be made to oscillate by providing a feedback loop from the output to

the input cavity, but a much simpler structure results if the electron beam direction is reversed by a negative electrode, termed the reflector.

Reflex Klystrons*: A schematic diagram of a reflex klystron is shown in Fig. 13. The velocity-modulation process takes place as before, but analysis shows that in the retarding field used to reverse the direction of electron motion, the phase of the current wave is exactly opposite to that in the 2-cavity klystron. When the bunched beam returns to the cavity gap, a positive field extracts maximum energy from the beam, since the direction of electron motion has now been reversed. Consideration of the phase conditions shows that for a fixed cavity potential, the reflex klystron will oscillate only near certain discrete values of reflector voltage for which the transit time measured from the gap center to the reflection point and back is given by

$$\omega\tau = 2\pi\left(N + \tfrac{3}{4}\right)$$

where N is an integer called the mode number.

By varying the reflector voltage around the value corresponding with the mode center, it is possible to vary the oscillation frequency by a small percentage. This fact is made use of in providing automatic frequency control or in frequency-modulation transmission.

Reflex-Klystron Performance Data: The performance data for a reflex klystron are usually given in the form of a reflector-characteristic chart. This

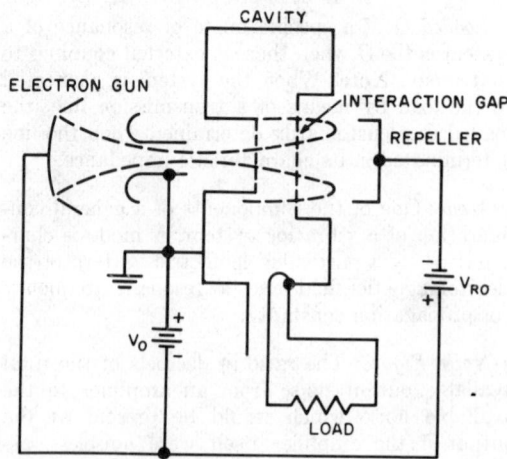

Fig. 13—Schematic of reflex klystron with power supply. *J. W. Gewartowski and H. A. Watson, "Principles of Electron Tubes," 1965: p. 311. Courtesy of D. Van Nostrand Company, Inc.*

* J. R. Pierce and W. G. Shepherd, "Reflex Oscillators," *Bell System Technical Journal,* Vol. 26, pp. 460–681; July 1947.

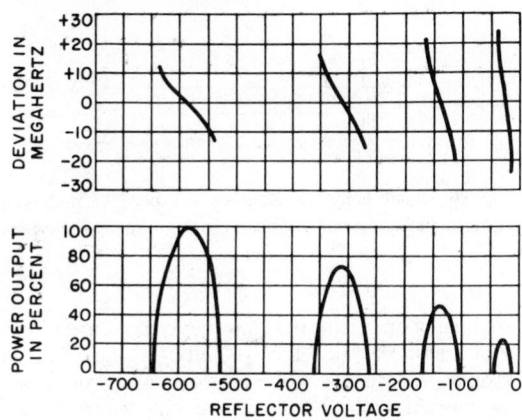

Fig. 14—Klystron reflector characteristic chart. *Courtesy of Sperry Gyroscope Company.*

chart displays power output and frequency deviation as a function of reflector voltage. Several modes are often displayed on the same chart. A typical chart is shown in Fig. 14.

There are two rather distinct classes of reflex klystrons in current large-scale manufacture (Table 11).

(A) Low-power tubes suitable for use in local-oscillator service, maser pumping, antenna-pattern testing, or similar applications. These tubes have power outputs in the range of 10 milliwatts to 1 to 2 watts. Typically, for local-oscillator service, a power of 10 to 100 milliwatts is necessary to operate crystal mixers with the required degree of isolation. For such applications as antenna testing or pumping of cryogenically cooled masers, power

of 500 milliwatts to 2 watts is usually required. The electronic tuning range required is about 50 megahertz independent of center frequency, but the linearity of the Δf versus ΔV_r characteristic is relatively unimportant.

(B) Tubes as frequency modulators in microwave links. These usually require considerably greater power, up to about 10 watts, and the linearity of the Δf versus ΔV_r characteristic over a limited (for example 10-megahertz) excursion is of primary importance as this parameter determines the harmonic margins in the system. Second-harmonic margins of -96 decibels for deviations of 125 kilohertz have been observed; the third-harmonic margins are about -120 decibels.

Multicavity Klystrons: Multicavity klystrons* have been perfected for use in two rather different fields of application: applications requiring extremely high pulse powers and continuous-wave systems in which moderate powers (tens of kilowatts) are required. An example of the first application is a power source for nuclear-particle acceleration, while ultra-high-frequency television and troposcatter transmitters are examples of the latter.

A multicavity klystron amplifier is shown schematically in Fig. 15. The example shown has 3 cavities all coupled to the same beam. The radio-frequency input modulates the beam as before. The bunched beam induces an amplified voltage across the second cavity, which is tuned to the operating frequency. This amplified voltage remodulates the beam with a certain phase shift and the now more-strongly bunched beam excites a highly amplified wave in the output circuit.

TABLE 11—CLASSES OF REFLEX KLYSTRONS.

Frequency (MHz)	Power Output (mW)	Useful Mode Width Δf_{3db} (MHz)	Operating Voltage
Local oscillators			
3 000	150	40	300
9 000	40	40	350
24 000	35	120	750
35 000	>15	50	2000
50 000	10–20	60–140	600
Maser pumps			
35 000	500–1500	70	2000
45 000	500–1000	80	2000
Frequency-modulation transmitters			
4 000	10 000	40	1100
7 000	10 000	37	750
9 000	6 000	60	500

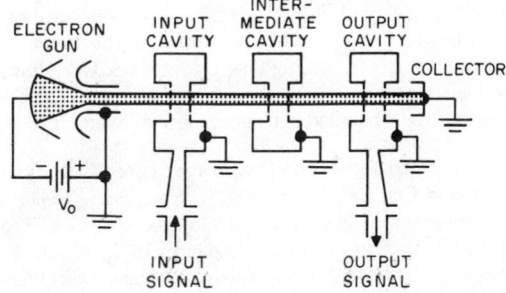

Fig. 15—Three-cavity klystron amplifier. *J. W. Gewartowski and H. A. Watson, "Principles of Electron Tubes," 1965: p. 340. Courtesy of D. Van Nostrand Company, Inc.*

* M. Chodorow, E. L. Ginzton, I. R. Neilson, and S. Sonkin, "Design and Performance of a High-Power Pulsed Klystron," *Proceedings of the IRE*, Vol. 41, pp. 1584–1602; November 1953. D. H. Priest, C. E. Murdock, and J. J. Woerner, "High-Power Klystrons at U.H.F.," *Proceedings of the IRE*, Vol. 41, pp. 20–25; January 1953.

It is found that the optimum power output is obtained when the second cavity is slightly de-tuned. Moreover, when increased bandwidth is required, the second cavity may be loaded with a resultant lowering in overall gain. Modern multi-cavity klystrons use magnetically focused high-perveance beams, and under these conditions high gains, large power outputs, and reasonable values of efficiency are readily obtained.

Continuous-wave multicavity klystrons are available with outputs of around 10 kilowatts at frequencies up to 5000 megahertz. The efficiencies are of the order of 30 percent and the gains vary between 20 and 50 decibels, according to the number of cavities, bandwidth, et cetera. Pulsed tubes have been designed for outputs of 30 megawatts and with efficiencies of over 40 percent at frequencies near 3000 megahertz.

Traveling-Wave Tubes

The traveling-wave tube* differs from the klystron in that the radio-frequency field is not confined to a limited region but is distributed along a wave-propagating structure. A longitudinal electron beam interacts continuously with the field of a wave traveling along this wave-propagating structure. In its most common form it is an amplifier, although there are related types of tubes that are basically oscillators.

The principle of operation may be understood by reference to Fig. 16. An electron stream is produced by an electron gun, travels along the axis of the tube, and is finally collected by a suitable electrode. Spaced closely around the beam is a circuit, in this case a helix, capable of propagating a slow wave. The circuit is proportioned so that the phase velocity of the wave is small with respect to the velocity of light. In typical low-power tubes, a value of the order of one-tenth of the velocity of light is used; for higher-power tubes the phase velocity may be two or three times higher. Suitable means are provided to couple an external radio-

* J. R. Pierce, "Traveling-Wave Tubes," D. Van Nostrand Co., New York; 1950. R. Kompfner, "Reports on Progress in Physics," Vol. 15, pp. 275–327, The Physical Society, London, England; 1952. R. G. E. Hutter, "Traveling-Wave Tubes," Advances in Electronics and Electron Physics, Vol. 6, Academic Press, New York; 1954. A bibliography is given in a survey paper by J. R. Pierce, "Some Recent Advances in Microwave Tubes," *Proceedings of the IRE*, Vol. 42, pp. 1735–1747; December 1954. S. Sensiper, "Electromagnetic Wave Propagation on Helical Structures," *Proceedings of the IRE*, Vol. 43, pp. 149–161; February 1955. A. H. W. Beck, "Space-Charge Waves and Slow Electromagnetic Waves," Pergamon Press, New York; 1958. D. A. Watkins, "Topics in Electromagnetic Theory," John Wiley & Sons, New York; 1958.

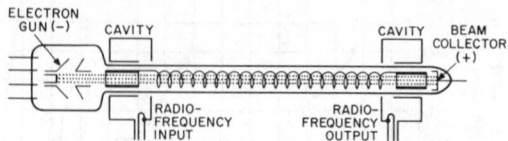

Fig. 16—Basic helix traveling-wave tube. The magnetic beam-focusing system between input and output cavities is not shown.

frequency circuit to the slow-wave structure at the input and output. The velocity of the electron stream is adjusted to be approximately the same as the axial phase velocity of the wave on the circuit.

When a wave is launched on the circuit, the longitudinal component of its field interacts with the electrons traveling in approximate synchronism with it. Some electrons will be accelerated and some decelerated, resulting in a progressive rearrangement in phase of the electrons with respect to the wave. The electron stream, thus modulated, in turn induces additional waves on the helix. This process of mutual interaction continues along the length of the tube with the net result that direct-current energy is given up by the electron stream to the circuit as radio-frequency energy, and the wave is thus amplified.

By virtue of the continuous interaction between a wave traveling on a broad-band circuit and an electron stream, traveling-wave tubes do not suffer the gain-bandwidth limitation of ordinary types of electron tubes. By proper circuit design, such tubes are made to have bandwidths of an octave in frequency, and even more in special cases.

The helix is an extremely useful form of slow-wave circuit because the impedance that it presents to the wave is relatively high and because, when properly proportioned, its phase velocity is almost independent of frequency over a wide range.

An essential feature of this type of tube is the approximate synchronism between the electron stream and the wave. For this reason, the traveling-wave tube will operate correctly over only a limited range in voltage. Practical considerations require that the operating voltages be kept as low as is consistent with obtaining the necessary beam input power; the voltage, in turn, dictates the phase velocity of the circuit. The electron velocity v in centimeters/second is determined by the accelerating voltage V in accordance with

$$v = 5.93 \times 10^7 V^{1/2}.$$

Figure 17 shows a typical relationship between gain and beam voltage. The gain G of a traveling-wave tube is given approximately by

$$G = A + BCN$$

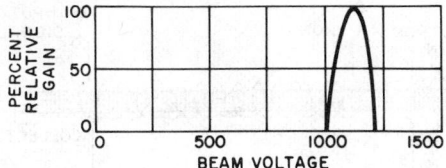

Fig. 17—Traveling-wave-tube gain versus accelerating voltage.

in decibels, where $A=$ the initial loss due to the establishment of the modes on the helix and lies in the range from -6 to -9 decibels, $B=$ a gain coefficient that accounts for the effect of circuit attenuation and space charge, $C=$ a gain parameter that depends on the impedances of the circuit and the electron stream $= [(E^2/(\omega/v)^2 P) \times (I_0/8V_0)]^{1/3}$, $I_0=$ beam current, $V_0=$ beam voltage, $N=$ number of active wavelengths in tube $= (l/\lambda_0)(c/v)$, $l=$ axial length of the helix, $\lambda_0=$ free-space wavelength, $v=$ phase velocity of wave along tube, and $c=$ velocity of light. The term $E^2/(\omega/v)^2 P$ is a normalized wave impedance that may be defined in a number of ways.

In practice, the attenuation of the circuit will vary along the tube, and consequently the gain per unit length will not be constant. The total gain will be a summation of the gains of various sections of the tube.

Commonly, C is of the order of 0.02 to 0.2 in helix traveling-wave tubes. The gain of low- and medium-power tubes varies from 20 to 70 decibels with 30 decibels being a common value. The gain in a tube designed to produce appreciable power will vary somewhat with signal level when the beam voltage is adjusted for optimum operation. Figure 18 shows a typical characteristic.

To restrain the physical size of the electron steam as it travels along the tube, it is necessary to provide a focusing field, either magnetic or electrostatic, of a strength appropriate to overcome the space-charge forces that would otherwise cause

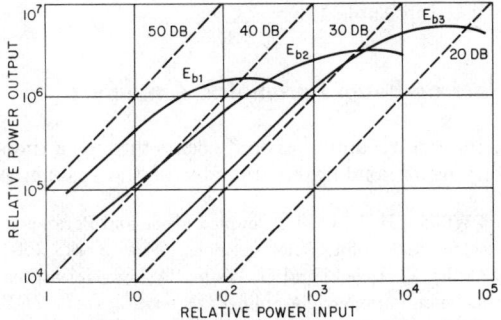

Fig. 18—Gain of traveling-wave tube as a function of input level and beam voltage. $E_{b1} < E_{b2} < E_{b3}$.

the beam to spread. Until fairly recently a longitudinal magnetic field supplied by a solenoid electromagnet was used for this purpose. Continuing demands for improved efficiency and reliability, and for weight and size reduction, however, have forced the development of permanent magnet-type focusing structures. At the present time, focusing by periodically reversing magnetic fields produced by permanent magnet structures is rapidly becoming predominant.

Several techniques for electrostatic containment of the electron stream have also been developed. Typical is the use of a bifilar helix slow-wave structure where an appropriate voltage difference between the helices provides, in effect, a distributed Einzel lens. Because of voltage breakdown problems as well as increased power supply requirements, electrostatic focusing has not yet proved practical for use in linear beam tubes.

Other types of slow-wave circuits in addition to the helix are possible, including a number of periodic structures. In general, such designs are capable of operation at higher power levels but at the expense of bandwidth.

Traveling-Wave-Tube Performance Data

Traveling-wave tubes are designed to emphasize particular inherent characteristics for specific applications. Three general classes are distinguished,

Low-Noise Amplifiers: Tubes of this class are intended for the first stage of a receiver and are proportioned to have the best possible noise figure. This requires that the random variations in the electron stream be minimized and that steps be taken also to minimize partition noise. Tubes have been made for commercial use with noise figures as low as 3 decibels in S-band and 6 decibels or less over the entire range from 1000 to 12 000 megahertz. Gains of from 20 to 35 decibels are typical. The maximum power output is generally not more than a few milliwatts. Performance of this order can be achieved with either permanent-magnet or electromagnet focusing structures. Recently a new class of tubes has been developed which offers medium-power performance (1 to 5 watts) with reasonable low-noise performance (noise figures of 10 to 14 decibels).

Intermediate Power Amplifiers: These tubes are intended to provide power gain under conditions where neither noise nor large values of power output are important. Gains of 30 or more decibels are customary and the maximum output power is usually in the range from 100 milliwatts to 1 watt.

Power Amplifiers: For this class of tubes, the application is usually the output stage of a transmitter; the power output, either continuous-wave or pulsed, is of primary importance. Much active development continues in this area and the values of power obtainable are steadily increasing. At present, continuous-wave powers range from tens of kilowatts in the ultra-high-frequency region to more than 100 watts at 10 000 megahertz. Tubes especially designed for pulsed operation provide considerably higher power. Several megawatts of peak power have been achieved at 3000 megahertz. Efficiencies in excess of 30 percent have been obtained and this may be further enhanced by recently developed collector-depression techniques. Power gains of 30 to 50 decibels are normal.

Backward-Wave Oscillators

A member of the traveling-wave-tube family, the O-type backward-wave oscillator,* makes use of the interaction of the electron stream with a radio-frequency circuit wave whose phase and group velocities are 180° apart. The group velocity, and thus the direction of energy flow, is directly opposed to the direction of electron motion. Figure 19 shows schematically a backward-wave tube with connection to both ends of the slow-wave structure, so that operation as either oscillator or amplifier could be achieved. An electron beam is produced by the electron gun, traverses the slow-wave structure, and is dissipated in the collector structure. During its transit the beam is confined by a longitudinal magnetic field. With a beam current of sufficient magnitude, the beam-structure interaction will produce oscillations and microwave power will be delivered from the end of the structure adjacent to the electron gun. At beam-current levels below the "start-oscillation" value, a radio-frequency signal may be introduced at the collector end of the device and the tube will operate as an amplifier.

To improve interaction efficiency, electron beams with hollow cross sections are usually used. This places all the electrons as close as possible to the slow-wave structure in the region of maximum radio-frequency field. The reason here is that the strength of the −1 space harmonic field goes to zero on the axis. To produce this hollow-cross-section beam it is necessary to use magnetically confined electron flow from the cathode, and thus the electron gun is entirely immersed in the magnetic field.

* H. Heffner, "Analysis of the Backward-Wave Traveling-Wave Tube," *Proceedings of the IRE,* Vol. 42, pp. 930–937; June 1954. A. H. W. Beck, "Space-Charge Waves and Slow Electromagnetic Waves," Pergamon Press, New York; 1958: pp. 241–255. R. Kompfner and N. T. Williams, "Backward-Wave Tubes," *Proceedings of the IRE,* Vol. 41, pp. 1602–1611; November 1953.

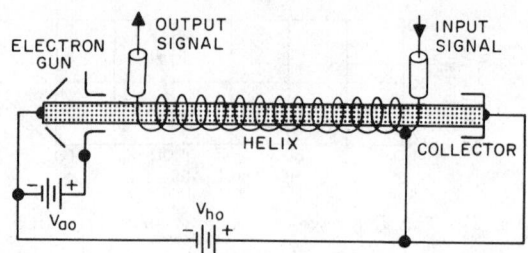

Fig. 19—A traveling-wave tube in operation as a backward-wave amplifier. A separate power supply connected to the anode permits beam-current control independent of the helix voltage. *J. W. Gewartowski and H. A. Watson, "Principles of Electron Tubes," 1965: p. 398. Courtesy of D. Van Nostrand Company, Inc.*

O-type backward-wave devices are voltage tunable, with the frequency being porportional to the $\frac{1}{2}$ power of the cathode-helix voltage as well as dependent on the dimensions of the structure. Typically, tuning over full octave range is possible and in special cases a range of 2 or more octaves can be achieved. However, where confined limits are desired on power variation or other special characteristics, more-restricted frequency ranges may be necessary. Where full octave tuning is used, power output variation of 6 to 10 decibels across the range is usual. In most cases a separate control element in the gun permits adjustment of beam-current amplitude and thus provides control of power output. Oscillators of this type have very low pulling figures but the pushing figure is often substantial. Frequency stability is generally excellent, with the achievable value normally depending on power supply capabilities rather than inherent tube limitations.

O-type backward-wave oscillators are generally low-power devices, with 10 to 50 milliwatts being typical. However, in the range from 1 to 4 gigahertz up to several hundred milliwatts is feasible, while in the range from 50 to 100 gigahertz 5 to 10 milliwatts is relatively difficult to achieve reliably. Typical performance for low-power helix-type permanent-magnet-focused backward-wave oscillators is listed in Table 12.

Electron-Beam Parametric Amplifiers

Parametric amplification* occurs through a time-varying or nonlinear parameter of the system. A

* William H. Louisell, "Coupled Mode and Parametric Electronics," John Wiley & Sons, New York; 1960. R. Adler, G. Hrbek, and G. Wade, "A Low-Noise Electron-Beam Parametric Amplifier," *Proceedings of the IRE,* Vol. 46, pp. 1756–1757; October 1958. T. J. Bridges and A. Ashkin, "A Microwave Adler Tube," *Proceedings of the IRE,* Vol. 48, pp. 361–363; March 1960.

TABLE 12—PERFORMANCE OF TYPICAL LOW-POWER BACKWARD-WAVE OSCILLATORS.

Frequency Range (GHz)	Tuning Voltage (V)	Cathode Current (mA)	Minimum Power Output (mW)
1.0–2.0	250–1150	15	100
2.0–4.0	300–1800	10	100
4.0–8.0	250–2400	12	25
5.3–11.0	245–2400	10	25
8.0–12.4	550–2400	10	25

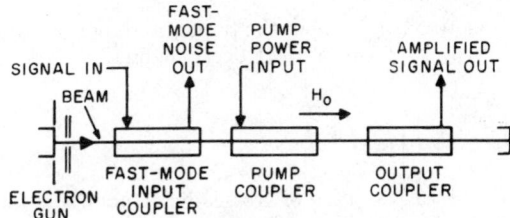

Fig. 20—Block diagram of O-type parametric amplifier. *R. Adler, G. Hrbek, and G. Wade, "A Low-Noise Electron-Beam Parametric Amplifier," Proceedings of the IRE, Vol. 46, p. 1756; October 1958.*

simple mechanical example is a child pumping up a swing. He gives energy to the swing amplitude by raising and lowering his center of gravity at twice the swing frequency. The time-varying parameter is the effective length of the swing.

Parametric amplifiers differ from usual amplifiers in the frequency of the energy source; for parametric amplifiers the energy comes from a high-frequency source whereas for the usual amplifier it comes from a direct-current source. There normally are 3 frequencies associated with parametric amplifiers: the signal frequency f_s; the energy source or pump frequency f_p; and the difference or idler frequency $f_i = f_p - f_s$. Although no energy is supplied at f_i, a circuit must be provided to support this frequency. Four-frequency parametric amplifiers, where the fourth frequency is a second idler or a second pump frequency, have been made. The advantage of this device is that the pump frequency (frequencies) can be lower than the signal frequency. The disadvantage is that a fourth circuit must be provided. Two-frequency (degenerate) parametric amplifiers, in which $f_p = 2f_s$ so that $f_i = f_s$, are the common form of the electron-beam devices. (The child in the swing is a version of the degenerate parametric amplifier.) The advantage of the degenerate parametric amplifier is that only 2 circuits need to be provided—1 for the signal frequency and 1 for the pump frequency. The disadvantage is that the phasing between the signal and pump must be adjusted precisely. In each of the foregoing cases, the ratios of the powers at the several frequencies are governed by the Manley–Rowe relations.*

The parametric amplifier's principal virtue is its low-noise behavior. It has this feature in common with the maser to which it bears a superficial resemblance.

Electron-beam parametric amplifiers are, for the most part, linear beam (O-type) devices, although crossed-field (M-type) devices have been built.* The usual O-type amplifier, such as the traveling-wave tube, uses the coupling between a circuit and the slow space-charge wave on the beam to obtain gain. Since this wave carries negative kinetic power, it is not possible to couple the noise on this wave out of the beam. For parametric amplification, it is necessary to use either the fast space-charge wave or the fast cyclotron wave. Because these waves carry positive kinetic power, it is possible (theoretically) to couple all the noise on these waves out of the beam.

A block diagram of an O-type parametric amplifier is shown in Fig. 20. A microwave tube version of this block diagram is shown in Fig. 21. This device has 20-decibel gain at 4140 megahertz with a bandwidth of 67 megahertz and a double-channel noise figure of 2.4 decibels. The tube uses the fast

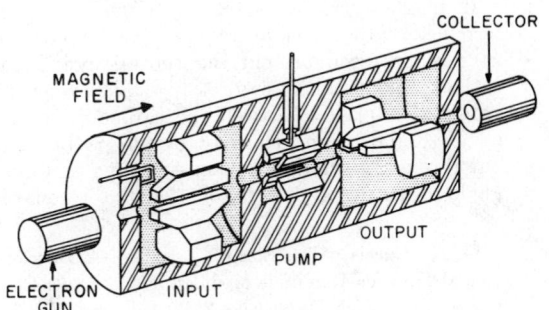

Fig. 21—Cross section of a microwave O-type parametric amplifier. *T. J. Bridges and A. Ashkin, "A Microwave Adler Tube," Proceedings of the IRE, Vol. 48, p. 362; March 1960.*

* J. M. Manley and H. E. Rowe, "Some General Properties of Nonlinear Elements—Part I, General Energy Relations," *Proceedings of the IRE*, Vol. 44, pp. 904–913; July 1956.

* J. Wilhelm Klüver, "A Low Noise M-Type Parametric Amplifier," *IEEE Transactions on Electron Devices*, Vol. ED-11, pp. 205–215; May 1964.

cyclotron wave, and the input and output are microwave forms of the Cuccia coupler.*

Crossed-Field Tubes

The earliest type of crossed-field tube† was the magnetron oscillator. The carcinotron oscillator and the crossed-field amplifier have been developed more recently. Crossed-field tubes generally operate with higher conversion efficiencies than linear beam devices, making them especially attractive for high-power applications.

Magnetrons

A magnetron‡ is a high-vacuum tube containing a cathode and an anode, the latter usually divided into two or more segments. A constant magnetic field modifies the space-charge distribution and the current-voltage relations. In modern usage, the term "magnetron" refers to the magnetron oscillator in which the interaction of the electronic space charge with the resonant system converts direct-current power into alternating-current power, usually at microwave frequencies.

Many forms of magnetrons have been made in the past and several kinds of operation have been employed. The type of tube that is now almost universally employed is the multicavity magnetron generating traveling-wave oscillations. It possesses the advantages of good efficiency at high frequencies, capability of high outputs either in pulsed or continuous-wave operation, moderate magnetic-field requirements, and good stability of operation. A section through the basic anode structure of a typical magnetron is shown in Fig. 22.

In magnetrons, the operating frequency is determined by the resonant frequency of the separate cavities arranged around the central cylindrical cathode and parallel to it. A high direct-current potential is placed between the cathode and the cavities, and radio-frequency output is brought out through a suitable transmission line or waveguide usually coupled to one of the resonator cavities. Under the action of the radio-frequency voltages across these resonators and the axial magnetic field, the electrons from the cathode form a bunched space-charge cloud that rotates around the tube axis, exciting the cavities and maintaining their radio-frequency voltages.

Magnetron Performance Data

The performance data for a magnetron are usually given in terms of two diagrams, the performance chart and the Rieke diagram.

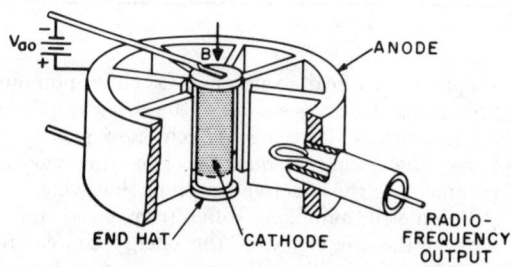

Fig. 22—Magnetron oscillator. *J. W. Gewartowski and H. A. Watson, "Principles of Electron Tubes," 1965: p. 428. Courtesy of D. Van Nostrand Company, Inc.*

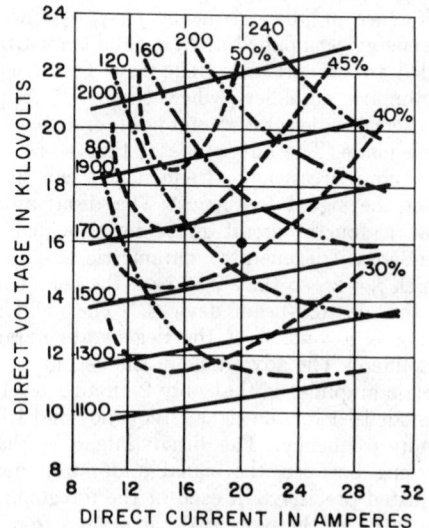

CONSTANT FIELD IN GAUSSES
CONSTANT OUTPUT IN KILOWATTS
CONSTANT OVERALL EFFICIENCY
TYPICAL OPERATING POINT

Fig. 23—Performance chart for pulsed magnetron. *Courtesy of Bell System Technical Journal.*

* C. L. Cuccia, "The Electron Coupler—A Developmental Tube for Amplitude Modulation and Power Control at Ultra-High Frequencies," *RCA Review*, Vol. 10, pp. 270–303; June 1949.

† "Crossed-Field Microwave Devices," E. Okress, Editor, Academic Press, New York; 1961.

‡ G. B. Collins, "Microwave Magnetrons," Vol. 6, Radiation Laboratory Series, 1st edition, McGraw-Hill Book Company, New York; 1948. J. B. Fisk, H. D. Hagstrum, and P. L. Hartman, "The Magnetron as a Generator of Centimeter Waves," *Bell System Technical Journal*, Vol. 25, pp. 167–348; April 1946.

Performance Chart: This is a plot of anode current along the abscissa and anode voltage along the ordinate of rectangular-coordinate paper. For a fixed typical tube load, pulse duration, pulse-repetition rate, and setting of the tuner of tunable tubes; lines of constant magnetic field, power output, efficiency, and frequency may be plotted over the complete operating range of the tube. Regions of unsatisfactory operation are indicated by cross hatching. For tunable tubes, it is customary to show performance charts for more than one setting of the tuner. In the case of magnetrons with attached magnets, curves showing the variation of anode voltage, efficiency, frequency, and power output with change in anode current are given. A typical chart for a magnetron having 8 resonators is given in Fig. 23.

Rieke Diagram: This shows the variation of power output, anode voltage, efficiency, and frequency with changes in the voltage standing-wave ratio and phase angle of the load for fixed typical operating conditions such as magnetic field, anode current, pulse duration, pulse-repetition rate, and the setting of the tuner for tunable tubes. The Rieke diagram is plotted on polar coordinates, the radial coordinate being the reflection coefficient measured in the line joining the tube to the load, and the angular coordinate being the angular distance of the voltage standing-wave minimum from a suitable reference plane on the output terminal. On the Rieke diagram, lines of constant frequency, anode voltage, efficiency, and output may be drawn (Fig. 24).

Magnetrons for pulsed operation have been built to deliver peak powers ranging from 3 megawatts at 3000 megahertz to 100 kilowatts at 30 000

megahertz. Continuous-wave magnetrons having outputs ranging from 1 kilowatt at 3000 megahertz to a few watts at 30 000 megahertz have been produced. Operation efficiencies up to 60 percent at 3000 megahertz are obtained, falling to 30 percent at 30 000 megahertz.

Carcinotron

The carcinotron is an M-type backward-wave oscillator in which the electron stream traverses the tube and interacts with the fields on the slow-wave structure under conditions where the electric and magnetic fields are perpendicular to each other. Figure 25 shows schematically a linear version of the carcinotron. In the electron gun, current is drawn from the cathode when the accelerator voltage is applied. Because of the presence of the magnetic field, directed as shown, the electron paths are curved approximately 90° so that they enter the interaction region between the slow-wave structure and the sole. If the voltages and the magnetic field strength are proper, the electrons will travel along a path approximately parallel to the structure until they reach the collector.

Although Fig. 25 shows a linear arrangement, carcinotrons are conventionally designed in a circular arrangement to conserve magnet size and weight. In this arrangement, the sole approximates the appearance of the cathode of a magnetron and the slow-wave structure is in the position of the magnetron anode, but neither the sole, nor the structure is re-entrant.

The carcinotron performance is similar to that of the O-type backward-wave oscillator but it offers several of the advantages of crossed-field devices. High-efficiency operation is possible with

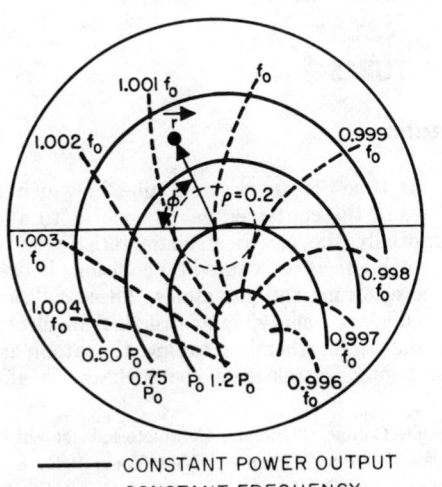

—— CONSTANT POWER OUTPUT
- - - - CONSTANT FREQUENCY

Fig. 24—Rieke diagram. *Courtesy of Bell System Technical Journal.*

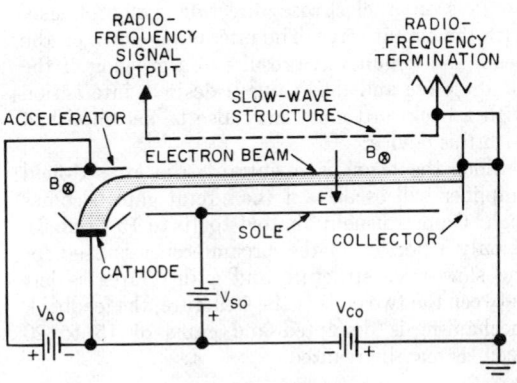

Fig. 25—Linear version of an M-carcinotron oscillator. *J. W. Gewartowski and H. A. Watson, "Principles of Electron Tubes," 1965: p. 459. Courtesy of D. Van Nostrand Company, Inc.*

values of 20 to 30 percent being readily obtained. This efficiency capability makes the carcinotron useful as a high-power device with continuous-wave capabilities of hundreds of watts through X-band. Its construction is such as to permit direct scaling to very-high frequencies, with several milliwatts of power having been achieved at frequencies beyond 300 gigahertz.

The carcinotron, like the O-type backward-wave oscillator, is voltage-tunable with the oscillation frequency being approximately directly proportional to the cathode slow-wave-structure voltage. This linear relationship simplifies the associated electronic tuning circuit considerably. Frequency pushing is also considerably lower than in O-type backward-wave oscillators. The M-type carcinotron has the disadvantage, however, that it is relatively noisy, with spurious power output often not more than 10 to 15 decibels below the main signal output.

In addition to obvious usage as high-power tunable signal sources, carcinotrons with their high noise may be used as electronic countermeasure jamming sources.

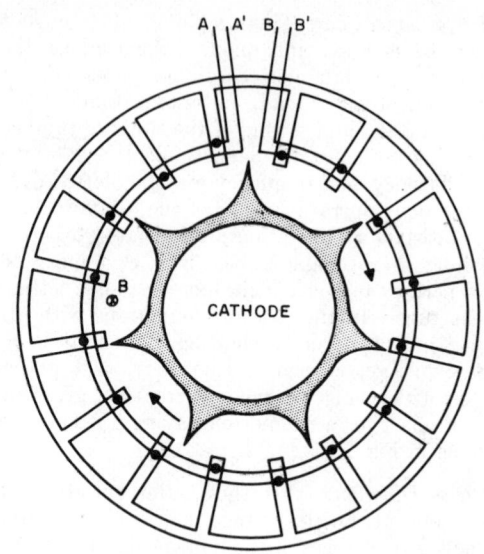

Fig. 26—Schematic drawing of a crossed-field amplifier. *J. W. Gewartowski and H. A. Watson, "Principles of Electron Tubes," 1965: p. 449. Courtesy of D. Van Nostrand Company, Inc.*

Crossed-Field Amplifiers

Crossed-field amplifiers* as a general class of tubes are mechanically quite similar to the crossed-field oscillator or magnetron. As may be seen from Fig. 26, the major difference is that the slow-wave structure is not re-entrant whereas in the magnetron both the beam and the circuit are re-entrant.

Referring to Fig. 26, voltage and magnetic field are applied as for the magnetron. A radio-frequency signal is applied to the structure and progresses in a clockwise direction toward the output terminal. Current spokes, produced in the cathode-circuit region by the radio-frequency electric fields, also progress in a clockwise direction synchronously with the circuit wave. The interaction between the beam and circuit wave results in a growing of the circuit wave and thus gain. If desired, interaction with a backward mode may also be accomplished with this device.

Since the beam is re-entrant, the crossed-field amplifier will oscillate if the circuit gain becomes high. Gain is usually limited to 10 to 15 decibels. If only a portion of the circumference is used for the slow-wave structure and a drift area is left between the two ends of the structure, the feedback mechanism is disrupted and gains of 15 to 20 decibels may be realized.

The power output of the crossed-field amplifier is essentially independent of the radio-frequency drive signal and it thus operates as a saturated amplifier. This characteristic makes it unsuitable for amplifying amplitude-modulated signals.

Crossed-field amplifiers offer the advantage of relatively high efficiency, 40 to 60 percent or even higher, and they may be designed to provide very high peak output powers. Their disadvantages are their low gain, limited bandwidth, high noise, and saturated-amplifier characteristic.

GAS TUBES

Ionization

A gas tube* is an electron tube in which the pressure of the contained gas is such as to affect substantially the electrical characteristics of the tube. Such effects are caused by collisions between moving electrons and gas atoms. These collisions, if of sufficient energy, may dislodge an electron from the atom, thereby leaving the atom as a positive ion. The electron space charge is effec-

---

* W. C. Brown, "Description and Operating Characteristics of the Platinotron—A New Microwave Tube Device," *Proceedings of the IRE*, Vol. 45, pp. 1209–1222; September 1957.

* J. D. Cobine, "Gaseous Conductors," 1st edition, McGraw-Hill Book Company, New York; 1941. E. H. Kennard, "Kinetic Theory of Gases," McGraw-Hill Book Company, New York; 1938: p. 149. L. B. Loeb, "Basic Processes of Gaseous Electronics," University of California Press, Berkeley, California; 1955.

TABLE 13—IONIZATION PROPERTIES OF GASES.

Gas	Ionization Energy (volts)	Collision Probability P_c
Helium	24.5	12.7
Neon	21.5	17.5
Nitrogen	16.7	37.0
Hydrogen (H_2)	15.9	20.0
Argon	15.7	34.5
Carbon monoxide	14.2	23.8
Oxygen	13.5	34.5
Krypton	13.3	45.4
Water vapor	13.2	55.2
Xenon	11.5	62.5
Mercury	10.4	67.0

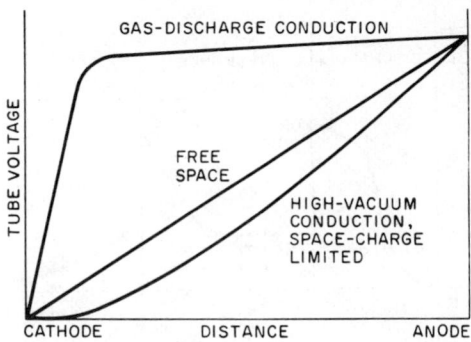

Fig. 27—Voltage distribution between plane parallel electrodes showing effect of space-charge neutralization in a hot-cathode gas tube.

tively neutralized by these positive ions and comparatively high free-electron densities are easily created.

Table 13 gives the energy in electronvolts necessary to produce ionization. The column P_c is the kinetic-theory collision probability per centimeter of path length for an electron in a gas at $15°$ Celsius at a pressure of 1 millimeter of mercury. The collision frequency is given by

$$f_c = vP_c\rho$$

where f_c = collisions per second, P_c = collision probability in collisions per centimeter per millimeter of pressure, and ρ = gas pressure in millimeters of mercury.

Characteristics of Gas Tubes

Gas tubes may be generally divided into two classes, depending on whether the cathode is hot or cold and thus on the mechanism by which electrons are supplied.

Hot-Cathode Gas Tubes: The electrons in the hot-cathode gas tube are produced thermionically. The voltage drop across such tubes is that required to produce ionization of the gas and is generally a few tens of volts. The current conducted by the tube depends primarily on the emission capability of the cathode. Figure 27 shows the effect of the ionized gas on the voltage distribution in a hot-cathode tube.

Cold-Cathode Gas Tubes: The electrons in a cold-cathode tube are produced by bombardment of the cathode by ions and/or by the action of a localized high electric field. The voltage drop across such a tube is higher than in the hot-cathode tube because of this mechanism of electron generation, and the current which can flow is limited. Figure 28 shows the effect of tube geometry and gas pressure on the voltage required to initiate the discharge.

Figure 29 shows a typical volt-ampere characteristic of a cold-cathode discharge. Cold-cathode gas tubes may be divided into two categories, depending on the region of this characteristic in which they operate.

Glow Discharge Tubes require a drop of several hundred volts across the tube and operate in region II. The current is of the order of tens of milliamperes.

Arc Discharge Tubes operate in region III. They are not, strictly speaking, cold-cathode tubes since the current is drawn from a localized spot on the cathode which is consequently heated and provides a large thermionic current. The voltage drop is thus lowered. Such a tube is capable of conducting currents of thousands of amperes at voltage drops of tens of volts. Mercury-pool cathodes are used

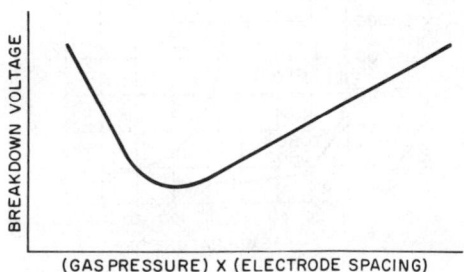

Fig. 28—Effect of gas pressure and tube geometry on gap voltage required for breakdown in a cold-cathode gas tube.

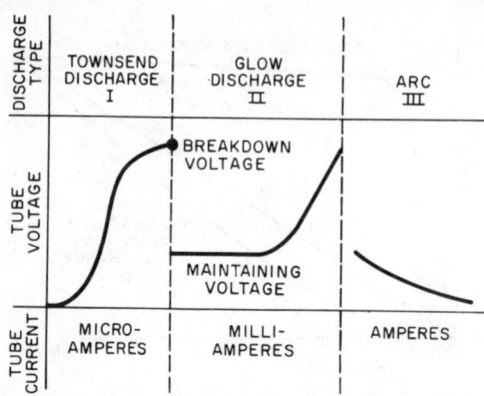

Fig. 29—Typical volt-ampere characteristic of cold-cathode gas discharge.

in one common form of arc discharge tube, supplying the electron current from an arc spot on the mercury-pool surface. The mercury vapor evaporated from the surface provides the gas atmosphere which is ionized.

Power Applications of Gas Tubes

Power Rectifier and Control Tubes: Mercury-vapor rectifiers, thyratrons, and ignitrons employ the very high current-carrying capacity of gas discharge tubes with low power losses for rectification and control in high-power equipment. The operation of mercury-vapor tubes depends on temperature insofar as tube voltage drop and peak inverse voltages are concerned (Fig. 30).

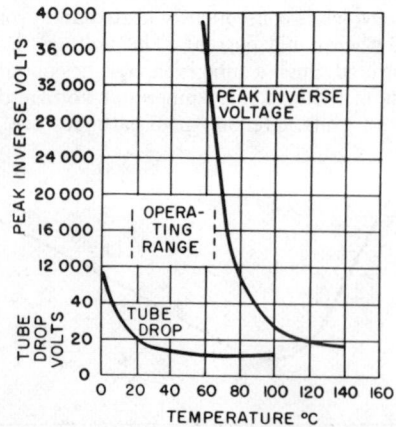

Fig. 30—Tube drop and arc-back voltages as a function of the condensed mercury temperature in a hot-cathode mercury-vapor tube. *Courtesy of McGraw-Hill Book Company.*

Hydrogen Thyratrons are hot-cathode hydrogen-filled triodes designed for use as electronic switching devices where short anode delay time is important. In pulsing service they are capable of switching tens of megawatts at voltages of tens of kilovolts. Anode delay time and time jitter are in the nanosecond range, and the tubes do not depend on ambient temperature for proper operation. Hydrogen thyratrons are also used in crowbar applications to protect other circuit components against fault voltages or currents and are capable of handling peak currents of several-thousand amperes.

Triggered Spark Gaps are cold-cathode gas tubes operating in the arc discharge region III. The gaps contain two high-power electrodes and a trigger electrode which is generally fired through a step-up pulse transformer by a simple low-energy pulse. The gaps are used as electronic switching devices for peak currents of tens of thousands of amperes and voltages of tens of kilovolts. They can discharge stored energies of several thousand joules and are used for energy transfer in exploding-bridge-wire circuits, gas plasma discharges, spark chambers, and Kerr cells. They are also used in crowbar applications for fast-acting protection of other circuit components against fault voltages and currents. Before conduction the gap presents a low capacitance and a very high impedance to the circuit. After triggering, when the gap is conducting, the impedance drops to a few ohms or less.

Voltage Regulators of the glow discharge type take advantage of the volt-ampere characteristic in region II, where the voltage is nearly independent of the current. They operate at milliamperes and up to a few hundred volts.

Voltage regulators of the corona discharge type operate at currents of less than a milliampere and at voltages up to several thousand volts.

Microwave Applications of Gas Tubes

Noise Sources: Gas discharge devices possess a highly stable and repeatable effective noise temperature when in the fired condition. This feature provides a convenient and accurate means for determining noise figure. The microwave energy radiated from a gas discharge plasma is coupled into a radio-frequency transmission line with which it is used. The amount of radio-frequency power available from a gas discharge tube depends mainly on the nature of the gas fill, the geometric characteristic of the discharge tube, and the electron temperature of the positive column or plasma. The design parameter which most strongly determines the noise temperature is the type of gas employed. Any of the noble gases may be used in a noise source. In practice, however, only two or three are normally used:

Gas	$F = \text{ENR (dB)}$
Helium	21.0
Neon	18.5
Argon	15.3

When referring to a noise source or generator, the ratio of its noise power output to thermal noise power is called the Excess Noise Ratio (ENR).

$$F = \text{ENR} = \frac{[(T_2/T_0)-1] - Y[(T_1/T_0)-1]}{Y-1}$$

where $Y =$ ratio of the noise output power of the receiver with the noise generator ON, to that with the noise generator OFF, $T_0 = 290$ K, $T_1 =$ temperature (in kelvins) of the termination, and $T_2 =$ effective noise temperature (in kelvins) of the noise generator in the fired condition. The expression $[(T_2/T_0)-1]$ is termed the excess noise power of the noise source. When $T_1 = T_0 = 290$ K

$$\text{ENR} = [(T_2/T_0)-1]/(Y-1)$$

$$\text{ENR (dB)} = 10 \log_{10} \text{ENR}.$$

The effective temperature of the noise source is equal to the temperature of the discharge only if the coupling of the transmission line to the discharge is complete. Otherwise there is a reduction in the noise power output which can best be determined by measuring the fired and unfired insertion loss of the unit at the frequency of interest. The relation between these factors is given by

$$[(T_e/T_0)-1]/[(T_2/T_0)-1] = 1-(L_u/L_f)$$

where $[(T_e/T_0)-1]$ is the effective excess noise power of the generator, $[(T_2/T_0)-1]$ is the excess noise power, and L_u and L_f are the insertion losses in the unfired and fired conditions, respectively. This correction should be subtracted from the apparent measured noise figure.

Noise figure is always measured with reference to a standard temperature of 290 K (T_0). If the ambient temperature (T_1) of the noise-generator termination differs from the standard temperature, the noise figure calculated must be corrected. To find the correction factor, substitute the ambient temperature of the noise-generator termination for T_1 in the following equation and add the temperature factor (F_T) to the noise figure calculated.

$$F_T = [Y/(Y-1)][(T_1/T_0)-1].$$

TR Tubes: Transmit–receive tubes are gas discharge devices designed to isolate the receiver section of radar equipment from the transmitter during the period of high power output. A typical TR tube and its circuit are illustrated in Fig. 31.

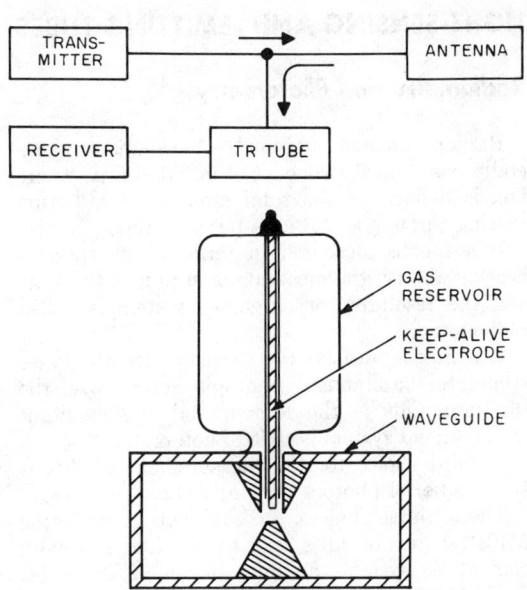

Fig. 31—Diagram of a TR tube and circuit.

The cones in the waveguide form a transmission cavity tuned to the transmitter frequency and the tube conducts received low-power-level signals from the antenna to the receiver. When the transmitter is operated, however, the high-power signal causes gas ionization between the cone tips, which detunes the structure and reflects all the transmitter power to the antenna. The receiver is protected from the destructively high level of power and all of the available transmitter power is useful output.

Microwave Gas Discharge Circuit Elements: Because of the high free-electron density, the plasmas of gas discharges are capable of strong interaction with electromagnetic waves in the microwave region. In general, microwave phase shift and/or absorption result. If used in conjunction with a magnetic field, these effects can be increased and made nonreciprocal. Phase shift is a result of the change in dielectric constant caused by the plasma according to

$$\epsilon_p/\epsilon_0 = 1 - (0.8 \times 10^{-4} N_0/f_s^2)$$

where $\epsilon_p =$ dielectric constant in plasma, $\epsilon_0 =$ dielectric constant in free space, $N_0 =$ electron density in electrons/centimeter^3, and $f_s =$ signal frequency in megahertz.

Absorption of microwave energy results when electrons, having gained energy from the electric field of the signal, lose this energy in collisions with the tube envelope or neutral gas molecules. This absorption is a maximum when the frequency of collisions is equal to the signal frequency and the absolute magnitude is proportional to the free-electron density.

LIGHT-SENSING AND -EMITTING TUBES

Radiometry and Photometry

Radiometric and photometric* systems are generally based on the concept of radiated flux, where flux is defined as the total amount of radiation passing through a unit area per unit time.

If a flux is measured in terms of its thermal heating ability, the most common unit is the watt and the resultant measurement system is called radiometry.

If a flux is measured in terms of its ability to stimulate the standard photopic human eye, the resultant unit is the lumen, and the resultant measurement system is called photometry.

A third choice for the measurement of flux is the number of photons per unit time.

These three choices, in conjunction with the MKS† system of units, lead to the three mutually compatible systems of units shown in Table 14. Table 15 gives equivalents between units in different photometric measurement systems.

Flux Units: The number of lumens dL_λ and the number of photons per second dN_λ associated with a monochromatic flux dW_λ in watts are given by

$$dL_\lambda = 680 E_\lambda \, dW_\lambda$$

and

$$dN_\lambda = (\lambda/hc) \, dW_\lambda$$

where 680 = number of lumens per watt of radiation at the peak photopic eye response, E_λ = normalized (to unity maximum) photopic human eye response (Fig. 32), λ = wavelength of the monochromatic radiation (meters), h = Planck's constant $\simeq 6.6 \times 10^{-34}$ (joule-second), and c = velocity of light $\simeq 3.0 \times 10^8$ (meters per second).

The number of lumens L and the number of photons per second N between the wavelength units of λ_3 to λ_4 associated with a distributed spectral radiation source having a wattage W between the wavelength limits, λ_1 and λ_2, are given by

$$L/W = 680 \int_0^\infty E_\lambda w_\lambda \, d\lambda \Big/ \int_{\lambda_1}^{\lambda_2} w_\lambda \, d\lambda$$

and

$$Nhc/W = \int_{\lambda_3}^{\lambda_4} \lambda w_\lambda \, d\lambda \Big/ \int_{\lambda_1}^{\lambda_2} w_\lambda \, d\lambda$$

where

$$W = w_{\lambda \, max} \int_{\lambda_1}^{\lambda_2} w_\lambda \, d\lambda$$

---

* J. W. T. Walsh, "Photometry," Constable and Co., Ltd., London; 1958.

† Meter, kilogram, second.

and where $w_{\lambda max}$ = maximum spectral density in watts per unit wavelength in the spectral band between λ_1 and λ_2, and w_λ = relative spectral distribution of the radiation source on thermal-energy basis, normalized to a maximum value of unity. Some typical w_λ spectral distributions are shown in Fig. 32.

Optical Imaging: In an optical lens system of flux-gathering diameter D_f in meters, focal length f in meters, and optical transmittance T, the ratio $f/D_f = n_f$ is called the f-number of the lens. If the surface of an object of radiance or luminance B in flux units per steradian per meter² is imaged by this system with a linear magnification m, and assuming Lambertian emittance characteristics over the solid angle subtended by the optical system, the image will be subjected to an irradiance or illuminance I_L in flux units per meter² given by

$$I_L = \pi BT/[4n_f^2(m+1)^2 + m^2].$$

For objects at infinity, $m = 0$, and

$$I_L(\text{object at infinity}) = \pi BT/4n_f^2.$$

If the irradiance (or illuminance) I_L in flux units per meter² is allowed to fall on a nonabsorbing Lambertian diffusing surface, the resultant image radiance (or luminance) B_i in flux units per steradian per meter² is given by

$$\Pi B_i = I_L.$$

Any desired method of measuring flux units, such as watts, lumens, or photons/second (Table 14), can be selected for expressing the object radiance (or luminance) B in flux units steradian⁻¹ meter⁻² and the irradiance (or illuminance) I_L in flux units meter⁻² in these relationships. Thus, a radiance B in watt steradian⁻¹ meter⁻² would be paired with an irradiance I_L in watt meter⁻², a luminance B in lumen steradian⁻¹ meter⁻² with an illuminance I_L in lumen meter⁻², and a radiance B in photon second⁻¹ steradian⁻¹ meter⁻² with an irradiance I_L in photon second⁻¹ meter⁻².

Any spectral distribution modifications, if present, would be included in the numerical magnitude of the lens transmission T, defined as the ratio of the total output flux from the optical system to the corresponding input flux.

Selection of appropriate alternative pairs of luminance and illuminance units when the flux units are not explicitly stated (first column of Table 15) must be made with care. Thus candle centimeter⁻² (or stilb) would be paired with phot, candle meter⁻² (or nit) with lux, and candle foot⁻² with footcandle. Even greater difficulty arises when the factor Π in the preceding relationships is absorbed or included in the units of luminance. Thus the product ΠB in apostilb would be paired with I_L in lux, the product ΠB in lambert with

TABLE 14—COMPATIBLE SYSTEMS OF RADIATION UNITS.

Parameter	Radiometric System	Photometric System	Photon System
Flux	watt	lumen	photon sec^{-1}
Source intensity	watt ster^{-1}	lumen ster^{-1}	photon sec^{-1} ster^{-1}
Incidence	watt m^{-2} (irradiance)	lumen m^{-2} (illuminance)	photon sec^{-1} m^{-2}
Exitance	watt m^{-2} (emittance)	lumen m^{-2} (emittance)	photon sec^{-1} m^{-2}
Sterance	watt ster^{-1} m^{-2} (radiance)	lumen ster^{-1} m^{-2} (luminance)	photon sec^{-1} ster^{-1} m^{-2}
Energy	watt sec	lumen sec	photon

Note: The terms in parentheses are often used to characterize a measurement as either radiometric or photometric.

TABLE 15—PHOTOMETRIC EQUIVALENTS.

Photometric Unit	Equivalent Unit Based on the Lumen (lm) as the Unit of Flux	Equivalent Lumen-MKS Unit
Source Intensity, C		
1 candle (Note 1)	1 lm ster^{-1} (Note 2)	1 lm ster^{-1} (Note 2)
1 candela	1 lm ster^{-1}	1 lm ster^{-1}
1 int. candle (Note 1)	1 lm ster^{-1} (Note 2)	1 lm ster^{-1} (Note 2)
1 Hefner candle	0.92 lm ster^{-1}	0.92 lm ster^{-1}
1 candlepower (Note 1)	1 lm ster^{-1} (Note 2)	1 lm ster^{-1} (Note 2)
Surface Luminance, B		
1 candle cm^{-2}	1 lm ster^{-1} cm^{-2}	10^4 lm ster^{-1} m^{-2}
1 candle m^{-2}	1 lm ster^{-1} m^{-2}	1 lm ster^{-1} m^{-2}
1 candle in^{-2}	1 lm ster^{-1} in^{-2}	1.55×10^3 lm ster^{-1} m^{-2}
1 candle ft^{-2}	1 lm ster^{-1} ft^{-2}	10.8 lm ster^{-1} m^{-2}
1 nit	10^{-4} lm ster^{-1} cm^{-2}	1 lm ster^{-1} m^{-2}
1 stilb	1 lm ster^{-1} cm^{-2}	10^4 lm ster^{-1} m^{-2}
1 apostilb	π^{-1} lm ster^{-1} m^{-2}	π^{-1} lm ster^{-1} m^{-2}
1 lambert	π^{-1} lm ster^{-1} cm^{-2}	$10^4 \pi^{-1}$ lm ster^{-1} m^{-2}
1 millilambert	$10^{-3} \pi^{-1}$ lm ster^{-1} cm^{-2}	$10\pi^{-1}$ lm ster^{-1} m^{-2}
1 footlambert	π^{-1} lm ster^{-1} ft^{-2}	$10.8\pi^{-1}$ lm ster^{-1} m^{-2}
Illuminance of a Surface, I_L		
1 lux	1 lm m^{-2}	1 lm m^{-2}
1 phot	1 lm cm^{-2}	10^4 lm m^{-2}
1 milliphot	10^{-3} lm cm^{-2}	10 lm m^{-2}
1 footcandle	1 lm ft^{-2}	10.8 lm m^{-2}
Energy, U		
1 talbot	1 lm sec	1 lm sec

Notes:
1. Unit becoming obsolete.
2. For a discussion of small differences in source intensity definitions and magnitudes a standard textbook on photometry should be consulted, such as J. W. T. Walsh, "Photometry," Constable and Co., Ltd., London; 1958.

I_L in phot, the product ΠB in millilambert with I_L in milliphot, and the product ΠB in footlambert with I_L in footcandle. These difficulties are avoided by the use of the compatible systems of radiation units shown in Table 14.

Typical Approximate Illumination Values at the

Earth's Surface:

Sun at zenith $\simeq 10^4$ footcandles

$\simeq 10^5$ lumen meter^{-2}

Full moon $\simeq 3 \times 10^{-2}$ footcandles

$\simeq 3 \times 10^{-1}$ lumen meter^{-2}.

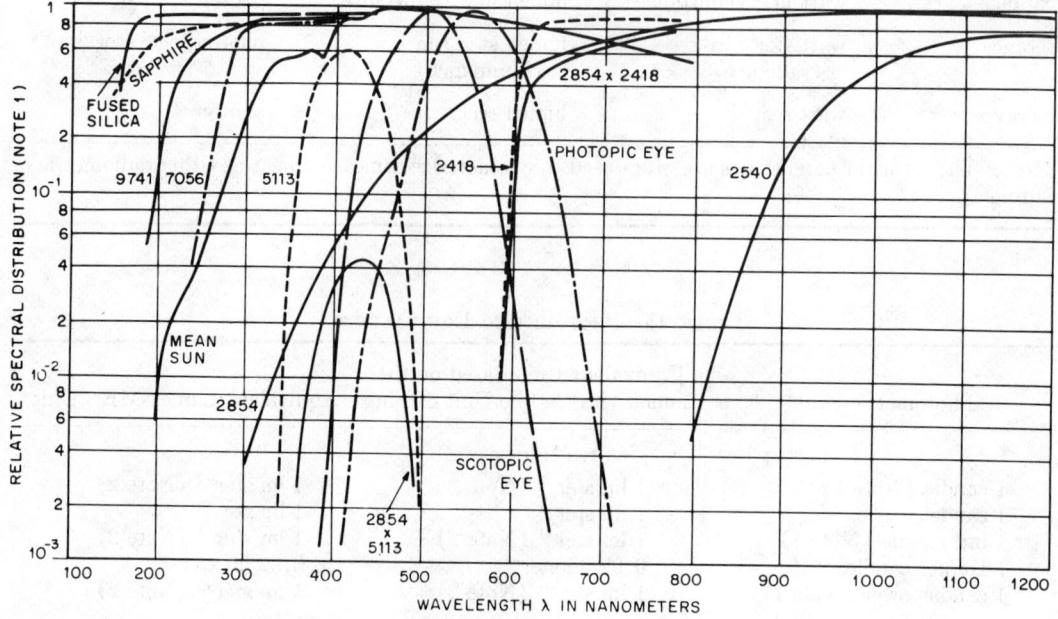

Fused silica: transmission through polished "Suprasil" (1 millimeter thick)

Sapphire: transmission through polished Sapphire (1 millimeter thick)

9741: transmission through polished Corning 9741 glass (1 millimeter thick)

7056: transmission through polished Corning 7056 glass (1 millimeter thick)

Mean sun: mean solar distribution at Earth's surface

5113: transmission through polished Corning 5113 filter (CS-5-58) (half-stock thickness)

2854×5113: product of 5113 curve and 2854 curve

2854: spectral density distribution of 2854 K color-temperature tungsten lamp

Scotopic eye: relative response of dark-adapted eye

Photopic eye: standard eye response

2418: transmission through polished Corning 2418 filter (CS-2-62) (stock thickness)

2854×2418: product of 2854 and 2418 curves

2540: transmission through polished Corning 2540 filter (CS-7-56) (stock thickness)

Note 1: "Relative spectral distribution" designates the relative radiant-energy density distribution w_λ for sources, the relative visual stimulation for equi-energy inputs for the eye response, and the spectral transmission t_λ for windows and filters. The transmission characteristics of individual filter and window samples can be expected to depart appreciably from these typical values.

Fig. 32—Useful spectral distributions.

Typical Approximate Brightness Values:

	footlamberts	lm ster^{-1}m^{-2}
Highlights, 35-milli-meter movie	$\simeq4$	$\simeq100$
Page brightness for reading fine print	$\simeq10$	$\simeq3\times10^2$
November football field	$\simeq50$	$\simeq1.5\times10^3$
Surface of moon seen from Earth	$\simeq1.5\times10^3$	$\simeq5\times10^4$
Summer baseball field	$\simeq3\times10^3$	$\simeq10^5$
Surface of 40-watt frosted lamp bulb	$\simeq8\times10^3$	$\simeq2.5\times10^5$
Crater of carbon arc	$\simeq4.5\times10^7$	$\simeq10^9$
Sun seen from Earth	$\simeq5.2\times10^8$	$\simeq1.5\times10^{10}$

Photoconductivity

Photoconductivity* is the increase in electrical conductivity of a material which takes place when the material is illuminated with infrared, visible, or ultraviolet light.

The absorption of light is a quantum process in which electrons are excited to higher energy levels. Ordinarily, the excited electrons are more mobile than unexcited electrons. Photoconductivity is commonly analyzed in terms of the number and mobility of the excited electrons in an electron conduction band and of electron vacancies or "holes" in a lower-energy valence band. To maintain a steady current, both types of current carriers must be generated in the volume of the material, or else charge carriers must enter the photoconductor at one of the electrodes. Many photoconductors make "ohmic" contacts with their electrodes. These serve as practically unlimited reservoirs of mobile electrons, free to enter the photoconductor volume. Even in these photoconductors the steady dark current is usually limited to a low value by a build-up of a space-charge-potential barrier in the photoconductor.

At the same time that mobile photoelectrons are excited (thermally or optically) in the interelectrode volume, positive charges must also be generated; these compensate the charge of the

* R. H. Bube, "Photoconductivity of Solids," John Wiley & Sons, New York; 1960. S. M. Ryokin, "Photoelectric Effects in Semiconductors," Consultants Bureau; 1964.

photoelectrons in such a way that a "photocurrent" can be superimposed on the small space-charge-limited "dark" current originating at the electrode. If the positive charges are immobile, then long after the photoelectron has passed through the photoconductor into the anode, the immobile positive charges may remain to support a photocurrent of electrons, drawing on the reservoir of electrons at the cathode. This will continue until the immobile "holes" or impurity centers are neutralized by recombination with some of the mobile electrons. Since the recombination lifetime may be much longer than the electron transit time between electrodes, the number of "photoelectrons" transported across the photoconductor may be much larger than the rate of generation of photoelectrons in the photoconductor volume. This ratio of photocurrent to generation rate is called the photoconductive gain.

The photoconductive gain of a pure material can often be greatly increased by addition of localized traps lying near the conducting band. Since these are in thermal equilibrium with the conducting band, they serve as an additional reservoir of the charge carriers. This can increase both the response time and the sensitivity by a large factor.

Practically all materials are photoconductors in the sense that light of the correct wavelengths will generate current carriers. However, in many materials the photoconductivity is not detectable by ordinary measurements, either because of very short carrier lifetimes or because of a large dark current. The useful photoconductors, characterized by comparatively long lifetimes and low dark currents, have most of their charge carriers immobile (in the dark). Light of the proper energy can excite these carriers through the forbidden energy regions into the conduction bands. The long-wavelength limit of photoconductivity at low temperatures is approximately given by

$$\lambda_{max}=hc/E_g$$

where E_g is the forbidden band gap, h is Planck's constant, and c is the velocity of light. For wavelengths longer than 5 micrometers, this equation gives a band gap smaller than $\frac{1}{4}$ volt. Photoconductors with such small energy gaps are usually cooled to reduce the dark conductivity due to thermal excitation of carriers across the gap.

A commonly used figure of merit for photoconductors is the detectivity D^*, defined as the signal-to-noise ratio at a given chopping frequency with an amplifier bandwidth of 1 hertz for a photodetector of 1 square centimeter, divided by the light flux in watts. Detectivities for several typical photoconductors at room temperature are shown in Fig. 33. The photoconductors with long-

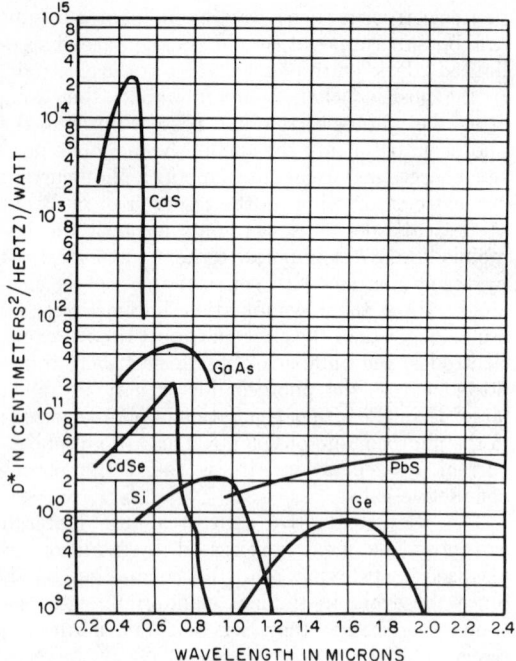

Fig. 33—Detectivity for some typical photoconductors at room temperature.

wavelength cutoffs will be considerably more sensitive if they are cooled below room temperature.

Phosphors

Fluorescent screens* are used in various electron devices such as image tubes, cathode-ray tubes, and storage cathode-ray tubes to convert electron energy into radiant energy. These viewing screens are comprised of many small-diameter (2 to 3 microns) phosphor crystals which emit light when bombarded by high-energy electrons. The spectral response of a phosphor screen is determined by its chemical and physical composition, deposition methods, and the tube processing procedures. Phosphor screens with given output characteristics have been categorized and assigned type numbers. Typical absolute spectral-response characteristics of aluminized phosphor screens are shown in Fig. 34.

Phosphor Efficiency: Over a rather wide range of magnitudes, the ratio U/q of total radiated energy

* "Optical Characteristics of Cathode-Ray Tube Screens," JEDEC Publication No. 16, (J6-C3-1). H. W. Leverenz, "Introduction to Luminescence of Solids," John Wiley & Sons, New York; 1950.

U into a 2π solid angle, to exciting electron charge q for typical aluminized phosphor screens, Fig. 35, is independent of excitation time, beam-current magnitude, and bombarded area.

Consequently the ratio W/I of average total radiated flux W to average exciting current I is also independent of the magnitude of the average current I, the peak current I_{max}, and the bombarding area, and may therefore be used to describe the flux-generating properties of raster-scanned as well as steady-state excited phosphor screens. Experimentally, it is found that the ratio W/I is not a linear function of electron beam voltage V, as might be expected if the phosphor converts the beam energy IV linearly into radiated flux, but behaves in general as shown in Fig. 36, with an offset energy component IV_k followed by an approximately linear dependence on the added energy $I(V-V_k)$ over the usual working voltage range, $V_{min} < V < V_{max}$.

The flux-generating properties of a phosphor screen behaving in the above manner are therefore approximately described by

$$W/I = \mathcal{E}_w(V-V_k), \qquad \text{for } V_{min} < V < V_{max}$$

where W = average output flux (watts), I = average exciting electron beam current (amperes), $\mathcal{E}_w$ = dimensionless phosphor efficiency ratio [radiated watts per exciting electron beam watts (watts watt⁻¹)], V = electron beam voltage, V_k = extrapolated offset knee voltage (Fig. 36), and V_{min}, V_{max} = limits of operating electron beam voltage.

The offset knee voltage V_k for typical aluminized phosphor screens has a magnitude between 1 and 3 kilovolts, and may be attributed primarily to electron beam power losses in penetrating the aluminizing layer plus possible inert-phosphor-particle coatings.

The average monochromatic radiant flux output per unit wavelength dw_λ from a phosphor screen at wavelength λ under these conditions is given by

$$dw_\lambda = \mathcal{E}_\lambda I(V-V_k)$$

where

$$\mathcal{E}_w = \int_0^\infty \mathcal{E}_\lambda \, d\lambda = \mathcal{E}_{\lambda \, max} \int_0^\infty w_\lambda \, d\lambda$$

$\mathcal{E}_\lambda$ = spectral density distribution of the phosphor efficiency or "spectral efficiency" [(watts meter⁻¹) watt⁻¹]

$\mathcal{E}_{\lambda \, max}$ = maximum spectral power density efficiency [(watts meter⁻¹) watt⁻¹]

w_λ = normalized relative power density of the radiant energy spectrum.

Typical values of the "spectral efficiency" $\mathcal{E}_\lambda$ are plotted in Fig. 34 for a number of commonly

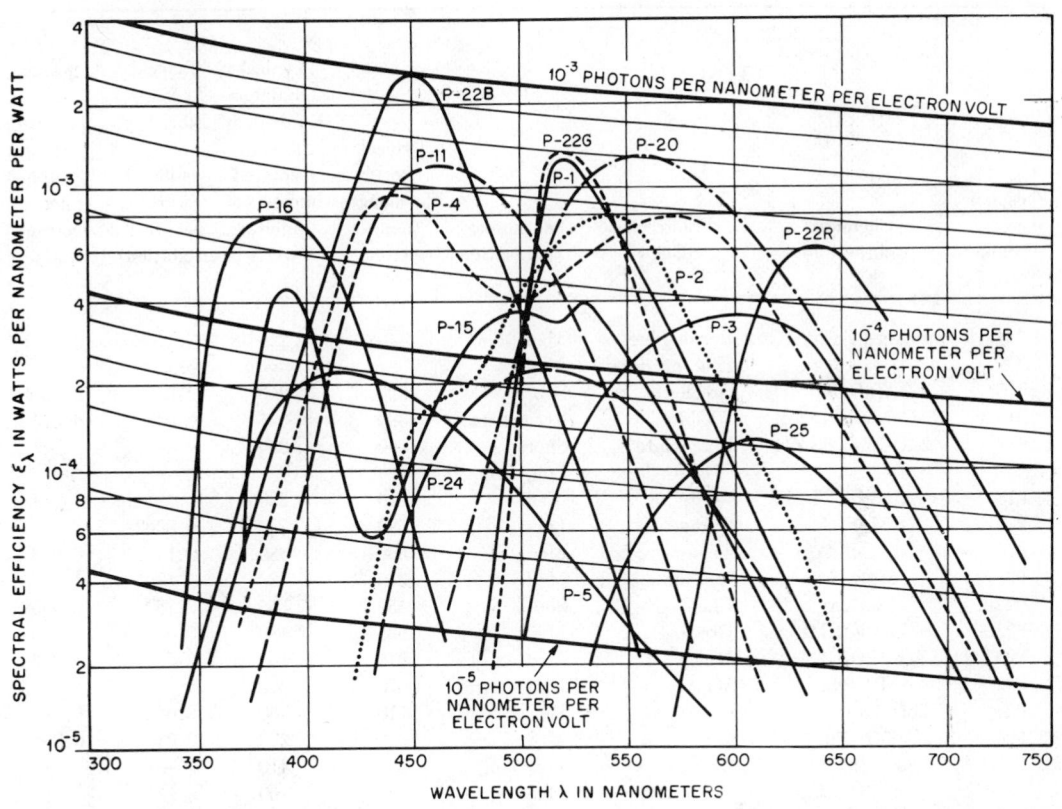

Phosphor Screen P Number	Chemical Composition	Fluorescent Color	Persistence Classification	Typical Peak Wavelength (nano-meters)	Typical Luminous Equivalent $\varepsilon_L/\varepsilon_w$ (radiated lumens per radiated watt)	Typical Absolute Efficiency ε_w (radiated watts per watt excitation)	Typical Quantum Yield Factor Y (photons per electron-volt)
P1	$Zn_2SiO_4:Mn$	Yellow-green	Med.	525	520	0.06	0.026
P2	$ZnS:Cu$	Yellow-green	Med.	533	460	0.07	0.03
P3	$Zn_8BeSi_5O_{19}:Mn$	Yellow-orange	Med.	603	380	0.041	0.02
P4	$ZnS:Ag+ZnCdS:Ag$ (all sulfide type)	White	Med. short	459	290	0.15	0.067
	(silicate-sulfide type)	White	Med.	450	290	—	—
	(silicate type)	White	Med.	410	240	—	—
P5	$CaWO_4:W$	Blue	Med. short	417	90	0.025	0.009
P6	$ZnS:Ag+ZnCdS:Ag$	White	Short	565	340	—	—
P7	$ZnS:Ag$ on $ZnCdS:Cu$	White	Med. short	440	280	—	—
		(White decay)	Long	—	—	—	—
P10	KCl	(dark trace)	Long	—	—	—	—
P11	$ZnS:Ag(Ni)$	Blue	Med. short	460	140	0.10	0.038
P12	$ZnMgF_2:Mn$	Orange	Long	590	410	—	—
P13	$MgSiO_3:Mn$	Red-orange	Med.	640	140	—	—

Fig. 34—Typical absolute spectral response characteristics of aluminized phosphor screens. (*cont on next page*)

Phosphor Screen P Number	Chemical Composition	Fluorescent Color	Persistence Classification	Typical Peak Wavelength (nano-meters)	Typical Luminous Equivalent $\mathcal{E}_L/\mathcal{E}_w$ (radiated lumens per radiated watt)	Typical Absolute Efficiency $\mathcal{E}_w$ (radiated watts per watt excitation)	Typical Quantum Yield Factor Y (photons per electron-volt)
P14	ZnS:Ag+ZnCdS:Cu	Purple-blue (Yel.-or. decay)	Med. short Med.	440 —	250 —	— —	— —
P15	ZnO:Zn	Green	Short	390	250	0.051	0.02
P16	CaMgSiO$_3$:Ce	UV-blue	Very short	380	25	0.049	0.015
P17	ZnO+ZnCdS:Cu	Blue-white (Yellow decay)	Short Long	550 —	350 —	— —	— —
P18	CaMgSiO$_3$:Ti+P3	White	Med.	410	230	—	—
P19	KMgF$_2$:Mn	Orange	Long	590	390	0.0002	—
P20	ZnCdS:Ag	Yellow-green	Med.	560	480	0.14	0.063
P21	MgF$_2$:Mn	Red-orange	Med.	610	360	—	—
P22B	ZnS:Ag	Blue	Short	450	55	0.15	0.055
P22G	Zn$_2$SiO$_4$:Mn	Green	Med.	525	530	0.06	0.025
P22R	Zn$_3$(PO$_4$)$_2$:Mn	Red	Med.	645	150	0.05	0.022
P23	P4 type	White	Med. short	570	320	—	—
P24	ZnO:Zn (Special)	Green	Short	510	360	0.026	0.011
P25	CaSiO$_3$:Pb, Mn	Orange	Med.	610	320	0.013	0.006
P26	ZnF	Orange	Very long	590	410	—	—
P27	Zn$_3$(PO$_4$)$_2$:Mn	Red-orange	Med.	635	60	—	—
P28	ZnS:Ag, Cu	Green-yellow	Long	550	500	—	—
P29	P2 & P25 type	—	—	—	—	—	—
P31	ZnS:Cu	Green	Med. short	522	230	0.22	—

Notes:

Since the response characteristics of phosphor screens depend on such variable parameters as chemical composition, particle size, deposition methods, and the tube processing procedures, considerable departure from the data given on this chart is to be expected for individual screen samples.

With the exception of efficiency and quantum yield factor, the data are based primarily on JEDEC Publication No. 16 entitled "Optical Characteristics of Cathode-Ray-Tube Screens."

Efficiency and quantum yield factor data on phosphor screen types shown in the figure are derived directly from experimental measurements. For the remaining phosphors, data from various published sources, primarily phosphor manufacturers, havebeen extrapolated to the stated units of measurement whenever possible.

Phosphor excitation is expressed in terms of the power dissipated by the electron beam in the phosphor layer proper, with corrections for power losses to the aluminum layer coating and to the glass substrate in the case where the electron beam completely penetrates the phosphor layer. Expressing the phosphor excitation in this manner minimizes the variations of efficiency ratings with accelerating potential. The ratings given are most accurate in the region from 8 kV to 12 kV. Typical phosphor screen dead voltages, as a result of the aluminum coating, can be expected to be on the order of 1 kV to 3 kV.

Radiant output is expressed in terms of the total flux leaving the phosphor exit window. Luminance characteristics may be calculated from the data given provided a radiating area is also specified and assuming a cosine-law radiance distribution (approximately valid only).

Input current levels are restricted to the linear response region for each phosphor with respect to both average and peak current densities.

Quantum yield factors are tabulated in terms of photons per electronvolt, making it possible to multiply the listed numerical factor by the selected effective excitation voltage to give the quantum yield in photons per electron.

Fig. 34—Continued.

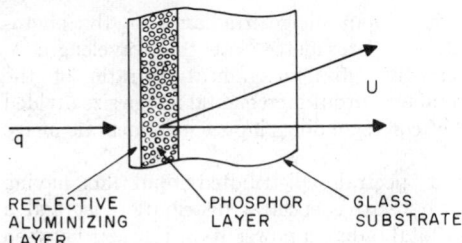

REFLECTIVE
ALUMINIZING
LAYER

PHOSPHOR
LAYER

GLASS
SUBSTRATE

Fig. 35—Aluminized phosphor-screen construction.

used JEDEC-registered phosphor materials. The apparently redundant designation (watts meter^{-1}) watt^{-1} (or the equivalent) is retained in Fig. 34 and in this text to emphasize the fact that watt watt^{-1} refers to ratios of radiated watts to effective exciting electron beam wattage. The spectral distributions shown in Fig. 34 refer only to the first fast component of phosphor emission and ignore the slower less-intense excitation levels occurring in some phosphors.

If the total flux output is measured in lumens L instead of watts, then the corresponding relationships are

$$L/I = \mathcal{E}_L(V - V_k), \qquad \text{for } V_{min} < V < V_{max}$$

$$dL_\lambda = \mathcal{E}_{L\lambda} I (V - V_k)$$

$$\mathcal{E}_L = \int_0^\infty \mathcal{E}_{L\lambda}\, d\lambda$$

where the subscript L indicates the use of lumens for measuring the flux. Phosphor efficiency $\mathcal{E}_L$ in lumens watt^{-1} and spectral efficiency $\mathcal{E}_{L\lambda}$ in (lumens meter^{-1}) watt^{-1} are related to their corresponding absolute values $\mathcal{E}$ and $\mathcal{E}_\lambda$ by

$$\mathcal{E}_{L\lambda} = 680 E_\lambda \mathcal{E}_\lambda$$

and

$$\mathcal{E}_L = 680 \mathcal{E}_w \int_0^\infty w_\lambda E_\lambda\, d\lambda \Big/ \int_0^\infty w_\lambda\, d\lambda$$

where 680 = luminous equivalent of monochromatic

radiation in lumen watt^{-1} at the peak eye-response wavelength, and E_λ = relative photopic eye response.

Some typical values of the dimensionless "spectral-matching-factor ratio"

$$\int_0^\infty w_\lambda E_\lambda\, d\lambda \Big/ \int_0^\infty w_\lambda\, d\lambda$$

appearing in these relationships are given in Table 4.

If the total flux output is measured in photons second^{-1} instead of watts or lumens, then the corresponding relationships are

$$N/n_e = Y(V - V_k)$$

$$dN_\lambda = y_\lambda n_e (V - V_k)$$

$$Y = \int_0^\infty y_\lambda\, d\lambda$$

where N = total number of output photons, n_e = total number of triggering input electrons, Y = "quantum yield factor" (photons electronvolt^{-1}), dN_λ = number of photons per unit wavelength (photons meter^{-1}), and y_λ = spectral efficiency (photons meter^{-1}) electronvolt^{-1} (shown on curved coordinate scales of Fig. 34).

The "quantum yield factor" Y, tabulated in Fig. 34 for the listed typical phosphor behavior, is useful in predicting the quantum yield N/n_e of a phosphor screen for an effective exciting electron beam voltage $V - V_k$ according to the above equation.

For a phosphor screen radiating according to Lambert's Cosine Law (often approximately, but not exactly, valid), the radiance B in (watts ster^{-1}) meter^{-2} and the luminance B_L in (lumens ster^{-1}) meter^{-2} is given by

$$\pi B = \mathcal{E}_w J (V - V_k)$$

$$\pi B_L = \mathcal{E}_L J (V - V_k)$$

where J = exciting current density (amperes meter^{-2}).

LIGHT-SENSING TUBES

Image Tubes and Image Intensifiers

An image tube* is an optical-image-in to optical-image-out electron tube device, combining an input

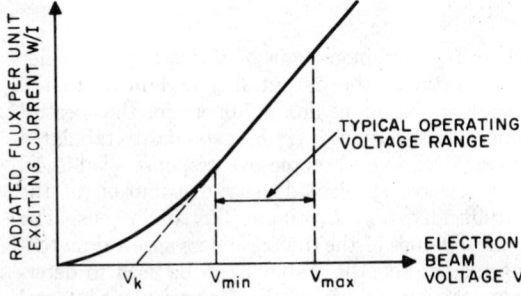

Fig. 36—Typical phosphor-screen behavior.

* "Photo-Electric Image Devices," Advances in Electronics and Electron Physics, Vols. 12, 16, 22A, and 22B, Academic Press, New York and London; 1960, 1962, and 1966. H. V. Soule, "Electro-Optical Photography at Low Illumination Levels," John Wiley & Sons, New York; 1968.

photocathode and an output phosphor screen such that photoelectrons emitted from each point on the photocathode subsequently excite a corresponding individual image "point" on the phosphor screen. Various focusing means, including magnetic and electrostatic electron lenses, may be used to assure maximum point-to-point correlation between the input and output images. The principal operating requirements are a lens to form the input image and a high-voltage supply, typically 5–25 kilovolts, to provide sufficient electron beam energy to excite the output phosphor screen.

If means are provided within the image tube to amplify the photoelectrons before they strike the output phosphor screen, or if the tube without such means produces a much brighter output image than the input image would produce on a diffusing screen, the tube is commonly called an image-intensifier tube.

Image-intensifier tubes are used to amplify the brightness of a faint input image for better visual or photographic viewing, whereas image tubes without amplification are used to convert radiation from one spectral region to another (image conversion) or to perform such control operations as optical shuttering by programing the applied high voltage.

The total output flux w_o in watts exiting (through 2Π steradians) from the phosphor-screen faceplate of an image intensifier tube for an input monochromatic flux dW_λ in watts at a wavelength λ is given by

$$dW_o = s_\lambda G \mathcal{E}_w (V - V_k) \, dW_\lambda$$

$$= G_\lambda \, dW_\lambda$$

where G_λ = monochromatic wattage gain of the image intensifier tube at a wavelength λ = ratio of the total output flux dW_o in watts to the input monochromatic flux dW_λ in watts, s_λ = radiant sensitivity of the input photocathode in amperes per watt (Fig. 5), G = internal current gain ratio of the image intensifier tube = ratio of the current bombarding the output phosphor screen to the corresponding photocurrent leaving the input photocathode, $\mathcal{E}_w$ = absolute phosphor screen efficiency = ratio of the total radiated flux in watts to the exciting electron beam power in watts dissipated in the particles of the output phosphor screen (Fig. 34), V = energy of the electron beam in volts bombarding the output phosphor screen, and V_k = extrapolated knee voltage of the output phosphor screen (Fig. 36).

If the phosphor screen radiates flux according to Lambert's Law (usually only approximately valid), the corresponding output image radiance R_o in watt steradian^{-1} meter^{-2} is given by

$$R_o = G_\lambda I_{\lambda_1} / \Pi m^2$$

where I_{λ_1} = input image irradiance on the photocathode in watt meter^{-2} at the wavelength λ_1 and m = differential magnification ratio of the image tube = output incremental image size divided by the corresponding input incremental image size.

For a spectrally distributed input flux having a known relative spectral distribution w_λ and a known total radiated power $W_{\lambda_1\lambda_2}$ in watts between the wavelength limits λ_1 and λ_2, the resulting total output flux W_o in watts exiting from the image tube is given by

$$W_o = s_{\lambda\,max} \left(\int_0^\infty \sigma_\lambda w_\lambda \, d\lambda \Big/ \int_{\lambda_1}^{\lambda_2} w_\lambda \, d\lambda \right)$$
$$\times G\mathcal{E}_w (V - V_k) W_{\lambda_1\lambda_2}$$

$$= G_{\lambda_1\lambda_2} W_{\lambda_1\lambda_2}$$

where $s_{\lambda\,max}$ = peak radiant sensitivity of the input photocathode in amperes per watt, σ_λ = relative radiant sensitivity of the input photocathode as a function of wavelength λ normalized to unity maximum, w_λ = relative spectral distribution of the power density spectrum of the input flux normalized to unity maximum, and $G_{\lambda_1\lambda_2}$ = wattage gain of the image tube for the relative spectral distribution w_λ and the wavelength limits λ_1 and λ_2. Typical values for the magnitude of the dimensionless spectral-matching-factor ratio

$$\int_0^\infty \sigma_\lambda w_\lambda \, d\lambda \Big/ \int_{\lambda_1}^{\lambda_2} w_\lambda \, d\lambda$$

are found in Table 4.

The total output flux L_o in lumens exiting from an image tube, corresponding to the total output flux W_o in watts, can be computed from the flux conversion relationships given in the section on Radiometry and Photometry, or from the following relationship

$$L_o = s_{\lambda\,max} \left(\int_0^\infty \sigma_\lambda w_\lambda \, d\lambda \Big/ \int_0^\infty E_\lambda w_\lambda \, d\lambda \right)$$
$$\times G\mathcal{E}_w \left(\int_0^\infty E_\lambda w_{o\lambda} \, d\lambda \Big/ \int_0^\infty w_{o\lambda} \, d\lambda \right)(V - V_k)L_i$$

$$= G_L L_i$$

where G_L = luminous gain of the image intensifier tube = ratio of the output flux in lumens to the corresponding input flux in lumens for the spectral input distribution w_λ, E_λ = standard tabulated average relative photopic eye response (Table 4), $w_{o\lambda}$ = relative spectral density distribution of the output flux, and L_i = input flux in lumens. The typical values of the dimensionless spectral matching factors given in Table 4 can be used to determine the magnitude of the dimensionless integral ratios appearing in these relationships.

For the special case where the input flux $L_i(2854)$ in lumens is generated by a 2854 K color-temperature tungsten-filament lamp, the output flux $L_o(2854)$ in lumens is given by

$$L_o(2854) = S(2854)G\mathcal{E}_w\left(\int_0^\infty w_{o\lambda}E_\lambda\, d\lambda \middle/ \int_0^\infty w_{o\lambda}\, d\lambda\right)$$
$$\times (V - V_k)L_i(2854)$$
$$= G_L(2854)L_i(2854)$$

where $G_L(2854)=$ luminous gain of the image intensifier for 2854 K tungsten-lamp radiation, and $S(2854)=$ luminous sensitivity of the input photocathode for 2854 K tungsten-lamp radiation. The magnitude of the luminous gain $G_L(2854)$ is commonly used to characterize the image intensification properties of an image intensifier tube.

If the output phosphor screen radiates flux according to Lambert's Law (usually only approximately valid), the output image luminance (or brightness) B_o in lumen steradian^{-1} meter^{-2} is given by

$$B_o = G_L I_i / \pi m^2$$

where $G_L=$ luminous gain of the image intensifier for the particular input spectral distribution w_λ, $I_i=$ input illuminance (or illumination) on the photocathode in lumen meter^{-2} for the spectral distribution w_λ, and $m=$ differential magnification ratio of the image tube= output incremental image size divided by the corresponding input incremental image size.

Internal current gain G of an image intensifier tube can be obtained by the use of an internal sandwich electrode, in which an auxiliary or sandwich photocathode is mounted in close proximity to and following an auxiliary or sandwich phosphor. Photoelectrons from the input photocathode of the tube are then imaged onto this sandwich phosphor screen and the flux from this screen is coupled to the sandwich photocathode, generating an enhanced photocurrent. The current gain ratio G of this sandwich phosphor–photocathode combination, defined as the ratio of output photocurrent to input photocurrent, is given by

$$G = s_{\lambda\, \max,\, \text{sand}}$$
$$\left(\int_0^\infty w_{\lambda\,(\text{sand})}\sigma_{\lambda\,(\text{sand})}\, d\lambda \middle/ \int_0^\infty w_{\lambda\,(\text{sand})}\, d\lambda\right)$$
$$\mathcal{E}_{w,\,\text{sand}}(V_{\text{sand}} - V_{k,\,\text{sand}})\gamma$$

where $s_{\lambda\,\max,\,\text{sand}}=$ peak monochromatic responsivity of the sandwich photocathode in ampere watt^{-1}, $w_{\lambda\,(\text{sand})}=$ relative spectral distribution of the flux emitted by the sandwich phosphor screen, $\sigma_{\lambda\,(\text{sand})}=$ the relative spectral distribution of the sandwich photocathode, $\mathcal{E}_{w,\,\text{sand}}=$ absolute efficiency of the sandwich phosphor screen in watt watt^{-1}, $V_{\text{sand}}=$ electron beam energy in

volts bombarding the sandwich phosphor screen, $V_{k,\,\text{sand}}=$ extrapolated knee voltage in volts for the sandwich phosphor screen, and $\gamma=$ optical coupling efficiency of the sandwich electrode= ratio of the flux falling onto the sandwich photocathode to the corresponding flux emitted by the sandwich phosphor screen. Typical values of the dimensionless ratio of the two integrals appearing in this relationship are given in Table 4.

The combination of a phosphor screen and a photocathode to produce current gain G can also be achieved by optically coupling the output flux from one image tube to the input of a second tube.

Resolution in image tubes and image intensifier tubes is a subjective parameter describing the number of pairs of equally spaced illuminated and unilluminated bars per unit distance at the photocathode imaged onto the input photocathode surface which can just be distinguished visually by a trained observer under stated test conditions.

Distortion is a parameter describing any change in the geometric shape of the output image compared with the input image. Radially increasing magnification leads to "pincushion" distortion, radially decreasing magnification leads to "barrel" distortion, and radially changing image rotation leads to "S" distortion.

Vacuum Photodiodes

The combination of a photocathode and an anode electrode for collecting the emitted photocurrent in an evacuated envelope is called a vacuum photodiode. A positive anode potential sufficient to assure collection of all emitted photoelectrons (that is, to "saturate" the diode phototube) is normally required, the tube then acting as a constant-current generator (Fig. 37). The power supply potential V_B must assure sufficient anode potential in the presence of a voltage drop in the load resistor R_L.

Under these conditions the total anode output current I_a, neglecting all noise fluctuations, is given by

$$I_a = I_s + I_b + I_d + I_L$$

where $I_s=$ emitted photocathode signal current, $I_b=$ emitted photocathode photocurrent due to stray background flux, $I_d=$ photocathode therm-

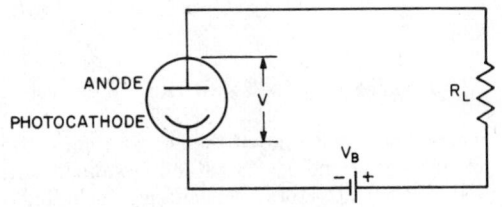

Fig. 37—Photodiode circuit.

ionic dark current, and I_L= residual dark current (leakage, etc.).

The instantaneous value of the signal current I_s will follow the instantaneous signal flux input magnitude from direct current up to an upper frequency limit (commonly 200–2000 megahertz) set by the transit-time spread of the electrons crossing the gap between cathode and anode, and including induced displacement currents during transit.

For steady-state or slowly varying input flux rates, the total noise current output i_n from the diode and load resistor is given by

$$i_n^2 = 2e\Delta f(I_s + I_b + I_d) + (4kT\Delta f/R_L) + i_L^2$$

where e= the electron charge=1.6×10^{-19} coulomb, Δf= noise-current measurement bandwidth, k= Boltzmann factor=1.38×10^{-23} joule per K, T= absolute temperature of the load resistor (K), R_L= load resistance (ohms), and i_L= residual dark noise current (from leakage, stray pickup, etc.).

To increase the absolute level of the noise voltage generated by the noise current i_n so that tube noise predominates over the noise voltage of the subsequent amplifiers or indicating circuits, and to suppress load-resistor noise relative to tube noise, large values of the load resistor (of the order of 10^7–10^9 ohms) are commonly used.

For a plane-parallel vacuum photodiode, the space-charge-limited output current $I_{a(max)}$ in amperes for a given applied cathode-to-anode potential difference V in volts is given by

$$I_{a(max)} = 2.33 \times 10^{-6} A V^{3/2}/d^2$$

where A= uniformly emitting emission area (meters^2), and d= anode–cathode spacing (meters).

In practice, linear output currents up to approximately half of this maximum limit can be obtained.

For a plane-parallel vacuum phototube, the output anode current I_A as a function of time t for an ultra-short exciting light pulse is given by

$$I_A = (2QRC/T^2)[(t/RC) + \exp(-t/RC) - 1]$$

$$\text{for } 0 < t < T$$

$$I_A = (2QRC/T^2)[(T/RC) + \exp(-I/RC) - 1]$$

$$\times \exp[-(t-T)/RC] \quad \text{for } t \geq T$$

and

$$T = d(2m/eV)^{1/2} = (3.37 \times 10^{-6}) d/(V)^{1/2} \text{ seconds}$$

where T= transit time of the charge from cathode to anode (seconds), V= cathode-to-anode potential (volts), C= total capacitance including external circuit capacitance, R= load resistance, and Q= total charge.

Gas Photodiodes

In diode phototubes not containing a high vacuum, ionization by collision of electrons with neutral molecules may occur so that more than one electron reaches the anode for each originally emitted photoelectron. This "gas amplification factor" has a value of between 3 and 5; a higher factor causes instabilities. Gas-tube operation is restricted to frequencies below about 10 000 hertz.

Photomultipliers

The combination of a photocathode and a secondary-emission electron multiplier is called a photomultiplier.[*] Emitted photoelectrons from the photocathode are directed under the influence of a suitable electrode, often called the "focus electrode," to the surface of a secondary-emitting electrode, the "first dynode." Subsequently emitted secondary electrons, increased in number by the effective secondary-emission ratio σ_1, are then directed to the secondary-emitting surface of a subsequent dynode by an appropriate electric field for further multiplication. Continuing this process for n successive dynodes and collecting the multiplier charge at an output electrode called the "anode" or "collector," leads to a charge or current amplification G given by

$$G = \sigma^n, \quad \text{for} \quad \sigma_1 = \sigma_2 = \sigma_3 \cdots = \sigma$$

where gains as high as 10^5–10^9 are commonly achieved in 10-stage to 16-stage multipliers.

Output Current: Disregarding all noise fluctuations of the output current, and assuming operation within the usual linear-response region, a photomultiplier acts as a constant-current source generating an output current I_o given by

$$I_o = I_s + I_b + I_d$$

$$= G\epsilon I_{ks} + G\epsilon I_{kb} + G\epsilon I_{kd} + I_{ad}$$

where I_s= anode signal current due to an incident signal flux to be detected, I_b= anode current due to any background flux simultaneously present on the photocathode, I_d= anode dark current, G= current gain of the electron multiplier, ϵ= collection efficiency (ratio of current entering the electron multiplier to emitted photocathode current), I_{ks}= photocathode signal current due to an incident signal flux to be detected, I_{kb}= photocathode current generated by any background flux simul-

* S. Rodda, "Photo-Electric Multipliers," Macdonald and Co., London; 1953.

taneously present on the photocathode, $I_{kd}=$ photocathode dark current, and $I_{ad}=$ component of anode dark current I_d not originating from the photocathode.

The output signal current I_s follows the instantaneous value of the input signal flux from direct current up to an upper frequency limit (typically 20–200 megahertz) established by the response time or "transit-time spread" (typically 1–10 nanoseconds).

The ratio of the output signal current I_s in amperes to the triggering input flux L in lumens is called the anode luminous sensitivity A in amperes/lumen and is given by

$$A = I_s/L = G\epsilon I_{ks}/L = G\epsilon S$$

where $S=$ photocathode luminous sensitivity in amperes/lumen (Table 3).

The value of the input flux L_D in lumens giving an output anode current just equal to the anode dark current I_d is called the "equivalent anode-dark-current input" or, more precisely, the luminous equivalent of the anode dark current and is given by

$$L_D = I_d/A = I_d/G\epsilon S.$$

Signal and Noise: Noise fluctuations of the output current in photomultipliers can be divided into two classes: *dark noise*, occurring in the absence of input flux; and *noise-in-signal*, including "quantum" noise resulting from the inherent quantum nature of the input flux as well as uncontrolled fluctuations of that flux. The presence of an appreciable, in fact often predominant, noise-in-signal current component in photomultipliers depending on the instantaneous signal current magnitude requires caution in applying noise concepts to photomultipliers, and may lead to erroneous conclusions regarding photomultiplier behavior, particularly for a modulated flux input.

For a steady-state unmodulated flux input, the total noise current i_n flowing in the load resistance R is given by

$$i_n^2 = 2eGK\Delta f(I_s + I_b + G\epsilon I_{kd}) + i_r^2 + (4kT\Delta f/R)$$

where $K=$ photomultiplier noise factor, $\Delta f=$ noise bandwidth of the noise-current measuring circuits (hertz), $i_r=$ residual photomultiplier anode dark noise current, excluding dark current emission from the photocathode (amperes), $(4kT\Delta f/R)^{1/2}=$ Johnson-Nyquist noise current in the load resistance R (amperes), $k=$ Boltzmann's constant, 1.38×10^{-23} joule/K, and $T=$ absolute temperature of the load resistance (K).

For photomultipliers with a constant gain per stage σ in the first few stages of the electron

multiplier, the noise factor K may be estimated from

$$K = \sigma/(\sigma-1).$$

The luminous equivalent of the anode dark noise current is called the equivalent noise input, abbreviated as ENI and defined as "the peak-to-peak value of a square-wave-chopped flux input which gives an rms value for the fundamental component of the output current just equal to the rms value of the dark noise current measured for a 1-hertz bandwidth." The ENI in lumens is given by

$$\text{ENI} = \frac{\pi(2)^{-1/2}(2eGK\epsilon I_{kd} + i_r^2(1\text{ hertz}) + 4kT/R)^{1/2}}{G\epsilon S}$$

where the factor $\pi(2)^{-1/2}$ converts peak-to-peak flux magnitude to equivalent rms magnitude of the fundamental component, and the notation i_r^2 (1 hertz) indicates that the residual dark noise current i_r excluding photocathode dark noise current is to be evaluated for a 1-hertz bandwidth.

Photomultipliers as Scintillation and Single-Electron Counters

In combination with suitable scintillating material, typically thallium-activated NaI crystals, photomultipliers are extensively used to detect the single flashes of light generated by the scintillating material on bombardment by a single triggering particle, typically a gamma ray from a nuclear disintegration process. If the scintillating material generates an average of N photons per disintegration incident on the effective photocathode of peak quantum efficiency Y_{max}, the resultant average charge pulse Q_A appearing in the anode circuit (disregarding all photons or electrons producing no output charge) will be given by

$$Q_A = NY_{max}\alpha Ge$$

where α is a spectral matching factor describing the relative match between the scintillator spectral output and the cathode sensitivity. (Refer to section on photoemission.)

Because of the random statistical fluctuations of cathode quantum efficiency Y_{max} and electron-multiplier gain G, as well as in the number of effective photons N generated by the scintillator, the anode charge Q_A will vary in magnitude from pulse to pulse, introducing ambiguity in the determination of the average magnitude of N, which in turn is used to determine the energy of the triggering input particle, for example the gamma ray. The ratio of the spread of the amplitude of individually observed values of the charge Q_A

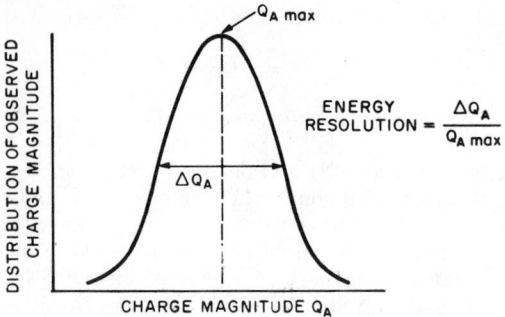

Fig. 38—Energy resolution of photomultiplier.

at half maximum to the most probable value $Q_{A(\max)}$ is called the "energy resolution" of the photomultiplier-plus-scintillator combination and is commonly 7–10% minimum. (See Fig. 38.)

If the input flux has no time-coherent groups of photons, as it does in scintillation detection, photoelectrons are emitted singly at random emission times from the photocathode and also generate an average output charge Q_A given by

$$Q_A = Ge$$

where all photons or electrons generating no output pulse are disregarded in measuring G and computing Q_A. Assuming sufficiently large gain G and sufficiently low generation of dark pulses of similar charge amplitude, the individual anode pulses of charge amplitude Ge can be detected and counted individually, the photomultiplier then acting as a single-electron counter.

Image Dissectors

Principle of Operation: The image dissector* is a specialized electronically deflectable photomultiplier. Inside its evacuated envelope, a photocathode emits electrons in proportion to the illumination from an optical image incident on it. These image electrons are accelerated and focused into a plane containing a defining aperture. By means of deflection coils or plates, the resulting electron image is moved across this plane and a sampling

---

* V. K. Zworykin and G. A. Morton, "Television," John Wiley & Sons, New York; 1940: p. 230 ff. K. R. Spangenberg, "Vacuum Tubes," 1st edition, McGraw-Hill Book Company, New York; 1948: p. 729. P. T. Farnsworth, "Television by Electron Image Scanning," *Journal of the Franklin Institute*, Vol. 218, pp. 411–444; October 1934. D. G. Fink, "Television Engineering," 2nd edition, McGraw-Hill Book Company, Inc., New York, N.Y.; 1952: pp. 95–99. C. C. Larson and B. C. Gardner, "The Image Dissector," *Electronics*, Vol. 12, No. 10, p. 24; October 1939.

of electrons passes through the aperture. These electrons impinge on the secondary-emitting surface of a dynode where, on the average, they give rise to 3 to 5 secondary electrons. These secondary electrons pass, in turn, to the next in a series of dynodes where their number is further multiplied. Depending on the number of the dynodes of the electron multiplier chain, their quality, and their applied potentials, a single electron passing through the defining aperture typically produces 10^5 to 10^7 electrons at the multiplier output anode. An example of a modern image dissector is shown schematically in Fig. 39.

To a first approximation, the resolution properties of an image dissector are determined solely by the geometric size and shape of the dissecting aperture. The predicted falloff in peak-to-peak signal modulation m for a dissector with various aperture sizes and shapes scanning a bar pattern input of increasing spatial density R in line pairs per meter is shown in Fig. 40.

A lower limit on the effective aperture size of an image dissector occurs because of the finite emission energy of photoelectrons. The combination of a finite average lateral emission energy component V_{ro} in volts with a finite average axial emission energy component V_{zo} leads to an electron image on the dissector aperture plate (for a point optical input image) which is blurred according to an approximately Gaussian current density distribution called the "point-spread function" of the electron lens of the image dissector. For a magnetically focused image dissector with a uniform electric accelerating field over a distance L_1 in meters between the photocathode and an electron-transmissive mesh followed by an electric-field-free drift and deflection space (Fig. 39) of length L_2 in meters prior to the aperture plate, the full width at half maximum W of the resulting point-spread function at the aperture plate in meters is given by

$$W \simeq 1.23 L_1 \frac{(V_{ro} V_{zo})^{1/2}}{V_a} \left[1 + 0.42 \frac{L_2 - 2L_1}{L_1} \left(\frac{\bar{V}_{zo}}{V_a} \right)^{1/2} \right]$$

where V_a is the beam energy in volts within the drift and deflection space. To maintain this minimum (focused) beam size at the aperture requires a solenoidal magnetic field strength B in weber meter^{-2} given by

$$B = \pi n (2m V_a)^{1/2} / L(e)^{1/2}$$

where $n =$ number of beam loops between cathode and aperture plate $= 1, 2, 3, \cdots$, $m =$ mass of the electron $= 9.1 \times 10^{-31}$ kilogram, and $e =$ electron charge $= 1.6 \times 10^{-19}$ coulomb.

Given a dissector with the minimum useful aperture size (approximately one beamwidth W in

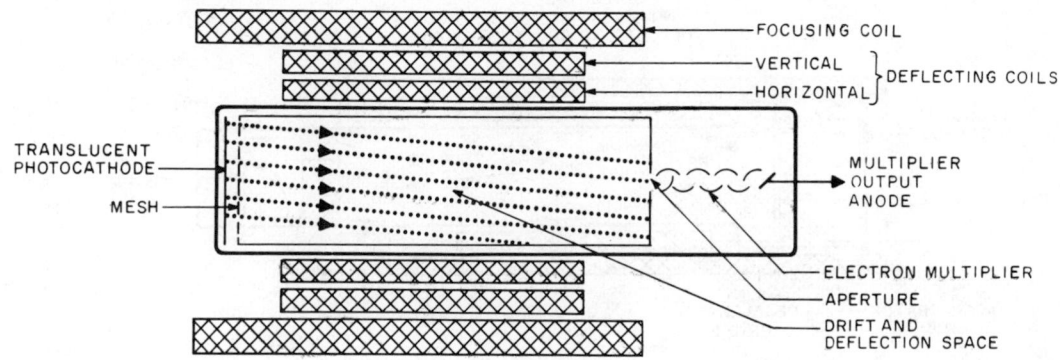

Fig. 39—Image dissector.

diameter), the resulting maximum value of the spatial pattern density $R_{\max}$ in line pairs per meter which can be detected at a modulation ratio $m \simeq 0.05$ is given by

$$R_{\max} \simeq 0.7/W.$$

Signal: For an input photocathode illuminance I_L in lumen meter^{-2}, the output signal current I_o from an image dissector is given by

$$I_o = StaGI_L$$

where $S =$ photocathode luminous sensitivity in ampere lumen^{-1}, $t =$ mesh transmission including collection efficiency losses between photocathode and first dynode $=$ ratio of the current transmitted by the mesh to the current incident on the mesh, $a =$ effective aperture area measured at the photo-

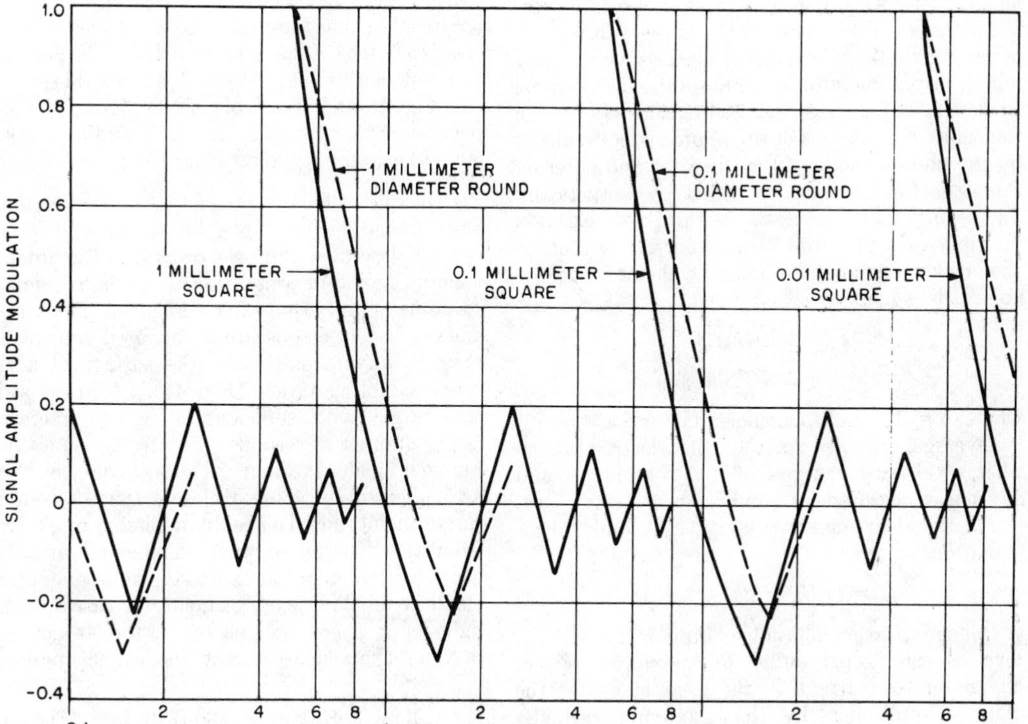

Fig. 40—Bar-pattern spatial density.

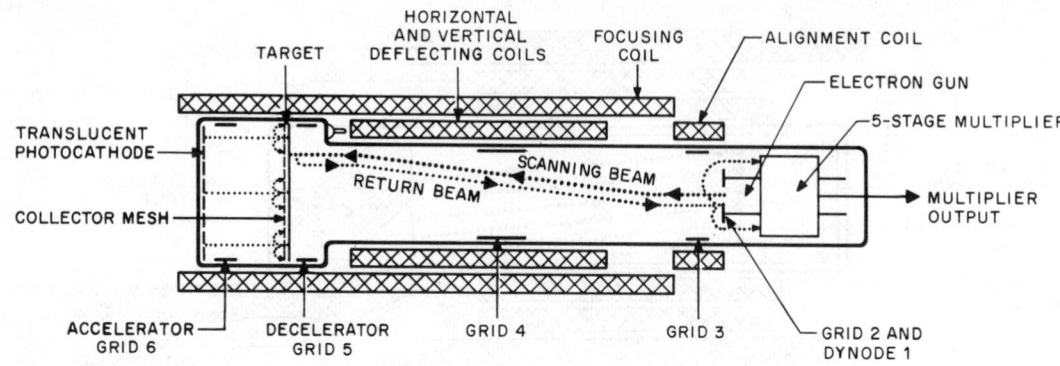

Fig. 41—Image orthicon. *By permission of RCA, copyright proprietor.*

cathode surface (and thus compensating for possible electron optical magnification changes within the dissector), and $G=$ electron multiplier current gain. The luminous sensitivity S pertains to the particular spectral distribution characteristic of the incident flux. If other units are used to measure the incident illuminance, for example watt meter^{-2} or photon second^{-1} meter^{-2}, the emitted photocathode current density SI_L in the above relationship may be computed according to the methods described in the section on Photoemission.

Noise: At the usual input illuminance levels necessary for satisfactory signal-to-noise-ratio performance of dissectors, dark noise (unlike all other conventional camera tubes) is completely negligible in magnitude. The total noise output current i_n in amperes is caused exclusively by the shot-noise fluctuations of the photocurrent emitted by the photocathode of the dissector and entering the dissector aperture, modified by statistically fluctuating gain processes within the electron multiplier. If the total output current is due to the incident signal illuminance I_L, the output noise current is given by

$$i_n{}^2 = 2eKGI_o\Delta f$$
$$= 2eKG^2 StaI_L\Delta f$$

where $K=$ the multiplier noise factor $\simeq \sigma/(\sigma-1)$, $\sigma=$ average gain per stage of the electron multiplier, $e=$ electron charge $= 1.6\times10^{-19}$ coulomb, and $\Delta f=$ noise measurement bandwidth in hertz.

The signal-to-noise power ratio S/N under these conditions is given by

$$S/N = I_o{}^2/i_n{}^2 = StaI_L/2eK\Delta f.$$

Dynamic Range: The upper limit to input light level is established either by maximum space-charge-limited current in the final stages of the electron multiplier, by the maximum cathode-current density consistent with desired cathode lifetime, or by excessive current in the multiplier

bleeder resistors. The lower limit is set usually by "shot noise" in the signal.

Image Orthicons

The image orthicon[*] is the camera tube most widely used for live commercial television. This fact derives from its high sensitivity, its close spectral-sensitivity match to the human eye, and its relatively fast response. Good-quality commercial television pictures can be generated by an image orthicon viewing a 5-to-20-footlambert ($\simeq$15–50 lumen steradian^{-1} meter^{-2}) scene through an F/5.6 lens. The image orthicon is generally available with either S-10 or S-20 spectral response (see Fig. 5), and is capable of 500 picture elements per raster height (9.9 line pairs/millimeter) at 30% video-amplitude response.

Principle of Operation: In the image section, a light image incident on the translucent photocathode liberates photoelectrons into the adjacent vacuum region in proportion to the light intensity (gamma is unity) on each element of the cathode. These photoelectrons are accelerated toward and magnetically focused onto the surface of a thin semiconducting target (Fig. 41). Electrons strike this target with sufficient energy to liberate a larger number of secondary electrons (typically 5) for each incident primary. The secondary electrons are collected by a mesh closely spaced from the target membrane. Hence, by depletion of electrons from the thin membrane, incremental areas become positive in proportion to the number of photoelectrons striking each element. In cases of highlight-level operation, parts of the target may become charged to target (collector) mesh po-

* A. Rose, P. K. Weimer, and H. B. Law, "The Image Orthicon—A Sensitive Television Pickup Tube," *Proceedings of the IRE*, Vol. 34, No. 7, pp. 424–432; July 1946.

tential, and saturation charge results. This phenomenon accounts for the so-called "knee" in the signal-vs-illumination transfer curve (Fig. 42).

Because the target membrane is very thin, of the order of microns, a charge distribution pattern formed on the image-section surface appears nearly simultaneously and identically on the scanning-section surface.

In the scanning section, an electron gun generates a highly apertured electron beam from a fraction to tens of microamperes in intensity. A solenoidal magnetic-focus coil and saddle-type deflection coils surrounding the scan section focus this beam on the insulator target and move it across the target. Scan-beam electrons impinge on the target at very low velocity, giving rise to relatively few secondary electrons. The target acts somewhat as a retarding field electrode and reflects a large number of the beam electrons that have less than average axial velocity. These two phenomena, small but finite secondary emission and reflection of slow beam electrons, limit scan-beam modulation to a maximum of about 30% at high light levels, and to 2 orders less at threshold. As will be shown later, the large unmodulated return beam current is the primary source of noise in the image orthicon.

Another problem created by the retarding-field aspect of low-velocity target scanning appears when the deflected beam does not strike the target normally. Since the entire beam-velocity component normal to the surface is now reduced by the cosine of the angle of incidence, the effective beam impedance is greatly increased. To overcome this problem, the decelerating field between grids 4 and 5 is shaped such that the electron beam always approaches normal to the plane of the target at a low velocity. If the elemental area on the target is positive, then electrons from the scanning

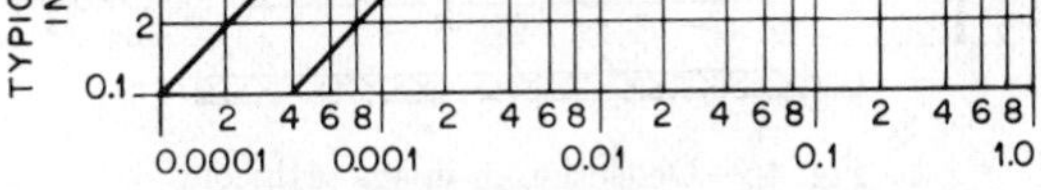

Fig. 42—Basic light-transfer characteristic for type *5820* and *5826* image orthicons. The curves are for small-area highlights illuminated by tungsten light, white fluorescent light, or daylight. *By permission of RCA, copyright proprietor.*

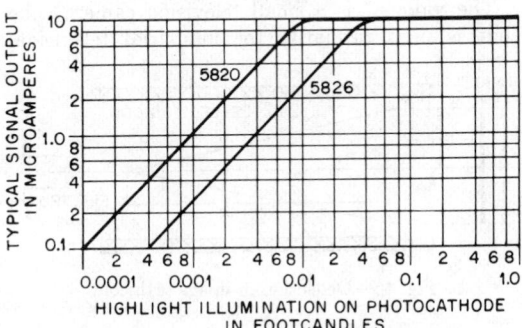

beam deposit until the charge is neutralized. If the elemental area is at cathode potential (corresponding to a dark picture area), no electrons are deposited. In both cases the excess beam electrons are turned back and focused into a 5-stage electron multiplier. The charges existing on either side of the semiconductive target membrane will, by conductivity, neutralize each other in less than one frame time. Electrons turned back at the target form a return beam that has been amplitude-modulated in accordance with the charge pattern of the target.

The return beam is redirected by the deflection and focus fields toward the electron gun where it originated. Atop the electron gun, and forming the final aperture for that gun, is a flat secondary-emitting surface comprising the first dynode of the electron multiplier. The return beam strikes this surface, generating secondary electrons in a ratio of approximately 4:1.

Grid 3 facilitates a more complete collection by dynode 2 of the secondary electrons emitted from dynode 1. The gain of the multiplier is high enough that in operation the limiting noise is the shot noise of the returned electron beam rather than the input noise of the video amplifier.

Signal and Noise: Typical signal output current for tube types *5820* and *5826* are shown in Fig. 42. The tubes should be operated so that the highlights on the photocathode bring the signal output slightly over the knee of the signal-output curve.

The spectral response of the types *5820* and *5826* is shown in Fig. 43. Note that when a Wratten 6 filter is used with the tube, a spectral curve closely approximating that of the human eye is obtained.

From the standpoint of noise, the total television system can be represented as shown in Fig. 44, where I_s= signal current, I_n= total image-orthicon noise current, E_{nt}= thermal noise in R_1, E_{ns}= shot noise in the input amplifier tube, R_1= input load, C_1= total input shunt capacitance, and R_t= shot-noise equivalent resistance of the input amplifier $= 2.5/g_m$ for triode or cascode input $= [I_b/(I_b+I_c)][(2.5/g_m)+(20I_{c2}/g_m^2)]$ for pentode input, with g_m= transconductance of input tube or cascode combination, I_b= amplifier direct plate current, and I_c= amplifier direct screen-grid current.

The noise added per stage is

$$\Delta n = [\sigma/(\sigma-1)]^{1/2}$$

where σ= stage gain in the multiplier. For a total multiplier noise figure to be directly usable, it must be referred to the first-dynode current, therefore, for 5 multiplier stages

$$\overline{\Delta N} = \Delta n^2 + \frac{\Delta n^2}{\sigma^2} + \frac{\Delta n^2}{\sigma^4} + \frac{\Delta n^2}{\sigma^6} + \frac{\Delta n^2}{\sigma^8}$$

where $\Delta N=$ electron-multiplier noise factor referred to multiplier input.

After combining all noise sources

$$\frac{S}{N}=\frac{I_s}{\left\{F\left[2eIk_m^2+4kT\left(\frac{1}{R_1}+\frac{R_t}{R_1^2}+\frac{\omega^2C_1^2R_t}{3}\right)\right]\right\}^{1/2}}$$

where $S/N=$ signal-to-noise ratio, $F=$ bandwidth in hertz, $e=$ electron charge $=1.6\times10^{-19}$ coulomb, $I=$ image-orthicon beam current, $k_m=$ electron-multiplier noise factor, referred to multiplier output $=m\Delta N$, $k=$ Boltzmann's constant $=1.38\times10^{-23}$ joule/kelvin, $T=$ absolute temperature in kelvins, and $\omega=2\pi f$ in hertz.

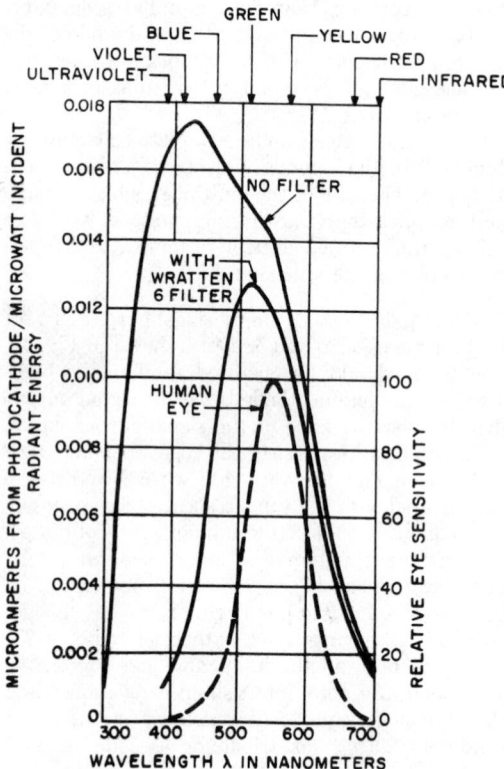

Fig. 43—Spectral sensitivity of image orthicon. *By permission of RCA, copyright proprietor.*

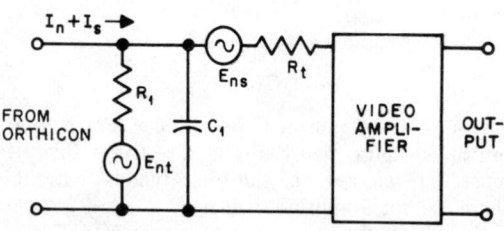

Fig. 44—Equivalent circuit for noise in orthicon and first amplifier stage.

The signal current is an alternating-current signal superimposed on a larger direct beam current. This can be thought of as a modulation of the beam current. Properly adjusted tubes obtain as much as 30-percent modulation.

$$I_s=mMI$$

where $m=$ multiplier gain and $M=$ percentage modulation. If S/N is now rewritten

$$\frac{S}{N}=\frac{I_s}{\left[4kTF\left(\frac{2eI_sm\overline{\Delta N}^2}{4kTM}+\frac{1}{R_1}+\frac{R_t}{R_1^2}+\frac{\omega^2C_1^2R_t}{3}\right)\right]^{1/2}}.$$

In typical television operation, the thermal noise of the load resistor and the shot noise of the first amplifier can be neglected.

Focusing and Scanning Fields: The electron optics of the scanning section of the tube are quite complicated and space does not permit the inclusion of the complete equations. A simple relationship between the strength of the magnetic focusing field and the magnetic deflection field is given below.

The image orthicon is usually operated with multiple-node focus in the scanning section. Working at a multiple-node focus not only demands more focus current but also more deflection current. Note the deflection path in Fig. 45. Let $H=$ horizontal dimension of scanned area or target, $L=$ effective length of horizontal deflection field, $H_d=$ horizontal deflection field (peak-to-peak value), and $H_f=$ focusing field. Then

$$H_d=H_fH/L$$

for the image orthicon, $H\approx1.25$ inches, and $L\approx4$ inches. Thus $H_f\approx75$ gauss, and $H_d\approx23$ gauss.

Vidicons

The vidicon* is a small television camera tube that is used primarily for industrial television,

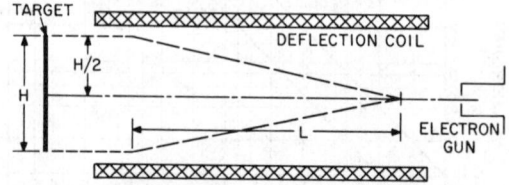

Fig. 45—Deflection in image orthicon.

* B. H. Vine, R. B. Janes, and F. S. Veith, "Performance of the Vidicon—A Small Developmental Camera Tube," *RCA Review*, Vol. 13, No. 1, pp. 3–10; March 1952. P. Weimer, S. Forgue, and R. Goodrich, "The Vidicon Photoconductive Camera Tube," *Electronics*, Vol. 23, No. 5, pp. 70–73; May 1950.

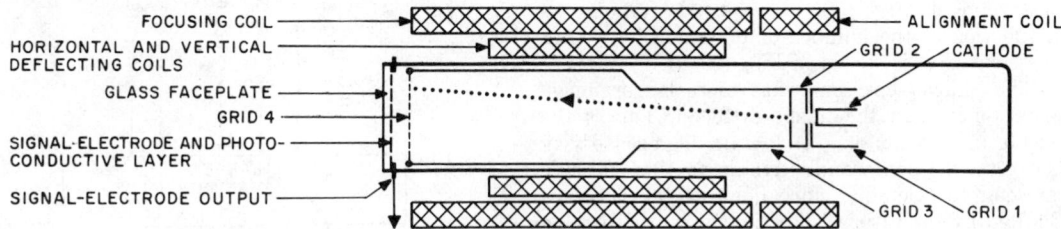

Fig. 46—Vidicon construction. *By permission of RCA, copyright proprietor.*

space applications, and studio film pickup because of its small size and simplicity.

As shown in Fig. 46, the tube consists of a signal electrode composed of a transparent conducting film on the inner surface of the faceplate; a thin layer (a few micrometers) of photoconductive material deposited on the signal electrode; a fine mesh screen (grid 4) located adjacent to the photoconductive layer; a focusing electrode (grid 3) connected to grid 4; and an electron gun.

Principle of Operation: Each elemental area of the photoconductor can be likened to a leaky capacitor with one plate electrically connected to the signal electrode that is at some positive voltage (usually about 20 volts) with respect to the thermionic cathode of the electron gun and the other plate floating except when commutated by the electron beam. Initially, the gun side of the photoconductive surface is charged to cathode potential by the electron gun, thus leaving a charge on each elemental capacitor. During the frame time, these capacitors discharge in accordance with the value of their leakage resistance, which is determined by the amount of light falling on that elemental area. Hence, there appears on the gun side of the photoconductive surface a positive-potential pattern corresponding to the pattern of light from the scene imaged on the opposite surface of the layer. Even those areas that are dark discharge slightly, since the dark resistivity of the material is not infinite.

The electron beam is focused at the surface of the photoconductive layer by the combined action of the uniform magnetic field and the electrostatic field of grid 3. Grid 4 serves to provide a uniform decelerating field between itself and the photoconductive layer such that the electron beam always approaches the surface normally and at a low velocity. When the beam scans the surface, it deposits electrons where the potential of the elemental area is more positive than that of the electron-gun cathode. At this moment the electrical circuit is completed through the signal-electrode circuit to ground. The amount of signal current which flows depends on the amount of discharge in the elemental capacitor, which in turn depends on the amount of light falling on this area.

Alignment of the beam is accomplished by a transverse magnetic field produced by external coils located at the base end of the focusing coil.

Deflection of the beam is accomplished by the transverse magnetic fields produced by external deflecting coils.

Signal and Noise: Since the vidicon acts as a constant-current generator as far as signal current is concerned, the value of the load resistor is determined by band-pass and noise considerations in the input circuit of the video amplifier. Unlike the image orthicon, vidicon signal current is removed at the target, and only that portion of the scan beam actually involved in the target discharge contributes shot noise. Moreover, electron-beam contributions to noise are minimal for low-light portions of the scene.

The primary noise associated with vidicon operation is seldom scan-beam shot noise. Where the signal current is less than 1 microampere and the band pass is relatively wide, the principal noise in the system is contributed by the input circuit and the first stage of the video amplifier. To minimize the thermal noise of the load resistor, its resistance is made much higher than flat-band-pass considerations would indicate, since signal voltage increases directly and noise voltage increases as the square root. To correct for attenuation of the signal with increasing frequency, the amplitude response of the video amplifier frequently employs high-frequency boost of the following form, where C_1 and R_1 refer to Fig. 47.

$$G = G_0 (1 + 4\pi^2 F^2 C_1^2 R_1^2)^{1/2} / R_1.$$

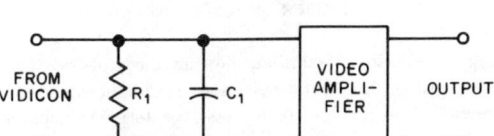

Fig. 47—Input circuit for first-stage amplifier in vidicon circuit.

A representative plot of amplitude response as a function of the number of television lines (per raster height) is shown in Fig. 48.

The vidicon has somewhat more lag or image persistence than the image orthicon. This is the result of two factors. To obtain high-sensitivity surfaces, the photoconductive decay time is made as long as tolerable, since quantum efficiency is limited by the ratio of effective carrier lifetime to carrier transit time across the photoconductor. A second source of lag is simply the *RC* time constant of the target recharging circuit; that is, the target capacitance and the beam impedance.

The spectral response of most commercial vidicons, designated S-18, is more actinic than the human eye. Figure 49 compares these responses with the spectrum of a 2854 K tungsten source.

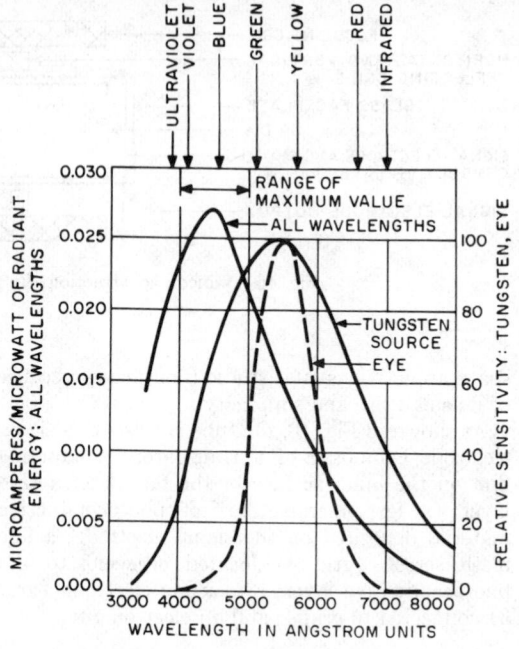

Fig. 49—Spectral response of vidicon. *By permission of RCA, copyright proprietor.*

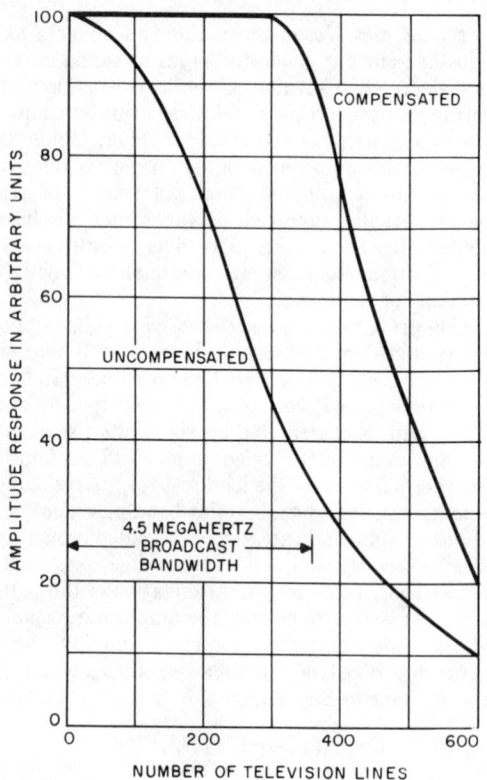

Fig. 48—Vidicon resolution, showing uncompensated and compensated horizontal responses. Highlight signal-electrode microamperes=0.35; test pattern=transparent square-wave resolution wedge; 80 television lines= 1-megahertz bandwidth. *By permission of RCA, copyright proprietor.*

Variations of the Vidicon

Interest in optical guidance and surveillance from air and spacecraft has given rise to a wide variety of vidicon camera tubes. To treat these variations in detail becomes encyclopedic, but the following gives some indication of the choices now available to the user.

Effective Sensitivity: True photoconductive tubes now offer sensitivities of 150–200 nanoamperes for $\frac{1}{2}$-footcandle illumination with 20 nanoamperes dark current. Improved methods of deposition of photoconductors have made possible higher voltage operation without objectionable dark shading. Special devices using junction effects promise even better sensitivity.

Spectral Response: Available photoconductors, taken as a whole, provide sensitivity over the entire visible range with usual (7056) glass windows. Quartz window tubes offer useful sensitivity to below 2000 angstrom units. Numerous applications of direct excitation of photoconductors by X-radiation have been reported. High-velocity electron excitation (bombardment-induced conductivity) is also in use.

Size and Deflection: Vidicons are available in sizes ranging from $\frac{1}{2}$ inch to 2 inches in diameter. Various combinations of deflection and focus are available.

Storage: A number of manufacturers have produced vidicons with long storage characteristics. Many are merely long-lag tubes; however, a few rely on high-resistivity materials or barrier layers to retain stored charge through minimal dark current. One such device, once exposed properly to a scene, regenerates the scene through readout over a period of the order of half an hour.

LIGHT-EMITTING TUBES

Cathode-Ray Tubes

A cathode-ray tube* is a vacuum tube in which an electron beam, deflected by applied electric and/or magnetic fields, indicates by a trace on a fluorescent screen the instantaneous value of the actuating voltages and/or currents. A typical high-intensity cathode-ray tube with post-deflection acceleration is shown in Fig. 50.

Principle of Operation: The function of the cathode-ray tube is to convert an electrical signal

into a visual display. The tube contains an electron-gun structure (to provide a narrow beam of electrons) and a phosphor screen (refer to section on Phosphor Screens). The electron beam is directed to the phosphor screen and strikes it, causing light to be emitted in a small area or spot in proportion to the intensity of the electron beam. The beam intensity varies as a function of the electrical signal which is applied to the control element in the electron gun.

The electrical signal that controls the beam intensity corresponds to the desired picture information; therefore, in modulating the electron beam the individual picture elements can be reproduced on the phosphor screen in the same degree of black or white as in the original picture. Although one element is not enough to reproduce a picture, the same process is carried out for all picture elements in successive order, and each element is positioned

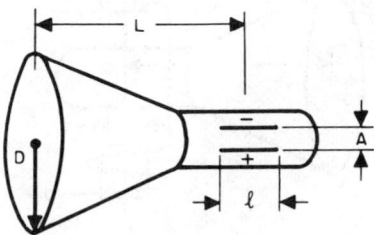

Fig. 51—Electrostatic deflection.

correctly on the phosphor screen to reproduce the entire picture. Means are provided either internally or externally of the tube for positioning or deflecting the electron beam over the phosphor-screen area in some systematic fashion to reproduce the entire picture as a visual output.

Electric-Field Deflection: Deflection is proportional to the deflection voltage, inversely proportional to the accelerating voltage, and in the direction of the applied field (Fig. 51). For structures using straight and parallel deflection plates, it is given by

$$D = E_d L l / 2 E_a A$$

where D = deflection in centimeters, E_a = accelerating voltage, E_d = deflection voltage, l = length of deflection plates or deflecting field in centimeters, L = length from center of deflecting field to screen in centimeters, and A = separation of plates.

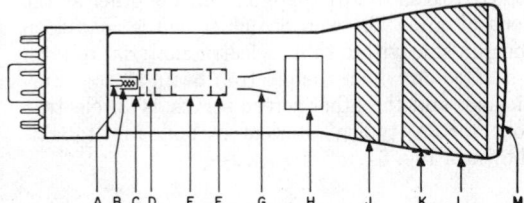

Fig. 50—Electrode arrangement of typical electrostatic focus and deflection cathode-ray tube. A—heater, B—cathode, C—control electrode, D—screen grid or pre-accelerator, E—focusing electrode, F—accelerating electrode, G—deflection-plate pair, H—deflection-plate pair, J—conductive coating connected to accelerating electrode, K—intensifier-electrode terminal, L—intensifier electrode (conductive coating on glass), M—fluorescent screen.

* K. R. Spangenberg, "Vacuum Tubes," 1st edition, McGraw-Hill Book Company, New York; 1948.

Magnetic-Field Deflection: Deflection is proportional to the flux or the current in the coil, inversely proportional to the square root of the accelerating voltage, and at right angles to the direction of the applied field (Fig. 52). Deflection is given by

$$D = 0.3LlH/(E_a)^{1/2}$$

where $H =$ flux density in gauss, and $l =$ length of deflecting field in centimeters.

Deflection Sensitivity: The deflection sensitivity is linear up to the frequency where the phase of the deflecting voltage begins to reverse before an electron has reached the end of the deflecting field.

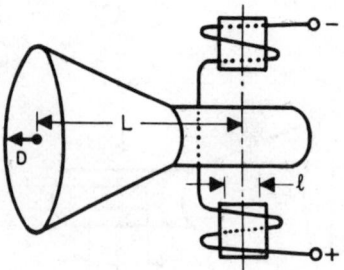

Fig. 52—Magnetic deflection.

Beyond this frequency, sensitivity drops off, reaching zero and then passing through a series of maxima and minima as $n = 1, 2, 3, \cdots$. Each succeeding maximum is of smaller magnitude.

$$D_{\text{zero}} = n\lambda(v/c)$$

$$D_{\text{max}} = (2n-1)(\lambda/2)(v/c)$$

where $D =$ deflection in centimeters, $v =$ electron velocity in centimeters/second, $c =$ velocity of light $\approx 3 \times 10^{10}$ centimeters/second, and $\lambda =$ free-space wavelength in centimeters.

Magnetic Focusing: There is more than one value of current that will focus, and best focus is at the minimum value. For an average coil $IN = 220 \ (V_0 d/f)^{1/2} =$ ampere-turns, $V_0 =$ accelerating voltage in kilovolts, $d =$ mean diameter of coil, $f =$ focal length, and d and f are in the same units.

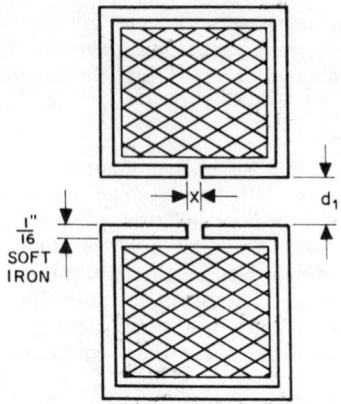

Fig. 53—Magnetic focusing.

A well-designed shielded coil will require fewer ampere-turns.

Figure 53 is an example of good shield design, where

$$x = d_1/20.$$

Storage Cathode-Ray Tubes

The storage cathode-ray tube* produces a visual display of controllable duration. The tube has two electron guns, a phosphor viewing screen, and two fine-mesh metal screens. One of the electron guns is referred to as the writing gun and the other as the flooding gun. The web of one screen is coated on the gun side with a thin dielectric material to form a surface on which the electron beam stores information, and the other screen serves as an electron collector. A typical storage cathode-ray tube is shown in Fig. 54.

Principle of Operation: The writing gun emits a pencil-like electron beam which is intensity modulated by the information to be stored. The information is in the form of an electrical input signal. The storage surface is scanned by this high-resolution beam which actually strikes this surface. A positive-charge image, corresponding in value to the input signal pattern, is imposed on the storage surface where it remains until it decays or is erased. The storage screen forms an array of ele-

* M. Knoll and B. Kazan, "Storage Tubes and their Basic Principles," John Wiley & Sons, New York; 1952.

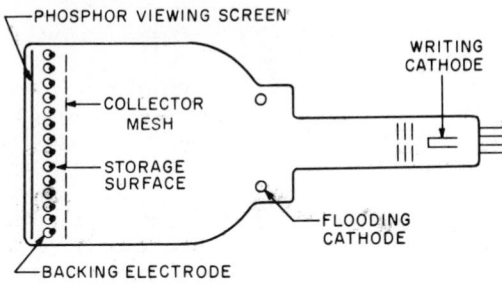

Fig. 54—Construction of storage cathode-ray tube.

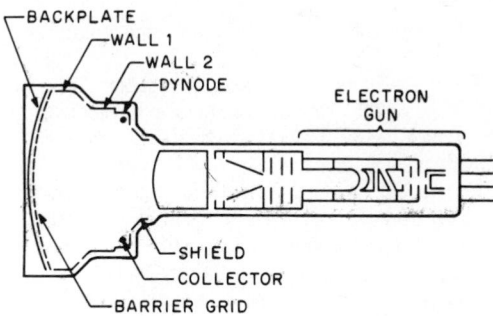

Fig. 55—Construction of barrier-grid storage tube.

mental electron guns, with each mesh hole considered as a control element of one of the guns. After the desired information has been stored on the storage mesh, the entire surface is flooded by an electron beam from the flooding gun. The value of positive charge deposited at each mesh aperture controls the amount of flooding beam current that can pass through the mesh aperture to the phosphor viewing screen. The current that passes through the mesh strikes the phosphor viewing screen, where a light output is observed in proportion to the bombarding-current density and the energy with which the electrons strike the phosphor. In other words, a grey scale is reproduced in the stored image. After the stored information has been observed or recorded, it is erased from the storage surface by flooding the storage surface with low-velocity electrons. Thus a net negative charge is deposited on each elemental area of the storage surface until flooding cathode potential is reached. The storage surface is then prepared for storing a new image.

Barrier-Grid Storage Tube

The barrier-grid storage tube is a form of cathode-ray tube in which information can be stored as an electrostatic charge. There is no visual display, since both the input and the output signals are electrical. The operation of this tube is closely similar to that of the storage cathode-ray tube. The tube (Fig. 55) consists of an electron gun for writing in and reading out the information, a curved target on which the information is stored, a collector electrode, and an associated electron optical system for focusing the signal electrons onto the collector electrode.

Principle of Operation: The target of the barrier-grid storage tube consists essentially of an array

of elemental capacitors, which are charged or discharged by the action of the electron beam from the electron gun. During the write operation, an electrical signal (such as a video signal) is applied to a control electrode of the electron gun, which electrode in turn modulates the electron beam generated by the gun. This modulated beam is scanned across the target, and the information is stored on the target in the form of a pattern of charged areas. During the read cycle an unmodulated electron beam is scanned across the target. As this reading beam approaches each elemental charged area on the target, electrons leave the area and are attracted to the collector. This current constitutes the output signal.

Scan Converter

The scan converter is a tube in which information can be stored as an electrostatic charge. In the scan converter there is no visual display; its output as well as its input are electrical signals. The tube (Fig. 56) consists of two electron guns on opposite ends of the tube on the same axis, a

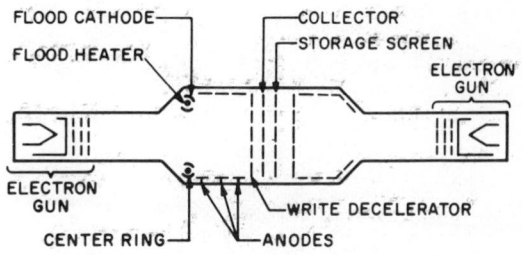

Fig. 56—Construction of scan converter.

collector mesh, and a storage screen comprised of a fine-mesh metal screen coated with a dielectric material serving as the storage surface. The holes of the screen remain open. Each electron gun directs a beam of electrons to the common storage mesh. Each beam is scanned over the storage surface by its own deflection system. One beam writes and the other reads, and it is possible for both modes of operation to occur simultaneously.

Principle of Operation: An electrical input signal, corresponding to the information to be stored, is applied to the writing gun. This signal controls the intensity of the electron beam which deposits or writes a pattern of electrostatic charge on the storage surface. This written information is read out in the form of an electrical output signal by the reading gun. The value of charge that has been deposited at each mesh aperture of the storage screen controls the amount of writing-beam current that can pass through the screen to the collector. The output signal thus varies as a function of the charge stored at each point on the storage screen. Either of the electron beams may be used to erase a stored picture after it has been read out thousands of times.

APPLICATIONS OF ELECTRON TUBES

The advent of transistors is limiting the use of vacuum tubes to special applications. The voltage-handling capabilities of electron tubes satisfy the requirements for high-power oscillator, amplifier, and certain pulse service applications. Tubes are used in the high-power stages of radio and similar transmitters, as modulators of high-power radio-frequency amplifiers, and for specific conditions as pulse generators for radar and other pulse service equipment. Transistors and other semiconductor devices have largely replaced tubes in low-power applications.

CLASSIFICATION

It is common practice to differentiate between types of vacuum-tube circuits, particularly amplifiers, on the basis of the operating regime of the tube.

Class-A: Grid bias and alternating grid voltages such that plate current flows continuously throughout electrical cycle ($\theta_p = 360$ degrees).

Class-AB: Grid bias and alternating grid voltages such that plate current flows appreciably more than half but less than entire electrical cycle ($360° > \theta_p > 180°$).

Class-B: Grid bias close to cutoff such that plate current flows only during approximately half of electrical cycle ($\theta_p \approx 180°$).

Class-C: Grid bias appreciably greater than cutoff so that plate current flows for appreciably less than half of electrical cycle ($\theta_p < 180°$).

A further classification between circuits in which positive grid current is conducted during some portion of the cycle, and those in which it is not, is denoted by subscripts 2 and 1, respectively. Thus a class-AB_2 amplifier operates with a positive swing of the alternating grid voltage such that positive electronic current is conducted and accordingly in-phase power is required to drive the tube.

GENERAL DESIGN

For quickly estimating the performance of a tube from catalog data, or for predicting the characteristics needed for a given application, the ratios given below may be used.

Table 1 gives correlating data for typical operation of tubes in the various amplifier classifications. From the table, knowing the maximum ratings of a tube, the maximum power output, currents, voltages, and corresponding load impedance may be estimated. Thus, taking for example a type F-124-A water-cooled transmitting tube as a class-C radio-frequency power amplifier and oscillator—the constant-current characteristics of which are shown in Fig. 1—published maximum ratings are as follows.

DC plate voltage:

$$E_b = 20\,000 \text{ volts}$$

DC grid voltage:

$$E_c = 3000 \text{ volts}$$

DC plate current:

$$I_b = 7 \text{ amperes}$$

RF grid current:

$$I_g = 50 \text{ amperes}$$

Plate input:

$$P_i = 135\,000 \text{ watts}$$

Plate dissipation:

$$P_p = 40\,000 \text{ watts.}$$

Maximum conditions may be estimated as follows. For $\eta = 75$ percent

$P_i = 135\,000$ watts

$E_b = 20\,000$ volts.

Power output $P_o = \eta P_i = 100\,000$ watts.
Average dc plate current $I_b = P_i/E_b = 6.7$ amperes.

Fig. 1—Constant-current characteristics with typical load lines AB—class C, CD—class B, EFG—class A, and HJK—class AB.

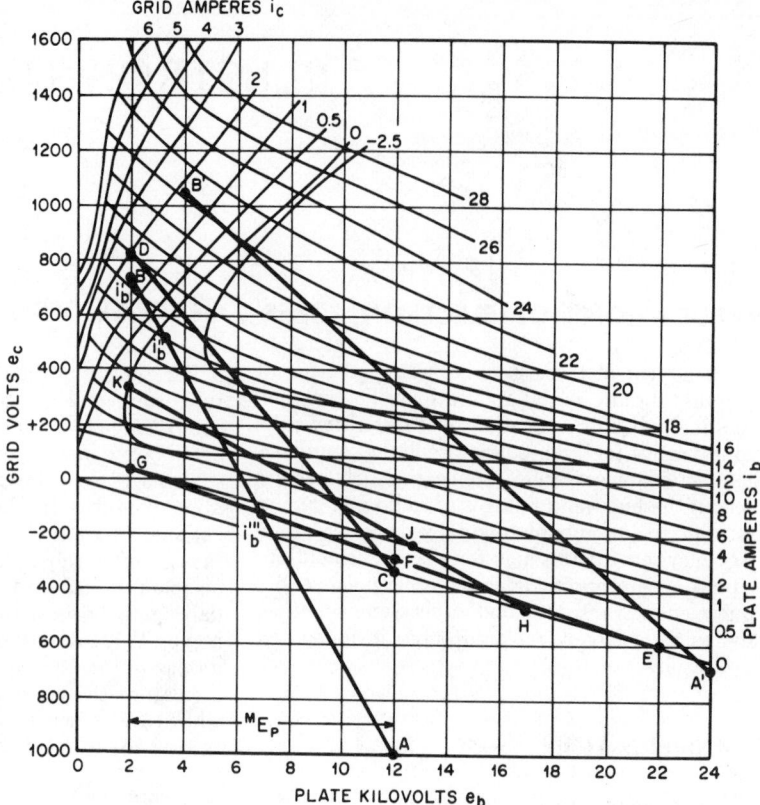

From tabulated typical ratio $^{M}i_b/I_b = 4$, instantaneous peak plate current $^{M}i_b = 4I_b = 27$ amperes.*

The rms plate alternating-current component, taking ratio $I_p/I_b = 1.2$

$$I_p = 1.2I_b = 8 \text{ amperes.}$$

The rms value of the plate alternating-voltage component from the ratio $E_p/E_b = 0.6$ is $E_p = 0.6E_b = 12\,000$ volts.

The approximate operating load resistance R_l is now found from

$$R_l = E_p/I_p = 1500 \text{ ohms.}$$

An estimate of the grid drive power required may be obtained by reference to the constant-current characteristics of the tube and determination of the peak instantaneous positive grid current $^{M}i_c$ and the corresponding instantaneous total grid voltage $^{M}e_c$. Taking the value of grid bias E_c for the given operating condition, the peak alternating grid drive voltage is

$$^{M}E_g = (^{M}e_c - E_c)$$

from which the peak instantaneous grid drive

*In this discussion, the superscript M indicates the use of the maximum or peak value of the varying component, i.e., $^{M}i_b =$ maximum or peak value of the alternating component of the plate current.

power is

$$^{M}P_c = {}^{M}E_g \, {}^{M}i_c.$$

An approximation to the average grid drive power P_g, necessarily rough due to neglect of negative grid current, is obtained from the typical ratio

$$I_c/^{M}i_c = 0.2$$

of dc to peak value of grid current, giving

$$P_g = I_c E_g = 0.2^{M}i_c E_g \text{ watt.}$$

Plate dissipation P_p may be checked with published values since

$$P_p = P_i - P_o.$$

It should be borne in mind that combinations of published maximum ratings as well as each individual maximum rating must be observed. Thus, for example in this case, the maximum dc plate operating voltage of 20 000 volts does not permit operation at the maximum dc plate current of 7 amperes since this exceeds the maximum plate input rating of 135 000 watts.

Plate load resistance R_l may be connected directly in the tube plate circuit, as in the resistance-coupled amplifier, through impedance-matching elements as in audio-frequency transformer coupling, or effectively represented by a loaded paral-

TABLE 1—TYPICAL AMPLIFIER OPERATING DATA. MAXIMUM SIGNAL CONDITIONS—PER TUBE.

Function	Class A	Class B af (p-p)	Class B rf	Class C rf
Plate efficiency η (percent)	20–30	35–65	60–70	65–85
Peak instantaneous to dc plate current ratio $^M i_b / I_b$	1.5–2	3.1	3.1	3.1–4.5
RMS alternating to dc plate current ratio I_p / I_b	0.5–0.7	1.1	1.1	1.1–1.2
RMS alternating to dc plate voltage ratio E_p / E_b	0.3–0.5	0.5–0.6	0.5–0.6	0.5–0.6
DC to peak instantaneous grid current $I_c / ^M i_c$		0.1–0.25	0.1–0.25	0.1–0.25

lel-resonant circuit as in most radio-frequency amplifiers. In any case, calculated values apply only to effectively resistive loads, such as are normally closely approximated in radio-frequency amplifiers. With appreciably reactive loads, operating currents and voltages will in general be quite different and their precise calculation is quite difficult.

The physical load resistance present in any given setup may be measured by audio-frequency or radio-frequency bridge methods. In many cases, the proper value of R_l is ascertained experimentally as in radio-frequency amplifiers that are tuned to the proper minimum dc plate current. Conversely, if the circuit is to be matched to the tube, R_l is determined directly as in a resistance-coupled amplifier or as

$$R_l = N^2 R_s$$

in the case of a transformer-coupled stage, where N is the primary-to-secondary voltage transformation ratio. In a parallel-resonant circuit in which the output resistance R_s is connected directly in one of the reactance legs

$$R_l = X^2 / R_s = L / C r_s = Q X$$

where X is the leg reactance at resonance (ohms), L and C are leg inductance in henries and capacitance in farads, respectively, and $Q = X / R_s$.

GRAPHIC DESIGN METHODS

When accurate operating data are required, more-precise methods must be used. Because of the nonlinear nature of tube characteristics, graphic methods usually are most convenient and rapid. Examples of such methods are given below.

A comparison of the operating regimes of class A, AB, B, and C amplifiers is given in the constant-current characteristics graph of Fig. 1. The lines corresponding to the different classes of operation

are each the locus of instantaneous grid e_c and plate e_b voltages, corresponding to their respective load impedances.

For radio-frequency amplifiers and oscillators having tuned circuits giving an effectively resistive load, plate and grid tube and load alternating voltages are sinusoidal and in phase (disregarding transit time), and the loci become straight lines.

For amplifiers having nonresonant resistive loads, the loci are in general nonlinear except in the distortionless case of linear tube characteristics (constant r_p), for which they are again straight lines.

Thus, for determination of radio-frequency performance, the constant-current chart is convenient. For solution of audio-frequency problems, however, it is more convenient to use the $(i_b - e_c)$ transfer characteristics of Fig. 2 on which a dynamic load line may be constructed.

Methods for calculation of the most important cases are given below.

Class-C Radio-Frequency Amplifier or Oscillator

Draw straight line from A to B (Fig. 1) corresponding to chosen dc operating plate and grid voltages, and to desired peak alternating plate and grid voltage excursions. The projection of AB on the horizontal axis thus corresponds to $^M E_p$. Using Chaffee's 11-point method of harmonic analysis, lay out on AB points

$$e_p' = {}^M E_p$$

$$e_p'' = 0.866 {}^M E_p$$

$$e_p''' = 0.5 {}^M E_p$$

to each of which correspond instantaneous plate currents i_b', i_b'', and i_b''' and instantaneous grid currents i_c', i_c'', and i_c'''. The operating currents

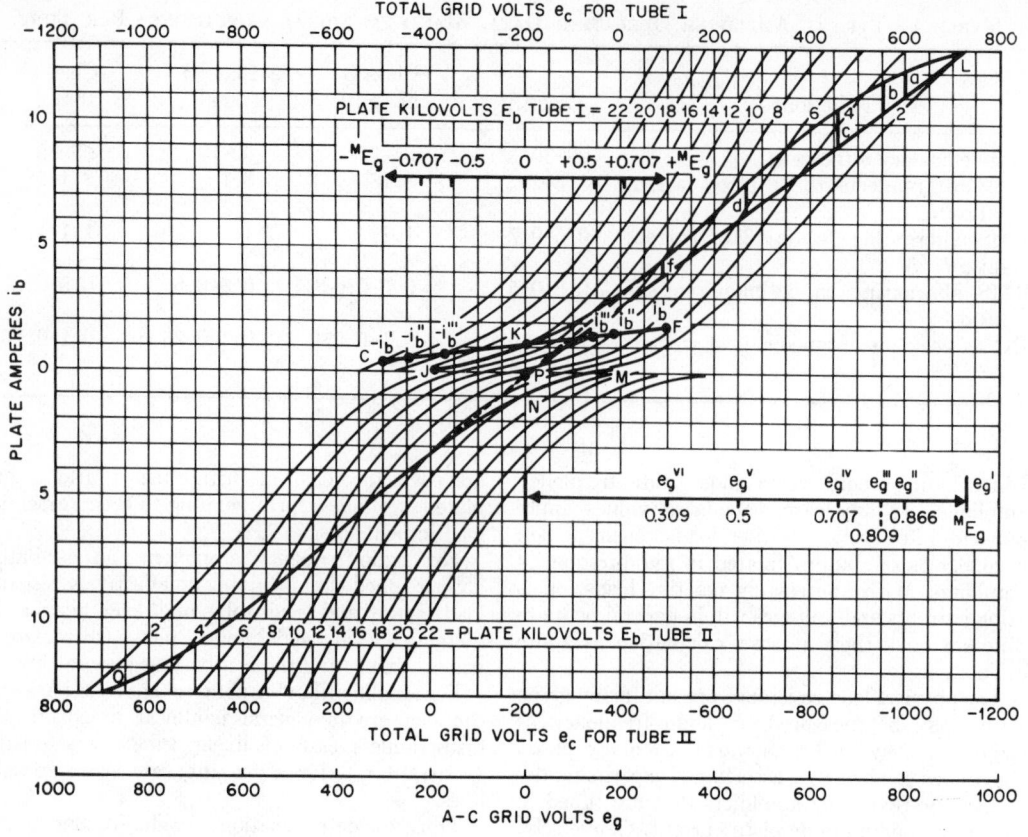

Fig. 2—Transfer characteristics i_b versus e_b with class A_2—CKF and class B—OPL load lines.

are obtained from

$$I_b = [i_b' + 2i_b'' + 2i_b''']/12$$

$$I_c = [i_c' + 2i_c'' + 2i_c''']/12$$

$$^M I_p = [i_b' + 1.73i_b'' + i_b''']/6$$

$$^M I_g = [i_c' + 1.73i_c'' + i_c''']/6.$$

Substitution of the above in the following give the desired operating data.

Power output $P_o = (^M E_p \, ^M I_p)/2$

Power input $P_i = E_b I_b$

Average grid excitation power $P_g = (^M E_g \, ^M I_g)/2$

Peak grid excitation power $^M P_c = ^M E_g i_c'$

Plate load resistance $R_l = ^M E_p / ^M I_p$

Grid bias resistance $R_c = E_c / I_c$

Plate efficiency $\eta = P_o / P_i$

Plate dissipation $P_p = P_i - P_o$.

The above procedure may also be applied to plate-modulated class-C amplifiers. Taking the above data as applying to carrier conditions, the analysis is repeated for $^{crest}E_b = 2E_b$ and $^{crest}P_o = 4P_o$

keeping R_l constant. After a cut-and-try method has given a peak solution, it will often be found that combination fixed and self grid biasing as well as grid modulation are indicated to obtain linear operation.

TABLE 2—CLASS-C RF AMPLIFIER DATA FOR 100-PERCENT PLATE MODULATION.

Symbol	Preliminary Carrier	Detailed	
		Carrier	Crest
E_b (volts)	12 000	12 000	24 000
$^M E_p$ (volts)	10 000	10 000	20 000
E_c (volts)	—	−1 000	−700
$^M E_g$ (volts)	—	1 740	1 740
I_b (amp)	2.9	2.8	6.4
$^M I_p$ (amp)	4.9	5.1	10.2
I_c (amp)	—	0.125	0.083
$^M I_g$ (amp)	—	0.255	0.183
P_i (watts)	35 000	33 600	154 000
P_o (watts)	25 000	25 500	102 000
P_g (watts)	—	220	160
η (percent)	75	76	66
R_l (ohms)	2 060	1 960	1 960
R_c (ohms)	—	7 100	7 100
E_{cc} (volts)	—	−110	−110

To illustrate the preceding exposition, a typical amplifier calculation is given below.

Operating requirements (carrier condition):

$E_b = 12\ 000$ volts

$P_o = 25\ 000$ watts

$\eta = 75$ percent.

Preliminary calculation (refer to Tables 1 and 2):

$E_p/E_b = 0.6$

$E_p = 0.6 \times 12\ 000 = 7200$ volts

$^M E_p = 1.41 \times 7200 = 10\ 000$ volts

$I_p = P_o/E_p$

$I_p = 25\ 000/7200 = 3.48$ amperes

$^M I_p = 4.9$ amperes

$I_p/I_b = 1.2$

$I_b = 3.48/1.2 = 2.9$ amperes

$P_i = 12\ 000 \times 2.9 = 35\ 000$ watts

$^M i_b/I_b = 4.5$

$^M i_b = 4.5 \times 2.9 = 13.0$ amperes

$R_l = E_p/I_p = 7200/3.48 = 2060$ ohms.

Complete Calculation: Lay out carrier operating line AB on constant-current graph, Fig. 1, using values of E_b, $^M E_p$, and $^M i_b$ from preliminary calculated data. Operating carrier bias voltage E_c is chosen somewhat greater than twice cutoff value (1000 volts) to locate point A.

The following data are taken along AB.

$i_b' = 13$ amp

$i_b'' = 10$ amp

$i_b''' = 0.3$ amp

$i_c' = 1.7$ amp

$i_c'' = -0.1$ amp

$i_c''' = 0$ amp

$E_c = -1000$ volts

$e_c' = 740$ volts

$^M E_p = 10\ 000$ volts.

From the equations, complete carrier data as follows are calculated.

$^M I_p = [13 + 1.73 \times 10 + 0.3]/6 = 5.1$ amp

$P_o = (10\ 000 \times 5.1)/2 = 25\ 500$ watts

$I_b = [13 + 2 \times 10 + 2 \times 0.3]/12 = 2.8$ amp

$P_i = 12\ 000 \times 2.8 = 33\ 600$ watts

$\eta = (25\ 500/33\ 600) \times 100 = 76$ percent

$R_l = (10\ 000/5.1) = 1960$ ohms

$I_c = [1.7 + 2(-0.1)]/12 = 0.125$ amp

$^M I_g = [1.7 + 1.7(-0.1)]/6 = 0.255$ amp

$P_g = (1740 \times 0.255)/2 = 220$ watts.

Operating data at 100-percent positive modulation crests are now calculated knowing that here

$E_b = 24\ 000$ volts $R_l = 1960$ ohms

and for undistorted operation

$P_o = 4 \times 25\ 500 = 102\ 000$ watts

$^M E_p = 20\ 000$ volts.

The crest operating line A′B′ is now located by trial so as to satisfy the above conditions, using the same equations and method as for the carrier condition.

It is seen that to obtain full-crest power output, in addition to doubling the alternating plate voltage, the peak plate current must be increased. This is accomplished by reducing the crest bias voltage with resultant increase of current conduction period but lower plate efficiency.

The effect of grid secondary emission to lower the crest grid current is taken advantage of to obtain the reduced grid-resistance voltage drop required. By use of combination fixed and grid resistance bias, proper variation of the total bias is obtained. The value of grid resistance required is given by

$$R_c = -(E_c - {}^{\text{crest}}E_c)/(I_c - {}^{\text{crest}}I_c)$$

and the value of fixed bias by

$$E_{cc} = E_c - (I_c R_c).$$

Calculations at carrier and positive crest together with the condition of zero output at negative crest give sufficiently complete data for most purposes. If accurate calculation of audio-frequency harmonic distortion is necessary, the above method may be applied to the additional points required.

Class-B Radio-Frequency Amplifiers

A rapid approximate method is to determine by inspection from the tube characteristics $(i_b - e_b)$ the instantaneous current, i_b' and voltage e_b' corresponding to peak alternating voltage swing from operating voltage E_b.

AC plate current:

$$^M I_p = i_b'/2$$

DC plate current:

$$I_b = i_b'/\pi$$

AC plate voltage:

$$^M E_p = E_b - e_b'$$

Power output:

$$P_o = [(E_b - e_b')i_b']/4$$

Power input:

$$P_i = E_b i_b'/\pi$$

Plate efficiency:

$$\eta = (\pi/4)[1 - (e_b'/E_b)].$$

Thus $\eta \approx 0.6$ for the usual crest value of $^ME_p \approx 0.8E_b$.

The same method of analysis used for the class-C amplifier may also be used in this case. The carrier and crest condition calculations, however, are now made from the same E_b, the carrier condition corresponding to an alternating-voltage amplitude of $^ME_p/2$ such as to give the desired carrier power output.

For greater accuracy than the simple check of carrier and crest conditions, the radio-frequency plate currents $^MI_p'$, $^MI_p''$, $^MI_p'''$, $^MI_p^0$, $-^MI_p'''$, $-^MI_p''$, and $-^MI_p'$ may be calculated for seven corresponding selected points of the audio-frequency modulation envelope $+^ME_g$, $+0.707^ME_g$, $+0.5^ME_g$, 0, -0.5^ME_g, -0.707^ME_g, and $-^ME_g$, where the negative signs denote values in the negative half of the modulation cycle. Designating

$$S' = {}^MI_p' - (-{}^MI_p')$$

$$D' = {}^MI_p' + (-{}^MI_p') - 2{}^MI_p^0$$

the fundamental and harmonic components of the output audio-frequency current are obtained as

$${}^MI_{p1} = (S'/4) + [S''/2(2)^{1/2}] \text{ (fundamental)}$$

$${}^MI_{p2} = (5D'/24) + (D''/4) - (D'''/3)$$

$${}^MI_{p3} = (S'/6) - (S'''/3)$$

$${}^MI_{p4} = (D'/8) - (D''/4)$$

$${}^MI_{p5} = (S'/12) - [S''/2(2)^{1/2}] + (S'''/3)$$

$${}^MI_{p6} = (D'/24) - (D''/4) + (D'''/3).$$

This detailed method of calculation of audio-frequency harmonic distortion may, of course, also be applied to calculation of the class-C modulated amplifier, as well as to the class-A modulated amplifier.

Class-A and AB Audio-Frequency Amplifiers

Approximate equations assuming linear tube characteristics:

Maximum undistorted power output

$${}^MP_o = ({}^ME_p \, {}^MI_p)/2$$

when plate load resistance

$$R_l = r_p \left[\frac{E_c}{({}^ME_p/\mu) - E_c} - 1 \right]$$

and negative grid bias

$$E_c = ({}^ME_p/\mu)[(R_l + r_p)/(R_l + 2r_p)]$$

giving maximum plate efficiency

$$\eta = {}^ME_p \, {}^MI_p / 8E_b I_b.$$

Maximum maximum undistorted power output

$${}^{MM}P_o = {}^ME_p^2/16r_p$$

when

$$R_l = 2r_p \qquad E_c = \tfrac{3}{4}({}^ME_p/\mu).$$

An exact analysis may be obtained by use of a dynamic load line laid out on the transfer characteristics of the tube. Such a line is CKF of Fig. 2, which is constructed about operating point K for a given load resistance r_l from

$$i_b^S = [(e_b^R - e_b^S)/R_l] + i_b^R$$

where R, S, etc., are successive conveniently spaced construction points.

Using the seven-point method of harmonic analysis, plot instantaneous plate currents i_b', i_b'', i_b''', i_b, $-i_b'''$, $-i_b''$, and $-i_b'$ corresponding to $+^ME_g$, $+0.707^ME_g$, $+0.5^ME_g$, 0, -0.5^ME_g, -0.707^ME_g, and $-^ME_g$, where 0 corresponds to the operating point K. In addition to the equations given under class-B radio-frequency amplifiers

$$I_b \text{ average} = I_b + (D'/8) + (D''/4)$$

from which complete data may be calculated.

Class-AB and B Audio-Frequency Amplifiers

Approximate equations assuming linear tube characteristics give (referring to Fig. 1, line CD) for a class-B audio-frequency amplifier

$${}^MI_p = i_b'$$

$$P_o = {}^ME_p \, {}^MI_p/2$$

$$P_i = (2/\pi) E_b \, {}^MI_p$$

$$\eta = (\pi/4)({}^ME_p/E_b)$$

$$R_{pp} = 4({}^ME_p/i_b') = 4R_l.$$

Again an exact solution may be derived by use of the dynamic load line JKL on the $(i_b - e_c)$ characteristic of Fig. 2. This line is calculated about the operating point K for the given R_l (in the same way as for the class-A case). However, since two tubes operate in phase opposition in this case, an identical dynamic load line MNO represents the other half cycle, laid out about the oper-

TABLE 3—DESIGN INFORMATION FOR 3 CLASSES OF AMPLIFIERS.

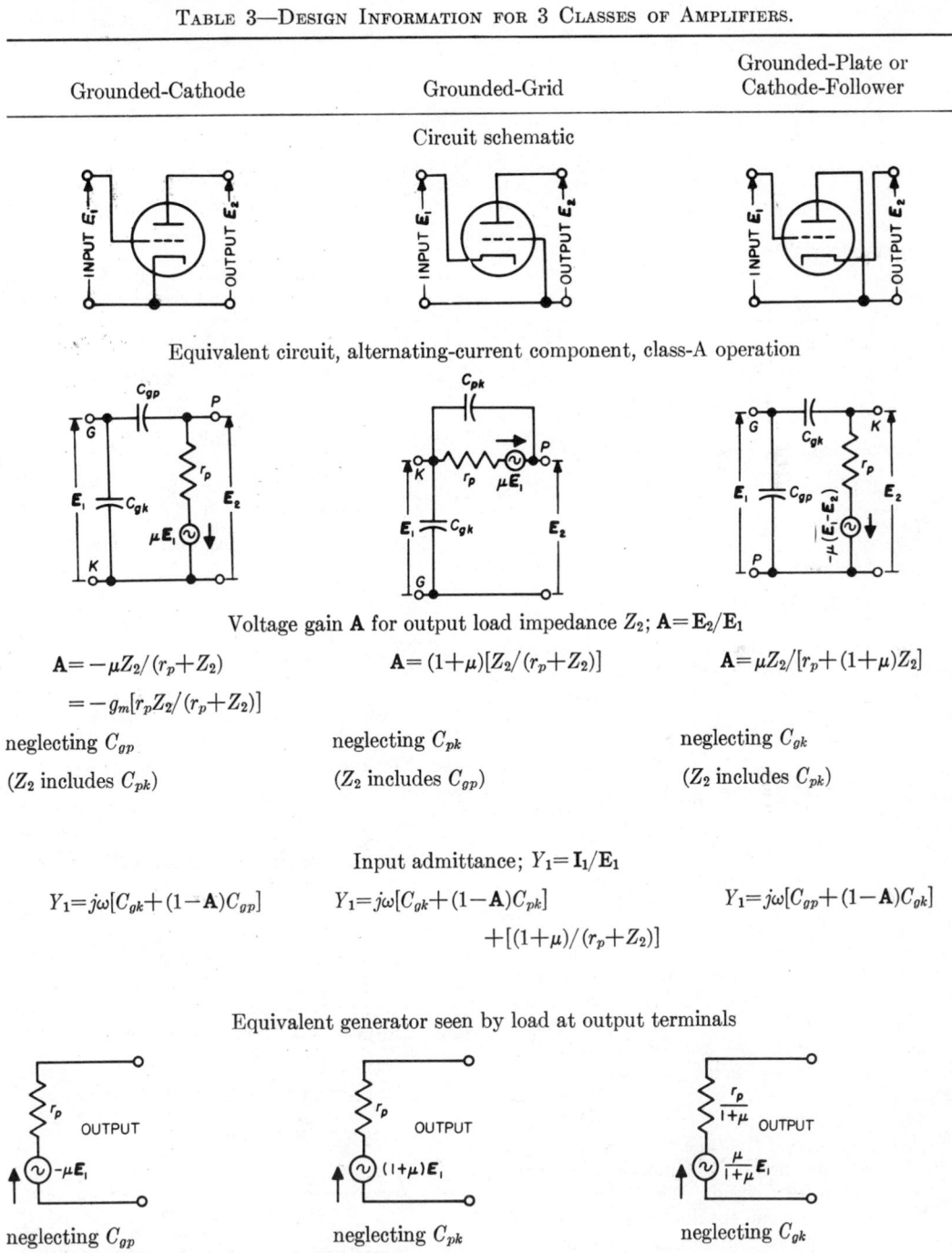

Grounded-Cathode	Grounded-Grid	Grounded-Plate or Cathode-Follower

Circuit schematic

Equivalent circuit, alternating-current component, class-A operation

Voltage gain A for output load impedance Z_2; $A = E_2/E_1$

$A = -\mu Z_2/(r_p + Z_2)$

$= -g_m[r_p Z_2/(r_p + Z_2)]$

neglecting C_{gp}

(Z_2 includes C_{pk})

$A = (1+\mu)[Z_2/(r_p + Z_2)]$

neglecting C_{pk}

(Z_2 includes C_{gp})

$A = \mu Z_2/[r_p + (1+\mu)Z_2]$

neglecting C_{gk}

(Z_2 includes C_{pk})

Input admittance; $Y_1 = I_1/E_1$

$Y_1 = j\omega[C_{gk} + (1-A)C_{gp}]$

$Y_1 = j\omega[C_{gk} + (1-A)C_{pk}]$

$+ [(1+\mu)/(r_p + Z_2)]$

$Y_1 = j\omega[C_{gp} + (1-A)C_{gk}]$

Equivalent generator seen by load at output terminals

neglecting C_{gp}

neglecting C_{pk}

neglecting C_{gk}

ating bias abscissa point but in the opposite direction (see Fig. 2).

Algebraic addition of instantaneous current values of the two tubes at each value of e_c gives the composite dynamic characteristic OPL for the two tubes. Inasmuch as this curve is symmetrical about point P, it may be analyzed for harmonics along a single half-curve PL by the Mouromtseff 5-point method. A straight line is drawn from P to L and ordinate plate-current differences a, b, c, d, f between this line and curve, corresponding to e_g'', e_g''', e_g^{IV}, e_g^{V}, and e_g^{VI}, are measured. Ordinate distances measured upward from curve PL are taken positive.

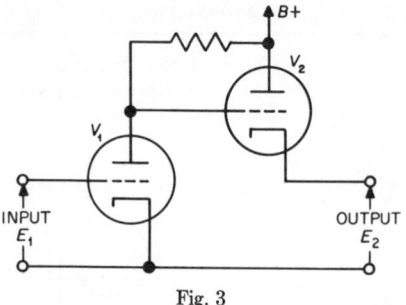

Fig. 3

Fundamental and harmonic current amplitudes and power are found from

$$^{M}I_{p1} = i_{b}' - {}^{M}I_{p3} + {}^{M}I_{p5} - {}^{M}I_{p7} + {}^{M}I_{p9} - {}^{M}I_{p11}$$

$$^{M}I_{p3} = 0.4475(b+f) + (d/3) - 0.578d - \tfrac{1}{2}{}^{M}I_{p5}$$

$$^{M}I_{p5} = 0.4(a-f)$$

$$^{M}I_{p7} = 0.4475(b+f) - {}^{M}I_{p3} + 0.5{}^{M}I_{p5}$$

$$^{M}I_{p9} = {}^{M}I_{p3} - \tfrac{2}{3}d$$

$$^{M}I_{p11} = 0.707c - {}^{M}I_{p3} + {}^{M}I_{p5}.$$

Even harmonics are not present due to dynamic characteristic symmetry. The direct-current and power-input values are found by the 7-point analysis from curve PL and doubled for two tubes.

CLASSIFICATION OF AMPLIFIER CIRCUITS

The classification of amplifiers in classes A, B, and C is based on the operating conditions of the tube. Another classification can be used, based on the type of circuits associated with the tube.

A tube can be considered as a four-terminal network with two input terminals and two output terminals. One of the input terminals and one of the output terminals are usually common; this common junction or point is usually called "ground."

When the common point is connected to the filament or cathode of the tube, we can speak of a grounded-cathode circuit: the most-conventional type of vacuum-tube circuit. When the common point is the grid, we can speak of a grounded-grid circuit, and when the common point is the plate or anode, we can speak of the grounded-anode circuit.

This last type of circuit is most commonly known by the name of *cathode-follower*.

A fourth and most-general class of circuit is obtained when the common point or ground is not directly connected to any of the three electrodes of the tube. This is the condition encountered at uhf where the series impedances of the internal tube leads make it impossible to ground any of them. It is also encountered in such special types of circuits as the *phase-splitter*, in which the impedance from plate to ground and the impedance from cathode to ground are made equal to obtain an output between plate and cathode balanced with respect to ground.

Design information for the first three classifications is given in Table 3, where

Z_2 = load impedance to which output terminals of amplifier are connected

E_1 = phasor input voltage to amplifier

E_2 = phasor output voltage across load impedance Z_2

A = voltage gain of amplifier = E_2/E_1

Y_1 = input admittance to input terminals of amplifier

$\omega = 2\pi \times$ (frequency of excitation voltage E_1)

$j = (-1)^{1/2}.$

AMPLIFIER PAIRS

The basic amplifier classes are often used in pairs or combination forms for special characteristics. The availability of dual triodes makes these combined forms especially useful.

Grounded-Cathode—Grounded-Plate

This pairing provides the gain and 180-degree phase reversal of a grounded-cathode stage (Fig. 3) with a low source impedance at the output terminals. It is especially useful in feedback circuits or for amplifiers driving a low or unknown load impedance. In tuned amplifiers, the possibility of oscil-

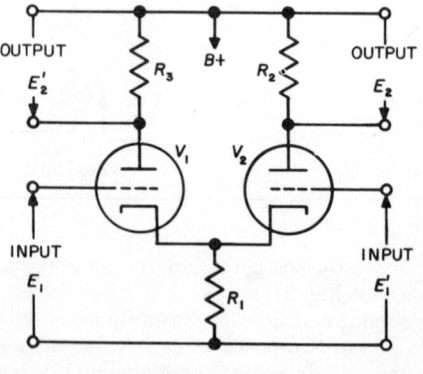

Fig. 4

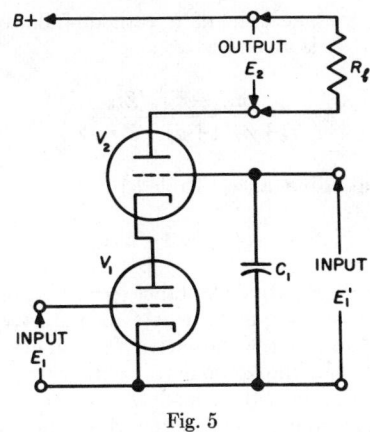

Fig. 5

lation must be considered. Direct coupling is useful for pulse work, permitting large positive input and negative output excursions.

Grounded-Plate—Grounded-Grid (Cathode-Coupled)

Direct coupling is usual, making a very simple structure. Several modified forms are possible with special characteristics.

Cathode-Coupled Amplifier: As a simple amplifier, R_3 and input E_1' (Fig. 4) are short-circuited. Output E_2 is in phase with input E_1. Gain (with $R_1 \gg 1/g_m$) is given by $\mathbf{A} \approx g_m R_2/2$. Even-harmonic distortion is reduced by symmetry, as in a push-pull stage. Due to the in-phase input and output relations, this circuit forms the basis for various R-C oscillators and the class of cathode-coupled multivibrators.

Symmetrical Clipper: With suitable bias adjustment, symmetrical clipping or limiting occurs between V_1 cutoff and V_2 cutoff, without drawing grid current.

Differential Amplifier: With input supplied to E_1 and E_1', the output E_2 responds (approximately) to the difference $E_1 - E_1'$. Balance is improved by constant-current supply to the cathode such as a high value of R_1 (preferably connected from a highly negative supply) or a constant-current pentode. The signal to E_1' should be slightly attenuated for precise adjustment of balance.

Phase Inverter: With R_3 and R_2 both used, approximately balanced (push-pull) outputs (E_2 and E_2') are obtained from either input E_1' or E_1. As a phase inverter (paraphase), one input (E_1) is used, the other being grounded, and R_3 is made slightly less than R_2 to provide exact balance.

Grounded-Cathode—Grounded-Grid (Cascode)

This circuit (Fig. 5) has characteristics somewhat resembling the pentode, with the advantage that no screen current is required. V_2 serves to isolate V_1 from the output load R_l, giving voltage gain equation

$$A = \frac{\mu_1 R_l}{r_{p1} + [(r_{p2} + R_l)/(\mu_2 + 1)]}.$$

For $R_l \ll \mu r_p$,

$$A \approx g_{m1} R_l.$$

For $R_l \gg \mu r_p$,

$$A \approx \mu_1 \mu_2.$$

As an rf amplifier, the grounded-grid stage V_2 drastically reduces capacitive feedback from output to input, without introducing partition noise (as produced by the screen current of a pentode). Shot noise contributed by V_2 is negligible due to the highly degenerative effect of r_{p1} in series with the cathode. The noise figure thus approaches the theoretical noise of V_1 used as a triode, without the undesirable effects of triode plate–grid capacitance.

Because of the 180° phase relation of input and output, this circuit is also valuable in audio feedback circuits, replacing a single stage with considerable increase in gain (for high values of R_l).

The grid of V_2 provides a second input connection E_1' useful for feedback or for gating. The voltage gain from E_1' to the output is considerably reduced, being given by

$$A = R_l \mu / (R_l + \mu r_p).$$

For $R_l \ll \mu r_p$,

$$A_2 \approx R_l / r_{p1}.$$

For $R_l \gg \mu r_p$,

$$A_2 \approx \mu.$$

CATHODE-FOLLOWER DATA

General Characteristics

(**A**) High-impedance input, low-impedance output.

(**B**) Input and output have one side grounded.

(**C**) Good wide-band frequency and phase response.

(**D**) Output is in phase with input.

(**E**) Voltage gain or transfer is always less than one.

(**F**) A power gain can be obtained.

(**G**) Input capacitance is reduced.

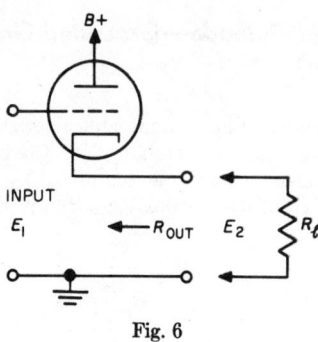

Fig. 6

General Case (Fig. 6)

$\text{Transfer} = E_{\text{out}}/E_{\text{in}} = g_m R_l/[g_m R_l + 1 + (R_l/r_p)]$

$R_{\text{out}} =$ output resistance

$\quad = r_p/(\mu+1)$ or approximately $1/g_m$

$g_m =$ transconductance in mhos
$\quad$ (1000 micromhos = 0.001 mho)

$R_l =$ total load resistance

Input capacitance $= C_{gp} + [C_{gk}/(1+g_m R_l)]$.

Specific Cases

(A) To match the characteristic impedance of the transmission line, R_{out} must equal Z_0 (Fig. 7).

(B) If R_{out} is less than Z_0, add resistor R_c' in series (Fig. 8) so that $R_c' = Z_0 - R_{\text{out}}$.

(C) If R_{out} is greater than Z_0, add resistor R_c in parallel (Fig. 9) so that

$$R_c = Z_0 R_{\text{out}}/(R_{\text{out}} - Z_0).$$

Note 1: Normal operating bias must be provided. To couple a high impedance into a low-impedance transmission line, for maximum transfer choose a tube with a high g_m.

Note 2: Oscillation may occur in a cathode-follower if the source becomes inductive and load capacitive at high frequencies. The general expression for voltage gain of a cathode-follower (including C_{gk}) is given by

$$A = \frac{\mu Z_2 + Z_2 r_p/Z_{gk}}{r_p + Z_2(1+\mu) + Z_2 r_p/Z_{gk}}.$$

The input admittance (Table 3)

$$Y_1 = j\omega[C_{gp} + (1-A)C_{gk}]$$

may contain negative-resistance terms causing oscillation at the frequency where an inductive grid circuit resonates the capacitive Y_1 component.

The use of a simple triode (or pentode) grounded-cathode circuit with a load resistor equal to Z_0 provides an equally good match with slightly higher gain $(g_m R_l)$, but will overload at a lower maximum voltage. The anode-follower provides output approximating the cathode-follower without the risk of oscillation.

NEGATIVE FEEDBACK

The following quantities are functions of frequency with respect to magnitude and phase.

$E, N, D =$ signal, noise, and distortion output voltage with feedback

$e, n, d =$ signal, noise, and distortion output voltage without feedback

$\quad A =$ voltage amplification magnitude of amplifier at a given frequency

$\quad \mathbf{A} =$ amplification including phase angle (complex quantity)

$\quad \beta =$ fraction of output voltage fed back (complex quantity); for usual negative feedback, β is negative

$\quad \phi =$ phase shift of amplifier and feedback circuit at a given frequency.

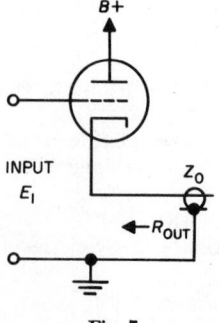

Fig. 7

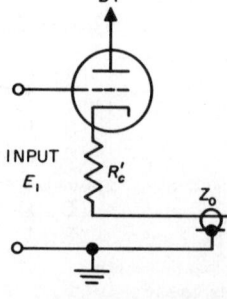

Fig. 8

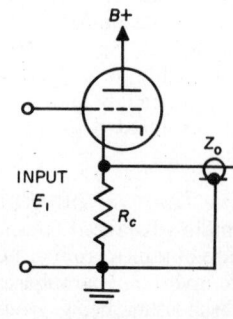

Fig. 9

Reduction in Gain Caused by Feedback (Fig. 10)

The total output voltage with feedback is

$$E+N+D=e+\frac{n}{1-\mathbf{A}\beta}+\frac{d}{1-\mathbf{A}\beta}.$$

It is assumed that the input signal to the amplifier is increased when negative feedback is applied, keeping $E=e$.

$(1-\mathbf{A}\beta)$ is a measure of the amount of feedback. By definition, the amount of feedback expressed in decibels is

$$20 \log_{10} | 1-\mathbf{A}\beta |$$

voltage gain with feedback $= \mathbf{A}/(1-\mathbf{A}\beta)$

change of gain $= 1/(1-\mathbf{A}\beta)$.

If the amount of feedback is large, i.e., $-\mathbf{A}\beta \gg 1$,

voltage gain becomes $-1/\beta$

and so is independent of $\mathbf{A}$.

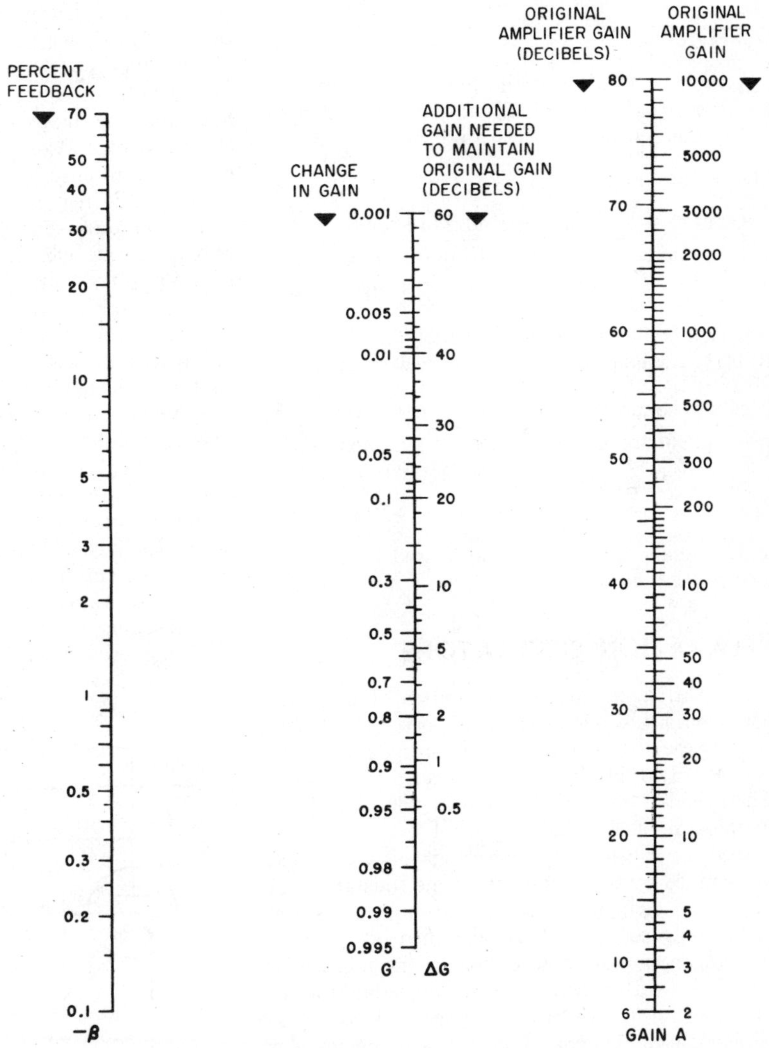

PERCENT FEEDBACK

ORIGINAL AMPLIFIER GAIN (DECIBELS)

ORIGINAL AMPLIFIER GAIN

ADDITIONAL GAIN NEEDED TO MAINTAIN ORIGINAL GAIN (DECIBELS)

CHANGE IN GAIN

Fig. 10—In negative-feedback amplifier considerations β, expressed as a percentage, has a negative value. A line across the β and A scales intersects the center scale to indicate change in gain. It also indicates the amount, in decibels, the input must be increased to maintain original output.

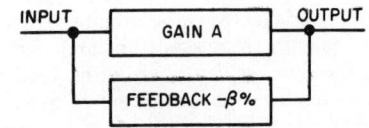

In the general case when ϕ is not restricted to 0 or π

voltage gain $= \mathbf{A}/(1+|\mathbf{A}\beta|^2-2|\mathbf{A}\beta|\cos\phi)^{1/2}$

change of gain $= (1+|\mathbf{A}\beta|^2-2|\mathbf{A}\beta|\cos\phi)^{-1/2}$.

Hence if $|\mathbf{A}\beta|\gg 1$, the expression is substantially independent of ϕ.

On the polar diagram relating $(\mathbf{A}\beta)$ and ϕ (Nyquist diagram), the system is unstable if the point $(1, 0)$ is enclosed by the curve. Examples of Nyquist diagrams for feedback amplifiers will be found in the chapter on Feedback Control Systems.

DISTORTION

A rapid indication of the harmonic content of an alternating source is given by the *distortion factor*, which is expressed as a percentage.

(Distortion factor)

$$= \left[\frac{\text{sum of squares of amplitudes of harmonics}}{\text{square of amplitude of fundamental}}\right]^{1/2}$$
$$\times 100 \text{ percent.}$$

If this factor is reasonably small, say less than 10 percent, the error involved in measuring it as

$$\left[\frac{\text{sum of squares of amplitudes of harmonics}}{\begin{array}{c}\text{sum of squares of amplitudes of fundamental}\\ \text{and harmonics}\end{array}}\right]^{1/2}$$
$$\times 100 \text{ percent}$$

is also small. This latter is measured by the *distortion-factor meter*.

RELAXATION OSCILLATORS

Relaxation oscillators are a class of oscillator characterized by a large excess of positive feedback, causing the circuit to operate in abrupt transitions between two blocked or overloaded end-states. These end-states may be stable, the circuit remaining in such condition until externally disturbed; or *quasi-stable*, recovering (after a period determined by coupling-circuit time constants and bias) and switching back to the opposite state. Relaxation oscillators are classified as *bistable*, *monostable*, or *astable* according to the number of stable end-states. Most circuits are adaptable to all three forms. Multistate devices are also possible. A wide variety of circuit arrangements is possible, including multivibrators, blocking oscillators, trigger circuits, counters, and circuits of the phantastron, sanotron, and sanophant class. Relaxation oscillators are often used for counting and frequency division, and to generate nonsinusoidal waveforms for timing, triggering, and similar applications.

Multivibrators

A number of multivibrator circuits are formed from three basic two-stage amplifiers (grounded-cathode–grounded-cathode, grounded-plate–grounded-grid, and grounded-cathode–grounded-grid, or combinations of these types), that readily provide the needed positive feedback with simple resistance or resistance–capacitance coupling. End-states may be any two of the four "blocked" conditions corresponding to cutoff or saturation in either stage. In general, the duration of a quasi-stable state will be determined by the exponential decay of charge stored in a coupling-circuit time constant, (the circuit switching back to the opposite state when the saturated or the cutoff tube recovers gain), while stable states are produced by direct coupling with bias sufficient to hold one tube inoperative. The memory effect of charge storage also operates in the case of stable end-states to ensure completion of transfer across the unstable region. The timing accuracy of an astable or quasi-stable multivibrator is considerably improved by supplying the grid resistors from a high positive voltage $(B+)$. The recovery from a cutoff condition thereby becomes an exponential towards a voltage much higher than the operating point, terminating in switchover when the cut off tube conducts. Grid conduction serves to clamp the capacitor voltage during the conducting state, erasing residual charge from the previous state. The starting condition for the next transition is thus more precisely determined and the linearity of the exponential recovery is improved by the more nearly constant-current discharge (since the range from cutoff to zero bias represents a smaller fraction of total charge). The grid-circuit time constant must be appropriately increased to obtain the same dwell time.

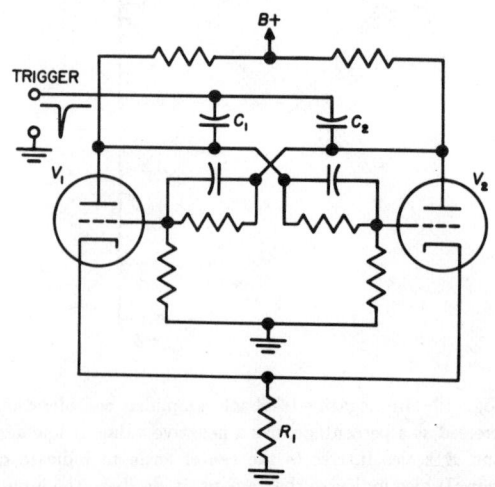

Fig. 11—Symmetrical bistable multivibrator (basic binary counter).

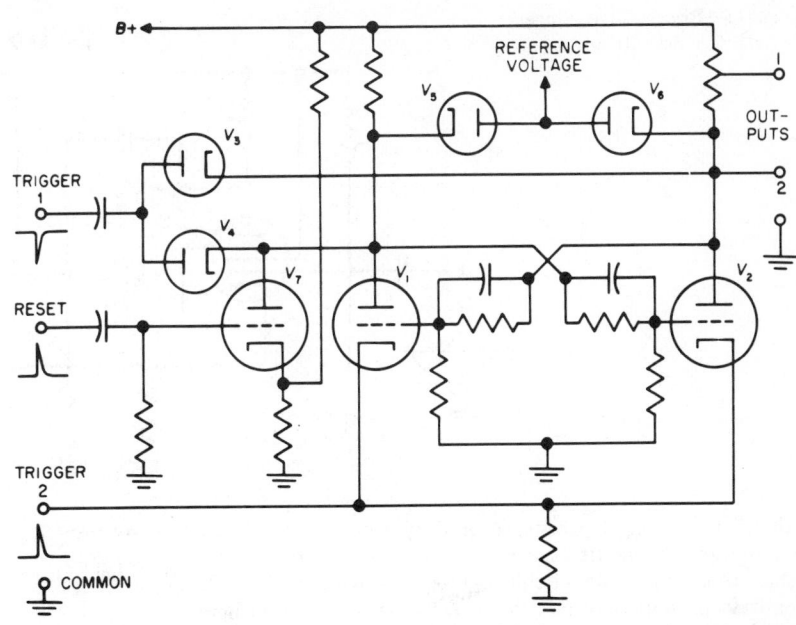

Fig. 12—Binary counter stage.

Bistable Circuits

Bistable circuits are especially suited for binary counters and frequency dividers and as trigger circuits to produce a step or pulse when an input signal passes above or below a selected amplitude.

Symmetrical Bistable Multivibrator: The circuit is shown in Fig. 11. Trigger signal may be applied to both plates, both grids, or if pentodes are used, to both suppressor grids.

Binary Counter Stage: An adaptation of the symmetrical bistable multivibrator is shown in Fig. 12. Alternative trigger inputs are shown with corresponding outputs to drive a following stage. The use of coupling diodes (V_3, V_4) reduces the tendency of C_1, C_2 in the circuit of Fig. 11 to cause misfiring by unbalanced stored charge. Tubes

V_5 and V_6 illustrate the application of clamping diodes, especially useful in high-speed circuits, to fix critical operating voltages. Pentodes with plate and grid clamping are suitable for very high speeds.

Schmitt Trigger: The circuit of Fig. 13 has the property that an output of constant peak value (a flat-topped pulse) is obtained for the period that the input waveform exceeds a specific voltage.

Monostable Circuits

Monostable multivibrators are useful for driven-sweep, pulse, and timing-wave generators. The absence of time constants and residual charge "memory" in the stable state reduces jitter when they are driven with irregularly spaced timing signals. Monostable versions may be derived from

Fig. 13—Basic Schmitt trigger.

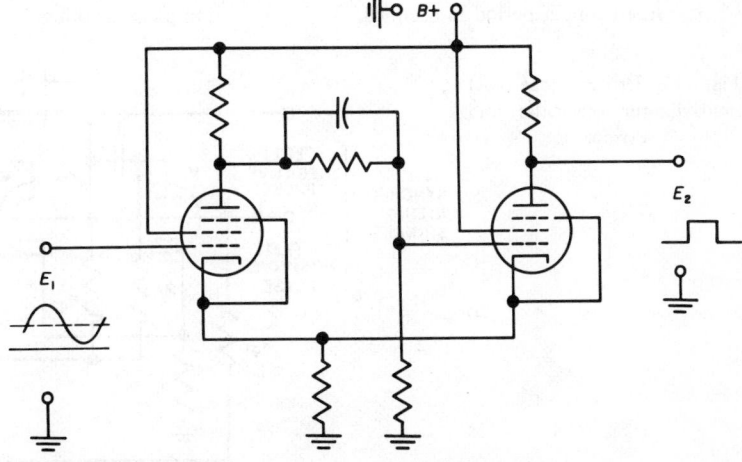

Fig. 14—Regenerative clipper (modified Schmitt trigger).

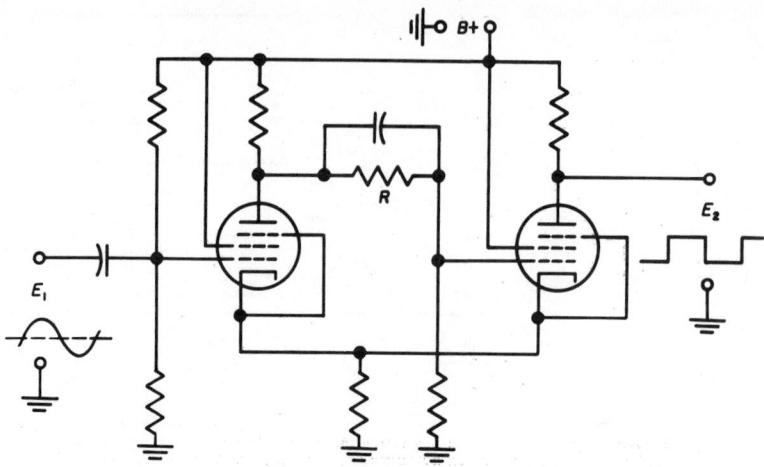

all of the foregoing bistable multivibrators by elimination of the direct (dc) coupling to one or the other grid. The circuit of Fig. 14 with R omitted is commonly used for pulse generation.

Most astable circuits can be made monostable by sufficient inequality of bias. The circuit of Fig. 17 is an example.

Sweep waveforms can be produced by integration of pulse outputs. The phantastron class of Miller sweep generators is also particularly useful for this purpose.

Driven (One-Shot) Multivibrator: The circuit is given in Fig. 15. Equations are

$$f_{mv}=f_s$$

f_{mv} = multivibrator frequency in hertz

f_s = synchronizing frequency in hertz.

Conditions of operation are

$$f_s > f_n \quad \text{or} \quad \tau_s < \tau_n.$$

where

f_n = free-running frequency in hertz

τ_s = synchronizing period in seconds

τ_n = free-running period in seconds.

At the control resistor R_c

$$\tau_{n2}=R_{g2}C \log_e[(E_{b1}-E_{m1}+E_{c2})/(E_{c2}+E_{x2})].$$

where

E_{b1} = plate-supply voltage V_1

E_{m1} = minimum ac voltage on plate of V_1

E_{c2} = bias voltage of V_2

E_{x2} = cutoff voltage corresponding to E_{b1}.

Regenerative Clipper: Bias on the first grid places the circuit of Fig. 14 in the center of the unstable region, giving regenerative clipping.

Phantastron: The phantastron circuit is a form of monostable multivibrator with similarities to the Miller sweep circuit. It is useful for generating very short pulses and linear sweeps. It uses a characteristic of pentodes: that while cathode current is determined mainly by control-grid potential, the screen-grid, suppressor-grid, and plate potentials determine the division of current between plate and screen. In certain tubes, such as the 6AS6, the transconductance from suppressor grid to plate is sufficiently high so that the plate current

Fig. 15—Driven (one-shot) multivibrator schematic and waveforms.

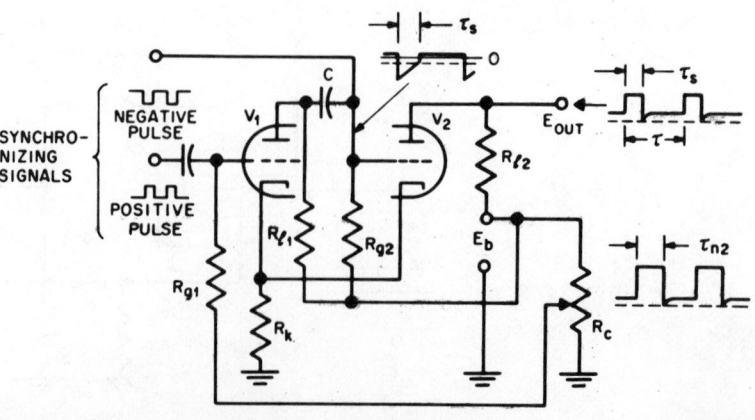

Fig. 16—Cathode-coupled
phantastron.

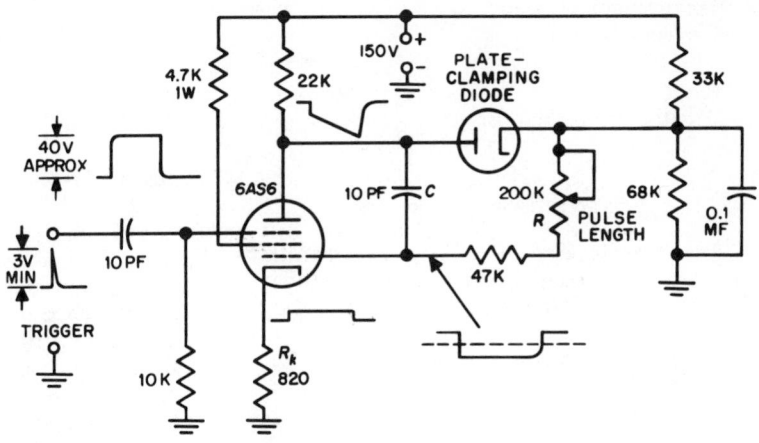

may be cut off completely with a small negative bias on the suppressor.

A typical phantastron circuit is shown in Fig. 16. During operation it switches between two states of interest.

(**A**) *Stable*: The control grid is slightly positive and draws current. Cathode current is maximum and the suppressor is biased negatively to plate-current cutoff by the cathode current in R_k. The plate is at a high potential determined by the clamping diode, and the screen potential is low.

(**B**) *Unstable*: When a positive trigger is applied to the suppressor grid (or a negative trigger to the control grid, cathode, or plate) the plate conducts, driving the control grid negative, reducing the cathode current, and taking most of the screen current. The plate potential then runs down linearly as in the Miller circuit.

The end of this period comes when the control grid goes positive again, resulting in increase of cathode current, suppressor cutoff, and heavy screen current.

In the circuit shown, the pulse width is adjustable from 0.3 to 0.6 microsecond. For longer pulses, it is possible to get a wide range of control both by varying R and C and by varying the plate-clamping potential.

Decreasing R_k results in astable operation.

Astable Circuits

The operating principles of the multivibrator and the exponential recovery from quasi-stable states are illustrated by the analysis of the free-running multivibrator.

Free-Running Zero-Bias Symmetrical Multivibrator: Exact equation for semiperiod (Figs. 17 and 18)

$$\tau_1 = \{R_{g1} + [R_{l2} r_p / (R_{l2} + r_p)]\} C_1$$
$$\times \log_e [(E_b - E_m)/E_x]$$

where

$$\tau = \tau_1 + \tau_2 = 1/f, \quad \tau_1 = \tau_2, \quad R_{g1} = R_{g2}, \quad C_1 = C_2.$$

f = repetition frequency in hertz

τ = period in seconds

τ_1 = semiperiod in seconds

r_p = plate resistance of tube in ohms

E_b = plate-supply voltage

E_m = minimum alternating voltage on plate

E_x = cutoff voltage corresponding to E_b

C = capacitance in farads.

Fig. 17—Schematic diagram of symmetrical multivibrator and voltage waveforms on tube elements.

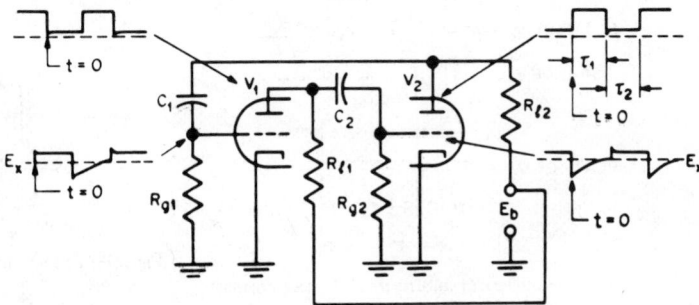

Approximate equation for semiperiod, where $R_{g1} \gg [R_{l2}r_p/(R_{l2}+r_p)]$, is

$$\tau_1 = R_{g1}C_1 \log_e[(E_b-E_m)/E_x].$$

Equation for buildup time is

$$\tau_B = 4(R_l+r_p)C$$

$$= 98 \text{ percent of peak value.}$$

Free-Running Zero-Bias Unsymmetrical Multivibrator: See symmetrical multivibrator (above) for circuit and terminology; the waveforms are given in Fig. 19.

Equations for fractional periods are

$$\tau_1 = \{R_{g1}+[R_{l2}r_p/(R_{l2}+r_p)]\}C_1$$
$$\times \log_e[(E_{b2}-E_{m2})/E_{x1}]$$
$$\tau_2 = \{R_{g2}+[R_{l1}r_p/(R_{l1}+r_p)]\}C_2$$
$$\times \log_e[(E_{b1}-E_{m1})/E_{x2}]$$
$$\tau = \tau_1+\tau_2 = 1/f.$$

Free-Running Positive-Bias Multivibrator: Equations for fractional period (Fig. 20) are

$$\tau_1 = \{R_{g1}+[R_{l2}r_p/(R_{l2}+r_p)]\}C_1$$
$$\times \log_e[(E_{b2}-E_{m2}+E_{c1})/(E_{c1}+E_{x1})]$$
$$\tau_2 = \{R_{g2}+[R_{l1}r_p/(R_{l1}+r_p)]\}C_2$$
$$\times \log_e[(E_{b1}-E_{m1}+E_{c2})/(E_{c2}+E_{x2})]$$

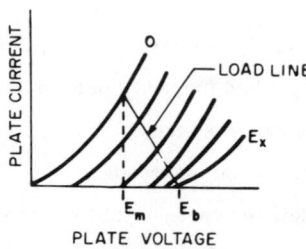

Fig. 18—Multivibrator potentials on plate-characteristic curve.

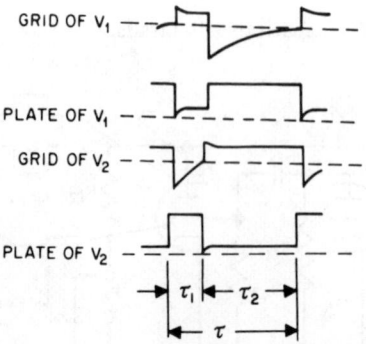

Fig. 19—Unsymmetrical multivibrator waveforms.

where

$$\tau = \tau_1+\tau_2 = 1/f$$

E_c = positive bias voltage.

Blocking Oscillators

The blocking oscillator (Figs. 21–23) is a single-tube relaxation oscillator using a close-coupled (current) transformer that imposes a fixed current ratio between grid current and plate current, while also providing the polarity reversal for positive feedback. There are, therefore, two end-states that satisfy the requirement i_p/i_g = turns ratio: one in the positive-grid region, with large grid current, and one at cutoff, with both currents zero. Astable and monostable forms are illustrated in the following discussion.

Astable Blocking Oscillator: Conditions for blocking are

$$E_1/E_0 < 1-\epsilon^{1/\alpha f - \theta}$$

where

E_0 = peak grid volts

E_1 = positive portion of grid swing in volts

f = frequency in hertz

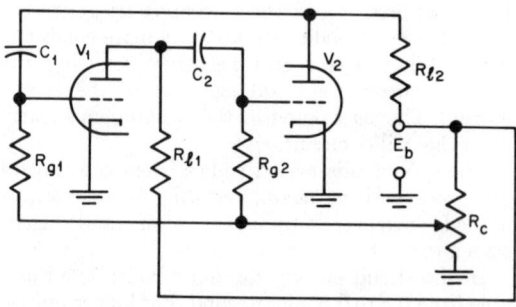

Fig. 20—Free-running positive-bias multivibrator.

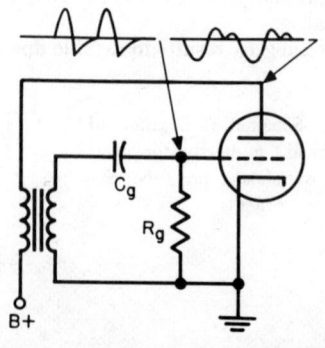

Fig. 21—Free-running blocking oscillator—schematic and waveforms.

$\alpha=$ grid time constant in seconds

$\epsilon=2.718=$ base of natural logs

$\theta=$ decrement of wave.

(A) Use strong feedback

$\qquad = E_0$ is high.

(B) Use large grid time constant

$\qquad = \alpha$ is large.

(C) Use high decrement (high losses)

$\qquad = \theta$ is high.

Pulse width is

$$\tau_1 \approx 2(LC)^{1/2}$$

where

$\tau_1=$ pulse width in seconds

$L=$ magnetizing inductance of transformer in henries

$C=$ interwinding capacitance of transformer in farads.

$$L = M(n_1/n_2)$$

where

$\qquad M=$ mutual inductance between windings

$n_1/n_2=$ turns ratio of transformer.

Repetition frequency

$$\tau_2 \approx 1/f \approx R_g C_g \log_e[(E_b+E_g)/(E_b+E_x)]$$

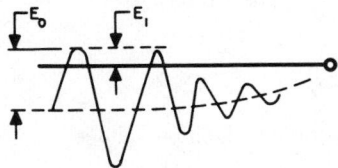

Fig. 22—Blocking-oscillator grid voltage.

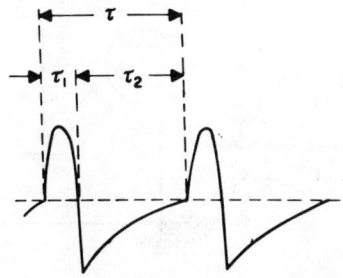

Fig. 23—Blocking-oscillator pulse waveform.

where

$\qquad \tau_2 \gg \tau_1$

$f=$ repetition frequency in hertz

$E_b=$ plate-supply voltage

$E_g=$ maximum negative grid voltage

$E_x=$ grid cutoff in volts

$\qquad \tau = \tau_1 + \tau_2 = 1/f.$

Astable Positive-Bias Wide-Frequency-Range Blocking Oscillator: Typical circuit values (Fig. 24) are

$R=0.5$ to 5 megohms

$C=50$ picofarads to 0.1 microfarad

$R_k=10$ to 200 ohms

$R_b=50\ 000$ to 250 000 ohms

$\Delta f=100$ hertz to 100 kilohertz.

Monostable Blocking Oscillator: Operating conditions (Fig. 25) are:

(A) Tube off unless positive voltage is applied to grid.

(B) Signal input controls repetition frequency.

(C) E_c is a high negative bias.

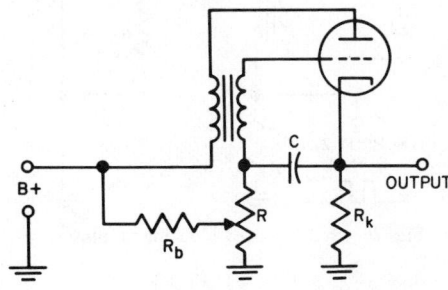

Fig. 24—Free-running positive-bias blocking oscillator.

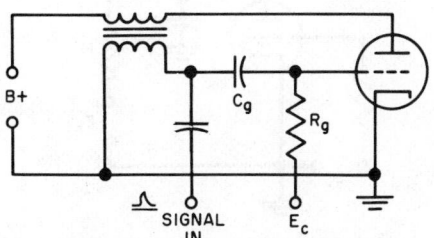

Fig. 25—Driven blocking oscillator.

Synchronized Astable Blocking Oscillator: Operating conditions (Fig. 26) are

$$f_n < f_s \quad \text{or} \quad \tau_n > \tau_s$$

where

f_n = free-running frequency in hertz

f_s = synchronizing frequency in hertz

τ_n = free-running period in seconds

τ_s = synchronizing period in seconds.

Gas-Tube Oscillators

A simple relaxation oscillator is based on the negative-resistance characteristic of a glow discharge, the two end-states corresponding to ignition and extinction potential of the discharge. Two astable forms are discussed. The circuit of Fig. 27 may also be used with a simple diode (neon lamp), omitting the grid resistor and bias. The circuit of Fig. 28 may be made monostable if the supply voltage is less than the ignition voltage at the selected bias.

Astable Gas-Tube Oscillator: This circuit is often used as a simple generator of the sawtooth waveform necessary for the horizontal deflection of a cathode-ray-oscilloscope beam. Equation for period (Fig. 27)

$$\tau = \alpha RC (1 + \alpha/2)$$

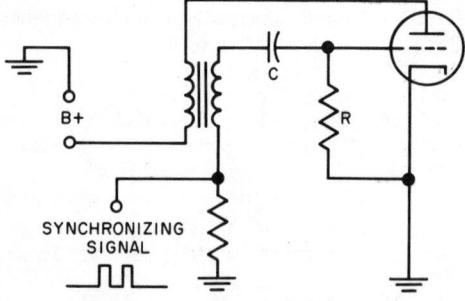

B+

SYNCHRONIZING
SIGNAL

Fig. 26—Synchronized blocking oscillator.

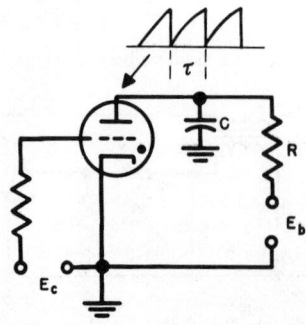

Fig. 27—Free-running gas-tube oscillator.

where

τ = period in hertz

$\alpha = (E_i - E_x)/(E - E_x)$

E_i = ignition voltage

E_x = extinction voltage

E = plate-supply voltage.

Velocity error

= change in velocity of cathode-ray-tube spot/trace period

Maximum percentage error = $\alpha \times 100$
if $\alpha \ll 1$.

Position error

= deviation of cathode-ray-tube trace from linearity.

Maximum percentage error = $(\alpha/8) \times 100$
if $\alpha \ll 1$.

Synchronized Astable Gas-Tube Oscillator: Conditions for synchronization (Fig. 28) are

$$f_s = Nf_n$$

where

f_n = free-running frequency in hertz

f_s = synchronizing frequency in hertz

N = an integer.

For $f_s \neq Nf_n$, the maximum δf_n before slipping is given by

$$(E_0/E_s)(\delta f_n/f_s) = 1$$

where

$\delta f_n = f_n - f_s$

E_0 = free-running ignition voltage

E_s = synchronizing voltage referred to plate circuit.

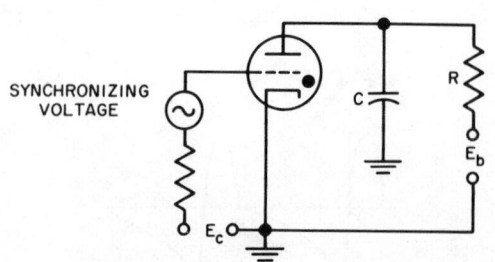

SYNCHRONIZING
VOLTAGE

Fig. 28—Synchronized gas-tube oscillator.

CHAPTER 19
SEMICONDUCTORS AND TRANSISTORS

GENERAL INFORMATION

This chapter describes the operating principles and chief characteristics of the most commonly used types of semiconductor devices. Under each topic are given a few key references which may be consulted for further information. For detailed characteristics and specifications, consult the comprehensive technical manuals published by the major semiconductor manufacturers, which present individual device specifications and such application information as typical circuits, heat-sink design, etc.

Complete tabulations of transistor, diode, and silicon-controlled-rectifier characteristics are published by D.A.T.A., Inc., 32 Lincoln Ave., Orange, N.J.

Relevant military specifications are:

MIL-S-19500D, Supplement 1; 11 May 1964—"General Specification for Semiconductor Devices."

MIL-STD-701E; 30 Dec. 1964—"Preferred and Guidance Lists of Semiconductor Devices."

QPL-19500-25; 21 Oct. 1964—"List of Products Qualified under MIL-S-19500."

MIL-E-IE, Supplement 1; 30 Dec. 1963—"General Specification for Electron Tubes" (includes some semiconductor devices).

QPL-1-45; 23 Oct. 1964. "List of Products Qualified under MIL-E-1."

MIL-S-38103, Supplement 1A; 15 Oct. 1963—"General Specification for Established Reliability Devices."

MIL-S-55191, Supplement 1; 7 May 1963—"General Specification for Semiconductors" (microelements).

Physical Constants

Electron charge

$$q = 1.60 \times 10^{-19} \text{ C}$$

Boltzmann's constant

$$k = 8.61 \times 10^{-5} \text{ eV/ K}$$

Dielectric constant of free space

$$\epsilon_v = 8.85 \times 10^{-14} \text{ F/cm}$$

Wavelength of 1 eV photon

$$= 1.24 \text{ } \mu\text{m} = 1.24 \times 10^{-4} \text{ cm}$$

Thermal voltage

$$kT/q = 0.026 \text{ V at } T = 300 \text{ K}$$

Principles of Semiconductors*

Table 1 shows some of the important properties of common semiconductors. Silicon and germanium are technologically the most important, particularly silicon with the advent of integrated circuits. Although germanium cannot be operated above about 100°C, large numbers of germanium transistors and diodes are still made. Microwave germanium transistors have superior frequency response because of the higher mobility. GaAs and InSb are the best known of the III-V compounds; GaAs is of interest because of its high electron mobility and good high-temperature capability, while InSb has a remarkably high electron mobility useful for Hall-effect devices.

The semiconductors of Table 1 crystallize in a diamond lattice, each atom having four nearest neighbors. The atoms are held together by covalent bonds, each involving a shared pair of electrons. These electrons occupy a band of energies known as the valence band and are not available for conducting current. Thermal agitation results in the breaking of occasional bonds, promoting electrons across the energy gap to a higher energy band (conduction band) where they can contribute to current flow. The remaining electrons in the incomplete valence band can act to produce an additional net current flow in a manner conveniently described by assigning a fictitious particle with positive charge, termed a hole, to each vacancy in the valence band.

The continuous thermal generation of hole-electron pairs is balanced by various recombination mechanisms. In an intrinsic semiconductor (no impurities), the equilibrium densities of holes and

* W. Shockley, "Electrons and Holes in Semiconductors," D. Van Nostrand Company, Princeton, N.J.; 1950. A. Nussbaum, "Semiconductor Device Physics," Prentice-Hall, Inc., Englewood Cliffs, N.J.; 1962.

TABLE 1—SEMICONDUCTOR PROPERTIES (AT $T = 290$ K).

	Si	Ge	GaAs	InSb
Energy gap, eV	1.106*	0.67*	1.35†	0.17†
Temp coeff of energy gap,‡ eV/ K×10⁴	−4	−4.5	−5	−2.7
Melting point, °C†	1412	958	1238	523
Thermal conductivity, W/cm-°C	1.42‖	0.52‖	0.44‖	0.17§
Thermal coeff of linear expansion, °C⁻¹×10⁶ ‖	4.2	5.5	5.7	
Lattice constant, Å†	5.42	5.65	5.65	6.48
Dielectric constant†	11.8	16.0	11.1	15.9
Electron mobility, cm²/volt-sec	1350*	3900*	6800‡	80 000‡
Temp dependence	$T^{-2.5}$	$T^{-1.66}$		$T^{-1.7}$
Hole mobility	480*	1900*	680‡	4000‡
Temp dependence	$T^{-2.7}$	$T^{-2.33}$		$T^{-2.1}$
n_i, cm⁻³	1.5×10¹⁰ *	2.4×10¹³ *		1.35×10¹⁶ §
Temp dependence of n_i †	T‡ exp$(-1.21/kT)$	T‡ exp$(-0.785/kT)$		

* E. M. Conwell, "Properties of Silicon and Germanium, II", *Proc. IRE*, Vol. 46, pp. 1281–1300; June 1958.
† "Semiconductors," N. B. Hannay, ed., Reinhold Publishing Co., New York; 1959.
‡ R. H. Bube, "Photoconductivity of Solids," John Wiley & Sons, New York; 1960.
§ R. A. Smith, "Semiconductors," Cambridge University Press, New York; 1959.
‖ "RADC Reliability Physics Notebook," First Edition, J. Vaccaro and H. C. Gorton, ed., Rome Air Development Center and Battelle Memorial Institute; 1965.

(conduction) electrons are each equal to the intrinsic density n_i, which is characteristic of the semiconductor and a strong function of temperature. If impurities such as phosphorus from group V of the periodic table are introduced during crystal growth or by diffusion, and substitute for some of the semiconductor atoms, four of their five valence electrons form bonds and the fifth is only weakly bound so it readily enters the conduction band. Such an impurity is termed a donor, since each atom donates one conduction electron, leaving behind a fixed positive charge. In equilibrium, the pn product is constant independent of the doping.

$$pn = n_i{}^2.$$

Thus, if n is increased by adding donors, p is proportionately decreased. In this case, the semiconductor is n type, the electrons are majority carriers, and the holes are minority carriers. The densities are determined by the above relation and by the condition for charge balance

$$n = N_D + p$$

where N_D is the donor density. Analogous remarks apply to group III impurities such as boron, termed acceptors, which introduce holes resulting in p type material. If both donors and acceptors are present, the conductivity type is determined by the net doping $N_D - N_A$.

Current flow takes place in a semiconductor by drift of carriers in an electric field and by the diffusion of carriers from a region of high concentration. In the former, $J = \sigma E$, where the conductivity is given by

$$\sigma = q(\mu_n n + \mu_p p).$$

The μ's are mobilities, describing the average drift velocity achieved by the carriers per unit electric field, and q is the electron charge. Diffusion current is described for electrons by $qD_n\, dn/dx$ and for the holes by $-qD_p\, dp/dx$. The diffusion constants D are related to the mobilities by

$$D/\mu = kT/q$$

where k is Boltzmann's constant and T the absolute temperature. At room temperature, the thermal voltage kT/q is about 26 millivolts.

At sufficiently high doping levels, the mobility decreases because of additional scattering from the impurity ions. At high electric fields, the mobility also decreases, until the carriers reach a limiting velocity almost independent of the field. This limiting velocity is of the order of 10⁶ to 10⁷ cm/s.

If equilibrium is disturbed by introducing excess carriers, e.g., by light or by injection from a pn junction, recombination mechanisms act to restore it. The simplest way to describe the recombination rate for excess electrons in p material, for example, is

$$dn/dt = -(n - n_0)/\tau_n$$

where n_0 is the equilibrium density and τ_n is called the lifetime of electrons. The excess density thus decays as exp $(-t/\tau_n)$. Lifetime is very sensitive to crystalline perfection as well as to various impurities, and in silicon can range from several microseconds down to a few nanoseconds.

Lifetime can be reduced for such applications as high-speed switching diodes and transistors by introducing impurity atoms (such as gold) which act as recombination centers.

pn JUNCTION DEVICES*

Diodes

General Principles. The boundary between p and n regions within a single semiconductor crystal introduces electrical properties of great importance. Figure 1 shows a pn junction having the p side more heavily doped $(N_A > N_D)$. A dipole charge layer composed of fixed donor and acceptor ions exists near the boundary, producing the potential barrier required to keep the mobile charges (holes and electrons) in place. This dipole layer is termed a depletion region, since within it the mobile carriers are largely swept out. A few holes generated by thermal excitation (or by illumination) in the n region near the barrier will fall into the p region and constitute a small reverse current I_{pr}. At equilibrium (no applied bias), I_{pr} is exactly balanced by a forward current I_{pf} consisting of the few holes in the p region which have enough energy to climb the barrier (similar remarks apply to the currents I_{nr} and I_{nf}).

Under forward external bias (V positive), the potential barrier is lowered, allowing much larger forward current to flow. Excess holes and electrons are injected into the n and p regions, respectively; under steady-state conditions the excess density decays with distance according to

$$\exp(-x/L), \qquad L^2 = D\tau$$

where the diffusion length L^2 is the product of the diffusion constant D and lifetime τ for the appropriate carrier. The injected minority density is greatest on the lightly doped side, and the corresponding current flow is determined by this density and by how rapidly carriers can be removed from the vicinity of the junction by recombination or by a nearby reverse-biased junction as in a transistor.

Under reverse bias (V negative), the potential barrier is increased, reducing the forward component and leaving only the small reverse current which does not depend on the barrier height. The

* W. Shockley, "The Theory of p–n Junctions in Semiconductors and p–n Junction Transistors," *Bell System Telephone Journal*, Vol. 28, p. 435; 1949. W. Shockley, "Problems Related to P–N Junctions in Silicon," *Solid-State Electronics*, Vol. 2, pp. 35–67; 1961. J. L. Moll, "The Evolution of the Theory for the Voltage-Current Characteristic of P–N Junctions," *Proc. IRE*, Vol. 46, pp. 1076–1083; 1958.

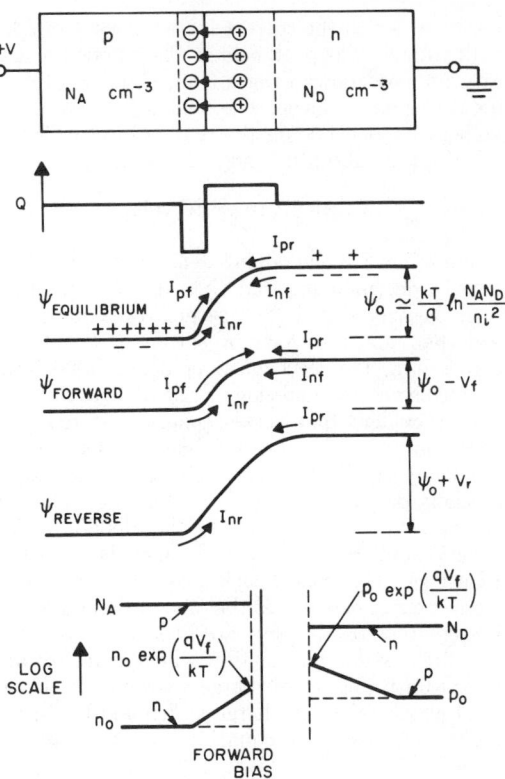

Fig. 1—Charge, potential, and carrier-density relationships in a pn junction.

device thus has rectifying properties, and the static VI characteristic is approximately

$$I = I_r[\exp(qV/nkT) - 1]$$

where the coefficient n varies from 1 to 2 depending on the current level and the nature of the recombination centers in the material near the junction. The reverse current I_r is very strongly temperature dependent, in theory increasing about 16% per °C for silicon and 10% per °C for germanium around room temperature. In practice, the reverse current increases more slowly, doubling every 8° to 10°C.

As the reverse voltage is increased, the depletion layer widens to provide the increased space charge in accordance with Poisson's equation. The capacitance of the junction for small voltage changes is that of a capacitor with parallel plates spaced by the depletion-layer width. Thus the capacitance depends on reverse bias (approximately as $V^{-1/2}$ for an abrupt junction and $V^{-1/3}$ for a graded junction), a property made use of in varactors.

In the high electric field of a reverse-biased junction, charge carriers can acquire enough energy to make hole–electron pairs by impact ionization. These new carriers can in turn create more, and if the field is large enough over a sufficiently long distance, a self-sustaining avalanche results in which the current increases very rapidly with voltage. The peak field in the junction necessary for breakdown ranges from about 3×10^5 to 5×10^5 V/cm in silicon. The avalanche breakdown voltage V_A depends on the doping level near the junction; for silicon it is approximated by

$$V_A \simeq 5.6 \times 10^{13} N^{-3/4} \text{ volts}$$

where N cm^{-3} is the doping density on the lightly doped side, assuming that the other side is much more heavily doped. For reverse breakdowns of less than about 6 volts in silicon or 3 volts in germanium, the mechanism is Zener breakdown (tunneling or field emission) rather than avalanche.

It is evident that excess charge is stored near the junction under forward bias. In some high-frequency applications the charging current in specially designed diodes is much greater than the steady conduction current, so the diode acts like a very nonlinear capacitor of much larger value than its reverse-bias capacitance. The presence of stored charge means that the diode will not recover immediately from a forward-bias low-impedance condition, since the junction will remain forward biased until the excess charge is swept out. If the total excess minority charge is denoted by Q, the basic (approximate) equation for the transient behavior for positive Q is

$$dQ/dt + Q/\tau = I$$

where I is the diode terminal current and τ is an effective lifetime. This equation says that Q is removed by recombination within the diode and by reverse current, and the steady-state Q for a forward current I_f is τI_f. A short reverse recovery time is obtained by reducing τ and by restricting the volume available for charge storage.

*Power Rectifiers.** Power rectifiers are large-area *pn* or *pin* junctions produced by alloying or diffusion techniques. They operate at current densities up to 1000 A/cm^2 and are usually silicon because of its high temperature capability. They are available in average current ratings from 400 mA to 1000 A and peak reverse voltages of over 1000 V. Peak single-cycle surge current ratings are typically 10 to 20 times I_{Av}, and reverse currents at 150°C are usually less than $2 \times 10^{-4} I_{Av}$. Maximum operating (junction) and storage temperatures are from 175° to 200°C, and power dissipation capabilities are derated to zero at this point from 100°–140°C case temperature.

For high-frequency rectification, diodes with low reverse recovery times (< 0.2 μs) and with up to 10 A average current are made by providing a

narrow base width or reducing lifetime in the silicon by gold doping.

*Zener Diodes.** Zener diodes (or reference diodes) are silicon *pn* junctions designed to be operated into the reverse-bias avalanche breakdown region (true Zener breakdown occurs only in diodes breaking down at less than about 6 volts), where they exhibit low dynamic resistance. They are widely used for clipping and voltage-regulation applications, and can be obtained in a very large number of breakdown (or working) voltage V_Z ratings from about 2 volts to 1500 volts. Typical temperature coefficients of working voltage at constant current range from about $-0.1\%/°C$ for 3-volt diodes and $+0.05\%/°C$ for 10-volt diodes, to $+0.1\%/°C$ for 100-volt diodes. The product of dynamic resistance and test current at which it is measured ranges from 100 to 500 mA-ohms for 10-volt diodes, and from 500 to 2500 mA-ohms for 100-volt diodes. Measured at constant power, the dynamic resistance varies approximately as V_Z^2. Some large-area types in large packages can dissipate up to 50 watts of power.

Varactors and Voltage-Variable Capacitors.† Varactors are *pn* junctions (generally of silicon, but gallium arsenide is used for highest frequency response) making use of the variation of junction capacitance with reverse bias. They may employ special impurity profiles to enhance the capacitance variation and to minimize series resistance losses. They are used as nearly lossless frequency multipliers in solid-state transmitters requiring more output power than is available from transistors. High-power varactors may be designed as charge-storage (or step-recovery) diodes† in which the charge injected during a short forward-bias pulse can be largely recovered on current reversal, greatly enhancing the capacitance-voltage nonlinearity. Low-power varactors are often used as voltage-variable capacitors for electronic tuning.

Varactors are characterized by their capacitance variation with voltage, breakdown voltage $V_{(BR)}$, and series resistance r_s. In an abrupt-junction varactor, the capacitance varies as $(V+0.7)^{-1/2}$. The Q is given by $1/(2\pi f r_s C)$, and the cutoff frequency f_{co} is Qf. Gallium arsenide leads to higher Q because of its high mobility and resulting lower resistance for a given doping level. To achieve reasonable efficiency when doubling or tripling,

* H. W. Henkel, "Germanium and Silicon Rectifiers," *Proc. IRE*, Vol. 47, pp. 1086–1099; June 1958. R. N. Hall, "Power Rectifiers and Transistors," *Proc. IRE*, Vol. 40, pp. 1512–1518; Nov. 1952.

* A. E. Garside and P. Harvey, "The Characteristics of Silicon Voltage-Reference Diodes," *Proc. IEE*, Vol. 106B, Suppl. No. 17, pp. 982–990; 1959. K. G. McKay, "Avalanche Breakdown in Silicon," *Phys. Rev.*, Vol. 94, pp. 877–884; 15 May 1954.

† P. Penfield and R. Rafuse, "Varactor Applications," MIT Technology Press, Cambridge, Mass.; 1962. L. A. Blackwell and K. L. Kotzebue, "Semiconductor-Diode Parametric Amplifiers," Prentice-Hall, Inc., Englewood Cliffs, N.J.; 1961.

the Q should be greater than 50–100 at the output frequency. The maximum power input is proportional to $f_iC_0V_{(BR)}^2$, where C_0 is the zero-bias capacitance. Series resistance increases with breakdown voltage, so that the higher-frequency units have lower voltage ratings. This leads to a relationship between power, frequency, and impedance level of the form $PZf^2=$ constant. Impedance level is important from the standpoint of ease of matching to typical source impedances, and in maintaining package reactance and losses proportionately low.

Varactors are available giving output power of 12 W at 1 GHz, 7 W at 2 GHz, 1 W at 5 GHz, and 50 mW at 20 GHz. Efficiencies range from 70%–80% at the lower frequencies when doubling, to 10%–20% for high-order multiplication.

Variable-capacitance diodes are also used in low-noise parametric amplifiers, up-converters, and down-converters, where high Q and maximum capacitance nonlinearity are important considerations. A figure-of-merit performance is given by γQ, where γ relates to the capacitance variation produced by the pump voltage expressed as $C(1+2\gamma\cos\omega t)$.

Charge-Storage (Snap-Off or Step-Recovery) Diodes. Charge-storage diodes are designed so that most of the minority carriers injected under forward bias are stored near the junction and are immediately available to contribute to reverse conduction. This is accomplished by a doping profile having a steep gradient near the junction, or by a *pin* profile with a narrow *i* region. When a reverse-bias voltage is applied, the stored charge flows out as reverse current, and until it is exhausted, the diode maintains a low impedance level. When the charge is depleted, the current decreases rapidly to zero and reverse voltage quickly builds up at a rate determined by the reverse-bias capacitance and the external circuit. The rapid current transition can serve either to initiate or terminate a fast pulse, or as a source of harmonics.

A charge-storage diode is characterized by its breakdown voltage, junction capacitance (zero bias), an effective lifetime τ describing the charge $Q_f=\tau I_f$ stored by a forward current I_f, and a transition time describing the rate of current fall under specified conditions. The latter time can be shorter than 0.1 ns (when driven from a 10-volt reverse bias and a load of 50 ohms) in a diode typically having a lifetime of 10 ns and a capacitance of 2 pF. High-order harmonic generation with efficiencies exceeding $1/n$ can be obtained for

output frequencies up to 10 GHz with output powers in the tens to hundreds of milliwatts.

pin Diodes. *pin* diodes consist of heavily doped p^+ and n^+ end regions separated by a lightly doped region which can usually be regarded as intrinsic. If this center region is thick $(10–100\,\mu m)$, the device is a useful high-voltage rectifier with a low forward drop at high currents because of conductivity modulation of the *i* region by the large numbers of carriers injected from the end regions.

The *pin* diode can also act as a variable resistance at microwave frequencies which are too high for rectification to take place because of the relatively large recovery time of the thick *i* layer. At zero or reverse bias, the *i* layer introduces a high resistance. Under forward bias, the injection and storage of carriers reduces the resistance of the *i* region according to $R_i\sim(20–50)/I^{0.87}$ ohms, where I is the forward-bias current in mA. These diodes are used as microwave switches when driven with abrupt bias changes, or as variable-resistance microwave amplitude modulators. Typical impedance levels are 1 ohm when "on" (50 mA bias) and 8 kilohms shunted by a 0.15-pF package capacitance when "off" (zero bias). The package will typically add 1–3 nH of series inductance.

pin diodes with thin *i* regions $(<5\,\mu m)$ can have recovery times fast enough to serve as useful charge-storage diodes for harmonic generation.

Point-Contact Diodes.† When a metal point contacts the surface of a semiconductor, a short electrical pulse of suitable voltage and duration can break through the oxide and form a metal-semiconductor junction. Under the proper conditions, this is a rectifying junction in which the semiconductor injects electrons into the metal, and no significant storage of carriers occurs. (See *Hot-Electron Diodes*, later in this chapter). On some types, it is thought that a small *pn* junction is formed if the metal alloys with the semiconductor. The area can be made extremely small with correspondingly small capacitance. These diodes are very fast and have commonly been used as microwave mixers or video detectors for many

* J. L. Moll, S. Krakauer, and R. Shea, "*P-N* Junction Charge Storage Diodes," *Proc. IRE*, Vol. 50, pp. 43–53; Jan. 1962. S. Krakauer, "Harmonic Generation, Rectification, and Lifetime Evaluation with the Step-Recovery Diodes," *Proc. IRE*, Vol. 50, pp. 1665–1676; July 1962.

* H. S. Veloric and M. B. Prince, "High-Voltage Conductivity-Modulated Silicon Rectifier," *Bell System Technical Journal*, Vol. 36, pp. 975–1004; July 1957. J. K. Hunton and A. G. Ryals, "Microwave Variable Attenuators and Modulators Using *P-I-N* Diodes," *IRE PGMTT Transactions*, Vol. 10, p. 262; July 1962.

† H. C. Torrey and C. A. Whitmer, "Crystal Rectifiers," McGraw-Hill Book Company, New York, N.Y.; 1948. A. Uhlir, "Two-Terminal *P-N* Junction Devices for Frequency Conversion and Computation," *Proc. IRE*, Vol. 44, pp. 1183–1191; Sept. 1956. M. Cutler, "Point Contact Rectifier Theory," *IRE Trans. on Electron Devices*, Vol. ED-4, pp. 201–207; 1957.

years. Noise figures range from around 6 dB at 3 GHz to 10 dB at 35 GHz with a 6–8-dB conversion loss. An rf power (cw) of about 350 mW or pulse energy of 1–10 ergs is sufficient to burn them out.

The gold-bonded diode used in computer logic circuit applications for moderately fast switching is made by pressing a preformed gold whisker against an n-type germanium chip, then alloying it into the chip with controlled pulses of current. A pn junction is thought to be formed.

Tunnel and Backward Diodes.* A tunnel diode has sufficiently heavy doping on both sides of the junction that the potential barrier becomes thin enough for the probability of quantum-mechanical tunneling through the barrier to be relatively large. Because of this tunneling, and as a result of the energy band structure of semiconductors, a volt-ampere characteristic having a voltage-controlled (single valued in voltage) negative resistance is obtained. In the reverse direction (p side negative), the device is a good conductor, and remains so in the forward direction up to a peak current I_P and corresponding voltage V_P. Between I_P and the valley point characterized by (I_V, V_V), a negative conductance $g = dI/dV$ is obtained. Beyond V_V, the current rises again because of injection current as in an ordinary pn junction diode, reaching a value of I_P at a voltage V_{PP}.

Since the tunneling phenomenon is very fast, tunnel diodes find applications as microwave negative-resistance amplifiers and in fast switching circuits. The response is generally limited by the inductance of the package and by the capacitance of the junction in combination with ohmic series resistance in the semiconductor. Above the resistive cutoff frequency (which can be 30 GHz or more), negative resistance no longer appears at the package terminals.

Most commercial tunnel diodes are made from germanium or gallium arsenide, since a high peak-to-valley ratio I_P/I_V is hard to obtain in silicon. Typical values of I_P/I_V are 3.5, 6, and 15 for Si, Ge, and GaAs, respectively; corresponding values of V_P are 65, 55, and 150 mV and of V_V are 420, 350, and 500 mV.

A backward diode is a tunnel diode having $I_P \simeq I_V$, used as a rectifier in which the "forward" direction of conduction (when the p side is negative) occurs without the usual voltage offset of a conventional diode, and with a lower temperature coefficient of current at constant voltage. The "reverse" direction corresponds to the conventional forward direction, so that the reverse breakdown voltage is very low (~ 600 mV for Si).

* I. Lesk, N. Holonyak, V. Davidsohn, and M. Aarons, "Germanium and Silicon Tunnel Diodes—Design, Operation and Application," *IRE Wescon Convention Record*, Part 3; 1959. R. N. Hall, "Tunnel Diodes," *IRE Trans. on Electron Devices*, Vol. ED-7, pp. 1–9; Jan. 1960.

The backward diode is useful when dealing with small-amplitude waveforms.

Photodiodes.* When photons irradiate a semiconductor having a band gap less than the energy of the photons, hole-electron pairs are produced. This phenomenon, the photovoltaic effect, is best exploited by forming a pn junction in a semiconductor such as silicon, germanium, or gallium arsenide. The resulting photodiode may have a quantum efficiency (ratio of photoelectrons to impinging photons) of from 0.2 to 0.7.

Photodiodes have two primary operating modes: photovoltaic and photoconductive. In the photovoltaic mode, the unbiased junction is illuminated to stimulate the production of hole-electron pairs. Charge separation then occurs in the field of the junction and a current flow results. Since photovoltaic cells produce a short-circuit photocurrent which is linear with respect to the radiation incident upon the pn junction and since no internal noise is present, they are well suited for very-low-level light detection and measurement. The major application for photovoltaic cells, however, is the conversion of solar radiation into electrical power. The operating specification of most interest in this application is the power conversion efficiency (ratio of the power produced by the cell to the power in the incident photon flux). Cell efficiency is affected by such factors as surface reflectance, absorption between the surface and the pn junction, and transmissivity of the junction region. Power efficiencies of commercial solar cells range from 5% to a maximum of about 15%. The best developed solar-cell material is silicon, but considerable work has been expended on gallium arsenide and cadmium sulfide. Fabrication advances now permit the production of long silicon ribbons and this should eventually result in significant cost reductions for silicon solar-energy converters.

In the photoconductive mode, a photodiode (usually of special construction) is reverse biased and exposed to optical radiation. The resulting hole-electron pairs along the junction separate and establish a current flow which consists of a small and constant reverse leakage (the dark current) and the photocurrent (the signal). The photocurrent is linear with respect to the incident radiation. The dark current establishes a minimum noise level below which operation in the presence of a dc signal is impractical. Pulsating signals having an amplitude less than the noise level can be detected by employing capacitive coupling to block the dc dark current. An important advantage of photoconductive operation over the photovoltaic mode is response time. A photodiode resembles a parallel-plate capacitor with a voltage-

* R. J. McIntyre, H.C. Sprigings, and P. P. Webb, "Solid-State Detectors for Laser Applications," *RCA Lasers*, pp. 52–59; 1972.

controlled dielectric. Therefore, unbiased operation is slowed by junction capacitance while biased operation reduces junction capacitance and enhances response time.

A photodiode of major importance is the avalanche photodiode. This device is a modified photodiode biased just below the avalanche breakdown region. Photons impinging upon the junction region create hole-electron pairs which initiate an avalanche multiplication of carriers across the junction. The resulting internal gain (which may be a few hundred), high signal-to-noise ratio, and rapid response time make avalanche photodiodes suitable for operation in applications formerly reserved for the photomultiplier tube (see Chapter 17).

*Electroluminescent Diodes.** The recombination of holes and electrons which occurs when current is injected into a *pn* junction requires the release of energy for equilibrium to be resumed. This energy, which corresponds to the band-gap separation of the minority carriers, may be in the form of a photon (radiative recombination), a series of phonons (lattice vibrations), or both. In some cases, the energy may be transferred to another electron. Electroluminescent diodes, which are often termed "light-emitting diodes" (LEDs), are those in which radiative recombination predominates over other equilibrium byproducts. Recombination radiation emitted by LEDs is peaked at or near the band-gap energy and in practical devices ranges from 0.55 μm to 34 μm. The spectral emission width of a representative device may be 25-30 nm and this provides sufficient monochromaticity for the production of discrete wavelength bands (including the visible colors green, yellow, orange, and red).

Direct band-gap semiconductors (those in which energy equilibrium at recombination is accompanied by emission of a photon) make more efficient LEDs than indirect band-gap semiconductors (those in which the resumption of equilibrium is accompanied by emission of a photon and phonons). Silicon and germanium, both indirect band-gap materials, produce far too little recombination radiation for practical use, but more specialized indirect or direct-indirect band-gap materials, particularly solid solution alloys of gallium and gallium arsenide (GaAs), can be tailored to produce reasonably efficient radiation.

The best developed material for electroluminescent diodes is GaAs, and suitably fabricated laboratory devices have exhibited an internal quantum efficiency (ratio of emitted photons per injected electrons) very near unity. Due to such factors as internal absorption, contact shadows, and re-

* F. M. Mims, III," Light Emitting Diodes," Howard W. Sams & Co., Inc., Indianapolis; 1973. H. Kressel, "Semiconductor Lasers: Devices," *Laser Handbook*, North-Holland Publ. Co., Amsterdam; 1972.

fractive-index-induced surface reflectance, the external quantum efficiency of practical GaAs LEDs is much lower (≤ 0.1). Both quantum efficiency and recombination radiation wavelength are affected by temperature; thus the external quantum efficiency of specially fabricated LEDs may exceed 0.4 at 20 kelvins. Wavelength is directly related to junction temperature and typically varies 0.25 nm per °C. Therefore changes in either the ambient temperature or junction heating will alter the emitted wavelength.

A wide range of economical, long-lived plastic and metal packaged LEDs is commercially available. It is common practice to install the diode chip inside a miniature directional reflector and encapsulate the entire assembly in an index-matching epoxy to enhance the radiation extraction efficiency. This fabrication technique improves the external efficiency by a factor of 2 or more. Visible emitters suitable for use as visual indicators and other display roles include: GaP and GaP:N (green—550 nm); $GaAs_{0.25}P_{0.75}$:N (yellow—610 nm); $GaAs_{0.6}P_{0.4}$ (red—660 nm); $Al_{0.3}Ga_{0.7}As$ (red—675 nm); and GaP:Zn,O (red—690 nm). Several of these materials, particularly $GaAs_{0.6}P_{0.4}$, are in widespread use as display devices for digital clocks, watches, and calculators. The most common infrared emitting materials are GaAs (905 nm) and GaAs:Si (940 nm). These materials offer much higher efficiencies than visible emitters and are useful in such applications as optical communications and ranging, position sensing, object detection, and electrooptical isolation. GaAs is characterized by a reasonable power output ($\simeq 1.5$ mW @ 100 mA I_F) and very fast turn-on time ($\simeq 1$ ns), while GaAs:Si is more efficient ($\simeq 10$ mW @ 100 mA I_F) but slower ($t_{on} \simeq 300$ ns).

Several fundamental modifications of the basic *pn*-junction electroluminescent diode exist, and chief among these is the semiconductor injection laser. In its simplest form, the injection laser is a direct band-gap LED having an exceptionally flat and uniform junction (the active region) bounded on facing sides by two parallel mirrors perpendicular to the plane of the junction, which provide a Fabry-Perot resonant cavity. The mirrors are usually produced by cleaving the semiconductor chip along parallel planes to produce perfectly parallel and flat surfaces. The remaining two sides of the chip perpendicular to the junction plane are intentionally roughened during the sawing process, which separates bars of material into individual chips. This surface roughening suppresses off-axis lasing modes. The high index of refraction of the semiconductor provides sufficient reflectance at the semiconductor-air interface for the optical feedback necessary for laser action.

Below threshold, the injection laser behaves like a conventional LED, but as the current injection is increased, a threshold point occurs where

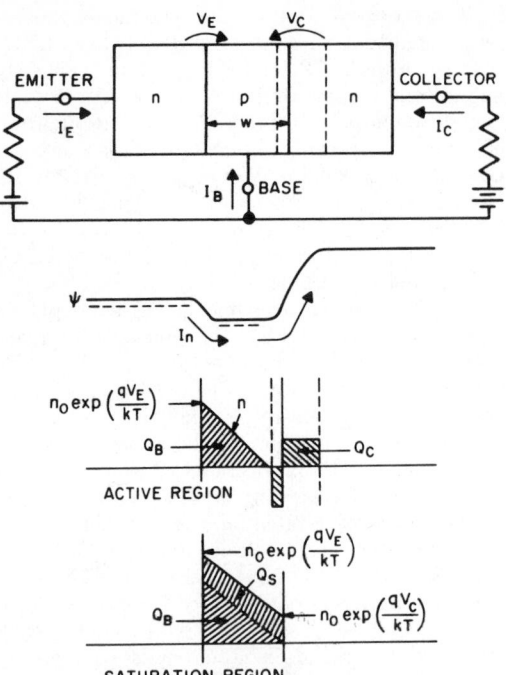

Fig. 2—Potential and charge relationships in a transistor.

heterojunction, which confines both the recombination region and the optical wave to a very thin layer, have been fabricated with a threshold current density of 1000 A/cm². These devices are now commercially available and can be operated continuously at 300 kelvins with appropriate heatsinking. Individual diodes may emit up to a few tens of milliwatts at 500 mA forward bias.

Injection laser recombination radiation is characterized by a narrow spectral bandwidth ($\simeq$0.3 nm), beam directionality, and moderate beam divergence. The beam directionality permits collection of up to 90% of the radiation by simple optics in most cases. The beam divergence, which may be 10–15° in the plane of the junction (half-power measurement) and 25–45° perpendicular to the plane of the junction, is a result of the laser's very narrow emission region. Essentially a diffraction limited slit, the emission region and internal lasing modes both contribute to the various diffraction patterns and structures seen in the far field pattern of the beam from most injection lasers. Injection laser applications, which include optical communications and ranging, are described elsewhere in this book.

Transistors[*]

General Principles. The basic transistor principles are illustrated in Fig. 2, which shows an *npn* unit. Similar remarks apply to the complementary *pnp* type. In normal operation, the forward-biased emitter junction injects electrons into the base region. The doping levels are chosen so that nearly all the emitter current is composed of these injected electrons, and only a small fraction consists of holes injected into the emitter. The base is thin enough that nearly all the injected electrons diffuse to the edge of the depletion region of the reverse-biased collector junction, where the field sweeps them across into the collector bulk. The fraction of the emitter current reaching the collector is denoted by α. The balance, $(1-\alpha)I_E$, is the base current consisting of holes recombining with some of the excess electrons in the base, or being injected into the emitter. Amplification is possible because current is transferred from the very-low-impedance emitter junction to the very-high-impedance collector junction.

In many applications, the base is considered as the control electrode. In the active region (emitter forward-biased, collector reverse-biased), the

the hole-electron population in the active region becomes inverted. Spontaneous recombination of holes and electrons then produces photons which stimulate in-phase recombination and photon emission by other holes and electrons, and lasing occurs as the optical gain in the active region overcomes absorption and other losses. The mirrors on either end of the active region provide the optical feedback necessary to sustain laser action, and a small fraction of the wave propagating between the mirrors emerges from each on each pass. One end facet on many commercial lasers is overcoated with a reflective Au film to cause all the radiation to emerge from only one end of the device and thus enhance collection efficiency.

The most common and best developed injection laser utilizes GaAs (905 nm), though many other semiconductors have been used to produce wavelengths ranging from 630 nm ($Al_xGa_{1-x}As$) to 34 μm (PbSnSe). The high current density required to achieve lasing in conventional injection lasers ($\geq$8000 A/cm²) precludes continuous operation above temperatures greater than about 77 kelvins. Room-temperature operation requires brief current pulses (10–100 amperes for typical diodes) no more than 200 ns wide and with a duty cycle of 0.1%. Peak pulse power outputs at 300 kelvins of up to 100 watts from single diodes and several kilowatts from diode arrays have been obtained from commercial devices. Injection lasers whose GaAs active region is sandwiched between two $Al_xGa_{1-x}As$ layers to produce a double

* J. G. Linvill and J. F. Gibbons, "Transistors and Active Circuits," McGraw-Hill Book Co., New York, N.Y.; 1961. W. W. Gartner, "Transistors: Principles, Design and Application," D. Van Nostrand, Princeton, N.J.; 1960. A. B. Phillips, "Transistor Engineering," McGraw-Hill Book Co., New York; 1962.

steady-state ratio of collector to base currents is

$$\alpha/(1-\alpha)=\beta$$

so that current gain as well as impedance transformation is obtained.

A transistor switch may be operated with both junctions reverse-biased ("open" or high-impedance state), with emitter forward-biased and collector reverse-biased (active region), or with both junctions forward-biased (saturation region or low-impedance "on" state). The latter case is usually reached by supplying a base current larger than I_{CM}/β, where I_{CM} is the maximum current the collector circuit will deliver. The transistor is usefully considered as operating in a normal mode, where the collector current obeys the standard diode equation plus a term for current transferred from the emitter

$$I_C= -\alpha_N I_E - I_{co}\exp\,(qV_C/kT-1)$$

and in an inverted mode, which defines an inverse alpha α_I

$$I_E= -\alpha_I I_C - I_{eo}\exp\,(qV_E/kT-1).$$

The current I_{co} is the reverse collector leakage current when the emitter is open, and I_{eo} is the reverse emitter current when the collector is open. These equations refer to the current reference directions shown in Fig. 2, and V_E and V_C are positive for forward bias. Both equations hold simultaneously in all regions, and may be manipulated for various purposes by making use of the relations $I_B= -(I_E+I_C)$ and $\alpha_N I_{eo}\simeq\alpha_I I_{co}$.

The above equations become inaccurate at high currents and high voltages. The effects occurring are illustrated by *npn* common-emitter–collector characteristics of typical shape shown in Fig. 3. At high currents, β decreases so that the curves crowd together. At high voltages, avalanche multiplication in the collector junction becomes important. The base-collector junction avalanche breakdown voltage is BV_{CBO}. Below this voltage, the multiplication factor M is given by

$$1/M=1-(V_{CB}/BV_{CBO})^n$$

where n ranges from 3 to 6. The effective current gain is $M\alpha$, which can become equal to 1 or greater at voltages where $M\geq 1/\alpha$. The breakdown voltage BV_{CEO} is defined for the base open ($I_B=0$) and is determined by the locus of points for which $M\alpha=1$. The voltage defined in Fig. 3 is the lowest value on the curve, which often has a negative resistance portion; this value is often denoted by LV_{CEO}. If the base is shorted to the emitter, the breakdown voltage BV_{CES}, closer to BV_{CBO}, is defined. Actual curve shapes for any given device may differ considerably from those of Fig. 3.

The static equations above do not predict the high-frequency or fast-switching properties of the transistor. These properties are conveniently

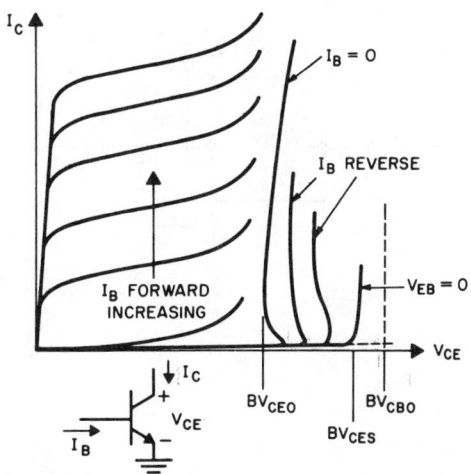

Fig. 3—Typical common-emitter static characteristics.

summarized by a charge-control model. When current flows in a transistor, excess charge (in the form of both minority and majority carriers) is stored in the base layer and, if the device is in saturation, in the collector. In addition, charge is stored in the capacitance associated with the reverse-biased collector junction. To change the voltage or current in the device requires charge flow to or from the control electrode (usually the base).

The excess minority charge in the base is denoted by Q_B (see Fig. 2). To a first approximation, this charge is proportional to the collector current

$$I_C=\omega_t Q_B$$

where ω_t is a characteristic frequency of the device depending principally on the base layer width. A narrow base results in a small charge stored per unit current and a high ω_t. To maintain charge neutrality in the base, an equal number of excess minority carriers must be furnished by the base current. Thus charge must be supplied or withdrawn to change the current through the device. The base current furnishes this charge and supplies recombination current to maintain the total charge. The basic charge-control equation is

$$I_B=dQ_B/dt+Q_B/\tau$$

where τ is an effective lifetime. This can also be written approximately as

$$\omega_t I_B=dI_C/dt+I_C/\tau$$

showing that the current gain I_C/I_B is $\beta=\omega_t\tau$ at low frequencies, and that at high frequencies it approaches the value $\omega_t/j\omega$. For that reason, ω_t is sometimes termed the current gain-bandwidth product. If the collector voltage V_{CB} is varying rapidly, an additional component of base current

$$I_B(C_c)= -dQ_c/dt= -C_c dV_{CB}/dt$$

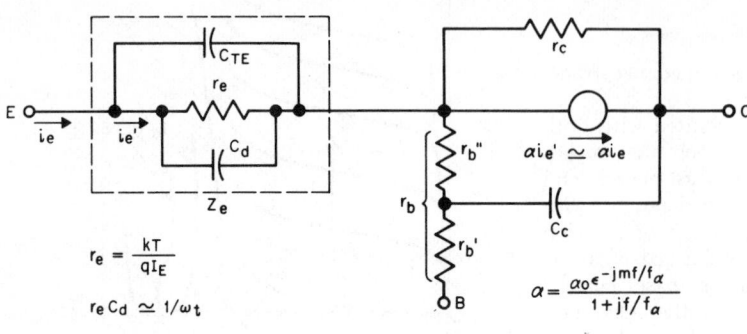

$$r_e = \frac{kT}{qI_E}$$

$$r_e C_d \simeq 1/\omega_t$$

$$\alpha = \frac{\alpha_0 \epsilon^{-jmf/f_\alpha}}{1 + jf/f_\alpha}$$

$$f_\alpha \simeq (1 + m) f_t$$

Fig. 4—Small-signal transistor equivalent circuit.

flows to charge the capacitance C_c of the collector-base junction. At high frequencies, a considerable total base current can flow through the resistance of the thin base layer. The resulting loss limits the high-frequency power gain, and the switching speed is limited by the amount of current the driving circuit can supply through the base resistance.

Silicon and germanium are used for all commercial transistors at the present time. Germanium is principally useful for some low-cost alloy types and for microwave transistors where the high carrier mobility is an advantage. Because of the smaller energy gap, germanium devices are useful to only about 100°C, whereas silicon devices can be used to approximately 200°C. Silicon technology is much more developed, the most significant feature being the relative ease with which a silicon-dioxide film can be thermally grown and subsequently selectively etched by a photolithographic process. Very precise and complex geometries can be formed in the film, which acts as diffusion mask so that impurities of p or n type can be introduced in selected areas. These areas can be as small as a few square micrometers.

*Small-Signal Transistors.** Devices are available with upper frequency limits from a few megahertz to over 3 GHz. Types intended for small-signal amplifiers and low-power oscillators rarely have dissipation ratings (or power output capabilities) greater than a few hundred milliwatts. Their electrical characteristics may be approximately described by the equivalent circuits of Fig. 4 and of Fig. 5, which is a high-frequency version of Fig. 4

* R. L. Pritchard, "High-Frequency Power Gain of Junction Transistors," *Proc. IRE*, Vol. 43, pp. 1075–1085; Sept. 1955. R. P. Abraham, "Transistor Behavior at High Frequencies," *IRE Trans. on Electron Devices*, Vol. ED-7, pp. 59–69; Jan. 1960. E. G. Nielson, "Behavior of Noise Figure in Junction Transistors," *Proc. IRE*, Vol. 45, pp. 957–963; July 1957. J. M. Rollett, "Stability and Power Gain Invariants of Linear Two-Ports," *IRE Trans. on Circuit Theory*, Vol. CT-9, pp. 29–32; March 1962.

transformed to be convenient for common-emitter use.

In Fig. 4, the impedance of the forward-biased emitter junction is represented by Z_e, with C_{Te} (sometimes denoted by C_{ib}) being the space-charge capacitance of the junction. The current generator αi_e represents the current transfer to the collector junction and has the frequency response shown, where the excess phase factor m ranges from about 0.2 to greater than 1, depending largely on the doping profile in the base. The resistance r_c represents the back resistance of the reverse-biased collector junction as well as effects caused by the collector space-charge layer widening into the base with increased voltage, thereby increasing the current gain. Another effect of this widening is to introduce a reverse transmission represented by the resistance r_b''. The element r_b' is the effective ohmic resistance of the base layer, and C_c is the collector junction capacitance.

The circuits of Figs. 4 and 5 do not show package impedances, which can exert a profound influence at very-high frequencies. To a first approximation, the package introduces lumped capacitances between the terminals and inductances in series with each terminal. For the common-emitter configuration, the most important of these elements are those providing coupling between output and input: the emitter inductance L_e which introduces an apparent resistance $\omega_t L_e$ into the input circuit,

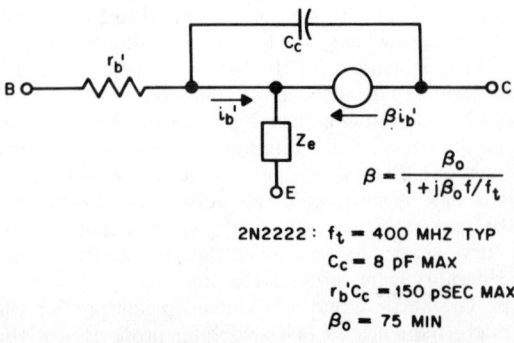

$$\beta = \frac{\beta_0}{1 + j\beta_0 f/f_t}$$

2N2222 : f_t = 400 MHZ TYP

C_c = 8 pF MAX

$r_b'C_c$ = 150 pSEC MAX

β_0 = 75 MIN

Fig. 5—High-frequency small-signal equivalent circuit.

and the collector–base capacitance which decreases the maximum stable gain.

Data sheets for small-signal applications frequently give values for the hybrid parameters, which are defined by the circuit of Fig. 6. Table 2 shows the low-frequency hybrid parameters for both common-emitter and common-base in terms of the elements of Fig. 4, sometimes termed T parameters. These relations are convenient for converting between common-base and common-emitter parameters. Table 3 shows equations for the high-frequency hybrid parameters in terms of the elements of Fig. 5. For illustration, numerical values are given for a type $2N2222$ general-purpose npn silicon vhf amplifier and switch.

The maximum frequency of oscillation is given by

$$f^2{}_{max} \simeq f_T/8\pi r_b C_c$$

and the power gain (common-emitter) at frequency f is about $f^2{}_{max}/f^2$ for a load admittance in the neighborhood of $\omega_t C_c - j\omega C_c$. The gain is of course limited at the low-frequency end by the resistive elements of Fig. 4. Transistors are available with f_{max} frequencies up to the order of 6 GHz.

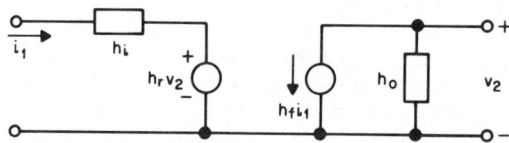

Fig. 6—Circuit defining hybrid parameters.

An important consideration is whether the transistor will oscillate with certain terminating impedances. The criterion for stability is that the stability factor k be greater than 1.

$$k = \frac{2\mathrm{Re}(h_i)\,\mathrm{Re}(h_o) - \mathrm{Re}(h_r h_f)}{|h_r h_f|}$$

where Re indicates the real part. Conversely, if $k<1$, the device will oscillate (because of its own internal feedback) with certain passive terminating impedances. In that case, the maximum available gain is unbounded, but a measure of the useable gain is given by

$$G_{MS} = |h_f/h_r|$$

where G_{MS} is termed the maximum stable gain. If $k>1$, the maximum available gain is

$$G_{MA} = G_{MS}[k - (k^2-1)^{1/2}].$$

Noise is generated in transistors by the "shot effect" in the bias current and by thermal noise in the base resistance. Surface recombination is thought to provide a source of noise which becomes significant at very-low frequencies. At frequencies above $(1-\alpha_0)f_\alpha$, the noise figure is

$$F = 1 + \frac{1 + 2r_b'/r_e + (f/f_\alpha)^2(1 + r_s/r_e + r_b'/r_e)^2}{r_s/r_e}$$

TABLE 2—LOW-FREQUENCY HYBRID PARAMETERS (SEE FIGS. 4 AND 6).

$$h_{fb} = -\alpha_0$$

$$h_{rb} = r_b/r_c$$

$$h_{ib} = r_e + r_b(1-\alpha_0)$$

$$h_{ob} = 1/r_c$$

$$h_{fe} = \alpha_0/(1-\alpha_0) = \beta_0[75\text{–}375]$$

$$h_{re} = r_b/(1-\alpha_0)\,r_c[4\times10^{-3}\text{ max}]$$

$$h_{ie} = r_b + r_e/(1-\alpha_0)[0.25\text{–}1.25\text{ k}\Omega]$$

$$h_{oe} = 1/(1-\alpha_0)\,r_c[25\text{–}200\ \mu\text{mhos}]$$

where r_s is the source resistance. It is evident that an optimum source resistance exists for minimum F. The minimum F is a function only of the ratios f/f_α and r_b'/r_e: for $f/f_\alpha = 0.5$ and $r_b'/r_e = 2$, $F_{min} \simeq 3.5$. Noise figure rises rapidly as f approaches f_α. Present attainable noise figures are less than 4 dB at 1 GHz.

*Switching Transistors.** Transistors intended for switching are used in the common-emitter connection, with base electrode controlling the impedance of the collector–emitter circuit. They are usually characterized by the following quantities in addition to the usual maximum voltage and leakage-current ratings. (Illustrative values are given for the $2N2369$ npn silicon high-speed switch):

$V_{CE(sat)}$: Collector-emitter saturation voltage at specified collector current I_c; usually $I_B = 0.1I_c$.

TABLE 3—HIGH-FREQUENCY HYBRID PARAMETERS (SEE FIGS. 5 AND 6).

$$h_{fb} \simeq -\alpha$$

$$h_{rb} \simeq j\omega r_b'C_c$$

$$h_{ib} \simeq Z_e + r_b'(1-\alpha)$$

$$h_{ob} \simeq j\omega C_c = j\omega C_{ob}$$

$$h_{fe} = \beta \simeq \omega_t/j\omega$$

$$h_{re} \simeq h_{oe}Z_e$$

$$h_{ie} \simeq r_b' + Z_e(1+\omega_t/j\omega)$$

$$h_{oe} \simeq j\omega C_c + \omega_t C_c$$

*J. L. Moll, "Large-Signal Transient Response of Junction Transistors," *Proc. IRE*, Vol. 42, pp. 1773–1784; Dec. 1954. J. G. Linvill, "Lumped Model of Transistors and Diodes," *Proc. IRE*, Vol. 46, pp. 1141–1152; June 1958. J. J. Sparks, "A Study of the Charge Control Parameters of Transistors," *Proc. IRE*, Vol. 48, pp. 1696–1705; Oct. 1960.

(0.25 V at $I_c = 10$ mA.) In terms of the forward and inverse current gain parameters α_N and α_I discussed in the general-principles section, the saturation voltage is

$$(q/kT) V_{CE \, (\text{sat})} = \ln \left[1 + (1-\alpha_I) I_C/I_B \right]$$
$$- \ln \left[1 - (1-\alpha_N) I_C/\alpha_N I_B \right] - \ln \alpha_I.$$

In practical devices, resistive voltage drops in various parts of the structure, particularly in the collector bulk, may be the controlling factor.

$V_{BE \, (\text{sat})}$: Base-emitter saturation voltage under above conditions (0.70 to 0.85 V).

h_{FE}: Direct-current signal current gain I_C/I_B at specified point in active region (40 to 120 at $I_C = 10$ mA, $V_{CE} = 1$ V).

h_{fe}: Small-signal current gain magnitude at specified frequency. Usually f is high enough that h_{fe} is behaving as f_T/f, so that f_T can be estimated by $f_T \sim f h_{fe}$. (5 min at 100 MHz; $f_T \sim 500$ MHz.)

C_{ob}: Output capacitance in common base, approximately equal to C_c in Fig. 4 (4 pF max at $V_{CB} = 5$ V).

t_{on}: Turn-on time= sum of delay and rise times to a specified current level from a specified reverse bias on the base, when the base is driven by a current step. The collector voltage swing is also usually given. If f_T is known, the rise time can be estimated from the charge-control equation given in the general-principles section ($t_{on} = 12$ ns max, $I_C = 10$ mA, $I_B = 3$ mA, $V_{EB} = 1.5$ V, $V_{CC} = 3$ V).

t_s: Storage time. When a transistor is driven into saturation, more base current is supplied than is needed to sustain the collector current. This results in a charge Q_s being stored in the base in addition to the Q_B shown in Fig. 2. Before the transistor will come out of saturation, Q_s must be removed, usually by an external reverse base current. The time needed to do this is t_s. (13 ns max for $I_C = I_{B \, on} = I_{B \, off} = 10$ mA).

t_{off}: The time required to reduce I_C to zero from a specified point in saturation. The storage time is one component of the off time. (18 ns max, $I_C = 10$ mA, $I_{B \, on} = 3$ mA, $I_{B \, off} = 1.5$ mA, $V_{CC} = 3$ V).

Q_T: Control charge. The charge which must be withdrawn from the base to turn the transistor off. This number is useful in estimating the value of a "speed-up" capacitor C where the transistor base is effectively driven by a voltage step V_{in} in series with C so that a charge $(V_{in} - V_{be}) \, C$ is extracted. (50 pC max for $I_C = 10$ mA, $I_B = 1$ mA).

Switching transistors cover an extremely wide range of current levels (a few mA to about 60 A) and speeds (a few nanoseconds to several microseconds) with the highest-speed units being relatively low-current low-voltage devices. High-current devices have very low impedance levels, and circuit inductance becomes an important limitation on switching time. Avalanche operation is possible with some types of switching transistors by biasing the collector-emitter circuit near the avalanche breakdown point and applying a small base signal to trigger avalanche breakdown. Very fast rise-time pulses of high current are available from remarkably simple circuits using the avalanche mode of operation.

Power Transistors. The term power transistor applies to devices with dissipation ratings higher than about 1 watt. Since they may be used as either switches or amplifiers, the discussion of the previous two sections applies. Special considerations apply to removing heat from the device and to avoiding thermal instability. The dissipation capabilities are usually expressed by a power derating curve showing average power dissipation as a function of case temperature. Controlling case temperature in a particular environment is a problem in heat-sink design. A typical derating curve is flat up to about 25°C and then decreases linearly to zero at 100°C for germanium or at 200°C for silicon devices.

Large-area transistors are particularly susceptible to a form of thermal instability in which a tendency for current to become nonuniformly distributed is reinforced by the temperature rise occurring in the high-current region. This can lead to an extreme concentration of current and a localized thermal runaway condition in which thermally generated current into the base stimulates further emission. An equilibrium situation is reached only by a sharp drop in collector voltage, termed "secondary breakdown." The localized heating frequently destroys the device. A transistor is much more sensitive to secondary breakdown at high voltage and low current than at low voltage and high current, and in the former condition the rated dissipation usually cannot be reached. Transistors are often characterized by a safe operating area on the I_C–V_{CE} plane, and different areas may be shown for direct-current and pulsed operation, since for short pulses the safe limits for dc can be exceeded. Some advanced types of silicon power transistors incorporate resistive stabilization in the emitter circuit to inhibit current concentration and extend the safe operating area.

For reliable operation, both power–temperature derating and safe operating areas must be considered. Switching transients cause additional temperature increases which depend on the thermal time constant of the transistor, short transients tending to be averaged over a thermal time constant.

* M. A. Clark, "Power Transistors," *Proc. IRE*, Vol. 46, pp. 1185–1204; June 1958. R. M. Scarlett and W. Shockley, "Secondary Breakdown and Hot Spots in Power Transistors," *IEEE Convention Record*, Vol. 11, Part 3, pp. 3–13; 1963. J. Tatum, "RF Large-Signal Power Amplifiers," *Electronic Design News*; May, June, July, 1965.

Power transistors range from cheap germanium-alloy devices for low-frequency power amplifiers and slow switches, to silicon rf power transistors which in the present state of the art can deliver 50 watts at 150 MHz, 15 watts at 500 MHz, and up to 2 watts at 2 GHz (pulsed). These types are characterized by their maximum power output and power gain as a function of frequency, usually in a specified circuit, and the intermodulation distortion performance is frequently quoted. Rf power transistors (and high-frequency devices in general) obey a law roughly of the form $P \cdot Z \cdot f^n = $ constant, where P is power output, Z is impedance level, and n ranges from 2 to 4. Thus, high-power high-frequency devices have extremely low impedance levels (for example, a 15-watt 500-MHz device has an input impedance of the order of 1 ohm), leading to difficulty with package impedances and circuit matching. Transistors are available with internally grounded emitters for lowest inductance, and with broad collector and base leads intended for strip-line circuits.

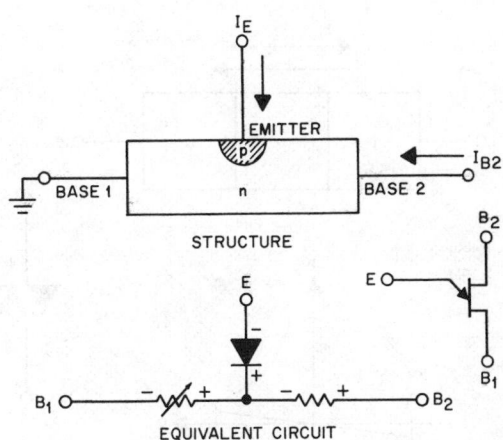

Fig. 7—Unijunction transistor structure and equivalent circuit.

Phototransistors.[*] Hole-electron pairs are produced in the vicinity of the collector-base junction when photons irradiate a bipolar transistor made from a semiconductor having a band gap less than the energy of the photons. Conventional transistors are housed in opaque packages to avoid noise and other adverse effects resultant from this optical radiation sensitivity, but phototransistors are installed in a housing having a transparent window or self-contained collection lens to better exploit this property. Since carrier generation occurs at the collector-base junction, the base region is made much larger than its counterpart in conventional transistors to enhance photon collection efficiency.

Phototransistors can be made in various configurations from a range of semiconductors, but the most common are *npn* units made from silicon. Since phototransistor action is controlled by an optical signal at the base, electrical connection to the base region is optional and therefore most commercial phototransistors have but two terminals. It is a popular misconception that application of electrical base bias via a base terminal increases phototransistor sensitivity by establishing a collector current which optimizes h_{FE}. In fact, the base bias causes some of the photocurrent to be shunted around the base-emitter junction with a resultant decrease in optical sensitivity. Optical bias from a small lamp or LED, however, can sometimes be used to enhance the lower sensitivity limit of phototransistors.

Unijunction Transistors.[†] The unijunction transistor (UJT) is a three-terminal, single-junction device which exhibits negative resistance and switching characteristics totally unlike those of conventional bipolar transistors. As shown in Fig. 7, the UJT consists of a bar of *n*-type silicon having ohmic contacts designated base 1 (B_1) and base 2 (B_2) on either side of a single *pn* junction designated the emitter. In operation, a positive voltage is applied to B_2, and B_1 is placed at ground potential. The B_2-E-B_1 junctions then act like a voltage divider which reverse-biases the emitter junction. An external voltage having a potential higher than this reverse bias will forward-bias the emitter and inject holes into the silicon bar which move toward B_1. The emitter-B_1 resistance then decreases, and this, in turn, causes the emitter voltage to decrease as the emitter current increases. The resultant negative resistance can be utilized in such applications as relaxation oscillators, voltage sensors, pulse generators, timers, and trigger generators.

Diac.[*] The diac is a two-terminal, transistor-like component which exhibits bistable switching for either polarity of a suitably high applied voltage. As shown in Fig. 8, the diac closely resembles a *pnp* transistor without an external base terminal. Doping concentrations at both junctions are very similar, and the net result is a bidirectional avalanche switch. The negative-resistance characteristic of the diac makes it useful for very simple relaxation oscillators and pulse generators, but its major application is in conjunction with a triac to produce ac phase-control circuits useful for motor-speed control, light dimming, and other ac power-control functions.

* J. Bliss, "Applications of Phototransistors in Electro-Optic Systems," Motorola Application Note AN-508; 1969.

† General Electric Co., "Transistor Manual," pp. 301–347; 1969.

* RCA, "Transistor, Thyristor, and Diode Manual," pp. 43–44; 1969.

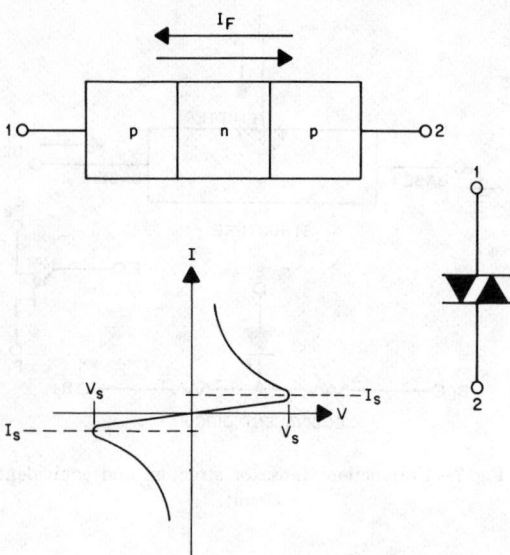

Fig. 8—Diac structure and characteristics.

pnpn Devices

The *pnpn* device is a three-junction semiconductor switch having an internal feedback-induced, bistable, negative-resistance characteristic similar to that of a gas-filled thyratron. A typical *pnpn* device consists of alternating p and n layers, at least three of which are arranged for suitable interaction. The family of *pnpn* devices includes the four-layer diode, the silicon controlled rectifier, and the triac, each of which is described in detail below.

*Four-Layer Diodes.** The four-layer or Shockley diode is the most elementary *pnpn* device. Figure 9 shows the volt-ampere characteristics of a typical four-layer diode. In the forward-blocking (off) state, the center junction is reverse-biased, and a small reverse current flows. Under forward bias, electrons are injected at junction J_1 into base layer p_1 from where they diffuse across to be collected at J_2, with current gain α_1. These electrons can be thought of as providing base current to the transistor formed by $p_2 n_2 p_1$. Similarly, holes are injected at J_3 and are collected at J_2 with current gain α_2. These holes provide base current for the transistor formed by $n_1 p_1 n_2$. If the terminal current is I, and the reverse current of the center junction J_2 is I_L, then a consideration of the total current across J_2 (which must equal I) gives $I = I_L + \alpha_1 I + \alpha_2 I$ or

$$I = I_L / (1 - \alpha_1 - \alpha_2).$$

* F. E. Gentry, F. W. Gutzwiller, N. Holonyak, Jr., and E. E. Van Zastrow, "Semiconductor Controlled Rectifiers: Principles and Applications of *pnpn* Devices," Prentice-Hall, Inc., Englewood Cliffs, N.J.; 1964.

Evidently, when the sum of the two alphas is 1 or more, an unstable regenerative situation exists. Therefore the only stable condition is when both internal transistors are saturated, thus forward-biasing center junction J_2 and switching the device into a very-low-impedance "on" state.

In silicon transistors, alpha (current gain) is a function of current, being small at very-low current levels and increasing with current. For a four-layer diode to switch into its low-impedance "on" state, the current through it must increase to the point where the sum of the alphas is 1. Accordingly, in a practical four-layer-diode switching circuit, triggering is achieved by momentarily raising the forward voltage until sufficient avalanche current (switching current I_s in Fig. 9) flows across the center junction J_2.

In any four-layer device, a rapidly increasing forward voltage in the blocking (off) state causes a displacement current $C\,dV/dt$ to flow in the capacitance C of the reverse-biased center junction. This current is carried by injected carriers at both emitters and contributes to an increase in the two alphas. If the sum of the alphas becomes 1, the device may unintentionally self-switch at a lower voltage than V_s. This phenomenon is termed the "rate-effect," and four-layer devices are often specified with respect to the dV/dt they can tolerate without premature switching.

Switching times of four-layer diodes are highly dependent upon circuit conditions. After the trigger pulse is applied, a brief delay follows while the anode current accumulates. This delay depends on the nature of the triggering pulse. The subsequent rise time to 90% of the final current depends partly on the load circuit. Total turn-on times

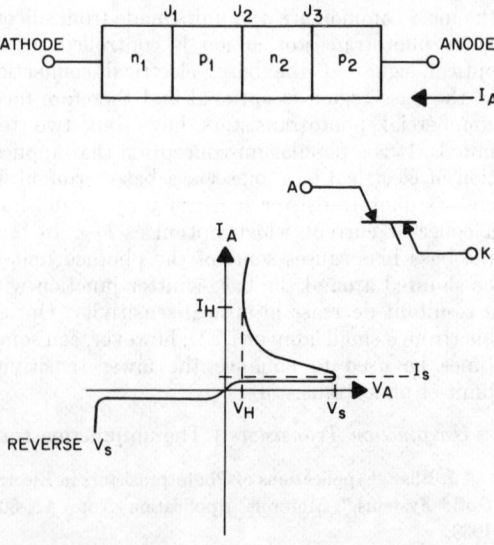

Fig. 9—Four-layer diode structure and characteristics.

range from about 10 ns for some devices to 100 ns. Turn-off is achieved by reducing the load current below holding current I_H, a result automatically accomplished when a capacitor in a charging circuit is the source of forward voltage. To assure complete turn-off, the excess carriers must be removed from both base layers by recombination or by reverse current flow, and turn-off time is therefore defined as the minimum time interval from removal or reversal of load current to application of a specified rate of voltage rise without the device switching on prematurely. Turn-off time is always larger then turn-on time and limits the peak pulse repetition rate which can be achieved in oscillatory circuits.

*Silicon Controlled Rectifiers.** The silicon (or semiconductor) controlled rectifier (SCR) is a four-layer diode with external electrical connection to the p_1 region (see Fig. 10). A small current applied to this third terminal, which is designated the gate, can control the switching level of the device. Therefore, the SCR is inherently more flexible than the basic self-switching four-layer diode and can be thought of as the solid-state analog of the conventional electromechanical relay.

The volt-ampere characteristics of the SCR, which closely resemble those of the four-layer diode, are shown for a variety of gate conditions in Fig. 10. In operation, a potential is applied to the gate to forward-bias J_1 and cause electrons to be injected into p_1. The appropriate equation for anode current then becomes $I_A = (\alpha_1 I_G + I_L)/(1 - \alpha_1 - \alpha_2)$. No gate bias is required after the

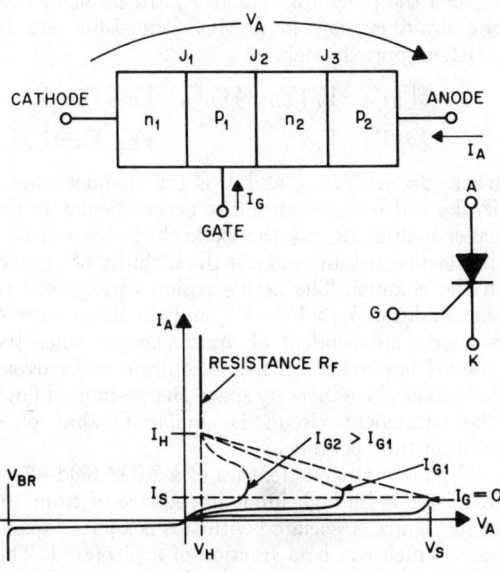

Fig. 10—SCR structure and characteristics.

* General Electric Co., "SCR Manual," pp. 12–13; 1967.

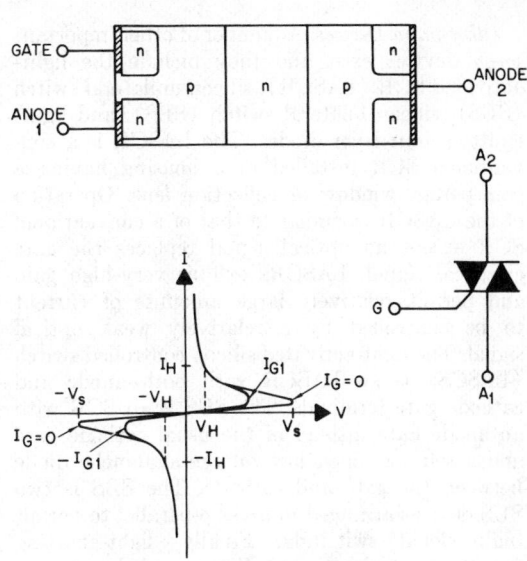

Fig. 11—Triac structure and characteristics.

device has begun to switch or has turned on, but a minimum forward-holding current I_H is required to keep the device in the low-impedance switch after it has been switched on. This means that an SCR connected across a constant load stays on once activated but can be turned off by simply disconnecting the load or momentarily bypassing the device with an anode-cathode short circuit.

Like four-layer diodes, SCRs may switch on prematurely or without a gate signal if a rapidly increasing forward voltage is applied to the device. Therefore the dV/dt becomes an important operating specification. Also, a maximum di/dt specification is sometimes given since a high-current SCR in a very-low-inductance circuit may suffer damage if the very rapid rate of current increase does not allow time for the entire junction area to be uniformly switched on. Total turn-on times for typical SCRs range from less than 50 ns for some small devices (1 A, 400 V) to about 10 μs for large-junction-area devices (400 A, 1200 V).

*Triacs.** The triac is a modified *pnpn* device having five layers and capable of switching on for either polarity of an applied voltage. As can be seen in Fig. 11, the triac has a single gate lead and is therefore the ac equivalent of the SCR. Since switching can be initiated by a positive or negative gate signal for either direction of current across the device, the triac has four basic operating modes. The triac exhibits I_H and temperature characteristics similar to those of the SCR. Applications for triacs include ac motor-speed control ac light dimmers and general ac power-control applications.

* General Electric Co., "SCR Manual," pp. 13–14; 1967.

Other pnpn Devices. A number of other important *pnpn* devices exist and they include the light-activated SCR (LASCR), silicon unilateral switch (SUS), silicon bilateral switch (SBS), and light-emitting four-layer diodes. The LASCR is a conventional SCR installed in a housing having a transparent window or collection lens. Operation of the LASCR is similar to that of a conventional SCR except an optical signal replaces the gate electrical signal. LASCRs exhibit very-high gain and permit relatively large amounts of current to be controlled by a relatively weak optical signal. The light-activated silicon controlled switch (LASCS) is an LASCR with both anode and cathode gate terminals. The SUS is an SCR with an anode gate instead of the usual cathode gate and a self-contained low-voltage avalanche diode between the gate and cathode. The SBS is two SUS devices arranged in inverse-parallel to permit bidirectional switching. Finally, light-emitting four-layer diodes are gallium-arsenide devices which emit recombination radiation when they are switched on. These devices are unique in that exceptionally simple optical pulse sources can be fabricated since the source is also the switching element. A particularly interesting device of this kind is the *pnpn* injection laser.

MAJORITY-CARRIER DEVICES

Field-Effect Transistors*

A field-effect transistor is a majority-carrier device in which the resistance of a current path from source to drain electrodes is modulated by the voltage applied to a gate electrode. In a junction-type device, the gate is a reverse-biased *pn* junction where the depletion layer extends into a conducting channel and effectively removes carriers from it. In the more recently developed and more important insulated-gate or metal-oxide-semiconductor (MOS) types, the field from a metal gate electrode extends through a thin insulator into the semiconductor layer between source and drain electrodes, modifying its conductivity.

Figure 12 shows a cross section of an *n*-channel insulated-gate field-effect MOS device (the complementary *p*-channel units are similar but opposite in polarity). Depending on the surface treatment, an *n* conducting channel may be formed and exist with the gate voltage $V_G = 0$. A negative gate voltage will then drive electrons out of the channel,

* J. T. Wallmark, "The Field-Effect Transistor—A Review," *RCA Review*, Vol. 24, pp. 641–660; Dec. 1963. S. R. Hofstein and F. P. Heiman, "The Silicon Insulated-Gate Field-Effect Transistor," *Proc. IEEE*, Vol. 51, pp. 1190–1202; Sept. 1963. "Field-Effect Transistors: Physics, Technology and Applications," J. T. Wallmark, ed., Prentice-Hall, Inc., Englewood Cliffs, N.J.; 1966.

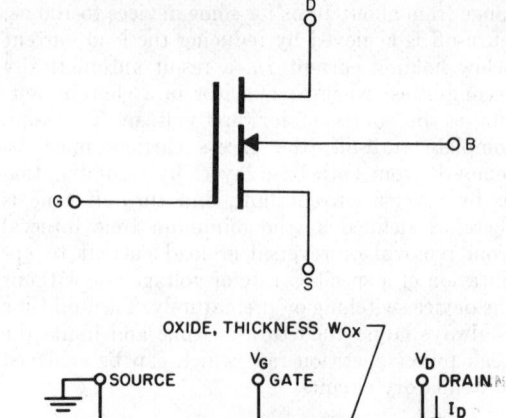

Fig. 12—Insulated-gate field-effect (MOS) transistor; *n*-channel enhancement type.

increasing the resistance from source to drain. This is termed depletion-mode operation. Conversely, if no channel exists with $V_G = 0$, one can be formed by applying positive V_G and attracting electrons to a thin surface layer. This is termed enhancement-mode operation (see Fig. 12).

Mobile charge exists in the channel for gate voltages greater than a characteristic voltage termed the pinch-off voltage V_P. In terms of this, the drain current in a MOS transistor can be written approximately as

$$I_D = \beta\left[(V_G - V_P)V_D - \tfrac{1}{2}V_D^2\right], \quad V_D \leq V_G - V_P$$
$$= \tfrac{1}{2}\beta(V_G - V_P)^2, \qquad\qquad V_D \geq V_G - V_P$$

where $\beta = \epsilon\mu W/Lw_{ox}$ and L is the channel length, W the width of the structure perpendicular to the paper in Fig. 12, w_{ox} the oxide thickness and ϵ its dielectric constant, and μ is the mobility of carriers in the channel. The active region corresponds to the condition $V_D > V_G - V_P$, and the drain current is nearly independent of drain voltages since the channel is pinched off near the drain and current flow takes place there by space-charge-limited flow. The equivalent circuit is similar to that of a vacuum-tube pentode.

The most obvious feature of a MOS field-effect transistor is its high input impedance of from 10^9 to 10^{15} ohms. Associated with this is a gate capacitance which can be a fraction of a picofarad. The gain–bandwidth figure $g_m/2\pi C_{gate}$ can easily be made greater than 100 MHz. Switching speed is limited by the charging time of the gate capacitance through the channel resistance, but can be faster than 10 ns.

Field-effect transistors can be made perfectly symmetrical, with source and drain interchangeable, so that they are useful as bilateral switches. Furthermore, they have no voltage offset, which is an advantage in chopper applications. A further advantage in many applications is their thermal stability and lack of thermal runaway. In contrast to bipolar transistors, their current decreases with increasing temperature.

Field-effect devices have found wide use in integrated circuits because they are naturally a plane-surface device well suited for production in large arrays by simple means.

Hot-Electron Diodes*

Hot-electron (hot-carrier or Schottky barrier) diodes consist of rectifying metal–semiconductor contacts in which current flows by means of majority carriers. Most are made on n-type silicon with a metal such as gold and, when forward-biased, inject electrons into the metal. Since these injected electrons have greater velocities than the thermal electrons, they are termed "hot." There is no minority-carrier storage, so switching times can be extremely fast. The voltage and current characteristics are described closely by the ideal diode equation (see pn diodes above), and the reverse characteristics and capacitance are similar to those of an abrupt pn junction.

Hot-electron diodes have advantages over point-contact diodes as microwave mixers, including less dependence of noise figure on local-oscillator power, more resistance to burnout, and more reproducible characteristics. These diodes are also appropriate for high-speed switching, harmonic generation, and parametric amplification.

Bulk-Effect and Transit-Time Devices†

A great deal of effort is being devoted to electromagnetic interactions in bulk semiconductors because of their promise of radio-frequency generation and amplification with higher power than any previous devices. Although bulk-effect devices are still in the early development stage, encouraging results have already been obtained. Two types seem to be of greatest technological importance; transferred-electron (or "Gunn") devices and avalanche transit-time devices.

The transferred-electron effect takes place in suitable "two-valley" semiconductors such as GaAs or InP. Electrons gain energy from an

applied electric field, and at fields of several kilovolts per centimeter enough energy is gained that the electrons begin to transfer to a conduction-band valley having a much smaller mobility. The result is that the current begins to decrease with increasing field until nearly all the electrons have been transferred. This bulk negative resistance leads to various types of electrical instability. One type occurs in relatively long heavily doped samples, in which narrow domains of high electric field are generated. These domains travel through the semiconductor at about the maximum electron velocity (about 10^7 cm/s), and the frequency of the resulting oscillation is determined principally by their transit time.

At the present time, powers over 100 watts pulsed and 100 milliwatts cw have been obtained with Gunn diodes at 2 GHz, and smaller powers at higher frequencies. Another type of behavior occurs in short and lightly doped samples which can act as negative-resistance devices producing stable small-signal amplification. The noise figure obtained is presently rather high (20 decibels).

The other type of device of potential importance is the avalanche transit-time (ATT) diode which consists of a pn junction designed with a high field avalanching region at one end of a relatively high-resistance lower field region across which the avalanche-generated carriers drift. The internally generated current is controlled by the terminal voltage, and the terminal current is delayed by the transit across the lower field region. ATT diodes can be operated as small-signal amplifiers (although present noise figures are very high) or as microwave oscillators and are tunable over wide frequency ranges. Cw power outputs of 30 mW at 14 GHz and 150 mW at 5.5 GHz have been obtained.

Hall-Effect Devices*

When a magnetic field B is applied perpendicularly to a current I flowing in a semiconductor bar, an electric field is produced perpendicular to both B and I and proportional to BI, and an open-circuit voltage V_H is consequently observed. The ratio V_H/V, where V is the applied voltage, is proportional to $\mu_H BI$, where μ_H is the Hall mobility. This is of the same order of magnitude as the conductivity mobility, so that materials such as InSb or InAs, because of their very high electron mobility, are very suitable for Hall devices. Applications include multipliers, measuring magnetic fields, and measuring electromagnetic power flow.

* D. Kahng and L. A. D'Asaro, "Gold-Epitaxial Silicon High-Frequency Diodes," *B.S.T.J.*, Vol. 43, pp. 225–232; Jan. 1964.

†Bulk-Effect and Transit-Time Devices—Special Issue on Semiconductor Bulk-Effect and Transit-Time Devices, *IEEE Transactions on Electron Devices*, Vol. ED-13; Jan. 1966.

* A. Clawson and H. Wieder, "Bibliography on the Hall-Effect Theory and Applications," *Solid-State Electronics*, Vol. 7, pp. 387–396; May 1964.

LETTER SYMBOLS FOR SEMICONDUCTOR DEVICES*

The letter symbols recommended by the IEC for semiconductor devices are given in the following. In general, these letter symbols agree with IEC publication No. 27, *Letter Symbols to be Used in Electrical Technology*, which is applicable. However, where different recommendations are given here, they should be used.

General

(*A*) *Current, Voltages, and Power.* The recommended basic letter symbols are included in Table 4. Where both upper-case and lower-case letters are shown, the upper-case letter should be used to represent: maximum (peak) values, average (mean) values, continuous (dc) values, and root-mean-square values.

TABLE 4—BASIC LETTER SYMBOLS FOR VOLTAGE, CURRENT, AND POWER.

Letter	Designates
I, i	current
V, v, or U, u^1	voltage
P, p	power

[1] While the letters U and u are recommended by the IEC for voltages and V and v are reserve symbols, V and v are so widely used in the field of semiconductors, they are given equal rank.

Lower-case basic letters are used to represent instantaneous values which vary with time.

The recommended general subscripts for voltage, current, and power are included in Table 5. Other subscripts are included in the specific lists later in this chapter.

Where both upper-case and lower-case letters are given in Table 5, the choice should be made using the following guidelines: If more than one subscript is used, and both style subscripts exist, the subscript should be either all lower-case or all upper-case. Upper-case subscripts should be used to indicate: continuous (dc) values (without signal) (I_B), instantaneous total values (i_B), average total values (I_{BAV}), and maximum (peak) total values (I_{BM}). Lower-case letters should be used to indicate values applying to varying components only: instantaneous values (i_b), root-

mean-square values (I_b), maximum (peak) values (I_{bm}), and average values (I_{bav}).

If it is necessary to indicate the terminal carrying the current, it is indicated by the first subscript (e.g.: I_B, i_B, i_b, I_b). If it is necessary to indicate the points between which a voltage is measured, it should be done by the first two subscripts. The first subscript indicates one terminal point of the device and the second the reference terminal or the circuit node. Where there is no possibility of confusion, the second subscript may be omitted (e.g.: $V_{BE}, v_{BE}, v_{be} V_{be}$, or $U_{BE}, u_{BE}, u_{be}, U_{be}$). Supply voltages or supply currents are indicated by repeating the appropriate terminal subscript (e.g.: V_{CC} or U_{CC}, I_{EE}). If it is also necessary to indicate a reference terminal, a third subscript is used (e.g.: V_{CCE} or U_{CCE}).

If a device has more than one terminal of the same kind, the subscript is formed by the appropriate letter for the terminal followed by a number; in the case of multiple subscripts, hyphens may be necessary to avoid misunderstanding.

Examples:

I_{B2}—continuous (dc) current flowing into the second base terminal

V_{B2-E} or U_{b2-E}—continuous (dc) voltage between the second base and the emitter terminals.

For multiple-unit devices, the subscripts are modified by a number preceding the letter subscript; in the case of multiple subscripts, hyphens may be necessary to avoid misunderstandings.

TABLE 5—GENERAL SUBSCRIPTS FOR VOLTAGE, CURRENT, AND POWER.

Subscript	Designates
Av, av	average
F, f	forward
M, m	maximum (peak) value
MIN, min	minimum value
O, o	open circuit
R, r	reverse or, as a second subscript, repetitive
S, s	short circuit or, as a second subscript, surge and/or nonrepetitive
(BR)	breakdown
(OV)	overload
tot	total

---

* Adopted from International Electrotechnical Commission (IEC) Publications No. 148, 1969, and No. 148A, 1974.

Examples:

I_{2c}—continuous (dc) current into the collector terminal of the second unit

V_{1c-2c} or U_{1c-2c}—continuous (dc) voltage between the collector terminals of the first and the second unit.

(*B*) *Electrical Parameters.* The most important basic letters used for electrical parameters of semiconductor devices are given in Table 6. Upper-case letters are used to represent electrical parameters of external circuits and of circuits in which the device forms only a part, and all inductances and capacitances. Lower-case letters are used to represent electrical parameters inherent in the device (with the exception of inductances and capacitances).

The recommended general subscripts used for the most important electrical parameters of semiconductor devices are included in Table 7. Other subscripts are included in specific lists later in this chapter.

Where both upper-case and lower-case letters are listed, the choice between the two styles shall be made according to the following: If more than one subscript is used, subscripts for which both styles exist shall either be all upper-case or all lower-case.

Example:

h_{FE}, y_{RE}, h_{fe}, but C_{Te} (T has no lower-case variant).

TABLE 6—BASIC LETTER SYMBOLS FOR ELECTRICAL PARAMETERS.

Letter	Designates
B, b	susceptance; imaginary part of an admittance (y) four-pole matrix parameter
C	capacitance
G, g	conductance; real part of an admittance (y) four-pole matrix parameter
H, h	hybrid (h) four-pole matrix parameter
L	inductance
R, r	resistance; real part of an impedance (z) four-pole matrix parameter
X, x	reactance; imaginary part of an impedance (z) four-pole matrix parameter
Y, y	admittance; admittance (y) four-pole matrix parameter
Z, z	impedance; impedance (z) four-pole matrix parameter

TABLE 7—GENERAL SUBSCRIPTS FOR ELECTRICAL PARAMETERS.

Subscript	Designates
F, f	forward; forward transfer
I, i	input
O, o	output
R, r	reverse; reverse transfer
T	depletion layer
1^1	input
2^1	output
11^2	input
12^2	reverse transfer
21^2	forward transfer
22^2	output

[1] Applicable to all electrical parameters except four-pole matrix parameters.
[2] Applicable to four-pole matrix parameters only.

The upper-case of a subscript shall be used for the designation of static (dc) values.

Examples:

h_{21E} or h_{FE}—static value of forward current transfer ratio in common-emitter configuration

R_E—dc value of the external emitter resistance.

The lower-case of a subscript shall be used for the designation of small-signal values.

Examples:

h_{21e} or h_{fe}—small-signal value of the short-circuit forward current transfer ratio in common-emitter configuration

$Z_e = R_e + jX_e$—small-signal value of the external impedance.

Each element of the four-pole matrix is identified according to the following rules:

The first letter subscript or double numeric subscript indicates input, output, forward, or reverse transfer, chosen from the list of subscripts given in Table 7.

Examples: h_{11} or h_i
h_{22} or h_o
h_{21} or h_f
h_{12} or h_r.

A further subscript is used for the identification of the circuit configuration. When no confusion is possible, this further subscript may be omitted.

Examples: h_{21e} or h_{fe}, h_{21E} or h_{FE}.

If only h_f is written, the circuit configuration must be understood. If only h_{21} is written, the circuit configuration must be understood as well as the kind of parameter (small-signal or static value).

If it is necessary to distinguish between real and imaginary parts of electrical parameters, no additional subscripts should be used. If basic symbols for the real and imaginary parts exist, these may be used.

Examples:

$$Z_c = R_e + jX_e$$
$$Y_{fe} = G_{fe} + jB_{fe}.$$

If such symbols do not exist or if they are not suitable, the following notation shall be used:

Re (h_{11b}) etc. for the real part
Im (h_{11b}) etc. for the imaginary part.

(*C*) *Other Quantities*. The basic letter symbol for time and duration is t.

Example: rise time t_r.

The basic letter symbol for temperature is t. When there is a possibility of confusion with t to represent "time," the letter theta (ϑ or θ) should be used instead. If this is not suitable and if there is no risk of confusion, the letter T (usually indicating kelvin temperature) may be used instead.

Examples: t_{amb}; ϑ_{amb}; $T_{amb} = 25°C$.

The following general subscripts are used with the basic letter symbol for temperature:

amb—ambient
case—case
J, j—junction
stg—storage.

The basic letter symbol for frequency is f.

The following quantities and their letter symbols are recommended for the quantities indicated:

thermal resistance—R_{th}*
thermal derating factor—K_t
virtual temperature
internal equivalent temperature $\left.\right\} t_{(vj)}$
transient thermal impedance—$Z_{(th)t}$
thermal impedance under pulse conditions—$Z_{(th)p}$.

* When a possibility of misinterpretation exists, because of combination of other letter symbols with the subscript of R_{th}, the subscript should be enclosed in parentheses, as $R_{(th)}$.

Low-Power Signal Diodes

Letter Symbols. The basic letter symbols given in Tables 4 and 6 are also used for low-power signal diodes. In addition to the general subscripts for voltage, current, and power listed in Table 5, the following special subscripts are used for low-power signal diodes:

A, a—anode
K, k—cathode
O—average output rectified.

In addition to the general subscripts listed in Table 7 for electrical parameters, the following special subscripts are used for low-power signal diodes:

δ, d—damping
r—recovery, recovered, rectification
S, s—storage, stored.

List of Letter Symbols. The symbols in Table 8 are recommended for use in the field of low-power signal diodes. They have been compiled in accordance with the general rules.

Mixer Diodes

Letter Symbols. The basic letter symbols given in Tables 4 and 6 are also used for mixer diodes. The general subscripts given in Tables 5 and 7 also apply.

List of Letter Symbols. The symbols listed in Table 9 are recommended for radar applications of mixer diodes.

Rectifier Diodes

Letter Symbols. The basic letter symbols given in Tables 4 and 6 are also used for rectifier diodes. In addition to the subscripts listed in Tables 5 and 7, the following are also used for rectifier diodes:

A, a—anode
K, k—cathode
O—average output rectified
(TO)—threshold
T—slope.

List of Symbols. Table 10 lists the symbols recommended for use in the field of rectifier diodes.

Voltage Reference and Voltage Regulator Diodes

Letter Symbols. The basic letter symbols given in Tables 4 and 6 are also used for voltage reference and regulator diodes. In addition to the subscripts listed in Tables 5 and 7, the symbols Z and z are used to designate *working*.

TABLE 8—LETTER SYMBOLS FOR LOW-POWER SIGNAL DIODES.

Name and designation	Letter symbol	Remarks
Voltages		
Forward continuous (direct) voltage	V_F	
Instantaneous total forward voltage	v_F	
Average forward voltage	$V_{F(AV)}$	
Reverse continuous (direct) voltage	V_R	
Instantaneous total reverse voltage	v_R	
Peak reverse voltage	V_{RM}	
Surge reverse voltage	V_{RSM}	
Breakdown voltage	$V_{(BR)}$	
Currents		
Forward continuous (direct) current	I_F	
Instantaneous forward current	i_F	
Peak forward current	I_{FM}	
Surge forward current	I_{FSM}	
Average output rectified current	I_O	
Reverse continuous (direct) current	I_R	
Instantaneous reverse current	i_R	
Peak reverse current	I_{RM}	
Powers		
Surge nonrepetitive power	P_{SM}	
Switching parameters		
Forward recovery time	t_{fr}	
Reverse recovery time	t_{rr}	
Reverse recovery current	i_{rr}	
Recovered charge (stored charge)	Q_s	

TABLE 8—CONTINUED.

Name and designation	Letter symbol	Remarks
Sundry quantities		
Differential resistance	r	
Damping coefficient	δ or d	
Damping resistance	r_δ or r_d	
Efficiency	η	
Rectification efficiency	η_r	

TABLE 9—LETTER SYMBOLS FOR MIXER DIODES.

Name and designation	Letter symbol	Remarks
Electrical power		
CW power	P_{CW}	
RF power	P_{RF}	
Average RF power	P_{RFAV}	
Pulse RF power	P_{RFP}	
Burn-out energy	$E_M; W_M; E_{HFM}; W_{HFM}$	
Other parameters		
Terminal capacitance	C_{tot}	
Diode (terminal) admittance	y_t	
IF terminal impedance	z_{if}	
Series inductance	L_s	
Voltage standing wave ratio	S_V	The abbreviations VSWR and vswr are in common use
Conversion loss	L_c	

TABLE 10—LETTER SYMBOLS FOR RECTIFIER DIODES.

Name and designation	Letter symbol	Remarks
Voltages		
Continuous (direct) forward voltage	V_F	
Crest (peak) forward voltage	V_{FM}	
Average forward voltage (with I_O specified)	$V_{F(AV)}$	
Continuous (direct) reverse voltage	V_R	
Crest (peak) working reverse voltage	V_{RWM}	
Repetitive peak reverse voltage (maximum recurrent reverse voltage)	V_{RRM}	
Nonrepetitive peak reverse voltage (peak transient reverse voltage)	V_{RSM}	
Breakdown voltage	$V_{(BR)}$	
Slope resistance	r_T	
Threshold voltage	$V_{(TO)}$	
Currents		
Continuous (direct) forward current	I_F	
Repetitive peak forward current	I_{FRM}	
Overload forward current	$I_{(OV)}$	
Surge (nonrepetitive) forward current	I_{FSM}	
Averaged output rectified current	I_O	
Continuous (direct) reverse current	I_R	
Average reverse current (with I_O specified)	$I_{R(AV)}$	

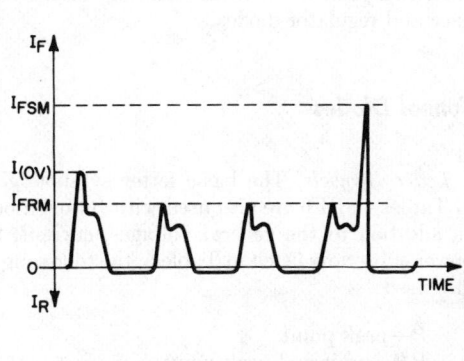

TABLE 11—LETTER SYMBOLS FOR VOLTAGE REFERENCE AND VOLTAGE REGULATOR DIODES.

Name and designation	Letter symbol	Remarks
Voltages		
Working voltage	V_Z	
Continuous (direct) reverse voltage below the working voltage range	V_R	
Noise voltage within the working voltage range	V_{nz}	The symbol V_n is also acceptable if no misunderstanding is possible
Currents		
Continuous (direct) reverse current within the working voltage range	I_Z	
Continuous (direct) reverse current at a voltage below the working voltage range	I_R	
Sundry quantities		
Differential resistance within the working voltage range	r_z	
Temperature coefficient of working voltage	α_{VZ}	Reserve symbol: S_Z

List of Symbols. Table 11 lists the letter symbols recommended for use in the field of voltage reference and regulator diodes.

Tunnel Diodes

Letter Symbols. The basic letter symbols given in Tables 4 and 6 are also used with tunnel diodes. In addition to the general voltages, current, and power subscripts listed in Table 5, the following are also used:

    P—peak point
    PP—projected peak point
    V—valley point.

In addition to the general subscripts listed in Table 7 for electrical parameters, the following are

also used:

    s—series
    p—parasitic (parallel)
    r—resistive.

List of Symbols. The symbols contained in Table 12 are recommended for use in the field of tunnel diodes.

Variable Capacitance Diodes

Letter Symbols. The basic letter symbols given in Tables 4 and 6 are also used for variable capacitance diodes (varactors). The general subscripts listed in Tables 5 and 7 also apply.

List of Letter Symbols. The recommended symbols for use in the field of variable capacitance diodes are given in Table 13.

TABLE 12—LETTER SYMBOLS FOR TUNNEL DIODES.

Name and designation	Letter symbol	Remarks
Voltages and currents		
Peak point current	I_P	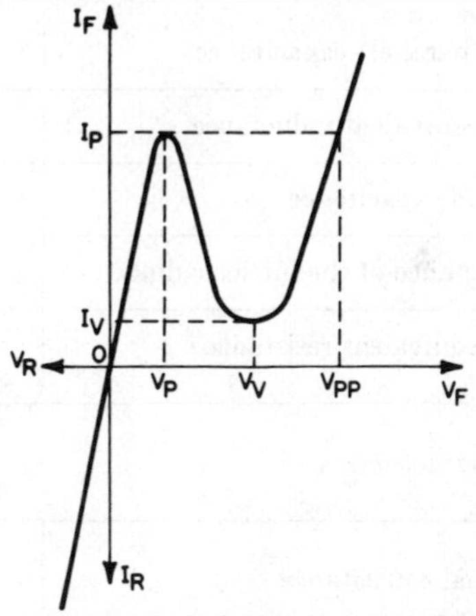
Valley point current	I_V	
Peak to valley point current ratio	I_P/I_V	
Reverse continuous (direct) current	I_R	
Peak point voltage	V_P	
Valley point voltage	V_V	
Projected peak point voltage	V_{PP}	
Reverse continuous (direct) voltage	V_R	
Equivalent circuit parameters		
Total series equivalent inductance	L_s	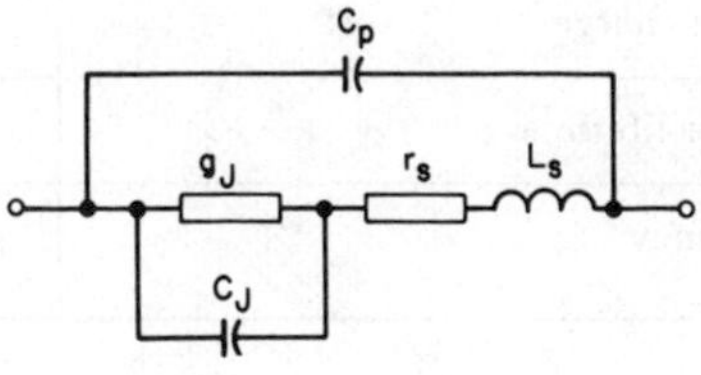
Total series equivalent resistance	r_s	
Junction capacitance of the intrinsic diode	C_j	
Negative conductance of the intrinsic diode	g_j	
Stray (parallel) capacitance	C_p	
Other small-signal parameters		
Resistive cut-off frequency	f_r	
Terminal capacitance	C_{tot}	

Thyristors

Letter Symbols. The basic letter symbols given in Tables 4 and 7 are also used for thyristors. In addition to the voltage, current, and power general subscripts listed in Table 5, the following special subscripts are used for thyristors:

A, a—anode
K, k—cathode
G, g—gate
D, d—off-state, nontrigger
T, t—on-state, trigger
H, h—holding
(BO)—breakover
Q, q—turn-off
(TO)—threshold.

In addition to the general subscripts listed in

TABLE 13—LETTER SYMBOLS FOR VARIABLE-CAPACITANCE DIODES.

Name and designation	Letter symbol	Remarks
Equivalent circuit parameters		
Stray (parallel) capacitance	C_p	
Series equivalent inductance	L_s	
Junction capacitance	C_j	
Conductance of the intrinsic diode	g_i	
Series equivalent resistance	r_s	
Other parameters		
Terminal capacitance	C_{tot}	
Effective Q value	Q_{eff}	
Cut-off frequency	f_{co}	
Transition frequency	f_T	
Stored charge	Q_s	
Carrier lifetime	$\tau_p; \tau_n$	
Efficiency	η	

Table 7 for electrical parameters, the following special subscripts are used for thyristors:

     t—turn-on
     q—turn-off
     T—slope.

List of Letter Symbols. The symbols in Table 14 are recommended for use in the field of thyristors.

Bipolar Transistors

Letter Symbols. The basic letters given in Tables 4 and 6 are used for bipolar transistors. The subscripts listed in Tables 5 and 7 also apply. In addition to the list of recommended general subscripts given in Table 5, the following special subscripts are recommended for voltages, currents, and powers in the field of bipolar transistors:

     B, b—base terminal
     C, c—collector terminal

     E, e—emitter terminal
     fl—floating
     pt—punch-through (penetration, reach through)
     R—(not as a first subscript) specified resistance
     sat—saturation
     X—specified circuit.

In addition to the list of recommended general subscripts given in Table 7, the following special subscripts are recommended for electrical parameters in the field of bipolar transistors:

     B, b—base; common-base configuration
     C, c—collector; common-collector configuration
     E, e—emitter; common-emitter configuration
     L—large signal
     sat—saturation
     S, s—storage
     T—transition.

TABLE 14—LETTER SYMBOLS FOR THYRISTORS.

Name and designation	Letter symbol	Remarks

Principal voltages: anode-cathode voltages

Name and designation	Letter symbol	Remarks
Continuous (direct) off-state voltage	V_D	
Peak off-state voltage	V_{DM}	
Crest (peak) working off-state voltage	V_{DWM}	
Repetitive peak off-state voltage	V_{DRM}	
Nonrepetitive peak off-state voltage	V_{DSM}	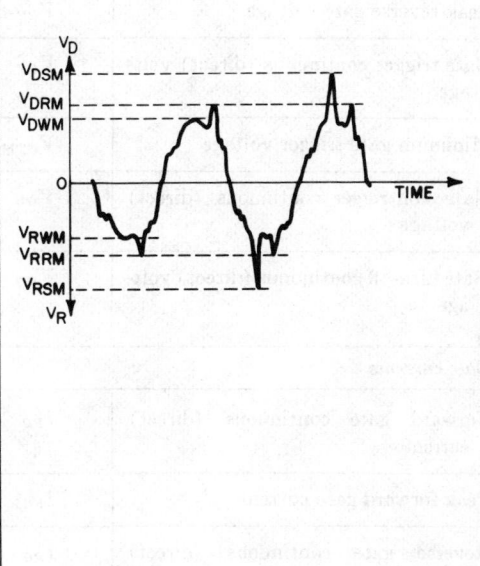
Breakover continuous (direct) voltage	$V_{(BO)}$	
Continuous (direct) on-state voltage	V_T	
Minimum on-state voltage	V_{TMIN}	
On-state threshold voltage	$V_{T(TO)}$	
Continuous (direct) reverse voltage	V_R	
Crest (peak) working reverse voltage	V_{RWM}	
Repetitive peak reverse voltage	V_{RRM}	
Nonrepetitive peak reverse voltage	V_{RSM}	
Reverse breakdown voltage	$V_{(BR)}$	

Principal currents: anode currents, cathode currents

Name and designation	Letter symbol	Remarks
Continuous (direct) off-state current	I_D	
Continuous (direct) breakover current	$I_{(BO)}$	
Continuous (direct) holding current	I_H	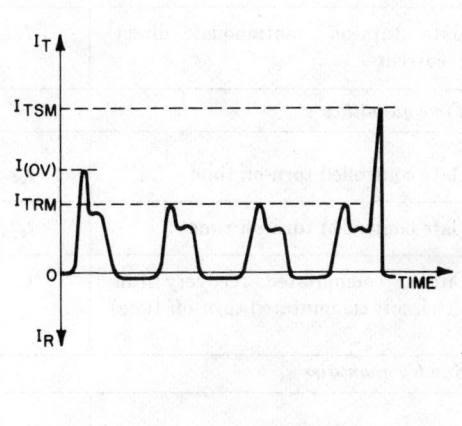
Continuous (direct) on-state current	I_T	
Overload on-state current	$I_{(OV)}$	
Repetitive peak on-state current	I_{TRM}	
Surge (nonrepetitive) on-state current	I_{TSM}	
Continuous (direct) reverse blocking current	I_R	
Repetitive peak reverse current	I_{RRM}	

TABLE 14—CONTINUED.

Name and designation	Letter symbol	Remarks
Gate voltages		
Forward gate continuous (direct) voltage	V_{FG}	
Peak forward gate voltage	V_{FGM}	
Reverse gate continuous (direct) voltage	V_{RG}	
Peak reverse gate voltage	V_{RGM}	
Gate trigger continuous (direct) voltage	V_{GT}	
Minimum gate trigger voltage	V_{GTMIN}	
Gate nontrigger continuous (direct) voltage	V_{GD}	
Gate turn-off continuous (direct) voltage	V_{GQ}	
Gate currents		
Forward gate continuous (direct) current	I_{FG}	
Peak forward gate current	I_{FGM}	
Reverse gate continuous (direct) current	I_{RG}	
Gate trigger continuous (direct) current	I_{GT}	
Gate nontrigger continuous (direct) current	I_{GD}	
Gate turn-off continuous (direct) current	I_{GQ}	
Time quantities		
Gate controlled turn-on time	t_{gt}	
Gate controlled turn-off time	t_{gq}	
Circuit commutated recovery time (circuit commutated turn-off time)	t_q	
Sundry quantities		
On-state slope resistance	r_T	

TABLE 15—LETTER SYMBOLS FOR BIPOLAR TRANSISTORS.

Name and designation	Letter symbol	Remarks
Voltages		
Collector-base (dc) voltage	V_{CB}	
Collector-emitter (dc) voltage	V_{CE}	
Emitter-base (dc) voltage	V_{EB}	
Base-emitter (dc) voltage	V_{BE}	
Collector-base (dc) voltage with $I_E=0$ I_C specified	V_{CBO}	
Emitter-base (dc) voltage with $I_C=0$ I_E specified	V_{EBO}	
Collector-emitter (dc) voltage with $I_B=0$ I_C specified	V_{CEO}	
Collector-emitter (dc) voltage with $R_{BE}=R$ I_C specified	V_{CER}	
Collector-emitter (dc) voltage with $V_{BE}=0$ I_C specified	V_{CES}	
Collector-emitter (dc) voltage with $V_{BE}=X$ specified (reverse biased emitter-base) I_C specified	V_{CEX}	
Breakdown voltages (open-circuit)	$V_{(BR)\dots O}$	The abbreviation *BV* is in common use for these quantities
Breakdown voltage, collector-base with $I_E=0$ I_C specified	$V_{(BR)CBO}$	
Breakdown voltage, emitter-base with $I_C=0$ I_E specified	$V_{(BR)EBO}$	
Breakdown voltage, collector-emitter with $I_B=0$ I_C specified	$V_{(BR)CEO}$	
Breakdown voltages (specified circuit)		The abbreviation *BV* is in common use for these quantities
Breakdown voltage, collector-emitter with $R_{BE}=R$ I_C specified	$V_{(BR)CER}$	
Breakdown voltage, collector-emitter with $V_{BE}=X$ specified I_C specified	$V_{(BR)CEX}$	
Breakdown voltage (short-circuit)	$V_{(BR)\dots S}$	The abbreviation *BV* is in common use for this quantity
Breakdown voltage, collector-emitter with $V_{BE}=0$ I_C specified	$V_{(BR)CES}$	

TABLE 15—CONTINUED.

Name and designation	Letter symbol	Remarks
Voltages—Con.		
Floating voltage, emitter-base with $I_E = 0$ V_{CB} specified	V_{EBfl}	
Punch-through (penetration) voltage	V_{pt}	
Saturation voltage, collector-emitter with I_B specified I_C specified	V_{CEsat}	
Saturation voltage, base-emitter with I_B specified I_C specified	V_{BEsat}	
Currents		
Base (dc) current	I_B	
Collector (dc) current	I_C	
Emitter (dc) current	I_E	
Collector cut-off current with $I_E = 0$ V_{CB} specified	I_{CBO}	
Collector cut-off current with $I_B = 0$ V_{CE} specified	I_{CEO}	
Emitter cut-off current with $I_C = 0$ V_{EB} specified	I_{EBO}	
Collector cut-off current with $R_{BE} = R$ V_{CE} specified	I_{CER}	
Collector cut-off current with $V_{BE} = 0$ V_{CE} specified	I_{CES}	
Collector cut-off current with $V_{BE} = X$ V_{CE} specified	I_{CEX}	
Base cut-off current with $V_{BE} = X$ V_{CE} specified	I_{BEX}	
Powers		
Collector power dissipation with t_{amb} or t_{case} specified	P_C	
Total input power (dc or average) to all electrodes with t_{amb} or t_{case} specified	P_{tot}	

TABLE 15—CONTINUED.

Name and designation	Letter symbol	Remarks

Static electrical parameters (specified for bias conditions)

Name and designation	Letter symbol	Remarks
Static value of the forward current transfer ratio (in common-emitter configuration)	h_{21E} or h_{FE}	$h_{21E} = \dfrac{I_C}{I_B} = \dfrac{I_E}{I_B} - 1$ with $V_{CE} = $ constant
Static value of the input resistance (in common-emitter configuration)	h_{11E} or h_{IE}	$h_{11E} = \dfrac{V_{BE}}{I_B}$ with $V_{CE} = $ constant
Inherent (large-signal) forward current transfer ratio	h_{21EL} or h_{FEL}	$h_{21EL} = \dfrac{I_C - I_{CBO}}{I_B + I_{CBO}}$ with $V_{CE} = $ constant

Small-signal electrical parameters (specified for bias and frequency conditions)

Name and designation	Letter symbol	Remarks
Small-signal value of the short-circuit input impedance:		
in common-emitter configuration	h_{11e} or h_{ie}	$h_{11e} = \dfrac{V_{be}}{I_b}$ with $V_{ce} = $ constant
in common-base configuration	h_{11b} or h_{ib}	$h_{11b} = \dfrac{V_{eb}}{I_e}$ with $V_{cb} = $ constant
Small-signal value of the open-circuit reverse voltage transfer ratio:		
in common-emitter configuration	h_{12e} or h_{re}	$h_{12e} = \dfrac{V_{be}}{V_{ce}}$ with $I_b = $ constant
in common-base configuration	h_{12b} or h_{rb}	$h_{12b} = \dfrac{V_{eb}}{V_{cb}}$ with $I_e = $ constant
Small-signal value of the short-circuit forward current transfer ratio:		
in common-emitter configuration	h_{21e} or h_{fe}	$h_{21e} = \dfrac{I_c}{I_b}$ with $V_{ce} = $ constant
in common-base configuration	h_{21b} or h_{fb}	$h_{21b} = \dfrac{I_c}{I_e}$ with $V_{cb} = $ constant
Small-signal value of the open-circuit output admittance:		
in common-emitter configuration	h_{22e} or h_{oe}	$h_{22e} = \dfrac{I_c}{V_{ce}}$ with $I_b = $ constant
in common-base configuration	h_{22b} or h_{ob}	$h_{22b} = \dfrac{I_c}{V_{cb}}$ with $I_e = $ constant

TABLE 15—CONTINUED.

Name and designation	Letter symbol	Remarks

Small-signal electrical parameters (specified for bias and frequency conditions)—Con.

Name and designation	Letter symbol	Remarks
Real part of the small-signal value of the short-circuit input impedance:		$h_{11e} = \mathrm{Re}(h_{11e}) + \mathrm{Im}(h_{11e})$ $h_{11b} = \mathrm{Re}(h_{11b}) + \mathrm{Im}(h_{11b})$
in common-emitter configuration	$\mathrm{Re}(h_{11e})$	
in common-base configuration	$\mathrm{Re}(h_{11b})$	
Imaginary part of the small-signal value of the short-circuit input impedance:		
in common-emitter configuration	$\mathrm{Im}(h_{11e})$	
in common-base configuration	$\mathrm{Im}(h_{11b})$	
Input capacitance (output short-circuited to ac):		
in common-emitter configuration	C_{11es} or C_{ies}	$h_{11e} \simeq \mathrm{Re}(h_{11e}) + \dfrac{1}{j\omega C_{11es}}$
in common-base configuration	C_{11bs} or C_{ibs}	$h_{11b} \simeq \mathrm{Re}(h_{11b}) + \dfrac{1}{j\omega C_{11bs}}$
Input capacitance (output open-circuited to ac):		
in common-emitter configuration	C_{11eo} or C_{ieo}	
in common-base configuration	C_{11bo} or C_{ibo}	
Output capacitance (input open-circuited to ac):		
in common-emitter configuration	C_{22eo} or C_{oeo}	$h_{22e} = \mathrm{Re}(h_{22e}) + j\omega C_{22eo}$
in common-base configuration	C_{22bo} or C_{obo}	$h_{22b} = \mathrm{Re}(h_{22b}) + j\omega C_{22bo}$
Output capacitance (input short-circuited to ac):		
in common-emitter configuration	C_{22es} or C_{oes}	$y_{22e} = \mathrm{Re}(y_{22e}) + j\omega C_{22es}$
in common-base configuration	C_{22bs} or C_{obs}	$y_{22b} = \mathrm{Re}(y_{22b}) + j\omega C_{22bs}$
Reverse transfer capacitance (input short-circuited to ac):		
in common-emitter configuration	C_{12es} or C_{res}	
in common-base configuration	C_{12bs} or C_{rbs}	

TABLE 15—CONTINUED.

Name and designation	Letter symbol	Remarks

Small-signal electrical parameters (specified for bias and frequency conditions)—Con.

Name and designation	Letter symbol	Remarks
Collector-base capacitance for transistors with isolated device terminals and a separate screen lead	$C_{c,cb}$	
Small-signal value of the short-circuit input admittance:		
in common-emitter configuration	y_{11e} or y_{ie}	$y_{11e} = \dfrac{I_b}{V_{be}}$ with V_{ce} = constant and $y_{11e} = \dfrac{1}{h_{11e}}$
in common-base configuration	y_{11b} or y_{ib}	$y_{11b} = \dfrac{I_e}{V_{eb}}$ with V_{cb} = constant and $y_{11b} = \dfrac{1}{h_{11b}}$
Small-signal value of the short-circuit reverse transfer admittance:		
in common-emitter configuration	y_{12e} or y_{re}	$y_{12e} = \dfrac{I_b}{V_{ce}}$ with V_{be} = constant
in common-base configuration	y_{12b} or y_{rb}	$y_{12b} = \dfrac{I_e}{V_{cb}}$ with V_{eb} = constant
Small-signal value of the short-circuit forward transfer admittance:		
in common-emitter configuration	y_{21e} or y_{fe}	$y_{21e} = \dfrac{I_c}{V_{be}}$ with V_{ce} = constant
in common-base configuration	y_{21b} or y_{fb}	$y_{21b} = \dfrac{I_c}{V_{eb}}$ with V_{cb} = constant
Small-signal value of the short-circuit output admittance:		
in common-emitter configuration	y_{22e} or y_{oe}	$y_{22e} = \dfrac{I_c}{V_{ce}}$ with V_{be} = constant
in common-base configuration	y_{22b} or y_{ob}	$y_{22b} = \dfrac{I_c}{V_{cb}}$ with V_{eb} = constant
Modulus of the short-circuit reverse transfer admittance:		
in common-emitter configuration	$\lvert y_{12e}\rvert$ or $\lvert y_{re}\rvert$	
in common-base configuration	$\lvert y_{12b}\rvert$ or $\lvert y_{rb}\rvert$	
Phase of the short-circuit reverse transfer admittance:		
in common-emitter configuration	φ_{y12e} or φ_{yre}	
in common-base configuration	φ_{y12b} or φ_{yrb}	

Im (y_{12e}) · y_{12e} · $\varphi\, y_{12e}$ · Re(y_{12e})

TABLE 15—CONTINUED.

Name and designation	Letter symbol	Remarks

Small-signal electrical parameters (specified for bias and frequency conditions)—Con.

Name and designation	Letter symbol	Remarks
Modulus of the short-circuit forward transfer admittance:		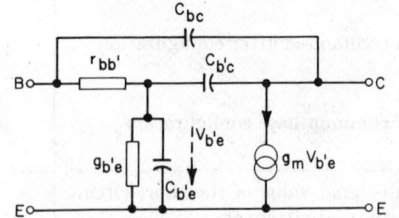
in common-emitter configuration	$\|y_{21e}\|$ or $\|y_{fe}\|$	
in common-base configuration	$\|y_{21b}\|$ or $\|y_{fb}\|$	
Phase of the short-circuit forward transfer admittance:		
in common-emitter configuration	φ_{y21e} or φ_{yfe}	
in common-base configuration	φ_{y21b} or φ_{yfb}	

Modified hybrid π equivalent circuit parameters

Name and designation	Letter symbol	Remarks
Base intrinsic resistance	$r_{bb'}$	This equivalent circuit is only a first order approximation, valid for most transistors over a certain frequency range.
Intrinsic base-emitter conductance	$g_{b'e}$	
Intrinsic base-emitter capacitance	$C_{b'e}$	
Intrinsic base-collector capacitance	$C_{b'c}$	
Intrinsic transconductance	g_m	
Base-collector capacitance	C_{bc}	

Frequency parameters

Name and designation	Letter symbol	Remarks
Cut-off frequency:		
in common-emitter configuration	f_{h21e} or f_{hfe}	
in common-base configuration	f_{h21b} or f_{hfb}	
in common-collector configuration	f_{h21c} or f_{hfc}	
Frequency of unity current transfer ratio	f_1	$f_1 = f$ for $\|h_{21e}\| = 1$
Transition frequency	f_T	$f_T = f \times \|h_{21e}\|$ (h_{21e} is measured in a region where the roll-off is 6 dB/octave)
Maximum frequency of oscillation	f_{max}	

TABLE 15—CONTINUED.

Name and designation	Letter symbol	Remarks
Switching parameters		
Pulse average time	t_w, t_{pav}	
Pulse time	t_p	
Duty cycle	D, δ	
Delay time	t_d	
Rise time	t_r	
Carrier storage time	t_s	
Fall time	t_f	
Turn-on time	t_{on}	$t_d + t_r$
Turn-off time	t_{off}	$t_s + t_f$
Emitter depletion layer capacitance	C_{Te}	

TABLE 15—CONTINUED.

Name and designation	Letter symbol	Remarks
Switching parameters—Con.		
Collector depletion layer capacitance	C_{Tc}	
Collector time coefficient	τ_C	
Rise time coefficient	τ_R	
Fall time coefficient	τ_F	
Stored charge	Q_s	
Transient current ratio in saturation	h_{21Esat} or h_{FEsat}	
Collector-emitter saturation resistance:		
small-signal value	r_{cesat}	
static value	r_{CEsat}	
Sundry quantities		
Noise	N, n	
Noise figure	F, F_n	
Noise current	I_n	
Noise voltage	V_n	
Noise power	P_n	
Effective noise bandwidth	B	
Amplification	A	
Current amplification	$A_I A_i$	
Voltage amplification	$A_V A_v$	
Gain	G	
Power gain	$G_P G_p$	
Insertion power gain	$G_I G_i$	
Transducer power gain	$G_T G_t$	
Available power gain	$G_A G_a$	
Efficiency	η	
Collector efficiency	η_C	

TABLE 15—CONTINUED.

Name and designation	Letter symbol	Remarks
External circuit parameters		
Emitter (dc) voltage supply	V_{EE}	
Base (dc) voltage supply	V_{BB}	
Collector (dc) voltage supply	V_{CC}	
External emitter resistance	R_E	
External base resistance	R_B	
External collector resistance	R_C	
External resistance connecting base to emitter	R_{BE}	
Generator resistance	R_G	
Load resistance	R_L	
Load capacitance	C_L	

TABLE 16—LETTER SYMBOLS FOR FIELD-EFFECT TRANSISTORS.

Name and designation	Letter symbol	Remarks
Currents		
Drain (dc) current	I_D	
Drain current, at a specified gate-source condition	I_{DSX}	
Drain current, at a specified (external) gate-source resistance	I_{DSR}	
Drain current, with gate short-circuited to source ($V_{GS}=0$)	I_{DSS}	
Source (dc) current	I_S	
Source current, at a specified gate-drain condition	I_{SDX}	
Source current, with gate short-circuited to drain ($V_{GD}=0$)	I_{SDS}	

TABLE 16—CONTINUED.

Name and designation	Letter symbol	Remarks
Currents—Con.		
Gate (dc) current	I_G	
Forward gate current	I_{GF}	
Gate cut-off current (of a junction field-effect transistor), with source open-circuited	I_{GDO}	
Gate cut-off current (of a junction field-effect transistor), with drain open-circuited	I_{GSO}	
Gate cut-off current (of a junction field-effect transistor), with drain short-circuited to source	I_{GSS}	
Gate leakage current (of an insulated-gate field-effect transistor), with drain short-circuited to source	I_{GSS}	
Gate cut-off current (of a junction field-effect transistor), with specified drain-source circuit conditions	I_{GSX}	
Substrate current	I_B; I_U	
Voltages		
Drain-source (dc) voltage	V_{DS}	
Gate-source (dc) voltage	V_{GS}	
Gate-source cut-off voltage (of a junction field-effect transistor and of a depletion type insulated-gate field-effect transistor)	$V_{GS(OFF)}$; V_{GSoff}	
Gate-source threshold voltage (of an enhancement type insulated-gate field-effect transistor)	V_{GST}; $V_{GS(th)}$; $V_{GS(TO)}$	
Forward gate-source (dc) voltage	V_{GSF}	
Reverse gate-source (dc) voltage	V_{GSR}	
Gate-drain (dc) voltage	V_{GD}	
Source-substrate (dc) voltage	V_{SB}; V_{SU}	

TABLE 16—CONTINUED.

Name and designation	Letter symbol	Remarks
Voltages—Con.		
Drain-substrate (dc) voltage	V_{DB}; V_{DU}	
Gate-substrate (dc) voltage	V_{GB}; V_{GU}	
Gate-gate voltage (for multigate devices	V_{G1-G2}	
Gate-source breakdown voltage with drain short-circuited to source	$V_{(BR)GSS}$	
Electrical power		
Drain-source (dc) power dissipation	P_{DS}	
Resistances (or conductances) and capacitances		
Drain-source resistance Gate-source resistance Gate-drain resistance Gate resistance (with $V_{DS}=0$ or $v_{ds}=0$)	r_{DS}; r_{ds} r_{GS}; r_{gs} r_{GD}; r_{gd} r_{GSS}; r_{gss}	
Drain-source on-state resistance	$r_{DS(ON)}$; $r_{ds(on)}$; r_{DSon}	
Drain-source off-state resistance	$r_{DS(OFF)}$; $r_{ds(off)}$; r_{DSoff}	
Open-circuit gate-source capacitance (drain-source and gate-drain open-circuited to ac)	C_{gso}	
Open-circuit gate-drain capacitance (drain-source and gate-source open-circuited to ac)	C_{gdo}	
Open-circuit drain-source capacitance (gate-drain and gate-source open-circuited to ac)	C_{dso}	
Short-circuit input capacitance in common-source configuration; gate-source capacitance (drain-source short-circuited to ac)	C_{iss}; C_{11ss}	
Short-circuit output capacitance in common-source configuration; drain-source capacitance (gate-source short-circuited to ac)	C_{oss}; C_{22ss}	

TABLE 16—CONTINUED.

Name and designation	Letter symbol	Remarks

Resistances (or conductances) and capacitances—Con.

Name and designation	Letter symbol	Remarks
Short-circuit input conductance in common-source configuration	g_{iss}	
Short-circuit output conductance in common-source configuration	g_{oss}	
Common-source reverse transfer capacitance with input short-circuited to a.c.	C_{rss}; C_{12ss}	
Short-circuit output capacitance in common-drain configuration (gate-drain short-circuited to ac)	C_{ods}; C_{22ds}	

Small-signal y parameters in common-source configuration and π equivalent circuit parameters

Name and designation	Letter symbol	Remarks
Short-circuit input admittance	$y_{is} = \mathrm{Re}_{(yis)} + j\omega C_{is}$ $y_{11s} = \mathrm{Re}_{(y11s)} + j\omega C_{11s}$	
Short-circuit reverse transfer admittance	$y_{rs} = \mathrm{Re}_{(yrs)} + j\omega C_{rs}$ $y_{12s} = \mathrm{Re}_{(y12s)} + j\omega C_{12s}$	
Short-circuit forward transfer admittance	$y_{fs} = \mathrm{Re}_{(yts)} + j\mathrm{Im}_{(yfs)}$ $y_{21s} = \mathrm{Re}_{(y21s)} + j\mathrm{Im}_{(y21s)}$	
Short-circuit output admittance	$y_{os} = \mathrm{Re}_{(yos)} + j\omega C_{os}$ $y_{22s} = \mathrm{Re}_{(y22s)} + j\omega C_{22s}$	
Modulus of the short-circuit reverse transfer admittance	$\mid y_{rs}\mid$; $\mid y_{12s}\mid$	
Phase of the short-circuit reverse transfer admittance	φ_{yrs}; φ_{y12s}	
Modulus of the short-circuit forward transfer admittance	$\mid y_{fs}\mid$; $\mid y_{21s}\mid$	
Phase of the short-circuit forward transfer admittance	φ_{yfs}; φ_{y21s}	

TABLE 16—CONTINUED.

Name and designation	Letter symbol	Remarks

Small-signal y parameters in common-source configuration and π equivalent circuit parameters—Con.

Name and designation	Letter symbol	Remarks
Gate-source conductance (in the π equivalent circuit)	g_{gs}	
Gate-drain conductance (in the π equivalent circuit)	g_{gd}	
Drain-source conductance (in the π equivalent circuit)	g_{ds}	
Forward transconductance (in the π equivalent circuit)	g_{ms}; g_m	
Gate-source capacitance (in the π equivalent circuit)	C_{gs}	
Gate-drain capacitance (in the π equivalent circuit)	C_{gd}	
Drain-source capacitance (in the π equivalent circuit)	C_{ds}	

Other parameters

Name and designation	Letter symbol	Remarks
Power gain	G_P; G_p	
Cut-off frequency (in the common-source configuration)	f_{yfs}	
Noise voltage	V_n	
Noise figure	F	
Temperature coefficient of drain current	α_{ID}	
Temperature coefficient of drain-source resistance	α_{rds}	

TABLE 16—CONTINUED.

Name and designation	Letter symbol	Remarks

Other parameters—Con.

Switching times:		
Turn-on delay time	$t_{d(on)}$	
Turn-off delay time	$t_{d(off)}$	
Rise time	t_r	
Fall time	t_f	
Turn-on time $(t_{d(on)} + t_r)$	t_{on}	
Turn-off time $(t_{d(on)} + t_f)$	t_{off}	

List of Letter Symbols. The symbols contained in Table 15 are recommended for use in the field of bipolar transistors.

Field-Effect Transistors

Letter Symbols. The basic symbols given in Tables 4 and 6 are also used for field-effect transistors. In addition to the subscripts listed in Tables 5 and 7, the following additional subscripts also apply to the field of field-effect transistors:

B, b; U, u—substrate
C, c—collector
D, d—drain
E, e—emitter
G, g—gate
S, s—source
T, th; (TO)—threshold.

List of Letter Symbols. The symbols recommended for use in the field of field-effect transistors are given in Table 16.

CHAPTER 20
TRANSISTOR CIRCUITS

BASIC CIRCUITS*

This chapter gives condensed descriptions of the various types of circuits in which transistors are operated together with design information enabling the determination of the circuit parameters. The following symbols are used:

A_i = current amplification
A_v = voltage amplification
$a = r_m/r_c$
G = power gain
i_b = base current
i_e = emitter current
i_c = collector current
i_{c0} = collector current with $i_e = 0$
i_l = load current
r_b = base resistance
r_c = collector resistance
r_e = emitter resistance
r_g = generator resistance
r_i = input resistance
r_l = load resistance
r_m = equivalent emitter–collector transresistance
r_o = output resistance
v_g = signal input voltage
y_l = load admittance
z_l = load impedance
α = short-circuit current multiplication factor
$\beta = \alpha/(1-\alpha)$
Δ = determinant.

The bipolar transistor is a 3-terminal device and is connected into a 4-terminal circuit in any of 3 possible methods, as illustrated by Tables 1–3.

TRANSISTOR NETWORKS

Figure 1 and Tables 4–6 give the characteristics of properly terminated 4-terminal networks. Table 7

* R. F. Shea et al., "Principles of Transistor Circuits," John Wiley & Sons, Inc., New York, N.Y.; 1953. Also, "The Transistor, Selected Reference Material," Bell Telephone Laboratories, New York, N.Y.; 1951. Also, W. H. Duerig et al., "Transistor Physics and Electronics," Applied Physics Laboratory, The Johns Hopkins University, Baltimore, Md.; 1953.

gives h-parameter equations for the three bipolar transistor configurations.

TYPICAL TRANSISTOR CHARACTERISTICS

Typical values of impedances and gains for junction-type bipolar transistors are given in Table 8. Gain calculations for transistor amplifiers are given in Table 9.

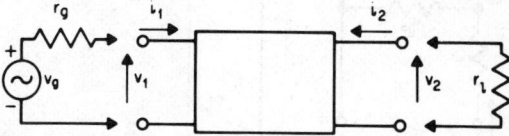

Fig. 1—Four-terminal network showing current and voltage conventions.

COMPARISON WITH ELECTRON TUBES

The bipolar transistor is current-operated, not voltage-operated. As a guide in circuit design, it is possible to replace the constant-voltage source of the electron tube with a current source. This principle (called duality) may be extended by replacing elements with given voltage characteristics by elements having equivalent current characteristics.*

It is sometimes possible, when consideration is given to loading effects, to convert electron-tube circuits directly to junction-transistor circuits by using the electron-tube analogy shown in Fig. 2.

SMALL-SIGNAL AMPLIFIERS

General

In designing small-signal amplifiers, it must be remembered that the transistor is a bilateral device;

* R. L. Wallace, Jr. and G. Raisbeck, "Duality as a Guide in Transistor Circuit Design," *Bell System Technical Journal*, vol. 30, pp. 381–417; April 1951.

TABLE 1—COMMON-BASE CIRCUIT.

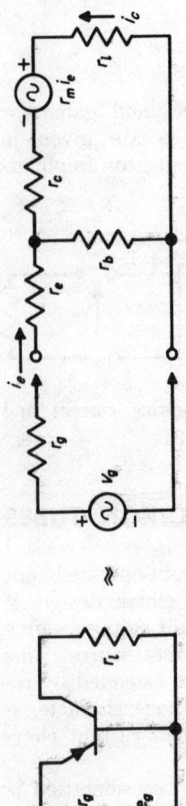

Conditions for validity	Exact Equation	Approximate Equations
		$r_e \ll r_c - r_m$, $r_b \ll r_c$; $r_e \ll r_c - r_m$, $r_b \ll r_c$, $r_l \ll r_c - r_m$
Input resistance $= r_i$	$r_e + r_b - \dfrac{r_b(r_b + r_m)}{r_l + r_c + r_b}$	$r_e + r_b \cdot \dfrac{r_c(1-a) + r_l}{r_c + r_l}$; $r_e + r_b(1-a) + r_g$
Output resistance $= r_o$	$r_c + r_b - \dfrac{r_b(r_b + r_m)}{r_g + r_e + r_b}$	$r_c \cdot \dfrac{r_c + r_b(1-a) + r_g}{r_e + r_b + r_g}$; $\dfrac{r_e + r_b(1-a) + r_g}{r_e + r_b + r_g} r_c$
Voltage amplification $= A_v$	$\dfrac{r_l(r_m + r_b)}{r_b(r_c - r_m + r_e + r_l) + r_e(r_c + r_l)}$	$\dfrac{a r_c r_l}{r_c[r_e + r_b(1-a)] + r_l(r_e + r_b)}$; $\dfrac{a r_l}{r_e + r_b(1-a)}$
Current amplification $= A_i$	$\dfrac{-(r_m + r_b)}{r_b + r_c + r_l}$	$\dfrac{-a}{1 + r_l/r_c}$; $-a$
Power gain $= G$	$\dfrac{r_l(r_m + r_b)^2}{(r_b + r_c + r_l)[r_b(r_c - r_m + r_e + r_l) + r_e(r_c + r_l)]}$	$\dfrac{a^2 r_c^2 r_l}{(r_c + r_l)r_c[r_e + r_b(1-a) + r_l(r_e + r_b)]}$; $\dfrac{a^2 r_l}{r_e + r_b(1-a)}$

TABLE 2—COMMON-EMITTER CIRCUIT.

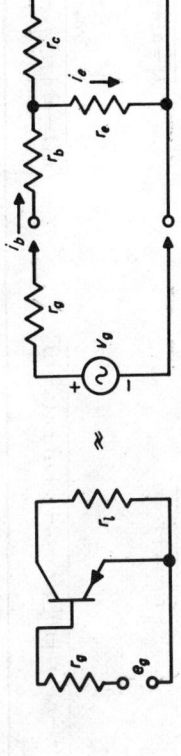

	Exact Equation	Approximate Equations	
Conditions for validity	—	$r_e \ll r_c - r_m$ $r_b \ll r_e$	$r_e \ll r_c - r_m$ $r_b \ll r_e$ $r_l \ll r_c - r_m$
Input resistance $= r_i$	$r_e + r_b + \dfrac{r_e(r_m - r_e)}{r_l + r_e + r_c - r_m}$	$r_b + r_e \cdot \dfrac{r_e + r_l}{r_c(1-a) + r_l}$	$r_b + \dfrac{r_e}{1-a}$
Output resistance $= r_o$	$r_c + r_e - r_m + \dfrac{r_e(r_m - r_e)}{r_g + r_b + r_e}$	$r_c(1-a) + r_e \cdot \dfrac{r_m + r_g}{r_e + r_b + r_g}$	$r_c(1-a) + r_e \cdot \dfrac{r_m + r_g}{r_e + r_b + r_g}$
Voltage amplification $= A_v$	$\dfrac{-r_l(r_m - r_e)}{r_b(r_c - r_m + r_e + r_l) + r_e(r_c + r_l)}$	$\dfrac{-a r_l}{r_c[r_e + r_b(1-a)] + r_l(r_e + r_b)}$	$\dfrac{-a r_l}{r_e + r_b(1-a)}$
Current amplification $= A_i$	$\dfrac{r_m - r_e}{r_c - r_m + r_e + r_l}$	$\dfrac{a}{1 - a + r_l/r_c}$	$\dfrac{a}{1-a}$
Power gain $= G$	$\dfrac{r_l(r_m - r_e)^2}{(r_c - r_m + r_e + r_l)[r_b(r_c - r_m + r_e + r_l) + r_e(r_c + r_l)]}$	$\dfrac{a^2 r_c^2 r_l}{[r_c(1-a) + r_l]r_c[r_e + r_b(1-a)] + r_l(r_e + r_b)}$	$\dfrac{a^2 r_l}{(1-a)[r_e + r_b(1-a)]}$

TABLE 3—COMMON-COLLECTOR CIRCUIT.

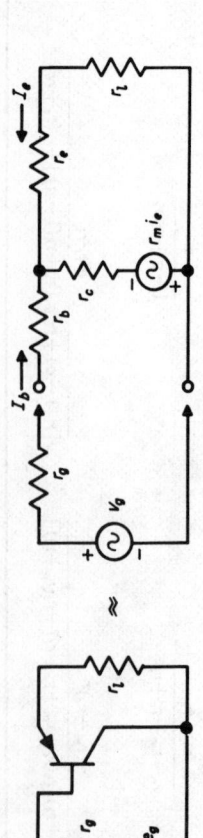

	Exact Equation	Approximate Equations	
Conditions for validity	—	$r_e \ll r_c - r_m$ $r_b \ll r_c$	$r_e \ll r_c - r_m$ $r_b \ll r_c$ $r_e \ll r_l \ll r_c - r_m$
Input resistance $= r_i$	$r_b + r_c + \dfrac{r_c(r_m - r_c)}{r_l + r_e + r_c - r_m}$	$r_b + r_c \cdot \dfrac{r_e + r_l}{r_c(1-a) + r_l}$	$\dfrac{r_l}{1-a}$
Output resistance $= r_o$	$r_e + r_c - r_m - \dfrac{r_c(r_c - r_m)}{r_g + r_b + r_c}$	$r_e + r_c(1-a) \cdot \dfrac{r_g + r_b}{r_g + r_c}$	$r_e + (r_b + r_g)(1-a)$
Voltage amplification $= A_v$	$\dfrac{r_c r_l}{r_b(r_c - r_m + r_e + r_l) + r_c(r_e + r_l)}$	$\dfrac{r_l}{r_e + r_b(1-a) + r_l}$	1
Current amplification $= A_i$	$\dfrac{r_c}{r_c - r_m + r_e + r_l}$	$\dfrac{1}{(1-a) + r_l/r_c}$	$(1-a)^{-1}$
Power gain $= G$	$\dfrac{r_c^2 r_l}{(r_c - r_m + r_e + r_l)\left[r_b(r_c - r_m + r_e + r_l) + r_c(r_e + r_l)\right]}$	$\dfrac{r_l r_c}{\left[r_c(1-a) + r_l\right]\left[r_e + r_b(1-a) + r_l\right]}$	$(1-a)^{-1}$

TABLE 4—DEFINITIONS OF EQUIVALENT-CIRCUIT PARAMETERS.

Parameter	Common Base	Common Emitter	Common Collector	Definition
z	z_{11}, z_{11b}, or z_{ib}	z_{11e} or z_{ie}	z_{11c} or z_{ic}	Input impedance with open-circuit output
	z_{12}, z_{12b}, or z_{rb}	z_{12e} or z_{re}	z_{12c} or z_{rc}	Reverse transfer impedance with open-circuit input
	z_{21}, z_{21b}, or z_{fb}	z_{21e} or z_{fe}	z_{21c} or z_{fc}	Forward transfer impedance with open-circuit output
	z_{22}, z_{22b}, or z_{ob}	z_{22e} or z_{oe}	z_{22c} or z_{oc}	Output impedance with open-circuit input
y	y_{11}, y_{11b}, or y_{ib}	y_{11e} or y_{ie}	y_{11c} or y_{ic}	Input admittance with short-circuit output
	y_{12}, y_{12b}, or y_{rb}	y_{12e} or y_{re}	y_{12c} or y_{rc}	Reverse transfer admittance with short-circuit input
	y_{21}, y_{21b}, or y_{fb}	y_{21e} or y_{fe}	y_{21c} or y_{fc}	Forward transfer admittance with short-circuit output
	y_{22}, y_{22b}, or y_{ob}	y_{22e} or y_{oe}	y_{22c} or y_{oc}	Output admittance with short-circuit input
h	h_{11}, h_{11b}, or h_{ib}	h_{11e} or h_{ie}	h_{11c} or h_{ic}	Input impedance with short-circuit output
	h_{12}, h_{12b}, or h_{rb}	h_{12e} or h_{re}	h_{12c} or h_{rc}	Reverse open-circuit voltage amplification factor
	h_{21}, h_{21b}, or h_{fb}	h_{21e} or h_{fe}	h_{21c} or h_{fc}	Forward short-circuit current amplification factor
	h_{22}, h_{22b}, or h_{ob}	h_{22e} or h_{oe}	h_{22c} or h_{oc}	Output admittance with open-circuit input

Note: $h_{11} = 1/y_{11}$ and $h_{22} = 1/z_{22}$.

any change in the output circuit will affect all preceding stages. Junction transistors have $\alpha < 1$ and, therefore, should not cause instability troubles at low frequencies.

Small-signal amplifiers are generally of the common-emitter amplifier configuration. A basic small-signal common-emitter circuit is shown in Fig. 3. Also shown are the equivalent circuits for the z, y, and h parameters.

Biasing Techniques

In Fig. 4A and 4B, battery polarity is shown for *pnp* transistors. The polarity is reversed for *npn* transistors.

In Fig. 4B

$$e_3 \equiv e_1 + e_2$$
$$e_1 \equiv e_3 r_2 / (r_3 + r_2)$$
$$e_2 \equiv e_3 r_3 / (r_3 + r_2).$$

The branch currents in Fig. 4B are

$$i_c = \frac{i_{c0}(1 + r_1/r_2 + r_1/r_3) + \alpha e/r_3}{1 - \alpha + r_1/r_2 + r_1/r_3}$$

$$i_e = (i_c - i_{c0})/\alpha$$

$$i_b = i_e(1 - \alpha) - i_{c0}$$

where i_{c0} = collector current when $i_e = 0$.

TABLE 5—EQUIVALENT CIRCUITS FOR 4-TERMINAL NETWORKS.

Param-eter	Circuit Equations	Equivalent Circuit
z	$v_1 = z_{11}i_1 + z_{12}i_2$ $v_2 = z_{21}i_1 + z_{22}i_2$	
y	$i_1 = y_{11}v_1 + y_{12}v_2$ $i_2 = y_{21}v_1 + y_{22}v_2$	
h	$v_1 = h_{11}i_1 + h_{12}v_2$ $i_2 = h_{21}i_1 + h_{22}v_2$	

TABLE 6—CHARACTERISTICS OF 4-TERMINAL NETWORKS.

Network Terminal Characteristic	z	y	h
Input impedance $=z_{in}$	$z_{11} - \dfrac{z_{12}z_{21}}{z_{22}+z_l}$	$\dfrac{y_{22}+y_l}{y_{11}(y_{22}+y_l)-y_{12}y_{21}}$	$h_{11} - \dfrac{h_{12}h_{21}}{h_{22}+y_l}$
Output impedance $=z_{out}$	$z_{22} - \dfrac{z_{12}z_{21}}{z_{11}+z_g}$	$\dfrac{y_{11}+y_g}{y_{22}(y_{11}+y_g)-y_{12}y_{21}}$	$\dfrac{h_{11}+z_g}{h_{22}(h_{11}+z_g)-h_{12}h_{21}}$
Voltage amplification $=A_v$	$\dfrac{z_{21}z_l}{z_{11}(z_{22}+z_l)-z_{12}z_{21}}$	$\dfrac{-y_{21}}{y_{22}+y_l}$	$\dfrac{-h_{21}}{h_{11}(h_{22}+y_l)-h_{12}h_{21}}$
Current amplification $=A_i$	$\dfrac{z_{21}}{z_{22}+z_l}$	$\dfrac{y_{21}y_l}{y_{11}(y_{22}+y_l)-y_{12}y_{21}}$	$\dfrac{h_{21}y_l}{h_{22}+y_l}$
Power gain $=G$	$\dfrac{z_{21}{}^2 z_l}{[z_{11}(z_{22}+z_l)-z_{12}z_{21}][z_{22}+z_l]}$	$\dfrac{y_{21}{}^2 y_l}{[y_{11}(y_{22}+y_l)-y_{12}y_{21}][y_{22}+y_l]}$	$\dfrac{h_{21}{}^2 y_l}{[h_{11}(h_{22}+y_l)-h_{12}h_{21}][h_{22}+y_l]}$

TABLE 7—h-PARAMETER EQUATIONS FOR THREE BIPOLAR TRANSISTOR CONFIGURATIONS.

Parameter	Grounded Base	Grounded Emitter	Grounded Collector
h_{11}	given as h_{11} or h_{ib}	$h_{ie} \approx h_{11}/(1+h_{21})$	$h_{ic} \approx h_{11}/(1+h_{21})$
h_{12}	given as h_{12} or h_{rb}	$h_{re} \approx (\Delta h^e - h_{12})/(1+h_{21})$	$h_{rc} \approx 1$
h_{21}	given as h_{21} or h_{fb}	$h_{fe} \approx -h_{21}/(1+h_{21})$	$h_{fc} \approx -(1+h_{21})^{-1}$
h_{22}	given as h_{22} or h_{ob}	$h_{oe} \approx h_{22}/(1+h_{21})$	$h_{oc} \approx h_{22}/(1+h_{21})$

$\Delta h^e = h_{ie}h_{oe} - h_{re}h_{fe}$.

Coupling Circuits (Fig. 5)

Bipolar transistors may be cascaded in much the same manner as electron tubes. Common-base, common-emitter, or common-collector configurations may be used. The stages are usually coupled by transformers or by R–C networks. Other means, such as direct coupling, can also be used.

Unlike the unilateral electron tube, the bipolar transistor is bilateral and essentially a current-operated device. In addition, the bipolar transistor (except in common-collector circuits) generally has an input impedance that is comparable to or lower than the output impedance. Care must be taken to match impedances between stages. The common-collector stage is a useful impedance-matching device and, in view of the efficiency of the bipolar transistor, it can be used for impedance matching in place of a transformer. The equations given in Tables 1–3 may be used to determine the interstage transformation ratios.

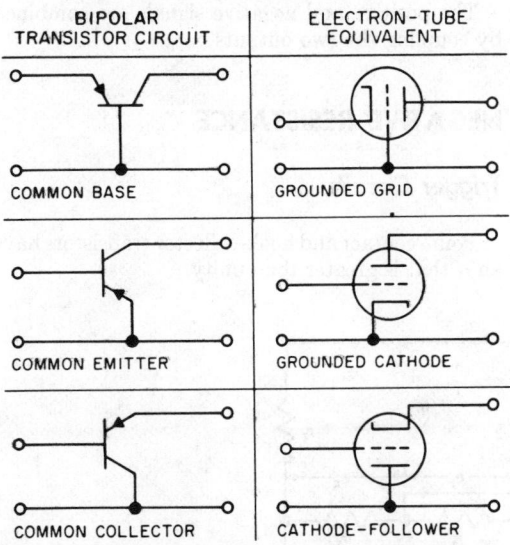

Fig. 2—The 3 basic bipolar transistor circuits are at the left; the electron-tube equivalent circuits at the right.

LARGE-SIGNAL OPERATION

Output Stage*

The bipolar transistor output stage has two power limitations:

(**A**) The maximum voltage that can be applied between the collector and base of the transistor.

(**B**) The temperature rise in the transistor.

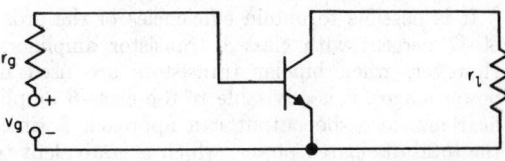

A. BASIC SMALL-SIGNAL COMMON-EMITTER CIRCUIT

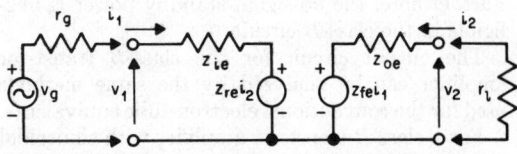

B. EQUIVALENT CIRCUIT FOR z PARAMETERS

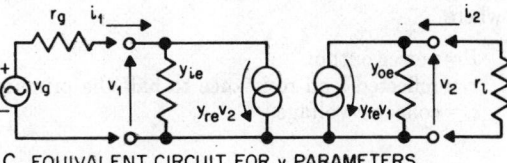

C. EQUIVALENT CIRCUIT FOR y PARAMETERS

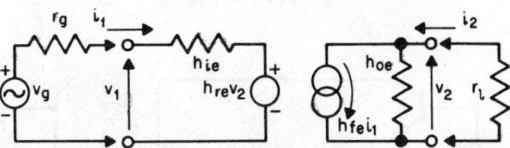

D. EQUIVALENT CIRCUIT FOR h PARAMETERS

Fig. 3—Basic small-signal amplifier using common-emitter circuit with equivalent circuits for z, y, and h parameters.

* P. I. Richards, "Power Transistors, Circuit Design and Data," Transistor Products, Inc., Waltham, Mass.

TABLE 8—JUNCTION-TYPE BIPOLAR TRANSISTOR CHARACTERISTICS.

	Common Base	Common Emitter	Common Collector
Maximum voltage amplification=A_v with $r_g=0$ and $r_l=\infty$	1.7×10^4	-1.7×10^5	1
Maximum current amplification=A_i with $r_l=0$	$+0.95$	-19	$+19$
Input resistance=r_i in ohms:			
$r_l=0$	35	750	120
$r_l=\infty$	270	270	5×10^6
Output resistance=r_o in ohms:			
$r_g=0$	6.8×10^5	7×10^5	37
$r_g=\infty$	5×10^6	2.5×10^3	2.5×10^5
Matched input resistance in ohms	100	450	6×10^4
Matched output resistance in ohms	2×10^6	4×10^5	3×10^3
Typical equivalent generator resistance=r_g in ohms	300	300	2×10^4
Small-signal power gain=G with typical r_g and r_l	25	40	12

The second limitation is especially important, because it can lead to a "runaway" effect. The higher the temperature, the higher the i_{c0}, which, in turn, leads to higher temperature and ultimately to failure of the transistor. (See Fig. 6.)

It is possible to obtain efficiencies of the order of 47 percent with class-A transistor amplifiers. However, when bipolar transistors are used in power stages, it is advisable to use class-B amplification, since the output can approach 3 times the total dissipated power, which is equivalent to 6 times the allowable dissipation of each unit. Furthermore, the no-signal standby power is negligible in the class-B circuit.

The output circuit for the class-B transistor amplifier can be analyzed by the same methods used for the conventional electron-tube equivalents.

For a class-B transistor amplifier with sinusoidal driving voltage

$$P=e_c^2/2r_l$$

where

$P=$ power output
$r_l=$ reflected load resistance to half the primary
$e_c=$ collector voltage.

$$\eta=\frac{\pi}{4(1+\pi r_l i_{c0}/e_c)}$$

where η is the efficiency at maximum power output levels. In actual cases η will be 65 to 75 percent.

The equivalent circuit for large-signal operation is given in Fig. 7.

Complementary Symmetry

A class-B transistor amplifier can be constructed without the need for a separate phase inverter or a push-pull output transformer. This can be done by using a pnp and an npn transistor as shown in Fig. 8.

The pnp unit will amplify the negative part of the input signal and the npn transistor will amplify the positive part. In this manner, phase inversion is automatically accomplished.

The positive and negative signals are combined by coupling the two outputs.

NEGATIVE RESISTANCE

Trigger Circuits

Point-contact and hook-collector transistors have an α that is greater than unity.

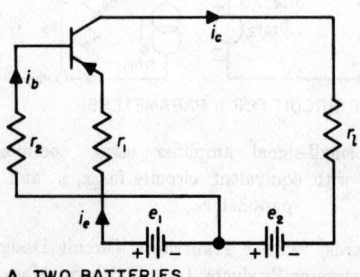

A. TWO BATTERIES

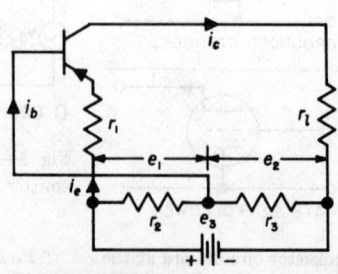

B. ONE BATTERY

Fig. 4—Bipolar transistor biasing methods.

TABLE 9—GAIN CALCULATIONS FOR BIPOLAR TRANSISTOR AMPLIFIERS.

	4-Terminal Network	Common Base		Common Emitter		Common Collector	
Conditions for validity		$R_l \ll \dfrac{1+h_{fe}}{h_{oe}}$ $R_g \gg \dfrac{h_{ie}}{1+h_{fe}}$		$R_l \ll \dfrac{1}{h_{oe}}$ $R_g \gg h_{ie}$		$h_{ie} \ll R_g \ll \dfrac{1+h_{fe}}{h_{oe}}$ $\dfrac{h_{ie}}{1+h_{fe}} \ll R \ll \dfrac{1}{h_{oe}}$	
Current amplification $= A_i$	$\dfrac{-h_{fe}}{1+h_{fe}}$	h_{fb}	h_{fe}	$\dfrac{-h_{fb}}{1+h_{fb}}$	$-(1+h_{fe})$	$\dfrac{-1}{1+h_{fb}}$	
Voltage amplification $= A_v$	$\dfrac{h_{fe}R_l}{h_{ie}}$	$\dfrac{-h_{fb}R_l}{h_{ib}}$	$\dfrac{-h_{fe}R_l}{h_{ie}}$	$\dfrac{h_{fb}R_l}{h_{ib}}$	1	1	
Input impedance $= z_{in}$	$\dfrac{h_{ie}}{1+h_{fe}}$	h_{ib}	h_{ie}	$\dfrac{h_{ib}}{1+h_{fb}}$	$(1+h_{fe})R_l$	$\dfrac{R_l}{1+h_{fb}}$	
Output impedance $= z_{out}$	$\dfrac{1+h_{fe}}{h_{oe}}$	$\dfrac{1}{h_{ob}}$	$\dfrac{1}{h_{oe}}$	$\dfrac{1+h_{fb}}{h_{ob}}$	$\dfrac{R_g}{1+h_{fe}}$	$(1+h_{fb})R_g$	

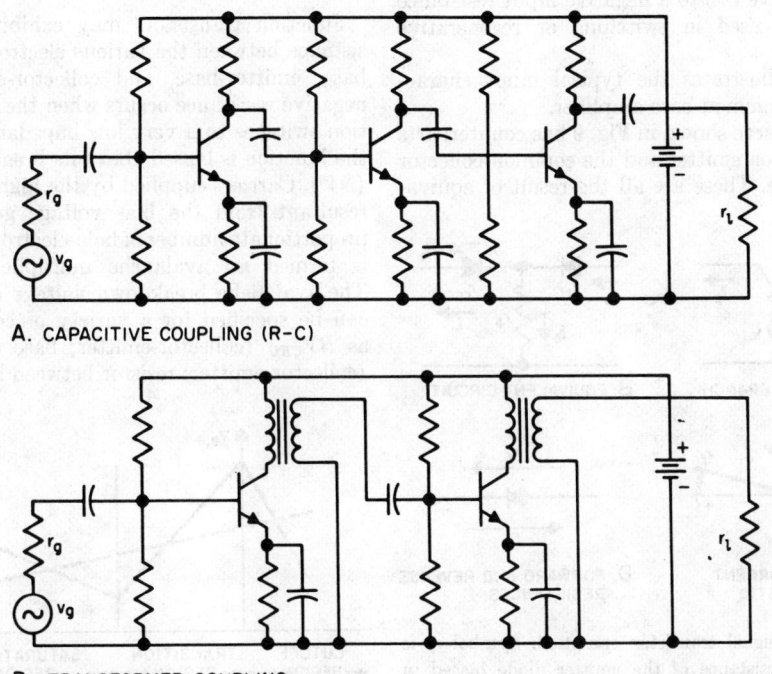

A. CAPACITIVE COUPLING (R-C)

B. TRANSFORMER COUPLING

Fig. 5—Multistage coupling.

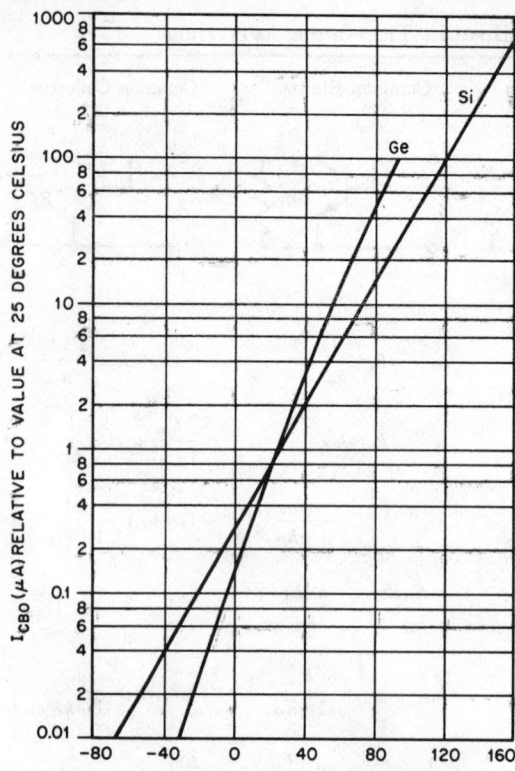

Fig. 6—Variation of I_{CB0} with temperature.

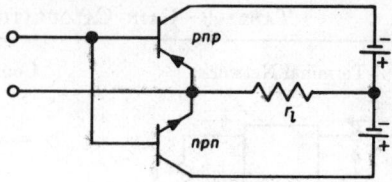

Fig. 8—Complementary symmetry for push-pull stage.

lent transistor properties, and only the common-base curve will be considered in this discussion.

Monostable operation is obtained if the load line intersects a positive-resistance portion only once, either in the saturation region or in the cutoff region.

Bistable operation is obtained when the load line intersects a positive-, a negative-, and again a positive-resistance region.

Astable operation is obtained when the load line intersects only the negative-resistance part of the characteristic.

A circuit that may be used as an astable or monostable trigger is shown in Fig. 10.

The emitter current is

$$i_e = \frac{r_c e_c}{r_l(r_b' + r_c + r_l)} \exp - \left[\frac{(r_b' + r_l)t}{r_b' r_l C} \right].$$

The period of the pulse is

$$t = \frac{r_b r_l C}{r_b' + r_l} \ln \frac{r_c [a(r_l + r_b') - r_b]}{r_l(r_b' + r_c + r_l)}.$$

Junction transistors may exhibit negative resistance between the various electrodes (collector-base, emitter-base, and collector-emitter). This negative resistance occurs when the specified junction switches to a very low impedance state when the junction is biased above its breakdown voltage (BV). Carriers supplied by the high field gradient resultant from the bias voltage generate a disproportionate number of hole-electron pairs in what is termed an avalanche multiplication process. The avalanche breakdown voltage of a transistor can be specified for a variety of conditions such as BV_{CEO} (collector-emitter; base open), BV_{CER} (collector-emitter; resistor between base and emit-

This can give rise to a negative input resistance that can be used in switching or regenerative circuits.

Figure 9 illustrates the typical input characteristic of a common-base amplifier.

The "N" curve shown in Fig. 9 has counterparts for the common-emitter and the common-collector configurations. These are all the result of equiva-

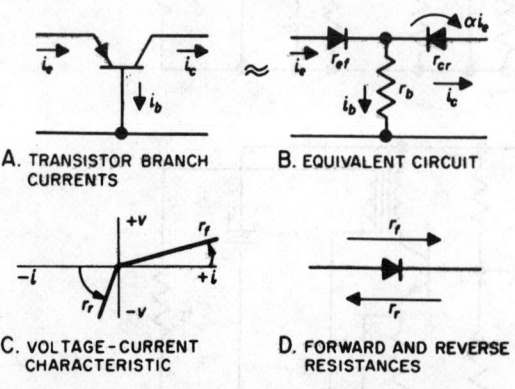

A. TRANSISTOR BRANCH CURRENTS

B. EQUIVALENT CIRCUIT

C. VOLTAGE-CURRENT CHARACTERISTIC

D. FORWARD AND REVERSE RESISTANCES

Fig. 7—Large-signal transistor operation. Symbol r_f is the dynamic resistance of the emitter diode biased in the forward conducting direction, and r_r is the dynamic resistance of the collector diode biased in the reverse direction.

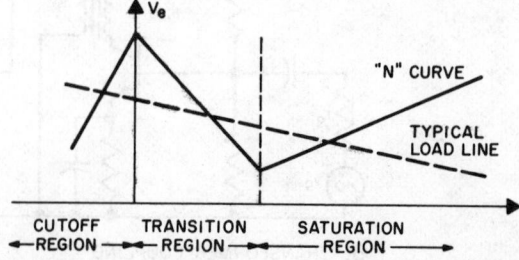

Fig. 9—Input resistance of a common-base amplifier using a point-contact transistor.

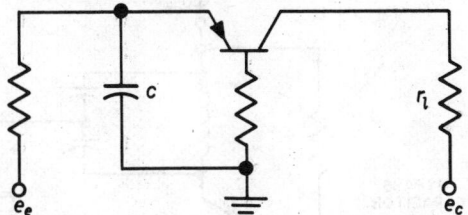

Fig. 10—Astable or monostable trigger circuit.

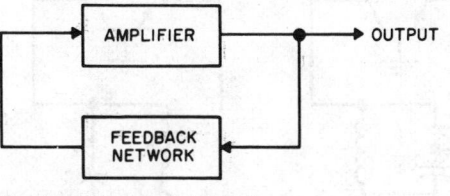

A. FEEDBACK OSCILLATOR

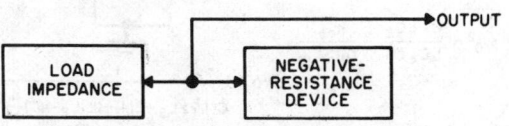

B. NEGATIVE-RESISTANCE OSCILLATOR

Fig. 12—Block diagrams of oscillator circuits.

ter), BV_{CEX} (collector-emitter; emitter-base reverse biased to yield highest possible BV_{CE}), etc.

Junction avalanche transistor capacitive discharge circuits, which can be triggered externally as in Fig. 11 or operated as self-switching relaxation oscillators as in Fig. 13, can produce very fast rise-time pulses ($t_{on} < 10$ nanoseconds) at high discharge currents ($I_M > 25$ amperes). These circuits are therefore useful for operating components such as injection lasers which require very brief, high-current drive pulses.

Oscillators

Oscillators may be grouped into two classes (Figs. 12 and 13).

(**A**) Four-terminal or feedback oscillators.
(**B**) Two-terminal or negative-resistance oscillators.

The point-contact or the hook-collector transistor can be used as a two-terminal oscillator by placing a resonant circuit in series with the base lead (Fig. 13A), or in parallel with the emitter resistance (Fig. 13B), or in parallel with the collector resistance (Fig. 13C).

Additional oscillator circuits are shown in Figs. 14 and 15.

VIDEO-FREQUENCY AMPLIFIERS

Low-Frequency Compensation

A transistor amplifier may be compensated to give improved low-frequency response by splitting

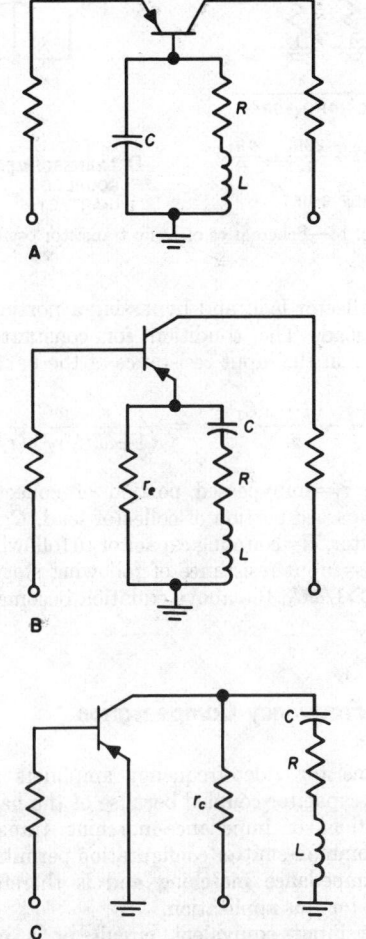

Fig. 13—Basic oscillator circuits using negative-resistance characteristics of bipolar transistors.

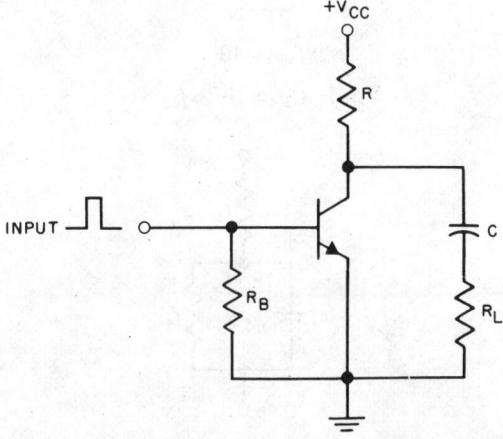

Fig. 11—Avalanche transistor pulse generator.

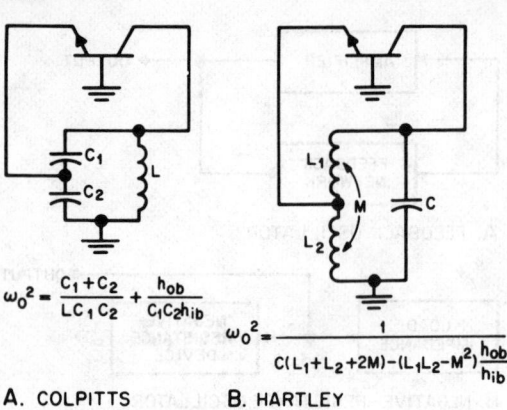

$$\omega_0^2 = \frac{C_1+C_2}{LC_1C_2} + \frac{h_{ob}}{C_1C_2h_{ib}}$$

A. COLPITTS

$$\omega_0^2 = \frac{1}{C(L_1+L_2+2M)-(L_1L_2-M^2)\frac{h_{ob}}{h_{ib}}}$$

B. HARTLEY

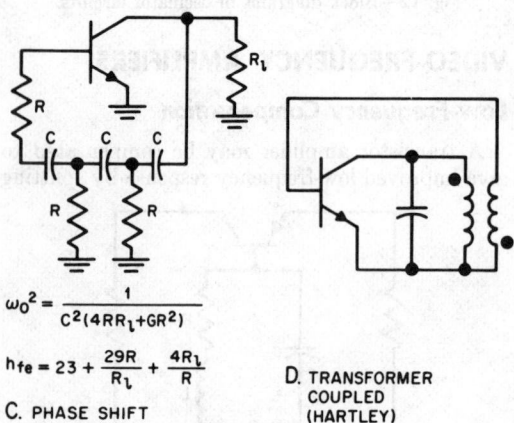

$$\omega_0^2 = \frac{1}{C^2(4RR_1+GR^2)}$$

$$h_{fe} = 23 + \frac{29R}{R_1} + \frac{4R_1}{R}$$

C. PHASE SHIFT

D. TRANSFORMER COUPLED (HARTLEY)

Fig. 14—Schematics of basic transistor oscillators.

the collector load and bypassing a portion of this split load. The condition for constant current flowing in the input resistance of the next stage is

$$\frac{r_1+r_2/(1+\omega^2C_1^2r_2^2)}{r_i} = \frac{\omega C}{(1+\omega^2C_1^2r_2^2)(r_2^2\omega C_1)}$$

where $r_1=$ unbypassed portion of collector load, $r_2=$ bypassed portion of collector load, $C_1=$ bypass capacitor, $C=$ coupling capacitor to following stage, and $r_i=$ input resistance of following stage. When $r_2 \approx r_1 \gg 1/\omega C_1$, the above equation becomes $r_1/r_i \approx C/C_1$.

High-Frequency Compensation

Transistor video-frequency amplifiers are generally capacitor-coupled because of the bandwidth limitations of impedance-matching transformers. The common-emitter configuration permits reasonable impedance matching and is therefore best suited for this application.

The input equivalent circuit of a common-emitter stage for high frequencies is shown in Fig. 16.

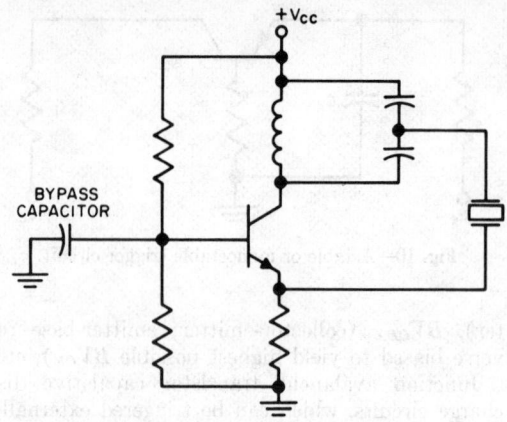

A. SERIES RESONANT

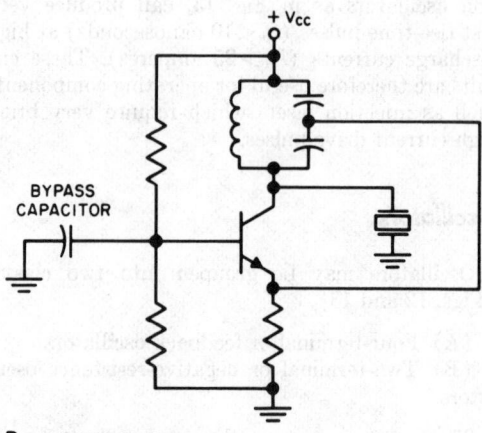

BYPASS CAPACITOR

B. ANTIRESONANT

Fig. 15—Crystal-controlled oscillator circuits.

The input impedance is approximately

$$z_i = r_3 + r_3/[1+j(10f/f_{\alpha0})]$$

where, for most transistors currently available for use as video amplifiers

$$r_3 = r_4$$

$$2\pi f_{\alpha0}r_4 = 10$$

$$C_3r_4 = 10/2\pi f_{\alpha0}.$$

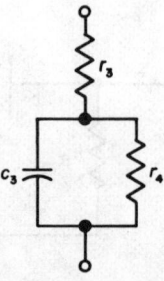

Fig. 16—Input equivalent circuit used for high-frequency compensation for a common-emitter amplifier.

High-frequency compensation may be obtained if an inductance L is placed in series with the collector load resistance r_1. The value of the compensating L may be obtained from the following equations.

$$|A_i| = \left(\frac{r_1^2 + \omega^2 L^2}{A^2 + B^2}\right)^{1/2}$$

$$\times \{[(1/\alpha_0) - 1]^2 + [(1/\alpha_0)(\omega/\omega_{\alpha 0})]^2\}^{-1/2}$$

where

$$A = r_1 + r_3[1 + (10\omega/\omega_{\alpha 0})^2]^{-1}$$

$$\times \left[2 - 2\omega^2 C_2 L + (1 - \omega^2 C_2 L)^2 \frac{10\omega^2}{\omega_{\alpha 0}} + r_1 C_2 \omega \frac{10\omega}{\omega_{\alpha 0}}\right]$$

and

$$B = \omega L + \omega r_3[1 + (10\omega/\omega_{\alpha 0})^2]^{-1}$$

$$\times \left[2 C_2 r_1 + C_2 r_1 \left(\frac{10\omega}{\omega_{\alpha 0}}\right)^2 + \omega C_2 L \frac{\omega 10}{\omega_{\alpha 0}} - \frac{10}{\omega_{\alpha 0}}\right].$$

If $\omega \ll \omega_2$

$$|A_i| = \frac{r_1}{r_1 + 2r_3} \frac{\alpha_0}{1 - \alpha_0}$$

where $\omega_2 =$ cutoff frequency of amplifier, $\alpha_0 =$ low-frequency alpha, and $C_2 =$ capacitance across L and r_1.

In addition to the compensation described above, series inductance can be used to resonate with the input capacitance.

Another method of high-frequency compensation is available. The emitter resistance may be only partly bypassed, resulting in degeneration at lower frequencies. The compensation conditions are similar to that of electron-tube cathode compensation.

INTERMEDIATE-FREQUENCY AMPLIFIERS

Series-Resonant Interstages

For the series-resonant coupling circuit (Fig. 17), the power gain per stage is

$$G \approx |\beta|^2 r_{i2}/r_{i1}.$$

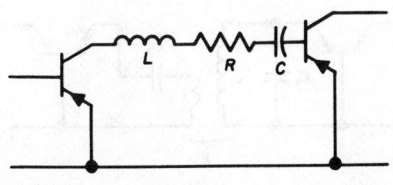

Fig. 17—Series-resonant interstage circuit.

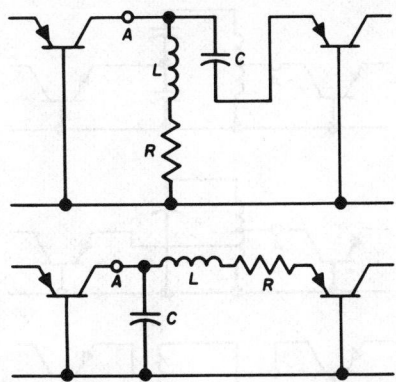

Fig. 18.—Parallel-resonant interstage circuits.

For iterated stages, $r_{i1} = r_{i2}$, and

$$G \approx |\beta|^2.$$

For common-base stages

$$G \approx |\alpha|^2 r_{i2}/r_{i1}$$

where $\alpha =$ common-base current gain, $\beta = \alpha/(1 - \alpha)$, $r_{i1} =$ input resistance of stage, and $r_{i2} =$ input resistance of following stage.

Junction transistors give less than unity gain using this coupling circuit with common-base or common-collector circuits. Point-contact transistors may be used in the common-base connection.

$$f_0/\Delta f_{3dB} = Q = \omega_0 L/(R + r_{i2})$$

where $f_0 =$ center frequency and $\Delta f_{3dB} =$ 3-decibel bandwidth.

Parallel-Resonant Interstages

If Q (> 10) includes the effect of the input impedance of the next stage for common-base stages (Fig. 18)

$$G \approx |\alpha|^2 Q^2 r_{i2}/r_{i1}.$$

For common-emitter stages

$$G \approx |\beta|^2 Q^2 r_{i2}/r_{i1}.$$

The equations below apply also.

Parallel-Resonant Interstage with Impedance Transformation: Power gain per stage:

$$G = A_i^2 (r_1/r_{i1})$$

$\times$ (fraction of output power delivered to load).

Let

$r_{i1} =$ input resistance of stage
$r_{i2} =$ input resistance of next stage
$g_i =$ conductance seen at A (Fig. 18) due to r_{i2}
$g_n =$ conductance seen at A due to network losses R
$g_o =$ output conductance of transistor

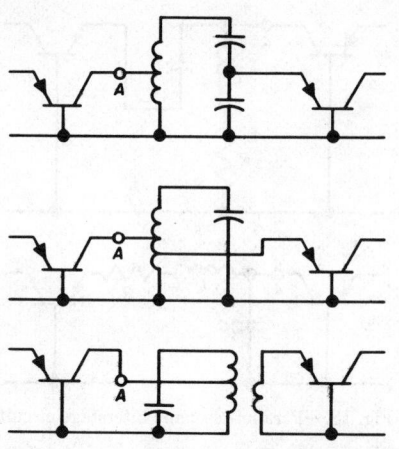

Fig. 19—Three types of single-tuned resonant interstage circuits.

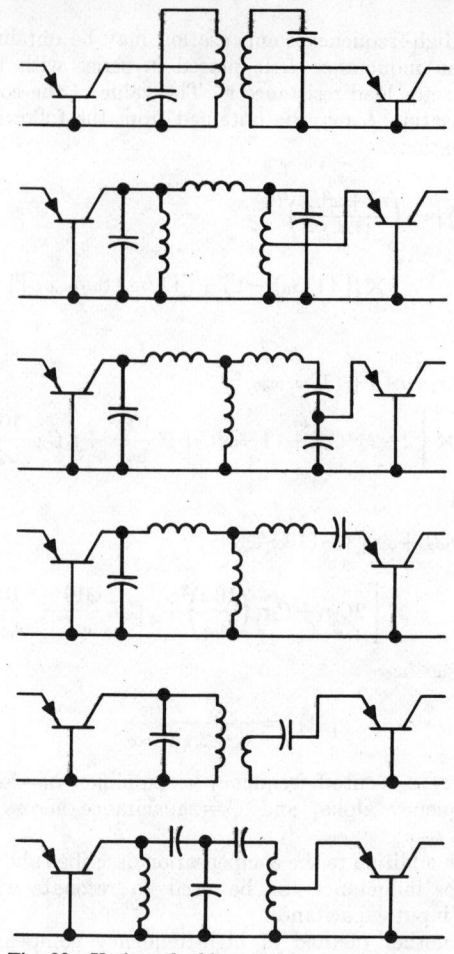

Fig. 20—Various double-tuned interstage circuits.

$p=$ ratio of equivalent series resistance seen at A to input resistance of next stage

$= r_1/r_{i2}$

$z_l = r_l + jx_l =$ total load impedance seen at A

$$z_c = \frac{r_c(1-j\omega r_c C_c)}{1+\omega^2 r_c^2 C_c^2} = \text{collector impedance.}$$

Then, for common-base stages, power gain is

$$G = \left|\frac{\alpha}{1+z_l/z_c}\right|^2 p\, \frac{r_{i2}}{r_{i1}}\left(\frac{g_i}{g_i+g_n}\right)^2.$$

For common-emitter connection

$$G = \left|\frac{\alpha}{1-\alpha+z_l/z_c}\right|^2 p\, \frac{r_{i2}}{r_{i1}}\left(\frac{g_i}{g_i+g_n}\right)^2.$$

For common-collector stages

$$G = \left|\frac{1}{1-\alpha+z_l/z_c}\right|^2 p\, \frac{r_{i2}}{r_{i1}}\left(\frac{g_i}{g_i+g_n}\right)^2$$

$$f_0/f_{3dB} = Q = \omega_0 C/(g_0+g_n+g_i)$$

where C is the total C seen at A (Fig. 18) due to the transistor output, the coupling network, and the following stage.

If $z_l \ll z_c$ and $g_i \gg g_n$ (load not matched, network losses low) and successive stages are identical ($r_{i1}=r_{i2}$):

For common-base stages

$$G = |\alpha|^2 p.$$

For common-emitter stages

$$G = |\beta|^2 p.$$

For common-collector stages

$$G = |\beta+1|^2 p.$$

Tuned-Circuit Interstages

Other configurations of single-tuned interstages are shown in Fig. 19. Any of the 3 bipolar transistor configurations may be used in these circuits.

Double-Tuned Interstages

For double-tuned interstages (Fig. 20), the same gain equations apply as for the single-tuned case. For a given bandwidth, however, p may be made larger in the double-tuned case.

The T and π equivalents of the transformers will not always be physically realizable.

For large bandwidth, the condition $Q_1 \gg Q_2$ is desirable, since then loading resistors are not required with their accompanying power loss.

For $Q_1 \gg Q_2$, for transitional coupling (Fig. 21)

$$\Delta f_{3dB}/f_0 = k = 1/Q_2\sqrt{2}$$

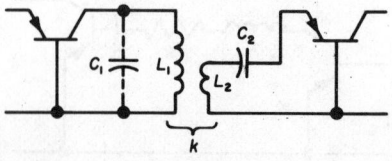

Fig. 21—Double-tuned interstage circuit using transitional coupling.

where $k=$ coefficient of coupling. If $z_i=r_i+jx_i=$ input impedance of next stage, then

$$Q_2=(\omega_0 L_2+x_i)/r_i$$

$$L_2=(Q_2 r_i-x_i)/\omega_0$$

$$=(r_i/2\pi\Delta f_{3dB})-(x_i/\omega_0)$$

L_2, C_2, and x_i are series-resonant at f_0

$$L_1 C_1=1/\omega_0^2$$

$$p\approx Q_2^2 C_2/C_1$$

C_1 includes the transistor output capacitance.

TEMPERATURE COMPENSATION

The i_c of a bipolar transistor may increase appreciably with temperature. This is objectionable since it increases the power dissipated in the transistor and so increases its temperature rise. Two possible methods for stabilizing i_c against temperature variations follow.

The circuit of Fig. 22A depends on negative feedback, similar to cathode bias in electron tubes, i_e being stabilized by the degeneration produced by R_1 at direct current. Capacitor C must bypass R_1 at the frequencies to be amplified.

For the circuit of Fig. 22A, with α being assumed constant over the operating range

$$i_c=\frac{i_{c0}(1+R_1/R_2+R_1/R_3)+\alpha e/R_3}{1-\alpha+R_1/R_2+R_1/R_3}.$$

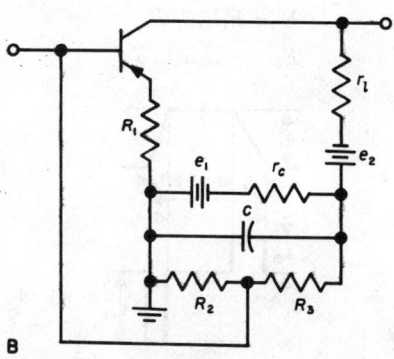

A

B

Fig. 22—Two types of temperature compensation for bipolar transistors.

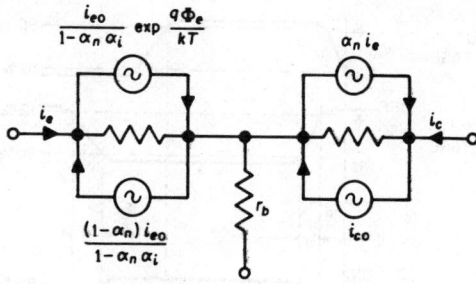

A. REGIONS I AND II

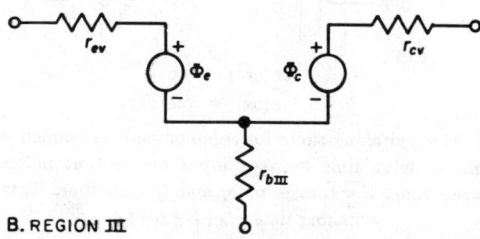

B. REGION III

Fig. 23—Ebers and Moll equivalent circuit for generating, amplifying, and shaping pulse waveforms.

When the variation with frequency of the phase shift resulting from R_1 and C is objectionable, or if C must be made inconveniently large, the circuit of Fig. 22B may be used. Since r_c and R_3 are higher resistances than R_1, a smaller C may be used for the same bypassing effect. Here stabilization is obtained by the drop in i_c influencing base potential, and R_1 is made small to minimize degeneration of signal frequencies.

If $R_3\gg R_1$ and $r_c\gg R_1$, then

$$i_c=\frac{i_{c0}[(r_c/R_3)(1+R_1/R_2)+1+R_1/R_2+R_1/R_3]+\alpha e_1/R_3}{1-\alpha+R_1/R_2+R_1/R_3+(r_c/R_3)(1+R_1/R_2)}.$$

PULSE CIRCUITS

Bipolar transistors may be used to generate, amplify, and shape pulse waveforms.

The Ebers and Moll* equivalent circuits of Fig. 23 give the large-signal transient response of a junction transistor. The parameters are:

$i_{e0}=$ saturation current of emitter junction with zero collector current

$i_{c0}=$ saturation current of collector junction with zero emitter current

$\alpha_n=$ transistor direct-current gain with the emitter functioning as an emitter and the collector functioning as a collector (normal α)

* J. J. Ebers and J. L. Moll, "Large-Signal Behavior of Junction Transistors," also, J. L. Moll, "Large-Signal Transient Response of Junction Transistors," *Proc. IRE*, vol. 42, pp. 1761–1772, 1773–1784; Dec. 1954.

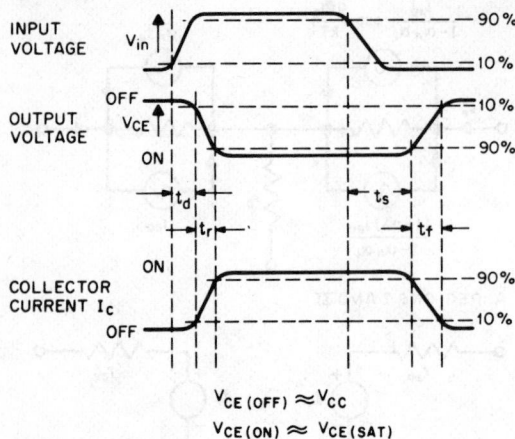

Fig. 24—Switching times for common-emitter configuration. t_d=delay time between input and output pulses, t_r=rise time, t_s=storage time, and t_f=fall time. Total switching time=$t_d+t_r+t_s+t_f$.

α_i= transistor direct-current gain with the collector functioning as an emitter and the emitter functioning as a collector (inverted α)

$$\Phi_e = (kT/q) \ln \left\{ -\left[(i_e+\alpha_i i_c)/i_{e0}\right]+1 \right\}$$

    = emitter-to-junction voltage

$$\Phi_c = (kT/q) \ln \left\{ -\left[(i_c+\alpha_n i_e)/i_{c0}\right]+1 \right\}$$

    = collector-to-junction voltage

k= Boltzmann's constant

T= absolute temperature

q= charge on electron.

The switching time (Fig. 24) can be calculated from the small-signal equivalent-circuit parameters; the turn-on time, from cutoff to saturation, depends on the frequency response of the transistor in the active region. The turn-off time, from saturation to cutoff, depends on minority-carrier storage time and decay time. Carrier storage time is that required for the operating point to move out of the saturation region into the active region on removal of the drive current and is a function of the frequency response of the transistors in the saturation region. Decay time follows the storage time and returns the transistor to cutoff; it depends on the frequency response in the active region. Switching time of the order of $3/\omega_n$ is realized if

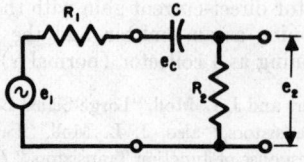

Fig. 25—Capacitive differentiation.

carrier storage is avoided.

$$\text{Turn-on time} = \omega_n^{-1} \frac{i_{e2}}{i_{e2}-0.9i_c/\alpha_n}$$

$$\text{Storage time} = \frac{\omega_n+\omega_i}{\omega_n\omega_i(1-\alpha_n\alpha_i)} \ln \frac{i_{e2}+i_{e1}}{i_c/\alpha_n+i_{e2}}$$

$$\text{Decay time} = \omega_n^{-1} \ln \frac{i_c+\alpha_n i_{e2}}{(i_c+\alpha_n i_{e2})/10}$$

where ω_n= cutoff frequency of normal alpha, ω_i= cutoff frequency of inverted alpha, i_{e1}, i_{e2}= emitter current before and after switching step is applied, and i_c= collector current in the saturation state.

CAPACITIVE-DIFFERENTIATION AMPLIFIERS

Capacitive-differentiation systems employ a series RC circuit (Fig. 25) with the output voltage e_2 taken across R_2. The latter includes the resistance of the load, which is assumed to have a negligible reactive component compared with R_2. In many applications the circuit time constant $RC \ll T$, where T is the period of the input pulse e_1. Thus, transients constitute a minor part of the response, which is essentially a steady-state phenomenon within the time domain of the pulse.

Differential Equation

$$e_1 = e_c + RC(de_c/dt)$$

where $R = R_1+R_2$. Then

$$e_2 = R_2C(de_c/dt) = (R_2/R)(e_1-e_c).$$

When the rise and decay times of the pulse are each $\gg RC$

$$e_2 \approx R_2C(de_1/dt).$$

Trapezoidal Input Pulse

When T_1, T_2, and T_3 are each much greater than RC, the output response e_2 is approximately rectangular, as shown in Fig. 26.

$$E_{21} = E_1R_2C/T_1$$
$$E_{23} = -E_1R_2C/T_3.$$

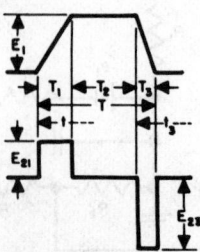

Fig. 26—Trapezoidal input pulse and principal output response.

More accurately, for any value of T, but for widely spaced input pulses, the following are obtained.

$0 < t < T_1$:

$$e_{21} = \frac{E_1 R_2 C}{T_1} [1 - \exp(-t/RC)].$$

$T_1 < t < (T_1 + T_2)$:

$$e_{22} = \frac{E_1 R_2 C}{T_1} [\exp(T_1/RC) - 1] \exp(-t/RC).$$

Note: $\exp(-t/RC) = \epsilon^{-t/RC}$.

$(T_1 + T_2) < t < T$:

$$e_{23} = -(E_1 R_2 C/T_3) \left(1 - \{T_3/T_1[\exp(T_1/RC) - 1]\right.$$
$$\left. + \exp[(T_1 + T_2)/RC]\} \exp(-t/RC)\right).$$

$t > T$:

$$e_{2x} = (E_1 R_2 C/T_3) \{(T_3/T_1)[\exp(T_1/RC) - 1]$$
$$+ \exp[(T_1 + T_2)/RC]$$
$$- \exp(T/RC)\} \exp(-t/RC)$$
$$= A \exp(-t/RC).$$

$T_2 \gg RC$:

$$e_{23} = -(E_1 R_2 C/T_3)[1 - \exp(-t_3/RC)].$$

For a long train of identical pulses repeated at regular intervals of T_r between starting points of adjacent pulses, add to each of the above (e_{21}, e_{22}, e_{23}, and e_{2x}) a term

$$e_{20} = [A/\exp(T_r/RC) - 1] \exp(-t/RC)$$

where A is defined in the expression for e_{2x} above.

Rectangular Input Pulse

Figure 27 is a special case of Fig. 26, with $T_1 = T_3 = 0$.

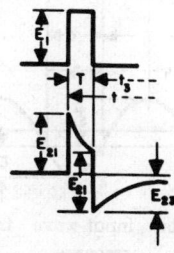

Fig. 27—Single rectangular input and output response.

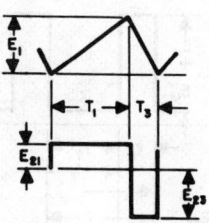

Fig. 28—Triangular input pulse and output response.

$0 < t < T$:

$$e_{21} = (R_2/R) E_1 \exp(-t/RC) = E_{21} \exp(-t/RC).$$

$t > T$:

$$e_{23} = -(R_2/R) E_1 [\exp(T/RC) - 1] \exp(-t/RC)$$
$$= E_{23} \exp(-t_3/RC)$$

where

$$E_{23} = -(R_2/R) E_1 [1 - \exp(-T/RC)].$$

Triangular Input Pulse

Figure 28 is a special case of the trapezoidal pulse, with $T_2 = 0$. The total output amplitude is approximately

$$|E_{21}| + |E_{23}| = |E_1| R_2 C[(T_1 + T_3)/T_1 T_3]$$

which is a maximum when $T_1 = T_3$.

CAPACITIVE-INTEGRATION AMPLIFIERS

Capacitive-integration circuits employ a series RC circuit (Fig. 29) with the output voltage e_2 taken across capacitor C. The load admittance is accounted for by including its capacitance in C, while its shunt resistance is combined with R_1 and R_2 to form a voltage divider treated by Thévenin's theorem. In contrast with capacitive differentiation, time constant $RC \gg T$ in many applications. Thus, the output voltage is composed mostly of the early part of a transient response to the input voltage wave. For a long repeated train of identical input pulses, this repeated transient response becomes steady state.

Circuit Equations

$$e_1 = e_2 + RC(de_2/dt)$$

where $R = R_1 + R_2$.

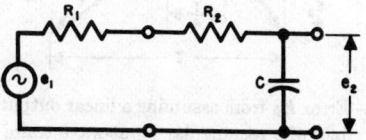

Fig. 29—Capacitive integration circuit.

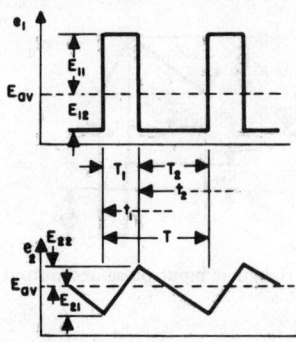

Fig. 30—Rectangular input-wave train and output response.

When $t \ll RC$ and E_{20} is very small compared with the amplitude of e_1

$$e_2 \approx E_{20} + (RC)^{-1} \int_0^t e_1 \, dt$$

where $E_{20} =$ value of e_2 at time $t = 0$.

Rectangular Input-Wave Train (Fig. 30)

$$E_{Av} = T^{-1} \int_0^T e_1 \, dt.$$

Then

$$E_{11}T_1 + E_{12}T_2 = 0.$$

After equilibrium or steady state has been established

$$e_{21} = E_{Av} + E_{11}[1 - \exp(-t_1/RC)]$$
$$+ E_{21} \exp(-t_1/RC)$$

$$e_{22} = E_{Av} + E_{12}[1 - \exp(-t_2/RC)]$$
$$+ E_{22} \exp(-t_2/RC).$$

If the steady state has not been established at time $t_1 = 0$, add to e_2 the term

$$(E_{20} - E_{Av} - E_{21}) \exp(-t_1/RC).$$

When $T_1 = T_2 = T/2$, then

$$E_{11} = -E_{12} = E_1$$

$$E_2 = E_{22} = -E_{21} = E_1 \tanh(T/4RC).$$

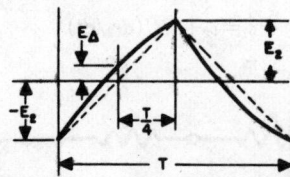

Fig. 31—Error E_Δ from assuming a linear output (dashed line) of a rectangular input-wave train.

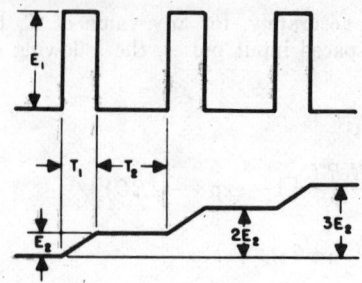

Fig. 32—Biased rectangular input-wave train and output response.

Approximately, for any T_1 and T_2, provided $T \ll RC$

$0 < t_1 < T_1$:

$$e_{21} = E_{Av} - E_2(1 - 2t_1/T_1).$$

$0 < t_2 < T_2$:

$$e_{22} = E_{Av} + E_2(1 - 2t_2/T_2)$$

where

$$E_2 = E_{22} = -E_{21} = E_{11}T_1/2RC$$
$$= -E_{12}T_2/2RC.$$

The error due to assuming a linear output voltage wave (Fig. 31) is

$$E_\Delta / E_2 \approx T/8RC$$

when $T_1 = T_2 = T/2$. The error in E_2 due to setting $\tanh(T/4RC) = T/4RC$ is comparatively negligible. When $T/RC = 0.7$, the approximate error in E_2 is only 1 percent. However, the error E_Δ is 1 percent of E_2 when $T/RC = 0.08$.

Biased Rectangular Input Wave

In Fig. 32, when $(T_1 + T_2) \ll RC$, and $E_{20} = 0$ at $t = 0$, the output voltage approximates a series of steps.

$$E_2 = E_1 T_1/RC.$$

Triangular Input Wave

In Fig. 33, when $(T_1 + T_2) \ll RC$, and after the steady state has been established, then (approxi-

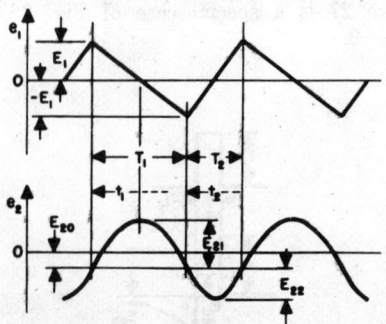

Fig. 33—Triangular input-wave train and output response.

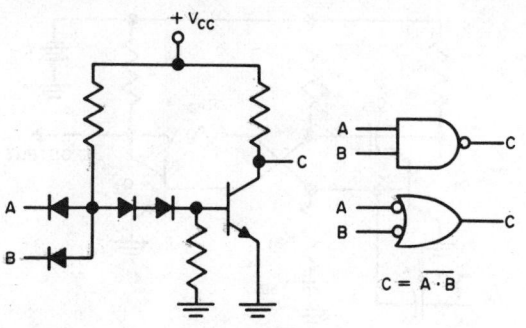

Fig. 34—NAND circuit for positive logic.

$$C = \overline{A \cdot B}$$

mately)

$0 < t_1 < T_1$:

$$e_{21} = E_{20} + E_{21} - 4E_{21}\left[(t_1/T_1) - \tfrac{1}{2}\right]^2.$$

$0 < t_2 < T_2$:

$$e_{22} = E_{20} + E_{22} - 4E_{22}\left[(t_2/T_2) - \tfrac{1}{2}\right]^2$$

where

$$E_{20} = E_1(T_2 - T_1)/6RC$$

$$E_{21} = E_1 T_1/4RC$$

$$E_{22} = -E_1 T_2/4RC.$$

MEASUREMENT OF SMALL-SIGNAL PARAMETERS

The small-signal parameters may be represented by ratios of small alternating voltages and currents if care is taken to keep the magnitudes of these signals small compared with the direct-current condition. For instance

$$z_{11} = r_e + r_b$$
$$= [\partial v_e/\partial i_e]_{i_c} \approx [\Delta v_e/\Delta i_e]_{i_c} \approx [v_e/i_e]_{i_c}.$$

Also

$z_{11} = e_1/i_1$ when $i_2 = 0$

$z_{12} = e_1/i_2$ when $i_1 = 0$

$z_{21} = e_2/i_1$ when $i_2 = 0$

$z_{22} = e_2/i_2$ when $i_1 = 0$

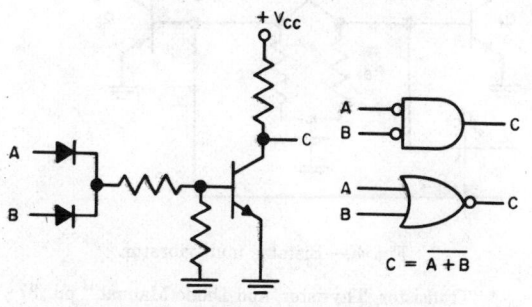

$$C = \overline{A + B}$$

Fig. 35—NOR circuit for positive logic.

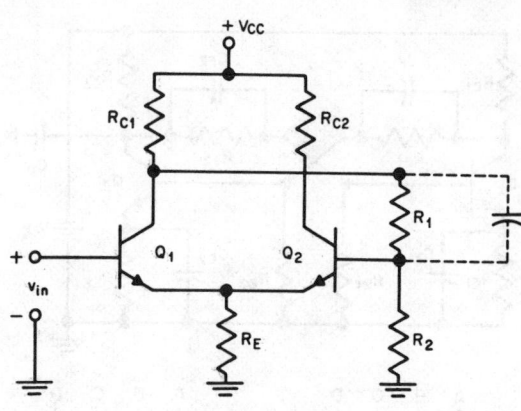

Fig. 36—Schmitt trigger.

and

$h_{11} = e_1/i_1$ when $e_2 = 0$

$h_{12} = e_1/e_2$ when $i_1 = 0$

$h_{21} = i_2/i_1$ when $e_2 = 0$

$h_{22} = i_2/e_2$ when $i_1 = 0.$

TYPICAL BIPOLAR TRANSISTOR CIRCUITS

Figures 34 and 35 show the NAND and NOR circuits, respectively, for positive logic, while Table 10 gives their relationship for positive and negative logic.

Figure 36 shows a Schmitt trigger circuit and Fig. 37 two blocking oscillators.

Figures 38, 39, and 40 show typical multivibrator circuits. For basic descriptions, refer to the section on relaxation oscillators in the chapter "Electron-Tube Circuits."

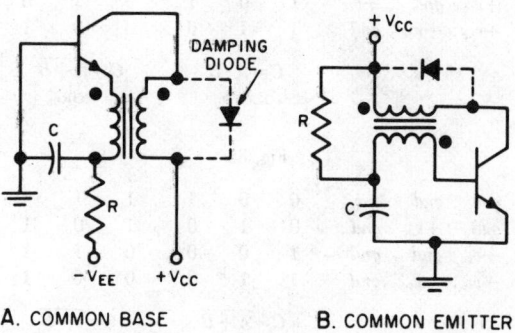

A. COMMON BASE   B. COMMON EMITTER

Fig. 37—Blocking oscillators.

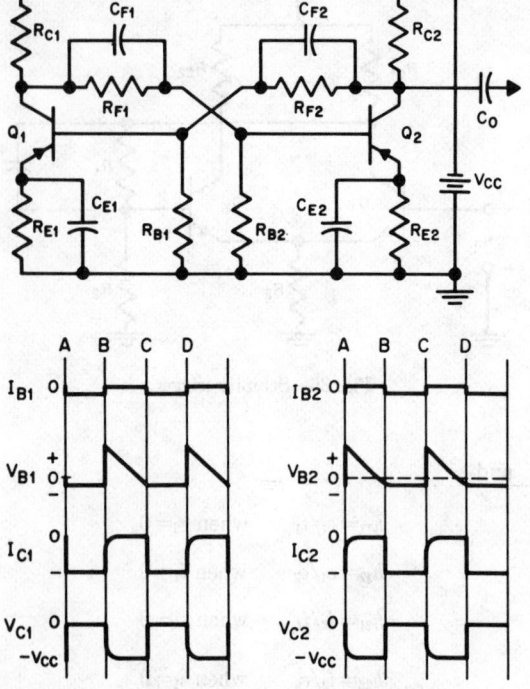

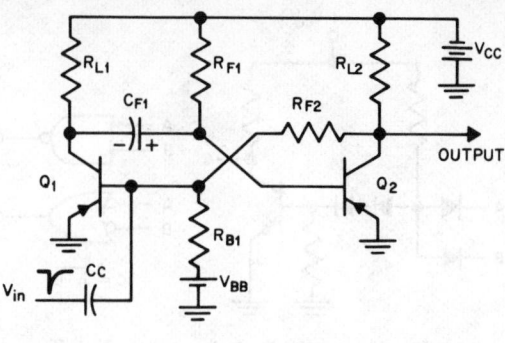

Fig. 39—Monostable multivibrator and waveforms of pertinent voltages.

Fig. 38—Astable multivibrator and waveforms of pertinent currents and voltages.

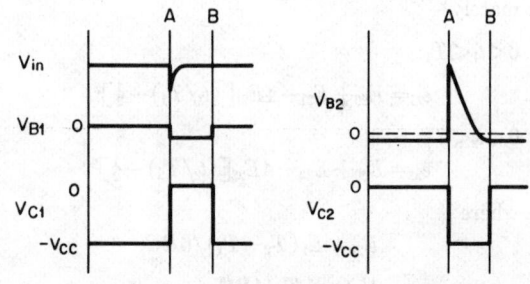

FET CIRCUIT CONSIDERATIONS

Small Signal Amplifiers*

The field-effect transistor has the three basic small-signal amplifier configurations shown in Fig. 41. The common-source configuration is the usual arrangement since it provides very high input impedance, moderate output impedance, and

TABLE 10—NAND–NOR RELATIONSHIP FOR POSITIVE AND NEGATIVE LOGIC.

Circuit Voltage States			Positive Logic $+v_{cc}=1$ $gnd=0$			Negative Logic $gnd=1$ $+v_{cc}=0$		
A	B	C	A	B	C	A	B	C
Fig. 34								
gnd	gnd	$+v_{cc}$	0	0	1	1	1	0
gnd	$+v_{cc}$	$+v_{cc}$	0	1	1	1	0	0
$+v_{cc}$	gnd	$+v_{cc}$	1	0	1	0	1	0
$+v_{cc}$	$+v_{cc}$	gnd	1	1	0	0	0	1
			$C=\overline{A \cdot B}$ NAND			$C=\overline{A+B}$ NOR		
Fig. 35								
gnd	gnd	$+v_{cc}$	0	0	1	1	1	0
gnd	$+v_{cc}$	gnd	0	1	0	1	0	1
$+v_{cc}$	gnd	gnd	1	0	0	0	1	1
$+v_{cc}$	$+v_{cc}$	gnd	1	1	0	0	0	1
			$C=\overline{A+B}$ NOR			$C=\overline{A \cdot B}$ NAND		

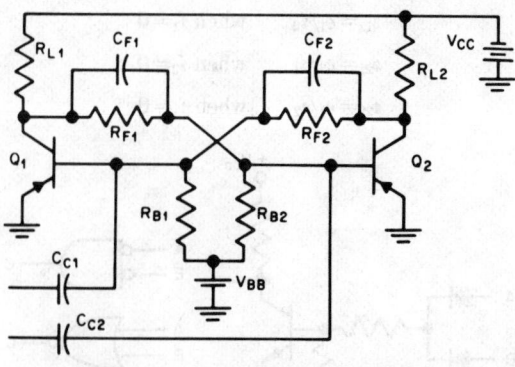

Fig. 40—Bistable multivibrator.

* "Transistor, Thyristor, and Diode Manual," pp. 37–39, RCA; 1969.

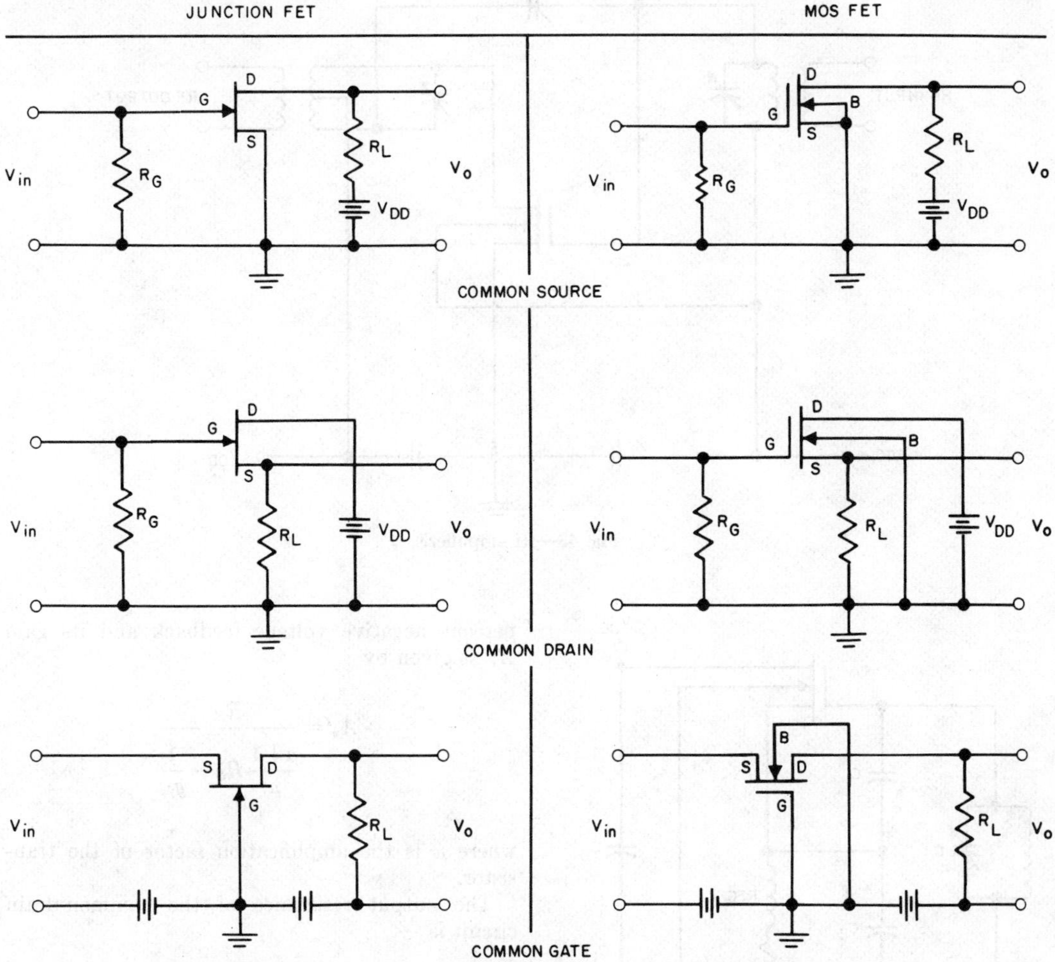

JUNCTION FET | MOS FET

COMMON SOURCE

COMMON DRAIN

COMMON GATE

Fig. 41—The 3 basic FET small-signal amplifier configurations.

a voltage gain greater than unity. The common-source voltage gain is given by

$$A_v = \frac{g_{fs} r_{os} R_L}{r_{os} + R_L}$$

where g_{fs} is the gate-to-drain forward transconductance, r_{os} is the common-source output resistance, and R_L is the load resistance.

The addition of a source resistor will reduce the voltage gain by producing a negative voltage feedback in proportion to the output current. The voltage gain then becomes

$$A_v' = \frac{g_{fs} r_{os} R_L}{r_{os} + (g_{fs} r_{os} + 1) R_S + R_L}$$

where R_S is the total source resistance in series with the source terminal.

The use of a source feedback resistor increases the common-source output impedance according to

$$Z_o = r_{os} + (g_{fs} r_{os} + 1) R_S$$

The common-drain configuration, often called a source-follower, has a very high input impedance, low output impedance, a voltage gain less than 1, and no output phase reversal. The circuit has 100

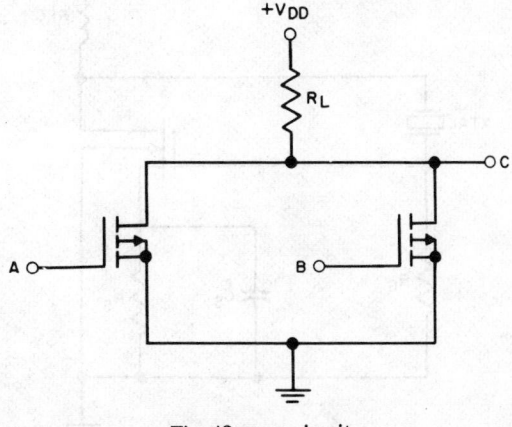

Fig. 42—NOR circuit.

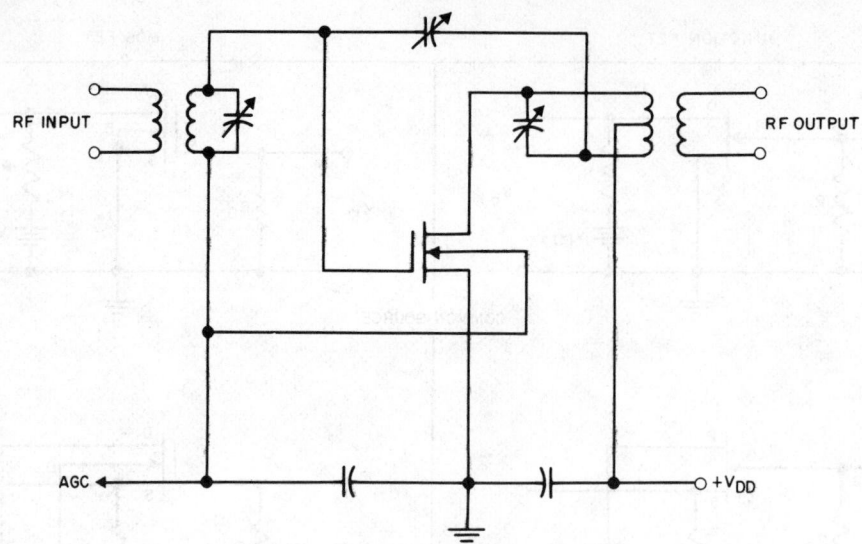

Fig. 43—Rf amplifier.

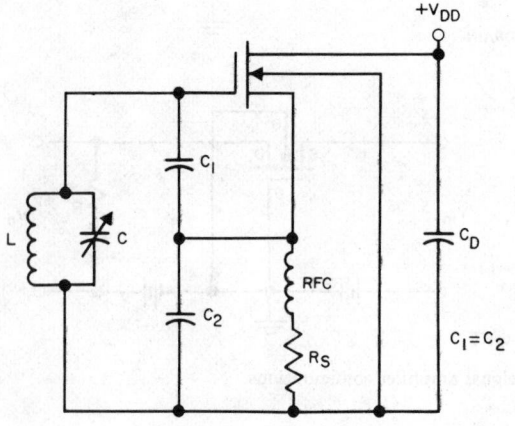

Fig. 44—Colpitts oscillator.

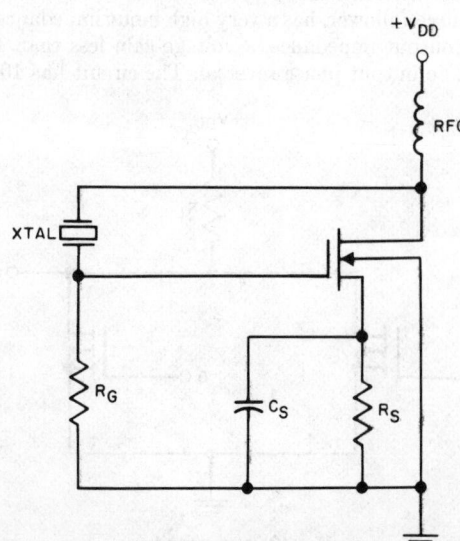

Fig. 45—Pierce-type crystal oscillator.

percent negative voltage feedback and its gain A_v' is given by

$$A_v'= \frac{R_S}{\dfrac{\mu+1}{\mu}R_S+\dfrac{1}{g_{fs}}}$$

where μ is the amplification factor of the transistor.

The output resistance of the common-drain circuit is

$$R_o'= \frac{r_{os}R_S}{(g_{fs}r_{os}+1)R_S+r_{os}}$$

where r_{os} is the common-source output resistance of the transistor.

The common-gate configuration has a low input impedance, high output impedance, low voltage gain, and relatively high frequency response. The common-gate voltage gain is

$$A_v= \frac{(g_{fs}r_{os}+1)R_L}{(g_{fs}r_{os}+1)R_G+r_{os}+R_L}$$

where R_G is the resistance of the input signal source.

TYPICAL FET CIRCUITS

Figure 42 shows a FET NOR circuit. Figure 43 shows an rf amplifier with a usable frequency response extending to beyond 100 megahertz. Figure 44 shows a Colpitts oscillator, and Fig. 45 shows a Pierce-type crystal controlled oscillator.

CHAPTER 21

MICROMINIATURE ELECTRONICS

DEFINITIONS AND TERMINOLOGY

Active Device: A device exhibiting transistance, e.g., gain or control.

Active Elements: Circuit components that have gain or that direct current flow: diodes, transistors.

Adder: Switching circuits that generate sum and carry bits.

Address: A code designating the location of information and instructions in the main storage unit of the computer.

Alumina (Aluminum Oxide): An inorganic insulator frequently used as a base substrate for hybrid film circuits because of its excellent electrical and physical properties.

Aluminum: Metal widely used as a conducting film interconnection medium in silicon and hybrid film microcircuits. Aluminum has a low melting and boiling point, is easily evaporated in vacuum, and is easily patterned by chemical etching.

Analog Computer: A continuously variable computer. A differential analyzer. It measures the effect of changes in one variable on all other variables in a system. Its operation is analogous to a slide rule.

AND: A Boolean logic operator analogous to multiplication. Of two variables, both must be true for the output to be true.

Å or Angstrom: A unit of linear measure equal to 10^{-10} meter.

Anodization: The formation of an insulating oxide layer by electrolytic action on metals such as tantalum, titanium, aluminum, and niobium (which are commonly known as valve metals and are used in thin-film work).

Artwork: Pattern of microcircuit or passive-element configurations formed of alternately clear and opaque regions in a dimensionally stable medium. Artwork is made some multiple of the final mask size, usually on a mylar laminate known as Ruby Studnite, which is clear mylar with a peelable opaque overlay. By cutting the overlay with a knife edge mounted on a coordinatograph, sections of the overlay may be removed to form clear areas in an opaque background.

Assembly: A number of parts or subassemblies (or any combination thereof) electrically and mechanically joined together to perform a specific function. *Examples*: audio-frequency amplifier, vacuum-tube voltmeter, distance meter, analog-to-digital converter, logic card.

Asynchronous: Operation of a switching network by a free-running signal that triggers successive instructions; the completion of one instruction triggers the next.

Autogenous (Self Grown): SiO_2 grown on the surface of a slice of silicon during the natural course of diffusion processes or oxidation.

Base Media: Materials employed in the preparation of artwork, such as Ruby Studnite.

Batch Process: A process that potentially yields a large number of products in a single or small number of steps. Batch processes are attractive because of their suitability for mass production with its attendant economies.

Beam Lead: A silicon chip connection lead formed in place chemically and suspended over a void or space. The cantilevered end can be formed on the silicon chip and then attached to the interconnecting pattern or it can be cantilevered from the interconnecting pattern and bonded to the chip.

Beryllium Oxide—Beryllia (BeO): Substrate material whose most interesting property is its high thermal conductivity, about half the value of copper. This makes it an excellent nonelectrically conducting heat sink. Also used in cases for semiconductor circuits and devices.

Binary: A system of numerical representation that uses only two symbols, 0 and 1.

Bipolar: Minority-carrier device in which current passes across junctions of p-type (holes are carriers) and n-type (electrons are carriers) semiconductor.

Bit: Abbreviation for binary digit; a unit of binary information.

Black Box: A useful mathematical approach to an electronic circuit that concerns itself only with the input and output and ignores the interior elements, discrete or integrated.

Bond(ing): The attachment of discrete wire or ribbon leads to an integrated or hybrid integrated circuit by such methods as soldering, welding, thermocompression bonding, etc.

Breadboard Model: An assembly of discrete components in rough form to prove the feasibility of a circuit, system, or design principle.

Buffer: A noninverting member of the digital family which may be used to handle a large fan-out or to convert input and output levels. Normally a buffer is an emitter-follower type of circuit.

Buried Layer: A high-conductivity diffused layer under the collector area of a transistor, used to lower collector resistance.

Byte: A fixed-length binary bit pattern (word).

Cathode Sputtering: When a low-pressure glow discharge is established between two electrodes, the cathode disintegrates slowly under bombardment by the ionized gas molecules; this phenomenon is termed cathode sputtering. See *Sputtering*.

Ceramic: A class of products composed of inorganic compounds formed through heat processing. Often used as a base substrate for thin- and thick-film circuits or for assembling chips and conductor patterns, e.g., aluminum oxide.

Cermet: Cermets are mixtures of insulating materials (ceramics such as refractory oxides, glasses) and highly conducting materials (metals). *Examples*: silicon monoxide and chromium (thin films) and silver plus palladium plus glass (thick films).

Chemofacture: A coined word indicating fabrication by chemical means rather than by manual means. Printed wiring is chemofactured.

Chip:
(**A**) A simple or complex semiconductor die.
(**B**) A single substrate on which all the active and passive elements of an electronic circuit have been fabricated using the semiconductor technologies of diffusion, passivation, masking, photoresist, and epitaxial growth. Normally, a chip is not ready for use until it is packaged and provided with terminals for connection to the outside world. Also called a die.

Chip Approach, Multiple: An approach to integrated circuits in which the circuit is distributed in pieces on two or more chips or substrates. All the active elements are diffused on one or more semiconductor chips. The passive elements are similarly grouped on one or more separate chips. In some cases passive components may be diffused into separate semiconductor chips or may be deposited on some other base substrate, typically aluminum oxide. The whole group of separate chips is interconnected by bonded wires, and external connections are also made using bonded leads.

Circuit: The combination of a number of electrical elements or parts to accomplish a desired function. *Examples*: filter, oscillator, amplifier—can be discrete or integrated.

Circuit Element: A basic constituent of a circuit, exclusive of interconnections.

Clear: To restore a memory or storage device to a standard state, usually the zero state. Also called reset.

Clock: A pulse generator which controls the timing of switching circuits and memory states and equals the speed at which the major portion of the computer operates.

Clock Rate: The speed (frequency) at which the major portion of a computer operates.

Coefficient of Capacitance, Temperature: Defined as $\text{TCC} = \Delta C / C \, \Delta T$, ΔC being the capacitance change within the temperature range ΔT, and C a reference value (normally room temperature capacitance). TCC is usually given as a multiple of $10^{-6}/°\text{C}$ or parts per million (ppm) per degree Celsius.

Coefficient of Resistance, Temperature: Defined as $\text{TCR} = \Delta R / R \, \Delta T$, ΔR being the resistance change within the temperature range ΔT, and R a reference value (normally room temperature resistance). TCR is usually given as a multiple of $10^{-6}/°\text{C}$ or parts per million (ppm) per degree Celsius.

Compatible: A process of depositing thin-film passive components on the passivated surface of a monolithic structure containing diffused active electrical elements. *Synonym*: compatible-hybrid.

Component:
(**A**) A discrete, packaged conventional electrical element that performs one electrical function (traditional use).
(**B**) A packaged electrical-circuit inseparable assembly (EIA).

See *Device* for primarily active electrical elements.

See *Component Part* for primarily passive electrical elements.

Component Part:
(**A**) The traditional passive electrical discrete part.

(B) The physical realization of an electrical property(ies) in a physically independent body that cannot practicably be further reduced or divided without destroying its function.

Conductive Pattern: A design formed from an electrically conductive material on an insulating base.

Conductor: A single conductive line forming an electrical connection between terminal areas.

Connector: A part or a constituent of a part. Its function is to provide separability between two circuits.

Coordinatograph: A precision drafting instrument used to provide master artwork. Rectilinear and radial instruments having plotting precision of ±0.0012 inch are available.

Cordwood: The technique of producing assembly modules by bundling parts as closely as possible and interconnecting them into circuits by welding or soldering leads together. (Part of the microminiature technology using discrete devices and components.)

Cordwood Modules: Form of packaging in which discrete active and passive components are assembled between two parallel printed-circuit boards with the axes of all cylindrical parts parallel to each other but orthogonal to the interconnecting printed-circuit boards. Leads protrude from common end planes for interconnection.

Counter: A device capable of changing states in a specified sequence on receiving appropriate input signals; a circuit which provides an output pulse or other indication after receiving a specified number of input pulses.

Counter, Binary: A flip-flop having a single input. Each time a pulse appears at the input, the flip-flop changes state (called a *T* flip-flop).

Counter, Ring: A loop or circuit of interconnected flip-flops so arranged that only one is "on" at any given time and that, as input signals are received, the position of the "on" state moves in sequence from one flip-flop to another around the loop.

Crossover: The point at which two conductors, insulated from each other, cross paths.

Cryotron: Thin-film cryogenic (very-low-temperature) switching device consisting of two crossing strips made of superconductive material with different critical field curves (e.g., tin and lead) separated by an insulating layer.

Current-Mode-Logic (*CML*): Operates in the nonsaturated mode as distinguished from all the other forms which operate in the saturated mode.

Decimal: A system of numerical representation which uses ten symbols, 0, 1, 2, ···, 9.

Delay: Undesirable delay effects are caused by rise time and fall time that reduce circuit speed, but delay units may be used deliberately to prevent inputs from changing while clock pulses are present. The delay time is always less than the clock-pulse interval.

Density: The number of logic gates or individual components which can be integrated onto a single substrate.

Deposition: The process of applying a material to a base (substrate or other layer) by means of vacuum, electrical, chemical, screening, or vapor methods.

Device:
(A) A combination of inseparable physical material to form a part containing one or more *active* elements. *Examples*: transistor, diode, integrated circuit.

(B) A component part capable of affecting the behavior of an electronic circuit. *Examples* of component parts that are not devices are: connectors, terminals, fuses.

Diced Electrical Elements: Microminiature component parts and device parts usually formed on or in a substrate in checkerboard fashion and then separated by a slicing process. Used in film components, diffused devices of both discrete and integrated form, etc.

Die: A single component or device sliced from a larger substrate. See *Chip*.

Die Bonding: The method of attaching a die or chip to a package for mechanical and thermal connection.

Dielectric: A nonconductor of electric current in which a steady electric field can be set up with a negligible current; an insulator.

Diffusion: A thermal process by which impurities are deliberately introduced into a material, for example a semiconductor. This thermal process introduces tiny amounts of impurities into the base material. Diffusion is a difficult process in solids though quite easy in fluids. Just drop a bit of coloring matter in a glass of water and the color will gradually distribute itself throughout the water, whereas it takes time and high temperature to diffuse phosphorus into silicon.

Digital Circuit: A circuit which operates like a switch, that is, it is either on or off.

Digital Computer: A discretely variable computer that counts separate units.

Diode-Transistor Logic (DTL): Logic performed by diodes. The transistor acts as an amplifier and the output is inverted.

Direct-Coupled-Transistor Logic (DCTL): Logic performed by directly coupled transistor stages.

Discrete:
(A) Electronic circuits built of separate, finished components.
(B) Separate, finished devices and component parts ready for use in electronic assemblies.

Distributed-Type Element: An element whose electrical function is not identifiable with a particular physical area.

Dope: To add chemical impurities to a semiconductor by a process such as diffusion or alloying. An impurity so added is called a *dopant*.

Electrical Properties: The concept of basic electrical characterization. Basic electrical properties are resistance, capacitance, inductance, transistance, and rectification.

Electroless Deposition: Plating based on controlled reduction of metals on certain catalytic surfaces.

Element:
(A) Active element—device or incremental volume of a part which exhibits transistance or gain. *Examples*: transistor, diode.
(B) Passive element—component part or incremental volume of a part not exhibiting useful transistance. *Examples*: resistor, capacitor.

Emitter-Coupled Logic (ECL): Same as *Current-Mode Logic*.

Epitaxial Growth: A chemical reaction in which silicon or other element is deposited from a gaseous mixture and grows in crystalline form on the surface of a silicon wafer or other crystalline substrate.

Etching: Chemical erosion with acids or other corrosive agents that produces a design on a surface by eating away selected exposed portions of the surface.

Eutectic Bond: A thermometallurgical bond generally used to provide contact between semiconductor chips or film assembly chips and substrates, e.g., a bond obtained by heating gold and silicon in intimate contact at their eutectic temperature.

Evaporation, Vacuum: Process of vaporizing metals and compounds in vacuum by the use of heated filaments, electron beams, etc. Vacuum evaporation involves heating a material in vacuum to such a temperature that its vapor pressure reaches 0.001 to 0.01 torr. Under these conditions, if the vacuum-chamber pressure is 10^{-4} to 10^{-6} torr, the evaporant material rapidly vaporizes, is propagated rectilinearly from the source, and ultimately condenses on cooler exposed surfaces in the vacuum environment. See *Thin Film*.

Evaporator: High-vacuum apparatus in which materials may be evaporated.

Exclusive OR: Logic expression in which the output is true if either of two variables is true but *not* if both are true.

Fall Time: A measure of the time required for a circuit to change its output from a high level (1) to a low level (0).

Fan-In: The number of inputs available on a gate.

Fan-Out: The number of gates that a given gate can drive. The term is applicable only within a given logic family.

Field Effect Transistor (FET): Majority-carrier unipolar 3-terminal device in which field modulation of current is the active process. May use either *pn* or metal-oxide semiconductor.

Film: General term applied to electronic coatings whose thickness dimension is minimal. Thin-film coatings range from 1 Å to 1 μm = 10^{-4} cm = 10^4 Å. Thick films range from 10 μm to 100 μm.

Flip-Chip: Standard planar silicon transistor, diode, or integrated circuit having metallic pellets deposited in small holes in oxide over lead areas. This permits face-down bonding after the die is flipped over and located on a suitable substrate.

Flip-Flop: An electronic circuit having two stable states and the ability to change from one state to the other on application of a signal in a specified manner. Specific types follow.

Flip-Flop, "D": D stands for delay. A flip-flop whose output is a function of the input which appeared one pulse earlier, that is, if a 1 appears at its input, the output a pulse later will be a 1.

Flip-Flop, "J–K": A flip-flop having two inputs designated J and K. At the application of a clock pulse, a 1 on the J input will set the flip-flop to the 1 or "on" state; a 1 on the K input will reset it to the 0 or "off" state, and 1's simultaneously on both inputs will cause it to change state regardless of the state it had been in.

Flip-Flop, "R–S": A flip-flop having two inputs designated R and S. At the application of a clock pulse, a 1 on the S input will set the flip-flop to the 1 or "on" state; and a 1 on the R input will reset it to the 0 or "off" state. It is assumed that 1's will never appear simultaneously at both inputs.

Flip-Flop, "R–S–T": A flip-flop having three inputs, R, S, and T. The R and S inputs produce states as described for the R–S flip-flop above; the T input causes the flip-flop to change states.

Flip-Flop, "T": A flip-flop having only one input. A pulse appearing on the input will cause the flip-flop to change states.

Functional Block or Molecular Circuit: A three-dimensional circuit in which the molecular arrangement of the material performs the desired electrical functions. It is not possible to relate areas within the block to functions performed. (Strictly, this type of functional operation is not yet available, but the term is sometimes used loosely to mean monolithic silicon integrated devices.) A piezoelectric delay line in some respects measures up to this definition.

Gate: A circuit having two or more inputs and one output, the output depending on the combination of logic signals at the inputs. There are four gates: AND, OR, NAND, NOR. The definitions below assume positive logic is used.

Gate, AND: All inputs must have 1-state signals to produce a 1-state output.

Gate, NAND: All inputs must have 1-state signals to produce a 0-state output.

Gate, NOR: Any one input or more having a 1-state signal will yield a 0-state output.

Gate, OR: Any one input or more having a 1-state signal is sufficient to produce a 1-state output.

Glassivation: Deposition and melting of vitreous material on a semiconductor die to passivate and protect underlying device junctions.

Half-Adder: A device which will accept two signals representing the augend and addend and produce output signals representing the sum and carry.

Half-Shift Register: Another name for flip-flop.

High-Density Packaging: The concept of decreasing radically (through miniaturization and other techniques) the volume needed to package a given set of components.

Hole: An electron vacancy; an unfilled state in a valence band of electrons; a positive charge.

Hybrid: A microelectronic circuit manufactured using a combination of techniques and methods, e.g., mixing monolithic integrated circuits with thin-film or thick-film techniques.

Hybrid Integrated Circuits: The arrangement consisting of one or more integrated circuits in combination with one or more discrete devices, or alternatively, the combination of more than one type of integrated circuit into a single device.

Impurities: Material added to silicon or germanium to create a p-type section or n-type section. *Examples* of impurities: boron (p), phosphorus (n), arsenic (n).

Integrated Circuit (IC): The physical realization of two or more circuit elements inseparably associated on or within a substrate to form an electrical network.

Interaction: The effects from two or more elements, parts, assemblies, or equipments on each other where each is performing a function.

Interconnections:
(A) All of the methods of providing electrical paths between any combination of metals, semiconductors, or other materials which make up a circuit.
(B) Connections external to the component part, device, subassembly, assembly, and all of those units that may be classed as functional items.

Intraconnections: Connections inseparably associated with circuit elements within a component part.

Inverter: The output is always in the opposite logic state as the input. Also called a NOT circuit.

Ion: When an electrically neutral atom acquires one or more additional electrons or loses one or more of its electrons, the resulting species with its state of charge is called an ion. In the first case it becomes a negative ion and in the second case a positive ion.

Ion Implantation: A doping technique wherein selected regions of a semiconductor substrate retain energetic ions of virtually any desired impurity.

Isolation: In silicon integrated circuits the problem of insulating components from each other is severe because of proximity of location and low silicon resistivity. Isolation in monolithic circuits is achieved by thermal oxidation, etching, epitaxial growth of intrinsic layers, use of isolation diffusions, and/or reverse-biased pn junctions.

Jumper: A direct electrical connection that is not a portion of the conductive pattern between two points in a printed circuit.

Junction: A boundary between a region of n-type to p-type semiconductor.

Kodak Metal Etch Resist (KMER): Organic photoresist compositions used in industry for selective protection (maskings) against etchants.

Kodak Photo Resist (KPR): See *Photoresist.*

Land: A metallic contact area.

Layout: Geometric arrangement of conductors and components in printed-wiring or integrated-circuit design.

Linear Circuit: A circuit whose output is an amplified version or a predetermined variation of its input.

Logic: A mathematical approach to the solution of complex situations by the use of symbols to define basic concepts, also called *symbolic logic.* The three basic logic symbols are AND, OR, and NOT.

Logic Forms: The following are some ways of performing logic with bipolar integrated circuits.

DCTL
    Direct-Coupled-Transistor Logic
DTL
    Diode-Transistor Logic
ECL (CML)
    Emitter-Coupled Logic
RCTL
    Resistor-Capacitor-Transistor Logic
RTL
    Resistor-Transistor Logic
TTL (or T^2L)
    Transistor-Transistor Logic
CTL
    Complementary-Transistor Logic
I^2L
    Integrated Injection Logic

Magnetic Integrated Circuit: The physical realization of two or more magnetic-circuit elements inseparably associated to perform all or at least a major portion of its intended function.

Masking: The process of coating certain areas of a surface so that they will be protected from treatment (etching, depositing, anodizing, etc.) the uncoated areas are to receive. Photoresist or metal masking are common examples.

Mean Time Before Failure (MTBF): Mean time before failure for an equipment as averaged over a stated time interval with a stated definition of failure.

Mean Time to Repair (MTTR): Mean time to repair an equipment by some scheme of testing, analysis, and replacement or repair averaged over a stated time interval.

Memory: A storage device into which information can be inserted and held for use at a later time.

Metallizing: Process used to make an insulating surface conductive. Processes include chemical reduction, electroless plating, vacuum evaporation, and cathode sputtering.

Metal-Oxide Semiconductor (MOS):
    (A) Capacitor—Metal-Oxide Semiconductor; the metal and semiconductor each act as plates of the capacitor while the silicon dioxide acts as the dielectric.
    (B) Transistor—Metal-Oxide Semiconductor; in an MOS transistor, the conductivity of a bulk semiconductor is modulated by a field impressed across an insulator layer deposited on the semiconductor. A metal film (gate electrode) lies on the insulator surface. A field applied to the gate modulates conductivity on the surface of the semiconductor.
    (C) Integrated circuits made from combinations of MOS transistors.

Microcircuit: Another name for integrated circuit.

Microelectronics:
    (A) Another name for integrated circuits.
    (B) That entire body of electronic art connected with or applied to the realization of electronic systems from extremely small electronic parts (Electronic Industries Association).
    (C) The art of electronic equipment design and its construction that uses any of the microminiaturization schemes. This art of electronics deals with microminiature parts, subassemblies, and assemblies. Electronic parts are replaced by active and passive elements through the use of fabrication processes such as screening, vapor deposition, diffusion, and photoetching to integrate as many elements as possible. Electronic assemblies are composed of devices or components showing radical departures from normal industry practice in minuteness of size and in form factors.

Micrometer: One-millionth of a meter or 10 000 angstroms.

Migration: Phenomenon encountered with some of the heavier metals such as silver. Under polarization and in the presence of moisture, growth of dendrites or "whiskers" may occur which can bridge small gaps and cause short-circuits. Can be inhibited by gold or rhodium flash coat, or by plastic overcoating.

Mil: One-thousandth of an inch or 250 000 angstroms.

Miniaturization: The technique of packaging that seeks to reduce the size and weight of electronic parts and supplant vacuum-tube circuits with transistor compatible parts.

Module: A subassembly in a packaging scheme displaying regularity and separable repetition. May or may not be separable from other modules after initial assembly.

Monolithic (Greek for "One-Stone"): A single flat-surfaced chip of silicon on which patterns may be etched, scribed, diffused, etc., the result being a single chip of material into whose surface have been formed transistors, diodes, resistors, capacitors, etc.

MOS: Acronym for Metal-Oxide-Semiconductor; MOS transistors were formerly called IGFETs (Insulated-Gate Field-Effect Transistors).

Morphology, Integrated: The structural characterization of an electronic component in which the identity of the current- or signal-modifying areas, patterns, or volumes has become lost in the integration of electronic materials, in contrast to an assembly of devices performing the same function.

Morphology, Translational: The structural characterization of an electronic component in which the areas or patterns of resistive, conductive, dielectric, and active materials (in or on the surface of the structure) can be identified in a one-to-one correspondence with devices assembled to perform an equivalent function.

Multiple Chip: Another name for hybrid-type circuit manufacture. See *Chip Approach, Multiple*.

NAND *Gate* $(D=\overline{A \cdot B \cdot C}$ *for Positive Inputs*): The simultaneous presence of all inputs in the positive state generates an inverted output.

Negative Logic: The more-negative voltage (or current) level represents the 1-state; the less-negative level represents the 0-state.

Nichrome: A nickel-chromium alloy used in making thin-film resistive components.

NOR *Gate* $(D=\overline{A+B+C}$ *for Positive Inputs*): The presence of one or more positive inputs generates an inverted output.

NOT: A Boolean logic operator indicating negation. A variable designated NOT will be the opposite of its AND or OR function. A switching function for only one variable.

n-Type, p-Type: A semiconductor may be doped with two types of impurities. *n*-type impurity means that an excess of negative charges are present in the semiconductor. A *p*-type semiconductor has an excess of positive charges. See *Hole*.

Ohmic Contact: Low-resistance nonvoltage-dependent contacts to semiconductor device.

Ohms per Square: The resistance of any square area, measured between parallel sides of thin or thick films of resistive materials or semiconductors.

OR: A Boolean operator analogous to addition (except that two truths will only add up to one truth). Of two variables, only one need be true for the output to be true.

Oxidation: A chemical reaction in which a thin portion of the surface of a silicon wafer is converted to silicon dioxide.

Oxide Masking: Use of autogenous SiO_2 as a barrier to permit selective area processing of silicon. The oxide formed by oxidation of silicon is selectively etched for exposure of the underlying silicon to diffusion or metallization.

Packaged Integrated Circuit: An integrated-circuit assembly within a container to permit usage without further protection or processing.

Packaging: The process of physically locating, connecting, and protecting elements, circuits, components, modules, subsystems, and junction effects.

Parallel Operation: Pertaining to the manipulation of information within computer circuits in which the digits of a word are transmitted simultaneously on separate lines. It is faster than serial operation but requires more equipment.

Parasitics: The undesired capacitance between elements in a monolithic structure which limits frequency response; these are minimized by zone isolation techniques.

Part: One unit or two or more pieces joined together which are not normally subject to disassembly without destruction of designed use. *Examples*: composition resistor, transistor, screw, pulse transformer, gear, substrate, or board with printed resistors or conductors.

Passivation: In semiconductor technology, protection (usually by silicon dioxide) of the junctions and surfaces of components and integrated circuits from harmful environments.

Passive Elements: Electrical elements not capable of performing the transistance function of amplifying or control functions. Resistance and capacitance are passive elements. Also, those components in a circuit which have no gain characteristics: capacitors, resistors, inductors.

Passive Substrate: A substrate that does not exhibit transistance, such as glass or ceramic.

Pellet Packaging: Technique using pellet-shaped components inserted into holes in a punchboard-type printed-wiring board. Used in discrete-component assemblies.

Photoresist: Photosensitive organic polymeric material used to selectively protect substrate surfaces against subsequent plating or etching.

Planar Diffusion: Diffusion through a two-dimensional area as in the fabrication of transistors and silicon monolithic integrated circuits.

Planar Technology: Technique of fabricating semiconductor or integrated-circuit devices in which the lateral diffusion of base dopant in the semiconductor is restricted from the start to the extent desired by the transistor requirements, thereby eliminating the need for mesa etching. Planar or surface passivation is defined as the growth of a chemical film layer on the surface of a semiconductor material which provides electrical stability of the surface; it isolates the surface from electrical and chemical conditions of the environment that degrade device characteristics and thereby adversely affect operating performance.

Positive Logic: The more positive voltage (or current) level represents the 1-state; the less positive level represents the 0-state.

Potted: Encapsulated by a material in the liquid state, e.g., plastic, which is then allowed to harden.

Printed Circuit: (In microelectronics) a pattern comprising printed wiring and printed elements, all formed in a predetermined design in (or attached to) the surface of a substrate.

Printed Contact: That portion of a printed circuit used to connect the circuit to a plug-in receptacle and to perform the function of a pin in a male plug.

Printed Element: An element in printed form, such as a printed inductor, resistor, capacitor, or transmission line.

Printed Wiring: (In microelectronics) a portion of a printed circuit consisting of a conductor pattern to provide point-to-point electrical connection only. *Examples*: screened conductive elements on a ceramic substrate; etched conductive lines on a printed-wiring film.

Printed-Wiring Board: A completely processed conductor pattern, usually formed on a stiff flat base. It serves as a means of electrical interconnection and physical attachment for discrete components.

Printing: Reproduction of a pattern on a surface by any of various processes, such as embossing, silk screening, transfer press, etc., where the tools are in contact with the substrate.

Probing: The method of making electrical connection to an integrated-circuit die to determine its electrical properties before it is packaged.

Production Model: A model in its final mechanical and electrical form of production design made by production tools, jigs, fixtures, and methods.

Propagation Delay: A measure of the time required for a change in logic level to spread through a chain of circuit elements.

Prototype: A model suitable for use in complete evaluation of form, design, and performance.

p-Type: See *n-Type*.

Pulse: A change of voltage or current of some finite duration and magnitude. The duration is called the pulse width or pulse length; the magnitude of the change is called the pulse amplitude or pulse height.

Purple Plague: Gold–aluminum intermetallic compound $AuAl_2$, formed at the bond point between gold connecting wire and aluminum contact lands on silicon integrated circuits. In its advanced stages of formation (accelerated by heat) the compound is brittle, resulting in a mechanically weak bond susceptible to catastrophic failure. $AuAl_2$ is purple in color, hence the descriptive term *purple plague*.

Register: A device used to store a certain number of digits in the computer circuits, often one word. Certain registers may also include provisions for shifting, circulating, or other operations.

Reliability: The ability of a device (circuit, component, element, etc.) to perform within the desired range over a measured period of time.

Resistor-Capacitor-Transistor Logic (RCTL): Same as *RTL* except that capacitors are used to enhance switching speed.

Resistor-Transistor Logic (RTL): Logic is performed by resistors. The transistor produces an inverted output from any positive input.

Rise Time: A measure of the time required for a circuit to change its output from a low level (0) to a high level (1).

Saturated Logic: A logic format wherein gate transistors are saturated or unsaturated to represent the two possible logic states.

Screening: Selective transfer of material through a metallic or plastic (silk) screen. Used in application of resist to printed-circuit boards and in deposition of thick-film circuits. The screen is generally masked by transfer of a photographically produced plastic pattern from its temporary film base support to the screen.

Scribing: Cutting or scratching an integrated-circuit wafer with a precision diamond scribe tool to permit dicing.

Semiconductor: Materials may be classified as one of 3 types by resistivity: Low-resistivity materials are called conductors; very-high-resistivity materials are called dielectrics; materials with intermediate resistivities are called semiconductors. *Examples*: germanium, silicon.

Semiconductor Integrated Circuit: The physical realization of two or more circuit elements inseparably associated on or in a semiconductor substrate to form an electrical network.

Serial Operation: The handling of information within computer circuits in which the digits of a word are transmitted one at a time along a single line. Though slower than parallel operation, its circuits are much less complex.

Shift Register: An element in the digital family which uses flip-flops to perform a displacement or movement of a set of digits one or more places to the right or left. If the digits are those of a numerical expression, a shift may be the equivalent of multiplying the number by a power of the base. See *Half-Shift Register*.

Silicon Dioxide (SiO₂): The result of oxidation on silicon; used in integrated circuits to protect or passivate the surface.

Silicon Monoxide (SiO): Capacitor dielectric, evaporated, usually used in the form of a thin film.

Skew: Time delay or offset between any two signals.

Slew Rate: Rate at which output can be driven from limit to limit over the dynamic range.

Specification: A detailed description of the characteristics of a product and of the criteria which must be used to determine whether the product is in conformity with the description.

Sputtering: Deposition of material in film form using a high-voltage glow discharge. See *Cathode Sputtering*.

Stability: Consistency of parameter values over the life of an item.

Subassembly: Two or more parts which form a portion of an assembly or of a unit replaceable as a whole but having a part or parts which are individually replaceable. *Example*: terminal board with mounted parts.

Substrate: The single body of material on or in which circuit elements are fabricated. Also, the base material used to make integrated circuits; may be either passive (thin-film, thick-film, hybrid) or active (monolithic, compatible).

Subsystem: An interconnected combination of a set of related circuits or integrated circuits to form a logical subdivision of an equipment or operational system.

Swiss Cheese: A technique of circuit fabrication using a thin board containing a lattice of holes into which pellet-type circuit elements are inserted and attached.

Synchronous Timing: Operation of a switching network by a clock pulse generator. Slower and more critical than asynchronous timing but requires fewer and simpler circuits.

Thermal Oxidation: Oxidation of silicon (or similar chemical element) by exposure to a combination of heat and oxidizing atmosphere.

Thermocompression Bonding: Direct bonding of two materials under heat and pressure without the presence of a third phase (such as solder) and without melting. Thermocompression bonding has been widely used for lead attachment in the fabrication of monolithic silicon integrated circuits.

Thick Film: A method of manufacturing integrated circuits by screen deposition of thick films; usually only passive elements are made this way.

Thin Film: A method of manufacturing integrated circuits by depositing thin layers of materials to perform electrical functions; usually evaporation or sputtering is used. Normally only passive elements are made this way.

Thin-Film Transistors (TFT): The TFT, sometimes called the insulated-gate field-effect transistor, is a classical field-effect device in which the conductivity of a semiconductor is modulated by impressing a field across an insulator overlying the conducting film channel.

Transistance: The electrical property which affects voltages or currents so as to accomplish gain or switching action. Examples of transistance occur in transistors, diodes, voltage-controlled rectifiers, and electron tubes.

Transistor-Transistor Logic (TTL): A modification of DTL which replaces the diode cluster with a multiple-emitter transistor.

Ultrasonic Bonding: Direct bonding of two materials through the use of pressure and lead agitation achieved by connection to an ultrasonic transducer. Widely used in making aluminum-to-aluminum connections in integrated circuits.

Ultrasonic Cleaning: Use of ultrasonic transducers to effect cavitations in conventional detergent media. Ultrasonically induced cavitation cleans very effectively on a microscopic basis.

Unipolar (Field-Effect) Transistor: Majority-carrier field-modulated current device. In one type (metal–oxide semiconductor) a field is impressed across an insulator. In another type (*pn junction*) a reverse-biased *pn* junction is used to supply the field.

Vacuum Evaporation: See *Evaporation, Vacuum*.

Vapor Plating: Deposition of films by thermal decomposition of compound vapor.

Weld: (*In microelectronics*) a method used to connect circuit leads to a device or part where the electrical bond is formed primarily by the application of a pulse of electrical energy and applied pressure.

Welded Circuit: A circuit made up of electronic parts which have their leads connected by welding techniques.

Wire Bonding: The method used to connect a silicon die to the leads of the device package; common methods are ball-, nailhead-, stitch-, and wedge-bonding. Various combinations of heat, pressure, and ultrasonic energy are used in the bonding process.

Word: An assemblage of bits considered as an entity in a computer.

MICROELECTRONICS

Microelectronics is the field of miniature electronics. Historically, the novelty of extremely small components and circuits stimulated the early development of microelectronics, but the field has now matured to the point where such practical benefits as low production costs, increased reliability, improved performance, and a host of new applications have become the predominant growth stimulants.

Originally, microelectronics utilized miniaturized discrete components and careful, often precise, layout and construction techniques to achieve miniaturized circuits. The cost of this space savings was often exorbitant and included high assembly expense, increased assembly time and difficulty of repair. The microelectronics art was advanced significantly by the advent of hybrid circuits, wherein such passive components as resistors, capacitors, and conducting paths are screened onto a common insulating substrate upon which are then installed microminiature active components such as transistors. In particular, both cost and assembly time were reduced while the miniaturization potential was greatly increased.

Discrete and hybrid microminiature construction techniques are still viable ways of achieving miniaturization, but the ultimate component density of each is dwarfed by the monolithic integrated circuit (IC). The IC exploits the fact that such components as resistors, diodes, transistors, and others can be fabricated from a semiconductor such as silicon. It was merely an extension of the solid-state art to fabricate entire circuits in a tiny silicon chip rather than to assemble discrete components made from separate bits of silicon into the same circuit.

The individual components of an IC are formed during precisely controlled chemical processing and temperature cycling steps. The various components, now buried in a silicon substrate, are then interconnected by chemically deposited conducting paths. The final circuit, which is usually a silicon chip less than 100 mils square, is then installed in a protective package, connected to the access pins by miniature wire leads, and covered with a protective seal.

Monolithic integrated circuits such as these can be combined with other ICs and discrete components on a common substrate to produce hybrid ICs. This construction method is often used when a circuit requires components such as inductors, transducers, and power devices which are difficult or even impossible to fabricate as part of a monolithic semiconductor substrate.

In recent years, microelectronics has been adapted to many traditional circuit applications as well as a host of new ones. Figure 1 shows how hybrid and monolithic technologies are employed to implement these applications, many of which are summarized along with their manufacturing processes in the sections that follow.

FABRICATION METHODS

Hybrid Integrated Circuits

There are two basic types of hybrid integrated circuits, multiple-chip and film hybrid; the latter type may be further divided into thin- and thick-film varieties (see Fig. 1).

Multiple-Chip Integrated Circuits: Multiple-chip integrated circuits consist of assemblies of active and passive silicon-based components or circuits, usually made by conventional semiconductor proc-

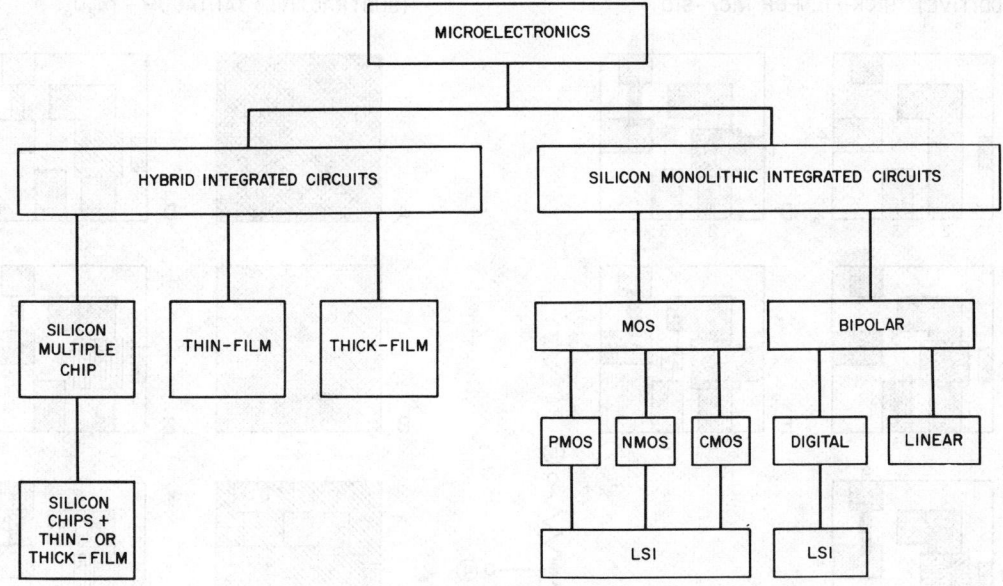

Fig. 1—Subdivisions of microelectronics.

esses. The components or circuits are individually mounted on an insulating substrate using metallized conductor patterns or wire bonding techniques.

Film Hybrid Integrated Circuits: Film hybrid integrated circuits are fabricated on inert substrates (glass, ceramic, etc.) by depositing films of conductive, resistive, and dielectric materials in the appropriate sequence. The films may be formed by vacuum evaporation, cathode sputtering, or electron-beam evaporation. Films deposited by these techniques have a thickness of less than 10 000 Å and are referred to as thin films (*tf*). A second class of film hybrid circuits is made by silk-screening conductive, resistive, and dielectric materials on ceramic substrates, followed by furnace firing and glazing. Films deposited by this technique are considerably thicker than 10 000 Å, more like 10 μm or greater, and are referred to as thick films (*TF*).

There are two classes of film deposition processes, additive and subtractive. In the additive process, electronic materials are deposited through selective mask, screen, or stencil only on areas of a substrate where they are meant to function. Different materials are deposited sequentially.

In the subtractive process, different electronic materials are sequentially deposited in layers which completely cover the substrate. Detailed geometric structure within each layer is subsequently determined by photomasking and etching each layer.

Additive processes tend to be less expensive than subtractive ones since fewer steps are usually involved. This is illustrated in Fig. 2.

Film hybrid circuits employ a range of standard substrate materials for thick- and thin-film techniques. Standard design equations govern the dimensional layout of resistors, capacitors, and inductors, regardless of materials and techniques chosen. All film hybrid circuits use microminiature active components. These components are attached at a terminal stage of fabrication, using welding or soldering techniques.

Note 1: Capacitors exceeding 10 000 picofarads and inductors exceeding 10 microhenries are obtained by the addition of discrete microminiature components. Thin-film components having these values offer poor area efficiency and compromise performance because of capacitor pinhole problems and inductor low Q factors. Thick-film capacitors show no pinhole problems but are subject to area efficiency considerations.

Note 2: Typical packaging configurations for transistors employed with film hybrid circuits are provided in the section on integrated-circuit packages.

Properties of Substrate Materials

The properties of glass and ceramic substrate materials commonly used in the fabrication of film hybrid integrated circuits are listed in Tables 1 and 2.

FILM MATERIALS AND PROCESSES

Table 3 summarizes the film materials and processes most widely used in industry.

(ADDITIVE) THICK-FILM OR NiCr–SiO   (SUBTRACTIVE) TANTALUM – Ta$_2$O$_5$

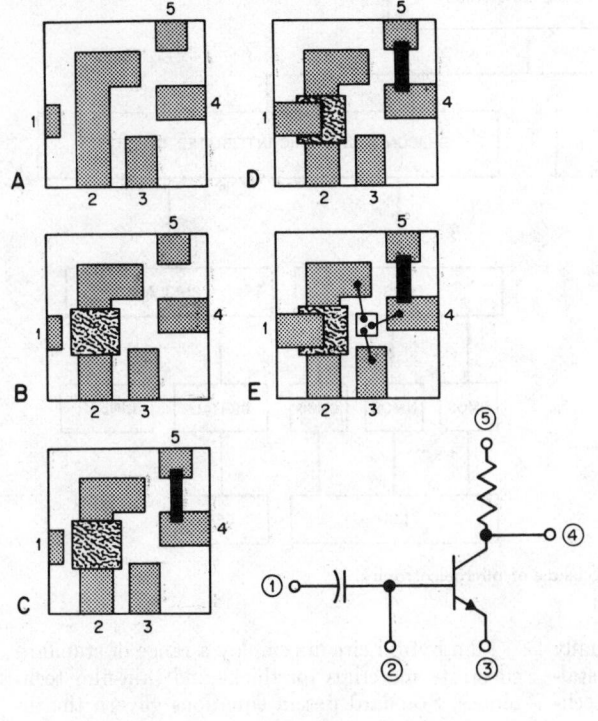

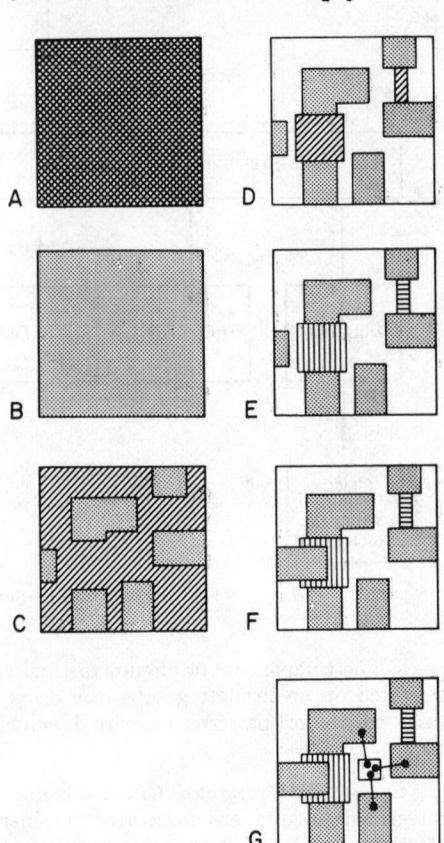

Additive

Thick-film or NiCr–SiO

A. Deposition of lands and interconnections on alumina
B. Deposition of capacitor dielectric
C. Deposition of resistive material
D. Counterelectrode capacitors
E. Attach add-on components, leads, and package

Subtractive

Tantalum–Ta$_2$O$_5$

A. Deposition of tantalum on glass or glazed aluminum
B. Deposition of aluminum or chromium–gold
C. Apply photoresist and etch aluminum or chromium–gold
D. Apply photoresist and etch tantalum
E. Anodize resistors and capacitors
F. Counterelectrode capacitors
G. Attach add-on components, leads, and package

Fig. 2—Process sequence for film hybrid circuits.

Film-Resistor Design

The resistance of a film resistor of the type shown in Fig. 3 is given by

$$R = \rho l / wt \text{ (ohm)} \qquad (1)$$

where R = resistance in ohms, ρ = resistivity in ohm-cm, l = length in cm, w = width in cm, and t = thickness in cm.

For a film of constant thickness, this expression reduces to

$$R = \rho'(l/w). \qquad (2)$$

For a unit square ($l = w$) of resistor, $R = \rho'$, where ρ' is the resistance per square (ohms/square) of the film in question. Such normalization reduces resistor design to a choice of appropriate l/w ratio.

The electrical characteristics of various film-resistor systems are listed in Table 4. Values for

TABLE 1—PROPERTIES OF COMMONLY USED GLASS SUBSTRATE MATERIALS.

Glass Type	Soda-Lime	Alkali Zinc Borosilicate (Microsheet)	Lime Alumino-Silicate (Alkali-Free)	Lime Alumino-Silicate (Alkali-Free)	Barium Alumino-Silicate (Alkali-Free)	Alkali Borosilicate	96% Silica	Fused Silica	Fotoceram*	Synthetic Sapphire†
Code number‡	0080	0211	1715	1723	7059	7740	7900	7940	—	—
Annealing point (°C)	512	542	866	710	650	565	910	1050	—	—
Softening point (°C)	696	720	1060	910	872	820	1500	1580	700	2040
Thermal exp. coef. (10^{-6}/°C)	9.2	7.2	3.5	4.6	4.5	3.25	0.8	0.56	10.4	5.8
Thermal conductivity (cal/cm/sec/°C at 25°C)	0.0023	—	—	0.0032	—	0.0027	0.0038	0.0034	0.005	0.098
Density (gm/cm³)	2.47	2.57	2.48	2.63	2.76	2.23	2.18	2.20	2.46	3.98
Dielectric constant (1 MHz) at 25°C	6.9	6.6	5.9	6.4	5.8	4.6	3.9	3.9	5.6	9.4 to 11.5
Loss tangent (1 MHz) at 25°C	0.01	0.0047	0.0024	0.0013	0.0011	0.0062	0.0006	0.00002	0.006	—
Log volume resistivity (ohms/cm³) at 25°C	6.4	8.3	13.6	14.1	13.5	8.1	9.7	11.8	14.0	12.0
Dielectric strength at 25°C (kilovolts rms)	0.35	2.0	>10	>10	>10	2.0	7.0	>10	—	—
Weatherability (g/cm²)	>5	0.05 to 0.25	<0.01	<0.01	<0.01	0.05 to 0.25	<0.01	<0.01	—	unaffected
Chemical durability										
5% HCl, 24 hr. (mg/cm²)	0.02	0.03	0.1	0.4	5.5	0.005	0.001	0.001	—	unaffected
5% NaOH, 6 hr. (mg/cm²)	0.5	2.0	1.2	0.3	3.7	1.1	1.1	0.7	—	unaffected
0.02N Na₂CO₃, 6 hr. (mg/cm²)	0.1	0.1	0.15	0.1	0.3	0.1	0.03	0.03	—	unaffected

*Trade name, Corning Glass Works, Corning, New York.
†Courtesy of Adolf Meller Company, Providence, R.I.
‡Code numbers of Corning Glass Works, Corning, New York.

TABLE 2—PROPERTIES OF COMMONLY USED CERAMIC SUBSTRATE MATERIALS.

Ceramic Type	Alumina (96% Al₂O₃)	Alumina (99.7% Al₂O₃)*	Dense Alumina (85% Al₂O₃)	Dense Alumina (94% Al₂O₃ +CaO +SiO₂)	Dense Alumina (96% Al₂O₃ +MgO +SiO₂)	Beryllia (BeO)	Dense Beryllia (98% BeO)	Dense Beryllia (99.5% BeO)	Titania (TiO₂)†	Barium Titanate	Magnesium Titanate*
Code number‡	—	—	576	719	614	—	735	754	192	—	—
Softening temp. (°C)	1650	>1650	1100	1500	1550	—	1600	1600	1600	1550	—
Melting point (°C)	2050	>2050	—	—	—	2550	—	—	1920	1700	—
Thermal exp. coef. (10^{-6}/°C)	6.4	5.0	6.5	6.2	6.4	7.5	6.1	6.0	8.3	8.1	7.5
Thermal conductivity cal/cm/sec/°C at 25°C	0.08	0.045	0.060	0.073	0.084	0.20	0.50	0.55	0.012	0.003	0.1
Density (gm/cm³)	3.7	3.9	3.4	3.58	3.7	3.08	2.9	2.88	4.0	5.5	3.6
Dielectric constant (1 MHz) at 25°C	10.0	9.5 (10 GHz)	8.3	8.9	9.3	7.1	6.3	6.4	85	10 to 10⁴	16 (10 GHz)
Loss tangent (1 MHz) at 25°C	0.0002	0.0001	0.0058	0.0018	0.0028	0.0006	0.0006	0.0006	0.0002	>0.02	0.0002
Log volume resistivity (ohms/cm³) at 30°C	14.0	>14	10.7	12.8	10.0	>14	13.8	>14	14.0	12.0	>14
Dielectric strength at 25°C and 60 Hz (volts/mil)	—	—	230	230	230	—	255	260	100	—	—
Surface smoothness (microinches)§	10 to 30	10 to 30	10 to 30	10 to 30	10 to 30	10 to 30	10 to 30	10 to 30	0.1 to 30	10 to 30	10 to 30
Chemical resistance	excellent	excellent	excellent	excellent	excellent	excellent	excellent	excellent	good	good	good

*Substrates used as dielectrics for microminiature microwave strip lines.
†Also known as Rutile.
‡Code numbers of ALSIMAG brand ceramics, American Lava Corp., Chattanooga, Tennessee.
§1 microinch=250 Å.

TABLE 3—FILM MATERIALS AND PROCESSES.

Application	Material	Process
Thin-film resistors	Nichrome (Ni–Cr)	Vacuum evaporation
	Tin oxide (SnO₂)	Vapor plating
	Tantalum–tantalum nitride (Ta–TaN)	Sputtering
	Cermet chromium-silicon monoxide (Cr–SiO)	Vacuum evaporation
Thick-film resistors	Palladium–silver Palladium oxide–silver and glass	Silk screen
Thin-film dielectrics	Silicon monoxide (SiO)	Vacuum evaporation
	Silicon dioxide (SiO₂)	Vapor plating, high temperature, steam oxidation, reactive sputtering
	Aluminum oxide (Al₂O₃)	Vapor plating, anodization of aluminum films
	Tantalum oxide (Ta₂O₅)	Anodization of tantalum, reactive sputtering
Thick-film dielectrics	BaTiO₃ or TiO₂ and glass mixtures	Silk screen
Thin-film conductors	Chromium–gold	Vacuum evaporation
	Chromium–copper	Vacuum evaporation
	Aluminum	Vacuum evaporation
	Nickel on ceramic	Electroless plating
Thick-film conductors	Gold–platinum–glass Gold–glass Silver–glass	Silk screen

integrated and molded carbon resistors are included for comparison.

Film-Capacitor Design

Film capacitors are generally made in flat-plate form, as shown in Fig. 4.

The capacitance of such a structure is given by

$$C = (0.225KA/t)(N-1)$$

$$= (0.0885KA'/t')(N-1)$$

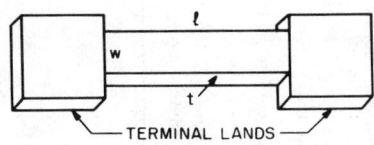

Fig. 3—Film resistor.

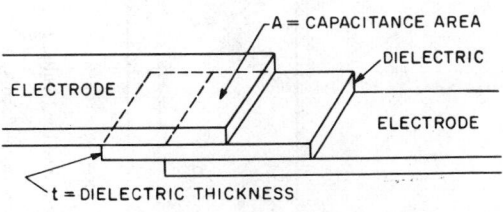

Fig. 4—Film capacitor.

TABLE 4—FILM, INTEGRATED, AND MOLDED CARBON RESISTOR CHARACTERISTICS.

Parameters	Nichrome (Ni–Cr)	Tin Oxide (SnO₂)	Tantalum–Tantalum Nitride (Ta–TaN)	Silver–Palladium Oxide (Ag–PdO) Cermet (Thick-Film)	Chromium–Silicon Monoxide (Cr–SiO) Cermet (Thin-Film)	Silicon (Si), p Type	Silicon (Si), n Type	Conventional Molded Carbon
Range (ohms/square)	40 to 400	50 to 200	20 to 200	10 to 10 000	100 to 1000	50 to 250	2.5	—
Maximum practical value (ohms)	1 Meg	1 Meg	1 Meg	1 Meg	1 Meg	25 000	250	22 Meg
Minimum value (ohms)	20	25	10	10	50	25	5	0.24
Tolerance as deposited (%)*	±5 to ±10	±15	±5.0	±10 to ±15	±10 to ±15	±10	±15	±5
Adjusted tolerance (%)	±1.0	±1.0	±0.05	±0.1	±1.0	—	—	—
Temperature coefficient (ppm/°C)	±100	—	−50 to −100	±10 to ±200	±500	+50 to +2500	+100	±1500
Maximum power dissipation (μW/sq/mil)†	2.0	2.0	2.0	5.0	2.0	3.0	3.0	2W total
Maximum voltage (volts)	±100	±100	±100	>100	±100	±20	±5	±750

*Untrimmed or unadjusted.
†Approximate figure; power dissipation depends greatly on substrate, age, and geometry.

TABLE 5—FILM- AND INTEGRATED-CAPACITOR CHARACTERISTICS.

Parameters	Silicon Dioxide (SiO₂)*	Aluminum Oxide (Al₂O₃)	Tantalum Oxide (Ta₂O₅)	Silicon Monoxide (SiO)	Thick Film	pn Junction
Design characteristic (pF/sq mil)†	0.25	0.3	2.5	0.01	0.32	1.2
Voltage coefficient	$v^{-1/2}$	0	0‡	0‡	0	0
Temperature coefficient (ppm/°C)	+100	+125	+400	±200	±300	—
Dissipation factor at 1 megahertz (%)	0.7	0.4	0.3	0.7	0.1	—
Dielectric constant	3.8	9.6	25	6	150 to 200	13.7
Maximum voltage§	<100	<100	<100	<100	≫100	<25
Tolerance (%)	±10 to ±20	±10 to ±20	±10 to ±20	±15 to ±20	±10 to ±20	±20

*Sputtered quantitatively.
†At practical thickness.
‡Anodized films show some polarization and voltage nonlinearity if not properly anodized.
§Function of thickness.

where C = capacitance (picofarads), K = dielectric constant, N = number of capacitor plates, A = area of electrode overlap (square inches), A' = area of electrode overlap (square centimeters), t = dielectric thickness (inches), and t' = dielectric thickness (centimeters).

The electrical characteristics of film capacitors are listed in Table 5. Values for integrated capacitors are included for comparison.

Film-Inductor Design*

Film inductors are limited to low inductance values (a few microhenries or less), relatively low Q, and exhibit coupling problems. For most hybrid applications, microminiature discrete inductors are used.

Two-dimensional spiral inductors can be made by patterning thin or thick conductive metallic films (chrome–copper, aluminum, or palladium–gold) followed by electroplating to increase cross-sectional area and reduce series resistance.

*From H. G. Dill, "Designing Inductors for Thin-Film Application," *Electronic Design*, Vol. 12, No. 4, pp. 52–59; 1964.

A typical film-inductor pattern is shown in Fig. 5. The following equations may be used in calculating the inductance for such a flat square spiral coil.

Equations for the Flat Square Spiral Coil of Fig. 5:

$$S = D^2$$

$$N = (D/2)(q+p)^{-1}$$

$$= (D/2_p)(1+r^{-1})^{-1}$$

$$l = 4(D/2)N$$

$$= (D^2/p)(1+r^{-1})^{-1}$$

$$L = 85 \times 10^{-10}\, DN^{5/3}$$

$$= 27 \times 10^{-10}\, (D^{8/3}/p^{5/3})(1+r^{-1})^{-5/3}$$

where D = dimension in cm as shown in Fig. 5, S = surface area of coil ($D_1 = 0$), N = number of turns, L = inductance in henries, l = length of spiral in cm, p = spiral width in cm, and q = distance between the outer edges of two adjacent spirals.

$$p/q = r.$$

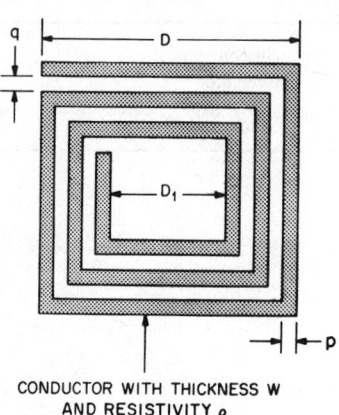

Fig. 5—Film inductor. This is a flat square spiral coil, with the spiral extending to the center ($D_1=0$; $q=p$).

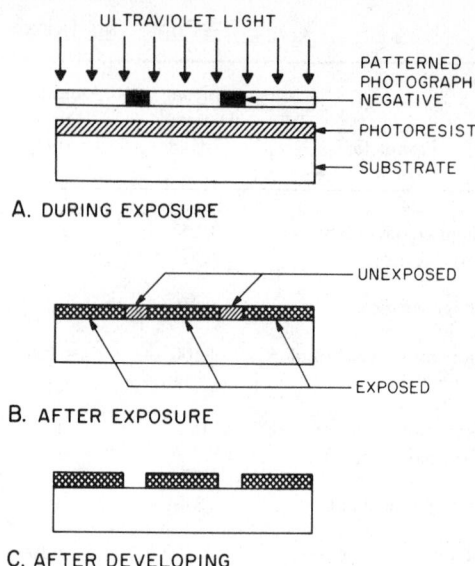

Fig. 6—Steps in the use of a photoresist.

PHOTORESIST APPLICATION

Photoresists are organic photosensitive polymeric liquids that have been used in the fabrication of printed circuits for several years. They are also widely used in the fabrication of both monolithic and film hybrid integrated circuits.

The steps involved in the use of a photoresist are shown in Fig. 6. The silicon or film substrate is thoroughly cleaned and then covered with a layer of photoresist by spincoating, spraying, or immersion. After the photoresist dries, the coated substrate is exposed to ultraviolet light through a patterned photographic negative (Fig. 6A). As a result, certain portions of the photoresist are exposed (Fig. 6B). The photoresist is then developed in a solvent such as trichloroethylene, washing away the unexposed photoresist (Fig. 6C). After developing, the substrate is left with its surface covered with a film in specific areas and is ready for the evaporation printing or deposition process. Chemical or electro-etching is used if it is desired to remove material from the unprotected areas of the underlying substrate. After these processes are completed, the photoresist is removed with a suitable solvent stripper.

The equipment necessary and procedures used in the photoresist process are similar to those used in contact printing. A large accurately drawn pattern (artwork) is prepared by manual drafting or coordinatograph and then photographically reduced. A vacuum copyholder is required to ensure intimate contact between the photographic negative and the substrate. Special mechanical jigs and fixtures are required to guarantee accurate registration.

The Eastman Kodak Company gives excellent information on the techniques of microphotography and photoresist applications in the following publications: "Kodak Photosensitive Resists for Industry," "Kodak Plates and Films for Science and Industry," and "Techniques of Microphotography." Tables* 6A and 6B list the etchants for materials commonly used in monolithic and film hybrid integrated circuits.

Monolithic Integrated Circuits

In this type of integrated circuit, both active and passive circuit components and then interconnections are fabricated in and on the surface of a single crystal of semiconductor material. Silicon is normally employed because of its inherent passivation properties, the ease with which silicon dioxide can be grown and selectively etched, and silicon's greater thermal range of use relative to germanium. There are two types of silicon monolithic integrated circuits, bipolar and *MOS* (metal-oxide-semiconductor).

Bipolar Integrated Circuits: These use transistor structures identical to conventional minority-carrier (bipolar) discrete transistors. This distinguishes them from the *MOS* type, which are majority-carrier or unipolar devices. The steps employed in the fabrication of bipolar silicon monolithic integrated circuits are summarized in Figs. 7 through 14.

*Adapted from T. D. Schlabach and D. K. Rider, "Printed and Integrated Circuitry, Materials and Processes," McGraw-Hill Book Company; 1963.

TABLE 6A—MATERIALS TO BE ETCHED USING ETCHANTS IN TABLE 6B.

Aluminum: (1) or (2). Tantalum, titanium, gold, and platinum are unaffected by these etchants.
Chromium: (3), (4), (5), or (6). 6 normal sulfuric acid will also work but is less desirable.
Constantan: (1).
Copper: (7). Tantalum and titanium are unaffected by this etchant. Best to bake photoresist.
Gold: (8), (9), or (10).
Gold–palladium: (11). For fired-on films. Will attack glass and organic resists in time.
Manganese: (3) or (4).
Molybdenum: (12). Can also be etched anodically in chromic acid (100 grains/liter).
Nickel: (13).
Nichrome: (1) or (14). Etchant (14) is also suitable for nickel and nickel-base magnetic alloys.
Palladium: (8) or (9).
Platinum: (8) or (9). Also for platinum alloys. Etchant (9) is for fired-on films.
Silicon: (15) or (16).
Silicon dioxide: (17), (18), or (19). Etchant (17) removes 1000 angstrom units of oxide per minute. (18) removes 3000 angstrom units of oxide per minute. (19) provides controlled slow etching.
Silver: (20). Dilute HNO_3 may also be used but is less desirable.
Steel: (1) or (21). Etchant (1) is also for stainless steel. (21) is for shim steel.
Tantalum: (22). Gold and nichrome are unaffected by this etchant, but on glass these may slough off if not protected.
Tellurium: (23) or (24).
Tin oxide: (5), then use (1) or (25). Baked resist needed. Nascent hydrogen reduces tin oxide to tin, which is soluble in chloride etchants.
Titanium: (26). Gold, nichrome, and tantalum are unaffected.
Tungsten: (22).

TABLE 6B—ETCHANTS FOR MATERIALS COMMONLY USED IN MONOLITHIC AND FILM HYBRID INTEGRATED CIRCUITS.*

(1) 2.25–3.75 molar ferric chloride
(2) 2–3 normal sodium hydroxide
(3) 6 normal hydrochloric acid (HCl)
(4) 2 parts etchant (1) and 1 part concentrated HCl
(5) Powdered zinc in dilute HCl
(6) 1 part 50% $K_3Fe(CN)_6$ and 1 part 10% KOH
(7) Saturated ferric chloride
(8) Aqua regia
(9) 1 part concentrated HCl, 1 part concentrated HNO_3, and 2 parts H_2O
(10) 35 grains potassium iodide, 10 grains iodine, and 100 milliliters H_2O
(11) 2 parts concentrated HF, 1 part concentrated HNO_3, and 3 parts H_2O
(12) 1 part concentrated H_2SO_4, 1 part concentrated HNO_3, and 3 parts H_2O
(13) Concentrated HCl (70°C)
(14) 4 parts concentrated HCl and 1 part H_2O
(15) 4 parts etchant (1), 4 parts concentrated HNO_3, and 1 part concentrated HF
(16) (CP-4) 5 parts concentrated HNO_3, 3 parts concentrated HF, 3 parts glacial acetic acid, and 10 drops (per 50 milliliters of acid mixture) liquid bromine
(17) Saturated aqueous solution of ammonium fluoride
(18) Saturated aqueous solution of ammonium bifluoride
(19) Etchall cream
(20) 2 molar ferric nitrate
(21) 3 parts concentrated HNO_3 and 7 parts H_2O
(22) 1 part concentrated HF, 1 part concentrated HNO_3, and 2 parts H_2O
(23) 6 normal HNO_3
(24) 1 molar ammonium persulfate
(25) 2 molar cupric chloride in 4 normal MHL
(26) 1 part concentrated HF and 20 parts H_2O

*Etchants are used at room temperature except for (13), which is used at 70 degrees Celsius. Parts are by volume.

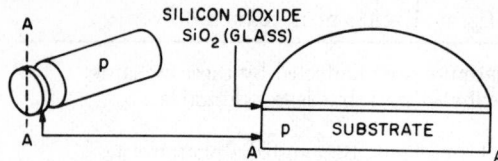

Fig. 7—Initial wafer.

(Fig. 7) A thin circular wafer of p-type silicon is cut from a rod. This material has high electrical resistivity and serves simply as a substrate or support on the surface of which other layers will be formed. The silicon wafer is first oxidized and a surface layer of silicon dioxide (SiO_2) is grown.

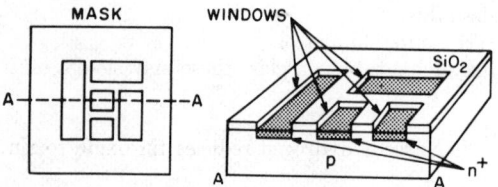

Fig. 8—Masking and etching of glass for diffusion of n^+ buried layer.

(Fig. 8) The entire glass surface is coated with a photolithographic material that resists acid when exposed to light. The surface is exposed to light through a photograph negative (mask) and the unexposed areas (windows) are etched away to permit a diffusion that produces n^+ material. These diffused areas, called buried layers, materially reduce the collector saturation resistance of the integrated-circuit transistors.

As the circular wafer will make many rectangular chips, on each of which individual integrated circuits will be formed, only a single chip is shown in this view. In the following steps only an edge-on view is shown.

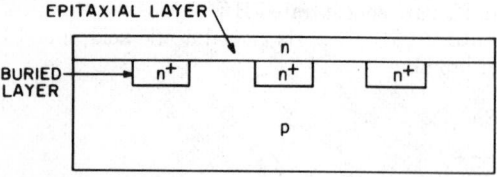

Fig. 9—Epitaxial deposition of n material.

(Fig. 9) During the n^+ diffusion, new SiO_2 grows over the wafer. This layer is entirely etched away, whereupon an epitaxial n layer is deposited over the entire surface of the wafer, resulting in islands of n^+ material as a buried layer.

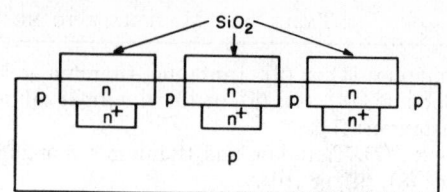

Fig. 10—p-type isolation diffusion.

(Fig. 10) The wafer is reoxidized and a new layer of silicon dioxide is grown over the top surface of the wafer. This is then etched away except over the n regions. A p dopant is then diffused through the exposed n epitaxial layer, forming isolated n islands that become sites for the subsequent formation of devices.

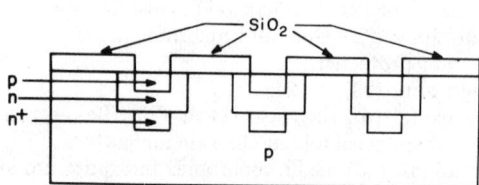

Fig. 11—p-type base diffusion.

(Fig. 11) Windows are again opened above the n areas in a new silicon-dioxide layer and p-type impurity is diffused for the formation of transistor bases, diode and junction-capacitor electrodes, and resistors.

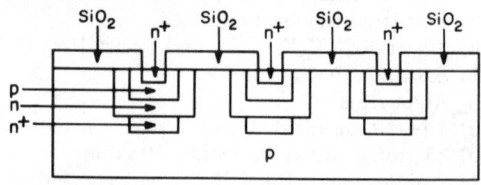

Fig. 12—n^+-type emitter diffusion.

(Fig. 12) After another SiO_2 regrowth, photoresist coating, masking, and etching of windows, an n^+ diffusion is carried out to form transistor emitters, additional diode and capacitor electrodes, plus degenerative layers for interconnections and crossovers.

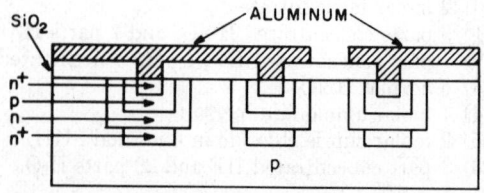

Fig. 13—Metallization to form connections.

(Fig. 13) A final oxide-etching step opens windows over emitter and other desired contact areas. The entire wafer is then covered with a layer of aluminum deposited by evaporation. The aluminum layer is transformed into the appropriate interconnection pattern by a final photolithographic step and etching.

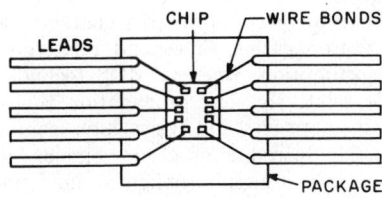

Fig. 14—Flatpack mounting of chip.

(Fig. 14) After metallization, all the integrated circuits on the wafer are electrically probed. Those that do not meet test specifications are marked with an ink dot for subsequent removal. After electrical test the wafer is scribed and broken into a number of chips or dice. Each die is a single integrated circuit.

Each integrated circuit is mounted in a suitable package. Aluminum or gold wire bonds are made between integrated-circuit lands and package terminals. The package is sealed and tested again electrically and environmentally before being shipped.

MOS Integrated Circuits: A major development in silicon integrated circuits has been the emergence of practical MOS monolithic ICs. These consist of intricate combinations of MOS transistors which are formed into a silicon substrate in a manner similar to that used to fabricate bipolar ICs.

The major difference between bipolar and MOS IC fabrication is that the latter requires only one diffusion step. A typical MOS IC structure in cross section is shown in Fig. 15. The structure in this figure, which shows two MOS transistors, is fabricated by diffusing heavily doped p-type islands into a lightly doped n-type silicon substrate. A thin layer of insulating silicon dioxide is then deposited or grown on the substrate. Next, appropriate contact holes are etched into the oxide layer, the substrate is metalized, and individual source, drain, and gate electrodes are formed by standard photolithographic processes.

MOS ICs are noted for their simplicity (hence very high component density) and very low power consumption. Conventional MOS ICs are subject to source-substrate and drain-substrate parasitic capacitance which impairs operating speed, but a new technique wherein individual MOS transistors are formed on an insulating sapphire substrate (silicon-on-sapphire, or SOS) significantly reduces parasitic capacitance and greatly improves operating speed.

MOS ICs require fewer fabrication steps than most bipolar ICs, and many of the steps require less control. Though conventional MOS ICs are slower than their bipolar counterparts, they are simpler and therefore have a much higher potential component density (with the possible exception of I²L bipolar ICs). Therefore, MOS technology has been admirably suitable for medium- and large-scale-integration (MSI and LSI). Applications for MOS LSI are diverse and include calculator, watch, and microprocessor chips.

PACKAGING

The requirements that an integrated-circuit package must satisfy are many and varied.

(**A**) It must be small enough to preserve the miniature character of the enclosed silicon die, yet large enough to permit interfacing with human or machine handling.

(**B**) It must be physically and mechanically durable enough to withstand stresses and shock during assembly and handling, to permit interconnection by various means, and to serve in a military environment.

(**C**) It must display reliable electrical integrity of connection between the inside of the circuit and external leads.

(**D**) It must be capable of being hermetically sealed, if necessary.

(**E**) It must provide low thermal resistance to its environment.

(**F**) It must provide adequate electrical and magnetic shielding.

(**G**) It should be suitable for inexpensive fabrication and assembly using automatic machine processes.

Microelectronic device outlines fall into a number of general types, or families. The most common are the flatpack, header, and dual-in-line (DIP) families. Examples of these are shown in Figs. 16, 17, and 18, respectively. Other families include the grid array, axial quad, quad header, and flange mounted types.

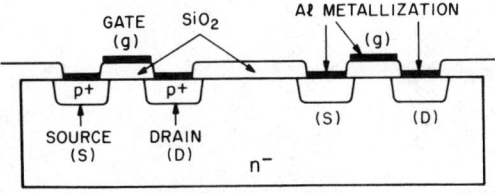

Fig. 15—*MOS* integrated circuit.

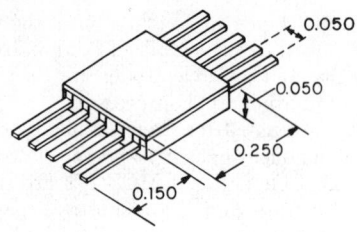

Fig. 16—Ten-lead flatpack.

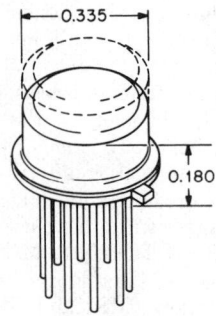

Fig. 17—Ten-lead header type package.

DIGITAL INTEGRATED CIRCUITS

Digital integrated circuits are those in which discrete voltage or current levels are used to represent the binary bits 0 and 1 so that logical decisions may be implemented. The earliest monolithic digital ICs were simple logic gates made using conventional bipolar technology. Though bipolar technology retains a highly significant role in digital ICs, rapid advances in the MOS field have provided a competing technology of major importance.

Since the early 1960s when digital ICs first became commercially available, numerous methods for implementing the various logic functions utilizing both bipolar and MOS technology have been developed. The most important of these logic families are described below.

Bipolar Logic Families

Bipolar technology is primarily characterized by very fast propagation times and relatively high power consumption, whereas MOS technology is noted for relatively slow propagation times, low power consumption, and very high component densities. The introduction of new bipolar fabrication methods and circuit configurations is rapidly blurring these considerations.

Direct-Coupled-Transistor Logic (DCTL): The earliest commercial IC family, DCTL has poor noise immunity, and its variation in input characteristics causes a base current-hogging problem. Original DCTL is now obsolete, but newer bipolar logic families have evolved which retain the simplicity of DCTL.

Resistor-Transistor Logic (RTL): The earliest IC logic family to achieve widespread commercial acceptance, RTL is noted for its economy and high speed-power product. Disadvantages of RTL are poor noise immunity and relatively low fanout capability.

Resistor-Capacitor-Transistor Logic (RCTL): RCTL is a variation of RTL wherein bypass capacitors are connected across gate input resistors to increase operating speed without increasing power consumption. Noise immunity is poor.

Diode-Transistor Logic (DTL): DTL is slower than other bipolar logic families, but it has lower

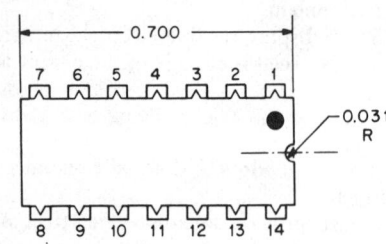

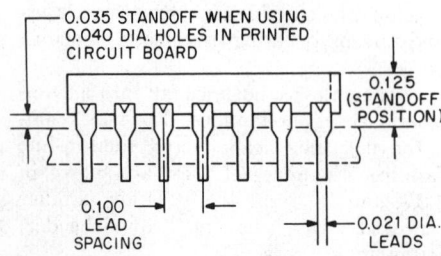

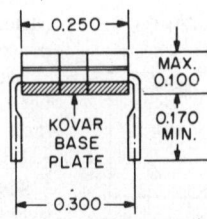

Fig. 18—Dual-in-line package with 14 leads.

power dissipation. It is voltage compatible with the popular TTL family, and both DTL and TTL can be mixed in a single circuit.

High-Threshold Logic (HTL): HTL improves the poor noise immunity of DTL with the addition of zener diodes at the gate inputs. Higher power dissipation and operating voltage are the result, but the subsequent noise immunity may be a desirable tradeoff.

Transistor-Transistor Logic (TTL): Higher noise immunity and faster operating speed than RTL, DTL, and HTL have contributed to the widespread acceptance of TTL among design engineers. Highly complex functions involving networks of interconnected gates, multivibrators, and registers are readily implemented with TTL logic. The TTL family is subdivided into standard types and Schottky types, which incorporate high-speed Schottky diodes to increase operating speed.

Complementary-Transistor Logic (CTL): A high-speed, medium-noise-immune logic family. Power consumption is higher than for TTL, and a more difficult manufacturing process results in higher cost.

Emitter-Coupled Logic (ECL): The logic families described thus far utilize saturated switching transistors. ECL, sometimes termed current-mode logic (CML), achieves switching by means of nonsaturated transistors; differing current levels are used to simulate the two logic status conditions. ECL offers exceptionally high speed, high input impedance, and low output impedance, but it has lower noise immunity than saturated logic and cannot be directly interfaced with other logic families.

Complementary TTL (CTTL): CTTL utilizes a multiemitter input transistor and complementary *pnp-npn* output transistors to achieve the same operating speeds as TTL at a tenth of the power consumption.

Integrated Injection Logic (I^2L): A logic family wherein conventional bipolar processing is used to achieve highly compact gate structures which operate at near TTL speeds while consuming a fraction of the power required for TTL. I^2L uses vertical *npn* transistors with multiple collectors and lateral *pnp* current sources as loads. An I^2L gate can be fabricated from a single complementary transistor pair and no resistors. I^2L technology is ideally suited for bipolar LSI at least equally as dense as MOS LSI and with faster operating speeds.

Current-Hogging Logic (CHL): CHL is a variation of I^2L but has higher noise immunity since saturated logic is implemented.

Complementary-Constant-Current Logic (C^3L): C^3L is a modification of DCTL which has the very high component density potential of I^2L but faster operating speeds.

Triple-Diffused Emitter-Follower Logic (3D-EFL): 3D-EFL is an unsaturated logic family which offers moderately fast operating speed and a simple triple-diffusion fabrication process. The tight component packing density of 3D-EFL coupled with the simple manufacturing process provides the potential for simpler production of bipolar LSI circuits.

Summary of Bipolar Logic: Table 7 summarizes the most important bipolar logic families and gives examples of simple gate circuits for each.

MOS Logic Families

Device fabrication technology rather than circuit format is the major design consideration for the MOS logic families. There are fewer MOS logic families than bipolar ones, but MOS technology offers several important advantages including simpler processing and the relative ease with which complex LSI can be implemented.

PMOS: The *p*-channel enhancement MOS (PMOS) technology resulted in the first commercial MOS ICs. PMOS remains the simplest MOS process and enjoys widespread commercial availability. Propagation delay (1 μs per gate) is considerably slower than for bipolar logic, but power consumption (0.1–1.0 mW per gate) is significantly lower.

NMOS: NMOS technology provides a faster logic family than PMOS, since electron majority carriers are inherently more mobile than holes. NMOS requires more careful processing than PMOS but gives a better speed-power product (100 ns propagation delay at 100 μW per gate).

CMOS: The CMOS logic family utilizes both PMOS and NMOS technology in a complementary arrangement on a common substrate. CMOS is almost immune to noise (1.5 volt noise margin), operates from a wide range of supply voltages (3–15 volts), and consumes less than a microwatt per gate while operating at moderately fast speeds (10 kHz) and no power at all during standby. Also, with a gate propagation delay of from 10 to 100 ns, CMOS is faster than either PMOS or NMOS. A wide range of CMOS logic circuits with pin-for-pin compatibility with the TTL family is available.

SOS: MOS logic circuits are inherently slower than bipolar types due to parasitic capacitance between the source and drain regions and the silicon substrate. A major advance in MOS IC fabrication

TABLE 7—TYPICAL CIRCUITS AND CHARACTERISTICS OF THE MAJOR BIPOLAR LOGIC FAMILIES. *Adapted from Electronic Design, Rene Colen (Microelectronics Editor), p. 171; 17 May 1966.* © *1966, Hayden Publishing Co., New York City.*

Symbol	Circuit Diagram	Speed*	Power*	Fan-out*	Noise Immunity*	Trade Name	Remarks
DCTL		Medium	Medium	Low	Low	Series 53	Variations in input characteristics result in base-current "hogging" problem. Proper operation not always guaranteed. More susceptible to noise because of low operating and signal voltages.
RTL		Low	Low	Low	Low	RTL	Very similar to DCTL. Resistors resolve current "hogging" problem and reduce power dissipation. However, operating speed is reduced.
RCTL		Low	Low	Low	Low	Series 51	Though capacitors can increase speed capability, noise immunity is affected by capacitive coupling of noise signals.
DTL		Medium	Medium	Medium	Medium to high	930 DTL	Use of pull-up resistor and charge-control technique improves speed capabilities. Many variations of this circuit exist, each having specific advantages.

Type		Description	Col 1	Col 2	Col 3	Col 4
TTL	SUHL Series 54/74	Very similar to DTL. Has lower parasitic capacity at inputs. With the many existing variations, this has become very popular.	High	Medium	Medium	Medium to high
CML (ECL)	MECL ECCSL	Similar to a differential amplifier, the reference voltage sets the threshold voltage. High-speed, high fan-out operation is possible with associated high power dissipation. Also known as emitter-coupled logic (ECL).	High	High	High	Medium to high
CTL	CTML	More-difficult manufacturing process results in compromises of active device characteristics and higher cost.	High	Medium	Medium	Medium
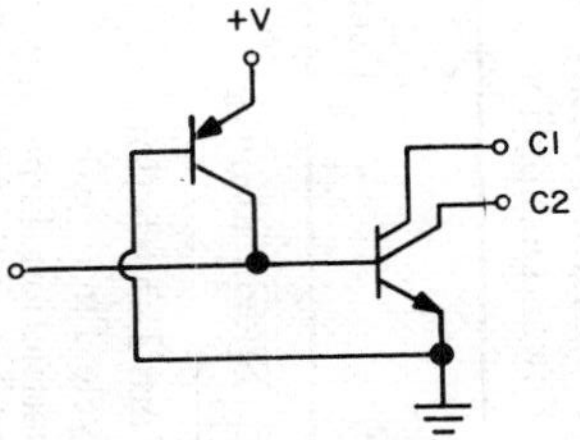 I²L	I²L	Provides smallest and most dense bipolar gate. Simple manufacturing process and higher component packing density than any MOS process. Also known as merged-transistor logic (MTL).	High	Low	High	Medium

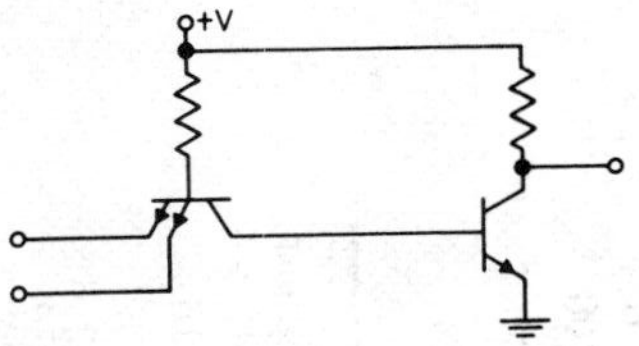

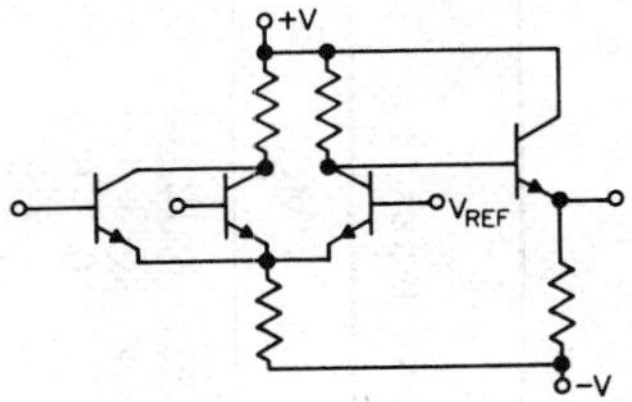

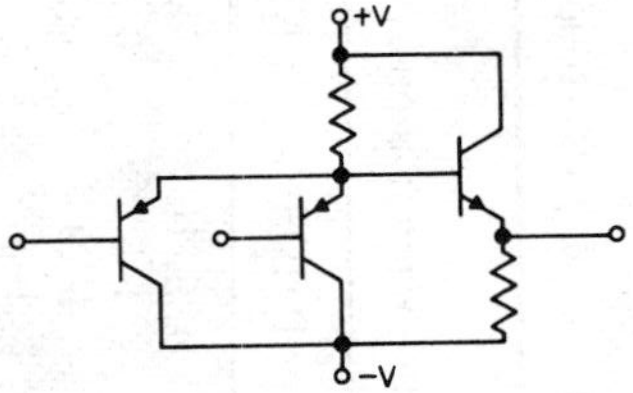

TTL

CML (ECL)

CTL

I²L

*Low =	<5 MHz	<5 mW	<5	<300 mV
Medium =	5 to 15 MHz	5 to 15 mW	5 to 10	300 to 500 mV
High =	>15 MHz	>15 mW	>10	>500 mV

reduces this capacitance to about 5% of its usual value by epitaxially depositing a 1-micrometer silicon film on an insulating sapphire substrate (silicon-on-sapphire, or SOS). Individual transistors are isolated from one another by an etching process, and interconnecting conduction paths are then deposited over them. CMOS logic circuits fabricated with SOS technology have exhibited gate propagation delays of a few nanoseconds and are expected to offer important competition to TTL logic in areas where the additional cost of the sapphire substrate is offset by high component density and fewer processing steps.

Summary of MOS Logic: Table 8 summarizes the most important MOS logic families.

Commercial Digital Integrated Circuits

A wide variety of digital integrated circuits is commercially available. Since the operation of a particular logic circuit is predictable according to truth table, most commercial devices are standardized, and many manufacturers second-source one another with pin-for-pin equivalent devices. Several manufacturers have developed product lines of logic functions implemented by a single logic family which have evolved into industry standards. The most successful of these is the series 54/74 bipolar TTL line. This line includes a diverse range of logic circuits ranging from simple arrays of gates (e.g., 7400 quadruple 2-input NAND gate) to sophisticated circuits capable of implementing relatively complex functions (e.g., 74192 up-down decade counter). Information and applications literature on this and other standardized IC lines is available from the various manufacturers of integrated circuits.

Special-Purpose Digital Logic Circuits

Rapid advances in both bipolar and MOS LSI technology have resulted in a variety of highly specialized complex logic circuits, including single-chip calculator, microprocessor, and digital clock and watch ICs. All these functions were originally implemented using MOS techniques, but recent advances in bipolar technology have stimulated interest in alternative approaches. For example, I^2L digital watch chips are in production as are both Schottky and I^2L microprocessors. The latter development is significant since bipolar technology permits faster operating speeds. SOS-MOS technology is also receiving considerable interest as a possible method to achieve complex logic functions with high operating speeds. Table 9 summarizes some of the operating capabilities for a variety of commercially available microprocessor chips.

New Developments

Though great advances have been achieved in component packing densities since 1960, further advances may be more in the direction of better yield, lower cost, and improved packaging. Component densities will continue to improve somewhat, but since the limits imposed by device physics are rapidly being approached, further advances may be less dramatic. This topic is discussed in detail in R. W. Keyes, "Physical Limits in Digital Electronics," *IEEE Proceedings*, May 1975, pp. 740–767.

MEMORY INTEGRATED CIRCUITS

A technology of vital supplementary importance to digital integrated circuits is memory integrated

TABLE 8—TYPICAL MOS INTEGRATED CIRCUIT CHARACTERISTICS.

Characteristic	PMOS	NMOS	CMOS
Propagation Delay (μs)	1	0.1	0.01–0.1
Power Dissipation (μW)	100–1000	100	1–100

TABLE 9—MICROPROCESSOR TECHNOLOGIES.*

Specification	PMOS	NMOS	CMOS	Bipolar
Word Length (bits)	4, 8, 16	8, 16	8, 12, 16	16
Cycle Time (μs)	1–5	0.5–2.5	0.3–3.0	0.1
Instruction Time (μs)	4–20	1.5–5.0	0.3–6.0	0.6
Chip Count	1–5	1	1–2	7

* For additional information, see L. Altman, Editor, *Microprocessors*, Electronics Book Series, McGraw-Hill Book Co., Inc., New York; 1975.

circuits. The successful development of practical semiconductor memories has been a major factor in the miniaturization of electronic calculators, computers, and other logic devices. Semiconductor memories may be classified as read only or read/write; both types, along with the appropriate fabrication technologies, are described below.

Read-Only Memories (ROMs)

Read-only memories store fixed data in the form of binary bit patterns. Access to the ROM is random, and any memory address can be found within a uniform time increment. The bit pattern of most ROMs is programmed during the manufacturing process (mask programming) according to the user's specifications. Such preprogrammed ROMs capable of storing up to 2^{14} bits are commercially available. Other ROMs may be field-programmed by the user (PROMs).

Field-programmable PROMs typically incorporate transistor arrays with a fusible emitter link which can be opened by a current pulse having appropriate width and amplitude. A 0 is stored by leaving the fusible link intact; a 1 is stored by electrically separating the link.

The electrically alterable or erasable PROM (EAPROM) can be reprogrammed by the user. One type of erasable PROM utilizes an array of floating-gate avalanche-injection transistors, each of which can store an electrically impressed charge whose amplitude is 40 volts or more. The stored charge can be removed (erased) by exposure to ultraviolet or X-radiation. Erasable PROMs are about ten times more costly than mask-programmed ROMs, and their storage capability is only 2^{13} bits per chip, but their operating flexibility makes them highly desirable for numerous applications.

Read/Write Memories (RAMs)

Calculators and computers require a temporary data storage capability during instruction process-ing and problem solving, and this is accomplished with read/write memories. Commercial RAMs have less storage capability than ROMs (2^{12} bits per chip) at a similar cost per bit. Unlike ROMs which store data permanently, RAMs have a volatile storage capability, and continuous application of power is required to retain stored data. This power requirement can be an excessive drain on some types of portable equipment, but newly developed CMOS RAMs operate for weeks or months from small dry cells. For example, a 1024-bit CMOS RAM has a standby power consumption of only 100 μW, whereas an equivalent NMOS RAM requires more than 100 mW.

Memory Technologies

Semiconductor memories can be implemented with variations of both bipolar and MOS technology. The most important are summarized in Table 10. As in the case of digital logic ICs, bipolar generally provides significantly faster access time than MOS, but MOS has significantly higher storage capacity.

The densest memory technology thus far is the MOS charge-coupled device (CCD), and 16 kilobit CCD memories (64 serial registers of 256 bits each) are commercially available. These memories have a maximum data transfer rate of 1.8 megabytes per second. CCD memories now offer a significant challenge to the traditional disc and drum storage methods as exemplified by the introduction of a commercial CCD memory system occupying a circuit board with a surface area of only 1160 cm^2 and having a storage capacity of more than 1.1 megabits. This memory has a worst-case access time of 193 μs (more than 40 times faster than conventional disc memories) and is expandable.

New Developments

As in the case of digital logic integrated circuits, memory integrated circuits are approaching the

TABLE 10—SEMICONDUCTOR MEMORY TECHNOLOGIES.*

Type	Access (ns)	Cycle Time (ns)	Power/Bit (μW)	Density (bits/chip)
PMOS	300–400	550	100	4096
NMOS	150–250	400	100	4096
CMOS	50–60	550	20	1024
CCD	80–250	500	10	16 384
TTL	50–60	90	500	1024
ECL	45–50	65	500	1024
I^{2}L	50	100	—†	—†

* Adapted from R. Allan, "Semiconductor Memories," *IEEE Spectrum*, August 1975, pp. 40–45.
† Under development.

miniaturization limits imposed by device physics. Therefore, further improvements can be expected to include emphasis on reliability, packaging, and economy. Also, it is not unlikely that important developments in optical and magnetic bubble technology eventually may be coupled with semiconductor logic and memory technology. Magnetic bubbles are miniature cylindrical magnetic domains embedded in a thin film of orthoferrite material such as $PbFe_{12}O_{19}$, $Y_3Fe_5O_{12}$, etc. Magnetic bubbles can be used to achieve memory functions at theoretical densities of millions of bits per square inch. Since the energy necessary to move a bubble is miniscule, magnetic bubble memories are of considerable interest to design engineers.

LINEAR INTEGRATED CIRCUITS

Unlike digital integrated circuits which respond to and produce two-state logic signals, linear ICs give an output signal which varies, often linearly, with respect to a varying input signal. Since linear ICs can be utilized in a wide variety of applications, standardization of circuit elements as in the case of most digital ICs is impractical. Nevertheless, several important families of linear ICs have evolved. They include the device categories described below.

Differential Amplifier

The differential amplifier is a building-block circuit which may be used alone or in conjunction with other similar amplifiers on a common substrate. As shown in Fig. 19, the basic differential amplifier consists of two identical transistors connected to respond to the difference between two input signals while simultaneously blocking identical input signals. The ability of the differential circuit to block identical input signals is very useful where noise is a problem.

The differential amplifier can be used as a linear amplifier across a very wide frequency range (dc to VHF) and in numerous other applications including mixers, product detectors, amplitude modulators, and frequency multipliers.

Operational Voltage Amplifier

The basic operational voltage amplifier, or op-amp, is a dual-input differential amplifier followed by one or more direct-coupled gain stages (Fig. 20). The typical op-amp exhibits very high voltage gain which can be easily adjusted downward with the application of external feedback (e.g., resistive coupling between the output and one of the inputs). In practice, op-amps often

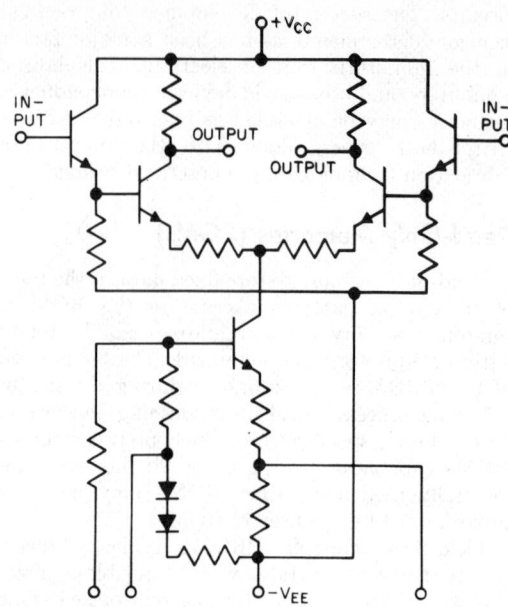

Fig. 19—Basic differential amplifier.

employ two direct-coupled balanced differential amplifiers in cascade with the first stage push-pull driving the second stage. An output stage provides additional gain. Since the op-amp employs direct interstage coupling, it is ideally suited for conventional bipolar IC fabrication methods.

Any op-amp can be connected as shown in Fig. 21 to provide a linear voltage gain equal to R_f/R_{in}, where R_f is the feedback resistance and R_{in} is the input resistor. The "ideal" op-amp would possess these additional characteristics:

1. Infinite gain
2. Infinite input impedance
3. Zero output impedance
4. Zero output for zero input
5. Infinite bandwidth

No existing op-amp meets these ideal characteristics, but many op-amps come close to meeting one or more of them. For example, op-amps with a CMOS input stage which gives an input impedance of 1.5×10^{12} ohms, an open-loop voltage gain of 110 dB (320,000 volts/volt), and a unity-gain bandwidth of 15 MHz are commercially available.

The op-amp was originally designed to perform such mathematical operations as integration, differentiation, summation, multiplication, comparison, and others. Though integrated op-amps are still used in these applications, the use of various types of external feedback networks permits a vast array of additional applications. Accordingly, the op-amp is the most versatile of linear ICs, and

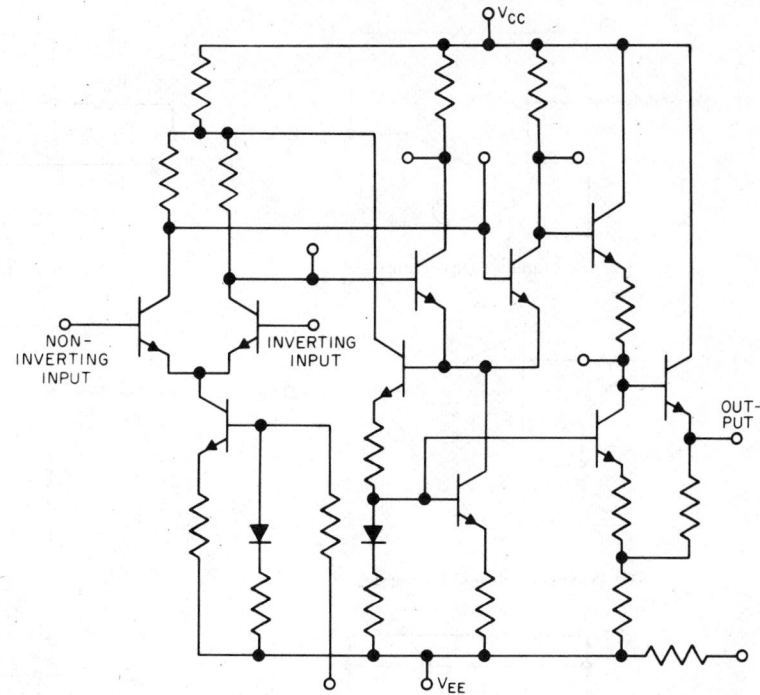

Fig. 20—Schematic diagram of the basic operational amplifier.

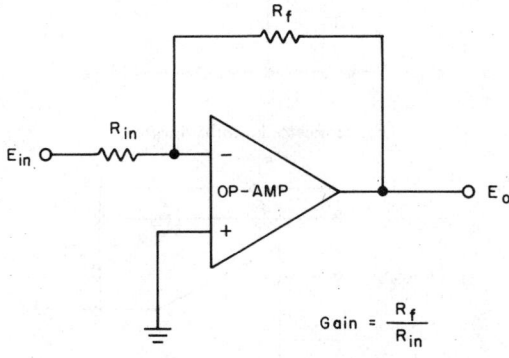

Fig. 21—Basic op-amp gain block.

its circuit applications include analog-digital conversion, peak detection, averaging, function generation, oscillator and pulse circuits, biological function amplification, voltage-controlled oscillator circuits, voltage comparison, active filters, and numerous others.

Figure 22 summarizes some of the more important op-amp circuit configurations and gives relevant formulas for calculating circuit performance. Detailed information regarding the many commercial op-amps, their operating specifications, pinout configurations, and typical application circuits is available from the various manufacturers.

Operational Transconductance Amplifier

The operational transconductance amplifier (OTA) provides a transconductance gain and a current output rather than a voltage gain and output. Since the output of the OTA is the product of the transconductance and the input voltage, the output circuit can be thought of as an infinite-impedance current generator. This contrasts with the output of the operational amplifier, which can be characterized as a zero-impedance voltage generator.

Considerable control over the performance of an OTA is possible. For example, varying the OTA bias current can completely control the open-loop gain of the device. The OTA bias current can also be used to control total power input. These operating characteristics combine to make the OTA usable in such widely diverse applications as gyrators, multipliers, amplifiers with AGC, sample and hold circuits, multiplexers, multivibrators, and many others.

Multipurpose Amplifiers

Numerous types of multipurpose amplifiers which may or may not incorporate differential inputs are commercially available as bipolar ICs. This family of linear circuits includes many different single-ended circuits which operate from

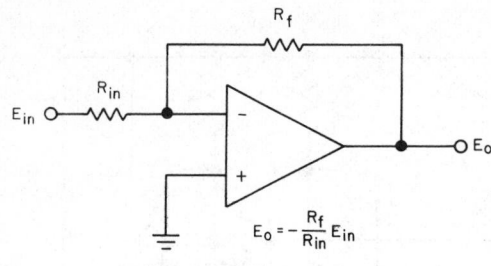

(A) Dc amplifier (inverting).

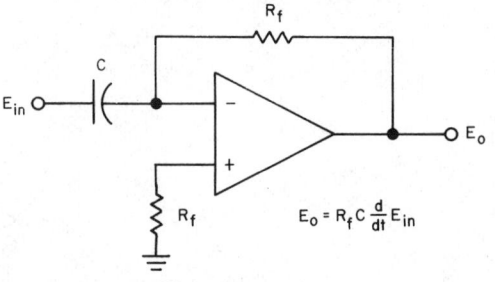

(B) Dc amplifier (noninverting).

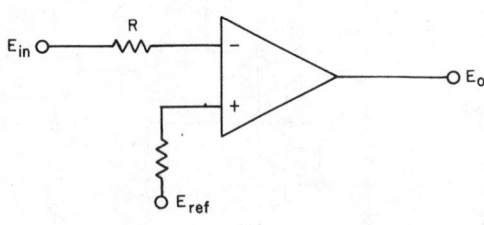

(C) Analog-to-ditigal converter.

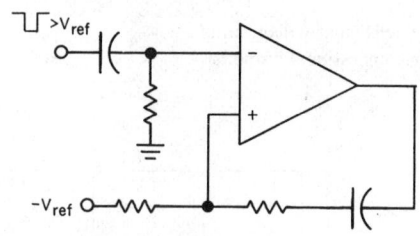

(D) Differentiator.

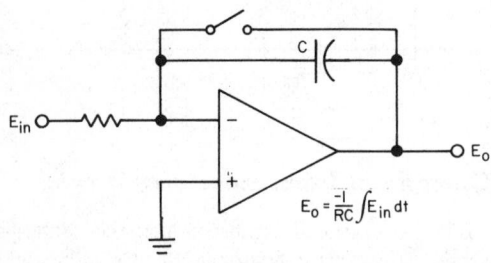

(E) Integrator.

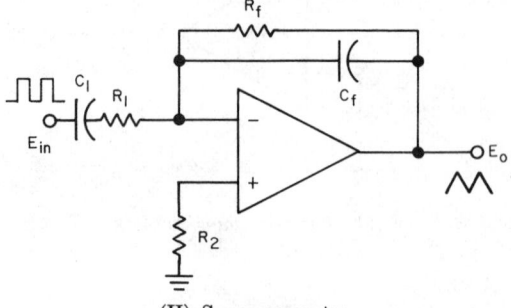

(F) Monostable multivibrator.

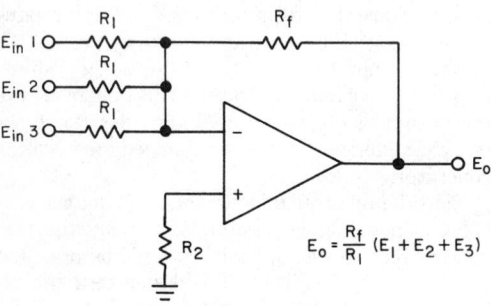

(G) Averaging amplifier.

(H) Sweep generator.

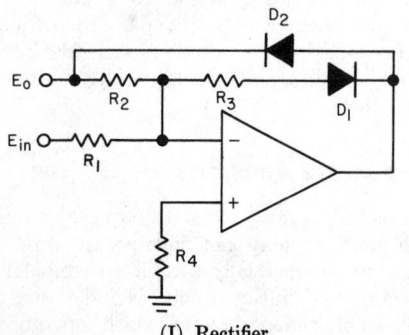

(I) Rectifier.

Fig. 22—Typical op-amp circuit applications.

single-polarity power supplies and are suitable for such applications as audio, i-f, rf, wideband, and video amplification. Many of these devices are utilized in commercial and industrial applications, and typical user products include miniature hearing aids, radios, stereo amplifiers, tape recorders, intrusion alarms, power amplifiers, televisions, and many others.

Special-Purpose Linear Integrated Circuits

Since any circuit which can be assembled from discrete semiconductor components can almost always be duplicated or simulated with monolithic IC technology, a great many special-purpose linear ICs exist. These include custom designs for use in various kinds of commercial products as well as a wide range of off-the-shelf numbered units. The commercial units include such devices as music synthesizers, phase-locked loops, tone decoders, function generators, FM demodulators, FM stereo demultiplexers, timers, and others. Though many of these linear circuits are dedicated to a specific application or function, the addition of a few external components can often result in a wide range of additional applications. The list of special-purpose linear ICs grows constantly, and the design engineer is well advised to consult current manufacturer's literature for information on new circuits and circuit application data.

BIBLIOGRAPHY

General Survey of Microelectronics

1. *Proceedings of the IEEE*, Microelectronics issue, Vol. 52, No. 12; Dec. 1964.
2. E. Keonjian (editor), "Microelectronics," McGraw-Hill Book Company, New York; 1963.
3. S. Levine, "Principles of Solid State Microelectronics," Holt, Rinehart, & Winston, New York; 1963.
4. G. Dummer and J. Granville, "Miniature and Microminiature Electronics," John Wiley & Sons, New York; 1961.

5. S. Weber (editor), "Large and Medium Scale Integration," McGraw-Hill Book Company, New York; 1974.

Silicon Monolithic Integrated Circuits

1. A. Khambata, "Introduction to Integrated Semiconductor Circuits," J. Wiley & Sons, New York; 1963.
2. R. Warner and J. Fordenwalt, "Integrated Circuits—Design Principles and Fabrication," Motorola Semiconductor Products Inc., 5005 East McDowell Road, Phoenix, Arizona; 1965.
3. "Microelectronics," G. Madland (editor), Integrated-Circuits Engineering, Inc., Phoenix, Arizona; 1964.
4. R. B. Sorkin, "Integrated Electronics," McGraw-Hill Book Company, New York; 1970.

Thin-Film Hybrid Integrated Circuits

1. L. Holland, "Thin-Film Microelectronics," John Wiley & Sons, New York; 1966.

Thick-Film Hybrid Integrated Circuits

1. E. Davis, W. Harding, and R. Schwartz, "An Approach to Low-Cost, High-Performance Microelectronics," *Wescon Proceedings*, San Francisco, California; August 1963.
2. J. O'Connell and E. Zaratkiewicz, "Thick-Film Technology," *Electrical Communication*, Vol. 41, No. 4, pp. 391–406; 1966.
3. R. Ilgenfritz, "Thick-Film Hybrid Microelectronic Circuit Technology," *Semiconductor Products and Solid State Technology*, vol. 9, pp. 35–42; June 1966.

Linear Integrated Circuits

1. E. R. Hnatek, "Applications of Linear Integrated Circuits," John Wiley & Sons, New York; 1975.
2. United Technology Publications, Inc., "Modern Applications of Linear ICs," Tab Books, Blue Ridge Summit, Pa.; 1974.

CHAPTER 22
OPTOELECTRONICS

INTRODUCTION

The technological marriage of the fields of optics and electronics is called optoelectronics. Optoelectronics is a remarkably diverse field involving such topics as the nature of optical radiation, the interaction of light with matter, radiometry, photometry, and the characteristics of various sources and sensors.

THE OPTICAL SPECTRUM

By convention, electromagnetic radiation is generally specified according to its metric wavelength. The frequency of a specific electromagnetic wavelength can be found from

$$f = \frac{3 \times 10^4}{\lambda}$$

where f is frequency in MHz and λ is wavelength in cm. The optical portion of the electromagnetic spectrum extends from 10 nm to 10^6 nm and is divided into three major categories: ultraviolet, visible, and infrared.

Ultraviolet

Those wavelengths falling below the visible spectrum and above x-rays are collectively designated ultraviolet (UV). UV is classified according to wavelength as extreme (10–200 nm), far (200–300 nm), or near (300–370 nm). UV is sometimes designated as short-wave or long-wave.

Visible

Those wavelengths between ~370–750 nm can be perceived by the human eye and are therefore collectively designated as visible light. Visible light is classified according to the various colors its wavelengths elicit in the mind of a standard observer. The major color categories are violet (370–455 nm), blue (456–492 nm), green (493–577 nm), yellow (578–597 nm), orange (598–622 nm), and red (623–750 nm).

Infrared

Those wavelengths falling above the visible spectrum and below microwaves are collectively designated infrared (IR). IR is classified according to its wavelength as near (750–1.5 × 10³ nm), middle (1.6×10³–6×10³ nm), far (6.1×10³–4×10⁴ nm) and far-far (4.1×10⁴–10⁶ nm).

BLACKBODY RADIATION*

Any surface having a temperature > 0 K (absolute zero) is a source of blackbody radiation. The term "blackbody" is actually a misnomer since a very hot surface will emit visible radiation. A more accurate term is universal radiator since a perfect blackbody is both a perfect emitter and a perfect absorber. Accordingly, Kirchoff's radiation law states

$$\frac{W}{\alpha} = W_{bb}$$

where W is radiant emittance (exitance), α is the absorptance of the blackbody, and W_{bb} is the radiant exitance from the blackbody.

Planck's law gives the blackbody spectral energy density for unpolarized radiation:

$$W_\lambda = 2\pi c^2 h \lambda^{-5} (e^{hc/\lambda kT} - 1)^{-1}$$

where c is the speed of light, h is Planck's constant, λ is the wavelength k is Boltzmann's constant, and T is the kelvin temperature of the blackbody.

Integrating Planck's law over all wavelengths gives the Stefan-Boltzmann law for total radiant exitance (W cm^{-2})

$$W = \sigma T^4$$

where σ is the Stefan-Boltzmann constant.

Wien's displacement law is the empirical expression which gives the peak wavelength of blackbody radiation as a function of temperature: $\lambda_{\max} T = 2897.9\mu$ (K).

* W. L. Wolfe, *Handbook of Military Infrared Technology*, Office of Naval Research, Washington, DC, 1965.

The temperature of an incandescent source can be used to assign the appropriate blackbody radiation curve. For example, radiation from the sun approximately corresponds to that from a blackbody at 5900 K. Fig. 1 shows the spectral-radiant exitance curves for blackbodies at temperatures from 200 to 6000 K.

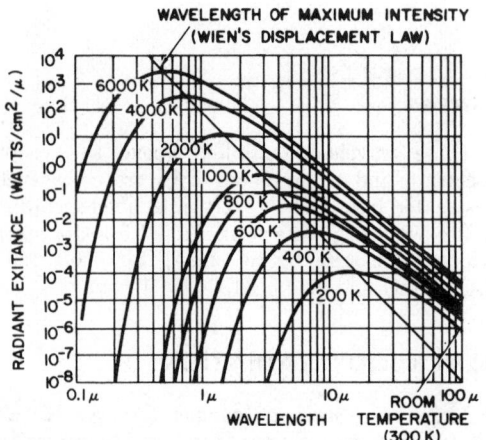

Fig. 1—Blackbody spectral radiant exitance curves from 200 K to 6000 K. Note the wavelength of maximum intensity as given by Wien's displacement law.

Since the blackbody spectral-radiant exitance curves for many common sources peak in the infrared, a popular misconception equates the infrared wavelengths with heat. In fact, any wavelength of radiation contains energy which, upon being absorbed by matter, is transformed into heat.

INTERACTION OF OPTICAL WAVES WITH MATTER

An optical wave may interact with matter by being reflected, refracted, absorbed, or transmitted. The interaction normally involves two or more of these effects.

Reflectance

Some of the optical radiation impinging upon any surface is reflected away from the surface. Reflectance varies according to the properties of the surface and the wavelength and in real circumstances may range from more than 98% (smoked MgO at visible wavelengths) to less than 1% (lampblack at visible wavelengths). Reflection from a surface may be either diffuse, specular, or both. A diffuse reflector has a surface which is rough when compared to the wavelength of the impinging radiation. Lambert's law specifies a

perfectly diffuse surface as one having a constant radiance independent of the viewing angle according to

$$L = \frac{M}{\pi}$$

where L is radiance in $W/m^2/sr$ and M is the radiant exitance in W/m^2. In practice, the reflectance of real reflectors varies with the cosine of the viewing angle. A specular reflector has a surface which is smooth when compared to the wavelength of the impinging radiation. A perfect specular reflector will reflect an oncoming beam without altering the divergence of the beam. A narrow beam of optical radiation impinging upon a specular reflector obeys two rules:

1. The angle of reflection is equal to the angle of incidence.
2. The incident ray and the reflected ray lie in the same plane as a normal line extending perpendicularly from the surface.

Fig. 2 illustrates both diffuse and specular reflection.

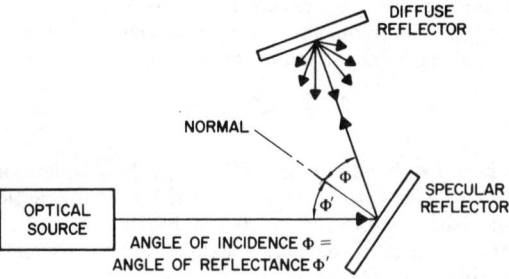

Fig. 2—Specular and diffuse (lambertian) reflectance. (*From F. M. Mims, III, Optoelectronics," p. 19, 1975, Howard W. Sams & Co., Inc.*)

Absorptance

Some of the optical radiation impinging upon any substance is absorbed by the substance. Absorptance varies according to the properties of the substance and the wavelength and in real circumstances may range from a low of <2 dB/km for certain ultrapure fused silica glasses to $>98\%$ for lampblack.

Transmittance

Some of the optical radiation impinging upon a substance is transmitted into the substance. The penetration depth may be slight in which case the transmittance is 0. Certain ultrapure silica glasses may have a transmittance $>75\%$ at certain wavelengths over a distance of 1 km. The reflectance, absorptance, and transmittance of a substance must correspond to

$$\rho + \alpha + \tau = 1$$

where ρ is reflectance, α is absorptance, and τ is transmittance.

Refraction

A ray of optical radiation passing from one medium to another is bent at the interface of the two mediums if the angle of incidence is $\neq 90°$. The index of refraction for a substance is the sine of the angle of incidence divided by the sine of the angle of refraction. Refractive index varies with wavelength and ranges from 1.0002914 (air at 656 nm) to 2.7 (crystalline titanium oxide).

RADIOMETRY

Radiometry is the science of measuring optical radiation at any wavelength. All fundamental radiometric measurements derive from a measure of optical energy, and optical energy is measured with a calorimeter. Since optical energy induces heat into an absorber, it follows that a thermally sensitive detector can be used to measure optical energy. Radiometric methodology is discussed more fully under the heading "Radiometric and Photometric Standards."

Radiometric Terms and Definitions

Five fundamental expressions for radiant power and its distribution in space have been developed and are defined below.

Radiant Flux: Radiant flux is the rate of flow of radiant energy per unit time. Therefore, radiant flux is equivalent to radiant power, and the unit of radiant flux is the watt.

Radiant Incidence: Radiant incidence is the density of radiant flux that irradiates a surface. Radiant incidence (formerly irradiance) is expressed in watts per unit of area.

Radiant Intensity: Radiant intensity is the measure of radiant power per unit of solid angle in watts per steradian.

Radiant Exitance: Radiant exitance is the measure of radiant power emitted by a surface. Radiant exitance (formerly emittance) is expressed in watts per unit area and is often used to describe the optical power reflected from a surface.

Radiant Sterance: Radiant sterance is the radiant intensity per unit of projected area. Radiant sterance (formerly radiance) is expressed in watts-steradian-square meter and is measured by dividing the radiant intensity from a source by the projected area of the source as viewed from a specified angle.

Visualizing the Radiometric Definitions: Fig. 3 summarizes the five fundamental radiometric terms and definitions.

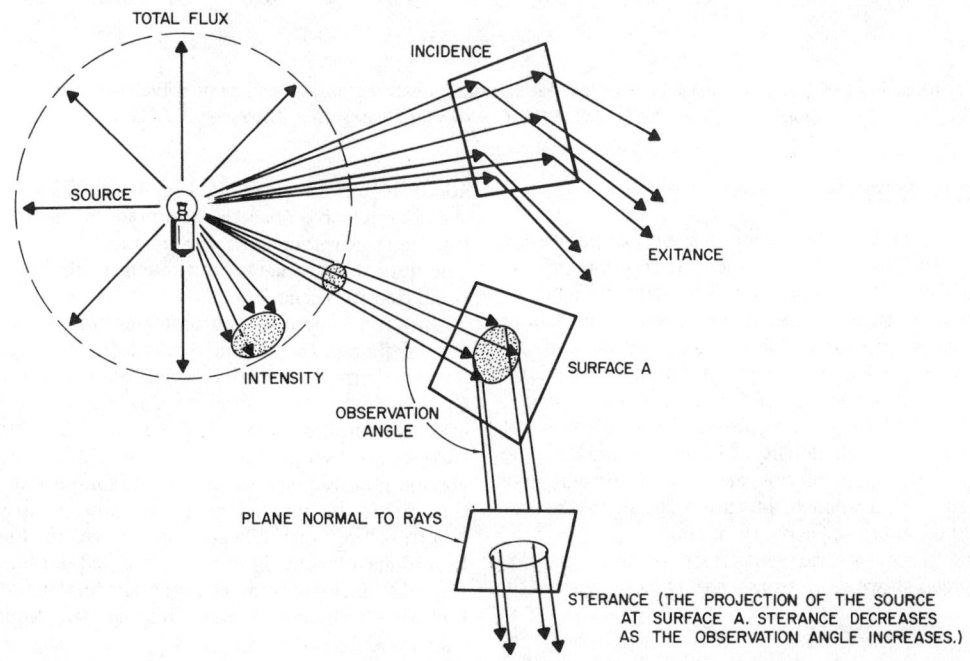

Fig. 3—A physical visualization of the radiometric and photometric terms. (*From F. M. Mims, III, "Optoelectronics," p. 25, 1975, Howard W. Sams & Co., Inc.*)

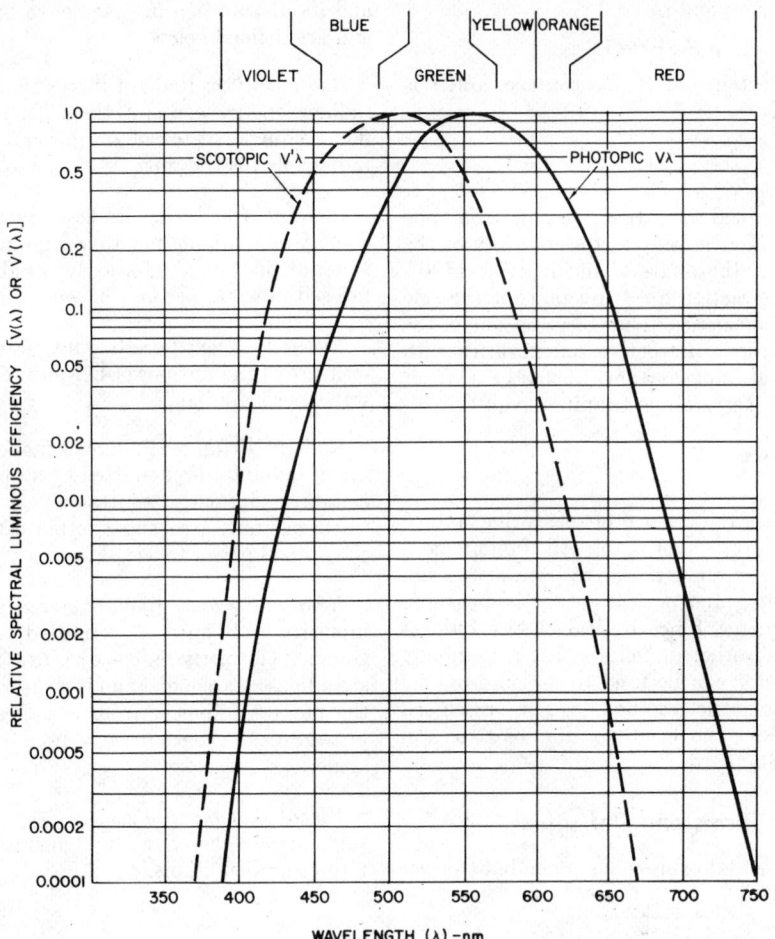

Fig. 4—Relative spectral luminous efficiency of the standard observer as a function of wavelength. (*From RCA Electro-Optics Handbook," Fig. 5-8, RCA Commercial Engineering, Harrison, NJ, 1974.*)

PHOTOMETRY*

Photometry is the science of measuring visible light with respect to the spectral response of the human eye. The retina of the human eye has two types of optical receptors, cones and rods. Cones are primarily responsible for color vision and are highly concentrated ($\sim 10^4$) in the 0.3 mm diameter spot at the center of the field of vision called the fovea. Rods are not present at the fovea but have a very high density ($1.5 \times 10^5 / \text{mm}^2$) in the peripheral regions of the retina. They do not give rise to color response, but in weak light are significantly more sensitive than cones.

The peak spectral sensitivity of rod vision is displaced downward from that of cone vision by

* R. K. Clayton, *Light and Living Matter, Vol. 2: The Biological Part*, McGraw-Hill Book Co., New York, 1971 and *RCA Electro-Optics Handbook*, RCA Commercial Engineering, Harrison, NJ, 1974.

about 45 nm as shown in Fig. 4, a pair of curves showing relative spectral luminous efficiency as a function of wavelength for a standard observer. The data used to generate the curves in Fig. 4 are tabulated in Table 1.

The fundamental photometric unit of visible optical flux is the lumen, and at the peak of the photopic luminosity curve (555 nm), 673 lumens correspond to one radiometric watt. At the peak of the scotopic luminosity curve, 1725 lumens correspond to one radiometric watt. These conversion factors permit the rod and cone response of the eye to be plotted as a function of absolute spectral luminous efficacy as shown in Fig. 5. Luminous efficacy is the ratio of total luminous (visible) flux to total radiometric flux in lm/W. Luminous efficiency η is related to luminous efficacy K according to

$$\eta = \left(\frac{K}{673}\right) 100\%$$

TABLE 1. RELATIVE SPECTRAL LUMINOUS EFFICIENCY VALUES FOR THE STANDARD OBSERVER (*RCA Electro-Optics Handbook, RCA Commercial Engineering, Harrison, NJ, 1974, Fig. 5-7.*)

Wavelength nm	Photopic V(λ) $L \geq 3$nt (cd m^{-2})	Scotopic V'(λ) (L$\leq 3 \times 10^{-5}$nt(cd m^{-2})
380	0.00004	0.00059
390	0.00012	0.00221
400	0.0004	0.00929
410	0.0012	0.03484
420	0.0040	0.0966
430	0.0116	0.1998
440	0.0230	0.3281
450	0.0380	0.4550
460	0.0600	0.5672
470	0.0910	0.6756
480	0.1390	0.7930
490	0.2080	0.9043
500	0.3230	0.9817
510	0.5030	0.9966
520	0.7100	0.9352
530	0.8620	0.8110
540	0.9540	0.6497
550	0.9950	0.4808
560	0.9950	0.3288
570	0.9520	0.2076
580	0.8700	0.1212
590	0.7570	0.0655
600	0.6310	0.03325
610	0.5030	0.01593
620	0.3810	0.00737
630	0.2650	0.003335
640	0.1750	0.001497
650	0.1070	0.000677
660	0.0610	0.0003129
670	0.0320	0.0001480
680	0.0170	0.0000716
690	0.0082	0.00003533
700	0.0041	0.00001780
710	0.0021	0.00000914
720	0.00105	0.00000478
730	0.00052	0.000002546
740	0.00025	0.000001379
750	0.00012	0.000000760
760	0.00006	0.000000425
770	0.00000	0.000000241
780	......	0.000000139

Infrared Visual Response*

The lower limits of visual response at the extremities of both the photopic and scotopic luminosity curves are not necessarily restricted to the range of wavelengths normally defined as

* F. J. Gardiner, "Effectiveness of IR Covert Illuminators," *RCA Lasers*, 1974, pp. 178–181.

visible light. For example, many observers can see the 905 nm radiation from a GaAs injection laser or the 1.06 μm radiation from a Nd^3:YAG laser if the intensity is sufficiently high. Fig. 6 is an extension of the absolute scotopic luminosity curve showing that a GaAs injection laser emitting at 905 nm will elicit a visual response in the mind of an observer whose eyes are sufficiently sensitive to respond to $\sim 10^{-6}$ lmW^{-1}. Since moderately high energy densities at the retina may be required to elicit visual response at near-infrared wavelengths, the safety factor must not be discounted.

Photometric Terms and Definitions

Photometric measurements are almost always based upon the photopic luminosity curve. For special purpose applications where night vision photometry is required, the scotopic luminosity curve is employed. Photometric terminology parallels radiometric terminology and Fig. 3 can be used to visualize the various definitions.

Luminous Flux: Luminous flux or luminous power is the rate of flow of light per unit of time. The unit of luminous flux is the lumen.

Incidence: Incidence is the density of luminous power that illuminates a surface. One lumen illuminating an area of one square meter gives an incidence of one lux. One lumen illuminating an area of one square foot gives an incidence of one footcandle.

Luminous Intensity: Luminous intensity is the measure of luminous flux per unit solid angle. The unit of luminous intensity is the candela, and one lumen of flux per steradian gives a luminous intensity of one candela.

Luminous Exitance: Luminous exitance is the measure of luminous flux emitted by a surface. Luminous exitance (formerly emittance) is expressed in lumens per unit area.

Luminous Sterance: Luminous sterance is a measure of brightness. Luminous sterance (formerly luminance) is measured by dividing the luminous intensity in a given plane by the projected area of the source at the plane. Luminous sterance is expressed in terms of lm/sr/unit area. A variety of terms describing various measures of luminous sterance (luminance) are in common use and are tabulated in Chapter 17 (Table 15). While luminous sterance is essentially a measure of photometric brightness, it can be misleading to use luminous sterance as a figure of merit for brightness due to the subjective nature of vision. For example, subtle differences in the packaging of two otherwise identical electroluminescent diodes may cause one to appear brighter than the other. Contrast is but one influencing factor, and an electrolumines-

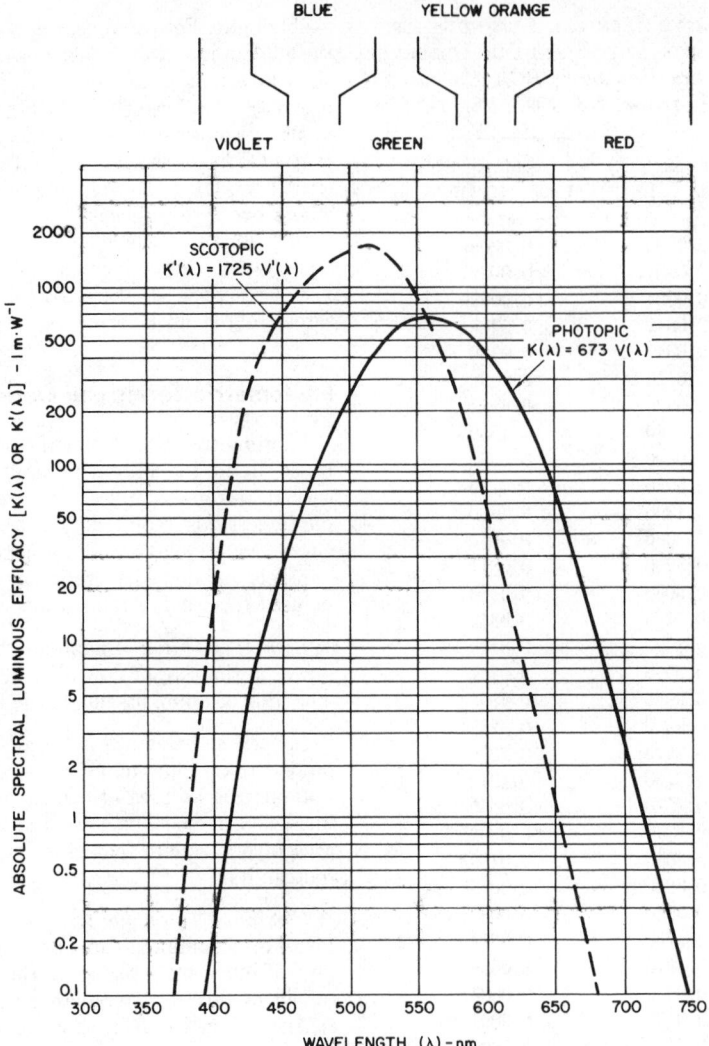

Fig. 5—Absolute spectral luminous efficacy of the standard observer as a function of wavelength. (*From RCA Electro-Optics Handbook," Fig. 5-9, 1974, RCA Commercial Engineering, Harrison, NJ.*)

cent diode mounted on a shiny header will appear dimmer than an identical diode mounted on a nonreflective header.

RADIOMETRIC AND PHOTOMETRIC STANDARDS

Measuring the output of a radiant source is one of the most challenging problems in optoelectronics. The National Bureau of Standards has invested many years in perfecting optical-power measurement standards and techniques, but even the best NBS calibrations may have an uncertainty of ±1%. This uncertainty arises in part from variations in the flux emitted by the standard lamp, slight variations in distance between the standard lamp and the calorimeter monitoring its output, the regulation of the standard lamp's power supply, and slight changes in ambient temperature. Fig. 7 illustrates the magnitude of the optical measurement problem and emphasizes the importance of careful radiometric procedures. The phrase "calibration traceable to NBS" is commonly found on the specification sheets for devices marketed by numerous optoelectronics firms, but the phrase is meaningless if information about the calibration procedure is not included.

OPTICAL SOURCES*

The most important optoelectronic sources are tungsten lamps, fluorescent lamps, glow discharge

* F. M. Mims, III, *Optoelectronics*, Howard W. Sams & Co., Inc., Indianapolis, 1975.

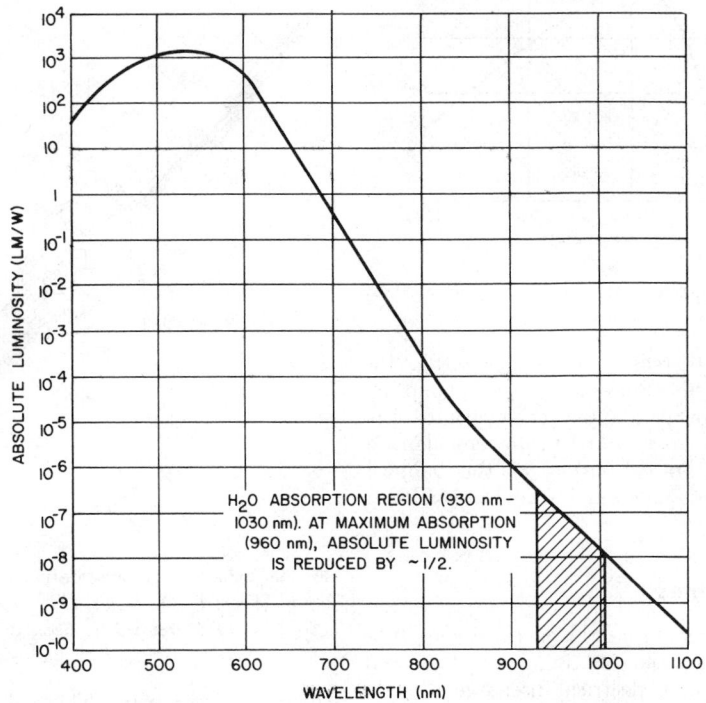

Fig. 6—Calculated extension of the standard scotopic luminosity curve neglecting H_2O absorption in the ocular media of the eye. (*Adapted from F. J. Gardiner, "Effectiveness of IR Covert Illuminators," pp. 178–181, 1974, RCA Lasers.*)

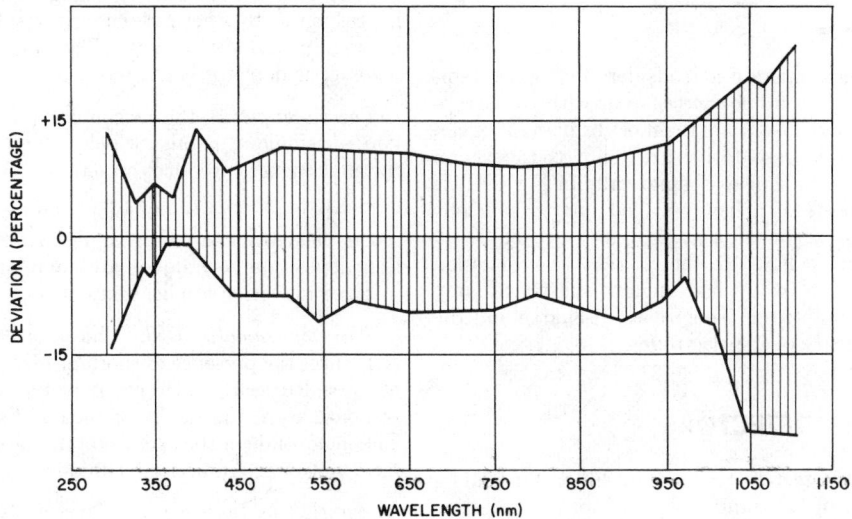

Fig. 7—Percentage deviation spread in the calibration of an identical photodetector by eleven independent measurement laboratories. (*Adapted from F. Grum and J. Cameron, "Detector Intercomparison Results," pp. 82–84, November 1974, Electro-Optical Systems Design.*)

lamps, electroluminescent diodes, and lasers. Electroluminescent diodes are described in Chapters 19 and 34. Lasers are described in Chapters 39 and 34. Natural sources are described in Chapters 17 and 34.

Tungsten Lamps

Fig. 8 shows the spectral output of a tungsten lamp at a temperature of 2800 K. A typical tungsten lamp is an efficient optical source, but

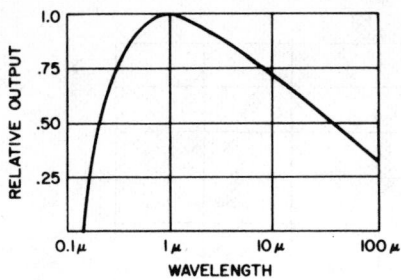

Fig. 8—Spectral output of a tungsten lamp at 2800 K.

only ~5% of its radiant flux falls within the visible wavelengths. Since a tungsten lamp requires a glass envelope to support the vacuum or to contain the inert gas required to prevent filament oxidation, the infrared output of the lamp is restricted to >3 μm unless an infrared transmitting window is employed.

Fluorescent Lamps

A typical fluorescent lamp is a sealed glass tube filled with argon gas and containing a small amount of mercury. When an electrical discharge is established in the tube, ultraviolet radiation is produced which causes a phosphor coating on the inside wall of the tube to fluoresce with a bright white glow.

Arc Lamps

Arc lamps operate at considerably higher temperatures than other sources and, with the exception of certain lasers, are the most brilliant artificial sources. A representative arc lamp is the xenon short-arc lamp shown schematically in Fig. 9. This lamp consists of a heavy-walled quartz envelope filled with xenon at a typical pressure of 20–40 atmospheres. Fig. 10 shows that the emission spectra from a typical xenon short-arc lamp originate both from the gaseous discharge and the incandescence of the electrodes.

OPTICAL DETECTORS

Optical detectors respond to either thermal or quantum effects induced by an optical radiation stimulus.

Thermal Detectors*

Thermal detectors respond to optical radiation-induced temperature variations and are therefore well-suited for broadband detection throughout

* W. L. Wolfe, *Handbook of Military Infrared Technology*, Office of Naval Research, Washington, DC, 1965.

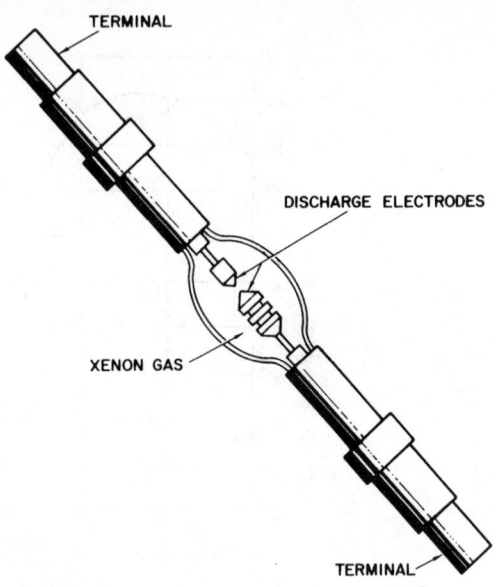

Fig. 9—Schematic representation of a xenon short-arc lamp. (*From F. M. Mims, III, "Optoelectronics," p. 43, 1975, Howard W. Sams & Co., Inc.*)

the optical spectrum. Thermal detectors include the bolometer, thermocouple, thermopile, thermopneumatic cell, and pyroelectric cell.

Bolometer: The bolometer changes its resistance in response to thermal energy resultant from impinging radiant energy. The most common bolometric detector is the termistor.

Thermocouple: A thermocouple is a junction of two dissimilar metals which, upon absorbing thermal energy, produces an emf.

Thermopile: The thermopile is an array of thermocouples. Miniature thermopiles made with thick film deposition techniques are commonly used in infrared detection applications.

Thermopneumatic Cell: The thermopneumatic cell senses the presence of thermal energy by means of a sealed cell which expands or contracts in response to variation in applied radiant energy. The magnitude of the expansion can be detected by means of interferometric techniques.

Pyroelectric Detector: The pyroelectric detector is a temperature sensitive current source. In a typical detector, a thin wafer of a ferroelectric crystal such as triglycine sulfate or lithium tantalate forms a capacitor whose capacitance is altered by thermal energy.

Quantum Detectors

Quantum detectors respond to variations in the number of incident photons. Quantum detectors

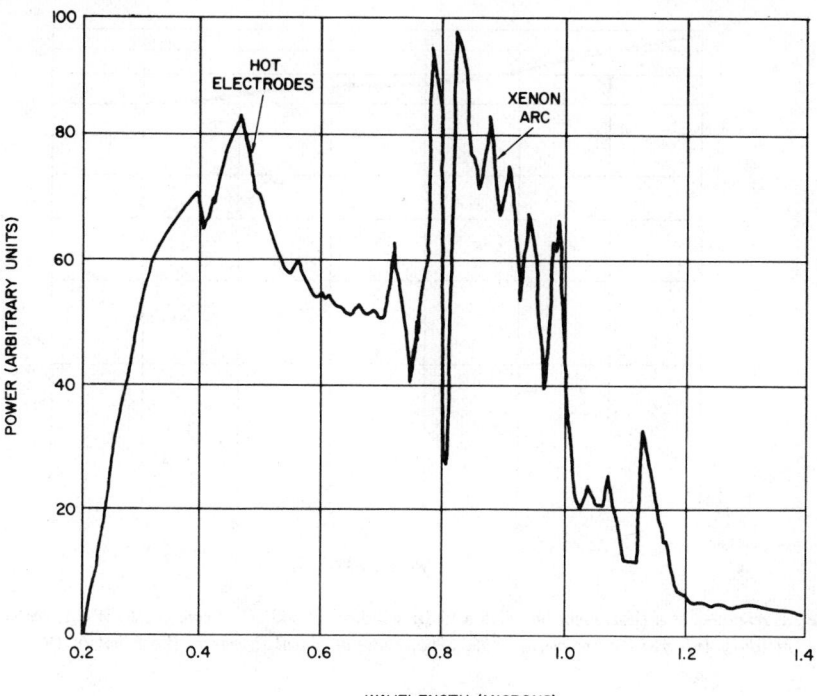

Fig. 10—Spectral output from a typical xenon short-arc lamp. (*Adapted from RCA Electro-Optics Handbook* (*EOH-11*), *p. 77, 1974, RCA Commercial Engineering, Harrison, NJ.*)

have a more limited spectral sensitivity range than thermal detectors but are characterized by relatively fast response time and high sensitivity. Quantum detectors include photovoltaic cells, photoconductive cells, photoelectromagnetic cells, and photoemissive devices. Photoemissive devices (photomultiplier tubes) are described in Chapters 17 and 34.

Photovoltaic Cells: The photovoltaic cell produces an emf when exposed to radiant energy of the appropriate wavelength. Photovoltaic cells are fabricated by producing a pn junction in such semiconductors as Si, Se, GaAs, InAs, and InSb. The quantum aspects of photovoltaic cells are described in Chapter 19.

Photoconductive Cells: The photoconductive cell responds to variations in radiant energy with a similar variation in the number of free charge carriers. Photoconductive detectors may be formed from pn junctions or unipolar semiconductors. Typical photoconductive semiconductors include CdS, CdSe, PbS, Si, PbSe, PbTe, Te, InSb, Ge:Au, Ge:Cu, Ge:Hg, Ge:Cd, Ge:Zn, Ge-Si:Zn, and Ge-Si:Au.

Photoelectromagnetic Cells: Photoelectromagnetic detectors produce an output voltage which varies in response to incident optical radiation. The voltage originates from separation of charge carriers by a magnetic field.

Spectral Response of Optical Detectors

A knowledge of detector spectral response is essential when matching a detector to a source. The spectral response of various detectors is summarized in Figs. 11 through 15.

OPTICAL COMPONENTS

Optical components are used both to manipulate and control optical radiation and to provide optical access to various sources and sensors.

Optical Materials

Glass is the most common optical material at visible and near-infrared wavelengths, but other wavelengths require more exotic materials. Fig. 16 shows the spectral transmission of several optical materials suitable for ultraviolet, visible, and near-infrared wavelengths. Fig. 17 shows the spectral transmission of several calcium aluminate glasses suitable for middle infrared wavelengths. Fig. 18 shows the spectral transmittance of a wide range of optical materials suitable for transmission throughout the optical spectrum.

The Thin Simple Lens

A thin simple lens refracts an optical ray passing through it as shown in Fig. 19. A lens may be

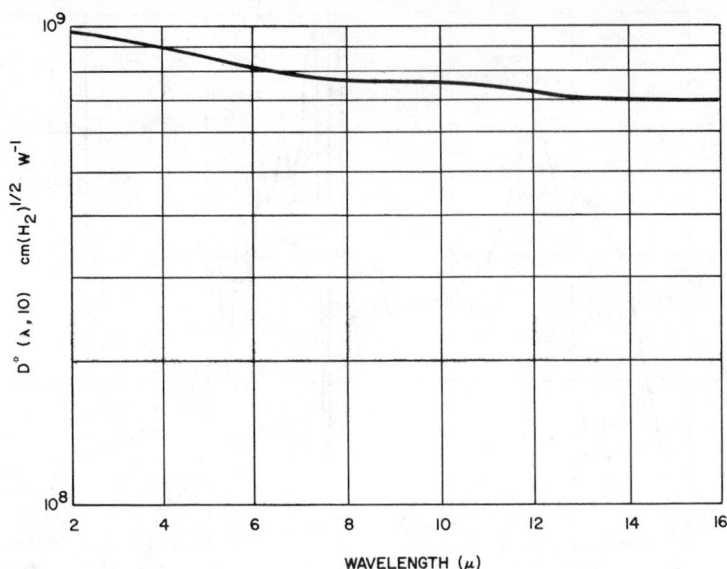

Fig. 11—Spectral response of a thermocouple with a CsBr window at 300 K. (*From W. L. Wolfe, editor, Handbook of Military Infrared Technology, p. 499, 1965, Office of Naval Research, Washington, DC.*)

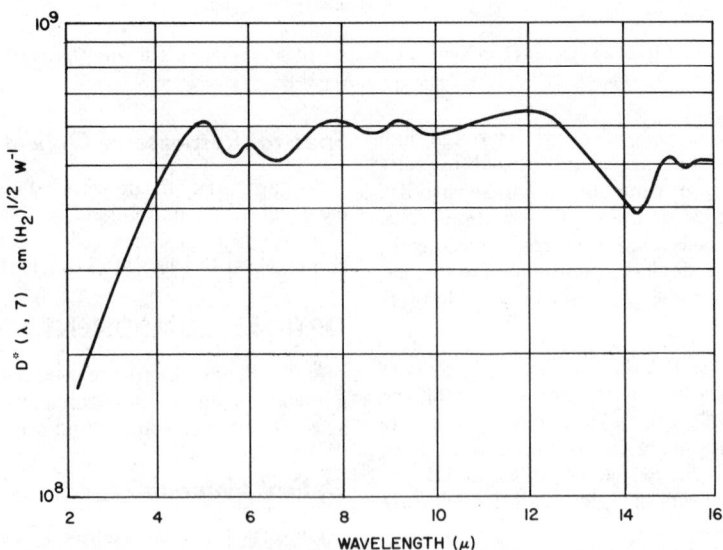

Fig. 12—Spectral response of a typical thermistor at 300 K. (*From W. L. Wolfe, editor, Handbook of Military Infrared Technology, p. 498, 1965, Office of Naval Research, Washington, DC.*)

either positive (converging) or negative (diverging). The focal point of a lens is that point at which the image of an infinitely distant point source is reproduced. Both the source and the focal point lie on the lens axis. The focal length is the distance between the lens and the focal point. The focal point of a converging lens is real and positive. The focal point of a diverging lens is virtual and negative. The relationship of the focal length (f) to the distances between the lens and the object being imaged (s) and the lens and the focused image (s') is given by the gaussian form of the thin lens equation

$$\frac{1}{s} + \frac{1}{s'} = \frac{1}{f}$$

The combined focal length for two thin lenses in contact or close proximity and having the same

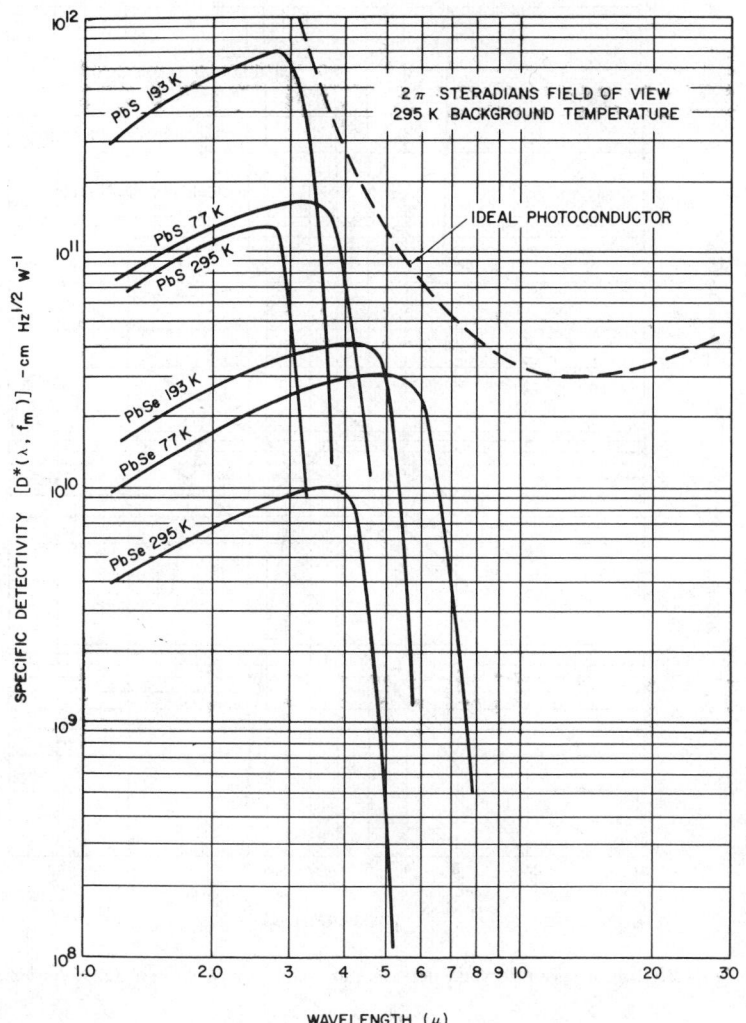

Fig. 13—Spectral detectivities of several thin film detectors at frequency f_m. (*From Santa Barbara Research Center.*)

optical axis is given by

$$\frac{1}{f} = \frac{1}{f_1} + \frac{1}{f_2}$$

The relationship between the focal length f and the refractive index (n) is

$$\frac{1}{f} = (n-1)\left(\frac{1}{r_1} - \frac{1}{r_2}\right)$$

where r_1 is the radius on the left lens surface and r_2 is the radius on the right lens surface.

The magnification (m) of a lens is given by

$$m = \frac{s'}{s} = \frac{\text{image size}}{\text{object size}}$$

The f/number of a lens defines its light collecting ability and is given by

$$f/\text{number} = \frac{f}{D}$$

where D is the diameter of the lens. A small f/number denotes a large lens diameter for a specified focal length and a higher light collecting ability or faster "speed" than a large f/number.

Numerical aperture (N.A.) is a measure of the acceptance angle of a lens and is given by

$$\text{N.A.} = n \sin\theta$$

where n is the refractive index of the object or image medium (air$\approx$1) and θ is half the maximum acceptance angle. When $n\approx1$ and θ is small, N.A.$\approx1/(2f/\text{no})$.

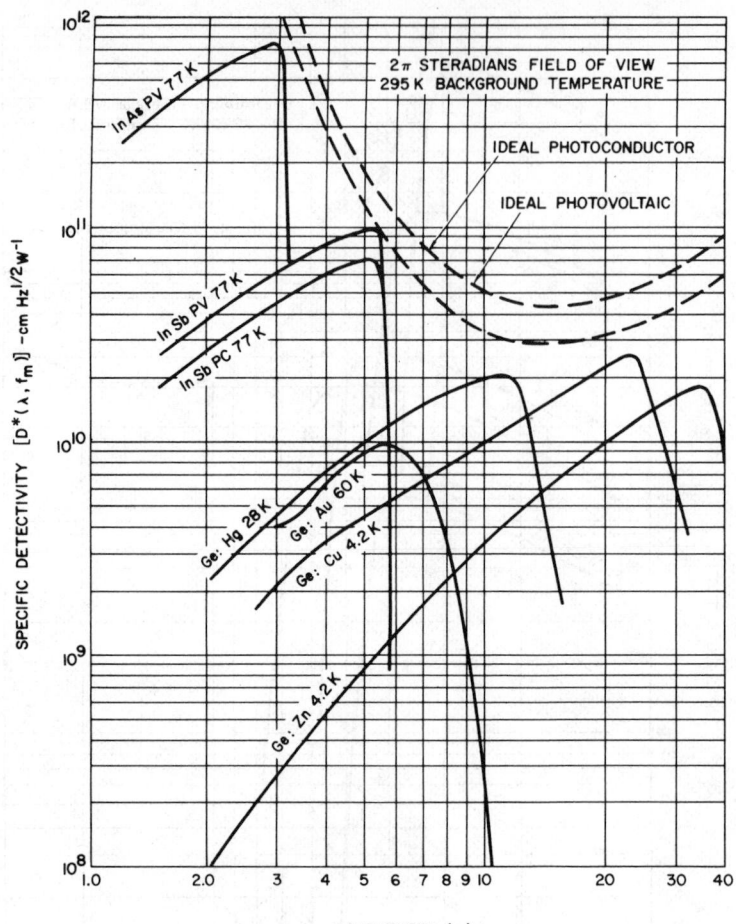

Fig. 14—Spectral detectivities of several crystal detectors at frequency f_m. (*From Santa Barbara Research Center.*)

Diffraction Limited Thin Lens

The performance of a lens or any other optical component or system of components which has no aberrations is limited only by aperture-induced diffraction, and such a lens or component is said to be diffraction limited. Monochromatic light of uniform transverse intensity collected by a diffraction limited spherical lens will be focused to a central spot called the Airy disc surrounded by several concentric and alternating light and dark rings. The diameter of the Airy disc is given by

$$d_u = 2.44\lambda(f/\text{no}).$$

Monochromatic light of gaussian transverse intensity distribution such as that emitted by many lasers will be focused to a spot having a diamter at the $1/e^2$ points of

$$d_g = 1.22\lambda(f/\text{no}).$$

Fig. 20 illustrates the performance of a diffraction limited lens.

Lens Aberrations*

Discounting the diffraction limit, the ability of a lens to produce a perfect image is limited by certain inherent aberrations. Lens aberrations result from the physical constraints of refraction as well as imperfect lens construction. The seven Seidel or primary aberrations are defined as follows:

Spherical Aberration: The variation of focus as a function of aperture wherein rays passing through the axis and peripheral regions of a lens are focused differently is spherical aberration.

Coma: The variation of focal length, hence magnification, with aperture is coma.

Astigmatism: The separate focusing of rays passing through the vertical and horizontal axes of a lens is astigmatism.

* W. L. Wolfe, *Handbook of Military Infrared Technology*, Office of Naval Research, Washington, DC, 1965.

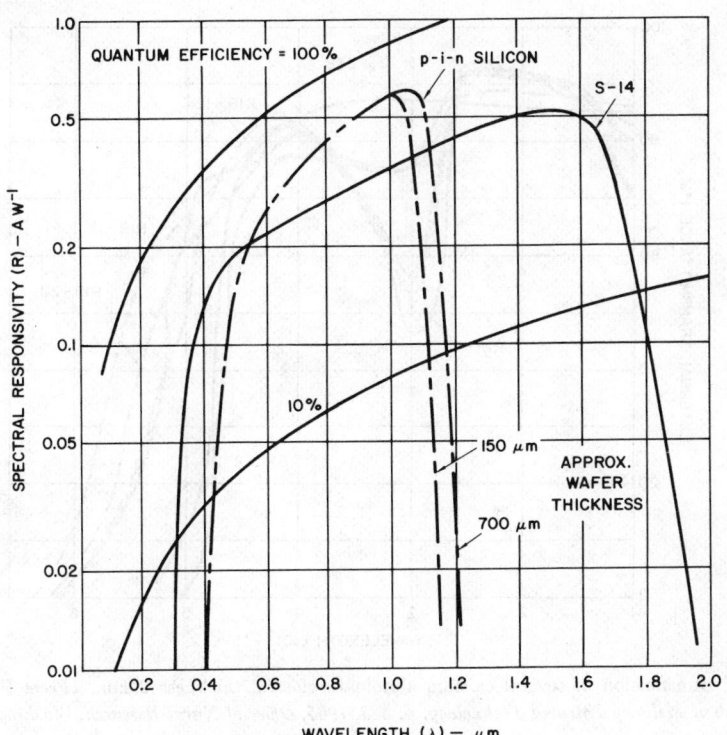

Fig. 15—Spectral response of pin Si and Ge photodetectors. (*From RCA Electro-Optics Handbook, p. 159, 1974, RCA Commercial Engineering, Harrison, NJ.*

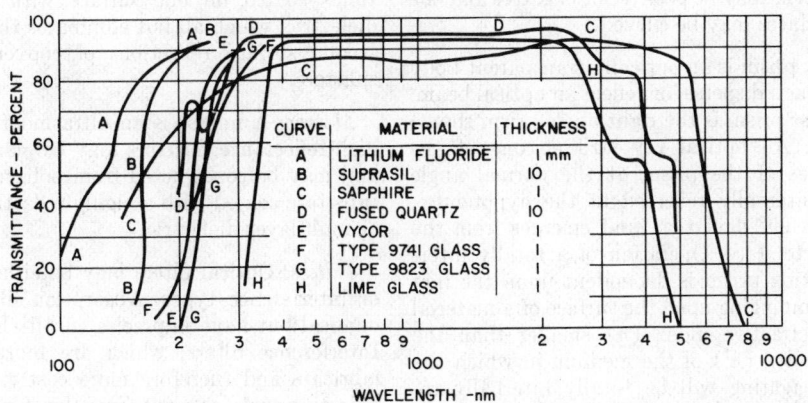

Fig. 16—Spectral transmission of several ultraviolet, visible, and near-infrared optical window materials. (*From RCA Electro-Optics Handbook, p. 220, 1974, RCA Commercial Engineering, Harrison, NJ.*)

Field Curvature: The appearance of an image focused by an astigmatic lens is distorted by field curvature.

Distortion: The variation of magnification with distance from the optical axis causes two types of distortion. Pincushion or positive distortion is when the image of a rectangle has concave sides. Barrel or negative distortion occurs when the image of a rectangle has convex sides.

Chromatic Distortion: The variation of focal length with wavelength is axial (longitudinal) chromatic distortion. Fig. 21 illustrates chromatic distortion and shows how a compound lens called an achromat can be used to correct it.

Optical Components

A variety of optical components is employed in optoelectronics and several of the more important components are described below.

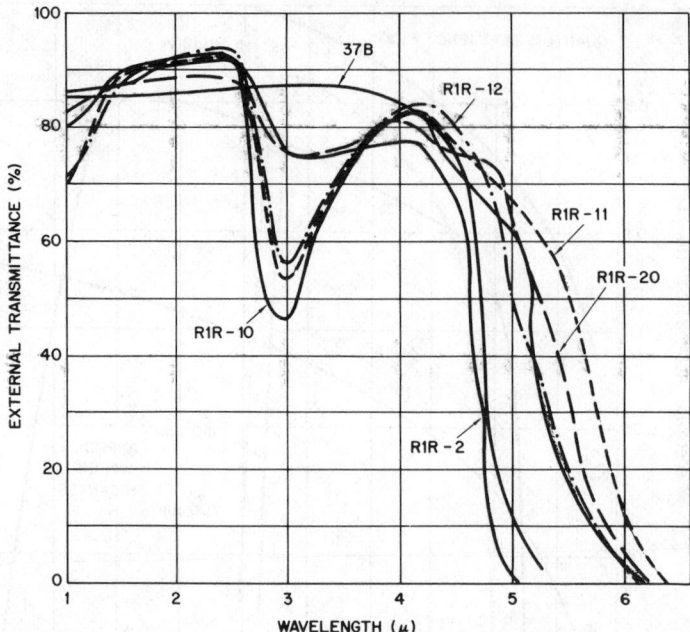

Fig. 17—Spectral transmission of several calcium aluminate glasses; thickness 2 mm. (*From W. L. Wolfe, editor, Handbook of Military Infrared Technology, p. 323, 1965, Office of Naval Research, Washington, DC.*)

Cylinder Lens: A cylinder lens is a section of a cylinder and therefore magnifies in only one plane. A cylinder lens may be positive or negative and one or both surfaces may be curved.

Prism: A prism is an optically transparent body used to refract, disperse, or reflect an optical beam. The simplest prism is the right angle prism shown in Fig. 22. An optical ray striking one of the shorter faces of the prism at the normal angle is totally internally reflected at the hypotenuse, undergoes a 90° deviation, and emerges from the second shorter face. Operation of a totally internally reflecting prism is dependent upon the fact that a ray impinging upon the surface of a material having a refractive index (n) smaller than the refractive index (n') of the medium in which the ray is propagating will be totally internally reflected when the angle of incidence is greater than a certain critical angle (ϕ) given by

$$\text{Sin}\phi_c = \frac{n'}{n}$$

Collimator (Beam Expander): The combination of two simple lenses shown in Fig. 23 is commonly used to increase the diameter of a laser beam while reducing its divergence.

Beam Splitter: Beam splitters are used to divide an optical beam and are of two basic types. A conventional beam splitter is ordinarily a thin plane parallel plate, one surface of which is usually coated with a partially reflecting film of thin

metal or a multilayer dielectric. A pellicle beam splitter is an ultrathin ($<10\,\mu$m) membrane, sometimes coated on one surface with a multilayer dielectric, which all but eliminates the undesirable second surface reflections of conventional beam splitters.

Mirror: A mirror is an ultrasmooth material of high reflectance. Mirrors may be planar or curved and may be constructed from polished metal or a substrate coated with a highly reflective metal film or multilayer dielectric.

Filters: Optical filters may be either absorption or interference types. Absorption filters are economical but have imprecise cutoff characteristics. Interference filters, which are more difficult to fabricate and therefore more costly, have highly reflecting surfaces made from thin films of metal or multilayer dielectrics which form a fixed Fabry-Perot interferometer which transmits a very narrow spectral band. Fig. 24 compares the spectral transmittance of a typical absorption and interference filter. Optical filters are designated according to their transmission characteristics as low-pass, bandpass, or high-pass.

Optical Waveguide Fibers: An optical waveguide fiber is a thin, flexible medium having a refractive index which exceeds that of the surrounding medium and is capable of conducting an optical ray by means of total internal reflection or guiding induced by a graded refractive index. Optical fibers are described in more detail in Chapter 34.

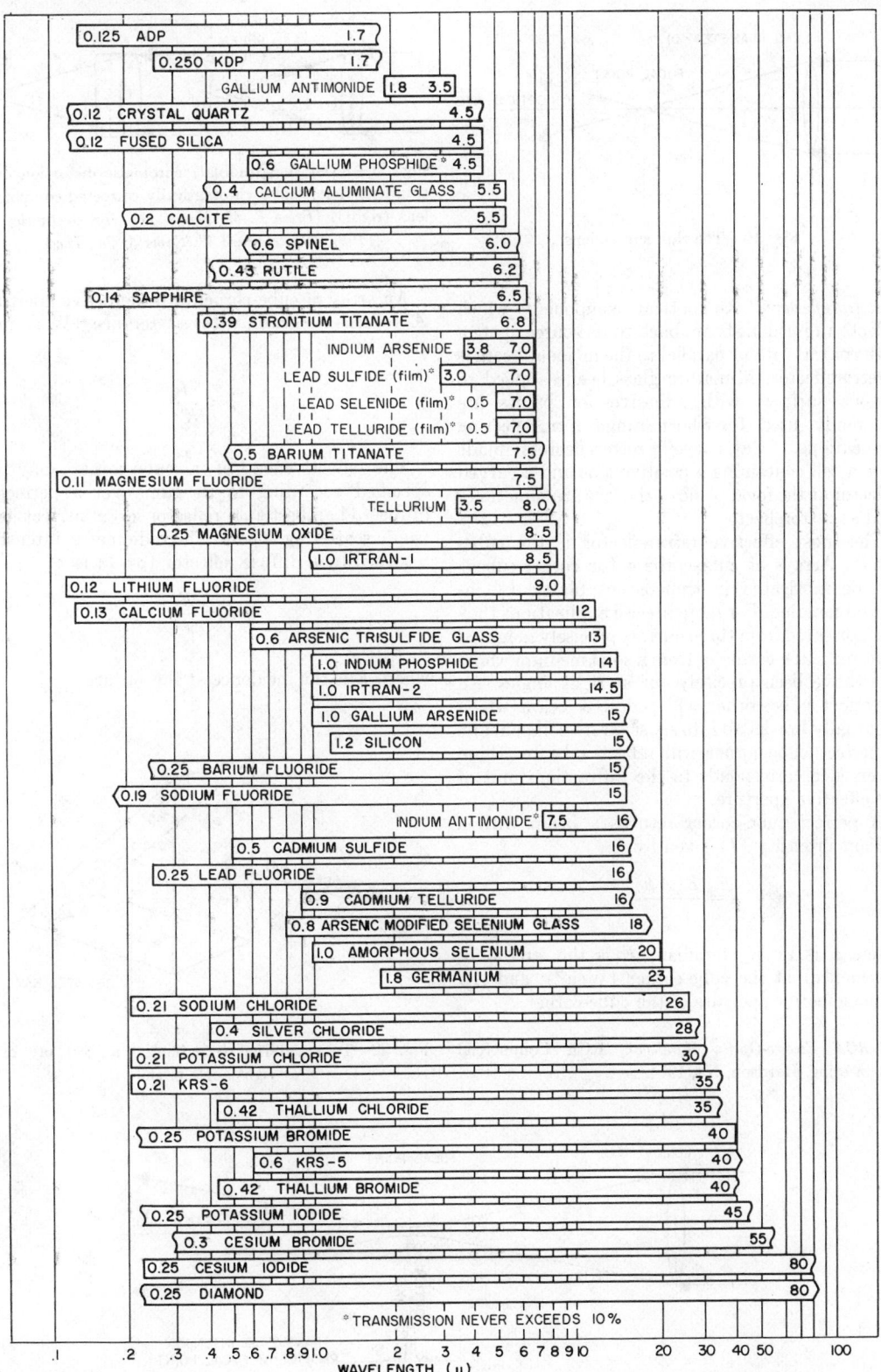

Fig. 18—Spectral transmission of a range of IR-transmitting materials; cutoff is 10% transmittance. (*From W. L. Wolfe, editor, Handbook of Military Infrared Technology, p. 327, 1965, Office of Naval Research, Washington, DC.*)

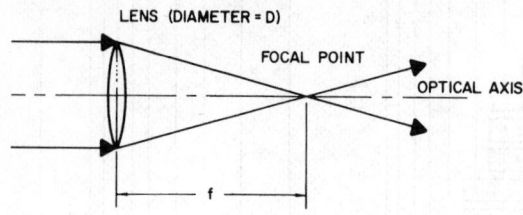

Fig. 19—The thin simple lens.

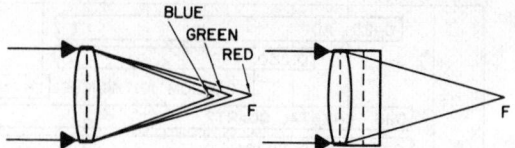

Fig. 21—Axial (longitudinal) chromatic distortion in a thin lens (left) and a chromatically corrected compound lens (right). (*From F. M. Mims, III, Optoelectronics, p. 109, 1975, Howard W. Sams & Co., Inc.*)

Retroreflector*: An optical component which reflects an incident beam back to its source along a path coaxial with or parallel to the incident beam is a retroreflector. Miniature glass beads applied to various surfaces with adhesives or paints are commonly used for short range retroreflecting applications. "Cat's eye" retroreflectors made from a cell containing a positive lens and a curved reflector at the focal plane of the lens are sometimes used as retroreflectors.

The most effective retroreflector is the cube-corner. Arrays of cube-corners (or corner-cubes) can be fabricated in transparent plastic by injection molding. For more precise applications they are fabricated from three mirrors precisely mounted to form a cube-corner or from a solid medium whose sides have been precisely cut at right angles. An imperfect cube-corner will reflect a beam which eventually breaks up into six separate components. A perfect cube-corner will reflect a beam whose divergence corresponds to the diffraction limit of the effective aperture.

A perfect cube-corner returns a beam with a radiant intensity (I) given by

$$I = \frac{E(A_e\theta)^2}{2}$$

where I is in W/steradian, E is the irradiance (incidence) at the cube-corner (W/m^2), and $A_e\theta$ is the effective aperture of the cube-corner.

* *RCA Electro-Optics Handbook*, RCA Commercial Engineering, Harrison, NJ, 1974.

An array of cube-corners with effective aperture A returns a beam with a sterance (W/sr/m^2) given by

$$I = \frac{I}{EA}$$

The effectiveness of a cube-corner may be specified according to its gain over a perfectly diffuse (Lambertian) reflector or a perfect isotropic scattering sphere. The radiometric intensity returned by a diffuse reflector ($\rho = 1$) is

$$I_d = \frac{E}{\pi}$$

where E is the incidence at the surface.

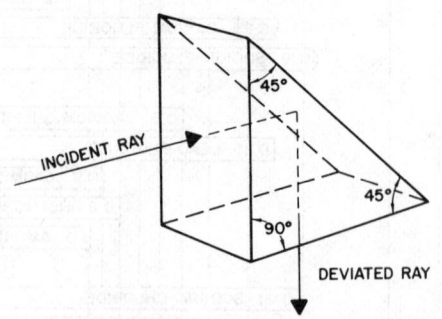

Fig. 22—Total internal reflection of a light ray in a right angle prism.

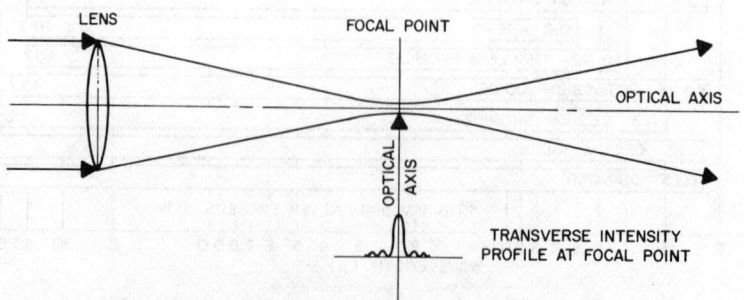

Fig. 20—The diffraction-limited thin simple lens.

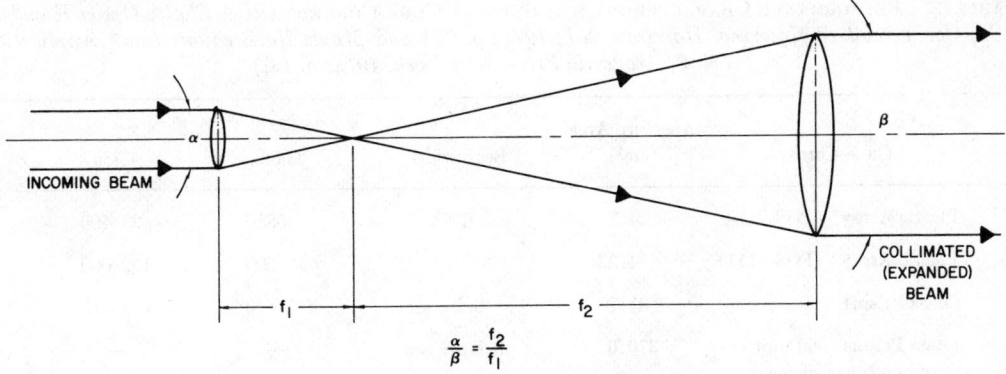

Fig. 23—Operation of a collimator (beam expander). (*Adapted from S. S. Charschan, editor, Lasers in Industry, p. 617, 1972, Van Nostrand Reinhold Co., NY.*)

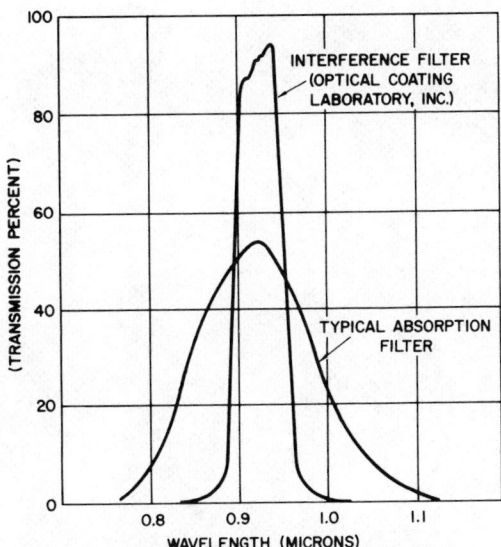

Fig. 24—Spectral transmission of a typical multilayer dielectric interference filter and a typical broadband absorption filter. (*From F. M. Mims, III, Light Beam Communications, p. 108, 1975, Howard W. Sams & Co., Inc.*)

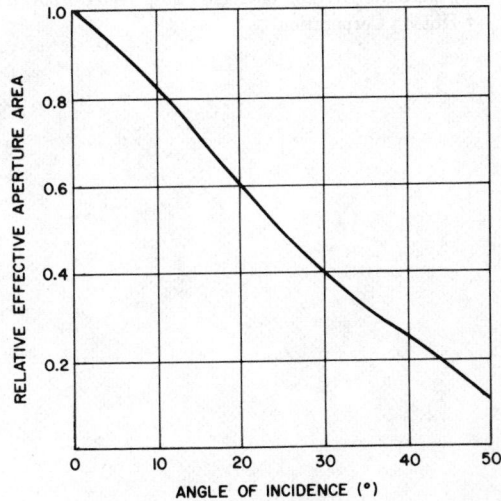

Fig. 25—Effective aperture area of a cube-corner retro-reflector as a function of the angle of incidence. (*From H. H. Plotkin, Geos-I Laser Retroreflector Design and Preliminary Signal Calculations, NASA Rep. No. X-524-64-205, 1964.*)

The radiometric intensity returned from an isotropic scattering sphere of area A is

$$E_i = \frac{EA}{4\pi}$$

The gain of a perfect cube-corner over an equal aperture perfectly diffuse reflector is then

$$\text{Gain}_d = \pi \frac{I}{EA}$$

and the gain of a perfect cube-corner over an equal area perfect isotropic scattering sphere is then

$$\text{Gain}_i = 4\pi \frac{I}{EA}$$

Retroreflectors are particularly useful in long range applications where precise optical alignment is difficult or impossible since they do not have to be aligned normal to the axis of the oncoming beam. However, the misalignment of a cube-corner with respect to the optical axis reduces its effective aperture as can be seen in Fig. 25. Table 2 tabulates the performance of several cube-corners and cube-corner arrays.

TABLE 2. PERFORMANCE CHARACTERISTICS OF SEVERAL CUBE-CORNERS (*RCA Electro-Optics Handbook, RCA Commercial Engineering, Harrison, NJ, 1974, p. 224 and Monte Ross, editor, Laser Applications, Vol. 2, Academic Press, New York, 1974, p. 14.*)

Cube-Corner	Aperture Area (cm²)	Beamwidth	Gain$_d$	Gain$_i$
Plastic Array* (FOS-21)	24.7	0.7°	5850	23 400
Plastic Array* (FOS-3111)	4.32	0.3°	40 600	162 000
Glass Prism†	31.7	6 arc sec	4.7×10^9	1.9×10^{10}
Glass Prisms (300 unit array on lunar surface)	370.0	3.5 arc sec	—	—
Glass Prisms (400 unit array on GEOS 2)	1100.0	20 arc sec	—	—

* Stimsonite Division of Elastic Stop Nut Corporation
† Hutson Corporation

CHAPTER 23
MODULATION

Modulation is a process whereby certain characteristics of a wave (often called a carrier) are varied or selected in accordance with a message signal. Modulation can be divided into continuous modulation in which the modulated wave is always present and pulsed modulation in which no signal is present between pulses.

PART 1—CONTINUOUS MODULATION

In continuous modulation* the modulated carrier can be given by the expression $s(t) = A(t) \cos\theta(t)$, where $A(t)$ is the *instantaneous amplitude* and $\theta(t)$ is the *instantaneous phase*. For a sinusoidal carrier of angular frequency ω_c, this expression reduces to $s(t) = A(t) \cos[\omega_c t + \phi(t)]$, where $\phi(t)$ is the carrier phase. When the instantaneous amplitude $A(t)$ is varied linearly by the message function and the carrier phase is constant, the process is called *amplitude modulation*; when the carrier phase angle $\phi(t)$ is modulated by the message function, the process is called *angular or phase modulation*.

The concept of rotating vector can be used to represent a sinusoidal vector modulated in both amplitude and phase as shown in Fig. 1, where $s(t)$ is represented as the projection of a rotating vector on a fixed reference axis.

$$s(t) = A(t) \cos[\omega_c t + \phi(t)]$$

$$= \mathrm{Re}(A(t) \exp\{j[\omega_c t + \phi(t)]\}).$$

$A(t)$ represents the envelope of the modulated

* P. F. Panter, "Modulation, Noise, and Spectral Analysis," Chapters 5 and 6, McGraw-Hill Book Co., New York, N. Y.; 1965.

carrier and $\phi(t)$ is the modulated phase. The vector rotates with an instantaneous angular frequency $\omega_i(t)$ which is given by

$$\omega_i(t) = \omega_c + [d\phi(t)/dt].$$

In amplitude modulation only the amplitude changes, and the general expression reduces to

$$s(t) = \mathrm{Re}[A(t) \exp(j\phi_0) \cdot \exp(j\omega_c t)]$$

while in phase modulation, only the phase changes so that

$$s(t) = \mathrm{Re}\{A_c \exp[j\phi(t)] \cdot \exp(j\omega_c t)\}$$

where A_c is constant.

ANALYTIC SIGNAL REPRESENTATION OF MODULATED WAVEFORMS

A real signal

$$s(t) = A(t) \cos[\omega_c t + \phi(t)]$$

may be expressed either as

$$s(t) = \mathrm{Re}(A(t) \exp\{j[\omega_c t + \phi(t)]\})$$

or as

$$s(t) = \mathrm{Re}[\psi(t)]$$

where $\psi(t)$ is the analytic signal defined by

$$\psi(t) = s(t) + j\hat{s}(t).$$

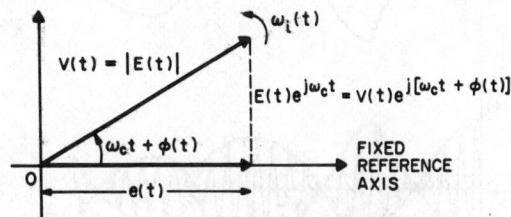

Fig. 1—Fixed-reference vector diagram. *From P. F. Panter, "Modulation, Noise, and Spectral Analysis," Fig. 2-7, © 1965, McGraw-Hill Book Company.*

The function $\hat{s}(t)$ is the Hilbert transform of $s(t)$, namely

$$\hat{s}(t) = \pi^{-1} \int \frac{s(t)}{t-\tau} \, d\tau.$$

Basically, the analytic signal $\psi(t)$ is a complex function of a real variable whose real and imaginary parts form a Hilbert pair. The analytic signal is simply a formalized version of the "rotating vector" discussed above. If $S(j\omega)$ is the Fourier transform of $s(t)$, then $\Psi(j\omega)$, the Fourier transform of $\psi(t)$, can be written in terms of $S(j\omega)$ as

$$\Psi(j\omega) = 2S(j\omega) \qquad \omega > 0$$
$$= S(j\omega) \qquad \omega = 0$$
$$= 0 \qquad \omega < 0.$$

Also, $\hat{S}(j\omega)$, the Fourier transform of $\hat{s}(t)$, is given by

$$\hat{S}(j\omega) = -j(\mathrm{sgn}\omega) S(j\omega)$$

where

$$\mathrm{sgn}x = 1 \qquad x > 0$$
$$= 0 \qquad x = 0$$
$$= -1 \qquad x < 0$$

and $\mathrm{sgn}x$ is the signum function.

AMPLITUDE MODULATION

In amplitude modulation, the frequency components of the modulating signal are translated to occupy a different position in the spectrum. It is essentially a multiplication process in which the time functions that describe the modulating signal and carrier are multiplied together. The following amplitude-modulation systems are discussed.

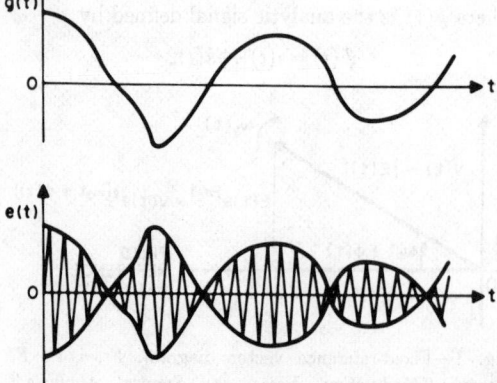

Fig. 2—Double-sideband waveforms. *From P. F. Panter, "Modulation, Noise, and Spectral Analysis," Fig. 5-3, © 1965, McGraw-Hill Book Company.*

(**A**) Double-sideband suppressed carrier (DSB–SC), also called DSB.

(**B**) Conventional amplitude modulation (AM).

(**C**) Vestigial sideband.

(**D**) Single sideband (SSB).

Double Sideband (DSB)

In DSB modulation the message signal $g(t)$, whose Fourier transform is $G(j\omega)$, is considered to have zero dc component. The product

$$e(t) = A_c g(t) \cos\omega_c t$$

represents a double-sideband suppressed-carrier signal and A_c = amplitude of unmodulated carrier. The radio-frequency envelope follows the wave-

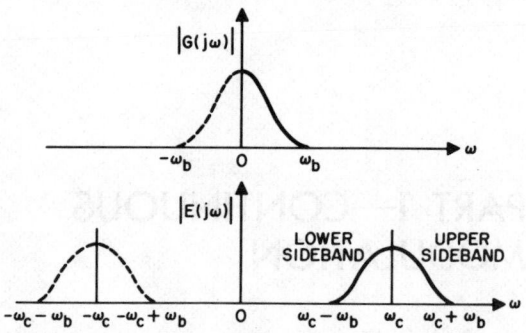

Fig. 3—Baseband signal and double-sideband spectra. *From P. F. Panter, "Modulation, Noise, and Spectral Analysis," Fig. 5-2, © 1965, McGraw-Hill Book Company.*

form of the modulating signal $g(t)$ as shown in Fig. 2. The spectral components of the DSB signal $e(t)$ are given by its Fourier transform

$$E(j\omega) = \tfrac{1}{2}G[j(\omega-\omega_c)] + \tfrac{1}{2}G[j(\omega+\omega_c)]$$

as shown in Fig. 3. Note that the upper and lower sidebands are translated symmetrically $\pm\omega_c$ about the origin.

Conventional Amplitude Modulation (AM)

In amplitude modulation a dc term is added to the modulating signal $g(t)$. The resulting waveform shown in Fig. 4 is given by

$$e(t) = [A_0 + as(t)] \cos\omega_c t = A_0[1 + m_a s(t)] \cos\omega_c t$$

where a = maximum amplitude of modulating function, $g(t) = as(t)$, $|s(t)| \le 1$; $m_a = a/A_0$ = modulation index or degree of modulation, $0 \le m_a \le 1$; A_0 = amplitude of unmodulated carrier; and $|m_a s(t)| \le 1$, to ensure an undistorted envelope.

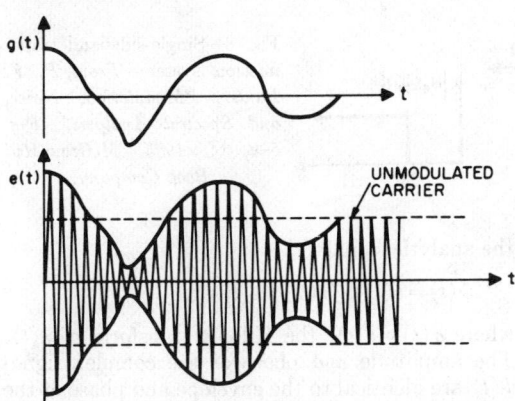

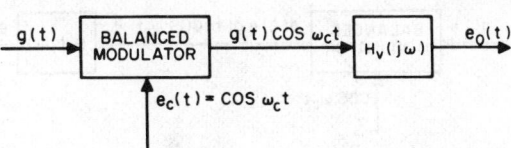

Fig. 6—Vestigial-sideband transmission system. *From P. F. Panter, "Modulation, Noise, and Spectral Analysis," Fig. 5-7, © 1965, McGraw-Hill Book Company.*

Fig. 4—Amplitude modulation of a carrier. The modulating signal is at top and the amplitude-modulated carrier at bottom. *From P. F. Panter, "Modulation, Noise, and Spectral Analysis," Fig. 5-4, © 1965, McGraw-Hill Book Company.*

For a signal $g(t)$ consisting of a sum of M sinusoidal components

$$g(t) = \sum_{k=1}^{M} a_k \cos(\omega_k t + \theta_k)$$

and

$$e(t) = A_0\left[1 + A_0^{-1}\sum_{k=1}^{M} a_k \cos(\omega_k t + \theta_k)\right]\cos\omega_c t$$

$$= \underbrace{A_0 \cos\omega_c t}_{\text{carrier}} + \underbrace{\tfrac{1}{2}a_1 \cos[(\omega_c+\omega_1)t+\theta_1]}_{\text{upper sideband}}$$

$$+ \underbrace{\tfrac{1}{2}a_1 \cos[(\omega_c-\omega_1)t-\theta_1]}_{\text{lower sideband}}+\cdots$$

$$+ \underbrace{\tfrac{1}{2}a_M \cos[(\omega_c+\omega_M)t+\theta_M]}_{\text{upper sideband}}$$

$$+ \underbrace{\tfrac{1}{2}a_M \cos[(\omega_c-\omega_M)t-\theta_M]}_{\text{lower sideband}}$$

where a_k is the amplitude and ω_k is the angular frequency of the kth component of the modulating signal, and θ_k is the constant-phase part of its phase. Each frequency component gives rise to a pair of sidebands $\omega_c \pm \omega_k$ symmetrically located about the carrier frequency ω_c (Fig. 5).

$$\text{Degree of peak modulation} = A_0^{-1}\sum_{k=1}^{M} a_k,$$

for ω_k not harmonically related.

$$\text{Degree of rms modulation} = A_0^{-1}\left(\sum_{k=1}^{M} a_k^2\right)^{1/2}.$$

Vestigial Sideband

Vestigial-sideband modulation is derived from a DSB signal by passing the output of the product modulator through a filter whose transfer function is $H_v(j\omega)$, as shown in Fig. 6. The transfer function $H_v(j\omega)$ of the filter treats the two sidebands of the DSB signal in such a manner as to attenuate one sideband differently from the other. The process of vestigial-sideband modulation by the use of the filter network $H_v(j\omega)$ may be replaced by an equivalent vestigial system shown in Fig. 7, where the transfer functions $H_i(j\omega)$ and $H_q(j\omega)$ are given by

$$H_i(j\omega) = \tfrac{1}{2}\{H_v[j(\omega-\omega_c)] + H_v[j(\omega+\omega_c)]\}$$

$$H_q(j\omega) = (1/2j)\{H_v[j(\omega-\omega_c)] - H_v[j(\omega+\omega_c)]\}.$$

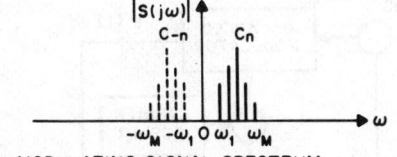

A. MODULATING SIGNAL SPECTRUM

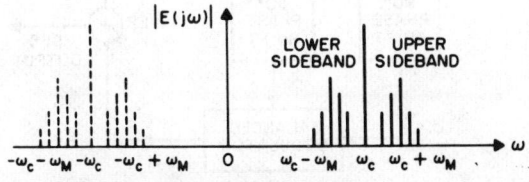

B. AMPLITUDE-MODULATED SPECTRUM

Fig. 5—AM spectrum—periodic modulating signal. *From P. F. Panter, "Modulation, Noise, and Spectral Analysis," Fig. 5-5, © 1965, McGraw-Hill Book Company.*

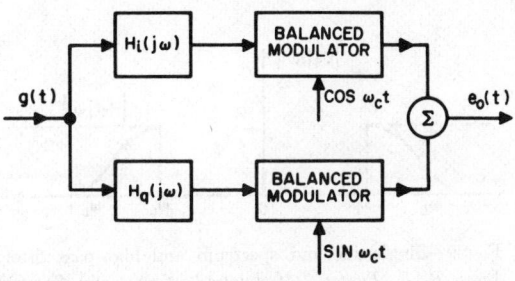

Fig. 7—Equivalent vestigial-sideband transmission system. *From P. F. Panter, "Modulation, Noise, and Spectral Analysis," Fig. 5-8, © 1965, McGraw-Hill Book Company.*

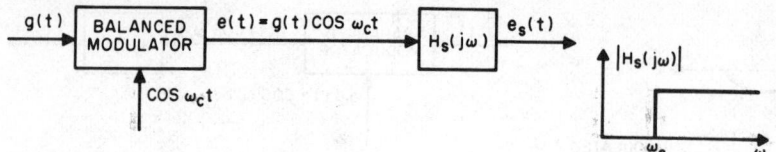

Fig. 8—Single-sideband transmission system. *From P. F. Panter, "Modulation, Noise, and Spectral Analysis," Fig. 5-9, © 1965, McGraw-Hill Book Company.*

Single Sideband (SSB)

Single-sideband transmission may be produced in the same manner as vestigial sideband by using a high-pass filter $H_s(j\omega)$ which completely eliminates all signals on one side of the carrier frequency, as shown in Fig. 8. The transfer function $H_s(j\omega)$ of the ideal high-pass filter is defined by

$$H_s(j\omega) = [\tfrac{1}{2} + \tfrac{1}{2} \operatorname{sgn}(\omega - \omega_c)] + [\tfrac{1}{2} - \tfrac{1}{2} \operatorname{sgn}(\omega + \omega_c)]$$

where $\operatorname{sgn}\omega$ is the signum function. The output spectrum $E_s(j\omega)$ is given by

$$E_s(j\omega) = H_s(j\omega) E(j\omega)$$

$$= \tfrac{1}{2} G[j(\omega - \omega_c)][\tfrac{1}{2} + \tfrac{1}{2} \operatorname{sgn}(\omega - \omega_c)]$$

$$+ \tfrac{1}{2} G[j(\omega + \omega_c)][\tfrac{1}{2} - \tfrac{1}{2} \operatorname{sgn}(\omega + \omega_c)]$$

and is shown in Fig. 9.

The SSB signal can also be regarded as the resultant of quadrature modulation of a carrier by a pair of signals in phase quadrature (Fig. 10). The modulated wave

$$e_s(t) = s(t) \cos\omega_c t - \sigma(t) \sin\omega_c t$$

represents an upper-sideband signal with no spectral components below the carrier angular frequency ω_c, where $s(t)$ is an arbitrary message function and $\sigma(t)$ its harmonic conjugate.

This equation can be written in the form

$$e_s(t) = [s^2(t) + \sigma^2(t)]^{1/2} \cos\{\omega_c t + \tan^{-1}[\sigma(t)/s(t)]\}$$

$$= \alpha(t) \cos[\omega_c t + \phi(t)]$$

regarding the single-sideband signal as a hybrid amplitude-modulated and phase-modulated wave. The envelope $\alpha(t)$ and phase $\phi(t)$ are related by

the analytic signal

$$\psi(t) = s(t) + j\sigma(t) = \alpha(t) \exp[j\phi(t)]$$

where $\sigma(t) = \hat{s}(t)$, the Hilbert transform of $s(t)$. The amplitude and phase of the complex signal $\psi(t)$ are identical to the envelope and phase of the single-sideband wave. The Fourier transform of the analytic signal $\psi(t)$ is

$$\Psi(j\omega) = S(j\omega) + jS(j\omega)$$

$$= S(j\omega) + S(j\omega) = 2S(j\omega), \qquad \omega > 0$$

$$= S(j\omega) - S(j\omega) = 0, \qquad \omega < 0.$$

Thus, a study of single sideband can be made through the analytic signal without reference to the arbitrary carrier frequency ω_c.

DEMODULATION OR DETECTION OF AMPLITUDE MODULATION

The process of separating the modulating signal from a modulated carrier is called demodulation or detection. In DSB or SSB detection, the detector must be supplied with a carrier wave that is synchronized with the wave used at the transmitter. This method of detection is called coherent or synchronous detection. In conventional amplitude-modulation systems, coherent detection is not necessary and the modulating signal may be recovered by the use of envelope detection, e.g., the modulated carrier is applied to a half-wave rectifier

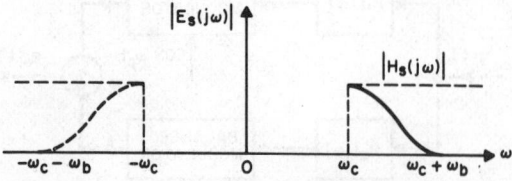

Fig. 9—Single-sideband spectrum and high-pass filter. *From P. F. Panter, "Modulation, Noise, and Spectral Analysis," Fig. 5-10, © 1965, McGraw-Hill Book Company.*

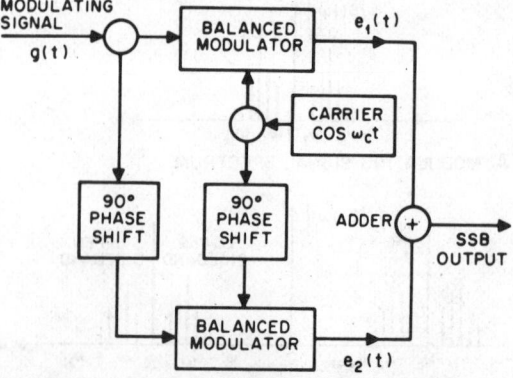

Fig. 10—Phase-shift method of generating SSB. *From P. F. Panter, "Modulation, Noise, and Spectral Analysis," Fig. 5-18, © 1965, McGraw-Hill Book Company.*

whose output is then filtered to provide the desired modulating signal.

DSB Detection

In DSB reception the incoming signal $e_r(t)$ is multiplied by a locally generated signal which is phase-synchronized with the carrier component of the received signal $e_r(t)$, as shown in Fig. 11. The detected output after filtering is given by

$$e_d(t) = kg(t) \cos(\phi_c - \phi_0), \qquad k = \text{constant}$$

where $(\phi_c - \phi_0)$ represents the phase difference between the transmitted carrier and the locally generated oscillator. When the local carrier is in phase with the incoming carrier, the detected signal is maximum. The output signal-to-noise ratio $(S/N)_o$ is related to the input signal-to-noise ratio $(S/N)_i$ by the expression

$$\frac{(S/N)_o}{(S/N)_i} = 2 \cos^2(\phi_c - \phi_0)$$

where the noise in each case is measured in a band occupied by the signal. This represents a maximum improvement of 3 decibels when the local oscillator is in phase with the incoming carrier.

AM Detection

Synchronous Detection:

$$\frac{(S/N)_o}{(S/N)_i} = \frac{2m_a^2 \langle g \rangle^2(t) \cos^2(\phi_c - \phi_0)}{1 + m_a^2 \langle g \rangle^2(t)}, \qquad |g(t)| \le 1$$

where $\langle g \rangle^2(t)$ equals the mean-square value of the message function, which is maximum for $m_a = 1$ and $\phi_c = \phi_0$.

Envelope Detection: In case of a carrier much stronger than the noise (high input carrier-to-noise ratio) we have

$$\frac{(S/N)_o}{(S/N)_i} = \frac{2m_a^2 \langle g \rangle^2(t)}{1 + m_a^2 \langle g \rangle^2(t)}$$

which is identical to the case of synchronous detection with $\phi_c = \phi_0$.

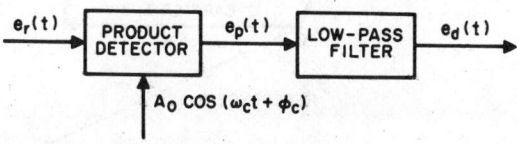

Fig. 11—Block diagram of double-sideband (DSB) receiver. *From P. F. Panter, "Modulation, Noise, and Spectral Analysis," Fig. 6-1, © 1965, McGraw-Hill Book Company.*

In case of poor input carrier-to-noise ratio, the message function $g(t)$ may be lost in the noise, which results in a threshold effect. This effect exists only in envelope detection and does not exist if synchronous or coherent detection is used.

SSB Detection

$$(S/N)_o/(S/N)_i = \cos^2(\phi_c - \phi_0)$$

where the signal component of the output is measured by the correlation of the detected output with the transmitted signal.

COMPARISON OF AMPLITUDE-MODULATION SYSTEMS

For equal power in the sidebands, the output signal-to-noise power ratios are identical.

For the same average total transmitted power, the following relations hold.

$$(S/N)_o(\text{DSB})/(S/N)_o(\text{AM}) = 1 + r^{-1}$$

where r equals the ratio of the mean-square power of the message function to its peak power, and

$$(S/N)_o(\text{DSB})/(S/N)_o(\text{SSB}) = 1.$$

For equal peak power

$$(S/N)_o(\text{DSB})/(S/N)_o(\text{AM}) = 4$$

for any waveform of the modulating signal.

To compare the merits of SSB versus DSB and AM on the basis of signal-to-noise ratio, the waveform of the modulating signal must be specified. This is illustrated in Fig. 12 for a modulating signal $\sin^\nu x$, $0 \le \nu \le 1$.

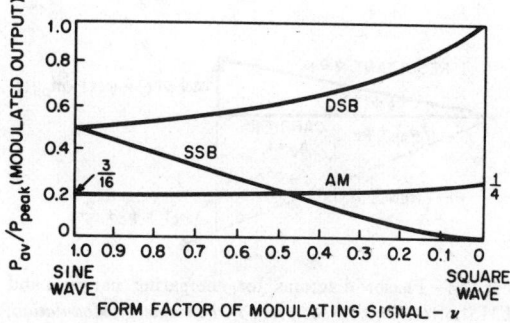

Fig. 12—Average-to-peak power relations as a function of modulating signal. *After W. K. Squires and E. Bedrosian, "The Computation of Single-Sideband Peak Power," Proceedings of the IRE, vol. 48, p. 124, Fig. 2; January 1960.*

EXPONENTIAL MODULATION

In exponential or angular modulation,* the carrier analytic signal $A_c \exp[j(\omega_c t+\phi_c)]$ is multiplied by the transformed message function $\exp[j\psi(t)]$ to produce an angle-modulated carrier analytic signal.

$$e(t) = \mathrm{Re}\{A_c \exp[j(\omega_c t+\phi_c)] \cdot \exp[j\psi(t)]\}$$
$$= \mathrm{Re}\{A_c \exp[j\phi(t)]\} \quad (1)$$

where

A_c = amplitude of unmodulated carrier

ω_c = angular frequency of unmodulated carrier

ϕ_c = carrier phase angle

$\phi(t) = [\omega_c t+\phi_c+\psi(t)]$

= instantaneous phase angle modulated by the message function $g(t)$.

Expanding equation (1) in powers of $\psi(t)$, we have

$$e(t) = \mathrm{Re}\{A_c \exp[j(\omega_c t+\phi_c)]$$
$$\times[1+j\psi(t)-(1/2!)\psi^2(t)-j(1/3!)\psi^3(t)+\cdots]\}.$$

When $|\psi(t)|_{\max} \gg 1$, we have nonlinear modulation since the carrier is multiplied by higher powers of $\psi(t)$. In case $|\psi(t)|_{\max} \ll 1$, the exponential modulation is approximately linear and is given by

$$e(t) \cong \mathrm{Re}\{A_c[1+j\psi(t)]\exp[j(\omega_c t+\phi_c)]\}.$$

Note that for amplitude modulation we have

$$e_{\mathrm{AM}}(t) = \mathrm{Re}\{A_c[1+m_a g(t)]\exp[j(\omega_c t+\phi_c)]\}.$$

The comparison of narrow-band angle modulation (small phase deviation) with AM is shown in Fig. 13. The general case when $|\psi(t)| \gg 1$ is illustrated in Fig. 14. Expressing equation (1) in the real form we obtain

$$e(t) = A_c[\cos\omega_c t+\phi_c+\psi(t)]$$

where for phase modulation

$$\psi(t) = m_p g(t), \ m_p = \text{constant}$$

and for frequency modulation

$$\psi(t) = m_f \int_0^t g(\tau)\,d\tau, \ m_f = \text{constant}.$$

The instantaneous frequency $\omega_i(t)$ is defined by

$$\omega_i(t) = \left[\frac{d\phi(t)}{dt}\right] = \left[\omega_c+\frac{d\psi(t)}{dt}\right].$$

In *phase modulation*, the instantaneous phase of the modulated signal varies proportionally with the modulating signal $g(t)$

$$e_{\mathrm{PM}}(t) = A_c \cos[\omega_c t+m_p g(t)]$$

where ϕ_c has arbitrarily been set to zero.

For single-tone sinusoidal modulation $g(t) = \cos\omega_m t$, we have

$$e_{\mathrm{PM}}(t) = A_c \cos(\omega_c t+m_p \cos\omega_m t)$$

where $m_p = \Delta\theta$, and the peak phase deviation is independent of ω_m.

The instantaneous frequency

$$\omega_i(t) = d\phi(t)/dt$$
$$= \omega_c - m_p \omega_m \sin\omega_m t$$

and the peak frequency deviation $\Delta\omega = m_p \omega_m$ is proportional to the modulating frequency ω_m.

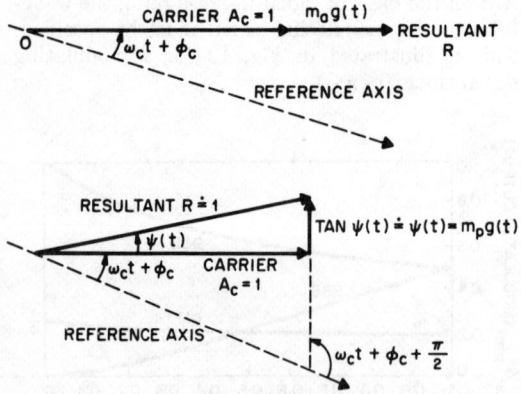

Fig. 13—Phasor diagrams for comparing narrow-band FM (bottom) with AM. *From P. F. Panter, "Modulation, Noise, and Spectral Analysis," Fig. 7-3, © 1965, McGraw-Hill Book Company.*

* P. F. Panter, "Modulation, Noise, and Spectral Analysis," Chapters 7, 11, 14, 15, and 16, McGraw-Hill Book Co., New York, N. Y.; 1965.

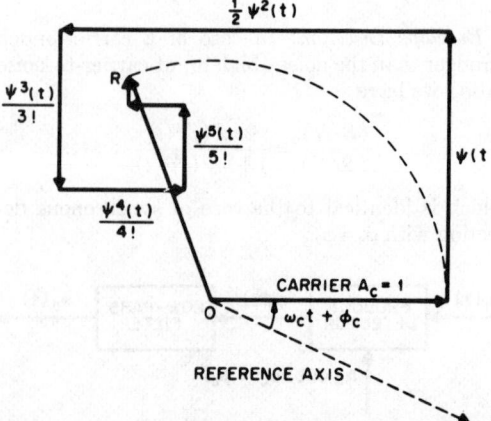

Fig. 14—Phasor diagram of exponentially modulated signal for large phase deviation. *From P. F. Panter, "Modulation, Noise, and Spectral Analysis," Fig. 7-4, © 1965, McGraw-Hill Book Company.*

In *frequency modulation*, the instantaneous frequency of the modulated signal is proportional to $g(t)$

$$\omega_i(t) = \omega_c + m_f g(t)$$

or

$$e_{FM}(t) = A_c \cos\left[\omega_c t + m_f \int_0^t g(\tau) d\tau\right].$$

For single-tone sinusoidal modulation

$$\omega_i(t) = \omega_c + \Delta\omega \cos\omega_m t$$

$$e_{FM}(t) = A_c \cos[\omega_c t + (m_f/\omega_m)\sin\omega_m t].$$

The peak frequency deviation $\Delta\omega \equiv m_f$ is independent of ω_m, while the peak phase deviation $\Delta\theta = \Delta\omega/\omega_m$ is inversely proportional to ω_m; $\Delta\theta$ (in radians) is the modulation index often denoted by β. For broad-band application $\Delta\omega \ll \omega_c$ and $\beta \gg 1$.

Frequency Spectrum of Single-Tone Angular Modulation

Small Phase Deviation (Narrow-Band PM):

$$e(t) = A_c \cos(\omega_c t + \beta \sin\omega_m t), \qquad \beta \ll 1$$

$$e(t) \cong A_c(\cos\omega_c t - \beta \sin\omega_m t \sin\omega_c t)$$

$$= \underbrace{A_c \cos\omega_c t}_{\text{carrier}} - \underbrace{\tfrac{1}{2}(A_c\beta)\cos(\omega_c - \omega_m)t}_{\text{lower sideband}}$$

$$+ \underbrace{\tfrac{1}{2}(A_c\beta)\cos(\omega_c + \omega_m)t.}_{\text{upper sideband}}$$

The corresponding equation for AM is

$$e_{AM}(t) = A_c \cos\omega_c t + \tfrac{1}{2}(A_c m_a)\cos(\omega_c - \omega_m)t$$

$$+ \tfrac{1}{2}(A_c m_a)\cos(\omega_c + \omega_m)t.$$

The vector representation of AM and narrow-band PM is illustrated in Fig. 15.

Large Phase Deviation (Wide-Band PM):

$$e(t) = A_c \cos(\omega_c t + \beta \sin\omega_m t), \qquad \beta \gg 1$$

$$= A_c[\cos\omega_c t \cos(\beta \sin\omega_m t)$$

$$- \sin\omega_c t \sin(\beta \sin\omega_m t)]$$

$$= A_c\Big[\cos\omega_c t \sum_{n=-\infty}^{\infty} J_n(\beta)\cos n\omega_m t$$

$$- \sin\omega_c t \sum_{n=-\infty}^{\infty} J_n(\beta)\sin n\omega_m t\Big].$$

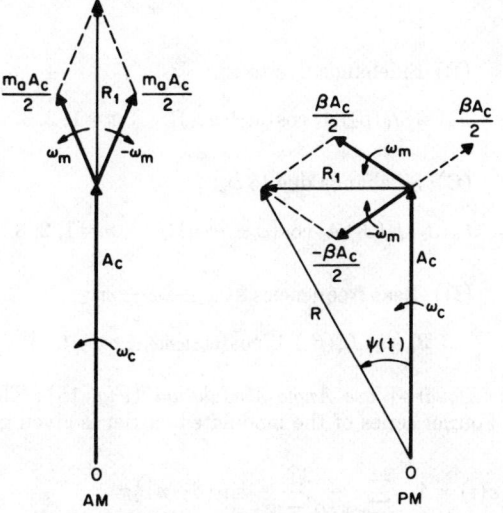

Fig. 15—Vector representation of AM and narrow-band PM. *From P. F. Panter, "Modulation, Noise, and Spectral Analysis," Fig. 7–5,* © *1965, McGraw-Hill Book Company.*

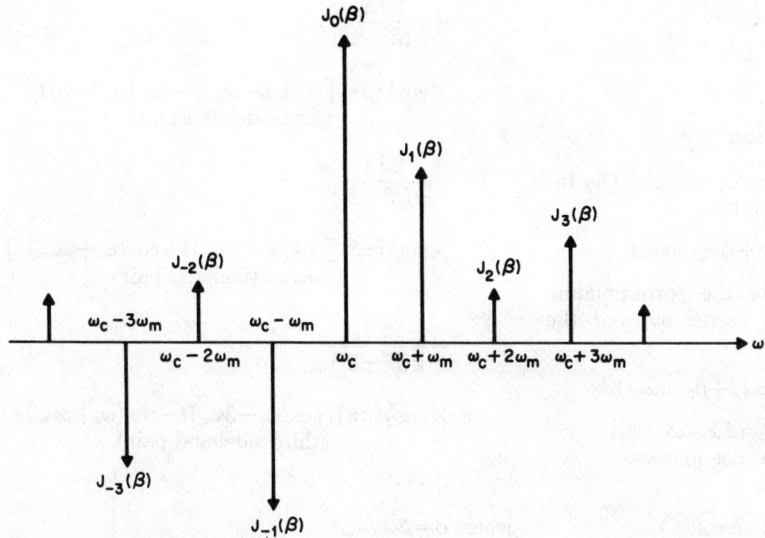

Fig. 16—Composition of FM wave into sidebands. *From P. F. Panter, "Modulation, Noise, and Spectral Analysis," Fig. 7–6,* © *1965, McGraw-Hill Book Company.*

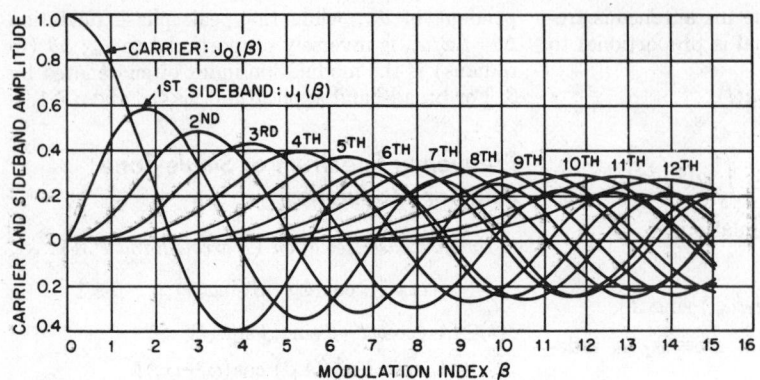

Fig. 17—Plot of Bessel functions of first kind as a function of argument β. From P. F. Panter, "Modulation, Noise, and Spectral Analysis," Fig. 7–8, © 1965, McGraw-Hill Book Company.

The waveform for wide-band modulation is given by

$$e(t) = A_c\{J_0(\beta)\ \cos\omega_c t$$
$$-J_1(\beta)\left[\cos(\omega_c-\omega_m)t-\cos(\omega_c+\omega_m)t\right]$$
$$+J_2(\beta)\left[\cos(\omega_c-2\omega_m)t+\cos(\omega_c+2\omega_m)t\right]$$
$$-J_3(\beta)\left[\cos(\omega_c-3\omega_m)t-\cos(\omega_c+3\omega_m)t\right]+\cdots\}$$
$$= A_c \sum_{n=-\infty}^{\infty} J_n(\beta)\ \cos(\omega_c+n\omega_m)t$$

as shown in Fig. 16.

In practical application the required bandwidth is finite, for—beyond a certain frequency range from the carrier, depending on the magnitude of β—the sideband amplitudes, which are proportional to $J_n(\beta)$, are negligibly small (see Fig. 17). Note that at $\beta=2.404$, $J_0(\beta)=0$ and the carrier amplitude is zero.

The average power in an angle modulated wave is constant

$$P = \tfrac{1}{2}(A_c^2) \sum_{-\infty}^{\infty} J_n^2(\beta)$$
$$= \tfrac{1}{2}(A_c^2).$$

Multitone Angle Modulation

Two-Tone Angle Modulation ω_1 and ω_2: The instantaneous frequency is given by

$$\omega_i(t) = \omega_c+\Delta\omega_{c1}\ \cos\omega_1 t+\Delta\omega_{c2}\ \cos\omega_2 t$$

where $\Delta\omega_{c1}$ and $\Delta\omega_{c2}$ denote the corresponding frequency deviations of the carrier ω_c, and the FM signal is

$$e(t) = A_c \cos(\omega_c t+\beta_1 \sin\omega_1 t+\beta_2 \sin\omega_2 t)$$

where $\beta_1=\Delta\omega_{c1}/\omega_1$ and $\beta_2=\Delta\omega_{c2}/\omega_2$.

The spectral components are as follows.

(A) Carrier:

$$J_0(\beta_1)J_0(\beta_2)A_c\ \cos\omega_c t.$$

(B) Sidebands due to ω_1:

$$J_n(\beta_1)J_0(\beta_2)A_c\ \cos(\omega_c\pm n\omega_1)t, \qquad n=1,2,3.$$

(C) Sidebands due to ω_2:

$$J_m(\beta_2)J_0(\beta_1)A_c\ \cos(\omega_c\pm m\omega_2)t, \qquad m=1,2,3.$$

(D) Beat frequencies at $\omega_c\pm n\omega_1\pm m\omega_2$:

$$J_n(\beta_1)J_m(\beta_2)A_c\ \cos(\omega_c\pm n\omega_1\pm m\omega_2)t.$$

Square-Wave Angle Modulation (Fig. 18): The Fourier series of the modulated carrier is given by

$$e(t) = A_c \sum_{n=-\infty}^{\infty} \frac{2\beta}{\pi(\beta^2-n^2)}\ \sin(\beta-n)\tfrac{1}{2}\pi$$
$$\times\cos(\omega_c+n\omega_m)t$$
$$= (2A_c/\pi\beta)\ \sin\tfrac{1}{2}(\pi\beta)\ \cos\omega_c t$$
$$\text{(carrier)}$$

$$+\frac{2\beta A_c}{\pi(\beta^2-1^2)}$$
$$\times\cos\tfrac{1}{2}(\beta\pi)\left[\cos(\omega_c-\omega_m)t-\cos(\omega_c+\omega_m)t\right]$$
$$\text{(first sideband pair)}$$

$$-\frac{2\beta A_c}{\pi(\beta^2-2^2)}$$
$$\times\sin\tfrac{1}{2}(\pi\beta)\left[\cos(\omega_c-2\omega_m)t+\cos(\omega_c+2\omega_m)t\right]$$
$$\text{(second sideband pair)}$$

$$-\frac{2\beta A_c}{\pi(\beta^2-3^2)}$$
$$\times\cos\tfrac{1}{2}(\beta\pi)\left[\cos(\omega_c-3\omega_m)t-\cos(\omega_c+3\omega_m)t\right]$$
$$\text{(third sideband pair)}$$

$$+\cdots$$

where $\beta=\Delta\omega_c/\omega_m$.

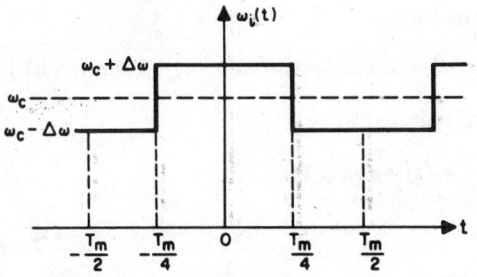

Fig. 18—Frequency modulation by square wave. *From P. F. Panter, "Modulation, Noise, and Spectral Analysis," Fig. 7-9, © 1965, McGraw-Hill Book Company.*

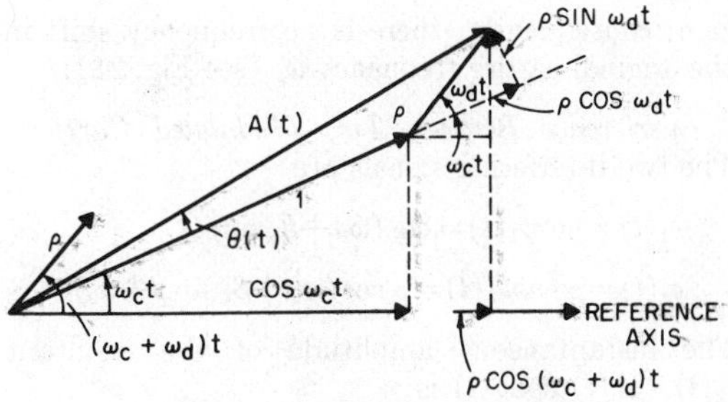

Fig. 20—Vectorial additions of unmodulated carriers. *From P. F. Panter, "Modulation, Noise, and Spectral Analysis," Fig. 11-1, © 1965, McGraw-Hill Book Company.*

Spectral Distribution of an FM/FM Signal

Let ω_c = carrier angular frequency, ω_s = subcarrier, and ω_m = modulating angular frequency. The instantaneous frequency of the carrier wave is

$$\omega_i(t) = \omega_c + \Delta\omega \cos[\omega_s t + \phi_s + \beta_s \sin(\omega_m t + \phi_m)]$$

where $\Delta\omega$ = peak frequency deviation of carrier, and $\beta_s = \Delta\omega_s/\omega_m$ = peak phase deviation of subcarrier. The spectral distribution is given by

$$e(t) = A_c \sum_{p=-\infty}^{\infty} \sum_{q=-\infty}^{\infty} J_p(\beta) J_q(p\beta_s)$$

$$\times \cos[(\omega_c + p\omega_s + q\omega_m)t + \phi_c + p\phi_s + q\phi_m]$$

where $\beta = \Delta\omega/\omega_s$ = peak phase deviation of carrier.

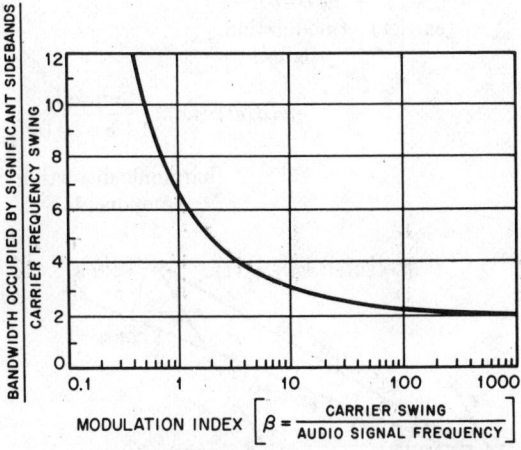

Fig. 19—Significant bandwidth (normalized) vs modulation index β. *From C. E. Tibbs and G. G. Johnstone, "Frequency Modulation Engineering," John Wiley & Sons, Inc., New York. Courtesy of Chapman & Hall, Ltd., London, England.*

Bandwidth Considerations in Multitone FM

An estimate of the IF bandwidth required for transmission of FM carrier by a complex modulating signal is given by

$$\beta_{\mathrm{IF}} = 2(\Delta F + 2f_m) = 2\Delta F(1 + 2/\beta)$$

where ΔF = peak frequency deviation for the system, and f_m = highest baseband frequency (see Fig. 19).

Interference in FM Reception

Interference Between Two Unmodulated Carriers: Let $\cos\omega_c t$ denote the desired signal and $\rho \cos(\omega_c + \omega_d)t$ denote the interfering signal, where $\rho < 1$ and $\omega_d \ll \omega_c$.

The vectorial addition of the unmodulated carriers, as shown in Fig. 20, is given by

$$e_r(t) = \cos\omega_c t + \rho \cos(\omega_c + \omega_d)t$$

$$= A(t) \cos[\omega_c t + \theta(t)]$$

where the envelope

$$A(t) = (1 + \rho^2 + 2\rho \cos\omega_d t)^{1/2}$$

$$\cong (1 + \rho \cos\omega_d t), \qquad \rho \ll 1$$

and the phase angle

$$\theta(t) = \tan^{-1}[\rho \sin\omega_d t/(1 + \rho \cos\omega_d t)].$$

The instantaneous frequency of the resultant is given by

$$\omega_i(t) = \omega_c + [d\theta(t)/dt]$$

$$= \omega_c + \omega_d \sum_{n=1}^{\infty} (-1)^{n+1} \rho^n \cos n\omega_d t.$$

Note that $d\theta(t)/dt$ has an average value equal to

zero; consequently, there is no frequency shift in the original carrier frequency ω_c (see Fig. 21).

Interference Between Two Modulated Carriers: The two interfering signals are

$$e_1(t) = \cos\psi_1(t) = \cos(\omega_1 t + \beta_1 \sin pt)$$

$$e_2(t) = \rho \cos\psi_2(t) = \rho \cos(\omega_2 t + \beta_2 \sin qt + \psi_0).$$

The instantaneous amplitude of the resultant $e_r(t) = A(t)\cos\phi(t)$ is

$$A(t) = [1 + \rho^2 + 2\rho \cos\psi(t)]^{1/2}$$

$$\psi(t) = \psi_2(t) - \psi_1(t).$$

The instantaneous phase angle of the resultant is

$$\phi(t) = \psi_1(t) + \tan^{-1}\frac{\rho \sin\psi(t)}{1 + \rho \cos\psi(t)}$$

$$= \psi_1(t) - \sum_{s=1}^{\infty} \frac{(-1)^s \rho^s}{s}$$

$$\times \sum_{n=-\infty}^{\infty} J_n(s\beta_2) \sum_{m=-\infty}^{\infty} J_m(s\beta_1)$$

$$\times \sin[(s\omega_2 - s\omega_1 - mp + nq)t + s\psi_0]$$

and the instantaneous frequency of the resultant of the two frequency-modulated signals is

$$\omega_i(t) = d\phi(t)/dt$$

$$= \omega_1 + \Delta\omega_1 \cos pt - \sum_{s=1}^{\infty} \frac{(-1)^s \rho^s}{s} \sum_{m=-\infty}^{\infty} J_m(s\beta_1)$$

$$\times \sum_{n=-\infty}^{\infty} J_n(s\beta_2)(s\omega_2 - s\omega_1 - mp + nq)$$

$$\times \cos[(s\omega_2 - s\omega_1 - mp + nq)t + s\psi_0].$$

Multipath Transmission Interference (Fig. 22):

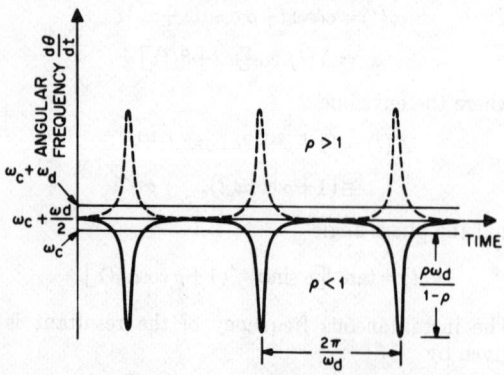

Fig. 21—Instantaneous frequency of resultant due to two-carrier interference: $\rho = 0.8$, solid curve; $\rho = 1/0.8$, dashed curve. *From P. F. Panter, "Modulation, Noise, and Spectral Analysis," Fig. 11-3, © 1965, McGraw-Hill Book Company.*

Direct wave

$$e_1(t) = \cos\psi_1(t) = \cos[\omega_c(t - t_1) + \beta \sin p(t - t_1)]$$

and reflected wave

$$e_2(t) = \rho \cos\psi_2(t)$$

$$= \rho \cos[\omega_c(t - t_2) + \beta \sin p(t - t_2) + \phi_0]$$

where

$t_1 =$ time required for the direct wave to reach the receiver

$t_2 =$ time required for the reflected wave to reach the receiver

$\phi_0 =$ angle of reflection of reflected signal

$\beta = \Delta\omega/p =$ modulation index

$p =$ angular frequency of modulating signal.

The resultant wave is $e_r(t) = A(t)\cos\phi(t)$, where

$$A(t) = [1 + \rho^2 + 2\rho \cos\psi(t)]^{1/2}$$

and

$$\phi(t) = \psi_1(t) + \tan^{-1}\{\rho \sin\psi(t)/[1 + \rho \cos\psi(t)]\}$$

where $\quad \psi(t) = \psi_2(t) - \psi_1(t).$

The instantaneous frequency $\omega_1(t)$ is given by

$$\omega_i(t) = d\phi(t)/dt$$

$$= \quad \omega_c \quad + \Delta\omega \cos p(t - t_1)$$
$$\text{(carrier)} \quad \text{(modulation signal)}$$

$$+ (d/dt)\ \tan^{-1}\frac{\rho\sin\psi(t)}{1 + \rho \cos\psi(t)}.$$

(harmonic distortion components)

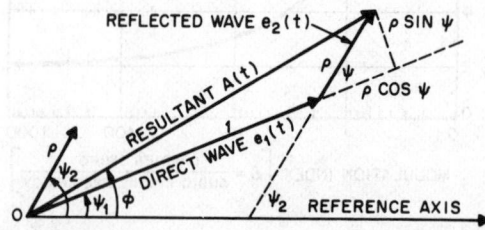

Fig. 22—Vector diagram showing direct wave, reflected wave, and resultant. *From P. F. Panter, "Modulation, Noise, and Spectral Analysis," Fig. 11-7, © 1965, McGraw-Hill Book Company.*

The harmonic distortion components are given by

$$2p \sum_{n=1}^{\infty} \frac{(-\rho)^n}{n} \sin(n\theta_0)$$

$$\times \sum_{m=1}^{\infty} (-1)^m 2m J_{2m}(nz) \sin 2mB$$

$$-2p \sum_{n=1}^{\infty} \frac{(-\rho)^n}{n} \cos(n\theta_0)$$

$$\times \sum_{m=0}^{\infty} (-1)^m (2m+1) J_{2m+1}(nz) \sin(2m+1)B$$

where

$$z = 2\beta \sin\tfrac{1}{2}(pt_0)$$

$$B = p[t - t_1 - \tfrac{1}{2}(t_0)]$$

$$\theta_0 = -\omega_c t_0 + \phi_0$$

$$t_0 = t_2 - t_1.$$

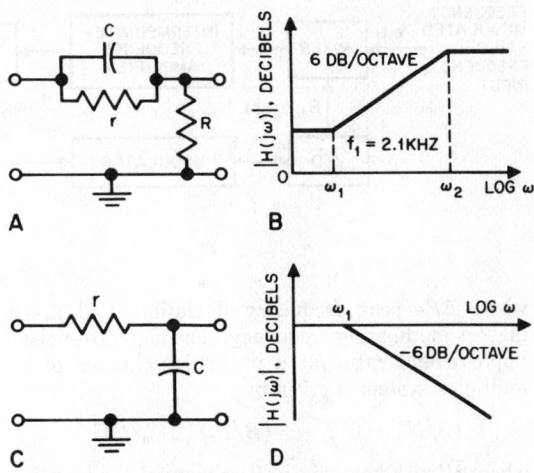

A   **B**

C   **D**

Fig. 24—Pre-emphasis and de-emphasis networks. A, Pre-emphasis network ($r \gg R$, $rC = 75$ μsec); B, Asymptotic response ($\omega_1 = 1/rC$, $\omega_2 \doteq 1/RC$); C, De-emphasis network ($rC = 75$ μsec); D, Asymptotic response ($f_1 = 2.1$ kHz). *From P. F. Panter, "Modulation, Noise, and Spectral Analysis," Figs. 14–6 and 14–7, © 1965, McGraw-Hill Book Company.*

Signal-to-Noise Improvement in FM Systems

The performance of a conventional FM receiver in the presence of random fluctuation noise is commonly judged on the basis of the variation of the output signal-to-noise $(S/N)_o$ power ratio as a function of the carrier-to-noise power ratio $(C/N)_i$ measured at the input to the limiter. This relationship is shown graphically in Fig. 23. The threshold of full improvement occurs when $(C/N)_i$ is about 12 decibels. For all values of the carrier greater than the threshold, the output $(S/N)_o$ is proportional to the input $(C/N)_i$. The signal-to-noise improvement ratio for a single channel FM system is given by

$$(S/N)_o / (C/N)_i = (\Delta\Phi)^2, \text{ using a phase detector}$$

where $\Delta\Phi$ = peak phase deviation, and

$$(S/N)_o / (C/N)_i = 3(\Delta F/f_m)^2$$
$$= 3\beta^2, \text{ using a frequency discriminator}$$

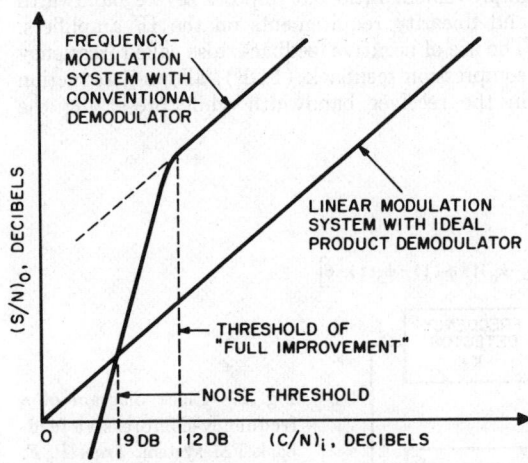

Fig. 23—Noise performance of conventional FM receiver. *From P. F. Panter, "Modulation, Noise, and Spectral Analysis," Fig. 14–2, © 1965, McGraw-Hill Book Company.*

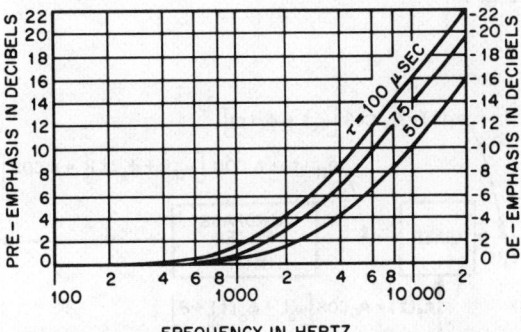

Fig. 25—Pre-emphasis and de-emphasis circuit response, for time constants of $\tau = 50$, 75, and 100 μsec. *From C. E. Tibbs and G. G. Johnstone, "Frequency Modulation Engineering," John Wiley & Sons, Inc., New York. Courtesy of Chapman & Hall, Ltd., London, England.*

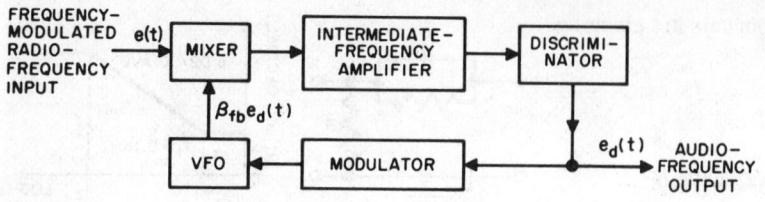

where ΔF = peak frequency deviation, and f_m = highest modulating frequency. The signal-to-noise improvement ratio for a particular channel of a multiplex system is given by

$$(S/N)_o/(C/N)_i = (B/B_c)(\Delta F_m/f_n)^2$$

where $2B$ = IF bandwidth, B_c = channel bandwidth, ΔF_m = peak channel frequency deviation, and f_n = midband channel frequency in the nth channel.

In the nonlinear region when the noise is larger than the carrier there exists a signal-suppression effect, the average amplitude of the discriminator output is reduced, and for $(C/N)_i \ll 1$ we have $(S/N)_o \propto (C/N)_i^2$.

Signal-to-Noise Improvement Through De-Emphasis

The $(S/N)_o$ ratio of the high-frequency end of the baseband can be increased by passing the modulating signal (at the transmitting end) through a pre-emphasis network (Fig. 24A and B) which emphasizes the higher signal frequencies, and then passing the output of the discriminator through a de-emphasis network (Fig. 24C and D) to restore the original signal-power distribution. Typical preemphasis and de-emphasis circuit responses for general time constants τ are shown in Fig. 25.

The improvement factor ρ_{FM} is given by

$$\rho_{FM} = \frac{(2\pi f_m \tau)^3}{3(2\pi f_m \tau - \tan^{-1} 2\pi f_m \tau)}$$

where f_m denotes the highest baseband frequency.

For narrow-band FM

$$\rho_{FM} \to 1.$$

For wide-band, f_m is large, and

$$\rho_{FM} \to (2\pi f_m \tau)^2/3.$$

The mean $(S/N)_o$ ratio for FM with pre-emphasis is given by

$$(S/N)_o = \rho_{FM} \cdot 3\beta^2 (C/N)_i.$$

Application of Negative Feedback to FM Systems

The use of a large modulation index in a FM system considerably increases the signal-to-noise improvement ratio but imposes severe bandwidth and linearity requirements on the IF amplifiers. The use of negative feedback, also called frequency compression feedback (FCF), allows a reduction in the receiver bandwidth while preserving the

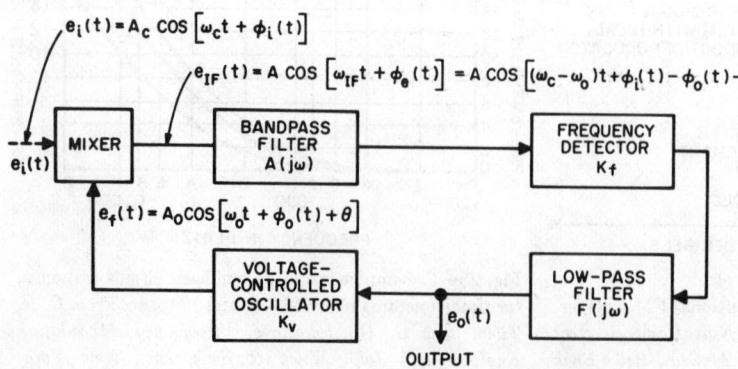

advantage of high modulation index. This is shown in Fig. 26. Assuming a noiseless incoming signal

$$e(t) = A_c \cos \left[\omega_c t + \Delta\omega \int_0^t g(u) \, du \right].$$

The instantaneous angular frequency is

$$\omega_i(t) = \omega_c + \Delta\omega g(t)$$

and the variable portion of the discriminator output is

$$e_d(t) = [k_d/(1+k_d\beta_{fb})]\Delta\omega g(t)$$
$$\cong (\Delta\omega/\beta_{fb})g(t), \; k_d\beta_{fb} \gg 1$$

where $\Delta\omega g(t)$ = instantaneous-frequency deviation of the incoming signal, k_d = discriminator constant, and β_{fb} = feed-back factor. Thus the effective index of modulation is reduced by β_{fb} and the IF bandwidth may be reduced to accept only one pair of sidebands.

Threshold Extension Using Frequency Compression Feedback

The threshold level of a FM receiver determines the maximum operating range of the FM communication system; hence any technique that lowers the threshold will enhance the system reliability. Frequency compression feedback may be used to lower the threshold (commonly referred to as threshold extension), as shown in Fig. 27. In this system, the threshold is approximately given by

$$\rho_T = (C/N)_i = 4.8[(F-1)/F]^2$$

where $F = 1 + K_v K_f$ is the feedback factor, and $K_v K_f$ is the loop gain.

PART 2—PULSE MODULATION

In pulse-modulation systems[*], the unmodulated carrier is usually a series of regularly recurrent

[*] P. F. Panter, "Modulation, Noise, and Spectral Analysis," Chapters 17, 18, 20, 21, and 22, McGraw-Hill Book Co., New York, N. Y.; 1965.

pulses; information is conveyed by modulating some parameter of the transmitted pulses such as the amplitude, duration, time of occurrence, or shape of pulse. This type of modulation is based on the "sampling principle," which states that a continuous message waveform that has a spectrum of finite width could be recovered from a set of discrete instantaneous samples whose rate is higher than twice the highest signal frequency. This discrete set of periodic samples of the message function is used to modulate some parameter of the carrier pulses. In *pulse-amplitude modulation* (PAM), the series of periodically recurring pulses is modulated in amplitude by the corresponding instantaneous samples of the message function. In *pulse-time modulation* (PTM), the instantaneous samples of the message function are used to vary the time of occurrence of some parameter of the pulsed carrier. Pulse-duration, pulse-position, and pulse-frequency modulation are particular forms of pulse-time modulation. In *pulse-duration modulation* (PDM), the time of occurrence of either the leading or trailing edge of each pulse (or both) is varied from its unmodulated position by the samples of the modulating wave. This is also called *pulse-length* or *pulse-width modulation* (PWM). In *pulse-position (or phase) modulation* (PPM), the samples of the modulating wave are used to vary the position in time of a pulse, relative to its unmodulated time of occurrence. Pulse-position modulation is essentially the same as PDM, except that the variable edge is now replaced by a short pulse. In *pulse-frequency modulation* (PFM), the samples of the message function are used to modulate the frequency of the series of carrier pulses.

Pulse modulation is used for time-division multiplexing (TDM). In TDM systems, each of a number of sampled messages is used to modulate a pulsed carrier. However, each pulsed carrier is allocated a different time interval for its transmission, and thus at each instant of time only one carrier is being transmitted, as shown in Fig. 28.

The pulse-modulation systems enumerated so far are examples of uncoded pulse systems. In *pulse-code modulation* (PCM), the modulating signal waveform is sampled at regular intervals as in conventional pulse modulation. However, in PCM, the samples are first quantized into discrete steps; i.e., within a specified range of expected sample values, only certain discrete levels are allowed and these are transmitted over the system by means of a code pattern of a series of pulses.

Another example of a code-modulation system is *delta modulation*. As in PCM, the range of signal amplitudes is quantized and binary pulses are produced at the sending end at regular intervals. However, in delta-modulation systems, instead of the absolute quantized signal amplitude being transmitted at each sampling, the transmitted

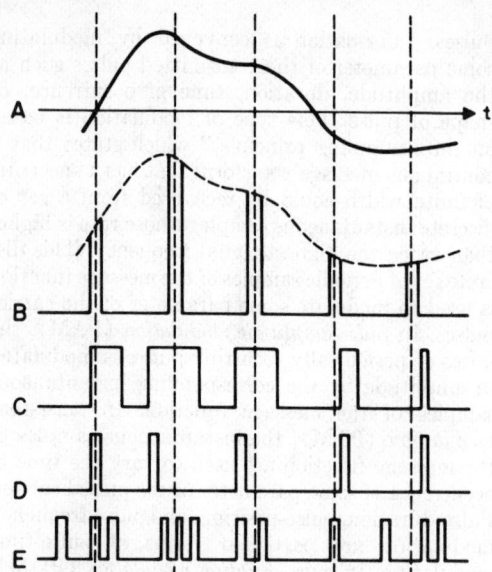

Fig. 28—Signal waveforms in TDM systems: A, Modulating signal waveform; B, Pulse-amplitude modulation; C, Pulse-length modulation; D, Pulse-position modulation; E, Pulse-code modulation. *From P. F. Panter, "Modulation, Noise, and Spectral Analysis," Fig. 18-2, © 1965, McGraw-Hill Book Company.*

pulses carry the information corresponding to the derivative of the amplitude of the modulating signal.

SAMPLING

Sampling in the Time Domain

If a signal $f(t)$ is sampled at regular intervals of time and at a rate higher than twice the highest significant signal frequency, then the samples contain all the information of the original signal. The function $f(t)$ may be reconstructed from these samples by the use of a low-pass filter. The reconstruction equation is

$$f(t) = \alpha \sum_{n=-\infty}^{\infty} f(n\alpha/2B) \frac{\sin 2\pi B(t-n\alpha/2B)}{2\pi B(t-n\alpha/2B)},$$

$$0 < \alpha \leq 1$$

where $f(t)$ is band-limited to B hertz, and the samples are taken at sampling intervals $\alpha/2B$ seconds apart.

Sampling in the Frequency Domain

A time-limited signal $f(t)$ which is zero outside the range $t_1 < t < t_2$ is completely determined by the values of the spectrum function $F(j\omega)$ at the angular-frequency sampling points given by

$$\omega_n = n[2/(t_2 - t_1)].$$

The function $f(t)$ expressed in terms of its sampling values in the frequency domain is given by the reconstruction equation

$$f(t) = \sum_{n=-\infty}^{\infty} (t_2 - t_1)^{-1} F\left(j \frac{2\pi n}{t_2 - t_1}\right)$$

$$\times \exp[j2\pi nt/(t_2 - t_1)].$$

Sampling of a Band-Pass Function (B_0, $B_0 + B$)

The reconstruction equation for $f(t)$ in terms of its sampled values is

$$f(t) = 2BT \sum_{n=-\infty}^{\infty} f(nT) \frac{\sin \pi B(t-nT)}{\pi B(t-nT)}$$

$$\times \cos 2\pi B_c(t-nT)$$

where $B_c = B_0 + (B/2)$, the center frequency of the band-pass signal, and the permissible values of T are given by

$$m/2B_0 \leq T \leq [(m+1)/2(B_0 + B)],$$

$$m = 0, 1, 2, \cdots$$

provided $B_0 \neq 0$.

The minimum sampling frequency for a band-limited signal of width B is illustrated in Fig. 29.

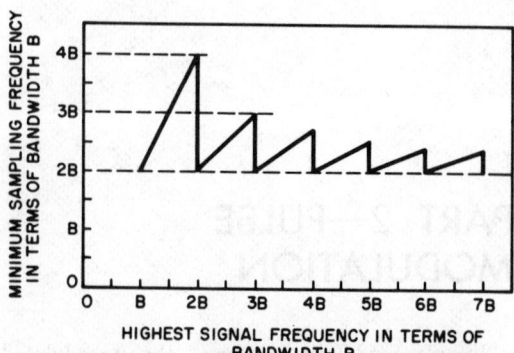

Fig. 29—Minimum sampling frequency for band of width B. *From P. F. Panter, "Modulation, Noise, and Spectral Analysis," Fig. 17-13, © 1965, McGraw-Hill Book Company.*

PULSE-AMPLITUDE MODULATION (PAM)

In PAM, the samples of the message function are used to amplitude modulate the successive carrier pulses. When the modulated pulses follow the amplitude variation of the sampled time function during the sampling interval, the process is called *natural sampling* or *top sampling*. In contrast with natural sampling we have instantaneous or square-topped sampling, where the amplitude of the pulses is determined by the instantaneous value of the sampled time function corresponding to a single instant (i.e., center or edge) of the sampling time interval. PAM can be instrumented by two distinct methods. The first produces a variation of the amplitude of a pulse sequence about a fixed nonzero value or pedestal and constitutes double-

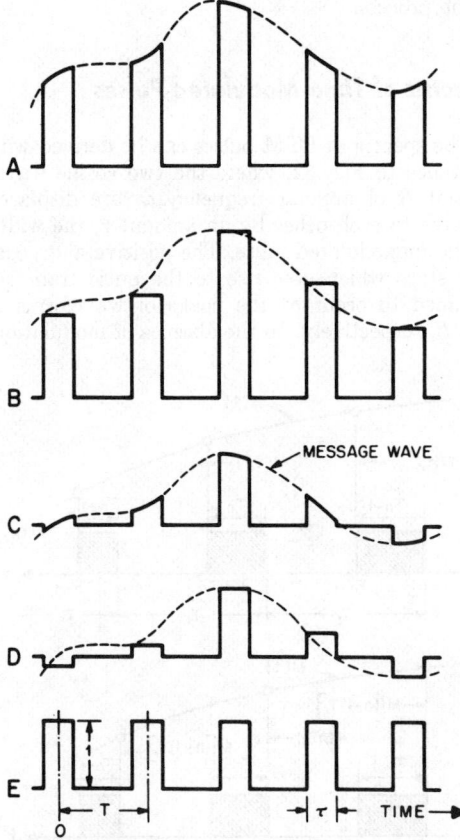

A

B

MESSAGE WAVE

C

D

E

$\leftarrow$ T $\rightarrow$ $\rightarrow$ τ $\leftarrow$ TIME $\rightarrow$

O

Fig. 30—Various shapes of amplitude-modulated pulses: A, Single-polarity pulses; B, Single-polarity flat-top pulses; C, Double-polarity pulses; D, Double-polarity flat-top pulses; E, Unit sampling function. *From H. S. Black, "Modulation Theory," courtesy of D. Van Nostrand Company, Inc., Princeton, N. J.*

sideband amplitude modulation (Fig. 30A and 30B). In the second method the pedestal is zero, and the output signal consists of double-polarity modulated pulses and constitutes double-sideband suppressed-carrier modulation (Fig. 30C and 30D).

Spectra of Amplitude-Modulated Pulses

Double-Polarity AM Pulses—Natural (or Top) Sampling: In the process of natural sampling (or exact scanning), the modulated pulses follow the sampled time function during the sampling interval. The unit sampling function (Fig. 30E) consists of a train of unmodulated periodic pulses of unit amplitude given by

$$p_T(t) = (\tau/T) \sum_{n=-\infty}^{\infty} \frac{\sin(n\pi\tau/T)}{n\pi\tau/T} \exp(jn\omega_0 t)$$

where $\omega_0 = 2\pi f_0 = 2\pi/T$ is the fundamental angular frequency of the pulse train, τ is the duration of the pulse, and τ/T is the duty cycle. Double-polarity AM pulses are obtained by multiplying the message signal $f(t)$ by the unit sampling function $p_T(t)$. In case of sinusoidal modulation, $f(t) = A \cos(\omega_m t + \phi)$, and the waveform of the AM pulses is given by

$$f_{s_1}(t) = f(t) p_T(t)$$

$$= (\tau/T) A \cos(\omega_m t + \phi)$$

$$+ (\tau/T) A \sum_{n=1}^{\infty} \frac{\sin(n\pi\tau/T)}{n\pi\tau/T}$$

$$\times \cos[(n\omega_0 \pm \omega_m) t \pm \phi].$$

In the general case, the message function $f(t)$ is band-limited, and its spectrum is $F(j\omega)$. The output spectrum is

$$F_{s_1}(j\omega) = (\tau/T) F(j\omega) + (\tau/T) \sum_{n=1}^{\infty} \frac{\sin(n\pi\tau/T)}{n\pi\tau/T}$$

$$\times \{F[j(\omega - n\omega_0)] + F[j(\omega + n\omega_0)]\}$$

$$= \frac{\tau}{T} \sum_{n=-\infty}^{\infty} \frac{\sin(n\omega_0\tau/2)}{n\omega_0\tau/2} F[j(\omega - n\omega_0)].$$

The spectrum of the double-polarity AM pulses consists of the original modulation spectrum and an infinite number of upper and lower sidebands around ω_0 and its harmonics.

Double-Polarity AM Pulses—Instantaneous (or Square-Top) Sampling: In case of sinusoidal modulation, the output waveform is given by

$$f_{s_2}(t) = (\tau/T) A \sum_{n=-\infty}^{\infty} \frac{\sin[\pi(\tau/T)(n\omega_0 + \omega_m)/\omega_0]}{[\pi(\tau/T)(n\omega_0 + \omega_m)/\omega_0]}$$

$$\times \cos[(n\omega_0 + \omega_m)(t - \tfrac{1}{2}\tau) + \phi].$$

In the general case, the output spectrum is

$$F_{s_2}(j\omega) = (\tau/T)\,\frac{\sin(\omega\tau/2)}{\omega\tau/2}\sum_{n=-\infty}^{\infty} F[j(\omega-n\omega_0)].$$

Single-Polarity AM Pulses—Natural Sampling:
For sinusoidal modulation

$$f_{s_3}(t) = [1+m_a\cos(\omega_m t+\phi)]$$

$$\times(\tau/T)\sum_{n=-\infty}^{\infty}\frac{\sin(n\pi\tau/T)}{n\pi\tau/T}\exp(jn\omega_0 t)$$

$$= p_T(t)+f_{s_1}(t)$$

where m_a is the modulation index. In the general case

$$F_{s_3}(j\omega) = P(j\omega)+F_{s_1}(j\omega)$$

where $P(j\omega)$ is the Fourier transform of $p_T(t)$.

Single-Polarity AM Pulses—Instantaneous Scanning: For sinusoidal modulation

$$f_{s_4}(t) = p_T(t)+f_{s_2}(t).$$

In the general case

$$F_{s_4}(j\omega) = P(j\omega)+F_{s_2}(j\omega).$$

Signal-to-Noise Ratio in PAM

$$(S/N)_i = \tfrac{1}{2}m_a^2 P/N_0 B$$

where $P=$ average power of unmodulated radio-frequency pulse train, $N_0=$ noise-power density in watts/hertz, and $B=$ channel (RF) bandwidth. Also

$$(S/N)_o = \frac{\tfrac{1}{2}m_a^2(\tau/T)P}{N_0 f_m}$$

where $f_m=$ top frequency of message function. By blocking the receiver between pulses to eliminate the noise in the interpulse period, the $(S/N)_o$ at the output of a low-pass filter is

$$(S/N)_o = \tfrac{1}{2}m_a^2 P/N_0 f_m$$

which is identical to the result obtained for conventional CW carrier amplitude modulation.

In practice, PAM provides a poorer signal-to-noise ratio than conventional AM, because the receiver is unblocked for rather longer than the pulse-duration time owing to the sloping sides of the pulse.

PULSE-TIME MODULATION (PTM)

The improvement in signal-to-noise ratio obtained by the use of time-modulated pulses of constant amplitude instead of amplitude-modulated pulses led to the development of systems using pulse-duration and pulse-position modulation. The sampling associated with pulse modulation may be either natural or uniform (periodic). Natural sampling may be defined as a process of sampling in which the time of sampling coincides with the time of appearance of the time-modulated pulse as shown in Fig. 31A. In the process of natural sampling, the pulse duration τ_n corresponds to the value of the modulating signal $M(t_n)$ at that instant, and consequently the sampling intervals t_n are not equal but depend on the modulation level. Uniform sampling may be defined as a process of sampling where the variation in the parameter of the pulse is proportional to the modulating signal at uniformly spaced sampling times. This is illustrated in Fig. 31B, where the width of the pulses is proportional to the modulating values $M(t_n)$ which are sampled at equal intervals $t_n=nT_r$ and are independent of the modulation process.

Spectra of Time-Modulated Pulses

The spectra of PTM pulses can be derived with reference to Fig. 32, where the two cosine waves A and B of angular frequency ω_r are displaced relative to each other by an amount τ, the width of the unmodulated pulse. The positive and negative steps which give rise to the pulse train are assumed to occur at the peaks of waveforms A and B, respectively. In the absence of modulation,

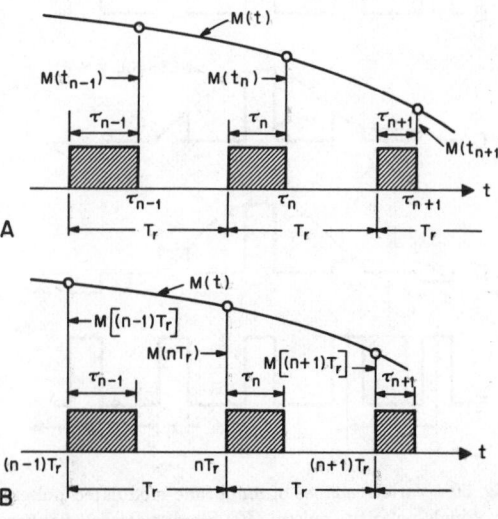

Fig. 31—PDM using natural and uniform sampling: A, Natural sampling; B, Uniform sampling. *From P. F. Panter, "Modulation, Noise, and Spectral Analysis," Fig. 18-14, © 1965, McGraw-Hill Book Company.*

the time of occurrence of the positive and negative steps is given by

$$\omega_r(t+\tau/2) = 2n\pi$$

and

$$\omega_r(t-\tau/2) = 2n\pi.$$

With natural modulation, the time of occurrence of the positive and negative steps is given by

$$\omega_r(t+\tau/2) + \beta \sin(\omega_m t + \phi) = 2n\pi$$

and

$$\omega_r(t-\tau/2) + \beta \sin(\omega_m t + \phi) = 2n\pi.$$

Similarly, with uniform modulation, the time of occurrence or the position of the leading and trailing edges of the pulses is determined by

$$\omega_r(t+\tau/2) + \beta \sin(\omega_m t + \phi) = 2n\pi$$

$$\omega_r(t-\tau/2) + \beta \sin(\omega_m \overline{t-\tau} + \phi) = 2n\pi$$

where ω_m is the modulating frequency and β is the modulation index. Pulses whose moments of occurrence satisfy these equations are said to be time modulated. In pulse-frequency modulation $\beta = \Delta\omega/\omega_m$, while in pulse-phase (or pulse-position) modulation β is constant independent of the modulating frequency.

Pulse-Frequency Modulation—Natural Sampling: A useful expression for an infinite train of unmodulated pulses is in the form

$$p_T(t) = (A/2\pi j) \sum_{k=-\infty}^{\infty} k^{-1}\{\exp[jk\omega_r(\tau/2)]$$

$$- \exp[-jk\omega_r(\tau/2)]\} \exp(jk\omega_r t)$$

where A is the amplitude of the pulses, and ω_r is

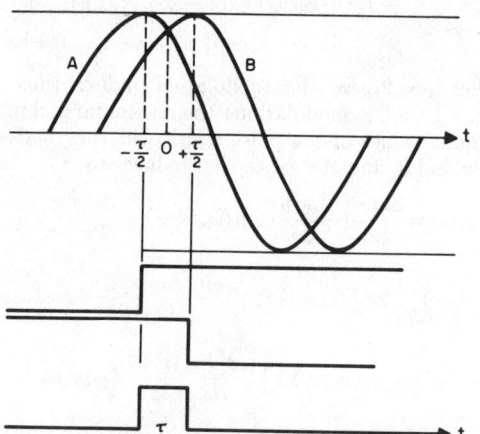

Fig. 32—Modulation process (modified). *From P. F. Panter, "Modulation, Noise, and Spectral Analysis," Fig. 17-14, © 1965, McGraw-Hill Book Company.*

the pulse repetition frequency. Frequency modulation can be taken into account by substituting for $\omega_r\tau/2$ in the expressions for the leading and trailing edges in the last equation, the expressions

$$\tfrac{1}{2}(\omega_r\tau) + \beta \sin(\omega_m t + \phi)$$

and

$$\tfrac{1}{2}(\omega_r\tau) - \beta \sin(\omega_m t + \phi).$$

The frequency-modulated pulse train is then

$$p_m(t) = (A/2\pi j)$$

$$\times \sum_{k=-\infty}^{\infty} k^{-1}\big(\exp\{j[k\omega_r(\tau/2) + k\beta \sin(\omega_m t + \phi)]\}$$

$$- \exp\{-j[k\omega_r(\tau/2) - k\beta \sin(\omega_m t + \phi)]\}\big)$$

$$\times \exp(jk\omega_r t)$$

$$= \frac{A\omega_r\tau}{2\pi} + \frac{A\omega_r\tau}{\pi} \sum_{k=1}^{\infty} \frac{\sin[k\omega_r(\tau/2)]}{k\omega_r(\tau/2)}$$

$$\times \Big(J_0(k\beta) \cos k\omega_r t + \sum_{n=1}^{\infty} J_n(k\beta)$$

$$\times \{\cos[(k\omega_r + n\omega_m)t + n\phi]$$

$$+ (-1)^n \cos[(k\omega_r - n\omega_m)t - n\phi]\}\Big).$$

This expression may be compared with that for the spectrum of a frequency-modulated continuous wave given by

$$e_{FM}(t) = AJ_0(\beta) \cos\omega_r t$$

$$+ A \sum_{n=1}^{\infty} J_n(\beta)\{\cos[(\omega_r + n\omega_m)t + n\phi]$$

$$+ (-1)^n \cos[(\omega_r - n\omega_m)t - n\phi]\}.$$

The conclusions reached are as follows.

(**A**) With pulse-frequency modulation using natural sampling, the direct-current component of the pulse spectrum has no sideband of the modulating frequency.

(**B**) The kth harmonic of the pulse repetition frequency is frequency modulated, the modulation index being $k\beta$.

Pulse-Frequency Modulation—Uniform Sampling: In this type of modulation, the displacement of waveform B of Fig. 32 from its unmodulated position at any instant of time t will depend on the value of the modulating voltage at $(t-\tau)$. The expression for the modulated pulse train becomes

$$p_m(t) = (A/2\pi j)$$

$$\times \sum_{k=1}^{\infty} k^{-1} (\exp\{ j[k\omega_r(\tau/2) + k\beta \sin(\omega_m t + \phi)]\}$$

$$- \exp\{ -j[k\omega_r(\tau/2) - k\beta \sin(\omega_m \overline{t-\tau} + \phi)]\})$$

$$\times \exp(jk\omega_r t)$$

$$= \frac{A\omega_r\tau}{2\pi} + A\left(\frac{\Delta\omega}{2\pi}\right)\tau \frac{\sin[\omega_m(\tau/2)]}{\omega_m(\tau/2)}$$

$$\times \cos[\omega_m t + \phi - (\omega_m\tau/2)]$$

$$+ \frac{A\omega_r\tau}{\pi} \sum_{k=1}^{\infty} \left(J_0(k\beta) \frac{\sin[k\omega_r(\tau/2)]}{k\omega_r(\tau/2)} \right)$$

$$\times \cos k\omega_r t + \sum_{n=1}^{\infty} J_n(k\beta) \left\{ \frac{\sin(k\omega_r + n\omega_m)(\tau/2)}{k\omega_r(\tau/2)} \right.$$

$$\times \cos[(k\omega_r + n\omega_m)t + n\phi - n\omega_m(\tau/2)]$$

$$\left. + (-1)^n \frac{\sin(k\omega_r - n\omega_m)(\tau/2)}{k\omega_r(\tau/2)} \right.$$

$$\left. \times \cos[(k\omega_r - n\omega_m)t - n\phi + n\omega_m(\tau/2)] \right\}).$$

The conclusions reached are as follows.

(A) The direct-current component of the pulse spectrum has a sideband of the modulating frequency of amplitude

$$(A\Delta\omega\tau/2\pi)\{\sin[\omega_m(\tau/2)]/\omega_m(\tau/2)\}.$$

Modulation can therefore be recovered by means of a low-pass filter.

(B) The upper and lower sidebands of the kth harmonic of the pulse repetition frequency are not equal in amplitude, whereas in the case of natural sampling they are equal.

Pulse-Position (or Pulse-Phase) Modulation: The waveform of pulse-phase modulation can be directly derived from that for pulse-frequency modulation by substituting $\omega_r\tau_d$ for β, where $\omega_r\tau_d$ represents the peak phase deviation of waveforms A and B which is constant independent of the modulation frequency ω_m. The resulting waveform is

Natural sampling:

$$p_m(t) = \frac{A\omega_r\tau}{2\pi} + \frac{A\omega_r\tau}{\pi} \sum_{k=1}^{\infty} \frac{\sin[k\omega_r(\tau/2)]}{k\omega_r(\tau/2)}$$

$$\times (J_0(k\omega_r\tau_d) \cos k\omega_r t + \sum_{n=1}^{\infty} J_n(k\omega_r\tau_d)$$

$$\times \{\cos[(k\omega_r + n\omega_m)t + n\phi]$$

$$+ (-1)^n \cos[(k\omega_r - n\omega_m)t - n\phi]\}).$$

We note that each pulse-repetition-frequency harmonic is phase-modulated, with peak deviation equal to $k\omega_r\tau_d$. Also, there is no sideband ac-

companying the direct-current component of the pulse spectrum, and hence modulation cannot be recovered by means of a low-pass filter.

Uniform sampling:

$$p_m(t) = \frac{A\omega_r\tau}{2\pi} + \frac{A\omega_r\omega_m\tau_d\tau}{2\pi} \frac{\sin[\omega_m(\tau/2)]}{\omega_m(\tau/2)}$$

$$\times \cos[\omega_m t + \phi - (\omega_m\tau/2)] + \frac{A\omega_r\tau}{\pi} \sum_{k=1}^{\infty} \left(J_0(k\omega_r\tau_d) \right.$$

$$\times \frac{\sin[(k\omega_r(\tau/2)]}{k\omega_r(\tau/2)} \cos k\omega_r t + \sum_{n=1}^{\infty} J_n(k\omega_r\tau_d)$$

$$\times \left\{ \frac{\sin(k\omega_r + n\omega_m)(\tau/2)}{k\omega_r(\tau/2)} \cos[(k\omega_r + n\omega_m)t \right.$$

$$\left. + n\phi - n\omega_m(\tau/2)] + (-1)^n \frac{\sin(k\omega_r - n\omega_m)(\tau/2)}{k\omega_r(\tau/2)} \right.$$

$$\left. \times \cos[(k\omega_r - n\omega_m)t - n\phi + n\omega_m(\tau/2)] \right\}).$$

This is an equation very similar to that for pulse-frequency modulation.

Pulse-Width Modulation: The spectrum for width-modulated pulses can be obtained from the spectrum of phase-modulated pulses. If the trailing edge, instead of being displaced in the same direction as the leading edge, is displaced in the opposite direction, pulse-width modulation will be produced.

Considering first the case of *symmetrical double-edge modulation*, the expression for the width-modulated pulse train becomes

$$p_m(t) = \frac{A}{2\pi j} \sum_{n,k=-\infty}^{\infty} k^{-1}\{J_n(k\omega_r\tau_d) \exp[jk\omega_r(\tau/2)]$$

$$- J_n(-k\omega_r\tau_d) \exp[-jk\omega_r(\tau/2)]\}$$

$$\times \exp\{ j[(k\omega_r + n\omega_m)t + n\phi]\}.$$

Let $m = 2\tau_d/\tau$, the modulation index; thus for $m = 1$ (100% modulation) the maximum and minimum values of the pulse width will vary between 2τ and 0, and the expression reduces to

$$p_m(t) = \frac{A\omega_r\tau}{2\pi} + \frac{A\omega_r m\tau}{2\pi} \sin(\omega_m t + \phi) + \frac{A\omega_r\tau}{2\pi}$$

$$\times \sum_{k=1}^{\infty} \left[2J_0[k\omega_r m(\tau/2)] \frac{\sin[k\omega_r(\tau/2)]}{k\omega_r(\tau/2)} \cos k\omega_r t \right.$$

$$+ 4J_1[k\omega_r m(\tau/2)] \frac{\cos[k\omega_r(\tau/2)]}{k\omega_r(\tau/2)} \cos(k\omega_r t)$$

$$\times \sin(\omega_m t + \phi) + 4J_2[k\omega_r m(\tau/2)] \frac{\sin[k\omega_r(\tau/2)]}{k\omega_r(\tau/2)}$$

$$\left. \times \cos(k\omega_r t) \cos(2\omega_m t + 2\phi) + \cdots \right].$$

We note that the direct-current component of the pulse spectrum has a sideband of the modulating frequency of amplitude $A\omega_r m\tau/2\pi$; therefore, modulation can be recovered by means of a low-pass filter.

In the case of *single-edge modulation*, only one edge is being modulated (e.g., the leading edge), and the resulting spectrum is given by

$$p_m(t) = \frac{A\omega_r\tau}{2\pi} + \frac{A\omega_r m\tau}{2\pi}\sin(\omega_m t + \phi) + \frac{A\omega_r\tau}{2\pi}$$

$$\times \sum_{k=1}^{\infty}\left[\frac{\sin k\omega_r(t-\tau/2)}{k\omega_r(\tau/2)} + \frac{J_0(k\omega_r m\tau)}{k\omega_r(\tau/2)}\sin k\omega_r(t+\tau/2)\right.$$

$$+\frac{2J_1(k\omega_r m\tau)}{k\omega_r(\tau/2)}\cos k\omega_r(t+\tau/2)\,\sin(\omega_m t+\phi)$$

$$+\frac{2J_2(k\omega_r m\tau)}{k\omega_r(\tau/2)}\sin k\omega_r(t+\tau/2)$$

$$\left.\times\cos(2\omega_m t+2\phi)+\cdots\right].$$

In this case also, the modulating signal can be extracted by means of a low-pass filter.

Signal-to-Noise Improvement Ratio in PTM

In PDM, the noise manifests itself as jitter in the leading and trailing edges of the recovered pulses, and the slopes of the pulse edges influence noise reduction. PPM systems are affected by noise in the same manner as PDM systems. Considering trapezoidal pulses (Figs. 33 and 34), the S/N power ratio at the demodulator output is

$$(S/N)_o = \tfrac{1}{2}(t_0/\tau_r)^2(A_c/\sigma)^2.$$

The ratio of peak pulse power to mean noise power is

$$(C/N)_i = (A_c/\sigma)^2.$$

Hence

$$(S/N)_o = \tfrac{1}{2}(t_0/\tau_r)^2(C/N)_i.$$

The $(S/N)_o$ can be improved by decreasing the

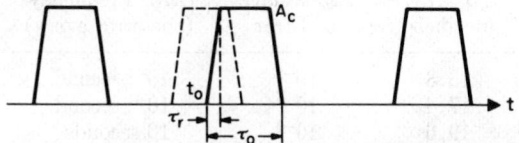

Fig. 33—Pulse-position modulation of trapezoidal pulses. *From P. F. Panter, "Modulation, Noise, and Spectral Analysis," Fig. 18–26, © 1965, McGraw-Hill Book Company.*

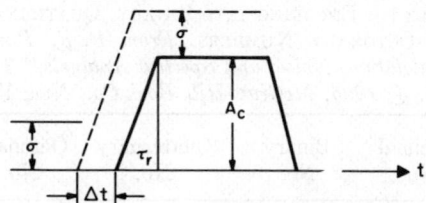

Fig. 34—Variation in pulse position due to noise or interference. *From P. F. Panter, "Modulation, Noise, and Spectral Analysis," Fig. 18–27, © 1965, McGraw-Hill Book Company.*

pulse rise τ_r or correspondingly by widening the transmission bandwidth. For $B \cong 1/\tau_r$

$$(S/N)_o = \tfrac{1}{2}t_0^2 B^2(C/N)_i.$$

For $B \cong 1/2\tau_r$

$$(S/N)_o = 2t_0^2 B^2(C/N)_i.$$

As in the case of FM, the $(S/N)_o$ ratio cannot be improved indefinitely by widening the bandwidth, because the noise power introduced at the receiver increases with bandwidth and eventually becomes comparable to the signal and "takes over" the system. A threshold level thus also exists just as in the FM case. This threshold level is usually taken as $A_c/\sigma = 2$, or $(C/N)_i = 4$ (6 dB).

PULSE-CODE MODULATION (PCM)

In PCM, several pulses are used as a code group to describe the quantized amplitude of a single sample. For example, a code group of n on–off pulses (binary code) can represent 2^n discrete amplitudes or levels, including zero level. In general, in an s-ary PCM system, the number of quantized amplitude levels the code group can express (including zero level) is given by

$$M = s^n.$$

If a stands for 0 or 1, the binary notation with n digits, $a_1, a_2, \cdots, a_n$, represents the number

$$a_1 2^0 + a_2 2^1 + a_3 2^2 + \cdots + a_n 2^{n-1}.$$

In the ternary number system, a stands for the pulse amplitude 0, 1, 2, and the code group of n digits represents the number

$$a_1 3^0 + a_2 3^1 + \cdots + a_n 3^{n-1}.$$

Table 1 shows how the 64 numbers from 0 through 63 are represented in binary, quaternary, and octonary notation.

TABLE 1.—ENCODING INTO BINARY, QUATERNARY, AND OCTONARY NUMBERS. *From P. F. Panter, "Modulation, Noise, and Spectral Analysis," Table 20–1, © 1965, McGraw-Hill Book Co., New York.*

Decimal No.	Binary No.	Quaternary No.	Octonary No.
0	000000	000	00
1	000001	001	01
2	000010	002	02
3	000011	003	03
4	000100	010	04
5	000101	011	05
6	000110	012	06
7	000111	013	07
8	001000	020	10
9	001001	021	11
10	001010	022	12
11	001011	023	13
12	001100	030	14
⋮	⋮	⋮	⋮
62	111110	332	76
63	111111	333	77

Transmission Requirements for PCM

Minimum Channel Bandwidth for No Intermodulation:

$$B = nf_m$$

where $n=$ number of pulses in the code group, and $f_m=$ highest frequency of message signal.

Threshold Power: The reliability of detecting the presence or absence of a pulse is a function of the signal-to-noise ratio. The probability of error using synchronous detection in the presence of white gaussian noise is given by

$$p = \tfrac{1}{2}[1 - \mathrm{erf}(V_0/2\sigma)]$$

where $V_0=$ peak signal pulse, $\sigma=$ rms noise amplitude, and erf$x$ is the error function

$$\mathrm{erf}x \equiv (2\pi)^{-1/2} \int_{-x}^{x} \exp(-u^2/2)\, du.$$

The rapid decrease of the probability of error as the signal-to-noise ratio is increased is shown in Table 2. Note that there exists in PCM a fairly definite threshold, at about 20 dB, above which the interference is negligible.

Average Signal Power: The average signal power of a code group with s possible discrete unipolar

pulse levels uniformly distributed is given by

$$P = (V_0^2/6)[(2s-1)/(s-1)].$$

For the binary code group with unipolar on–off pulses, $s=2$ and $P=V_0^2/2$. The average signal power of a bipolar s-ary system is

$$P = (V_0^2/12)[(s+1)/(s-1)]$$

which for a bipolar binary system reduces to

$$P = V_0^2/4.$$

Channel Capacity of a PCM System

$$C = B \log_2 s^2 = B \log_2[1 + (12P/K^2N)] \text{ bit/second}$$

where $N=$ average channel noise power, and $K=$ constant (relating the spacing of levels of code pulses to the rms noise voltage σ) to attain the required probability of error $V_0 = (s-1)K\sigma$. Comparing the channel capacity of a PCM system with the capacity $C = B \log_2(1+P)$ of Shannon's ideal system, it follows that a PCM system requires $K^2/12$ times the power theoretically required to realize a given channel capacity for a given bandwidth.

Quantization Noise in a PCM System

Representing the message signal by certain discrete allowed levels or steps is called quantizing. It inherently introduces an initial error in the amplitude of the samples, giving rise to quantization noise.

Uniform Spacing of Levels: In this case, the quantizing interval or step $\Delta v=$ constant, and the

TABLE 2.—RELATIONSHIP OF PROBABILITY OF ERROR TO SIGNAL-TO-NOISE RATIO. *From P. F. Panter, "Modulation, Noise, and Spectral Analysis," Table 20–2, © 1965, McGraw-Hill Book Co., New York.*

P/N (decibels)	Probability of Error	Error Frequency (one error every)
13.3	10^{-2}	10^{-3} second
17.4	10^{-4}	10^{-1} second
19.6	10^{-6}	10 seconds
21.0	10^{-8}	20 minutes
22.0	10^{-10}	1 day
23.0	10^{-12}	3 months

quantizing noise power is given by

$$N_q = (\Delta v)^2/12$$

assuming that the quantization noise is uniformly distributed between $\pm\Delta v/2$. Assuming that the amplitudes of the samples are uniformly distributed, the signal power recovered from the quantized samples is

$$S_q = [(M^2-1)/12](\Delta v)^2$$

where $M =$ number of discrete levels assigned to message signal. The ratio of the signal power to the quantizing noise power is

$$S_q/N_q = M^2-1 \cong M^2, \qquad M \gg 1.$$

Nonuniform Spacing of Levels: Quantization noise can be reduced by the use of nonuniform spacing of levels, to provide smaller steps for weaker signals and coarser quantization near the peak of large signals. Quantization noise can be minimized by an optimum level distribution which is a function of the probability density of the signal. The optimum level spacing Δv_k is given by

$$[p(v_k)]^{1/3}\Delta v_k = k/M, \qquad k = \text{constant}.$$

With optimum level spacing the total minimum error power is

$$(N_q)_{min} = (2/3M^2)\left\{\int_0^V [p(v)]^{1/3}\,dv\right\}^3$$

where $p(v) =$ probability density of the message signal, and the nonuniform levels are symmetri-

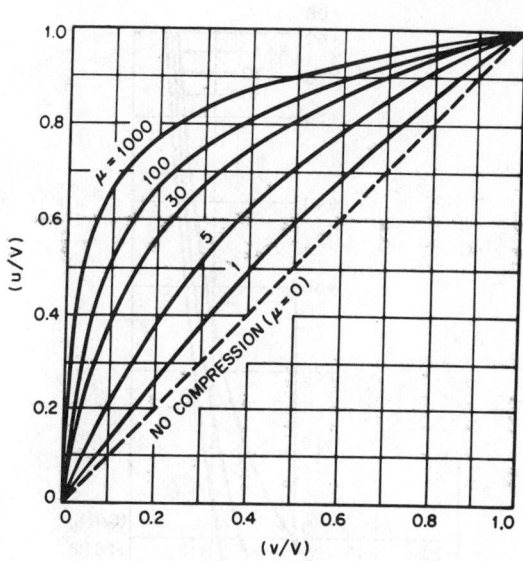

Fig. 36—Logarithmic compression characteristics. *From P. F. Panter, "Modulation, Noise, and Spectral Analysis," Fig. 20–11, © 1965, McGraw-Hill Book Company.*

cally disposed about zero level in the amplitude range $(-V, V)$.

In practice, nonuniform quantization is realized by compression, followed by uniform quantization as in Fig. 35. The logarithmic compression curve shown in Fig. 36 renders the distortion largely independent of the signal and is relatively easy to obtain in practice.

$$u = k \log[1+(\mu v/V)]$$

where $v =$ input voltage, $u =$ output voltage, $\mu =$ compression parameter, and $k =$ undetermined constant. By adjusting the maximum values of the input and the compressed signals to be equal, this gives

$$u = \frac{V \log(1+\mu v/V)}{\log(1+\mu)}, \qquad 0 \leq v \leq V$$

and

$$u = \frac{-V \log(1-\mu v/V)}{\log(1+\mu)}, \qquad -V \leq v \leq 0.$$

The quantizing noise power using logarithmic compression is

$$N_q = (\alpha^2/12)(V^2+\mu^2 S)$$

where

$$\alpha = \frac{2 \log(1+\mu)}{\mu M}$$

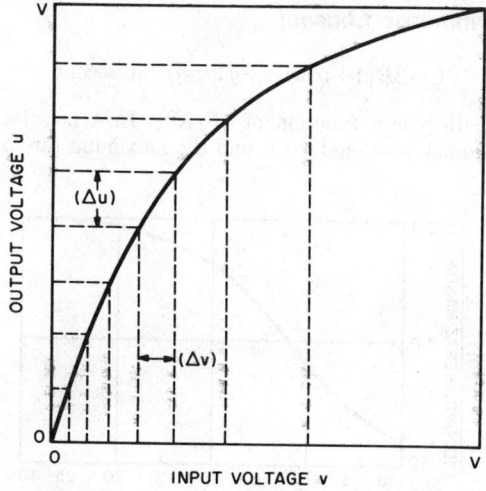

Fig. 35—Compression characteristic of "compressor". *From P. F. Panter, "Modulation, Noise, and Spectral Analysis," Fig. 20–10, © 1965, McGraw-Hill Book Company.*

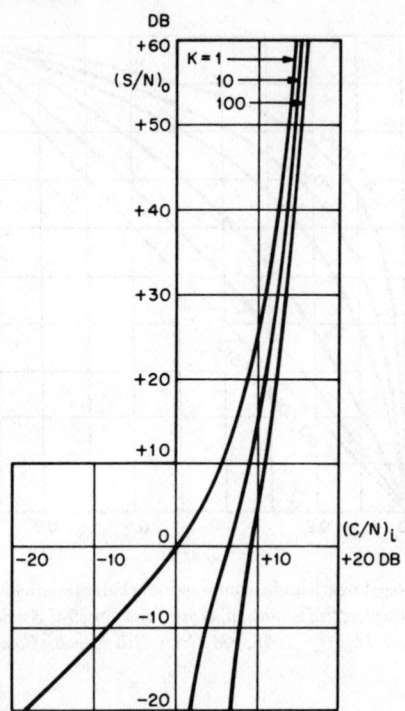

Fig. 37—Output signal-to-noise ratio for PCM. *From H. F. Mayer, "Principles of Pulse Code Modulation," Advan. Electron., vol. 3, © 1951, Academic Press, Inc., New York.*

and

$$S = \int_{-V}^{V} v^2 p(v) \, dv = \text{average signal power.}$$

False-Pulse Noise in a PCM System

In addition to quantization noise, a PCM system is characterized by *false-pulse noise*, which originates primarily at the receiving end of the system and is caused by noise spikes breaking through the threshold. This type of noise decreases rapidly as the signal power is increased above threshold. The effect of the false pulses introduced in the code group is to introduce an error in the decoded samples. The mean-square error introduced in the decoded signal is defined as the false-pulse noise. The output signal-to-noise ratio at the decoder is

$$(S/N)_o = (1/4pq) - 1$$

where p is the probability of sending out state one and receiving state zero and vice versa, and q is the probability that no transmission fault occurs ($p+q=1$). The output signal-to-noise ratio drops from infinity with a noiseless channel ($p=0$, $q=1$) to zero in the case of an infinitely large channel noise ($p=q=1/2$).

The output signal-to-noise ratio for K links in tandem is given by

$$(S/N)_{o,(K)} = \frac{(q-p)^{2K}}{1 - (q-p)^{2K}}.$$

These expressions for the $(S/N)_o$ in a PCM system are given in terms of the probability of false pulses in the code group due to channel noise. The following expressions relate the output signal-to-noise ratio to the input carrier-to-noise ratio for one link.

$$(S/N)_o = (\pi/8)^{1/2} (V_0/2\sigma) \, \exp[\tfrac{1}{2}(V_0/2\sigma)^2].$$

For unipolar or on–off binary system

$$(S/N)_o = (\pi/16)^{1/2} [(C/N)_i]^{1/2} \exp[\tfrac{1}{4}(C/N)_i].$$

For bipolar binary system

$$(S/N)_o = (\pi/8)^{1/2} [(C/N)_i]^{1/2} \exp[\tfrac{1}{2}(C/N)_i].$$

For K links in tandem

$$(S/N)_{o(K)} = (\text{erf}x)^{2K} / [1 - (\text{erf}x)^{2K}]$$

where $x = V_0/2\sigma$. For high $(C/N)_i$, $x \gg 1$, and

$$\text{erf}x \cong 1 - (2/\pi)^{1/2} \frac{\exp(-x^2/2)}{x}.$$

The rapid improvement in $(S/N)_o$ for small increases in $(C/N)_i$ is illustrated in Fig. 37 for various links in tandem.

Capacity or Maximum Rate of Transmission Over a Noisy Binary Symmetric Channel

$$C = 2B(1 + p \log_2 p + q \log_2 q) \text{ bit/second}$$

where p is a function of $(C/N)_i$. In a noiseless channel, $p=0$ and $q=1$, and the maximum rate of

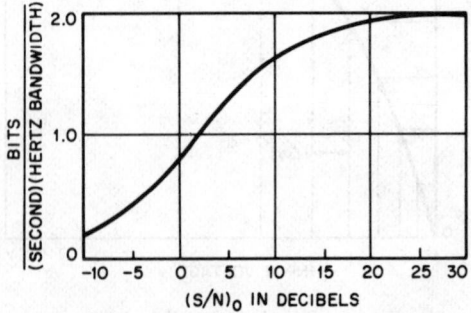

Fig. 38—Rate of information and output signal-to-noise ratio. *From P. F. Panter, "Modulation, Noise, and Spectral Analysis," Fig. 21–8, © 1965, McGraw-Hill Book Company.*

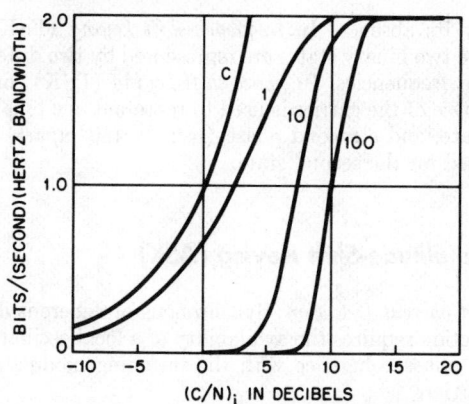

Fig. 39—Rate of transmission and channel carrier-to-noise ratio. *From H. F. Mayer, "Principles of Pulse Code Modulation," Advan. Electron., vol. 3, © 1951, Academic Press, Inc., New York.*

transmission is 2 bits/(second) (hertz bandwidth) for binary orthogonal pulses. The maximum rate of transmission as a function of $(S/N)_o$ is shown in Fig. 38. The rate of transmission as a function of $(C/N)_i$, with K the number of links in tandem, is shown in Fig. 39.

DELTA MODULATION (DM)

In a DM system, instead of the absolute signal amplitude being transmitted at each sampling, only the changes in signal amplitude from sampling instant to sampling instant are transmitted. As shown in Fig. 40, the transmitted pulse train $e_2(t)$

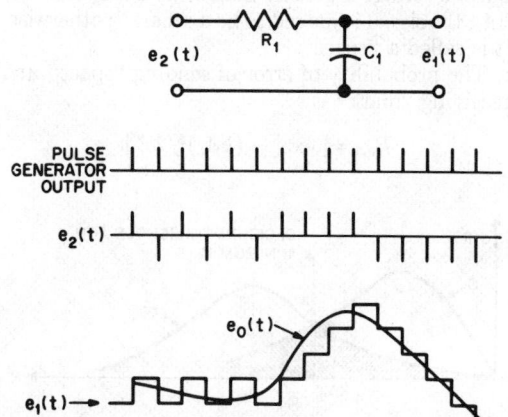

Fig. 40—Delta modulation waveforms using single integration. *From P. F. Panter, "Modulation, Noise, and Spectral Analysis," Fig. 22-2, © 1965, McGraw-Hill Book Company.*

of positive and negative pulses at the output of the encoder can be assumed to be generated at a constant clock rate. The transmitted pulses from the pulse generator are positive if the change in signal amplitude is positive, otherwise the transmitted pulses are negative. In the decoder, the delta-modulated pulse train $e_2(t)$ is integrated into the voltage $e_1(t)$, which consists of the original message function plus noise components due to sampling. These are eliminated by a low-pass filter, so that the reconstructed signal of the final output is a close replica of the original modulating signal $e_0(t)$.

Signal-to-Noise Ratio in DM

The difference between the original and reconstructed signals gives rise to a "quantizing noise" that can be decreased by increasing the "sampling frequency," which in DM is made equal to the pulse frequency. The quantized noise power using single integration is given by

$$N_0 = \tfrac{2}{3}(f_m/f_s)(\Delta v)^2$$

where f_m = highest modulating frequency, and Δv = height of unit step in volts.

A DM system has no fixed maximum signal amplitude limitation but overloads when the slope of the signal is too large. The largest slope the system can reproduce is one that changes by one level or step every pulse interval, so that the maximum signal power depends on the type of signal. The signal power in the calculation of signal-to-noise ratio is taken as the power of the sinusoidal tone, which is just below the overload point. The maximum amplitude at such a sinusoidal signal of frequency f that can be transmitted with single integration without overloading is

$$A = f_s(\Delta v)/2\pi f.$$

The average signal power is

$$S_0 = f_s^2(\Delta v)^2/4\pi^2 f^2$$

so that the signal-to-noise ratio for single integration is

$$(S/N)_o = \tfrac{3}{2}r^3(f_m/\pi f)^2$$

where $r = f_s/2f_m$ = bandwidth expansion factor. The signal-to-noise ratio using double integration is

$$(S/N)_o = \tfrac{3}{2}r^5(f_m/\pi f)^4.$$

Thus, the improvement in signal-to-noise ratio varies with f_s^3 for the system with single integration, whereas it varies with f_s^5 for double integration.

DIGITAL DATA MODULATION SYSTEMS

In a digital data communication system*, the information source consists of a finite number of discrete messages which are coded into a sequence of waveforms or symbols; each waveform is selected from a finite alphabet of signal waveforms. Thus, the problem of transmitting information is reduced to the problem of transmitting a sequence of waveforms, each one selected from a specified and finite set. This is in contrast to the problem of transmitting analog information where the resulting set of waveforms is infinite. However, as in analog modulation, the information-carrying digital waveforms are used to modulate a sinusoidal carrier to place the relatively low-frequency energy of the video signals into the higher-frequency band. In the detection process, either *coherent detection* is used (where the receiver is phase-locked with the transmitter) or *noncoherent detection* (where the receiver is not phase-locked with the transmitter). However, the receiver will be assumed to be time synchronized in all the digital modulation systems under discussion.

At the receiver, the problem of reception reduces itself to the problem of deciding between the signal waveforms. Since the decision of the discrete-signal receiver is either right or wrong, the criterion of performance of a digital communication system is ordinarily based on the probability of error, i.e., the probability of choosing an incorrect message from a finite set of possible transmitted messages.

REVIEW OF DIGITAL MODULATION METHODS

This section outlines the most common modulation methods of producing discrete binary carrier-modulated signals. In binary modulation systems, the digital information to be transmitted is assumed to be coded in binary form using two elementary signals. These two signals are called "mark" and "space" or "one" and "zero"; they are of equal and finite duration and occur with equal probability. The two signals are generated by modulating a sinusoidal carrier in amplitude, frequency, or phase in a time sequence of two mutually exclusive states. In *amplitude-shift keying* (ASK), the sinusoidal carrier is pulsed so that one of the binary states is represented by the presence of the carrier while the other state is represented

* P. F. Panter, "Modulation, Noise, and Spectral Analysis," Chapter 23, McGraw-Hill Book Co., New York, N. Y.; 1965.

by its absence. In *frequency-shift keying* (FSK), the two binary states are represented by two different frequencies. In *phase-shift keying* (PSK), one phase of the carrier is used to represent one binary state, and a second phase (usually 180° apart) is used for the second state.

Amplitude-Shift Keying (ASK)

Coherent Detection: Synchronous or coherent detection requires the availability of a local oscillator in phase coherence with the incoming modulated carrier.

The probability of error P_e or the fraction of the total number of incorrect decisions made by the receiver is

$$P_e = \tfrac{1}{2}[1 - \mathrm{erf}(A_c/2\sigma)]$$

where A_c = amplitude of carrier, and $\sigma^2 = N =$ average channel noise power. For high input-carrier-to-noise ratio $A_c \gg \sigma$ and

$$\mathrm{erf}\, t \doteq 1 - (2/\pi)^{1/2} \frac{\exp(-t^2/2)}{t}, \quad t = A_c/2\sigma$$

so that for large $(C/N)_i$

$$P_e = [\pi(C/N)_i]^{-1/2} \exp[-\tfrac{1}{4}(C/N)_i]$$

where $(C/N)_i = A_c/2\sigma^2$.

Envelope Detection: In envelope detection, signal phase coherence is not required in the detection process, and a simple envelope detector is used following the IF amplifier. The decision at the output is based on a threshold kA_c; if the detector output voltage is greater than some fixed threshold kA_c, the signal is judged to be a "mark"; otherwise, it is called a "space".

The probability of error of sending "space" and receiving "mark" is

$$P_{e,s} = \tfrac{1}{2} \exp[-(kA_c)^2/2\sigma^2].$$

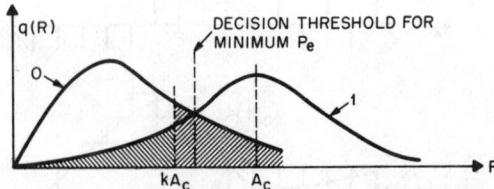

Fig. 41—Decision threshold for envelope detection of binary ASK. $q(R)$ is the probability density of carrier plus noise. *From P. F. Panter, "Modulation, Noise, and Spectral Analysis," Fig. 23-2, © 1965, McGraw-Hill Book Company.*

The probability of sending "mark" and receiving "space" is

$$P_{e,m}=\tfrac{1}{2}\exp[-(1+k^2)A_c{}^2/2\sigma^2]\sum_{n=1}^{\infty}(k)^nI_n(A_c{}^2/\sigma^2)$$

where I_n= Bessel function of nth order and imaginary argument.

The two types of error, namely $P_{e,s}$ and $P_{e,m}$, are not in general equiprobable, but can be made so for a given $(C/N)_i$ by a proper choice of the decision threshold kA_c. The results are plotted in Fig. 41, where it is seen that the decision factor k approaches the value of 0.5 for high $(C/N)_i$.

Frequency-Shift Keying (FK)

In a multiple-coded FSK system, digital information is transmitted by using as a code the sequential transmission of carrier pulses of constant amplitude and several different frequencies. In a binary system, only two carriers are used, with the code block being any desired length.

Coherent Detection: The probability of error is given by

$$P_e=\tfrac{1}{2}[1-\mathrm{erf}(A_c/2\sigma)]=\tfrac{1}{2}[1-\mathrm{erf}(E/2N_0)^{1/2}]$$

where E= energy content of signal waveform, and N_0= noise power density. Hence

$$(C/N)_i=E/N_0.$$

Noncoherent or Envelope Detection: A noncoherent FSK system uses envelope detection, and the decision as to whether a "mark" or a "space" was transmitted is made on the basis of which detector output has the highest amplitude at the sampling time. The probability of error is

$$P_e=\tfrac{1}{2}\exp(-E/2N_0)=\tfrac{1}{2}\exp[-\tfrac{1}{2}(C/N)_i].$$

A plot of error probability as a function of $(C/N)_i$ is shown in Fig. 42 for both coherent and non-

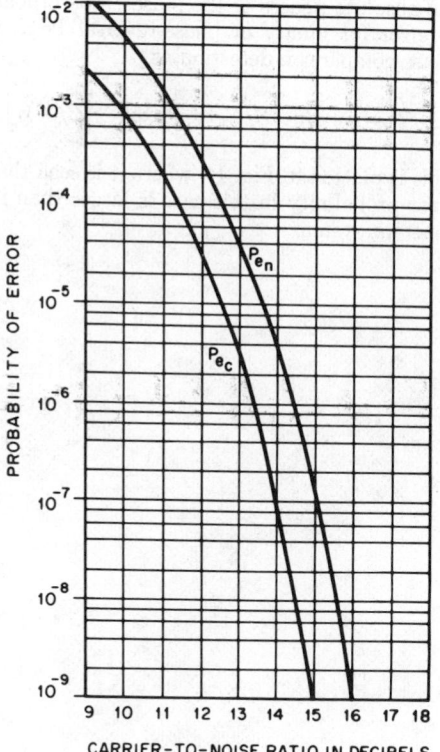

Fig. 42—Probability of error in binary FSK system: P_{e_n}=probability of error for a noncoherent FSK system; P_{e_c}=probability of error for a coherent FSK system. *From P. F. Panter, "Modulation, Noise, and Spectral Analysis," Fig. 23-5, © 1965, McGraw-Hill Book Company.*

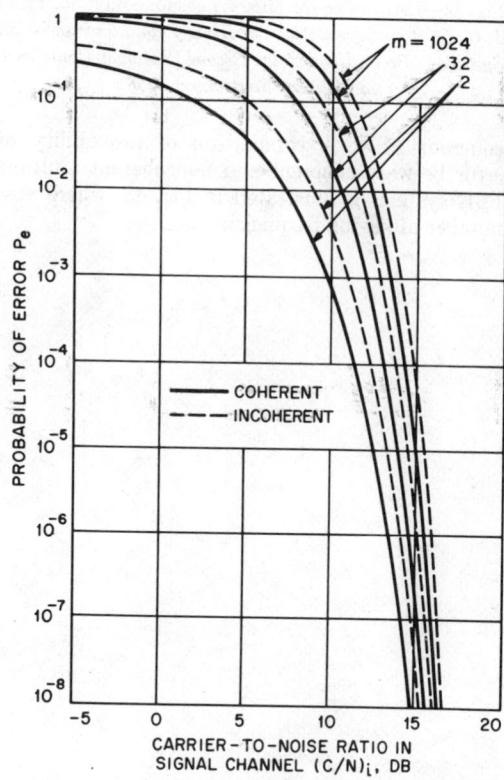

Fig. 43—A comparison of probability of error between coherent and noncoherent multiple FSK systems. *From H. Akima, "The Error Rates in Multiple FSK Systems, National Bureau of Standards Technical Note 167; March 1963.*

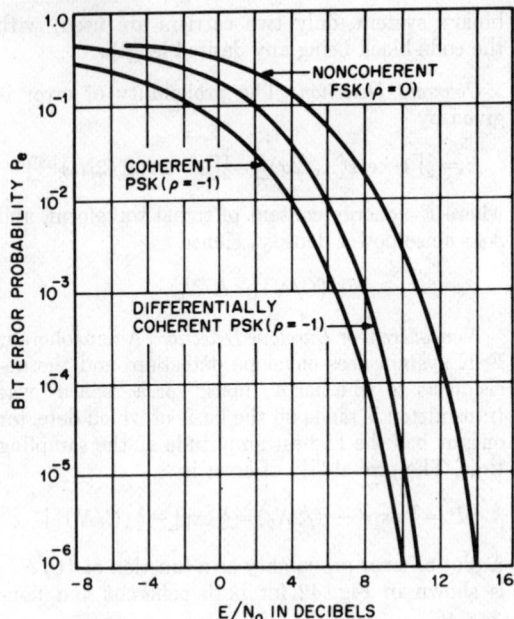

Fig. 44—Error rates for binary phase modulation. *From J. G. Lawton, "Comparison of Binary Data Transmission Systems," Proceedings of the Second National Conference on Military Electronics; 1958.*

coherent FSK. A comparison of probability of error between coherent and noncoherent multiple FSK systems is illustrated in Fig. 43, where $m=$ number of keying frequencies.

Phase-Shift Keying (PSK)

In digital phase-shift keying or digital phase modulation, digital information is transmitted by using as a code the sequential transmission of carrier pulses of constant amplitude, angular frequency, and duration, but of different relative phase.

Coherent Detection: In coherent detection, a phase reference is provided in the receiver, permitting the receiver to be phase-synchronized with the transmitter. For binary PSK, the probability of error is

$$P_e = \tfrac{1}{2}[1 - \text{erf}(E/N_0)^{1/2}].$$

Phase-Comparison or Differentially Coherent Detection: In this system phase comparison of successive samples is used in the detection process, and the information is conveyed by the phase transitions between carrier pulses rather than by the absolute phases of the pulses. The probability of error for binary or phase-reversal PSK, using phase-comparison detection, is

$$P_e = \tfrac{1}{2}\exp(-E/N_0) = \tfrac{1}{2}\exp[-(C/N)_i].$$

This is plotted in Fig. 44, where it is seen that the error probability in this case is larger than in the coherent system.

GENERAL

The equations and charts of this chapter are for transmission lines operating in the TEM mode.* At the beginning of several of the sections (e.g., "Fundamental Quantities," "Voltage and Current," "Impedance and Admittance," "Reflection Coefficient") there are accurate equations, according to conventional transmission-line theory. These are applicable from the lowest power and communication frequencies, including direct current, up to the frequency where a higher mode begins to appear on the line.

Following the accurate equations are others that are specially adapted for use in radio-frequency problems. In cases of small attenuation, the terms $\alpha^2 x^2$ and higher powers in the expansion of $\exp\alpha x$, etc., are neglected. Thus, when $\alpha x = (\alpha/\beta)\theta = 0.1$ neper (or about 1 decibel), the error in the approximate equations is of the order of 1 percent.

Much of the information is useful also in connection with special lines, such as those with spiral (helical) inner conductors, which function in a quasi-TEM mode; likewise for microstrip.

It should be observed that Z_0 and Y_0 are complex quantities and the imaginary part cannot be neglected in the accurate equations, unless preliminary examination of the problems indicates the contrary. Even when attenuation is small, $Z_0 = 1/Y_0$ must often be taken at its complex value, especially when the standing-wave ratio is high. In the first few pages of equations, the symbol R_0 is used frequently. However, in later charts and special applications, the conventional symbol Z_0 is used where the context indicates that the quadrature component need not be considered for the moment.

Rule of Subscripts and Sign Conventions

The equations for voltage, impedance, etc., are generally for the quantities at the input terminals of the line in terms of those at the output terminals

---

* The information on pp. 24-1–24-18 is valid for single-mode waveguides in general, except for equations where the symbols R, L, G, or C per unit length are involved.

(Fig. 1). In case it is desired to find the quantities at the output in terms of those at the input, it is simply necessary to interchange the subscripts 1 and 2 in the equations and to place a minus sign before x or θ. The minus sign may then be cleared through the hyperbolic or circular functions; thus

$$\sinh(-\gamma x) = -\sinh\gamma x, \quad \text{etc.}$$

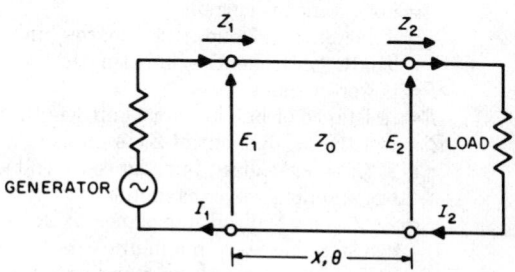

Fig. 1—Transmission line with generator, load.

SYMBOLS

Voltage and current symbols usually represent the alternating-current complex sinusoid, with magnitude equal to the root-mean-square value of the quantity.

Certain quantities, namely C, c, f, L, T, v, and ω are shown with an optional set of units in parentheses. Either the standard units or the optional units may be used, provided the same set is used throughout.

$A = 10 \log_{10}(1/\eta) = $ dissipation loss in a length of line in decibels

$A_0 = 8.686\alpha x = $ normal or matched-line attenuation of a length of line in decibels.

$B_0 = $ susceptive component of Y_0 in mhos

$C = $ capacitance of line in farads/unit length (microfarads/unit length)

$c = $ velocity of light in vacuum in units of length/second (units of length/microsecond). See page 3-11.

$E = $ voltage (root-mean-square complex sinusoid) in volts

$_fE = $ voltage of forward wave, traveling toward load

$_rE = $ voltage of reflected wave

$|E_{flat}|$ = root-mean-square voltage when standing-wave ratio = 1.0

$|E_{max}|$ = root-mean-square voltage at crest of standing wave

$|E_{min}|$ = root-mean-square voltage at trough of standing wave

e = instantaneous voltage

$F_p = G/\omega C$ = power factor of dielectric

f = frequency in hertz (megahertz)

G = conductance of line in mhos/unit length

G_0 = conductive component of Y_0 in mhos

$g_a = Y_a/Y_0$ = normalized admittance at voltage standing-wave maximum

$g_b = Y_b/Y_0$ = normalized admittance at voltage standing-wave minimum

I = current (root-mean-square complex sinusoid) in amperes

$_fI$ = current of forward wave, traveling toward load

$_rI$ = current of reflected wave

i = instantaneous current

L = inductance of line in henries/unit length (microhenries/unit length)

P = power in watts

R = resistance of line in ohms/unit length

R_0 = resistive component of Z_0 in ohms

$r_a = Z_a/Z_0$ = normalized impedance at voltage standing-wave maximum

$r_b = Z_b/Z_0$ = normalized impedance at voltage standing-wave minimum

$S = |E_{max}/E_{min}|$ = voltage standing-wave ratio

T = delay of line in seconds/unit length (microseconds/unit length)

v = phase velocity of propagation in units of length/second (units of length/microsecond)

X_0 = reactive component of Z_0 in ohms

x = distance between points 1 and 2 in units of length (also used for normalized reactance = X/Z_0)

$Y_1 = G_1 + jB_1 = 1/Z_1$ = admittance in mhos looking toward load from point 1

$Y_0 = G_0 + jB_0 = 1/Z_0$ = characteristic admittance of line in mhos

$Z_1 = R_1 + jX_1$ = impedance in ohms looking toward load from point 1

$Z_0 = R_0 + jX_0$ = characteristic impedance of line in ohms

Z_{oc} = input impedance of a line open-circuited at the far end

Z_{sc} = input impedance of a line short-circuited at the far end

α = attenuation constant = nepers/unit length = $0.1151 \times$ decibels/unit length

β = phase constant in radians/unit length

$\gamma = \alpha + j\beta$ = propagation constant

ϵ = base of natural logarithms = 2.718; or dielectric constant of medium (relative to air), according to context

$\eta = P_2/P_1$ = efficiency (fractional)

$\theta = \beta x$ = electrical length or angle of line in radians

$\theta° = 57.3\theta$ = electrical angle of line in degrees

λ = wavelength in units of length

λ_0 = wavelength in free space

$\rho = |\rho| \angle 2\psi$ = voltage reflection coefficient

$\rho_{dB} = -20 \log_{10}(1/\rho)$ = voltage reflection coefficient in decibels

ϕ = time phase angle of complex voltage at voltage standing-wave maximum

ψ = half the angle of the reflection coefficient = electrical angle to nearest voltage standing-wave maximum on the generator side

$\omega = 2\pi f$ = angular velocity in radians/second (radians/microsecond).

FUNDAMENTAL QUANTITIES AND LINE PARAMETERS

$$dE/dx = -(R + j\omega L)I$$

$$d^2E/dx^2 = \gamma^2 E$$

$$dI/dx = -(G + j\omega C)E$$

$$d^2I/dx^2 = \gamma^2 I$$

$$\gamma = \alpha + j\beta = [(R + j\omega L)(G + j\omega C)]^{1/2}$$

$$= j\omega(LC)^{1/2}$$

$$\times [(1 - jR/\omega L)(1 - jG/\omega C)]^{1/2}$$

$$\alpha = (\tfrac{1}{2}\{[(R^2 + \omega^2 L^2)(G^2 + \omega^2 C^2)]^{1/2} + RG - \omega^2 LC\})^{1/2}$$

$$\beta = (\tfrac{1}{2}\{[(R^2 + \omega^2 L^2)(G^2 + \omega^2 C^2)]^{1/2} - RG + \omega^2 LC\})^{1/2}$$

$$Z_0 = 1/Y_0 = [(R + j\omega L)/(G + j\omega C)]^{1/2}$$

$$= (L/C)^{1/2}[(1 - jR/\omega L)/(1 - jG/\omega C]^{1/2}$$

$$= R_0(1 + jX_0/R_0)$$

$$Y_0 = 1/Z_0 = G_0(1 + j\,B_0/G_0)$$

$$\alpha = \tfrac{1}{2}(R/R_0 + G/G_0)$$

$$\beta\,B_0/G_0 = \tfrac{1}{2}(R/R_0 - G/G_0)$$

$$R_0 = [M/2(G^2 + \omega^2 C^2)]^{1/2}$$

$$G_0 = [M/2(R^2 + \omega^2 L^2)]^{1/2}$$

$$B_0/G_0 = -X_0/R_0 = (\omega RC - \omega LG)/M$$

where

$$M=[(R^2+\omega^2L^2)(G^2+\omega^2C^2)]^{1/2}+RG$$
$$+\omega^2LC$$

$$1/T=v=f\lambda=\omega/\beta$$

$$\beta=\omega/v=\omega T=2\pi/\lambda$$

$$\gamma x=\alpha x+j\beta x=(\alpha/\beta)\theta+j\theta$$

$$\theta=\beta x=2\pi\,x/\lambda=2\pi\,fTx$$

$$\theta\degree=57.3\theta=360x/\lambda=360fTx.$$

(**A**) Special case—distortionless line: When $R/L=G/C$, the quantities Z_0 and α are independent of frequency.

$$X_0=0$$

$$\alpha=R/R_0$$

$$Z_0=R_0+j0=(L/C)^{1/2}$$

$$\beta=\omega(LC)^{1/2}.$$

(**B**) For small attenuation: $R/\omega L$ and $G/\omega C$ are small.

$$\gamma=j\omega(LC)^{1/2}\{1-j[(R/2\omega L)+(G/2\omega C)]\}$$

$$=j\beta(1-j\alpha/\beta)$$

$$\beta=\omega(LC)^{1/2}=\omega L/R_0=\omega CR_0$$

$$T=1/v=(LC)^{1/2}=R_0C$$

$$\alpha/\beta=(R/2\omega L)+(G/2\omega C)=(R/2\omega L)+\tfrac{1}{2}F_p$$

$$=(Rv/2\omega R_0)+\tfrac{1}{2}F_p$$

$$=\text{attenuation in nepers/radian}$$

$$=\frac{\text{(decibels per 100 feet) (wavelength in line in meters)}}{1663}$$

$$\alpha=\tfrac{1}{2}R(C/L)^{1/2}+\tfrac{1}{2}G(L/C)^{1/2}$$

$$=(R/2R_0)+\pi(F_p/\lambda)$$

$$=(R/2R_0)+\tfrac{1}{2}(F_p\beta)$$

where R and G vary with frequency, while L and C are nearly independent of frequency.

$$Z_0=1/Y_0$$

$$=(L/C)^{1/2}\{1-j[(R/2\omega L)-(G/2\omega C)]\}$$

$$=R_0(1+jX_0/R_0)$$

$$=1/[G_0(1+j\,B_0/G_0)]$$

$$=(1/G_0)(1-jB_0/G_0)$$

$$R_0=1/G_0=(L/C)^{1/2}$$

$$B_0/G_0=-(X_0/R_0)=(R/2\omega L)-\tfrac{1}{2}F_p=(\alpha/\beta)-F_p$$

$$X_0=-[R/2\omega(LC)^{1/2}]+(G/2\omega C)(L/C)^{1/2}$$

$$=-(R\lambda/4\pi)+(\tfrac{1}{2}F_p)R_0.$$

(**C**) With certain exceptions, the following few equations are for ordinary lines (e.g., not spiral delay lines) with the field totally immersed in a uniform dielectric of dielectric constant ϵ (relative to air). The exceptions are all the quantities not including the symbol ϵ, these being good also for special types such as spiral delay lines, microstrip, etc.

$$L=1.016R_0(\epsilon^{1/2})\times10^{-3}\text{ microhenries/foot}$$

$$=\tfrac{1}{3}R_0(\epsilon^{1/2})\times10^{-4}\text{ microhenries/centimeter}$$

$$C=1.016[(\epsilon^{1/2})/R_0]\times10^{-3}\text{ microfarads/foot}$$

$$=[(\epsilon^{1/2})/3R_0]\times10^{-4}\text{ microfarads/centimeter}$$

$$v/c=1016/R_0C'=\epsilon^{-1/2}$$

$$=\text{velocity factor (with capacitance }C'\text{ in picofarads/foot)}$$

$$\lambda=\lambda_0v/c=c/f(\epsilon^{1/2})=\lambda_0/(\epsilon^{1/2})$$

$$T=1/v=R_0C'\times10^{-6}=1.016\times10^{-3}/(v/c)$$

$$=1.016\times10^{-3}\epsilon^{1/2}\text{ microseconds/foot (with capacitance }C'\text{ in picofarads/foot)}.$$

The line length is

$$x/\lambda=xf(\epsilon^{1/2})/984\text{ wavelengths}$$

$$\theta=2\pi x/\lambda=xf(\epsilon^{1/2})/156.5\text{ radians}$$

where xf is the product of feet times megahertz.

VOLTAGE AND CURRENT

$$E_1={}_fE_1+{}_rE_1={}_fE_2\epsilon^{\gamma x}+{}_rE_2\epsilon^{-\gamma x}$$

$$=E_2\{[(Z_2+Z_0)/2Z_2]\epsilon^{\gamma x}+[(Z_2-Z_0)/2Z_2]\epsilon^{-\gamma x}\}$$

$$=\tfrac{1}{2}(E_2+I_2Z_0)\epsilon^{\gamma x}+\tfrac{1}{2}(E_2-I_2Z_0)\epsilon^{-\gamma x}$$

$$=E_2[\cosh\gamma x+(Z_0/Z_2)\sinh\gamma x]$$

$$=E_2\cosh\gamma x+I_2Z_0\sinh\gamma x$$

$$=[E_2/(1+\rho_2)](\epsilon^{\gamma x}+\rho_2\epsilon^{-\gamma x})$$

$$I_1={}_fI_1+{}_rI_1={}_fI_2\epsilon^{\gamma x}+{}_rI_2\epsilon^{-\gamma x}$$

$$=Y_0({}_fE_2\epsilon^{\gamma x}-{}_rE_2\epsilon^{-\gamma x})$$

$$=I_2\{[(Z_0-Z_2)/2Z_0]\epsilon^{\gamma x}+[(Z_0+Z_2)/2Z_0]\epsilon^{-\gamma x}\}$$

$$=\tfrac{1}{2}(I_2+E_2Y_0)\epsilon^{\gamma x}+\tfrac{1}{2}(I_2-E_2Y_0)\epsilon^{-\gamma x}$$

$$=I_2[\cosh\gamma x+(Z_2/Z_0)\sinh\gamma x]$$

$$=I_2\cosh\gamma x+E_2Y_0\sinh\gamma x$$

$$=[I_2/(1-\rho_2)](\epsilon^{\gamma x}-\rho_2\epsilon^{-\gamma x})$$

$$E_1=AE_2+BI_2$$

$$I_1=CE_2+DI_2$$

where the general circuit parameters are $A=\cosh\gamma x$, $B=Z_0\sinh\gamma x$, $C=Y_0\sinh\gamma x$, and $D=\cosh\gamma x$.

Refer to section on "Matrix Algebra."

(**A**) When point 2 is at a voltage maximum or minimum, x' is measured from voltage maximum and x'' from voltage minimum (similarly for currents)

$$E_1=E_{max}(\cosh\gamma x'+S^{-1}\sinh\gamma x')$$
$$=E_{min}(\cosh\gamma x''+S\sinh\gamma x'')$$
$$I_1=I_{max}(\cosh\gamma x'+S^{-1}\sinh\gamma x')$$
$$=I_{min}(\cosh\gamma x''+S\sinh\gamma x'').$$

When attenuation is neglected

$$E_1=E_{max}(\cos\theta'+j\,S^{-1}\sin\theta')$$
$$=E_{min}(\cos\theta''+j\,S\sin\theta'').$$

(**B**) Letting $Z_l=$ impedance of load, $l=$ distance from load to point 2, and $x_l=$ distance from load to point 1

$$E_1=E_2\frac{\cosh\gamma x_l+(Z_0/Z_l)\sinh\gamma x_l}{\cosh\gamma l+(Z_0/Z_l)\sinh\gamma l}$$

$$I_1=I_2\frac{\cosh\gamma x_l+(Z_l/Z_0)\sinh\gamma x_l}{\cosh\gamma l+(Z_l/Z_0)\sinh\gamma l}.$$

(**C**) $e_1=\sqrt{2}\,|\,_fE_2\,|\,\epsilon^{\alpha x}\sin[\omega t+2\pi(x/\lambda)-\psi_2+\phi]$
$$+\sqrt{2}\,|\,_rE_2\,|\,\epsilon^{-\alpha x}\sin[\omega t-2\pi(x/\lambda)+\psi_2+\phi]$$
$i_1=\sqrt{2}\,|\,_fI_2\,|\,\epsilon^{\alpha x}$
$$\times\sin[\omega t+2\pi(x/\lambda)-\psi_2+\phi+\tan^{-1}(B_0/G_0)]$$
$$+\sqrt{2}\,|\,_rI_2\,|\,\epsilon^{-\alpha x}$$
$$\times\sin[\omega t-2\pi(x/\lambda)+\psi_2+\phi+\tan^{-1}(B_0/G_0)].$$

(**D**) For small attenuation

$$E_1=E_2\{[1+(Z_0/Z_2)\alpha x]\cos\theta$$
$$+j[(Z_0/Z_2)+\alpha x]\sin\theta\}$$
$$I_1=I_2\{[1+(Z_2/Z_0)\alpha x]\cos\theta$$
$$+j[(Z_2/Z_0)+\alpha x]\sin\theta\}.$$

(**E**) When attenuation is neglected

$$E_1=E_2\cos\theta+jI_2Z_0\sin\theta$$
$$=E_2[\cos\theta+j(Y_2/Y_0)\sin\theta]$$
$$=\,_fE_2\epsilon^{j\theta}+\,_rE_2\epsilon^{-j\theta}$$
$$I_1=I_2\cos\theta+jE_2Y_0\sin\theta$$
$$=I_2[\cos\theta+j(Z_2/Z_0)\sin\theta]$$
$$=Y_0(\,_fE_2\epsilon^{j\theta}-\,_rE_2\epsilon^{-j\theta}).$$

General circuit parameters are

$$A=\cos\theta$$
$$B=jZ_0\sin\theta$$
$$C=jY_0\sin\theta$$
$$D=\cos\theta.$$

IMPEDANCE AND ADMITTANCE

$$\frac{Z_1}{Z_0}=\frac{Z_2\cosh\gamma x+Z_0\sinh\gamma x}{Z_0\cosh\gamma x+Z_2\sinh\gamma x}$$

$$\frac{Y_1}{Y_0}=\frac{Y_2\cosh\gamma x+Y_0\sinh\gamma x}{Y_0\cosh\gamma x+Y_2\sinh\gamma x}.$$

(**A**) By interchange of subscripts and change of signs (see p. 24-1), the load impedance is

$$\frac{Z_2}{Z_0}=\frac{Z_1\cosh\gamma x-Z_0\sinh\gamma x}{Z_0\cosh\gamma x-Z_1\sinh\gamma x}.$$

For a length of uniform line or a symmetrical network

$$Z_1=Z_{oc}(Z_2+Z_{sc})/(Z_2+Z_{oc})$$
$$Z_2=Z_{oc}(Z_1-Z_{sc})/(Z_{oc}-Z_1).$$

(**B**) The input impedance of a line at a position of maximum or minimum voltage has the same phase angle as the characteristic impedance.

$$Z_1/Z_0=Z_b/Z_0=Y_0/Y_b=r_b+j0=S^{-1}$$

at a voltage minimum (current maximum).

$$Y_1/Y_0=Y_a/Y_0=Z_0/Z_a=g_a+j0=S^{-1}$$

at a voltage maximum (current minimum).

(**C**) When attenuation is small

$$\frac{Z_1}{Z_0}=\frac{[(Z_2/Z_0)+\alpha x]+j[1+(Z_2/Z_0)\alpha x]\tan\theta}{[1+(Z_2/Z_0)\alpha x]+j[(Z_2/Z_0)+\alpha x]\tan\theta}.$$

For admittances, replace Z_0, Z_1, and Z_2 by Y_0, Y_1, and Y_2, respectively.

When A and B are real

$$\frac{A\pm jB\tan\theta}{B\pm jA\tan\theta}=\frac{2AB\pm j(B^2-A^2)\sin2\theta}{(B^2+A^2)+(B^2-A^2)\cos2\theta}.$$

(**D**) When attenuation is neglected

$$\frac{Z_1}{Z_0}=\frac{Z_2/Z_0+j\tan\theta}{1+j(Z_2/Z_0)\tan\theta}=\frac{1-j(Z_2/Z_0)\cot\theta}{Z_2/Z_0-j\cot\theta}$$

and similarly for admittances.

(**E**) When attenuation $\alpha x=\theta\alpha/\beta$ is small and standing-wave ratio is large (say >10) (*Note:* The complex value of Z_0 or Y_0 must be used in computing the resistive component of Z_1 or Y_1.

For θ measured from a voltage minimum

$$Z_1/Z_0 = [r_b + (\alpha/\beta)\theta](1 + \tan^2\theta) + j\tan\theta$$

$$= [r_b + (\alpha/\beta)\theta](\cos^2\theta)^{-1} + j\tan\theta$$

(See Note 1)

$$Z_0/Z_1 = Y_1/Y_0$$

$$\left.\begin{array}{l} = [r_b + (\alpha/\beta)\theta](1 + \cot^2\theta) - j\cot\theta \\[4pt] = [r_b + (\alpha/\beta)\theta](\sin^2\theta)^{-1} - j\cot\theta. \end{array}\right\}$$

(See Note 2)

For θ measured from a voltage maximum

$$Z_0/Z_1 = Y_1/Y_0 = [g_a + (\alpha/\beta)\theta](1 + \tan^2\theta) + j\tan\theta$$

(See Note 1)

$$Z_1/Z_0 = [g_a + (\alpha/\beta)\theta](1 + \cot^2\theta) - j\cot\theta.$$

(See Note 2)

Note 1: Not valid when $\theta \approx \pi/2$, $3\pi/2$, etc., due to approximation in denominator $1 + (r_b + \theta\alpha/\beta)^2 \tan^2\theta = 1$ (or with g_a in place of r_b).

Note 2: Not valid when $\theta \approx 0$, π, 2π, etc., due to approximation in denominator $1 + (r_b + \theta\alpha/\beta)^2 \cot^2\theta = 1$ (or with g_a in place of r_b). For open- or short-circuited line, valid at $\theta = 0$.

(**F**) When x is an integral multiple of $\lambda/2$ or $\lambda/4$: For $x = n\lambda/2$, or $\theta = n\pi$

$$\frac{Z_1}{Z_0} = \frac{(Z_2/Z_0) + \tanh n\pi(\alpha/\beta)}{1 + (Z_2/Z_0)\tanh n\pi(\alpha/\beta)}.$$

For $x = n\lambda/2 + \lambda/4$, or $\theta = (n+\tfrac{1}{2})\pi$

$$\frac{Z_1}{Z_0} = \frac{1 + (Z_2/Z_0)\tanh(n+\tfrac{1}{2})\pi(\alpha/\beta)}{(Z_2/Z_0) + \tanh(n+\tfrac{1}{2})\pi(\alpha/\beta)}.$$

(**G**) For small attenuation, with any standing-wave ratio: For $x = n\lambda/2$, or $\theta = n\pi$, where n is an integer

$$\frac{Z_1}{Z_0} = \frac{(Z_2/Z_0) + n\pi(\alpha/\beta)}{1 + (Z_2/Z_0)n\pi(\alpha/\beta)}$$

$$g_{a1} = \frac{g_{a2} + \alpha n\lambda/2}{1 + g_{a2}\alpha n\lambda/2} = S_1^{-1}.$$

For $x = (n+\tfrac{1}{2})\lambda/2$, or $\theta = (n+\tfrac{1}{2})\pi$, where n is an integer or zero

$$\frac{Z_1}{Z_0} = \frac{1 + (Z_2/Z_0)(n+\tfrac{1}{2})\alpha(\lambda/2)}{(Z_2/Z_0) + (n+\tfrac{1}{2})\alpha(\lambda/2)}$$

$$g_{b1} = \frac{1 + g_{a2}(n+\tfrac{1}{2})(\alpha/\beta)\pi}{g_{a2} + (n+\tfrac{1}{2})(\alpha/\beta)\pi} = S_1.$$

Subscript a refers to the voltage-maximum point and b to the voltage minimum. In the above equations, subscripts a and b may be interchanged, and/or r may be substituted in place of g, except for the relationships to standing-wave ratio.

LINES OPEN- OR SHORT-CIRCUITED AT THE FAR END

Point 2 is the open- or short-circuited end of the line, from which x and θ are measured.

(**A**) Voltages and Currents: Use equations of "Voltages and Currents" section (p. 24-3), with the following conditions.

Open-circuited line:

$$\rho_2 = 1.00\angle 0° = 1.00$$

$$_rE_2 = {}_fE_2 = E_2/2$$

$$_rI_2 = -{}_fI_2$$

$$I_2 = 0$$

$$Z_2 = \infty.$$

Short-circuited line:

$$\rho_2 = 1.00\angle 180° = -1.00$$

$$_rE_2 = -{}_fE_2$$

$$E_2 = 0$$

$$_rI_2 = {}_fI_2 = I_2/2$$

$$Z_2 = 0.$$

(**B**) Impedances and admittances:

$$Z_{oc} = Z_0\coth\gamma x$$

$$Z_{sc} = Z_0\tanh\gamma x$$

$$Y_{oc} = Y_0\tanh\gamma x$$

$$Y_{sc} = Y_0\coth\gamma x.$$

(**C**) For small attenuation: Use equations for large swr in (**E**), p. 24-4, with the following conditions.

Open-circuited line:

$$g_a = 0.$$

Short-circuited line:

$$r_b = 0.$$

(**D**) When attenuation is neglected:

$$Z_{oc} = -jR_0\cot\theta$$

$$Z_{sc} = jR_0\tan\theta$$

$$Y_{oc} = jG_0\tan\theta$$

$$Y_{sc} = -jG_0\cot\theta.$$

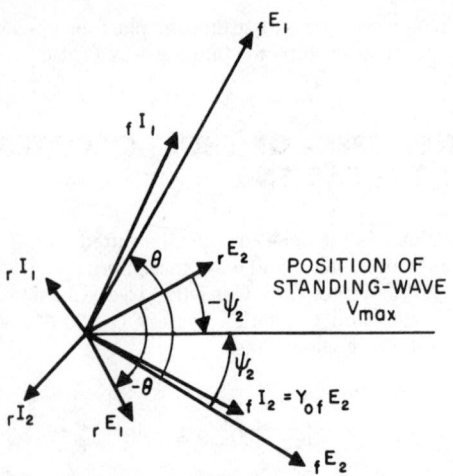

Fig. 2—Diagram of complex voltages and currents at two fixed points on a line with considerable attenuation. (Diagram rotates counterclockwise with time.)

(E) Relationships between Z_{oc} and Z_{sc}:

$$(Z_{oc}Z_{sc})^{1/2}=Z_0$$

$$\pm(Z_{sc}/Z_{oc})^{1/2}=\tanh\gamma x\approx(\alpha/\beta)\theta(1+\tan^2\theta)+j\tan\theta$$

$$=\frac{\alpha\theta}{\beta\cos^2\theta}+j\tan\theta$$

$$\approx j\tan\theta[1-j(\alpha/\beta)\theta(\tan\theta+\cot\theta)]$$

$$=j\tan\theta[1-j(\alpha/\beta)(2\theta/\sin2\theta)].$$

Note: Above approximations not valid for $\theta\approx\pi/2$, $3\pi/2$, etc.

$$\pm(Z_{oc}/Z_{sc})^{1/2}=\coth\gamma x\approx(\alpha/\beta)\theta(1+\cot^2\theta)-j\cot\theta$$

$$=\frac{\alpha\theta}{\beta\sin^2\theta}-j\cot\theta$$

$$\approx -j\cot\theta[1+j(\alpha/\beta)\theta(\tan\theta+\cot\theta)]$$

$$=-j\cot\theta[1+j(\alpha/\beta)(2\theta/\sin2\theta)].$$

Note: Above approximations not valid for $\theta\approx\pi$, 2π, etc.

(F) When attenuation is small (except for $\theta\approx n\pi/2$, $n=1, 2, 3, \cdots$)

$$\pm(Z_{sc}/Z_{oc})^{1/2}=\pm(Y_{oc}/Y_{sc})^{1/2}$$

$$=\pm j[-(C_{oc}/C_{sc})]^{1/2}$$

$$\times[1-j\tfrac{1}{2}(G_{oc}/\omega C_{oc}-G_{sc}/\omega C_{sc})]$$

where $Y_{oc}=G_{oc}+j\omega C_{oc}$ and $Y_{sc}=G_{sc}+j\omega C_{sc}$. The $+$ sign is to be used before the radical when C_{oc} is positive, and the $-$ sign when C_{oc} is negative.

(G) $R/|X|$ component of input impedance of low-attenuation nonresonant line:

Short-circuited line (except when $\theta\approx\pi/2$, $3\pi/2$, etc.)

$$R_1/|X_1|=G_1/|B_1|$$

$$=|(\alpha/\beta)\theta(\tan\theta+\cot\theta)+(B_0/G_0)|$$

$$=|(\alpha/\beta)(2\theta/\sin2\theta)+(B_0/G_0)|.$$

Open-circuited line (except when $\theta\approx\pi$, 2π, etc.)

$$R_1/|X_1|=G_1/|B_1|$$

$$=|(\alpha/\beta)\theta(\tan\theta+\cot\theta)-(B_0/G_0)|$$

$$=|(\alpha/\beta)(2\theta/\sin2\theta)-(B_0/G_0)|.$$

VOLTAGE REFLECTION COEFFICIENT AND STANDING-WAVE RATIO

$$\rho={}_rE/{}_fE=-{}_rI/{}_fI=(Z-Z_0)/(Z+Z_0)$$

$$=(Y_0-Y)/(Y_0+Y)=|\rho|\angle2\psi$$

where ψ is the electrical angle to the nearest voltage maximum on the generator side of point where ρ is measured (Figs. 2, 3, and 4).

$$\rho_1=\rho_2\epsilon^{-2\alpha x}\angle-2\theta$$

$$|\rho_1|=|\rho_2|/10^{A_0/10}.$$

Voltage reflection coefficient in decibels

$$\rho_{dB}=-20\log_{10}|1/\rho|.$$

The minus sign is frequently omitted.

$$|\rho_{dB}\text{ at input}|=|\rho_{dB}\text{ at load}|+2A_0.$$

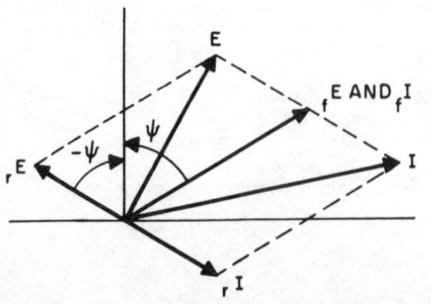

Fig. 3—Voltages and currents at time $t=0$ at a point ψ electrical degrees toward the load from a voltage standing-wave maximum.

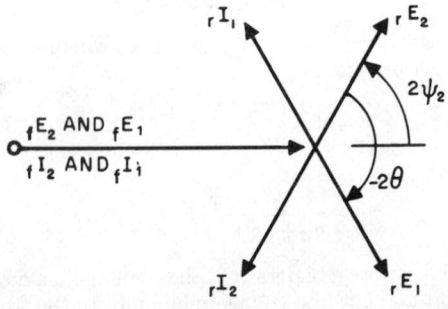

Fig. 4—Abbreviated diagram of a line with zero attenuation.

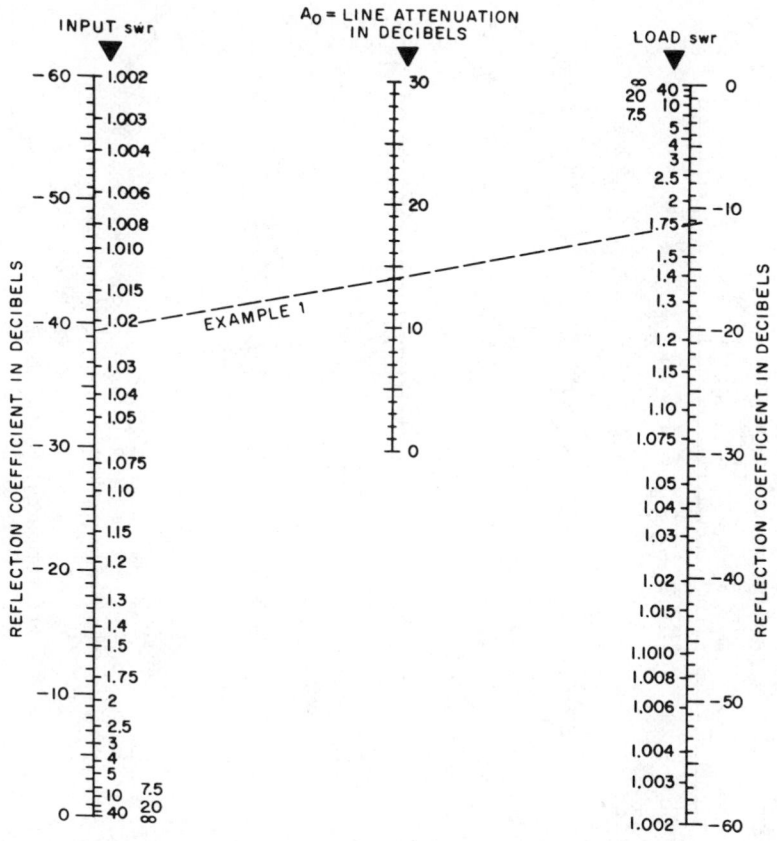

Fig. 5—Line attenuation and voltage reflection coefficient for low swr.

These two relationships and standing-wave ratio versus reflection coefficient in decibels are shown in Figs. 5 and 6.

$$Z = E/I = (_fE +_rE)/(_fI +_rI)$$

$$= Z_0[(1+\rho)/(1-\rho)]$$

$$Z/Z_0 = (1+\rho)/(1-\rho)$$

$$= \frac{1 + jS \cot\psi}{S + j \cot\psi}$$

$$S = |E_{max}/E_{min}| = |I_{max}/I_{min}|$$

$$= \frac{|_fE| + |_rE|}{|_fE| - |_rE|} = \frac{|_fI| + |_rI|}{|_fI| - |_rI|}$$

$$= \frac{1 + |\rho|}{1 - |\rho|} = r_a = g_a^{-1} = g_b = r_b^{-1}$$

$$|\rho| = (S-1)/(S+1)$$

$$1/S_1 = \tanh[\alpha x + \tanh^{-1}(1/S_2)]$$

$$= \tanh[0.1151 A_0 + \tanh^{-1}(1/S_2)].$$

(A) For high standing-wave ratio. When the ratio S_1 is greater than 6/1, then with 1 percent

accuracy or better

$$1/S_1 = 1/S_2 + \alpha x = 1/S_2 + 0.115 A_0$$

$$|\rho_{dB}| = 17.4/S.$$

Subject to the conditions below, the standing-wave ratio is given by one or the other of

$$S \approx (1+x^2)/r$$

$$S \approx (1+b^2)/g$$

where

$$r + jx = Z/Z_0 = (1/R_0)[R - (B_0/G_0)X + jX]$$

$$g + jb = Y/Y_0 = (1/G_0)[G + (B_0/G_0)B + jB].$$

Conditions, for 1-percent accuracy

$$r < 0.1 |x + 1/x| \quad \text{when } |x| > 0.3$$

$$g < 0.1 |b + 1/b| \quad \text{when } |b| > 0.3.$$

The boundary of the 1-percent-error region can be plotted on the Smith chart (Fig. 7) by use of the equation (for impedances)

$$|\cot\psi| = 0.1 S^2/(S^2 - 1)^{1/2}.$$

The same boundary line on the chart holds when

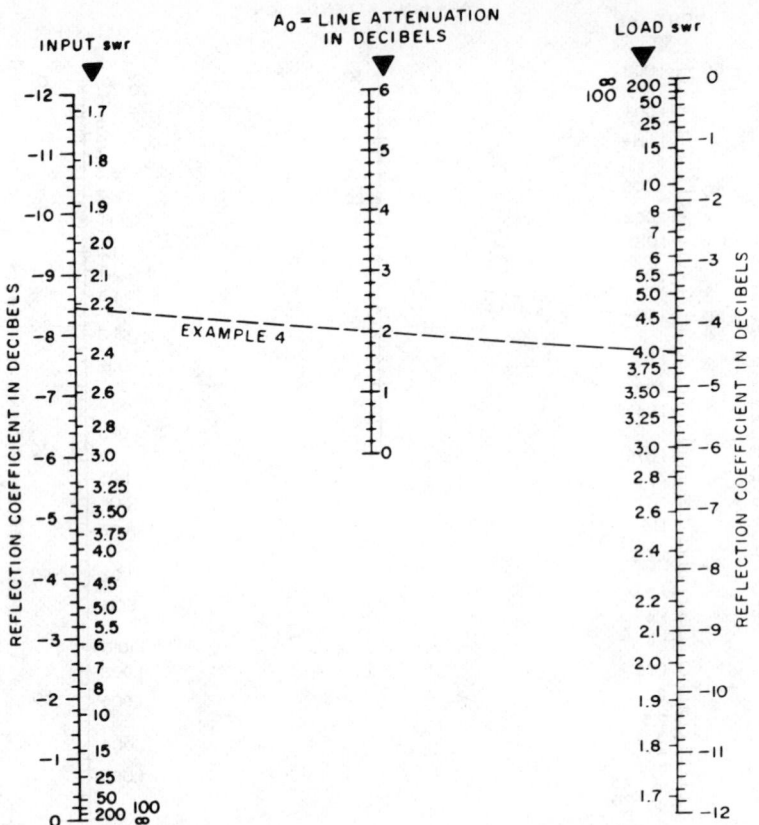

Fig. 6—Line attenuation and voltage reflection coefficient for high swr.

reading admittances. The area outside the solid heart-shaped curve is where the swr equation is accurate to within 1 percent. The area outside the dashed curve is where the reciprocal of $r+jx$ lies in the permitted region.

POWER AND EFFICIENCY

The net power flowing toward the load is

$$P= |{}_fE|^2G_0[1-|\rho|^2+2|\rho|(B_0/G_0)\sin2\psi]$$

where $|E|$ is the root-mean-square voltage.

Example: Derive the power equation

$$P=(\text{real})EI^*.$$

When the following expressions are substituted in this equation, the power equation results.

$$E={}_fE(1+\rho)$$
$$I={}_fEY_0(1-\rho)$$
$$I^*={}_fE^*Y_0^*(1-\rho^*)$$
$$Y_0^*=G_0(1-jB_0/G_0)$$
$$\rho=|\rho|\exp j2\psi$$
$$\rho^*=|\rho|\exp{-j2\psi}.$$

(A) When the angle B_0/G_0 of the characteristic admittance is negligibly small, the net power flowing toward the load is given by

$$P=G_0(|{}_fE|^2-|{}_rE|^2)$$
$$=|{}_fE|^2G_0(1-|\rho|^2)$$
$$=|E_{\max}E_{\min}|/R_0$$
$$P_1=|{}_fE_2|^2G_0(\epsilon^{2(\alpha/\beta)\theta}-|\rho_2|^2\epsilon^{-2(\alpha/\beta)\theta}).$$

(B) Efficiency, when B_0/G_0 is negligibly small:

$$\eta=P_2/P_1=\frac{1-|\rho_2|^2}{\epsilon^{2(\alpha/\beta)\theta}-|\rho_2|^2\epsilon^{-2(\alpha/\beta)\theta}}$$

$$=\eta_{\max}\frac{1-|\rho_2|^2}{1-|\rho_2|^2\eta^2_{\max}}=\frac{1-|\rho_2|^2}{1-|\rho_1|^2}\epsilon^{-2\alpha x}$$

$$=\frac{1/|\rho_2|-|\rho_2|}{1/|\rho_1|-|\rho_1|}=\frac{S_1-1/S_1}{S_2-1/S_2}.$$

The maximum error in the above expressions is

$$\pm100(S_2-1/S_2)B_0/G_0 \quad\text{percent}$$
$$\pm4.34(S_2-1/S_2)B_0/G_0 \quad\text{decibels.}$$

When the ratio S_1 is greater than 6/1:

$$\eta\approx S_1/S_2\approx(1+0.115A_0S_2)^{-1}.$$

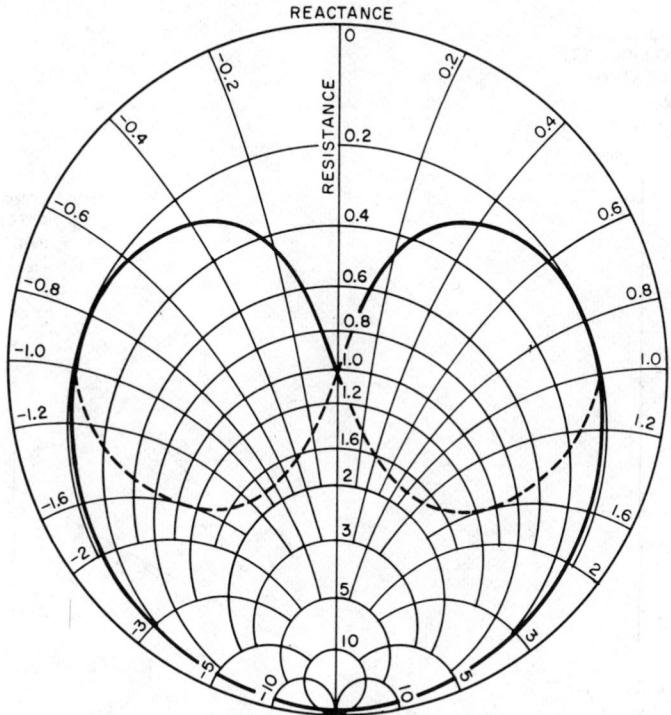

Fig. 7—Permitted region for use of equation $S \approx (1+x^2)/r$. *W. W. Macalpine, "Computation of Impedance and Efficiency of Transmission Line with High Standing-Wave Ratio," Trans. of the AIEE, vol. 72, part 1, p. 336, Fig. 2; July 1953.*

When the load matches the line, $\rho_2=0$ and the efficiency is accurately

$$\eta_{\max}= \exp[-2(\alpha/\beta)\theta]= \exp(-2\alpha x)=10^{-A_0/10}$$

$$A-A_0= 10 \log_{10}(\eta_{\max}/\eta).$$

Figure 8 is drawn from the expressions in this paragraph.

(**C**) Efficiency, when swr is high:

$$\eta= \frac{P_2}{P_1}= \frac{R_2}{R_1}\left(\frac{1+x_1^2}{1+x_2^2}\right)= \frac{G_2}{G_1}\left(\frac{1+b_1^2}{1+b_2^2}\right)$$

$$= \frac{R_2}{R_0^2 G_1}\left(\frac{1+b_1^2}{1+x_2^2}\right)= \frac{R_0^2 G_2}{R_1}\left(\frac{1+x_1^2}{1+b_2^2}\right)$$

where R is the ohmic resistance while x is the normalized reactance and similarly for G and b. It is important that the R's and G's be computed properly, using equations in the section on "Transformation of Impedance on Lines with High SWR," page 24-10. Note the identity of the efficiency equations with the left-hand terms of the impedance equations. The conditions for accuracy are the same as stated for the impedance equations for high standing-wave ratio.

Example: Physical significance of the equation for efficiency at high standing-wave ratio: Subject to stated conditions, approximately, $x= \cot\psi$ and $I=I_{\max}\sin\psi$. $I_{\max}=$ current standing-wave maximum, practically constant along line when stand-

ing-wave ratio >6. Then

$$P=I^2R=I_{\max}^2R/(1+x^2).$$

When line length is greater than $\frac{1}{3}$ wavelength, then

$$\eta\approx[1+0.115A_0(1+x_2^2)(R_0/R_2)]^{-1}.$$

(**D**) Loss in nepers $=\frac{1}{2}\log_\epsilon(P_1/P_2)=0.1151\times$ (loss in decibels).

For a matched line, loss $=$ attenuation $=(\alpha/\beta)\theta= \alpha x$ nepers.

Loss in decibels $= 10 \log_{10}(P_1/P_2)=8.686\times$ (loss in nepers).

When $2(\alpha/\beta)\theta$ is small

$$P_1/P_2=1+2(\alpha/\beta)\theta\frac{1+|\rho_2|^2}{1-|\rho_2|^2}$$

and

decibels/wavelength

$$= 10 \log_{10}\left(1+4\pi(\alpha/\beta)\frac{1+|\rho_2|^2}{1-|\rho_2|^2}\right).$$

(**E**) For the same power flowing in a line with standing waves as in a matched or "flat" line:

$$P=|E_{\text{flat}}|^2/R_0$$

$$|E_{\max}|=|E_{\text{flat}}|S^{1/2}$$

$$|E_{\min}|=|E_{\text{flat}}|/S^{1/2}$$

$$|_fE|=\frac{1}{2}|E_{\text{flat}}|(S^{1/2}+S^{-1/2})$$

$$|_rE|=\frac{1}{2}|E_{\text{flat}}|(S^{1/2}-S^{-1/2}).$$

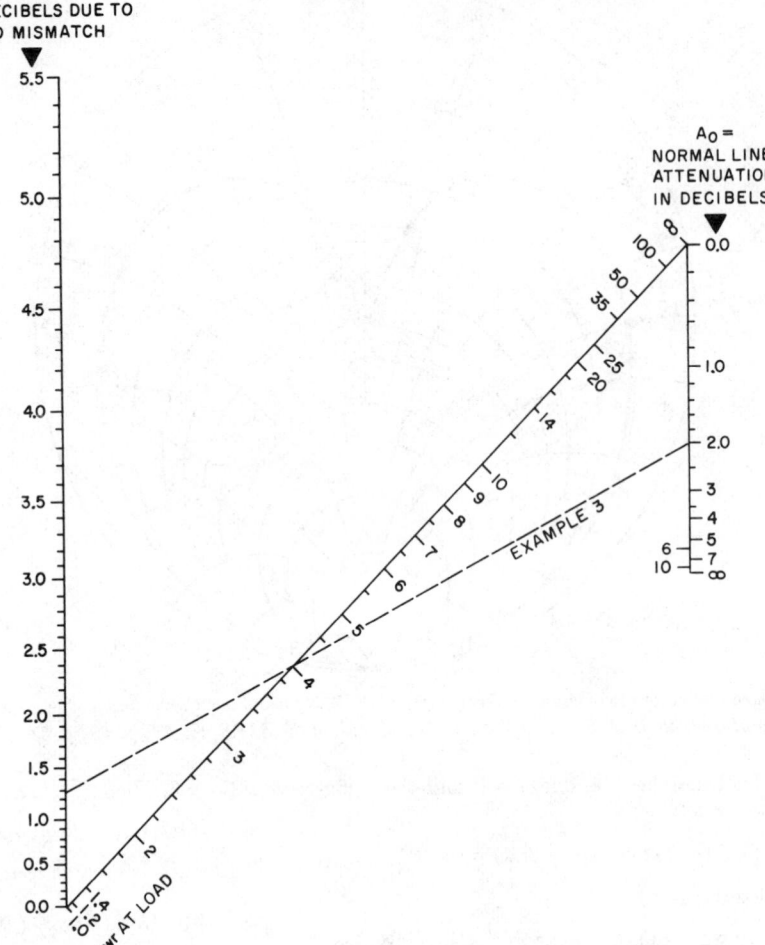

Fig. 8—Standing-wave loss factor. Due to load mismatch, an increase of loss in decibels as read from this figure must be added to normal line attenuation to give total dissipation loss in the line. This does not include mismatch loss due to any difference of line *input* impedance from the conjugate of the generator impedance [(**B**), p. 24–8].

When the loss is small, so that S is nearly constant over the entire length, then per half wavelength

(power loss)/(loss for flat line) $\approx \frac{1}{2}(S+1/S)$.

(**F**) The power dissipation per unit length, for unity standing-wave ratio, is

$$\Delta P_d/\Delta x = 2\alpha P$$

$$\frac{\text{(dissipation in watts/foot)}}{\text{(line power in kilowatts)}}$$

$$= 2.30 \text{ (decibels/100 feet)}$$

where the decibels/100 feet is the normal attenuation for a matched line.

When swr > 1, the dissipation at a current maximum is S times that for swr$=1$, assuming the attenuation to be due to conductor loss only. The multiplying factor for local heating reaches a minimum value of $(S+1/S)/2$ all along the line when conductor loss and dielectric loss are equal.

(**G**) Further considerations on power and efficiency are given in "Mismatch and Transducer Loss," p. 24–11.

TRANSFORMATION OF IMPEDANCE ON LINES WITH HIGH SWR*

When standing-wave ratio is greater than 10 or 20, resistance cannot be read accurately on the Smith chart, although it is satisfactory for reactance.

* W. W. Macalpine, "Computation of Impedance and Efficiency of Transmission Lines with High Standing-Wave Ratio," *Transactions of the AIEE*, vol. 72, part I, pp. 334–339; July 1953; also *Electrical Communication*, vol. 30, pp. 238–246; September 1953.

Use the equation (Fig. 9)

$$R_1 = R_2\frac{1+x_1^2}{1+x_2^2} + R_0(1+x_1^2)$$

$$\times\left[(\alpha/\beta)\theta + (B_0/G_0)\left(\frac{x_1}{1+x_1^2} - \frac{x_2}{1+x_2^2}\right)\right]$$

where $R=$ ohmic resistance, $x=X/R_0=$ normalized reactance.

When admittance is given or required, similar equations can be written with the aid of the following tabulation. The top row shows the terms in the above equation.

R_1	R_2	x_1^2	x_2^2	R_0	x_1	$-x_2$
G_1	G_2	b_1^2	b_2^2	$1/R_0$	$-b_1$	b_2
R_1	$G_2R_0^2$	x_1^2	b_2^2	R_0	x_1	b_2
G_1	R_2/R_0^2	b_1^2	x_2^2	$1/R_0$	$-b_1$	$-x_2$

For transforming R to G or vice versa:

$$R = R_0^2G\,|\,x/b\,|$$

where x and b are read on the Smith chart in the usual manner for transforming impedances to admittances.

The conditions for roughly 1-percent accuracy of the equations are:

Standing-wave ratio greater than 6/1 at input; $|B_0/G_0|<0.1$; $r+jx$ or $g+jb$ (whichever is used, at each end of line) meet the requirements stipulated in paragraph (**A**) ("For high standing-wave ratio") on p. 24–7; and the line parameters and given impedance be known to 1-percent accuracy.

When line length is greater than $\frac{1}{3}$ wavelength, then

$$R_1 \approx R_2[(1+x_1^2)/(1+x_2^2)] + 0.115A_0R_0(1+x_1^2)$$

$$\frac{R_1/R_0}{1+x_1^2} \approx \frac{R_2/R_0}{1+x_2^2} + (\alpha/\beta)\theta.$$

The equation for resistance transformation is derived from expressions for high swr in paragraph (**A**), just referred to.

Example: A load of $0.4-j2000$ ohms is fed through a length of RG–218/U cable at a frequency of 2.0 megahertz. What are the input impedance and the efficiency for a 24-foot length of cable and for a 124-foot length?

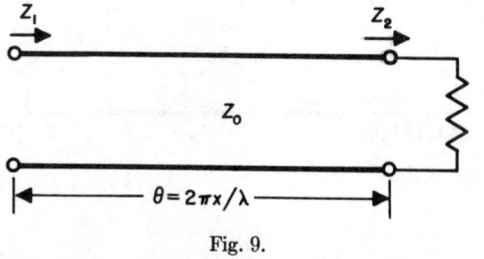

Fig. 9.

For RG–218/U, the attenuation at 2.0 megahertz is 0.095 decibel/100 feet (see Fig. 40). The dielectric constant $\epsilon=2.26$ and F_p is negligibly small. Then, by equations in (**B**) and (**C**), p. 24–3:

$$B_0/G_0 = \alpha/\beta = (\text{dB/100 ft})(\lambda_{\text{meters}})/1663$$

$$= [0.095\times150/(2.26)^{1/2}]/1663 = 0.0057$$

$$x/\lambda = xf\epsilon^{1/2}/984 = 24\times2.0\times1.5/984 = 0.073$$

$$\theta = 2\pi x/\lambda = 0.46 \text{ radian for 24-foot length}$$

while

$$x/\lambda = 0.38 \text{ and } \theta = 2.4 \text{ for 124-foot length.}$$

$$Z_2/Z_0 \approx (0.4-j2000)/50 = 0.008-j40.$$

For the 24-foot length, by the Smith chart

$$x_1 = X_1/Z_0 = -1.9, \quad \text{or } X_1 = -95 \text{ ohms.}$$

The conditions for accuracy of the resistance transformation equation are satisfied. Now

$$1+x_1^2 = 1+(1.9)^2 = 4.6$$

$$1+x_2^2 = 1+(40)^2 = 1600$$

$$R_1 = 0.4(4.6/1600) + 50\times4.6\times0.0057$$

$$\times[0.46-(1.9/4.6)+(40/1600)]$$

$$= 0.0012+0.105 = 0.106 \text{ ohm.}$$

The efficiency equation in paragraph (**C**) on p. 24–9, gives

$$\eta = 0.0012/0.106 = 0.0113, \text{ or 1.1 percent}$$

where the 0.0012 figure is taken directly from the first quantity on the right-hand side of the computation of R_1.

Similarly, for the 124-foot length, $x_1=1.1$, $X_1=55$ ohms, $1+x_1^2=2.21$, $R_1=0.00055+1.83=1.83$ ohms

$$\eta = 0.00055/1.83 = 3.1\times10^{-4}, \text{ or 0.03 percent.}$$

Tabulating the results,

Length (feet)	Input Impedance (ohms)	Efficiency (%)	Loss (dB)
24	$0.106-j95$	1.1	19.6
124	$1.8\ +j55$	0.03	35

The considerably greater loss for 124 feet compared with 24 feet is because the transmission passes through a current maximum where the loss per unit length is much higher than at a current minimum.

MISMATCH AND TRANSDUCER LOSS

Figures 5, 6, and 8, plus the equations in this section, permit the calculation of loss when imped-

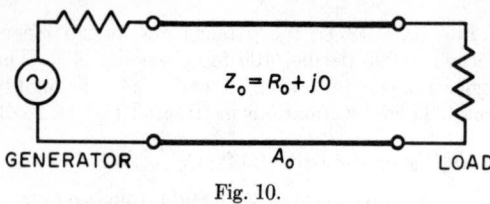

GENERATOR A_0 LOAD

Fig. 10.

ance mismatch exists in a transmission-line system; also the change in standing-wave ratio along a line due to attenuation.

One End Mismatched

When either generator or load impedance is mismatched to the Z_0 of the line and the other is matched (Fig. 10)

$(\text{mismatch loss}) = P_m/P$

$$= 1/(1- |\rho|^2) = (S+1)^2/4S \quad (1)$$

where $P =$ power delivered to load, $P_m =$ power that would be delivered were system matched, and $S =$ standing-wave ratio of mismatched impedance referred to Z_0.

Compared with an ideal transducer (ideal matching network between generator and load)

$(\text{transducer loss}) = A_0 + 10 \log_{10}(P_m/P)$ decibels

$$(2)$$

where $A_0 =$ normal attenuation of line.

Generator and Load Mismatched

$$|X_0/R_0| \ll 1.$$

When mismatches exist at both ends of the system (Fig. 11)

$(\text{mismatch loss at input})$

$= P_m/P$

$$= [(R_g+R_1)^2 + (X_g+X_1)^2]/4R_gR_1 \quad (3)$$

$(\text{transducer loss}) = (A-A_0) + A_0$

$$+ 10 \log_{10}(P_m/P) \text{ decibels} \quad (4)$$

where $(A-A_0) =$ standing-wave loss factor obtained from Fig. 8 for $S =$ standing-wave ratio at load.

Notes on Equation (3)

This equation reduces to Eq. (1) when X_g and/or X_1 is zero.

In Eq. (3), the impedances can be either ohmic or normalized with respect to any convenient Z_0.

When determining input impedance R_1+jX_1 on

Smith chart, adjust radius arm for S at input, determined from that at output by aid of Figs. 5 and 6.

For junction of two admittances, use Eq. (3) with G and B substituted for R and X, respectively.

Equation (3) is valid for a junction in any linear passive network. Likewise Eq. (1) when at least one of the impedances concerned is purely resistive. Determine S as if one impedance were that of a line.

Examples

Example 1: The swr at the load is 1.75 and the line has an attenuation of 14 decibels. What is the input swr?

Using Fig. 5, set a straightedge through the 1.75 division on the "load swr" scale and the 14-decibel point on the middle scale. Read the answer on the "input swr" scale, which the straightedge intersects at 1.022.

Example 2: Readings on a reflectometer show the reflected wave to be 4.4 decibels below the incident wave. What is the swr?

Using Fig. 6, locate the reflection coefficient 4.4 (or −4.4) decibels on either outside scale. Beside it, on the same horizontal line, read swr=4.0+.

Example 3: A 50-ohm line is terminated with a load of $200+j0$ ohms. The normal attenuation of the line is 2.00 decibels. What is the loss in the line?

Using Fig. 8, align a straightedge through the points $A_0 = 2.0$ and swr=4.0. Read $A-A_0 = 1.27$ decibels on the left-hand scale. Then the transmission loss in the line is

$$A = 1.27 + 2.00 = 3.27 \text{ decibels.}$$

This is the dissipation or heat loss as opposed to the mismatch loss at the input (example 4).

Example 4: In the preceding example, suppose the generator impedance is $100+j0$ ohms, and the line is 5.35 wavelengths long. What is the mismatch loss between the generator and the line?

According to example 3, the load swr=4.0 and the line attenuation is 2.0 decibels. Then, using Fig. 6, the input swr is found to be 2.22. On the

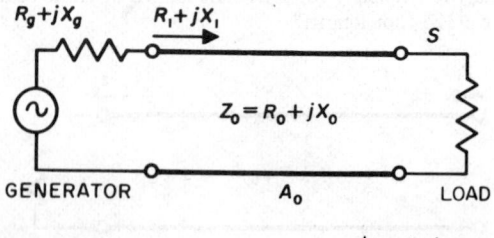

Fig. 11.

Smith chart, locate the point corresponding to 0.35 wavelength toward the generator from a voltage maximum, and swr= 2.22. Read the input normalized impedance as $0.62+j0.53$ with respect to $Z_0 = 50$ ohms. Now the mismatch loss at the input can be determined by use of (3). However, since the generator impedance is nonreactive, (1) can be used if desired. Refer to the following paragraph and to the "Notes on Equation (3)" above.

With respect to $100+j0$ ohms, the normalized impedance at the line input is $0.31+j0.265$, which gives swr= 3.5 according to the Smith chart. Then by (1), $P_m/P = 1.45$, giving a mismatch loss of 1.62 decibels. The transducer loss is found by using the results of examples 3 and 4 in (4). This is

$$1.27+2.00+1.62 = 4.9 \text{ decibels.}$$

ATTENUATION AND RESISTANCE OF TRANSMISSION LINES AT ULTRA-HIGH FREQUENCIES

The normal or matched-line attenuation in decibels/100 feet is

$$A_{100} = 4.34 R_t/Z_0 + 2.78 f \epsilon^{1/2} F_p$$

where the total line resistance/100 feet (for perfect surface conditions of the conductors) is, for copper coaxial line

$$R_t = 0.1(1/d+1/D)f^{1/2}$$

and for copper 2-wire open line

$$R_t = (0.2/d)f^{1/2}$$

where $D =$ diameter of inner surface of outer coaxial conductor in inches, $d =$ diameter of conductors (coaxial-line center conductor) in inches, $f =$ frequency in megahertz, $\epsilon =$ dielectric constant relative to air, and $F_p =$ power factor of dielectric at frequency f.

For other conductor materials, the resistance of conductor of diameter d (and similarly for D) is

$$0.1(1/d)(f\mu_r\rho/\rho_{Cu})^{1/2} \text{ ohms/100 feet.}$$

Refer to section on "Skin Effect," in Chapter 6.

RESONANT LINES

Symbols

$f_0 =$ resonance frequency in megahertz
$G_a =$ conductance load in mhos at voltage standing-wave maximum, equivalent to some or all of the actual loads
$k =$ coefficient of coupling
$n =$ integral number of quarter wavelengths
$p = k^2 Q_{1s} Q_{2s} =$ load transfer coefficient or matching factor

$P_c =$ power converted into heat in resonator
$P_m =$ power available from generator in watts
$\quad = E_{oc}^2/4R_{gen}$
$P_x =$ power transferred when load is directly connected to generator (for single resonators); or an analogous hypothetical power (for two coupled resonators)
$Q =$ figure of merit of a resonator as it exists, whether loaded or unloaded
$Q_d =$ doubly loaded Q (all loads being included)
$Q_s =$ singly loaded Q (all loads included except one). For a pair of coupled resonators, Q_{1s} is the value for the first resonator when isolated from the other. (Similarly for Q_{2s})
$Q_u =$ unloaded Q
$R_b =$ resistance load in ohms at voltage standing-wave minimum, equivalent to some or all of the actual loads
$R_u =$ resistance similar to R_b except for unloaded resonator
$R_1 =$ generator resistance, referred to short-circuited end
$R_2 =$ load resistance
$S_x = R_1/R_2$ or $R_2/R_1 =$ mismatch factor between generator and load
$Z_{10} =$ characteristic impedance of the first of a pair of resonators
$\theta_1 =$ electrical angle from a voltage standing-wave minimum point

(**A**) Q of a resonator (electrical, mechanical or any other) is

$$Q = 2\pi \frac{\text{(energy stored)}}{\text{(energy dissipated per cycle)}}$$

$$= 2\pi f \frac{\text{(energy stored)}}{\text{(power dissipation)}}.$$

In a freely oscillating system, the amplitude decays exponentially.

$$I = I_0 \exp(-\pi ft/Q).$$

(**B**) Unloaded Q of a resonant line:

$$Q_u = \beta/2\alpha$$

the line length being n quarter-wavelengths, where n is a small integer. The losses in the line are equivalent to those in a hypothetical resistor at the short-circuited end (**E**), p. 24-4:

$$R_u = n\pi Z_0/4Q_u.$$

(**C**) Loaded Q of a resonant line (Fig. 12):

$$Q^{-1} = Q_u^{-1} + (4R_b/n\pi Z_0) + (4G_a/n\pi Y_0)$$

$$= (4/n\pi Z_0)(R_u+R_b+G_a/Y_0^2).$$

All external loads can be referred to one end and represented by either R_b or G_a as in Fig. 13.

The total loading is the sum of all the individual loadings.

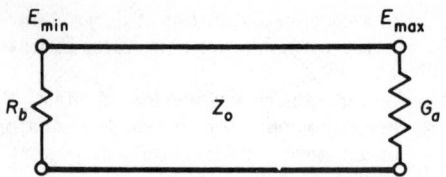

Fig. 12—Quarter-wave line with loadings at nominal short-circuit and open-circuit points.

General conditions:

$$R_b/Z_0 = G_a/Y_0 \ll 1.0$$

or, roughly, $Q > 5$.

(**D**) Input admittance and impedance:

The converse of the equations for Fig. 13 can be used at the resonance frequency. Then R or G is the input impedance or admittance, while

$$R_b = n\pi Z_0/4Q_s$$

where $Q_s =$ singly loaded Q with the losses and all the loads considered except that at the terminals where input R or G is being measured.

In the vicinity of the resonance frequency, the input admittance when looking into a line at a tap point θ_1 in Fig. 14 is approximately

$$Y = G + jB = \frac{n\pi Y_0}{4\sin^2\theta_1}\left(Q_s^{-1} + j2\frac{f-f_0}{f_0}\right)$$

provided

$$|f - f_0|/f_0 \ll 1.0$$

and

$$\left|\left[\theta(f-f_0)/f_0\right]\cot\theta_1\right| \ll 1.0$$

where $\theta = n\pi/2 =$ length of line at f_0. It is not valid when $\theta_1 \approx 0$, π, 2π, etc., except that it is good near the short-circuited end when $f - f_0 \approx 0$.

Such a resonant line is approximately equivalent to a lumped LCG parallel circuit, where

$$\omega_0^2 L_1 C_1 = (2\pi f_0)^2 L_1 C_1 = 1.$$

Admittance of the equivalent circuit is

$$Y = G + j\left[\omega C_1 - (1/\omega L_1)\right]$$
$$\approx \omega_0 C_1 \{Q_s^{-1} + j2[(f-f_0)/f_0]\}.$$

Then, subject to the conditions stated above

$$L_1 = (4\sin^2\theta_1)/n\pi\omega_0 Y_0$$

$$C_1 = n\pi Y_0/(4\omega_0\sin^2\theta_1) = nY_0/(8f_0\sin^2\theta_1)$$

$$G = n\pi Y_0/(4Q_s\sin^2\theta_1)$$

$$Q_s = \omega_0 C_1/G = 1/\omega_0 L_1 G.$$

Similarly, the input impedance at a point in

series with the line (Fig. 13C and D) is

$$Z = R + jX = \frac{n\pi Z_0}{4\cos^2\theta_1}\left(Q_s^{-1} + j2\frac{f-f_0}{f_0}\right)$$

provided

$$|f - f_0|/f_0 \ll 1.0$$

and

$$\left|\theta\left[(f-f_0)/f_0\right]\tan\theta_1\right| \ll 1.0.$$

It is not valid when $\theta_1 \approx \pi/2$, $3\pi/2$, etc.

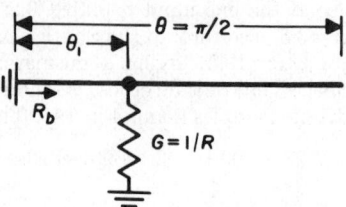

A. SHUNT OR TAPPED LOAD.

$$R_b = (Z_0^2/R)\sin^2\theta_1$$
OR
$$G_a = G\sin^2\theta_1 = R_b/Z_0^2$$

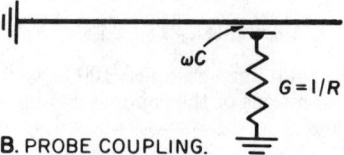

B. PROBE COUPLING.

$$G_a = (\omega^2 C^2/G)\sin^2\theta_1$$
OR
$$R_b = Z_0^2\omega^2 C^2 R\sin^2\theta_1$$
PROVIDED $G \gg \omega^2 C^2$

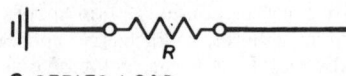

C. SERIES LOAD.

$$R_b = R\cos^2\theta_1$$

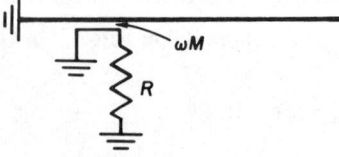

D. LOOP COUPLING.

$$R_b = (\omega^2 M^2/R)\cos^2\theta_1$$
PROVIDED $X_{loop} \ll R$

Fig. 13—Typical loaded quarter-wave sections with apparent R_b equivalent to the loading at distance θ_1 from voltage-minimum point of the line. Outer conductor not shown.

The voltage standing-wave ratio at resonance, on the generator (Fig. 15) is

$$S = (R_2 + R_u)/R_1$$

$$= \frac{(R_2/R_1)Q_u + Q_d}{Q_u - Q_d}.$$

When $R_1 = R_2$

$$S = \frac{1 + Q_d/Q_u}{1 - Q_d/Q_u}$$

$$\rho = Q_d/Q_u.$$

(E) Insertion loss (Fig. 15): At resonance, for either a distributed or a lumped-constant device

(dissipation loss)

$$= 10 \log_{10}(P_x/P_{out})$$

$$= 20 \log_{10}[1/(1 - Q_d/Q_u)]$$

$$\approx 20 \log_{10}(1 + Q_d/Q_u)$$

$$\approx 8.7 Q_d/Q_u \text{ decibels}$$

(mismatch loss)

$$= 10 \log_{10}(P_m/P_x)$$

$$= 10 \log_{10}[(1 + S_x)^2/4S_x] \text{ decibels.}$$

The dissipation loss also includes a small additional mismatch loss due to the presence of the resonator. The error in the form $20 \log_{10}(1 + Q_d/Q_u)$ is about twice that of the form $8.7 Q_d/Q_u$. The last expression $(8.7 Q_d/Q_u)$ is in error compared with the first, $20 \log_{10}[1/(1 - Q_d/Q_u)]$, by roughly $-50(Q_d/Q_u)$ percent for $(Q_d/Q_u) < 0.2$.

The selectivity is given on page 9–7, where $Q = Q_d$. That equation is accurate over a smaller range of $(f - f_0)$ for a resonant line than it is for a single tuned circuit.

At resonance*

$$P_{out}/P_{in} = R_2/(R_u + R_2)$$

$$= \frac{Q_u - Q_d}{Q_u + (R_1/R_2)Q_d} = 1 - Q_s/Q_u$$

where Q_s is for the resonator loaded with R_2 only.

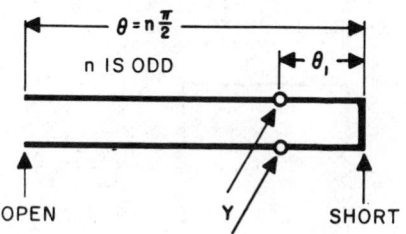

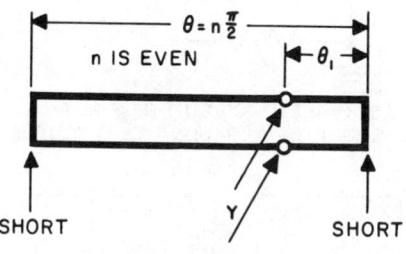

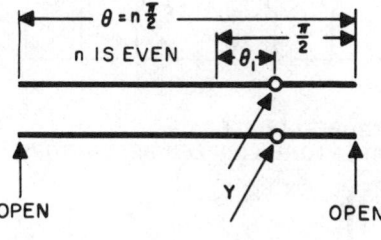

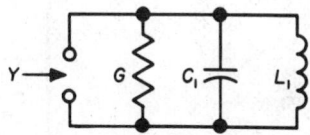

Fig. 14—Resonant transmission lines and their equivalent lumped circuit.

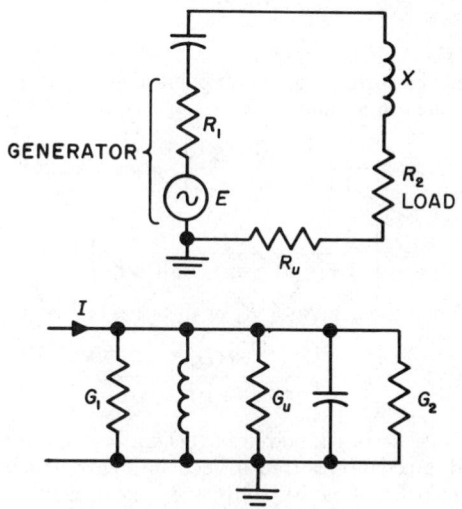

Fig. 15—Equivalent circuits of a resonant line (or a lumped tuned circuit) as seen at the short-circuited and open-circuited ends. All the power equations are good for either lumped or distributed parameters.

* When the line is resonated by a reactive load ($\theta \neq n\pi/2$), it is frequently preferable to use the resistance form of the equation. Compute R_u by the method given on p. 24–11, or on p. 24–5, where $Z_0 = R_0(1 - jB_0/G_0)$.

The maximum power transfer, for fixed Q_u, Q_d, and Z_0 occurs when $R_1 = R_2$. Then

$$P_{out}/P_{in} = (Q_u - Q_d)/(Q_u + Q_d) = 1 - Q_s/Q_u$$

$$P_{out}/P_m = (1 - Q_d/Q_u)^2$$

$$P_{in}/P_m = 1 - (Q_d/Q_u)^2.$$

When the generator R_1 or G_1 is negligibly small (then $Q = Q_s = Q_d$)

$$(P_{in}/P_{out})_s = Q_u/(Q_u - Q)$$

(F) Power dissipation $(= P_c)$:

$$P_c/P_m = \frac{4(Q_d/Q_u)(1 - Q_d/Q_u)}{1 + R_2/R_1}.$$

For matched input and output $(R_1 = R_2)$

$$P_c/P_m = 2(Q_d/Q_u)(1 - Q_d/Q_u)$$

$$\approx 2Q_d/Q_u \quad (\text{for } Q_d \ll Q_u)$$

$$P_c/P_{out} = 2Q_d/(Q_u - Q_d)$$

$$P_c/P_{in} = 2Q_d/(Q_u + Q_d).$$

For generator matched by load plus cavity

$$P_c/P_m = 2Q_d/Q_u.$$

When the generator R_1 or G_1 is negligibly small

$$(P_c/P_{out})_s = Q/(Q_u - Q)$$

$$(P_c/P_{in})_s = Q_s/Q_u.$$

(G) Voltage and current:

At the current-maximum point of an n-quarter-wavelength resonant line

$$I_{sc} = 4\left[\frac{P_m Q_d(1 - Q_d/Q_u)}{(1 + R_2/R_1)n\pi Z_0}\right]^{1/2}, \text{ rms amperes}$$

$$= 4\left[\frac{P_m Q_d}{\{1 + [(R_2 + R_u)/R_1]\}n\pi Z_0}\right]^{1/2}.$$

When the generator R_1 or G_1 is negligibly small

$$I_{sc} = 2\left[\frac{P_s Q_s}{n\pi Z_0(R_2 + R_u)/R_s}\right]^{1/2}$$

where P_s = rated power of generator, R_s = rated load impedance as transformed into current-maximum point of cavity, and $I = I_{sc}\cos\theta_1$, while $E = Z_0 I_{sc}\sin\theta_1$.

The voltage and current are in quadrature time phase.

When $R_1 = R_2 + R_u$ and $n = 1$

$$I_{sc} \approx (8P_m Q_d/\pi Z_0)^{1/2}.$$

In a lumped-constant tuned circuit

$$I = 2\left[\frac{P_m Q_d(1 - Q_d/Q_u)}{(1 + R_2/R_1)X}\right]^{1/2}.$$

(H) Pair of coupled resonators (Fig. 16):

With inductive coupling near the short-circuited end of a pair of quarter-wave resonant lines

$$k = (4/\pi)\omega M/(Z_{10}Z_{20})^{1/2}.$$

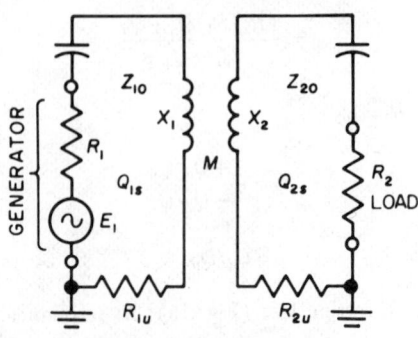

A. EQUIVALENT CIRCUIT WITH RESISTANCES AS SEEN AT THE SHORT-CIRCUITED END.

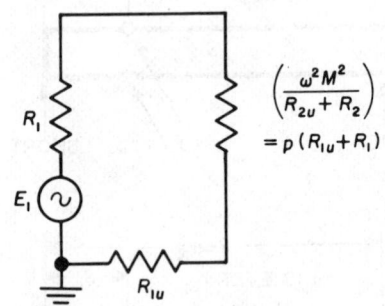

B. EQUIVALENT CIRCUIT OF FIRST RESONATOR AT RESONANCE FREQUENCY.

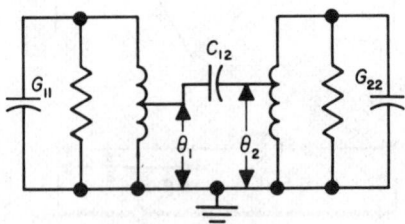

C. PROBE-COUPLED OR APERTURE-COUPLED RESONATORS.

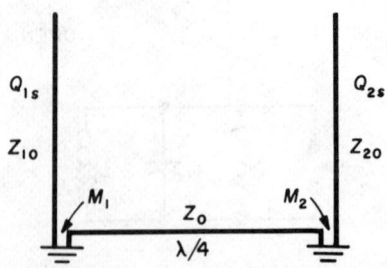

D. QUARTER-WAVELENGTH LINE COUPLING.

Fig. 16—Two coupled resonators.

For coupling through a lossless quarter-wavelength line, inductively coupled near the short-circuited ends of the resonators (Fig. 16D):

$$k = \frac{4\omega^2 M_1 M_2}{\pi Z_0 (Z_{10} Z_{20})^{1/2}}.$$

Probe coupling near top (Fig. 16C):

$$k = (4/\pi)\omega C_{12}(Z_{10} Z_{20})^{1/2} \sin\theta_1 \sin\theta_2.$$

For lumped-constant coupled circuits, p and k are defined on pp. 9-7 and 9-1.

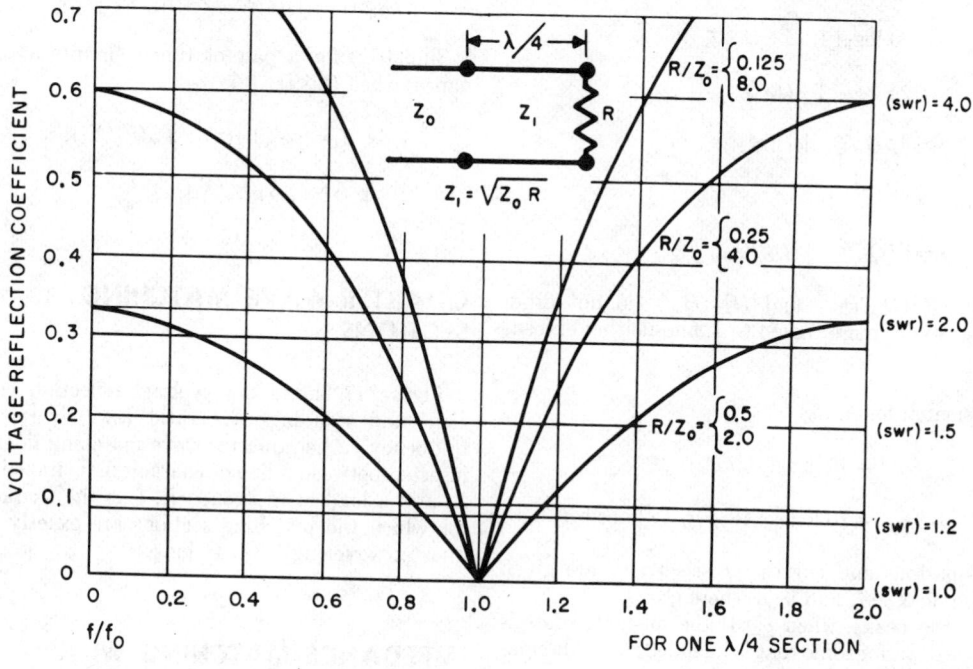

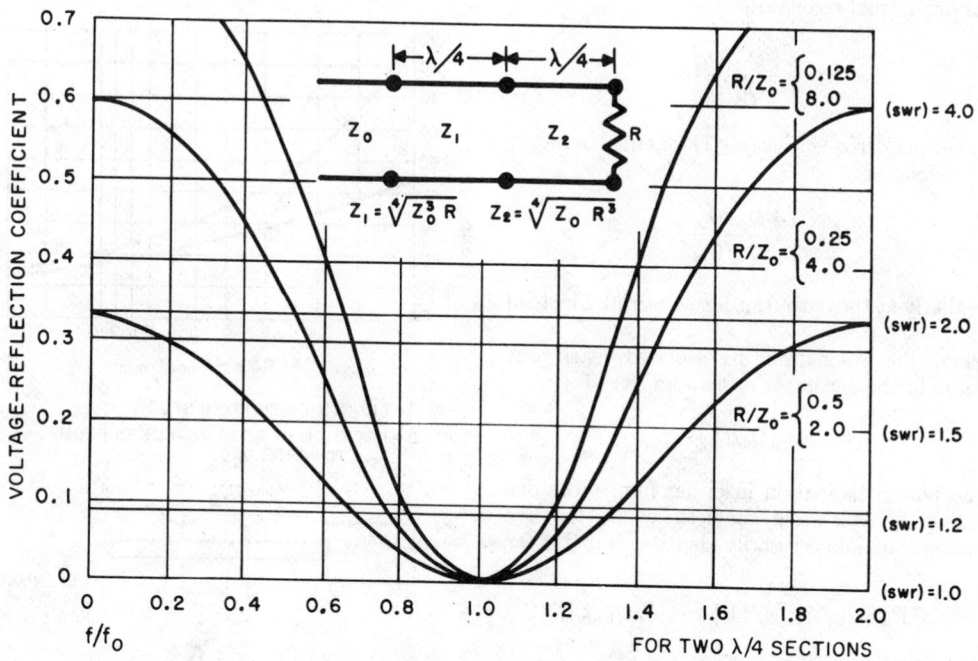

Fig. 17—Quarter-wave matching sections.

In either lumped or distributed resonators

(dissipation loss)

$$= 10 \log_{10}(P_x/P_{out})$$

$$= 10 \log_{10}[1/(1-Q_{1s}/Q_{1u})(1-Q_{2s}/Q_{2u})]$$

$$\approx 20 \log_{10}[1/(1-Q_s/Q_u)]$$

$$\approx 20 \log_{10}(1+Q_s/Q_u)$$

$$\approx 8.7 Q_s/Q_u \text{ decibels}$$

where

$$Q_s/Q_u = [(Q_{1s}/Q_{1u})(Q_{2s}/Q_{2u})]^{1/2}$$

provided (Q_{1s}/Q_{1u}) and (Q_{2s}/Q_{2u}) do not differ by a ratio of more than 4 to 1, and neither exceeds 0.2.

(mismatch loss at f_0)

$$= 10 \log_{10}(P_m/P_x)$$

$$= 10 \log_{10}[(1+p)^2/4p] \text{ decibels.}$$

Equations and curves for selectivity are given on pp. 9–6, 9–7, and 9–9, where $Q=Q_s$.

At the peaks, when $p \geq 1$, the mismatch loss is zero, except for some that is included in the dissipation loss.

Input voltage standing-wave ratio at f_0 for equal or unequal resonators

$$S = \frac{p+Q_{1s}/Q_{1u}}{1-Q_{1s}/Q_{1u}}.$$

At the peak frequencies $(p \geq 1)$ for equal or nearly equal resonators

$$S = \frac{1+Q_{1s}/Q_{1u}}{1-Q_{1s}/Q_{1u}}.$$

Similarly at the output, using subscript 2 instead of 1.

When the resonators are isolated, each one presents to the generator or load an swr of

$$S = (Q_u/Q_s) - 1.$$

The power dissipation in either lumped or distributed (quarter-wave) devices, where the two resonators are not necessarily identical, but $Q_s \ll Q_u$, is

$$P_{1c} = I_{1sc}^2 R_{1u}[4/(1+p)^2] P_m Q_{1s}/Q_{1u}$$

$$P_{2c} = [4p/(1+p)^2] P_m Q_{2s}/Q_{2u}.$$

These equations and those below for the currents assume that P_m is concentrated at f_0.

The currents in quarter-wave resonant lines, when $Q_s \ll Q_u$

$$I_{1sc} = [4/(1+p)](P_m Q_{1s}/\pi Z_{10})^{1/2}$$

$$I_{2sc}/I_{1sc} = (p Z_{10} Q_{2s}/Z_{20} Q_{1s})^{1/2}.$$

Similarly, for a pair of tuned circuits at resonance, when $Q_s \ll Q_u$

$$I_1 = [2/(1+p)](P_m Q_{1s}/X_1)^{1/2}$$

$$I_2/I_1 = (p X_1 Q_{2s}/X_2 Q_{1s})^{1/2}.$$

QUARTER-WAVE MATCHING SECTIONS

Figure 17 shows how voltage-reflection coefficient and standing-wave ratio (swr) vary with frequency f when quarter-wave matching lines are inserted between a line of characteristic impedance Z_0 and a load of resistance R. f_0 is the frequency for which the matching sections are exactly one-quarter wavelength $(\lambda/4)$ long.

IMPEDANCE MATCHING WITH SHORTED STUB (Fig. 18)

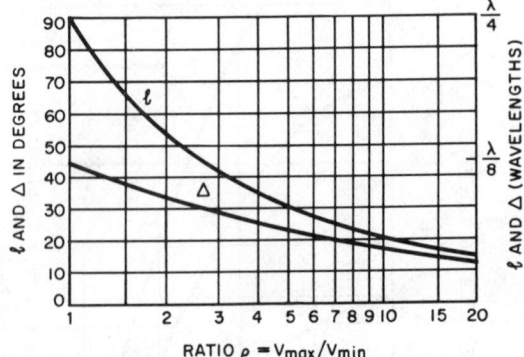

ℓ = LENGTH OF SHORTED STUB

Δ = LOCATION OF STUB MEASURED FROM V_{min} TOWARD LOAD

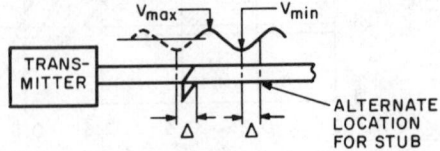

Fig. 18.

IMPEDANCE MATCHING WITH OPEN STUB *(Fig. 19)*

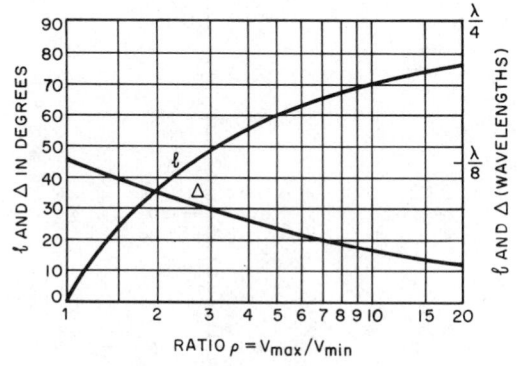

ℓ = LENGTH OF OPEN STUB

Δ = LOCATION OF STUB MEASURED FROM V_{min} TOWARD TRANSMITTER

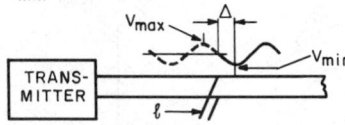

Fig. 19.

LENGTH OF TRANSMISSION LINE

Figure 20 gives the actual length of line in centimeters and inches when given the length in electrical degrees and the frequency, provided that the velocity of propagation on the transmission line is equal to that in free space. The length is equal to that in free space. The length is given on the L-scale intersection by a line between λ and l°, where

$$l^\circ = \frac{360L \text{ in centimeters}}{\lambda \text{ in centimeters}}.$$

Example: $f=600$ megahertz, $l^\circ=30$, length $L=1.64$ inches or 4.2 centimeters.

MEASUREMENT OF IMPEDANCE WITH SLOTTED LINE

Symbols

Z_0= characteristic impedance of line
Z= impedance of load (the unknown)
Z_1= impedance at first V_{min}
λ= wavelength on line
χ= distance from load to first V_{min}
(swr) $= V_{max}/V_{min}$

$$\theta^\circ = 180 \; (\chi/\tfrac{1}{2}\lambda) = 0.012f\chi/k$$
$k=$ velocity factor
$\quad=$ (velocity on line)/(velocity in free space)

where f is in megahertz and χ in centimeters.

Procedure *(Fig. 21)*

Measure $\lambda/2$, χ, V_{max}, and V_{min}.
Determine

$$Z_1/Z_0 = 1/(\text{swr}) = V_{min}/V_{max}$$

(wavelengths toward load) $=\chi/\lambda=0.5\chi/(\lambda/2)$.

Then Z/Z_0 may be found on an impedance chart. For example, suppose

$$V_{min}/V_{max}=0.60 \quad \text{and} \quad \chi/\lambda=0.40.$$

Refer to the chart, such as Fig. 22 reproduced in part here. Lay off with slider or dividers the distance on the R/Z_0 axis from the center point (marked 1.0) to 0.60. Pass around the circumference of the chart in a counterclockwise direction from the starting point 0 to the position 0.40, toward the load. Read off the resistance and reactance components of the normalized load impedance Z/Z_0 at the point of the dividers. Then it is found that

$$Z= Z_0(0.77+j0.39).$$

Similarly, there may be found the admittance of the load. Determine

$$Y_1/Y_0 = V_{max}/V_{min}=1.67$$

in the above example. Now pass around the chart counterclockwise through $\chi/\lambda=0.40$, starting at 0.25 and ending at 0.15. Read the components of the normalized admittance.

$$Y=1/Z= (1/Z_0)(1.03-j0.53).$$

Alternatively, these results may be computed by

$$Z= R_s+jX_s$$

$$= Z_0 \frac{1-j(\text{swr}) \tan\theta}{(\text{swr})-j \tan\theta}$$

$$= Z_0 \frac{2(\text{swr})-j[(\text{swr})^2-1] \sin2\theta}{[(\text{swr})^2+1]+[(\text{swr})^2-1] \cos2\theta}$$

$$Y=G+jB=1/Z= (1/R_p)-j(1/X_p)$$

$$= Y_0 \frac{2(\text{swr})+j[(\text{swr})^2-1] \sin2\theta}{[(\text{swr})^2+1]-[(\text{swr})^2-1] \cos2\theta}$$

where R_s and X_s are the series components of Z, while R_p and X_p are the parallel components.

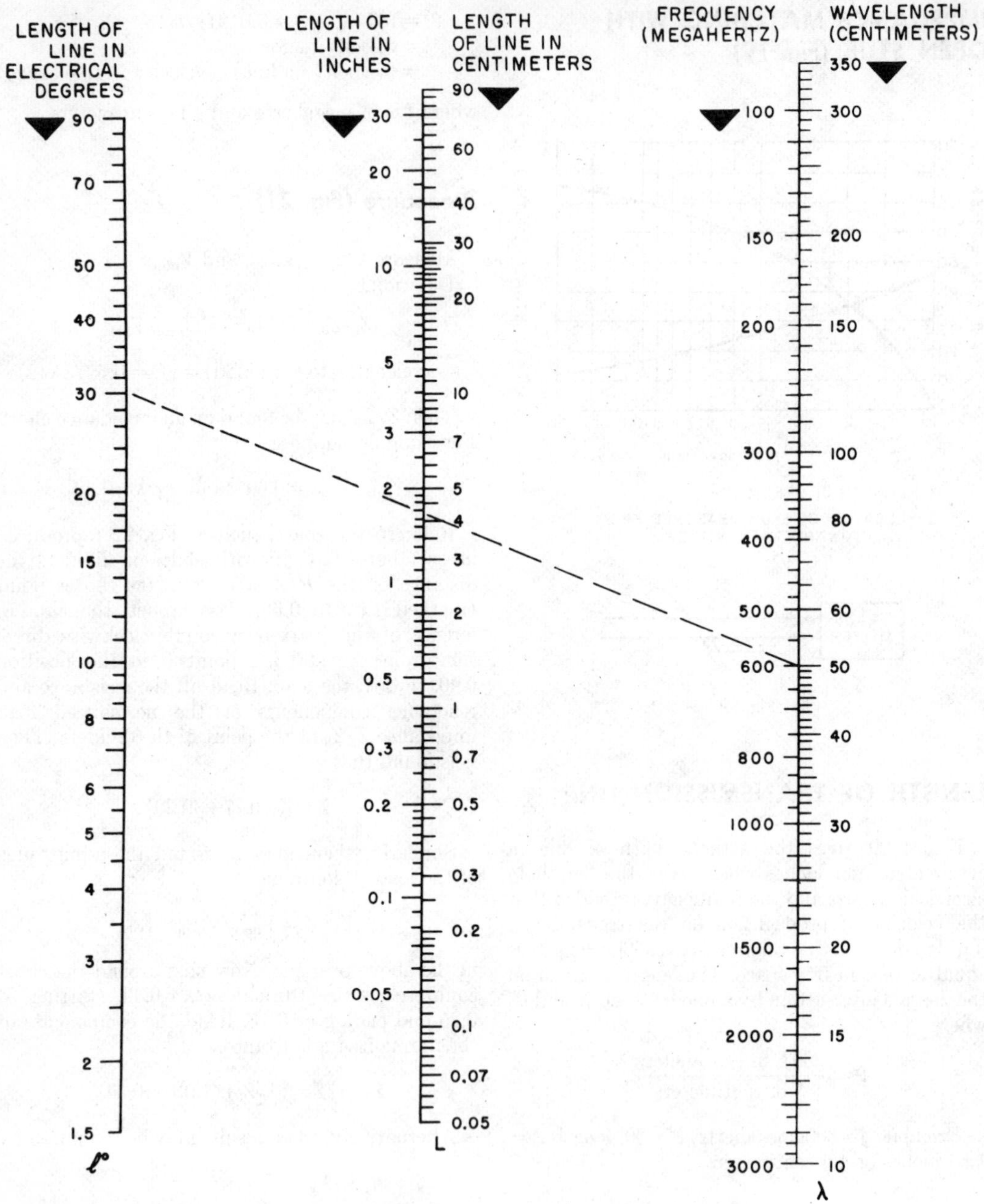

LENGTH OF LINE IN ELECTRICAL DEGREES
LENGTH OF LINE IN INCHES
LENGTH OF LINE IN CENTIMETERS
FREQUENCY (MEGAHERTZ)
WAVELENGTH (CENTIMETERS)
ℓ°
L
λ

Fig. 20—Determination of length of transmission line.

DETECTOR
PROBE
Vmax
Z1
LOAD Z
λ/2
x, θ
SECOND Vmin
FIRST Vmin

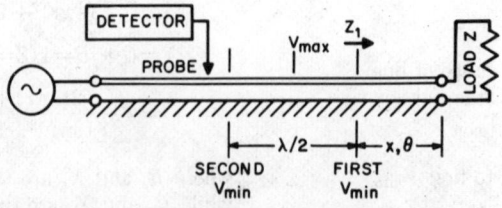

Fig. 21—Measurements on line.

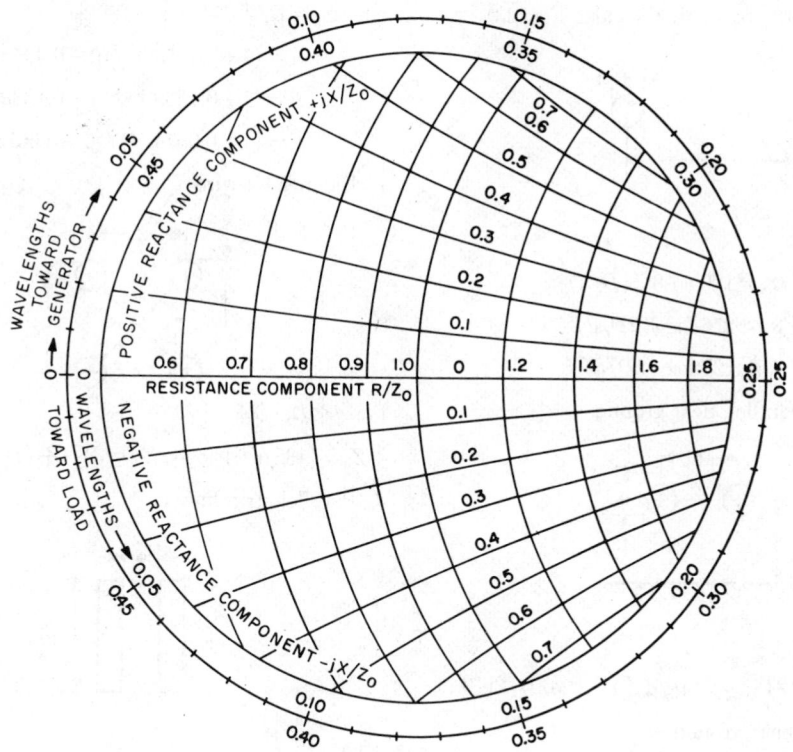

Fig. 22—Smith chart, center portion.

CHARACTERISTIC IMPEDANCE OF LINES

A. Single coaxial line (See also Fig. 23).

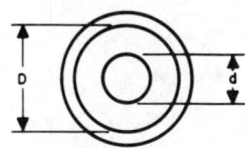

$$Z_0 = (138/\epsilon^{1/2}) \log_{10} (D/d)$$
$$= (60/\epsilon^{1/2}) \log_e (D/d)$$

ϵ = dielectric constant

$$= 1 \text{ in air}$$

B. Balanced shielded line

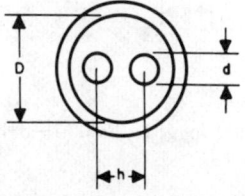

For $D \gg d$, $h \gg d$,

$$Z_0 = (276/\epsilon^{1/2}) \log_{10}\{2v[(1-\sigma^2)/(1+\sigma^2)]\}$$
$$= (120/\epsilon^{1/2}) \log_e\{2v[(1-\sigma^2)/(1+\sigma^2)]\}$$

$$v = h/d \qquad \sigma = h/D$$

C. Beads—dielectric ϵ_1

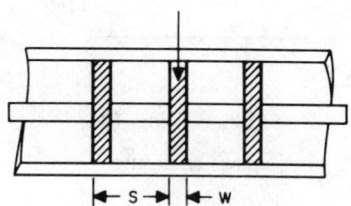

For lines *A.* and *B.*, if insulating beads are used at frequent intervals—call new characteristic impedance Z_0'

$$Z_0' = Z_0/\{1 + [(\epsilon_1/\epsilon) - 1](W/S)\}^{1/2}$$

$$W \ll S \ll \lambda/4$$

D. Open 2-wire line in air (See also Fig. 23).

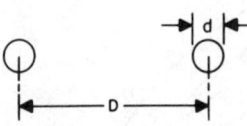

$$Z_0 = 120 \cosh^{-1}(D/d)$$
$$\approx 276 \log_{10}(2D/d)$$
$$\approx 120 \log_e(2D/d)$$

E. Wires in parallel, near ground

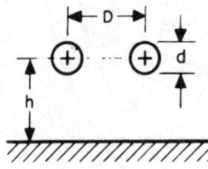

For $d \ll D, h,$

$$Z_0 = (69/\epsilon^{1/2}) \log_{10}\{(4h/d)[1+(2h/D)^2]^{1/2}\}$$

F. Balanced, near ground

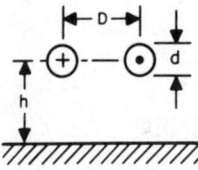

For $d \ll D, h,$

$$Z_0 = (276/\epsilon^{1/2}) \log_{10}\{(2D/d)[1+(D/2h)^2]^{-1/2}\}$$

G. Single wire, near ground

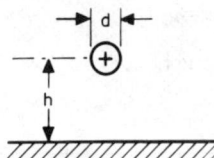

For $d \ll h,$

$$Z_0 = (138/\epsilon^{1/2}) \log_{10}(4h/d)$$

H. Single wire, square enclosure

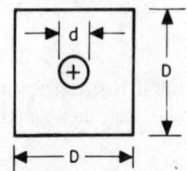

$$Z_0 \approx [138 \log_{10}\rho + 6.48 - 2.34A - 0.48B - 0.12C]\epsilon^{-1/2}$$

where $\rho = D/d$

$$A = (1+0.405\rho^{-4})/(1-0.405\rho^{-4})$$
$$B = (1+0.163\rho^{-8})/(1-0.163\rho^{-8})$$
$$C = (1+0.067\rho^{-12})/(1-0.067\rho^{-12})$$

I. Balanced 4-wire

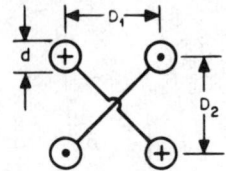

For $d \ll D_1, D_2,$

$$Z_0 = (138/\epsilon^{1/2}) \log_{10}\{(2D_2/d)[1+(D_2/D_1)^2]^{-1/2}\}$$

J. Parallel-strip line

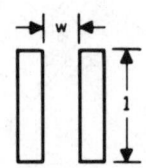

$$w/l < 0.1$$
$$Z_0 \approx 377 (w/l)$$

K. Five-wire line

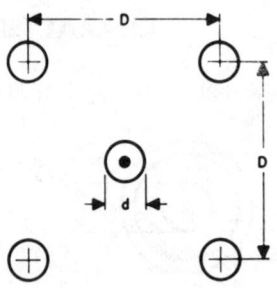

For $d \ll D,$

$$Z_0 = (173/\epsilon^{1/2}) \log_{10}(D/0.933d)$$

L. Wires in parallel—sheath return

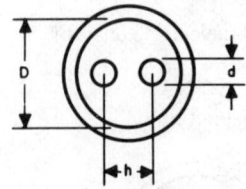

For $d \ll D, h,$

$$Z_0 = (69/\epsilon^{1/2}) \log_{10}[(\nu/2\sigma^2)(1-\sigma^4)]$$
$$\sigma = h/D$$
$$\nu = h/d$$

M. Air coaxial with dielectric supporting wedge

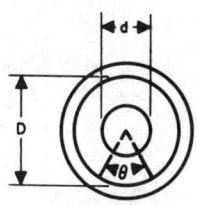

$$Z_0 \approx \frac{138 \log_{10}(D/d)}{[1+(\epsilon-1)(\theta/360)]^{1/2}}$$

ϵ = dielectric constant of wedge

θ = wedge angle in degrees

N. Balanced 2-wire—unequal diameters

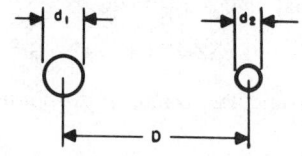

$$Z_0 = (60/\epsilon^{1/2}) \cosh^{-1} N$$

$$N = \tfrac{1}{2}[(4D^2/d_1 d_2) - (d_1/d_2) - (d_2/d_1)]$$

O. Balanced 2-wire near ground

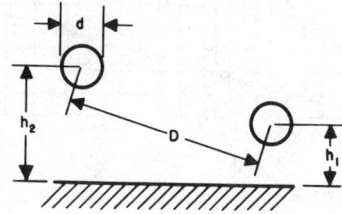

For $d \ll D, h_1, h_2$,

$$Z_0 = (276/\epsilon^{1/2}) \log_{10}\{(2D/d)[1+(D^2/4h_1 h_2)]^{-1/2}\}$$

Holds also in either of the following special cases:

$$D = \pm(h_2 - h_1)$$

or

$$h_1 = h_2 \text{ (see } F. \text{ above)}$$

P. Single wire between grounded parallel planes—
ground return

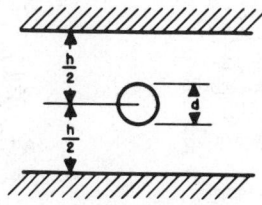

For $d/h < 0.75$,

$$Z_0 = (138/\epsilon^{1/2}) \log_{10}(4h/\pi d)$$

Q. Balanced line between grounded parallel planes

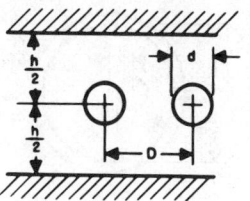

For $d \ll D, h$,

$$Z_0 = (276/\epsilon^{1/2}) \log_{10}\left(\frac{4h \tanh(\pi D/2h)}{\pi d}\right)$$

R. Balanced line between grounded parallel planes

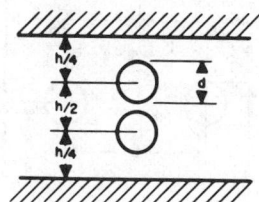

For $d \ll h$,

$$Z_0 = (276/\epsilon^{1/2}) \log_{10}(2h/\pi d)$$

S. Single wire in trough

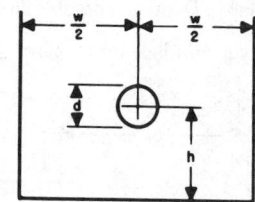

For $d \ll h, w$,

$$Z_0 = (138/\epsilon^{1/2}) \log_{10}\left(\frac{4w \tanh(\pi h/w)}{\pi d}\right)$$

T. Balanced 2-wire line in rectangular enclosure

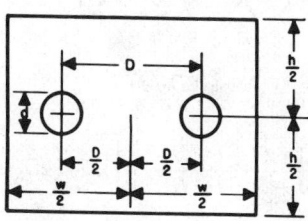

For $d \ll D, w, h$,

$$Z_0 = (276/\epsilon^{1/2})\left[\log_{10}\left(\frac{4h \tanh(\pi D/2h)}{\pi d}\right) \right.$$
$$\left. - \sum_{m=1}^{\infty} \log_{10}\left(\frac{1+u_m^2}{1-v_m^2}\right)\right]$$

where

$$u_m = \frac{\sinh(\pi D/2h)}{\cosh(m\pi w/2h)} \qquad v_m = \frac{\sinh(\pi D/2h)}{\sinh(m\pi w/2h)}$$

U. Eccentric line

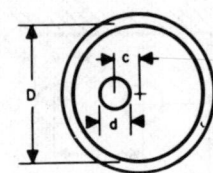

$$Z_0 = (60/\epsilon^{1/2}) \cosh^{-1} U$$
$$U = \tfrac{1}{2}[(D/d) + (d/D) - (4c^2/dD)]$$

V. Balanced 2-wire line in semi-infinite enclosure

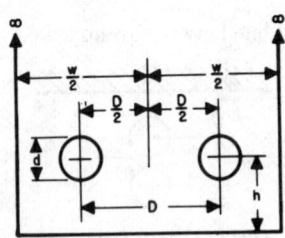

For $d \ll D, w, h$,
$$Z_0 = (276/\epsilon^{1/2}) \log_{10}[2w/\pi d (A^{1/2})]$$

where
$$A = \operatorname{cosec}^2(\pi D/w) + \operatorname{cosech}^2(2\pi h/w)$$

W. Outer wires grounded, inner wires balanced to ground

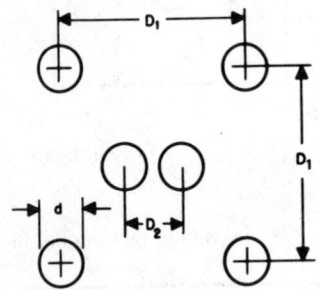

$$Z_0 \approx (276/\epsilon^{1/2}) \left\{ \log_{10}(2D_2/d) \right.$$
$$\left. - \left[\log_{10} \frac{1 + (1 + D_2/D_1)^2}{1 + (1 - D_2/D_1)^2} \right]^2 \left[\log_{10}(2D_1\sqrt{2}/d) \right]^{-1} \right\}$$

X. Split thin-walled cylinder

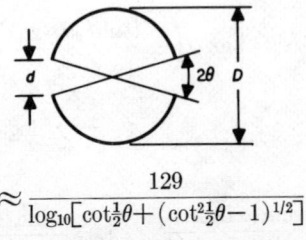

$$Z_0 \approx \frac{129}{\log_{10}[\cot \tfrac{1}{2}\theta + (\cot^2 \tfrac{1}{2}\theta - 1)^{1/2}]}$$

For θ small:
$$Z_0 \approx 129/\log_{10}(4D/d)$$

Courtesy of Electronic Engineering

Y. Slotted air line

When a slot is introduced into an air coaxial line for measuring purposes, the increase in characteristic impedance in ohms, compared with a normal coaxial line, is less than

$$\Delta Z = 0.03\theta^2$$

where θ is the angular opening of the slot in radians.

0 TO 220 OHMS

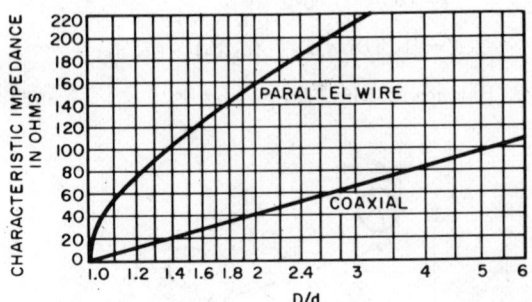

0 TO 700 OHMS

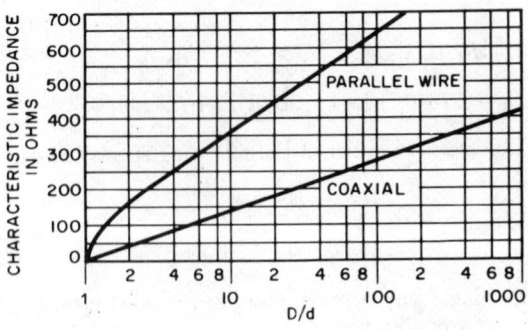

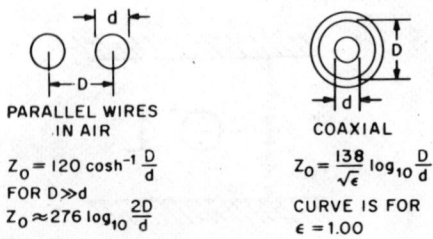

PARALLEL WIRES IN AIR

$$Z_0 = 120 \cosh^{-1} \frac{D}{d}$$
FOR $D \gg d$
$$Z_0 \approx 276 \log_{10} \frac{2D}{d}$$

COAXIAL

$$Z_0 = \frac{138}{\sqrt{\epsilon}} \log_{10} \frac{D}{d}$$
CURVE IS FOR $\epsilon = 1.00$

Fig. 23.

VOLTAGE GRADIENT IN A COAXIAL LINE *(Fig. 24)*

C' = capacitance in picofarads/foot

D = diameter of inner surface of outer conductor in same units as d

d = diameter of inner conductor

E = total voltage across line (E and ΔE both rms or both peak)

r = radius (r and Δr both in same units)

ϵ = net effective dielectric constant (= 1 for air); $1/\epsilon^{1/2}$ = velocity factor

$$\Delta E/\Delta r = \frac{0.434E}{r \log_{10}(D/d)}$$

$$= 0.059 E C'/r\epsilon$$

$$= 60E/rZ_0\epsilon^{1/2}$$

$$= (6.10\times10^4 E)/rZ_0^2 C'.$$

At the voltage standing-wave maximum

(gradient at surface of inner conductor)

$$= (5.37/d)(SP_{kW}/Z_0\epsilon)^{1/2}$$

$$= 5450(SP_{kW})^{1/2}/dC'Z_0^{3/2} \text{ peak volts/mil}$$

where d is in inches (1 mil = 0.001 inch). For amplitude or pulse modulation, let P_{kW} be the power in kilowatts at the crest of the modulation cycle. Thus, if the carrier is 1 kilowatt and modulation 100 percent, set

$$P_{kW} = 4 \text{ kilowatts.}$$

Example: What is the voltage gradient at inner conductor of a $6\frac{1}{8}$-inch rigid 50-ohm line with 500 kilowatts continuous-wave power, unity swr? Let $\epsilon = 1.00$ and $d = 2.60$ inches.

$$(\text{gradient}) = (5.37/2.60)(500/50)^{1/2}$$

$$= 6.55 \text{ peak volts/mil.}$$

The breakdown strength of air at atmospheric pressure is 29 000 peak volts/centimeter, or 74

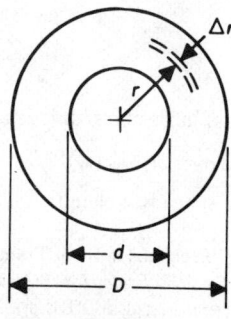

Fig. 24.

peak volts/mil (experimental value, before derating).

MICROSTRIP*

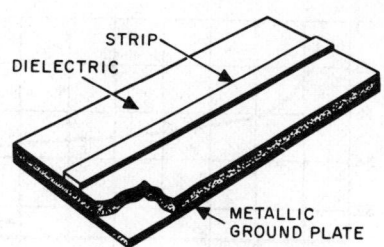

Fig. 25—Microstrip line.

Microstrip consists of a wire above a ground plane, being analogous to a 2-wire line in which one of the wires is represented by the image in the ground plane of the wire that is physically present. Figure 25 illustrates a short length of microstrip line, showing the metallic-strip conductor bonded to a dielectric sheet, to the other side of which is bonded a metallic ground plate.

Phase Velocity

Theoretically, for the TEM mode with conductors completely immersed in the dielectric, the velocity of propagation is

$$v = c/(\epsilon_r)^{1/2}$$

where c = velocity of light in vacuum and ϵ_r = dielectric constant relative to air.

For Teflon-impregnated Fibreglas dielectric, this gives 604 feet per microsecond. Experimental measurements on a line with $\frac{7}{32}$-inch strip width and dielectric sheet $\frac{1}{16}$-inch thick give

$$v = 655 \text{ feet/microsecond.}$$

Typical measurements together with the theoretical TEM wavelength are plotted in Fig. 26.

Characteristic Impedance

If it were not for fringing and leakage flux, the theoretical characteristic impedance would be

$$Z_0 = (h/w)(\mu/\epsilon)^{1/2}$$

$$= 377(h/w)(1/\epsilon_r)^{1/2}$$

* D. D. Grieg and H. F. Engelmann, "Microstrip—A New Transmission Technique for the Kilomegacycle Range," and two accompanying papers in *Proceedings of the IRE*, vol. 40, pp. 1644–1663; December 1952; also *Electrical Communication*, vol. 30, pp. 26–54; March 1953.

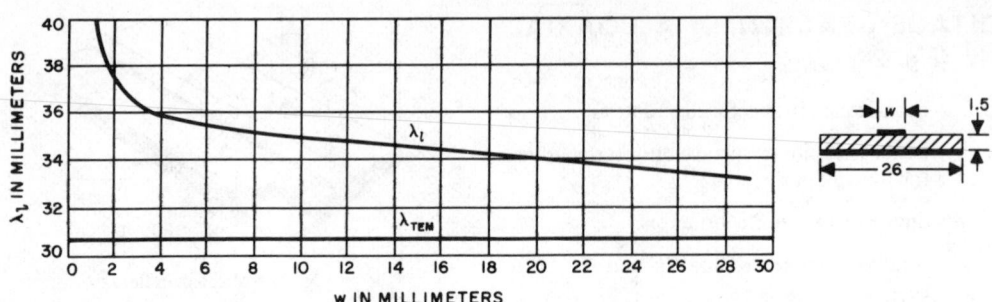

Fig. 26—Wavelength in microstrip versus width of strip conductor. The dimensions in the sketch at right are in millimeters. Dielectric was Fibreglas G–6. Measurements were taken at 4770 megahertz.

where $h=$ thickness of dielectric, $w=$ width of strip conductor, $\epsilon=$ dielectric constant in farads/meter, and $\mu=$ permeability in henries/meter.

Figure 27 shows the experimentally determined Z_0 for typical microstrip lines.

Attenuation

Conductor loss for copper, in decibels/foot

$$\alpha_{\mathrm{Cu}}=7.25\times10^{-5}(1/h)\,(f_{\mathrm{MHz}}\epsilon_r)^{1/2}.$$

Dielectric loss in decibels/foot

$$\alpha_d=2.78\times10^{-2}f_{\mathrm{MHz}}F_p(\epsilon_r)^{1/2}$$

where $F_p=$ power factor or loss angle, and $h=$ dielectric thickness in inches.

A correction factor for conductor attenuation is shown in Fig. 28 for use in the equation

$$\alpha_c=\alpha_0\times\Delta$$

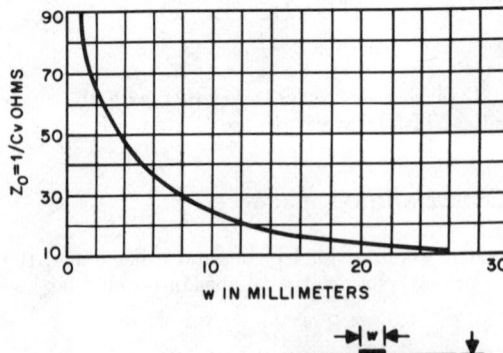

Fig. 27—Characteristic impedance for microstrip with Fibreglas G–6 dielectric. Dimensions in sketch are in millimeters. C is the measured electrostatic capacitance in farads per unit length and v is the phase velocity in units of length per second.

where α_0 is, for copper conductors, given by α_{Cu} above.

$$\alpha_0=\alpha_{\mathrm{Cu}}(\mu_r\rho/\rho_{\mathrm{Cu}})^{1/2}$$

where $\mu_r=$ relative permeability, and $\rho/\rho_{\mathrm{Cu}}=$ resistivity relative to copper.

The measured attenuation of a typical microstrip line is shown in Fig. 40. The relatively high attenuation is due to the small physical size of the line.

Power-Handling Capacity

For a microstrip line composed of a strip $\frac{7}{32}$-inch wide on a Teflon-impregnated Fibreglas base $\frac{1}{16}$-inch thick:

(**A**) At 3000 megahertz with 300 watts cw, the temperature under the strip conductor has been measured at 50° Celsius rise above 20° Celsius ambient.

(**B**) Under pulse conditions, corona effects appear at the edge of the strip conductor for pulse power of roughly 10 kilowatts at 9000 megahertz.

STRIP TRANSMISSION LINES*

Strip transmission lines differ from microstrip in that a second ground plane is placed above the conductor strip (Fig. 29). The characteristic impedance is shown in Fig. 30 and the attenuation in Fig. 31.

Attenuation

Dielectric loss in decibels/unit length

$$\alpha_d=27.3F_p(\epsilon_r)^{1/2}/\lambda_0$$

where $\lambda_0=$ free-space wavelength.

* S. B. Cohn, "Problems in Strip Transmission Lines," *Transactions of the IRE Professional Group on Microwave Theory and Techniques*, vol. MTT3, pp. 119–126; March 1955. Other papers on strip-type lines also appear in that issue of the journal.

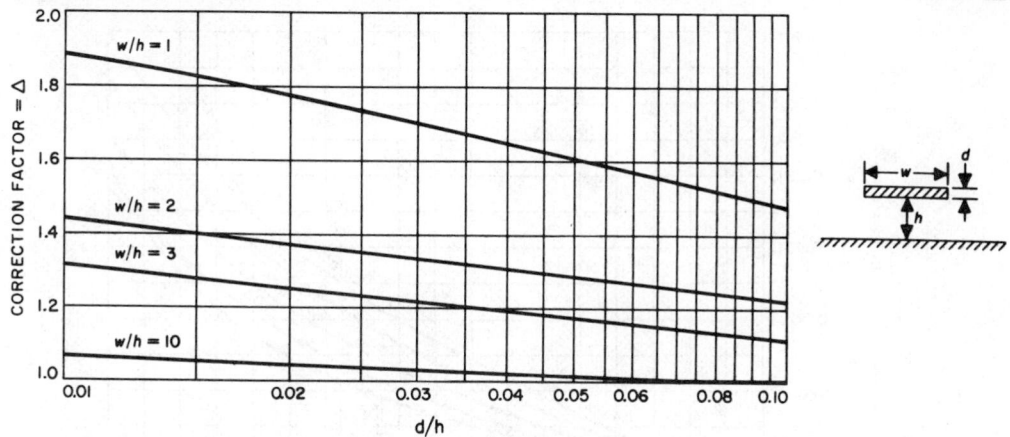

Fig. 28—Correction factor Δ for conductor attenuation (case of wide strip of small thickness above infinite ground plane).

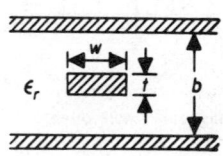

Fig. 29.

Conductor loss in decibels/unit length

$$\alpha_c = (y/b) \left(f_{GHz} \epsilon_r \mu_r \rho / \rho_{Cu} \right)^{1/2}$$

where $y=$ ordinate from Fig. 31, and $\rho/\rho_{Cu}=$ resistivity relative to copper.

The unit of length in α_d is that of λ_0, and in α_c it is that of b.

LINES WITH HELICAL INNER CONDUCTOR

Spiral Delay Line

For a transmission line with helical inner conductor (spiral delay line) where axial wavelength and length of line are both long compared with line diameter (similar to Fig. 32 in dimensional symbols):

$$L' = 0.30 n^2 d^2 \left[1 - (d/D)^2 \right]$$

microhenries/axial foot, where d is in inches and $n = 1/\tau =$ turns/inch.

$$C' = 7.4 \epsilon_r / \log_{10}(D/d)$$

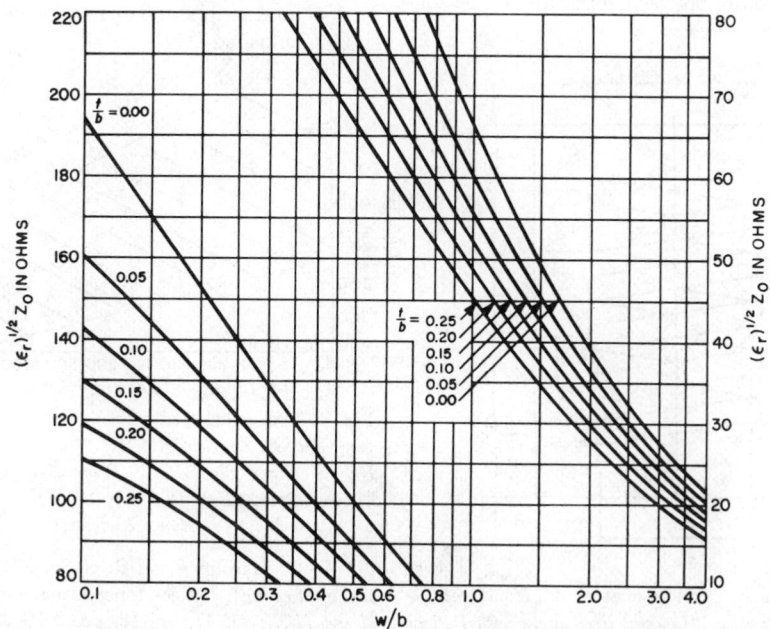

Fig. 30—Plot of strip-transmission-line Z_0 versus w/b for various values of t/b. For lower-left family of curves, refer to left-hand ordinate values; for upper-right curves, use right-hand scale. *Courtesy of Transactions of the IRE Professional Group on Microwave Theory and Techniques.*

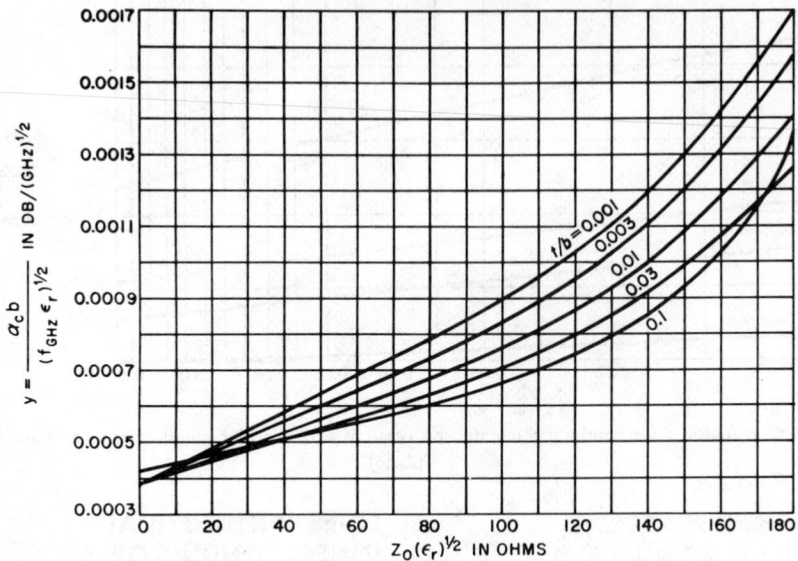

Fig. 31—Theoretical attenuation of copper-shielded strip transmission line in dielectric medium ϵ_r. *Courtesy of Transactions of the IRE Professional Group on Microwave Theory and Techniques.*

picofarads/axial foot.

$$Z_0 = (L'/C')^{1/2} \times 10^3 \text{ ohms}$$

$$T = (L'C')^{1/2} \times 10^{-3}$$

microseconds/axial foot.

$$\alpha_{dB} = 4.34 R/Z_0 + 27.3 F_p f T$$

decibels/axial foot where $R=$ total conductor resistance in ohms/axial foot, $f=$ frequency in megahertz, $F_p=$ power factor, and $\epsilon_r=$ relative dielectric constant of medium between spiral and outer conductor.

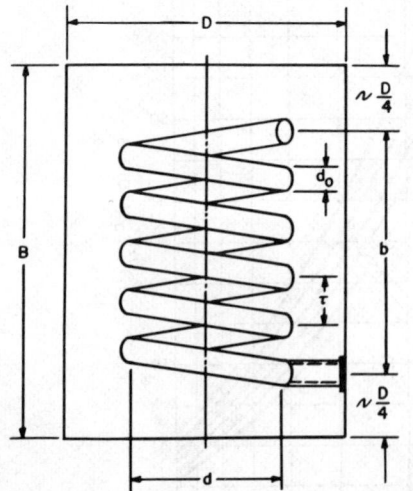

Fig. 32—Outline sketch of resonator. *W. W. Macalpine and R. O. Schildknecht, "Coaxial Resonators with Helical Inner Conductor," Proceedings of the IRE, vol. 47, no. 12, p. 2100; December 1959. © 1959 Institute of Radio Engineers.*

HELICAL RESONATOR*

Symbols

$b=$ axial length of coil, inches
$B=$ inside length of shield, inches
$d=$ mean diameter of turns, inches
$D=$ inside diameter of shield, inches
$d_0=$ diameter of conductor, inches
$f_0=$ frequency of resonance, megahertz
$n=$ turns per inch
$N=$ total number of turns of winding
$Q_u=$ unloaded Q

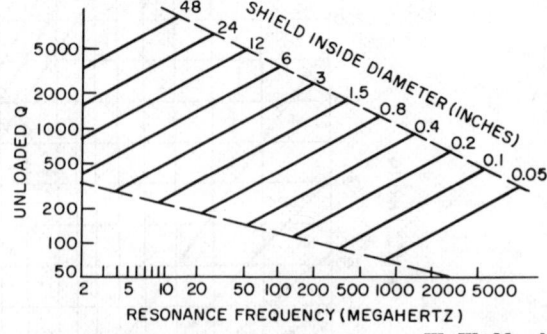

Fig. 33—Unloaded Q of helical resonator. *W. W. Macalpine and R. O. Schildknecht, "Coaxial Resonators with Helical Inner Conductor," Proceedings of the IRE, vol. 47, no. 12, p. 2100; December 1959. © 1959 Institute of Radio Engineers.*

* W. W. Macalpine and R. O. Schildknecht, "Coaxial Resonators with Helical Inner Conductor," *Proceedings of the IRE*, vol. 47, no. 12, pp. 2099–2105; December 1959. W. W. Macalpine and R. O. Schildknecht, "Helical Resonator Design Chart," *Electronics*, p. 140; 12 August 1960.

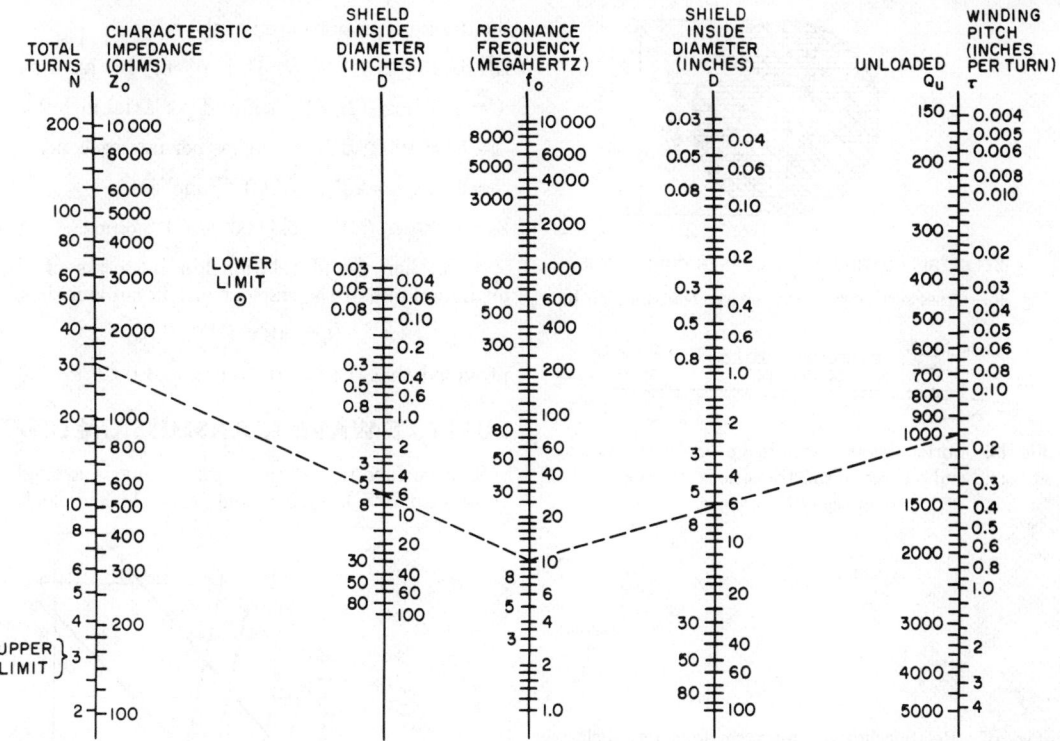

Fig. 34—Design chart for quarter-wave helical resonators. *W. W. Macalpine and R. O. Schildknecht, "Coaxial Resonators with Helical Inner Conductor," Proceedings of the IRE, vol. 47, no. 12, p. 2101; December 1959.* © 1959 Institute of Radio Engineers.

δ = skin depth, inch

$\tau = 1/n$ = pitch of winding, inches.

Other symbols have the usual meanings, page 24-1.

The helical resonator shown in Fig. 32 consists of a shield (outer conductor) enclosing a coil (inner conductor). One end of the coil is solidly connected to the shield and the other end open-circuited, except possibly for a trimming capacitor. It operates as a distributed-parameter system equivalent to a quarter-wave coaxial transmission-line resonator. Probe, loop, or aperture coupling can be used for input and output circuits, or between adjacent coupled resonators.

Unloaded Q versus resonance frequency and shield diameter is shown in Fig. 33. The region plotted is where the helical resonator gives better Q for a given volume than other types. For higher Q and frequency, a conventional coaxial-line resonator is preferable. When the Q and frequency are lower than the plotted region, a lumped LC resonant circuit is indicated. These conditions are marked on Fig. 34 as "upper limit" (3 turns) and "lower limit," respectively.

Figure 34 is usually accurate to within ± 10 percent. It is plotted from the following equations and is limited by the conditions noted.

Unloaded Q is given for a resonator that consists of a single-layer coil of copper conductor on a low-loss form and is enclosed in a shield of seamless copper tubing. A shielded coil below its resonance

frequency also has ideally a true Q predicted by this equation. The equation gives a practical working value of Q somewhat below the theoretical maximum.

$$Q_u = 50 D f_0^{1/2}$$

provided:

$$0.45 < d/D < 0.6$$

$$b/d > 1.0$$

$$0.4 < d_0/\tau < 0.6 \text{ at } b/d = 1.5$$

$$0.5 < d_0/\tau < 0.7 \text{ at } b/d = 4.0$$

$$d_0 > 5\delta.$$

Total number of turns

$$N = 1900/(f_0 D) \text{ turns}$$

for $d/D = 0.55$, and $b/d > 1.0$.

Pitch of winding and characteristic impedance:

$$\tau = 1/n = (f_0 D^2)/2300 \text{ inches per turn}$$

$$Z_0 = 98\,000/(f_0 D) \text{ ohms}$$

for $d/D = 0.55$, and $b/d = 1.5$.

General conditions for all equations:

$$B \approx (b + D/2)$$

$$\tau < d/2.$$

Simplified and empirical equations from which

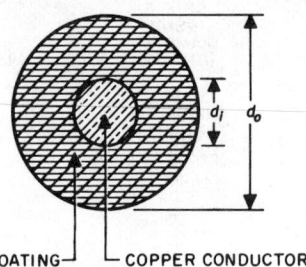

Fig. 35—Cross section of surface-wave transmission line.

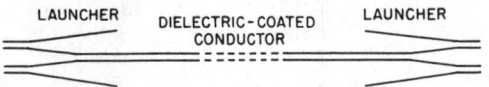

Fig. 36—Surface-wave transmission line with launchers at each end. These form transitions to coaxial line. *Courtesy of Electronics.*

the design equations are developed:

$$L = 0.025 n^2 d^2 [1 - (d/D)^2] \text{ microhenry per axial inch}$$
$$C = 0.75 / \log_{10}(D/d) \text{ picofarad per axial inch}$$
$$v = f_0 \lambda = 1000 (LC)^{-1/2} \text{ inches per microsecond}$$
$$b = 0.94 \lambda / 4 = 235 f_0^{-1} (LC)^{-1/2} \text{ inches}$$
$$Z_0 = 1000 (L/C)^{1/2} = 235\,000 (b f_0 C)^{-1} \text{ ohms.}$$

A further useful relationship in terms of the inside volume of the shield (vol) in cubic inches is

$$Q_u = 50 (\text{vol})^{1/3} f_0^{1/2}$$

provided that $0.4 < d/D < 0.6$, and $1.0 < b/d < 3.0$.

SURFACE-WAVE TRANSMISSION LINE*

The surface-wave transmission line is a single-conductor line having a relatively thick dielectric

Fig. 37—Relationship among wire diameter, dielectric layer, phase-velocity reduction, and impedance (for brown polyethylene). *Courtesy of Electronics.*

WAVELENGTH-TO-DIAMETER RATIO = λ/d_i

RATIO OF COATED-TO-UNCOATED WIRE = d_0/d_i

PHASE-VELOCITY REDUCTION = $100\, \delta v/c$ PERCENT

IMPEDANCE = Z IN OHMS (FOR $\epsilon_r = 2.3$)

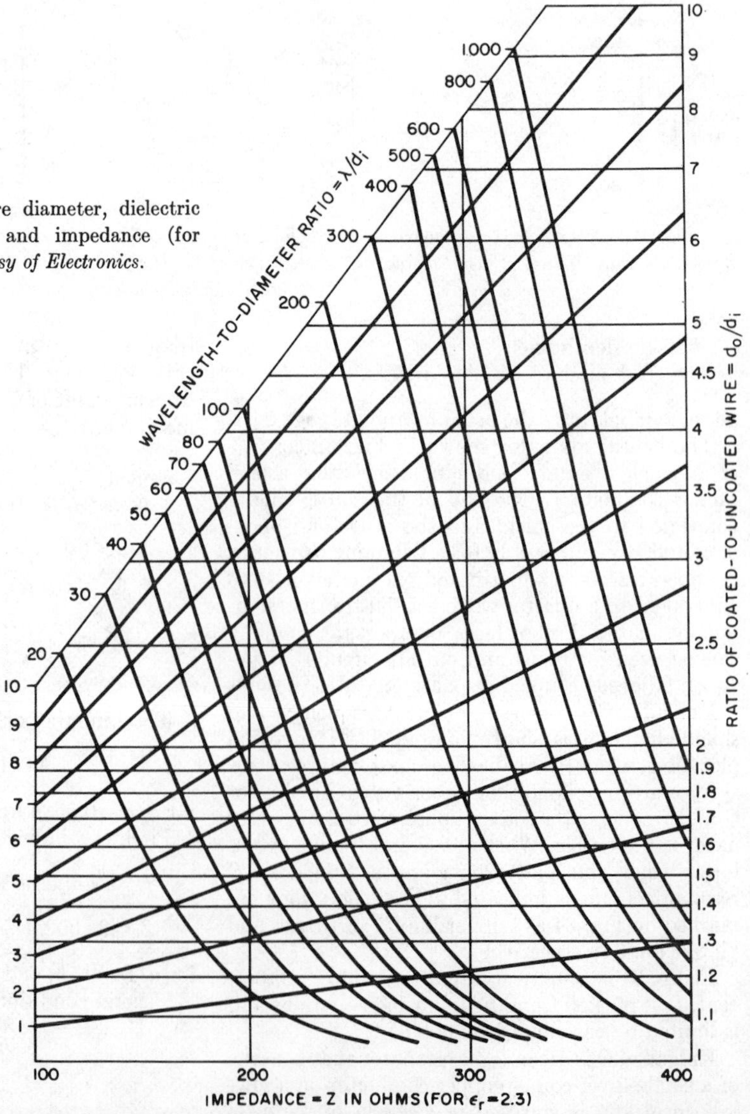

* G. Goubau, "Designing Surface-Wave Transmission Lines," *Electronics*, vol. 27, pp. 180–184; April 1954.

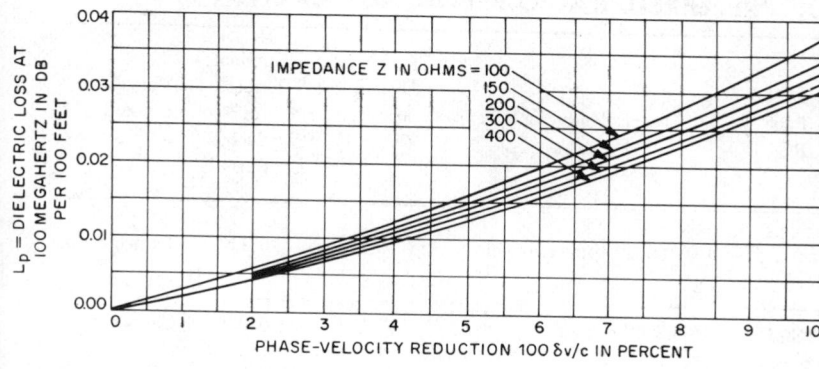

Fig. 38—Dielectric loss at 100 megahertz for brown polyethylene ($\epsilon_r = 2.3$ and $F_p = 5 \times 10^{-4}$). *Courtesy of Electronics.*

sheath (Fig. 35). The sheath diameter is often 3 or more times the conductor diameter. A mode of propagation that is practically nonradiating is excited on the line by means of a conical horn at each end as shown in Fig. 36. The mouth of the horn is roughly one-quarter to one wavelength in diameter. Losses are about half those of a 2-wire line, but the surface-wave line has a practical lower frequency limit of about 50 megahertz.

Design charts are given in Figs. 37 to 39 together with equations for attenuation losses.

The losses in the two launchers combined vary from less than 0.5 decibel to a little more than 1.0 decibel, according to their design.

Conductor loss L_c by the equation below is 5 percent over the theoretical value for pure copper. Dielectric loss L_p for polyethylene at 100 megahertz is shown in Fig. 38. For other dielectrics and frequencies, find L_i by the equation.

$$L_c = 0.455 f^{1/2}/Z d_i \text{ decibels}/100 \text{ feet}$$

$$L_i = 26 f F_p L_p/(\epsilon_r - 1) \text{ decibels}/100 \text{ feet}$$

$$L_i = L_p f/100$$

for brown polyethylene (Fig. 38).

Symbols

c = velocity of propagation in free space
d_i = diameter of the conductor (inches in equation for L_c)
d_o = outside diameter of the dielectric coating
f = frequency in megahertz
F_p = power factor of dielectric
L_c = conductor loss in decibels/100 feet
L_i = dielectric loss in decibels/100 feet
L_p = dielectric loss shown in Fig. 38
Z = waveguide impedance in ohms
δv = reduction in phase velocity
ϵ_r = dielectric constant relative to air
λ = free-space wavelength.

Example: At 900 megahertz ($\lambda = 0.333$ meter), a 200-foot line is required having a permissible loss of 1.0 decibel/100 feet (not including the launcher losses). What are its dimensions?

Allowing 20 percent for dielectric loss, the conductor loss would be $L_c = 0.8$ decibel/100 feet. Assuming $Z = 250$ ohms as a first approximation, the formula for L_c gives $d_i = 0.068$ inch. Use No. 14 AWG wire ($d_i = 0.064$ and $\lambda/d_i = 204$). Now going to Fig. 37 and assuming that $100\delta v/c = 6$ percent is adequate, we find that $d_o/d_i = 3$ and $Z = 270$ ohms.

Recomputing, $L_c = 0.79$ decibel/100 feet. By Fig. 38, $L_p = 0.017$ at 100 megahertz for brown polyethylene. Using the same material at 900 megahertz, the loss is $L_i = 0.15$ decibel/100 feet.

For 200 feet, the combined conductor and dielectric loss is 1.9 decibels, to which must be added the loss of 0.5 to 1.0 decibel total for the two launchers.

Dielectric Other Than Polyethylene (Fig. 39)

Determine Z and $\delta v/c$ for polyethylene ($\epsilon_r = 2.3$) from Fig. 37. Then use Fig. 39 to find the value of d_o/d_i required for the same performance with actual dielectric constant ϵ_r. Make computation of new dielectric loss, using Fig. 38 and equation for L_i.

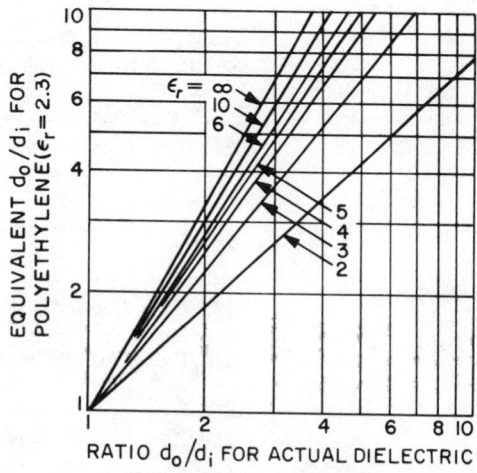

Fig. 39—Conversion chart for dielectric other than polyethylene. *Courtesy of Electronics.*

ARMY–NAVY LIST OF PREFERRED RADIO-FREQUENCY CABLES*

Class of Cables		JAN Type RG–	Inner Conductor†	Dielectric Material (Note 1)	Nominal Diameter of Dielectric (in.)	Shielding Braid
50 ohms	Single braid	58C/U	19/0.0071″ tinned copper	A	0.116	Tinned copper
		213/U	7/0.0296″ copper	A	0.285	Copper
		215/U	7/0.0296″ copper	A	0.285	Copper
		218/U	0.195″ copper	A	0.680	Copper
		219/U	0.195″ copper	A	0.680	Copper
		220/U	0.260″ copper	A	0.910	Copper
		221/U	0.260″ copper	A	0.910	Copper
	Double braid	55B/U	0.032″ silvered copper	A	0.116	Silvered copper
		212/U	0.0556″ silvered copper	A	0.185	Silvered copper
		214/U	7/0.0296″ silvered copper	A	0.285	Silvered copper
		217/U	0.106″ copper	A	0.370	Copper
		223/U	0.035″ silvered copper	A	0.116	Silvered copper
		224/U	0.106″ copper	A	0.370	Copper
75 ohms	Single braid	11A/U	7/26 AWG tinned copper	A	0.285	Copper
		12A/U	7/26 AWG tinned copper	A	0.285	Copper
		34B/U	7/0.0249″ copper	A	0.460	Copper
		35B/U	0.1045″ copper	A	0.680	Copper

Protective Covering (Note 2)	Nominal Overall Diameter (in.)	Weight (lb/ft)	Nominal Capacitance (pF/ft)	Maximum Operating Voltage (rms)	Remarks
IIa	0.195	0.029	28.5	1 900	Small-size flexible cable
IIa	0.405	0.120	29.5	5 000	Medium-size flexible cable (formerly RG–8A/U)
IIa, with armor	0.475 max.	0.160	29.5	5 000	Same as RG–213/U, but with armor (formerly RG–10A/U)
IIa	0.870	0.491	29.5	11 000	Large-size low-attenuation high-power transmission line (formerly RG–17A/U)
IIa, with armor	0.945 max.	0.603	29.5	11 000	Same as RG–218/U, but with armor (formerly RG–18A/U)
IIa	1.120	0.745	29.5	14 000	Very-large low-attenuation high-power transmission cable (formerly RG–19A/U)
IIa, with armor	1.195 max.	0.925	29.5	14 000	Same as RG–220/U, but with armor (formerly RG–20A/U)
IIIa	0.206	0.032	28.5	1 900	Small-size flexible cable
IIa	0.332	0.093	28.5	3 000	Small-size microwave cable (formerly RG–5B/U)
IIa	0.425	0.158	30.0	5 000	Special medium-size flexible cable (formerly RG–9B/U)
IIa	0.545	0.236	29.5	7 000	Medium-size power transmission line (formerly RG–14A/U)
IIa	0.216	0.036	28.5	1 900	Small-size flexible cable (formerly RG–55A/U)
IIa, with armor	0.615 max.	0.282	29.5	7 000	Same as RG–217/U, but with armor (formerly RG–74A/U)
IIa	0.412	—	20.5	5 000	Medium-size flexible video cable
IIa, with armor	0.475	—	20.5	5 000	Similar to RG–11A/U, but with armor
IIa	0.630	0.231	21.5	6 500	Large-size high-power low-attenuation flexible cable
IIa, with armor	0.945 max.	0.480	21.5	10 000	Large-size high-power low-attenuation video and communication cable

Class of Cables		JAN Type RG–	Inner Conductor†	Dielectric Material (Note 1)	Nominal Diameter of Dielectric (in.)	Shielding Braid
75 ohms Continued	Single braid	59B/U	0.0230″ copper-covered steel	A	0.146	Copper
		84A/U	0.1045″ copper	A	0.680	Copper
		85A/U	0.1045″ copper	A	0.680	Copper
		164/U	0.1045″ copper	A	0.680	Copper
		307A/U	17/0.0058″ silvered copper	A Foamed	0.029	Silvered copper
	Double braid	6A/U	21 AWG copper-covered steel	A	0.185	Inner: silvered copper. Outer: copper
		216/U	7/0.0159″ tinned copper	A	0.285	Copper
High temperature	Single braid	144/U	7/0.0179″ silvered copper-covered steel	F1	0.285	Silvered copper
		178B/U	7/0.004″ silvered copper-covered steel	F1	0.034	Silvered copper
		179B/U	Same as above	F1	0.063	Silvered copper
		180B/U	Same as above	F1	0.102	Silvered copper
		187A/U	7/0.004″ annealed silvered copper-covered steel	F1	0.060	Silvered copper
		195A/U	Same as RG–187A/U	F1	0.102	Silvered copper
		196A/U	Same as RG–187A/U	F1	0.034	Silvered copper
		211A/U	0.190″ copper	F1	0.620	Copper

Protective Covering (Note 2)	Nominal Overall Diameter (in.)	Weight (lb/ft)	Nominal Capacitance (pF/ft)	Maximum Operating Voltage (rms)	Remarks
IIa	0.242	—	21.0	2 300	General-purpose small-size video cable
IIa, with lead sheath	1.000	1.325	21.5	10 000	Same as RG–35B/U, except lead sheath instead of armor for underground installations
IIa, with lead sheath and special armor	1.565 max.	2.910	21.5	10 000	Same as RG–84A/U, with special armor for underground installations
IIa	0.870	0.490	21.5	10 000	Same as RG–35B/U, except without armor
IIIa	0.270	—	20	400	
IIa	0.332	—	20.0	2 700	Small-size video and communication cable
IIa	0.425	0.121	20.5	5 000	Medium-size flexible video and communication cable (formerly RG–13A/U)
Teflon-tape moisture seal with double-braid type-V jacket	0.410	0.120	20.5	5 000	Similar to RG–11A/U, except cable core is teflon. $Z = 75$ ohms
IX	0.075 max.	—	29.0	1 000	$Z = 50$ ohms
IX	0.105	—	20.0	1 200	
IX	0.145	—	15.5	1 500	$Z = 95$ ohms
VII	0.110	—	—	1 200	Miniaturized cable. $Z = 75$ ohms
VII	0.155	—	—	1 500	Miniaturized cable. $Z = 95$ ohms
VII	0.080	—	—	1 000	Miniaturized cable. $Z = 50$ ohms
Same as RG–144/U	0.730	0.450	29.0	7 000	Semiflexible cable operating at $-55°C$ to $+200°C$ (formerly RG–117A/U). $Z = 50$ ohms

Class of Cables		JAN Type RG–	Inner Conductor†	Dielectric Material (Note 1)	Nominal Diameter of Dielectric (in.)	Shielding Braid
High temperature *Continued*	Single braid	228A/U	0.190″ copper	F1	0.620	Copper
		302/U	0.025″ silvered copper-covered steel	F1	0.146	Silvered copper
		303/U	0.039″ silvered copper-covered steel	F1	0.116	Silvered copper
		304/U	0.059″ silvered copper-covered steel	F1	0.185	Silvered copper
		316/U	7/0.0067″ annealed silvered copper-covered steel	F1	0.060	Silvered copper
	Double braid	115/U	7/0.028″ silvered copper	F2	0.250	Silvered copper
		142B/U	0.039″ silvered copper-covered steel	F1	0.116	Silvered copper
		225/U	7/0.0312″ silvered copper	F1	0.285	Silvered copper
		226/U	19/0.0254″ silvered copper wire	F2	0.370	Copper
		227/U	7/0.0312″ silvered copper	F1	0.285	Silvered copper
Pulse	Single braid	26A/U	19/0.0117″ tinned copper	E	0.288	Tinned copper
		27A/U	19/0.0185″ tinned copper	D	0.455	Tinned copper
	Double braid	25A/U	19/0.0117″ tinned copper	E	0.288	Tinned copper
		28B/U	19/0.0185″ tinned copper	D	0.455	Inner: tinned copper. Outer: galvanized steel
		64A/U	19/0.0117″ tinned copper	E	0.288	Tinned copper

Protective Covering (Note 2)	Nominal Overall Diameter (in.)	Weight (lb/ft)	Nominal Capacitance (pF/ft)	Maximum Operating Voltage (rms)	Remarks
Teflon-tape moisture seal with double-braid type-V jacket, with armor	0.795	0.600	29.0	7 000	Same as RG–211A/U, but with armor (formerly RG–118A/U). $Z = 50$ ohms
IX	0.206	—	21.0	2 300	$Z = 75$ ohms
IX	0.170	—	28.5	1 900	$Z = 50$ ohms
IX	0.280	—	28.5	3 000	$Z = 50$ ohms
IX	0.102	—	—	1 200	Miniaturized cable. $Z = 50$ ohms
Same as RG–144/U	0.375	—	29.5	5 000	Medium-size cable for use where expansion and contraction are a major problem. $Z = 50$ ohms
IX	0.195	—	28.5	1 900	Small-size flexible cable. $Z = 50$ ohms
Same as RG–144/U	0.430	0.176	29.5	5 000	Semiflexible cable operating at −55°C to +200°C (formerly RG–87A/U). $Z = 50$ ohms
Same as RG–144/U	0.500	0.247	29.0	7 000	Medium-size cable for use where expansion and contraction are a major problem (formerly RG–94A/U). $Z = 50$ ohms
Same as RG–228A/U	0.490	0.224	29.5	5 000	Same as RG–225/U, but with armor (formerly RG–116/U). $Z = 50$ ohms
IV, with armor	0.505	0.189	50.0	10 000	High-voltage cable. $Z = 48$ ohms
IV, with armor	0.670	0.304	50.0	15 000 peak	Large-size cable. $Z = 48$ ohms
IV	0.505	0.205	50.0	10 000	High-voltage cable. $Z = 48$ ohms
IV	0.750	0.370	50.0	15 000 peak	Large-size cable. $Z = 48$ ohms
IV	0.475 max.	0.205	50.0	10 000	Medium-size cable. $Z = 48$ ohms

Class of Cables		JAN Type RG–	Inner Conductor†	Dielectric Material (Note 1)	Nominal Diameter of Dielectric (in.)	Shielding Braid
Pulse *Continued*	Double braid	156/U	7/21 AWG tinned copper	First layer A; second layer H	0.285	Inner: tinned copper. Outer: galvanized steel. Tinned copper outer shield
		157/U	19/24 AWG tinned copper	First layer H; second layer A; third layer H	0.455	
		158/U	37/21 AWG tinned copper		0.455	
		190/U	19/0.0117″ tinned copper		0.380	Same as above
		191/U	30 AWG tinned copper; single braid over supporting elements; 0.485″ max.	First layer H; second layer J; third layer H	1.065	Same as above
	Four braids	88/U	19/0.0117″ tinned copper	E	0.288	Tinned copper
Low capacitance	Single braid	62A/U	0.0253″ solid copper-covered steel	A	0.146	Copper
		63B/U	0.0253″ copper-covered steel	A	0.285	Copper
		79B/U	0.0253″ copper-covered steel	A	0.285	Copper
	Double braid	71B/U	0.0253″ copper-covered steel	A	0.146	Tinned copper
High attenuation	Single braid	301/U	7/0.0203″ Karma wire	F1	0.185	Karma wire
High delay	Single braid	65A/U	No. 32 Formex F. Helix diameter 0.128″	A	0.285	Copper
Twin conductor	Single braid	57A/U	Each conductor 7/0.0285″ plain copper	A	0.472	Tinned copper
		130/U		A	0.472	Tinned copper
		131/U		A	0.472	Tinned copper

Protective Covering (Note 2)	Nominal Overall Diameter (in.)	Weight (lb/ft)	Nominal Capacitance (pF/ft)	Maximum Operating Voltage (rms)	Remarks
IIa	0.540	0.211	30.0	10 000	Taped inner layers, first layer type K and second layer type A-1R, between the outer braid of the outer conductor and the tinned copper shield. Triaxial pulse cables. $Z = 50$ ohms
IIa	0.725	0.317	38.0	15 000	
IIa	0.725	0.380	78.0	15 000	Same as above, except $Z = 25$ ohms
VIII over one wrap of type K	0.700	0.353	50.0	15 000	Taped inner layers, two wraps of type K and two wraps of type L between the outer braid and the tinned copper shield. Pulse cable. $Z = 50$ ohms
Same as above	1.460	1.469	85.0	15 000	Same as RG–190/U, except $Z = 25$ ohms
IIa	0.515	—	50.0	10 000	Medium-size multishielded high-voltage cable. $Z = 48$ ohms
IIa	0.242	0.382	14.5	750	$Z = 93$ ohms
IIa	0.405	0.082	10.0	1 000	Medium-size low-capacitance air-spaced cable. $Z = 125$ ohms
IIa, with armor	0.475 max.	0.138	10.0	1 000	Same as RG–63B/U, but with armor. $Z = 125$ ohms
IIIa	0.250 max.	—	14.5	750	Low-capacitance cable. $Z = 93$ ohms
IX	0.245	—	29.0	3 000	High-attenuation cable. $Z = 50$ ohms
IIa	0.405	0.096	44.0	1 000	High-impedance video cable; high-delay line. $Z = 950$ ohms. (Refer to Note 3.)
IIa	0.625	0.225	17.0	3 000	$Z = 95$ ohms
I	0.625	0.220	17.0	8 000	Same as RG–57A/U, except inner conductors are twisted to improve flexibility. $Z = 95$ ohms
I, with aluminum armor	0.710	0.295	17.0	8 000	Same as RG–130/U, but with armor. $Z = 95$ ohms

Class of Cables	JAN Type RG–	Inner Conductor†	Dielectric Material (Note 1)	Nominal Diameter of Dielectric (in.)	Shielding Braid
Double braid	22B/U	Each conductor 7/0.0152″ copper	A	0.285	Tinned copper
	111A/U		A	0.285	Tinned copper
Twin coaxial	181/U	Each conductor 7/26 AWG copper	A	0.210	Copper inner braids and common braid

* From "RF Transmission Lines and Fittings," MIL–HDBK–216, 4 January 1962, revised 18 May 1965. Requirements for listed cables are in Specification MIL–C–17.

† Diameter of strands given in inches. As, 7/0.0296″ = 7 strands, each 0.0296-inch diameter.

Note 1—Dielectric materials: A = Polyethylene, D = Layer of synthetic rubber between two layers of conducting rubber, E = Layer of conducting rubber plus two layers of synthetic rubber, F1 = Solid polytetrafluoroethylene (Teflon), F2 = Semisolid or taped polytetrafluoroethylene (Teflon), H = Conducting synthetic rubber, and J = Insulating butyl rubber.

ATTENUATION AND POWER RATING OF LINES AND CABLES

Attenuation

Figure 40 illustrates the attenuation of general-purpose radio-frequency lines and cables up to their practical upper frequency limit. Most of these are coaxial-type lines, but waveguide and microstrip are included for comparison.

The following notes are applicable to this figure.

(**A**) For the RG-type cables, only the number is given (for instance, the curve for RG-218/U is labeled 218. Refer to table on pages 24-32–24-41.) The data on RG-type cables are taken mostly from, "RF Transmission Lines and Fittings," MIL–HDBK–216, 4 January 1962, revised 18 May 1965, and from "Solid Dielectric Transmission Lines," Electronic Industries Association Standard RS-199, December 1957.

Some approximation is involved in order to simplify the figure. Thus, where a single curve is labeled with several type numbers, the actual attenuation of each individual type may be slightly different from that shown by the curve.

(**B**) The curves for rigid copper coaxial lines are labeled with the diameter of the line only, as $7/8″C$. These have been computed for the lines listed in "Rigid Coaxial Transmission Lines, 50

Ohms," Electronic Industries Association Standard RS-225, August 1959. The computations considered the copper losses only, on the basis of a resistivity $\rho = 1.724$ microhm-centimeters; a derating of 20 percent has been applied to allow for imperfect surface, presence of fittings, etc., in long installed lengths. Relative attenuations of the different sizes are

$$A_{6\frac{1}{8}″} \approx 0.13 A_{\frac{7}{8}″}$$

$$A_{3\frac{1}{8}″} \approx 0.26 A_{\frac{7}{8}″}$$

$$A_{1\frac{5}{8}″} \approx 0.51 A_{\frac{7}{8}″}.$$

(**C**) Typical curves are shown for three sizes of 50-ohm semirigid cables such as Styroflex, Spiroline, Heliax, Alumispline, etc. These are labeled by size in inches, as $7/8″S$.

(**D**) The microstrip curve is for Teflon-impregnated Fibreglas dielectric $1/16$-inch thick and conductor strip $7/32$-inch wide.

(**E**) Shown for comparison is the attenuation in the $TE_{1,0}$ mode of 5 sizes of brass waveguide. The resistivity of brass was taken as $\rho = 6.9$ microhm-centimeters, and no derating was applied. For copper or silver, attenuation is about half that for brass. For aluminum, attenuation is about two-thirds that for brass.

Protective Covering (Note 2)	Nominal Overall Diameter (in.)	Weight (lb/ft)	Nominal Capacitance (pF/ft)	Maximum Operating Voltage (rms)	Remarks
IIa	0.420	0.116	16.0	1 000	Small-size balanced twin-conductor cable. $Z=95$ ohms
IIa, with armor	0.490 max.	0.145	16.0	1 000	Same as RG–22B/U, but with armor. $Z=95$ ohms
IIa	0.640	—	12	3 500	Filled-to-round, unbalanced transmission cable. Twin coaxial. $Z=125$ ohms

Note 2—Jacket types: I = Polyvinyl chloride (colored black), IIa = Noncontaminating synthetic resin, IIIa = Noncontaminating synthetic resin (colored black), IV = Chloroprene, V = Fibreglas, silicone-impregnated varnish, VII = Polytetrafluoroethylene, VIII = Polychloroprene, and IX = Fluorinated ethylene propylene.

Note 3—For RG–65A/U, delay = 0.042 microsecond per foot at 5 megahertz; dc resistance = 7.0 ohms/foot.

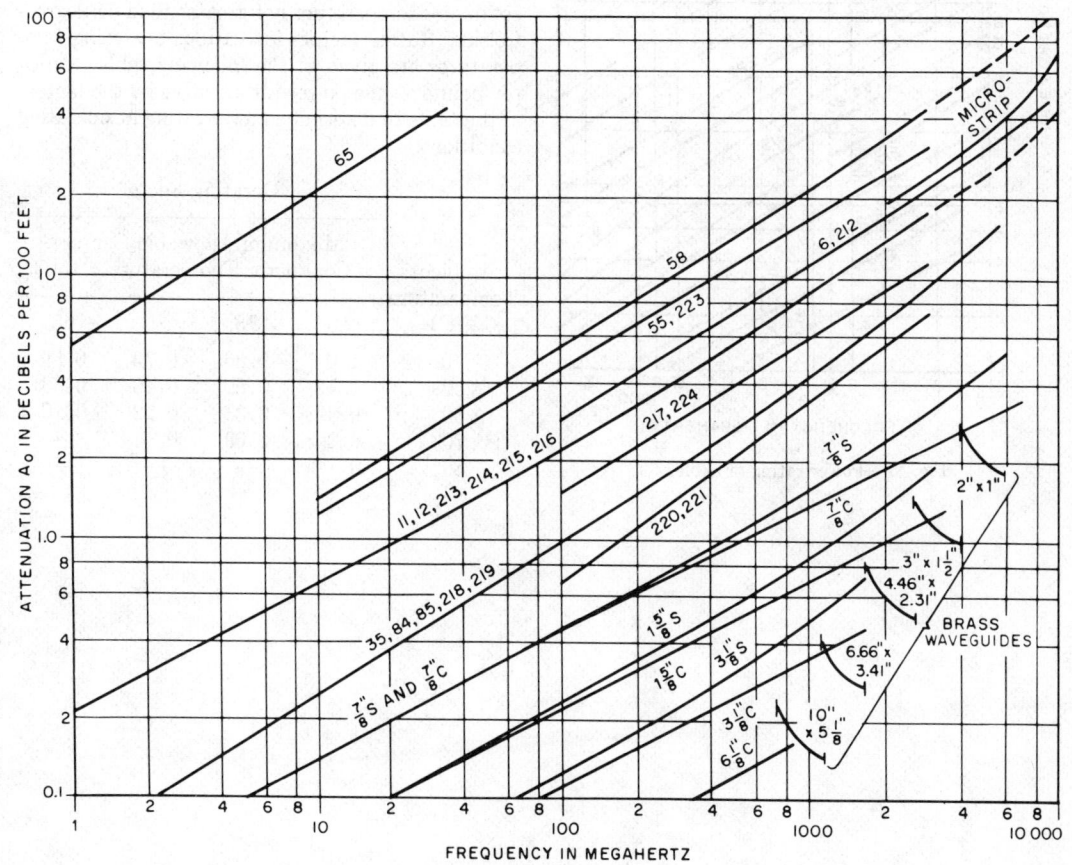

Fig. 40—Attenuation of cables.

Power Rating

Figure 41 shows the approximate power transmitting capabilities of various coaxial-type lines. The following notes are applicable.

(A) Identification of the curves for the RG-type cables is the same as in Fig. 40. The data for these cables are from the same sources. For polyethylene cables, an inner-conductor maximum temperature of 80 degrees Celsius is specified (G). For high-temperature cables (types 211, 228; 225,

227) the inner-conductor temperature is 250 degrees Celsius.

(B) The curves for 50-ohm rigid coaxial line are labeled with the diameter of the line only, as $\frac{7}{8}''C$. These are rough estimates based largely on miscellaneous charts published in catalogs.

(C) For Styroflex, Spiroline, Heliax, Alumispline, etc., cables, refer to (C) above.

(D) The curves are for unity voltage standing-wave ratio. Safe operating power is inversely proportional to swr expressed as a numerical ratio greater than unity. Do not exceed maximum operating voltage (see pp. 24-25 and 24-32–24-40).

(E) An ambient temperature of 40 degrees Celsius is assumed.

(F) The 4 curves meeting the 100-watt ordinate may be extrapolated: at 3000 megahertz for 55, 58, power is 28 watts; for 59, power is 44 watts; and for 6, 212, power is 58 watts.

(G) Electronic Industries Association Standard RS–199 states that operation of a polyethylene dielectric cable at a center-conductor temperature in excess of 80 degrees Celsius is likely to cause permanent damage to the cable. Where practicable, and particularly where continuous flexing is required, it is recommended that a cable be selected which, in regular operation, will produce a center-conductor temperature not greater than 65 degrees Celsius. Rating factors for various operating temperatures are given in the following table. Multiply points on the power-rating curve by the factors in the table to determine power rating at operating conditions.

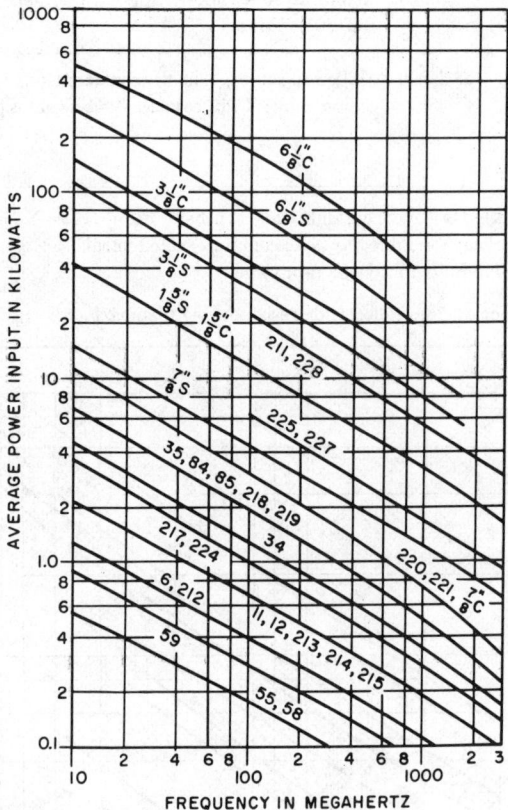

Fig. 41—Power rating of cables.

Derating Factor

Ambient Temperature (°C)	Maximum Allowable Center-Conductor Temperature (°C)			
	80	75	70	65
40	1.0	0.86	0.72	0.59
50	0.72	0.59	0.46	0.33
60	0.46	0.33	0.22	0.10
70	0.20	0.09	0	—
80	0	—	—	—

CHAPTER 25

WAVEGUIDES AND RESONATORS

PROPAGATION OF ELECTROMAGNETIC WAVES IN HOLLOW WAVEGUIDES

For propagation of energy at microwave frequencies through a hollow metal tube under fixed conditions, the following different types of waves are available.

TE Waves: Transverse-electric waves, sometimes called H waves, characterized by the fact that the electric vector (E vector) is always perpendicular to the direction of propagation. This means that

$$E_z \equiv 0$$

where z is the direction of propagation.

TM Waves: Transverse-magnetic waves, also called E waves, characterized by the fact that the magnetic vector (H vector) is always perpendicular to the direction of propagation. This means that

$$H_z \equiv 0$$

where z is the direction of propagation.

Note—TEM Waves: Transverse-electromagnetic waves. These waves are characterized by the fact that both the electric vector (E vector) and the magnetic vector (H vector) are perpendicular to the direction of propagation. This means that

$$E_z = H_z = 0$$

where z is the direction of propagation. This is the mode commonly excited in coaxial and open-wire lines. It cannot be propagated in a waveguide.

The solutions for the field configurations in waveguides are characterized by the presence of the integers m and n, which can take on separate values from 0 or 1 to infinity. Only a limited number of these different m,n modes can be propagated, depending on the dimensions of the guide and the frequency of excitation. For each mode there is a definite lower limit or cutoff frequency below which the wave is incapable of being propagated. Thus, a waveguide is seen to exhibit definite properties of a high-pass filter.

The propagation constant $\gamma_{m,n}$ determines the amplitude and phase of each component of the wave as it is propagated along the length of the guide. With $z =$ (direction of propagation) and $\omega = 2\pi \times$ (frequency), the factor for each component is

$$\exp[\,j\omega t - \gamma_{m,n} z\,].$$

Thus, if $\gamma_{m,n}$ is real, the phase of each component is constant, but the amplitude decreases exponentially with z. When $\gamma_{m,n}$ is real, it is said that no propagation takes place. The frequency is considered below cutoff. Actually, propagation with high attenuation does take place for a small distance, and a short length of guide below cutoff is often used as a calibrated attenuator.

When $\gamma_{m,n}$ is imaginary, the amplitude of each component remains constant, but the phase varies with z. Hence, propagation takes place. $\gamma_{m,n}$ is purely imaginary only in a lossless guide. In the practical case, $\gamma_{m,n}$ usually has both a real part $\alpha_{m,n}$, which is the attenuation constant, and an imaginary part $\beta_{m,n}$, which is the phase propagation constant. Then $\gamma_{m,n} = \alpha_{m,n} + j\beta_{m,n}$.

RECTANGULAR WAVEGUIDES

Figure 1 shows a rectangular waveguide and a rectangular system of coordinates, disposed so that the origin falls on one of the corners of the waveguide; z is the direction of propagation along the guide, and the cross-sectional dimensions are y_0 and x_0.

For the case of perfect conductivity of the guide walls with a nonconducting interior dielectric (usu-

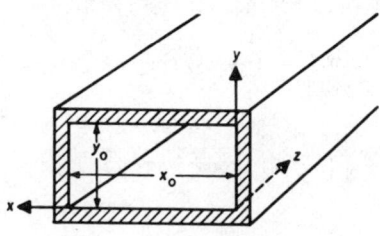

Fig. 1—Rectangular waveguide.

ally air), the equations for the $TM_{m,n}$ or $E_{m,n}$ waves in the dielectric are

$$E_x = -A \frac{\gamma_{m,n}}{\gamma_{m,n}z + \omega^2\mu\epsilon} (m\pi/x_0) \sin[(n\pi/y_0)y]$$

$$\times \cos[(m\pi/x_0)x] \exp(j\omega t - \gamma_{m,n}z)$$

$$E_y = -A \frac{\gamma_{m,n}}{\gamma_{m,n}z + \omega^2\mu\epsilon} (n\pi/y_0) \cos[(n\pi/y_0)y]$$

$$\times \sin[(m\pi/x_0)x] \exp(j\omega t - \gamma_{m,n}z)$$

$$E_z = A \sin[(n\pi/y_0)y] \sin[(m\pi/x_0)x]$$

$$\times \exp(j\omega t - \gamma_{m,n}z)$$

$$H_x = -A \frac{j\omega\epsilon}{\gamma_{m,n}z + \omega^2\mu\epsilon} (n\pi/y_0) \cos[(n\pi/y_0)y]$$

$$\times \sin[(m\pi/x_0)x] \exp(j\omega t - \gamma_{m,n}z)$$

$$H_y = A \frac{j\omega\epsilon}{\gamma_{m,n}z + \omega^2\mu\epsilon} (m\pi/x_0) \sin[(n\pi/y_0)y]$$

$$\times \cos[(m\pi/x_0)x] \exp(j\omega t - \gamma_{m,n}z)$$

$$H_z \equiv 0$$

where ϵ is the dielectric constant and μ the permeability of the dielectric material in meter-kilogram-second (rationalized) units.

Constant A is determined solely by the exciting voltage. It has both amplitude and phase. Integers m and n may individually take values from 1 to infinity. No TM waves of the 0,0 type or 1,0 type are possible in a rectangular guide so that neither m nor n may be 0.

Equations for the $TE_{m,n}$ waves or $H_{m,n}$ waves in a dielectric are

$$E_x = -B \frac{j\omega\mu}{\gamma_{m,n}z + \omega^2\mu\epsilon} (n\pi/y_0) \sin[(n\pi/y_0)y]$$

$$\times \cos[(m\pi/x_0)x] \exp(j\omega t - \gamma_{m,n}z)$$

$$E_y = B \frac{j\omega\mu}{\gamma_{m,n}z + \omega^2\mu\epsilon} (m\pi/x_0) \cos[(n\pi/y_0)y]$$

$$\times \sin[(m\pi/x_0)x] \exp(j\omega t - \gamma_{m,n}z)$$

$$E_z \equiv 0$$

$$H_x = B \frac{\gamma_{m,n}}{\gamma_{m,n}z + \omega^2\mu\epsilon} (m\pi/x_0) \cos[(n\pi/y_0)y]$$

$$\times \sin[(m\pi/x_0)x] \exp(j\omega t - \gamma_{m,n}z)$$

$$H_y = B \frac{\gamma_{m,n}}{\gamma_{m,n}z + \omega^2\mu\epsilon} (n\pi/y_0) \sin[(n\pi/y_0)y]$$

$$\times \cos[(m\pi/x_0)x] \exp(j\omega t - \gamma_{m,n}z)$$

$$H_z = B \cos[(n\pi/y_0)y] \cos[(m\pi/x_0)x]$$

$$\times \exp(j\omega t - \gamma_{m,n}z).$$

Constant B depends only on the original exciting voltage and has both magnitude and phase; m and n individually may assume any integer value from

0 to infinity. The 0,0 type of wave where both m and n are 0 is not possible, but all other combinations are.

As stated previously, propagation only takes place when the propagation constant $\gamma_{m,n}$ is imaginary.

$$\gamma_{m,n} = [(m\pi/x_0)^2 + (n\pi/y_0)^2 - \omega^2\mu\epsilon]^{1/2}.$$

This means, for any m,n mode, propagation takes place when

$$\omega^2\mu\epsilon > (m\pi/x_0)^2 + (n\pi/y_0)^2$$

or, in terms of frequency f and velocity of light c, when

$$f > \frac{c}{2\pi(\mu_1\epsilon_1)^{1/2}} [(m\pi/x_0)^2 + (n\pi/y_0)^2]^{1/2}$$

where μ_1 and ϵ_1 are the relative permeability and relative dielectric constant, respectively, of the dielectric material with respect to free space.

The wavelength in the air-filled waveguide is always greater than the wavelength in free space. The wavelength in the dielectric-filled waveguide may be less than the wavelength in free space. If λ is the wavelength in free space and the medium filling the waveguide has a relative dielectric constant ϵ

$$\lambda_{g(m,n)} = \frac{\lambda}{[\epsilon - (m\lambda/2x_0)^2 - (n\lambda/2y_0)^2]^{1/2}}$$

$$= \frac{\lambda}{[\epsilon - (\lambda/\lambda_c)^2]^{1/2}}$$

where $(1/\lambda_c)^2 = (m/2x_0)^2 + (n/2y_0)^2$.

The phase velocity within the guide is also always greater than in an unbounded medium. The phase velocity v and group velocity u are related by

$$u = c^2/v$$

where the phase velocity is given by $v = c\lambda_g/\lambda$, and the group velocity is the velocity of propagation of the energy.

To couple energy into waveguides, it is necessary to understand the configuration of the characteristic electric and magnetic lines. Figure 2 shows the field configuration for a $TE_{1,0}$ wave. Figure 3 shows the instantaneous field configuration for a higher mode, a $TE_{2,1}$ wave.

In Fig. 4 are shown only the characteristic E lines for the $TE_{1,0}$, $TE_{2,0}$, $TE_{1,1}$, and $TE_{2,1}$ waves. The arrows on the lines indicate their instantaneous relative directions. To excite a TE wave, it is necessary to insert a probe to coincide with the direction of the E lines. Thus, for a $TE_{1,0}$ wave, a single probe projecting from the side of the guide parallel to the E lines would be sufficient to couple into it. Two ways of coupling from a coaxial line

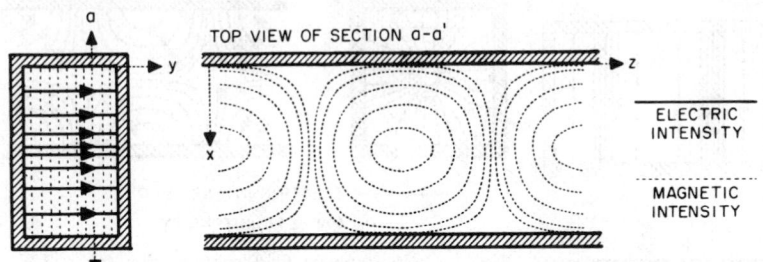

Fig. 2—Field configuration for a $TE_{1,0}$ wave.

to a rectangular waveguide to excite the $TE_{1,0}$ mode are shown in Fig. 5. With structures such as these, it is possible to make the standing-wave ratio due to the junction less than 1.15 over a 10- to 15-percent frequency band.

Figure 6 shows the instantaneous configuration of a $TM_{1,1}$ wave; Figure 7 shows the instantaneous field configuration for a $TM_{2,1}$ wave. Coupling to this type of wave may be accomplished by inserting a probe, which is parallel to the E lines, or by means of a loop oriented to link the lines of flux.

CIRCULAR WAVEGUIDES

The usual coordinate system is ρ, θ, z, where ρ is the radial direction; θ is the angle; z is in the longitudinal direction.

TM Waves (E Waves): $H_z \equiv 0$

$$E_\rho = H_\theta \eta (\lambda/\lambda_{g(m,n)}) \exp(j\omega t - \gamma_{m,n}z)$$

$$E_\theta = -H_\rho \eta (\lambda/\lambda_{g(m,n)}) \exp(j\omega t - \gamma_{m,n}z)$$

$$E_z = AJ_n(k_{m,n}\rho) \cos n\theta \exp(j\omega t - \gamma_{m,n}z)$$

$$H_\rho = -jA(2\pi n/\lambda k_{m,n}{}^2 \eta \rho)J_n(k_{m,n}\rho) \sin n\theta$$
$$\times \exp(j\omega t - \gamma_{m,n}z)$$

$$H_\theta = -jA(2\pi/\lambda k_{m,n}\eta)J_n'(k_{m,n}\rho) \cos n\theta$$
$$\times \exp(j\omega t - \gamma_{m,n}z)$$

where $\eta = (\mu/\epsilon)^{1/2}$, with μ and ϵ in absolute units.

By the boundary conditions, $E_z = 0$ when $\rho = a$ the radius of the guide. Thus, the only permissible values of k are those for which $J_n(k_{m,n}a) = 0$, because E_z must be zero at the boundary.

The numbers m,n take on all integral values from zero to infinity. The waves are seen to be characterized by the numbers m and n, where n gives the order of the bessel functions, and m gives the order of the root of $J_n(k_{m,n}a)$. The bessel function has an infinite number of roots, so that there are an infinite number of k's that make $J_n(k_{m,n}a) = 0$.

TE Waves (H Waves): $E_z \equiv 0$

$$E_\rho = jB(2\pi n\eta/\lambda k_{m,n}{}^2\rho)J_n(k_{m,n}\rho) \sin n\theta$$
$$\times \exp(j\omega t - \gamma_{m,n}z)$$

$$E_\theta = jB(2\pi\eta/\lambda k_{m,n})J_n'(k_{m,n}\rho) \cos n\theta$$
$$\times \exp(j\omega t - \gamma_{m,n}z)$$

$$H_\rho = -E_\theta(\lambda_{g(m,n)}/\eta\lambda) \exp(j\omega t - \gamma_{m,n}z)$$

$$H_\theta = E_\rho(\lambda_{g(m,n)}/\eta\lambda) \exp(j\omega t - \gamma_{m,n}z)$$

$$H_z = BJ_n(k_{m,n}\rho) \cos n\theta \exp(j\omega t - \gamma_{m,n}z).$$

Again n takes on integral values from zero to infinity. The boundary condition $E_\theta = 0$ when $\rho = a$ still applies. To satisfy this condition, k must be such as to make $J_n'(k_{m,n}a)$ equal to zero [where the superscript indicates the derivative of $J_n(k_{m,n}a)$]. It is seen that m takes on values from

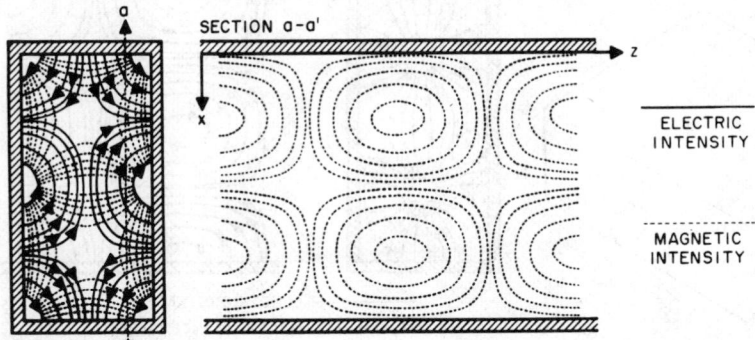

Fig. 3—Field configuration for a $TE_{2,1}$ wave.

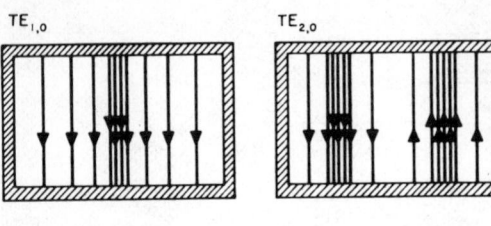

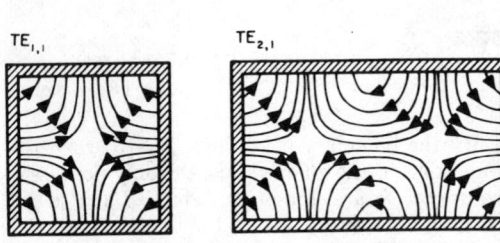

Fig. 4—Characteristic E lines for TE waves.

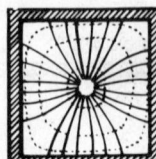

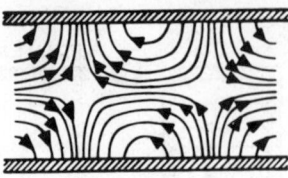

——— ELECTRIC INTENSITY

--------- MAGNETIC INTENSITY

Fig. 6—Instantaneous field configuration for a $TM_{1,1}$ wave.

1 to infinity since there are an infinite number of roots of $J_n{}'(k_{m,n}a)$.

For circular waveguides, the cutoff frequency for the m,n mode is

$$f_{c(m,n)} = ck_{m,n}/2\pi$$

where c = velocity of light, and $k_{m,n}$ is evaluated from the roots of the bessel functions.

$$k_{m,n} = U_{m,n}/a \quad \text{or} \quad U_{m,n}{}'/a$$

where a = radius of guide or pipe, and $U_{m,n}$ is the

root of the particular bessel function of interest (or its derivative).

The wavelength in any guide filled with a homogeneous dielectric ϵ (relative) is

$$\lambda_g = \lambda_0/[\epsilon - (\lambda_0/\lambda_c)^2]^{1/2}$$

where λ_0 is the wavelength in free space, and λ_c is the free-space cutoff wavelength for any mode under consideration.

Tables 1 and 2 are useful in determining the values of k. For TE waves the cutoff wavelengths are given in Table 1, and for TM waves the cutoff wavelengths are given in Table 2, where n is the order of the bessel function and m is the order of the root.

Figure 8 shows λ_0/λ_g as a function of λ_0/λ_c. From this, λ_g may be determined when λ_0 and λ_c are known.

The pattern of magnetic force of TM waves in a circular waveguide is shown in Fig. 9. Only the maximum lines are indicated. To excite this type of pattern, it is necessary to insert a probe along the length of the waveguide and concentric with the H lines. For instance, in the $TM_{0,1}$ type of wave, a probe extending down the length of the waveguide at its very center would provide the proper excitation. This method of excitation is shown in Fig. 10. Corresponding methods of excitation may be used for the other types of TM waves shown in Fig. 9.

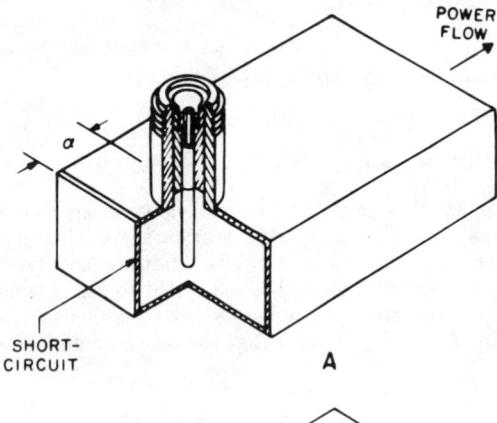

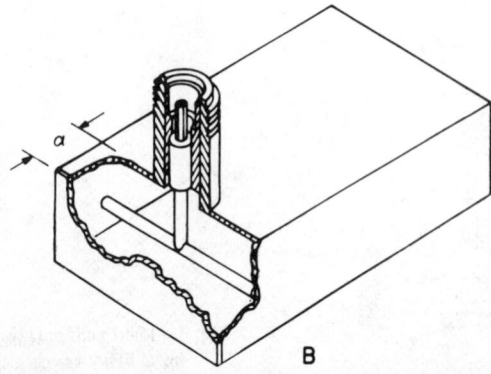

Fig. 5—Methods of coupling to $TE_{1,0}$ mode ($\alpha \approx \lambda_g/4$).

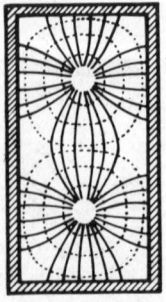

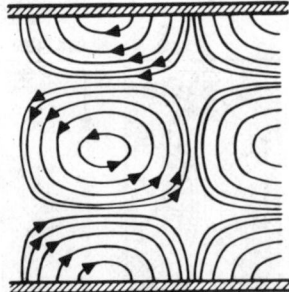

——— ELECTRIC INTENSITY

--------- MAGNETIC INTENSITY

Fig. 7—Instantaneous field configuration for a $TM_{2,1}$ wave.

TABLE 1—TE WAVES, VALUES OF λ_c/a
(where a = radius of guide).

n \ m	0	1	2
1	1.640	3.414	2.057
2	0.896	1.178	0.937
3	0.618	0.736	0.631

TABLE 2—TM WAVES, VALUES OF λ_c/a.

n \ m	0	1	2
1	2.619	1.640	1.224
2	1.139	0.896	0.747
3	0.726	0.618	0.541

Figure 11 shows the patterns of electric force for TE waves. Again only the maximum lines are indicated. This type of wave may be excited by an antenna that is parallel to the electric lines of force. The $TE_{1,1}$ wave may be excited by means of an antenna extending across the waveguide. This is illustrated in Fig. 12.

Propagating E waves have a minimum attenuation at $(3)^{1/2}f_c$. The $H_{1,1}$ wave has minimum attenuation at the frequency $2.6\,(3)^{1/2}f_c$.

The $H_{0,1}$ wave has the interesting and useful property that attenuation decreases as the frequency increases. This has made the $H_{0,1}$ mode exceedingly useful in the transmission of microwave signals over long distances.

Table 3 presents some of the important equations for various guides.

SQUARE WAVEGUIDES

Waveguide having interior dimensions $x_0 = y_0$ (Fig. 1) has found increasingly important application in dual polarized horn feeds and waveguide multiplexers. Usually these involve simultaneous propagation of the orthogonally oriented dominant modes, $TE_{1,0}$ and $TE_{0,1}$. These modes are theoretically capable of propagation without cross-coupling, at the same frequency, in lossless waveguide of square cross section. In practice, wall

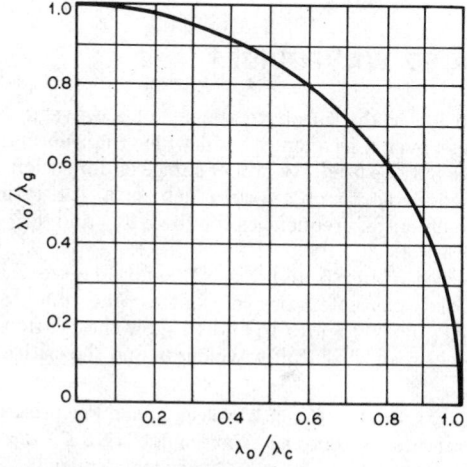

Fig. 8—Chart for determining guide wavelength.

losses, surface irregularities, and unequal transverse interior dimensions give rise to $TE_{1,0}$ and $TE_{0,1}$ mode cross-conversion. This occurs continually along the waveguide so that unless special care is taken, long lengths of dual-polarized waveguide exhibit a deteriorated mode isolation as a function of length of guide.

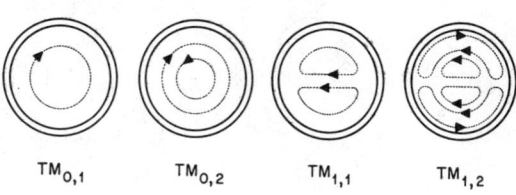

Fig. 9—Patterns of magnetic force of TM waves in circular waveguides.

Most important in establishing the initial mode isolation is proximity of the operating frequency to the cutoff frequency of the $TE_{1,0}$ mode in the square waveguide so that the total operating frequency band is above $TE_{1,0}$ cutoff and well below $TE_{1,1}$ cutoff. The lowest operating frequency should be approximately 25% above the $TE_{1,0}$ cutoff frequency. Thus, a dual-polarized feed propagating a 4400-MHz signal should use a square waveguide having internal dimensions of about 1.68 inches. If the internal dimensions are arrived at by using the 1.87 internal dimension of WR 187 waveguide, the dual mode isolation will probably not exceed 35 dB. By operating about 25% above the $TE_{1,0}$ mode cutoff frequency and well below $TE_{1,1}$ cutoff, the isolation can exceed 50 dB.*

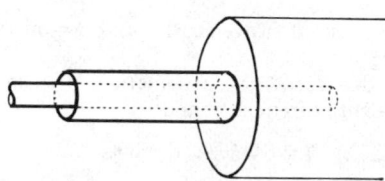

Fig. 10—Method of coupling to circular waveguide for $TM_{0,1}$ wave.

* D. J. LeVine and W. Sichak, "Dual-Mode Horn Feed for Microwave Multiplexing," *Electronics*, McGraw-Hill Book Co., New York, N. Y.; September 1954.

TABLE 3—CUTOFF WAVELENGTHS AND ATTENUATION FACTORS.*

Type of Guide (copper) †	Cutoff Wavelength λ_c $f_c = (c/\lambda_c)$	Attenuation Constant α (dB/100 ft)
Coaxial line‡ TEM	0	$\dfrac{3.58 \times 10^{-6}(f)^{1/2}(1/b)[1+(b/a)]}{\ln(b/a)}$ NOTE: The figure on p. 22-41 includes a derating factor of 20% applied to calculated α.
Rectangular pipe $TE_{m,0}$ or $H_{m,0}$	$2a/m$	$\dfrac{1.107}{a^{3/2}} \times \dfrac{\frac{1}{2}(a/b)(f/f_c)^{3/2}+(f/f_c)^{-1/2}}{[(f/f_c)^2-1]^{1/2}}$
Circular pipe:		
$TM_{0,1}$ or $E_{0,1}$	$2.613a$	$\dfrac{0.485}{a^{3/2}} \times \dfrac{(f/f_c)^{3/2}}{[(f/f_c)^2-1]^{1/2}}$
$TE_{1,1}$ or $H_{0,1}$	$3.412a$	$\dfrac{0.423}{a^{3/2}} \times \dfrac{(f/f_c)^{-1/2}+(1/2.38)(f/f_c)^{3/2}}{[(f/f_c)^2-1]^{1/2}}$
$TE_{0,1}$ or $H_{0,1}$	$1.640a$	$\dfrac{0.611}{a^{3/2}} \times \dfrac{(f/f_c)^{-1/2}}{[(f/f_c)^2-1]^{1/2}}$

* Dimensions are in inches and frequency in hertz; vacuum dielectric.
† For other metals multiply α by the square root of ratio of resistivity relative to that of copper.
‡ Inner and outer conductors same material.

ATTENUATION IN A WAVEGUIDE BEYOND CUTOFF

When a waveguide is used at a wavelength greater than the cutoff wavelength, there is no real propagation and the fields are attenuated exponentially. The attenuation L in a length d is given by

$$L = 54.5(d/\lambda_c)[1-(\lambda_c/\lambda)^2]^{1/2} \text{ decibels}$$

where $\lambda_c =$ cutoff wavelength, and $\lambda =$ operating wavelength.

Note that for $\lambda \gg \lambda_c$, attenuation is essentially independent of frequency and

$$L = 54.5d/\lambda_c \text{ decibels}$$

where λ_c is a function of geometry.

STANDARD WAVEGUIDES

Table 4 lists some properties and dimensions of standard rectangular waveguides. For other than

theoretical vacuum performance, consider the relative value of ϵ for sea level, 20°C air, as approximately 1.0006. Rounded inner corners also modify performance slightly.*

RIDGED WAVEGUIDES

To lower the cutoff frequency of a waveguide for use over a frequency band wider than normal, ridges may be used. By proper choice of dimensions, it is possible to obtain as much as a 4:1 ratio between cutoff frequencies for the $TE_{2,0}$ and $TE_{1,0}$ modes.

Tables 5 and 6 and Figs. 13 and 14 give the essential characteristics of single- and double-ridged guides. Figures 15 and 16 show the relationship between the cutoff wavelength and the critical

* M. M. Brady, "Cutoff Wavelengths and Frequencies of Standard Rectangular Waveguides," *I.E.E. Electronics Letters*, vol. 5, no. 17, London, England; 21 August 1969.

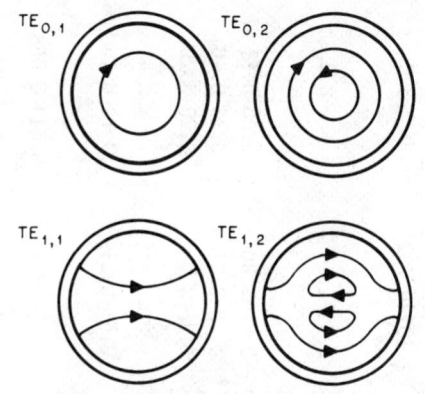

Fig. 11—Patterns of electric force of TE waves in circular waveguides.

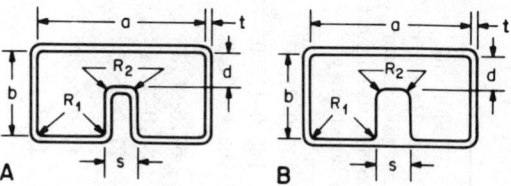

Fig. 13—Single-ridged waveguides. Refer to Table 5.

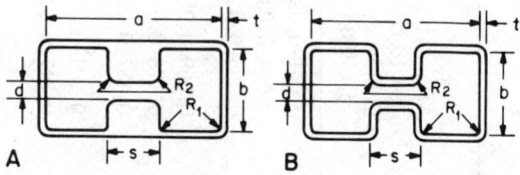

Fig. 14—Double-ridged waveguides. Refer to Table 6.

dimensions. Figures 17 and 18 describe the bandwidth (ratio of cutoff wavelengths of the 0,1 and 0,2 modes). The price paid for increased bandwidth is an increase in attenuation relative to the equivalent rectangular guide and is shown in Figs. 19 and 20.

Coaxial line can be coupled to either the single- or double-ridged guide by partly inserting the probe into the open space between the ridge and the opposite wall or ridge (Fig. 5A) or connecting it directly across the gap.

FLEXIBLE WAVEGUIDES

Flexible waveguide is used to join rigid sections or components that cannot be accurately dimensioned, positioned, or rendered immobile in space. Thus, rather than attempt to precisely specify and hold every bend, twist, and straight section of a long waveguide run between fixed points, it is often adequate to leave a short run or bend to be filled by a flexible guide insert. It is also used to permit thermally induced relative movement and to insulate portions of a waveguide run from shock and vibration. Flexible waveguide should not be treated as a link between a cabinet and its frequently opened doors and drawers, unless the

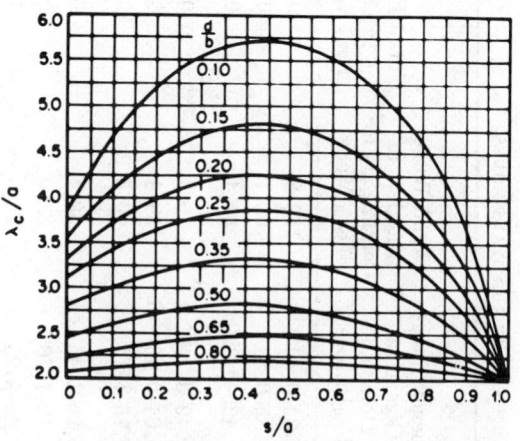

Fig. 15—Cutoff wavelength, single-ridged guide. *From S. Hopfer, "The Design of Ridged Waveguides," Transactions of the I.R.E., vol. MTT-3, no. 5, fig. 5; 1955.*

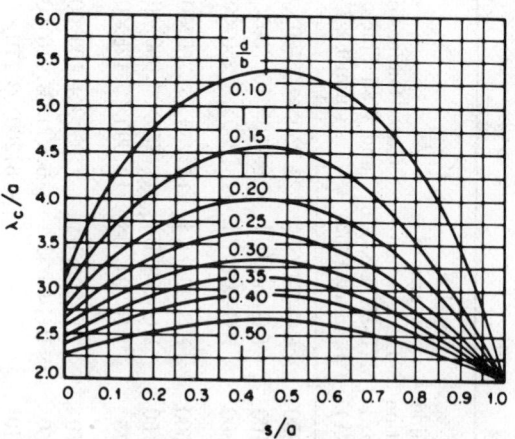

Fig. 16—Cutoff wavelength, double-ridged guide. *From S. Hopfer, "The Design of Ridged Waveguides," Transactions of the I.R.E., vol. MTT-3, no. 5, fig. 2; 1955.*

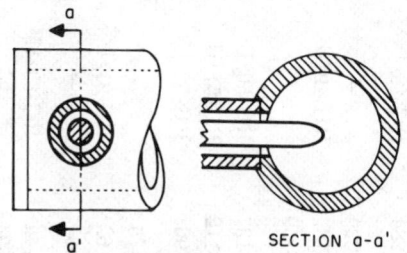

Fig. 12—Method of coupling to circular waveguide for $TE_{1,1}$ wave.

TABLE 4—STANDARD WAVEGUIDES.

EIA Waveguide Designation (Standard RS-261-A)	JAN Waveguide Designation (MIL-HDBK-216, 4 January 1962)	Outer Dimensions and Wall Thickness (in inches)	Frequency Range in Gigahertz for Dominant ($TE_{1,0}$) Mode	Cutoff Wavelength λ_c in Centimeters for $TE_{1,0}$ Mode	Cutoff Frequency f_c in Gigahertz for $TE_{1,0}$ Mode	Theoretical Attenuation, Lowest to Highest Frequency in dB/100 ft	Theoretical Power Rating in Megawatts for Lowest to Highest Frequency*
WR-2300	RG-290/U†	23.250×11.750×0.125	0.32–0.49	116.8	0.256	0.051–0.031	153.0–212.0
WR-2100	RG-291/U†	21.250×10.750×0.125	0.35–0.53	106.7	0.281	0.054–0.034	120.0–173.0
WR-1800	RG-201/U†	18.250×9.250×0.125	0.425–0.620	91.4	0.328	0.056–0.038	93.4–131.9
WR-1500	RG-202/U†	15.250×7.750×0.125	0.49–0.740	76.3	0.393	0.069–0.050	67.6–93.3
WR-1150	RG-203/U†	11.750×6.000×0.125	0.64–0.96	58.4	0.514	0.128–0.075	35.0–53.8
WR-975	RG-204/U†	10.000×5.125×0.125	0.75–1.12	49.6	0.605	0.137–0.095	27.0–38.5
WR-770	RG-205/U†	7.950×4.100×0.125	0.96–1.45	39.1	0.767	0.201–0.136	17.2–24.1
WR-650	RG-69/U	6.660×3.410×0.080	1.12–1.70	33.0	0.908	0.317–0.212	11.9–17.2
WR-510		5.260×2.710×0.080	1.45–2.20	25.9	1.16	—	—
WR-430	RG-104/U	4.460×2.310×0.080	1.70–2.60	21.8	1.375	0.588–0.385	5.2–7.5
WR-340	RG-112/U	3.560×1.860×0.080	2.20–3.30	17.3	1.735	0.877–0.572	—
WR-284	RG-48/U	3.000×1.500×0.080	2.60–3.95	14.2	2.08	1.102–0.752	2.2–3.2
WR-229	—	2.418×1.273×0.064	3.30–4.90	11.6	2.59	—	—
WR-187	RG-49/U	2.000×1.000×0.064	3.95–5.85	9.50	3.16	2.08–1.44	1.4–2.0
WR-159	—	1.718×0.923×0.064	4.90–7.05	8.09	3.71	—	—
WR-137	RG-50/U	1.500×0.750×0.064	5.85–8.20	6.98	4.29	2.87–2.30	0.56–0.71
WR-112	RG-51/U	1.250×0.625×0.064	7.05–10.00	5.70	5.26	4.12–3.21	0.35–0.46
WR-90	RG-52/U	1.000×0.500×0.050	8.20–12.40	4.57	6.56	6.45–4.48	0.20–0.29
WR-75	—	0.850×0.475×0.050	10.00–15.00	3.81	7.88	—	—
WR-62	RG-91/U	0.702×0.391×0.040	12.40–18.00	3.16	9.49	9.51–8.31	0.12–0.16
WR-51	—	0.590×0.335×0.040	15.00–22.00	2.59	11.6	—	—
WR-42	RG-53/U	0.500×0.250×0.040	18.00–26.50	2.13	14.1	20.7–14.8	0.043–0.058
WR-34	—	0.420×0.250×0.040	22.00–33.00	1.73	17.3	—	—
WR-28	RG-96/U‡	0.360×0.220×0.040	26.50–40.00	1.42	21.1	21.9–15.0	0.022–0.031

WR-22	RG-97/U‡	0.304×0.192×0.040	33.00–50.00	1.14	26.35	31.0–20.9	0.014–0.020
WR-19	—	0.268×0.174×0.040	40.00–60.00	0.955	31.4	—	—
WR-15	RG-98/U‡	0.228×0.154×0.040	50.00–75.00	0.753	39.9	52.9–39.1	0.0063–0.0090
WR-12	RG-99/U‡	0.202×0.141×0.040	60.00–90.00	0.620	48.4	93.3–52.2	0.0042–0.0060
WR-10	—	0.180×0.130×0.040	75.00–110.00	0.509	59.0	—	—
WR-8	RG-138/U§	0.140×0.100×0.030	90.00–140.00	0.406	73.84	152–99	0.0018–0.0026
WR-7	RG-136/U§	0.125×0.0925×0.030	110.00–170.00	0.330	90.84	163–137	0.0012–0.0017
WR-5	RG-135/U§	0.111×0.0855×0.030	140.00–220.00	0.259	115.75	308–193	0.00071–0.00107
WR-4	RG-137/U§	0.103×0.0815×0.030	170.00–260.00	0.218	137.52	384–254	0.00052–0.00075
WR-3	RG-139/U§	0.094×0.0770×0.030	220.00–325.00	0.173	173.28	512–348	0.00035–0.00047

* For these computations, the breakdown strength of air was taken as 15 000 volts per centimeter. A safety factor of approximately 2 at sea level has been allowed.

† Aluminum, 2.83×10^{-6} ohm-cm resistivity. ‡ Silver, 1.62×10^{-6} ohm-cm resistivity. § JAN types are silver, with a circular outer diameter of 0.156 inch and a rectangular bore matching EIA types. All other types are of a Cu-Zn alloy, 3.9×10^{-6} ohm-cm resistivity.

Note: Equivalent designations of waveguides follow.

EIA	British	IEC	EIA	British	IEC	EIA	British	IEC
WR-2300	00	-R3	WR-340	9A	-R26	WR-51	19	-R180
WR-2100	0	-R4	WR-284	10	-R32			
WR-1800	1	-R5				WR-42	20	-R220
WR-1500	2	-R6	WR-229	11A	-R40	WR-34	21	-R260
WR-1150	3	-R8	WR-187	12	-R48	WR-28	22	-R320
WR-975	4	-R9	WR-159	13	-R58			
WR-770	5	-R12	WR-137	14	-R70	WR-22	23	-R400
WR-650	6	-R14	WR-112	15	-R84	WR-19	24	-R500
WR-510	7	-R18	WR-90	16	-R100	WR-15	25	-R620
WR-430	8	-R22	WR-75	17	-R120	WR-12	26	-R740
			WR-62	18	-R140	WR-10	27	-R900
						WR-8	28	-R1200

TABLE 5—CHARACTERISTICS OF SINGLE-RIDGED WAVEGUIDES. *From MIL-HDBK-216, "RF Transmission Lines and Fittings," 4 January 1962.*

| Frequency Range (GHz) | $f_{c1.0}$ (GHz) | $\lambda_{c1.0}$ (in.) | $f_{c2.0}$ (GHz) | Dimensions in Inches | | | | | | | At $f=(3)^{1/2} f_{c1.0}$ | |
				a	b	d	s	t	R_1 (max)	R_2	Atten* (dB/ft)	Power Rating† (kW)
							Bandwidth 2.4:1					
0.175–0.42	0.148	79.803	0.431	28.129	12.658	5.278	4.360	—	—	1.056	0.00024	32 870.
0.267–0.64	0.226	52.260	0.658	18.421	8.289	3.457	2.855	—	—	0.691	0.00045	14 100.
0.42–1.0	0.356	33.177	1.036	11.695	5.263	2.195	1.813	0.125	0.047	0.439	0.00087	5 682.
0.64–1.53	0.542	21.792	1.577	7.682	3.457	1.442	1.191	0.125	0.047	0.288	0.00164	2 451.
0.84–2.0	0.712	16.588	2.072	5.847	2.631	1.097	0.906	0.080	0.047	0.219	0.00248	1 421.
1.5–3.6	1.271	9.293	3.699	3.276	1.474	0.615	0.508	0.080	0.047	0.123	0.00591	445.8
2.0–4.8	1.695	6.968	4.933	2.456	1.105	0.461	0.381	0.080	0.047	0.092	0.00908	250.6
3.5–8.2	2.966	3.982	8.632	1.404	0.632	0.264	0.218	0.064	0.031	0.053	0.0212	81.87
4.75–11.0	4.025	2.934	11.714	1.034	0.465	0.194	0.160	0.050	0.031	0.039	0.0333	44.43
7.5–18.0‡	6.356	1.858	18.498	0.655	0.295	0.123	0.1015	0.050	0.015	0.025	0.0661	17.82
11.0–26.5‡	9.322	1.267	27.130	0.4466	0.2010	0.0838	0.0692	0.040	0.015	0.017	0.117	8.285
18.0–40.0‡	15.254	0.7743	44.393	0.2729	0.1228	0.0512	0.0423	0.040	0.015	0.010	0.246	3.035
							Bandwidth 3.6:1					
0.108–0.39	0.092	128.37	0.404	31.218	14.048	2.402	5.307	—	—	0.480	0.0016	14 550.
0.27–0.97	0.229	51.572	1.006	12.542	5.644	0.965	2.132	—	—	0.193	0.0065	2 348.
0.39–1.4	0.331	35.680	1.454	8.677	3.905	0.668	1.475	0.125	0.047	0.134	0.0112	1 124.
0.97–3.5	0.822	14.367	3.611	3.494	1.572	0.269	0.594	0.080	0.047	0.054	0.0438	182.2
1.4–5.0	1.186	9.958	5.210	2.422	1.090	0.186	0.412	0.080	0.047	0.037	0.0758	87.56
3.5–12.4	2.966	3.982	13.030	0.968	0.436	0.075	0.165	0.050	0.031	0.015	0.300	13.99
5.0–18.0‡	4.237	2.787	18.613	0.678	0.305	0.052	0.115	0.050	0.015	0.010	0.513	6.857
12.4–40.0‡	10.508	1.124	46.162	0.273	0.123	0.021	0.046	0.040	0.015	0.004	2.008	1.115

* Copper.
† Based on breakdown of air—15 000 volts per cm (safety factor of approx 2 at sea level). Corner radii considered.
‡ Figure 13B in these frequency ranges only.

TABLE 6—CHARACTERISTICS OF DOUBLE-RIDGED WAVEGUIDES. *From MIL–HDBK–216, "RF Transmission Lines and Fittings," 4 January 1962.*

Frequency Range (GHz)	$f_{c1,0}$ (GHz)	$\lambda_{c1,0}$ (in.)	$f_{c2,0}$ (GHz)	Dimensions in Inches							At $f=(3)^{1/2}f_{c1,0}$	
				a	b	d	s	t	R_1 (max)	R_2	Atten* (dB/ft)	Power Rating† (kW)
Bandwidth 2.4:1												
0.175–0.42				29.667	13.795	5.863	7.417	—	—	1.173		
0.267–0.64				19.428	9.034	3.839	4.857	—	—	0.768		
0.42–1.0				12.333	5.737	2.437	3.083	0.125	0.050	0.487		
0.64–1.53				8.100	3.767	1.601	2.025	0.125	0.050	0.320		
0.84–2.0				6.167	2.868	1.219	1.542	0.125	0.050	0.244		
1.5–3.6				3.455	1.607	0.683	0.864	0.080	0.050	0.137		
2.0–4.8				2.590	1.205	0.512	0.648	0.080	0.050	0.102		
3.5–8.2				1.480	0.688	0.292	0.370	0.064	0.030	0.058		
4.75–11.0				1.090	0.506	0.215	0.272	0.050	0.030	0.043		
7.5–18.0				0.691	0.321	0.136	0.173	0.050	0.020	0.027		
11.0–26.5‡				0.471	0.219	0.093	0.118	0.040	0.015	0.019		
18.0–40.0‡				0.288	0.134	0.057	0.072	0.040	0.015	0.011		
Bandwidth 3.6:1												
0.108–0.39	0.092	128.37	0.401	34.638	14.894	2.904	8.660	—	—	0.581	0.0014	28 830.
0.27–0.97	0.229	51.572	0.999	13.916	5.984	1.167	3.479	—	—	0.233	0.0055	4 653.
0.39–1.4	0.331	35.680	1.444	9.628	4.140	0.807	2.407	0.125	0.050	0.161	0.0097	2 227.
0.97–3.5	0.822	14.367	3.587	3.877	1.667	0.325	0.969	0.080	0.050	0.065	0.0378	361.2
1.4–5.0	1.186	9.958	5.176	2.687	1.155	0.225	0.672	0.080	0.050	0.045	0.0656	173.5
3.5–12.4	2.966	3.982	12.944	1.074	0.462	0.090	0.269	0.050	0.030	0.018	0.259	27.74
5.0–18.0	4.237	2.787	18.490	0.752	0.323	0.063	0.188	0.050	0.020	0.013	0.443	13.59
12.4–40.0‡	10.508	1.124	45.857	0.303	0.130	0.025	0.076	0.040	0.015	0.005	1.730	2.210

* Copper.

† Based on breakdown of air—15 000 volts per cm (safety factor of approx 2 at sea level). Corner radii considered.

‡ Figure 14B in these frequency ranges only.

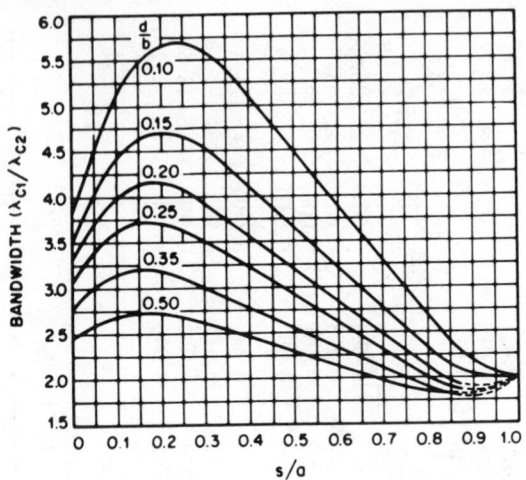

Fig. 17—Bandwidth curves, single-ridged guide. *From S. Hopfer, "The Design of Ridged Waveguides," Transactions of the I.R.E., vol. MTT-3, no. 5, fig. 9; 1955.*

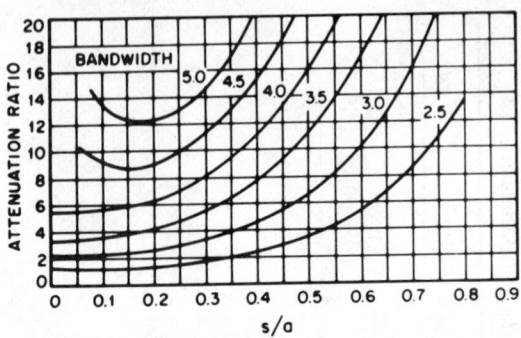

Fig. 19—Attenuation ratio, parametric in bandwidth, single-ridged guide. *From S. Hopfer, "The Design of Ridged Waveguides," Transactions of the I.R.E., vol. MTT-3, no. 5, fig. 11; 1955.*

cabinet is specified and/or designed for that type of service. Most flexible-waveguide structures are susceptible to cracking under these conditions if flexure is often repeated.

Flexible waveguide is available in many different forms. It may be made from flat ribbons, wound on a rectangular mandrel, with the edges convoluted or folded in and interlocked. The convoluted guide may be soldered or unsoldered, since the bending and twisting results from a flexure of each turn and not from a relative sliding as in the case of the interlocked guide. If soldered, it is more difficult to flex and essentially loses twist capability.

Corrugated flexible guide may be made by properly shaping thin-wall seamless rectangular tubing, or by bending and soldering corrugated sheet metal (with due consideration of current flow so that a low-loss joint results).

A bellows-type guide is produced from a group of radial chokes in tandem configuration and made of a flexible alloy.

Vertebral guide is made from a tandem chain of choke–cover sections contained within a neoprene or rubber jacket.

In general, all types except the seamless corrugated waveguide should be jacketed with neoprene or rubber. The unsoldered convoluted, interlocking, and vertebral guide must be jacketed to be pressurized.

Table 7 gives the properties of soldered convoluted flexible waveguide.

The wide variety of manufacturing techniques, the jacketing material and thickness, the length dependence, and other characteristics make it impossible to limit the ($\pm$) stretch, twist, and centerline displacement ranges of flexible waveguide. These are usually described in terms of maximum acceptable VSWR or loss of the section as a function of the ($\pm$) stretch, twist, or dis-

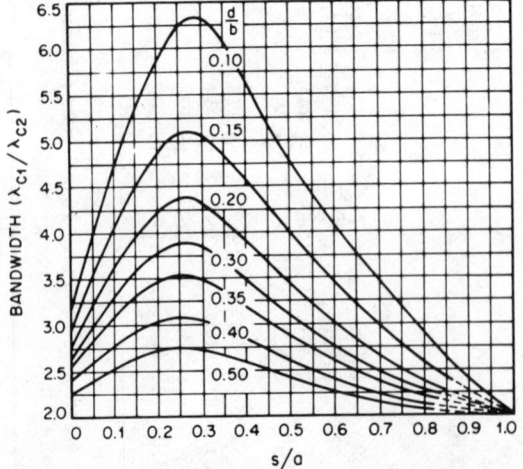

Fig. 18—Bandwidth curves, double-ridged guide. *From S. Hopfer, "The Design of Ridged Waveguides," Transactions of the I.R.E., vol. MTT-3, no. 5, fig. 10; 1955.*

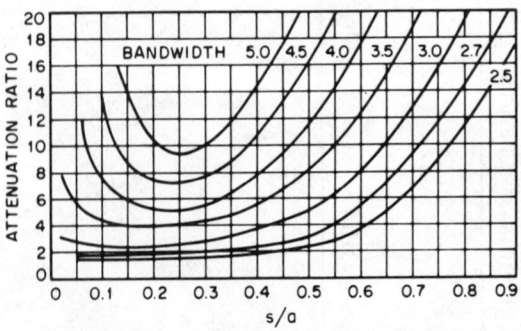

Fig. 20—Attenuation ratio, parametric in bandwidth, double-ridged guide. *From S. Hopfer, "The Design of Ridged Waveguides," Transactions of the I.R.E., vol. MTT-3, no. 5, fig. 12; 1955.*

TABLE 7—PROPERTIES OF SOLDERED CONVOLUTED FLEXIBLE WAVEGUIDES. *From MIL–HDBK–216, "RF Transmission Lines and Fittings," 4 January 1962.*

Dimensions (inches)		Minimum Bending Radii (inches)				Equivalent Rectangular Waveguide Type	Weight (lb/ft)	Nominal Attenuation (dB/100 ft)	Nominal Power Rating (MW)	Maximum Operating Pressure (psi)
		Standard Molded Assembly		Unjacketed or Special Molded Assembly						
Inside	Outside	H Plane	E Plane	H Plane	E Plane					
6.500×3.250	6.660×3.410	27	13	17	$8\frac{1}{2}$	RG-69/U	2.88	0.50	10	15
4.300×2.150	4.460×2.310	18	9	$11\frac{1}{2}$	$5\frac{3}{4}$	RG-104/U	1.46	0.80	8.0	20
2.840×1.340	3.000×1.500	14	7	9	$4\frac{1}{2}$	RG-48/U	0.530	1.50	2.0	30
1.872×0.872	2.000×1.000	8	4	5	$2\frac{1}{2}$	RG-49/U	0.332	3.0	1.0	30
1.372×0.622	1.500×0.750	5	$2\frac{1}{2}$	$3\frac{1}{4}$	$1\frac{5}{8}$	RG-50/U	0.266	4.7	0.50	30
1.122×0.497	1.250×0.625	$3\frac{1}{2}$	$1\frac{3}{4}$	$2\frac{1}{4}$	$1\frac{1}{8}$	RG-51/U	0.200	5.7	0.40	45
0.900×0.400	1.000×0.500	3	$1\frac{1}{2}$	2	1	RG-52/U	0.112	9.0	0.25	60
0.622×0.311	0.702×0.391	3	$1\frac{1}{2}$	2	1	RG-91/U	0.085	15.0	0.20	60
0.420×0.170	0.500×0.250	$2\frac{1}{2}$	$1\frac{1}{4}$	$1\frac{1}{2}$	$\frac{3}{4}$	RG-53/U	0.050	29.0	0.10	60
0.280×0.140	0.360×0.220	$2\frac{1}{2}$	$1\frac{1}{4}$	$1\frac{1}{2}$	$\frac{3}{4}$	RG-96/U	0.039	35.0	0.05	60

TABLE 8—COMPOSITION AND CONDUCTIVITY OF WAVEGUIDE MATERIAL. *From R. M. Cox and W. E. Rupp, "Fight Waveguide Losses 5 Ways," Microwaves, vol. 5, no. 8, p. 34, Hayden Microwave Corp.; August 1966.*

| Material | Composition (%) | | | | | | % Conductivity* |
	Cu	Zn	P	Ag	Al	Mg	
Copper (oxygen free)	99.95† min	—	—	—	—	—	97.6 min
Copper DLP (deoxidized, low phosphorus)	99.90† min	—	0.004–0.012	—	—	—	96.1 min
Commercial bronze	89–91	9–11	—	—	—	—	44.2 min
Silver (fine)	0.08 max	—	—	99.90 min	—	—	100.0 min
Coin silver	9–10.4	0.06	—	89.6–91.0	—	—	82.0 min
Aluminum 1100	0.2	0.10	—	—	99.0 min	—	59.5 min
Aluminum 6061	0.15–0.40	0.25	—	—	95	0.8–1.2	40.0 min
Magnesium	—	0.05	0.6–1.4	—	2.5–3.5	94.0	37.5‡

* International Annealed Copper Standard.

† Any silver present is counted in the copper content.

‡ MIL-HDBK-216, Military Standardization Handbook, "RF Transmission Lines and Fittings," 4 January 1962.

placement, and are best established on advice from the manufacturer selected.

DIELECTRIC-ROD WAVEGUIDES

This type of waveguide has applications in antenna structures, laser devices, fiber optics, and millimetric-wave techniques.

The field structures for the nonradiating modes fall into two classes—circularly symmetric and nonsymmetric modes. The cutoff wavelengths (λ_c) for these modes are (for the $E_{0,m}$ and $H_{0,m}$ modes)*

$$\lambda_c = \pi d (\epsilon - 1)^{1/2} / j_{0,m}$$

where d = rod diameter, ϵ = relative dielectric constant, and $j_{0,m}$ = mth root of $J_0(X)$.

Analysis of the field equations reveals the necessary coexistence of an E wave with an H wave to obtain a nonsymmetric field structure.†

These modes are described as HE if the H mode is predominant, and as EH if the E mode predominates. The special case of the $HE_{1,1}$ mode, referred to as the "dipole" mode because of the resemblance of the transverse-electric-field pattern to that of the electrostatic dipole, is of special interest because it has zero cutoff frequency.‡

* H. M. Barlow and J. Brown, "Radio Surface Waves," Oxford at the Clarendon Press; 1962: p. 71.

† D. G. Kieley, "Dielectric Aerials," Methuen's Monographs on Physical Subjects, John Wiley, New York; 1953: pp. 7–29.

‡ H. M. Barlow and J. Brown, "Radio Surface Waves," Oxford at the Clarendon Press; 1962: p. 69.

Figure 21 describes the relation between λ/λ_0 and d/λ_0 for rods of different ϵ (λ = operating wavelength, λ_0 = free-space wavelength, d = rod diameter), and the field structure is shown in Fig. 22.

The attenuation of the $HE_{1,1}$ mode, for material having relatively low loss,* is found from

$$\alpha = 27.3 (\epsilon/\lambda_0) R \tan\delta \text{ (dB/cm)}$$

where ϵ = relative dielectric constant, $\tan\delta$ = loss tangent of dielectric, and R = attenuation factor (dimensionless). Note that this closely resembles the expression for TEM mode propagation in a low-loss dielectric medium,† given by

$$\alpha = [27.3 (\epsilon)^{1/2}/\lambda] \tan\delta \text{ (dB/cm)}.$$

For d/λ_0 larger than 0.8, $R \approx [1/(\epsilon)^{1/2}]$. For other values see Elsasser*.

The $HE_{1,1}$ mode waveguide is impractical for many frequencies in the UHF and lower bands because of size of rod and field spread adjacent to the rod outside of the dielectric. At much higher frequencies, there is another order of difficulty in obtaining long fibers of suitable material. Thus these guides, used as unsupported optical fibers, are not suited for transmission over long distances but may be found useful as imaged lines, wherein the rod cross section is a semicircle, with the flat

* W. M. Elsasser, "Attenuation in a Dielectric Rod," *Journal of Applied Physics*, vol. 20, pp. 1193–1196; December 1949.

† G. L. Ragan, "Microwave Transmission Circuits," First Edition, McGraw-Hill Book Co., New York; 1948: p. 29.

side resting on a highly conductive, large copper sheet image plane.*

WAVEGUIDE LOSSES

Hollow, enclosed single-conductor waveguides, propagating in the interior space, exhibit losses via dissipation in the waveguide walls and the dielectric material filling the space, leakage through the walls and connections to the guide, and localized power absorption (and heating) at the connections (flanges) because of poor contact or fabrication. The following discussion assumes that the dielectric is air, with zero loss tangent, and that the depth of penetration into the walls is very much less than the wall thickness, so that no appreciable wall leakage occurs.

Waveguide Material and Modes

Figure 23 shows attenuation as a function of percent conductivity for WR–112 waveguide at

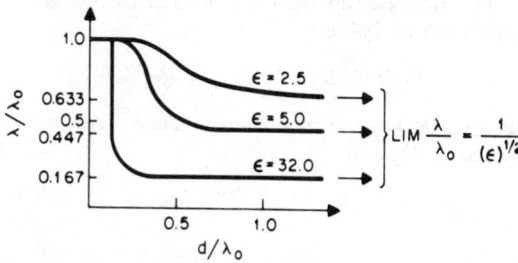

Fig. 21—Wavelength of $HE_{1,1}$ mode as a function of d/λ_0. *From D. G. Kieley, "Dielectric Aerials," Methuen's Monographs on Physical Subjects, p. 27, John Wiley & Sons, New York; 1953.*

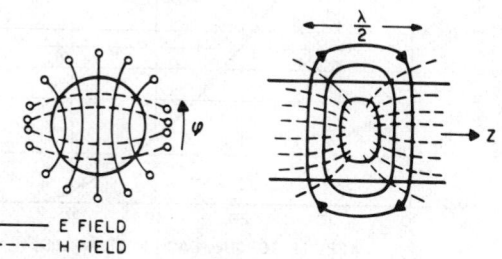

—— E FIELD
---- H FIELD

Fig. 22—$HE_{1,1}$ mode, field distribution. *From D. G. Kieley, "Dielectric Aerials," Methuen's Monographs on Physical Subjects, p. 27, John Wiley & Sons, New York; 1953.*

* "Advances in Microwaves," editor L. Young, vol. 1, Academic Press, New York; 1966: pp. 75–113 ("Optical Waveguides," by A. E. Karbowiak). Also E. Sobel, "Quasi-Optical Transmission Methods for Millimeter Waves," Northeast Electronic Research and Engineering Meeting—1962 (*NEREM Record*), First Edition, published by Louis Winner, New York; November 1962: pp. 152–153.

9.0 GHz.* Table 8 relates material composition and percent conductivity.

To obtain the lowest attenuation where a choice of waveguide size and cross section is possible, consideration should be given to mode selection. Figure 24 shows the relation between various waveguides for different modes and cross sections over a broad frequency band. Figure 25 relates the attenuation rates of several circular and square guides (in the $TE_{0,1}$, $TE_{1,1}$, and $TE_{1,0}$ modes) over a limited frequency band.*

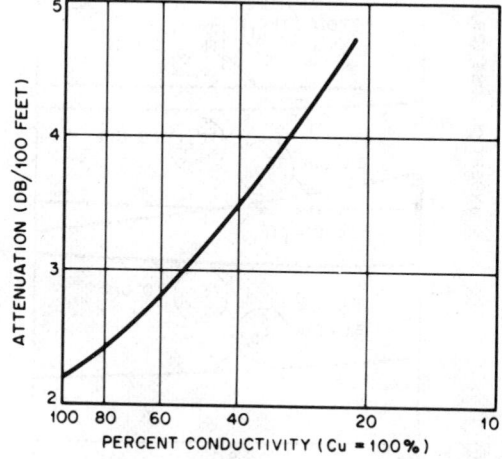

Fig. 23—Attenuation as a function of percent conductivity for WR–112 waveguide at 9.0 GHz.

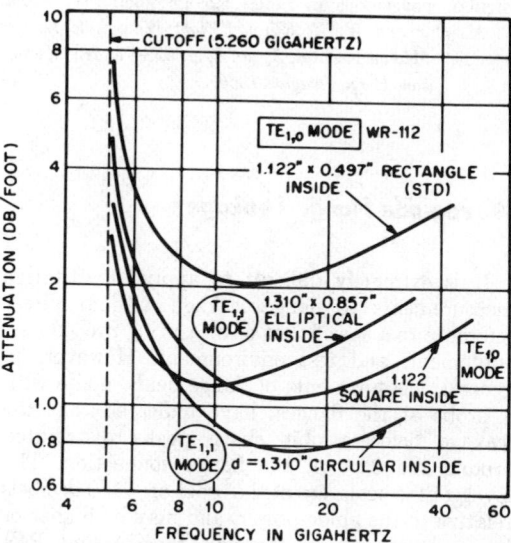

Fig. 24—Attenuation as a function of frequency for various waveguides. *From R. M. Cox and W. E. Rupp, "Fight Waveguide Losses 5 Ways," Microwaves, vol. 5, no. 8, p. 36, Hayden Microwave Corp.; August 1966.*

* R. M. Cox and W. E. Rupp, "Fight Waveguide Losses 5 Ways," *Microwaves*, vol. 5, no. 8, pp. 32–40, Hayden Microwave Corp.; August 1966.

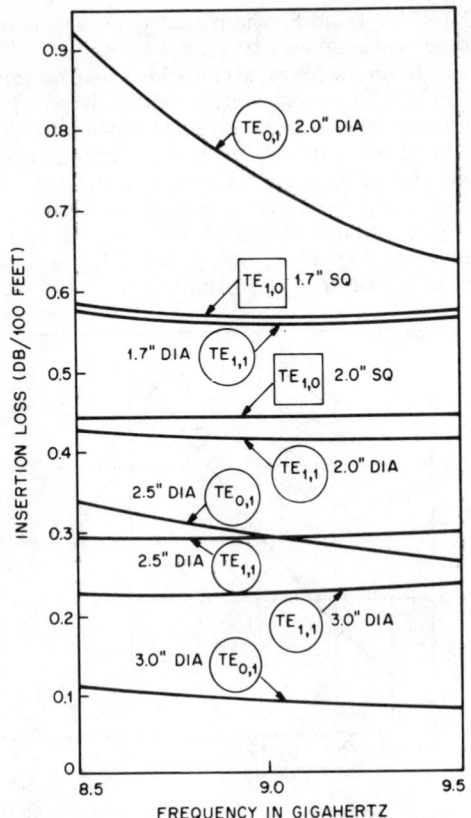

Fig. 25—Attenuation curves for various square and circular waveguides in range 8.5–9.5 gigahertz. *From R. M. Cox and W. E. Rupp, "Fight Waveguide Losses 5 ways," Microwaves, vol. 5, no. 8, p. 37, Hayden Microwave Corp.; August 1966.*

Waveguide Flange Leakage*

It is extremely difficult to apply quantitative measurements to flange leakage without direct reference to a specific set of measuring procedures, equipment, and test environment. However, in general, measurements of flange fields, made with a probe at the flanges, have indicated that the leakage fields exhibit sharp peaks distributed around the edge of the flange connection. The levels of the peaks are of the order of -130 decibels relative to the guide power, and may be higher or lower depending on the bolt tension and RFI gaskets employed.

*S. Galagan, "Electrical Characteristics of Waveguide Seals with EMI Supplement," unpublished report prepared for the Parker Seal Co., Culver City, Cal.; January 1964.

Flange Resistance and Bolt Torque†

The equivalent series resistance of a die-cast sealed rectangular aluminum S-band flange pair as a function of bolt torque (each bolt) is shown in Fig. 26, as an indication of the importance of proper bolt tightening at a flange connection. Table 9 gives the bolt torque for several bolt sizes to meet the recommended value of about 1000 lb/linear inch of flange connection, which is estimated to give a satisfactory waveguide seal for high-power applications.

Figure 27 shows typical flange resistance as a function of frequency for a UG-53/U plus UG-54/U choke–cover combination, with and without a mesh gasket seal. A combination of two cover flanges, without seals, may have about 10 times the choke–cover resistance.

Flange Insertion Loss*

The relationship between the flange resistance and insertion loss is

$$L(\text{decibels}) = 10 \log (1 - R_F/Z_0)$$

where $R_F =$ radio-frequency flange resistance (measured), and

$$Z_0 = 593b/a \left[1 - (f_c/f)^2 \right]^{-1/2} \text{ (ohms).‡}$$

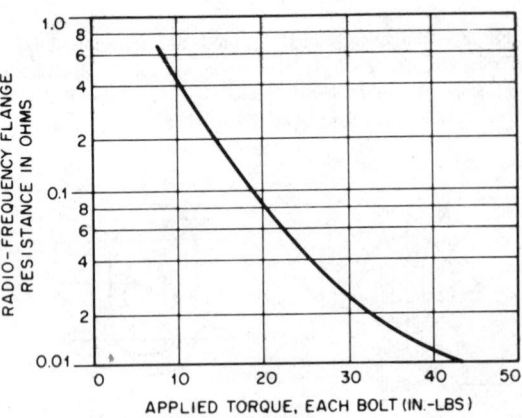

Fig. 26—Typical flange radio-frequency resistance as a function of bolt torque. *From Handbook Catalog W5460, "Waveguide Flange and EMI Sealing," © 1967, Parker Seal Co., Culver City, California.*

† Handbook Catalog No. W5460, "Waveguide Flange and EMI Sealing," Parker Seal Company, Culver City, California; Copyright 1967.

‡ This is the characteristic impedance, defined as the maximum transverse voltage divided by the total longitudinal current.

The approximate value of flange insertion loss calculated from the above equation may be scaled to other flange sizes and frequencies by

$$R_{F1}/R_{F2} \sim (A_1/A_2)(f_1/f_2)^{1/2}$$

where A = flange area, f = frequency, and the subscripts refer to the two conditions.

Losses and Noise Temperature

In radio telescopes, satellite ground antennas, and other loss-sensitive waveguide systems, the noise-temperature contributions must be controlled and accounted for in the system design. The waveguide losses may be converted to noise temperature* by

$$T = T_R(1 - a)$$

where T = temperature in degrees Kelvin, T_R = temperature of the lossy insert in degrees Kelvin, and a = power transmission coefficient. For A (decibels) = $-10 \log_{10} a$

$$T = T_R(1 - 10^{-A/10}).$$

Figure 28 relates the noise temperature added to a lossless waveguide system at 290°K by inserting a 290°K pad having A decibels of loss between the measuring point (or input to the receiver) and a 0°K load, as a function of the insertion loss A (decibels).

Some typical values of insertion loss measured for a specific configuration are:

23-decibel directional coupler*	0.03 decibel
Flexible waveguide*	0.023 decibel
Flanges (UG-53/U plus UG-53/U)†	0.0017 decibel

WAVEGUIDE CIRCUIT ELEMENTS‡

Just as at low frequencies, it is possible to shape metallic or dielectric pieces to produce local concentrations of magnetic or electric energy within a waveguide and thus produce what are, essentially,

* A. J. Giger, S. Pardee, Jr., and P. R. Wickliffe, Jr., "The Ground Transmitter and Receiver," *Bell System Technical Journal*, vol. 42, no. 4, part 1, p. 1096; July 1963.

† Handbook Catalog No. W5460, "Waveguide Flange and EMI Sealing," Parker Seal Company, Culver City, California; Copyright 1967.

‡ C. G. Montgomery, R. H. Dicke, and E. M. Purcell, "Principles of Microwave Circuits," McGraw-Hill Book Company, New York; 1948: Chapters 1 and 6. Also N. Marcuvitz, "Waveguide Handbook," McGraw-Hill Book Company, New York; 1951.

TABLE 9—RECOMMENDED TORQUE TABLE. *From Handbook Catalog No. W5460, "Waveguide Flange and EMI Sealing,"* © *1967, Parker Seal Co., Culver City, Cal.*

Screw Size	Threads per Inch	Recommended Torque (inch-lb)	Tension* (lb)
No. 4	40	4.5	235
	80	5.5	280
No. 6	32	8.5	360
	40	10	410
No. 8	32	18	625
	36	20	685
No. 10	24	23	705
	32	32	940
$\frac{1}{4}''$	20	80	1800
	28	100	2200
$\frac{5}{16}''$	18	140	2540
	24	150	2620
$\frac{3}{8}''$	16	250	3740
	24	275	3950
$\frac{7}{16}''$	14	400	4675
	20	425	4700
$\frac{1}{2}''$	13	550	6110
	20	575	6140

* Tension (lb) = torque (inch-lb)/0.2 × diameter of bolt (in.).

lumped inductances or capacitances over a limited frequency bandwidth.

This behavior as a lumped element will be evident only at some distance from the obstacle in the guide, since the fields in the immediate vicinity are disturbed.

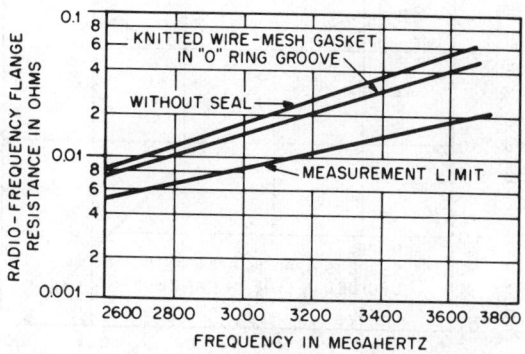

Fig. 27—Typical flange resistance as a function of frequency for choke–cover combination (UG–53/U plus UG–54/U, measured). *From Handbook Catalog W5460, "Waveguide Flange and EMI Sealing,"* © *1967, Parker Seal Co., Culver City, California.*

Capacitive elements are formed from electric-field concentrating devices, such as screws or thin diaphragms inserted partially along electric-field lines. These are susceptible to breakdown under high power. Figure 29 shows the relative susceptance B/Y_0 for symmetrical and asymmetrical diaphragms for small b/λ_g.

A common form of shunted lumped inductance is the diaphragm. Figures 30 and 31 show the relative susceptance B/Y_0 for symmetrical and asymmetrical diaphragms in rectangular waveguides. These are computed for infinitely thin diaphragms. Finite thicknesses result in an increase in B/Y_0.

Another form of shunt inductance that is useful because of mechanical simplicity is a round post completely across the narrow dimension of a rectangular guide (for $TE_{1,0}$ mode). Figure 32 gives the normalized values of the elements of the equivalent 4-terminal network for several post diameters.

Frequency dependence of waveguide susceptances may be given approximately as

Inductive $= B/Y_0 \propto \lambda_g$

Capacitive $= B/Y_0 \propto 1/\lambda_g$ (distributed)

$$= B/Y_0 \propto \lambda_g/\lambda^2 \text{ (lumped)}.$$

Distributed capacitances are found in junctions and slits, whereas tuning screws act as lumped capacitances.

HYBRID JUNCTIONS*

The hybrid junction is illustrated in various forms in Fig. 33. An ideal junction is characterized by the fact that there is no direct coupling between arms 1 and 4 or between 2 and 3. Power flows

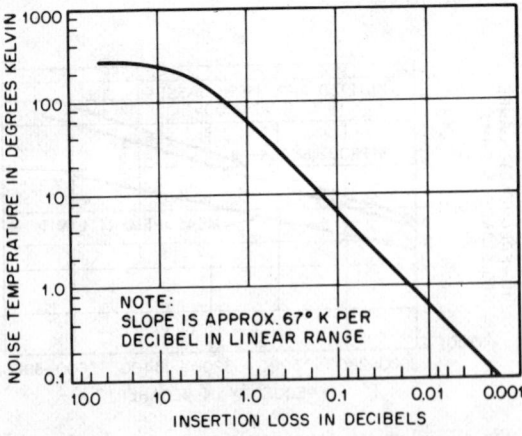

Fig. 28—Noise temperature as a function of insertion loss (290°K ambient).

* C. G. Montgomery, R. H. Dicke, and E. M. Purcell, "Principles of Microwave Circuits," McGraw-Hill Book Company, New York; 1948: Chapter 9.

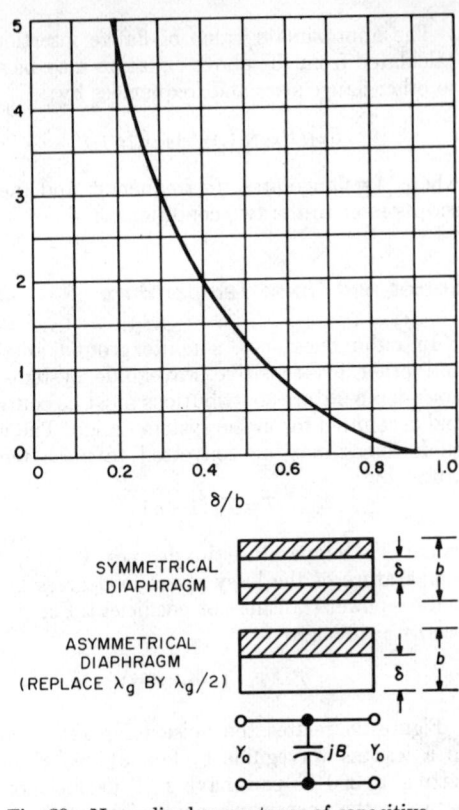

Fig. 29—Normalized susceptance of capacitive diaphragms.

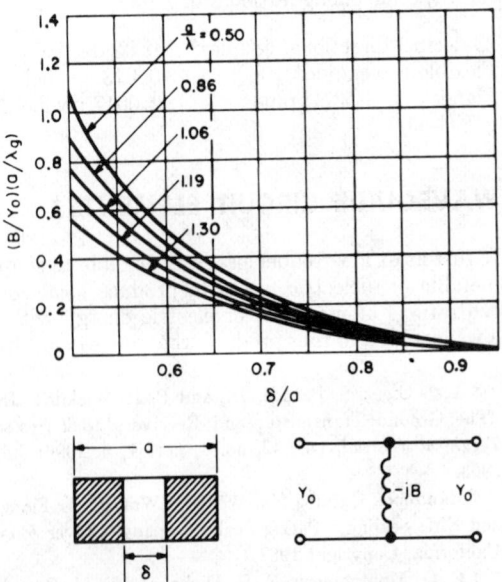

Fig. 30—Normalized susceptance of a symmetrical inductive diaphragm. *Reprinted from "Microwave Transmission Circuits," by George L. Ragan, 1st ed., 1948; by permission, McGraw-Hill Book Co., N.Y.*

TABLE 10—EQUATIONS FOR A RIGHT-CIRCULAR-CYLINDRICAL CAVITY.

Mode	λ_0 Resonant Wavelength	Q (all dimensions in same units)
$TM_{0,1,1}(E_0)$	$\dfrac{4}{[(1/h)^2+(2.35/a^2)]^{1/2}}$	$(\lambda_0/\delta)\,(a/\lambda_0)[1+(a/2h)]^{-1}$
$TE_{0,1,1}(H_0)$	$\dfrac{4}{[(1/h)^2+(5.93/a^2)]^{1/2}}$	$(\lambda_0/\delta)\,(a/\lambda_0)\left[\dfrac{1+0.168(a/h)^2}{1+0.168(a/h)^3}\right]$
$TE_{1,1,1}(H_1)$	$\dfrac{4}{[(1/h)^2+(1.17/a^2)]^{1/2}}$	$(\lambda_0/\delta)\,(h/\lambda_0)\left[\dfrac{2.39h^2+1.73a^2}{3.39(h^3/a)+0.73ah+1.73a^2}\right]$

from 1 to 4 only by virtue of reflections in arms 2 and 3. Thus, if arm 1 is excited, the voltage arriving at arm 4 is

$$E_4=\tfrac{1}{2}E_1[\Gamma_2\exp(j2\theta_2)-\Gamma_3\exp(j2\theta_3)]$$

and the reflected voltage in arm 1 is

$$E_{r1}=\tfrac{1}{2}E_1[\Gamma_2\exp(j2\theta_2)+\Gamma_3\exp(j2\theta_3)]$$

where E_1 is the amplitude of the incident wave, Γ_2 and Γ_3 are the reflection coefficients of the terminations of arms 2 and 3, and θ_2 and θ_3 are the respective distances of the terminations from the junctions. In the case of the rings, θ is the distance between the arm-and-ring junction and the termination.

If the decoupled arms of the hybrid junction are independently matched and the other arms are terminated in their characteristic impedances, then all four arms are matched at their inputs.

RESONANT CAVITIES

A cavity enclosed by metal walls has an infinite number of natural frequencies at which resonance will occur. One of the more common types of

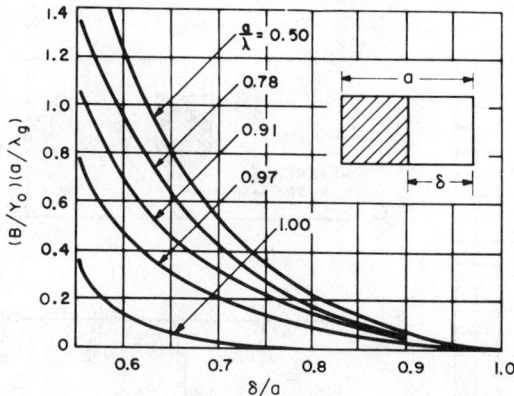

Fig. 31—Normalized susceptance of an asymmetrical inductive diaphragm. *Reprinted from "Microwave Transmission Circuits," by George L. Ragan, 1st ed., 1948; by permission, McGraw-Hill Book Co., N.Y.*

cavity resonators is a length of transmission line (coaxial or waveguide) short-circuited at both ends.

Resonance occurs when

$$2h=l(\lambda_g/2)$$

where l is an integer and

$2h=$ length of the resonator

$\lambda_g=$ guide wavelength in resonator

$$=\lambda/[\epsilon-(\lambda/\lambda_c)^2]^{1/2}$$

where $\lambda=$ free-space wavelength, $\lambda_c=$ guide cutoff wavelength, and $\epsilon=$ relative dielectric constant of medium in cavity.

For $TE_{m,n}$ or $TM_{m,n}$ waves in a rectangular cavity with cross section a, b

$$\lambda_c=2/[(m/a)^2+(n/b)^2]^{1/2}$$

where m and n are integers.

For $TE_{m,n}$ waves in a cylindrical cavity

$$\lambda_c=2\pi a/U_{m,n}'$$

where a is the guide radius, and $U_{m,n}'$ is the mth root of the equation $J_n'(U)=0$.

For $TM_{m,n}$ waves in a cylindrical cavity

$$\lambda_c=2\pi a/U_{m,n}$$

where a is the guide radius, and $U_{m,n}$ is the mth root of the equation $J_n(U)=0$.

For TM waves, $l=0, 1, 2, \cdots$.

For TE waves, $l=1, 2, \cdots$, but not 0.

Rectangular Cavity of Dimensions a, b, 2h

$$\lambda=2/[(l/2h)^2+(m/a)^2+(n/b)^2]^{1/2}$$

where only one of l, m, n may be zero.

Cylindrical Cavities of Radius a and Length 2h

$$\lambda=1/[(l/4h)^2+(1/\lambda_c)^2]^{1/2}$$

where λ_c is the guide cutoff wavelength.

Spherical Resonators of Radius a

$$\lambda = 2\pi a / U_{m,n} \text{ for a TE wave}$$

$$\lambda = 2\pi a / U_{m,n}' \text{ for a TM wave.}$$

Values of $U_{m,n}$:

$$U_{1,1} = 4.5, \quad U_{2,1} = 5.8, \quad U_{1,2} = 7.64.$$

Values of $U_{m,n}'$:

$$U_{1,1}' = 2.75 = \text{lowest-order root.}$$

Additional Cavity Equations

Note that resonant modes are characterized by three subscripts in the mode designations of Tables 10 and 11 and Fig. 34.

Figure 34 is a mode chart for a right-circular-cylindrical resonator, showing the distribution of resonant modes with frequency as a function of cavity shape. With the aid of such a chart, the various possible resonances can be predicted as the length $(2h)$ of the cavity is varied by a movable piston.

Effect of Temperature and Humidity on Cavity Tuning

The resonant frequency of a cavity changes with temperature and humidity (due to changes in dielectric constant of the atmosphere) and with thermal expansion of the cavity. A homogeneous cavity made of one kind of metal will have a thermal-tuning coefficient equal to the linear coefficient of expansion of the metal, since the fre-

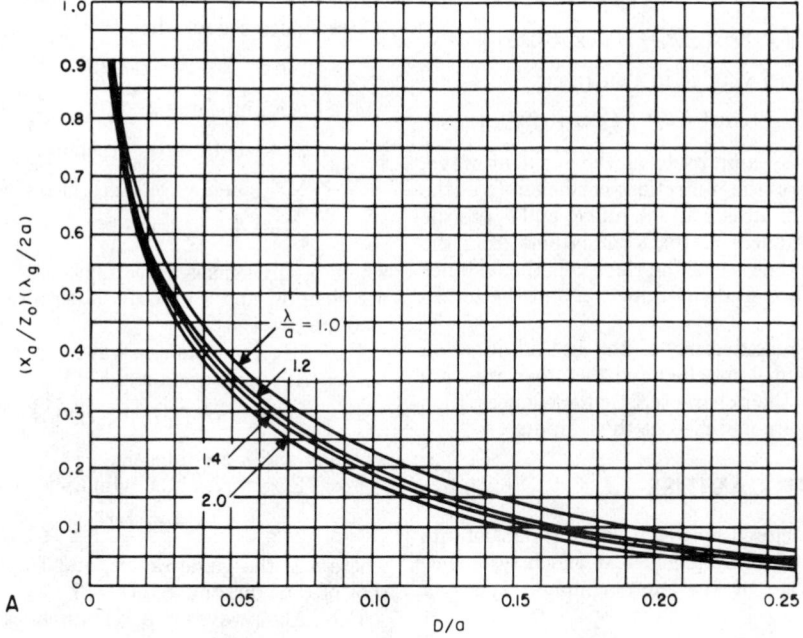

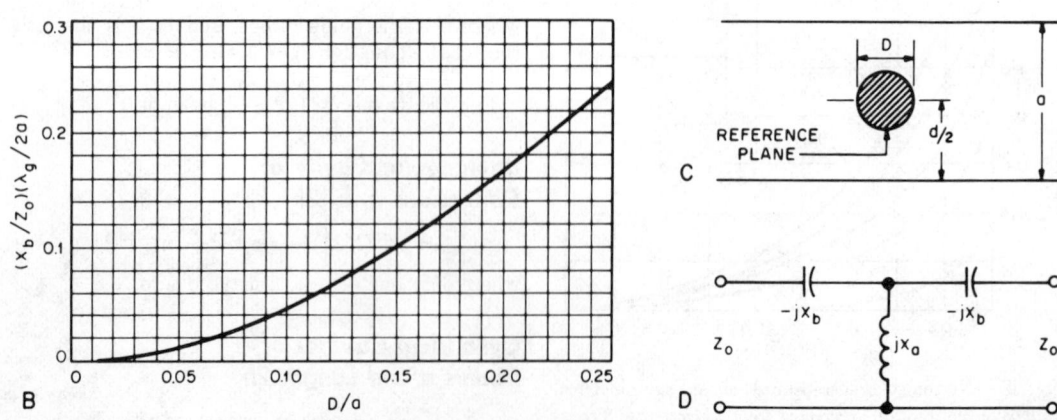

Fig. 32—Equivalent circuit for inductive cylindrical post. A—Shunt reactance characteristic. B—Series reactance characteristic.

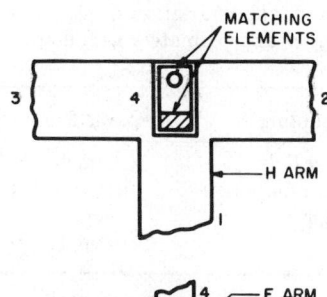

MATCHING ELEMENTS

H ARM

E ARM

WAVEGUIDE HYBRID JUNCTION (MAGIC T)

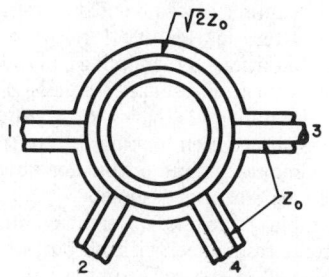

SHUNT COAXIAL HYBRID RING

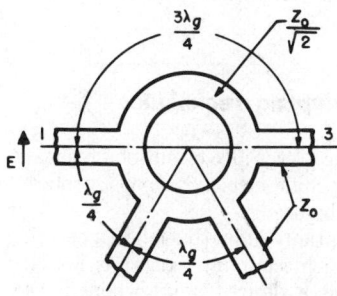

E-PLANE WAVEGUIDE HYBRID RING

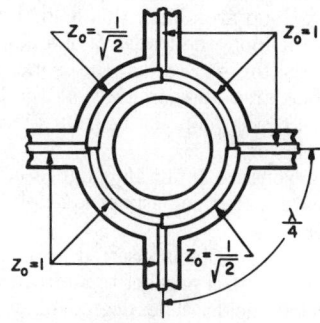

SYMMETRICAL COAXIAL HYBRID

Fig. 33—Hybrid junctions.

quency is inversely proportional to the linear dimension of the cavity.

Metal	Linear Coefficient of Expansion/°C
Yellow brass	20
Copper	17.6
Mild steel	12
Invar	1.1

$\times 10^{-6}$

The relative dielectric constant of air (vacuum = 1) is given by

$$k_e = 1 + 210 \times 10^{-6}(P_a/T) + 180$$
$$\times 10^{-6}[1 + (5580/T)](P_w/T)$$

where P_a and P_w are partial pressures of air and water vapor in millimeters of mercury, and T is the absolute temperature. Figure 35 is a nomograph showing change of cavity tuning relative to conditions at 25 degrees Celsius and 60 percent relative humidity (expansion is not included).

Coupling to Cavities and Loaded Q

Near resonance, a cavity may be represented as a simple shunt-resonant circuit, characterized by a loaded $Q = Q_l$, where $1/Q_l = (1/Q_0) + (1/Q_{\text{ext}})$, Q_0 is the unloaded Q characteristic of the cavity

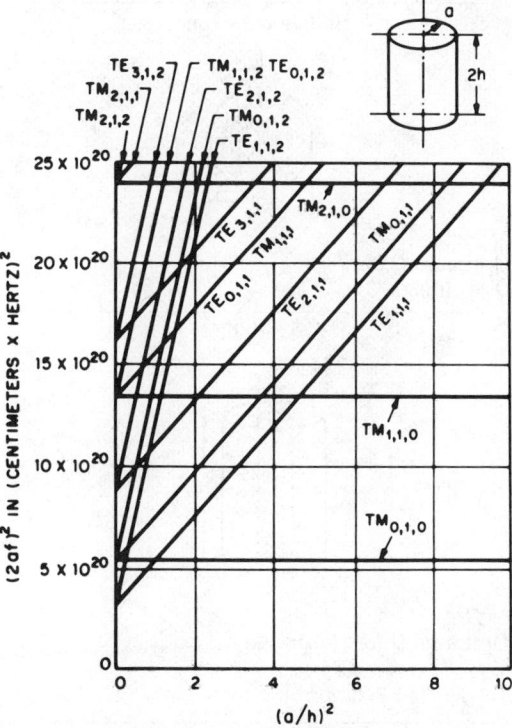

Fig. 34—Mode chart for right-circular-cylindrical cavity.
Reprinted from "Techniques of Microwave Measurements," by Carol G. Montgomery, 1st ed., 1947; by permission, McGraw-Hill Book Co., N.Y.

Table 11—Characteristics of Various Types of Resonators.

Square prism $TE_{1,0,1}$

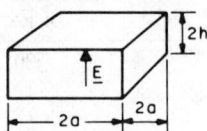

$\lambda_0 = 2.83a$
$Q = (0.353\lambda/\delta)[1 + (0.177\lambda/h)]^{-1}$

Circular cylinder $TM_{0,1,0}$

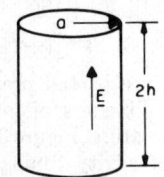

$\lambda_0 = 2.61a$
$Q = (0.383\lambda/\delta)[1 + (0.192\lambda/h)]^{-1}$

Sphere

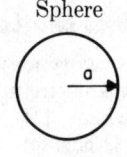

$\lambda_0 = 2.28a$
$Q = 0.318(\lambda/\delta)$

Sphere with cones

$\lambda_0 = 4a$
Optimum Q for $\theta = 34°$
$Q = 0.1095(\lambda/\delta)$

Coaxial TEM

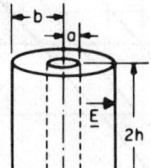

$\lambda_0 = 4h$
Optimum Q for $(b/a) = 3.6$
　　$(Z_0 = 77$ ohms$)$
　　$\lambda/[4\delta + 7.2(h\delta/b)]$

Skin depth in meters $= \delta = (10^7/2\pi\omega\sigma)^{1/2}$, where $\sigma =$ conductivity of wall in mhos/meter, and $\omega = 2\pi \times$ frequency.

itself, and $1/Q_{ext}$ is the loading due to the external circuits. The variation of Q_{ext} with size of the coupling is approximately as follows:

Coupling	$1/Q_{ext}$ is Proportional to
Small round hole	$(\text{diameter})^6$
Symmetrical inductive diaphragm	$(\delta)^4$ (see Fig. 30)
Small loop	$(\text{diameter})^4$

Equations for Coupling Through a Cavity

Table 12 summarizes some of the useful relationships in a 4-terminal cavity (transmission type) for three conditions of coupling: matched input (input resistance at resonance equals Z_0 of input line), equal coupling ($1/Q_{in} = 1/Q_{out}$), and matched output (resistance seen looking into output terminals at resonance equals output-load resistance). A matched generator is assumed.

In the table, g_c' is the apparent conductance of the cavity at resonance, with no output load; the transmission T is the ratio of the actual output-circuit power delivered to the available power from the matched generator. The loaded Q is Q_l and unloaded Q is Q_0.

Cavity Coupling Techniques*

To couple power into or out of a resonant cavity, either waveguide or coaxial loops, probes, or apertures may be used.

The essentially inductive loop (a certain amount of electric-field coupling exists) is inserted in the resonator at a desired point where it can couple to a strong magnetic field. The degree of coupling may be controlled by rotating the loop so that more or less loop area links this field. For a fixed location of the loop, the loaded Q of a loop-coupled coaxial resonator varies as the square of the effective loop area and inversely as the square of the distance of the loop center from the resonator axis of revolution.

The off-resonance input impedance of the loop is low, a feature that sometimes is helpful in series connections.

The capacitive probe is inserted in the resonator at a point where it is parallel to and can couple to strong electric fields. The degree of coupling is controlled by adjusting the length of the probe relative to the electric field.

* C. G. Montgomery, R. H. Dicke, and E. M. Purcell, "Principles of Microwave Circuits," McGraw-Hill Book Company, New York; 1948: Chapter 7.

TABLE 12—COUPLING THROUGH A CAVITY.

	Matched Input	Equal Coupling	Matched Output
Input standing-wave ratio	1	$1+g_c'=2(T^{-1/2}-1)$	$1+2g_c'$
Transmission ratio$=T$	$1-g_c'=1-2\rho$	$(1+g_c'/2)^{-2}=(1-\rho)^2$	$(1+g_c')^{-1}=1-2\rho$
$Q_l/Q_0=\rho$	$\frac{1}{2}g_c'=\frac{1}{2}(1-T)$	$[g_c'/(2+g_c')]=1-(T^{1/2})$	$[g_c'/2(1+g_c')]=\frac{1}{2}(1-T)$

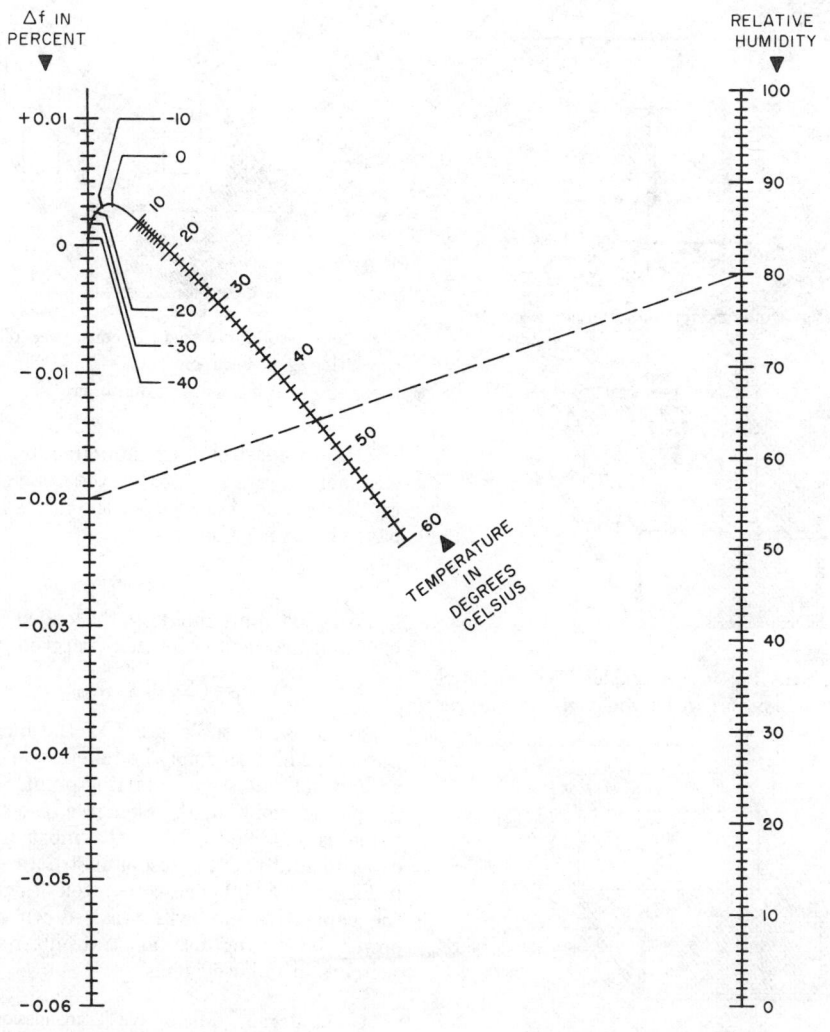

Fig. 35—Effect of temperature and humidity on cavity tuning. *Reprinted from "Techniques of Microwave Measurements," by Carol G. Montgomery, 1st ed., 1947; by permission, McGraw-Hill Book Co., N.Y.*

The off-resonance input impedance of the probe-coupled resonator is high; this property is useful in parallel connections.

Aperture coupling is suitable when coupling waveguides to resonators or in coupling resonators together. In this case, the aperture must be located and shaped to excite the proper propagating modes.

For all means of coupling, the input impedance at resonance and the loaded Q may be adjusted by proper selection of the point of coupling and the degree of coupling.

Simple Waveguide Cavity*

A cavity may be made by enclosing a section of waveguide between a pair of large shunt susceptances, as shown in Fig. 36. Its loaded Q is given by

$$Q_l = \tfrac{1}{4}(\lambda_g/\lambda)^2(b^4+4b^2)^{1/2}\tan^{-1}(2/b)$$

and the resonant guide wavelength λ_{g0} is obtained from

$$2\pi l/\lambda_{g0} = \tan^{-1}(2/b).$$

Resonant Irises

Resonant irises may be used to obtain low values of loaded Q (<30). The simplest type is shown in

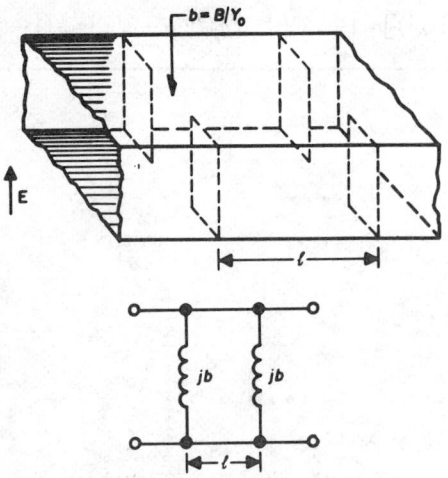

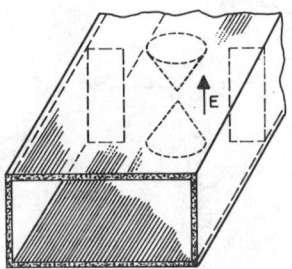

Fig. 39—Resonant structure consisting of cones with capacitive gap between apexes and thin symmetrical inductive diaphragm.

Fig. 37. It consists of an inductive diaphragm and a capacitive screw located in the same plane across the waveguide. For $Q_l<50$, the losses in the resonant circuit may be ignored and

$$1/Q_l \approx 1/Q_{\text{ext}}.$$

To a good approximation, the loaded Q (matched load and matched generator) is given by

$$Q_l = (B_l/2Y_0)(\lambda_{g0}/\lambda)^2$$

where B_l is the susceptance of the inductive diaphragm. This value may be taken from charts such as Figs. 30 and 31 as a starting point, but because of the proximity of the elements, the susceptance value is modified. Exact Q's must be obtained experimentally. Other resonant structures are given in Figs. 38 and 39. These are often designed so that the capacitive gap will break down under high power levels for use as transmit–receive (tr) switches in radar systems.

Fig. 36—Waveguide cavity and equivalent circuit.

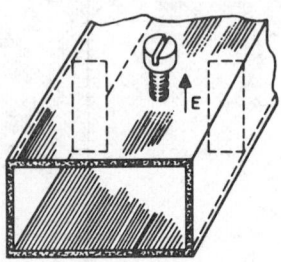

Fig. 37—Resonant iris in waveguide. The capacitive screw is tuned to resonance with the inductive diaphragm.

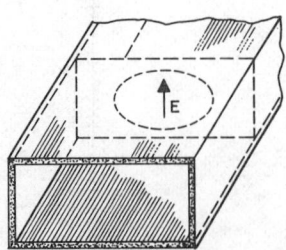

Fig. 38—Resonant element consisting of an oblong aperture in a thin transverse diaphragm.

* G. L. Ragan, "Microwave Transmission Circuits," McGraw-Hill Book Company, New York; 1948: Chapter 10.

Microwave structures are characterized by dimensions that are of the order of the wavelength of the propagated signal. The notions of current, voltage, and impedance, useful at lower frequencies, have been successfully extended to these structures, but these quantities are not as directly available for measurement; there are no voltmeters or ammeters and no apparent "terminal pair" between which to connect them. The electromagnetic field itself, distributed throughout a region, becomes the relevant quantity.

Within uniform structures, which are the usual form of waveguides, the *power flow* and the *phase* of the field at a cross section are the quantities of importance. The most usual form of measurement, that of the standing-wave pattern in a slotted section, is easily interpreted in terms of *travelling* waves and gives directly the *reflection coefficient*. The scattering description of waveguide junctions was introduced* to express this point of view. It is not, however, restricted to microwaves; a low-frequency network can be considered as a "waveguide junction" between transmission lines† connected to its terminal pairs, and the scattering matrix is a useful complement to the impedance and admittance descriptions.

AMPLITUDE OF A TRAVELLING WAVE

In a uniform waveguide, a travelling wave is characterized, for a given mode and frequency, by the electromagnetic-field distribution in a transverse cross section and by a propagation constant h. The field in any other cross section, at a distance z in the direction of propagation, has the same pattern but is multiplied by $\exp(-jhz)$. A wave propagating in the opposite direction, for the same mode and frequency, varies with z as $\exp(jhz)$. When losses are negligible, h is real.

The *amplitude* of a travelling wave, at a given cross section in the waveguide, is a complex number a defined as follows. The square $|a|^2$ of the magnitude of a is the power flow,* that is, the integral of the Poynting vector over the waveguide cross section. The phase angle of a is that of the transverse field in the cross section.†

The amplitude of a given travelling wave varies with z as $\exp(-jhz)$.

The wave amplitude has the dimensions of the square root of a power. The meter–kilogram–second unit is therefore the $(\text{watt})^{1/2}$.

REFLECTION COEFFICIENT

Definition

At a cross section in a waveguide, the reflection coefficient is the ratio of the amplitudes of the waves travelling respectively in the negative and the positive directions.

The positive direction must be specified and is usually taken as toward the load. To give a definite phase to the reflection coefficient, a convention is necessary that describes how the phases of waves travelling in opposite directions are to be compared. The usual convention is to compare in the two waves the phases of the transverse electric-field vectors.‡

For a short-circuit, produced, for instance, by a perfect conducting plane placed across the waveguide, the reflection coefficient is $W = -1$. For an open circuit, it is $W = +1$ and for a matched load, $W = 0$.

---

* C. G. Montgomery, R. H. Dicke, and E. M. Purcell, "Principles of Microwave Circuits," McGraw-Hill Book Company, Inc., New York, N. Y.; 1948.

† Transmission lines are in fact considered as special cases of waveguides: See "IRE Standards on Antennas and Waveguides: Definitions of Terms, 1953," The Institute of Radio Engineers, Inc.; New York, N.Y.: 1953. Published in *Proceedings of the IRE*, vol. 41, pp. 1721–1728; December 1953.

* The amplitude is sometimes defined to make the power flow equal to $\frac{1}{2}|a|^2$ rather than to $|a|^2$. This would correspond to the use of peak values instead of root-mean-square values.

† This phase is well defined for a pure mode, since the field has the same phase everywhere in the cross section.

‡ The dual convention, based on the magnetic-field vector, would give the "current" reflection coefficient, equal to minus the "voltage" reflection coefficient. The latter is used almost exclusively and the "voltage" is implicit.

When the cross section is displaced by z in the positive direction, the reflection coefficient W becomes

$$W' = W \exp (2jhz). \qquad (1)$$

Measurement

In a slotted waveguide equipped with a sliding voltage probe,* the position of a maximum is one where the phase of the reflection coefficient is zero.

The ratio of the maximum to the minimum (the standing-wave ratio or swr) is

$$(\mathrm{swr}) = (1 + |W|)/(1 - |W|).$$

Therefore

$$W = [(\mathrm{swr}) - 1]/[(\mathrm{swr}) + 1] \qquad (2)$$

is the value of W at the position of a maximum. At the position of a minimum, which is easier to locate in practice, the reflection coefficient is $[1 - (\mathrm{swr})]/[1 + (\mathrm{swr})]$.

At any other position, the value of W is obtained by applying (1). If the reflection coefficient is wanted in some waveguide connected to the slotted section, a good match must obtain at the transition or a correction must be applied as explained in problems **a** and **b** on pages 26-5 to 26-7.

SCATTERING MATRIX OF A JUNCTION

Definition

To define accurately the waves incident on a waveguide junction and those reflected (or scattered) from it, some reference locations must be chosen in the waveguides. These locations are called the ports† of the junction. In a waveguide that can support several propagating modes, there should be as many ports as there are modes. (These ports may or may not have the same physical location in the multimode waveguide.)

At each port i of a junction, consider the amplitude a_i of the incident wave, travelling toward the junction, and the amplitude b_i of the scattered wave, travelling away from it. As a consequence of Maxwell's equations, there exists a linear relation between the b_i and the a_i. Considering the a_i (where i varies from 1 to n) as the components of a vector **a**, and the b_i as the components of a vector **b**, this relation can be expressed by

$$\mathbf{b} = \mathbf{S}\mathbf{a}$$

* A probe that gives a reading proportional to the electric field.

† At lower frequencies, for a network connecting transmission lines, a port is a terminal pair.

where $\mathbf{S} = (s_{ij})$ is an $n \times n$ matrix called the *scattering matrix* of the junction.

The s_{ii} is the *reflection coefficient* looking into port i and s_{ij} is the *transmission coefficient* from j to i, all other ports being terminated in matching impedances.

Properties

For a *reciprocal* junction, the transmission coefficient from i to j equals that from j to i; the matrix **S** is symmetrical.

$$\mathbf{S} = \tilde{\mathbf{S}}$$

where $\tilde{\mathbf{S}}$ denotes the transpose of **S**.

The total power incident on the junction is

$$|\mathbf{a}|^2 = \sum_{i=1}^{i=n} |a_i|^2.$$

The total power reflected is

$$|\mathbf{b}|^2 = \sum_{i=1}^{i=n} |b_i|^2.$$

For a lossless junction, these two powers are equal

$$|\mathbf{a}|^2 = |\mathbf{b}|^2.$$

This implies that the matrix **S** is unitary (see page 46-29).

$$\mathbf{S}\dagger = \mathbf{S}^{-1}.$$

For a *passive* junction with losses, $|\mathbf{b}|^2 < |\mathbf{a}|^2$, hence the matrix $1 - \mathbf{S}\mathbf{S}\dagger$ is definite positive.

Change of Terminal Plane

If the port in arm i is moved away from the junction by ϕ_i electrical radians, the scattering matrix becomes

$$\mathbf{S}' = \Phi \mathbf{S} \Phi \qquad (3)$$

where

$$\Phi = \begin{bmatrix} e^{-j\phi_1} & 0 & 0 & 0 & \cdots \\ 0 & e^{-j\phi_2} & 0 & 0 & \cdots \\ 0 & 0 & e^{-j\phi_3} & 0 & \cdots \\ \cdot & \cdot & \cdot & \cdot & \cdots \\ \cdot & \cdot & \cdot & \cdot & \cdots \\ \cdot & \cdot & \cdot & \cdot & \cdots \end{bmatrix}. \qquad (4)$$

TWO-PORT JUNCTIONS

The 2-port junction includes the case of an obstacle or discontinuity placed in a waveguide as well as that of two essentially different waveguides connected to each other.

If reciprocity applies, the scattering matrix

$$S = \begin{bmatrix} s_{11} & s_{12} \\ s_{21} & s_{22} \end{bmatrix} \qquad (5)$$

is symmetrical

$$s_{21} = s_{12}.$$

For a lossless junction, the scattering coefficients can be expressed by

$$s_{11} = + \tanh (u/2) \exp (-2j\alpha)$$
$$s_{22} = - \tanh (u/2) \exp (-2j\beta)$$
$$s_{12} = + \operatorname{sech} (u/2) \exp [-j(\alpha+\beta)] \qquad (6)$$

in terms of three parameters, u, α, and β.

This corresponds to the representation of the junction by an ideal transformer with transformer ratio $n = \exp (-u/2)$, of hyperbolic amplitude u, placed between two sections of transmission line with electrical lengths α and β, respectively.

The quantity $-20 \log_{10} | s_{12} |$ is the insertion loss.

TRANSFORMATION MATRIX

To find the effect of successive obstacles in a waveguide or to combine 2-port junctions placed in cascade, it is convenient to introduce the wave transformation matrix T.

This matrix T relates the travelling waves on one side of the junction to those on the other side. Using the notations of Fig. 1

$$\begin{bmatrix} A_1 \\ B_1 \end{bmatrix} = T \begin{bmatrix} A_2 \\ B_2 \end{bmatrix}. \qquad (7)$$

The 2×2 transformation matrix T may be deduced from the scattering matrix S

$$T = s_{21}^{-1} \begin{bmatrix} 1 & -s_{22} \\ s_n & -\det S \end{bmatrix}. \qquad (8)$$

Conversely, if $T = (t_{ij})$, the scattering matrix is

$$S = t_{11}^{-1} \begin{bmatrix} t_{21} & \det T \\ 1 & -t_{12} \end{bmatrix}. \qquad (9)$$

When reciprocity applies to the junction

$$\det T = s_{12}/s_{21} \qquad (10)$$

becomes unity.

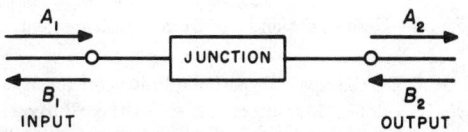

Fig. 1—Convention for wave transformation matrix T.

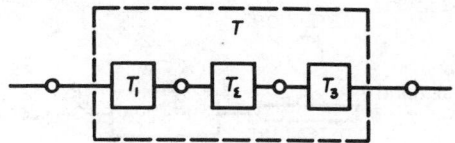

Fig. 2—Junctions in cascade.

The input reflection coefficient $W' = B_1/A_1$ is related to the load reflection coefficient $W = B_2/A_2$ by

$$W' = (t_{21} + t_{22}W)/(t_{11} + t_{12}W) \qquad (11)$$
$$= s_{11} + [s_{12}^2 W/(1 - s_{22}W)]. \qquad (12)$$

When a number of junctions 1, 2, 3, are placed in cascade (Fig. 2), the output port of each of them being the input port of the following one, the resulting junction has the transformation matrix

$$T = T_1 T_2 T_3.$$

If n similar junctions with transformation matrix T are cascaded, the resulting transformation matrix is T^n.

Letting trace $T = t_{11} + t_{22} = 2 \cos \theta$

$$T^n = (\sin n\theta/\sin \theta) T - [\sin (n-1)\theta/\sin \theta]. \qquad (13)$$

MEASUREMENT OF THE SCATTERING MATRIX*

A slotted line is placed on side 1 of the junction (see Fig. 3). For any load with reflection coefficient W, placed on side 2, the input reflection coefficient W' can be measured. W' is called the *image* of W. The images of various known loads can be plotted on a reflection chart and the scattering coefficients deduced by the following procedures.

(A) With a matched load, one obtains directly s_{11} plotted as O' on Fig. 4. O' is called the iconocenter.

(B) With a sliding short-circuit on side 2, or any variable reactive load, the input reflection coefficient describes a circle Γ', image of the unit circle Γ. This circle can be deduced from 3 or more measurements. Let C be its center and R its radius (Fig. 4). The magnitudes of the scattering coefficients result:

$$| s_{11} | = OO'$$
$$| s_{22} | = O'C/R$$
$$| s_{12} |^2 = R(1 - | s_{22} |^2). \qquad (14)$$

* G. A. Deschamps, "Determination of the Reflection Coefficients and Insertion Loss of a Waveguide Junction," *Journal of Applied Physics*, vol. 24, pp. 1046–1050; August 1953: Also, *Electrical Communication*, vol. 31, pp. 57–62; March 1954.

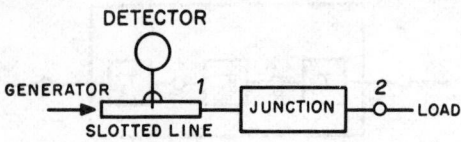

Fig. 3—Slotted-line setup for scattering-matrix measurement.

The phases of these coefficients all follow from one more measurement.

(**C**) The input reflection coefficient is measured with an open-circuit load placed at port 2, or for a short-circuit placed a quarter wave away from it. This may be one of the measurements taken in step (**B**). It gives (Fig. 5) the point P', image of the point $P(W=+1)$.

A point P'' is constructed by projecting P' through O' onto Q on Γ', then Q through C onto P'' on Γ' (Fig. 5). Then

$$\text{Phase of } s_{11} = \text{angle } (OP,\ OO')$$

$$\text{Phase of } s_{22} = \text{angle } (O'C,\ CP'')$$

$$\text{Phase of } s_{12} = \tfrac{1}{2} \text{ angle } (OP,\ CP''). \quad (15)$$

(**D**) When no matched load is available, as was assumed in (**A**), the iconocenter O' may be obtained as in Fig. 6. Let P_1, P_2, P_3, P_4 represent the input reflection coefficients when a short circuit is placed successively at port 2 and at distances $\lambda/8$, $\lambda/4$, and $3\lambda/8$ from it. These points define the circle Γ' [(as in (**B**)] and the intersection I (the crossover point) of P_1P_3 and P_2P_4 may be used to find O': draw perpendiculars to CI at points C and I up to their intersections with Γ' and get C' and I'; then O' is the intersection of CI and $C'I'$.

The point P_3 is identical to P' in (**C**) above, hence the four measurements give the complete scattering matrix by constructing P'' and applying (14) and (15).

(**E**) The construction of O' in (**D**) above is valid with any sliding load not necessarily reactive.

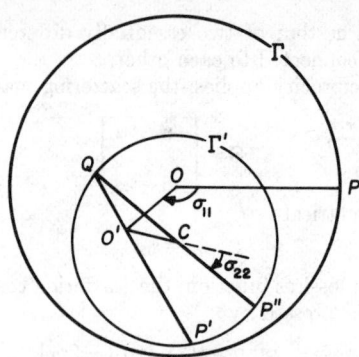

Fig. 5—Construction for the phases of the scattering coefficients.

Taking a load with small standing-wave ratio increases the accuracy of the construction.

(**F**) When exact measurements of the displacements of the sliding load are difficult to make, for instance if the wavelength is very short, the point O' may be obtained as follows. Using a reactive load, construct the circle Γ' as in (**B**) above, then using a sliding load as in (**E**) above, construct a circle Γ'' (see Fig. 7). The iconocenter O' is the hyperbolic midpoint of the diameter of Γ'' (through C) with respect to Γ'. It may be constructed by means of the hyperbolic protractor* (page 26-6), or by means of the dotted-line construction (Fig. 7).

GEOMETRY OF REFLECTION CHARTS

The following brief outline is complemented by the section on hyperbolic trigonometry on pages 46-10 to 46-13.

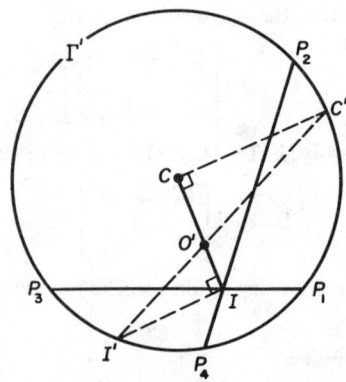

Fig. 6—Determination of O' from 4 measurements.

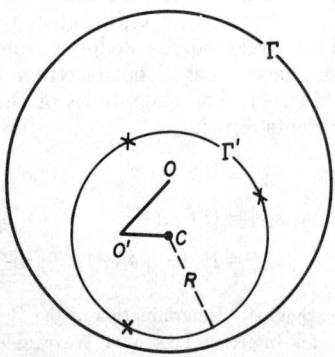

Fig. 4—Construction for the magnitudes of the scattering coefficients.

* G. A. Deschamps, "Hyperbolic Protractor for Microwave Impedance Measurements and Other Purposes," International Telephone and Telegraph Corporation, 320 Park Avenue, New York, New York 10022; 1953.

Conformal Charts

A reflection coefficient can be represented by a point in a plane just as any complex number is represented on the Argand diagram.

The passive loads, $|W| \leq 1$, are represented by points inside a unit circle Γ. Inside this circle, the lines of constant resistance and reactance may be drawn (Smith chart) or the lines of constant magnitude and phase of the impedance (Carter chart).

The transformation from a load reflection coefficient W to its image W' through a 2-port junction is bilinear as in (11) or (12). On the reflection chart, this transformation maps circles into circles and preserves the angle between curves and the cross ratio of 4 points; if

$$[W_1, W_2, W_3, W_4] = \frac{W_1 - W_3}{W_1 - W_4} : \frac{W_2 - W_3}{W_2 - W_4}$$

denotes the cross ratio of 4 reflection coefficients W_1, W_2, W_3, and W_4, then

$$[W_1', W_2', W_3', W_4'] = [W_1, W_2, W_3, W_4].$$

The transformation through a lossless junction preserves also the unit circle Γ and therefore leaves invariant the *hyperbolic distance* defined on page 44-11. The hyperbolic distance to the origin of the chart is the *mismatch*, that is, the standing-wave ratio expressed in decibels: It may be evaluated by means of the proper graduation on the radial arm of the Smith chart. For two arbitrary points W_1, W_2, the hyperbolic distance between them may be interpreted as the mismatch that results from the load W_2 seen through a lossless network that matches W_1 to the input waveguide.

Projective Chart

The reflection coefficient W is represented by the point $\bar{W}$ (Fig. 8) on the same radius of the circle

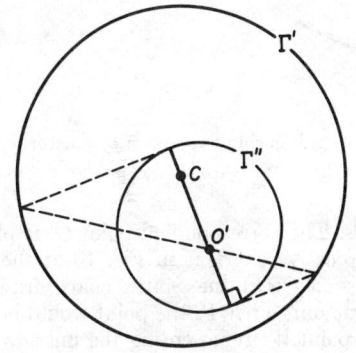

Fig. 7—Use of circles Γ'' and Γ' for determination of O'.

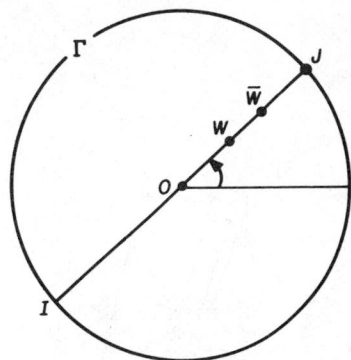

Fig. 8—Representation of a reflection coefficient by W on a Smith chart and $\bar{W}$ on the projective chart.

Γ but at a distance

$$O\bar{W} = (2OW/(1+OW^2)) \qquad (16)$$

from the origin.

This is equivalent to using the standing-wave ratio squared instead of the direct ratio:

$$\bar{W}J/\bar{W}I = (WJ/WI)^2. \qquad (17)$$

The transformation (11), (12), when the junction is lossless, is represented on this chart by a projective transformation, that is, one that maps straight lines into straight lines and preserves the cross ratio of four points on a straight line. It therefore preserves the hyperbolic distance defined on page 46-11.

EVALUATION OF HYPERBOLIC DISTANCE

On the projective chart, the hyperbolic distance $\langle AB \rangle$ between two points A and B inside the circle Γ can be evaluated by means of a hyperbolic protractor as shown in Fig. 9. The line AB is extended to its intersections I and J with Γ. The protractor is placed so that the sides OX, OY of the right angle go through I and J. (This can be done in many ways but does not affect the result.) The numbers read on the radial lines of the protractor going through A and B, respectively, are added if A and B are on opposite sides of the radial line marked O; subtracted otherwise: This result divided by 2 is the distance $\langle AB \rangle$. In Fig. 9, for instance

$$\langle AB \rangle = \tfrac{1}{2}(12+4) = 8 \text{ decibels.}$$

Problem a

A slotted line with 100-ohm characteristic impedance is used to make measurements on a 60-ohm coaxial line. The transition acts as an ideal transformer. Find the reflection coefficient W of an

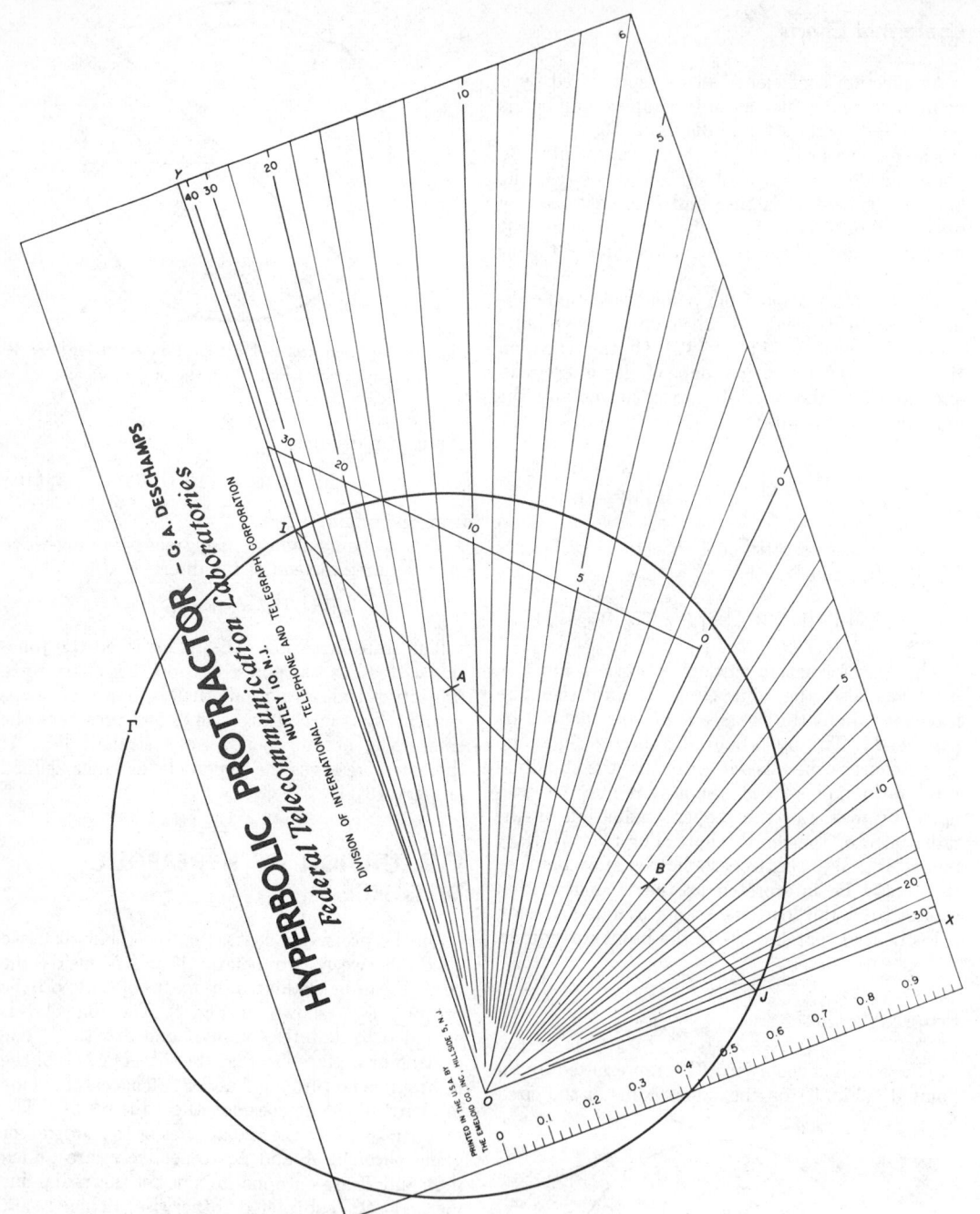

Fig. 9—Definition and evaluation of hyperbolic distance $\langle AB \rangle$ using hyperbolic protractor.

obstacle placed in the coaxial line, knowing that it produces a reflection coefficient

$$W' = 0.5 \exp\ (j\pi/2)$$

in the slotted line.

A match in the coaxial line appears in the slotted line as a normalized impedance of 0.6, hence the mismatch (standing-wave ratio in decibels) is 4.5 decibels. The corresponding point $\bar{O}'$ is plotted on the projective chart as in Fig. 10 at the distance $\langle O\bar{O}' \rangle = 4.5$. (On the Smith chart drawn inside the same unit circle Γ, the point would be O'.)

The point $\bar{W}'$ representing the unknown load is plotted at the hyperbolic distance

$$20 \log_{10} \left[(1+0.5)/(1-0.5) \right] = 9.5 \text{ decibels}$$

from the origin in the direction $+90°$. The hyperbolic distance

$$\langle \bar{O}'\bar{W}' \rangle = 11 \text{ decibels}$$

is measured with the protractor. This is the mismatch produced by the obstacle in the coaxial line. It corresponds to a magnitude of the reflection coefficient of 0.56.

The phase of this reflection coefficient is the elliptic angle $\langle \bar{O}'P, \bar{O}'\bar{W}' \rangle$.

It is evaluated as explained on page 46-11: extend QO' up to R on Γ and measure the arc

$$PR = 56°.$$

The answer is

$$W = 0.56 \underline{/56°}.$$

Problem b

If the transition between the slotted line and the waveguide is not an ideal transformer as in problem **a**, its properties may be found by the method described on page 26-3. In particular, if the transition has no losses (the circle Γ' coincides with Γ), the point O' may be found as in (**A**), (**D**), (**E**), or (**F**), above, the point P' as in (**C**) or (**D**), above, and this completes the calibration.

For any load placed in the waveguide and producing the reflection coefficient W' in the slotted line, the corrected standing-wave ratio in decibels is the hyperbolic distance $[O'W']$. This is evaluated by constructing $\bar{O}'$, $\bar{W}'$ on the projective chart and measuring $\langle \bar{O}'\bar{W}' \rangle$ with the protractor. The phase angle is the elliptic angle $\langle \bar{O}'P', \bar{O}'\bar{W}' \rangle$ (see page 46-11).

Problem c

A section of coaxial line 90 electrical degrees in length and with 100-ohm characteristic impedance

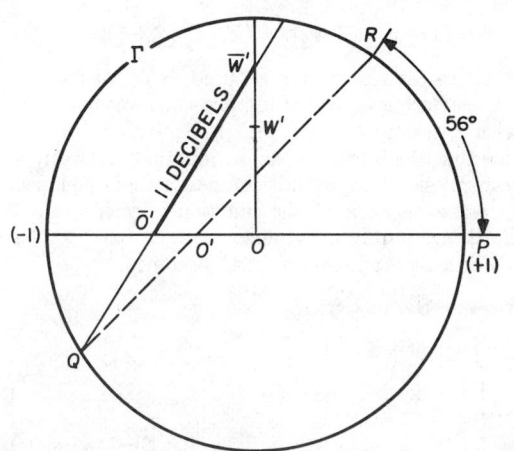

Fig. 10—Measurement of reflection coefficient with a mismatched slotted line.

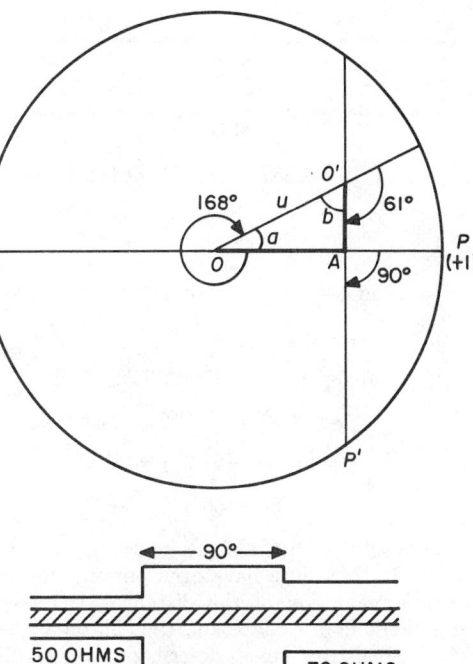

Fig. 11—Solution for transformation in transmission line.

is inserted between a 50-ohm coaxial line on one side and a 70-ohm coaxial line on the other (Fig. 11). Find the transformer ratio $n = \exp(-u/2)$ and the electrical lengths α, β, of the representation (6).

The two discontinuities are assumed to act as ideal transformers with hyperbolic amplitudes

$$20 \log_{10} \frac{100}{50} = 6 \text{ decibels} = 0.67 \text{ neper}$$

and

$$20 \log_{10} \frac{70}{100} = -3.1 \text{ decibels} = -0.36 \text{ neper}.$$

The characteristic polygon* on the projective chart is a triangle OAO' with right angle A; hence, $u = \langle OO' \rangle$ is given by

$$\cosh u = \cosh 0.69 \cosh 0.36$$

$$u = 0.78 \text{ neper} = 6.8 \text{ decibels}$$

$$n = \exp(-u/2) = 1/1.48.$$

The length of line α and β can be deduced from evaluating the elliptic angles $\langle OA, OO' \rangle = a$ and

* G. A. Deschamps, "Hyperbolic Protractor for Microwave Impedance Measurements and Other Purposes," International Telephone and Telegraph Corporation, 320 Park Avenue, New York, New York 10022; 1953: pp. 15–16 and p. 41.

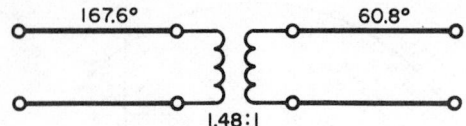

Fig. 12—Equivalent circuit for Fig. 11.

$\langle O'A, O'O \rangle = b$

$\tan a = \tanh 0.36/\sinh 0.69 = 0.46$

$a = 24.7°$

$\tan b = \tanh 0.69/\sinh 0.36 = 1.62$

$b = 58.4°$

$\alpha = \frac{1}{2}(360° - 24.7°) = 167.6°$

$\beta = \frac{1}{2}(180° - 58.4°) = 60.8°.$

The resulting equivalent network is shown in Fig. 12. It could also have been obtained by geometrical evaluation of the distance $\langle OO' \rangle$ with the hyperbolic protractor and the elliptic angles a and b by constructions as described on pages 26-5 and 46-11.

CORRESPONDENCES WITH CURRENT, VOLTAGE, AND IMPEDANCE VIEWPOINTS

Normalized Current and Voltage

In a waveguide, at a point where the amplitudes of the waves travelling in the positive and negative directions are, respectively, a and b, the normalized voltage v and the normalized current i are defined by

$$v = a + b$$

$$i = a - b. \tag{18}$$

The *net power flow* at that point in the positive direction is

$$|a|^2 - |b|^2 = \operatorname{Re} vi^*. \tag{19}$$

Current and Voltage Not Normalized

A more-general definition for current and voltage becomes possible when a meaning has been assigned to the characteristic impedance Z_0 of the waveguide

$$V = vZ_0^{1/2}$$

$$I = iY_0^{1/2} \tag{20}$$

where $Y_0 = 1/Z_0$ is the characteristic admittance and v and i are the normalized values defined above.

Conversely, if by some convention the voltage (or the current) has been defined, a characteristic impedance will result from (20). This is the case for a 2-conductor waveguide supporting the TEM mode; the characteristic impedance is the ratio of voltage to current in a travelling wave.

If V and I are the voltage and the current at a point in a waveguide of characteristic impedance $Z_0 = 1/Y_0$, the amplitudes of the waves travelling in both directions at that point are

$$a = \frac{1}{2}(VY_0^{1/2} + IZ_0^{1/2})$$

$$b = \frac{1}{2}(VY_0^{1/2} - IZ_0^{1/2}). \tag{21}$$

Normalized Impedance and Admittance

At a point in a waveguide, the normalized impedance is $Z = v/i$ and the normalized admittance is the inverse, $Y = 1/Z$.

They are related to the reflection coefficient $W = b/a$ by

$$Z = (1+W)/(1-W)$$

$$Y = (1-W)/(1+W) \tag{22}$$

hence

$$W = (1-Y)/(1+Y) = (Z-1)/(Z+1). \tag{23}$$

Impedance and Admittance Matrix of a Junction

The **Z** and **Y** matrices of a junction are defined in terms of the scattering matrix **S** by

$$\mathbf{Y} = (1-\mathbf{S})(1+\mathbf{S})^{-1}$$

$$\mathbf{Z} = (1+\mathbf{S})(1-\mathbf{S})^{-1}. \tag{24}$$

The matrices **Y** and **Z** do not always exist since **S** may have eigenvalues $+1$ or -1, which means that $\det(1-\mathbf{S})$ or $\det(1+\mathbf{S})$ may be zero. Conversely

$$\mathbf{S} = (1-\mathbf{Y})(1+\mathbf{Y})^{-1} = (\mathbf{Z}-1)(\mathbf{Z}+1)^{-1}. \tag{25}$$

These equations may be used as definitions for the scattering matrix of lumped-constant networks with n terminal pairs. This is equivalent to considering the network as a junction between n transmission lines of unit characteristic impedance.

If the network or the junction is reciprocal, **Y** and **Z** are purely imaginary.

For a 2-port junction, (24) becomes

$$\mathbf{Y} = (1-\mathbf{S})/(1+\mathbf{S})$$

$$= [\det(1+\mathbf{S})]^{-1}$$

$$\times \begin{bmatrix} 1 - \det\mathbf{S} + (s_{22}-s_{11}) & -2s_{12} \\ -2s_{21} & 1 - \det\mathbf{S} - (s_{22}-s_{11}) \end{bmatrix}$$

$$\tag{26}$$

and

$$\mathbf{Z}=(1+\mathbf{S})/(1-\mathbf{S})$$

$$=[\det{(1-\mathbf{S})}]^{-1}$$

$$\times\begin{bmatrix} 1-\det{\mathbf{S}}-(s_{22}-s_{11}) & 2s_{12} \\ 2s_{21} & 1-\det{\mathbf{S}}+(s_{22}-s_{11}) \end{bmatrix}$$

$$\tag{27}$$

$$\det{(1+\mathbf{S})}=1+\operatorname{tr}\mathbf{S}+\det{\mathbf{S}}$$

$$=1+(s_{11}+s_{22})+(s_{11}s_{22}-s_{12}{}^2)$$

$$\det{(1-\mathbf{S})}=1-\operatorname{tr}\mathbf{S}+\det{\mathbf{S}}$$

$$=1-(s_{11}+s_{22})+(s_{11}s_{22}-s_{12}{}^2).$$

The matrices **Y** and **Z** relate normalized voltages and currents at both ports (Fig. 13) as follows:

$$\begin{bmatrix} v_1 \\ v_2 \end{bmatrix}=\mathbf{Z}\begin{bmatrix} i_1 \\ i_2 \end{bmatrix}$$

$$\begin{bmatrix} i_1 \\ i_2 \end{bmatrix}=\mathbf{Y}\begin{bmatrix} v_1 \\ v_2 \end{bmatrix}.$$

Fig. 13—Sign convention for defining the impedance and admittance of a 2-port junction.

TRANSFORMATION MATRIX

A transformation matrix useful for composing 2-port junctions in cascade relates the voltage and current on one side of the junction to the same quantities on the other side. With the notation in Fig. 14

$$\begin{bmatrix} v' \\ i' \end{bmatrix}=\mathbf{U}\begin{bmatrix} v \\ i \end{bmatrix}.\tag{28}$$

The matrix **U**, sometimes called the *ABCD* matrix, has the same properties as **T** described earlier.

For a series element with normalized impedance Z

$$\mathbf{U}=\begin{bmatrix} 1 & Z \\ 0 & 1 \end{bmatrix}$$

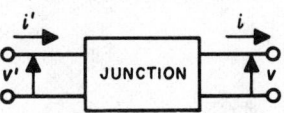

Fig. 14—Sign convention for voltages and currents related by the transformation matrix.

and for a shunt element with normalized admittance Y

$$\mathbf{U}=\begin{bmatrix} 1 & 0 \\ Y & 1 \end{bmatrix}.$$

A product of matrices of these types gives the transformation matrix for any ladder network.

For the shunt element Y, the scattering matrix is

$$\mathbf{S}=(2+Y)^{-1}\begin{bmatrix} -Y & 2 \\ 2 & -Y \end{bmatrix}\tag{29}$$

hence

$$s_{11}=s_{22}$$

$$s_{12}=1+s_{11}.\tag{30}$$

For the series element Z, the scattering matrix is

$$\mathbf{S}=(2+Z)^{-1}\begin{bmatrix} Z & 2 \\ 2 & Z \end{bmatrix}\tag{31}$$

hence

$$s_{11}=s_{22}$$

$$s_{12}=1-s_{11}.\tag{32}$$

Relations (30) and (32) are characteristic, respectively, of a shunt and a series obstacle in a waveguide.

The matrix **T** can be deduced from **U** and vice versa:

$$\mathbf{T}=\frac{1}{2}\begin{bmatrix} 1 & 1 \\ 1 & -1 \end{bmatrix}\mathbf{U}\begin{bmatrix} 1 & 1 \\ 1 & -1 \end{bmatrix}$$

$$=\frac{1}{2}\begin{bmatrix} u_{11}+u_{12}+u_{21}+u_{22} & u_{11}-u_{12}+u_{21}-u_{22} \\ u_{11}+u_{12}-u_{21}-u_{22} & u_{11}-u_{12}-u_{21}+u_{22} \end{bmatrix}.$$

$$\tag{33}$$

A similar equation will transform **T** into **U**, since

$$\mathbf{U}=\frac{1}{2}\begin{bmatrix} 1 & 1 \\ 1 & -1 \end{bmatrix}\mathbf{T}\begin{bmatrix} 1 & 1 \\ 1 & -1 \end{bmatrix}.\tag{34}$$

ELEMENTARY DIPOLE

Field Strength*

The elementary dipole forms the basis for many antenna computations. Since dipole theory assumes an antenna with current of constant magnitude and phase throughout its length, approximations to the elementary dipole are realized in practice only for antennas shorter than one-tenth wavelength. The theory can be applied directly to a loop whose circumference is less than one-tenth wavelength, thus forming a magnetic dipole. For larger antennas, the theory is applied by assuming the antenna to consist of a large number of infinitesimal dipoles with differences between individual dipoles of space position, polarization, current magnitude, and phase corresponding to the distribution of these parameters in the actual antenna. Field-strength equations for large antennas are then developed by integrating or otherwise summing the field vectors of the many elementary dipoles.

The outline below concerns electric dipoles. It also can be applied to magnetic dipoles by installing the loop perpendicular to the PO line at the center of the sphere in Fig. 1. In this case, vector H becomes E, the electric field; E_t becomes the magnetic tangential field; and E_r becomes the radial magnetic field.

In the case of a magnetic dipole, Table 1, showing variations of the field in the vicinity of the dipole, can also be used.

For electric dipoles, Fig. 1 indicates the electric and magnetic field components in spherical coordinates with positive values shown by the arrows.

r = distance OM
θ = angle POM measured from P toward M
I = current in dipole
λ = wavelength
f = frequency

$\omega = 2\pi f$
$\alpha = 2\pi/\lambda$
c = velocity of light (see page 3–12)
$v = \omega t - \alpha r$
l = length of dipole

* Based on R. Mesny, "Radio-Électricité Générale," Étienne Chiron, Paris, France; 1935.

The following equations expressed in meter-kilogram–second units (in vacuum) result:

$$E_r = - (30l\lambda I/\pi)(\cos\theta/r^3)(\cos v - \alpha r \sin v)$$

$$E_t = + (30l\lambda I/2\pi)(\sin\theta/r^3)$$
$$\times (\cos v - \alpha r \sin v - \alpha^2 r^2 \cos v)$$

$$H = + (1/4\pi)lI(\sin\theta/r^2)(\sin v - \alpha r \cos v). \quad (1)$$

These equations are valid for the elementary dipole at distances that are large compared with

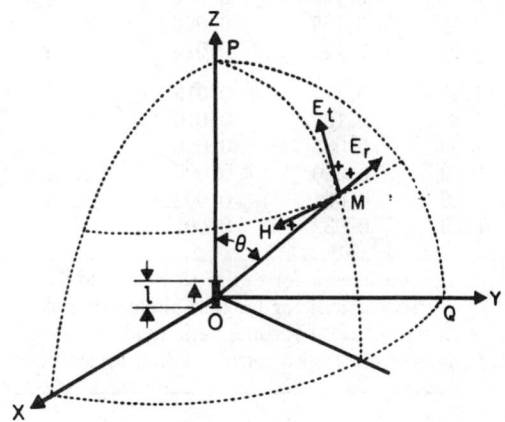

Fig. 1—Electric and magnetic components in spherical coordinates for electric dipoles.

the dimensions of the dipole. Length of the dipole must be small with respect to the wavelength, say $l/\lambda < 0.1$. The equations are for a dipole in free space. If the dipole is placed vertically on a plane of infinite conductivity, its image should be taken into account, thus doubling the above values.

Field at Great Distance

When distance r exceeds 5 wavelengths, as is generally the case in radio applications, the radial electric field E_r becomes negligible with respect

TABLE 1—VARIATIONS OF FIELD IN THE VICINITY OF A DIPOLE.

r/λ	$1/\alpha r$	A_r	ϕ_r	A_t	ϕ_t	A_H	ϕ_H
0.01	15.9	4028	3°.6	4012	3°.6	253	93°.6
0.02	7.96	508	7°.2	500	7°.3	64.2	97°.2
0.04	3.98	65	14°.1	61	15°.0	16.4	104°.1
0.06	2.65	19.9	20°.7	17.5	23°.8	7.67	110°.7
0.08	1.99	8.86	26°.7	7.12	33°.9	4.45	116°.7
0.10	1.59	4.76	32°.1	3.52	45°.1	2.99	122°.1
0.15	1.06	1.66	42°.3	1.14	83°.1	1.56	132°.3
0.20	0.80	0.81	51°.5	0.70	114°.0	1.02	141°.5
0.25	0.64	0.47	57°.5	0.55	133°.1	0.75	147°.5
0.30	0.56	0.32	62°.0	0.48	143°.0	0.60	152°.0
0.35	0.45	0.23	65°.3	0.42	150°.1	0.50	155°.3
0.40	0.40	0.17	68°.3	0.37	154°.7	0.43	158°.3
0.45	0.35	0.134	70°.5	0.34	158°.0	0.38	160°.5
0.50	0.33	0.106	72°.3	0.30	160°.4	0.334	162°.3
0.60	0.265	0.073	75°.1	0.26	164°.1	0.275	165°.1
0.70	0.228	0.053	77°.1	0.22	166°.5	0.234	167°.1
0.80	0.199	0.041	78°.7	0.196	168°.3	0.203	168°.7
0.90	0.177	0.032	80°.0	0.175	169°.7	0.180	170°.0
1.00	0.159	0.026	80°.9	0.157	170°.7	0.161	170°.9
1.20	0.133	0.018	82°.4	0.132	172°.3	0.134	172°.4
1.40	0.114	0.013	83°.5	0.114	173°.5	0.114	173°.5
1.60	0.100	0.010	84°.3	0.100	174°.3	0.100	174°.3
1.80	0.088	0.008	84°.9	0.088	174°.9	0.088	174°.9
2.00	0.080	0.006	85°.4	0.080	175°.4	0.080	175°.4
2.50	0.064	0.004	86°.4	0.064	176°.4	0.064	176°.4
5.00	0.032	0.001	88°.2	0.032	178°.2	0.032	178°.2

A_r = coefficient for radial electric field
A_t = coefficient for tangential electric field
A_H = coefficient for magnetic field
ϕ_r, ϕ_t, ϕ_H = phase angles corresponding to coefficients.

to the tangential field and

$$E_r = 0$$

$$E_t = -(60\pi lI/\lambda r) \sin\theta \cos(\omega t - \alpha r)$$

$$H = +E_t/120\pi. \qquad (2)$$

Field at Short Distance

In the vicinity of the dipole $(r/\lambda < 0.01)$, αr is very small and the terms in (1) containing αr become negligible. The ratio of the radial and tangential field is then

$$E_r/E_t = -2\cot\theta.$$

Hence, the radial field at short distance has a magnitude of the same order as the tangential field. These two fields are in opposition. Further,

the ratio of the magnetic and electric tangential fields is

$$H/E_t = r \tan v/60\lambda.$$

The magnitude of the magnetic field at short distances is, therefore, extremely small with respect to that of the tangential electric field, relative to their relationship at great distances. The two fields are in quadrature. Thus, at short distances, the effect of the dipole on an open circuit is much greater than on a closed circuit as compared with the effect at remote points.

Field at Intermediate Distance

At intermediate distance, say between 0.01 and 5.0 wavelengths, one should take into account all the terms of equations (1). This case occurs, for

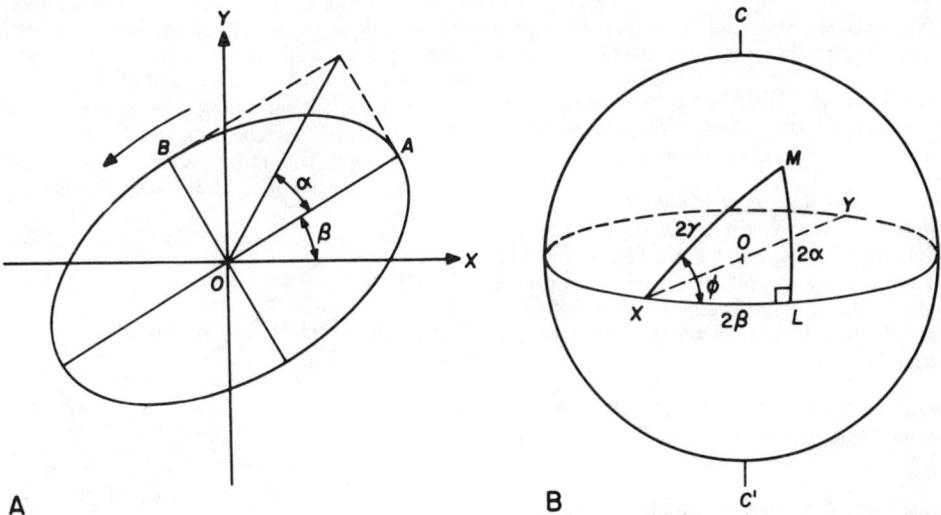

Fig. 2—Polarization ellipse at A and representation at B of a state of polarization by a point on a sphere.

instance, when studying reactions between adjacent antennas. To calculate the fields, it is convenient to transform the equations as follows:

$$E_r = -60\alpha^2 lI \cos\theta A_r \cos(v+\phi_r)$$

$$E_t = +30\alpha^2 lI \sin\theta A_t \cos(v+\phi_t)$$

$$H = -(1/4\pi)\alpha^2 lI \sin\theta A_H \cos(v+\phi_H) \quad (3)$$

where

$$A_r = [1+(\alpha r)^2]^{1/2}/(\alpha r)^3, \tan\phi_r = \alpha r$$

$$A_t = [1-(\alpha r)^2+(\alpha r)^4]^{1/2}/(\alpha r)^3, \cot\phi_t = (1/\alpha r)-\alpha r$$

$$A_H = [1+(\alpha r)^2]^{1/2}/(\alpha r)^2, \cot\phi_H = -\alpha r. \quad (4)$$

Values of A's and ϕ's are given in Table 1 as a function of the ratio between the distance r and the wavelength λ. The second column contains values of $1/\alpha r$ that would apply if the fields E_t and H behaved as at great distances.

LINEAR POLARIZATION

An electromagnetic wave is linearly polarized when the electric field lies wholly in one plane containing the direction of propagation.

Horizontal Polarization: The case where the electric field lies in a plane parallel to the earth's surface.

Vertical Polarization: The case where the electric field lies in a plane perpendicular to the earth's surface.

E Plane of an antenna is the plane in which the electric field lies. The principal E plane of an antenna is the E plane that also contains the direction of maximum radiation.

H Plane of an antenna is the plane in which the magnetic field lies. The H plane is normal to the E plane. The principal H plane of an antenna is the H plane that also contains the direction of maximum radiation.

ELLIPTICAL AND CIRCULAR POLARIZATION

A plane electromagnetic wave, at a given frequency, is elliptically polarized when the extremity of the electric vector describes an ellipse in a plane perpendicular to the direction of propagation, making one complete revolution during one period of the wave. More generally, any field vector, electric, magnetic, or other, is elliptically polarized if its extremity describes an ellipse.

Two perpendicular axes OX and OY are chosen for reference in the plane of the polarization ellipse, Fig. 2A. This plane is usually perpendicular to the direction of propagation. At a given frequency, the field components along these axes are represented by two complex numbers

$$X = |X| \exp j\phi_1$$

$$Y = |Y| \exp j\phi_2. \quad (5)$$

Amplitude of Elliptically Polarized Field: $E^2 = |X|^2 + |Y|^2$, so that the power density in free space for a plane wave is $E^2/240\pi$.

Axial Ratio: The ratio r of the minor axis to the major axis of the polarization ellipse$= OB/OA$.

Ellipticity Angle: $\alpha = \pm \tan^{-1} r$, where the sign is taken according to the sense of rotation.

Orientation Angle: The angle β between OX and the major axis of the polarization ellipse (indeterminate for circular polarization).

Polarization of Receiving Antenna: For plane waves incident in a given direction, the polarization of the incident wave that, for a given amplitude, induces the maximum voltage across the antenna terminals. If this voltage is expressed as hE, then h is the effective length of the antenna for the given direction.

Polarization Ratio: The ratio $P = Y/X$, a complex number with phase $\phi = \phi_2 - \phi_1$ and magnitude $\tan\gamma = |Y|/|X|$.

Relative Power received by an elliptically polarized receiving antenna as it is rotated in a plane normal to the direction of propagation of an elliptically polarized wave is given by

$$P_r = K \frac{(1 \pm r_1 r_2)^2 + (r_1 \pm r_2)^2 + (1 - r_1^2)(1 - r_2^2)\cos 2\theta}{(1 + r_1^2)(1 + r_2^2)}$$

$$(6)$$

where $K = $ constant, $r_1 = $ axial ratio of elliptically polarized wave, $r_2 = $ axial ratio of elliptically polarized antenna, and $\theta = $ angle between the direction of maximum amplitude in the incident wave and the direction of maximum amplitude of the elliptically polarized antenna.

The $(+)$ sign is to be used if both the receiving and transmitting antennas produce the same hand of polarization. The $(-)$ sign is to be used when one is left-handed and the other right-handed.

State of Polarization is specified either by the polarization ratio P (angles γ and ϕ) or by the shape, orientation, and sense of the polarization ellipse (angles α and β).

Polarization Charts

Problems on polarization can be solved by means of charts similar to those used for reflection coefficients and impedances.* These charts may be related to the representation introduced in optics by H. Poincaré: The angles 2α and 2β are taken as the latitude and longitude of a point on a sphere, Fig. 2B. Each state of polarization is thus represented by a single point on the sphere and vice

* V. H. Rumsey, G. A. Deschamps, M. L. Kales, and J. I. Bohnert, "Techniques for Handling Elliptically Polarized Waves with Special Reference to Antennas," *Proceedings of the IRE*, vol. 39, pp. 533–552; May 1951.

versa. Linear polarizations correspond to points on the equator and the two circular polarizations respectively to the poles C and C'. If X represents linear polarization along the reference axis, M some arbitrary polarization, and L the linear polarization along the major axis of the ellipse, the spherical triangle XLM has the following properties

$$XL = 2\beta \qquad\qquad LM = 2\alpha \qquad\qquad XM = 2\gamma$$

$$L = 90° \qquad\qquad\qquad\qquad\qquad\qquad X = \phi$$

From these come the following relations

$$\tan 2\beta = \tan 2\gamma \cos\phi$$

$$\sin 2\alpha = \sin 2\gamma \sin\phi$$

and

$$\cos 2\gamma = \cos 2\alpha \cos 2\beta$$

$$\tan\phi = \tan 2\alpha \csc 2\beta \qquad\qquad (7)$$

which convert from γ, ϕ (polarization ratio) to α, β (ellipse parameters) or vice versa.

These relations can be solved graphically on a chart (Fig. 3) that is a map of the sphere obtained by projection from pole C' on the plane of the equator.* The circles for constant ϕ and constant γ are shown. β is read on the rim and α can be obtained by rotating the point about the center of the chart to bring it on the γ scale on the vertical diameter. A radial arm bearing the same graduations (standing-wave ratio and decibels) as on the Smith chart can also be used. Figure 3 shows only the map of one hemisphere. Polarizations of the opposite sense can be plotted by considering the projection as taken from the pole C.

Example: Assume an axial ratio of 0.5 is measured with an angle of 15° between the maximum field and the reference axis. The intersection M of the radial line $\beta = 15°$ and a circle corresponding to $\alpha = 26.5°$ (since $\tan 26.5° = 0.5$) represents the measured polarization. This polarization can be considered to be produced by two similar radiators normal to each other, the ratio of whose currents is $\tan\gamma = 0.56$ (since the point lies on the $\gamma = 29°$ arc); the current in the radiator along the reference axis is larger and $\phi = 69°$ ahead of the current in the other radiator.

Voltage Induced by Wave of Arbitrary Polarization: If the polarization of the antenna is represented by the point M on the Poincaré sphere and that of the incident wave by N, the voltage induced

* This is a standard geographic projection. Chart H. O. Misc., No. 7736–1 having a 20-centimeter radius, may be obtained at nominal charge from the United States Navy Department Hydrographic Office, Washington D. C., 20301.

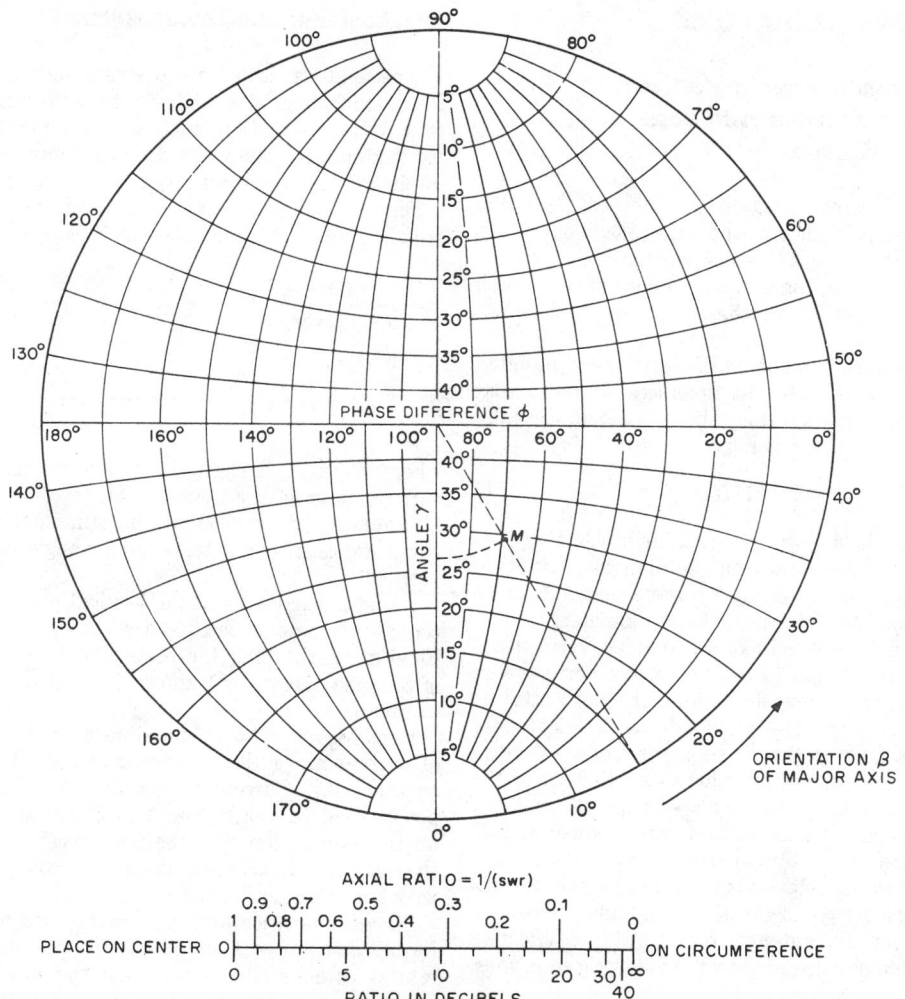

Fig. 3—Projection used in solving polarization problems. The dashed lines and point M are the construction for the example given in the text.

is

$$hE \cos\delta \qquad (8)$$

where 2δ is the angular distance MN. On Fig. 3, the angle 2δ can be obtained by the following construction. Plot the points M and N on a transparent overlay, rotate the overlay about the center 0 until the points M and N fall on the same ϕ circle, and read the difference between the γ's.

Measurement of Wave Polarization

By comparing the signals received by a dipole oriented successively in the directions X and Y, the ratio $|Y|/|X|$ representing the polarization of the wave is found. On Fig. 3, the point M is

on a known γ circle. To obtain another locus, compare the signals received with the same dipole oriented at 45° then 135° from OX. This gives a second circle that can be constructed as the first one with respect to points XY, then rotated by 90° by means of an overlay.

If many measurements are to be taken, the two systems of γ circles could be drawn in advance. This measurement leaves a sense ambiguity that can be resolved only by using receiving antennas with nonlinear polarization.*

* Other methods using the projective chart are described by G. A. Deschamps in "Hyperbolic Protractor for Microwave Impedance Measurements and other Purposes," International Telephone and Telegraph Corporation, 320 Park Ave., New York, New York, 10022; 1953.

VERTICAL RADIATORS

Field Strength From a Vertically Polarized Antenna With Base Close to Ground

The following equation is obtained from elementary-dipole theory and is applicable to low-frequency antennas. It assumes that the earth is a perfect reflector, the antenna dimensions are small compared with λ, and the actual height does not exceed $\lambda/4$.

The vertical component of electric field radiated in the ground plane, at distances so short that ground attenuation may be neglected (usually when $D<10\lambda$), is given by

$$E = 377 I h_e / \lambda D \qquad (9)$$

where $E=$ field strength in millivolts/meter, $I=$ current at base of antenna in amperes, $h_e=$ effective height of antenna, $\lambda=$ wavelength in same units as h_e, and $D=$ distance in kilometers.

The effective height of a grounded vertical antenna is equivalent to the height of a vertical wire producing the same field along the horizontal as the actual antenna, provided the vertical wire carries a current that is constant along its entire length and of the same value as at the base of the actual antenna. Effective height depends upon the geometry of the antenna and varies slowly with λ. For types of antennas normally used at low and medium frequencies, it is roughly one-half to two-thirds the actual height of the antenna.

For certain antenna configurations effective height can be calculated by the following equations.

Straight Vertical Antenna: $h \leq \lambda/4$

$$h_e = \frac{\lambda}{\pi \sin(2\pi h/\lambda)} \sin^2(\pi h/\lambda)$$

where $h=$ actual height.

Loop Antenna: $A < 0.001\lambda^2$

$$h_e = 2\pi n A / \lambda$$

where $A=$ mean area per turn of loop and $n=$ number of turns.

Adcock Antenna:

$$h_e = 2\pi a b / \lambda,$$

where $a=$ height of antenna and $b=$ spacing between antennas.

In the above equations, if h_e is desired in meters or feet, all dimensions h, A, a, b, and λ must be in meters or feet, respectively.

Practical Vertical-Tower Antennas

The field strength from a single vertical tower insulated from ground and either of self-supporting or guyed construction, such as is commonly used for medium-frequency broadcasting, may be calculated by the following equation. This is more accurate than (9). Near ground level the equation is valid within the range $2\lambda < D < 10\lambda$.

$$E = \frac{60I}{D \sin(2\pi h/\lambda)}$$

$$\times \left[\frac{\cos[2\pi(h/\lambda)\cos\theta] - \cos 2\pi(h/\lambda)}{\sin\theta} \right] \qquad (10)$$

where $E=$ field strength in millivolts/meter, $I=$ current at base of antenna in amperes, $h=$ height of antenna, $\lambda=$ wavelength in same units as h, $D=$ distance in kilometers, and $\theta=$ angle from the vertical.

Radiation patterns in the vertical plane for antennas of various heights are shown in Fig. 4. Field strength along the horizontal as a function of antenna height for 1 kilowatt radiated is shown in Fig. 5.

Both Figs. 4 and 5 assume sinusoidal distribution of current along the antenna and perfect ground conductivity. Current magnitudes for 1-kilowatt power used in calculating Fig. 5 are also based on the assumption that the only resistance is the theoretical radiation resistance of a vertical wire with sinusoidal current.

Since inductance and capacitance are not uniformly distributed along the tower and since current is attenuated in traversing the tower, it is impossible to obtain sinusoidal current distribution

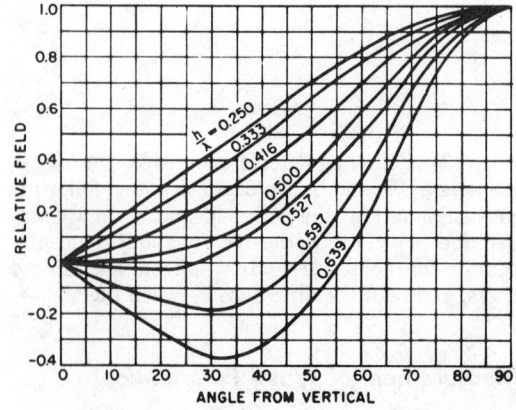

Fig. 4—Field strength as a function of angle of elevation for vertical radiators of different heights.

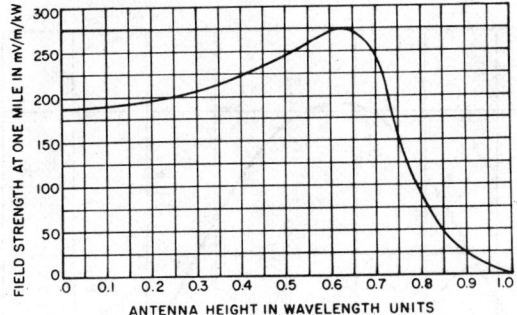

Fig. 5—Field strength along the horizontal as a function of antenna height for a vertical grounded radiator with 1 kilowatt of radiated power.

in practice. Consequently actual radiation patterns and field strengths differ from Figs. 4 and 5.* The closest approximation to sinusoidal current is found on constant-cross-section towers.

In addition, antenna efficiencies vary from about 70 percent for 0.15-wavelength physical height to over 95 percent for 0.6-wavelength height. The antenna input power must be multiplied by the efficiency to obtain the power radiated.

Average results of measurements of impedance at the base of several actual vertical radiators, as given by Chamberlain and Lodge,† are shown in Fig. 6.

For design purposes when actual resistance and current of the projected radiator are unknown, resistance values may be selected from Fig. 6 and the resulting effective current obtained from

$$I_e = (W\eta/R)^{1/2} \qquad (11)$$

where I_e = current effective in producing radiation in amperes, W = watts input, η = antenna efficiency, varying from 0.70 at $h/\lambda = 0.15$ to 0.95 at $h/\lambda = 0.6$, and R = resistance at base of antenna in ohms.

If I_e from (11) is substituted in (10), reasonable approximations to the field strength at unit distances, such as 1 kilometer or 1 mile, will be obtained.

The practical equivalent of a higher tower may be secured by adding a capacitance "hat" with or without tuning inductance at the top of a lower tower.*

A good ground system is important with vertical-radiator antennas. It should consist of at least 120 radial wires, each $\frac{1}{2}$ wavelength or longer, buried 6 to 12 inches below the surface of the soil. A ground screen of high-conductivity metal mesh, bonded to the ground system, should be used on or above the surface of the ground adjacent to the tower.

FIELD STRENGTH AND RADIATED POWER FROM ANTENNAS IN FREE SPACE

Isotropic Radiator

The power density P at a point due to the power P_t radiated by an isotropic radiator is

$$P = P_t/4\pi R^2 \text{ watt/meter}^2 \qquad (12)$$

where R = distance in meters and P_t = transmitted power in watts.

The electric-field strength E in volts/meter and power density P in watts/meter^2 at any point are related by

$$P = E^2/120\pi$$

where 120π is known as the resistance of free space. From this

$$E = (120\pi P)^{1/2} = (30P_t)^{1/2}/R, \text{ volt/meter}. \qquad (13)$$

Half-Wave Dipole

For a half-wave dipole in the direction of maximum radiation

$$P = 1.64P_t/4\pi R^2 \qquad (14)$$

$$E = (49.2P_t)^{1/2}/R. \qquad (15)$$

These relations are shown in Fig. 7.

---

* For information on the effect of some practical current distributions on field strength, see H. E. Gihring and G. H. Brown, "General Considerations of Tower Antennas for Broadcast Use," *Proceedings of the IRE*, vol. 23, pp. 311–356; April 1935.

† A. B. Chamberlain and W. B. Lodge, "The Broadcast Antenna," *Proceedings of the IRE*, vol. 24, pp. 11–35; January 1936.

* For additional information see G. H. Brown, "A Critical Study of the Characteristics of Broadcast Antennas as Affected by Antenna Current Distribution," *Proceedings of the IRE*, vol. 24, pp. 48–81; January 1936. G. H. Brown and J. G. Leitch, "The Fading Characteristics of the Top-Loaded WCAU Antenna," *Proceedings of the IRE*, vol. 25, pp. 583–611; May 1937. Also, C. E. Smith and E. M. Johnson, "Performance of Short Antennas," *Proceedings of the IRE*, vol. 35, pp. 1026–1038; October 1947.

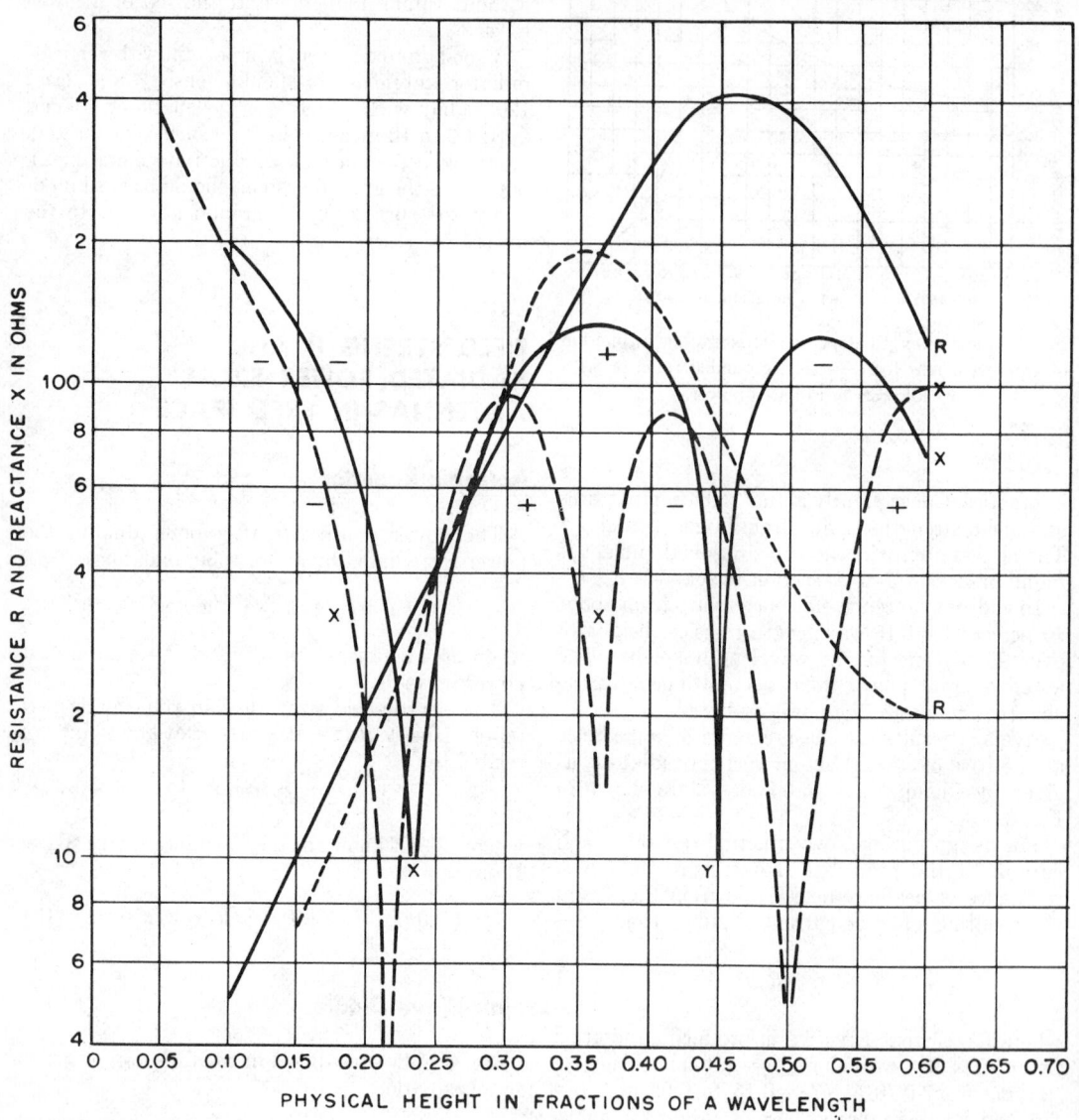

Fig. 6—Resistance and reactance components of impedance between tower base and ground of vertical radiators as given by Chamberlain and Lodge. Solid lines show average results for 5 guyed towers; dashed lines show average results for 3 self-supporting towers. *Courtesy of Proceedings of the IRE.*

Received Power

To determine the power intercepted by a receiving antenna, multiply the power density from Fig. 7 by the receiving area. The receiving area is

$$\text{area} = G\lambda^2/4\pi$$

where G = gain of receiving antenna and λ = wavelength in meters.

The receiving areas and gains of common antennas are given in Table 4 on page 25–44.

Equation (16) can be used to determine the power received by an antenna of gain G_r when the transmitted power P_t is radiated by an antenna of gain G_t.

$$P_r = P_t G_r G_t \lambda^2 / (4\pi R)^2. \qquad (16)$$

G_t and G_r are the gains over an isotropic radiator. If the gains over a dipole are known, instead of gain over isotropic radiator, multiply each gain by 1.64 before inserting in (16).

RADIATION FROM AN END-FED CONDUCTOR OF ANY LENGTH

Expressions for field strength $F(\theta)$ for several radiator configurations are as follows.

(**A**) Half-wave, resonant:

$$F(\theta) = \cos(90° \sin\theta)/\cos\theta$$

(**B**) Any odd number of half waves, resonant:

$$F(\theta) = \cos[(l°/2) \sin\theta]/\cos\theta$$

(**C**) Any even number of half waves, resonant:

$$F(\theta) = \sin[(l°/2) \sin\theta]/\cos\theta$$

(**D**) Any length, resonant:

$$F(\theta) = (1/\cos\theta)[1 + \cos^2 l° + \sin^2\theta \sin^2 l°$$
$$- 2\cos(l° \sin\theta)\cos l° - 2\sin\theta\sin(l° \sin\theta)\sin l°]^{1/2}$$

(**E**) Any length, nonresonant:

$$F(\theta) = \tan(\theta/2)\sin(l°/2)(1 - \sin\theta)$$

where $l° = 360l/\lambda$ = length of radiator in electrical degrees, energy to flow from left-hand end of radiator, l = length of radiator in same units as λ, θ = angle from the normal to the radiator, and λ = wavelength.

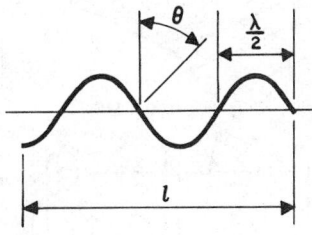

See also Fig. 8.

RHOMBIC ANTENNAS

Linear radiators may be combined in various ways to form antennas such as the horizontal vee, inverted vee, etc. The type most commonly used at high frequencies is the horizontal terminated rhombic shown in Fig. 9.

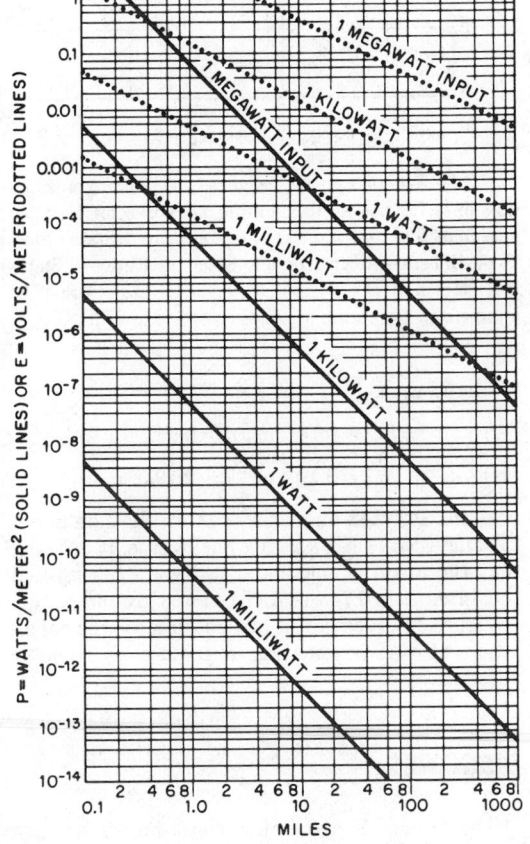

Fig. 7—Power density at various distances from a half-wave dipole.

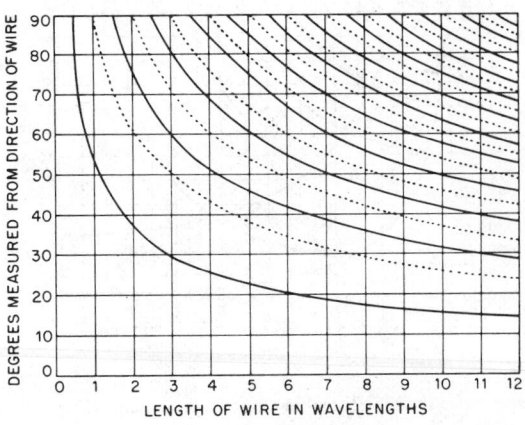

Fig. 8—Directions of maximum (solid lines) and minimum (dotted lines) radiation from a single-wire radiator. Direction given here is $(90° - \theta)$.

In designing rhombic antennas* for high-frequency radio circuits, the desired vertical angle Δ of radiation above the horizon must be known or assumed. When the antenna is to operate over a wide range of radiation angles or is to operate on several frequencies, compromise values of h, L, and φ must be selected. Gain of the antenna increases as the length L of each side is increased; however, to avoid too-sharp directivity in the vertical plane, it is usual to limit L to less than 6 wavelengths.

Knowing the side length and radiation angle desired, the height h above ground and the tilt angle φ can be obtained from Fig. 10.

Example: Find h and φ if Δ = 20 degrees and L = 4λ. On Fig. 10 draw a vertical line from Δ = 20 degrees to meet L/λ = 4 curve and h/λ curves, From intersection at L/λ = 4, read on the right-hand scale φ = 71.5 degrees. From intersection on h/λ curves, there are two possible values on the

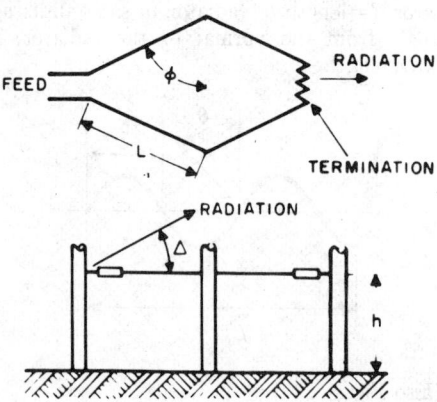

Fig. 9—Dimensions and radiation angles for rhombic antenna.

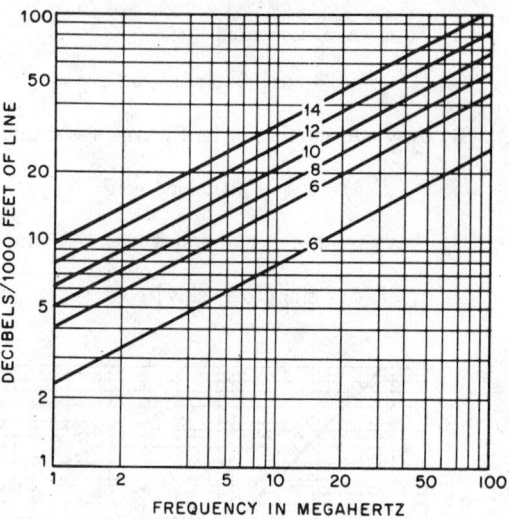

Fig. 11—Attenuation of balanced 600-ohm transmission lines for use as terminating networks for rhombic antennas. The top 5 curves are for United States Steel Type 12 or American Iron and Steel Institute 410 Stainless Steel and the bottom curve is for iron wire. The numbers on the curves are sizes in American Wire Gauge.

left-hand scale

(**A**) $h/\lambda = 0.74$ or $h = 0.74\lambda$

(**B**) $h/\lambda = 2.19$ or $h = 2.19\lambda$.

Similarly, with an antenna 4λ on the side and a tilt angle φ = 71.5°, working backwards, it is found that the angle of maximum radiation Δ is 20°, if the antenna is 0.74λ or 2.19λ above ground.

Figure 11 gives useful information for the calculation of the terminating resistance of rhombic antennas.

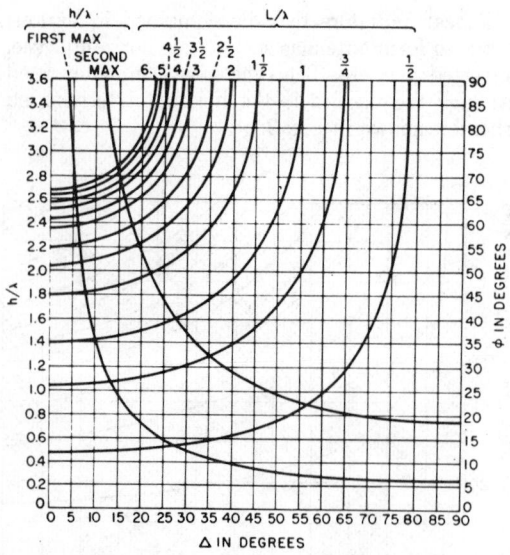

Fig. 10—Rhombic-antenna design chart.

* For more complete information see A. E. Harper, "Rhombic Antenna Design," D. Van Nostrand Company, New York; 1941.

DISCONES

The discone is a radiator whose impedance can be directly matched to a 50-ohm coaxial transmission line over a wide frequency band. The outer conductor of the transmission line is connected to the cone at the gap and the inner conductor to the center of the disc. The dimensions shown in Fig. 12 give the best impedance match over a wide band.* Since the bandwidth is inversely proportional to C_{min}, that dimension is usually made only slightly larger than the diameter of the coaxial transmission line. Dimensions S and D are determined from $S = 0.3C_{min}$ and $D = 0.7C_{max}$. L and ϕ determine how the standing-wave ratio varies with frequency at the low edge of the band, as shown in Fig. 13. A discone with $\phi = 60°$ and $C_{min}/L = 1/22$ had a standing-wave ratio of less than 1.5 over at least a 7/1 frequency range and a standing-wave ratio of less than 2 over at least a 9/1 range in frequency.

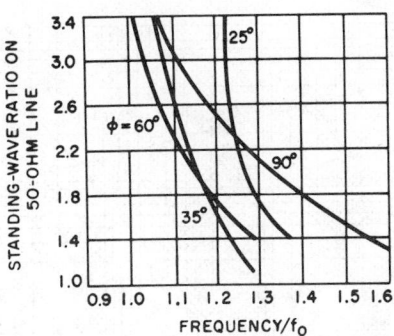

Fig. 13—Standing-wave ratio versus ratio of frequency to the frequency at which slant height is $\lambda/4$. *Courtesy of Electronics.*

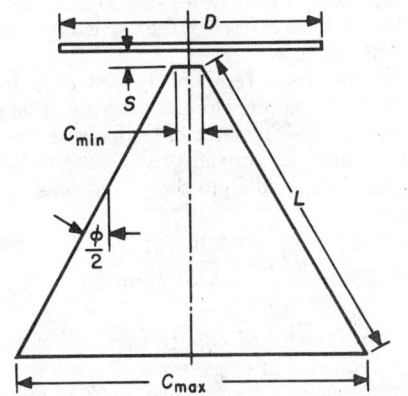

Fig. 12—Optimum discone dimensions. *Courtesy of Electronics.*

The pattern is omnidirectional in the H plane, while the E-plane pattern varies somewhat with frequency as shown in Fig. 14.

HELICAL ANTENNAS

Helical antennas can be classified either as to shape (such as cylindrical, flat, or conical) or as to type of pattern produced (such as normal or axial mode). Data will be given here only for the cylindrical helix radiating in the normal and axial modes.

Normal-Mode Helix

When the diameter is considerably less than a wavelength and the electrical length less than a

wavelength, the helix radiates in the normal mode (peak of the pattern normal to the helix axis). In contrast with the ordinary dipole, where the radiating electromagnetic wave appears to travel on the dipole with the velocity of light in the surrounding medium, the velocity of the wave along the axis of the helix is lower and depends on the frequency, diameter, and number of turns per unit length. The velocity can be decreased by large factors with a corresponding decrease in axial length for quarter-wave or half-wave resonance.

Velocity of Propagation: The phase velocity along the helix axis is

$$(c/v)^2 = 1 + (M\lambda/\pi D)^2 \qquad (17)$$

where c = velocity of light in surrounding medium, v = axial velocity, λ = wavelength in surrounding medium, D = mean helix diameter (same units as λ), and M = value obtained from Fig. 15.

The apparent phase velocity in the direction of the wire is equal to the axial velocity divided by the sine of the pitch angle, or

$$\left(\frac{V_w}{c}\right)^2 = \frac{1 + (N\pi D)^2}{1 + (M\lambda/\pi D)^2} \qquad (18)$$

where N is the number of turns per unit length. Figure 15 shows the variation of V_w/c when the

* J. J. Nail, "Designing Discone Antennas," *Electronics*, vol. 26, pp. 167–169; August 1953.

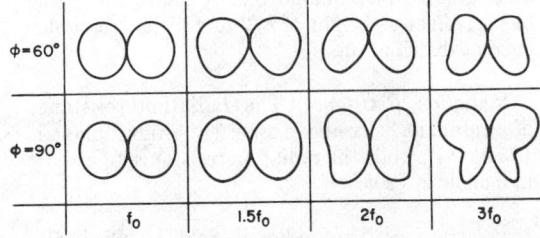

Fig. 14—Discone E-plane patterns. *Courtesy of Electronics.*

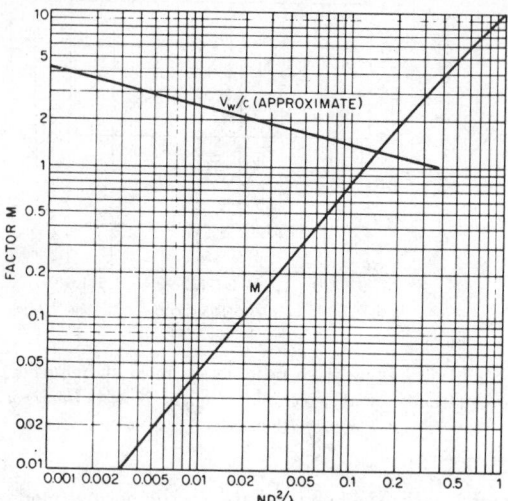

Fig. 15—Chart giving M for (17) and (18) and also showing apparent phase velocity V_w/c.

terms in (18) are much greater than unity. Figure 16 shows, for a particular case, how the frequency for quarter-wave resonance varies with the number of turns per unit length for constant wire length. When $ND \geq 1$ and $ND^2/\lambda \leq 1/5$, (18) reduces to

$$V_w/c \approx 1.25(h/D)^{1/5} \qquad (18A)$$

where $h=$ height of the quarter-wavelength helix.

To obtain a real input impedance (resonance), each half of the helical antenna must be a quarter-wavelength long at the velocity given above or for $ND^2/\lambda < 1/5$

$$h/\lambda = V/4c$$

$$= \tfrac{1}{4}[1+20(ND)^{5/2}(D/\lambda)^{1/2}]^{-1/2} \qquad (19)$$

where h is the length of each half.

Effective Height: The effective height of a resonant helix above a perfect ground plane is $2h/\pi$ because the current distribution is similar to that of a quarter-wave monopole. A short monopole has an effective height of $h/2$ due to its triangular current distribution.

Radiation Resistance: The radiation resistance of a resonant helix above a perfect ground plane is $(25.3h/\lambda)^2$, while the radiation resistance of a short monopole is $(20h/\lambda)^2$.

Polarization: The radiated field is elliptically polarized and the ratio of the horizontally polarized

field E_h to the vertically polarized field E_v is

$$E_h/E_v = \frac{(N\pi D)J_1(\pi D/\lambda)}{J_0(\pi D/\lambda)}$$

$$\approx 5ND^2/\lambda \qquad (20)$$

where $J_0, J_1=$ Bessel functions* of the first kind.

The approximation is valid for diameters less than 0.1 wavelength. Circular polarization is obtained with a resonant helix when the height is about 0.9 times the diameter.

The horizontal polarization is decreased considerably when the helix is used with a ground plane. The vertical pattern of the horizontally polarized field then varies as $2(h/\lambda)\sin\theta\cos\theta$, while the vertical pattern of the vertically polarized field varies as $\cos\theta$.

Losses for short resonant helices may be appreciable because the wire diameter must be much smaller than the diameter of a dipole of the same height. Neglecting proximity effects, the ratio of the power dissipated P_l to the power radiated P_r is

$$P_l/P_r = \frac{2 \times 10^{-4}(V_w/c)}{d(h/\lambda)^2 F_{\mathrm{MHz}}^{1/2}} \qquad (21)$$

where $d=$ diameter of copper wire in inches, and $F_{\mathrm{MHz}}=$ frequency in megahertz.

The efficiency is thus $1/(1+P_l/P_r)$. Figure 17 is a plot of height versus resonant frequency for 3 wire diameters for 50-percent efficiency, assuming that $V_w/c=1$.

Q and Tap Point: The Q factor† can be calcu-

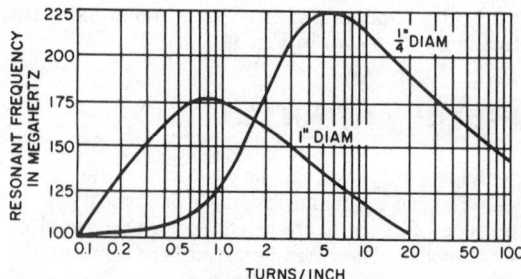

Fig. 16—Resonant frequency for various helix configurations with same length of wire.

---

* Table of Bessel functions is given in Chapter 47.

† Unloaded Q. When the antenna is driven by a zero-resistance generator, the 3-decibel bandwidth is f_0/Q. When driven by a generator whose resistance matches the resonant resistance of the antenna, the 3-decibel bandwidth is $2f_0/Q$.

lated* approximately

$$Q = \pi Z_0 / 4R_{\text{base}} \qquad (22)$$

where

$Z_0 =$ characteristic impedance

$$= 60(c/V)[\ln(4h/D) - 1]$$

$R_{\text{base}} =$ radiation resistance plus wire resistance

$$= (25.3h/\lambda)^2 + 0.125(V_w/c)/dF_{\text{MHz}}^{1/2}$$

where $d =$ wire diameter in inches.

The input resonant resistance R_{tap} with one end of the resonant helix connected to a perfectly

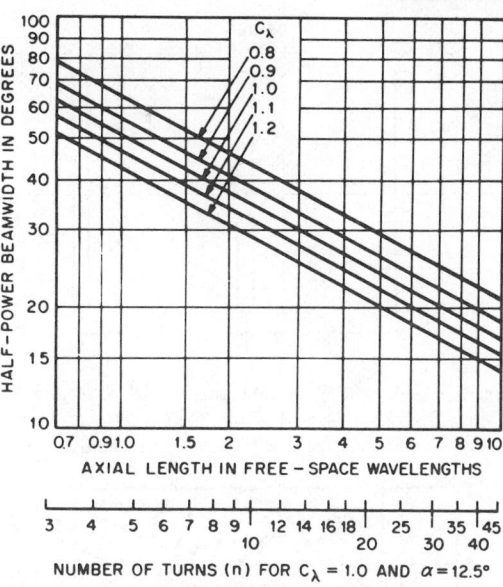

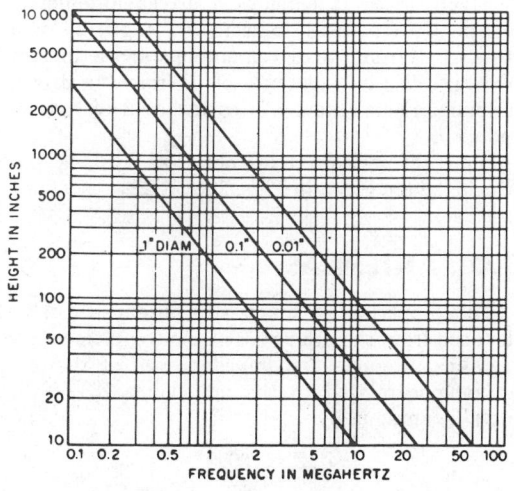

Fig. 17—Helix height versus frequency for 50-percent efficiency, assuming $V_w/c = 1$.

conducting ground plane is

$$R_{\text{tap}} = (4/\pi)QZ_0 \sin^2\theta \qquad (23)$$

where $\theta =$ angular distance between tap point and the ground plane.

Axial-Mode Helix

When the helix circumference is of the order of a wavelength, an end-fire circularly polarized pattern (axial ratio less than 6 decibels) is obtained.†

* A. G. Kandoian and W. Sichak, "Wide-Frequency-Range Tuned Helical Antennas and Circuits," *Electrical Communication*, vol. 30, pp. 294–299; December 1953; also, *Convention Record of the IRE 1953 National Convention*, Part 2—Antennas and Communication; pp. 42–47.

† J. D. Kraus, "Antennas," McGraw-Hill Book Co. Inc., New York; 1950: p. 213.

Fig. 18—Axial-mode helical-antenna beamwidth versus length. $\theta_{3dB} = 115/C_\lambda(nS_\lambda)^{1/2}$, degrees; directivity $G = 15C_\lambda^2 nS_\lambda$, where $nS_\lambda =$ length, $n =$ number of turns, $S_\lambda =$ spacing in air wavelengths, $C_\lambda =$ circumference, and $\alpha =$ pitch angle. J. D. Kraus, "Antennas," Fig. 7-22, © 1950, McGraw-Hill Book Company.

The half-power beamwidth versus axial length in free-space wavelengths is given in Fig. 18. C_λ, the circumference in air wavelengths, varies between 0.8 and 1.2.

PARASITIC ARRAY

*Multielement Yagi-Uda Array**

Closely coupled parasitic arrays where the parasitic element may function as either a director or

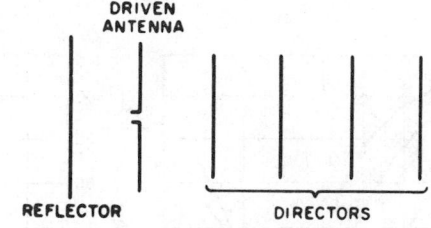

Fig. 19—Yagi-Uda array.

* S. Uda and Y. Mushiake, "Yagi-Uda Antenna," Sasaki Printing and Publishing Co., Ltd., Sendai, Japan; 1954.

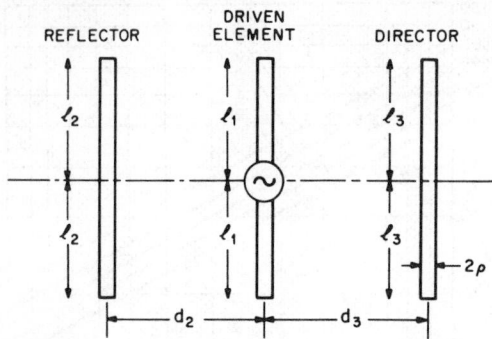

Fig. 20—3-element Yagi–Uda antenna. *S. Uda and Y. Mushiake, "Yagi–Uda Antenna," Fig. 9–1, © 1954, Sasaki Printing and Publishing Company.*

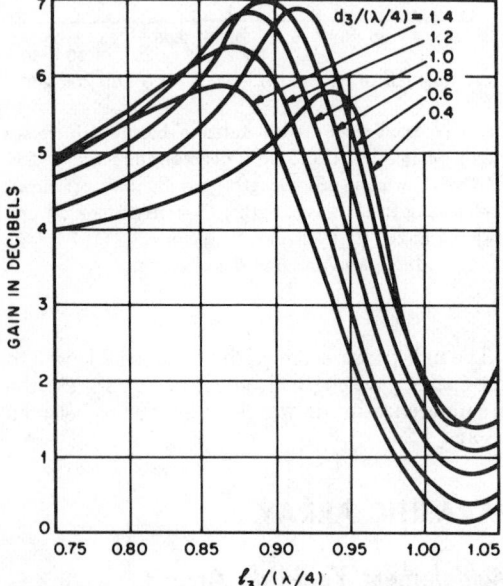

Fig. 21—Calculated gain of 3-element Yagi–Uda antenna for indicated values of $d_3/(\lambda/4)$. $l_1=l_2=d_2=\lambda/4$ and $\rho=\lambda/200$. *S. Uda and Y. Mushiake, "Yagi–Uda Antenna," Fig. 9–3, © 1954, Sasaki Printing and Publishing Company.*

reflector are often used in practice, particularly between 100 and 1000 megahertz. These were first described by S. Uda in Japanese and subsequently in English by H. Yagi. They are often referred to as Yagi arrays but lately have been known as Yagi–Uda arrays.

Figure 19 shows a multielement Yagi–Uda array. Little is gained from using more than 1 reflector but a worthwhile gain is obtained by using many directors.

3-Element Yagi–Uda Array

The 3-element array, shown in Fig. 20, is extensively used. It employs a driven antenna of length $2l_1$, a parasitic reflector of length $2l_2$ at a distance d_2 from the driven antenna and a director of length $2l_3$ at a distance of d_3 from the driven antenna. Each antenna is constructed of rods of radius ρ.

Some theoretical curves obtained by Uda and Mushiake are given in Figs. 21 and 22.

SLOT ANTENNAS

The properties of many slot antennas can be deduced from the properties of the complementary metallic antenna. The impedance Z_s of the slot antenna is related to the impedance Z_m of the metallic antenna by

$$Z_m Z_s = (60\pi)^2. \qquad (24)$$

The magnitude of the electric field E_s produced by the slot is proportional to the magnitude of the magnetic field H_m of the metallic antenna, and H_s is proportional to E_m. The electric- and magnetic-plane patterns of the slot are similar to the magnetic- and electric-plane patterns, respectively, of the metallic antenna.

Example: Slot antenna in an infinite metallic plane, Fig. 23. The complementary metallic antenna is a dipole. For a narrow slot a half-wavelength long, fed at the center, the impedance is

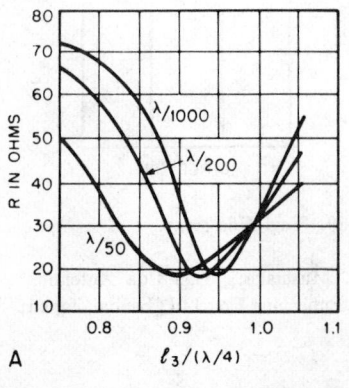

A

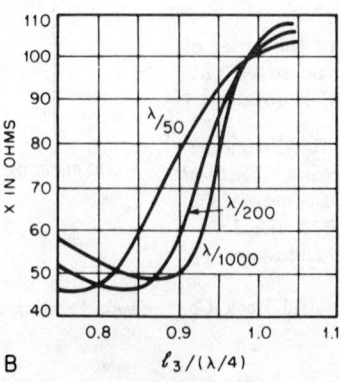

B

Fig. 22—Calculated input impedance of 3-element Yagi–Uda antenna for indicated values of ρ. In both curves $l_1=l_2=d_2=\lambda/4$, and $d_3/(\lambda/4)=0.8$. *S. Uda and Y. Mushiake, "Yagi–Uda Antenna," Fig. 9–2, © 1954, Sasaki Printing and Publishing Company.*

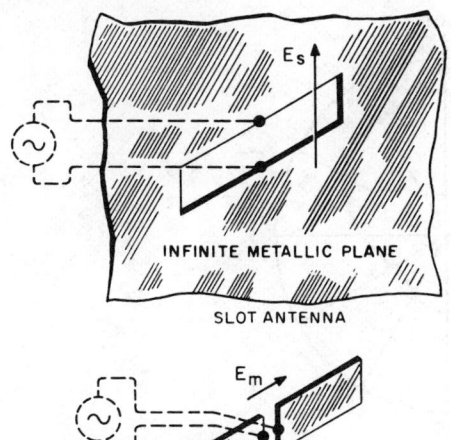

Fig. 23—Slot antenna and its metallic counterpart.

$(60\pi)^2/73 = 494$ ohms if the slot radiates on both sides. (If a cavity is added to suppress radiation on one side, the impedance doubles.) The E-plane pattern of the slot and the H-plane pattern of the dipole are omnidirectional, while the slot H-plane pattern is the same as the dipole E-plane pattern.

Impedance of Thin Rectangular Slot (Fig. 24)

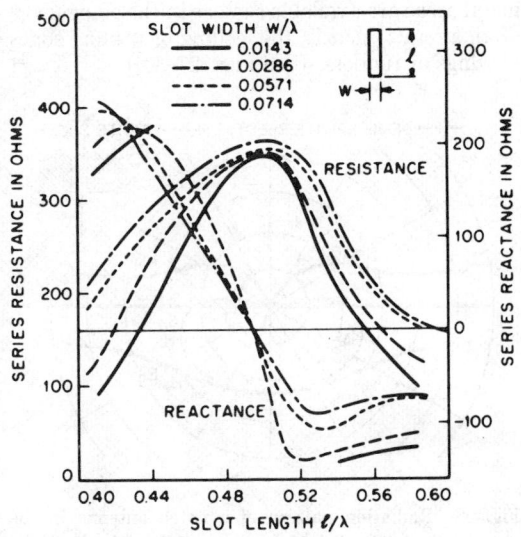

Fig. 24—Impedance of thin rectangular slot. *H. Jasik, "Antenna Engineering Handbook," Fig. 8-16,* © *1961, McGraw-Hill Book Company.*

Impedance of Small Annular Slots

The annular-slot antenna, the complement of a loop, is often used as a flush-mounted antenna to produce a pattern and polarization similar to that of a short dipole mounted on a large ground plane. When the outer diameter is less than about a tenth

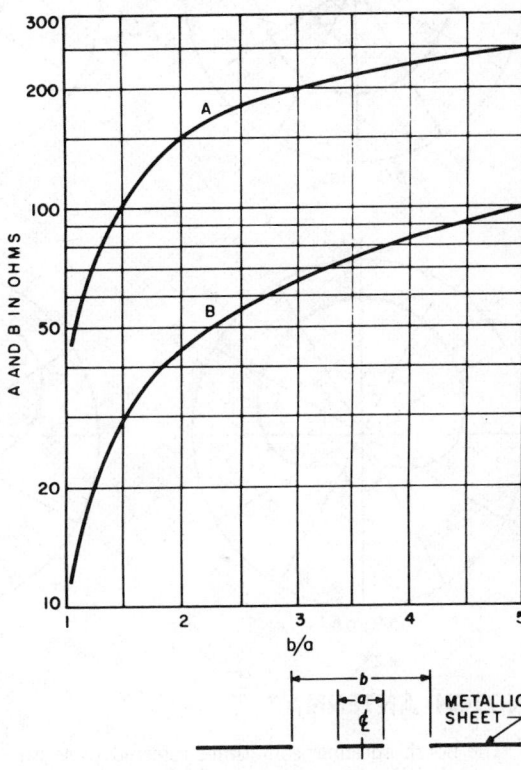

Fig. 25—Impedance of annular-slot antenna. $R = A(b/\lambda)^2$ and $X = B(\lambda/b)$ (capacitive).

of a wavelength, the impedance* is given by Fig. 25.

Axial Slots on Cylinders

Figure 26 shows how the E-plane pattern† of an axial slot in the surface of a cylinder varies with diameter and wavelength.

* H. Levine and C. H. Papas, "Theory of the Circular Diffraction Antenna," *Journal of Applied Physics*, vol. 22, pp. 29–43; January 1951.
† G. Sinclair, "Patterns of Slotted-Cylinder Antennas," *Proceedings of the IRE*, vol. 36, pp. 1487–1492; December 1948.

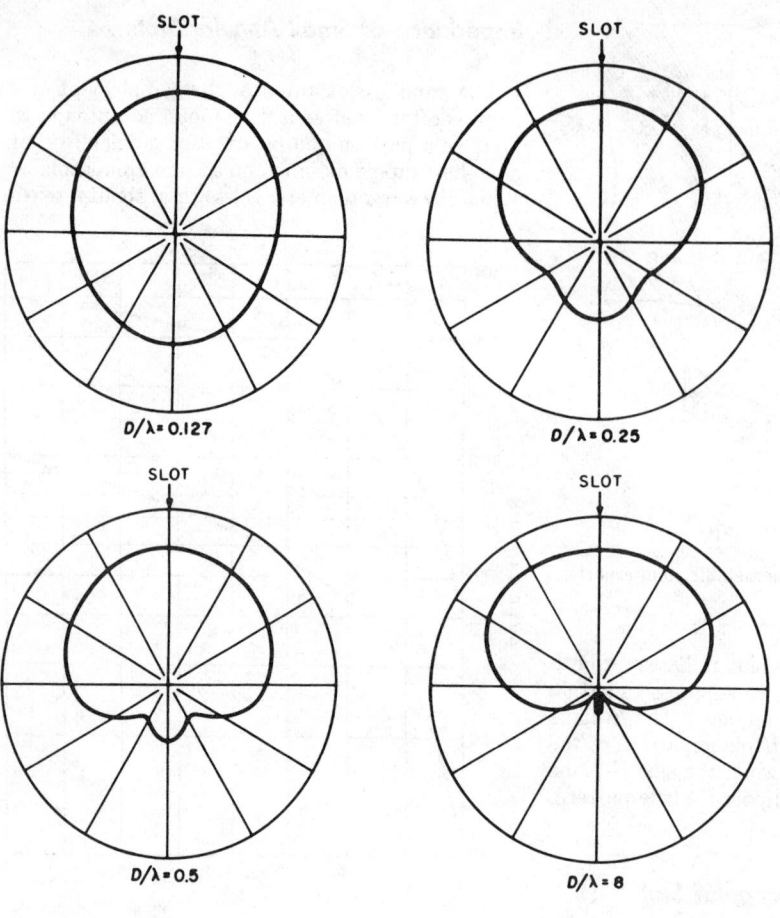

Fig. 26—Radiation pattern for single axially slotted cylindrical antenna of diameter *D. Courtesy of Proceedings of the IRE.*

NOTCH ANTENNA*

The notch antenna, sometimes referred to as an open-ended slot antenna, is used in cases where a broad-band radiator is necessary and pseudo-sheet-metal areas are available such as in the empennage of an aircraft, that is, the leading or trailing edges of wings or rudders. (See Figs. 27–30.)

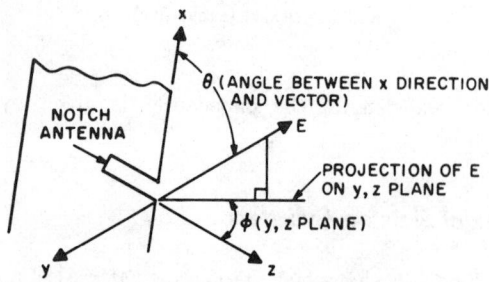

Fig. 27—Notch antenna. *Adapted from H. Jasik, "Antenna Engineering Handbook," Fig. 8–12, © 1961, McGraw-Hill Book Company.*

* R. H. J. Cary, "The Slot Aerial and Its Applications to Aircraft," *Proceedings of the Institution of Electrical Engineers,* Part III, vol. 99, pp. 187–196, 210–213; July 1952.

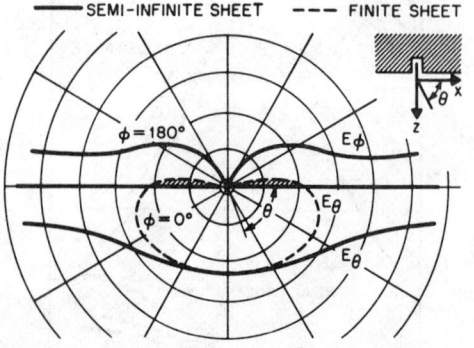

Fig. 28—Radiation pattern of a notch antenna in the plane of the notched sheet (*xz* plane of Fig. 27). *H. Jasik, "Antenna Engineering Handbook," Fig. 8–13, © 1961, McGraw-Hill Book Company.*

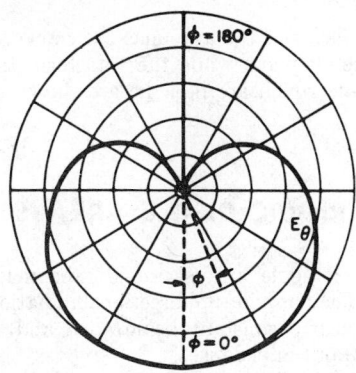

Fig. 29—Radiation pattern of a notch antenna in the plane perpendicular to the notched sheet through the centerline of the notch (*yz* plane of Fig. 27). *H. Jasik, "Antenna Engineering Handbook," Fig. 8–14, © 1961, McGraw-Hill Book Company.*

FREQUENCY-INDEPENDENT ANTENNAS*

The performance of a lossless antenna is independent of frequency if its dimensions, when measured in wavelengths, are held constant. If the shape of the antenna can be specified entirely in angles, its performance is independent of frequency. An example is the infinite biconical antenna.

Figures 31 and 32 are examples of frequency-independent antennas.

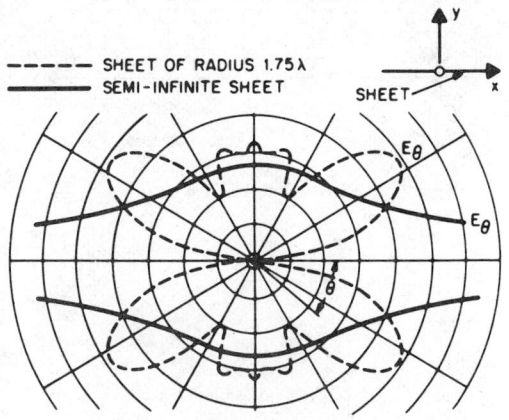

Fig. 30—Radiation pattern of a notch antenna in a plane perpendicular to both the notched sheet and the notch (*xy* plane of Fig. 27). *H. Jasik, "Antenna Engineering Handbook," Fig. 8–15, © 1961, McGraw-Hill Book Company.*

* V. H. Rumsey, "Frequency-Independent Antennas," IRE National Convention Record, Part 1; pp. 114–118; 1957.

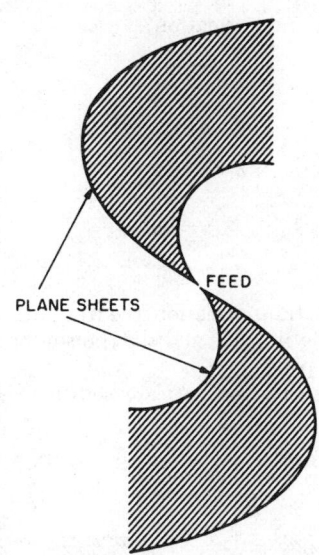

Fig. 31—Antenna formed by four planar curves that remain essentially the same when scaled to a different unit of length. *V. H. Rumsey, "Frequency–Independent Antennas," IRE National Convention Record, Part 1, Fig. 2; 1957.*

Logarithmic Periodic Antennas

If an antenna is designed so that its characteristics are periodic with the logarithm of the frequency, and the variation of the characteristics are negligible or small over a single period, then a practical frequency-independent antenna is essentially obtained.

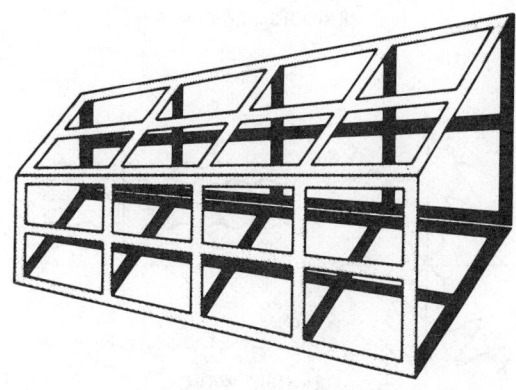

Fig. 32—Horn antenna with holes of uniformly expanding size. *V. H. Rumsey, "Frequency–Independent Antennas," IRE National Convention Record, Part 1, Fig. 7; 1957.*

Using the transformation

$$Z = \ln W$$

or if

$$W = \rho \exp(j\theta)$$

$$Z = x + jy$$

then

$$x = \ln \rho$$

$$y = \theta.$$

Using this transformation, the true antenna form will be in the W plane and the transformation will be in the Z plane.

Examples of structures are shown in Fig. 33.

The true forms of the antennas are shown on the left in the W plane while the transformations in the Z plane which are periodic are shown on the right.

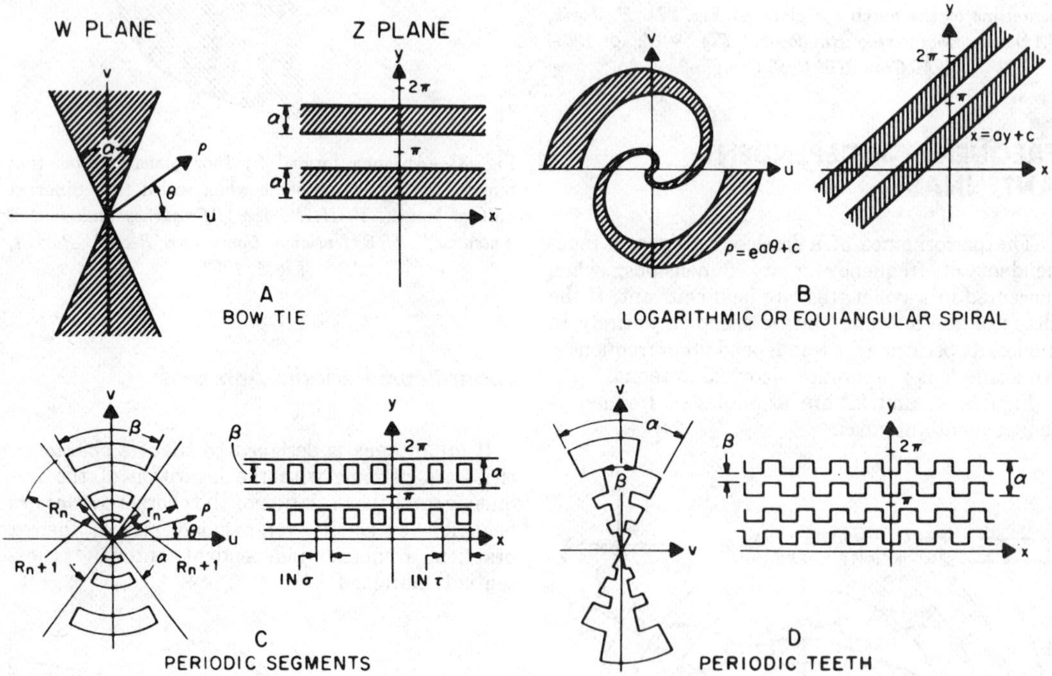

LOG PERIODIC DIPOLE ARRAYS*

Coplanar dipole arrays can also be made log periodic. They provide unidirectional radiation patterns of nearly constant beamwidth and nearly constant input impedance.

A schematic diagram of such an array is shown in Fig. 34, where a balanced feed is employed.

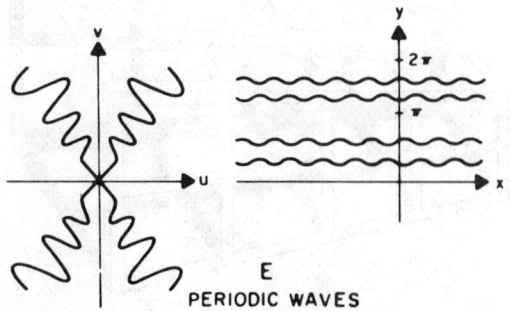

Fig. 33—Logarithmically periodic structures. *R. H. DuHamel and D. E. Isbell, "Broad-Band Logarithmically Periodic Antenna Structures," IRE 1957 National Convention Record, Part 1, pp. 119–128; 1957.*

---

* D. E. Isbell, "Log Periodic Dipole Arrays," IRE Transactions on Antennas and Propagation, vol. AP-8, no. 3, pp. 260–267; May 1960.

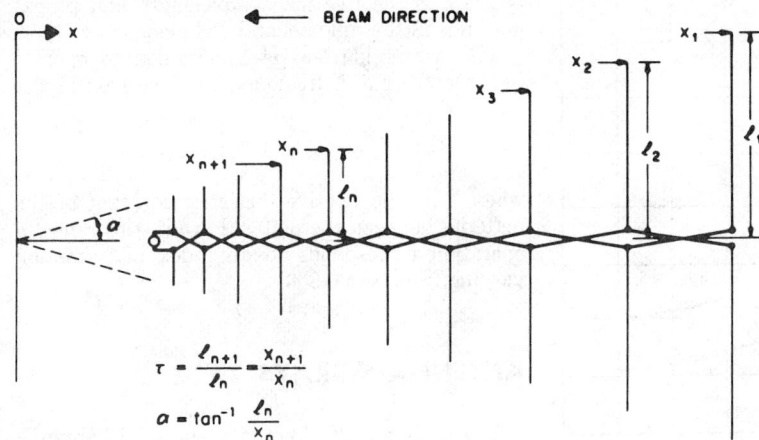

Fig. 34—Schematic diagram of a log periodic dipole array. *D. E. Isbell, "Log Periodic Dipole Arrays," IRE Transactions on Antennas and Propagation, vol. AP-8, no. 3, Fig. 3; 1960.*

τ is the geometric ratio and α the so-called angle of the array.

The radiating efficiency versus antenna size L at its widest point, in wavelengths, is shown for an antenna array with a τ of 0.89 and an α of 45° in Fig. 35.

The measured input impedance of τ's of 0.81, 0.89, and 0.95 are shown in Fig. 36.

The measured directivity plotted against the antenna angle α for the same three values of τ are shown in Fig. 37.

LENS-TYPE RADIATORS

Even though they may be more complicated to design, lenses have the following useful characteristics.

(**A**) Ability to scan over a wide angle compared with the beamwidth.

(**B**) Lower tolerances.

(**C**) Less rearward radiation.

(**D**) Lower construction costs in many cases.

Simple types of rotational lenses for pencil-beam operation are shown in Fig. 38.

Lens shape design equations for rotational lenses are shown in Fig. 39.

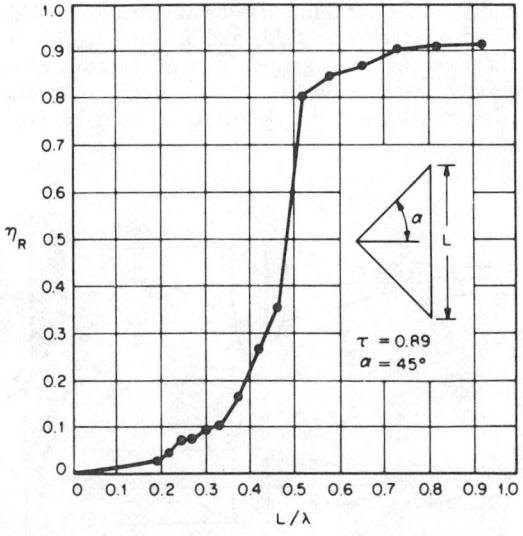

Fig. 35—Radiating efficiency versus antenna size of a log periodic dipole array. *D. E. Isbell, "Log Periodic Dipole Arrays," IRE Transactions on Antennas and Propagation, vol. AP-8, no. 3, Fig. 7; 1960.*

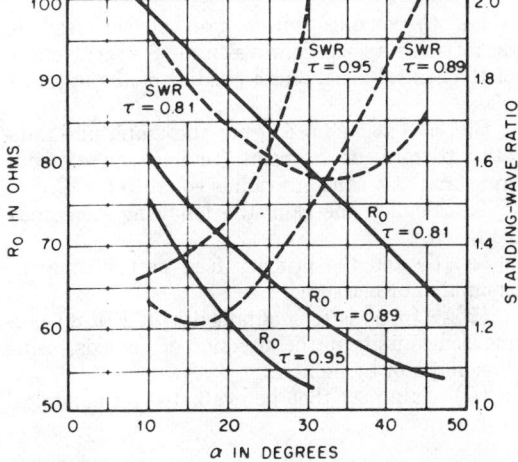

Fig. 36—Measured input impedance and standing-wave ratios for 3 values of τ for a log periodic dipole array. *Adapted from D. E. Isbell, "Log Periodic Dipole Arrays," IRE Transactions on Antennas and Propagation, vol. AP-8, no. 3, Figs. 8, 9, 10; 1960.*

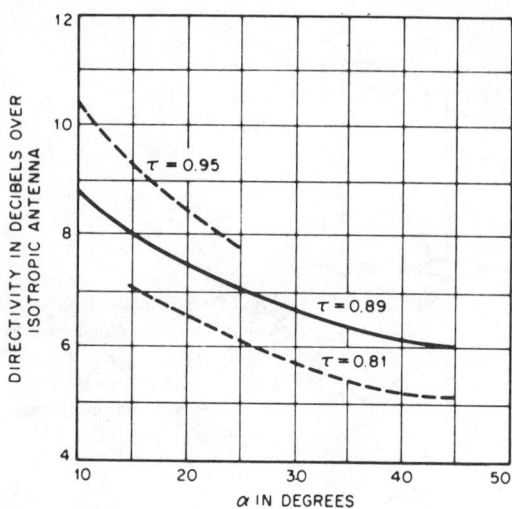

Fig. 37—Directivity versus τ and α of a log periodic dipole array. *D. E. Isbell, "Log Periodic Dipole Arrays," IRE Transactions on Antennas and Propagation, vol. AP-8, no. 3, Fig. 14; 1960.*

The gain of a lens antenna depends on the aperture illumination. For the case where the lens is either a figure of rotation or a cylindrical surface, the aperture illumination ratio is given in Fig. 40.

The maximum power gain G in decibels of a lossless single-refracting-surface rotational lens fed at its focus is given by

$$G = 20 \log_{10}\left[(\lambda f_1)^{-1} \int_{\text{aperture}} \rho A(\rho, \phi)\, d\rho\, d\phi\right] + G_f$$

(25)

where G_f = on-axis power gain of the feed in decibels, λ = free-space wavelength, f_1 = focal length of the central zone, and ρ = radius of the aperture plane.

$F(\theta, \phi)$ [and $F(\theta, Z)$] are the amplitude radiation patterns of the point (or line source) feeds measured at a constant radius equal to f_1.

To calculate the gain, the following steps may be taken.

(**A**) G_f and the primary feed pattern may be computed or measured.

(**B**) $A(\rho, \phi)$ is then obtained with $F(\theta, \phi)$ normalized to unity in the direction of the axis, using Fig. 40 point by point.

(**C**) Gain may then be evaluated by numerical integration.

Metal-Plate Waveguide Lenses

For plates spaced a distance d apart, the index of refraction n is

$$n = [1 - (\lambda/2d)^2]^{1/2}$$

(26)

where λ is the free-space wavelength and propagation is in the fundamental *TE* mode.

The waveguide lens is constrained to operate between the cutoff frequency and the next higher *TE* mode or

$$0.5 < [a(\epsilon_r)^{1/2}]/\lambda < 1.0$$

where ϵ_r is the relative dielectric constant of the material between the plates. For air, this range of operation corresponds to an index of refraction varying from 0 to 0.866.

ANTENNA ARRAYS

The basis for all directivity control in antenna arrays is wave interference. By providing a large number of sources of radiation, it is possible with a fixed amount of power to greatly reinforce radiation in a desired direction while suppressing it in undesired directions. The individual sources may be any type of antenna.

Distant Field

The distant magnetic field is related to the distant electric field of an array by

$$H = (\epsilon/\mu)^{1/2}(i_p \times E)$$

where ϵ = permittivity of the medium, μ = permeability of the medium, H = distant magnetic field, E = distant electric field, and i_p = unit vector in direction of propagation. $(i_p \times E)$ = vector cross product (see p. 46–31). Consequently, only one distant field, say the electric field, need be computed.

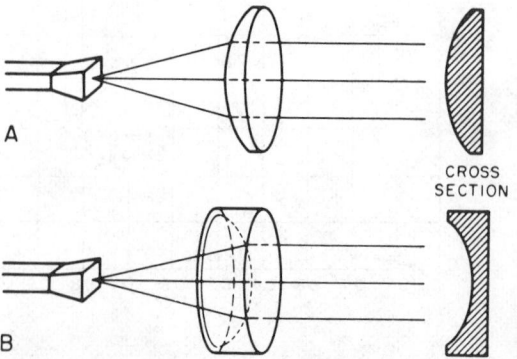

Fig. 38—Rotational lenses using horn feed for pencil-beam operation. In *A* the index of refraction $n > 1$; in *B* the index of refraction $n < 1$.

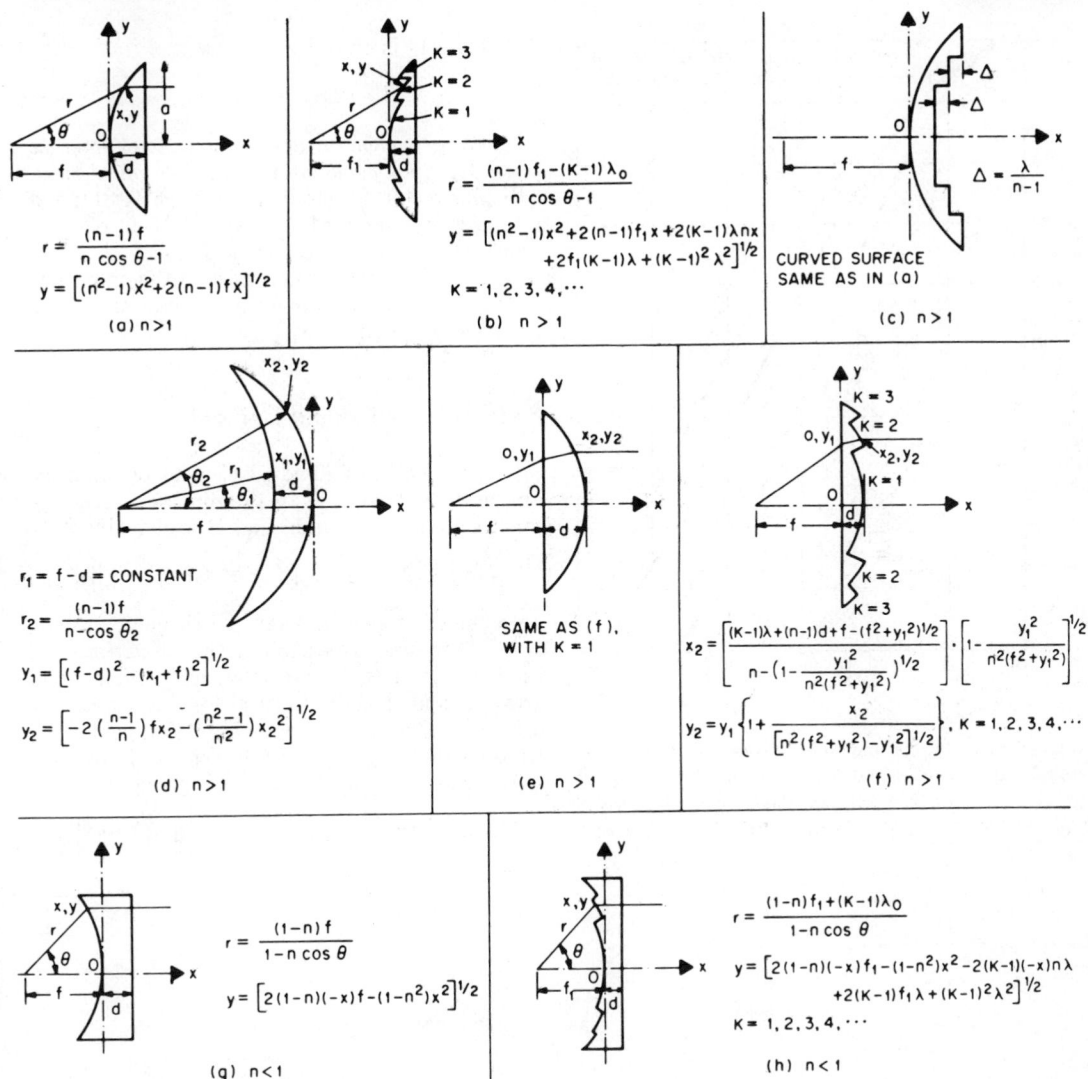

Fig. 39—Design equations for rotational lenses. *Adapted from H. Jasik, "Antenna Engineering Handbook," Figs. 14-2 and 14-5,* © *1961, McGraw-Hill Book Company.*

The distant electric field is given by the vector sum of the contribution from all the sources, with proper consideration for phase.

Amplitude of Individual Elements

The element of an array is a small antenna which has a pattern of electric field given by $A(\theta, \phi)$ which expresses the relative intensity radiated in a given direction θ, ϕ symbolized by a unit direction vector i_p. The magnitude e_p of the field contributed by a single element in the direction θ, ϕ of unit

vector i_p is given by

$$e_p = aA(\theta, \phi)$$

where $a =$ the excitation current with which the element is driven (may be a complex number), and $A(\theta, \phi) =$ the relative pattern strength of the element in the direction θ, ϕ.

Phase of Individual Elements

The phase ψ of the field contributed by an element of an array at the distant field point P

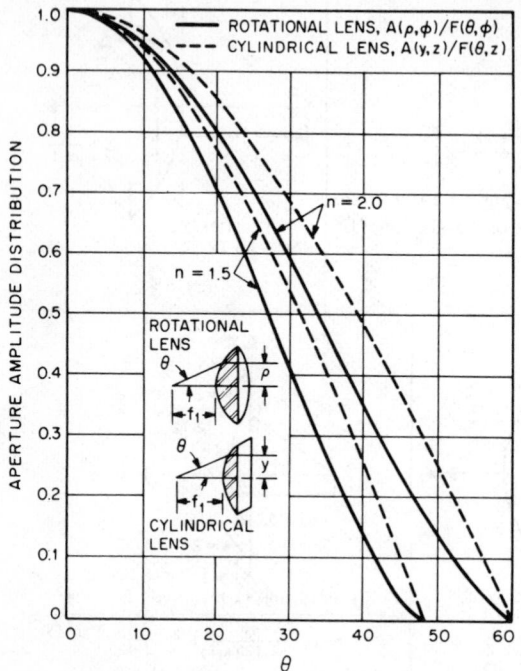

Fig. 40—Aperture illumination ratios for the simple cylindrical and rotational lenses of Fig. 39(a) and (g) for $n>1$. H. Jasik, "Antenna Engineering Handbook," Fig. 14-7(a), © 1961, McGraw-Hill Book Company.

(see Fig. 41) is given by

$$\psi = (2\pi/\lambda)(R \cdot i_p) - q$$

where λ = wavelength, R = position vector *from the phase reference point* for the array *to the element*, $(R \cdot i_p)$ = vector dot product (see page 44-31), and q = phase shift applied to the field of an element (most often in the transmission line).

The phase is made to operate on the amplitude of the electric field of an element by means of the exponential function, $\exp(j[\psi])$.

Field of an Individual Element

The field of an individual element of an array at the distant point P in the direction (θ, ϕ) of unit vector i_p as shown in Fig. 41 is given by

$$e_p = aA(\theta, \phi) \exp\{ j[(2\pi/\lambda)(R \cdot i_p) - q]\}.$$

In the most general array the elements are not alike. Therefore, the excitations a, the patterns A, the radii from the phase reference point for the array R, and the additional phase shifts q, are all different. One must distinguish the field contributed to P, e_{pk} by the kth element from that of the other elements. This field is given by

$$e_{pk} = a_k A_k(\theta, \phi) \exp\{ j[(2\pi/\lambda)(R_k \cdot i_p) - q_k]\}.$$

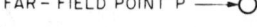

FAR − FIELD POINT P

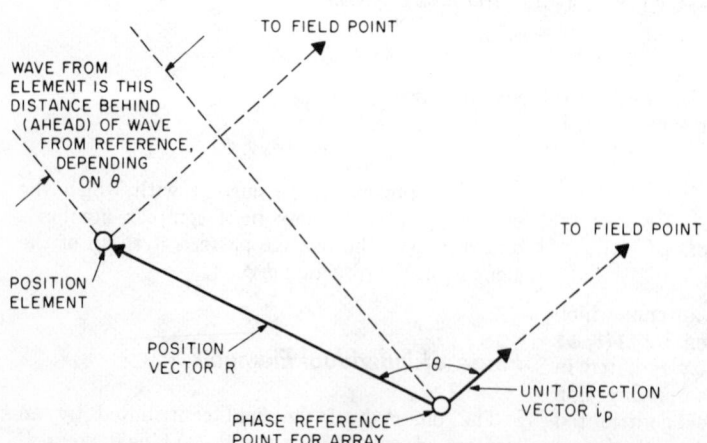

Fig. 41—Diagram for determining phase at far-field point P of the wave from an array element.

Field of the Array

The distant field E_p of an array at a point P is given by the vector sum of the fields contributed by all the elements.

$$E_p = \sum_k r_{pk}.$$

Therefore

$$E_p = \sum_k a_k A_k(\theta, \phi) \exp\{ j[(2\pi/\lambda)(R_k \cdot i_p) - q_k]\}.$$

$$(27)$$

Types of Arrays

    Linear
    Planar
    Circular
    Spherical
    Volumetric.

As can be seen, arrays are classified on the basis of the curve or surface on which the elements are placed. Unequal spacings of elements are useful in linear and planar arrays for beam scanning applications.

Arrays are also classified according to the behavior of their main beams as

    Ordinary (fixed beam)
    Phased (movable beam)
    Switched (stepped movable beam)
    Multiple (several beams)
    Adaptive (beam follows received signal)
    Combinations of the above.

Array elements are most usually equally spaced and parallel to each other in the linear, planar, and circular arrays. They cannot be equally spaced and are usually not parallel if mounted on a sphere.

Equally Spaced Linear Array: In this very important type elements are: parallel, similar, equally spaced (spacing $= s$), located on a straight line, and m in number.

The equation for the field at P is

$$E_p = A(\theta, \phi) \sum_{k=0}^{m-1} a_k \exp jk[(2\pi/\lambda) s \cos\theta - q] \quad (28)$$

where, because of straight-line geometry, R becomes ks, $(- \cdot i_p)$ becomes $\cos\theta$, q_k becomes kq, and θ becomes angle between line and i_p.

Sin θ' may be used in place of $\cos\theta$, where θ' is the angle between i_p and the normal to the line of the array. In most practical cases of interest, the equation for E_p must be used. In the special case of equally excited elements, simplified equations

(A)

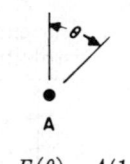

$$F(\theta) = A \quad (1)$$

(B)

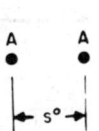

$$F(\theta) = 2A[\cos(\tfrac{1}{2}s^\circ \sin\theta)]$$

(C)

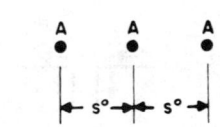

$$F(\theta) = A + 2A[\cos(s^\circ \sin\theta)]$$

(D)

$$F(\theta) = 4A[\cos(s^\circ \sin\theta) \cos(\tfrac{1}{2}s^\circ \sin\theta)]$$

(E)

$$m \text{ radiators (general case)}$$

$$F(\theta) = A \sin(m\tfrac{1}{2}s^\circ \sin\theta)/\sin(\tfrac{1}{2}s^\circ \sin\theta)$$

Fig. 42—Equal excitation field patterns. See Fig. 43 to compare A for common antenna types.

are given in Fig. 42. Element patterns of some common driven wire array elements are shown in Fig. 43.

Product Pattern Theorem

Equation (28) shows that the field of an array of identical parallel elements is given by the product of the pattern of one element and the pattern of an array of isotropic point sources located at the positions of the elements. That is

$$E_p = A(\theta, \phi) F_p$$

Directivity

Type of Radiator	Current Distribution	Horizontal E Plane $A(\theta)$	Vertical H Plane $A(\phi)$
(A) Half-wave dipole	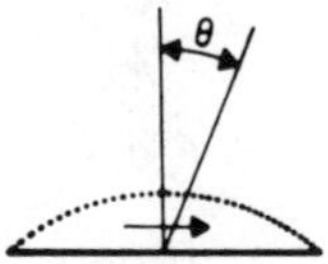	$A(\theta) = K \dfrac{\cos(\frac{1}{2}\pi \sin\theta)}{\cos\theta}$ $\approx K \cos\theta$	$A(\phi) = K(1)$
(B) Shortened dipole	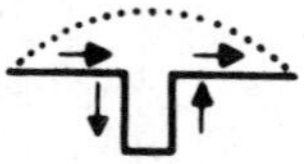	$A(\theta) \approx K \cos\theta$	$A(\phi) = K(1)$
(C) Lengthened dipole	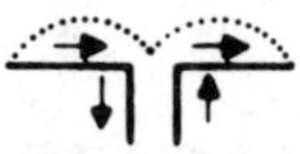	$A(\theta) = K\left[\dfrac{\cos[(\pi l/\lambda)\sin\theta] - \cos(\pi l/\lambda)}{\cos\theta}\right]$	$A(\phi) = K(1)$
(D) Horizontal loop	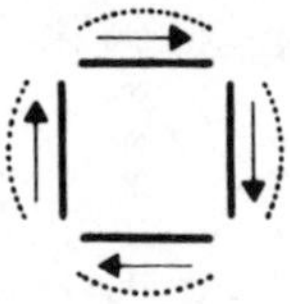	$A(\theta) \approx K(1)$	$A(\phi) = K\cos\phi$
(E) Horizontal turnstile	 i_1 and i_2 phased 90°	$A(\theta) \approx K'(1)$	$A(\phi) \approx K'(1)$

Fig. 43—Radiation patterns of several common types of antennas. $\theta =$ horizontal angle measured from perpendicular bisecting plane, $\phi =$ vertical angle measured from horizon, and K and K' are constants, with $K' \approx 0.7K$.

where F_p is the pattern of the array of isotropic sources.

$$F_p = \sum_{k=0}^{m-1} a_k \exp\{jk[(2\pi/\lambda)s\cos\theta - q]\}.$$

The element pattern $A(\theta, \phi)$ may be used to suppress side lobes arising from the array term F_p.

Associated Polynomial

The phase of an array element is

$$(2\pi/\lambda)s\cos\theta - q = \psi \quad \text{(see page 27-21)}.$$

Then the array pattern F_p can be expressed as

$$F_p = a_0 + a_1 \exp(j\psi) + a_2 \exp(j2\psi)$$
$$+ a_3 \exp(j3\psi) + \cdots + a_{m-1}\exp[j(m-1)\psi]$$

$$F_p = \sum_{k=0}^{m-1} a_k \exp(jk\psi).$$

That is, F_p is a polynomial in powers of $\exp(j\psi)$. F_p is called the associated polynomial for the array.

$$F_p = a_0 + a_1 Z + a_2 Z^2 + \cdots + a_{m-1}Z^{m-1}$$

where $Z = \exp(j\psi)$.

The variable Z and all its powers are restricted to the unit circle in the complex plane.

Array Computations of Power Patterns

Array power patterns are computed by multiplying the equation for the absolute value squared of the associated polynomial by the absolute value squared of the element pattern. The words "pattern of an antenna" usually refer to power patterns. The array power pattern $P(\theta, \phi)$ is given by

$$P(\theta, \phi) = |A(\theta, \phi)|^2 \left| \sum_{k=0}^{m-1} a_k \exp(jk\psi) \right|^2$$

where

$$|A(\theta, \phi)|^2 = \text{element factor}$$

$$\left| \sum_{k=0}^{m-1} a_k \exp(jk\psi) \right|^2 = \text{array factor}.$$

Rules for Linear Arrays

(A) Main beam is formed whenever $\psi = 0$.

(B) Gain drops if more than one main beam is formed.

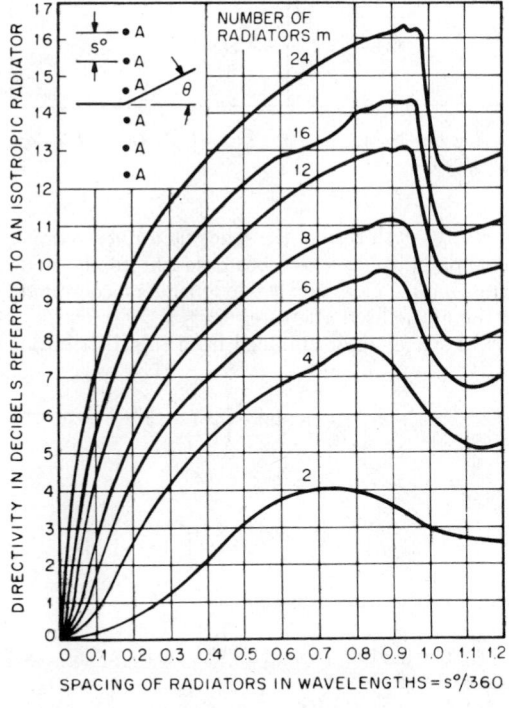

Fig. 44—Uniform broadside array of isotropic radiators.
Adapted from H. E. King, "The Microwave Engineers' Handbook and Buyers' Guide," p. 193, © *1966, Horizon House, Inc.*

(C) Only one main beam will form if

$$s/\lambda < (1 + \sin\theta_0)^{-1}$$

where s = element spacing, λ = wavelength, and θ_0 = angle first main beam makes with the normal to the array (angle of scan).

(D) Main beam is scanned by adjusting q in ψ.

(E) Spacing s which produces highest array directivity is as shown in Fig. 44.

(F) Pattern of uniform array of m elements (all a's equal, all q's = 0) is

$$P(\theta, \phi) = [\sin^2\tfrac{1}{2}(n\psi)] / [n^2 \sin^2\tfrac{1}{2}\psi].$$

(G) Side lobes of in-phase arrays are controlled by adjusting coefficients (a's), as discussed below. The binomial adjustment produces no side lobes, but the beam is very broad and the gain is low.

(H) Binomial array ($s = \tfrac{1}{2}\lambda$) has associated polynomial $F_p = (1+Z)^r$ for an $r+1$ element array

$$F_p = a_{r0} + a_{r1}Z + a_{r2}Z^2 + a_{r3}Z^3 + \cdots + a_rZ^r$$

$$A_{rk} = r!/k!(r-k)!$$

$$\text{Array factor} = |F_p|^2 = \cos2r(\tfrac{1}{2}\pi \cos\theta).$$

Distributions suitable for use with arrays follow.

LINE-SOURCE DISTRIBUTIONS*

For line sources, the current distribution is considered to be a function of only a single coordinate. The directivity pattern $E(u)$ resulting from a given distribution is simply related to the distribution by a finite Fourier transform, as given below.

$$E(u) = \tfrac{1}{2}l \int_{-1}^{+1} f(x) \exp(jux) \, dx \qquad (29)$$

where $f(x)$ = relative shape of field distribution over aperture as a function of x, $u = (\pi l/\lambda) \sin\phi$, l = overall length of aperture, ϕ = angle measured from normal to aperture, and x = normalized distance along aperture ($-1 \leq x \leq 1$). See Fig. 45.

The simplest type of aperture distribution is the uniform distribution where $f(x) = 1$ along the aperture and is zero outside of the aperture. The directivity pattern is given as

$$E(u) = l(\sin u/u) = l \frac{\sin[(\pi l/\lambda) \sin\phi]}{(\pi l/\lambda) \sin\phi}. \qquad (30)$$

* Text and Fig. 45 are by permission from H. Jasik, "Antenna Engineering Handbook," © 1961, McGraw-Hill Book Company.

TYPE OF DISTRIBUTION $-1 \leq x \leq 1$	DIRECTIVITY PATTERN $E(u)$		HALF-POWER BEAMWIDTH IN DEGREES 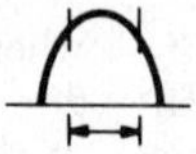	ANGULAR DISTANCE TO FIRST ZERO 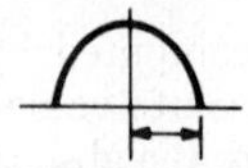	1ST SIDE LOBE IN DECIBELS BELOW MAXIMUM	GAIN FACTOR
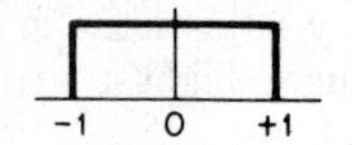 $f(x)=1$	$\ell \dfrac{\sin u}{u}$		$50.8 \dfrac{\lambda}{\ell}$	$57.3 \dfrac{\lambda}{\ell}$	13.2	1.0
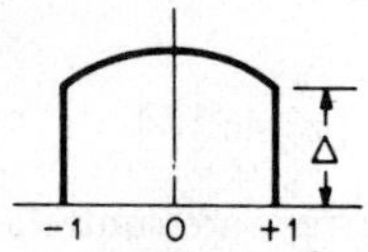 $f(x)=1-(1-\Delta)x^2$	$\ell(1+L) \dfrac{\sin u}{u}$ $L=(1-\Delta)\dfrac{d^2}{du^2}$	$\Delta=$ 1.0	$50.8 \dfrac{\lambda}{\ell}$	$57.3 \dfrac{\lambda}{\ell}$	13.2	1.0
		0.8	$52.7 \dfrac{\lambda}{\ell}$	$60.7 \dfrac{\lambda}{\ell}$	15.8	0.994
		0.5	$55.6 \dfrac{\lambda}{\ell}$	$65.3 \dfrac{\lambda}{\ell}$	17.1	0.970
		0	$65.9 \dfrac{\lambda}{\ell}$	$81.9 \dfrac{\lambda}{\ell}$	20.6	0.833
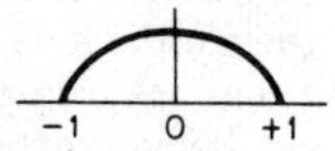 $\cos \dfrac{\pi x}{2}$	$\dfrac{\pi \ell}{2} \dfrac{\cos u}{(\frac{\pi}{2})^2 - u^2}$		$68.8 \dfrac{\lambda}{\ell}$	$85.9 \dfrac{\lambda}{\ell}$	23	0.810
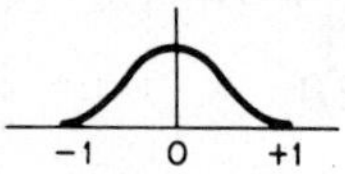 $\cos^2 \dfrac{\pi x}{2}$	$\dfrac{\ell}{2} \dfrac{\sin u}{u} \dfrac{\pi^2}{\pi^2-u^2}$		$83.2 \dfrac{\lambda}{\ell}$	$114.6 \dfrac{\lambda}{\ell}$	32	0.667
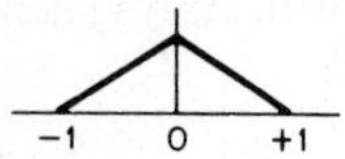 $f(x)=1-|x|$	$\dfrac{\ell}{2}\left(\dfrac{\sin \frac{u}{2}}{\frac{u}{2}}\right)^2$		$73.4 \dfrac{\lambda}{\ell}$	$114.6 \dfrac{\lambda}{\ell}$	26.4	0.75

Fig. 45—Line-source distributions. H. Jasik, "Antenna Engineering Handbook," p. 2–26, © 1961, McGraw-Hill Book Company.

Taylor Distribution*

The Taylor aperture distribution is designed to produce equal side lobes out to a point beyond which the side-lobe amplitude decreases.† The excitation peaks at the aperture end are adjusted on moving this transition point.

The Taylor aperture distribution is given by

$$g(p, A, \bar{n}) = 1 + 2 \sum_{n=1}^{\bar{n}-1} F(n, A, \bar{n}) \cos np \quad (31)$$

with the coefficient $F(n, A, \bar{n})$ given by

$$F(n, A, \bar{n}) = \frac{[(n-1)!]^2 \prod_{m=1}^{\bar{n}-1} 1-(n^2/z_m^2)}{(\bar{n}-1+n)!(\bar{n}-1-n)!} \quad (32)$$

* This section is reprinted from R. J. Spellmire, "Tables of Taylor Aperture Distributions," Memorandum 581, Hughes Aircraft Company; October 1958.

† T. T. Taylor, "Design of Line Sources for Narrow Beamwidth and Low Side Lobes," Technical Memorandum 316, Hughes Aircraft Company, Culver City, California; 1953.

where $z_m = m$th zero of the space factor, $\bar{n} =$ number of equiamplitude side lobes adjacent to the main beam on one side, and $p =$ independent coordinate of the normalized aperture, $-\pi \leq p \leq \pi$.

The space factor obtained from this distribution is given by

$$F(z, A, \bar{n}) = \frac{\sin \pi z}{\pi z} \prod_{n=1}^{\bar{n}-1} \frac{1-(z/z_n)^2}{1-(z/n)^2}$$

where

$$z = (2a/\lambda) \sin\theta, \qquad (\theta=0° \text{ is broadside})$$

which has zeros z_n given by

$$z_n = \pm\sigma[A^2+(n-\tfrac{1}{2})^2]^{1/2}, \qquad 1 \leq n \leq \bar{n}$$

$$z_n = \pm n, \qquad \bar{n} \leq n < \infty \quad (33)$$

where σ (the measure of deviation from the ideal space factor) is a number slightly greater than 1 given by

$$\sigma = \eta/[A^2+(\bar{n}-\tfrac{1}{2})^2]^{1/2}. \quad (34)$$

TABLE 2. GENERAL DESIGN PARAMETERS, TAYLOR DISTRIBUTION.

Design Side-Lobe Level (dB)	Side-Lobe Voltage Ratio η	β_0 (degrees)	A^2	Values of Parameter σ							
				$\bar{n}=3$	$\bar{n}=4$	$\bar{n}=5$	$\bar{n}=6$	$\bar{n}=7$	$\bar{n}=8$	$\bar{n}=9$	$\bar{n}=10$
15	5.62341	47.26	0.58950	1.14712	1.11631	1.09528	1.08043	1.06949	1.06112	1.05453	1.04921
16	6.30957	48.43	0.64798	1.14225	1.11378	1.09375	1.07941	1.06876	1.06058	1.05411	1.04887
17	7.07946	49.61	0.70901	1.13723	1.11115	1.09216	1.07835	1.06800	1.06001	1.05367	1.04852
18	7.94328	50.77	0.77266	1.13206	1.10843	1.09050	1.07724	1.06721	1.05942	1.05321	1.04815
19	8.91251	51.90	0.83891	1.12676	1.10563	1.08879	1.07609	1.06639	1.05880	1.05273	1.04777
20	10.00000	53.05	0.90777	1.12133	1.10273	1.08701	1.07490	1.06554	1.05816	1.05223	1.04738
21	11.2202	54.16	0.97927	1.11577	1.09974	1.08518	1.07367	1.06465	1.05750	1.05172	1.04697
22	12.5893	55.27	1.05341	1.11009	1.09668	1.08329	1.07240	1.06374	1.05682	1.05119	1.04654
23	14.1254	56.46	1.13020	1.10430	1.09352	1.08135	1.07108	1.06280	1.05611	1.05064	1.04610
24	15.8489	57.46	1.20965	1.09840	1.09029	1.07934	1.06973	1.06183	1.05538	1.05007	1.04565
25	17.7828	58.54	1.29177	1.09241	1.08698	1.07728	1.06834	1.06083	1.05463	1.04948	1.04518
26	19.9526	59.66	1.37654	1.08632	1.08360	1.07517	1.06690	1.05980	1.05385	1.04888	1.04469
27	22.3872	60.64	1.46395	1.08015	1.08014	1.07300	1.06543	1.05874	1.05305	1.04826	1.04420
28	25.1189	61.72	1.55406		1.07661	1.07078	1.06392	1.05765	1.05223	1.04762	1.04368
29	28.1838	62.77	1.64683		1.07300	1.06851	1.06237	1.05653	1.05139	1.04696	1.04316
30	31.6228	63.80	1.74229		1.06934	1.06619	1.06079	1.05538	1.05052	1.04628	1.04262
31	35.4813	64.83	1.84044		1.06561	1.06382	1.05916	1.05421	1.04963	1.04559	1.04206
32	39.8107	65.83	1.94126		1.06182	1.06140	1.05751	1.05300	1.04872	1.04488	1.04149
33	44.6684	66.83	2.04473			1.05893	1.05581	1.05177	1.04779	1.04415	1.04091
34	50.1187	67.85	2.15092			1.05642	1.05408	1.05051	1.04684	1.04341	1.04031
35	56.2341	68.84	2.25976			1.05386	1.05231	1.04923	1.04587	1.04264	1.03970
36	63.0957	69.82	2.37129			1.05126	1.05051	1.04792	1.04487	1.04186	1.03907
37	70.7946	70.82	2.48551				1.04868	1.04658	1.04385	1.04107	1.03843
38	79.4328	71.81	2.60241				1.04681	1.04521	1.04282	1.04025	1.03777
39	89.1251	72.73	2.72201				1.04491	1.04382	1.04176	1.03942	1.03711
40	100.0000	73.74	2.84428				1.04298	1.04241	1.04068	1.03858	1.03643

The parameter A is related to the side-lobe voltage ratio η through the equation

$$\pi A = \cosh^{-1}\eta. \tag{35}$$

The members of the pattern family $F(z, A, \bar{n})$ have two independent characteristics.

(A) Design side-lobe ratio η.

(B) Outer boundary of the region of uniform side lobes $\bar{n}$, an integer.

Once these two parameters have been chosen, the pattern, the distribution, and all other relevant data may be calculated.

The designer of a line source is usually given the following three conditions to be met.

(A) Side-lobe level in decibels $10 \log_{10}\eta^2$.

(B) Half-power beamwidth β_d.

(C) Ratio of total length to wavelength $2a/\lambda$.

To apply the Taylor distribution to a line source, the quantity $2a\beta_d/\lambda\beta_0$ is calculated for the given side-lobe level η.

β_0 is the beamwidth of the ideal pattern in terms of a standard beamwidth and is given by*

$$\beta_0 = 2 \sin^{-1}(1/\pi)\left[(\cosh^{-1}\eta)^2 - (\cosh^{-1}\eta/\sqrt{2})^2\right]^{1/2}. \tag{36}$$

Using the condition

$$\sigma \leq 2a\beta_d/\lambda\beta_0 \tag{37}$$

the minimum value of $\bar{n}$ is obtained, thus determining the distribution and the pattern.

Table 2 consists of the following.

Side-lobe level in decibels
Side-lobe voltage ratio η
Ideal pattern beamwidth β_0 in degrees
Parameter A^2
Parameter σ for $\bar{n}$.

The equations and Table 2 enable one to compute a Taylor line source distribution from which the array coefficients may be found by using the ordinates at corresponding points. Table 3 lists many such aperture distributions. Only half of the symmetrical distribution is listed in Table 3. The half-aperture is divided into 20 intervals.

PROBABILISTIC DESIGN

It is possible to determine the critical number of antenna elements randomly spaced to give a specified side-lobe level with a required probability from a probabilistic viewpoint. One such relationship is shown in Fig. 46.

DENSITY TAPERED ARRAY*

Unequal-spacing technique can also be used to shape the main beam and close-in side lobes with no amplitude tapering, and is referred to as density tapering or space tapering.†

The far-field pattern is shaped by the use of a variation in element density. The main-beam and close-in side-lobe shapes of a density-tapered array are essentially indistinguishable from those produced by an array of the same size using an amplitude taper generated by the same curve. However, the average side-lobe level far out in the pattern of a density-tapered array is approximately

$$S_A \approx 10 \log_{10} 1/N \tag{38}$$

for low first-side-lobe tapers. Note that the far-out average side-lobe level essentially depends only on the number of elements used; only the uniformity of the side-lobe level—staying close to the average everywhere—depends on the element placement. By contrast, the far-out side-lobe level of a perfect amplitude-tapered regularly spaced array is approximately $20 \log_{10} 1/N$. The difference in practice is not so great as these idealized results would indicate—for example, average side-lobe level of -30 decibels for a 1000-element density-tapered array, and -60 decibels for the same array with amplitude taper only—because of error effects.

For practical reasons—such as the control of interaction between real elements, ease of design of feed networks, and the recognition of the finite physical dimensions of real antennas—the elements cannot be placed in completely arbitrary locations, and a compromise technique is frequently used in which the elements are placed on a regularly spaced grid whose spacing was chosen to prevent grating lobe occurrence in visible space. Figure 47A shows a grid with space for 4000 elements in which only 900 have been placed, as indicated by the darkened squares. Figure 47B shows a cut of the resulting pattern with the predicted average side-lobe level of (38) indicated.

Ideally, all peaks would be 3 decibels above S_A if all of the peaks were equal; the performance that is indicated comes close to this ideal.

---

* The following three paragraphs and Fig. 47 are reprinted from J. L. Allen, "Array Antennas, New Applications for an Old Technique," *IEEE Spectrum*, vol. 1, no. 11, pp. 115–130; November 1964.

† R. E. Willey, "Space Tapering of Linear and Planar Arrays," *IRE Transactions on Antennas and Propagation*, vol. AP-10, pp. 369–377; July 1962.

---

* T. T. Taylor uses the equation $\beta_0 = (2/\pi)\left[(\cosh^{-1}\eta)^2 - (\cosh^{-1}\eta/\sqrt{2})^2\right]^{1/2}$. This approximation is valid for long line sources, but serious error can occur for short apertures and high side-lobe ratios.

TABLE 3—APERTURE DISTRIBUTIONS.

$g(p, A, \bar{n})$ for 20-Decibel Side-Lobe Level

p	$\bar{n}=3$	$\bar{n}=4$	$\bar{n}=5$	$\bar{n}=6$	$\bar{n}=7$	$\bar{n}=8$	$\bar{n}=9$	$\bar{n}=10$
0	1.316624	1.284708	1.280816	1.256022	1.256066	1.238560	1.222626	1.209990
1	1.312568	1.282674	1.275806	1.255240	1.250348	1.238698	1.228526	1.220218
2	1.300514	1.276194	1.261814	1.251098	1.236128	1.234926	1.235116	1.234768
3	1.280802	1.264236	1.241448	1.239582	1.219236	1.220296	1.222924	1.224840
4	1.253990	1.245446	1.217620	1.217600	1.202352	1.194684	1.188794	1.184462
5	1.220828	1.218576	1.192092	1.185666	1.182032	1.165674	1.150800	1.139114
6	1.182228	1.182956	1.164510	1.148018	1.152116	1.138874	1.126362	1.115688
7	1.139238	1.138858	1.132466	1.109906	1.110516	1.109988	1.109746	1.108592
8	1.093006	1.087706	1.092622	1.073698	1.062692	1.069794	1.077696	1.084078
9	1.044740	1.032014	1.042580	1.036550	1.017432	1.017304	1.018894	1.021336
10	0.995674	0.975106	0.982698	0.991774	0.977980	0.964048	0.951756	0.942524
11	0.947032	0.920634	0.917092	0.933634	0.936888	0.921076	0.905254	0.891292
12	0.899994	0.872014	0.853278	0.863114	0.881248	0.882102	0.881064	0.878270
13	0.855678	0.831878	0.800374	0.790958	0.806460	0.825952	0.843714	0.858020
14	0.815100	0.801660	0.776426	0.735362	0.727876	0.743354	0.760142	0.776336
15	0.779172	0.781424	0.755790	0.714574	0.678868	0.663434	0.653280	0.647634
16	0.748682	0.769938	0.766972	0.737604	0.692584	0.646576	0.607118	0.574474
17	0.724282	0.765028	0.793448	0.797898	0.777820	0.738148	0.701240	0.667820
18	0.706486	0.764086	0.824550	0.873746	0.905900	0.918794	0.926628	0.930684
19	0.695660	0.764678	0.848994	0.935992	1.020404	1.099434	1.165630	1.221138
20	0.692028	0.765080	0.858204	0.959950	1.066174	1.175128	1.268022	1.347450

$g(p, A, \bar{n})$ for 25-Decibel Side-Lobe Level

p	$\bar{n}=3$	$\bar{n}=4$	$\bar{n}=5$	$\bar{n}=6$	$\bar{n}=7$	$\bar{n}=8$	$\bar{n}=9$	$\bar{n}=10$
0	1.443664	1.435116	1.428806	1.412802	1.408658	1.396330	1.386904	1.378638
1	1.438970	1.430836	1.423444	1.409406	1.402928	1.393576	1.385930	1.379446
2	1.424930	1.417950	1.407840	1.398436	1.387230	1.383250	1.379052	1.375816
3	1.401670	1.396318	1.383252	1.378160	1.364582	1.361962	1.359168	1.356912
4	1.369422	1.365818	1.351176	1.347364	1.336548	1.329852	1.324894	1.320450
5	1.328554	1.326434	1.312746	1.306584	1.301792	1.291064	1.283032	1.276164
6	1.279596	1.278392	1.268318	1.258184	1.257664	1.248848	1.241172	1.234958
7	1.223276	1.222264	1.217448	1.205132	1.203558	1.201530	1.198530	1.196054
8	1.160558	1.159044	1.159320	1.149174	1.142616	1.144958	1.146474	1.147388
9	1.092642	1.090168	1.093476	1.089704	1.079752	1.079092	1.079764	1.080214
10	1.020982	1.017452	1.020574	1.024282	1.017324	1.010272	1.005582	1.002130
11	0.947270	0.943028	0.942864	0.950788	0.952492	0.944832	0.937604	0.931766
12	0.873390	0.869192	0.864132	0.870058	0.879560	0.880494	0.879146	0.877232
13	0.801388	0.798252	0.789100	0.787436	0.796558	0.807262	0.815690	0.821946
14	0.733372	0.732392	0.722406	0.711972	0.711080	0.720596	0.730868	0.740194
15	0.671446	0.673566	0.667586	0.653300	0.639788	0.634932	0.633464	0.633964
16	0.617610	0.623398	0.626276	0.617568	0.600180	0.580960	0.564338	0.550774
17	0.573666	0.583166	0.598048	0.604480	0.599198	0.583510	0.566924	0.551192
18	0.541122	0.553814	0.580860	0.607124	0.626698	0.637034	0.643320	0.646900
19	0.521118	0.535968	0.572040	0.615014	0.659356	0.702244	0.739894	0.772342
20	0.514372	0.529980	0.569382	0.618866	0.673534	0.731134	0.783388	0.829678

TABLE 3—CONTINUED

$g(p, A, \bar{n})$ for 30-Decibel Side-Lobe Level

p	$\bar{n}=4$	$\bar{n}=5$	$\bar{n}=6$	$\bar{n}=7$	$\bar{n}=8$	$\bar{n}=9$	$\bar{n}=10$
0	1.558106	1.555218	1.546266	1.541348	1.532636	1.525586	1.519102
1	1.551970	1.548884	1.540712	1.534816	1.527552	1.521304	1.515818
2	1.533.88	1.530116	1.523810	1.515970	1.511376	1.506748	1.503046
3	1.503642	1.499574	1.495108	1.486430	1.482698	1.478924	1.475738
4	1.462424	1.458178	1.454498	1.447314	1.441934	1.437510	1.433340
5	1.410794	1.406922	1.402732	1.398552	1.391580	1.386074	1.381110
6	1.349646	1.346722	1.341448	1.339588	1.333860	1.328776	1.324638
7	1.279974	1.278368	1.272682	1.270860	1.268740	1.266122	1.264042
8	1.202884	1.202580	1.198110	1.194598	1.195026	1.194970	1.194600
9	1.119604	1.120214	1.118530	1.113974	1.113484	1.113594	1.113366
10	1.031568	1.032430	1.033986	1.031120	1.028052	1.026092	1.024654
11	0.940440	0.940914	0.944534	0.945830	0.942792	0.939882	0.937742
12	0.848196	0.847920	0.851322	0.856462	0.857730	0.857782	0.857512
13	0.757138	0.756224	0.757298	0.762792	0.769000	0.774002	0.777678
14	0.669866	0.668914	0.667124	0.668788	0.674920	0.681264	0.686910
15	0.589206	0.589138	0.586148	0.582744	0.582660	0.583846	0.585850
16	0.518064	0.519810	0.518938	0.514046	0.507606	0.501802	0.497152
17	0.459246	0.463402	0.468040	0.468358	0.463642	0.457952	0.452054
18	0.415232	0.421816	0.433630	0.444530	0.451788	0.457116	0.460754
19	0.387986	0.396364	0.414216	0.435644	0.457852	0.478300	0.496602
20	0.378758	0.387802	0.408002	0.433820	0.462780	0.490278	0.515686

$g(p, A, \bar{n})$ for 35-Decibel Side-Lobe Level

p	$\bar{n}=5$	$\bar{n}=6$	$\bar{n}=7$	$\bar{n}=8$	$\bar{n}=9$	$\bar{n}=10$
0	1.665394	1.662912	1.659392	1.653666	1.649974	1.646654
1	1.657752	1.655422	1.651590	1.646510	1.642130	1.638124
2	1.634978	1.633034	1.628604	1.624752	1.619452	1.614516
3	1.597562	1.596024	1.591438	1.588120	1.583374	1.578932
4	1.546318	1.545026	1.541152	1.537368	1.534300	1.531500
5	1.482400	1.481134	1.478550	1.474360	1.472158	1.470218
6	1.407248	1.405914	1.404426	1.401050	1.398258	1.395702
7	1.322528	1.321254	1.320086	1.318544	1.315428	1.312452
8	1.230028	1.229104	1.227650	1.227522	1.225894	1.224280
9	1.131566	1.131270	1.129728	1.129576	1.130108	1.130638
10	1.028922	1.029340	1.028628	1.027710	1.028724	1.029782
11	0.923862	0.924786	0.925752	0.925088	0.924876	0.924670
12	0.818236	0.819244	0.821778	0.823114	0.822698	0.822108
13	0.714104	0.714806	0.717698	0.721296	0.723272	0.724944
14	0.613912	0.614174	0.615938	0.619738	0.624242	0.628522
15	0.520536	0.520614	0.520750	0.522180	0.525638	0.529094
16	0.437274	0.437718	0.437348	0.436244	0.435648	0.435124
17	0.367622	0.368996	0.370346	0.369904	0.367646	0.365272
18	0.314942	0.317538	0.322400	0.326748	0.329324	0.331480
19	0.282068	0.285690	0.294066	0.304426	0.315366	0.325656
20	0.270890	0.274912	0.284752	0.297834	0.312938	0.327318

TABLE 3—CONTINUED

$g(p, A, \bar{n})$ for 40-Decibel Side-Lobe Level

p	$\bar{n}=6$	$\bar{n}=7$	$\bar{n}=8$	$\bar{n}=9$	$\bar{n}=10$
0	1.766566	1.765820	1.762932	1.760116	1.757184
1	1.757258	1.756486	1.753812	1.750754	1.747584
2	1.729594	1.728770	1.726548	1.723204	1.719746
3	1.684334	1.683520	1.681570	1.678558	1.675430
4	1.622656	1.621966	1.620004	1.617818	1.615526
5	1.546122	1.545628	1.543692	1.542046	1.540340
6	1.456596	1.456288	1.454778	1.453118	1.451422
7	1.356264	1.356054	1.355302	1.353742	1.352118
8	1.247580	1.247398	1.247284	1.246526	1.245692
9	1.133172	1.133034	1.133096	1.133420	1.133718
10	1.015682	1.015686	1.015670	1.016398	1.017182
11	0.897572	0.897806	0.898000	0.898428	0.898898
12	0.781016	0.781466	0.782380	0.782850	0.783284
13	0.667952	0.668466	0.670174	0.671666	0.673122
14	0.560300	0.560716	0.562562	0.565110	0.567766
15	0.460322	0.460564	0.461720	0.463942	0.466340
16	0.370872	0.371046	0.371416	0.372098	0.372830
17	0.295434	0.295784	0.296298	0.296274	0.296162
18	0.237826	0.238542	0.240444	0.242172	0.243886
19	0.201574	0.202658	0.206332	0.211174	0.216288
20	0.189182	0.190424	0.194904	0.201304	0.208148

APPROXIMATIONS FOR RAPID EVALUATIONS

Some useful rules of approximate evaluation are given by Allen for estimating array parameters for in-phase linear arrays.

Half-Power-Beamwidth Normal to Array

$$\theta_L \simeq K\lambda/L \qquad (39)$$

where K depends on the highest side-lobe levels, λ is the wavelength used, and L is the length of the array in the plane parallel to θ in the same units used for λ.

Parameter K for Beamwidth Calculation

$$K_L \simeq 50 + (S-13) \text{ degrees}, \; S>13 \qquad (40)$$

where S is the side-lobe level in decibels. ($S = 13$ decibels for uniform illumination.)

Aperture Efficiency of In-Phase Array

$$\eta_L \simeq 1 - [(S-13)/100] \qquad (41)$$

for S as defined above, compared with the gain of an array all of whose elements are excited equally and in phase.

General Aperture Efficiency of Any Array

$$\eta = |\sum a_m|^2 / r \sum |a_m|^2 \qquad (42)$$

where r is the number of radiators.

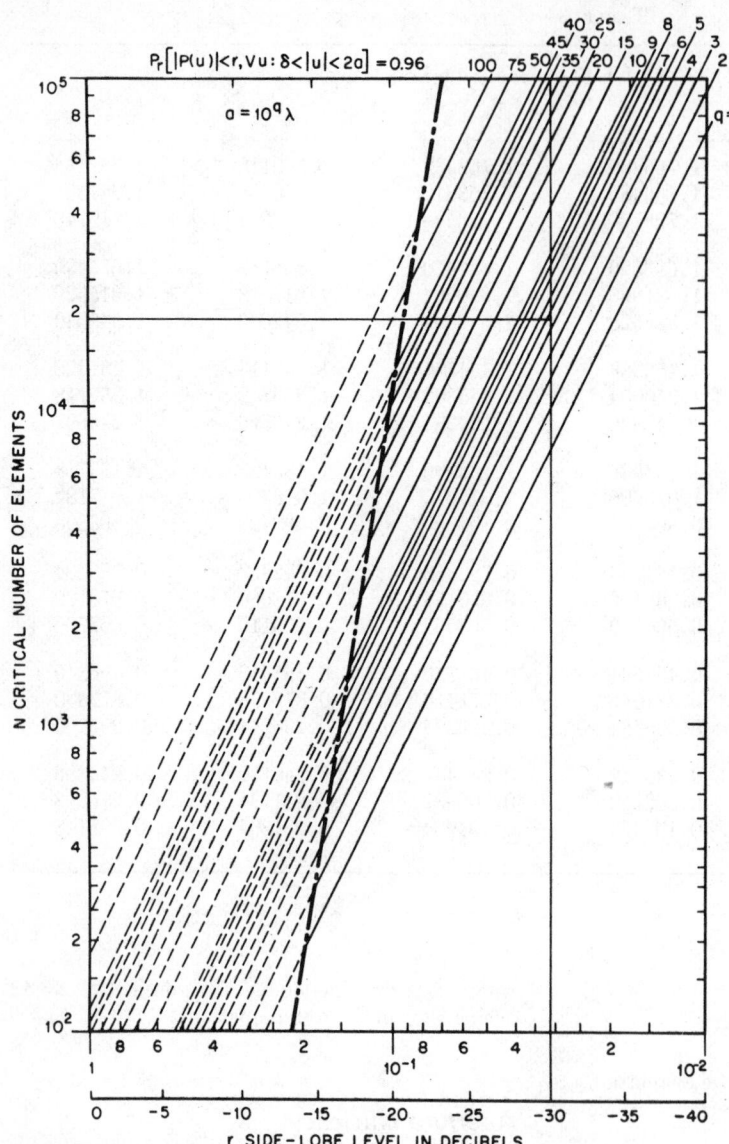

Fig. 46—Critical number of randomly spaced antenna elements required as a function of array size in wavelengths to produce the indicated side-lobe level with a probability of 96 percent. *Y. T. Lo, "Probabilistic Approach to the Design of Large Antenna Arrays," IEEE Transactions on Antennas and Propagation, vol. AP-11, no. 1, p. 96; January 1963.*

Directivity of Planar Array Radiating into Half-Space U

$$U = 4\pi\eta r\,(a/\lambda^2) \qquad (43)$$

where a is the area allotted each element.

Beamwidth of Rectangular Planar Array

$$\theta_p \simeq \theta_L. \qquad (44)$$

Efficiency of Rectangular Planar Array*

$$\eta_p \simeq \eta_L^2. \qquad (45)$$

Beamwidth Constant K for Circular Planar Array

$$K_{cp} \simeq 58 + 0.8\,(S-17). \qquad (46)$$

* Estimated independently of Allen.

Note 1: $S = 17$ decibels for uniformly illuminated circular aperture.

Note 2: Circularly symmetric distribution is assumed.

Efficiency of Circularly Symmetric Planar Array

$$\eta_{cp} = 1 - 1.5[(S - 17)/100]. \qquad (47)$$

As applied to an array, the half-power beamwidth* in the plane of the array θ may be used with the half-power beamwidth ϕ in the plane orthogonal to the line of the array.

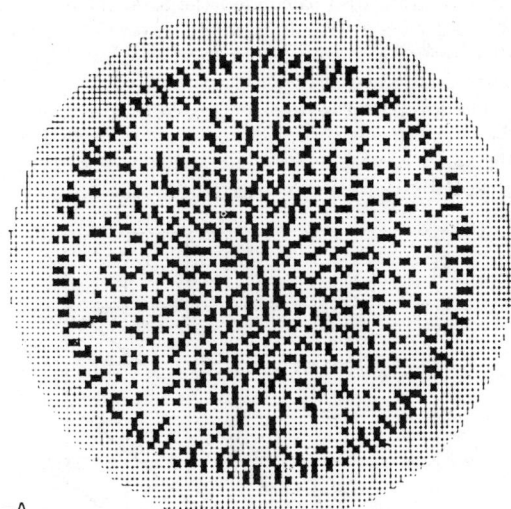

A

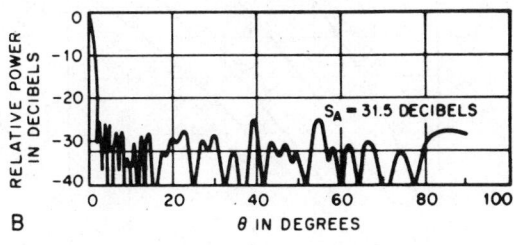

B

Fig. 47—Density-tapered array with 4000-element grid containing 900 active elements (darkened squares). Typical pattern cut for density-tapered array. *J. L. Allen, "Array Antennas, New Applications for an Old Technique," IEEE Spectrum, vol. 1, no. 11, Fig. 8; November 1964.*

* Kraus provides estimates for directivity based on beamwidth.

Directivity of an Array D

$$D \simeq 41\,253/\theta\phi, \qquad \theta \text{ and } \phi \text{ in degrees.} \quad (48)$$

Gain of an Array G

$$G \simeq \mathcal{E}D \qquad (49)$$

where $\mathcal{E}$ is a number less than 1 that takes all losses into account other than those due to η. The aperture efficiency has already operated to broaden θ and so its effect is included in D. The number of square degrees subtended by all space is 41 253.

It should be pointed out that these equations apply to arrays that produce only a single main beam. This includes a linear array of point sources whose main beam in the ϕ plane has a beam angle of 360°, and it includes arrays that radiate into a half space such as a planar array of dipoles over a ground plane. If the ground plane is removed so the planar array radiates a narrow beam in both directions, then D is divided by 2. In general, D is divided by the number of equal multilobes (main beams).

Gain of the In-Phase Planar Array Radiating into a Half Space

Combining the approximating equations above

$$G \simeq \mathcal{E} \frac{41\,253}{(K_\theta\lambda/L_\theta) \times (K_\phi\lambda/L_\phi)} \simeq \mathcal{E} \frac{4\pi A\,(57.3)^2}{\lambda^2 K_\theta K_\phi} \quad (50)$$

where $K_\theta =$ beam parameter in the θ plane, $K_\phi =$ beam parameter in the ϕ plane, $L_\theta =$ array length parallel to the θ plane, $L_\phi =$ array length parallel to the ϕ plane, and $A = L_\theta L_\phi$.

Gain and Beamwidth of a Beam Phased to Scan (Tilt) Off the Normal by an Angle When no Multilobes are Formed

$$G_\alpha \simeq G \cos\alpha$$

$$\theta_\alpha \simeq \theta \times (\cos\alpha)^{-1}$$

$$\theta_\alpha = \text{half-power beamwidth at } \alpha. \quad (51)$$

This equation holds good to α^* degrees† where $\alpha^* = 90° - \theta_0$ and $\sin\theta_0 = 1.06(\lambda/S)^{1/2}$.

† M. J. King and R. K. Thomas, "Gain of Large Scanned Arrays," *IRE Transactions on Antennas and Propagation*, vol. AP-8, no. 6, p. 635; November 1960.

RANDOM ERRORS IN ARRAYS

When an array antenna is built, the measured array pattern differs from that theoretically expected. This is due to systematic and random errors. Systematic errors can be caused by the presence of a radome, by line attenuation that was not properly accounted for, or by mutual coupling, for example. Systematic errors can ultimately be corrected for.

Random errors are those that arise in the construction or adjustment of an array, due to limitations of accuracy in fabrication or measurement. Errors will therefore be present in the phases and amplitudes of the radiation from the elements due to the accumulation of tolerances. These errors in the net outgoing waves will:

Raise the side lobes above the designed level
Reduce the gain
Change the pointing direction of the main beam.

If many units of the same design were manufactured, the errors in each unit would be different because of the differing circumstances of its construction. Therefore the description of the antenna performance in terms of the errors that arise when construction is within permitted tolerances must be on a statistical basis. The description of the gains and side lobes of arrays in the presence of random errors has been investigated.* The average pattern of many units of the same design was found with the assumption that the amplitude errors Δ_{mn} and phase errors δ_{mn} are distributed in a Gaussian manner and that the mean errors $\langle\Delta_{mn}\rangle$ and $\langle\delta_{mn}\rangle$ are zero.

Under this condition the average pattern or "mean power" $\langle P(\theta,\phi)\rangle$ is given by

$$\langle P(\theta,\phi)\rangle = P_0(\theta,\phi)$$

$$+S(\theta,\phi)\langle\mathcal{E}^2\rangle[\sum_{}^{M}\sum_{}^{N}I_{MN}{}^2/(\sum_{}^{M}\sum_{}^{N}I_{MN})^2] \quad (52)$$

where $P_0(\theta,\phi)=$ no-error pattern, $S(\theta,\phi)=\cos\theta(\cos^2\theta\cos^2\phi+\sin^2\phi)$, the element factor due to current flowing in the x direction, $\langle\mathcal{E}^2\rangle=\langle\Delta^2\rangle+\langle\delta^2\rangle$, $\langle\Delta^2\rangle=$ mean-square amplitude error, $\langle\delta^2\rangle=$ mean-square phase error, and $I_{MN}=$ current in element in Mth row and Nth column of a planar array located in the XY plane that is part of a right-handed spherical polar-coordinate system.

The statistical distribution about the mean power was shown by Ruze to be given by a modified Rayleigh distribution

$$W(r)=(2r/\sigma^2)\exp[-(a^2+r^2)/\sigma^2]I_0(2ar/\sigma^2)_{\text{arg}}$$

$$(53)$$

* J. Ruze, "Physical Limitations on Antennas," Technical Report 248, 30 October 1952, Research Laboratory of Electronics, Massachusetts Institute of Technology, ASTIA AD–62351.

where

$r=$ field strength

$a^2=P_0(\theta,\phi)$

$\sigma^2=S(\theta,\phi)\langle\mathcal{E}^2\rangle[\sum_{}^{M}\sum_{}^{N}I_{MN}{}^2/(\sum_{}^{M}\sum_{}^{N}I_{MN})^2]$

$I_0(x)_{\text{arg}}=$ modified Bessel function of the first kind of argument x.

The percentage of the cases where the side-lobe levels are exceeded by a given field r is calculated by integrating the probability distribution $W(r)$ from $-\infty$ up to r for a given root-mean-square error $\langle\mathcal{E}^2\rangle^{1/2}$. This was done by Ruze for a Dolph Chebishev array of 25 elements designed for a 29-decibel side-lobe level. Ruze found the theoretical field distributions shown in Fig. 48 for the positions of the side-lobe maxima in the zero-error pattern. Ruze also calculated the array factor for the 29-decibel distribution caused by error currents 40 percent in magnitude at random phase. This pattern is shown in Fig. 49.

The levels at the no-error side-lobe positions of the calculated pattern were determined from Fig. 49, and the probability that the side-lobe levels exceeded r decibels was determined. This was com-

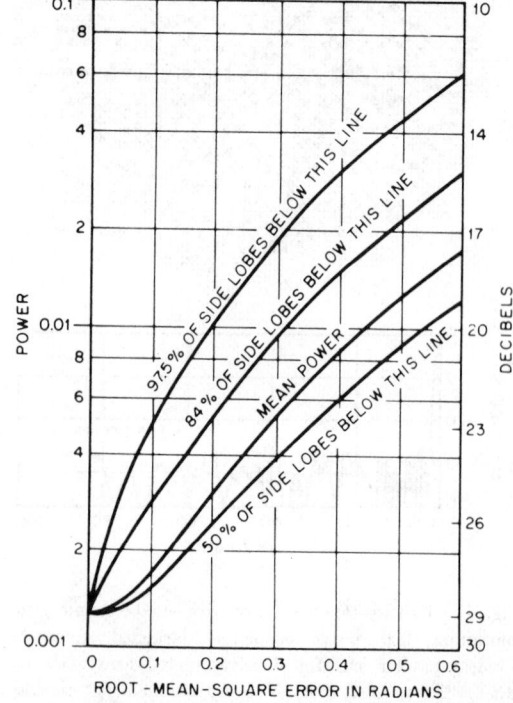

Fig. 48—Side-lobe distribution for 25-element broadside array. Designed for 29-decibel side-lobe suppression. Computed at design-lobe maxima. *H. Jasik, "Antenna Engineering Handbook," Fig. 2-23, © 1961, McGraw-Hill Book Company.*

Fig. 49—Theoretical effect of error currents on radiation pattern of 25-element broadside array designed for 29-decibel side-lobe suppression: $A =$ no error; $B = 0.40$ root-mean-square error in each element at random phase. H. Jasik, "Antenna Engineering Handbook," Fig. 2-24, © 1961, McGraw-Hill Book Company.

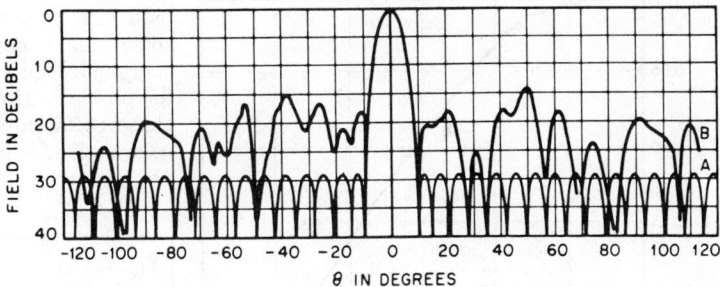

pared with the probability computed from the equation for $W(r)$. The comparison of the results is given in Fig. 50. The agreement shown is excellent.

Reduction in Gain with Random Errors

The gain G in the presence of random errors is estimated by

$$G \approx G_0 [1 + \tfrac{3}{4}\pi (d/\lambda)^2 \langle \mathcal{E}^2 \rangle]^{-1} \qquad (54)$$

where $d =$ element spacing and $\lambda =$ wavelength.

Beam Pointing with Random Errors

The pointing of the beam in the presence of random errors was investigated by Leichter* who found the general pointing-direction solution and then applied it to the case of uniform and Taylor amplitude distributions possessing random errors.

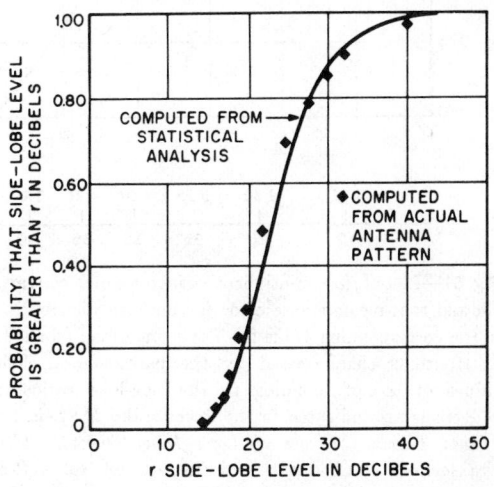

Fig. 50—Comparison of probabilities (computed in two ways) that side-lobe levels exceed r decibels. J. Ruze, "Physical Limitations on Antennas," Technical Report 248, Fig. 19, Research Laboratory for Electronics, MIT, ASTIA No. AD-62351; 30 October 1952.

Pointing Errors with Constant Amplitude Distribution

Maximum phase error σ_δ allowable to insure that beam pointing direction lies within angular region $\pm \gamma$ with given probability P in terms of the beamwidth γ_0 is given by

$$\sigma_\delta (P, \gamma) = (\pi/a_p)(L/3\rho)^{1/2}(\gamma/\gamma_0) \qquad (55)$$

where $\sigma_\delta(P, \gamma) =$ maximum root-mean-square error allowable for probability P and angular pointing error γ, $a_p =$ argument of cumulative normal distribution function corresponding to centered cumulative probability τ where $\tau = 1 - [(1-P)/2]$ (double-tailing effect), $L =$ half the length of the array, $\gamma_0 =$ beamwidth $= 2/2L$, $\rho =$ correlation length for phase error (taken to be the element spacing in an array), and $\gamma =$ beam pointing error in radians.

*M. Leichter, "Beam Pointing Errors of Long Line Sources," *IRE Transactions on Antennas and Propagation*, vol. AP-8, no. 3, pp. 268–275; May 1960.

Pointing Error for Taylor Distribution

$$\sigma_\delta (P, \gamma, \alpha) = f(\alpha)\sigma_\delta (P, \gamma) \qquad (56)$$

where $\alpha =$ parameter derived from maximum side lobes expected in Taylor design given in Fig. 51, and $f(\alpha) =$ function relating Taylor and constant-amplitude distributions given in Fig. 51.

In Fig. 51, $f(\alpha)$ is plotted in the range $(0, 1.75\pi)$ corresponding to side-lobe ratios from 13.2 to 40 decibels.

For α, beyond the range plotted in Fig. 51, $f(\alpha)$ is given approximately by

$$f(\alpha) \simeq 2\pi^{1/4}\alpha^{-3/4}, \qquad \alpha > 6.$$

Pointing Error for Scanned Beam

$$\sigma_{\delta s} (P, \gamma) = \cos\theta_s \sigma_\delta (P, \gamma) \qquad (57)$$

where $\theta_s =$ angle of scan off normal.

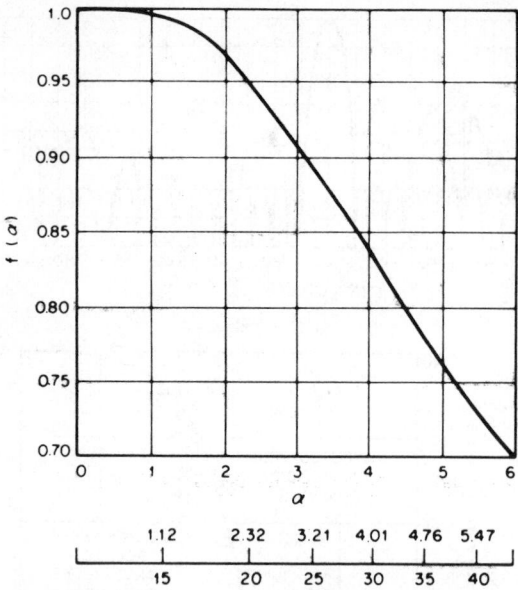

Fig. 51—Plot of $f(\alpha)$, multiplicative factor which converts allowed root-mean-square error for uniform distribution to the corresponding value for Taylor modified $(\sin x)/x$ distributions characterized by the parameter α. The values of α corresponding to the side-lobe ratios in −decibels are indicated in the lower scale. *M. Leichter, "Beam Pointing Errors of Long Line Sources," IRE Transactions on Antennas and Propagation, vol. AP-8, no. 3, p. 273; May 1960.*

Relationship Between RMS Pointing Error and RMS Phase Error

RMS phase error

$$\sigma_\delta = (\pi\sigma_p/\gamma_0)(L/3\Delta)^{1/2} \qquad (58)$$

because rms pointing error

$$\sigma_p \equiv \gamma/a_p.$$

EFFECT OF GROUND ON ANTENNA RADIATION AT VERY-HIGH AND ULTRA-HIGH FREQUENCIES

The behavior of the earth as a reflecting surface is considerably different for horizontal than for vertical polarization. For horizontal polarization the earth may be considered a perfect conductor, that is, the reflected wave at all vertical angles β is substantially equal to the incident wave and 180 degrees out of phase with it. $F(\beta)$ in Fig. 52B was derived on this basis. The approximation is good for practically all types of ground.

For vertical polarization, however, the problem is much more complex, as both the relative ampli-

tude K and relative phase ϕ change with vertical angle β and vary considerably with different types of ground. Figure 53 is a set of curves that illustrate the problem. The subscripts to the amplitude and phase coefficients K and ϕ refer to the type of polarization, H for horizontal and V for vertical.

Fig. 52—Several array directivity problems that do not fall into any of the preceding classes. $s° =$ spacing in electrical degrees, $h_1° =$ height of radiator in electrical degrees, and $d° =$ spacing of radiator from screen in electrical degrees.

(A) Two radiators any phase ϕ

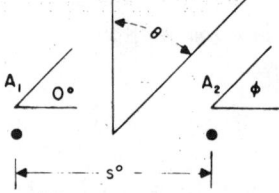

$$F(\theta) = [A_1^2 + A_2^2 + 2A_1A_2 \cos(s° \sin\theta + \phi)]^{1/2}.$$

When $A_1 = A_2$

$$F(\theta) = 2A \cos(\tfrac{1}{2}s° \sin\theta + \tfrac{1}{2}\phi).$$

(B) Radiator above ground (horizontal polarization)

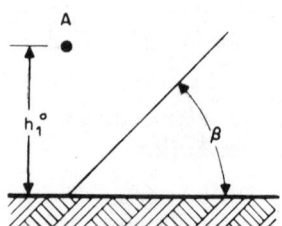

$$F(\beta) = 2A \sin(h_1° \sin\beta).$$

(C) Radiator parallel to screen

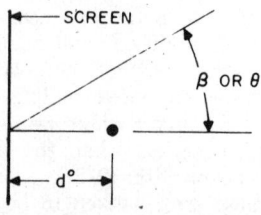

$$F(\beta) = 2A \sin(d° \cos\beta)$$

or

$$F(\theta) = 2A \sin(d° \cos\theta).$$

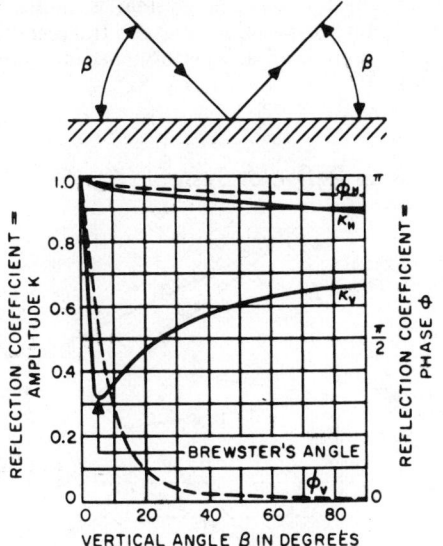

It is to be noted particularly that at grazing incidence ($\beta=0$) the reflection coefficient is the same for vertical and horizontal polarization. This is substantially true for practically all ground conditions.

Fig. 53—Typical ground-reflection coefficients for horizontal and vertical polarizations.

ELECTROMAGNETIC HORNS

Radiation from a waveguide may be obtained by placing an electromagnetic horn of a particular size at the end of the waveguide.

Figure 54 gives data for designing a horn to have a specified gain with the shortest length possible. The length L_1 is given by

$$L_1 = L[1 - (a/2A) - (b/2B)] \qquad (59)$$

where a = wide dimension of waveguide in the H plane and b = narrow dimension of waveguide in the E plane.

If $L \geq A^2/\lambda$, where A = longer dimension of aperture, the gain is given by

$$G = 10AB/\lambda^2. \qquad (60)$$

The half-power width in the E plane is given by

$$51\lambda/B \text{ degrees} \qquad (61)$$

and the half-power width in the H plane is given by

$$70\lambda/A \text{ degrees} \qquad (62)$$

where E = electric vector and H = magnetic vector.

Figure 55 shows how the angle between 10-decibel points varies with aperture.

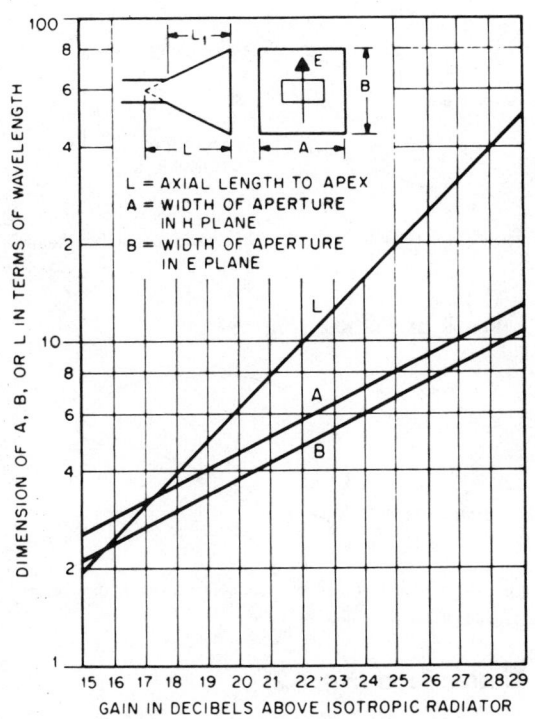

Fig. 54—Design of electromagnetic-horn radiator.

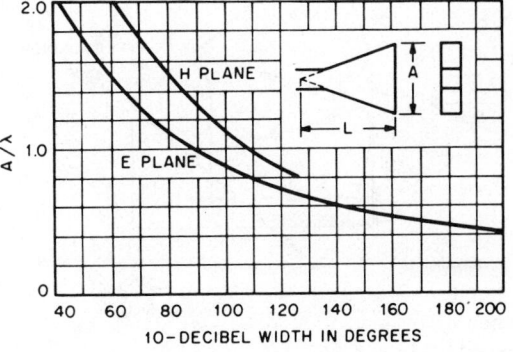

Fig. 55—10-decibel widths of horns. $L \geq A^2/\lambda$.

PARABOLOIDS

Paraboloidal Reflector (Fig. 56)

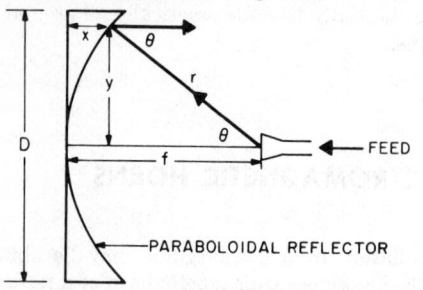

Fig. 56—Paraboloidal reflector design. f=focal length, $r = f \sec^2 (\theta/2)$, $x = f \tan^2 (\theta/2)$, and $y = 2f \tan (\theta/2)$.

Horn Parabolic Reflector (Fig. 57)

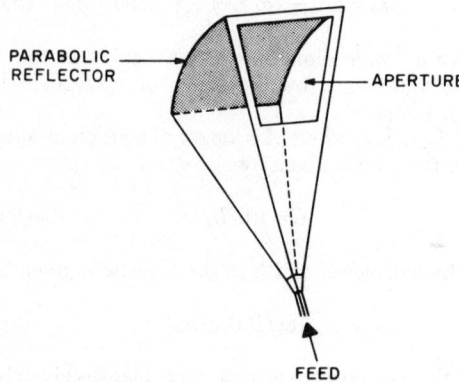

Fig. 57—Horn parabolic reflector achieves low noise by screening feed radiation.

Cassegrain Reflector System (Fig. 58)

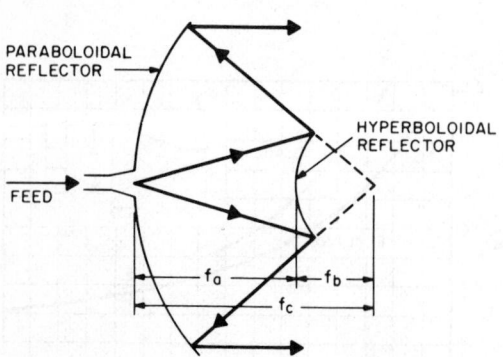

Fig. 58—Cassegrain reflector system.

The effective focal length $f_e = (f_a/f_b)f_c$. The usual range of magnification (f_a/f_b) is between 2 and 6.

Note: The Shwarzschild system is similar but uses reflector cross sections that are transcendental curves to improve off-axis performance for scanning purposes.

Gregorian Reflector System (Fig. 59)

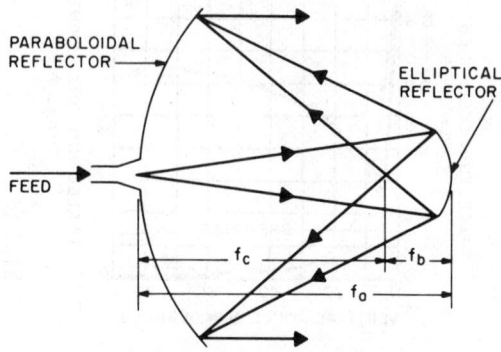

Fig. 59—Gregorian reflector system. Effective focal length $f_e = (f_a/f_b)f_c$ = magnification.

Parabolic Cylinder (Fig. 60)

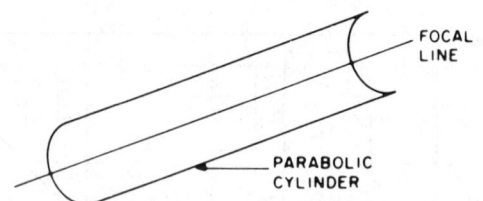

Fig. 60—Parabolic cylinder that produces a fan beam.

Cheese or Pillbox (Fig. 61)

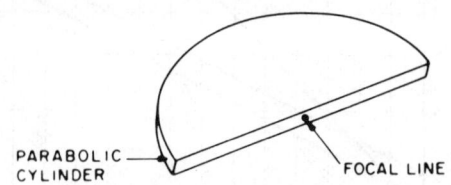

Fig. 61—Cheese or pillbox that produces a fan beam.

Cluster Feed (Fig. 62)

Feed horns are located on a spherical surface around the focal point of the parabola. Various distributions and scanning systems may be synthe-

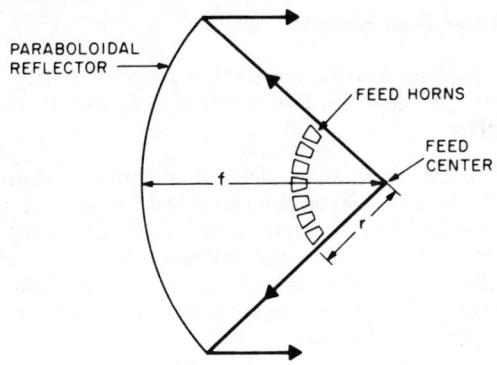

Fig. 62—Cluster feed.

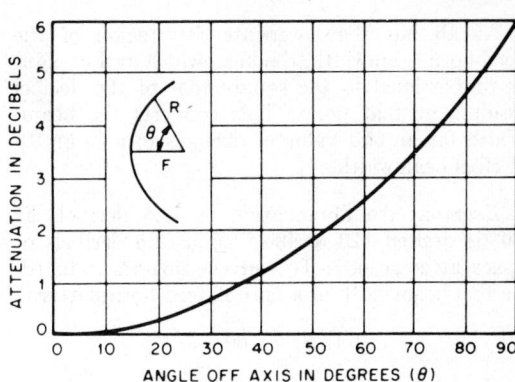

Fig. 64—Space attenuation as a function of feed angle for a parabola. $A = 20 \log (R/F) = 20 \log \sec^2 (\theta/2)$. *K. S. Kelleher, Aero Geo Astro Corp., The Microwave Engineers' Handbook and Buyers' Guide, p. 128, © 1964, Horizon House, Inc.*

sized using phased-array methods of feeding the cluster.

FEED DESIGN FOR PARABOLOIDAL ANTENNA

f = focal length

D = aperture diameter

θ = angle at feed point at focus from centerline of paraboloidal reflector to a line to any point on the reflector.

(**A**) Determine reflector edge illumination desired. For the normal feed-horn pattern, it has been determined that the reflected energy should be distributed so that the energy at the reflector edges is about 10 decibels below that at the center; if side-lobe performance is more important, the energy at the edge should be in the region of 20

decibels down. A compromise would be 15 decibels down at the edge.

(**B**) Determine f/D from reflector geometry.

(**C**) From the geometry it is now possible to determine the subtended beamwidth of the feed, which is the subtended angle at the focus. This is shown as a curve in Fig. 63.

(**D**) The space attenuation should be subtracted from the desired edge illumination. The space attenuation at the edge is given by

$$\text{Space Attenuation} = 20 \log[\sec^2(\theta/2)] \quad (63)$$

where θ is the angle to the edge. The attenuation curve is plotted in Fig. 64.

(**E**) Using the resultant (desired edge attenuation minus the space attenuation), determine the feed-horn-pattern beamwidth. Figure 65 may be used.

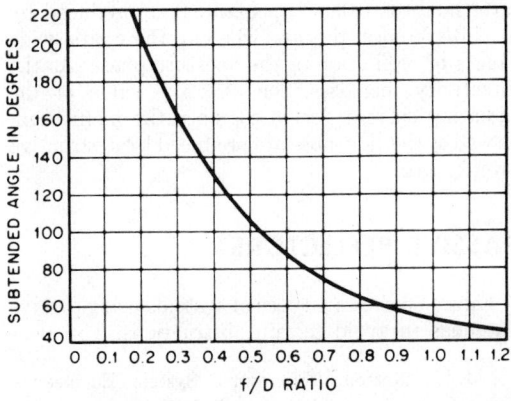

Fig. 63—Relationship of f/D ratio of a parabola with the subtended angle. *K. S. Kelleher, Aero Geo Astro Corp., The Microwave Engineers' Handbook and Buyers' Guide, p. 128, © 1964, Horizon House, Inc.*

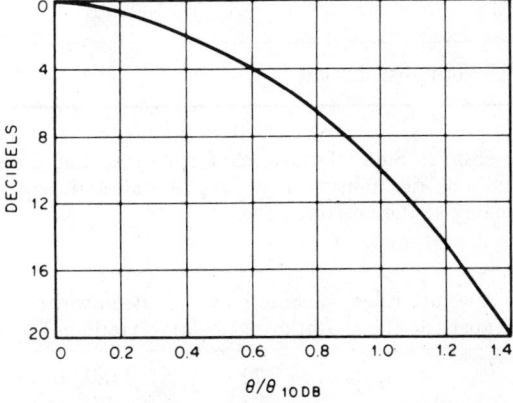

Fig. 65—Universal feed-horn pattern for $E_{dB} = 10(\theta/\theta_{10dB})^2$. *Adapted from K. S. Kelleher, Aero Geo Astro Corp., The Microwave Engineers' Handbook and Buyers' Guide, p. 128, © 1964, Horizon House, Inc.*

At all but the lower-intensity region of the feed-horn beamwidth, the beamwidth at any point is proportional to the square root of the decibel reading at that point. This converts the beamwidth for an odd value of decibels to a 3- or 10-decibel beamwidth.

Example: An illumination of 17.5 decibels at 60° is desired (20 decibels minus 2.5 decibels of space attenuation). To convert this value to 10-decibel beamwidth of a normal feed-horn pattern

$$(10/17.5)^{1/2} \times 60 = 45.4°$$

or the 10-decibel beamwidth is 90.8°.

Aperture Distributions with Resultant Side-Lobe Levels and Beamwidths

$L =$ source length

$\theta =$ angle measured from broadside

$x =$ distance measured along the source from the center

$p = 2\pi x/L$

$\lambda =$ wavelength.

$\cos^n p/2$: Cosine distribution with no pedestal and with zero energy at aperture edge.

n	Side-Lobe Ratio (decibels)	Beamwidth (radians)
0*	13.2	$0.88\lambda/L$
1	23	$1.2\lambda/L$
2	32	$1.45\lambda/L$
3	40	$1.66\lambda/L$

*Uniform distribution.

$\cos p/2$: Sum of patterns for cosine and for uniform distribution with a pedestal and with energy at the aperture edge.

Aperture Edge Taper (decibels)	Side-Lobe Ratio (decibels)	Beamwidth (radians)
10	20	$1.06\lambda/L$
15	22	$1.13\lambda/L$
∞*	23	$1.20\lambda/L$

*Cosine alone.

Other Distributions

Gaussian distribution yields a radiation pattern with no side lobes. It is a rather inefficient distribution.

Taylor distribution yields a radiation pattern with all side lobes of equal level and the narrowest beamwidth of all constant-phase distributions with side lobes at or below the specified level.

These distributions result in the same planar cross-section patterns as for linear arrays and are illustrated in Fig. 45.

Surface Errors

Surface errors are introduced into the reflector surface by manufacturing tolerances, wind, unexpected weight loading with changing position, snow, rain, and error factors such as malpositioning of feeds, changes in frequency, etc.

Regular errors are nonrandom errors introduced by surface deformations or feed-horn movement. These affect only phase. Linear variations rotate the entire pattern. Quadratic errors cause raised and blurred side lobes and reduce the directivity by broadening the main beam. Cubic errors cause a beam tilt, and the main beam and side-lobe patterns become asymmetric.

Random errors are statistical. These increase the side-lobe energy and decrease the main-beam energy while decreasing the directivity of the antenna.

Supergain

Directivity higher than that obtained with uniform phase is called supergain. It is produced by an interference process wherein there are wide ranges of oscillation of the aperture phase. As the directivity increases, the effective value of the radiating current decreases, since the main beam as well as the side lobes are produced by destructive interference.

PASSIVE REFLECTORS*

Figure 66 shows 4 types of systems using passive antennas to avoid terrain obstructions. To calcu-

* M. L. Norton, "Microwave System Engineering Using Large Passive Reflectors," *IRE Transactions on Communication Systems*, vol. CS-10, no. 3, pp. 304–311; September 1962. W. C. Jakes, Jr., "Theoretical Study of an Antenna–Reflector Problem," *Proceedings of the IRE*, vol. 41, pp. 272–274; February 1953.

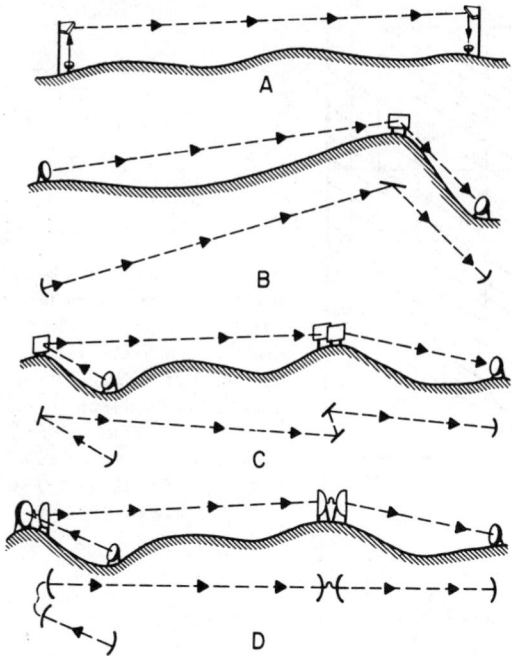

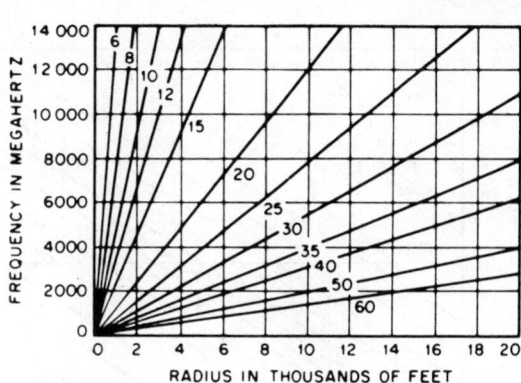

Fig. 67—Radius of the near field for the indicated values of antenna diameter D or reflector side dimension a in feet. *M. L. Norton, "Microwave System Engineering Using Large Passive Reflectors," IRE Transactions on Communication Systems, vol. CS-10, no. 3, Fig. 3; September 1962.*

Fig. 66—Types of passive antenna systems. A = periscope type. B = 2-hop system using single passive reflector. C = 3-hop system using single and double passive reflectors. D = 3-hop system using back-to-back antennas. *M. L. Norton, "Microwave System Engineering Using Large Passive Reflectors," IRE Transactions on Communication Systems, vol. CS-10, no. 3, Fig. 1; September 1962.*

late the loss of a microwave system using passive reflectors, it has to be determined whether the reflectors are in the near or far fields. A commonly accepted definition of the far field is the space in the field of an antenna where, over a given area, the spherical wave varies from a plane wave by less than $\lambda/16$, where λ is the wavelength. If the variation is more than $\lambda/16$, the given area is normally considered to be in the near field. The radius r of the near field plotted in Fig. 67 is given by

$$r = 2D^2/\lambda \quad \text{or} \quad 2a^2/\lambda \qquad (64)$$

where (all in the same units) D = antenna diameter, a = reflector side, and λ = wavelength.

The projected area (or effective area) a^2 of a passive reflector is given by

$$a^2 = A^2 \sin\tfrac{1}{2}\theta \qquad (65)$$

where A^2 = actual area and θ = deflection angle.

Figures 68 through 71 show performance characteristics of various passive systems.

Corner Reflectors

The corner reflector* is a simple directive antenna. The dimensions given in Fig. 72 will give a gain of 8 to 10 decibels over a dipole alone. If λ = wavelength

$$0.25\lambda \leq S \leq 0.7\lambda$$

length of reflector $\geq \lambda$
height of reflector $\geq 5\lambda/8$.

ANTENNA GAIN AND EFFECTIVE AREA

The gain of an antenna is a measure of how well the antenna concentrates its radiated power in a given direction. It is the ratio of the power radiated in a given direction to the power radiated in the same direction by a standard antenna (a dipole or isotropic radiator), keeping the input power constant. If the pattern of the antenna is known and there are no ohmic losses in the system, the gain G is defined by

$$G = \left(\frac{\text{maximum power intensity}}{\text{average power intensity}}\right)$$

$$= 4\pi \, |E_0|^2 \Big/ \iint_{\text{all angles}} |E|^2 d\Omega \qquad (66)$$

* J. D. Kraus, "Corner Reflector Antenna," *Proceedings of the IRE*, vol. 28, pp. 513–519; November, 1940.

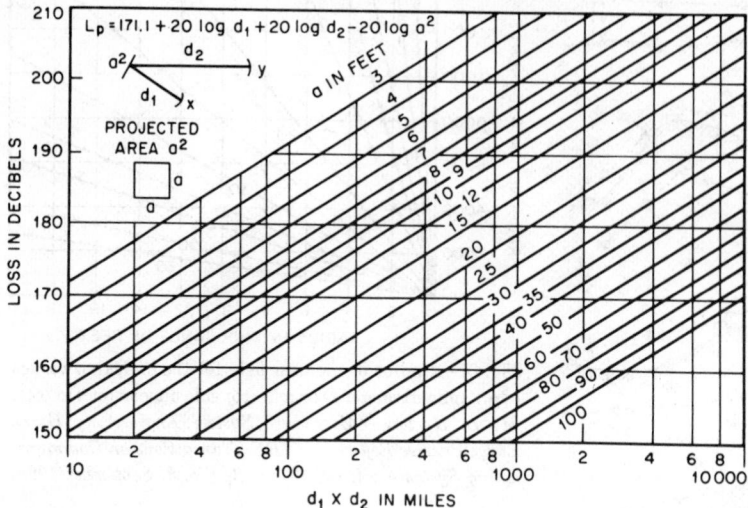

Fig. 68—Space loss on a 2-hop passive reflector system using passive reflectors in the far field for the indicated values of a in feet. M. L. Norton, "*Microwave System Engineering Using Large Passive Reflectors,*" *IRE Transactions on Communication Systems, vol. CS-10, no. 3, Fig. 8; September 1962.*

where $|E_0|$ = magnitude of the field at the maximum of the radiation pattern and $|E|$ = magnitude of the field in any direction.

The effective area A_r of an antenna is defined by

$$A_r = G\lambda^2/4\pi \qquad (67)$$

where G = gain of the antenna and λ = wavelength.

The power delivered by a matched antenna to a matched load connected to its terminals is PA_r, where P is the power density in watts/meter² at the antenna and A_r is the effective area in meters². The gains and receiving areas of some typical antennas are given in Table 4.

The gains and effective areas given in Table 4 apply in the receiving case only; when the polarizations are not the same, the gain is given by

$$G_\theta = G \cos^2\theta \qquad (68)$$

where G = gain of the antenna and θ = angle between plane of polarization of the antenna and the incident field.

Equation (68) applies only to linear polarization. Equation (6) gives the variation for circular or elliptical polarization. If a circularly polarized antenna is used to receive power from an incident wave of the same screw sense, the gains and receiving areas in Table 4 are correct. If a circularly polarized antenna is used to receive power from a linearly polarized wave (or vice versa) the gain or receiving area will be half those of Table 4.

If the half-power widths of a narrow-beam antenna are known, the approximate gain above an isotropic radiator may be computed from

$$G = 30\ 000/W_E W_H \qquad (69)$$

where W_E = E-plane half-power width in degrees, and W_H = H-plane half-power width in degrees.

Equation (69) is not accurate if the half-power

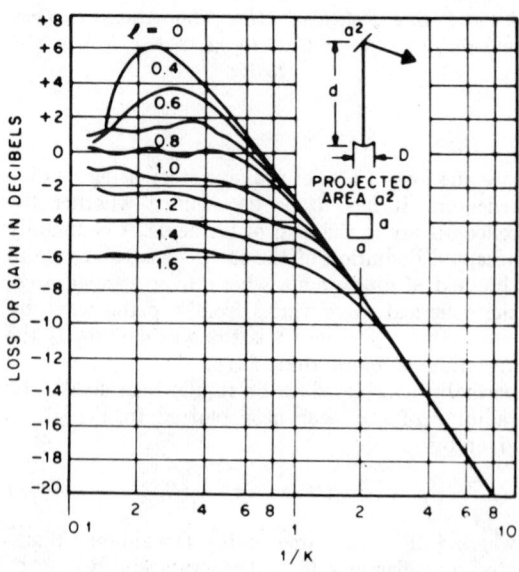

Fig. 69—Efficiency of passive reflectors in the near field. $1/K = \pi\lambda d/4a^2 = 772\alpha/fa^2$, and $l = D(\pi/4a^2)^{1/2} = 0.886D/a$. *After W. C. Jakes, Jr., "Theoretical Study of an Antenna-Reflector Problem," Proceedings of the IRE, vol. 41, no. 2, pp. 272–274; February 1953.*

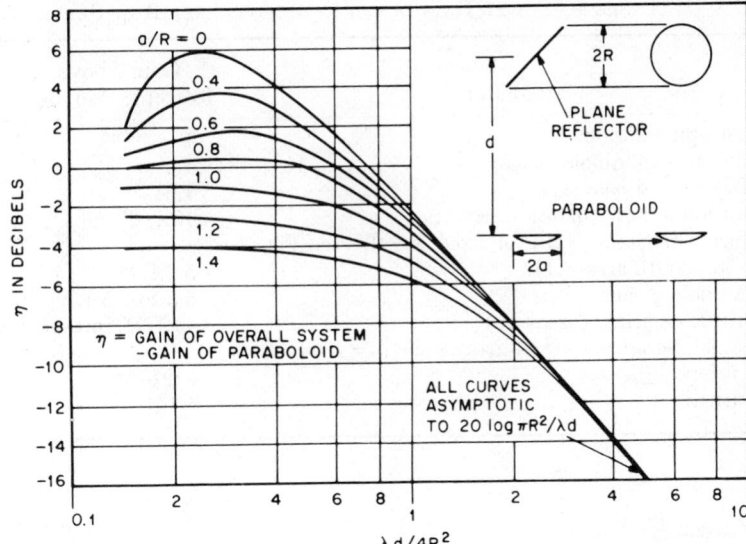

Fig. 70—Relative gain of periscope antenna system employing plane elevated reflector. *H. Jasik, "Antenna Engineering Handbook," Fig. 13-6,* © *1961, McGraw-Hill Book Company.*

widths are greater than about 20 degrees, or if there are many large side lobes.

SCANNING SYSTEMS

Conical Scanning

In conical scanning the radiation pattern of the antenna is pointed slightly off center and rotated around the boresight, keeping the angle offset constant. When the target is on boresight a continuous-wave signal results, and when the target is off boresight the signal is amplitude modulated at the rotation frequency. The magnitude and phase of the modulation determine the direction of the target.

This may be done mechanically by rotating the feed mechanically around the focal point. It may be done electronically by a phase-comparison conical-scan system as shown in Fig. 73.

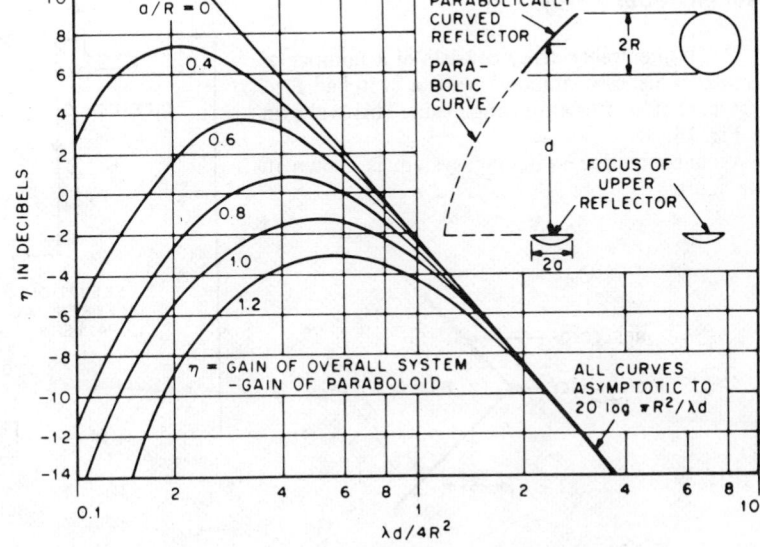

Fig. 71—Relative gain of periscope antenna system employing curved elevated reflector. *H. Jasik, "Antenna Engineering Handbook," Fig. 13-7,* © *1961, McGraw-Hill Book Company.*

TABLE 4—POWER GAIN G AND EFFECTIVE AREA A OF SEVERAL COMMON ANTENNAS.

Radiator	Gain Above Isotropic Radiator	Effective Area
Isotropic radiator	1	$\lambda^2/4\pi$
Infinitesimal dipole or loop	1.5	$1.5\lambda^2/4\pi$
Half-wave dipole	1.64	$1.64\lambda^2/4\pi$
Optimum horn (mouth area$=A$)	$10A/\lambda^2$	$0.81A$
Horn (maximum gain for fixed length—see Fig. 55, mouth area$=A$)	$5.6A/\lambda^2$	$0.45A$
Parabola or metal lens	6.3 to $7.5A/\lambda^2$	0.5 to $0.6A$
Broadside array (area$=A$)	$4\pi A/\lambda^2$ (max)	A (max)
Omnidirectional stacked array (length$=L$, stack interval$\leq\lambda$)	$\approx2L/\lambda$	$\approx L\lambda/2\pi$
Turnstile	1.15	$1.15\lambda^2/4\pi$

Monopulse

Monopulse systems differ from conical-scan systems inasmuch as they obtain tracking information by fixed beams. Four horns or other similar prime radiators may be clustered around the feed point as shown in Fig. 74.

Hybrids are used to obtain sums and differences, which are combined to generate an elevation differential pattern and an azimuth differential pattern. These are patterns that are zero along the boresight and maximum on either side of the boresight. The error signal is generated by comparison of the sum signal with each of the differential patterns.

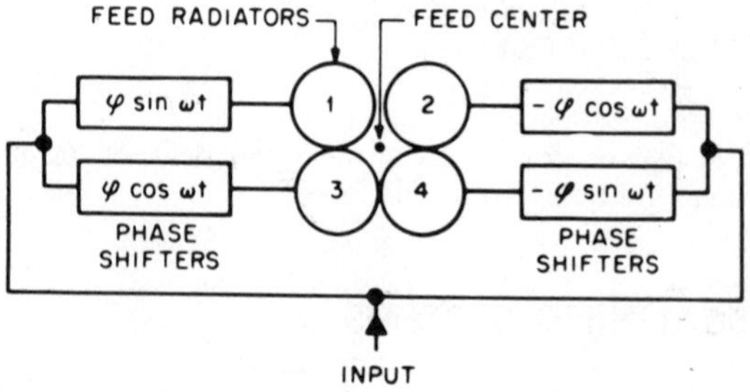

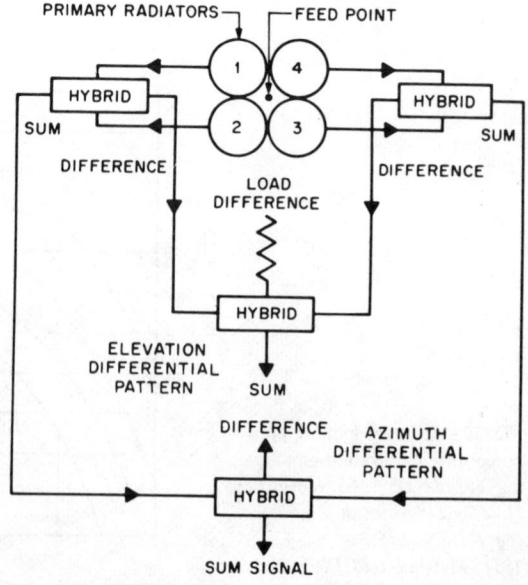

Fig. 73—Conical-scan system using phase shifters.

Wullenweber Array

The Wullenweber array consists of a number of circularly disposed antennas with a switched feed system that creates a rotating beam. This is shown in Fig. 75.

A noncontacting switching system is shown in the figure.

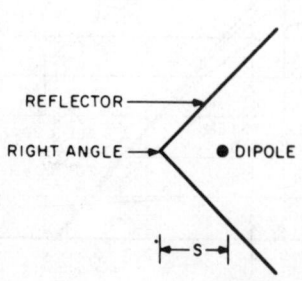

Fig. 72—Corner-reflector antenna.

Fig. 74—Monopulse system.

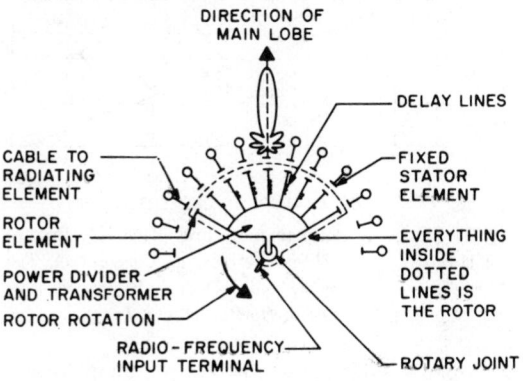

Fig. 75—Wullenweber array. R. C. Hansen, "Microwave Scanning Antennas," p. 219, © 1964, Academic Press Inc.

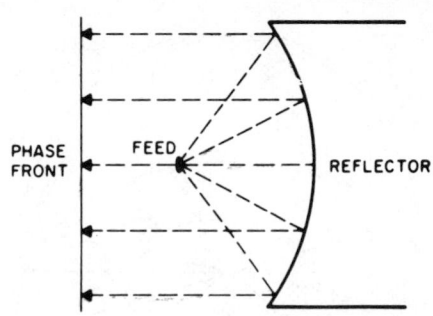

Fig. 77—Sectional view of hourglass scanner. After R. C. Hansen, "Microwave Scanning Antennas," © 1964, Academic Press Inc.

Hourglass Scanning System

Since the Wullenweber creates a beam in only one plane, the plane of the array, another method is used to obtain a beam at right angles to the plane of the array to create an effective pencil beam. This is done in the hourglass scanner, which combines a Wullenweber with a reflector. This is illustrated in Fig. 76.

Figure 77 is a sectional view of the reflecting surface and one of the feeds, showing how the vertical beam is created.

Luneberg Lens

The Luneberg lens is a spherically symmetric refracting structure which forms perfectly geometric images of two given concentric spheres on each other. In its most-well-known form for radio applications, one of the spheres has an infinite radius and the other sphere is at the surface of the lens. A beam of parallel rays that are incident from any direction will be focused at the surface of the lens diametrically opposite from the direction of incidence. This is illustrated in a planar case in Fig. 78.

The Luneberg lens that accomplishes this concentration of energy has a refractive index n that varies with the normalized radial coordinate ρ according to the relationship

$$n = (Z - \rho^2)^{1/2}. \qquad (70)$$

At the surface, $n = 1$, matching free space.

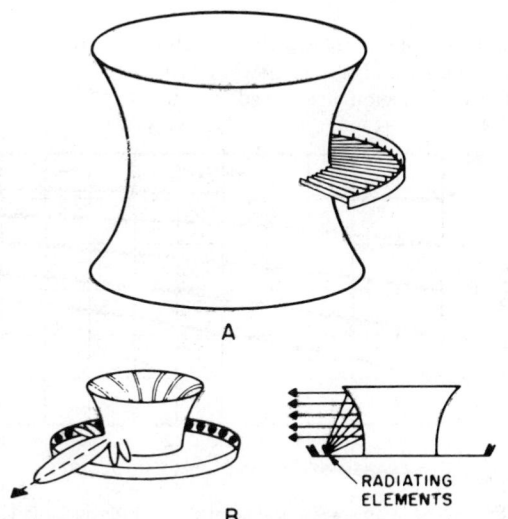

Fig. 76—Hourglass scanner. A pictorial view is at A; B is a half-parabolic hourglass system showing scanning lobe and ray paths. R. C. Hansen, "Microwave Scanning Antennas," pp. 218, 219, © 1964, Academic Press Inc.

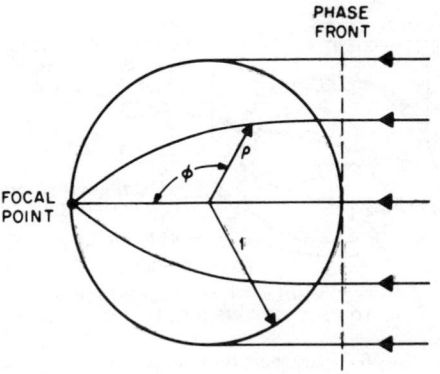

Fig. 78—Typical ray paths in a Luneberg lens. R. C. Hansen, "Microwave Scanning Antennas," p. 214, © 1964, Academic Press Inc.

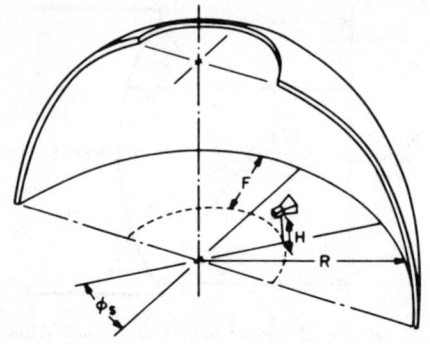

Fig. 79—180° solid-reflector torus. *R. C. Hansen, "Microwave Scanning Antennas," p. 220, © 1964, Academic Press Inc.*

Torus

The torus has a surface that is generated by rotating a section of a parabola about an axis parallel to the latus rectum. A 180° solid-reflector torus is shown in Fig. 79.

Scanning is limited in this 180° solid-reflector torus by spillover at the edges as well as blocking by the opposite edge as the angles approach ±90°. A 360° torus using 45° polarization to avoid blocking is shown in Fig. 80.

In this type of 360° reflector, a 45° wave reflected from one side passes through the other side without blocking since it is perpendicular to the reflecting wires.

NOISE TEMPERATURE (FIGS. 81–83)

MICROWAVE RADIATION HAZARDS*

Experiments have shown that hazards to personnel could exist from microwave radiation if

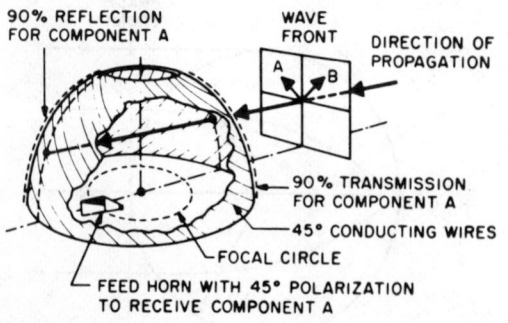

Fig. 80—360° parabolic torus antenna. *R. C. Hansen, "Microwave Scanning Antennas," p. 221, © 1964, Academic Press Inc.*

* W. W. Mumford, "Some Technical Aspects of Microwave Radiation Hazards," *Proceedings of the IRE,* vol. 49, no. 2, pp. 427–447; February 1961.

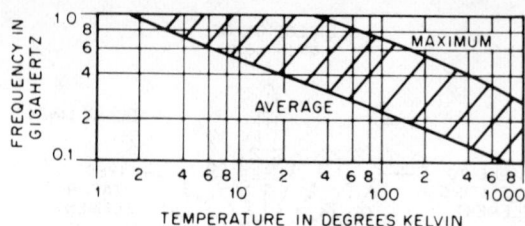

Fig. 81—Galactic noise temperature (after Ko, Brown, and Hazard). *R. C. Hansen, Aerospace Corporation, The Microwave Engineers' Handbook and Buyers' Guide, p. 207, © 1966, Horizon House, Inc.*

appropriate safety measures are not adopted and observed. From present information it is believed that exposure to a power density of 10 milliwatts per square centimeter (10 mW/cm²) should not be allowed for more than 1 hour, and that 1 mW/cm² is safe indefinitely. The damage to personnel is caused by heat generated by the microwave radiation, but it should be pointed out that serious damage can be caused to the eyes and by deep penetration through the skin where pain is felt only after the damage has occurred.

Radiation From an Antenna

The field in front of a paraboloidal antenna may be divided into 3 main regions as follows:

   Fresnel or Near Field
   Fraunhofer or Far Field
   Transition Zone.

There are no sharp dividing lines between the 3 regions, and the somewhat arbitrary limits set for each region are based on the way in which

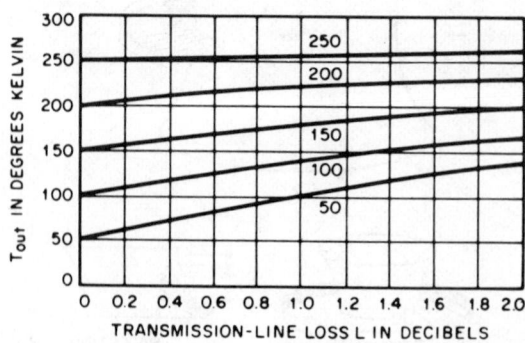

Fig. 82—Antenna noise-temperature degradation due to transmission-line loss for an ambient temperature of 290° Kelvin and for indicated antenna temperatures T_{ant} in degrees Kelvin. $T_{out} = [290(L-1) + T_{ant}]/L$. *R. C. Hansen, Aerospace Corporation, The Microwave Engineers' Handbook and Buyers' Guide, p. 207, © 1966, Horizon House, Inc.*

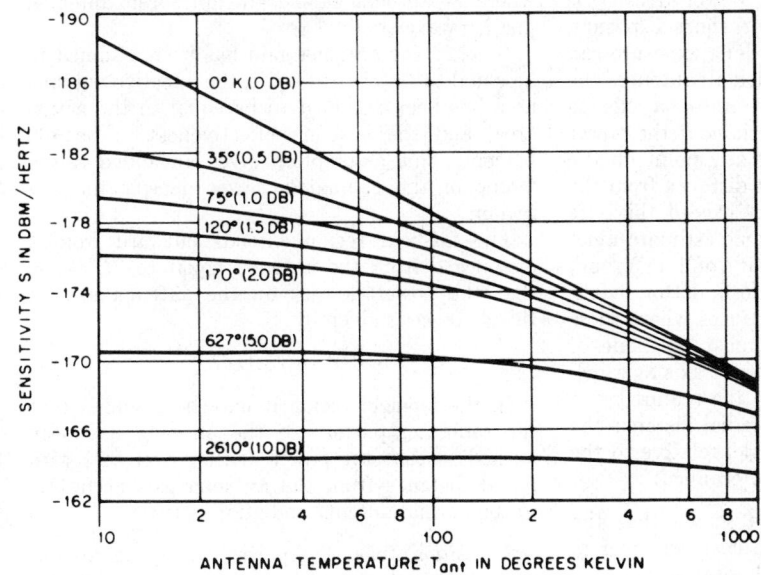

Fig. 83—System sensitivity versus antenna and receiver noise temperatures. Receiver temperatures in degrees Kelvin and noise figures in decibels are given for each curve. Absolute sensitivity $S = \alpha K(T_{ant} + T_{rec})$, where K = Boltzmann's constant = 198.6/dBm/hertz/degree. R. C. Hansen, Aerospace Corporation, *The Microwave Engineers' Handbook and Buyers' Guide*, p. 204, © 1966, Horizon House, Inc.

energy spreads as the distance from a paraboloidal antenna increases.

Fresnel or Near-Field Region: In this region the radiation is substantially confined within a cylindrical pattern having the same diameter as the antenna. This region may be considered to extend out from the antenna to a distance of

$$\pi D^2/8\lambda = A/2\lambda \qquad (71)$$

where D = diameter of antenna, A = area of antenna aperture, and λ = wavelength.

The energy is not uniform across the antenna in the Fresnel region and does not have a fixed value with distance from the antenna but varies about a mean value. For tapered antenna illumination the maximum power density on the antenna axis occurs at a distance from the antenna of $0.2D^2/\lambda$, the power density at this distance being about 2 decibels above the average near-field value. The maximum power density with uniform illumination is about the same as with tapered illumination, but the variations in power density along the antenna axis are of much greater amplitude. The power density just below the near-field cylindrical pattern is about 10 decibels less than the density on the antenna axis.

Fraunhofer or Far-Field Region: At a substantial distance from an antenna, the power density begins to decrease in proportion to the inverse square of the distance from the antenna. This may be said to occur at a distance from a paraboloidal antenna where the difference in path length between a ray on the axis of the beam, and a ray from the edge of the antenna to a given point on the beam axis, is less than about 1/16 of a wavelength. This distance, which can be shown to be $2D^2/\lambda$, is considered to be the beginning of the far-field region.

Transition Zone: Since the distance to the far-field region is about 5 times the length of the near-field region, there is a transition zone between the two. The decrease in power density in the transition zone varies along the antenna axis, and the decrease in density with increased radial distance can best be determined from power density contours as shown in Fig. 84, where the variation in power density is given for a paraboloidal-antenna aperture.

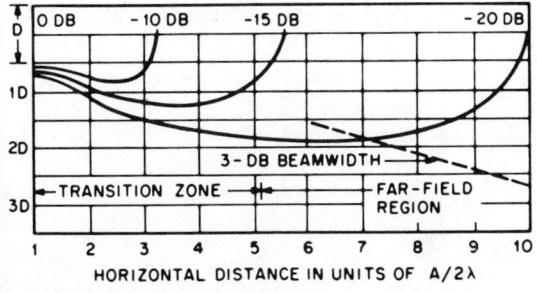

Fig. 84—Power density contours for paraboloidal antenna.

Determination of Power Densities

In estimating the radiation hazards that may exist in front of an antenna, it is necessary to determine in which areas the power density is greater than the safe limit for short exposure, and in which areas indefinitely long exposure can be permitted. With a paraboloidal antenna the power density is greatest on the antenna axis, so that it is first necessary to determine if the power density exceeds 10 mW/cm² at any point on the antenna axis, and if so, to what distance from the antenna does the power density exceed this safe limit for short exposure. The same estimate must then be made for a power density of 1 mW/cm², which is assumed to be the safe limit for indefinitely long exposure. At all distances where these limits are exceeded, estimates must be made of the radial distance from the antenna axis at which the power density is reduced to the required safe limit. From the estimated safe radial distance the minimum antenna elevation angle, relative to the terrain and buildings, may be determined.

Near-Field Region: It can be shown that the maximum power density in the near-field region, assuming a circular paraboloidal antenna, is

$$W = 16P/\pi D^2 = 4P/A \qquad (72)$$

where P = average transmitter power, A = area of antenna aperture, and D = antenna diameter.

This maximum power density occurs on the antenna axis at a distance of $0.2D^2/\lambda$ from the antenna. However, in considering radiation hazards it is usual to assume that the power density on the antenna axis in the near field is at its maximum value throughout the length of this region.

Most of the energy in the near-field region is confined to a cylinder of diameter equal to the antenna diameter, the power density decreasing very rapidly with radial distance outside this cylindrical region. With normal tapered antenna illumination it may be assumed that the power density at the edge of the near-field cylinder is 6 decibels below the maximum level at the antenna axis, and at a radial distance of $0.6D$ from the axis the power density is at least 10 decibels below the maximum level of $4P/A$.

Far-Field Region: In the far field the free-space power density W on the antenna axis is given by

$$W = GP/4\pi d^2 = AP/\lambda^2 d^2 \qquad (73)$$

where G = antenna gain, d = distance from antenna, and λ = wavelength.

In (72) the antenna gain has been assumed to be equal to $(\pi D/\lambda)^2$; that is, the effective antenna area has been assumed to be equal to the actual area, and the antenna effectiveness to be 100 percent. This assumption gives an added safety factor in the estimated power density for this region.

The far-field region extends outward from a distance d from the antenna equal to $2D^2/\lambda$, so that the power density on the antenna axis at this distance is given by

$$W = P/5.1D^2. \qquad (74)$$

In the far-field region it may be assumed that the radiation pattern of the antenna has been formed so that the power density decreases with radial distance from the antenna axis according to the normal antenna radiation pattern.

Transition Zone: The power density on the antenna axis in the transition zone decreases approximately as the square of the distance from the near-field region. Thus when estimating the variation in power density with distance along the antenna axis, only two regions need be considered. The first is the near field, in which the power density may be considered to be constant at the maximum value of $4P/A$. In the second region, which includes all distances beyond the near-field region, the power density on the antenna axis may be considered to decrease as the square of the distance from the outer limit of the near-field region.

In the transition zone the power density is a maximum on the antenna axis, decreasing rapidly as the radial distance from this axis increases. However, the falling off in power density with radial distance is somewhat less rapid than would be the case had the far-field antenna radiation pattern already been formed. The decrease in power density with radial distance within this area can best be estimated from the power density contours shown in Fig. 84. In this figure, contours of -10, -15, and -20 decibels, relative to the maximum near-field power density, are given for the transition zone and for the far-field region out to 10 times the near-field distance of $A/2\lambda$.

CHAPTER 28

ELECTROMAGNETIC-WAVE PROPAGATION

Radio waves may be propagated* from the transmitting antenna to the receiving antenna through or along the surface of the earth, through the atmosphere, or by reflection or scattering from natural or artificial reflectors. The conductivity and dielectric constant of the ground vary considerably from those of the atmosphere. At very-low frequencies, ground waves may be satisfactorily propagated for distances of several thousand kilometers. At high frequencies, however, the losses are so great that signals can be propagated for only a few hundred kilometers by ground wave. Propagation in the medium- and high-frequency bands is chiefly by ground wave and by reflection from the ionosphere, and severe fading is caused in these frequency bands by the interference between ground and ionospheric waves.

The refractive index of the atmosphere is an important factor in radio propagation. At frequencies between about 100 and 8000 megahertz, scattering of radio waves by inhomogeneities in the electromagnetic characteristics of the atmosphere can be used to provide satisfactory wideband communication up to several times the line-of-sight distance. New techniques are being developed for generating coherent high-power waves in the optical spectrum. Atmospheric absorption at these frequencies is high, but the large bandwidths and small antenna beamwidths may make such frequencies practical for certain applications.

VERY-LOW FREQUENCIES—UP TO 30 KILOHERTZ

The propagation of long radio waves is of considerable importance in reliable communication, long-range navigation, and the detection of nuclear explosions. Considerable progress has been made in recent years in understanding the propagation of such waves in the earth-ionosphere waveguide.†

At short distances from a transmitter the received signal is chiefly by a ground or surface wave, and at very-low frequencies its intensity is essentially inversely proportional to distance. At greater distances the field intensity falls at a higher rate

because of losses in the ground and because of the curvature of the earth. These losses increase with frequency. At sufficiently great distances the received level is chiefly due to sky waves reflected from the ionosphere. At intermediate distances the field is a combination of sky waves and ground waves that result in an interference pattern. The total field at the receiver may be obtained in two distinct ways. The first method, which leads to the geometric-optics theory, directly sums the contributions at the receiver from the primary source and each of its images. The second method treats the source and its images as self-illuminating diffraction gratings, one above the earth and one below, and leads to the waveguide mode theory. The advantages of mode theory are restricted to very-low frequencies, where relatively few modes can be supported in the earth-ionosphere waveguide. For example, when the height of the ionosphere is 80 kilometers and the wavelength is 20 kilometers, only the first 8 modes can be supported. When the wavelength is 2 kilometers, however, all modes up to the 80th can be supported.

Thus at very-low frequencies, and for distances greater than say 3000 kilometers, it is simpler to use the mode of lowest order to obtain the received field. At distances less than about 1000 kilometers, it is simpler to use ray theory. The above is based on an idealized condition, since it has been assumed that the earth is flat, that the ionosphere is sharply bounded, and the effect of the earth's magnetic field can be ignored. Even with these simplifying assumptions the results are useful. For a full treatment of the general case see Wait, Budden, and Johler.*

The results of calculations made by Wait and Spies, taking into account the curvature of the earth and the conductivities of the earth and the ionosphere, are shown in Figs. 1 and 2, where $\sigma_g=$ ground conductivity, $\omega_r=$ ionospheric conductivity parameter, and $n=$ mode order. It is seen that the lowest attenuation of mode 1 in the daytime ($h=70$ km) is at about 18 kilohertz and at night ($h=90$ km) at about 15 kilohertz.

* CCIR XIth Plenary Assembly, Oslo; 1966: Vol. II, Propagation.

† A. D. Watt, "VLF Radio Engineering," vol. 14, International Series of Monographs in Electromagnetic Waves, Pergamon Press, New York; 1967.

* J. R. Wait, "Electromagnetic Waves in Stratified Media," Pergamon Press, Long Island City, N.Y.; 1962. K. G. Budden, "The Wave-Guide Mode Theory of Wave Propagation," Prentice-Hall, Inc., New York, N.Y.; 1962. J. R. Johler, "Propagation of the Low-Frequency Radio Signal," *Proceedings of the IRE*, vol. 50, no. 4, pp. 404–427; 1962.

The phase velocity ratio of mode 1 in the daytime is greater than 1 for frequencies less than about 13 kilohertz and less than 1 for higher frequencies. At night the crossover frequency is about 9 kilohertz.

In daylight the attenuation of mode 1 is always less than that of mode 2. At night the attenuation for the two modes may be of the same order.

LOW AND MEDIUM FREQUENCIES— 30 TO 3000 KILOHERTZ*

For low and medium frequencies of approximately 30 to 3000 kilohertz with a short vertical antenna over perfectly reflecting ground

$E = 186.4(P_r)^{1/2}$ millivolts rms/meter at 1 mile
$E = 300(P_r)^{1/2}$ millivolts rms/meter at 1 kilometer

where P_r = radiated power in kilowatts.

Actual inverse-distance fields at 1 kilometer for a given transmitter output power depend on the height and power radiation efficiency of the antenna and associated circuit losses.

Typical values found in practice for well-designed stations are:

Small L or T antennas as on ships:

$40(P_t)^{1/2}$ millivolts/meter at 1 kilometer

Vertical radiators 0.15 to 0.25λ high:

$290(P_t)^{1/2}$ millivolts/meter at 1 kilometer

Vertical radiators 0.25 to 0.40λ high:

$322(P_t)^{1/2}$ millivolts/meter at 1 kilometer

Vertical radiators 0.40 to 0.60λ high or top-loaded vertical radiators:

$386(P_t)^{1/2}$ millivolts/meter at 1 kilometer

where P_t = transmitter output power in kilowatts. These values can be increased by directive antenna systems.

It has been found that the concept of basic transmission loss, also called path loss, is convenient for the analysis of radio communication systems. Basic transmission loss is the dimensionless ratio P_R/P_A, where P_R is the power radiated from a lossless, isotropic transmitting antenna and P_A is the power available from a lossless, isotropic receiving antenna in a matched load. The isotropic antennas are at the same physical locations and operate in the same band of frequencies as the actual antennas.

Surface-wave (commonly called ground-wave) basic transmission loss is plotted in Fig. 3 for

* "Radio Spectrum Utilization," Joint Technical Advisory Committee (IEEE and EIA), New York.; 1964. CCIR XIth Plenary Assembly, Oslo; 1966: Vol. II, Propagation.

vertically polarized propagation over land having a representative conductivity and dielectric constant and in Fig. 4 for vertically polarized propagation over sea water. Both antennas are 30 feet above the surface in both figures.

In the low-frequency and medium-frequency ranges, propagation losses for horizontally polarized transmission between antennas on the surface of the earth are impractically high. Ground constants typical of various terrain types are listed in Table 1.

Under the conditions used in Figs. 3 and 4, the earth's surface behaves like a nearly perfect reflector for the isotropic antennas that are only a small fraction of a wavelength from it and that are used to calculate the basic transmission loss. As a result, each isotropic antenna together with the surface of the earth has a gain of nearly 3.01 dB in the general direction of the horizon. By contrast, a lossless quarter-wave monopole erected over a good ground screen would have a gain of 5.16 dB, and a

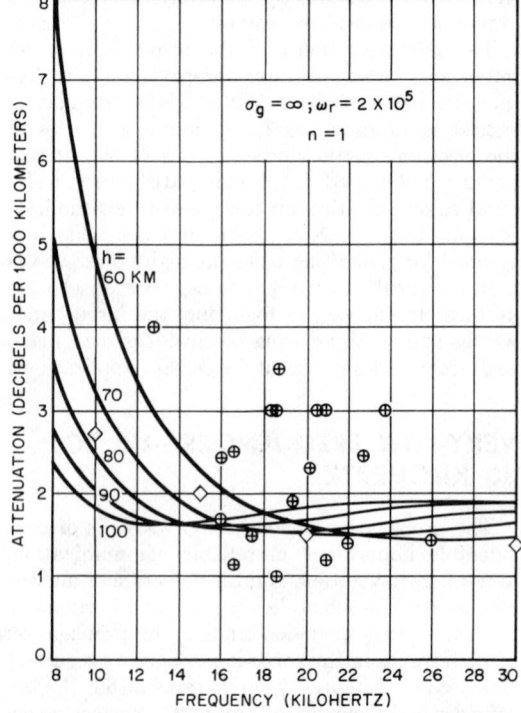

Fig. 1—Relation between attenuation, frequency, and height of the ionosphere. The diamonds represent some experimental observations by Taylor on the average daytime attenuation for west-to-east propagation over sea water. Attenuation rate in the opposite direction is greater by about 1 decibel per 1000 kilometers. *From "Radio Spectrum Utilization," Joint Technical Advisory Committee (IEEE and EIA), IEEE, New York; 1964: page 104.*

short lossless monopole would have a gain of 4.77 dB. It follows that the transmission loss between quarter-wave monopoles on the surface of the earth is very nearly $2\times(5.16-3.01)=4.30$ dB less than the basic transmission loss given in Figs. 3 and 4, and the transmission loss between short monopoles is $2\times(4.77-3.01)=3.52$ dB less.

Figures 3 and 4 do not include the effect of sky waves reflected from the ionosphere. Sky waves cause fading at medium distances and produce higher field strengths than the surface wave at longer distances, particularly at night. Sky-wave field strength is subject to diurnal, seasonal, and irregular variations due to changing properties of the ionosphere.

Figure 5 shows a family of propagation curves for F_0 computed from

$$F_0=80.2-10\log D-0.00176f^{0.26}D$$

where $D=$ distance in kilometers, $f=$ frequency in kilohertz, and F_0 is the annual median received

TABLE 1—GROUND CONDUCTIVITY AND DIELECTRIC CONSTANT FOR MEDIUM- AND LONG-WAVE PROPAGATION TO BE USED WITH NORTON, BURROWS, BREMMER, OR OTHER DEVELOPMENTS OF SOMMERFELD PROPAGATION EQUATIONS.

Terrain	Conductivity σ (mhos/meter)	Dielectric Constant ϵ (esu)
Sea water	5	80
Fresh water	8×10^{-3}	80
Dry, sandy, flat coastal land	2×10^{-3}	10
Marshy, forested flat land	8×10^{-3}	12
Rich agricultural land, low hills	1×10^{-2}	15
Pastoral land, medium hills and forestation	5×10^{-3}	13
Rocky land, steep hills	2×10^{-3}	10
Mountainous (hills up to 3000 feet)	1×10^{-3}	5
Cities, residential areas	2×10^{-3}	5
Cities, industrial areas	1×10^{-4}	3

field strength in decibels above 1 microvolt per meter that would be produced by a short, vertical transmitting dipole at or near the earth's surface and *radiating* 1 kilowatt. The empirical equation is based on measured data. Figure 5 therefore includes the effects of both sky-wave and surface-wave propagation.

Penetration of Waves

The extent to which the lower strata influence the effective ground constants depends on the depth of penetration of the radio energy. This in turn depends on the value of the constants and the frequency. If the depth of penetration is defined as that depth in which the wave has been attenuated to $1/e$ (37%) of its value at the surface, then over the frequency range from 10 kHz to 10 MHz, δ has the values shown in Table 2. It will be seen that, at frequencies of 10 MHz and above, only the surface of the ground need be considered, but at lower frequencies, strata down to a depth of 100 meters or more must be taken into account. It is particularly important to take account of the lower strata when the upper strata are of lower conductivity, since more energy penetrates to the lower levels than happens with an upper layer of higher conductivity.

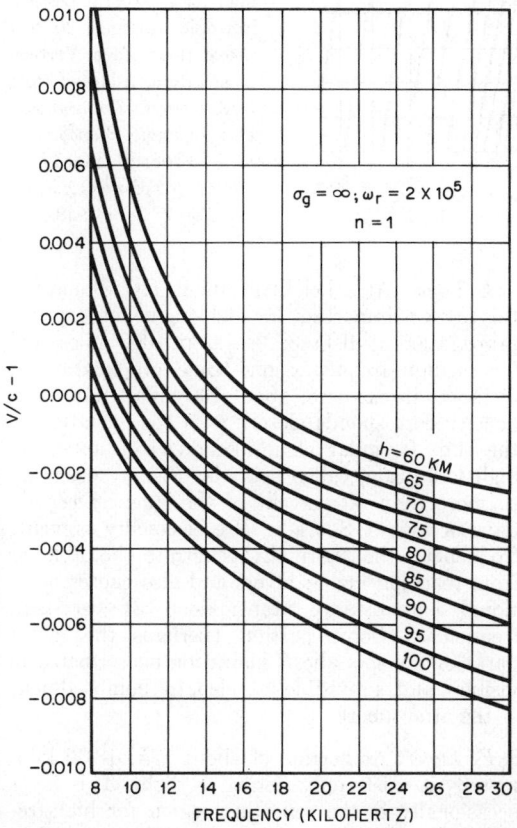

Fig. 2—Phase velocity V as a function of ionospheric height h and frequency, relative to the velocity in free space c. From "*Radio Spectrum Utilization*," *Joint Technical Advisory Committee (IEEE and EIA)*, IEEE, New York; 1964: page 105.

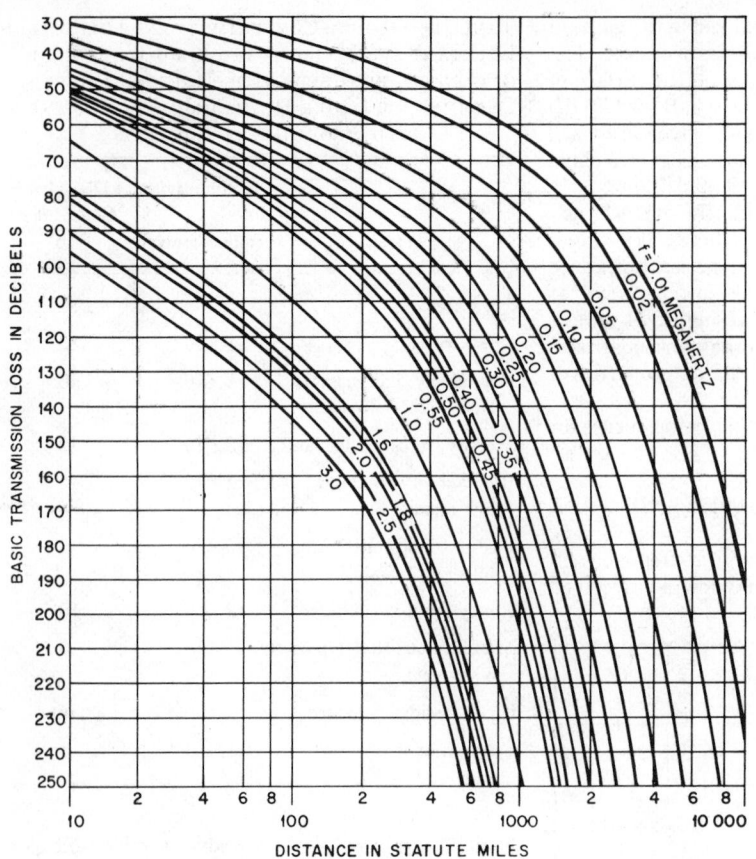

Fig. 3—Basic transmission loss expected for surface waves propagated over a smooth spherical earth. Over land: $\sigma = 0.005$ mho/meter, $\epsilon = 15$. Lossless isotropic antennas 30 feet above the surface. Vertical polarization. Adapted from K. A. Norton, "Transmission Loss in Radio Propagation: II," National Bureau of Standards Technical Note 12, Fig. 7; June 1959.

HIGH FREQUENCIES—3 TO 30 MEGAHERTZ*

At frequencies between about 3 and 25 megahertz and distances greater than about 100 miles, transmission depends chiefly on sky waves reflected from the ionosphere. This is a region high above the earth's surface where the rarefied air is sufficiently ionized (primarily by ultraviolet sunlight) to reflect or absorb radio waves, such effects being controlled almost exclusively by the free-electron density. The ionosphere is usually considered as consisting of the following layers.

D Layer: At heights from about 50 to 90 kilometers, it exists only during daylight hours, and ionization density corresponds with the elevation of the sun.

This layer reflects very-low- and low-frequency waves, absorbs medium-frequency waves, and weakens high-frequency waves through partial absorption.

E Layer: At a height of about 110 kilometers, this layer is important for high-frequency daytime propagation at distances less than 1000 miles, and for medium-frequency nighttime propagation at distances in excess of about 100 miles. Ionization density corresponds closely with the elevation of the sun. Irregular cloud-like areas of unusually high ionization, called sporadic E, may occur up to more than 50 percent of the time on certain days or nights. Sporadic E occasionally prevents frequencies that normally penetrate the E layer from reaching higher layers and also causes occasional long-distance transmission at very-high frequencies. Some portion (perhaps the major part) of the sporadic-E ionization is ascribable to visible- and subvisible-wavelength bombardment of the atmosphere.

F1 Layer: At heights of about 175 to 250 kilometers, it exists only during daylight. This layer occasionally is the reflecting region for high-frequency transmission, but usually oblique-incidence waves that penetrate the E layer also penetrate the F_1 layer and are reflected by the F_2 layer. The F_1 layer introduces additional absorption of such waves.

F2 Layer: At heights of about 250 to 400 kilo-

* K. Davies, "Ionospheric Radio Propagation," Monograph 80, National Bureau of Standards, Washington, D.C.; 7 April 1965. CCIR XIth Plenary Assembly, Oslo; 1966: Vol. II, Propagation.

TABLE 2—DEPTH OF PENETRATION OF WAVES INTO THE GROUND.

Frequency	Depth δ(m)		
	$\sigma=4$ mho/m $\epsilon=80$	$\sigma=10^{-2}$ mho/m $\epsilon=10$	$\sigma=10^{-3}$ mho/m $\epsilon=5$
10 kHz	2.5	50	150
100 kHz	0.80	15	50
3 MHz	0.14	5	17
10 MHz	0.08	2	9

meters, F_2 is the principal reflecting region for long-distance high-frequency communication. Height and ionization density vary diurnally, seasonally, and over the sunspot cycle. Ionization does not follow the elevation of the sun in any fashion, since (at such extremely low air densities and molecular-collision rates) the medium can store received solar energy for many hours, and, by energy transformation, can even detach electrons during the night. At night, the F_1 layer merges with the F_2 layer at a height of about 300 kilometers. The absence of the F_1 layer, and reduction in absorption of the E layer, causes nighttime field intensities and noise to be generally higher than during daylight.

As indicated to the right on Fig. 6, these layers are contained in a thick region throughout which ionization generally increases with height. The layers are said to exist where the ionization gradient is capable of refracting waves back to earth. Obliquely incident waves follow a curved path through the ionosphere due to gradual refraction or bending of the wave front. When attention need be given only to the end result, the process can be assimilated to a reflection.

Depending on the ionization density at each layer, there is a critical or highest frequency f_c at which the layer reflects a vertically incident wave. Frequencies higher than f_c pass through the layer at vertical incidence. At oblique incidence, and distances such that the curvature of the earth and ionosphere can be neglected, the maximum usable frequency is given by

$$MUF = f_c \sec \phi$$

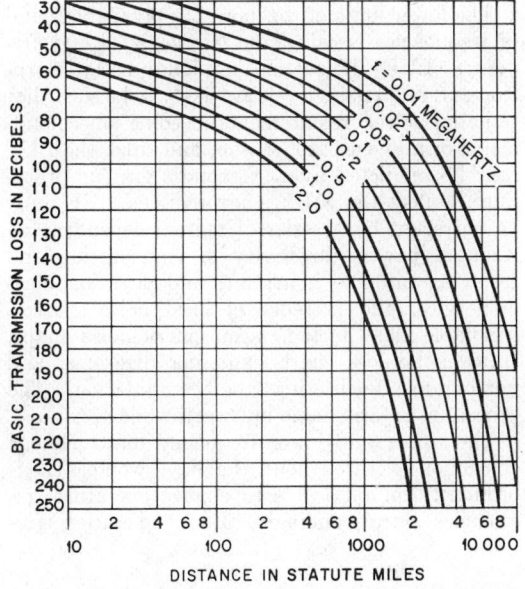

Fig. 4—Basic transmission loss expected for surface waves propagated over a smooth spherical earth. Over sea water: $\sigma=5$ mhos/meter, $\epsilon=80$. Lossless isotropic antenna 30 feet above the surface. Vertical polarization. Adapted from *K. A. Norton, "Transmission Loss in Radio Propagation: II," National Bureau of Standards Technical Note 12, Fig. 8; June 1959.*

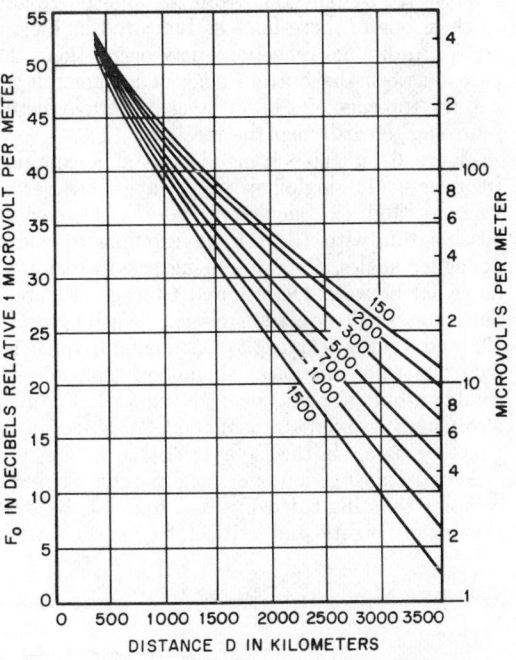

Fig. 5—Family of basic curves of F_0 to be used to determine the annual median value of the field strength for the frequencies (in kilohertz) indicated on the curves. *CCIR Thirteenth Plenary Assembly, vol. VI, Report 264-3, p. 108, Geneva; 1974.*

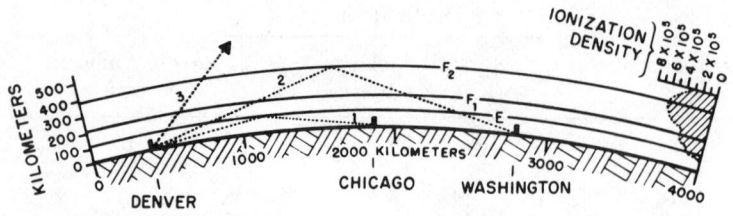

Fig. 6—Schematic explanation of skip-signal zones.

where MUF=maximum usable frequency for the particular layer and distance, and ϕ=angle of incidence at reflecting layer.

At greater distances, curvature is taken into account by the modification

$$\mathrm{MUF}=kf_c \sec \phi$$

where k is a correction factor that is a function of distance and vertical distribution of ionization.

f_c and height, and hence ϕ for a given distance, vary for each layer with local time of day, season, latitude, and throughout the 11-year sunspot cycle. The various layers change in different ways with these parameters. In addition, ionization is subject to frequent abnormal variations.

Ionospheric losses are a minimum near the maximum usable frequency and increase rapidly for lower frequencies during daylight.

High frequencies travel from the transmitter to the receiver by reflection from the ionosphere and earth in one or more hops as indicated in Figs. 6 and 7. Additional reflections may occur along the path between the bottom edge of a higher layer and the top edge of a lower layer, the wave finally returning to earth near the receiver.

Figure 6 indicates transmission on a common frequency, (1) single-hop via E layer, Denver to Chicago, and (2) single-hop via F_2, Denver to Washington, with (3) the wave failing to reflect at higher angles, thus producing a skip region of no signal between Denver and Chicago. Figure 7 illustrates single-hop transmission, Washington to Chicago, via the E layer (ϕ_1). At higher frequencies over the same distance, single-hop transmission would be obtained via the F_2 layer (ϕ_2). Figure 7 also shows two-hop transmission, Washington to San Francisco, via the F_2 layer (ϕ_3).

Actual transmission over long distances is more complex than indicated by Figs. 6 and 7, because the layer heights and critical frequencies differ with time (and hence longitude) and with latitude. Further, scattered reflections occur at the various surfaces.

Typical values of critical frequency for Washington, D.C., are shown in Fig. 8.

Preferably, operating frequencies should be selected from a specific frequency band that is bounded above and below by limits that are systematically determinable for the transmission path under consideration. The recommended upper limit is called the *optimum working frequency* (FOT) and is selected below the MUF to provide some margin for ionospheric irregularities and turbulence, as well as for the statistical deviation of day-to-day ionospheric characteristics from the predicted monthly median value. So far as may be consistent with available frequency assignments, operation in reasonable proximity to the upper frequency limit is preferable, in order to reduce absorption loss.

The lower limit of the normally available band of frequencies is called the *lowest useful high frequency* (LUF). Below this limit ionospheric absorption and radio noise levels are likely to be such that radiated-power requirements become uneconomical. For a given path, season, and time, the LUF may be predicted by a systematic graphic procedure. Unlike the MUF, the predicted LUF must be corrected by a series of factors dependent on radiated power, directivity of transmitting and receiving antennas in azimuth and elevation, class of service, and presence of local noise sources. Available data include atmospheric-noise maps, transmission-loss charts, antenna diagrams, and nomograms facilitating the computation. The procedure is formidable but worthwhile.

The upper and lower frequency limits change continuously throughout the day, whereas it is ordinarily impracticable to change operating frequencies correspondingly. Each operating fre-

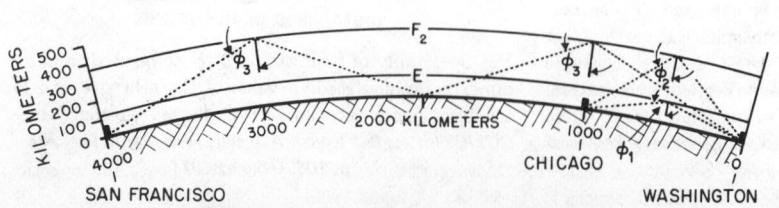

Fig. 7—Single-hop and two-hop transmission paths due to E and F_2 layers.

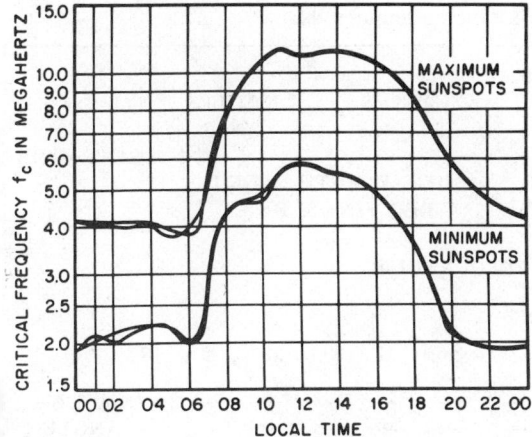

Fig. 8—Critical frequency for Washington, D.C. *National Bureau of Standards Circular 462.*

quency, therefore, should be selected to fall within the above limits for a substantial portion of the daily operating period.

Angles of Departure and Arrival

Angles of departure and arrival are of importance in the design of high-frequency antenna systems. These angles, for single-hop transmission, are obtained from the geometry of a triangular path over a curved earth with the apex of the triangle placed at the virtual height assumed for the altitude of the reflection. Figure 9 is a family of curves showing radiation angle for different distances.

$D =$ great-circle distance in statute miles
$H =$ virtual height of ionosphere layer in kilometers

$\Delta =$ radiation angle in degrees
$\phi =$ semiangle of reflection at ionosphere.

Forecasts of High-Frequency Propagation

In addition to forecasts for ionospheric disturbances, the Institute of Telecommunication Sciences of the Environmental Sciences Services Administration (ESSA) issues monthly predictions 3 months in advance, used to determine the optimum working frequencies for high-frequency communication.

In designing a high-frequency communication circuit, it is necessary to determine the optimum traffic frequencies, system loss, signal-to-noise ratio, angle of arrival, and circuit reliability. Manual methods for calculating the values of these factors are described,* as is the use of electronic computers for predicting the performance of high-frequency sky-wave communication circuits.†

Table 3 is a typical performance prediction prepared by computer. A general description of the circuit parameters used in the calculations is shown in the heading of the computer printout. Starting at the top of the page and reading from left to right, the heading may be described as follows.

The first line contains the month, the solar activity level in 12-month moving average Zurich sunspot number, and a circuit identification number. The second and third lines contain the transmitter and receiver locations, the bearings, and the distance. The fourth and fifth lines contain the antenna system for each terminal and their orientation relative to the great-circle path. The minimum angle indicates the lowest vertical angle considered in the mode selection.

The sixth line is the power delivered to the

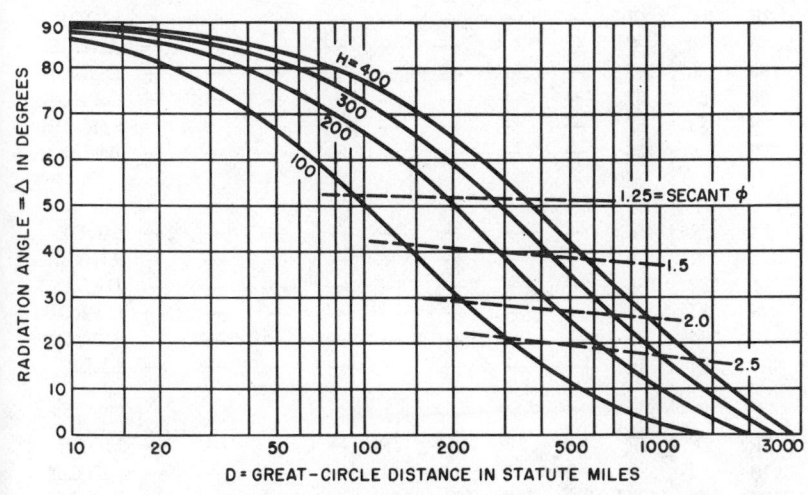
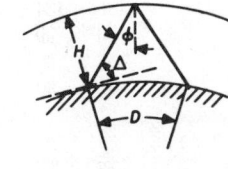

Fig. 9—Single-reflection radiation angle and great-circle distance.

* K. Davies, "Ionospheric Radio Propagation," National Bureau of Standards Monograph 80; 1 April 1965.
† D. L. Lucas and G. W. Haydon, "Predicting Statistical Performance Indexes for High-Frequency Ionospheric Telecommunication Systems," ESSA Technical Report IER1-ITSA; 1 August 1966.

TABLE 3—SYSTEM PERFORMANCE PREDICTIONS.

1	JAN		SSN=20	CH 5.029			
LONDONDERRY		TO	CHELTENHAM		AZIMUTHS		N.MILES
55.00N — 7.31W			38.75N — 76.85W		280.5 46.3		2880.3

RHOMBIC 50H 168L 70 DEG.   ANT=0 DB
OFF AZIMUTH 0 DEG.   MIN. ANGLE=0 DEG.   OFF AZIMUTH 0 DEG.
PWR=200.00 KW   3 MHZ MAN. NOISE=−154 DBW   REQ. S/N=61 DB

OPERATING FREQUENCIES

GMT	MUF	3	4	5	6	7	8	10	12	15	17	20	22	25	27	
2	8.6															
		2F	2F	2F	2F	2F	2F	2F	—	—	—	—	—	—	—	MODE
		7	5	5	5	5	6	7	—	—	—	—	—	—	—	ANGLE
		187	185	185	185	185	186	187	—	—	—	—	—	—	—	DELAY
		50	99	99	97	87	66	20	—	—	—	—	—	—	—	C.PROB.
		89	59	71	78	83	86	94	—	—	—	—	—	—	—	S/N..DB
		50	42	81	91	85	66	20	—	—	—	—	—	—	—	REL.
4	8.4															
		2F	2F	2F	2F	2F	2F	2F	—	—	—	—	—	—	—	MODE
		8	5	5	5	6	7	7	—	—	—	—	—	—	—	ANGLE
		188	185	185	185	186	187	188	—	—	—	—	—	—	—	DELAY
		50	99	99	96	84	61	16	—	—	—	—	—	—	—	C.PROB.
		89	58	69	77	82	87	95	—	—	—	—	—	—	—	S/N..DB
		50	34	77	89	82	60	16	—	—	—	—	—	—	—	REL.
6	8.3															
		2F	2F	2F	2F	2F	2F	2F	—	—	—	—	—	—	—	MODE
		7	5	5	5	6	7	7	—	—	—	—	—	—	—	ANGLE
		187	185	185	185	186	187	187	—	—	—	—	—	—	—	DELAY
		50	99	99	94	81	57	10	—	—	—	—	—	—	—	C.PROB.
		89	56	68	76	82	87	97	—	—	—	—	—	—	—	S/N..DB
		50	28	76	89	80	57	10	—	—	—	—	—	—	—	REL.
8	7.3															
		2F	2F	2F	2F	2F	2F	—	—	—	—	—	—	—	—	MODE
		7	5	5	5	6	7	—	—	—	—	—	—	—	—	ANGLE
		187	185	185	185	186	187	—	—	—	—	—	—	—	—	DELAY
		50	99	97	85	59	27	—	—	—	—	—	—	—	—	C.PROB.
		84	56	69	78	83	87	—	—	—	—	—	—	—	—	S/N..DB
		48	29	71	76	56	26	—	—	—	—	—	—	—	—	REL.
10	7.9															
		2F	2F	2F	2X	2X	2F	—	—	—	—	—	—	—	—	MODE
		6	4	4	1	1	6	—	—	—	—	—	—	—	—	ANGLE
		186	184	184	181	182	186	—	—	—	—	—	—	—	—	DELAY
		50	99	99	98	87	47	—	—	—	—	—	—	—	—	C.PROB.
		87	56	70	77	84	87	—	—	—	—	—	—	—	—	S/N..DB
		49	30	75	86	84	46	—	—	—	—	—	—	—	—	REL.
12	14.5															
		2F	3E	4F	3F	2F	2F	2F	2F	2F	2F	—	—	—	—	MODE
		4	3	15	10	4	3	3	3	4	4	—	—	—	—	ANGLE
		184	181	190	187	184	183	183	183	184	184	—	—	—	—	DELAY
		50	99	99	99	99	99	99	90	39	8	—	—	—	—	C.PROB.
		101	5	53	67	75	81	91	93	101	103	—	—	—	—	S/N..DB
		50	0	14	67	84	93	98	90	39	8	—	—	—	—	REL.

transmitting antenna, the man-made noise level assumed for the area in dBW in a 1-hertz bandwidth at 3 megahertz, and the hourly median signal-to-noise ratio required to provide the service requested. The signal is in the same units as the transmitter power and the noise is in a 1-hertz bandwidth. The seventh line contains the heading for the operating frequencies, which are given in megahertz in the eighth line. In addition to the operating frequencies, the eighth line also contains the time heading (GMT) and the classically defined maximum-usable-frequency heading (MUF), i.e., frequency which has a 50-percent probability of having a sky-wave path.

For each time and operating frequency the body of the tabulation contains: (A) the mode having the greatest probability (MODE), (B) the median vertical angle associated with this mode (ANGLE), (C) the propagation time in tenths of milliseconds (DELAY), (D) the percentage of days that any sky-wave mode is expected to exist, circuit probability (C.PROB.), (E) the median of the hourly median signal-to-noise ratios for the days sky-wave modes exist (S/N..DB), and (F) the percentage of days within the month that the median required signal-to-noise ratio is expected to be equalled or exceeded (REL.).

Bandwidth Limitations*

In high-frequency transmission, the communication bandwidth is limited by multipath propagation. The greatest limitation occurs when two or more paths exist with a different number of hops. The bandwidth may then be as small as 100 hertz, but such multipath may be minimized by operating near the MUF. Operation at a frequency within approximately 10% of the MUF is necessary for paths less than about 600 kilometers to obtain bandwidths greater than, say, 1 kilohertz. The

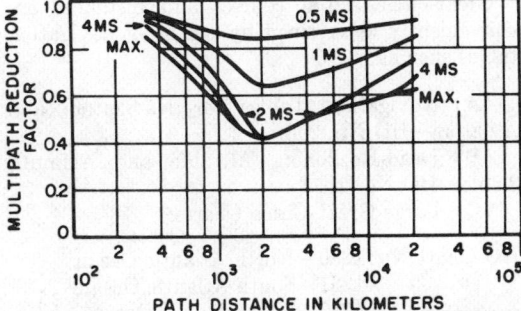

Fig. 10—Multipath reduction factor as a function of path distance. *R. K. Salaman, "A New Ionospheric Multipath Reduction Factor (MRF)," © 1962, Institute of Radio Engineers.*

* "Multipath Propagation Over High-Frequency Radio Circuits," CCIR Report 203, Vol. III, Geneva, 1974.

multipath reduction factor (MRF) is defined as the smallest ratio of MUF to operating frequency for which the range of multipath propagation time difference is less than a specified value. The MRF thus defines the frequency above which a specified minimum protection against multipath is provided. Figure 10 shows the MRF for various lengths of path.*

Diversity†

It has been shown that if two or more high-frequency radio channels are sufficiently separated in space, frequency, angle of arrival, time, or polarization, the fading on the various channels is more or less independent. Diversity systems make use of this fact to improve the overall performance, combining or selecting separate radio channels on a single high-frequency circuit.

Satisfactory diversity improvement can be obtained if the correlation coefficient of the fading on the various channels does not exceed about 0.6, and experiments have indicated that a frequency separation of the order of 400 hertz gives satisfactory diversity performance on long high-frequency paths. Spacing between antennas at right angles to the direction of propagation should be about 10 wavelengths. Polarization diversity has been found to be about equivalent to space diversity in the high-frequency band. Measurements have indicated that times varying from 0.05 to 95 seconds may be necessary to obtain fading correlation coefficients as low as 0.6 in high-frequency time-diversity systems. Angle-of-arrival diversity requires the use of large antennas so as to obtain the required vertical directive characteristics. Differences in the angle of arrival of 2° have been shown to give satisfactory diversity improvement on high-frequency circuits.

GREAT-CIRCLE CALCULATIONS

Referring to Fig. 11, A and B are two places on the earth's surface the latitudes and longitudes of which are known. The angles X and Y at A and B of the great circle passing through the two places and the distance Z between A and B along the

* R. K. Salaman, "A New Ionospheric Multipath Reduction Factor (MRF)," *IRE Transactions on Communication Systems*, vol. CS-10, pp. 220–222; June 1962.

† "Bandwidth and Signal-to-Noise Ratios in Complete Systems," CCIR Report 195, Vol. III, Geneva; 1974. G. L. Grisdale, J. H. Morriss, and D. S. Palmer, "Fading of Long-Distance Radio Signals and a Comparison of Space and Polarization Diversity Reception in the 6–18 Mc Range," *Proceedings of the IEE*, Part B, No. 13, pp. 39–51; January 1957.

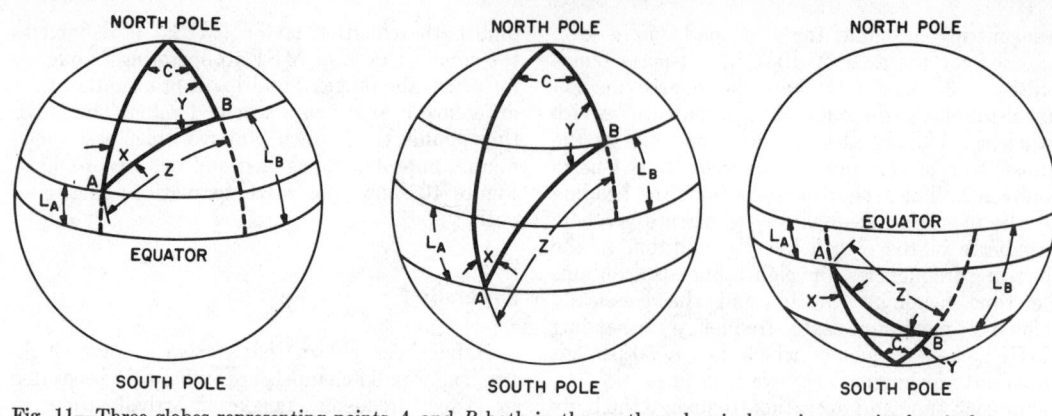

Fig. 11—Three globes representing points A and B both in the northern hemisphere, in opposite hemispheres, and both in the southern hemisphere. In all cases, L_A=latitude of A. L_B=latitude of B. C=difference of longitude.

great circle can be calculated as follows: B=place of greater latitude (nearer the pole), L_A=latitude of A, L_B=latitude of B, and C=difference of longitude between A and B.

Then

$$\tan \tfrac{1}{2}(Y-X) = \cot \tfrac{1}{2}C \frac{sin \tfrac{1}{2}(L_B-L_A)}{\cos \tfrac{1}{2}(L_B+L_A)}$$

and

$$\tan \tfrac{1}{2}(Y+X) = \cot \tfrac{1}{2}C \frac{\cos \tfrac{1}{2}(L_B-L_A)}{sin \tfrac{1}{2}(L_B+L_A)}$$

give the values of $\tfrac{1}{2}(Y-X)$ and $\tfrac{1}{2}(Y+X)$ from which

$$\tfrac{1}{2}(Y+X)+\tfrac{1}{2}(Y-X)=Y$$

and

$$\tfrac{1}{2}(Y+X)-\tfrac{1}{2}(Y-X)=X.$$

In the above equations, north latitudes are taken as positive and south latitudes as negative. For example, if B is latitude 60° N and A is latitude 20° S,

$$\frac{L_B+L_A}{2} = \frac{60+(-20)}{2} = \frac{60-20}{2} = \frac{40}{2} = 20°$$

$$\frac{L_B-L_A}{2} = \frac{60-(-20)}{2} = \frac{60+20}{2} = \frac{80}{2} = 40°.$$

If both places are in the southern hemisphere and L_B+L_A is negative, it is simpler to call the place of greater south latitude B and to use the above method for calculating bearings from true south and to convert the results afterward to bearings east of north.

The distance Z (in degrees) along the great circle between A and B is given by the following:

$$\tan \tfrac{1}{2}Z = \tan \tfrac{1}{2}(L_B-L_A)$$

$$\times [\sin \tfrac{1}{2}(Y+X)]/[\sin \tfrac{1}{2}(Y-X)].$$

The angular distance Z (in degrees) between A and B may be converted to linear distance as follows:

Z (in degrees) $\times 111.12$ = kilometers
Z (in degrees) $\times 69.05$ = statute miles
Z (in degrees) $\times 60.00$ = nautical miles.

In multiplying, the minutes and seconds of arc must be expressed in decimals of a degree. For example, $Z = 37°45'36''$ becomes $37.755°$.

Example: Find the great-circle bearings at Brentwood, Long Island, longitude 73°15'10''W, latitude 40°48'40''N, and at Rio de Janeiro, Brazil, longitude 43°22'07''W, latitude 22°57'09''S; and the great-circle distance in statute miles between the two points. Refer to Table 4.

Available Maps and Tables

Great-circle initial courses and distances are conveniently determined by means of navigation tables such as:

(A) Navigation Tables for Navigators and Aviators—HO No. 206.
(B) Dead-Reckoning Altitude and Azimuth Table—HO No. 211.
(C) Large Great-Circle Charts:

HO Chart No. 1280—North Atlantic Ocean
1281—South Atlantic Ocean
1282—North Pacific Ocean
1283—South Pacific Ocean
1284—Indian Ocean

The above tables and charts may be obtained at a nominal charge from the United States Navy Department Hydrographic Office, Washington, D.C.

TABLE 4—EXAMPLE OF GREAT-CIRCLE CALCULATIONS.

	Longitude	Latitude	
Brentwood	73°15′10″W	40°48′40″N	L_B
Rio de Janeiro	43°22′07″W	(−)22°57′09″S	L_A
	29°53′03″		C
		17°51′31″	L_B+L_A
		63°45′49″	L_B-L_A

$\frac{1}{2}C=14°56′31″$ $\frac{1}{2}(L_B+L_A)=8°55′45″$ $\frac{1}{2}(L_B-L_A)=31°52′54″$

$\log \cot 14°56′31″=10.57371$	$\log \cot 14°56′31″=10.57371$
plus $\log \cos 31°52′54″= \ \ 9.92898$	plus $\log \sin 31°52′54″= \ \ 9.72277$
$\overline{0.50269}$	$\overline{0.29648}$
minus $\log \sin 8°55′45″= \ \ 9.19093$	minus $\log \cos 8°55′45″= \ \ 9.99471$
$\log \tan \frac{1}{2}(Y+X)= \ \ 1.31176$	$\log \tan \frac{1}{2}(Y-X)= \ \ 0.30177$
$\frac{1}{2}(Y+X)=87°12′26″$	$\frac{1}{2}(Y-X)=63°28′26″$

Bearing at Brentwood $=\frac{1}{2}(Y+X)+\frac{1}{2}(Y-X)=Y=150°40′52″$ East of North
Bearing at Rio de Janeiro $=\frac{1}{2}(Y+X)-\frac{1}{2}(Y-X)=X=23°44′00″$ West of North

$\frac{1}{2}(L_B-L_A)=31°52′54″$	$\log \tan 31°52′54″=9.79379$
$\frac{1}{2}(Y+X)=87°12′26″$	plus $\log \sin 87°12′26″=9.99948$
	$\overline{9.79327}$
$\frac{1}{2}(Y-X)=63°28′26″$	minus $\log \sin 63°28′26″=9.95170$
	$\log \tan \frac{1}{2}Z=9.84157$
	$\frac{1}{2}Z=34°46′24″$ $Z=69°32′48″$

$69°32′48″=69.547°.$
Linear distance $=69.547\times69.05=4802$ statute miles.

EFFECT OF NUCLEAR EXPLOSIONS ON RADIO PROPAGATION*

Nuclear explosions below an altitude of about 15 kilometers have little effect on radio transmission. However, a detonation occurring at an altitude between 15 and 60 kilometers can produce blackout in the low-frequency, medium-frequency, and high-frequency bands over a radius of several hundred kilometers. This effect lasts only for a few minutes except in an area close to the site of the explosion. In general, it can be said that the effect of nuclear explosions is greatest near the site of the detonation, but the effects of ionization and shock waves do not last longer than a few minutes at distances greater than a few hundred kilometers from the site of the explosion.

* S. Glasstone, "The Effects of Nuclear Weapons," US Government Printing Office, Washington, D.C.; 1962.

IONOSPHERIC SCATTER PROPAGATION*

This type of transmission permits communication in the frequency range from approximately 30 to 60 megahertz and over distances from about 1000 to 2000 kilometers. It is believed that this type of propagation is due to scattering from the lower D region of the ionosphere and that the useful bandwidth is restricted to less than 10 kilohertz. The greatest use for this type of transmission has been for printing-telegraph channels, particularly in the auroral regions where conventional high-frequency ionospheric transmission is often unreliable.

The median attenuation over paths between 800 and 1000 miles in length is about 80 decibels

* "Ionospheric Scatter Transmission," *Proceedings of the IRE*, vol. 48, no. 1, pp. 5–29; 1960. CCIR XIIIth Plenary Assembly, Vol. VI, Report 260-2, and Vol. III, Report 109-2, Geneva, 1974.

greater than the free-space path attenuation at 30 megahertz and about 90 decibels greater than the free-space value at 50 megahertz.

METEOR-BURST PROPAGATION*

Frequencies in the very-high- and ultra-high-frequency bands may be propagated by reflection from columns of ionization produced by meteors entering the lower E region. Experimental single-channel 2-way telegraph circuits have been operated in the frequency range from 30 to 40 megahertz over distances of 600 to 1300 kilometers with transmitter powers of 1 to 3 kilowatts. One-way transmission of voice and facsimile have also been made with transmitter powers of 1 kilowatt and 20 kilowatts, respectively.

The frequency range from about 50 to 80 megahertz has been found best suited for meteor-burst transmission.

PROPAGATION ABOVE 30 MEGAHERTZ LINE-OF-SIGHT CONDITIONS†

Radio Refraction ‡

Under normal propagation conditions, the refractive index of the atmosphere decreases with height so that radio rays travel more slowly near the ground than at higher altitudes. This variation in velocity with height results in bending of the radio rays. Uniform bending may be represented by straight-line propagation, but with the radius of the earth modified so that the relative curvature between the ray and the earth remains unchanged. The new radius of the earth is known as the effective earth radius, and the ratio of the effective earth radius to true earth radius is usually denoted by K. The average value of K in temperate

---

* "Intermittent Communication by Meteor-Burst Propagation," CCIR XIII Plenary Assembly, Vol. VI, Report 251-1, Geneva, 1974. P. A. Forsyth, E. L. Vogan, D. R. Hansen, and C. O. Hines, "The Principles of JANET—A Meteor-Burst Communication System," *Proceedings of the IRE*, Vol. 45, pp. 1642–1657; December 1957.

† K. Bullington, "Radio Propagation at Frequencies Above 30 Megacycles," *Proceedings of the IRE*, vol. 35, pp. 1122–1136; October 1947. D. E. Kerr, "Propagation of Short Radio Waves," McGraw–Hill Book Company, New York; 1951. CCIR XIth Plenary Assembly, Vol. II, Oslo; 1966.

‡ B. R. Bean and E. J. Dutton, "Radio Meteorology," Monograph 92, Institute for Telecommunication Sciences and Aeronomy, Environmental Science Services Administration (ESSA), Superintendent of Documents, Washington, D.C.; 1966.

climates is about 1.33; however, values from about 0.6 to 5.0 are to be expected.

The decrease in the refractive index with height may at times be so great that the ray is bent down with a radius equal to that of the earth so that the earth may then be considered to be flat. A further increase in the refractive-index gradient results in the radio ray being bent down sufficiently to be reflected from the earth. The ray then appears to be trapped in a duct between the earth and the maximum height of the radio path.

Under certain atmospheric conditions the refractive index may increase with height, causing the radio rays to bend upward. Such inverse bending results in a decrease in path clearance on line-of-sight paths.

The distance to the radio horizon over smooth earth, when the height h is very small compared with the earth's radius, is given with a good approximation by

$$d = (3Kh/2)^{1/2}$$

where h = height in feet above the earth, d = distance to radio horizon in miles, and K = ratio of the effective to the true radius of the earth.

Over a smooth earth, a transmitter antenna at height h_t (feet) and a receiving antenna at height h_r (feet) are in radio line-of-sight provided the spacing in miles is less than $(2h_t)^{1/2} + (2h_r)^{1/2}$ (assuming $K = 1.33$).

The nomogram in Fig. 12 gives the radio-horizon distance between a transmitter at height h_t and a receiver at height h_r. Figure 13 extends the first nomogram to give the maximum radio-path length between two airplanes whose altitudes are known. Both figures assume a value of $K = 1.33$.

Path Plotting and Profile-Chart Construction

Path Plotting: When laying out a microwave system, it is usually convenient to plot the path on a profile chart. Such charts are scaled to indicate the departure of the curvature of the earth from a straight line. Referring to Fig. 14

$$D^2 + R^2 = (h+R)^2 = h^2 + 2Rh + R^2$$

$$D^2 = h^2 + 2Rh$$

where D = distance, R = radius of earth (3960 miles), and h = altitude. Since $h \ll R$, $D = (2Rh)^{1/2}$, and inserting the true earth's radius with R and D in statute miles and h in feet

$$D = \left(\frac{2 \times 3960}{5280} h\right)^{1/2}$$

$$D = [(3/2)h]^{1/2}$$

$$h = (2/3)D^2$$

for true earth, where D is in miles and h in feet.

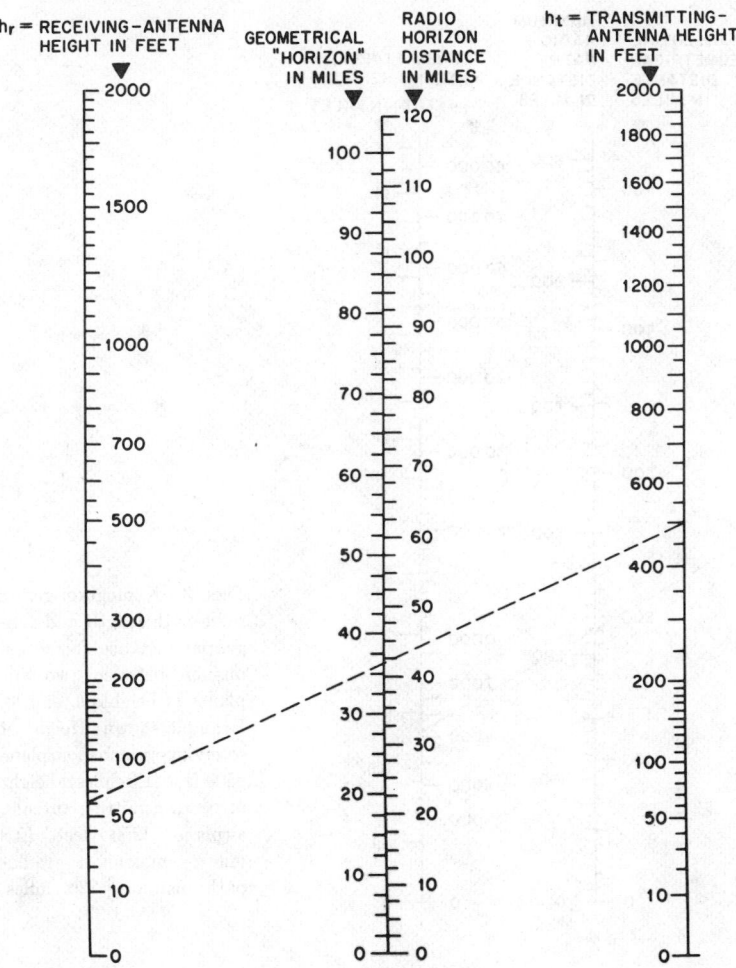

Fig. 12—Nomogram giving radio-horizon distance in miles when h_r and h_t are known. Example shown: Height of receiving antenna 60 feet; height of transmitting antenna 500 feet; maximum radio-path length = 41.5 miles. (K = 1.33)

Using a value of $K = 1.33$

$$D = [(3/2)h]^{1/2}(4/3)^{1/2} = (2h)^{1/2}$$

$$h = D^2/2.$$

Or more generally

$$h = 2D^2/3K.$$

Profile Paper: Using a 4/3 effective-radius factor, the departure from a horizontal tangent line is

$$h = D^2/2$$

where symbols are as above. Using this equation, a template can be made for convenient drawing of profile paper (Fig. 15). For instance, if the horizontal scale is 10 miles/inch, the vertical scale 100 feet/inch, and a width corresponding to 40 miles is desired, the following points may be plotted.

Distance from Center (horizontal)		Distance from Horizontal (vertical)
0 miles = 0 inches	and	0 feet = 0 inches
5 miles = $\frac{1}{2}$ inch	and	$12\frac{1}{2}$ feet = $\frac{1}{8}$ inch
10 miles = 1 inch	and	50 feet = $\frac{1}{2}$ inch
15 miles = $1\frac{1}{2}$ inches	and	$112\frac{1}{2}$ feet = $1\frac{1}{8}$ inches
20 miles = 2 inches	and	200 feet = 2 inches

A typical example of a template constructed according to these figures is given in Fig. 16. If a different scale is desired than is provided on available profile-chart paper (for example, if a 50-mile hop is to be plotted on 30-mile paper), then the scale of miles may be doubled to extend the range of the paper to 60 miles. The vertical scale in feet must then be quadrupled; i.e., 100-foot divisions become 400-foot divisions (Fig. 15).

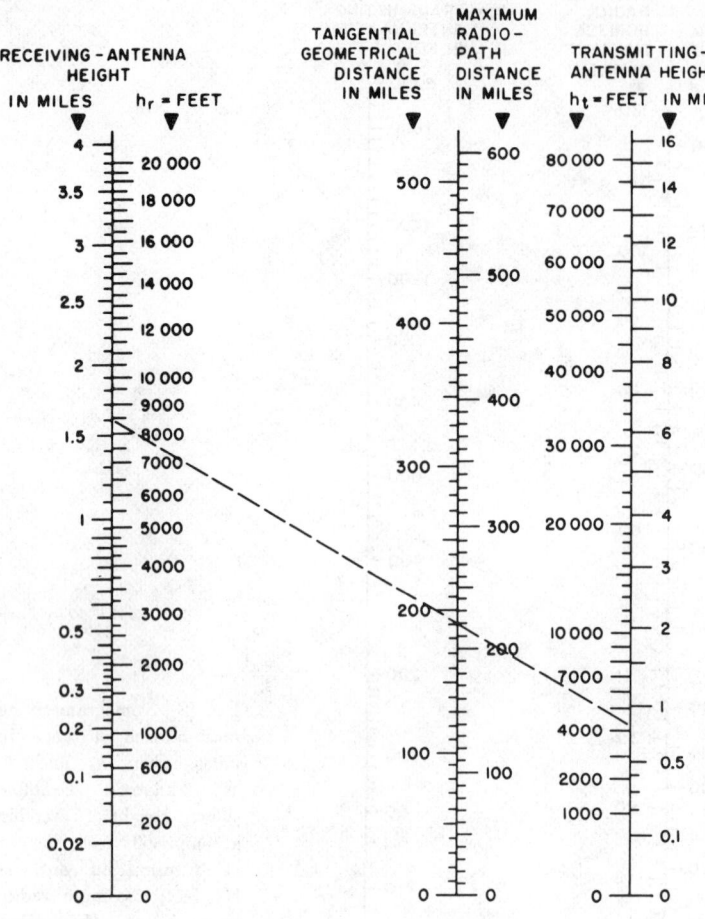

Fig. 13—Nomogram giving radio-path length and tangential distance for transmission between two airplanes at heights h_r and h_t. Example shown: Height of receiving-antenna airplane 8500 feet (1.6 miles); height of transmitting-antenna airplane 4250 feet (0.8 mile); maximum radio-path distance = 220 miles. ($K=1.33$)

Fresnel Zones

The Fresnel–Kirchhoff theory was originally developed to account for the diffraction of light when obstructed by diaphragms, and when transmitting through apertures of various shapes and sizes. This theory may be applied to radio and sound waves and is based on the concept that any small element of space in the path of a wave may be considered as the source of a secondary wavelet, and that the radiated field can be built up by the superposition of all these wavelets (Huygens principle).

Consider a transparent screen between a distant transmitter T and a receiver R, with the distance from screen to transmitter being at least 10 times the distance from screen to receiver, and with the plane of the screen perpendicular to the direction T–R. Concentric circles may be drawn on this screen, with the centers at the point where the line T–R intersects the screen at O, the radius of the first circle being such that the difference in length between the path O–R and the path from the circumference of this circle to R is $\frac{1}{2}$ wavelength (λ). The radii of the other circles are such that the corresponding path length differences are integral multiples of $\frac{1}{2}\lambda$. The radius of the first circle is $(d\lambda)^{1/2}$, where d is the distance from O to R, and the radius of the second circle is $(2d\lambda)^{1/2}$, of the third $(3d\lambda)^{1/2}$, etc. The area within the first circle is called the first Fresnel zone, and the other ring-shaped areas are the second, third, etc.,

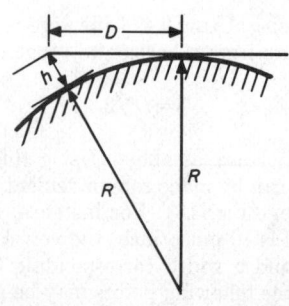

Fig. 14—Straight line tangent to earth's surface.

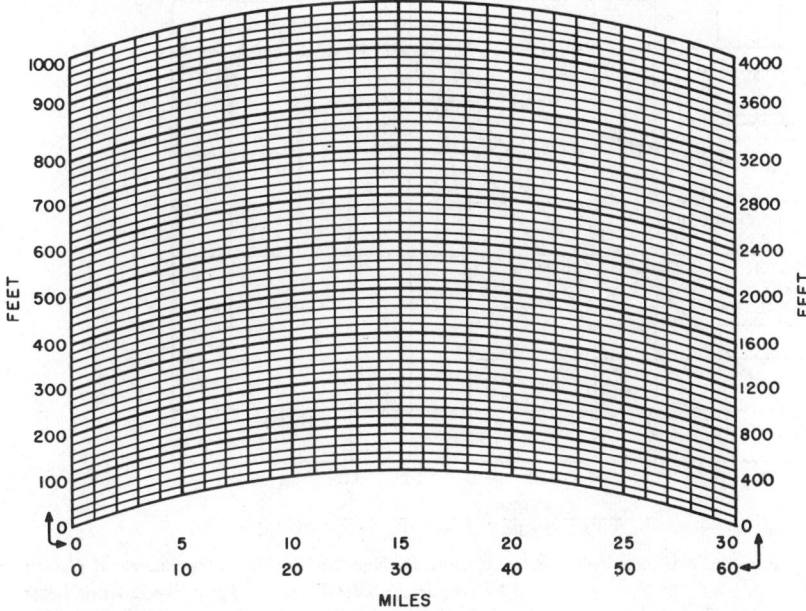

Fig. 15—Typical 4/3-earth profile paper, 1000-foot scale.

Fresnel zones. The fields from the odd-number zones are in phase at R, and the fields from the even-number zones are also in phase at R but are opposite in phase to the fields from the odd-number zones. It can be shown that the effect at R of each zone is nearly equal. If an infinitely absorbing screen is provided with an aperture of the same diameter as the first Fresnel zone, it will be found that the field at R is twice as great as the unobstructed or free-space field. If the aperture is increased to include the second zone, the field at R will then be nearly zero, since the fields from zones 1 and 2 are nearly equal in amplitude and opposite in phase. With a continued increase in the diameter of the aperture, further maxima and minima appear; the amplitude of these oscillations decreases very gradually until eventually the field

at R approaches the free-space value, which is half that due to the first Fresnel zone. If the distance from the screen to the transmitter is d_1, and from the screen to the receiver is d_2, then the general expression for the radius of the nth Fresnel zone is

$$\{n\lambda[(d_1 \times d_2)/(d_1+d_2)]\}^{1/2}.$$

Required Path Clearance

A criterion to determine whether the earth is sufficiently removed from the radio line-of-sight ray to allow mean free-space propagation conditions to apply is to have the first Fresnel zone clear all obstacles in the path of the rays. This first zone

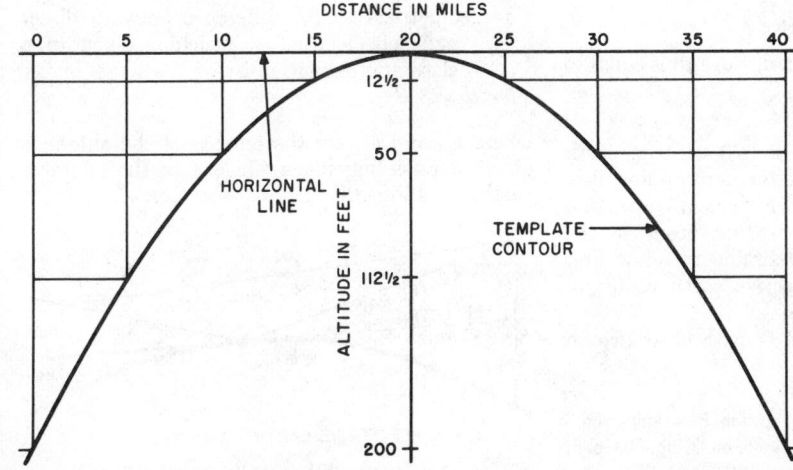

Fig. 16—Construction of a template for profile charts. Drawing is actual size.

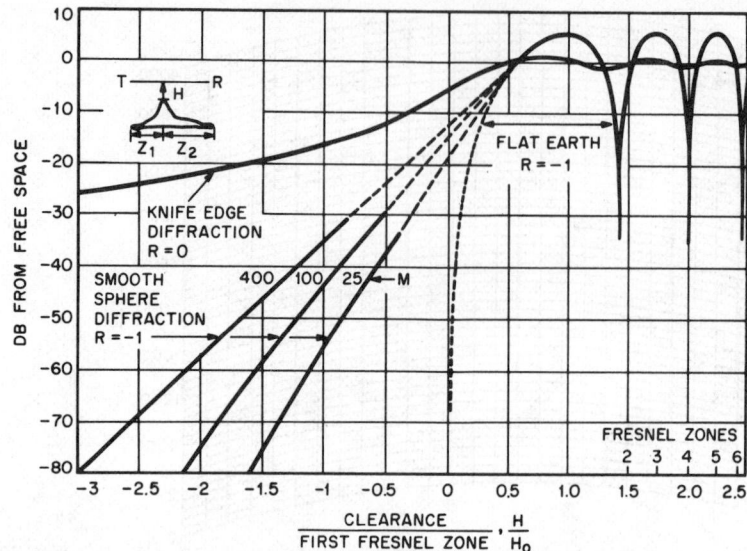

Fig. 17—Effect of path clearance on radio transmission. R = reflection coefficient of surface; H = clearance; H_0 = first Fresnel zone radius = $[(\lambda Z_1, Z_2)/(Z_1 + Z_2)]^{1/2}$; $M = (H_1/K^{1/3})\{[1 + (H_2/H_1)^{1/2}]/2\}^2(F/4000)^{2/3}$; H_1, H_2 = antenna height in feet above a smooth sphere; F = frequency in megahertz; K = (effective earth radius)/(true earth radius). *K. Bullington, "Radio Propagation Fundamentals," Bell System Technical Journal, vol. 36, no. 3, Fig. 8, © 1957 American Telephone and Telegraph Company.*

is bounded by points for which the transmission path from transmitter to receiver is greater by one-half wavelength than the direct path. Let d be the length of the direct path and d_1 and d_2 be the distances to transmitter and receiver from a point P. The radius of the first Fresnel zone at P is approximately given by

$$R_1{}^2 = \lambda\,(d_1 d_2/d)$$

where all quantities are expressed in the same units.

Expressing d in miles and frequency F in megahertz, the first Fresnel-zone radius in feet at P is given by

$$R_{1m} = 2280\,(d_1 d_2/Fd)^{1/2}.$$

The maximum occurs when $d_1 = d_2$ and is equal to

$$R_{1m} = 1140\,(d/F)^{1/2}.$$

While a fictitious earth of 4/3 of true earth radius is generally accepted for determining first Fresnel-zone clearance under normal refraction condition, unusual conditions that occur in the atmosphere may make it desirable to allow first Fresnel clearance of an effective earth radius of 0.7 to 0.5 of the true radius.

Figure 17 shows the effect of path clearance on radio transmission.*

* K. Bullington, "Radio Propagation Fundamentals," *Bell System Technical Journal*, vol. 36, no. 3, pp. 593–626; 1957.

Interference Between Direct and Reflected Rays

Where there is one reflected ray combining with the direct ray at the receiving point (Fig. 18), the resulting field strength (neglecting the difference in angles of arrival, and assuming perfect reflection at T) is related to the free-space intensity, irrespective of the polarization, by

$$E = 2E_d \sin 2\pi\,(\delta/2\lambda)$$

where

E = resulting field strength $\rbrace$ same
E_d = direct-ray field strength $\rbrace$ units
δ = geometrical length difference between direct and reflected paths, which is given to a close approximation by
$\delta = 2h_{at}h_{ar}/d$

where h_{at} and h_{ar} are the heights of the antennas above a reflecting plane tangent to the effective earth. (See Fig. 18).

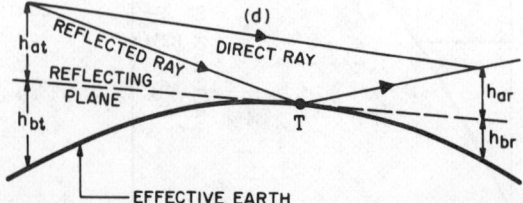

Fig. 18—Interference between direct and reflected rays.

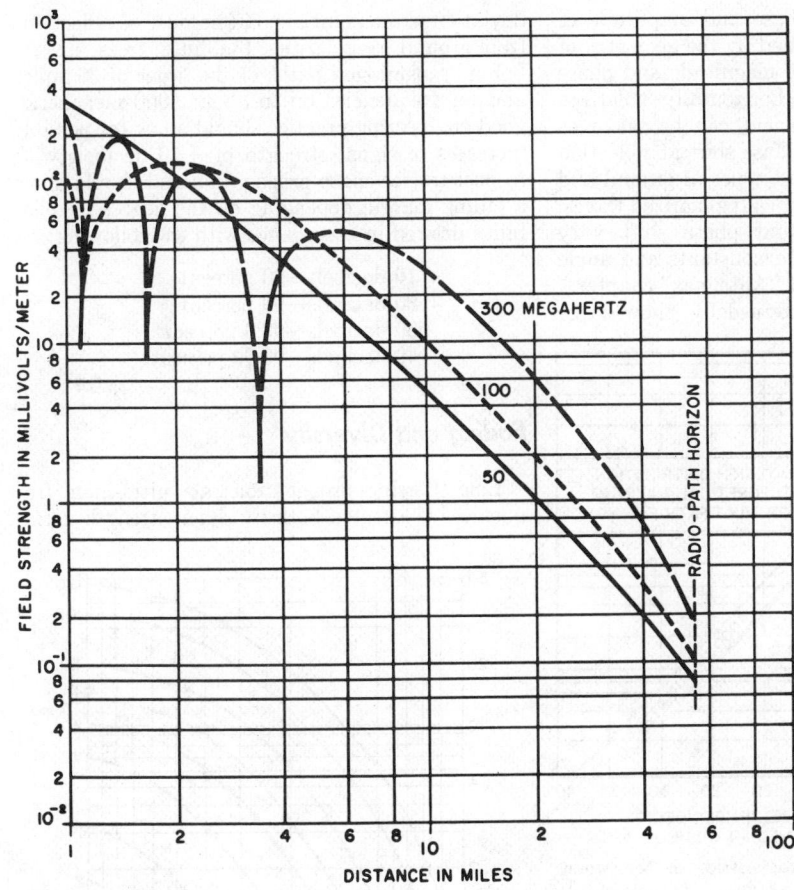

Fig. 19—Variation of resultant field strength with distance and frequency. Antenna heights: 1000 feet, 30 feet; power: 1 kilowatt; ground constants: $\sigma = 5 \times 10^{-14}$ emu, $\epsilon = 15$ esu; polarization: horizontal.

The following cases are of interest.

$$E = 0 \quad \text{for } h_{at}h_{ar} = d\lambda/2$$

$$E = 2E_d \quad \text{for } h_{at}h_{ar} = d\lambda/4$$

$$E = E_d \quad \text{for } h_{at}h_{ar} = d\lambda/12.$$

In case $h_{at} = h_{ar} = h$,

$$E = 0 \quad \text{for } h = (d\lambda/2)^{1/2}$$

$$E = 2E_d \quad \text{for } h = (d\lambda/4)^{1/2}$$

$$E = E_d \quad \text{for } h = (d\lambda/12)^{1/2}.$$

All these equations are written with the same units for all quantities.

Space-Diversity Reception

When h_{ar} is varied, the field strength at the receiver varies approximately according to the preceding equation. The use of two antennas at different heights provides a means of compensating to a certain extent for changes in electrical-path differences between direct and reflected rays (space-diversity reception).

The antenna spacing at the receiver should be approximately such as to give a $\lambda/2$ variation between geometrical-path differences in the two cases. An approximate value of the spacing is given by $\lambda d/4h_{at}$ when all quantities are in the same units.

The spacing in feet for d in miles, h_{at} in feet, λ in centimeters, and f in megahertz, is given by

$$\text{spacing} = 43.4\lambda d/h_{at}$$

$$= 1.3 \times 10^6 d/fh_{at}.$$

Example: $\lambda = 3$ centimeters, $d = 20$ miles, and $h_{at} = 50$ feet; therefore spacing = 52 feet.

Variation of Field Strength with Distance

Figure 19 shows the variation of resultant field strength with distance and frequency; this effect is due to interference between the free-space wave and the ground-reflected wave as these two components arrive in or out of phase.

To compute the field accurately under these conditions, it is necessary to calculate the two components separately and to add them in correct phase relationship. The phase and amplitude of the reflected ray is determined by the geometry of the path and the change in magnitude and phase at ground reflection. For horizontally polarized waves, the reflection coefficient can be taken as approximately 1, and the phase shift at reflection as 180 degrees, for nearly all types of ground and angles of incidence. For vertically polarized waves, the reflection coefficient and phase shift vary appreciably with the ground constants and angle of incidence. (See Fig. 53 of "Antennas" chapter.)

Measured field strengths usually show large deviations from point to point because of reflections from ground irregularities, buildings, trees, etc.

For transmission paths of the order of 30 miles and for frequencies up to about 6000 megahertz, good engineering practice should allow for possible increases of signal strength of +10 decibels with respect to free-space propagation and should allow a fading margin depending on the degree of reliability desired in accordance with the following:

     10 decibels—90 percent
     20 decibels—99 percent
     30 decibels—99.9 percent
     40 decibels—99.99 percent.

Fading and Diversity*

Line-of-sight propagation at ultra-high frequencies is affected both by signal-strength varia-

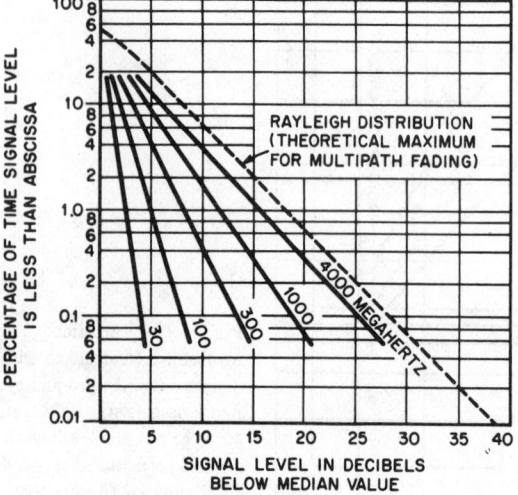

Fig. 20—Typical fading characteristics in the worst month on line-of-sight paths of 30 to 40 miles with clearance of 50 to 100 feet. *K. Bullington, "Radio Propagation Fundamentals," Bell System Technical Journal, vol. 36, no. 3, Fig. 4, © 1957 American Telephone and Telegraph Company.*

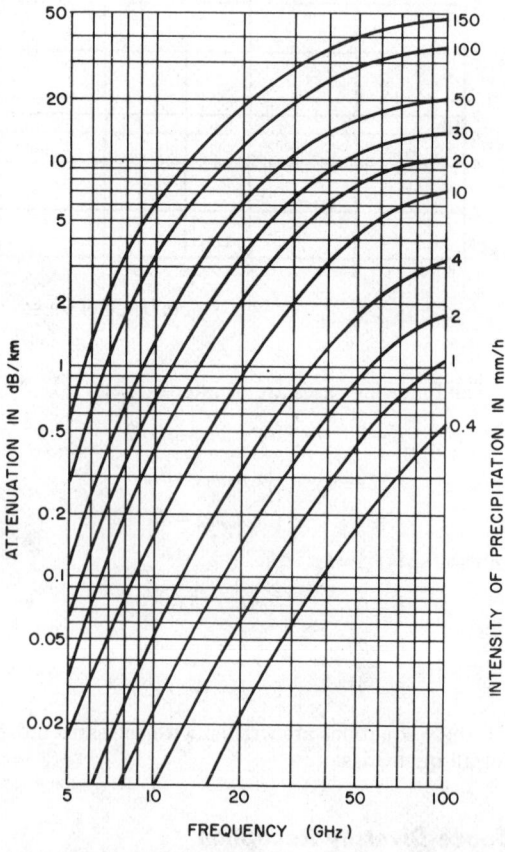

Fig. 22—Attenuation due to precipitation. *CCIR XIIIth Plenary Assembly, Vol. V, Report 233-3, Geneva, 1974.*

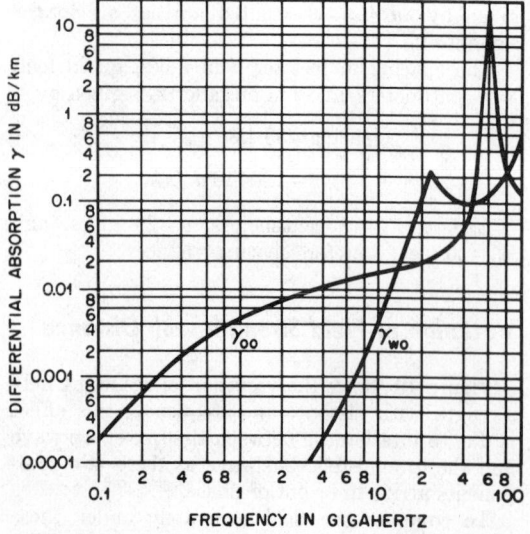

Fig. 21—Atmospheric absorption versus wavelength. *CCIR XIIIth Plenary Assembly, Vol. V, Report 233-3, Geneva, 1974.*

* K. Bullington, "Radio Propagation Fundamentals," *Bell System Technical Journal*, vol. 36, no. 3, pp. 593–626; 1957. K. W. Pearson, "Method for the Prediction of the Fading Performance of a Multisection Microwave Link," *Proceedings of the IEE*, vol. 112, no. 7, pp. 1291–1300; July 1965.

tions due to multipath transmission and by bending of the beam due to abnormal variation of refractive index with height in the lower atmosphere.

At frequencies below about 8000 megahertz, and on paths having adequate clearance, the fading on line-of-sight paths is due to multipath transmission. Multipath fading may be divided into two main types; the first is relatively rapid and is caused by interference between two or more rays arriving by slightly different paths; this is known as *atmospheric-multipath*. The second type of fading is less rapid and is due to interference between direct and reflected rays; this is referred to as *reflection-multipath*. In general, the number of fades per unit time due to atmospheric-multipath increases with path length; however, the duration of a fade of a given depth tends to decrease with increasing path length. Figure 20 shows the typical fading characteristics of a line-of-sight path.

Either frequency or space diversity may be used to reduce the amplitude of multipath fading. In the case of atmospheric-multipath fading on line-of-sight paths, it has been found that considerable diversity improvement can usually be obtained with a frequency difference of 100 to 200 megahertz or with a vertical antenna spacing of between 100 and 200 wavelengths.

Atmospheric Absorption*

Oxygen and water vapor may absorb energy from a radio wave by virtue of the permanent electric dipole moment of the water molecule and the permanent magnetic dipole moment of the oxygen molecule. Figure 21 shows water-vapor absorption γ_{wo} and oxygen absorption γ_{oo} as a function of frequency.

The attenuation due to rain increases with frequency and with increasing rate of precipitation. Figure 22 shows the frequency dependence of attenuation due to precipitation. Typical rainfall rates in a temperate climate are shown in Fig. 23. In temperate climates rainfall rates exceeding 1 inch (25.4 millimeters) per hour are unlikely to occur over an area larger than about 4 miles in diameter.

Free-Space Transmission Equations

If the incoming wave is a plane wave having a power flow per unit area equal to P_0, the available power at the output terminals of a receiving antenna may be expressed as

$$P_r = A_r P_0$$

where A_r is the effective area of the receiving antenna.

* "Transmission Loss Predictions for Tropospheric Communication Circuits," National Bureau of Standards Technical Note No. 101.

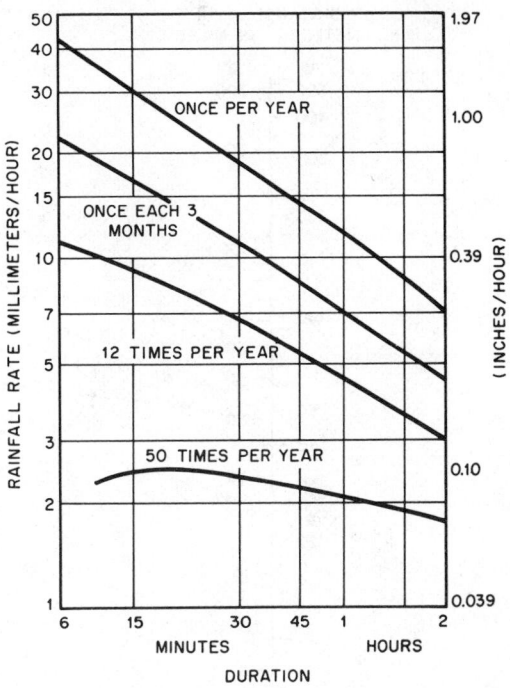

Fig. 23—Rainfall duration in England. *From E. G. Bilham, "Climate of British Isles," Macmillan Company, Toronto, Canada; 1938.*

The free-space path attenuation is given by

$$\text{attenuation} = 10 \log (P_t/P_r)$$

where P_t is the power radiated from the transmitting antenna (same units as for P_r). Then

$$P_r/P_t = A_r A_t / d^2\lambda^2$$

where A_r = effective area of receiving antenna, A_t = effective area of transmitting antenna, λ = wavelength, and d = distance between antennas.

The length and surface units in the equation should be consistent. This is valid provided $d \gg 2a^2/\lambda$, where a is the largest linear dimension of either of the antennas.

Path Attenuation Between Isotropic Antennas:

$$P_t/P_r = 4.56 \times 10^3 f^2 d^2$$

where f is in megahertz and d is in miles.
Path attenuation α (in decibels) is

$$\alpha = 36.6 + 20 \log f + 20 \log d.$$

A nomogram for the solution of α is given in Fig. 24.

Effective Areas of Typical Antennas (Refer to Antennas Chapter)

Hypothetical isotropic antenna (no heat loss)

$$A = (1/4\pi)\lambda^2 \approx 0.08\lambda^2.$$

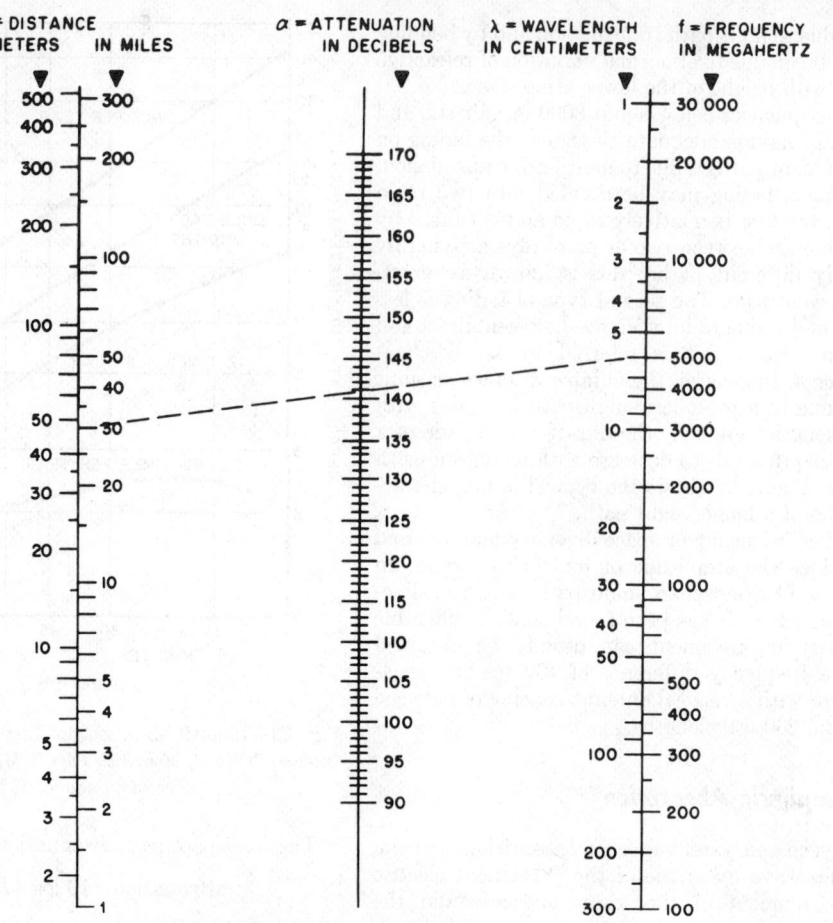

$$\alpha = 36.6 + 20 \ \mathrm{LOG} \ f(\mathrm{MHZ}) + 20 \ \mathrm{LOG} \ d(\mathrm{MILES}) \ \mathrm{DECIBELS}$$

Fig. 24—Nomogram for solution of free-space path attenuation α between isotropic antennas. Example shown: distance 30 miles; frequency 5000 megahertz; attenuation = 141 decibels.

Small uniform-current dipole, short compared with wavelength (no heat loss)

$$A = (3/8\pi)\lambda^2 \approx 0.12\lambda^2.$$

Half-wavelength dipole (no heat loss)

$$A \approx 0.13\lambda^2.$$

Parabolic reflector of aperture area S (here, the factor 0.54 is due to nonuniform illumination of the reflector)

$$A \approx 0.54S.$$

Very long horn with small aperture dimensions compared with length

$$A = 0.81S.$$

Horn producing maximum field for given horn length

$$A = 0.45S.$$

The aperture sides of the horn are assumed to be large compared with the wavelength.

Antenna Gain Relative to Hypothetical Isotropic Antennas

If directive antennas are used in place of isotropic antennas, the transmission equation becomes

$$P_r/P_t = G_t G_r [P_r/P_t]_{\mathrm{isotropic}}$$

where G_t and G_r are the power gains due to the directivity of the transmitting and receiving antennas, respectively.

The apparent power gain is equal to the ratio of the effective area of the antenna to the effective area of the isotropic antenna (which is equal to $\lambda^2/4\pi \approx 0.08\lambda^2$).

The apparent power gain due to a paraboloidal reflector is thus

$$G = 0.54(\pi D/\lambda)^2$$

where D is the aperture diameter, and an illumination factor of 0.54 is assumed. In decibels, this becomes

$$G_{\mathrm{dB}} = 20 \log f + 20 \log D - 52.6$$

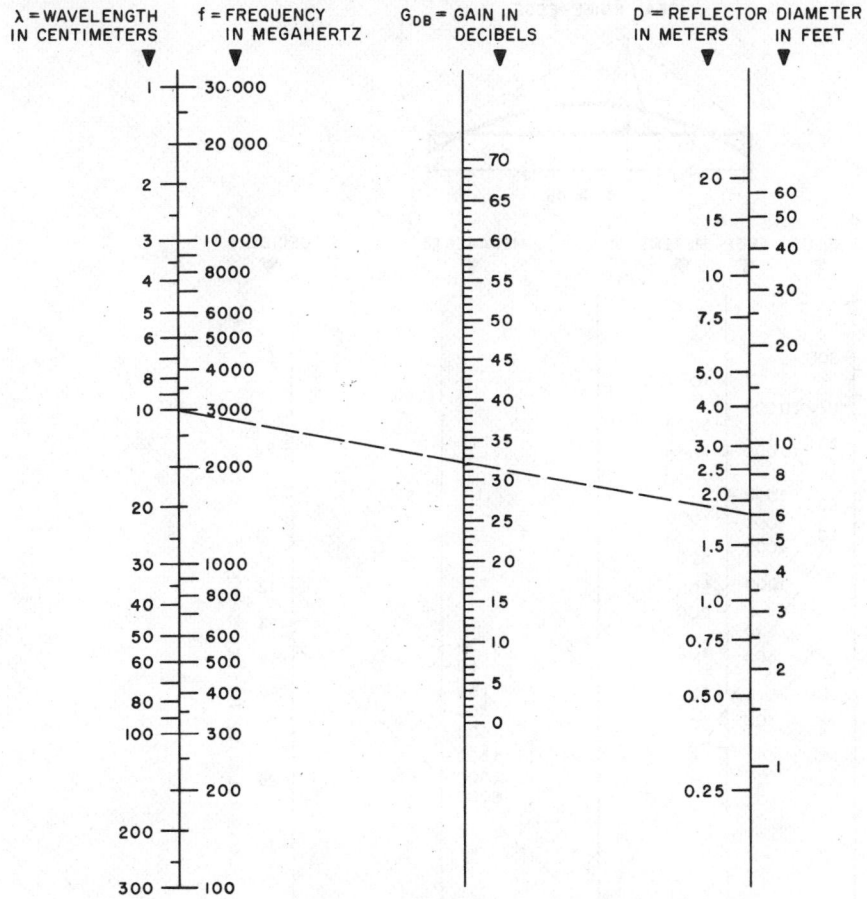

10 LOG G = 20 LOG f(MHZ) + 20 LOG D(FEET) - 52.6

Fig. 25—Nomogram for determination of apparent power gain G_{dB} (in decibels) of a paraboloidal reflector. Example shown: frequency 3000 megahertz; diameter 6 feet; gain=32 decibels.

where f= frequency in megahertz, and D= aperture diameter in feet.

The solution for G_{dB} may be found in Fig. 25.

Antenna Beam Angle

The beam angle θ in degrees is related to the apparent power gain G of a paraboloidal reflector with respect to isotropic antennas approximately by

$$\theta^2 \approx 27\,000/G.$$

Since $G=5.5\times 10^{-6}D^2f^2$, the beam angle becomes

$$\theta \approx (7\times 10^4)/fD$$

where θ= beam angle between 3-decibel points in degrees, f= frequency in megahertz, and D= diameter of paraboloid in feet.

Transmitter Power for a Required Output Signal/Noise Ratio

Using the above expressions for path attenuation and reflector gain, the ratio of transmitted power to theoretical receiver noise, in decibels, is given by

$$10 \log (P_t/P_n)$$
$$= A_p + (S/N) + (nf) - G_t - G_r - (\bar{n}\bar{i}\bar{f})$$

where S/N= required signal/noise ratio at receiver in decibels, (nf)= noise figure of receiver in decibels (see chapter "Radio Noise and Interference" for definition), $(\bar{n}\bar{i}\bar{f})$= noise improvement factor in decibels due to modulation methods where extra bandwidth is used to gain noise reduction (see Modulation chapter for definition), P_n= theoretical noise power in receiver, P_t= radiated transmitter power, G_t= gain of transmitting antenna in decibels, G_r= gain of receiving antenna in decibels, and A_p= path attenuation in decibels.

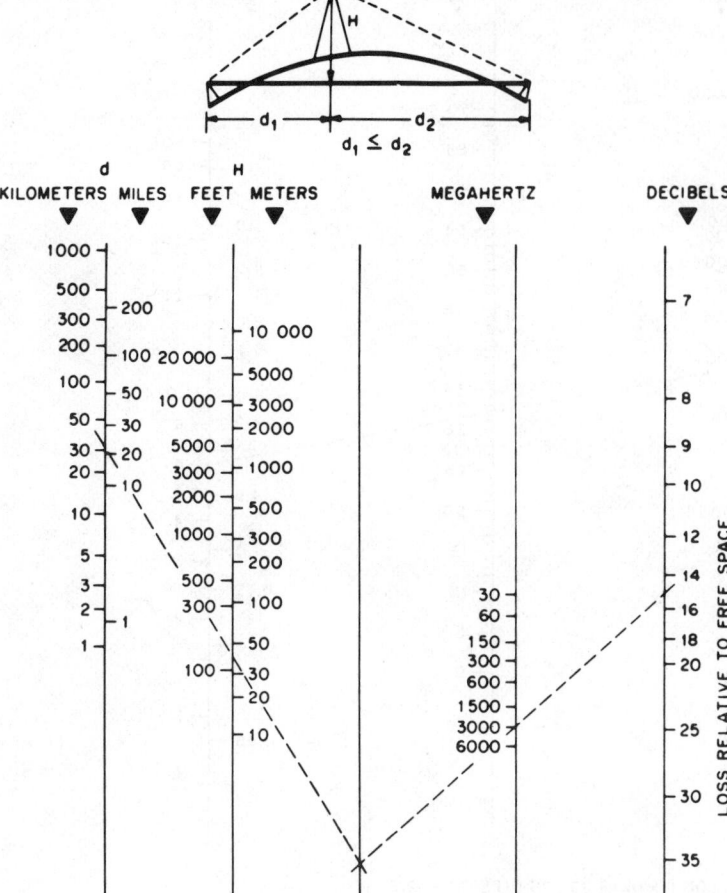

Fig. 26—Knife-edge diffraction loss relative to free space. *K. Bullington, "Radio Propagation Fundamentals," Bell System Technical Journal, vol. 36, no. 3,* Fig. 7, © 1957 *American Telephone and Telegraph Company.*

An equivalent way to compute the transmitter power for a required output signal/noise ratio is given below directly in terms of reflector dimensions and system parameters.

(A) Normal free-space propagation

$$P_t = \frac{\beta_1\beta_2}{40}\frac{BL^2}{f^2r^4}\frac{F}{K}\frac{S}{N}.$$

(B) With allowance for fading

$$P_t = \frac{\beta_1\beta_2}{40}\frac{BL^2}{f^2r^4}\frac{F}{K}\sigma\left(\frac{S}{N}\right)_m.$$

(C) For multirelay transmission in n equal hops

$$P_t = \frac{\beta_1\beta_2}{40}\frac{BL^2}{f^2r^4}\frac{F}{K}\sigma\left(\frac{S}{N}\right)_{nm}.$$

(D) Signal/noise ratio for nonsimultaneous fading is

$$10 \log (S/N)_n = 10 \log \sigma (S/N)_{1m} - 10 \log \bar{n}$$

where

P_t = power in watts available at transmitter output terminals (kept constant at each repeater point)

β_1 = loss power ratio (numerical) due to transmission line at transmitter

β_2 = same as β_1 at receiver

B = root-mean-square bandwidth (generally approximated to bandwidth between 3-decibel attenuation points) in megahertz

L = total length of transmission in miles

f = carrier frequency in megahertz

r= radius of paraboloidal reflectors in feet

F= power-ratio noise figure of receiver (a numerical factor; see chapter "Radio Noise and Interference")

K= improvement in signal/noise ratio due to the modulation used. (For instance, $K=3m^2$ for frequency modulation, where m is the ratio of maximum frequency deviation to maximum modulating frequency. Note that this is the numerical power ratio.)

σ= numerical ratio between available signal power in case of normal propagation to available signal power in case of maximum expected fading

S/N= required signal/noise power ratio at receiver

$(S/N)_m$= minimum required signal/noise power ratio in case of maximum expected fading

$(S/N)_{nm}$= same as above in case of n hops, at repeater number n

$(S/N)_{1m}$= same as above at first repeater

$(S/N)_n$= same as above at end of n hops

n= number of equal hops

m= number of hops where fading occurs

$$\bar{n}=n-m+\sum_1^m \sigma_k$$

σ_k= ratio of available signal power for normal conditions to available signal power in case of actual fading in hop number k (equation holds in case signal power is increased instead of decreased by abnormal propagation or reduced hop distance).

KNIFE-EDGE DIFFRACTION PROPAGATION*

Diffraction loss at an ideal knife-edge can be estimated from Fig. 26. However, the transmission loss over a practical knife-edge diffraction path depends critically on the shape of the diffracting edge. Since a natural obstacle, such as a mountain ridge, may depart considerably from an ideal knife-edge, the diffraction loss in practice is usually 10 to 20 decibels greater than that estimated for the ideal case.

A nonuniform transverse profile of the diffracting edge, or reflections on the transmission paths each side of the diffracting edge, may result in multipath transmission causing variations in the received level as a function of frequency, space, and time. The amplitude of such variations may be

* K. Bullington, "Radio Propagation Fundamentals," *Bell System Technical Journal*, vol. 36, no. 3, pp. 593–626; 1957.

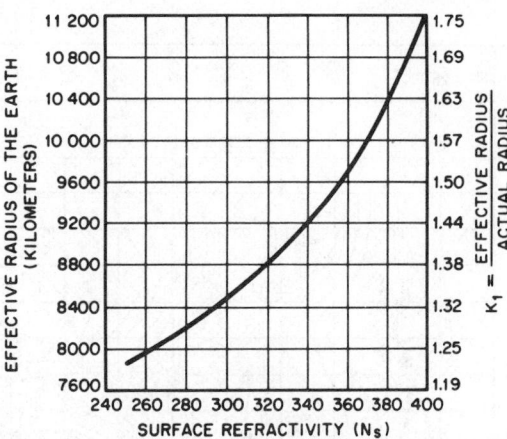

Fig. 27—Variation of the effective radius of the earth a as a function of the surface refractivity N_s. *Report no. 338-2 of vol. V—Propagation, CCIR XIIIth Plenary Assembly, Geneva; 1974.*

reduced by either space or frequency diversity and by the use of narrow-beamwidth antennas.

TROPOSPHERIC SCATTER PROPAGATION*

Weak but reliable fields are propagated several hundred miles beyond the horizon in the very-high-, ultra-high-, and super-high-frequency bands. An important parameter in scatter propagation is the scatter angle or angle of intersection of the transmitting and receiving antenna beams. This angle θ in radians is given by

$$\theta=\frac{2d-d_t-d_v}{2R}+\frac{h_t-H_t}{d_t}+\frac{h_v-H_v}{d_v}$$

where

d= great-circle distance between transmitting and receiving antennas

d_t= distance to the horizon from the transmitting antenna

d_v= distance to the horizon from the receiving antenna

h_t= height above sea level of the transmitting horizon

h_v= height above sea level of the receiving horizon

H_t= height above sea level of the transmitting antenna

H_v= height above sea level of the receiving antenna

R= effective radius of the earth.

The same units are used for distances and heights.

* "Estimation of Tropospheric-Wave Transmission Loss," CCIR XIIIth Plenary Assembly, Vol. V, Report 425-1, Geneva, 1974. A. F. Harvey, "Microwave Engineering," Academic Press, Inc., New York, N.Y.; 1963.

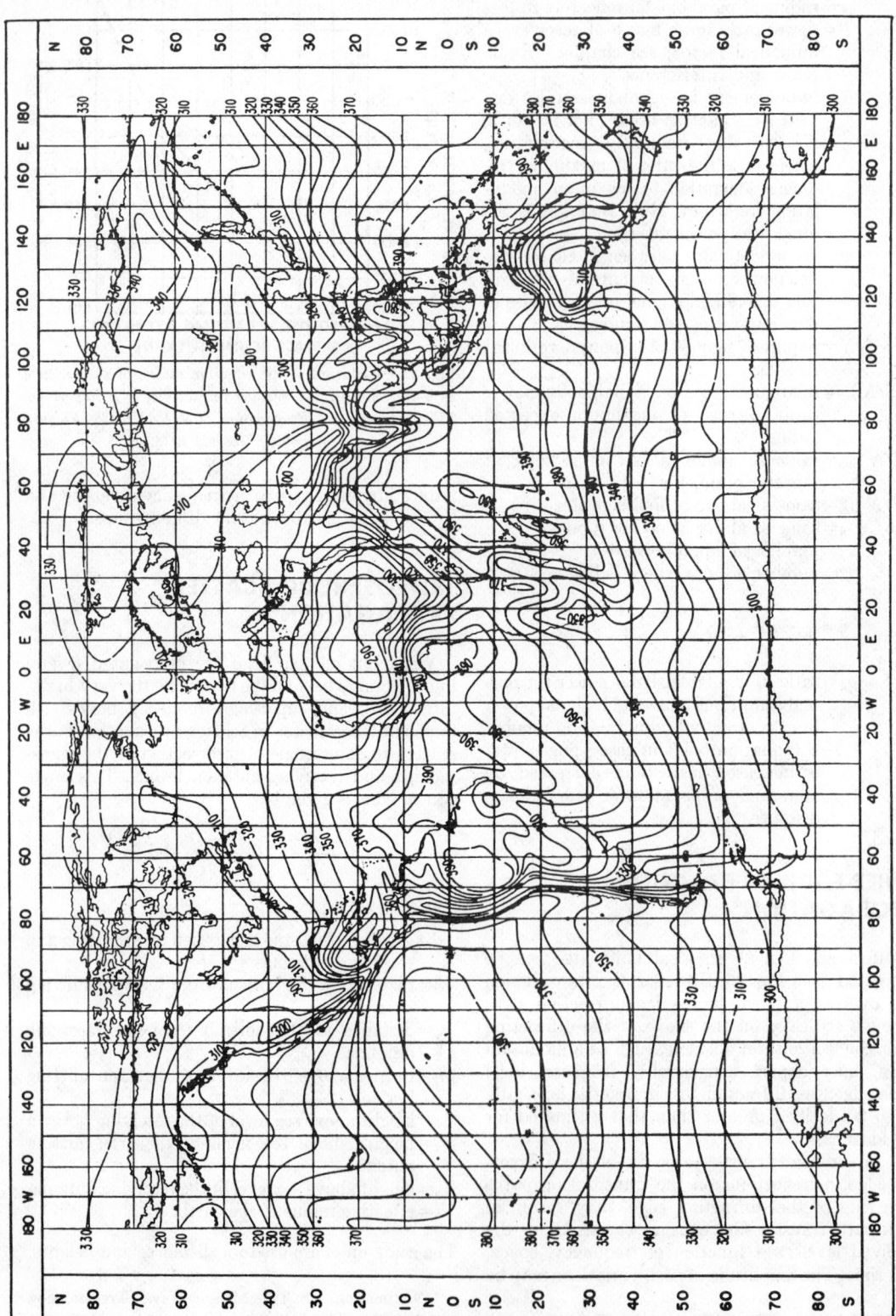

Fig. 28—Worldwide mean value of N_0 for February. *"Influence of the Atmosphere on Wave Propagation,"* Report no. 563 of vol. V—Propagation, CCIR XIIIth Plenary Assembly, Geneva, 1974.

The effective radius of the earth is a function of the refractive index gradient and may be estimated from Fig. 27. This curve is based on the correlation found between the decrease in the refractive index in the first kilometer of altitude above the earth's surface and the surface value of the refractive index. Figure 28 shows typical mean values of the refractive index at sea level.

The long-term median transmission loss due to forward scatter is approximately

$$L(50) = 30 \log f - 20 \log d + F(\theta d) - G_p - V(d_e) \text{ dB}$$

where $F(\theta d)$ is shown in Fig. 29 as a function of the product θd. The angular distance θ is the angle between radio horizon rays in the great-circle plane containing the antennas and d is the distance between antennas.

A semi-empirical estimate of the path antenna gain G_p is provided by

$$G_p = G_t + G_r - 0.07 \exp[0.055(G_t + G_r)] \text{dB}$$

for values of G_t and G_r each less than 50 dB.

$V(d_e)$, shown in Fig. 30, is an adjustment for the indicated types of climate.

This division is, of course, rather crude and local geographical conditions may require serious modifications. A brief description of these climates is given in Annex 1 of CCIR Report 238-2, Geneva, 1974.

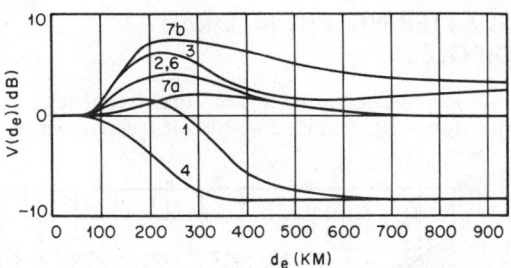

Fig. 30—The function $V(d_e)$ for the types of climate indicated on the curves.

1. Equatorial (data from Congo and Ivory Coast).
2. Continental subtropical (Sudan).
3. Maritime subtropical (data from West Coast of Africa).
4. Desert (Sahara).
5. Mediterranean (no curves available).
6. Continental temperate (data from France, Federal Republic of Germany, and U.S.A.).
7a. Maritime temperate, over land (data from U.K.).
7b. Maritime temperate, over sea (data from U.K.).
8. Polar (no curves available).

CCIR XIIIth Plenary Assembly, Vol. V, Report 238-2, Geneva, 1974.

Fast and slow fading is experienced on tropospheric scatter paths. Fast fading is due to multipath transmission, is in general Rayleigh distributed, and can be considerably reduced by diversity, an antenna spacing of 60 wavelengths usually being adequate. Slow fading, with periods of hours or days, is caused by changes in the gradient of the refractive index of the atmosphere along the transmission path and is little affected by diversity.

The plane-wave gains of large antennas are not fully realized on tropospheric scatter paths. The power on such a path is received, not from a single point source, but from a volume in the atmosphere that subtends a solid angle at the receiving antenna. If the antenna beam angles are such as to limit the available scattering volume, then the received power will be correspondingly limited and the antennas are said to suffer an antenna-to-medium coupling loss. The resulting median loss of received power is likely to be about 5 decibels for two 40-decibel-gain plane-wave antennas, and 17 decibels for two 50-decibel-gain antennas. The extent to which the path antenna gain is a function of the scatter angle θ or the height of the scatter volume has not yet been established.

Multipath transmission limits the communication bandwidth that can be used on a single carrier; however, useful bandwidths of several megahertz have been shown to be available on some 200-mile scatter paths. Narrow-beam antennas and diversity reduce the effects of multipath transmission.

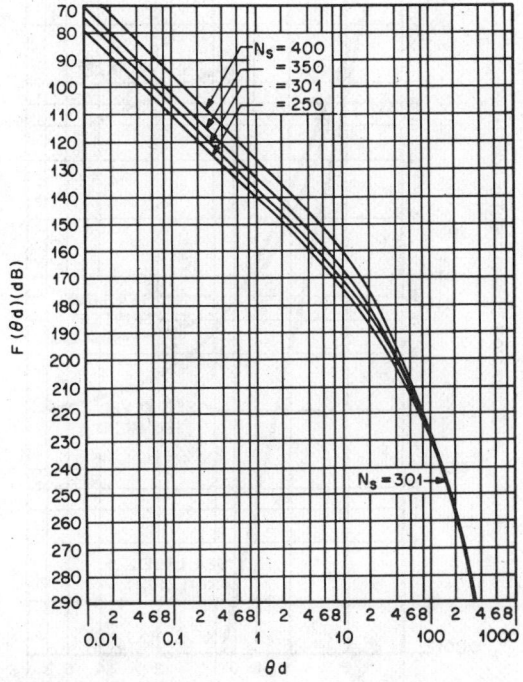

Fig. 29—Attenuation function $F(\theta d)$, where d is in kilometers and θ is in radians, for indicated values of surface refractivity N_S. *CCIR XIIIth Plenary Assembly, Vol. V, Report 238-2, Geneva, 1974.*

SCATTERING FROM ORBITAL DIPOLES*

It was demonstrated in 1963 that frequencies in the ultra-high- and super-high-frequency bands may be propagated by scattering from fine microwave dipoles, about 1.8 centimeters in length, launched into orbit around the earth at a mean height of 3650 kilometers. Tests of this system were known as the West Ford experiment. A band of frequencies near 8000 megahertz was used during the tests, and digital data were transmitted at rates from 20 000 bits per second, initially, down to around 100 bits per second. A multipath time-delay spread of about 100 microseconds and a doppler frequency smear of 1 to 2 kilohertz were observed during these experiments.

EARTH–SPACE COMMUNICATION

Communication between earth and outer space (see chapter on Space Communication) must pass through the earth's atmosphere, so that the optimum frequencies for this service are those which pass through the atmosphere with minimum attenuation. A range of frequency little attenuated by the atmosphere is known as a window; one such window occurs between the critical frequency of

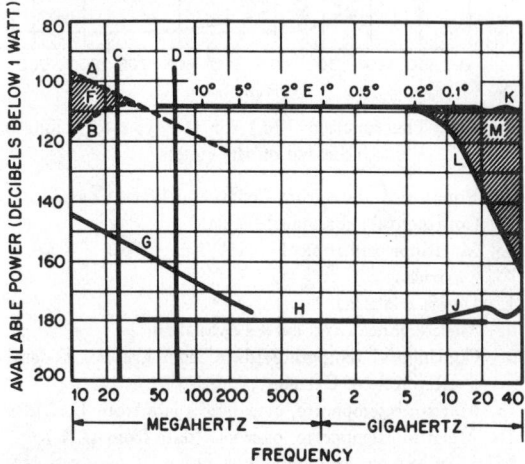

Fig. 31—General frequency limits in a simple earth-to-spacecraft communication system. (Spacecraft: isotropic antenna; transmitter power, 1 watt; bandwidth, 1 kilohertz; distance, 1000 kilometers. Earth station antenna: paraboloid; diameter, 20 meters; efficiency, 55% ——. 15-decibel gain above isotropic antenna – – –).

A: Signal level during ideal nighttime conditions (no absorption).

B: Typical signal level during daytime conditions, assuming an angle of elevation of 5°.

C: Minimum frequency to assure penetration of earth's ionosphere: polar region, oblique path; tropical region, vertical path.

D: Minimum frequency to assure penetration of earth's ionosphere: tropical region, oblique path.

E: Beamwidth of paraboloid between half-power points.

F: Effect of ionospheric absorption.

G: Minimum cosmic noise. Maximum value will be found to be higher by about 15 decibels.

H: Noise level corresponding to a temperature of 70° Kelvin.

J: Noise due to absorption in a clear atmosphere, assuming an elevation angle of 5°.

K: Typical signal level for a vertical path in a clear atmosphere.

L: Typical signal level in heavy rain (16 millimeters/hour), vertical depth 1 kilometer, assuming an elevation angle of 5°.

M: Effect of varying atmospheric conditions and elevation angles.

CCIR Tenth Plenary Assembly, vol. IV, Report 205, p. 194, Geneva; 1963.

* "Project West Ford Issue," *Proceedings of the IEEE*; May 1964.

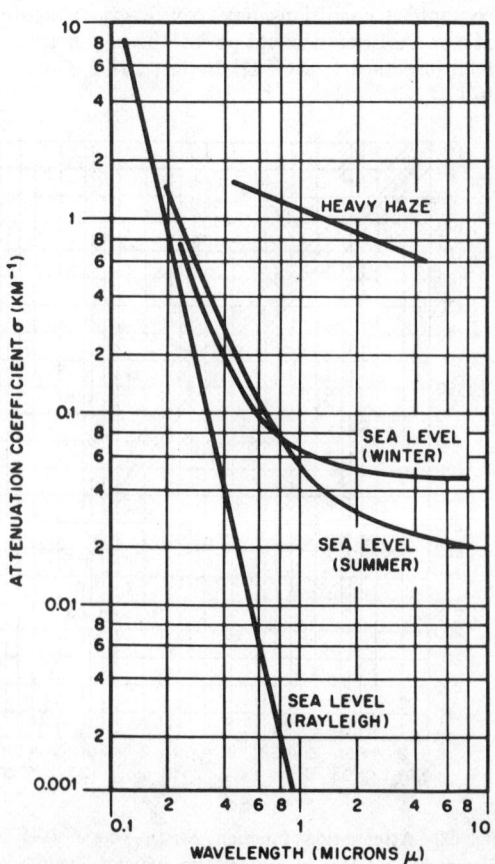

Fig. 32—Atmospheric attenuation coefficients for several conditions as a function of wavelength for window regions of the spectrum.

TABLE 5—INTERNATIONAL VISIBILITY CODE.

Code No.	Description	Daylight Visual Range (meters)		Exponential Attenuation Coefficient (km⁻¹)	
		From	To	From	To
0	Dense fog		<50		>86
1	Thick fog	50	200	86	21
2	Moderate fog	200	500	21	8.5
3	Light fog	500	1 000	8.5	4.3
4	Thin fog	1 000	2 000	4.3	2.1
5	Haze	2 000	4 000	2.1	1.1
6	Light haze	4 000	10 000	1.1	0.43
7	Clear	10 000	20 000	0.43	0.21
8	Very clear	20 000	50 000	0.21	0.07
9	Exceptionally clear	$>50 000$		<0.07	

the ionosphere and the frequency absorbed by rainfall and oxygen. This frequency range extends from about 10 to 10 000 megahertz. Another window exists in the optical and infrared region of 10^6 to 10^9 megahertz. Figure 31 shows the general frequency limits for earth–space communication.

LINE-OF-SIGHT PROPAGATION AT OPTICAL FREQUENCY*

The scattering and absorption by gases, vapors, and suspended matter in the earth's atmosphere leads to attenuation of a collimated beam of light. Where the attenuation coefficient σ varies slowly as a function of wavelength, it is defined by

$$H = H_0 \exp (-\sigma x)$$

where H and H_0 are the irradiance values (in power per unit area) measured at two points separated by a distance x. Attenuation due to scattering from a purely gaseous atmosphere is termed Rayleigh scattering, after that early investigator; it is proportional to the inverse fourth power of wavelength. For this reason, Rayleigh scattering is more important for the shorter wavelengths, the blue regions of the spectrum and the ultraviolet.

At longer visible wavelengths and in the infrared, scattering from the particulate matter (the aerosols in the atmosphere) becomes more important. Since the size of these particles may be comparable to the wavelengths, a different scattering law is followed, with attenuation being less dependent on frequency. The Naval Research Laboratories in their reports Numbers 4031 and 5453 give measured attenuation coefficients for several atmospheric conditions. Referring to Fig. 32, sea-level attenuation due to scattering is shown for the Rayleigh atmosphere, a very clear summer and winter atmosphere, and a heavy-haze condition.

The International Visibility Code is frequently used to describe the scattering characteristics of the atmosphere for visible light, and is given in Table 5, together with corresponding attenuation coefficients. It should be noted that, because of the various types of haze and fog conditions experienced in the atmosphere, it is not possible to extrapolate reliable ultraviolet or infrared attenuation from these visibility values.

Figure 32 is to be used for attenuation coefficients in atmospheric windows well removed from absorption bands. Absorption is due primarily to water vapor, carbon dioxide, and ozone in the atmosphere, although oxygen and nitrogen absorption is experienced at wavelengths shorter than $0.2\ \mu$. The position of the primary H_2O and CO_2 bands in the infrared is shown in Fig. 33, which indicates the approximate transmission for a 0.3-kilometer path at sea level containing 5.7 millimeters of precipitable water. Window regions are shown shaded. The simple exponential transmission law is not applicable in absorbing regions, because of the rapid change of attenuation coefficient with wavelength.*

Broad-band attenuation coefficients are suffi-

* B. Cooper, "Optical Communications in the Earth's Atmosphere," *IEEE Spectrum*, pp. 83–88; July 1966. D. G. C. Luck, "Some Factors Affecting Applicability of Optical-Band Radio (Coherent Light) to Communication," *RCA Review*, pp. 359–409; September 1961.

* J. N. Howard, "The Transmission of the Atmosphere in the Infrared," *Proceedings of the IRE*, vol. 47, no. 9, pp. 1451–1457; 1959.

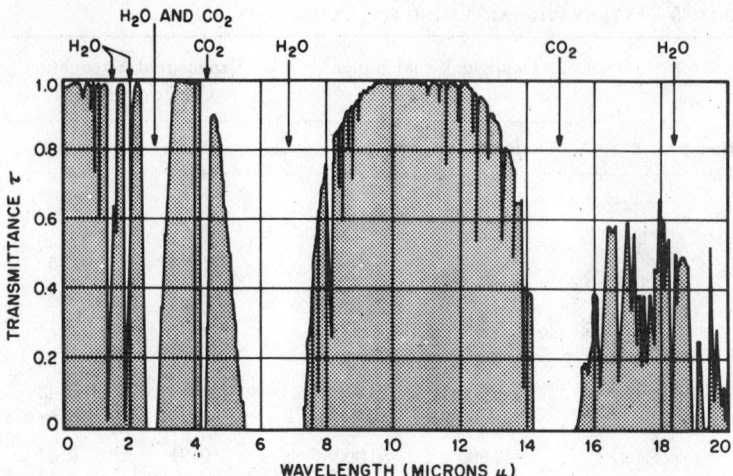

Fig. 33—Atmospheric transmission over a 0.3-kilometer path at sea level containing 5.7 millimeters of precipitable water.

ciently accurate to predict performance with incoherent light for such applications as photometry or radiometry. Because of the narrow spectral width of coherent laser light, it is necessary to use highly resolved transmission data to predict laser system performance. This is to ensure that the laser emission does not coincide with one of the molecular transition lines which may be strongly absorbing, even in wings of the main absorption bands. High-resolution solar spectrum atlases are a convenient source for locating telluric spectral lines.

Small-scale variations in the refractive index of the atmosphere serve to decrease the spatial coherence of a propagating light wavefront. This decrease may be shown to be related to the mean-square angular deflection of the beam as it passes through the atmosphere.

CHAPTER 29

RADIO NOISE AND INTERFERENCE

NOISE AND ITS SOURCES

Noise and interference from other communication systems are two factors limiting the useful operating range of all radio equipment.* There are a number of different sources of radio noise—the most important being atmospheric noise, galactic noise, man-made noise, and receiver noise. The levels of noise may be expressed in various ways; perhaps the most convenient is to refer the received noise power to the thermal noise power at a reference temperature of 290 Kelvins.

In estimating the noise level at the receiver due to external sources, the gain and orientation of the receiving antenna must be considered. Since, in general, the available noise power is proportional to bandwidth, it may be expressed in terms of an effective antenna noise factor f_a, which is defined by

$$f_a = P_n/kT_0B = T_a/T_0$$

where P_n = noise power available from an equivalent loss-free antenna (watts), k = Boltzmann's constant = 1.38×10^{-23} joules per Kelvin, T_0 = reference temperature, taken as 290 Kelvins, B = effective receiver noise bandwidth (hertz), and T_a = effective antenna temperature in the presence of external noise (Kelvins).

Figure 1 shows noise level F_a in decibels above kT_0B and T_a in Kelvins as a function of frequency for the more important sources of radio noise. Atmospheric noise curves obtained from CCIR Report 322 are shown for New York City for the summer nighttime and winter daytime. These two curves represent the extremes in expected levels of atmospheric radio noise at this location. Galactic noise, also taken from CCIR Report 322, is shown. Within a ±2-decibel temporal variation (neglecting ionospheric shielding), the values shown will be the upper limit of galactic noise, but in any given situation, the received noise should be calculated considering critical frequencies and any directional properties of the antenna. Two curves of man-made noise are shown, the upper curve being

the expected value in urban areas and the lower curve the expected value in a suburban area. These curves are composite curves obtained from measurements made by Work Group 3 of the FCC Advisory Committee, Land Mobile Radio Service, and the Institute for Telecommunication Sciences and Aeronomy, Environmental Science Services Administration, in Washington, D.C., plus measurements made in New York City and reported in the *Bell System Technical Journal* of November 1952. Since fairly good agreement was obtained for urban noise from these three independent sets of measurements, the values given can probably be used for any fair-sized urban area. The suburban curve is shown as 16 decibels below the urban curve, which is about the average difference found. The frequency range of the measurements shown are from approximately 20 megahertz through 450 megahertz, with the dashed curves showing the extrapolation above this frequency. In a quiet rural receiving site chosen with care, the man-made noise will normally be below galactic noise at 20 megahertz and above. Typical receiver noise temperature is shown as varying from about 100 K at 10 megahertz to 2000 K at 10 000 megahertz, corresponding to a noise figure** ranging from −1.3 decibels to 9.0 decibels.

A source of noise not shown in Fig. 1 is that due to atmospheric absorption that is relatively low, but can be of importance at frequencies above 1000 megahertz when low-noise amplifiers are employed at the receiver. Blake's† calculation shows that for antennas directed about 5 degrees above the horizontal, the effective antenna-noise temperature is about 90 K at 1000 megahertz and about 60 K at 3000 megahertz, being essentially constant at 60 K from 3000 to 10 000 megahertz. These values include an irreducible 36 K term caused by ground temperature.

---

* A. D. Watt, R. M. Coon, E. L. Maxwell, and R. W. Plush, "Performance of Some Radio Systems in the Presence of Thermal and Atmospheric Noise," *Proceedings of the IRE*, vol. 46, pp. 1914–1923; December 1958.

** H. T. Friis, "Noise Figures of Radio Receivers," *Proceedings of the IRE*, vol. 32, pp. 419–422, July 1944, and "Standards of Measuring Noise in Linear 2-ports," *Proceedings of the IRE*, vol. 48, pp. 60–68, January 1960.

† L. V. Blake, "Antenna and Receiving-System Noise-Temperature Calculation," *NRL Report 5668*, U.S. Naval Research Laboratory, Washington, D.C., September 19, 1961.

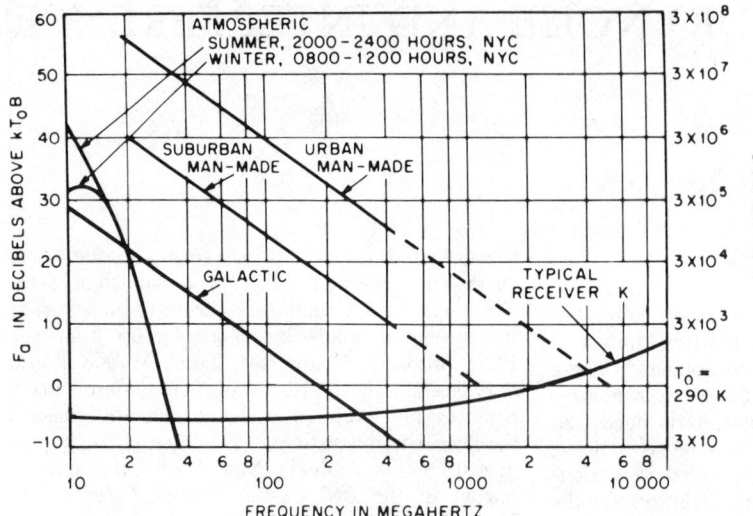

Fig. 1—Median values of average noise power expected from various sources (omnidirectional antenna near surface).

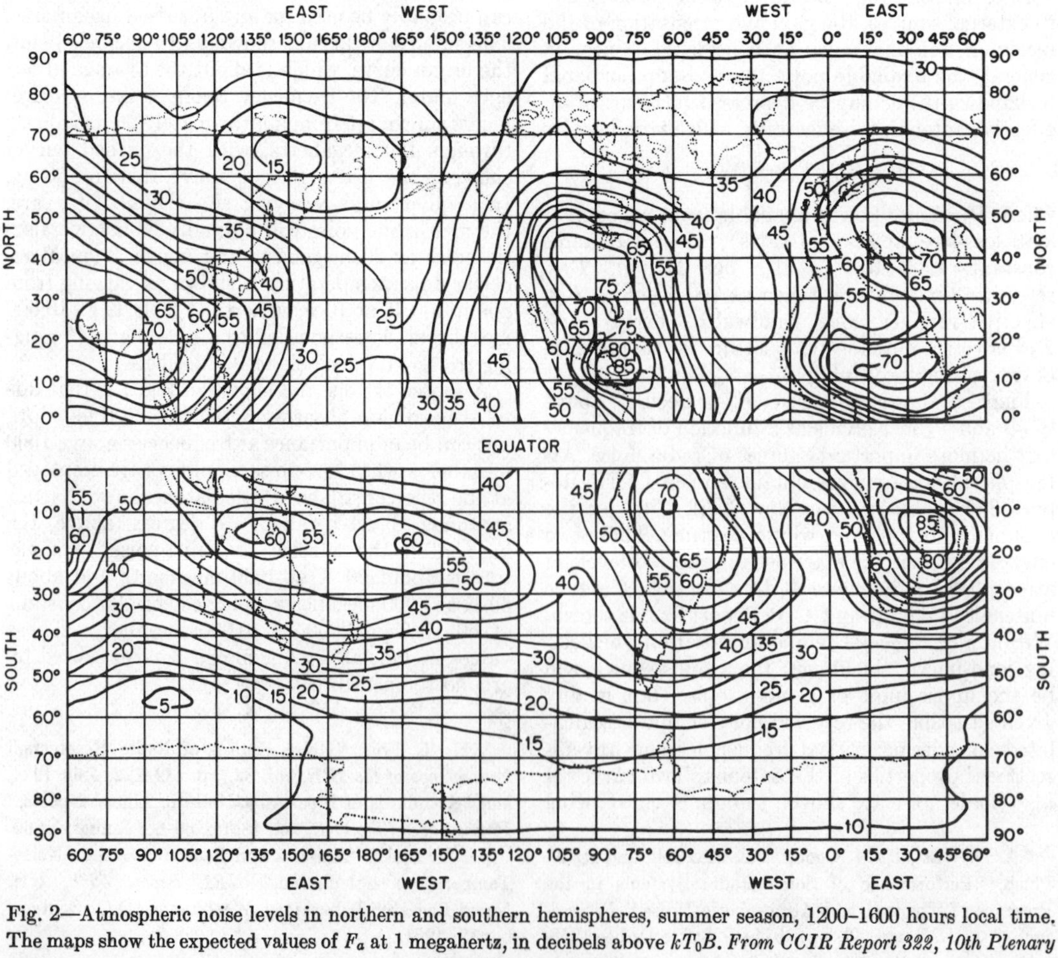

Fig. 2—Atmospheric noise levels in northern and southern hemispheres, summer season, 1200–1600 hours local time. The maps show the expected values of F_a at 1 megahertz, in decibels above kT_0B. From CCIR Report 322, 10th Plenary Assembly, Geneva; 1963.

Atmospheric Noise

This noise is produced mostly by lightning discharges in thunderstorms. The noise level thus depends on frequency, time of day, weather, season of the year, and geographical location.

Subject to variations due to local stormy areas, noise generally decreases with increasing latitude on the surface of the globe. Noise is particularly severe during the rainy seasons in areas such as Caribbean, East Indies, equatorial Africa, northern India, et cetera.

Atmospheric noise usually predominates in quiet locations at frequencies below about 20 megahertz. Maps may be used to show the atmospheric noise level at the receiver for various parts of the world; such a map is shown in Fig. 2 for 1200–1600 local time in the summer. This map gives the median noise level in decibels above kT_0B at a frequency of 1 megahertz, as received on a short vertical grounded dipole (k is Boltzmann's constant, T_0 is 290 Kelvins, and B is the receiver bandwidth in hertz). This parameter is related to the rms noise field strength by

$$E_n = F_a + 20 \log_{10} f_{MHz} - 65.5$$

where E_n = rms noise field for a 1-kilohertz bandwidth in decibels above 1 microvolt per meter, F_a = noise level in decibels above kT_0B, and f_{MHz} = frequency in megahertz.

The frequency dependence of atmospheric noise for the same season and time block is shown in Fig. 3. CCIR Report 322* gives the distribution of radio noise throughout the world at frequencies between 0.01 and 100 megahertz. In this report, the expected value of radio noise is based on measurements made with a short vertical antenna over a perfectly conducting ground. The use of directive receiving antennas may modify the level of received noise considerably.

Galactic Noise†

Galactic noise at radio frequencies is caused by disturbances originating outside the earth or its atmosphere. The chief sources of such radio noise are the sun, celestial radio-sky background radiation concentrated along the galactic plane, and a large number of discrete, cosmic sources distributed chiefly among the galactic plane. Galactic noise reaching the surface of the earth extends from about 15 to 100 000 megahertz, being limited at

* "World Distribution and Characteristics of Atmospheric Radio Noise," CCIR Report 322, 10th Plenary Assembly, Geneva; 1963.

† J. L. Steinberg and J. Lequeux, "Radio Astronomy," McGraw-Hill Book Company, New York, N.Y.; 1963. J. L. Powsey and R. N. Bracewell, "Radio Astronomy," Clarendon Press, Oxford, England; 1955.

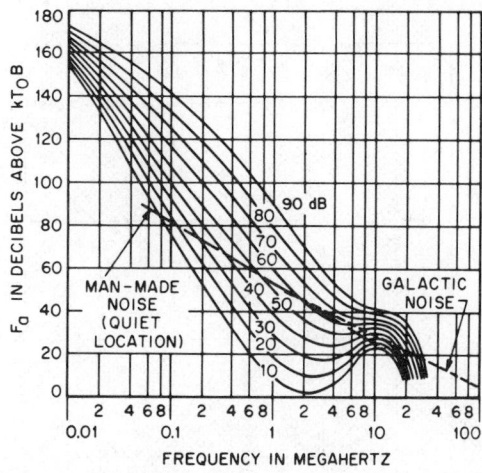

Fig. 3—Variation of radio noise with frequency, for data given in Fig. 2 legend. *From CCIR Report 322, 10th Plenary Assembly, Geneva; 1963.*

the low end of the spectrum by ionospheric absorption and at the high end by atmospheric absorption.

Galactic-noise temperature is a dominant factor for a typical radio receiver operating within the frequency range from about 40 to 250 megahertz, as shown in Fig. 1. Above about 250 megahertz, internal receiver noise predominates. In practice, the importance of galactic noise is restricted by atmospheric noise to frequencies not lower than about 18 megahertz with an upper limit at about 1000 megahertz where galactic noise reduces to low values corresponding to about 2 K (minimum), and 100 K (maximum), remaining essentially constant at these values up to 10 000 megahertz.

The low-noise region from about 1000 to 10 000 megahertz is most amenable to application of special low-noise antennas. In the tropospheric-noise region from 10 000 to 100 000 megahertz, radio-sky noise temperature increases mainly due to water-vapor resonance and oxygen resonance in the atmosphere. However, high-gain receiving antennas directed at the sun can result in excessive antenna-noise temperature. For example, an 85-foot-diameter, parabolic (dish) antenna directed at the sun will experience an antenna-noise temperature rise of about 40 000 K, at 400 megahertz, with the "quiet sun" centered on the antenna main lobe, compared to only about 600 K rise for Cassiopeia A for similar conditions.

Radio-Sky Maps: Published in 1973 by Taylor**, Figs. 4 and 5 show detailed radio-sky maps of the whole celestial sphere for the 136-megahertz and 400-megahertz space research satellite frequency bands. Also shown are the well-known discrete

** R. F. Taylor, "136MHz/400MHz Radio-Sky Maps" *Proceedings of the IEEE*, vol. 61, no. 4, *Proceedings Letters*, pp. 469–472, April 1973.

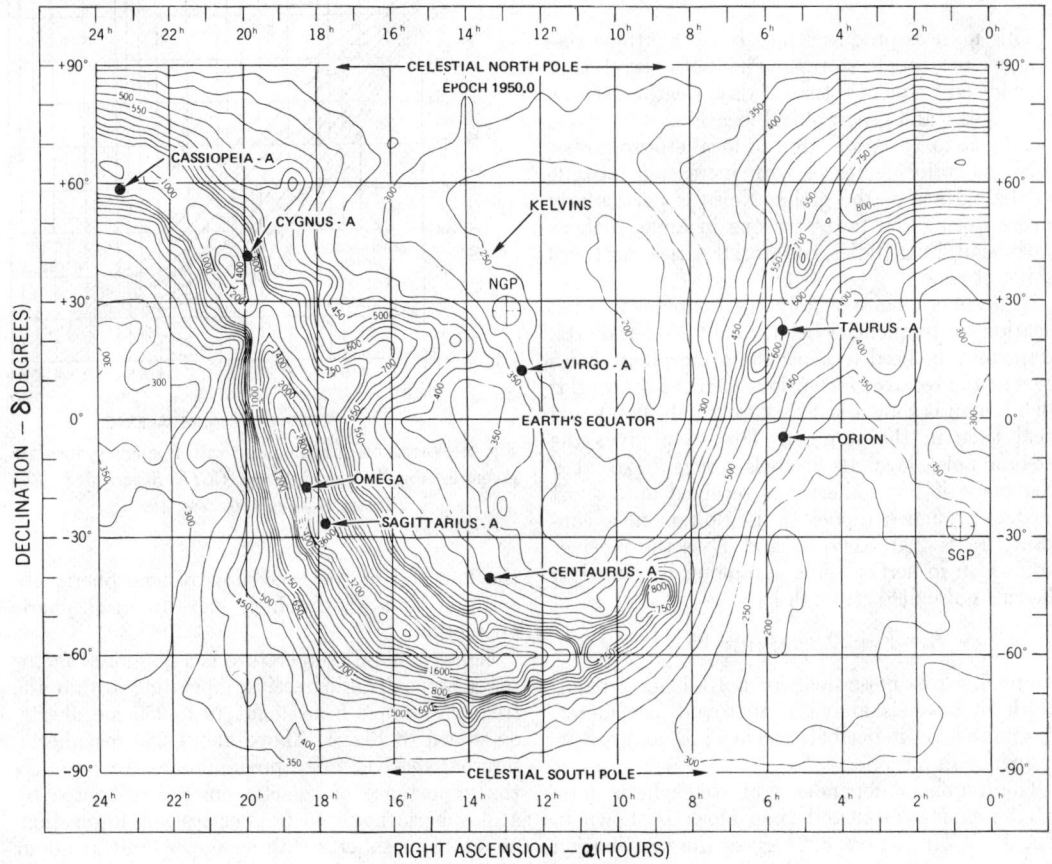

Fig. 4—Radio-sky map. 136-megahertz brightness temperature (Kelvins).

sources of cosmic radio waves, known as radio stars, including the intense source, Cassiopeia A. The 136-megahertz and 400-megahertz radio-sky maps in Figs. 4 and 5 are a composite of data obtained using high-gain antennas with solid-angle beams ranging in size from 2° to 5° half-power beamwidth (HPBW) at 136 megahertz, and 7° to 16° HPBW at 400 megahertz.

Figure 6 shows the level of galactic noise in decibels relative to a noise temperature of 290 K when receiving on a half-wave dipole. The noise levels shown in this figure assume no atmospheric absorption, and refer to the following sources of galactic noise.

Galactic Plane: Galactic noise from the galactic plane in the direction of the center of the galaxy. The noise levels from other parts of the galactic plane can be as much as 12 to 15 decibels below the levels given in Fig. 6.

Quiet Sun: Noise from the "quiet" sun; that is, solar noise at times when there is little or no sunspot activity.

Disturbed Sun: Noise from the "disturbed" sun. The term disturbed refers to times of sunspot and solar-flare activity.

Cassiopeia A: Noise from a high-intensity discrete source of cosmic noise known as Cassiopeia A. This is one of more than a hundred known discrete sources. Cassiopeia A subtends a solid angle at the earth's surface of only about 5 arc minutes.

The levels of cosmic noise received by a highly directive antenna with mainlobe pointed along the galactic plane can be obtained from equations given by Kraus[***] for the antenna-noise temperature (T_A) at the output terminals of an ideal, loss-free, antenna as

$$T_A = \frac{\int_0^{\theta=90°-\theta_0} \int_0^{\phi=2\pi} T(\theta,\phi)\, G(\theta\phi)\ \sin\theta d\theta d\phi}{\int_0^{\theta=90°-\theta_0} \int_0^{\phi=2\pi} G(\theta,\phi)\ \sin\theta d\theta d\phi}\ K$$

[***] J. D. Kraus, "Radio Astronomy," McGraw-Hill Book Company, 1966.

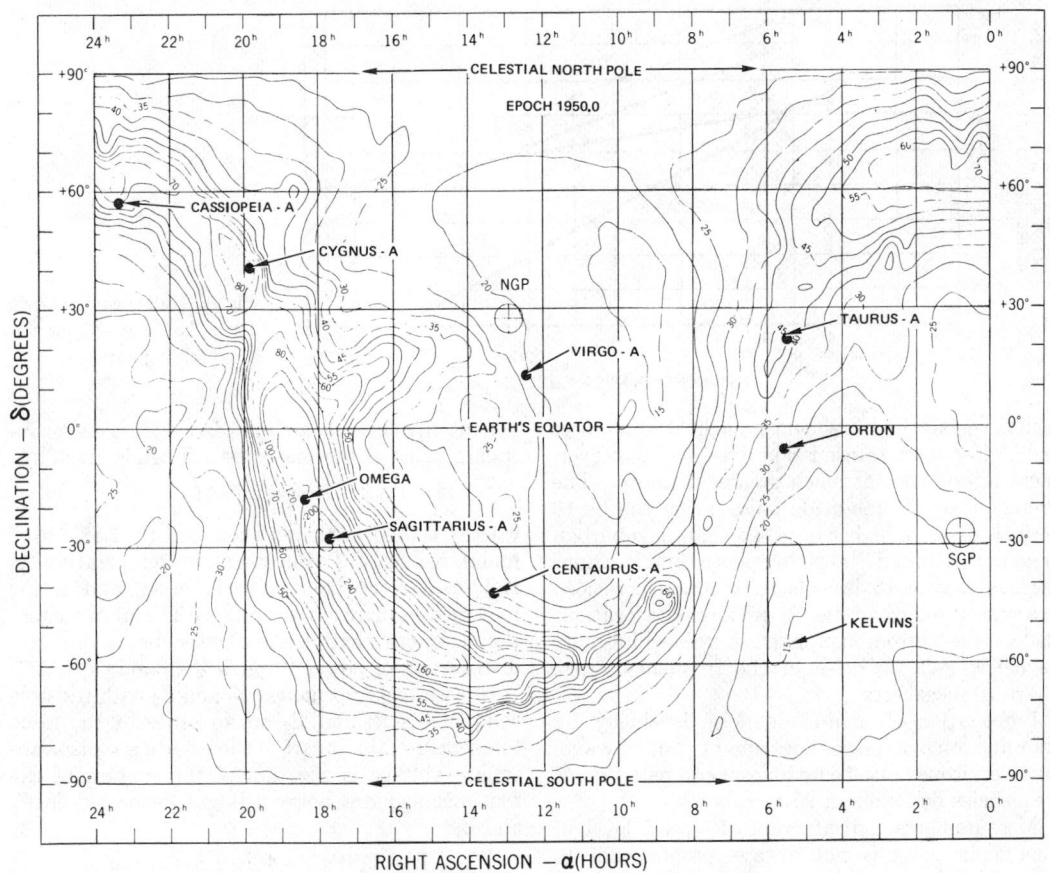

Fig. 5—Radio-sky map. 400-megahertz brightness temperature (Kelvins).

where

$\theta = 0°$ at zenith

$\phi = 360°$ azimuth angle

$T(\theta,\phi) =$ brightness-noise temperature distribution from Radio-Sky Map, Kelvins

$G(\theta,\phi) =$ antenna radiation pattern gain distribution, assumed symmetrical

$\theta_0 =$ minimum elevation angle between antenna's mainlobe axis and the horizon, degrees.

However, for a practical antenna, Taylor and Stocklin† give a simplified approximation for T_A including contributions from the main lobe, side lobes and back lobe as,

$$T_A \simeq 0.82\ T_{sky} + 0.13\,(\bar{T}_{sky} + T_E)\ K,\ \text{for a solid-}$$
angle beam, $\theta_{HPBW} = \phi_{HPBW} \leq 25°$

† R. F. Taylor and F. J. Stocklin, "VHF/UHF Stellar Calibration Error Analysis," *Proceedings International Telemetering Conference*, Washington, D.C., vol. VII pp. 553–566, September 27–29, 1971.

where

$T_{sky} =$ mean value of sky-brightness temperature within mainlobe HPBW, in Kelvins

$\bar{T}_{sky} =$ mean value of sky-brightness temperature within antenna side lobes, in Kelvins

$T_E \simeq T_0 = 290$ K, effective noise temperature of earth.

For example, a 136-megahertz, phase-array, directive antenna with main lobe HPBW equal to 12°, pointed near Cassiopeia A, has a value of T_A equal to approximately 870 K, for T_{sky} equal to 950 K and $\bar{T}_{sky}$ equal to 400 K obtained from Fig. 4.

Man-Made Noise

The amplitude of man-made noise decreases with increasing frequency and varies considerably with location. It is chiefly due to electric motors, neon signs, power lines, and ignition systems located within a few hundred yards of the receiving an-

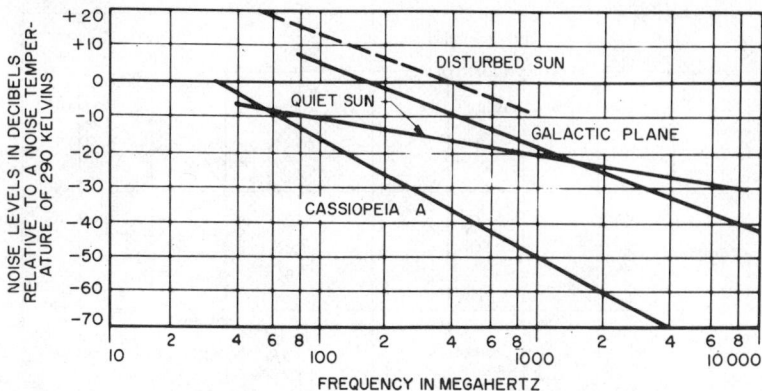

Fig. 6—Galactic noise levels for a half-wave-dipole receiving antenna.

tenna; certain high-frequency medical appliances and high-voltage transmission lines may, however, cause interference at much greater distances. The average level of man-made noise power can be 16 decibels or more higher in urban than in suburban areas in the United States; in remote rural locations the level may be 15 decibels below that experienced in a typical suburban site. In quiet remote locations the noise level from man-made sources will usually be below galactic noise in the frequency range above 10 megahertz.

Propagation of man-made noise is chiefly by transmission over power lines and by ground wave; however, it may also be by ionospheric reflection at frequencies below about 20 megahertz.

Measurements indicate that the peak level of man-made noise is not always proportional to bandwidth for bandwidths greater than about 10 kilohertz. According to the best available information, the peak field strengths of man-made noise (except diathermy and other narrow-band noise) increase as the receiver bandwidth is increased, substantially as shown in Fig. 7 for bandwidths greater than about 10 kilohertz.

Precipitation Static*

Precipitation static is produced by rain, hail, snow, or dust storms in the vicinity of the receiving antenna and is important chiefly at frequencies below 10 megahertz. This form of interference can be reduced by eliminating sharp points from the antenna and its surroundings, and also by providing means for dissipating the charges which build up on an antenna and on its surroundings during electrical storms.

Thermal Noise

Thermal noise is caused by the thermal agitation of electrons in resistances. Let R = resistive com-

---

* R. L. Tanner and J. E. Nanevicz, "An Analysis of Corona-Generated Interference in Aircraft," *Proceedings of the IEEE*, vol. 52, pp. 44–52; January 1964.

ponent in ohms of an impedance Z. The mean-square value of thermal-noise voltage is given by

$$E^2 = 4RkT \cdot \Delta f$$

where k is the Boltzmann's constant ($= 1.38 \times 10^{-23}$ joules/Kelvin), T the absolute temperature in Kelvins, Δf the bandwidth in hertz, and E the root-mean-square noise voltage. The above equation assumes that thermal noise has a uniform distribution of power through the bandwidth Δf.

In case two impedances Z_1 and Z_2 with resistive components R_1 and R_2 are in series at the same temperature, the square of the resulting root-mean-square voltage is the sum of the squares of the root-mean-square noise voltages generated in Z_1 and Z_2:

$$E^2 = E_1{}^2 + E_2{}^2 = 4(R_1 + R_2)kT \cdot \Delta f.$$

In case the same impedances are in parallel at the same temperature, the resulting impedance Z is calculated as is usually done for alternating-current circuits, and the resistive component R of Z is then determined. The root-mean-square noise voltage is the same as it would be for a pure resistance R.

It is customary in temperate climates to assign to T a value such that $1.38T = 400$, corresponding to about 17 degrees Celsius or 63 degrees Fahrenheit. Then $E^2 = 1.6 \times 10^{-20} R \cdot \Delta f$.

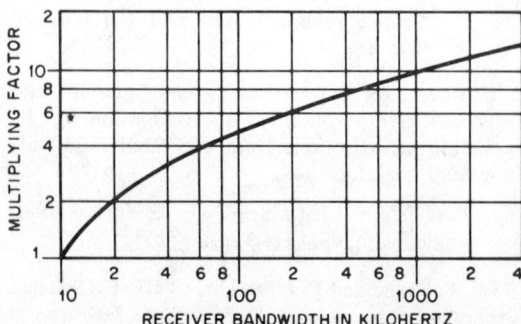

Fig. 7—Bandwidth factor for man-made noise.

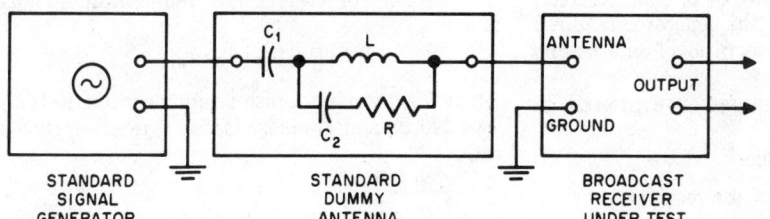

Fig. 8—Measurement of equivalent noise sideband input of a broadcast receiver.

Noise in Amplifiers

The ultimate sensitivity of an amplifier is set by the noise inherent to its input stage. For discussions of the noise produced in electron tubes and in transistors, refer to the pertinent chapters.

NOISE MEASUREMENTS

Measurement for Broadcast Receivers*

For standard broadcast receivers, the noise properties are determined by means of the equivalent noise sideband input (ensi). The receiver is connected as shown in Fig. 8. Components of the standard dummy antenna are $C_1 = 200$ picofarads, $C_2 = 400$ picofarads, $L = 20$ microhenries, and $R = 400$ ohms.

The equivalent noise sideband input

$$(\text{ensi}) = mE_s(P'_n/P'_s)^{1/2}$$

where E_s = root-mean-square unmodulated carrier-input voltage, m = degree of modulation of signal carrier at 400 hertz, P'_s = root-mean-square signal-power output when signal is applied, and P'_n = root-mean-square noise-power output when signal input is reduced to zero. It is assumed that no appreciable noise is transferred from the signal generator to the receiver, and that m is small enough for the receiver to operate without distortion.

Noise Factor of a Receiver

A more precise evaluation of the quality of a receiver as far as noise is concerned is obtained by means of its noise factor.**

It should be clearly realized that the noise factor evaluates only the linear part of the receiver, i.e., up to the demodulator.

The equipment used for measuring noise factor is shown in Fig. 9. The incoming signal (applied to the receiver) is replaced by an unmodulated signal generator with R_0 = internal resistive component, E_i = root-mean-square open-circuit carrier voltage, and E_n = root-mean-square open-circuit noise voltage produced in signal generator. Then

$$E_n^2 = 4kT_0R_0\Delta f'$$

where k is the Boltzmann's constant ($= 1.38 \times 10^{-23}$ joules/Kelvin), T_0 the temperature in Kelvins, and $\Delta f'$ the effective bandwidth of receiver (determined as below).

If the receiver does not include any other source of noise, the ratio E_i^2/E_n^2 is equal to the power carrier/noise ratio measured by the indicator:

$$E_i^2/E_n^2 = (E_i^2/4R_0)/kT_0\Delta f' = P_i/N_i.$$

The quantities $E_i^2/4R_0$ and $kT_0\Delta f'$ are called the *available* carrier and noise powers, respectively.

The output carrier/noise power ratio measured in a resistance R may be considered as the ratio of an available carrier-output power P_0 to an available noise-output power N_0.

The noise factor F of the receiver is defined by

$$P_0/N_0 = F^{-1}(P_i/N_i)$$

$$F = (N_0/N_i)(P_0/P_i)^{-1}$$

$$= E^2_{i1:1}/4kT_0R_0\Delta f' = P_{i1:1}/kT_0\Delta f'$$

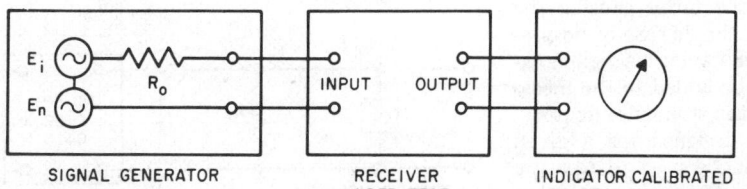

Fig. 9—Measurement of the noise factor of a receiver. The receiver is considered as a 4-terminal network. Output refers to last intermediate-frequency stage.

---

* "Standards on Radio Receivers: Methods of Testing Broadcast Radio Receivers, 1938," published by The Institute of Radio Engineers; 1942.

** The definition of the noise factor was first given by H. T. Friis, "Noise Figures of Radio Receivers," *Proceedings of the IRE*, vol. 32, pp. 419–422; July 1944.

where $P_0/P_i =$ available gain G of the receiver, $P_{i1:1} =$ available power from the generator required to produce a carrier-to-noise ratio of one at the receiver output.

Noise figure is the noise factor expressed in decibels:

$$F_{dB} = 10 \log_{10} F.$$

Effective bandwidth $\Delta f'$ of the receiver is

$$\Delta f' = G^{-1} \int G_f \, df$$

where G_f is the differential available gain. $\Delta f'$ is generally approximated to the bandwidth of the receiver between those points of the response showing a 3-decibel attenuation with respect to the center frequency.

Measurement of Noise Figure with a Thermal Noise Source

For the case where the spurious responses of the receiver are negligible, receiver noise figure can be conveniently measured by using the noise output of a thermal noise source having an equivalent generator resistance equal to that specified for use with the receiver.

With the noise source off, but still possessing the correct output resistance, receiver gain is adjusted for a convenient amount of noise power output; then with the noise source on, and still possessing the correct output resistance, the noise power output is increased by a convenient power ratio (N_2/N_1). The measured noise figure is then given by

$$NF = (\text{excess})_{dB} - 10 \log[(N_2/N_1) - 1].$$

For a thermal diode operating in the temperature limited emission mode

$$(\text{excess})_{dB} = 10 \log(20 R_d I_d)$$

where R_d is the noise source output resistance, and I_d is the diode current in amperes.

When the receiver has appreciable spurious responses, the correction factor which must be used with the above simple equation is a complex function of the spurious response ratios, and of the percentage of total internal receiver noise produced by the circuits preceding the mixer causing the spurious responses. For the simple case of no preselection, and a diode mixer having negligible excess noise, 3 decibels must be added to the measured noise figure to obtain the true noise figure.

A thermal noise source designed for a given generator impedance R_1 can be used to measure the noise figure of a receiver designed for a higher generator R_2 by adding a resistor $(R_2 - R_1)$ between noise source and receiver input and using

$$NF = NF_{read} - 10 \log(R_2/R_1).$$

Conversion of receiver noise temperature to noise factor:

$$F = 1 + (T_R/T_0)$$

where $T_R =$ receiver noise temperature in Kelvins, $T_0 = 290$ K, and $F =$ noise factor of receiver (power ratio).

Conversely,

$$T_R = (F - 1) T_0.$$

Determination of effective noise temperature of receiving system (i.e. antenna, transmission line, and receiver):

$$T_E = T_A + (LF - 1) T_0$$

where $T_E =$ effective noise temperature of receiving system, $T_A =$ antenna noise temperature, $L =$ transmission line loss (power ratio), $F =$ noise factor of receiver (power ratio), and $T_0 = 290$ K.

Determination of the effective input noise power of the receiving system:

$$N_i = kBT_E$$

where $N_i =$ effective input noise power of the receiving system, $k =$ Boltzmann's constant (1.38×10^{-23} joules/Kelvin), $B =$ bandwidth in hertz, and $T_E =$ effective noise temperature in Kelvins; or

$$dBm_i = -198.6 + 10 \log B + 10 \log T_E.$$

Calculation of Noise Figure

The active device can be defined for noise figure calculations as in Fig. 10.

R_{eq} can be experimentally obtained by measuring, with a "zero impedance" generator, the equivalent microvolts of noise V_{sc}, in a bandwidth B, in series with the input terminals, with an almost-short-circuit on the output terminals. R_{eq} is then given by

$$R_{eq} = |V_{sc}|^2 / 1.64 \times 10^{-20} \langle BW \rangle. \qquad (1)$$

R_e is straightforwardly obtained by input impedance measurements with a short-circuit on the output terminals.

ρ can be experimentally obtained by approximately open-circuiting the input terminals at the

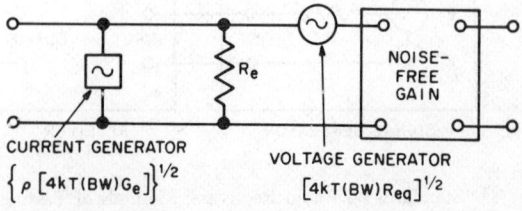

Fig. 10—Calculation of the noise figure of a receiver.

frequency of interest with a tuned circuit of parallel resonant resistance R_0, and measuring the total equivalent microvolts of noise produced across the input terminals, with an almost-short-circuit on the output terminals. Then, assuming negligible correlation

$$\rho=[1+(R_e/R_0)]^2\{[\mid V_{oc}\mid^2/(1.64\times10^{-20}\langle BW\rangle R_e)] \\ -(R_{eq}/R_e)\}-(R_e/R_0). \quad (2)$$

When the above characterized device is used with an input transforming circuit of parallel resonant resistance R_r, the resulting noise factor can be calculated as follows: First calculate R_1 and β from

$$R_1^{-1}=R_r^{-1}+R_e^{-1} \quad (3)$$

$$\beta=[1+\rho(R_r/R_e)]/[1+(R_r/R_e)]. \quad (4)$$

In terms of the above quantities and the transformed generator resistance R_s seen by the input terminals of the active device, the resulting noise factor is given by

$$F=1+2(R_{eq}/R_1)+(R_{eq}/R_s)+(R_s/R_1) \\ \times[\beta+(R_{eq}/R_1)]. \quad (5)$$

It should be noted that to minimize noise figure the input circuit should always be tuned so as to null any part of the noise due to βR_1, which is correlated with the noise due to R_{eq}. Equation (5) can be applied to this best noise figure tuning case if ρ is obtained, by some method, from only the uncorrelated part of the βR_1 noise.

This resulting noise figure is minimized when the transformed generator resistance has the value

$$R_{s\ opt}=\{(R_1R_{eq}/\beta)/[1+(R_{eq}/\beta R_1)]\}^{1/2} \quad (6)$$

and with this optimum source resistance

$$F_{opt}=1+2\beta(R_{eq}/R_1)^{1/2} \\ \times\{[1+(R_{eq}/R_1)]^{1/2}+(R_{eq}/R_1)^{1/2}\}. \quad (7)$$

Noise Factor of Cascaded Networks

The overall noise factor of two networks a and b in cascade (Fig. 11) is

$$F_{ab}=F_a+[(F_b-1)/G_a]$$

provided $\Delta f_b'\leq\Delta f_a'$.

The additional noise due to external sources influencing real antennas (such as galactic noise), may be accounted for by an apparent antenna temperature, bringing the available noise-power input to $kT_a\Delta f'$ instead of $N_i=kT_0\Delta f'$ (the physical

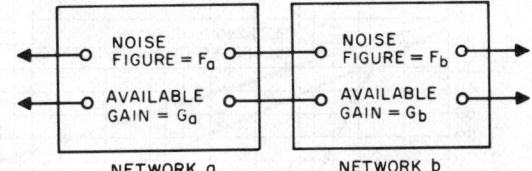

Fig. 11—Overall noise figure F_{ab} of two networks, a and b, in cascade.

antenna resistance at temperature T_0 is generally negligible in high-frequency systems). The internal noise sources contribute $(F-1)N_i$ as before, so that the new noise factor is given by

$$F'N_i=(F-1)N_i+kT_0\Delta f'$$

$$F'=(F-1)+(T_a/T_0).$$

The average temperature of the antenna for a 6-megahertz equipment is found to be 3000 Kelvins, approximately. The contribution of external sources is thus of the order of 10, compared with a value of $(F-1)$ equal to 1 or 2, and becomes the limiting factor of reception. At 3000 megahertz, however, values of T_a may fall below T_0.

Noise Improvement Factor

In case the receiver includes demodulation processes that produce a signal/noise ratio improvement (nif), this improvement ratio must, of course, be considered when evaluating the carrier required to produce a desired output signal/noise ratio. Noise improvement factor is also discussed in the chapter "Modulation."

MEASUREMENT OF EXTERNAL RADIO NOISE*

External noise fields, such as atmospheric, galactic, and man-made, are measured in the same way as radio-wave field strengths, with the exception that peak, rather than average, values of noise are usually of interest, and also that the overall band-pass action of the measuring apparatus must be accurately known in measuring noise. When measuring noise that varies over wide limits with time, such as atmospheric noise, it is generally best to use automatic recorders.

* W. Q. Crichlow, et al., "Special Report on Characteristics of Terrestrial Radio Noise," International Scientific Radio Union (URSI), Commission IV; August 1960.

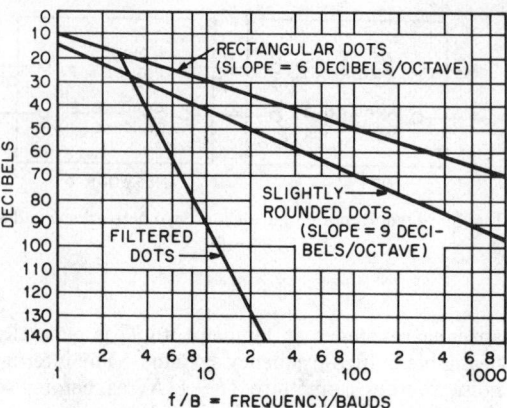

Fig. 12—Curves giving the envelopes for Fourier spectra of the emission resulting from several shapes of a single telegraph dot. For the upper curve the dot is taken to be rectangular and its length is $\frac{1}{2}$ of the period T corresponding to the fundamental dotting frequency. The dotting speed in bauds is $B = 1/t = 2/T$. The bottom curve would result from the insertion of a filter with a pass band equal to 5 units on the f/B scale, and having a slope of 30 decibels/octave outside of the pass band.

INTERFERENCE EFFECTS IN VARIOUS SYSTEMS*

Besides noise, the efficiency of radio communication systems can be limited by the interference produced by other radio communication systems. The amount of tolerable signal/interference ratio, and the determination of conditions for entirely satisfactory service, are necessary for the specification of the amount of harmonic and spurious frequencies that can be allowed in transmitter equipments, as well as for the correct spacing of adjacent channels.

Simple Telegraphy

It is considered that satisfactory radiotelegraph service is provided when the radio-frequency interference power available in the receiver, averaged over a cycle when the amplitude of the interfering wave is at a maximum, is at least 10 decibels below the available power of the desired signal averaged in the same manner, at the time when the desired signal is a minimum.

To determine the amount of interference produced by one telegraph channel on another, Figs. 12 and 13 will be found useful.

* "Handbook of Radio-Frequency Interference," vols. 1 to 4, Fredrick Research Corp., Wheaton, Maryland; 1962.

Frequency-Shift Telegraphy and Facsimile

It is estimated that the interference level of −10 decibels as recommended in the previous case will also be suitable for frequency-shift telegraphy and facsimile.

Double-Sideband Telephony

The multiplying factor for frequency separation between carriers as required for various ratios of signal/interference is given in the following table. This factor should be multiplied by the highest modulation frequency.

The acceptance band of the receiving filters in hertz is assumed to be 2× (highest modulation frequency) and the cutoff characteristic is assumed to have a slope of 30 decibels/octave.

Ratio of Desired to Interfering Carriers in Decibels	Multiplying Factor for Various Ratios of Signal/Interference			
	20 dB	30 dB	40 dB	50 dB
60	0	0	0	0
50	0	0	0	0.60
40	0	0	0.60	1.55
30	0	0.60	1.55	1.85
20	0.60	1.55	1.85	1.96
10	1.55	1.85	1.96	2.00
0	1.85	1.96	2.00	2.55
−10	1.96	2.00	2.55	2.85
−20	2.00	2.55	2.85	3.2
−30	2.55	2.85	3.2	3.6
−40	2.85	3.2	3.6	4.0
−50	3.2	3.6	4.0	4.5
−60	3.6	4.0	4.5	5.1
−70	4.0	4.5	5.1	5.7
−80	4.5	5.1	5.7	6.4
−90	5.1	5.7	6.4	7.2
−100	5.7	6.4	7.2	8.0

Broadcasting

As a result of a number of experiments, it is possible to set down the following results for carrier frequencies between 150 and 285 kilohertz and between 525 and 1560 kilohertz.

Frequency Separation Between Carriers in Kilohertz	Minimum Ratio of Desired and Interfering Carriers in Decibels
11	0*
10	6†
9	14†
8	26‡
5 (or less)	60†

* Extrapolated.
† Experimental.
‡ Interpolated.

These experimental results agree reasonably well with the theoretical results of the preceding table with a highest modulation frequency of about 4500 hertz, and with a signal/interference ratio of 50 decibels.

Single-Sideband Telephony

Experience shows that the separation between adjacent channels need be only great enough to assure that the nearest frequency of the interfering signal is 40 decibels down on the receiver filter characteristic when due allowance has been made for the frequency instability of the carrier wave.

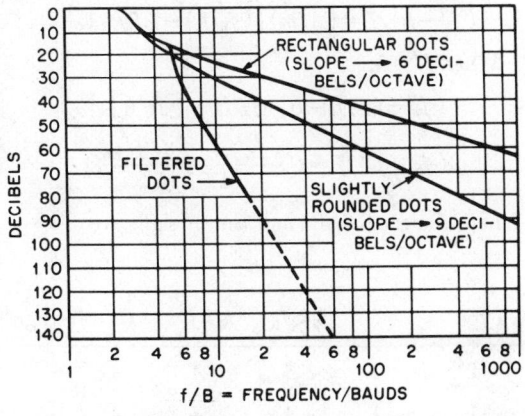

Fig. 13—Received power as a function of frequency separation between transmitter frequency and midband frequency of the receiver.

SPURIOUS RESPONSES

In superheterodyne receivers, where a nonlinear element is used to get a desired intermediate-frequency signal from the mixing of the incoming signal and a local-oscillator signal, interference from spurious external signals results in a number of undesired frequencies that may fall within the intermediate-frequency band. Likewise, when two local oscillators are mixed in a transmitter or receiver to produce a desired output frequency, several unwanted components are produced at the same time due to the imperfections of the mixer characteristic. The following tables show how the location of the spurious frequencies can be determined.

Defining and Coincidence Equations

Mixing for Difference Frequency

Defining Equations	Coincidence
Type I:	
$f_x = \pm(f_1 - f_2)$	$[f_2/f_1]_\infty = (m+1)/(n+1)$
$f_x' = \pm(nf_2 - mf_1')$	
Type II:	
$f_x = \pm(f_1 - f_2)$	$[f_2/f_1]_\infty = (m-1)/(n-1)$
$f_x' = \pm(mf_1' - nf_2)$	
Type III:	
$f_x = f_1 - f_2$	$[f_2/f_1]_\infty = (1-m)/(n+1)$
$f_x' = mf_1' + nf_2$	

Mixing for Sum Frequency

Type IV:	
$f_x = f_1 + f_2$	$[f_2/f_1]_\infty = (m-1)/(n+1)$
$f_x' = mf_1' - nf_2$	
Type V:	
$f_x = f_1 + f_2$	$[f_2/f_1]_\infty = (m+1)/(n-1)$
$f_x' = nf_2 - mf_1'$	
Type VI:	
$f_x = f_1 + f_2$	$[f_2/f_1]_\infty = (1-m)/(n-1)$
$f_x' = mf_1' + nf_2$	

In types I and II, both f_x and f_x' must use the same sign throughout. Types III and VI are relatively unimportant except when $m = n = 1$.

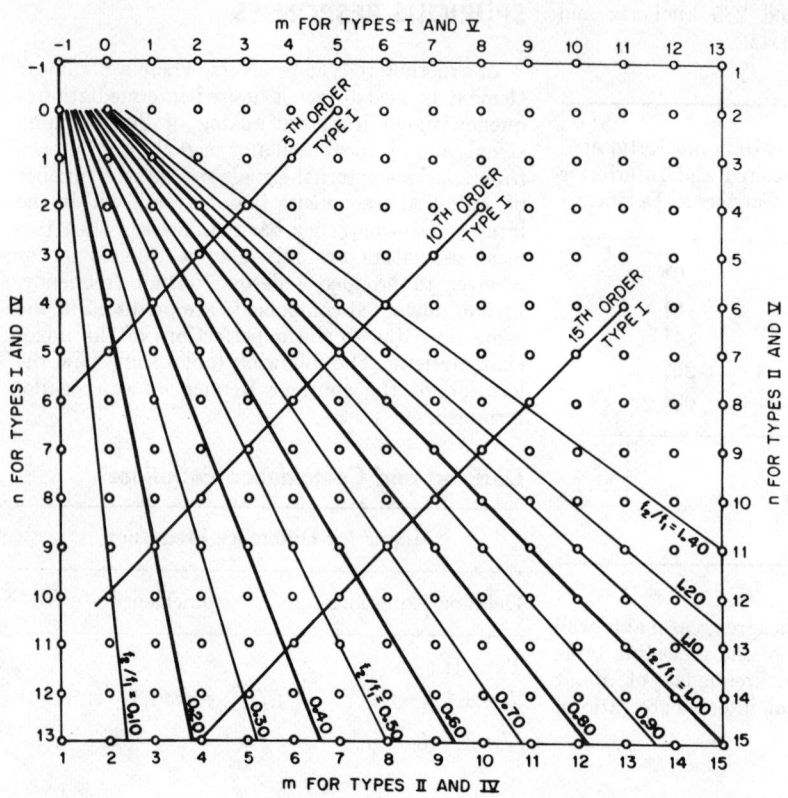

Fig. 14—Chart of spurious responses. Each circle represents a spurious-response coincidence, where $f_1'=f_1$ and $f_x'=f_x$.

Symbols

f_1 = signal frequency (or first source)

f_1' = spurious signal ($f_1'=f_1$ for mixing local sources, but when dealing with a receiver, usually $f_1'\neq f_1$)

f_2 = local-injection frequency (or second source)

f_x = desired mixer-output frequency

f_x' = spurious mixer-output frequency

$k=m+n$ = order of response, where m and n are positive integers.

Coincidence is where $f_1'=f_1$ and $f_x'=f_x$.

Image (m = n = 1)

Kind of Mixing	Receiver ($f_x'=f_x$)	Two Local Sources ($f_1'=f_1$)
Difference	$f_1'=\pm(2f_2-f_1)$	
	$=\pm(f_1-2f_2)$ $f_2<f_1$	$f_x'=f_1+f_2$
	$=f_1+2f_2$ $f_2>f_1$	
Sum	$f_1'=f_1+2f_2$	$f_x'=\pm(f_1-f_2)$
	$=2f_x-f_1$	

Intermediate-frequency rejection must be provided for spurious signal $f_1'=f_x$ where $m=1$, $n=0$.

Selectivity Equations (For Types I, II, IV, and V only.)

When $f_x'=f_x$,

$$(f_1'-f_1)/f_1=(A/m)\{(f_2/f_1)-[f_2/f_1]_\infty\}.$$

When $f_1'=f_1$,

$$(f_x'-f_x)/f_1=B\{(f_2/f_1)-[f_2/f_1]_\infty\}$$

$$(f_x'-f_x)/f_x=C\frac{(f_2/f_1)-[f_2/f_1]_\infty}{1\mp f_2/f_1}.$$

Where the coefficients and the $\mp$ signs are

Type	A	B ($f_2<f_1$)	B ($f_2>f_1$)	C	$\mp$ Sign
I	$n+1$	A	$-A$	A	$-$
II	$n-1$	$-A$	A	$-A$	$-$
IV	$n+1$	$-A$	$-A$	$-A$	$+$
V	$n-1$	A	A	A	$+$

Variation of Output Frequency vs Input-Signal Deviation

For any type $\Delta f_x' = \pm m \, \Delta f_1'$:

Use the $+$ or the $-$ sign according to defining equation for type in question.

Table of Spurious Responses (Below)

Type I Coincidences:

$$[f_2/f_1]_{co} = (m+1)/(n+1)$$

where $f_x' = f_x$ and $f_1' = f_1$.

Chart of Spurious Responses (Fig. 14)

Example: Suppose two frequencies whose ratio is $f_2/f_1 = 0.12$ are mixed to obtain the sum frequency. The spurious responses are found by laying a transparent straightedge on the chart, passing through the circle $-1, -1$ and lying a little to the right of the line marked $f_2/f_1 = 0.10$. It is observed that the straightedge passes near circles indicating the responses

$$\text{Type IV:} \begin{cases} m=1 \\ n=0 \end{cases} \begin{cases} =2 \\ =7 \end{cases} \begin{cases} =2 \\ =8 \end{cases}$$

$$\text{Type V:} \begin{cases} m=0 \\ n=9 \end{cases} \begin{cases} =0 \\ =10 \end{cases}$$

The actual frequencies of the responses f_x' or f_1' can be determined by substituting these coefficients m and n in the defining equations.

Frequency Ratio $=[f_2/f_1]_{co}$			Lowest Order			
Fraction	Decimal	Reciprocal	k_I	m_I	n_I	Highest Orders
1/1	1.000	1.000	2	1	1	All even orders $m=$ (See note 2.)
8/9	0.889	1.125	15	7	8	
7/8	0.875	1.143	13	6	7	
6/7	0.857	1.167	11	5	6	
5/6	0.833	1.200	9	4	5	
4/5	0.800	1.250	7	3	4	
7/9	0.778	1.286	14	6	8	$\begin{cases} m_I=5 \\ n_I=7 \end{cases}$
3/4	0.750	1.333	5	2	3	
5/7	0.714	1.400	10	4	6	
7/10	0.700	1.429	15	6	9	$\begin{cases} m_I=3 \\ n_I=5 \end{cases} \begin{cases} =5 \\ =8 \end{cases}$
2/3	0.667	1.500	3	1	2	
5/8	0.625	1.600	11	4	7	
3/5	0.600	1.667	6	2	4	$\begin{cases} m_I=5 \\ n_I=9 \end{cases}$
4/7	0.571	1.750	9	3	6	
5/9	0.556	1.800	12	4	8	
6/11	0.545	1.833	15	5	10	$\begin{cases} m_I=1 \\ n_I=3 \end{cases} \begin{cases} =2 \\ =5 \end{cases} \begin{cases} =3 \\ =7 \end{cases} \begin{cases} =4 \\ =9 \end{cases}$
1/2	0.500	2.000	1	0	1	

Types II, IV, and V Coincidences: For each ratio $[f_2/f_1]_{co}$ there are also the following responses:

Type	k	m	n
II	$k_{II}=k_I+4$	$m_{II}=m_I+2$	$n_{II}=n_I+2$
IV	$k_{IV}=k_I+2$	$m_{IV}=m_I+2$	$n_{IV}=n_I$
V	$k_V=k_I+2$	$m_V=m_I$	$n_V=n_I+2$

Notes:

1. When $f_2 > f_1$, use reciprocal column and interchange the values of m and n.

2. At $[f_2/f_1]_{co} = 1/1$, additional important responses are

type II: $m=n=2$

type IV: $m=2$, $n=0$

type V: $m=0$, $n=2$.

CHAPTER 30

BROADCASTING AND RECORDING

INTRODUCTION

In the United States, broadcasting to the general public is regulated by the Federal Communications Commission (FCC), which allocates frequencies and establishes technical standards.* Three general classes of broadcast stations have been established. These are *standard broadcast stations* (amplitude modulation in the band 535–1605 kHz), *FM broadcast stations* (frequency modulation in the band 88–108 MHz), and *television broadcast stations* (operating in the bands 54–72, 74–88, 174–216, and 470–890 MHz with vestigial-sideband amplitude modulation of the visual carrier and frequency modulation of the aural carrier.)

This chapter also discusses certain aspects of international broadcasting between 5950 and 26 100 kilohertz in accordance with international agreements. Other technical information related to broadcasting is covered under auxiliary services, intercity transmission, recording, and terminal facilities.

STANDARD BROADCASTING†

Standard-broadcast stations are licensed for operation on channels spaced by 10 kilohertz and occupying the band from 535 to 1605 kilohertz. The major classifications are clear channel, regional channel, and local channel. A clear-channel station renders primary and secondary service over wide areas and is protected against objectionable interference. A regional-channel station renders primary service to larger cities and the surrounding rural areas; a channel may be occupied by several stations, and the primary service area may be limited by interference. A local station is designed to render service primarily to a city or town and its nearby surburban or rural areas. Its primary service area may also be limited by interference.

* Federal Communications Commission Rules and Regulations, Volume III, Part 73, Sept. 1972.

† FCC Rules and Regulations, Part 73, Subpart A; 1972.

Field-Strength Requirements

Primary Service:

City business, factory areas, 10 to 50 millivolts/meter, ground wave

City residential areas, 2 to 10 millivolts/meter, ground wave

Rural—all areas during winter or northern areas during summer, 0.1 to 0.5 millivolt/meter, ground wave

Rural—southern areas during summer, 0.25 to 1.0 millivolt/meter, ground wave.

Secondary Service: All areas having sky-wave field strength equal to or greater than 500 microvolts/meter for 50 percent or more of the time.

Table 1 outlines generally the protected contours and permissible interference for the various classes of stations. There are additional details and some exceptions in paragraphs 73.21–73.29 and 73.181–73.190 of Part 73 of the FCC Rules and Regulations, September 1972.

Coverage Data

Figures 1, 2, and 3 show computed values of ground-wave field strength as a function of the distance from the transmitting antenna. These are used to determine coverage and interference. They were computed for the frequencies indicated, a dielectric constant equal to 15 for ground and 80 for sea water (referred to air as unity), and for the surface conductivities noted. The curves are for radiation from a short vertical antenna at the surface of a uniformly conductive spherical earth, with an antenna power and efficiency such that the inverse-distance field is 100 millivolts/meter at one mile. (Twenty such charts, for frequencies at intervals throughout the standard broadcast band, are contained in Section 73.184 of the FCC Rules and Regulations.) Figure 4 shows the estimated effective field for vertical omnidirectional antennas of various heights. Figures 5 and 6 show the effective ground conductivity for various parts of the U.S. and Canada, and Fig. 7 shows the sky-wave fields for 10 percent and 50 percent of the time.

TABLE 1—CLASSIFICATION OF STANDARD-BROADCAST STATIONS.

Class of Channel	Class of Station	Permissible Power (kW)	Signal-Intensity Contour of Area Protected from Objectionable Interference (microvolts/meter)		Permissible Interfering Signal on Same Channel (microvolts/meter)	
			Day[1]	Night	Day[1]	Night[3]
Clear	I-A	50	Sc=100 Ac=500	Sc=500[2] Ac=500[1]	5	25
	I-B	10–50	Sc=100 Ac=500	Sc=500[2] Ac=500[1]	5	25
	II-A	0.25–50 day 10–50 night	500	500[1]	25	25
	(II-B II-D)	0.25–50	500	2500[1]	25	125
Regional	III-A	1–5	500	2500[1]	25	125
	III-B	0.5–5 day 0.5–1 night	500	4000[1]	25	200
Local	IV	0.25–1 day 0.25 night	500	not prescribed	25	not prescribed

Notes:

Sc—same channel, Ac—adjacent channel.

[1] Ground wave.

[2] 50% sky wave.

[3] 10% sky wave.

Station Performance Requirements*

Modulation: 85% to 95%.

Audio-Frequency Distortion: Harmonics less than 5% arithmetical sum or root-sum-square amplitude up to 84% modulation; less than 7.5% for 85% to 95% modulation.

Audio-Frequency Response: Transmission characteristic flat between 100 and 5000 hertz to within 2 decibels, referred to 1000 hertz.

Noise: At least 45 decibels, unweighted, below 100% modulation for the frequency band from 30 to 20 000 hertz.

Carrier-Frequency Stability: Within 20 hertz of assigned frequency.

* Refer to Section 73.40 of the FCC Rules and Regulations.

FREQUENCY-MODULATION BROADCASTING*

Frequency-modulation broadcasting stations are authorized for operation on 100 allocated channels, each 200 kilohertz wide, extending consecutively from channel 201 on 88.1 megahertz to channel 300 on 107.9 megahertz. Commercial broadcasting is authorized on channels 221 (92.1 megahertz) through 300. Noncommercial educational broadcasting is licensed on channels 201 through 220 (91.9 megahertz). (Certain exceptions are listed in Section 73.204 of the FCC Rules and Regulations.)

Station Service Classification

Class-A Stations: Render service primarily to a relatively small community, city, or town and the rural surroundings. The coverage will not exceed

* FCC Rules and Regulations, Part 73, Subparts B and C; 1972.

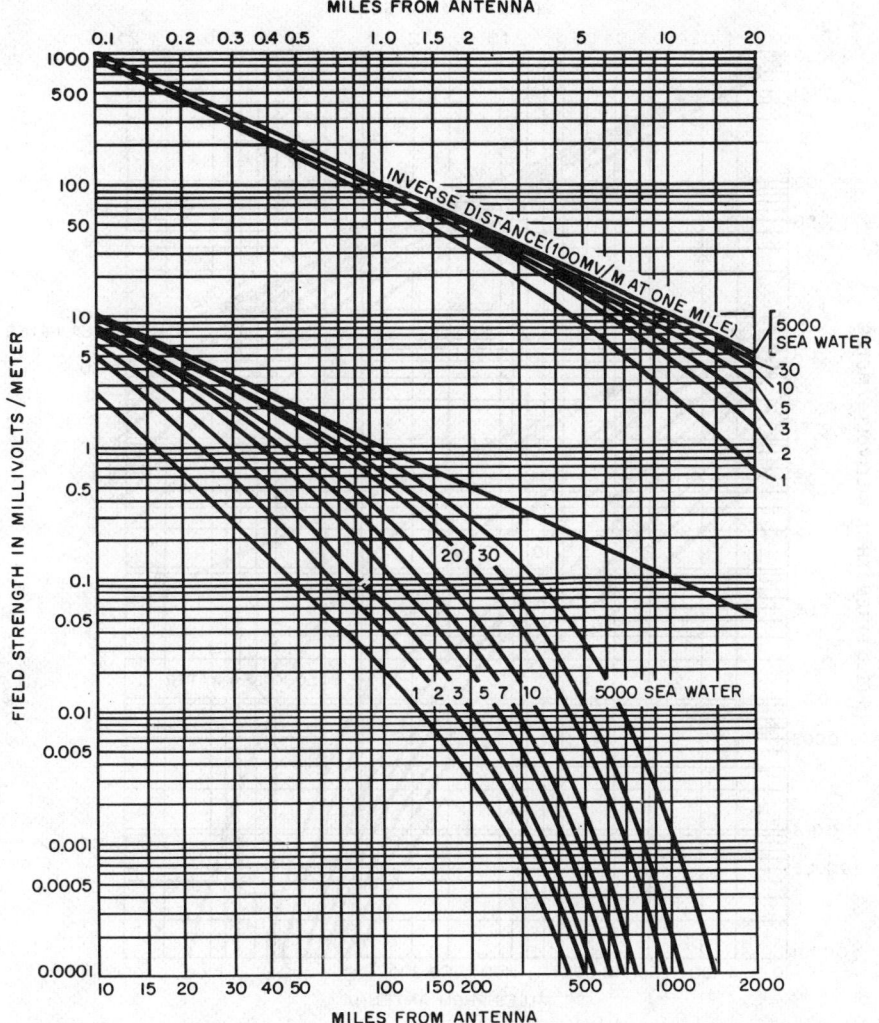

Fig. 1—Ground-wave field strength plotted against distance. Computed for 550 kilohertz. Dielectric constant = 15. Ground-conductivity values above are in millimhos/meter.

the equivalent of 3 kilowatts effective radiated power* at an antenna height above average terrain† of 300 feet. Minimum effective radiated power is 100 watts. Class-*A* channels are 221, 224, 228, 232, 237, 240, 244, 249, 252, 257, 261, 265, 269, 272, 276, 280, 285, 288, 292, and 296.

Class-B Stations: Render service to a large community. These stations operate in Zone I or Zone IA,* and their coverage will not exceed the equivalent of an effective radiated power of 50 kilowatts at an antenna height 500 feet above average terrain. Minimum effective radiated power is 5 kilowatts.

Class-C Stations: Render service to a large community. These stations operate in Zone II,†

* Effective radiated power is the product of antenna gain and antenna input power. Antenna input power is transmitter power minus transmission-line loss.

† Average terrain is defined as the average of the elevations between 2 and 10 miles from the antenna along eight radials evenly spaced by 45°.

* Generally speaking, Zone I is the northeastern part of the U.S., and Zone IA is Puerto Rico, the Virgin Islands, and California south of 40° latitude. For exact boundaries, refer to paragraph 73.205 of the FCC Rules and Regulations.

† Zone II includes Alaska, Hawaii, and other parts of the U.S. not in Zone I or IA.

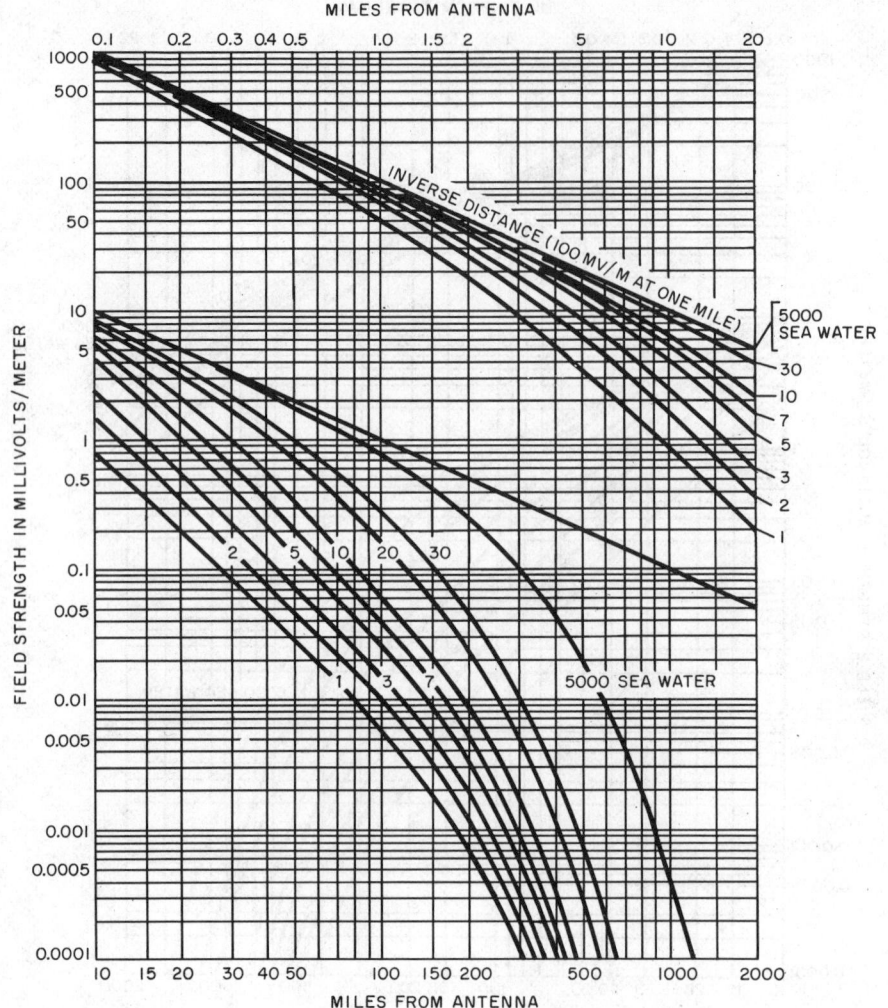

Fig. 2—Ground-wave field strength plotted against distance. Computed for 1000 kilohertz. Dielectric constant = 15. Ground-conductivity values above are in millimhos/meter.

and their coverage will not exceed the equivalent of an effective radiated power of 100 kilowatts at an antenna height of 2000 feet above average terrain. Minimum effective radiated power is 25 kilowatts.

Class-D Stations: Operate on a noncommercial educational channel with transmitter power output no greater than 10 watts.

If antenna heights for Class-*A*, -*B*, or -*C* stations are in excess of those noted above, effective radiated powers will be decreased as shown in Fig. 8.

Channel Availability

Commercial FM channels have been set forth in a table of assignments which lists communities and available channels.* The Table of Assignments (paragraph 73.202 of the FCC Rules and Regulations) is based on maximum service within the limits imposed by co-channel and adjacent-channel interference. A listed channel not in use may be applied for if the unlisted community is within 10 miles for Class-*A* channels, 15 miles for Class *B*/*C* channels, of the listed community and provided other requirements, including the minimum mileage separations in Table 2, are met.

Coverage

The extent of coverage of an FM station is determined by two field-strength contours: 3.16 millivolts/meter is the minimum field strength

* A table of assignments for noncommercial educational FM stations is contained in Section 73.507 of the FCC Rules and Regulations.

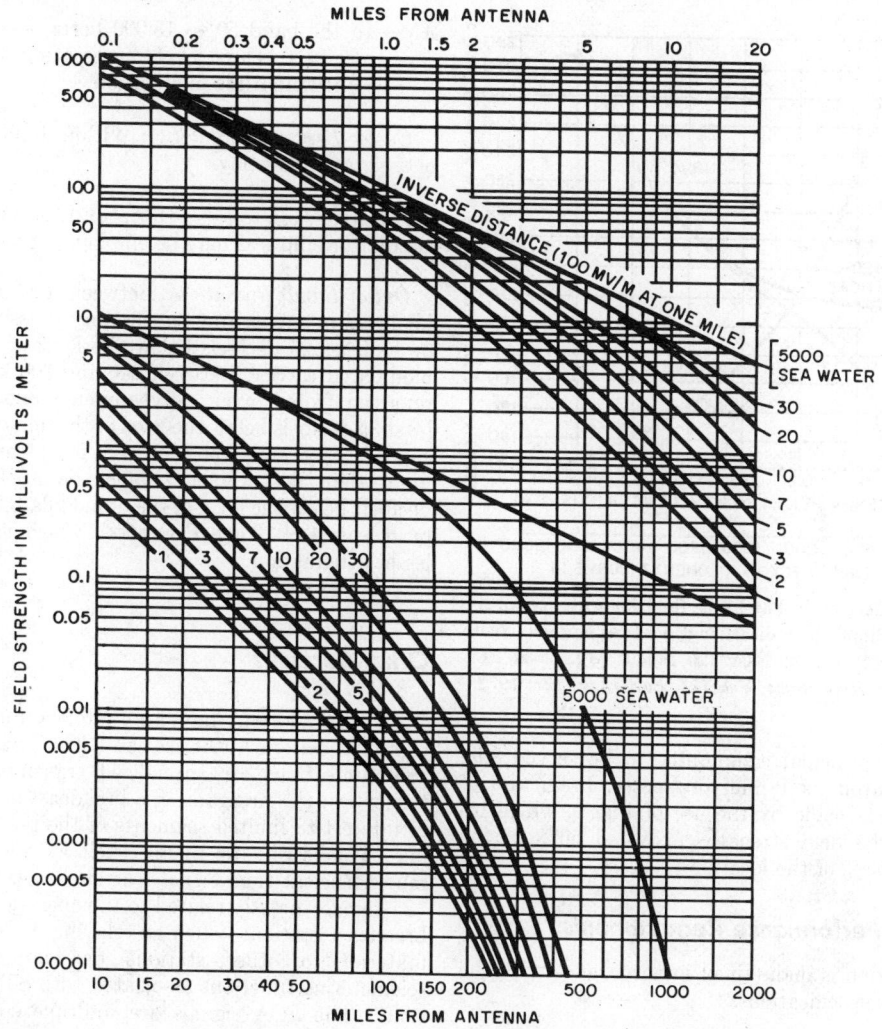

Fig. 3—Ground-wave field strength plotted against distance. Computed for 1600 kilohertz. Dielectric constant=15. Ground-conductivity values are in millimhos/meter.

TABLE 2—MINIMUM MILEAGE SEPARATIONS BETWEEN FM STATIONS.

	Class A				Class B				Class C			
	Separation in Kilohertz				Separation in Kilohertz				Separation in Kilohertz			
	Co-channel	200	400	600	Co-channel	200	400	600	Co-channel	200	400	600
Class					Minimum Mileage Separations							
A	65	40	15	15	—	65	40	40	—	105	65	65
B	—	65	40	40	150	105	40	40	170	135	65	65
C	—	105	65	65	170	135	65	65	180	150	65	65

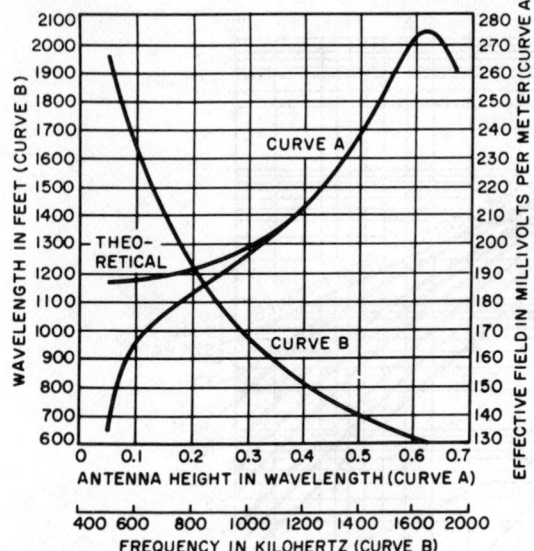

WAVELENGTH IN FEET (CURVE B)
EFFECTIVE FIELD IN MILLIVOLTS PER METER (CURVE A)

CURVE A

THEO-
RETICAL

CURVE B

ANTENNA HEIGHT IN WAVELENGTH (CURVE A)

FREQUENCY IN KILOHERTZ (CURVE B)

Fig. 4—Effective field at 1 mile for 1 kilowatt (curve A). Use for simple omnidirectional vertical antenna with ground system of at least 120 radials $\lambda/4$. *From FCC Rules and Regulations, Vol. III, Part 73, p. 127; 1972.*

over the principal community to be served; the outer contour is 1 millivolt/meter. Prediction of coverage is made by the use of Fig. 9, which indicates the field strengths exceeded 50% of the time at 50% of the locations.

Station Performance Requirements

Operation is maintained in accordance with the following specifications.

Audio-Frequency Response: Transmitting system capable of transmitting the band of frequencies 50 to 15 000 hertz. Pre-emphasis employed and response maintained within limits shown by curves of Fig. 10.

Audio-Frequency Distortion: Maximum combined audio-frequency harmonic root-mean-square voltage in system output less than as shown below.

Modulating Frequency (hertz)	Percent Harmonic
50–100	<3.5
100–7500	<2.5
7500–15 000	<3.0

Modulation: Frequency modulation with a modulating capability of 100 percent corresponding to a frequency swing of ±75 kilohertz.

Noise:

FM—In the band 50 to 15 000 hertz, at least 60 decibels below 100-percent modulation at 400-hertz modulating frequency.

AM—In the band 50 to 15 000 hertz, at least 50 decibels below level representing 100-percent amplitude modulation.

Center-Frequency Stability: Within ±2000 hertz of assigned frequency.

Antenna Polarization: Horizontal required, but circular or elliptical may be employed if desired.

Out-of-Band Radiation: Between 120 and 240 kilohertz removed from carrier, any emission must be at least 25 decibels below the level of the unmodulated carrier. Between 240 and 600 kilohertz removed from carrier, any emission must be at least 35 decibels below the level of the unmodulated carrier. Any emission removed from carrier by more than 600 kilohertz must be at least 80 decibels below the level of the unmodulated carrier, or at least $43+10 \log_{10}P$ (watts), whichever is the lesser attenuation.

Other Services

An FM station may apply for a Subsidiary Communications Authorization (SCA). Subsidiary communications are specialized transmissions of two types: (1) programs of a broadcast nature but of interest to limited segments of the public wishing to subscribe to them (such as background music, special time signals, or storecasting), and (2) signals directly related to the operation of FM broadcast stations (such as relaying of broadcast material to other stations or remote-control telemetering functions associated with STL operation). The SCA signals are multiplexed on the main channel by frequency modulation of subcarriers in the range of 20 and 75 kilohertz for a monophonic station, and in the range of 53 and 75 kilohertz for a stereophonic station. Detailed standards are in Section 73.319 of the FCC Rules and Regulations; 1972. (SCA transmissions can include "visual transmission," which is the transmission of signal which, when used with appropriate receiving apparatus, permits visual presentation of the information transmitted, as on a viewing screen or a graphic record.)

Stereophonic transmission by FM stations uses the main channel and a stereophonic subchannel. The standards as set forth by the FCC in paragraph 73.322 of the Rules and Regulations are abstracted below. It is important to recognize that a requirement of the standard is that monophonic receivers must receive monophonic or stereophonic programs without degradation; that is, the systems must be compatible.

The main channel refers to the band of frequencies from 50 to 15 000 hertz that frequency modulate the main carrier. The stereophonic sub-

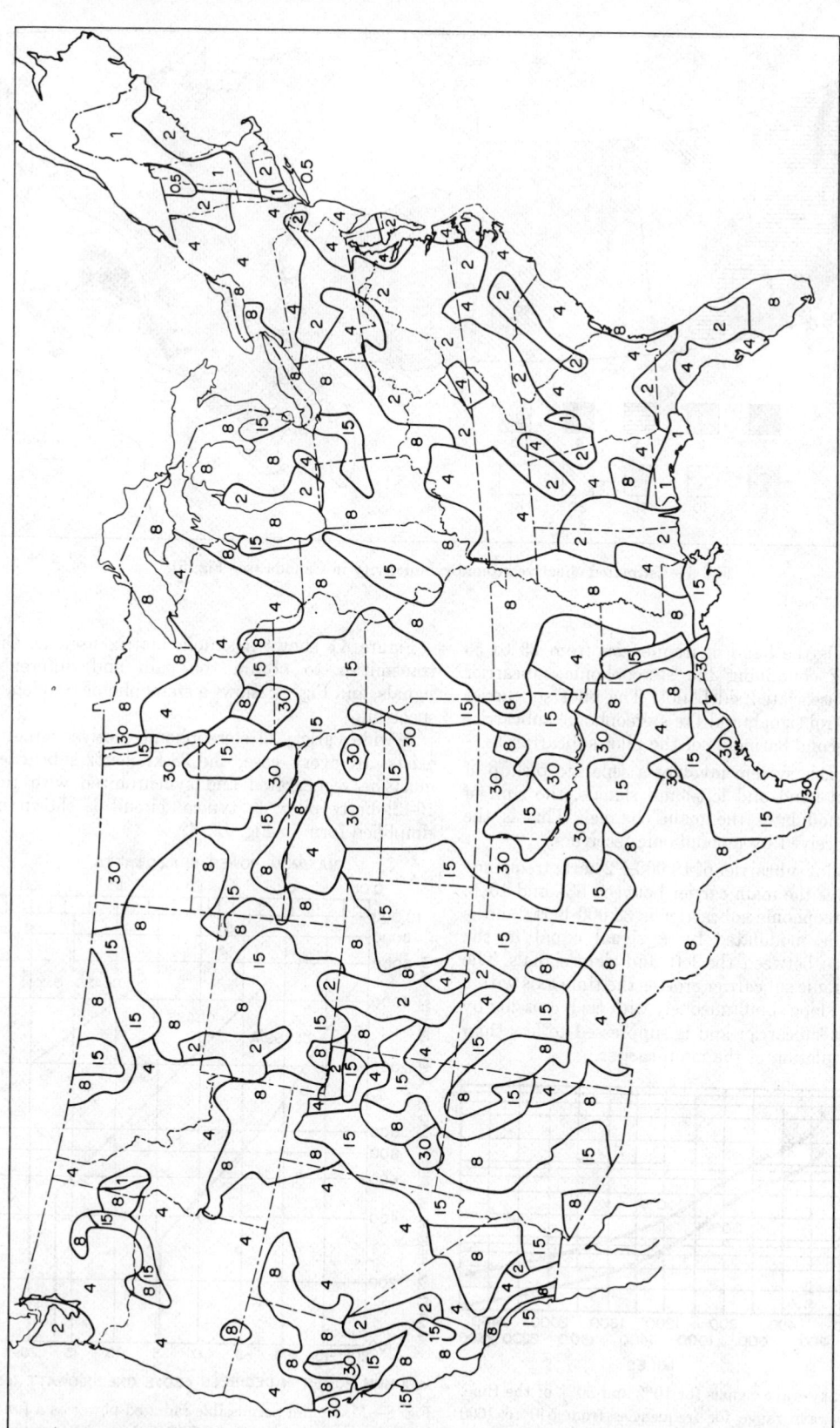

Fig. 5—Estimated effective ground conductivity in the United States. The numbers are in millimhos/meter. The conductivity of sea water (not shown) is assumed to be 5000 millimhos/meter.

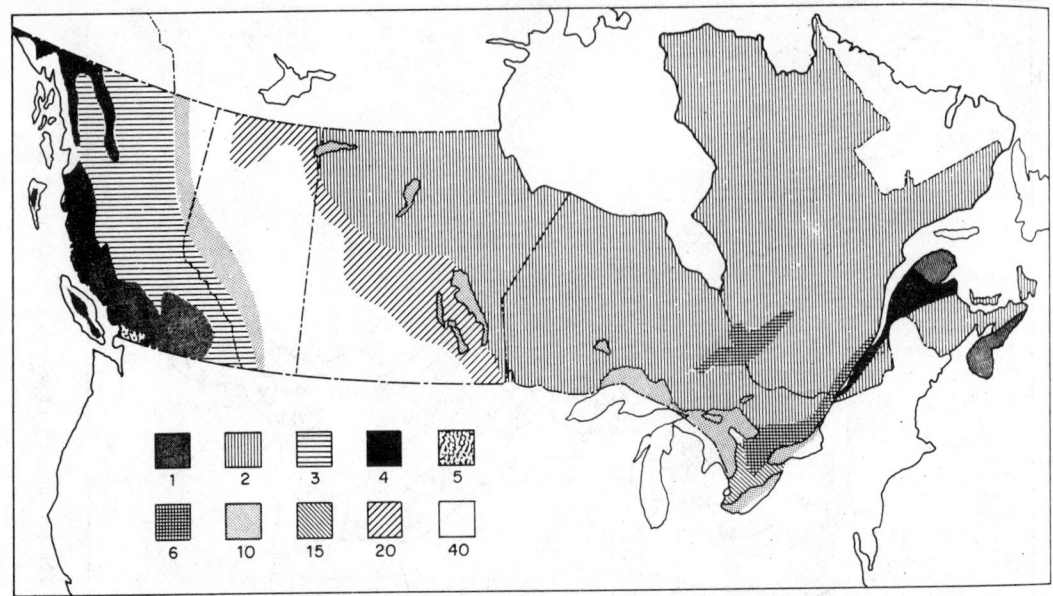

Fig. 6—Estimated effective ground conductivity in Canada (see Fig. 5).

channel is the band of frequencies from 23 to 53 kilohertz containing the stereophonic subcarrier and its associated sidebands. The pilot subcarrier is a control signal, and the stereophonic subcarrier is the second harmonic of the pilot subcarrier.

The basic system involves a separate pickup of the right-hand and left-hand signals, the sum of which modulates the main channel. This is the signal received by monophonic receivers.

The pilot subcarrier of $19\,000 \pm 2$ hertz frequency modulates the main carrier between 8% and 10%. The stereophonic subcarrier is 38 000 hertz and is amplitude modulated by a signal equal to the difference between the left and right signals. The stereophonic subcarrier crosses the time axis with a positive slope simultaneously with each crossing by the pilot subcarrier and is suppressed to less than 1% modulation of the main carrier.

Figure 11 shows a simple matrix used at the transmitter to obtain the sum and difference signals, and Fig. 12 shows a stereophonic frequency spectrum.

A wide variety of stereophonic receiver circuits exists. In every case, the 38-kilohertz subcarrier must be regenerated and synchronized with the 19-kilohertz pilot. A typical circuit is shown in simplified form in Fig. 13.

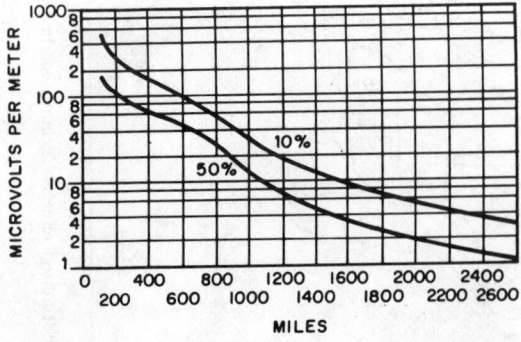

Fig. 7—Sky-wave signals for 10% and 50% of the time. The sky-wave range for frequencies from 540 to 1600 kilohertz is based on a radiated field of 100 millivolts/meter at 1 mile at the pertinent vertical angle.

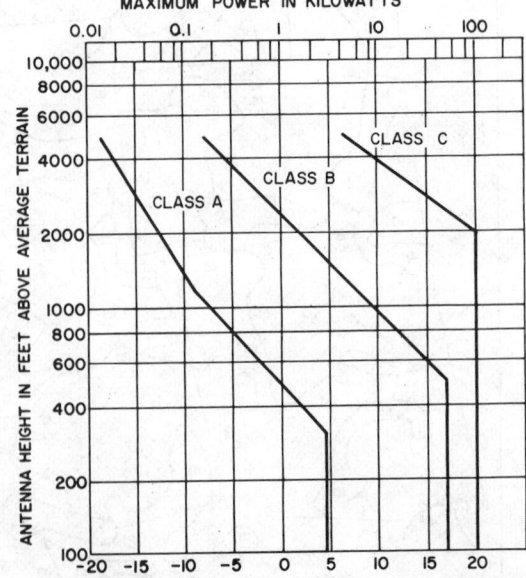

Fig. 8—Maximum permissible radiated power as a function of antenna height. *From FCC Rules and Regulations, Vol. III, Part 73, p. 197; 1972.*

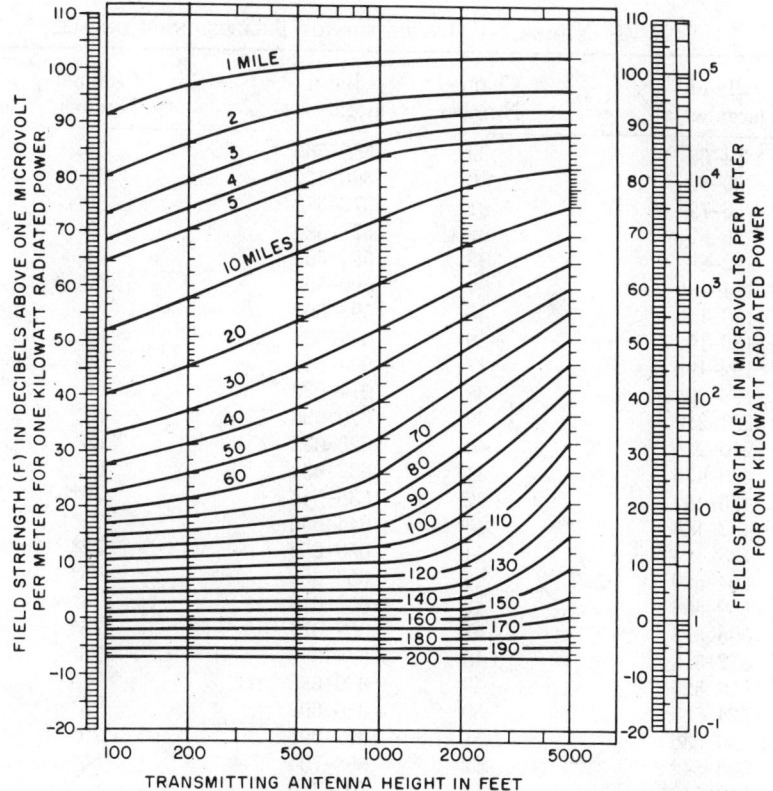

Fig. 9—F (50,50) FM channels, showing estimated field strength exceeded at 50% of the potential receiver locations for at least 50% of the time at a receiving antenna height of 30 feet. *From FCC Rules and Regulations, Vol. III, Part 73, p. 191; 1972.*

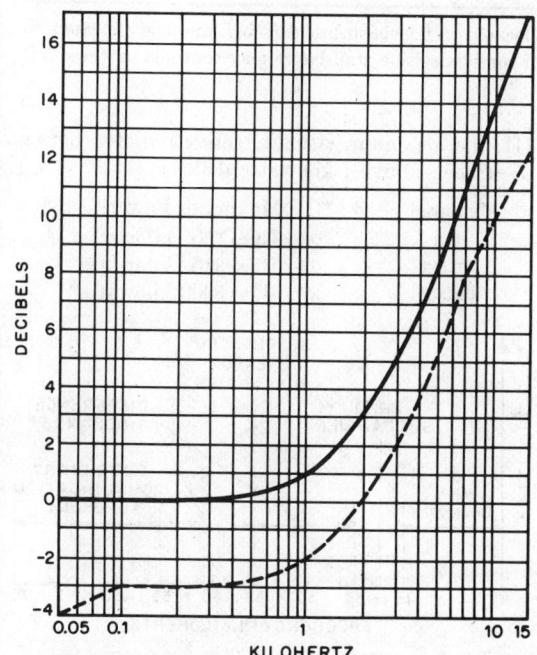

Fig. 10—Standard pre-emphasis curve for frequency-modulation and television aural broadcasting. Time constant = 75 microseconds (solid line). Frequency-response limits are set by the two lines.

TELEVISION BROADCASTING*

Channel Designations

Television broadcast stations are authorized for commercial and educational operation on the channels shown in Table 3. Assignment of channels to specific communities is made by the FCC, and the channels are designated as commercial or educational.

Coverage Data

The channel assignments have been made in such a manner as to facilitate maximum interference-free coverage in the available frequency bands. The radiated power of a particular station is fixed by several considerations.

Minimum Power is 100 watts effective visual radiated power. No minimum antenna height is specified.

* FCC Rules and Regulations, Part 73, Subpart E; 1972.

TABLE 3—NUMERICAL DESIGNATION OF TELEVISION CHANNELS.

Channel Number	Band (megahertz)	Channel Number	Band (megahertz)	Channel Number	Band (megahertz)
2	54–60	29	560–566	57	728–734
3	60–66	30	566–572	58	734–740
4	66–72	31	572–578	59	740–746
5	76–82	32	578–584	60	746–752
6	82–88	33	584–590	61	752–758
7	174–180	34	590–596	62	758–764
8	180–186	35	596–602	63	764–770
9	186–192	36	602–608	64	770–776
10	192–198	37	608–614	65	776–782
11	198–204	38	614–620	66	782–788
12	204–210	39	620–626	67	788–794
13	210–216	40	626–632	68	794–800
14	470–476	41	632–638	69	800–806
15	476–482	42	638–644	70*	806–812
16	482–488	43	644–650	71*	812–818
17	488–494	44	650–656	72*	818–824
18	494–500	45	656–662	73*	824–830
19	500–506	46	662–668	74*	830–836
20	506–512	47	668–674	75*	836–842
21	512–518	48	674–680	76*	842–848
22	518–524	49	680–686	77*	848–854
23	524–530	50	686–692	78*	854–860
24	530–536	51	692–698	79*	860–866
25	536–542	52	698–704	80*	866–872
26	542–548	53	704–710	81*	872–878
27	548–554	54	710–716	82*	878–884
28	554–560	55	716–722	83*	884–890
		56	722–728		

* The frequencies between 806 and 890 MHz, formerly allocated to television broadcasting, are now allocated to the land mobile services. Operation, on a secondary basis, of some television translators may continue on these frequencies.

Maximum Power: (See Figs. 14 and 15.) Except as limited by antenna heights in excess of 1000 feet (2000 feet for channels 14-83) in Zone I and antenna heights in excess of 2000 feet in Zones II and III, the maximum visual effective radiated power in decibels above 1 kilowatt (dBk) is:

Channel	Maximum Power
2–6	20 dBk = 100 kilowatts
7–13	25 dBk = 316 kilowatts
14–83	37 dBk = 5000 kilowatts*

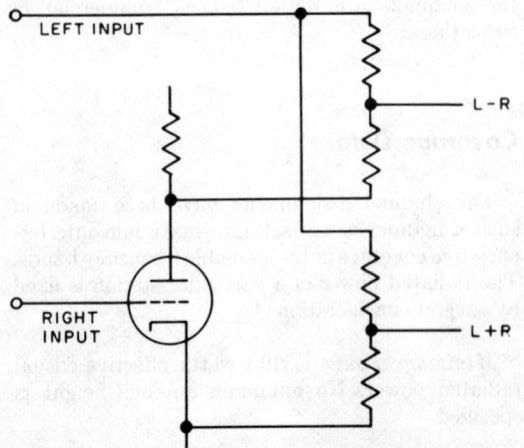

Fig. 11—Matrix for sum and difference signals.

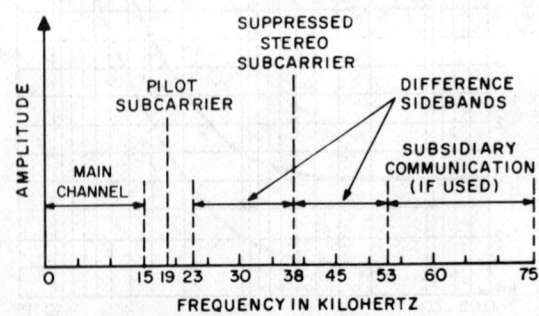

Fig. 12—Resulting stereophonic frequency spectrum.

* Limited to 1000 kilowatts if station is within 250 miles of Canadian border.

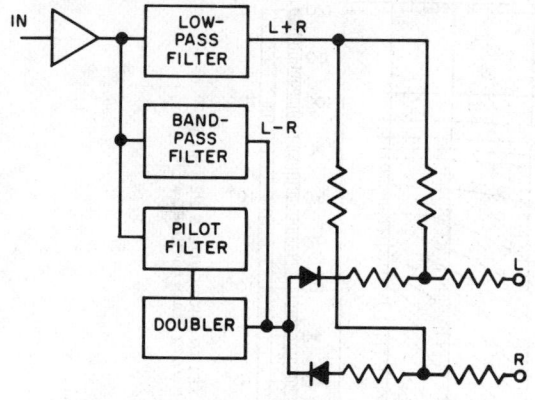

Fig. 13—Stereophonic receiver circuit.

Zone I is the same as Zone I for FM allocations. Zone II includes Puerto Rico, Alaska, the Hawaiian Islands, the Virgin Islands, and other parts of the US not in Zones I and III. Zone III is essentially a strip along the southeastern border of the US from Florida to Texas. Detailed descriptions of the zones are in the FCC Rules and Regulations, Vol. III, Section 73.609.

Grade of Service: Grades of service are designated Grade *A* and Grade *B*. The signal strength in

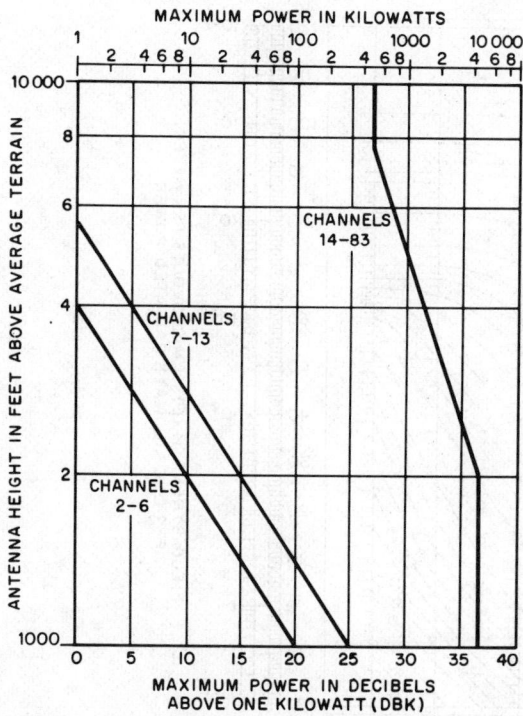

Fig. 14—Maximum television-station power versus antenna height for Zone I. *From FCC Rules and Regulations, Vol. III, Part 73, p. 275; 1972.*

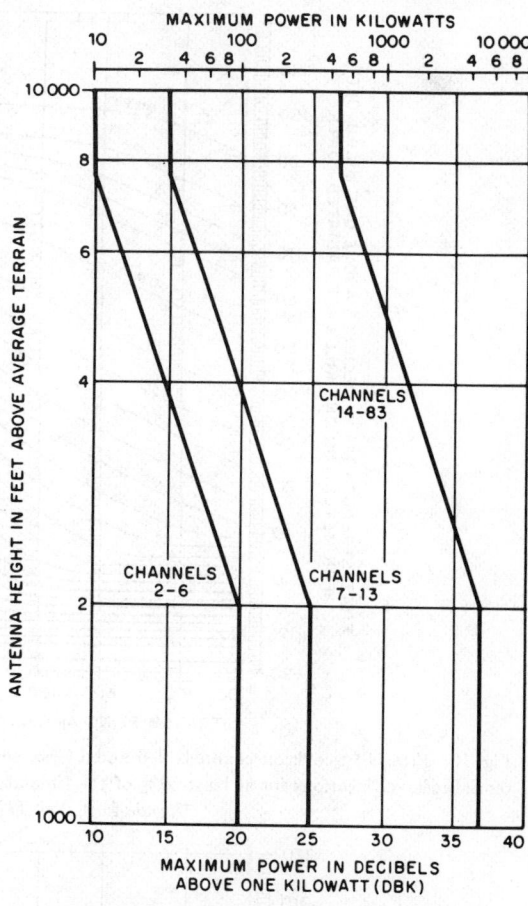

Fig. 15—Maximum television-station power versus antenna height for Zones II and III. *From FCC Rules and Regulations, Vol. III, Part 73, p. 277; 1972.*

decibels above 1 microvolt/meter (dBμ) specified for each service is:

Channel	Grade A	Grade B
2–6	68 dBμ = 2510 μV/m	47 dBμ = 224 μV/m
7–13	71 dBμ = 3550 μV/m	56 dBμ = 631 μV/m
14–83	74 dBμ = 5010 μV/m	64 dBμ = 1585 μV/m

Transmitter Location: The transmitter location must be so chosen that on the basis of effective radiated power and antenna height, the following minimum field strength in decibels above 1 microvolt/meter will be provided over the principal community to be served.

Channel	Signal
2–6	74 dBμ = 5010 μV/m
7–13	77 dBμ = 7080 μV/m
14–83	80 dBμ = 10 000 μV/m

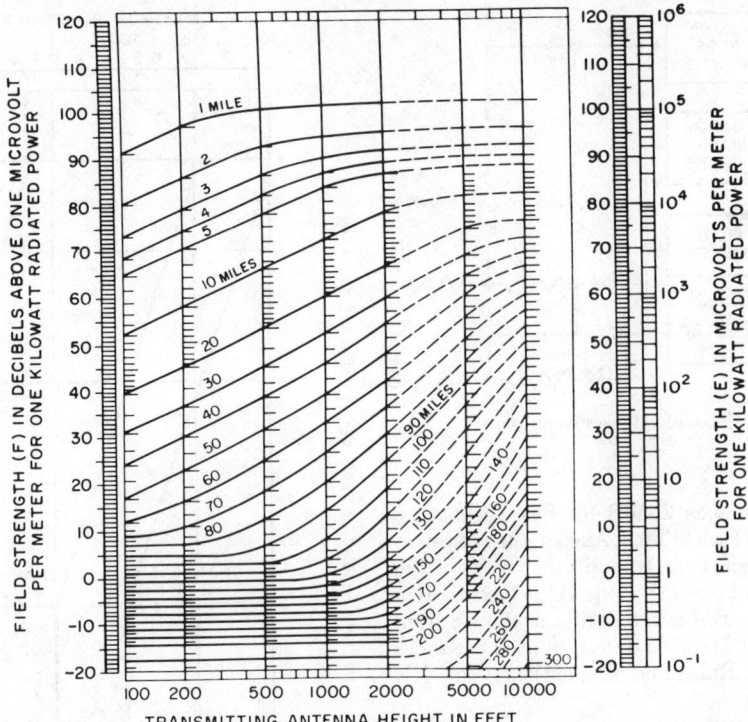

Fig. 16—$F(50,50)$ television channels 2–6 and 14–83, showing estimated field strength exceeded at 50% of the potential receiver locations for at least 50% of the time at a receiving antenna height of 30 feet. *From FCC Rules and Regulations, Vol. III, Part 73, p. 289; 1972.*

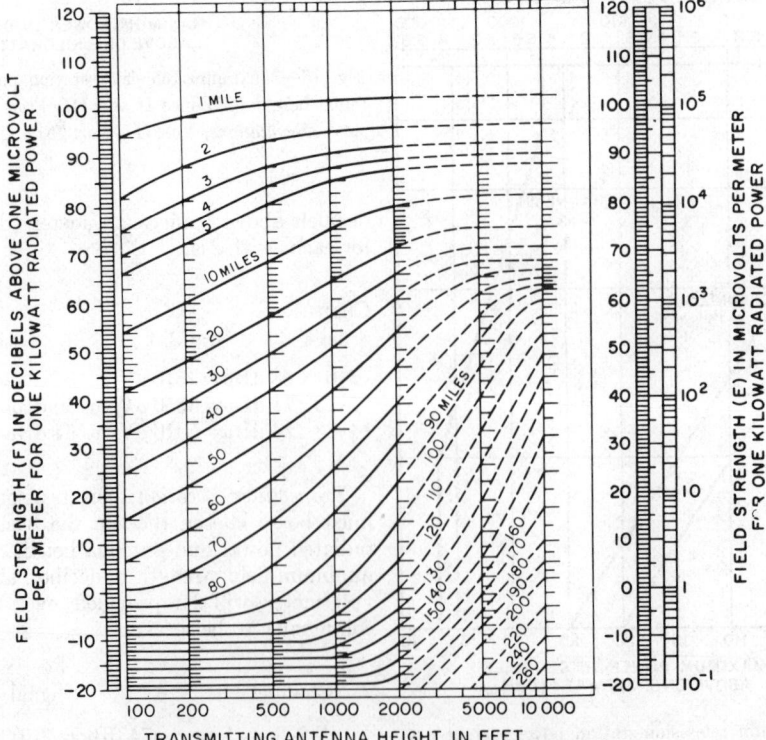

Fig. 17—$F(50,50)$ television channels 7–13, showing estimated field strength exceeded at 50% of the potential receiver locations for at least 50% of the time at a receiving antenna height of 30 feet. *From FCC Rules and Regulations, Vol. III, Part 73, p. 291; 1972.*

The curves of Figs. 16 and 17 give estimated field strengths for the television channels at different heights and powers. The antenna height is the height of the radiation center of the antenna above average terrain. Average terrain is determined by the elevations between 2 and 10 miles from the antenna site, taken along 8 radials separated by 45° in azimuth. Effective radiated power is the product of the antenna gain and the antenna input power. Antenna input power is the peak visual output power of the transmitter less transmission-line and diplexer losses. The procedures to be followed in determining the effective radiated power to be used in the prediction of coverage are detailed in Section 73.684 of the FCC Rules and Regulations.

Directional antennas may be employed to improve service. The ratio of maximum to minimum radiation in the horizontal plane shall not exceed 10 decibels for channels 2–13, and 15 decibels for channels 14–83 if the transmitter power is more than 1 kilowatt. There is no restriction for channels 14–83 if the transmitted power is 1 kilowatt or less.

Transmission Standards

The standards for television transmission in the US, as defined by the FCC, are:

Channel Width: 6 megahertz.

Picture Carrier Location: 1.25 megahertz ±1000 hertz above lower boundary of the channel.*

Aural Center Frequency: 4.5 megahertz ±1000 hertz above visual carrier.

Polarization of Radiation: Horizontal.

* The table of assignments specifies that certain stations operate with carrier frequencies offset 10 kilohertz above or below the normal carrier frequencies. This is done to minimize co-channel interference.

Modulation: Amplitude-modulated composite picture and synchronizing signal on visual carrier, together with frequency-modulated audio signal on aural carrier. Figure 18 shows the channel frequency spectrum, and Fig. 19 shows the video waveform for color transmission. (Monochrome waveform is similar; the principal difference is the absence of burst.)

Scanning Lines: 525 lines/frame interlaced two to one.

Scanning Sequence: Horizontally from left to right, vertically from top to bottom.

Horizontal Scanning Frequency: 2/455 times chrominance subcarrier frequency (15 734.264± 0.044 hertz). (For monochrome only, nominal value of 15 750 hertz may be used.)

Vertical Scanning Frequency: 2/525 times the horizontal scanning frequency (59.94 hertz). (For monochrome only, the nominal value of 60 hertz may be used.)

Aspect Ratio: 4 units horizontal, 3 units vertical.

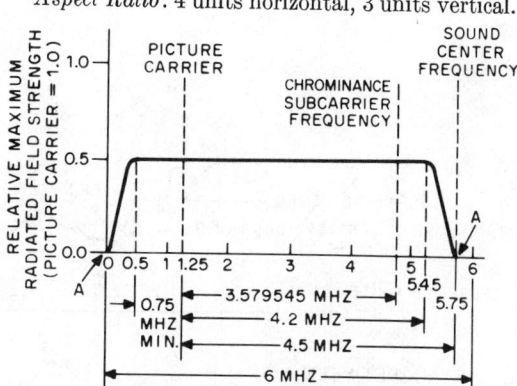

CHANNEL FREQUENCY SPECTRUM IN MEGAHERTZ
REFERRED TO LOWER FREQUENCY LIMIT OF CHANNEL

(A) FIELD STRENGTH AT POINTS A SHALL NOT BE GREATER THAN –20 dB.

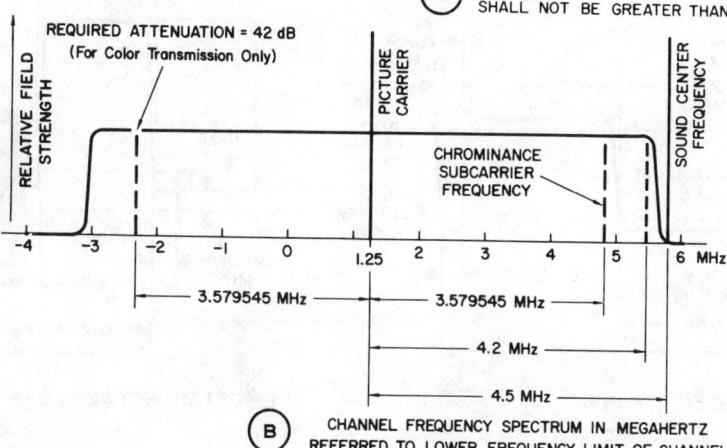

CHANNEL FREQUENCY SPECTRUM IN MEGAHERTZ
REFERRED TO LOWER FREQUENCY LIMIT OF CHANNEL

Fig. 18—Radio-frequency amplitude characteristics of television picture transmission. Drawings not to scale. (A) Stations other than those for which B applies. (B) Stations on channels 15–83 with transmitter peak visual power of 1 kilowatt or less.

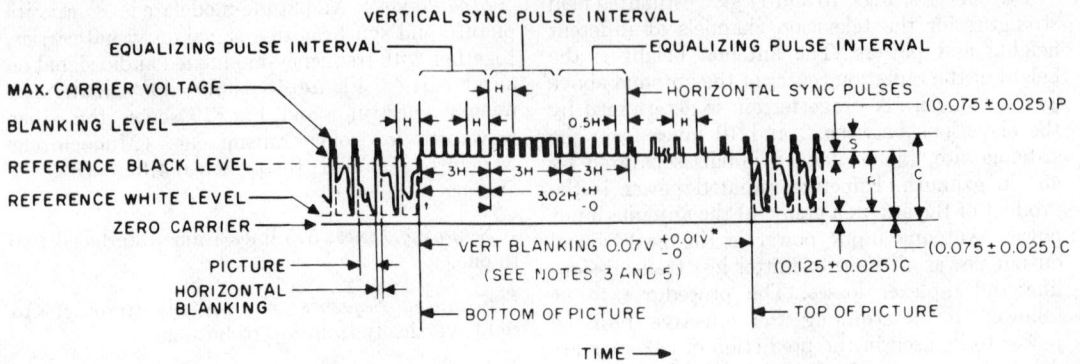

A   FIELD I

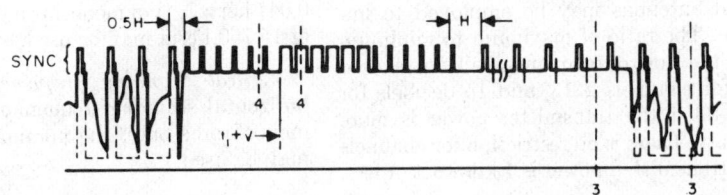

B   FIELD 2

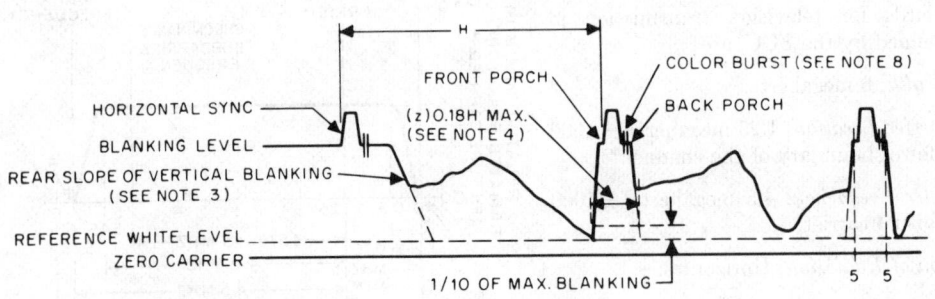

C – DETAIL BETWEEN 3-3 IN B

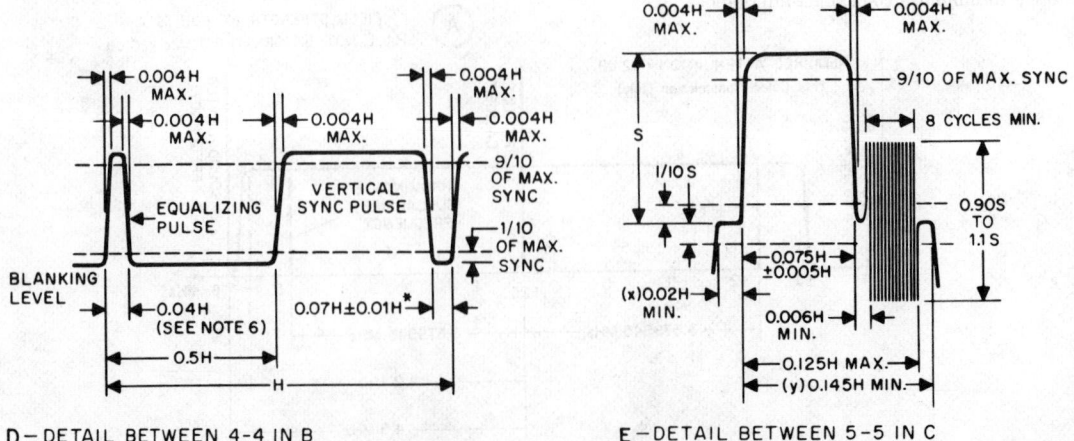

D – DETAIL BETWEEN 4-4 IN B   E – DETAIL BETWEEN 5-5 IN C

Fig. 19—Television composite-signal waveform data. (See notes at right).

Chrominance Subcarrier Frequency: 3.579545 megahertz ±10 hertz.

Blanking Level: Shall be transmitted at 75±2.5 percent of the peak carrier level.

Reference Black Level: Black level is separated from the blanking level by 7.5±2.5 percent of the video range from blanking level to reference white level.

Reference White Level: Luminance signal of reference white is 12.5±2.5 percent of peak carrier.

Peak-to-Peak Variation: Peak-to-peak variation of transmitter output in one frame due to all causes must not exceed 5 percent.

Polarity of Transmission: Negative—a decrease in initial light intensity causes an increase in radiated power.

Transmitter Brightness Response: For monochrome transmission, radio-frequency output varies in an inverse logarithmic relation to the brightness of the scene.

Aural-Transmitter Power: Maximum radiated power is 20 percent (minimum, 10 percent) of peak visual transmitter power.

For color transmission the luminance component shall be transmitted as amplitude modulation of the picture carrier and the chrominance components as amplitude-modulation sidebands of a pair of suppressed subcarriers in quadrature.*

* See also J. W. Wentworth, "Color Television Engineering," McGraw-Hill Book Company, New York; 1955.

Color Signal: The equation of the complete color signal is

$$E_M = E_Y'$$
$$+ [E_Q' \sin(\omega t + 33°) + E_I' \cos(\omega t + 33°)]$$

where

$$E_Q' = +0.41(E_B' - E_Y') + 0.48(E_R' - E_Y')$$

$$E_I' = -0.27(E_B' - E_Y') + 0.74(E_R' - E_Y')$$

$$E_Y' = +0.30E_R' + 0.59E_G' + 0.11E_B'.$$

For color-difference frequencies below 500 kilohertz, the signal can be represented by

$$E_M = E_Y' + \left\{ \frac{1}{1.14} \left[\frac{1}{1.78}(E_B' - E_Y') \sin\omega t \right. \right.$$
$$\left. \left. + (E_R' - E_Y') \cos\omega t \right] \right\}.$$

The symbols have the following significance.

E_M = total video voltage, corresponding to the scanning of a particular picture element, applied to the modulator of the picture transmitter

E_Y' = gamma-corrected voltage of the monochrome (black-and-white) portion of the color-picture signal, corresponding to the given picture element

E_Q', E_I' = amplitudes of two orthogonal components of the chrominance signal, corresponding respectively to narrow-band and wide-band axes

---

Notes:

1. H = time from start of one line to start of next line.
2. V = time from start of one field to start of next field.
3. Leading and trailing edges of vertical blanking should be complete in less than $0.1H$.
4. Leading and trailing slopes of horizontal blanking must be steep enough to preserve minimum and maximum values of $(x+y)$ and z under all conditions of picture content.
5. Dimensions marked with an asterisk indicate that tolerances given are permitted only for long-time variations, and not for successive cycles.
6. Equalizing pulse area shall be between 0.45 and 0.5 of the area of a horizontal synchronizing pulse.
7. Color burst follows each horizontal pulse, but is omitted following the equalizing pulses and during the broad vertical pulses.
8. Color bursts to be omitted during monochrome transmission.
9. The burst frequency shall be 3.579545 megahertz. The tolerance on the frequency shall be ±10 hertz with a maximum rate of change of frequency not to exceed 1/10 hertz per second.
10. The horizontal scanning frequency shall be 2/455 times the burst frequency.
11. The dimensions specified for the burst determine the times of starting and stopping the burst but not its phase. The color burst consists of amplitude modulation of a continuous sine wave.
12. Dimension P represents the peak excursion of the luminance signal from blanking level but does not include the chrominance signal. Dimension S is the synchronizing amplitude above blanking level. Dimension C is the peak carrier amplitude.
13. Refer to FCC standards for further explanations and tolerances.
14. Horizontal dimensions not to scale in A, B, and C.

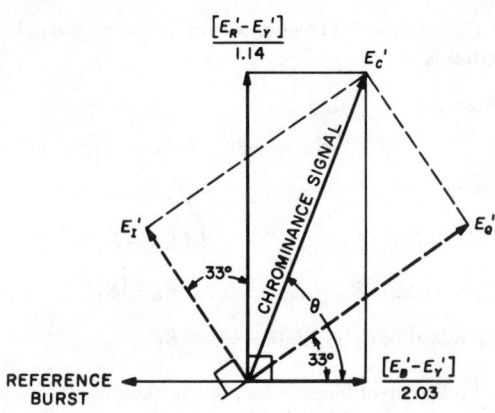

Fig. 20—Phases of color signal.

E_R', E_G', E_B' = gamma-corrected voltages corresponding to red, green, and blue signals during the scanning of the given picture element

ω = angular frequency = 2π times the frequency of the chrominance subcarrier.

The portion of each expression between brackets represents the chrominance subcarrier signal that carries the chrominance information.

The phase reference in the E_M equation is the phase of the burst +180°, as shown in Fig. 20. The burst corresponds to amplitude modulation of a continuous sine wave.

The equivalent bandwidth assigned before modulation to the color difference signals E_Q' and E_I' are as follows:

Q-channel bandwidth:

At 400 kilohertz, less than 2 decibels down.
At 500 kilohertz, less than 6 decibels down.
At 600 kilohertz, at least 6 decibels down.

I-channel bandwidth:

At 1.3 megahertz, less than 2 decibels down.
At 3.6 megahertz, at least 20 decibels down.

The gamma-corrected voltages E_R', E_G', and E_B' are suitable for a color-picture tube having primary colors with the chromaticities listed below in the CIE (Commission Internationale de l'Éclairage) system of specification, and having a transfer gradient (gamma exponent) of 2.2 associated with each primary color. The voltages E_R', E_G', and E_B' may be, respectively, of the form $E_R^{1/\gamma}$, $E_G^{1/\gamma}$, and $E_B^{1/\gamma}$, although other forms may be used with advances in the state of the art.

Color	x	y
Red (R)	0.67	0.33
Green (G)	0.21	0.71
Blue (B)	0.14	0.08

The radiated chrominance subcarrier vanishes on the reference white of the scene. The numerical values of the signal specification assume that this condition will be produced as CIE Illuminant C ($x = 0.310$, $y = 0.316$).

E_Y', E_Q', E_I', and the components of these signals shall match each other in time to 0.05 microsecond.

The angles of the subcarrier measured with respect to the burst phase, when reproducing saturated primaries and their complements at 75 percent of full amplitude, shall be within ±10 degrees and the amplitudes within ±20 percent of the values specified above. The ratios of the measured amplitudes of the subcarrier to the luminance signal for the same saturated primaries and their complements must fall between the limits of 0.8 and 1.2 of the values specified for their ratios.

The interval beginning with line 17 and continuing through line 20 of the vertical blanking interval of each field may be used for the transmission of test signals and cue and control signals. Test signals may include signals used to supply reference modulation levels so that variations in light intensity of the scene viewed by the camera will be faithfully transmitted, and signals designed to check the performance of the overall transmission system or its individual components. Cue and control signals shall be related to the operation of the television broadcast station. (Field 1 line numbers start with first equalizing pulse in field 1; field 2 line numbers start with second equalizing pulse in field 2.) Modulation of the television transmitter by such signals shall be confined to the area between the reference white level and the blanking level except where test signals include chrominance subcarrier frequencies, in which case positive excursions of chrominance components may exceed reference white and negative excursions may extend into the synchronizing area. In no case may the modulation excursions produced by test signals extend beyond peak-of-sync or to zero carrier level. Test signals or cue and control signals may not be transmitted during that portion of each line devoted to horizontal blanking. Line 19 in each field may be used only for transmission of the standard vertical interval reference (VIR) signal.

The intervals within the first and last 10 microseconds of lines 22 through 24 and 260 through 262 (on a "field" basis) may contain coded patterns for the purpose of electronic identification of television broadcast programs and spot announcements. No single transmission of such coded patterns shall exceed one second in duration.

Multiplexing of the aural carrier may be employed for the purpose of transmitting telemetry and alerting signals from the transmitter site to the control point of a television broadcast station. Multiplexing is limited to the use of a single subcarrier in the range 20 to 50 kHz. The maximum

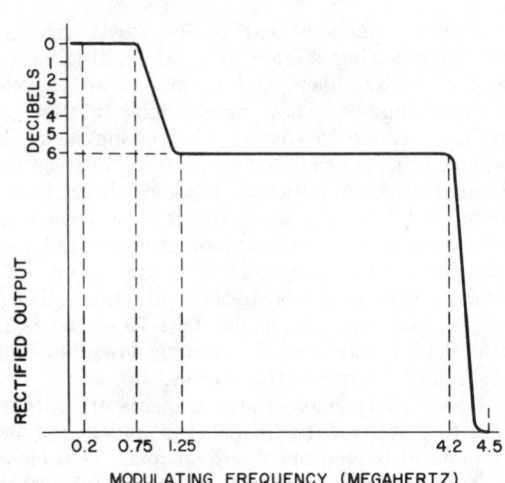

Fig. 21—Ideal demodulated amplitude characteristic of television transmitter.

modulation of the aural carrier by the subcarrier shall not exceed 10 percent of the maximum permissible degree of modulation.

Visual Transmitter Design*

Overall Frequency Response: For monochrome, the output measured in the antenna transmission line after vestigial-sideband filters shall be within limits of +0 and

- −2 decibels at 0.5 megahertz
- −2 decibels at 1.25 megahertz
- −3 decibels at 2.0 megahertz
- −6 decibels at 3.0 megahertz
- −12 decibels at 3.5 megahertz

with respect to the video amplitude characteristic of Fig. 21.

For color transmission, additional requirements are that a 3.58-megahertz sine wave shall be 6±2 decibels below a 200-kilohertz sine wave, between 2.1 and 4.1 megahertz the amplitude shall be within ±2 decibels of its value at 3.58 megahertz, and the amplitude shall be no more than 4 decibels below that value at 4.18 megahertz.

For modulating frequencies of 1.25 megahertz or greater, lower-sideband radiation must be 20 decibels below carrier level. In addition, the radiation of the lower sideband due to modulation by the color subcarrier (3.579545 megahertz) must be attenuated by a minimum of 42 decibels. For monochrome and color, the field strength of the upper sideband for a modulating frequency of 4.75

---

* See also EIA standard RS240, "Electrical Performance Standards for Television Broadcast Transmitters."

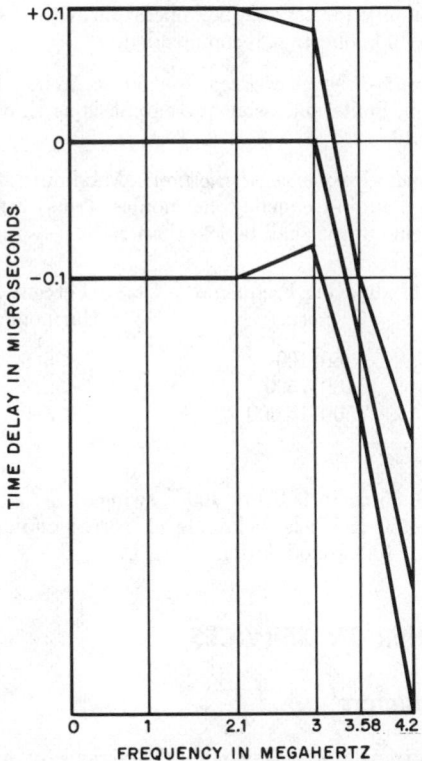

Fig. 22—Envelope-delay curve for television transmitter.

megahertz or greater shall be attenuated at least 20 decibels.

Envelope Delay: The modulated radiated signal shall have an envelope delay relative to the average envelope delay between 0.05 and 0.2 megahertz of zero microseconds up to a frequency of 3.0 megahertz; then linearly decreasing to 4.18 megahertz, being 0.17 microsecond at 3.58 megahertz. The tolerance on the envelope delay is ±0.05 microsecond at 3.58 megahertz, linearly increasing to ±0.1 microsecond down to 2.1 megahertz and up to 4.18 megahertz, and remaining at ±0.1 microsecond down to 0.2 megahertz. See Fig. 22.

Horizontal Pulse-Timing Variations: Variation of the time interval between successive pulse leading edges shall be less than 0.5 percent of the average interval.

Horizontal Pulse-Repetition Stability: Rate of change of leading-edge recurrence frequency shall not exceed 0.15 percent/second.

Aural Transmitter

Modulation: Satisfactory operation must be maintained with a frequency swing of ±25 kilohertz, which is defined as 100% modulation. A capability for satisfactory operation with a swing of ±40 kilohertz is recommended.

Audio-Frequency Response: 50 to 15 000 hertz within limits and using pre-emphasis as shown in Fig. 10.

Audio-Frequency Distortion: Maximum combined audio-frequency harmonics (rms) at the system output shall be less than

Modulating Frequency (hertz)	Percent Harmonic
50–100	<3.5
100–7500	<2.5
7500–15 000	<3.0

Noise:

FM—55 decibels below 100% swing.
AM—50 decibels below level corresponding to 100% modulation.

OTHER TV SERVICES

Translators*

A translator retransmits the signals of a TV broadcast station by frequency conversion and amplification, without significantly altering any characteristic except the amplitude and frequency.

Translators may be authorized to operate in the very-high-frequency TV band provided that no interference is caused to the direct reception of any station operating on the same or an adjacent channel. Ultra-high-frequency translators may be authorized on channels 55 through 69 provided certain mileage separations are maintained.

Very-high-frequency translators may not exceed 1 watt peak visual power (10 watts west of the Mississippi River) except where authorized in a community listed in the table of assignments, in which case they must be operated with a power of 100 watts. Ultra-high-frequency translators shall be limited to 100 watts peak visual power output, except that on channels listed in the table of assignments, the authorized peak visual output is either 100 watts or 1000 watts.

No limit is placed on the effective radiated power that can be obtained by using directive antennas.

Any emissions more than 3 megahertz outside the channel shall be −30 decibels for transmitters rated at 1 watt or less, −50 decibels for transmitters rated at more than 1 watt, and −60 decibels for transmitters rated at more than 100 watts.

* FCC Rules and Regulations, Part 74, Subpart G; 1972.

Cable Television (CATV) Systems

These systems consist of an advantageously located receiving antenna array and a distribution system of amplifiers and coaxial cable to subscribers' homes. They were originally used to provide service to communities having poor TV service or none at all, but today many of them are located in communities having good direct reception. CATV systems may provide improved reception without the use of outdoor antennas, and may also provide a greater choice of programs (up to 20 or more VHF channels) to their subscribers. These systems are regulated under Part 76 of the FCC Rules and Regulations. The technical requirements of Part 76 are summarized below.

Unless other channel arrangements are authorized to a system, the frequency boundaries of the channels delivered to subscribers must be as listed in Table 3. The frequency of the visual carrier shall be 1.25 MHz ±25 kHz above the lower boundary of the channel, except that, in those systems that supply subscribers with a converter in order to facilitate delivery of cable television channels, the frequency of the visual carrier at the output of the converter shall be 1.25 MHz ±250 kHz above the lower frequency boundary of the channel. The frequency of the aural carrier shall be 4.5 MHz ±1 kHz above the frequency of the visual carrier.

The visual signal level, across a terminating impedance which correctly matches the internal impedance of the cable system as viewed from the subscriber terminals, shall be not less than 1 millivolt across 75 ohms or 2 millivolts across 300 ohms. (At other impedance values, the minimum visual signal level shall be $(0.0133 \, Z)^{1/2}$ millivolts, where Z is the appropriate impedance value.)

The visual signal level on each channel shall not vary more than 12 decibels within any 24-hour period. It shall be maintained within 3 decibels of the visual signal level of any visual carrier within 6-MHz nominal frequency separation, and within 12 decibels of the visual signal level on any other channel. The signal level must not be great enough to cause signal degradation due to overload in the subscriber's receiver.

The peak-to-peak variation in visual signal level caused by undesired low-frequency disturbances (hum or repetitive transients) generated within the system, or by inadequate low-frequency response, shall not exceed 5% of the visual signal level.

The rms voltage of the aural signal shall be maintained between 13 and 17 decibels below the associated visual signal level.

Between 0.5 and 5.25 MHz above the lower channel limit, the amplitude characteristic shall be within a range of ±2 decibels relative to the level at 1.25 MHz above the lower channel limit.

For each signal which is delivered by a cable television system to subscribers within the pre-

TABLE 4—RADIATION LIMITS FOR CATV SYSTEMS

Frequencies	Radiation limit (microvolts/ meter)	Distance (feet)
Up to and including 54 MHz	15	100
Over 54 up to and including 216 MHz	20	10
Over 216 MHz	15	100

dicted Grade B contour for that signal, or each signal which is first picked up within its predicted Grade B contour, the ratio of visual signal level to system noise, and of visual signal level to any undesired co-channel television signal operating on proper offset assignment, shall be not less than 36 decibels.

The ratio of visual signal level to the rms amplitude of any coherent disturbances such as intermodulation products or discrete-frequency interfering signals not operating on proper offset assignments shall not be less than 46 decibels.

The terminal isolation provided each subscriber shall be not less than 18 decibels and shall be sufficient to prevent reflections caused by open-circuited or short-circuited subscriber terminals from producing visible picture impairments at any other subscriber terminal.

Radiation from a cable television system shall be limited as in Table 4.

In many cases, signals are transmitted from a distant reception point to a CATV system by microwave relay stations operating in the Cable Television Relay Service.* Channel assignments are in the band 12.7-12.95 GHz.

The transmitter peak output power shall not be greater than necessary, and in any event, shall not exceed 5 watts on any channel; except that stations using frequency modulation to transmit a baseband of frequency-division multiplexed standard television signals may be authorized to use peak power of 15 watts, 30 watts, or 60 watts, depending on the frequency assignments. Local distribution service (LDS) stations shall use for the visual signal either vestigial sideband AM transmission or frequency-division multiplexed FM transmission. When vestigial sideband AM transmission is used, the peak power of the visual signal on all channels shall be maintained within 2 decibels of equality. The mean power of the aural signals on each channel shall not exceed a level 7 decibels below the peak power of the visual signal. In no case shall the power delivered by a transmitter to the antenna system in the frequency band between 12.70 and 12.75 GHz exceed $+10$ dBW. Additionally, the

maximum equivalent isotropically radiated power of a station shall not exceed $+55$ dBW.

A cable television relay station may be authorized to employ any type of emission (if included in Part 78) suitable for the simultaneous transmission of visual and aural television signals. Any emission appearing on a frequency outside of the channel authorized for a transmitter shall be attenuated below the power of the emission in accordance with the following limits. For stations using FM or double sideband AM transmission: on any frequency outside the channel limits by between zero and 50% of the authorized channel width, at least 25 decibels below the mean power of the emission; on any frequency outside the channel limits by more than 50% and up to 150% of the authorized channel width, at least 35 decibels below the mean power of the emission; and on any frequency outside the channel limits by more than 150% of the authorized channel width, at least $43+10 \log_{10}$ (power in watts) decibels below the mean power of the emission. For stations using vestigial sideband AM transmission, at least 50 decibels below the peak power of the emission.

Cable television relay stations shall use directive transmitting antennas. Normally, the maximum beamwidth in the horizontal plane between half power points of the major lobe shall not exceed 3°. However, if a need to serve a larger sector, or more than a single sector, can be shown, greater beamwidth or multiple antennas may be authorized for LDS stations. Either vertical, horizontal, or elliptical polarization may be employed. (The Commission reserves the right to specify the polarization of the transmitted signal.) The choice of receiving antennas is left to the discretion of the licensee. However, licensees will not be protected from interference which results from the lack of adequate antenna discrimination against unwanted signals. The transmitting antenna system of stations employing maximum equivalent isotropically radiated power exceeding $+45$ dBW in the frequency band between 12.70 and 12.75 GHz shall be oriented so that the direction of maximum radiation of any antenna shall be at least 1.5° away from the geostationary satellite orbit, taking into account the effect of atmospheric refraction.

Frequency tolerances are listed in Table 5.

Instructional Television†

Instructional television, in addition to using the very- and ultra-high-frequency channels in the table of assignments, may operate on a series of channels for Instructional Television Fixed Stations. The primary purpose of these stations, rather than broadcasting to conventional TV receivers, is to direct instructional material to specified

---

* FCC Rules and Regulations, Part 78; 1972.

† FCC Rules and Regulations, Part 74, Subpart I; 1972.

receiving locations. The assigned frequencies are given in Table 6.

It is intended that an applicant may have up to four transmitters in a single area of operation and will use the channels in one of the groups listed. Directive transmitting and receiving antennas may be used to minimize interference. Power is limited to that required to perform the desired service, and a special showing is required if the proposed peak visual power exceeds 10 watts. The aural power may be between 10% and 70% of the peak visual power. Radio-frequency harmonics shall be at least 60 decibels below the peak visual output power; all other emissions more than 3 megahertz outside of the channel shall be −30 decibels for transmitters rated less than 10 watts, and −40 decibels for transmitters rated 10 watts or more. The standards for broadcast transmission shall apply except for relaxed specifications for lower-sideband attenuation, and the allowable use of horizontal, vertical, or circular polarization of antennas.

Additional Applications

An important aspect of television in recent years has been the marked increase in the use of TV equipment for "closed-circuit" applications. Here the picture and sound signals are wired directly

TABLE 5—FREQUENCY TOLERANCES FOR CABLE TELEVISION RELAY SERVICE

Type of transmission:	Tolerance
FM, including modulation by a frequency-division-multiplexed baseband of standard television signals, using 25 MHz or greater authorized bandwidth.	0.02 percent
Vestigial sideband AM:	
Visual carrier	0.0005 percent
Aural carrier	4.5 MHz±1 kHz above visual carrier frequency
FM and double sideband AM using 12.5 MHz or less authorized bandwidth.	0.005 percent [1]

[1] This tolerance shall apply to stations authorized on or after July 1, 1974, and to equipment type-accepted on or after January 1, 1974. A frequency tolerance of 0.02 percent shall apply to all other equipment in this category. Except that upon suitable showing that the occupied bandwidth (Ref. § 2.202) is sufficiently less than the channel width desired, accompanying any application for authorization, special tolerances may be specified in said authorization.

from the point of pickup to the point of reception. Two methods of transmission are used: video transmission, wherein the video signal itself is passed over coaxial lines and received on picture

TABLE 6—FREQUENCY ASSIGNMENTS OF INSTRUCTIONAL TELEVISION FIXED STATIONS.

Channel	Band Limits (megahertz)
Group A	
A–1	2500–2506
A–2	2512–2518
A–3	2524–2530
A–4	2536–2542
Group B	
B–1	2506–2512
B–2	2518–2524
B–3	2530–2536
B–4	2542–2548
Group C	
C–1	2548–2554
C–2	2560–2566
C–3	2572–2578
C–4	2584–2590
Group D	
D–1	2554–2560
D–2	2566–2572
D–3	2578–2584
D–4	2590–2596
Group E	
E–1	2596–2602
E–2	2608–2614
E–3	2620–2626
E–4	2632–2638
Group F	
F–1	2602–2608
F–2	2614–2620
F–3	2626–2632
F–4	2638–2644
Group G	
G–1	2644–2650
G–2	2656–2662
G–3	2668–2674
G–4	2680–2686
Group H*	
H–1	2650–2656
H–2	2662–2668
H–3	2674–2680

* These frequencies shared with other stations.

monitors as used in a broadcast studio (the full 6-megahertz bandwidth is available) and the sound is fed on a separate pair; or radio-frequency transmission, wherein picture and sound are modulated on a radio-frequency carrier and distributed to conventional TV receivers, as is done in a CATV system. Since the equipment used is often the same as that used in broadcasting, the standards are the same or similar. However, since no broadcasting is involved, and no licensing is required, there are no required technical standards other than those necessary for compatibility between the pickup devices and the receiving devices. Where broadcast standards have been modified, the modifications are usually in the scanning (the number of lines and frames) and in the synchronizing signal.

Some interest exists in what is known as "Pay-TV." The subscriber pays a fee to receive special or selected programs not available on broadcast channels. In some schemes these programs have been wired directly into the home using techniques similar to those of CATV. In other schemes they have been broadcast, using an experimental FCC license. In these latter cases the broadcast signal must be "scrambled" so that only paying subscribers with special devices can receive the program.

Auxiliary Broadcast Services*

There are three principal types of auxiliary broadcast services: Remote Pickup Broadcast Stations, Aural Broadcast STL (Studio Transmitter Link) and Intercity Relay Stations, and Television Auxiliary Broadcast Stations.

Remote Pickup Broadcast Stations are used to transmit aural program material and operational communications in connection with broadcasts not originating in a studio. Groups of frequencies are available in the regions of 1600 kilohertz and 26, 152, and 450 megahertz. Amplitude modulation is used below 25 megahertz; above this frequency either amplitude or frequency modulation is used.

Aural Broadcast STL and Intercity Relay Stations are used to transmit aural program material and may be licensed in any one of ten 500-kilohertz channels in the frequency band 947–952 megahertz. Frequency modulation with a swing of ±200 kilohertz is usually employed, and directional antennas are required.

Television Auxiliary Broadcast Stations include mobile stations, studio-to-transmitter links, intercity relay stations, and translator relay stations. For the first three types, channels may be assigned

* FCC Rules and Regulations, Part 74, Subparts D, E, and F; 1972.

for simultaneous transmission of picture and sound in Band *A*, Band *B*, or Band *D*. Band *A* consists of 7 channels between 1990 and 2110 megahertz, plus 3 channels between 2450 and 2500 megahertz. Band *B* consists of ten 25-megahertz channels between 6875 and 7125 megahertz. Band *D* consists of twenty-two 25-megahertz channels between 12 700 and 13 250 megahertz.

Frequencies in the bands 17 700–19 300, 19 400–19 700, 27 525–31 300, and 38 600–40 000 MHz are available for assignment on a case-by-case basis for television pickup, STL, and intercity relay purposes. Certain frequencies in the band 1990-2110 MHz may be used by translator relay stations on a secondary basis (i.e., translator relay stations must not interfere with television pickup, STL, or intercity relay stations, but must accept any interference caused by such stations).

Any suitable type of emission may be employed, and power is limited to that necessary to render satisfactory service. (Television translator relay stations are limited to amplitude modulation for picture and frequency modulation for sound, obtained by heterodyne conversion from a television broadcast station.) Out-of-band emissions, based on percentage of channel width, are limited to levels below the unmodulated carrier as follows:

Upper Channel Limit (%)	Lower Channel Limit (%)	Attenuation
0 to +50	0 to −50	25 dB
+50 to +150	−50 to −150	35 dB
>+150	>−150	$43+10 \log_{10}(W)$

where W = power in watts.

POINT-TO-POINT TRANSMISSION

In the United States, broadcasters generally use the facilities of communication common carriers for the transmission of aural and visual program material between cities, and often to connect various locations within the same city.

For audio transmission, the most frequently used service is amplitude-equalized and delay-equalized and covers the approximate frequency range from 100 to 5000 hertz. Higher-grade service up to 15 kilohertz is available.

The transmission objectives for local channels are amplitude ±1 decibel over the frequency range specified, plus a signal-to-noise ratio of 62 decibels at transmission level of +8 volume units (VU).

Intercity audio facilities may use a *B-22* cable system, which is capable of transmitting a frequency band from 35 to 8000 hertz, or may use two voice channels on type-*K* carrier systems.

Video transmission is either by coaxial cable or radio. Service is available for transmission of either monochrome or NTSC color signals. The requirements for the two are similar except that for color transmission the frequency response at 3.58 megahertz must be closely controlled (± 0.1 decibel) and there are required tolerances on differential gain (± 0.4 decibel) and differential phase ($\pm 0.5°$).

TABLE 7—TYPICAL SYSTEMS FOR
POINT-TO-POINT TRANSMISSION.

Cable			
Designation	Use	Video Band (megahertz)	Repeater Spacing (miles)
A2A	local	4.5	3–4
L1	intercity	2.7	7.8
L3	intercity	4.0	4

Radio			
Designation	Use	Frequency Band (megahertz)	Video Channels (2-way)
TD-2	intercity	3700–4200	6
TH	intercity	5925–6425	8
TJ	short-haul	10 700–11 700	6
TL*	short-haul	10 700–11 700	6

* All solid state except klystrons.

Table 7 outlines some of the systems that have been developed.

INTERNATIONAL BROADCASTING SERVICE IN THE UNITED STATES*

International broadcasting stations employ frequencies in allocated bands between 5950 and 26 100 kilohertz, and are intended to serve the general public in foreign countries. The area in which reception is desired is known as the "target area." Frequencies assigned to US stations by the FCC are as follows:

Band	Frequency (kilohertz)
A	5 950–6 200
B	9 500–9 775
C	11 700–11 975
D	15 100–15 450
E	17 700–17 900
F	21 450–21 750
G	25 600–26 100

* FCC Rules and Regulations, Part 73, Subpart F; 1972.

Assignments are made for use of specific frequencies, at specified hours, and for transmission to specified target areas.

Frequencies assigned are as close as possible to the optimum working frequency, which has been defined as that frequency which is returned to the surface of the earth for a specific transmission path and time of day on 90% of the days of the month. In no case will frequencies exceed the maximum usable frequency for more than 15 minutes during any transmission. The maximum usable frequency is the highest frequency which is returned to the earth for a specific path and time of day for 50% of the days of the month.

The minimum transmitter power is 50 kilowatts, the antenna power gain toward the target area is at least 10, and the field strength incident on the target area, either measured or calculated, shall exceed 150 microvolts/meter for 50% of the time. This value of field strength is intended to provide protection for at least 90% of each hour, during 90% of the days of each month, against atmospheric and industrial noise. Assignments in the US will provide 40 decibels of co-channel protection and 11 decibels of adjacent-channel protection at reference points in the target area.

PROGRAM SOURCES FOR BROADCASTING

If picture and/or sound are broadcast in "real time," such broadcasts are known as live broadcasts, and the pickup equipment (television cameras, microphones, and associated units) is designed to provide signals meeting industry and government standards. If picture and/or sound are recorded for replay at another time, and often at another location, industry standards have been developed to permit interchangeability. Sound signals may be recorded on disc or tape, while pictures and sound may be recorded on film or tape. Pertinent sections of the standards used in the US for such devices are abstracted below.

Disc Recording and Reproducing*

Speed: It shall be standard that the mean speed of the recording turntable (rpm) be either 33⅓ or 45±0.1%, and the mean speed of the reproducing turntable be either 33⅓ or 45±0.3%. It is recognized that 78.26-rpm discs are still in existence, but this speed is no longer considered a standard.

Rotation: It shall be standard that discs intended for broadcast applications be rotated in a clockwise direction as viewed from the side being reproduced and that the direction of feed shall be outside-in.

* See also P. J. Grey, "Disc Recording and Reproduction," American Photographic Book Publishing Company, New York, New York; March 1964.

Size: It shall be standard that the outer disc diameter fall within the following specified limits:

Nominal (inches)	Finished Discs (inches)
12	$11\frac{7}{8} \pm \frac{1}{32}$
10	$9\frac{7}{8} \pm \frac{1}{32}$
7	$6\frac{7}{8} \pm \frac{1}{32}$

It is recognized that 16-inch transcriptions are still in limited use, but this size is no longer considered a standard.

Flutter (Wow): In recording or reproducing, flutter is the deviation in frequency or pitch that results from minor periodic or random changes in the motion of the medium. The term "flutter" usually refers to cyclic deviations occurring at a relatively high rate (for example, 10 hertz). The term "wow" usually refers to cyclic deviations occurring at a relatively low rate (for example, a once-per-revolution speed variation of a turntable).

It shall be standard that the average deviation (measured over the range 0.5–200 hertz) from the mean speed of the recording turntable, when making the recording, shall not exceed 0.04% of the mean speed. The average deviation above shall be measured by a meter the dynamics of which shall be the same as those of the VU meter as specified in ANSI Standard C16.5-1961.

It shall be standard that the average deviation from the mean speed of the reproducing turntable when reproducing shall not exceed 0.1% of the mean speed.

Starting Time: It shall be standard that a reproducing turntable platen shall attain its mean speed in not more than 120 degrees rotation.

The reproducing characteristic for both monophonic and stereophonic discs shall be as shown in Table 8.

Grooves: It shall be standard that the groove shape for finished monophonic discs shall have an included angle of $90 \pm 5°$, a top width of not less than 0.0022 inch, and a bottom radius not greater than 0.00025 inch. It is recommended that discs with these groove shape characteristics be reproduced with a stylus having a tip radius of 0.001 inch ($+0.0001$ inch, -0.0002 inch) and an included angle of 40–55°.

It shall be standard that the groove shape for finished stereophonic discs shall have an included angle of $90 \pm 2°$, a top width of not less than 0.001 inch, and a bottom radius of not greater than 0.0002 inch. It is recommended that discs with these groove shape characteristics be reproduced with a stylus having a tip radius of 0.0005 to 0.0007 inch and an included angle of 40–55°. In stereophonic recording both channels are recorded in a single groove.

TABLE 8—RELATIVE OUTPUT VERSUS FREQUENCY FOR CONSTANT-VELOCITY INPUT.

Frequency (hertz)	Reproducing Characteristic (decibels)
15 000	-17.17
14 000	-16.64
13 000	-15.95
12 000	-15.28
11 000	-14.55
10 000	-13.75
9 000	-12.88
8 000	-11.91
7 000	-10.85
6 000	-9.62
5 000	-8.23
4 000	-6.64
3 000	-4.76
2 000	-2.61
1 000	0
700	$+1.23$
400	$+3.81$
300	$+5.53$
200	$+8.22$
100	$+13.11$
70	$+15.31$
50	$+16.96$
30	$+18.61$

Planes of modulation—It shall be standard that in a 45°—45° stereophonic disc the groove shall have orthogonal modulation planes inclined at 45° to a radial line on the surface of the disc, and the intersection of the modulation planes shall be normal to said radial lines.

Channel orientation—It shall be standard that the outer groove wall of the disc shall contain the right-hand channel information and the inner wall shall contain the left-hand channel information.

Phase—It shall be standard that the phase relationship between channels shall be such as to result in lateral groove displacement when the stereo recording system is driven with equal amplitude and in-phase signals, and the groove displacement shall be vertical when the stereophonic recording system is driven by equal-amplitude signals in antiphase (180°).

Noise: The following standards apply.

Low-frequency noise or rumble—It shall be standard that for a monophonic disc reproducing system the low-frequency noise voltage generated by the turntable, its associated pickup, and equalizer or equalized preamplifier when playing an essentially rumble-free silent groove, shall be at least 40 decibels below a reference level of 1.4

centimeters per second peak velocity at 100 hertz. The reference level of 1.4 centimeters per second peak velocity approximates the expected program level at 100 hertz and corresponds in amplitude to 7 centimeters per second peak velocity at 500 hertz. For stereophonic records the low-frequency noise shall be -35 decibels in each channel for a reference level of 1 centimeter per second peak velocity.

High-frequency noise—It shall be standard that the noise level measured with a standard volume indicator (ANSI Standard C16.5—1961) when reproducing a disc on a flat velocity basis over a frequency range between 500 and 15 000 hertz shall be at least 55 decibels below the level obtained under the same conditions of reproduction using a 1000-hertz tone recorded at a peak velocity of 7 centimeters per second. For stereophonic records the high-frequency noise shall be -50 decibels in each channel.

Recorded Level: The reference recorded program level for monophonic recordings shall produce the same deflection on a standard volume indicator as that produced by a 1000-hertz tone recorded at a peak velocity of 7 centimeters per second. For stereophonic recordings each channel shall produce the same deflection as that produced by a 1000-hertz tone recorded at a peak velocity of 5.0 centimeters per second.

Tape Recording and Reproducing of Sound*

The recording medium consists of finely divided ferrous-oxide particles deposited on a plastic backing known as the base. The recording or reproducing magnet is called a head, and it consists of soft iron pole pieces wound with coils of wire. During recording the head is "biased" with an alternating current in the 60–100-kilohertz range; the value of the bias current affects the frequency response, distortion, and signal-to-noise ratio. During recording the tape is moved through the field created at

* See also NAB standards, "Magnetic Tape Recording and Reproducing (Reel to Reel)," April 1965; and "Cartridge Tape Recording and Reproducing," October 1964.

the gap (a good recorder may have a gap size of 0.25 mil) in the recording head, and the flux pattern representing the magnitude and direction of the signal at the moment the tape leaves the gap is recorded. (See Fig. 23.)

During reproduction the process is reversed, and the flux from the tape will induce a corresponding voltage in the windings of the head. Commercial recorders usually use separate heads for record, reproduce, and erase.

Two types of audio tape devices are used by broadcasters: reel to reel and cartridge tape. Either may be used for both recording and reproducing.

Reel-to-Reel Standards:

Speed—The preferred tape speed is 7½ inches per second ±0.2%. Supplementary tape speeds are 15 and 3¾ inches per second ±0.2%. Special-purpose limited-performance systems operate at 7½, 3¾, and 1⅞ inches per second ±2%.

Tape wind—Recorded tape should normally be wound so that the start of the program material is at the outside of the reel, with the oxide-coated surface of the tape facing the hub of the reel.

Tape dimensions—Width 0.246±0.002 inch (nominal ¼″ tape). Thickness not more than 0.0022 inch.

Lengths—

Reel Diam.	Hub	1.5-mil Base	1.0-mil Base	0.5-mil Base
3″	1.75″	125′	200′	300′
5″	1.75″	600′	900′	—
7″	2.25″	1200′	1800′	—
10.5″	4.5″	2500′	3600′	—
14″	4.5″	5000′	7200′	—

Tracks—On multitrack recordings, with the tape moving left to right and the oxide-coated side facing away from the observer, track 1 shall be the top track, track 2 the next lower, etc. Track dimensions are as follows:

	Width	Center to Center	Remarks
Full track	0.238″	—	—
Two-track mono	0.082″	0.156″	—
Two-track stereo	0.082″	0.156″	Track 1 left channel, track 2 right channel*
Four-track mono	0.043″	0.134″	Recording sequence: 1—4—3—2
Four-track stereo	0.043″	0.134″	1—3 one direction, 2—4 other direction*
			1—4 left channel, 2—3 right channel

* Head gaps are so placed that when full track tape is reproduced the signals in the two output channels are in phase.

Electrical Performance:

Output level—A given reel shall have an average output level that is uniform within ±0.5 decibel at 400 hertz at 7½ inches per second. Any specified type of tape shall be uniform ±1.0 decibel from reel to reel.

Signal-to-noise ratio—

Speed (ips)	Unweighted (20–20 000 hertz)			Weighted (Fig. 24)		
	Full Track (decibels)	2 Track (decibels)	4 Track (decibels)	Full Track (decibels)	2 Track (decibels)	4 Track (decibels)
15	50	45	not used	58	53	not used
7½	50	45	45	60	55	52
3¾	46	46	45	57	54	52
Special-purpose	46	43	40	—	—	—

Flutter (Measured from 0.5 to 200 hertz using a 3-kilohertz signal)—

Speed (ips)	Unweighted (rms)	Weighted (rms) (Fig. 25)
15	0.15%	0.05%
7½	0.20%	0.07%
3¾	0.25%	0.10%
Special purpose	0.5%	—

Crosstalk—Adjacent-track signal-to-crosstalk ratio shall be not less than 60 decibels from 200 hertz to 10 kilohertz. Stereophonic channel separation shall be not less than 40 decibels from 100 hertz to 10 kilohertz.

Distortion—Shall be less than 3% rms for a 400-hertz sine wave recorded at 6 decibels above the standard reference level.

Recorded level—Standard reference level is a 400-hertz sine wave recorded at 7½ inches per second, adjusted to an output level 8 decibels below that which will produce 3% third-harmonic distortion. The standard reproducing characteristic and the standard system response limits are shown, respectively, in Figs. 26 and 27.

Cartridge Tape Standards:

Speed—7½ ips ±0.4%.

Tape dimensions—Width 0.246±0.002 inch (nominal ¼″ tape). Thickness not more than 0.0016 inch.

Cartridge sizes—Shown in Fig. 28.

Monophonic tape has a program track and a cue track, as shown in Fig. 29. Stereophonic tape has two program tracks and a cue track, as shown in Fig. 30.

Cue tones—

	Primary	Secondary	Tertiary
Frequency (Hz)	1000±75	150±30	8000±1000
Level (dB)*	+8±3	−2±3	+8±3
Duration (ms)	500±250	500±250	500±250
Purpose	Stop	End of message	Auxiliary

* Referred to the standard reference level.

Electrical performance—The standard reference level is the same as that for reel-to-reel tape. The reproduce characteristic is the same as the equivalent reel-to-reel characteristic shown in curve A

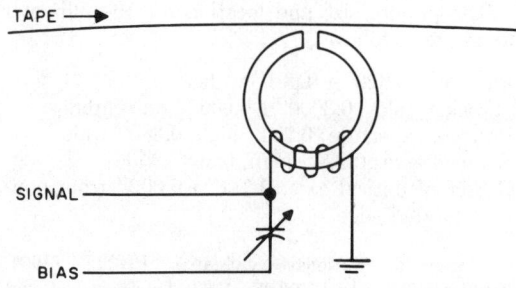

TAPE →

SIGNAL

BIAS

Fig. 23—Tape recording.

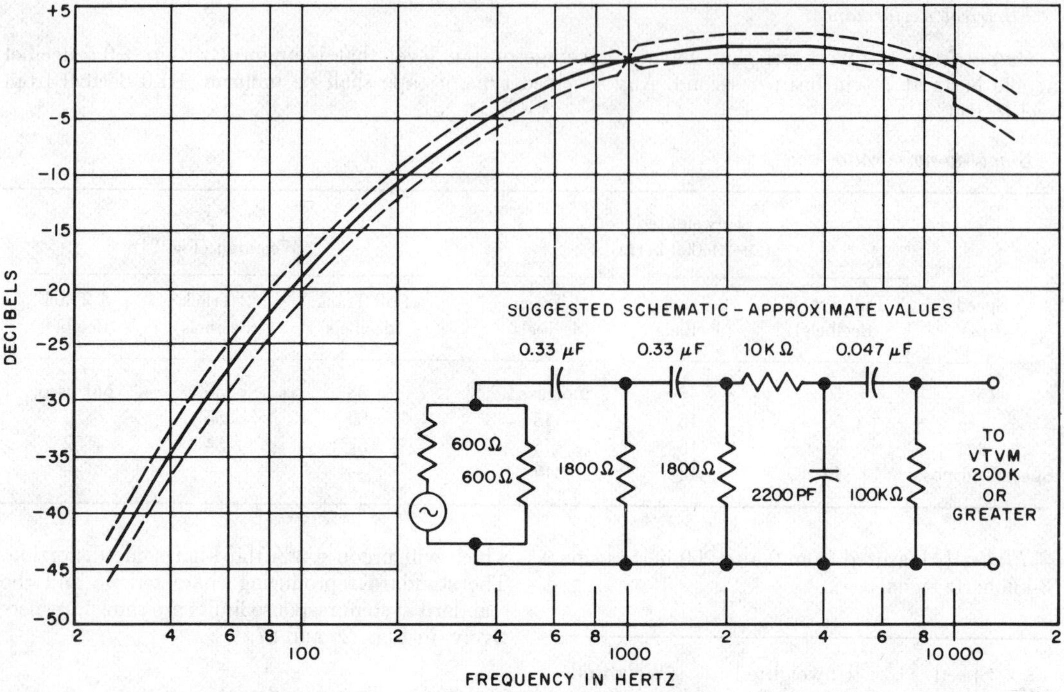

Fig. 24—Weighting curve for weighted noise measurements. The dashed lines represent tolerances.

of Fig. 26, and the system response limits are the same as shown in Fig. 27 for 7½ ips tape, limited to frequencies between 50 and 12 000 hertz.

Distortion—Same as for reel-to-reel tape.

Flutter (unweighted)—Same as for reel-to-reel tape 7½ ips.

Signal-to-noise ratio (unweighted)—Monophonic, 45 decibels; stereophonic, 42 decibels.

Crosstalk (cue tone to program channel)—

Monophonic:
    150 hertz   −50 decibels
    1000 hertz   −55 decibels
    8000 hertz   −50 decibels
Stereophonic:
    ≥ −50 decibels

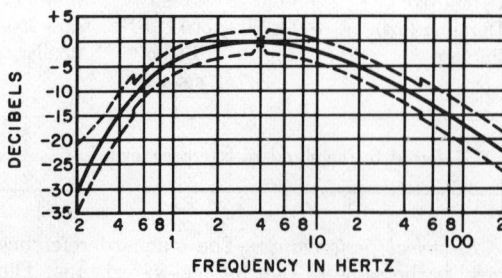

Fig. 25—Weighting curve for weighted flutter measurements.

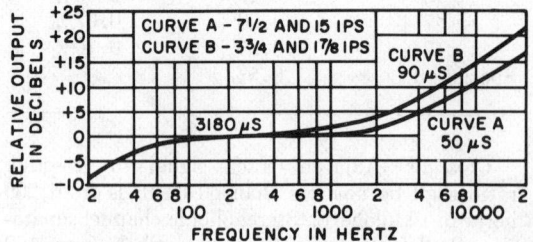

Fig. 26—NAB standard reproducing characteristic, reproducing amplifier output for constant flux in the core of an ideal reproducing head.

FILM FOR TELEVISION

Television broadcasters usually use standard 16-millimeter sound films and 35-millimeter slides; in some metropolitan centers 35-millimeter film and 3¼″×4″ slides are used.

The picture size and location for 16-millimeter film* are as follows.

Film size: 0.629″±0.001″ wide
Sprocket holes: 0.3000″±0.0005″ on centers
Projector aperture: 0.284″ high, 0.380″ wide
Scanned area: 0.276″ high, 0.368″ wide
Center of optical axis: 0.314″±0.002″ from non-perforated edge.

* American National Standards PH22.12—1964, PH22.8—1957, and PH22.96—1963; American National Standards Institute (ANSI).

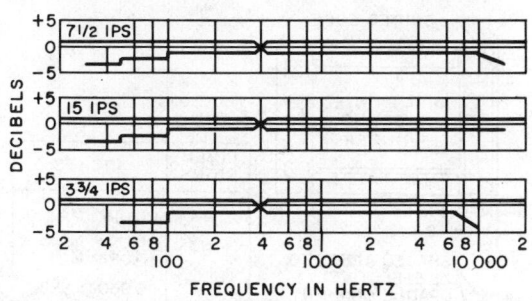

Fig. 27—NAB standard response limits of reproducing systems.

If the sound is optically recorded, the location of the sound track is as shown in Fig. 31. Magnetic sound striping is as shown in Fig. 32, and both optical and magnetic are shown in Fig. 33.†

A typical TV 16-millimeter film projector has the following capability.

Reel capacity: 2000 feet
Speed: 24 frames/second
Vertical jump: 0.2% picture width

† American National Standards PH22.41—1957 and PH22.87—1958.

Horizontal weave: 0.2% picture width
Shutter frequency: 120 cycles/second
Light application time: 30%
Light output: 50 lumens
Uniformity of illumination: Not less than 80%
Sound reproduction: Optical and magnetic
Sound output: +6 dBm at 600 ohms
Frequency response: 80–8000 hertz, ±1½ decibels
Wow and flutter: Less than 0.25%
Signal-to-noise ratio: Optical, 60 decibels below rated output
Magnetic, 40 decibels below rated output
Harmonics: Less than 2% from 100-7000 hertz
Starting time: Less than 5 seconds
Stopping distance: 3 feet of film

Television 2×2 slides* have been standardized as shown in Fig. 34.

Certain recommended practices have been established to assure that important information will be visible on the majority of home TV receivers, which are not necessarily in correct scanning

* American National Standard PH22.144—1965.

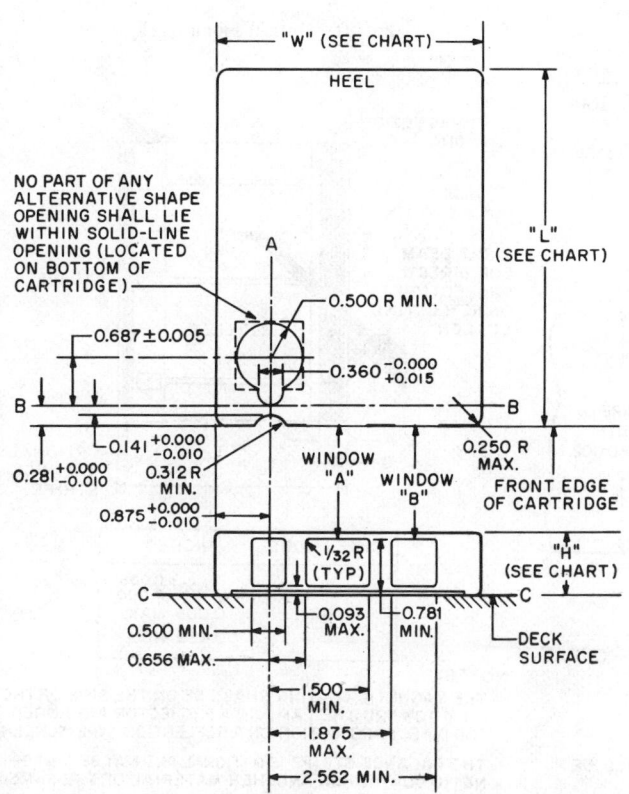

1. DIMENSIONS

ALL IMPORTANT OPERATING DIMENSIONS (IN INCHES) ARE REFERENCED FROM TWO IMAGINARY MUTUALLY PERPENDICULAR VERTICAL PLANES MARKED A-A AND B-B, AND A THIRD HORIZONTAL PLANE C-C, REPRESENTING THE DECK SURFACE OF THE CARTRIDGE TAPE PLAYER.

WHERE APPLICABLE, DIMENSIONS INCLUDE DRAFT ALLOWANCES.

2. MATERIALS

ALL MATERIALS USED IN THE CARTRIDGE CONSTRUCTION SHALL BE NONMAGNETIC.

CARTRIDGE NAB TYPE	WIDTH "W"±1/64"	LENGTH "L" MAX.	HEIGHT "H" MAX.
A	4"	5¼"	0.9375"
B	6"	7"	0.9375"
C	7⅝"	8½"	0.9375"

Fig. 28—NAB cartridge standard.

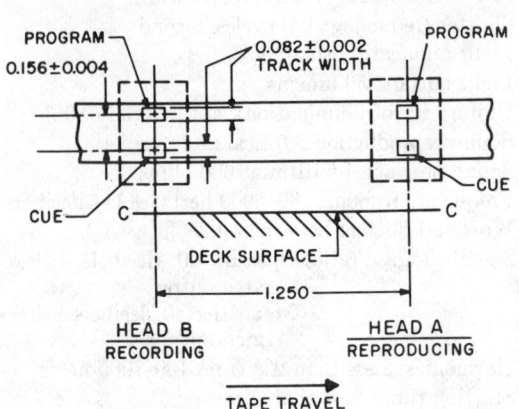

Fig. 29—Monophonic 2-track recorded track dimensions.

adjustment. These are known as the "safe action" area and the "safe title" area and are defined below.

	Height (in.)	Width (in.)	Corner Radius
For 16-mm film*:			
Scanned area	0.276	0.368	—
Safe action	0.248	0.331	0.066
Safe title	0.221	0.294	0.058
For 2×2 slides:			
Scanned area	0.844	1.125	—
Safe action	0.759	1.013	0.203
Safe title	0.675	0.900	0.180

* Society of Motion Picture and Television Engineers (SMPTE) Recommended Practices RP13—1963, RP8—1961, and RP7—1962.

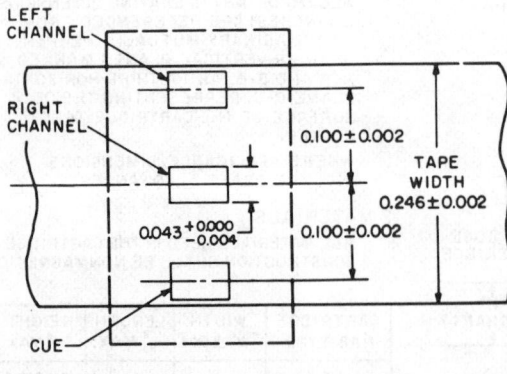
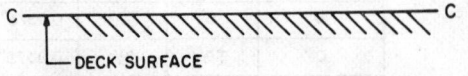

NOTES:

1. IF SECOND STEREO HEAD IS USED, IT SHOULD BE PLACED 1.250 INCHES FROM FIRST HEAD.
2. TRACK WIDTH: ALL TRACKS SHALL BE 0.043 +0.000 −0.004 INCH.

Fig. 30—Stereophonic 3-track recorded track dimensions.

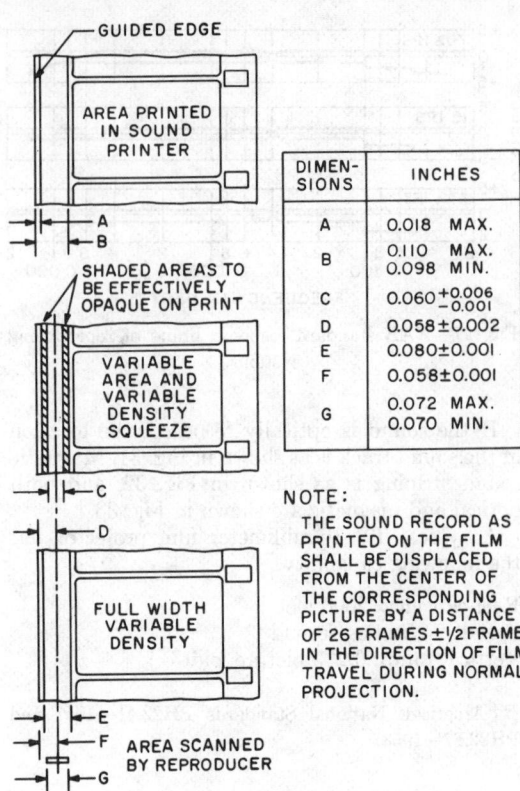

DIMEN-SIONS	INCHES
A	0.018 MAX.
B	0.110 MAX. 0.098 MIN.
C	0.060 +0.006 −0.001
D	0.058±0.002
E	0.080±0.001
F	0.058±0.001
G	0.072 MAX. 0.070 MIN.

NOTE:

THE SOUND RECORD AS PRINTED ON THE FILM SHALL BE DISPLACED FROM THE CENTER OF THE CORRESPONDING PICTURE BY A DISTANCE OF 26 FRAMES ±½ FRAME IN THE DIRECTION OF FILM TRAVEL DURING NORMAL PROJECTION.

Fig. 31—Optical sound track.

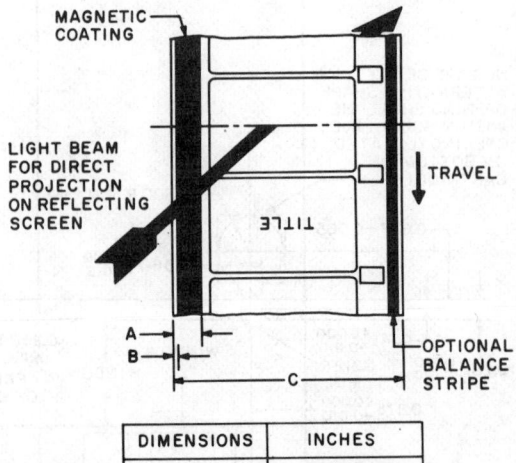

DIMENSIONS	INCHES
A	0.100 +0.005 −0.000
B	0.005 MAX.
C	0.628 NOM.

NOTES:

THE MAGNETIC COATING SHALL BE ON THE SIDE OF THE FILM TOWARD THE LAMP ON A PROJECTOR ARRANGED FOR DIRECT PROJECTION ON A REFLECTION-TYPE SCREEN.

THE BALANCE STRIPE IS OPTIONAL AND MAY BE A MAGNETIC COATING OR ANOTHER MATERIAL OF THE SAME THICKNESS.

Fig. 32—Magnetic sound stripe.

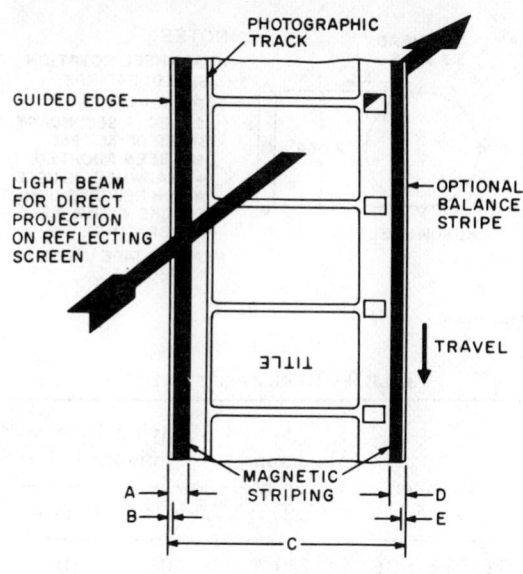

Fig. 33—Magnetic–photographic sound tracks.

DIMENSIONS	INCHES
A	0.058 MAX. 0.053 MIN.
B	0.005 MAX.
C	0.628 NOM.
D	0.031 MAX. 0.028 MIN.
E	0.002 MAX.

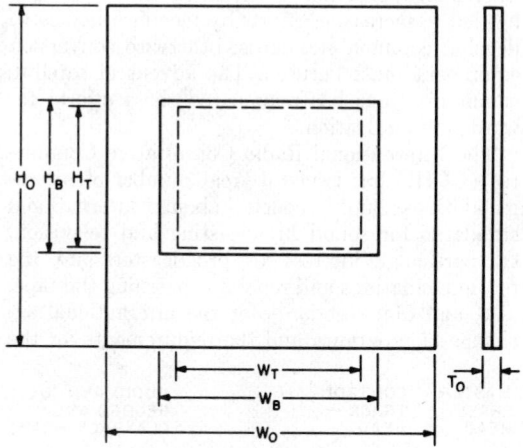

DIMENSIONS IN INCHES			
NOMINAL	OVERALL		
	H_O	W_O	T_O MAX.
2 x 2 SLIDE (DOUBLE 35)	$2 \, ^{+0}_{-1/32}$	$2 \, ^{+0}_{-1/32}$	1/8

TRANSMITTED PICTURE		PICTURE BACKGROUND		CENTERING TOLERANCE
H_T MAX.	W_T MAX.	H_B MIN.	W_B MIN.	
27/32	1 1/8	29/32	1 11/32	1/64

Fig. 34—Television slide standards.

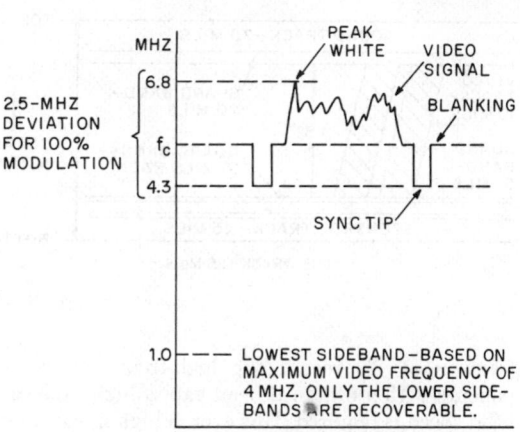

Fig. 35—Modulation system.

Another significant practice relates to the density and contrast range of films and slides, which must be accommodated to the available dynamic range of the television system. The recommendations are:

The minimum diffuse density of highlight areas shall have a normal value of 0.4 to 0.3 but not less than 0.3 for optimum reproduction in the television system. This value is not intended to apply to glint, specular highlights, or other small areas where details need not be reproduced.

The maximum diffuse density of lowlight areas shall have a normal value of 1.9 to 2.0 but not greater than 2.0 for optimum reproduction in the television system. This value is not intended to apply to small areas where details need not be reproduced.

The density of human faces, usually observed more intently than other picture areas, shall be greater than the measured minimum density by a value not less than 0.15 or more than 0.5 unless special effects are desired. These density values are important to preserve the proper density relationships between face tones and highlights.

The SMPTE has available a series of test films to check system performance.

TAPE FOR VIDEO AND SOUND REPRODUCTION*

With the development and widespread use of the video tape recorder, magnetic tape has replaced motion-picture film as a program storage medium in a great many applications. The standard system employed by broadcasters was developed as a compromise to overcome a number of conflicting requirements for the recording of a wide range of

---

* See also Harold E. Ennes, *Television Broadcasting: Tape and Disc Recording Systems,* Howard W. Sams & Co., Inc., Indianapolis, Indiana; 1973.

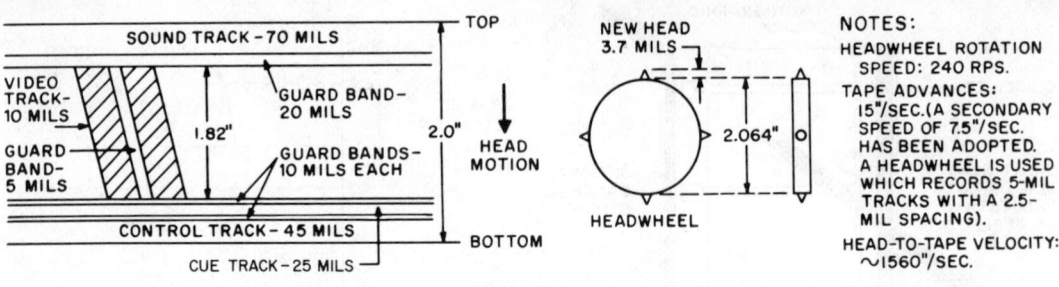

NOTES:

HEADWHEEL ROTATION SPEED: 240 RPS.

TAPE ADVANCES: 15"/SEC.(A SECONDARY SPEED OF 7.5"/SEC. HAS BEEN ADOPTED. A HEADWHEEL IS USED WHICH RECORDS 5-MIL TRACKS WITH A 2.5-MIL SPACING).

HEAD-TO-TAPE VELOCITY: ~1560"/SEC.

Fig. 36—Tape tracks.

frequencies. To record the high-frequency components with a practical head gap, a high tape-to-head speed is required. However, a high linear tape speed would pose tape-handling problems and would require reels of impractical size for any reasonable playing time. In addition, the output decreases as the recorded frequency decreases, and an unacceptable signal-to-noise ratio would result at the lower frequencies of the video signal.

To overcome these problems and accommodate the wide band of TV frequencies, a frequency-modulated radio-frequency carrier is employed. The modulated carrier is fed to four recording heads mounted in quadrature on the periphery of a 2-inch-diameter drum rotating at 14 400 rpm. A series of transverse tracks 10 mils wide are recorded across a 2-inch-wide tape moving at 15 inches per second. Sound is recorded on a longitudinal track at 15 inches per second.

Figures 35 through 39 illustrate the basic operation.

Standard reels are an important aspect of tape interchangeability. The specifications are outlined in Fig. 40 and Table 9.

A simpler type of video tape recorder, the helical-scan type, has found wide usage in nonbroadcast applications. The basic principle is for the tape to pass in a slanted path around a drum; the recording heads rotate on a disc inside the drum and trace slanted tracks across the tape, as shown in Fig. 41. The high degree of standardization that exists in quad-head machines is not found in helical-scan recorders, and there are a number of design variations. Differences among machines include the number of heads (one or two), tape wrap around the drum (full wrap or half wrap), and tape width (one inch or half inch).*

INTERNATIONAL RECOMMENDATIONS AND STANDARDS FOR SOUND AND TELEVISION BROADCASTING

The various broadcast standards and practices described on preceding pages apply in the United States and in many other countries. Other standards apply elsewhere. In recent years the need to interchange program material on film and tape has led to increasing efforts by technical bodies to develop common standards; otherwise conversion equipment must be used. The advent of satellite communications has augmented these efforts toward standardization.

The International Radio Consultative Committee (CCIR) has issued a great number of recommendations and reports about international standards for sound broadcasting and television. Of particular interest to broadcasters are the recommendations and reports concerning the tape, disc, and film standards for the international exchange of programs, and the requirements for the

* See Harry Kybett, *Video Tape Recorders*, Howard W. Sams & Co., Inc., Indianapolis, Indiana; 1974.

TABLE 9—REEL SPECIFICATIONS.

Diameter (inches)	Maximum Capacity (feet)	Playing Time (minutes)	
		7.5-in/s	15-in/s
6.50±0.01	750	20	10
8.00±0.01	1650	44	22
10.50±0.01	3600	96	48
12.50±0.01	5540	148	74
14.00±0.01	7230	192	96

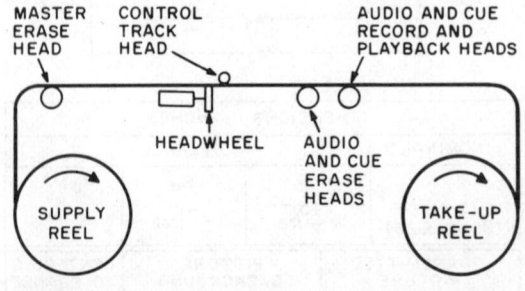

NOTE:

THE AUDIO-VIDEO SEPARATION SHALL BE 9.250 INCHES, WITH THE AUDIO RECORD PRECEDING THE CORRESPONDING PICTURE RECORD.

Fig. 37—Simplified diagram of tape operation.

transmission of television signals that have different standards. Also of interest are the technical aspects of three color systems proposed for international standardization, plus the technical aspects of broadcasting directly from satellites. Portions of these recommendations and reports follow.

Single-track sound recording on magnetic tape is essentially the same as in the United States,

except that the European spool is different. An adapter has been specified which permits the US type spool to fit the European type hub. Similarly, there is good agreement on the standards used for 2-track stereophonic tape recordings and for disc recordings.*

The international exchange of television programs on monochrome and color (types B and C) film† is effected by means of the following types:

35 COMOPT—35-millimeter film with combined picture and sound and an optical sound track.

16 COMOPT—16-millimeter film with combined picture and sound and an optical sound track.

35 COMMAG—35-millimeter film with combined picture and sound and a magnetic sound track.

16 COMMAG—16-millimeter film with combined picture and sound and a magnetic sound track.

35 SEPMAG—35-millimeter film for the picture with a magnetic sound track on a separate 35-millimeter film.

16 SEPMAG—16-millimeter film for the picture with a magnetic sound track on a separate 16-millimeter film.

35 MUTE—35-millimeter film without sound.

16 MUTE—16-millimeter film without sound.

If the picture and the sound films do not have the same width, two numbers are used. The first

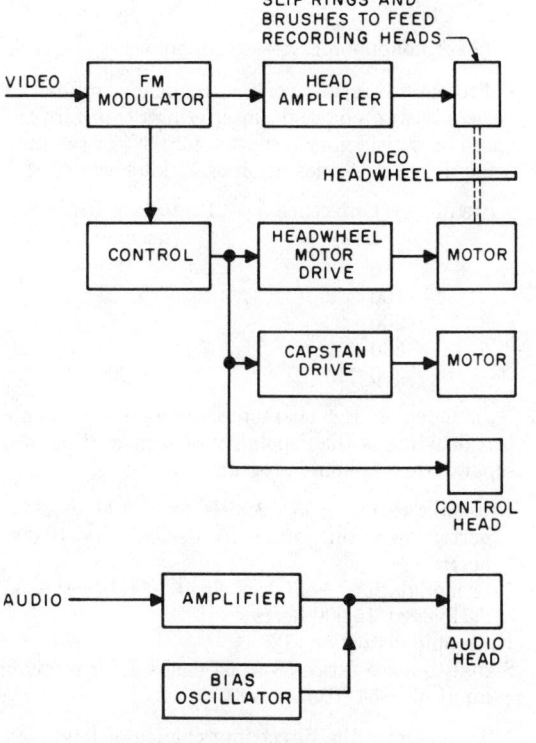

Fig. 38—Simplified diagram of record.

* CCIR Recommendation 408-3, Geneva; 1974.
† CCIR Recommendation 265-3, Geneva; 1974.

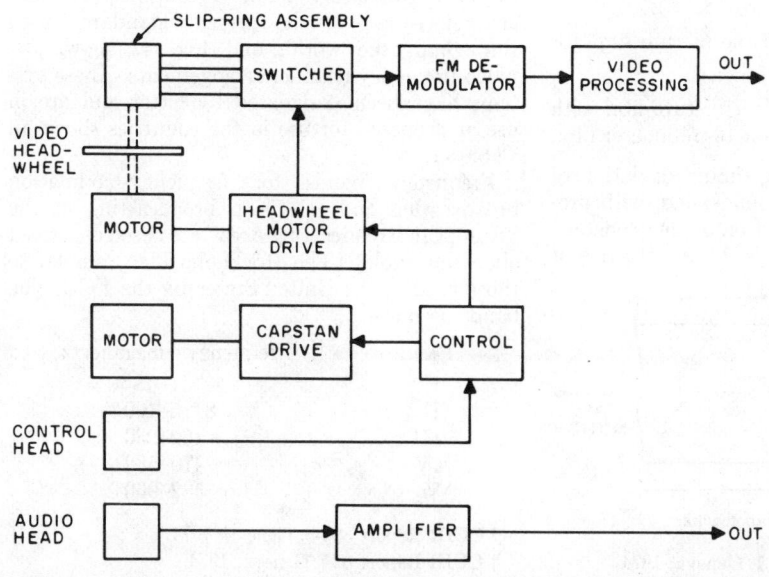

Fig. 39—Simplified diagram of playback.

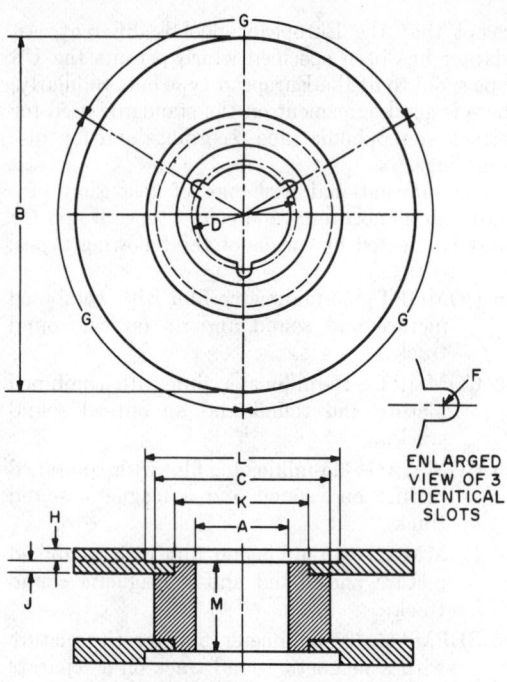

ENLARGED VIEW OF 3 IDENTICAL SLOTS

DIMENSIONS	INCHES
A	3.000 +0.004,−0
B	REFER TO TABLE 7
C	4.500 ±0.100
D	3.250 ±0.002
E	NOT USED
F	0.109 +0.003,−0
G	120° ±0.1°
H	0.025 MAX.
J	0.099 MAX.
K	3.600 MIN.
L	6.000 MIN.
M	2.212 ±0.003

Fig. 40—Reel specifications.

number indicates the width of the picture film, for example:

35/16 SEPMAG—35-millimeter picture film with magnetic sound track on 16-millimeter film.

In FM broadcasting* a maximum deviation of ±75 or ±50 kilohertz is recommended, with pre-emphasis characteristics based on a time constant of either 50 or 75 microseconds. In the absence of

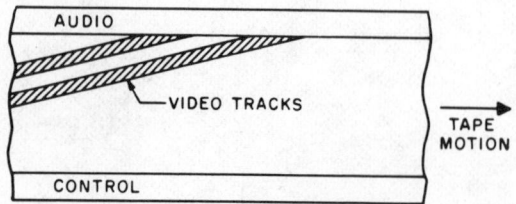

Fig. 41—Helical-scan tracks.

* CCIR Recommendation 412-1, Geneva; 1974.

industrial interference, a field strength of 50 microvolts/meter at 10 meters above the ground is considered to give acceptable service. In the presence of interference, satisfactory service requires:

Median Field Strength (microvolts/meter)		Area
M	S	
0.25	0.50	Rural
1	2	Urban
3	5	Large cities

(M = monophonic, S = stereophonic)

Protection ratios versus frequency difference between wanted and interfering transmitters, based on satisfactory reception for 99% of the time using a frequency deviation of 75 kilohertz, are:

Frequency Difference (kilohertz)	Protection Ratio (decibels)
0	28
100	12
200	6
300	−7
400	−20

Included in the characteristics of stereophonic broadcasting is the capability of transmitting two separate monophonic programs:*

Crosstalk ratio: −35 decibels from 100 to 3000 hertz, increasing to −15 decibels at 10 000 hertz

Intermodulation: −45 decibels at 1000 hertz; −30 decibels at 15 000 hertz

Harmonic distortion: 1%

Signal-to-noise ratio: 58 to 64 decibels for receiver input of −54 dBm.

To facilitate the direct interchange of television programs, efforts have been made to achieve common standards. Table 10 outlines pertinent characteristics† of the present standards used throughout the world, and Fig. 42 shows the video-frequency-channel arrangements. These systems have been designated by letter and are in use or proposed for use in the countries shown in Table 11.

Frequency bands for frequency-modulation broadcasting and television broadcasting in the "European Broadcasting Area" are based on agreed allocation tables (The Stockholm Plan) similar to those used in the United States by the FCC. The bands used are:

Band	Frequency (megahertz)
I	41–68
II	87.5–100
III	162–230
IV	470–582
V	582–960

* CCIR Report 300-3, Geneva; 1974.
† CCIR Report 624, Geneva; 1974.

TABLE 10—INTERNATIONAL TELEVISION STANDARDS.

	A	M	N	B	C	G	H	I	D, K	L	K1	E
Lines/frame	405	525	625	625	625	625	625	625	625	625	625	819
Fields/second	50	60	50	50	50	50	50	50	50	50	50	50
Interlace	2/1	2/1	2/1	2/1	2/1	2/1	2/1	2/1	2/1	2/1	2/1	2/1
Frames/second	25	30	—	25	25	25	25	25	25	25	25	25
Lines/second	10 125	15 750	15 625	15 625	15 625	15 625	15 625	15 625	15 625	15 625	15 625	20 475
Aspect ratio[1]	4/3	4/3	4/3	4/3	4/3	4/3	4/3	4/3	4/3	4/3	4/3	4/3
Video band (MHz)	3	4.2	4.2	5	5	5	5	5.5	6	6	6	10
RF band (MHz)	5	6	6	7	7	8	8	8	8	8	8	14
Visual polarity[2]	+	−	−	−	+	−	−	−	−	+	+	+
Sound modulation	A3	F3	F3	F3	A3	F3	F3	F3	F3	F3	F3	A3
Pre-emphasis in microseconds	—	75	75	50	50	50	50	50	50	—	50	—
Deviation (kHz)	—	25	25	50	—	50	50	50	50	—	50	—
Gamma of display device	2.8	2.2	2.2	—	—	—	—	2.8	—	—	—	—

Notes:

[1] In all systems the scanning sequence is from left to right and top to bottom.

[2] All visual carriers are amplitude modulated. Positive polarity indicates that an increase in light intensity causes an increase in radiated power. Negative polarity (as used in the US—Standard M) means that a decrease in light intensity causes an increase in radiated power.

Bands I and II contain television stations as well as frequency-modulation broadcast stations. Bands III, IV, and V contain only television stations.

Comparable bands in the United States are the low VHF television band, channels 2–6 (54–88 megahertz); the frequency-modulation band for sound broadcasting (88–108 megahertz); the high VHF television band, channels 7–13 (174–216 megahertz); and the UHF television band, channels 14–83 (470–960 megahertz).

The television standards proposed or used (and described in Fig. 42) are:

Bands I, II, and III
A	405-line
B	625-line
C	625-line (Belgian)
D	625-line
E	819-line
K1	625-line

Bands IV and V:
G	625-line
H	625-line
I	625-line
K	625-line
L	625-line.

Because of the different characteristics of the systems used in the very-high-frequency channels, there are differences in the channel width and in the specific frequencies used for picture and sound carriers. In Bands IV and V a uniform channel width of 8 megahertz provides 61 channels numbered 21 through 81. The visual carrier frequency is a nominal 1.25 megahertz above the lower limit of the channel.

The 625-line color or monochrome standard* proposed is:

Number of lines per picture	625
Field frequency (fields/second)	50
Interlace	2/1
Picture-frame frequency (pictures/second)	25
Line frequency and tolerance when operated nonsynchronously (lines/second)	15 625 ±0.02% (B) ±0.0001% (C)
Aspect ratio	4/3
Scanning sequence:	
Lines	left to right
Fields	top to bottom
Systems capable of operating independently of power-supply frequency	yes
Approximate gamma of picture signal	0.4
Nominal video bandwidth in megahertz	5; 5.5 or 6

In connection with the recording of monochrome television signals on magnetic tape,† Table 12 outlines differences between systems for 525 lines/60 fields and 625 lines/50 fields.

* CCIR Recommendation 472-1, Geneva; 1974.

† CCIR Recommendation 469-1, Geneva; 1974.

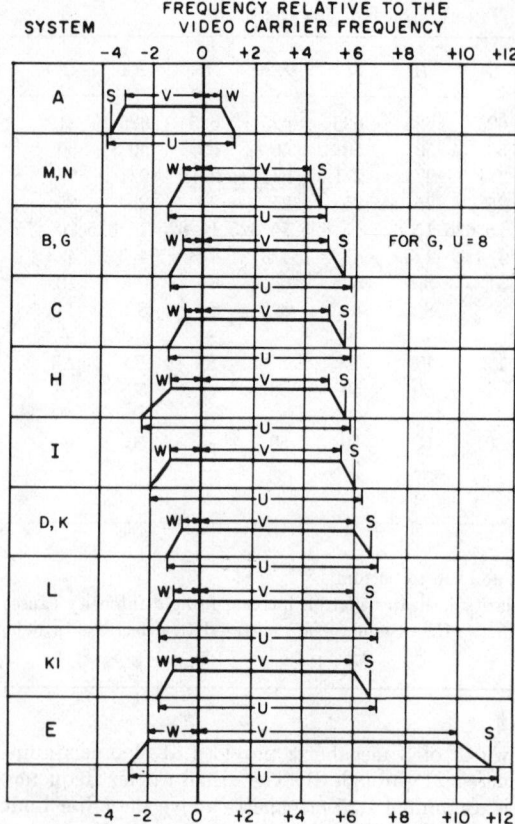

S = SOUND CARRIER
U = LIMITS OF RADIO-FREQUENCY CHANNEL
V = NOMINAL WIDTH OF MAIN SIDEBAND
W = NOMINAL WIDTH OF VESTIGIAL SIDEBAND

Fig. 42—International television transmission standards.

Color Television

Three color-television systems have been proposed for a European color standard: The present NTSC (National Television System Committee) system used in the United States, but modified for 625 lines/50 fields, the PAL (Phase Alternation Line) system, and the SECAM (Sequential With Memory) system. The systems are similar in that they separate the luminance and chrominance information, and transmit the chrominance information in the form of two color difference signals which modulate a color subcarrier transmitted within the video band of the luminance signal. The systems are different in the processing of the chrominance information. Figures 43 and 44 are simplified diagrams of the NTSC system.

In the NTSC system, the color difference signals I and Q amplitude-modulate subcarriers that are displaced by $\pi/2$, giving a suppressed-carrier output. A burst of the subcarrier frequency is transmitted during the horizontal back porch to synchronize the color demodulator.

TABLE 11—USE OF TELEVISION STANDARDS DESCRIBED IN TABLE 10 AND FIG. 42.

Country	Standard Used
Argentina	N
Australia	B
Austria	B,G
Belgium	C,B,H
Bulgaria	D,K
Canada	M
Czechoslovakia	D,K
Denmark	B,G
Finland	B,G
France	E,L
Hungary	D,K
India	B
Iran	B,G
Ireland	A,I
Israel	B,G
Italy	B,G
Japan	M
Korea	M
Luxembourg	C,L
Mexico	M
Monaco	E,L
Morocco	B,H
Netherlands Antilles	M
New Zealand	B
Nigeria	B,I
Norway	B,G
Pakistan	B
Panama	M
Peru	M
Poland	D,K
Portugal	B,G
Rhodesia	B,G
Romania	D,K
Saudi Arabia	B
Spain	B,G
Sweden	B,G
Switzerland	B,G
The Netherlands	B,G
United Kingdom	A,I
United States of America	M
Union of Soviet Socialist Republics	D,K
West Germany	B,G

In the PAL system, the phase of the subcarrier is changed from line to line, which requires the transmission of a line switching signal as well as a color burst.

In the SECAM system, the color subcarrier is frequency modulated, alternately, by the color difference signals. This is accomplished by an electronic line-to-line switch; the switching information is transmitted as a line-switching signal.

The long-distance transmission of television

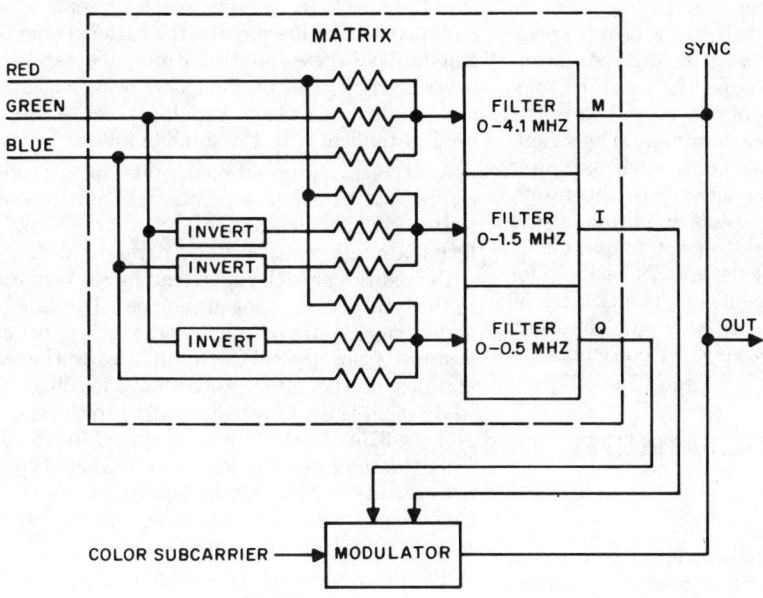

$$E_M = 0.30E_R + 0.59E_G + 0.11E_B$$
$$E_I = 0.60E_R - 0.28E_G - 0.32E_B$$
$$E_Q = 0.21E_R - 0.52E_G + 0.31E_B$$

Fig. 43—NTSC transmitted signal.

signals has been facilitated by standardizing certain system characteristics.*

(**A**) Interconnection at video frequencies

Impedance—75 ohms unbalanced
Return loss—at least 24 decibels
Amplitude:
    Input—1 volt peak-to-peak
    Output—1 volt peak-to-peak
Polarity—Black-to-white transitions, positive going

(**B**) Interconnection at intermediate frequencies

Impedance—75 ohms unbalanced
Amplitude:
    Input—0.3 volt root-mean-square
    Output—0.5 volt root-mean-square

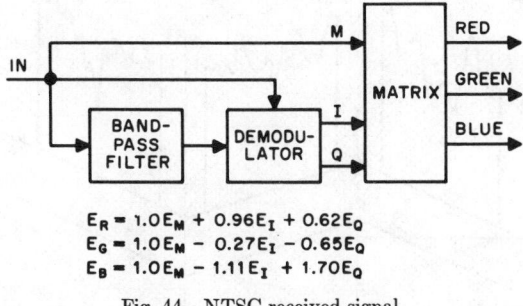

$$E_R = 1.0E_M + 0.96E_I + 0.62E_Q$$
$$E_G = 1.0E_M - 0.27E_I - 0.65E_Q$$
$$E_B = 1.0E_M - 1.11E_I + 1.70E_Q$$

Fig. 44—NTSC received signal.

* CCIR Recommendations 421-3 and 403-2, Geneva; 1974.

Intermediate frequency:
    Up to 1000 megahertz—35 megahertz
    Above 1000 megahertz—70 megahertz

In a hypothetical reference circuit of about 1600 miles the signal-to-noise ratios for different systems are as follows:

System (lines)	405	525	625	625	819	819
Video band (MHz)	3	4	5	6	5	10
Signal-to-weighted-noise ratio (dB)	50	56	52	57	52	50

TABLE 12—DIFFERENCES IN TAPE RECORDING BETWEEN 525-LINE AND 625-LINE SYSTEMS.

	525-Line	625-Line
Tape speed (inches/second)	15	15⅝
Head rotation (rotations/second)	240	250
Tracks/field	16	20
Tracks/second	960	1000
Control-track frequency (hertz)	240	250
Frame pulses (per second)	30	25
Frequency at sync level (MHz)	4.3	5.0
Frequency at blanking (MHz)	5	5.5
Frequency at white level (MHz)	6.8	6.8
Tape width (inches)	2	2
Tape thickness (inches)	0.0015	0.0015

The signal-to-noise ratio for random noise is defined as the ratio, in decibels, of the peak-to-peak amplitude of the picture signal to the root-mean-square amplitude of the noise, between 10 kilohertz and the upper limit of the video band. The weighting is specified for each system. The signal-to-noise ratios for periodic noise, such as power supply hum and for pulse noise, are the ratios, in decibels, of the peak-to-peak amplitude of the picture signal to the peak-to-peak amplitude of the noise. Recommendations are 35 decibels for the 525-line system (except in Japan) and 30 decibels for all other systems for periodic noise, and 11 and 25 decibels, respectively, for pulse noise.

BROADCASTING VIA SATELLITES

Direct Broadcasting

The feasibility of direct broadcasting of sound and television to the public by means of satellites is being actively considered.* The requirements for an active synchronous satellite to produce a field of 50 microvolts/meter in the frequency-modulation band over North America have been calculated.

Required predetection S/N ratio (dB)	26
Required signal power (dBW)	−115
Receiving-antenna gain (dB)	6
Transmitting-antenna gain (dB)	23.6
Transmitter power (W)	205

The requirements to cover about one-third of the earth's surface with a television broadcast in the ultra-high-frequency band have been calculated.

Frequency (MHz)	650
Peak carrier-to-root-mean-square noise (dB)	30
Required signal power (dBW)	−94.1
Receiving-antenna gain (dB)	15.2
Transmitting-antenna gain (dB)	19.3
Visual transmitter power (kW)	427
Sound transmitter power (kW)	107

Program Relays†

Operated by the International Telecommunications Satellite Consortium (INTELSAT), communication satellites in earth-synchronous orbits provide television service in the Atlantic, Pacific,

* CCIR Report 215-3, Geneva; 1974.

† From L. S. Golding and J. E. D. Ball, "Satellite Television Covers the World." Reprinted by permission from *IEEE Spectrum*, Vol. 10, No. 8, Aug. 1973. Copyright 1973 by the Institute of Electrical and Electronics Engineers, Inc.

and Indian Ocean regions, reaching some 47 different countries. Frequently, the national television standards of these countries differ, and as a result, particularly if color programs are being exchanged, high-quality television standards converters are used in tandem with the satellite link.

A typical international satellite television circuit starts from an international television center (ITC), such as the one in New York, and terminates at another ITC, perhaps in London, with facilities at each end extending the connection to the local broadcasting authorities. The video and audio signals travel over radio relay or cable systems from the ITCs to link with the earth stations of the satellite system. The 5925–6425-MHz band is used for transmitting to the satellite, and the 3700–4200-MHz band is used for receiving from the satellite. The television channel typically uses about 35 MHz of the 500 MHz available in each band, with the remaining bandwidth used mainly for telephone service.

To protect earthbound communications from interference, the CCIR recommends a limit on the power flux density in any radio frequency interval of 4 kHz within the total satellite link band of 500 MHz. The resulting limit on effective isotropic radiated power (eirp) in the direction of the earth's surface is approximately +12 dBW.

To allow a total eirp larger than +12 dBW for satellite television transmission, a dispersal waveform [1] is commonly added to the video signal at the transmitting earth station. The power dispersal waveform insures that even in the absence of the television signal, when the carrier is unmodulated, the RF carrier power will not exceed the +12-dBW limit in any 4-kHz band. As shown in Fig. 45, the

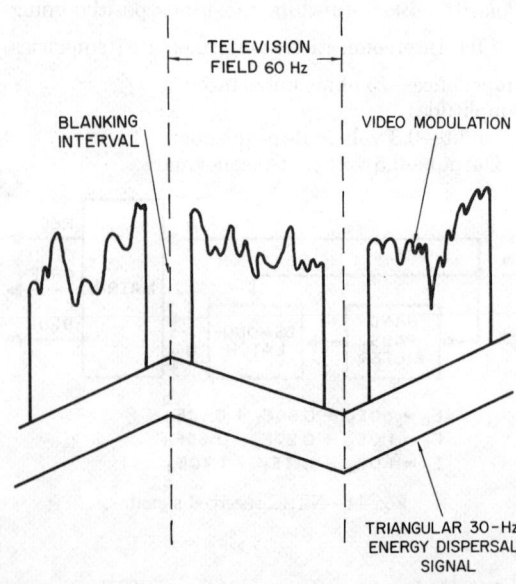

Fig. 45—Use of energy-dispersal signal.

dispersal waveform is a low-frequency symmetrical triangular wave having a fundamental frequency of 30 Hz and locked in phase with the 60-Hz rate of the television field waveform.

The points of inflection of the triangular waveform are made to coincide with the television field blanking interval so as to minimize the possibility of interfering with the picture. This technique provides adequate power dispersion regardless of the amplitude of video modulation and thus permits very high eirp values to be assigned to the satellite television RF carrier. The triangular waveform is removed at the receiving earth station before the video signal is passed on to the terrestrial link.

Quality objectives for television are derived from the judgments of human observers as they view television pictures. Using existing data on such human judgments together with information on the characteristics of TV channels, it is possible to formulate a set of objectives for a satellite TV channel which would provide high levels of picture quality.

Current CCIR recommendations have been aimed toward adequate performance requirements. A satellite hypothetical reference circuit (SHRC) provides the framework for these recommendations. The SHRC is defined as starting at the transmitting earth station and terminating at the receiving earth station [2].

Presently, recommendations by the CCIR are a nominal bandwidth of 4.2 MHz (for 525-line transmissions) and a 56-dB ratio of peak-to-peak picture signal to weighted random noise [3]. However, a number of other performance requirements also play a role in determining television quality.

A near-complete list of these requirements is given in Table 13. If the satellite power and bandwidth needed to meet these objectives are not available, a lower set of objectives must be established. It is important that proper balance be maintained between these lower objectives so that their relative subjective impairment of the transmitted picture signal remains reasonably constant.

In this context, the four most important parameters to consider are: the attenuation-frequency response of the channel, the signal to weighted-random-noise ratio in the channel, the differential gain of the channel, and the differential phase distortion introduced by the channel.

Generally speaking, the other transmission parameters can be satisfied consistent with the requirements set for these four. The attenuation-frequency response of the channel, for instance, may be relaxed if there is no requirement to transmit color signals, and in that case, a direct trade can be made between channel noise and channel bandwidth. However, the channel must usually be capable of transmitting color signals and, as a result, must have a video bandwidth of 4.2 MHz

TABLE 13—SATELLITE TELEVISION PERFORMANCE
REQUIREMENTS*

Transmission parameter	Performance requirement
Insertion gain†	±0.5 dB
Insertion gain variations	
Short term (1 second)†	±0.2 dB
Long term (1 hour)†	±1.0 dB
Noise	
Random (weighted)† ‡	56 dB
Impulsive†	25 dB
Periodic, below 1 kHz** ††	50 dB
Periodic, 1 kHz to 0.2 MHz††	55 dB
Attenuation frequency‡‡	±0.5 dB
Envelope delay†	±63 ns
Differential gain†	1.2 dB
Differential phase†	3 degrees
Linear distortion	
Field time† **	±1 percent
Line time†	±1 percent
Short time††	±1 percent
Luminance-chrominance inequalities	
Gain††	0.5 dB
Delay††	±50 ns
Synchronizing signal	
Nonlinearity†	−10 percent
Distortion†	+5 percent

* These requirements are for a CCIR-type satellite hypothetical reference circuit (SHRC).

† Based on CCIR Recommendation 421-2.

‡ Present provisional CCIR recommended value for the SHRC given in Recommendation 354-1.

** Measured after clamping. Clamping is used in satellite links to attenuate the energy dispersal waveform upon reception.

†† Based on CCIR Recommendation 451-1.

‡‡ Based on CCIR Recommendation 421-2 and Report 407-1.

in order to permit the color information to pass satisfactorily.

Present satellite transmission systems for television use frequency modulation. For FM, the RF bandwidth B_{RF} and the video bandwidth f_v are related by Carson's Rule, so that $B_{\text{RF}} = 2\Delta f + 2f_v$, where Δf is the peak FM deviation. Adherence to the Carson's Rule bandwidth will ensure a good-quality picture with tolerable signal distortion levels. However, if the resulting signal-to-random noise level is unsatisfactory, it may be desirable to increase the peak deviation of the FM carrier.

The consequences of increased deviation are increased signal distortion, primarily differential gain and differential phase. Furthermore, overdeviation can produce impulsive noise when the picture signal contains high-amplitude, high-frequency

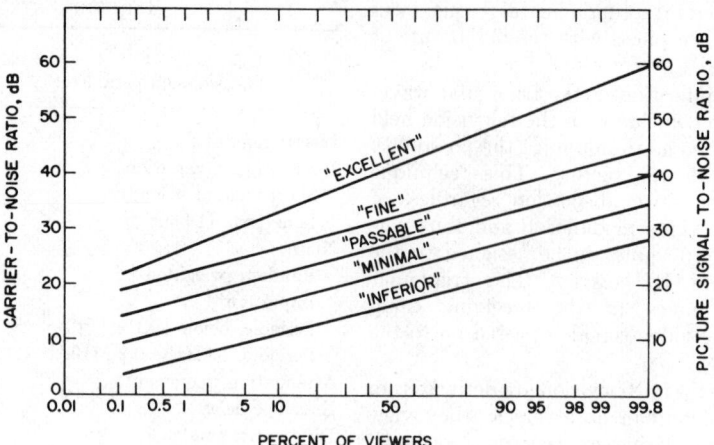

Fig. 46—Effect of random noise on subjective picture quality.

components as in highly saturated color information. To minimize these undesirable consequences, the practical upper limit on overdeviation for color signals is about 4 dB. Although such overdeviation can improve the signal-to-noise ratio, it should only be used when absolutely necessary, and it should not be considered a permanent feature of any satellite circuit.

For television transmission, noise is specified as a weighted parameter [4], with the weighting determined by subjective testing. A weighting network attenuates noise at the upper end of the video band relative to noise at the lower end, since the upper-frequency noise has been found more annoying to viewers. This principle holds true for color television signals as well as monochrome, although it is modified somewhat due to a need to maintain a higher signal-to-noise ratio at or near the color subcarrier frequency.

Subjective tests reported by Barstow and Christopher [5] quantify this difference between monochrome and color systems. Their results also show the level at which the "median" practiced observer will see noise in color TV pictures. They used white noise in a 4.2-MHz band, obtaining a "just perceptible" response when the noise was added to the viewed picture at 44–45 dB below the peak-to-peak value of the composite video signal. In CCIR terms, their "just perceptible" level would be 47–48 dB below the picture signal, excluding the synchronizing pulse. The Bell System 4000-mile user-to-user requirement is 50 dB below the picture signal.

Similar subjective tests were conducted by the Television Allocations Study Organization (TASO) [6]. Results of these tests are shown in Fig. 46. The left ordinate shows TASO signal-to-noise measurements, and the right ordinate shows CCIR figures for the ratio of peak-to-peak pic-

ture signal to weighted random noise. The Barstow and Christopher median-observer just-perceptible rating corresponds roughly to a point on the "excellent" rating line representing 75 percent of the TASO viewers. Note that CCIR's requirement of 56 dB for the SHRC reference channel would approximate a rating of "excellent" by about 99 percent of TASO viewers.

Differential gain (DG) is defined in terms of a small-amplitude 3.58-MHz sine wave. It is the amplitude change of the sine wave as a function of luminance amplitude (picture brightness).

As can be seen in Fig. 47, the subjective effect of DG on critical observers has a reasonably symmetrical distribution across the dB scale. The Bell

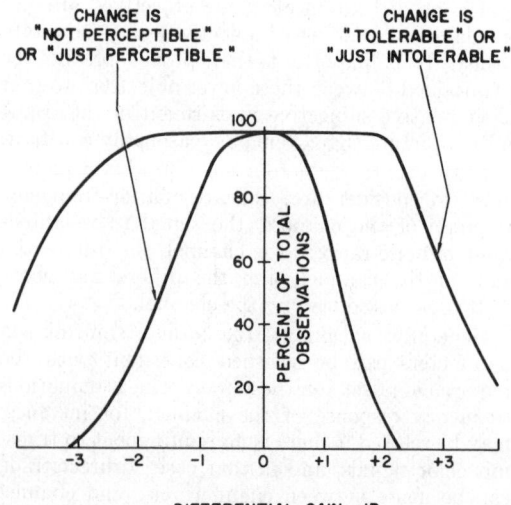

Fig. 47—Effect of differential gain on subjective picture quality.

System requirement for a 4000-mile intercity land link is 2 dB, and the CCIR recommendation for a 2500-mile terrestrial link is 1.2 dB [7].

Like DG, differential phase (DP) is also defined for a small-amplitude 3.58-MHz sine wave. It is the change in phase of the wave, as a function of luminance amplitude, relative to its phase at blanking level; DP is always given in degrees.

Shown in Fig. 48 is the subjective effect of DP on some critical observers. The Bell System 4000-mile user-to-user requirement for DP is 5 degrees, corresponding to a rating of "just perceptible" or better by about 50 percent of the critical observers in Fig. 48. Although no CCIR 2500-mile link recommendation is given for DP, comparison with the DG requirement suggests a value of three degrees.

For FM transmission via satellite, three development areas appear to hold some promise for more efficient operation in the future. These areas are: improved pre-emphasis and de-emphasis techniques [8], threshold extension demodulators, and time and frequency diplexed audio channels [9]. Each of these appears to offer some hope of reducing the bandwidth or power, or both, needed to provide a given output signal quality.

In the video spectrum of the composite color signal and in some of the subjective weighting curves for color signals, the interval near the color subcarrier has a local maximum. More optimal pre-emphasis networks can be expected to reflect details such as this in the shape of the video spectrum and weighting network. This consideration is particularly important in wideband FM systems.

One of the difficulties in FM transmission of television signals is the great variation in the signal as a function of picture material. Variations in chrominance signal amplitude can range from values as high as 40 percent of the luminance signal image, for highly saturated colors, to small values near zero for nearly monochromatic scenes. Luminance signal amplitudes and dc levels also vary over wide ranges. These variations suggest the use of time-varying or adaptive pre-emphasis which would monitor the signal energy in various frequency intervals of the video spectrum. This would allow the adjustment of the pre-emphasis characteristic so as to maintain a more constant signal-to-noise ratio with changes in picture material.

Use of FM demodulators with feedback would allow reduction of the FM threshold carrier-to-noise ratio and permit satisfactory operation at lower ratios. The difficulty in developing such threshold extension demodulators is the requirement that they operate with large (30–40 MHz) bandwidths and meet linearity requirements stringent enough to avoid crosstalk between the luminance and chrominance components of the signal.

Time diplexed transmission of sound signals is receiving widespread attention. Systems proposed in France, Britain, the U.S.S.R., and the U.S. all use various types of pulse-modulated sound signals transmitted during the video line-blanking period.

Standards Conversion: Signals of three principal color television standards are transmitted over satellite television links. These standards are: the 525-line NTSC (National Television Systems Committee) system, used in the U.S., Canada, and Japan; the 625-line SECAM (sequential color with memory) system, used in France, Eastern Europe, and the U.S.S.R.; and the 625-line PAL (phase alternation line) system, used in West Germany and most of the other countries of Western Europe.

INTELSAT recommends that, where possible, conversion should take place after the signal has been transmitted over the satellite link. This minimizes the possibility of double-conversions which could take place on multiple-destination transmissions, and which would unnecessarily degrade the quality of the transmission.

Conversion is a process that involves one or more of the following steps:

1. Change in number of lines per frame including interpolation between lines to obtain a smooth reproduction in the vertical direction of a frame.

2. Change in number of fields per second including interpolation between fields to obtain a smooth portrayal of motion.

3. Adjustments to account for the different durations of the scanning lines in the input and output standards.

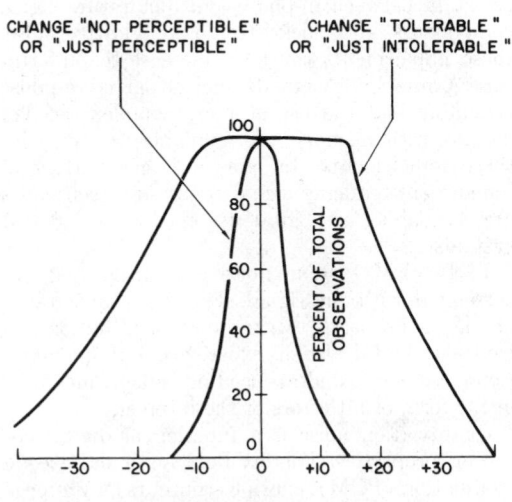

Fig. 48—Effect of differential phase on subjective picture quality.

4. Compensation for timing errors that occur in the waveform of the output standard.

5. Transcoding, to change the coding of the chrominance information from that of the input standard to that of the output standard.

The earliest type of standards converters were monochrome optical converters in which the original image was displayed on a cathode-ray tube (CRT) and then rescanned at the new standard. Optical conversion has been improved by work such as that done in the Federal Republic of Germany [10]. One such optical converter separates the luminance and chrominance signals using filters. After separation, the chrominance subcarrier frequency is changed, and CRT displays of the luminance and chrominance signals are provided. Plumbicon tubes are then used to convert these signals to the output standard [11].

Work in the United Kingdom and Japan [12] has led to the development of all-electronic converters using field and line delays. These converters are said to have improved performance with regard to movement blurring, geometric distortion, brevity of response, resolution, and flicker, and to require less adjustment than optical converters.

Future standards converters are likely to convert the video signal into digital form, using high-speed digital processors to convert the signal to the required line standard. The key device in such a digital converter would be a memory capable of storing at least one complete television field and of handling data at rates in excess of 100 Mb/s.

Digital Transmission Systems: Digital transmission techniques for commercial broadcast television signals over satellite links, based on circuits capable of reliably operating at data rates of 30–120 Mb/s, are feasible. Advantages cited for digital techniques include use of less RF bandwidth, less power, or both; easier handling of some transmission impairments and of interference between satellite and ground communications; and full compatibility with highly efficient time-division multiple-access satellite systems as well as with data transmission networks.

Shown in Fig. 49 is the required carrier-to-noise ratio versus required RF bandwidth needed to transmit an NTSC 525-line color TV signal and provide an output signal-to-noise ratio of 56 dB. This figure shows that for bit rates of 50 Mb/s or less, significant gains in power and bandwidth can be realized when compared to FM transmission. Each of the five digital-system operating points shown in the figure assumes a different combination of data rate and error-correcting code. The codes used in these calculations are forward-acting types that increase the transmission rate by a factor of either 2 or ⅗ and provide bit error-rate

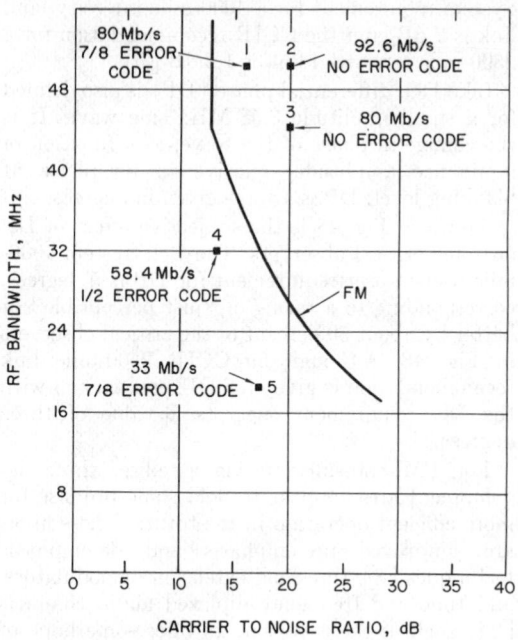

Fig. 49—Digital system performance.

reduction by a factor of 10^{-4} to 10^{-8}. Points 1 through 3 in Fig. 49 correspond to conventional pulse code modulation and, as noted in the figure, do not offer more efficient operation than FM. Points 4 and 5 correspond to proposed digital transmission systems utilizing source encoding techniques [13] and capable of providing improved bandwidth or power savings in comparison to an FM system.

In a digital system, transmission impairments manifest themselves as intersymbol interference, crosstalk between in-phase and quadrature signal components, and clock jitter. Like additive noise, these impairments cause an increase in bit error rate. A number of methods, such as partial response signalling and the use of more complex receiver designs with memory, are available to cope with these impairments. In fact, a larger variety of impairment-reducing signal processing techniques are available for digital systems than for FM systems.

In the FM system, noise is additive and uncorrelated with the signal. The usual appearance of the noise is as background scintillation. In contrast, digital systems experience both quantization noise due to digital encoding and channel noise in the form of bit errors at the receiver.

Quantization noise is a function of digital encoding properties. For example, with pulse code modulation (PCM), spurious contours may appear in the picture due to large step changes in gray level as the video signal crosses quantization boundaries. To make these contours imperceptible, a minimum of 8 bits per sample are necessary.

Differential PCM (DPCM) is a technique which quantizes amplitude changes from one sample to the next. With this technique, the quantization noise appears as a "busyness" around edges in the picture. Other types of digital encoding produce still different sorts of quantization noise. Subjective testing is the only available way to discover the bit rate needed to reduce quantization noise to acceptable levels. Here, parameters that significantly affect picture quality are varied until impairments are "just perceptible" to a practiced observer.

Channel noise due to bit errors can also be evaluated by subjective tests. In this case, the error rate is increased until the observer notices "just perceptible" degradation in the picture. For PCM, this occurs at an error rate of about 10^{-6}; for DPCM, it occurs at about 10^{-8}. The PCM noise appears as isolated picture-element errors, and DPCM noise is seen as streaks in the picture.

In FM transmission systems, the signal-to-noise ratio is measured with reference to a constant-level video signal. But in a digital transmission system, such a signal is not suitable since the quantization noise being measured is a function of the signal. To measure digital system signal-to-noise ratio, a gaussian noise loading signal can be used, with the measured ratio compared to a theoretically determined value. Another measurement technique is the noise-power ratio measurement, which has been used extensively for evaluating signal-to-noise ratio in voice channels.

Other objective tests involve the use of test patterns such as the pulse-and-bar, multiburst, and color-bar. These patterns permit oscilloscope measurements to be carried out to determine resolution or frequency response, transient response, crosstalk, and noise. In evaluating a digital transmission, system test patterns are also useful for obtaining objective measures of performance. However, it is likely that a different set of test patterns than those used to evaluate FM systems will prove more useful for digital systems. For example, a continuous ramp function is a useful pattern for showing up contouring present in the analog-to-digital conversion operation. But this pattern is not used a great deal in evaluating FM systems. Most present test patterns are single frame patterns, which would not be useful in evaluating frame-to-frame video encoding techniques.

References:

1. CCIR Report 384-1.
2. CCIR Recommendation 354-1.
3. CCIR Recommendation 352-1.
4. J. R. Cavanaugh, "A Single Weighting Characteristic for Random Noise in Monochrome and NTSC Color Television," *J. SMPTE*, Vol. 79, pp. 105–109, Feb. 1970.
5. J. M. Barstow and H. N. Christopher, "The Measurement of Random Video Interference to Monochrome and Color TV Pictures," *AIEE Transactions*, Part 1, pp. 313–320, 1962.
6. Television Allocations Study Organization, *TASO Report to the FCC*, Mar. 16, 1969.
7. CCIR Requirement 421-2.
8. R. A. Bruce, "Optimum Pre-Emphasis and De-Emphasis Networks for Transmission of Television by PCM," *IEEE Trans Communications Technology*, pp. 91–96, Sept. 1964.
9. CMTT Report CMTT/1035, Oct. 31, 1969.
10. F. Faeschke, "Methods for NTSC-PAL Colour Standards Conversion Between Television Standards With Different Vertical Frequencies," *NTZ*, pp. 177–180, Apr. 1968.
11. H. Wendt, "An Electro-Optical Standards Converter for Colour Television Signals," *NTZ*, Vol. 5, pp. 281–285, May 1969.
12. E. R. Rout and R. E. Davies, "Electronic Standards Conversion for Transatlantic Colour Television," *J. SMPTE*, Vol. 77, pp. 12–14, Jan. 1, 1969.
13. CCITT, Committee Special D Report No. 110.

Radar (*RA*dio *D*etection *A*nd *R*anging) is a method for detecting and measuring the range to a distant target with beamed microwaves. With the advent of a variety of efficient lasers and laser detectors, the term is sometimes now applied to detection and ranging at optical wavelengths.

GENERAL*

A simplified diagram of a typical, basic system for radio direction and range finding is shown in Fig. 1. A pulsed high-power transmitter emits radio waves for a short time interval through a directive antenna to illuminate the target. The returned echo is picked up, usually by the same antenna, is amplified by a high-gain wide-band receiver, and is displayed on an indicator. Direction of a target is usually indicated by noting the direction of the narrow-beam antenna at the time the echo is received. The range is measured in terms of time because the radar pulse travels with the speed of light, 300 meters one way per microsecond, or approximately 12 microseconds per round-trip radar mile. Chart 1 gives several convenient values of range and echo time.

* M. I. Skolnik, "Introduction to Radar Systems," McGraw-Hill Book Company, New York; 1962.

See also:

R. S. Berkowitz, "Modern Radar," John Wiley & Sons, Inc., New York; 1965.

D. P. Meyer and H. A. Mayer, "Radar Target Detection," Academic Press, Inc., New York; 1973.

Alternative systems may employ pulses of longer duration and radiation duty cycle up to 100 percent (cw). Measurement of radio-frequency Doppler velocity in the receiver circuits, then, permits a target's radial velocity to be computed.

CHART 1—RADAR RANGE/TIME RELATIONSHIPS.*

149.9	meters per microsecond
491.8	feet per microsecond
0.5	foot per nanosecond
163.9	yards per microsecond
0.08088	nautical mile per microsecond
0.09313	statute mile per microsecond
15	kilometers per 100 microseconds
8	nautical miles per 100 microseconds
1	nautical mile per 12.36 microseconds
1000	yards per 6.10 microseconds
2000	yards (1 "radar mile") per 12.2 microseconds

* Values given are approximately equal to one-half the velocity of light (c), where $c = 2.99776 \times 10^8$ meters per second.

The factors characterizing the operation of each component are shown in Fig. 1. These are discussed below in turn and combined into the free-space range equation. The propagation factors modifying free-space range are presented.

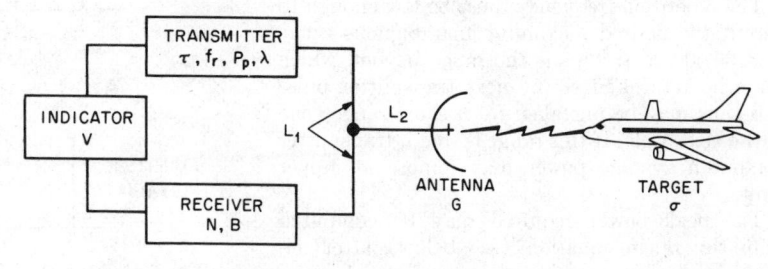

Fig. 1—Simplified diagram of a radar set.

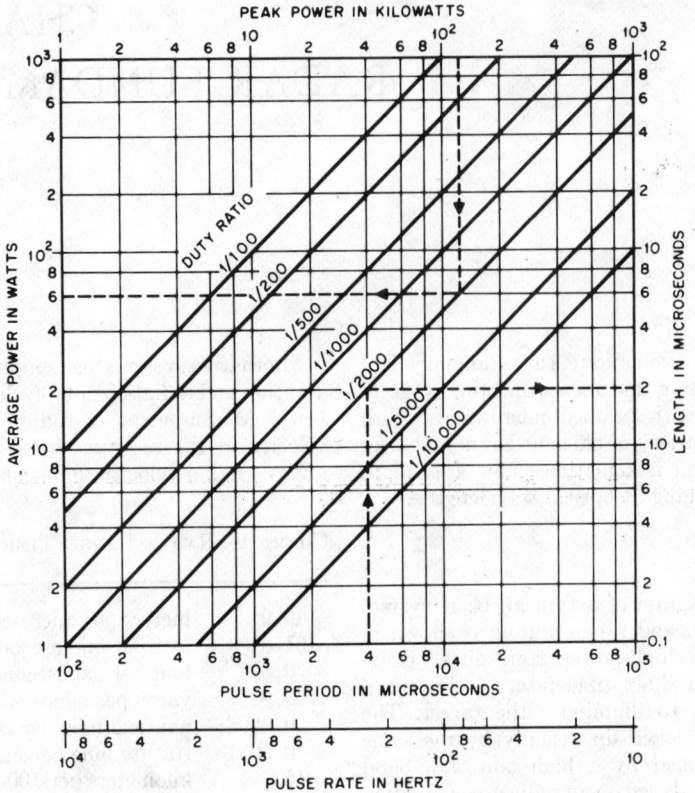

Fig. 2—Power-time relationships.

TRANSMITTER

Important transmitter factors are: $\tau =$ pulse length in microseconds, $f_r =$ pulse rate in hertz, $d =$ duty ratio $= \tau f_r \times 10^{-6} = P_a/P_p$, $P_a =$ average power in kilowatts, $P_p =$ peak power in kilowatts, and $\lambda =$ carrier wavelength in centimeters.

Pulse length is generally in the microsecond region. Longer pulses may be used for greater range if the transmitter power capacity permits. On the other hand, if a range resolution of ΔR feet is required, the pulse cannot be longer than $\Delta R/500$ microseconds unless pulse compression is used.

The repetition frequency must be low enough to permit the desired maximum unambiguous range $(f_r < 90\,000/R_u)$. This is the range beyond which the echo returns after the next transmitter pulse and thus may be mistaken for a short-range echo of the next cycle. If this range is small, transmitter maximum average power may impose an upper limit.

The peak power required may be computed from the range equation (see below) after determination or assumption of the remaining factors. Peak and average power may be interconverted by use of Fig. 2. Pulse energy is $P_p \tau \times 10^{-3}$ joules.

To improve range resolution when performance is limited by transmitter peak power, pulse compression techniques may be used.

Pulse compression waveforms permit achievement of range resolutions shorter than that corresponding to the radiated pulse width. The most common type of pulse compression is the linear fm type. The radio frequency of the transmitted pulse is made to change linearly during the pulse interval τ (Fig. 3A) by the frequency B. In the receiving

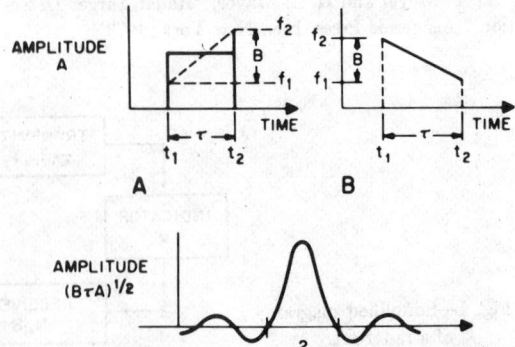

Fig. 3—Pulse compression signal.

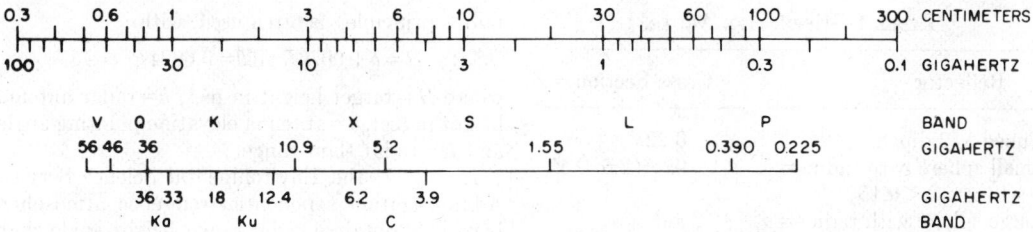

Fig. 4—Correlation between frequency, wavelength, and band nomenclature for radar.

circuits, the return signal is passed through a compression filter. Time delay versus frequency response of this filter is shown in Fig. 3B. The filter then has an output as shown in Fig. 3C.

The factor $B\tau$ (the pulse compression ratio) is a measure of the improvement in range resolution.

The choice of carrier frequency is a complex one, determined by a variety of interdependent factors —available transmitter power sources, antenna size, angular accuracy, aerial coverage requirements, and propagation considerations. Frequency–wavelength conversions are facilitated by Fig. 4, which also defines commonly employed band nomenclature.

ANTENNA

Antenna gain (G) and beamwidth (θ_1, θ_2) are directly related. For well designed antennas of conventional beam shape, as used in typical radar applications, the gain may be estimated from

$$G = K_1/(\theta_1, \theta_2)$$

where θ_1 and θ_2 are the 3-decibel beamwidths in orthogonal planes, and the constant K depends on type and side-lobe-level design. When θ_1 and θ_2 are expressed in degrees, K can be chosen to be about 27 000 for horn-fed paraboloidal reflectors, 30 000 for linear-array-fed parabolic cylinders, and 37 000 for two-dimensional arrays of radiators.

Antenna gain for large "plane" aperture antennas can also be estimated on the basis of antenna area (A) and wavelength (λ) with

$$G = K_2(4\pi A/\lambda^2)$$

where A and λ are in the same units, and K_2 is a constant depending on type and design, ranging between 0.6 and 0.9.

For large "plane" aperture antennas, the beamwidth θ_1 to the aperture dimensions d_1 along the plane of measurement is given by

$$\theta_1 = K_3(\lambda/d_1)$$

where λ and d_1 are in the same units, and K_3 is a constant which typically ranges between 55 for a 20-decibel side-lobe-level design and 75 for a 35-decibel side-lobe-level design.

Beamwidth for a horn-fed paraboloidal reflector is given in Fig. 5.

TARGET ECHOING AREA

The radar cross section (σ) is a figure of merit which describes the microwave reflection efficiency of a target of interest. Radar cross section is usually defined more specifically as that area of a lossless isotropic reflecting sphere which gives the same signal at the receiver as the target of interest. Since an isotropic sphere scatters an oncoming beam uniformly in all directions (4π steradians),

$$\sigma = 4\pi J/H$$

where J is the radiation intensity from the target and H is the irradiance at the target. For large complex structures and short wavelengths, the values vary rapidly with aspect angle. The effec-

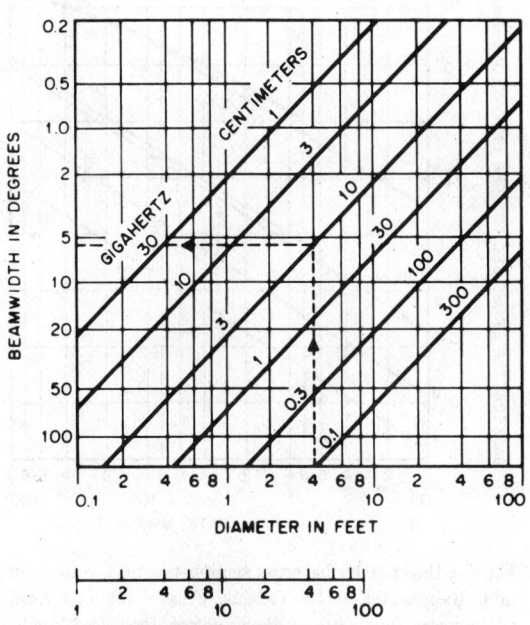

Fig. 5. Beamwidth of a paraboloidal reflector.

TABLE 1—EFFECTIVE AREAS.

Reflector	Cross Section $= \sigma$
Tuned $\lambda/2$ dipole	$0.22\lambda^2$
Small sphere with radius $= a$, where $a/\lambda < 0.15$	$9\pi a^2 (2\pi a/\lambda)^4$
Large sphere with radius $= a$, where $a/\lambda > 1$	πa^2
Corner reflector with one edge $= a$ (maximum)	$4\pi a^4/3\lambda^2$
Flat plate with area $= A$ (normal incidence)	$4\pi A^2/\lambda^2$
Cylinder with radius $= a$, length $= L$ (normal incidence)	$2\pi L^2 a/\lambda$
Small airplane	5 m²
Medium-size airplane	10 m²
Small ship	10^4 m²
Large ship	10^6 m²

tive areas of several important configurations* are listed in Table 1.

The target echoing area of rain is of interest because the rain return may obscure the radar return from desired targets; alternately, the rain return is of direct interest in radars designed for weather detection or analysis.

The radar cross section of rain can be found as

$$V\sigma_r$$

where V is the volume of the radar beam occupied by rain, and σ_r is the backscatter cross section per unit volume.

When the radar beam is within the rain region, then

$$V \approx (\pi/4) R^2 \theta_1, \theta_2 (\tau c/2)$$

where R = radar range, θ_1, θ_2 = effective antenna beamwidths, τ = pulse width, and c = velocity of propagation.

The value of σ_r is given in Fig. 6 for various wavelengths.

TARGET HEIGHT

The height of a target can be obtained from radar target range and antenna elevation pointing angle (θ). The height computation must account for earth curvature and for atmospheric refraction. A simplified model for refractive index ($\frac{4}{3}$ earth

* L. N. Ridenour, "Radar System Engineering," Vol. 1, Radiation Laboratory Series, McGraw-Hill Book Company, New York, New York; 1947: See pp. 64–68, 78, 80.

radius principle) is often used, with

$$H = h + 6076R \sin\theta = 0.6624R^2 \cos^2\theta$$

where H = target height in feet, h = radar antenna height in feet, θ = antenna elevation pointing angle, and R = radar slant range.

In Fig. 7, the Environmental Science Services Administration exponential reference atmosphere is used to obtain a radar range–height angle chart useful at longer ranges.

RECEIVER

The ratio of signal-to-noise power obtained from the antenna is reduced by noise generated within the radar receiver. The noise figure F is the measure of degradation in available signal-to-noise power between receiver input and output terminals. Equivalently

$$F = N_{out}/KT_0 B_n G$$

where N_{out} = available output noise power in watts, k = Boltzmann's constant (1.38×10^{-23} joule/kelvin), T_0 = standard temperature (290 kelvins), B_n = effective noise bandwidth in hertz, and G = available power gain.

For cascaded noisy networks with individual noise figures F_n and gains G_n, the overall noise figure is given by

$$F_0 = F_1 + (F_{2-1}/G_1) + \cdots + [F_{n-1}/(G_1 \cdots G_{n-1})].$$

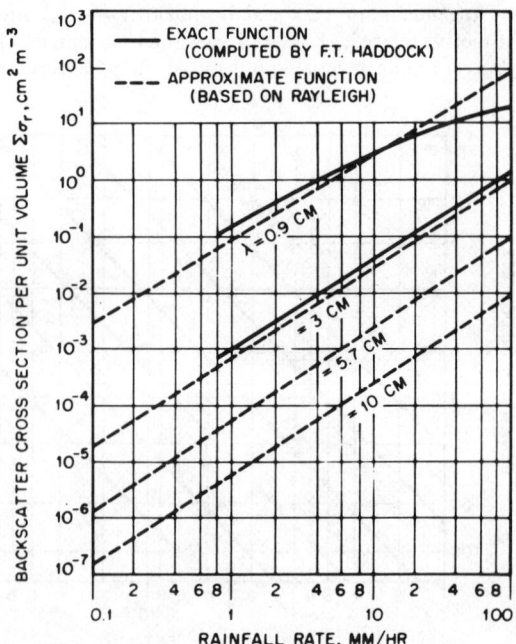

Fig. 6—Backscattering cross section per unit volume of rain. Rain water at 18° Celsius. *From Gunn and East, "Microwave Properties of Precipitation Particles," Quarterly Journal, Royal Meteorological Society, Vol. 80, pp. 522–545; October 1954.*

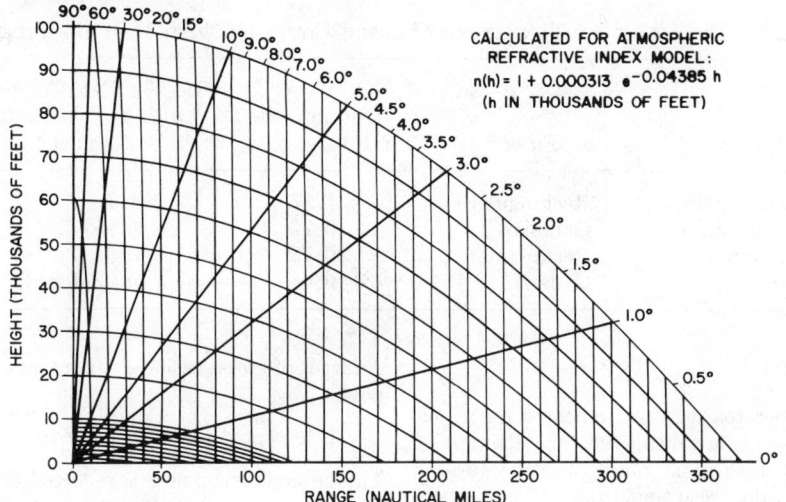

Fig. 7—Radar range-height-angle chart, calculated for an exponential model of the atmospheric refractive index. "Radar range" means distance along the ray path. Elevation angles are angles of rays with respect to horizontal at the radar antenna. *From L. V. Blake, "A Guide to Basic Pulse-Radar Maximum-Range Calculation," Part 1—Equations, Definitions, and Aids to Calculation, U.S. Naval Research Laboratory, Washington, D.C.; December 1969.*

The bandwidth of the receiver is chosen to optimize the detection of the radar return signal. The filter network whose output maximizes the output peak-signal-to-mean-noise-power ratio is called a matched filter. The transfer function of this filter is given by

$$H(\omega) = aF^*(\omega) \exp(-j\omega t_d)$$

where a = normalizing filter gain constant, $F^*(\omega)$ = complex conjugate of $F(\omega)$, the spectrum of the radar signal $f(t)$, and t_d = filter delay constant.

The impulse response of the matched filter is given by

$$h(t) = af(t_d - t).$$

It is usually not feasible to provide a matched filter. For simple types of signals, relatively simple filters can be used. Table 2 gives the loss in signal-to-noise ratio compared with a matched filter when band-pass filters of the types shown are used. B and τ are the 3-decibel bandwidth of the filter and 3-decibel pulsewidth of the signal, respectively.

When the optimum $B\tau$ is not used, signal detection is degraded. Figure 8 gives experimentally obtained data for several types of pulse and filter shapes.

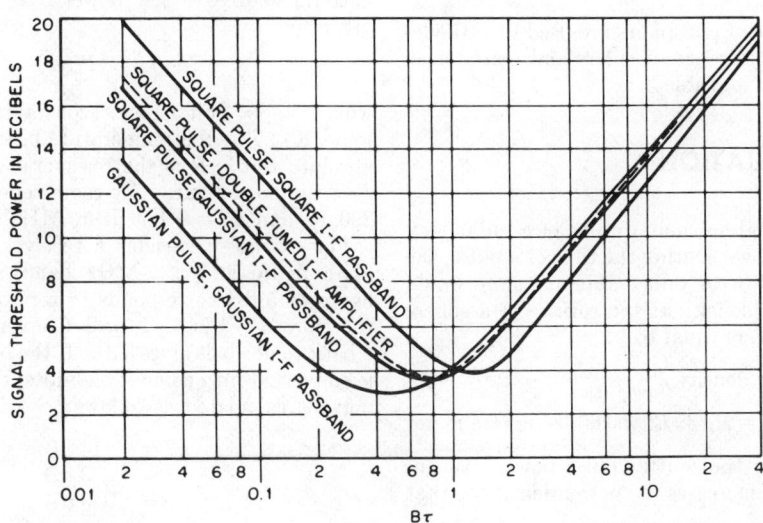

Fig. 8—Signal threshold power as a function of $B\tau$ for different shapes of pulses and of i–f passband. *From J. L. Lawson and G. E. Uhlenbeck, "Threshold Signals," Fig. 8–11, p. 208, Boston Technical Publishers, Inc., Cambridge, Mass.; 1964.*

TABLE 2—EFFICIENCY OF NONMATCHED FILTERS COMPARED WITH MATCHED FILTER.

Input Signal	Filter	Optimum $B\tau$ (IF band pass)	Loss in SNR Compared with Matched Filter (decibels)
Rectangular pulse	Rectangular	1.37	1.7
Rectangular pulse	Gaussian	0.72	0.98
Gaussian pulse	Rectangular	0.72	0.98
Gaussian pulse	Gaussian	0.44	0 (matched)

INDICATOR

When an operator uses an indicator on which radar video signals are displayed for detection of targets, signal integration takes place because of light intensification resulting from phosphor persistence and because of perception of adjacent intensified spots. This integration affects the signal-to-noise ratio required for detection. For a search radar using a typical Plan Position Indicator (PPI), experimental data indicate that ideal power integration is approached when the number of hits is small, but becomes less efficient as the number of hits increases. Figure 9 gives a PPI indicator visibility factor V. This is the required signal-to-noise ratio for a linear detector as a function of the number of pulses integrated, for a 50 percent probability of detection. The curves are calculated for a nonfluctuating signal for the indicated values of false-alarm probability (P_{fa}).

Many types of radar indicators other than PPI are used for display of radar video. These are shown in Fig. 10. Type A was the first type used and is the best example of a deflection-modulated display. The PPI is the most common intensity-modulated type.

Various types of phosphors are used in cathode-ray tubes, with a range of colors and persistence. These are listed in Table 3.

RANGE EQUATION

The theoretical maximum range of a radar may be found as follows. During the time of a pulse, the power radiated with unity antenna gain would result in power density at the surface of a sphere of radius R meters equal to

Radiated power density

$$= P_t/4\pi R^2 \text{ watts per square meter}$$

where P_t is the transmitter pulse power. An antenna of gain G increases the power density so that at the target

Power density

$$= P_t G/4\pi R^2 \text{ watts per square meter.}$$

A target of cross section σ square meters intercepts power

$$\text{Intercepted power} = (P_t G/4\pi R^2)\sigma \text{ watts.}$$

This intercepted power is reflected and reradiated so that at the radar after the return trip over the distance of R meters, the power density at the radar is

$$\text{Power density} = (P_t G/4\pi R^2)(\sigma/4\pi R^2).$$

The antenna of gain G has an effective aperture

$$\text{Effective aperture} = G\lambda^2/4\pi$$

so that at the antenna terminal the received power S_1 is

$$S_1 = (P_t G/4\pi R^2)(\sigma/4\pi R^2)(G\lambda^2/4\pi)$$

$$= P_t G^2 \sigma \lambda^2/(4\pi)^3 R^4.$$

A loss factor L may be placed in the denominator to account for all losses (transmission-line, atmospheric, and others) lumped together. Also at the input is the receiver noise (plus noise contributions from the antenna, transmission line, and galactic sources), the power (N) of which is given by

$$N = kTB(F-1)$$

where k is Boltzmann's constant $(1.38 \times 10^{-23}$ joule/K), T is the temperature in kelvins (above absolute zero), B is the receiver bandwidth, and F is the noise figure. For room temperature, $T = 290$ K, and $kT = -114$ dBm/MHz.

For example, consider a receiver with a 5-dB noise figure and a 2-MHz bandwidth. A noise figure of 5 dB corresponds to a power ratio of 3; therefore, $F-1$ is $3-1$, or 2. Converting the ratio 2 back to decibels gives 3 dB. If the bandwidth of 2 is converted into compatible units, the value of N may be calculated as follows:

$$
\begin{aligned}
kT &= -114 \text{ dBm} \\
F-1 &= 3 \text{ dB} \\
B &= 3 \text{ dB} \\
\hline
N &= -107 \text{ dBm.}
\end{aligned}
$$

The signal-to-noise ratio at the receiver input

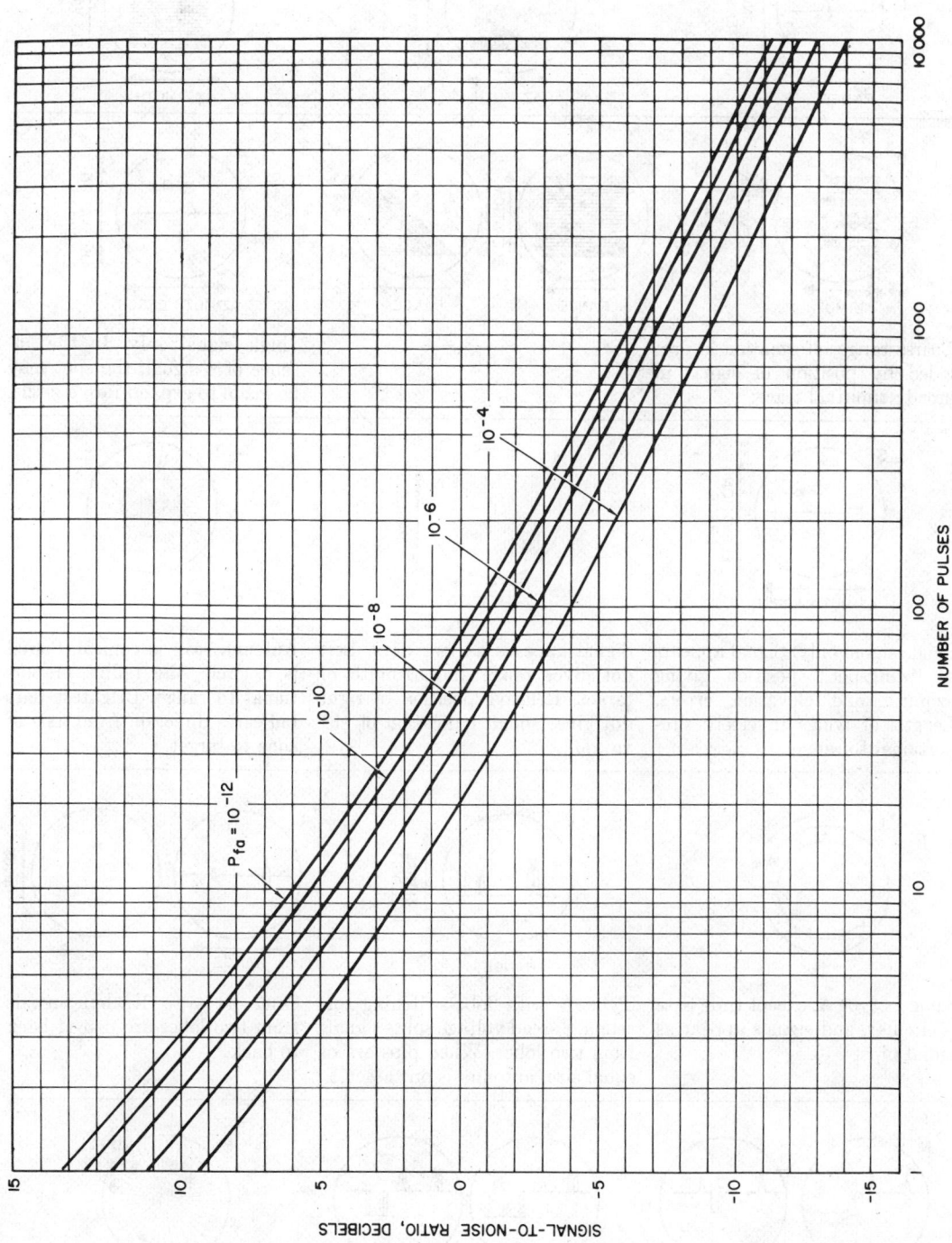

Fig. 9—Visibility factor $V_{0(50)dB}$ for PPI cathode-ray-tube display. The curve is applicable to intensity-modulated displays generally. *From L. V. Blake, "A Guide to Basic Pulse-Radar Maximum-Range Calculation," Part 1—Equations, Definitions, and Aids to Calculation, U.S. Naval Research Laboratory, Washington, D.C.; December 1969.*

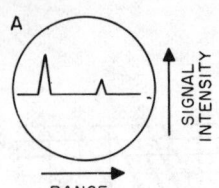

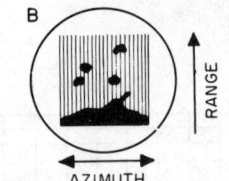

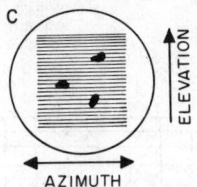

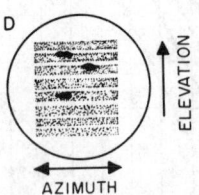

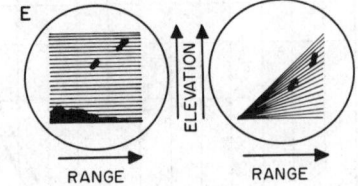

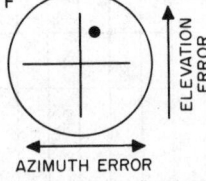

Coarse range information is provided by position of signal in broad azimuthal trace.

Single signal only. In the absence of a signal, the spot may be made to expand into a circle.

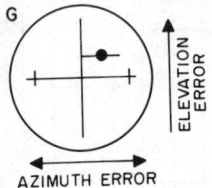

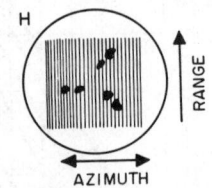

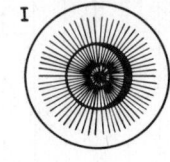

Single signal only. Signal appears as "wingspot," position giving azimuth and elevation errors. Length of wings inversely proportional to range.

Signal appears as two dots. Left dot gives range and azimuth of target. Relative position of right dot gives rough indication of elevation.

Antenna scan is conical. Signal is a circle, the radius proportional to range. Brightest part indicates direction from axis of cone to target.

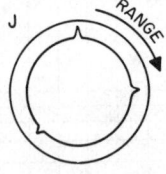

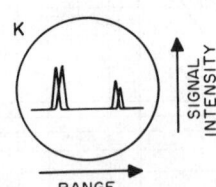

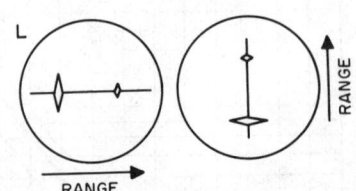

Same as type A, except time base is circular, and signals appear as radial pips.

Type A with lobe-switching antenna. Spread voltage splits signals from two lobes. When pips are of equal size, antenna is on target.

Same as type K, but signals from two lobes are placed back to back.

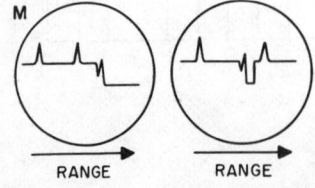

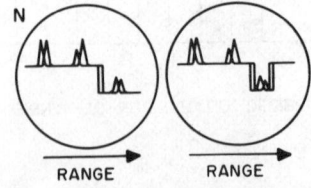

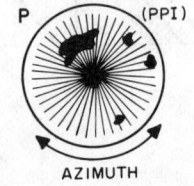

Type A with range step or range notch. When pip is aligned with step or notch, range can be read from a dial or counter.

A combination of type K and type M.

Range is measured radially from the center.

Fig. 10—Types of radar indicators. *Courtesy of McGraw-Hill Book Company.*

TABLE 3—CHARACTERISTICS OF EIA-REGISTERED STANDARD PHOSPHORS. *Adapted from R. K. Gessford, Sr., W. A. Dickinson, and J. H. Loughlin, "Cathode-Ray Tubes," Table II, p. 83, Electronics; 29 April 1960.* © *1960, McGraw-Hill Publishing Company.*

EIA Phosphor	Emission Color		Persistence	Application
	Fluorescence	Phosphorescence		
P-1	Yellowish green	Yellowish green	Medium	Used in cathode-ray oscillographs and radar.
P-2	Yellowish green	Yellowish green	Medium	Used in cathode-ray oscillographs.
P-3	Yellowish orange	Yellowish orange	Medium	
P-4	White	White	Medium to medium short	Used in monochrome television picture tubes.
P-5	Blue	Blue	Medium short	Photograph recording.
P-6	White	White	Short	Obsolete—originally used in television receivers.
P-7	White	Yellowish green	Blue—medium short Yellowish green—long	Used for radar.
P-8	Obsolete	replaced by P-7		
P-9	Obsolete			
P-10			Dark trace—very long	Outside source of light is used for observation. Persistance from seconds to several months.
P-11	Blue	Blue	Medium short	Photographic recording.
P-12	Orange	Orange	Long	Used for radar.
P-13	Reddish orange	Reddish orange	Medium	
P-14	Purplish blue	Yellowish orange	Blue—medium short Orange—medium	Used for military displays where repetition rate is 2 to 4 seconds after excitation is removed.
P-15	Green	Green	Visible—short Ultraviolet—very short	Television pickup of photographs by flying-spot scanning.
P-16	Bluish purple	Bluish purple	Very short	Television pickup of photographs by flying-spot scanning.
P-17	Yellow white to blue white	Yellow	Blue—short Yellow—long	Used for military displays.
P-18	White	White	Medium to medium short	Low-frame-rate television.
P-19	Orange	Orange	Long	Radar indicators.
P-20	Yellow green	Yellow green	Medium to medium short	High-visibility displays.
P-21	Reddish orange	Reddish orange	Medium	
P-22	Tricolor phosphor screen		Medium short	Used for color television.
P-23	White	White	Medium short	Low temperature—(sepia). Interchangeable with P-4.
P-24	Green	Green	Short	Used in flying-spot scanner tubes.
P-25	Orange	Orange	Medium	Used for military displays where repetition rate is 10 seconds to 2 minutes after excitation is removed.

TABLE 3—*Continued*

EIA Phosphor	Emission Color		Persistence	Application
	Fluorescence	Phosphorescence		
P-26	Orange	Orange	Very long	Used for radar display.
P-27	Reddish orange	Reddish orange	Medium	Color television monitor service.
P-28	Yellow green	Yellow green	Long	Used for radar display.
P-29	Two-color phosphor screen		Medium	Used as indicator in aircraft instruments.
P-30*	—	—	—	—
P-31	Green	Green	Medium short	Used in cathode-ray oscillographs.
P-32	Purple blue	Yellow green	Long	Used for radar display.
P-33	Orange	Orange	Very long	Used for radar display.

* No data available—not registered as yet with EIA.

during the time of the pulse is

$$S_1/N = (P_t G^2 \sigma \lambda^2)/[(4\pi)^3 R^4 LkTB(F-1)].$$

One scan past the target produces a sequence of n pulse returns where

$$n = \overline{PRF}\theta_{az}/6\overline{RPM}$$

where θ_{az} is the azimuth beamwidth in degrees between 3-dB points; $\overline{PRF}$ is pulse repetition frequency in pulses per second, and $\overline{RPM}$ is the antenna rotation rate in revolutions per minute. For a sequence of n pulses, there is an improvement in signal-to-noise ratio over that of a single pulse

$$S/N|_n = [I(n)](S_1/N)$$

where $I(n)$, the integration improvement factor, is a function of n which lies between n and $\sqrt{n}$ depending on the type of processing. The value of S_1/N required for a sequence of n pulses to produce a given probability of detection and probability of false alarm is called the visibility factor, V_0, which is a function of the desired probabilities of detection and false alarm and of n ($I(n)$ is included in V_0). The desired detection will take place if

$$S_1/N \geq V_0.$$

Figure 9 shows values for V_0. For example, for 10 pulses and a P_{fa} of 10^{-6}, 50% probability of detection requires $S/N = 3.4$ dB.

Combining the above equations gives the following expression for the detection range:

$$R = \{(P_t G^2 \sigma \lambda^2)/[(4\pi)^3 LkTB(F-1)V_0]\}^{1/4}.$$

SIGNAL ATTENUATION BY WEATHER

Atmospheric absorption causes attenuation of the rf signal. Using a standard value of oxygen

and water-vapor constant, Blake* has computed 2-way path attenuation for a radar at sea level as functions of range, frequency, and antenna pointing angle. Where significant (see Fig. 12), this absorption attenuation must be iteratively applied to obtain the actual range performance prediction. Precipitation can increase path attenuation significantly, particularly at the higher frequency, and hence must be considered in range performance prediction. A realistic environmental model does

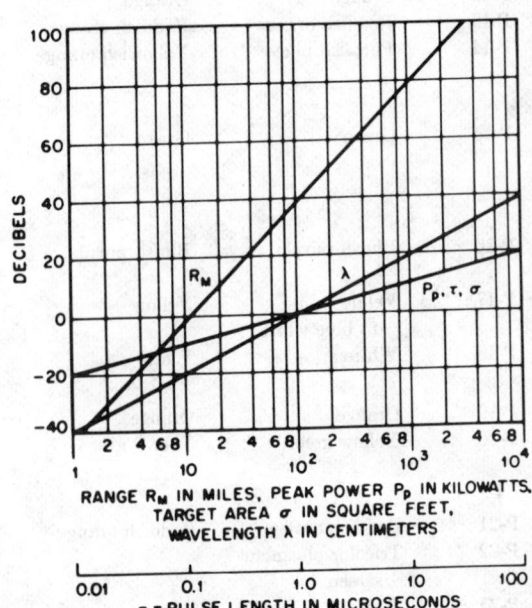

Fig. 11—Radar range equation.

* L. V. Blake, "A Guide to Basic Pulse-Radar Maximum-Range Calculation," Part 1—Equations, Definitions, and Aids to Calculation, U.S. Naval Research Laboratory, Washington, D.C.; December 1962.

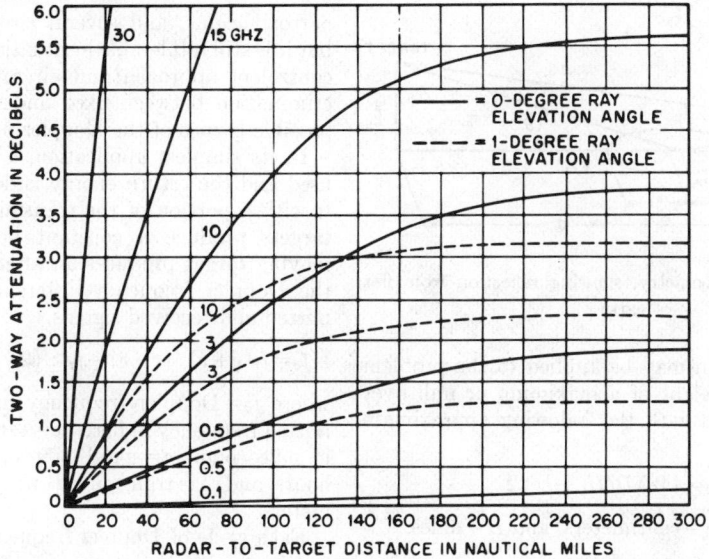

Fig. 12—Radar atmospheric attenuation. *From L. V. Blake,"A Guide to Basic Pulse-Radar Maximum-Range Calculation," Part 1—Equations, Definitions, and Aids to Calculation, U.S. Naval Research Laboratory, Washington, D.C.; December 1962.*

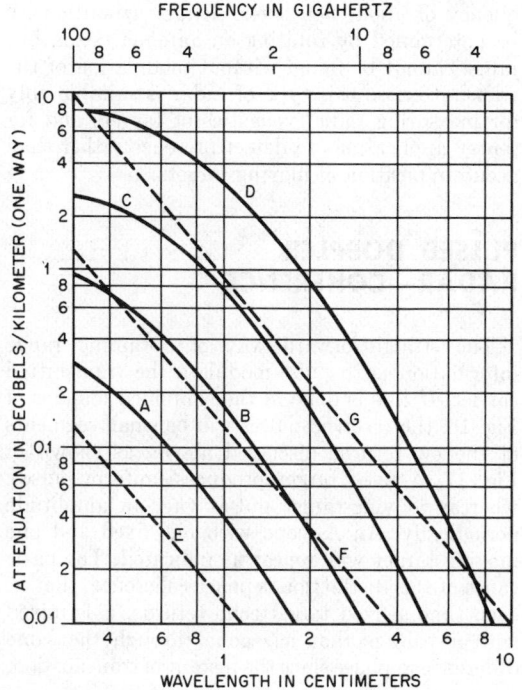

Fig. 13—Weather attenuation. $A = 0.25$ mm/hr (drizzle); $B = 1$ mm/hr (light rain); $C = 4$ mm/hr (moderate rain); $D = 16$ mm/hr (heavy rain). The broken curves show attenuation in fog or cloud. $E = 0.032$ g/m³ (visibility about 2000 ft); $F = 0.32$ g/m³ (visibility about 400 ft); $G = 2.3$ g/m³ (visibility about 100 ft). *From H. Goldstein, "Attenuation by Condensed Water," pp. 671–692 of "Propagation of Short Radio Waves," editor D. E. Kerr, McGraw-Hill Book Company, New York; 1951.*

not necessarily assume that precipitation occurs throughout the path of interest. Figure 13 provides attenuation per unit of path length versus frequency, for various types of weather, which must then be considered through the portion of model path in which precipitation is assumed.*

REFLECTION LOBES

The maximum theoretical free-space range of a radar is often appreciably modified, especially for low-frequency sets, by reflections from the earth's surface. For low angles and a flat earth, the modifying factor is

$$F = 2 \sin\left[(2\pi h_1 h_2)/\lambda R \right]$$

where h_1, h_2, and R are defined in Fig. 14, all in the same units as λ. The resulting vertical pattern is shown in Fig. 15 for a typical case. The angles of the maxima of the lobes and the minima, or nulls, may be found from

$$\theta_m = h_2/R = n\lambda/4h_1$$

where

$\theta_m =$ angle of maximum in radians, when

$$n = 1, 3, 5 \ldots$$

$=$ angle of minimum in radians, when

$$n = 0, 2, 4 \ldots$$

* H. Goldstein, "Attenuation by Condensed Water," pp. 671–692 of "Propagation of Short Radio Waves," McGraw-Hill Book Company, New York; 1951.

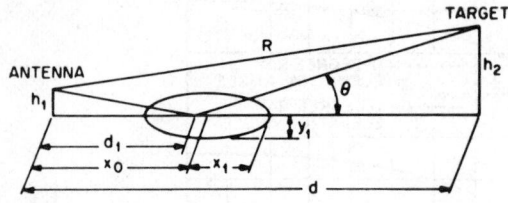

Fig. 14—Radar geometry, showing reflection from flat earth.

This expression may be applied to the problem of finding the height of a maximum or null over the curved earth with the following approximate result.

$$H_2 = 44n\lambda D/H_1 + D^2/2$$

where H = feet, λ = centimeters, and D = miles.

REFLECTION ZONE

The reflection from the ground occurs not at a point, but over an elliptical area, essentially the first Fresnel zone. The center of the ellipse and its dimensions may be found from

$$x_0 = d_1(1+2a)$$

$$x_1 = 2d_1[a(1+a)]^{1/2}$$

$$y_1 = 2h_1[a(1+a)]^{1/2}$$

where x_0, x_1, y_1, and d are shown in Fig. 14, and

$$d_1 = h_1/\sin\theta = h_1 d/(h_1+h_2)$$

$$a = \lambda/4h_1 \sin\theta.$$

In the maximum of the first lobe, $a = 1$, and the distances to the nearest and farthest points are

$$x_0 - x_1 = 0.7h_1^2/\lambda$$

$$x_0 + x_1 = 23.3h_1^2/\lambda$$

$$y_1 = 2(2h_1)^{1/2}.$$

These dimensions determine the extent of flat ground required to double the free-space range of a radar as above. The height limit of any large irregularity in the area is $h_1/4$. If the same area is available on a sloping site of angle ϕ, double range may be obtained on a target on the horizon. In this case

$$x_0 + x_1 = 1.46\lambda/\sin^2\phi.$$

CONTINUOUS-WAVE DOPPLER RADAR

Echoes from stationary objects confuse or mask those from aircraft, especially on PPI scopes. This effect may be minimized by use of short pulses,

narrow beams, and several circuit modifications, but it is still intolerable in situations such as ground control of approach and aircraft detection. Discrimination between fixed and moving targets is possible by use of the Doppler principle.

In its simplest application, a cw transmitter is used and the return energy is detected by mixing it with a portion of the transmitter power. Fixed targets produce a constant voltage, whereas a moving target produces an alternating voltage at the Doppler frequency difference between transmitted and received signals.

$$f_d = f_t(c+v)/(c-v) - f_t \approx (2v/c)f_t = 89.4(v/\lambda)$$

where f_d = Doppler frequency in hertz, f_t = transmitted frequency in hertz, v = target radial velocity in miles/hour, c = speed of propagation in miles/hour, and λ = transmitted wavelength in centimeters.

Each cycle of Doppler frequency corresponds to a target radial motion of one-half transmitted wavelength. Thus, a target moving with a radial velocity of 300 miles/hour = 440 feet/second will move about 880 half-waves per second at 1000 megahertz ($\lambda \approx 1$ foot), resulting in a Doppler frequency of about 880 hertz. Target azimuth may be determined by rotating an antenna beam, but range cannot be found without modulation of the transmitter, so this type of radar is suitable only for measuring radial velocities of targets and for sentry applications to detect presence rather than accurate position of moving targets.

PULSED DOPPLER RADAR—COHERENCE

The straightforward way of obtaining range information is to pulse-modulate the transmitted carrier. If this is done in the simplified manner of Fig. 16, the received pulses will be small segments of the cw returns discussed above, as shown in Fig. 17. A fixed target produces uniform pulses, whereas moving-target pulses vary in amplitude periodically. An A-scope with one fixed and one moving target will appear as indicated. The basic cause of this distinction is phase coherence; that is, each time a fixed target echo returns, it is mixed with a voltage that has gone through the same difference in phase since the instant of transmission.

MOVING-TARGET-INDICATOR RADAR

In most Doppler radars, the objective is to differentiate targets on the basis of their speed in order to eliminate the display of stationary targets. In this case only moving target indication (MTI) is provided, typically, on a PPI scope. To maintain the various other objectives of the radar, the same

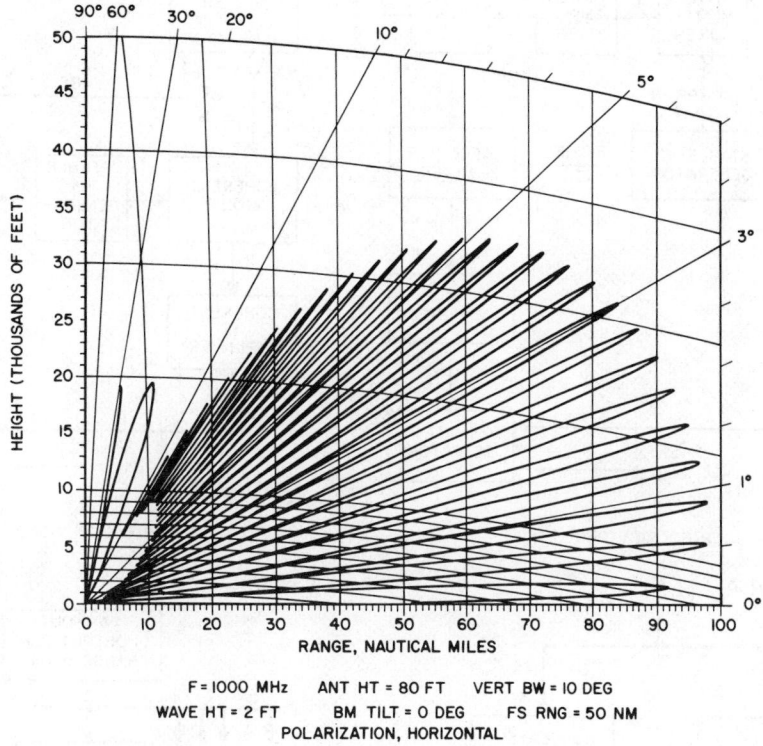

F = 1000 MHz   ANT HT = 80 FT   VERT BW = 10 DEG
WAVE HT = 2 FT   BM TILT = 0 DEG   FS RNG = 50 NM
POLARIZATION, HORIZONTAL

Fig. 15—Vertical-lobe pattern resulting from reflections from earth. *From L. V. Blake, "Machine Plotting of Radio/ Radar Vertical Coverage Diagrams," U.S. Naval Research Laboratory, Washington, D.C.; June 25, 1970.*

performance parameters are usually required—data rate, aerial coverage and range, range and angle resolution, etc. Discrimination is obtained by determining the change in target signal phase delay (from transmission to detection) in sequential signals. This method requires that the various parts of the system, in addition to providing the characteristics necessary to achieve the other system objectives, also be able to generate and process phase-stable signals. Typical MTI radars are shown in the block diagrams of Fig. 18. Radar A employs a fully coherent frequency generation system in which transmit, receive, and 1st and 2nd local-oscillator frequencies are all fully cohered. The output amplitude of the phase detector is directly related to the phase difference between the received signal and a reference signal which results directly from the propagation delay, and by design, is usually made bipolar. In radar B, the rf pulse transmitter (often a magnetron) has a random starting phase. To provide phase coherency, the i-f oscillator (2nd local oscillator) starts each transmission with its starting phase "locked" to a sample of the transmitter rf signal (heterodyned down in a signal mixer by the same signal as used for the receiver 1st-local-oscillator signal).

Discrimination against signals from stationary or slowly moving targets is obtained in the MTI canceller. Canceller A is a type used historically, particularly with search radars. The bipolar video signal train is injected into a delay channel which provides a delay exactly equal to the transmission

Fig. 16—Simple pulsed Doppler radar.

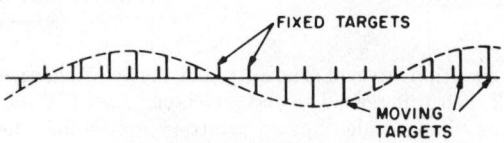

Fig. 17—Pulsed Doppler radar video signal.

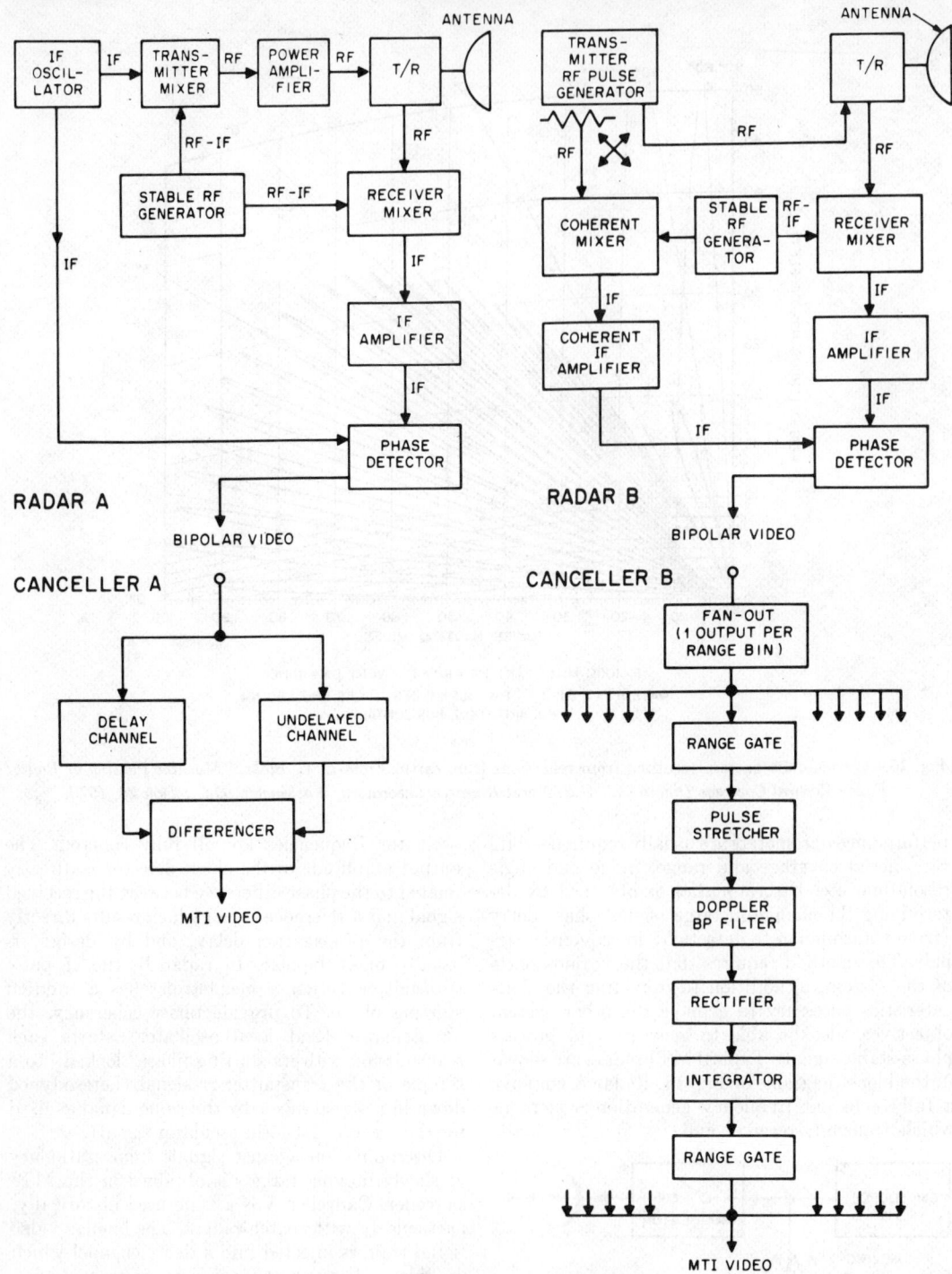

Fig. 18—MTI radar block diagrams.

repetition interval. The output of the delay channel is then differenced to the undelayed signal. While the bipolar video has an arbitrary amplitude, the change in this amplitude is related to target velocity. For low target velocity the difference in

target phase from pulse to pulse, and consequently in phase-detector output signal, is small and hence the output from the differencing circuit will be below a detectable level.

The delay channel can employ a variety of delay

media. Acoustic propagation properties of such materials as quartz, mercury, and alloys of magnetic materials are used with piezoelectric and magneto-electric transducers. For quartz and mercury lines, both am and fm modulation of the video signal have been used with carrier frequencies in the range of 10 to 100 megahertz. Alternative means have included the use of storage tubes (in either analog or quantized data format) and of ferrite core or similar binary storage memories with the bipolar video stored in a quantized format obtained by analog-to-digital conversion. The MTI video in these latter systems is developed by digital-to-analog conversion of the differenced digital data. Quantization errors must be kept well below the bipolar-video noise level. The sampling interval must be shorter than the signal pulse width.

Canceller B employs range-gated Doppler filters. The bipolar MTI video is applied to pulse stretchers through video range-gating circuits that pass video during a time interval corresponding to one pulse width. Thus, in a search radar the number of range-gated channels usually corresponds to the total number of range resolution cells. The pulse stretcher output signal has as its fundamental frequency the Doppler signal. This signal may be passed either through a single shaped filter designed to optimally reject clutter noise, or through a bank of narrow-band filters. The filter outputs are each rectified, smoothed (with the time constant dependent on the time on target), and after range gating the resulting video appears at the range of the initial input pulse, but delayed in angle corresponding to the smoothing time.

The output signal from the canceller (assuming received signals sufficiently large for phase-detector bipolar-video output to be related only to signal phase) is related to the Doppler frequency. For a single delay canceller, the voltage response is given by

$$E = E_0 \sin \pi f_d T$$

where E_0 = reference voltage output, f_d = Doppler frequency, and T = interpulse period.

When two single-delay cancellers are cascaded to provide "double" cancellation, the voltage response is given by

$$E = E_0 \sin^2 \pi f_d T.$$

E approaches zero not only as f_d approaches zero (the stationary-target case) but also when

$$f_d = n/T = n f_r$$

where $n = 1, 2, \ldots$, and f_r = pulse repetition frequency.

The blind speeds

$$V_n = \lambda n f_r / 89.4.$$

Blind speeds can be avoided by changing the interpulse period or the operating frequency. When

either of these is staggered on a repetitive pulse-to-pulse basis, with sequential deviation from the mean by $\pm \delta$, then the response of a single canceller is the average of the response given above, modified with factors $(1+\delta)$ and $(1-\delta)$. Usually, the power response of a system is of concern, hence the rms voltage response is then given by

$$E = [(E^2_{+\delta} + E^2_{-\delta})/2]^{1/2}$$

where:

$$E_{+\delta} = E_0 \sin \pi f_d T (1+\delta)$$
$$E_{-\delta} = E_0 \sin \pi f_d T (1-\delta).$$

For a double canceller, the rms voltage response

$$E = E_0 \sin^2 \pi f_d T (1+\delta)$$
$$+ \sin^2 \pi f_d T (1-\delta) - \tfrac{1}{2} \sin^2 2\pi f_d T.$$

MTI Limitations

Various factors, external and internal to the radar, limit the ability to see targets in clutter.

Clutter Motion: Because of internal fluctuations, the attenuation of the clutter signal (CA) is limited.

$$CA \approx a/[2(\pi f_0/f_r)^2] \quad \text{(single canceller)}$$
$$\approx a/[12(\pi f_0/f_r)^4] \quad \text{(double canceller)}$$

where a = factor depending on type of clutter and f_0 = radio frequency.

Typical values of a are as follows:

Rain clutter	2×10^{15}
Sea echo (windy)	1×10^{16}
Heavily wooded hills (20 mph wind)	2×10^{17}
Sparsely wooded hills (calm)	4×10^{19}

Antenna Rotation: Clutter attenuation is also limited by modulation of the clutter signal by the pattern of the antenna as it scans by the clutter element. Assuming a Gaussian antenna pattern, the limiting value of clutter attenuation is given by

$$CA = n^2/1.39 \quad \text{(single canceller)}$$
$$= n^4/3.84 \quad \text{(double canceller)}$$

where n = number of hits in 1-way 3-decibel antenna pattern.

Equipment Instabilities: The various radar subsystems suffer from instabilities in timing, amplitudes, and frequencies of the various signals involved. Usually these can be designed to be sufficiently low compared with the two previous limitations so as to not degrade MTI performance significantly. The various factors are usually unrelated, and hence their effects can be added on a root-mean-square basis.

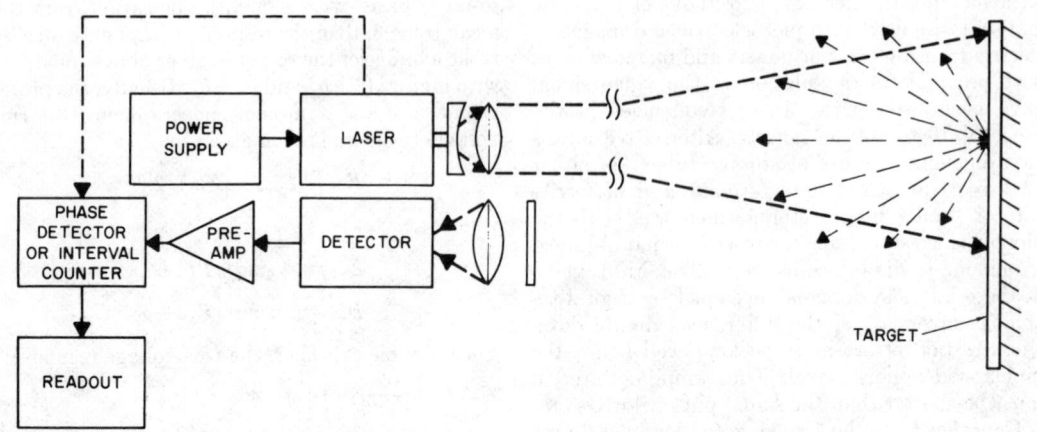

Fig. 19—Essential elements of a basic optical radar.

LASER RADAR

The development of various kinds of practical laser systems and detectors has made possible radar-like devices which operate at optical wavelengths. Most laser radars are used as short- to medium-range point-to-point rangefinders rather than area scanning devices, and they are ideal for this role since the very narrow beam of the laser permits the target to be irradiated with a high power density.

Laser radar has several advantages over its microwave counterpart for short-range detection. These include:

1. Very narrow beam width (less than 1 mrad)
2. Very small transmitter antenna (less than 10 cm diameter)
3. Very small receiver antenna (less than 50 cm diameter)
4. In the case of semiconductor injection laser systems, very simple drive circuitry for the transmitter.

Several different laser categories, including gas, solid state, and semiconductor, have been operated as both experimental and operational radars. The best developed gas laser rangefinders utilize the helium–neon (He–Ne) laser (P_0 less than 5 mW) and a photomultiplier receiver. The laser is intensity modulated to cause the phase of the received signal to be a sine function of the distance to the target. Helium–neon laser rangefinders with detection ranges against cooperative targets (e.g., arrays of precision glass corner cube retroreflectors) of more than 25 km with a measurement error of as little as a few millimeters per kilometer are commercially available.

For applications where the target is a noncooperative quasilambertian reflector (e.g., bare terrain, vegetation, structures, etc.), high-power pulsed solid-state lasers are employed. These rangefinders usually employ time of flight as a measurement of target distance. Suitable lasers for high-power optical rangefinders include ruby and both neodymium doped glass and YAG.

Semiconductor injection lasers are useful for detecting cooperative targets at ranges up to and even greater than 10 km and diffuse (lambertian), noncooperative targets at up to several hundred meters. A detailed description of the operation of an experimental injection laser rangefinder is given in W. Koechner, "Optical Ranging System Employing a High Power Injection Laser Diode," *IEEE Transactions on Aerospace and Electronic Systems*, pp. 81–91; January 1968.

Design Considerations

Figure 19 is a functional diagram of an elementary laser radar. The transmitter includes a pulsed or intensity-modulated laser, a power supply, and appropriate collimating optics. The receiver includes collection optics, a narrow-band-pass optical filter to reduce the noise effects of sunlight and other unwanted optical sources, a detector, a preamplifier, either a phase detection or a time-interval circuit, and a readout device.

The ultimate detection range of an optical radar is limited by the target reflectance coefficient, the nature of the target reflecting surface (i.e., diffuse lambertian or specular), background radiation at the target, atmospheric attenuation, and atmospheric backscatter of the transmitted beam. Figure

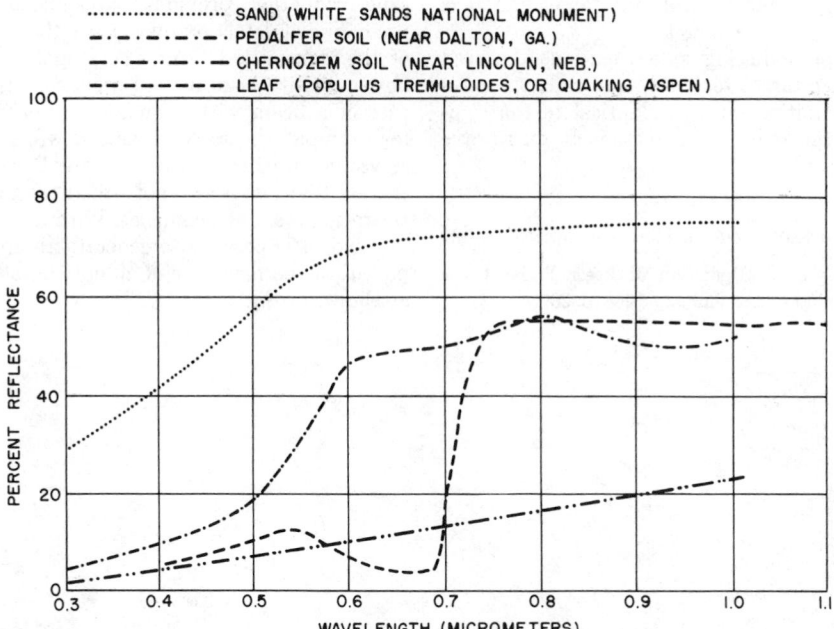

Fig. 20—Spectral reflectance of some natural terrain features. (*Soils from H. R. Condit, "The Spectral Reflectance of American Soils," Photogrammetric Engineering, pp. 955–966; September 1970. Vegetation from W. D. Billings and R. J. Morris, "Reflection of Visible and Infrared Radiation From Leaves of Different Ecological Groups," American Journal of Botany, pp. 327–331; May 1951.*)

20 gives the reflectance of several natural terrain features at a range of wavelengths and shows that the reflectance of many materials is enhanced in the near infrared. Atmospheric extinction coefficients are given for a range of visibility conditions in the chapter on optical communications, and target reflectance considerations are described in the next section.

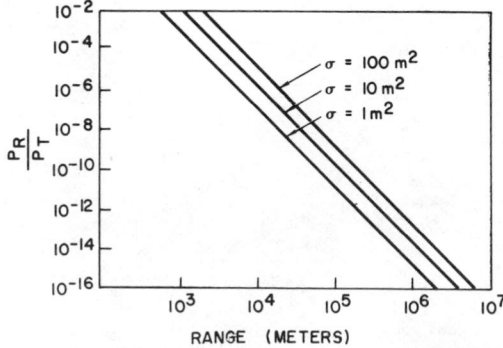

Fig. 21—Ratio of received power to transmitted power (P_R/P_T) versus detection range for different target cross sections. (*After M. Ross, "Laser Receivers," p. 342; John Wiley and Sons, Inc., 1966.*)

Optical Radar Range Equation

A standard form of the optical radar range equation is[*]

$$R^4 = P_T \sigma A_R \eta / P_R \Omega_T 4\pi$$

where P_T is the peak transmitted power, σ is the optical cross section of the target, A_R is the area of the receiver aperture, P_R is the received power, Ω_T is the transmitter beam divergence, and η is the overall optical transmission efficiency factor.

Since the parameters of an optical radar are well known to its operator, the variables of most interest are η and σ. The variable η is discussed for the atmosphere in the chapter on optical communications; σ is a figure of merit which describes target reflectance efficiency and is the product of the target area and the reflection efficiency of the target. Since at optical wavelengths the accepted reference target for σ is a lossless diffuse reflector, σ is that area of a perfectly lossless and diffuse flat plate in a plane perpendicular to the oncoming beam which gives the same reflected intensity as the target in question when both are irradiated under identical conditions. Therefore, σ may be calculated by[*]

$$\sigma = \pi J / H$$

[*] M. Ross, *Laser Receivers*, John Wiley & Sons, Inc., New York, pp. 340–343; 1966.

where J is the intensity from the target and H is the irradiance at the target.

An isotropic radiating sphere may also be used as a reference target for an optical radar. In this case, the definition for σ is identical to that employed for the radar cross section at microwave frequencies, or[*]

$$\sigma = 4\pi J/H.$$

It is important to note that the optical radar

[*] P. W. Wyman, "Definition of Laser Radar Cross Section," *Applied Optics*, January 1968, p. 207.

range equation provides only an approximate prediction of detection range since the cross section of the transmitted beam has a spatially variable power density in nearly all practical cases. In the case of a beam with a guassian intensity profile, for example, the range equation will give a conservative prediction for a target significantly smaller than the beam spot size at the target when the target is at the beam axis. With this precaution in mind, it is possible to generate an approximate plot of the performance of an optical radar system as shown in Fig. 21.

INTRODUCTION

The speed of propagation of radio waves (300 000 kilometers per second) permits measurement of *distance* as a function of *time* and of *direction* as a function of *differential distance* to two or more known points. In free space, radio navigation is capable of considerable accuracy. Along the surface of the earth, however, accuracy is reduced by the effects of multiple propagation paths between transmitter and receiver. Most navigation systems are, therefore, a compromise between service area, accuracy, and convenience of use. In general, complexity is minimized on the vehicle at the expense of greater complexity at the fixed station. Since aircraft and ships may move over large areas, systems which involve cooperation between a vehicle and a ground station have required a high degree of international standardization. These standards, once established, change slowly. Within each country, additional military systems are in use; some of these are compatible with the international civil systems.

MAJOR STANDARDIZING AGENCIES

International Telecommunication Union (ITU), Geneva, Switzerland. An agency of the United Nations. Allocates frequencies for best use of the radio spectrum.

International Civil Aviation Organization (ICAO), Montreal, Canada. An agency of the United Nations. Formulates standards and recommended practices, including navigation aids, for all civil aviation.

International Air Transport Association (IATA), Montreal, Canada. An association of scheduled airlines.

Federal Communications Commission (FCC), Washington, D.C. Licenses transmitters in the United States and aboard US registered ships and aircraft.

Federal Aviation Administration (FAA), Washington, D.C. Formerly *Federal Aviation Agency*.

Operates navigation aids and traffic control systems, for both military and civil aircraft, in the US and its possessions. (Not to be confused with *Civil Aeronautics Board (CAB)*, which regulates routes and fares of interstate and foreign airlines operating in the US.)

United States Coast Guard (USCG), Washington, D.C. Operates navigation aids for shipping.

Radio Technical Commission for Aeronautics (RTCA), Washington, D.C. Supported by contributions from industry and from government agencies. Seeks participation by manufacturers, users, and others in the generation of recommended standards for aviation electronics. Some of these standards have been adopted, at least in part, by the ICAO and by the FAA.

Radio Technical Commission for Marine (RTCM), Washington, D.C. Similar to RTCA, but for shipping.

Airlines Electronic Engineering Committee (AEEC), Annapolis Science Center, Annapolis, Maryland 21401. A division of Aeronautical Radio, Inc. *(ARINC)*, owned by the scheduled US airlines. Holds frequent meetings with manufacturers, issues newsletters, and publishes technical recommendations on avionics hardware purchased by the scheduled airlines. Has worldwide influence on airborne-equipment design.

REDUCTION OF ERRORS CAUSED BY MULTIPATH PROPAGATION

In most radio navigation systems, the desired path of the signal is the shortest one between transmitter and receiver; errors result from the admixture with signals which have traveled by longer, often variable, paths. To reduce such multipath effects, the following techniques are commonly used.

(A) *Pulse Transmission.* With suitable pulse duration and repetition rate, plus means at the receiver to recognize the leading edge of the pulses, the direct signal may be separated from that which

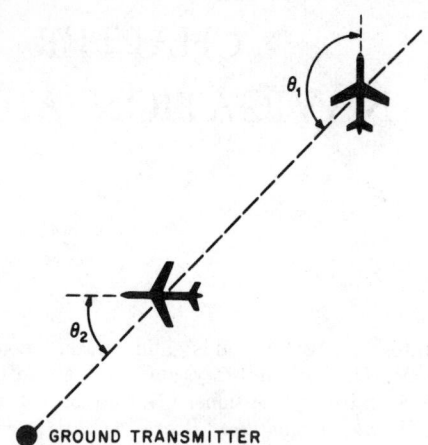

Fig. 1—Airborne direction finder.

has traveled a longer path. Effectively used in radar, DME, Loran-A, Loran-C.

(**B**) *Space Diversity*. The larger the aperture of an antenna system, the greater the statistical probability that the desired signals will add linearly while the multipath signals add randomly. Effectively used in doppler VOR, doppler DF. (Antenna directivity is very frequently used as an equivalent to space diversity. By proper shaping of the antenna pattern, energy may be increased along the direct path and reduced along undesired potentially interfering paths. In directional systems, such as ILS, horizontal directivity may be employed for this purpose; in nondirectional or omnidirectional systems, such as Tacan and DME, vertical directivity is used.)

(**C**) *Frequency Diversity*. While the line-of-sight path remains the same at all radio frequencies, indirect paths may vary with frequency. In such cases, spectrum-spreading techniques may achieve the same result as space diversity.

MAJOR RADIO NAVIGATION AIDS

For the specific extent to which these aids are currently implemented throughout the world, see the navigational facility maps issued at frequent intervals by the US Coast and Geodetic Survey, Washington Science Center, Rockville, Maryland 20852. For US statistics, see "The ATS Fact Book," issued annually by the FAA.

ADF (Airborne Direction Finder)

Rotatable-loop-antenna system on aircraft measures angle between aircraft axis and direction of arrival of radio wave from ground transmitter (Fig. 1). Not the most accurate bearing measuring system but very widely used, since it will operate on almost any type of station in the 200–1600-kilohertz band. Not used at high frequency because of ionospheric contamination. Some use at ultrahigh frequency but subject to severe vehicular site error. A ground station specifically intended for this service is called a Non-Directional Beacon (NDB) that operates in the 200 to 415 kHz band and is an ICAO standard.

AEROSAT*

The Aeronautical Satellite Program (AEROSAT) is an experimentation and evaluation program proposed by ESRO, Canada, and the United States. The total program will span almost ten years and plans call for use of satellites to provide improved communication and surveillance capability for oceanic air traffic control. Information from this program will support ICAO in defining an operational satellite system for the mid-1980's. A first generation of L-Band avionics based upon the reference avionics interface parameters of the AEROSAT system is being evaluated.

ATCRBS (Air Traffic Control Radar Beacon System)

Rotating directional ground-based interrogator at 1030 megahertz transmits approximately 400 pulse-pairs per second and receives replies from airborne transponder (Fig. 2) at 1090 megahertz, pulse-coded with identity, altitude, etc. There are 4096 codes available (Fig. 3). Decoded replies are displayed on a PPI indicator, often with those of

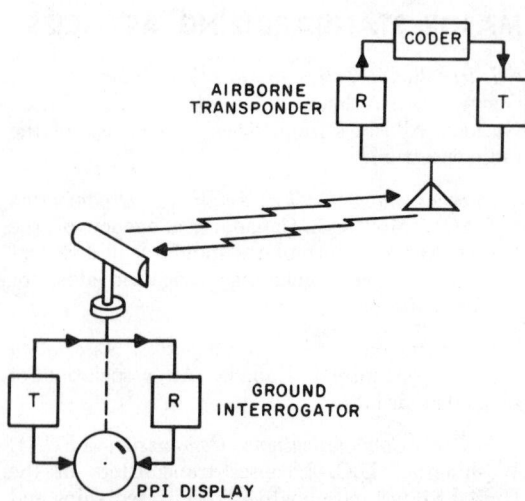

Fig. 2—Secondary surveillance radar (ATCRBS).

* W. A. Fried, "Aircraft Navigation: Recent Developments in Ground-Based and Satellite-Based Systems," IEEE, T-AES p. 27; January 1974.

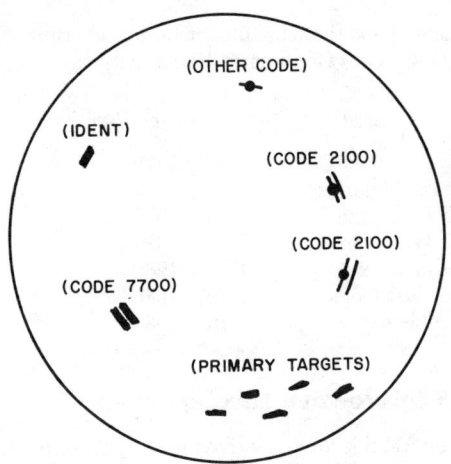

Fig. 3—Radar scope with various targets and codes displayed.

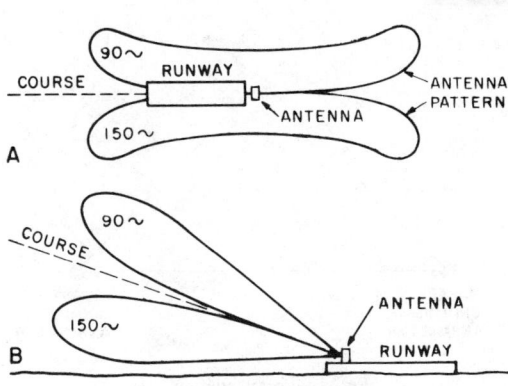

Fig. 5—Guidance systems.

associated primary radar. To suppress unwanted side lobes an omnidirectional pulsed pattern is also radiated from the ground, and the transponder is arranged to reply only when the directional signal exceeds the omnidirectional signal. This system is often referred to as Secondary Surveillance Radar (SSR).

DME (Distance Measuring Equipment—ICAO Standard)*

Airborne interrogater transmits about 30 pulse-pairs per second on one of 126 channels between 1025 and 1150 megahertz. Ground transponder replies on one of 126 channels at 962–1024 and 1151–1213 megahertz. Airborne indicator reads transmit-to-receive time on meter calibrated in nautical miles (see Fig. 4). (DME plus VOR

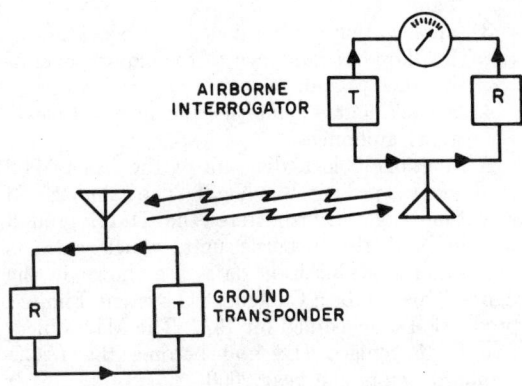

Fig. 4—Distance measurement.

* *IEEE Transactions on Aerospace and Navigational Electronics*, vol. ANE-12, no. 1; March 1965.

constitute the ICAO standard rho-theta system when both ground stations are co-located.)

ILS (Instrument Low-Approach System—ICAO Standard)

Lateral guidance (Fig. 5A) is provided by the *localizer* located at the far end of the runway. 40 channels from 108 to 112 megahertz. Two identical antenna patterns, the left-hand one modulated by 90 hertz and the right-hand one by 150 hertz. Vertical needle of airborne display, driven right by 90 hertz and left by 150 hertz, centers when aircraft is on course.

Vertical guidance (Fig. 5B) is provided by the *glide slope*, located to the side of the near end of the runway. 40 channels from 329 to 335 megahertz, paired one-for-one with localizer. Amplitude modulation at 150 hertz below course and 90 hertz above course. Horizontal needle of airborne display, driven up by 150 hertz and down by 90 hertz, is horizontal when the aircraft is on course. (The "crosspointer" is a display combining vertical localizer needle and horizontal glide-slope needle.)

Along-course guidance (Fig. 6) is by *fan markers*, all at 75 megahertz. The inner marker is close to the runway; the middle marker about 3500 feet out; the outer marker about 5 miles out.

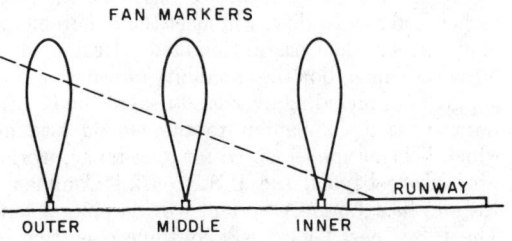

Fig. 6—Along-course guidance.

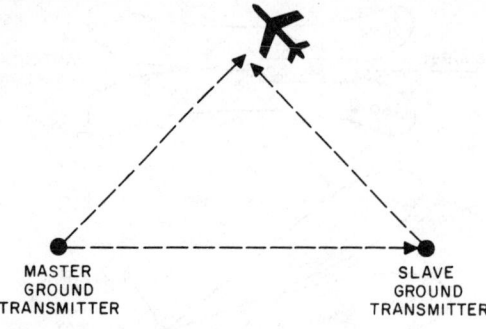

Fig. 7—Loran system.

DME may also be used for this purpose, 20 DME channels being paired with localizer frequencies.

The quality of ILS installations varies with equipment, terrain, etc. ICAO has established the following minimum-condition categories (assuming properly equipped aircraft and properly trained air crew):

Category	I	II	III_A	III_B	III_C
Minimum ceiling	200'	100'	50'	35'	0
Forward visibility	2600'	1200'	700'	150'	0

LORAN-A, C (Long Range Navigation)

Forty-five-microsecond pulses are transmitted 20 to 34 times per second by master station and repeated by slave station 300 miles away (Fig. 7). Aircraft, using oscilloscope display, notes difference in time of arrival and computes its position by use of two or more master-slave pairs. Lines of constant time difference are hyperbolic, with stations as foci. Loran-A pairs are distinguished by frequency: 1850, 1900, 1950 kilohertz, and by slight variations in pulse repetition frequency. (Loran-C uses same basic principle, but at 100 kilohertz, where longer ground-wave range is possible.)

Loran-C has been selected by the Federal Government as the radio-navigation system for the Coastal Confluence Region, Harbors and Estuaries of the United States. Loran-C, with its high accuracy and availability, will meet the requirements of all users in the Coastal Confluence Region while providing the nation the capability of wider system application, including over-land use, in the future. Loran-C is a companion to the *Omega* system, which is being implemented for long-range, worldwide usage beyond the U.S. Coastal Confluence Region. The Loran-A system will be phased out. The U.S. Coast Guard will continue operation of the Loran-A system for a period of two years subsequent to completion of each phase of the Loran-C

system. Thus the schedule for Loran-A termination in the Coastal Confluence Region is:

Area	Date of Termination
Hawaiian Islands	1 July 1979
Aleutian Islands	1 July 1979
Gulf of Alaska	1 July 1979
U.S. West Coast	1 July 1979
Gulf of Mexico	1 July 1980
U.S. East Coast	1 July 1980
Caribbean	1 July 1980

MLS (Microwave Landing System)*

The MLS is an air-derived data system in which the airborne unit obtains precision azimuth angle, elevation angle, and range data referenced to the runway. Angular position of the aircraft is measured by reference to ground transmitters that generate angle-encoded signals throughout the coverage sector in both azimuth and elevation. The airborne unit extracts the modulated angle data that corresponds to the line-of-sight angle from the ground antenna. Range measurements are made via airborne interrogation of a ground transponder. The replies from the ground beacon are tracked to extract range data from the round-trip time delay.

The system is capable of transmitting auxiliary data such as runway identity, equipment status, weather data, and siting constants to airborne units. The airborne unit computes position data or flight path deviation data suitable for inputs to the flight control system and/or for display to the pilot.

The basic MLS system is comprised of the following functions or elements:

1) a basic *C*-band elevation and azimuth guidance element.
2) a DME operating in a separate portion of *C* band.
3) an elevation guidance element (elevation 2) for flare-out guidance to touchdown operating in K_u band.
4) a back course azimuth (and optional elevation) guidance.

A functional block diagram of the basic MLS equipment is presented in Fig. 8. The azimuth and elevation angle transmitters, the DME ground beacon, and the airborne unit, which extracts range and angle position data, are shown in the figure. This will be a Category III system. Limited production is scheduled for 1977. The MLS objective is to replace ILS and become the IACO standard before the year 2000.

* V. J. Fritch, Jr. and L. J. Sanders, "Instrument Landing Systems," *IEEE Communications*, pp. 435–454, May, 1973.

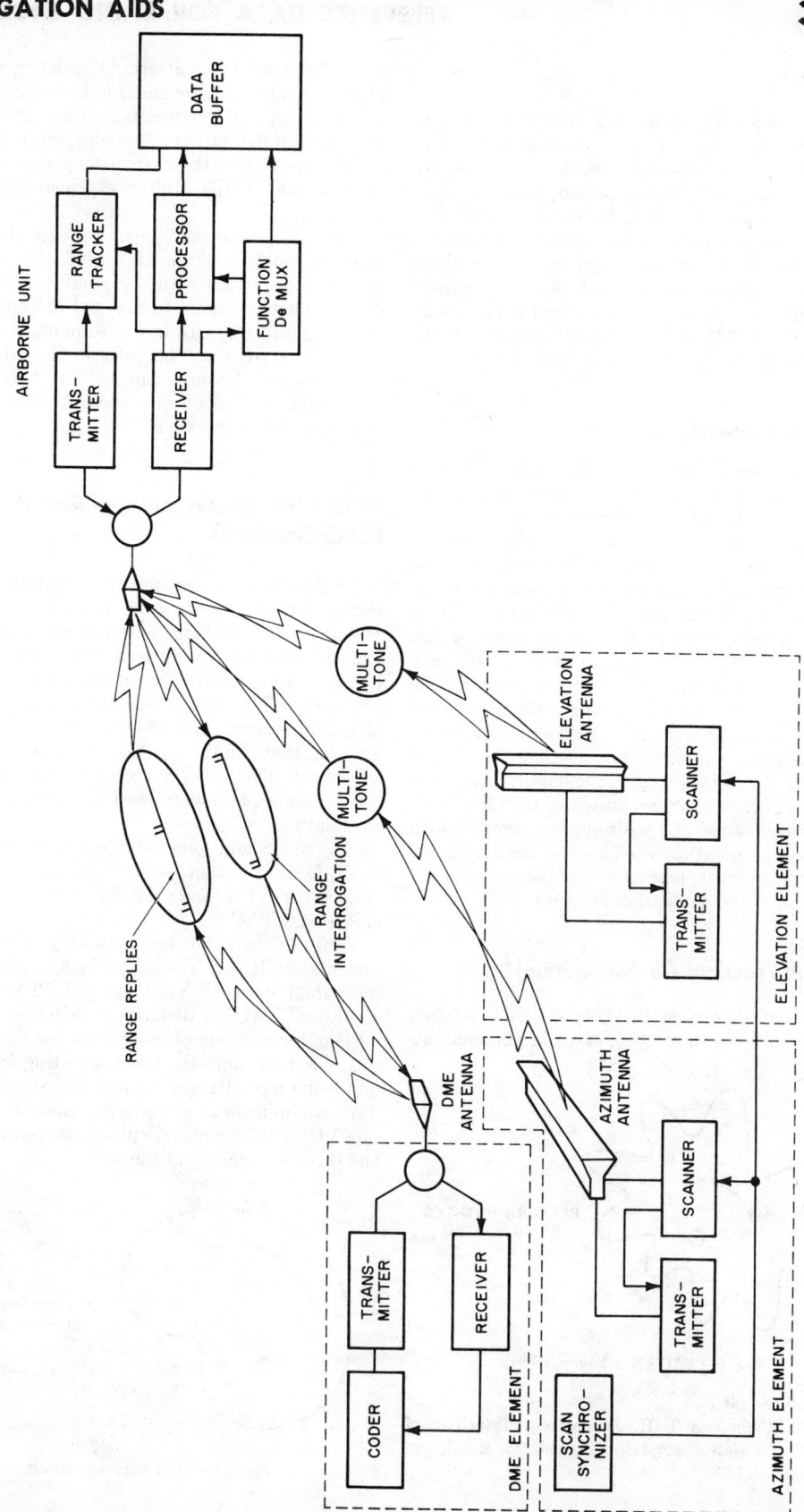

Fig. 8—Functional block diagram of MLS equipment.

Omega*

Omega is a world-wide VLF navigation system. Eight stations are required to provide world-wide coverage. Each of the eight stations sequentially transmits long, but precisely timed, pulses at three different frequencies, 10.2, 11.33, and 13.6 kilohertz. The precise timing permits automatic acquisition of the stations, and use of the three frequencies reduces the residual "lane ambiguity" of the phase measurement to 72 miles for hyperbolic operation and 144 miles for direct ranging operation. The position error is of the order of 1 mile.

Radar (Airborne)*

Recent techniques permit the obtaining of a semiautomatic position fix by means of airborne radar. The geographic coordinates of the fix point, or check point (CP), have previously been inserted into the navigation computer and are stored there. The computer calculates the aircraft position from its normal navigation sensor inputs. From these two data, it generates bearing and distance from the aircraft to the CP. The latter are converted into the equivalent radar coordinates, bearing and slant range. These are then displayed as cursors on the navigator's radar indicator. When the navigator observes the predetermined CP target on the radar image, he manually places the range and bearing cursors on top of the CP target. This process causes any position error signals which exist to be automatically fed back to the navigation computer, thereby fix-correcting the position information of the navigation system.

TACAN (Tactical Air Navigation)†

Constant-duty-cycle ICAO-type DME beacon, to which is added a rotating cardioid antenna pattern plus a 9-lobed pattern (Fig. 9), generating 15-hertz coarse bearing and 135-hertz fine bearing in the aircraft. Reference signals are transmitted by coded pulse bursts. Provides rho-theta in a single equipment. Higher frequency allows smaller antenna than VOR. Multilobe principle improves accuracy.

TACAN ground equipment consists of either a fixed or mobile transmitting unit. The airborne unit in conjunction with the ground unit reduces the transmitted signal to a visual presentation of both azimuth and distance information. TACAN is a pulse system and operates in the UHF band of frequencies. Its use requires TACAN airborne equipment and does not operate through conventional VOR equipment.

VOR (VHF Omnidirectional Range— ICAO Standard)

Ground transmitter radiates continuous-wave signals on one of the even frequencies between 108 and 118 megahertz (interleaved with 40 odd localizer frequencies between 108 and 112 megahertz). Cardioid antenna pattern rotates 30 times per second (Fig. 10), generating 30 hertz (AM) in airborne receiver. The 30-hertz reference signal is also radiated, (FM) ±480 hertz on a 9960-hertz subcarrier. The airborne receiver reads bearing as a function of phase between FM and AM 30-hertz modulations.

In the US and some other countries, the ICAO VOR/DME rho-theta system is implemented by co-locating VOR and TACAN stations. This is called VORTAC.

VORTAC is a facility consisting of two components, VOR and TACAN, which provides three individual services: VOR azimuth, TACAN azimuth and TACAN distance (DME) at one site. Although consisting of more than one component, incorporating more than one operating frequency, and using more than one antenna system, a VORTAC is a unified navigational aid. Both components of a VORTAC operate simultaneously and provide the three services at all times.

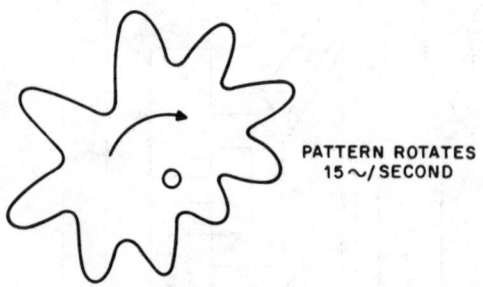

PATTERN ROTATES
15 ∿/SECOND

Fig. 9—TACAN 9-lobe pattern.

* Fried, *loc. cit.*

† R. I. Colin and S. H. Dodington, "Principles of Tacan," *Electrical Communication,* vol. 33, no. 1, pp. 11–25; March 1956.

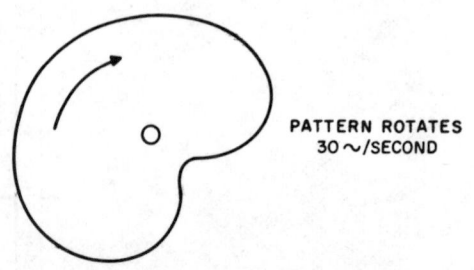

PATTERN ROTATES
30 ∿/SECOND

Fig. 10—VOR antenna pattern.

SYSTEM CHARACTERISTICS*

Inertial Navigation

Operates by double integration of measured acceleration.
Self-contained.
Unlimited coverage.
Passive operation.
Unlimited number of users.
Accuracy—2 nmi/h or better.
Ambiguities—none.
Error characteristics—errors among users are uncorrelated, aircraft separation is affected; absolute error diverges with time.
Reliability assured through triple redundancy.

Doppler

Operates by integration of measured velocity.
Self-contained.
Unlimited coverage.
Radiates RF energy in the 10-GHz region.
Unlimited number of users.
Accuracy—position error less than 1.4 percent of distance traveled since calibrating.
Ambiguities—none.
Error characteristics—errors are uncorrelated among users; aircraft separation is affected; absolute error diverges with time.

VOR/DME

Operates by angle-range position determination.
Frequency, VOR 108–118 MHz, passive operation DME 962–1213 MHz, active (air-ground-air) operation.
Range—50–200 mi (depends upon transmitter power and altitude).
DME function estimated to accommodate 100 users/station.
Accuracy ±4 nmi at 50 mi range (includes airborne equipment and pilotage errors); error diverges with range.
Ambiguities—VOR signal has ambiguity 180° from the true null; ambiguity resolved with a TO-FROM flag in the airborne receiver.
Error characteristics—most of the quoted error is uncorrelated among users and affects aircraft separation; some of the error, the part due to radiation anomalies, is correlated among users and does not affect separation.
Reliability—assured through redundancy, multiplicity of ground stations and duplication of airborne equipment.

* Graham W. Turner, "Survey of Aviation Navigation Systems," *IEEE Communications*, pp. 427–434; May 1973.

Time/Frequency

Operates by one-way ranging with a precision clock, coverage depends upon frequency, power, etc.
Requires at least two transmitters.
Passive operation.
Unlimited number of users.
Accuracy—potentially to a few feet.
Ambiguities—the mirror image of the true position with respect to the two transmitters; resolved by an approximate knowledge of position.
Error characteristics—errors associated with transmitters (location, timing, etc.) are correlated among users and do not influence separation; errors in airborne equipment are uncorrelated and do influence separation.

Loran A

Operates by measurement of two range-differences.
Frequency—1.8–2.0 MHz.
Coverage to a range of about 1000 mi.
Requires at least three transmitters.
Passive operation.
Unlimited number of users.
Accuracy.
Ambiguities—close to transmitting sites; geometric degeneration of accuracy along baseline extensions.
Error characteristics—errors associated with transmitters (location, timing, etc.) are correlated and do not influence separation; errors in airborne equipment are uncorrelated among users and do influence separation.

Loran C

Operates by measurement of two-range differences.
Frequency—LF.
Coverage to a range of about 2000 mi.
Requires at least three transmitters.
Passive operation.
Unlimited number of users.
Accuracy.
Ambiguities—close to transmitting sites; geometric degeneration along base line extensions.
Error characteristics—errors associated with transmitter sites (location, timing, etc.) are correlated among users and do not influence separation; errors in airborne equipment are uncorrelated among users and do influence separation.

Decca

Operates by measurement of two range differences.

Frequency—LF.

Coverage to a range of 200–300 mi.

Requires at least three transmitters.

Passive operation.

Unlimited number of users.

Accuracy.

Ambiguities—phase measurement ambiguity from lane to lane; requires land identification; geometric degeneration along base line extensions.

Error characteristics—errors associated with transmitter sites (location, timing, etc.) are correlated among users and do not influence separation; errors in airborne equipment are uncorrelated among users and do influence separation.

Omega

Operates by measurement of two range differences.

Frequency—VLF.

Coverage, global with 6–8 ground stations.

Requires at least three stations.

Passive operation.

Unlimited number of users.

Accuracy—1–2 nmi.

Ambiguities—phase measurement ambiguity from lane to lane; requires lane identification; geometric degeneration along base line extensions.

Error characteristics—errors associated with transmitter sites (location, synchronization, propagation effects, etc.) are correlated among adjacent users and do not influence separation; errors in airborne equipment are uncorrelated among users and do influence separation.

Consol

Operates by determining bearing angle from ground sites (a ground site consists of three in-line antennas).

Frequency LF/MF (300-kHz region).

Coverage—1000–1500 nmi from each station.

Requires two stations to provide a fix.

Passive operation.

Unlimited number of users.

Accuracy—a few nautical miles (depends upon range, diurnal effect, sea–land and propagation, etc).

Ambiguities—radial bearing line is ambiguous from one angular sector to the next; geometric degeneration along the line of the three antennas (baseline extensions).

Error characteristics information is derived by counting dots and dashes and correcting the counts for symbols lost due to cancellation; error tends to be quantized.

GLOSSARY OF NAVIGATION TERMS

Absolute Altimeter. Reads height above terrain, using radio reflections, as distinct from a barometric altimeter, which determines height by sensing local air pressure.

Accelerometer. Senses force per unit mass along a given axis, due to acceleration of vehicle which carries accelerometer.

ADF. Airborne Direction Finder.

Angle of Cut. Angle at which two lines of position intersect. Preferably a right angle.

A–N Range. An obsolescent system in which one side of the course is defined aurally by the morse-code letter *A* and the other side by the letter *N*, the two blending into a steady tone on course. Used in the 4-course range.

Approach Path. The portion of the flight path between start of descent and touchdown.

ARSR. FAA term for Air-Route Surveillance Radar.

ARTCC. FAA term for Air-Route Traffic Control Center. About 20 cover the United States, each connected to numerous radars and radio communication sets by microwave links and landlines.

ASDE. FAA term for Airport Surface Detection Equipment, a radar for observing vehicles on the airport surface.

ASR. FAA term for Airport Surveillance Radar.

ATCRBS. FAA term for Air Traffic Control Radar Beacon System. Same as ICAO's Secondary Surveillance Radar (SSR).

Azimuth. The angle in the horizontal plane with respect to a fixed reference, usually true North, measured clockwise (refer to Bearing).

Back-Course. In ILS, the course located on the opposite end of the runway behind the localizer.

Baseline. The line joining two points between which electrical time is compared. Large baselines produce high instrument accuracy but may also introduce instrument ambiguities.

Bearing. An angle in the horizontal plane with respect to a reference. Usually expressed in degrees measured clockwise from the reference. "Relative" bearing is to some arbitrary reference; "absolute" bearing is to North—usually magnetic North in navigation systems.

Bend. A departure of a course line from a straight line. It is usually oscillatory and generally caused by interference between direct and multipath signals (refer to Scalloping).

Boresighting. The process of aligning a directional antenna system, often by optical means.

Boundary Marker. In ILS, the innermost fan marker, near the approach end of the runway. (Sometimes called the inner marker.)

CADF. Commutated-Antenna Direction Finder. Uses a circular array of antennas, the receiver connected to them sequentially through a commutating device. A means of increasing the antenna aperture, so as to reduce site error.

Capture Effect. The tendency of a receiver to suppress the weaker of two simultaneously received signals.

CEP. Circular Error Probability. In a two-dimensional error distribution, the radius of a circle encompassing half the errors.

Chain. A network of stations operating as a group.

Clearance Sector. In ILS, the sector from the course to the back course, in which sector it is desirable to maintain the left-right needle off scale.

Coherent Pulses. Pulses used in navigation systems in which the phase of the radio-frequency cycles within the pulse is retained for measurement purposes (as in Loran-C, as contrasted with Loran-A, which uses only the envelope).

Compass Locator. Colloquial term for an LF/MF nondirectional beacon. Same as H-Beacon and NDB.

Cone of Ambiguity. In VOR and TACAN, the conical volume of airspace above the beacon in which bearing information is unreliable.

Cone of Silence. The conical volume above an antenna where field strength is relatively low.

Consol. A keyed continuous-wave short-baseline system operating in the LF/MF band and useful to about 1500 miles. Uses 3 radiators to generate two daisy-shaped patterns which are slowly rotated and also switched alternately. Can be read by simple receivers. Developed originally under the name Sonne. Still in limited use, mainly by shipping.

Consolan. A form of Consol using two radiators instead of three.

Course. The intended direction of travel. Also, the direction defined by a navigation aid.

Course-Line Computer. A vehicle-carried device which converts navigation signals into courses not generated directly by the signals themselves (for example, hyperbolic to straight-line, rhotheta to straight-line).

Course Softening. Intentional decrease in course sensitivity as the navigation aid is approached.

CPE. Circular Probable Error. Same as CEP.

Crab Angle. Correction angle to compensate for wind drift. The angular difference between course and heading.

Dead Reckoning. Determination of position at one time with respect to known position at a previous time, by the application of course and distance information derived without reference to external aids.

Decca. A navigation system that transmits continuous-wave signals on several related frequencies around 100 kilohertz. It comprises a master station, with slaves about 70 miles away. Lines of position are hyperbolic. Because of ionospheric contamination, its range is limited to about 180 miles. Decca is widely used by coastal shipping in Europe.

Decision Gate. In ILS, the point at which the pilot must decide to land or to execute a missed-approach procedure.

Dectra. An adaptation of the Decca system, using two stations at each end of a great-circle path to form a multiplicity of tracks. In experimental use in the North Atlantic.

Directional Localizer. In ILS, a localizer using directivity to maximize the signal along the runway and to minimize reflections from hangers, etc. May require a subsidiary low-directivity localizer to provide clearance.

DME. Distance Measuring Equipment. Provides distance information by time of round-trip transmission of pulses from an interrogator to a transponder. ICAO standard.

Doppler Navigator. A self-contained aid, transmitting two or more beams of electromagnetic or acoustic energy downward toward a reflecting surface and using the change in returned frequency, due to motion of the vehicle (Doppler effect), to measure the speed of the vehicle with respect to the reflecting surface. In wide use by transoceanic aircraft, using radio, and, to a lesser extent, by ships, using acoustic energy.

Doppler VOR. A wide-aperture VOR, to reduce site errors. Uses a 44-foot-diameter circle of antennas to generate the variable phase. The reference phase is transmitted by a single central antenna. Compatible with standard VOR receivers.

Drift Angle. Angle between heading and track, due to effect of wind or water currents.

Error, Attitude. Varies with attitude of vehicle. Often related to polarization error.

Error, Instrument. Error caused by the equipment itself.

Error, Polarization. Varies with polarization of antenna at one or both ends of signal path. Often related to attitude error.

Error, Propagation. Error caused by variations in the propagation medium.

Error, Readout. Error caused by failure of navigator to read his instrument properly.

Error, Site. Error caused by reflections from obstructions close to the site of the navigation aid. Of major concern in directional systems.

Fan Marker. A 75-megahertz ground-based transmitter having a fan-shaped beam pointing vertically and across the flight path to provide a fix. Obsolescent on enroute airways, but still a part of ILS.

Fix. A position determined without reference to a former position.

Flag Alarm. An indicator on a navigation instrument to warn when a reading is unreliable.

Flare-out. That part of the approach path which rapidly decreases the glide angle by nosing up the aircraft at touchdown.

GCA. Ground-Controlled Approach. Control of the approach of an aircraft by a ground-based human controller. The aircraft's position is determined by ground-based radar, and instructions are transmitted to the pilot, generally by voice. Used by the military, but less popular with civil authorities because of the division of responsibility between controller and pilot.

Geoid. The shape of the earth as defined by the hypothetical extension of mean sea level through all land masses.

Glide Slope. The facility that provides vertical guidance in the ILS. Usually established at about 3° above the horizontal.

Goniometer. An inductive or capacitive device having a stator connected to a fixed antenna system and a rotor connected to a receiver or transmitter, allowing the pattern of the fixed array to be rotated as the rotor is rotated. Extensively used in direction-finding and in omnirange beacons.

H-Beacon. An LF/MF nondirectional beacon, used by airborne direction finders. Same as NDB.

Heading. The horizontal direction in which a vehicle is pointed with respect to a reference, often magnetic North. Usually expressed in degrees, clockwise from the reference.

Homing. The process of approaching a desired point by directing the vehicle toward that point.

Hyperbolic System. A generic term for navigation systems (Decca, Loran, Omega) deriving position by measurement of differential distance to several stations. Two stations provide a hyperbolic LOP. Three or more stations provide a fix.

ILS. Instrument Low-Approach System. An ICAO standard, involving lateral guidance by a 108–112-megahertz localizer, vertical guidance by a 330–335-megahertz glide slope, and distance along the path by 75-megahertz fan markers.

ILS Reference Point. The optimum contact point for a landing. ICAO standards place it 500 to 1000 feet from the approach end of the runway.

Inertial Navigator. A self-contained dead-reckoning system that depends on the sensing of accelerations in three planes and double integration of them to obtain distance traveled. Accuracy decreases as time of travel increases, therefore it is most suited to high-speed vehicles.

Initial Conditions. The value set into an inertial navigator at the start of travel. May be "updated" during the journey by radio or visual fixes.

Inner Marker. Refer to boundary marker.

Instrument Approach. An approach using navigation instruments rather than direct visual reference to the terrain.

Interrogator. Pulse transmitter-receiver used in DME and secondary surveillance radar to elicit a reply from transponder.

Leader Cable System. An aid in which the path to be followed is defined by the magnetic field of an electric cable installed on the ground or under water.

Localizer. The facility that provides lateral guidance in ILS.

LOP. Line of Position. A line plotted on the earth's surface representing the locus of constant indication of navigational information (in VOR, straight radial lines; in DME, circles; in Decca, Loran, and Omega, hyperbolas).

Loran-A. A hyperbolic system using pulse envelopes at 2 megahertz with a baseline of about 300 miles; usable to several hundred miles.

Loran-C. A hyperbolic system using 100-kilohertz coherent pulses with baselines up to 500 miles; usable to over 1000 miles.

Loran-D. A shorter-baseline lower-power version of Loran-C, for tactical applications.

M-Array Glide Slope. A modification to obtain a greater degree of energy cancellation at low elevation angles, reducing site error due to hills on the approach path. (Called "M" because it was the 13th in a series of designs.)

MLS. Microwave Landing System. An experimental system that is planned to replace ILS before year 2000.

Middle Marker. A fan marker on an ILS approach path, approximately 3500 feet from the approach end of the runway.

Most Probable Position. A computed position based on several lines of position, all adjusted to a common time and weighted in accordance with their estimated probable errors.

NDB. Non-Directional Beacon. An LF/MF station used with airborne direction finders. Same as H-beacon. (Refer to ADF.)

Night Effect. An error occurring mainly at night, when ionospheric reflection is at a maximum. A major limitation to the useful range of continuous-wave systems in the LF/MF bands.

Octantal Error. A bearing error, usually due to departure of an antenna pattern from an ideal shape. It varies sinusoidally throughout the 360 degrees and has four positive and four negative maxima.

Omega. A hyperbolic system using high-power transmitters with time-shared continuous-wave signals at around 10 kilohertz, and having baselines and service range of 5000 miles.

Omnirange. Colloquialism for omnidirectional range, a facility providing bearings equally well in all directions. The best-known example is VOR.

Outer Marker. In ILS, the fan marker farthest from the end of the runway, usually about 5 miles out.

PAR. An FAA term for Precision Approach Radar. This X-band radar, which scans limited angles of elevation and azimuth, is the main component of the GCA system.

Pitch. The angular displacement between the longitudinal axis of the vehicle and the horizontal.

Quadrantal Error. An error in measured bearing, frequently due to antenna or goniometer characteristics, which varies sinusoidally throughout

the 360 degrees and has two positive and two negative maxima.

Reciprocal Bearing. The opposite direction to a bearing (bearing $\pm 180°$).

Rho-Rho. A generic term for navigation systems that derive position by measurement of distance to two stations (DME/DME).

Rho-Theta. A generic term for navigation systems that derive position by measurement of distance and bearing from a single station (VOR/DME, TACAN).

RMI. Radio Magnetic Indicator. A cockpit display combining omnirange bearing and vehicle magnetic heading, to generate an ADF-type indication.

Roll. The angular displacement between the transverse axis of the vehicle and the horizontal.

RVR. Runway Visual Range. The forward distance visible along the runway during a landing approach.

SAFI. Semi-Automatic Flight Inspection. An FAA system for systematically checking many ground-based navigation aids. Uses airborne recorders and ground-based computers.

Scalloping. Oscillatory course bends occurring at a rate higher than can be followed by the vehicle.

Self-Contained Aid. A navigation aid that does not require cooperating radio equipment external to the vehicle (doppler radar, inertial, star tracking).

Short-Distance Navigation. Navigation using aids of limited range; bounded at one end by approach navigation and at the other end by long-distance navigation, there being no universally accepted demarcation.

Slant Distance. Distance between two points not at same elevation. Also called slant range.

Sonar. A general name for sonic and ultrasonic underwater ranging.

Sonne. Refer to Consol.

Space-Referenced Aid. An aid that depends on reference to relatively "fixed" stars (a star tracker).

SSR. Secondary Surveillance Radar.

TACAN. Tactical Air Navigation. A rho-theta system in the 960–1215-megahertz band combining DME with a multilobe omnirange, the latter providing fine and coarse bearing information.

Theta-Theta. Generic term for navigation systems that derive position by measurement of bearing from two stations (VOR/VOR).

Track. Actual path traveled.

Transit. An experimental system in which the position of a surface vehicle is determined by the doppler shift in continuous-wave signals received from a low-orbit earth satellite.

Transponder. A transmitter-receiver triggered by an interrogator. Used chiefly in secondary surveillance radar and DME.

VOR. VHF Omnidirectional Range.

VORTAC. Co-located VOR and TACAN stations. This is the preferred method in the United States and some other countries of implementing the VOR/DME system of the ICAO. It allows aircraft equipped only with TACAN to use the same facilities and procedures as aircraft equipped only with VOR/DME.

Yaw. The angular displacement between the normal axis of the vehicle and the course line.

GENERAL REFERENCES

1. "Advanced Navigation Techniques," *AGARD Conf. Proc. 28*, W. T. Blackband, Ed. Slough, England: Technivision Services, 1970. Comprehensive treatment of long-range navigation systems containing many additional references.

2. Avionics Research Group, Ohio University, Athens, Ohio, "All-weather, Low-Level Navigation," Final Report, U.S. Army ECOM Contract DAA-68-C-0084, November 1972.

3. G. E. Beck, "Navigation Systems—A Survey of Modern Electronic Aids." London: Van-Nostrand-Reinhold, 1971.

4. N. Bowditch, "American Practical Navigator," U.S. Navy Hydrographic Office, Publ. 9, 1958.

5. T. W. Brogden, "The Omega Navigation System." *Navigation, J. Inst. Navigation*, vol. 15, no. 2, Summer 1968.

6. C. Broxmeyer, *Inertial Navigation Systems.* New York: McGraw-Hill, 1964.

7. W. R. Fried, "An Integrated Satellite Radio Navigation, Surveillance and Communication System," Proceedings of the National Radio Navigation Symposium, Institute of Navigation, Washington, D.C., Nov. 13–15, 1973.

8. W. R. Fried, "New Developments in Radar and Radio Sensors for Aircraft Navigation," *IEEE, T-AES*, pp. 25–33; January 1974.

9. P. H. Garrett, "Advances in Low-Frequency Radio Navigation Methods," *IEEE, T-AES*, pp. 562–574; July 1975.

10. ———, Handbook: VOR/VORTAC Siting Criteria, FAA, Department of Transportation, 1968.

11. L. Horowitz, "Direct-Ranging Loran," *NAECON Proc.* (Dayton, Ohio), May 1969.

12. *IEEE Trans. Aerosp. Navig. Electron.* (Special Issue of the VOR/DME Navigation System—Its Present Capabilities and Future Potential), vol. ANE-12, Mar. 1965 (entire issue).

13. M. Kayton and W. R. Fried, *Avionics Navigation Systems.* New York: Wiley, 1969.

14. R. H. Mayer, "Doppler Navigation for Commercial Aircraft in the Domestic Environment," *IEEE Trans. Aerosp. Navig. Electron*, vol. ANE-11, pp. 8–15, Mar. 1964.

15. B. Miller, "The C-5 Navigation System—An Application of Digital Synergistic Stochastic Hybrid Navigation Technology," *Proc. AGARD-NATO 9th Meeting of Guidance and Control Panel*, September 22–26, 1969.

16. "National Plan for Navigation," Dep. Transp., Washington, D.C., Apr. 1972. Contains a good review of prominent navigation aids from an operational viewpoint.

17. J. A. Pierce, "OMEGA," *IEEE Trans. Aerosp. Electron. Syst.*, vol. AES-1, pp. 206–215, Dec. 1965.

18. C. Powell, "The Decca Navigator System for Ship and Aircraft Use," *Proc. Inst. Elec. Eng.* (London), Paper 2567R, Mar. 1968.

19. T. A. Stansell, Jr., "The Navy Navigation Satellite System; Description and Status," *Navigation, J. Inst. Navigation*, vol. 15, no. 3, Fall 1968.

20. R. C. Stow and B. Danik, "Integrated Hybrid Inertial Navigation Systems," *Proc. AGARD-NATO 9th Meeting of Guidance and Control Panel*, September 22–26, 1969.

21. A. B. Winnick and D. M. Brandewie, "Recent VOR/DME System Improvements." *Proc. IEEE*, vol. 58, March 1970.

22. W. Zimmerman, "Optimum Integration of Aircraft Navigation Systems," *IEEE Trans. Aerospace and Electronic Systems*, vol. AES-5, Sept. 1969.

CHAPTER 33

SPACE COMMUNICATION

INTRODUCTION

The design of space communication systems is not inherently different from that of terrestrial microwave systems. However, the unique environment of space causes a change of emphasis in the design process. The major differences follow.

(**A**) There is no appreciable fading in space communication systems because of the conditions to which terrestiral microwave systems are subject. The only exception to this is rainfall, which affects systems above 4 GHz. For this reason it is common to calculate space communication systems with considerably more precision than terrestrial systems and to design them with considerably less margin. It is not unusual for system designers to argue about discrepancies of the order of 0.1 dB in systems which may have net losses of the order of 100 dB.

(**B**) In general, spacecraft equipment power output and power supply are extremely expensive, as is planetary receiver sensitivity. Thus, the question of appropriate economic tradeoffs is a basic one of space-system design and will ordinarily control the design of the system.

(**C**) Because of the above, advantage is usually taken of advances in detection and signal design as quickly as possible.

SYSTEM DESIGN

There are three basic types of space communication systems, operating in different environments and with decidedly different characteristics. On the following pages are developed the characteristics, principles, and basic information for design of each of these types.

Planet-to-Spacecraft Type: A communication system working in the planet-to-spacecraft direction is characterized, at least on Earth, by the relatively easy availability of electrical power and supporting structures for large antennas. Such a path will therefore have a relatively large available EIRP (Effective Isotropic Radiated Power); a relatively high noise background (both from the warm planet and from the man-made noise generated on the planet), and a requirement for precise aiming information for the planetary antenna.

Spacecraft-to-Planet Type: In the spacecraft-to-planet direction, the receiving antenna looks into the relative cold of space; it thus enjoys a relatively quiet background. Power generation in the spacecraft is extremely expensive, because of both the weight that must be put into orbit and the difficulty of disposing of waste heat. Antenna gain is also very expensive, since increasingly large antennas require increasingly precise stabilization of the spacecraft, which eventually results in an increase in the amount of fuel necessary for adjusting the position of the antenna and holding it within the prescribed tolerance. Reliability is of surpassing importance.

Spacecraft-to Spacecraft Type: Spacecraft-to-spacecraft links enjoy tremendous freedom because almost all frequencies are available for use, including optical frequencies. Electrical design of these links is relatively straightforward, the principal difficulty involving the maintenance of track between the two spacecraft.

CASE 1—THE PLANET-TO-SPACECRAFT LINK

System Noise

The antenna on a spacecraft receiver will "see" an environment that contains a number of noise sources besides its intended signal [1]. There is a background comprising the broad noise contribution of the galaxy as shown in Fig. 1, with concentrated sources of noise. The most conspicuous of these concentrated sources are the sun and any nearby planetary bodies in the antenna's field of view. The sun's noise temperature is shown in Fig. 2.

The net background noise temperature can be found in a practical case by simply averaging the noise temperature per unit solid angle, as follows.

When in the vicinity of a planet, the antenna

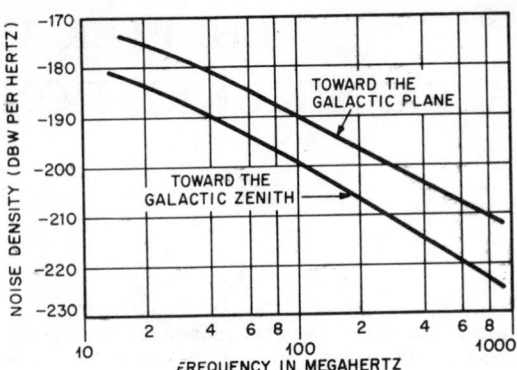

Fig. 1—Levels of galactic noise as a function of frequency.
After Perlman, et al. [1].

of the spacecraft "sees" the planet as a noise source having a noise temperature of T kelvins (where T is the approximate average surface temperature of the planet). The planet radiates radio noise at a level corresponding to this temperature. Table 1 lists approximate radio blackbody temperatures of bodies of planetary size, plus other physical constants of interest.

The effective noise temperature of a spacecraft antenna's field of view is approximately the blackbody radio temperature of the planet, so long as the planet (as "seen" by the spacecraft antenna) subtends an angle greater than the beamwidth of the satellite antenna.

When the angle subtended by the planet is smaller than the beamwidth of the satellite antenna, the satellite sees a combination of galactic noise and planetary radiation proportional to the relative areas of the planetary cross section and the beamwidth at the planetary distance. If the spacecraft is at a distance h from the surface of a planet of radius R (in consistent units), the planet will subtend an angle γ at the spacecraft

$$\gamma = 2 \arcsin\{1/[1+(h/R)]\}. \quad (1)$$

The effective distance of the planet is $h+R$, and the area of the zone bounded by the pattern of the spacecraft-antenna beam at that distance is

$$4\pi(h+R)^2/G \quad (2)$$

where G is the gain of the spacecraft antenna over an isotropic antenna.

The effective noise temperature the antenna sees is

$$T_e = T_p \left\{ \frac{\arcsin\{1/[1+(h/R)]\}}{\arccos[1-(2/G)]} \right\}^2$$

$$+ T_c \left\{ \frac{\arcsin\{1/[1+(h/R)]\} - \arccos[1-(2/G)]}{\arcsin\{1/[1+(h/R)]\}} \right\}^2$$

$$(3)$$

where T_p is the planetary blackbody temperature, and T_c is the average cosmic temperature in the

direction of look, which is the exact expression. At frequencies above 1 GHz the cosmic noise temperature (except for discrete sources) is usually ignored, and the expression reduces to

$$T_e = T_p \left\{ \frac{\arcsin\{1/[1+(h/R)]\}}{\arccos[1-(2/G)]} \right\}^2. \quad (4)$$

At distances of three or more planetary radii, $[(h/R)=3]$, and with gains of 3 or more, a further simplification can be made to

$$T_e = GT_p/4[(h/R)+1]^2 \quad (5)$$

with an error of less than 10 percent at $(h/R)=3$. The error decreases rapidly with increasing G or h/R. If the spacecraft antenna sees only part of the planetary surface (as in communication satellites with high-gain antennas) the noise contribution ideally should be found by vector integration over the visible surface of the planet or, more practically, by estimating the ratios of areas involved and their effective distances.

The noise density at the spacecraft receiver (in dBW of noise per hertz of receiver bandwidth) is $10 \log k(T_e+T_R)$, where T_e is the effective external noise temperature, T_R is the noise temperature of the receiver without antenna, and k is Boltzmann's constant. This can be stated as

$$n = -228.60 + 10 \log(T_e+T_R), \text{ in dBW/Hz.} \quad (6)$$

(In the path calculation equations, decibels are used throughout where possible.)

Because of the relatively noisy receivers often provided in spacecraft—especially when intended for use close to a planet, as in communication satellite service, it is sometimes more convenient to use the receiver noise figure. If F is the receiver noise figure (a number; the noise figure expressed in dB is $N=10 \log F$), it may be converted to noise temperature by the relation

$$T_R = 290(F-1) \quad (7)$$

which assumes the conventional 290 K reference temperature. The use of the noise figure is declining in space applications, since the standard temperature of 290 K is essentially meaningless

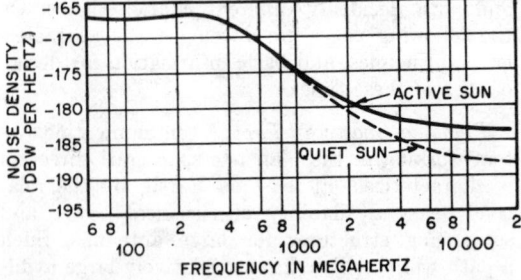

Fig. 2—Experimental and theoretical values of noise from quiet sun and active sun. Values were observed at the terminals of an antenna having a beam equal to or smaller than the solar disk. *After Perlman, et al.* [1].

away from a terrestrial environment. In addition, low-noise receivers produce negative standard noise figures when expressed in decibels, leading to confusion. If a spaceborne receiver had a "5-dB noise figure," its effective noise temperature by (7) would be about 627 K.

Path Losses

If the planet has a substantial atmosphere, there will be losses in signal strength from absorption by molecular-dipole resonances; in addition, there will be attenuation from precipitation [2]. The amount of this attenuation varies with the climate of the area to which communication is intended, as well as with the angle at which the transmitted signal leaves the surface of the planet (thus determining the distance the signal must travel through the atmosphere). Because of the resonant nature of some of the absorption, the attenuation is strongly frequency-dependent. Chapter 28, Electromagnetic-Wave Propagation, treats this subject in detail.

Figure 3 gives terrestrial measurements of clear-air attenuation at 30° and at the zenith. Figure 4 gives information on millimeter-wave attenuation caused by rainfall, about which information is particularly sparse. However, measurements are continuously being taken, and designers should examine the latest literature [8, 9] for statistical data if in a terrestrial path. The use of frequencies above 8 GHz is contemplated.

In the following equations the atmospheric attenuation is given simply as L_a (in dB). It is assumed that the designer of space communication systems will use the available data from space probes for the composition of the atmospheres of unfamiliar planets. The free-space loss is

$$L_{FS} = 92.45 + 20 \log f + 20 \log d \qquad (8A)$$

where L_{FS} is in decibels, d in kilometers, and f in gigahertz. Alternatively

$$L_{FS} = 96.58 + 20 \log f + 20 \log d \qquad (8B)$$

where d is in statute miles.

The total path loss is

$$L_p = 92.45 + 20 \log f + 20 \log d + L_a \qquad (9)$$

where L_p and L_a are in decibels, d in kilometers, and f in gigahertz.

The rf carrier-to-noise ratio in the receiver is then

$$C/N = P - L_{FS} - L_a + G - n - 10 \log B \qquad (10)$$

where P is the transmit EIRP in dBW, and B is the receiver noise bandwidth in Hz.

The gain [3] of the receiving antenna is G (in dB over isotropic). Expressions for the gains of various types of antennas are given in Table 2.

C/N Calculation

A complete equation for C/N (in dB) is

$$C/N = P + 136.15 - 20 \log f - 20 \log d - L_a$$
$$+ G - 10 \log(T_e + T_R) - 10 \log B. \qquad (11A)$$

Because the frequency term cancels, it is sometimes simpler to work with the effective area A of the spacecraft antenna. Equation (11A) then becomes

$$C/N = P + 157.61 + 10 \log A - 20 \log d - L_a$$
$$- 10 \log(T_e + T_R) - 10 \log B \qquad (11B)$$

where A is in square meters.

CASE 2—THE PLANETARY RECEIVER

General Considerations

Except for initial probes, planetary stations are characterized by high available transmitting powers and relatively high-gain antennas. Heavy cryogenic systems can be provided for extremely low-noise receivers, and the relatively low cosmic noise temperature makes their use practical. The effect is to reduce the power requirements of the spacecraft transmitter. Also, the stable platform afforded by the planetary surface allows precise antenna positioning.

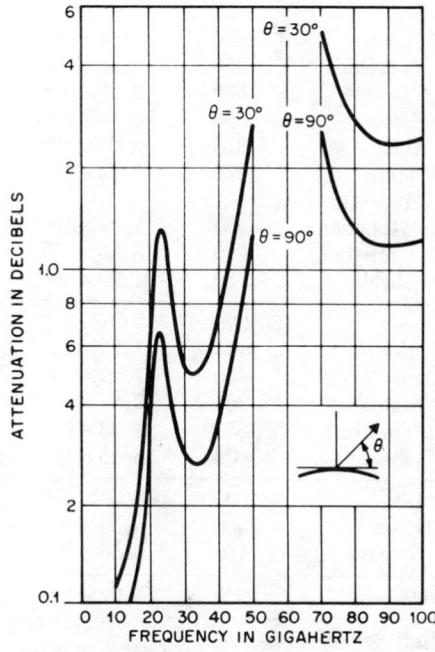

Fig. 3—Calculated values of attenuation at frequencies in the millimeter-wave band for a standard atmosphere, using a zenith path and a path with an elevation angle of 30 degrees. *From Hogg* [2].

TABLE 1—CHARACTERISTICS OF

	Radius (km)	Mean Distance from Principal (km)	Mass (kg)	Sidereal Period	Surface Temp (K)	Orbital Velocity (km/sec)
Sun	696 000		1.99×10^{30}		4500	
Mercury	2420	5.79×10^{7}	3.16×10^{23}	87d23h16m48s	616	47.844 489
Venus	6100	1.08×10^{8}	4.87×10^{24}	224d16h48m	577	35.002 275
Earth	6378	1.50×10^{8}	5.98×10^{24}	365d6h14m24s	288	29.772 050
Moon	1736.7	3.84×10^{5}	7.35×10^{22}	27d7h43m41s		
Mars	3410	2.27×10^{8}	6.39×10^{23}	686d23h31m12s	208 to 299	24.107 314
I Phobos	7.5	9.38×10^{3}	Unknown	7h39m14s*		
II Deimos	4	2.35×10^{4}	Unknown	1d6h17m55s		
Jupiter	70 400	7.78×10^{8}	1.90×10^{27}	4332d14h24m	180	13.051 423
V Amalthea	80	1.81×10^{5}		11h57m23s		
I Io	1867.5	4.22×10^{5}		1d18h27m34s		
II Europa	1575	6.71×10^{5}		3d3h13m42s		
III Ganymede	2575	1.071×10^{6}		7d3h42m33s		
IV Callisto	2590	1.88×10^{6}		16d16h32m9s		
VI	70	1.145×10^{7}		250d15h		
VII	20	1.17×10^{7}		260d1h		
X	10	1.17×10^{7}		260d		
XII	10	2.12×10^{7}		600d*		
XI	12.5	2.25×10^{7}		692d*		
VIII	15	2.35×10^{7}		739d*		
IX	10	2.37×10^{7}		745d*		
Saturn	59 700	1.43×10^{9}	5.69×10^{26}	10 759d	141	9.639 707
Rings:						
I Mimas	250	1.87×10^{5}		22h37m5s		
II Encelade	250	2.35×10^{5}		1d8h53m7s		
III Tethys	500	2.96×10^{5}		1d21h18m26s		
IV Dione	500	3.80×10^{5}		2d17h41m9s		
V Rhea	675	5.25×10^{5}		4d12h25m12s		
VI Titan	2475	1.22×10^{6}		15d22h41m25s		
X Themis				20d20h		
VII Hyperion	200	1.47×10^{6}		21d6h38m20s		
VIII Japetus	600	3.55×10^{6}		79d7h55m25s		
IX Phoebe	150	1.29×10^{7}		550d11h*		
Uranus	26 000	2.87×10^{9}	8.70×10^{25}	30 688d	102	6.791 246
V Miranda	100	1.23×10^{5}		1d9h56m		
I Ariel	450	1.92×10^{5}		2d12h29m*		
II Umbriel	350	2.67×10^{5}		4d3h28m*		
III Titania	850	4.38×10^{5}		8d16h57m*		
IV Oberon	800	5.86×10^{5}		13d11h7m*		
Neptune	24 000	4.50×10^{9}	1.03×10^{26}	60 181d	91	5.471 620
I Triton	2500	3.53×10^{5}		5d21h3m*		
II Nereid	150	5.56×10^{6}		359d9h36m		
Pluto	7000	5.91×10^{9}	5.38×10^{24}	90 737d	97	4.827 900

* Retrograde.

† Jupiter radiates a considerable amount of nonthermal energy by synchrotron radiation of relativistic electrons. The apparent noise temperature of Jupiter from a point outside its electron belts rises rapidly below about 4 GHz, reaching 2000 K at $\lambda = 21$ cm, 5000 K at 31 cm, and 20 000 K at 68 cm.

‡ Estimate.

Major Solar System Bodies.

Approximate Distance from Earth in km (Max)	(Min)	Period of Rotation	Orbital Eccentricity See Eq. (28)	Inclination of Plane of Orbit to Ecliptic or for Satellite to Planet's Orbit	Radio Blackbody Temp (K)
147 250 000	132 078 000	24d16h			See Fig. 2
218 864 800	80 465 000	58d15h55m	0.206	7°0'10''	
259 097 300	40 232 500	224d16h48m*	0.007	3°23'37''	580
		23h56m4s	0.017	0	254
					140 to 220
399 106 400	56 325 500	24h37m23s	0.093	1°51'1''	211
965 580 000	590 613 100	9h50m to 9h55m	0.048	1°18'31''	144†
			0.0028	3°6'9''	
			0.000	3°6'7''	
			0.003	3°5'8''	
			0.0015	3°2'3''	
			0.0075	2°42'7''	
			0.155	29°	
			0.207	28°	
			0.08	28°	
			0.21	163°	
			0.38	148°	
			0.27	157°	
1 654 360 400	1 197 319 200	10h14m	0.055	2°29'33''	106
			0.0190	26°44'7''	
			0.0001	26°44'7''	
			0.0000	26°44'7''	
			0.0020	26°44'7''	
			0.0009	26°41'9''	
			0.0289	26°7'1''	
			0.1043	26°0'0''	
			0.0284	16°18'1''	
			0.1659	174°42'	
3 154 228 000	2 584 535 800	10h45m*	0.047	46'21''	100‡
				98°	
				98°	
				98°	
				98°	
4 683 063 000	4 308 096 100	15h29m	0.009	1°46'45''	85
				40°	
7 563 710 000	4 296 831 000	6d9h	0.247	17°8'44''	80

Sources: 1967 World Almanac, p. 545; Flammarion Book of Astronomy, pp. 661–664; Handbook of Astronautical Engineering, pp. 1–41 and 2–42; IEEE Trans. on Aerospace and Electronic Systems, Vol. AES–3; No. 5, p. 759; IEEE Trans. on Antennas and Propagation, Vol. AT–12, pp. 908–913; Encyclopedia Brittanica.

TABLE 2—POWER GAIN G AND EFFECTIVE AREA A FOR SEVERAL COMMON ANTENNAS*. *From Northrop [3]*.

Antenna	Gain Above Isotropic Antenna G	Maximum Effective Area A
Isotropic (hypothetical)	1	$\lambda^2/4\pi = 0.079\lambda^2$
Infinitesimal dipole or loop	1.5	$(3/8\pi)\lambda^2 = 0.119\lambda^2$
Linear half-wavelength dipole	1.64	$(30/73\pi)\lambda^2 = 0.13\lambda^2$
Optimum horn (mouth area $= S$)	$10(S/\lambda^2)$	$0.81S$
Parabolic reflector (aperture $= S$, $\eta \approx 0.5$ to 0.6)	$4\pi\eta S/\lambda^2$	ηS
Broadside array (area $= S$)	$4\pi[S(\max)/\lambda^2]$	$S(\max)$
Turnstile	1.15	$1.15(\lambda^2/2\pi) = 0.0915\lambda^2$

* No heat loss.

Conventions of Planetary Station Design

Because of the very large antennas often used, antenna gains are often hard to measure precisely, and in any event these gains may vary with dish position because of deformation. Similarly, because of the extremely low noise temperature of the receivers, precise in-place measurements of their noise temperatures are very difficult to make. It has therefore become customary to measure and to specify the signal-to-noise ratio (which is relatively easier to measure) with a known received signal strength. A convenient way to express this is the ratio G/T, usually expressed in dBW/K [this is dimensionally misleading—see Equation (12)], which is a general rf figure of merit of the receiving portion of the station. It is the antenna

gain divided by the system noise temperature, or

$$G/T = G - 10 \log T \qquad (12)$$

where G is expressed in decibels, and T in K.

Similarly, the received carrier-to-noise ratio is often expressed as C/T (in dB), which is

$$C/T = 10 \log P_R - 10 \log T \qquad (13)$$

and is related to the true carrier-to-noise ratio by

$$C/N = C/T + 228.60 - 10 \log B \qquad (14)$$

where B is the receiver noise bandwidth in hertz, and the other values are in decibels.

Because of the relatively broad patterns of spacecraft antennas, and for convenience in comparing systems, the available signal is often expressed as a "flux density" ψ in watts/square meter or more usually in dBW/m². This flux density is

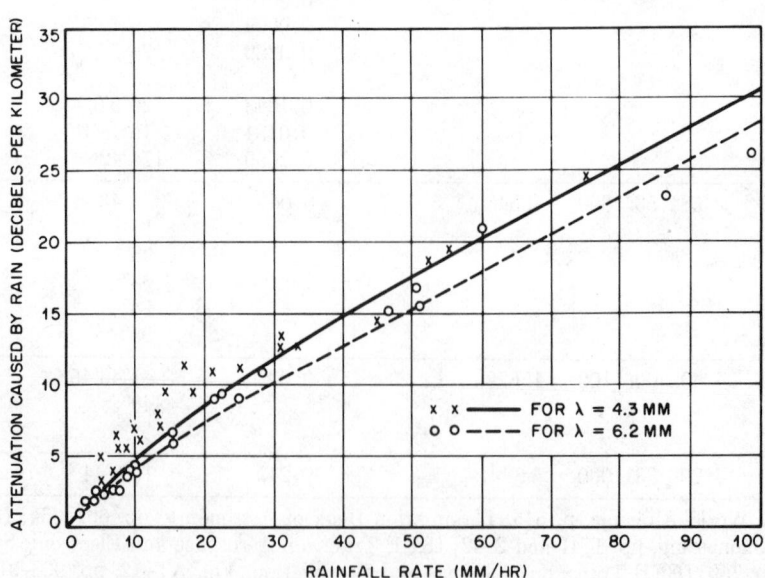

Fig. 4—Measurements at wavelengths of 6.2 and 4.3 millimeters (made by Bell Telephone Laboratories) of attenuation caused by rain, plus corresponding calculated values (solid and broken lines). *From Hogg [2]*.

given by

$$\psi = P_T - 70.99 - 20 \log d - L_a \qquad (15)$$

where P_T is the EIRP of the spacecraft in dBW, ψ is in dBW/m², and L_a is in dB.

With a known flux density and (G/T), the (C/T) can easily be found by

$$C/T = \psi + G/T - 20 \log f - 21.46 \qquad (16)$$

where f is in GHz, ψ in dBW/m², and the other values are in dB.

Noise Sources at a Planetary Station

It is much more complex to calculate the effective noise temperature of a planetary station than that of a spacecraft, because of the greater variety of noise sources and the greater sensitivity of the receiving system. The galactic noise contribution must be considered, along with noise radiation from the atmosphere and surrounding surfaces. In the region between 1 and 10 GHz, *all* noise sources must be considered; the greatest noise contribution is usually radiation from the planetary surface. To some extent the design of the antenna must be compromised, in that side-lobe levels must be kept lower than would otherwise be desirable, to avoid pickup from the ground. This often results in a compromise of the optimum illumination for the highest gain.

The general way to arrange a noise budget for such a station is to consider the lineup of equipment between the antenna feed and the low-noise preamplifier. This might look like the diagram shown in Fig. 5. The system noise temperature is obtained by adding the noise contributions listed in Table 3.

T_F, T_D, and T_{FL} are the actual physical temperatures of the respective devices shown in Fig. 5. The blackbody noise contributions of these devices will be significant if high transmit powers and cooled preamplifiers are involved. Usually it is helpful to cool as much of these assemblies as is possible.

CASE 3—THE SPACECRAFT-TO-SPACECRAFT LINK

At the present state of knowledge, it appears that the planner of space communication systems has almost complete design freedom so long as he is concerned with communication between two points that are permanently in space. The net path loss (in dB) is simply

$$L_p = 49.53 - 20 \log f + 20 \log d - 10 \log A_T - 10 \log A_R \qquad (17)$$

where f is in GHz, d in km, L_P in dB, and A_T and A_R (the areas of the transmitting and receiving antennas, respectively) in m².

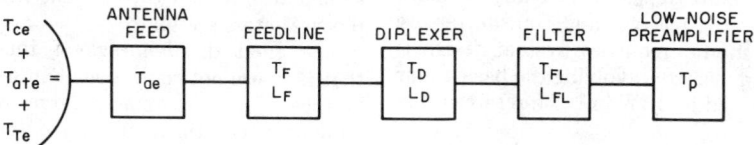

Fig. 5—Noise sources affecting a typical planetary receiver. The T's are temperatures in K; L's are losses expressed as numerical ratios. Refer to Table 3.

TABLE 3—NOISE BUDGET FOR PLANETARY STATION (SEE FIG. 5).

Cosmic noise	$\dfrac{\text{Cosmic (sky) temperature}}{\text{Atmospheric loss} = \alpha_a}$	$= T_{ce}$
Atmospheric noise	$[(\alpha_a - 1)/\alpha_a]$(atmospheric temp)	$= T_{ate}$
Terrain pickup (side lobes and main lobes at low look angles)		$= T_{Te}$
	Subtotal—antenna effective temperature	$= T_{ae}$
Feedline	$T_F(L_F - 1)$	$= T_{Fe}$
Diplexer	$T_D L_F(L_D - 1)$	$= T_{De}$
Filter	$T_{FL} L_F L_D(L_{FL} - 1)$	$= T_{FLe}$
Preamplifier	$T_P L_F L_D L_{FL}$	$= T_{Pe}$
	Total	$= T_{Se}$

As frequency is increased, the transmit power or the antenna sizes can be reduced; it is clearly better to use the highest frequency for which generators and receivers are available. Optical frequencies are a possible choice; the basic difficulty of this approach is to maintain the precise track and platform stability necessary for use of the narrow beamwidths available with an optical system.

CASE 4—SATELLITE REPEATERS

General

The earliest example of space communication involved a satellite repeater used for communication between two terrestrial points. Of course, satellite repeaters can be provided for communication between any two points at which the satellite is mutually visible, and they can be used in complex arrangements involving switching, store-and-forward, and the like. They are quite useful in space exploration. The most practical arrangements have involved active satellites, for reasons explained below, although a passive satellite's relative immunity from jamming makes it attractive for military applications.

Passive Satellite Repeaters

A passive satellite repeater is really a radar system in which the receiver and transmitter are not co-located. Radio-frequency system design is relatively uncomplicated, involving the basic radar equation. Expressed in dBW/m², the received flux density is

$$\psi = P_T - 141.98 - 20\log d_1 - 20\log d_2$$
$$+ 10\log\sigma - L_{aU} - L_{aD} \quad (18)$$

where P_T is the transmitted power in dBW, d_1 and d_2 are the slant ranges of the satellite to the transmitter and the receiver in km, σ is the radar cross section of the satellite in m², and L_{aU} and L_{aD} are the up- and down-path atmospheric losses, respectively, in dB. It is obvious that a passive satellite system using a relatively high altitude requires enormous transmitter EIRP for more than a few channels. For a carrier-to-noise ratio of 20 dB in a telephone channel ($kTB = -141.7$ dBm), an 85-foot-diameter antenna and a noiseless receiver, a 6000-km orbit, and a 100-foot-diameter spherical reflector, (and ignoring atmospheric losses), the transmitter power at 6 GHz must be about 150 watts per channel. More-efficient reflectors can be made, of course. The most effective in terms of reflecting cross section per unit weight is the type made of orbiting thin wire dipoles (needles).

It has been determined [4] that the dipoles have an average reflecting cross section of $0.158\lambda^2$ (per

dipole), including the effects of random dipole orientation. Exact calculations are complex, since, to be precise, a way must be found to sum the contributions of each dipole in the scattering volume.

The common scattering volume can be found geometrically. If the belt is much narrower than the antenna beamwidth, the scattering cross section is

$$\chi = N_l L (0.158\lambda^2) \quad (19)$$

where N_l is the number of dipoles per unit length and L is the length of the common cylinder.

If the belt is considerably wider than the beamwidth, the common volume will be an ellipsoid, the scattering cross section of which is

$$\chi = N_v V (0.158\lambda^2) \quad (20)$$

where N_v is the number of dipoles per unit volume and V is the volume of the ellipsoid.

Natural satellites are occasionally used for passive repeaters. They have the characteristics of large cross-sectional areas and relatively poor reflectivity. In general, these rough bodies are far from specular reflectors. The Moon, for instance, has an effective reflecting cross section of 9.48×10^{11} m², about $\frac{1}{10}$ of its physical cross section. Similar information is not now available for other natural satellites; it is reasonable to assume, however, that essentially airless satellites have similar terrain and thus similar ratios of radar to physical cross section.

The great disadvantage of these satellites is that they are not continuously in usable positions; however, their extremely large reflecting cross sections make them attractive for moving low-priority record traffic and for planetary exploration and development.

Active Satellite Repeaters

The simplest form of an active satellite is one in which a signal is received, amplified, and reradiated. The design constraints on such a device are set by the following considerations.

(A) When the satellite is used as a relay between planetary stations, up-link power is usually adequate to permit the use of uncooled receiver preamplifiers.

(B) The system performance is therefore set by the satellite EIRP. This is, in the final analysis, a function of the orbited weight of the satellite; it is determined by the specific weight of available primary power sources and by the lifting capacity and cost of available boosters.

(C) The operating cost of the satellite itself is simply the cost of replacing the satellite divided by the satellite's life expectancy.

$$\text{operating cost/year} = \frac{\text{launch cost}}{(\text{MTBF}) \times P} \quad (21)$$

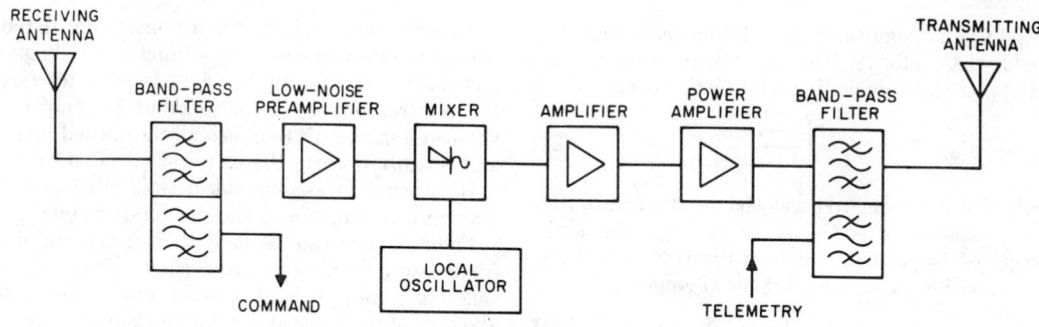

Fig. 6—Block diagram of a simple satellite repeater.

where MTBF is the mean time before failure, and P is the probability that the launch will be successful and that the satellite will work properly when in place.

For systems involving relatively small numbers of satellites, such as synchronous systems, conservative accounting demands that at least one launch failure be assumed at the outset. In-place standby equipment is not unusual in such systems. Therefore, reliability is of almost overwhelming importance.

The block diagram of a simple satellite repeater is given in Fig. 6. Because efficiency is paramount, systems of this type are usually operated quite close to the level at which limiting occurs. The limiting characteristics of most practical, efficient amplifiers are rather "hard." This characteristic is helpful when the repeater is used to amplify pulsed signals, and the amplifier in these cases is normally run in a limiting condition.

Some systems, however, use arrangements in which each of a number of rf carriers handles multichannel telephony, frequency modulated on the carrier. As carriers are added, each shares the available output power of the satellite. However, when the satellite's amplifiers are operated close to the limiting point, some of the power is dissipated as distortion products. This reduces the

expected output power slightly and also causes cross modulation and distortion among the carriers.

Measurements have been made [5] on a satellite using traveling-wave tubes. Table 4 gives the effect of added carriers on the available output power of the satellite, relative to its saturated power out. Figure 7 gives the intermodulation performance under the same conditions, using equipment for 240 voice channels. The referenced article presents a mathematical model that is suitable for more-general calculations using a computer.

This type of fm multiple-access system is quite susceptible to jamming and interference, since a signal strong enough to drive the amplifier into hard limiting will monopolize essentially all the available power of the satellite transponder.

Effects of Motion

Relative velocities of spacecraft are often great enough to produce Doppler frequency shifts of

TABLE 4—EFFECT OF ADDITIONAL CARRIERS ON SATELLITE OUTPUT POWER. *After Berman and Podraczky* [5].

Number of Equal Carriers	Loss of Useful Power Relative to the Power Obtained from a Single Carrier at Saturation (dB)	Available Output Power Per Carrier Relative to a Single Carrier at Saturation (dB)
1	0	0
2	1.20	−4.21
4	1.31	−7.33
6	1.37	−9.15
8	1.46	−10.49
100	1.50	−21.50

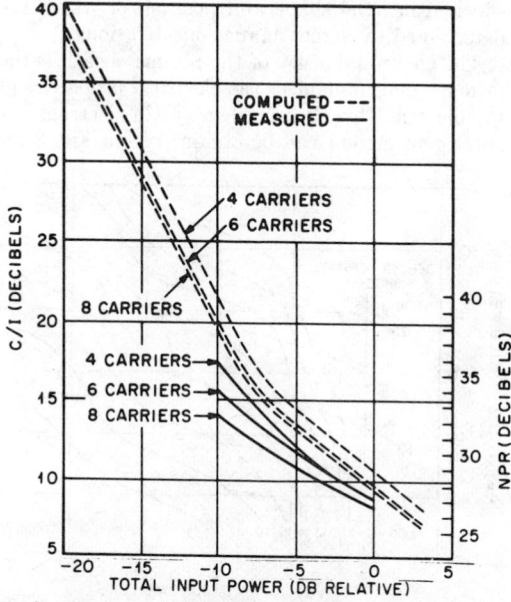

Fig. 7—Carrier-to-intermodulation ratio and noise power ratio vs input drive level for the middle radio-frequency channel. *From Berman and Podraczky* [5].

significant magnitude. In extreme cases there are relativistic effects. The exact expression for the Doppler shift (including relativistic effects) is

$$f_R/f_T = \left[\frac{1-(v/c)}{1+(v/c)}\right]^{1/2} \quad (22)$$

where v is the relative velocity of the transmitter and receiver, c is the speed of light, f_R is the received frequency, and f_T is the transmitted frequency. For $v \ll c$, the more usual equation

$$f_R/f_T = 1-(v/c) \quad (23)$$

is valid. Considerations of conservation of energy require that the received power shall decrease under these circumstances by the ratio

$$P_R/P_T = \frac{[1-(v/c)]^2}{1+(v/c)} \quad (24)$$

which will show a 1-dB reduction in power at $v = 0.075c$. The rate of transfer of information is correspondingly reduced, so that signal integration techniques can be used to restore the original signal-to-noise ratio.

SPACECRAFT IN AN ATMOSPHERE [6]

If a spacecraft is to pass through an atmosphere at high speed during some part of its flight, it may generate a plasma sheath that has drastic effects on both the attenuation of the atmosphere and on the patterns and impedances of the spacecraft antennas. The plasma is formed behind a shock front, the shape and position of which are determined by aerodynamic considerations.

The electron density of the plasma sheath is the primary determinant of the electrical properties of the plasma. This electron density (in electrons per cubic centimeter) can be obtained from Fig. 8 for the Earth's atmosphere, for a point immediately behind the normal shock, as a function of altitude and spacecraft velocity. For cylindrical spacecraft, this electron density drops to about 1/100 of this value at 3 spacecraft radii behind the normal shock, and roughly linearly thereafter at $\frac{1}{10}$ per 10 radii.

It is usual to assume the plasma thickness L, above the antenna, as $\frac{1}{2}$ the spacecraft radius.

Figure 8 also can be used to find the collision frequency g (collisions per second), which, at 3 spacecraft radii behind normal shock, drops to about $\frac{1}{10}$ of the value at normal shock and decreases linearly thereafter at $\frac{1}{10}$ per 10 radii.

The resulting plasma will behave much like a high-pass filter, with a definite cutoff at approximately the "plasma frequency" ω_p.

$$\omega_p = 5.642 \times 10^{-2}(N_e)^{1/2} \quad (25)$$

where N_e is the electron density from Fig. 8. Above cutoff, there will be several absorption lines from dipole resonances. These absorption lines usually extend well into the infrared region.

A general idea of the resulting behavior of the plasma can be gained from Fig. 9, where the absorption coefficient and refractive index of the plasma are plotted for an electron density of 10^{12} electrons per cubic centimeter.

More-precise information on the absorption coefficient α and the index of refraction n can be obtained from Figs. 10 and 11, respectively, for frequencies of $10\omega_p$ and below.

Mitchell's method [6] can be used as follows to calculate the loss in the plasma sheath at a particular point in the flight plan:

(**A**) Determine the sheath electron density N_e and collision frequency g at the antenna's location on the spacecraft. For the Earth's atmosphere, use Fig. 8.

(**B**) Calculate ω_p from (25).

(**C**) Using Figs. 10 and 11 for the frequency of interest, find α and n.

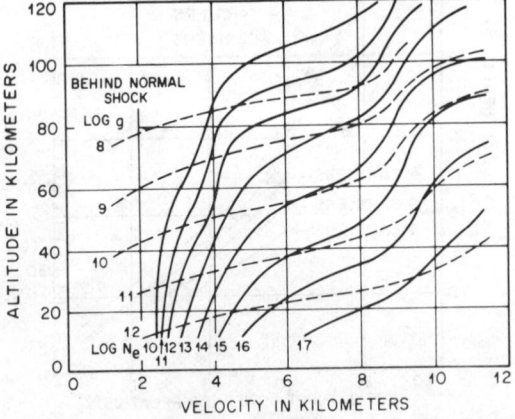

Fig. 8—Sheath electron density N_e and collision frequency g. From Mitchell [6].

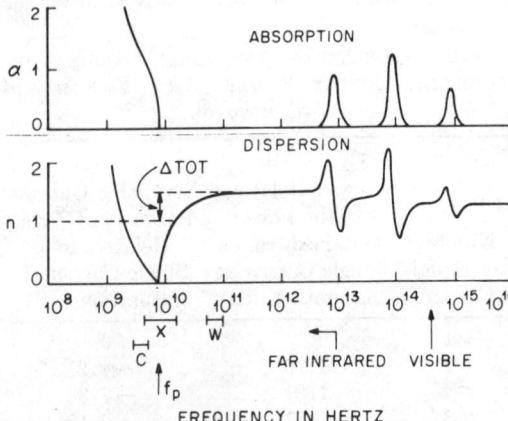

Fig. 9—Spectral view of α and n for a plasma of $N_e = 10^{12}$ electrons/cm³. After Mitchell [6].

The plasma loss L_p (in dB) can now be calculated from

$$L_p = -8.685\alpha k_0 L + 10 \log \frac{(|1-R|^2)}{[|1-R\exp(i\phi)|^2]} \quad (26)$$

where α = absorption coefficient (from Fig. 10), $k_0 = 2\pi/\lambda = 2.094 \times 10^{-2} f$, f = frequency of interest in MHz, L = thickness of the plasma sheath, usually taken as $\frac{1}{2}$ the spacecraft radius, in the same units as λ, $R = (k_r - 1)/(k_r + 1)$, $k_r = k/k_0 = n + i\alpha$, n = index of refraction (from Fig. 11), $\phi = 2kT$, $k = k_r k_0$, $T = 2/(k_r + 1)$, and $i = (-1)^{1/2}$.

The first term of (26) is the loss due to absorption, and the second term is that due to reflection. The equation is valid when $(k_0) \times$ (spacecraft radius) $\gg 1$. This is usually the case for frequencies above 500 megahertz. Shane and Fante [7] have pointed out that the above equation ignores the effect of the antenna mismatch when the antenna has high Q. These effects have been calculated for only a small number of antenna types; the reader is referred to the literature for the effect on specific antenna types when high-Q antenna systems are considered.

ORBIT MECHANICS

To calculate the orbit of a spacecraft around a planet, in the simplest case where only these two

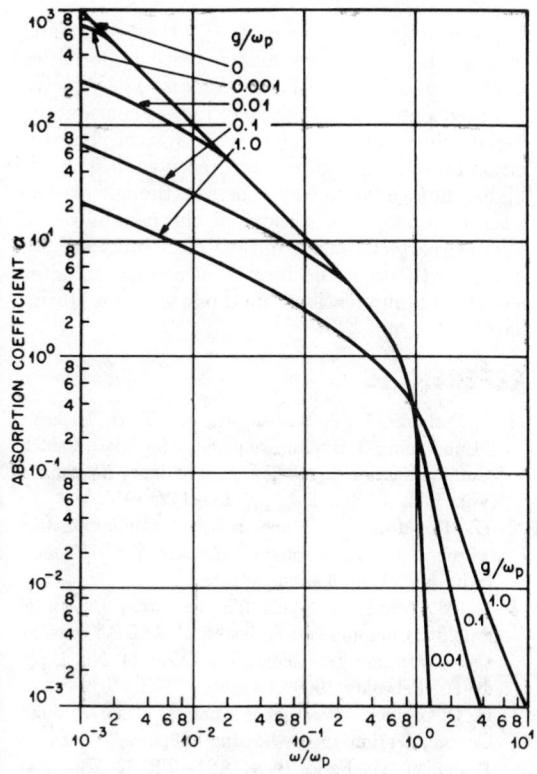

Fig. 10—Plasma absorption as a function of ω/ω_p and g/ω_p. From Mitchell [6].

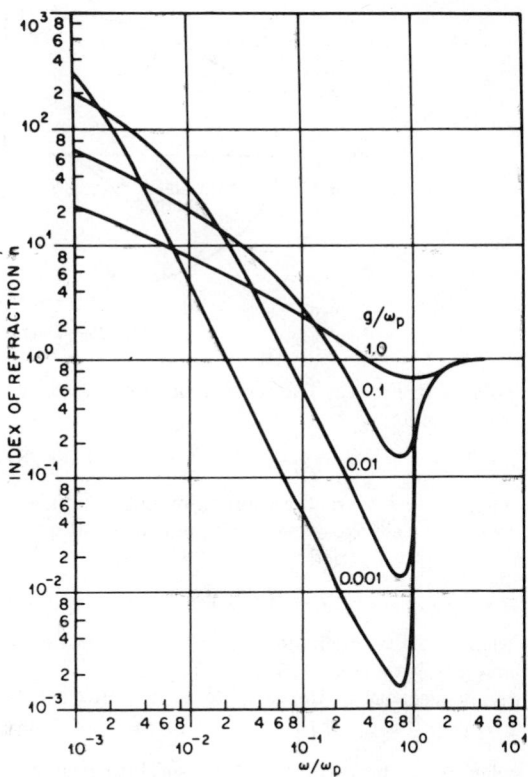

Fig. 11—Plasma index of refraction as a function of ω/ω_p and g/ω_p. From Mitchell [6].

bodies are considered (the two-body problem), the following three independent differential equations must be solved. (Relativistic effects can be ignored, and the satellite's mass is negligible compared with that of the planet.)

$$d^2x/dt^2 = -\mu(x/r^3)$$
$$d^2y/dt^2 = -\mu(y/r^3)$$
$$d^2z/dt^2 = -\mu(z/r^3) \quad (27)$$

where $r = (x^2 + y^2 + z^2)^{1/2}$, $\mu = Gm_p$ (for Earth, $\mu = 5.17 \times 10^{12}$, km^3/hr^2), $G = 8.64 \times 10^{-13}$, (km)^3/kg(hr)^2 (the gravitation constant), and m_p = the planetary mass.

The result is a hyperbola (if the satellite exceeds escape velocity) or an ellipse; specification of an orbit requires that six constants of integration be mentioned.

When such complications as the effect of perturbing bodies and nonsphericities of the planet are involved, the mathematics of exact orbit specification becomes hair-raising; very large computers are used for precise numerical orbit calculations.

Idealized Case

For an elliptical orbit, the planet is at one of the foci of the ellipse (the barycenter). This is

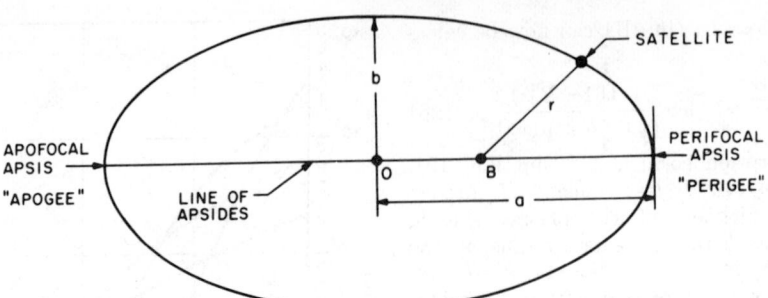

Fig. 12—Diagram of satellite orbit.

point B in Fig. 12. The eccentricity ϵ of the ellipse is defined as

$$\epsilon = [1-(b^2/a^2)]^{1/2} \qquad (28)$$

where a and b are the semimajor and semiminor axes of the ellipse, respectively, in consistent units.

The orbital period is

$$T = 2\pi a^{3/2}/\mu^{1/2} \qquad (29)$$

where a is the semimajor axis in km, T is in hours, and μ is as defined above.

The velocity of the satellite is, instantaneously

$$v = \{\mu[(2/r)-(1/a)]\}^{1/2} \qquad (30)$$

where r is the distance of the satellite from the barycenter. The total energy of the satellite (the sum of its potential and kinetic energies at any instant) is

$$E = -m\mu/2a \qquad (31)$$

where m is the mass of the satellite.

LASER COMMUNICATIONS [10]

The prospect of earth-satellite and deep-space laser communications has been studied by the United States Air Force and the National Aeronautics and Space Administration. The potential advantages of a laser communications link include exceptionally high antenna gain, very high radiated power, and compact transmitting optics. The major obvious disadvantage of a laser communications link is the atmospheric attenuation at optical wavelengths; even a thin layer of overcast could effectively eliminate the reception of laser signals from a space probe. Some microwave-laser communication tradeoff studies have assumed that the optical communications approach will require an earth-synchronous satellite to house the receiver and a microwave link to relay the received signals on to earth. Other studies have assumed that the receiver for a laser system could be located in an arid region with low overcast probability, but, as in the case of a microwave link, more than one receiver would be required to assure continuous reception. A NASA-funded study by Bell Laboratories in the late 1960s found that the requirement of a synchronous

satellite relay station and other factors made microwave links operating in the 5–10 GHz range more advantageous than laser communications.

TABLE 5—COMPARISON IN DECIBELS OF A 2-GHz MICROWAVE SYSTEM WITH TWO PROPOSED LASER SYSTEMS. *After Cook [10].*

Parameter	0.53 μm System	10.6 μm System
Effective Radiated Power	+60	+51
Receiving Area	−33	−33
Noise Temperature	−30	−17
Receiver Losses	−7	−4
Net Comparison	−10	−3

For example, Table 5 compares the performance of a 2-GHz microwave system with hypothetical 0.53 μm and 10.6 μm laser systems. The effective radiated power of either laser link is considerably higher than for the microwave system, but this advantage is mitigated by receiver losses and higher noise levels. Nevertheless, though existing laser technology falls short of the capabilities of microwave technology, optical data links are still in a growth stage and may offer a practical alternative communications method at some future date.

REFERENCES

1. S. Perlman, W. J. Russell, Jr., and F. H. Dickson, "Concerning Optimum Frequencies for Space Vehicle Communication," *IRE Trans. on Military Electronics*, Vol. MIL-4, Nos. 2–3, pp. 184–192; 1960.
2. D. C. Hogg, "Millimeter-Wave Communication Through the Atmosphere," *Science*, Vol. 159, No. 3810, pp. 39–46; 5 January 1968.
3. G. M. Northrop, "Aids for the Gross Design of Satellite Communication Systems," *IEEE Trans. on Communication Technology*, Vol. COM-14, No. 1, pp. 46–56; February 1966.
4. C. C. Gander, "Parametric Analysis of Air/Ground Communication via Orbiting Dipoles," Wright-Patterson Air Force Base, ASD TR 62-579, (AD 289 879); September 1962.
5. A. L. Berman and E. I. Podraczky, "Experimental

Determinations of Intermodulation Distortion Produced in a Wideband Communications Repeater," *IEEE International Convention Record*, Part II, pp. 69–88; 1967.

6. F. H. Mitchell, Jr., "Communication-System Blackout During Reentry of Large Vehicles," *Proc. IEEE*, Vol. 55, No. 5, pp, 619–626; May 1967.

7. E. D. Shane and R. L. Fante, "Comment on 'Communication-System Blackout During Reentry of Large Vehicles,'" *Proc. IEEE*, Vol. 55, No. 11, p. 2055; November 1967.

8. O. E. DeLange, A. F. Dietrich, and D. C. Hogg, "An Experiment on Propagation of 60-GHz Waves Through Rain," *Bell System Technical Journal*, Vol. 54, No. 1, p. 165 ff; January 1975.

9. S. H. Lin, "A Method for Calculating Rain Attenuation Distributions on Microwave Paths," *Bell System Technical Journal*, Vol. 54, No. 6, p. 1051 ff; July–August 1975.

10. J. S. Cook, "Deep Space Communications," *Bell Laboratories Record*, pp, 213–218; August 1970.

BIBLIOGRAPHY

R. F. Filipowsky and E. I. Muehldorf, "Space Communications Systems," Prentice-Hall, Englewood Cliffs, New Jersey; 1965.

"Space Communications," editor A. V. Balakrishnan, McGraw-Hill Book Company, New York; 1963.

G. N. Krassner and J. V. Michaels, "Introduction to Space Communication Systems," McGraw-Hill Book Company, New York; 1964.

Edited Lectures, United States Seminar on Communication Satellite Earth Station Technology, Washington, D. C., 16–27 May 1966, Communications Satellite Corporation; 1967.

E. J. Baghdady and K. W. Kruse, "Signal Design for Space Communication and Tracking Systems," *IEEE Trans. on Communication Technology*, Vol. COM–13, No. 4, pp. 484–498; December 1965.

R. C. Barthle and R. D. Briskman, "Trends in Design of Communications Satellite Earth Stations," *Microwave Journal*, Vol. 10, No. 11, pp. 26–30, 100–108; October 1967.

R. C. Chapman, Jr. and J. B. Millard, "Intelligible Crosstalk Between Frequency Modulated Carriers Through AM–PM Conversion," *IEEE Trans. on Communication Systems*, Vol. CS–12, No. 2, pp. 160–166; June 1964.

J. Dimeff, W. D. Gunter, Jr., and R. J. Hruby, "Spectral Dependence of Deep-Space Communications Capability," *IEEE Spectrum*, Vol. 4, No. 9, pp. 98–104; September 1967.

CHAPTER 34
OPTICAL COMMUNICATIONS

INTRODUCTION

An optical communications system requires an optical source, a means for modulating the source, and an optical receiver. Frequently the optical coupling between the source and receiver is enhanced by means of external optical antennas or an optical fiber waveguide. Optical communication links have exhibited the potential for supplementing conventional communication methods and offer the advantages and disadvantages outlined in Table 1.

OPTICAL SOURCES

A great variety of optical sources has been used in both operational and developmental optical communication links, but by far the most successful are certain electroluminescent diodes and lasers. Both these sources offer relatively high modulation bandwidths and a range of emission wavelengths.

Electroluminescent Diodes*

The operation of electroluminescent diodes is dependent upon the radiative recombination of hole-electron pairs at a forward-biased pn junction, a process described in detail in Chapter 19. Since the radiant emission from electroluminescent diodes is generally linear with respect to applied current below the saturation region, they can be easily and directly pulse or analog modulated by simply controlling the forward bias. Depending upon its constituent bandgap, an electroluminescent diode normally emits a relatively narrow spectral bandwidth (approximately 300–400 nm) between

---

*S. E. Miller, T. Li, and E. A. J. Marcatilli, "Optical-Fiber Transmission Systems, Part II: Devices and Systems Considerations," *Proceedings of the IEEE*, December 1973, pp. 1726–1751.

about 0.5 μm and 30 μm. Thus far, electroluminescent diodes of GaAs, GaAs:Si, GaAs-Al$_x$Ga$_{1-x}$As, and related materials emitting in the near infrared (approximately 750–950 nm) have demonstrated best success in optical communication applications.

Electroluminescent diodes normally emit radiation into a relatively broad pattern and can therefore not be coupled to optical waveguide fibers as efficiently as laser sources. Nevertheless, reasonable coupling efficiency can be obtained by utilizing modified structures such as the one shown in Fig. 1. This structure has a typical emission area of 2×10^{-5} cm^2 and a sterance of 100 W/sr/cm^2 at a forward bias of 150 mA. The output wavelength can be peaked anywhere between 750 and 950 nm by adjusting the aluminum concentration in the Al$_x$Ga$_{1-x}$As alloy making up the diode. About 2 mW can be launched into a suitable optical waveguide fiber using this structure. This and most other electroluminescent diodes exhibit operating lifetimes of more than 10^4 hours at the relatively high current densities required to elicit significant optical output.

Numerous electroluminescent diode configurations have been utilized in atmospheric optical communication links, and a typical device structure is shown schematically in Fig. 2. Typically, approximately 20% of the radiation emitted by such a structure can be collected and collimated by a simple fl lens. A typical cw power output of about 7 mW at 100 mA forward bias and a pulsed power output of more than 90 mW at several amperes forward bias can be obtained from economical commercial devices. These relatively high power levels are obtained from silicon compensated GaAs diodes having a peak spectral wavelength of about 940 nm. The high peak powers are obtained at the expense of modulation bandwidth since the rise and fall times of GaAs:Si diodes are approximately 300 and 200 ns, respectively, as compared to about 10 ns for typical GaAs devices. Somewhat lower optical powers (e.g. a few milliwatts at 100 mA forward bias) can be obtained from GaAs electroluminescent diodes.

TABLE 1—ADVANTAGES AND DISADVANTAGES OF OPTICAL COMMUNICATION SYSTEMS.

Optical Link	Advantages	Disadvantages
Atmosphere	Moderately Secure Absence of RFI Moderately High Data Rates No Licensing Requirements	Requires Critical Alignment Restricted to Line-of-Sight Subject to Atmospheric Interference Subject to Interference from Ambient Illumination
Optical Fiber Waveguide	Secure Very High Data Rates Compact Size Low Cost Potential Absence of Crosstalk and EM Interference	Requires Special Handling Requires Special Splicing and Termination Methods

Superluminescent Diodes

Specially fabricated GaAs-Al$_x$Ga$_{1-x}$As double heterostructure and other electroluminescent diodes can be made to emit both spontaneous (quantum noise) and stimulated (amplified quantum noise) radiation. As in the case of a laser, stimulated emission implies a narrower spectral emission width and higher radiance than spontaneous emission. Fig. 3 shows the construction of a typical superluminescent diode (SLD). The structure of the device in the figure is virtually identical to that of a stripe geometry injection laser with the

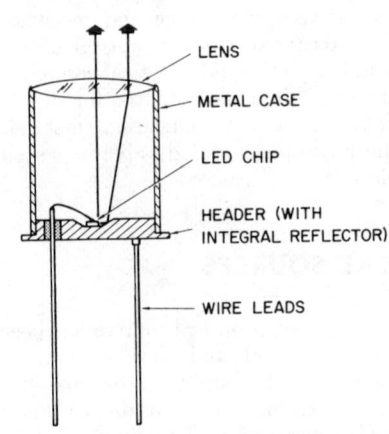

Fig. 2—Construction of a typical GaAs electroluminescent diode of the type employed in atmospheric communication links.

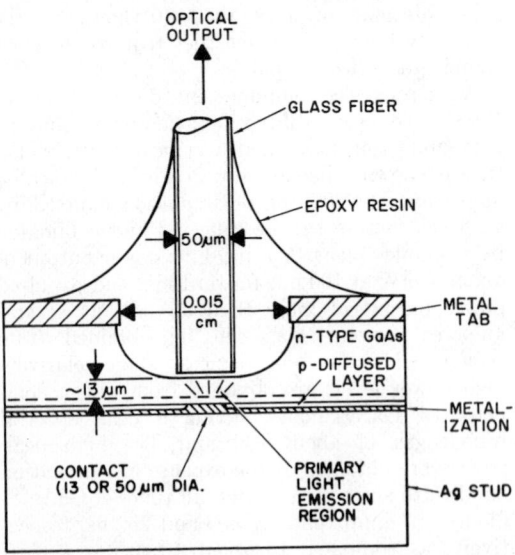

Fig. 1—GaAs electroluminescent diode specifically designed for direct coupling to an optical-fiber waveguide. (*After C. A. Burrus and R. W. Dawson, "Small-Area High-Current-Density GaAs Electroluminescent Diodes and a Method of Operation for Improved Degradation Characteristics," Applied Physics Letters, 1 August 1970, pp. 97–99.*)

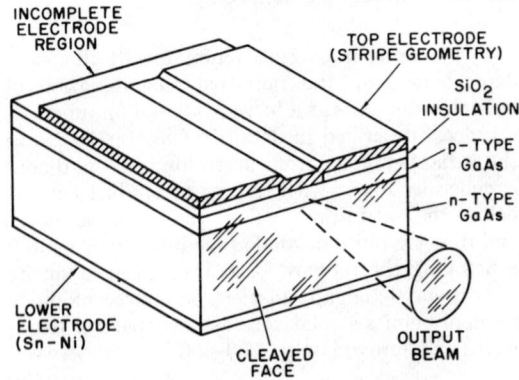

Fig. 3—Stripe-geometry semiconductor injection laser modified with an incomplete upper contact electrode to achieve superluminescent emission. (*After T. Lee, C. A. Burrus, Jr., and B. I. Miller, "A Stripe-Geometry Double-Heterostructure Amplified-Spontaneous-Emission [Superluminescent] Diode," IEEE Journal of Quantum Electronics, August 1973, pp. 820–828.*)

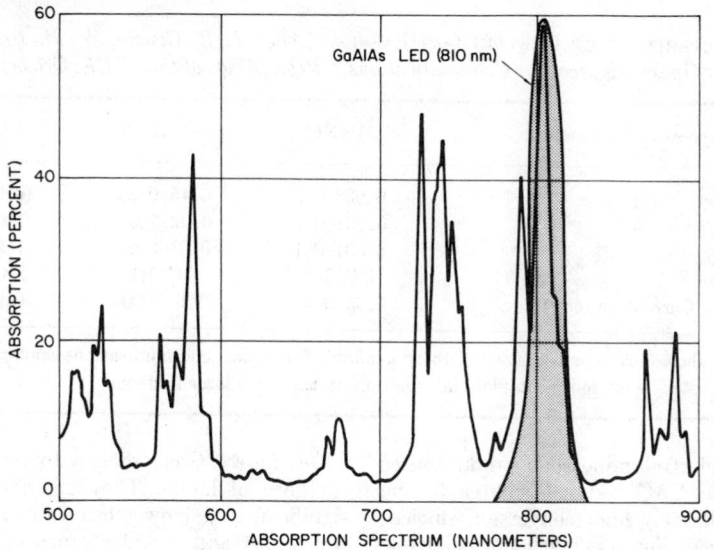

Fig. 4—Absorption spectrum of Nd³:YAG versus peak spectral emission of a typical GaAlAs electroluminescent diode. (*From F. M. Mims, III, "Light Beam Communications," p. 76, 1975, Howard W. Sams & Co., Inc.*)

major exception being an incomplete upper electrode. The incomplete electrode eliminates current injection near one of the mirrors, and this effectively isolates the mirror from the optical wave propagating along the plane of the junction and suppresses laser action. A similar effect can be had by angling one mirror of an injection laser a few degrees away from the normal or by coating one of the end mirrors of an injection laser with an antireflective film. The SLD is less efficient than most other electroluminescent sources, but its narrow spectral width and high radiance make it well suited for optical waveguide links. Pulsed radiation of 50 mW has been launched into an optical fiber from an SLD source.

Lasers*

The laser is an optical amplifier provided with feedback necessary to initiate and sustain laser oscillations. An abbreviated theory of the laser is treated separately in Chapter 39. Briefly, the optical emissions which initiate laser action are quantum processes resulting from the excited atoms, ions, or molecules of an active medium resuming equilibrium and, in the process, giving off a portion of the previously absorbed excitation energy as optical photons. When more atoms, ions, or molecules of the active medium are in an excited rather than unexcited state, optical gain exceeds optical loss, and stimulated emission occurs as photons emitted by atoms, ions, or molecules

* J. E. Geusic, W. B. Bridges, and J. I. Pankove, "Coherent Sources for Communications," *Proceedings of the IEEE*, October 1970, pp. 1419–1439.

spontaneously returning to the ground state stimulate the emission of still other photons. Facing mirrors on either side of the active medium provide a resonant cavity which causes regenerative gain to occur for photons travelling coaxially along the cavity axis. Oscillation occurs if one or more cavity resonances fall within the laser transitions for which gain exceeds losses.

Laser action has been observed in literally thousands of active mediums representing solid-state ion systems, liquids, gases, and semiconductors. Lasers made from each of these active medium categories have some potential as optical communication sources, but several different laser systems are exceptionally well suited for optical communications. These lasers, which are characterized by narrow spectral width, narrow beamwidth, and ease of modulation, are described below.

Solid-State Ion Lasers: The most important solid-state ion laser for communication applications is Nd-doped YAlG (more commonly designated YAG). YAG lasers have emitted up to 10^3 watts cw and are more efficient than HeNe and Ar gas lasers with power efficiencies ranging from 2.9% for a 100 watt system to 1.7% for a 750 watt system. Nd:YAG has the relatively broad absorption spectrum shown in Fig. 4 and has been traditionally pumped with such broad-spectrum sources as incandescent lamps and flash lamps. Recently, considerable attention has been given a novel and efficient pumping scheme wherein electroluminescent diodes or injection lasers are metallurgically tailored to emit a narrow spectral width at one or more of the Nd ion's peak absorption bands and utilized to pump a miniature YAG rod. Since virtually all of the radiation

TABLE 2—CHARACTERISTICS OF CERTAIN GAS LASERS. (*After J. E. Geusic, W. B. Bridges, and J. I. Pankove, "Coherent Optical Sources for Communications," Proceedings of the IEEE, October 1970, p. 1423.*)

Operating Parameter	HeNe	Ar	CO_2
Wavelength (μm)	0.633	0.45–0.53	10.6
Output Power (W)	0.001–0.1	0.02–100	1.0–100
Efficiency (%)	0.001–0.1	0.01–0.2	1–20
Excitation	DC, RF	DC, RF	DC, RF
Discharge Current (A/cm²)	0.05–0.5	100–2000	0.01–0.1

Note: The values shown are representative of those available for optical communications applications and are not all-inclusive. See Chapter 39 for more complete information on these gas laser systems.

emitted by the electroluminescent pump source which reaches the YAG rod contributes to the lasing process, relatively high conversion efficiencies can be achieved. Fig. 5 is a schematic outline of one such laser wherein a 0.45 mm×5.00 mm Nd:YAG rod is end-pumped by a single electroluminescent diode. This configuration delivers a few milliwatts in the lowest order, single frequency mode (TEM_{00}). Direct modulation of electroluminescent diode-pumped YAG lasers is not feasible at high modulation rates due to the long (approximately 230 μs) fluorescent lifetime of the Nd ion. However, a variety of external modulation methods employing electro-optical crystals may be employed.

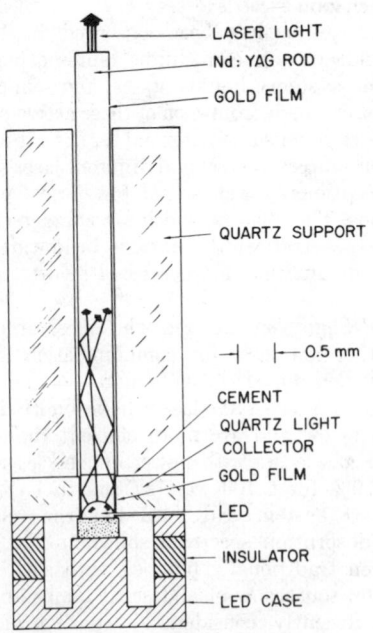

Fig. 5—Construction of a miniature Nd³: YAG laser end-pumped by an electroluminescent diode. (*After D. A. Draegert, "Single-Diode End-Pumped Nd:YAG Laser," IEEE Journal of Quantum Electronics, December 1973, pp. 1146–1149.*)

Gas Lasers: Gas lasers are by far the most varied category of lasers. They are characterized by a significantly narrower beam divergence and spectral width and a wider range of emission wavelengths than any other laser category. Many gas laser systems are suitable for communications applications, but the three best-studied systems are helium-neon, argon, and carbon dioxide. Table 2 compares the operating performance of each of these lasers.

The HeNe type is by far the most advanced gas laser, and commercial units have exhibited operating lifetimes of more than 2×10^4 hours. A HeNe laser can be directly modulated by varying the discharge current or externally modulated with a variety of electro-optic crystals. Practical HeNe lasers are limited to a maximum power output of 100 mW, and for applications requiring higher output power, argon or CO_2 lasers may be employed. Like HeNe, argon provides a highly visible output beam (see Table 2), but at a power level of several watts or more. Carbon dioxide lasers provide still higher output powers (more than 100 watts), but their 10.6 μm operating wavelength requires exotic optical materials and specialized detectors.

Semiconductor Injection Lasers: A specially modified electroluminescent diode with two facing end mirrors and a highly uniform and planar junction region, the semiconductor injection laser is characterized by a very fast risetime (about 1 ns), efficient and straightforward direct modulation, and a peak spectral output which can be tailored over a range of from 750 to 900 nm. Stripe-geometry double heterostructure injection lasers can deliver about 10 mW cw over a spectral width of about 1 nm. These lasers can be operated cw at above 300 K with appropriate heat sinking and can be both pulse and amplitude modulated. Single heterostructure injection lasers can deliver more than 10 watts peak pulsed power at 300 K ($t_{on} \leq 200$ ns). Low-power cw injection lasers are well suited as sources for optical-fiber waveguide links, while high-power pulsed lasers are well suited for atmospheric communications.

OPTICAL DETECTORS

Both solid-state and electron-multiplier phototubes (photomultiplier tubes) are employed as detectors in optical communications receivers. Since both natural (solar, stellar, lunar, etc.) and artificial background radiation may impose an undesirable noise level in the detector, narrow-bandpass optical filters are often placed before the detector. This noise blocking technique is especially effective with narrow spectral bandwidth sources such as many lasers.

Photomultiplier Tubes*

The operation of the photomultiplier tube is described in detail in Chapter 17. Briefly, a photomultiplier converts an optical signal into an electrical signal by both photoemission and secondary emission. Radiation impinging upon a photocathode generates photoelectrons which are accelerated by an electric field toward the first of a series of secondary-emission electrodes termed *dynodes*. Secondary emission from the first dynode is accelerated toward the second dynode, and the process continues until the secondary emission from the final dynode is collected by the anode. Since the ratio of input to output electrons at each dynode may be on the order of 4, a 12-stage photomultiplier may have a gain of 4^{12}. If the input pulse has a width of 5 ns, the anode output current will be about 1 mA.

Several parameters affecting the performance of photomultiplier tubes as detectors for optical communication receivers are described below:

Spectral Sensitivity: The spectral sensitivities of the most important photoemissive surfaces are shown in Chapter 17 (Fig. 5). By selecting an appropriate photoemissive surface, peak spectral sensitivity may be centered anywhere over a range of from about 250 nm to 1000 nm.

Radiant Sensitivity: Radiant sensitivity is the ratio of anode output current to incident optical power at a specified wavelength. Radiant sensitivity is expressed in A/W and is strongly dependent upon the spectral content of the incident radiation. The peak radiant sensitivity of commercial photomultiplier tubes ranges from about 10^{-1} A/W (116 bialkali at 380 nm) to 3×10^{-3} A/W (S1 at 800 nm). With careful attention to operating

conditions (low-noise circuits, carefully designed optics, absence of thermal gradients, etc.), certain high quantum efficiency photomultiplier tubes can be used to detect individual photons.

Response Time: The emission of both photoelectrons and secondary electrons requires a finite time interval. Practical photomultiplier tubes have pulse rise times ranging from about 0.8 to 20 ns. The rise time decreases slightly as the anode-cathode voltage is increased.

Stability: Photomultiplier tubes are subject to fatigue and resultant loss of anode sensitivity as a consequence of operating history. Operational lifetimes of several thousand hours can be obtained if the tube is operated within its specified ratings.

Operational Requirements: Photomultiplier tubes require a high-voltage (600–3000 volts) power supply for operation. This voltage must be staircased so that each dynode is operated at its specified voltage level, and this is normally accomplished by a voltage divider network connected directly to the socket of the tube.

Mechanical Geometry: Photomultiplier tubes are normally housed within a glass envelope, and they are consequently mechanically fragile. Typical photomultiplier tubes are cylindrical in shape and measure several centimeters in diameter and up to 10 or more centimeters in length.

Solid-State Detectors*

A wide variety of solid-state detectors have been employed in experimental, developmental, and operational optical communication links including phototransistors, photo-FETs, both selenium and silicon solar cells, pn photodiodes, pin photodiodes, and avalanche photodiodes. The most important of these detectors for practical optical communication purposes are pin and avalanche silicon photodiodes. Both these detector configurations may have a quantum efficiency of 0.8–0.9 at their regions of peak spectral sensitivity (600–900 nm). Several parameters concerning the performance of pin photodiodes are described below:

Spectral sensitivity—The spectral sensitivity of typical silicon pin photodiodes ranges from less than 300 nm to more than 1100 nm, as shown in Fig. 6. The spectral sensitivity may be restricted to a narrow spectral band with a narrow-bandpass optical filter.

* *RCA Photomultiplier Manual*, RCA Electronic Components, Harrison, NJ, 1970.

* S. E. Miller, T. Li, and E. A. J. Marcatilli, "Optical-Fiber Transmission Systems, Part II: Devices and Systems Considerations," *Proceedings of the IEEE*, December 1973, pp. 1726–1751.

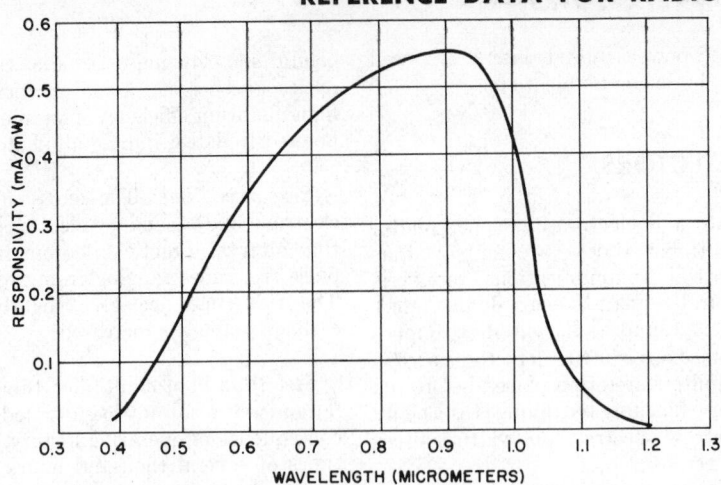

Fig. 6—Spectral response of a typical silicon photodiode. (*From F. M. Mims, III, "Optoelectronics,"* *p. 84, 1975, Howard W. Sams & Co., Inc.*)

Radiant sensitivity—The maximum radiant sensitivity (i.e. responsivity) for silicon pin photodiodes ranges from 0.3 to 0.6 A/W at the region of peak spectral sensitivity (850 to 900 nm).

Response time—Properly biased pin photodiodes exhibit pulse rise times of less than 5 ns.

Stability—Under most operating conditions, silicon pin photodiodes exhibit no significant change in performance with time.

Operating requirements—A typical silicon pin photodiode requires a reverse bias of from several volts to 90 volts DC. Conventional discrete and integrated-circuit amplifier circuits can be used to amplify the detected signal.

Mechanical geometry—Silicon pin photodiodes can be made with a wide variety of active area configurations. Optical communications receivers normally employ diodes with an active area of a few square millimeters to 1 cm². Diodes such as these are usually installed in compact metal packages with a glass window (or self-contained interference filter) and are mechanically robust.

Silicon avalanche photodiodes are distinguished by internal gain resulting from an avalanche multiplication process (see Chapter 19). They share certain of the operating characteristics of silicon pin photodiodes, including spectral sensitivity, stability, and mechanical geometry. Other operating characteristics are described below:

Radiant sensitivity—At the peak spectral sensitivity of 900 nm, silicon avalanche photodiodes may exhibit a sensitivity of more than 100 A/W. The radiant sensitivity (i.e. responsivity) varies with the reverse bias, a parameter unique to each diode. Thus a diode with a reverse breakdown voltage (V_{BR}) of 450 volts may have a responsivity several times greater than a diode with a V_{BR} of 300 volts.

Response time—Typical avalanche photodiodes have response times of about 2 ns (10% to 90% points).

Operating requirements—Avalanche photodiodes must be reverse biased to within less than 1% of V_{BR} for optimum gain. Since V_{BR} varies with temperature, a temperature-tracking bias supply is required for optimum results. Typical V_{BR} ratings of avalanche photodiodes range from 100 to 500 volts.

Table 3 compares the primary operating characteristics of a representative solid-state photodetector and the photomultiplier tube.

TABLE 3—COMPARISON OF SOLID-STATE PIN PHOTODIODE WITH PHOTOMULTIPLIER. (*After P. H. Wendland, "Solid State Combo Senses Light Well Enough to Vie With Tubes," Electronics, May 24, 1971, pp. 50–54.*)

Characteristic	PIN Photodiode	Photomultiplier
Spectral Range (nm)	200–1100	200–1000 (photocathode dependent)
Response Time (ns)	5	5
Sensitivity (A/W)	0.3–0.6	0.003–0.1
Minimum Detectable Light Level	5×10^{-15} W	Single Photon
Operating Voltage (V)	3–90	600–3000

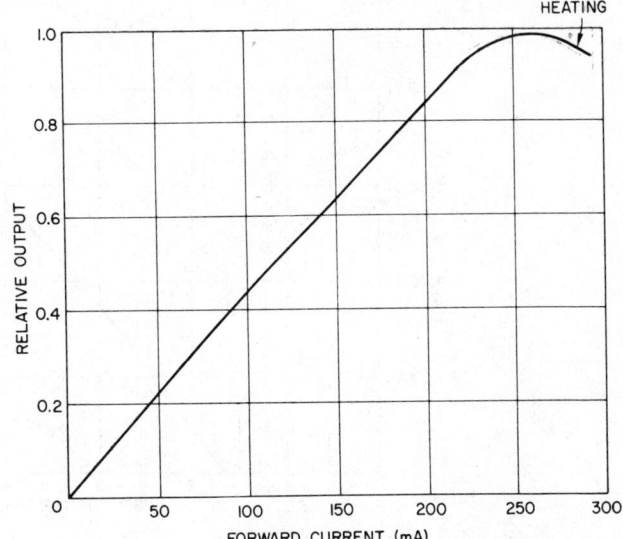

Fig. 7—Optical power output versus forward current for a typical GaAs electroluminescent diode. (*From F. M. Mims, III, "Light Emitting Diodes," p. 19, 1973, Howard W. Sams & Co., Inc.*)

MODULATION TECHNIQUES

Optical communications systems may employ various configurations of the continuous (analog) or pulsed modulation formats described in detail in Chapter 23. Electroluminescent diode sources can be directly amplitude or pulse modulated by varying their forward bias, and HeNe lasers may be directly amplitude or pulse modulated by varying their discharge current. All optical communication sources can be externally modulated by means of an electro-optical crystal.

Direct Modulation

The most important optical sources suitable for direct modulation are all classes of electroluminescent diodes and the HeNe laser.

Electroluminescent Diodes: Fig. 7 is a typical plot of optical power output as a function of forward current for a GaAs electroluminescent diode, the linear nature of which illustrates the suitability of these diodes for analog modulation. Since a GaAs electroluminescent diode may have a turn-on time of approximately 10^{-9} second, suitably designed diodes may have a modulation bandwidth greater than 100 MHz.

Injection Laser: Like the electroluminescent diode, the injection laser exhibits an output which is linear with respect to forward current over a portion of its operating range (see Fig. 40, Chapter 39), but higher current driving circuitry is required (more than 350 mA for cw double heterostructure lasers and more than 5 A for single heterostructure pulsed lasers). The high resonance frequency of the injection laser gives rise to very rapid fluctuations in power output and results in a very large modulation bandwidth. The resonance frequency ν_r for a single mode is

$$\nu_r \cong \frac{m}{2\pi}\left\{\frac{1}{\tau_s \tau_p}\left(\frac{I}{I_t} - 1\right)\right\}^{1/2}$$

where τ_s is the electron lifetime for spontaneous recombination, τ_p is the photon lifetime (inversely proportional to cavity losses), I is the forward current, I_t is the threshold current, and m is the gain exponent dependent upon the band-tail structure.* Double heterostructure injection lasers may have a ν_r of up to several GHz, and prebiased lasers have been pulse modulated at 2.3 Gb/s. Prebiasing the laser to just below its threshold reduces the time required to populate the upper levels and achieve population inversion.

HeNe Laser: Fig. 8 shows the dependence of the optical output of a HeNe laser tube on the discharge current of the tube. These lasers can be 100% pulse modulated (full off-full on) at more than 160 kHz and 15% analog modulated at more than 600 kHz. Helium-neon lasers are commonly modulated externally also.

External Modulation

The most important external modulators are of the reactive type, although absorptive modulators may have potential value. External modulators are characterized by such factors as spectral transmissivity, insertion loss, temperature variations,

* S. E. Miller, T. Li, and E. A. J. Marcatilli, "Optical-Fiber Transmission Systems, Part II: Devices and Systems Considerations," *Proceedings of the IEEE*, December 1973, p. 1732.

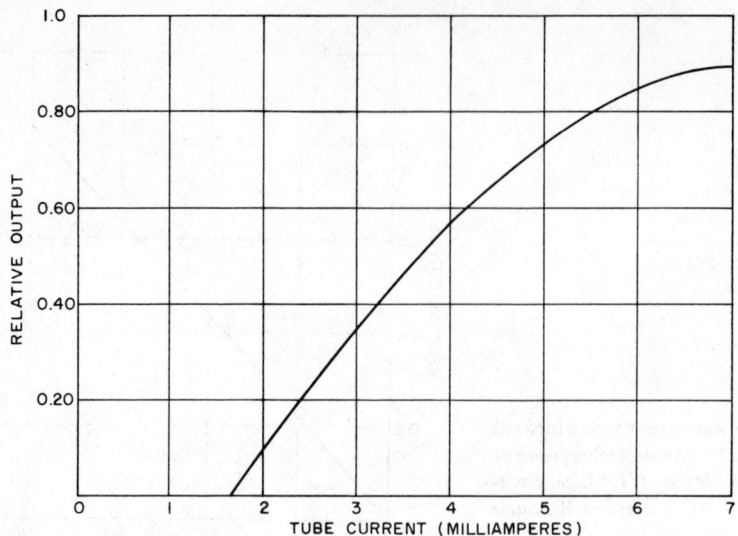

Fig. 8—Output power of a HeNe laser as a function of plasma tube discharge current.

bandwidth, and drive requirements. The most important operating characteristic is the power dissipation (milliwatts) per unit bandwidth (megahertz) required for a specific modulation rate $(P/\Delta f)$.

Electro-Optic Modulation: An electric field applied to a suitable transparent medium may alter the index of refraction of the medium and cause an optical beam passing through the medium to experience a phase shift. This phenomenon, which is elaborated upon in Chapter 39 ("Nonlinear Optical Effects Produced by Laser Beams"), gives rise to two distinct modulation types. The

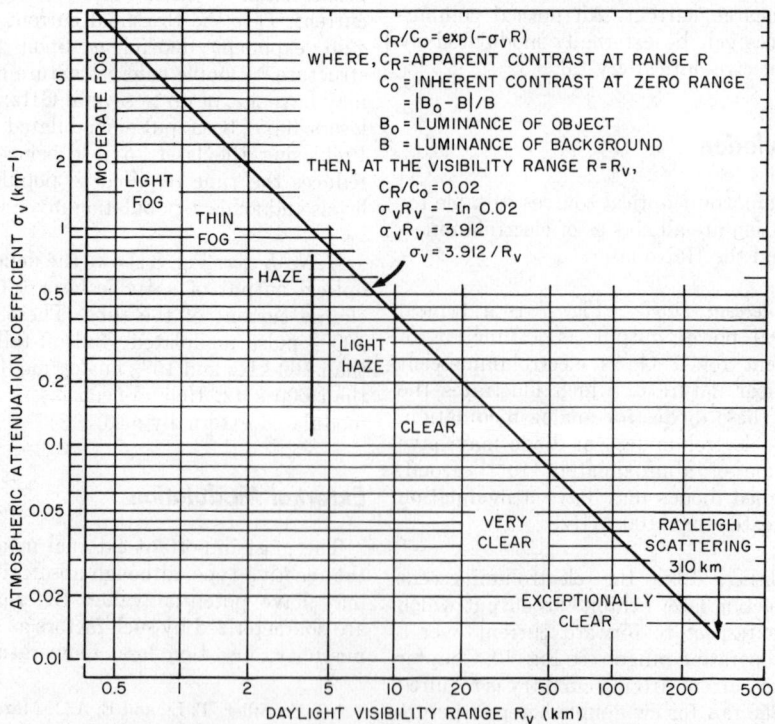

Fig. 9—Atmospheric attenuation coefficient for visible light as a function of daylight visibility range. (*From "Electro-Optics Handbook," p. 7-7, 1968, RCA Aerospace Systems Division.*)

first is a phase modulation wherein an optical beam passing through the medium is polarized along a principal plane so that the emerging beam retains its original polarization and amplitude but the plane of the polarization is rotated (phase modulated). In the second, the incident beam is prepolarized, and an analyzer placed opposite the modulator causes the beam to be intensity or amplitude modulated.

Typical electro-optic modulators require operating voltages of more than 100 volts and have a typical $P/\Delta f$ of several milliwatts per megahertz. Two of the most practical electro-optic modulator materials in the 0.4–4.0 μm band are lithium niobate ($LiNbO_3$) and lithium tantalate ($LiTaO_3$). A $LiTaO_3$ modulator with a $P/\Delta f$ of 0.1 mW/MHz has modulated a 496 nm beam at 1.5 GHz (30% modulation depth).

Acousto-Optic Modulation: Intensity modulation over bandwidths of greater than 300 MHz is

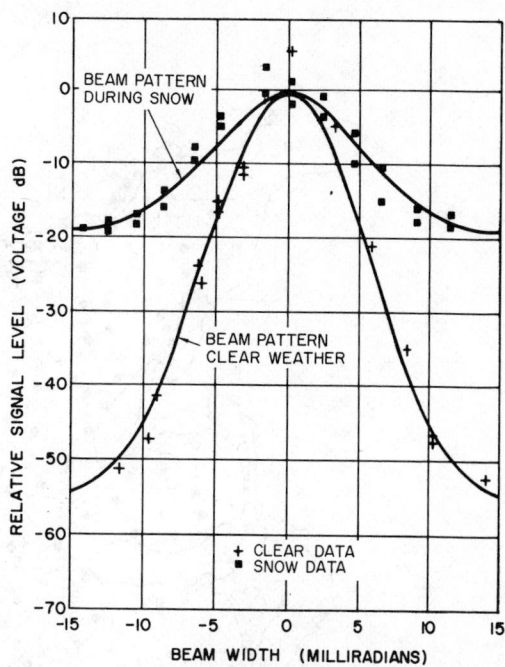

Fig. 11—Spatial intensity distribution of an optical beam in a clear atmosphere and in the presence of snow. (*From L. E. Wood and S. Murahata, "Optical Communication Systems for Short-Haul Applications," U.S. Department of Commerce Office of Telecommunications Report 73-3, March 1973, p. 74.*)

possible when an optical beam is diffracted by the phase grating created by the photoelastic effect resulting from the propagation of an acoustic wave in certain materials. Acousto-optic deflection and modulation has been observed in As_2Se_3 glasses, GaAs, and $LiNbO_3$.

Magneto-Optic Modulation: Certain magnetic iron garnets have demonstrated Faraday rotation and optical absorption with a $P/\Delta f$ of as little as 0.01 mW/MHz. These crystals currently have high attenuation at the important laser communications wavelength of 1.06 μm (Nd:YAG).

TRANSMISSION MEDIA

Experimental and operational optical communications links have utilized both the atmosphere and optical-fiber waveguides as transmission media, and those factors influencing these transmission media are described below.

Atmospheric Links*

An optical beam propagating through the earth's atmosphere is altered by atmospheric absorption,

* W. K. Pratt, *Laser Communication Systems*, John Wiley & Sons, Inc., New York, 1969.

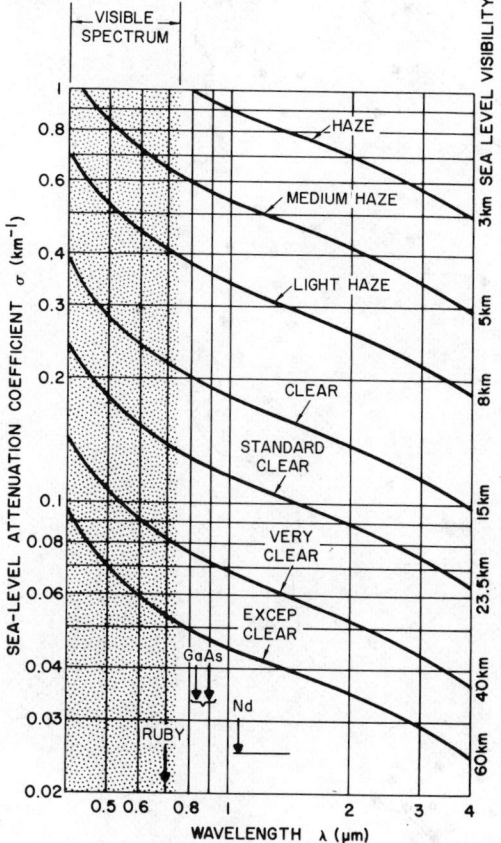

Fig. 10—Approximate variation of attenuation coefficients with wavelength at sea level for various atmospheric conditions (neglects absorption by water vapor and carbon dioxide). (*From "Electro-Optics Handbook," p. 7-8, 1968, RCA Aerospace Systems Division.*)

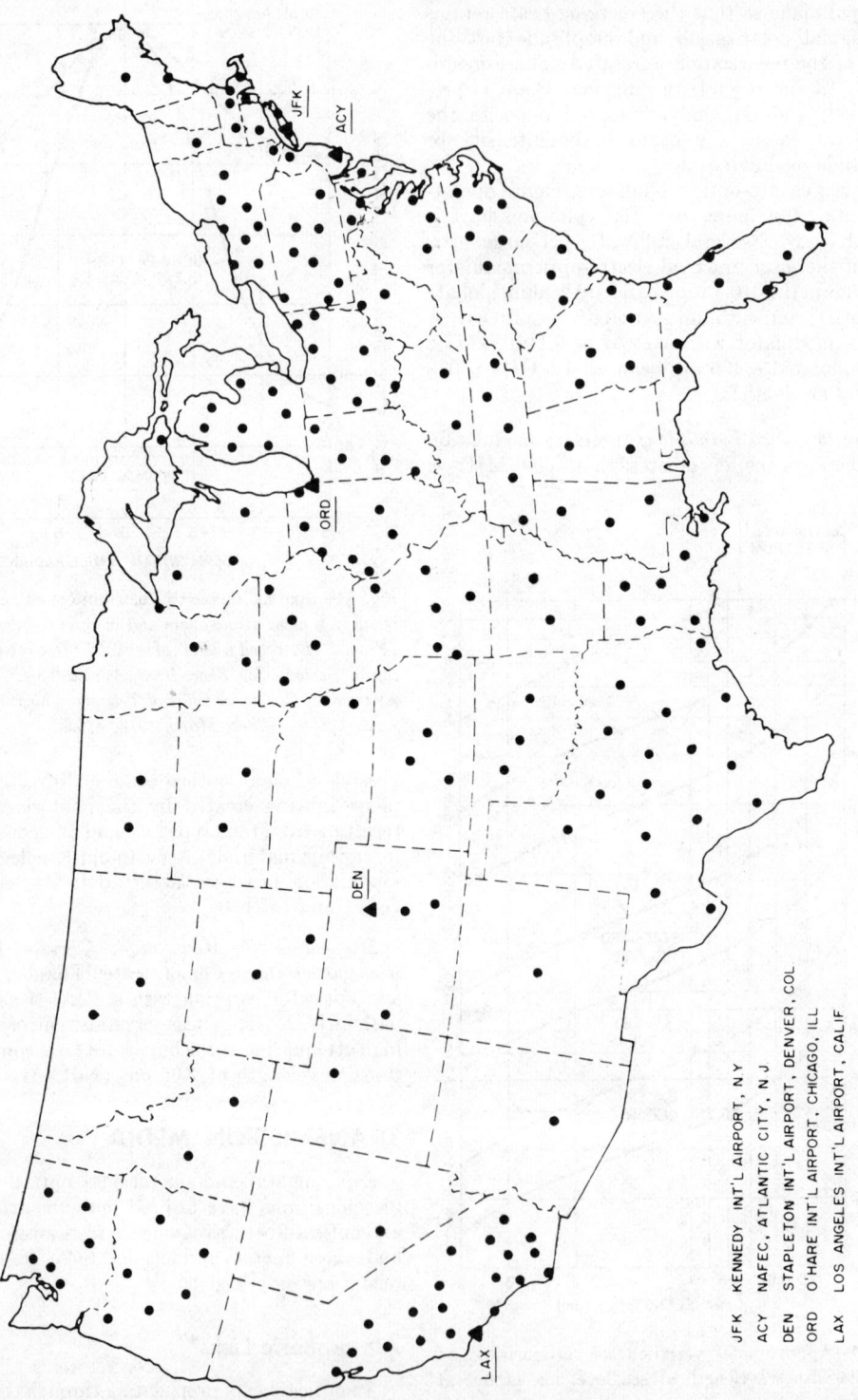

JFK KENNEDY INT'L AIRPORT, N Y
ACY NAFEC, ATLANTIC CITY, N J
DEN STAPLETON INT'L AIRPORT, DENVER, COL.
ORD O'HARE INT'L AIRPORT, CHICAGO, ILL
LAX LOS ANGELES INT'L AIRPORT, CALIF

Fig. 12—Geographical locations of transmissometer-equipped airports in the United States. (*From L. E. Wood and S. Murahata, "Optical Communication Systems for Short-Haul Applications," U.S. Department of Commerce, Office of Telecommunications Report 73-3, March 1973.*)

scattering, and turbulence. These factors may act alone or collectively to increase, often adversely, both the spatial and temporal intensity distribution of an optical beam. The transmission of the atmosphere may be expressed as

$$\tau_a = \exp(-\alpha_a L)$$

where L is the atmospheric range, and the attenuation or extinction coefficient α_a is the sum of the absorption and scattering coefficients. Atmospheric absorption is a spectrally selective phenomenon when the absorbant is gaseous in nature, and the principal gaseous absorbers are water vapor, carbon dioxide, and ozone. Since these gases absorb at narrow spectral lines rather than bands, the detailed transmissivity of the atmosphere as a function of wavelength is highly complex. Fig. 33 in Chapter 28 shows the general atmospheric transmission over a 3 km path at sea level over a spectral range of 1 to 20 μm and shows the existence of several high-transmissivity windows. Since this figure is only general, a careful consideration of the transmissivity near the edges of windows should be made if a narrow-bandwidth source (e.g., a laser) is being considered as an optical source in a communicator.

Fig. 32 in Chapter 28 shows the atmospheric attenuation coefficients for several atmospheric conditions as a function of wavelength, and Table 5 in Chapter 28 gives the International Visibility Code and specifies the visibility range for various atmospheric conditions. These data are expanded upon in Figs. 9 and 10 which show, respectively, the atmospheric attenuation coefficients for visible light as a function of daylight visibility range and the variation of these coefficients with wavelength.

Broadband atmospheric absorption can be attributed to precipitation, aerosols, and other particulates in the atmosphere. Particulates are also a major cause of scattering. Atmospheric scattering may be molecular (Rayleigh) or particulate (Mie) in origin, with the latter contributing far more loss than the former. Rayleigh scattering predominates at shorter wavelengths (consider the blue color of the sky) and has the attenuation coefficient given by

$$\alpha_{SR} = 0.827 N A_p^3 \lambda_c^{-4}$$

where N is the particle count per unit volume, A_p is the cross-sectional area (in square centimeters) of a typical scattering particle, and λ is given in centimeters.

Mie scattering is given by

$$\alpha_{SM} = \frac{3.91}{V} \left[\frac{\lambda_c}{0.55} \right]^{-0.585 V^{1/3}}$$

where V is the visual range (kilometers) and λ is expressed in micrometers. Mie scattering causes the spatial power distribution of an optical beam to be spread outward as shown in Fig. 11.

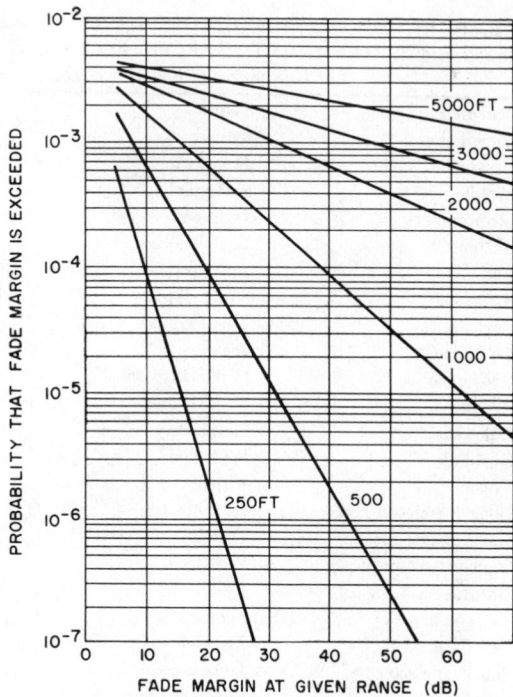

Fig. 13—Optical availability for Denver. (*From L. E. Wood and S. Murahata, "Optical Communication Systems for Short-Haul Applications," U.S. Department of Commerce, Office of Telecommunications Report 73-3, March 1973.*)

The temporal intensity distribution of an optical beam is altered by atmospheric turbulence. Since temperature changes the refractive index of the atmosphere, temperature gradients over an optical path will cause deviations in a ray as it passes between regions of differing refractive index. Intensity fluctuations are most pronounced when the optical beam passes near landforms which are good thermal absorbers and therefore good thermal emitters. Asphalt, roof tops, and other terrain features can induce severe intensity fluctuations upon an optical beam propagating nearby.

The Office of Telecommunications of the United States Department of Commerce has studied the prospects for successful operation of short-haul atmospheric optical communication links using a variety of experimental and analytical methods, including the evaluation of National Weather Service Runway Visual Range (RVR) and Runway Visibility Value (RVV) measurements taken from transmissometer equipped airports. Fig. 12 is a map of the United States showing the location of airports equipped with transmissometers, and Chart 1 lists those cities nearest the transmissometer equipped airports. Figs. 13 through 17 show the optical channel availability for Denver, Los Angeles, New York City, Atlantic City, and Chicago, respectively. With the exception of

CHART 1—CITIES IN THE UNITED STATES NEAREST TRANSMISSOMETER-EQUIPPED AIRPORTS. (*L. E. Wood and S. Murahata, "Optical Communication Systems for Short-Haul Applications," U.S. Department of Commerce Office of Telecommunications Report 73-3, March 1973.*)

Alabama	Peoria	St. Louis	Philadelphia
Birmingham	Quincy	Springfield	Pittsburgh (2)
Huntsville	Rockford	*Montana*	Reading
Mobile	Springfield	Billings	Wilkes-Barre
Montgomery	*Indiana*	Great Falls	*Rhode Island*
Arizona	Fort Wayne	*Nebraska*	Providence
Phoenix	Indianapolis	Lincoln	*South Carolina*
Arkansas	South Bend	Omaha	Charleston
Fort Smith	Terre Haute	*Nevada*	Columbia
Little Rock	*Iowa*	Reno	Greer
California	Cedar Rapids	*New Jersey*	*South Dakota*
Arcata	Des Moines	Newark	Huron
Bakersfield	Sioux City	*New Mexico*	Sioux Falls
Burbank	Waterloo	Albuquerque	*Tennessee*
Fresno	*Kansas*	*New York*	Bristol
Long Beach	Hutchinson	Albany	Chattanooga
Los Angeles (2)	Topeka	Binghamton	Knoxville
Oakland	Wichita	Buffalo	Memphis
Ontario	*Kentucky*	Elmira	Nashville
Sacramento (2)	Lexington	Islip	*Texas*
San Diego	Standeford	New York City (2)	Amarillo
San Francisco	*Louisiana*	Niagara Falls	Abilene
San Jose	Baton Rouge	Rochester	Austin
Santa Ana	Lafayette	Syracuse	Beaumont
Santa Barbara	Lake Charles	Utica	Corpus Christi
Stockton	Monroe	*North Carolina*	Dallas
Colorado	New Orleans	Asheville	El Paso
Colorado Springs	Shreveport	Fayetteville	Ft. Worth
Denver	*Maine*	Greensboro	Houston
Grand Junction	Bangor	Raleigh	Longview
Pueblo	Portland	Wilmington	Lubbock
Connecticut	*Maryland*	Winston-Salem	Midland
Windsor Locks	Baltimore	*North Dakota*	San Angelo
Delaware	*Massachusetts*	Bismarck	San Antonio
Wilmington	Bedford	Fargo	Waco
District of Columbia	Boston	*Ohio*	Wichita Falls
Washington (2)	Nantucket	Akron-Canton	*Utah*
Florida	Worcester	Cincinnati	Salt Lake City
Jacksonville	*Michigan*	Cleveland	*Vermont*
Miami	Alpena	Dayton	Burlington
Orlando	Detroit	Mansfield	*Virginia*
Pensacola	Flint	Port Columbus	Lynchburg
St. Petersburg	Grand Rapids	Toledo	Newport News
Tallahassee	Kalamazoo	Youngstown	Richmond
Tampa	Lansing	*Oklahoma*	*Washington*
West Palm Beach	Muskegon Co.	Clinton	Moses Lake
Georgia	Saginaw	Oklahoma City	Seattle (2)
Atlanta	Willow Run	Tulsa	Spokane
Augusta	*Minnesota*	*Oregon*	Yakima
Columbus	Duluth	Eugene	*West Virginia*
Macon	Minneapolis-St. Paul	Klamath Falls	Charleston
Savannah	Rochester	Medford	Wheeling
Idaho	*Mississippi*	Pendleton	*Wisconsin*
Boise	Jackson	Portland	Green Bay
Pocatello	Meridian	Troutdale	Madison
Illinois	*Missouri*	*Pennsylvania*	Milwaukee
Champaign	Joplin	Allentown	*Wyoming*
Chicago (2)	Kansas City (2)	Erie	Casper
Moline	St. Joseph	Harrisburg	Cheyenne

Atlantic City, the availability for the longest path exceeds 99% in all cases. Figs. 13 through 17 are based on a model which gives the probability that the atmospheric attenuation will exceed a fade margin A before the error rate falls below a specified threshold:

$$p = p_0 10^{-AL_f/10L_p}$$

where A is given in decibels, L_p is the optical range, p_o is related to the frequency of fog and snow, and L_f describes the intensity of fog.* Both p_o and L_f are independently determined for each site.

Though atmospheric optical communication links can be expected to provide reasonable service in short-haul applications (shorter than 2 km), longer-range links in areas having frequent precipitation can be expected to yield poor service. Fig. 18, for example, summarizes the performance over a 9-month period of a 24-channel HeNe laser telephone link between the city of Yerevan and the Burakan Astrophysical Observatory in the USSR. This system employs an external optical

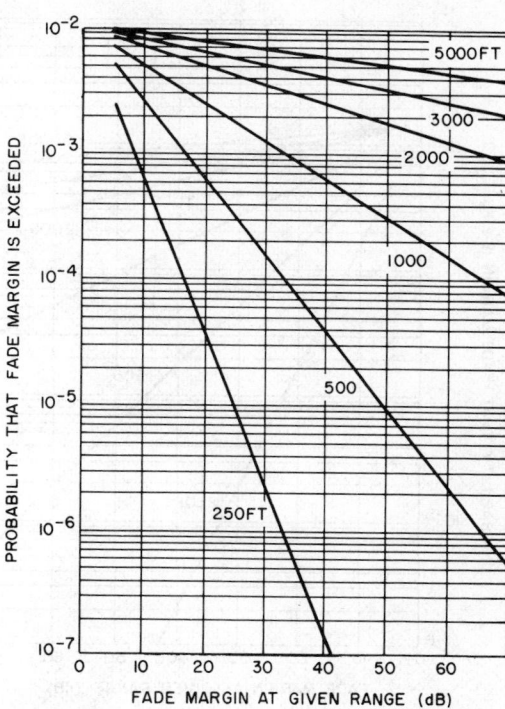

Fig. 15—Optical availability for New York. (*From L. E. Wood and S. Murahata, "Optical Communication Systems for Short-Haul Applications," U.S. Department of Commerce, Office of Telecommunications Report 73-3, March 1973.*)

modulator which impresses a 192-kHz stream of 0.5-μs pulses onto the 40-mW beam of a HeNe laser.

Optical-Fiber Waveguide Links*

Many of the limitations of atmospheric optical communication links can be readily overcome by employing an optical-fiber waveguide. Indeed, this communications method compares favorably with more conventional methods (see Table 4) and offers the potential of very high bandwidth. Currently, very low loss optical-fiber waveguides are more costly than conventional transmission means and require both careful handling and specialized methods for making connections and splices. However, optical-fiber waveguides offer several important advantages which may outweigh their disadvantages in many applications.

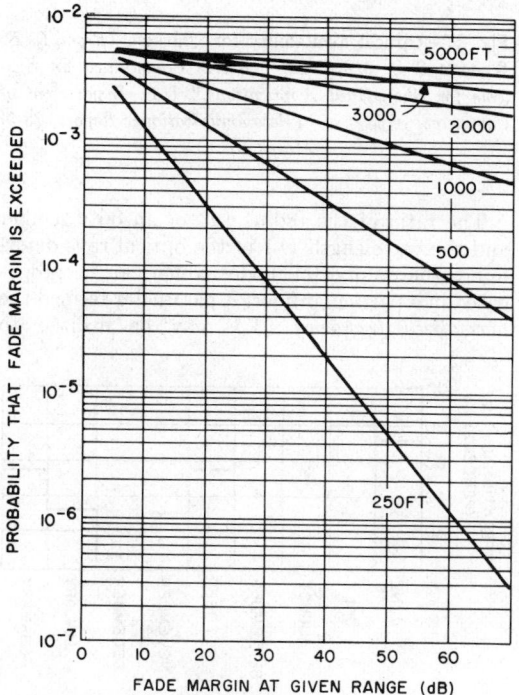

Fig. 14—Optical availability for Los Angeles. (*From L. E. Wood and S. Murahata, "Optical Communication Systems for Short-Haul Applications," U.S. Department of Commerce, Office of Telecommunications Report 73-3, March 1973.*)

* L. E. Wood and S. Murahata, "Optical Communication Systems for Short-Haul Applications," U.S. Department of Commerce, Office of Telecommunications Report 73-3; March 1973, p. 11.

* R. L. Gallawa, "Optical Fiber Links for Telecommunications, Part Two," U.S. Army Strategic Communications Command Technical Report No. SCC-ACO-1-73; July 1973. D. Gloge, "Optical Fibers for Communication," *Applied Optics*, February 1974, pp. 249–254. J. S. Cook, "Communication by Optical Fiber," *Scientific American*, November 1973, pp. 28–35.

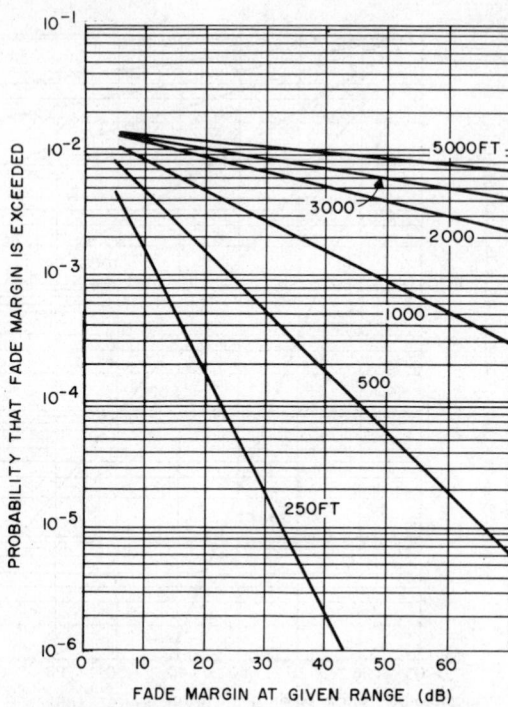

Fig. 16—Optical availability for Atlantic City. (*From L. E. Wood and S. Murahata, "Optical Communication Systems for Short-Haul Applications," U.S. Department of Commerce, Office of Telecommunications Report 73-3, March 1973.*)

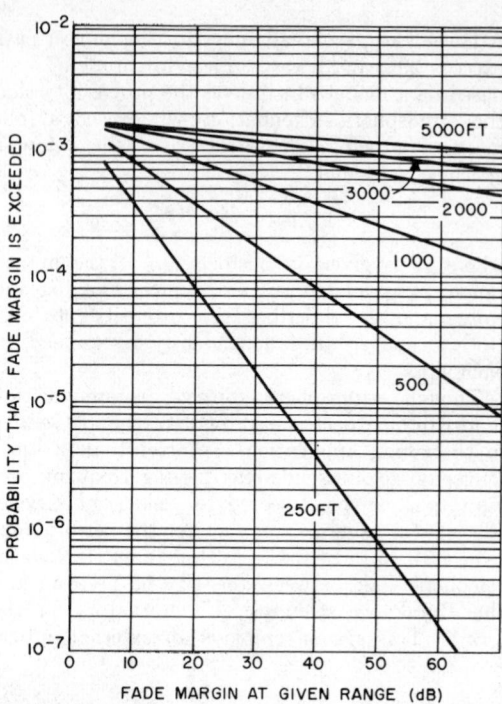

Fig. 17—Optical availability for Chicago. (*From L. E. Wood and S. Murahata, "Optical Communication Systems for Short-Haul Applications," U.S. Department of Commerce, Office of Telecommunications Report 73-3, March 1973.*)

They include:

1. Very high data rate capability
2. Compact dimensions
3. Potential low cost
4. Immunity to lightning, inductive cross talk, and the electromagnetic pulse effects induced by nuclear explosions
5. General immunity to moisture and temperature variabilities
6. Potential upgrading of bandwidth and information-carrying capacity as new modulation and demodulation methods are developed

Fig. 19 shows how a fiber conducts an optical ray by means of total internal reflection. Fig. 20 is a schematic outline of a typical optical fiber showing a low-loss core having a refractive index n_1 and a cladding having a refractive index n_2. The maximum acceptance angle of this fiber with respect to its axis is given by

$$\text{Sin}\theta \, \text{max} = (n_1{}^2 - n_2{}^2)^{1/2}$$

where $\sin\theta$ is designated the numerical aperture (N.A.) of the fiber. It should be noted that an optical ray will propagate through an unclad fiber. In this case, n_2 is 1.

The ratio of the radius (a) of an optical fiber and the wavelength (λ) of the optical rays determines the number of modes which can be propagated in a particular fiber. A parameter termed the *normalized frequency* (V) may be utilized to

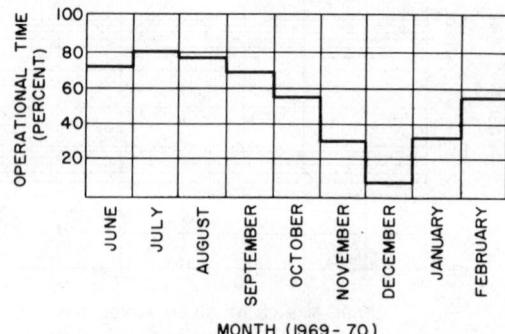

Fig. 18—Operational time of a 24-channel 40-mW HeNe laser communications link between the city of Yerevan and the Burakan Astrophysical Observatory (USSR). (*After R. A. Kazarian, R. G. Manucharian, E. S. Vartanian, and E. G. Vartanian, "Experimental Operation of a Multichannel Laser Communication Link," IEEE Transactions on Aerospace and Electronic Systems, January 1971, pp. 111–113.*)

TABLE 4—COMPARISON OF CONVENTIONAL AND OPTICAL TRANSMISSION SYSTEMS. (*From R. L. Gallawa, "Telecommunication via Certain Guided Wave Structures," U.S. Department of Commerce, Department of Telecommunications Report 73-4, March 1973, p. 27.*)

Transmission Medium	Frequency	Channel Capacity (Telephone Circuits)	Typical Repeater Spacing or Attenuation
Wire Pairs	1–140 kHz	Up to 240 (16 pairs)	0.1–0.3 dB/km
Multipair Cable	12–250 kHz	Up to 1400 (24 pairs)	3 dB/km
Coaxial Cable			Repeater Spacing
L-1	64–3096 kHz	1800	12.8 km
L-3	308–8320 kHz	16 740	6.4 km
L-4	0.5–17 MHz	32 400	3.2 km
L-5	3–51 MHz	90 000	1.6 km
Circular Waveguide (Over-Moded)	50 GHz	$>10^5$	0.5–4 dB/km
Optical Fiber	10^{14} Hz	$>10^5$	2.0–800 dB/km

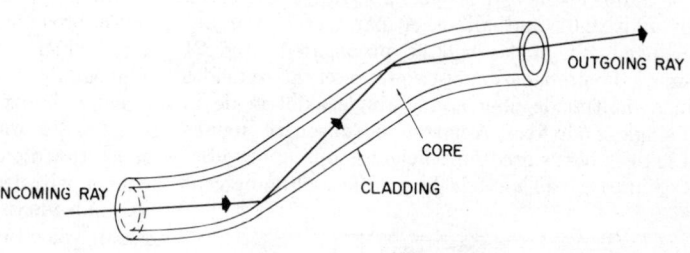

Fig. 19—Propagation of an optical ray through an optical-fiber waveguide via total internal reflection.

calculate the number of modes which will propagate in a specific fiber:

$$V = \frac{2\pi a}{\lambda}(n_1^2 - n_2^2)^{1/2}$$

Only a single mode will propagate when $V \leq 2.405$, but as V increases, the number (N) of modes increases according to

$$N = \frac{V^2}{2}$$

Subtle changes in some of the parameters can have a disproportionate effect on the number of modes propagating in a fiber. For example,

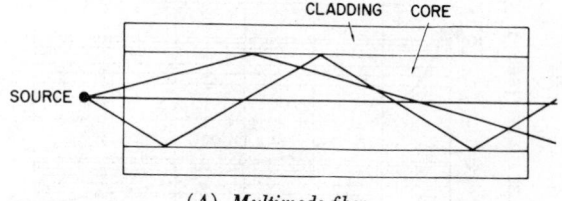

(A) Multimode fiber.

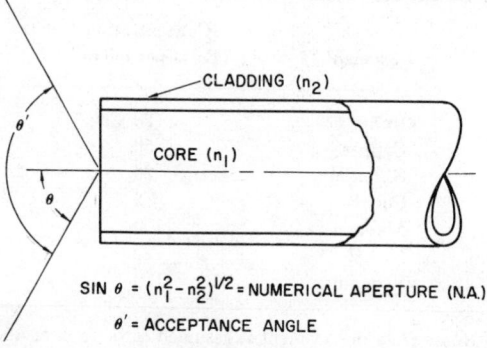

SIN θ = $(n_1^2 - n_2^2)^{1/2}$ = NUMERICAL APERTURE (N.A.)

θ' = ACCEPTANCE ANGLE

Fig. 20—The acceptance angle and numerical aperture of an optical-fiber waveguide.

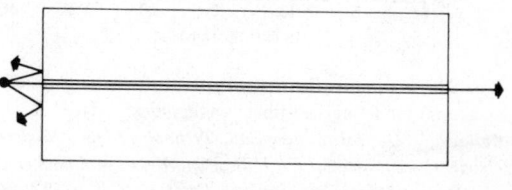

(B) Single-mode fiber.

Fig. 21—Comparison of ray propagation in multimode and single-mode optical waveguides. (*From F. M. Mims, III, "Light Beam Communications," p. 117, 1975, Howard W. Sams & Co., Inc.*)

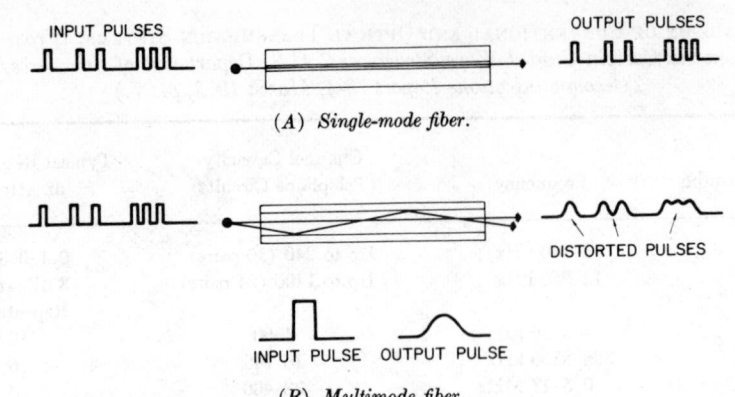

(A) Single-mode fiber.

(B) Multimode fiber.

Fig. 22—Pulse distortion in a multimode optical-fiber waveguide. (*From F. M. Mims, III, "Light Beam Communications," p. 117, 1975, Howard W. Sams & Co., Inc.*)

Gallawa* has shown that for $\lambda = 1$ μm, $n_1 = 1.51$, and $n_2 = 1.5$, a single axial mode will propagate if the radius of the core is ≥ 2.2 μm. If n_1 is increased by only 0.01 and all other parameters are unchanged, six modes will be propagated. Fig. 21 shows the propagation of two representative modes in a multimode fiber and a single axial mode in a single-mode fiber. A narrow beam source such as a laser is best suited for efficient coupling of radiation into a single-mode fiber. Electroluminescent diodes are best suited for use with multimode fibers.

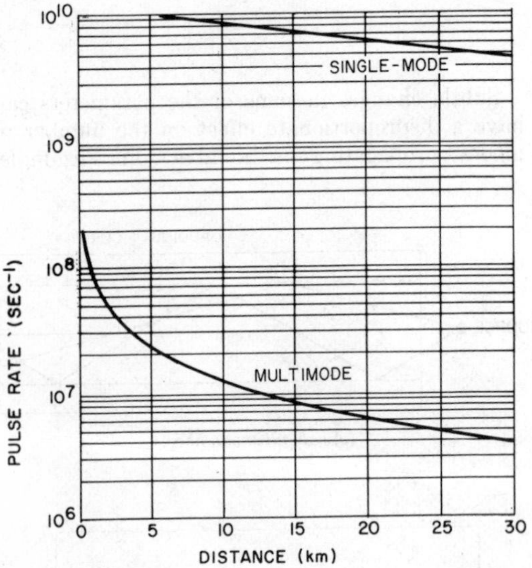

Three types of fiber-induced dispersion lead to pulse broadening and limit the maximum data rates which can be transmitted through a fiber. Waveguide dispersion occurs as the optical frequency changes and, in turn, alters the λ dimensions of the waveguide. Material dispersion occurs since the dielectric properties of the waveguide change with frequency. Finally, multimode propagation leads to dispersion as a result of the varying group velocities of various modes. Of these three factors, material dispersion is the dominant cause of pulse spreading in single-mode fibers, and multimode delays are the dominant cause of pulse spreading in multimode fibers. Fig. 22 illustrates pulse broadening and shows how it limits the maximum data rate by blurring adjacent pulses.

Fig. 23—Maximum theoretical pulse rate for single-mode and multimode optical-fiber waveguides. (*After R. L. Gallawa, "Optical Waveguide Technology for Modern Urban Communications," U.S. Department of Commerce, Office of Telecommunications, October 1974, p. 18.*)

* R. L. Gallawa, "Optical Fiber Links for Telecommunications, Part Two," U.S. Army Strategic Communications Command Technical Report No. SCC-ACO-1-73; July 1973.

TABLE 5—MAXIMUM TOLERABLE CONCENTRATIONS FOR VARIOUS IMPURITY IONS IN A LOW-LOSS OPTICAL FIBER. (*From A. D. Pearson and W. G. French, "Low-Loss Glass Fibers for Optical Transmission," Bell Laboratories Record, April 1972, pp. 103–109.*)

Element	Concentration (Parts per billion)
Iron	20
Copper	50
Chromium	20
Cobalt	2
Manganese	100
Nickel	20
Vanadium	100

Note: Only one element is assumed to be present and in its worst valence state. The maximum tolerable loss is assumed to be 20 dB/km.

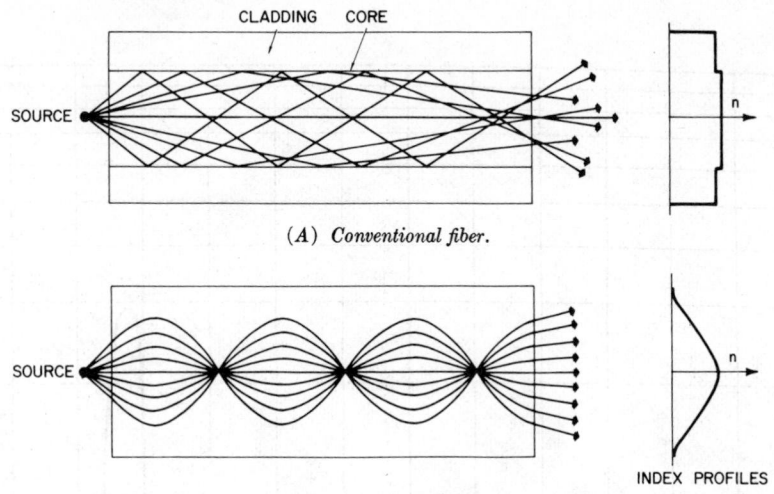

(A) *Conventional fiber.*

(B) *Fiber with graded-index cladding.*

Fig. 24—Optical ray propagation in conventional and graded refractive index fibers. (*From F. M. Mims, III, "Light Beam Communications," p. 115, 1975, Howard W. Sams & Co., Inc.*)

Fig. 23 compares the maximum data rates for single- and multimode fibers.

Pulse broadening in multimode fibers can be reduced considerably by employing a fiber with a parabolically graded refractive-index profile to impart a similar velocity to each of the propagating modes. Fig. 24 compares mode propagation in conventional and graded-index fibers.

Various glasses and plastics can be used to make optical waveguide fibers. The lowest losses thus far (about 2 dB/km at 1.05 μm) have been obtained from fibers of ultrapure fused silica. Contaminants severely limit the transmissivity of an optical fiber as can be seen in Table 5. Ultrapure fibers present difficult processing requirements, and multicomponent glasses having much higher losses (about 800 dB/km) are more readily manufactured and offer a more economical alternative to the relatively costly (at present) very low loss fibers.

Plastic optical fibers are available as an economical supplement to glass for short-haul links wherein moderately high losses (up to 500 dB/km at 656 nm) are acceptable. The transmission attenuation of DuPont PFX Fiber Optic Cable is shown in Fig. 25.

DESIGN CONSIDERATIONS

The functional diagram of a typical optical communications system operating in the atmosphere is shown in Fig. 26. The design considerations for a system such as this involve such engineering trade-offs as source and detector selection, modulation format, transmitter beamwidth, and receiver aperture size. Some of these factors are determined in advance by system requirements (e.g. modulation format), while others are largely determined by engineering constraints. A careful consideration of the optical communications range equation, the desired receiver signal-to-noise ratio (SNR), and the noise contribution of major background sources will greatly assist in the design of a practical system and the prediction of its performance.

Optical Communications Range Equation*

In the usual case, the receiver in an optical communications system operating in the atmosphere intercepts only a small fraction of the transmitted beam. Both the spatial intensity distribution of the transmitted beam and the probability distribution of the transmitter pointing error are usually assumed to be gaussian. For the case where the receiver intercepts a portion of the transmitted beam having an intensity only half that at the beam axis, the approximate received power level is

$$P_R \approx \frac{\pi^2 \tau_a P_A d_T^2 d_R^2}{32 R^2 \lambda_c^2}$$

where τ_a is the atmospheric transmission, P_A is the peak transmitter power, d_T is the diameter of the transmitter aperture, d_R is the diameter of the receiver aperture, R is the range, and λ_c is the wavelength of the transmitted beam. Since pointing errors tend to be gaussian in nature, the actual P_R is typically more than 70% greater than that given by the range equation.

---

* W. K. Pratt, "Laser Communication Systems," John Wiley & Sons, Inc., New York; 1969.

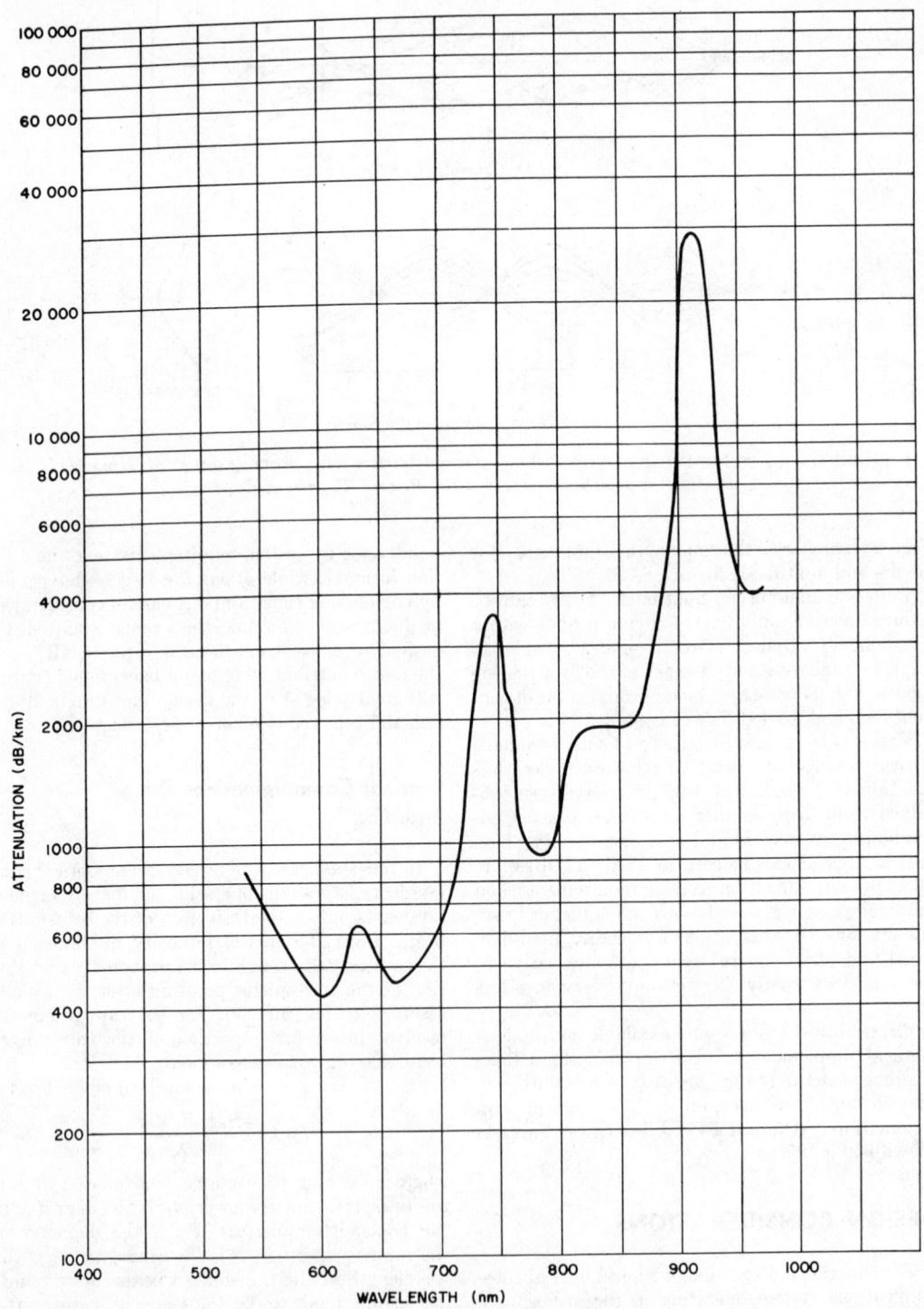

Fig. 25—Attenuation as a function of wavelength for DuPont PFX low-loss fiber-optic cable.

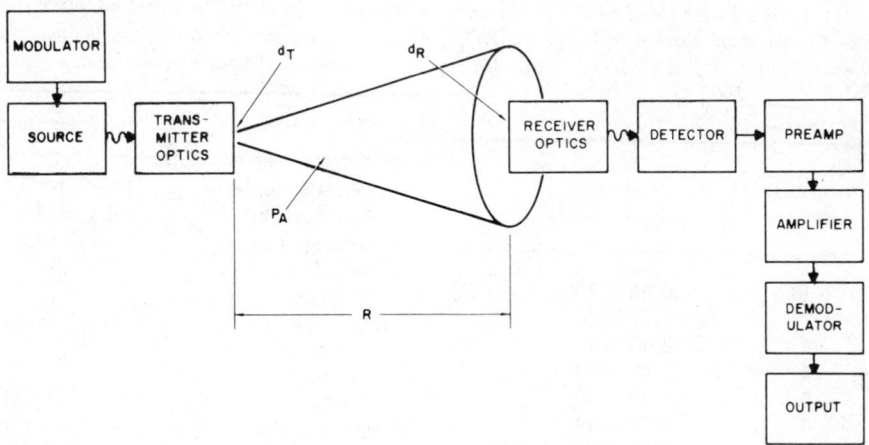

Fig. 26—Functional block diagram of a typical optical communications system designed to operate in the atmosphere.

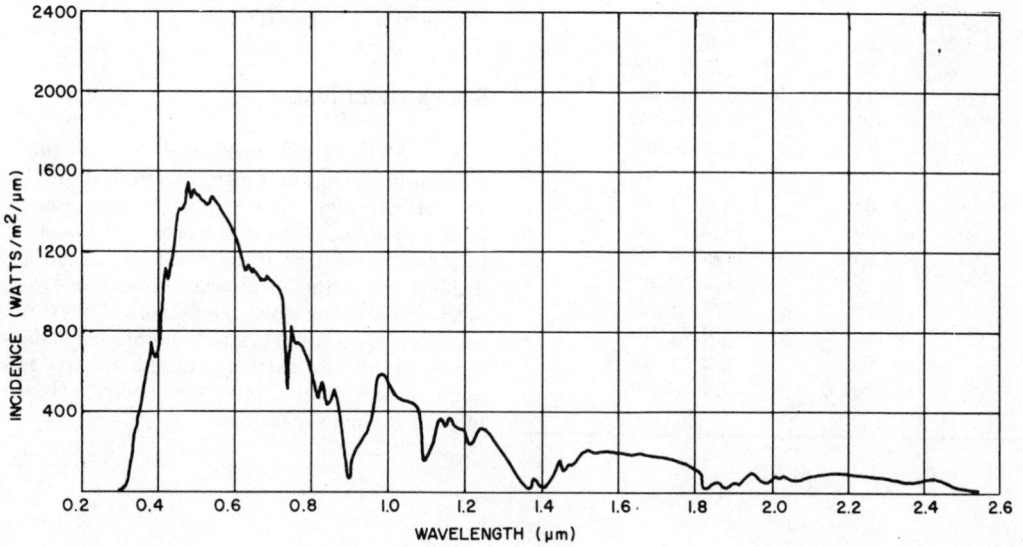

Fig. 27—The solar constant at sea level for an air mass of 1 (solar angle = 90°). (*Adapted from D. M. Gates, "Spectral Distribution of Solar Radiation at the Earth's Surface," Science, February 1966, pp. 523–529.*)

Signal-to-Noise Ratio*

The maximum range of an optical communication system is limited by the SNR of the receiver. One definition of SNR is

$$\frac{S}{N} = \frac{\rho^2 P_R^2 R_1 G^2}{2eB(\rho P_R + \rho P_b + I_d)R_1 G^2 + 2FkTB}$$

where ρ is the responsivity of the detector (am-

peres/watt), P_R is the received power (watts), P_b is the background illumination (watts), R_1 is the load resistance, e is the charge on an electron, B is the bandwidth, F is the noise factor of the preamplifier, k is Boltzmann's constant, T is the temperature (kelvins), I_d is the detector dark current (amperes), and G is the internal gain of the photodetector. The optimum SNR of a receiver occurs when the receiver bandwidth is the reciprocal of 2.5 times the pulse width $(B = 0.4\tau)$. This establishes a peak-to-rms signal ratio of $\sqrt{2}$. Since the noise in an optical receiver has a gaussian intensity distribution, and since the ratio of peak

* W. J. Hannan, J. Bordogna, and D. Karlsons, "Practical Aspects of Injection Laser Communication Systems," *RCA Review*, December 1967, pp. 609–619.

TABLE 6—SOLAR INCIDENCE AS A FUNCTION OF SOLAR ANGLE. (*From D. S. Bond and F. P. Henderson, "The Conquest of Darkness," AD 346297, Defense Documentation Center, Alexandria, Va.; 1963.*)

Solar Angle (degrees)	Incidence (lumens/m²)
−18	6.51×10^{-4}
−12	8.31×10^{-3}
−6	3.40×10^{0}
−5	1.08×10^{1}
−0.8	4.53×10^{2}
0	7.32×10^{2}
5	4.76×10^{3}
10	1.09×10^{4}
15	1.86×10^{4}
20	2.73×10^{4}
25	3.67×10^{4}
30	4.70×10^{4}
35	5.70×10^{4}
40	6.67×10^{4}
45	7.59×10^{4}
50	8.50×10^{4}
55	9.40×10^{4}
60	10.2×10^{4}
65	10.8×10^{4}
70	11.3×10^{4}
75	11.7×10^{4}
80	12.0×10^{4}
85	12.2×10^{4}
90	12.4×10^{4}

TABLE 7—INCIDENCE FROM SEVERAL NATURAL SOURCES. (*From "RCA Electro-Optics Handbook," RCA Aerospace Systems Division, Burlington, Massachusetts; 1968.*)

Source	Incidence (lumens/m²)
Direct Sunlight	$1–1.3 \times 10^{5}$
Full Daylight	$1–2 \times 10^{4}$
Overcast Day	10^{3}
Very Dark Day	10^{2}
Twilight	10^{1}
Deep Twilight	10^{0}
Full Moon	10^{-1}
Quarter Moon	10^{-2}
Starlight	10^{-3}
Overcast Starlight	10^{-4}

to rms noise is 4:1, the SNR at the detection threshold is 32 (15 dB).

Background Noise

The SNR of an optical receiver, hence its maximum detection range, is reduced by background radiation. The major daytime source of background radiation is the sun, and the spectral incidence of sunlight at sea level is shown in Fig. 27. Table 6 gives the incidence (lumens/m²) at sea level for various solar elevations. Other major sources of background radiation are solar radiation scattered by the earth's atmosphere, the moon, and the stars. Table 7 gives the incidence (lumens/m²) for these noise sources.

CHAPTER 35
WIRE TRANSMISSION

DEFINITIONS OF COMMONLY USED TERMS

System Reference Level Point: A point in a communication circuit arbitrarily chosen as a reference point for signal level measurements. Common equivalent terms are "0 dB transmission level point," "zero level," "zero level point," "0 dB TL," and "0 TLP." Previous practice has been to consider the transmitting toll switchboard as the 0 dB TLP.

Relative Level: The relative level at any point in the circuit is a measure of the power gain or loss between the 0 TLP and the point under consideration. Signal powers and interference levels may be referred to the 0 TLP as "a signal level of −16 dBm0," which indicates the power level the signal would have registered had it been measured at the 0 TLP. Note that the 0 TLP may not be accessible for measurement and, in fact, need not even exist in a given system. Reference of signal and interference powers to the 0 TLP is convenient in system design.

Volume: A method of expressing the amplitude of complex nonperiodic signals such as speech. Volume is expressed in volume units (VU) and is defined as the reading obtained on a specified meter when read in a prescribed manner. The volume indicator is not frequency weighted in its response.

Noise: Noise, in its broadest definition, consists of any undesired signal in a communication circuit. Noise may be classified as thermal or white noise, impulse noise, crosstalk, tone interference, and miscellaneous.
Noise is measured on voice communication channels through weighting networks which simulate the interfering effect of noise on human observers using modern telephone sets.

Thermal Noise: A form of noise occurring on all transmission media and in all communications apparatus arising from random electron motion. It is characterized by uniform energy distribution over the frequency spectrum and a normal or Gaussian distribution of levels.

Impulse Noise: Noncontinuous noise consisting of irregular pulses of short duration and relatively high amplitude. Some sources of impulse noise in voice communication channels are: induced interference by transients due to relay and switch operation, transients due to switching or lightning in adjacent power circuits, and crosstalk from high-level telegraph circuits.

Crosstalk: Interference from other communication channels is called crosstalk. It is classified as near-end and far-end crosstalk and as intelligible and unintelligible crosstalk. Near-end crosstalk is measured at the sending terminal while far-end crosstalk is measured at the receiving terminal.
Intelligible crosstalk can be understood by the listener and, because it diverts his attention, it has more interfering effect than unintelligible crosstalk. Crosstalk into a voice-frequency circuit from adjacent voice-frequency circuits or between groups and supergroups in carrier systems is generally intelligible. Crosstalk due to incomplete suppression of sidebands, to intermodulation of two or more carrier channels, or to otherwise intelligible crosstalk between carrier channels having offset frequency spectra is generally unintelligible. Such crosstalk is often classed as miscellaneous noise. Intermodulation crosstalk in wide-band carrier systems approaches thermal noise in its spectral distribution.

Tone Interference: Interference due to single tones or complex periodic waveforms.

Miscellaneous Noise: Interferences that cannot readily be placed in any of the preceding categories.

Reference Noise: 1 picowatt (10^{-12} watt) of 1000-hertz power. Also commonly stated as −90 dBm. This reference noise level is used in the Western Electric 3A Noise Measuring Set. The previous standard 2B Noise Measuring Set used a reference noise level of −85 dBm at 1000 hertz.

dBrn: Decibels above reference noise is the unit of measurement of noise power used in both the original Western Electric Noise Measuring Set and the current standard 3A. The original set contained a frequency-weighting network corresponding to the 144-type receiver while the present set has a

TABLE 1—COMPARISON OF VARIOUS NOISE
MEASURING SETS.

Noise Measuring Set	Reading Due to 0 dBm of	
	1000 Hertz	White Noise Limited to 0–3-Kilohertz Band
Western Electric 2B (144 line weighting)	90 dBrn	82 dBrn
Western Electric 2B (F1A line weighting)	85 dBa	82 dBa
Western Electric 3A (C message weighting)	90 dBrn	88 dBrn
CCITT psophometer	+1 dBm*	−2 dBm*

* The psophometer is defined as measuring the internal (open-circuit) voltage of an equivalent noise generator having an impedance of 600 ohms and delivering noise power to a 600-ohm load. For convenience in comparison, the psophometric electromotive force has been converted to dBm.

frequency-weighting network corresponding to the 500-type telephone set receiver. Due to the differences in network response, different readings would be obtained from the two instruments if the same signal were measured. Refer to Table 1.

dBa: Decibels above *adjusted* reference noise (−85 dBm). The weighting network of the 2B Noise Measuring Set was adjusted for the response of the 302-type telephone set (F1A weighting). So that similar readings would be obtained on a noise

signal consisting of thermal noise limited in frequency spectrum from 0 to 3 kilohertz, the reference was adjusted to −85 dBm. The dBa as a noise unit appears in earlier technical literature, but is being supplanted by the dBrn.

pWp: Picowatts of noise psophometrically weighted. Units of noise power derived from measurements with the CCITT recommended psophometer. The psophometer is frequency weighted by a curve having a shape similar to the F1A weighting curve. The reference noise is −90 dBm (1 picowatt) at 800 hertz.

Net Loss (Equivalent): The net loss of a transmission system is the difference between the relative levels at the input and the output of the system. By custom the net loss is understood to be measured at 1000 hertz in the American and Canadian plant and at 800 hertz in European practice (CCITT).

Singing Margin: The singing margin of a circuit is defined as the amount by which the net loss of the two directions of transmission may be reduced before oscillation (singing) occurs. Inadequate singing margin results in distorted transmission and a hollow or "rain-barrel" sounding transmission due to prolonged echoes.

Return Loss: A measure of the match between the two impedances on either side of a junction point, defined by

$$\text{RL (dB)} = 20 \log_{10} \left| \frac{Z_1 + Z_2}{Z_1 - Z_2} \right|$$

where Z_1 and Z_2 are the complex impedances of the two halves of the circuit.

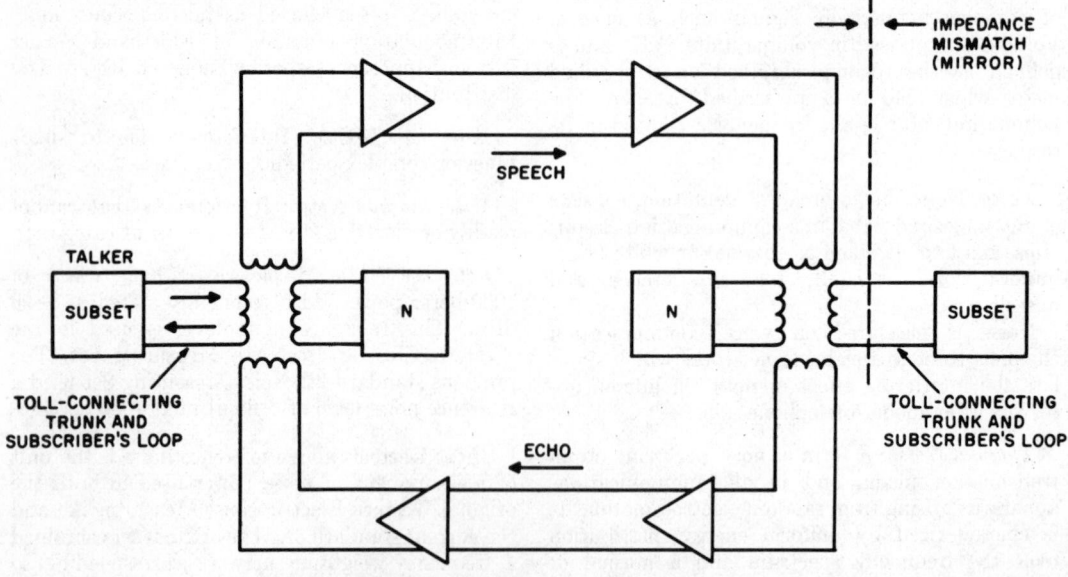

N = COMPROMISE BALANCING NETWORK OF 4-WIRE SET

Fig. 1—Echo in a toll connection. *"Transmission Systems for Communications," revised third edition, p. 29,* © *1964 Bell Telephone Laboratories, Inc.*

Insertion Loss—4-Wire Terminating Sets: In hybrid circuits or 4-wire terminating sets, the insertion loss between the 4-wire transmitting and receiving circuits is the sum of the input and output losses of the hybrid plus the return loss between the 2-wire drop and the balancing network.

Echo: A signal returned to the talker after making one or more round trips between talker and listener (see Fig. 1). The first talker echo is generally the most important. Echo return loss is defined as the average return loss at the 4-wire terminating set over the frequency band from 500 to 2500 hertz, as this range is the most important for echo control.

MESSAGE-CHANNEL OBJECTIVES— THE EXCHANGE PLANT

Subscriber Sets

The 500-type subscriber set provides improved transmission performance with improved frequency response, particularly on long subscriber loops. Figures 2, 3, and 4 compare the performance of the 500-type set with the older 302-type set.

Resistance Design of Subscriber Loops

Present practice is to design subscriber loops such that the maximum loop resistance is limited to a value dependent on requirements for supervisory signaling or transmission, whichever governs. Resistance objectives in current use are:

For Step-by-Step and Panel Offices: Maximum loop resistance of 850 and 1200 ohms, respectively, limited by supervisory signaling and dial-pulsing requirements.

For No. 5 Crossbar and ESS Offices: Maximum loop resistance of 1300 ohms, based primarily on transmission considerations. The 1300-ohm limit applies provided (A) that 500-type sets are always used on loops greater than 10 000 feet in length (302-type sets may still be used in shorter loops) and (B) that loops of 18 000 feet and greater are

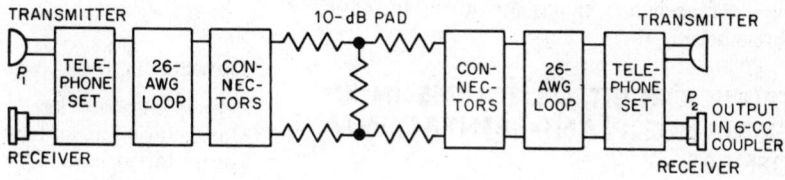

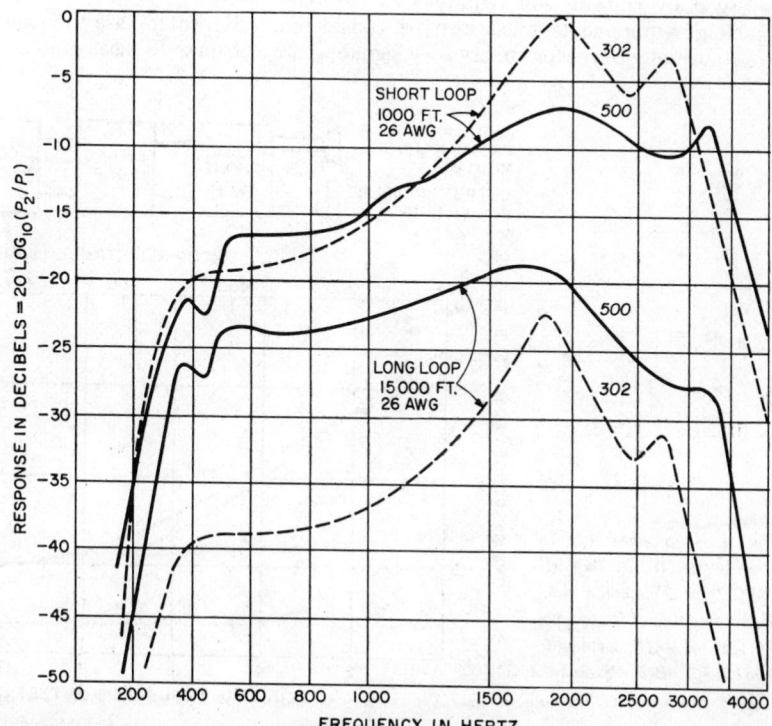

Fig. 2—Comparison of overall response. *W. F. Tuffnell, "500-Type Telephone Set," Bell Laboratories Record, vol. 29, pp. 414-418; September 1951.* © *Bell Telephone Laboratories, Inc., 1951.*

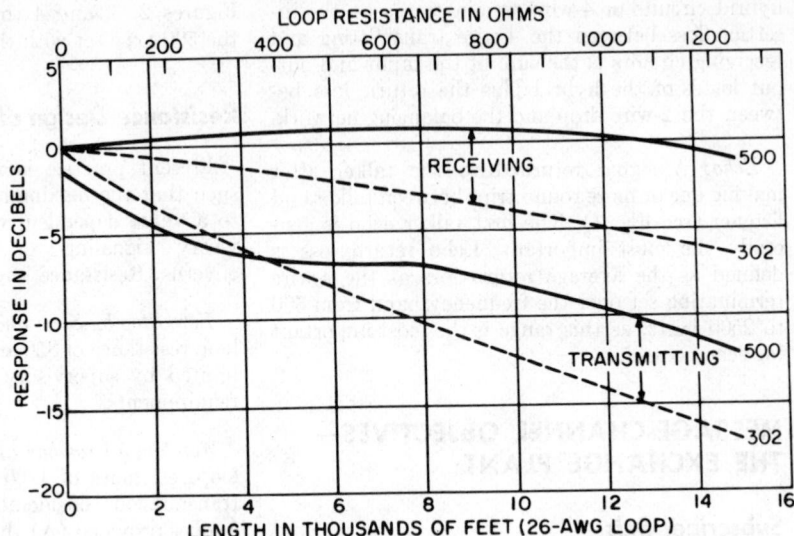

Fig. 3—Relative volume levels. *W. F. Tuffnell, "500-Type Telephone Set," Bell Laboratories Record, vol. 29, pp. 414–418; September 1951.* © *Bell Telephone Laboratories, Inc., 1951.*

loaded to reduce total attenuation and flatten the amplitude–frequency response.

Typical values of electrical properties of various types of exchange cables are given in Tables 2 through 10.

TRUNK-CIRCUIT OBJECTIVES IN THE EXCHANGE PLANT—PENTACONTA OFFICES

Interoffice trunk-circuit design is primarily governed by transmission requirements although signaling requirements must also be considered. Resistance objectives for supervisory signaling are

Crossbar systems: 3000 ohms maximum
Panel systems: 1300 ohms maximum
Step-by-step systems: 2000 ohms maximum—
                        anticipated increase
                        to 5000 ohms.

Transmission objectives are

Average insertion loss: 4 decibels
Maximum insertion loss: 6 decibels
Future target loss: Nearly 0 decibels with
                          loading or equaliza-
                          tion.

Repeaters are frequently required in interoffice trunks to meet present transmission objectives.

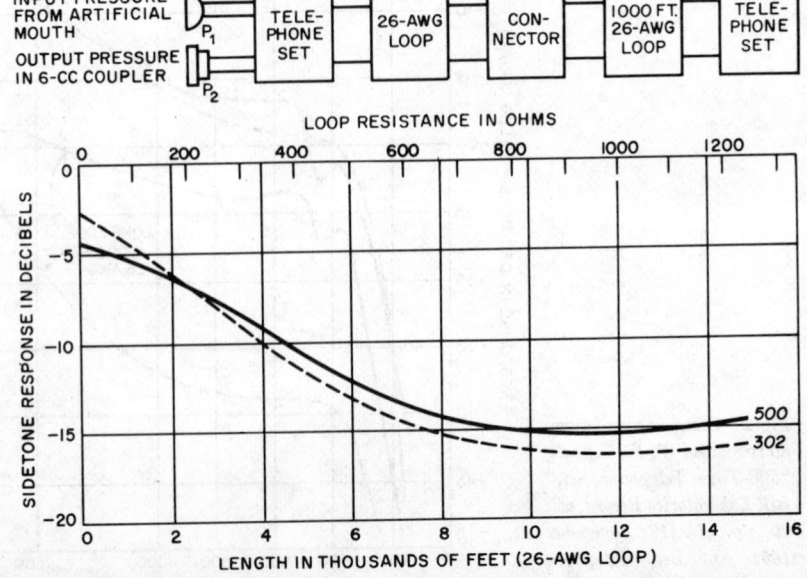

Fig. 4—Comparative side-tone levels. *W. F. Tuffnell, "500-Type Telephone Set," Bell Laboratories Record, vol. 29, pp. 414–418; September 1951.* © *Bell Telephone Laboratories, Inc., 1951.*

TABLE 2—LINE CONSTANTS OF COPPER OPEN-WIRE PAIRS.

	Resistance in Ohms/Loop Mile						Inductance in Millihenries/Loop Mile					
	165 Mil		128 Mil		104 Mil		165 Mil		128 Mil		104 Mil	
Freq (kHz)	12″ DP	8″ CS	12″ DP	8″ CS	12″ DP	8″ CS	12″ DP	8″ CS	12″ DP	8″ CS	12″ DP	8″ CS
0.1	4.10	4.10	6.82	6.82	10.33	10.33	3.37	3.11	3.53	3.27	3.66	3.40
0.5	4.13	4.13	6.83	6.83	10.34	10.34	3.37	3.10	3.53	3.27	3.66	3.40
1.0	4.19	4.19	6.87	6.87	10.36	10.36	3.37	3.10	3.53	3.27	3.66	3.40
1.5	4.29	4.29	6.94	6.94	10.41	10.41	3.37	3.10	3.53	3.26	3.66	3.40
2.0	4.42	4.42	7.02	7.02	10.47	10.47	3.36	3.10	3.53	3.26	3.66	3.40
3.0	4.76	4.76	7.24	7.24	10.62	10.62	3.35	3.09	3.52	3.26	3.66	3.40
5.0	5.61	5.61	7.92	7.92	11.11	11.11	3.34	3.08	3.52	3.25	3.66	3.40
10	7.56	7.56	10.05	10.05	12.98	12.98	3.31	3.04	3.49	3.23	3.64	3.38
20	10.23	10.23	13.63	13.63	17.14	17.14	3.28	3.02	3.46	3.20	3.61	3.35
30	12.26	12.26	16.26	16.26	20.55	20.55	3.26	3.00	3.44	3.17	3.58	3.33
50	15.50	15.50	20.41	20.41	25.67	25.67	3.25	2.99	3.43	3.16	3.57	3.31
100	21.45	21.45	28.09	28.09	35.10	35.10	3.24	2.98	3.42	3.15	3.55	3.29
150	26.03	26.03	33.96	33.96	42.42	42.42	3.23	2.97	3.41	3.14	3.54	3.28
200	29.89	29.89	38.93	38.93	48.43	48.43	3.23	2.97	3.40	3.14	3.54	3.28
500	46.62	46.62	60.53	60.53	74.98	74.98	3.22	2.96	3.39	3.13	3.53	3.27
1000	65.54	65.54	84.84	84.84	104.9	104.9	3.22	2.96	3.38	3.12	3.52	3.26

	Leakage Conductance in Micromhos/Loop Mile					Capacitance in Microfarads/Loop Mile	
	Dry—All Gauges		Wet—All Gauges				
Freq (kHz)	12″—DP	8″—CS	12″—DP	8″—CS	Wire Size	12″	8″
0.1	0.04	0.04	2.5	2.0	In space:		
0.5	0.15	0.06	3.0	2.3	165 mil	0.00898	0.00978
1.0	0.29	0.11	3.5	2.6	128 mil	0.00855	0.00928
1.5	0.43	0.15	4.0	2.9	104 mil	0.00822	0.00888
2.0	0.57	0.20	4.5	3.2	On 40-wire line, dry:		
3.0	0.85	0.30	5.5	3.7	165 mil	0.00915	0.01000
5.0	1.4	0.49	7.5	4.6	128 mil	0.00871	0.00948
10	2.8	0.97	12.1	6.6	104 mil	0.00857	0.00908
20	5.6	1.9	20.5	9.6	On 40-wire line, wet:		
30	8.4	2.9	28.0	12.1	165 mil	0.0093	0.0102
50	14.0	4.8	41.1	15.7	128 mil	0.0089	0.0097
					104 mil	0.0085	0.0093

Wires spaced 8 and 12 inches.
Insulators:
  40 pairs toll and double-petticoat (DP) per mile.
  53 pairs Pyrex glass (CS) per mile.
Temperature 68° Fahrenheit.

TABLE 3—LINE CONSTANTS OF 40% COPPER-CLAD STEEL OPEN-WIRE PAIRS.

Freq (kHz)	Resistance in Ohms/Loop Mile						Inductance in Millihenries/Loop Mile					
	165 Mil		128 Mil		104 Mil		165 Mil		128 Mil		104 Mil	
	12" DP	8" CS	12" DP	8" CS	12" DP	8" CS	12" DP	8" CS	12" DP	8" CS	12" DP	8" CS
0.0	9.8	9.8	16.2	16.2	24.6	24.6	—	—	—	—	—	—
0.1	10.0	10.0	16.3	16.3	24.6	24.6	3.37	3.11	3.53	3.27	3.66	3.40
0.5	10.0	10.0	16.4	16.4	24.7	24.7	3.37	3.10	3.53	3.27	3.66	3.40
1.0	10.1	10.1	16.6	16.6	24.8	24.8	3.37	3.10	3.53	3.27	3.66	3.40
1.5	10.1	10.1	16.7	16.7	24.9	24.9	3.37	3.10	3.53	3.26	3.66	3.40
2.0	10.2	10.2	16.8	16.8	25.2	25.2	3.36	3.10	3.53	3.26	3.66	3.40
3.0	10.4	10.4	17.1	17.1	25.4	25.4	3.35	3.09	3.52	3.26	3.66	3.40
5.0	10.6	10.6	17.4	17.4	26.0	26.0	3.34	3.08	3.52	3.25	3.66	3.40
10	10.8	10.8	17.7	17.7	26.5	26.5	3.31	3.04	3.49	3.23	3.64	3.38
20	11.4	11.4	18.2	18.2	27.1	27.1	3.28	3.02	3.46	3.20	3.61	3.35
30	12.3	12.3	18.8	18.8	27.5	27.5	3.26	3.00	3.44	3.17	3.58	3.33
50	14.5	14.5	20.4	20.4	28.7	28.7	3.25	2.99	3.43	3.16	3.57	3.31
100	20.8	20.8	26.5	26.5	33.3	33.3	3.24	2.98	3.42	3.15	3.55	3.29
150	25.9	25.9	32.5	32.5	39.6	39.6	3.23	2.97	3.41	3.14	3.54	3.28

Freq (kHz)	Leakage Conductance in Micromhos/Loop Mile				Capacitance in Microfarads/Loop Mile		
	Dry—All Gauges		Wet—All Gauges				
	12"—DP	8"—CS	12"—DP	8"—CS	Wire Size	12"	8"
0.1	0.04	0.04	2.5	2.0	In space:		
0.5	0.15	0.06	3.0	2.3	165 mil	0.00898	0.00978
1.0	0.29	0.11	3.5	2.6	128 mil	0.00855	0.00928
1.5	0.43	0.15	4.0	2.9	104 mil	0.00822	0.00888
2.0	0.57	0.20	4.5	3.2	On 40-wire line, dry:		
3.0	0.85	0.30	5.5	3.7	165 mil	0.00915	0.01000
5.0	1.4	0.49	7.5	4.6	128 mil	0.00871	0.00948
10	2.8	0.97	12.1	6.6	104 mil	0.00857	0.00908
20	5.6	1.9	20.5	9.6	On 40-wire line, wet:		
30	8.4	2.9	28.0	12.1	165 mil	0.0093	0.0102
50	14.0	4.8	41.1	15.7	128 mil	0.0089	0.0097
					104 mil	0.0085	0.0093

Wires spaced 8 and 12 inches.
Insulators:
    40 pairs toll and double-petticoat (DP) per mile.
    53 pairs Pyrex glass (CS) per mile.
Temperature 68° Fahrenheit.

TABLE 4—ATTENUATION OF COPPER OPEN-WIRE PAIRS.

Freq (kHz)	165 Mil			128 Mil			104 Mil		
	12″ DP	12″ CS	8″ CS	12″ DP	12″ CS	8″ CS	12″ DP	12″ CS	8″ CS
Dry Weather									
0.1	0.023	0.023	0.025	0.032	0.032	0.034	0.041	0.041	0.0425
0.5	0.029	0.029	0.0315	0.045	0.045	0.048	0.063	0.063	0.067
1.0	0.030	0.030	0.0325	0.047	0.047	0.0505	0.067	0.067	0.072
1.5	0.031	0.031	0.0335	0.048	0.048	0.051	0.068	0.068	0.073
2.0	0.0325	0.032	0.035	0.0485	0.048	0.052	0.069	0.069	0.074
3.0	0.036	0.034	0.038	0.051	0.050	0.054	0.071	0.070	0.076
5.0	0.044	0.041	0.0445	0.057	0.055	0.0595	0.076	0.074	0.080
10	0.061	0.056	0.0605	0.076	0.070	0.076	0.093	0.087	0.094
20	0.088	0.076	0.083	0.108	0.096	0.104	0.129	0.116	0.125
30	0.110	0.092	0.100	0.135	0.116	0.125	0.159	0.140	0.151
50	0.148	0.118	0.127	0.179	0.147	0.158	0.209	0.176	0.189
100	—	0.165	0.178	—	0.204	0.220	—	0.244	0.262
150	—	0.203	0.218	—	0.249	0.268	—	0.296	0.317
200	—	0.235	0.25	—	—	—	—	—	—
500	—	—	0.42±	—	—	—	—	—	—
1000	—	—	0.7±	—	—	—	—	—	—
Wet Weather									
0.1	0.032	0.029	0.030	0.043	0.039	0.040	0.054	0.049	0.0505
0.5	0.037	0.034	0.036	0.053	0.050	0.053	0.072	0.069	0.0705
1.0	0.039	0.035	0.037	0.056	0.052	0.055	0.076	0.073	0.0775
1.5	0.041	0.037	0.0385	0.058	0.0535	0.0565	0.078	0.0745	0.0795
2.0	0.043	0.038	0.040	0.060	0.0545	0.058	0.0805	0.076	0.0805
3.0	0.0485	0.041	0.044	0.064	0.0575	0.061	0.0845	0.078	0.083
5.0	0.060	0.050	0.0525	0.075	0.0645	0.068	0.094	0.084	0.089
10	0.085	0.068	0.072	0.102	0.083	0.0885	0.120	0.101	0.106
20	0.127	0.095	0.101	0.150	0.116	0.123	0.173	0.137	0.144
30	0.161	0.118	0.124	0.188	0.142	0.150	0.216	0.168	0.176
50	0.220	0.154	0.162	0.253	0.185	0.195	0.287	0.217	0.227
100	—	0.228	0.237	—	0.271	0.283	—	0.313	0.326
150	—	0.288	0.299	—	0.339	0.353	—	0.390	0.405

Wires spaced 8 and 12 inches.

Insulators:

    40 pairs toll and double-petticoat (DP) per mile.

    53 pairs Pyrex glass (CS) per mile.

Temperature 68° Fahrenheit.

TABLE 5—ATTENUATION OF 40% COPPER-CLAD STEEL OPEN-WIRE PAIRS.

Freq (kHz)	165 Mil			128 Mil			104 Mil		
	12″ DP	12″ CS	8″ CS	12″ DP	12″ CS	8″ CS	12″ DP	12″ CS	8″ CS

Attenuation in Decibels per Mile

Dry Weather

Freq (kHz)	12″ DP	12″ CS	8″ CS	12″ DP	12″ CS	8″ CS	12″ DP	12″ CS	8″ CS
0.2	0.054	0.054	0.057	0.073	0.073	0.077	0.091	0.091	0.096
0.5	0.067	0.067	0.071	0.097	0.097	0.103	0.127	0.127	0.134
1.0	0.073	0.073	0.078	0.112	0.112	0.120	0.152	0.152	0.162
1.5	0.076	0.076	0.082	0.118	0.118	0.127	0.162	0.162	0.174
2.0	0.077	0.077	0.083	0.120	0.120	0.130	0.168	0.168	0.180
3.0	0.079	0.079	0.085	0.124	0.124	0.134	0.174	0.174	0.188
5.0	0.082	0.082	0.088	0.127	0.127	0.138	0.179	0.179	0.195
10	0.085	0.085	0.092	0.131	0.131	0.142	0.186	0.186	0.201
20	0.088	0.088	0.096	0.135	0.135	0.147	0.191	0.191	0.207
30	0.095	0.095	0.103	0.139	0.139	0.152	0.195	0.195	0.211
50	0.110	0.110	0.119	0.150	0.150	0.163	0.206	0.206	0.221
100	0.156	0.156	0.168	0.188	0.188	0.203	0.234	0.234	0.252
150	0.199	0.199	0.214	0.233	0.233	0.251	0.273	0.273	0.293

Wet Weather

Freq (kHz)	12″ DP	12″ CS	8″ CS	12″ DP	12″ CS	8″ CS	12″ DP	12″ CS	8″ CS
0.2	0.066	0.060	0.063	0.089	0.081	0.084	0.111	0.101	0.105
0.5	0.077	0.072	0.076	0.111	0.104	0.110	0.145	0.136	0.142
1.0	0.083	0.078	0.084	0.126	0.119	0.126	0.168	0.160	0.169
1.5	0.088	0.082	0.087	0.130	0.124	0.133	0.178	0.170	0.181
2.0	0.089	0.083	0.089	0.136	0.128	0.137	0.184	0.176	0.188
3.0	0.093	0.086	0.092	0.140	0.132	0.142	0.192	0.183	0.196
5.0	0.100	0.091	0.097	0.147	0.137	0.148	0.201	0.190	0.205
10	0.111	0.098	0.104	0.159	0.145	0.155	0.214	0.200	0.215
20	0.126	0.107	0.115	0.175	0.155	0.166	0.233	0.212	0.228
30	0.145	0.120	0.127	0.197	0.168	0.177	0.253	0.224	0.238
50	0.184	0.147	0.153	0.230	0.190	0.199	0.288	0.247	0.261
100	0.282	0.219	0.227	0.314	0.254	0.265	0.372	0.303	0.317
150	0.370	0.285	0.295	0.415	0.324	0.336	0.461	0.367	0.382

Wires spaced 8 and 12 inches.
Insulators:
    40 pairs toll and double-petticoat (DP) per mile.
    53 pairs Pyrex glass (CS) per mile.
Temperature 68° Fahrenheit.

TABLE 6—CHARACTERISTICS OF STANDARD TYPES OF AERIAL COPPER-WIRE TELEPHONE CIRCUITS.

Type of Circuit	Gauge of Wires (mils)	Spacing of Wires (inches)	Primary Constants per Loop Mile				Propagation Constant				Line Impedance				Wavelength (miles)	Velocity (miles per second)	Attenuation (dB per mile)
							Polar		Rectangular		Polar		Rectangular				
			R (ohms)	L (henries)	C (μF)	G (μmho)	Magnitude	Angle (deg +)	α	β	Magnitude	Angle (deg −)	R (ohms)	X (ohms −)			
Nonpole pair phys	165	8	4.11	0.00311	0.01000	0.11	0.0353	83.99	0.00370	0.0351	565	5.88	562	58	179.0	179 000	0.0325
Nonpole pair side	165	12	4.11	0.00337	0.00915	0.29	0.0352	84.36	0.00346	0.0350	612	5.35	610	57	179.5	179 500	0.030
Pole pair side	165	18	4.11	0.00364	0.00863	0.29	0.0355	84.75	0.00325	0.0353	653	5.00	651	57	178.0	178 000	0.028
Nonpole pair phan	165	12	2.06	0.00208	0.01514	0.58	0.0355	85.34	0.00288	0.0354	373	4.30	372	28	177.5	177 500	0.025
Nonpole pair phys	128	8	6.74	0.00327	0.00948	0.11	0.0358	80.85	0.00569	0.0353	603	8.97	596	94	178.0	178 000	0.0505
Nonpole pair side	128	12	6.74	0.00353	0.00871	0.29	0.0356	81.39	0.00533	0.0352	650	8.32	643	94	178.5	178 500	0.047
Pole pair side	128	18	6.74	0.00380	0.00825	0.29	0.0358	81.95	0.00502	0.0355	693	7.72	686	93	177.0	177 000	0.044
Nonpole pair phan	128	12	3.37	0.00216	0.01454	0.58	0.0357	82.84	0.00445	0.0355	401	6.73	398	47	177.0	177 000	0.039
Nonpole pair phys	104	8	10.15	0.00340	0.00908	0.11	0.0367	77.22	0.00811	0.0358	644	12.63	629	141	175.5	175 500	0.072
Nonpole pair side	104	12	10.15	0.00366	0.00837	0.29	0.0363	77.93	0.00760	0.0355	692	11.75	677	141	177.0	177 000	0.067
Pole pair side	104	18	10.15	0.00393	0.00797	0.29	0.0365	78.66	0.00718	0.0358	730	10.97	717	139	175.5	175 500	0.063
Nonpole pair phan	104	12	5.08	0.00223	0.01409	0.58	0.0363	79.84	0.00640	0.0357	421	9.70	415	71	176.0	176 000	0.056

1000 hertz.
DP (double petticoat) insulators for all 12- and 18-inch spaced wires.
CS (special glass with steel pin) insulators for all 8-inch spaced wires.

Notes: 1. All values are for dry-weather conditions.
    2. All capacitance values assume a line carrying 40 wires.
    3. Resistance values are for temperature of 20°C (68°F).

TABLE 7—REPRESENTATIVE VALUES OF TOLL-CABLE LINE AND PROPAGATION CONSTANTS.

Freq (kHz)	Resistance (ohms/mile)			Inductance (millihenries/mile)			Conductance (micromhos/mile)			Capacitance (µF/mile)	Characteristic Impedance (ohms)			Phase Shift (radians/mile)			Attenuation (decibels/mile)		
	13	16	19	13	16	19	13	16	19	13, 16, or 19	13	16	19	13	16	19	13	16	19
0	20.7	41.8	83.8	1.070	1.100	1.112	—	—	—	0.0610	—	—	—	—	—	—	—	—	—
0.1	20.7	41.8	83.8	1.069	1.100	1.112	0.40	0.25	0.10	0.0610	530−j505	745−j730	1050−j1040	0.020	0.027	0.040	0.17	0.24	0.35
0.5	20.7	41.9	83.9	1.065	1.099	1.112	1.4	0.75	0.40	0.0609	250−j210	345−j315	480−j460	0.050	0.064	0.092	0.36	0.51	0.77
1.0	20.8	42.0	84.0	1.060	1.098	1.111	2.5	1.5	1.0	0.0609	195−j140	255−j215	345−j319	0.075	0.092	0.133	0.47	0.69	1.06
1.5	20.9	42.1	84.1	1.057	1.097	1.111	3.5	2.0	1.6	0.0608	170−j105	225−j175	290−j255	0.100	0.116	0.17	0.53	0.79	1.27
2.0	21.0	42.2	84.2	1.053	1.096	1.110	4.5	2.65	2.35	0.0608	160−j85	205−j150	255−j215	0.120	0.140	0.20	0.58	0.87	1.44
3.0	21.3	42.4	84.3	1.046	1.095	1.110	6.5	4.15	4.05	0.0607	145−j63	180−j115	217−j170	0.170	0.189	0.25	0.63	1.00	1.68
5.0	22.0	43.0	84.5	1.035	1.093	1.109	10.5	7.6	8.0	0.0606	135−j42	155−j72	182−j120	0.26	0.28	0.35	0.70	1.16	2.03
10	24.0	44.5	85.3	1.007	1.085	1.105	21.0	18.5	20.0	0.0605	131−j23	142−j40	155−j73	0.50	0.52	0.59	0.80	1.32	2.43
20	29.1	49.5	89.0	0.968	1.066	1.095	47.0	46.2	50.0	0.0604	128−j15	137−j25	141−j41	0.97	1.00	1.07	1.04	1.55	2.77
30	35.5	55.4	94.0	0.945	1.047	1.085	78.0	80.5	87.5	0.0602	126−j12	135−j18	137−j30	1.43	1.48	1.57	1.27	1.78	3.02
50	47.5	67.0	105.5	0.910	1.015	1.065	150.	160.	180.	0.0600	124−j10	133−j13	134−j20	2.34	2.42	2.60	1.75	2.24	3.53
100	71.3	91.7	137.0	0.870	0.963	1.017	350.	400.	450.	0.0598	121−j7.3	130−j9	131−j13	4.54	4.71	5.00	2.72	3.31	4.80
150	90.0	111.2	165.0	0.850	0.935	0.980	600.	700.	800.	0.0595	119−j6.0	127−j7	129−j11	6.73	6.94	7.25	3.60	4.27	6.00
200	—	—	—	—	—	—	—	—	—	—	—	—	—	—	—	—	—	—	7.00
500	—	—	—	—	—	—	—	—	—	—	—	—	—	—	—	—	—	—	12±
1000	—	—	—	—	—	—	—	—	—	—	—	—	—	—	—	—	—	—	18±
For 0° F:																			
Increase by	—	—	—	—	—	—	—	—	—	—	—	—	—	—	—	—	—	—	—
Decrease by	9%	9%	9%	0.5%	0.5%	0.5%	50%	50%	50%	2%	—	—	—	2%	2%	2%	9%	9%	9%
For 110° F:																			
Increase by	8%	8%	8%	0.4%	0.4%	0.4%	50%	50%	50%	2%	—	—	—	2%	2%	2%	9%	9%	9%
Decrease by	—	—	—	—	—	—	—	—	—	—	—	—	—	—	—	—	—	—	—

13, 16, and 19 AWG quadded toll cable.
Nonloaded.
All figures for loop-mile basis.
Temperature 55° Fahrenheit.

TABLE 8—APPROXIMATE CHARACTERISTICS OF STANDARD TYPES OF PAPER-INSULATED TOLL TELEPHONE CABLE CIRCUITS.

| Wire Gauge (AWG) | Type of Loading* | Spacing of Load Coils (miles) | Constants Assumed to be Distributed per Loop Mile | | | | Propagation Constant | | | | Line Impedance | | | | Wavelength (miles) | Velocity (miles per second) | Cutoff Frequency f_c (hertz) | Attenuation (dB per mile) |
			R (ohms)	L (henries)	C (μF)	G (μmho)	Polar Magnitude	Polar Angle (deg+)	Rect. α	Rect. β	Polar Magnitude	Polar Angle (deg−)	Rect. R (ohms)	Rect. X (ohms)				
							Side Circuit											
19	N.L.S.	—	84.0	0.001	0.061	1.0	0.183	47.0	0.1249	0.134	470	42.8	345	319.4	46.9	46900	—	1.06
19	H-31-S	1.135	87.2	0.028	0.061	1.0	0.277	76.6	0.0643	0.269	710	13.2	691	162.2	23.3	23300	6700	0.56
19	H-44-S	1.135	88.4	0.039	0.061	1.0	0.319	79.9	0.0561	0.314	818	9.9	806	140.8	20.0	20000	5700	0.49
19	H-88-S	1.135	91.2	0.078	0.061	1.0	0.441	84.6	0.0418	0.439	1131	5.2	1126	102.8	14.3	14300	4000	0.36
19	H-172-S	1.135	96.3	0.151	0.061	1.0	0.610	87.0	0.0323	0.609	1565	2.8	1563	76.9	10.3	10300	2900	0.28
19	B-88-S	0.568	97.7	0.156	0.061	1.0	0.620	87.0	0.0322	0.619	1590	2.8	1588	76.7	10.2	10200	5700	0.28
16	N.L.S.	—	42.1	0.001	0.061	1.5	0.129	49.1	0.0842	0.097	331	40.7	255	215.4	64.5	64500	—	0.69
16	H-31-S	1.135	44.5	0.028	0.061	1.5	0.266	82.8	0.0334	0.264	683	7.0	677	83.0	23.8	23800	6700	0.29
16	H-44-S	1.135	45.7	0.039	0.061	1.5	0.315	84.6	0.0296	0.313	808	5.2	805	72.8	20.1	20000	5700	0.26
16	H-88-S	1.135	48.5	0.078	0.061	1.5	0.438	87.6	0.0224	0.437	1124	2.7	1123	53.1	14.4	14400	4000	0.19
16	H-172-S	1.135	53.6	0.151	0.061	1.5	0.608	88.3	0.0183	0.608	1562	1.5	1562	41.1	10.3	10300	2900	0.16
16	B-88-S	0.568	54.9	0.156	0.061	1.5	0.618	88.3	0.0185	0.618	1587	1.5	1587	41.4	10.2	10200	5700	0.16
13	N.L.S.	—	20.8	0.001	0.061	2.5	0.094	52.9	0.0568	0.075	242	36.9	195	140.0	83.6	83600	—	0.47
							Phantom Circuit											
19	N.L.P.	—	42.0	0.0007	0.100	1.5	0.165	47.8	0.1106	0.122	262	42.0	195	175.2	51.5	51500	—	0.96
19	H-18-P	1.135	43.5	0.017	0.100	1.5	0.270	78.7	0.0529	0.264	429	11.1	421	82.6	23.8	23800	7000	0.46
19	H-25-P	1.135	44.2	0.023	0.100	1.5	0.308	81.3	0.0466	0.305	491	8.5	485	72.4	20.6	20600	5900	0.40
19	H-50-P	1.135	45.7	0.045	0.100	1.5	0.424	85.3	0.0351	0.423	675	4.5	673	53.3	14.9	14900	4200	0.30
19	H-63-P	1.135	47.8	0.056	0.100	1.5	0.472	86.0	0.0331	0.471	752	3.8	750	49.8	13.3	13300	3700	0.29
19	B-50-P	0.568	49.0	0.089	0.100	1.5	0.594	87.4	0.0273	0.593	945	2.4	944	39.8	10.6	10600	5900	0.24
16	N.L.P.	—	21.0	0.0007	0.100	2.4	0.116	50.0	0.0746	0.089	185	39.0	144	116.3	70.6	70600	—	0.65
16	H-18-P	1.135	22.2	0.017	0.100	2.4	0.262	84.0	0.0273	0.260	417	5.8	415	41.8	24.1	24100	7000	0.24
16	H-25-P	1.135	22.8	0.023	0.100	2.4	0.303	85.4	0.0243	0.302	483	4.4	481	36.8	20.8	20800	5900	0.21
							Physical Circuit											
16	H-50-P	1.135	24.3	0.045	0.100	2.4	0.422	87.4	0.0189	0.422	672	2.4	672	27.5	14.9	14900	4200	0.16
16	H-63-P	1.135	26.4	0.056	0.100	2.4	0.471	87.7	0.0185	0.471	749	2.0	749	26.6	13.4	13400	3700	0.16
16	B-50-P	0.568	27.5	0.089	0.100	2.4	0.593	88.5	0.0157	0.593	944	1.3	944	21.4	10.6	10600	5900	0.14
13	N.L.P.	—	10.4	0.0007	0.100	2.4	0.086	55.1	0.0442	0.071	137	33.9	114	76.3	89.1	89100	—	0.43
16	B-22	0.568	43.1	0.040	0.061	1.5	0.315	85.0	0.0273	0.314	809	4.8	806	67.1	20.0	20000	11300	0.24

1000 hertz.

* The letters H and B indicate loading-coil spacings of 6000 and 3000 feet, respectively, and the figures show the inductance in millihenries of the loading coils used. NL indicates no loading and P and S are for phantom and side circuits, respectively.

TABLE 9—APPROXIMATE CHARACTERISTICS OF STANDARD TYPES OF PAPER-INSULATED EXCHANGE TELEPHONE CABLE CIRCUITS.

Wire Gauge (AWG)	Code No.	Type of Loading	Loop Mile Constants C (μF)	Loop Mile Constants G (μmho)	Prop. Const. Polar Mag	Prop. Const. Polar Angle (deg)	Prop. Const. Rect. α	Prop. Const. Rect. β	Mid-Section Char. Imp. Polar Mag	Mid-Section Char. Imp. Polar Angle (deg)	Mid-Section Char. Imp. Rect. Z_{01}	Mid-Section Char. Imp. Rect. Z_{02}	Wave-length (miles)	Velocity (miles per second)	Cut-off Freq (hertz)	Attenuation (dB per mile)
26	BST	NL	0.083	1.6	—	—	—	—	910	—	—	—	—	—	—	2.9
	ST	NL	0.069	1.6	0.439	45.30	0.307	0.310	1007	44.5	719	706	20.4	20 400	—	2.67
24	DSM	NL	0.085	1.9	—	—	—	—	725	—	—	—	—	—	—	2.3
	ASM	NL	0.075	1.9	0.355	45.53	0.247	0.251	778	44.2	558	543	25.0	25 000	—	2.15
		M88	0.075	1.9	0.448	70.25	0.151	0.421	987	23.7	904	396	14.9	14 900	3100	1.31
		H88	0.075	1.9	0.512	75.28	0.130	0.495	1160	14.6	1122	292	12.7	12 700	3700	1.13
		B88	0.075	1.9	0.684	81.70	0.099	0.677	1532	8.1	1515	215	9.3	9 270	5300	0.86
22	CSA	NL	0.083	2.1	0.297	45.92	0.207	0.213	576	43.8	416	399	29.4	29 400	—	1.80
		M88	0.083	2.1	0.447	76.27	0.106	0.434	905	13.7	880	214	14.5	14 500	2900	0.92
		H88	0.083	2.1	0.526	80.11	0.0904	0.519	1051	9.7	1040	177	12.1	12 100	3500	0.79
		H135	0.083	2.1	0.644	83.50	0.0729	0.640	1306	6.3	1300	144	9.8	9 800	2800	0.63
		B88	0.083	2.1	0.718	84.50	0.0689	0.718	1420	5.3	1410	130	8.75	8 750	5000	0.60
		B135	0.083	2.1	0.890	86.50	0.0549	0.890	1765	3.3	1770	102	7.05	7 050	4000	0.48
19	CNB	NL	0.085	1.6	0.188	47.00	0.128	0.138	400	—	—	—	—	—	—	1.23
	DNB	NL	0.066	1.6	0.383	82.42	0.0505	0.380	453	42.8	333	308	45.7	45 700	—	1.12
		M88	0.066	1.6	0.459	84.60	0.0432	0.459	950	8.9	939	146	16.6	16 600	3200	0.44
		H88	0.066	1.6	0.569	86.53	0.0345	0.570	1137	5.2	1130	103	13.7	13 700	3900	0.38
		H135	0.066	1.6	0.651	87.23	0.0315	0.651	1413	4.0	1410	99	11.0	11 000	3200	0.30
		H175	0.066	1.6					1643	3.3	1640	95	9.7	9 700	2800	0.27
		B88	0.066	1.6	0.641	86.94	0.0342	0.641	1565	2.8	1560	77	9.8	9 800	5500	0.30
16	NH	NL	0.064	1.5	0.133	49.10	0.0868	0.1004	320	40.6	243	208	62.6	62 600	—	0.76
		M88	0.064	1.5	0.377	85.88	0.0271	0.377	937	4.6	934	76	16.7	16 700	3200	0.24
		H88	0.064	1.5	0.458	87.14	0.0238	0.458	1130	2.8	1130	55	13.7	13 700	3900	0.21

1000 hertz.

In the third column of the above table the letters M, H, and B indicate loading-coil spacings of 9000 feet, 6000 feet, and 3000 feet, respectively, and the figures show the inductance in millihenries of the loading coils used. NL indicates no loading.

TABLE 10—REPRESENTATIVE VALUES OF LINE AND PROPAGATION CONSTANTS OF MISCELLANEOUS CABLES.

Freq (kHz)	Resistance (ohms/mile)	Inductance (mH/mile)	Conductance (μmhos/mile)	Capacitance (μF/mile)	Characteristic Impedance (ohms)	Phase Shift (radians/ mile)	Attenuation (dB/mile)
\multicolumn{8}{c}{16-Gauge Spiral-Four (Disc-Insulated) Toll-Entrance Cable}
0.1	42.4	2.00	0.042	0.02491	—	0.024	0.18
0.5	42.9	1.98	0.053	0.02491	540–j460	0.045	0.32
1.0	43.4	1.94	0.074	0.02491	428–j324	0.067	0.44
1.5	43.9	1.89	0.102	0.02491	380–j275	0.085	0.49
2.0	44.4	1.82	0.127	0.02491	350–j230	0.101	0.55
3.0	45.5	1.74	0.186	0.02490	307–j157	0.145	0.64
5.0	47.5	1.64	0.320	0.02490	279–j107	0.218	0.74
10	50.8	1.56	0.72	0.02489	258–j63	0.405	0.85
20	56.9	1.53	1.95	0.02488	226–j36	0.78	0.99
30	63.0	1.52	3.54	0.02488	248–j26	1.15	1.10
50	73.0	1.51	7.1	0.02488	245–j19	1.90	1.31
100	94.8	1.46	16.9	0.02488	243–j13	3.80	1.71
150	113.5	1.44	27.1	0.02488	240–j10	5.65	2.08
200	130.0	1.43	38.0	0.02487	—	—	2.35

22 AWG Emergency Cable

	Resistance (ohms/mile)	Inductance (mH/mile)	Conductance (μmhos/mile)	Capacitance (μF/mile)	Characteristic Impedance (ohms)	Phase Shift (radians/ mile)	Attenuation (dB/mile)
Side:							
0	166	1.00	—	—	—	—	—
1	—	—	1.3	0.063	468–j449	—	1.53
Phant:							
0	83	0.69	—	—	—	—	—
1	—	—	2.1	0.100	265–j250	—	1.37

19 AWG CL Emergency Cable

	Resistance (ohms/mile)	Inductance (mH/mile)	Conductance (μmhos/mile)	Capacitance (μF/mile)	Characteristic Impedance (ohms)	Phase Shift (radians/ mile)	Attenuation (dB/mile)
Side:							
dry 0	92	1.39	negligible	—	—	—	—
wet 0	92	1.39	negligible	—	—	—	—
dry 1	—	—	negligible	0.110	272–j244	—	1.48
wet 1	—	—	negligible	0.14	239–j214	—	1.69
Phant:							
dry 0	46	0.5	negligible	—	—	—	—
wet 0	46	0.5	negligible	—	—	—	—
dry 1	—	—	negligible	0.25	124–j116	—	1.58
wet 1	—	—	negligible	0.28	117–j109	—	1.69

TABLE 10—(*Continued*)

Freq (kHz)	Resistance (ohms/mile)	Inductance (mH/mile)	Conductance (μmhos/mile)	Capacitance (μF/mile)	Characteristic Impedance (ohms)	Phase Shift (radians/ mile)	Attenuation (dB/mile)
\multicolumn{8}{c}{Coaxial Cable 0.27-Inch Diam (New York–Philadelphia 1936 Type)}							
\multicolumn{8}{l}{Temperature 68° Fahrenheit}							
50	24	0.48	23	0.0773	78.5	—	1.3
100	32	0.47	46	0.0773	78	—	1.9
300	56	0.445	156	0.0772	76	—	3.2
1000	100±	0.43	570	0.0771	74.5	—	6.1
\multicolumn{8}{c}{Coaxial Cable 0.27-Inch Diam (Stevens Point–Minneapolis Type)}							
\multicolumn{8}{l}{Temperature 68° Fahrenheit}							
10	—	—	—	—	—	—	0.75
20	—	—	—	—	—	—	0.92
30	—	—	—	—	—	—	1.10
50	—	—	—	—	$79{-}j6$	—	1.38
100	—	—	—	—	$77.8{-}j4$	—	1.70
300	—	—	—	—	$76.1{-}j2$	—	3.00
1000	—	—	—	—	$75{-}j1.3$	—	5.6
3000	—	—	—	—	$74.5{-}j1.1$	—	10
10000	—	—	—	—	—	—	18
\multicolumn{8}{c}{Coaxial Cable 0.375-Inch Diam (Polyethylene Discs)}							
10	—	—	—	—	—	—	0.53
20	—	—	—	—	—	—	0.65
30	—	—	—	—	—	—	0.72
50	—	—	—	—	50±	—	0.90
100	—	—	—	—	—	—	1.18
300	—	—	—	—	—	—	2.1
1000	—	—	—	—	—	—	4.0
3000	—	—	—	—	—	—	7
10000	—	—	—	—	—	—	13

All figures for loop-mile basis.
Nonloaded.
Temperature 55° Fahrenheit.

Toll Connecting Trunks

Present transmission design objectives require a minimum loss of 2 decibels in each toll connecting trunk to help mask the generally poor impedances of the exchange plant. When necessary, impedance-correcting networks are added. Maximum loss of toll connecting trunks is set at 4 decibels. These transmission objectives may be obtained in the following ways.

(**A**) If the normal loss of a trunk is less than 2 decibels, a 2-decibel pad is inserted in the trunk at the toll office.

Series Type	Shunt Type

Generation of negative Z and Y

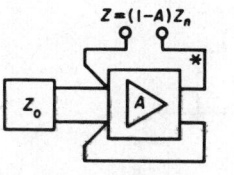 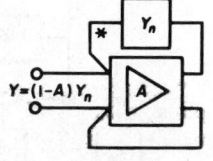

Typical $-Z$ generator	Typical $-Y$ generator
*Positive feedback	*Positive feedback

Insertion gain between line A and line B

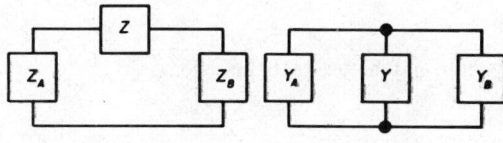

Gain $= 20 \log_{10}$	Gain $= -20 \log_{10}$
$\times \lvert 1 + [Z/(Z_A + Z_B)] \rvert$ dB	$\times \lvert 1 + [Y/(Y_A + Y_B)] \rvert$ dB

Stability conditions

$Z_A + Z_B + Z > 0$	$Y_A + Y_B + Y > 0$

Typical network configurations for telephone lines

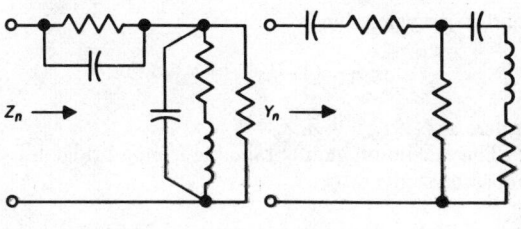

Z network for loaded cable	Y network for nonloaded cable and open wire

---

Maximum practical gain for a $-Z$ or $-Y$ repeater

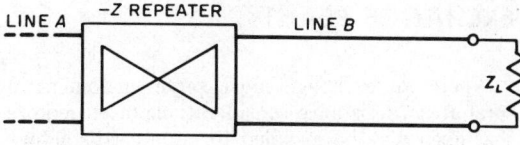

Characteristic impedance $= Z_0$
Propagation constant $= \gamma = \alpha + j\beta$ per unit length l

---

For a series ($-Z$-type) repeater:

Maximum gain

$$= -20 \log_{10} \left| 1 - M \left(\frac{N_A Z_{0,A} + N_B Z_{0,B}}{Z_{0,A} + Z_{0,B}} \right) \right| \text{ dB}$$

where

$$N = \frac{1 - \lvert \Gamma \rvert}{1 + \lvert \Gamma \rvert}$$

= minimum normalized impedance seen by repeater

$$\Gamma = \left(\frac{Z_L - Z_0}{Z_L + Z_0} \right) \exp(-2\gamma l)$$

= load reflection coefficient plus twice line loss

$M =$ stability factor, usually 0.9 (stability margin $= 1 - M$).

For a shunt ($-Y$-type) repeater:

Substitute $Y_{0,A}$ for $Z_{0,A}$, and $Y_{0,B}$ for $Z_{0,B}$

Fig. 5—Negative-impedance telephone repeaters.

(**B**) If the normal loss of a trunk is between 2 and 4 decibels, no pad is inserted but impedance-correcting networks may be added to improve the return loss.

(**C**) If the trunk loss exceeds 4 decibels, repeaters are added and the trunk insertion loss adjusted to be equal to the quantity $(VNL+2)$ decibels, where VNL represents the via net loss computed as though the toll connecting trunk were an intertoll trunk. Impedance correction may also be necessary.

USE OF OPEN WIRE IN THE EXCHANGE PLANT

Where an exchange must serve customers in rural areas, resistance considerations often require that open wire be installed to permit satisfactory supervisory signaling. If the number of subscribers to be served warrants their use, subscriber carrier systems are frequently employed to increase the capacity of the open wire.

REPEATERS IN THE EXCHANGE PLANT

If trunk losses exceed the allowable maximum value, it is necessary to add gain to the circuit. Amplifiers designed for this purpose are termed voice-frequency repeaters. While the advent of carrier transmission systems has made long voice-frequency circuits obsolete, the improved requirements for the interoffice and toll-connecting-trunk losses have resulted in the use of voice-frequency repeatered trunks in the exchange plant. Two types of repeaters are typically employed, negative-impedance and 4-wire repeaters.

Negative-Impedance Repeaters—2-Wire

A negative-impedance telephone repeater is a voice-frequency repeater that provides effective gain by inserting a negative impedance into the line to cancel out the line impedances that cause transmission losses.

It is possible to generate two distinct types of negative impedances. The series type is stable when it is terminated in an open circuit and oscillates when connected to a low impedance. The shunt type is stable when short-circuited but will oscillate when terminated in a high impedance.

The shunt type may be regarded as a negative admittance.

Because they represent lumped impedance discontinuities, series or shunt negative-impedance repeaters cause reflections at the point of insertion. These reflections produce echoes and limit the gain obtainable. To overcome these objections, series and shunt repeaters are used in combination.

Figure 5 illustrates the characteristics of the two types of repeater, and uniform lines are assumed. For nonuniform lines, reflections at all junctions must be computed and referred to the repeater location. In switched telephone trunks, Z_L is generally taken as zero or infinity.

Between lines having reasonably similar impedances, the bridged-T-configuration combination

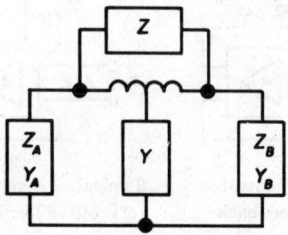

Fig. 6—Series-shunt repeater.

repeater may be used. Its insertion gain is

$$G_T = 20 \log_{10} \left| \frac{1 - ZY/4}{1 + \dfrac{ZY}{4} + \dfrac{Z}{Z_A + Z_B} + \dfrac{Y}{Y_A + Y_B}} \right| \text{ dB.}$$

The characteristic impedance of the series-shunt repeater (Fig. 6) is

$$Z_0 = (Z/Y)^{1/2}$$

and its transmission is

$$\exp\gamma = (1 - \tfrac{1}{2}x)/(1 + \tfrac{1}{2}x)$$

where $x = (ZY)^{1/2} = Z/Z_0 = Y/Y_0$.

The maximum gain obtainable from a bridged-T repeater is given by

$$20 \log_{10}(\exp\gamma) < (RL_A/2) + (RL_B/2)$$

where RL_A and RL_B are the minimum return losses of the two lines relative to the characteristic impedance of the repeater. For best results, the characteristic impedance of the repeater should be

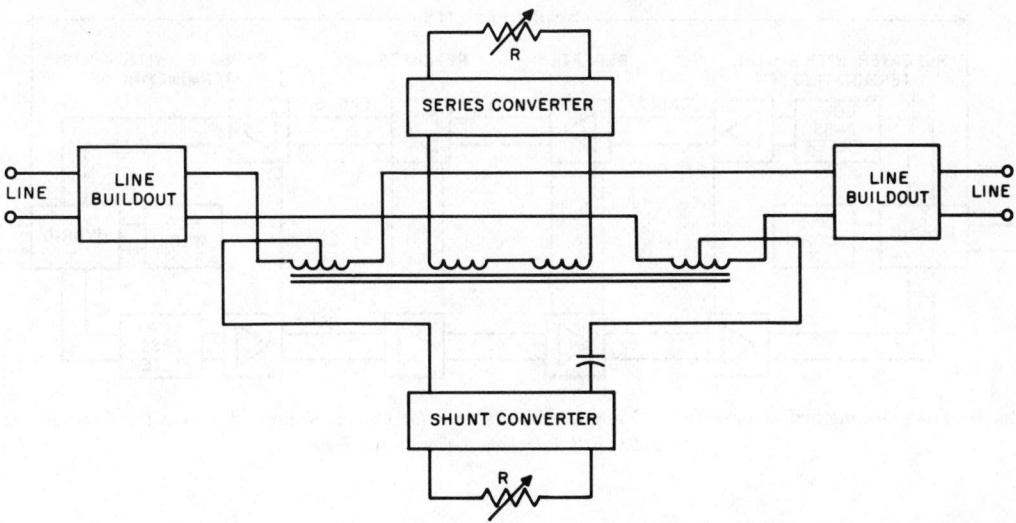

Fig. 7—Block schematic of E6 repeater. *"Transmission Systems for Communications,"* revised third edition, p. 71, © 1964, Bell Telephone Laboratories, Inc.

matched to that of the line having the higher return loss.

In practice, the above gain must be reduced somewhat to allow a margin of stability.

In cases where the combination repeater is inserted between lines whose impedances differ by 3:1 or more, an *"L"* configuration (with the Z-type toward the higher impedance) may prove advantageous because of its impedance-matching properties.

One design (Fig. 7) avoids the difficulty of adjusting the negative Z or Y network by building out the line impedances to appear as fixed impedances of 900 ohms in series with capacitance of 2 microfarads. The converter or gain unit is designed to operate with this impedance. This arrangement permits the repeater to be used between cable circuits having different impedances. Since good return losses between the cable and converter can be obtained, echoes are negligible with this repeater configuration. Cable characteristic changes due to variations in ambient temperature and variations in end terminations during signaling limit the maximum usable gain to about 12 decibels.

Four-Wire Voice-Frequency Repeaters

If voice-frequency trunks must include more gain than can be obtained from 2-wire repeaters, 4-wire operation is employed (Fig. 8). Repeaters for 4-wire circuits are conventional 1-way ampli-

fiers. The facility is converted from 2-wire to 4-wire at the terminations by the use of 4-wire terminating sets consisting of resistance or transformer hybrids with appropriate balancing networks for the connecting office facilities. Low-pass filters may be included in the terminating sets to limit the possibility of the circuit singing outside the voice band. Four-wire voice-frequency trunks, because of the length of the trunk, are generally operated at via net loss (VNL) as discussed below.

MESSAGE-CHANNEL OBJECTIVES— THE TOLL TRANSMISSION PLANT

Delay: Delay, by itself, is seldom annoying in speech communication until the delay has reached a value of approximately 600 milliseconds. Delays encountered in the modern toll plant seldom reach this value but delay of this magnitude can be expected in circuits operating via synchronous-orbit satellites.

Echoes (Fig. 1) without delay appear as sidetone and are not detrimental unless quite excessive. Delayed echoes, on the other hand, are particularly annoying. Table 11 gives the round-trip echo loss tolerated by the average subscriber as a function of delay.

Terminal Net Loss: Terminal net loss is defined as the total of all losses in a toll connection, including the via net losses of the links and the pads in the toll connecting trunks.

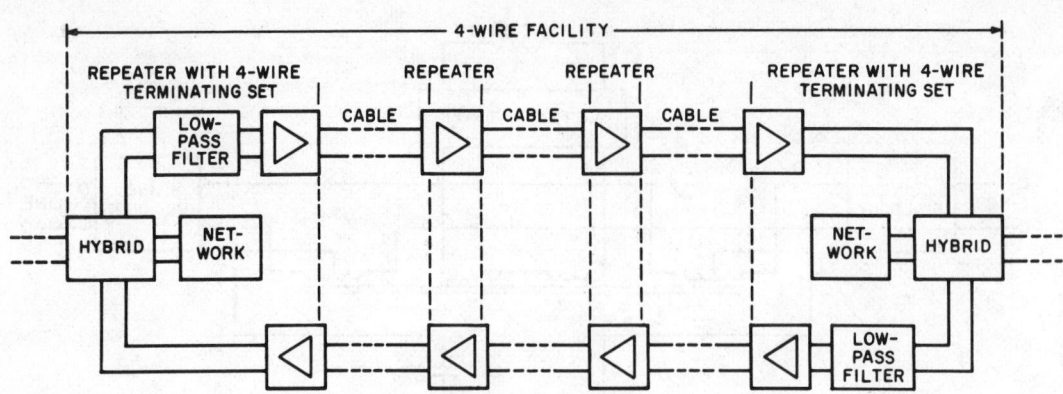

Fig. 8—Four-wire method of operation. *"Transmission Systems for Communications,"* revised third edition, p. 74, © 1964, Bell Telephone Laboratories, Inc.

Via Net Loss: The via net loss of a trunk is the insertion loss at which the trunk is operated for purposes of echo control. The via net loss of a trunk is given by

$$VNL = \frac{0.2L}{V} + 0.4 \text{ dB}$$

where V is the velocity of propagation and L is the length of the circuit. The quantity $0.2/V$ is defined as the via-net-loss factor (VNLF). Table 12 lists VNLF's for some of the more-common facility types. The required via net loss for any trunk may be computed by

VNL = VNLF

×Length of Circuit in Miles + 0.4 dB.

The via net losses thus determined should result in an echo design acceptable to the user for 99 percent of the toll calls.

Implicit in the via-net-loss design of trunk transmission is the return loss at the 4-wire terminating sets, particularly in the band of frequencies from 500 to 2500 hertz. The present objective for echo return loss is a mean value of 11 decibels with a standard deviation of 3 decibels. In some non-commercial systems, it may not be economical to attempt to meet this requirement.

Net-Loss Variations: The net loss of any link will vary randomly from its design value as a result of equipment and transmission-medium variations. Present design objectives call for the distribution of deviations from the computed value of net loss to have a mean of 0 decibels and a standard deviation of 1.41 decibels. This variation is taken into consideration in computing the VNL.

The present maintenance objective for net loss

Table 11—Subscriber Reaction to Echo Delay.

"Transmission Systems for Communications," revised third edition, p. 34, © 1964 Bell Telephone Laboratories, Inc.

Round-Trip Delay (milliseconds)	Mean Required Round-Trip Loss L_0 (decibels)
0	1.4
20	11.1
40	17.7
60	22.7
80	27.2
100	30.9

Table 12—Via-Net-Loss Factors.

"Transmission Systems for Communications," revised third edition, p. 76, © 1964 Bell Telephone Laboratories, Inc.

Facility	VNLF (decibels per mile)
Two-wire open wire (all wire sizes)	0.01
Two-wire 19H88-50	0.03
Four-wire 19H44-25	0.01
Carrier systems (all types)	0.0015

TABLE 13—CIRCUIT NET LOSS IN DECIBELS.

Circuit Length		CCITT Rec. G.131 (decibels)	VNL Design (decibels)
(kilometers)	(statute miles)		
0	0	0.5	0.40
250	155.3	0.5	0.63
500	310.7	0.5	0.87
750	466.0	1.0	1.10
1000	621.4	1.0	1.33
1250	776.7	1.5	1.56
1500	932.1	1.5	1.80
1750	1087.4	2.0	2.03
2000	1242.8	2.0	2.26
2250	1398.1	2.5	2.50
2500	1553.5	2.5	2.73

as given by the Bell System* is that the distribution of deviations of trunk losses from their prescribed values shall have a bias of not more than ±0.25 decibel and a distribution grade of not more than 1.0 decibel. Bias and distribution grade are computed using the following equations.

(A) Computation of bias:

$$\text{Bias} = B = n^{-1} \sum_1^n \Delta_k$$

where n is the total number of trunks terminating at a switching center, and Δ_k is the difference between the measured and design loss of the kth trunk. The sign of Δ_k is positive if the measured loss is greater than the design loss and negative if it is less.

(B) Computation of distribution grade:

$$\text{Distribution Grade} = D = \left[n^{-1} \sum_1^n (\Delta_k)^2 - B^2 \right]^{1/2}$$

where the terms are as defined in (A) above. Note that when the bias is zero, D is equal to the standard deviation.

NET-LOSS OBJECTIVES—CCITT RECOMMENDATIONS

The CCITT currently recommends† that international toll circuits in the new transmission plan be given an insertion loss of 0.5 decibel for each

* "Notes on Distance Dialing—1961", American Telephone and Telegraph Company, New York, N. Y.

† CCITT Recommendation G.131, Green Book, Vol. III-1, Fifth Plenary Assembly, Geneva; December 1972.

500 kilometers length or fraction thereof, assuming that the international circuit is derived from carrier systems in coaxial cable or radio relay.

Table 13 compares the losses of long-haul switching trunks designed in accordance with CCITT Recommendation G.131 and with VNL. For the VNL design, a VNLF of 0.0015 decibel per statute mile is used. The 0.4-decibel term is included.

Subscriber Echo Tolerance: Table 11 gives the tolerable echo loss for the average subscriber. The loss presently required to satisfy individual subscribers is normally distributed with a standard deviation of approximately 2.5 decibels.

Echo Suppressors: If the trunk requires a VNL of more than 2.5 decibels or if the delay exceeds 45 milliseconds, American practice is to insert echo suppressors and then set the net loss of the trunk to 0.5 decibel. Echo suppressors are voice-operated devices which, when one party is talking, insert a high loss in the opposite direction of transmission. Echo suppressors are generally applied only on final routes between regional centers in direct distance dialing or on high-usage groups extending between switching centers in widely separated regions.

Echo suppressors must be used carefully as they can cause clipping of the start and finish phonemes of words. If two or more circuits containing echo suppressors are switched in tandem, a phenomenon known as *lockout* may occur if both parties attempt to talk simultaneously. In this case neither party will be heard by the other.

The CCITT recommends* that the overall loss of a connection may be adjusted so that echo currents are sufficiently attenuated or, alternatively, an echo suppressor may be fitted if the loss adjustment results in an excessive insertion loss.

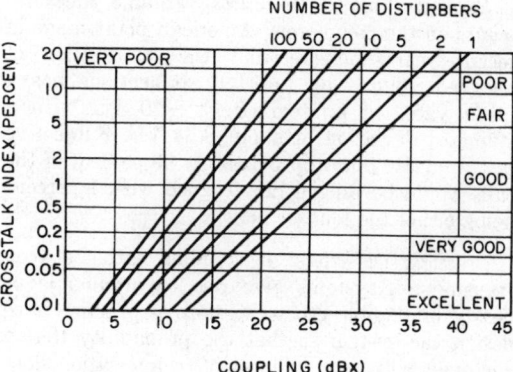

Fig. 9—Crosstalk judgment curves. *"Transmission Systems for Communications,"* revised third edition, p. 48, © 1964 Bell Telephone Laboratories, Inc.

* CCITT Recommendation G.131, Green Book, Vol. III-1, Fifth Plenary Assembly, Geneva; December 1972.

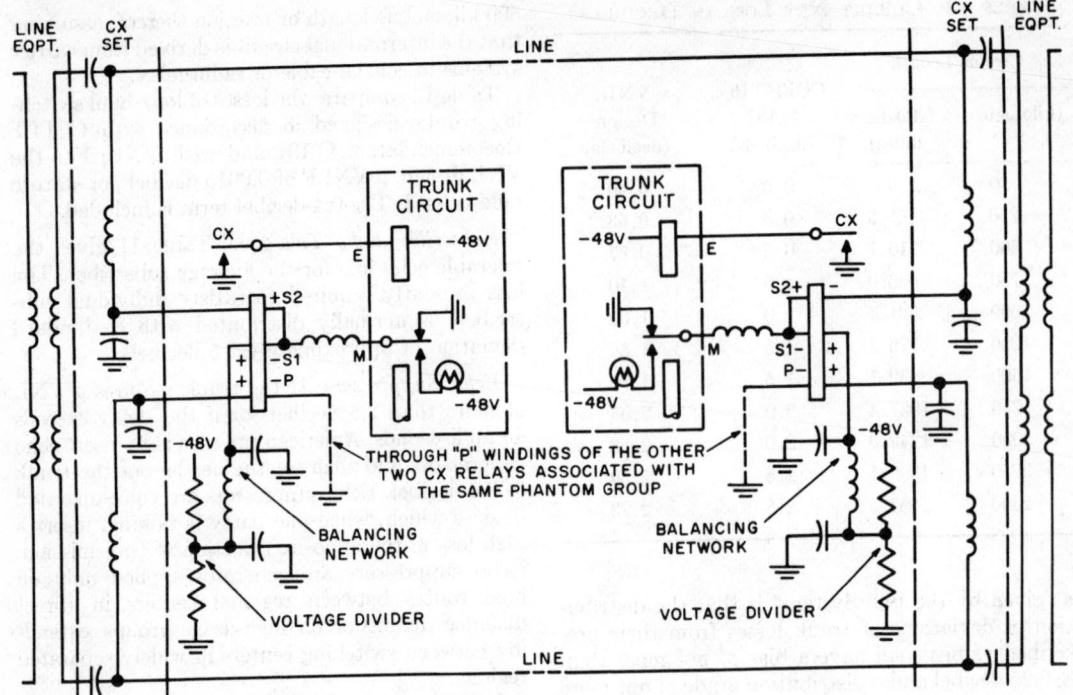

Fig. 10—Composite signaling for 1 voice channel. *"Notes on Distance Dialing—1961," p. 16,* © *1961 American Telephone and Telegraph Company.*

Noise Objectives: Noise objectives in present use are:

Long-haul circuits: 38 dBa or 44 dBrn (*C* message weighting)

Short- and medium-haul circuits: 32 dBa or 38 dBrn (*C* message weighting) at the 0 TLP.

Received Volume Objectives: Volume measurements on the commercial American plant made in recent years indicate that the distribution of received volume on intertoll connections has a mean value of approximately −30 VU with a standard deviation of about 8.34 VU. Subjective tests indicate that approximately 99 percent of the calls would be judged fair to good with 1 percent being either too loud or poor.

Crosstalk Objectives: If coupling paths between transmission systems give rise to intelligible or nearly intelligible crosstalk, normal practice is to design the system so that the probability that a customer will hear a "foreign" conversation does not exceed 1 percent. In measuring crosstalk, a commonly used unit is the dBx, which is defined as the difference between 90 decibels of loss and the transmission of the coupling path. Figure 9 illustrates the relationship between the crosstalk coupling in dBx and the crosstalk index, which is defined as the chance of encountering intelligible crosstalk.

OVERALL SYSTEM DESIGN OBJECTIVES—CCITT

A summary of the CCITT recommendations for system objectives and design criteria for circuits used in the international telephone service is given in Chapter 2.

Interoffice Signaling

If repeatered or nonrepeatered voice-frequency circuits are used as interoffice trunks, direct-current signaling is frequently used to provide both supervisory and numeric signals. The common circuits follow.

Composite (CX) Signaling: Composite signaling (Fig. 10) stems from the use of composite sets to superimpose signaling on the line wires along with the speech. CX signaling uses each of the line wires separately to provide a signaling channel in each direction of transmission. The composite sets are effectively combinations of high- and low-pass filters designed to prevent mutual interference between speech and signal circuits. Additional wires must be used if compensation is required for differences in earth potential at the circuit terminals. E and M control leads are used to interconnect the CX set and the trunk relay circuit. Suitable bypass circuits must be provided at repeaters.

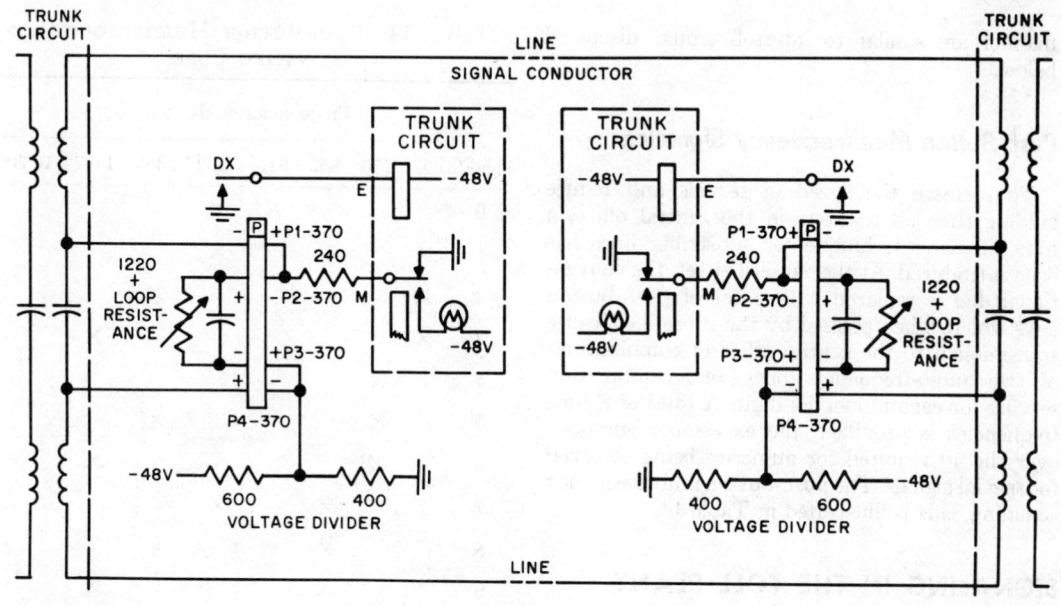

Fig. 11—DX signaling circuit. *"Notes on Distance Dialing—1961," p. 18, © 1961 American Telephone and Telegraph Company.*

DX Signaling: DX signaling (Fig. 11) resembles the differential full-duplex telegraph circuit, using one line wire to transmit the signals and the other for earth potential compensation. The full-duplex operation permits independent 2-way signal transmission. DX signaling works through negative-impedance repeaters without special bypass arrangements, but signaling bypasses are required at conventional-type repeaters. E and M control is used between the DX circuit and the trunk relays.

E and M Lead Control

Most signaling systems other than loop signaling are separated from the trunk relay circuit and generally are introduced between the trunk circuit and the line. The name "E and M" historically stems from conventional designations of the interconnecting leads on circuit drawings. Signaling between the trunk relay circuit and a separate signaling unit is accomplished over two leads, an M lead which transmits signals outgoing from the trunk unit to the line and the E lead which transmits incoming signals from the line to the trunk unit. Figure 6 of Chapter 2 illustrates the E and M lead signals as conventionally used.

SIGNALING IN THE EXCHANGE PLANT

Subscriber Loop Signaling

Subscriber loop signaling generally uses control of direct current in the subscriber loop to provide both supervisory and numeric signals. Supervisory on-hook (open-loop) and off-hook (closed-loop) signals are used to detect when the calling subscriber is demanding service and when a called subscriber answers. Numerics are transmitted as dial pulses obtained by opening and closing the loop at a rate of the order of 10 to 12 pulses per second, the number of pulses in a train representing the dialed digit. Ringing signals for summoning the called subscriber to the telephone are commonly transmitted using a high-voltage low-frequency signal (about 90 to 100 volts at $16\frac{2}{3}$ to 25 hertz) to directly actuate a bell in the subscriber set.

The use of such a high-voltage signal for ringing presents some difficulties in electronic switching exchanges now being introduced in the plant, particularly those employing solid-state devices. One approach has been to transmit a tone at normal speech level, using this tone to actuate an audible device.

Alternating-Current Signaling Techniques in the Exchange Plant

To reduce the holding time on common-control equipment in switching centers, multifrequency pulsing is being used more and more for numeric transmission. This is combined in some cases with direct-current (CX or DX) circuits for transmission of supervisory signals. Where interoffice trunks are provided via carrier multiplex systems, no direct-current path is available and alternating-current techniques must be used to transmit supervisory or line signals. Trunks operated in this manner are similar to intertoll trunks discussed below.

Push-Button Multifrequency Signaling

To increase the speed of service and reduce holding time on registers in the central office, a tone signaling technique for subscriber lines has been introduced. At the subscriber set, the conventional dial is replaced with a set of push-button keys which, when pressed by the subscriber, cause transmission to the central office of combinations of two audio-frequency tones, one combination serving for each numerical digit. A total of 8 tone frequencies is provided, the excess combinations over the 10 required for numerics being reserved for special signals. The push-button multifrequency signaling code is illustrated in Table 14.

SIGNALING IN THE TOLL PLANT

Signaling over long-haul intertoll trunks uses techniques whereby signaling information is transmitted at audio frequencies within the voice channel (in-band signaling) or just above the voice-channel spectrum (out-of-band signaling). Levels are comparable to speech levels. Use of alternating-current techniques is mandatory in frequency-division-multiplex single-sideband carrier systems, as no direct-current path exists through the equipment. Some short-haul systems have been designed that make use of frequency shift of the channel carrier to transmit supervisory and numeric signals, but difficulties in applying this technique to wide-band carrier systems has limited the frequency-shift keying to systems having only a few channels. E and M control of alternating-current signaling is almost universally used in the United States and Canada. Separate signaling systems may be used to transmit supervisory and numeric signals.

In-Band Signaling Systems—American Telephone and Telegraph Company

Present designs for application to toll trunks have the following characteristics:

Frequency:
    For 4-wire trunks: 2600 hertz
    For 2-wire trunks: 2600 hertz
                          2400 hertz
Receiver sensitivity:
    −29 dBm at 0 TLP for 4-wire circuits
    −32 dBm at 0 TLP for 2-wire circuits
(The transmit level is normally adjusted to provide a receive level of −19 dBm at the 0 TLP for 4-wire circuits and −22 dBm at the 0 TLP for 2-wire circuits.)

TABLE 14—PUSH-BUTTON MULTIFREQUENCY SIGNALING CODE.

Signal	Frequencies in Hertz							
	697	770	852	941	1209	1336	1477	1633*
0				×	×			
1	×				×			
2	×					×		
3	×						×	
4		×			×			
5		×				×		
6		×					×	
7			×		×			
8			×			×		
9			×				×	

* 1633 hertz is used in combination with the other 7 frequencies for special-category signals.

Pulsing characteristics:
    8 to 12 pps with break of 46 to 76 percent

Provision is made to minimize mutual interference between signal and speech circuits by the use of a guard circuit which inhibits operation of the receiver when frequencies other than signal tone are present. Additional arrangements insert a tone blocking filter in the receiving speech path when calls are made to lines which do not return answer supervision.

Multifrequency Trunk Signaling

To increase the speed of setting up interoffice connections, multifrequency signaling is often applied to trunk circuits for transmission of switching information. Digital information is transmitted by combinations of 2 of the following 5 audio frequencies: 700, 900, 1100, 1300, and 1500 hertz. A sixth frequency of 1700 hertz is used in combination with the 1100-hertz frequency as a "priming" signal and in combination with the 1500-hertz frequency as a "start" signal. Table 15 gives the standard multifrequency signaling code. Each tone is customarily transmitted at a level of −6 dBm0.

In-Band Signaling Systems—CCITT

Two systems are found in the CCITT Recommendations (refer to Chapter 2).

TABLE 15—MULTIFREQUENCY TRUNK SIGNALING CODE.

Signal	Frequencies in Hertz					
	700	900	1100	1300	1500	1700
Priming			×			×
Start					×	×
0				×	×	
1	×	×				
2	×		×			
3		×	×			
4	×			×		
5		×		×		
6			×	×		
7	×				×	
8		×			×	
9			×		×	

1 VF System: This system uses an in-band tone of 2280 hertz with a level of −6 dBm at the 0 TLP. Supervisory signals use various pulse lengths for coding, while numerics are transmitted by a 6-digit start-stop code consisting of 1 start bit, 4 information bits, and 1 stop bit. The 4 information bits are transmitted high-order bit first and form a set of 16 numeric signals, the first 10 of which correspond to the numerics 1 through 0 (10). Of the remainder, codes 11 and 12 are used for operator call-in, codes 13, 14, and 16 are spares, and code 15 is used to signal "end of pulsing."

2 VF System: The 2 VF system uses 2 frequencies, 2040 and 2400 hertz, either singly or in combination to provide the required supervisory and numeric signals. Levels transmitted for each frequency are −9 dBm at the 0 TLP. Supervisory signals are transmitted using combination tone bursts for coding. Numerics are transmitted using a 4-digit binary code. The allocation of numeric codes is similar to the 1 VF system.

Out-of-Band Signaling—American Telephone and Telegraph Company

Type N, O, and ON carrier systems have built-in options for using an out-of-band signaling channel at 3700 hertz. Since the signaling path is completely outside the voice band, it is unnecessary to provide protection against false operation on speech signals or to insert and remove filters in the speech path.

Out-of-Band Systems—CCITT

Recommendation Q.21 gives 4 recommended out-of-band systems, 3 of which are recommended for use with carrier systems having 12 channels per group, and the other for those systems having 8 channels per group.

Systems for Use with 12-Channel Groups:

Type I—Compatible with only those group and supergroup reference pilots having a displacement from the virtual carrier of 140 hertz.

Frequency   Virtual carrier (zero frequency)
Level   High (−3 dBm0 approximately)
Signals   Discontinuous

Type II—Compatible with only those group and supergroup reference pilots having a displacement from the virtual carrier of 80 hertz.

IIa—For use with discontinuous signals

Frequency   3825 hertz
Level   High (−5 dBm0 approximately)

IIb—For use with semicontinuous signals

Frequency   3825 hertz
Level   Low (−20 dBm0 approximately)

Systems for Use with 8-Channel Groups:

Frequency   4400 hertz

Level:
For discontinuous signals   −6 dBm0
For semicontinuous signals   −17.4 dBm0 to −20 dBm0

MULTIPLEX TRANSMISSION IN CARRIER TELEPHONY

Multiplexing Techniques

Two basic techniques are used for the transmission of a plurality of telephone channels over a single transmission medium.

(A) Frequency-division systems, in which a unique band of frequencies within the wide-band frequency spectrum of the transmission medium is allotted to each communication channel on a continuous time basis.

(B) Time-division systems, in which each communication channel is allotted a discrete time slot within a basic sampling frame, with a theoretical occupancy of the entire wide-band frequency spectrum for the allotted time.

Several types of modulation techniques may be employed with each of the multiplexing techniques.

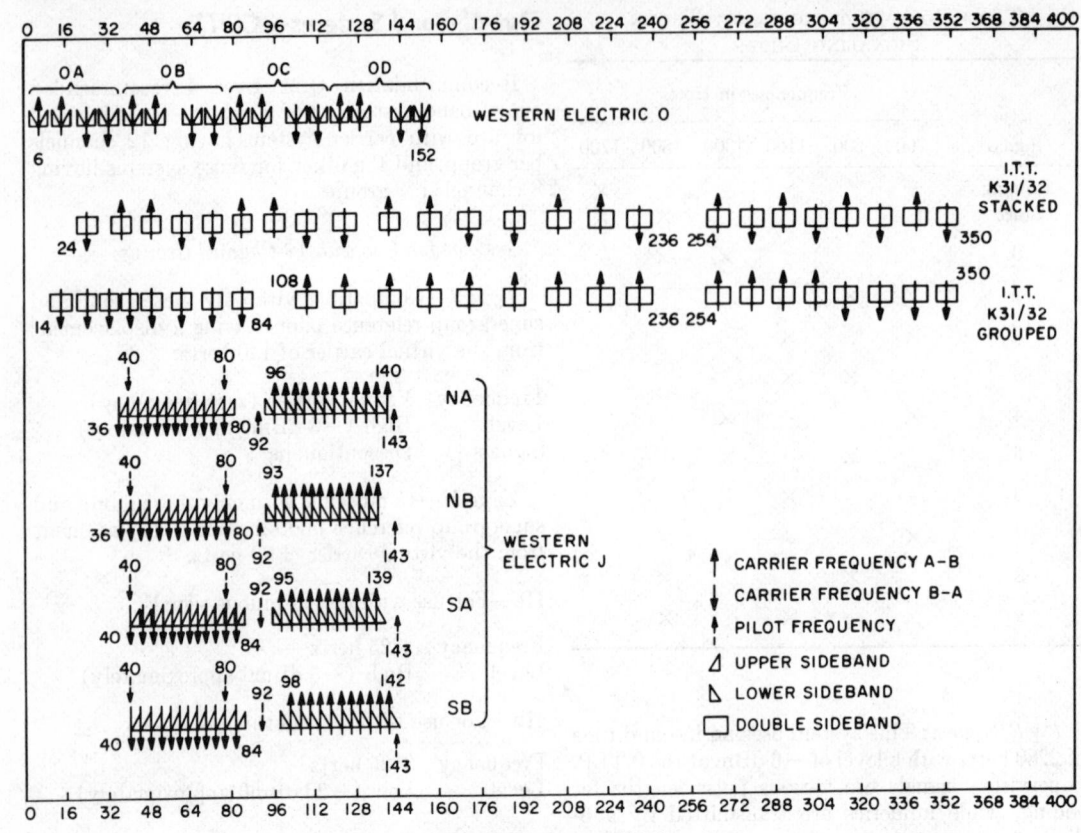

Fig. 12.—Open-wire carrier systems (4-wire). The numbers are in kilohertz.

Modulation Techniques

Modulation techniques suitable for use with time- and frequency-domain systems are listed below. For further details on the various modulation techniques, refer to Chapter 23.

Time-Division Systems:

> Pulse-Amplitude Modulation (PAM)
> Pulse-Duration or Pulse-Width Modulation (PDM or PWM)
> Pulse-Position or Pulse-Time Modulation (PPM or PTM)
> Pulsed-Frequency Modulation (PFM)
> Pulse-Code Modulation (PCM)
> Delta Modulation.

Frequency-Division Systems:

> Single-Sideband Suppressed Carrier (SSSC)
> Single-Sideband Transmitted Carrier (SSTC)
> Double-Sideband Suppressed Carrier (DSSC)
> Double-Sideband Transmitted Carrier (DSTC).

Carrier Telephony in the Communications Plant

The majority of systems in the communications plant use frequency-division multiplex (FDM) with single-sideband suppressed carrier, although some short- to medium-haul systems use other modulation techniques.

Subscriber Carrier

Subscriber carrier systems are used to provide service to customers at considerable distances from the central office exchange. Essentially, the terminating and signaling circuits are arranged to work from a subset on the customer's premises to a subscriber line circuit in the exchange, with the signaling requirements those of the subscriber line.

Trunk Carrier

Trunk carrier systems operate between trunk terminations at switching offices. Trunk systems may be further categorized as short- or medium-haul systems for use between central offices within

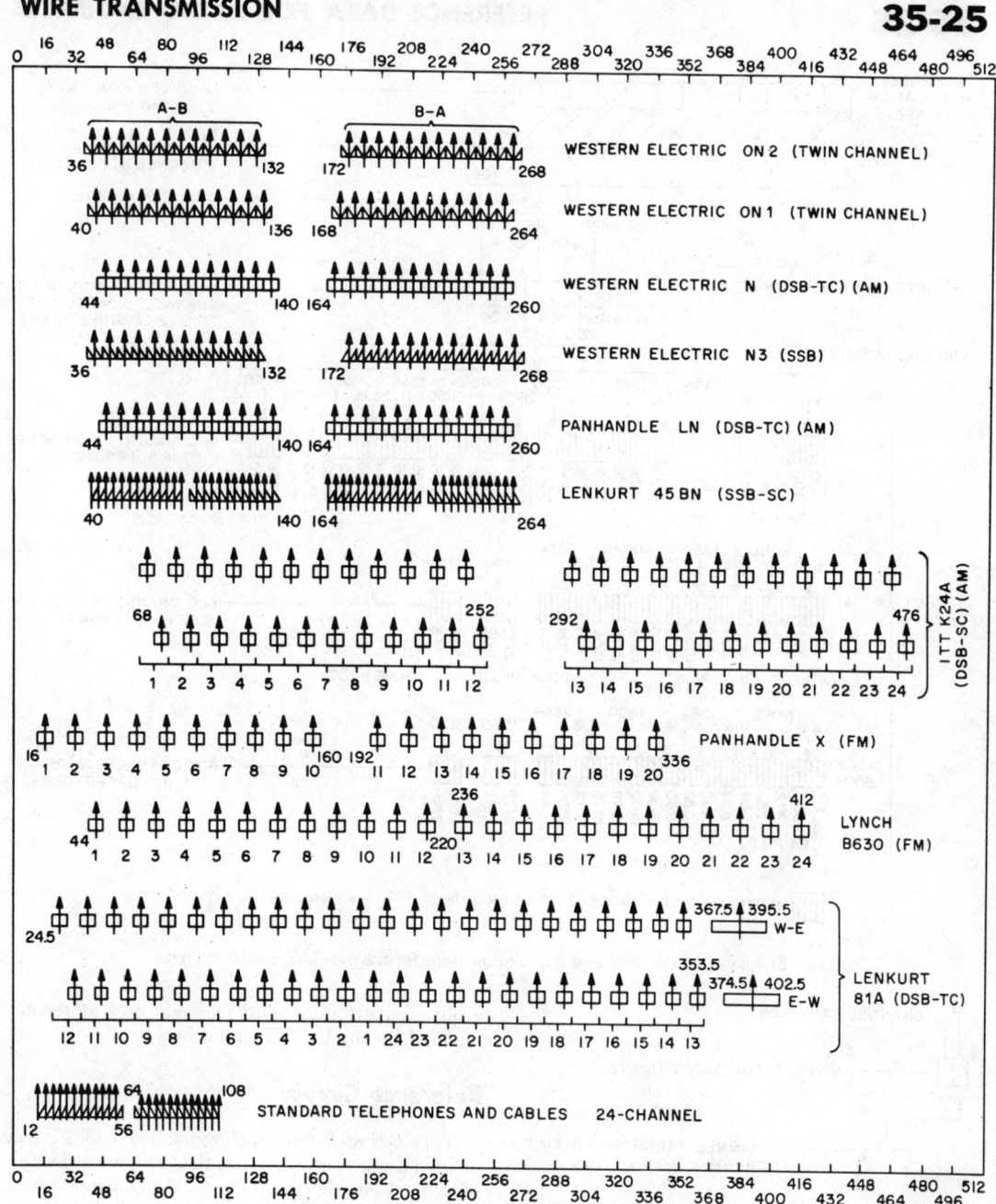

Fig. 13—Cable carrier systems (2-wire).

a local area, and long-haul systems having performance specifications suitable for transcontinental or intercontinental connections.

Modulation Plans and Frequency Allocations

Low- to Medium-Channel-Density Systems: Figures 12 and 13 illustrate the frequency allocations of channels in some of the commonly used low-

and medium-channel-density systems. The majority of these systems would also be classified as short-to-medium-haul systems, although the Western Electric Type J system was originally designed to provide transcontinental circuit performance capability.

High-Channel-Density Systems: Modulation plans and frequency allocations for high-channel-density systems are given in Figs. 14 through 17, which show frequency bands occupied by basic

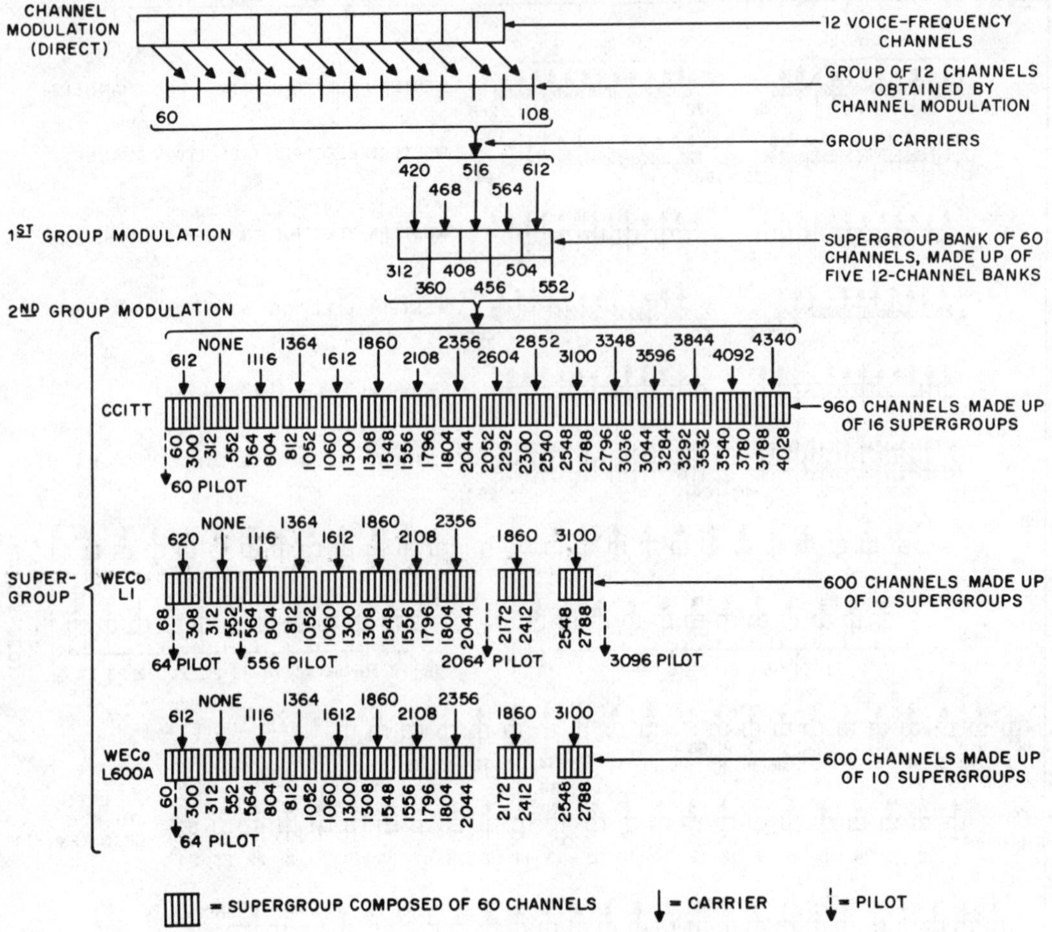

Fig. 14—Frequency allocations and modulation steps for coaxial-cable carrier systems.

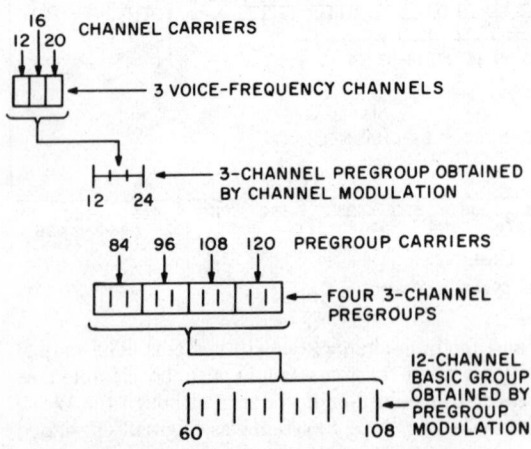

60–108 KILOHERTZ 12-CHANNEL BASIC GROUP
IS END TO END COMPATIBLE WITH
DIRECT CHANNEL MODULATION OF FIG. 14

Fig. 15—Frequency allocations for the translation of 12 audio-frequency channels into a basic group by pregroup modulation (double modulation).

groups, supergroups, and mastergroups as recommended by the CCITT and others.

Reference Circuits

Hypothetical Reference Circuits: The CCITT has defined a number of hypothetical reference circuits of defined length (generally 2500 kilometers) and with a specified number of terminal and intermediate equipments. (Refer to Chapter 2.)

Bell System Reference Circuit: The Bell Telephone System long-haul circuit design for circuits of 1000 to 4000 statute miles in length is based on a 38-dBa noise objective including all sources of noise. This 38-dBa objective for 4000-mile circuits is reduced by a factor of 10 log $4000/L$, where L is the length of the circuit in statute miles. The minimum value, however, is 31 dBa, applying to circuits less than 800 miles in length.

DCS Reference Circuit: The United States Defense Communications System (DCS) has

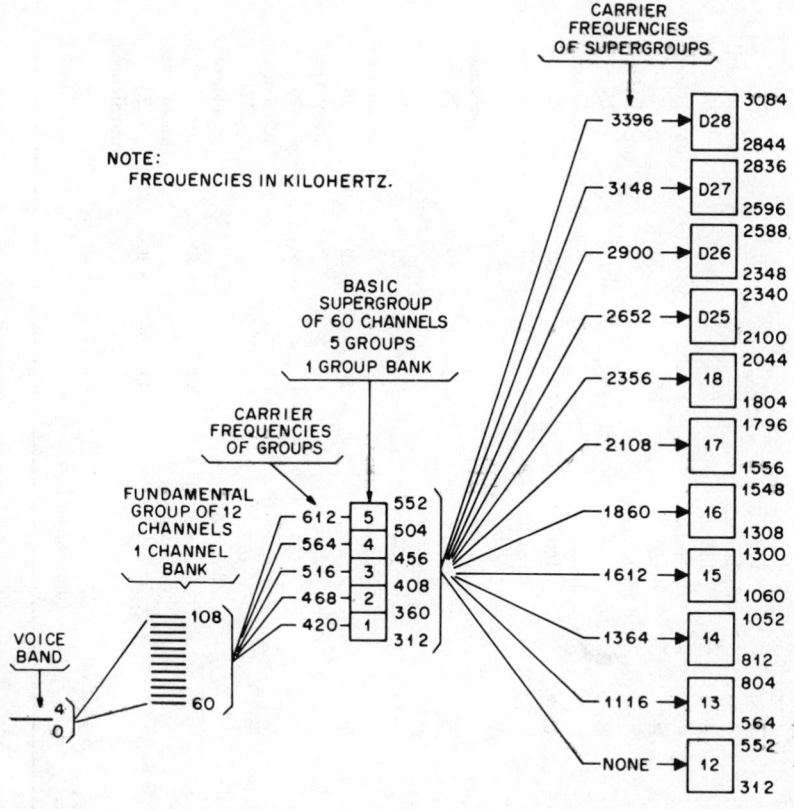

NOTE:
FREQUENCIES IN KILOHERTZ.

Fig. 16—Frequency allocation, basic mastergroup. © *1967 American Telephone and Telegraph Company.*

---

defined a reference circuit of 6000 nautical miles in length having 6 homogeneous links and having a 38-dBa0 (25 000 pWp0) noise objective. Linear proration of noise is used for links exceeding 333 nautical miles.

OVERALL PARAMETERS FOR CARRIER TRANSMISSION SYSTEMS

Noise Objectives—CCITT

The CCITT recommended noise objective should apply to any telephone channel having the same makeup as the hypothetical reference circuit. Noise is measured at the 0 TLP. (Refer to Chapter 2 for specific details.)

Accuracy of Carrier Frequencies

Accuracy of virtual carrier frequencies should be sufficient to keep the difference between the transmitted and received audio frequency to 2 hertz or less, including any intermediate modulating and demodulating processes. Master oscillators of current design are generally capable of maintaining the base frequency to within ± 1 part in 10^7 of its assigned value over a 30-day period.

TRANSLATING-EQUIPMENT INTERFACES

Channel-Translating Equipment

Figure 18 illustrates the limits of channel-translator frequency response recommended by the CCITT, and Fig. 19 illustrates the limits of frequency response used by the Bell System.

Impedance: The most commonly used impedance specifications for the 4-wire voice-frequency input and output impedances are:

Impedance: nominally 600 ohms balanced to ground
Return loss: at least 26 decibels against a 600-ohm resistance over the frequency spectrum from 300 to 3400 hertz.

Levels: Commonly used levels at the 4-wire voice termination are:

Transmitting: -16 dBr.
Receiving: $+7$ dBr.

Group and Supergroup Levels and Impedances

No recommendations are made by CCITT for levels at group and supergroup distribution frames,

Table 16—Relative Power Levels at Group and Supergroup Distribution Frames of Carrier Systems in Different Countries. *G.233 Geneva; 1972.*

Country	Relative Power Level at Group Distribution Frame		Basic Group at Distribution Frame	Impedance at Group Distribution Frame (ohms)	Relative Power Level at Supergroup Distribution Frame		Impedance at Supergroup Distribution Frame (ohms)
	Transmit (dBr)	Receive (dBr)			Transmit (dBr)	Receive (dBr)	
Australia:							
System 1	−36.5	−30.5	B	150 (balanced)	−35	−30.5	75 (unbalanced)
System 2	−42	−5	B	135 (balanced)	−35	−30	75 (unbalanced)
Belgium	−37	−8	B	150 (balanced)	−35	−30	75 (unbalanced)
France	−52 / −33*	−17 / −15*	B	150 (balanced)	−45	−35	75 (unbalanced)
Democratic German Republic	−36	−30	B	150 (balanced)	−35	−30	75 (unbalanced)
Germany (Federal Republic)	−36 / −36	−23* / −30	B	150 (balanced)	−36* / −35	−23* / −30	75 (unbalanced)
India	−36.5	−30.4	B	150 (balanced)	−34.8	−30.4	75 (unbalanced)
Hungary, Italy, Netherlands	−37	−30	B	150 (balanced)	−35	−30	75 (unbalanced)
Japan (Nippon Telegraph and Telephone Public Corporation)	−36	−18	B	75 (balanced)	−29	−29	75 (unbalanced)
Mexico (Teléfonos de México)	−47	−10	B	150 (balanced)	−47	−24	75 (unbalanced)
People's Republic of Poland	−36	−23*	B	150 (balanced)	−36	−23	75 (unbalanced)
Spain, Denmark, Ireland, New Zealand, Norway, United Kingdom	−37	−8	B	75 (unbalanced)	−35	−30	75 (unbalanced)
Sweden					−35	−30	75 (unbalanced)
Switzerland	−41 / −36.5*	−7.8 / −30.5*	A or B	75 (unbalanced)	−35	−26	75 (unbalanced)

TABLE 16—*Continued*

Country	Relative Power Level at Group Distribution Frame		Basic Group at Distribution Frame	Impedance at Group Distribution Frame (ohms)	Relative Power Level at Supergroup Distribution Frame		Impedance at Supergroup Distribution Frame (ohms)
	Transmit (dBr)	Receive (dBr)			Transmit (dBr)	Receive (dBr)	
U.S.S.R.	−36	−23	B	150 (balanced)	−36	−23	75 (unbalanced)
United States of America (American Telephone and Telegraph Co.)	−42	−5	B	135 (balanced)	−25	−28	75 (unbalanced)

Basic group A = 12–60 kilohertz, and basic group B = 60–180 kilohertz.
* Values proposed for new equipments.

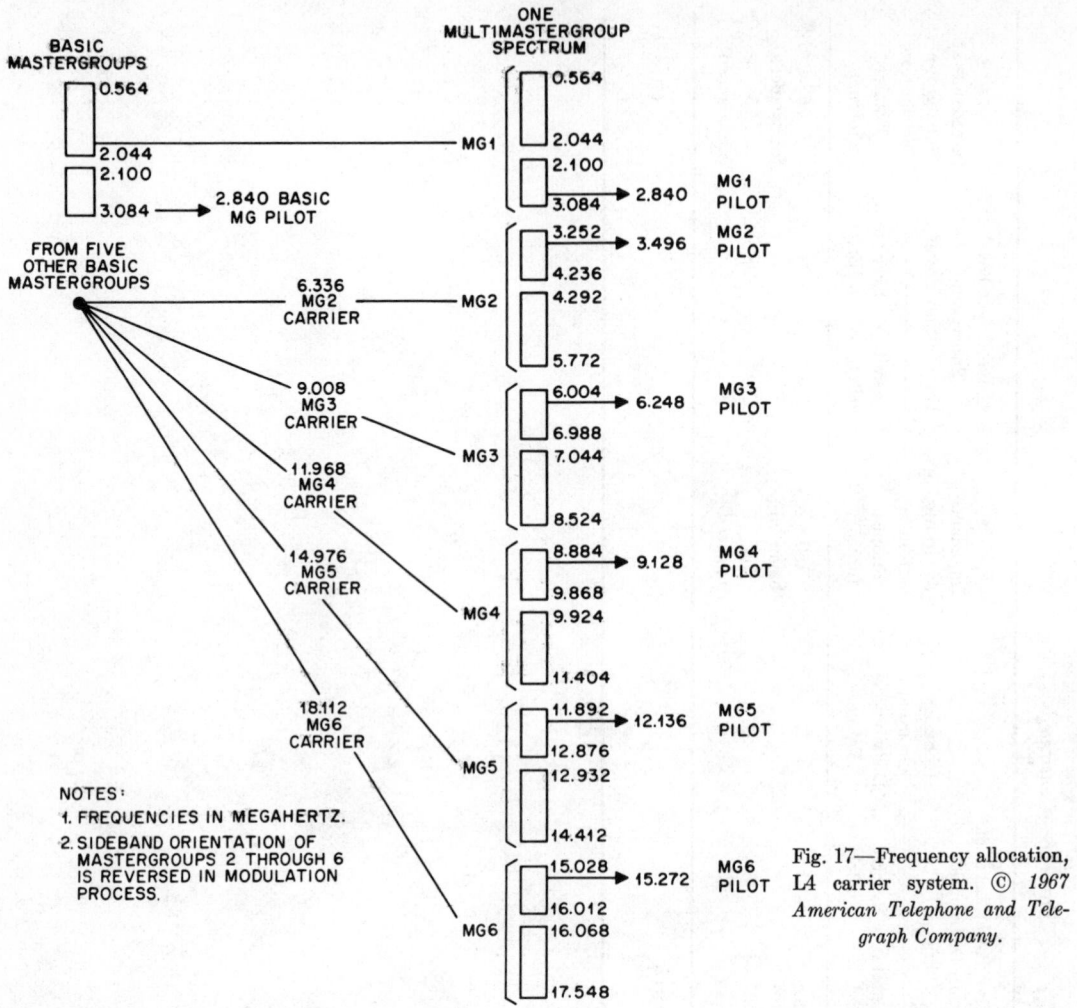

Fig. 17—Frequency allocation, L4 carrier system. © 1967 American Telephone and Telegraph Company.

as various administrations and operating companies have adopted widely different values. Table 16 gives values used in various operating administrations and companies throughout the world.

Return loss of the group or supergroup translating equipment is generally specified as not less than 20 decibels against a resistive impedance of the nominal value over the frequency band of the basic group or supergroup.

Mastergroup Levels and Impedances

CCITT recommended relative levels and impedances at mastergroup distribution frames are:

Transmit: -36 decibels or -4.1 nepers
Receive: -23 decibels or -2.6 nepers

across a 75-ohm impedance unbalanced to ground, with a return loss of not less than 20 decibels against a 75-ohm resistance over the basic mastergroup frequency band.

Supermastergroup Levels and Impedances

CCITT recommended relative levels and impedances at supermastergroup distribution frames are:

Transmit: -33 decibels or -3.8 nepers
Receive: -25 decibels or -2.9 nepers

across a 75-ohm impedance unbalanced to ground, with a return loss of not less than 20 decibels against a 75-ohm resistance over the basic supermastergroup frequency band.

Through Connection of Groups, Supergroups, Mastergroups, and Supermastergroups

If it is desired to carry blocks of channels through a station without resorting to demodulation to audio, through equipment is used consisting basically of filters plus any pads or amplifiers required for level conditioning. The CCITT recommends that the through-connection equipment should provide the following relative suppression of unwanted components (these values apply provisionally to supermastergroups).

Intelligible crosstalk components	70 dB or 8.0 nepers
Unintelligible crosstalk components	70 dB or 8.0 nepers
Possible crosstalk components	35 dB or 4.0 nepers (wherever possible components appear)
Harmful out-of-band components	40 dB or 4.6 nepers
Harmless out-of-band components	17 dB or 2.0 nepers.

The various components are defined in Recommendation G.242.

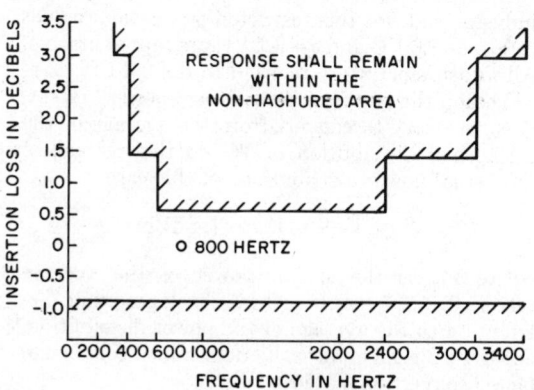

Fig. 18—Limits of the insertion loss for CCITT channel translating equipment.

Pilots on Groups, Supergroups, Mastergroups, and Supermastergroups

CCITT recommended values for pilot frequencies are given in Table 17.

Frequency Accuracy of Pilots

Pilot frequency 84.080 and 411.920 kHz	±1	Hz
Pilot frequency 84.140 and 411.860 kHz	±3	Hz
Pilot frequency 1552 kHz	±2	Hz
Pilot frequency 11 096 kHz	±10	Hz

The above pilots are generally used for regulating or continuity pilots. Other system pilots are *line*-regulating and continuity pilots and frequency-comparison pilots. These generally are specific to

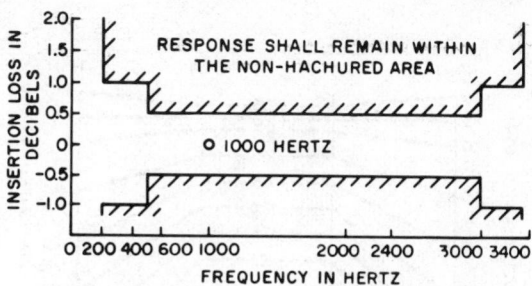

Fig. 19—Limits of the insertion loss for Bell System channel translating equipment.

the particular transmission medium. Recommended frequencies for frequency pilots are 60 and 308 kilohertz.

Bell System pilot frequencies are generally 92 kilohertz for group-regulating pilots, and 64 or 308 kilohertz for synchronizing or line pilots.

COMPANDORS IN CARRIER TELEPHONY

The use of compandors can produce substantial subjective improvement in the noise and crosstalk performance of a voice channel used for speech. The typical compandor consists of a volume compressor at the transmitting end of the 4-wire path and a volume expandor at the receiving end. The compressor acts to compress the transmitted volume range to approximately half the applied range, generally by providing gain for low-level

TABLE 17—FREQUENCY AND LEVEL OF GROUP, SUPERGROUP, MASTERGROUP, AND SUPER-MASTERGROUP PILOTS. G.241

Pilot for	Frequency (kilohertz)	Absolute Power Level at a Zero-Relative-Level Point
Basic group B	84.080	—20 dB (—2.3 nepers)
	84.140	—25 dB (—2.9 nepers)
	104.080	—20 dB (—2.3 nepers)
Basic supergroup	411.860	—25 dB (—2.9 nepers)
	411.920	—20 dB (—2.3 nepers)
	547.920	—20 dB (—2.3 nepers)
Basic mastergroup	1552	—20 dB (—2.3 nepers)
Basic supermaster-group	11096	—20 dB (—2.3 nepers)
Basic 15-super-group assembly (No. 1)	1552	—20 dB (—2.3 nepers)

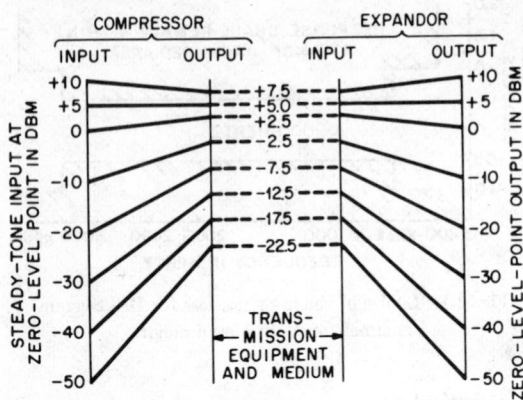

Fig. 20—Compandor function.

speech phonemes and less gain or none for high-level phonemes. Low-level phonemes which convey much of the actual intelligence are thus transmitted on the line at a higher average level, preventing the line noise from masking them. The receiving volume expandor performs the reverse function so that the dynamic level response is maintained nearly linear over the system. The expandor substantially reduces received noise between syllables and words.

Figure 20 illustrates the level modifications produced by the compandor. The noise and crosstalk improvement generally credited to the use of compandors is of the order of 20 to 22 decibels. These figures hold for improvement on marginally noisy and quiet circuits but deteriorate rapidly on circuits having high noise, so that on circuits having a signal-to-noise ratio of 20 decibels or less without compandors, the improvement disappears and the compandor may actually degrade circuit performance further.

Compandors will accentuate level variations on the circuit, thus circuit level stability must be held within close limits.

LOAD CARRYING CAPACITY OF MULTICHANNEL SYSTEMS

The data from many previous measurements indicate that talker volumes on civilian telephone

TABLE 18—NUMBER OF ACTIVE CHANNELS.

N	$N_a = N/4$	Exceeded 1% of Time
12	3	7
40	10	17
100	25	36
600	150	170
1200	300	330
1800	450	490

networks are distributed log-normally with a median value (V_0) of -12.5 VU and a standard deviation (σ) of 5.0 VU. More-recent tests indicate that the trend is toward lower median talker volumes and larger standard deviations. Values representative of the present population are $V_0 = -16.8$ VU and $\sigma = 6.4$ VU. The values of $V_0 = -12.5$ VU and $\sigma = 5.0$ VU continue to be used. For design purposes, Fig. 21 illustrates the currently used curve of talker volume distribution.

Independent tests made on military systems indicate that, for this restricted population, values of $V_0 = -8.6$ VU and $\sigma = 3.7$ VU are representative. All measurements are referred to the 0 TLP.

To find the power of an "average-power" talker, it is necessary to convert from the average of the log-normal distribution in VU to the average of the actual power distribution. In this case

$$P_{op} = V_0 + 0.115\sigma^2 - 1.4 \text{ dBm}$$

where P_{op} is the speech power of the average talker, $0.115\sigma^2$ converts from the log-normal distribution to the average of the power distribution, and 1.4 dB is an empirically determined conversion factor converting VU into dBm.

The load carrying capacity of a multichannel system is conveniently defined as the power of a

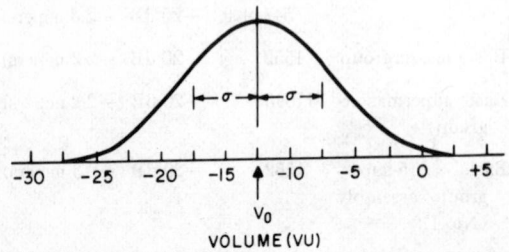

Fig. 21—Talker volume distribution.

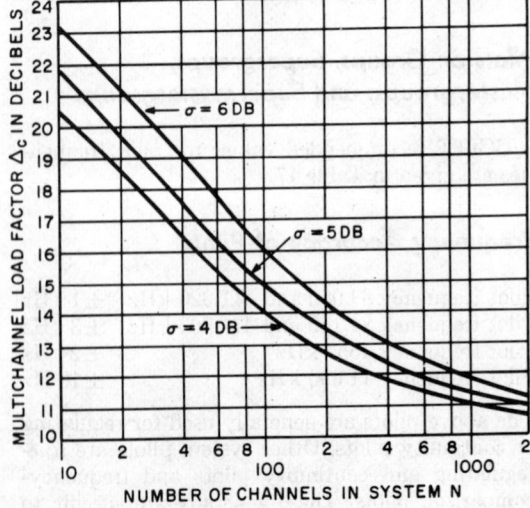

Fig. 22—Determination of load capacity.

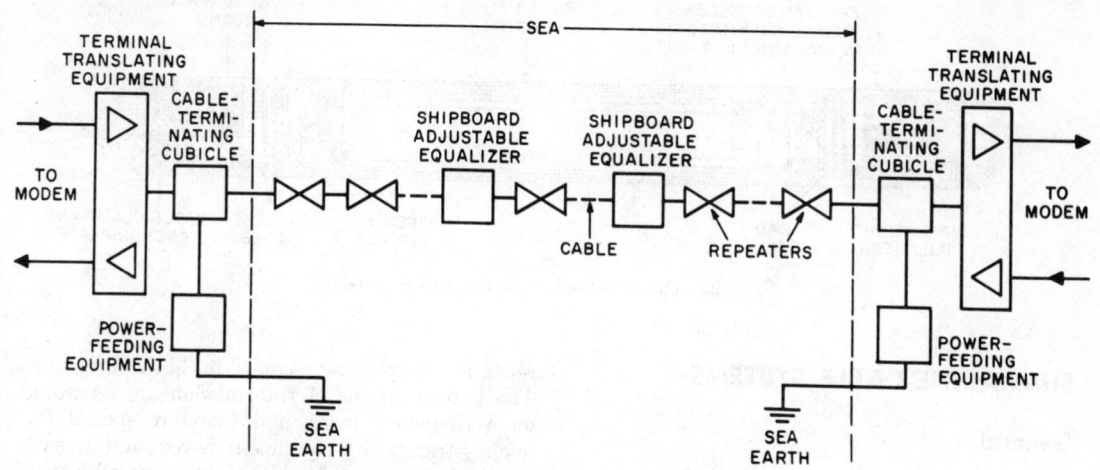

Fig. 23—Block schematic of submarine-cable system.

sine-wave test tone that the system must transmit without overload. The power of this signal at the 0 TLP will be defined as P_s.

In any multichannel system, if the number of channels is large, the average number of active channels (N_a) is approximately 25 percent of the total number of channels. Table 18 gives values more nearly in accord with the expected values.

The use of TASI or other types of channel-sharing equipment will change this value. For small numbers of channels the departure from the 25-percent figure is greater.

The required load capacity of a system may be expressed in terms of the average talker power, the number of channels in the system, and a multichannel load factor Δ_c which expresses the expected ratio of the peak power to average power of the multichannel signal. A curve showing the relation between Δ_c and the total number of channels is given in Fig. 22. Curves for 3 values of σ are shown. The complete expression for P_s is given as

$$P_s = V_0 + 0.155\sigma^2 - 1.4 + 10 \log_{10} N_a + \Delta_c \text{ dBm0}$$

or

$$P_s = P_{op} + 10 \log_{10} N_a + \Delta_c \text{ dBm0}$$

where N_a is the number of active channels, and Δ_c is obtained from the curves.

This determination of P_s is specifically limited to those systems where the baseband is transmitted "flat." If pre-emphasis is used to obtain improved signal-to-noise ratios, the effects of this pre-emphasis must be included.

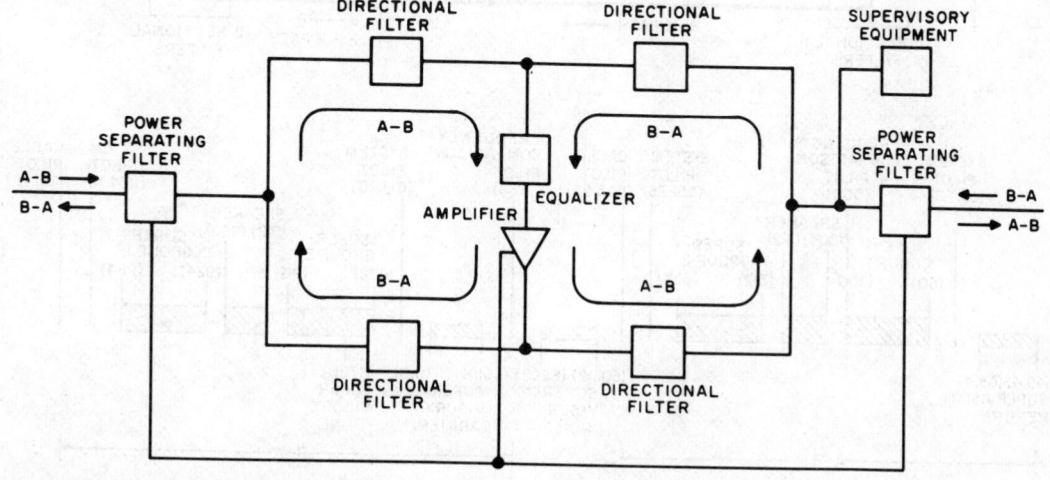

Fig. 24—Block schematic of a two-way submarine repeater.

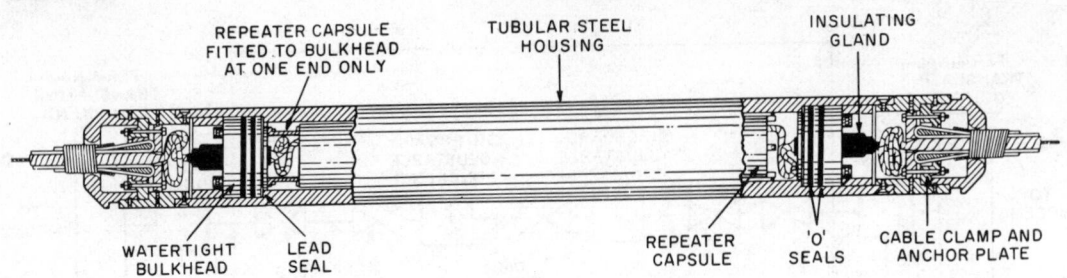

Fig. 25—Cross section of a submarine repeater.

SUBMARINE-CABLE SYSTEMS

General

The transmission of multichannel speech over a system which includes a cable involves the provision of submerged repeaters and their associated special terminal and power-feeding equipment.

The most usual arrangement is shown in Fig. 23 where a single cable is equipped with submerged repeaters amplifying signals in both directions. The two directions of transmission are separated on a frequency basis, and therefore special frequency-translating equipment is required at each terminal to translate the signals from the submarine cable to a position in the frequency spectrum suitable for transmission to the local network.

Early cables such as TAT-1, TAT-2, and Florida–Puerto Rico employed 1-way repeaters and required two cables, one for each direction of

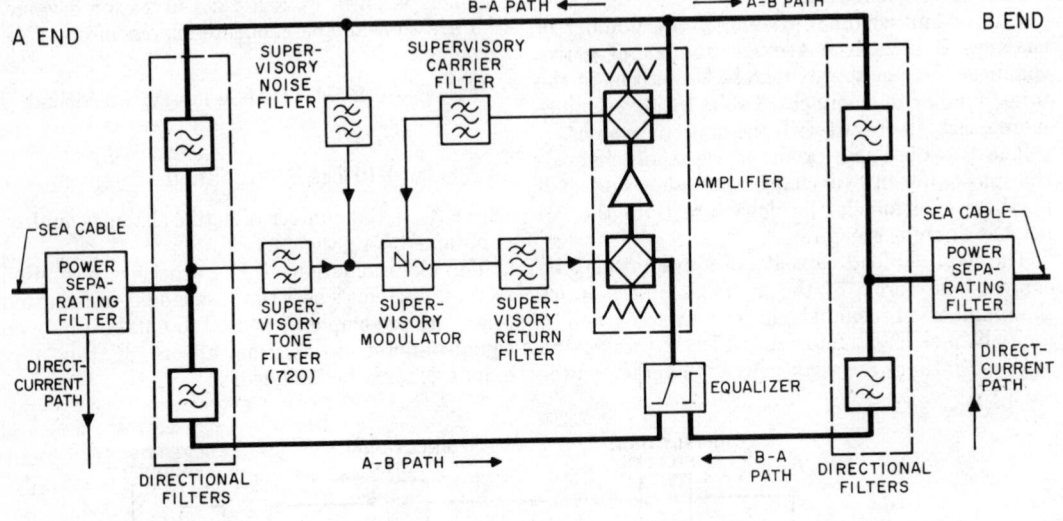

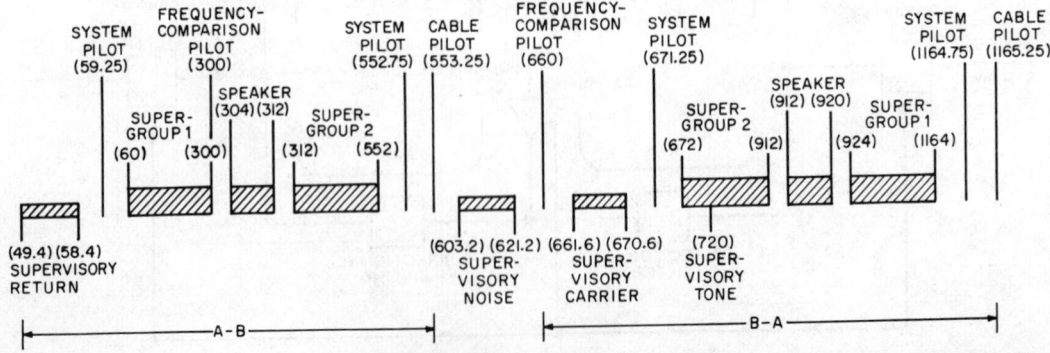

Fig. 26—Simplified block diagram of 120/160-channel submarine repeater and the system frequency spectrum. Frequencies in kilohertz are in parentheses.

TABLE 19—DATA FOR REPRESENTATIVE TYPES OF SUBMARINE-CABLE SYSTEM.

Number of 3-Kilohertz Channels	Traffic Bands		Type of Cable	Repeater Spacing (nautical miles)	Approximate Repeater Gain (decibels)	System Type
	A–B (kilohertz)	B–A (kilohertz)				
24	24–96	120–192	0.46-inch armored	26.8	61.5	—
80	20–260	312–552	0.62-inch armored	21	60.5	Short-haul
80	60–300	360–608	0.99-inch lightweight*	26.3	55	Long-haul
80	60–300	360–608	1.00-inch lightweight†	29.5	55	Long-haul
160	60–552	672–1164	0.62-inch armored	11.9	50	Short-haul
160	60–552	672–1164	0.99-inch lightweight*	17.3	50	Long-haul
160	60–552	672–1164	1.00-inch lightweight†	19.4	50	Long-haul
360	312–1428	1848–2964	1.00-inch lightweight†	9.5	40	Long-haul
640	312–2292	2792–4772	0.935-inch armored	7.5	43	Short-haul

* British Post Office.
† Standard Telephones and Cables.

transmission. This design has been superseded by the 2-way repeater with single cable.

Special power-feeding equipment must also be provided. The vacuum tubes in all repeaters are connected in tandem and fed by a constant-current source which is controlled to ±1 percent to prevent damage to the tubes. Double-end power feeding is normally employed, thereby limiting the maximum voltage to which a repeater is subjected.

Submerged Repeaters

Figure 24 shows a generalized block schematic of a repeater. The directions of transmission are separated on a frequency basis by the directional filters, a single amplifier being employed for both directions. The amplifier is usually a 3-stage negative-feedback amplifier employing either vacuum tubes or transistors. The construction of the repeater is such that maximum reliability is provided, and the electrical circuits are enclosed in a high-

tensile-steel pressure-resisting housing fitted with suitable glands for cable entry and exit (Fig. 25).

Types of Systems

Various types of system are available depending on the number of voice circuits required and the type of cable. Table 19 shows the repeater spacing, signal bands, type of cable, etc., for representative types of system.

Figure 26 shows in greater detail the frequency spectrum and the repeater block schematic for a typical long-distance 160-channel system.

Equalization

On long submarine repeatered systems the build-up of residual differences between the cable loss characteristic and the repeater gain characteristic necessitates periodic correction. This is done by the insertion during installation of shipboard ad-

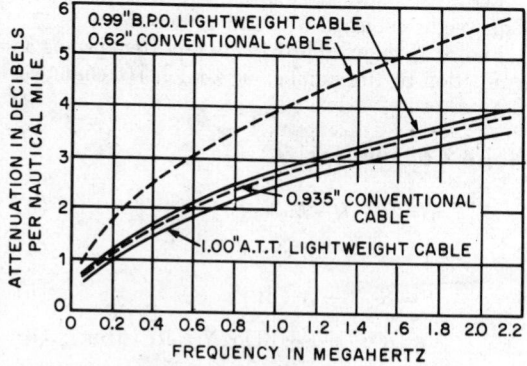

Fig. 27—Attenuation of conventional and lightweight submarine cables at 3°C and 1500 fathoms.

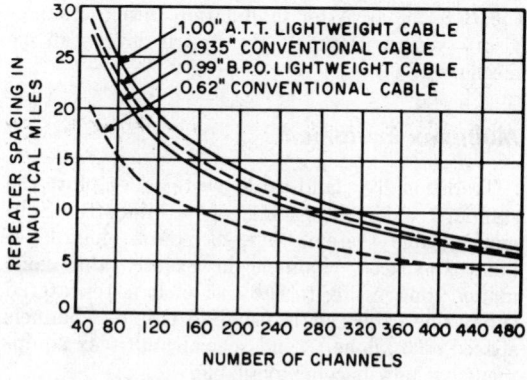

Fig. 28—Approximate repeater spacing on conventional and lightweight submarine cables.

TABLE 20—POWER FEED DATA.

System	Length (nautical miles)	Cable	Repeater Voltage (volts)	Power Feed Current (amperes)	Power Feed Voltage (volts)
TAT–1 (36)	1942	0.62-inch conventional	62.8	0.250	2300
TAT–1 (60)	326	0.62-inch conventional	124	0.316	1200
Cantat (60)	1518	0.99-inch lightweight	73.5	0.430	4900
	+554	0.62-inch conventional			
TAT–3 (96)	3518	1.00-inch lightweight	50	0.389	5500

justable equalizers (see Fig. 23) usually at intervals of 10 to 12 repeaters. The necessity for adjustment on board ship requires a demountable pressure-resisting housing for the equalizer.

Supervisory Arrangements

All submerged repeatered systems provide supervisory arrangements whereby a faulty repeater may be located from the terminal stations. The complexity of the arrangement varies, depending on the type of system, from simple fault location to the provision for measuring noise and intermodulation on each repeater.

As an example of the latter, see Fig. 26. Here each repeater is equipped with a different supervisory carrier filter in the band from 661 to 670 kilohertz, so that by sending the appropriate frequency from the B terminal the supervisory modulator in any repeater can be made operative. In this condition the noise generated from the repeater can be picked off and will be translated to the supervisory return path and returned to the B terminal, where its level can be measured. Similarly, by sending further test frequencies intermodulation tests can be made. Gain measurements can be made by transmitting 720 kilohertz from the B terminal. After modulation into the supervisory return channel, its return level can be measured.

Multiplex Equipment

Traditionally, land-cable systems employ the standard channel spacing of 4 kilohertz. Submarine-cable systems can, of course, handle 4-kilohertz-spaced channels in exactly the same manner but, owing to the cost of long repeatered submarine cables, it is usual to employ channels spaced at 3 kilohertz and special multiplex equipment has now become available.

Because of a significant reduction in the guard space between channels made possible by the terminal-equipment design and the absence of demodulations to voice over the long span of submarine-cable circuits, the channel response of a long-haul submarine-cable system differs little from that of a land line coaxial system of comparable length having a number of internal demodulations and remodulations.

Another method of increasing the utilization of the submarine cable is the employment of Time Assignment Speech Interpolation (TASI). This is is a high-speed switching system whereby talkers seize a circuit only when they begin to speak. Such an arrangement approximately doubles the circuit capacity.

Submarine-Cable-System Equations

$$f_m = bN + B \qquad (1)$$

$$G = A\ L/N \qquad (2)$$

where $f_m =$ top working frequency, $b =$ channel bandwidth in kHz, $B =$ unused bandwidth below lowest channel and between channels, $N =$ number of channels, $G =$ gain per repeater at f_m, $A =$ cable attenuation in decibels per nautical mile at f_m, and $L =$ total length of cable.

Figure 27 shows attenuation of conventional and lightweight cables at 3°C and 1500 fathoms.

Figure 28 shows approximate repeater spacing as a function of the number of 4-kilohertz channels.

Noise Calculations

$$S = W - H - M_2 \ (\text{dBr}) \qquad (3)$$

$$I = S - G \ (\text{dBr}) \qquad (4)$$

$$I^1 = S - G - A \ (\text{dBr}) \qquad (5)$$

$$N_1 = KBT + F + 10 \log N + M_1 \ (\text{dBm}) \qquad (6)$$

$$N_0 = KBT + F + 10 \log N + M_1 - I^1. \qquad (7)$$

Substituting values for KBT and for I^1

TABLE 21—CODE COMBINATIONS.

Character	International Morse	International Cable	Character	International Morse	International Cable
A	·—	+−	.	·—·—·—	Punctuations
B	—···	−+++	;	—·—·—·	not
C	—·—·	−+−+	(Comma) ,	— —··— —	shown
D	—··	−++	:	— — —···	because
E	·	+	?	··— —··	of
F	··—·	++−+	(Apos) '	·— — — —·	many
G	— —·	− —+	-	—····—	variations
H	····	++++	/	—··—·	among
I	··	++	Ä	·—·—	systems.
J	·— — —	+− — —	Á or Å	·— —·—	
K	—·—	−+−	É	··—··	
L	·—··	+−++	CH	— — — —	
M	— —	− —	Ñ	— —·— —	
N	—·	−+	Ö	— — —·	
O	— — —	− — —	Ü	··— —	
P	·— —·	+− —+	(or)	—·— —·—	
Q	— —·—	− —+−	"	·—··—·	
R	·—·	+−+	—	—···—	
S	···	+++	=	—··—	
T	—	−	SOS	··· — — — ···	
U	··—	++−	Attention	—·—·—	
V	···—	+++−	CQ	—·—· — —·—	
W	·— —	+− —	DE	—·· ·	
X	—··—	−++−	Go ahead	—·—	
Y	—·— —	−+− —	Wait	·—···	
Z	— —··	− —++	Break	—···—·—	
1	·— — — —	+− — — —	Understand	···—·	
2	··— — —	++− — —	Error	········	
3	···— —	+++− —	OK	·—·	
4	····—	++++−	End of message	·—·—·	+−+−+
5	·····	+++++	End of work	···—·—	
6	—····	−++++			
7	— —···	− —+++			
8	— — —··	− — —++			
9	— — — —·	− — — —+			
0	— — — — —	− — — — —			

$$T \text{ (dBm0)} = 139.5 + F + 10 \log N + G - W + H + A + M_1 + M_2 \quad (8)$$

where W = power-handling capacity in dBm, H = test-tone load capacity in dBm, G = repeater gain at top frequency, F = amplifier noise figure in dB, N_1 = noise at final-amplifier input in dBm, N_0 = noise at receiver 0 dB point, I = repeater input relative level in dBr, I^1 = amplifier input relative level in dBr, S = repeater output relative level in dBr, M_1 = repeater input planning margin in dB, M_2 = repeater output planning margin in dB, A = attenuation in repeater input network in dB, KBT = available noise power from matched source in dBm, N = number of repeaters, and T = target noise level (dBm0).

Power Feeding a Chain of Repeaters

The power feed of submarine cables is a constant-current source with voltages of opposite polarity applied between the center conductor and sea earth at each end of the cable.

Consider a chain of n repeater sections and L nautical miles.

$$V_t = irL + (N-1)V_r \quad (9)$$

where V_t = total power feed voltage, i = power feed current, r = direct-current resistance of the cable center conductor, N = number of repeaters, and V_r = voltage drop at each repeater.

Table 20 shows typical systems and their power feed data.

TABLE 22—TIMING OF FIVE-UNIT START-STOP TELEPRINTER CODES. *From F. W. Smith, "Transmission Speeds and Pulse Lengths of Commonly Used Five-Unit Start-Stop Printing Telegraph Codes," Western Union Technical Review, p. 140; October 1957.*

Code Pattern (Total No. of Pulses per Character)	Nominal Transmitting Speeds			Nominal Pulse Lengths (milliseconds)		Receiving Shaft Speed (rpm)	Milliseconds per Character	Where Used
	Operations per Minute	Average Words per Minute	Bauds	Start and 5 Code Pulses	Rest Pulse			
7.42	368*	61.33	45.45	22	31	420	163	Bell System—US
7	390*	65	45.45	22	22	420	154	Western Union—US
7.5	400†	66.67	50	20	30	461.5	150	ITU standard—Europe
7.42	404†	67.33	50	20	28.4	461.7	148.4	US military, for interoperation with allies
7	428.6†	71.43	50	20	20	461.5	140	Former CCIT standard—Europe
7.42	460	76.67	56.88	17.57	25	525.7	130.43	US—all commercial and military users
7.42	600‡	100	74.2	13.47	19.18	685	100	US—all users
7	636‡	106	74.2	13.47	13.47	685	94.3	US military—limited use

* These two codes are compatible.
† These three codes are compatible.
‡ These two codes are compatible.

TELEGRAPH FACILITIES

International Morse and Cable Codes (Table 21)

International Morse Code is determined by combinations of unipolar current pulses of short and long ($\approx 1:3$, occasionally $1:2$) durations.

$$A = {}^{+}_{0} \; \square\!\square\!\square \;\; \square\!\square\!\square$$

International Cable Code is determined by combinations of bipolar current pulses of the same length.

$$A = {}^{+}_{0} \!\!\! {}_{-} \; \square\!\square$$

Teleprinters

The most widely used teleprinters operate at 60 to 100 words per minute and use the start-stop mode of operation. Each character is composed of a 5-bit code group preceded by a start pulse and followed by a stop pulse. All bit intervals are equal

TABLE 23—FIVE-UNIT TELEPRINTER-CODE ALPHABETS.

Bit Numbers 5 4 3 2 1	Letters Case	Figures Case				
	International Alphabet #2	International Alphabet #2	US Alphabets Military Std	Weather	TWX	Telex
0 0 0 0 0	Blank*	Blank*	Blank*	—	Blank*	Blank*
0 0 0 0 1	E	3	3	3	3	3
0 0 0 1 0	Line Feed	Line Feed	Line Feed	Line Feed	Line Feed	Line Feed
0 0 0 1 1	A	—	—	↑	—	—
0 0 1 0 0	Space	Space	Space	Space	Space	Space
0 0 1 0 1	S	(Apos) '	Bell	Bell	Bell	(Apos) '
0 0 1 1 0	I	8	8	8	8	8
0 0 1 1 1	U	7	7	7	7	7
0 1 0 0 0	Car. Ret	Car. Ret	Car. Ret	Car. Ret	Car. Ret	Car. Ret
0 1 0 0 1	D	WRU	$	↗	$	WRU
0 1 0 1 0	R	4	4	4	4	4
0 1 0 1 1	J	Aud Sig	(Apos) '	↙	(Comma) ,	Bell
0 1 1 0 0	N	(Comma) ,	(Comma) ,	○	¼	(Comma) ,
0 1 1 0 1	F	†	!	→	WRU	$
0 1 1 1 0	C	:	:	○	½	:
0 1 1 1 1	K	(	(	←		(
1 0 0 0 0	T	5	5	5	5	5
1 0 0 0 1	Z	+	"	+	"	"
1 0 0 1 0	L	)	)	↖	¾	)
1 0 0 1 1	W	2	2	2	2	2
1 0 1 0 0	H	†	Stop	↓		#
1 0 1 0 1	Y	6	6	6	6	6
1 0 1 1 0	P	0	0	Ø	0	0
1 0 1 1 1	Q	1	1	1	1	1
1 1 0 0 0	O	9	9	9	9	9
1 1 0 0 1	B	?	?	⊕	⅝	?
1 1 0 1 0	G	†	&	↘	&	&
1 1 0 1 1	Figures	Figures	Figures	Figures	Figures	Figures
1 1 1 0 0	M	.	.	.	.	.
1 1 1 0 1	X	/	/	/	/	/
1 1 1 1 0	V	=	;	①	⅜	;
1 1 1 1 1	Letters	Letters	Letters	Letters	Letters	Letters

Notes: Transmission Order: Bit 1→Bit 5.

* "Blank" in US; "No Action" in International Alphabet #2.

† Unassigned (domestic variation, not used internationally).

except for the stop pulse, which is somewhat longer. Table 22 gives the time and speed relationships of various commonly used codes.

Teleprinter Codes: Letter and figure assignments for teleprinter codes are given in Tables 23, 24, and 25.

TABLE 24—MOORE ARQ CODE (COMPARED WITH 5-UNIT TELEPRINTER CODE).

Code Assignments		Moore ARQ Code	Five-Unit TTY Code
Letters Case	Figures Case	Bit Numbers 7 6 5 4 3 2 1	Bit Numbers 5 4 3 2 1
Blank	Blank	1 1 1 0 0 0 0	0 0 0 0 0
E	3	0 0 0 1 1 1 0	0 0 0 0 1
Line Feed	Line Feed	0 0 0 1 1 0 1	0 0 0 1 0
A	—	0 1 0 1 1 0 0	0 0 0 1 1
Space	Space	0 0 0 1 0 1 1	0 0 1 0 0
S	(Apos) '	0 1 0 1 0 1 0	0 0 1 0 1
I	8	0 0 0 0 1 1 1	0 0 1 1 0
U	7	0 1 0 0 1 1 0	0 0 1 1 1
Car. Ret	Car. Ret	1 1 0 0 0 0 1	0 1 0 0 0
D	⊕	0 0 1 1 1 0 0	0 1 0 0 1
R	4	0 0 1 0 0 1 1	0 1 0 1 0
J	Bell	1 1 0 0 0 1 0	0 1 0 1 1
N	(Comma) ,	0 0 1 0 1 0 1	0 1 1 0 0
F	□	1 1 0 0 1 0 0	0 1 1 0 1
C	:	0 0 1 1 0 0 1	0 1 1 1 0
K	(	1 1 0 1 0 0 0	0 1 1 1 1
T	5	1 0 1 0 0 0 1	1 0 0 0 0
Z	+	1 0 0 0 1 1 0	1 0 0 0 1
L	)	0 1 0 0 0 1 1	1 0 0 1 0
W	2	1 0 1 0 0 1 0	1 0 0 1 1
H	□	0 1 0 0 1 0 1	1 0 1 0 0
Y	6	1 0 1 0 1 0 0	1 0 1 0 1
P	0	0 1 0 1 0 0 1	1 0 1 1 0
Q	1	1 0 1 1 0 0 0	1 0 1 1 1
O	9	0 1 1 0 0 0 1	1 1 0 0 0
B	?	1 0 0 1 1 0 0	1 1 0 0 1
G	□	1 0 0 0 0 1 1	1 1 0 1 0
Figures	Figures	0 1 1 0 0 1 0	1 1 0 1 1
M	.	1 0 0 0 1 0 1	1 1 1 0 0
X	/	0 1 1 0 1 0 0	1 1 1 0 1
V	=	1 0 0 1 0 0 1	1 1 1 1 0
Letters	Letters	0 1 1 1 0 0 0	1 1 1 1 1
Signal I	Signal I	0 0 1 0 1 1 0	
Idle α	Idle α	1 0 0 1 0 1 0	
Idle β	Idle β	0 0 1 1 0 1 0	

Note: Transmission Order: Bit 1→Bit 7.

Carrier Telegraph Systems

Standard carrier telegraph systems multiplex up to 20 or more telegraph circuits into one standard 4-kilohertz voice channel employing frequency-division techniques. The most prevalent form of modulation is frequency-shift keying (FSK); however, amplitude modulation is also used.

The channel spacing for FSK systems is commonly 120 or 170 hertz. The 120-hertz spacing is used for systems operating at 60 words per minute or less, while 170-hertz spacing is used for 100 words per minute or less. (Operation at 100 words per minute on a 120-hertz spaced channel is possible for single links or if some degradation in performance can be tolerated.) The usual frequency shift for 120-hertz-spaced channels is ±30 hertz and the frequency shift for 170-hertz-spaced channels is ±42.5 hertz. Higher-speed channels for special application are normally derived by replacing the terminal equipment of two or more adjacent channels with a single wide-band channel terminal.

The channel frequency allocations for carrier telegraph systems are as follows.

(**A**) 170-hertz spacing commencing with 425 hertz as the center frequency of the lowest frequency channel.

(**B**) 120-hertz spacing commencing with 420 hertz as the center frequency of the lowest frequency channel.

Recent developments in modulation techniques for the transmission of teleprinter data have, among others, included the Kineplex and Duobinary techniques. These techniques represent approaches that have as an objective increasing the bit rate that can be effectively transmitted within a given bandwidth. In these techniques, increased transmission speed is obtained at a small sacrifice in immunity to noise.

Kineplex: In the Kineplex technique, two bits at a time are transmitted by phase modulation of an audio tone burst with the information represented by the difference in phase between the tone burst and the previously transmitted tone burst; for example, $(0,0) = 0$; $(0,1) = \pi/2$; $(1,1) = \pi$; and $(1,0) = 3\pi/2$. Thus, the Kineplex is basically a quaternary phase-modulation system.

Duobinary: In the Duobinary technique, the input bit stream is so processed that the output consists of a 3-level signal occupying half of the bandwidth of the original signal, and having the special property that the output signal may change level by only one step between successive bits. The 3-level output signal may be used to modulate an audio tone either by frequency modulation or amplitude modulation.

TABLE 25—BELL SYSTEM INFORMATION INTERCHANGE CODE AND ASSOCIATED 4-ROW
KEYBOARD ARRANGEMENT.

Bit Numbers	Code Assignments

Code Assignments rows:

b_7	0	0	0	0	1	1	1	1
b_6	0	0	1	1	0	0	1	1
b_5	0	1	0	1	0	1	0	1

b_4 b_3 b_2 b_1	Controls		Graphics					
0 0 0 0	NULL	DLE						
0 0 0 1	SOM	X ON						
0 0 1 0	EOA	TAPE						
0 0 1 1	EOM	X OFF						
0 1 0 0	EOT	~~TAPE~~						
0 1 0 1	WRU	ERROR						
0 1 1 0	RU	SYN	Graphic					
0 1 1 1	BELL	LEM	subset					
			identical					
1 0 0 0	①	S_0	with					
1 0 0 1	TAB	S_1	Fig. 30					
1 0 1 0	LINE FEED	S_2						
1 0 1 1	VT	S_3						
1 1 0 0	FORM	S_4						
1 1 0 1	RETURN	S_5					CNFM	
1 1 1 0	SO	S_6					ALT MODE	
1 1 1 1	SI	S_7					ESC	
							RUB OUT	

① Unassigned

CTRL	Control key	ALT MODE	Alternate mode	RU	"Are you···?"
SO	Shift out	ESC	Escape	BELL	Audible signal
SI	Shift in	①	Unassigned	TAB	Shift to next tab point
DLE	Data link escape	SYN	Synchronous idle	LINE FEED	Advances platen to next line
X ON	Transmitter on	NULL	Null/idle		
TAPE	Turn on tape	SOM	Start of message	VT	Vertical tab control
X OFF	Transmitter off	EOA	End of address	FORM	Form feedout operation
~~TAPE~~	Turn off tape	EOM	End of message	RETURN	Starts new line
LEM	Logical end of media	EOT	End of transmission	RUB OUT	Deletes character
CNFM	Confirm	WRU	"Who are you?"	ERROR	Error

Keyboard arrangement:

Row 1: ! 1   " 2   # 3 $ 4   % 5   & 6   ' 7   (8   ) 9   0   * :   = —

Row 2: ALT MODE   Q   W   WRU E   TAPE R   ~~TAPE~~ T   Y   U   TAB I   ← O   @ P   LINE FEED   RE-TURN

Row 3: CTRL   SOM A   X OFF S   EOT D   RU F   BELL G   H   J   VT K   FORM L   + ;   RUB OUT

Row 4: SHIFT   Z   X   EOM C   V   EOA B   ↑ N   M   < ,   > .   ? /   SHIFT

TABLE 26—IBM DATA-TRANSCEIVER CODE.

Bit Number (R 7 4 2 1)	000	001	010	011	100	101	110	111
0 0 0 0 0								
0 0 0 0 1								TPH/TGH
0 0 0 1 0								@
0 0 0 1 1				(NA)		(NA)	Space	
0 0 1 0 0								#
0 0 1 0 1				(NA)		(NA)	9	
0 0 1 1 0				(NA)		(NA)	8	
0 0 1 1 1	G	P			X			
0 1 0 0 0								(NA)
0 1 0 0 1				(NA)		(NA)	6	
0 1 0 1 0				(NA)		(NA)	5	
0 1 0 1 1	D	M			U			
0 1 1 0 0				(NA)		(NA)	3	
0 1 1 0 1	B	K			S			
0 1 1 1 0	A	J			/			
0 1 1 1 1	$\bar{0}$							
1 0 0 0 0								Restart
1 0 0 0 1				SOC/EOC		$\overset{+}{0}$	EOT	
1 0 0 1 0				*		%	□	
1 0 0 1 1	&	—			Ø			
1 0 1 0 0				$	,			
1 0 1 0 1	I	R			Z			
1 0 1 1 0	H	Q			Y			
1 0 1 1 1	7							
1 1 0 0 0				(NA)		(NA)	(NA)	
1 1 0 0 1	F	O			W			
1 1 0 1 0	E	N			V			
1 1 0 1 1	4							
1 1 1 0 0	C	L			T			
1 1 1 0 1	2							
1 1 1 1 0	1							
1 1 1 1 1								

Transmission Order: Bit X→Bit 1.

Legend:
TPH/TGH	Telephone/telegraph	$\overset{+}{0}$ Plus zero
SOC/EOC	Start or end of card	$\bar{0}$ Minus zero
EOT	End of transmission	□ "Lozenge" (special symbol)
(NA)	Valid but not assigned	

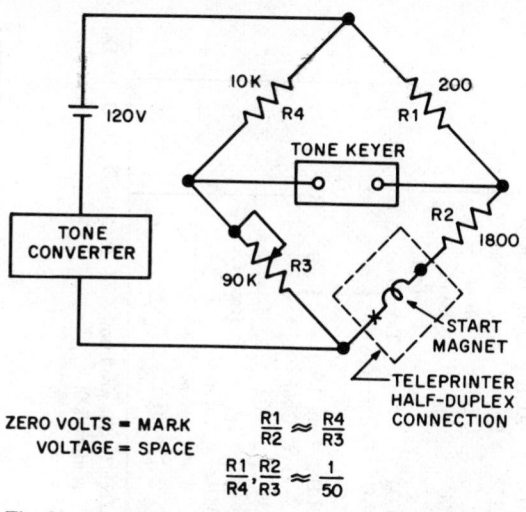

$$\frac{R1}{R2} \approx \frac{R4}{R3}$$

$$\frac{R1}{R4}, \frac{R2}{R3} \approx \frac{1}{50}$$

ZERO VOLTS = MARK
VOLTAGE = SPACE

Fig. 29—Typical bridge hybrid (half duplex to full duplex).

In both the Kineplex and Duobinary modulation techniques, a plurality of tone channels may be multiplexed into a standard 4-kilohertz voice channel using frequency-division techniques.

Definitions

Arrhythmic: Refers to the random generation of characters as in the case of those generated by a teleprinter keyboard. The antonym—rhythmic—applies to outputs such as those generated by automatic tape senders.

Break: To break, in a communication circuit, is for the receiving user to interrupt the sending user and to take control of the circuit. This term is generally used in connection with half-duplex and simplex telegraph circuits.

TABLE 27—US DEPARTMENT OF DEFENSE 8-UNIT CODE.

Bit Numbers								Code Assignments							
								0	0	0	0	1	1	1	1
								0	0	1	1	0	0	1	1
								0	1	0	1	0	1	0	1
P	C	I_2	I_1	D_3	D_2	D_1	D_0	Control Functions				Alphanumerics			
				0	0	0	0					MS	K	)	0
				0	0	0	1					UC	L	-	1
				0	0	1	0					LC	M	+	2
				0	0	1	1					LF	N	<	3
				0	1	0	0					CR	O	=	4
				0	1	0	1					Space	P	>	5
				0	1	1	0					A	Q	—	6
				0	1	1	1	①	①	①	①	B	R	$	7
				1	0	0	0					C	S	*	8
				1	0	0	1					D	T	(	9
				1	0	1	0					E	U	"	(Apos) '
				1	0	1	1					F	V	:	;
				1	1	0	0					G	W	?	/
				1	1	0	1					H	X	!	.
				1	1	1	0					I	Y	(Comma) ,	□
				1	1	1	1					J	Z	⊕	Idle

Notes:
1. Transmission order: Bit $D_0 \rightarrow$ Bit P.
2. Parity-bit P provides *odd* parity for transmission.
3. When punched in perforated tape, bit patterns have *even* parity.
4. Alphanumerics have 1, control functions 0, in bit-position C.

Legend:

MS	Master space	CR	Carriage return
UC	Upper case	⊕	Stop
LC	Lower case	□	"Lozenge" (special symbol)
LF	Line feed	①	Unassigned

TABLE 28—CONTROL-FUNCTION ASSIGNMENTS IN DEPARTMENT OF DEFENSE CODE ALPHABETS.

Control Function Assignments[1,2]

Bit Numbers[3,4,5] $D_3\,D_2\,D_1\,D_0$	Fieldata Code (Standard Form) $I_2 I_1$=00	=01	=10	=11	COMLOGNET Common Language Code =10	=11	SACCOMNET (465L) Control Code =01	=10	=11
0 0 0 0	Blank/Idle	Cont K	Dial 0	RTT	β	①	①	①	AA
0 0 0 1	TST	Cont L	Dial 1	RTR	#	ACK₁	Line Good Mess.	①	
0 0 1 0	TCL	Cont M	Dial 2	NRR	t	ACK₂	Main Alarm	Null/Idle	
0 0 1 1	TAB	Cont N	Dial 3	EOBK	OWD	REQ	①	EOD	
0 1 0 0	Cont CR	Cont O	Dial 4	EOB	①	WBT		EOA	
0 1 0 1	Cont SP	Cont P	Dial 5	EOF	@ %	REP	Display	SOM	KM
0 1 1 0	Cont A	Cont Q	Dial 6	EOC	¢	SOML	Alert Loop	ACK Begin Interr	
0 1 1 1	Cont B	Cont R	Dial 7	AKR		ER	Last Older Mess.		
1 0 0 0	Cont C	Cont S	Dial 8	RPB	Beil	DM		Send # Mess.	①
1 0 0 1	Cont D	Cont T	Dial 9	ISN	&	EOM	Take Over	D* End Interr	AL
1 0 1 0	Cont E	Cont U	SOC	NISN	Σ	SOLB	①	①	①
1 0 1 1	Cont F	Cont V	SOB	CWF	≠	EDB		Line Good Char.	①
1 1 0 0	Cont G	Cont W	SOD	①	≢	EOLB	Last Resort		RR
1 1 0 1	Cont H	Cont X	①	SAC	°	RM	①	①	
1 1 1 0	Cont I	Cont Y	①	SPC	+₀	SOMH	①	Interr Mess.	①
1 1 1 1	Cont J	Cont Z	Stop	DEL	⁻₀	①	①	①	

Notes:

1. For alphanumeric assignments, refer to Table 27.
2. Bit patterns marked ① are unassigned.
3. In COMLOGNET, bits are numbered: P-C-I_1-I_2-I_3-D_2-D_1-D_0.
4. In SACCOMNET, bits are numbered: P-C-D_5-D_4-D_3-D_2-D_1-D_0.
5. Control bit C is always 0.

BIT NUMBERS							COLUMN → ROW ↓	$\begin{smallmatrix}0&0\\0\end{smallmatrix}$ 0	$\begin{smallmatrix}0&0\\1\end{smallmatrix}$ 1	$\begin{smallmatrix}0&1\\0\end{smallmatrix}$ 2	$\begin{smallmatrix}0&1\\1\end{smallmatrix}$ 3	$\begin{smallmatrix}1&0\\0\end{smallmatrix}$ 4	$\begin{smallmatrix}1&0\\1\end{smallmatrix}$ 5	$\begin{smallmatrix}1&1\\0\end{smallmatrix}$ 6	$\begin{smallmatrix}1&1\\1\end{smallmatrix}$ 7
b_7	b_6	b_5	b_4	b_3	b_2	b_1									
			0	0	0	0	0	NUL	DLE	SP	0	@	P	`	p
			0	0	0	1	1	SOH	DC1	!	1	A	Q	a	q
			0	0	1	0	2	STX	DC2	"	2	B	R	b	r
			0	0	1	1	3	ETX	DC3	#	3	C	S	c	s
			0	1	0	0	4	EOT	DC4	$	4	D	T	d	t
			0	1	0	1	5	ENQ	NAK	%	5	E	U	e	u
			0	1	1	0	6	ACK	SYN	&	6	F	V	f	v
			0	1	1	1	7	BEL	ETB	'	7	G	W	g	w
			1	0	0	0	8	BS	CAN	(	8	H	X	h	x
			1	0	0	1	9	HT	EM	)	9	I	Y	i	y
			1	0	1	0	10	LF	SUB	*	:	J	Z	j	z
			1	0	1	1	11	VT	ESC	+	;	K	[	k	{
			1	1	0	0	12	FF	FS	,	<	L	\	l	l
			1	1	0	1	13	CR	GS	—	=	M	]	m	}
			1	1	1	0	14	SO	RS	.	>	N	^	n	~
			1	1	1	1	15	SI	US	/	?	O	_	o	DEL

NUL	Null, or all zeros	DC1	Device control 1
SOH	Start of heading	DC2	Device control 2
STX	Start of text	DC3	Device control 3
ETX	End of text	DC4	Device control 4
EOT	End of transmission	NAK	Negative acknowledge
ENQ	Enquiry	SYN	Synchronous idle
ACK	Acknowledge	ETB	End of transmission block
BEL	Bell, or alarm	CAN	Cancel
BS	Backspace	EM	End of medium
HT	Horizontal tabulation	SUB	Substitute
LF	Line feed	ESC	Escape
VT	Vertical tabulation	FS	File separator
FF	Form feed	GS	Group separator
CR	Carriage return	RS	Record separator
SO	Shift out	US	Unit separator
SI	Shift in	SP	Space
DLE	Data link escape	DEL	Delete

Fig. 30—Recommended USA Standard Code for Information Interchange (USASCII) X3.4—1967 for use on tape.
© 1967 USA Standards Institute. Reprinted by permission.

Bias Distortion: Distortion affecting a binary modulation system wherein all intervals of one state are uniformly longer or shorter than the duration of the corresponding theoretical interval.

Characteristic Distortion: Distortion caused by transients which, as a result of modulation, are present in the transmission channel and depend on its transmission qualities.

Fortuitous Distortion: Distortion resulting from random causes.

Cyclic Distortion: Distortion which is neither characteristic, bias, nor fortuitous and which in general has a periodic character.

End Distortion (As related to start-stop teleprinter signals): The shifting of the end of all marking pulses from their proper positions in relation to the beginning of the start pulse.

Full Duplex: A type of operation that permits simultaneous communication in both directions between the called and the calling parties.

Half Duplex: A circuit designed for duplex operation but which can only be operated in one direction at a time because of the terminal equipment. Figure 29 shows a typical hybrid circuit. The transmission facility permits full-duplex operation. The terminal facility is arranged for half-duplex operation.

Isochronous Modulation: Modulation (or demodulation) in which the time interval separating any two significant instants is theoretically equal to the unit interval or a multiple of this.

Simplex: The type of operation which permits the transmission of signals in either direction alternately.

Speech Plus Duplex: A method of operation that permits the simultaneous transmission of speech and telegraphy over a single voice channel.

Telegraph-Signal Distortion: The shifting of the transition points of the signal pulses from their proper position relative to the beginning of the start pulse. The magnitude of the distortion is expressed in percent of the perfect unit pulse length.

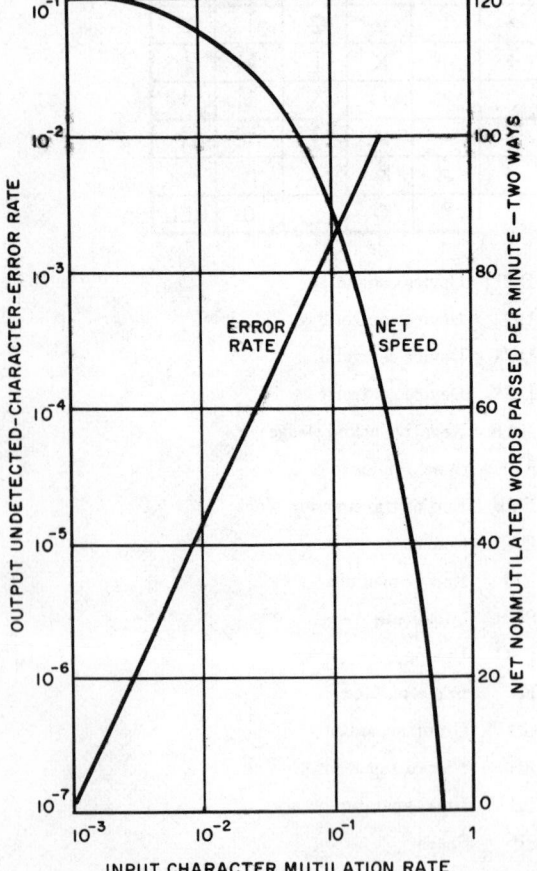

Fig. 31—Net speed and improvement of error rate using ARQ equipment. (Normal transmission in each direction is 60 words per minute.)

COMMERCIAL STANDARDS FOR ACCEPTABLE ERROR RATES

Teleprinter Transmission

Unprotected high-frequency radio circuits:	1 error in 10^3 characters
ARQ high-frequency radio or landline or submarine-cable circuits:	1 error in 10^4 characters

Data Transmission

For circuits which have been conditioned for 600 or 1200 or 2400 bits per second, 1 error in 10^5 bits.

For CCITT Recommendations on Telegraph, refer to Chapter 2.

COMMUNICATION CODES

Requirements for synchronization, error detection, standardization, compatibility, and card readers have resulted in the development of a number of specialized codes. Some of these codes are presented in Fig. 30 and Tables 26 through 28. The Hollerith 12-unit card code is given in Chapter 40.

ARQ

Automatic request-repeat (ARQ) systems may be implemented with any code having error-detecting properties. This mode of error control is generally used in conjunction with high-frequency radio transmission. In ARQ codes, the receiving terminal is arranged to detect a transmission error and automatically transmit the RQ (request-repeat) signal to the transmitting terminal. The transmitting terminal then retransmits the character or code block involved until it is correctly received. Implementation of an ARQ system requires a buffer storage of the transmitted characters that has sufficient capacity for a transmitted character to be held in store long enough to assure that it was correctly received. Commercial radiotelegraph links commonly use the Moore ARQ code given in Table 24. In this code each permissible code character contains exactly 3 ones and 4 zeros. Thus any other combination of ones and zeros is recognizable as an error and a request for retransmission is effected. The code fails in the case of error bursts that convert the received character into another having 3 ones and 4 zeros. All error-detecting codes fail when the error pattern is such as to convert the code character into another permissible character.

Figure 31 shows the improvement in character-error rate and the net speed obtained by use of ARQ equipment.

CHAPTER 36

SWITCHING NETWORKS AND TRAFFIC CONCEPTS

PART 1—COORDINATE SWITCHING NETWORKS

DEFINITIONS OF TERMS

Switching Network: That part of a switching system that establishes transmission paths between pairs of terminals.*

Space-Division Switching Network: A switching network in which the transmission paths are physically distinct.*

Coordinate Switch: A rectangular array of crosspoints in which one side of the crosspoints is multipled in rows and the other side in columns.*

Switching Stage: Those switches in a switching network that have identical parallel functions.*

Crosspoint: A two-state switching device having a low transmission impedance in one state and a very high one in the other.*

Concentration: The function associated with a switching network having fewer outlet than inlet terminals.

Expansion: The function associated with a switching network having more outlet than inlet terminals.

Full Availability (Switch): A switch (or switching network) capable of providing a path from every inlet terminal to every outlet terminal in the absence of traffic.

Single-Linkage Array: The mesh or spread of interconnections between the stages of a switching network whereby every switch of one stage has one connection to every switch of the adjacent stage. Sometimes known as a "spider web".

Nonfolded Network: A network in which each line or trunk is connected to an inlet terminal, an

* Source: *Bell System Technical Journal*; September 1964.

outlet terminal, or to both, and which is capable of completing a path between any two lines from the inlet terminal of one to the outlet terminal of the other.

Folded Network: A network in which each line or trunk is connected to both an inlet terminal and an outlet terminal, and which is capable of completing a path between any two lines from either terminal of one to a terminal of the other.

Nonblocking Network: A network in which there is at all times at least one available path between any pair of idle lines or trunks connected thereto, regardless of the number of paths already occupied.

Blocking: The inability to interconnect two idle lines connected to a network because all possible paths between them are already in use.

PROPERTIES OF COORDINATE SWITCHING NETWORKS

A coordinate switch for interconnecting a number of inlet terminals and a number of outlet terminals is one wherein every inlet and outlet is associated with a row or column, at the intersections (crosspoints) of which connecting devices are provided.

In electromechanical coordinate switches the connecting devices may be individual contact-making relays, in which case the number of complete relays (coil and contacts) required is the product of the numbers of inlets and outlets. Alternatively, the whole crosspoint array may be provided by a crossbar switch, in which a single relay coil is associated with each row and column of the switch and the concurrent energizing of a row coil and a column coil closes an individual set of crosspoint contacts.

More generally, the connecting device may in principle be any two-state device that exhibits a low-impedance state and a high-impedance state.

When the number of inlets or outlets, or both, is large, the number of crosspoints can be reduced significantly by replacing a single coordinate switch

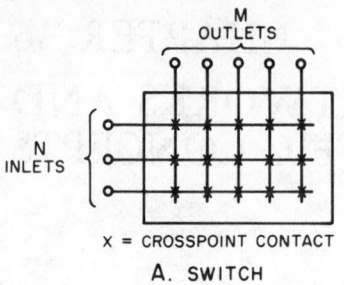

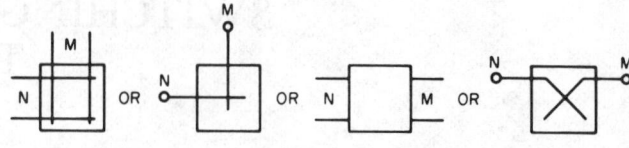

A. SWITCH   **B. CONVENTIONAL EQUIVALENT SYMBOLS**

Fig. 1—Single-stage "rectangular" coordinate switch (full availability).

by a number of interconnected smaller coordinate switches, arranged in a multistage coordinate switching network in one of many possible ways, the properties of some of which are described hereafter.

SINGLE-STAGE COORDINATE SWITCHES

Figure 1 shows a rectangular coordinate switch (switching matrix, matrix) interconnecting inlets from N lines and outlets to M other lines. When interconnection is possible at every crosspoint, the switch provides *full availability* and is inherently *nonblocking* (see definitions).

Number of crosspoint contacts required$=NM$.

When full availability of all outlets is not a requirement, an economy of crosspoint contacts can be obtained by use of a *limited-availability* (restricted-access) coordinate switch. Figure 2 shows a square switch with 5 inlets and 5 outlets, in which every inlet has access to only 3 of the outlets. The grading is said to be homogeneous when, as shown, each set of 3 outlets is unique.

When the outlets from a switch are connected to the *same* lines or trunks as are the inlets to the switch, a triangular "folded" arrangement may be used. Figure 3 shows a triangular switch for N inlets-outlets. The switch provides full availability and is nonblocking.

Number of crosspoint contacts required

$$= N(N-1)/2.$$

MULTISTAGE COORDINATE SWITCHING NETWORKS

3-Stage Networks—Configuration

General Case: Figure 4 shows a 3-stage network having N/n primary switches with n inlets each, S secondary switches, and M/m tertiary switches with m outlets each, to interconnect a total of N inlets to M outlets.

To satisfy the requirements for single-linkage spread connections (see definition) between stages

$$a=d=S; \quad b=N/n; \quad c=M/m.$$

The number of crosspoints X required is given by

$$X=S[N+M+(NM/nm)].$$

The relationship of the number of secondary switches S to the number of inlets and outlets, n and m, determines the degree of internal blocking presented by the network.

Symmetrical Case: Figure 5 shows part of a network for the case where $N=M$, $n=m$. The number of crosspoints required is given by

$$X=S[2N+(N^2/n^2)].$$

Nonblocking 3-Stage Networks

(A) *General Case*: Figure 6 shows part of a network that presents no internal blocking (see definition). When $N\neq M$, the condition for nonblocking is given by $S=n+m-1$.

The number of crosspoints required is given by

$$X=(n+m-1)[N+M+(NM/nm)].$$

A minimum number of crosspoints is obtained

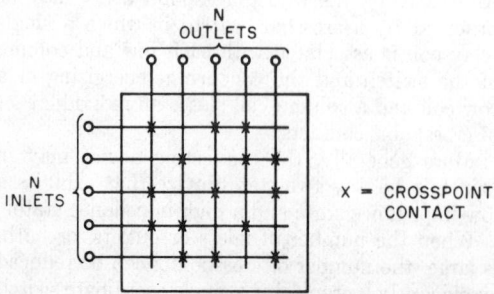

Fig. 2—Single-stage "square" coordinate switch (limited availability).

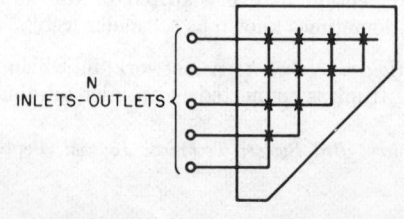

Fig. 3—Single-stage "folded" (triangular) switch.

when $m=n$ and when n satisfies the equation

$$NM/(N+M)=n^3/(n-1). \qquad (1)$$

Columns 1 and 2 of Table 1 show corresponding values of n and $NM/(N+M)$ satisfying equation (1), from which the optimum value of n for given values of N and M may be selected. Figure 7 shows part of the network.

The number of crosspoints required is then given by

$$X_{min}= (2n-1)[N+M+(NM/n^2)].$$

(**B**) *Symmetrical Case*: Figure 8 shows part of a symmetrical nonblocking network, where $M=N$, $m=n$. The condition for nonblocking is given by

$$S=2n-1.$$

The number of crosspoints required is given by

$$X=N(2n-1)[2+(N/n^2)].$$

A minimum number of crosspoints is obtained when n satisfies the equation

$$N=2n^3/(n-1). \qquad (2)$$

Columns 1 and 3 of Table 1 show corresponding values of n and N satisfying equation (2), from which the optimum value of n may be selected.

For large values of n, beyond the range of practical 3-stage networks (see (**C**) below)

$$2n^2 \rightarrow N.$$

Columns 1 and 4 of Table 1 show corresponding values of n and N satisfying the equation

$$N=2n^2. \qquad (3)$$

In no case is the optimum value of n selected by using column 3 more than unity less than the value indicated by column 4; in many cases, the same value is indicated by both columns.

(**C**) *Comparison of 3-Stage Network and Single-Stage Switch*: A single-stage rectangular coordinate switch having N inlets, M outlets, and NM crosspoints is inherently nonblocking. The most favorable nonblocking 3-stage network requires fewer than NM crosspoints if

$$NM/(N+M)>n^2(2n-1)/(n-1)^2.$$

Table 2 shows, for some practical values of n, the limiting value of $NM/(N+M)$ below which a single-stage switch requires fewer crosspoints.

Table 3 compares single-stage switches and 3-stage symmetrical nonblocking networks for typical values of N and n, where N/n is integral (see (**D**) below), to illustrate the trends of the design choices.

(**D**) *Practical Nonblocking 3-Stage Network with Minimum Crosspoints*: When a nonblocking net-

TABLE 1—RELATIONSHIP OF n, N, AND M IN NONBLOCKING 3-STAGE NETWORKS.

n	$NM/(N+M)$ Eq. (1)	N Eq. (2)	N Eq. (3)
2	8.0	16*	8
3	13.5	27*	18
4	21.3	42.7	32
5	32.2	62.5	50
6	43.2	86.4	72
7	57.2	114.3	98
8	73.1	146.3	128
9	91.1	182.2	162
10	111.1	222.2	200
11	133.1	266.2	242
12	157.1	314.2	288

* The only two integral solutions of Equation (2).

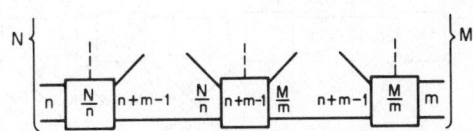

Fig. 6—Nonblocking 3-stage network.

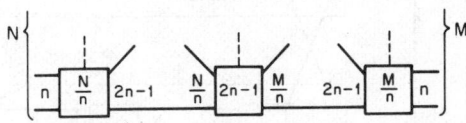

Fig. 7—Nonblocking 3-stage network with minimum crosspoints.

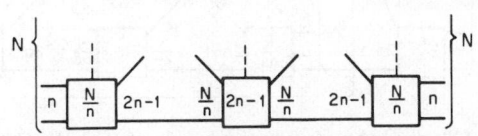

Fig. 8—Symmetrical nonblocking network with minimum crosspoints.

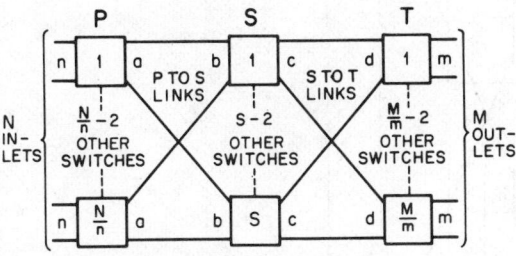

Fig. 4—Three-stage network—general case.

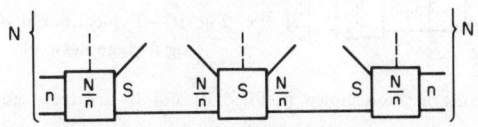

Fig. 5—Symmetrical 3-stage network.

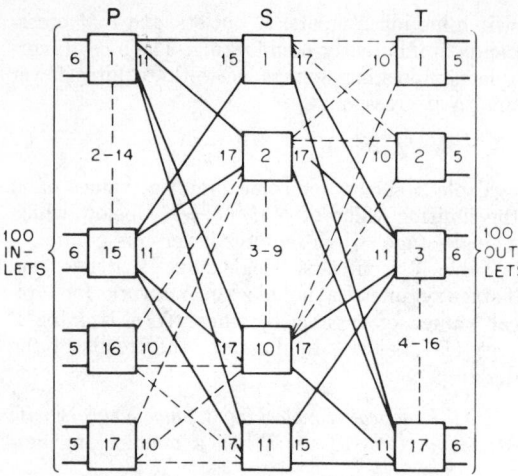

Fig. 9—Nonblocking 3-stage network for 100 inlets, 100 outlets.

TABLE 2—LOWER LIMITS OF $NM/(N+M)$ FOR NONBLOCKING 3-STAGE NETWORKS.

n	2	3	4	5	6
$NM/(N+M)>$	12	11.25	12.45	14.06	15.84

primary-stage switches are required, 15 with 6 inlets each and 2 with 5 inlets. The larger switches require $2n-1=11$ outlets each, cross-linked to 11 secondary-stage switches (links shown by continuous lines). The smaller switches require $2n-2=10$ outlets each, cross-linked to 10 only of the secondary-stage switches (links shown by dashed lines). Thus, the secondary-stage switches also are of two sizes. The total number of crosspoints is 5291 (as compared with 5423 for a nonblocking network for 102 inlets and outlets, with $n=6$).

The method may readily be extended to a network in which the numbers of inlets N and outlets M are unequal, by superimposing $(2n-1)$ links from each larger primary-stage and tertiary-stage switch on $(2n-2)$ links from each smaller switch, both sets spread over $(2n-1)$ secondary-stage switches.

(E) *Extension to 5-Stage and 7-Stage Networks*: If the number of inlets and outlets on each secondary-stage switch of a 3-stage network is large, it is advantageous to use a 5-stage network. One possible arrangement is shown in Fig. 10,* in which each secondary switch of a symmetrical 3-stage network conforming to Fig. 8 is expanded into a nonblocking 3-stage subnetwork.

work with a minimum number of crosspoints is sought, the indicated optimum value of inlets n per primary-stage switch may be such that N/n is not integral. The desired result may be achieved by providing some of the primary-stage and tertiary-stage switches with $(n-1)$ inlets and outlets, respectively, by adjusting the sizes of the secondary-stage switches, and by superimposing two sets of interstage linkages.

The method is illustrated in Fig. 9, a nonblocking network for 100 inlets and 100 outlets requiring a minimum number of crosspoints. The value $n=6$ is selected from column 3 of Table 1. The nearest multiple of 6 that exceeds 100 is 102. Thus, 17

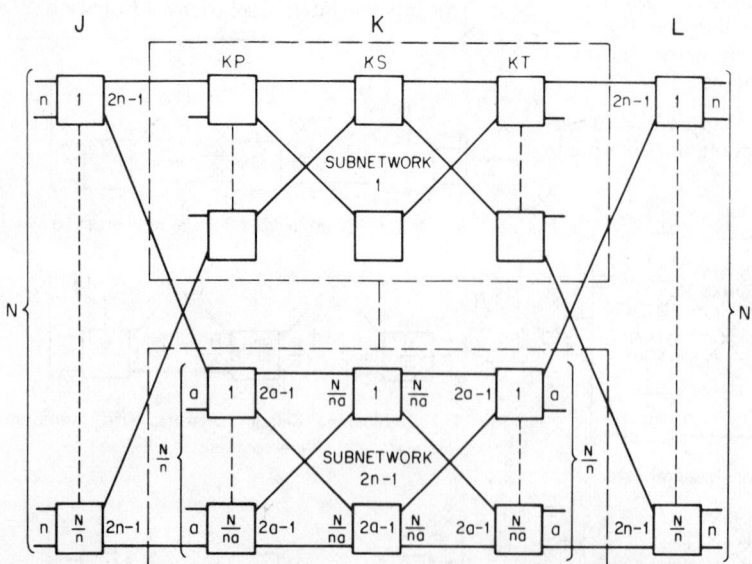

Fig. 10—Typical nonblocking 5-stage network.

* To our knowledge, no rigorous proof exists that this network or those shown in Figs. 12 and 13 are truly nonblocking.

TABLE 3—COMPARISON OF TYPICAL 1-STAGE AND 3-STAGE NONBLOCKING COORDINATE NETWORKS ($M=N$, $m=n$, N/n INTEGRAL).

	1-Stage	3-Stage			1-Stage	3-Stage			1-Stage	3-Stage	
N	X	n	X	N	X	n	X	N	X	n	X
8*	64	2	96	50*	2500	5	1800	100	10 000	5	5 400
15	225	3	275	54	2916	6	2079	100	10 000	10	5 700
16	256	2	288	55	3025	5	2079	105	11 025	7	5 655
16	256	4	336	56	3136	4	2156	108	11 664	6	5 940
18*	324	2	351	56	3136	7	2236	108	11 664	9	6 120
18	324	3	360	60	3600	5	2376	110	12 100	5	6 336
20	400	2	420	60	3600	6	2420	110	12 100	10	6 479
20	400	4	455	64	4096	4	2688	120	14 400	6	7 040
24	576	3	560	64	4096	8	2880	120	14 400	8	6 975
24	576	4	588	70	4900	5	3024	128*	16 384	8	7 680
25	625	5	675	70	4900	7	3120	130	16 900	10	8 151
27	729	3	675	72*	5184	6	3168	140	19 600	7	8 840
30	900	3	800	75	5625	5	3375	140	19 600	10	9 044
30	900	5	864	80	6400	5	3744	144	20 736	8	9 180
32*	1024	4	896	80	6400	8	3900	144	20 736	9	9 248
35	1225	5	1071	81	6561	9	4131	150	22 500	10	9 975
36	1296	4	1071	84	7056	6	4004	160	25 600	8	10 800
36	1296	6	1188	84	7056	7	4056	160	25 600	10	10 944
40	1600	4	1260	90	8100	6	4455	162*	26 244	9	11 016
40	1600	5	1296	90	8100	9	4760	170	28 900	10	11 951
45	2025	5	1539	91	8281	7	4563	180	32 400	9	12 920
48	2304	4	1680	96	9216	6	4928	180	32 400	10	12 996
48	2304	6	1760	96	9216	8	5040	190	36 100	10	14 089
49	2401	7	1911	98*	9604	7	5096	200*	40 000	10	15 200

Legend:

$N=$ Inputs, outputs per network.
$n=$ Inputs per P-switch, outputs per T-switch.
$X=$ Crosspoints required.
*= Optimum configuration (where $N=2n^2$, except when $n=3$).

If $N>160$, a 5-stage nonblocking network can be designed—by judicious selection of the parameters n and a—which requires fewer crosspoints than the most favorable nonblocking 3-stage network. The advantage increases slowly with N; at $N=240$ the advantage is less than 5%.

The number of crosspoints is given by

$$X=N(2n-1)\{2+(2a-1)[(2/n)+(N/n^2a^2)]\}.$$

A minimum number of crosspoints is obtained when n and a satisfy the equations

$$N=2na^3/(a-1)$$
$$=[na^2(2n^2+2a-1)]/[(2a-1)(n-1)].$$

When the number of inlets and outlets N is such that the switches of the center stage of a 5-stage network become large, each center-stage switch can with advantage be expanded in like manner into a nonblocking 3-stage assembly. A 7-stage network results that is nonblocking overall.

(F) "Folded" Nonblocking 3-Stage Networks: Figure 11 shows the case where the outlets from the tertiary-stage switches of a symmetrical 3-stage network are connected to the same lines or trunks as are the inlets to the primary-stage switches. The number of crosspoints required to ensure no internal blocking is significantly less than in the case described in (B), provided the control circuits permit a path through the network to be established either in the direction primary-to-secondary-to-tertiary-stage or in the reverse direction. Such a network is said to function as a "folded" network.

The condition for nonblocking is given by

$$S = n.$$

The number of crosspoints required is given by

$$X = N[2n + (N/n)].$$

A minimum number of crosspoints is obtained when n satisfies $2n^2 = N$. Corresponding values of n and N may be selected from columns 1 and 4 of Table 1. The number of crosspoints is then given by

$$X_{min} = 2N(2N)^{1/2}.$$

When N/n is not integral for the optimum value of n, some of the primary-stage and tertiary-stage switches may be provided with $(n-1)$ inlets and outlets each, their numbers being determined by the method indicated in (D).

(G) "Folded" Nonblocking 5-Stage Networks: When N is large (> 350), it is advantageous to use a 5-stage folded network. Each secondary-stage switch may be expanded into a nonfolded nonblocking 3-stage subnetwork by the method described in (E). Figure 12 shows part of the resulting network (compare Fig. 10). The number of crosspoints is given by

$$X = N\{2n + (2a-1)[2 + (N/na^2)]\}.$$

A minimum number of crosspoints is obtained when n and a satisfy the equations

$$N = 4na^3/(4a-1) = 2n^2a^2/(2a-1).$$

Since $a \leq n$, approximate values for n and a are given by

$$N = n^3 = a^3.$$

Another method of expanding a "folded" nonblocking 3-stage network into a 5-stage network, when the secondary-stage switches become uneconomically large, is shown in Fig. 13. The secondary stage is divided into a number of groups of smaller switches, and the primary and tertiary stages are each divided into a number of 2-stage subnetworks. The primary- and tertiary-stage subnetworks and secondary-stage groups are cross-connected by a single-linkage spread in such a way that one switch in every primary-stage subnetwork has one connection to one switch in every secondary-stage group, and vice-versa. The number of crosspoints is given by

$$X = N\{2n + (2a-1)[2 + (N/na^2)]\}.$$

(H) Source Reference: Further information for the design of nonblocking multistage coordinate switching networks may be obtained from C. Clos, "A Study of Non-Blocking Switching Networks," Bell System Tech. J.; March 1953.

Networks with Internal Blocking

General Case: When a network presenting no internal blocking is not a requirement, a multistage network providing full availability between the inlets and the outlets—but introducing a measure of blocking—can be designed that requires fewer crosspoints. The number of secondary-stage switches S required in a 3-stage network to interconnect N inlets and M outlets, with a selected value of n, may be represented in a practical case as shown in Fig. 14 by $S = kn$, where k (constant) $< (2n-1)/n$ determines the blocking.

The number of crosspoints required is given by

$$X = kn[N + M + (NM/n^2)].$$

A minimum number of crosspoints is obtained if n satisfies

$$n^2 = NM/(N+M).$$

The number of crosspoints is then given by

$$X_{min} = 2k[NM(N+M)]^{1/2}.$$

Symmetrical Case—"Folded" Network Preferred: When the network is symmetrical, $M = N$ and, if $n = \frac{1}{2}(2N)^{1/2}$

$$X_{min} = 2kN(2N)^{1/2}.$$

When the inlets and outlets of a symmetrical network are connected to the same lines or trunks, it is advantageous to arrange the network to function as a "folded" network (see (F) above). The internal blocking for a given value of k is thereby significantly reduced.

Traffic-Carrying Capability of Blocking Networks: Equations and tables for determining the blocking in 3-stage networks, and the switch dimensions required to carry a given density of traffic, may be found in the following publications.

M. Van Den Bossche and R. G. Knight, "Traffic Problems and Blocking in a Three-Link Switching System," Bell Telephone Manufacturing Company, Antwerp; February 1957.

T. L. Bowers, "Blocking in 3-Stage 'Folded' Switching Arrays," IEEE Paper No. CP 63-1461.

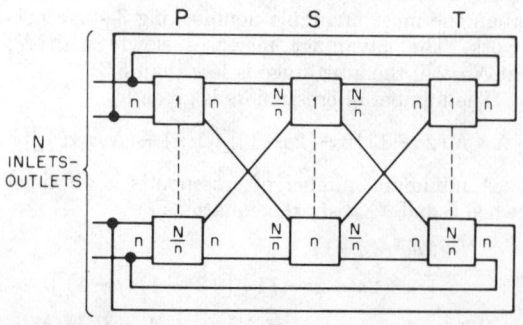

Fig. 11—"Folded" nonblocking 3-stage network.

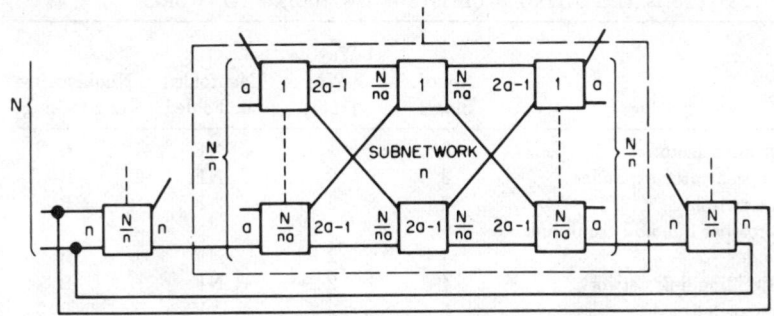

Fig. 12—"Folded" non blocking 5-stage network.

T. L. Bowers, "Derivation of Blocking Formulae for 3-Stage 'Folded' Switching Arrays," IEEE Paper No. CP 63-1462.

Switching Systems Employing Multistage Networks

Civilian and military switching systems in the US, which employ multistage coordinate switching networks satisfying various combinations of the elective parameters, include the typical examples shown in Table 4.

COMPARISON OF CROSSPOINT DEVICES

Table 5 lists comparative information relating to crosspoint devices currently in use (1967) in space-division switching systems for telephone, telegraph, and digital data communications. The table shows estimates of selected properties of three

of the most widely adopted devices. The figures refer to crosspoints in coordinate switching networks used in military switching centers, which normally carry heavier traffic between fewer lines than do civilian switching centers.

PART 2—TRAFFIC CONCEPTS

DEFINITIONS OF TERMS

Lost Calls Cleared: The assumption that calls not immediately satisfied at the first attempt are cleared from the system and do not reappear during the period under consideration. Used in the Erlang *B* loss-probability equation.

Lost Calls Delayed: The assumption that calls not immediately satisfied are held in the system until satisfied. Used in the Erlang *C* delay-probability equation.

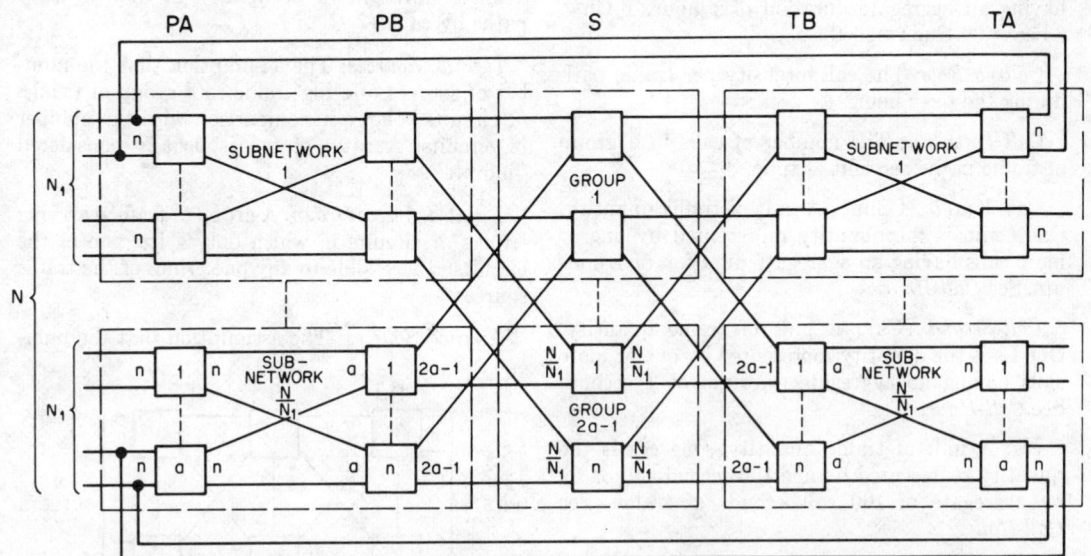

Fig. 13—"Folded" nonblocking 5-stage network (alternative arrangement).

TABLE 4—TYPICAL SYSTEMS EMPLOYING MULTISTAGE SWITCHING NETWORKS.

Company	System	No. of Stages	1-Way or 2-Way Traffic	Nonfolded or Folded	Nonblocking or Blocking
AECo	EAX group selector	3	1	NF	B
ITT	Telex concentrator-expander	3	2	NF	B
ITT	Autovon switch	3	2	F	NB
AT&T	No. 5 crossbar line and trunk link network	4	2	NF	B
AT&T	No. 1 ESS line link network	4	2	NF	B
AECo	Autovon switch	5	2	F	NB
ITT	Telex directional matrix	3 or 5	1	NF	B
AT&T	No. 1 ESS line and trunk link network	8	2	NF	B

Lost Calls Held: The assumption that calls not immediately satisfied at the first attempt are held in the system for a period not exceeding the average holding time of all calls and are thereafter cleared from the system. Used in the Poisson loss-probability equation.

Busy Hour: The continuous one-hour period which has the maximum average traffic intensity.

Busy-Hour Call (BHC): A European unit of call intensity. One BHC is one call during the busy hour.

Call: A discrete engagement or occupation of a traffic path.

Call Concentration: The average ratio of calls during the busy hour (BHC) to calls during the day.

Call-Hour (Ch): A unit of traffic quantity. One Ch is the quantity represented by one or more calls having an aggregate duration of 1 hour. 1 Ch≡ 36 ccs≡ 60 Cmin≡ 3600 Cs.

Calling Rate: The call intensity per traffic path during the busy hour.

Call Intensity: The number of calls in a group of traffic paths per unit of time.

Call-Minute (Cmin): A unit of traffic quantity. One Cmin is the quantity represented by one or more calls having an aggregate duration of 1 minute. See *Call-Hour*.

Call-Second (Cs): A unit of traffic quantity. One Cs is the quantity represented by one or more calls having an aggregate duration of 1 second. See *Call-Hour*.

ccs: A unit of traffic quantity. One ccs is the quantity represented by one 100-second call or by an aggregate of 100 call-seconds of traffic. See *Call-Hour*.

Equated Busy-Hour Call (EBHC): A European unit of traffic intensity. One EBHC is the average intensity in one or more traffic paths occupied in the busy hour by one 2-minute call or for an aggregate duration of 2 minutes. 30 EBHC=1 E (numerically).

Erlang (E): The international dimensionless unit of traffic intensity. One E is the intensity in a traffic path continuously occupied, or in one or more paths carrying an aggregate traffic of 1 call-hour per hour, 1 call-minute per minute, etc.

Full-Availability Group: A group of traffic-carrying trunks or circuits in which every circuit is accessible to all the traffic sources.

Grade of Service: A measure of the probability that, during a specified period of peak traffic, a call offered to a group of trunks or circuits will fail to find an idle circuit at the first attempt. Usually applied to the busy hour of traffic.

Holding Time: The duration of occupancy of a traffic path by a call. Sometimes used to mean the average duration of occupancy of one or more paths by calls.

Infinite Sources: The assumption that the number of sources offering traffic to a group of trunks or circuits is large in comparison with the number of circuits. A ratio of ten is usually considered "infinite".

Limited-Access Group: A group of traffic-carrying trunks or circuits in which only a fraction of the circuits is accessible to any one group of the traffic sources.

Limited Sources: The assumption that the num-

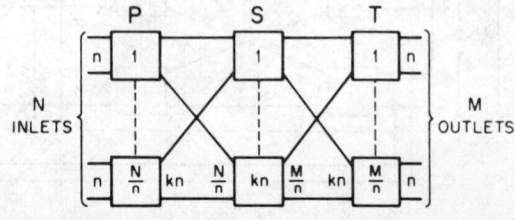

Fig. 14—Three-stage network with internal blocking.

TABLE 5—PROPERTIES OF CROSSPOINT DEVICES.

Parameter	Crossbar Switch	Reed Relay	*pnpn* Diode
Bandwidth limit ($\not>$1000-line network)	> 100 kHz	150 kHz*	> 50 kHz
Switching speed (3-stage network)	600–800 ms	3 ms	1 μs
Size (crosspoints/cu. ft.)	20	100†	> 10 000
Crosstalk (at 4 kHz, $\not>$1000-line network)	> 70 dB	88 dB	> 70 dB
Reliability (per crosspoint)	> 10^7 operations	10^8 operations	MTBF: 4%/10^5 hrs
Cost (per 2-way crosspoint)	$3–4 (6-wire)	$8–10 (6-wire)	$2 (2 diodes)

* Miniature type (in coaxial environment): > 50 MHz.

† Miniature type: 500.

MTBF = mean time before failure.

ber of sources offering traffic to a group of trunks or circuits does not exceed a specified number.

Occupancy: The traffic intensity in one or more traffic paths. 100% occupancy of one path≡1 E.

Traffic Concentration: The average ratio of the traffic quantity during the busy hour to the traffic quantity during the day.

Traffic Density: See *Traffic Intensity*.

Traffic Flow: See *Traffic Intensity*.

Traffic Intensity: The traffic quantity in one or more traffic paths per unit of time.

Traffic Load: See *Traffic Intensity*.

Traffic Path: A channel, time slot, frequency band, line, trunk, switch, or circuit over which individual communications pass in sequence.

Traffic (Quantity): The aggregate engagement time or occupancy time of one or more traffic paths.

Traffic Rate: The traffic intensity per traffic path during the busy hour.

Traffic Unit (TU): A unit of traffic intensity. One TU is the average intensity in one or more traffic paths carrying an aggregate traffic of 1 call-hour in one hour (the busy hour unless otherwise specified). 1 TU=1 E (numerically).

Unit Call (UC): A unit of traffic intensity. One UC is the average intensity in one or more traffic paths occupied during one hour by one 100-second call, or for an aggregate duration of 100 seconds. 36 UC≡1 E.

TRAFFIC UNITS AND THEIR RELATIONSHIP

In the terminology of traffic, in the context of communications, two expressions in common use cause much confusion and misunderstanding, "traffic quantity" and "traffic intensity". The following usage is well established in the telephone industry and is not in conflict (except where noted) with the authorities listed below. The traffic units in common use, and their relationship, are summarized in Table 6.

Traffic Quantity

The measure of "traffic quantity" is the aggregate engagement time or occupancy time of a traffic path or a group of paths [1]. It is expressed in hours (more often, call-hours), minutes, or seconds and does not, of itself, refer to any particular period of time. The dimension of the unit of traffic quantity is t.

A more practical expression, derived from the above definition, is used in some U.S. papers, namely, the ccs, defined [1, 2] (strictly) by

1 ccs= one 100-second call or

$$100 \text{ call-seconds of traffic.}$$

It follows that

$$1 \text{ call-hour} = 36 \text{ ccs (strictly).}$$

Call Intensity

The expression "busy-hour call" (BHC), in common use in Europe [1], means one call *of unspecified duration* in the busy hour and is *not* a measure of "traffic quantity". Its dimension is t^{-1}.

Traffic Intensity

The measure of traffic intensity [1, 2, 3] (sometimes called traffic density [4], traffic load [5], traffic flow [6, 7], or just traffic) is the traffic quantity in a traffic path or a group of paths *per unit of time*. It is expressed in erlangs (E) or traffic units (TU) [6], defined by

1 E=1 TU=1 call-hour per hour

$$= 1 \text{ call-minute per minute, etc}$$

TABLE 6—TRAFFIC UNITS.

Term	Unit	Abbreviation	Dimension
Busy hour	Hour	BH	t
Call	Call	C	None
Call intensity	Call/hour, etc.	C/h, BHC*	t^{-1}
Holding time	Minute, second, etc.	min, s	t
Traffic (quantity):	Call-hour	Ch	t
1 Ch = 60 Cmin	Call-minute	Cmin	t
= 3600 Cs	Call-second	Cs	t
= 36 ccs	Hundred call-seconds	ccs	t
Traffic intensity:	Erlang	E	None
1 E = 1 TU†	Traffic unit	TU†	None
= 30 EBHC*	Equated busy-hour call	EBHC*	None
= 36 ccs/h‡ = 36 UC	Unit call	UC	None
= 60 Cmin/h			
= 3600 Cs/h			

* Applied to the busy hour only.
† "In the busy hour" implied, unless otherwise stated.
‡ Often abbreviated "ccs", with "per hour" implied.

The unit of traffic intensity is a pure number, a ratio without dimension. One erlang of traffic intensity in one traffic path means continuous occupancy of that path, *irrespective of the duration* of that occupancy. The erlang was adopted in 1946 by the CCIF (now CCITT) as the international unit of traffic intensity [3, 4].

Another expression, in common use in Europe, is the "equated busy-hour call" (EBHC), which is defined as the average traffic intensity in the busy hour resulting from one 2-minute call or from several calls having an aggregate duration of two minutes [1].

In the US, the accepted unit of traffic intensity is the "ccs per hour" or the "unit call" (UC) that is synonymous with it [2, 5, 8, 9]. However, some tables [9] use ccs to mean the *intensity* "ccs per hour" rather than the aggregate *quantity* of 100-call-second units that this expression literally means. Although it has no inherent connotation of traffic intensity, its use in that sense *must be accepted as conventional practice*. It follows that

$$1 \text{ E} = 36 \text{ UC} = 36 \text{ ccs per hour (strictly)}$$

$$= 36 \text{ ccs (conventionally)}.$$

Only the EBHC has any inherent connotation of the busy hour, or of any particular time interval. But the TU, as used by the British Post Office and most European administrations, has an *implied* association with the busy hour, unless otherwise stated [6].

The use of the term "traffic" to mean "traffic intensity" is deprecated.

Ambiguity of "ccs"

The dual use of the ccs unit in American practice has been explained [2] as follows:

"Traffic intensity may also be expressed in 'ccs per hour' and many engineering tables are given in these units. The relation to the 'erlang' is: 1 erlang = 36 ccs per hour."

"In engineering practice, units of 'hundred call-seconds per hour' are often abbreviated 'ccs', with the 'per hour' *implied but not stated*."

References

1. C. F. J. Boehlen, "Telephone Traffic Definitions," *Telephone Engineer and Management*, para. 2b, 2g; Feb.–March 1944.
2. "Switching Systems," AT&T Company; 1961: page 13.
3. Brockmeyer, Halstrom, and Jensen, "Life and Works of A. K. Erlang," *Transactions of the Danish Academy of Technical Sciences*, vol. 2, p. 21, Copenhagen; 1948.

TABLE 7—TRUNK LOADING CAPACITY—FULL AVAILABILITY. ERLANG B EQUATION. *From Bulletin 485,* © *1953, Automatic Electric Company.*

Trunks	Grade of Service 1 in 1000		Grade of Service 1 in 500		Grade of Service 1 in 200		Grade of Service 1 in 100		Grade of Service 1 in 50		Grade of Service 1 in 20	
	UC	TU	UC	TU	UC	TU	UC	TU	UC	TU	UC	TU
1	0.04	0.001	0.07	0.002	0.2	0.005	0.4	0.01	0.7	0.02	1.8	0.05
2	1.8	0.05	2.5	0.07	4	0.11	5.4	0.15	7.9	0.22	14	0.38
3	6.8	0.19	9	0.25	13	0.35	17	0.46	22	0.60	32	0.90
4	16	0.44	19	0.53	25	0.70	31	0.87	39	1.09	55	1.52
5	27	0.76	32	0.90	41	1.13	49	1.36	60	1.66	80	2.22
6	41	1.15	48	1.33	58	1.62	69	1.91	82	2.28	107	2.96
7	57	1.58	65	1.80	78	2.16	90	2.50	106	2.94	135	3.74
8	74	2.05	83	2.31	98	2.73	113	3.13	131	3.63	163	4.54
9	92	2.56	103	2.85	120	3.33	136	3.78	156	4.34	193	5.37
10	111	3.09	123	3.43	143	3.96	161	4.46	183	5.08	224	6.22
11	131	3.65	145	4.02	166	4.61	186	5.16	210	5.84	255	7.08
12	152	4.23	167	4.64	190	5.28	212	5.88	238	6.62	286	7.95
13	174	4.83	190	5.27	215	5.96	238	6.61	267	7.41	318	8.83
14	196	5.45	213	5.92	240	6.66	265	7.35	295	8.20	350	9.73
15	219	6.08	237	6.58	266	7.38	292	8.11	324	9.01	383	10.63
16	242	6.72	261	7.26	292	8.10	319	8.87	354	9.83	415	11.54
17	266	7.38	286	7.95	318	8.83	347	9.65	384	10.66	449	12.46
18	290	8.05	311	8.64	345	9.58	376	10.44	414	11.49	482	13.38
19	314	8.72	337	9.35	372	10.33	404	11.23	444	12.33	515	14.31
20	339	9.41	363	10.07	399	11.09	433	12.03	474	13.18	549	15.25
21	364	10.11	388	10.79	427	11.86	462	12.84	505	14.04	583	16.19
22	389	10.81	415	11.53	455	12.63	491	13.65	536	14.90	617	17.13
23	415	11.52	442	12.27	483	13.42	521	14.47	567	15.76	651	18.08
24	441	12.24	468	13.01	511	14.20	550	15.29	599	16.63	685	19.03
25	467	12.97	495	13.76	540	15.00	580	16.12	630	17.50	720	19.99
26	493	13.70	523	14.52	569	15.80	611	16.96	662	18.38	754	20.94
27	520	14.44	550	15.28	598	16.60	641	17.80	693	19.26	788	21.90
28	546	15.18	578	16.05	627	17.41	671	18.64	725	20.15	823	22.87
29	573	15.93	606	16.83	656	18.22	702	19.49	757	21.04	858	23.83
30	600	16.68	634	17.61	685	19.03	732	20.34	789	21.93	893	24.80
31	628	17.44	662	18.39	715	19.85	763	21.19	822	22.83	928	25.77
32	655	18.20	690	19.18	744	20.68	794	22.05	854	23.73	963	26.75
33	683	18.97	719	19.97	774	21.51	825	22.91	887	24.63	998	27.72
34	711	19.74	747	20.76	804	22.34	856	23.77	919	25.53	1033	28.70
35	739	20.52	776	21.56	834	23.17	887	24.64	951	26.43	1068	29.68
36	767	21.30	805	22.36	864	24.01	918	25.51	984	27.34	1104	30.66
37	795	22.03	834	23.17	895	24.85	950	26.38	1017	28.25	1139	31.64
38	823	22.86	863	23.97	925	25.69	981	27.25	1050	29.17	1175	32.63
39	851	23.65	892	24.78	955	26.53	1013	28.13	1083	30.08	1210	33.61
40	880	24.44	922	25.60	986	27.38	1044	29.01	1116	31.00	1246	34.60
41	909	25.24	951	26.42	1016	28.23	1076	29.89	1149	31.92	1281	35.59
42	937	26.04	981	27.24	1047	29.08	1108	30.77	1182	32.84	1317	36.58
43	966	26.84	1010	28.06	1078	29.94	1140	31.66	1215	33.76	1353	37.57
44	995	27.64	1040	28.88	1109	30.80	1171	32.54	1248	34.68	1388	38.56
45	1024	28.45	1070	29.71	1140	31.66	1203	33.43	1282	35.61	1424	39.55
46	1053	29.26	1099	30.54	1171	32.52	1236	34.32	1315	36.53	1459	40.54
47	1083	30.07	1129	31.37	1202	33.38	1268	35.21	1349	37.46	1495	41.54
48	1111	30.88	1159	32.20	1233	34.25	1300	36.11	1382	38.39	1531	42.54
49	1141	31.69	1189	33.04	1264	35.11	1332	37.00	1415	39.32	1567	43.54
50	1170	32.51	1220	33.88	1295	35.98	1364	37.90	1449	40.25	1603	44.53

TABLE 7—*Continued*

Trunks	Grade of Service 1 in 1000		Grade of Service 1 in 500		Grade of Service 1 in 200		Grade of Service 1 in 100	
	UC	TU	UC	TU	UC	TU	UC	TU
51	1200	33.33	1250	34.72	1327	36.85	1397	38.80
52	1229	34.15	1280	35.56	1358	37.72	1429	39.70
53	1259	34.98	1310	36.40	1390	38.60	1462	40.60
54	1289	35.80	1341	37.25	1421	39.47	1494	41.50
55	1319	36.63	1371	38.09	1453	40.35	1527	42.41
56	1349	37.46	1402	38.94	1484	41.23	1559	43.31
57	1378	38.29	1432	39.79	1516	42.11	1592	44.22
58	1408	39.12	1463	40.64	1548	42.99	1625	45.13
59	1439	39.96	1494	41.50	1579	43.87	1657	46.04
60	1468	40.79	1525	42.35	1611	44.76	1690	46.95
61	1499	41.63	1556	43.21	1643	45.64	1723	47.86
62	1529	42.47	1587	44.07	1675	46.53	1756	48.77
63	1559	43.31	1617	44.93	1707	47.42	1789	49.69
64	1590	44.16	1648	45.79	1739	48.31	1822	50.60
65	1620	45.00	1679	46.65	1771	49.20	1855	51.52
66	1650	45.84	1710	47.51	1803	50.09	1888	52.44
67	1681	46.69	1742	48.38	1835	50.98	1921	53.35
68	1711	47.54	1773	49.24	1867	51.87	1954	54.27
69	1742	48.39	1804	50.11	1900	52.77	1987	55.19
70	1773	49.24	1835	50.98	1932	53.66	2020	56.11
71	1803	50.09	1867	51.85	1964	54.56	2053	57.03
72	1834	50.94	1898	52.72	1996	55.45	2087	57.96
73	1865	51.80	1929	53.59	2029	56.35	2120	58.88
74	1895	52.65	1960	54.46	2061	57.25	2153	59.80
75	1926	53.51	1992	55.34	2093	58.15	2186	60.73
76	1957	54.37	2024	56.21	2126	59.05	2219	61.65
77	1988	55.23	2055	57.09	2159	59.96	2253	62.58
78	2019	56.09	2087	57.96	2191	60.86	2286	63.51
79	2050	56.95	2118	58.84	2223	61.76	2319	64.43
80	2081	57.81	2150	59.72	2256	62.67	2353	65.36
81	2112	58.67	2182	60.60	2289	63.57	2386	66.29
82	2143	59.54	2213	61.48	2321	64.48	2420	67.22
83	2174	60.40	2245	62.36	2354	65.38	2453	68.15
84	2206	61.27	2277	63.24	2386	66.29	2487	69.08
85	2237	62.14	2308	64.13	2419	67.20	2521	70.02
86	2268	63.00	2340	65.01	2452	68.11	2554	70.95
87	2299	63.87	2372	65.90	2485	69.02	2588	71.88
88	2330	64.74	2404	66.78	2517	69.93	2621	72.81
89	2362	65.61	2436	67.67	2550	70.84	2655	73.75
90	2393	66.48	2468	68.56	2583	71.76	2688	74.68
91	2425	67.36	2500	69.44	2616	72.67	2722	75.62
92	2456	68.23	2532	70.33	2650	73.58	2756	76.56
93	2488	69.10	2564	71.22	2682	74.49	2790	77.49
94	2519	69.98	2596	72.11	2715	75.41	2823	78.43
95	2551	70.85	2628	73.00	2748	76.32	2857	79.37
96	2582	71.73	2660	73.90	2781	77.24	2891	80.31
97	2614	72.61	2692	74.79	2814	78.16	2925	81.24
98	2645	73.48	2724	75.68	2847	79.07	2958	82.18
99	2677	74.36	2757	76.57	2880	79.99	2992	83.12
100	2709	75.24	2789	77.47	2913	80.91	3026	84.06

TABLE 7—*Continued*

Trunks	Grade of Service 1 in 1000		Grade of Service 1 in 500		Grade of Service 1 in 200		Grade of Service 1 in 100	
	UC	TU	UC	TU	UC	TU	UC	TU
101	2740	76.12	2821	78.36	2946	81.83	3060	85.00
102	2772	77.00	2853	79.26	2979	82.75	3094	85.95
103	2804	77.88	2886	80.16	3012	83.67	3128	86.89
104	2836	78.77	2918	81.05	3045	84.59	3162	87.83
105	2867	79.65	2950	81.95	3078	85.51	3196	88.77
106	2899	80.53	2983	82.85	3111	86.43	3230	89.72
107	2931	81.42	3015	83.75	3145	87.35	3264	90.66
108	2963	82.30	3047	84.65	3178	88.27	3298	91.60
109	2995	83.19	3080	85.55	3211	89.20	3332	92.55
110	3027	84.07	3112	86.45	3244	90.12	3366	93.49
111	3059	84.96	3145	87.35	3277	91.04	3400	94.44
112	3091	85.85	3177	88.25	3311	91.97	3434	95.38
113	3122	86.73	3209	89.15	3344	92.89	3468	96.33
114	3154	87.62	3242	90.06	3378	93.82	3502	97.28
115	3186	88.51	3275	90.96	3411	94.74	3536	98.22
116	3218	89.40	3307	91.86	3444	95.67	3570	99.17
117	3250	90.29	3340	92.77	3478	96.60	3604	100.12
118	3282	91.18	3372	93.67	3511	97.53	3639	101.07
119	3315	92.07	3405	94.58	3544	98.45	3673	102.02
120	3347	92.96	3437	95.48	3578	99.38	3707	102.96
121	3379	93.86	3470	96.39	3611	100.31	3741	103.91
122	3411	94.75	3503	97.30	3645	101.24	3775	104.86
123	3443	95.64	3535	98.20	3678	102.17	3809	105.81
124	3475	96.54	3568	99.11	3712	103.10	3843	106.76
125	3507	97.43	3601	100.02	3745	104.03	3878	107.71
126	3540	98.33	3633	100.93	3779	104.96	3912	108.66
127	3572	99.22	3666	101.84	3812	105.89	3946	109.62
128	3604	100.12	3699	102.75	3846	106.82	3981	110.57
129	3636	101.01	3732	103.66	3879	107.75	4015	111.52
130	3669	101.91	3765	104.57	3912	108.68	4049	112.47
131	3701	102.81	3797	105.48	3946	109.62	4083	113.42
132	3733	103.70	3830	106.39	3980	110.55	4118	114.38
133	3766	104.60	3863	107.30	4013	111.48	4152	115.33
134	3798	105.50	3896	108.22	4047	112.42	4186	116.28
135	3830	106.40	3929	109.13	4081	113.35	4221	117.24
136	3863	107.30	3961	110.04	4114	114.28	4255	118.19
137	3895	108.20	3994	110.95	4148	115.22	4289	119.14
138	3928	109.10	4027	111.87	4181	116.15	4324	120.10
139	3960	110.00	4060	112.78	4215	117.09	4358	121.05
140	3992	110.90	4093	113.70	4249	118.02	4392	122.01
141	4025	111.81	4126	114.61	4283	118.96	4427	122.96
142	4058	112.71	4159	115.53	4316	119.90	4461	123.92
143	4090	113.61	4192	116.44	4350	120.83	4496	124.88
144	4122	114.51	4225	117.36	4384	121.77	4530	125.83
145	4155	115.42	4258	118.28	4418	122.71	4564	126.79
146	4188	116.32	4291	119.19	4451	123.64	4599	127.74
147	4220	117.22	4324	120.11	4485	124.58	4633	128.70
148	4253	118.13	4357	121.03	4519	125.52	4668	129.66
149	4285	119.03	4390	121.95	4552	126.46	4702	130.62
150	4318	119.94	4423	122.86	4586	127.40	4737	131.58

Table 8—Trunk Loading Capacity—Full Availability. Poisson Equation.
From Bulletin 485, © 1953, Automatic Electric Company.

Trunks	Grade of Service 1 in 1000		Grade of Service 1 in 100		Grade of Service 1 in 50		Grade of Service 1 in 20		Grade of Service 1 in 10	
	UC	TU	UC	TU	UC	TU	UC	TU	UC	TU
1	0.1	0.003	0.4	0.01	0.7	0.02	1.9	0.05	3.8	0.10
2	1.6	0.05	5.4	0.15	7.9	0.20	12.9	0.35	19.1	0.55
3	6.9	0.20	16	0.45	20	0.55	29.4	0.80	39.6	1.10
4	15	0.40	30	0.85	37	1.05	49	1.35	63	1.75
5	27	0.75	46	1.30	56	1.55	71	1.95	88	2.45
6	40	1.10	64	1.80	76	2.10	94	2.60	113	3.15
7	55	1.55	84	2.35	97	2.70	118	3.25	140	3.90
8	71	1.95	105	2.90	119	3.30	143	3.95	168	4.65
9	88	2.45	126	3.50	142	3.95	169	4.70	195	5.40
10	107	2.95	149	4.15	166	4.60	195	5.40	224	6.20
11	126	3.50	172	4.80	191	5.30	222	6.15	253	7.05
12	145	4.05	195	5.40	216	6.00	249	6.90	282	7.85
13	166	4.60	220	6.10	241	6.70	277	7.70	311	8.65
14	187	5.20	244	6.80	267	7.40	305	8.45	341	9.45
15	208	5.80	269	7.45	293	8.15	333	9.25	370	10.30
16	231	6.40	294	8.15	320	8.90	362	10.05	401	11.15
17	253	7.05	320	8.90	347	9.65	390	10.85	431	11.95
18	276	7.65	346	9.60	374	10.40	419	11.65	462	12.85
19	299	8.30	373	10.35	401	11.15	448	12.45	492	13.65
20	323	8.95	399	11.10	429	11.90	477	13.25	523	14.55
21	346	9.60	426	11.85	458	12.70	507	14.10	554	15.40
22	370	10.30	453	12.60	486	13.50	536	14.90	585	16.25
23	395	10.95	480	13.35	514	14.30	566	15.70	616	17.10
24	419	11.65	507	14.10	542	15.05	596	16.55	647	17.95
25	444	12.35	535	14.85	572	15.90	626	17.40	678	18.85
26	469	13.05	562	15.60	599	16.65	656	18.20	710	19.70
27	495	13.75	590	16.40	627	17.40	686	19.05	741	20.60
28	520	14.45	618	17.15	656	18.20	717	19.90	773	21.45
29	545	15.15	647	17.95	685	19.05	747	20.75	805	22.35
30	571	15.85	675	18.75	715	19.85	778	21.60	836	23.20
31	597	16.60	703	19.55	744	20.65	809	22.45	868	24.10
32	624	17.35	732	20.35	773	21.45	840	23.35	900	25.00
33	650	18.05	760	21.10	803	22.30	871	24.20	932	25.90
34	676	18.80	789	21.90	832	23.10	902	25.05	964	26.80
35	703	19.55	818	22.70	862	23.95	933	25.90	996	27.65
36	729	20.25	847	23.55	892	24.80	964	26.80	1028	28.55
37	756	21.00	876	24.35	922	25.60	995	27.65	1060	29.45
38	783	21.75	905	25.15	951	26.40	1026	28.50	1092	30.35
39	810	22.50	935	25.95	982	27.30	1057	29.35	1125	31.25
40	837	23.25	964	26.80	1012	28,10	1088	30.20	1157	32.14
41	865	24.05	993	27.60	1042	28.95	1120	31.10	1190	33.05
42	892	24.80	1023	28.40	1072	29.80	1151	31.95	1222	33.95
43	919	25.55	1052	29.20	1103	30.65	1183	32.85	1255	34.85
44	947	26.30	1082	30.05	1133	31.45	1214	33.70	1287	35.75
45	975	27.10	1112	30.90	1164	32.35	1246	34.60	1320	36.65
46	1003	27.85	1142	31.70	1194	33.15	1277	35.45	1352	37.55
47	1030	28.60	1171	32.55	1225	34.05	1309	36.35	1385	38.45
48	1058	29.40	1201	33.35	1255	34.85	1340	37.20	1417	39.35
49	1086	30.15	1231	34.20	1286	35.70	1372	38.10	1450	40.30
50	1115	30.95	1261	35.05	1317	36.60	1403	38.95	1482	41.15

TABLE 8—*Continued*

Trunks	Grade of Service 1 in 1000		Grade of Service 1 in 100		Grade of Service 1 in 50	
	UC	TU	UC	TU	UC	TU
51	1143	31.75	1291	35.85	1349	37.45
52	1171	32.55	1322	36.70	1380	38.35
53	1200	33.35	1352	37.55	1410	39.15
54	1228	34.10	1382	38.40	1441	40.05
55	1256	34.90	1412	39.20	1472	40.90
56	1285	35.70	1443	40.10	1503	41.75
57	1313	36.45	1473	40.90	1534	42.60
58	1342	37.30	1504	41.80	1565	43.45
59	1371	38.10	1534	42.60	1596	44.35
60	1400	38.90	1565	43.45	1627	45.20
61	1428	39.65	1595	44.30	1659	46.10
62	1457	40.45	1626	45.15	1690	46.95
63	1486	41.30	1657	46.05	1722	47.85
64	1516	42.10	1687	46.85	1752	48.65
65	1544	42.90	1718	47.70	1784	49.55
66	1574	43.70	1749	48.60	1816	50.45
67	1603	44.55	1780	49.45	1847	51.30
68	1632	45.35	1811	50.30	1878	52.15
69	1661	46.15	1842	51.15	1910	53.05
70	1691	46.95	1873	52.05	1941	53.90
71	1720	47.80	1904	52.90	1973	54.80
72	1750	48.60	1935	53.75	2004	55.65
73	1779	49.40	1966	54.60	2036	56.55
74	1809	50.25	1997	55.45	2067	57.40
75	1838	51.05	2028	56.35	2099	58.30
76	1868	51.90	2059	57.20	2130	59.15
77	1898	52.70	2091	58.10	2162	60.05
78	1927	53.55	2122	58.95	2194	60.95
79	1957	54.35	2153	59.80	2226	61.85
80	1986	55.15	2184	60.65	2258	62.70
81	2016	56.00	2215	61.55	2290	63.60
82	2046	56.85	2247	62.40	2321	64.45
83	2076	57.65	2278	63.30	2354	65.40
84	2106	58.50	2310	64.15	2386	66.30
85	2136	59.35	2341	65.05	2418	67.15
86	2166	60.15	2373	65.90	2451	68.10
87	2196	61.00	2404	66.80	2483	68.95
88	2226	61.85	2436	67.65	2515	69.85
89	2256	62.65	2467	68.55	2547	70.75
90	2286	63.50	2499	69.40	2579	71.65
91	2317	64.35	2530	70.30	2611	72.55
92	2346	65.15	2562	71.15	2643	73.40
93	2377	66.05	2594	72.05	2674	74.30
94	2407	66.85	2625	72.90	2706	75.15
95	2437	67.70	2657	73.80	2739	76.10
96	2468	68.55	2689	74.70	2771	76.95
97	2498	69.40	2721	75.60	2803	77.85
98	2528	70.20	2752	76.45	2836	78.80
99	2559	71.10	2784	77.35	2868	79.65
100	2589	71.90	2816	78.20	2900	80.55

4. J. Kruithof, "Telephone Traffic Calculus," Bell Telephone Manufacturing Company, Antwerp; January 1952.

5. A. L. Gracey, "Basic Theory Underlying Bell System Facilities Capacity Tables," *Transactions of the AIEE*, vol. 69, p. 238; 1950.

6. G. S. Berkeley, "Traffic and Trunking Principles in Automatic Telephony," Ernest Benn Ltd, London; 1949: pages 18 and 315.

7. CCITT Blue Book, vol. II, Geneva; 1965: page 70.

8. "Traffic Engineering Practices—Trunk Facilities—Basic Trunk Tables," AT&T Co.; March 1960: Div. G, Section 5-a.

9. "Method of Calculating Trunking and Switch Quantities for Strowger Automatic Telephone Exchanges," Bulletin 485, Automatic Electric Company, Chicago, Illinois; 1953.

TRAFFIC EQUATIONS AND TABLES

Grade of Service

The overall "grade of service" of a switching system refers to the anticipated ratio of calls "lost" at the first attempt to the total number of attempts to establish a connection through the system during a specified period of time, usually the busy hour. The grades of service contributed by each constituent switch, switching network, or trunk group are usually added to arrive at the overall grade of service.

The grade of service provided by a particular group of trunks or circuits of specified size and carrying a specified traffic intensity, or the probability that a call offered to the group will find all available trunks already occupied, depends on a number of factors, the most important of which follow.

(**A**) The distribution in time and duration of the offered traffic; for example, random or periodic arrival, and constant or exponentially distributed holding time.

(**B**) The number of traffic sources; for example, high ("infinite") or limited.

(**C**) The availability of the trunks in the group to the traffic sources; for example, full availability or restricted access.

(**D**) The conditions under which calls are "lost" or blocked (see definitions).

Traffic Equations

The two most commonly used equations for determining the probability of blocked calls are known as the Erlang B and the Poisson equations. Both assume "infinite sources".

The Erlang B equation is based on the further assumption that lost-calls-cleared conditions apply. It has been standardized by the CCITT and is used throughout Europe.

The Poisson equation, also known as the Molina equation, is based on the alternative assumption that lost-calls-held conditions apply and is generally preferred in the U.S.

The Erlang C equation is also based on "infinite sources" and assumes lost-calls-delayed conditions.

If the number of sources is limited, the less frequently used Engset equation corresponds to the Erlang B, and the Binomial equation corresponds to the Poisson.

The Erlang B equation is given by

$$P = (y^n/n!)/(\sum_0^n y^n/n!).$$

The Poisson equation is given by

$$P = e^{-y} \sum_{n+1}^{\infty} (y^n/n!)$$

where, in both cases, P = probability of loss or blocking, n = number of trunks, and y = traffic offered (in erlangs).

Traffic Tables

Table 7 shows the maximum traffic intensity that a group of trunks will carry under full-availability conditions, with Erlang B assumptions, for several required grades of service. The traffic intensities are tabulated both in unit calls UC (ccs per hour) and in traffic units TU (erlangs). A similar table, based on grades of service of 1, 3, 5, and 7 in 100, is to be found in the CCITT Blue Book [10].

Table 8 shows corresponding trunk loading capabilities based on the use of the Poisson equation. This table is the familiar "American Table."

Other tables and curves based on the Erlang B, Poisson, Erlang C, Engset, and Binomial equations are in the sources listed below.

References

In addition to references [8] and [9] listed earlier, traffic information may be obtained from the following.

10. CCITT Blue Book, vol. II, Geneva; 1965: page 239.

11. "Switching Systems," AT&T Company; 1961: Chapter 3.

THEORY OF SOUND WAVES*

Sound (or a sound wave) is an alteration in pressure, stress, particle displacement, or particle velocity that is propagated in an elastic material; or the superposition of such propagated alterations. Sound (or sound sensation) is also the sensation produced through the ear by the above alterations.

Wave Equation

Behavior of sound waves is given by the wave equation

$$\nabla^2 p = (1/c^2)(\partial^2 p/\partial t^2) \qquad (1)$$

where p is the instantaneous pressure increment above and below a steady pressure (dynes/centimeter^2); p is a function of time and of the three coordinates of space. Also, $t =$ time in seconds, $c =$ velocity of propagation in centimeters/second, and $\nabla^2 =$ the Laplacian, which for the particular case of rectangular coordinates x, y, and z (in centimeters), is given by

$$\nabla^2 \equiv (\partial^2/\partial x^2) + (\partial^2/\partial y^2) + (\partial^2/\partial z^2). \qquad (2)$$

Plane Waves: For a plane wave of sound, where variations with respect to y and z are zero, $\nabla^2 p = \partial^2 p/\partial x^2 = d^2 p/dx^2$; the latter is approximately equal to the curvature of the plot of p versus x at some instant. Equation (1) states simply that, for variations in x only, the acceleration in pressure p (the second time derivative of p) is proportional to the curvature in p (the second space derivative of p).

Sinusoidal variations in time are usually of interest. For this case the standard procedure is to put $p = $ (real part of $\bar{p}\epsilon^{j\omega t}$), where the phasor $\bar{p}$ now satisfies the equation

$$\nabla^2 \bar{p} + (\omega/c)^2 \bar{p} = 0. \qquad (3)$$

* Lord Rayleigh, "Theory of Sound," vols. 1 and 2, Dover Publications, New York; 1945. P. M. Morse, "Vibration and Sound," 2nd edition, McGraw-Hill Book Company, New York; 1948.

Velocity phasor $\bar{v}$ of the sound wave in the medium is related to the complex pressure phasor $\bar{p}$ by

$$\bar{v} = -(1/j\omega\rho_0) \text{ grad } \bar{p}. \qquad (4)$$

Specific acoustic impedance $\bar{Z}$ at any point in the medium is the ratio of the pressure phasor to the velocity phasor, or

$$\bar{Z} = \bar{p}/\bar{v}. \qquad (5)$$

Spherical Waves: The solutions of (1) and (3) take particularly simple and instructive forms for the case of one dimensional plane and spherical waves in one direction. Table 1 summarizes the pertinent information.

For example, the acoustic impedance for spherical waves has an equivalent electrical circuit comprising a resistance shunted by an inductance. In this form, it is obvious that a small spherical source (r is small) cannot radiate efficiently since the radiation resistance $\rho_0 c$ is shunted by a small inductance $\rho_0 r$. Efficient radiation begins approximately at the frequency where the resistance $\rho_0 r$ equals the inductive (mass) reactance $\rho_0 c$. This is the frequency at which the period ($= 1/f$) equals the time required for the sound wave to travel the peripheral distance $2\pi r$.

Sound Intensity

The sound intensity is the average rate of sound energy transmitted in a specified direction through a unit area normal to this direction at the point considered. In the case of a plane or spherical wave, the intensity in the direction of propagation is given by

$$I = p^2/\rho c \text{ ergs/second/centimeter}^2 \qquad (6)$$

where $p =$ pressure (dynes/centimeter^2), $\rho =$ density of the medium (grams/centimeter^3), and $c =$ velocity of propagation (centimeters/second).

The sound intensity is usually measured in decibels, in which case it is known as the intensity

TABLE 1—SOLUTIONS FOR VARIOUS PARAMETERS.

Factor	Type of Sound Wave	
	Plane Wave	Spherical Wave
Equation for p	$\partial^2 p/\partial x^2 = (1/c^2)(\partial^2 p/\partial t^2)$	$(\partial^2 p/\partial x^2) + (2/r)(\partial p/\partial r) = (1/c^2)(\partial^2 p/\partial t^2)$
Equation for $\bar{p}$	$(d^2\bar{p}/dx^2) + (\omega/c)^2\bar{p} = 0$	$(d^2\bar{p}/dx^2) + (2/r)(d\bar{p}/dt) + (\omega/c)^2\bar{p} = 0$
Solution for p	$p = F[t-(x/c)]$	$p = (1/r)F[t-(x/c)]$
Solution for $\bar{p}$	$\bar{p} = \bar{A}\exp(-j\omega x/c)$	$\bar{p} = (1/r)\bar{A}\exp(-j\omega r/c)$
Solution for $\bar{v}$	$\bar{v} = (\bar{A}/\rho_0 c)\exp(-j\omega x/c)$	$\bar{v} = (\bar{A}/\rho_0 cr)[1+(c/j\omega r)]\exp(-j\omega r/c)$
$\bar{Z}$	$\bar{Z} = \rho_0 c$	$\bar{Z} = \rho_0 c/[1+(c/j\omega r)]$
Equivalent electrical circuit for $\bar{Z}$		

where

    $p =$ excess pressure in dynes/centimeter^2
    $\bar{p} =$ complex excess pressure in dynes/centimeter^2
    $t =$ time in seconds
    $x =$ space coordinate for plane wave in centimeters
    $r =$ space coordinate for spherical wave in centimeters
    $\bar{v} =$ complex velocity in centimeters/second
    $\bar{Z} =$ specific acoustic impedance in dyne-seconds/centimeter^3
    $c =$ velocity of propagation in centimeters/second
    $\omega = 2\pi f; f =$ frequency in hertz
    $F =$ an arbitrary function
    $\bar{A} =$ complex constant
    $\rho_0 =$ density of medium in grams/centimeter^3.

level and is equal to 10 times the logarithm (to the base 10) of the ratio of the sound intensity (expressed in watts/centimeter^2) to the reference level of 10^{-16} watt/centimeter^2. Table 2 shows the intensity levels of some familiar sounds.

SOUND IN GASES

The acoustic behavior of a medium is determined by its physical characteristics and, in the case of gases, by the density, pressure, temperature, specific heat, coefficients of viscosity, and the amount of heat exchange at the boundary surfaces.

The velocity of propagation in a gas is a function of the equation of state ($PV = RT$ plus higher-order terms), the molecular weight, and the specific heat.*

For small displacements relative to the wavelength of sound, the velocity is given by

$$c = (\gamma p_0/\rho_0)^{1/2} \qquad (7)$$

* H. C. Hardy, D. Telfair, and W. H. Pielemeier, "The Velocity of Sound in Air," *Journal of the Acoustical Society of America*, vol. 13, pp. 226–233; January 1942. See also L. Beranek, "Acoustic Measurements," John Wiley & Sons, New York; 1949: p. 46.

TABLE 2—INTENSITY LEVELS.

Type of Sound	Intensity Level (decibels above 10^{-16} watt/centimeter^2)	Intensity (microwatts/centimeter^2)	Root-Mean-Square Sound Pressure (dynes/centimeter^2)	Root-Mean-Square Particle Velocity (centimeters/second)	Peak-to-Peak Particle Displacement for Sinusoidal Tone at 1000 Hertz (centimeters)
Threshold of painful sound	130	1000	645	15.5	6.98×10^{-3}
Airplane, 1600 rpm, 18 feet	121	126	228	5.5	2.47×10^{-3}
Subway, local station, express passing	102	1.58	25.5	0.98	4.40×10^{-4}
Noisiest spot at Niagara Falls	92	0.158	8.08	0.31	1.38×10^{-4}
Average automobile, 15 feet	70	10^{-3}	0.645	15.5×10^{-3}	6.98×10^{-6}
Average conversational speech, 3.25 feet	70	10^{-3}	0.645	15.5×10^{-3}	6.98×10^{-6}
Average office	55	3.16×10^{-5}	0.114	2.75×10^{-3}	1.24×10^{-6}
Average residence	40	10^{-6}	20.4×10^{-3}	4.9×10^{-4}	2.21×10^{-7}
Quiet whisper, 5 feet	18	6.3×10^{-9}	1.62×10^{-3}	3.9×10^{-5}	1.75×10^{-8}
Reference level	0	10^{-10}	2.04×10^{-4}	4.9×10^{-6}	2.21×10^{-9}

where γ = ratio of the specific heat at constant pressure to that at constant volume, p_0 = the steady pressure of the gas in dynes/centimeter^2, and ρ_0 = the steady or average density of the gas in grams/centimeter^3.

The values of the velocity in a few gases are given in Table 3 for 0 degrees Celsius and 760 millimeters of mercury barometric pressure.

The velocity of sound c in dry air is given by the following experimentally verified equation.

$$c = 33\ 145 \pm 5 \text{ centimeters/second}$$

$$= 1087.42 \pm 0.16 \text{ feet/second}$$

for the audible-frequency range, at 0 degrees Celsius

and 760 millimeters of mercury with 0.03-mole-percent content of CO_2.

The velocity in air for a range of about 20 degrees Celsius change in temperature is given by

$$c = 33\ 145 + 60.7 T_c \text{ centimeters/second}$$

$$= 1052.03 + 1.106 T_f \text{ feet/second}$$

where T_c is the temperature in degrees Celsius and T_f in degrees Fahrenheit. For values of T_c greater than 20 degrees, the following equation may be used.

$$c = 33\ 145 \times (T_k/273)^{1/2} \text{ centimeters/second}$$

where T_k is the temperature in degrees Kelvin.

TABLE 3—VELOCITY OF SOUND IN VARIOUS GASES.*

| Gas | Symbol | Velocity | |
		meters/second	feet/second
Air	—	331.45	1087.42
Ammonia	NH_3	415	1361
Argon	A	319	1046
Carbon monoxide	CO	337.1	1106
Carbon dioxide	CO_2	268.6	881 (above 100 hertz)
Carbon disulfide	CS_2	189	606
Chlorine	Cl	205.3	674
Ethylene	C_2H_4	317	1040
Helium	He	970	3182
Hydrogen	H_2	1269.5	4165
Illuminating gas	—	490.4	1609
Methane	CH_4	432	1417
Neon	Ne	435	1427
Nitric oxide	NO	325	1066
Nitrous oxide	N_2O	261.8	859
Nitrogen	N_2	334	1096
Oxygen	O_2	317.2	1041
Steam (100°C)	H_2O	404.8	1328

* From "Handbook of Chemistry and Physics," "International Critical Tables," and *Journal of the Acoustical Society of America.*

For other corrections, if extreme accuracy is desired, reference should be made to the literature.*

From (5) and Table 1, characteristic impedance is equal to the ratio of the sound pressure to the particle velocity.

$$\bar{Z} = \bar{p}/\bar{v} = \rho_0 c \cos\phi$$

where for plane waves, $\phi = 0$ and $\cos\phi = 1$; and for spherical waves, $\tan\phi = \lambda/2\pi r$. λ = wavelength of acoustic wave, and r = distance from sound source. For r greater than a few wavelengths, $\cos\phi \approx 1$.

Characteristic impedance $\rho_0 c$ in dyne-seconds/centimeter³ (rayls) for several gases at 0 degrees Celsius and 760 millimeters of mercury is given in Table 4.

* H. C. Hardy, D. Telfair, and W. H. Pielemeier, "The Velocity of Sound in Air," *Journal of the Acoustical Society of America*, vol. 13, pp. 226–233; January 1942.

TABLE 4—CHARACTERISTIC IMPEDANCE $\rho_0 c$ FOR GASES.

Gas	Symbol	$\rho_0 c$
Air	—	42.86
Argon	A	56.9
Carbon dioxide	CO_2	51.1
Carbon monoxide	CO	42.1
Helium	He	17.32
Hydrogen	H_2	11.40
Neon	Ne	38.3
Nitric oxide	NO	43.5
Nitrous oxide	N_2O	51.8
Nitrogen	N_2	41.8
Oxygen	O_2	45.3

SOUND IN LIQUIDS

In liquids, the velocity of sound is given by

$$c = (1/K\rho_0)^{1/2} \text{ centimeters/second}$$

where K = compressibility in centimeters/second²/ gram and may be regarded as constant.

$$K = (47 \times 10^{-9})/981$$

for most liquids.

Figures for the velocity of sound in centimeters/ second through some liquids are given in Table 5.

SOUND IN SOLIDS

The velocity of sound in solids is determined by the shape and size of the bounded medium as compared with the wavelength of the excitation. For rods or square bars with unconstrained sides, the velocity of propagation varies with the ratio of thickness to wavelength, being, for 1 wavelength in diameter, about 0.65 times the zero-diameter-to-wavelength ratio.

Some experimental values are given in Table 6.

TABLE 5—VELOCITY OF SOUND IN LIQUIDS.

Liquid	Temp in °C	Velocity in (cm/sec) $\times 10^5$
Alcohol, ethyl	12.5	1.24
	20	1.17
Benzene	20	1.32
Carbon disulfide	20	1.16
Chloroform	20	1.00
Ether, ethyl	20	1.01
Glycerin	20	1.92
Mercury	20	1.45
Pentaine	18	1.05
	20	1.02
Petroleum	15	1.33
Turpentine	3.5	1.37
	27	1.28
Water, fresh	17	1.43
Water, sea (36 parts/thousand salinity)	15	1.505

ACOUSTIC AND MECHANICAL NETWORKS AND THEIR ELECTRICAL ANALOGS*

The present advanced state of the art of electrical network theory suggests its advantageous application, by analogy, to equivalent acoustic and mechanical networks. Actually, Maxwell's initial work on electrical networks was based on the previous work of Lagrange in dynamic systems. The following is a brief summary showing some of the network parameters available in acoustic and mechanical systems and their analysis using Lagrange's equations.

Table 7 shows the analogous behavior of electrical, acoustic, and mechanical systems. These are analogous in the sense that the equations (usually differential equations) formulating the various physical laws are alike.

Lagrange's Equations

The Lagrangian equations are partial differential equations describing the stored and dissipated energy and the generalized coordinates of the system. They are

$$\frac{d}{dt}\left(\frac{\partial T}{\partial \dot{q}_v}\right) + \frac{\partial F}{\partial \dot{q}_v} + \frac{\partial V}{\partial q_v} = Q_v, \qquad (v = 1, 2, \cdots, n)$$

(8)

where T and V are, as in Table 7, the system's total kinetic and potential energy (in ergs), F is $\frac{1}{2}$ the rate of energy dissipation (in ergs/second, Rayleigh's dissipation function), Q_v the generalized forces (dynes), and q_v the generalized coordinates (which may be angles in radians, or displacements in centimeters). For most systems (and those considered herein) the generalized coordinates are equal in number to the number of degrees of freedom in the systems required to determine uniquely the values of T, V, and F.

Example

As an example of the application of these equations toward the design of electroacoustic trans-

* E. G. Keller, "Mathematics of Modern Engineering," vol. 2, 1st ed., John Wiley & Sons, New York; 1942. Also, H. F. Olson, "Dynamical Analogies," 1st ed., D. Van Nostrand, New York; 1943.

ducers, consider the idealized crystal microphone in Fig. 1.

This system has 2 degrees of freedom since only 2 motions, namely the diaphragm displacement x_d and the crystal displacement x_c, are needed to specify the system's total energy and dissipation.

A sound wave impinging on the microphone's diaphragm creates an excess pressure p (dynes/centimeter²). The force on the diaphragm is then pA (dynes), where A is the effective area of the diaphragm. The diaphragm has an effective mass m_d, in the sense that the kinetic energy of all the parts associated with the diaphragm velocity $\dot{x}_d$ $(=dx_d/dt)$ is given by $m_d\dot{x}_d^2/2$. The diaphragm is supported in place by the stiffness S_d. It is coupled to the crystal via the stiffness S_o. The crystal has a stiffness S_c, an effective mass of m_c (to be computed below), and is damped by the mechanical resistance R_c. The only other remaining parameter is the acoustic stiffness S_a introduced by compression of the air-tight pocket enclosed by the diaphragm and the case of the microphone.

The total potential energy V stored in the system for displacements x_d and x_c from equilibrium position, is

$$V=\tfrac{1}{2}S_d x_d^2+\tfrac{1}{2}S_a(x_d A)^2+\tfrac{1}{2}S_c x_c^2+\tfrac{1}{2}S_o(x_d-x_c)^2.$$

$$(9)$$

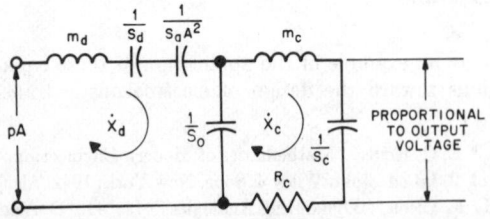

OUTPUT VOLTAGE

Fig. 1—Crystal microphone analyzed by use of Lagrange's equations.

The total kinetic energy T due to velocities $\dot{x}_d$ and $\dot{x}_c$ is

$$T=\tfrac{1}{2}m_c\dot{x}_c^2+\tfrac{1}{2}m_d\dot{x}_d^2. \qquad (10)$$

(This neglects the small kinetic energy due to motion of the air and that due to the motion of the spring S_o). If the total weight of the unclamped part of the crystal is w_c (grams), one can find the effective mass m_c of the crystal as soon as some assumption is made as to movement of the rest of the crystal when its end moves with velocity $\dot{x}_c$. Actually, the crystal is like a transmission line and has an infinite number of degrees of freedom. Practically, the crystal is usually designed so that its first resonant frequency is the highest passed by the microphone. In that case, the end of the crystal moves in phase with the rest, and in a manner that, for simplicity, is here taken as parabolically. Thus it is assumed that an element of the crystal located y centimeters away from its clamped end moves by the amount $(y/h)^2 x_c$, where h is the length of the crystal. The kinetic energy of a length dy of the crystal due to its velocity of $(y/h)^2\dot{x}_c$ and its mass of $(dy/h)w_c$ is $\tfrac{1}{2}(dy/h)\times w_c(y/h)^4\dot{x}_c^2$. The kinetic energy of the whole crystal is the integral of the latter expression as y varies from 0 to h. The result is $\tfrac{1}{2}(w_c/5)\dot{x}_c^2$. This shows at once that the effective mass of the crystal is $m_c=w_c/5$, i.e., $\tfrac{1}{5}$ its actual weight.

The dissipation function is $F=\tfrac{1}{2}R_c\dot{x}_c^2$. Finally, the driving force associated with displacement x_d of the diaphragm is pA. Substitution of these expressions and (9) and (10) in Lagrange's equations (8) results in the force equations

$$m_d\ddot{x}_d+S_d x_d+S_a A^2 x_d+S_o(x_d-x_c)=pA$$

$$m_c\ddot{x}_c+S_o(x_c-x_d)+S_c x_c+R_c\dot{x}_c=0. \qquad (11)$$

These are the mechanical version of Kirchhoff's law that the sum of all the resisting forces (rather than voltages) are equal to the applied force. The equivalent electrical circuit giving these same differential equations is shown in Fig. 1. The crystal produces, by its piezoelectric effect, an open-circuit voltage proportional to the displacement x_c. By means of this equivalent circuit, it is now easy, by using the usual electrical-circuit techniques, to find the voltage generated by this microphone per unit of sound-pressure input, and also its amplitude- and phase-response characteristic as a function of frequency.

It is important to note that this process of analysis not only results in the equivalent electrical circuit, but also determines the effective values of the parameters in that circuit.

PRINCIPLE OF RECIPROCITY

A network having two pairs of terminals as shown in Fig. 2 has a voltage and a current value

TABLE 6—VELOCITY c OF SOUND IN LONGITUDINAL DIRECTION FOR BAR-SHAPED SOLIDS.*

Material	Velocity in (cm/sec)$\times 10^5$	Material	Velocity in (cm/sec)$\times 10^5$
Aluminum	5.24	Crystals:	
Antimony	3.40	Quartz X-cut	5.44
Bismuth	1.79	Ammonium dihydrogen phosphate ($NH_4H_2PO_4$) 45° Z-cut	3.28
Brass	3.42	Rochelle salt (sodium potassium tartrate, $KNaC_4H_4O_6 \cdot 4H_2O$)	
Cadmium	2.40	45° Y-cut	2.47
Constantan	4.30	45° X-cut	2.47
Copper	3.58	Calcium fluoride (CaF_2, fluorite) X-cut	6.74
German silver	3.58	Sodium chloride (NaCl, rock salt) X-cut	4.51
Gold	2.03	Sodium bromide (NaBr) X-cut	2.79
Iridium	4.79	Potassium chloride (KCl, sylvite) X-cut	4.14
Iron	5.17	Potassium bromide (KBr) X-cut	3.38
Lead	1.25		
Magnesium	4.90	Glasses:	
Manganese	3.83	Heavy flint	3.49
Nickel	4.76	Extra-light flint	4.55
Platinum	2.80	Crown	5.30
Silver	2.64	Heaviest crown	4.71
		Quartz	5.37
Steel	5.05	Granite	3.95
Tantalum	3.35	Ivory	3.01
Tin	2.73	Marble	3.81
Tungsten	4.31	Slate	4.51
Zinc	3.81	Woods:	
		Elm	1.01
Cork	0.50	Oak	4.10

* B. W. Henvis, "Wavelengths of Sound," *Electronics*, vol. 20, pp. 134, 136; March 1947.

associated with each terminal, making four quantities.

If any two are specified, the others may be found by the two linear relations

$$V_1 = aI_1 + bI_2$$

$$V_2 = cI_1 + dI_2$$

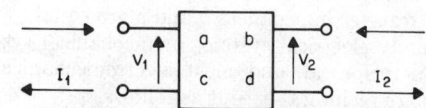

Fig. 2—Network showing principle of reciprocity.

TABLE 7A—ANALOGOUS BEHAVIOR OF SYSTEMS—PARAMETER OF ENERGY DISSIPATION
(OR RADIATION).

Electrical	Mechanical	Acoustic

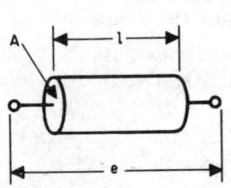

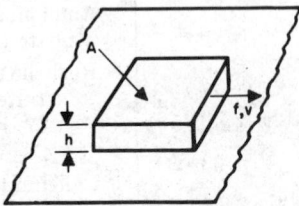

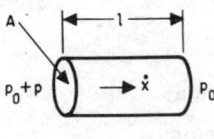

current in wire	viscous damping vane	gas flow in small pipe

$$P = Ri^2$$

$$i = e/R = dq/dt = \dot{q}$$

$$R = \rho l/A$$

$$P = R_m v^2$$

$$v = f/R_m = dx/dt = \dot{x}$$

$$R_m = \mu A/h$$

$$P = R_a \dot{X}^2$$

$$\dot{X} = p/R_a = dX/dt$$

$$R_a = 8\mu\pi l/A^2$$

where

i = current in amperes
e = voltage in volts
q = charge in coulombs
t = time in seconds
R = resistance in ohms
ρ = resistivity in ohm-centimeters
l = length in centimeters
A = cross-sectional area of wire in centimeters^2
P = power in watts

where

v = velocity in centimeters/second
f = force in dynes
x = displacement in centimeters
t = time in seconds
R_m = mechanical resistance in dyne-seconds/centimeter
μ = coefficient of viscosity in poises
h = height of damping vane in centimeters
A = area of vane in centimeters^2
P = power in ergs/second

where

$\dot{X}$ = volume velocity in centimeters^3/second
p = excess pressure in dynes/centimeter^2
X = volume displacement in centimeters^3
t = time in seconds
R_a = acoustic resistance in dyne-seconds/centimeter^5
μ = coefficient of viscosity in poises
l = length of tube in centimeters
A = area of tube in centimeters^2
P = power in ergs/second

where a, b, c, d are complex impedances. One or both of the terminal pairs may be replaced by a mechanical system. With MKS (meter, kilogram, second) units:

A system satisfies the principle of reciprocity if the two transfer impedances b and c are equal.

All purely electrical systems or mechanical systems are reciprocal, and most electromechanical systems are reciprocal except as follows.

In the case of crystal or electrostatic transducers the principle holds true, but in the case of magnetic or electrodynamic transducers it is true in magnitude but not in sign. Combinations of the two types of transducers can be made to violate reciprocity in magnitude.*

However, for the calibration of pressure-gradient, dynamic, ribbon, and capacitor type microphones, the reciprocity technique is accurate and simple to apply.

* E. M. McMillan, "Violation of the Reciprocity Theorem in Linear Passive Electromechanical Systems," *Journal of the Acoustical Society of America*, vol. 18, pp. 344–347; October 1946.

TABLE 7B—ANALOGOUS BEHAVIOR OF SYSTEMS—PARAMETER OF ENERGY STORAGE
(ELECTROSTATIC OR POTENTIAL ENERGY).

Electrical	Mechanical	Acoustic

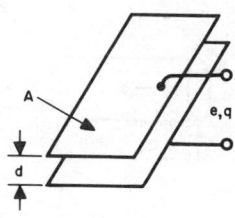

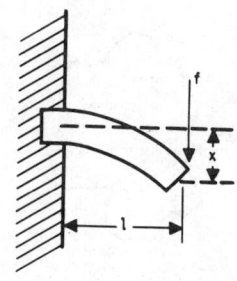

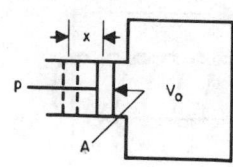

capacitor with closely spaced plates	clamped-free (cantilever beam)	piston acoustic compliance (at audio frequencies, adiabatic expansion)

$$W_e = q^2/2C = Sq^2/2$$

$$q = Ce = e/S$$

$$C = (kA/36\pi d) \times 10^{-11}$$

$$V = x^2/2C_m = S_m x^2/2$$

$$x = C_m f = f/S_m$$

$$C_m = l^3/3EI$$

$$V = X^2/2C_a = S_a X^2/2$$

$$X = C_a p = p/S_a = xA$$

$$C_a = V_0/c^2\rho$$

where

C = capacitance in farads
S = stiffness = $1/C$
W_e = energy in watt-seconds
k = relative dielectric constant (= 1 for air, numeric)
A = area of plates in centimeters^2
d = separation of plates in centimeters

where

C_m = mechanical compliance in centimeters/dyne
S_m = mechanical stiffness = $1/C_m$
V = potential energy in ergs
E = Young's modulus of elasticity in dynes/centimeter^2
I = moment of inertia of cross-section in centimeters^4
l = length of beam in centimeters

where

C_a = acoustic compliance in centimeters^5/dyne
S_a = acoustic stiffness = $1/C_a$
V = potential energy in ergs
c = velocity of sound in enclosed gas in centimeters/second
ρ = density of enclosed gas in grams/centimeter^3
V_0 = enclosed volume in centimeters^3
A = area of piston in centimeters^2

For the calibration of a microphone T^x, two operations are necessary.

(**A**) Place a sound source producing a spherical field at the location d centimeters from the microphone to be calibrated. The distance d should be much larger than the largest dimension L of any of the transducers used and should be much larger than L^2/λ, where λ is the wavelength of the sound in centimeters.

At this distance, determine the open-circuit voltage e_{oc}^x of the microphone T^x in abvolts

(volts $\times 10^{-8}$) and of a second auxiliary microphone T' put in its place (e_{oc}').

These voltages form the ratio

$$M_0^x/M_0' = e_{oc}^x/e_{oc}'$$

where $M_0^x = e_{oc}^x/P_{ff}$ = ratio of the open-circuit voltage produced by a transducer in a plane-wave sound field to the sound pressure present before the microphone was inserted, and M_0' a similar ratio for the second auxiliary microphone. M_0^x is the required calibration.

TABLE 7C—ANALOGOUS BEHAVIOR OF SYSTEMS—PARAMETER OF ENERGY STORAGE
(MAGNETOSTATIC OR KINETIC ENERGY).

Electrical	Mechanical	Acoustic

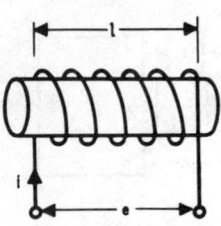

for a very long solenoid	for translational motion in one direction; m is the actual weight in grams	gas flow in a pipe
$W_m = Li^2/2$	$T = mv^2/2$	$T = M\dot{X}^2/2$
$e = L(di/dt) = L(d^2q/dt^2) = L\ddot{q}$	$f = m(dv/dt) = m(d^2x/dt^2) = m\ddot{x}$	$p = M(d\dot{X}/dt) = M(d^2X/dt^2)$
$L = 4\pi ln^2 Ak \times 10^{-9}$		$= M\ddot{X}$
		$M = \rho l/A$

where	where	where
L = inductance in henries W_m = energy in watt-seconds l = length of solenoid in centimeters A = area of solenoid in centimeters^2 n = number of turns of wire/centimeter k = relative permeability of core ($=1$ for air, numeric)	m = mass in grams T = kinetic energy in ergs v = velocity in centimeters/second	M = inertance in grams/centimeter^4 T = kinetic energy in ergs l = length of pipe in centimeters A = area of pipe in centimeters^2 ρ = density of gas in grams/centimeter^3

(**B**) Drive the microphone to be calibrated with a constant current $i_T{}^x$ and, placing the secondary auxiliary microphone at the same distance d, measure the open-circuit voltage e_{oc}' across its output.

The value of $M_0{}^x$ in practical units is obtained from

$$M_0{}^x = [(e_{oc}'/i_T{}^x) \cdot (e_{oc}{}^x/e_{oc}')(2d\lambda/pc) \times 10^{-7}]^{1/2}$$

where p is the density of the gas and c is the velocity of propagation of sound in that medium, the product pc being called the characteristic impedance of the medium.*

* W. R. MacLean, "Absolute Measurement of Sound Without a Primary Standard," *Journal of the Acoustical Society of America*, vol. 12, pp. 140–146; July 1940. Also L. Beranek, "Acoustical Measurements," John Wiley & Sons, New York; 1949: pp. 116–121.

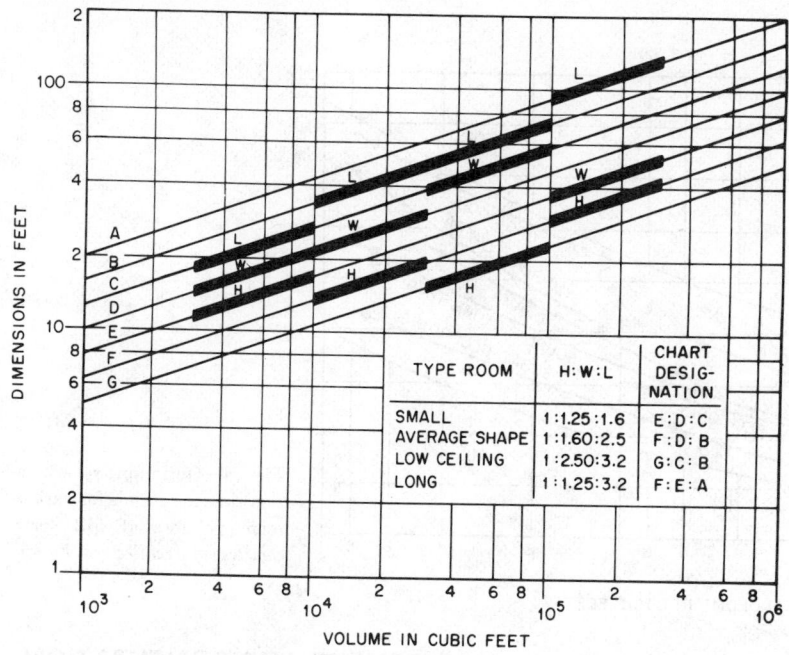

TYPE ROOM	H:W:L	CHART DESIGNATION
SMALL	1:1.25:1.6	E:D:C
AVERAGE SHAPE	1:1.60:2.5	F:D:B
LOW CEILING	1:2.50:3.2	G:C:B
LONG	1:1.25:3.2	F:E:A

VOLUME IN CUBIC FEET

Fig. 3—Preferred room dimensions based on $2^{1/3}$ ratio. Permissible deviation is ±5 percent. *Courtesy of Acoustical Society of America and RCA.*

SOUND IN ENCLOSED ROOMS*

Good Acoustics—Governing Factors

Reverberation Time or Amount of Reverberation: This varies with frequency and is measured by the time required for a sound, when suddenly interrupted, to die away or decay to a level 60 decibels below the original sound.

The reverberation time and the shape of the reverberation-time/frequency curve can be controlled by selecting the proper amounts and varieties of sound-absorbent materials and by the methods of application. Room occupants must be considered inasmuch as each person present contributes a fairly definite amount of sound absorption.

Standing Sound Waves: Resonant conditions in sound studios cause standing waves by reflections from opposing parallel surfaces, such as ceiling-floor and parallel walls, resulting in serious peaks in the reverberation-time/frequency curve. Standing sound waves in a room can be considered comparable to standing electrical waves in an improperly terminated transmission line where the transmitted power is not fully absorbed by the load.

* F. R. Watson, "Acoustics of Building," 3rd ed., John Wiley & Sons, New York; 1941.

Room Sizes and Proportions for Good Acoustics

The frequency of standing waves is dependent on room sizes: frequency decreases with increase of distance between walls and between floor and ceiling. In rooms with two equal dimensions, the two sets of standing waves occur at the same frequency with resultant increase of reverberation time at resonant frequency. In a room with walls and ceilings of cubical contour this effect is tripled, and elimination of standing waves is practically impossible.

The most advantageous ratio for height:width: length is in the proportion of $1:2^{1/3}:2^{2/3}$ or separated by $\frac{1}{3}$ or $\frac{2}{3}$ of an octave.

In properly proportioned rooms, resonant conditions can be effectively reduced and standing waves practically eliminated by introducing numerous surfaces disposed obliquely. Thus, large-order reflections can be avoided by breaking them up into numerous smaller reflections. The object is to prevent sound reflection back to the point of origin until after several reflections.

Most desirable ratios of dimensions for broadcast studios are given in Fig. 3.

Optimum Reverberation Time

Optimum, or most desirable reverberation time, varies with (A) room size, and (B) use, such as music, speech, etc. (see Figs. 4 and 5).

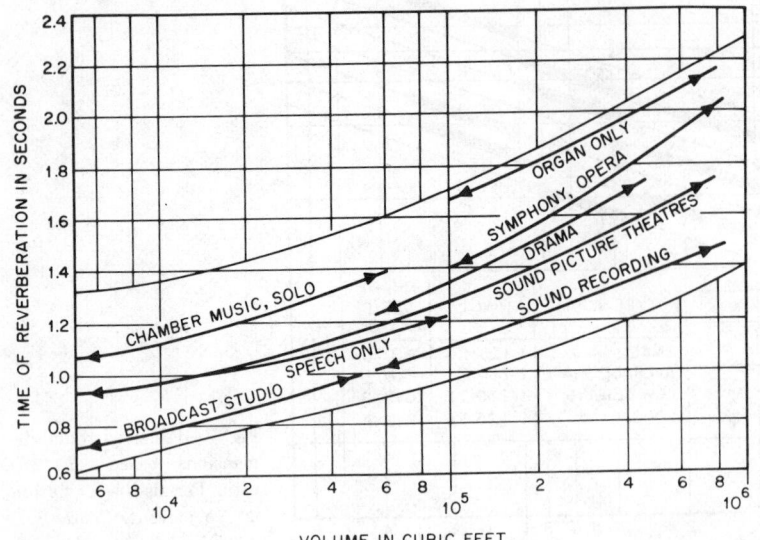

Fig. 4—Optimum reverberation time in seconds for various room volumes at 512 hertz. *Courtesy of Architectural Forum.*

Figure 5 shows the desirable ratio of the reverbation time for various frequencies to the reverberation time for 512 hertz. The desirable reverberation time for any frequency between 60 and 8000 hertz may be found by multiplying the reverberation time at 512 hertz (from Fig. 4) by the desirable ratio in Fig. 5 which corresponds to the frequency chosen.

The reverberation time affects the intelligibility of speech unless suitable speech cadences are developed. The intelligibility at a sound intensity of about 1 dyne/cm² is shown in Fig. 6.

MEASUREMENT AND COMPUTATION OF REVERBERATION TIME

The reverberation time of an enclosed space that already exists is an important quantity that is relatively easy to measure. Such measurements can yield invaluable information about the space in a more accurate and easier-to-process form than can calculation only. When the enclosed space does not exist (new construction, for example) except on the architect's drawing board, it is then necessary to use the most accurate method of calculation available.

The Measurement of Reverberation Time

The degree of accuracy required is determined by the use to which the data will be put. If the purpose is merely to compare the reverberation time to one of the existing criteria charts, the

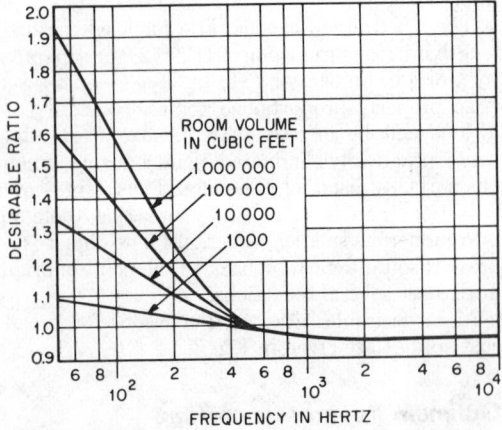

Fig. 5—Desirable relative reverberation time versus frequency for various structures and auditoriums. *Courtesy of Western Electric Company.*

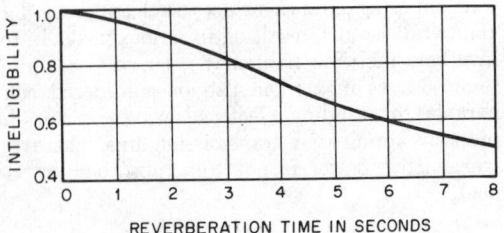

Fig. 6—Intelligibility as a function of reverberation time.

by-ear method using a simple stopwatch will often thoroughly satisfy the requirement. If the enclosed space is a concert hall of significance, the Schroeder-Kuttruff method is almost mandatory. In addition, in a concert hall, the early decay time (EDT) is of great interest. The majority of day-to-day reverberation measurements are taken with quite satisfactory results by using the interrupted-noise method.

Using the Ears-and-Stopwatch Method

The ears-and-stopwatch method of measuring reverberation time works best with reverberation times in excess of 2 seconds. In auditoriums where speech intelligibility is of importance, the design engineer needs to take careful note of the potential articulation losses whenever the RT_{60} exceeds 1.6 seconds. (The RT_{60} is the time required for the reverberant sound to decay 60 dB.) Below that value, the reverberation time will not detrimentally affect the intelligibility, although making it too low may require excessive acoustic power to be generated at the sound source.

The measurement procedure is to excite the space with some steady noise of wide spectral content. (A random-noise generator, interstation noise from an fm tuner, a bursting balloon, or a small yachting cannon all have been employed at different times.) Care must be exercised to insure that the noise source exceeds the ambient noise by at least 30 dB and preferably by the full 60 dB. It is usually wise to band-limit the noise in some fashion, if possible, especially at the lower frequencies, as it is the region around 1000 Hz and 2000 Hz that affects intelligibility.

The listener stands in the reverberant field, at least 2/3 of the room distance away from the sound source (Fig. 7). The sound source is usually turned toward a corner in order to excite the maximum number of room modes. When the noise is shut off, the stopwatch is started, and when the listener judges that the sound has dropped to the ambient noise level, the stopwatch is halted. (Practicing in spaces where the RT_{60} is already known will rapidly "calibrate" the ears to a surprisingly accurate degree.) If only 30 dB of decay is available, the time recorded by the stopwatch is doubled. If 60 dB of decay is available, the time is read directly.

In defense of this deceptively simple method, let it be said that data containing as high as a 50% error is vastly more useful than no data at all in engineering a sound system into an existing acoustic environment.

The Interrupted-Noise Method

The interrupted-noise method of measuring reverberation time is illustrated in Fig. 8. This is the most widely used method and yields excellent field results; only in the most critical concert halls is this method found wanting. It has been stated by Atal that the subjective assessment of

Fig. 7—The ears-and-stopwatch method.

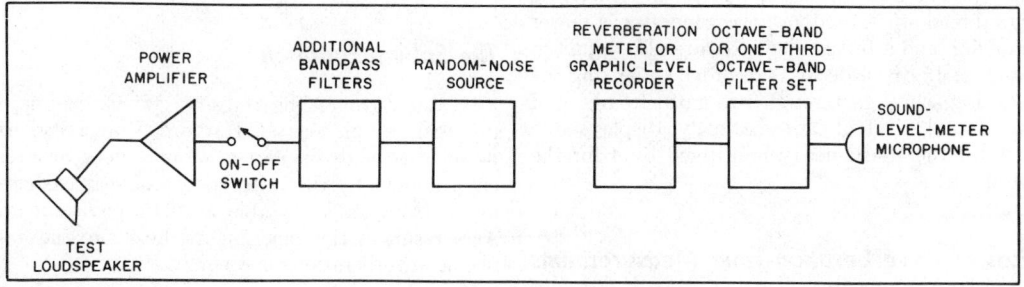

Fig. 8—The interrupted-noise method.

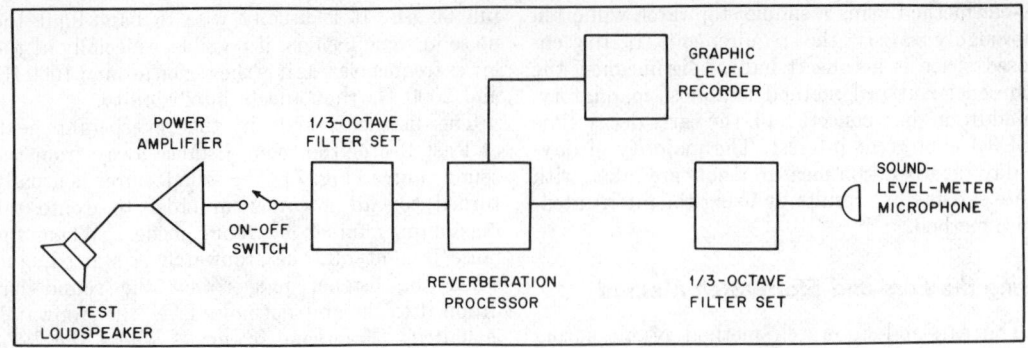

Fig. 9—The Schroeder-Kuttruff method.

reverberation time is governed by the early decay time (EDT, which is the time it takes the early decay to change by 15 dB). Good agreement between this EDT and subjective estimates of the presence or absence of excessive reverberation has been obtained in a number of concert halls. This first 15 dB of decay is difficult to obtain with the interrupted-noise method because of the statistical variations that occur in the test signal over short time periods.

The Schroeder-Kuttruff Method

The Schroeder-Kuttruff method (Fig. 9) employs a 2.7-ms rectangular pulse that is used to excite a standard 1/3-octave bandpass filter. The resulting "ringing" of the filter provides a statistically reliable signal that allows highly repeatable decay recordings and excellent resolution of the EDT.

Reverberation-Time Meters

At the present time, at least one manufacturer provides an RT_{60} meter for measurement of reverberation time. This device contains in a hand-held carrying case the measuring microphone, octave-band filters, and a direct-reading digital readout. A random-noise generator, a power amplifier, and a test loudspeaker are also required. The first 15 dB of decay is recorded by letting the decaying sound go through two gate circuits, and the digital readout automatically displays the time for the 15-dB decay multiplied by 4 for the full 60 dB.

Uses of Reverberation-Time Measurements

The main reason for such extensive efforts to determine RT_{60} accurately is that this is the easiest

way to find the average absorption coefficient $(\bar{a})$ and its associated value of reflectivity $(1-\bar{a})$ in an acoustic environment. Traditionally, the way to increase intelligibility in an acoustic environment is to increase $\bar{a}$.

When an electroacoustic system is employed in a given environment, an increase in the directivity factor, Q, of the sound source (see section on public-address systems) can be substituted for increased $\bar{a}$ over a surprisingly wide range of applications. Therefore, the contemporary electroacoustic designer needs to have an accurate $\bar{a}$ figure in order to calculate the minimum Q that he requires in the sound sources. A further use of the evaluation of both RT_{60} and Q is their role in the establishment of an acceptable articulation loss of consonants in speech. (See section on public-address systems.)

Of greatly increased current interest is the role of $\bar{a}$ in helping to reduce the reverberant sound field in an acoustic environment, thereby making the overall sound level in the space lower than it was before the absorption was applied. It should be noted, however, that absorption will not lower the sound level for any listener in the direct sound field of the sound source.

For a better understanding of the beneficial engineering uses of these measurements, it is necessary to examine in detail the commonly used equations associated with them.

The Sabine Equation

At the turn of the century, W. C. Sabine, a professor of physics at Harvard University, experimented with the correction of a poor acoustic environment by the introduction of seat cushions taken from an acceptable acoustic environment. As a result of the experiments, he wrote the first usable reverberation-time equation:

$$RT_{60} = \frac{0.049V}{S\bar{a}}$$

where RT_{60} is the time in seconds required for a sound to decay 60 dB, V is the volume of the room in cubic feet, S is the boundary surface area in square feet, and $\bar{a}$ is the *average* absorption coefficient. The value of $\bar{a}$ is:

$$\bar{a}=\frac{s_1a_1+s_2a_2+\cdots s_na_n}{S}$$

where s_1, s_2, etc., are boundary surface areas; a_1, a_2, etc., are the absorption values for the boundary areas with which they are associated; and s_na_n is the total absorption of the people, furniture, etc., present in the room. For metric use, the constant 0.049 becomes 0.161, V is in cubic meters, and S is in square meters.

The variations of the reverberation-time equation are:

$$\bar{a}=\frac{0.049V}{S \cdot RT_{60}}$$

$$V=\frac{S\bar{a} \cdot RT_{60}}{0.049}$$

$$S=\frac{0.049V}{\bar{a} \cdot RT_{60}}$$

The Norris-Eyring Equation

By 1930, R. F. Norris recognized that, in the limiting case, the Sabine equation predicted a finite reverberation time in a room with 100% absorption present. He further demonstrated that for true absorption values in excess of 0.63, this equation could give $\bar{a}$ values in excess of 1.0 (100% absorption). In conjunction with C. F. Eyring, Norris wrote an equation that gave $\bar{a}$ values from 1.0 to 0 for true absorption values when cal-

culated from actual RT_{60} measurements. The Norris-Eyring equation is:

$$RT_{60}=\frac{0.049V}{-S\ln(1-\bar{a})}$$

$$\bar{a}=1-\exp(-0.049V/S \cdot RT_{60})$$

$$V=\frac{RT_{60}[-S\ln(1-\bar{a})]}{0.049}$$

$$S=\frac{0.049V}{-RT_{60}\ln(1-\bar{a})}$$

If the value of RT_{60} is measured and the corresponding value of $\bar{a}$ is calculated, insertion of this value of $\bar{a}$ into the expression $-\ln(1-\bar{a})$ converts the $\bar{a}$ value into a Sabine $\bar{a}$. For example, $-\ln(1-0.63)=0.99$.

When tables of absorption values are examined, it is of vital importance to know which formula was used in determining the numerical values. (See Fig. 10 for a typical method of obtaining $\bar{a}$ values in a reverberation chamber.) If the values were obtained using the Sabine equation, be sure to use the Sabine equation variations consistently for any further manipulations of the data. If the Norris-Eyring equation was used, remain consistent in its use for any further manipulations of the data.

In the method illustrated in Fig. 10, the absorption coefficient of the test material is calculated from the equation

$$a=\frac{0.049V}{A_1+A_2+A_3}\left(\frac{1}{T_M}-\frac{1}{T_E}\right)$$

where a is the absorption coefficient of the test material, V is the internal volume of the test room in cubic feet, $A_1+A_2+A_3$ is the surface area of the test material, T_M is the measured RT_{60} of the

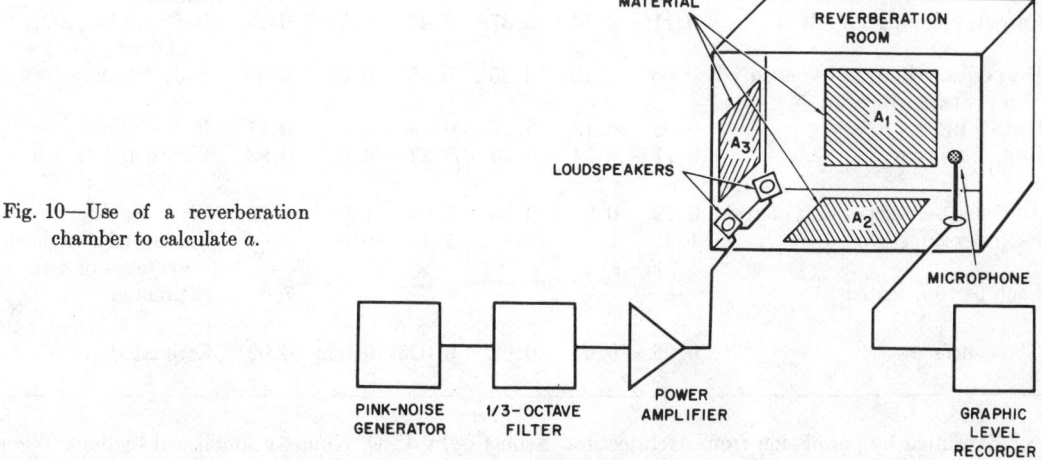

Fig. 10—Use of a reverberation chamber to calculate a.

test room with the test material installed as shown, and T_E is the measured RT_{60} of the empty test room. If S_m is defined as $A_1+A_2+A_3$, the equation may be written in the following alternative forms:

$$S_m = \frac{0.049V}{a}\left(\frac{1}{T_M}-\frac{1}{T_E}\right)$$

$$T_M = \frac{1}{\dfrac{aS_m}{0.049V}+\dfrac{1}{T_E}}$$

$$T_E = \frac{1}{\dfrac{1}{T_M}-\dfrac{aS_m}{0.049V}}$$

$$V = \frac{1}{0.049\dfrac{}{aS_m}\left(\dfrac{1}{T_M}-\dfrac{1}{T_E}\right)}$$

The total $S\bar{a}$ is the sum of $S\bar{a}$ with the room empty plus Sa of the test sample.

As an example, let $S_m = 75$ ft^2, $V = 8000$ ft^3, $S = 2400$ ft^2, $T_M = 6.5$, $T_E = 9.0$, and $\bar{a} = 0.018$ for the empty room. Then:

$$a = \frac{0.049 \times 8000}{75}\left(\frac{1}{6.5}-\frac{1}{9.0}\right) = 0.223$$

$$RT_{60} = \frac{0.049 \times 8000}{(2400 \times 0.018)+(75 \times 0.223)} = 6.5$$

Table 8 gives absorption coefficients of some typical building materials.

TABLE 8—ACOUSTIC COEFFICIENTS OF MATERIALS AND PERSONS.*

Description	Sound Absorption Coefficients (hertz)						Authority
	128	256	512	1024	2048	4096	
Brick wall unpainted	0.024	0.025	0.031	0.042	0.049	0.07	W. C. Sabine
Brick wall painted	0.012	0.013	0.017	0.02	0.023	0.025	W. C. Sabine
Plaster+finish on wood lath— wood studs	0.020	0.022	0.032	0.039	0.039	0.028	P. E. Sabine
Plaster+finish coat on metal lath	0.038	0.049	0.060	0.085	0.043	0.056	V. O. Knudsen
Poured concrete unpainted	0.010	0.012	0.016	0.019	0.023	0.035	V. O. Knudsen
Poured concrete painted and varnished	0.009	0.011	0.014	0.016	0.017	0.018	V. O. Knudsen
Carpet, pile on concrete	0.09	0.08	0.21	0.26	0.27	0.37	Building Research Station
Carpet, pile on $\frac{1}{8}$ in. felt	0.11	0.14	0.37	0.43	0.27	0.25	Building Research Station
Draperies, velour, 18 oz per sq yd in contact with wall	0.05	0.12	0.35	0.45	0.38	0.36	P. E. Sabine
Ozite $\frac{3}{8}$ in.	0.051	0.12	0.17	0.33	0.45	0.47	P. E. Sabine
Rug, axminster	0.11	0.14	0.20	0.33	0.52	0.82	Wente and Bedell
Audience, seated per sq ft of area	0.72	0.89	0.95	0.99 ·	1.00	1.00	W. C. Sabine
Each person, seated	1.4	2.25	3.8	5.4	6.6	—	Bureau of Standards, averages of 4 tests
Each person, seated	—	—	—	—	—	7.0	Estimated
Glass surfaces	0.05	0.04	0.03	0.025	0.022	0.02	Estimated

* Reprinted by permission from Architectural Acoustics by V. O. Knudsen, published by John Wiley and Sons.

The Fitzroy Equation

The next major improvement in the use of these basic equations came in 1959 with the publication of Dariel Fitzroy's paper "Reverberation Formula Which Seems To Be More Accurate With Nonuniform Distribution of Absorption" in the *Journal of the Acoustical Society of America*. Fitzroy had become concerned with the discrepancies that occurred between the calculated and measured RT_{60} when the absorption was not uniformly distributed about the acoustic environment—for example, if the total ceiling is covered with highly absorptive acoustic tile, but all other boundary surfaces are hard and reflective. The Fitzroy equation is the same as the Norris-Eyring equation *if* the absorption is evenly distributed, but it yields much more accurate answers when uneven distribution is encountered.

$$RT_{60} = \frac{0.049V}{S^2}\left[\frac{2xy}{-\ln(1-\bar{a}_{xy})}\right.$$

$$\left. + \frac{2xz}{-\ln(1-\bar{a}_{xz})} + \frac{2yz}{-\ln(1-\bar{a}_{yz})}\right]$$

where x and y are the height and width of the room, and $\bar{a}_{xy}$ is the average absorption coefficient for the two end walls; x and z are the height and length, and $\bar{a}_{xz}$ is the average absorption coefficient for the two side walls; and y and z are the width and length, and $\bar{a}_{yz}$ is the average absorption coefficient for the floor and ceiling.

This equation is used only for calculating the expected RT_{60} from the drawings of the acoustic environment. Once the expected RT_{60} is calculated, that value is inserted into the standard Norris-Eyring equation variation to obtain the true $\bar{a}$ value.

The Fitzroy equation is an invaluable tool in the study of how and where to place absorbent materials for optimum control of RT_{60}.

Basic Considerations in the Measurement, Calculation, and Application of Acoustic Treatment

While the measurements and/or calculations of RT_{60}, $\bar{a}$, etc., can be made today with acceptable accuracy, these techniques and equations supply only a few hints of the variations in application of the material itself to achieve the optimum results with the minimum cost. A few of the more basic rules are listed:

1. Diffusion is highly desirable, and both absorption and room geometry should be employed to enhance it.

2. Every effort should be made to preserve useful reflecting surfaces (those within 30 to 50 feet of a sound source).

3. Rarely should absorption be placed on ceilings. Preferred choices include the floor—carpets also lower noise levels at the source as well as providing absorption—rear walls, etc.

4. It should be considered that too high an RT_{60} will detrimentally affect intelligibility, and an RT_{60} that is too low requires much higher power output to overcome the excessive absorption.

5. Low-frequency absorption is usually controlled by diaphragmatic action and high-frequency absorption by soft, fuzzy materials.

6. Materials useful as absorbers are almost never useful as isolaters. Absorbers are intended to control the reverberant field *within* an acoustic environment. Isolators are intended to keep sound inside a given environment or to keep sounds in other environments outside of the given environment. Good isolators are characterized by mass and rigidity. Good absorbers are characterized by porousness and nonrigidity.

These room parameters directly interact with the directivity factor, Q, of the sound system, and they should be adjusted to optimize the overall room–sound-system performance.

PUBLIC ADDRESS SYSTEMS

Successful speech and music reinforcement systems require a threefold design solution:

1. The reconciliation of the reverberation time, the directivity factor of the loudspeaker, and the distance from the sound source to the farthest listener so that an acceptable articulation loss for consonants in speech is obtained.

2. The adjustment of the system parameters, within the limits set forth as necessary to achieve good articulation, to insure the required acoustic gain. See Fig. 11.

3. The determination from the first two steps of the electrical power required at the input of the transducers in order to produce the acoustic power needed at the listener's ears.

Basic Definitions

$D_1 =$ the distance in feet from the sound source (loudspeaker) to the microphone.

$D_2 =$ the distance in feet from the sound

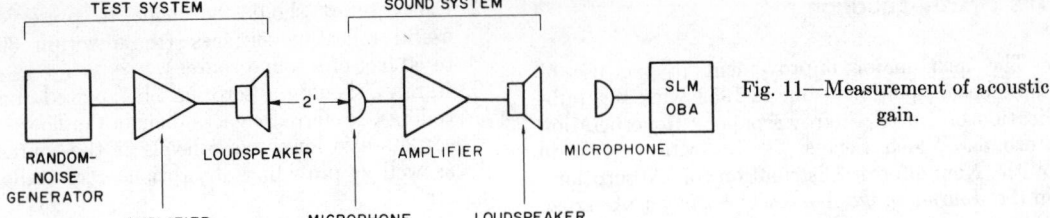

Fig. 11—Measurement of acoustic gain.

source (loudspeaker) to the most distant listener.

$D_s=$ the distance in feet from the talker (performer) to the microphone.

$D_0=$ the distance in feet from the talker (performer) to the most distant listener. See Fig. 12.

$EAD=$ the equivalent acoustic distance in feet. This is the maximum distance from a talker that a listener can comfortably stand and hear clearly without a sound system. See Fig. 13.

$EPR=$ the electrical power in watts required at the input of a loudspeaker to achieve the specified acoustic level.

$r=$ the distance from the sound source for calculations.

$n+1=$ the number of loudspeaker groups; n is the number of groups of the same acoustic power output as the "1" group (which supplies direct sound to the listener) not supplying direct sound to a given single point of observation.

$\%AL_{CONS}=$ the percentage of articulation loss for consonants in speech. A successful speech system should not exceed a maximum of 15% AL_{CONS}.

$V=$ the volume of the room in cubic feet.

$RT_{60}=$ the reverberation time in seconds for 60 dB of decay.

$Q=$ the directivity factor. Also called R_θ.

$R=$ the room constant in square feet. $R=S\bar{a}/(1-\bar{a})$

$D_c=$ the critical distance, the distance at which the direct sound level equals the reverberant sound level.

$L_{SENS}=$ the output of the loudspeaker in dB-SPL at 4 feet for an electrical input power of 1 watt.

$dB\text{-}SPL_D=$ the desired acoustic program level at the listener's ears in dB-SPL.

$NOM=$ the number of open microphones contributing equal input level to the sound system.

$FSM=$ the feedback stability margin; 6 dB is the minimum considered adequate for speech purposes.

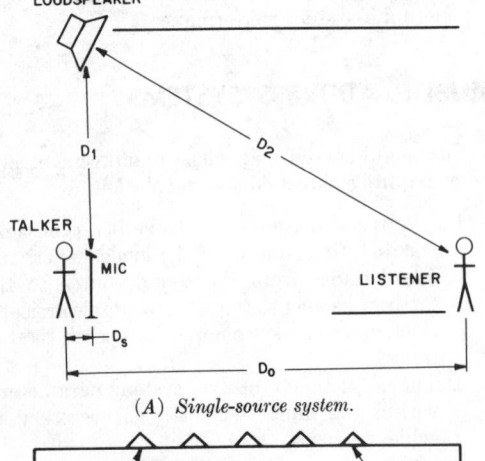

(A) Single-source system.

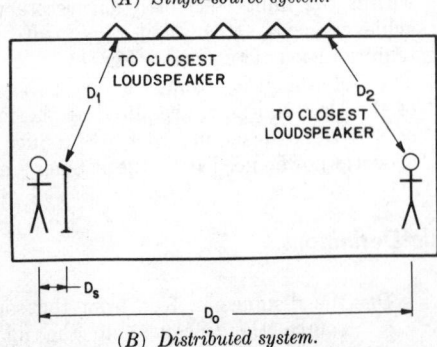

(B) Distributed system.

Fig. 12—Basic parameters of sound systems.

Key Equations

$$\Delta D_x=-10\log\left(\frac{Q}{4\pi r^2}+\frac{4(n+1)}{R}\right)$$

$$r=\left\{Q\,\Big/\,4\pi\left[10^{\pm\Delta D_x/10}-\frac{4(n+1)}{R}\right]\right\}^{1/2}$$

$$D_c=0.141\left(\frac{QR}{(n+1)}\right)^{1/2}$$

$$\Delta D_1+\Delta EAD-\Delta D_s-\Delta D_2-10\log NOM$$
$$-6\,\text{dB}\,FSM=0$$

$$\Delta D_s+\Delta D_2-\Delta EAD+10\log NOM+6\,\text{dB}\,FSM$$
$$=\text{Min}\,\Delta D_1$$

$$\Delta D_s+\Delta D_2-\Delta D_1+10\log NOM+6\,\text{dB}\,FSM$$
$$=\text{Min}\,\Delta EAD$$

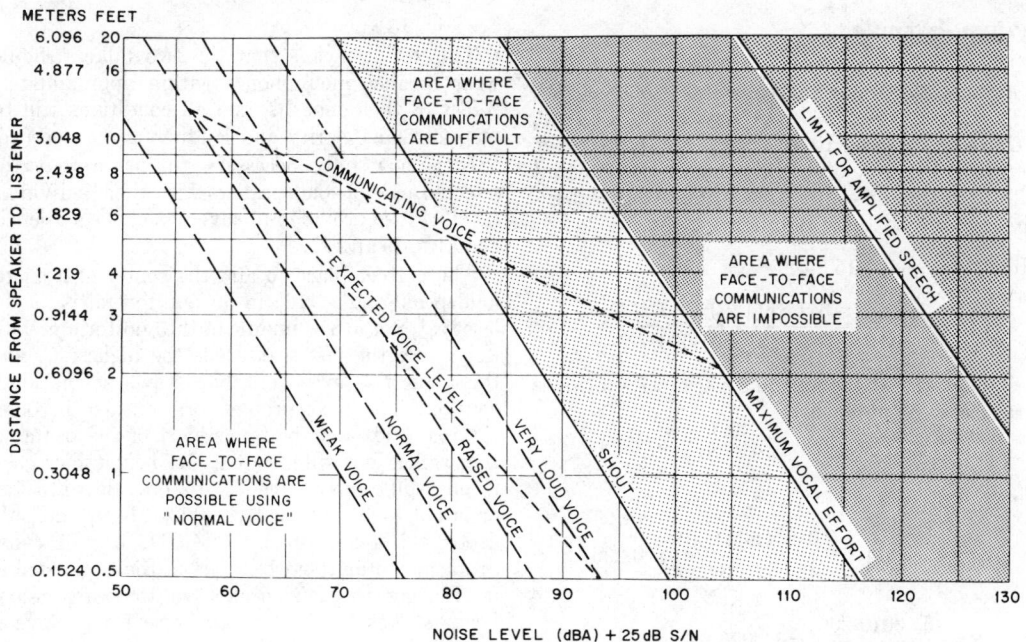

Fig. 13—Nomograph for finding EAD.

$\Delta D_1 + \Delta EAD - \Delta D_2 - 10 \log NOM - 6 \text{ dB } FSM$
$$= \text{Max } \Delta D_s$$

$\Delta D_1 + \Delta EAD - \Delta D_s - 10 \log NOM - 6 \text{ dB } FSM$
$$= \text{Max } \Delta D_2$$

$$10^{(\Delta D_1 + \Delta EAD - \Delta D_s - \Delta D_2 - 6 \text{ dB } FSM)/10} = \text{Max } NOM$$

$$\% AL_{\text{CONS}} = \frac{641.81 \, (D_2)^2 \, (RT_{60})^2 \, (n+1)}{VQ}$$

Maximum D_2 for AL_{CONS} of 15%
$$= \left(\frac{15VQ}{641.81 \, (RT_{60})^2 \, (n+1)} \right)^{1/2}$$

Maximum RT_{60} for AL_{CONS} of 15%
$$= \left(\frac{15VQ}{641.81 \, (D_2)^2 \, (n+1)} \right)^{1/2}$$

Minimum V for AL_{CONS} of 15%
$$= \frac{641.81 \, (D_2)^2 \, (RT_{60})^2 \, (n+1)}{15Q}$$

Minimum Q for AL_{CONS} of 15%
$$= \frac{641.81 \, (D_2)^2 \, (RT_{60})^2 \, (n+1)}{15V}$$

Minimum Norris–Eyring $S\bar{a}$ for AL_{CONS} of 15%
$$= S\{1 - \exp[-1.24 D_2 V / S \, (15VQ)^{1/2}]\}$$

Maximum $(n+1)$ for AL_{CONS} of 15%
$$= \frac{15VQ}{641.81 \, (D_2)^2 \, (RT_{60})^2}$$

$$EPR = 10^{\,[(\text{dB-SPL}_\text{D}+10)+(\Delta D_2 - \Delta 4') - L_{\text{SENS}}]/10}$$

Note: For $(\Delta D_2 - \Delta 4')$, see Fig. 14.

Max Prog Level in dB-SPL from
Available Electrical Power
$$= \frac{\text{Watts avail}}{10} - (\Delta D_2 - \Delta 4') + L_{\text{SENS}}$$

LISTENER AT
DISTANT POSITION

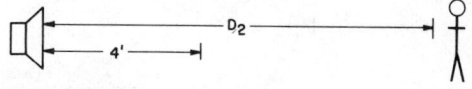

SOUND-SYSTEM
LOUDSPEAKER

ATTENUATION OF SOUND IN dB FROM 4' IN FRONT OF
LOUDSPEAKER TO LISTENER $= \Delta D_2 - \Delta 4'$

WHERE, $\Delta D_2 = -10 \text{ LOG}_{10} \left(\dfrac{Q}{4\pi (D_2)^2} + \dfrac{4}{R} \right)$

$\Delta 4' = -10 \text{ LOG}_{10} \left(\dfrac{Q}{4\pi 16} + \dfrac{4}{R} \right)$

Fig. 14—Calculation of D_2 attenuation.

System Example

Assume the following values: $V = 500\,000$ ft^3, $RT_{60} = 2.5$ s, $S = 42\,500$ ft^2, $D_2 = 125$ ft, $n+1 = 2$, $NOM = 2$, $L_{SENS} = 110$ dB-SPL, EAD desired $= 8$ ft, dB-SPL$_D = 85$ db-SPL. Design a system that will fulfill these requirements in an acoustical environment with these parameters.

Min Q that allows 15 $\% AL_{CONS}$

$$= \frac{641.81\,(125)^2\,(2.5)^2\,(2)}{15\,(500\,000)} = 16.71$$

A standard available unit has a Q of 17, so this is a realizable requirement.

$$\bar{a} = 1 - \exp(-0.049 V / S \cdot RT_{60})$$

Therefore:

$$\bar{a} = 1 - \exp\left(-\frac{0.049 \times 500\,000}{42\,500 \times 2.5}\right) = 0.206$$

$$R = \frac{42\,500\,(0.206)}{1 - 0.206} = 11\,026 \text{ ft}^2$$

$$D_c = 0.141 \left(\frac{17\,(11\,026)}{2}\right)^{1/2} = 43 \text{ ft}$$

Distance D_1 should be $\geq D_c < 45$ ft. This is to insure maximum available acoustic gain while avoiding time-dely interference. A good D_1 selection would be 40 feet.

See Fig. 15.

This result means that the two talkers should wear lavalier microphones within approximately 2 feet of their mouths, and all conditions will be met so far as clarity and loudness are concerned.

It is now only necessary to insure that the selected pair of loudspeakers with their individual Qs of 17 also provide full coverage of the audience. See Figs. 16 and 17.

The correct place to aim the center of a single loudspeaker (Fig. 16) in an auditorium is at the last seat, not at the middle of the auditorium. This is because the last seat needs the highest Q, and the highest Q exists on the 0° axis of the loudspeaker. The next problem to consider involves the rear wall and the upper part of the beam. If the wall is not tilted out as its height increases, is not sufficiently irregular to provide diffusion, or is not highly absorptive (99% absorption only drops the reflection by 20 dB), it will cause problems no matter what part of the beam strikes it. Placing the loudspeaker so that its on-axis beam strikes the wall at an angle from above is usually the remedy. Always try to avoid aiming a loudspeaker directly at a wall in such a way that the on-axis beam is perpendicular to the flat surface.

Fig. 16 illustrates a typical case. Here, the Q is chosen for the last seat, and the coverage pattern of the loudspeaker is used to insure smooth coverage from the rear to the front of the auditorium. After the coverage is assured, a quick

	One Source	Two Sources
$\Delta 40' = -10 \log\left[\dfrac{17}{4\pi\,(40)^2} + \dfrac{4\,(n+1)}{11\,026}\right] =$	29.18 dB	28.04 dB
$\Delta 8' = -10 \log\left[\dfrac{17}{4\pi\,(8)^2} + \dfrac{4\,(n+1)}{11\,026}\right] =$	16.68 dB	16.6 dB
$\Delta 125' = -10 \log\left[\dfrac{17}{4\pi\,(125)^2} + \dfrac{4\,(n+1)}{11\,026}\right] =$	33.47 dB	30.9 dB
$10 \log 2 =$	3.01 dB	3.01 dB
$\Delta 4' = -10 \log\left[\dfrac{17}{4\pi\,(4)^2} + \dfrac{4\,(n+1)}{11\,026}\right] =$	10.71 dB	10.71 dB

$\Delta D_s = 29.18 + 16.68 - 33.47 - 3.01 - 6 = 3.38$ dB for one source

$\Delta D_s = 28.04 + 16.6 - 30.9 - 3.01 - 6 = 4.73$ dB for two sources

$$\text{Max } D_s = r = \left[17 \Big/ 4\pi\left(10^{-3.38/10} - \frac{4}{11\,026}\right)\right]^{1/2} = 1.72 \text{ ft for one source}$$

$$\text{Max } D_s = r = \left[17 \Big/ 4\pi\left(10^{-4.73/10} - \frac{4 \times 2}{11\,026}\right)\right]^{1/2} = 2.01 \text{ ft for two sources}$$

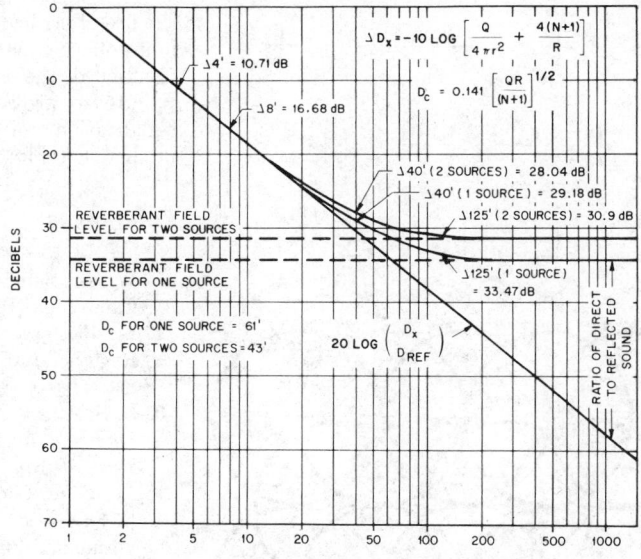

Fig. 15—Decibel changes with distance from source.

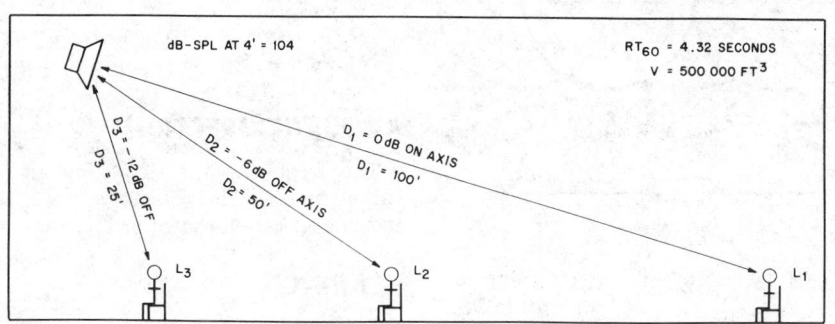

Fig. 16—Loudspeaker orientation (ideal case).

Rel Q at $L_3 = 1$

   D_I at $L_3 = 0$ dB

dB-SPL at L_3

$$= 104 + 20 \log \frac{4}{25} - 12$$

$$= 76 \text{ dB-SPL}$$

Max D_3 That Allows an
AL_{CONS} of 15%

$$= \left(\frac{15VQ}{641.81(RT_{60})^2}\right)^{1/2}$$

$$= \left(\frac{15 \times 500\,000 \times 1}{641.81 \times 18.66}\right)^{1/2}$$

$$= 25 \text{ ft}$$

Rel Q at $L_2 = 4$

   D_I at $L_2 = 6$ dB

dB-SPL at L_2

$$= 104 + 20 \log \frac{4}{50} - 6$$

$$= 76 \text{ dB-SPL}$$

Max D_2 That Allows an
AL_{CONS} of 15%

$$= \left(\frac{15VQ}{641.81(RT_{60})^2}\right)^{1/2}$$

$$= \left(\frac{15 \times 500\,000 \times 4}{641.81 \times 18.66}\right)^{1/2}$$

$$= 50 \text{ ft}$$

Axial Q at $L_1 = 16$

   D_I at $L_1 = 12$ dB

dB-SPL at L_1

$$= 104 + 20 \log \frac{4}{100}$$

$$= 76 \text{ dB-SPL}$$

Max D_1 That Allows an
AL_{CONS} of 15%

$$= \left(\frac{15VQ}{641.81(RT_{60})^2}\right)^{1/2}$$

$$= \left(\frac{15 \times 500\,000 \times 16}{641.81 \times 18.66}\right)^{1/2}$$

$$= 100 \text{ ft}$$

$$D_I = 10 \log Q; \quad Q = 10^{D_I/10}$$

NOTE: Level changes with distance are for "direct sound."

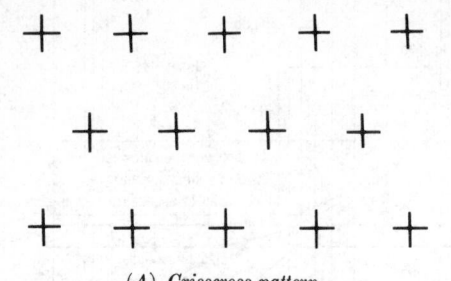

(A) Crisscross pattern.

(B) Crisscross, 50% overlap.

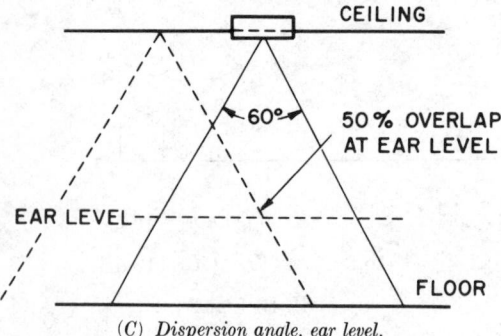

(C) Dispersion angle, ear level.

Fig. 17—Proper distribution density in an overhead distributed sound system.

calculation of the desired Q for 15% AL_{CONS} at each closer location clearly reveals that determining the Q for the last seat also does so for all nearer seats, provided smooth coverage is achieved (± 2 dB).

The electrical power required is:

$$EPR = 10^{\,[(85+10)+(33.47-10.71)-110]/10} = 5.97 \text{ watts}$$

This is 5.97 watts per loudspeaker; therefore, since two loudspeakers are involved (for example, a front speaker and a rear speaker with time-delay correction), the total EPR is 2(5.97), or approximately 12 watts.

It should be kept in mind that the accuracy of sound-system design equations is predicated on the use of critical-bandwidth band-rejection, minimum-phase equalizers to adjust the total transfer function of the sound system to the environment. Failure to provide for such sound-system–room equalization can lead to large variations from the predicted results.

References

1. W. C. Sabine, *Collected Papers on Acoustics*, Cambridge (USA); 1923.
2. M. R. Schroeder, "New Method of Measuring Reverberation Time," *Journal of the Acoustical Society of America*, Vol. 37, page 409; 1965.
3. B. S. Atal, M. R. Schroeder, and G. M. Sessler, *Subjective Reverberation Time and Its Relation to Time Decay*, 5th International Congress on Acoustics, Liege, Paper G32; 1965.
4. R. F. Norris, "A Derivation of the Reverberation Formula," Published in Appendix II of V. O. Knudsen's *Architectural Acoustics*, John Wiley & Sons, Inc., New York; 1932.
5. D. B. Davis, *Acoustical Tests and Measurements*, Howard W. Sams & Co., Inc., Indianapolis; 1965.

ACOUSTIC SPECTRUM

The frequency ranges of human voices and various musical instruments are compared in the acoustic spectrum shown in Fig. 18.

HEARING*

The auditory system consists of the periphery sensors, acoustic neurological transducers, the ears, the eighth cranial nerve leading to a programming and priority switching center at various levels of the brain stem, and finally to the auditory area of the cortex located near the Sylvian fissure of the frontal-lobe convolution.

The auditory system does much more than detect minute sounds. Among other functions, it preferentially places more weight on certain pre-programmed characteristic sounds, localizes the direction of most sounds by a variety of ingenious techniques, and initiates involuntary actions for visual acquisition of the source.

The hearing mechanism was probably evolved to help man survive in a hostile environment, and not for linguistic communication or for musical entertainment.

The ear (Fig. 19) is divided for convenience into 3 sections, the external, the middle, and the inner ear.

* J. L. Flanagan, "Speech Analysis, Synthesis and Perception," Academic Press, New York; 1965. Also "Technical Aspects of Sound," E. G. Richardson, ed., Elsevier Press, New York; 1953.

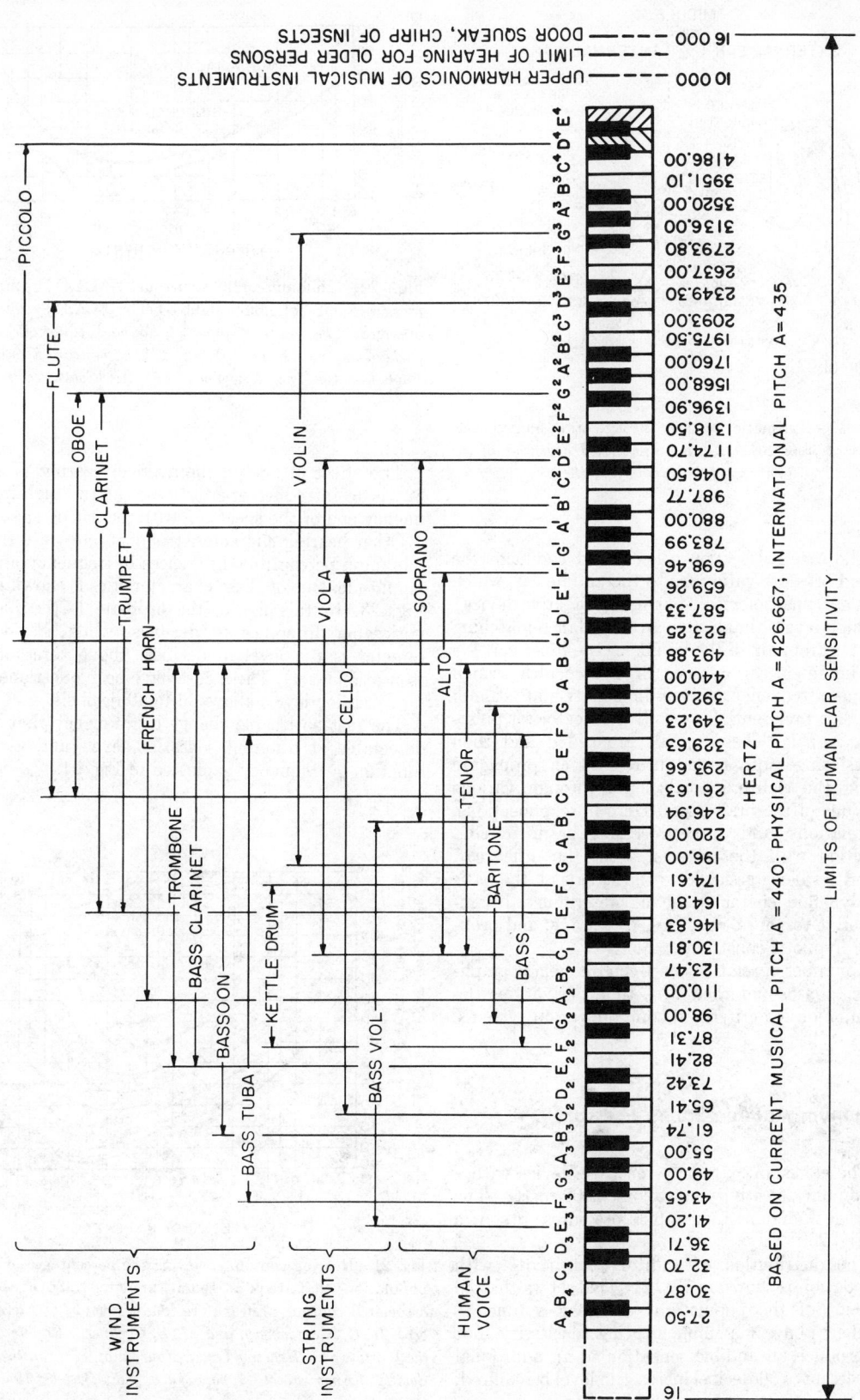

Fig. 18—Acoustic spectrum in hertz.

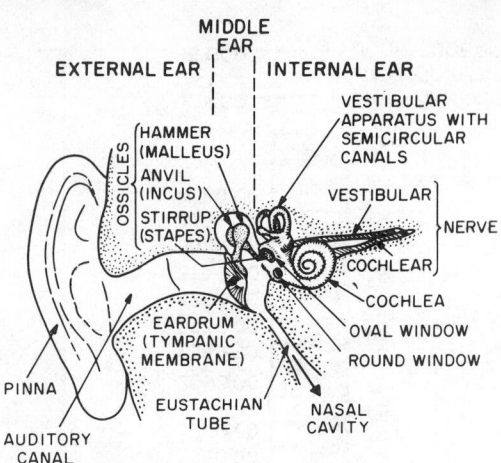

Fig. 19—Schematic section through the human ear. *Adapted from Brödel, three unpublished drawings of the anatomy of the human ear; 1946.*

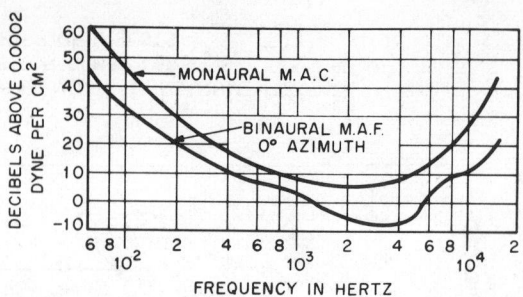

Fig. 20—Minimum-audible-ear-canal (M.A.C.) sound pressure and minimum-audible-field (M.A.F.) sound pressure. *"Technical Aspects of Sound," editor E. G. Richardson, vol. 1, fig. 106, p. 248, American Elsevier Publishing Co., Inc., New York; 1953. All Rights Reserved.*

The external section collects and conducts the sound pressure variations to the middle ear, which acts as a "pathological" barrier (protective device) to high-level stimuli and an impedance-matching section from air to the liquids of the inner ear.

The inner ear consists of the cochlea, which acts as a frequency analyzer for steady-state stimuli and as a preferentially sensitive detector for pulse-type signals. These signals are coded and then transformed into neural firings which propagate along the eighth cranial nerve through various sections of the brain stem. Here interconnections, correlations, and data processing occur in conjunction with the response from the other ear, priorities being assigned to the signals on the basis of dynamic characteristic as determined by instinctive response, conditioned reaction, and probably hypnotic constraint.

A number of relations between frequency, amplitude, phase, and their first time derivations determine design criteria in communication systems.

Minimum Audible Sound Pressure

The ear is an extremely sensitive device with a maximum sensitivity at about 2000 hertz. The average minimum-audible-field pressures are shown in Fig. 20.

The distribution of auditory sensitivity with population is shown in Fig. 21. This shows that 50 percent of the population will hear a tone 18 decibels above maximum hearing sensitivity, but to produce an audible sensation in an additional 49 percent, a 50-decibel increase in level is required.

The effect of age on the average hearing (Fig. 22) is to introduce greater losses at the high-frequency end of the spectrum with increasing age.

Other hearing characteristics of interest are the minimum perceptible differences in frequency and amplitude. One of these characteristics is shown in Fig. 23. In this figure, the ordinate is the just noticeable difference in decibels, while the parameter is the level in decibels above threshold (sensation level). These data have been questioned for sensation levels above 30 to 40 decibels.

The just noticeable change in frequency that is detectable at different sensation levels and as a function of frequency is plotted in Fig. 24.

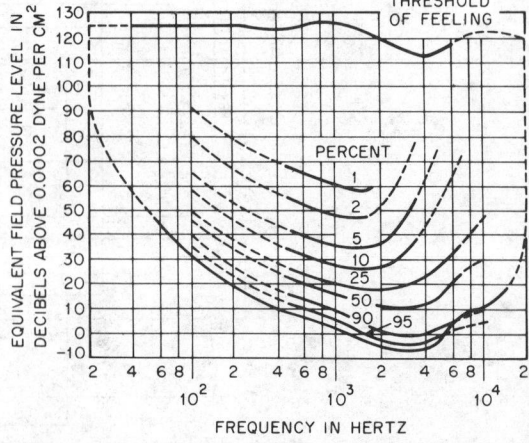

Fig. 21—Hearing contours showing the percentages of a population that require sound pressures above the associated contour to hear pure tones. *From J. C. Steinberg, H. C. Montgomery, and M. B. Gardner, "Results of the World's Fair Hearing Tests," Journal of the Acoustical Society of America, vol. 12, no. 2, fig. 8, p. 300; October 1940.*

The ordinate is in hertz and not in percent change. At a sensation level of 5 decibels above threshold and at 30 hertz, a 4/30 or 15-percent change is necessary for detection, whereas at 1000 hertz, less than 1-percent change will be noticeable. This means that the low-frequency instruments of an orchestra can be permitted larger tolerances in tuning.

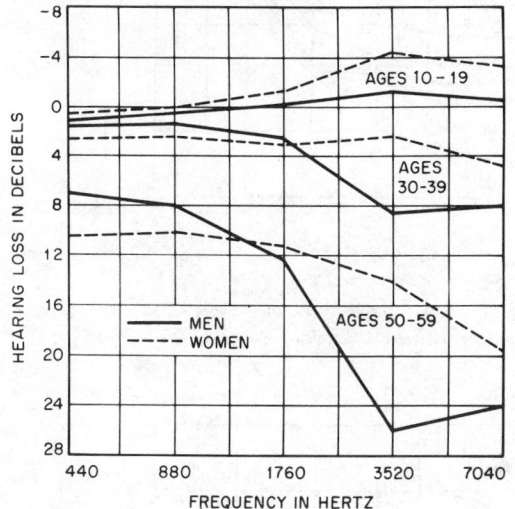

Fig. 22—Effect of age on the average hearing in a population. *From J. C. Steinberg, H. C. Montgomery, and M. B. Gardner, "Results of the World's Fair Hearing Tests," Journal of the Acoustical Society of America, vol. 12, no. 2, fig. 3, p. 293; October 1940.*

When a wide-band Gaussian noise is present, other sounds are masked; that is, their loudness apparently decreases.

The precise effect of noise on complex sounds is difficult to measure since it depends on the waveform and the power density spectrum of the sounds, there being less masking for spike-like stimuli.

Loudness level is the intensity level of a 1000-hertz tone which is equal in loudness to the given tone. The relation between loudness and frequency is given by the equal-loudness contours of Fig. 25. These indicate that at low listening levels, the bass and high frequencies must be boosted in strength to preserve the same relative loudness between different tones.

Figure 26 shows the deterioration in word articulation with noise for various signal-to-noise ratios.

The loudness of a signal (sensation of loudness) depends on the magnitude of the sound as expressed in terms of energy (intensity) and on the frequency of the signal. The intensity is measured by the number of decibels that the sound is above the arbitrary reference level such as

$$0 \text{ dB} = 10^{-16} \text{ watt per cm}^2$$

$$= 0.0002 \text{ dyne per cm}^2$$

$$= 73.8 \text{ dB below 1 dyne per cm}^2.$$

The sensation level indicates the number of decibels that the sound is above the threshold of hearing at a given frequency.

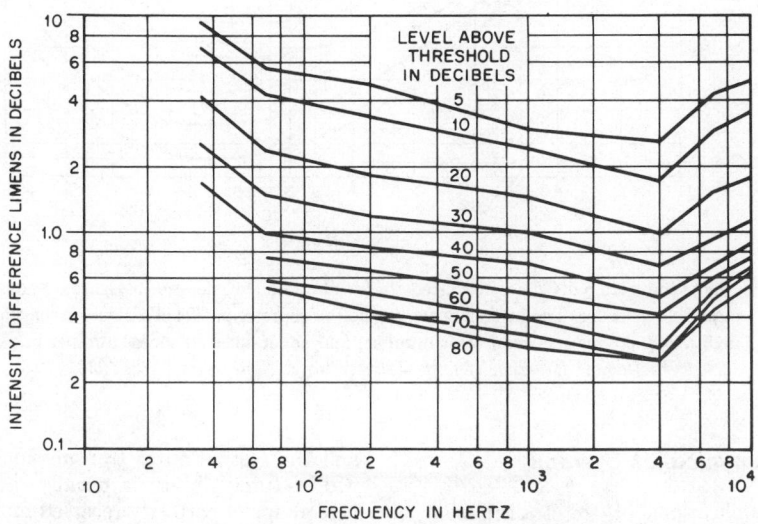

Fig. 23— Intensity difference limens for pure tones at different levels above threshold. *From "Technical Aspects of Sound," editor E. G. Richardson, vol. 1, fig. 110, p. 252, American Elsevier Publishing Co., Inc., New York; 1953. All Rights Reserved.*

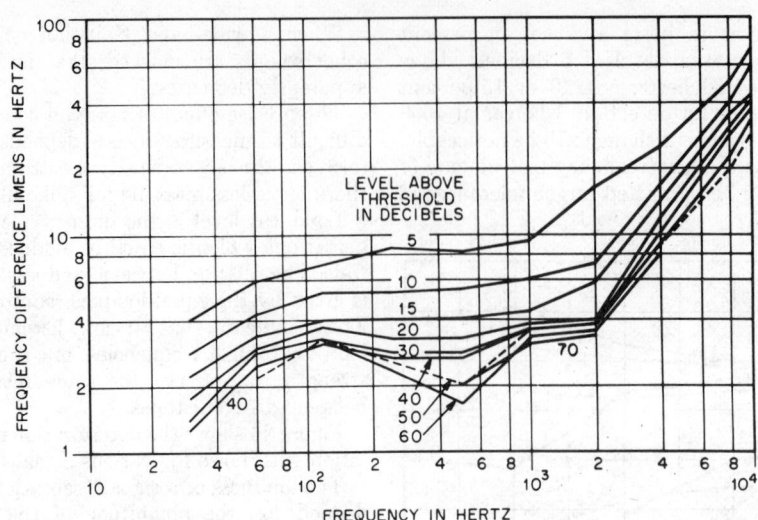

Fig. 24—Frequency difference limens for pure tones at different levels above threshold. *From "Technical Aspects of Sound," editor E. G. Richardson, vol. 1, fig. 111, p. 253, American Elsevier Publishing Co., Inc., New York; 1953. All Rights Reserved.*

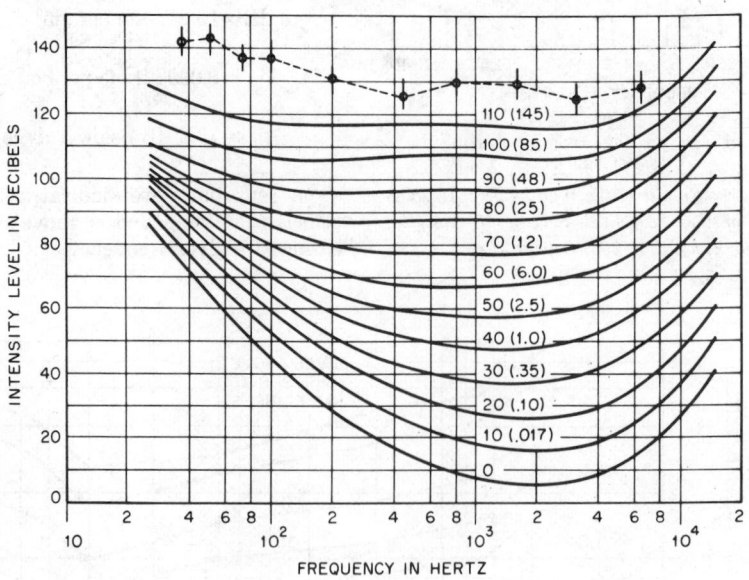

Fig. 25—Equal-loudness contours plotted against intensity (decibels above reference pressure). The zero contour is the curve for minimum audible pressure. The broken curve at the top represents Wegel's data for the threshold of feeling. The parameter is designated as loudness level (first number) and as loudness in sones (number in parentheses). *From H. Davis, "Hearing," fig. 45, John Wiley & Sons, New York; 1938.*

Speech-Communication Systems

In many applications of the transmission of information by speech, a premium is placed on intelligibility rather than on flawless reproduction. Especially important is the reduction of intelligibility as a function of both the background noise and the restriction of transmission-channel bandwidth. Intelligibility is usually measured by the percentage of correctly received monosyllabic nonsense words uttered in an uncorrelated sequence. This score is known as syllable articulation. Because the sounds are nonsense syllables, one part of the word is entirely uncorrelated with the re-

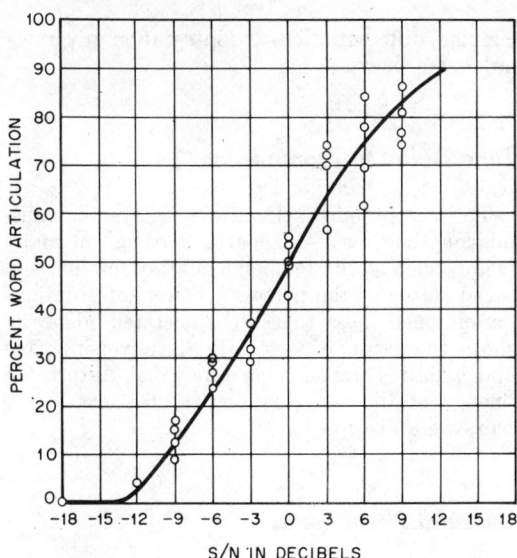

Fig. 26—Word articulation for continuous speech heard in the presence of continuous noise, plotted as a function of the signal-to-noise ratio in decibels. The average level of the speech was held constant at approximately 90 decibels above 0.0002 dyne per square centimeter. *From G. A. Miller and J. C. Licklider, "The Intelligibility of Interrupted Speech," Journal of the Acoustical Society of America, vol. 22, fig. 9, p. 171; March 1950.*

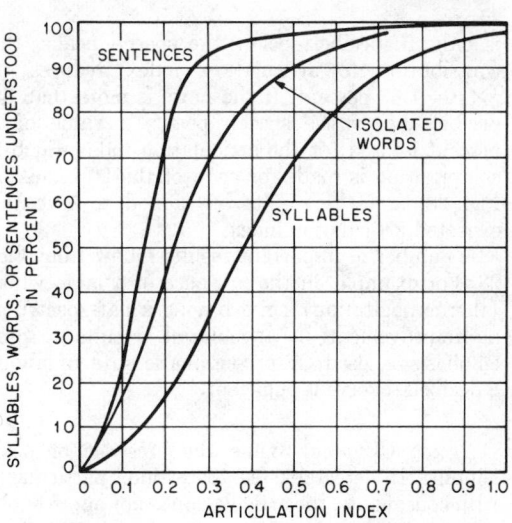

Fig. 27—Relations between various measures of speech intelligibility. Relations are approximate; they depend on the type of material and the skill of the talkers and listeners.

mainder, so it is not consistently possible to guess the whole word correctly if only part of it is received intelligibly. Obviously, if the test speech were a commonly used word, or say a whole sentence with commonly used word sequences, the score would increase because of correct guessing from the context. Figure 27 shows the interrelationship between syllable, word, and sentence articulation. Also given is a quantity known as articulation index.

Articulation Index: The concept and use of articulation index is obtained from Fig. 28. The abscissa is divided into 20 bandwidths of unequal frequency intervals. Each of these bands will contribute 5 percent to the articulation index when the speech spectrum is not masked by noise and is sufficiently loud to be above the threshold of audibility. The ordinates give the root-mean-square peaks and minimums (in $\frac{1}{8}$-second intervals) and the average sound pressures created at 1 meter from a speaker's mouth in an anechoic (echo-free) chamber. The units are in decibels pressure per hertz relative to a pressure of 0.0002 dyne/centimeter². (For example, for a bandwidth of 100 hertz, rather than 1 hertz, the pressure would be that indicated plus 20 decibels; the latter figure is obtained by taking 10 times logarithm [to the base 10] of the ratio of the 100-hertz band to the indicated band of 1 hertz.)

An articulation index of 5 percent results in any of the 20 bands when a full 30-decibel range of speech-pressure peaks to speech-pressure minimums is obtained in that band. If the speech minimums are masked by noise of a higher pressure, the contribution to articulation is accordingly reduced to a value given by $\frac{1}{6}$ [(decibels level of speech peaks) − (decibels level of average noise)]. Thus, if the average noise is 30 decibels under the speech peaks, this expression gives 5 percent. If the noise

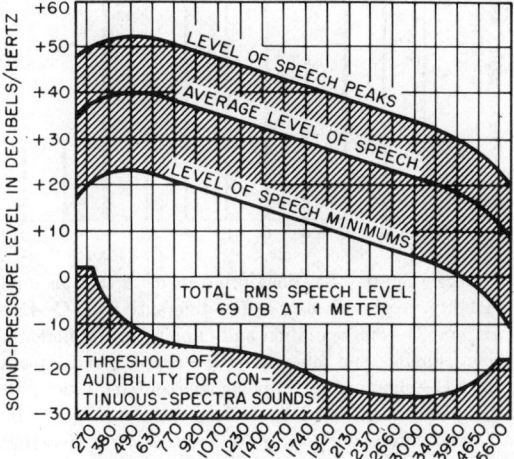

Fig. 28—Bands of equal articulation index. 0 decibels = 0.0002 dyne/centimeter². *Courtesy of Proceedings of the I.R.E.*

is only 10 decibels below the speech peaks, the contribution to articulation index reduces to $\frac{1}{6} \times 10 = 1.67$ percent. If the noise is more than 30 decibels below the speech peaks, a value of 5 percent is used for the articulation index. Such a computation is made for each of the 20 bands of Fig. 28, and the results are added to give the expected articulation index.

A number of important results follow from Fig. 28. For example, in the presence of a large white (thermal-agitation) noise having a flat spectrum, an improvement in articulation results if pre-emphasis is used. A pre-emphasis rate of about 8 decibels/octave is sufficient.

Speech Clipping: While the presence of peak clipping is detectable as distortion, particularly with consonants, the articulation is not appreciably affected by even large amounts of peak clipping.* The deterioration from clipping is determined apparently by the masking and smearing caused by the intermodulation frequencies produced by the nonlinear clipping circuit. Consequently, the articulation after clipping depends on whether the higher frequencies are preferentially amplified (differentiation) or attenuated (similar to integration) before clipping.

The articulation resulting from sequences of clipping, differentiation, and integration in various orders are shown in Fig. 29.

Time Delay in Transmission

The acceptability of various delays in transmission time over long paths depends on many factors such as the temperament, occupation, and social status of the partners. Fewer interruptions permit longer delay times to be tolerated. Figure 30 shows the results of tests with 22 conversing pairs and indicates transmission times that disturb the fluency of the conversations (at 0.4 second, 10 pairs were disturbed).

Binaural Phenomena

A number of hearing characteristics form the basis for stereophonic listening.

Effect of Intensity: When two tone sources, differing only in intensity, are impressed separately on each ear, the source seems to be located toward the side with the greater intensity. Figure 31 shows the effect of moving a tone source from the right

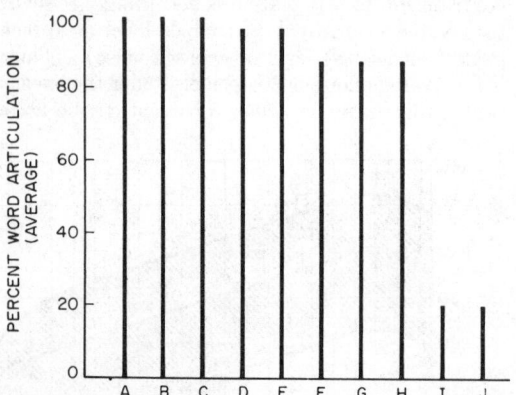

Fig. 29—Effects of various types of distortion on intelligibility. A—no distortion; B—differentiation; C—integration; D—differentiation and clipping; E—differentiation, clipping, and integration; F—clipping and integration; G—clipping; H—clipping and differentiation; I—integration and clipping; J—integration, clipping, and differentiation. *Courtesy of the Journal of the Acoustical Society of America.*

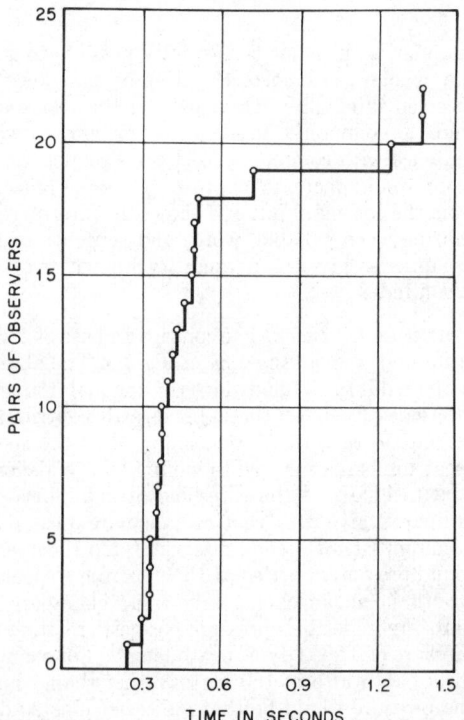

Fig. 30—Transmission delay times for 22 conversing pairs.

* J. C. R. Licklider and I. Pollack, "Effects of Differentiation, Integration, and Infinite Peak Clipping upon the Intelligibility of Speech," *Journal of the Acoustical Society of America*, vol. 20, pp. 42–51; 1948.

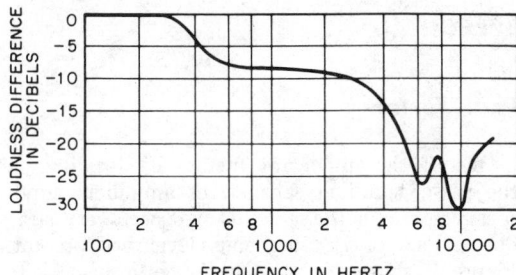

Fig. 31—The difference in loudness level produced in the right ear when a source of pure tone is moved from the right to the left of an observer. The curve shows how marked is the sound shadow produced, at different frequencies, by the head. *From J. C. Steinberg and W. B. Snow, "Physical Factors," Bell System Technical Journal, vol. 13, no. 2, fig. 4, p. 253;* © *1934, American Telephone and Telegraph Co., reprinted by permission.*

to the left of an observer. Below 300 hertz there is no shift in the apparent source.

Figure 32 shows the amount of increase of a tone in one ear over the other produced by a displacement in azimuth.

Effect of Phase: When the phase of a tone is different at each ear, the listener imagines the source to be located toward the side of the leading phase. The phase effect is most noticeable at frequencies below about 1000 hertz.

Effect of Latency of Arrival: When transient sounds such as clicks are heard, localization is made on the basis of time of arrival, and the sound source shifts to the side of the first arrival.

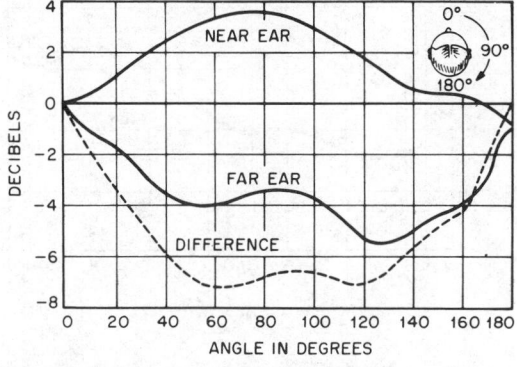

Fig. 32—The variation in loudness level as a speech source is rotated in a horizontal plane around the head. The broken curve shows the difference in loudness level between the two ears for various azimuths. *From J. C. Steinberg and W. B. Snow, "Physical Factors," Bell System Technical Journal, vol. 13, no. 2, fig. 3, p. 252;* © *1934, American Telephone and Telegraph Co., reprinted by permission.*

The minimum perceptible latency is of the order of 0.1 millisecond. The maximum latency beyond which the sounds are resolved as two separate ones is about 2 milliseconds. Between these values the apparent displacement is roughly proportional to the time difference.

Factors Determining Localization

In practical situations, closed rooms with sound-reflecting walls reduce our ability to localize sources, particularly at the lower frequencies where the phase of the field at the ear is determined to a large extent by the room dimensions and where no phase changes can exist within the distance between the two ears. The best cues to azimuth localization occur on transient-like sounds and noises. The phase of the reproducing system is

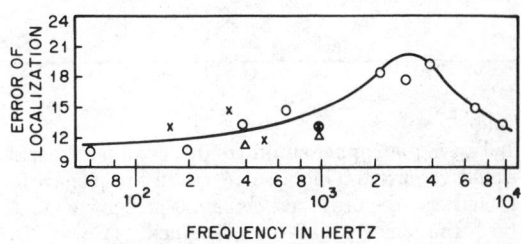

Fig. 33—The average of the errors, in degrees, made by two observers in localizing a source of tone at various frequencies. Circles and crosses are for two different series of observations. Triangles are for impure tones. *From H. Davis, "Hearing," fig. 74, John Wiley & Sons, New York, New York; 1938.*

important for these types of sounds. For sharp clicks, the echoes from the room are generally resolved in time and produce an awareness of the room geometry.

The sound reaching the observer should reach the ears directly rather than through a reflected path, since otherwise the intensities will be distorted and the integrity of the leading or transient edges will be lost. Figure 33 shows the errors made in localizing a source of tone at various frequencies.

Minimum-Discernible-Bandwidth Changes

Table 9 gives the increase in high-frequency bandwidth required to produce a minimum discernible change in the output quality of speech and music.

These bandwidths are known as difference-limen units. For example, a system transmitting music

TABLE 9—BANDWIDTH INCREASES NECESSARY TO GIVE AN EVEN CHANCE OF QUALITY IMPROVEMENT BEING NOTICEABLE. ALL FIGURES ARE IN KILOHERTZ.

Minus One Limen		Reference Frequency	Plus One Limen	
Speech	Music		Music	Speech
—	—	3	3.6	3.3
3.4	3.3	4	4.8	4.8
4.1	4.1	5	6.0	6.9
4.6	5.0	6	7.4	9.4
5.1	5.8	7	9.3	12.8
5.5	6.4	8	11.0	—
5.8	6.9	9	12.2	—
6.2	7.4	10	13.4	—
6.4	8.0	11	15.0	—
7.0	9.8	13	—	—
7.6	11.0	15	—	—

and having an upper cutoff frequency of 6000 hertz would require an increase in cutoff frequency to 7400 hertz before there is a 50-percent chance that the change can be discerned. (Curve B, Fig. 34.)

Figure 34 is based on the data of Table 9. For any high-frequency cutoff along the abscissa, the ordinates give the next higher and next lower cutoff frequencies for which there is an even chance of discernment. As expected, one observes that, for frequencies beyond about 4000 hertz, restriction of upper cutoff affects music more than speech.

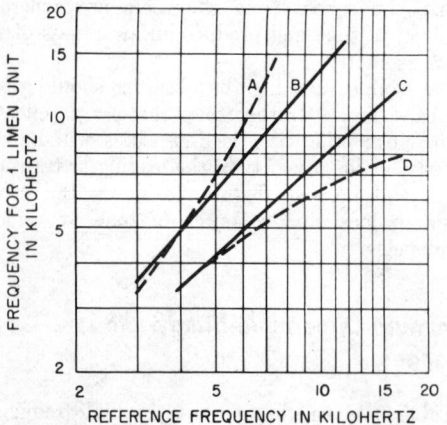

Fig. 34—Minimum-discernible-bandwidth changes. A—plus 1 limen for speech; B—plus 1 limen for music; C—minus 1 limen for music; D—minus 1 limen for speech.
Courtesy of Bell System Technical Journal.

SPEECH*

Peak Factor

One of the important factors in deciding on the power-handling capacity of amplifiers, loudspeakers, etc., is the fact that in speech very large fluctuations of instantaneous level are present. Figure 35 shows the peak factor (ratio of peak to root-mean-square pressure) for unfiltered (or wideband) speech, for separate octave bandwidths below 500 hertz, and for separate $\frac{1}{2}$-octave bandwidths above 500 hertz. The peak values for sound pressure of unfiltered speech, for example, rise 10 decibels higher than the averaged root-mean-square value over an interval of $\frac{1}{8}$ second, which corresponds roughly to a syllabic period. However, for a much longer interval of time, say the time duration of one sentence, the peak value reached by the sound pressure for unfiltered speech is about 20 decibels higher than the root-mean-square value averaged for the entire sentence.

Thus, if the required sound-pressure output demands a long-time average of, say, 1 watt of electrical power from an amplifier, then, to take care of the instantaneous peaks in speech, a maximum peak-handling capacity of 100 watts is needed. If the amplifier is tested for amplitude distortion with a sine wave, 100 watts of peak instantaneous power exists when the average power

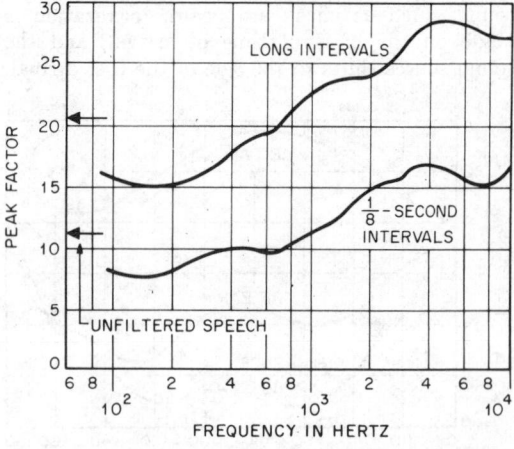

Fig. 35—Peak factor (ratio of peak to root-mean-square pressures) in decibels for speech in 1- and in $\frac{1}{2}$-octave frequency bands, for $\frac{1}{8}$- and for 75-second time intervals.
Courtesy of the Journal of the Acoustical Society of America.

* J. L. Flanagan, "Speech Analysis, Synthesis and Perception," Academic Press, New York; 1965. Also "Technical Aspects of Sound," E. G. Richardson, ed., Elsevier Press, New York; 1953.

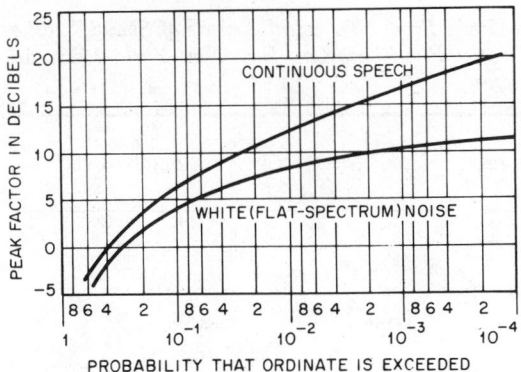

Fig. 36—Statistical properties of the peak factor in speech. The abscissa gives the probability (ratio of the time) that the peak factor in the uninterrupted speech of one person exceeds the ordinate value. Peak factor = (decibels instantaneous peak value) − (decibels root-mean-square long-time average).

of the sine-wave output is 50 watts. This shows that if no amplitude distortion is permitted at the peak pressures in speech sounds, the amplifier should give no distortion when tested by a sine wave of an average power 50 times greater than that required to give the desired long-time-average root-mean-square pressure.

The foregoing puts a very stringent requirement on the amplifier peak power. In relaxing this specification, one of the important questions is what percentage of the time will speech overload an amplifier of lower power than that necessary to take care of all speech peaks. This is answered in Fig. 36; the abscissa gives the probability of the (peak)/(long-time-average) powers exceeding the ordinates for continuous speech and white noise. When multiplied by 100, this probability gives the expected percent of time during which peak distortion occurs. If 1 percent is taken as a suitable criterion, then a 12-decibel ratio of (peak)/(long-time-average) powers is sufficient. Thus, the amplifier should be designed with a power reserve of 16 in order that peak clipping may occur not more than about 1 percent of the time.

Characteristics of Speech

The production of speech involves the generation of a complex Fourier spectrum, the preferential selection of certain components, and the radiation of the resulting acoustic energy.

Vowels and semivowels are produced by causing expirating air to excite the vocal cords, which generate a wide-band spectrum of sound energy consisting of a fundamental frequency and many harmonics of this frequency. The fundamental frequency has an average value of 130 hertz for the male voice and about 205 hertz for the female voice. The fundamental frequency varies with the individual and is caused to change during the production of the vowels and semivowels.

The preferential selective process of certain components is imposed by the resonances and constrictions involving the tongue and lips and the cavities of the vocal tract (Fig. 37). The vocal tract has resonances and selectively transmits certain bands of frequencies. Harmonics occurring near these resonant points will be selectively transmitted and are called formants of the vowel.

The consonants usually involve greater constrictions than the vowels. There are four general types of articulatory formations used in generating consonants. These are shown in Table 10.

The following 4 formations are indicated by the headings across the top of the table.

(**A**) *Plosives* (or stops) are formed by blocking the speech path so that no air flows and suddenly creating an opening somewhere along the tract so as to form a puff of air.

(**B**) *Fricatives* are formed by blowing air through a narrow orifice or across the cutting edge of the teeth.

(**C**) The *nasal consonants* are formed by lowering the soft palate and blocking the mouth so that air must pass through the nose.

(**D**) The *glides or semivowels* are like the vowels in that their formation is relatively open.

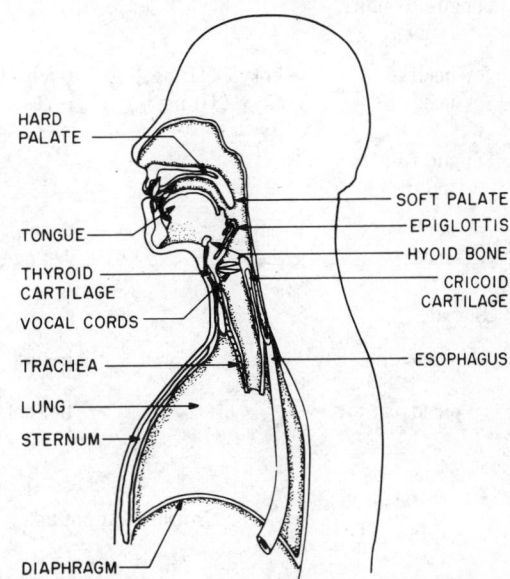

Fig. 37—Schematic median section of the head and thorax. *From "Technical Aspects of Sound," editor E. G. Richardson, vol. 1, fig. 85, p. 200, American Elsevier Publishing Co., Inc., New York; 1953. All Rights Reserved.*

TABLE 10—THE CONSONANT CHART (MODIFIED FROM IPA). *From "Technical Aspects of Sound," editor E. G. Richardson, vol. 1, table 2, p. 205, American Elsevier Publishing Co., Inc., New York; 1953. All Rights Reserved.*

	Plosives (Stops)	Fricatives	Nasals	Semivowels
Two lips				
Voiceless	p–pit	φ–Schwester (Ger.)		
Voiced	b–bit	β–saber (Sp.)	m–me	w–was
Lip to teeth				
Voiceless		f–fill		
Voiced		v–vest	ɱ–invento (Sp.)	ʋ–westland (Dutch)
Tongue to teeth or gum ridge				
Voiceless	t–to	θ–thick s–see		
Voiced	d–do	ð–this z–zoo	n–no	ɹ–red l–lateral
Tongue to gum ridge and hard palate				
Voiceless		ʃ–shed		
Voiced		ʒ–pleasure		
Tongue to hard palate				
Voiceless	c–kotya (Hung.)	ç–ich (Ger.)		
Voiced	ɟ–nagy (Hung.)	j–yes	ɲ–ogni (It.)	λ–egli (It.)
Tongue to soft palate				
Voiceless	k–cap	x–ach (Ger.)		
Voiced	g–gap	ɣ̥–vagen (Ger.)	ŋ–sing	
Glottal				
Voiceless	ʔ–verein (North Ger.)	h–hat		
Voiced		ɦ–ahead		

Initial Compounds		Final Compounds	
pr	pl	nt	ld
hw	kw	nd	rz
st	bl	st	ks
tr	sp	ts	kt
fr	kl	nk	rd

Speech Compression Characteristics

With proper coding and for a 30-decibel signal-to-noise power ratio, the conventional 3-kilohertz channel should transmit information at about 30 000 bits per second. From Table 11 it can be calculated that, with probabilities of occurrence in English as shown, 4.9 bits per phoneme are involved. In conversational speech about 10 phonemes per second are produced, so that speech covers less than 50 bits per second.

Caution is necessary to avoid erroneous conclusions as to the possible economies which could be effected. The auditory apparatus is able to scan the multidimensional attributes of the acoustic image and concentrate on minute details which may have emotional content or information as to the physical environment of the source.

The compression of speech, in view of the possible economic gains, has attracted considerable attention. H. W. Dudley's dedicated efforts in this area during his 40 years at Bell Telephone Laboratories and his contributions form the basis for most subsequent work.

His Voder, diagrammed in Fig. 38, contained 10 contiguous band-pass filters which covered the speech band. These filters were activated by volume controls operated by finger keys. Three ad-

TABLE 11—RELATIVE FREQUENCIES OF ENGLISH SPEECH SOUNDS IN STANDARD PROSE (*after Dewey*).

Vowels and Diphthongs			Consonants		
Phoneme	Relative Frequency of Occurrence (%)	$-P(x_i) \log_2 P(x_i)$	Phoneme	Relative Frequency of Occurrence (%)	$-P(x_i) \log_2 P(x_i)$
I	8.53	0.3029	n	7.24	0.2742
a	4.63	0.2052	t	7.13	0.2716
æ	3.95	0.1841	r	6.88	0.2657
ɛ	3.44	0.1672	s	4.55	0.2028
ɒ	2.81	0.1448	d	4.31	0.1955
ʌ	2.33	0.1264	l	3.74	0.1773
i	2.12	0.1179	ð	3.43	0.1669
e, eɪ	1.84	0.1061	z	2.97	0.1507
u	1.60	0.0955	m	2.78	0.1437
aɪ	1.59	0.0950	k	2.71	0.1411
oʊ	1.30	0.0815	v	2.28	0.1244
ɔ	1.26	0.0795	w	2.08	0.1162
ʊ	0.69	0.0495	p	2.04	0.1146
aʊ	0.59	0.0437	f	1.84	0.1061
ɑ	0.49	0.0376	h	1.81	0.1048
o	0.33	0.0272	b	1.81	0.1048
ju	0.31	0.0258	ŋ	0.96	0.0644
ɔɪ	0.09	0.0091	ʃ	0.82	0.0568
			g	0.74	0.0524
			j	0.60	0.0443
			tʃ	0.52	0.0395
			dʒ	0.44	0.0344
			θ	0.37	0.0299
			ʒ	0.05	0.0055
Totals	38%			62%	

$H(X) = -\Sigma_i P(x_i) \log_2 P(x_i) = 4.9$ bits. If all phonemes were equiprobable, then $H(X) = \log_2 42 = 5.4$ bits.

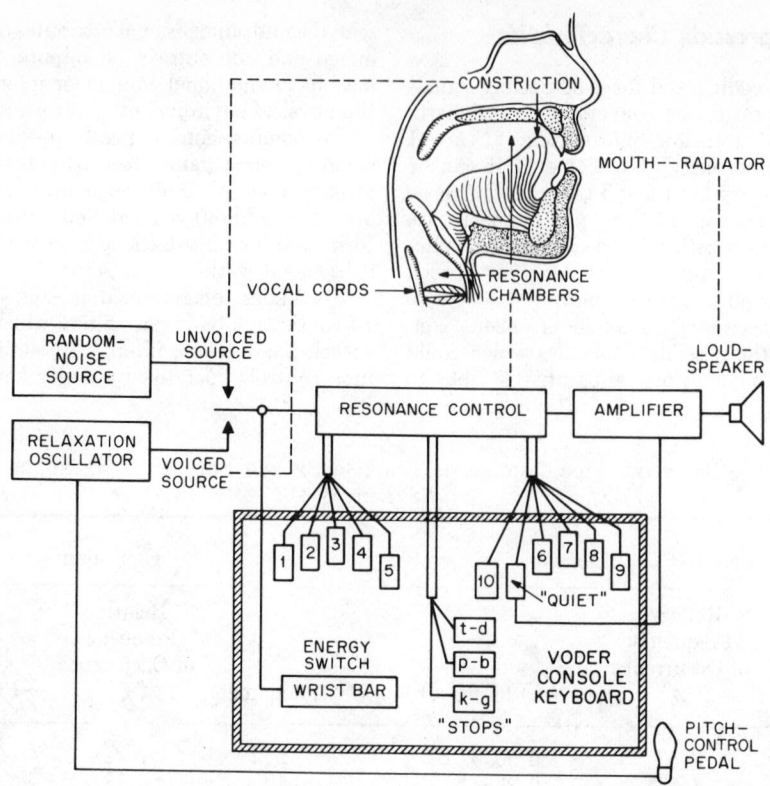

Fig. 38—Schematic diagram of the Voder synthesizer. *From H. W. Dudley, R. R. Riesz, and S. A. Watkins, "A Synthetic Speaker," Journal of the Franklin Institute, vol. 227, no. 6, fig. 6, p. 748; June 1939.*

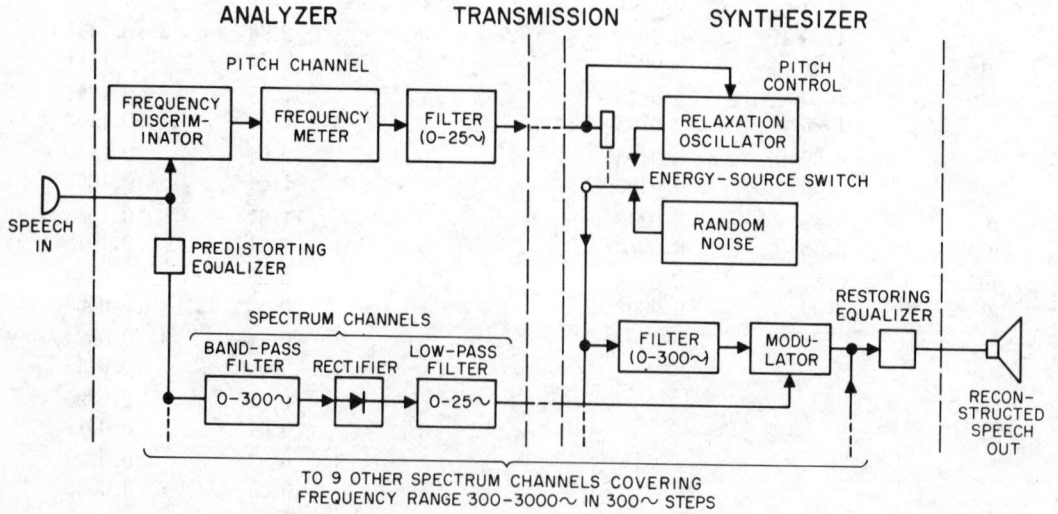

Fig. 39—Block diagram of the original spectrum channel vocoder. *From H. W. Dudley, "The Vocoder," Bell Laboratories Record, vol. 17, fig. 1, p. 123; 1939.*

ditional finger keys provided transient excitation of the selected filters for stop-consonant sounds.

The filter was supplied with either a noise or a buzz oscillator. A pedal controlled the pitch of the buzz oscillator.

The information generated by such a device is

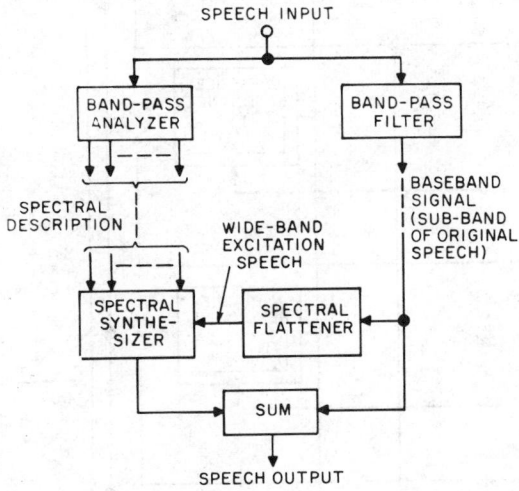

Fig. 40—Diagram of speech-excited vocoder.

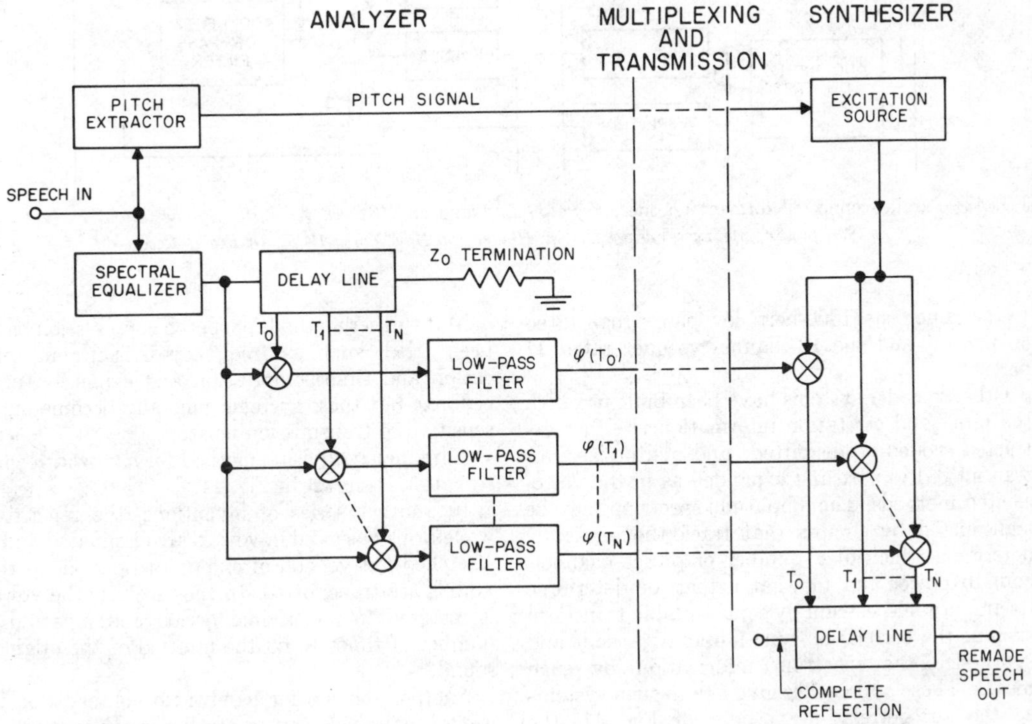

of the order of 10–20 bits per second, requiring a transmission bandwidth of a few hertz. Enormous economies in bandwidth might be realizable if (A) the deterioration in quality were acceptable, (B) pre- and post-processing equipment expenses justified the reduction in transmission costs, and (C) further demand for a service were restricted by the spectrum available.

The band-compression speech system based on analysis–synthesis experiments of Dudley was called vocoder (voice coder) and is now known as the spectrum channel vocoder. This is shown in Fig. 39.

There have been many variations of the spectrum channel vocoder mainly directed toward reducing the redundancy of the various channels. Considerable progress has been made by so-called voice-excited techniques. In these techniques, the excitation information is transmitted by selecting only a portion of the band and expanding this portion by intermodulation produced by a nonlinear device. This is then used as a source of excitation for the usual vocoder channels. A representative system is shown in Fig. 40.

Intelligibility and speech-quality tests indicate that for overall bandwidth compressions of about 3 to 1, the voice-excited vocoder was rated in 72 percent of the tests as good as normal telephone.

Fig. 41—Autocorrelation vocoder. *From M. R. Schroeder, "Correlation Techniques for Speech Bandwidth Compression," Journal of the Audio Engineering Society, vol. 10, no. 2, fig. 2, p. 164; April 1962.*

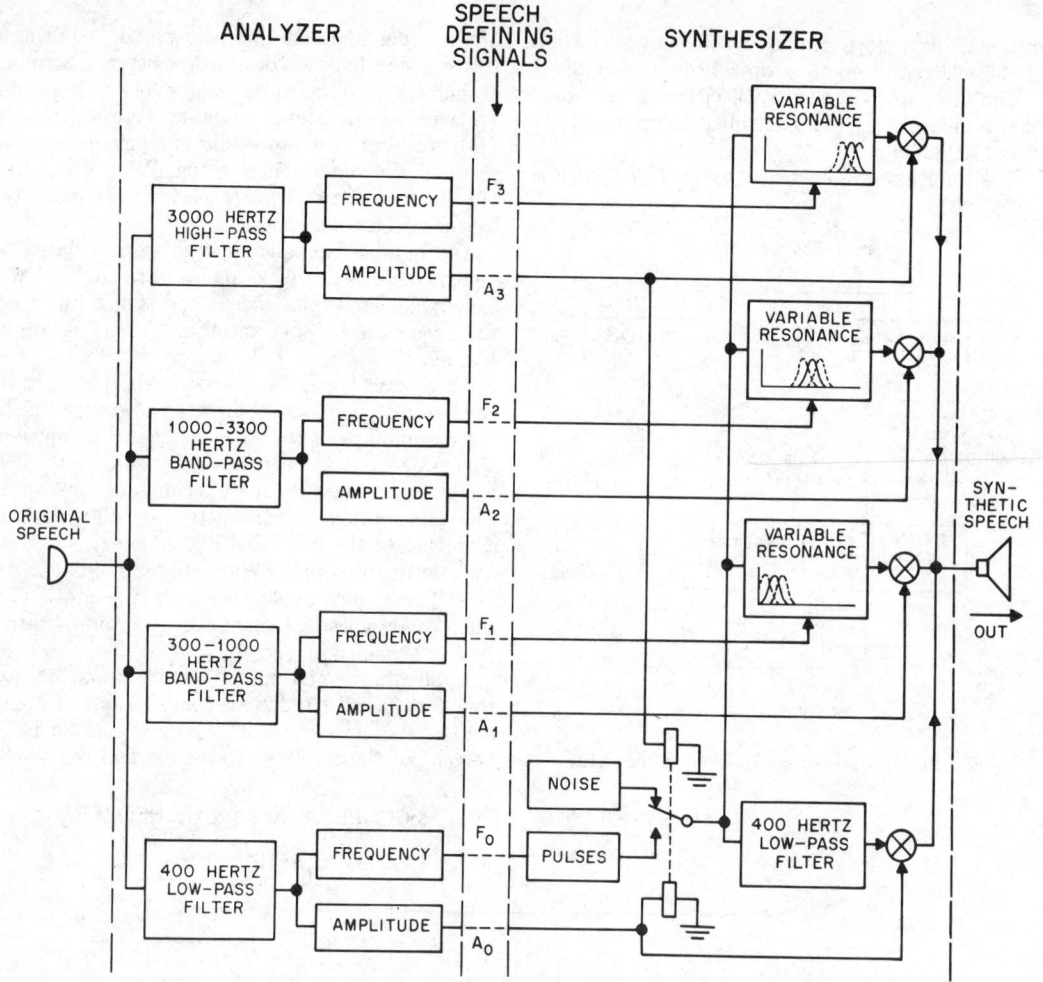

Fig. 42—Parallel-connected formant vocoder. *From J. L. Flanagan, "Speech Analysis, Synthesis and Perception," fig. 8.9, p. 262, Springer-Verlag, Berlin, Heidelberg, New York; 1965. All Rights Reserved.*

In the same tests, 1800-hertz low-pass signals rated 36 percent and the 18-channel vocoder rated 17 percent.

Other vocoder systems have been built in which the pitch and excitation information is either extracted, coded, transmitted, and synthesized, or transmitted in part and expanded as in the voice-excited methods. The amplitude spectrum may be transmitted by circuits that track the formants, determine which of a number of preset channels contain power and to what extent, or determine its amplitude spectrum by some suitable transform such as the correlation function and transmit and synthesize the spectrum information by such means. These approaches give rise to such systems as the autocorrelation vocoder of Fig. 41, the formant vocoder of Fig. 42, and the voice-excited formant vocoder of Fig. 43.

Many other methods of speech compression have been tried, such as frequency-division multiplication and time compression and expansion procedures, but these systems generally become more sensitive to transmission noise.

A frequency-division method for bandwidth conservation is shown in Fig. 44.

In another series of inventions, the repetitive waveforms observed in vowels are eliminated and a stretched-out version of one or two periods of the vowels are transmitted. In the receiver, the vowel is restored to its original period and repeated a number of times to fill the interval of the original sound.

Another scheme for bandwidth conservation in use for many years on transatlantic cables is called TASI (Time Assignment Speech Interpolation). In this system, use is made of the fact that on a

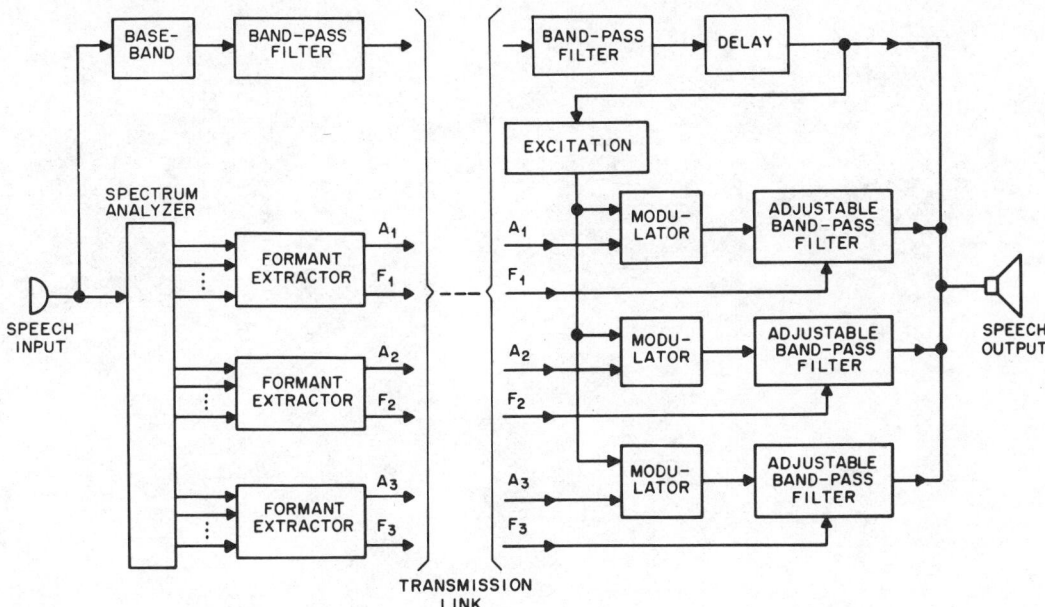

Fig. 43—Voice-excited formant vocoder. *From J. L. Flanagan, "Resonance Vocoder and Baseband Complement," IRE Transactions on Audio, vol. AU-8, no. 3, fig. 10, p. 100; 1960.*

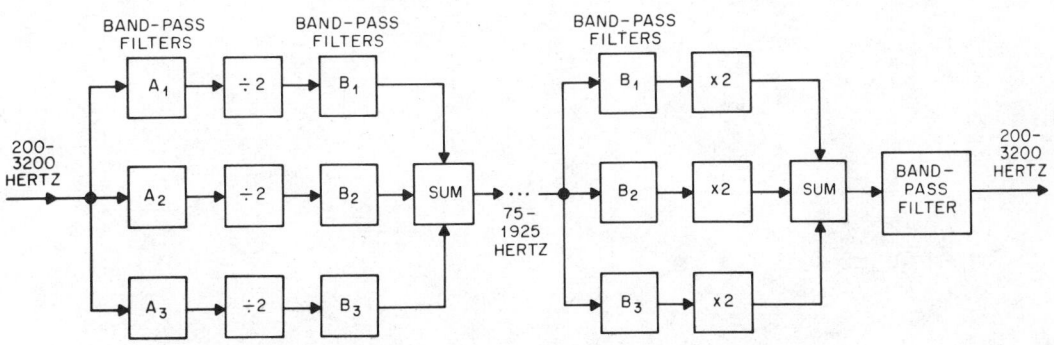

Fig. 44—Block diagram of the Vobanc frequency-division-multiplication system. *From J. L. Flanagan, "Speech Analysis, Synthesis and Perception," fig. 8.14, p. 274, Springer-Verlag, Berlin, Heidelberg, New York; 1965. All Rights Reserved.*

4-wire circuit one party is normally quiet and listening about half the time, and there are also many hesitations, pauses, and other silent periods even in the transmitting channel. During these silent intervals, the talker loses his channel. He is reassigned a channel, usually a different one, when he decides to talk again. The identification of the receiver to whom the talker was connected and subsequent re-addressing is done by the logic of the equipment through either a short identification signal before each talk period or through an auxiliary channel.

TASI increases the utility of 36-channel cable to 2 or 3 times the normal operation.

GENERAL

An atom consists of a positively-charged dense core or nucleus surrounded by a cloud of negatively charged electrons. The nucleus comprises 99.95% of the mass of the atom within a radius of 10^{-13} centimeter as compared with 10^{-8} centimeter for the atomic radius. Nuclear structure is determined by a balance between repulsive electrical forces among the positively charged protons and the attractive nuclear force. The nuclear force has a finite range of influence so that a nucleus, once broken up, is in unstable equilibrium. Nuclear fission energy is released in an event such as this.

The investigation of nuclei necessarily involves very short distances and very high energies. It includes observation of the behavior of particles scattered by nuclei or of breakup constituents. Bombarding particles having these characteristics are produced by accelerators. The observations of the products of subsequent nuclear reactions lead to our state of knowledge of the nucleus and of nuclear forces.

FUNDAMENTAL PARTICLES

The groupings are of particles with similar specific interactions and decay modes. Table 1 lists fundamental particles with mass <2 nucleon mass. A description of each follows.

Photons

Phenomena involving energy transfer by electromagnetic disturbances are conveniently explained by considering the energy to be transferred in discrete bundles called photons γ. The energy carried by a photon is proportional to the frequency of the associated wave. For E in ergs and γ in second^{-1}, the relation is $E = h\gamma$. The constant h (Planck's constant) $= 6.62 \times 10^{-27}$ erg-second. Photons emitted in radioactive decay are called gamma rays. Photons emitted in the acceleration (deceleration) of charged particles are called x-rays.

Leptons

Electrons are particles occurring in nature as negatively charged extranuclear atomic constituents. Certain radioactive materials may emit fast electrons called positrons $(+)$ or beta rays $(-)$ (also symbolized $\beta\pm$). $\beta+$ ultimately annihilate atomic electrons and emit electromagnetic radiation.

Neutrinos are particles of zero charge and zero mass. The neutrino ν is emitted accompanying the radioactive emission of $\beta\pm$ (beta decay). This emission of leptons in pairs is characteristic of all lepton-producing reactions.

Muons: The radioactive decay of mesons and hyperons produced in very high energy reactions often results in the emission of a lepton pair one of whose constituents is a muon. The muon μ subsequently decays into another lepton as it emits a lepton pair. Neutrinos created in muon reactions are not identical to those that accompany beta decay. Characteristically, the muon interacts rather weakly with matter. In slowing down after emission, a μ^- is often captured into "atomic" states of nuclei just as is an electron.

Mesons and Hyperons

Pions: The particle whose role relative to the nuclear force field may be analogous to that of the photon to the electromagnetic field is the pion. It is produced in very high energy reactions and interacts strongly with matter. Charged mesons decay into μ-ν or e-ν pairs while neutral pions decay into photon pairs.

Kaons and *hyperons* are unstable particles produced in nuclear reactions with bombarding energy > several gigaelectron-volts. Their lifetimes are short and they have various decay modes. These particles are sometimes called *strange* particles because the relationship between their production probability and lifetime is very unlike that of other elementary particles.

38-1

TABLE 1—ELEMENTARY PARTICLES (1965). *From A. H. Rosenfeld, A. Barbaro-Galtieri, W. H. Barkas, P. L. Bastien, J. Kirz, and M. Roos, Review of Modern Physics, volume 37, page 633; 1965.*

General Classification	Particle	Symbol	Charge	Mass in m_e	Rest Energy in MeV	Mean Life in Seconds
Photon	Photon	γ	0	0	0	Stable
Lepton	Neutrino	ν	0	0	0	Stable
	Electron	e	$\pm$	1	0.51	Stable
	Muon	μ	$\pm$	207	105.7	2.2×10^{-6}
Meson	Pion (charged)	II	$\pm$	273	139.6	2.6×10^{-8}
	Pion (neutral)	II°	0	264	135	1.8×10^{-16}
	Kaon (charged)	K	$\pm$	967	494	1.2×10^{-8}
	Kaon (neutral)	K°	0	973	497	$(K_1°) 0.8 \times 10^{-10}$
						$(K_2°) 6 \times 10^{-8}$
Baryon:						
Nucleon	Proton	p	$\pm$	1836	938.3	Stable
	Neutron	n	0	1838	939.6	1.0×10^{3}
Hyperon	Lambda	$\Lambda°$	0	2182	1115	2.6×10^{-10}
	Sigma (charged)	Σ	$+$	2327	1189	0.8×10^{-10}
			$-$	2342	1197	1.6×10^{-10}
	Sigma (neutral)	$\Sigma°$	0	2333	1192	$<1 \times 10^{-14}$
	Xi (charged)	Ξ	$-$	2585	1321	1.8×10^{-10}
	Xi (neutral)	$\Xi°$	0	2571	1314	3×10^{-10}

Nucleons

Proton is the positively charged constituent of all stable nuclei. In particular, it is the nucleus of the hydrogen atom. At very high energies antiprotons (p with *negative* charge) are produced. Subsequently the antiproton annihilates with an ordinary proton with the emission of electromagnetic radiation.

Neutron is the neutral constituent of all stable nuclei. Free neutrons interact strongly with nuclei and are usually captured with the emission of electromagnetic energy. A nucleus of charge Z and mass A is made up of Z protons and $A-Z=N$ neutrons.

Light Nuclei

Certain light nuclei have been widely used as bombarding particles. These include nuclei of deuterium (deutron), tritium (triton), and helium (helium-3 and alpha particle). Alpha particles α are emitted from certain heavy radioisotopes in a process called α decay.

TERMINOLOGY

Atomic Nucleus: Consists of protons and neutrons, Z and N in number. The number of protons Z is referred to as the atomic number.

Nuclear Charge: Carried by the protons, each of which has charge $e = 1.6 \times 10^{-19}$ coulomb.

Mass Number: An integer A equal to the total number of neutrons and protons in the nucleus. $A = N + Z$. The symbolic representation of a nucleus is ^{A}X, where X is the appropriate chemical symbol: Carbon, with 6 protons and 6 neutrons, is written ^{12}C.

Atomic Mass Unit (amu): A unit of mass equal to 1.660×10^{-24} gram and equivalent to the mass of each of the particles of a fictitious substance whose molecular weight is 1 gram. One atomic mass unit is approximately the mass of the neutron or proton.

Isotopes: Nuclei with common Z. Isotopes are chemically indistinguishable: The three naturally occurring isotopes of oxygen are ^{16}O, ^{17}O, and ^{18}O. Nuclei with common A are called isobars; with common N, isotones.

Binding Energy: The energy required to separate all of the component neutrons and protons of the nucleus is called the total nuclear binding energy B. Binding energy and mass defect are equivalent according to the relativistic mass–energy relation. The fraction B/A is approximately 8×10^6 electron-volts for all but extremely light nuclei and represents on the average the energy required to remove a single neutron or proton from a nucleus.

Electron-Volt (eV): A unit convenient for representing the energy of charged particles accelerated by electric fields. The electron-volt is equal to 1.6×10^{-19} joule and is the kinetic energy acquired by a particle bearing one unit of electric charge (1.6×10^{-19} coulomb) that has been accelerated through a potential difference of 1 volt. According to the relativistic mass–energy equation, 1 amu equals 931 megaelectron-volts (MeV), where 1 MeV equals 10^6 eV.

Fission; Fusion: The breakup of nuclei into nuclear fragments that are themselves nuclei is fission. The coalescing of two nuclei to form a heavier one is fusion. The mass defect for medium-weight nuclei is greater than that for light or heavy nuclei. Light and heavy nuclei in general both have an average total weight greater than that of medium-weight nuclei into which they might fission or fuse. Thus, when uranium breaks into its fission fragments, or two deuterium nuclei fuse to form helium, there is a net loss in mass. The mass lost appears as an equivalent amount of kinetic energy of the nuclei or their decay products. In the fission of ^{235}U, for example, each fissioning nucleus releases approximately 200 MeV$\approx 10^{-4}$ erg of energy or 3×10^{10} fissions per second≈ 1 watt.

Nuclear radius of a nucleus of mass number A is given approximately by R equals $r_0 A^{1/3}$. Experimental values quoted for r_0 range from 1.1 to 1.5×10^{-13} centimeter. The unit of length, 10^{-13} centimeter, is called the fermi.

Nuclear Reaction: A process in which a nucleus struck by a fast-moving particle combines with it to form an energy aggregate. This briefly formed compound nucleus breaks up almost immediately either into the original nucleus and particle or into a different nucleus and one or more secondary particles, effecting a nuclear transmutation in the second case. A typical reaction represented in detail is

$$^1H + {}^7Li \rightarrow {}^8Be \rightarrow {}^7Be + {}^1n$$

or in abbreviated form, $^7Li(p, n)^7Be$. The bombarding and emitted particles in this reaction are a proton and a neutron, respectively.

Cross section of a nuclear reaction is a measure of the probability of its occurrence. Quantitatively, the total cross section σ is the inverse of the number of particles that must strike 1 centimeter^2 of target material to induce a nuclear reaction in 1 nucleus of the target. If the number of target nuclei/centimeter$^2 = N$, and there are F bombarding particles incident on each centimeter^2 of the target/unit time, the number of nuclear events n (per centimeter^2/unit time) is given by $n = NF\sigma$. The barn$= 10^{-24}$ centimeter^2 is commonly used to express cross-section values.

Stable Nucleus: One that retains its identity indefinitely unless disturbed by external forces.

Radioactive Nucleus or Unstable Nucleus: One which ultimately transforms spontaneously into a nucleus of a different kind. The transformation occurs through the emission of beta particles, alpha particles, or gamma rays (radioactive decay); through the breakup of the nucleus into one or more nuclear fragments (spontaneous nuclear fission); or through the absorption or capture of an extranuclear electron from the atomic shell (electron capture).

Radioactive Decay Constant λ: The fraction of nuclei of a radioactive material disintegrating in unit time. The radioactive nuclei remaining after time t in a material consisting originally of N_0 nuclei is given by $N = N_0 \exp(-\lambda t)$.

Half-life τ of a radioactive material is the interval in which its original activity is reduced by half and, given in terms of the decay constant, is $\tau = 0.693/\lambda$.

Relativistic Concepts: Two concepts fundamental to the explanation of nuclear and atomic phenomena stem from the special theory of relativity. These are:

Relativistic Mass: The behavior of bodies moving at an appreciable fraction of the velocity of light can be explained in the same manner as that for low velocities if they are assumed to have a mass that increases with velocity. The relativistic velocity-dependent mass

$$m = m_0 / (1 - v^2/c^2)^{1/2}$$

where $m_0 =$ mass of body at rest, $v =$ velocity of body, and $c =$ velocity of light (all in consistent units), must be used in all accurate calculations of the behavior of energy nuclear and atomic phenomena. The relativistic-mass increase is important in the design of high-energy particle accelerators.

Mass-Energy Equivalence: The kinetic energy of a moving body is given accurately by $(m - m_0)c^2$. (The familiar expression $m_0 v^2/2$ is an approximation applicable only at low velocities.) By inference, a body at rest has associated with it the so-called rest energy $E = m_0 c^2$. A striking example is the tremendous quantity of energy released during nuclear fission.

HIGH-ENERGY-PARTICLE ACCELERATORS

Accelerators are used to provide beams of various particles for the study of nuclear structure and nuclear reactions. For this purpose accelerators with energies from 100 KeV to ~ 200 GeV have been built or are proposed. These accelerate stable charged particles from electrons to ^{238}U. In addition, the production of X rays by the bombardment of heavy targets by electrons has led to the use of electron accelerators for biological and industrial radiographic purposes.

When sufficiently high bombarding energies are attained, it is possible for part of the energy to be converted into matter with the creation of particles. Energies of several hundred MeV are sufficient to create pions. Antiprotons and neutrons have been

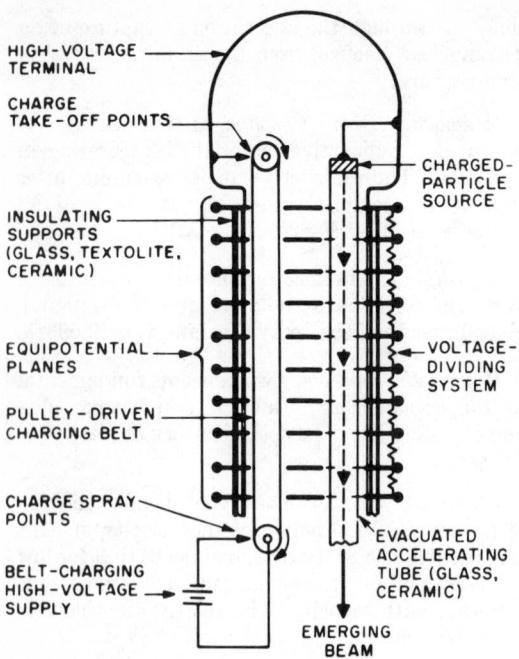

Fig. 1—Van de Graaff generator.

created with bombarding energies ~5–6 gigaelectron-volts (GeV) and particles as heavy as antideuterons have been produced. The production of antiparticles, hyperons, and K particles for the purpose of understanding their behavior has been a major impetus to the construction of accelerators with energies as high as 200 GeV.

Direct-Voltage Accelerators

Historically, the first accelerators consisted of an ion source and a target held at a high potential difference. This arrangement has the advantage that a direct current of accelerated ions may be attained. It has the disadvantage that the maximum energy is severely restricted by the potential difference and gradient that can be maintained between the ion source and target.

Van de Graaff Accelerator

Electric charge is sprayed on a traveling insulated belt in Fig. 1 and is transported against the voltage gradient to a rounded terminal supported by an insulating column. The energy is generated from the mechanical power source driving the belt. Ions are injected from an ion source in the terminal into an evacuated tube. An appropriate gradient is maintained down the tube by voltage dividers to accelerate and focus the ion beam. The beam is deflected by a magnet through an orbit fixed by slits. Feedback signals can be derived from these

slits to control a corona discharge to the terminal. In this way voltage stability of 0.01% has been attained. By pressurizing the atmosphere around the generator, a compact arrangement is obtained capable of quite high voltages. Terminal potentials greater than 13 MV have been reached. Ion currents of several hundred μA and electron currents of 1 mA can be produced at voltages ~5 MV.

A multistage version of the Van de Graaff accelerator called the tandem Van de Graaff has been commercially produced. The terminal is located at the center of the accelerator and is maintained at a potential V. Negative ions are accelerated from ground to $+V$, then pass through a stripping foil which removes electrons to form ions of positive charge. These are accelerated further from $+V$ to ground. Protons have been accelerated to ~20 MeV and heavy ions to >100 MeV in these devices.

Rectified-Voltage Accelerators

Higher currents may be attained using cascaded rectifier circuits than with electrostatic generators. The Cockcroft-Walton generator is the oldest and most commonly used device of this type. Voltages up to 2 MV and accelerated currents of 1 mA have been attained. The beam-handling and insulation problems are similar to those of the Van de Graaff accelerator. The voltage limitation is set by the available rectifiers and capacitors. Circuits with a large number of stages have poor voltage regulation. The insulated-core-transformer generator uses a transformer with many secondaries. The core sections are insulated from the secondaries, each other, and the primary.

The Dynamitron generator charges multiple secondaries in parallel by capacitive coupling. In each of these the secondaries are connected in series to generate the final voltage. The current capabilities of Dynamitron generators are >10 mA at voltages ~4 MV.

Cyclic Accelerators

The maximum energy for a direct-voltage accelerator is ultimately limited by the attainable voltage. Cyclic accelerators bypass this limitation by applying an alternating field to the particle over an extended spatial region in such a way that

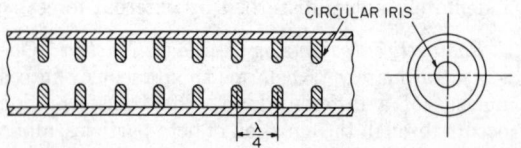

Fig. 2—Traveling-wave-type iris-loaded linear electron accelerator.

the field at the position of the particle is always either accelerating or zero.

This basic condition may be satisfied by a hollow waveguide structure for which the phase velocity of the radio-frequency traveling wave matches the particle velocity. Such an arrangement is most satisfactory for electrons whose velocity is nearly constant (the speed for a 2.0-MeV electron is $>0.98c$, where c is the speed of light). A segmented structure consisting of hollow metal cylinders separated by gaps also satisfies the conditions. The lengths of cylinders and gaps and the frequency of the applied voltage are chosen so that the field in the gap is accelerating in synchronism with the passage of the accelerated particle through the gap.

The orbit for the particle may be a closed circle (synchrotron), an open spiral (cyclotron), or a straight line (linear accelerator). The beam is pulsed with a microstructure at the frequency of the applied voltage. This may be superposed on a gross macrostructure because of some feature of the accelerator operation.

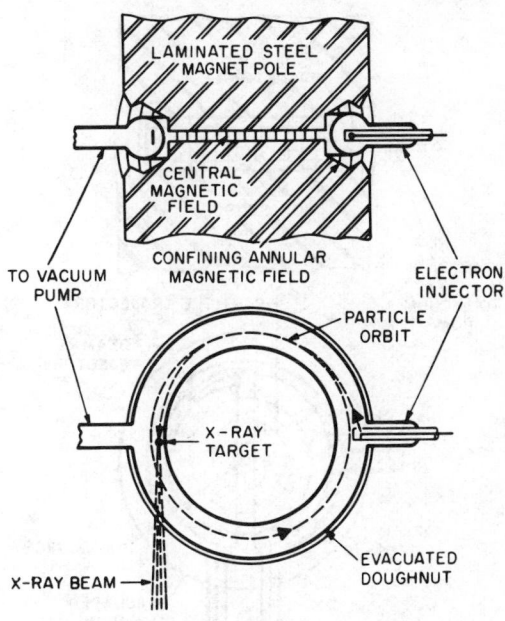

Fig. 3—Betatron.

Linear Accelerators

The *traveling-wave linear accelerator* (Fig. 2) moves electrons along a straight path by means of a radio-frequency wave. The accelerating tube is a long waveguide longitudinally loaded with field-perturbing obstacles (usually irises placed at $\lambda/4$ intervals). High-power radio-frequency energy passes into the waveguide and builds up an oscillating radio-frequency excitation. The dimensions and loading are chosen such that a traveling wave of high amplitude is established with its E vector parallel to the axis and traveling with a phase velocity equal to the local velocity of the electrons being accelerated. Since the accelerating-tube dimensions are proportional to the wavelength of the oscillator, operating frequencies in the very-high-frequency and microwave regions have been used. For example, accelerators designed to produce long pulses (<50 microseconds) operate with L-band excitation ($\sim$1250 megahertz), while those designed for extreme compactness have recently been built at X-band frequencies (10 000 megahertz). Most accelerators built until very recently have used S band (2850 megahertz) to produce $\sim$1-microsecond pulses. To a large extent the exact choice has depended on the parallel development of high-power klystrons and magnetrons.

Standing-Wave Accelerators

The standing-wave accelerator uses a single right-circular-cylindrical cavity with a series of axial drift tubes. The length L of a drift tube corresponding to energy E for a particle of mass m and a driving frequency f is $L = (1/f)(2E/m)^{1/2}$. The

cavity is excited in a mode such that an axial electric field is established. Considerations of phase stability restrict the operation to that portion of the cycle for which the accelerating field increases with time. Considerations of radial focusing, on the other hand, restrict operation to that portion of the cycle in which the acceleration field decreases with time. Accordingly additional focusing structures have been included in practical accelerators. Accelerators have been built for protons up to 68 MeV and for heavy ions up to 10 MeV/nucleon. Proposals have been made for proton linear accelerators with energies up to 1 GeV for injection into high-energy synchrotrons and for the production of mesons.

Betatron

The only induction accelerator is the betatron (Fig. 3). An electron beam is deflected into a circular orbit by a magnetic field B. The flux linking the orbit is caused to increase with time. The induced circumferential electric field accelerates the electrons to a final energy limited by maximum field, orbit radius, and radiation losses by the accelerated electrons (synchrotron radiation). The condition for which a particle moves in a circle of radius r results from $p = eBr$, where p and e are the momentum and charge of the electron, respectively. A rate of change of flux $d\Phi/dt$ gives rise to a field $E = -(1/2\pi r)(d\Phi/dt)$. By applying the laws of motion we find that the condition that particles move at a fixed radius r during acceleration is that $B_{orbit} = 0.5(\Delta\Phi/\pi r^2)$. To attain maximum energy, the flux linking the orbit is caused

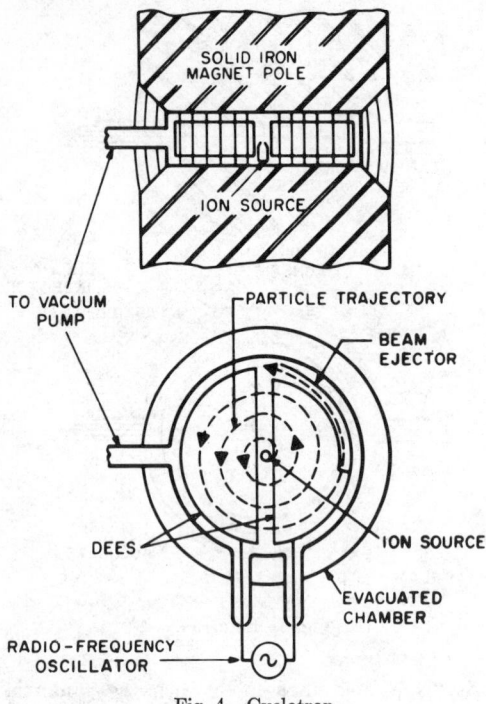

Fig. 4—Cyclotron.

to change from $-\Phi_{max}$ to $+\Phi_{max}$ while B_{orbit} goes from B_{min} to B_{max}. The maximum energy attained is 300 MeV although the usually encountered betatrons designed for medical and radiographic applications have energies ~15–25 MeV.

Cyclotron

The cyclotron (Fig. 4) in many ways is similar to the standing-wave linear accelerator. The machine uses hollow metal electrodes called "dees" in place of the cylindrical drift tubes. A unipolar magnetic field causes the particle to execute circular orbits while within a dee. Radio-frequency voltage is applied to the dees and the radius of the orbit increases as the particle is accelerated during the passage through the gap.

The time necessary to complete a semicircular portion of the orbit is Be/mc. Thus, to the extent that the mass m remains constant, the synchronous time is constant.

The acceleration process is repeated until the ion reaches the outer radius of the dee or its relativistic mass change causes it to fall out of synchronism with the accelerating voltage. This takes place at ~15 MeV/nucleon.

The limitation on maximum energy due to relativistic change in mass may be overcome in two ways. Either the frequency may be modulated in such a way as to track the circulation frequency of the accelerated particles (frequency-modulation cyclotron) or the circulation time can be made isochronous (isochronous cyclotron) by causing B to increase with radius in the proper manner. Modern cyclotrons have been built on both principles.

It can be shown, however, that the condition for a vertical focusing beam is that the field shall fall off with radius. The vertical focusing is introduced in isochronous cyclotrons by introducing azimuthal variations in the magnetic field ("Thomas shims" or a "spiral ridge").

Synchrotron

The ultimate limits on energy attainable with a cyclotron are associated with the economics of the very large magnet and the difficulty in achieving the required radio-frequency structure. Both of these problems are bypassed if the orbit is kept at constant radius. Then the magnetic field is required only along a fixed relatively small volume in space, and the construction of and access to the accelerating electrodes becomes relatively straightforward. As the particle energy increases, any change in particle rotation frequency must be matched by modulating the frequency of the accelerating voltage. Injection takes place at as high a velocity as possible to minimize the range of modulation.

Electron Synchrotron: If injection takes place at several million volts, the frequency modulation required is only a few percent. In practice, operation at constant frequency is attainable by allowing the radial extent of the magnetic field to be several percent so that the particle will change radius slightly as it is accelerated. Electron synchrotrons with energy as high as 10 GeV have been built. The ultimate limitation is loss of energy to X radiation as the electron is accelerated in a circular path.

Proton Synchrotron: These devices are usually constructed in "racetrack" form, with deflecting magnets at the curves and accelerating, focusing, and beam-ejecting structures in the straight sections. In machines involving energy up to several GeV, injection at several MeV is sufficient to allow a small range of frequency modulation. This injection takes place from a Van de Graaff accelerator. For larger machines, a linear accelerator is used to attain injection energies of tens of MeV. In these accelerators, the magnetic field increases linearly in time while the radio-frequency voltage is appropriately frequency modulated.

For the very largest machines, economic constraints require that the magnet cross section be kept as small as possible. For this reason little if any radial or vertical variations in orbit can be tolerated. The use of alternate-gradient focusing magnets permits focusing both radially and verti-

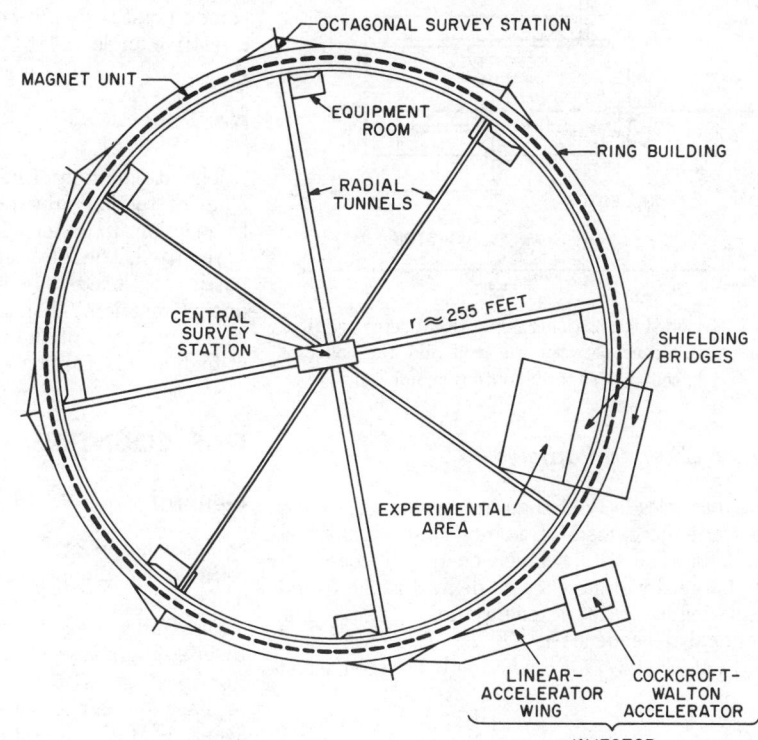

Fig. 5—CERN 32-GeV synchrotron.

cally; this has enabled the construction of accelerators of about 30 GeV maximum energy and 840 feet in diameter, but with a vacuum-chamber aperture only 3×6 inches. The CERN proton synchrotron appears in Fig. 5.

REACTOR

The basic components of a reactor are (A) fissionable fuel that releases energy and neutrons, and (B) a moderator that slows emitted neutrons to energies at which it is probable that they will cause further fissions. Monitoring of the power level is accomplished by instrumentation that measures temperature and radiation level. Appropriate power levels are maintained by control rods that are efficient neutron absorbers. Research activities with reactors include neutron spectroscopy as well as effects on solid-state properties. Thermal fluxes and fast fluxes $> 10^{14} n/\mathrm{cm}^2/\mathrm{second}$ have been attained. Figure 6 shows a modern research reactor.

NUCLEAR INSTRUMENTATION

Interaction of Radiations with Matter

Nuclear studies are carried out by observing the properties (number and kind, energy, time, angular distributions, and correlations) of radiations emitted in nuclear reactions or by radioactive nuclei. If the radiation is charged, it will lose energy as it traverses any medium. It may excite or ionize the atoms of the medium along its trajectory. Measurements of the excitation or ionization may then be used to infer the properties of the radiations. When uncharged radiations, such as gamma rays or neutrons, interact with matter, they produce secondary charged particles. The detection of these secondaries enables detection and spectroscopy of the primaries.

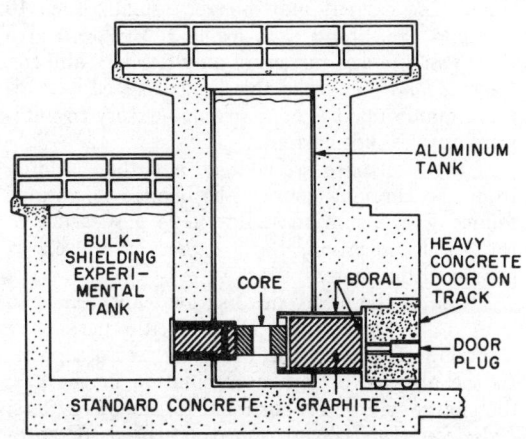

Fig. 6—Swimming-pool reactor.

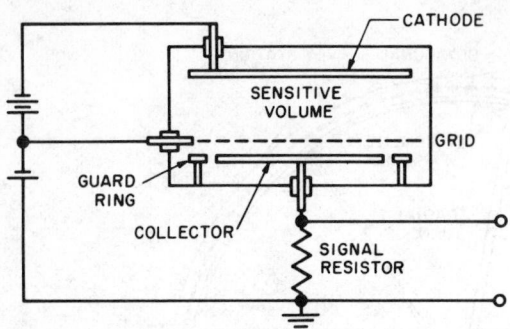

Fig. 7—Gridded ion chamber. Only that energy required to move electrons between the grid and the collector contributes to the output signal.

Heavy Charged Particles

For particles heavier than an electron, the most important mechanism of energy loss is ionization loss. A particle with a given energy E, mass M, and charge Z will have a well-defined range R and a well-defined rate of energy loss dE/dx. The functional dependence is $dE/dx \sim f(MZ^2/E)$ and $R \sim g[(M/Z^2)(E/M)^h]$, where h is between 1 and 2.

Electrons

Electrons lose energy not only by ionization but also by bremsstrahlung due to sudden deceleration of the electron in the field of the nucleus. For electrons moving in a medium of atomic number Z, loss by radiation dominates for $E > 800/Z$ MeV. As a result there is considerable straggling in the range of electrons of a given energy.

Gamma Rays

It is characteristic of uncharged radiations that they interact in single encounters rather than continuously. As a result, for a beam of gamma rays of energy E passing through matter, there is an exponential attenuation of intensity on traversing a thickness x. The three dominant interaction mechanisms are as follows.

(**A**) *Photoelectric effect* in which atoms ejected are ionized by the photon. The energy of the electron $\sim E_\gamma$.

(**B**) *Compton scattering* in which the photon is elastically scattered by a free electron. The energy transfer depends on the angle of scattering so that it is continuously distributed to a maximum energy $\sim E_\gamma - 0.25$ MeV.

(**C**) *Pair production* in which the photon creates an electron-positron pair in the field of the nucleus. The initial kinetic energy of the electron and positron is $E_\gamma - 1.02$ MeV. The positron ultimately

interacts with a free electron usually after it has come to rest and emits two 0.511-MeV photons at a relative angle of 180°.

Neutrons

The attenuation of neutrons is also exponential. The interaction may be by elastic scattering or by reaction. In either case, heavy charged particles are emitted. The total energy transferred to charged particles is fixed only for a charged-particle-producing reaction. Thermal neutrons are often captured by nuclei of the medium with the emission of high-energy gamma rays.

GAS COUNTERS

General

Gas counters are devices that collect the charge released by the passage of an ionizing particle. They may be used to detect the presence of radiation. Under special conditions they may be used to infer the energy lost by the ionizing particles. For most gases, the energy required to form a primary ion pair is ~ 30–35 eV. Thus, one particle losing 1 MeV will release a 3×10^4 primary ion pair or $\sim 5 \times 10^{-15}$ coulomb of charge of either sign. Electrodes placed in the counter maintain an electric field that tends to separate negative and positive ions. These are collected at the anode and cathode, respectively. The motion of the ions induces currents and charges on the electrodes.

Ion Chambers

A gas counter in which the primary ions create no further ionization is called an ion chamber (see Fig. 7). The current flowing between the electrodes is just the primary ionization per unit time. This current may be very small, since 10 particles per second each losing 1 MeV will give rise to an average current of only 5×10^{-14} ampere. *Current ion chambers* of this type are used as radiation monitors. It is of course mandatory to minimize any leakage currents.

If the electrodes are charged and then isolated from the charging source, the change in voltage following exposure to radiation is a measure of the total ionization. Such devices are used as personnel dosimeters.

If one of the electrodes has low capacitance to ground and to the other electrode, the passage of a single ionizing particle will give rise to a pulse. In the absence of ion recombination in the gas, the charge will be proportional to the energy loss. *Pulse ion chambers* are sometimes used for spectroscopy. The resolution (spread in collected charge

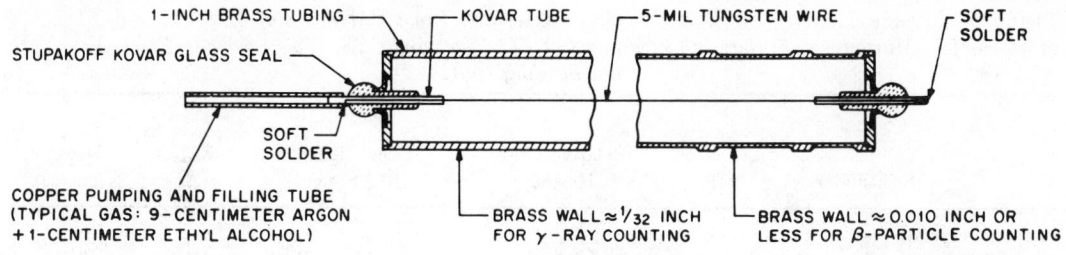

1-INCH BRASS TUBING — — KOVAR TUBE — 5-MIL TUNGSTEN WIRE — SOFT SOLDER

STUPAKOFF KOVAR GLASS SEAL

SOFT SOLDER

COPPER PUMPING AND FILLING TUBE
(TYPICAL GAS: 9-CENTIMETER ARGON
+ 1-CENTIMETER ETHYL ALCOHOL)

BRASS WALL ≈ 1/32 INCH
FOR γ-RAY COUNTING

BRASS WALL ≈ 0.010 INCH OR
LESS FOR β-PARTICLE COUNTING

Fig. 8—Typical Geiger-Müller counter.

for a fixed incident particle energy) may be limited by fluctuations in the number of ion pairs as well as by noise. Resolutions as good as 1% have been reported.

Proportional Counters

As the field strength increases in a gas counter, the electrons freed by ionization begin to gain enough energy between collisions to cause ionizations of their own. If the probability that an electron will cause an ionization is p (<1), the number of ion pairs created per primary ion pair will be $1+p+p^2+p^3+\cdots=1/(1-p)$. Thus there is a multiplication of charge by the counter. If the density of ionization remains so slow that recombination is not probable, the final charge will be proportional to the number of primary ion pairs. A counter designed to use gas multiplication as a linear amplifying device is called a proportional counter. The considerations previously applied for pulse height resolution continue to be appropriate. In addition, there will be an additional spread due to any nonuniform multiplications. Ordinarily, resolutions attained with a proportional counter are not as good as those obtained with a pulse ion chamber ($>4\%$). Rise times are 10^{-6} second and counting rates $\sim10^5$/second are possible.

Geiger Counters

When $p\to1$, gas multiplication $=1/(1-p)\to\infty$. An avalanche takes place that is eventually limited by a change in the operating conditions. A geiger

counter is such a counter (Fig. 8). The slowly moving positive-ion space charge accumulates around the anode (region of highest field) and then modifies the field so that $p<1$. The discharge must then be quenched to avoid relaxation oscillations. Quenching is ordinarily done by including a polyatomic gas that absorbs energy by dissociating rather than by ionizing. A charge of $\sim10^9$ ions is released per count. The size is entirely independent of the nature of the initial ionizing event. *Dead time* of ~200 microseconds typically exists in the counter following a discharge. Useful counting rates typically are $<2\times10^3$/second.

Scintillation Counters

In Fig. 9, scintillation counters use photomultipliers actuated by fluorescence produced when charged particles strike certain materials. The method is particularly advantageous for materials that are transparent to their own fluorescent radiation, since in that case large detectors with correspondingly high γ efficiency may be used. Inorganic crystals of materials such as sodium iodide are dense and contain heavy elements. They respond to gamma rays with a pulse height spectrum that is characteristic of the energy of the particular gamma ray. A typical pulse height spectrum is shown in Fig. 10. The rate at which light is emitted at the beginning of the pulse depends on the scintillation decay time, the relative light yield for a given energy radiation, and the energy of the radiation. This rate is sufficiently high that timing of events occurring as close together as 10^{-11} second is possible with organic scintillators and as close together as 10^{-9} second with sodium

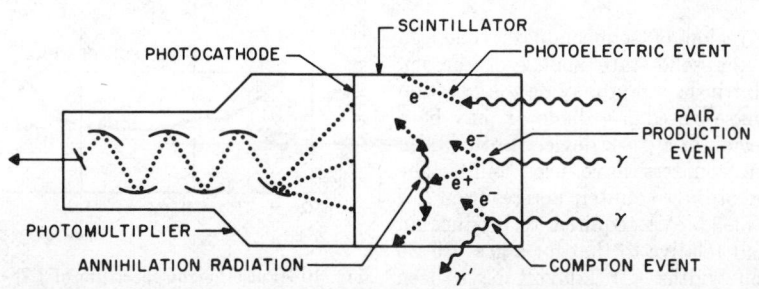

SCINTILLATOR

PHOTOCATHODE

PHOTOELECTRIC EVENT

PAIR PRODUCTION EVENT

PHOTOMULTIPLIER

ANNIHILATION RADIATION

COMPTON EVENT

Fig. 9—Scintillation counter showing typical events that contribute to the pulse height spectrum.

TABLE 2—CHARACTERISTICS OF SELECTED SCINTILLATORS. *From "Methods of Experimental Physics," ed.-in-chief L. Marton, vol. 5, chapter 14, editors L. C. L. Yuan and C. S. Wu, Academic Press, New York and London; 1961.*

Scintillator	Relative Pulse Height	Decay Time ($\times 10^{-9}$ Second)	λ_{max} (Å)	Density (gm/cc)
Inorganic crystal:				
NaI(Te)	2.1	250	4100	3.7
CsI(Te)	0.6	1500	4500	4.5
LiI(Eu)	0.7	2000	4400	4.1
Organic crystal:				
Anthracene	1.0	32*	4400	1.25
p, p'—Quaterphenyl	0.85	7*	4200	1.2
Trans-Stilbene	0.60	6.4*	4100	1.16
Liquid scintillators:				
Toluene POPOP†	0.67	<3*	4300	0.87
Plastic scintillators:				
Polystyrene TP‡	0.28	≤3	3550	0.9
PVT POPOP§	0.47	<3	4300	0.9

* Also exhibit ∼370-nanosecond component when irradiated with neutrons.

† 1, 4-bis-2-(5-phenyloxazolyl)-benzene; 0.15 gm/liter.

‡ *p*-Terphenyl; 36 gm/liter.

§ POPOP; 1 gm/liter.

iodide. Properties of some common scintillators are given in Table 2.

Cerenkov Counters

Cerenkov counters make use of the visible light emitted by relativistic charged particles moving in a medium of index of refraction n with a velocity $v > c/n$. A fast electron or proton entering a clear material will emit visible light in a narrow cone of angle $\theta \times \cos^{-1}(c/vn)$. The light pulse is detected by a photomultiplier. Arranging optical coupling so that only light moving in a narrow range of θ is collected makes possible a velocity-sensitive, fast ($< 10^{-9}$ second) counter. (See Fig. 11.)

Semiconductor Detectors

An intrinsic region of semiconductor between p-n regions is the solid-state analog of the ion chamber. The intrinsic region may be the junction region of a reverse-biased p-n diode or may be a compensated region in a p-i-n device. Both silicon and germanium counters have been built. The great advantage of these counters derives from low energy loss (∼3–3.5 eV) required to produce an electron-hole pair relative to that for a gas (30–35 eV). Resolution widths <1 kilovolt have been obtained for electrons on silicon and gamma rays on germanium. Widths of 10–15 keV have been obtained for heavy particles on silicon.

Thick detectors are typically made by lithium drift compensation to form the intrinsic layer. Thicknesses ∼1 centimeter have been obtained with volumes >50 cm³.

The use of germanium is advantageous for gamma-ray spectroscopy because of its relatively

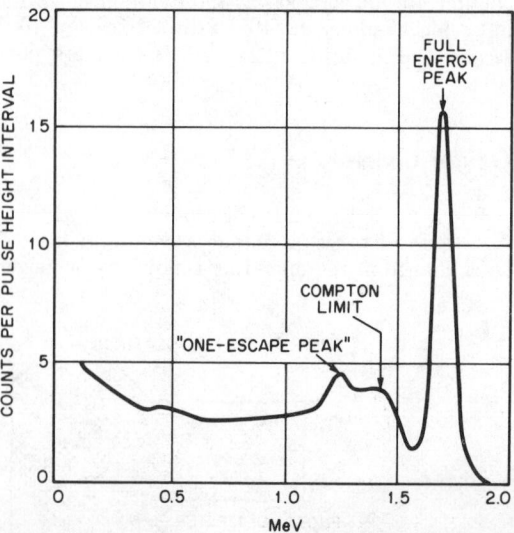

Fig. 10—Pulse height spectrum of 1.78-MeV gamma as observed with 3″×3″ No. 1 scintillation spectrometer.

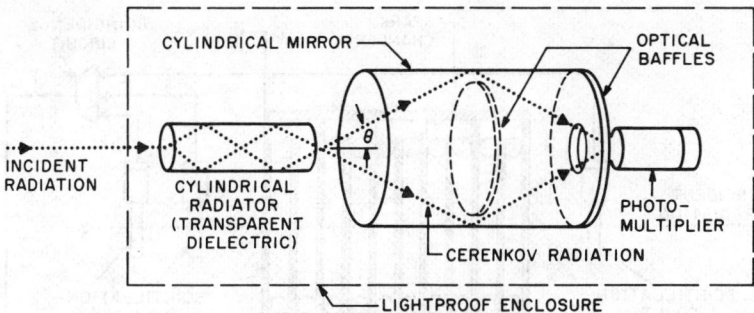

Fig. 11—Cerenkov counter. Position of baffle restricts detection to a narrow range of θ, hence to a narrow range of velocity of incident radiation.

high atomic number. However, its large energy gap makes it necessary to operate germanium counters at low temperatures (liquid nitrogen boiling point of 78° Kelvin is convenient). A diagram of a germanium spectrometer and a pulse height spectrum are shown in Figs. 12 and 13.

Bubble Chamber

A superheated liquid will boil at nucleation centers and ionization centers caused by the passage of an ionizing particle. The liquid in the bubble chamber (liquid hydrogen is often used) is superheated by a sudden reduction of pressure. Stereo photographs taken during the sensitive period may be studied to reconstruct the events that caused the tracks. High magnetic fields are used to confine the charged particles and to enable a determination

of their momenta. The use of such chambers is especially important with low-duty-cycle accelerators because the efficiency of detecting any event taking place in the bubble chamber is essentially 100 percent.

Spark Chamber

A discharge chamber shown in Fig. 14 consisting of a series of parallel plane avalanche counters will respond to the passage of an ionizing particle by discharging along the trajectory. A photograph will enable the trajectory to be reconstructed in a way similar to the bubble chamber. However, the spark chamber may be made sensitive in a time $\ll 1$ microsecond, remaining sensitive for a few microseconds. Coincidence telescopes may be used to sensitize the counter only for interesting events. A high degree of spatial and time sensitivity is possible. Recently chambers have been built using sonic detection of pulses using parallel wires to form the planes so that electric detection is now possible.

Neutron Detectors

Neutrons are detected by the charged secondaries they produce. The $^6\mathrm{Li}(n, \alpha)^3\mathrm{H}$ and $^{10}\mathrm{B}(n, \alpha)^7\mathrm{Li}$ reactions are often used for the detection of thermal neutrons by incorporating $^{10}\mathrm{B}$ or $^6\mathrm{Li}$ in a gas or solid counter. Fast neutrons may

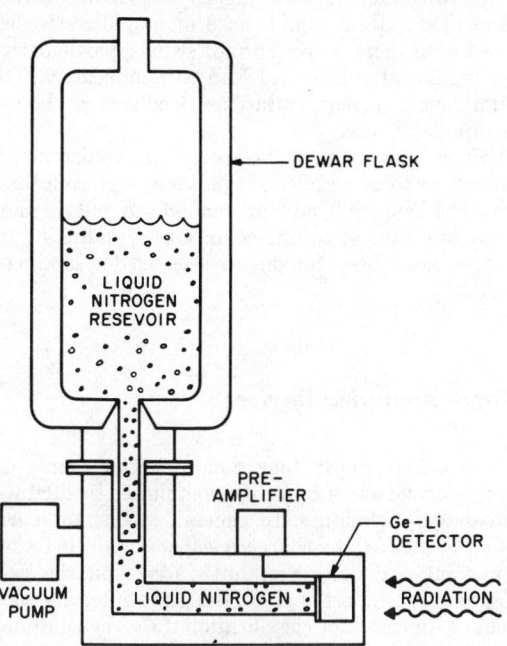

Fig. 12—Germanium-lithium gamma-ray spectrometer.

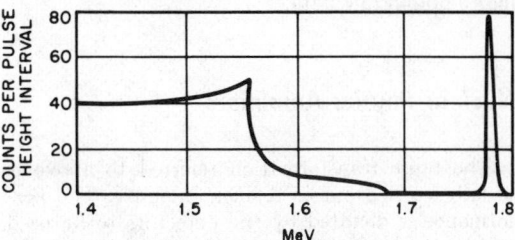

Fig. 13—Pulse height spectrum of 1.78-MeV gamma with Ge-Li spectrometer. Note suppressed zero on abscissa.

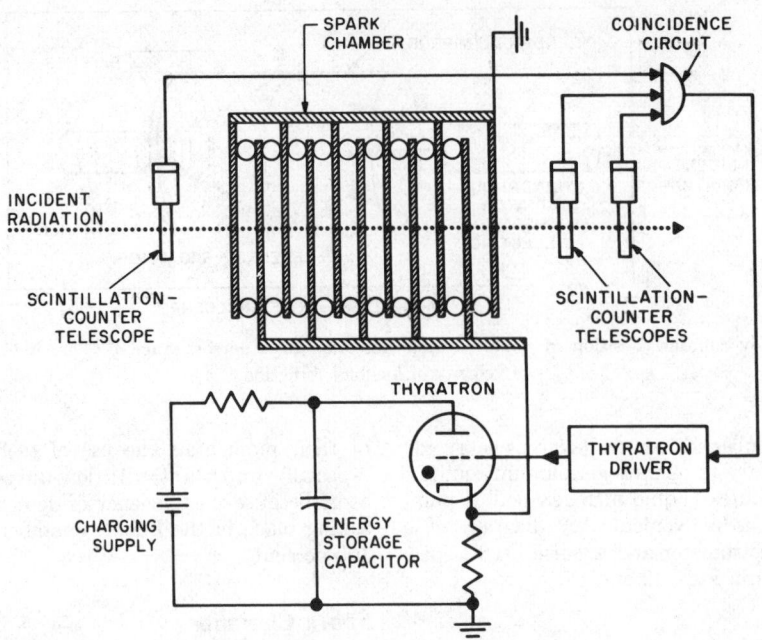

Fig. 14—Spark chamber with coincidence telescope.

be detected by providing a hydrogenous moderator surrounding a slow-neutron detector. Spectroscopy of fast neutrons may be carried out by observing the recoil protons following a $^1H(n, n)^1H$ reaction.

INSTRUMENTATION

An event in a nuclear-physics detector appears as a current pulse. The information content usually is the total charge in the pulse and its time of occurrence. The charge is at low level, typically ranging from 10^4–10^7 electrons. Before the pulse can be processed to extract the information, it must be transformed by amplification and pulse shaping. Subsequently time-measuring devices and analog-to-digital converters complete the processing to obtain raw data.

Nuclear-Physics Amplifiers

The pulse transformation referred to above is usually carried out in a linear manner. The performance is dictated by the detectors to be used and by the information desired.

Durations of current pulses range from about 1 nanosecond to a few microseconds for most de-

tectors. Timing resolution of the order of 1% of the current duration is sometimes desired. Accordingly, the amplifiers must be fast. Semiconductor detectors may attain line widths of 0.1% and have exhibited no measurable nonlinearity. This capability places stringent restrictions on the linearity, stability, and noise of amplifiers to be used with them. Large current swings, particularly in high-level stages, will lead to nonlinearity. To limit such current variations, feedback or bootstrapping is used.

Pulse shaping must be used to minimize noise as well as to avoid pileup of pulses at high counting rates. Pileup will add at random to pulses and may also shift operating conditions. Finally, input stages must use components selected for minimal noise.

Time-Measuring Devices

Measurements in time ranges shorter than ∼1 microsecond are ordinarily accomplished by digital-to-analog techniques. In general, a constant current is allowed to charge a capacitor during the time interval. The resultant voltage on the capacitor is proportional to the time interval and can be digitized for classification if the distribution of time intervals is of interest. The same device is known as a coincidence circuit if used in conjunction with a differential voltage discriminator. A standard output is produced if the voltage on the capacitor is within a predetermined range.

Completely digital measurements are possible for extended times to an accuracy of about 10 nanoseconds using 100-megahertz counters. This limit will be surpassed as faster semiconductor devices become available on a production basis. Completely digital devices with time resolution <1 nanosecond and time ranges of several hundred nanoseconds have also been built using vernier techniques.

Analog-to-Digital Converters

Analog-to-digital converters are used to divide an amplitude range into denumerable intervals. The characteristics of interest for nuclear-physics applications include resolution, stability, dynamic range, dead time, and integral and differential linearity. Although satisfactory performance in all characteristics is important, the occurrence of narrow peaks in pulse height spectra places special emphasis on differential linearity. The variation in size of the intervals must either be negligible or at least small and smooth. As a result, most analog-to-digital converters in use are of the counter type. The pulse to be digitized charges a capacitor to its peak value. The capacitor is discharged by a constant current in a time measured by a periodic clock oscillator and counter. A single threshold device marks the end of discharge, and the same current is used for all pulses. Therefore the widths of adjacent channels remain constant or vary smoothly with any imperfections in the discharging current. Devices with dynamic ranges of 100:1, 2^{12} channels, differential linearity <1%, integral linearity <0.1%, and clock frequencies of 8–50 megahertz have been described.

Successive approximation counters in which the pulse is successively classified to the nearest $\frac{1}{2}$, $\frac{1}{4}$, $\frac{1}{8}$, etc., of full scale have been proposed on several occasions. These have the advantage of speed since they require only N comparisons rather than 2^N comparisons for 2^N channels. Until recently it has been considered that the differential nonlinearities associated with the necessary component tolerances would preclude their use for large numbers of channels. A variation in the device, in which a pulse step (random in size relative to the incoming pulse) is added before digitization and its digital value subtracted subsequently, has the effect that any nonlinearity is averaged over an interval equal to the range of added steps. The success of this method seems to have removed the earlier objections to this type of analog-to-digital converter.

Data Acquisition Systems

Complete systems using computers with fixed programs have been available commercially for over a decade. These systems include amplifiers, conditional gates, and analog-to-digital converters. The number generated by the latter provides an address for a magnetic core memory in which the frequency of occurrence of each address is totalized. The numbers stored in the memory may be displayed as a histogram on an oscilloscope, or may be outputs to a listing device such as a typewriter or on computer-compatible media such as magnetic or paper tape. The controls are quite flexible and allow considerable automation of operation.

A number of systems based on small stored-program computers have been described in the literature. These latter systems are particularly useful in a class of experiments known as multiparameter experiments. In such experiments, events are characterized by simultaneous values of several pulse heights and detector times. Were a totalizing system to be used, millions of storage locations would be required. Instead, the digital numbers or descriptors are stored sequentially in a temporary buffer and subsequently transferred to magnetic tape for later analysis. Contact with the experiment in real time is maintained by totalizing the frequency of occurrence of certain meaningful classes of descriptors.

The computing power available makes possible the totalization of quantities derived by reducing raw data. Under some conditions, the reduced data may be more immediately meaningful to the experimenter than would the raw data. This is particularly true when the output relation between sets of detectors is a function of only a few parameters such as nature of particle (proton, alpha particle, etc.). Then the event may be totalized with the few values of that parameter as part of the descriptor rather than the large ranges of outputs of counters that enabled the identification.

In general, the prime virtue of such a system is its flexibility of function and its unlimited capacity. Disadvantages are its cost, its somewhat slower response, and the need to program each new experiment. However, the changes in the economics of small computers, their increased speed, and the exchange of programing techniques should lead to a more widespread use. A block diagram of a typical small system is shown in Fig. 15.

RADIOLOGICAL SAFETY

Radiological Health

The dissipation of energy in body tissue by radiation may affect humans by damaging cells with resultant biological changes. Two classes of change

TABLE 3—MAXIMUM PERMISSIBLE EXPOSURE TO EXTERNAL RADIATION. *From "Protection against Neutron Radiation up to 30 Million Electron Volts," National Bureau of Standards Handbook 63, U.S. Government Printing Office, Washington, D.C.; 1957.*

		Ionizing Radiation of Any Type	
	Type of Exposure	Region of Exposure	Magnitude
Radiation worker	Long-term average maximum weekly dose	Whole body	0.1 rem
		Local (hands, forearms, feet, ankles)	1.5 rem
		skin	0.6 rem
	Accidental or emergency exposure (once in lifetime)	Whole body	25 rem*
		Local (hands, forearms, feet, ankles) in addition to whole body dose	100 rem*
	Accumulated dose	The maximum dose for any 13-week period is 3 rem. The maximum accumulated dose is $5(N-18)$ rem, where N is the age in years (>18).†	
Nonradiation worker	Long-term average maximum yearly dose	Critical organs	0.5 rem
Neutrons, of energy:	40-hour week		
Thermal			670 n/cm²/sec‡
1–10 MeV			17 n/cm²/sec
10–30 MeV			10 n/cm²/sec

* "Permissible Dose from External Sources of Ionizing Radiation," National Bureau of Standards Handbook 59, U.S. Government Printing Office, Washington, D. C.; 1959.

† 10 Code of Federal Regulations, Part 20, Commerce Clearing House, Inc.; New York.

‡ H. Blatz, "Introduction to Radiological Health," © McGraw-Hill Book Company, New York, N.Y.; 1964.

must be considered. For somatic changes, damage appears in the individual who has been irradiated. The second class, genetic, may appear in future generations. The levels of radiation, the exposure rate, and the length of time over which exposure occurs are closely connected with the nature and extent of any damage. It is important to establish levels of radiation exposure below which an undue health hazard is not encountered. Further, it is necessary to exercise administrative controls and provide monitoring so that these limits are actually observed.

The establishment of tolerance limits is based on limited experience with the relatively few cases of human exposure and on theoretical considerations. For instance, there is probably no threshold below which no genetic damage results. In addition, the net effect on the human race depends also on the number of individuals affected. The exposure level that can be tolerated by any individual without somatic effects is high compared with the exposure level that can cause genetic effects. Accordingly, exposure limits for the general population are set at a much lower level than are those

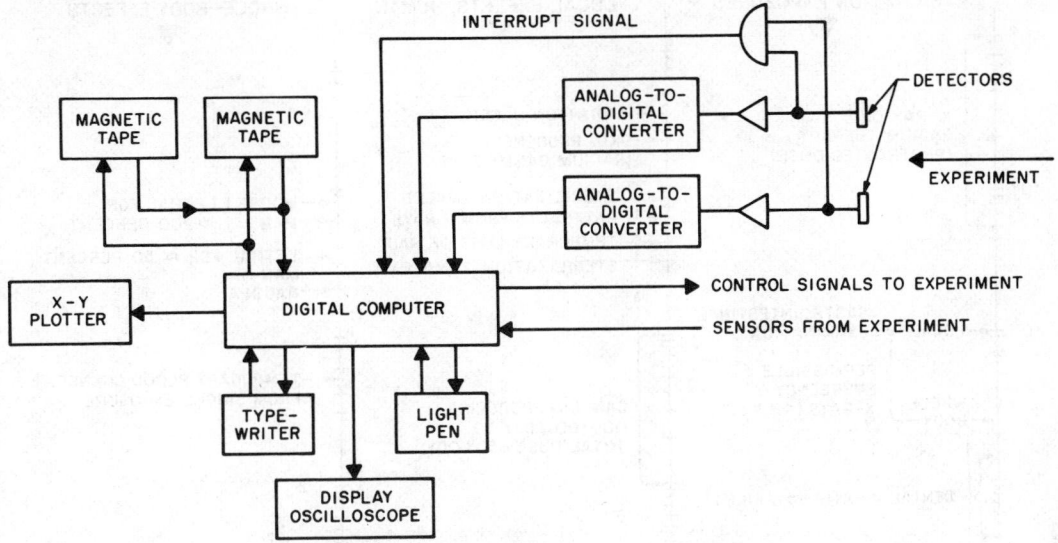

Fig. 15—Computer-based data acquisition system.

for the relatively small number of radiation workers. In each case a balance is attempted between possible damage to the individual and benefits he may gain by the use of the radiation source causing the damage.

Biological changes tend to be linearly related to the exposure, whether it is severe and short term or smaller and occurring over an extended period of time. Some biological effects are subject to recovery. Critical somatic damage takes place either when the rate of continuing damage exceeds the rate at which repair takes place or when a sufficient percentage of cells are so damaged that the function of vital organs is impaired. Human tolerances to external radiation exposure are indicated in Table 3.

Radiation Quantities and Units

Activity: The quantity of radioactive material is defined in terms of a transformation rate or the number of α or β disintegrations (nuclear transformations) that take place per unit time. The curie is defined as 3.7×10^{10} transformations/second.

Exposure: The effect of ionizing radiation on matter is to release charge either by direct ionization or by the liberation of ionizing particles. For X rays or γ rays, an exposure that releases 2.58×10^{-4} coulomb/kilogram of dry air is defined as one roentgen. The exposure R due to a source of C curies of activity emitting γ rays of total energy

TABLE 4—QUALITY FACTOR.

Particle	QF
X or γ ray, β particle	1
Proton	5
α particle	20
Slow neutron	5
Fast neutron	10

TABLE 5—MAXIMUM PERMISSIBLE AMOUNTS OF RADIOISOTOPES IN TOTAL BODY. *From "Maximum Permissible Body Burdens and Maximum Permissible Concentrations of Radionuclides in Air and in Water for Occupational Exposure," National Bureau of Standards Handbook No. 69, U.S. Government Printing Office, Washington, D.C.; 5 June 1959.*

Radio-isotope	Where Concentrated	Maximum Permissible Burden in Total Body in Microcuries
^{226}Ra	bone	0.2
^{239}Pu	bone	0.4
^{90}Sr	bone	20
^{32}P	bone	30
^{45}Ca	bone	200
^{137}Cs	liver	30

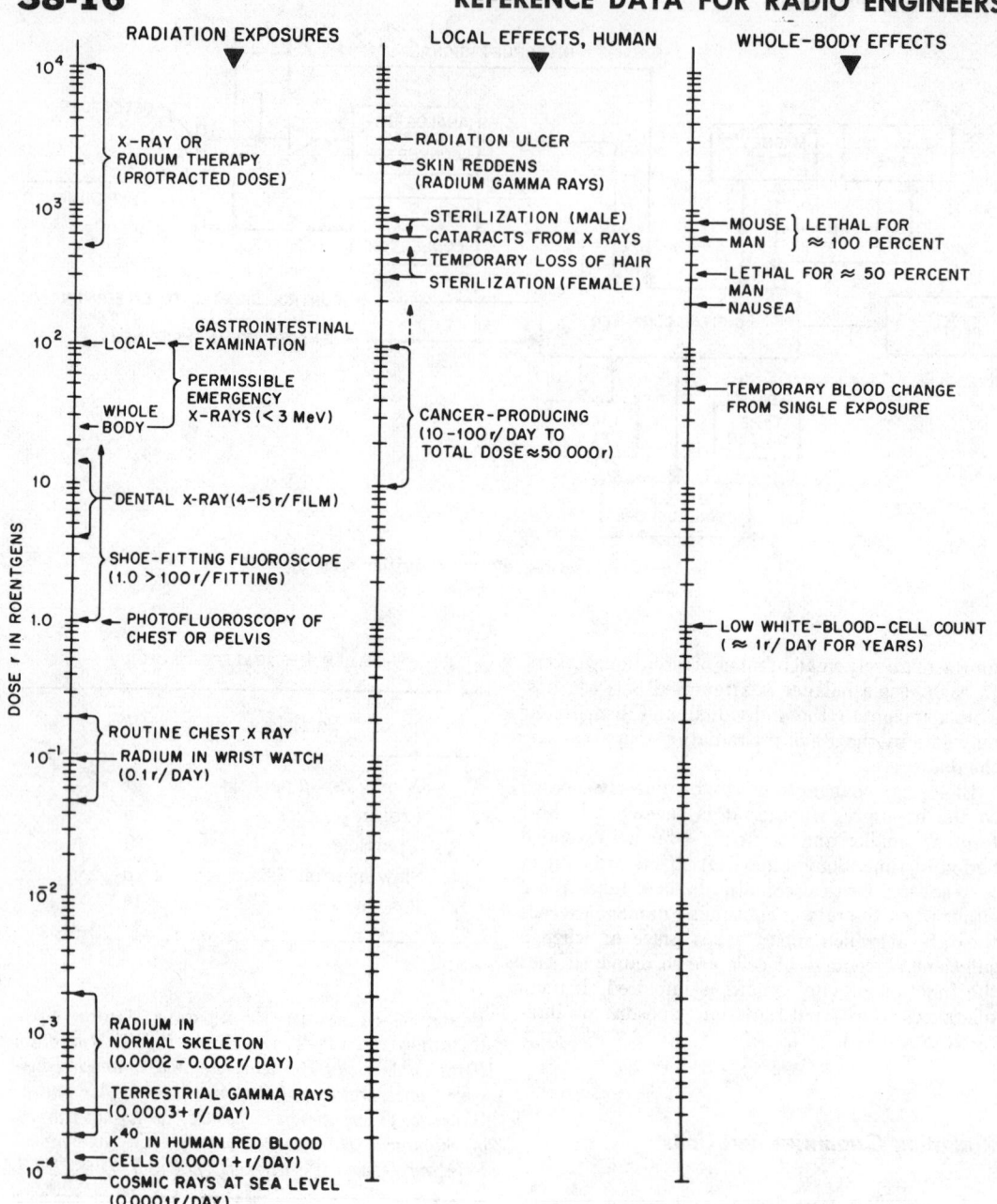

Fig. 16—Chart of radiation effects. *After R. D. Evans and C. R. Williams.*

E MeV per disintegration at a distance of F feet is

$$R \approx 6CE/F^2 \text{ roentgen.}$$

Absorbed Dose: Radiation damage is dependent on the energy absorbed per volume of the object under irradiation. An object that has absorbed $\frac{1}{100}$ joule/kilogram of ionization energy has absorbed a dose of 1 rad.

Quality Factor: Most cell damage is indirect; that is, the environment of the cell is altered by the radiation. Poisons generated in biological fluids cause the cell damage. For this reason, a particle that ionizes densely is more likely to cause cell damage than is a lightly ionizing particle. The biological damage accompanying a particular absorbed dose, due to a particular radiation, relative to the damage for the same dose due to electrons, is called the quality factor (QF).

Dose Equivalent: The dose equivalent is defined as the product of (absorbed dose)$\times (QF)\times$other possible modifying factors such as possible non-uniform distribution of an internally deposited isotope. The unit of dose equivalent is the rem= rad$\times QF$. Some current values of dose equivalents appear in Table 4.

Radiation Dosimetry

Instruments for the measurement of dose are usually based on standard particle detectors in conjunction with count integrating and count rate circuits. Particular devices are designed for specific applications. Thin-window ion chambers and scintillation counters are used for α dosimetry. Geiger-Müller counters, ion chambers, and scintillation counters are used for β and γ rays.

Proportional counters and scintillation counters based on ^{6}Li or ^{10}B are used for dosimetry of slow neutrons. The incorporation of a suitable moderating shield permits these same counters to be used for fast neutrons. Pocket dosimeters and photographic film worn on the body are widely used for personnel monitoring. The dosimeters consist of ion chambers that are charged. The state of charge may be read on an internal calibrated electrometer so that the amount of charge lost because of radiation may be measured. The blackening of film also permits the radiation exposure to be estimated. By interposing various metal foils, differential estimates of the exposure due to γ rays, to β rays, and to neutrons may be made.

Radioisotopes

In addition to possible hazards due to radiations from radioisotopes, an additional hazard exists because they may be taken into the body through inhalation, ingestion, or breaks in the skin. Radiations are then very efficiently absorbed by the body and are particularly damaging. In this case, the lifetime, chemical character, and form of the isotope are important. These properties determine the extent to which it is absorbed, the organs to which it preferentially migrates, the ease with which it is excreted by the body, and its effective lifetime in the body. Long-lived isotopes, which are retained in and around bone marrow and which emit high-energy α rays and β rays, are particularly hazardous. Radium, strontium, and plutonium are included in this latter class.

Although it is imperative that all radioisotopes be handled with utmost care, short-lived isotopes generally constitute a lesser hazard than do long-lived isotopes for the same initial activity. Untrained personnel should not attempt to handle unsealed radioactive materials. Tolerance levels for several internally absorbed radioactive materials appear in Table 5. Figure 16 shows the general biological effects of radiation.

BIBLIOGRAPHY

Elementary Nuclear Physics

I. Kaplan, "Nuclear Physics," 2nd ed., Addison-Wesley International Division, Reading, Mass.; 1963.

R. Resnick and D. Halliday, "Physics for Students of Science and Engineering," 2nd ed., John Wiley & Sons, New York, N.Y.; 1962.

Advanced Nuclear Physics

S. DeBenedetti, "Nuclear Interactions," John Wiley & Sons, New York, N.Y.; 1964.

J. M. Blatt and V. Weisskopf, "Theoretical Nuclear Physics," John Wiley & Sons, New York, N.Y.; 1952.

Accelerators

M. S. Livingston and J. P. Blewett, "Particle Accelerators," International Series in Pure and Applied Physics, McGraw-Hill Book Company, New York, N.Y.; 1962.

"200-BEV Accelerator Design Study," Report No. 16 000, University of California Radiation Laboratory, Berkeley.

Continuing literature appears in "Proceedings of Accelerator Conference," printed biennially in *IEEE Transactions on Nuclear Science*.

Nuclear Instrumentation

R. L. Chase, "Nuclear Pulse Spectrometry," McGraw-Hill Book Company, New York, N.Y.; 1961.

E. Gatti and V. Svelto, "Coincidence Circuits and Time-sorters," *Nucleonics*, vol. 23, no. 7, p. 62; July 1965.

E. Fairstein and J. Hahn, "Nuclear Pulse Amplifiers," (5 parts), *Nucleonics*, vol. 23, no. 7, p. 56; July 1965 through vol. 24, no. 3, p. 68; March 1966.

Radiation Detectors

"Nuclear Physics: Volume 5—Methods of Experimental Physics," Editor-in-Chief L. Marton, Editors L. C. L. Yuan and C. S. Wu, Academic Press, New York, N.Y.; 1961.

Continuing literature appears in "Proceedings of Scintillation Counter Symposium," printed biennially in *IEEE Transactions on Nuclear Science*.

Radiological Safety

D. E. Barnes and D. Taylor, "Radiation Hazards and Protection," 2nd ed., Pitman Publishing Corp., New York, N.Y.; 1963.

H. Blatz, "Introduction to Radiological Health," McGraw-Hill Book Company, New York, N.Y.; 1964.

Data Acquisition Systems

"Automatic Acquisition and Reduction of Nuclear Data," Report No. E53S, European–American Nuclear Data Committee; 1964.

CHAPTER 39

QUANTUM ELECTRONICS

INTRODUCTION

Maser* is an acronym for *Microwave Amplification* by *Stimulated Emission* of *Radiation*. A laser is a maser operating at optical frequencies (*Light Amplification by Stimulated Emission of Radiation*). Since the operation of masers and lasers is dependent upon quantum processes and interactions, they are known as quantum electronic devices.

MASERS

Introduction

The maser is a quantum amplifier for microwave frequencies. Masers may utilize a solid or gaseous active medium, and the operation of both is dependent on the following definitions.

Planck's Law:

$$E_2 - E_1 = hf. \qquad (1)$$

An atom can make discontinuous jumps or transitions from one allowed energy level to another, accompanied by either the emission or absorption of a photon of electromagnetic radiation at frequency f. (Refer to Table 1.)

Spontaneous Transitions consist of the spontaneous falling of an atom from a higher energy level to a lower, accompanied by the emission of a photon at the frequency given by (1). This is a dominant process at high frequencies (optical range).

Stimulated Transitions between energy levels are caused or stimulated by externally applied electromagnetic radiation. Upward transitions

---

* J. P. Gordon, H. J. Zeiger, and C. H. Townes, "The Maser—New Type of Microwave Amplifier, Frequency Standard, and Spectrometer," *Physical Review*, Vol. 99, No. 4, pp. 1264–1274; 15 August 1955. A. E. Siegman, "Microwave Solid-State Masers," McGraw-Hill Book Company, New York, N.Y.; 1964.

---

represent an attenuation process; downward transitions represent an amplification process.

W_{12}: stimulated transition probability per unit time = the number of stimulated jumps out of each level per unit time

$W_{12} = W_{21}$, and this probability is proportional to the strength of the applied radiation.

Boltzmann's Distribution: At thermal equilibrium, the ratio of the populations of the two energy states is always given by

$$N_2/N_1 = \exp[-(E_2 - E_1)/kT]. \qquad (2)$$

Net Power Absorption:

$$P_{abs} = hf_{21}(W_{12}N_1 - W_{21}N_2) = hf_{21}W_{21}(N_1 - N_2). \qquad (3)$$

If $(N_1 - N_2) < 0$, then $P_{abs} < 0$ (see Fig. 1). Negative power absorption means that the net power flow is from the atoms to the signal, that is, the signal is amplified. This is the basic maser process.

Relaxation Processes, such as spin-lattice relaxation in solid-state masers, or gas collisions or spin-exchange in gas masers, tend to destroy the population inversion and return the populations to thermal equilibrium after some relaxation time.

Gas Masers

Because of the low density of atoms in gases, the bandwidth of gas masers as amplifiers is very narrow and practically untunable. The power saturation level is also low, of the order of 10^{-10} to 10^{-12} watt.

As oscillators, however, gas masers have exceedingly high spectral purity, and some are used as frequency standards.

Gas masers have been developed in two types: beam masers and optically pumped gas-cell masers.

*Condition for Oscillations**: The condition for oscillations in terms of the loaded Q of the resonant

---

* K. Shimoda, T. C. Wang, and C. H. Townes, "Further Aspects of the Theory of the Maser," *Physical Review*, Vol. 102, No. 5, pp. 1308–1321; 1 June 1956.

TABLE 1—PHOTON ENERGIES AS A FUNCTION OF WAVELENGTH.

Wavelength (millimeters)	Frequency (hertz)	Energy of a Photon (hf)		Minimum Number of Photons per Second for 1 Watt of Power
		(joules)	(electronvolts)	
10	3×10^{10}	1.98×10^{-23}	1.24×10^{-4}	5×10^{22}
1	3×10^{11}	1.98×10^{-22}	1.24×10^{-3}	5×10^{21}
0.1	3×10^{12}	1.98×10^{-21}	1.24×10^{-2}	5×10^{20}
6.9×10^{-4}	4.35×10^{14}	2.87×10^{-19}	1.8	3.5×10^{18}

structure, its volume V, the matrix element μ, and interaction time with the radiation field t, is given by

$$N_2-N_1\geq (3hV/2\pi\mu^2Q)\left[(\sin^2\delta/\delta^2)\,t^2\right]^{-1} \quad (4)$$

where $\delta = (\omega-\omega_0)\,t/2$, ω is the frequency of oscillation, and ω_0 is the natural transition frequency.

For the case of a beam maser at resonance, (4) reduces to

$$N_2-N_1\geq (3hV/(4\pi^2\mu^2QT_0^2)) \quad (5)$$

where T_0 is the flight time of a molecule through the resonant structure.

For a radiation pumped gas maser, (4) reduces to

$$N_2-N_1\geq (3h\,\Delta fV)/(2\pi^2\mu^2Q) \quad (6)$$

where Δf is the natural linewidth of the transition.

Linewidth of Maser Oscillator: The half-width of the maser oscillator is approximately

$$\Delta\nu_{\rm osc}=4\pi kT(\Delta f)^2/P_B \quad (7)$$

where P_B is the output power. For the typical case of an ammonia maser, $P_B=10^{-10}$ watt, $\Delta f=2000$ hertz, and $\Delta\nu_{\rm osc}=4\times10^{-3}$ hertz, which is 2×10^{-13} the frequency ν_0 and shows the high degree of monochromaticity of the output radiation of a maser oscillator.

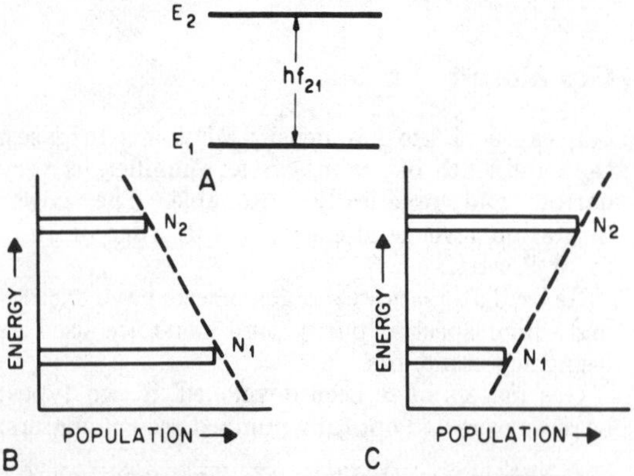

Frequency Pulling: If the cavity frequency tuning ν_c is not set equal to the atomic or molecular frequency ν_0, the actual frequency of the resonance ν will be pulled by

$$\nu-\nu_0= (\nu_c-\nu_0)\,(Q/Q_L) \quad (8)$$

where Q is the cavity Q, and Q_L is the molecular line Q.

Beam Masers: The types mostly used for frequency standards are the ammonia maser* (Fig. 2) and the hydrogen maser† (Fig. 3).

The ammonia molecule has electric dipole transitions and the inhomogeneous electric field focuser concentrates the upper-energy molecules in the center of the beam. After entering the cavity these molecules release their energy by stimulated emission.

The frequency of the 3–3 line of $^{14}NH_3$ is 23 870.129 megahertz. The frequency of the 3–3

Fig. 1—(A) Two energy levels and the corresponding transition frequency. (B) Energy-level populations at thermal equilibrium. (C) A negative population difference, or population inversion. *A. E. Siegman, "Microwave Solid-State Masers," p. 4, © 1964, McGraw-Hill Book Company.*

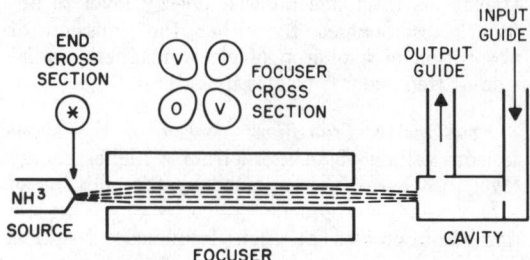

Fig. 2—Block diagram of the molecular beam spectrometer and oscillator. *J. P. Gordon, H. T. Zeiger, and C. H. Townes, "Microwave Molecular Oscillator and New Hyperfine Structure in the Microwave Spectrum of NH₃," Physical Review, Vol. 95, No. 1, pp. 282–284; 1 July 1954.*

* J. P. Gordon, H. J. Zeiger, and C. H. Townes, "Microwave Molecular Oscillator and New Hyperfine Structure in the Microwave Spectrum of NH₃," *Physical Review*, Vol. 95, No. 1, pp. 282–284; 1 July 1954.

† D. Kleppner, H. M. Goldenberg, and N. F. Ramsey, "Theory of the Hydrogen Maser," *Physical Review*, Vol. 126, No. 2, pp. 603–615; 15 April 1962.

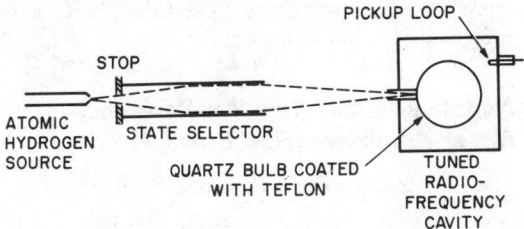

Fig. 3—Schematic diagram of atomic hydrogen maser. *D. Kleppner, H. M. Goldenberg, and N. F. Ramsey, "Theory of the Hydrogen Maser," Physical Review, Vol. 126, No. 2, pp. 603–615; 15 April 1962.*

line of $^{15}NH_3$, measured* with an accuracy of 5 parts in 10^{11}, is 22 789 421 701±1 hertz (A_1 time).

The atomic hydrogen type has magnetic dipole transitions and the inhomogeneous magnetic field focuser concentrates the upper-energy atoms in the center of the beam. After entering the coated quartz bulb, the atoms release their energy to the cavity by stimulated emission.

The frequency of the $F=1$ $m_F=0$ to $F=0$ $m_F=0$ transition is† 1 420 405 751.800±0.028 hertz (A_1 time).

Optically Pumped Gas-Cell Rubidium Maser (*Fig. 4*): An increase in the population of the upper level of the hyperfine transition in the ground state of Rb^{87} can be obtained by "optical pumping"‡ and proper light filtering. With sufficient gain, oscillations can be induced in the cavity.§ The frequency of the oscillation ≈6 834 682 600

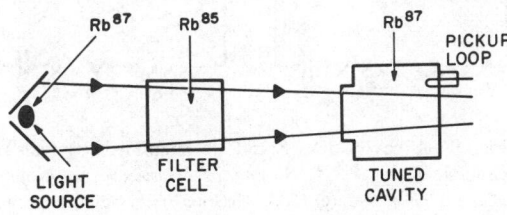

Fig. 4—Schematic diagram of rubidium maser.

* J. De Prins, "$^{15}NH_3$ Double-Beam Maser as a Primary Frequency Standard," *IRE Transactions on Instrumentation*, Vol. I-11, No. 3–4, pp. 200–203; December 1962.

† S. B. Crampton, D. Kleppner, and N. F. Ramsey, "Hyperfine Separation of Ground-State Atomic Hydrogen," *Physical Review Letters*, Vol. 11, No. 7, pp. 338–340; 1 October 1963. W. Markowitz et al, *Frequency*, Vol. 1, p. 46; July–August 1963.

‡ A. Kastler, "Optical Methods of Atomic Orientation and of Magnetic Resonance," *Journal of the Optical Society of America*, Vol. 47, pp. 460–465; June 1957.

§ P. Davidovits and W. A. Stern, "A Field-Independent Optically Pumped Rb^{87} Maser Oscillator," *Applied Physics Letters*, Vol. 6, No. 1, pp. 20–21; 1 January 1965.

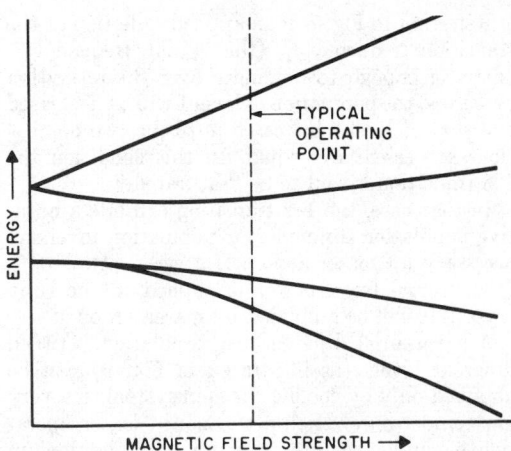

Fig. 5—The four Zeeman energy levels of the Cr^{3+} ion in ruby for a direct-current magnetic field oriented at 90° to the symmetry axis of ruby. *From A. E. Siegman, "Microwave Solid-State Masers," Fig. 1–4, © 1964, McGraw-Hill Book Company.*

hertz; it depends on gas collisions and is affected by the pressure of the buffer gas. However, the device can be used as a secondary frequency standard of good long-time stability and extremely good short-time stability.

Solid-State Masers

Three-Level Masers: Paramagnetic materials such as ruby, with their multiple energy levels and microwave transitions, are used in the so-called "3-level-maser" scheme* (Figs. 5 and 6).

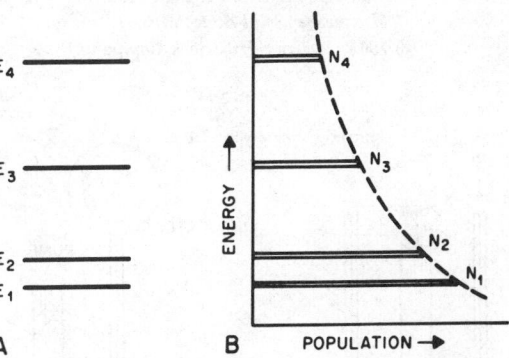

Fig. 6—(A) The four Zeeman levels for a particular operating point in ruby, and (B) the thermal-equilibrium distribution of spins among the four levels at a particular (low) temperature. *From A. E. Siegman, "Microwave Solid-State Masers," Fig. 1–5, © 1964, McGraw-Hill Book Company.*

* N. Bloembergen, "Proposal for a New Type Solid-State Maser," *Physical Review*, Vol. 104, pp. 324–327; 15 October 1956.

Referring to Fig. 7, if an applied radiation at the transition frequency f_{13} (the "pump frequency") is strong enough to dominate over the relaxation processes, the population of level 1 will be decreased and that of level 3 increased until the two populations are essentially equal. In this condition the 1–3 transition is said to be "saturated."

In that case, the 1–2 transition exhibits a negative population difference or population inversion necessary for maser action. If a weak signal at f_{21} (the "signal frequency") is applied to the ruby crystal, it will be amplified by maser action.

A substantial difference in population between different levels (as illustrated in Fig. 6) can be obtained only by cooling the spin system to a very low temperature. High-performance maser operation generally requires the use of liquid helium (4.2 K or colder), although maser operation with considerably reduced performance has been ob-

served at liquid-nitrogen temperature (77 K) and even at dry-ice temperature (195 K).

Negative-Q and Negative-Resistance Maser Amplifiers (Figs. 8 and 9)

$$\text{Magnetic } Q = -Q_m = \omega_0 L/(-R_m)$$
$$= \omega_0 W_s/(-P_m) \quad (9)$$

where W_s is the stored energy in the cavity and $-P_m$ is the power absorbed by the maser crystal; that is, P_m is the power emitted by the spin system due to the signal applied to it.

$$1/Q_m = (\gamma^2 \hbar/\pi\mu_0)(I \, \Delta N \sigma^2 \eta/\Delta f_L) \quad (10)$$

where $\gamma = g\beta\mu_0/\hbar$, $I =$ inversion ratio, $\Delta f_L =$ magnetic resonance linewidth in frequency unit, $\sigma^2 =$ maximum value of transition probability tensor, and $\eta =$ filling factor for cavity.

In the usual microwave maser, $hf/kT \ll 1$ and $\Delta N \approx (hf/kT)N/n$, with $N =$ total number of spins

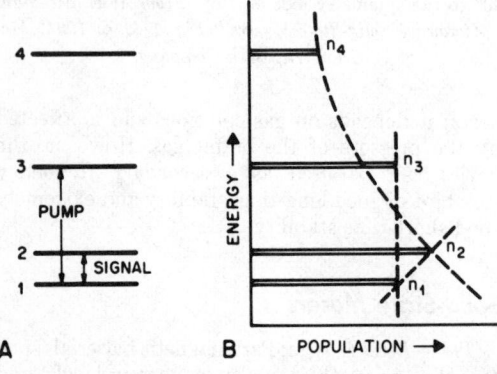

Fig. 7—(A) The pump and signal frequencies in a typical 3-level maser, and (B) the population distribution which results when the 1–3 transition is saturated. *From A. E. Siegman, "Microwave Solid-State Masers," Fig. 1–7, © 1964, McGraw-Hill Book Company.*

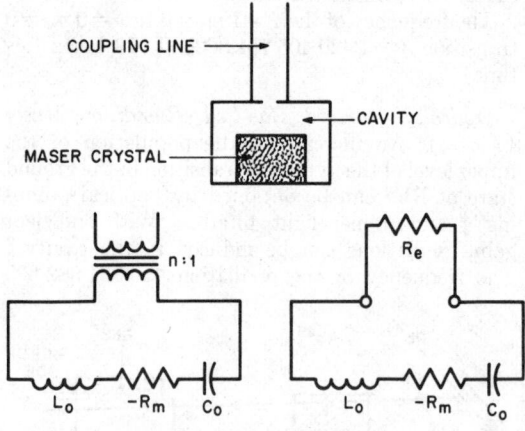

Fig. 9—A cavity maser and its simplified equivalent circuits. *From A. E. Siegman, "Microwave Solid-State Masers," Fig. 6–3, © 1964, McGraw-Hill Book Company.*

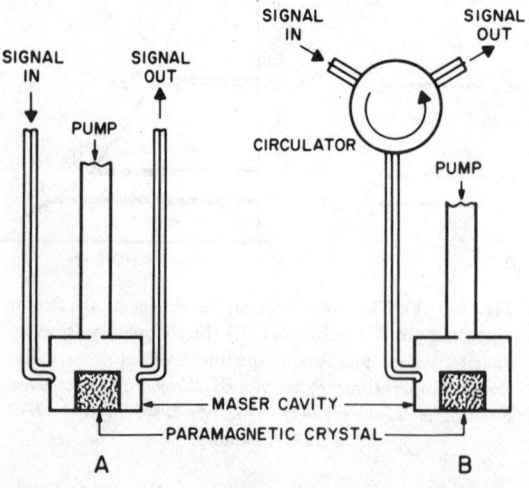

Fig. 8—Cavity maser amplifiers: (A) two-port or transmission type; (B) circulator type. *From A. E. Siegman, "Microwave Solid-State Masers," Fig. 1–8, © 1964, McGraw-Hill Book Company.*

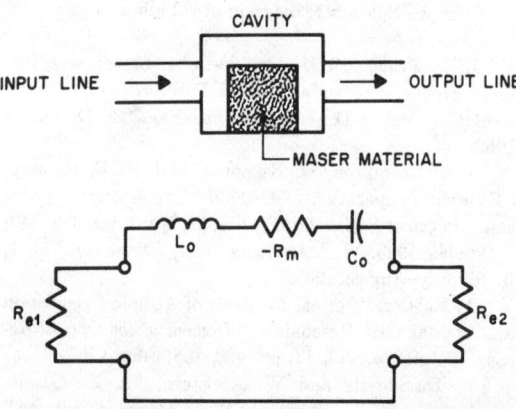

Fig. 10—A 2-port cavity maser and its equivalent circuit. *From A. E. Siegman, "Microwave Solid-State Masers," Fig. 6–7, © 1964, McGraw-Hill Book Company.*

per unit volume, $n=$ number of energy levels, and

$$\frac{1}{Q_m}=\frac{2g^2\beta^2\mu_0}{h}\frac{hf}{kT}\frac{N}{n}\frac{I\sigma^2\eta}{\Delta f_L}$$

$$\approx10^{-18}\frac{hf}{kT}\frac{N}{n}\frac{I\sigma^2\eta}{\Delta f_L}. \qquad (11)$$

Example: Maser material: pink ruby, 0.1% chromium concentration.

$$N=5\times10^{19}\text{ spins/cm}^3$$

$$n=4\text{ levels}$$

$$T=4.2\text{ K}$$

$$\sigma^2=1$$

$$\Delta f_L=100\text{ megahertz}$$

$$\eta=1$$

$$I=2$$

then $Q_m\approx100$.

Cavity-Maser Gain and Gain-Bandwidth Product

Reflection Cavity Maser (Fig. 9): Voltage gain

$$g(\omega)=\frac{1/Q_e+1/Q_m}{1/Q_e-1/Q_m}\cdot(1+4Q_{\text{tot}}^2\delta^2)^{-1} \qquad (12)$$

where external $Q=Q_e=\omega_0L_0/R_e$ and $\delta\equiv(\omega-\omega_0)/\omega_0$. $1/Q_{\text{tot}}=1/Q_e-1/Q_m$ at high gain, $Q_e\approx Q_m$, and the mid-band power gain G_0 of the maser is

$$G_0=(2Q_{\text{tot}}/Q_m)^2. \qquad (13)$$

The full amplifier bandwidth Δf between the 3-decibel points is

$$\Delta f=f_0/Q_{\text{tot}}=f_0(1/Q_e-1/Q_m) \qquad (14)$$

and the gain-bandwidth product is

$$g_0(\Delta f/f_0)=[g_0/(g_0-1)]\cdot(2/Q_m)$$

or

$$g_0(\Delta f/f_0)\approx2/Q_m \qquad (15)$$

if $g_0\gg1$. For example, if $Q_m=100$, $\Delta f/f_0\approx2\%/g_0$, which is obviously small for any reasonable voltage gain.

Two-Port Cavity Maser (Fig. 10): Mid-band

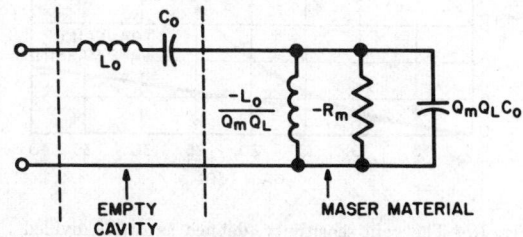

Fig. 11—Equivalent circuit for a single-tuned maser cavity including the effects of the magnetic-resonance linewidth. *From A. E. Siegman, "Microwave Solid-State Masers," Fig. 6–9, © 1964, McGraw-Hill Book Company.*

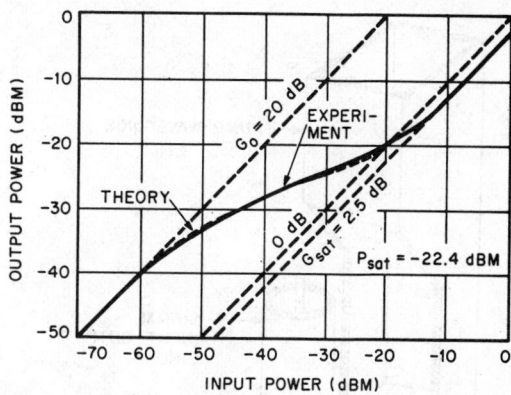

Fig. 12—Experimental cavity maser saturation curve measured by Arams and Okwit, plus a theoretical curve fitted to the experimental results. *From A. E. Siegman, "Microwave Solid-State Masers," Fig. 6–35, © 1964, McGraw-Hill Book Company.*

power gain g_0

$$g_0=\frac{4/Q_{e1}Q_{e2}}{(1/Q_{e1}+1/Q_{e2}-1/Q_m)^2} \qquad (16)$$

with $Q_{e1}=\omega_0L_0/R_{e1}$ and $Q_{e2}=\omega_0L_0/R_{e2}$. If $\epsilon=Q_m/Q_{e1}$ and $1-\epsilon=Q_m/Q_{e2}$, then

$$g_0(\Delta f/f_0)\approx(2/Q_m)[\epsilon(1-\epsilon)]^{1/2} \qquad (17)$$

with the same maximum gain-bandwidth product of $\frac{1}{2}$ for $\epsilon=\frac{1}{2}$ as the 1-port maser using the same cavity (for $\epsilon=\frac{1}{2}$, the coupling is equally divided between input and output).

More-Exact Analysis of the Cavity Maser (Fig. 11): If the magnetic Q_m is not large compared with the linewidth Q defined by $Q_L=f_0/\Delta f_L$, then

$$g(\Delta f/f_0)\approx2/(Q_m+Q_L). \qquad (18)$$

The linewidth of $Q=Q_L=\omega_0/\Delta\omega_L=\omega_0T_2/2$, where T_2 is the spin-temperature relaxation time. Improving Q_m will not improve the gain-bandwidth product of a simple cavity maser once Q_m is significantly less than Q_L.

Power Saturation: At sufficiently high signal levels the signal-frequency population difference of a 3-level maser is reduced, and hence the maser gain is also reduced, ultimately to zero. Saturation effects appear in typical cavity masers at amplifier output of about -30 to -50 dBm. A saturation power P_{sat} can be defined as

$$P_{\text{sat}}=\omega W_{\text{sat}}/Q_{m(0)}$$

$$P_{\text{in}}/P_{\text{sat}}=(1+g_{\text{sat}})^2/2(g_0-g_{\text{sat}})\frac{g_0-g}{(g+1)^2(g-g_{\text{sat}})}. \qquad (19)$$

The saturation-power parameter P_{sat} is the maximum power output from an oscillating maser under optimum conditions (Fig. 12).

Practical Realizations of Cavity Masers: Typical

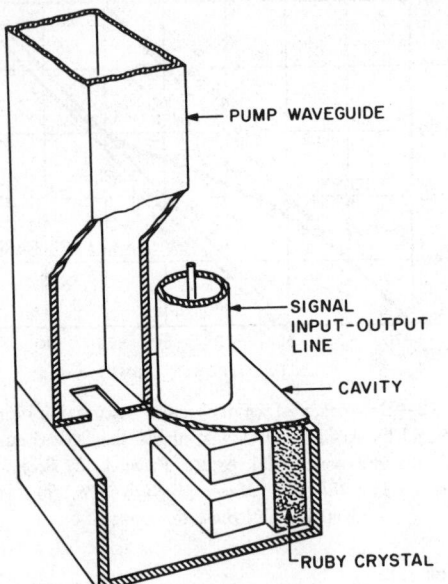

Fig. 13—High-efficiency 3000-megahertz cavity maser constructed at Stanford University. *From A. E. Siegman, "Microwave Solid-State Masers," Fig. 1–10, © 1964, McGraw-Hill Book Company.*

examples of the present state of the art might be bandwidths of 2 megahertz at 1500 megahertz, 6 at 3000 megahertz, and 20 at 9000 megahertz, with a gain of 20 decibels in each instance. Maser amplifiers have so far been successfully operated at signal frequencies from 300 to 75 000 megahertz, thus covering the entire microwave range. The pump power required in typical cases is a few tens of milliwatts or less (Fig. 13).

Traveling-Wave Maser (Fig. 14)

Gain:

$$G_{dB} = 27 (SN/Q_m) \qquad (20)$$

where S = slowing factor of circuit. $S = C/V_g$, where C = velocity of light, V_g = group velocity of the wave on the line, and $N = l/\lambda_0$ (the length of the circuit expressed in free-space wavelengths).

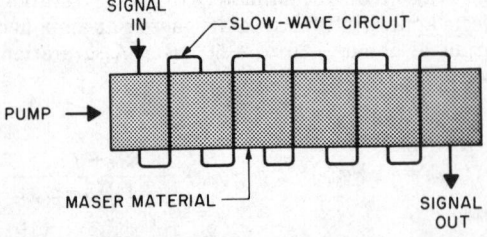

Fig. 14—Schematic diagram of a traveling-wave maser. *From A. E. Siegman, "Microwave Solid-State Masers," Fig. 7–1, © 1964, McGraw-Hill Book Company.*

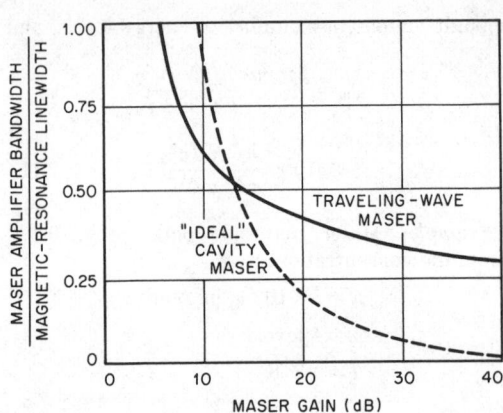

Fig. 15—Ratio of maser bandwidth to magnetic-resonance linewidth for a traveling-wave maser and for an "ideal" cavity maser (in which $Q_m \ll Q_L$). *From A. E. Siegman, "Microwave Solid-State Masers," Fig. 7–2, © 1964, McGraw-Hill Book Company.*

Example: Typical values for ruby maser at S band: At 3000 megahertz, magnetic $Q_m = 100$, $S = 100$, circuit length $l = 10$ cm = ($N = 1$), and gain $G_{dB} = 27$ dB.

Bandwidth:

$$\Delta f = \Delta f_L \{3/[G_{dB}(f_0) - 3]\}^{1/2} \qquad (21)$$

where $\Delta f_L = 1/\pi T_2$ is the full 3-decibel bandwidth of the magnetic-resonance line (see Fig. 15).

The traveling-wave maser can easily have both nonreciprocal forward gain and nonreciprocal reverse isolation or attenuation. In addition, the traveling-wave maser has much better stability against pump power fluctuations and fluctuations in magnetic Q_m than does the cavity maser (see Fig. 16).

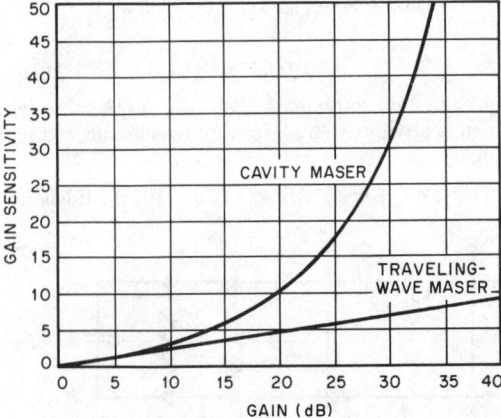

Fig. 16—The gain sensitivity, defined as $\Delta G/G$ divided by $\Delta Q_m/Q_m$, for cavity and traveling-wave masers. *From A. E. Siegman, "Microwave Solid-State Masers," Fig. 7–4, © 1964, McGraw-Hill Book Company.*

Saturation (Fig. 17):

Practical Realizations of Traveling-Wave Masers (Figs. 18–21):

Maser Noise Figure

Noise Figure:

$$F = 1 + \frac{T_a + (|\rho_2|^2/G) T_L}{T_g} \qquad (22)$$

where $T_a =$ noise temperature of maser amplifier, $T_g =$ noise temperature of generator, $T_L =$ noise temperature of the load, and $\rho_2 =$ reflection coefficient into output.

For the standard definition of noise figure,[*] $T_g =$ room temperature, and the load noise term

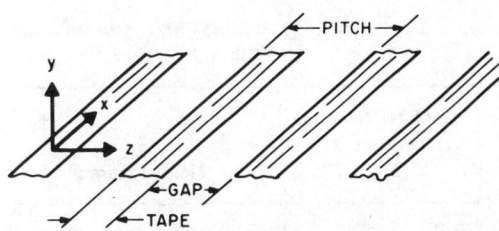

Fig. 18—The basic transverse-tape or ladder-line array. *From A. E. Siegman, "Microwave Solid-State Masers," Fig. 7-15, © 1964, McGraw-Hill Book Company.*

is either not included or is hidden in the T_a term. With these restrictions the noise figure (Table 2) reduces to

$$F = 1 + (T_a/290 \text{ K}). \qquad (23)$$

Maser Noise Output (Fig. 22):

$$\text{Maser noise output} = (G-1)\frac{hfB}{1-\exp(-hf/kT_m)}. \qquad (24)$$

Assuming $hf/kT \ll 1$, then for a 1-port (reflection) cavity maser

$$T_a \approx T_m + (Q_m/Q_0) T_0 \approx T_m \text{ if } Q_0 \to \infty.$$

For a 2-port (transmission) cavity maser

$$T_a = (1-\epsilon)^{-1} T_m$$

where T_m is the magnitude of the negative spin temperature, and $\epsilon \equiv R_L/R_m$ is a parameter that

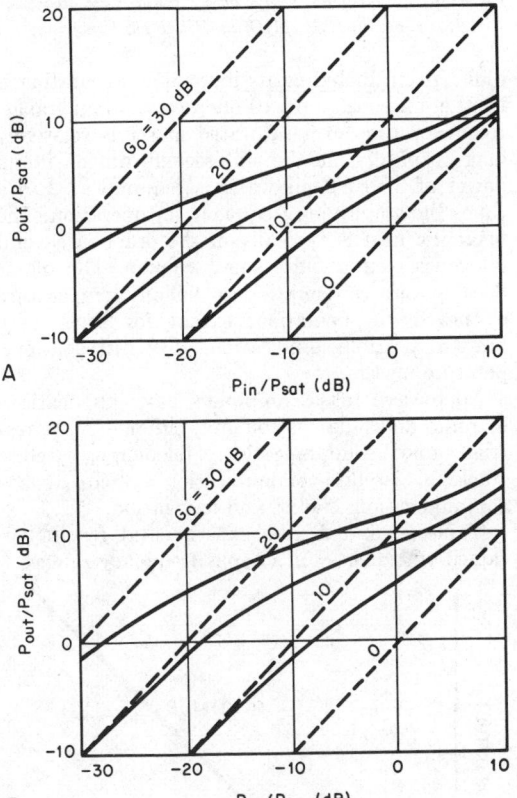

Fig. 17—Saturation characteristics of the traveling-wave maser: (A) uniform field distribution; (B) exponential field distribution. The quantity P_{sat} is defined differently for the two cases. *From A. E. Siegman, "Microwave Solid-State Masers," Fig. 7-5, © 1964, McGraw-Hill Book Company.*

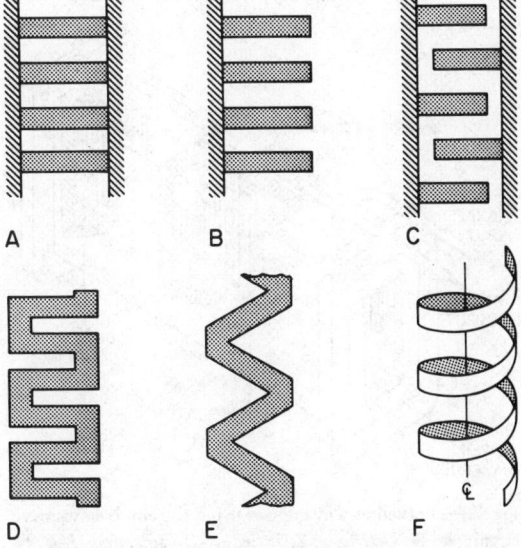

Fig. 19—Various slow-wave circuits derived from the transverse-tape array: (A) easitron; (B) comb; (C) interdigital line; (D) meander line; (E) zig-zag line; (F) tape helix. *From A. E. Siegman, "Microwave Solid-State Masers," Fig. 7-16, © 1964, McGraw-Hill Book Company.*

[*] Institute of Radio Engineers, "IRE Standards on Methods of Measuring Noise in Linear Two-Ports," 1958. Also, "Representation of Noise in Linear Two-Ports," *Proceedings of the IRE*, Vol. 48, pp. 60 and 69; January 1960.

TABLE 2—NOISE TEMPERATURE AND NOISE FIGURE.

Amplifier Noise Temperature T_a (K)	Noise Figure F
1	1.003 = 0.014 dB
7	1.023 = 0.1 dB
35	1.12 = 0.5 dB
75	1.26 = 1 dB
290	2 = 3 dB
870	4 = 6 dB
2 600	10 = 10 dB
29 000	100 = 20 dB

describes how the total external loading is divided between the source R_g and the load R_L. For the maximum gain-bandwidth condition, which is $\epsilon = \frac{1}{2}$, this gives $T_a = 2T_m$, or twice as noisy as a 1-port maser using the same maser cavity (see Fig. 23).

For the traveling-wave maser

$$T_a \approx T_m + (L_{\text{dB}} \text{ ohmic}/G_{\text{dB}})\, T_0$$

$$T_a \approx T_m \quad \text{if } L_{\text{dB}} \ll G_{\text{dB}}.$$

Microwave Maser Experimental Results*

Three-level masers have now been operated over a frequency range of more than two decades from 300 to 76 000 megahertz. Tuning ranges are within 5 to 30 percent. Equivalent input temperatures of 5 to 10 K are relatively easy for the maser am-

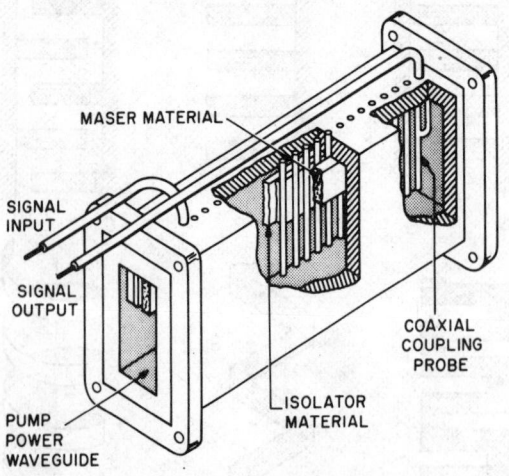

Fig. 20—Traveling-wave maser using the comb slow-wave circuit. *R. W. DeGrasse, E.O. Schulz-DuBois, and H. E. D. Scovil, "The Three-Level Solid-State Traveling-Wave Maser," Bell System Technical Journal, Vol. 38, No. 2, Fig. 1–13; March 1959.*

* A. E. Siegman, "Microwave Solid-State Masers," McGraw-Hill Book Company, New York, N.Y.; 1964.

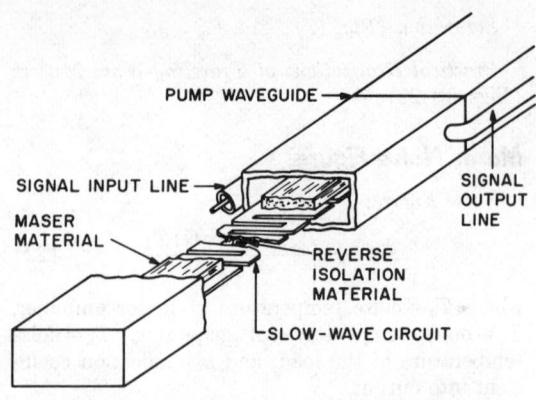

Fig. 21—Traveling-wave maser using a meander-line slow-wave circuit developed at Stanford University. *From A. E. Siegman, "Microwave Solid-State Masers," Fig. 1–12, © 1964, McGraw-Hill Book Company.*

plifier itself, including its input line. Saturation is in the microwatt or 10-microwatt input range. Recovery time of a saturated maser is relatively long, typically in the millisecond range. Pump power of 5 to 50 milliwatts is required at 2 to 4 times the amplifying frequency for operation. The magnetic field is typically of the order of several kilogauss over a quite small volume. The maser does require refrigeration to helium temperature or nearby in operation, except for some very limited possible applications for nitrogen-temperature masers.

Microwave maser amplifiers have applications in radio and radar astronomy, radiometry, terrestrial radar, communications, telemetry, satellite tracking, satellite-communication systems, space communications, radar, and navigation.

Tables 3 and 4 give experimental results obtained with masers in various frequency ranges.

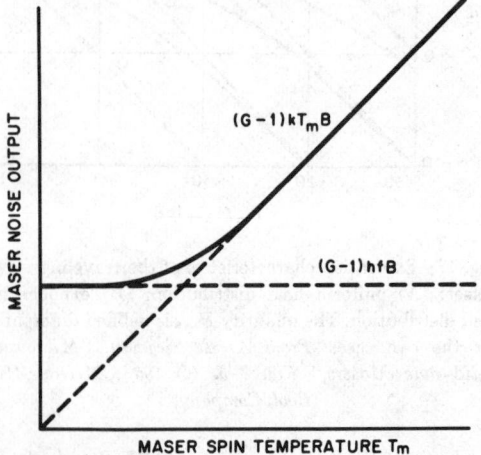

Fig. 22—Noise output of a maser amplifier as a function of the maser spin temperature T_m. *From A. E. Siegman, "Microwave Solid-State Masers," Fig. 8–22, © 1964, McGraw-Hill Book Company.*

TABLE 3—EXPERIMENTAL RESULTS WITH CAVITY MASERS. A. E. Siegman, "Microwave Solid-State Masers," pp. 440–444, © McGraw-Hill Book Company, 1964.

Signal Frequency (GHz)	Pump Frequency (GHz)	Gain-Bandwidth Product (MHz)	Maser Crystal	Operating Point	Bath Temperature (K)	Pump Power (mW)	References and Remarks
0.30	5.40	0.32	Cr^{3+} in $K_3Co(CN)_6$	a-c plane; ~80 g	1.6		Kingston, Lincoln Laboratory. Used in Venus radar echo experiments.
0.45	5.40	0.5	Cr^{3+} in $K_3Co(CN)_6$	a-c plane; ~80 g	1.25		
0.38–0.45	~11.8	0.56	Ruby	90°; 70 g	1.7		Wessel, General Electric.
1.2	10.84	11 (19)	Ruby	90°; 1740 g	4.2 (1.5)		Arams and Okwit, Airborne Instruments Laboratory. Maser cavity tunable 0.8–2.0 gigahertz.
1.75	11.80	37.5	Ruby	90°; 2100 g	1.5		
1.82	11.86	20	Ruby	90°; 2150 g	4.2		
1.37	8.0	>1.5	Cr^{3+} in $K_3Co(CN)_6$	a-c plane	<2		Artman, Bloembergen, and Shapiro, Harvard University.
1.38	9.07	1.85	Cr^{3+} in $K_3Co(CN)_6$	a-c plane; 1200 g	1.25	28	Autler and McAvoy, Lincoln Laboratory.
1.42	3.85	2.7	Cr^{3+} in $K_3Co(CN)_6$	a-c plane; 480 g	1.4	~100	Bolger, Robinson, and Ubbink, Kamerlingh Onnes Laboratory, Leiden, The Netherlands.
2.2	12.7	12	Ruby	90°	~2	2	Hansen and Rowley, Lockheed Missile Systems Division. Untuned pump.
2.39	13.0	25	Ruby	90°	1.4		Higa, Jet Propulsion Laboratory [unpublished].
2.54	23.70	21	Ruby	86°; 2800 g	1.6	~30	Chang, Cromack, and Siegman, Stanford University. Large gain-bandwidths attributed to efficient cavity design.
2.70	13.40	16.5	Ruby	88°; 2800 g	1.6	~5	
2.74	13.60	>55	Ruby	90°; 2800 g	1.6	~5	
2.97	10.30	5	Ruby	23.5°; 2650 g	1.6	~5	

TABLE 3—EXPERIMENTAL RESULTS WITH CAVITY MASERS. (*Continued*)

Signal Frequency (GHz)	Pump Frequency (GHz)	Gain-Bandwidth product (MHz)	Maser Crystal	Operating Point	Bath temperature (K)	Pump power (mW)	References and Remarks
2.8	9.4	1.8	Cr^{3+} in $K_3Co(CN)_6$	a-c plane; ~2000 g	1.25	~10	McWhorter and Meyer, Lincoln Laboratory. An important early maser experiment.
~3.0	~15.0		Ruby		2		Zverev, Kornienko, Manenkov, and Prokhorov, Lebedev Institute, Moscow.
6.2	11.5	~1	Gd^{3+} and Cd^{3+} in lanthanum ethyl sulfate	90°; 1800 g	1.2	38	Scovil, Bell Telephone Laboratory.
9.06	17.52		Gd^{3+} and Ce^{3+} in lanthanum ethyl sulfate	17°; 2850 g	1.2	>250	Scovil, Feher, and Seidel, Bell Telephone Laboratory. First solid-state maser operation.
8.4-9.7	~23.0	40-65	Ruby	55°	4.2-1.6	3-15	Morris, Kyhl, and Strandberg, Massachusetts Institute of Technology.
8.5-9.9	~24.0	40-250	Ruby	48°-55°	4.2		Kikuchi, Lambe, Makhov, Terhune, et al., University of Michigan. First use of ruby.
9.5	23.5	50-100	Ruby	55°; 3740 g	1.4	30	Giordmaine, Alsop, Mayer, and Townes, Columbia.
9.0	24.0	35-105	Ruby	55°-60°; 4000 g	4.2	10	Forward, Goodwin, and Kiefer, Hughes Research Laboratory.
9.0	25.0	220	Ruby		4.2		Inaba, Research Institute of Electrical Communication, Tohoku University, Sendai, Japan [unpublished].

9.4–9.61	22.85–23.85	100–230	Ruby	55°; 3900–4300 g	1.35	30–60	Gianino and Dominick, Ewen-Knight Corporation.
8.2–10.6	~35.0	>>25	Cr^{3+} in TiO_2 (rutile)	$\theta=80°$; $\phi=90°$; 3300 g	4.2	~10	Gerritsen and Lewis, RCA Laboratory (unpublished).
~9.0	~34.0	>>8	Fe^{3+} in TiO_2		4.2	~10	
10.4	58.4	126	Cr^{3+} in beryl (emerald)	90°; 1900 g	4.2		Goodwin, Hughes Research Laboratory.
9.54	10.85	4	Ruby	32°; small H_0	4.2	100	Arams, Airborne Instruments Laboratory. Use of cross-relaxation effects for low f_p/f_s ratio. *Ibid.* Harmonic pumping.
10.59	9.60		Ruby	90°; 1675 g	4.2	8	
9.52	24.0	3.8	Ruby	55°	56	>50	Ditchfield and Forrester, Royal Radar Establishment, Malvern, England. Liquid O_2 cooling.
9.14	23.1	14	Ruby	55°; 4000 g	77		Maiman, Hughes Research Laboratory. Highest-temperature maser operation.
9.14	23.1		Ruby	55°; 4000 g	195		
9.4	25.0		Fe^{3+} in Al_2O_3	1200 g	1.8		Kornienko and Prokhorov, Lebedev Institute, Moscow.
12.3	31.8	15	Fe^{3+} in Al_2O_3	120 g	4.2	10	King and Terhune, University of Michigan. Near-zero-field operation.
22.3	49.9		Cr^{3+} in TiO_2	$\theta=30°$; $\phi=90°$; 4500 g	4.2		Gerritsen and Lewis, RCA Laboratory.
33.7–36.1	70.4	10–50	Fe^{3+} in TiO_2	(100) plane; 3000 g	4.2	2–10	Foner and Momo, Lincoln Laboratory.
49.0–57.0	78.1–78.8		Fe^{3+} in TiO_2	73° from [110] in(001); 7200–5500 g (push-pull operation)	1.6	50	Carter, Columbia University.
~76.0	~150.0		Fe^{3+} in TiO_2	Many; ~25 000 g			Heller, Lincoln Laboratory.
96.3	65.2		Fe^{3+} in TiO_2	$\theta=53°$; $\phi=28.5°$; 7350 g	2.1		Hughes, Westinghouse. Five-energy-level "push-pull-push" pumping scheme.

Table 4—Experimental Results with Traveling-Wave Masers. A. E. Siegman, "Microwave Solid-State Masers," p. 445, © McGraw-Hill Book Company, 1964.

Laboratory	Signal Frequency (GHz)	Pump Frequency (GHz)	Circuit (length; slowing factor)	Material	Operating Point	Operating Temp. (K)	Electronic Gain,[a] dB/in. (F/B ratio[b])	Net Gain[c] (dB)	Instant Bandwidth (MHz)	Tuning range (MHz)	Reverse Isolation (F/B ratio[b])
Airborne Instr. Lab.	1.12–1.37	10.9	Comb[g] (1.9"; 98)	Ruby	90°; 1700 g	1.6	3.5[h] (1:1)	4.5	25[f]	250	
Airborne Instr. Lab.	2.15–2.35	12.8	Comb[g] (6.5"; 72–80)	Ruby	90°; 2400 g	1.8	5[h] (1:1)	30	25[j]	200	YIG disks (30:1)
Bell Tel. Lab.	2.39	13.0	Comb[g] (5.3"; 190)	Ruby	90°; 2500 g	1.8	4.5 (1:1)	36	13	10	YIG disks (>10:1)
Stanford Univ.	2.4–3.1	14.0	Meander line (3"; ~100)	Ruby	90°; 2800 g	1.6	8 (3.5:1)	20	25–40	700	YIG slab
MELabs, Palo Alto, Calif.	3.0	14.0	Modified combs[f]	Ruby	90°; 2800 g	4.2	4–8 (2:1)		20–40		YIG slab
Mullard Res. Lab., England	3.0	14.0	Comb (4.7"; ~160)	Ruby	90°; 2900 g	1.37	7.5 (—)	21.5	17		YIG disks (>20:1)
Bell Tel. Lab.	6.0–6.3	11.7–12.3	Comb[d] (1"; 40)	Gadolinium ethyl sulfate	1800 g	1.6	12 (10:1)	25	25	300	YIG spheres (30:1)

Bell Tel. Lab.[l]	5.75–6.1	18.9–19.5	Comb[d] (5"; 40)	Ruby	90°; 4000 g	1.5	6 (3.5:1)	23–13[e]	25–65[e]	350	(a) Dark ruby (8:1) (b) YIG spheres (60:1)
Lincoln Lab.	8.2	25.6	Meander line	Ruby	55°	4.2	12 (—)	18	25	400	YIG slab
Univ. of Mich.	9.45–9.85	24.0	Karp circuit[i] (3.5")	Ruby	55°; 4100 g	1.6	8.5[h,k] (—)	22[k]	35–40[j]	400	YIG slab
Airborne Instr. Lab.	23–41	43–82	Dielectric-filled Cr^{3+} in rutile waveguide		Various	1.5	26 (—)				Ferrite

[a] Defined as output with pump on ÷ output with pump off, expressed in decibels.
[b] Ratio of forward gain (or reverse loss) in decibels to reverse gain (or forward loss) in decibels.
[c] Electronic gain of entire circuit, minus total losses.
[d] Twelve fingers per inch.
[e] Broadband low-gain operation resulting from misalignment of two ruby samples in structure.
[f] Thirty to fifty fingers per inch.
[g] Ruby placed on both sides of slow-wave circuit.
[h] Electronic gain was reduced to ~40% of this value when maser was operated at 4.2 K.
[i] Ridged-waveguide easitron structure.
[j] Predicted value if gain were increased to 30 decibels.
[k] Predicted value. Because of accidental misalignment of two ruby samples in the structure, the resonance line was split. Actual gain in each line was approximately half the tabulated value
[l] This maser has most recently been operated with superconducting electromagnet.

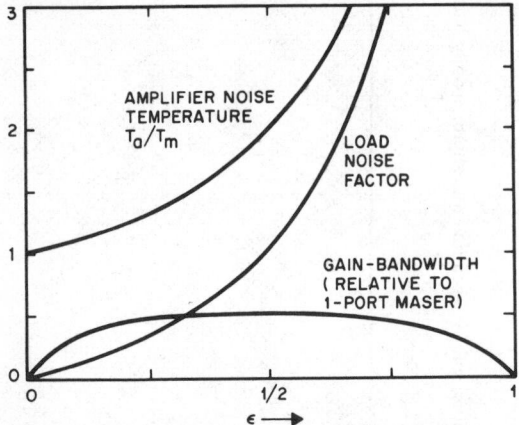

Fig. 23—Various significant quantities for the 2-port (transmission) cavity maser, as a function of the input/output coupling ratio ϵ. *From A. E. Siegman, "Microwave Solid-State Masers," Fig. 8–21, © 1964, McGraw-Hill Book Company.*

LASERS*

Introduction

The fundamental quantum aspects of laser operation are essentially identical to those of a maser. A suitable active material is excited by an external stimulus to raise a population of its constituent atoms, ions, or molecules from the ground state to higher energy levels. After an interval called the lifetime, an excited atom will resume equilibrium by spontaneously falling back to the ground state, usually through several intermediate levels. The downward transitions between levels above the ground state will be accompanied by emissions of photons, lattice vibrations (heat), or both.

If the stimulus input is made sufficiently intense, the number of excited atoms will exceed the number of atoms at the ground state, and a population inversion will be established. Now, an applied signal in the form of spontaneously emitted radiation (quantum noise) will be amplified since more of its energy will be expended on stimulating downward transitions with a resultant in-phase energy contribution than on upward transitions with a net signal loss.

The system described thus far comprises a true optical amplifier capable of producing a superradiant beam having a relatively narrow bandwidth. If the active material is enclosed in a lossy cylindrical housing with a transparent end window, the system will emit a narrow beam with a di-

---

* A. L. Schawlow and C. H. Townes, "Infrared and Optical Masers," *Physical Review*, Vol. 112, pp. 1940–1949; Dec. 1958. A. E. Siegman, "An Introduction to Lasers and Masers," McGraw-Hill Book Company, New York; 1971.

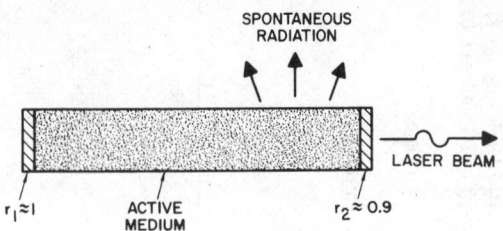

Fig. 24—The essential elements of a laser cavity.

vergence corresponding to the field of view defined by the housing.

The superradiant emission of an optical amplifier may be described as "incoherent" laser radiation since no single-frequency oscillation occurs and the output consists of amplified quantum noise. To produce quasi-coherent laser radiation, the active material must be placed in an optical resonating cavity made from two facing mirrors, one of which is 100% reflecting and the other usually more than 90% reflecting to permit extraction of a portion of the intercavity beam (Fig. 24). Now, amplified superradiant radiation is reflected back into the active medium whereupon additional in-phase amplification occurs on each pass, and the laser output, which emerges from one or both end mirrors, includes a narrow linewidth, amplitude stabilized oscillation. The oscillation has an amplitude orders of magnitude greater than the still-present quantum noise.

Three- and Four-Level Lasers

In a three-level laser medium such as ruby, the terminal state is separated from the ground state by an intermediate metastable state (Fig. 25). Atoms raised to the terminal level by the pumping stimulus almost immediately collapse to the metastable state where they remain for a time called the fluorescent lifetime. They then resume equilibrium spontaneously or upon stimulation by a photon of appropriate energy.

Operation of a four-level laser medium such as neodymium is characterized by a low pumping threshold since there are two intermediate levels between the ground and terminal states and radiative emission occurs during downward transitions between them (Fig. 26). Since the population of the second level is initially negligible, a population inversion between the second and third levels is much more readily achieved than in a three-level laser.

Laser Operating Conditions and Parameters

Laser action has been observed in a remarkably diverse variety of active mediums with spectral

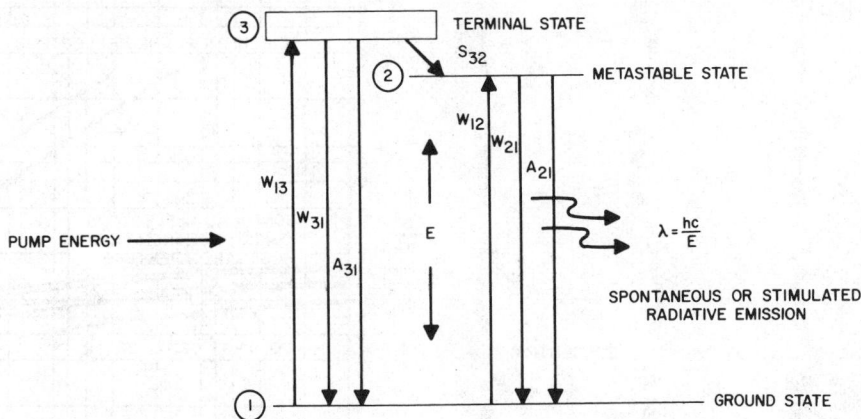

Fig. 25—The three-level laser showing major energy transition probabilities.

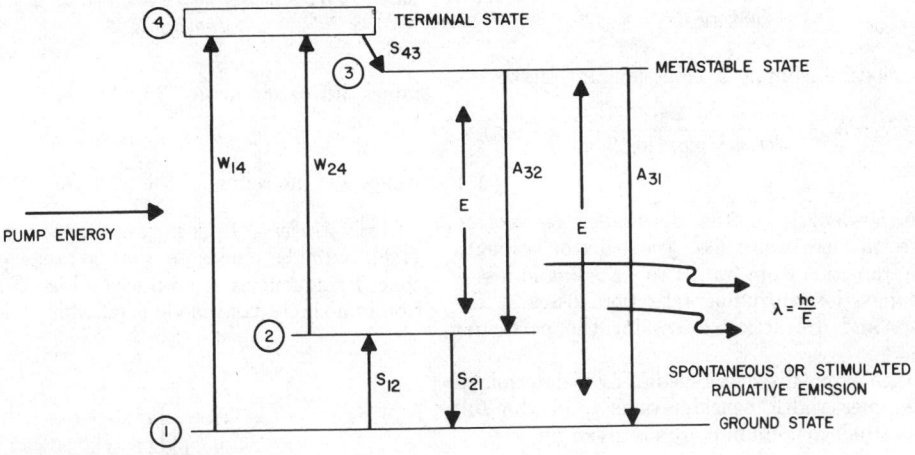

Fig. 26—The four-level laser showing major energy transition probabilities.

outputs ranging from 0.3 μm in the ultraviolet to 800 μm in the far infrared. Several important operating conditions and other parameters characterize practical laser systems, including the following.

Cavity Q: The Q of a laser resonating cavity may be defined as

$$Q \equiv \frac{n_0\omega}{2\alpha_{0c} + (c/L)\ln(1/r_1 r_2)} \quad (25)$$

where α_{0c} is the cavity loss, L is the cavity length, r_1 and r_2 are the mirror reflectance coefficients, n_0 is the refractive index of the laser medium, and ω is the angular velocity $(2\pi f)$.

Threshold: The onset of laser oscillation is an abrupt, nonlinear process whereupon a narrow linewidth, amplitude stabilized oscillation is superimposed over the amplified quantum noise. The gain of the travelling wave in the amplifier must equal absorption in the laser medium and losses at each mirror (collectively termed R), and this may be expressed

$$R^{\alpha l_A} = 1 \quad (26)$$

where α is gain and l_A is the active length of the amplifier. If the mirror losses are unequal, R becomes $(R_1 R_2)^{1/2}$. The condition for maser oscillation can be modified for application at optical

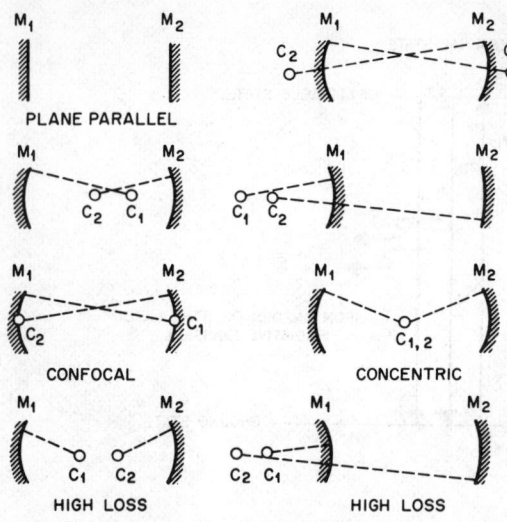

Fig. 27—Examples of mirror configurations for optical masers. All except the bottom two exhibit low-loss resonant modes. *From A. Yariv and J. P. Gordon, "The Laser," Proceedings of the IEEE, Vol. 51, No. 1, Fig. 10; January 1963.*

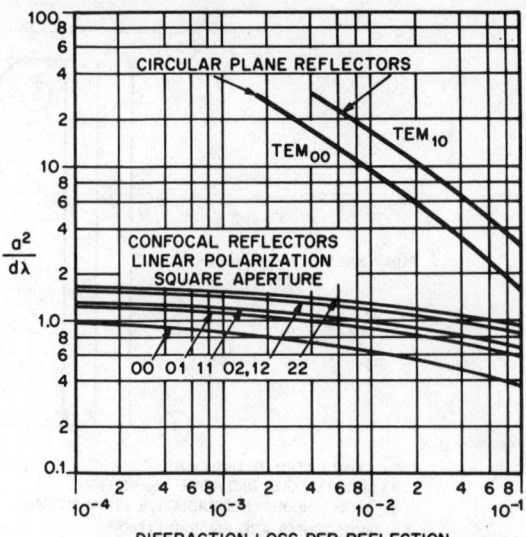

Fig. 28—Diffraction losses for plane-parallel and confocal resonators. *From A. Yariv and J. P. Gordon, "The Laser," Proceedings of the IEEE, Vol. 51, No. 1, Fig. 13; January 1963.*

frequencies. Assuming a Gaussian line shape

$$\Delta N_c = N_2 - N_1 \geq \frac{\Delta \nu_D}{2fD(\pi e^2/mc)(\log_e 2/\pi)^{1/2}} \text{ (loss)} \quad (27)$$

where D = length of the discharge, e = electron charge, m = electron mass, f = oscillator strength of the transition from state 1 to state 2, and loss = single pass loss including reflection losses at the mirrors and diffraction losses for the particular mode.

In case of gas-discharge media, $\Delta \nu$ is determined by Doppler width considerations, and the full width at half-maximum points is given by

$$\Delta \nu_D = (2\nu/c)\left[(2kT/M)\log_e 2\right]^{1/2} \quad (28)$$

where k = Boltzmann's constant, c = velocity of light, T = average atomic temperature, and M = atomic mass.

Combining (27) and (28) gives the oscillation condition as

$$N_2 - N_1 \geq \frac{\nu}{f(\pi e^2/mc^2)}(2\pi kT/M)^{1/2}\text{(losses)}. \quad (29)$$

Optical Resonator Configurations: Fig. 27 shows several laser resonator configurations. For low losses, the dimensions of the reflectors must satisfy the relation

$$a_1 a_2/d\lambda \gtrsim 1 \quad (30)$$

where a_1 and a_2 are half the widths of the two reflectors, respectively, in any arbitrary direction perpendicular to the resonator axis, and d is the distance between the reflectors.

Diffraction Losses: The parameter that deter-

mines diffraction losses (Fig. 28) is

$$N = a^2/d\lambda \quad (31)$$

where a is the radius of the mirrors.

Axial Modes: A laser may oscillate in many axial (longitudinal) modes so that a range of closely spaced frequencies is produced (Fig. 29). Operation in a single axial mode is possible in some laser

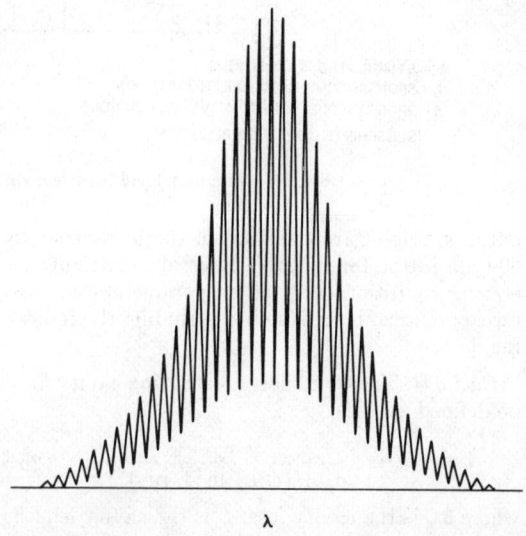

Fig. 29—Typical axial modes from a mode-locked Nd^{3+}: $Y_3Al_5O_{12}$ laser.

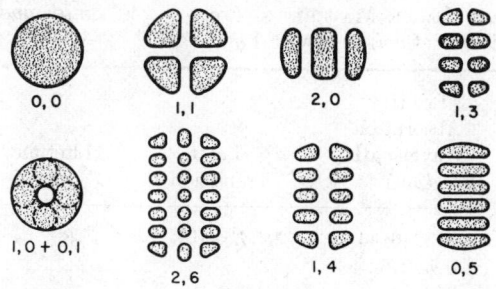

Fig. 30—Examples of Hermite-Gaussian transverse modes. (*After A. E. Siegman, "An Introduction to Lasers and Masers," p. 331; McGraw-Hill Book Company, 1971.*)

systems, and the result is a highly monochromatic output.

Transverse Modes: A laser may oscillate in many transverse modes with a corresponding increase in beam spreading. Fig. 30 shows several examples of low- and high-order Hermite-gaussian transverse modes. These modes are designated TEM_{mn} where m and n are integers which specify the various modes. Operation in the lowest order transverse electromagnetic mode (TEM_{00}) is possible in certain laser systems.

Spot Size: A laser oscillating in the TEM_{00} mode produces a beam with the gaussian transverse amplitude distribution $\exp[-(x^2+y^2)/w^2]$, where w is the spot size. The spot size is defined as the radius at which the field intensity of the lowest-order (TEM_{00}) mode falls to $1/e^2$ of its on-axis value. The spot size of a TEM_{00} beam contains approximately 86 percent of the total energy within the beam. A confocal resonator oscillating in the TEM_{00} mode has a spot size given by

$$w = (d\lambda/\pi)^{1/2} \qquad (32)$$

where d is the confocal resonator spacing.

Coherence: The optical properties of a laser beam may be very generally described as quasi-coherent; *i.e.* the monochromaticity of the beam may be described as temporal coherence and the highly directional nature of the beam as spatial coherence. These definitions are very general since coherence theory provides a far deeper and very specific definition of coherence.

Spiking: Many pulsed and cw lasers exhibit a spiked output (Fig. 31). Spiking is a form of relaxation oscillation caused by the momentary loss of population inversion during laser oscillation and the reattainment of population inversion as a consequence of continued pumping.

Q-Switching: If the Q of a laser cavity is temporarily spoiled by misaligning or blocking one of the resonator mirrors, virtually all the active atoms can be raised to the terminal state. If the cavity Q is quickly restored at the proper instant, a very brief, high-power "giant" pulse will be produced as a consequence of the available stored energy and the very high internal gain. Very large Q-switched Nd:glass lasers have delivered more than 10^{12} watts in pulses ranging from 10^{-9} to 10^{-10} second in length.

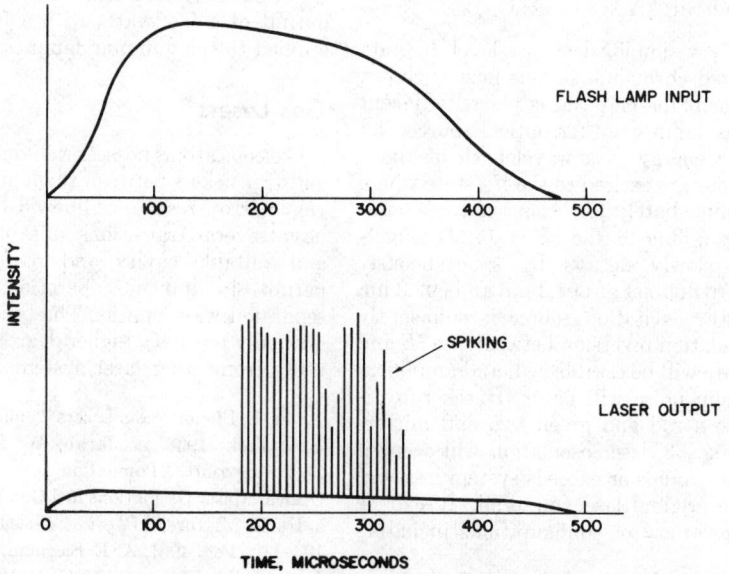

Fig. 31—Spiking behavior during a ruby laser pulse.

TABLE 5—PROPERTIES OF SEVERAL MAJOR SOLID-STATE LASER MATERIALS. *From W. V. Smith and P. P. Sorokin, "The Laser," McGraw-Hill Book Company, New York; 1966.*

Ion	Host	Laser Wavelength (μm)	Pulse Threshold (joules at 300 K)	Primary Absorption Wavelength (μm)	Laser Transition	Lifetime (ms)
Cr^{3+}	Al_2O_3	0.6943	2000	0.32–0.44 0.5–0.6	$^2E(\bar{E}) \rightarrow {}^4A_2$	3
U^{3+}	CaF_2	2.613	1200	0.9	$^4I_{11/2} \rightarrow {}^4I_{9/2}$	—
Nd^{3+}	Gd_2O_3	1.0789	9	0.57–0.6	$^4F_{3/2} \rightarrow {}^4I_{11/2}$	0.12
Nd^{3+}	$Y_3Al_5O_{12}$	1.0648	2	0.57–0.6	$^4F_{3/2} \rightarrow {}^4I_{11/2}$	0.20
Nd^{3+}	$CaWO_4$	1.0582	1.6	0.57–0.6	$^4F_{3/2} \rightarrow {}^4I_{11/2}$	—
Nd^{3+}	$CaWO_4$	1.3392	3.6	0.57–0.6	$^4F_{3/2} \rightarrow {}^4I_{11/2}$	—
Nd^{3+}	Glass	1.06	13	0.78–0.86	$^4F_{3/2} \rightarrow {}^4I_{11/2}$	—

Practical Laser Systems

Laser action has been observed in hundreds of different solids, liquids, gases, and semiconductors at thousands of discrete wavelengths ranging from 0.3 μm to 800 μm. The various laser systems employ strikingly different quantum processes, excitation methods, and mechanical construction and range in size from microminiature injection laser diodes to enormous gas dynamic lasers.

Solid-State Lasers

The most common solid-state lasers employ optical pumping of a host crystal or glass doped with an appropriate ion. Table 5 lists the properties of some of the more important optically-pumped solid-state laser materials. The best developed optically-pumped solid-state laser materials are ruby, Nd^{3+}:glass, and Nd^{3+}:YAG ($Y_3Al_5O_{12}$).

Ruby Laser[*]: A simplified energy-level diagram for triply ionized chromium in this host is shown in Fig. 32. When this material is placed adjacent to a xenon flashlamp or other optical source and irradiated with energy at a wavelength of about 550 nm, Cr^{3+} ions are excited to the $^4F_2$ state where they almost immediately lose some of the excitation energy by falling to the ^2E state. This metastable state slowly decays by spontaneously emitting a sharp doublet at 694.3 nm and 692.9 nm (300 K). If the excitation source is sufficiently intense, a population inversion between the ^2E and the ground state will be established, and amplification of quantum noise will occur. If the ruby is fabricated into a rod and given two end mirrors as shown in Fig. 33, laser oscillation will occur if the system gain equals or exceeds system losses.

Ruby is the original laser material. It can be operated in a variety of configurations including cw, conventional "long-pulse," and Q-switched. Very-high-energy ruby laser systems employ water or cryogenic cooling to remove thermal energy from the rod. Fig. 34 shows an apparatus used for obtaining cw laser action in ruby.

Nd^{3+}:Glass Laser: The operation of a Nd^{3+}:glass laser is similar to that of ruby with the major exceptions being that Nd^{3+} is a four-level system and the predominant radiative emission is 1.06 μm in the near infrared. Glass lasers are easier to fabricate than ruby lasers, but they are more sensitive to thermal and beam-induced damage.

Nd^{3+}: $Y_3Al_5O_{12}$ Laser: Operation of this system is similar to Nd^{3+}:glass, but lower thresholds are possible and YAG is a more robust host than glass. Quality YAG crystals are difficult to grow, particularly with usable diameters of greater than 5 cm, but Nd:YAG lasers can produce a cw power output of a few watts with a few hundred watts applied to the pumping lamp.

Gas Lasers[*]

Gaseous atoms possess well-defined energy levels, and transitions between them are characterized by very narrow resonance linewidths. Often a gas will have several transitions suitable for laser action, and suitable cavity and resonator design may permit simultaneous operation at several widely separated wavelengths. The beams from gas lasers normally possess a higher degree of coherence than those from other laser systems since they have a

[*] T. H. Maiman, "Stimulated Optical Radiation in Ruby," *Nature*, Vol. 187, pp. 493–494; August 1960.

[*] A. L. Bloom, "Gas Lasers," John Wiley & Sons, Inc., New York; 1968. A. Javan, W. B. Bennett, Jr., and D. R. Herriott, "Population Inversion and Continuous Optical Maser Oscillations in a Gas Discharge Containing a HeNe Mixture," *Physical Review Letters*, Vol. 6, pp. 106–110; Feb. 1961. A. E. Siegman, "An Introduction to Lasers and Masers," McGraw-Hill Book Company, New York; 1971.

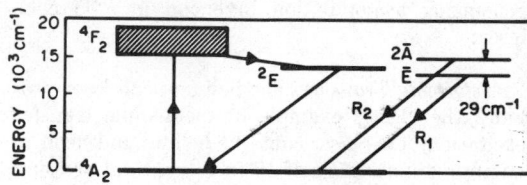

Fig. 32—Energy-level diagram of Cr^{3+} in corundum, showing pertinent processes. *From T. H. Maiman, "Stimulated Optical Radiation in Ruby," Nature, Vol. 187, Fig. 1; August 1960.*

very narrow linewidth and are easily operated in the lowest order mode (TEM_{00}). For example, a gas laser utilizing a plane-parallel Fabry-Perot resonating cavity may produce a beam having a divergence of 3×10^{-6} radian. A confocal resonator will yield a smaller spot size at the cavity aperture but a larger divergence (approximately 10^{-3} radian). This is not a limiting factor, since a simple telescope can be used to produce a considerably more parallel beam (with the additional benefit of transverse mode control by means of a suitable aperture stop within the telescope).

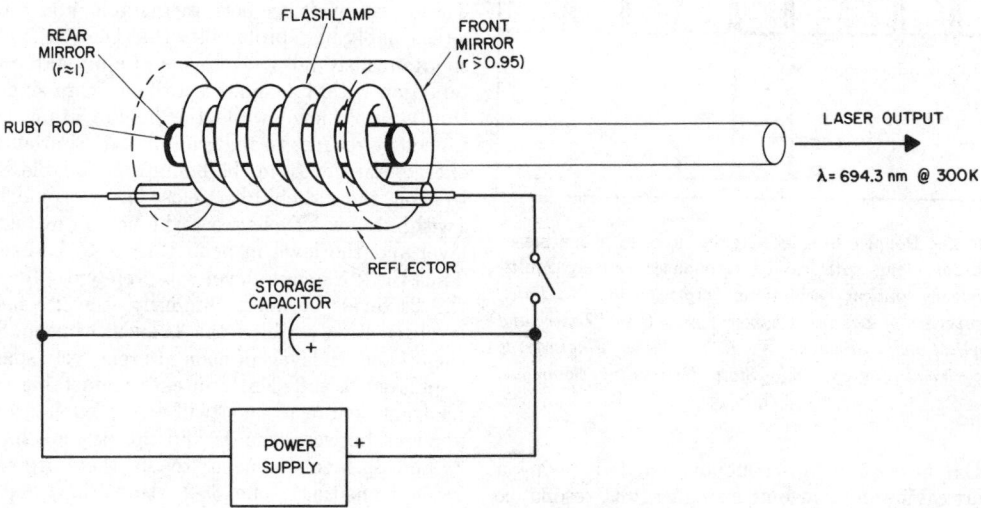

Fig. 33—Basic construction of a ruby laser.

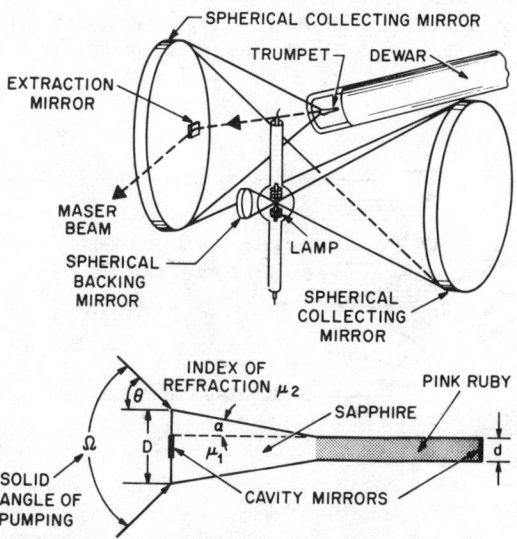

Fig. 34—Ruby continuous-wave laser setup. *From A. Yariv and J. P. Gordon, "The Laser," Proceedings of the IEEE, Vol. 51, No. 1, Fig. 9; January 1963.*

As noted earlier, the gain curve in a gaseous laser medium has a Doppler width given by (28). For a helium-neon laser oscillating at 1.153 μm, one obtains a $\Delta\nu_D$ of 800 MHz. The longitudinal separation of the cavity modes is given by $c/2L$, where L is the reflector separation. For a practical cavity length of about 1 meter, the axial mode separation is about 150 MHz; thus several of these resonances fall within the Doppler-broadened gain curve. Since, in general, there will be several modes above threshold for oscillation, the output from such a laser will generally have discrete frequencies separated by $c/2L$ (see Fig. 35). The theoretical limit of the frequency width, $\Delta\nu_{osc}$, is determined by the spontaneous emission into the oscillating mode or

$$\Delta\nu_{osc} = 8\pi h\nu(\Delta\nu_c)^2/p \qquad (33)$$

For an optical power output (p) of 10^{-3} W and a cavity resonance width of $\Delta\nu_c \simeq 0.3$ MHz, one obtains a theoretical linewidth of $\Delta\nu_{osc} \simeq 3\times10^{-3}$ Hz. This degree of line narrowing has not yet been observed, primarily because of experimental difficulties.

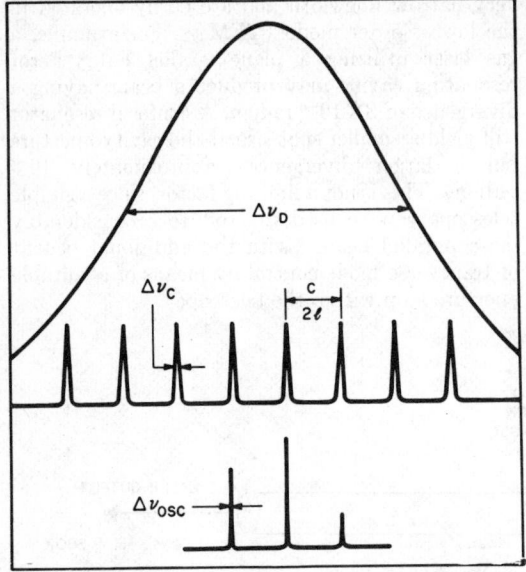

Fig. 35—Doppler broadened gain curve of a gas maser medium along with cavity resonances causing multifrequency optical oscillations. *From C. K. N. Patel, "Properties of Lasers," Chapter 2, Fig. 6 in "Lasers and Applications," edited by W. S. C. Chang, Engineering Experiment Station, Ohio State University; Columbus, Ohio.*

Gas lasers can be frequency tuned by using a short cavity and sweeping a single cavity resonance across the Doppler-broadened gain curve by varying the cavity spacing. Frequency tuning can also be obtained in some systems by incorporating a wavelength-selective device such as a prism or grating in the cavity.

There are three major quantum processes for generating a population inversion in a gaseous laser:

Metastable Transfer: The helium-neon laser provides the classic example of metastable transfer excitation. This laser contains helium and neon in an approximate ratio of 10 to 1 at a total pressure of about 1 torr. Fig. 36, a simplified energy-level diagram showing the pertinent states of both gases, reveals that the helium 2^3S and 2^1S levels have almost the same energies as the neon $3s$ and $2s$ levels. A dc or rf discharge established in the gas mixture will pump helium atoms to the 2^3S and 2^1S levels as a result of energetic electron impact. These two levels are both metastable, and there is a reasonably high probability that before the atoms spontaneously fall back to the ground state an interaction with a neon atom will occur, raising it to the $3s$ or $2s$ level while the helium atom simultaneously returns to the ground state. For efficient energy transfer from the helium metastable states to neon, there should be a close energy coincidence (within a few kT) between the helium metastable level and the level in neon that is to be excited. Thus the 2^3S helium level will preferentially excite the $2s$ levels of neon. Similarly, the 2^1S helium level selectively excites the $3s$ levels of neon. Since the $3p$ and $2p$ levels of neon are relatively sparsely populated, a sufficiently intense metastable transfer from helium to neon will create population inversions between the $2s$ and $2p$, $3s$ and $2p$, and $3s$ and $3p$ levels in neon. In all, there are eleven $2s$-$2p$ transitions, one $3s$-$2p$ transition, and one $3s$-$3p$ transition. These transitions and their respective wavelength equivalents are shown in Table 6. The $3s$-$3p$ transition is well suited for laser operation, since it is characterized by very high gain, but the 3.39 μm wavelength is inconvenient for many applications. The most commonly used

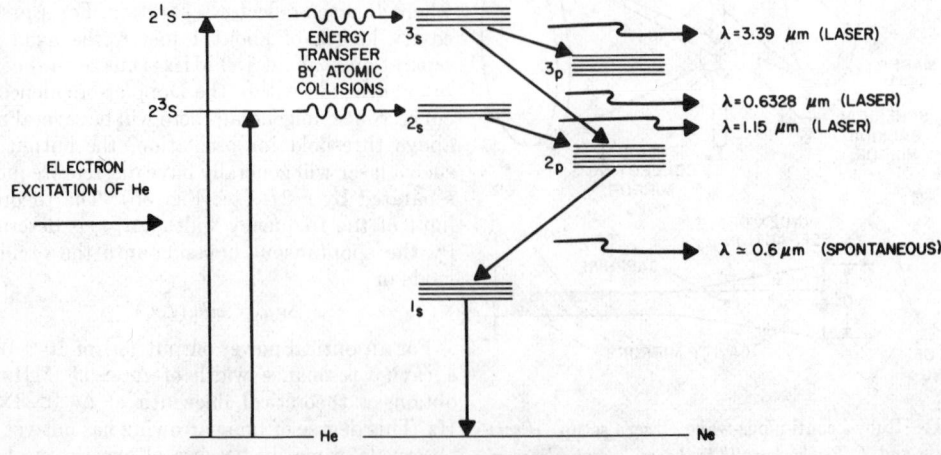

Fig. 36—Metastable transfer in the helium-neon laser.

TABLE 6—OBSERVED TRANSITIONS IN THE HENE LASER (PASCHEN NOTATION). *A. Yariv and J. P. Gordon, "The Laser," Proceedings of the IEEE, Vol. 51, No. 1, p. 15; January 1963.*

λ (micrometers)	Transition
1.0801	$2S_3-2p_7$
1.08475	$2S_2-2p_6$
1.11461	$2S_4-2p_8$
1.11806	$2S_5-2p_9$
1.13936	$2S_5-2p_8$
1.14123	$2S_2-2p_5$
1.15259	$2S_2-2p_4$
1.16047	$2S_2-2p_3$
1.16173	$2S_3-2p_5$
1.17700	$2S_2-2p_2$
1.19882	$2S_3-2p_2$
1.20696	$2S_5-2p_6$
1.52349	$2S_2-2p_1$
3.39	$3S-2p$
	Visible Transition
0.6328	$3S_2-2p_4$

helium-neon transition is the $3s-2p$, which yields a visible wavelength of 632.8 nm.

Electron Impact: Excitation in the noble-gas lasers (He, Ne, A, Kr, and Xe) takes place via inelastic collisions with the discharge electrons. The collision cross section for excitation of a given level is proportional to the dipole matrix element connecting the excited level with the ground state.

Consequently, the electron-collision process provides a discriminant mechanism for level excitation. The best developed of these lasers is argon (A), and continuous outputs of several watts are obtainable from commercial devices (as compared to up to 0.05 W from HeNe). The output wavelength of an argon laser can be adjusted to one or more discrete frequencies (e.g., 488 nm and 514 nm). These wavelengths happen to be useful in a great variety of applications.

Molecular Vibrational-Rotational Excitation: The laser systems described thus far are dependent upon atomic processes, but very efficient operation may be obtained in some gases via molecular vibrational-rotational excitation. The principal example is the carbon-dioxide laser. When an electrical discharge is passed through a mixture of CO_2 and such other gases as N_2, He, and H_2O (which serve as catalysts for increasing efficiency), the CO_2 molecules are excited to higher than normal vibrational-rotational states by electron collisions. This fulfills the requirement of a population inversion, and oscillation will occur if the gas is placed between two optical resonators. The principal wavelength emission is 10.6 μm in the infrared, a region where conventional optical materials are unsuitable. Though exotic optical materials must be used in order to couple the laser beam from the cavity to the outside, the longer wavelength simplifies optical alignment over those lasers operating at shorter wavelengths. Lasers using CO_2 are well developed, and a variety of practical configurations have evolved. Certain of these configurations are unique and will be mentioned later.

Fig. 37 shows the construction of a basic CO_2

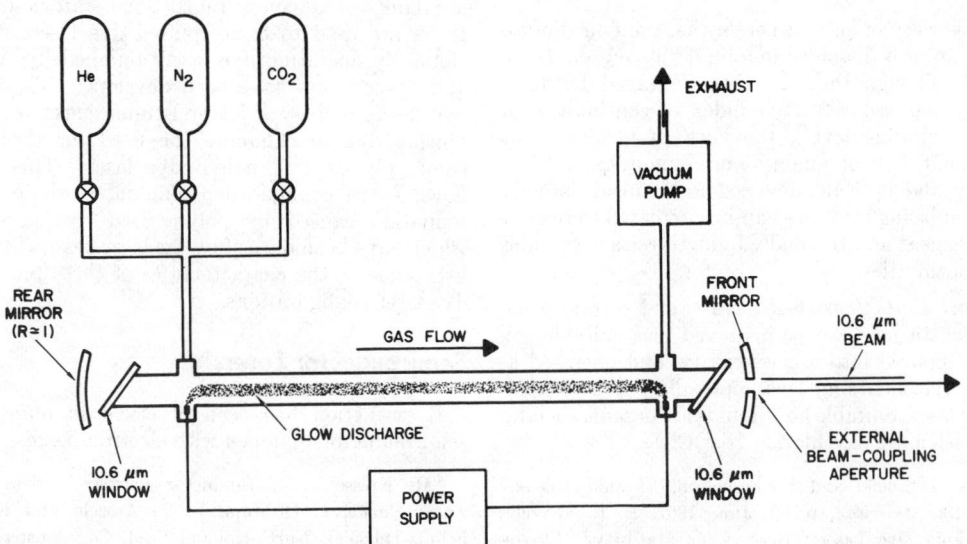

Fig. 37—Basic schematic for a CO_2 gas laser. (*After F. Mims, III, Light-Beam Communications, Fig. 3–19; Howard W. Sams & Co., Inc., 1975.*)

TABLE 7—CHARACTERISTICS OF THE PRIMARY ORGANIC DYE LASERS. *From B. B. Snavely, "Organic Dye Lasers: Headed Toward Maturity," Electro-Optical System Design, pp. 30–36; April 1975.*

Characteristic	Laser-Pumped Pulsed Dye Laser	Flash Lamp-Pumped Pulsed Dye Laser	CW Dye Laser
Peak Power	10^7–10^8 W	Several MW	2 W (approx)
Energy per Pulse	Several Joules or Less	Several Joules or Less	—
Pulse Width	5–30 ns	50 ns-1 ms	—
Mode-Locked Pulse Width	—	10 ps	1.5 ps
Tuning Range	340 nm–1200 nm (Total) 375 nm–550 nm (Representative Single Dye)	340 nm–700 nm (Total)	525 nm–710 nm (Total)
Linewidth	40 MHz (Tuned)	40 MHz (Tuned)	10 MHz (Tuned) (approx)
Efficiency	0.5* (approx)	10^{-2}† (approx)	0.35* (approx)

* Optical Output/Optical Input   † Optical Output/Electrical Input

laser. Since the energy extracted from a laser such as this is directly related to the volume of excited gas, considerable output power may be obtained by increasing the length of the discharge tube or, more conveniently, flowing the gas through the discharge tube at a high velocity. Continuous outputs in excess of 10^4 watts at power efficiencies of 20–30 percent are achievable. These power levels, which can be quite hazardous, are useful in a variety of applications (e.g., cutting, drilling, welding, etc.).

Liquid Lasers*

A variety of lasers wherein the lasing medium is a liquid or is dissolved in a liquid have been developed. Though they are disadvantaged by thermally induced refractive-index discontinuities at high pumping levels, they are characterized by excellent optical quality and immunity to thermally and optically induced mechanical damage. Liquid lasing mediums can be circulated to remove excess heat and thermally induced refractive-index discontinuities.

Rare-Earth Ions: Eu^{3+}, Nd^{3+}. and certain other rare-earth ions can be dissolved in a suitable solvent poured into a glass cavity and operated as a conventional four-level, optically-pumped, solid-state laser. Suitable host liquids are organic chelates and selenium oxychloride ($SeOCl_2$).

* A. Lempicki and H. Samelson, "Liquid Lasers," *Scientific American*, p. 80; June 1967. B. B. Snavely, "Organic Dye Lasers: Headed for Maturity," *Electro-Optical System Design*, pp. 30–36; April 1973. R. C. Cunningham, "Dye Lasers Today—And Tomorrow?" *Electro-Optical System Design*, pp. 13–18; Dec. 1974.

Organic Dyes: A diverse family of organic-dye lasers has been developed. The most efficient laser dyes are conjugated carbon atom chains having alternating single and double bonds, the length of which largely determines the fluorescent wavelength. The conjugated chain has broadly dispersed energy levels, which make frequency tuning over a typical range of 100 nm or more feasible for a given dye laser by varying the wavelength reflectance of the output resonator. Dye lasers are characterized by very brief lifetimes, which means very rapid, high-power optical excitation pulses are required to achieve a population inversion. Special arc lamps and conventional solid-state and gas lasers are used to excite pulsed dye lasers. Continuously operating dye lasers pumped by A, N, and other lasers have been developed. These cw dye lasers employ a jet or laminar-sheet of free-flowing dye to eliminate the need for the cells commonly used in pulsed dye lasers. This free-flowing dye configuration eliminates window attenuation caused by polymerized dye particles which have been burned by the laser beam. Table 7 lists some of the characteristics of three practical dye laser configurations.

Semiconductor Lasers*

Laser action has been observed in numerous semiconductors pumped with electron beams, opti-

* H. Kressel, "Semiconductor Lasers: Devices," in *Laser Handbook*, (Editors: F. T. Arecchi and E. O. Schulz-DuBois), North-Holland Publ. Co., Amsterdam; 1972. R. Glicksman, "Recent Progress in Injection Lasers," *Proceedings of the Technical Program of the Electro-Optical System Design Conference*, 1970.

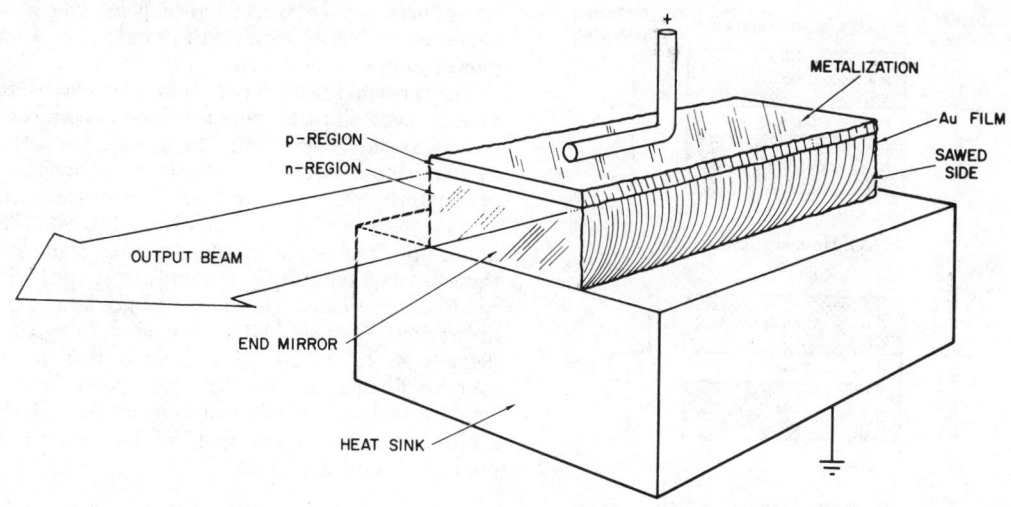

Fig. 38—Construction of typical injection laser. (*After R. Campbell and F. Mims, III, Semiconductor Diode Lasers, Fig. 1-8; Howard W. Sams & Co., Inc., 1972.*)

cal beams, and direct-current injection into a pn junction. The most important semiconductor lasers are the pn-junction injection type.

Injection Lasers: Minority carriers recombine at a forward biased pn junction and, upon resuming equilibrium, remove the energy which separates the carriers (equal to or greater then the band gap) by emitting a photon, a series of phonons (thermal inducing lattice vibrations), or both. The radiative process is exceptionally efficient (quantum efficiency approximately unity) in some semiconductors, the most notable being gallium arsenide (GaAs) and its alloys. Nonlaser electroluminescent diodes made from GaAs ($\lambda = 900$ nm), GaAsP ($\lambda = 660$ nm), and GaP ($\lambda = 550$ nm) are routinely employed in a great variety of optoelectronic applications including communications, indicators, numeric displays, and isolators. The external efficiency of these diodes is considerably less than the internal efficiency due to losses caused by internal absorption, internal reflection, contact shadows, and other factors.

An efficient injection laser requires a semiconductor in which radiative recombination predominates over other recombination byproducts, and this is best accomplished with a direct band gap material such as GaAs or GaAlAs. Fig. 38 shows the construction of a typical homojunction injection laser. Early homojunction lasers were made by diffusing an acceptor, usually Zn, into n-type GaAs having a carrier concentration in the 10^{17}–10^{18} cm^{-3} range. Considerable improvement in both operating efficiency and usable life are made possible by fabricating the junction using a carefully controlled epitaxial process wherein a p-type layer is grown upon a smoothly polished n-type sub-strate. The hole concentration in the p-type region is typically 1–3×10^{19} cm^{-3}, and the electron concentration in the n-type region is typically 2–4×10^{18} cm^{-3}. The wafer is mechanically cleaved along parallel and facing crystallographic planes to produce bars which are sawed into individual laser diodes. The high index of refraction of GaAs (approximately 3.6) causes the cleaved facets to have a reflectance of about 35 percent each, and the high gain of the material permits lasing to occur without adding a reflective coating. However, it is common practice to coat the rear facet with a reflective gold film to permit all the laser radiation to be extracted from one end of the diode.

The homojunction laser requires a very high threshold current density (greater than 25,000 A/cm^2 at 300 K), but a variety of modified junction geometries have been developed which significantly lower the threshold current density. The single-heterojunction (SH) and double-heterojunction (DH) structures (see Fig. 39) are the most important of these modified structures. Lasers of the SH type incorporate an (AlGa)As-GaAs-GaAs junction geometry, and the refractive index discontinuity at the (AlGa)As-GaAs interface provides a degree of electron and photon confinement which reduces J_{th} to about 8000 A/cm^2. Lasers of the DH type incorporate an (AlGa)As-GaAs-(AlGa)As junction geometry, and the refractive index discontinuities at the (AlGa)As-GaAs and GaAs-(AlGa)As interfaces provide an even higher degree of electron and photon confinement with a significant lowering of J_{th} (less than 1500 A/cm^2 at 300 K). By isolating all the active region from the uppermost metallization layer except for a narrow (approximately 1 μm) stripe perpendicular to the end facets, the thermal

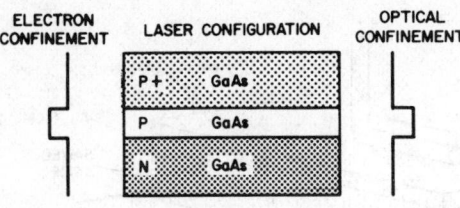

(A) Homostructure.

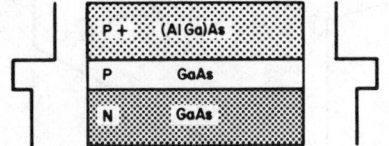

(B) Single heterostructure.

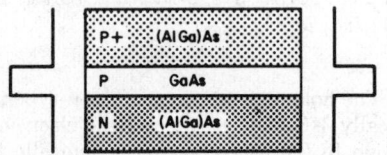

(C) Double heterostructure.

Fig. 39—The three major injection laser structures and their respective refractive index discontinuity induced electron and optical confinement. (*After R. Campbell and F. Mims, III, Semiconductor Diode Lasers, Fig. 2–9; Howard W. Sams & Co., Inc., 1972.*)

byproducts can be reduced to a point where cw operation at 300 K with a power output of a few tens of milliwatts is feasible.

The threshold behavior of an injection laser provides a classic illustration of the onset of laser oscillations, as can be seen in Fig. 40. The optical output from injection lasers is not nearly as directional or spectrally pure as that from most other lasers, as shown by a typical example in Fig. 41. The miniature dimensions of the emitting regions of these lasers (typically a few mils $\times$ 1 μm) give rise to diffraction effects which produce large beam divergences (typical half angles range from a few degrees $\times$ 25° to 15° $\times$ 30°, depending on the junction dimensions; see Fig. 42). Special fabrication geometries or external mirrors and mode-limiting apertures can be used to enhance the beam quality of injection lasers.

Metal-Vapor Lasers*

The quantum aspects of metal-vapor lasers closely resemble those of gas lasers which achieve excitation by metastable transfer. In a typical metal-vapor laser, for example, energy stored in the metastable levels of helium is transferred to atoms of vapors of Cs, Pb, Se, Cd, or any of a variety of other metals. Since these metals have a number of intermediate energy states, a population inversion can be readily established between the terminal state and one or more intermediate states. Metal-vapor lasers require special discharge

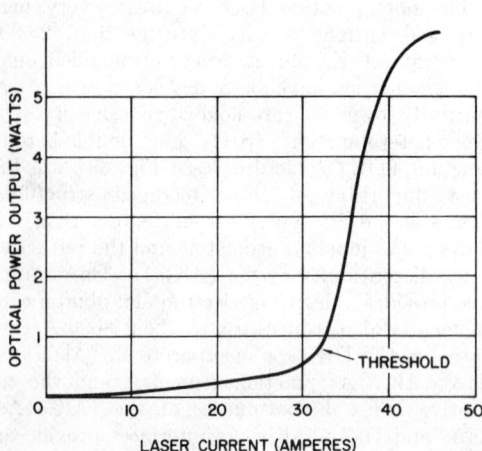

Fig. 40—Optical power output as a function of input current into a typical SH injection laser. Note the abrupt threshold region. (*After R. Campbell and F. Mims, III, Semiconductor Diode Lasers, Fig. 2–15; Howard W. Sams & Co., Inc., 1972.*)

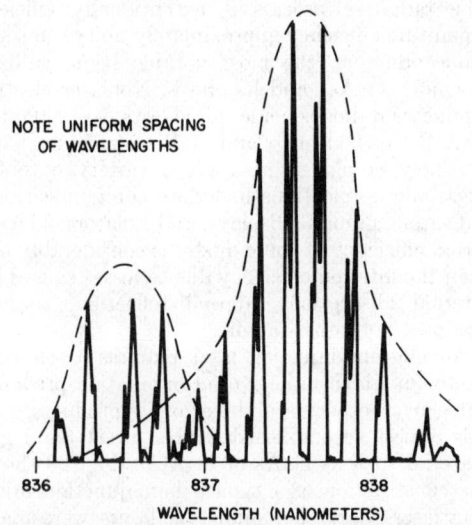

Fig. 41—Axial modes of a typical DH stripe geometry injection laser. (*After R. Campbell and F. Mims, III, Semiconductor Diode Lasers, Fig. 2–20; Howard W. Sams & Co., Inc., 1972.*)

* W. T. Silfvast, "Metal-Vapor Lasers," *Scientific American,*" pp. 88–97; Feb. 1973.

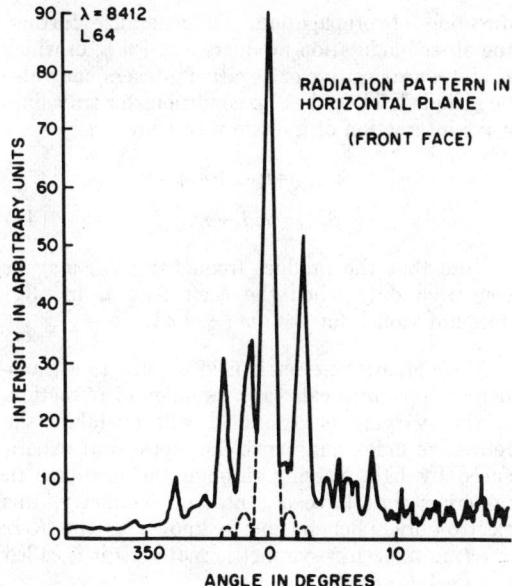

Fig. 42—Radiation pattern for one of the emission wavelengths measured in the junction plane. *From Quantum Electronics: Proceedings of the Third International Congress, Dunod Ed. Paris and Columbia University Press, New York, Vol. 2, Fig. 4, p. 1886; 1964.*

tubes for uniform distribution of the vapor and heating of the constituent metal to its vapor pressure. Metal-vapor lasers have been operated at a wide range of wavelengths including the ultraviolet, and many metals are suitable for this type of laser system.

Chemical Lasers*

A variety of chemically pumped lasers have been fabricated wherein a chemical reaction produced optical radiation which was used to pump an appropriate laser material. True chemical lasers are more elegant in that the chemical reaction becomes an inherent part of the laser system. One of the most successful chemical lasers thus far utilizes a fuel mixture of H_2 or D_2 and SF_6 with a reaction initiated by a transverse electrical discharge. Peak pulse powers of 25 MW or more have been achieved (multiline). Conventional chemical lasers produce highly toxic byproducts which are expelled into the atmosphere. An HF/DF chemical laser which recirculates its exhaust and has a nominal efficiency of about 2 percent has been developed.

Other Lasers

A great many exotic laser systems, many of which incorporate technologies described thus far,

* R. J. Freiberg, P. P. Chenausky, and D. W. Fradin, "Self-Contained Recirculating Chemical Laser," *Proceedings of the 1974 International Electron Devices Meeting.*

exist. Among these systems are high-power gas dynamic lasers (cf. CO_2 molecular vibrational-rotational lasers) which produce cw beams with a multimode power output of greater than 1×10^5 watts. There is abundant laser literature, and new devices and applications are reported on a frequent basis.

Modulation and Demodulation of Laser Beams

Several laser systems are highly compatible with optical communication technology. A number of practical modulation and demodulation methods for laser beams have been developed, and this topic is covered in the chapter on optical communications.

Nonlinear Optical Effects Produced by Laser Beams*

The unique optical nature of the beam produced by many lasers is responsible for a variety of interesting and unusual nonlinear optical effects.

Traveling-Wave Interaction: The polarization of the nonlinear medium may be expanded in powers of E.

$$P = a_1E + a_2E^2 + a_3E^3 + a_4E^4 + \cdots + a_kE^k + \cdots \tag{34}$$

In general, the coefficients a_k depend on the frequencies present in E and P. In centrosymmetric materials, P is an odd function of E, $P(E) = -P(-E)$, and all even coefficients must vanish. In noncentrosymmetric materials, P is neither an odd nor an even function of E. Thus, only materials that lack a center of symmetry can generate even harmonics, or sum and difference frequencies, when two signals are present.

Consider the case of two traveling waves passing through a dielectric medium of length L as shown in Fig. 43A. The exciting field, in complex form, is

$$E(t, z) = E_a \exp[j(\omega_a t - \beta_a z)] + E_b \exp[j(\omega_b t - \beta_b z)] \tag{35}$$

where the phase velocities of the two waves in the medium are given by

$$v_a = \omega_a/\beta_a \quad \text{and} \quad v_b = \omega_b/\beta_b \tag{36}$$

respectively. For the sake of illustration, concen-

* P. A. Franken and J. F. Ward, "Optical Harmonics and Nonlinear Phenomena," *Reviews of Modern Physics*, Vol. 35, pp. 23–29; January 1963. I. P. Kaminow, "Parametric Principles in Optics," *IEEE Spectrum*, Vol. 2, pp. 35–43; April 1965.

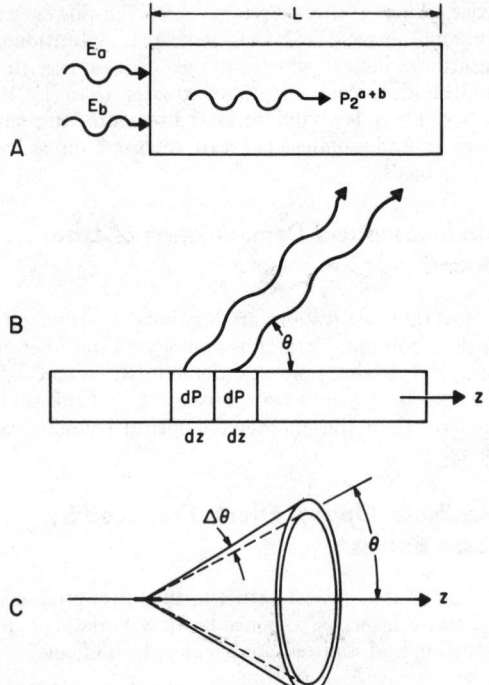

Fig. 43—(A) Polarization wave P^{a+b} at the sum frequency is produced by two incident waves E_a and E_b. (B) Energy radiated by differential elements dz adds in phase at angle θ with beam width $\Delta\theta$. (C) Cone of radiation produced. *I. P. Kaminow, "Parametric Principles in Optics," IEEE Spectrum, Vol. 2, Fig. 3, p. 37; April 1965.*

trate on only one term in the nonlinear response, a_2E^2, and only one of the modulation products it produces, the sum frequency

$$\omega_{a+b} \equiv \omega_a + \omega_b. \qquad (37)$$

The corresponding complex polarization obtained by substituting (35) into (34) is

$$P_2{}^{a+b}(t,z) = a_2 E_a E_b \exp\{ j[(\omega_a+\omega_b)t - (\beta_a+\beta_b)z]\} \qquad (38)$$

where the actual polarization is the real part of (38).

To collimate appreciable energy at ω_{a+b}, it is necessary to satisfy the condition

$$|\cos\theta - 1| < \lambda/2L \qquad (39)$$

or

$$|\beta_a+\beta_b-\beta_{a+b}| < \pi/L \qquad (40)$$

where it has been assumed that $\lambda/L \ll 1$.

Equation (40) holds for two interacting waves, E_a and E_b, traveling in the same direction. If the two waves are not parallel, then (40) may be written in vector form with β_a and β_b directed normal to the wavefronts of E_a and E_b in the

direction of propagation. To generalize further, the other modulation products $m\omega_a + n\omega_b$, in which m and n are positive or negative integers, may also be included. Then the ω, β conditions for traveling-wave interaction of the two waves are

$$\omega_{m+n} = m\omega_a + n\omega_b$$
$$\beta_{m+n} = m\beta_a + n\beta_b. \qquad (41)$$

Note that the product frequency ω_{m+n} may be generated only when the coefficient a_k in (34) does not vanish for $k = |m| + |n|$.

Modulation: An electric field applied to a transparent substance can vary its index of refraction. A time-varying electric field will modulate the refractive index and hence the phase shift experienced by light passing through the medium. In materials that have a center of symmetry, this electro-optic phenomenon is known as the *Kerr effect*; in noncentrosymmetric materials it is called the *Pockels effect*.

Light modulation by these electro-optic effects may be viewed as parametric mixing of an optical field and a radio-frequency modulating field. Taking ω_a to be the optical carrier frequency and ω_b the modulating frequency, the first sidebands occurring in connection with the Kerr effect are at $\omega_a \pm 2\omega_b$, corresponding to $m=1$, $n=\pm2$ in (41), since $a_2=0$ and $a_3\neq0$. For modulation by the Pockels effect, however, the first sidebands appear at $\omega_a \pm \omega_b$ and are displaced from the carrier by the fundamental modulating frequency rather than the second harmonic. The process corresponds to $m=1$, $n=\pm1$, with $a_2\neq0$.

The crystal most widely employed for its Pockels effect is potassium dihydrogen phosphate (KDP).

The efficiency of the electro-optic effect is not reduced appreciably when the modulating field and optical propagation vector are not collinear, provided the light beam is in a crystallographic (110) plane and the modulating field is along the optic axis in the (001) direction. One may conceive of two types of modulator. In the first, a pure phase modulator, the incident light would be polarized in a principal plane; for example, along the (110) direction normal to the plane of incidence. The emergent light would retain its original polarization and amplitude but would be phase modulated. In the other type, the incident light would have components of polarization in both principal planes. The emergent light beam, when passed through an analyzer, will be intensity modulated.

To reduce the field strength and power required to produce a given modulation index, it is desirable that the interaction length L be large. Then, for optical and radio-frequency waves traveling in the same direction, the radio-frequency modulating and optical carrier velocities must be matched. With a radio-frequency dielectric constant of 20 and an optical dielectric constant of 2.25 in KDP, one method for achieving this match is to partly fill the cross section of a parallel-plate transmission line as shown in Fig. 44. At low enough frequencies, the effective radio-frequency dielectric constant of the structure will be determined by the ratio of air to KDP in the cross section of the transmission line as indicated in the figure.

Harmonic Generation, Optical Mixing, Heterodyning, and Rectification: Consider now the case in which ω_a and ω_b are both optical frequencies. The sum and difference frequencies may be produced by a noncentrosymmetric crystal $(a_2 \neq 0)$. Both a sum frequency (mixing) and a difference frequency (heterodyning) have been observed for $\omega_a \neq \omega_b$. For $\omega_a = \omega_b$, the sum frequency becomes the second harmonic and the difference frequency a direct-current polarization of the medium. Both effects have been observed. Third-harmonic generation of light has also been observed using a centrosymmetric crystal $(a_2 = 0, a_3 \neq 0)$.

Much of this work was accomplished using high-power pulsed optical masers focused inside the crystals. With the resultant short interaction length and high field strength, it was not essential that the β conditions be satisfied precisely. However, when these conditions are satisfied, the interaction becomes much more efficient.

Parametric Amplification by the Raman Effect*: When mechanical atomic vibrations occur, they influence the electronic polarizability (or refractive

* E. Garmire, F. Pandarese, and C. H. Townes, "Coherently Driven Molecular Vibrations and Light Modulation," *Physical Review Letters*, Vol. 11, pp. 160–163; 15 August 1963. H. J. Zeiger, P. E. Tannenwald, S. Kern, and R. Herendeen, "Two-Step Raman Scattering in Nitrobenzene," *Physical Review Letters*, Vol. 11, pp. 419–422; 1 November 1963. R. Chiao and B. P. Stoicheff, "Angular Dependence of Maser-Stimulated Raman Radiation in Calcite," *Physical Review Letters*, Vol. 12, p. 290; 16 March 1964. L. W. Davis, S. L. McCall, and A. P. Rodgers, "Raman Maser Study of Optical Difference Frequency Production," *Journal of Applied Physics*, Vol. 35, pp. 2289–2290; August 1964. N. Bloembergen and Y. R. Shen, "Coupling Between Vibrations and Light Waves in Raman Laser Media," *Physical Review Letters*, Vol. 12, p. 504; 4 May 1964. N. Bloembergen and Y. R. Shen, "Multimode Effects in Stimulated Raman Emission," *Physical Review Letters*, Vol. 13, p. 720; 4 December 1964.

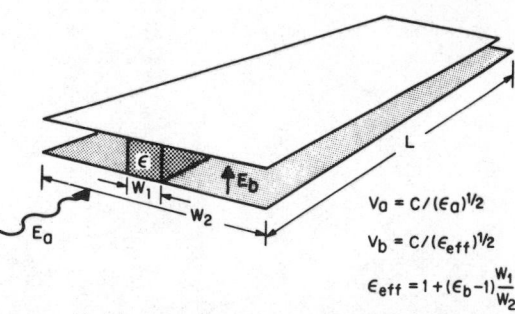

Fig. 44—Parallel-plate structure containing a KDP rod with optical dielectric constant ϵ_a and radio-frequency dielectric constant ϵ_b. *I. P. Kaminow, "Parametric Principles in Optics," IEEE Spectrum, Vol. 2, Fig. 5, p. 40; April 1965.*

index) of the medium and thereby modulate transmitted light at the vibration frequency ω_ν. If the atomic displacement in the normal coordinate corresponding to the ω_ν vibrational mode is called x_ν, then a term $b_\nu E x_\nu$ must be added to the expansion for P in (34) to account for the vibrational-optical interaction. The coefficient b_ν is very large only when x_ν is driven near the resonant frequency ω_ν. If E is an optical field at ω_p and x_ν is oscillating at ω_ν, then the transmitted light will have components at ω_+ and ω_-. Sidebands produced by atomic vibrations in this fashion are called *Raman frequencies*. Those vibrational modes with symmetry such that $b_\nu \neq 0$ are said to be Raman active. (A center of symmetry or lack of it in a substance does not enter directly into determining Raman activity of a vibrational mode.) For historical reasons, the lower sideband ω_- is called the Stokes frequency and the upper one, ω_+, the anti-Stokes frequency.

The response function of a Raman-active medium has the form

$$P = a_1 E + a_2 E^2 + a_3 E^3 + \cdots + b_\nu E x_\nu + \cdots \quad (42)$$

where, as before, a_2 vanishes when a center of symmetry is present. The a_2 term is capable of mixing two electromagnetic waves (optical or radio-frequency) to produce a third, the a_3 term mixes three electromagnetic waves to produce a fourth electromagnetic wave, and the b_ν term mixes a vibrational wave and an electromagnetic wave to produce another electromagnetic wave.

CHAPTER 40
DIGITAL COMPUTERS

INTRODUCTION

A digital computer is a complex assembly of electronic logic circuits and memory devices which, when instructed by external commands to execute internally stored instructions, performs calculations or compiles, analyzes, and correlates data. Because of this, a digital computer might better be termed a digital processor, but through convention the term computer is more commonly used. Digital computers are highly complex devices which may incorporate considerable support equipment such as mass data storage and retrieval systems and a great variety of input and output devices. Since all operations of a computer are ultimately based upon number relationships, most notably patterns of binary bits, this chapter begins with a treatment of number systems and their relationship to digital computing systems.

NUMBERS

Numbers describe quantities and may be complex or describe a continuum. The following discussion concentrates on the subsets of numbers and on the manipulations of the sets of numbers that are commonly associated with the internal arithmetic of digital computers.

Natural numbers are the numbers 1, 2, 3, $\cdots$. Integers are the numbers 0, 1, 2, 3, $\cdots$; -1, -2, -3, $\cdots$. Proper fractions are real numbers whose magnitude is less than one.

Number System Base

Numbers are commonly represented by means of a base (or radix) b and an ordered set of symbols A_m, A_{m-1}, A_{m-2}, $\cdots$; A_1, A_0, A_{-1}, $\cdots$, A_{-r}. The base b is an integer larger than 1 and each of the symbols A_k is one of the integers 0, 1, 2, $\cdots$, $b-1$. The relation between a number N, the symbols A_k, and the base b is expressed by

$$N = A_m b^m + A_{m-1} b^{m-1} + \cdots + A_1 b$$
$$+ A_0 + A_{-1} b^{-1} + \cdots + A_{-r} b^{-r}$$
$$= \sum_{k=-r}^{m} A_k b^k.$$

Conventionally, digits of a number are written in decreasing order of significance from left to right. A decimal point or radix point is usually placed between A_0 and A_{-1}. If no decimal point is used, the rightmost symbol is A_0. The most common bases are *decimal* (base 10), *binary* (base 2), *octal* (base 8), *hexadecimal* (base 16), and *duodecimal* (base 12). In writing binary and octal numbers, we conventionally use the appropriate subset of the 10 decimal digits; in writing hexadecimal and duodecimal numbers, alphabetical letters are used in addition to the 10 decimal digits. A common convention for hexadecimal is to use the letters A, B, C, D, E, F to represent the values 10, 11, 12, 13, 14, 15, respectively.

Sometimes a subscript is added to a number to indicate the base. Thus for example, the decimal number 71 may be written as 71_{10}, and its octal equivalent as 107_8. Some examples of numbers follow. Systems are given in Table 1.

$$1756_{10} = (1 \times 10^3) + (7 \times 10^2) + (5 \times 10^1)$$
$$+ (6 \times 10^0).$$

$$2.19_{10} = (2 \times 10^0) + (1 \times 10^{-1}) + (9 \times 10^{-2}).$$

$$603.17_8 = (6 \times 8^2) + (0 \times 8^1) + (3 \times 8^0)$$
$$+ (1 \times 8^{-1}) + (7 \times 8^{-2}) = 387_{10} + 15/64.$$

$$1756_8 = (1 \times 8^3) + (7 \times 8^2) + (5 \times 8^1) + (6 \times 8^0)$$
$$= 1006_{10}.$$

$$10110_2 = (1 \times 2^4) + (0 \times 2^3) + (1 \times 2^2)$$
$$+ (1 \times 2^1) + (0 \times 2^0) = 22_{10}.$$

$$101.10_2 = (1 \times 2^2) + (0 \times 2^1) + (1 \times 2^0)$$
$$+ (1 \times 2^{-1}) + (0 \times 2^{-2}) = 5_{10} + \tfrac{1}{2}.$$

TABLE 1—EXPRESSION OF A NUMBER IN DIFFERENT CODES.

System	Code
Decimal	347
Binary (base 2)	101011011
Binary Coded Decimal	0011 0100 0111
Binary Coded Octal	101 011 011
Octal (base 8)	533

CONVERSION BETWEEN NUMBER SYSTEMS

Often there is a need to express a number in one base, given the number expressed in another base.

Octal-to-Binary Conversion

A binary digit will henceforth be called a bit. To convert from octal to binary, replace each octal digit by the equivalent three bits. To convert from binary to octal, group the bits into groups of three starting from the radix point and convert each group of three bits into the equivalent octal digit. See Table 2. Thus

$$3051.126_8 = 11\ 000\ 101\ 001.\ 001\ 010\ 110$$

$$3\quad 0\quad 5\quad 1.\ 1\quad 2\quad 6$$

The conversion from binary to hexadecimal and back is similar to the conversions involving binary and octal, except that the hexadecimal conversions involve groups of four bits instead of groups of three bits.

Arbitrary Conversion

The conversion problem has a natural division between integers and fractions. Each base can describe any integer in a finite number of digits, but some fractions—irrational fractions—cannot be described in any base by a finite number of digits. In fact, it may be impossible to convert a fraction expressed in finite form in one base, into a finite form in some other given base without error.

The following describes three of the many algorithms for converting a number from one base to another. In principle, each of these three algorithms can apply to the conversion from any base to any other base. In practice, the choice of a conversion algorithm depends on whether the arithmetic is performed in the base in which the number is originally expressed or whether the arithmetic is performed in the base into which the number is

to be converted. The first procedure applies to conversion into the base in which calculations are made; the second and third procedures apply to conversions from the base in which calculations are made.

First Conversion Procedure

This conversion procedure applies the definition of the value of a number expressed in some base. Powers of the base being converted from are useful. See powers of 2 and 8 in Table 3. The following example illustrates the conversion from octal to decimal of the number 347.265_8 and two ways of organizing the calculation.

$$347.265_8 = (3 \times 8^2) + (4 \times 8^1) + (7 \times 8^0)$$
$$+ (2 \times 8^{-1}) + (6 \times 8^{-2}) + (5 \times 8^{-3})$$
$$= \{[(3 \times 8) + 4] \times 8\} + 7$$
$$+ \{\{\{[(5 \times \tfrac{1}{8}) + 6] \times \tfrac{1}{8}\} + 2\} \times \tfrac{1}{8}\}$$
$$= 231.353\ 515\ 625_{10}.$$

Second Conversion Procedure

This procedure consists of a sequence of steps which get better and better approximations in the new base to the number. In each step we find the integral power of the new base b^m such that the difference between the latest approximation and the number either equals b^m or lies between b^m and b^{m+1}. The new approximation is the sum of the latest approximation and b^m. In practice, we just calculate the differences as shown in the example below of the conversion of the decimal number 6.13 to binary. A table of powers of 2 such as Table 3 is necessary for this calculation. The first

TABLE 2—EQUIVALENT NUMERICAL REPRESENTATIONS (BINARY, OCTAL, AND EXCESS 3).

Decimal	Binary	Octal	Binary Coded Decimal Excess 3	Reflected Binary Excess 3
0	0000	0	0011	0010
1	0001	1	0100	0110
2	0010	2	0101	0111
3	0011	3	0110	0101
4	0100	4	0111	0100
5	0101	5	1000	1100
6	0110	6	1001	1101
7	0111	7	1010	1111
8	1000	10	1011	1110
9	1001	11	1100	1010

term in the approximation to 6.13 is $2^2=4$ which gives a difference of 2.13. The second term is $2^1=2$ as it is less than 2.13 while 2^2 is more than 2.13. The next approximations are $2^{-3}=0.125$ and $2^{-8}=0.003\ 906\ 25$. If this approximation is good enough, we can stop with 6.13_{10} approximated by

$$2^2+2^1+2^{-3}+2^{-8}=110.00100001_2.$$

For conversions to bases other than binary, a given integral power may appear a number of times. Thus if b^m appears k times this means that the digit corresponding to b^m in the answer has the value k.

$$
\begin{array}{r}
6.13 \\
-4 \\
\hline
2.13 \\
-2 \\
\hline
0.13 \\
-0.125 \\
\hline
0.005 \\
-0.00390625 \\
\hline
0.00109375 \ \text{(error)}.
\end{array}
$$

Third Conversion Procedure

This procedure converts integers in one way and proper fractions in another way. A number such as 6.13, which is neither an integer nor a proper fraction, will be considered as the sum of an integer and a proper fraction which are converted separately and then combined.

Integers are converted by repeated division and extraction of the remainder. In the first step of the process, the number is divided by the new base b to get an integral quotient and an integer remainder less than b. This first remainder is the digit to the left of the decimal point in the new base. In each succeeding step the latest quotient is divided by the new base to get an integral quotient and a remainder less than the new base b. The remainders taken in order form the converted number. The example below shows the conversion of 637_{10} into 1175_8 and 13_{10} into 1101_2. The first four steps of the conversion of 637_{10} into octal have quotients 79, 9, 1, 0, and remainders 5, 7, 1, 1, respectively.

$$
\begin{array}{ll}
0+1 & 0+1 \\
8\,\overline{|1}+1 & 2\,\overline{|1}+1 \\
8\,\overline{|9}+7 & 2\,\overline{|3}+0 \\
8\,\overline{|79}+5 & 2\,\overline{|6}+1 \\
8\,\overline{|637} & 2\,\overline{|13}
\end{array}
$$

A proper fraction is converted into a new base by repeated multiplication. One starts with the

fraction to be converted. In each step, the latest fraction is multiplied by the new base b. The integer part of the resulting number is extracted, leaving a new proper fraction. The integer extracted at the nth step is the coefficient of b^{-n}. The examples below show the conversion of 0.13_{10} into the octal number $0.1024\cdots$, and the binary number $0.00100001\cdots$

$$
\begin{array}{lll}
0.13 & 0.13 & \rightarrow 0.08 \\
\times 8 & \times 2 & \times 2 \\
\hline
1.04 & 0.26 & 0.16 \\
\times 8 & \times 2 & \times 2 \\
\hline
0.32 & 0.52 & 0.32 \\
\times 8 & \times 2 & \times 2 \\
\hline
2.56 & 1.04 & 0.64 \\
\times 8 & \times 2 & \times 2 \\
\hline
4.48 & 0.08 & 1.28
\end{array}
$$

Binary Coding

A binary code is a rule for representing a set of symbols by a second set of symbols, where each symbol of the second set consists of a number of binary digits (i.e., each digit is either a one or a zero). It is common practice to represent decimal numbers or the alphabet in some binary code, because the physical apparatus of digital computers is inherently binary or works best in binary fashion. Of the many possible binary codes for a decimal number, one common code expresses the decimal number as a base-2 number as described above. However, because of the complexity of pure base-2 encoding and the problems of encoding fractions, there has arisen another set of binary codes which encodes each input decimal digit separately into binary. These codes are used in binary coded decimal (BCD) computers whose number representation and arithmetic are decimal in nature and in the input and output devices of binary computers. Since there are more than 2^3 possibilities for a decimal digit, a binary code for representing decimal digits requires a minimum of four bits.

Some binary codes for decimal digits use more than four bits for reasons of hardware simplicity or error correction and detection. The simplest of these codes is the one which encodes decimal digit k into ten bits with the kth bit equal to one and the other bits equal to zero. There are a number of binary codes whose input symbols include the integers zero through nine, the alphabet, and other symbols. These codes are widely used in digital computer manipulations, in communications, punched cards, etc. and are described in the section on input and output equipment.

TABLE 3—POWERS OF 2, 8, AND 16.

n	m	k	$2^n=8^m=16^k$	$2^{-n}=8^{-m}=16^{-k}$
0	0	0	1	1
1			2	0.5
2			4	0.25
3	1		8	0.125
4		1	16	0.062 5
5			32	0.031 25
6	2		64	0.015 625
7			128	0.007 812 5
8		2	256	0.003 906 25
9	3		512	0.001 953 125
10			1 024	0.000 976 562 5
11			2 048	$0.^3$ 488 281 25
12	4	3	4 096	$0.^3$ 244 140 625
13			8 192	$0.^3$ 122 070 312 5
14			16 384	$0.^3$ 061 035 156 25
15	5		32 768	$0.^3$ 030 517 578 125
16		4	65 536	$0.^3$ 015 258 789 062 5
17			131 072	$0.^3$ 007 629 394 531 25
18	6		262 144	$0.^3$ 003 814 697 265 625
19			524 288	$0.^3$ 001 907 348 632 812 5
20		5	1 048 576	$0.^6$ 953 674 316 406 25
21	7		2 097 152	$0.^6$ 476 837 158 203 125

Note: Superscripts appearing in the right-hand column indicate a group of that many zeros.

The simplest form of binary coding encodes each digit into the 4-bit base-2 equivalent. See Tables 1 and 4.

Excess-3 Codes

In applications where the binary zero is represented by the absence of a signal, it is not desirable to have the decimal symbol 0 represented by 0000, since it cannot then be distinguished from no signal. This is avoided by choosing ten of the possible 4-bit representations that do not include the position 0000. Such a code is given in Table 2. This code uses the binary notation for 3 as the representation for 0. Each of the other 9 symbols is represented by the binary equivalent of the symbol plus 3. For that reason, it is known as an excess-3 code. It has the further property that it is "self-complementing"; that is, the 9's complement of the decimal symbol is formed by changing 1's to 0's and the 0's to 1's in the coded representation of the symbol. This property is useful in performing many of the arithmetic operations within the computer.

Gray or Reflected Binary Codes

A certain group of codes are frequently used when mechanical analogs (position, shaft recording, etc.) are converted into digital form for computer input processes or for recording. This type of code gets its usefulness from the property that one and only one digit of the code changes in proceeding to or from the next higher or next lower number. Thus there is no intermediate instant which can be interpreted as a number which is in error by more than one unit in the least significant position.

A common code of this type is the reflected code or Gray code in Table 4. This code is useful because its conversion to base-2 binary or vice versa is easy. The two rules for this conversion are the following. We define b_n and r_n to be the nth position measured from the right of the base-2 binary number and the reflected binary number, respectively. The first rule is that the most significant digit of both numbers are equal. The second rule is that the sum of $b_{n+1}+b_n+r_n$ is even. Thus a given reflected binary digit is a zero or a one as the equivalent base-2 binary digit and the base-2 binary digit to its left are the same or different. In the conversion from reflected binary to base-2 binary, the leftmost digit is determined by the first rule and then the second rule is used to determine the other digits in descending order of importance.

The 10 decimal symbols can be represented in a reflected binary-3 representation by converting the original number first into excess-3 notation and then using the reflected code transformation. It has the useful property that only one digit change is required in advancing from the 9 to 0 representation and that change occurs in the most significant position. Refer to Table 2.

COMPUTER REPRESENTATION OF NUMBERS

Computers and other digital devices use a variety of number systems. Binary number systems and binary coded decimal systems are the most common. Certain operations use the natural numbers and others use signed numbers. A register with n symbols and b levels per symbol representing the natural numbers will typically represent the numbers $0, 1, 2, \cdots, b^n-1$ rather than the numbers $1, 2, \cdots, b^n$ (i.e., the first item indicated is numbered 0 rather than 1). Since a number is represented inside a computer by a sequence of digital devices, conventions are needed by which these digital devices can represent the information

TABLE 4—EQUIVALENT NUMERICAL REPRESENTATIONS (BINARY, REFLECTED BINARY, AND HEXADECIMAL).

Decimal	Binary	Reflected Binary	Hexadecimal
0	0000	0000	0
1	0001	0001	1
2	0010	0011	2
3	0011	0010	3
4	0100	0110	4
5	0101	0111	5
6	0110	0101	6
7	0111	0100	7
8	1000	1100	8
9	1001	1101	9
10	1010	1111	A
11	1011	1110	B
12	1100	1010	C
13	1101	1011	D
14	1110	1001	E
15	1111	1000	F

TABLE 5—DECIMAL AND BINARY NUMBERS IN VARIOUS CONVENTIONS.

Number	Sign and Magnitude	Ones or Nines Complement	Twos or Tens Complement
$+62_{10}$	062	062	062
-62_{10}	162	937	938
$+1101_2$	0 1101	0 1101	0 1101
-1101_2	1 1101	1 0010	1 0011
-1000_2	1 1000	1 0111	1 1000
-0000_2	1 0000	1 1111	0 0000

carried by the plus sign, the minus sign, and the decimal point in ordinary notation.

Signed Numbers

The conventions given below for the expression of positive and negative numbers apply to numbers expressed in any base. These procedures all use the standard notation for positive numbers and a chosen digit (conventionally the leftmost digit) to represent the sign. See Table 5.

In the simplest convention (called *Sign and Magnitude*) the magnitude of the number is expressed in the usual fashion, and a special digit called the sign digit is 0 for positive numbers and 1 for negative numbers. Another convention obtains the negative of a number N in the base b by replacing each digit of N including the sign digit by its complemented value (i.e., by $b-1$-the value of the digit). This convention is called the nines complement for decimal numbers and the ones complement for binary numbers. Thus the negative of the decimal number $+62$ (which is stored in the computer as 0 62) is 937, and the negative of the binary number $+011_2$ (which is stored as 0 011) is 1100. This procedure has the feature that one can multiply any number which is in binary notation or in excess-3 decimal notation, be it positive or negative, by minus one just by complementing each digit (replacing each 0 by a 1 and each 1 by a 0).

A similar convention, called the tens complement for decimal numbers and the twos complement for binary numbers, obtains the negative of a nonzero number N by adding one to the nines complement of N in decimal notation or the ones complement of N in binary notation. The negative of zero is zero. Thus in this convention the negative of the decimal number $+62$ (which is stored in the computer as 062) is 938 and the negative of the binary number $+011$ (which is stored as 0011) is 1101.

The sign and magnitude convention is easiest to understand. The complement notations have hardware advantages, since one adder can simply add two numbers with any combination of signs or relative magnitudes. A source of much confusion for both the sign and magnitude convention and the ones or nines complement convention is the existence of both a positive and a negative zero, while the twos or tens complement has only a positive zero.

Fixed Point and Floating Point

A number stored in a given number of digits has many possible interpretations. One common interpretation is to assume a fixed-point number with the radix point always to the left of the most significant numerical digit. Thus, for example, assuming a sign and magnitude convention, a 36-bit number will go from $-1+2^{-35}$ to $+1-2^{-35}$ in steps of 2^{-35}. In this case the rightmost bit has the weight 2^{-35}, the number to the left of this bit has the weight 2^{-34}, the leftmost bit is the sign, and the bit to the right of the sign bit is the most significant bit with a weight of 2^{-1}. An alternative interpretation that is also used is to assume the radix point to be at the right of the least significant digit, which would result in a range of all the integers from $-2^{35}+1$ to $+2^{35}-1$. In fact, the radix point could be assumed anyplace. For each choice of radix position, 36 bits would represent 2^{36} or so numbers, each with a given meaning. The determining factor in the choice of a radix point is the extent of problems associated with the results of multiplication and division. The contents of the register containing the results of fixed-point addition and subtraction are not affected by the placement of the radix point. If we choose the radix point to the left of the most significant digit, no multiplication will result in overflow (i.e., a number outside of the computer range), but division will sometimes result in such a number. If all numbers are integers, multiplication sometimes results in overflow but division will not except for division by zero. Intermediate choices of radix point cause problems with both multiplication and division.

In floating-point notation with base b, the digits of a number are divided into two pieces to make two subnumbers N_1 and N_2, where N_1 is an integer and N_2 is a fixed-point fraction with the resulting number interpreted as $N_2 \times b^{N_1}$. N_2 is called the mantissa and N_1 the exponent. Thus the 36-bit number of the previous example could be broken into a 10-bit exponent and a 26-bit mantissa. Thus N_2 would run from $-1+2^{-25}$ to $+1+2^{-25}$ in steps of 2^{-25} and N_1 would run from -511 to $+511$. (2^{511} is approximately 6.5×10^{153}). Certain numbers such as one half can be represented in a number of ways such as $N_2 = 2^{-1}$, $N_1 = 0$; $N_2 = 2^{-2}$, $N_1 = 1$, $N_2 = 2^{-3}$, $N_1 = 2$, etc. The number is usually kept in normalized form (i.e., for any nonzero number

one uses the form in which the most significant digit of the mantissa is nonzero) to avoid the many problems associated with multiple notations for the same number.

To summarize, there are many possible interpretations of the stored digits composing a number. Fixed-point notation is usually applied when dealing with numbers with known magnitudes in a narrow range such as in business calculations. Floating-point notation is used when the numbers either occupy a wide range of magnitudes or the magnitudes are unknown. This occurs in many scientific calculations. In principle, with enough storage and time we can program either a fixed-point or a floating-point machine to do any given calculation. However, the performance of a fixed-point calculation with floating-point arithmetic or vice versa is often grossly inefficient in programming effort, storage, and the time required for the calculation.

Overflow and Underflow

A variety of problems are associated with the fact that legitimate arithmetic operations on numbers within the range of a machine may result in numbers outside the range of the machine. Thus, in fixed-point notation in which all numbers are interpreted as fractions, the addition of two large numbers of like sign or the subtraction of two large numbers with unlike sign or a division where the result is larger than one, result in an overflow. In floating-point notation, addition, subtraction, multiplication, and division can cause either overflow (a number whose magnitude is larger than the largest number the computer can hold) or underflow (a number whose exponent is less than the algebraically smallest legal exponent). For example, if the smallest legal exponent is -511 and the exponent of the result is -753, an underflow has occurred. There are a number of possibilities for computer action in such cases. One is to stop the computer immediately. This is ordinarily done only when testing the computer, as this takes too much time. A common alternative is to perform the operation with the usual rules, which results in a number that must be interpreted in an appropriate way, as the usual interpretation is misleading; in addition, an overflow alarm is set.

A third possibility (often used for division by zero) is to refuse to do the operation and set an alarm. A fourth possibility (useful for floating-point arithmetic) is to replace an overflow result by the largest representable magnitude possible and to replace an underflow result by the smallest possible nonzero number and set an alarm. Once the alarm is set, the program can then take the appropriate action. This might be to (A) stop the calculation, (B) repeat the calculation in a way that overflow does not occur, (C) ignore the overflow (in many logical operations overflows are meaningless), or (D) interpret the overflow (multiple-precision arithmetic uses the overflow equipment as an essential part of the process).

Roundoff

Multiplication in fixed point and all operations in floating point can produce results with more digits than can fit inside a standard register. There are several methods for removing the excess digits. The different methods require different amounts of effort and for a uniform distribution of remainders produce different distributions of error. In extensive calculations the presence of a bias in this distribution is significant, as the total error with numbers of one sign tends to grow in proportion to the number of errors in the presence of a bias, while the total error tends to grow in proportion to the square root of the number of errors in the absence of a bias. One common operation in this case is truncation—as many digits as fit are stored and the remaining digits of lower significance are forgotten. This operation is simplest but produces a strong bias in the distribution of the resulting errors. An alternative operation is roundoff—one is added to the least significant digit of the magnitude of the stored number if the remainder is equal to or larger than half of that digit, and nothing is added otherwise. There is no bias or only a small bias in this case. Another roundoff procedure (applicable to a binary computer) is to always set the least significant digit to a 1. This procedure has no bias or only a small bias, a variance which is four times the variance of the previous procedure, and requires little effort. An additional possibility is to randomly (with probability of 50 percent) add one to the least significant digit of the magnitude of the truncated number. This introduces no bias but requires a source of random digits.

Multiple-Precision Arithmetic

If greater precision is desired than that obtainable in a single machine register, we can program the computer so that two or more registers contain a given number. The arithmetic operations become much more complicated. In adding two numbers, each stored in n registers, we must perform n ordinary additions and then appropriately add numbers and subtract numbers from various registers to correct for overflow conditions or for the condition when the intermediate set of result registers are not all of the same sign. In multiplication there are n^2 possible single register multiplications and $n^2 - 2n + 1$ single register additions, of which only the more significant operations are usually performed. Multiple-precision division requires an iteration procedure.

TABLE 6—ADDITION TABLE FOR ONE DIGIT.

Carry In	0	0	0	0	1	1	1	1
Addend Digit	0	0	1	1	0	0	1	1
Augend Digit	0	1	0	1	0	1	0	1
Sum Digit	0	1	1	0	1	0	0	1
Carry Out	0	0	0	1	0	1	1	1

Addition

The rules of addition inside a computer depend on the notation and on the base. Usually the calculations are done in one of the complement notations, and even computers using the sign and magnitude notation have intermediate results in complement form. We described twos complement arithmetic for binary numbers. Tens complement arithmetic has essentially the same rules. The only difference between twos complement arithmetic and ones complement arithmetic is that, in the former, carries out of the sign position are discarded while, in the latter, carries out of the sign position are added to the least significant digit (such a carry is called an end-around carry).

Addition is performed in the usual way from right to left with carries out of one position added to the next digit position on the left. In the calculation of the nth digit position of the sum, there is one digit from the nth digit position of each of the two numbers to be added and a possible carry from the previous stage. The sum digit and the carry out take on the value 0 or 1 as shown in Table 6.

The same procedure applies to the addition of two numbers in twos complement whether the numbers are both positive, both negative, or of different signs. In performing the addition, the sign digit and the numerical digits are lumped together as one composite number with the sign digit being the most significant digit of the composite number. Thus, carries out of the most significant numerical digit are added to the sign position. The carry out of the sign position has no effect. These rules are illustrated by the following examples using twos complement integers.

Carry	1	Carry	111	Carry	11 1
+5	0 0101	+7	0 0111	−3	1 1101
+6	0 0110	+5	0 0101	−4	1 1100
+11	0 1011	+12	0 1100	−7	1 1001

Carry 11	111	Carry 11	111	Carry	11
+7	0 0111	+13	0 1101	−15	1 0001
−5	1 1011	−1	1 1111	+3	0 0011
+2	0 0010	+12	0 1100	−12	1 0100

TABLE 7—SUBTRACTION TABLE FOR ONE DIGIT.

Borrow In	0	0	0	0	1	1	1	1
Subtrahend Digit	0	0	1	1	0	0	1	1
Minuend Digit	0	1	0	1	0	1	0	1
Difference Digit	0	1	1	0	1	0	0	1
Borrow Out	0	1	0	0	1	1	0	1

Subtraction

Subtraction as such is not commonly performed in complement notation. Instead if $A-B$ must be computed, we first form the negative of B, namely $(-B)$, and then add A and $(-B)$ to form $A+(-B)$, which is the answer. This trick means that, with one adder and simple complementing equipment, a computer can perform both addition and subtraction. If desired, we can build a subtractor (i.e., a device which finds the difference of two numbers). The rules for subtraction are similar to those for addition except that instead of a carry from the previous stage, there is a borrow that represents a value of -1, and the algebraic sum of the digit values (subtrahend—minuend—borrow in) is either $+1$, 0, -1, or -2. The nth digit of the result is one if the sum is $+1$ or -1, and there is a borrow out of the nth digit if the algebraic sum is negative. See Table 7.

Multiplication and Division

Multiplication and division of signed numbers are usually done by first calculating the product or ratio of the magnitude of the input numbers, and then taking either this result or its negative depending on whether the input numbers have the same or different signs. Thus the problem is reduced to the case of positive numbers. The multiplication of two numbers can be carried out as in ordinary arithmetic by multiplying each digit of the multiplier by the multiplicand in turn, writing the products in staggered positions, and then summing all the terms. In binary multiplication the multiplier digits are either 0 or 1, and thus the partial products are either a shifted version of the multiplicand or all zeros. The addition of the partial sums is a little difficult, but the following rule is sufficient. Let s_n be the sum of the carries into the nth column and of the partial-product digits in the nth column.

If s_n is odd, the nth full product digit is one and the carry into the next stage equals $\frac{1}{2}(s_n-1)$; if s_n is even, the nth full product digit is zero and the carry into the next stage equals $\frac{1}{2}s_n$. See example. The carry in and sums s_n are given in decimal notation.

```
Multiplicand   1 1 0 1 1

Multiplier     1 0 1 1 1
               ─────────
                 1 1 0 1 1
               1 1 0 1 1
             1 1 0 1 1          Partial Products
           0 0 0 0 0
         1 1 0 1 1

         1 1 2 2 1 1 1 0 0  Carry In
         2 2 3 5 4 3 3 2 1  Sums sₙ

       1 0 0 1 1 0 1 1 0 1  Product
```

In digital computers it is usually convenient to sum the partial products at each step so that only three arithmetic registers are needed: one for the partial sum, one for the multiplicand, and one for the multiplier. The multiplicand becomes larger by one digit in each step until it is almost two words long, but we can discard one digit of the multiplier at each step with the result that at every step the multiplier and the partial sum fit into two words.

There are several methods of performing division. Many of these involve repeated subtraction in one form or another. Basically, the divisor is successively subtracted from the partial remainder to get a smaller but nonnegative partial remainder, and a 1 is placed in the quotient corresponding to the position from which the rightmost digit of the divisor was subtracted. See example.

```
                1 0 1 1  Quotient
      1 0 0 1 │ 1 1 0 1 0 1 0
              1 0 0 1
              ─────────
                1 0 0 0 1
                1 0 0 1
                ─────────
                1 0 0 0 0
                1 0 0 1
                ─────────
                1 1 1   Final Remainder
```

Floating-Point Arithmetic

Floating-point arithmetic requires several operations in addition to those of fixed-point operation. Assume two floating-point numbers $N_2 \times b^{N_1}$ and $M_2 \times b^{M_1}$ where $M_1 \geq N_1$. Before adding or subtracting the numbers N_2 and M_2, we would first have to shift the number N_2 to the right $M_1 - N_1$ places so that corresponding inputs to the adder would have the same weight. For multiplication and division, the mantissas are multiplied or divided and the exponents are summed or differenced as shown by

$$(N_2 \times b^{N_1}) \times (M_2 \times b^{M_1}) = N_2 M_2 \times b^{N_1 + M_1}$$

$$(N_2 \times b^{N_1}) / (M_2 \times b^{M_1}) = (N_2 / M_2) \times b^{N_1 - M_1}$$

$$(M_2 \times b^{M_1}) / (N_2 \times b^{N_1}) = (M_2 / N_2) \times b^{M_1 - N_1}.$$

A normalization procedure is usually performed as the last step of a floating-point operation, as the intermediate results are not always normalized.

SWITCHING ALGEBRA

Switching algebra provides techniques for the design and analysis of digital equipments in terms of two-state logic.

Digital computers and other digital equipments are predominantly composed of binary devices—devices which have only two states. In practice, each state is associated with a range of values, and the binary devices are designed so that the signal values for properly operating devices are either in one state or the other. In most of the design of digital equipments, the individual digital devices are considered to be ideal two-state devices with no reference to their analog nature. Switching algebra allows us to manipulate a description of the behavior of a collection of ideal binary devices. There are a large number of notations used in switching algebra. This chapter uses notation similar to that of reference [3].

Binary States

Because of the many forms of binary logic devices and binary storage devices, there are many names for the two states. Thus the two states may be associated with a binary digit being 0 or 1, a signal voltage being high or low, two terminals being open-circuited or short-circuited, a relay being operated or not, the direction of saturation of a magnetic circuit, the presence or absence of a hole in a card, etc. Because much of the development of switching algebra was done in the related field of propositional calculus, the two states are often associated with true and false. In the following, the two states are associated with the binary digits 0, 1.

The assignment of the values 0,1 to two physical states is arbitrary. When the assignment of the states 0 and 1 to a physical equipment is changed, the description of the equipment by switching algebra changes with all the binary state values being interchanged, with AND circuits becoming OR circuits and vice versa. Thus, assuming that the inputs have either a high or a low voltage, the diode circuit of Fig. 1 is an OR circuit if the high voltage represents the one state, and an AND circuit if the lower voltage represents the one state. (Note that the return voltage $-V$ is chosen such that at least one diode is always conducting.) This duality in the assignment of states has a strong theoretical interest, and on occasion the analysis or synthesis of a function is easier in a single form than in the dual form. However, this duality of approach tends to generate confusion, and many engineers use one of the interpretations exclusively.

Switching Variables

A switching variable will be described either by a capital letter alone or by a capital letter with a subscript. Thus X, X_1, Y, Z_n are four examples of variables.

Basic Postulates*

The first postulate of switching algebra is that, at any one instant, a variable will either be in the zero state or in the one state but not both.

1A. If $X = 1$, then $X \neq 0$

1B. If $X = 0$, then $X \neq 1$

Switching algebra is based on three basic functions: the complement function, the AND function, and the OR function.

The complement of a variable or a constant is indicated by a prime. Thus the complements of X, X', 0, 1 are written X', $(X')'$, $0'$, $1'$, respectively. The complement of one state is equal to the other state.

2A. If $X = 0$ then $X' = 0' = 1$

2B. If $X = 1$ then $X' = 1' = 0$

X	X'
0	1
1	0

* S. H. Caldwell, "Switching Circuits and Logical Design," John Wiley & Sons, New York; 1958: pp. 36–44. E. J. McCluskey, "Introduction to the Theory of Switching Circuits," McGraw-Hill Book Co., New York; 1965: pp. 85–87, 112.

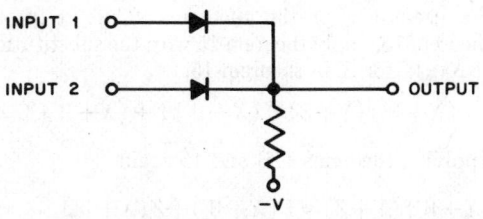

Fig. 1—Diode circuit.

The AND function of two or more variables is 1 if and only if all the variables are 1. The AND function is denoted by the usual algebraic notation for a product. Thus, X AND Y is denoted by $X \cdot Y$ or XY; 0 AND 1 is denoted by $0 \cdot 1$. The values of the function X AND Y follow.

X	Y	XY
0	0	0
0	1	0
1	0	0
1	1	1

The OR function (sometimes called the INCLUSIVE OR function) of two or more variables is 1 if and only if one or more of the variables is 1. The OR function is denoted by the usual algebraic notation for a sum. Thus X OR Y is denoted by $X+Y$; 0 OR 1 is denoted by $0+1$. The values of the function X OR Y follow.

X	Y	X+Y
0	0	0
0	1	1
1	0	1
1	1	1

Note that while the AND function obeys the conventional arithmetic rules for a product, the OR function does not obey the conventional arithmetic rules for the sum of $1+1=1$ in switching algebra.

3A. $0 \cdot 0 = 0$ 3B. $1+1=1$

4A. $1 \cdot 1 = 1$ 4B. $0+0=0$

5A. $1 \cdot 0 = 0 \cdot 1 = 0$ 5B. $0+1=1+0=1$

In the absence of parentheses, the order of performing the operations is that conventionally used in algebra. Thus $X \cdot Y + Z$ equals $(X \cdot Y) + Z$ and not $X \cdot (Y+Z)$.

Theorems

A large number of useful theorems follow from the basic definitions.

6A. $X+0=X$ 6B. $X \cdot 1 = X$

7A. $X+1=1$ 7B. $X \cdot 0 = 0$

8A. $X+X=X$ 8B. $X \cdot X = X$

9. $(X')'=X$

10A. $X+X'=1$ 10B. $X \cdot X'=0$

11A. $X+Y=Y+X$ 11B. $XY=YX$

12. $X+XY=X$ 13. $XY'+Y=X+Y$

14A. $X+Y+Z=(X+Y)+Z=X+(Y+Z)$

14B. $XYZ=(XY)Z=X(YZ)$

15. $X(Y+Z)=XY+XZ$

16. $(X+W)(Y+Z)=XY+XZ+WY+WZ$

17A. $(X+Y)(X'+Z)(Y+Z)$

$$= (X+Y)(X'+Z)$$

17B. $XY+X'Z+YZ=XY+X'Z$

18A. $(X_1+X_2+\cdots+X_n)'=X_1'X_2'\cdots X_n'$

18B. $(X_1 X_2 \cdots X_n)'=X_1'+X_2'+\cdots+X_n'$

19. $f(X_1, X_2, \cdots, X_n, +, \cdot)'$

$$=f(X_1', X_2', \cdots, X_n', \cdot, +)$$

20. $f(X_1, X_2, \cdots, X_n)$

$$= X_1 f(1, X_2, \cdots, X_n)+X_1'f(0, X_2, \cdots, X_n)$$

21. $f(X_1, X_2, \cdots, X_n)=[X_1+f(0, X_2, \cdots, X_n)]$

$$\cdot [X_1'+f(1, X_2, \cdots, X_n)]$$

Theorems 18A and 18B show how to take the complement of a sum or a product. Theorem 19 provides a general rule for the complement of an arbitrary expression. It says that to derive the complement of an expression, we need only replace each variable by its complement, each OR function by an AND function, each AND function by an OR function, and maintain the order of the operations. The application of theorem 19 requires care because of the possibility of losing implicit products or the order of the operations. Mistakes can be avoided by initially filling in each product dot and each parentheses before applying theorem 19. Thus $XY+WZ$ becomes $(X \cdot Y)+(W \cdot Z)$ and the complement is clearly $(X'+Y') \cdot (W'+Z')$. Careless application of theorem 19 to $XY+WZ$ can lead to erroneous forms such as $X'Y' \cdot W'Z'$ or $X'+Y' \cdot W'+Z'$. The theorems labeled by the same number (such as 6A and 6B) are duals of each other. Each theorem of the pair can be derived from the other by applying theorem 19 and then complementing each of the alphabetically labeled variables. Thus the result of applying theorem 19 to equation 11A is $X'Y'=Y'X'$ which becomes $XY=YX$ by changing X' into X and Y' into Y.

In these theorems, each of the variables can represent an arbitrary function. This greatly widens the usefulness of the theorems. Thus to prove theorem 16, apply theorem 15 with the substitution of $X+W$ for X in theorem 15.

$$(X+W)(Y+Z)=(X+W)Y+(X+W)Z.$$

Applying theorems 11B and 15 again

$$(X+W)(Y+Z)=Y(X+W)+Z(X+W)$$

$$=YX+YW+ZX+ZW.$$

TABLE 8—TABLE OF COMBINATIONS PROVING
$XY+X'Z=XY+X'Z+YZ$.

Case No.	X	Y	Z	YZ	$XY+X'Z$	$XY+X'Z+YZ$
0	0	0	0	0	0	0
1	0	0	1	0	1	1
2	0	1	0	0	0	0
3	0	1	1	1	1	1
4	1	0	0	0	0	0
5	1	0	1	0	0	0
6	1	1	0	0	1	1
7	1	1	1	1	1	1

Rearrange terms using theorems 11A and 11B

$$(X+W)(Y+Z)=XY+XZ+WY+WZ.$$

The theorems, of course, can be applied in treating any function as well as other theorems. Also, with practice, the application of the theorems and the processes of factoring, expanding, and rearranging terms become automatic and take little time.

There are three common approaches to proving theorems or manipulating forms in switching algebra. One approach is to use known theorems as was done in the previous paragraph. A second approach is to use a map, which is discussed later in this chapter. A third approach is called perfect induction.

Perfect Induction

Any theorem or equation of switching algebra with a finite number of variables can be derived by perfect induction—i.e., examining all the possible cases and showing that the result is valid for every case. An equation with n variables will have 2^n cases as each variable has two possibilities; namely, 0, 1. A systematic method of applying perfect induction is to form a table of combinations (also called a truth table) listing all the possible cases and both sides of the equation. See Table 8. An n-digit binary number is conventionally associated with each case where the mth bit of the number has the same value as the mth variable for that case. The cases are listed in the order of their associated numbers. Table 8 shows the proof of theorem 17B. Perfect induction is important for proving the basic theorems and dealing with functions of a few variables. By itself it becomes unwieldy for functions with many variables, since these functions have too many cases.

Canonical Expressions

An arbitrary switching function can be described by a table of combinations which gives the value of the function for every possible case. The canonical expressions using fundamental products and sums provide a simple method for deriving an algebraic expression for a function, given the table of combinations for the function. Fundamental products and sums both contain all the input variables. A fundamental product for a given case has the value 1 for that case and the value 0 for all other cases. Each input variable in a fundamental product is primed if the variable is 0 for the given case and unprimed if the variable is 1 for the given case. See Table 9. A fundamental sum for a given case has the value 0 for that case and the value 1 for all other cases. Each input variable in a fundamental sum is primed if the variable is 1 for the given case and unprimed if the variable is 0 for the given case.

The canonical sum of a function is a sum of the fundamental products corresponding to the cases for which the function has a 1. The canonical product of a function is a product of the fundamental sums corresponding to the cases for which the function has a zero. Thus the expression $XY+X'Z$ (which has 1 for cases 1, 3, 6, 7 and zero for cases 0, 2, 4, 5) has the canonical expres-

TABLE 9—FUNDAMENTAL PRODUCTS AND SUMS.

Case No.	X	Y	Z	Fundamental Product	Fundamental Sum
0	0	0	0	$X'Y'Z'$	$X+Y+Z$
1	0	0	1	$X'Y'Z$	$X+Y+Z'$
2	0	1	0	$X'YZ'$	$X+Y'+Z$
3	0	1	1	$X'YZ$	$X+Y'+Z'$
4	1	0	0	$XY'Z'$	$X'+Y+Z$
5	1	0	1	$XY'Z$	$X'+Y+Z'$
6	1	1	0	XYZ'	$X'+Y'+Z$
7	1	1	1	XYZ	$X'+Y'+Z'$

sions

$$XY+X'Z=X'Y'Z+X'YZ+XYZ'+XYZ$$
$$=(X+Y+Z)(X+Y'+Z)$$
$$\times(X'+Y+Z)(X'+Y+Z').$$

If the function has very few ones, the canonical sum is simpler; if the function has very few zeros, the canonical product is simpler.

Minimization

If a digital equipment design has been reduced to algebraic expressions, the next step is to implement these expressions using the minimum hardware. This section describes procedures for simplifying the expressions and thus minimizing the required hardware. Procedures commonly used are algebraic reduction, map reduction, and Quine–McCluskey reduction. Algebraic reduction is automatically applied to simple apparent cases by the designer; map reduction is the standard procedure for more-complicated cases (with as many as 6 variables); and Quine–McCluskey reduction is essential for large multivariable redundant functions or for automated reduction procedures.

Certain general comments can be made about minimization procedures. First, often there is no expression for a function that is the best for all hardware types. Differing limitations on delay or driving capability, differing costs of components, the fact that a variable is easily available in one form but not in its complement, etc., will make a different form preferable in different cases. Second, minimization of the hardware required to implement each algebraic function is only part of the overall minimization process, although it is the best analyzed part of this process. Hardware can also be minimized by time sharing so that one piece serves several functions, or by generating certain auxiliary functions which are used in many places. Often we can make significant savings compared with a given design by changing the overall approach to the system design or the choice of basic variables and thus reducing the number and complexity of the expressions to be implemented. Third, the minimization process is not a goal in itself, and the savings due to minimization must be balanced against the design cost. Usually, the preliminary expressions can be drastically reduced with little effort. Eventually, the possible savings become smaller and the effort required to reduce these savings so large that the minimization effort becomes uneconomical. The design should minimize parts that are repeated many times, as the savings per part are multiplied by the number of parts.

In summary, minimization procedures for algebraic expressions offer significant savings in system hardware. Also, for most cases and for all types of

hardware, the standard minimization procedures are valid. For example, the reduction of the expression $X+XY$ to X is useful for all types of hardware. The novice designer can safely ignore the special cases in which one form is best for one type of hardware and another form is best for another type of hardware.

Algebraic Minimization

There are several common definitions of a minimum expression of a function. The minimum expression can be defined as the expression containing the smallest number of *literals*. A literal is defined to be each appearance of a variable or its complement. Thus the expressions $X+X'$, $XY'Z'$, and $XY+X'Z$ have 2, 3, and 4 literals, respectively. Other minimization procedures attempt to obtain the minimal sum (the sum of products) with the fewest terms and literals, or the minimal product (the product of sums) with the fewest terms and literals.

One approach is to use the identities, which we have previously listed, to minimize a form. This is illustrated in the following examples.

The expression $XYZ+XYZ'+X'YZ+X'YZ'$ can be simplified by repeated application of the identities

$$X(Y+Z)=XY+XZ,\ X+X'=1,\ X\cdot1=X.$$
$$XYZ+XYZ'+X'YZ+X'YZ'$$
$$=XY(Z+Z')+X'Y(Z+Z')=XY\cdot1+X'Y\cdot1$$
$$=XY+X'Y=(X+X')Y=1\cdot Y=Y.$$

The expression $X'Y+YW'+Z+(Y'+XW)Z'$ can be simplified by using the identities $X(Y+Z)=XY+XZ$, $XY'+Y=X+Y$, $(XY)'=X'+Y'$, $X+1=1$, and $X+X'=1$.

$$X'Y+YW'+Z+(Y'+XW)Z'$$
$$=Y(X'+W')+Z+Y'+XW$$
$$=Z+Y'+XW+(X'+W')$$
$$=Z+Y'+XW+(XW)'=Z+Y'+1=1.$$

Another example is

$$XW+XWY+X'=XW(1+Y)+X'$$
$$=XW\cdot1+X'=XW+X'=W+X'.$$

Some functions are especially difficult to minimize. This is illustrated by the following example.*

$$T=ABE+ABF+ACDE+BCDF.$$

* W. H. Burkhart, "Theorem Minimization," *Proceedings of the Association for Computing Machinery*, Pittsburgh, Pa.; May 1952.

Straightforward factoring of

$$T = ABE + ABF + ACDE + BCDF$$

yields

$$T = AB(E+F) + CD(AE+BF)$$

with 10 literals while special effort is needed to find the simpler equivalent form

$$T = (AB+CD)(AE+BF).$$

There is a systematic procedure called the Quine–McCluskey method for finding the minimal sum.

Map Method

The map method is the most common method for minimizing functions of four or less variables. It is also useful for functions with five or six variables.

The map is a pictorial description of a table of combinations. See Fig. 2.* A map of n variables contains 2^n squares, with each square representing one of the fundamental products of literals. Thus, square number 0 in each of the maps represents the case of "all variables equal zero." The combination of variables of a given square can be determined by the notation on the row and column corresponding to that square. Thus the first through the fourth columns of Figs. 2C, 2D, and 2E correspond to the cases of $X=Y=0$; $X=0$, $Y=1$; $X=Y=1$; and $X=1$, $Y=0$, respectively. Likewise, the first row of Fig. 2D corresponds to $Z=0$ and

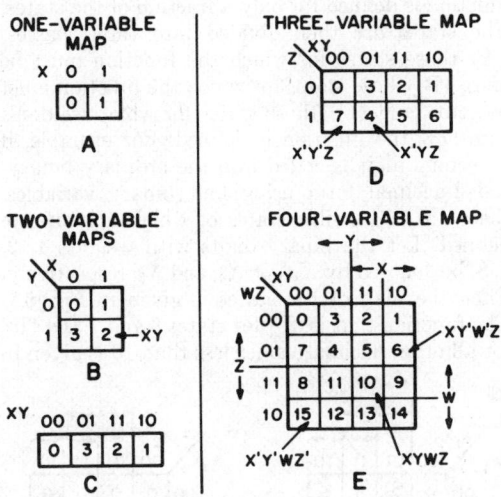

Fig. 2—Switching algebra maps.

* The numbering in Fig. 2 corresponds to a Gray code of the variables. More commonly, the map numbering uses the number associated with the variables in the table of combinations. Thus it is easy to associate a table of combinations with a map and vice versa.

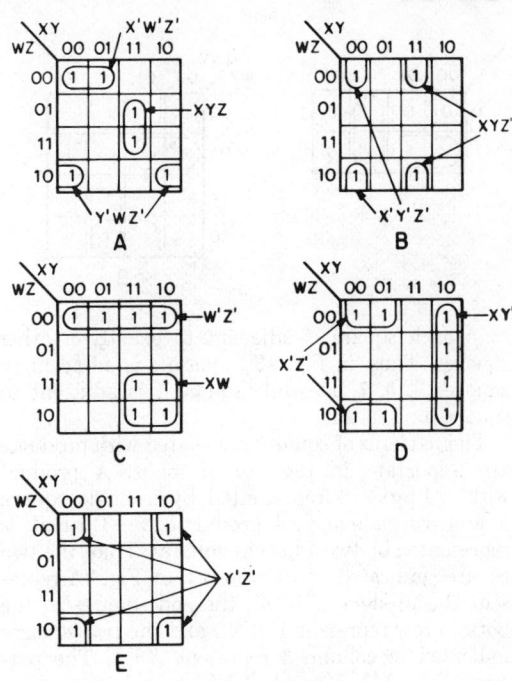

Fig. 3—Map interpretation.

the second row of Fig. 2D corresponds to $Z=1$. Also the first through the fourth rows of Fig. 2E correspond to the cases of $W=Z=0$; $W=0$, $Z=1$; $W=Z=1$; and $W=1$, $Z=0$.

Thus, square 3 in all of the maps corresponds to $Y=1$ and the other variables equal to zero; square 6 in Fig. 2E corresponds to $X=Z=1$, $Y=W=0$ or the product term $XY'W'Z$; and square 15 corresponds to $X=Y=0$, $W=1$, $Z=0$ or the product term $X'Y'WZ'$. There are several notations for the values of the rows and columns. In one notation each row (column) is described by the binary values of the variables. Thus in Fig. 2E the notation 10 for row 4 denotes that $W=1$ and $Z=0$. An alternative notation is to indicate the rows or columns in which each of the variables is a one. Thus in Fig. 2E an "X" is shown above the line separating the last two columns to show that in these columns $X=1$.

A function can be described using a map by placing "ones" in the squares corresponding to the combinations for which the function is a one and "zeros" in the other squares.

In a map, adjacent squares differ in only one variable. Thus, in Fig. 2D, square 2 corresponding to the term XYZ' is adjacent to square 3 corresponding to $X'YZ'$. In these maps the first column is considered to be adjacent to both the second and the last column, while the top row is adjacent to both the row below it and to the bottom row. (Thus in all the maps of Fig. 2, since a Gray code numbering is used, each numbered square is adjacent to a square with either the preceding or the succeeding number.) Also, in an n-variable

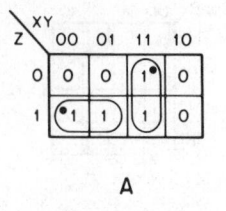

A

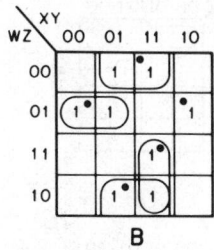

B

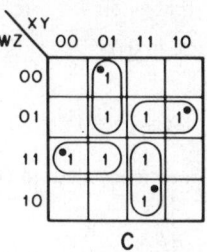

C

Fig. 4—Formation of a minimum sum of products by map method.

map each square is adjacent to exactly n other squares. Thus, in Fig. 2E, square 4 is adjacent to squares 3, 5, 7, 11 while square 1 is adjacent to squares 0, 2, 6, 14.

The patterns of squares associated with products are important in the use of maps. A product with n literals is represented by a single square of an n-variable map. A product of $n-1$ literals is represented by two adjacent squares. Thus, the two squares indicated in the top row of Fig. 3A represent the product $X'W'Z'$, the end squares in the bottom row represent $Y'WZ'$, and the two squares indicated in column 3 represent XYZ. The patterns for $X'Y'Z'$ and XYZ' are indicated in Fig. 3B. A product of $n-2$ literals represents a pattern of four squares. Thus the patterns for $W'Z'$ and for XW are indicated in Fig. 3C, the patterns for XY' and $X'Z$ are indicated in Fig. 3D, and the pattern for $Y'Z'$ is indicated in Fig. 3E. Similarly, products with $n-3$ literals have 8 squares, products with $n-4$ literals have 16 squares, etc. If we visualize the end squares of the rows and columns being adjacent, then the pattern of a product in every case is a rectangle with the number of squares along each side being a power of 2. Thus the products covering two squares are either 1×2 or 2×1, and the products covering four squares are either 1×4 or 2×2 or 4×1.

The map method is more complicated for functions of 5 and 6 variables (which require 2 or 4 maps of four variables, respectively, to describe the function).

The formation of a minimum sum of products by the map method requires two steps.

(A) The function is described by placing ones and zeros in appropriate squares of the map.

(B) All the ones of the map are covered using a minimum number of terms.

Example 1: $XY+YZ+X'Z$. The map for this function is given in Fig. 4A. Clearly the two terms $XY+X'Z$ cover all the ones and the term YZ is redundant.

Example 2: $X'W'Z+X'YW'+XYZ'+YWZ'+XYW+XY'W'Z$. The map for this function is given in Fig. 4B. The best sum for this map is $X'W'Z+YZ'+XYW+XY'W'Z$.

Example 3: An interesting function is indicated by the map in Fig. 4C, where the best sum

$X'YW'+XW'Z+X'WZ+XYW$ does not include the *"obvious"* term YZ.

There are a number of rules for forming the best sum.

(A) Essential products must be used. For an essential product we can find a square such that any other product, included in the function and containing the selected square, is totally included in the essential product. Thus in examples 1, 2, and 3, all the terms in the minimal sum are essential products, and squares which could be used to prove the products to be essential are indicated by dots.

(B) If there are several patterns to cover a given square, the largest pattern (the product with the fewest literals) is preferred.

(C) After the essential products have been used, the remaining squares containing ones are covered.

Don't-Care States

In many designs of digital equipment, a logic function is defined for only a fraction of the states. The states are then divided into three classes: (A) those states for which the function must be ones, (B) those states for which the function must be zeros, and (C) those states for which we don't care how the function is defined. For example, if a decimal digit is coded into the ordinary binary-coded-decimal form using four binary variables, there are 16 possible states of which only 10 are defined. Let the binary digits with weights 1, 2, 4, 8, be denoted by X_1, X_2, X_4, and X_8, respectively. Then the map of the states is given in Fig. 5A. The function that is "1" for states 6 and 7 and "0" for all other decimal values less than 10 is given in

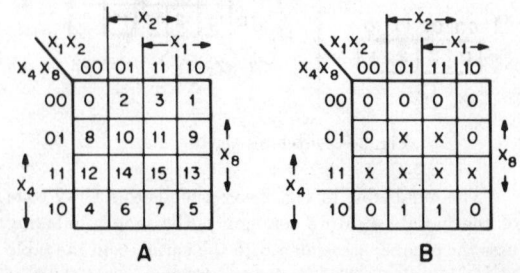

A

B

Fig. 5—Don't-care states.

TABLE 10—TYPICAL LOGIC DEVICES.

Major Class	Subclass	Gates	Flip-Flops	Inverters	Amplifiers	Speed
Electronic	Diode logic	×				Medium
	Transistor logic (resistor-transistor logic, diode-transistor logic, transistor–transistor logic, current switching, etc.)	×	×	×	×	High
	Integrated circuits	×	×	×	×	Med.-high
	Tunnel diode	×	×			Very high
	Cryotron	×	×			Medium
	Neuristor	×	×		×	Medium
	Vacuum-tube circuits	×	×	×	×	Med.-high
Relay	—	×	×	×	×	Slow
Magnetic	Magnetic-core logic	×	×	×		Medium
Mechanical		×	×	×	×	Slow
Pneumatic or fluidic	—	×	×	×	×	Slow

Fig. 5B where X denotes a don't-care state. This function is given by $X_2 X_4 X_8'$ if don't-care states are assumed equal to zero. However, this function can be simplified to $X_2 X_4$ if the don't-care states 14 and 15 are defined to be one and the other don't-care states are defined to be zero. In general, significant savings can be introduced by the appropriate definition of don't-care states.

LOGIC HARDWARE

A complete data processing system is composed of a large number of elementary units, such as gates, flip-flops, storage devices, and input/output devices, arranged to perform the necessary logic operations. These elementary units can take many different forms. They may consist of semiconductor circuits, integrated circuits, magnetic circuits, relays, vacuum-tube circuits, mechanical devices, or more-unusual devices. In any case, the essential characteristic which a digital logic element must possess is that it operate in two or more well-defined and distinct states. A transistor is usually operated in a saturated or a cutoff condition. A relay has two states—open and closed. A mechanical lever may have two states—say up and down. A magnetic core has two possible states, saturated in one direction or in the other. Many devices actually have a continuum of states, but in practice are restricted to operation in just two states to obtain adequate protection against noise, component variations, and other degrading influences.

The same considerations have generally, with a few exceptions, precluded the use of multistate devices and led to the almost exclusive use of two-state (binary) devices.

Table 10 lists a number of devices used to perform logic operations. It does not include devices used solely for storage purposes, such as magnetic-core memories, magnetic tape, magnetic drums, paper tape, punched cards, photographs, and plug boards. Very often equipment uses a combination of different devices: a low-speed portion might use diode logic, while a high-speed portion might use transistor logic. Some devices are used because of certain desirable characteristics but must be supplemented by others because they do not perform all necessary functions.

Generally speaking, equipment requires all four types of devices: gates, flip-flops, inverters, and amplifiers. If a given logic device is unsuitable for some, it must be supplemented by another. Many of the devices are unsuitable for inverters or amplifiers, and are usually supplemented by transistor circuits for these purposes.

The most common building blocks used to construct digital computers are semiconductor logic circuits. A typical example is the diode–transistor logic (*DTL*) circuit shown in Fig. 6A. The operating conditions are chosen so that the transistor is either nonconducting or is in the saturated state. Two logic signals, A and B, are connected as inputs, and the output signal F is a function of the inputs. The input signals would be derived from an identi-

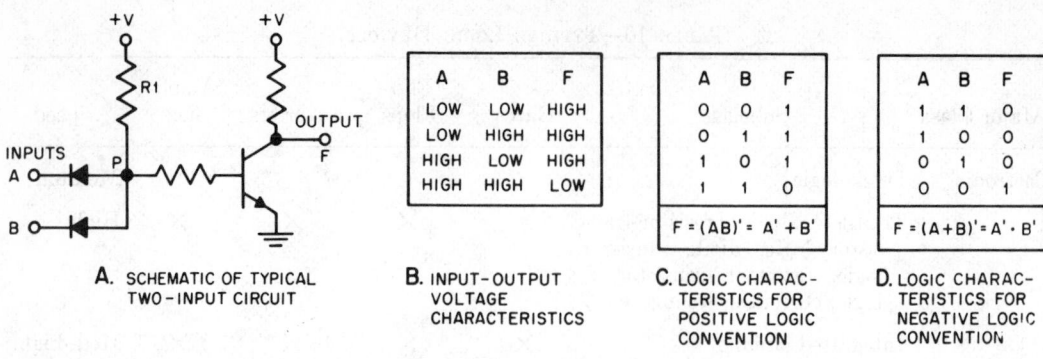

A	B	F
LOW	LOW	HIGH
LOW	HIGH	HIGH
HIGH	LOW	HIGH
HIGH	HIGH	LOW

A	B	F
0	0	1
0	1	1
1	0	1
1	1	0

$F = (AB)' = A' + B'$

A	B	F
1	1	0
1	0	0
0	1	0
0	0	1

$F = (A+B)' = A' \cdot B'$

A. SCHEMATIC OF TYPICAL TWO-INPUT CIRCUIT

B. INPUT-OUTPUT VOLTAGE CHARACTERISTICS

C. LOGIC CHARACTERISTICS FOR POSITIVE LOGIC CONVENTION

D. LOGIC CHARACTERISTICS FOR NEGATIVE LOGIC CONVENTION

Fig. 6—Diode–transistor logic (DTL).

cal or compatible circuit and therefore would be restricted to two states; one state is approximately at ground potential and the other is several volts positive. Resistor $R1$ is made small compared with the back resistance of the diodes and large compared with the forward resistance of the diodes. As a result, the voltage at point P is close to ground if one or more of the inputs is low and several volts positive if all of the inputs are high. The transistor stage is a nonlinear inverter designed to cut off when any input is low and to saturate when all inputs are high. With the use of additional diodes, the circuit is readily extended to accommodate additional input signals.

The logic function performed by this circuit depends on the way in which the logic levels are defined, which is an arbitrary choice. The majority of computers use *positive* logic, in which the logic 1 is chosen as the more positive of the levels. Some use *inverted* logic or *negative* logic, in which the logic 1 is chosen as the more negative of the levels. Fig. 6B tabulates the input/output voltage characteristics of the *DTL* circuit; while Fig. 6C and Fig. 6D show what these correspond to for each polarity convention. The function provided when the positive logic convention is used is $F = (AB)' = A' + B'$. This is usually referred to as the NAND function (Not AND). When the inverted convention is used, the function $F = (A+B)' = (A \cdot B)'$ results. It is referred to as the NOR function (Not OR).

Either convention can be used, provided it is followed consistently. Since it seems more natural to say that a flip-flop is triggered by a logic one rather than a zero, a common practice is to choose the polarity convention to suit the triggering characteristics of the flip-flop.

A flip-flop is a bistable device which is either in a zero or a one state and is used to remember the state of a digital system. A flip-flop has a true or 1's output which corresponds to the actual flip-flop state and a complement or 0's output which corresponds to the complement of the flip-flop state. A flip-flop can change state when actuated by its inputs. If a flip-flop is reset, its state becomes zero

independent of the past state; if a flip-flop is set, its state becomes one independent of its past state; if the flip-flop is triggered or complemented, its state is complemented.

A large variety of logic circuits, or building blocks, have been used to construct digital equipment. Table 11 is a list of typical logic circuits.

Most electronic circuits are *special purpose*; that is, each circuit is designed for one specific application. On the other hand, digital logic circuits are usually *general purpose*; each circuit is designed for operation over a wide range of applications. The applications differ in the number and type of load circuits connected to the driving circuit, the length and arrangement of the wiring between the driving and load circuits, the logic states of other signals into the load circuits, etc.

To be able to readily interconnect a group of building blocks to form some desired logic configuration, and to have assurance that it will work correctly without adjustment, it is general practice to use a set of application rules covering all the factors of significance. For each circuit type, the rules would typically specify the fan-in (the number of input signals which may be connected), the fan-out (the number of load circuits which may be connected, or, if they do not represent equal loading, the various combinations of load types which may be connected), the limits which must be observed regarding the length and routing of interconnecting wires, and the stage delay (which may be related to the other parameters). Depending on the exact characteristics of the circuits, the rules may include a number of other topics, such as how the clock signal is distributed, limitations on the physical locations of certain circuits, etc.

Of course, the rules will be peculiar to the particular set of logic circuits considered. Moreover, they will not necessarily cover every possible application. The set of rules is somewhat peculiar to the application, since they are always based on assumptions regarding the environment in which the circuits are to function: the ambient noise level,

TABLE 11—TYPICAL LOGIC BUILDING BLOCKS.

Circuit Type	Circuit Symbol*	Typical Variations	Remarks
Flip-flop		S–R type	Sets and resets (clears) in response to whichever of two input signals is 1.
		J–K type	Same as above, except if both inputs are 1, the circuit will complement.
		T type	Complements whenever triggered.
Gate		AND	$F = ABC \cdots N$, where F is the output signal and $A, B, C, \cdots, N$ are the inputs.
		OR	$F = A + B + C + \cdots + N$
		NAND	$F = (ABC \cdots N)'$
		NOR	$F = (A + B + C + \cdots + N)'$
		Exclusive OR	$F = AB' + A'B$
	Above symbol if applicable or: (FUNCTION)	Threshold	$F = 1$ if the arithmetic sum of the inputs exceeds a fixed value (the threshold); $F = 0$ otherwise. In the most general case, the various logic inputs can be weighted nonuniformly when summing them.
Inverter	A ▷o– F		$F = A'$
Amplifier	HFO	High fan-out	For driving heavy logic loads.
	CD	Cable driver	For driving low-impedance cables.
	LC	Interface	For changing signal levels at the interface with other equipment.
	CLK	Clock	High timing accuracy, for distributing clock signals.
Miscellaneous	X MS	Delay	For delaying a signal.
	OSC X MHZ	Oscillator	For generating clock or other reference timing signals.
	SS X MS	Pulse generator	For reshaping of signals (e.g., "single-shot" or monostable multivibrator).

* Consistent with MIL-STD-806B. Assumes higher voltage level corresponds to a logic "one".

and the characteristics of the noise. The circuits have some fixed noise margin which, if exceeded, may produce errors. The total disturbance is a composite of a number of factors: ground noise, power supply noise, cross-coupling between signal lines, noise coupled from external sources, etc. In a small unit these disturbances may be insignificant, and the rules may impose few restrictions. In a large unit, particularly a high-speed one, these disturbances may be very substantial, and therefore the rules may be extremely restrictive. In fact, the circuits might be unusable in some difficult applications.

Logic Subassemblies

Certain logic blocks, or subunits, recur frequently in digital equipment. The following briefly describes some of the most common ones.

Register: A generic term for any group of flip-flops used to store or operate on a set of related bits of information. Specific types are described below.

Buffer Register: A register used as a temporary store in transferring information between two units, usually because the two units are asynchronous, are operating at different speeds, or are performing independent tasks.

Storage Register: A register used to temporarily store data for later use.

Shift Register: A register having the capability of serially shifting the data from each stage of the register to the adjacent one. It may be a unidirectional or a bidirectional shift register, depending on whether it can shift only from left to right, only from right to left, or both.

Serial–Parallel Converter: A register having the capability of accepting input data in serial form, shifting the data to accumulate a full "word," and then transferring the word out in parallel form. A parallel–serial converter operates in the reverse way.

Adder: A device which accepts two words and produces a third word that is the sum of the two inputs. An adder is designated as a parallel adder if it operates on words in parallel form, and as a serial adder if it operates on them in serial form. Adders are also distinguished according to the numerical code in which the operation is carried out (binary, binary coded decimal, excess-three coded decimal, etc.) and the method of representing negative numbers (one's complement, two's complement, nine's complement, etc.).

Accumulator: A register capable of accepting an input word, then forming and storing the sum of that input and the word previously stored in the register. The register includes the addition capabilities of an adder; although sometimes the word "accumulator" is used to designate a simple storage register fed from an adder.

Counter: A storage device capable of changing from one state to the next in succession on receipt of a trigger signal. Counters differ in the number system they use (binary, decimal, etc.). In addition to the usual arrangement, several special arrangements have useful characteristics. In a ring counter, only one flip-flop in the counter is set at any instant. In a shift-register counter, the amount of equipment is often minimized and very high counting rates are possible. In a Gray code counter, only one flip-flop changes state at any time.

Read/Write Memory (RAM): A generic term for either a volatile or nonvolatile memory the contents of which may be altered at will or read out without alteration. RAMs are randomly addressable and are readily implemented using various semiconductor technologies.

Read-Only Memory (ROM): A generic term for a nonvolatile memory whose contents are mask-programmed during the semiconductor processing procedure and whose contents may not be altered thereafter. ROMs are randomly addressable and are utilized for the storage of reference data, microinstructions, and various codes.

Programmable Read-Only Memory (PROM): A ROM whose contents may be field-programmed by the end user. PROMs offer an economical solution to the preparation of test programs and other bit patterns without having to resort to the expense of first-time mask programming of custom ROMs. PROMs are also useful for one-of-a-kind circuits and applications.

Electrically-Alterable Read-Only Memory (EAROM): The EAROM is a ROM whose contents may be altered (*i.e.* erased) by application of an external stimulus. The most common erase mechanisms are direct electronic erase signals and application of short-wave ultraviolet or X-radiation. After erasure, an EAROM may be reprogrammed by the user.

Clock: In a synchronous machine, the signal distributed throughout the machine to synchronize the operations. The term "clock" is also used to designate the oscillator or other device which acts as the source of the clock signal. The signal is distributed from the clock generator throughout the machine by means of the clock distributor, a tree-like arrangement of amplifiers, usually having well-controlled delay characteristics to ensure that the pulses trigger the various parts of the machine simultaneously. In a single-phase clock machine, all circuits receive the same signal. In a multiphase clock machine, two or more clock phases are distributed throughout the equipment, with each circuit connected to one of the available phases. This is usually done either to prevent "race conditions," in which the relative time delays through different logic paths is of importance, or to control the direction of information flow in circuits using two-terminal devices (e.g., tunnel diodes).

Synchronous Logic: Digital circuits in which the system changes state only at certain instants determined by a clock.

Asynchronous Logic: Digital circuits in which the system changes state at arbitrary instants of time determined by the propagation delays through the system components.

Comparator: An arrangement of gates which accepts two corresponding words or characters, and generates a signal to indicate the relationship between the two inputs. An equal comparator is one which indicates whether the two inputs are identically equal. A magnitude comparator indicates which of the two inputs is numerically larger.

Control: The portion of the machine which exercises overall control over the operation of the other portions. It normally contains counters to keep track of the steps required to carry out the operations. Based on the counter states and on the conditions of various status signals from other sections, the control section generates the appropriate control signals for the other sections.

Coder, Decoder, Translator, Matrix: An arrangement of gates used to convert data from one code to another. For example, it might change a character in one alphabet (say Hollerith) to another alphabet (say binary). It might convert a binary code of n bits to enable one out of 2^n lines. The various possible names for the devices are used almost interchangeably.

Checking Device: A device which either generates checking information to be appended to some data, tests the accuracy of existing checking information, or both. The most common type uses only a single parity bit as the checking information; it is usually called a parity checker. A wide variety of other types of checking arrangements can be used (e.g., cyclic code, Hamming code, and other multiple parity-bit techniques, modulo sum check, 2-out-of-5 code, etc.).

Microprocessors

Rapid advances in semiconductor large-scale integration (LSI) technology have permitted the combination and interconnection of an unprecedented number of logic gates, logic subassemblies, and memory arrays on a single silicon substrate chip. The end result of this major technology advance has been the fabrication of a new class of microminiature processors collectively termed microprocessors. The term microprocessor describes the physical size of the device and does not imply any inherent operating limitations. Indeed microprocessors routinely incorporate features (e.g. operational register stack) not found in more advanced conventional computers.

Since microprocessors can be designed to duplicate the central processing unit of a conventional computer, their main advantages are not operational but are instead low cost, compact size, and limited power consumption. For example, a typical microcomputer might consist of a single microprocessor chip, a random-access memory for logic control, a read-only memory for storing the computer instruction set, appropriate input and output chips, a crystal-controlled clock, and a power supply. All this circuitry can be placed on a single compact printed circuit card.

Due to their low cost, microprocessors are well-suited for such dedicated applications as traffic-signal controllers, coin changers, electronic games and amusement devices, meter readers, credit checkers, appliance controllers, gas-pump monitoring, electronic scales, cash registers, and "smart" test equipment and instruments. The compatibility of the microprocessor with a dedicated application is well illustrated by the single-chip calculator, a device which was the immediate forerunner of the first commercial microprocessors.

Fig. 7 shows the architecture of the first 8-bit microprocessor (Intel 8080). All the circuits in Fig. 7 are contained on a single 165×191 mil silicon n-channel MOS chip comprising about 5000 transistors and installed in a 40-pin dual in-line package (DIP). The chip has 78 basic machine language instructions and can execute assembly language instructions in 2 to 9 microseconds at a maximum clock frequency of 2 MHz. An internal 3-bit stack pointer permits seven levels of subroutine nesting, and a 16-bit stack pointer can address up to 65 kilobytes of external stack memory. Additional internal memory includes a 16-bit program counter, an 8-bit accumulator, an 8-bit status register, and a stack of six 8-bit registers. The 8080 can be connected as a computer with the addition of only six external TTL packages.

Fig. 8 shows how another single chip monolithic microprocessor, the Motorola XC6800, can be connected to external circuitry to form a complete microcomputer. This microprocessor has 72 basic machine language instructions, a 1-MHz maximum clock frequency, 8-bit parallel processing capability, 16-bit addressing of up to 65 kilobytes of memory, and six internal registers.

The first commercial microprocessors used PMOS and NMOS technology, but newer chips employing low-power CMOS have since appeared.

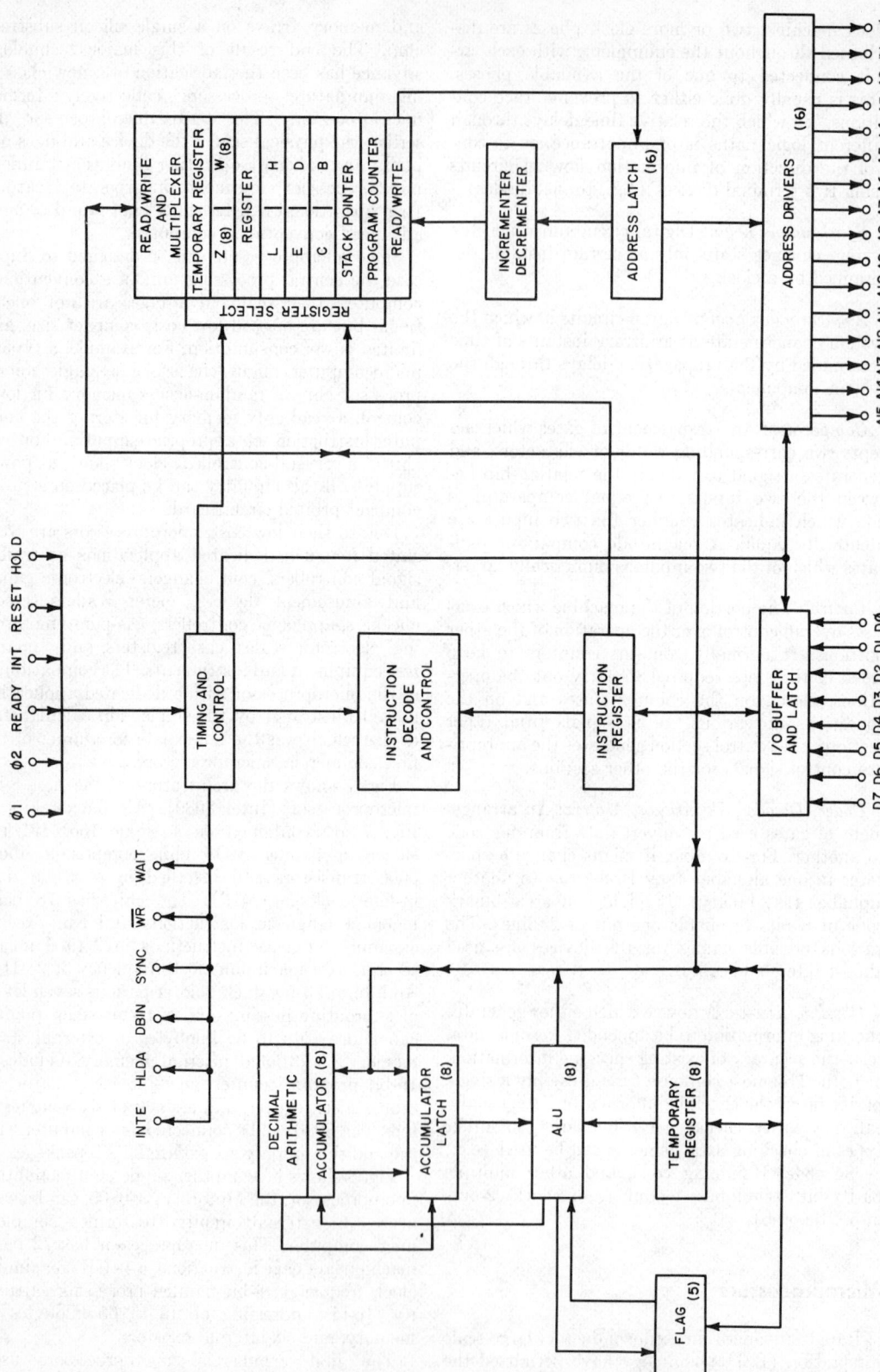

Fig. 7—Organization of 8080 microprocessor. (Numbers in parentheses are bit count.)

More recently, bipolar technology, which is inherently five to ten times faster than NMOS, has been used to produce I²L and Schottky microprocessor chips. It is expected that future improvements in LSI technology will permit a processor, memory, and input/output circuitry to be combined on a single monolithic substrate as a complete single-chip microcomputer.

Advances in microprocessor and related technologies are occurring quite rapidly, and the reader is advised to consult current technical periodicals for new developments. In addition to the excellent technical information available from manufacturers of microprocessors, two helpful references are A. G. Vacroux, "Microcomputers," *Scientific American*, May 1975, pp. 32–40 and L. Altman, Editor, *Microprocessors*, Electronics Book Series, McGraw-Hill, New York; 1975.

COMPUTER ORGANIZATION

When we refer to a digital computer or a data processor, we usually mean what is known as a stored-program machine. A program is a series of instructions to the machine (e.g., add the number in memory location X to the present contents of the accumulator, shift the accumulator contents

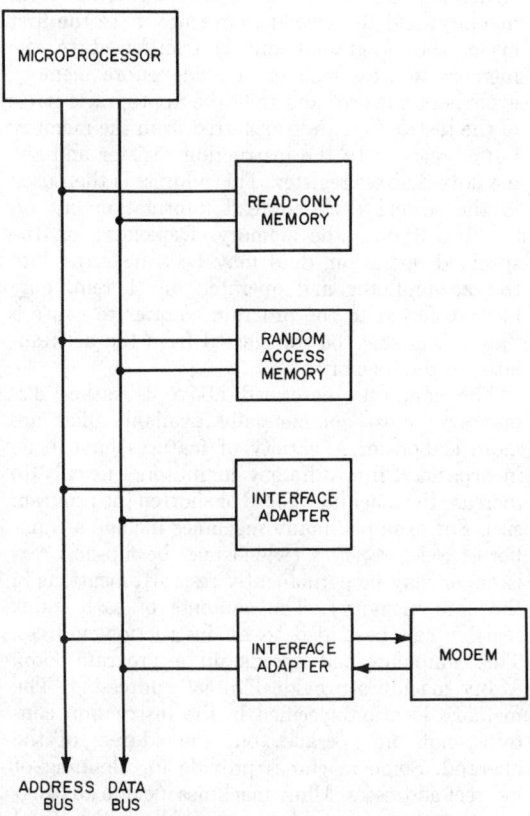

ADDRESS DATA
BUS   BUS

Fig. 8—Microprocessor microcomputer architecture.

right by 3-bit places, etc.). The execution of these instructions in the proper sequence causes the machine to perform the desired overall task. If a machine is designed to do only a simple or fixed series of steps, the sequence can be wired into the machine; this is known as a wired-program machine. In a stored-program machine, the series of instructions are stored in a memory within the computer. Each instruction is read from memory, in the appropriate order, and then decoded and executed. To change the task the machine performs, all that is necessary is to change the stored instructions (although it should be noted that in practice the preparation and checking of the programs often is very time consuming and expensive).

The accomplishment of a given task typically might involve performing a million instructions in sequence. If this had to be prepared beforehand as one long string of a million instructions, and fed into the machine in that form, the cost of program preparation would be prohibitive and the computer speed would be held down to that of the input device feeding the program in. Instead, advantage is taken of the fact that the total task can be broken into a number of short subroutines, which recur frequently throughout the overall task. Some subroutines can be used in exactly the same form each time; others require only that the memory addresses involved in the instructions be modified each time the subroutine is used. The actual program fed into the machine might consist of only a thousand instructions, which include the subroutines, plus possibly some instructions for modifying the addresses used in the subroutines (many computers have special hardware for this purpose), plus other "branch" or "jump" type instructions. A jump instruction can be used to cause the machine to end the present subroutine, and jump to another or back to the beginning of the one just finished. The latter is a "program loop", which can be traversed repeatedly until terminated by a "conditional jump" instruction when some specified condition is met.

The actual program of a thousand instructions, with appropriate branch instructions and conditions for branching, causes the computer to perform the instructions in a complicated order, skipping over some and repeating others until the task is completed.

As mentioned above, the machine can modify the address of an instruction. Normal practice is to store both data and instruction words in the same memory; the machine makes no distinction between them and can perform an operation on an instruction word just as well as on a data word.

The complete set of different instructions a given machine is capable of performing is its instruction repertoire. The number and variety of instructions included in different machines cover a wide range. It has been shown that any digital computation can be performed by a combination of a

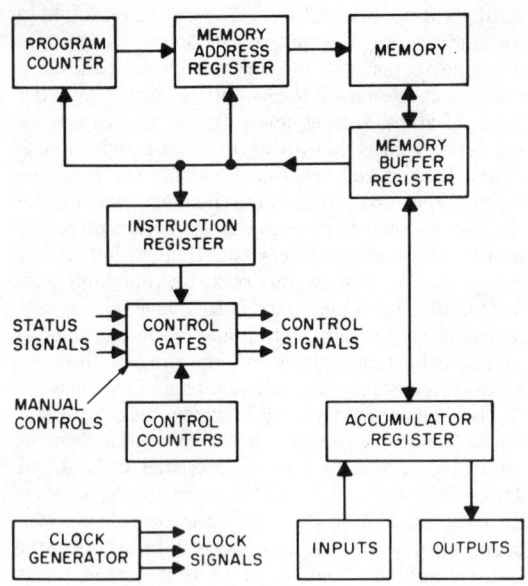

Fig. 9—Diagram of elementary single-address digital computer.

few very basic instructions. However, the processing time required by such a machine would be impractically long. Today's computers have a repertoire of from about 15 to over 100 different instructions.

An instruction always contains an operation code, which specifies the particular instruction (e.g., add, store in memory, etc.), and ordinarily at least one memory address. Most machines are single-address machines. The address usually specifies either an address in memory from which the operand is to be read and then used in performing the specified operation, or the address in which data are to be stored. A single-address machine, after executing an instruction, normally obtains the next instruction from the memory location immediately following; although in a jump-type instruction, the address contained in the instruction is used to cause it to go to the specified address to obtain the next instruction. Some machines are of the two-address, three-address, or four-address type. The additional addresses contained in an instruction can be used in many ways; e.g., to specify an additional operand address, the address of the next instruction, or where the results of the operation are to be stored.

Figure 9 is a block diagram of the principal portions of a simple single-address computer. It consists of the following sections.

Memory: Used to store all information, including both data and the program.

Memory Address Register: Used to hold the memory address for the current memory cycle.

Memory Buffer Register: Used as a buffer for information transfers between the core memory and the other sections.

Program Counter: Used to hold the address from which the current instruction is read.

Accumulator Register: A register and associated logic used to store a word of information and to perform a variety of operations thereon; arithmetic, logic, testing, or comparison operations are usually included, plus operations which shift or move the word or a part of the word.

Instruction Register: Used to hold the operation code portion of the current information.

Control Counters: The counter or counters used to keep track of the sequence of steps required to perform the instruction.

Control Gates: The logic which generates all signals required to control all of the other sections.

Clock Generator: Used in a synchronous machine to provide a clock or synchronizing signal to all parts of the machine.

Two memory cycles are required for most instructions: the first to fetch the instruction from memory, and the second to execute it. In the first cycle, the program count is transferred to the memory address register, a read–restore memory cycle is performed, and then the appropriate parts of the instruction are transferred from the memory buffer register to the instruction register and the memory address register. This address is then used in the second cycle to read information out of, or store it into, the memory. Depending on the specified operation, data may be transferred into the accumulator and operated on therein, may be transferred to the program counter to cause a "jump", or may be transferred from the accumulator to the memory.

The computer described above is rather elementary; most commercially available ones are more elaborate. A variety of features have been incorporated into different computers—usually to increase the machine's speed or shorten the program size. For example, many machines include a number of *index registers* (which may be flip-flop registers or may be permanently reserved locations in the core memory). The contents of each index register can be added to an instruction address. This simplifies and speeds up a program loop. Many machines provide indirect addressing. The memory location specified in the instruction contains, not the operand, but the address of the operand. Some machines provide for chaining of indirect addresses. Many machines include floating-point arithmetic operations as well as the usual fixed-point operations.

There are even greater variations between different machines in the input/output section, which is the equipment used to interconnect with other devices, such as magnetic tape and magnetic drum storage devices, paper tape and punched card readers and punches, analog-to-digital and digital-to-analog devices, communication equipment, and other computers. The input/output equipment also includes means for communication with the operator of the processor: indicator lights and control switches. Since most of the peripheral equipment is asynchronous with respect to the computer, most computers will include (A) one or more buffer registers to hold data or change its format during a transfer, and (B) an interrupt subsystem to cause the computer to respond to a signal from a peripheral device by interrupting its current task and entering a subroutine to handle the data transfer. If the number of devices is small and frequency of interruptions is low, a simple interrupt arrangement suffices. On the other hand, if the number of devices is large and the frequency of interruptions is high, a machine may be designed with an elaborate interrupt subsystem which automatically and quickly handles the "housekeeping" steps required in changing from one task to another and which provides for different priorities of interrupts, with low-priority subroutines being interrupted if a high-priority request occurs.

In many machines certain input/output operations are actually controlled by the peripheral device. Instead of the computer's program performing the data transfer, it merely pauses momentarily and permits the peripheral device to transfer data directly to or from the core memory, after which the program proceeds. This arrangement is common for data transfers to or from magnetic tape, drum, and disk memories. It requires that the peripheral equipment include a controller which keeps track of the pertinent addresses and counts off the characters or words in the block transfer.

Sometimes more than one computer is used for a given application. It may be that one acts as a satellite to relieve the main computer of some task, such as performing input/output work. Where high reliability is a requirement, a spare computer and spare peripheral equipment can be used, with provisions for switching in a spare unit to replace a defective one. In some applications, the spare computer is off-line, either continuously testing itself or doing some low-priority work. In other applications, the spare computer carries out exactly the same tasks as the on-line machine to minimize the switchover time and also to act as a check on the on-line machine. This is known as the *shadow technique*. Where extremely high reliability is required, three machines can perform the same task simultaneously; if a discrepancy occurs, the majority are assumed to be correct.

Another technique is to have two or more computers on-line and sharing the work, with all having access to the same memories and peripheral equipment. Control equipment prevents two computers from using the same device at the same time.

A variety of techniques have been used for multiprocessor applications. They generally are costly in equipment; not only are two or more computers required, but also substantial amounts of switching and control equipment. The programs may be extremely complex, particularly if the switchover is to be fully or almost fully automatic. In most instances to date, it has been necessary to depend on human intervention to determine which machine is defective and to initiate the switchover to the spare processor.

INPUT/OUTPUT EQUIPMENT

There are many types of computer input and output equipment. The data may be read directly into or out of the computer, or into or out of intermediate storage. There are also devices to transfer data between different input or output media.

Digital Magnetic Tape

Magnetic tape is used for computer input/output operations and as computer on-line and off-line bulk storage. Magnetic tape is a cheap permanent storage. Several hundred million bits can be stored on a 2400-foot reel (10.5-inch diameter) of 0.5-inch tape. The disadvantages of magnetic tape are a tendency toward errors (error rates from 10^{-5} to 10^{-9} per bit) and slow access. There are a number of parity checking codes and elaborate procedures to detect and correct almost all magnetic tape errors. Even if the tape is mounted on a transport, it takes several minutes to go from one end of the tape to the other.

The tape itself is composed of magnetic material deposited on a paper-thin plastic ribbon. The ribbon is usually wound on reels much as movie film. Although the tape can be stored off-line from the computer, it can be read into the computer only when it is mounted on a tape transport which passes it from the feed reel onto a takeup reel over a magnetic read–write head. The computer writes on the tape in a similar manner.

There are a number of ways to record data on tape. Figure 10 shows the magnetic flux recorded on tape as a function of distance along the tape.

For *RZ* (return-to-zero) and pulse recording, the tape is normally unmagnetized and pulses corresponding to the data are recorded on the tape. The *RB* (return-to-bias) modulation is the same as pulse recording except that the tape is nominally biased or magnetized in one direction.

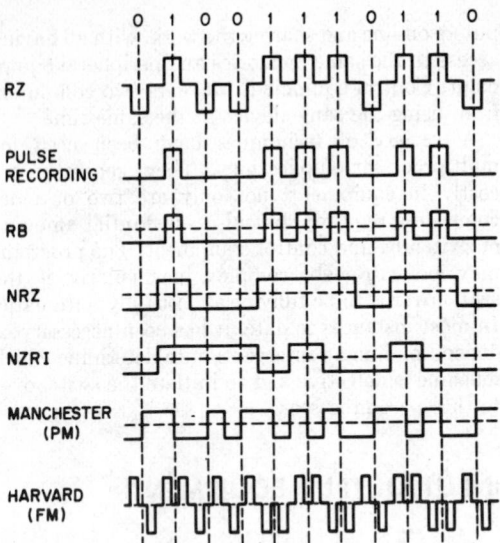

Fig. 10—Recording techniques.

RZ · PULSE RECORDING · RB · NRZ · NZRI · MANCHESTER (PM) · HARVARD (FM)

For *NRZ* (nonreturn-to-zero), *NRZI* (nonreturn-to-zero inverted), and *PM* (phase modulation), the tape is actually fully magnetized in one direction or the other. These three methods are the most common and they differ on how data are recorded on tape. For *NRZ* a 1 corresponds to positive magnetization; for *NRZI* a 1 corresponds to a transition in either direction; for *PM* a 1 corresponds to a negative transition at data time with transitions at intermediate times being ignored. *NRZ* and *NRZI* are equivalent and put the smallest frequency response requirements on the tape and on the tape transport electronics. *PM* is effective against varying oxide thickness and signal strength. For *FM* (frequency modulation) a 1 corresponds to two pulses in the same direction.

Magnetic tape comes in many widths, bit densities, formats, etc. Tape widths vary from 0.25 inch to several inches, but 0.5-inch and 1-inch tapes are most common. A given tape system has a number of channels across the width of the tape on which data are written or read in parallel. There may be one or many channels on a tape, but 7 and 9 channels are most common. The set of bits written in parallel usually represents one or possibly two characters with associated parity bits (also clock, if any).

Data on tape are written in blocks (Fig. 11)

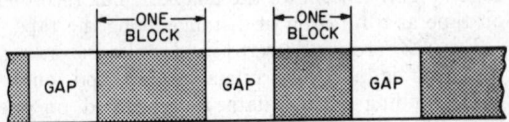

Fig. 11—Tape blocks.

with uniform spacing between the bits of a block in each channel. The length of a block is not fixed. The gaps between blocks, in which no data are written, are typically of the order of 0.75 inch. A group of blocks is called a file, and the end of a

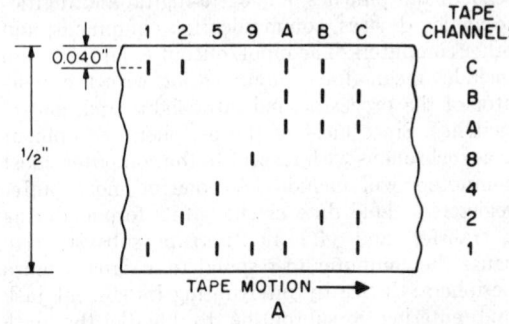

A

TAPE CHANNELS: C B A 8 4 2 1 — columns 1 2 5 9 A B C

TAPE MOTION →

	--	-A	B-	BA
----	bl	ƀ	-	⅋+
---1	1	/	J	A
--2-	2	S	K	B
--21	3	T	L	C
-4--	4	U	M	D
-4-1	5	V	N	E
-42-	6	W	O	F
-421	7	X	P	G
8---	8	Y	Q	H
8--1	9	Z	R	I
8-2-	ø	‡	!	?
8-21	#=	,	$	⧧
84--	@'	%(	*	¤)
84-1	:	v	]	[
842-	>	\	;	<
8421	√	+++	△	‡

B

SYM-BOL	NAME
bl	Blank (Space)
ƀ	Substitute Blank
-	Minus Sign, Hyphen
⅋	Ampersand
/	Slash
\	Backslash
<	Less-Than Sign
>	Greater-Than Sign
#	Number Sign
@	At Sign
*	Asterisk
¤	Lozenge
'	Prime, Apostrophe
(	Left Parenthesis
)	Right Parenthesis
[	Left Bracket
]	Right Bracket
‡	Record Mark
⧧	Group Mark
△	Mode Change
+++	Segment Mark
√	Tape Mark (Radical)
v	Word Separator
=	Equal Sign

C

Fig. 12—Seven-track data format. In *A* (shown oxide side down), the vertical lines on the tape represent ones that are written 0.048 inch long and are scanned over 0.030 inch for reading to allow for misalignment with the read/write head. In *B*, the standard 7-track binary-coded-decimal code and graphics are given (parity bit not shown). Figure 12*C* gives symbol names.

file is indicated by a very long gap and/or special characters written on the tape. Means must also be provided to recognize the beginning and end of the tape. Some systems have parity check characters associated with each block and methods for indicating unusable tape areas.

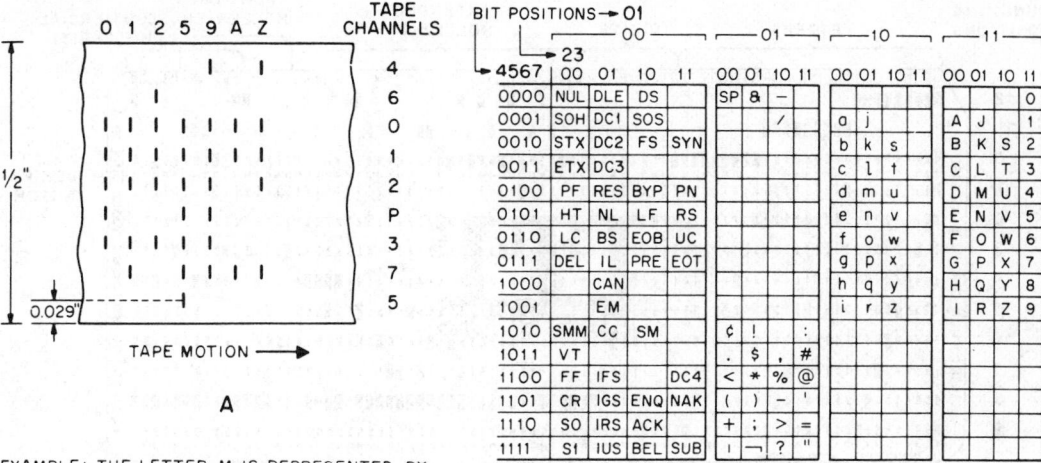

4567	\[00] 00	01	10	11	\[01] 00	01	10	11	\[10] 00	01	10	11	\[11] 00	01	10	11
0000	NUL	DLE	DS		SP	&	−									0
0001	SOH	DC1	SOS				/		a	j			A	J		1
0010	STX	DC2	FS	SYN					b	k	s		B	K	S	2
0011	ETX	DC3							c	l	t		C	L	T	3
0100	PF	RES	BYP	PN					d	m	u		D	M	U	4
0101	HT	NL	LF	RS					e	n	v		E	N	V	5
0110	LC	BS	EOB	UC					f	o	w		F	O	W	6
0111	DEL	IL	PRE	EOT					g	p	x		G	P	X	7
1000		CAN							h	q	y		H	Q	Y	8
1001		EM							i	r	z		I	R	Z	9
1010	SMM	CC	SM		¢	!		:								
1011	VT				.	$	,	#								
1100	FF	IFS		DC4	<	*	%	@								
1101	CR	IGS	ENQ	NAK	(	)	_	'								
1110	SO	IRS	ACK		+	;	>	=								
1111	SI	IUS	BEL	SUB	|	¬	?	"								

(BIT POSITIONS → 01; header groups 00, 01, 10, 11 each subdivided 23 = 00 01 10 11. Tape channels diagram A shown oxide side down; example: THE LETTER M IS REPRESENTED BY BIT CONFIGURATION 11 01 0100. NOTE: EOB = ETB AND PRE = ESC ARE DUPLICATE ASSIGNMENTS.)

Fig. 13—Nine-track data format. In *A* (shown oxide side down), the ones on the tape are written 0.044 inch long and are scanned over 0.040 inch for reading. In *B*, the 9-track Extended Binary Coded Decimal Interchange Code (EBCDIC) is given.

Most tape transports move the data area of the tape over the read–write head at a fixed speed and stop and start in the inter-record gap. Typical tape speeds are 37.5, 75, 112.5, 120, and 150 inches per second. Many different speeds are available from as low as 1 inch per second. Incremental tape transports are available which write or read standard tape formats and stop and start after each character. Incremental transports can read or write on tape at any speed up to 650 steps per second and at tape densities of up to 556 bits per inch.

Figures 12 through 14 illustrate IBM compatible tape formats. The most common bit densities are 200, 556, 800, or 1600 bits per inch.

Disc Memories

Magnetic tape has an exceptionally high storage capability but very long access time. The disc memory is a competitive system wherein data are stored on concentric tracks on a magnetic disk. A typical high-density disc (e.g., the IBM 3330) has 200 tracks per inch with a storage capacity of 4000 bits per inch.

An important innovation in disc memory technology is the so-called floppy disc. Made of flexible plastic and resembling a 45-rpm record, the floppy disc is considerably more economical and easier to ship and store than the metal discs used in conventional disc memory systems. Commercial floppy disc systems with a storage capacity of 2 million bits per disc and a 250 kHz bit-serial access rate are available. Random access times of less than

500 milliseconds make these systems far superior to less expensive but considerably slower magnetic tape storage systems.

Punched Cards

The punched card is a widely used input/output medium for digital computers. The standard IBM card (Fig. 15) is 3.25″×7.375″×0.0067″ in size. It is divided into 12 rows and 80 columns, and therefore has 12×80 = 960 hole positions.

The Remington Rand card (Fig. 16) has the same dimensions as the IBM card. It consists of two parts with 45 columns each. A column (in each part) has 6 hole sites.

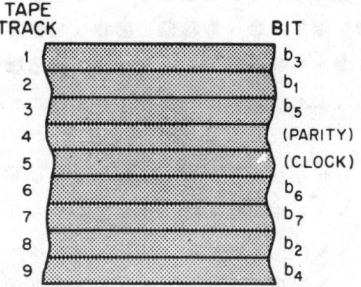

Fig. 14—Tape for USA Standard Code for Information Interchange (USASCII) X3.4—1967. © *1967 USA Standards Institute. Reprinted by permission.* See Fig. 30 of Chapter 35 for code table.

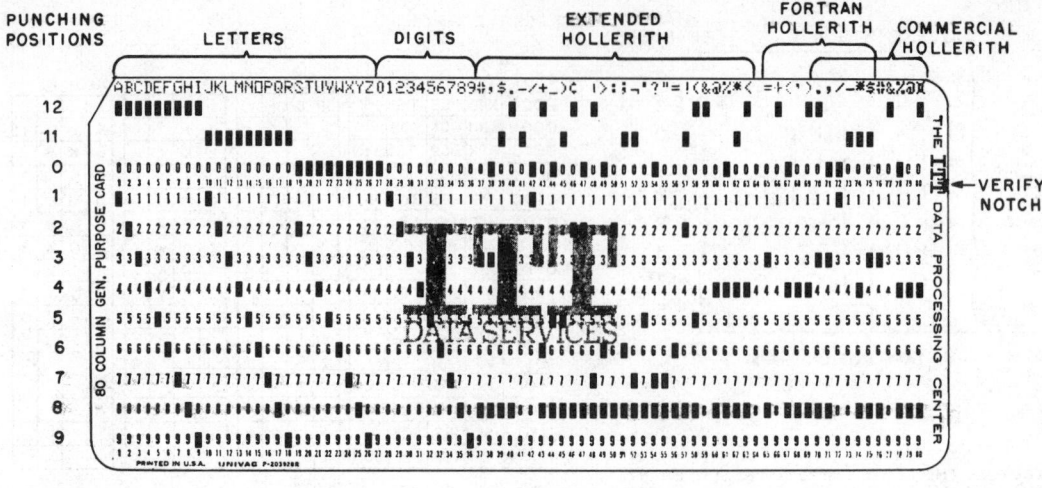

Fig. 15—IBM card with Hollerith code. *Courtesy of IBM Corporation.*

Card Readers

Reading can take place either serially (column by column), in parallel (row by row), or all at once. Most readers handle an entire card at a time, with all positions on the card being read simultaneously. Readers have speeds of from 100 to 2000 cards per minute. Reading is accomplished either mechanically with brushes or photoelectrically.

Card Punches

Card-punch on-line equipment normally operates at a nominal speed of 100 to 240 cards per minute. Manual punches are available for off-line punching.

Paper-Tape Readers

The most common paper tape is either 5-level (5 holes per row; Fig. 17) or 8-level (8 holes per row; Fig. 18). Information appears at *hole sites* where holes may or may not be punched. A character consists of a row of hole sites across the tape. A hole exists where a bit is a 1. Characters are placed along the length of the tape. The tape and hence the characters are scanned sequentially. A character row has punched in it a small hole defined as a *location hole*. It serves to indicate the presence of a character and acts as a mechanical means for grasping and moving the paper tape.

The tape is moved through the read station of the paper-tape reader, and all hole sites of a given character are examined simultaneously.

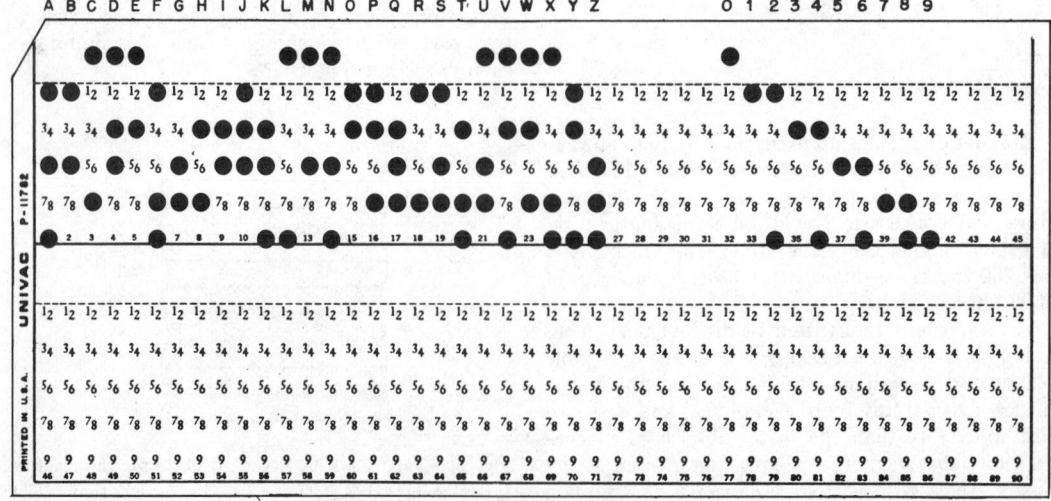

Fig. 16—Remington Rand card with 90 columns. *Courtesy of Sperry Rand Corporation.*

Reading is accomplished mechanically by wire brush or photoelectrically. Mechanical reading of paper tape takes place at low speeds in the range of 60 to 100 characters per second. Photoelectric paper-tape reading equipment can read at speeds in the range of 2000 characters per second, either unidirectionally or bidirectionally.

Paper-Tape Punches

A paper-tape punch punches the proper information and moves the tape one character width after each punch cycle. The punch is a slow device, operating at speeds up to 200 characters per second or greater.

Printers

The printer prints either a character at a time or can store and print a line at a time. Speeds vary from one or two characters per second up to 1000 lines or more per minute. Printing is done by mechanical (hammer and type wheel), cathode-ray tube and photosensitive paper, or electrostatic means.

Visual Displays

Cathode-ray tubes are commonly used for visual output displays. Figure 19 shows the CHARAC-TRON* tube, which can read out up to about 100 000 computer words per minute. The electron

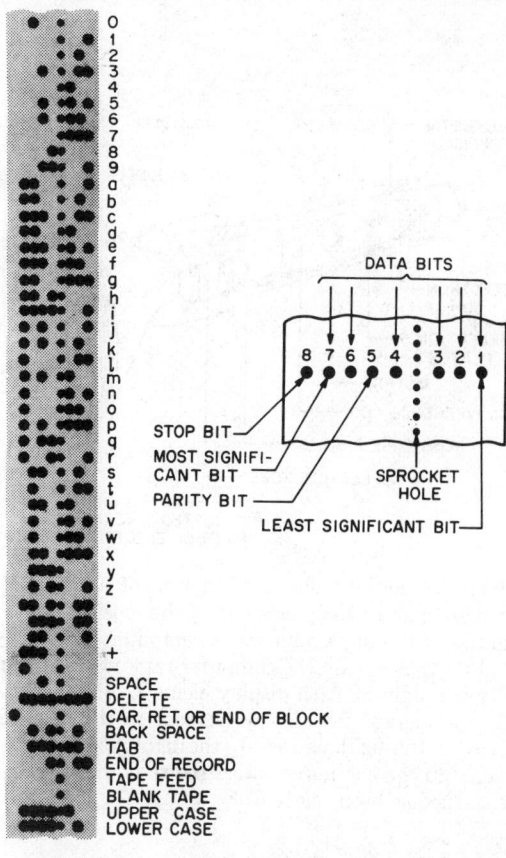

Fig. 18—Eight-channel hole code standard as proposed in EIA Standard RS–244. The typical format for 8-channel tape is also shown. *Courtesy of Electronics Industries Association.*

beam is deflected by the selection plates and then passes through the proper character in the matrix. A 6-bit code selects the desired character from 64 within the matrix, and coils keep the beam from scattering. A 20-bit signal then selects the position on the screen where the character is to be shown. The display rate in this example is 20 000 characters per second. Characters can also be formed either by a series of dots or by approximate deflection signals.

In some devices the light from the cathode-ray tube is used to drive a xerographic printer. The combination of a crt and a xerographic printer is capable of speeds as high as 5 to 10 pages per second. Alternatively, the crt may be photographed and the film negatives used to produce prints or for displays.

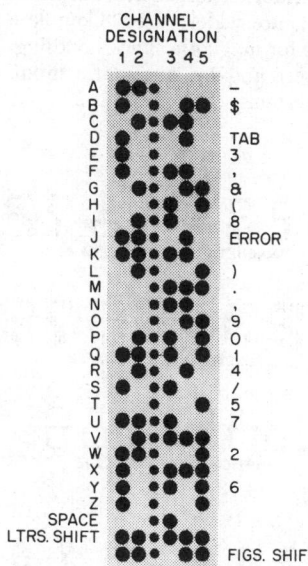

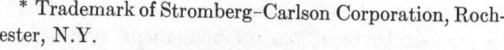

Fig. 17—Five-channel 29-code paper tape.

* Trademark of Stromberg–Carlson Corporation, Rochester, N.Y.

Light Pens

A light pen is a light-sensitive device by which a person can select some portion of the display for

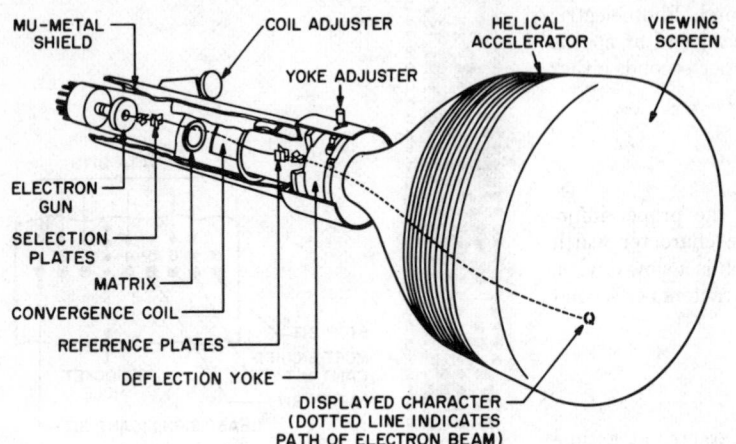

MU-METAL SHIELD
COIL ADJUSTER
YOKE ADJUSTER
HELICAL ACCELERATOR
VIEWING SCREEN
ELECTRON GUN
SELECTION PLATES
MATRIX
CONVERGENCE COIL
REFERENCE PLATES
DEFLECTION YOKE
DISPLAYED CHARACTER (DOTTED LINE INDICATES PATH OF ELECTRON BEAM)

Fig. 19—Construction of CHARACTRON® tube (Registered Trademark of Stromberg–Carlson Corporation).

computer action. The combination of a pictorial display and a light pen is one of the most effective means of linking a man and a computer.

In a typical case the computer rapidly scans the display, lighting each display element in sequence. At the instant the display element in the field of view of the light pen is lit, the light pen sends a signal to the computer indicating that this display element has been selected.

Keyboards

A keyboard is an input device which has a number of push buttons or keys. When a key is pressed, coded electrical pulses are presented to the processor. One commonly used keyboard is the typewriter, which can produce hard copy and coded pulses, and can also be used as an output device.

Reading Original Documents

Equipment is available that handles and reads original paper or card documents at high rates of speed. The source documents containing alphabetic and numeric printed characters (Fig. 20) may be printed in printer's ink or magnetic ink. Automatic reading of source documents eliminates the large amount of clerical work (typist and punch-card operations, etc.) otherwise needed to transform the original document into a computer input medium.

Printer's ink is most often used. However, the reading of normal printer's ink is based on black-and-white recognition principles and is susceptible to mutilation or strikeovers. Photoelectric scanning techniques have been developed to read and decipher printed characters in almost any style of printing. Each character is scanned a number of

times and each character produces its individual pattern of pulses. Line readers have also been developed.

Extraneous markings over magnetic ink have much less mutilation effect than the same markings over printer's ink. Magnetic-ink characters are specially formed to give a unique character signal indication when the document is passed under a magnetic reading head. In magnetic character recognition, the magnetic head integrates within a given time interval the total magnetic ink that constitutes the character.

Checks preprinted in magnetic ink and containing such information as account number, bank number, etc., are widely used. Complete systems are available for magnetic check handling (sorting, scanning, etc.) and for automatic input to computer data systems.

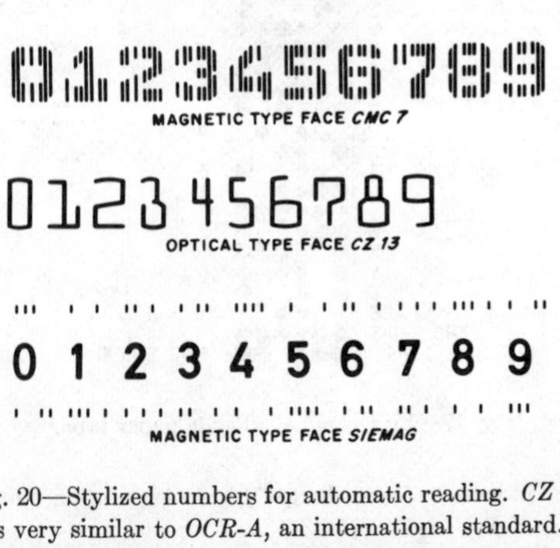

0 1 2 3 4 5 6 7 8 9
MAGNETIC TYPE FACE E 13 B

0 1 2 3 4 5 6 7 8 9
MAGNETIC TYPE FACE CMC 7

0 1 2 3 4 5 6 7 8 9
OPTICAL TYPE FACE CZ 13

0 1 2 3 4 5 6 7 8 9
MAGNETIC TYPE FACE SIEMAG

Fig. 20—Stylized numbers for automatic reading. CZ 13 is very similar to OCR-A, an international standard.

COMPUTER PROGRAMMING

Programming is the process of preparing a set of coded instructions which, when executed by a digital computer, yield the solution of a specific problem or perform specific functions. This section introduces some basic programming concepts and various levels of programming languages. Each computer and each programming language has its own unique instruction repertoire, method of operation, etc. These should be studied and understood before preparing a program on a specific language for execution by a specific computer.

A computer has the ability to automatically execute a program stored within itself. During execution of the program, the computer performs various digital operations (adding two numbers, moving data in and out of storage, reading in or printing out data, etc.). If the stored program is changed, the actions of the computer change. Thus, the computer actions depend on both the configuration of the computer hardware (the physical computer equipment) and the software (the programs stored within the computer). A given computer capability can be provided either by hardware alone or by a combination of hardware and software. For example, a square-root capability can be added to a computer either by additional equipment for a square-root instruction or by a program subroutine that computes a square root using the existing equipment. In the following, techniques are described by which a computer under control of a special processor program will accept programs written in a language much richer and easier to use than the language provided by the hardware. The choice of a given mixture of hardware and software depends on factors such as cost, speed, ease of maintenance, and flexibility.

There are three nominal levels of programming language: machine language, assembler language, and compiler language. The following sections briefly describe the general properties, advantages, and disadvantages of each level. Succeeding sections describe these levels in greater detail. (These levels are idealizations, and many systems permit programming at several levels.)

Machine Language

Machine language, the most elementary programming level, bears little resemblance to higher level languages customarily employed by programmers. Machine language consists of binary bit patterns comprising instructions or data which directly control the operation of each functional element of the processor. A specific bit pattern is termed a microinstruction, one or more of which may be required to implement a macroinstruction (i.e., a more conventional computer instruction). Very little programming is done at the machine language level since it is extremely tedious, time-consuming, and difficult to correct when errors are found. However, machine language programming is vitally important to the development of dedicated applications for microprocessors, initial operation of new computers for which higher languages have not yet been developed, repair of faulty computers, and debugging of programs when more conventional methods fail and the program must be examined at the microinstruction level.

Assembler Language

In the assembler-language level, the programmer generally writes one program instruction or one program constant for each memory location occupied by the program. Thus, the programmer has the same direct control over the computer operation that he would have if he wrote in machine language. However, the assembler language offers the programmer many aids such as the following.

(A) The form of the assembler language is easier for humans to use. For example, each instruction in assembler language is divided into a few easily understood fields; each instruction in machine language is a string of digits not easily understood. Thus, the addition of one number to another might have the form AD X3 in assembler language and 34071245 in machine language.

(B) In assembler language, much of the detail of the machine language can be ignored and it will be treated by standard conventions.

(C) Automatic allocation of storage. In machine language we must decide where each instruction and each constant is placed, and each reference to an instruction or constant must use the appropriate storage location. This is tedious in itself. If we now wish to insert several instructions or constants in the middle of a program, the part of the program beyond this point must be relocated and all references to this part of the program must also be adjusted. This makes changes difficult. In an assembler language, locations are referred to by a name such as AC27 or JOBNUM, and the computer—under the control of an assembler processor program—does all the bookkeeping. If the program is changed, all the addresses are automatically recomputed.

(D) Easy incorporation of existing subroutines into the program.

(E) Ability to insert comments into the program. Without comments, which describe the program and the meaning of the variables, it is very difficult to understand most programs.

(F) The assembler processor program finds many clerical or syntactical errors. Examples of such errors include the use of illegal codes, giving two statements the same label, or branching to a nonexistent address.

Compiler Language

The compiler language is an intermediate step between common human language and machine language, and is an attempt to satisfy both human and computer requirements. Thus, humans should find it easy and effective to program a computer task in compiler language, while means must be provided to automatically and quickly transform that program unambiguously into machine language.

At the compiler-language level, there is usually no simple relation between a compiler statement and a machine order instruction or constant. One compiler statement usually corresponds to many machine-language instructions, but a compiler statement can correspond to one or to zero machine-language instructions. Most programming is done in compiler language as it is easiest to learn and use. The time and effort to write a typical program in compiler language is a small fraction of the time to write the equivalent program in assembler language. Compilers allow us to easily generate or use subroutines for computation or input and output. The machine time to do a computation and the storage requirements are usually larger for a machine program produced by a compiler than for a machine program produced by an assembler. However, if the programmer is a novice or if the program and the computer are extremely complex, the opposite will be true and the speed/storage characteristics of compiler generated code will be superior to those of assembler generated code. Programs in commonly used compiler languages can be processed and executed on a large variety of computers, while assembler languages are usually limited to a given machine or family of machines. Assembler-language programs are preferred over compiler-language programs if computation time or computer storage is very critical, or if special techniques not easily available from a compiler are needed.

Programming a Problem

This section describes typical steps in the computation of a result using a compiler language. A similar description applies if an assembler language is used. In practice there may be deviations from this outline. Certain steps may be combined, certain steps may be repeated because of errors, and additional outputs may be produced.

(**A**) *Program Definition*. A decision is made to solve a given program on a computer and the mathematical techniques to be used are chosen.

(**B**) *Program Outline*. The program is outlined and the interfaces between different parts of the program are defined. For a simple program this step requires little effort; for large complex programs this step may require man-years of effort to produce the needed flow charts and documentation.

(**C**) *Coding*. The compiler-language program is written. The output of this step is a set of handwritten sheets.

(**D**) *Machine Input Preparation*. The handwritten sheets are converted into a form suitable for machine input, e.g., punched cards or tape.

(**E**) *Compilation*. At this step two programs are involved—the compiler-language program (also called a source program) for performing the desired computation, and the compiler processor program which processes the source program. The computer, under control of the compiler processor program, processes the source program and produces as output an object program. The object program is a machine-language program corresponding to the operations indicated by the source program.

(**F**) *Execution*. The object program is read into and executed by the computer, which reads in appropriate data as needed, performs the desired computation, and produces the desired output.

Clearly a key factor in this procedure is the compiler processor program, which translates from a source-language program a human easily understands to a machine-language program the computer understands. In general, compilers are complex, expensive programs which can be written only by highly trained and skilled programmers.

Program Libraries

Given a problem to be solved by a computer, there may be no need to write a program. Often similar problems have arisen before and a program already exists which can solve the given problem. In this case the existing program, along with data describing the specific problem parameters, is read into the computer and the desired results are obtained. For example, programs are available which can find the roots of an arbitrary polynomial given the polynomial coefficients. Large libraries of routines cover such problems as matrix inversion, statistical regression, the design of mechanical structures, the analysis of electrical networks, etc., and are available from computer manufacturers or from associations of computer users.

Even if no program is available which solves the given problem to the desired accuracy, there may be existing routines which, with a few changes, can solve the problem. The programming for a given problem often raises a question whether we should write a simple program to solve the immediate problem or a more complicated program which will be applicable to a larger class of future problems.

MACHINE LANGUAGE

Both instructions and data are stored inside the computer as strings of digits. Most computers do not have special symbols denoting the placement of the decimal point, the end of a word, or determining what are data and what is an instruction. The interpretation of a string of digits is determined by conventions built into the computer and by the position of the digits. If the digits are in the instruction register, they are interpreted as an instruction; if the digits are in the arithmetic section, they are interpreted to be a fixed-point or floating-point number depending on whether the instruction being performed is a fixed-point or floating-point instruction. This approach has two advantages: (A) There is no fixed allocation of memory between instructions and data, so that some programs can fill most of storage with numbers and other programs can fill most of storage with instructions. (B) The program can modify some of its own instructions as necessary. This technique is very useful, especially for modifying the address part of an instruction.

The following is a typical set of conventions.

Fixed-Point Numbers:

Ordinary notation $+.3116527$

Machine notation 03116527

The sign is represented by the most significant digit, with 0 representing *plus* and 1 representing *minus*. The machine is arranged to act as if the decimal point is as shown in the diagrams.

Floating-Point Numbers:

Ordinary notation $-.6527 \times 10^{+31}$

Machine notation $0\ 3\ 1\ 1\ 6\ 5\ 2\ 7$

Sign of Exponent   Magnitude of Exponent   Sign of Mantissa   Magnitude of Mantissa

A floating-point number has an exponent and a mantissa. The exponent consists of one digit representing the sign of the exponent and several digits representing the exponent magnitude. The exponent magnitude is an integer. The mantissa part consists of one digit representing the sign of the mantissa and a number of digits representing the mantissa magnitude. The mantissa is always a proper fraction as shown.

Instructions:

Order code	Special digits	Address or addresses

An instruction has several fields. The order code describes the operation to be performed, such as adding two numbers, multiplying two numbers, or storing a number in memory. Internally, each possible operation is given a number and the order-code field of the instruction contains the number of the operation to be performed.

The address part of the instruction contains a number or numbers describing a location or locations in storage. In a single-address computer, this location is that of the operand for the order code. In a multiple-address computer, the other addresses can refer to other operands or to the location of the next instruction to be performed. The address portion is often given a nonstandard interpretation for instructions that have no operand address.

The special-digits field of the instruction selects one of a set of permissible variations of the order. The use of this field differs greatly from machine to machine. Typical uses of this field are: to determine whether or not the instruction address is to be modified by an index register, or to determine the length of the fields to be processed by the order.

For example, assume that the addition order was indicated by the digits 34, and a special-digits section of 0 indicated that the order was to be performed in the nominal mode. Then the instruction

34	0	71245
Order Code	Special Digits	Address

would mean that the contents of memory location 71245 were to be added in the nominal mode.

ASSEMBLER LANGUAGE

A typical machine language microinstruction is a binary bit pattern such as 01 111 010. Since a processor might have an instruction set consisting of 75 or more basic instructions and several dozen variations, machine language programming at the binary bit pattern level is exceptionally tedious and prone to human error. An assembler language is a form of shorthand in which mnemonics are used in place of bit patterns to implement the machine language microinstructions. For example, the bit pattern 11 001 010 is the machine language "Jump if Zero" for a commercial microprocessor (Intel 8080), and the machine language mnemonic for this instruction is "JZ." Since different computers utilize different machine language instruction sets, a sample assembler language and some sample programs for a hypothetical computer are developed below.

The hypothetical computer has several components that directly concern the programmer. The *accumulator* retains the results of arithmetic operations. The *memory* has a number of locations that are numbered sequentially. In the assembler program, the locations can be described by symbols that contain a sequence of letters or numbers. The first character of the symbol sequence is always alphabetical. In the conversion from the source program (i.e., the assembler-language program) to the object program, the assembler processor replaces the symbols by appropriate numbers.

Typical Arithmetic Instructions

AD M   This instruction adds the contents of memory location M to the "old" contents of the accumulator and stores the sum in the accumulator. The old contents of the accumulator are erased. In this instruction and in every other instruction of our repertoire except store (denoted by the mnemonic ST), the contents of the memory are unchanged. In a program, the symbol M denoting a memory location can be replaced by any other symbol.

SU M   This instruction subtracts the contents of memory location M from the old contents of the accumulator and stores the difference in the accumulator.

MU M   This instruction stores in the accumulator the product of the contents of memory location M and the old contents of the accumulator.

Many computers have other arithmetic instructions, such as divide, square-root, and floating-point instructions. The hypothetical computer has only fixed-point operations.

Data Transfer Instructions

A large number of instructions transfer data from one location to another without modifying the data. The following are a few simplified data transfer instructions. In practice, the data transfer instructions for input, output, and mass-storage devices are very complicated due to considerations of timing, errors, and the provision of mass data transfer with a minimum of program control.

CA M   This instruction erases or clears the accumulator and places the contents of memory location M in the accumulator.

ST M   This instruction stores the contents of the accumulator in memory location M. The old contents of memory location M are lost. The accumulator contents are unchanged by this instruction.

SA M   This instruction transfers that part of the accumulator contents corresponding to an instruction address into the address field of the instruction in memory location M. Modification of the address field of instructions is an important programming concept.

IN M   This instruction reads one character from the input device denoted by M into the accumulator.

OU M   This instruction places one character from the accumulator into the output device denoted by M.

The assembler processor program translates the symbolic addresses of the input and output devices into numerical equivalents that are recognized by the hardware.

Simple Arithmetic Program

Problem: Given four memory locations denoted by X, A, B, C whose contents are denoted by x, a, b, and c, respectively, form $ax^2 + bx + c$ and place this number in location X.

Note that X (the number of the memory location) is, in general, different from its contents. For example, the tenth location (i.e., $X = 10$) might have contents .12345 (i.e., $x = .12345$). Note also that the program changes the value of x, and thus any description of the program must distinguish between initial values and final values.

The following program accomplishes the above task. To help the reader, the value of the accumulator for each step of this program is given in the comments column. In an assembler the use of the comments section is optional. Program execution starts at the top instruction and proceeds in order to the bottom instruction.

Instruction	Comments
CA X	x
MU A	ax
AD B	$ax + b$
MU X	$ax^2 + bx$
AD C	$ax^2 + bx + c$
ST X	$ax^2 + bx + c$

Branch Instructions

Although instructions are normally performed in order, means must be provided for changing the order. This capability for changing the order of instructions permits among other things the writing of (A) programs of moderate length to perform repetitive tasks, (B) programs to deal with functions that are defined one way for part of the range of the input variable and in another way for some other part of the range, and (C) programs that respond to external inputs.

The machine instructions used to change the order in which instructions are performed are most easily explained as follows. The computer instructions are stored in memory locations. The computer remembers which instruction is to be performed next by means of an "instruction counter" which contains the location in memory of the next instruction to be performed. Thus, the instruction counter is in effect a pointer which shows which instruction is next to be performed. In normal operation, as soon as one instruction is completed, the instruction counter advances to the next memory location so that orders are performed in sequence. The *branch instructions* are instructions which can modify the contents of the instruction counter, thereby changing the order in which instructions are performed.

TR M This instruction places the contents of the address part of the instruction into the instruction counter. Thus, since M is the address part of the instruction, the next instruction to be performed is in memory location M. This instruction is an unconditional transfer of the instruction sequence.

TN M The behavior of this instruction depends on the contents of the accumulator. If the contents of the accumulator are negative, the next instruction to be executed is in memory location X. If the contents are positive or zero, the next instruction to be executed is in the regular sequence. This instruction is called a conditional transfer.

There are as many possible conditional transfer instructions as there are conditions on which a transfer could be based. For example, the transfer condition might be that the accumulator was zero, that a certain console switch was on, etc. The following instruction is a practical way to implement a large number of conditional transfer instructions, with each instruction dependent on a different condition.

SO M, N This instruction has three fields: An order field represented by the letters SO; an address field represented by the letter M; and a special-digits field represented by the letter N. The special-digits field contains an integer which indicates one of many conditions. Thus, if the Nth condition is fulfilled when this order is performed, the next instruction performed is in location M. Otherwise, the next instruction to be performed is in the regular sequence.

Similar instructions can be used to turn on or turn off one of N indicators.

HALT This instruction halts the computer. Although this instruction does not modify the instruction counter, it is included here because it does change the order in which instructions are performed by terminating them.

Function with a Break Point

Problem: Given locations X, Y, XM, $A1$, $B1$, $A2$, $B2$ with contents x, y, x_m, a_1, b_1, a_2, b_2; place a_1x+b_1 in Y if $x<x_m$. Otherwise place a_2x+b_2 in Y.

The assembler notation must now be expanded to include a column in which the location of each instruction can be labeled by a symbol. This labeling is optional, and a programmer may label as many or as few locations as he wishes as long as the references are labeled correctly.

The following program performs the task described above. The comments column contains notes for the programmer, such as describing the contents of the accumulator.

Instruction Location	Instruction	Comments
	CA X	x
	SU XM	$x-x_m$
	TN RR	Branch if $x<x_m$
	CA A2	a_2
	MU X	a_2x
	AD B2	a_2x+b_2
	TR RS	Unconditional transfer
RR	CA A1	a_1
	MU X	a_1x
	AD B1	a_1x+b_1
RS	ST Y	
	HALT	

Data Words

The assembler language must provide means to specify constants and to reserve blocks of storage. These means can be provided by pseudo-operations such as the following.

DW X This pseudo-operation places the decimal constant X into storage. In a program, X will be a number rather than a symbol. Thus possible pseudo-operations are "DW 105329" or DW-69991. In a binary computer there will be pseudo-operations for binary and octal constants as well.

BS N This pseudo-operation reserves a block of storage N words long. In a program, N will be a number rather than a symbol. A possible pseudo-operation is BS 250.

Problem with a Loop

This section describes a key programming idea—the use of a program loop to do a repetitive task.

Problem: Given (1) a set of 100 constants x_1, x_2, $\cdots$, x_{100}, stored in consecutive locations starting at location X; (2) three constants a, b, c stored in locations A, B, C; and (3) a block of 100 locations starting at Y.

Place in location Y	$ax_1^2 + bx_1 + c$
Place in the location after Y	$ax_2^2 + bx_2 + c$
$\vdots$	$\vdots$
Place in the 99th location after Y	$ax_{100}^2 + bx_{100} + c$

This problem can be solved by placing in a loop the simple arithmetic program previously described. The solution shows how instructions can be manipulated in the arithmetic section as if they were numbers. Since the address portion of an instruction is the least significant part of the instruction, adding a constant k to the instruction makes the address field of the instruction larger by k and leaves the other parts unchanged.

Instruction Location	Instruction	Comments
R1	CA X	CA $(X+N-1)$ on the Nth loop
	MU A	ax
	AD B	$ax + b$
R2	MU X	MU $(X+N-1)$ on the Nth loop
	AD C	$ax^2 + bx + c$
R3	ST Y	$ST(Y+N-1)$ on the Nth loop
	CA R1	Add one to address field
	AD ONE	of the instruction in location R1
	ST R1	Store new instruction into R1
	SA R2	Copy this address into R2
	CA R3	Add one to address field
	AD ONE	of the instruction in location R3
	ST R3	Store new instruction into R3
	CA N1	
	AD ONE	
	ST N1	$N1 = -100 + N$ on Nth loop
	TN R1	
	HALT	Exit after 100th loop
X	BS 100	Data storage
Y	BS 100	
N1	DW -100	
ONE	DW 1	

Other Assembler Orders

Indexing Orders: The simple loop program just described performed a repetitive operation using the same instructions over and over again. While the loop program required fewer memory locations and less programming than straightforward hundredfold repetition with appropriate addresses of the six orders (from location R1 to location R3) needed to treat one x_i, the loop program also required about twice the number of executed computer steps as the straightforward repetition program. The extra computer steps are loop bookkeeping instructions. By means of indexing, loop programs can be written which at the expense of additional hardware take fewer memory locations than the simple loop program and run almost as fast as the straightforward repetition program. Indexing requires the following:

(**A**) An index register to store an integer which can be used to modify the effective address of an instruction. Thus, if the index register contains the number N, the instruction CA M will operate on memory location M under normal operation and on $M+N$ when the instruction is indexed.

(**B**) Instructions to transfer data into and out of the index register.

(**C**) A special branch instruction which increments the index register, thereby modifying the effective address of the indexed instructions; counts the number of loops; and also acts as a branch instruction which either continues the loop or terminates the loop. This special instruction requires about one machine operation time and performs all the loop bookkeeping.

Logic Orders: There are instructions which, instead of performing arithmetic operations on two computer words, perform Boolean operations such as AND, INCLUSIVE OR, EXCLUSIVE OR, etc. An example of such an instruction is the following.

AN M　This instruction places a 1 in the ith digit of the accumulator if both the ith digit of memory location M and ith digit of the old contents of the accumulator are 1. Otherwise, a 0 is placed in the ith digit of the accumulator ($i = 1, 2, 3, \cdots, n$ for an n-digit machine).

Shift Orders: There are instructions for shifting the data stored in the accumulator. Shift operations in which the data moved off the end of the accumulator are lost are distinguished from cycling operations in which the data moved off the end of the accumulator are recycled to the other end.

Initial contents of accumulator	012345678
Contents of accumulator shifted right one place	001234567

Contents of accumulator cycled   801234567
    right one place

Macroinstructions and Subroutines

Although most assembler instructions generate only one machine instruction, some assembler instructions called macroinstructions generate more than one machine instruction. Macroinstructions are used to incorporate subroutines into an assembler program.

Subroutine: A subroutine is a set of instructions that directs the computer to perform a function which is expected to be used in different programs or in different parts of a single program. For example, a subroutine which computes a square root can be used in a wide variety of problems. Subroutines may be classified in two categories: open and closed.

(**A**) *Open Subroutine*. An open subroutine is inserted in a program at the location where it is to be used. The open subroutine is normally used (A) if the main program requires it only once and (B) if it is short.

(**B**) *Closed Subroutine*. When the main program uses a subroutine several times, the subroutine need appear only once, with a calling sequence placed in each position of the program where the subroutine is used. The calling sequence provides the subroutine with the set of parameters and/or addresses needed to execute the subroutine and transfers control to the subroutine. An example of a parameter is the number whose square root will be calculated in the square-root routine for use by the main program. The subroutine then performs its calculation, communicates the results, and transfers control back to the main body of the program. Ordinarily the subroutine returns to the location following the calling sequence.

COMPILER LANGUAGE

Compilers are used for scientific calculation, business data processing, artificial language processing, and real-time control. The following describes some basic language forms taken from Algol, Fortran, and PL-1, which are popular compilers for scientific processing. This description illustrates the capability of compiler languages in a fraction of the space needed to formulate a complete and consistent system.

The choice of a compiler language depends heavily on a number of characteristics other than the ease with which a problem can be programmed in the compiler language. The following questions are pertinent. How expensive is it to write the compiler processor program? Even if it is very easy to program problems in a given compiler language (for example, an arbitrary mixture of English and mathematical equations), it may be prohibitively expensive to write the compiler processor program which converts from the compiler-language program to a machine-language program. How long does it take the compiler processor program to convert the compiler-language program (i.e., source program) into a machine-language program (i.e., object program)? How fast is the object program and how large is it? What debugging aids are offered by the compiler? The fact that a compiler language is a standard used by many organizations and computers may be important for the following reasons. First, an extensive library of routines written in this language will be available. Second, if programs are written in this language and the user changes computers, it is likely the new computer will accept the same compiler-language programs with little or no reprogramming. Third, there will be many programmers who know this language.

Syntax

A compiler language, as do most natural languages, possesses a syntax which describes how elements of the language may be combined. Punctuation marks and certain special symbols are essential parts of the syntax. A basic operation in the compiler language has the form

Label: Generalized Statement;

The *label* is a sequence of symbols identifying the basic operation. The *generalized statement* is used in many ways. For example, the generalized statement can tell the program to compute a new value for a variable, to transfer to a different part of the program, or can declare that certain variables represent integer quantities. The colon is used to terminate the label while the semicolon is used to terminate the basic operation. If no label is needed, the label and the colon may be omitted.

A generalized statement consisting of the word PROCEDURE is used to start a program or subroutine. A generalized statement consisting of the word END is used to end a program or subroutine. Thus, if S_1, S_2, S_3, S_4 represent generalized statements and L_1, L_2, L_3 represent labels, a program in computer language may have the form

L1:PROCEDURE;L2:S1;S2;S3;L3:S4;END;

In Fortran, positional information is used in place of some of the punctuation in the syntax described above. Therefore, certain columns of a card are reserved for labels and other columns for generalized statements.

Numbers

The compiler accepts numbers written in the ordinary form using the ten decimal digits and the special symbols $+$ $-$.

Examples:

3	$+3$	3.	27.416
83500	2750	.263	$-.264$

A number can also be written in a special form to indicate a number times a power of 10.

Compiler Notation	Ordinary Notation
5.E3	$5. \times 10^3$
3.14E–6	3.14×10^{-6}

Identifiers

A sequence of letters and numbers starting with a letter constitutes an identifier.

Examples:

A   B2   JOB   A16B43ZY

Identifiers are used to represent variables or labels. Subscripted variables can be represented by enclosing the subscripts in parentheses and separating the subscripts by commas.

Compiler Notation	Ordinary Notation
A (3)	A_3
BA (1, 2)	$BA_{1,2}$

Identifiers can be chosen freely, except that they cannot be the same as certain sequences reserved for special use.

The program describes which variables represent integers, real numbers, complex numbers, or arrays, by declarative statements. Thus, if the three variables A, B, and D are integers and CM is a two-dimensional array of real numbers with 6 rows and 10 columns, the following declarations are made.

INTEGER A, B, D; REAL ARRAY CM (6, 10);

Arithmetic Expressions

Arithmetic expressions can be formed using constants, variables, arithmetic operation symbols,

blanks, and parentheses. The operation symbols are $+$ $-$ $*$ $/$ $\uparrow$. The sum, difference, product, quotient of A and B are written as A+B A−B A*B A/B. A raised to the power B is written A ↑ B.

In contrast to ordinary notation in which various-size letters, subscripts, and superscripts are used, compiler-language expressions are composed of a single sequence of equal-size letters. In an expression, constants or variables may not be adjacent to other constants or variables but must be separated by an operation symbol. Also, any two operation symbols must be separated by either a constant or a variable.

Compiler Notation	Ordinary Notation
3.*X+B	$3X+B$
2.*X↑2−Y*(6+Y*(3+Y))	
	$2X^2 - Y(6+Y(3+Y))$

Although parentheses can always be used to clarify the meaning of an expression, certain rules define the order in which operations are performed when parentheses are absent. Thus, it is not *a priori* clear whether 3.*X+B means 3.*(X+B) or (3.*X)+B and whether 2.*X↑2 means 2.*(X↑2) or (2.*X)↑2. The compiler has the following order assigned to arithmetic operations; first the lowest-order operations are performed, then the next-order operations are performed, etc.

Order	Operations
1	$\uparrow$
2	*/
3	+ −

Thus 2.*X ↑ 2 − Y*(6+Y*(3+Y)) is equivalent to (2.*(X↑2))−(Y*(6+(Y*(3+Y)))).

If operations are of the same order, they are performed from left to right. Thus A*B/C*D= ((A*B)/C)*D.

Arithmetic Statements

An arithmetic statement has the general form

$$A = B$$

where A is either a subscripted or unsubscripted variable, and B represents an arbitrary arithmetic expression.

The arithmetic statement means that the computer will compute the value of the expression B, *using the existing values for the variables*, and then give the variable A this newly computed value. If the expression B does not use the variable A, the arithmetic statement is equivalent to an ordinary equality.

Examples:

$$Y=3.*X+R; Z=2.\uparrow Y;$$

However, the arithmetic statement is not an equality if the expression B does use the variable A. Thus

$$X=X+1$$

means increase the value of X by 1, i.e., the computer first computes the value of the expression $X+1$ using the old value of X and then replaces X with the value of the just-computed expression.

Note that if the operations represented by an arithmetic statement were written in assembler language, the assembler-language form would be both longer and less easily understood.

Branch Statements

Ordinarily, statements are executed in sequence. At any statement one can change the sequence of executed instructions with a GO TO *LABEL* statement which associates the next executed statement with the label. Thus

$$\text{GO TO TALLY};$$

means that the next executed statement is the one labeled with the identifier TALLY.

Often a branch is dependent on a condition. Conditions are of the form

$$(\text{expression})\rho(\text{expression})$$

where ρ is one of the symbols

$$< \leq = > \geq \neq$$

with the conventional meaning for these symbols. Thus, if the program branches to TALLY only if the product $X*Y$ is greater than 1, the following statement is written

$$\text{IF } X*Y>1 \text{ THEN GO TO TALLY};$$

If the condition is not fulfilled, the program continues in sequence.

There is also a conditional statement that covers both outcomes of the condition. This statement has the form

$$\text{IF B THEN S1 ELSE S2};$$

which means that S1 is performed if B is true and S2 is performed if B is false. Thus, if Y is defined in terms of X as

$$Y=\begin{cases}0 & X\leq 6 \\ 2X^4 & X>6\end{cases}$$

the following statement suffices.

$$\text{IF } X\leq 6 \text{ THEN } Y=0 \text{ ELSE } Y=2*X\uparrow 4;$$

Loops

The compiler language has special forms to describe loops in which a certain statement or set of statements is performed repeatedly. For example, assume that we are given two arrays, Y and X, with 100 elements and that $Y_j=3/X_j$ and $Z_j=2Y_j^2$ for $J=3, 5, 7, \cdots, 25$. Then the loop performing this operation can be written as a DO loop as follows.

```
           ARRAY X(100), Y(100);
MAR:       DO J=3 TO 25 BY 2;
           Y(J)=3/X(J);
           Z(J)=2*Y(J)↑2;
END MAR;
```

The statements

$$Y(J)=3/X(J); Z(J)=2*Y(J)\uparrow 2$$

will be performed 12 times as J goes through the values 3, 5, 7, $\cdots$, 25. Each loop is associated with a label. The loop label is the label of the DO statement which starts the loop and the loop label follows the END which terminates the loop. The initial value, the final value, and the increment of the loop index variable can be described by arithmetic expressions as well as by constants. Thus the following statement is valid.

$$\text{DO } J=2*X/Y \text{ TO DELTA}/3 \text{ BY} -M(X,Y)+R;$$

A loop need not run through all the values of the index variable indicated in the DO statement. The loop may be terminated by a conditional IF statement which transfers control out of the loop and terminates the loop.

A program loop may be placed inside another program loop, which may in turn be inside a third program loop, etc. The only restriction on loop placement is that no loop may be partly inside and partly outside of some other loop. Thus, the following program calculates Z, where

$$Z = \sum_{i=1}^{10} \sum_{j=6}^{25} A_{i,j} X_j Y_i$$

```
         Z=0;
Loop 1: DO I=1 TO 10;
Loop 2: DO J=6 TO 25;
         Z=Z+A(I, J)*X(J)*Y(I)
         END LOOP 2; END LOOP 1;
```

Functions

There are certain standard functions such as sine, cosine, and logarithm which are commonly used in arithmetic expressions. The compiler accepts mathematical expressions using these functions. Thus in compiler language $X+\text{SIN}(X)$ is equivalent to $X+\sin X$ in ordinary notation, and $\text{EXP}(X)$ is equivalent to $\exp(X)$ in ordinary notation. The argument of a function is always enclosed in parentheses. Also, the argument of a function can be an arbitrary expression. Thus if $Z = \exp(\sin X + 2)$ it can be calculated either by

$$Y = \text{SIN}(X) + 2;$$

$$Z = \text{EXP}(Y);$$

or by

$$Z = \text{EXP}(\text{SIN}(X) + 2);$$

The compiler automatically generates whatever functions are stored in its library without any effort on the part of the programmer. In addition, the programmer can form his own functions by writing function procedures. The procedure is labeled with the name of the function. The arguments of the function are placed in parentheses following the word PROCEDURE. The last executed statement of the procedure is a RETURN statement which contains the answer in parentheses. Thus the form of a function procedure is

FUNCTION NAME: PROCEDURE (List of Arguments);

Program to Find Value of Function;

RETURN (FUNCTION VALUE);

END;

Example: A square-root procedure usually has two parts; one part finds an approximation to the square root, and a second part improves the accuracy of the approximation. The following procedure improves the accuracy of an approximation to a square root.

The function is named SRT and has two arguments X_0 and Y. Define

$$X_{n+1} = \tfrac{1}{2}\left[X_n + (Y/X_n)\right] \quad \text{for } n = 0, 1, 2, \cdots, 5.$$

SRT (X0, Y) equals X_6 if

$$|(X_5 - X_4)/X_5| \geq 10^{-4}$$

and the first X_n for which

$$|(X_n - X_{n-1})/X_n| < 10^{-4}$$

if $(X_5 - X_4)/X_5 < 10^{-4}$. SRT(X0, Y) lies between $Y^{1/2}$ and $(1+5.1\times10^{-9})Y^{1/2}$ if X_0 differs from $Y^{1/2}$ by less than a factor of 2. A function procedure for SRT is the following. The function ABS is such that $\text{ABS}(X) = |X|$.

```
SRT: PROCEDURE (X0, Y);
    LOOP: DO J=1 TO 6 BY 1
        Z=X0; X0=(X0+Y/X0)/2;
        IF (ABS((X0-Z)/X0)<1.E-4) THEN GO
            TO RET;
    END LOOP;
RET: RETURN (X0);
END;
```

Subroutines

Function procedures are basically limited to the calculation of single-valued functions. Subroutine procedures are more general. Thus a subroutine can invert a matrix or process a table. The compiler language calls a subroutine by giving the name of the subroutine and listing the input and output information needed for the subroutine. This information includes certain constants, variables, arrays, and functions to be operated on, as well as the labels of the statements to which the subroutine will exit on various contingencies.

Input and Output Statement

Input and output operations require that we specify the data to be read in or out and the form of the data. The following describes one of the many ways to perform this task.

The form of the data can be described by a FORMAT statement with the general form of

LABEL: FORMAT (Description of Data Form);

where the label is an identifier for the FORMAT statement.

Some simple examples of data descriptions are:

I6 is a six-digit integer such as +12345 or 123456

F8.4 is a fixed-point number using 8 character spaces with 4 digits to the right of the decimal point such as −21.4503 or +10.0000

3X means 3 blank spaces

3HAGE means the word AGE is placed in the output. The H description is composed of three parts; (A) an integer describing the length of the output phrase, (B) the letter H, and (C) the output phrase.

SPACE means go to the next output line.

Plus signs and leading zeros are conventionally omitted in numerical output. Commas are used to separate different data descriptions.

The input or output statement has the form

Verb Label, List

The verb is a word such as READ or PRINT describing the action. The label identifies the pertinent FORMAT statement. The List is a list of variables which are to be accepted by the computer for an input statement and a list of the output variables for an output statement.

For example, output of the form

XbbbAREAbbbbVOLUME
12bbb7.3bbbbbb81.6

where b denotes a blank space, can be produced by the following program statements if the three variables X1, AR, VOL have the values 12, 7.3, and 81.6, respectively.

LA1: FORMAT (1HX, 3X, 4HAREA, 4X,
        6HVOLUME, SPACE, I2, 2X, F4.1, 4X,
        F6.1);
    PRINT LA1, X1, AR, VOL;

Other Compiler Forms

An actual compiler will allow more-complicated versions of the forms described above and will also allow additional forms to treat programming problems not described above. Moreover, a compiler will have a strict set of rules describing what is allowed, not allowed, and what may either be rejected or cause erroneous operation.

The forms described above strongly resemble mathematical notation. Compilers such as Cobol use a language resembling English. Thus a Cobol program uses statements such as

ADD INSURANCE, RETIREMENT, INCOME TAX, FICA GIVING TOTAL DEDUCTION THEN SUBTRACT TOTAL DEDUCTION FROM GROSS INCOME GIVING NET INCOME. IF NET INCOME EXCEEDS 2000.00 GO TO PAYROLL-ERROR-ROUTINE.

Other compilers designed for specific fields such as mechanical structure design or surveying accept forms chosen to meet the needs of those fields. Compiler programs may accept input or produce output in forms other than ordinary printing. Thus, compiler programs may produce output in graphic form or as oscilloscope displays.

Speed and Storage

Compilers provide techniques for minimizing storage requirements or execution time. Techniques for saving storage space include the storage of data so that they are accessible by several procedures, and the use of a given block of storage for different data at different times. Techniques for saving time include performing an operation once instead of doing it repeatedly, and choosing the fastest method to perform a given operation.

Example:

$$W = SIN(X)*SIN(Y); \quad Z = SIN(X)*COS(Y);$$

is equivalent to

$$U = SIN(X); \quad W = U*SIN(Y); \quad Z = U*COS(Y);$$

but the first method obtains $SIN(X)$ twice while the second method obtains $SIN(X)$ once.

Example $U = X*X$; is mathematically equivalent to $U = X \uparrow 2$, but the execution times can differ by as much as a factor of 10, as the first operation is compiled as a simple multiplication while the second operation may be compiled using the mathematical identity $A^B = \exp[B \log_e(A)]$. (Note that the mathematical equivalent of X*X and X↑2 is true only if factors such as roundoff and truncation errors are ignored.)

Often the computer is so fast and the memory so large that speed and storage considerations can be safely ignored. If speed or storage is a problem, we may (A) rewrite the program so as to correct the problems, (B) use a different language or compiler, or (C) use a larger or faster computer.

Debugging

Because programs tend to be complicated, most of them initially have errors and must undergo a debugging phase in which the errors are corrected and the programmer gains confidence in the program. Because debugging can take much effort, we should exert ourselves to write simple programs which will have fewer errors and be easier to debug.

In the debugging phase, the program works on test problems for which the answers or certain properties of the answers are known. Much debugging involves very simple test problems, as these can reveal most errors and require the least effort. A common beginner's fault is to use extremely complicated test problems to correct errors that could be corrected more easily by simple test problems. After the problems discovered by the simple test problems are corrected, only then should we use test programs with full complexity. The problems should fully test each feature of the program.

Common debugging aids include the following.

Error Printouts and Indications: The compiler and the computer both can provide indications of errors. These are very useful in debugging.

Dump Programs: Dump programs are developed to dump (i.e., print out) various areas within memory units as a debugging aid. By the use of parameters, the dump programs may be actuated to dump various areas of the memory units (e.g., fifth through sixteenth records on tape, locations 1000 to 1126 of memory, entire drum area, etc.). These dumps are generally performed when an error occurs within a program. The programmer is then able to determine the exact status of the computer at the time of error. By the use of dumps, the programmer may be able to determine the cause of the error and a solution to the problem.

Trace Programs: In normal computer programs, the programmer has access to almost none of the details of the computer operation. To do otherwise would drastically slow down the computer because of the effort of providing all this output. However, in debugging, the programmer may want to know the details of certain parts of program operation. The trace programs provide this facility. The programmer can compile the program so that, as the program is run, extensive details of the computation of selected portions of the program are printed out. At the conclusion of the test run, the programmer can study the output of the trace program to determine if the object program is operating correctly.

REFERENCES

1. "Handbook of Automation, Computation and Control," edited by E. M. Grabbe, S. Ramo, and D. E. Wooldridge, John Wiley & Sons, New York; 1959.
2. M. Phister, Jr., "Logical Design of Digital Computers," John Wiley & Sons, New York; 1958.
3. E. J. McCluskey, "Introduction to the Theory of Switching Circuits," McGraw-Hill Book Co., New York; 1965.
4. T. C. Bartee, J. L. Lebow, and I. S. Reed, "Theory and Design of Digital Machines," McGraw-Hill Book Co., New York; 1962.
5. D.D. McCracken, "Digital Computer Programming," John Wiley & Sons, New York; 1957.
6. R. S. Ledley, "Programming and Utilizing Digital Computers," McGraw-Hill Book Co., New York; 1962.
7. "Introduction to System Programming," edited by P. Wegner, Academic Press, New York; 1964.
8. E. I. Organick, "A Fortran IV Primer," Addison-Wesley Publishing Co., Inc., Reading, Mass.; 1966.
9. F. L. Bauer, R. Baumann, M. Feliciano, and K. Samelson, "Introduction to Algol," Prentice-Hall, Inc., Englewood Cliffs, New Jersey; 1964.
10. G. A. Korn and T. M. Korn, "Electronic Analog and Hybrid Computers," McGraw-Hill Book Co., New York; 1964.

INTRODUCTION

Since the late 1940's communication theory has developed along two main lines. These lines have their origins in the work of Wiener [1, 2] and Shannon [3, 4] and are essentially statistical in nature. The branch of communication theory that has come to be associated with the name of Shannon is commonly termed *Information Theory*.

Both Wiener and Shannon were concerned ultimately with the problem of the accurate reproduction of signals after their transmission over unreliable communication links. An important difference in their work is that while Wiener assumed that the signal could be processed only after it had been perturbed by noise, Shannon assumed that it could be processed both before and after perturbation. Wiener's name has come to be associated with the problem of extracting the signals belonging to a given ensemble from noise of a known type* whereas Shannon's name is associated more with the problem of encoding signals selected from a given ensemble in such a way as to make possible their accurate reproduction after transmission over a noisy communication system. This chapter discusses that part of communication theory arising from the work of Shannon.

Information theory as developed by Shannon is a "theory of measure" in the sense that it provides the communication engineer with methods for determining the limits of achievement when transmitting information over noisy communication links. Shannon has shown [3] that it is possible, even with a noisy communication link, to transmit information at a certain finite rate, determined by the link, with a probability of error that can be made as small as desired. This is a major result of the theory.

Shannon's theory is concerned with the statistical properties of symbols selected from suitably defined alphabets (ensembles) and is not concerned with the meaning associated with the

* For a very readable account of Wiener's work on statistical filtering and prediction, reference should be made to: Y. W. Lee, "Statistical Theory of Communication," John Wiley & Sons, Inc., New York; 1960.

selected symbols. As Shannon [3] states, "These semantic aspects of communication are irrelevant to the engineering problem."

Although the classical information theory of Shannon provides the engineer with methods for determining the performance limits of a system working under fixed physical conditions, it provides him with only vague indications as to how a data transmission system should be designed to achieve errorfree information transmission at a finite transmission rate. Some of the main developments in information theory during recent years have been concerned with the refinement and extension of the original theory [5–11] and with the vitally important practical problem of encoding and decoding messages so as to achieve errorfree transmission at a finite rate.

MODEL OF THE COMMUNICATION PROCESS

A block diagram of a rather general communication system is shown in Fig. 1 and the equivalent binary (on/off) system is shown in Fig. 2. The various parts of the system are discussed briefly in this section and in more detail in the sections which follow.

Information Source

The information source selects symbols (letters, numbers, words, sounds, etc.) from an alphabet (or ensemble) of possible symbols. The alphabet from which the symbols are selected is fixed and is independent of the communication process. Combinations of successively (sequentially) selected symbols form the messages that are to be transmitted over the communication system. The selective and statistical nature of the source is a main feature of modern communication theory.

Source-to-Signal Encoder

The source-to-signal encoder transforms the successively selected symbols into distinct physical

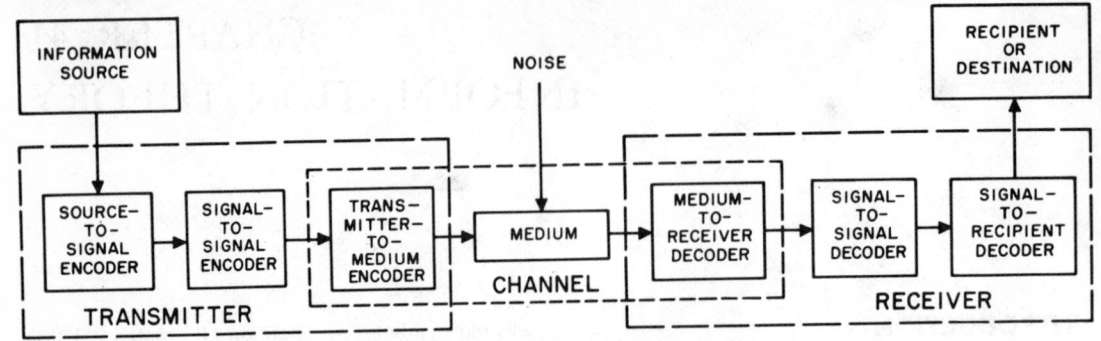

Fig. 1—Diagram of a communication system.

signals. These signals may, for example, take the form of voltage pulses as in telegraphic systems or more-continuous voltage/time functions as in radio and telephone systems. It is important to note the distinction between the symbols (which are selections from some predetermined alphabet) and the signals (which are physical representations of the selected symbols).

Signal-to-Recipient Decoder

The signal-to-recipient decoder operates inversely to the source-to-signal encoder. It converts physical signals into symbols suitable for use by the recipient. Typical of the outputs from signal-to-recipient decoders are the outputs from teleprinters and from radio and telephone systems. It is important to note that the signals which constitute the input to the signal-to-recipient decoder are dependent on any previous decisions made at the medium-to-receiver decoder.

Signal-to-Signal Encoder

The signal-to-signal encoder converts the signal representing a symbol into another of generally more complex form. The conversion process generally involves adding redundancy to the signals and is that part of the system which implements the encoding necessary when employing error-detecting or error-correcting codes.

Signal-to-Signal Decoder

The signal-to-signal decoder operates inversely to the signal-to-signal encoder. It converts from a coded signal (which, due to transmission impairments, may differ from that at the output of the signal-to-signal encoder) and produces a signal which, ideally, should correspond directly to the signal output from the source-to-signal encoder.

Transmitter-to-Medium Encoder

The transmitter-to-medium encoder (or modulator) operates on the encoded signals that represent information symbols and converts them into a form suitable for transmission over the medium connecting the transmitter to the receiver. Generally, there are restrictions on the signals that can be sent over the transmission medium. These restrictions may take the form of limitations on the power, bandwidth, and duration of the electrical signals used, and the transmitter-to-medium encoder must be designed so as to produce suitable signals.

Medium-to-Receiver Decoder

The medium-to-receiver decoder (or detector) operates inversely to the transmitter-to-medium encoder. It converts the modulated signals that are received into signals similar to those at the output of the signal-to-signal encoder. The device often acts as a primary decision maker and, in a binary system, may decide whether the received pulse is a binary 1 or a binary 0. The output signals from the medium-to-receiver decoder are used in the remaining decoding part of the receiver.

Channel

The channel is the medium and the fixed terminal equipment linking the transmitter and receiver. The term "fixed terminal equipment" needs elaboration since the application of information theory requires a careful definition of what constitutes the channel. Figures 1 and 2 show the transmitter-to-medium encoder and medium-to-receiver decoder as parts of the transmitter and receiver and also as parts of the channel. Convention has it that they be regarded as belonging to the transmitter and receiver. However, if the modulation and demodulation processes are fixed in the sense that

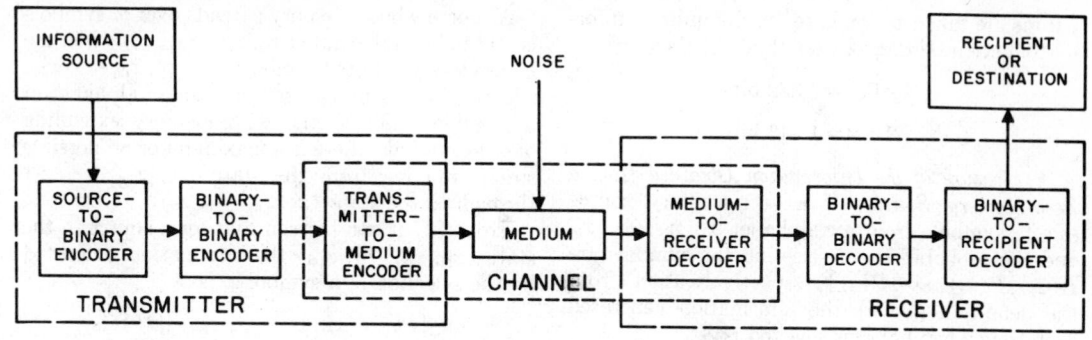

Fig. 2—Diagram of a binary communication system.

the designer is either unwilling or unable to change them, then they should be regarded as part of the channel. In general, in the application of Shannon's theorems, the channel represents that part of the system that the designer cannot or will not change [12] and includes the decision processes leading to the demodulator output.

DISCRETE-INFORMATION SOURCE AND BINARY ENCODING OF ITS OUTPUT*

Classification of Information Sources

The information source generates messages by making successive selections from an alphabet of possible symbols. The source may be either *discrete* or *continuous*. The latter system is considered in a following section.

A discrete-information source is one that selects symbols from a finite set $x_1, x_2, \cdots, x_n$ according to some probability rule. Telegraphy is a simple example of a discrete source and transmission system.

A continuous-information source is one that makes selections from an alphabet that is continuous within its range. An example of the output of a continuous source is the position taken by the pointer of an instrument used to measure the amplitude of some variable that may take any value within certain range limits.

This chapter considers only those sources known mathematically as *ergodic sources*. An ergodic source is one in which every sequence of symbols produced by the source is the same in statistical properties. If it is observed long enough, such a source will produce, with probability approaching

* For a detailed discussion of the topics treated in this section consult: R. M. Fano, "Transmission of Information," Chapters 2, 3, and 4, MIT Press and John Wiley & Sons, Inc., New York; 1961.

unity, a sequence of symbols that is "typical." In simple terms, this means that if a sequence is sufficiently long it will almost certainly contain numbers of symbols and symbol combinations that are independent of the particular sequence.

An information source is said to be *memoryless* or to have *zero memory* if successive symbols generated by the source are statistically independent. That is, a source has zero memory if each symbol is selected without influence from all previous selections. If previously selected symbols influence the selection of a symbol then the source is said to possess memory. If the selection of a symbol is influenced only by the immediately preceding symbol, the source is known mathematically as a *Markov* source. If the selection is influenced by the m previously selected symbols, the source possesses memory and is sometimes called an *mth-order* Markov source.

A Measure of Information, The Entropy Function

Definition*: If an event x_i occurs with probability $P(x_i)$ then the amount of information associated with the *known occurrence* of the event is defined to be

$$I(x_i) = \log_a[P(x_i)]^{-1}.$$

If, in the definition, logarithms are to the base 2, the units of information are in *bits* (a shortened form of binary digits). If logarithms are taken to the base e, the units of information are in *nats* (a shortened form of natural units). And if loga-

* The use of a logarithm in the definition of information is justified by the fact that information is required to have an additive property. Further justification for the use of this definition is given in this section under the heading "Properties and Interpretation of the Entropy Function." For a detailed justification of the definition consult: R. M. Fano, "Transmission of Information," Chapter 2, MIT Press and John Wiley & Sons, Inc., New York; 1961.

rithms are taken to the base 10, the units of information are in *hartleys* (after R. V. L. Hartley).

$$1 \text{ hartley} = 3.322 \text{ bits}$$

$$1 \text{ nat} = 1.443 \text{ bits.}$$

A Measure of the Information Obtained from a Zero-Memory Source: If a zero-memory source selects symbols from an alphabet $x_1, x_2, \cdots, x_n$, and the probabilities of selecting the symbols are $P(x_1), P(x_2), \cdots, P(x_n)$, respectively, then (from the definition above) the information generated each time a symbol x_i is selected is

$$\log_2[P(x_i)]^{-1} \text{ bits.} \qquad (1)$$

Since the symbol x_i will, *on average*, be selected $N \cdot P(x_i)$ times in a total of N selections, the *average amount* of information H' obtained from N selections is

$$H' = NP(x_1) \log_2[P(x_1)]^{-1} + \cdots$$
$$+ NP(x_n) \log_2[P(x_n)]^{-1} \text{ bits.}$$

Therefore, the average amount of information per symbol selection is

$$H'/N = H = P(x_1) \log_2[P(x_1)]^{-1} + \cdots$$
$$+ P(x_n) \log_2[P(x_n)]^{-1}$$

that is

$$H = \sum_{i=1}^{n} P(x_i) \log_2[P(x_i)]^{-1}$$

$$= -\sum_{i=1}^{n} P(x_i) \log P(x_i) \text{ bits/symbol.} \qquad (2)$$

The quantity H given by (2) is called the *Entropy Function*. This term is used since the form of (2) is the same as that derived in statistical mechanics [13] for the thermodynamic quantity entropy.

Note: The information associated with N selections from the statistically independent set is, on average, equal to N times the information per single selection.

A Measure of the Information Obtained from a Source with Memory: For a source whose memory extends over m symbols, the dependence on previous selections can be expressed mathematically in terms of a conditional probability. This gives the probability that the source will select x_i given that the m previous selections were $x_{l1}, x_{l2}, \cdots, x_{lm}$, where x_{lm} is the symbol selected immediately before the selection of x_i, and x_{l1} is the symbol selected m symbols before the selection of x_i. This conditional probability may be written

$$P(x_i/x_{l1}, x_{l2}, \cdots, x_{lm}).$$

It should be understood here that x_{li}; $i = 1$, $2, \cdots, m$; may be any one of the n possible source symbols; $x_1, x_2, \cdots, x_n$.

A source whose memory extends over m symbols is said to be in the state $(x_{l1}, x_{l2}, \cdots, x_{lm})$ when the m previously selected symbols were $x_{l1}, x_{l2}, \cdots, x_{lm}$. Clearly, for a source selecting from an alphabet of n possible symbols, and with memory extending over m symbols, there is a maximum of n^m possible states, ranging from the state $(x_1, x_1, \cdots, x_1)$ through to the state $(x_n, x_n, \cdots, x_n)$.

From (1) it can be seen that for a source in the state $(x_{l1}, x_{l2}, \cdots, x_{lm})$ the information generated by the selection of a symbol x_i is

$$\log_2\{P[x_i/(x_{l1}, x_{l2}, \cdots, x_{lm})]\}^{-1} \text{ bits}$$

and since the source can select any one of the symbols $x_1, x_2, \cdots, x_n$, the *average amount* of information generated per selection when the source is in the state $(x_{l1}, x_{l2}, \cdots, x_{lm})$ is

$$H[X/(x_{l1}, x_{l2}, \cdots, x_{lm})]$$
$$= \sum_{i=1}^{n} P[x_i/(x_{l1}, \cdots, x_{lm})]$$
$$\times \log_2\{P[x_i/(x_{l1}, \cdots, x_{lm})]\}^{-1} \text{ bits.} \qquad (3)$$

The function $H[X/(x_{l1}, x_{l2}, \cdots, x_{lm})]$ is called the *conditional entropy* and is a measure of the average amount of information generated by a source in state $(x_{l1}, x_{l2}, \cdots, x_{lm})$ when selecting a source symbol.

Since the source can be in any one of n^m possible states, it follows that if the probability of it being in the lth state is denoted by $P(x_{l1}, x_{l2}, \cdots, x_{lm})$, then the *average amount of information* generated by the source in selecting a symbol is

$$H = \sum_{l=1}^{n^m} P(x_{l1}, \cdots, x_{lm}) \cdot \sum_{i=1}^{n} P[x_i/(x_{l1}, \cdots, x_{lm})]$$
$$\times \log_2\{P[x_i/(x_{l1}, \cdots, x_{lm})]\}^{-1} \text{ bits.}$$

Therefore, using Bayes's Theorem, this can be rewritten as

$$H = \sum_{l=1}^{n^m} \sum_{i=1}^{n} P(x_i, x_{l1}, \cdots, x_{lm})$$
$$\times \log_2\{P[x_i/(x_{l1}, \cdots, x_{lm})]\}^{-1} \text{ bits/symbol.} \qquad (4)$$

The information generated by the source in selecting N symbols is $H' = NH$.

Properties and Interpretation of the Entropy Function

The entropy function has a number of properties which substantiate it as a reasonable measure of information. Some of these properties follow.

(**A**)

$$H = -\sum_{i=1}^{n} P(x_i) \log P(x_i)$$

is continuous in $P(x_i)$.

(**B**) If the probabilities $P(x_i)$ are equal $[P(x_i)=1/n]$, then $H=\log n$ and is, therefore, a function which increases with increasing n. This is a reasonable property of a measure of information since the more symbols available for selection, the larger the initial uncertainty and hence the greater the change in going from the uncertain to the certain state associated with the selection of a particular symbol.*

(**C**) $H=0$ if and only if all the $P(x_i)$ are zero except one which is unity. This again is a reasonable property of a measure of information, since if the outcome of a selection is known before the selection is made, then when it is made nothing is learned from it.

(**D**) For a given n, that is, a given number of source symbols, H is a maximum and equal to $\log n$ when all the $P(x_i)$'s are equal $[P(x_i)=1/n]$. This too is a reasonable property since it is the situation which intuitively has the most choice or uncertainty associated with it.

If an information source selects from an alphabet of only two symbols, x_1 and x_2, then it is said to be a *binary source*. If the probability of the symbols occurring are P and $Q(=1-P)$, respectively, the entropy function for a zero-memory source is

$$H=-P\log_2 P-(1-P)\log_2(1-P).$$

This function is shown in Fig. 3.

The output of a binary source is in binary digits or *binits*. The distinction between the bit, which is a measure of information, and the binit, which is an output binary symbol, should be carefully noted. Figure 3 shows that on average the amount of information provided by a binary source is always equal to or less than 1 bit/binit. The binary source provides one bit of information for each selected symbol only when the two symbols are equiprobable.

Regardless of whether a source possesses memory or not, the entropy function can be interpreted as the average amount of information provided by the source per symbol selected or, alternatively, as the average amount of information necessary to specify which symbol has been selected. If a source is allowed to select n symbols, where n is a very large number, then it will, with high probability, select only 2^{nH} different sequences of symbols, each having a probability of occurrence equal to $1/2^{nH}$ [5]. This is a direct physical interpretation of H. It means that, theoretically, any very long sequence of n symbols selected by the source can

* It should be noted that since, for equiprobable events, $H=\log n$; 1 bit can be defined as the amount of information obtained when the selection is made from 2 equiprobable events, and 1 hartley as the amount of information obtained when the selection is made from 10 equiprobable events. It is not possible to define the nat meaningfully in the same way since a physical selection cannot be made from e equiprobable events.

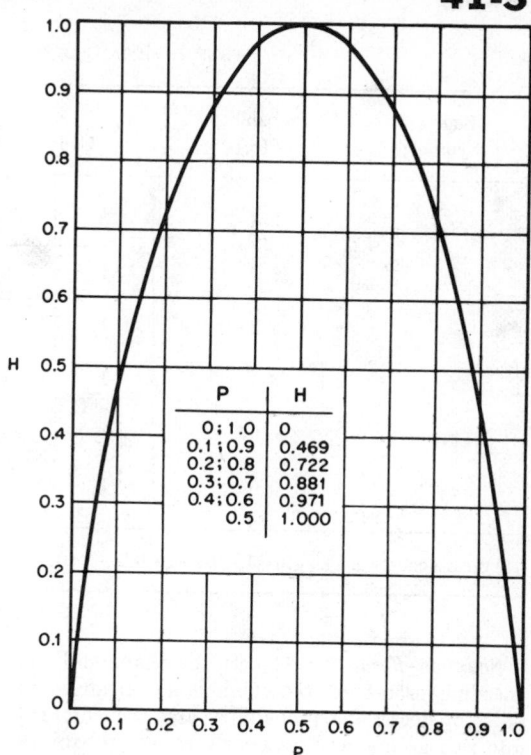

P	H
0 ; 1.0	0
0.1 ; 0.9	0.469
0.2 ; 0.8	0.722
0.3 ; 0.7	0.881
0.4 ; 0.6	0.971
0.5	1.000

Fig. 3—The entropy function; $H=-P\log_2 P-(1-P)\log_2(1-P)$.

be encoded and retransmitted using only nH binary digits, each carrying one bit of information.

Binary Encoding of an Information Source

When a symbol is selected by an information source, an average amount of information equal to H is produced. This implies that it should be possible to use the source-to-binary encoder in such a way as to transmit the selected symbol using, *on average*, only H binary digits (H being the lower limit). The lower limit can, in general, only be attained by encoding long blocks of source symbols. Often, in practice, many more digits are used than are theoretically necessary. In this section two methods are discussed for encoding the output of a source so as to represent selections unambiguously when using a reduced number of binary digits. The practical importance of encoding of this type is somewhat limited, since in general redundancy ("unnecessary" binary digits) is of considerable value in combatting interfering noise. In fact, error-detecting and error-correcting codes are designed to add redundancy. However, there are circumstances, particularly if occasional errors are not too serious or if the interfering noise is slight, where it may be to advantage to use as few binary digits as possible to specify and transmit a selected symbol.

TABLE 1—EXAMPLE OF SHANNON–FANO ENCODING.

Source Symbol	Probability $P(x_i)$	Code Words Representing Each Symbol					
x_1	0.4	0					code word 1
x_2	0.2	1	0				code word 2
x_3	0.2	1	1	0			code word 3
x_4	0.1	1	1	1	0		code word 4
x_5	0.07	1	1	1	1	0	code word 5
x_6	0.03	1	1	1	1	1	code word 6

Average code-word length $= 1 \times 0.4 + 2 \times 0.2 + 3 \times 0.2 + 4 \times 0.1 + 5 \times 0.07 + 5 \times 0.03 = 2.3$ binary digits/symbol.

Shannon–Fano Encoding: In the Shannon–Fano encoding procedure, the symbols are arranged in order of decreasing probability and then divided into two groups with as nearly equal probability as possible. The binary digit zero is assigned to each symbol in the upper group and the binary one to each symbol in the lower group. The process is repeated by dividing each of the two groups into two subgroups of nearly equal probability. A binary zero is then assigned to each symbol in the upper subgroup of each group and a binary one to each symbol in the lower subgroup of each group. The process is continued until each subgroup contains only one symbol.

This encoding procedure (Table 1) has the important properties of being economical in the use of binary digits and permitting unambiguous decoding on a symbol-by-symbol basis.

Table 2 is an alternative method of constructing the code words.

Huffman Encoding: Although the Shannon–Fano method of encoding is generally satisfactory, there is no guarantee that the average number of binary

TABLE 2—ALTERNATIVE METHOD OF SHANNON–FANO ENCODING.

Source Symbol	Probability	Code Words Representing Each Symbol				
x_1	0.4	0	0			code word 1
x_2	0.2	0	1			code word 2
x_3	0.2	1	0			code word 3
x_4	0.1	1	1	0		code word 4
x_5	0.07	1	1	1	0	code word 5
x_6	0.03	1	1	1	1	code word 6

Average code word length $= 2 \times 0.4 + 2 \times 0.2 + 2 \times 0.2 + 3 \times 0.1 + 4 \times 0.07 + 4 \times 0.03 = 2.3$ binary digits/symbol.

The entropy of this zero-memory source is

$H = -0.4 \log 0.4 + 0.2 \log 0.2 + 0.2 \log 0.2 + 0.1 \log 0.1 + 0.07 \log 0.07 + 0.03 \log 0.03 = 2.21$ bits/symbol.

TABLE 3—EXAMPLE OF HUFFMAN ENCODING.

Source Symbol	1st $P(x_i)$	2nd	3rd	4th	5th
					Arrangement

Source Symbol:	Code Words:
x_1	1
x_2	00
x_3	010
x_4	0111
x_5	01100
x_6	01101

Average code word length = 2.06 binary digits/symbol.

The entropy of the source = 1.999 bits/symbol.

digits used to represent a source symbol will be as small as or smaller than the average number used when encoding by some other scheme. An encoding procedure developed by Huffman [14] is optimum in the sense that no other encoding scheme uses, on average, a smaller number of binary digits to represent a symbol.

The Huffman encoding procedure (Table 3) is as follows.

Stage 1. The symbols are arranged in order of decreasing probability. (First arrangement.)

Stage 2. The two symbols of lowest probability are combined to form a single symbol whose probability is the sum of the two constituent symbols.

Stage 3. A new symbol set is formed from the original set, with the combined symbol replacing its two constituent symbols in the list. The new

symbol set is then arranged in order of decreasing probability. (Second arrangement.)

Stage 4. Stage 2 is repeated.

Stage 5. Stage 3 is repeated.

Stage 6. Stages 1 through 5 are repeated until a single symbol of unit probability is obtained.

Stage 7. Whenever two symbols are combined to form a new symbol, a binary zero is assigned to the upper symbol and a binary one to the lower symbol in the combination. The complete code word for a particular source symbol is the sequence of binary digits leading from the final unit-probability symbol back through the various symbol junctions to the source symbol in question.

Note: The *average* number of binary digits necessary to represent a source symbol can be reduced towards the entropy limit, H, if either the Shannon–Fano or the Huffman technique is used to encode blocks of source symbols rather than individual source symbols.

Relative Entropy and Redundancy

The ratio of the entropy of a source to the maximum value the entropy could take for the same set of source symbols is called the relative entropy.

Redundancy (R) is defined as being equal to 1 minus the relative entropy.

$$R = 1 - H/H_{max}$$

where H is the entropy and H_{max} the maximum value of the entropy.

COMMUNICATION CHANNEL

Classification of Communication Channels

Communication channels are generally classified according to the nature of their inputs and outputs, and the nature of the conditional probabilities relating these inputs and outputs.

If the input to a channel is discrete and the output is also discrete, the channel is said to be *discrete*. If the input and output are both continuous, the channel is said to be *continuous*. If the input is discrete and the output continuous the channel is said to be *discrete-to-continuous*. The channel is said to be *continuous-to-discrete* if the input is continuous and the output discrete. If the conditional probabilities relating the input symbols and the output symbols remain unaltered as successive symbols are transmitted, the channel is said to be *constant* or *memoryless*. If these probabilities depend on previously occurring input and output events, the channel is said to possess *memory*. This section considers only those channels that are discrete, memoryless, and one-way.*

Representation of a Channel

After a symbol or message has been selected by an information source and, possibly, suitably encoded (either by a technique such as that due to Huffman, or so as to implement some error-correcting technique) it is fed to the channel for transmission. At the receiver end of the channel a decision is made as to the symbol or message that was transmitted; this constitutes the output from the channel. Because of various forms of inter-

* For a discussion of systems involving feedback and two-way transmission consult: E. J. Baghdady, "Lectures on Communication System Theory," Chapter 14, McGraw-Hill Book Co. Inc., New York; 1961. L. S. Schwartz, "Principles of Coding, Filtering, and Information Theory," Chapter 10, Spartan Books, Inc., Baltimore, and Cleaner-Hume Press Ltd., London; 1963.

ference, incorrect decisions will be made from time to time and the output of the channel will differ from its input. The decisions made by the detector or decision-making part of the channel can be related to the channel input symbols by a set of conditional probabilities.

If the set of n input symbols is denoted by $x_1, x_2, \cdots, x_n$, and the set of k output symbols by $y_1, y_2, \cdots, y_k$, then the channel, which includes the transmitter and the decision-making process at the receiver end, can be represented by a diagram as shown in Fig. 4 or by a channel matrix as shown beneath the figure.

In the diagram and matrix representations of the channel, the $P(y_i/x_j)$'s are called the *forward probabilities*. $P(y_i/x_j)$ is the probability that a decision will be made that results in an output symbol y_i when, in fact, the transmitted symbol was x_j. Clearly, since for a particular input symbol a decision must be reached as regards an output symbol

$$\sum_{i=1}^{k} P(y_i/x_j) = 1.$$

The probability of obtaining a symbol y_i as output from the channel is

$$P(y_i) = \sum_{j=1}^{n} p(x_j) \cdot P(y_i/x_j).$$

From Bayes's rule it follows that the probability that a symbol x_j was transmitted, given that the output from the channel is y_i, is

$$P(x_j/y_i) = \frac{P(y_i/x_j) \cdot P(x_j)}{P(y_i)}$$

and, therefore

$$P(x_j/y_i) = \frac{P(y_i/x_j) \cdot P(x_j)}{\sum_{j=1}^{n} P(x_j) \cdot P(y_i/x_j)}.$$

$P(x_j/y_i)$ is called the *reverse* or *backward probability*.

A Measure of the Information Transmitted Over a Channel

Before an output is obtained from the communication channel, the probability that symbol x_j is the channel input is $P(x_j)$. The entropy associated with the input symbols is therefore

$$H(X) = \sum_{j=1}^{n} P(x_j) \log_2 [P(x_j)]^{-1} \text{ bits/symbol.}$$

(5)

This *a priori entropy* can be interpreted as the average number of bits of information carried by

Fig. 4—Diagram of a discrete channel.

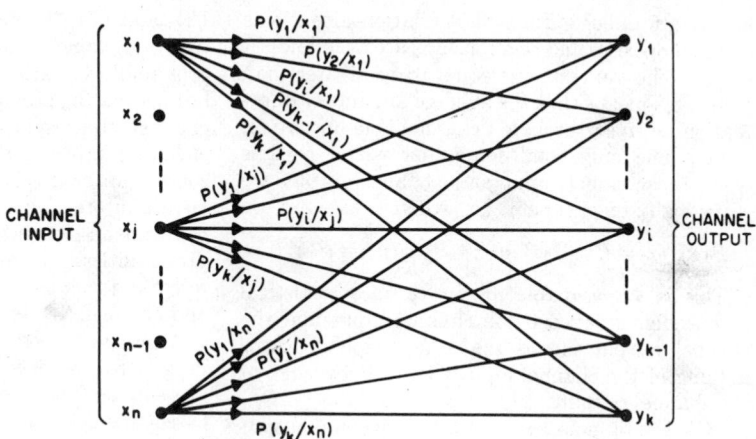

$$
\begin{array}{c}
\text{Input} \\
\text{Symbols}
\end{array}
\quad
\begin{array}{ccccc}
 & & \text{Output Symbols} & & \\
y_1 & y_2 & y_i & & y_k
\end{array}
$$

	y_1	y_2		y_i		y_k
x_1	$P(y_1/x_1)$	$P(y_2/x_1)$	$\cdots$	$P(y_i/x_1)$	$\cdots$	$P(y_k/x_1)$
x_2	$P(y_1/x_2)$	$P(y_2/x_2)$	$\cdots$	$P(y_i/x_2)$	$\cdots$	$P(y_k/x_2)$
x_j	$P(y_1/x_j)$	$P(y_2/x_j)$	$\cdots$	$P(y_i/x_j)$	$\cdots$	$P(y_k/x_j)$
x_n	$P(y_1/x_n)$	$P(y_2/x_n)$	$\cdots$	$P(y_i/x_n)$	$\cdots$	$P(y_k/x_n)$

an input symbol or as the average number of bits necessary to specify an input symbol.

After reception of an output symbol y_i, the probabilities associated with the input symbols are

$$P(x_1/y_i),\ P(x_2/y_i),\ \cdots,\ (P(x_n/y_i))$$

and the entropy associated with the set of input symbols $x_1, x_2, \cdots, x_n$ is

$$H(X/y_i) = \sum_{j=1}^{n} P(x_j/y_i)\ \log_2[P(x_j/y_i)]^{-1} \text{ bits.}$$

Taking the average over all possible output symbols gives

$$
\begin{aligned}
& H(X/Y) \\
&= \sum_{i=1}^{k} P(y_i) \cdot H(X/y_i) \\
&= \sum_{i=1}^{k} P(y_i) \cdot \sum_{j=1}^{n} P(x_j/y_i) \cdot \log_2[P(x_j/y_i)]^{-1} \\
&= \sum_{i=1}^{k} \sum_{j=1}^{n} P(y_i) \cdot P(x_j/y_i) \cdot \log_2[P(x_j/y_i)]^{-1} \\
&= \sum_{i=1}^{k} \sum_{j=1}^{n} P(y_i, x_j)\ \log_2[P(x_j/y_i)]^{-1} \text{ bits/symbol.}
\end{aligned}
\tag{6}
$$

$H(X/Y)$ is called the *a posteriori entropy* or *equivocation* and can be interpreted as the average number of bits of information carried by an input symbol after a symbol has been received at the output of the channel, or as the average number of bits necessary to specify an input symbol after a symbol has been received at the output of the channel. $H(X/Y)$ is a measure of the uncertainty associated with the input after the output has been received. This uncertainty is caused by the channel noise.

The difference between the *a priori* and *a posteriori* entropies, $I = H(X) - H(X/Y)$, is sometimes called the *mutual information* and, more frequently, the *information rate*. It follows from the interpretations of $H(X)$ and $H(X/Y)$ that I is a measure of the amount of information gained by the recipient as a result of observing the symbol at the output of the channel.

$$I = H(X) - H(X/Y) \text{ bits/symbol.} \tag{7}$$

Some Properties of Mutual Information and the Associated Entropies

The mutual information I has a number of important properties, and the associated entropies satisfy a number of important relationships. Some of the properties and relationships are as follows.

(**A**) The value of I is equal to or greater than zero. This means that the average amount of information received through a channel is nonnegative.

(**B**) The only condition under which $I=0$ is when the channel input and channel output are statistically independent, i.e., when

$$P(x_j, y_i) = P(x_j) \cdot P(y_i); \quad P(x_j/y_i) = P(x_j).$$

This is a reasonable property, since statistical independence between the channel input and the channel output means that the recipient learns nothing of the channel input from a knowledge of the channel output.

(**C**) For a noiseless channel, once an output symbol has been observed there is no uncertainty as to which input symbol was transmitted; hence $H(X/Y)=0$ and $I=H(X)$, the entropy of the channel input.

The following important relationships* can be shown to be true.

(**A**)

$$I = H(X) - H(X/Y) = H(Y) - H(Y/X)$$

where

$$H(Y) = \sum_{i=1}^{k} P(y_i) \log[P(y_i)]^{-1}$$

and

$$H(Y/X) = \sum_{j=1}^{n} \sum_{i=1}^{k} P(y_i, x_j) \log_2[P(y_i/x_j)]^{-1}.$$

(**B**)

$$H(X, Y) = H(X) + H(Y/X) = H(Y) + H(X/Y)$$

where

$$H(X, Y) = \sum_{i=1}^{k} \sum_{j=1}^{n} P(x_j, y_i) \log[P(x_j, y_i)]^{-1}.$$

(**C**)

$$H(X, Y) \leq H(X) + H(Y)$$

$$H(X) \geq H(X/Y)$$

$$H(Y) \geq H(Y/X)$$

with equality in each case if, and only if, X and Y are statistically independent.

Channel Capacity

The channel capacity C is defined as the maximum rate at which information can be transmitted over a channel. As can be seen from (5),

* R. M. Fano, "Transmission of Information," Chapter 5, MIT Press and John Wiley & Sons, Inc., New York; 1961. Also Reza [15] and Goldman [16].

(6), and (7), the mutual information, or information rate, depends not only on the *fixed* conditional probabilities relating the channel input and output but also on the probabilities with which the various channel input symbols are chosen. By a suitable encoding process, the output symbols from the source can be used in such a way that the $P(x_j)$'s governing the channel input symbols maximize the transmission rate for a fixed set of conditional probabilities. The encoding process is sometimes referred to as a *statistical matching* of the source and channel.

Although the calculation of channel capacity is, in general, somewhat involved algebraically, it presents no fundamental difficulties* and in certain cases† the calculation becomes relatively simple.

$$C = \max(I) = \max[H(X) - H(X/Y)]. \quad (8)$$

Capacity of Some Simple Channels

Binary Symmetric Channel: The channel shown in Fig. 5 is known as a binary symmetric channel. The channel input and output are binary and the probabilities are symmetrical.

$$C = 1 + p \log_2 p + (1-p) \log_2(1-p) \text{ bits/symbol.}$$

$$(9)$$

Erasure Channel: The channel shown in Fig. 6 is known as an erasure channel. It can be interpreted as the model of a channel in which the decision maker at the receiver end of the channel prints out an erasure if the ratio of the *a posteriori* probabilities associated with the channel input symbols is not sufficiently large.‡

* The general procedure for evaluating the capacity of a discrete memoryless channel is discussed in: R. M. Fano, "Transmission of Information," MIT Press and John Wiley & Sons, Inc., New York; 1961: pages 136–141.

† If the probability elements in the matrix representation of a channel satisfy certain conditions, the calculation of channel capacity becomes relatively simple. If each row in the matrix is a permutation of some set of conditional probabilities $p_1, p_2, \cdots, p_k$, the channel is said to be *uniform from the input*. If the columns of the matrix are permutations of some set of conditional probabilities $p_{11}, p_{12}, \cdots, p_{1n}$, then the channel is said to be *uniform from the output*. Under these conditions the calculation of channel capacity is relatively simple. The reasons for the simplicity of the calculation and method of calculation are discussed in Fano, *op cit*, pages 127–130.

‡ For a discussion of statistical decision theory and decision processes involving rejections (or erasures) consult: D. Middleton, "Introduction to Statistical Communication Theory," Chapter 23, McGraw-Hill Book Co., New York; 1960.

The capacity of this channel is

$$C = (1-q)[1 - \log_2(1-q)]$$
$$+ (1-p-q) \log_2(1-p-q)$$
$$+ p \log_2 p \text{ bits/symbol.} \quad (10)$$

By increasing the erasure rate, the probability of an incorrect decision can be reduced to a negligible value and the channel represented as shown in Fig. 7. This channel is known as a binary erasure channel and has a capacity of

$$C = 1-q \text{ bits/symbol.} \quad (11)$$

Rayleigh Fading Channel: Pierce [17] has considered a channel in which binary information is transmitted by frequency-shift keying (FSK) in the presence of Rayleigh fading and Gaussian noise. In the system considered by Pierce there are two receivers, one for each transmitted symbol, and envelope detection is used. The transmitted signal is assumed to be that associated with the receiver giving the largest output. The fading of

successive bauds is assumed to be statistically independent, as is the additive noise in the two receivers. It is assumed also that no phase or amplitude changes occur during each baud. Pierce has shown that a system possessing these properties, and satisfying the assumptions, can be represented as a binary symmetric channel with crossover probability of

$$p = [2 + (S_0 T/N_0)]^{-1}$$

where S_0 is the average transmitter power, T the baud duration, and N_0 the noise power/hertz at each receiver. The channel is shown diagrammatically in Fig. 8 and the function

$$p = [2 + (S_0 T/N_0)]^{-1}$$

is given in Fig. 9 for various values of signal-to-noise ratio. The capacity of the channel, for various values of S_0, T, and N_0, can be derived from Eq. (9).

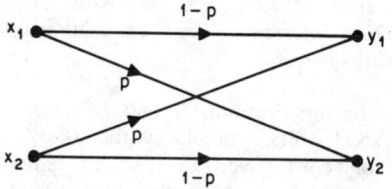

Fig. 5—Binary symmetric channel.

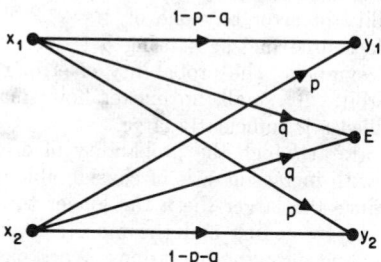

Fig. 6—Erasure channel.

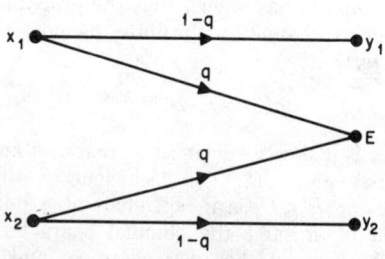

Fig. 7—Binary erasure channel.

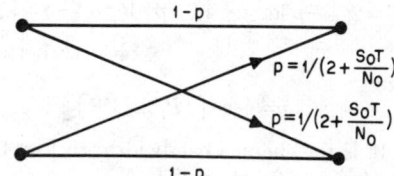

Fig. 8—Rayleigh fading channel with additive Gaussian noise.

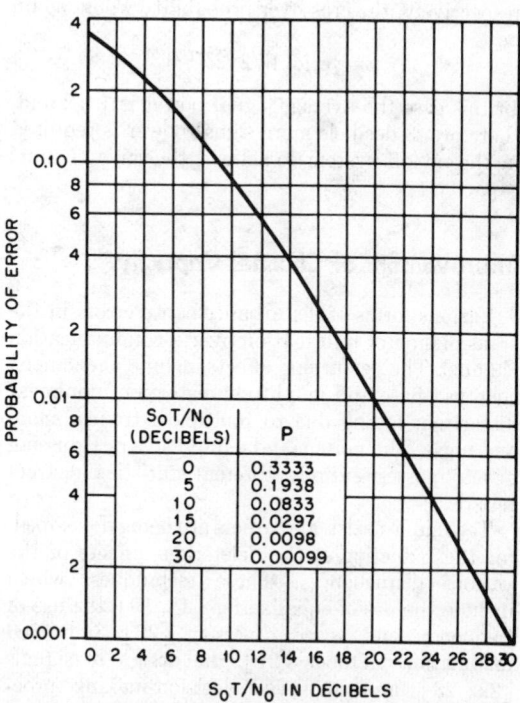

$S_0 T/N_0$ (DECIBELS)	P
0	0.3333
5	0.1938
10	0.0833
15	0.0297
20	0.0098
30	0.00099

Fig. 9—Crossover error probability in a Rayleigh fading channel.

Binary Channel with Gaussian Noise: If information is transmitted over a binary channel as a series of positive and negative pulses of amplitude $V/2$ and the channel is perturbed by additive Gaussian noise of average power $N(=\sigma^2)$, the crossover error probability is

$$p = \int_{-\infty}^{0} [\sigma(2\pi)^{1/2}]^{-1} \exp[x-(V/2)^2/2\sigma^2]dx$$

$$= \tfrac{1}{2}[1- \operatorname{erf}(V/2\sqrt{2}\sigma)]$$

$$= \tfrac{1}{2} \operatorname{erfc}(V/2\sqrt{2}\sigma)$$

$$= \tfrac{1}{2} \operatorname{erfc}(V/2\sqrt{2}N^{1/2}).$$

For pulses transmitted at the Nyquist rate and of duration $(1/2W)$ seconds, $(V/2)^2$ is equal to P, the average signal power. The crossover probability can then be written as

$$p = \tfrac{1}{2}\operatorname{erfc}[(P/2N)^{1/2}]$$

and the channel capacity as

$$C/W = 2[1+p\log_2 p+(1-p)\log_2(1-p)]$$

$$\text{bits/sec/hertz} \quad (12)$$

where

$$p = \tfrac{1}{2}\operatorname{erfc}[(P/2N)^{1/2}]$$

and W is the channel bandwidth. In Fig. 10, C/W is plotted as a function of P/N.

Note: If the information were transmitted as a series of on/off pulses, of amplitude V and zero, respectively, the crossover probability would again be

$$p = \tfrac{1}{2}\operatorname{erfc}(V/2\sqrt{2}N^{1/2}).$$

In this case the average signal power is $V^2/2$ and, therefore, 3 decibels more signal power is required in the on/off system to achieve the same channel capacity.

Improvement of Channel Capacity

Various forms of disturbance cause errors in the transmission of information over a communication channel. The perturbing effects include phenomena such as phase and amplitude distortion, nonlinear distortion, fading due to multipath transmission, and noise. The noise may be impulsive or Gaussian or may possess entirely different statistical characteristics.

Techniques exist, and others are being developed, for the reduction of the deleterious effects of the various disturbances. These techniques, which include the use of equalization [18, 19], the use of frequency and space diversity [20], improved modulation methods [21], the design of signals [22, 23], and improved decision-making processes [24], are channel modifications leading to a reduction in the error probabilities and a consequent increase in channel capacity.

Fundamental Theorem of Information Theory

The mutual information, $I=H(X)-H(X/Y)$, is a measure of the average amount of information transmitted over a channel. This is not to say, however, that the channel output is free from error or that a recipient could be certain of the channel input from a knowledge of the channel output. Knowing the channel output simply means that the channel input could be encoded using $H(X)-H(X/Y)$ fewer binary digits. The measure I, and more particularly its maximum value C (the channel capacity) has, however, been given definite significance in terms of errorfree transmission in a theorem due to Shannon [3]. The theorem, which is variously known as *Shannon's Second Theorem*, the *Noisy Channel Coding Theorem*, or the *Fundamental Theorem*, can be stated approximately as:

If an information source has an entropy H and a noisy channel a capacity C, then, provided $H \leq C$, the output from the source can be transmitted over the channel and recovered with an arbitrarily small probability of error. If $H > C$, it is not possible to transmit and recover information with an arbitrarily small probability of error.

Note: In the theorem, H and C are measured in bits/sec $(=$ bits/symbol $\times$ symbols/sec$)$.

To achieve errorfree transmission it is necessary that messages from the source be encoded using long sequences of n channel symbols. Shannon's theorem means that, with a channel of capacity C, it is possible to transmit with an arbitrarily small probability of error any one of $M=2^{n(C-\lambda)}$ equiprobable source messages using a sequence of n channel symbols. The probability of error can be made arbitrarily small, no matter how small the λ, provided n is sufficiently large.

The rate at which the probability of error decreases with increasing n is of considerable importance, since the larger the n the longer the delay introduced by coding and the more complex the encoding and decoding operations. Work has been done [9, 25, 26] which shows that for various channels the probability of error decreases exponentially (or almost exponentially) with increasing n. Fano, *op cit*, has shown that the probability of error for a channel with finite memory has a *general form*

$$p(e) = K \cdot 2^{(-na/R')}$$

where K is a slowly varying function of n and the transmission rate R'. The coefficient a, which is positive for $R' < C$, is independent of n but is a function of R' and the channel characteristics. Also, Shannon [9] has derived upper and lower bounds for the error probability in channels with

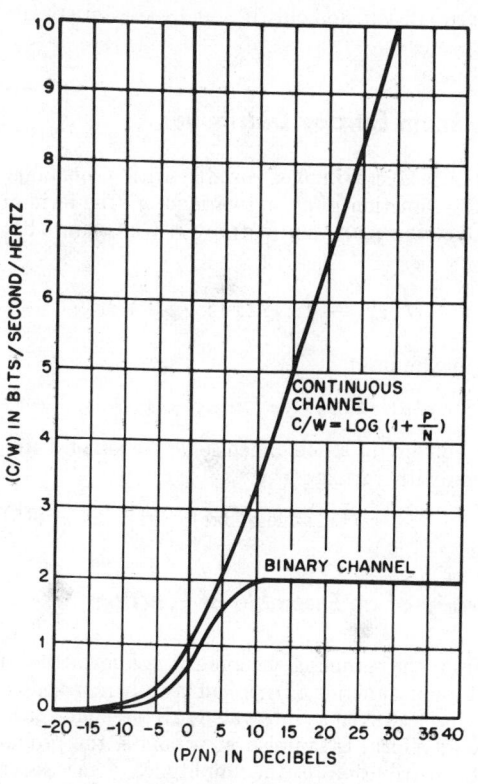

P/N (DECIBELS)	C/W (BINARY CHANNEL)	C/W (CONTINUOUS CHANNEL)
-20	0.0092	0.01
-15	0.0288	0.04
-10	0.0897	0.13
-5	0.271	0.40
0	0.738	1.00
5	1.537	2.06
10	1.982	3.46
15	1.999	5.03
20	2.000	6.66
25	2.000	8.31
30	2.000	9.97
35	2.000	11.63
40	2.000	13.29

Fig. 10—Channel capacity in bits/sec/hertz with Gaussian noise.

additive Gaussian noise in which optimal coding and decoding are used.

CONTINUOUS-INFORMATION SYSTEMS

Continuous-Information Source

A discrete-information source generates information at a finite rate, and the entropy function is a measure of the information generated. In the case of a continuous-information source the situation is, however, more involved. Complications arise because a continuously variable quantity can assume any one of an infinite number of values and requires, therefore, an infinite number of binary digits for its exact specification. An immediate consequence of this is that in order to transmit the output of a continuous-information source and recover it exactly, a channel of infinite capacity is required. Since, in practice, a continuous channel is perturbed by noise and therefore has a finite capacity (as will be shown later), it is not possible to transmit the output of a continuous source over a noisy channel and recover it exactly.

The fundamental difficulties associated with continuous sources can be avoided in practice, since it is the transmission and exact recovery of information (where the information represents the source output as accurately as desired) that is important and not the transmission of exact continuous information. Shannon [3] has shown that, if the output of a continuous source is specified to within certain tolerance limits, it is possible—in very general cases—to assign a definite value to the rate at which information is generated by the source. This information can be transmitted over a channel and the probability of error in recovery made arbitrarily small, provided the rate of generation is less than the channel capacity.

Sampling Theorem

The Sampling Theorem [4] is an important aid in the study and analysis of communication systems involving the use of continuous-time functions that are of limited bandwidth.

The theorem states that, *if a function of time $f(t)$ contains no frequencies higher than W hertz, it is completely determined by giving the value of the function at a series of points spaced $1/2W$ seconds apart.*

If $f(t)$ contains no frequencies greater than W hertz, then it can be expressed as

$$f(t) = \sum_{n=-\infty}^{n=+\infty} X_n[\sin\pi(2Wt-n)/\pi(2Wt-n)]$$

where

$$X_n = f(n/2W).$$

It is important to understand that the theorem

makes no mention of the time origin of the samples. The time origin is unimportant; it is only the spacing of the samples which matters.

If the function $f(t)$ is substantially zero outside a time interval T and contains no frequencies higher than W hertz, it can be specified by $2TW$ ordinates.

Entropy of a Continuous Distribution*

The entropy of a continuous variable x with probability density function $p(x)$ is defined as

$$H(x) = -\int_{-\infty}^{+\infty} p(x) \, \log_2 p(x) \, dx. \qquad (13)$$

With an n-dimensional density function $p(x_1, x_2, \cdots, x_n)$ the entropy is defined as

$$H(X) = -\int_{-\infty}^{+\infty} \cdots \int_{-\infty}^{+\infty} p(x_1, x_2, \cdots, x_n)$$
$$\times \log_2 p(x_1, x_2, \cdots, x_n) \, dx_1, \cdots, dx_n. \qquad (14)$$

In the case of two variables x and y, the joint and conditional entropies are defined to be

$$H(x, y)$$
$$= -\int_{-\infty}^{+\infty} \int_{-\infty}^{+\infty} p(x, y) \, \log_2 p(x, y) \, dxdy$$

$$H(x/y)$$
$$= -\int_{-\infty}^{+\infty} \int_{-\infty}^{+\infty} p(x, y) \, \log_2 [p(x, y)/p(y)] \, dxdy$$

$$H(y/x)$$
$$= -\int_{-\infty}^{+\infty} \int_{-\infty}^{+\infty} p(x, y) \, \log_2 [p(x, y)/p(x)] \, dxdy.$$

As in the discrete case

$$H(x, y) \leq H(x) + H(y)$$
$$H(x/y) \leq H(x)$$
$$H(y/x) \leq H(y)$$

---

* The difficulties involved in regarding

$$H(x) = \int_{-\infty}^{+\infty} p(x) \, \log p(x) \, dx$$

as the limiting case of

$$\lim_{\delta x_i \to 0} \left[-\sum_i p(x_i) \delta x_i \, \log p(x_i) \delta x_i \right]$$

are discussed in Goldman [16], p. 127. The entropy function, in the continuous case, is dependent on the coordinate system (see Shannon [3], p. 57) and any change in the coordinates will give rise to a change in the entropy function. The entropy function is as important in the continuous as in the discrete case, since the concepts of mutual information and channel capacity depend on the difference of two entropies and the difference is absolute and independent of the coordinate systems.

with equality if, and only if, x and y are statistically independent.

Maximum Entropy Distribution

If x is a continuous variable with probability density function $p(x)$ and variance σ^2, the form of $p(x)$ giving maximum entropy is Gaussian. That is

$$H(x) = -\int_{-\infty}^{+\infty} p(x) \, \log_2 p(x) \, dx$$

is a maximum if

$$p(x) = [(2\pi)^{1/2}\sigma]^{-1} \cdot \exp - (x^2/2\sigma^2).$$

The entropy of a one-dimensional Gaussian distribution with variance σ^2 is

$$H(x) = \log_2 (2\pi e)^{1/2} \cdot \sigma. \qquad (15)$$

Entropy of an Ensemble of Functions

From the sampling theorem, it is known that a continuous function of time can be fully represented by samples taken at intervals $1/2W$ seconds apart. If a waveform is sampled at n points, the probability distribution for the amplitudes of successive samples is of the general form $p(x_1, x_2, \cdots, x_n)$ and the entropy of the set of possible time functions is given by (14).

The entropy *per sample* is defined to be

$$H_1(X) = -\lim_{n \to \infty} n^{-1} \int_{-\infty}^{+\infty} \cdots \int_{-\infty}^{+\infty} p(x_1, x_2, \cdots, x_n)$$
$$\times \log_2 p(x_1, x_2, \cdots, x_n) \, dx_1, \cdots, dx_n.$$

The entropy *per second* is

$$H(X) = -\lim_{T \to \infty} T^{-1} \int_{-\infty}^{+\infty} \cdots \int_{-\infty}^{+\infty} p(x_1, \cdots, x_{2WT})$$
$$\times \log_2 p(x_1, \cdots, x_{2WT}) \, dx_1, \cdots, dx_{2WT}$$

and, since $n = 2WT$, it follows that $H(X) = 2WH_1(X)$.

If the set of possible waveforms has the characteristics of white Gaussian noise of average power $N \, (=\sigma^2)$, then samples are independent, and

$$H_1(X) = \log_2 (2\pi e N)^{1/2} \text{ bits/sample.}$$

The entropy per second is

$$H(x) = 2W \log_2 (2\pi e N)^{1/2}$$
$$= W \log_2 2\pi e N \text{ bits/second.}$$

Entropy Power

An important concept in continuous-information systems is that of entropy power. The entropy power of a given signal set (ensemble) is defined to be the power of white noise limited to the same bandwidth as the original signals and having the same entropy as the signals.

If a set of signals has an entropy H_1, the power of white noise having the same entropy is given by

$$N_1 = (2\pi e)^{-1} \cdot \exp(2H_1).$$

The power N_1 is the entropy power of the signals.

It should be noted that since white noise has the maximum entropy for a given power, *the entropy power of any noise is less than or equal to its actual power.*

Capacity of a Continuous Channel*

If the input to a continuous channel is in the form of continuous time functions, the output will be a perturbed version of these signals, and the input and output signals—being limited to a bandwidth W—can be represented during a time interval T by $n = 2TW$ samples. The probability density functions for the input, for the output, and for the conditional relationship between input and output, are

$$P(x_1, x_2, \cdots, x_n) = P(X)$$

$$P(y_1, y_2, \cdots, y_n) = P(Y)$$

and

$$P[(y_1, y_2, \cdots, y_n)/(x_1, x_2, \cdots, x_n)] = P(Y/X)$$

respectively.

The rate of transmission I of information over the continuous channel is defined in a way analogous to that for the discrete case.

$$I = H(X) - H(X/Y) = H(Y) - H(Y/X)$$

$$= -\int_{-\infty}^{+\infty} P(X) \log_2 P(X) \, dx$$

$$+ \int_{-\infty}^{+\infty} \int_{-\infty}^{+\infty} P(X, Y)$$

$$\times \log_2[P(X, Y)/P(Y)] \, dX dY \text{ bits}/n \text{ samples.}$$

The capacity of the channel is defined as the maximum value of I with respect to all possible sets of input signals.

$$C = \lim_{T \to \infty} \max_{P(X)} T^{-1}[H(X) - H(X/Y)] \text{ bits/second.}$$

$$(16)$$

* For a detailed discussion of the topics of this section the reader should consult Shannon [3] and Fano, *op cit*, Chapter 5. For an elegant geometrical interpretation of communication in the presence of noise, the reader should consult Shannon [4].

Capacity of a Channel in Which the Noise is Additive and Independent of the Input

The rate at which information is transmitted over the channel is

$$I = H(X) - H(X/Y) = H(Y) - H(Y/X).$$

Since the output Y is related to the input X by $Y = X + n$, where n is the noise, and since X and n are statistically independent, $H(Y/X)$ can be shown to be equal to $H(n)$, the entropy of the noise.

The rate of transmission of information therefore is

$$I = H(Y) - H(n)$$

and the capacity is found by maximizing $H(Y)$ with respect to the input.

$$C = \max_{P(X)} [H(Y)] - H(n). \tag{17}$$

Capacity of a Continuous Channel Perturbed by Additive White Gaussian Noise

From (17) the capacity of the channel is seen to be given by

$$C = \max_{P(X)} [H(Y)] - H(n).$$

If the noise is white Gaussian noise, the entropy of the noise is given by

$$H(n) = W \log_2 2\pi e N \text{ bits/second}$$

where W is the channel bandwidth, and N the average noise power.

If the average transmitter power is limited to P, the average receiver signal power is $P + N$. From before, the distribution $P(X)$, having maximum entropy for a given power $P + N (= \sigma^2)$, is Gaussian and has entropy

$$H(Y) = W \log_2 2\pi e (P + N) \text{ bits/second.}$$

The channel capacity therefore is

$$C = W \log_2 2\pi e (P + N) - W \log_2 2\pi e N$$

$$= W \log_2 (1 + P/N) \text{ bits/second.} \tag{18}$$

This means that, by using sufficiently long signals coded to have the properties of white Gaussian noise, it is possible to transmit information over the channel at a rate equal to or less than C, with an arbitrarily small probability of error.

The function $C/W = \log_2(1 + P/N)$ is plotted in Fig. 10 for various values of P/N.

Capacity of a Channel Perturbed by an Arbitrary Type of Noise

When dealing with an arbitrary perturbing noise, the maximizing problem associated with the determination of channel capacity cannot be solved explicitly. However, upper and lower bounds can be determined for C in terms of the channel bandwidth, the average transmitter power, the average noise power, and the entropy power of the noise.

The capacity C, in bits/second, is bounded by the inequalities

$$W \log_2[(P+N_1)/N_1] \leq C \leq W \log[(P+N)N_1] \tag{19}$$

where W = bandwidth, P = average transmitter power, N = average noise power, and N_1 = entropy power of the noise.

ERROR-CORRECTING AND ERROR-DETECTING CODES*

Concept of Coding; A Fundamental Difficulty

As stated earlier, the fundamental theorem of information theory implies that it is possible to transmit any one of $M = 2^{nR}$ equiprobable source messages over a binary system by using sequences of n binary digits, and that if R is less than the channel capacity C, then the probability of error can be made arbitrarily small provided n is sufficiently large. This means that of the n binary digits transmitted, the equivalent of only nR are message carrying digits, the remaining $n(1-R)$ digits being redundant in the sense that they carry no message information. The ratio nR/n is called the *rate of information transmission*, or simply the *rate*, and is measured in bits/binary digit.

In his proof of the fundamental theorem, Shannon avoided the difficult and as yet unsolved problem of specifying a code that satisfied the conditions of the theorem; he considered the *average* probability of error over all randomly chosen codes of length n and demonstrated that this average tends to zero as n tends to infinity. It is the problem of *systematically* producing a code that satisfies the conditions of the fundamental theorem, rather than selecting one at random and hoping that it will be a good one, that has been the subject of considerable attention since the initial publication of Shannon's paper.

Basically, the concept of coding information for transmission consists of two operations. The first is an *encoding operation* in which nR information

digits are converted to, and represented by, a larger block* of n digits. The n binary digits are transmitted over the channel, and at the receiver the second operation (a decoding operation) is performed. In decoding, the n received digits are used in the receiver and a decision is made as to the original nR information digits that were transmitted from the source.

If the block coding concept is considered a little further, a fundamental difficulty encountered in practical attempts to transmit in accordance with Shannon's theorem becomes apparent. To perform the encoding operation, facilities must be available so that the nR-bit information sequence can be converted to an n-binit sequence, with a one-to-one correspondence between the two sequences. It would seem at first sight that the encoder would have to store each of the 2^{nR} possible n-binit sequences and select the appropriate sequence on reception of the nR information digits. As regards decoding, a similar amount of storage would seem to be necessary. During transmission, errors occur so that the received n-binit sequence may be any one of a set of 2^n, and the receiver has the task of comparing each of the 2^{nR} possible transmitted sequences with the received sequence before making a decision as to which was the most likely transmitted sequence. The prospect of having encoding and decoding equipment whose complexity grows exponentially with n is extremely prohibitive in practice, and attempts have been made to ease the storage problem. One approach adopted is that in which algebraic structure, and Group Theory in particular, is employed. It can be shown [29] that encoding can be performed with equipment whose complexity grows only linearly with n, and that the storage necessary at the decoder is $2^{n(1-R)}$ n-binit sequences [30]. In a second approach, which is essentially probabilistic, a sequential technique has been employed [12] in an attempt to reduce the storage necessary for the decoding operation. Algebraic coding is considered in more detail later in this section and examples of some important codes are given. Also, the problem of the systematic synthesis of efficient multiple-error correcting codes is mentioned and an important class of these codes considered and illustrated with examples.

Code Words, Block Codes, Linear Codes, Group Codes, Parity-Check Codes

As stated above, information can be transmitted by using blocks of n binary digits. In any effective coding system, not all of the possible 2^n n-binit

---

* In this section only binary codes will be considered. The reader interested in a deeper understanding of coding, and in coding for more-general signal alphabets, should consult W. W. Peterson, "Error-Correcting Codes," MIT Press; 1961.

* Codes of this type are called *block* codes. Alternative methods of encoding and decoding can be found in [12], [27], [28], and others.

sequences are used. The subset of sequences that are used is the *code*, and each member of the subset is called a *code word*.

If encoding and decoding are carried out using distinct n-binit sequences, then—as mentioned earlier—the code is said to be a *block code*. In certain cases the alternative terms *linear code*, *group code*, and *parity-check code* are used. The term linear code is used since, in a code, the subset of sequences that form the code words generally satisfies the conditions of a linear associative algebra. The term group code is used since the study of block codes can be developed to advantage by using group theory. The term parity-check code is used since the code words generally consist of information digits and redundant digits that are referred to as parity-check digits.

Information Digits, Parity-Check Digits, Systematic Codes

Code words are generally constructed so that, in addition to the information digits, they contain a number of "redundant" digits. The "redundant" digits are generally formed as linear combinations of the *information digits* and are called *parity-check digits*.

If, within a code, each code word is such that the first k digits are information digits and the next $m(=n-k)$ digits are check digits, then the code is said to be a *systematic code*.

Error-Detecting Codes, Error-Correcting Codes

The parity checks mentioned above form the basis of *error-detecting* and *error-correcting* codes. An error-detecting code is one whose code-word structure is such that the presence of an error or errors in the received sequence can be *detected but not corrected*. An error-correcting code is one whose code-word structure is such that the presence of an error or errors can be detected, the position or positions located, and necessary *corrections* made.

Example: An example of an elementary error-detecting code is that in which a single parity-check digit is used to detect the presence of an *odd* number of errors in a received sequence. In this code a single digit is added to the information digits, the additional digit being chosen to make the total number of 1's in the word an even number. This kind of check is called an *even-parity check** and can be illustrated as follows: If the

---

* It should be noted that an odd-parity check, in which a check digit is added to make the total number of 1's in the transmitted sequence an *odd* number, would be equally satisfactory.

information digits are 01011, and an even-parity check is to be used, then the transmitted sequence will be 010111, and the presence of 1, 3, or 5 errors can be *detected but not corrected*.

Modulo 2 Arithmetic

Modulo 2 arithmetic plays an important role in the study of binary codes. The rules of modulo 2 arithmetic are as follows.

$$0+0=0$$
$$1+0=1$$
$$0+1=1$$
$$1+1=0$$
$$0\cdot0=0$$
$$0\cdot1=0$$
$$1\cdot0=0$$
$$1\cdot1=1$$

The sign $\oplus$ is sometimes used to denote modulo 2 addition.

Error Patterns

If the transmitted sequence is V and the received sequence is U, then the sequence $U-V$ is called the *error pattern*. Clearly, the error pattern is that pattern which, when added to the transmitted code word, results in the received sequence.

Example: If 011011 is transmitted and 101101 is received then the error pattern is 011011−101101, which is equal to 011011+101101 in modulo 2 arithmetic. The error pattern is seen to be 110110.

Hamming Distance

The Hamming distance between two n-digit binary sequences is the number of digits in which they differ. For example, if the sequences are 1010110 and 1001010, then the Hamming distance is 3.

Minimum-Distance Decoding

In minimum-distance decoding, a received sequence is compared with all possible transmitted sequences and the decision made that the transmitted sequence is that sequence whose Hamming distance from the received sequence is a minimum. For errors that are independent from binary digit to binary digit, minimum-distance decoding leads to the smallest overall probability of error, and is equivalent to *maximum-likelihood decoding*.

Relationship Between Hamming Distance and Error-Detecting* and Error-Correcting Properties of a Code

If the Hamming distance between *any two* words of a code is equal to $e+1$, then it is possible to *detect* the presence of any e or fewer errors in a received sequence.

If the Hamming distance between any two words of a code is equal to $2e+1$, then it is possible to *correct* any e or fewer errors occurring in a received sequence.

Elements of Parity-Check Coding†

A parity-check code, or group code, can be defined uniquely in terms of a *parity-check matrix*. A sequence $v(=v_1, v_2, \cdots, v_n)$ is a code word if, and only if, it satisfies the matrix equation $H \cdot v^T = 0$, where H is the parity-check matrix, and v^T is the *transpose* of the row matrix $v = v_1, v_2, \cdots, v_n$. If the parity-check matrix is taken to be of the general form

$$H = \begin{bmatrix} a_{11} & a_{12} & \cdots & a_{1n} \\ a_{21} & a_{22} & \cdots & a_{2n} \\ \cdot & \cdot & & \cdot \\ \cdot & \cdot & & \cdot \\ \cdot & \cdot & & \cdot \\ a_{m1} & a_{m2} & \cdots & a_{mn} \end{bmatrix}$$

then the requirement that a code word satisfies the above matrix equation is seen to be equivalent to the requirement that the word satisfies the following set of m simultaneous equations

$$a_{11} \cdot v_1 + a_{12} \cdot v_2 + \cdots + a_{1n} \cdot v_n = 0$$
$$\vdots \qquad\qquad \vdots$$
$$a_{m1} \cdot v_1 + a_{m2} \cdot v_2 + \cdots + a_{mn} \cdot v_n = 0. \qquad (20)$$

If the row rank of the parity-check matrix is m, this means that m rows of the matrix are linearly independent and hence, in solving the equations, that $n-m$ of the elements $v_1, v_2, \cdots, v_n$ of the code word can be chosen arbitrarily; the remaining m digits are determined, in terms of these chosen

*Until comparatively recently error-detection techniques, with a request for retransmission on detection of an error, have been used more extensively than automatic error correction on the grounds of equipment economy. However, with the present trend towards the miniaturization of electronic equipment, automatic error correction is increasing in practical importance and it, rather than error detection, forms the main topic of this section.

† Arithmetic modulo 2 will be used throughout the remainder of this chapter.

digits, as the solution of (20). The $n-m$ arbitrarily chosen digits are the information digits, and the remaining m digits—determined as the solution of the set of simultaneous equations—are the parity-check digits.

The parity-check matrix is used in both encoding and decoding operations and must be stored, in some form, in both the encoder and decoder.

The encoding operation can be illustrated by an example in which the parity-check matrix is taken as

$$H = \begin{bmatrix} 100011 \\ 010001 \\ 001010 \\ 000110 \end{bmatrix}.$$

Since this matrix has a row rank of 4, the code words are seen to contain 4 parity-check digits and 2 information digits. The parity-check digits C_1, C_2, C_3, C_4 can be determined from the information digits I_1 and I_2 by using the matrix equation above. If the code word v is arbitrarily chosen to be of the form $C_1, C_2, C_3, C_4, I_1, I_2$, then the parity-check digits must satisfy the set of simultaneous equations

$$C_1 + I_1 + I_2 = 0$$
$$C_2 + I_2 \quad\;\; = 0$$
$$C_3 + I_1 \quad\;\; = 0$$
$$C_4 + I_1 \quad\;\; = 0.$$

The resulting code words for this example are thus seen to be 000000, 011111, 101110, and 110001.

In decoding, the parity-check matrix is multiplied by the transpose of the received sequence $v'(=v_1', v_2', \cdots, v_n')$ and an m-digit sequence called the *corrector* or *syndrome* obtained. Following the determination of the syndrome, a correction can then be made by assuming that a particular syndrome always comes about as a result of the presence of one particular error pattern. The syndrome c is related to the received sequence and the parity-check matrix by the matrix equation

$$c = H \cdot [v']^T.$$

Clearly, if the received sequence v' is the same as a possible transmitted sequence, then the syndrome is zero and the received sequence must be assumed correct. If, however, errors occur during transmission, and they are not such as to convert the transmitted sequence into a sequence that corresponds to another permissible transmission sequence, then the syndrome will be nonzero. In this case the received sequence v' is equal to the sum of the transmitted sequence v and the error pattern x, and the syndrome is

$$c = H \cdot [v+x]^T = Hv^T + Hx^T = Hx^T.$$

It can be seen that this syndrome is, in fact, equal to the modulo 2 sum of those columns of the parity-check matrix whose positions correspond to the positions of 1's in the error pattern x. Since it is possible for a number of error patterns to result in the same syndrome, it is clear that any practical decoder cannot correct all error patterns. The decoder which examines all the error patterns that result in a particular syndrome, and selects as the transmission error that error pattern containing the least number of 1's, is a minimum-distance decoder.

The following example illustrates minimum-distance decoding based on the above ideas. Let it again be assumed that the parity-check matrix

is of the form

$$H = \begin{bmatrix} 100011 \\ 010001 \\ 001010 \\ 000110 \end{bmatrix}$$

Table 4 shows the code words, error patterns, and received sequences, together with the syndromes calculated using the matrix H. It can be seen from the table that all single errors and some double errors can be corrected, but that no patterns of 3 or more errors can be corrected. It should be noted that to perform the minimum-distance decoding operation for a group code it is necessary to store only the parity-check matrix, together with the 2^m syndromes and their associated error patterns.

From the ideas and example presented above, it should be clear that if a code is to be such that any pattern of e or fewer errors is to be correctible, then each such error pattern must give rise to a distinct syndrome. This means that no two sets of e columns of the parity-check matrix should have the same modulo 2 sum, or, expressed alternatively, that *every set of 2e columns of the parity-check matrix should be linearly independent if the code is to be able to correct any pattern of e or fewer errors.*

For a given word length n, the problem of systematically producing a parity-check matrix with every set of $2e$ columns linearly independent is one of the difficult problems of coding theory. A general method of synthesizing such a matrix is the method of Sacks [31]. This method, which may also be used as a proof of the Varsharmov–Gilbert–Sacks bound,* is very laborious and is

TABLE 4—MINIMUM-DISTANCE DECODING.

Error Patterns	Code Words				Transposed Syndrome for Each Received Sequence
	000000	011111	101110	110001	
000000	000000	011111	101110	110001	0000*
100000	100000	111111	001110	010001	1000*
010000	010000	001111	111110	100001	0100*
001000	001000	010111	100110	111001	0010*
000100	000100	011011	101010	110101	0001*
000010	000010	011101	101100	110011	1011*
000001	000001	011110	101111	110000	1100*
110000	110000	101111	011110	000001	1100
101000	101000	110111	000110	011001	1010*
100100	100100	111011	001010	010101	1001*
100010	100010	111101	001100	010011	0011*
100001	100001	111110	001111	010000	0100
011000	011000	000111	110110	101001	0110*
010100	010100	001011	111010	100101	0101*
010010	010010	001101	111100	100011	1111*
010001	010001	001110	111111	100000	1000
001100	001100	010011	100010	111101	0011
001010	001010	010101	100100	111011	1001
001001	001001	010110	100111	111000	1110*
000110	000110	011001	101000	110111	1010
000101	000101	011010	101011	110100	1101*
000011	000011	011100	101101	110010	0111*

* Indicates a correctible error pattern. Note that the error pattern 110000 cannot be corrected since it gives the same syndrome as the more probable error pattern 000001. Note also that no patterns of 3 or more errors can be corrected since the 16 four-digit syndromes have all been allocated to the correction of single and double errors.

* The Varsharmov–Gilbert–Sacks bound is a lower bound in the sense that a parity-check code capable of correcting any e or fewer errors, and having code words of length n, can always be constructed if the number of check digits is equal to or greater than m, where m is the smallest integer satisfying the condition

$$2^m > \sum_{i=0}^{2e-1} {}^{n-1}c_i.$$

The Varsharmov–Gilbert–Sacks bound is a sufficient but *not* necessary condition, since if $m = m'$ is the smallest integer for which the condition

$$2^m > \sum_{i=0}^{2e-1} {}^{n-1}c_i.$$

is satisfied then it is certainly possible to construct a code (with words of length n) capable of correcting any pattern of e or fewer errors. However, it is also possible in many cases to construct a code capable of correcting any e or fewer errors with less than m' check digits.

also inefficient since the rates (the ratio of information digits to word length) are not as high as those obtainable by other methods of synthesis. A number of very important synthesis procedures follow and are illustrated with examples.

Hamming Single-Error-Correcting Code

From the preceding theory, it can be seen that if we wish to correct all single errors that might occur in an n-digit sequence, we need only arrange the parity-check matrix so that its n columns are nonzero and distinct. It thus follows that a *single-error-correcting* binary code, with code words of length n, can be constructed if it contains m check digits, where m is the smallest integer satisfying the condition $2^m \geq n+1$. If the parity-check matrix is arranged so that the binary content of each column (when converted to its decimal equivalent) indicates the position of the column in the matrix, and the positions of the check digits in the code word are arranged to coincide with those columns in the matrix that contain only a single one, then the code is known as a Hamming single-error-correcting code (after the pioneer work of Hamming [32]). This particular arrangement of the parity-check matrix, while possessing no additional error-correcting properties as compared with any other arrangement of the same set of columns, has the following practical advantages.

(A) Each check digit can be determined directly from the information digits independent of the other check digits.

(B) The position of an error can be determined simply by converting the resulting syndrome to its decimal equivalent, this number being the location of the error.

Example: Consider the construction of a Hamming single-error-correcting code for words of length $n=15$. In this case the condition $2^m \geq 15+1$ must be satisfied and a single-error-correcting code can, therefore, be constructed with words containing 11 information digits and 4 check digits. The parity-check matrix H is

$$H = \begin{bmatrix} 000000011111111 \\ 000111100001111 \\ 011001100110011 \\ 101010101010101 \end{bmatrix}$$

and the code word structure is

$$C_1 C_2 I_1 C_3 I_2 I_3 I_4 C_4 I_5 I_6 I_7 I_8 I_9 I_{10} I_{11}$$

where C_i is the ith check digit and I_j is the jth information digit. For this code the check digits

can be seen to be determined from

$$C_1 + I_1 + I_2 + I_4 + I_5 + I_7 + I_9 + I_{11} = 0$$
$$C_2 + I_1 + I_3 + I_4 + I_6 + I_7 + I_{10} + I_{11} = 0$$
$$C_3 + I_2 + I_3 + I_4 + I_8 + I_9 + I_{10} + I_{11} = 0$$
$$C_4 + I_5 + I_6 + I_7 + I_8 + I_9 + I_{10} + I_{11} = 0.$$

If it is desired to transmit the information digits 10101010101, then the check digits (which can be determined from the above equations) are found to be $C_1 = 1$, $C_2 = 0$, $C_3 = 1$, $C_4 = 0$ and the transmitted code word which results is seen to be 101101001010101.

As an illustration of decoding, let it be assumed that the received sequence is 100101001010101. For this received sequence the syndrome is found to be

$$\begin{matrix} 0 \\ 0 \\ 1 \\ 1 \end{matrix}$$

which has a decimal equivalent of $1 \times 2^0 + 1 \times 2^1 + 0 \times 2^2 + 0 \times 2^3 = 3$, indicating that the error is in the third digit of the received sequence. If the received sequence is assumed to be 101101001010100 then the syndrome is

$$\begin{matrix} 1 \\ 1 \\ 1 \\ 1 \end{matrix}$$

which has a decimal equivalent of 15, indicating that the error is in the fifteenth digit of the received sequence.

Reed-Muller Error-Correcting Codes [33, 34]

Reed–Muller codes are a class of *multiple-error-correcting* codes that have a wide range of information rates and error-correcting ability. These codes are such that for any integers r and s, where r is *less* than s, there is a code with words of length

$$n = 2^s \text{ that contains } m = 1 + {}^s c_1 + {}^s c_2 + \cdots + {}^s c_{s-r-1}$$

check digits and is capable of correcting any pattern of $2^{s-r-1} - 1$ or fewer errors.

Encoding Process: In the encoding operation, the transmitted sequence $f(= f_0, f_1, \cdots, f_{n-1})$ is obtained from the $n-m$ information digits by using an rth-degree encoding expression of the following general form

$$f(= f_0, f_1, \cdots, f_{n-1}) = g_0 \cdot x_0 + g_1 \cdot x_1 + \cdots + g_s \cdot x_s$$
$$+ g_{12} \cdot x_1 \cdot x_2 + \cdots + g_{s-1,s} \cdot x_{s-1} \cdot x_s$$
$$+ \cdots + g_{1,2,\cdots,r} \cdot x_1 \cdot x_2 \cdots x_r$$
$$+ \cdots + g_{s-r+1,\cdots,s} \cdot x_{s-r+1} \cdots x_s. \quad (21)$$

In this expression, the coefficients g_0, g_1, $\cdots$, $g_{1,2,3\cdots r}$, etc., are information digits, and the sequences x_1, x_2, $\cdots$, x_s are basis "vectors" of length n having the form

$$x_0 = 111111 \cdots\cdots\cdots\cdots 1111$$

$$x_1 = 010101 \cdots\cdots\cdots\cdots 0101$$

$$x_2 = 00110011 \cdots\cdots\cdots 00110011$$

$$x_3 = 000011110000 \cdots 00001111$$

$$x_4 = 0000000011111111 \cdots\cdots$$

$$\vdots \qquad\qquad \vdots$$

$$x_s = \underbrace{00\cdots 0000}_{2^{s-1}\ 0\text{'s}}\ \underbrace{1111\cdots\cdots 1111}_{2^{s-1}\ 1\text{'s}} \ .$$

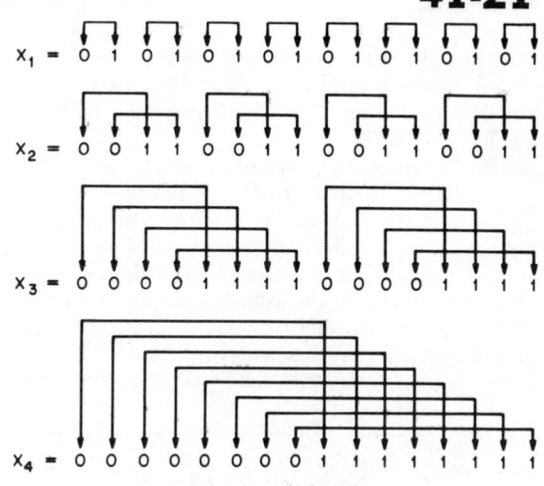

Fig. 11—Symbol pairing scheme for use in decoding Reed-Muller codes.

Examples Illustrating Encoding Process:—

Case 1. Consider the case where $s=4$ and $r=1$. Under these circumstances the general encoding expression (21) becomes

$$f = g_0 \cdot x_0 + g_1 \cdot x_1 + g_2 \cdot x_2 + g_3 \cdot x_3 + g_4 \cdot x_4$$

and code words are generated by using the g_i's as information digits. The words of this code are of length $n = 2^s = 2^4 = 16$, and the code is capable of correcting any pattern of $2^{s-r-1} - 1 = 3$ or fewer errors.

Suppose, by way of illustration, that it is desired to transmit the information digits 10100. For this sequence the transmitted sequence $f (= f_0, f_1, \cdots, f_{15})$ is seen to be

$$f = 1 \cdot 1111111111111111$$

$$+ 0 \cdot 0101010101010101$$

$$+ 1 \cdot 0011001100110011$$

$$+ 0 \cdot 0000111100001111$$

$$+ 0 \cdot 0000000011111111$$

$$= 1100110011001100.$$

Case 2. As a second illustration, consider the case where $s=4$ and $r=2$. Under these circumstances (21) becomes

$$f = g_0 \cdot x_0 + g_1 \cdot x_1 + g_2 \cdot x_2 + g_3 \cdot x_3 + g_4 \cdot x_4$$

$$+ g_{12} \cdot x_1 \cdot x_2 + g_{13} \cdot x_1 \cdot x_3 + g_{14} \cdot x_1 \cdot x_4$$

$$+ g_{23} \cdot x_2 \cdot x_3 + g_{24} \cdot x_2 \cdot x_4 + g_{34} \cdot x_3 \cdot x_4.$$

This code, which has words of length $2^s = 2^4 = 16$, contains $1 + {}^4c_1 = 5$ check digits and 11 information digits, and is capable of correcting any single error that occurs in a received sequence.

From the above encoding expression, the transmitted sequence corresponding to the information sequence 01000100001 is

$$f (= f_0, f_1, \cdots, f_{15}) = 0 \cdot x_0 + 1 \cdot x_1 + 0 \cdot x_2 + 0 \cdot x_3 + 0 \cdot x_4$$

$$+ 1 \cdot x_1 \cdot x_2 + 0 \cdot x_1 \cdot x_3 + 0 \cdot x_1 \cdot x_4$$

$$+ 0 \cdot x_2 \cdot x_3 + 0 \cdot x_2 \cdot x_4 + 1 \cdot x_3 \cdot x_4$$

and, as

$$x_1 = 0101010101010101$$

$$x_1 \cdot x_2 = 0001000100010001$$

and

$$x_3 \cdot x_4 = 0000000000001111$$

the transmitted sequence is seen to be

$$0100010001001011.$$

Decoding Process: A general decoding algorithm for these codes has been devised by Reed [34]. The algorithm enables any pattern of $2^{s-r-1} - 1$ or fewer errors to be corrected.

In the decoding operation, each information digit is computed a number of times in terms of certain selected subsets of the elements $f_0, f_1, \cdots, f_{n-1}$ of the received sequence, and a majority decision is made as to whether the information digit in question is a one or a zero. In decoding, the rth-degree coefficients ($g_{12}, g_{13}, \cdots, g_{34}$ in Case 2, above) are first obtained and then a new "received" sequence is computed by adding the newly found r'th-order terms ($g_{12} \cdot x_1 \cdot x_2, \cdots, g_{34} \cdot x_3 \cdot x_4$ in Case 2) to the original received sequence. This new "received" sequence is then used and the $(r-1)$th-degree coefficients extracted in the same way as the r'th-degree coefficients. The process is repeated until either the message is extracted or an indeterminacy occurs.

A general scheme for determining which subsets of the elements $f_0, f_1, \cdots, f_{n-1}$ should be used in checking the information digits $g_i, \cdots, g_{ij}, \cdots, g_{ijk}$, etc., is as follows.

Arrange the basis vectors in order as shown in Fig. 11, and for each vector x_i associate the jth zero with the jth one as indicated. Each pair of associated elements is termed a matching pair (refer to W. W. Peterson, "Error-Correcting Codes," MIT Press; 1961).

The 2^{s-1} subsets of 2 elements used to determine g_i are the 2^{s-1} matching pairs in the basis vector x_i. Each of the 2^{s-2} subsets of 4 elements used to determine g_{ij} is obtained by taking a matching pair of components in x_i together with the associated pair in x_j. In the same manner, each of the 2^{s-3} subsets of 8 elements used to determine g_{ijk} is obtained by taking a matching pair in x_i and associating with it a matching pair in x_j and four

matching components in x_k. The scheme can be extended in a straightforward manner to obtain check relations for higher-order coefficients.

Example Illustrating Decoding Process:—Let us consider the case where $s=4$ and $r=2$ as above and assume that it is desired to transmit the information sequence 10000000001. The transmitted sequence for this particular information sequence is 1111111111110000. Let us suppose that the received sequence is 0111111111110000.

Using the scheme described above and Fig. 11, it can be seen that the check relations are

$$g_1=f_0+f_1=f_2+f_3=f_4+f_5=f_6+f_7=f_8+f_9=f_{10}+f_{11}=f_{12}+f_{13}=f_{14}+f_{15}$$

$$g_2=f_0+f_2=f_1+f_3=f_4+f_6=f_5+f_7=f_8+f_{10}=f_9+f_{11}=f_{12}+f_{14}=f_{13}+f_{15}$$

$$g_3=f_0+f_4=f_1+f_5=f_2+f_6=f_3+f_7=f_8+f_{12}=f_9+f_{13}=f_{10}+f_{14}=f_{11}+f_{15}$$

$$g_4=f_0+f_8=f_1+f_9=f_2+f_{10}=f_3+f_{11}=f_4+f_{12}=f_5+f_{13}=f_6+f_{14}=f_7+f_{15}$$

$$g_{12}=f_0+f_1+f_2+f_3=f_4+f_5+f_6+f_7=f_8+f_9+f_{10}+f_{11}=f_{12}+f_{13}+f_{14}+f_{15}$$

$$g_{13}=f_0+f_1+f_4+f_5=f_2+f_3+f_6+f_7=f_8+f_9+f_{12}+f_{13}=f_{10}+f_{11}+f_{14}+f_{15}$$

$$g_{14}=f_0+f_1+f_8+f_9=f_2+f_3+f_{10}+f_{11}=f_4+f_5+f_{12}+f_{13}=f_6+f_7+f_{14}+f_{15}$$

$$g_{23}=f_0+f_2+f_4+f_6=f_1+f_3+f_5+f_7=f_8+f_{10}+f_{12}+f_{14}=f_9+f_{11}+f_{13}+f_{15}$$

$$g_{24}=f_0+f_2+f_8+f_{10}=f_1+f_3+f_9+f_{11}=f_4+f_6+f_{12}+f_{14}=f_5+f_7+f_{13}+f_{15}$$

$$g_{34}=f_0+f_4+f_8+f_{12}=f_1+f_5+f_9+f_{13}=f_2+f_6+f_{10}+f_{14}=f_3+f_7+f_{11}+f_{15}$$

and, on substituting the received element values into the check relations for the coefficients g_{ij}, the following values are obtained.

$$g_{12}=1; \quad g_{12}=0; \quad g_{12}=0; \quad g_{12}=0 \qquad \therefore \ g_{12}=0 \text{ by majority decision}$$

$$g_{13}=1; \quad g_{13}=0; \quad g_{13}=0; \quad g_{13}=0 \qquad \therefore \ g_{13}=0 \text{ by majority decision}$$

$$g_{14}=1; \quad g_{14}=0; \quad g_{14}=0; \quad g_{14}=0 \qquad \therefore \ g_{14}=0 \text{ by majority decision}$$

$$g_{23}=1; \quad g_{23}=0; \quad g_{23}=0; \quad g_{23}=0 \qquad \therefore \ g_{23}=0 \text{ by majority decision}$$

$$g_{24}=1; \quad g_{24}=0; \quad g_{24}=0; \quad g_{24}=0 \qquad \therefore \ g_{24}=0 \text{ by majority decision}$$

$$g_{34}=0; \quad g_{34}=1; \quad g_{34}=1; \quad g_{34}=1 \qquad \therefore \ g_{34}=1 \text{ by majority decision.}$$

At this stage six information digits have been decoded.

The new "received" sequence, $f'=g_0 \cdot x_0+g_1 \cdot x_1+g_2 \cdot x_2+g_3 \cdot x_3+g_4 \cdot x_4$, can now be computed by adding the sequence

$$g_{12} \cdot x_1 \cdot x_2+g_{13} \cdot x_1 \cdot x_3+g_{14} \cdot x_1 \cdot x_4+g_{23} \cdot x_2 \cdot x_3+g_{24} \cdot x_2 \cdot x_4+g_{34} \cdot x_3 \cdot x_4 \text{ to } f.$$

The sequence f' is found to be 0111111111111111 and, on using these new elements in the check relations for g_1, g_2, g_3, and g_4, the following values are obtained for the information digits g_1, g_2, g_3, and g_4.

$$g_1=1; \quad g_1=0; \quad g_1=0; \quad g_1=0; \quad g_1=0; \quad g_1=0; \quad g_1=0; \quad g_1=0$$

$$g_2=1; \quad g_2=0; \quad g_2=0; \quad g_2=0; \quad g_2=0; \quad g_2=0; \quad g_2=0; \quad g_2=0$$

$$g_3=1; \quad g_3=0; \quad g_3=0; \quad g_3=0; \quad g_3=0; \quad g_3=0; \quad g_3=0; \quad g_3=0$$

$$g_4=1; \quad g_4=0; \quad g_4=0; \quad g_4=0; \quad g_4=0; \quad g_4=0; \quad g_4=0; \quad g_4=0.$$

By majority decisions the values of g_1, g_2, g_3, and g_4 are taken as 0, 0, 0, and 0, respectively.

On adding $g_1 \cdot x_1 + g_2 \cdot x_2 + g_3 \cdot x_3 + g_4 \cdot x_4$ to the sequence f', the sequence corresponding to $g_0 x_0$ is obtained. This sequence is found to be 0111111111111111 and, as x_0 is 1111111111111111, g_0 must equal 1 by majority decision. The decoded information sequence is thus 10000000001, which is correct.

Iterated (or Product) Codes*

It is possible to use simple systematic codes to produce more-powerful codes with increased error-correcting ability. These codes are called *iterated* or *product* codes.

As an example of an iterated code, consider the code formed from a simple systematic code in which a single check digit is added as a means of *detecting* an odd number of errors in a code word. The information digits are arranged in a two-dimensional (or higher-dimensional) array as shown in Fig. 12, and an even-parity-check digit is added to each row and each column. In addition, checks are also carried out on check digits.

The specific code shown in Fig. 12 is clearly more powerful than the original codes from which it was constructed, since it is able to *correct* any single error that might occur. The position of an error is located as the common element of the row and the column whose parity checks fail. Iterated codes may be generalized in that the rows of the array may be taken from one type of systematic code while the columns are taken from a different type of systematic code.

Bose-Chaudhuri Codes

In recent years, some of the most important advances in the development of multiple-error-correcting codes have been concerned with a large class of codes known as *cyclic codes*.† These codes are of extreme practical importance because of the ease with which they can be synthesized, and the ease with which they can be encoded and decoded by using simple feedback shift registers.

For a thorough understanding of cyclic codes it is necessary to have a knowledge of abstract

* For a more detailed discussion of these codes, consult Peterson, *op cit*, Chapter 5.

† A code is defined to be a cyclic code if, for any code word $V(=V_1, V_2, \cdots, V_n)$ in the code, the vector $V'(=V_n, V_1, V_2, \cdots, V_{n-1})$, obtained by cyclically shifting V one unit to the right, is also a word of the code.

algebra that is beyond the scope and nature of this chapter. It is therefore proposed to consider only briefly the important class of codes that are known as Bose–Chaudhuri codes.* The properties of Bose-Chaudhuri codes are outlined and a method of constructing the parity-check matrix for the correction of multiple errors is given. Also, practical methods for the encoding and decoding of these codes are discussed.

The codes of Bose–Chaudhuri [35, 36] are a class of cyclic codes that are particularly effective in the detection and correction of randomly occurring multiple errors. These codes have the property that for any positive integers m and e there is a code with words of length $n = 2^m - 1$; this code contains no more than $m \cdot e$ parity-check digits, and is capable of correcting any pattern of e or fewer errors.

The parity-check matrix, H, for a Bose-Chaudhuri code that has words of length $n = 2^m - 1$, and is capable of correcting any pattern of e or fewer errors, can be derived as follows.

(A) Take an m by m matrix, Z, of the form

$$Z = \begin{bmatrix} 010000 & \cdots & 000 \\ 001000 & \cdots & 000 \\ 000100 & \cdots & 000 \\ \vdots & & \vdots \\ 000000 & \cdots & 010 \\ 000000 & \cdots & 001 \\ \alpha_0 \alpha_1 \alpha_2 & \cdots & \alpha_{m-1} \end{bmatrix}$$

and select the binary digits α_0, α_1, $\cdots$, α_{m-1} so that the polynomial

$$C(x) = \alpha_0 + \alpha_1 \cdot x + \alpha_2 \cdot x^2 + \cdots + \alpha_{m-1} \cdot x^{m-1} + x^m$$

is irreducible† and does not divide $x^k - 1$ for any k less than $2^m - 1$.‡

* For a detailed discussion of the theory and properties of cyclic codes, including codes for the correction of random errors and codes for the correction of bursts of errors, consult Peterson, *op cit*, Chapters 6 through 10.

† This means that the polynomial cannot be factorized into factors with the coefficients 0 and 1. Tables of irreducible polynomials are given in Peterson, *op cit*, Appendix C.

‡ These conditions ensure that the matrices

$$Z, Z^2(=Z \cdot Z), \cdots, Z^n$$

are distinct, that is, that the matrix Z has a maximum period.

(B) Take any nonzero vector X of m elements.

(C) Form the parity-check matrix, H, as follows.

$$H = \begin{bmatrix} X & Z \cdot X & Z^2 \cdot X & Z^3 \cdot X & \cdots & Z^{n-1} \cdot X \\ X & Z^3 \cdot X & Z^6 \cdot X & Z^9 \cdot X & \cdots & Z^{3(n-1)} \cdot X \\ X & Z^5 \cdot X & Z^{10} \cdot X & Z^{15} \cdot X & \cdots & Z^{5(n-1)} \cdot X \\ & & \vdots & & & \\ X & Z^{2e-1} \cdot X & \cdots & \cdots & \cdots & Z^{(2e-1)(n-1)} \cdot X \end{bmatrix}$$

In this matrix, Z^i is the matrix Z multiplied by itself i times, and $Z^i \cdot X$ is the matrix obtained by multiplying the matrix Z^i by the matrix X.

The matrix H may contain a number of rows that are all zeros and may also contain a number of repeated rows. Clearly, rows of this type are of no value in parity checking and should be removed from H. After the removal of these rows, the matrix contains rows that are linearly independent. The number of independent rows is equal to the row rank of the matrix and, as was explained earlier, this is equal to the number of check digits. The rank of H can be determined directly as follows. If $f_i(x)$, for $i = 1, 3, \cdots, 2e-1$, is a polynomial with 0's and 1's as coefficients, and is such that it is the *minimum* degree polynomial for which the matrix equation $f_i(x = Z^i) = 0$ is satisfied; then the polynomial $f(x)$, which is the *least common multiple* of the $f_i(x)$'s [that is, is the polynomial of lowest degree which is a multiple of each $f_i(x)$] is known as the *generator polynomial* of the Bose–Chaudhuri code. The degree of the generator polynomial is equal to the rank of H.

Example: Consider the synthesis of a triple-error-correcting Bose–Chaudhuri code for which $m = 4$ and $e = 3$. This code has words of length $n = 2^4 - 1 = 15$. Let it be assumed that

$$X = \begin{bmatrix} 1 \\ 0 \\ 0 \\ 0 \end{bmatrix}$$

and that the matrix Z is

$$Z = \begin{bmatrix} 0100 \\ 0010 \\ 0001 \\ 1100 \end{bmatrix}$$

For this choice of Z the characteristic polynomial

$$C(x) = \alpha_0 + \alpha_1 x + \cdots + \alpha_{m-1} x^{m-1} + x^m$$

becomes

$$C(x) = 1 + x + x^4$$

which is irreducible and does not divide $x^k - 1$ for any k less than 15.

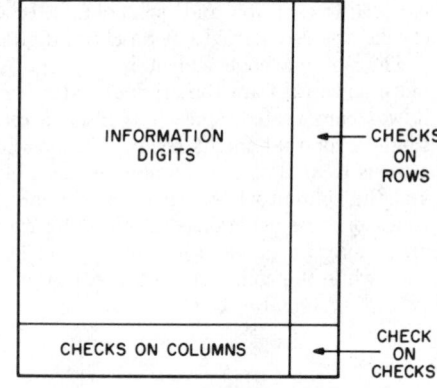

A. GENERAL EXAMPLE OF AN ITERATED CODE

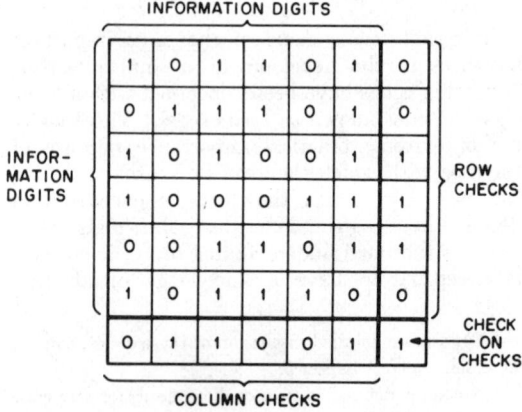

B. SPECIFIC EXAMPLE OF AN ITERATED CODE

Fig. 12—Examples of iterated codes.

By simple matrix multiplication it can be shown that

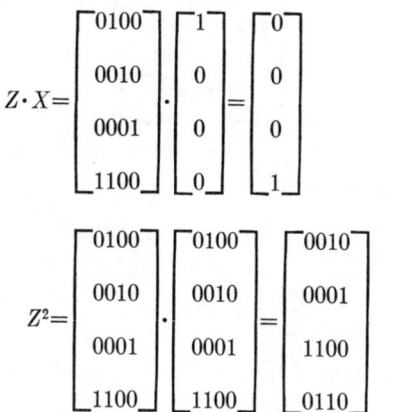

$$Z \cdot X = \begin{bmatrix} 0100 \\ 0010 \\ 0001 \\ 1100 \end{bmatrix} \cdot \begin{bmatrix} 1 \\ 0 \\ 0 \\ 0 \end{bmatrix} = \begin{bmatrix} 0 \\ 0 \\ 0 \\ 1 \end{bmatrix}$$

$$Z^2 = \begin{bmatrix} 0100 \\ 0010 \\ 0001 \\ 1100 \end{bmatrix} \cdot \begin{bmatrix} 0100 \\ 0010 \\ 0001 \\ 1100 \end{bmatrix} = \begin{bmatrix} 0010 \\ 0001 \\ 1100 \\ 0110 \end{bmatrix}$$

and, therefore, that

$$Z^2 \cdot X = \begin{bmatrix} 0 \\ 0 \\ 1 \\ 0 \end{bmatrix}$$

By continuing the matrix multiplication, it can be seen that the parity-check matrix H is

$$H = \begin{bmatrix} 1000100110101111 \\ 0001001101011111 \\ 0010011010111110 \\ 0100110101111100 \\ 1000110001100001 \\ 0111101111011111 \\ 0010100101000101 \\ 0001100011000011 \\ 1011011011011011 \\ 0000000000000000 \\ 0110110110110111 \\ 0110110110110111 \end{bmatrix}$$

This matrix has a row of all zeros and two identical rows. If the row of zeros is removed, together with one of the two identical rows, the

following matrix is obtained as parity-check matrix for the code.

$$H = \begin{bmatrix} 1000100110101111 \\ 0001001101011111 \\ 0010011010111110 \\ 0100110101111100 \\ 1000110001100001 \\ 0111101111011111 \\ 0010100101000101 \\ 0001100011000011 \\ 1011011011011011 \\ 0110110110110111 \end{bmatrix}$$

This matrix has 10 independent rows and is, therefore, of rank 10. Also, it can be seen that every set of 6 columns of this matrix is linearly independent, and the code can thus correct all patterns of 3 or fewer errors.

The rank of the matrix above could have been obtained directly from the fact that the minimal polynomials $f_1(x)$, $f_3(x)$, and $f_5(x)$ are $1+x+x^4$, $1+x+x^2+x^3+x^4$, and $1+x+x^2$, respectively. Therefore

$$f(x) = f_1(x) \cdot f_3(x) \cdot f_5(x)$$
$$= g(x)$$
$$= 1+x+x^2+x^4+x^5+x^8+x^{10}$$

has degree 10, and the rank of H is thus 10. The polynomial $g(x)$ is the generator polynomial for a Bose–Chaudhuri code with 5 information and 10 check digits.

Encoding and Decoding of Bose–Chaudhuri Codes: The encoding and decoding of Bose–Chaudhuri (and other) cyclic codes can be easily mechanized by using feedback shift registers. Details of the general encoding procedure and the decoding procedure for the *detection* of errors can be found in Peterson, *op cit*, Chapter 8, and in [37]. The techniques of decoding for the *correction* of errors are more complicated and are discussed in Chapter 9 of Peterson's book and also in [38–42]. This chapter discusses only the encoding and the decoding for the *detection* of errors.

The encoding and decoding procedures can best be discussed in terms of code polynomials. Any code word can be expressed as a polynomial, the coefficients of which are the elements of the code

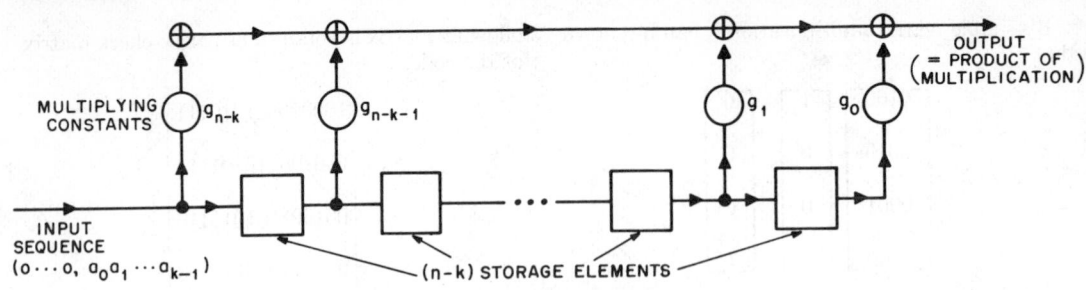

Fig. 13—Circuit for multiplying fixed polynomial $g(x)=g_0+g_1 \cdot x+\cdots+g_{n-k}\cdot x^{n-k}$ by the polynomial $P(x)=a_0+a_1 \cdot x+\cdots+a_{k-1}\cdot x^{k-1}$.

word. For example, if a code word is 1011001, it can be represented by a polynomial

$$1 \cdot x^0+0 \cdot x^1+1 \cdot x^2+1 \cdot x^3+0 \cdot x^4+0 \cdot x^5+1 \cdot x^6$$
$$=1+x^2+x^3+x^6.$$

Cyclic codes have the important property that any code word is a multiple of the generator polynomial. That is, any code word polynomial $T(x)$ is related to the generator polynomial $g(x)$ by the expression $T(x)=P(x)\cdot g(x)$, where $P(x)$ is some multiplying polynomial. If the code words are of length n, and the number of check digits is $m=n-k$, then the polynomial $T(x)$ has degree equal or less than $n-1$ and, as $g(x)$ has degree m, the encoding operation can be regarded as the simple multiplication of the generator polynomial by the polynomial $P(x)$, which is of degree $k-1$ or less and has as coefficients the k information digits.

A feedback-shift-register scheme for the multiplication of a fixed polynomial

$$g(x)=g_0+g_1 \cdot x+\cdots+g_{n-a}\cdot x^{n-k}$$

by any polynomial

$$P(x)=a_0+a_1 \cdot x+\cdots+a_{k-1}\cdot x^{k-1}$$

is shown in Fig. 13.* The system consists of $n-k$ stages of shift register, the initial contents of which are set to zero. The operation of multiplication can be performed with this circuit by feeding an n-digit sequence into the register and observing the output at the position indicated in Fig. 13. The input sequence consists of k information digits, which are fed into the register high-order first, and $n-k$ zeros, which are fed into the register after the information digits.

After transmission over the channel, an encoded message may contain errors. If the error pattern is represented by a polynomial $E(x)$, in the same manner as the code words, then the received se-

---

*Peterson, in Chapter 8 of his book, has discussed schemes for encoding so that the resulting code words are the words of a *systematic* code.

quence $R(x)$ is

$$R(x)=T(x)+E(x)=P(x)\cdot g(x)+E(x)$$

and, provided $E(x)$ is not a multiple of $g(x)$, the presence of errors in $R(x)$ can be detected by simply dividing $R(x)$ by $g(x)$. In the absence of errors, $g(x)$ will divide $R(x)$ exactly, but if errors have occurred $g(x)$ will not divide $R(x)$ exactly and a remainder will exist. The presence or absence of a remainder can be used to determine whether errors have occurred. It should be noted that if $E(x)=q(x)\cdot g(x)$, then $g(x)$ will divide $R(x)$ exactly and the presence of errors will not be detected. This is not surprising since

$$R(x)=P(x)\cdot g(x)+q(x)\cdot g(x)=S(x)\cdot g(x)$$

which corresponds to a permissible code word sequence, and the decoder must assume it to be such.

Figure 14 shows a general circuit for dividing a polynomial

$$R(x)=r_0+r_1 \cdot x+\cdots+r_{n-1}\cdot x^{n-1}$$

by a polynomial

$$g(x)=g_0+g_1 \cdot x+\cdots+g_{n-k}\cdot x^{n-k}.$$

In dividing, the register contents are set to zero initially and the digits $r_{n-1}, \cdots, r_1$ then shifted into the register in the order in which they are received. After a total of $n-1$ shifts, the content of the registers is equal to the remainder after dividing $R(x)$ by $g(x)$.

Some Concluding Remarks

In addition to the algebraic approach to coding, some of the main aspects of which have been introduced above, the probabilistic approach has also been investigated [12, 43–46] and holds promise of being an economical method for transmitting information, with a negligibly small probability of error, at rates close to the channel capacity. For reasons of space, this highly important aspect of modern coding theory is not discussed in this chapter, other than to mention that sequential decision

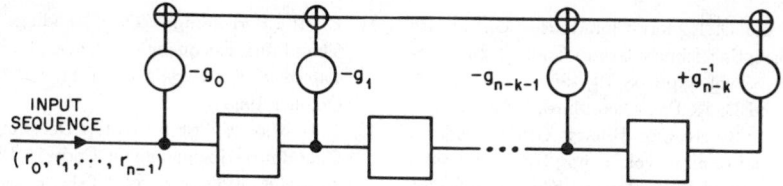

Fig. 14—Circuit for dividing polynomial $R(x) = r_0 + r_1 \cdot x + \cdots + r_{n-1} \cdot x^{n-1}$ by the polynomial $g(x) = g_0 + g_1 \cdot x + \cdots + g_{n-k} \cdot x^{n-k}$.

procedures are used to overcome the exponentially growing complexity of normal decoding equipment. For a discussion of sequential decoding, the reader should consult Wozencraft [12] and the somewhat simpler explanation given by Fano [43].

REFERENCES

1. N. Wiener, "Cybernetics," MIT Press and John Wiley & Sons, New York; 1948.

2. N. Wiener, "Extrapolation, Interpolation and Smoothing of Stationary Time Series," MIT Press and John Wiley & Sons, New York; 1949.

3. C. E. Shannon, "A Mathematical Theory of Communication," *BSTJ*, vol. 27, p. 379 and p. 623; 1948.

4. C. E. Shannon, "Communication in the Presence of Noise," *Proc. IRE*, vol. 37, p. 10; 1949.

5. B. McMillan, "The Basic Theorems of Information Theory," *Ann. Math. Statistics*, vol. 24, p. 196; 1953.

6. A. Feinstein, "Foundations of Information Theory," McGraw-Hill Book Co., New York; 1957.

7. A. I. Khinchin, "Mathematical Foundations of Information Theory," Dover Publications, Inc., New York; 1957.

8. A. Feinstein, "On the Coding Theorem and its Converse for Finite-Memory Channels," *Information and Control*, vol. 2, p. 25; 1959.

9. C. E. Shannon, "Probability of Error for Optimal Codes in a Gaussian Channel," *BSTJ*, vol. 38, no. 3, p. 611; 1959.

10. C. E. Shannon, "Coding Theorems for Discrete Source with a Fidelity Criterion," Information and Decision Process, Editor R. E. Machol, McGraw-Hill Book Co., New York; 1960: p. 93.

11. J. Wolfowitz, "Coding Theorems of Information Theory," Second Edition, Springer-Verlag; Berlin & New York; 1964.

12. J. M. Wozencraft and B. Reiffen, "Sequential Decoding," MIT Press and John Wiley & Sons, Inc., New York; 1961.

13. R. C. Tolman, "The Principles of Statistical Mechanics," Oxford University Press; 1938.

14. D. A. Huffman, "A Method for the Construction of Minimum-Redundancy Codes," *Proc. IRE*, vol. 40, p. 1098; Sept. 1952.

15. F. M. Reza, "An Introduction to Information Theory," Chapter 3, McGraw-Hill Book Co., New York; 1961.

16. S. Goldman, "Information Theory," Prentice-Hall, Inc., Englewood Cliffs, N. J.; 1955: p. 348.

17. J. N. Pierce, "Theoretical Diversity Improvement in Frequency-Shift Keying," *Proc. IRE*, vol. 46, no. 5, part 1, pp. 903–910; May 1958.

18. W. R. Bennett and J. R. Davey, "Data Transmission," Chapter 15, McGraw-Hill Book Co., New York; 1965.

19. R. W. Lucky, "Automatic Equalization for Digital Communication," *BSTJ*, vol. 44, no. 4; April 1965.

20. E. J. Baghdady, "Lectures on Communication System Theory," McGraw-Hill Book Co., New York; 1961: Chapters 5, 6, and 7.

21. Chapter 19 of Reference 20.

22. Chapters 10 and 11 of Reference 20.

23. A. V. Balakrishnan, ed., "Advance in Communication Systems," vol. 1, Chapter 1, Academic Press, New York; 1965.

24. D. Middleton, "Introductions to Statistical Communication Theory," McGraw-Hill Book Co., New York; 1960: Chapter 23.

25. A. Feinstein, "Error Bounds in Noisy Channel without Memory," *IRE Transactions on Information Theory*, vol. IT-1, no. 2, pp. 13–14; Sept. 1955.

26. D. Blackwell *et al*, "The Capacity of a Class of Channels," *Ann. Math. Statistics*, vol. 30, pp. 1229–1241; Dec. 1959.

27. D. W. Hagelbarger, "Recurrent Codes: Easily Mechanized, Burst-Correcting, Binary Codes," *BSTJ*, vol. 38, pp. 969–984; July 1959.

28. J. L. Massey, "Threshold Decoding," MIT Press, Cambridge, Mass.; 1963.

29. P. Elias, "Coding for Noisy Channels," *IRE Convention Record*, part 4, pp. 37–46; 1955.

30. D. Slepian, "A Class of Binary Signalling Alphabets," *BSTJ*, vol. 35, pp. 203–234; Jan. 1956.

31. G. E. Sacks, "Multiple Error Correction by Means of Parity Checks," *IRE Trans. on Information Theory*, vol. IT-4, no. 4; December 1958.

32. R. W. Hamming, "Error Detecting and Error Correcting Codes," *BSTJ*, vol. 29, pp. 147–160; 1950.

33. D. E. Muller, "Application of Boolean Algebra to Switching Circuit Design and to Error Detection," *IRE Trans. on Electronic Computers*, vol. EC-3, pp. 6–12; 1954.

34. I. S. Reed, "A Class of Multiple-Error-Correcting Codes and the Decoding Scheme," *IRE Trans. on Information Theory*, vol. PGIT-4, pp. 38–49; 1954.

35. R. C. Bose and D. K. Ray-Chaudhuri, "On a Class of Error Correcting Binary Group Codes," *Information and Control*, vol. 3, pp. 68–79; 1960.

36. R. C. Bose and D. K. Ray-Chaudhuri, "Further Results on Error Correcting Binary Group Codes," *Information and Control*, vol. 3, pp. 279–290; 1960.

37. W. W. Peterson and D. T. Brown, "Cyclic Codes for Error Detection," *Proc. IRE*, vol. 49, pp. 228–235; January 1961.

38. W. W. Peterson, "Encoding and Error-Correction Procedures for Bose-Chaudhuri Codes," *IRE Trans. on Information Theory*, vol. IT-6, no. 4, pp. 459–470; September 1960.

39. R. T. Chien, "Cyclic Decoding Procedures for Bose–Chaudhuri–Hocquenghem Codes," *IEEE Trans. on Information Theory*, vol. IT-10, no. 4, pp. 357–363; October 1964.

40. G. D. Forney, Jr., "On Decoding BCH Codes," *IEEE Trans. on Information Theory*, vol. IT-11, no. 4, pp. 549–557; October 1965.

41. E. R. Berlekamp, "On Decoding Binary Bose–Chaudhuri–Hocquenghem Codes," *IEEE Trans. on Information Theory*, vol. IT-11, no. 4, pp. 577–579; October 1965.

42. J. L. Massey, "Step-by-Step Decoding of the Bose–Chaudhuri–Hocquenghem Codes," *IEEE Trans. on Information Theory*, vol. IT-11, no. 4, pp. 580–585; October 1965.

43. R. M. Fano, "A Heuristic Discussion of Probabilistic Decoding," *IEEE Trans. on Information Theory*, vol. IT-9, no. 2, pp. 64–74; April 1963.

44. K. M. Perry and J. M. Wozencraft, "SECO: A Self-Regulating Error Correcting Coder-Decoder," *IRE Trans. on Information Theory*, vol. IT-8, pp. 128–135; September 1962.

45. R. G. Gallager, "Low-Density Parity-Check Codes," *IRE Trans. on Information Theory*, vol. IT-8, pp. 21–28; January 1962.

46. R. G. Gallager, "Low-Density Parity-Check Codes," MIT Press, Cambridge, Mass.; 1963.

CHAPTER 42

PROBABILITY AND STATISTICS

GENERAL

A *random experiment* is one that can be repeated a large number of times, under similar circumstances, but which yields unpredictable results at each trial. For example, rolling of a die is a random experiment where the result is one of the numbers 1, 2, 3, 4, 5, or 6. Observing the noise voltage across a resistor is another random experiment that gives a number V dependent on the instant of observation. A random experiment may consist of observing or measuring elements taken from a set that is then known as a *population*. A real number associated with the result of a random experiment is called a *random variable* or *variate*. A variate may be *discrete*, as in the case of the die, or *continuous* as in the case of a noise voltage.

Fluctuations of the result of a random experiment are due to causes that cannot be entirely controlled. However, if the conditions of the experiment are sufficiently uniform (for instance, if the same die is used in successive throws or if the resistor is at a constant temperature), some statistical regularity will be observed when results of a large number of experiments are considered. This regularity is expressed by the law that gives the probability of obtaining a given result or a result falling within a given range of values. The law of probability is assumed to be the same for each performance of the experiment, independently of the results of other trials.

A discrete variate, which may take values x_1, x_2, $\cdots$, x_n from a finite or denumerable set is described by p_k, its probability function. p_k is the probability of obtaining x_k as the result of one trial.

$$0 \leq p_k \leq 1$$

$$\sum_{\text{all } k} p_k = 1.$$

The *cumulative probability function*

$$P(x) = \sum_{x_k \leq x} p_k$$

also describes the variate. The p_k are the jumps of this function.

For a discrete variate of more than one dimension, several possibilities arise. For example, in two dimensions let (x_j, y_k) be the coordinates of a point in the (x, y) plane. $p(x_j, y_k)$ is the *joint* probability that $x = x_j$ and $y = y_k$.

$$p_1(x_j) = \sum_{\text{all } k} p(x_j, y_k)$$

and

$$p_2(y_k) = \sum_{\text{all } j} p(x_j, y_k)$$

are, respectively, the probabilities that $x = x_j$ independent of y, and $y = y_k$ independent of x.

$$p(x_j \mid y_k) = p(x_j, y_k) / p_2(y_k), \quad p_2(y_k) > 0$$

and

$$p(y_k \mid x_j) = p(x_j, y_k) / p_1(x_j), \quad p_1(x_j) > 0$$

are, respectively, the probabilities that $x = x_j$ given that $y = y_k$ has already occurred, and that $y = y_k$ given that $x = x_j$ has already occurred.

x_j and y_k are said to be *independent* if $p(x_j, y_k) = p_1(x_j) p_2(y_k)$ for all x_j and y_k.

A continuous variate is one that takes values from a nondenumerable set. The probability that one trial of the experiment gives a result between x and $x+dx$ is $p(x)dx$, where $p(x)$ is known as the *probability density function*. The *cumulative distribution function* is

$$P(x) = \int_{-\infty}^{x} p(s) \, ds.$$

$P(x)$ is the probability that the variate is less than or equal to x.

$$P(x_1) \geq P(x_2) \quad \text{if} \quad x_1 \geq x_2$$

$$P(-\infty) = 0$$

$$P(+\infty) = \int_{-\infty}^{+\infty} p(s) \, ds = 1$$

$$p(x) \geq 0$$

$$p(x) = dP/dx.$$

For a continuous random variable of more than one dimension or *multivariate*, the probability density function p and the cumulative distribution function P can also be defined. For instance, if (x, y) are the coordinates of a point in the plane, then $p(x, y)\,dx\,dy$ is the probability that the multivariate has its abscissa between x and $x+dx$ and its ordinate between y and $y+dy$.

The *marginal* probability density functions are

$$p_1(x) = \int_{-\infty}^{\infty} p(x, y)\ dy$$

$$p_2(y) = \int_{-\infty}^{\infty} p(x, y)\ dx.$$

For example, $p_1(x)\ dx$ is the probability that the variate x lies between x and $x+dx$ independent of y.

The *conditional* probability functions are

$$p(x \mid y) = p(x, y)/p_2(y), \quad \text{for} \quad p_2(y) > 0$$

and

$$p(y \mid x) = p(x, y)/p_1(x), \quad \text{for} \quad p_1(x) > 0.$$

For example, $p(x \mid y_0)\ dx$ is the probability that the variate x lies between x and $x+dx$, knowing that $y = y_0$.

Two variates x and y are said to be independent if

$$p(x, y) = p_1(x)\,p_2(y)$$

for all x and y.

The cumulative distribution function is

$$P(x, y) = \int_{-\infty}^{x} ds \int_{-\infty}^{y} p(s, t)\,dt.$$

$P(x, y)$ is the probability that the variates are less than or equal to x and y, respectively.

DEFINITIONS

Quantities used to describe the main features of a distribution are listed below. The mean and median locate the center. Geometrically the mean is the abscissa of the center of gravity of the probability (density) function and the median divides that function into two equal parts. The rms, variance, standard deviation, and mean absolute deviation are measures of the spread about the mean. The moments of order three and four about the mean describe asymmetry and peakedness, respectively, of the probability (density) function.

The equations containing $\sum$ and $\int$ refer to the discrete and continuous cases, respectively. Some authors combine these cases by means of a Stieltjes integral, for example

$$\mu = \int_{-\infty}^{\infty} x\,dP(x).$$

Expected Value or Mathematical Expectation: For any variable y equal to a given function $g(x)$ of the random variable x, the expected value is

$$E(y) = \sum_{\text{all } k} g(x_k)\,p_k$$

$$= \int_{-\infty}^{+\infty} g(x)\,p(x)\ dx.$$

Average or Mean:

$$\mu = E(x) = \sum_{\text{all } k} p_k x_k$$

$$= \int_{-\infty}^{+\infty} x\,p(x)\ dx.$$

Root-Mean-Square:

$$r = [E(x^2)]^{1/2} = \Big[\sum_{\text{all } k} p_k x_k^2\Big]^{1/2}$$

$$= \Big[\int_{-\infty}^{\infty} x^2 p(x)\ dx\Big]^{1/2}.$$

Moment of Order r, About the Origin:

$$\nu_r = E(x^r) = \sum_{\text{all } k} p_k x_k^r$$

$$= \int_{-\infty}^{+\infty} x^r p(x)\ dx$$

$$\nu_1 = \mu.$$

Moment of Order r, About the Mean:

$$\mu_r = E[(x-\mu)^r] = \sum_{\text{all } k} p_k (x_k - \mu)^r$$

$$= \int_{-\infty}^{+\infty} (x-\mu)^r p(x)\ dx.$$

Variance:

$$\sigma^2 = \mu_2 = E[(x-\mu)^2] = \sum_{\text{all } k} p_k (x_k - \mu)^2$$

$$= \int_{-\infty}^{+\infty} (x-\mu)^2 p(x)\ dx.$$

Standard Deviation or RMS Deviation from the Mean:

$$\sigma = \{E[(x-\mu)^2]\}^{1/2} = \left[\sum_{\text{all } k} p_k(x_k-\mu)^2\right]^{1/2}$$

$$= \left[\int_{-\infty}^{+\infty} (x-\mu)^2 p(x)\ dx\right]^{1/2}.$$

Mean Absolute Deviation or Mean Absolute Error:

$$\text{mae} = E(|\ x-\mu\ |) = \sum_{\text{all } k} p_k\ |\ x_k-\mu\ |$$

$$= \int_{-\infty}^{+\infty} |\ x-\mu\ |\ p(x)\ dx.$$

Median: A value m such that the variate x_k (or x) has equal probabilities of being larger or smaller than m. For the continuous case

$$\int_{-\infty}^{m} p(x)\ dx = \int_{m}^{+\infty} p(x)\ dx = \tfrac{1}{2}.$$

Mode: A value of x (or x_k) where the probability density $p(x)$ (or p_k) is largest. There may be more than one mode.

p-Percent Value or Percentile: A value of x exceeded only p percent of the time; that is, with probability $p/100$. For continuous distributions the p-percent value denoted by x_p satisfies

$$1 - P(x_p) = \int_{x_p}^{+\infty} p(x)\ dx = p/100.$$

The median is the 50-percent value.

Quartile: The 25- and the 75-percent values are called the lower and upper quartiles, respectively.

Chebishev Inequality:

$$\text{Prob}(|\ x-\mu\ | \geq \epsilon) \leq \sigma^2/\epsilon^2.$$

CHARACTERISTIC FUNCTION

The characteristic function $C(u)$ for a distribution defined by its probability (density) function p or by its cumulative distribution function P is

$$C(u) = E[\exp(jux)] = \int_{-\infty}^{+\infty} \exp(jux)\ dP(x)$$

$$= \int_{-\infty}^{+\infty} \exp(jux) p(x)\ dx$$

$$= \sum_{\text{all } k} p_k \exp(jux_k).$$

Properties

$$C(0) = 1$$

$$|\ C(u)\ | \leq 1, \quad \text{for } u \text{ real}$$

$$\nu_r = \frac{j^{-r} d^r C(u)}{du^r}\bigg|_{u=0}$$

$$C(-u) = C^*(u), \quad \text{for } u \text{ real}$$

where the asterisk denotes the complex conjugate. $C(u)$ can be expanded in terms of the moments ν_r

$$C(u) = 1 + \sum_{r=1}^{\infty} \nu_r(ju)^r/r!$$

The function C is the Fourier transform of p in the continuous case, hence

$$p(x) = (1/2\pi) \int_{-\infty}^{\infty} \exp(-jux) C(u)\ du.$$

The moment generating function is

$$C(-jt) = E(e^{tx}).$$

For a multivariate $\mathbf{x} = (x_1, x_2, \cdots, x_n)$, the characteristic function is

$$C(u_1, u_2, \cdots, u_n)$$

$$= E\{\exp[\ j(u_1 x_1 + u_2 x_2 + \cdots + u_n x_n)]\}$$

or

$$C(\mathbf{u}) = E[\exp(\ j\mathbf{u}\tilde{\mathbf{x}})].$$

ADDITION OF STATISTICALLY INDEPENDENT VARIABLES

If two continuous independent variates x_1, x_2 have probability densities $p_1(x_1)$ and $p_2(x_2)$, the probability density function for their sum $x = x_1 + x_2$ is the convolution integral

$$p(x) = \int_{-\infty}^{\infty} p_1(x-\xi) p_2(\xi)\ d\xi$$

or in shortened form

$$p = p_1 * p_2.$$

Similarly, the cumulative distribution function for the sum is

$$P(x) = P_1 * p_2 = \int_{-\infty}^{+\infty} P_1(x-\xi)\ dP_2(\xi)$$

$$= \int_{-\infty}^{+\infty} P_1(x-\xi) p_2(\xi)\ d\xi.$$

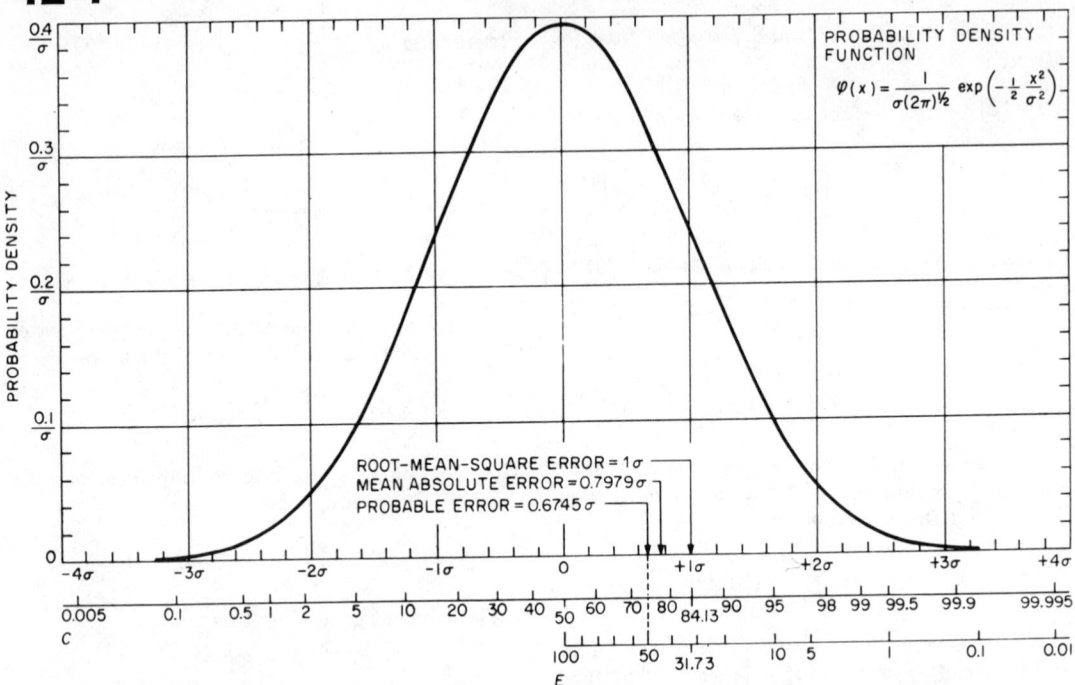

Fig. 1—The normal distribution. σ is the standard deviation. Scale C is the cumulative distribution function in percent $= 100\ \Phi(x)$. For example, the probability of finding x between $-\sigma$ and $+2\sigma$ is $97-16=81$ percent. Scale E is the probability that the error (absolute deviation) exceeds the value read on the axis. For example, if the deviation is larger than 2σ in either direction, probability is 4.5 percent.

Instead of computing these convolutions, it is sometimes simpler to use the corresponding property of the characteristic functions

$$C(u) = C_1(u)\,C_2(u)$$

and to deduce $p(x)$ as the Fourier transform of $C(u)$. This property extends to the sum of n independent variates.

DISTRIBUTIONS

Normal Distribution

The *normal* or *Gaussian* distribution is often found in practice because it occurs whenever (1) a large number of independent random causes produces additive effects and (2) an appreciable fraction of the causes produces effects of nearly maximum variance. This is known as the *Central Limit Theorem*: If the variates x_i are independent, have a mean μ and a standard deviation σ, then the sum $x_1 + x_2 + \cdots + x_n$ has a distribution that is

approximately normal with mean $n\mu$ and standard deviation $\sigma n^{1/2}$. (The x_i may be discrete or continuous.)

The normal probability density function, for a mean of zero and a standard deviation σ, is

$$\varphi_\sigma(x) = [1/\sigma\,(2\pi)^{1/2}]\exp[-\tfrac{1}{2}(x/\sigma)^2].$$

(See Fig. 1 and the table on p. 47–19.) When the mean value is μ instead of 0, the probability density becomes $\varphi_\sigma(x-\mu)$.

The cumulative distribution function

$$\Phi(x) = \int_{-\infty}^{x} \varphi_\sigma(x)\ dx$$

is given by scale C on Fig. 1 and more accurately by the table on p. 47–19. Related to Φ are the error function erf t and the complementary error function erfc t.

$$\text{erf } t = (2/\pi^{1/2})\int_{0}^{t}\exp(-t^2)\ dt = 2\Phi[t2^{1/2}]-1$$

erfc $t = 1 - \text{erf } t$.

The distribution of the absolute deviation from

the mean $|x-\mu|$, sometimes called the error, is given by scale E on Fig. 1. The median value of the error, equal to 0.6745σ, is called the *probable error*. It is exceeded 50 percent of the time. The average of $|x-\mu|$, equal to 0.7979σ, is the *mean absolute error*. The 3σ error is exceeded with probability of about 0.3 percent.

Additive Property: The linear combination, with constant coefficients of n normal independent random variables, is also a normal random variable. If

$$y=c_1x_1+c_2x_2+\cdots+c_nx_n$$

where x_i has mean μ_i and variance σ_i^2, then y has a mean

$$\mu=\sum_{i=1}^{n}c_i\mu_i$$

and a variance

$$\sigma^2=\sum_{i=1}^{n}c_i^2\sigma_i^2.$$

Poisson Distribution

A random experiment that leads to the Poisson distribution might consist of counting during a given time T the electrons emitted by a cathode, the telephone calls received at a central office, or the noise pulses exceeding a threshold value. In all these cases the events are, in general, independent of each other and there is a constant probability νdt that one of them will occur during a short interval dt.

The probability that exactly k events will occur during time interval T is given by the Poisson frequency function

$$p_k=(m^k/k!)\exp(-m),\quad k=0,1,2,\cdots$$

where the parameter $m=\nu T$ is the average number of events during interval T.

The variance of k is

$$E[(k-\nu T)^2]=m.$$

The standard deviation is

$$m^{1/2}.$$

The characteristic function $C(u)$ is

$$\exp\{m[\exp(ju)-1]\}.$$

Binomial Distribution

If the result of a random experiment is one of two alternatives, the statistics are completely de-

fined by the probability p of one of the alternatives. The trial may be the flipping of a coin or the testing of a transistor taken at random. The preferred alternative or "success" could be a head in the first case, an acceptable transistor in the second case. The probability of failure in one trial is

$$q=1-p.$$

In n independent trials, the probability of exactly k "successes" is given by

$$p_k=C_k^n p^k(1-p)^{n-k}$$

$$=[n!/k!(n-k)!]p^k(1-p)^{n-k},\quad 0\leq k\leq n.$$

This is called the binomial distribution because p_k is the kth term in the development of the binomial $(p+q)^n$.

The average of k is np and the variance is

$$E[(k-np)^2]=npq.$$

The standard deviation is

$$(npq)^{1/2}.$$

The probability of at least one success in n independent trials is

$$1-(1-p)^n.$$

The characteristic function $C(u)$ is

$$[p\exp(ju)+1-p]^n.$$

Application: If 15 percent of the components from a given lot are defective, the probability of finding exactly 3 bad ones in a set of 10 is

$$C_3^{10}(0.15)^3(0.85)^7=\frac{10\times9\times8}{1\times2\times3}15^3\,85^7\,10^{-20}$$

$$=13\text{ percent.}$$

The probability of finding at least one good component in a set of 3 is

$$1-(0.15)^3=99.7\text{ percent.}$$

If n is large, the binomial distribution may be approximated by a normal distribution with mean np and standard deviation $(npq)^{1/2}$. If the product np is small and n is large, a better approximation is the Poisson distribution with parameter $m=np$.

Exponential Distribution

An important case where this distribution arises is the following. In a Poisson process, the probability that the interval between two consecutive events lies between t and $t+dt$ is

$$\nu \exp(-\nu t)\ dt = d[1 - \exp(-\nu t)]$$

with $t \geq 0$. The average interval is

$$E[t] = 1/\nu.$$

The probability density function is

$$p(t) = \nu \exp(-\nu t), \quad t \geq 0.$$

The root-mean-square is

$$[E(t^2)]^{1/2} = \sqrt{2}/\nu.$$

The standard deviation is

$$\{E[(t-1/\nu)^2]\}^{1/2} = 1/\nu.$$

The median is

$$(\log_e 2)/\nu = 0.6931/\nu.$$

The cumulative distribution function is

$$1 - \exp(-\nu t).$$

The probability that an interval is larger than t is

$$\exp(-\nu t).$$

Problem: Pulses of noise, above a certain level, occur with an average density of 2 per millisecond. A device is triggered every time two pulses occur within the same 5-microsecond interval. How often does this happen? Since $\nu t = 0.01$, then

$$\exp(-0.01) = 0.990$$

(from table on p. 47–18) is the probability that one interval will exceed 5 microseconds. The device is triggered by 1 percent of the pairs of consecutive pulses, hence 20 times per second.

Multivariate Normal Distribution

The multivariate $\mathbf{x} = (x_1, x_2, \cdots, x_n)$ is normally distributed about the origin if the probability density function is

$$\varphi_M(\mathbf{x}) = [(2\pi)^n \det M]^{-1/2} \exp[-\tfrac{1}{2}(\mathbf{x}M^{-1}\tilde{\mathbf{x}})]$$

where the *moment* or *covariance matrix* $M = (\mu_{ij})$ is of order n and $\tilde{\mathbf{x}}$ denotes the transpose of $\mathbf{x}$. The coefficients μ_{ij} are the second-order moments

$$\mu_{ij} = E[x_i x_j].$$

Sometimes μ_{ii}, the variance of x_i, is denoted by σ_i^2 and μ_{ij}, the covariance of x_i and x_j, is expressed by $\sigma_i \sigma_j r_{ij}$. The r_{ij} are correlation coefficients.

Any linear function of $\tilde{\mathbf{x}}$, say $\tilde{\mathbf{y}} = L\tilde{\mathbf{x}}$, where L is a matrix of order $m \times n$, is normally distributed with the moment matrix

$$N = LML.$$

The characteristic function of the multivariate normal distribution is

$$C(\mathbf{u}) = E[\exp(j\mathbf{u}\tilde{\mathbf{x}})] = \exp[-\tfrac{1}{2}(\mathbf{u}M\tilde{\mathbf{u}})].$$

The sum of two independent, normally distributed multivariates $\mathbf{x}$ and $\mathbf{y}$ with covariance matrices M and N, respectively, is normally distributed with covariance matrix $M + N$

$$\varphi_M * \varphi_N = \varphi_{M+N}.$$

Normal Distribution in Two Dimensions: The probability density function at the point (x, y) is

$$\varphi(x, y) = [2\pi\sigma_1\sigma_2(1-\rho^2)^{1/2}]^{-1}$$

$$\times \exp\left[-[2(1-\rho^2)]^{-1}\left(\frac{x^2}{\sigma_1^2} - \frac{2\rho xy}{\sigma_1\sigma_2} + \frac{y^2}{\sigma_2^2}\right)\right]$$

where σ_1^2 and σ_2^2 are the variances of x and y, and ρ is their correlation coefficient.

Circular Case—Rayleigh Distribution: When the two normally distributed variates have the same variance $(\sigma_1 = \sigma_2 = \sigma)$ and are not correlated $(\rho = 0)$

$$\varphi(x, y) = (2\pi\sigma^2)^{-1} \exp[-\tfrac{1}{2}(x^2 + y^2)/\sigma^2].$$

The distance R to the origin, $R = (x^2 + y^2)^{1/2}$, is distributed according to the probability density function

$$p(R) = (R/\sigma^2) \exp(-R^2/2\sigma^2), \quad \text{for} \quad R \geq 0.$$

This is sometimes called the Rayleigh distribution. When a large number of small independent random phasors with equiprobable phases are added, the extremity of the vector sum is distributed according to the circular normal bivariate distribution. The magnitude R of the sum has therefore the probability density $p(R)$. This applies to the electromagnetic field scattered by a large number of small scatterers. It also describes the distribution of the envelope of a narrow band of Gaussian noise.

Figure 2 shows the function $p(R)$, and scale C gives the probability that some given level will be exceeded. The rms of R is $\sigma(2)^{1/2}$. The average

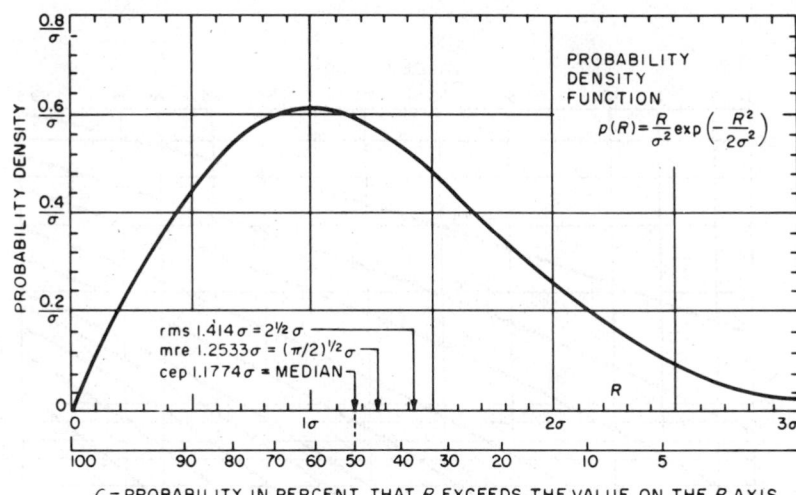

PROBABILITY
DENSITY
FUNCTION

$$p(R) = \frac{R}{\sigma^2} \exp\left(-\frac{R^2}{2\sigma^2}\right)$$

rms $1.414\sigma = 2^{1/2}\sigma$
mre $1.2533\sigma = (\pi/2)^{1/2}\sigma$
cep 1.1774σ = MEDIAN

R

C = PROBABILITY IN PERCENT THAT R EXCEEDS THE VALUE ON THE R AXIS

Fig. 2—Rayleigh distribution. R is the distance to the origin in a bivariate normal distribution. σ is the standard deviation for either component of the normal distribution.

$\sigma(\pi/2)^{1/2} = 1.2533\sigma$ is the *mean radial error*. The *median* or 50-percent value, 1.1774σ, is also called cep (circular error probable), because it is the radius of the 50-percent-probability circle in the x, y plane.

Using $X = \frac{1}{2}R^2$ (power) as the variable

$$p(R)\ dR = [\exp(-X/X_0)]\ d(X/X_0)$$

with $X_0 = \sigma^2$ (an exponential distribution).

When the circular normal distribution has its center at a distance S from the origin, the distance R to the origin is distributed according to

$$p(R)\ dR = (R/\sigma^2)$$

$$\times \exp[-(R^2 + S^2)/2\sigma^2] I_0(RS/\sigma^2)\ dR$$

where I_0 = Bessel function of zero order with imaginary argument. This is the distribution of the envelope of a sine wave plus some Gaussian noise. It also represents the distribution of the amplitude of a field that results from the addition of a fixed vector and a random component obtained, for instance, by scattering from a large number of small independent scatterers. See Fig. 3.

Chi-Square Distribution

The distribution of the sum of the squares of n independent normal variates, each having mean zero and variance unity, is called the chi-square distribution of n degrees of freedom.

The probability density function for this sum x is

$$k_n(x) = \frac{x^{(n/2)-1}}{2^{n/2}\Gamma(n/2)} \exp(-x/2).$$

(x, being the sum of n squares, is nonnegative.) The parameter n is called the *number of degrees of freedom*. The mean of x is n, its variance is $2n$, and its characteristic function is $(1 - j2t)^{-n/2}$.

The p-percent value of x (exceeded p percent of the time) is denoted, for n degrees of freedom, by $\chi_p^2(n)$:

$$\int_{\chi_p^2}^{\infty} k_n(x)\ dx = p/100.$$

Curves of $\chi_p^2(n)$ versus p are shown in Fig. 4. For values of n greater than 30, $(2x)^{1/2}$ is approxi-

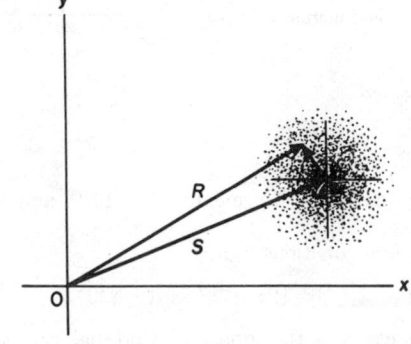

Fig. 3.

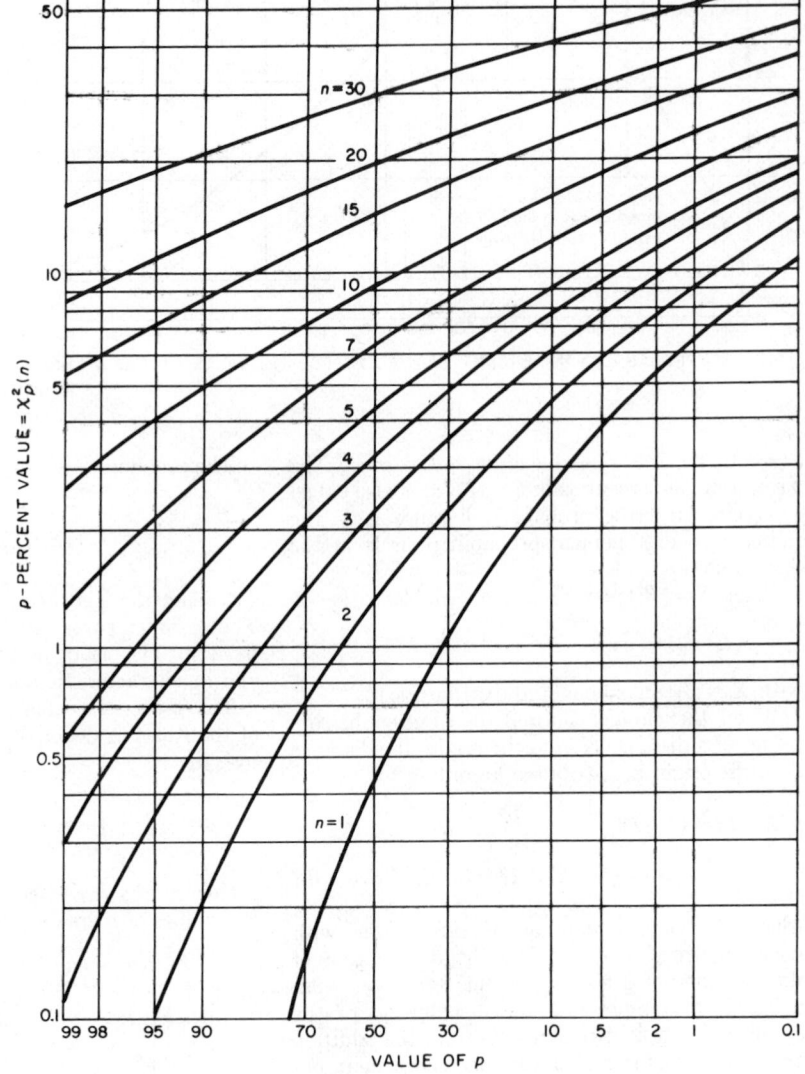

Fig. 4—Chi-square distribution. The function $\chi_p{}^2(n)$ vs p.

mately normal with mean $(2n-1)^{1/2}$ and unit variance.

The first functions k_n are:

$$k_1(x) = (2\pi x)^{-1/2} \exp(-x/2).$$

In this case x is the square of a normal variate.

$$k_2(x) = \tfrac{1}{2} \exp(-x/2).$$

In this case x corresponds to R^2/σ^2 in the Rayleigh distribution.

$$k_3(x) = (x/2\pi)^{1/2} \exp(-x/2).$$

In this case x is the square of the distance to the origin of a point in space having a normal distribution with spherical symmetry.

t Distribution

The t distribution of n degrees of freedom arises from the ratio

$$t = x/(y/n)^{1/2}.$$

x is a normal variate with zero mean and unit variance; y has a chi-square distribution with n degrees of freedom.

The probability density function is

$$p(t) = (n\pi)^{-1/2} \frac{\Gamma[(n+1)/2]}{\Gamma(n/2)} (1+t^2/n)^{-(n+1)/2},$$

$$\text{for} \quad -\infty < t < \infty$$

where Γ refers to the gamma function. The mean of t is zero and its variance is $n/(n-2)$ for $n \geq 3$.

Values of the cumulative distribution of $|t|$ may be read from Fig. 5.

SAMPLING

If a random experiment is repeated independently n times, the set of variates $x_1, x_2, \cdots, x_n$ is called a random sample of size n. The distribution of x from which the sample is drawn is known as the *parent distribution*.

The numbers $x_1, \cdots, x_n$ may not all be different and may form a smaller set $x_1, \cdots, x_k, \cdots, x_m$, where x_k occurs n_k times. The definitions given earlier can be applied to a sample (or to an arbi-

trary set of numbers) by using the *relative frequencies* n_k/n in place of the probabilities p_k.

The *sample mean* is defined to be

$$\bar{x} = (1/n)(x_1 + x_2 + \cdots + x_n).$$

This estimate is unbiased: $E(\bar{x}) = \mu$. It usually has the least variance among unbiased estimators.

The *sample variance* is defined to be

$$s^2 = (n-1)^{-1} \sum_{i=1}^{n} (x_i - \bar{x})^2.$$

The factor $(n-1)$ makes s^2 unbiased: $E(s^2) = \sigma^2$.

For a normal variate the standard deviation σ can also be deduced from the sample range; that is, from the difference between the largest number and the smallest number in the sample. For a sample of size n, σ is obtained by dividing the range by the number c_n in the table.*

n	c_n
5	2.33
10	3.08
20	3.73
30	4.09
100	5.02

Usually this estimator will have a larger variance than the sample variance.

Let the x_k be in such order that

$$x_1 \leq x_2 \leq x_3 \leq \cdots \leq x_n.$$

The sample median is defined to be

$$\xi = x_{(n+1)/2}$$

if n is odd and

$$\xi = \tfrac{1}{2}[x_{n/2} + x_{(n/2)+1}]$$

if n is even. A sample may always be so ordered.

Interval Estimation of the Mean and Variance of a Normal Variate

A p-percent *confidence interval* is such that the quantity estimated falls within that interval p percent of the time. Intervals of this type can be

* From E. S. Pearson, "Percentage Limits for the Distribution of Range in Samples for a Normal Population," *Biometrika*, vol. 24, pp. 404–417; November 1932: see p. 416. See also, E. S. Pearson and H. O. Hartley, "Biometrika Tables for Statisticians," vol. 1, Cambridge University Press, London, England; 1954: see Table 22.

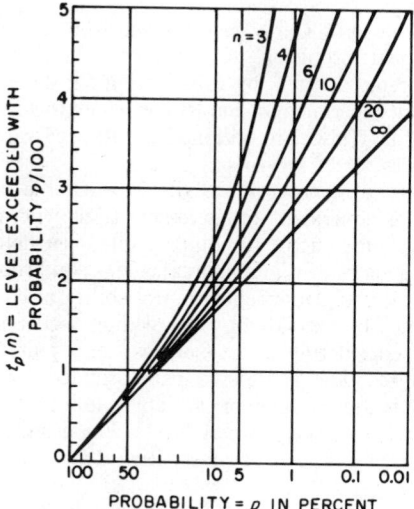

Fig. 5—Student's t distribution. For n degrees of freedom, the ordinate on the curve labeled n is the value $t_p(n)$ exceeded, in either direction, with a probability $p/100$.

deduced from a given sample for the mean μ and for the variance σ of the parent population.

Mean, σ known

$$\bar{x}-a\sigma/n^{1/2}<\mu<\bar{x}+a\sigma/n^{1/2}$$

where a is chosen so that $\phi(a)=0.5+p/200$.

Mean, σ unknown

$$\bar{x}-(s/n^{1/2})t_{100-p}(n-1)$$
$$<\mu<\bar{x}+(s/n^{1/2})t_{100-p}(n-1).$$

Example: For a sample of size 5, a 99% confidence interval for μ, σ unknown, is

$$\bar{x}-2.06s<\mu<\bar{x}+2.06s$$

referring to Fig. 5 for the value of $t_{100-p}(4)$.

Variance, μ known

$$\sum_{i=1}^{n}(x_i-\mu)^2/\chi_{(100-p)/2}^2(n)$$

$$<\sigma^2<\sum_{i=1}^{n}(x_i-\mu)^2/\chi_{(100+p)/2}^2(n).$$

Variance, μ unknown

$$(n-1)s^2/\chi_{(100-p)/2}^2(n-1)$$
$$<\sigma^2<(n-1)s^2/\chi_{(100+p)/2}^2(n-1).$$

Example: For a random sample of size 5 a 90% confidence interval for σ^2, μ unknown, is

$$0.42s^2<\sigma^2<5.7s^2$$

referring to Fig. 4 for the values of $\chi_5^2(4)$ and $\chi_{95}^2(4)$.

CHI-SQUARE TEST

The problem is to find how well a sample taken from a population agrees with some distribution function assumed for that population.

Let the random sample be divided into m disjoint sets and let the number of sample points falling within the ith set be f_i. From the assumed distribution and the size of the sample, the expected number g_i of points in the ith set is computed. The deviation between this and the actual result is expressed by

$$D=\sum_{i=1}^{m}(f_i-g_i)^2/g_i.$$

If the f_i are sufficiently large, this deviation is distributed according to the chi-square distribution

with $m-1$ degrees of freedom, approximately. The curves of Fig. 4 can be used to evaluate in percent the significance of a given deviation.

If the assumed parent distribution is not completely known and r parameters defining it have been determined to fit the sample, the number of degrees of freedom is reduced to $m-1-r$. For example, in the following application ν is unknown; hence $r=1$.

Application: During three successive one-hour periods the number of telephone calls received at a station was 11, 15, and 23, while during two nonoverlapping two-hour periods it was 40 and 37. How does this agree with a Poisson process?

Since the density ν (the number of calls per hour) has not been specified, it is deduced from the sample

$$\nu=(11+15+23+40+37)/7=18.$$

The deviation from the expected number is

$$7^2/18+3^2/18+5^2/18+4^2/36+1^2/36=5.1.$$

For $5-2=3$ degrees of freedom, this deviation is exceeded about 15 percent of the time. The assumption of a Poisson process is therefore very good. It would have been significantly doubtful only if the deviation obtained was exceeded as rarely as 5 percent or less of the time, which corresponds to D larger than 7.8 in this application.

MONTE CARLO METHOD

The Monte Carlo method consists of solving statistical problems, or problems that can be interpreted as such, by substituting for the actual random experiment a simpler one where the desired probability laws are obtained by drawing random numbers.

Reading in order the digits in the table on p. 47-14 is equivalent to successive trials where the result is one out of 10 equiprobable eventualities. Taking pairs of digits simulates 100 equiprobable eventualities. An event with probability of 63 percent may be simulated by the reading of successive pairs, considering as a "success" any pair from 00 to 62. The successive pairs divided by 100 approximate a random variable uniformly distributed over the interval (0, 1). For a smoother approximation, 3 or 4 consecutive digits could be used.

Any continuous variate x with cumulative distribution function $P(x)$ can be transformed into a new variate r that is uniformly distributed between zero and one by means of $r=P(x)$. Conversely, the variate x can be simulated by solving $P(x)=r_i$

for x, where the r_i are successive random numbers. For example, from p. 47–14 we have 0.49, 0.31, 0.97, 0.45, 0.80, etc. The table on p. 47–19 gives the corresponding values of x that will be normally distributed with mean zero and variance one: 0.0, -0.5, 1.9, -0.1, 0.8. This simulates the result of successive shots aimed at the point $x=0$.

To obtain accurate numerical results by the Monte Carlo method, a large number of trials N may be necessary since the accuracy increases roughly as the square root of N. There are cases, however, where only a crude evaluation is needed and it may be obtained even with a short table such as that on p. 47–14.

Problem: Airplanes arrive over an airfield at random, independently of each other, at the average rate of one per minute. The landing operation takes $\frac{3}{4}$ minute and only one airplane can be handled at a time. Will many airplanes have to wait before landing? The cumulative distribution function for the interval t minutes between arrival of successive airplanes is $1-\exp(-t)$ (see p. 47–18). The successive intervals, during an imaginary experiment, may therefore be taken as $t_i=-\log_e(1-r_i)$, where r_i are the random numbers uniformly distributed between 0 and 1. This is equivalent to $t_i=-\log_e r_i$. Starting at the top left of the table on p. 47–14 gives 0.71, 1.17, 0.03, 0.80, 0.22, 0.13, 0.25, 0.40, 0.37, 0.46, 0.17, 0.15, 0.37, 0.65, 3.91, 2.21, 0.17, $\cdots$ for the successive intervals in minutes. It is apparent that after a few minutes airplanes will be waiting. A few other trials using other parts of the table show that this situation is not exceptional. The traffic density is too high. The problem could be made more realistic by assuming a normal distribution of the landing times, simulated for instance as explained above.

RANDOM OR STOCHASTIC PROCESSES

A random or stochastic process is a family of random variables $\{y_t\}$ where the index t is usually considered to be time. If t assumes discrete values (for example, all the positive integers), the process is called *discrete*. If t takes on all the values from an interval, the process is said to be *continuous*.

The noise voltage y_t across a resistor is an example of a stochastic process. Suppose the general shape of a typical recording of this noise voltage does not appear to change: The oscillations show no tendency to become larger or smaller, do not change in rapidity, etc. This type of process is called *stationary*. The probability density functions of stationary stochastic processes are independent of time:

$$p(y_{t_1})=p(y_{t_2}), \quad \text{for any } t_1 \text{ and } t_2.$$

The process is called *Gaussian* or *normal* if the joint distribution

$$p(y_{t_1}, y_{t_2}, y_{t_3}, \cdots), \quad \text{for any } t_1, t_2, t_3, \cdots$$

is a multidimensional normal distribution.

Suppose the random variables in the family exhibit similar statistical properties and one member could be taken as representative of the entire family. This type of process is called *ergodic*. For ergodic processes, time averages are equivalent to ensemble averages, e.g.

$$\lim_{T\to\infty} T^{-1}\int_0^T y_t \, dt = \int_{-\infty}^{\infty} y_t p(y_t) \, dy_t.$$

Power Density Spectrum

In arriving at the power density spectrum of a stationary random function, let

$$F_T(\nu)=\int_0^T f(t)\exp(-2\pi j\nu t)\,dt$$

be the Fourier transform of the given random function $f(t)$ limited to the interval 0 to T.

The power density spectrum is defined by

$$W(\nu)=\lim_{T\to\infty} T^{-1}\,|\,F_T(\nu)\,|^2.$$

The function W is an even function of frequency

$$W(-\nu)=W(\nu)$$

since for a real function f

$$F_T(-\nu)=F_T{}^*(\nu).$$

Sometimes the spectrum is limited to positive frequencies by considering

$$W'(\nu)=2W(\nu) \quad \text{for} \quad \nu>0$$

$$=0 \qquad \text{for} \quad \nu<0.$$

The power in a band extending from ν_1 to ν_2 is

$$\int_{\nu_1}^{\nu_2} W'(\nu)\,d\nu.$$

Correlation Functions

The autocorrelation function is defined by

$$\varphi(\tau)=\lim_{T\to\infty} T^{-1}\int_0^T f(t)f(t+\tau)\,dt.$$

It is particularly useful because its Fourier transform is the power density spectrum

$$W(\nu) = \int_{-\infty}^{+\infty} \varphi(t) \exp(-j2\pi\nu t) \, dt$$

$$\varphi(t) = \int_{-\infty}^{+\infty} W(\nu) \exp(j2\pi\nu t) \, d\nu$$

or also

$$W'(\nu) = 4 \int_{0}^{\infty} \varphi(t) \cos(2\pi\nu t) \, dt$$

$$\varphi(t) = \int_{0}^{\infty} W'(\nu) \cos(2\pi\nu t) \, d\nu.$$

The cross-correlation function is defined by

$$\varphi_{fg} = \lim_{T \to \infty} T^{-1} \int_{0}^{T} f(t) g(t+\tau) \, dt.$$

The mean square of $f(t)$ is

$$\varphi(0) = \int_{-\infty}^{+\infty} W(\nu) \, d\nu = \int_{0}^{\infty} W'(\nu) \, d\nu.$$

If the process is Gaussian it is entirely specified by its second-order properties: power spectrum or autocorrelation function. For instance $p(y_t, y_{t+\tau})$ is a bivariate normal probability density function with $\mu_{11} = \mu_{22} = \varphi(0)$, and $\mu_{12} = \varphi(\tau)$.

Effect of a Linear Filter

A linear filter is defined by its impulse response $h(t)$ or by its transfer function $H(\nu)$, which is the Fourier transform of $h(t)$.

If the input to the filter is the random function $f_1(t)$, the output is the random function

$$f_2(t) = h*f_1$$

$$= \int_{-\infty}^{+\infty} h(t-\tau) f_1(\tau) \, d\tau.$$

Introducing the power gain

$$G(\nu) = |H(\nu)|^2$$

the power density spectrum of f_2 is

$$W_2 = GW_1.$$

The autocorrelation function of f_2 is

$$\varphi_2 = g*\varphi_1$$

where g is the Fourier transform of G or

$$g(t) = h(t)*h(-t) = \int_{-\infty}^{+\infty} h(\tau) h(\tau+t) \, d\tau.$$

CHAPTER 43

RELIABILITY AND LIFE TESTING

DEFINITIONS AND TERMINOLOGY

Average Life: The mean value for a normal distribution of lives. Generally applied to mechanical failures resulting from "wear-out".

Component: (Normally used interchangeably with the term "unit"). A component is defined as an article which is normally a combination of parts, subassemblies, or assemblies, and is a self-contained element of a complete operating equipment and performs a function necessary to the operation of that equipment. *Examples*: indicator unit, power unit, receiver, transmitter, rotating antenna, modulator unit, amplifier unit.

Confidence Level (Coefficient): The degree of desired trust or assurance in a given result. A confidence level is always associated with some assertion and measures the probability that a given assertion is true. For example, it could be the probability that a particular characteristic will fall within specified limits, i.e., the chance that the true value of P lies between $P=a$ and $P=b$.

Configuration Management: Management of and knowledge of where all specifications, procedures, and associated test results are located and assigned, so that it is possible to produce these controlled items and all reliability evaluations and predictions pertaining to the system.

Cumulative Distribution Function: If x is a random variable, then the cumulative distribution function of x is defined to be the function F such that for every real number t, $F(t)$ is the probability that a given outcome of x will not exceed t; in symbols: $F(t) = \text{Pr}, (-\infty \leq t)$.

Defect: Any deviation of a unit of product from specified requirements. A unit of product may contain more than one defect.

Degradation Failure: A failure that results from a gradual change in performance characteristics of an equipment or part with time.

Design Reviews:
(A) Preliminary design review: As soon as possible after a contract has been signed, a bread-board model should be built and its reliability estimated. Reliability engineering shall re-evaluate all parts and components and determine and recommend improvements in the design.
(B) Intermediate design reviews: While developing the system, conduct design reviews on a formal basis at all suppliers as well as at the prime contractor. This program should be coordinated with the reliability growth program. Account must be taken of the contract requirements for reliability goals at scheduled points in the production schedule.
(C) Critical design review: When engineering believes the design is ready to be "frozen" and also when a satisfactory prototype has met the qualification and other reliability tests, a final design review shall be scheduled. This formal review takes into account all contract demands as modified by the most recent changes in the contract. If the product is adjudged to be satisfactory, the final design may be approved for production. The block system should be used and authorization should be given to production to make x units per the specifications and blueprints without any change.

Downtime: Time during which equipments are not capable of doing useful work because of malfunction. This does not include preventive-maintenance time. In other words, downtime is measured from the occurrence of a malfunction to the correction of that malfunction.

Equipment: Material or articles (such as sets) comprising an outfit. The term "equipment" sometimes is used instead of "set", i.e., one or more assemblies or a combination of items capable of independently performing a complete function. *Examples*: Radio receiver, digital computer, automobile. An equipment may contain several sets as components. An example would be two radio receivers assembled for dual-diversity reception. The combination would constitute the equipment.

Failure: A failure is a detected cessation of ability to perform a specified function or functions within previously established limits on the area of interest. It is beyond adjustment by the operator by means of controls normally accessible to him during the routine operation of the device. This requires that measurable limits be established to define satisfactory performance of the function.

Failure Mode Analysis: Research, development, and production engineers as well as the reliability engineers analyze the basic design and determine by simulation and logistics what possible failures might occur. Corrective measures for eliminating and preventing failures are built into the basic design. Standard forms for the failure mode analysis are made available and the results must be given to the prime contractors and production engineers for evaluation.

Failure Rate and Hazard Rate: Failure rate is generally the rate at which failure occurs during an interval of time (given that it has not occurred before the start of that interval) as a function of the total interval length. Hazard rate is an instantaneous failure rate and is defined as the limit of the failure rate as the time interval approaches 0. An example might be: A family takes an automobile trip of 120 miles and completes the trip in three hours. The average rate was 40 mph, although they drove at various rates of speed. The rate at any given instance could be determined by reading the speedometer at that instance. Therefore, the 40 mph is equivalent to the failure rate and the speed at any instant of time equals the hazard rate.

Inherent Reliability: The basic or generic failure rates of components have often been compiled by several companies as well as by some government agencies. A library covering failure rates should be part of a good reliability program. Such a listing gives concretely the reliability that can be guaranteed.

Lot Size: A specific quantity of similar material or collection of similar units from a common source; in inspection work, the quantity offered for inspection and acceptance at any one time. It may be a collection of raw material, parts, subassemblies inspected during production, or a consignment of finished products to be sent out for service.

Maintainability: The maintainability of an equipment in a specified maintenance environment is the probability that a failure will be repaired within a specified time after the failure occurs.

Mean Time Before Failure (MTBF): The total measured operating time of a population of equipments divided by the total number of failures of a repairable equipment is defined as the ratio of the total operating time to the total number of failures. The measured operating time of the equipments of the population that did not fail must be included. This measurement is normally made during that period of time between the early life and wear-out failures. In the case of exponentially dis-

tributed times before failure, this ratio is the reciprocal of the failure rate.

MTBF can be determined by dividing the product of the number of equipments tested N and the test time t by the number of failures f which occur during that time, i.e., mtbf or often just $m = Nt/f$. "m" is the reciprocal of λ, i.e., $m = 1/\lambda$ and is related to the probability of survival by the exponential law $P_s = e(-t/m)$. The figure of merit m (sometimes expressed as $\bar{t}$) is convenient for use in determining if the reliability of an equipment is likely to be adequate for missions of specific lengths.

Mean Time to Failure (MTTF): The measured operating time of a single piece of equipment divided by the total number of failures of the equipment during the measured period of time. This measurement is normally made during that period of time between the early life and wear-out failures.

Mean Time to Repair (MTTR): The measured repair time divided by the total number of failures of the equipment.

Mission Success Rate: That percentage of the total missions uninterrupted by failure of the equipment. This figure of merit is more closely dependent on the reliability of the parts included in the system and on the design of the system than are either maintenance ratio or in-commission rate. However, this measure of reliability is valuable primarily to a using agency that has a regular schedule of missions. A mission success rate obtained by one agency is not typical of the equipment in general and will not necessarily apply for other agencies with different operating schedules.

Mode of Failure: The physical description of the manner in which a failure occurs and the operating condition of the equipment or part at the time of the failure.

Part Failure Rate: The rate at which a part fails to perform its intended function.

Probability: The limiting relative frequency in an infinite random series. If an event can occur in n ways and its failure in m ways, and if these $m + n$ ways are equally likely, then the mathematical probability that the event will occur in any one trial is the ratio $n/(n+m)$.

In other words, the probability of an event is the theoretical relative frequency with which it will occur, such relative frequency being the ratio of the number of times the event is observed under experimental conditions to the total of a great number of observations made under those conditions.

Probability of Survival: A numerical expression of reliability with the accepted nomenclature of

P_s and a range from 0 to 1.0 indicating the extremes of "impossibility" and "certainty".

In other words, the probability of a given equipment performing its intended function or the given use cycle is

$$R(t; x) = 1 - F(t; x)$$
$$= R(x+t)/R(x), \quad t \geq 0.$$

Product Effectiveness: The entire reliability program must be tied in with the quality engineering programs for securing the most effective operation possible. Product effectiveness includes all the elements for securing at minimum cost a product with maximum effectiveness. Programs for quality assurance contain schedules and procedures that encompass within them reliability, preventive maintenance, value engineering, human engineering, quality control, inspection, and tests that result in systems and products that will prove most effective when in operation.

Reliability Allocations: With an overall system reliability goal and where the confidence level is known, reliability values may be allocated to every component in the system by the use of available failure rates and weighting factors.

Reliability Demonstration: Critical tests must be programmed to provide sufficient valid data to determine the reliability of the system and all critical component parts. Provision should be made for processing all such data as expeditiously as possible to speed up all phases of the program with truly reliable materials and parts.

Reliability Evaluations and Predictions—Summary: It is impossible in the space allotted to this chapter to detail all the important features of reliability. Only a few of the tables required for detailed reliability estimations are listed. At the end of the chapter a short list of references is given. These contain more-complete lists of references. Also, many yearly conferences on reliability have been held at which the more important techniques are covered in the technical papers usually printed as Transactions. These should be added to a good library on reliability and will supply many missing steps in reliability engineering.

Reliability Goals: Requests for bids, specifications, and contracts requiring quality assurance and reliability generally describe completely the reliability goals. The reliability of the system with specified confidence levels is the principal goal. Values to be secured at specified points in time are listed on the growth curves. A reliability demonstration program is detailed to show that the reliability goals have been attained. How are such reliability goals expressed? The simplest statement covers only an expected value. The quality assurance paragraph in the specification will state simply: "The desired reliability is 99.7%." This is too simple, as the length of the mission or number of cycles of operation has not been detailed. The statement should be, "For 100 hours of operation the specified reliability is 99.7%."

Many contracts also introduce confidence levels. For example, after the initial design review, the reliability of the system must be 99% with a confidence level of 0.60. After qualifying, the system must have a reliability of 99.7% for a mission time of 10 hours with a confidence level of 0.99. Thus expressed, this goal has within it a final goal plus some information concerning the desired growth curve for reliability. In many programs, provision for the reliability demonstration program has not been made, or in making final arrangements for the finalized contract it is cancelled because of lack of funds. The reliability engineer should establish a very modest program to check the achievement of the reliability goals by means of an economic reliability demonstration program. Thus, it must be emphasized that the simplest possible reliability demonstration test should be made a part of the final program covered by contract and funds. This provides vital evidence that the reliability goals have been achieved and that the customer reliability requirements have been met.

Reliability Growth Curves: Periodic reports, such as monthly or quarterly, should be prepared containing up-to-date predictions of the system's reliability. These should be presented graphically on the reliability growth curves for this system and its various components and parts.

Reliability Predictions: Many agencies and companies have compiled failure rates for parts, components, subassemblies, assemblies, and systems. These generic failure rates are used as basic data to predict a value for reliability.

Sample: One or more sample units selected at random from a quantity of product to represent that quality of product for inspection purposes.

Sequential Sampling: Sampling inspection in which, after each unit is inspected, the decision is made to accept, to reject, or to inspect another unit. *Note*: Sequential sampling as defined here is sometimes called "unit sequential sampling" or "multiple sampling." Multiple sampling is preferred, to differentiate from sequential testing.

Specification Limits: The specification limit(s) is the requirement that a quality characteristic should meet. This requirement may be expressed as an upper or a lower specification limit (herein called a single specification limit), or both upper and lower specification limits (herein called a double specification limit).

System (General): A combination of parts, assemblies, and sets joined together to perform a specific operational function or functions. *Examples*: Piping system, refrigeration system, air conditioning system.

Test to Failure: Testing conducted on one or more items until a predetermined number of failures have been observed. Failures are induced by increasing electrical, mechanical, and/or environmental stress levels, usually in contrast to life tests in which failures occur after extended exposure to predetermined stress levels. A life test can be considered a test to failure using age as the stress.

Value Engineering: One feature of a good reliability program is a concurrent value engineering program. Many programs include incentive provisions. If an operation is unusually expensive, provisions should be made for a series of improvements that fall in with the value engineering and incentive programs. Provisions for successive cost reduction programs as well as basic value engineering improvements should be added to each contract. It may be made a part of the incentive programs that are now usually included in military and government contracts for procurement.

Weapon System: A missile and all the necessary support equipment (either ground or airborne) necessary to launch and properly operate a missile.

Wear-Out Period: The wear-out period of an equipment is that period of equipment life, following the normal operating period, during which the equipment failure rate increases above the normal rate.

RELIABILITY PREDICTIONS

The handbook approach to reliability predictions has much merit where bona fide failure-rate values (λ) and mean-time-before-failure values (MTBF) for components, parts, and systems are available. When applied in the usual reliability relations, the results may be used to make fairly valid predictions concerning the reliability of a product or system. Likewise these data will provide fairly good evidence for evaluating its quality, dependability, maintainability, and availability. A comprehensive design review will show whether the design has merit. If the approved design is followed exactly, production will be able to manufacture the type and kind of desired product that will also meet the reliability demands of service.

Potential growth in reliability will be a definite phase of the program. When carried to fruition it provides at little additional cost a valid demonstration of the reliability of the product and the effect of improvement in design scheduled as part of the reliability and quality program. These are the ideal conditions under which well-trained engineers may use, with a high degree of belief and with little danger of being wrong, the usual tools and techniques of reliability. Under true statistical control of the parts that make up the whole, valid reliability predictions are possible.

RELIABILITY DEFINITIONS

The following three definitions of reliability cover the different concepts concerning reliability that permeate the contracts and purchase orders now being prepared and given to prime and subcontractors.

(**A**) Reliability is the probability that a given product, system, or action will achieve its designated goal successfully under the specified environmental conditions and for a prescribed period of time or for the number of cycles of operation required for the mission or task.

(**B**) Reliability is the ratio of the number of successes to the total number of trials for a well-defined program, that is

$$R = S/N$$

where R designates reliability, $S =$ number of successes, and $N =$ number of trials.

(**C**) Reliability measured in terms of mean time before failure (MTBF) can be determined only for N units in operation when the total number of hours in operation or total number of cycles operated for all N units is known when the last unit has failed; it is given as

$$\mathrm{MTBF} = m = \sum t_i/N$$

or

$$m = \sum c_i/N$$

where $t_i =$ time to failure in hours for the ith unit, and/or $c_i =$ time to failure in cycles for the ith unit.

The failure rate then is taken as

$$\lambda = 1/m.$$

For mission time T and assuming an exponential distribution for the mission

$$R = \exp(-\lambda T)$$

where the density function is the exponential

$$F(t) = \lambda \exp(-\lambda t).$$

FUNCTION OF RELIABILITY

Stress on reliability is manifested at the present time for products of all kinds, for instrumentation and functional performance, and in the dependability of the individuals assigned to specific tasks. In dealing with coworkers and employees it is hoped that tasks assigned will be accomplished per schedule and as well or better than anticipated.

When a particular product is purchased, such as an automobile, it is expected to provide satisfactory transportation immediately and for many thousands of miles before breakdown, wear-out, or failure. By scheduled maintenance programs, manufacturers of these automobiles are able to guarantee a minimum of 50 000 miles of satisfactory performance within specified time limits. The term "reliability" is not usually applied to a new home just purchased, but it does apply for the more reputable and experienced builders and contractors. The electrical outlets, heating and water systems, windows, and roofs are supposed to perform their functions satisfactorily for years. The same may be said for all the elements, parts, and functions of such a home. The reliability, maintainability, and availability of any element, product, article, and system may be estimated in terms of known information concerning actual field performance. If such field information is not obtainable, then tests may be simulated to provide these measures.

A good reliability program sets up measures necessary to provide adequate assurance that the desired reliability goal has been achieved. The nature of these goals for both commercial and military applications is given in succeeding sections. Reliability is applied to the review of designs, to failure mode analysis, and to the use of production processes to secure absolute adherence to specification requirements as given on the engineering drawings. What part does reliability play in the quality-assurance program and how much is totally independent? What changes may be made, such as the introduction of redundancy if necessary to secure the specified reliability goal? When are characteristics truly independent? What tradeoffs are required to attain desired results? To solve many of these reliability problems, it is necessary to set up an applicable mathematical model. Their derivations will not be given but how to use them will.

RELIABILITY AND QUALITY: THE ADMINISTRATIVE MIX

Reliability, quality control, and engineering have been attempting to put "reliability" in a proper position in the management spectrum. Some feel that it belongs in engineering, but most organization charts (Fig. 1) do not reflect this expectation. Organizations such as the Institute of Electrical and Electronics Engineers and American Society for Quality Control have defined this proposed location as noted in the chart.

Today, reliability seems to fit most naturally into the product-assurance or product-effectiveness section of major space equipment manufacturers.

The most important organizational aspect is the fact that reliability and well-versed quality personnel use one tool which, when shared jointly,

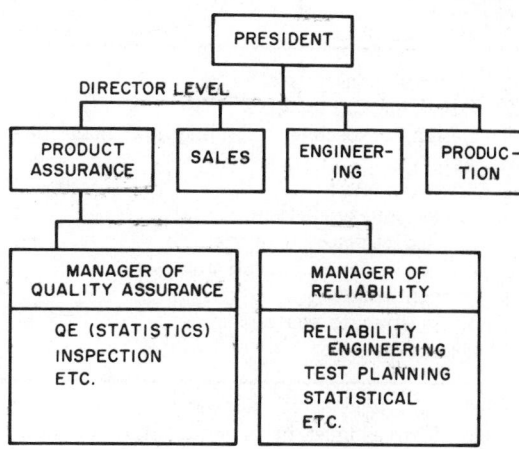

Fig. 1—Typical organization chart showing reliability chain of responsibility.

adds to the "experience-retention" capability; that tool is statistics. If the intent of any organization is to get the most out of the "fall of data," the organization of Fig. 1 is effective. Because of the problem of decision making, reliability is best placed at the staff level.

In a small organization, reliability is usually shared by engineering and by quality assurance. Some firms, to better satisfy the requirements of the Department of Defense, put reliability in the quality organization spectrum of operations, viz.,

> Quality and reliability assurance tasks.
> Data reduction.
> Planning.
> Inspection, etc.
> Test plan.
> Vendor survey.
> Quality-control engineering.

It may be noted that, early in the growth of reliability, the group or person assigned to that responsibility was given the task as part of engineering, much as is being accomplished today with the advent of "value engineering," "configuration control management," etc. In many firms a reliability "person" is assigned from the product-effectiveness organization to an engineering group for "task-force" duty. Reliability-oriented personnel can do most good when they are attached to engineering operations. Decisions must be made abruptly because of the short design–production cycle, and it is best to have all the tools for this type of decision-making at hand for immediate use. Reliability training can be effectively used.

RELIABILITY TASKS

The most important part of reliability planning lies in the relationship between reliability and engineering or the program manager. Reliability,

TABLE 1—TASK DELINEATION CHART.

1	Program plan update	11	Change and configuration control
2	Education and manuals	12	Reports and project review
3	Design to specified reliability and maintainability	13	Corrective action control
4	Apportionment	14	Supplier control
5	Models and prediction	15	Manufacturing reliability and maintainability control
6	Cost-effectiveness analysis	16	Failure diagnosis
7	Failure modes and effects analysis	17	Data acquisition and reduction
8	Human factors	18	Verification
9	Stress/strength analysis	19	Summary
10	Design review	20	References

to be effective, must fit into the total program. One of the best ways to gain acceptance of the reliability or maintainability effort is to establish a proper sequence for its insertion into a master operating plan. For companies without a particular project this plan usually is discussed with the engineering head and the production head. For corporations working on a program that requires reliability by specified intent, the customer invariably requires a reliability plan. The major items (or task delineators) in a reliability plan are noted in Table 1.

In the plan itself, the time relationship of the tasks to reliability and overall project tasks are noted as milestones. Some of these milestones are then placed on more-sophisticated PERT* or CPM† charts for management review of major effort.

DATA PROCESSING—COMPUTERS AND RELIABILITY

Data processing has become an important reliability tool. Historically, the computer (a form of data processor) was built to calculate on an iterative basis. Then the machine's capability was expanded to allow it to do iterative calculations over long periods of time. This resulted in accumulations of vast amounts of knowledge, which was stored in the memory of the computer.

The reliability engineer became the champion of the computer many years ago, because he had large calculations with many variables. Today's engineer, although not the reliability type, is becoming more cognizant of the machine's capability.

The reliability engineer uses the computer for the following purposes.

(**A**) Data reduction.
(**B**) System reliability predictions.
(**C**) Calculation of reliability tables.
(**D**) Plotting of data.

* Program Evaluation and Review Technique.
† Critical Path Method.

(**E**) Control of experiments (i.e., preplanning and preprogramming to execute an experiment by machine control). Logic for control of equipment (turn on—turn off).
(**F**) Modelling.
(**G**) Information retrieval.
(**H**) Reliability summary of parameter behavior.
(**I**) On-line computation of experiments.

One area that needs expansion is information retrieval. The reliability engineer definitely defers to history as a tool. Prior test results or experimental results are part of the balance sheet that helps decide on capability or design changes. When dealing with the placement of a part in an unknown environment, it is best to draw on earlier, similar experiments to predict behavior. American industry is spewing out tons of test data on products, processes, materials, etc. Some format must be used to collect the results of the report. This is where information retrieval (IR) enters the picture.

Statistical programs commonly used by reliability engineers (and available as standard computer programs) follow.

Frequency plots (and skewness calculations).
Range test.
Dispersion.
Central tendency.
Regression analysis.
F test.
t test.
Confidence-factor estimation.
Multiple-range test.
Line correlations.
Chi-square test.
Analysis of variance.
Analysis of distributions, (binomial, exponential, Poisson, etc.).
Availability, estimation, and prediction models.
Sampling statistics (directed towards sequential life testing).
Response surface experimentation.
Curve fitting.
Worse-case analysis (circuit design).

STANDARDS FOR RELIABILITY AND QUALITY CONTROL

Measures of reliability are derived from mathematical theory and mathematical models and are covered in later sections. Their basis is good-quality products made to established standards under statistical control. Economic quality standards are established from past satisfactory results, eliminating uncontrolled units of product. Satisfactory components made to established standards lay the foundations for reliable products. Reliability standards are covered under reliability goals. Reliability uses many of the features of operations research, particularly the use of mathematical models. The inherent reliability of each part, article, or system may be expressed numerically. These measures are tied in with probability theory.

RELIABILITY AND PROBABILITY

Reliability is considered as one form of probability. If a reliability of 0.99 is desired, this means that on the average the function will be satisfactorily performed 99 times in 100 trials. It may be stated that a reliability R has been secured if the probability of success P under the conditions and environment stipulated is such that $R=P$. Such a reliability value is considered an *expected* value and may be expressed as

$$E(R)=P.$$

This form of reliability is often considered as a point probability. Sets of such point probabilities are contrasted with distribution probabilities, where according to the location in the distribution, confidence levels may be associated with such probability values, whereas the expected values are usually associated with a level of 0.50 since expectancy is associated with a 50–50 chance. Successive experiments under which more data are accumulated improve the worth of the expected value. Under distribution theory, with the accumulation of more data, better confidence limits may be associated with any selected reliability value in the improved distribution. When a series of expected values are obtained by successive tests, the variability of these values may be determined; a probability may then be established as the confidence associated with a range within which the expected reliability values may lie in the past and in the future if statistical control is maintained. All results may be combined to provide a single expected value, used as the average or mean of the distribution of probabilities. In the simplest case the set of reliability or probability values will consist of points of equal weight. Each value is based on the same number of tests, n_i, with $n=n_1=n_2=n_3=\cdots=n_m$ for m tests with equal sample sizes. If, however, the n_i values are different from test to test, then the number of units used will be different and may be used as weighting factors. It is best to use relative weights, i.e., w_i, which will be secured from the relation

$$w_i=n_i/\sum_{i=1}^{i=m} n_i.$$

The distribution of reliability values actually has two weighting factors, (A) the frequency of occurrence of a particular test value, f_j, $j=1$ to c, and (B) the relative weight w_i derived from the relation above. When the n_i values are equal, use $w_j=w_i$ for each distribution value x_j and

$$\sum_{i=1}^{i=c} w_i= \sum_{j=1}^{j=c} f_j w_j$$

where $c=$ number of cells for x_j values.

Example: For the ith test, where n_i units are placed on test and d_i units in 1000 hours of test are classed as defective, then the measure of the reliability of the test is

$$R_i= (n_i-d_i)/n_i=s_i/n_i$$

where $s_i=n_i-d_i$, and the ratio failure rate λ is secured from the relation

$$\lambda_i=d_i/n_i$$

where this particular form of reliability is based on each unit being tested for 1000 hours with each occurrence of a defective unit being recorded regardless of time, if such failure occurs at 1000 hours or less of operating time. Later, reliability will be tied in with the number of hours of operation and $R(t)$ for time operated, while $R(c)$ for number of cycles operated will be used to measure the reliability of products and systems. The following two sets of data were secured for this example. Set 1 consists of results obtained where all n_i values are identical $(n=100)$ whereas in set 2 the n_i values vary in number from test to test. Table 2 lists these two sets of data for $m=20$ tests each.

Test Results

When summarized, the R_i values are noted as R_j values. The best estimate of the total reliability is the sum of these parts and is termed R. For set 1 the number of different cells is $c=7$, whereas for set 2 the number of different cells is $c=8$.

Summary of Test Results

Table 3 summarizes the results for the two sets of test results. It determines the arithmetic average (the mean) for each distribution, and also its variance and standard deviation. Using these values, rough estimates of the limits for confidence bands for both one-tailed and two-tailed distributions are given on the assumption that the distributions are essentially normal.

For set 2, giving each value the same weight regardless of the number of tests gives a first estimate of the average reliability as $\bar{R} = 19.486/20 = 0.9743$. The best value is determined by weighting values by the number of tests, giving $\bar{R}' = 0.9750$. The variance σ_R^2 is determined from

$$\sigma_R^2 = \left(\sum f_j w_j R_j^2 / \sum f_j w_j \right) - \bar{R}^2$$

$$= (0.950795250/1.0000) - 0.950625$$

$$= 0.00017025$$

$$\sigma_R = (0.0001702500)^{1/2} = 0.013048.$$

Analysis of Test Results—2 Sets of Weighting Factors (Tables 4 and 5)

For the twenty tests in set 1, the number of units on test is 100 in all cases, giving a total of 2000 units tested. For the 20 tests in set 2, the number of units on test varies from 100 to 2000, giving a total of 8000 units tested. If the number of units on test is used as the weighting factor, it will be used in the summary for the total distribution of the 10 000 units tested.

Observed vs. Theoretical Probabilities for Distributions

The average values for sets 1 and 2 are identical. Set 1 has a larger variance. With a failure rate of

TABLE 2—DETAILED TEST RESULTS FOR 2 SETS OF TEST DATA.

	Set 1				Set 2			
Test	Units on Test n_i	Failures d_i	Successes s_i	Reliability $R_i = s_i/n_i$	Units on Test n_i	Failures d_i	Successes s_i	Reliability $R_i = s_i/n_i$
1	100	0	100	1.000	200	2	198	0.990
2	100	1	99	0.990	500	0	500	1.000
3	100	3	97	0.970	700	7	693	0.990
4	100	0	100	1.000	100	4	96	0.960
5	100	2	98	0.980	1000	27	973	0.973
6	100	4	96	0.960	400	6	394	0.985
7	100	3	97	0.970	2000	54	1946	0.973
8	100	5	95	0.950	200	6	194	0.970
9	100	1	99	0.990	300	6	294	0.980
10	100	2	98	0.980	100	0	100	1.000
11	100	0	100	1.000	200	3	197	0.985
12	100	0	100	1.000	400	16	384	0.960
13	100	3	97	0.970	100	5	95	0.950
14	100	6	94	0.940	200	8	192	0.960
15	100	4	96	0.960	100	3	97	0.970
16	100	5	95	0.950	100	4	96	0.960
17	100	2	98	0.980	300	6	294	0.980
18	100	2	98	0.980	200	2	198	0.990
19	100	3	97	0.970	400	16	384	0.960
20	100	4	96	0.960	500	25	475	0.950
Total	2000	50	1950	(0.975)	8000	200	7800	(0.975)

$m = 20$, $n = n_i = 100$, $\sum d_i = 50$, $\sum n_i = 2000$.

$$R = (\sum n_i - \sum d_i)/\sum n_i = \sum s_i / \sum n_i$$
$$= 1950/2000 = 0.975$$

$m = 20$, $n_1 = 200$, $n_2 = 500$, $n_3 = 700$, $n_4 = 100$, etc.; $\sum n_i = 8000$.

$$R = (\sum n_i - \sum d_i)/\sum n_i = (8000 - 200)/8000$$
$$= 7800/8000 = 0.975$$

or

$$R = \Sigma s_i / \Sigma n_i = 7800/8000 = 0.975$$

TABLE 3—SUMMARY OF TABLE 2 TEST RESULTS.

Set 1

R_j	f_j	f_jR_j	w_j	w_jf_j	$w_jf_jR_j$	$w_jf_jR_j^2$
0.940	1	0.940	0.05	0.05	0.0470	0.04418
0.950	2	1.900	0.05	0.10	0.0950	0.09025
0.960	3	2.880	0.05	0.15	0.1440	0.13824
0.970	4	3.880	0.05	0.20	0.1940	0.18818
0.980	4	3.920	0.05	0.20	0.1960	0.19208
0.990	2	1.980	0.05	0.10	0.0990	0.09801
1.000	4	4.000	0.05	0.20	0.2000	0.20000
Sum=	20	19.500	—	1.00	0.9750	0.95094

R_j	R_j^2	$f_jR_j^2$	ΔR_j	$\langle\Delta R_j\rangle^2$	$w_jf_j\langle\Delta R_j\rangle^2$
0.940	0.8836	0.8836	−0.035	0.001225	0.00006125
0.950	0.9025	1.8050	−0.025	0.000625	0.00006250
0.960	0.9216	2.7648	−0.015	0.000225	0.00003375
0.970	0.9409	3.7636	−0.005	0.000025	0.00000500
0.980	0.9604	3.8416	+0.005	0.000025	0.00000500
0.990	0.9801	1.9602	+0.015	0.000225	0.00002250
1.000	1.0000	4.0000	+0.025	0.000625	0.00012500
Sum=	—	19.0188	—	—	0.00031500

Average: $\bar{R} = 19.500/20 = 0.9750$

$\sigma_R^2 = (\sum f_jR_j^2/\sum f_j) - \bar{R}^2 = (19.0188/20) - (0.975)^2 = 0.950940 - 0.950625 = 0.000315$.

Also $\sigma_R^2 = \sum f_j\langle\Delta R_j\rangle^2/\sum f_j$ (where $\Delta R_j = R_j - \bar{R}_j$) and $\sigma_R = 0.017748$; hence
$\sigma_R^2 = \frac{1}{20}(0.001225 + 0.001250 + 0.000675 + 0.000100 + 0.000100 + 0.000450 + 0.002500)$
$= \frac{1}{20}(0.006300) = 0.000315$, or using relative weights, $w_jf_j = 1.00$, $\sigma_R^2 = w_jf_j\langle\Delta R_j\rangle^2 = 0.000315$,
given in last column above. Hence $\sigma_R = (0.000315)^{1/2} = 0.017748$.

Set 2

R_j	f_j	f_jR_j	w_i values			$f_jw_j = \sum w_i$	$f_jw_jR_j$	$f_jw_jR_j^2$
0.950	2	1.900	0.0125,	0.0625		0.0750	0.071250	0.067687500
0.960	5	4.800	(2) 0.0125,	(2) 0.05,	0.025	0.1500	0.144000	0.138240000
0.970	2	1.940	0.0250,	0.0125		0.0375	0.036375	0.035283750
0.973	2	1.946	0.1250,	0.2500		0.3750	0.364875	0.355023375
0.980	2	1.960	0.0375,	0.0375		0.0750	0.073500	0.072030000
0.985	2	1.970	0.0500,	0.0250		0.0750	0.073875	0.072766875
0.990	3	2.970	0.0250,	0.0875,	0.0250	0.1375	0.136125	0.134763750
1.000	2	2.000	0.0625,	0.0125		0.0750	0.075000	0.075000000
Sum=	20	19.486	—			1.0000	0.975000	0.950795250

R_j	R_j^2	ΔR_j	$\langle\Delta R_j\rangle^2$	$f_jw_j\langle\Delta R_j\rangle^2$
0.950	0.902500	−0.025	0.000625	0.0000468750
0.960	0.921600	−0.015	0.000225	0.0000337500
0.970	0.940900	−0.005	0.000025	0.0000009375
0.973	0.946729	−0.002	0.000004	0.0000015000
0.980	0.960400	+0.005	0.000025	0.0000018750
0.985	0.970225	+0.010	0.000100	0.0000075000
0.990	0.980100	+0.015	0.000225	0.0000309375
1.000	1.000000	+0.025	0.000625	0.0000468750
Sum=	—	—	—	0.0001702500

TABLE 4—DETERMINATION OF WEIGHTING FACTORS FOR SETS 1 AND 2 COMBINED.

Units Tested	Test Occurrences	Total Tested	Weighting Factors as Determined	
			w_i per Total Tested	w_i per Single Test
100	25	2 500	0.25	0.01
200	5	1 000	0.10	0.02
300	2	600	0.06	0.03
400	3	1 200	0.12	0.04
500	2	1 000	0.10	0.05
700	1	700	0.07	0.07
1000	1	1 000	0.10	0.10
2000	1	2 000	0.20	0.20
Sum=	40	10 000	1.00	—

TABLE 5—DISTRIBUTION OF SETS 1 AND 2 COMBINED.

R_j	Frequency f_j	Failures d_j	Units Tested n_j	Relative Weight w_j	$w_j R_j$	$w_j R_j^2$
0.940	1	6	100	0.01	0.0094	0.00883600
0.950	4	40	800	0.08	0.0760	0.07220000
0.960	8	60	1 500	0.15	0.1440	0.13824000
0.970	6	21	700	0.07	0.0679	0.06586300
0.973	2	81	3 000	0.30	0.2919	0.28401870
0.980	6	20	1 000	0.10	0.0980	0.09604000
0.985	2	9	600	0.06	0.0591	0.05821350
0.990	5	13	1 300	0.13	0.1287	0.12741300
1.000	6	0	1 000	0.10	0.1000	0.10000000
Sum=	40	250	10 000	1.00	0.9750	0.95082420

R_j	R_j^2	ΔR_j	$\langle \Delta R_j \rangle^2$	$w_j \langle \Delta R_j \rangle^2$
0.940	0.883600	−0.035	0.001225	0.00001225
0.950	0.902500	−0.025	0.000625	0.00005000
0.960	0.921600	−0.015	0.000225	0.00003375
0.970	0.940900	−0.005	0.000025	0.00000175
0.973	0.946728	−0.002	0.000004	0.00000120
0.980	0.960400	+0.005	0.000025	0.00000250
0.985	0.970225	+0.010	0.000100	0.00000600
0.990	0.980100	+0.015	0.000225	0.00002925
1.000	1.000000	+0.025	0.000625	0.00006250
Sum=	—	—	—	0.00019920

The average is given directly in column $w_j R_j$ as 0.9750 since its divisor is 1.00 per column w_j. However, it may also be obtained from the relation

$$\bar{R}_j = \left(\sum n_i - \sum d_i \right) / \sum n_i = (10\ 000 - 250)/10\ 000 = 9750/10\ 000 = 0.975.$$

The variance is given directly as the sum of the terms in column $w_j \langle \Delta R_j \rangle^2$. It may also be obtained by subtracting $\bar{R}_j^2$ from the sum of column $w_j R_j^2$, giving

$$\sigma_R^2 = 0.95082420 - (0.975)^2 = 0.95082420 - 0.95062500 = 0.00019920.$$

$$\sigma_R = (0.00019920)^{1/2} = 0.014114.$$

0.025, for n units tested using the exponential distribution, the theoretical probabilities are given in Table 6.

Figure 2 shows the distribution for this example for the 20 sets of values of failures found when $n=100$ as observed in set 1, for 1000 hours of test for each unit unless it failed earlier. The time of failure for each unit was not recorded. Theoretical values for the Exponential Law where the failure rate is 2.5%, and for the Normal Law where $\bar{R}=0.975$ and $\sigma_R=0.01775$, have been determined and are shown on Fig. 2 for comparison. The Normal Law appears to be a better fit than the Exponential for these data.

Individual and cumulated observed results are given in Fig. 3 for sets 1 and 2, modified by weighting to give the best unbiased results. Theoretical results for these cases are given for both the Exponential Law and the Normal Law.

DERIVATION OF THEORETICAL DISTRIBUTIONS

Statistical techniques are necessary to derive various theoretical distributions that fit the observed reliability data. Which distribution best fits the data and can be used for making valid reliability predictions can then be determined by visual comparisons or by the use of the chi-square (χ^2) test for goodness of fit. The following procedures were used to derive frequencies comparable with set 1 data where $n=100$ units for each test. Set 2 includes a large number of different values from $n=100$ to $n=2000$. An approximate method is to use the number of tests as the weighting factor and to assume all values to be observed for $n=100$ with an assumed frequency equal to the number of sets of $n=100$ that were actually tested. For example, if $R=0.96$ is observed for $n=500$,

TABLE 6—EXACT AND CUMULATED PROBABILITIES ASSOCIATED WITH EXPONENTIAL FOR UNITS TESTED AND TEST OCCURRENCES IN TABLE 4.

$n=100, 200, 300, 400, 500, 700, 1000,$ and 2000. Failure rate $r=0.025$.

	$n=100$						$n=200$					
	Observed in 25 Tests		Exact Theor. Prob.*	Cumulated			Observed in 5 Tests		Exact Theor. Prob.*	Cumulated		
				Observed		Theor. Prob.*				Observed		Theor. Prob.*
Failures	n	P		n	P		n	P		n	P	
0	5	0.20	0.082	5	0.20	0.082	0	0	0.007	0	0	0.007
1	2	0.08	0.205	7	0.28	0.287	0	0	0.034	0	0	0.041
2	4	0.16	0.257	11	0.44	0.544	2	0.40	0.084	2	0.40	0.125
3	5	0.20	0.214	16	0.64	0.758	1	0.20	0.140	3	0.60	0.265
4	5	0.20	0.134	21	0.84	0.892	0	0	0.175	3	0.60	0.440
5	3	0.12	0.067	24	0.96	0.959	0	0	0.175	3	0.60	0.615
6	1	0.04	0.028	25	1.00	0.987	1	0.20	0.146	4	0.80	0.761
7	0	0	0.010			0.997	0	0	0.104	4	0.80	0.865
8	0	0	0.003			1.000	1	0.20	0.065	5	1.00	0.930

Cumulated

	$n=300$			$n=400$			$n=500$			$n=700$			$n=1000$		
	Obsvd.		Theor. Prob.*	Obsvd.		Theor. Prob.*	Obsvd.		Theor. Prob.*	Obsvd.		Theor. Prob.*	Obsvd.		Theor. Prob.*
Failures	n	P		n	P		n	P		n	P		n	P	
0	0	0	0.001	0	0	0.000	1	0.500	0.000	0	0	0.000	0	0	0.000
6	2	1.00	0.378	1	0.333	0.130	1	0.500	0.035	0	0	0.002	0	0	0.000
7			0.525	1	0.333	0.220	1	0.500	0.070	1	1.00	0.004	0	0	0.000
16			0.998	3	1.00	0.973	1	0.500	0.869			0.421	0	0	0.038
25			1.000			0.999	2	1.00	0.999			0.965	0	0	0.553
27			1.000			1.000			1.000			0.987	1	1.00	0.700

For $n=2000$ during the test 54 failures were observed. For $r=0.025$, then $rn=(0.025)(2000)=50.000$. For this value for $rn=50$ in an exponential table, the probability of 54 or less failures is 0.742.

* Based on Poisson exponential probability values (Molina tables).

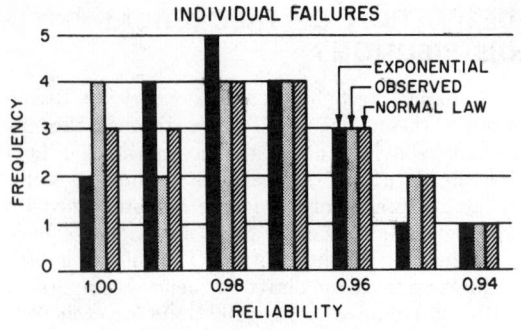

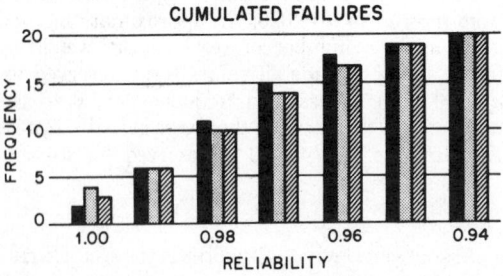

Fig. 2—Observed reliability distributions of individual and cumulated failures for set 1, consisting of 20 sets with $n=100$ on test for 1000 hours each. The corresponding theoretical frequencies for Exponential and Normal Law are also shown. Failure rate $r=0.025$ and $rn=2.5$ for Exponential; $\bar{R}=0.975$ and $\sigma_R=0.01775$ for Normal Law.

then $R=0.96$ has a frequency of 5 for $n=100$ instead of a frequency of 1 for $n=500$. The feature neglected is the distribution of failures that occurred in the five 100-unit groups since all are assumed to be at the average value. This treatment has been applied to sets 1 and 2 combined.

Exponential Law

For the exponential the average of the distribution is $\langle a \rangle$ or $\langle x \rangle = \langle rn \rangle$ and its average and variance are equal. Tables are available that give exact individual terms and cumulated terms for the probabilities for the exponential, where $c=$ the number of failures, $r=$ failure rate, $n=$ sample size $=$ number of units tested, $rn=$ expected number of failures, and $P_{m,n,rn}$ is the probability of the occurrence of exactly m failures in a sample of n units with the expected value rn. Mathematically it is derived from

Individual term:

$$P_{m,n,rn} = \exp(-rn)\,(rn)^m/m!$$

Cumulated term:

$$P_{c\text{ or less},n,rn} = \sum_{m=0}^{m=c} \left[\exp(-rn)\,(rn)^m/m!\right].$$

Some tables give cumulated values for $c+1$ to n

rather than from 0 to c, so enter such tables at $c+1$.

Normal Law

For those cases where the terms are variable, the area under the Normal Law curve between cell boundaries may be found in terms of probability values. These cell boundaries, measured in units of standard deviation, are determined as z values evaluated in standard-deviation units measured from the average. These z values are used to determine corresponding Normal Law probabilities from Normal Law probability tables giving areas. Do not use ordinate probabilities, also given in many of these tables. These Normal Law probability tables give integral values from various

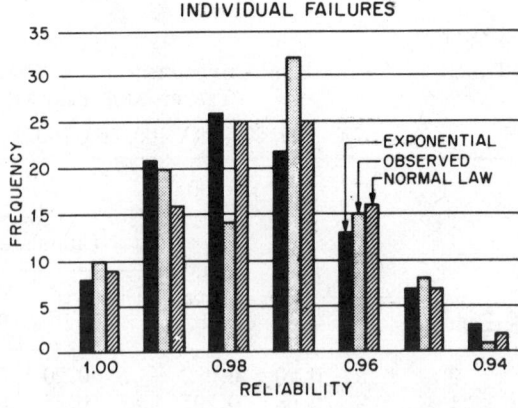

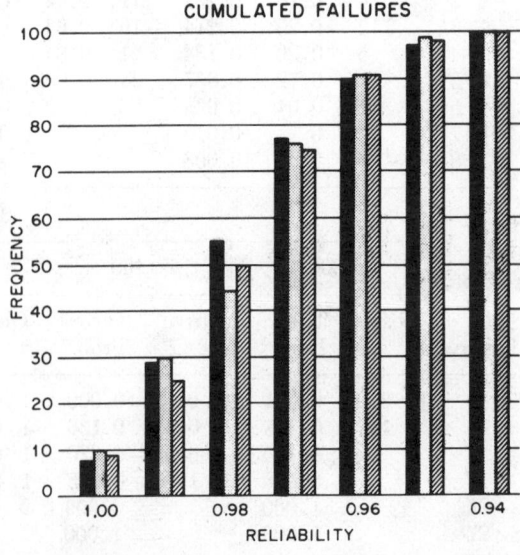

Fig. 3—Observed vs. theoretical failure distributions for sets 1 and 2 combined, consisting of 40 sets with $n=100$–2000 on test for 1000 hours each. Failure rate $r=0.025$ for Exponential; $\bar{R}=0.975$ and $\sigma_R=0.01466$ for Normal Law.

limits such as from $-\infty$ to x, or from $-x$ to $+x$, etc. The differences between these probabilities corresponding to the boundaries give the expected frequency in the cell defined by the boundaries used. The example below gives the steps required to secure these theoretical frequencies.

Statistical Techniques: Checks for Goodness of Fit. The end cells at each end of the distribution must not be empty and must contain at least one observed value. It is better to have equal cell widths except for the end cells. It is necessary to check the theoretical distribution to see if it is a close fit to the observed distribution. This is done by checking the theoretical values for each cell against the observed frequency for the same cell, i.e., f_i' vs. f_i for the ith cell. The check for goodness of fit is made by means of the chi-square (χ^2) distribution

$$\chi^2 = \sum \left[(f-f')^2/f' \right].$$

The probability showing how good the theoretical distribution fits is determined from the χ^2 value computed per the above equation. It is secured by entering a χ^2 table at the number of degrees of freedom (d.f.) derived from the number of cells used in the observed distribution less the number of statistics used in deriving the theoretical distribution. For an exponential curve, d.f. $= c-1$, where $c =$ number of cells, and the expected value rn is known and was used to derive the theoretical frequencies for each cell. For a Normal Law distribution, where the average and standard deviation are used to derive the theoretical frequencies, d.f. $= c-2$. The probability value (from the χ^2 table) indicates that, in repeated tests, fits as good or better than the one observed will occur with the frequency given by the probability found for goodness of fit.

This same chi-square distribution is used also to determine reliability probability values, as noted in this Chapter. It is desirable to have available the most extensive chi-square tables available to expedite computations of reliability predictions. Use those tables having a wide range of both probability values and also degrees of freedom (d.f.). Many tables, such as those first computed, only cover d.f. values to 30. Later tables cover up to d.f. $= 100$.

Determination of Theoretical Distributions For Set 1 Data Where $n = 100$

Table 7 gives the observed and two theoretical distributions for set 1 data. It gives the steps used in deriving theoretical frequencies for the Normal Law distribution using the average $\bar{R}$ and σ_R for the observed distribution. In this computation, σ based on n units rather than s based on $n-1$

units has been used as the estimate of the variability.

Reliability Values for Sets 1 and 2 Combined ($n = 100$ to 2000)

Table 8 lists the observed reliability values modified by weighting by the respective sample sizes for each group to be equivalent to the single sample size $n = 100$. The two reliability values that cause difficulty in this determination of equivalence are 0.985 and 0.973. These two values were observed values for larger sample sizes. The value 0.985 was found when $n = 200$ and $n = 400$; also the value 0.973 was secured when $n = 1000$ and $n = 2000$. Failures are reported in integral values, so the reliability values 0.985 and 0.973 cannot exist for $n = 100$ test units. For the observed distribution and also for the theoretical distributions for the Exponential and the Normal Law, frequencies for 0.985 and 0.973 are prorated to add into the 0.990, 0.980, and 0.970 cells for the observed distribution to permit better comparisons for the case $n = 100$. This is method 2 as noted in Table 8. Method 1 uses the observed distribution for all observed cell values, but for the exponential prorates the theoretical values for $rn = 2.5$. In this method theoretical exponential probabilities for 1, 2, and 3 failures for $n = 100$ are computed for the nonstandard 0.985 and 0.975 cells. The upper half of Table 8 shows the details for this exponential distribution, in which the probabilities for the frequencies of occurrence are read from a Poisson exponential table for $rn = (0.025)(100)$, $rn = 2.5$.

Normal Law (Gaussian) Probabilities

With $\bar{R}_j = 0.9750$ and $\sigma_R = 0.014114$, based on a Normal Law distribution (often termed a Gaussian or Laplacian distribution) for a confidence level of 0.90,

$$R_{0.90} = 0.9750 - 1.282(0.0141)$$

$$= 0.9750 - 0.0181 = 0.9569.$$

For a confidence level of 0.95, then

$$R_{0.95} = 0.9750 - 1.645(0.0141)$$

$$= 0.9750 - 0.0232 = 0.9518.$$

This means that the results of the tests on these 10 000 units, assuming a Normal Law distribution for a confidence level of 0.90, show that the reliability will be 95.69% or better. Also these results show a reliability of 95.18% or better for a confidence level of 0.95. Table 9 lists a series of these confidence levels with the observed fractional occurrences compared with the theoretical values based on the assumption of a Normal Law distribution for the average and standard deviation values listed above, i.e., $\bar{R}_j = 0.9750$ and $\sigma_R = 0.014114$.

MEASURES OF RELIABILITY

Failure rate is used as one measure of reliability. It is defined as the ratio of the total number of failures

$$\sum_{i=1}^{i=n} f_i$$

observed during total test time

$$\sum_{i=1}^{i=n} t_i$$

where f_i is the number of failures observed in test time t_i for the ith sample. If there are n units on test, then (where r denotes the failure rate)

$$r = \sum_{i=1}^{i=n} f_i \Big/ \sum_{i=1}^{i=n} t_i, \quad (i = 1 \text{ to } n)$$

or if

$$f = \sum_{i=1}^{i=n} f_i, \quad \text{and} \quad t = \sum_{i=1}^{i=n} t_i$$

then

$$r = f/t.$$

Time is expressed usually in hours, but may be expressed in milliseconds, seconds, minutes, days, etc. A common unit of time is 1 hour. However, for failure rates 1000 hours or 1 000 000 hours is a common measure.

The reciprocal of the failure rate is the Mean Time Before Failure $(MTBF) = m$, where $m = 1/r$, hence

$$m = t/f.$$

TABLE 7—DATA FOR SET 1: 20 GROUPS OF RELIABILITY RESULTS FOR $n = 100$.

Derivation of Normal Law Theoretical Probabilities.

Boundary Values	Deviations from $\bar{R}$ Numerical	Deviations from $\bar{R}$ z Values σ_R Units	Normal Law Probabilities Corresponding to z	Reliability Values	Theoretical Frequency Individual Prob.	Theoretical Frequency Individual No.	Cumulated No.
0.995	+0.020	+1.127	0.3701	1.000	0.1299	3	3
0.985	+0.010	+0.5634	0.2134	0.990	0.1567	3	6
0.975	0	0	0	0.980	0.2134	4	10
0.965	−0.010	−0.5634	0.2134	0.970	0.2134	4	14
0.955	−0.020	−1.127	0.3701	0.960	0.1567	3	17
0.945	−0.030	−1.690	0.4545	0.950	0.0844	2	19
				0.940	0.0455	1	20

Observed and Theoretical Individual and Cumulated Values for Failure Rate $r = 0.025$. Exponential and Normal Laws.

			$m = 20$ Sets					
Reliability Values	Observed Values Ind.	Observed Values Cum.	Exponential, $rn = 2.5$ Probabilities Ind.	Exponential, $rn = 2.5$ Probabilities Cum.	Exponential, $rn = 2.5$ Number Ind.	Exponential, $rn = 2.5$ Number Cum.	Normal Law $\bar{R} = 0.975$, $\sigma_R = 0.01775$ Individual	Normal Law $\bar{R} = 0.975$, $\sigma_R = 0.01775$ Cumulated
1.00	4	4	0.082	0.082	2	2	3	3
0.99	2	6	0.205	0.287	4	6	3	6
0.98	4	10	0.257	0.544	5	11	4	10
0.97	4	14	0.214	0.758	4	15	4	14
0.96	3	17	0.134	0.892	3	18	3	17
0.95	2	19	0.067	0.959	1	19	2	19
0.94	1	20	0.041	1.000	1	20	1	20
Sum=	20				20		20	

The Greek letter lambda (λ) is used to represent the ratio failure rate. If $f=$ total failures during a given interval of time as before

$$\lambda = f/n.$$

Another measure of reliability is the probability of survival P_s. It may be applied to an individual part or to a system. Calabro* gives a theorem covering group survival as derived from part survival, stating on page 64: "The probability of survival for a group of identical parts expressed as a percentage is equal to the probability of survival of each individual part composing the group."

* S. R. Calabro, "Reliability Principles and Practices," McGraw-Hill Book Company, New York; 1962.

In many field trials it has been found that the failure rate after a given wear-in period has a growth pattern that matures quite rapidly, giving a constant failure rate until the parts start to wear out, at which time the failure rate tends to increase rapidly with use. Most texts give a curve for these three phases sometimes called the "bathtub" curve. The functions usually associated with such curves are: the exponential density function $\lambda \exp(-\lambda t)$, which has associated with it the exponential $\exp(-\lambda t)$, where the failure rate λ is usually considered constant in phase 2 and also may have associated with it many other functions, such as the *negative binomial*, the *positive binomial*, and possibly others, where the failure rate is essentially constant. In practice the exponential is assumed in the majority of cases.

TABLE 8—DATA FOR SETS 1 AND 2 COMBINED:

$n=100$ TO 2000, $r=0.025$, $\bar{R}=0.975$, $\sigma_R=0.0141$.

The upper half of the table shows R values as observed for a series of different n values. The lower half shows R values modified to fit cells for $n=100$ in steps of 0.01.

Reliability Values	Individual Observed Frequencies		Frequencies Modified by Weighting Sets 1+2	Individual			Cumulated		
				Theoretical				Theoretical	
	Set 1	Set 2		Exponential	Normal Law	Observed	Observed	Exponential	Normal Law
1.000	4	2	10	8	8	10		8	8
0.990	2	3	13	14	11	23		22	19
0.985	—	2	6	16	11	29		38	30
0.980	4	2	10	17	16	39		55	46
0.973	—	2	30	6	14	69		61	60
0.970	4	2	7	15	16	76		76	76
0.960	3	5	15	13	16	91		89	92
0.950	2	2	8	7	6	99		96	98
0.940	1	0	1	4	2	100		100	100
Sum=	20	20	100	100	100				

$n=2000; 8000$

Reliability Values	Modified Observed Frequencies	Individual		Modified Observed Frequencies	Cumulated	
		Theoretical			Theoretical	
		Exponential	Normal Law		Exponential	Normal Law
1.00	10	8	9	10	8	9
0.99	20	21	16	30	29	25
0.98	14	26	25	44	55	50
0.97	32	22	25	76	77	75
0.96	15	13	16	91	90	91
0.95	8	7	7	99	97	98
0.94	1	3	2	100	100	100
Sum=	100	100	100			

$\bar{R}=0.9750$; $\sigma_R^2=0.0002150$; $\sigma_R=0.01466$

TABLE 9—RELIABILITIES ASSOCIATED WITH ONE-TAILED CONFIDENCE LEVELS.

Confidence Level	Multiplying Factor for Normal Law, z	$n = 10\,000$ units		Normal Law Theoretical Reliability
		Observed Reliability		
0.90	1.282	95.07%		95.69%
0.95	1.645	94.50%		95.18%
0.96	1.751	94.38%		95.03%
0.97	1.881	94.25%		94.85%
0.98	2.054	94.12%		94.60%
0.99	2.326	94.00%		94.22%

DISTRIBUTION FUNCTIONS USED IN RELIABILITY

As noted above, the exponential distribution is used in a large percentage of cases as the best reliability function for solving reliability problems. The Normal Law or Gaussian probability distribution is used in many other cases. The Weibull distribution, of which the exponential is a special case, is used in many refined studies. The Weibull cumulative distribution function (or failure distribution) is defined as

$$F(t) = 1 - \exp-(t-\nu)^{\beta/\alpha}, \quad \text{for } t \geq \nu; \, \alpha, \beta > 0$$

$$= 0, \text{ for other } t, \, \alpha, \, \beta \text{ values} \quad (1)$$

otherwise $F(t)$ is the probability that an item will fail at or before time t. The three unknown parameters are

$\alpha =$ scale parameter

$\beta =$ shape parameter

$\nu =$ location parameter.

On setting $\beta = 1$, Eq. (1) becomes the exponential distribution with delay, i.e., ν can be thought of as a guarantee period within which no failures can occur, hence may be thought of as a minimum life. When the location parameter $\nu = 0$ and Eq. (1) reduces to the more common form, then

$$F(t) = 1 - \exp(-t^{\beta/\alpha}), \quad \text{for } t \geq 0; \, \alpha, \beta \geq 0. \quad (2)$$

For Eq. (2) when $\beta = 1$, the exponential distribution results with $\alpha =$ mean life. Do not use α as the mean life unless $\beta = 1$.

The reliability function $R(t) = 1 - F(t)$ for the two-parameter Weibull distribution is

$$R(t) = \exp(-t^{\beta/\alpha})$$

where $R(t)$ is the probability that no item will fail at or before time t.

Note especially that the density function is secured by taking the first derivative of $F(t)$. Hence the Weibull probability density function is the first derivative of Eq. (2), the Weibull failure distribution. Thus

$$f(t) = \beta t^{\beta-1}/\alpha \exp(-t^{\beta/\alpha})$$

which for $\beta = 1$ is identical to the exponential density function with $\alpha = 1/\lambda = m$; this gives $f(t) = \lambda \exp(-\lambda t)$, given in previous sections.

The Normal Law or Gaussian density function is given by

$$F(t) = [1/\sigma(2\pi)^{1/2}] \exp[-(t-m)^2/2\sigma^2]$$

where t is the component age. The normal distribution depends on age, whereas the exponential does not. An exponential type of universe of components, when placed in a test in which the failed components are not replaced, suffers its greatest losses in the test period before m, the mean life, whereas for a normal distribution of components the greatest losses occur around m, the mean life. The reliability $R(t)$ is given by

$$R(t) = [1/\sigma(2\pi)^{1/2}] \int_t^\infty \exp[-(t-m)^2/2\sigma^2] dt.$$

The logarithmic-normal or log-normal is used when the failure rate is known not to be constant. The log-normal is the same as the normal distribution given above except that the variable is replaced by its logarithm. Many skew distributions become normal when this replacement of x by $\log x$ is applied.

For the gamma distribution, the density function is given as

$$f(t) = (\alpha! \beta^{\alpha+1})^{-1} t^\alpha \exp(-t/\beta)$$

where $\alpha! = 1 \cdot 2 \cdot 3 \cdots (\alpha-1)\alpha$, termed a factorial. The corresponding reliability function is given by the relation below derived from the first derivative

$$R(t) = \int_t^\infty (\alpha! \beta^{\alpha+1})^{-1} t^\alpha \exp(-t/\beta) dt.$$

The hazard rate is secured for all distributions from

$$z(t) = f(t)/R(t).$$

The hazard rate for the exponential is

$$z(t) = 1/m = \lambda$$

whereas for the Weibull distribution it is

$$z(t) = \beta t^{\beta-1}/\alpha.$$

The density function for the rectangular distribution is

$$f(t) = t/\alpha.$$

The corresponding reliability function for the rectangular distribution is

$$R(t) = 1 - t/\alpha.$$

Since the hazard rate is $f(t)/R(t)$, then for the rectangular distribution

$$z(t) = t/(\alpha - t).$$

Another distribution that is often confused with the exponential is the Poisson exponential, which covers the sum of the probabilities for integral values denoting the number of failures observed. Its density function is

$$f(x) = \alpha^x e^{-\alpha}/x!.$$

Its reliability function is expressed as a summation rather than an integral. It is expressed as

$$R(x) = \sum_{i=x}^{i=\infty} \alpha^i e^{-\alpha}/i!.$$

The hazard rate is not applicable to this distribution, nor is it applicable to the positive binomial and the hypergeometric functions that follow. For these distributions only the density functions are given. The corresponding reliability functions are secured by taking the applicable sums over the designated range of values covered by each individual case.

The binomial density function is

$$f(x) = C_x^n (1-p)^x p^{n-x}$$

where p is the failure rate and might be replaced by r or λ, as applicable. The nonparametric case is covered by the hypergeometric density function

$$f(x) = (1/C_n^N) C_{n-x}^{N-pN} C_x^{pN}$$

where N is the total number of units in the lot or universe under consideration, n is the sample size or number of units on test, and x is the number of failures postulated in the sample where there are pN failures in the lot of N units with p the failure rate.

Note that reliability values may readily be obtained for the exponential. These same probability values are secured from the chi-square (χ^2) distribution. Life-test theorems developed by Benjamin Epstein also show that the same probabilities may be secured from the use of Fisher's F ratio test. For a pictorial and also mathematical presentation of most of the distributions given above, see p. 13 of reference [4] and pp. 72–73 of reference [1].

RELIABILITY DEMANDS AND CHECKS IN COMMERCIAL AND MILITARY PRODUCTS

Guarantees are given to cover the performance of most commercial items. These guarantees must be realistic and in line with the costs of the products purchased, and therefore the manufacturer must make reliability and life studies to determine what are reasonable guarantees. Many companies have accumulated sufficient evidence about the life characteristics to be able to give guarantees closely in line with the quality and reliability of their products. For example, car batteries are guaranteed for 12, 15, 18, 24, 36, and even 48 months. Large quantities of data evidently have provided sufficient evidence to justify the time guarantee for each type of battery.

Practically all military contracts contain clauses under Quality Assurance requiring reliability programs. Many contracts include provisions for a preliminary design review and also a critical design review before qualification tests. In addition, some contracts require a reliability demonstration test. Use is made in this latter case of all the engineering data that might be obtained in the qualification tests, if they precede the reliability demonstration test.

Reliability Analysis of Qualification Test Results

Qualification test results cover only a short period of actual operation, hence often are ignored in evaluating the reliability of the product being qualified for release. Good reliability engineering demands that all test results be compiled and added to the meager life test results to provide maximum reliability evidence. At times such qualification tests and prior bench tests provide the only available reliability data for making a valid reliability evaluation. It is obvious that these results provide low confidence levels. However, as noted under the Bayesian approach, all these data and somewhat similar data from other products provide fairly good evidence concerning the reliability of the product in question.

Design of Experiments for Reliability

To meet commercial and military demands for reliability, it is desirable to establish critical tests for all the critical items and characteristics. This determines whether the so-called economic quality standards are too loose or too tight to achieve the economic reliability value established in the contract with the confidence level stipulated, if any. The number of tests must be minimized to reduce costs and also to secure maximum information at minimum costs. All the knowledge and skill of the most experienced and trained mathematical statisticians are required to make use of the most efficient design of experiments for the most critical characteristics for the product involved. The data secured from these tests will be used as part of the data required for the reliability demonstration test.

Levels for the Confidence Limits

Reliability magnitudes required depend on the use to be made of the product or system under consideration. A set of values for many commercial products might be at a 95%, 98%, or 99% reliability value at a confidence level of 0.90. For the first runs the confidence level often used is 0.60, requested initially by many industrial associations as one of the first values on the learning curve. For military uses or commercially where life or property can be endangered, reliability values of 0.997 or even 0.999 are desired with confidence levels of 0.95 to 0.99. For the most dangerous missions in space, reliability values of 0.999 or even 0.9999 are desired with confidence levels of 0.99 or 0.999, if possible. To determine these values requires a large amount of data; hence in many cases it is impossible to secure actual data for the product involved. It is necessary to secure similar data from somewhat similar products, if possible. This is known as the Bayesian approach. With meager data for the product involved, use is made of other information from other tests. The data are weighted according to the confidence felt in the various sets of data that have been accumulated as being usable.

Bayesian Probabilities

Since 1964 an increased emphasis has been placed on the use of Bayesian probabilities. Products have been qualified for immediate operation or application because of similarity to other products which have already been qualified. The Bayesian approach makes use of all similar data to provide fairly valid estimates of the reliability of the new product. Chapter 13 of reference [2] covers the use of Bayes' theorem in reliability. It compares on pages 121 and 122 the compound probabilities obtained for a series–parallel group of components per the usual practice as contrasted with the application of Bayes' theorem. The mathematical relations are different in line with the great difference in the paths assumed to be combined and generally give different numerical values. The use of Bayes' theorem tends to provide more-optimistic reliability predictions.

SYSTEMS RELIABILITY

When preparing a proposal it is necessary to make a first estimate of the possible reliability of the system with the available component parts. This estimate is usually very crude. The design has not been developed and is not firm. Some parts may not even have been developed, so that the first reliability value indicates that it is very feasible to bid on the potential contract. It may indicate that it will be very difficult to secure the reliability demanded, so that ample additional funds must be provided for original research in many new areas.

When a new system is being produced, the majority of the subsystems usually have been produced for other purposes. It is possible to use the prior estimates of reliability secured from these other projects to evaluate the new system. It must be shown that the system will achieve its mission, say 999 times out of 1000 attempts, and the available data are sufficient to give a confidence level of 0.99 that this desired reliability will be achieved.

COMPONENT RELIABILITY

Initially it is essential for the reliability engineer to tabulate as far as possible all the expected failure rates for the component parts in each subsystem. If all the products made by a company are component parts, then data must be accumulated giving under the full range of environmental conditions the successful operation of the part for its entire life. The test results will provide the necessary reliability data to set up failure rates that are meaningful. The equations for determining these reliability values in terms of failure rates (such as percentage failures per 1000 hours or per 1000 cycles of operation) or in terms of mean time before failure, are given in the referenced texts, and also earlier in this chapter.

Of importance in component reliability are the relationships of reliability specifications covering the development of failure rates for components. A good example is the issuance of MIL–R–38000 military specifications by the U.S. Air Force. These specifications are an outgrowth of the Minuteman effort and detail the methods for design acceptance and qualification levels for components. An example is included below.

Usually a number of components are placed in a life test. The number depends on the failure rates required and the number of failures decided on to allow test completion. For example, let the following requirements apply.

(A) Failure rate level to be met: 0.1% per 1000 hours.

(B) Confidence factor: 90%.

(C) Number of failures allowed: 0.

(D) Total test time required to meet the above conditions: 2.3×10^6 unit hours.

In other words, in a test of 230 resistors, the test duration would be 10 000 hours. To simplify the statistics and the calculation, a nomogram (Fig. 4) abstracted from MIL–R–38100 is included. The statistics herein are generally based on exponential-wear-out curves.

Part Failure Rates

Failure rates for component parts are shown in Table 10. These rates reflect improvements in parts since the advent of the Minuteman and Polaris programs. The growth is of the order of 5 to 100 times better than the failure rates of parts used in 1950 generation systems.

Generic Failure Rates

Many references provide generic failure rates for basic units. Such basic values are taken as the standards and then are multiplied by one or more weighting factors as noted previously. Weighting factors must include whether the part is used with or without periodic preventive maintenance; whether used in space, aircraft, or on the ground; and must also include the allotted safety factors. The generic failure rates also may be increased or decreased to take account of new engineering information, differences in temperatures, power ratings, altitudes, pressures, and primarily applications. Parts used in outer space must meet much more rigid requirements than those used for ground equipment where preventive maintenance is part of the program.

The first allocation of reliability values for a proposal will be based on the best estimate of the engineering design for the new product. This makes it possible to note the component parts, materials, and hardware that the designer expects to use. Generic failure rates are selected from tables for these parts, modified by proper weighting factors. These provide an expected value for the new product covered by the proposal.

ALLOCATION OF RELIABILITIES

Management and engineering provide an overall reliability goal for the system. Such a goal might be a reliability of 0.997. A confidence level may also be associated with this reliability value. At

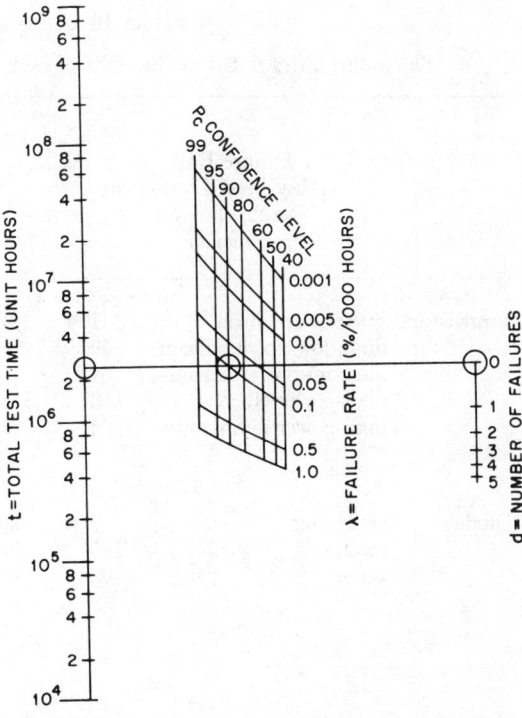

Fig. 4—Failure rate and confidence level for total test time. The example shown is for the Exponential Law $P_C = 1 - \exp(-2.3) = 0.90$. As indicated, $\lambda = 10^{-6}$, $t = 2.3 \times 10^6$, $\lambda t = 2.3$, and $P_o = \exp(-\lambda t)$.

the beginning of the reliability growth pattern, the confidence level may be 0.50, or just an expected value, or perhaps a minimum of 0.60. This will be followed by a series of probability values through 0.90, 0.95, 0.99, and even as large as 0.9999. (How these are evaluated is covered in the section on confidence levels.) Assuming that initially we are dealing with expected values, the allocation of reliability probability values among the component parts should be related to their respective generic probability values.

Where P_s (probability of survival) is the reliability for the system, then for two parts $P_1 P_2 = P_s$. A first estimate would be secured by assuming that the values will be equal for these parts, so that $P_1 = P_2 = P_s^{1/2} = P$, or $P = P_s^{1/2}$. However, the better way to handle the allocation problem is to determine what failure rates may be allocated to the component parts so that the final sum will provide the reliability value specified for the system.

$$R_s(t) = \exp(-\lambda_s t).$$

For n components, then $\lambda_s = \lambda_1 + \lambda_2 + \lambda_3 + \lambda_4 + \cdots + \lambda_n$, and for whatever weights that should be assigned to each of the n components corresponding failure-rate (λ) values may be obtained to give λ_s for the system. Page 190 of reference [1] gives

TABLE 10—SUMMARY OF FAILURE RATES.

The failure rates noted at the right of each column are in "bits" (1 failure in 10^8 hours).

Part	Failure Rates (Raytheon Equipment on Apollo and Polaris)		Failure Rates Chosen for Spacecraft Life/Reliability Estimates		MIL-HDBK-217A (1966)	
Transistors	low-power silicon	1.4	low-power:		0.2 junction	
	dual, low-power silicon	49	digital	0.5	temperature	47
	medium-power silicon	61	analog	1	0.1 junction	
	high-power silicon	153			temperature	30
	high-power germanium	194	high-power	10		
			weighted average	1		
Diodes	switching	2.6	low-power digital	0.2		
	rectifier	135	low-power analog	0.6	0.2 junction	
	zener	93	zener diode <1 watt	2	temperature	20
			zener diode >1 watt	20		
			weighted average	0.6	0.1 junction	
					temperature	14
Capacitors	ceramic	1.6	solid tantalum	2	derated to 40%:	
	paper (mylar)	31	all others	0.4	ceramic	2
	solid tantalum	4.3	weighted average	0.6	paper	12
					high-reliability	
					paper	3
Resistors	film	3.3	wirewound	1.0	derated to 40%:	
	carbon	0.2	all others	0.2	film	270
	wirewound	10	weighted average	0.3	carbon	0.35
					wirewound	1.5
Integrated circuits	NOR gate	1.4		1.0		

details for this method. The simplest case might be as follows: $\lambda_s = 0.010$. When $n = 4$, then for components A, B, C, and D the failure rates might be $\lambda_1 = 0.002$, $\lambda_2 = 0.004$, $\lambda_3 = 0.001$, and $\lambda_4 = 0.003$. If the weights are not right, a redetermination might give $\lambda_2 = 0.0035$ and $\lambda_3 = 0.0015$. This is the usual method used for allocation, making use of available failure rates for the components involved. It provides a basis for comparing tabulated values with allocated values.

The AGREE report [10] uses a more complex method, detailed on pages 192–195 of reference [1]. The AGREE report cautions against the use of its method for items of low importance. Its basis is the complexity of the units rather than their failure rates. Consideration is given to the number of hours the jth unit will be required to operate in T system hours and the total number of modules in the system. The importance factor of each unit must also be considered. All units do not operate the same length of time, hence these times of operation must be considered and many of these t_j values will be less than T hours. The AGREE report gives the allocated rate of the jth unit as

$$\lambda_j = n_j [-\log R^*(T)]/E_j t_j N.$$

This is used in many instances to allocate reliabilities among the components.

TABLE 11—CORRELATION OF FAILURE RATES.

	Fits	Bits	Failures/10^6 Hrs	%/1000 Hrs	Failures/10^3 Hrs	Failures/Hr
Fits	1	10^{-1}	10^{-3}	10^{-4}	10^{-6}	10^{-9}
Bits	10	1	10^{-2}	10^{-3}	10^{-5}	10^{-8}
Failures/10^6 hrs	10^3	10^2	1	10^{-1}	10^{-3}	10^{-6}
%/1000 hrs	10^4	10^3	10	1	10^{-2}	10^{-5}
Failures/10^3 hrs	10^6	10^5	10^3	10^2	1	10^{-3}
Failures/hr	10^9	10^8	10^6	10^5	10^3	1

Illustrating the simpler method, if an assembly is made up of two components of equal weight and is to have an end reliability of 0.99, then each component should have a reliability of at least 0.995. The product $0.995 \times 0.995 = 0.990025$. If the assembly is made up of three units of equal weight, then each should have a reliability of 0.997, since $(0.997)^3 = 0.991026973$. The same procedure is used even if the parts have widely different importance and hence extremely different reliability values. These preliminary estimates of the reliability values required are used in determining costs for proposals to secure new business. These values become even more stringent when the confidence level is increased.

RELIABILITY AND PROBABILITY— CONFIDENCE LEVELS AND LIMITS

When expected values of reliability and/or failure rates are used as measures of the reliability values given in the contract, such probabilities are called point probabilities. They are derived from the data by the use of the applicable equations in this chapter. The "method of attributes" is generally used since reliability is closely related to failure rates derived directly from attributes data. In the method of attributes, units on test are classified by the test into two categories—failures or successes, defectives or nondefectives, unsatisfactory or satisfactory—all twofold classifications. If a multifold classification is used, such as the exact measurement in ohms of a resistor, a distribution is obtained. Reliability distributions for $R(t)$ are a combination of the above methods, since the values are based on failure rates and also on mean-time-before-failure values, which lie in a multifold classification. Confidence limits are provided when multifold classifications are given.

Measures of Reliability

Reliability may be measured in many ways. Failures are recorded with respect to time of occurrence, plus type and cause of failure. For example, 5 failures have been observed in 20 units on test for 100 000 hours, where some failures occurred before 20 000 hours each and others occurred slightly after 20 000 hours. Fifteen units operated more than 20 000 hours without failing. Thus the failure rate is found by dividing 5 by 100 000, giving a failure rate of 50%/10 000 hours, or 50/1 000 000 hours, or 50/10^6 hours, or 5%/1000 hours. The MTBF = m value is found to be $m = 100\,000/5 = 20\,000$ hours.

The unit of measure of failure rates is often considered to be a bit, which is 0.01 of a failure every million hours. The failure rates for parts are decreasing so rapidly that the industry talks about failure rates in "fits," which are one-tenth of a bit. Table 11 covers the various terms used in these calculations. To convert a failure rate term in the first column to any term in the heading of succeeding columns, multiply by the number under the new term heading, i.e., to convert bits to %/1000 hours, multiply by 10^{-3}.

Two basic methods are used in securing reliability data. In the first method the 20 units above are placed on test with the stipulation that each unit be operated 5000 hours, which requires over 200 days of continuous operation. During this period of operation the time at which each of the 5 failures occur must be recorded. If they occur at 4200, 4350, 4400, 4750, and 4900 hours, and if the test is stopped at the end of 5000 hours, the total time of operation is $100\,000 - (800 + 650 + 600 + 250 + 100) = 97\,600$ hours. If the technician making the test wishes to obtain the total operating time of 100 000 hours, then the remaining 15 units must operate $2400/15 = 160$ hours each more than the original 5000 hours specified with no

TABLE 12—TIME IN HOURS TO FAILURE FOR RELIABILITY TEST ($n=20$ UNITS).

	6 500		8 000		10 200		5 500	
	4 200		7 000		4 750		8 900	
	9 400		4 350		4 900		10 500	
	4 400		9 100		8 750		9 150	
	7 800		8 100		9 200		12 500	
Sum=	32 300	+	36 550	+	37 800	+	46 550	=153 200

additional failures. Otherwise the estimate of the failure rate should be $r=5/97\ 600=5.12/100\ 000=5.12\%/1000$ hours. The value of $m=$ MTBF is $m=97\ 600/5=19\ 520$ hours.

The second basic method for determining the failure rate is to operate all 20 units until they fail. This makes it difficult to plan a testing program, as some equipment may be tied up for two or even several times the period designated for the original test. However, failure rates found by the second method should be more representative of actual field failure rates. The data secured under these conditions for the 20 units might be as given in Table 12.

This gives a failure rate much larger than the first estimate, i.e., $r=20/153\ 200=13.05\%/1000$ hours. If the 5% value is near the actual central value, then the total operating time should be about 400 000 hours, or 20 000 hours on the average for each unit=833 days.

It is desirable to establish confidence levels for reliability values. A distribution of probability values is used to give either a one-sided confidence value or a two-sided confidence value with both upper and lower limits. When determining failure rates a one-sided level applies. A confidence level of 0.95 might be stipulated for a reliability of 0.95 or better (the goal stipulated in the contract). In the first set of data the rate of 5%/1000 hours may be expressed as a reliability of 95% where the mission time is 1000 hours. If the mission time is 10 hours then the corresponding reliability is 99.95%. These values are associated with expected values rather than having a range of values associated with a designated confidence-level value. Where a confidence limit is used, a limit is established which is associated with this limit. It will be expressed as a failure rate or an MTBF value.

Confidence Levels or Limits

As noted previously, to obtain valid confidence limits it is necessary to have a distribution. The meaning of the terms confidence level, degree of belief, and confidence limits is important. Is there adequate information so that the observer is almost certain that the reliability forecast will be met? Or is the sample size so small that the observer may secure widely different results with a subsequent test of the same nature? What confidence does the observer have that the data are good and that the results are adequate to justify the reliability prediction made? What degree of belief P_b does one have that the data are good, are sufficient to make the desired predictions, and are sufficiently uniform to make it possible to establish a band within which it is expected future values of the same kind will fall? When dealing with minimum or maximum values, the confidence level or limit is associated with one limit only and the distribution taken is considered a one-tailed probability associated with limits from $-\infty$ to x, or from x to $+\infty$, where x is the limiting value. If both the minimum and maximum values are specified or if bands of approximately equal probability widths are desired, two-tailed limits must be computed. The limits associated with some given confidence level (such as 0.90) might be from x_a to x_b, which might be expressed as from $x=a$ to $x=b$. Use is often made of the Gaussian or Normal Law distributions in determining such confidence limits.

One-Sided Confidence Limit

When it is desired to specify that the true mean time between failures must exceed a given minimum value with a confidence level of $(1-\alpha)$, the procedure for a one-sided confidence limit is applied. This provides a tail area α and means there is a probability α that the m value actually observed by test will be smaller than the specified minimum and a probability of $1-\alpha$ that it will be larger. Reference [2] denotes this one-sided confidence limit by the notation C_L to distinguish it from the two-sided lower limit L. Its value is given by

$$C_L=(2r/\chi^2_{\alpha;2r})\widehat{m}=2T/\chi^2_{\alpha;2r}$$

where tests are continued until the rth failure occurs with $r=1, 2, \cdots, d$; $T=$ accumulated test time$= \sum t_i$, $m = T/r =$ an estimate of the mean time between failures, and $1-\alpha =$ confidence level prescribed.

Note that in this case $2r =$ degrees of freedom (d.f.).

However, a test can also be terminated at some preselected test time without a failure occurring exactly at that time. For such a case Epstein* has shown that for the accumulated hours of operating time $T = \sum t_i$, then

$$m \geq 2T/\chi^2_{\alpha; 2r+2}$$

where d.f.$= 2r+2$ and the case where $r=0$ is covered. For $r=0$, then

$$C_L = 2T/\chi^2_{\alpha; 2}.$$

In the percent survival method, the accumulated operating time T is not measured and only the straight test duration time t_d is known, at which time r failures of n units on test are counted. In this method confusion may exist between chance failures and failures due to actual wear-out. The time to wear-out must be known, and it is necessary to design and select parts from manufacturers that can be made so that their respective wear-out time is many hours past the time of the mission. Again referring to both reference [2] and to Epstein, for a one-sided confidence level of $1-\alpha$ the lower-limit estimated reliability for t_d hours is given by

$$\hat{R}(t_d) = \frac{1}{1 + [(r+1)/(n-r)]F_{\alpha; 2r+2; 2n-2r}}$$

where F is the upper α percentage point of the Fisher distribution (termed the F distribution) with the two corresponding degrees of freedom, $2r+2$ and $2n-2r$. For this estimate of reliability there is a probability of $1-\alpha$ that the true reliability for t_d hours is equal to or larger than $R(t_d)$. It must be noted that this reliability estimate is nonparametric and is valid for the exponential as well as the nonexponential case.

A general mathematical approach is used in many cases to determine the confidence levels for either one-sided or two-sided distributions for various density functions. Confidence levels and reliability values are related by the two following general relations, where $P_b =$ degree of belief, equivalent to the confidence level. One relation covers continuous distributions and makes use of the area under the density function secured by integration, while the second relation covers sum-

mations for integral values. These relations are

$$P_b = \int_0^{x^*} f(x)\, dx \Big/ \int_0^{x=\infty} f(x)\, dx$$

$$P_b = \sum_0^{x^*} F(x) \Big/ \sum_0^{x=\infty} F(x).$$

For the exponential density function, use of these relations gives

$$\begin{aligned} P_b &= \int_0^{t^*} \lambda \exp(-\lambda t)\, dt \Big/ \int_0^\infty \lambda \exp(-\lambda t)\, dt \\ &= -\exp(-\lambda t)_0{}^{t*} \Big/ -\exp(-\lambda t)_0{}^\infty \\ &= [-\exp(-\lambda t^*) + 1]/(0+1) \\ &= 1 - \exp(-\lambda t^*). \end{aligned}$$

The reliability $R(t)$ for time t for P_b (the one-sided confidence level) is derived from the term λt^*, where no failures have been observed in time $T = \sum t_i$. From an exponential table determine $\lambda t^* = a$, corresponding to $1 - P_b$. Then $\lambda = a/t^* = a/T$, since t^* corresponds to the total time required for the test. The final reliability value is determined from

$$R(t) = \exp(-\lambda t) = \exp(-at/T).$$

Two-Sided Confidence Limit

If a test is terminated when the rth failure has occurred, the ratio $2r(\hat{m}/m)$ has a chi-square distribution with $2r$ degrees of freedom. The two-sided confidence interval at a confidence level of $(1-\alpha)$ is

$$\hat{m}(2r/\chi^2_{\alpha/2; 2r}) \leq m \leq \hat{m}(2r/\chi^2_{1-\alpha/2; 2r}).$$

Here $\hat{m}$ represents the estimate of m derived from the samples tested and is the MTBF. The lower limit L is given by

$$L = (2r/\chi^2_{\alpha/2; 2r})\, \hat{m} = 2T/\chi^2_{\alpha/2; 2r}$$

while the upper confidence limit is given by

$$U = (2r/\chi^2_{1-\alpha/2; 2r})\, \hat{m} = 2T/\chi^2_{1-\alpha/2; 2r}.$$

Herein $\hat{m} = T/r$ and can be derived from either a replacement or a nonreplacement test, while $T = \sum t_i$, the sum of the operating times accumulated by all the components during the test. When the test is terminated at time t_d without a failure occurring exactly at that time, then the degrees of freedom for the lower limit are changed from $2r$ to $2r+2$. The upper and lower limits are given by

$$2T/\chi^2_{\alpha/2; 2r+2} \leq m \leq 2T/\chi^2_{1-\alpha/2; 2r}.$$

---

* Ben Epstein, "Estimation From Life Test Data," *IRE Transactions on Reliability and Quality Control,*" vol. RQC-9; April 1960.

TABLE 13—RELIABILITY VALUES FOR A MISSION OF $t=10$ HOURS FOR $\lambda=2.5\%/1000$ HOURS FOR 6 ONE-TAILED CONFIDENCE LEVELS FOR EXPONENTIAL: $R(\lambda t)=R[\lambda t+z(\lambda t)^{1/2}]$, z GIVEN IN NORMAL LAW (GAUSSIAN) TABLES FOR P_z TABULATED FOR CONFIDENCE LEVEL.

Confidence Level	Normal Law z Values	Upper Limit for $(\lambda t)_z=0.00025+z(0.00025)^{1/2}$	$R(\lambda t)_z=\exp[-(\lambda t)_z]$
0.90	1.282	0.02052	0.97968
0.95	1.645	0.02626	0.97408
0.96	1.751	0.02794	0.97245
0.97	1.881	0.02999	0.97045
0.98	2.054	0.03272	0.96780
0.99	2.326	0.03702	0.96366

From these limits giving lower and upper limits, L and U, in terms of mean time before failure, for any mission time t, then lower and upper limiting values for the reliability $R(t)$ may be readily computed from

$$L::R_L:R_L(t)=\exp(-t/L)$$

$$U::R_U:R_U(t)=\exp(-t/U).$$

When the Gaussian (Normal Law) distribution applies or is used as a means of determining upper and lower limits for either m or $R(t)$, where $\bar{m}$, σ_m and $\bar{R}$, σ_R are known, either symmetric or non-symmetric confidence limits may be determined from

$$L:\bar{m}-z_\alpha\sigma_m;\ \bar{R}-z_\alpha\sigma_R$$

$$U:\bar{m}+z_\beta\sigma_m;\ \bar{R}+z_\beta\sigma_R$$

where $\alpha+\beta=\gamma=$ probability for the specified confidence band.

As an aid to the calculation of confidence bands under certain stated conditions, several of the military specifications listed in Table 14 allow an easy calculation of these limits. One specification to note is MIL-R-22973.

RELIABILITY AND LIFE EXPECTANCIES—USE OF NORMAL LAW THEORY

In most cases reliability values are associated with life usage or with the time of storage. A mission may require t hours to be accomplished. For example, it may require 10 hours to drive an automobile from Los Angeles to San Francisco, a distance of approximately 420 miles. What is $R(10)$, the reliability of accomplishing this mission in 10 hours at any time? In a prior example, $p=2.5\%$ where it was assumed that each unit was tested 1000 hours. The failure rate then may be expressed as $2.5\%/1000$ hours. If the mission time is 10 hours the reliability $R(t)$, assumed to be based on the exponential, is determined as follows. For $t=10$ hours, $v=2.5\%/1000$ hours $=0.000025$/hr, and for this case

$$R(10)=\exp[-0.000025(10)]$$

$$=\exp(-0.00025)=0.99975.$$

This value of reliability is based on the expected value. For the exponential the variance is equal to the expected value. Hence, since for this 10-hour mission $t=0.00025$, $\sigma_t=(0.00025)^{1/2}=0.01581$. For a 90% confidence level using the proper multiplying factor based on the Normal Law

$$t_{0.90}=0.00025+1.282(0.01581)$$

$$=0.00025+0.02026842=0.02051842.$$

The corresponding reliability for $t=10$ hours is $R_{0.90}(t=10)=\exp(-0.02052)$, $R(10)_{0.90}=0.97968$. For a 95% confidence level based on the Normal Law:

$$(\lambda t)_{0.95}=0.00025+1.645(0.01581)$$

$$=0.00025+0.0260075=0.0262575.$$

For this expected value of λt with $t=10$ hours, $R(10)=\exp(-0.02626)=0.97408$. For a 99% confidence level based on the Normal Law:

$$(\lambda t)_{0.99}=0.00025+2.326(0.01581)$$

$$=0.00025+0.03677406=0.03702406.$$

Hence $R(t=10)_{0.99}=\exp-0.03702=0.96366$. For the six confidence levels often used, the reliability values for a one-tailed confidence level are given in Table 13.

Reverting to the data in Table 8, there existed only one set of 100 units $=n$ out of 10 000 that had a reliability observed of 0.94. Associated with this value is a confidence level of $(10\,000-100)/$

$10\,000 = 9900/10\,000 = 0.99$ (Table 9). This reliability value is determined for an assumed operating period of 1000 hours. Hence this gives $R(1000) = 0.94$ when the confidence level $P_C = 0.99$. Hence the value of λ is thus computed

$$R(1000) = 0.94 = \exp[-\lambda(1000)] = \exp-0.0619.$$

Then $1000\lambda = 0.0619$, and $\lambda = 6.19\%/1000$ hours. For $t = 10$ hours, then

$$R(10) = \exp(-\lambda t) = \exp[-0.0619(10)/1000]$$

$$= \exp(-0.000619) = 0.99938.$$

Thus the actual data provide more-optimistic estimates of the reliability based on field test results. Since the distribution as graphed appears to be almost rectangular in Fig. 2, the assumption of normality is pessimistic.

In these life tests each failure must be carefully analyzed to determine whether it is a *chance* failure or a *wear-out* failure. These results must be fed back to the design engineers to make certain that corrective measures for improving the life characteristics are taken and established as standard procedures.

DESIGN IMPROVEMENTS—REDUNDANCY

If it is necessary to improve a system's reliability markedly, it is necessary to change the design. Different materials and parts can be used that have better reliability histories. Parts are considered to operate in series where independent, or they may be operated in parallel or in a series-parallel combination. As noted previously, the Bayesian theorem, together with additional data, provides better estimates of the final reliability. All these techniques and statistical tools should be used to improve the reliability of the system. The use of redundant circuits and parts may be the best and cheapest way to achieve the established contract reliability goals.

Series Circuits and Assemblies

When parts are assembled essentially in series, the system reliability is determined from the product of its component parts. It is determined from

$$R_S = \prod_{i=1}^{i=h} R_i \ \text{(for } h \text{ component parts).}$$

One of the problems is to determine the confidence level for a group of h components in series when the confidence level for each component

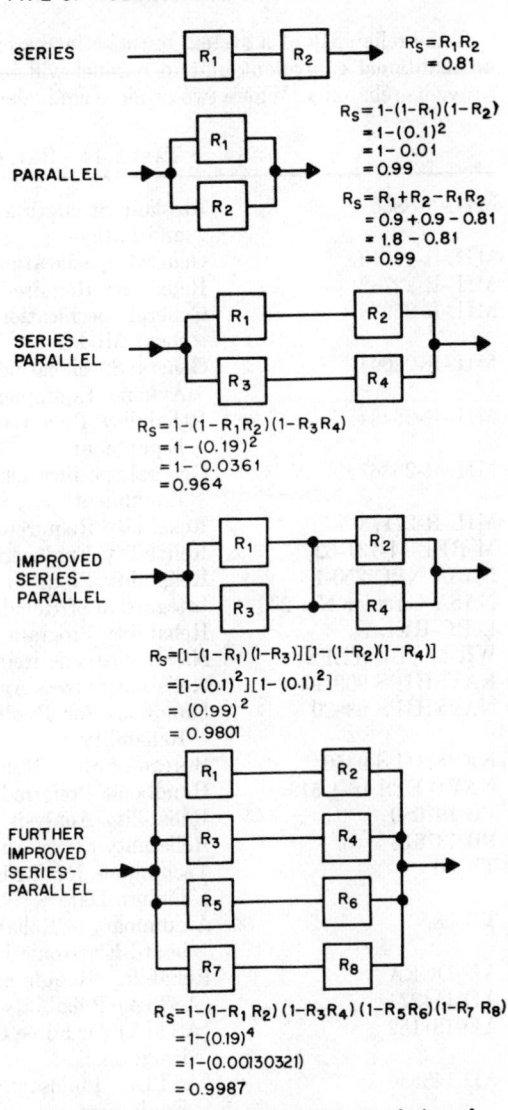

Fig. 5—Diagrams of redundant systems—relations for computing system reliabilities. All parts are assumed to have equal reliability ($R_C = 0.90$).

reliability value is known. The usual technique is to express the reliabilities of all components in probabilities associated with the designated confidence level, such as $P_C = 0.90$. Then the confidence level for the system will not be greater than P_C and will not be less than $(P_C)^h$ for the h components. Recent developments in mathematical inequalities have indicated that the confidence level for the system is generally greater than the mean of the spread from $(P_C)^h$ to P_C. For example, for $h = 2$, $P_C = 0.90$, the range is 0.81 to 0.90. Experiment and field test results indicate that P_{Cs} is closer to P_C.

Parallel Circuits and Assemblies

If the reliability of a system must be increased, an additional equivalent unit in parallel will improve its reliability. Where two or more equivalent units are in parallel, then (where $Q_c = 1 - R_c$ is the unreliability or lack of reliability) the reliability of the system R_S is given by

$$R_S = 1 - (1 - P_{C_1})(1 - P_{C_2}) \cdots (1 - P_{C_h}).$$

TABLE 14—RELIABILITY SPECIFICATIONS.

MIL-A-8866	Airplane Strength and Rigidity Reliability Requirements, Repeated Loads and Fatigue
MIL-R-19610	General Specifications for Reliability of Production Electronic Equipment
MIL-R-22732	Reliability Requirements for Shipboard and Ground Electronic Equipment
MIL-R-22973	General Specification for Reliability Index Determination for Avionic Equipment Models
MIL-R-23094	General Specification for Reliability Assurance for Production Acceptance of Avionic Equipment
MIL-R-26484	Reliability Requirements for Development of Electronic Subsystems for Equipment
MIL-R-26667	General Specification for Reliability and Longevity Requirements, Electronic Equipment
MIL-R-27173	Reliability Requirements for Electronic Ground Checkout Equipment
M-REL-M-131-62	Reliability Engineering Program Provisions for Space System Contractors
NASA NPC 250-1	Reliability Program Provisions for Space System Contractors
NASA Circular No. 293	Integration of Reliability Requirements into NASA Procurements
LeRC-REL-1	Reliability Program Provisions for Research and Development Contracts
WR-41 (BUWEPS)	Naval Weapons Requirements, Reliability Evaluation
NAVSHIPS 900193	Reliability Stress Analysis for Electronic Equipment
NAVSHIPS 93820	Handbook for Prediction of Shipboard and Shore Electronic Equipment Reliability
NAVSHIPS 94501	Bureau of Ships Reliability Design Handbook
NAVWEPS 16-1-519	Handbook Preferred Circuits—Naval Aeronautical Electronic Equipment
PB 181080	Reliability Analysis Data for Systems and Components Design Engineers
PB 131678	Reliability Stress Analysis for Electronic Equipment, TR-1100
TR-80	Techniques for Reliability Measurement and Prediction Based on Field Failure Data
TR-98	A Summary of Reliability Prediction and Measurement Guidelines for Shipboard Electronic Equipment
AD-DCEA	Reliability Requirements for Production Ground Electronic Equipment
AD 114274	(ASTIA) Reliability Factors for Ground Electronic Equipment
AD 131152	(ASTIA) Air Force Ground Electronic Equipment-Reliability Improvement Program
AD 148556	(ASTIA) Philosophy and Guidelines—Prediction on Ground Electronic Equipment
AD 148801	(ASTIA) Methods of Field Data Acquisition, Reduction and Analysis
AD 148977	(ASTIA) Prediction and Measurement of Air Force Ground Electronic Reliability
MIL-HDBK-217	Reliability Stress and Failure Rate Data for Electronic Equipment
RADC 2623	Reliability Requirements for Ground Electronic Equipment
USAF BLTN 2629	Reliability Requirements for Ground Electronic Equipment
AR-705-25	Reliability Program for Material and Equipment
OP 400	General Instructions: Design, Manufacture and Inspection of Naval Ordnance Equipment
MIL-STD-721	Definitions for Reliability Engineering
MIL-STD-756	Procedures for Prediction and Reporting Prediction of Reliability of Weapon Systems
MIL-STD-781	Test Levels and Accept/Reject Criteria for Reliability of Nonexpendable Electronic Equipment
MIL-STD-785	Requirements for Reliability Program (for Systems and Equipments)
DOD H-108	Sampling Procedure and Table for Life and Reliability Testing

For example, if $P_C = 0.90$ for one part, $R_C = P_C$. Then, when $h = 3$, $R_S = 1 - (0.10)^3 = 1 - 0.001 = 0.999$, which is much better than $P_C = 0.90$ established for one component part. These relations hold for all values of R_C. For example, if $R_1 = 0.90$, $R_2 = 0.80$, and $R_3 = 0.70$, then $R_S = 1 - (0.1)(0.2) \times (0.3) = 1 - 0.006 = 0.994$.

Series-Parallel Circuits and Assemblies

Figure 5 shows series, parallel, and series-parallel assemblies, plus system reliability values for each assembly.

Tradeoffs

There are often limitations on the weight and volume of the overall system. If reliability is too low for the prescribed requirements, a tradeoff must be made between (A) adding too much redundancy to improve the reliability and (B) increasing the components, weight, and volume of the system. Stringent design reviews should provide the best basis for making these decisions.

STATISTICS AND RELIABILITY

The basic science of reliability is interleaved with statistical concepts. For more-detailed information on reliability, consult the references at the end of this chapter. Chapters 42 and 46 of this Handbook give additional statistical and mathematical information.

AVAILABILITY AND MAINTAINABILITY

Availability and maintainability are used today to describe a relationship to measure product effectiveness in the field.

Product availability is defined as the probability that the system will operate satisfactorily at any point in time where time includes not only operating life, but also active repair time and administrative and logistic time. An equation for availability is

$$A = \text{MTBF}/(\text{MTBF}+\text{MTTR})$$

where A = availability, MTBF = mean time before failure, and MTTR = mean time to repair. The calculation of MTTR is related to repair hours, while the calculation of MTBF is related to component operating hours. Figure 6 is a graph of the above equation.

It is evident from the figure that the effect of maintainability on availability increases as the ratio of MTBF to MTTR decreases. If an item has an inherently low MTBF, the MTTR must be very low to sustain a good level of availability.

In the design of any complex system, an optimum condition should be established between reliability and maintainability, so that reliability is not increased beyond the point where very little availability gain is obtained because of lack of consideration of the effect of maximum maintainability.

To look at this another way, manufacturers of microelectronic devices claim that MTBF is very high and repair time is nil or very low. Look again at Fig. 6 and the ratio of MTBF/MTTR in the region to the right of 100. Little is to be gained by designing a module that can be repaired extremely quickly, if to do the job special tools and costs are involved. In other words, the throwaway concept in this case is clearly justifiable.

RELIABILITY SPECIFICATION INDEX

Table 14 presents a family tree of U.S. Government documents establishing and supporting reliability requirements. New specifications are added frequently to build up the reliability factors and requirements. One area (not listed in the table) that is expanding steadily is special parts reliability specifications such as the MIL-R-38000 series. These specifications cover the acceptance and qualification testing of high-reliability parts.

REFERENCES

1. "Reliability Engineering," Edited by William H Von Alven, ARINC Research Corporation, Prentice-Hall, Inc., Englewood Cliffs, N.J.; 1964 (23 contributors).
2. Igor Bazovsky, "Reliability Theory and Practice," Prentice-Hall, Inc., Englewood Cliffs, N.J.; 1961.
3. S. R. Calabro, "Reliability Principles and Practices," McGraw-Hill Book Company, Inc., New York, N.Y.; 1962.

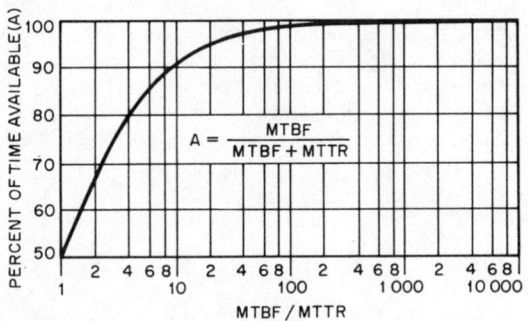

Fig. 6—Product availability.

4. Frank M. Gryna, Jr., Naomi J. McAfee, Clifford M. Ryerson, and Stanley Zwerling, Editors, "Reliability Training Text," sponsored by ASQC and IEEE, Second Edition; March 1960.

5. Norman L. Johnson and Fred C. Leone, "Statistics and Experimental Design in Engineering and the Physical Sciences," Volumes 1 and 2, John Wiley & Sons, Inc., New York, N.Y.; 1964.

6. C. G. Lambe, "Elements of Statistics," Longmans, Green and Co., London and New York; 1952.

7. Richard R. Landers, "Reliability and Product Assurance, A Manual for Engineering and Management," Prentice-Hall, Inc., Englewood Cliffs, N.J.; 1963.

8. David K. Lloyd and Myron Lipow, "Reliability: Management Methods and Mathematics," Prentice-Hall, Inc., Englewood Cliffs, N.J.; 1962.

9. "Statistical Theory of Reliability," Marvin Zelen, Editor: Proceedings of an Advanced Seminar Conducted by the Mathematics Research, United States Army, at the University of Wisconsin, Madison, 8–10 May 1962, The University of Wisconsin Press, Madison, Wisconsin; 1963.

10. Advisory Group on Reliability of Electronic Equipment (AGREE), Office of the Assistant Secretary of Defense (Research and Engineering), "Reliability of Military Electronic Equipment," Supt. of Documents, U.S. Government Printing Office, Washington, D.C.; 4 June 1957.

11. Ad Hoc Study Group on Parts Specification Management for Reliability, Office of the Director of Defense Research and Engineering and Office of the Assistant Secretary of Defense Supply and Logistics, "Parts Specification Management for Reliability," Volumes 1 and 2, PSMR-1, Supt. of Documents, U.S. Government Printing Office, Washington D.C.; May 1960.

12. Milton Abramowitz and Irene A. Stegun, "Applied Mathematics Series AMS 55, Handbook of Mathematical Functions with Formulas, Graphs and Mathematical Tables," National Bureau of Standards, Supt. of Documents, U.S. Government Printing Office, Washington, D.C., third printing with corrections; 1965.

13. A. Hald, "Statistical Tables and Formulas," John Wiley & Sons, Inc., New York, N.Y.; 1952.

14. E. C. Molina, "Poisson's Exponential Binomial Limit, Table 1—Individual Terms; Table 2—Cumulated Terms," D. Van Nostrand Co., Inc., New York, N.Y.; 1949.

15. Arnold N. Lowan and Staff, "Applied Mathematics Series AMS 23, Tables of Normal Probability Function," National Bureau of Standards, U.S. Govt. Printing Office, Supt. of Documents, Washington, D.C.; 1953.

16. Arnold N. Lowan and Staff, "Applied Mathematics Series AMS 14, Tables of the Exponential Function, e^x," National Bureau of Standards, 4th Edition, U.S. Govt. Printing Office, Supt. of Documents, Washington, D.C.; 1961.

17. Staff of the Computation Laboratory, "Tables of the Error Function and of its First Twenty Derivatives," Harvard University Press, Cambridge, Mass.; 1952.

18. A. S. Goldman and T. B. Slattery, "Maintainability," John Wiley & Sons, Inc., New York, 1964, (Contributions by S. Firstman, Rand Corporation, and J. Rigney, University of Southern California): Chapter on General Electric Company "TEMPO".

19. "Reliability Handbook," edited by W. Grant Ireson, Editor-in-Chief, Executive Head, Department of Industrial Engineering, Stanford University, McGraw-Hill Book Company, Inc.; 1966 (19 contributors).

CHAPTER 44

FOURIER WAVEFORM ANALYSIS

FOURIER TRANSFORM OF A FUNCTION

The Fourier transform $F(y)$ of a function of real variable $f(x)$ [where $f(x)$ may be real or complex] is defined by the integral

$$F(y) = \int_{-\infty}^{\infty} f(x) \, \exp(-j2\pi xy) \, dx$$

provided this integral exists for every real value of x.

A sufficient, but not necessary, existence condition is that $f(x)$ be absolutely integrable; that is

$$\int_{-\infty}^{\infty} |f(x)| \, dx < \infty.$$

$(\sin x)/x$ is an important example of a function which has a Fourier transform even though it is not absolutely integrable.

In general, $F(y)$ is complex; letting $f(x) = f_r(x) + jf_i(x)$, where $f_r(x)$ and $f_i(x)$ are real valued, one has

$$F(y) = \int_{-\infty}^{\infty} [f_r(x) \, \cos 2\pi yx + f_i(x) \, \sin 2\pi yx] \, dx$$

$$-j \int_{-\infty}^{\infty} [f_r(x) \, \sin 2\pi yx - f_i(x) \, \cos 2\pi yx] \, dx.$$

Conversely, the function of real variable $f(x)$, whose Fourier transform is a given function $F(y)$, is given by the integral (inverse Fourier transform)

$$f(x) = \int_{-\infty}^{\infty} F(y) \, \exp(j2\pi xy) \, dy$$

where it is assumed that, at points of discontinuity of the integral (if any), the function $f(x)$ is given the value

$$f(x) = [f(x^+) + f(x^-)]/2.$$

$f(x^+)$ and $f(x^-)$ are the limits of $f(x-t)$ as t approaches 0 through positive and negative values, respectively.

Letting $F(y) = F_r(y) + jF_i(y)$, one has

$$f(x) = \int_{-\infty}^{\infty} [F_r(y) \, \cos 2\pi yx - F_i(y) \, \sin 2\pi yx] \, dy$$

$$+j \int_{-\infty}^{\infty} [F_r(y) \, \sin 2\pi yx + F_i(y) \, \cos 2\pi yx] \, dy.$$

In many engineering applications it is customary to denote the variable y as "frequency"; in most cases x represents time or space.

Introducing the radian frequency $\omega = 2\pi y$ as a variable, the definitions of the Fourier transform and of its inverse are written as

$$F(\omega/2\pi) = F_1(\omega)$$

$$= \int_{-\infty}^{\infty} f(x) \, \exp(-j\omega x) \, dx$$

$$f(x) = (2\pi)^{-1} \int_{-\infty}^{\infty} F_1(\omega) \, \exp(j\omega x) \, d\omega.$$

The properties of the Fourier transform are listed in Table 1; see also page 42-11 for the case of random functions.

44-1

TABLE 1—PROPERTIES OF FOURIER TRANSFORM.*

		Function	Fourier Transform
1.	Definition	$f(x)$	$F(y) = \int_{-\infty}^{+\infty} f(x)\, \exp(-2\pi jxy)\, dx$
2.	Inverse transform	$f(x) = \int_{-\infty}^{+\infty} F(y)\, \exp(2\pi jxy)\, dy$	$F(y)$
3.	Linearity	$a\, f(x)$ $f_1(x) \pm f_2(x)$	$a\, F(y)$ $F_1(y) \pm F_2(y)$
4.	Translation or shifting theorem	$g(x) = f(x - x_0)$, x_0 = real const. $g(x) = \exp(2\pi jy_0 x) f(x)$, y_0 = real const.	$G(y) = \exp(-2\pi jx_0 y)\, F(x)$ $G(y) = F(y - y_0)$
5.	Change of scale	$g(x) = f(x/a)$, a = real const.	$G(y) = \lvert a \rvert\, F(ay)$
6.	Frequency shifting and change of scale	$g(x) = \exp(2\pi jy_0 x) f(x/a)$, y_0 and a = real const.	$G(y) = \lvert a \rvert\, F\big[a(y - y_0)\big]$
7.	Interchange of function and transform	$g(x) = F(x)$	$G(y) = f(-y)$
8.	Convolution in x-space (product of Fourier transforms)	$h = f*g = g*f$ i.e., $h(x) = \int_{-\infty}^{+\infty} f(x - \tau)\, g(\tau)\, d\tau = \int_{-\infty}^{\infty} f(\tau)\, g(x - \tau)\, d\tau$	$H = F \cdot G$
8a.	Convolution in y-space (product of inverse Fourier transforms)	$h = f \cdot g$	$H = F * G$
9.	Unit pulse (or Dirac function)	$\delta(x)$ $f(x) = 1$ (for all x)	$F(y) = 1$ (for all y) $\delta(y)$

10. Periodic train of equal pulses

$$A \sum_{n=-\infty}^{n=+\infty} \delta(x-nT)$$

$$(A/T) \sum_{n=-\infty}^{n=+\infty} \delta(y-n/T)$$

11. Derivative in x-space

$$g(x) = d^n f/dx^n$$

$$G(y) = (2\pi j y)^n F(y), \text{ if } G(y) \text{ exists}$$

11a. Derivative in y-space

$$g(x) = (-2\pi j x)^n f(x)$$

$$G(y) = d^n F/dy^n$$

12. Integral in x-space

$$g(x) = \int_{-\infty}^{x} f(x)\,dx$$

$$G(y) = [1/(2\pi j y)]F(y) + F(0)\delta(y)$$

$$\text{where } F(0) = \int_{-\infty}^{\infty} f(x)\,dx$$

12a. Integral in y-space

$$g(x) = -[1/(2\pi j x)]f(x)$$

$$G(y) = \int_{-\infty}^{y} F(y)\,dy$$

13. Symmetry

$$g(x) = f(-x)$$

$$G(y) = F(-y)$$

$$f \text{ even}: f(x) = f(-x)$$

$$F \text{ even}: F = 2 \int_0^\infty f(x) \cos(2\pi x y)\,dx$$

$$f \text{ odd}: f(x) = -f(-x)$$

$$F \text{ odd}: F = -2j \int_0^\infty f(x) \sin(2\pi x y)\,dx$$

14. Complex conjugate

$$g(x) = f^*(x)$$

$$G(y) = F^*(-y)$$

Hence, if $f(x)$ is real

$$F(-y) = F^*(y)$$

15. Area under the curve

$$\int_{-\infty}^{+\infty} f(x)\,dx = F(0)$$

$$\int_{-\infty}^{+\infty} F(y)\,dy = f(0)$$

16. Parseval's theorem

$$\int_{-\infty}^{+\infty} f^*(x) g(x)\,dx$$

$$= \int_{-\infty}^{+\infty} F^*(y) G(y)\,dy$$

TABLE 1—*Continued*

	Function	Fourier Transform
16a. Alternative forms	$\int_{-\infty}^{+\infty} f(x)g(x)\,dx$	$=\int_{-\infty}^{+\infty} F(-y)G(y)\,dy$
	$\int_{-\infty}^{+\infty} f(u)G(u)\,du$	$=\int_{-\infty}^{+\infty} F(u)g(u)\,du$
16b. "Energy" relation	$\int_{-\infty}^{+\infty} \|f(x)\|^2\,dx$	$=\int_{-\infty}^{+\infty} \|F(y)\|^2\,dy$
17. Initial value theorem	If $f(x)=0$ for $x<0$, is real, and contains no pulses	
	$f(0^+)=2\int_{-\infty}^{\infty} F_r(y)\,dy$	
	where $F_r(y)=\text{Re}[F(y)]$	
18. Relationships between $F_r(y)$ and $F_i(y)$	a) If $f(x)=0$ for $x<0$, is real, and contains no pulses	
	$\int_{-\infty}^{\infty} F_r^2(y)\,dy=\int_{-\infty}^{\infty} F_i^2(y)\,dy$	
	where $F_i(y)=\text{Im}[F(y)]$	
	b) The following integral relationships apply (Hilbert transforms)	
	$F_r(y)=2\int_{-\infty}^{\infty} [F_i(\tau)/2\pi(y-\tau)]\,d\tau$	
	$F_i(y)=-2\int_{-\infty}^{\infty} [F_r(\tau)/2\pi(y-\tau)]\,d\tau$	
	in which the Cauchy principal values of the integrals are taken	

* In the table, functions of x are denoted by lower-case letters and their transforms by the corresponding capital letters.

COMMON PULSE FORMS AND SPECTRA* (TABLE 2)

TABLE 2—TIME AND FREQUENCY FUNCTIONS FOR COMMONLY ENCOUNTERED PULSE SHAPES.

Time Function	Frequency Function

A. Rectangular pulse

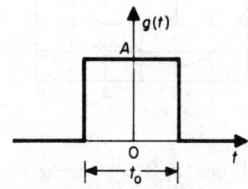

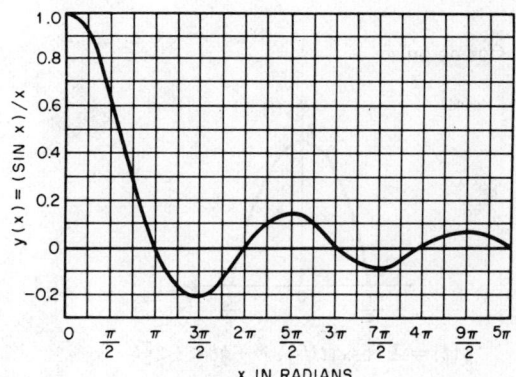

x IN RADIANS

$$g(t) = A \text{ for } -\tfrac{1}{2}t_0 < t < \tfrac{1}{2}t_0$$

$$= 0 \text{ otherwise}$$

Area $\mathcal{A} = At_0$

$$G(f) = \mathcal{A}(\sin\alpha)/\alpha$$

where $\alpha = \pi t_0 f$

[See curve $(\sin x)/x$ above.]

B. Isosceles-triangle pulse

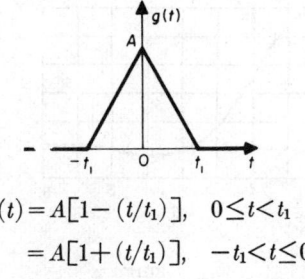

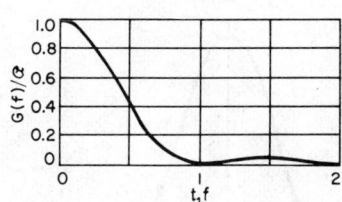

$$g(t) = A[1 - (t/t_1)], \quad 0 \le t < t_1$$

$$= A[1 + (t/t_1)], \quad -t_1 < t \le 0$$

$$= 0, \text{ otherwise}$$

Area $\mathcal{A} = At_1$

$$G(f) = \mathcal{A}[(\sin\alpha)/\alpha]^2$$

where $\alpha = \pi t_1 f$

C. Sawtooth pulse

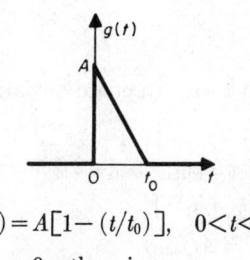

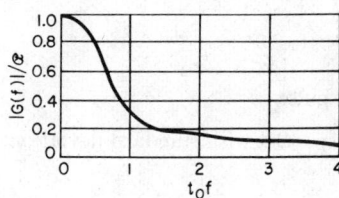

$$g(t) = A[1 - (t/t_0)], \quad 0 < t < t_0$$

$$= 0, \text{ otherwise}$$

Area $\mathcal{A} = \tfrac{1}{2}At_0$

$$G(f) = \mathcal{A}(j/\alpha)\{[(\sin\alpha)/\alpha]\exp(-j\alpha) - 1\}$$

$$= \mathcal{A}\,\frac{1 - \exp(-2j\alpha) - 2j\alpha}{2\alpha^2}$$

where $\alpha = \pi t_0 f$

TABLE 2—*Continued*

Time Function	Frequency Function

D. Any pulse of polygonal form may be represented as a linear combination of waveforms such as **A**, **B**, and **C** above eventually after some shifts in time. The pulse spectrum is the same linear combination of the corresponding spectra (eventually modified according to property 4, Table 1).

E. Cosine pulse

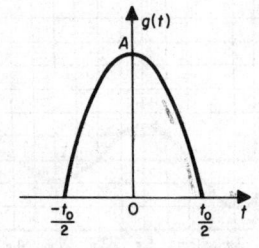

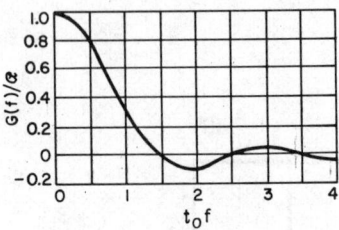

$g(t) = A \cos\pi(t/t_0), \quad -\tfrac{1}{2}t_0 < t < \tfrac{1}{2}t_0$

$= 0$, otherwise

Area $\mathcal{C} = (2/\pi) A t_0$

$G(f) = \mathcal{C}\{[\cos(\pi/2)\alpha]/(1-\alpha^2)\}$

where $\alpha = 2t_0 f$

For $\alpha = 1$, $G(f) = \mathcal{C}\pi/4$

F. Cosine-squared pulse

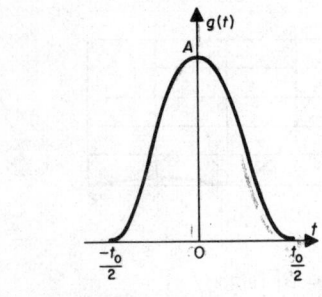

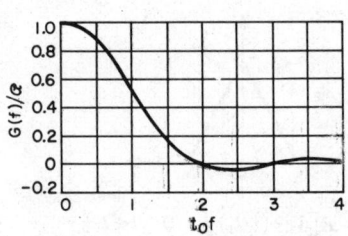

$g(t) = A \cos^2\pi(t/t_0)$
$\left. = \tfrac{1}{2}A[1+\cos2\pi(t/t_0)] \right\}$ $\quad -\tfrac{1}{2}t_0 < t < \tfrac{1}{2}t_0$

$= 0$, otherwise

Area $\mathcal{C} = \tfrac{1}{2}A t_0$

$G(f) = \mathcal{C} \dfrac{\sin\pi\alpha}{\pi\alpha(1-\alpha^2)}$

where $\alpha = t_0 f$

For $\alpha = 1$, $G(f) = \tfrac{1}{2}\mathcal{C}$

G. Gaussian pulse

Use curve on p. 42-4 with standard deviation

$\sigma = t_1$

$\quad = (2\ln2)^{-1/2}t_{6dB}; \quad \Delta t_{6dB} \equiv 2t_{6dB}$

$g(t) = A \exp[-\tfrac{1}{2}(t/t_1)^2]$

$\quad = A \exp[-(\ln2)(t/t_{6dB})^2]$

Area $\mathcal{C} = (2\pi)^{1/2} A t_1$

$\quad = \tfrac{1}{2}(\pi/\ln2)^{1/2} A \Delta t_{6dB}$

Use curve on p. 42-4 with standard deviation

$\sigma = f_1 = 1/2\pi t_1$

$\quad = (2\ln2)^{1/2}/\pi\Delta t_{6dB}; \quad f_{3dB} = 2^{1/2}\ln2/\pi\Delta t_{6dB}$

$G(f) = \mathcal{C} \exp[-\tfrac{1}{2}(f/f_1)^2]$

$\quad = (\ln2/2\pi)^{1/2}(A/f_{3dB})$

$\qquad\qquad \times \exp[-\tfrac{1}{2}(\ln2)(f/f_{3dB})^2]$

$\quad = \tfrac{1}{2}(\pi/\ln2)^{1/2} A \Delta t_{6dB}$

$\qquad\qquad \times \exp[-(\pi^2/4\ln2)(f\Delta t_{6dB})^2]$

TABLE 2—*Continued*

Time Function	Frequency Function

H. Critically damped exponential pulse

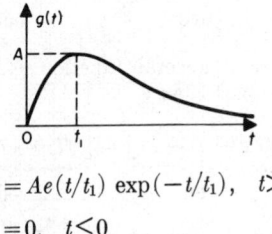

$$g(t) = Ae(t/t_1) \exp(-t/t_1), \quad t>0$$

$$= 0, \quad t \leq 0$$

$$e = 2.71828\cdots$$

Area $\mathcal{Q} = Aet_1$

$$G(f) = \mathcal{Q}[1/(1+j\alpha)^2]$$

where $\alpha = 2\pi t_1 f$

---

* For an extensive tabulation of the Fourier transform and its inverse, see G. A. Campbell and R. M. Foster, "Fourier Integrals for Practical Applications," D. Van Nostrand Co., Inc., New York; 1948. See also "Tables of Integral Transforms," vol. 1, Bateman Manuscript Project, editor A. Erdélyi, McGraw-Hill Book Co., New York; 1954.

FOURIER SERIES

Real Form of Fourier Series

For a periodic function with period 2π, defined by its values in the interval $-\pi$ to $+\pi$ or 0 to 2π (as illustrated in Fig. 1)

$$f(x) = \tfrac{1}{2}A_0 + \sum_{n=1}^{n=\infty} (A_n \cos nx + B_n \sin nx)$$

x in radians

$$= \tfrac{1}{2}C_0 + \sum_{n=1}^{n=\infty} C_n \cos(nx + \phi_n)$$

where

$$C_0 = A_0$$

$$C_n = (A_n{}^2 + B_n{}^2)^{1/2}$$

$$\cos\phi_n = A_n/C_n$$

$$\sin\phi_n = -B_n/C_n.$$

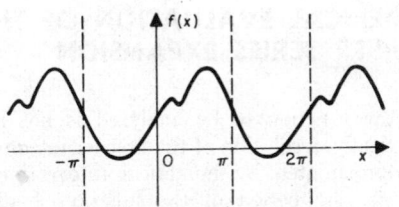

Fig. 1—Periodic wave.

The coefficients A_0, A_n, and B_n are determined by

$$A_n = \pi^{-1} \int_{-\pi}^{\pi} f(x) \cos nx \, dx = \pi^{-1} \int_0^{2\pi} f(x) \cos nx \, dx$$

$$B_n = \pi^{-1} \int_{-\pi}^{\pi} f(x) \sin nx \, dx = \pi^{-1} \int_0^{2\pi} f(x) \sin nx \, dx$$

for $n = 0, 1, 2, \cdots$.

Arbitrary Period

For a periodic function with period T, defined by its values in the intervals $-T/2$ to $+T/2$ or from 0 to T instead of from $-\pi$ to $+\pi$ or 0 to 2π, the Fourier expansion is given by

$$f(x) = \tfrac{1}{2}A_0$$

$$+ \sum_{n=1}^{n=\infty} [A_n \cos 2n(\pi/T)x + B_n \sin 2n(\pi/T)x]$$

and the coefficients by

$$A_n = (2/T) \int_{-T/2}^{T/2} f(x) \cos(2n\pi x/T) \, dx$$

$$= (2/T) \int_0^T f(x) \cos(2n\pi x/T) \, dx$$

$$B_n = (2/T) \int_{-T/2}^{T/2} f(x) \sin(2n\pi x/T) \, dx$$

$$= (2/T) \int_0^T f(x) \sin(2n\pi x/T) \, dx$$

for $n = 0, 1, 2, \cdots$.

Complex Form of Fourier Series

For Functions with Period 2π

$$f(x) = \sum_{n=-\infty}^{n=+\infty} D_n \exp(jnx)$$

where

$$D_n = (2\pi)^{-1} \int_{-\pi}^{+\pi} f(x) \exp(-jnx) dx$$

and n takes on all positive and negative integral values including zero.

For real functions

$$D_n = \tfrac{1}{2}(A_n - jB_n) = \tfrac{1}{2}C_n \exp(j\phi_n)$$

$$D_{-n} = \tfrac{1}{2}(A_n + jB_n) = \tfrac{1}{2}C_n \exp(-j\phi_n) = D_n{}^*$$

$$D_0 = \tfrac{1}{2}A_0 = \tfrac{1}{2}C_0.$$

For Functions with Period T

$$f(x) = \sum_{n=-\infty}^{n=+\infty} D_n \exp[j(2n\pi x/T)]$$

$$D_n = T^{-1} \int_0^T f(x) \exp[-j(2n\pi x/T)] dx.$$

Average Power

The average power of the periodic waveform $f(x)$ is

$$T^{-1} \int_0^T |f(x)|^2 \, dx = \sum_{n=-\infty}^{n=+\infty} |D_n|^2$$

$$= \tfrac{1}{4}C_0{}^2 + \tfrac{1}{2}\sum_{n=1}^{n=\infty} C_n{}^2$$

$$= \tfrac{1}{4}A_0{}^2 + \tfrac{1}{2}\sum_{n=1}^{n=\infty}(A_n{}^2 + B_n{}^2).$$

Odd and Even Functions

If $f(x)$ is an odd function, i.e.,

$$f(x) = -f(-x)$$

then all the coefficients of the cosine terms (A_n) vanish and the Fourier series consists of sine terms alone.

If $f(x)$ is an even function, i.e.,

$$f(x) = f(-x)$$

then all the coefficients of the sine terms (B_n) vanish and the Fourier series consists of cosine terms alone, and a possible constant.

The Fourier expansions of functions in general include both cosine and sine terms. Every function capable of Fourier expansion consists of the sum of an even and an odd part.

$$f(x) = \tfrac{1}{2}A_0 + \underbrace{\sum_{n=1}^{n=\infty} A_n \cos nx}_{\text{even}} + \underbrace{\sum_{n=1}^{n=\infty} B_n \sin nx}_{\text{odd}}.$$

To separate a general function $f(x)$ into its odd and even parts, use

$$f(x) \equiv \underbrace{\tfrac{1}{2}[f(x) + f(-x)]}_{\text{even}} + \underbrace{\tfrac{1}{2}[f(x) - f(-x)]}_{\text{odd}}.$$

In some cases by suitable selection of the origin, the function may be made either odd or even, thus simplifying the expansion.

Odd or Even Harmonics

An odd or even function may contain odd or even harmonics. A condition that causes a function $f(x)$ of period 2π to have only odd harmonics in its Fourier expansion is

$$f(x) = -f(x+\pi).$$

A condition that causes a function $f(x)$ of period 2π to have only even harmonics in the Fourier expansion is

$$f(x) = f(x+\pi).$$

To separate a general function $f(x)$ into its odd and even harmonics, use

$$f(x) \equiv \underbrace{\tfrac{1}{2}[f(x) + f(x+\pi)]}_{\text{even harmonics}} + \underbrace{\tfrac{1}{2}[f(x) - f(x+\pi)]}_{\text{odd harmonics}}.$$

A periodic function may sometimes be changed from odd to even (and vice versa) by a shift of the origin, but the presence of particular odd or even harmonics is unchanged by such a shift.

NUMERICAL EVALUATION OF THE FOURIER SERIES EXPANSION

If the function to be analyzed is not known analytically, a solution of the Fourier integral may be approximated by numerical integration. For instance, the period of the function is divided into 12 equal parts as indicated by Fig. 2.

The values of the ordinates at these 12 points are recorded and the following computations made.

	Y_0	Y_1	Y_2	Y_3	Y_4	Y_5	Y_6
		Y_{11}	Y_{10}	Y_9	Y_8	Y_7	
Sum	S_0	S_1	S_2	S_3	S_4	S_5	S_6
Difference		d_1	d_2	d_3	d_4	d_5	

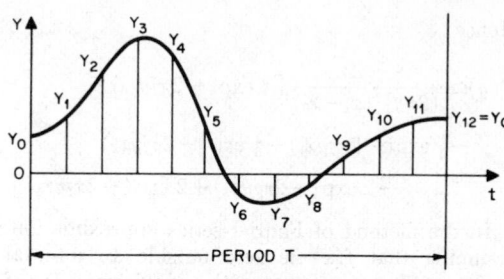

Fig. 2—Division of the period of the function for numerical solution.

The sum terms are arranged as follows.

	S_0	S_1	S_2	S_3		$\bar{S}_0$	$\bar{S}_1$
	S_6	S_5	S_4			$\bar{S}_2$	$\bar{S}_3$
Sum	$\bar{S}_0$	$\bar{S}_1$	$\bar{S}_2$	$\bar{S}_3$		$\bar{S}_7$	$\bar{S}_8$
Difference	D_0	D_1	D_2				

The difference terms are as follows.

	d_1	d_2	d_3		
	d_5	d_4		$\bar{S}_4$	D_0
Sum	$\bar{S}_4$	$\bar{S}_5$	$\bar{S}_6$	$\bar{S}_6$	D_2
Difference	D_3	D_4		D_5	D_6

The coefficients of the Fourier series are now obtained as follows, where $\frac{1}{2}A_0$ equals the average value, the $A_1 \cdots A_n$ expressions represent the coefficients of the cosine terms, and the $B_1 \cdots B_n$ expressions represent the coefficients of the sine terms.

$$\frac{1}{2}A_0 = \frac{\bar{S}_7 + \bar{S}_8}{12}$$

$$A_1 = \frac{D_0 + 0.866 D_1 + 0.5 D_2}{6}$$

$$A_2 = \frac{\bar{S}_0 + 0.5\bar{S}_1 - 0.5\bar{S}_2 - \bar{S}_3}{6}$$

$$A_3 = D_6/6$$

$$A_4 = \frac{\bar{S}_0 - 0.5\bar{S}_1 - 0.5\bar{S}_2 + \bar{S}_3}{6}$$

$$A_5 = \frac{D_0 - 0.866 D_1 + 0.5 D_2}{6}$$

$$A_6 = \frac{\bar{S}_7 - \bar{S}_8}{12}.$$

Also

$$B_1 = \frac{0.5\bar{S}_4 + 0.866\bar{S}_5 + \bar{S}_6}{6}$$

$$B_2 = \frac{0.866(D_3 + D_4)}{6}$$

$$B_3 = D_5/6$$

$$B_4 = \frac{0.866(D_3 - D_4)}{6}$$

$$B_5 = \frac{0.5\bar{S}_4 - 0.866\bar{S}_5 + \bar{S}_6}{6}.$$

NUMERICAL EVALUATION OF THE FOURIER TRANSFORM

The Fourier transform of a function $f(x)$ may be computed numerically by a polynomial approximation or by a Fourier-series evaluation.

In the first method, the function $f(x)$ is approximated by a polygon $p(x)$; indicating with x_i the values of x in correspondence of the vertices of the polygon, one differentiates twice $p(x)$, obtaining a series of pulses of area σ_i

$$d^2 p/dx^2 = \sum \sigma_i \delta(x - x_i).$$

Hence

$$P(y) \simeq F(y) = \frac{-1}{(2\pi y)^2} \sum \sigma_i \exp(-2\pi j y x_i).$$

An example is given by the function shown in Fig. 3, where the magnitudes of the pulse areas are, respectively

$$\sigma_0 = 2A/3(x_1 - x_0)$$

$$\sigma_1 = \sigma_2 = -A/3(x_1 - x_0)$$

$$\sigma_3 = -\sigma_4 = -2A/(x_1 - x_0).$$

Hence

$$F(y) \simeq \frac{-A}{(2\pi y)^2 (x_1 - x_0)} \left[\tfrac{2}{3} \exp(-2\pi jyx_0) \right.$$

$$-\tfrac{1}{3} \exp(-2\pi jyx_1) - \tfrac{1}{3} \exp(-2\pi jyx_2)$$

$$\left. -2 \exp(-2\pi jyx_3) + 2 \exp(-2\pi jyx_4) \right].$$

In the method of Fourier-series approximation, assuming that $f(x)$ is zero outside an interval $-T < x < T$, one evaluates the coefficients D_n of the expansion in complex form; i.e.,

$$f(x) = \sum_{n=-\infty}^{n=+\infty} D_n \exp(jn\pi x/T).$$

In this case the Fourier transform may be expressed as

$$F(y) = \sum_{n=-\infty}^{n=+\infty} 2TD_n \frac{\sin(2\pi yT - n\pi)}{2\pi yT - n\pi}.$$

PULSE-TRAIN ANALYSIS

If the pulse defined by the function $g(t)$ is repeated every interval T, a periodic waveform

$$y(t) = \sum_{n=-\infty}^{n=+\infty} g(t - nT)$$

results with period T and repetition frequency $F = 1/T$ (see Table 3A and 3B).

This pulse train may be expressed as a convolution product

$$y(t) = \left[\sum_{n=-\infty}^{n=+\infty} \delta(t - nT) \right] g(t)$$

and, applying properties 8 and 10 (Table 1), its Fourier transform is

$$Y(f) = (1/T) \left[\sum_{n=-\infty}^{n=+\infty} \delta(f - nF) \right] G(f).$$

TABLE 3—THE SPECTRUM FOR PULSE TRAINS. SPECTRA ARE IN GENERAL COMPLEX FUNCTIONS. THEY ARE REPRESENTED HERE BY REAL CURVES ONLY TO SIMPLIFY THE ILLUSTRATION.

Waveform	Spectrum
A. Single pulse	
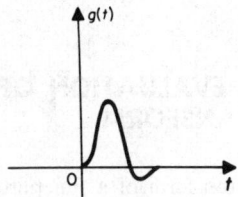	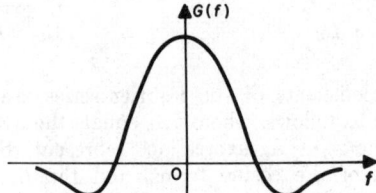
B. Infinite periodic pulse train	
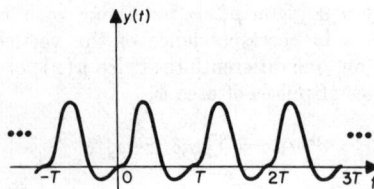	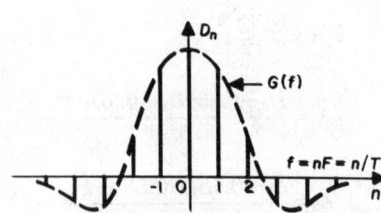
C. Limited pulse train	
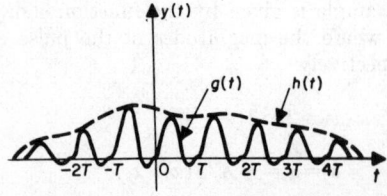	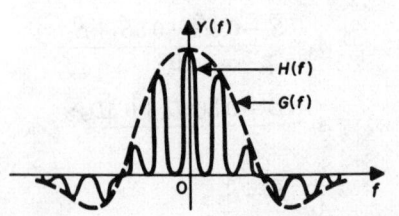

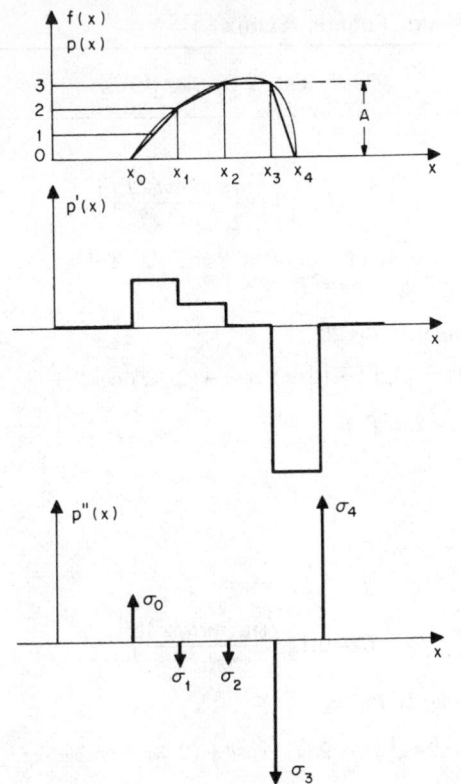

Fig. 3—Polynomial approximation for evolution of the Fourier transform.

The function $y(t)$ is represented by the Fourier series

$$y(t) = \sum_{n=-\infty}^{n=+\infty} D_n \exp(jnt)$$

where

$$D_n = (1/T) G(nF).$$

The coefficients D_n are obtained by sampling the pulse spectrum at frequencies that are multiples of the repetition frequency.

The amplitude C_n of the nth harmonic in the real representation (see p. 44-7) is

$$C_n = 2 | D_n | = (2/T) | G(nF) |.$$

By a translation τ of the time origin, the D_n are multiplied by the factor $\exp(-2\pi jn\tau/T)$; the C_n are not changed.

The constant term of the series

$$D_0 = A_0/2 = C_0/2$$

is the average amplitude

$$A_{av} = \mathcal{Q}/T = G(0)/T$$

where

$$\mathcal{Q} = \int_0^T g(t)\, dt$$

is the area under one pulse.

If the pulses do not overlap (i.e., if the function $g(t)$ is zero outside of some period a to $a+T$) the energy in a pulse is

$$E = \int_a^{a+T} g^2(t)\, dt = \int_{-\infty}^{+\infty} | G(f) |^2\, df.$$

The root-mean-square amplitude is

$$A_{rms} = (E/T)^{1/2}.$$

The average power of the pulse train is

$$E/T = A_{rms}^2 = \sum_{n=-\infty}^{n=+\infty} | D_n |^2 = \tfrac{1}{4} C_0^2 + \tfrac{1}{2} \sum_{n=1}^{n=+\infty} C_n^2.$$

A pulse train of finite extent, where all the pulses have the same shape and are spaced periodically, may be represented as a product

$$y(t) = h(t) \sum_{n=-\infty}^{n=+\infty} g(t-nT).$$

The function $h(t)$ defines the envelope of the pulse train.

The Fourier transform

$$Y(f) = (1/T) G(f) \sum_{n=-\infty}^{n=+\infty} H(f-nF)$$

may be interpreted, in the frequency domain, as a train of pulses having $G(f)$ as an envelope and a form defined by $H(f)$. See Table 3C.

When $h(t) = 1$, then $H(f)$ is the δ function. The pulse train is a periodic waveform having a line spectrum as explained above. See Table 3B.

The Fourier series coefficients for a number of commonly encountered pulse trains are given in Table 4.

When the pulse train is derived from a pulse listed in Table 4, the coefficients can also be read off the corresponding spectrum curve by sampling at values n/T of the frequency.

TABLE 4—PERIODIC WAVEFORMS AND FOURIER SERIES.

Waveform	Coefficient of Fourier Series

A. Rectangular wave

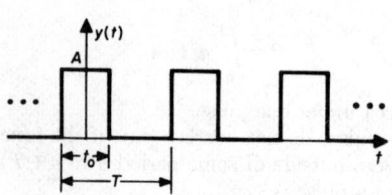

Derived from rectangular pulse, Table 2A

$$A_{\text{Av}} = A\,(t_0/T) \qquad A_{\text{rms}} = A\,(t_0/T)^{1/2}$$

$$C_n = 2D_n = 2A_{\text{Av}} \left| \frac{\sin\,(n\pi t_0/T)}{n\pi t_0/T} \right|$$

Can be read off curve of $(\sin x)/x$, Table 2A, by sampling at $n\pi t_0/T$

Example: If $T = 2t_0$

$$y(t) = 2A_{\text{Av}}[\tfrac{1}{2} + (2/\pi)\,\cos\theta - (2/3\pi)\,\cos3\theta + \cdots]$$

with $\theta = 2\pi t/T$

B. Isosceles-triangle wave

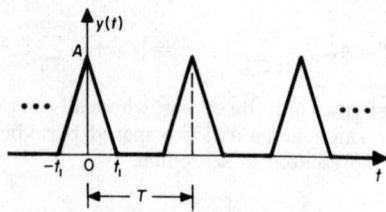

Derived from triangular pulse, Table 2B

$$A_{\text{Av}} = A\,(t_1/T) \qquad A_{\text{rms}} = A\,(2t_1/3T)^{1/2}$$

$$C_n = 2A_{\text{Av}} \left(\frac{\sin\,(n\pi t_1/T)}{n\pi t_1/T} \right)^2$$

Example: If $T = 2t_1$

$$y(t) = 2A_{\text{Av}}[\tfrac{1}{2} + (2/\pi)^2\,\cos\theta + (2/3\pi)^2\,\cos3\theta + \cdots]$$

with $\theta = 2\pi t/T$

C. Sawtooth wave

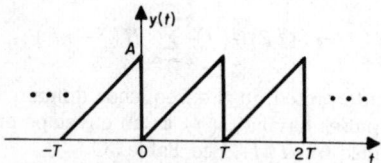

Derived from triangular pulse, Table 2C

$$A_{\text{Av}} = A/2 \qquad A_{\text{rms}} = A\,(3^{-1/2})$$

$$C_n = 2A_{\text{Av}}\,(1/\pi n)$$

$$y(t) = 2A_{\text{Av}}[\tfrac{1}{2} - (1/\pi)\,\sin\theta - (1/2\pi)\,\sin2\theta - \cdots]$$

D. Clipped sawtooth wave

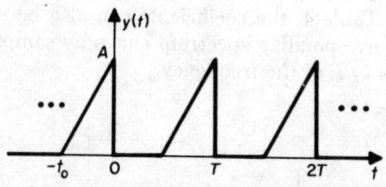

Derived from triangular pulse, Table 2C

$$A_{\text{Av}} = A\,(t_0/2T) \qquad A_{\text{rms}} = A\,(t_0/3T)^{1/2}$$

$$C_n = 2A_{\text{Av}}\,(1/\alpha^2)[\sin^2\alpha + \alpha(\alpha - \sin2\alpha)]^{1/2}$$

with $\alpha = n\pi t_0/T$

TABLE 4—Continued

Waveform	Coefficient of Fourier Series

E. Sawtooth wave

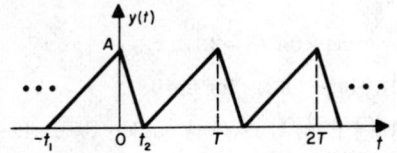

Derived from the sum of two triangular pulses, Table 2C

$$A_{Av}= A/2 \qquad A_{rms}= A\,(3^{-1/2})$$

$$C_n = 2A_{Av}\,(T^2/\pi^2 n^2 t_1 t_2)\,|\sin{(n\pi t_1/T)}|$$

with $t_1 + t_2 = T$

F. Symmetrical trapezoidal wave

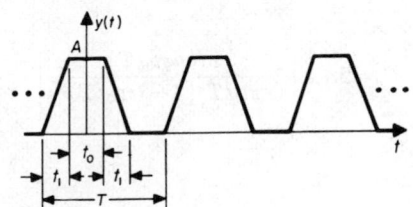

Derived as in Table 2D

$$A_{Av}= A[(t_0+t_1)/T]$$

$$A_{rms}= A[(3t_0+2t_1)/3T]^{1/2}$$

$$D_n = A_{Av}\,\frac{\sin{(\pi n t_1/T)}}{\pi n t_1/T}\,\frac{\sin[\pi n(t_1+t_0)/T]}{\pi n(t_1+t_0)/T}$$

$$C_n = 2\,|D_n|$$

G. Train of cosine pulses

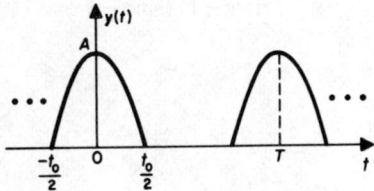

Derived from cosine pulse, Table 2E

$$A_{Av}= (2/\pi)\,A\,(t_0/T) \qquad A_{rms}= A\,(t_0/2T)^{1/2}$$

$$C_n = 2A_{Av}\left|\frac{\cos{(n\pi t_0/T)}}{1-(2nt_0/T)^2}\right|$$

For $nt_0/T = 1/2$, this becomes $\pi A_{Av}/2$

H. Full-wave-rectified sine wave

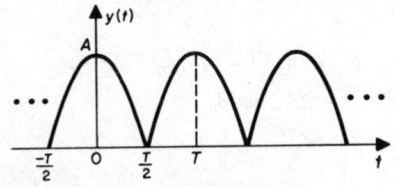

Derived from cosine pulse, Table 2E (same as Table 4G with $t_0 = T$)

$$A_{Av}= (2/\pi)\,A \qquad A_{rms}= A/(2^{1/2})$$

$$C_0 = 2A_{Av}$$

$$C_n = 2A_{Av}(4n^2-1)^{-1}, \quad \text{for } n\neq 0$$

$$y(t) = 2A_{Av}[\tfrac{1}{2}+\tfrac{1}{3}\cos\theta-\tfrac{1}{15}\cos2\theta+\tfrac{1}{35}\cos3\theta\cdots$$
$$-(-1)^n(4n^2-1)^{-1}\cos n\theta\cdots]$$

with $\theta = 2\pi t/T$

TABLE 4—*Continued*

Waveform	Coefficient of Fourier Series

I. Half-wave-rectified sine wave

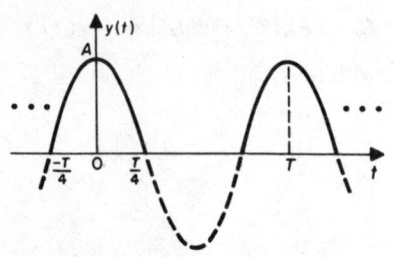

Derived from cosine pulse, Table 2E (same as Table 4G with $t_0 = T/2$)

$$A_{Av} = (1/\pi) A \qquad A_{rms} = A/2$$

$$C_0 = 2A_{Av}$$

$$C_{2n+1} = 0, \quad \text{except for } C_1 = 2A_{Av}(\pi/4)$$

$$C_{2n} = 2A_{Av}(4n^2 - 1)^{-1}, \quad \text{for } n \neq 0$$

$$y(t) = 2A_{Av}\left[\tfrac{1}{2} + (\pi/4)\cos\theta + \tfrac{1}{3}\cos 2\theta - \tfrac{1}{15}\cos 4\theta + \cdots\right.$$
$$\left. -(-1)^n (4n^2 - 1)^{-1}\cos 2n\theta \cdots\right]$$

J. Train of cosine-squared pulses

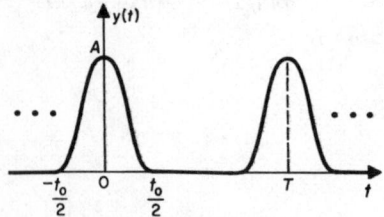

Derived from cosine-squared pulse, Table 2F

$$A_{Av} = \tfrac{1}{2}A(t_0/T) \qquad A_{rms} = \tfrac{1}{2}A(3t_0/2T)^{1/2}$$

$$C_n = 2A_{Av}\left| \frac{\sin(n\pi t_0/T)}{(n\pi t_0/T)[1 - (nt_0/T)^2]} \right|$$

K. Fractional sine wave

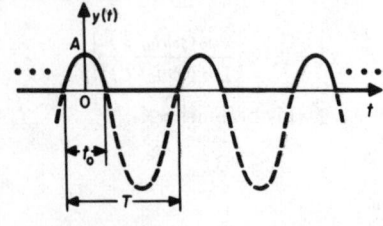

$$A_{Av} = \frac{A}{\pi}\frac{\sin\alpha - \alpha\cos\alpha}{1 - \cos\alpha}$$

$$A_{rms} = \frac{A}{(2\pi)^{1/2}}\frac{[2\alpha + \alpha\cos 2\alpha - (3/2)\sin 2\alpha]^{1/2}}{1 - \cos\alpha}$$

with $\alpha = \pi t_0/T$

$$C_n = 2A_{Av}\left| \frac{\sin n\alpha \cos\alpha - n\sin\alpha \cos n\alpha}{n(n^2 - 1)(\sin\alpha - \alpha\cos\alpha)} \right|$$

L. Critically damped exponential wave

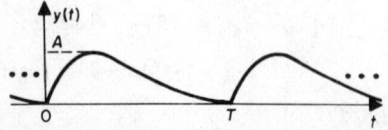

Derived from exponential pulse, Table 2H (period $T \gg$ period t_1 to make overlap negligible)

$$A_{Av} = Ae(t_1/T) \qquad A_{rms} = (Ae/2)(t_1/T)^{1/2}$$

$$e = 2.71828\cdots$$

$$C_n = 2A_{Av}[1 + (2\pi nt_1/T)^2]^{-1}$$
$$= 2A_{Av}\cos^2\theta_n$$

with $\tan\theta_n = 2\pi nt_1/T$

CHAPTER 45
MAXWELL'S EQUATIONS

GENERAL*

The following four basic laws of electro-magnetism for bodies at rest are derived from the fundamental, experimental, and theoretical work of Ampère and Faraday, and are valid for quantities determined by their average values in volumes that contain a very great number of molecules (macroscopic electromagnetism).

STATEMENT OF FOUR BASIC LAWS (RATIONALIZED MKS UNITS)

(A) The work required to carry a unit magnetic pole around a closed path is equal to the total current linking that path, that is, the total current passing through any surface that has the path for its periphery. This total current is the sum of the conduction current and the displacement current, the latter being equal to the derivative with respect to time of the electric flux ϕ_D passing through any surface that has the above closed path for its periphery.

(B) The electromotive force (emf) induced in any fixed closed loop is equal to minus the time rate of change of the magnetic flux ϕ_B through that loop. By electromotive force is meant the work required to carry a unit positive charge around the loop.

(C) The total electric flux diverging from a charge Q is equal to Q in magnitude.

(D) Magnetic flux lines are continuous (closed) loops. There are no sources or sinks of magnetic flux.

EXPRESSION OF BASIC LAWS IN INTEGRAL FORM

(A) $\quad \oint \mathbf{H} \cdot \mathbf{dl} = I_{\text{total}} = I_{\text{conduction}} + \dfrac{\partial \phi_D}{\partial t}$

* Developed from: J. E. Hill, "Maxwell's Four Basic Equations," *Westinghouse Engineer*, vol. 6, p. 135; September 1946.

where $\oint$ = a line integral around a closed path, $\mathbf{dl}$ = vector element of length along path, $\mathbf{H}$ = magnetic field vector, and ϕ_D = electric flux linking the path. The time rate of change of ϕ_D is written as a partial derivative to indicate that the loop does not move (the coordinates of each point of the loop remain fixed during integration).

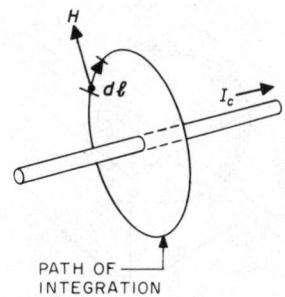

PATH OF INTEGRATION

(B) $\quad \oint \mathbf{E} \cdot \mathbf{dl} = -\dfrac{\partial \phi_B}{\partial t}$

where $\mathbf{E}$ = electric field vector and ϕ_B = magnetic flux linking the path of integration.

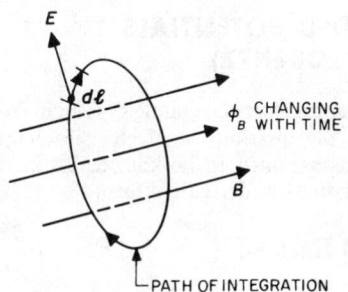

PATH OF INTEGRATION

(C) $\quad \displaystyle\int_S \mathbf{D} \cdot \mathbf{dS} = Q$

where S = any closed surface, $\mathbf{dS}$ = vector element of S, directed outward and normal to the element of surface, $\mathbf{D}=$ vector electric flux density, and $Q=$ the net electric charge within S.

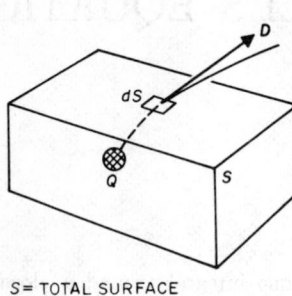

S = TOTAL SURFACE
Q = TOTAL CHARGE INSIDE S

$$(\mathbf{D}) \quad \int_S \mathbf{B} \cdot d\mathbf{S} = 0$$

where $\mathbf{B}=$ vector magnetic flux density.

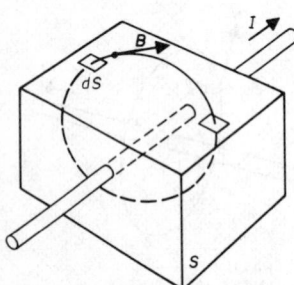

B LINES ARE CLOSED CURVES; AS MANY ENTER REGION AS LEAVE IT.

BASIC LAWS IN DERIVATIVE FORM (PAGE 45-4)

RETARDED POTENTIALS (H. A. LORENTZ)

Consider an electromagnetic system in free space in which the distribution of electric charges and currents is assumed to be known. From the four basic equations in derivative form

$$\text{curl } \mathbf{H} = \mathbf{j}_c + \epsilon_0 \frac{\partial \mathbf{E}}{\partial t} \qquad \text{curl } \mathbf{E} = -\mu_0 \frac{\partial \mathbf{H}}{\partial t}$$

$$\text{div } \mathbf{H} = 0 \qquad \text{div } \mathbf{E} = \frac{\rho}{\epsilon_0}$$

two retarded potentials can be determined:

one scalar,

$$\phi = \frac{1}{4\pi\epsilon_0} \int \frac{\rho^\dagger dV}{r}$$

one vector,

$$\mathbf{A} = \frac{\mu_0}{4\pi} \int \frac{\mathbf{j}_c^\dagger dV}{r} .$$

The integrals extend over the volume containing the charges and currents. The daggers mean that the values of the quantities are taken at time $t-r/c$, where r is the distance from the location of the charge or current to the point P considered, and $c=1/\sqrt{\epsilon_0\mu_0}=$ speed of light in free space.

The electric and magnetic fields at point P are expressed by

$$\mathbf{H} = \frac{1}{\mu_0} \text{curl } \mathbf{A}$$

$$\mathbf{E} = -\text{grad } \phi - \frac{\partial \mathbf{A}}{\partial t} .$$

FIELDS IN TERMS OF ONE VECTOR ONLY (HERTZ VECTOR)

The previous expressions imply a relation between ϕ and $\mathbf{A}$:

$$\text{div } \mathbf{A} = -\epsilon_0\mu_0 \frac{\partial \phi}{\partial t} .$$

Consider a vector $\mathbf{\Pi}$ such that $\mathbf{A} = \epsilon_0\mu_0 \, \partial\mathbf{\Pi}/\partial t$. Then for all variable fields

$$\phi = -\text{div } \mathbf{\Pi}.$$

The electric and magnetic fields can thus be expressed in terms of the vector $\mathbf{\Pi}$ only:

$$\mathbf{H} = \epsilon_0 \text{curl } \frac{\partial \mathbf{\Pi}}{\partial t}$$

$$\mathbf{E} = \text{grad div } \mathbf{\Pi} - \epsilon_0\mu_0 \frac{\partial^2 \mathbf{\Pi}}{\partial t^2} .$$

POYNTING VECTOR

Consider any volume V of the previous electromagnetic system enclosed in a surface S. It can be shown that

$$-\int_V \mathbf{E} \cdot \mathbf{j}_c dV = \frac{\partial}{\partial t} \int_V (\tfrac{1}{2}\epsilon_0 E^2 + \tfrac{1}{2}\mu_0 H^2) \, dV$$

$$+ \int_S (\mathbf{E} \times \mathbf{H}) \cdot d\mathbf{S}$$

where E and H are the magnitudes of the electric and magnetic field vectors, respectively, and $d\mathbf{S}$ is a vector element of S, directed outward from V and normal to the element of surface. The current densities j_c are delivered by sources contained in V. The interpretation of the equation is

(power delivered by sources in V) = (rate of

increase of energy stored in V) + (power

flow through S away from V).

From this equation and its interpretation, the densities of electric and magnetic stored energy (in joules per cubic meter) are taken to be $\frac{1}{2}\epsilon_0 E^2$ and $\frac{1}{2}\mu_0 H^2$, respectively. The vector product $\mathbf{E} \times \mathbf{H}$ is called the Poynting vector and is interpreted as the density of power flow (in watts per square meter) at a point.

In the single-frequency case, a complex Poynting vector $\frac{1}{2}\mathbf{E} \times \mathbf{H}^*$ is often used. Its real and imaginary parts are the densities of real and reactive power flow, respectively. The components of $\mathbf{E}$ and $\mathbf{H}$ are the complex amplitudes of the corresponding field components, and the asterisk means that the complex conjugate is to be taken. It can be shown that

$$-\int_V \frac{1}{2}(\mathbf{E}\cdot\mathbf{j}_c{}^*)\,dV$$

$$=j2\omega\int_V\left[\frac{1}{4}\mu_0(HH^*)-\frac{1}{4}\epsilon_0(EE^*)\right]dV$$

$$+\int_S \frac{1}{2}(\mathbf{E}\times\mathbf{H}^*)\cdot d\mathbf{S}.$$

This shows that in case there are no sources inside V and no net complex power leaves V, then the mean electric and magnetic energies stored inside V are equal.

SUPERPOSITION

The electromagnetic status of a system can be described by its fields and flux densities. When all of the equations that determine the status are linear, superposition applies to the fields and to the flux, current, and charge densities, since all of these quantities are then linearly interrelated. If $(\mathbf{E}, \mathbf{H}, \mathbf{D}, \mathbf{B}, \mathbf{j}_c, \rho)$ represents one set of sources and responses and $(\mathbf{E}', \mathbf{H}', \mathbf{D}', \mathbf{B}', \mathbf{j}_c', \rho')$ another set, the set resulting when the two sets of sources are applied together is $(\mathbf{E}+\mathbf{E}', \mathbf{H}+\mathbf{H}', \mathbf{D}+\mathbf{D}', \mathbf{B}+\mathbf{B}', \mathbf{j}_c+\mathbf{j}_c', \rho+\rho')$.

The equations that determine electromagnetic status consist of Maxwell's equations and the relations between the fields and the flux densities that are imposed by the materials and the free space of the system. Maxwell's equations themselves are linear, as are the free-space relations. Superposition applies, then, if the materials in the

system are linear; it can apply if the materials are inhomogeneous, anisotropic, or time-varying; still more complicated material relations are admissible; but superposition does not apply if one of the material relations is nonlinear. A linear relation between cause and effect is one for which multiplying cause by any number x produces x times the effect.

It should be noted that superposition does *not* apply to the material elements of a system; when one source is turned off so that the response to another source alone can be determined, the material elements of the dead source must remain in position. The source consisting of a paraboloidal dish, feed horn, feeder, and transmitter is turned off by disconnecting the transmitter from the feeder and terminating the feeder in the incremental output impedance of the transmitter; the dish and feed horn must remain in position. A source consisting of specified current and charge densities satisfying the law of conservation of electric charges has no material elements; when it is turned off, only the medium containing it remains. On the other hand, it is often useful to specify as a source the tangential electric or magnetic field on a surface. These sources have significant material elements that must be reckoned with when the sources are turned off. When the electric field source is turned off, the tangential electric field must vanish on the surface, regardless of the presence of other sources: The surface becomes a perfect short-circuit. It is convenient to regard the surface as containing a sheet of dielectric material of infinite permittivity, an "electric wall". When the magnetic field source is turned off, vanishing of the tangential magnetic field requires the surface to behave as a perfect open-circuit. The surface may be regarded as containing a sheet of magnetic material of infinite permeability, a "magnetic wall". Either wall totally reflects electromagnetic waves incident on it.

RECIPROCITY

A reciprocal multiport system is one for which the response at port n to an excitation at port m agrees with the response at port m to the same excitation at port n. The source and measuring device must have equal incremental impedances. Although certain nonlinear and time-varying systems may be reciprocal in accordance with this definition, it is usual to confine consideration of reciprocity to linear, time-invariant systems.

For an electromagnetic system for which ports have not been defined, let $\mathbf{E}$ and $\mathbf{H}$ be the fields in a volume V that result from the application of one set of sources external to V, and let $\mathbf{E}'$ and $\mathbf{H}'$ be the fields in V resulting from another set of external sources. If the portion of the system within V is

General Form	Static Case	Steady-State

(A)　$\left.\begin{array}{l}\text{curl }\mathbf{H}\\ \nabla\times\mathbf{H}\end{array}\right\}=\mathbf{j}_c+\dfrac{\partial\mathbf{D}}{\partial t}$　　　　$\left.\begin{array}{l}\text{curl }\mathbf{H}\\ \nabla\times\mathbf{H}\end{array}\right\}=0$　　　　$\left.\begin{array}{l}\text{curl }\mathbf{H}\\ \nabla\times\mathbf{H}\end{array}\right\}=\mathbf{j}_c$

$\mathbf{j}_c=0$

$\mathbf{j}_c=$ conduction current density　　　　$\dfrac{\partial\mathbf{D}}{\partial t}=0$　　Conduction current exists but time derivatives are zero

(B)　$\left.\begin{array}{l}\text{curl }\mathbf{E}\\ \nabla\times\mathbf{E}\end{array}\right\}=-\dfrac{\partial\mathbf{B}}{\partial t}$　　　　$\left.\begin{array}{l}\text{curl }\mathbf{E}\\ \nabla\times\mathbf{E}\end{array}\right\}=0$　　　　$\left.\begin{array}{l}\text{curl }\mathbf{E}\\ \nabla\times\mathbf{E}\end{array}\right\}=0$

(C)　$\left.\begin{array}{l}\text{div }\mathbf{D}\\ \nabla\cdot\mathbf{D}\end{array}\right\}=\rho$　　　　$\left.\begin{array}{l}\text{div }\mathbf{D}\\ \nabla\cdot\mathbf{D}\end{array}\right\}=\rho$　　　　$\left.\begin{array}{l}\text{div }\mathbf{D}\\ \nabla\cdot\mathbf{D}\end{array}\right\}=\rho$

$\rho=$ charge density

$=$ charge per unit volume

(D)　$\left.\begin{array}{l}\text{div }\mathbf{B}\\ \nabla\cdot\mathbf{B}\end{array}\right\}=0$　　　　$\left.\begin{array}{l}\text{div }\mathbf{B}\\ \nabla\cdot\mathbf{B}\end{array}\right\}=0$　　　　$\left.\begin{array}{l}\text{div }\mathbf{B}\\ \nabla\cdot\mathbf{B}\end{array}\right\}=0$

Notes:

The time dependence $e^{+j\omega t}$ has been used in the free-space single-frequency column.

For an explanation of the operator ∇ (del) and the associated vector operations, refer to the vector-analysis equations in the chapter "Mathematical Equations."

$\left.\begin{array}{l}\mu_0=4\pi\times10^{-7}\text{ henry/meter}\\[4pt] \epsilon_0=\dfrac{10^7}{4\pi c^2}\approx\dfrac{10^{-9}}{36\pi}\text{ farad/meter}\\[6pt] c=\text{speed of light in free space}\\[4pt] =\dfrac{1}{\sqrt{\epsilon_0\mu_0}}\approx3\times10^8\text{ meter/second}\end{array}\right\}$ in the rationalized meter-kilogram-second (MKS) system of units.

Maxwell's equations result in the law of conservation of electric charges, the integral form of which is

$$I=-\frac{\partial Q_i}{\partial t}$$

$Q_i=$ net sum of all electric charges within a closed surface S

$I=$ outgoing conduction current

and the derivative form

$$\text{div }\mathbf{j}_c=-\frac{\partial\rho}{\partial t}$$

DERIVATIVE FORM

Quasi-Steady-State	Free-Space	Free-Space Single-Frequency
$\left.\begin{array}{c}\text{curl } \mathbf{H} \\ \nabla \times \mathbf{H}\end{array}\right\} \approx \mathbf{j}_c$	$\left.\begin{array}{c}\text{curl } \mathbf{H} \\ \nabla \times \mathbf{H}\end{array}\right\} = \dfrac{\partial \mathbf{D}}{\partial t}$ $= \epsilon_0 \dfrac{\partial \mathbf{E}}{\partial t}$	$\left.\begin{array}{c}\text{curl } \mathbf{H} \\ \nabla \times \mathbf{H}\end{array}\right\} = j\omega\epsilon_0 \mathbf{E}$
$\partial \mathbf{D}/\partial t$ can be neglected except in capacitors (ac at industrial power frequencies)	$\mathbf{j}_c = 0$ and ϵ_0 is the dielectric constant of free space	$\omega = 2\pi f =$ angular frequency, $f =$ the frequency considered, and $j = \sqrt{-1}$
$\left.\begin{array}{c}\text{curl } \mathbf{E} \\ \nabla \times \mathbf{E}\end{array}\right\} = -\dfrac{\partial \mathbf{B}}{\partial t}$	$\left.\begin{array}{c}\text{curl } \mathbf{E} \\ \nabla \times \mathbf{E}\end{array}\right\} = -\dfrac{\partial \mathbf{B}}{\partial t}$ $= -\mu_0 \dfrac{\partial \mathbf{H}}{\partial t}$	$\left.\begin{array}{c}\text{curl } \mathbf{E} \\ \nabla \times \mathbf{E}\end{array}\right\} = -j\omega\mu_0 \mathbf{H}$
	$\mu_0 =$ magnetic permeability of free space	
$\left.\begin{array}{c}\text{div } \mathbf{D} \\ \nabla \cdot \mathbf{D}\end{array}\right\} = \rho$	$\left.\begin{array}{c}\text{div } \mathbf{E} \\ \nabla \cdot \mathbf{E}\end{array}\right\} = 0$	$\left.\begin{array}{c}\text{div } \mathbf{E} \\ \nabla \cdot \mathbf{E}\end{array}\right\} = 0$
$\left.\begin{array}{c}\text{div } \mathbf{B} \\ \nabla \cdot \mathbf{B}\end{array}\right\} = 0$	$\left.\begin{array}{c}\text{div } \mathbf{H} \\ \nabla \cdot \mathbf{H}\end{array}\right\} = 0$	$\left.\begin{array}{c}\text{div } \mathbf{H} \\ \nabla \cdot \mathbf{H}\end{array}\right\} = 0$

Boundary conditions at the surface of separation between two media 1 and 2 are

$$\mathbf{H}_{2T} - \mathbf{H}_{1T} = \mathbf{j}_s \times \mathbf{N}$$

$$\mathbf{E}_{2T} - \mathbf{E}_{1T} = 0$$

$$\mathbf{B}_{2N} - \mathbf{B}_{1N} = 0$$

$$\mathbf{D}_{2N} - \mathbf{D}_{1N} = \sigma.$$

Subscript T denotes a tangential and subscript N a normal component.

$\mathbf{N} =$ unit normal vector from medium 1 to medium 2, which is the positive direction for normal vectors

$\mathbf{j}_s =$ current density on the surface, if any

$\sigma =$ density of electric charge on the surface of separation.

reciprocal

$$\text{div } (\mathbf{E} \times \mathbf{H}' - \mathbf{E}' \times \mathbf{H}) = 0$$

throughout V. If ports are defined for the sub-system within V, the definition given first can be derived from this differential relation.

The materials in a reciprocal system can be inhomogeneous, anisotropic, and even more complex, but if they are anisotropic, the relations between the fields and the flux densities in the materials must involve symmetric tensors. Systems containing charged particles in a dc magnetic field, ferroelectric or ferromagnetic materials, or active devices are often not reciprocal.

MAXWELL'S EQUATIONS IN DIFFERENT SYSTEMS OF COORDINATES

When a particular system of coordinates is advantageously used, such as cylindrical, spherical, etc., the components are derived from the vector equations by means of the vector-analysis equations in the chapter "Mathematical Equations."

ELECTROMAGNETIC FORCES ON CHARGES AND CURRENTS

The vectors $\mathbf{E}$, $\mathbf{B}$ of a given electromagnetic field are defined as forces on unit charge and on unit current, respectively. More generally, if the charge density is ρ and the current density is $\mathbf{j}_c$, the mechanical forces exerted on a volume V of the medium are

$$\mathbf{F}_e = \int_V \rho \mathbf{E} \, dV$$

$$\mathbf{F}_m = \int_V \mathbf{j}_c \times \mathbf{B} \, dV.$$

In differential form, the mechanical forces are written as

$$d\mathbf{F}_e = \mathbf{E} dq$$

where

$$dq = \int_V \rho \, dV, \qquad V \to 0$$

and

$$d\mathbf{F}_m = I \, d\mathbf{l} \times \mathbf{B}$$

where

$$I \, d\mathbf{l} = \int_V \mathbf{j}_c \, dV, \qquad V \to 0.$$

The total force exerted on the charged and current-carrying material of volume V is

$$\mathbf{F}_e + \mathbf{F}_m = \int_V (\rho \mathbf{E} + \mathbf{j}_c \times \mathbf{B}) \, dV.$$

CHAPTER 46
MATHEMATICAL EQUATIONS

MENSURATION EQUATIONS

Areas and Lengths Associated with Plane Figures

Parallelogram

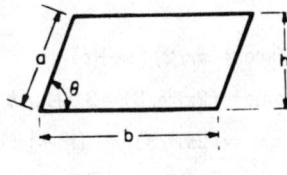

$$\text{Area} = bh$$

$$= ab \sin\theta$$

Trapezoid

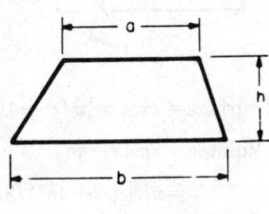

$$\text{Area} = \tfrac{1}{2}h(a+b)$$

Triangle

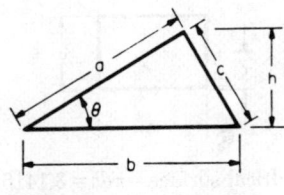

$$\text{Area} = \tfrac{1}{2}bh$$

$$= \tfrac{1}{2}ab \sin\theta$$

$$= [s(s-a)(s-b)(s-c)]^{1/2}$$

where $s = \tfrac{1}{2}(a+b+c)$. See p. 46-8.

Regular polygon

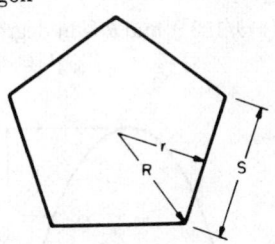

$$\text{Area} = \tfrac{1}{2}nrS$$

$$= nr^2 \tan(180°/n)$$

$$= \tfrac{1}{4}nS^2 \cot(180°/n)$$

$$= \tfrac{1}{2}nR^2 \sin(360°/n)$$

where n = number of sides, r = short radius, S = length of one side, R = long radius, and $r = R \cos(180°/n) = \tfrac{1}{2}S \cot(180°/n)$.

Circle

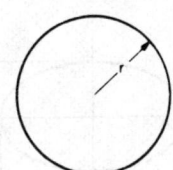

$$\text{Area} = \pi r^2$$

$$\text{Circumference} = 2\pi r$$

where r = radius and π = 3.1416.

Segment of circle

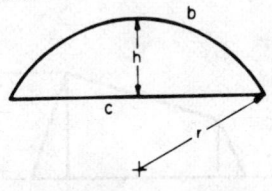

$$\text{Area} = \tfrac{1}{2}[br - c(r-h)]$$

where b = length of arc, and c = length of chord = $[4(2hr - h^2)]^{1/2}$.

46-1

Sector of circle

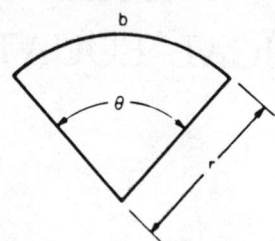

$$\text{Area} = br/2 = \pi r^2 (\theta/360°)$$

where $b = (\pi r \theta/180°)$ and θ is in degrees.

Parabola

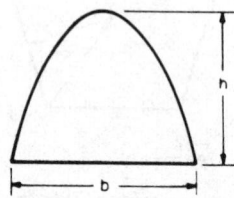

$$\text{Area} = \tfrac{2}{3}bh$$

$$\text{Arc length} = [4h^2 + (b^2/4)]^{1/2}$$
$$+ (b^2/8h)\ln\{4h + 2[4h^2 + (b^2/4)]^{1/2}/b\}$$

Ellipse

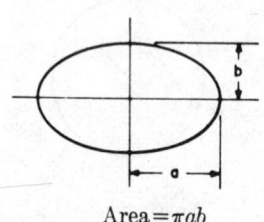

$$\text{Area} = \pi ab$$

$$\text{Circumference} = 4aE(k)$$

where $E(k)$ is a complete elliptic integral with $k = (a^2 - b^2)^{1/2}/a$ of the second kind, $a > b$.

Trapezium

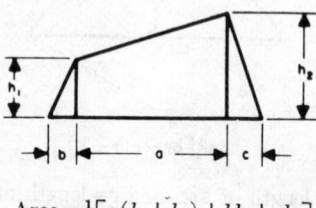

$$\text{Area} = \tfrac{1}{2}[a(h_1 + h_2) + bh_1 + ch_2]$$

Surface Areas and Volumes of Solid Figures

Sphere

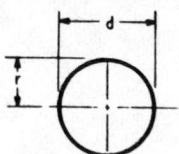

$$\text{Surface} = 4\pi r^2 = 12.5664r^2 = \pi d^2$$
$$\text{Volume} = (4\pi r^3/3) = 4.1888r^3$$

Sector of sphere

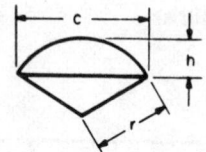

$$\text{Total surface} = (\pi r/2)(4h + c)$$
$$\text{Volume} = (2\pi r^2 h/3) = 2.0944r^2 h$$
$$= (2\pi r^2/3)[r - (r^2 - \tfrac{1}{4}c^2)^{1/2}]$$
$$c = [4(2hr - h^2)]^{1/2}$$

Segment of sphere

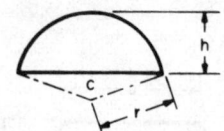

$$\text{Spherical surface} = 2\pi rh = \tfrac{1}{4}\pi(c^2 + 4h^2)$$
$$\text{Volume} = \pi h^2(r - \tfrac{1}{3}h)$$
$$= \pi h^2[(c^2 + 4h^2)/8h - \tfrac{1}{3}h]$$
$$c = [4(2hr - h^2)]^{1/2}$$

Cylinder

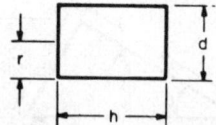

$$\text{Cylindrical surface} = \pi dh = 3.1416dh$$
$$\text{Total surface} = 2\pi r(r + h)$$
$$\text{Volume} = \pi r^2 h = 0.7854d^2 h$$
$$= c^2 h/4\pi = 0.0796c^2 h$$

$c = $ circumference

Torus or ring of circular cross-section

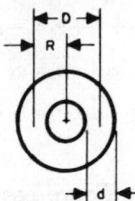

Surface $=4\pi^2 Rr=39.4784Rr=9.8696Dd$

Volume $=2\pi^2 Rr^2=19.74Rr^2=2.467Dd^2$

where $D=2R=$ diameter to centers of cross-section of torus and $r=d/2$.

Pyramid

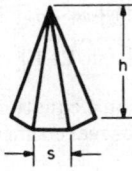

Volume $=Ah/3$.

When base is a regular polygon

$$\text{Volume}=\tfrac{1}{3}h\{nr^2[\tan(360°/2n)]\}$$

$$=\tfrac{1}{3}h\{\tfrac{1}{4}(ns^2)[\cot(360°/2n)]\}$$

where $A=$ area of base, $n=$ number of sides, and $r=$ short radius of base. See p. 46-1.

Pyramidal frustum

Volume $=\tfrac{1}{3}h[a+A+(aA)^{1/2}]$

where $A=$ area of base and $a=$ area of top.

Cone with circular base

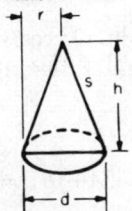

Conical area $=\pi rs=\pi r(r^2+h^2)^{1/2}$

Volume $=\pi r^2 h/3=1.047r^2h=0.2618d^2h$

where $s=$ slant height.

Conic frustum

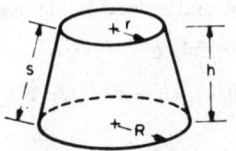

Volume $=(\pi h/3)(R^2+Rr+r^2)$

Area of conic surface $=\pi s(R+r)$

Development of conic surface

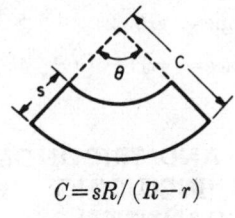

$$C=sR/(R-r)$$

$$\theta=360R/C$$

where θ is in degrees.

Wedge frustum

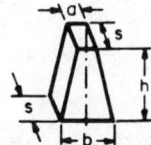

Volume $=\tfrac{1}{2}hs(a+b)$

where $h=$ height between parallel bases.

Ellipsoid

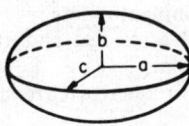

Volume $=(4\pi abc/3)$

$$\text{Surface}=2\pi\{c^2+[b/(a^2-c^2)^{1/2}][c^2F(\phi,k)$$

$$+(a^2-c^2)E(\phi,k)]\}\quad\text{if } a>b>c;$$

$$=2\pi a\{a+[c^2/(a^2-c^2)^{1/2}]$$

$$\times\ln[a+(a^2-c^2)^{1/2}]/c\}$$

if $a=b>c$ (oblate ellipsoid);

$$=2\pi c\{c+a^2(a^2-c^2)^{-1/2}\text{ arc sin}[(a^2-c^2)^{1/2}/a]\}$$

if $a>b=c$ (prolate ellipsoid).

$F(\phi, k)$ and $E(\phi, k)$ are incomplete elliptic integrals of the first and second kinds, respectively.

$$\phi = \arcsin[(a^2 - c^2)^{1/2}/a]$$

$$k = (a/b)[(b^2 - c^2)/(a^2 - c^2)]^{1/2}$$

Paraboloid

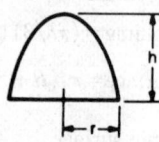

$$\text{Volume} = (\pi r^2 h/2)$$

$$\text{Curved surface} = (\pi r/6h^2)[(r^2 + 4h^2)^{3/2} - r^3]$$

ALGEBRAIC AND TRIGONOMETRIC EQUATIONS (INCLUDING COMPLEX QUANTITIES)

Quadratic Equation

If $ax^2 + bx + c = 0$, then

$$x = [-b \pm (b^2 - 4ac)^{1/2}]/2a$$

$$= 2c/[-b \mp (b^2 - 4ac)^{1/2}].$$

Solution of Cubic Equations*

Given $z^3 + a_2 z^2 + a_1 z + a_0 = 0$, let

$$q = \tfrac{1}{3}a_1 - \tfrac{1}{9}a_2^2; \quad r = \tfrac{1}{6}(a_1 a_2 - 3a_0) - \tfrac{1}{27}a_2^3.$$

If $q^3 + r^2 > 0$, one real root and a pair of complex conjugate roots

$q^3 + r^2 = 0$, all roots real and at least two are equal

$q^3 + r^2 < 0$, all roots real (irreducible case).

Let

$$s_1 = [r + (q^3 + r^2)^{1/2}]^{1/3}$$

$$s_2 = [r - (q^3 + r^2)^{1/2}]^{1/3}$$

then

$$z_1 = (s_1 + s_2) - (a_2/3)$$

$$z_2 = -\tfrac{1}{2}(s_1 + s_2) - (a_2/3) + (j\sqrt{3}/2)(s_1 - s_2)$$

$$z_3 = -\tfrac{1}{2}(s_1 + s_2) - (a_2/3) - (j\sqrt{3}/2)(s_1 - s_2).$$

* M. Abramovitz and I. A. Stegun, "Handbook of Mathematical Functions," National Bureau of Standards, Washington, D. C.: p. 17.

If z_1, z_2, z_3 are the roots of the cubic equation

$$z_1 + z_2 + z_3 = -a_2$$

$$z_1 z_2 + z_1 z_3 + z_2 z_3 = a_1$$

$$z_1 z_2 z_3 = -a_0.$$

Solution of Quartic Equations*

Given $z^4 + a_3 z^3 + a_2 z^2 + a_1 z + a_0 = 0$, find the real root u_1 of the cubic equation

$$u^3 - a_2 u^2 + (a_1 a_3 - 4a_0)u - (a_1^2 + a_0 a_3^2 - 4a_0 a_2) = 0$$

and determine the four roots of the quartic as solutions of the two quadratic equations

$$v^2 + \{(a_3/2) \mp [(a_3^2/4) + u_1 - a_2]^{1/2}\}v$$

$$+ (u_1/2) \mp [(u_1/2)^2 - a_0]^{1/2} = 0.$$

If all roots of the cubic equation are real, use the value of u_1 which gives real coefficients in the quadratic equation.

Complex Quantities

In the following equations all quantities are real except $j = (-1)^{1/2}$.

$$(A + jB) + (C + jD) = (A + C) + j(B + D)$$

$$(A + jB)(C + jD) = (AC - BD) + j(BC + AD)$$

$$\frac{A + jB}{C + jD} = \frac{AC + BD}{C^2 + D^2} + j\frac{BC - AD}{C^2 + D^2}$$

$$\frac{1}{A + jB} = \frac{A}{A^2 + B^2} - j\frac{B}{A^2 + B^2}.$$

Polar Form:

$$A + jB = \rho(\cos\theta + j\sin\theta) = \rho e^{j\theta}$$

De Moivre's Equation:

$$(A + jB)^v = \rho^v(\cos v\theta + j\sin v\theta)$$

where $\rho = (A^2 + B^2)^{1/2} > 0$, $\cos\theta = A/\rho$, and $\sin\theta = B/\rho$. For nonintegral v this quantity is many-valued.

Complex Conjugate:

$$\langle A + jB \rangle = (A + jB)^* = A - jB.$$

Analytic Function: Let $f(z)$ be a function of the

* M. Abramovitz and I. A. Stegun, "Handbook of Mathematical Functions," National Bureau of Standards, Washington, D. C.: p. 17.

complex variable $z=x+jy$. $f(z)$ is analytic at a point $z=z_0$, if

$$\lim_{\Delta z \to 0} [f(z_0+\Delta z)-f(z_0)]/\Delta z$$

exists independent of the manner in which Δz approaches zero. Analyticity is equivalent to the existence of a Taylor series about the point in question. In addition, $f(z)$ is analytic if and only if the Cauchy–Riemann equations hold:

$$\partial u/\partial x=\partial v/\partial y$$

$$\partial v/\partial x=-\partial u/\partial y$$

where $u=\mathrm{Re}f(z)$ and $v=\mathrm{Im}f(z)$. If $f(z)$ is analytic for all z, it is called an entire function.

Properties of e

$$e=\lim_{n\to\infty}(1+n^{-1})^n=\sum_{k=0}^{\infty}(k!)^{-1}=2.71828$$

$$e^{\pm jx}=\cos x \pm j \sin x=\exp(\pm jx).$$

Properties of Logarithms

If $\log_a x=N$, then $a^N=x$

$\log_a x=\log_a b \log_b x$

$\log_a xy=\log_a x+\log_a y$

$\log_a x/y=\log_a x-\log_a y$

$\log_a x^y=y \log_a x$

$\log_a b=1/\log_b a$

$a^{\log_a x}=x$

$\log_a 1=0$

$\log_a a=1$

$\log_e x=\ln x=\log_e 10 \log_{10} x=2.30259 \log_{10} x$

$\log_{10} x=\log_{10} e \log_e x=0.43429 \log_e x.$

Sums

In this section the following symbols will be used.

$\Gamma(\alpha)=$ Gamma function of α.

$\binom{n}{k}=\dfrac{n!}{k!(n-k)!}$ in which n and k are positive integers, $n\geq k$, and

$\binom{n}{0}=1.$

Arithmetic Progression:

$$\sum_{k=0}^{n-1}(a+kd)=a+(a+d)+(a+2d)+\cdots$$
$$+[a+(n-1)d]$$
$$=\tfrac{1}{2}n[2a+(n-1)d].$$

Geometric Progression:

$$\sum_{k=0}^{n-1}ar^k=a+ar+ar^2+\cdots+ar^{n-1}$$
$$=[a(r^n-1)/(r-1)], \quad \text{for} \quad r\neq 1$$
$$=na, \quad \text{for} \quad r=1.$$

Sums of Powers of Integers:

$$\sum_{k=1}^{n}k^2=1^2+2^2+3^2+\cdots+n^2$$
$$=[n(n+1)(2n+1)/6]$$

$$\sum_{k=1}^{n}k^3=1^3+2^3+3^3+\cdots+n^3$$
$$=[n^2(n+1)^2/4]$$

$$\sum_{k=1}^{n}k^4=1^4+2^4+3^4+\cdots+n^4$$
$$=\tfrac{1}{30}n(n+1)(2n+1)(3n^2+3n-1)$$

$$\sum_{k=1}^{n}k^r=1^r+2^r+3^r+\cdots+n^r$$
$$=\frac{n^{r+1}}{r+1}+\tfrac{1}{2}n^r+\sum_{k=1}^{[r/2]}(2k)^{-1}\binom{r}{2k-1}B_{2k}n^{r-2k+1}$$

where r is a positive integer, $[r/2]$ is the largest integer less than or equal to $r/2$, and B_{2k} is the $2k$th Bernoulli number. These sums are tabulated for $n=1, 2, 3, \cdots, 100$ and for $r=1, 2, 3, \cdots, 10$.[*]

Sums of integral powers of odd integers may be obtained from the above. For example,

$$\sum_{k=0}^{n}(2k+1)^2=1^2+3^2+5^2+\cdots+(2n+1)^2$$
$$=1^2+2^2+3^2+\cdots+(2n+1)^2$$
$$-2^2-4^2-6^2-\cdots-(2n)^2$$
$$=\sum_{k=1}^{2n+1}k^2-2^2\sum_{k=1}^{n}k^2$$
$$=\tfrac{1}{3}(n+1)(2n+1)(2n+3).$$

[*] M. Abramovitz and I. A. Stegun, "Handbook of Mathematical Functions," National Bureau of Standards, Washington, D. C.

Sums of Powers of Reciprocals of Integers:

$$\zeta(z) = \sum_{k=1}^{\infty} (1/k^z) = (1/1^z) + (1/2^z) + (1/3^z) + \cdots$$

for Re$z > 1$, $\zeta(z)$ is the Riemann zeta function of z

$$\sum_{k=1}^{\infty} (1/k^2) = (1/1^2) + (1/2^2) + (1/3^2) + \cdots$$

$$= \pi^2/6 = 1.64493$$

$$\sum_{k=1}^{\infty} (1/k^3) = (1/1^3) + (1/2^3) + (1/3^3) + \cdots$$

$$= 1.20206$$

$$\sum_{k=1}^{\infty} (1/k^4) = (1/1^4) + (1/2^4) + (1/3^4) + \cdots$$

$$= \pi^4/90 = 1.08232$$

$$\sum_{k=1}^{\infty} (1/k^{2r}) = (1/1^{2r}) + (1/2^{2r})$$

$$+ (1/3^{2r}) + \cdots$$

$$= [2^{2r-1}\pi^{2r} \mid B_{2r} \mid / (2r)!],$$

r integral

where B_{2r} is the $2r$th Bernoulli number. $\zeta(z)$ is tabulated for $z = 2, 3, 4, \cdots, 42$, to 20 decimal places in the "Handbook of Mathematical Functions." The truncated sums

$$\sum_{k=1}^{n} (1/k^r)$$

are related to the polygamma functions.

Finite Sums of Binomial Coefficients:

$$\sum_{k=0}^{n} \binom{m+k}{m} = \binom{m}{m} + \binom{m+1}{m} + \cdots + \binom{m+n}{m}$$

$$= \binom{m+n+1}{m+1}$$

$$\sum_{k=0}^{n} \binom{r}{k}\binom{s}{n-k} = \binom{r}{0}\binom{s}{n}$$

$$+ \binom{r}{1}\binom{s}{n-1} + \cdots + \binom{r}{n}\binom{s}{0}$$

$$= \binom{r+s}{n} \quad (r \geq n \text{ and } s \geq n).$$

Binomial Theorem:

Nonnegative integral exponent

$$(a+b)^n = a^n + na^{n-1}b + \tfrac{1}{2}[n(n-1)]a^{n-2}b^2 + \cdots + b^n$$

$$= \sum_{k=0}^{n} \binom{n}{k} a^{n-k}b^k.$$

For other n, integral negative and nonintegral, the series will be infinite. For convergence it is assumed $\mid a \mid > \mid b \mid$. (If $\mid a \mid < \mid b \mid$, interchange a and b.)

Negative integral exponent

$$(a+b)^{-n} = a^{-n} - na^{-n-1}b + \tfrac{1}{2}[n(n+1)]a^{-n-2}b^2 - \cdots$$

$$= \sum_{k=0}^{\infty} (-1)^k \binom{n+k-1}{k} a^{-n-k}b^k.$$

Nonintegral exponent

$$(a+b)^\alpha = a^\alpha + \alpha a^{\alpha-1}b + \tfrac{1}{2}\alpha(\alpha-1)a^{\alpha-2}b^2 + \cdots$$

$$= \sum_{k=0}^{\infty} \frac{\Gamma(\alpha+1)}{k!\Gamma(\alpha-k+1)} a^{\alpha-k}b^k.$$

Multinomial Series:

$$(x_1 + x_2 + \cdots + x_r)^n$$

$$= \sum_{n_1=0}^{n} \sum_{n_2=0}^{n-n_1} \sum_{n_3=0}^{n-n_1-n_2} \cdots \sum_{n_{r-1}=0}^{n-n_1-n_2-\cdots-n_{r-2}} n! \prod_{k=1}^{r} \frac{x_k^{n_k}}{n_k!}$$

where the interpretation $n_r = n - n_1 - n_2 - \cdots - n_{r-1}$ is to be used in the final product.

Combinations and Permutations

A combination is a selection from a number of things in which the order of the selected objects is disregarded, whereas a permutation is a selection in which the order is taken into consideration. For example, if from the letters a, b, and c a group of two is selected, then ab, bc, ac are the combinations and ab, ba, bc, cb, ac, ca are the permutations.

The number of different combinations of n (dissimilar) things taken r at a time is

$$\binom{n}{r} = C_r^n = \frac{n!}{r!(n-r)!}.$$

The number of different permutations of n (dissimilar) things taken r at a time is

$$P_r^n = n!/(n-r)! = n \times (n-1) \times \cdots \times (n-r+1).$$

Bernoulli Numbers

Definition:

$$B_n = (d^n/dx^n)[x/(e^x-1)]\big|_{x=0}.$$

Values for Small n:

$$B_0=1,\quad B_1=-\tfrac{1}{2},\quad B_2=\tfrac{1}{6},\quad B_4=-\tfrac{1}{30},$$

$$B_6=\tfrac{1}{42},\quad B_8=-\tfrac{1}{30},\quad B_{10}=\tfrac{5}{66},$$

$$B_{2n+1}=0\quad\text{for all integral }\ n>0..$$

Identities:

$$B_{2n}=(-1)^{n-1}\frac{2\cdot(2n)!}{(2\pi)^{2n}}\sum_{k=1}^{\infty}\frac{1}{k^{2n}},\ n\geq1$$

$$B_n=\sum_{k=0}^{n}\binom{n}{k}B_k,\quad n>1.$$

The latter identity may be used for recursive calculation of the B_n's.

Trigonometric Identities

$$1=\sin^2A+\cos^2A=\sin A\ \csc A=\tan A\ \cot A$$
$$=\cos A\ \sec A$$
$$\sin A=\cos A/\cot A=1/\csc A=\cos A\ \tan A$$
$$=\pm(1-\cos^2A)^{1/2}$$
$$\cos A=\sin A/\tan A=1/\sec A=\sin A\ \cot A$$
$$=\pm(1-\sin^2A)^{1/2}$$
$$\tan A=\sin A/\cos A=1/\cot A=\sin A\ \sec A$$
$$\sin A=(e^{jA}-e^{-jA})/2j$$
$$\cos A=(e^{jA}+e^{-jA})/2$$

$$\sin(A\pm B)=\sin A\ \cos B\pm\cos A\ \sin B$$
$$\cos(A\pm B)=\cos A\ \cos B\mp\sin A\ \sin B$$
$$\tan(A\pm B)=(\tan A\pm\tan B)/(1\mp\tan A\ \tan B)$$
$$=(\tan A\ \cot B\pm1)/(\cot B\mp\tan A)$$
$$\cot(A\pm B)=(\cot A\ \cot B\mp1)/(\cot B\pm\cot A)$$
$$=(\cot A\mp\tan B)/(1\pm\cot A\ \tan B)$$

$$\sin2A=2\ \sin A\ \cos A$$
$$\cos2A=\cos^2A-\sin^2A$$
$$\tan2A=(2\ \tan A)/(1-\tan^2A)$$
$$\sin3A=3\ \sin A-4\ \sin^3A$$
$$\cos3A=-3\ \cos A+4\ \cos^3A$$

$$\tan3A=(3\ \tan A-\tan^3A)/(1-3\ \tan^2A)$$
$$\cos nA=\mathrm{Re}(\cos A+j\ \sin A)^n$$
$$\sin nA=\mathrm{Im}(\cos A+j\ \sin A)^n$$
$$\sin\tfrac{1}{2}A=\pm[(1-\cos A)/2]^{1/2}$$
$$\cos\tfrac{1}{2}A=\pm[(1+\cos A)/2]^{1/2}$$
$$\tan\tfrac{1}{2}A=\sin A/(1+\cos A)=(1-\cos A)/\sin A$$
$$\sin A\pm\sin B=2\ \sin\tfrac{1}{2}(A\pm B)\ \cos\tfrac{1}{2}(A\mp B)$$
$$\cos A+\cos B=2\ \cos\tfrac{1}{2}(A+B)\ \cos\tfrac{1}{2}(A-B)$$
$$\cos B-\cos A=2\ \sin\tfrac{1}{2}(A+B)\ \sin\tfrac{1}{2}(A-B)$$
$$\tan A\pm\tan B=[\sin(A\pm B)/\cos A\ \cos B]$$
$$\cot A\pm\cot B=[\sin(B\pm A)/\sin A\ \sin B]$$
$$\sin^2A-\sin^2B=\sin(A+B)\ \sin(A-B)$$
$$\cos^2A-\sin^2B=\cos(A+B)\ \cos(A-B)$$
$$\tan\tfrac{1}{2}(A\pm B)=(\sin A\pm\sin B)/(\cos A+\cos B)$$
$$\cot\tfrac{1}{2}(A\mp B)=(\sin A\pm\sin B)/(\cos B-\cos A)$$
$$\cos^2A=\tfrac{1}{2}(\cos2A+1)$$
$$\cos^3A=\tfrac{1}{4}(\cos3A+3\ \cos A)$$
$$\cos^4A=\tfrac{1}{8}(\cos4A+4\ \cos2A+3)$$
$$\sin^2A=\tfrac{1}{2}(-\cos2A+1)$$
$$\sin^3A=\tfrac{1}{4}(-\sin3A+3\ \sin A)$$
$$\sin^4A=\tfrac{1}{8}(\cos4A-4\ \cos2A+3)$$
$$\sin A\ \cos B=\tfrac{1}{2}[\sin(A+B)+\sin(A-B)]$$
$$\cos A\ \cos B=\tfrac{1}{2}[\cos(A+B)+\cos(A-B)]$$
$$\sin A\ \sin B=\tfrac{1}{2}[\cos(A-B)-\cos(A+B)].$$

$$\sin A+m\ \sin B=\rho\ \sin C$$

with

$$\rho^2=1+m^2+2m\ \cos(B-A)$$

and

$$\tan(C-A)=[m\ \sin(B-A)]/[1+m\ \cos(B-A)].$$

$$\sum_i A_i\exp(j\theta_i)=\rho e^{j\psi}$$

with

$$\tan\psi=\left(\sum_i A_i\ \sin\theta_i/\sum_i A_i\ \cos\theta_i\right)$$

and

$$\rho=\left[\sum_i A_i^2+\sum_{i<j}\sum A_iA_j\ \cos(\theta_i-\theta_j)\right]^{1/2}.$$

In the previous notation

$$\sum_i A_i \cos\theta_i = \rho \cos\psi$$

$$\sum_i A_i \sin\theta_i = \rho \sin\psi$$

apply.

$$\sin x + \sin 2x + \sin 3x + \cdots + \sin mx$$
$$= [\sin\tfrac{1}{2}mx \, \sin\tfrac{1}{2}(m+1)x / \sin\tfrac{1}{2}x]$$

$$\cos x + \cos 2x + \cos 3x + \cdots + \cos mx$$
$$= [\sin\tfrac{1}{2}mx \, \cos\tfrac{1}{2}(m+1)x / \sin\tfrac{1}{2}x]$$

$$\sin x + \sin 3x + \sin 5x + \cdots + \sin(2m-1)x$$
$$= (\sin^2 mx / \sin x)$$

$$\cos x + \cos 3x + \cos 5x + \cdots + \cos(2m-1)x$$
$$= (\sin 2mx / 2 \sin x)$$

$$\tfrac{1}{2} + \cos x + \cos 2x + \cdots + \cos mx$$
$$= [\sin(m+\tfrac{1}{2})x / 2 \sin\tfrac{1}{2}x]$$

Angle (degrees)	Sine	Cosine	Tangent
0	0	1	0
30	$\frac{1}{2}$	$\frac{1}{2}\sqrt{3}$	$\frac{1}{3}\sqrt{3}$
45	$\frac{1}{2}\sqrt{2}$	$\frac{1}{2}\sqrt{2}$	1
60	$\frac{1}{2}\sqrt{3}$	$\frac{1}{2}$	$\sqrt{3}$
90	1	0	$\pm\infty$
180	0	-1	0
270	-1	0	$\pm\infty$
360	0	1	0
0–90	$+$	$+$	$+$
90–180	$+$	$-$	$-$
180–270	$-$	$-$	$+$
270–360	$-$	$+$	$-$

versine: $\text{vers } \theta = 1 - \cos\theta$

haversine: $\text{hav } \theta = \tfrac{1}{2}(1 - \cos\theta) = \sin^2\tfrac{1}{2}\theta$.

Approximations for Small Angles

$$\left.\begin{array}{l} \sin\theta = (\theta - \theta^3/6 \cdots) \\ \tan\theta = (\theta + \theta^3/3 \cdots) \\ \cos\theta = (1 - \theta^2/2 \cdots) \end{array}\right\} \theta \text{ in radians}$$

$\sin\theta = \theta$

with less than 1-percent error up to
$\theta = 0.24$ radian $= 14.0°$

with less than 10-percent error up to
$\theta = 0.78$ radian $= 44.5°$

$\tan\theta = \theta$

with less than 1-percent error up to
$\theta = 0.17$ radian $= 10.0°$

with less than 10-percent error up to
$\theta = 0.54$ radian $= 31.0°$.

Inequalities

$$\sin x \leq x < \tan x, \quad \text{for} \quad 0 \leq x < \pi/2$$

$$\sin x \geq (2/\pi) x, \quad \text{for} \quad 0 \leq x \leq \pi/2$$

$$\cos x < \sin x / x \leq 1, \quad \text{for} \quad 0 < x \leq \pi$$

where x is in radians.

PLANE TRIGONOMETRY

Right Triangles (Fig. 1)

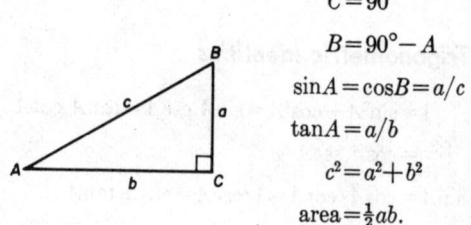

$$C = 90°$$
$$B = 90° - A$$
$$\sin A = \cos B = a/c$$
$$\tan A = a/b$$
$$c^2 = a^2 + b^2$$
$$\text{area} = \tfrac{1}{2}ab.$$

Fig. 1.

Oblique Triangles (Fig. 2)

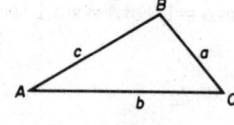

Fig. 2.

Sum of Angles:

$$A + B + C = 180°. \tag{1}$$

Law of Cosines:

$$a^2 = b^2 + c^2 - 2bc \cos A. \tag{2}$$

Law of Sines:

$$a/\sin A = b/\sin B = c/\sin C. \tag{3}$$

Law of Tangents:

$$\frac{a-b}{a+b} = \frac{\tan\frac{1}{2}(A-B)}{\tan\frac{1}{2}(A+B)}. \tag{4}$$

Half-Angle Equation (Fig. 3):

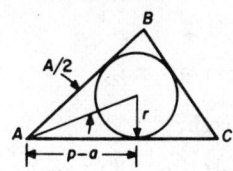

Fig. 3.

$$\tan\tfrac{1}{2}A = r/(p-a) \qquad (5)$$

where

$$2p = a+b+c$$

$$r = [(p-a)(p-b)(p-c)/p]^{1/2}.$$

Solving an Oblique Triangle:

Given	Use	To Obtain	
aBC	(1)	A	
	(3)	bc	
Abc	(1)	$B+C$	hence
	(4)	$B-C$	B, C
abc	(5) or (2)	ABC	
abA ambiguous case	(3) and (1)	BCc	

SPHERICAL TRIGONOMETRY

Spherical triangles are bounded by the arcs of great circles. These are circles formed by the intersection of a sphere with planes passing through the sphere's center. In the following equations α, β, γ are the angles and a, b, c are the corresponding opposite sides, respectively. The sides are measured by the angles subtended by the arcs; for example, a side extending from the Equator to the North Pole is a $90°$ side.

Right Spherical Triangles ($\gamma = 90°$) (Fig. 4)

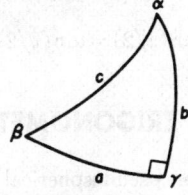

Fig. 4.

$$\cos c = \cos a \, \cos b = \cot\alpha \, \cot\beta$$
$$\cos\alpha = \sin\beta \, \cos a = \tan b \, \cot c$$
$$\cos\beta = \sin\alpha \, \cos b = \tan a \, \cot c$$
$$\sin a = \sin c \, \sin\alpha = \tan b \, \cot\beta$$
$$\sin b = \sin c \, \sin\beta = \tan a \, \cot\alpha. \qquad (6)$$

Oblique Triangles (Fig. 5)

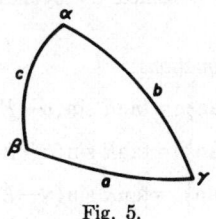

Fig. 5.

Law of Cosines for Sides:

$$\cos a = \cos b \, \cos c + \sin b \, \sin c \, \cos\alpha$$
$$\cos b = \cos c \, \cos a + \sin c \, \sin a \, \cos\beta$$
$$\cos c = \cos a \, \cos b + \sin a \, \sin b \, \cos\gamma. \qquad (7A)$$

Law of Cosines for Angles:

$$\cos\alpha = -\cos\beta \, \cos\gamma + \sin\beta \, \sin\gamma \, \cos a$$
$$\cos\beta = -\cos\gamma \, \cos\alpha + \sin\gamma \, \sin\alpha \, \cos b$$
$$\cos\gamma = -\cos\alpha \, \cos\beta + \sin\alpha \, \sin\beta \, \cos c. \qquad (7B)$$

Law of Sines:

$$\sin a/\sin\alpha = \sin b/\sin\beta = \sin c/\sin\gamma. \qquad (8)$$

Napier's Analogies:

$$\frac{\sin\tfrac{1}{2}(\alpha-\beta)}{\sin\tfrac{1}{2}(\alpha+\beta)} = \frac{\tan\tfrac{1}{2}(a-b)}{\tan\tfrac{1}{2}c} \qquad (9A)$$

$$\frac{\cos\tfrac{1}{2}(\alpha-\beta)}{\cos\tfrac{1}{2}(\alpha+\beta)} = \frac{\tan\tfrac{1}{2}(a+b)}{\tan\tfrac{1}{2}c} \qquad (9B)$$

$$\frac{\sin\tfrac{1}{2}(a-b)}{\sin\tfrac{1}{2}(a+b)} = \frac{\tan\tfrac{1}{2}(\alpha-\beta)}{\cot\tfrac{1}{2}\gamma} \qquad (9C)$$

$$\frac{\cos\tfrac{1}{2}(a-b)}{\cos\tfrac{1}{2}(a+b)} = \frac{\tan\tfrac{1}{2}(\alpha+\beta)}{\cot\tfrac{1}{2}\gamma} \qquad (9D)$$

Half-Angle Equations:

$$\tan\tfrac{1}{2}\alpha = \tan r/\sin(p-a)$$
$$\tan\tfrac{1}{2}\beta = \tan r/\sin(p-b)$$
$$\tan\tfrac{1}{2}\gamma = \tan r/\sin(p-c) \qquad (10A)$$

where $2p = a + b + c$ and

$$\tan^2 r = \frac{\sin(p-a) \, \sin(p-b) \, \sin(p-c)}{\sin p}.$$

$$\sin^2\tfrac{1}{2}\alpha = [\sin(p-b) \, \sin(p-c)/\sin b \, \sin c]$$

$$\cos^2\tfrac{1}{2}\alpha = [\sin p \, \sin(p-a)/\sin b \, \sin c]$$

$$\tan^2\tfrac{1}{2}\alpha = [\sin(p-b) \, \sin(p-c)/\sin p \, \sin(p-a)]$$

$$\tag{10B}$$

and equations obtained by cyclical permutation for β and γ.

Half-Side Equations:

$$\tan\tfrac{1}{2}a = \tan R \, \sin(\alpha - E)$$

$$\tan\tfrac{1}{2}b = \tan R \, \sin(\beta - E)$$

$$\tan\tfrac{1}{2}c = \tan R \, \sin(\gamma - E) \tag{11A}$$

where $2E = \alpha + \beta + \gamma - \pi$ is the spherical excess and

$$\tan^2 R = \frac{\sin E}{\sin(\alpha - E) \, \sin(\beta - E) \, \sin(\gamma - E)}.$$

$$\sin^2\tfrac{1}{2}a = -[\sin E \, \sin(E - \alpha)/\sin\beta \, \sin\gamma]$$

$$\cos^2\tfrac{1}{2}a = [\sin(E - \beta) \, \sin(E - \gamma)/\sin\beta \, \sin\gamma]$$

$$\tan^2\tfrac{1}{2}a = -[\sin E \, \sin(E - \alpha)/\sin(E - \beta)\sin(E - \gamma)]$$

$$\tag{11B}$$

and equations obtained by cyclical permutation for b and c.

Area: On a sphere of radius one, the area of a triangle is equal to the spherical excess

$$2E = \alpha + \beta + \gamma - \pi.$$

L'Huilier's Theorem:

$$\tan^2\tfrac{1}{2}E = \tan\tfrac{1}{2}p \, \tan\tfrac{1}{2}(p-a)$$

$$\times \tan\tfrac{1}{2}(p-b) \, \tan\tfrac{1}{2}(p-c). \tag{12}$$

Solving an Oblique Triangle*:

Given	Use	To Obtain
abc	(10)	$\alpha\beta\gamma$
$\alpha\beta\gamma$	(11)	abc
$ab\gamma$	(9)	$\alpha \pm \beta$, hence α, β, then c
$\alpha\beta c$	(9)	$a \pm b$, hence a, b, then γ
$ab\alpha$ ambiguous case	(8) (9)	β $c\gamma$
$\alpha\beta a$ ambiguous case	(8) (9)	b $c\gamma$

* See also great-circle calculations in Chapter 28.

HYPERBOLIC FUNCTIONS*

$$\sinh x = (e^x - e^{-x})/2$$

$$\cosh x = (e^x + e^{-x})/2$$

$$\tanh x = \sinh x/\cosh x$$

$$= [1 - \exp(-2x)]/[1 + \exp(-2x)]$$

$$= 1/\coth x$$

$$\operatorname{sech} x = 1/\cosh x$$

$$\operatorname{csch} x = 1/\sinh x$$

$$\sinh(-x) = -\sinh x$$

$$\cosh(-x) = \cosh x$$

$$\tanh(-x) = -\tanh x$$

$$\coth(-x) = -\coth x$$

$$\sinh jx = j \sin x$$

$$\cosh jx = \cos x$$

$$\tanh jx = j \tan x$$

$$\coth jx = -j \cot x$$

$$\cosh^2 x - \sinh^2 x = 1$$

$$1 - \tanh^2 x = 1/\cosh^2 x$$

$$\coth^2 x - 1 = 1/\sinh^2 x$$

$$\sinh 2x = 2 \sinh x \, \cosh x$$

$$\cosh 2x = \cosh^2 x + \sinh^2 x$$

$$\sinh(x \pm jy) = \sinh x \, \cos y \pm j \, \cosh x \, \sin y$$

$$\cosh(x \pm jy) = \cosh x \, \cos y \pm j \, \sinh x \, \sin y$$

$$\tanh(x \pm y) = (\tanh x \pm \tanh y)/(1 \pm \tanh x \, \tanh y).$$

If $y = \operatorname{gd} x$ (gudermannian of x) is defined by

$$x = \log_e \tan(\tfrac{1}{4}\pi + \tfrac{1}{2}y)$$

then

$$\sinh x = \tan y$$

$$\cosh x = \sec y$$

$$\tanh x = \sin y$$

$$\tanh(x/2) = \tan(y/2).$$

HYPERBOLIC TRIGONOMETRY

Hyperbolic (or pseudospherical) trigonometry applies to triangles drawn in the hyperbolic type

* Tables of hyperbolic functions appear in Chapter 47.

of non-Euclidean space. Reflection charts, used in transmission-line theory and waveguide analysis, are models of this hyperbolic space.*

Conformal Model (Fig. 6)

The space is limited to the inside of a unit circle Γ. Geodesics (or "straight lines" for the model) are arcs of circle orthogonal to Γ as shown in sketch of conformal model. The hyperbolic distance between two points A and B is defined by

$$[AB] = \log_e(BI/BJ):(AI/AJ)$$

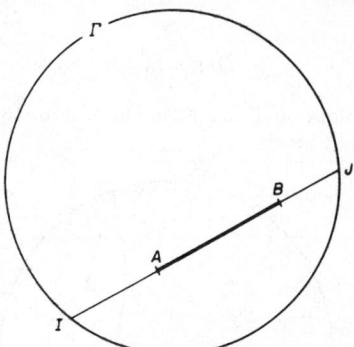

Fig. 7—Projective model.

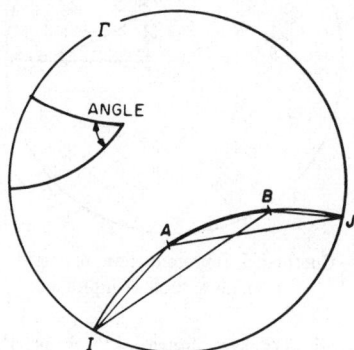

Fig. 6—Conformal model.

where I and J are the intersections with Γ of the geodesic AB. The distance $[AB]$ is expressed in nepers. For engineering purposes, a unit, corresponding to the decibel and equal to $1/8.686$ neper, is sometimes used.

As this model is conformal, the angle between two lines is the ordinary angle between the tangents at their common point.

Projective Model (Fig. 7)

The space is again composed of the points inside of a circle Γ. Geodesics are straight-line segments limited to the inside of Γ. (JI in sketch of projective model.)

The hyperbolic distance $\langle AB \rangle$ is defined by

$$\langle AB \rangle = \tfrac{1}{2}\log_e\big[(BI/BJ):(AI/AJ)\big]$$

* G. A. Deschamps, "Hyperbolic Protractor for Microwave Impedance Measurements and Other Purposes," International Telephone and Telegraph Corporation, 320 Park Ave., New York, N. Y. 10022; 1953.

and can be measured directly by means of a hyperbolic protractor. The angles for this model do not appear in true size except when at the center of Γ. An angle such as BAC, when it is considered in reference to the projective model, will be called an *elliptic* angle. It can be evaluated, as shown in Fig. 8, by projecting B and C through the hyperbolic midpoint of OA onto B' and C' on the circle Γ, then measuring $B'OC'$ as in Euclidean geometry.

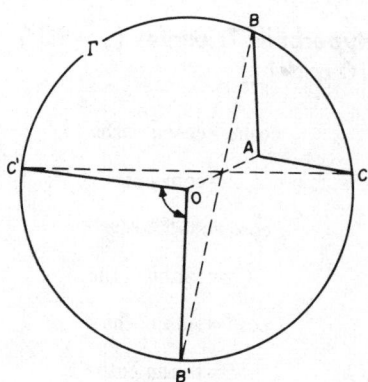

Fig. 8—Construction of angle on projective model.

The two models in Fig. 9 drawn inside the same circle Γ can be set into a distance-preserving correspondence by the transformation: $\mathfrak{B}(M) = M'$ defined by

$$[OM] = \langle OM' \rangle$$

or in terms of ordinary distances

$$OM' = 2OM/(1+OM^2).$$

The hyperbolic distance to the center O being denoted by u

$$OM = \tanh(u/2)$$

and

$$OM' = \tanh u.$$

The points on Γ are at an infinite distance from any point inside Γ.

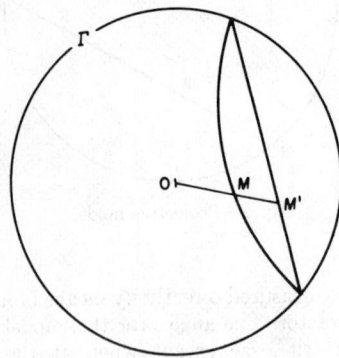

Fig. 9—Correspondence between the two models.

In the following equations, the sides are expressed in nepers, the angles in radians. The three points A,B,C are assumed to be inside the circle Γ.

Right Hyperbolic Triangles ($\gamma = 90°$) (Figs. 10 and 11)

$$\cosh c = \cosh a \cosh b$$

$$= \cot\alpha \cot\beta$$

$$\cos\alpha = \sin\beta \cosh a$$

$$= \tanh b \coth c$$

$$\cos\beta = \sin\alpha \cosh b$$

$$= \tanh a \coth c.$$

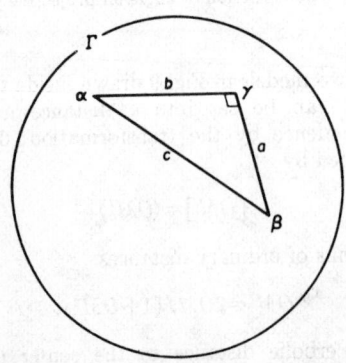

Fig. 10—Projective representation of right hyperbolic triangle.

When B is at infinity, i.e., on Γ

$$\cos A = \tanh b$$

$$\cot A = \sinh b$$

$$\csc A = \cosh b$$

$$\tan\tfrac{1}{2}A = \exp b$$

or

$$(\pi/2) - A = \mathrm{gd}\, b$$

(See definition of gd on p. 46-10.)

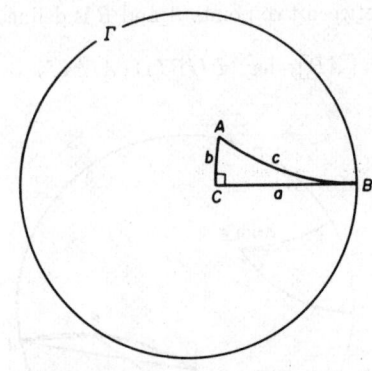

Fig. 11—Conformal representation of right hyperbolic triangle with B at infinity.

CB and AB are "parallel." A is also called angle of parallelism and is noted by

$$A = \sqcap (b)$$

$$= \pi/2 - \mathrm{gd}\, b.$$

Oblique Hyperbolic Triangles

Law of Cosines:

$$\cosh a = \cosh b \cosh c - \sinh b \sinh c \cos\alpha$$

$$\text{and permutations} \quad (13A)$$

$$\cos\alpha = -\cos\beta \cos\gamma + \sin\beta \sin\gamma \cosh a$$

$$\text{and permutations.} \quad (13B)$$

Law of Sines (Fig. 12):

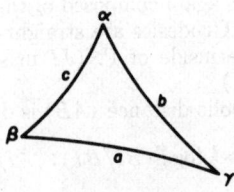

Fig. 12.

$$\sinh a/\sin\alpha = \sinh b/\sin\beta = \sinh c/\sin\gamma. \quad (14)$$

Napier's Analogies:

$$\frac{\sin\tfrac{1}{2}(\alpha-\beta)}{\sin\tfrac{1}{2}(\alpha+\beta)}=\frac{\tanh\tfrac{1}{2}(a-b)}{\tanh\tfrac{1}{2}c} \qquad (15A)$$

$$\frac{\cos\tfrac{1}{2}(\alpha-\beta)}{\cos\tfrac{1}{2}(\alpha+\beta)}=\frac{\tanh\tfrac{1}{2}(a+b)}{\tanh\tfrac{1}{2}c} \qquad (15B)$$

$$\frac{\sinh\tfrac{1}{2}(a-b)}{\sinh\tfrac{1}{2}(a+b)}=\frac{\tan\tfrac{1}{2}(\alpha-\beta)}{\cot\tfrac{1}{2}\gamma} \qquad (15C)$$

$$\frac{\cosh\tfrac{1}{2}(a-b)}{\cosh\tfrac{1}{2}(a+b)}=\frac{\tan\tfrac{1}{2}(\alpha+\beta)}{\cot\tfrac{1}{2}\gamma}. \qquad (15D)$$

Half-Angle Equations:

$$\tan\tfrac{1}{2}\alpha=\tanh r/\sinh(p-\alpha)$$

and permutations where $2p=a+b+c$ and

$$\tanh^2 r=\frac{\sinh(p-a)\,\sinh(p-b)\,\sinh(p-c)}{\sinh p}. \qquad (16A)$$

$$\sin^2\tfrac{1}{2}\alpha=\frac{\sinh(p-b)\,\sinh(p-c)}{\sinh b\,\sinh c}$$

$$\cos^2\tfrac{1}{2}\alpha=\frac{\sinh p\,\sinh(p-a)}{\sinh b\,\sinh c}$$

$$\tan^2\tfrac{1}{2}\alpha=\frac{\sinh(p-b)\,\sinh(p-c)}{\sinh p\,\sinh(p-a)}. \qquad (16B)$$

Half-Side Equations:

$$\coth\tfrac{1}{2}a=\coth R/\sin(\Delta+\alpha)$$

and permutations where $2\Delta=\pi-\alpha-\beta-\gamma$ is the hyperbolic defect and

$$\tanh^2 R=\frac{\sin\Delta}{\sin(\Delta+\alpha)\,\sin(\Delta+\beta)\,\sin(\Delta+\gamma)}. \qquad (17A)$$

$$\sinh^2\tfrac{1}{2}a=\frac{\sin\Delta\,\sin(\Delta+\alpha)}{\sin\beta\,\sin\gamma}$$

$$\cosh^2\tfrac{1}{2}a=\frac{\sin(\Delta+\beta)\,\sin(\Delta+\gamma)}{\sin\beta\,\sin\gamma}$$

$$\tanh^2\tfrac{1}{2}a=\frac{\sin\Delta\,\sin(\Delta+\alpha)}{\sin(\Delta+\beta)\,\sin(\Delta+\gamma)}. \qquad (17B)$$

Area: The hyperbolic area of a triangle is equal to the hyperbolic defect.

$$2\Delta=\pi-(\alpha+\beta+\gamma). \qquad (18)$$

Solving an Oblique Hyperbolic Triangle: Solution of an oblique hyperbolic triangle is analogous to that for an oblique spherical triangle, as follows.

Given	Use	To Obtain
abc	(16)	$\alpha\beta\gamma$
$\alpha\beta\gamma$	(17)	abc
$ab\gamma$	(15)	$\alpha\pm\beta$, hence α, β, then c
$\alpha\beta c$	(15)	$a\pm b$, hence a, b, then γ
$ab\alpha$ ambiguous case	(14)	β
	(15)	$c\gamma$
$\alpha\beta a$ ambiguous case	(14)	b
	(15)	$c\gamma$

PLANE ANALYTIC GEOMETRY

In the following, x and y are coordinates of a variable point in a rectangular-coordinate system.

Straight Line

General Equation:

$$Ax+By+C=0$$

A, B, and C are constants.

Slope-Intercept Form (Fig. 13):

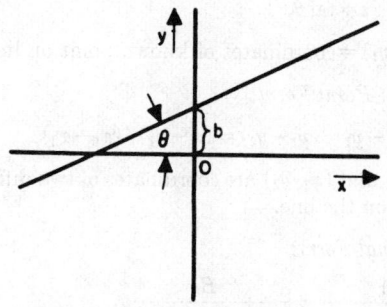

Fig. 13—Slope-intercept.

$$y=sx+b$$

$$b=y\text{-intercept}$$

$$s=\tan\theta$$

$$=\text{slope}.$$

Intercept-Intercept Form (Fig. 14):

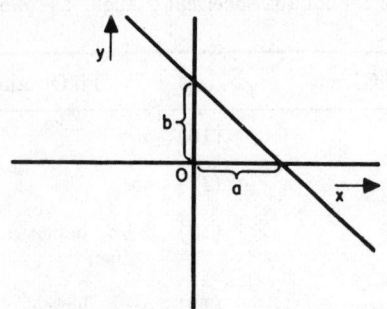

Fig. 14—Intercept-intercept.

$$(x/a)+(y/b)=1$$

$$a=x\text{-intercept}$$

$$b=y\text{-intercept}.$$

Point-Slope Form (Fig. 15):

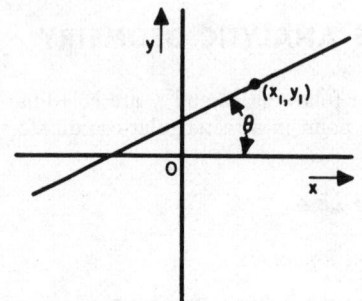

Fig. 15—Point-slope.

$$y-y_1=s(x-x_1)$$

$$s=\tan\theta$$

(x_1, y_1) = coordinates of known point on line.

Point-Point Form:

$$(y-y_1)/(y_1-y_2)=(x-x_1)/(x_1-x_2).$$

(x_1, y_1) and (x_2, y_2) are coordinates of two different points on the line.

Normal Form:

$$\frac{A}{\pm(A^2+B^2)^{1/2}}x+\frac{B}{\pm(A^2+B^2)^{1/2}}y$$

$$+\frac{C}{\pm(A^2+B^2)^{1/2}}=0.$$

The sign of the radical is chosen so that

$$\frac{C}{\pm(A^2+B^2)^{1/2}}<0.$$

Distance from Point (x_1, y_1) *to a Line:* Substitute coordinates of the point in the normal form of the line. Thus

$$\text{distance}=\frac{A}{\pm(A^2+B^2)^{1/2}}x_1+\frac{B}{\pm(A^2+B^2)^{1/2}}y_1$$

$$+\frac{C}{\pm(A^2+B^2)^{1/2}}.$$

Angle Between Two Lines:

$$\tan\phi=(s_1-s_2)/(1+s_1s_2)$$

where ϕ = angle between the lines, s_1 = slope of one line, and s_2 = slope of other line. When the lines are mutually perpendicular, $\tan\phi=\pm\infty$, whence

$$s_1=-1/s_2.$$

Transformation of Rectangular Coordinates

Translation:

$$x_1=h+x_2$$

$$y_1=k+y_2$$

$$x_2=x_1-h$$

$$y_2=y_1-k$$

(h, k) = coordinates of new origin referred to old origin.

Rotation (Fig. 16):

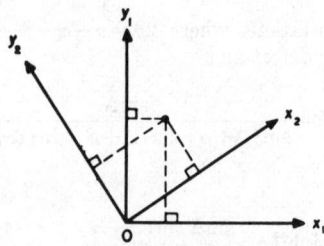

Fig. 16.

$$x_1=x_2\cos\theta-y_2\sin\theta$$

$$y_1=x_2\sin\theta+y_2\cos\theta$$

$$x_2=x_1\cos\theta+y_1\sin\theta$$

$$y_2=-x_1\sin\theta+y_1\cos\theta$$

(x_1, y_1) = "old" coordinates

$(x_2, y_2) =$ "new" coordinates

$\theta =$ counterclockwise angle of rotation of axes.

Circle

The equation of a circle of radius r with center at (m, n) is

$$(x-m)^2 + (y-n)^2 = r^2.$$

Tangent Line to a Circle: At (x_1, y_1) is

$$y-y_1 = -[(x_1-m)/(y_1-n)](x-x_1).$$

Normal Line to a Circle: At (x_1, y_1) is

$$y-y_1 = [(y_1-n)/(x_1-m)](x-x_1).$$

Parabola

Figure 17 shows an x-parabola centered at the origin open to the right.

Focus: F
Directrix: D
Vertex: O
Latus rectum: AA'
$e =$ eccentricity $= 1$
$MP = FP = 1$ for any point P on the parabola.

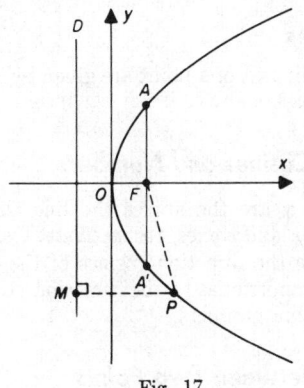

Fig. 17.

x-Parabola:

$$(y-k)^2 = \pm 2p(x-h)$$

where (h, k) are the coordinates of the vertex, and the sign used is plus or minus when the parabola is open to the right or to the left, respectively. The semilatus rectum is p.

y-Parabola:

$$(x-h)^2 = \pm 2p(y-k)$$

where (h, k) are the coordinates of the vertex. Use plus sign if parabola is open above, and minus sign if open below.

Tangent Lines to a Parabola:

$(x_1, y_1) =$ point of tangency.

For x-parabola

$$y-y_1 = \pm [p/(y_1-k)](x-x_1).$$

Use plus sign if parabola is open to the right, minus sign if open to the left.
For y-parabola

$$y-y_1 = \pm [(x_1-h)/p](x-x_1).$$

Use plus sign if parabola is open above, minus sign if open below.

Normal Lines to a Parabola:

$(x_1, y_1) =$ point of contact.

For x-parabola

$$y-y_1 = \mp [(y_1-k)/p](x-x_1).$$

Use minus sign if parabola is open to the right, plus sign if open to the left.
For y-parabola

$$y-y_1 = \mp [p/(x_1-h)](x-x_1).$$

Use minus sign if parabola is open above, plus sign if open below.

Ellipse

Figure 18 shows ellipse centered at origin. If the ellipse is centered at (h, k) instead, the equations that follow must be modified by replacing x, x_1, y, y_1 by $x-h, x_1-h, y-k, y_1-k$, respectively.

Foci: F, F'
Directrices: D, D'

$e =$ eccentricity < 1

$2a = A'A =$ major axis

$2b = BB' =$ minor axis

$2c = FF' =$ focal distance.

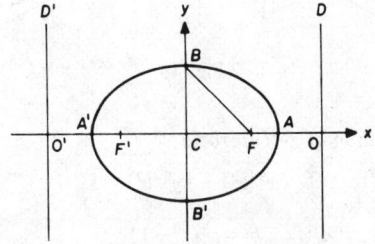

Fig. 18.

Then

$$OC = a/e$$

$$BF = a$$

$$FC = ae$$

$$1 - e^2 = b^2/a^2.$$

Equation of Ellipse:

$$(x^2/a^2) + (y^2/b^2) = 1.$$

Sum of the Focal Radii:

To any point on ellipse $= 2a$.

Equation of Tangent Line to Ellipse:

$(x_1, y_1) =$ point of tangency

$$(xx_1/a^2) + (yy_1/b^2) = 1.$$

Equation of Normal Line to an Ellipse:

$$y - y_1 = (a^2 y_1/b^2 x_1)(x - x_1).$$

Hyperbola

Figure 19 shows x-hyperbola centered at origin. If the hyperbola is centered at (h, k) instead, the equations that follow must be modified by replacing x, x_1, y, y_1 by $x-h$, x_1-h, $y-k$, y_1-k, respectively.

Foci: F, F'
Directrices: D, D'

$$e = \text{eccentricity} > 1$$

$$2a = \text{transverse axis} = A'A$$

$$CO = a/e$$

$$CF = ae.$$

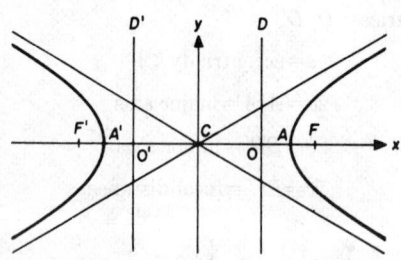

Fig. 19.

Equation of x-Hyperbola:

$$(x^2/a^2) - (y^2/b^2) = 1$$

where

$$b^2 = a^2(e^2 - 1).$$

Equation of Conjugate $(y\text{-})$ Hyperbola:

$$(y^2/b^2) - (x^2/a^2) = 1.$$

Tangent Line to x-Hyperbola:

$(x_1, y_1) =$ point of tangency

$$a^2 y_1 y - b^2 x_1 x = -a^2 b^2.$$

Normal Line to x-Hyperbola:

$$y - y_1 = -(a^2 y_1/b^2 x_1)(x - x_1).$$

Asymptotes to Hyperbola:

$$y = \pm (b/a) x.$$

SOLID ANALYTIC GEOMETRY

In the following, x, y, and z are the coordinates of a variable point in space in a right-handed rectangular-coordinate system. See Fig. 20.

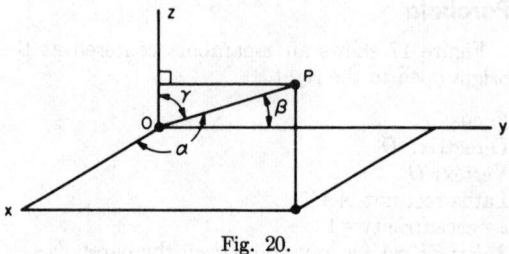

Fig. 20.

Coordinates

The coordinates of a point are given by a triplet, e.g., (x_0, y_0, z_0).

Direction Cosines and Numbers

α, β, and γ are the angles the line OP makes with the x, y, and z axes, respectively. $\cos\alpha$, $\cos\beta$, and $\cos\gamma$ are the direction cosines of the line OP. Numbers proportional to the direction cosines are called direction numbers.

Distance Between Two Points

$$d = [(x_1 - x_0)^2 + (y_1 - y_0)^2 + (z_1 - z_0)^2]^{1/2}.$$

Equations of a Plane

General Form:

$$Ax + By + Cz + D = 0.$$

Three-Point Form:

$$\begin{vmatrix} x - x_0 & y - y_0 & z - z_0 \\ x_1 - x_0 & y_1 - y_0 & z_1 - z_0 \\ x_2 - x_0 & y_2 - y_0 & z_2 - z_0 \end{vmatrix} = 0$$

where (x_0, y_0, z_0), (x_1, y_1, z_1), and (x_2, y_2, z_2) are three points on the plane.

Intercept Form:

$$(x/a)+(y/b)+(z/c)=1$$

where a, b, and c are the intercepts on the x, y, and z axes, respectively.

Point-Direction Form:

$$A(x-x_1)+B(y-y_1)+C(z-z_1)=0$$

where A, B, and C are the direction numbers of a normal to the plane and (x_1, y_1, z_1) is a point on the plane.

Normal Form:

$$x\cos\alpha+y\cos\beta+z\cos\gamma-p=0$$

where $\cos\alpha$, $\cos\beta$, and $\cos\gamma$ are the direction cosines of a normal to the plane, and p is the distance of the plane to the origin.

Equations of the Straight Line

Point-Direction Form:

$$(x-x_1)/A=(y-y_1)/B=(z-z_1)/C$$

or

$$x=At+x_1$$
$$y=Bt+y_1$$
$$z=Ct+z_1 \quad (parametric)$$

where A, B, and C are direction numbers of the line and (x_1, y_1, z_1) is a point on the line.

Two-Point Form:

$$(x-x_1)/(x_1-x_0)=(y-y_1)/(y_1-y_0)$$
$$=(z-z_1)/(z_1-z_0)$$

or

$$x=x_0+(x_1-x_0)t$$
$$y=y_0+(y_1-y_0)t$$
$$z=z_0+(z_1-z_0)t \quad (parametric)$$

where (x_0, y_0, z_0) and (x_1, y_1, z_1) are two points on the line.

Ellipsoid

$$(x^2/a^2)+(y^2/b^2)+(z^2/c^2)=1$$

where a, b, c are the semiaxes of the ellipsoid or the intercepts on the x, y, and z axes, respectively.

Prolate Spheroid

$$a^2(y^2+z^2)+b^2x^2=a^2b^2$$

where $a>b$, and x-axis=axis of revolution.

Oblate Spheroid

$$b^2(x^2+z^2)+a^2y^2=a^2b^2$$

where $a>b$, and y-axis=axis of revolution.

Paraboloid of Revolution

$$y^2+z^2=2px$$

x-axis=axis of revolution.

Hyperboloid of Revolution

Revolving an x-hyperbola about the x-axis results in the hyperboloid of two sheets

$$a^2(y^2+z^2)-b^2x^2=-a^2b^2.$$

Revolving an x-hyperbola about the y-axis results in the hyperboloid of one sheet

$$b^2(x^2+z^2)-a^2y^2=a^2b^2.$$

DIFFERENTIAL CALCULUS

List of Derivatives

In the following u, v, w are differentiable functions of x, and c is a constant.

General Equations:

$$dc/dx=0$$
$$dx/dx=1$$
$$(d/dx)(u+v-w)=(du/dx)+(dv/dx)-(dw/dx)$$
$$(d/dx)(cv)=c(dv/dx)$$
$$(d/dx)(uv)=u(dv/dx)+v(du/dx)$$
$$(d/dx)(v^c)=cv^{c-1}(dv/dx)$$
$$\frac{d}{dx}\left(\frac{u}{v}\right)=\frac{v(du/dx)-u(dv/dx)}{v^2}$$
$$dy/dx=(dy/dv)(dv/dx), \quad \text{if} \quad y=y(v)$$
$$dy/dx=(dx/dy)^{-1}, \quad \text{if} \quad dx/dy\neq0.$$

Transcendental Functions:

$$(d/dx)\ln v=v^{-1}(dv/dx)$$
$$(d/dx)(c^v)=c^v\ln c(dv/dx)$$
$$(d/dx)(e^v)=e^v(dv/dx)$$
$$(d/dx)(u^v)=vu^{v-1}(du/dx)+(\ln u)u^v(dv/dx)$$

$(d/dx)(\sin v) = \cos v(dv/dx)$

$(d/dx)(\cos v = -\sin v(dv/dx)$

$(d/dx)(\tan v) = \sec^2 v(dv/dx)$

$(d/dx)(\cot v) = -\csc^2 v(dv/dx)$

$(d/dx)(\sec v) = \sec v \tan v(dv/dx)$

$(d/dx)(\csc v) = -\csc v \cot v(dv/dx)$

$(d/dx)(\arc \sin v) = (1-v^2)^{-1/2}(dv/dx)$

$(d/dx)(\arc \cos v) = -(1-v^2)^{-1/2}(dv/dx)$

$(d/dx)(\arc \tan v) = (1+v^2)^{-1}(dv/dx)$

$(d/dx)(\arc \cot v) = -(1+v^2)^{-1}(dv/dx)$

$(d/dx)(\arc \sec v) = [v(v^2-1)^{1/2}]^{-1}(dv/dx)$

$(d/dx)(\arc \csc v) = -[v(v^2-1)^{1/2}]^{-1}(dv/dx)$

$(d/dx)(\sinh v) = \cosh v(dv/dx)$

$(d/dx)(\cosh v) = \sinh v(dv/dx)$

$(d/dx)(\tanh v) = \mathrm{sech}^2 v(dv/dx).$

TABLE OF INTEGRALS

Indefinite Integrals

General Equations:

$$\int af(x)\,dx = a\int f(x)\,dx$$

$$\int [f(x)+g(x)]dx = \int f(x)\,dx + \int g(x)\,dx$$

$$(d/dx)\int f(x)\,dx = f(x)$$

$$(d/dx)\int_u^v f(y,x)\,dy = f(v,x)\cdot(dv/dx)$$

$$-f(u,x)\cdot(du/dx) + \int_u^v \frac{\partial f(y,x)}{\partial x}\,dy$$

$$\int f'(x)g(x)\,dx = f(x)g(x) - \int f(x)g'(x)\,dx$$

$$\int f(x)\,dx = \int f[h(y)]h'(y)\,dy$$

$$\int_{x_1}^{x_2} f(x)\,dx = \int_{-x_2}^{-x_1} f(-x)\,dx$$

$$\int f(\sin x,\cos x)\,dx = \int f\left(\frac{2z}{1+z^2},\frac{1-z^2}{1+z^2}\right)\frac{dz}{1+z^2}.$$

Elementary Forms:

$$\int x^m dx = x^{m+1}/(m+1), \quad m \neq -1$$

$$\int (dx/x) = \ln|x|$$

$$\int e^x dx = e^x$$

$$\int \ln x\,dx = x\ln x - x$$

$$\int \sin x\,dx = -\cos x$$

$$\int \cos x\,dx = \sin x$$

$$\int \tan x\,dx = -\ln|\cos x|$$

$$\int \cot x\,dx = \ln|\sin x|$$

$$\int \csc x\,dx = \ln|\tan \tfrac{1}{2}x|$$

$$\int \sec x\,dx = \ln|\sec x + \tan x|$$

$$\int \sinh x\,dx = \cosh x$$

$$\int \cosh x\,dx = \sinh x$$

$$\int \tanh x\,dx = \ln|\cosh x|$$

$$\int \coth x\,dx = \ln|\sinh x|$$

$$\int \mathrm{sech}\,x\,dx = 2\tan^{-1}e^x$$

$$\int \mathrm{csch}\,x\,dx = \ln|\tanh \tfrac{1}{2}x|$$

$$\int dx/(1+x^2) = \tan^{-1}x$$

$$\int dx/(1-x^2) = \tfrac{1}{2}\ln[(1+x)/(1-x)]$$

$$\int dx/(x^2-1) = \tfrac{1}{2}\ln[(x-1)/(x+1)]$$

$$\int dx/(1-x^2)^{1/2} = \sin^{-1}x$$

$$\int dx/(x^2\pm1)^{1/2} = \ln|x+(x^2\pm1)^{1/2}|$$

$$\int (x^2\pm1)^{1/2}dx = \tfrac{1}{2}[x(x^2\pm1)^{1/2}\pm\ln|x+(x^2\pm1)^{1/2}|]$$

$$\int (1-x^2)^{1/2}dx = \tfrac{1}{2}[x(1-x^2)^{1/2}+\sin^{-1}x].$$

Forms Containing $(ax+b)$, $a{\neq}0$, $b{\neq}0$:

$$\int (ax+b)^n dx = [(ax+b)^{n+1}/a(n+1)], \quad n{\neq}-1$$

$$\int (ax+b)^{-1} dx = (1/a) \ln |ax+b|$$

$$\int x\,dx/(ax+b) = a^{-2}[ax+b-b\ln(ax+b)]$$

$$\int x\,dx/(ax+b)^2 = a^{-2}[b/(ax+b)+\ln(ax+b)]$$

$$\int dx/[x(ax+b)] = b^{-1}\ln[x/(ax+b)]$$

$$\int dx/[x(ax+b)^2] = [b(ax+b)]^{-1}+b^{-2}\ln[x/(ax+b)]$$

$$\int dx/[x^2(ax+b)] = -(bx)^{-1}+(a/b^2)\ln[(ax+b)/x]$$

$$\int dx/[x^2(ax+b)^2] = -(2ax+b)/[b^2x(ax+b)]+(2a/b^3)\ln[(ax+b)/x]$$

$$\int x^m(ax+b)^n dx = [x^{m+1}(ax+b)^n/(m+n+1)]+[bn/(m+n+1)]\int x^m(ax+b)^{n-1}dx$$

$$\int \frac{x^m dx}{(ax+b)^n} = [b(1-n)]^{-1}\left[\frac{-x^{m+1}}{(ax+b)^{n-1}}+(m-n+2)\int \frac{x^m}{(ax+b)^{n-1}}\,dx\right], \quad n{\neq}1$$

$$\int \frac{x^m dx}{ax+b} = \frac{x^m}{am}+(b/a)\int \frac{x^{m-1}}{ax+b}\,dx, \quad m{\neq}0.$$

Forms Containing $(ax+b)^{1/2}$, $a{\neq}0$, $b{\neq}0$:

$$\int x(ax+b)^{1/2}dx = \frac{2(3ax-2b)[(ax+b)^3]^{1/2}}{15a^2}$$

$$\int x^m(ax+b)^{1/2}dx = [2/a(2m+3)]\left\{x^m[(ax+b)^3]^{1/2}-mb\int x^{m-1}(ax+b)^{1/2}dx\right\}$$

$$\int \frac{(ax+b)^{1/2}}{x}\,dx = 2(ax+b)^{1/2}+(b)^{1/2}\ln\left|\frac{(ax+b)^{1/2}-(b)^{1/2}}{(ax+b)^{1/2}+(b)^{1/2}}\right|, \quad b>0$$

$$= 2(ax+b)^{1/2}-2(-b)^{1/2}\tan^{-1}[-(ax+b)/b]^{1/2}, \quad b<0$$

$$\int \frac{(ax+b)^{1/2}}{x^m}\,dx = -[(m-1)b]^{-1}\frac{[(ax+b)^3]^{1/2}}{x^{m-1}}+\tfrac{1}{2}[(2m-5)a]\int \frac{(ax+b)^{1/2}dx}{x^{m-1}}, \quad m{\neq}1$$

$$\int \frac{x\,dx}{(ax+b)^{1/2}} = [2(ax-2b)/3a^2](ax+b)^{1/2}$$

$$\int \frac{x^m dx}{(ax+b)^{1/2}} = \frac{2x^m(ax+b)^{1/2}}{(2m+1)a}-\frac{2bm}{(2m+1)a}\int \frac{x^{m-1}dx}{(ax+b)^{1/2}}$$

$$\int \frac{dx}{x(ax+b)^{1/2}} = (b)^{-1/2} \ln \left| \frac{(ax+b)^{1/2}-(b)^{1/2}}{(ax+b)^{1/2}+(b)^{1/2}} \right|, \quad b>0$$

$$= [2/(-b)^{1/2}] \tan^{-1}[(ax+b)/-b]^{1/2}, \quad b<0$$

$$\int \frac{dx}{x^n(ax+b)^{1/2}} = -\frac{(ax+b)^{1/2}}{(n-1)bx^{n-1}} - \frac{(2n-3)a}{(2n-2)b} \int \frac{dx}{x^{n-1}(ax+b)^{1/2}}, \quad n\neq 1.$$

Forms Containing $R=ax^2+bx+c$, $a\neq 0$, $x>0$: Let $q=4ac-b^2$.

$$\int \frac{dx}{R} = [2/(q)^{1/2}] \tan^{-1}[(2ax+b)/(q)^{1/2}], \quad q>0$$

$$= (-q)^{-1/2} \ln \frac{2ax+b-(-q)^{1/2}}{2ax+b+(-q)^{1/2}}, \quad q<0.$$

(If $q=0$, R is a perfect square.)

$$\int \frac{dx}{R^n} = \frac{2ax+b}{(n-1)qR^{n-1}} + \frac{2(2n-3)a}{q(n-1)} \int \frac{dx}{R^{n-1}}$$

$$\int \frac{x^m}{R^n} dx = \frac{x^{m-1}}{(2n-m-1)aR^{n-1}} - \frac{n-m}{2n-m-1} (b/a) \int \frac{x^{m-1}dx}{R^n} + \frac{m-1}{2n-m-1} (c/a) \int \frac{x^{m-2}dx}{R^n}$$

$$\int \frac{dx}{x^m R^n} = -[(m-1)cx^{m-1}R^{n-1}]^{-1} - \frac{m+n-2}{m-1} (b/c) \int \frac{dx}{x^{m-1}R^n} - \frac{m+2n-3}{m-1} (a/c) \int \frac{dx}{x^{m-2}R^n}.$$

Forms Containing $(R)^{1/2}=(ax^2+bx+c)^{1/2}$, $a\neq 0$: Let $q=4ac-b^2$.

$$\int \frac{dx}{(R)^{1/2}} = (a)^{-1/2} \ln \left[(R)^{1/2}+x(a)^{1/2}+\frac{b}{2(a)^{1/2}} \right], \quad a>0$$

$$= \frac{-1}{(-a)^{1/2}} \sin^{-1}\left[\frac{2ax+b}{(-q)^{1/2}} \right], \quad a<0$$

$$\int (R)^{1/2}dx = \frac{(2ax+b)(R)^{1/2}}{4a} + \frac{q}{8a} \int \frac{dx}{(R)^{1/2}}$$

$$\int \frac{dx}{R^n(R)^{1/2}} = \frac{2(2ax+b)(R)^{1/2}}{(2n-1)qR^n} + \frac{8a(n-1)}{(2n-1)q} \int \frac{dx}{R^{n-1}(R)^{1/2}}$$

$$\int R^n(R)^{1/2}dx = \frac{(2ax+b)R^n(R)^{1/2}}{4(n+1)a} + \frac{(2n+1)q}{8(n+1)a} \int \frac{R^ndx}{(R)^{1/2}}$$

$$\int \frac{xdx}{(R)^{1/2}} = \frac{(R)^{1/2}}{a} - \frac{b}{2a} \int \frac{dx}{(R)^{1/2}}$$

$$\int x(R)^{1/2}dx = \frac{R(R)^{1/2}}{3a} - \frac{b}{2a} \int (R)^{1/2}dx$$

$$\int \frac{x^mdx}{R^n(R)^{1/2}} dx = a^{-1} \int \frac{x^{m-2}dx}{R^{n-1}(R)^{1/2}} - (b/a) \int \frac{x^{m-1}dx}{R^n(R)^{1/2}} - (c/a) \int \frac{x^{m-2}dx}{R^n(R)^{1/2}}$$

$$\int \frac{x^mR^n}{(R)^{1/2}} dx = \frac{x^{m-1}R^n(R)^{1/2}}{(2n+m)a} - \frac{(2n+2m+1)b}{2a(2n+m)} \int \frac{x^{m-1}R^ndx}{(R)^{1/2}} - \frac{(m-1)c}{(2n+m)a} \int \frac{x^{m-2}R^ndx}{(R)^{1/2}}$$

$$\int \frac{dx}{x^m R^n (R)^{1/2}} = -\frac{(R)^{1/2}}{(m-1)cx^{m-1}R^n} - \frac{(2n+2m-3)b}{2c(m-1)} \int \frac{dx}{x^{m-1}R^n(R)^{1/2}} - \frac{(2n+m-2)a}{(m-1)c} \int \frac{dx}{x^{m-2}R^n(R)^{1/2}}$$

$$\int \frac{R^n dx}{x^m (R)^{1/2}} = -\frac{R^{n-1}(R)^{1/2}}{(m-1)x^{m-1}} + \frac{(2n-1)b}{2(m-1)} \int \frac{R^{n-1}dx}{x^{m-1}(R)^{1/2}} + \frac{(2n-1)a}{m-1} \int \frac{R^{n-1}dx}{x^{m-2}(R)^{1/2}}.$$

Logarithmic Integrands

$$\int \ln ax\, dx = x(\ln ax - 1)$$

$$\int \log_b x\, dx = \log_b e(\ln x - 1)x = [(\ln x - 1)x/\ln b]$$

$$\int (\ln x)^n dx = x(\ln x)^n - n \int (\ln x)^{n-1}dx$$

$$\int x^m \ln x\, dx = x^{m+1}[(m+1)^{-1}\ln x - (m+1)^{-2}], \quad m \neq -1$$

$$\int \frac{\ln x}{x}\, dx = \tfrac{1}{2}(\ln x)^2$$

$$\int x^m(\ln x)^n dx = [x^{m+1}(\ln x)^n/(m+1)] - [n/(m+1)] \int x^m(\ln x)^{n-1}dx, \quad m \neq -1$$

$$\int [(\ln x)^n/x]dx = (\ln x)^{n+1}/(n+1), \quad n \neq -1$$

$$\int dx/(x \ln x) = \ln | \ln x |$$

$$\int \frac{x^m dx}{(\ln x)^n} = -\frac{x^{m+1}}{(n-1)(\ln x)^{n-1}} + \frac{m+1}{n-1} \int \frac{x^m dx}{(\ln x)^{n-1}}, \quad n \neq 1$$

$$\int (x^m/\ln x)\, dx = \text{Ei}[(n+1)\ln x], \quad m \neq -1.$$

Exponential Integrands

$$\int a^{bx} dx = a^{bx}/(b \ln a)$$

$$\int x^m e^x dx = x^m e^x - m \int x^{m-1}e^x dx$$

$$\int dx/(a+be^{mx}) = (x/a) - (am)^{-1}\ln | a+be^{mx} |, \quad a \text{ and } m \neq 0$$

$$\int \frac{dx}{(a+be^{mx})^{1/2}} = [m(a)^{1/2}]^{-1} \ln \frac{(a+be^{mx})^{1/2} - (a)^{1/2}}{(a+be^{mx})^{1/2} + (a)^{1/2}}, \quad a > 0$$

$$= \{2/[m(-a)^{1/2}]\} \arctan[(a+be^{mx})^{1/2}/(-a)^{1/2}], \quad a < 0.$$

Trigonometric Integrands

$$\int \sin^2 x \, dx = \tfrac{1}{2}(x - \sin x \cos x)$$

$$\int \sin^n x \, dx = -[(\sin^{n-1} x \cos x)/n] + (n-1)/n \int \sin^{n-2} x \, dx$$

$$\int \cos^2 x \, dx = \tfrac{1}{2}(x + \sin x \cos x)$$

$$\int \cos^n x \, dx = [(\cos^{n-1} x \sin x)/n] + (n-1)/n \int \cos^{n-2} x \, dx$$

$$\int \sin x \cos^m x \, dx = -\cos^{m+1} x/(m+1)$$

$$\int \sin^m x \cos x \, dx = \sin^{m+1} x/(m+1)$$

$$\int \sin^n x \cos^m x = \frac{\cos^{m-1} x \sin^{n+1} x}{m+n} + \frac{m-1}{m+n} \int \cos^{m-2} x \sin^n x \, dx$$

$$= -\frac{\sin^{n-1} x \cos^{m+1} x}{m+n} + \frac{n-1}{m+n} \int \cos^m x \sin^{n-2} x \, dx$$

$$\int \frac{\sin^n x \, dx}{\cos^m x} = (m-1)^{-1} \left[\frac{\sin^{n-1} x}{\cos^{m-1} x} - (n-1) \int \frac{\sin^{n-2} x \, dx}{\cos^{m-2} x} \right]$$

$$\int \frac{\cos^m x}{\sin^n x} \, dx = -(n-1)^{-1} \left[\frac{\cos^{m-1} x}{\sin^{n-1} x} + (m-1) \int \frac{\cos^{m-2} x \, dx}{\sin^{n-2} x} \right]$$

$$\int \frac{dx}{\sin^m x \cos^n x} = [(n-1) \sin^{m-1} x \cos^{n-1} x]^{-1} + \frac{m+n-2}{n-1} \int \frac{dx}{\sin^m x \cos^{n-2} x}$$

$$= -[(m-1) \sin^{m-1} x \cos^{n-1} x]^{-1} + \frac{m+n-2}{m-1} \int \frac{dx}{\sin^{m-2} x \cos^n x}$$

$$\int \tan^n x \, dx = [\tan^{n-1} x/(n-1)] - \int \tan^{n-2} x \, dx$$

$$\int \cot^n x \, dx = -[\cot^{n-1} x/(n-1)] - \int \cot^{n-2} x \, dx$$

$$\int \sec^2 x \, dx = \tan x$$

$$\int \sec^n x \, dx = [\sin x/(n-1) \cos^{n-1} x] + [(n-2)/(n-1)] \int \sec^{n-2} x \, dx, \quad n \neq 1$$

$$\int \frac{dx}{a + b \cos x + c \sin x} = \frac{2}{(a^2 - b^2 - c^2)^{1/2}} \tan^{-1} \frac{(a-b) \tan(x/2) + c}{(a^2 - b^2 - c^2)^{1/2}}, \quad a^2 > b^2 + c^2$$

$$= (b^2 + c^2 - a^2)^{-1/2} \ln \left| \frac{(a-b) \tan(x/2) + c - (b^2 + c^2 - a^2)^{1/2}}{(a-b) \tan(x/2) + c + (b^2 + c^2 - a^2)^{1/2}} \right|, \quad a^2 < b^2 + c^2, \; a \neq b$$

$$= c^{-1} \ln |a+c \tan(x/2)|, \quad a=b$$

$$= \frac{-2}{c+(a-b)\tan(x/2)}, \quad a^2=b^2+c^2$$

$$\int x^n \sin ax\, dx = -\sum_{k=0}^{n} k! \binom{n}{k} \frac{x^{n-k}}{a^{k+1}} \cos[ax+(k\pi/2)], \quad n \text{ nonnegative integer}$$

$$\int x^n \cos ax\, dx = \sum_{k=0}^{n} k! \binom{n}{k} \frac{x^{n-k}}{a^{k+1}} \sin[ax+(k\pi/2)], \quad n \text{ nonnegative integer.}$$

Inverse Trigonometric Integrals, x and a>0:

$$\int \sin^{-1}(x/a)\, dx = x \sin^{-1}(x/a) + (a^2-x^2)^{1/2}$$

$$\int \cos^{-1}(x/a)\, dx = x \cos^{-1}(x/a) - (a^2-x^2)^{1/2}$$

$$\int \tan^{-1}(x/a)\, dx = x \tan^{-1}(x/a) - \tfrac{1}{2}a \ln(a^2+x^2)$$

$$\int \cot^{-1}(x/a)\, dx = x \cot^{-1}(x/a) + \tfrac{1}{2}a \ln(a^2+x^2)$$

$$\int \sec^{-1}(x/a)\, dx = x \sec^{-1}(x/a) - a \ln[x+(x^2-a^2)^{1/2}]$$

$$\int \csc^{-1}(x/a)\, dx = x \csc^{-1}(x/a) + a \ln[x+(x^2-a^2)^{1/2}]$$

$$\int x \sin^{-1}x\, dx = \tfrac{1}{4}[(2x^2-1)\sin^{-1}x + x(1-x^2)^{1/2}]$$

$$\int x \cos^{-1}x\, dx = \tfrac{1}{4}[(2x^2-1)\cos^{-1}x - x(1-x^2)^{1/2}]$$

$$\int x^n \sin^{-1}x\, dx = [x^{n+1}\sin^{-1}x/(n+1)] - (n+1)^{-1}\int [x^{n+1}/(1-x^2)^{1/2}]dx$$

$$\int x^n \cos^{-1}x\, dx = [x^{n+1}\cos^{-1}x/(n+1)] + (n+1)^{-1}\int [x^{n+1}/(1-x^2)^{1/2}]dx.$$

Miscellaneous Integrals:

$$\int \frac{dx}{ax^3+b} = \frac{p}{3b}\left[\tfrac{1}{2}\ln \frac{(x+p)^2}{x^2-px+p^2} + \sqrt{3}\ \text{arc tan}\ \frac{x\sqrt{3}}{2p-x}\right], \quad p=(b/a)^{1/3}$$

$$\int \frac{dx}{x^4+a^4} = (4a^3\sqrt{2})^{-1}\left[\ln\left(\frac{x^2+ax\sqrt{2}+a^2}{x^2-ax\sqrt{2}+a^2}\right) + 2\tan^{-1}\left(\frac{ax\sqrt{2}}{a^2-x^2}\right)\right]$$

$$\int \frac{dx}{x^4-a^4} = (1/4a^3)\left[\ln\left(\frac{x-a}{x+a}\right) - 2\tan^{-1}(x/a)\right].$$

Definite Integrals

$$\int_0^1 x^u(1-x)^v dx = [\Gamma(u+1)\Gamma(v+1)/\Gamma(u+v+2)], \quad u>-1, v>-1$$

$$= [u!v!/(u+v+1)!], \quad \text{if } u, v \text{ nonnegative integers}$$

$$\int_0^\infty x^u dx/(1+x^v) = (\pi/v) \csc[(u+1)\pi/v], \quad 0<u+1<v$$

$$\int_0^1 \frac{x^{2n+1}dx}{(1-x^2)^{1/2}} = \frac{(2n)(2n-2)\cdots 6\cdot 4\cdot 2}{(2n+1)(2n-1)\cdots 5\cdot 3\cdot 1}, \quad n \text{ positive integer}$$

$$\int_0^1 \frac{x^{2n}dx}{(1-x^2)^{1/2}} = \frac{(2n-1)(2n-3)\cdots 5\cdot 3\cdot 1}{(2n)(2n-2)\cdots 6\cdot 4\cdot 2}\cdot\tfrac{1}{2}\pi, \quad n \text{ positive integer}$$

$$\int_0^\infty x^b e^{-ax}dx = \Gamma(b+1)/a^{b+1}, \quad b>-1, a>0$$

$$= b!/a^{b+1}, \quad b \text{ nonnegative integer}$$

$$\int_0^\infty e^{-ax^2}dx = \tfrac{1}{2}(\pi/a)^{1/2}, \quad a>0$$

$$\int_0^\infty x^b e^{-ax^2}dx = \frac{\Gamma[(b+1)/2]}{2a^{(b+1)/2}}, \quad b>-1$$

$$\int_0^\infty \exp[-(ax^2+bx+c)]dx = \tfrac{1}{2}(\pi/a)^{1/2}\exp[(b^2-4ac)/4a][1-\mathrm{erf}(\tfrac{1}{2}ba^{1/2})]$$

$$\int_0^\infty \exp\{-[x^2+(a^2/x^2)]\}dx = \tfrac{1}{2}\exp(-2|a|)(\pi)^{1/2}$$

$$\int_0^{\pi/2} \sin^n x dx = \int_0^{\pi/2} \cos^n x dx = \tfrac{1}{2}(\pi)^{1/2}\frac{\Gamma[\tfrac{1}{2}(n+1)]}{\Gamma(\tfrac{1}{2}n+1)}, \quad n>-1$$

$$\int_0^\infty (\sin mx/x)\,dx = \tfrac{1}{2}\pi, \quad m>0$$

$$= 0, \quad m=0$$

$$= -\tfrac{1}{2}\pi, \quad m<0$$

$$\int_0^\infty (\sin x \cdot \cos mx/x)\,dx = 0, \quad |m|>1$$

$$= \tfrac{1}{4}\pi, \quad m=\pm 1$$

$$= \tfrac{1}{2}\pi, \quad -1<m<1$$

$$\int_0^\infty (\sin x/x)^2 dx = \tfrac{1}{2}\pi$$

$$\int_0^\infty \cos(x^2)\,dx = \int_0^\infty \sin(x^2)\,dx = \tfrac{1}{2}(\tfrac{1}{2}\pi)^{1/2}$$

$$\int_0^\infty [\cos mx/(1+x^2)]dx = \tfrac{1}{2}\pi e^{-|m|}$$

$$\int_0^\infty (\cos x/x^{1/2})\,dx = \int_0^\infty (\sin x/x^{1/2})\,dx = (\tfrac{1}{2}\pi)^{1/2}$$

$$\int_0^\infty \exp(-a^2x^2)\,\cos bx\,dx = [(\pi)^{1/2}/2\,|\,a\,|]\exp(-b^2/4a^2)$$

$$\int_0^1 [\ln x/(1-x)]\,dx = -\tfrac{1}{6}\pi^2$$

$$\int_0^1 [\ln x/(1+x)]\,dx = -\tfrac{1}{12}\pi^2$$

$$\int_0^1 [\ln x/(1-x^2)]\,dx = -\tfrac{1}{8}\pi^2$$

$$\int_0^1 [\ln x/(1-x^2)^{1/2}]\,dx = -\tfrac{1}{2}\pi\ln 2$$

$$\int_0^1 (\ln x)^n\,dx = (-1)^n n!$$

$$\int_0^1 x^m(\ln x^{-1})^n\,dx = [\Gamma(n+1)/(m+1)^{n+1}], \quad m>-1, n>-1$$

$$\int_0^{\pi/2} \ln \sin x\,dx = -\tfrac{1}{2}\pi\ln 2$$

$$\int_0^{\pi/2} \ln \cos x\,dx = -\tfrac{1}{2}\pi\ln 2$$

$$\int_0^\pi x \ln \sin x\,dx = -\tfrac{1}{2}\pi^2\ln 2$$

$$\int_0^\pi \ln(a+b\cos x)\,dx = \pi\ln\tfrac{1}{2}[a+(a^2-b^2)^{1/2}], \quad a\geq b.$$

SERIES

Taylor's Series for a Single Variable

$$f(z) = f(a)+f'(a)(z-a)+\tfrac{1}{2}[f''(a)](z-a)^2+\cdots+[f^{(n)}(a)/n!](z-a)^n+R_n$$

$$= \sum_{k=0}^n (k!)^{-1}f^{(k)}(a)(z-a)^k+R_n$$

where the remainder is bounded by $|R_n|\leq[M/(n+1)!]|\,z-a\,|^{n+1}$ in which

$$M = \max_{0\leq\theta\leq1} |f^{(n+1)}[a+\theta(z-a)]|.$$

If, a, z, and f are real, then there exists a real θ, $0<\theta<1$, such that the remainder is

$$R_n = \{f^{(n+1)}[a+\theta(z-a)]/(n+1)!\}(z-a)^{n+1}.$$

When $a=0$, this series is often called MacLaurin's series.

Taylor's Series for Two Variables

$$f(x,y) = f(a,b) + \left[\frac{\partial f(x,y)}{\partial x} \bigg|_{x=a,y=b} (x-a) + \frac{\partial f(x,y)}{\partial y} \bigg|_{x=a,y=b} (y-b) \right]$$

$$+ \tfrac{1}{2} \left[\frac{\partial^2 f(x,y)}{\partial x^2} \bigg|_{x=a,y=b} (x-a)^2 + 2 \frac{\partial^2 f(x,y)}{\partial x \partial y} \bigg|_{x=a,y=b} (x-a)(y-b) + \frac{\partial^2 f(x,y)}{\partial y^2} \bigg|_{x=a,y=b} (y-b)^2 \right] + \cdots$$

$$= \sum_{k=0}^{n} (k!)^{-1} \{ [(x-a)(\partial/\partial\xi) + (y-b)(\partial/\partial\eta)]^k f(\xi,\eta) \}_{\xi=a,\eta=b} + R_n$$

where the remainder is bounded by $|R_n| \leq [M/(n+1)!](|x-a| + |y-b|)^{n+1}$ in which

$$M = \max_{\substack{k=0,1,2,\cdots,n+1 \\ 0 \leq \theta \leq 1}} |\partial^{n+1} f(\xi,\eta)/\partial^k \xi \partial^{n+1-k}\eta|, \qquad \xi = a + \theta(x-a), \eta = b + \theta(y-b).$$

If $f(x,y)$ is a real function of two variables, there exists a number θ, $0 < \theta < 1$ such that the remainder is

$$R_n = [(n+1)!]^{-1} \{ [(x-a)(\partial/\partial\xi) + (y-b)(\partial/\partial\eta)]^{n+1} f(\xi,\eta) \}, \qquad \xi = a + \theta(x-a), \eta = b + \theta(y-b).$$

Miscellaneous Series

$$\ln(1+x) = x - \tfrac{1}{2}x^2 + \tfrac{1}{3}x^3 - \tfrac{1}{4}x^4 + \cdots = -\sum_{k=1}^{\infty} (-1)^k (x^k/k), \qquad |x| < 1$$

$$e^x = 1 + x + (x^2/2!) + (x^3/3!) + \cdots = \sum_{k=0}^{\infty} (x^k/k!), \qquad |x| < \infty$$

$$\sin x = x - \frac{x^3}{3!} + \frac{x^5}{5!} - \frac{x^7}{7!} + \cdots = \sum_{k=0}^{\infty} (-1)^k \frac{x^{2k+1}}{(2k+1)!}, \qquad |x| < \infty$$

$$\cos x = 1 - \frac{x^2}{2!} + \frac{x^4}{4!} - \frac{x^6}{6!} + \cdots = \sum_{k=0}^{\infty} (-1)^k \frac{x^{2k}}{(2k)!}, \qquad |x| < \infty$$

$$\sinh x = x + \frac{x^3}{3!} + \frac{x^5}{5!} + \frac{x^7}{7!} + \cdots = \sum_{k=0}^{\infty} \frac{x^{2k+1}}{(2k+1)!}, \qquad |x| < \infty$$

$$\cosh x = 1 + \frac{x^2}{2!} + \frac{x^4}{4!} + \frac{x^6}{6!} + \cdots = \sum_{k=0}^{\infty} \frac{x^{2k}}{(2k)!}, \qquad |x| < \infty$$

$$\tan x = x + \tfrac{1}{3}x^3 + (2x^5/15) + (17x^7/315) + (62x^9/2835) + \cdots, \qquad |x| < \tfrac{1}{2}\pi$$

$$\cot x = \frac{1}{x} - \frac{x}{3} - \frac{x^3}{45} - \frac{2x^5}{945} - \frac{x^7}{4725} - \cdots, \qquad |x| < \pi$$

$$\text{arc } \sin x = x + \frac{1}{2}\frac{x^3}{3} + \frac{1 \cdot 3}{2 \cdot 4}\frac{x^5}{5} + \frac{1 \cdot 3 \cdot 5}{2 \cdot 4 \cdot 6}\frac{x^7}{7} + \cdots, \qquad |x| < 1$$

$$\text{arc } \tan x = x - \tfrac{1}{3}x^3 + \tfrac{1}{5}x^5 - \tfrac{1}{7}x^7 + \cdots, \qquad |x| < 1$$

$$\text{arc } \sinh x = x - \frac{1}{2}\frac{x^3}{3} + \frac{1 \cdot 3}{2 \cdot 4}\frac{x^5}{5} - \frac{1 \cdot 3 \cdot 5}{2 \cdot 4 \cdot 6}\frac{x^7}{7} + \cdots, \qquad |x| < 1$$

$$\text{arc } \tanh x = x + \tfrac{1}{3}x^3 + \tfrac{1}{5}x^5 + \tfrac{1}{7}x^7 + \cdots, \qquad |x| < 1.$$

MATRIX ALGEBRA

Notation

A matrix of order $m \times n$ is a rectangular array of numbers, real or complex, consisting of m rows and n columns:

$$\mathbf{A} = \begin{bmatrix} a_{11} & a_{12} & a_{13} & \cdots & a_{1n} \\ a_{21} & a_{22} & a_{23} & \cdots & a_{2n} \\ \cdots & \cdots & \cdots & \cdots & \cdots \\ a_{m1} & a_{m2} & a_{m3} & \cdots & a_{mn} \end{bmatrix} = [a_{ij}].$$

A row (column) vector is a $1 \times n$ ($n \times 1$) matrix. An $n \times n$ matrix is called a square matrix. A matrix with all entries equal to zero is called a zero matrix; it is denoted by $\mathbf{0}$. A square zero matrix with the elements on the main diagonal, that is, the diagonal extending from the upper left corner to the lower right corner, replaced by ones is called an identity matrix. It is denoted by $\mathbf{I}_n$ or $\mathbf{I}$. A square matrix with all entries zero above or below the main diagonal is called triangular. For example,

$$(1, 2, 5), \quad \begin{bmatrix} 2 \\ 3 \end{bmatrix}, \quad \begin{bmatrix} 0 & 0 \\ 0 & 0 \\ 0 & 0 \end{bmatrix},$$

$$\begin{bmatrix} 1 & 4 & 3 & 4 \\ 0 & 3 & -1 & 5 \\ 0 & 0 & 4 & -2 \\ 0 & 0 & 0 & 2 \end{bmatrix}, \quad \begin{bmatrix} 1 & 0 & 0 \\ 0 & 1 & 0 \\ 0 & 0 & 1 \end{bmatrix}$$

are row, column, zero, triangular, and identity matrices, respectively.

Operations

Addition and Subtraction: If $\mathbf{A}$ and $\mathbf{B}$ are matrices of the same order with elements a_{ij} and b_{ij}, respectively, the matrix

$$\mathbf{C} = \mathbf{A} \pm \mathbf{B}$$

has elements

$$c_{ij} = a_{ij} \pm b_{ij}, \quad i = 1, 2, \cdots, m; j = 1, 2, \cdots, n.$$

Multiplication by a Number: If k is a number, real or complex, the matrix

$$\mathbf{C} = k\mathbf{A}$$

has elements

$$c_{ij} = ka_{ij}, \quad i = 1, 2, \cdots, m; j = 1, 2, \cdots, n.$$

Multiplication of Two Matrices: Let $\mathbf{A}$ and $\mathbf{B}$ be two matrices of orders $m \times n$ and $n \times p$, respectively. The matrix

$$\mathbf{C} = \mathbf{AB}$$

will have elements

$$c_{ij} = \sum_{k=1}^{n} a_{ik} b_{kj}, \quad i = 1, 2, \cdots, m; j = 1, 2, \cdots, p.$$

Matrix $\mathbf{C}$ has order $m \times p$. Note that the product $\mathbf{BA}$ is defined only when $m = p$. In general $\mathbf{AB} \neq \mathbf{BA}$ even when $m = n = p$, so it is necessary to distinguish between premultiplication and postmultiplication.

For a square matrix $\mathbf{A}$ of order n, powers are defined

$$\mathbf{A}^0 = \mathbf{I}, \ \mathbf{A}^1 = \mathbf{A}, \ \mathbf{A}^2 = \mathbf{A} \cdot \mathbf{A}, \ \mathbf{A}^3 = \mathbf{A} \cdot \mathbf{A}^2, \text{ etc.}$$

A polynomial function of $\mathbf{A}$ is a square matrix of order n given by

$$P(\mathbf{A}) = a_n \mathbf{A}^n + a_{n-1} \mathbf{A}^{n-1} + \cdots + a_1 \mathbf{A}^1 + a_0 \mathbf{A}^0$$

where the a_i are real or complex numbers.

Division of Two Matrices: Not defined.

Determinant

Definition: The determinant of a square matrix $\mathbf{A}$ of order n is usually defined

$$|\mathbf{A}| = \sum \pm a_{1i} a_{2j} \cdots a_{nr}$$

where the second subscripts $i, j, \cdots, r$ form a permutation (rearrangement) of the integers 1, 2, $\cdots$, n. The sum is taken over all permutations with a plus (minus) sign if the permutation is even (odd). A permutation is called even (odd) if an even (odd) number of inversions is necessary to attain the natural or ascending order. For example, $4132 \rightarrow 1432 \rightarrow 1342 \rightarrow 1324 \rightarrow 1234$; therefore, 4132 is an even permutation.

A square matrix is said to be singular if its determinant is zero and nonsingular otherwise.

Laplace's Development: By an expansion known as Laplace's development, the determinant of a matrix $\mathbf{A}$ of order n can be expressed in terms of determinants of matrices of order $n-1$.

$$|\mathbf{A}| = \sum_{j=1}^{n} (-1)^{i+j} a_{ij} M_{ij}$$

(expansion by the ith row)

$$= \sum_{i=1}^{n} (-1)^{i+j} a_{ij} M_{ij}$$

(expansion by the jth column).

M_{ij} is the determinant of the matrix formed by deleting the ith row and jth column of $\mathbf{A}$. Expanding by any row or column will lead to the same value of $|\mathbf{A}|$. The M_{ij} are in turn evaluated in terms of determinants of order $n-2$. This process is continued until, say, second-order determinants are obtained.

$$\begin{vmatrix} b_{11} & b_{12} \\ b_{21} & b_{22} \end{vmatrix} = b_{11}b_{22} - b_{12}b_{21}.$$

A first-order determinant has a value equal to its only entry. M_{ij} is called the minor and $(-1)^{i+j} \times M_{ij} = A_{ij}$ is called the cofactor of the element a_{ij}.

Laplace's development is valuable for a literal expansion. For numerical evaluation of determinants of large order, say greater than four, the Gauss algorithm described in the next section requires less effort.

Linear Transformations

The linear transformation or set of equations

$$a_{11}x_1 + a_{12}x_2 + \cdots + a_{1n}x_n = y_1$$
$$a_{21}x_1 + a_{22}x_2 + \cdots + a_{2n}x_n = y_2$$
$$\vdots$$
$$a_{m1}x_1 + a_{m2}x_2 + \cdots + a_{mn}x_n = y_m$$

may be compactly written in matrix form

$$\mathbf{AX} = \mathbf{Y}$$

where $\mathbf{A} = [a_{ij}]$ is a matrix of order $m \times n$, and $\mathbf{X}$ and $\mathbf{Y}$ are column vectors.

Inverse Matrix: If $m=n$ and $|\mathbf{A}| \neq 0$, then there exists an inverse matrix denoted by $\mathbf{A}^{-1}$ such that $\mathbf{AA}^{-1} = \mathbf{A}^{-1}\mathbf{A} = \mathbf{I}$. The inverse transformation expressing the x's in terms of the y's may then be compactly written

$$\mathbf{X} = \mathbf{A}^{-1}\mathbf{Y}.$$

This inverse transformation may be effected using *Cramer's rule*

$$x_i = (1/|\mathbf{A}|) \sum_{k=1}^{n} A_{ki}y_k, \quad i=1, 2, \cdots, n$$

where A_{ki} denotes the cofactor associated with the element a_{ki} in the original matrix $\mathbf{A}$. Cramer's rule provides a useful literal expansion of the solution. However, numerical evaluation of the determinants involved, say Laplace's development, requires of the order of $n!$ operations.

The following *Gauss algorithm* requires only of the order of n^3 operations and is therefore preferred for numerical evaluation when n is large: Renumber the x_k's if necessary to make $a_{11} \neq 0$. Normalize the first equation by dividing it by a_{11}. If $a_{21}=0$, leave the second equation intact. If $a_{21} \neq 0$, eliminate x_1 by subtracting the normalized first equation multiplied by a_{21} from the second equation. Similarly, eliminate x_1 from the remaining $n-2$ equations. The result is

$$x_1 + (a_{12}/a_{11})x_2 + (a_{13}/a_{11})x_3 + \cdots + (a_{1n}/a_{11})x_n = (y_1/a_{11})$$

$$[a_{22} - a_{21}(a_{12}/a_{11})]x_2 + [a_{23} - a_{21}(a_{13}/a_{11})]x_3 + \cdots + [a_{2n} - a_{21}(a_{1n}/a_{11})]x_n = y_2 - a_{21}(y_1/a_{11})$$

$$\vdots$$

$$[a_{n2} - a_{n1}(a_{12}/a_{11})]x_2 + [a_{n3} - a_{n1}(a_{13}/a_{11})]x_3 + \cdots + [a_{nn} - a_{31}(a_{1n}/a_{11})]x_n = y_n - a_{31}(y_1/a_{11}).$$

The entire process is now repeated with the first equation omitted. There then results a set of the form

$$x_1 + b_{12}x_2 + b_{13}x_3 + \cdots + b_{1n}x_n = c_{11}y_1$$

$$x_2 + b_{23}x_3 + \cdots + b_{2n}x_n = c_{21}y_1 + c_{22}y_2$$

$$b_{33}x_3 + \cdots + b_{3n}x_n = c_{31}y_1 + c_{32}y_2 + c_{33}y_3$$

$$\vdots$$

$$b_{n3}x_3 + \cdots + b_{nn}x_n = c_{n1}y_1 + c_{n2}y_2 + c_{nn}y_n.$$

Again the process is repeated with the first two equations omitted. Continuing in this manner yields a triangular form

$$x_1 + b_{12}x_2 + b_{13}x_3 + b_{14}x_4 + \cdots + b_{1n}x_n = c_{11}y_1$$

$$x_2 + b_{23}x_3 + b_{24}x_4 + \cdots + b_{2n}x_n = c_{21}y_1 + c_{22}y_2$$

$$x_3 + d_{34}x_4 + \cdots + d_{3n}x_n = e_{31}y_1 + e_{32}y_2 + e_{33}y_3$$

$$x_4 + \cdots + d_{4n}x_n = e_{41}y_1 + e_{42}y_2 + e_{43}y_3 + e_{44}y_4$$

$$\vdots$$

$$x_n = e_{n1}y_1 + e_{n2}y_2 + e_{n3}y_3 + \cdots + e_{n4}y_n.$$

Note that the last equation gives the value of x_n. This may be substituted in the next to the last equation to obtain x_{n-1} and so on. If the y's are literal as shown above, the process will yield the inverse transformation

$$X = A^{-1}Y$$

where A^{-1} is the inverse of the original matrix. If the y's are numerical, labor is saved by combining the values on the right side of each equation at each step of the algorithm.

Since the determinant of a triangular matrix is equal to the product of the elements on the main diagonal

$$|A| = a_{11}a_{22}'a_{33}' \cdots a_{nn}'$$

where a_{kk}' is the quantity the kth equation is divided by in the above Gauss algorithm. This is useful for evaluating determinants of large order since it requires only of the order of n^3 operations.

A matrix A may be viewed as consisting of column or row vectors. The largest number of linearly independent column vectors (which is the same as the largest number of linearly independent row vectors) is called the *rank* of the matrix, $\rho(A)$. A set of vectors V_i is linearly independent if

$$\sum_i a_i V_i = 0$$

implies that $a_i = 0$ for $i = 1, 2, \cdots$.

The rank is equal to the order of the largest nonvanishing determinant of the submatrix by deleting rows and columns of the original matrix. Consider the matrices

$$A = \begin{bmatrix} a_{11} & a_{12} & \cdots & a_{1n} \\ a_{21} & a_{22} & \cdots & a_{2n} \\ & & \vdots & \\ a_{m1} & a_{m2} & \cdots & a_{mn} \end{bmatrix}$$

and

$$B = \begin{bmatrix} a_{11} & a_{12} & \cdots & a_{1n} & y_1 \\ a_{21} & a_{22} & \cdots & a_{2n} & y_2 \\ & & \vdots & & \\ a_{m1} & a_{m2} & \cdots & a_{mn} & y_m \end{bmatrix}$$

The equations

$$AX = Y$$

have a solution if and only if

$$\rho(A) = \rho(B)$$

in which case the equations are said to be consistent.

If $\rho(A) < n = m$, that is, $|A| = 0$, and if the equations are consistent, then the Gauss algorithm will terminate before n steps. That is, the coeffi-

cients of all the x_k's will be zero in the remaining $n - \rho(A)$ equations. Therefore, among the x_k's there will be certain ones, $n - \rho(A)$ in number, which may be assigned arbitrary values. Similarly, if $m \neq n$, the Gauss algorithm will yield an equivalent set of $\rho(A)$ equations that has the same solution as the original set. Again $n - \rho(A)$ (possibly zero) of the x_k's may be assigned arbitrary values.

It is not necessary to know beforehand whether the equations are inconsistent. If they are inconsistent, the algorithm will yield an "equation" in which the coefficients of the x_k's on the left side are zero but there is a nonzero combination of the y_k's on the right side. Since the right side of the "equation" may contain accumulated round-off errors, an analysis of the error propagation in the Gauss algorithm may be necessary to determine whether a small right side is caused by inconsistent equations or by round-off errors.

Eigenvectors and Eigenvalues

An eigenvector of the square matrix A of order n is a nonzero vector X such that

$$AX = \lambda X.$$

The scalar λ is called an eigenvalue of A and X is called an eigenvector corresponding to or associated with λ. The eigenvalues may be determined from the characteristic equation

$$|A - \lambda I| = 0.$$

The corresponding eigenvectors X may then be found by solving

$$(A - \lambda I)X = 0.$$

The solution may be obtained by the Gauss algorithm. If the eigenvalues λ are distinct, an explicit solution may be obtained by taking a nontrivial row of cofactors from $A - \lambda_i I$. This is possible since the rank of $A - \lambda_i I$ is $n - 1$ and therefore, there exists a nonvanishing subdeterminant of order $n - 1$.

Note that eigenvectors are determined only to within a multiplicative constant.

Further Definitions and Properties

The matrix whose elements are a_{ji} is called the conjugate transpose of $A = [a_{ij}]$; it is denoted by $A^\dagger$. The conjugate transpose of a product is

$$(AB)^\dagger = B^\dagger A^\dagger.$$

If $A = A^\dagger$ ($A = -A^\dagger$), A is said to be Hermitian (skew-Hermitian). If $A^\dagger A = I$, A is called unitary.

If $A^\dagger A = AA^\dagger$, A is called normal. Diagonal, unitary, Hermitian, and skew-Hermitian matrices are special cases of normal matrices. If the matrix A is real (that is, all entries are real) the terms Hermitian, skew-Hermitian, and unitary are usually replaced by symmetric, skew-symmetric, and unitary, respectively. The eigenvalues of Hermitian, skew-Hermitian, and unitary matrices are real, pure imaginary, and of unit absolute value, respectively.

The inner product of vectors x and y is

$$X^\dagger Y = \sum_{i=1}^{n} x_i^* y_i$$

where x_i^* is the complex conjugate of x_i. (A square matrix of order one is considered here as a scalar.)

The length of a vector is given by

$$\| X \| = (X^\dagger X)^{1/2} = \left(\sum_{i=1}^{n} | x_i |^2 \right)^{1/2}.$$

Two vectors are said to be orthogonal if

$$X^\dagger Y = 0.$$

Orthogonality corresponds to perpendicularity in two and three dimensions. Inner products obey the following inequalities

$$| X^\dagger Y | \leq \| X \| \, \| Y \| \qquad \text{(Schwarz)}$$

$$\| X + Y \| \leq \| X \| + \| Y \| \qquad \text{(Triangle)}.$$

Two matrices A and B are called similar if there exists a nonsingular matrix C such that

$$B = CAC^{-1}.$$

Of particular interest is the case in which matrix B is diagonal

$$B = \begin{bmatrix} \lambda_1 & 0 & 0 & \cdots & 0 \\ 0 & \lambda_2 & 0 & \cdots & 0 \\ 0 & 0 & \lambda_3 & \cdots & 0 \\ & & & \vdots & \\ 0 & 0 & 0 & \cdots & \lambda_n \end{bmatrix}.$$

The diagonal elements λ_i are the eigenvalues of the matrix A. Not all matrices are similar to diagonal ones. However, if a matrix of order n has n linearly independent eigenvectors, as is the case with normal matrices or when all its eigenvalues are distinct, it can be diagonalized. This may be done as follows: Let

$$A \begin{bmatrix} x_{1i} \\ x_{2i} \\ \vdots \\ x_{ni} \end{bmatrix} = \lambda_i \begin{bmatrix} x_{1i} \\ x_{2i} \\ \vdots \\ x_{ni} \end{bmatrix}$$

Then

$$A \begin{bmatrix} x_{11} & x_{12} & \cdots & x_{1n} \\ x_{21} & x_{22} & \cdots & x_{2n} \\ & & \vdots & \\ x_{n1} & x_{n2} & \cdots & x_{nn} \end{bmatrix}$$

$$= \begin{bmatrix} x_{11} & x_{12} & \cdots & x_{1n} \\ x_{21} & x_{22} & \cdots & x_{2n} \\ & & \vdots & \\ x_{n1} & x_{n2} & \cdots & x_{nn} \end{bmatrix} \begin{bmatrix} \lambda_1 & 0 & 0 & \cdots & 0 \\ 0 & \lambda_2 & 0 & \cdots & 0 \\ & & & \vdots & \\ 0 & 0 & 0 & \cdots & \lambda_n \end{bmatrix}.$$

The matrix formed from the eigenvectors is nonsingular since the eigenvectors are linearly independent. Hence

$$C^{-1} = \begin{bmatrix} x_{11} & x_{12} & \cdots & x_{1n} \\ x_{21} & x_{22} & \cdots & x_{2n} \\ & & \vdots & \\ x_{n1} & x_{n2} & \cdots & x_{nn} \end{bmatrix}.$$

If a matrix A is reducible to diagonal form, polynomial functions of A are readily calculated

$$f(A) = f(C^{-1}BC)$$

$$= C^{-1} \begin{bmatrix} f(\lambda_1) & 0 & 0 & \cdots & 0 \\ 0 & f(\lambda_2) & 0 & \cdots & 0 \\ & & & \vdots & \\ 0 & 0 & 0 & \cdots & f(\lambda_n) \end{bmatrix} C.$$

Matrices can always be reduced to the Jordan canonical form, in which the eigenvalues are on the main diagonal and there are ones in certain places just above the main diagonal, and zeros elsewhere. In this form operations such as polynomial functions are simplified.

Hermitian Forms

A Hermitian form is a polynomial

$$X^\dagger A X = \sum_{j=1}^{n} \sum_{i=1}^{n} a_{ij} x_i^* x_j$$

where A is a Hermitian matrix. Hermitian forms are real-valued and satisfy the inequality

$$\lambda_n \| X \|^2 \leq X^\dagger A X \leq \lambda_1 \| X \|^2$$

in which λ_1 and λ_n are the largest and smallest eigenvalues of A. By a suitable change of variable

$$Y = TX$$

any Hermitian form $X^\dagger A X$ may be reduced.

$$X^\dagger A X = Y^\dagger T^\dagger A T Y = \sum_{i=1}^{n} \lambda_i | y_i |^2$$

by choosing T to be that unitary matrix which diagonalizes A.

The maximum value of $X^\dagger AX$ for all unit vectors X, that is, $\|X\|=1$, is the largest eigenvalue of A, say λ_1. A vector yielding this largest value will be a corresponding eigenvector, say X_1. The maximum of $X^\dagger AX$ overall unit vectors orthogonal to X_1 will be another eigenvalue, say λ_2. A vector yielding λ_2 will be a corresponding eigenvector, say X_2. The process is repeated considering unit vectors orthogonal to both X_1 and X_2. In this way the eigenvalues of any Hermitian matrix may be found.

VECTOR-ANALYSIS EQUATIONS

Rectangular Coordinates

(In the following, vectors are indicated in bold-faced type.)

Notation:

$$a = a\hat{a}$$

a = magnitude of a

$\hat{a}$ = unit vector in direction of a.

Associative Law: For addition

$$a + (b+c) = (a+b) + c = a+b+c.$$

Commutative Law: For addition

$$a + b = b + a.$$

Scalar or "Dot" Product (Fig. 21):

$$a \cdot b = b \cdot a$$

$$= ab \cos\theta$$

where θ = angle included by a and b.

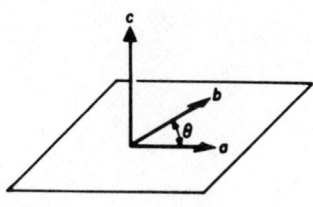

Fig. 21.

Vector or "Cross" Product:

$$a \times b = -b \times a$$

$$= ab \sin\theta \hat{c}$$

where

θ = smallest angle swept in rotating a into b

$\hat{c}$ = unit vector perpendicular to plane of a and b, and directed in the sense of travel of a right-hand screw rotating from a to b through the angle θ.

Distributive Law for Scalar Multiplication:

$$a \cdot (b+c) = a \cdot b + a \cdot c.$$

Distributive Law for Vector Multiplication:

$$a \times (b+c) = a \times b + a \times c.$$

Scalar Triple Product:

$$a \cdot (b \times c) = (a \times b) \cdot c = c \cdot (a \times b) = b \cdot (c \times a).$$

Vector Triple Product:

$$a \times (b \times c) = (a \cdot c)b - (a \cdot b)c$$

$$(a \times b) \cdot (c \times d) = (a \cdot c)(b \cdot d) - (a \cdot d)(b \cdot c)$$

$$(a \times b) \times (c \times d) = (a \times b \cdot d)c - (a \times b \cdot c)d.$$

Del Operator:

$$\nabla \equiv i(\partial/\partial x) + j(\partial/\partial y) + k(\partial/\partial z)$$

where i, j, k are unit vectors in the directions of the x, y, z coordinate axes, respectively.

Gradient:

$$\mathrm{grad}\phi = \nabla\phi$$

$$= i(\partial\phi/\partial x) + j(\partial\phi/\partial y) + k(\partial\phi/\partial z),$$

in Cartesian coordinates

$$\mathrm{grad}(\phi + \psi) = \mathrm{grad}\phi + \mathrm{grad}\psi$$

$$\mathrm{grad}(\phi\psi) = \phi\,\mathrm{grad}\psi + \psi\,\mathrm{grad}\phi.$$

Divergence:

$$\mathrm{div}\,a = \nabla \cdot a = (\partial a_x/\partial x) + (\partial a_y/\partial y) + (\partial a_z/\partial z),$$

in Cartesian coordinates

where a_x, a_y, a_z are components of a in the directions of the x, y, z coordinate axes, respectively.

$$\mathrm{div}(a+b) = \mathrm{div}\,a + \mathrm{div}\,b$$

$$\mathrm{div}(\phi a) = \phi\,\mathrm{div}\,a + a \cdot \mathrm{grad}\phi.$$

Curl:

$$\mathrm{curl}\,a = \nabla \times a$$

$$= i\left(\frac{\partial a_z}{\partial y} - \frac{\partial a_y}{\partial z}\right) + j\left(\frac{\partial a_x}{\partial z} - \frac{\partial a_z}{\partial x}\right)$$

$$+ k\left(\frac{\partial a_y}{\partial x} - \frac{\partial a_x}{\partial y}\right)$$

$$= \begin{vmatrix} \mathbf{i} & \mathbf{j} & \mathbf{k} \\ \partial/\partial x & \partial/\partial y & \partial/\partial z \\ a_x & a_y & a_z \end{vmatrix}, \quad \text{in Cartesian coordinates}$$

$\mathrm{curl}(\mathbf{a}+\mathbf{b})=\mathrm{curl}\,\mathbf{a}+\mathrm{curl}\,\mathbf{b}$

$\mathrm{curl}(\phi\mathbf{a})=\mathrm{grad}\phi\times\mathbf{a}+\phi\,\mathrm{curl}\,\mathbf{a}$

$\mathrm{curl}\,\mathrm{grad}\phi=0$

$\mathrm{div}\,\mathrm{curl}\,\mathbf{a}=0$

$\mathrm{div}(\mathbf{a}\times\mathbf{b})=\mathbf{b}\cdot\mathrm{curl}\,\mathbf{a}-\mathbf{a}\cdot\mathrm{curl}\,\mathbf{b}$

Laplacian $\equiv\nabla^2=\nabla\cdot\nabla$

$$\nabla^2\phi=(\partial^2\phi/\partial x^2)+(\partial^2\phi/\partial y^2)+(\partial^2\phi/\partial z^2),$$

in Cartesian coordinates

$\mathrm{curl}\,\mathrm{curl}\,\mathbf{a}=\mathrm{grad}\,\mathrm{div}\,\mathbf{a}-(\mathbf{i}\nabla^2 a_x+\mathbf{j}\nabla^2 a_y+\mathbf{k}\nabla^2 a_z)$

$\qquad =\nabla(\nabla\cdot\mathbf{a})-\nabla^2\mathbf{a}.$

Directional Derivative: Derivative of ϕ in the direction of $\mathbf{s}$

$$d\phi/ds=\hat{\mathbf{s}}\cdot\nabla\phi.$$

Integral Relations: In the following equations τ is a volume bounded by a closed surface S. The unit vector $\mathbf{n}$ is normal to the surface and is directed outwards. dS is an element of surface area. If the surface is represented by $z=f(x,y)$

then

$$dS=[1+(\partial f/\partial x)^2+(\partial f/\partial y)^2]^{1/2}dx\,dy.$$

$$\int_\tau \nabla\phi\,d\tau=\int_S \phi\mathbf{n}\,dS$$

$$\int_\tau \nabla\cdot\mathbf{a}\,d\tau=\int_S \mathbf{a}\cdot\mathbf{n}\,dS \quad \text{(Gauss' theorem)}$$

$$\int_\tau \nabla\times\mathbf{a}\,d\tau=\int_S \mathbf{n}\times\mathbf{a}\,dS$$

$$\int_\tau (\psi\nabla^2\phi-\phi\nabla^2\psi)\,d\tau=\int_S [\psi(\partial\phi/\partial n)-\phi(\partial\psi/\partial n)]dS$$

where $\partial/\partial n$ is the derivative in the direction of $\mathbf{n}$ (Green's theorem).

In the two following equations S is an open surface bounded by a contour C, with distance along C represented by s.

$$\int_S \mathbf{n}\times\nabla\phi\,dS=\int_C \phi\,d\mathbf{s}$$

$$\int_S (\nabla\times\mathbf{a})\cdot\mathbf{n}\,dS=\int_C \mathbf{a}\cdot d\mathbf{s} \quad \text{(Stokes' theorem)}$$

where $\mathbf{s}=s\hat{\mathbf{s}}$ and $\hat{\mathbf{s}}$ is the unit tangent vector along C.

Gradient, Divergence, Curl, and Laplacian in Coordinate Systems Other than Rectangular

Cylindrical Coordinates: (ρ,ϕ,z), unit vectors $\hat{\boldsymbol{\varrho}}$, $\hat{\boldsymbol{\phi}}$, $\mathbf{k}$, respectively

$\mathrm{grad}\psi=\nabla\psi=(\partial\psi/\partial\rho)\hat{\boldsymbol{\varrho}}+\rho^{-1}(\partial\psi/\partial\phi)\hat{\boldsymbol{\phi}}+(\partial\psi/\partial z)\mathbf{k}.$

Let $\mathbf{a}=a_\rho\hat{\boldsymbol{\varrho}}+a_\phi\hat{\boldsymbol{\phi}}+a_z\mathbf{k}$. Then

$\mathrm{div}\,\mathbf{a}=\nabla\cdot\mathbf{a}=\rho^{-1}(\partial/\partial\rho)(\rho a_\rho)+\rho^{-1}(\partial a_\phi/\partial\phi)+(\partial a_z/\partial z)$

$\mathrm{curl}\,\mathbf{a}=\nabla\times\mathbf{a}=[\rho^{-1}(\partial a_z/\partial\phi)-(\partial a_\phi/\partial z)]\hat{\boldsymbol{\varrho}}+[(\partial a_\rho/\partial z)-(\partial a_z/\partial\rho)]\hat{\boldsymbol{\phi}}+[\rho^{-1}(\partial/\partial\rho)(\rho a_\phi)-\rho^{-1}(\partial a_\rho/\partial\phi)]\mathbf{k}$

$\nabla^2\psi=\rho^{-1}(\partial/\partial\rho)[\rho(\partial\psi/\partial\rho)]+\rho^{-2}(\partial^2\psi/\partial\phi^2)+(\partial^2\psi/\partial z^2).$

Spherical Coordinates: (r,θ,ϕ), unit vectors $\hat{\mathbf{r}}$, $\hat{\boldsymbol{\theta}}$, $\hat{\boldsymbol{\phi}}$

$$r=\text{distance to origin}$$

$$\theta=\text{polar angle}$$

$$\phi=\text{azimuthal angle}$$

$\mathrm{grad}\psi=\nabla\psi=(\partial\psi/\partial r)\hat{\mathbf{r}}+r^{-1}(\partial\psi/\partial\theta)\hat{\boldsymbol{\theta}}+(r\sin\theta)^{-1}(\partial\psi/\partial\phi)\hat{\boldsymbol{\phi}}.$

Let $\mathbf{a}=a_r\hat{\mathbf{r}}+a_\theta\hat{\boldsymbol{\theta}}+a_\phi\hat{\boldsymbol{\phi}}$. Then

$\mathrm{div}\,\mathbf{a}=\nabla\cdot\mathbf{a}=r^{-2}(\partial/\partial r)(r^2 a_r)+(r\sin\theta)^{-1}(\partial/\partial\theta)(a_\theta\sin\theta)+(r\sin\theta)^{-1}(\partial a_\phi/\partial\phi)$

$\mathrm{curl}\,\mathbf{a}=\nabla\times\mathbf{a}=(r\sin\theta)^{-1}[(\partial/\partial\theta)(a_\phi\sin\theta)-(\partial a_\theta/\partial\phi)]\hat{\mathbf{r}}$

$\qquad\qquad +r^{-1}[(\sin\theta)^{-1}(\partial a_r/\partial\phi)-(\partial/\partial r)(ra_\phi)]\hat{\boldsymbol{\theta}}+r^{-1}[(\partial/\partial r)(ra_\theta)-(\partial a_r/\partial\theta)]\hat{\boldsymbol{\phi}}$

$\nabla^2\psi=r^{-2}(\partial/\partial r)[r^2(\partial\psi/\partial r)]+(r^2\sin\theta)^{-1}(\partial/\partial\theta)[\sin\theta(\partial\psi/\partial\theta)]+(r^2\sin^2\theta)^{-1}(\partial^2\psi/\partial\phi^2).$

Orthogonal Curvilinear Coordinates:

Coordinates: u_1, u_2, u_3

Metric coefficients: h_1, h_2, h_3 $(ds^2 = h_1^2 du_1^2 + h_2^2 du_2^2 + h_3^2 du_3^2)$

Unit vectors: i_1, i_2, i_3 $(ds = i_1 h_1 du_1 + i_2 h_2 du_2 + i_3 h_3 du_3)$

$$\text{grad}\psi = \nabla\psi = h_1^{-1}(\partial\psi/\partial u_1)i_1 + h_2^{-1}(\partial\psi/\partial u_2)i_2 + h_3^{-1}(\partial\psi/\partial u_3)i_3$$

$$\text{diva} = \nabla\cdot a = (h_1 h_2 h_3)^{-1}[(\partial/\partial u_1)(h_2 h_3 a_1) + (\partial/\partial u_2)(h_3 h_1 a_2) + (\partial/\partial u_3)(h_1 h_2 a_3)]$$

$$\text{curla} = \nabla\times a = (h_2 h_3)^{-1}[(\partial/\partial u_2)(h_3 a_3) - (\partial/\partial u_3)(h_2 a_2)]i_1 + (h_3 h_1)^{-1}[(\partial/\partial u_3)(h_1 a_1) - (\partial/\partial u_1)(h_3 a_3)]i_2$$
$$+ (h_1 h_2)^{-1}[(\partial/\partial u_1)(h_2 a_2) - (\partial/\partial u_2)(h_1 a_1)]i_3$$

$$= (h_1 h_2 h_3)^{-1}\begin{vmatrix} h_1 i_1 & h_2 i_2 & h_3 i_3 \\ \partial/\partial u_1 & \partial/\partial u_2 & \partial/\partial u_3 \\ h_1 a_1 & h_2 a_2 & h_3 a_3 \end{vmatrix}$$

$$\nabla^2\psi = (h_1 h_2 h_3)^{-1}\left[\frac{\partial}{\partial u_1}\left(\frac{h_2 h_3}{h_1}\frac{\partial\psi}{\partial u_1}\right) + \frac{\partial}{\partial u_2}\left(\frac{h_3 h_1}{h_2}\frac{\partial\psi}{\partial u_2}\right) + \frac{\partial}{\partial u_3}\left(\frac{h_1 h_2}{h_3}\frac{\partial\psi}{\partial u_3}\right)\right].$$

Space Curves

A curve may be represented vectorially as $r = r(s)$. See Fig. 22.

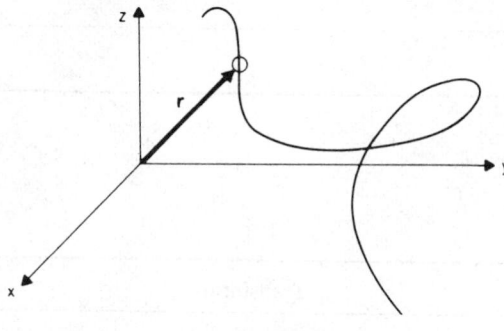

Fig. 22.

A unit tangent t is then given by

$$t = dr/ds.$$

The principal normal n is given by

$$n = (1/k)(dt/ds)$$

where k is the curvature. The radius of curvature $R = 1/k$. For a plane curve $y = f(x)$, the curvature may be computed from

$$k = |y''|/[1 + (y')^2]^{3/2}.$$

The binormal is defined by

$$b = t \times n.$$

These vectors satisfy Frenet's equations

$$dn/ds = -kt + \tau b$$

$$db/ds = -\tau n$$

where τ is the torsion. The torsion is zero everywhere if and only if the curve lies in a plane.

LAPLACE TRANSFORM

The Laplace transform of a function $f(t)$ is defined as

$$F(p) = \int_0^\infty f(t)e^{-pt}dt.$$

If this integral converges for some $p = p_0$, real or complex, then it will converge for all p such that $\text{Re}p > \text{Re}p_0$.

The inverse transform may be found by

$$f(t) = (j2\pi)^{-1}\int_{c-j\infty}^{c+j\infty} F(z)e^{tz}dz, \quad t > 0$$

where there are no singularities to the right of the path of integration.

TABLE OF LAPLACE TRANSFORMS

General Equations

Function	Transform*
Shifting theorem $f(t-a)$, $f(t)=0$, $t<0$	$e^{-ap}F(p)$, $a>0$
Convolution $\int_0^t f_1(\lambda)f_2(t-\lambda)\,d\lambda$	$F_1(p)F_2(p)$
Linearity $a_1 f_1(t)+a_2 f_2(t)$, $(a_1, a_2 \text{ const})$	$a_1F_1(p)+a_2F_2(p)$
Derivative $df(t)/dt$	$-f(0)+pF(p)$
Integral $\int f(t)\,dt$	$p^{-1}\left[\int f(t)\,dt\right]_{t=0}+[F(p)/p]$, $\operatorname{Re}p>0$
Periodic function $f(t)=f(t+r)$	$\int_0^r f(\lambda)e^{-p\lambda}d\lambda/(1-e^{-pr})$, $r>0$
$f(t)=-f(t+r)$	$\int_0^r f(\lambda)e^{-p\lambda}d\lambda/(1+e^{-pr})$, $r>0$
$f(at)$, $a>0$	$F(p/a)/a$
$e^{at}f(t)$	$F(p-a)$, $\operatorname{Re}p>\operatorname{Re}a$
$t^n f(t)$	$(-1)^n[d^nF(p)/dp^n]$
Final-value theorem $f(\infty)$	$\lim_{p\to 0} pF(p)$
Initial-value theorem $f(0+)$	$\lim_{p\to\infty} pF(p)$

* $F(p)$ denotes the Laplace transform of $f(t)$.

Miscellaneous Functions*

Function	Transform
Step $\quad u(t-a)=0$, $0\leq t<a$ $\qquad\qquad =1$, $t\geq a$	e^{-ap}/p
Impulse $\quad \delta(t)$	1
t^a, $\operatorname{Re}a>-1$	$\Gamma(a+1)/p^{a+1}$
e^{at}	$1/(p-a)$, $\operatorname{Re}p>\operatorname{Re}a$
$t^a e^{bt}$, $\operatorname{Re}a>-1$	$\Gamma(a+1)/(p-b)^{a+1}$, $\operatorname{Re}p>\operatorname{Re}b$
$\cos at$	$p/(p^2+a^2)$ $\Big\}$ $\operatorname{Re}p>\lvert\operatorname{Im}a\rvert$
$\sin at$	$a/(p^2+a^2)$
$\cosh at$	$p/(p^2-a^2)$ $\Big\}$ $\operatorname{Re}p>\lvert\operatorname{Re}a\rvert$
$\sinh at$	$1/(p^2-a^2)$

Miscellaneous Functions—(Continued)

Function	Transform
$\ln t$	$-(\gamma+\ln p)/p$, γ is Euler's constant $=0.57722$
$1/(t+a)$, $a>0$	$e^{ap}E_1(ap)$
e^{-at^2}	$\frac{1}{2}(\pi/a)^{1/2}e^{p^2/4a}\,\mathrm{erfc}[p/2(a)^{1/2}]$
Bessel function $J_v(at)$, $\mathrm{Re}v>-1$	$r^{-1}[(r-p)/a]^v$, $r=(p^2+a^2)^{1/2}$, $\qquad$ $\mathrm{Re}p>|\mathrm{Re}a|$
Bessel function $I_v(at)$, $\mathrm{Re}v>-1$	$R^{-1}[(R-p)/a]^v$, $R=(p^2-a^2)^{1/2}$, $\qquad$ $\mathrm{Re}p>|\mathrm{Re}a|$

* For an extensive listing, refer to "Tables of Integral Transforms," vol. 1, Bateman Manuscript Project, editor A. Erdéyli, McGraw-Hill Book Co., New York; 1954.

Inverse Transforms*

Transform	Function
1	$\delta(t)$
$1/(p+a)$	e^{-at}
$1/(p+a)^v$, $\mathrm{Re}v>0$	$t^{v-1}e^{-at}/\Gamma(v)$
$1/[(p+a)(p+b)]$	$(e^{-at}-e^{-bt})/(b-a)$
$p/[(p+a)(p+b)]$	$(ae^{-at}-be^{-bt})/(a-b)$
$1/(p^2+a^2)$	$a^{-1}\sin at$
$1/(p^2-a^2)$	$a^{-1}\sinh at$
$p/(p^2+a^2)$	$\cos at$
$p/(p^2-a^2)$	$\cosh at$
$1/(p^2+a^2)^{1/2}$	$J_0(at)$
e^{-ap}/p	$u(t-a)$
e^{-ap}/p^v, $\mathrm{Re}v>0$	$(t-a)^{v-1}u(t-a)/\Gamma(v)$
$(1/p)e^{-a/p}$	$J_0[2(at)^{1/2}]$
$(1/p^v)e^{-a/p}$ $\}$ $\mathrm{Re}v>0$	$(t/a)^{(v-1)/2}J_{v-1}[2(at)^{1/2}]$
$(1/p^v)e^{a/p}$	$(t/a)^{(v-1)/2}I_{v-1}[2(at)^{1/2}]$
$(1/p)\ln p$	$-\gamma-\ln t$, $\gamma=0.57722$

* Refer to "Tables of Integral Transforms," vol. 1, Bateman Manuscript Project.

SELECTED MATHEMATICAL FUNCTIONS

Exponential Integrals

Definitions:

$$E_1(z)=\int_z^{\infty}(e^{-t}/t)\,dt, \quad |\arg z|<\pi$$

in which the path of integration does not cross the negative t-axis and also excludes the origin. For $\arg z=\pi$ the following function is used.

$$Ei(x)=-\mathrm{pv}\int_{-x}^{\infty}(e^{-t}/t)\,dt, \quad x>0$$

where pv stands for the Cauchy principal value.

$$E_1(-x\pm j0)=Ei(x)\mp j\pi.$$

Series Expansions:

$$E_1(z)=-\gamma-\ln z-\sum_{n=1}^{\infty}[(-1)^n z^n/(n\cdot n!)],$$

$$|\arg z|<\pi$$

$$Ei(x)=\gamma+\ln x+\sum_{n=1}^{\infty}x^n/(n\cdot n!)$$

in which γ is Euler's constant, $\gamma=0.57722$.

Asymptotic Expansions:

$$E_1(z)\sim(e^{-z}/z)[1-z^{-1}+(2!/z^2)-(3!/z^3)+\cdots],$$

$$|\arg z|<\pi$$

$$Ei(x)\sim(e^x/x)[1+x^{-1}+(2!/x^2)+(3!/x^3)+\cdots].$$

Cosine and Sine Integrals

Definitions:

$$Si(z)=\int_0^z(\sin t/t)\,dt$$

$$Ci(z)=\gamma+\ln z+\int_0^z[(\cos t-1)/t]\,dt, \quad |\arg z|<\pi$$

where γ is Euler's constant.

Series Expansions:

$$Si(z)=\sum_{n=0}^{\infty}\frac{(-1)^n z^{2n+1}}{(2n+1)(2n+1)!}$$

$$Ci(z)=\gamma+\ln z+\sum_{n=1}^{\infty}\frac{(-1)^n z^{2n}}{2n(2n)!}.$$

Asymptotic Expansions:

$$Si(z) \sim \tfrac{1}{2}\pi - \left(1 - \frac{2!}{z^2} + \frac{4!}{z^4} - \frac{6!}{z^6} + \cdots\right)\frac{\cos z}{z}$$

$$- \left(1 - \frac{3!}{z^2} + \frac{5!}{z^4} - \frac{7!}{z^6} + \cdots\right)\frac{\sin z}{z^2}$$

$$Ci(z) \sim \left(1 - \frac{2!}{z^2} + \frac{4!}{z^4} - \frac{6!}{z^6} + \cdots\right)\frac{\sin z}{z}$$

$$- \left(1 - \frac{3!}{z^2} + \frac{5!}{z^4} - \frac{7!}{z^6} + \cdots\right)\frac{\cos z}{z^2}.$$

Psi and Polygamma Functions

Definitions:

$$\psi^{(n)}(z) = \frac{d^{n+1}}{dz^{n+1}}\left[\ln\Gamma(z)\right] \quad n=0, 1, 2, \cdots$$

$\psi^{(0)}(z) = \psi(z)$ is known as the psi function and $\psi^{(n)}(z)$ as the polygamma function of order n, $n=1, 2, \cdots$

Special Values:

$$\psi^{(n)}(1) = (-1)^{n+1}n!\zeta(n+1), \quad n=1, 2, \cdots$$

$$\psi(1) = -\gamma$$

$$\psi^{(n)}(\tfrac{1}{2}) = (-1)^{n+1}n!(2^{n+1}-1)\zeta(n+1),$$
$$n=1, 2, \cdots$$

$$\psi(\tfrac{1}{2}) = -\gamma - 2\ln 2$$

in which $\zeta(n+1)$ is the Riemann zeta function of $n+1$.

Series Expansions:

$$\psi^{(n)}(1+z) = \sum_{k=0}^{\infty} \frac{(-1)^{n+k+1}(n+k)!\zeta(n+k+1)z^k}{k!},$$
$$|z|<1, n=1, 2, \cdots$$

$$\psi(1+z) = -\gamma - \sum_{k=1}^{\infty}(-1)^k\zeta(k+1)z^k, \quad |z|<1.$$

Gamma Function

Definition:

$$\Gamma(z) = \int_0^{\infty} t^{z-1}e^{-t}dt$$

$$\Gamma(-z) = -\pi/[\Gamma(z+1)\sin\pi z], \quad \text{Re} z > 0.$$

$\Gamma(z)$ is an analytic function everywhere except at the negative integers.

Identities:

$$z\Gamma(z) = \Gamma(z+1)$$

$$\Gamma(\tfrac{1}{2}+z)\Gamma(\tfrac{1}{2}-z) = \pi \sec\pi z.$$

Special Values:

$$\Gamma(z) = (z-1)! \quad z=1, 2, 3, \cdots$$

$$\Gamma(z+\tfrac{1}{2}) = [(2z)!/2^{2z}z!](\pi)^{1/2}$$

n	1	2	3	4	5	6	7	8	9	10
$\Gamma(n)=(n-1)!$	1	1	2	6	24	120	720	5040	40 320	362 880.

Asymptotic Expansion (Stirling's Equation):

$$\Gamma(z) \sim e^{-z}z^{z-1/2}(2\pi)^{1/2}[1 + (1/12z) + (1/288z^2) + \cdots]$$

$$\text{Note } z! = z\Gamma(z) \sim e^{-z}z^{z+1/2}(2\pi)^{1/2}[1 + (1/12z) + (1/288z^2) + \cdots]$$

Asymptotic Expansions:

$$\psi^{(n)}(z) \sim [(-1)^n(n-1)!/z^n][1 + (n/2z) + \cdots],$$
$$\text{as } z \to \infty, |\arg z| < \pi, n=1, 2, \cdots$$

$$\psi(z) \sim \ln z - (1/2z) + \cdots.$$

Error Function

Definitions:

$$\text{erf} z = (2/\pi^{1/2})\int_0^z \exp(-t^2)dt$$

$$\text{erfc} z = (2/\pi^{1/2})\int_z^{\infty} \exp(-t^2)dt$$

$$= 1 - \text{erf} z.$$

The path of integration for large t must remain within $|\arg t| < \pi/4$ in the latter integral.

Series Expansion:

$$\text{erf} z = (2/\pi^{1/2})\sum_{n=0}^{\infty}[(-1)^n z^{2n+1}/n!(2n+1)],$$
$$|z| < \infty.$$

Asymptotic Expansion:

$$\text{erf} z \sim 1 - [\exp(-z^2)/z\pi^{1/2}]$$
$$\times [1 - (1/2z^2) + (3/4z^4) - \cdots],$$
$$\text{as } z \to \infty, |\arg z| < 3\pi/4.$$

Inequality:

$$\frac{(2/\pi^{1/2})\exp(-x^2)}{x+(x^2+2)^{1/2}}<\text{erfc}\,x\le\frac{(2/\pi^{1/2})\exp(-x^2)}{x+[x^2+(4/\pi)]^{1/2}},$$

$$x\ge 0.$$

Derivatives:

$$(d^{(n+1)}/dz^{(n+1)})\,\text{erf}\,z=(-1)^n(2/\pi^{1/2})H_n(z)e^{-z^2},$$

$$n=0,1,2,\cdots$$

where $H_n(z)$ is the Hermite polynomial of order n.

Relation to Gaussian Distribution:

$$\text{Prob}(X\le x)=[\sigma(2\pi)^{1/2}]^{-1}\int_{-\infty}^{x}\exp\left(-\frac{(t-\mu)^2}{2\sigma^2}\right)dt$$

$$=\frac{1}{2}\left[1+\text{erf}\left(\frac{x-\mu}{\sigma\sqrt{2}}\right)\right].$$

Fresnel Integrals

Definitions:

$$C(z)=\int_0^z\cos[(\pi/2)t^2]dt$$

$$S(z)=\int_0^z\sin[(\pi/2)t^2]dt.$$

Series Expansions:

$$C(z)=\sum_{n=0}^{\infty}\frac{(-1)^n(\pi/2)^{2n}}{(2n)!(4n+1)}z^{4n+1}$$

$$S(z)=\sum_{n=0}^{\infty}\frac{(-1)^n(\pi/2)^{2n+1}}{(2n+1)!(4n+3)}z^{4n+3},\quad |z|<\infty.$$

Asymptotic Expansions:

$$C(z)\sim\frac{1}{2}+\left(1-\frac{1\cdot3}{(\pi z^2)^2}+\frac{1\cdot3\cdot5\cdot7}{(\pi z^2)^4}-\cdots\right)\frac{\sin\frac{1}{2}\pi z^2}{\pi z}$$

$$-\left[(1/\pi z^2)-\frac{1\cdot3\cdot5}{(\pi z^2)^3}+\frac{1\cdot3\cdot5\cdot7\cdot9}{(\pi z^2)^5}-\cdots\right]\frac{\cos\frac{1}{2}\pi z^2}{\pi z}$$

$$S(z)\sim\frac{1}{2}-\left[1-\frac{1\cdot3}{(\pi z^2)^2}+\frac{1\cdot3\cdot5\cdot7}{(\pi z^2)^4}-\cdots\right]\frac{\cos\frac{1}{2}\pi z^2}{\pi z}$$

$$-\left[(1/\pi z^2)-\frac{1\cdot3\cdot5}{(\pi z^2)^3}+\frac{1\cdot3\cdot5\cdot7\cdot9}{(\pi z^2)^5}-\cdots\right]\frac{\sin\frac{1}{2}\pi z^2}{\pi z}.$$

Elliptic Integrals

First Kind:

$$F(\phi,k)=\int_0^\phi\frac{d\theta}{(1-k^2\sin^2\theta)^{1/2}}.$$

Second Kind:

$$E(\phi,k)=\int_0^\phi(1-k^2\sin^2\theta)^{1/2}d\theta.$$

If $\phi=\pi/2$, the elliptic integrals are said to be complete and they are denoted by K or $K(k)$ and E or $E(k)$.

Bessel Functions* (Fig. 23)

Definitions: Bessel functions are solutions to Bessel's differential equation

$$z^2(d^2w/dz^2)+z(dw/dz)+(z^2-\nu^2)w=0.$$

They are divided into

First kind

$$J_\nu(z)=(z/2)^\nu\sum_{k=0}^{\infty}\frac{(-1)^k(z/2)^{2k}}{k!\Gamma(k+\nu+1)}.$$

Second kind

$$Y_\nu(z)=\frac{J_\nu(z)\cos\nu\pi-J_{-\nu}(z)}{\sin\pi\nu},\quad \nu\text{ not an integer}$$

$$Y_n(z)=\lim_{\nu\to n}Y_\nu(z),\quad n\text{ integral}.$$

Third kind

$$H_\nu^{(1)}(z)=J_\nu(z)+jY_\nu(z)$$

$$H_\nu^{(2)}(z)=J_\nu(z)-jY_\nu(z).$$

The second and third kinds are sometimes called Neumann and Hankel functions, respectively.

The modified Bessel functions are solutions to Bessel's differential equation with z replaced by jz.

$$I_\nu(z)=\exp(-j\pi\nu/2)J_\nu[z\exp(j\pi/2)],$$

$$-\pi<\arg z\le\pi/2$$

$$=\exp(j3\pi\nu/2)J_\nu[z\exp(-j3\pi/2)],$$

$$\pi/2<\arg z\le\pi$$

$$K_\nu(z)=(j\pi/2)\exp(j\pi\nu/2)H_\nu^{(1)}[z\exp(j\pi/2)],$$

$$-\pi<\arg z\le\pi/2$$

$$=-(j\pi/2)\exp(-j\pi\nu/2)$$

$$\times H_\nu^{(2)}[z\exp(-j\pi/2)],\quad -\pi/2<\arg z\le\pi.$$

Integral Representations:

$$J_\nu(z)=\frac{2(z/2)^\nu}{\Gamma(\nu+\frac{1}{2})\pi^{1/2}}\int_0^{\pi/2}\cos(z\sin\phi)(\cos\phi)^{2\nu}d\phi,$$

$$\text{Re}\,\nu>-\frac{1}{2}$$

$$H_\nu^{(1)}(z)=-(j/\pi)e^{-j\pi\nu/2}\int_{-\infty}^{\infty}\exp(jz\cosh t-\nu t)\,dt,$$

$$0<\arg z<\pi$$

$$H_\nu^{(2)}(z)=(j/\pi)e^{j\pi\nu/2}\int_{-\infty}^{\infty}\exp(-jz\cosh t-\nu t)\,dt,$$

$$-\pi<\arg z<0.$$

* For an extensive treatment of Bessel functions, see G. N. Watson, "A Treatise on the Theory of Bessel Functions," Cambridge University Press, New York; 1943.

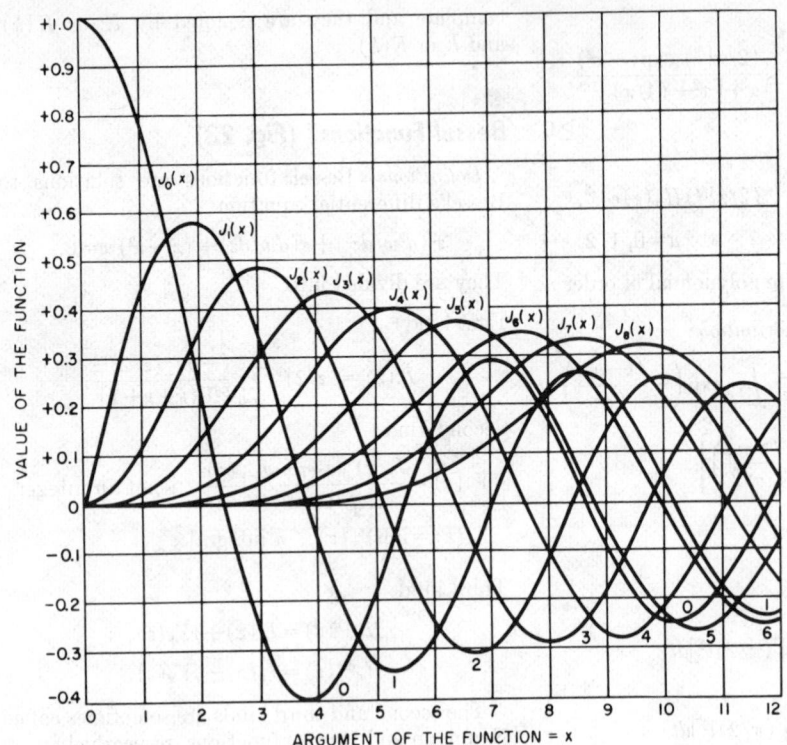

Fig. 23—Bessel functions for the first 8 orders.

Asymptotic Expressions: For $|z| \to \infty$

$$J_\nu(z) \sim [(2/\pi z)]^{1/2} \cos(z - \nu\pi/2 - \pi/4),$$
$$|\arg z| < \pi$$

$$Y_\nu(z) \sim [(2/\pi z)]^{1/2} \sin(z - \nu\pi/2 - \pi/4),$$
$$|\arg z| < \pi$$

$$H_\nu^{(1)}(z) \sim [(2/\pi z)]^{1/2} \exp[j(z - \nu\pi/2 - \pi/4)],$$
$$-\pi < \arg z < 2\pi$$

$$H_\nu^{(2)}(z) \sim [(2/\pi z)]^{1/2} \exp[-j(z - \nu\pi/2 - \pi/4)],$$
$$-2\pi < \arg z < \pi.$$

Recurrence Relations:

$$C_{\nu-1}(z) + C_{\nu+1}(z) = (2\nu/z) C_\nu(z)$$

$$C_{\nu-1}(z) - C_{\nu+1}(z) = 2C_\nu'(z)$$

where C denotes J, Y, $H^{(1)}$ or $H^{(2)}$.

Series Containing Bessel Functions:

$$\exp(-ju \sin x) = \sum_{n=-\infty}^{\infty} J_n(u) \exp(-jnx)$$

$$\cos(u \sin x) = J_0(u) + 2 \sum_{n=1}^{\infty} J_{2n}(u) \cos 2nx$$

$$\sin(u \sin x) = 2 \sum_{n=1}^{\infty} J_{2n-1}(u) \sin(2n-1)x$$

$$\cos(u \cos x) = J_0(u) + 2 \sum_{n=1}^{\infty} (-1)^n J_{2n}(u) \cos 2nx$$

$$\sin(u \cos x) = 2 \sum_{n=1}^{\infty} (-1)^{n+1} J_{2n-1}(u)$$
$$\times \cos(2n-1)x$$

$$1 = \sum_{n=-\infty}^{\infty} J_n^2(x).$$

Orthogonal Polynomials

Any set of polynomials $\{f_n(x)\}$ with the property

$$\int_a^b w(x) f_n(x) f_m(x) \, dx = 0, \quad \text{for } m \neq n$$

$$= h_n, \quad \text{for } m = n$$

is called a set of orthogonal polynomials on the interval (a, b) with respect to the weight function $w(x)$. These functions occur in the Gauss quadrature equations among other places. Chebishev polynomials are involved in the theory of the Chebishev filter; Hermite polynomials arise in the refinements of the Central Limit Theorem, the so-called Edgeworth series, etc.

The important properties are summarized in the following table.

$f_n(x)$	Name	a	b	$w(x)$	h_n	Explicit Expression
$T_n(x)$	Chebishev	-1	$+1$	$(1-x^2)^{-1/2}$	$\frac{1}{2}\pi,\ n\neq 0;\ \pi,\ n=0$	$\frac{1}{2}n\sum_{m=0}^{[n/2]}(-1)^m\frac{(n-m-1)!}{m!(n-2m)!}(2x)^{n-2m}$
$H_n(x)$	Hermite	$-\infty$	∞	e^{-x^2}	$2^n n!\pi^{1/2}$	$n!\sum_{m=0}^{[n/2]}(-1)^m\frac{(2x)^{n-2m}}{m!(n-2m)!}$
$L_n(x)$	Laguerre	0	∞	e^{-x}	1	$\sum_{m=0}^{n}(-1)^m\binom{n}{m}\frac{x^m}{m!}$
$P_n(x)$	Legendre	-1	1	1	$2/(2n+1)$	$2^{-n}\sum_{m=0}^{[n/2]}(-1)^m\binom{n}{m}\binom{2n-2m}{n}x^{n-2m}$

where $[n/2]$ denotes the largest integer less than or equal to $n/2$.

NUMERICAL ANALYSIS

Algorithms for Solving F(x) =0

Bisection Method and Regula Falsi (*Rule of False Position*): First determine x_1 and x_2 such that $F(x_1)F(x_2)<0$, i.e., x_1 and x_2 are points at which the function has opposite signs.

Bisection Method: Calculate

$$x_3 = (x_1+x_2)/2.$$

Regula Falsi: Calculate

$$x_3 = [x_1 F(x_2) - x_2 F(x_1)]/[F(x_2)-F(x_1)].$$

To obtain the next approximation, take x_3 and x_i, $i=1$ or 2, such that $F(x_3)F(x_i)<0$, and repeat the procedure.

Newton-Raphson: Take some initial value x_1 and calculate successively

$$x_{n+1} = x_n - [F(x_n)/F'(x_n)], \quad n=1, 2, 3, \cdots$$

This method may not converge. When it converges, the rate of convergence is generally faster than the bisection method or the regula falsi.

Algorithm for Solving F(x, y) =G(x, y) =0

The following is an extension of the Newton-Raphson method described above. Take some initial values x_1 and y_1 and calculate successively

$$x_{n+1} = x_n + \left[\left(\frac{\partial F}{\partial y}G - F\frac{\partial G}{\partial y}\right)\Big/\left(\frac{\partial F}{\partial x}\frac{\partial G}{\partial y}-\frac{\partial F}{\partial y}\frac{\partial G}{\partial x}\right)\right]\Big|_{(x,y)=(x_n,y_n)}$$

$$y_{n+1} = y_n + \left[\left(F\frac{\partial G}{\partial x} - \frac{\partial F}{\partial x}G\right)\Big/\left(\frac{\partial F}{\partial x}\frac{\partial G}{\partial y}-\frac{\partial F}{\partial y}\frac{\partial G}{\partial x}\right)\right]\Big|_{(x,y)=(x_n,y_n)}$$

for $n=1, 2, 3, \cdots$.

By using Taylor's series to first derivatives and Cramer's rule, this algorithm may be further extended to the case of m simultaneous equations in m variables.

Interpolation Polynomial

The polynomial of lowest degree which passes through n points (x_i, y_i), $i=1, 2, \cdots, n$, is given by

$$P(x) = \sum_{i=1}^{n}\left(y_i\prod_{k=1\ k\neq i}^{n}\frac{x-x_k}{x_i-x_k}\right)$$

where $x_i\neq x_k$ for $i\neq k$.

Interpolation at Equidistant Points

Let

$$f_i = f(x_0+ih)$$
$$g_{ij} = g(x+ih, y+jk).$$

Then $f(x_0+ph)$ may be approximated by

$$(1-p)f_0+pf_1, \quad \text{given two points}$$

$$[p(p-1)/2]f_{-1}+(1-p^2)f_0+[p(p+1)/2]f_1,$$

given three points

$$[-p(p-1)(p-2)/6]f_{-1}+[(p^2-1)(p-2)/2]f_0$$
$$-[p(p+1)(p+2)/2]f_1+[p(p^2-1)/6]f_2,$$

given four points.

$g(x_0+ph, y_0+qk)$ may be approximated by

$$(1-p-q)g_{00}+pg_{10}+qg_{01}, \quad \text{given three points}$$

$$(1-p)(1-q)g_{00}+p(1-q)g_{10}+q(1-p)g_{01}+pqg_{11},$$

given four points.

Integration (Fig. 24)

Let

$$f_i=f(x_0+ih)$$

$$\int_{x_0}^{x_0+mh} f(x)\,dx \approx h(\tfrac{1}{2}f_0+f_1+f_2+\cdots+f_{n-1}+\tfrac{1}{2}f_n),$$

(Trapezoidal Rule)

$$\int_{x_0}^{x_0+2nh} f(x)\,dx \approx \tfrac{1}{3}h(f_0+4f_1+2f_2+4f_3+2f_4+\cdots$$
$$+2f_{2n-2}+4f_{2n-1}+f_{2n}) \quad \text{(Simpson's Rule)}.$$

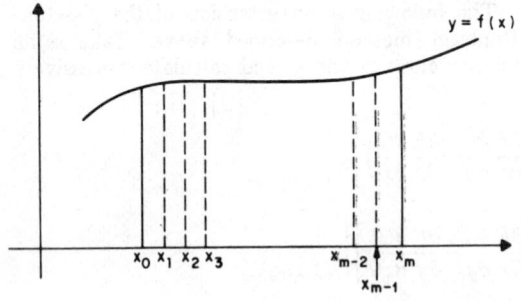

FOR SIMPSON'S RULE m MUST BE EVEN.

Fig. 24.

Differentiation

If a function $f(x)$ is known at two points a and b, then

$$f'(a) \approx [f(b)-f(a)]/(b-a).$$

If $f(x)$ is given at a discrete set of points, one may differentiate the interpolation polynomial stated above. The equation simplifies when the derivative is calculated at interpolation points which are equidistant, for example

$$f_{-1}' \approx (1/2h)(-3f_{-1}+4f_0-f_1)$$

$$f_0' \approx (1/2h)(-f_{-1}+f_1)$$

$$f_1' \approx (1/2h)(f_{-1}-4f_0+3f_1)$$

where

$$f_i=f(x_0+ih), \quad i=-1, 0, 1.$$

Error in Arithmetic Operations

Let the quantities A and B be known to within errors of a and b, respectively, where a and b are small relative to A and B. Then

Operation	Maximum Error (to first order in a and b)
$A \pm B$	$a+b$
$A \cdot B$	$\|AB\|(\|a/A\|+\|b/B\|)$
A/B	$\|A/B\|(\|a/A\|+\|b/B\|)$

The magnitude q of small random errors often has a normal distribution

$$(h/\pi^{1/2})\exp(-h^2q^2)$$

where h is called the index of precision. The rms of q denoted by σ is in terms of h

$$\sigma=1/h\sqrt{2}.$$

Let $f(x_1, x_2, \cdots, x_n)$ be a function of n variables x_i, which have normally distributed independent errors q_i with indices of precision h_i. Then $f(x_1, x_2, \cdots, x_n)$ will have an error Q, which is also normally distributed with index of precision H. For small errors

$$H \approx \{[h_1^{-1}(\partial f/\partial q_1)]^2+[h_2^{-1}(\partial f/\partial q_2)]^2+\cdots$$
$$+[h_n^{-1}(\partial f/\partial q_n)]^2\}^{-1/2}.$$

LOGARITHMS TO BASE 2 AND POWERS OF 2

x	$\log_2 x$
0.1	−3.32193
0.2	−2.32193
0.3	−1.73697
0.4	−1.32193
0.5	−1.00000
0.6	−0.73697
0.7	−0.51457
0.8	−0.32193
0.9	−0.15200
1.0	0.00000
1.1	0.13750
1.2	0.26303
1.3	0.37851
1.4	0.48543
1.5'	0.58496
1.6	0.67807
1.7	0.76553
1.8	0.84800
1.9	0.92600
2.0	1.00000
10	3.32193
100	6.64386
1000	9.96578

2^y	y

$$\log_2 x = \log_2 10 \, \log_{10} x = \log_2 e \, \log_e x$$

$$2^y = e^{y \log_e 2} = 10^{y \log_{10} 2}$$

$$\log_2 10 = 3.32193 = 1/\log_{10} 2$$

$$\log_{10} 2 = 0.30103 = 1/\log_2 10$$

$$\log_2 e = 1.44269 = 1/\log_e 2$$

$$\log_e 2 = 0.69315 = 1/\log_2 e$$

y	2^y
0.1	1.072
0.2	1.149
0.3	1.231
0.4	1.320
0.5	1.414
0.6	1.515
0.7	1.625
0.8	1.741
0.9	1.866
1	2
2	4
3	8
4	16
5	32
6	64
7	128
8	256
9	512
10	1 024
11	2 048
12	4 096
13	8 192
14	16 384
15	32 768
16	65 536
17	131 072
18	262 144
19	524 288
20	1 048 576
21	2 097 152
22	4 194 304
23	8 388 608
24	16 777 216
25	33 554 432
26	67 108 864
27	134 217 728
28	268 435 456
29	536 870 912
30	1 073 741 824
31	2 147 483 648
32	4 294 967 296

$\log_2 x$	x

47-1

COMMON LOGARITHMS OF NUMBERS AND PROPORTIONAL PARTS

	0	1	2	3	4	5	6	7	8	9	proportional parts								
											1	2	3	4	5	6	7	8	9
10	0000	0043	0086	0128	0170	0212	0253	0294	0334	0374	4	8	12	17	21	25	29	33	37
11	0414	0453	0492	0531	0569	0607	0645	0682	0719	0755	4	8	11	15	19	23	26	30	34
12	0792	0828	0864	0899	0934	0969	1004	1038	1072	1106	3	7	10	14	17	21	24	28	31
13	1139	1173	1206	1239	1271	1303	1335	1367	1399	1430	3	6	10	13	16	19	23	26	29
14	1461	1492	1523	1553	1584	1614	1644	1673	1703	1732	3	6	9	12	15	18	21	24	27
15	1761	1790	1818	1847	1875	1903	1931	1959	1987	2014	3	6	8	11	14	17	20	22	25
16	2041	2068	2095	2122	2148	2175	2201	2227	2253	2279	3	5	8	11	13	16	18	21	24
17	2304	2330	2355	2380	2405	2430	2455	2480	2504	2529	2	5	7	10	12	15	17	20	22
18	2553	2577	2601	2625	2648	2672	2695	2718	2742	2765	2	5	7	9	12	14	16	19	21
19	2788	2810	2833	2856	2878	2900	2923	2945	2967	2989	2	4	7	9	11	13	16	18	20
20	3010	3032	3054	3075	3096	3118	3139	3160	3181	3201	2	4	6	8	11	13	15	17	19
21	3222	3243	3263	3284	3304	3324	3345	3365	3385	3404	2	4	6	8	10	12	14	16	18
22	3424	3444	3464	3483	3502	3522	3541	3560	3579	3598	2	4	6	8	10	12	14	15	17
23	3617	3636	3655	3674	3692	3711	3729	3747	3766	3784	2	4	6	7	9	11	13	15	17
24	3802	3820	3838	3856	3874	3892	3909	3927	3945	3962	2	4	5	7	9	11	12	14	16
25	3979	3997	4014	4031	4048	4065	4082	4099	4116	4133	2	3	5	7	9	10	12	14	15
26	4150	4166	4183	4200	4216	4232	4249	4265	4281	4298	2	3	5	7	8	10	11	13	15
27	4314	4330	4346	4362	4378	4393	4409	4425	4440	4456	2	3	5	6	8	9	11	13	14
28	4472	4487	4502	4518	4533	4548	4564	4579	4594	4609	2	3	5	6	8	9	11	12	14
29	4624	4639	4654	4669	4683	4698	4713	4728	4742	4757	1	3	4	6	7	9	10	12	13
30	4771	4786	4800	4814	4829	4843	4857	4871	4886	4900	1	3	4	6	7	9	10	11	13
31	4914	4928	4942	4955	4969	4983	4997	5011	5024	5038	1	3	4	6	7	8	10	11	12
32	5051	5065	5079	5092	5105	5119	5132	5145	5159	5172	1	3	4	5	7	8	9	11	12
33	5185	5198	5211	5224	5237	5250	5263	5276	5289	5302	1	3	4	5	6	8	9	10	12
34	5315	5328	5340	5353	5366	5378	5391	5403	5416	5428	1	3	4	5	6	8	9	10	11
35	5441	5453	5465	5478	5490	5502	5514	5527	5539	5551	1	2	4	5	6	7	9	10	11
36	5563	5575	5587	5599	5611	5623	5635	5647	5658	5670	1	2	4	5	6	7	8	10	11
37	5682	5694	5705	5717	5729	5740	5752	5763	5775	5786	1	2	3	5	6	7	8	9	10
38	5798	5809	5821	5832	5843	5855	5866	5877	5888	5899	1	2	3	5	6	7	8	9	10
39	5911	5922	5933	5944	5955	5966	5977	5988	5999	6010	1	2	3	4	5	7	8	9	10
40	6021	6031	6042	6053	6064	6075	6085	6096	6107	6117	1	2	3	4	5	6	8	9	10
41	6128	6138	6149	6160	6170	6180	6191	6201	6212	6222	1	2	3	4	5	6	7	8	9
42	6232	6243	6253	6263	6274	6284	6294	6304	6314	6325	1	2	3	4	5	6	7	8	9
43	6335	6345	6355	6365	6375	6385	6395	6405	6415	6425	1	2	3	4	5	6	7	8	9
44	6435	6444	6454	6464	6474	6484	6493	6503	6513	6522	1	2	3	4	5	6	7	8	9
45	6532	6542	6551	6561	6571	6580	6590	6599	6609	6618	1	2	3	4	5	6	7	8	9
46	6628	6637	6646	6656	6665	6675	6684	6693	6702	6712	1	2	3	4	5	6	7	7	8
47	6721	6730	6739	6749	6758	6767	6776	6785	6794	6803	1	2	3	4	5	5	6	7	8
48	6812	6821	6830	6839	6848	6857	6866	6875	6884	6893	1	2	3	4	4	5	6	7	8
49	6902	6911	6920	6928	6937	6946	6955	6964	6972	6981	1	2	3	4	4	5	6	7	8
50	6990	6998	7007	7016	7024	7033	7042	7050	7059	7067	1	2	3	3	4	5	6	7	8
51	7076	7084	7093	7101	7110	7118	7126	7135	7143	7152	1	2	3	3	4	5	6	7	8
52	7160	7168	7177	7185	7193	7202	7210	7218	7226	7235	1	2	2	3	4	5	6	7	7
53	7243	7251	7259	7267	7275	7284	7292	7300	7308	7316	1	2	2	3	4	5	6	6	7
54	7324	7332	7340	7348	7356	7364	7372	7380	7388	7396	1	2	2	3	4	5	6	6	7

COMMON LOGARITHMS OF NUMBERS AND PROPORTIONAL PARTS (continued)

	0	1	2	3	4	5	6	7	8	9	1	2	3	4	5	6	7	8	9
55	7404	7412	7419	7427	7435	7443	7451	7459	7466	7474	1	2	2	3	4	5	5	6	7
56	7482	7490	7497	7505	7513	7520	7528	7536	7543	7551	1	2	2	3	4	5	5	6	7
57	7559	7566	7574	7582	7589	7597	7604	7612	7619	7627	1	2	2	3	4	5	5	6	7
58	7634	7642	7649	7657	7664	7672	7679	7686	7694	7701	1	1	2	3	4	4	5	6	7
59	7709	7716	7723	7731	7738	7745	7752	7760	7767	7774	1	1	2	3	4	4	5	6	7
60	7782	7789	7796	7803	7810	7818	7825	7832	7839	7846	1	1	2	3	4	4	5	6	6
61	7853	7860	7868	7875	7882	7889	7896	7903	7910	7917	1	1	2	3	4	4	5	6	6
62	7924	7931	7938	7945	7952	7959	7966	7973	7980	7987	1	1	2	3	3	4	5	6	6
63	7993	8000	8007	8014	8021	8028	8035	8041	8048	8055	1	1	2	3	3	4	5	5	6
64	8062	8069	8075	8082	8089	8096	8102	8109	8116	8122	1	1	2	3	3	4	5	5	6
65	8129	8136	8142	8149	8156	8162	8169	8176	8182	8189	1	1	2	3	3	4	5	5	6
66	8195	8202	8209	8215	8222	8228	8235	8241	8248	8254	1	1	2	3	3	4	5	5	6
67	8261	8267	8274	8280	8287	8293	8299	8306	8312	8319	1	1	2	3	3	4	5	5	6
68	8325	8331	8338	8344	8351	8357	8363	8370	8376	8382	1	1	2	3	3	4	4	5	6
69	8388	8395	8401	8407	8414	8420	8426	8432	8439	8445	1	1	2	3	3	4	4	5	6
70	8451	8457	8463	8470	8476	8482	8488	8494	8500	8506	1	1	2	2	3	4	4	5	6
71	8513	8519	8525	8531	8537	8543	8549	8555	8561	8567	1	1	2	2	3	4	4	5	5
72	8573	8579	8585	8591	8597	8603	8609	8615	8621	8627	1	1	2	2	3	4	4	5	5
73	8633	8639	8645	8651	8657	8663	8669	8675	8681	8686	1	1	2	2	3	4	4	5	5
74	8692	8698	8704	8710	8716	8722	8727	8733	8739	8745	1	1	2	2	3	4	4	5	5
75	8751	8756	8762	8768	8774	8779	8785	8791	8797	8802	1	1	2	2	3	3	4	5	5
76	8808	8814	8820	8825	8831	8837	8842	8848	8854	8859	1	1	2	2	3	3	4	5	5
77	8865	8871	8876	8882	8887	8893	8899	8904	8910	8915	1	1	2	2	3	3	4	4	5
78	8921	8927	8932	8938	8943	8949	8954	8960	8965	8971	1	1	2	2	3	3	4	4	5
79	8976	8982	8987	8993	8998	9004	9009	9015	9020	9025	1	1	2	2	3	3	4	4	5
80	9031	9036	9042	9047	9053	9058	9063	9069	9074	9079	1	1	2	2	3	3	4	4	5
81	9085	9090	9096	9101	9106	9112	9117	9122	9128	9133	1	1	2	2	3	3	4	4	5
82	9138	9143	9149	9154	9159	9165	9170	9175	9180	9186	1	1	2	2	3	3	4	4	5
83	9191	9196	9201	9206	9212	9217	9222	9227	9232	9238	1	1	2	2	3	3	4	4	5
84	9243	9248	9253	9258	9263	9269	9274	9279	9284	9289	1	1	2	2	3	3	4	4	5
85	9294	9299	9304	9309	9315	9320	9325	9330	9335	9340	1	1	2	2	3	3	4	4	5
86	9345	9350	9355	9360	9365	9370	9375	9380	9385	9390	1	1	2	2	3	3	4	4	5
87	9395	9400	9405	9410	9415	9420	9425	9430	9435	9440	0	1	1	2	2	3	3	4	4
88	9445	9450	9455	9460	9465	9469	9474	9479	9484	9489	0	1	1	2	2	3	3	4	4
89	9494	9499	9504	9509	9513	9518	9523	9528	9533	9538	0	1	1	2	2	3	3	4	4
90	9542	9547	9552	9557	9562	9566	9571	9576	9581	9586	0	1	1	2	2	3	3	4	4
91	9590	9595	9600	9605	9609	9614	9619	9624	9628	9633	0	1	1	2	2	3	3	4	4
92	9638	9643	9647	9652	9657	9661	9666	9671	9675	9680	0	1	1	2	2	3	3	4	4
93	9685	9689	9694	9699	9703	9708	9713	9717	9722	9727	0	1	1	2	2	3	3	4	4
94	9731	9736	9741	9745	9750	9754	9759	9763	9768	9773	0	1	1	2	2	3	3	4	4
95	9777	9782	9786	9791	9795	9800	9805	9809	9814	9818	0	1	1	2	2	3	3	4	4
96	9823	9827	9832	9836	9841	9845	9850	9854	9859	9863	0	1	1	2	2	3	3	4	4
97	9868	9872	9877	9881	9886	9890	9894	9899	9903	9908	0	1	1	2	2	3	3	4	4
98	9912	9917	9921	9926	9930	9934	9939	9943	9948	9952	0	1	1	2	2	3	3	4	4
99	9956	9961	9965	9969	9974	9978	9983	9987	9991	9996	0	1	1	2	2	3	3	3	4

proportional parts: 1 2 3 | 4 5 6 | 7 8 9

NATURAL TRIGONOMETRIC FUNCTIONS FOR DECIMAL FRACTIONS OF A DEGREE

deg	sin	cos	tan	cot	
0.0	.00000	1.0000	.00000	∞	90.0
.1	.00175	1.0000	.00175	573.0	.9
.2	.00349	1.0000	.00349	286.5	.8
.3	.00524	1.0000	.00524	191.0	.7
.4	.00698	1.0000	.00698	143.24	.6
.5	.00873	1.0000	.00873	114.59	.5
.6	.01047	0.9999	.01047	95.49	.4
.7	.01222	.9999	.01222	81.85	.3
.8	.01396	.9999	.01396	71.62	.2
.9	.01571	.9999	.01571	63.66	.1
1.0	.01745	0.9998	.01746	57.29	89.0
.1	.01920	.9998	.01920	52.08	.9
.2	.02094	.9998	.02095	47.74	.8
.3	.02269	.9997	.02269	44.07	.7
.4	.02443	.9997	.02444	40.92	.6
.5	.02618	.9997	.02619	38.19	.5
.6	.02792	.9996	.02793	35.80	.4
.7	.02967	.9996	.02968	33.69	.3
.8	.03141	.9995	.03143	31.82	.2
.9	.03316	.9995	.03317	30.14	.1
2.0	.03490	0.9994	.03492	28.64	88.0
.1	.03664	.9993	.03667	27.27	.9
.2	.03839	.9993	.03842	26.03	.8
.3	.04013	.9992	.04016	24.90	.7
.4	.04188	.9991	.04191	23.86	.6
.5	.04362	.9990	.04366	22.90	.5
.6	.04536	.9990	.04541	22.02	.4
.7	.04711	.9989	.04716	21.20	.3
.8	.04885	.9988	.04891	20.45	.2
.9	.05059	.9987	.05066	19.74	.1
3.0	.05234	0.9986	.05241	19.081	87.0
.1	.05408	.9985	.05416	18.464	.9
.2	.05582	.9984	.05591	17.886	.8
.3	.05756	.9983	.05766	17.343	.7
.4	.05931	.9982	.05941	16.832	.6
.5	.06105	.9981	.06116	16.350	.5
.6	.06279	.9980	.06291	15.895	.4
.7	.06453	.9979	.06467	15.464	.3
.8	.06627	.9978	.06642	15.056	.2
.9	.06802	.9977	.06817	14.669	.1
4.0	.06976	0.9976	.06993	14.301	86.0
.1	.07150	.9974	.07168	13.951	.9
.2	.07324	.9973	.07344	13.617	.8
.3	.07498	.9972	.07519	13.300	.7
.4	.07672	.9971	.07695	12.996	.6
.5	.07846	.9969	.07870	12.706	.5
.6	.08020	.9968	.08046	12.429	.4
.7	.08194	.9966	.08221	12.163	.3
.8	.08368	.9965	.08397	11.909	.2
.9	.08542	.9963	.08573	11.664	.1
5.0	.08716	0.9962	.08749	11.430	85.0
.1	.08889	.9960	.08925	11.205	.9
.2	.09063	.9959	.09101	10.988	.8
.3	.09237	.9957	.09277	10.780	.7
.4	.09411	.9956	.09453	10.579	.6
.5	.09585	.9954	.09629	10.385	.5
.6	.09758	.9952	.09805	10.199	.4
.7	.09932	.9951	.09981	10.019	.3
.8	.10106	.9949	.10158	9.845	.2
.9	.10279	.9947	.10334	9.677	.1
6.0	.10453	0.9945	.10510	9.514	84.0
	cos	sin	cot	tan	deg

deg	sin	cos	tan	cot	deg
6.0	.10453	0.9945	.10510	9.514	84.0
.1	.10626	.9943	.10687	9.357	.9
.2	.10800	.9942	.10863	9.205	.8
.3	.10973	.9940	.11040	9.058	.7
.4	.11147	.9938	.11217	8.915	.6
.5	.11320	9936	.11394	8.777	.5
.6	.11494	.9934	.11570	8.643	.4
.7	.11667	.9932	.11747	8.513	.3
.8	.11840	.9930	.11924	8.386	.2
.9	.12014	.9928	.12101	8.264	.1
7.0	.12187	0.9925	.12278	8.144	83.0
.1	.12360	.9923	.12456	8.028	.9
.2	.12533	.9921	.12633	7.916	.8
.3	.12706	.9919	.12810	7.806	.7
.4	.12880	.9917	.12988	7.700	.6
.5	.13053	.9914	.13165	7.596	.5
.6	.13226	.9912	.13343	7.495	.4
.7	.13399	.9910	.13521	7.396	.3
.8	.13572	.9907	.13698	7.300	.2
.9	.13744	.9905	.13876	7.207	.1
8.0	.13917	0.9903	.14054	7.115	82.0
.1	.14090	.9900	.14232	7.026	.9
.2	.14263	.9898	.14410	6.940	.8
.3	.14436	.9895	.14588	6.855	.7
.4	.14608	.9893	.14767	6.772	.6
.5	.14781	.9890	.14945	6.691	.5
.6	.14954	.9888	.15124	6.612	.4
.7	.15126	.9885	.15302	6.535	.3
.8	.15299	.9882	.15481	6.460	.2
.9	.15471	.9880	.15660	6.386	.1
9.0	.15643	0.9877	.15838	6.314	81.0
.1	.15816	.9874	.16017	6.243	.9
.2	.15988	.9871	.16196	6.174	.8
.3	.16160	.9869	.16376	6.107	.7
.4	.16333	.9866	.16555	6.041	.6
.5	.16505	.9863	.16734	5.976	.5
.6	.16677	.9860	.16914	5.912	.4
.7	.16849	.9857	.17093	5.850	.3
.8	.17021	.9854	.17273	5.789	.2
.9	.17193	.9851	.17453	5.730	.1
10.0	.1736	0.9848	.1763	5.671	80.0
.1	.1754	.9845	.1781	5.614	.9
.2	.1771	.9842	.1799	5.558	.8
.3	.1788	.9839	.1817	5.503	.7
.4	.1805	.9836	.1835	5.449	.6
.5	.1822	.9833	.1853	5.396	.5
.6	.1840	.9829	.1871	5.343	.4
.7	.1857	.9826	.1890	5.292	.3
.8	.1874	.9823	.1908	5.242	.2
.9	.1891	.9820	.1926	5.193	.1
11.0	.1908	0.9816	.1944	5.145	79.0
.1	.1925	.9813	.1962	5.097	.9
.2	.1942	.9810	.1980	5.050	.8
.3	.1959	.9806	.1998	5.005	.7
.4	.1977	.9803	.2016	4.959	.6
.5	.1994	.9799	.2035	4.915	.5
.6	.2011	.9796	.2053	4.872	.4
.7	.2028	.9792	.2071	4.829	.3
.8	.2045	.9789	.2089	4.787	.2
.9	.2062	.9785	.2107	4.745	.1
12.0	.2079	0.9781	.2126	4.705	78.0
	cos	sin	cot	tan	deg

NATURAL TRIGONOMETRIC FUNCTIONS FOR DECIMAL FRACTIONS OF A DEGREE (continued)

deg	sin	cos	tan	cot	deg
12.0	0.2079	0.9781	0.2126	4.705	78.0
.1	.2096	.9778	.2144	4.665	.9
.2	.2113	.9774	.2162	4.625	.8
.3	.2130	.9770	.2180	4.586	.7
.4	.2147	.9767	.2199	4.548	.6
.5	.2164	.9763	.2217	4.511	.5
.6	.2181	.9759	.2235	4.474	.4
.7	.2198	.9755	.2254	4.437	.3
.8	.2215	.9751	.2272	4.402	.2
.9	.2233	.9748	.2290	4.366	.1
13.0	0.2250	0.9744	0.2309	4.331	77.0
.1	.2267	.9740	.2327	4.297	.9
.2	.2284	.9736	.2345	4.264	.8
.3	.2300	.9732	.2364	4.230	.7
.4	.2317	.9728	.2382	4.198	.6
.5	.2334	.9724	.2401	4.165	.5
.6	.2351	.9720	.2419	4.134	.4
.7	.2368	.9715	.2438	4.102	.3
.8	.2385	.9711	.2456	4.071	.2
.9	.2402	.9707	.2475	4.041	.1
14.0	0.2419	0.9703	0.2493	4.011	76.0
.1	.2436	.9699	.2512	3.981	.9
.2	.2453	.9694	.2530	3.952	.8
.3	.2470	.9690	.2549	3.923	.7
.4	.2487	.9686	.2568	3.895	.6
.5	.2504	.9681	.2586	3.867	.5
.6	.2521	.9677	.2605	3.839	.4
.7	.2538	.9673	.2623	3.812	.3
.8	.2554	.9668	.2642	3.785	.2
.9	.2571	.9664	.2661	3.758	.1
15.0	0.2588	0.9659	0.2679	3.732	75.0
.1	.2605	.9655	.2698	3.706	.9
.2	.2622	.9650	.2717	3.681	.8
.3	.2639	.9646	.2736	3.655	.7
.4	.2656	.9641	.2754	3.630	.6
.5	.2672	.9636	.2773	3.606	.5
.6	.2689	.9632	.2792	3.582	.4
.7	.2706	.9627	.2811	3.558	.3
.8	.2723	.9622	.2830	3.534	.2
.9	.2740	.9617	.2849	3.511	.1
16.0	0.2756	0.9613	0.2867	3.487	74.0
.1	.2773	.9608	.2886	3.465	.9
.2	.2790	.9603	.2905	3.442	.8
.3	.2807	.9598	.2924	3.420	.7
.4	.2823	.9593	.2943	3.398	.6
.5	.2840	.9588	.2962	3.376	.5
.6	.2857	.9583	.2981	3.354	.4
.7	.2874	.9578	.3000	3.333	.3
.8	.2890	.9573	.3019	3.312	.2
.9	.2907	.9568	.3038	3.291	.1
17.0	0.2924	0.9563	0.3057	3.271	73.0
.1	.2940	.9558	.3076	3.251	.9
.2	.2957	.9553	.3096	3.230	.8
.3	.2974	.9548	.3115	3.211	.7
.4	.2990	.9542	.3134	3.191	.6
.5	.3007	.9537	.3153	3.172	.5
.6	.3024	.9532	.3172	3.152	.4
.7	.3040	.9527	.3191	3.133	.3
.8	.3057	.9521	.3211	3.115	.2
.9	.3074	.9516	.3230	3.096	.1
18.0	0.3090	0.9511	0.3249	3.078	72.0

cos	sin	cot	tan	deg

deg	sin	cos	tan	cot	deg
18.0	0.3090	0.9511	0.3249	3.078	72.0
.1	.3107	.9505	.3269	3.060	.9
.2	.3123	.9500	.3288	3.042	.8
.3	.3140	.9494	.3307	3.024	.7
.4	.3156	.9489	.3327	3.006	.6
.5	.3173	.9483	.3346	2.989	.5
.6	.3190	.9478	.3365	2.971	.4
.7	.3206	.9472	.3385	2.954	.3
.8	.3223	.9466	.3404	2.937	.2
.9	.3239	.9461	.3424	2.921	.1
19.0	0.3256	0.9455	0.3443	2.904	71.0
.1	.3272	.9449	.3463	2.888	.9
.2	.3289	.9444	.3482	2.872	.8
.3	.3305	.9438	.3502	2.856	.7
.4	.3322	.9432	.3522	2.840	.6
.5	.3338	.9426	.3541	2.824	.5
.6	.3355	.9421	.3561	2.808	.4
.7	.3371	.9415	.3581	2.793	.3
.8	.3387	.9409	.3600	2.778	.2
.9	.3404	.9403	.3620	2.762	.1
20.0	0.3420	0.9397	0.3640	2.747	70.0
.1	.3437	.9391	.3659	2.733	.9
.2	.3453	.9385	.3679	2.718	.8
.3	.3469	.9379	.3699	2.703	.7
.4	.3486	.9373	.3719	2.689	.6
.5	.3502	.9367	.3739	2.675	.5
.6	.3518	.9361	.3759	2.660	.4
.7	.3535	.9354	.3779	2.646	.3
.8	.3551	.9348	.3799	2.633	.2
.9	.3567	.9342	.3819	2.619	.1
21.0	0.3584	0.9336	0.3839	2.605	69.0
.1	.3600	.9330	.3859	2.592	.9
.2	.3616	.9323	.3879	2.578	.8
.3	.3633	.9317	.3899	2.565	.7
.4	.3649	.9311	.3919	2.552	.6
.5	.3665	.9304	.3939	2.539	.5
.6	.3681	.9298	.3959	2.526	.4
.7	.3697	.9291	.3979	2.513	.3
.8	.3714	.9285	.4000	2.500	.2
.9	.3730	.9278	.4020	2.488	.1
22.0	0.3746	0.9272	0.4040	2.475	68.0
.1	.3762	.9265	.4061	2.463	.9
.2	.3778	.9259	.4081	2.450	.8
.3	.3795	.9252	.4101	2.438	.7
.4	.3811	.9245	.4122	2.426	.6
.5	.3827	.9239	.4142	2.414	.5
.6	.3843	.9232	.4163	2.402	.4
.7	.3859	.9225	.4183	2.391	.3
.8	.3875	.9219	.4204	2.379	.2
.9	.3891	.9212	.4224	2.367	.1
23.0	0.3907	0.9205	0.4245	2.356	67.0
.1	.3923	.9198	.4265	2.344	.9
.2	.3939	.9191	.4286	2.333	.8
.3	.3955	.9184	.4307	2.322	.7
.4	.3971	.9178	.4327	2.311	.6
.5	.3987	.9171	.4348	2.300	.5
.6	.4003	.9164	.4369	2.289	.4
.7	.4019	.9157	.4390	2.278	.3
.8	.4035	.9150	.4411	2.267	.2
.9	.4051	.9143	.4431	2.257	.1
24.0	0.4067	0.9135	0.4452	2.246	66.0

cos	sin	cot	tan	deg

NATURAL TRIGONOMETRIC FUNCTIONS FOR DECIMAL FRACTIONS OF A DEGREE (continued)

deg	sin	cos	tan	cot		deg	sin	cos	tan	cot	
24.0	0.4067	0.9135	0.4452	2.246	**66.0**	**30.0**	0.5000	0.8660	0.5774	1.7321	**60.0**
.1	.4083	.9128	.4473	2.236	.9	.1	.5015	.8652	.5797	1.7251	.9
.2	.4099	.9121	.4494	2.225	.8	.2	.5030	.8643	.5820	1.7182	.8
.3	.4115	.9114	.4515	2.215	.7	.3	.5045	.8634	.5844	1.7113	.7
.4	.4131	.9107	.4536	2.204	.6	.4	.5060	.8625	.5867	1.7045	.6
.5	.4147	.9100	.4557	2.194	.5	.5	.5075	.8616	.5890	1.6977	.5
.6	.4163	.9092	.4578	2.184	.4	.6	.5090	.8607	.5914	1.6909	.4
.7	.4179	.9085	.4599	2.174	.3	.7	.5105	.8599	.5938	1.6842	.3
.8	.4195	.9078	.4621	2.164	.2	.8	.5120	.8590	.5961	1.6775	.2
.9	.4210	.9070	.4642	2.154	.1	.9	.5135	.8581	.5985	1.6709	.1
25.0	0.4226	0.9063	0.4663	2.145	**65.0**	**31.0**	0.5150	0.8572	0.6009	1.6643	**59.0**
.1	.4242	.9056	.4684	2.135	.9	.1	.5165	.8563	.6032	1.6577	.9
.2	.4258	.9048	.4706	2.125	.8	.2	.5180	.8554	.6056	1.6512	.8
.3	.4274	.9041	.4727	2.116	.7	.3	.5195	.8545	.6080	1.6447	.7
.4	.4289	.9033	.4748	2.106	.6	.4	.5210	.8536	.6104	1.6383	.6
.5	.4305	.9026	.4770	2.097	.5	.5	.5225	.8526	.6128	1.6319	.5
.6	.4321	.9018	.4791	2.087	.4	.6	.5240	.8517	.6152	1.6255	.4
.7	.4337	.9011	.4813	2.078	.3	.7	.5255	.8508	.6176	1.6191	.3
.8	.4352	.9003	.4834	2.069	.2	.8	.5270	.8499	.6200	1.6128	.2
.9	.4368	.8996	.4856	2.059	.1	.9	.5284	.8490	.6224	1.6066	.1
26.0	0.4384	0.8988	0.4877	2.050	**64.0**	**32.0**	0.5299	0.8480	0.6249	1.6003	**58.0**
.1	.4399	.8980	.4899	2.041	.9	.1	.5314	.8471	.6273	1.5941	.9
.2	.4415	.8973	.4921	2.032	.8	.2	.5329	.8462	.6297	1.5880	.8
.3	.4431	.8965	.4942	2.023	.7	.3	.5344	.8453	.6322	1.5818	.7
.4	.4446	.8957	.4964	2.014	.6	.4	.5358	.8443	.6346	1.5757	.6
.5	.4462	.8949	.4986	2.006	.5	.5	.5373	.8434	.6371	1.5697	.5
.6	.4478	.8942	.5008	1.997	.4	.6	.5388	.8425	.6395	1.5637	.4
.7	.4493	.8934	.5029	1.988	.3	.7	.5402	.8415	.6420	1.5577	.3
.8	.4509	.8926	.5051	1.980	.2	.8	.5417	.8406	.6445	1.5517	.2
.9	.4524	.8918	.5073	1.971	.1	.9	.5432	.8396	.6469	1.5458	.1
27.0	0.4540	0.8910	0.5095	1.963	**63.0**	**33.0**	0.5446	0.8387	0.6494	1.5399	**57.0**
.1	.4555	.8902	.5117	1.954	.9	.1	.5461	.8377	.6519	1.5340	.9
.2	.4571	.8894	.5139	1.946	.8	.2	.5476	.8368	.6544	1.5282	.8
.3	.4586	.8886	.5161	1.937	.7	.3	.5490	.8358	.6569	1.5224	.7
.4	.4602	.8878	.5184	1.929	.6	.4	.5505	.8348	.6594	1.5166	.6
.5	.4617	.8870	.5206	1.921	.5	.5	.5519	.8339	.6619	1.5108	.5
.6	.4633	.8862	.5228	1.913	.4	.6	.5534	.8329	.6644	1.5051	.4
.7	.4648	.8854	.5250	1.905	.3	.7	.5548	.8320	.6669	1.4994	.3
.8	.4664	.8846	.5272	1.897	.2	.8	.5563	.8310	.6694	1.4938	.2
.9	.4679	.8838	.5295	1.889	.1	.9	.5577	.8300	.6720	1.4882	.1
28.0	0.4695	0.8829	0.5317	1.881	**62.0**	**34.0**	0.5592	0.8290	0.6745	1.4826	**56.0**
.1	.4710	.8821	.5340	1.873	.9	.1	.5606	.8281	.6771	1.4770	.9
.2	.4726	.8813	.5362	1.865	.8	.2	.5621	.8271	.6796	1.4715	.8
.3	.4741	.8805	.5384	1.857	.7	.3	.5635	.8261	.6822	1.4659	.7
.4	.4756	.8796	.5407	1.849	.6	.4	.5650	.8251	.6847	1.4605	.6
.5	.4772	.8788	.5430	1.842	.5	.5	.5664	.8241	.6873	1.4550	.5
.6	.4787	.8780	.5452	1.834	.4	.6	.5678	.8231	.6899	1.4496	.4
.7	.4802	.8771	.5475	1.827	.3	.7	.5693	.8221	.6924	1.4442	.3
.8	.4818	.8763	.5498	1.819	.2	.8	.5707	.8211	.6950	1.4388	.2
.9	.4833	.8755	.5520	1.811	.1	.9	.5721	.8202	.6976	1.4335	.1
29.0	0.4848	0.8746	0.5543	1.804	**61.0**	**35.0**	0.5736	0.8192	0.7002	1.4281	**55.0**
.1	.4863	.8738	.5566	1.797	.9	.1	.5750	.8181	.7028	1.4229	.9
.2	.4879	.8729	.5589	1.789	.8	.2	.5764	.8171	.7054	1.4176	.8
.3	.4894	.8721	.5612	1.782	.7	.3	.5779	.8161	.7080	1.4124	.7
.4	.4909	.8712	.5635	1.775	.6	.4	.5793	.8151	.7107	1.4071	.6
.5	.4924	.8704	.5658	1.767	.5	.5	.5807	.8141	.7133	1.4019	.5
.6	.4939	.8695	.5681	1.760	.4	.6	.5821	.8131	.7159	1.3968	.4
.7	.4955	.8686	.5704	1.753	.3	.7	.5835	.8121	.7186	1.3916	.3
.8	.4970	.8678	.5727	1.746	.2	.8	.5850	.8111	.7212	1.3865	.2
.9	.4985	.8669	.5750	1.739	.1	.9	.5864	.8100	.7239	1.3814	.1
30.0	0.5000	0.8660	0.5774	1.732	**60.0**	**36.0**	0.5878	0.8090	0.7265	1.3764	**54.0**
	cos	sin	cot	tan	deg		cos	sin	cot	tan	deg

NATURAL TRIGONOMETRIC FUNCTIONS FOR DECIMAL FRACTIONS OF A DEGREE (continued)

deg	sin	cos	tan	cot	
36.0	0.5878	0.8090	0.7265	1.3764	54.0
.1	.5892	.8080	.7292	1.3713	.9
.2	.5906	.8070	.7319	1.3663	.8
.3	.5920	.8059	.7346	1.3613	.7
.4	.5934	.8049	.7373	1.3564	.6
.5	.5948	.8039	.7400	1.3514	.5
.6	.5962	.8028	.7427	1.3465	.4
.7	.5976	.8018	.7454	1.3416	.3
.8	.5990	.8007	.7481	1.3367	.2
.9	.6004	.7997	.7508	1.3319	.1
37.0	0.6018	0.7986	0.7536	1.3270	53.0
.1	.6032	.7976	.7563	1.3222	.9
.2	.6046	.7965	.7590	1.3175	.8
.3	.6060	.7955	.7618	1.3127	.7
.4	.6074	.7944	.7646	1.3079	.6
.5	.6088	.7934	.7673	1.3032	.5
.6	.6101	.7923	.7701	1.2985	.4
.7	.6115	.7912	.7729	1.2938	.3
.8	.6129	.7902	.7757	1.2892	.2
.9	.6143	.7891	.7785	1.2846	.1
38.0	0 6157	0.7880	0.7813	1.2799	52.0
.1	.6170	.7869	.7841	1.2753	.9
.2	.6184	.7859	.7869	1.2708	.8
.3	.6198	.7848	.7898	1.2662	.7
.4	.6211	.7837	.7926	1.2617	.6
.5	.6225	.7826	.7954	1.2572	.5
.6	.6239	.7815	.7983	1.2527	.4
.7	.6252	.7804	.8012	1.2482	.3
.8	.6266	.7793	.8040	1.2437	.2
.9	.6280	.7782	.8069	1.2393	.1
39.0	0 6293	0.7771	0.8098	1.2349	51.0
.1	.6307	.7760	.8127	1.2305	.9
.2	.6320	.7749	.8156	1.2261	.8
.3	.6334	.7738	.8185	1.2218	.7
.4	.6347	.7727	.8214	1.2174	.6
.5	.6361	.7716	.8243	1.2131	.5
.6	.6374	.7705	.8273	1.2088	.4
.7	.6388	.7694	.8302	1.2045	.3
.8	.6401	.7683	.8332	1.2002	.2
.9	.6414	.7672	.8361	1.1960	.1
40.0	0.6428	0.7660	0.8391	1.1918	50.0
.1	.6441	.7649	.8421	1.1875	.9
.2	.6455	.7638	.8451	1.1833	.8
.3	.6468	.7627	.8481	1.1792	.7
.4	.6481	.7615	.8511	1.1750	.6
40.5	0.6494	0.7604	0.8541	1.1708	49.5
	cos	sin	cot	tan	deg

deg	sin	cos	tan	cot	
40.5	0.6494	0.7604	0.8541	1.1708	49.5
.6	.6508	.7593	.8571	1.1667	.4
.7	.6521	.7581	.8601	1.1626	.3
.8	.6534	.7570	.8632	1.1585	.2
.9	.6547	.7559	.8662	1.1544	.1
41.0	0.6561	0.7547	0.8693	1.1504	49.0
.1	.6574	.7536	.8724	1.1463	.9
.2	.6587	.7524	.8754	1.1423	.8
.3	.6600	.7513	.8785	1.1383	.7
.4	.6613	.7501	.8816	1.1343	.6
.5	.6626	.7490	.8847	1.1303	.5
.6	.6639	.7478	.8878	1.1263	.4
.7	.6652	.7466	.8910	1.1224	.3
.8	.6665	.7455	.8941	1.1184	.2
.9	.6678	.7443	.8972	1.1145	.1
42.0	0.6691	0.7431	0.9004	1.1106	48.0
.1	.6704	.7420	.9036	1.1067	.9
.2	.6717	.7408	.9067	1.1028	.8
.3	.6730	.7396	.9099	1.0990	.7
.4	.6743	.7385	.9131	1.0951	.6
.5	.6756	.7373	.9163	1.0913	.5
.6	.6769	.7361	.9195	1.0875	.4
.7	.6782	.7349	.9228	1.0837	.3
.8	.6794	.7337	.9260	1.0799	.2
.9	.6807	.7325	.9293	1.0761	.1
43.0	0.6820	0.7314	0.9325	1.0724	47.0
.1	.6833	.7302	.9358	1.0686	.9
.2	.6845	.7290	.9391	1.0649	.8
.3	.6858	.7278	.9424	1.0612	.7
.4	.6871	.7266	.9457	1.0575	.6
.5	.6884	.7254	.9490	1.0538	.5
.6	.6896	.7242	.9523	1.0501	.4
.7	.6909	.7230	.9556	1.0464	.3
.8	.6921	.7218	.9590	1.0428	.2
.9	.6934	.7206	.9623	1.0392	.1
44.0	0.6947	0.7193	0.9657	1.0355	46.0
.1	.6959	.7181	.9691	1.0319	.9
.2	.6972	.7169	.9725	1.0283	.8
.3	.6984	.7157	.9759	1.0247	.7
.4	.6997	.7145	.9793	1.0212	.6
.5	.7009	.7133	.9827	1.0176	.5
.6	.7022	.7120	.9861	1.0141	.4
.7	.7034	.7108	.9896	1.0105	.3
.8	.7046	.7096	.9930	1.0070	.2
.9	.7059	.7083	.9965	1.0035	.1
45.0	0.7071	0.7071	1.0000	1.0000	45.0
	cos	sin	cot	tan	deg

LOGARITHMS OF TRIGONOMETRIC FUNCTIONS FOR DECIMAL FRACTIONS OF A DEGREE

deg	L sin	L cos	L tan	L cot		deg	L sin	L cos	L tan	L cot	
0.0	$-\infty$	0.0000	$-\infty$	∞	90.0	6.0	9.0192	9.9976	9.0216	0.9784	84.0
.1	7.2419	0.0000	7.2419	2.7581	.9	.1	9.0264	9.9975	9.0289	0.9711	.9
.2	7.5429	0.0000	7.5429	2.4571	.8	.2	9.0334	9.9975	9.0360	0.9640	.8
.3	7.7190	0.0000	7.7190	2.2810	.7	.3	9.0403	9.9974	9.0430	0.9570	.7
.4	7.8439	0.0000	7.8439	2.1561	.6	.4	9.0472	9.9973	9.0499	0.9501	.6
.5	7.9408	0.0000	7.9409	2.0591	.5	.5	9.0539	9.9972	9.0567	0.9433	.5
.6	8.0200	0.0000	8.0200	1.9800	.4	.6	9.0605	9.9971	9.0633	0.9367	.4
.7	8.0870	0.0000	8.0870	1.9130	.3	.7	9.0670	9.9970	9.0699	0.9301	.3
.8	8.1450	0.0000	8.1450	1.8550	.2	.8	9.0734	9.9969	9.0764	0.9236	.2
.9	8.1961	9.9999	8.1962	1.8038	.1	.9	9.0797	9.9968	9.0828	0.9172	.1
1.0	8.2419	9.9999	8.2419	1.7581	89.0	7.0	9.0859	9.9968	9.0891	0.9109	83.0
.1	8.2832	9.9999	8.2833	1.7167	.9	.1	9.0920	9.9967	9.0954	0.9046	.9
.2	8.3210	9.9999	8.3211	1.6789	.8	.2	9.0981	9.9966	9.1015	0.8985	.8
.3	8.3558	9.9999	8.3559	1.6441	.7	.3	9.1040	9.9965	9.1076	0.8924	.7
.4	8.3880	9.9999	8.3881	1.6119	.6	.4	9.1099	9.9964	9.1135	0.8865	.6
.5	8.4179	9.9999	8.4181	1.5819	.5	.5	9.1157	9.9963	9.1194	0.8806	.5
.6	8.4459	9.9998	8.4461	1.5539	.4	.6	9.1214	9.9962	9.1252	0.8748	.4
.7	8.4723	9.9998	8.4725	1.5275	.3	.7	9.1271	9.9961	9.1310	0.8690	.3
.8	8.4971	9.9998	8.4973	1.5027	.2	.8	9.1326	9.9960	9.1367	0.8633	.2
.9	8.5206	9.9998	8.5208	1.4792	.1	.9	9.1381	9.9959	9.1423	0.8577	.1
2.0	8.5428	9.9997	8.5431	1.4569	88.0	8.0	9.1436	9.9958	9.1478	0.8522	82.0
.1	8.5640	9.9997	8.5643	1.4357	.9	.1	9.1489	9.9956	9.1533	0.8467	.9
.2	8.5842	9.9997	8.5845	1.4155	.8	.2	9.1542	9.9955	9.1587	0.8413	.8
.3	8.6035	9.9996	8.6038	1.3962	.7	.3	9.1594	9.9954	9.1640	0.8360	.7
.4	8.6220	9.9996	8.6223	1.3777	.6	.4	9.1646	9.9953	9.1693	0.8307	.6
.5	8.6397	9.9996	8.6401	1.3599	.5	.5	9.1697	9.9952	9.1745	0.8255	.5
.6	8.6567	9.9996	8.6571	1.3429	.4	.6	9.1747	9.9951	9.1797	0.8203	.4
.7	8.6731	9.9995	8.6736	1.3264	.3	.7	9.1797	9.9950	9.1848	0.8152	.3
.8	8.6889	9.9995	8.6894	1.3106	.2	.8	9.1847	9.9949	9.1898	0.8102	.2
.9	8.7041	9.9994	8.7046	1.2954	.1	.9	9.1895	9.9947	9.1948	0.8052	.1
3.0	8.7188	9.9994	8.7194	1.2806	87.0	9.0	9.1943	9.9946	9.1997	0.8003	81.0
.1	8.7330	9.9994	8.7337	1.2663	.9	.1	9.1991	9.9945	9.2046	0.7954	.9
.2	8.7468	9.9993	8.7475	1.2525	.8	.2	9.2038	9.9944	9.2094	0.7906	.8
.3	8.7602	9.9993	8.7609	1.2391	.7	.3	9.2085	9.9943	9.2142	0.7858	.7
.4	8.7731	9.9992	8.7739	1.2261	.6	.4	9.2131	9.9941	9.2189	0.7811	.6
.5	8.7857	9.9992	8.7865	1.2135	.5	.5	9.2176	9.9940	9.2236	0.7764	.5
.6	8.7979	9.9991	8.7988	1.2012	.4	.6	9.2221	9.9939	9.2282	0.7718	.4
.7	8.8098	9.9991	8.8107	1.1893	.3	.7	9.2266	9.9937	9.2328	0.7672	.3
.8	8.8213	9.9990	8.8223	1.1777	.2	.8	9.2310	9.9936	9.2374	0.7626	.2
.9	8.8326	9.9990	8.8336	1.1664	.1	.9	9.2353	9.9935	9.2419	0.7581	.1
4.0	8.8436	9.9989	8.8446	1.1554	86.0	10.0	9.2397	9.9934	9.2463	0.7537	80.0
.1	8.8543	9.9989	8.8554	1.1446	.9	.1	9.2439	9.9932	9.2507	0.7493	.9
.2	8.8647	9.9988	8.8659	1.1341	.8	.2	9.2482	9.9931	9.2551	0.7449	.8
.3	8.8749	9.9988	8.8762	1.1238	.7	.3	9.2524	9.9929	9.2594	0.7406	.7
.4	8.8849	9.9987	8.8862	1.1138	.6	.4	9.2565	9.9928	9.2637	0.7363	.6
.5	8.8946	9.9987	8.8960	1.1040	.5	.5	9.2606	9.9927	9.2680	0.7320	.5
.6	8.9042	9.9986	8.9056	1.0944	.4	.6	9.2647	9.9925	9.2722	0.7278	.4
.7	8.9135	9.9985	8.9150	1.0850	.3	.7	9.2687	9.9924	9.2764	0.7236	.3
.8	8.9226	9.9985	8.9241	1.0759	.2	.8	9.2727	9.9922	9.2805	0.7195	.2
.9	8.9315	9.9984	8.9331	1.0669	.1	.9	9.2767	9.9921	9.2846	0.7154	.1
5.0	8.9403	9.9983	8.9420	1.0580	85.0	11.0	9.2806	9.9919	9.2887	0.7113	79.0
.1	8.9489	9.9983	8.9506	1.0494	.9	.1	9.2845	9.9918	9.2927	0.7073	.9
.2	8.9573	9.9982	8.9591	1.0409	.8	.2	9.2883	9.9916	9.2967	0.7033	.8
.3	8.9655	9.9981	8.9674	1.0326	.7	.3	9.2921	9.9915	9.3006	0.6994	.7
.4	8.9736	9.9981	8.9756	1.0244	.6	.4	9.2959	9.9913	9.3046	0.6954	.6
.5	8.9816	9.9980	8.9836	1.0164	.5	.5	9.2997	9.9912	9.3085	0.6915	.5
.6	8.9894	9.9979	8.9915	1.0085	.4	.6	9.3034	9.9910	9.3123	0.6877	.4
.7	8.9970	9.9978	8.9992	1.0008	.3	.7	9.3070	9.9909	9.3162	0.6838	.3
.8	9.0046	9.9978	9.0068	0.9932	.2	.8	9.3107	9.9907	9.3200	0.6800	.2
.9	9.0120	9.9977	9.0143	0.9857	.1	.9	9.3143	9.9906	9.3237	0.6763	.1
6.0	9.0192	9.9976	9.0216	0.9784	84.0	12.0	9.3179	9.9904	9.3275	0.6725	78.0
	L cos	L sin	L cot	L tan	deg		L cos	L sin	L cot	L tan	deg

LOGARITHMS OF TRIGONOMETRIC FUNCTIONS FOR DECIMAL FRACTIONS OF A DEGREE (continued)

deg	L sin	L cos	L tan	L cot	deg
12.0	9.3179	9.9904	9.3275	0.6725	78.0
.1	9.3214	9.9902	9.3312	0.6688	.9
.2	9.3250	9.9901	9.3349	0.6651	.8
.3	9.3284	9.9899	9.3385	0.6615	.7
.4	9.3319	9.9897	9.3422	0.6578	.6
.5	9.3353	9.9896	9.3458	0.6542	.5
.6	9.3387	9.9894	9.3493	0.6507	.4
.7	9.3421	9.9892	9.3529	0.6471	.3
.8	9.3455	9.9891	9.3564	0.6436	.2
.9	9.3488	9.9889	9.3599	0.6401	.1
13.0	9.3521	9.9887	9.3634	0.6366	77.0
.1	9.3554	9.9885	9.3668	0.6332	.9
.2	9.3586	9.9884	9.3702	0.6298	.8
.3	9.3618	9.9882	9.3736	0.6264	.7
.4	9.3650	9.9880	9.3770	0.6230	.6
.5	9.3682	9.9878	9.3804	0.6196	.5
.6	9.3713	9.9876	9.3837	0.6163	.4
.7	9.3745	9.9875	9.3870	0.6130	.3
.8	9.3775	9.9873	9.3903	0.6097	.2
.9	9.3806	9.9871	9.3935	0.6065	.1
14.0	9.3837	9.9869	9.3968	0.6032	76.0
.1	9.3867	9.9867	9.4000	0.6000	.9
.2	9.3897	9.9865	9.4032	0.5968	.8
.3	9.3927	9.9863	9.4064	0.5936	.7
.4	9.3957	9.9861	9.4095	0.5905	.6
.5	9.3986	9.9859	9.4127	0.5873	.5
.6	9.4015	9.9857	9.4158	0.5842	.4
.7	9.4044	9.9855	9.4189	0.5811	.3
.8	9.4073	9.9853	9.4220	0.5780	.2
.9	9.4102	9.9851	9.4250	0.5750	.1
15.0	9.4130	9.9849	9.4281	0.5719	75.0
.1	9.4158	9.9847	9.4311	0.5689	.9
.2	9.4186	9.9845	9.4341	0.5659	.8
.3	9.4214	9.9843	9.4371	0.5629	.7
.4	9.4242	9.9841	9.4400	0.5600	.6
.5	9.4269	9.9839	9.4430	0.5570	.5
.6	9.4296	9.9837	9.4459	0.5541	.4
.7	9.4323	9.9835	9.4488	0.5512	.3
.8	9.4350	9.9833	9.4517	0.5483	.2
.9	9.4377	9.9831	9.4546	0.5454	.1
16.0	9.4403	9.9828	9.4575	0.5425	74.0
.1	9.4430	9.9826	9.4603	0.5397	.9
.2	9.4456	9.9824	9.4632	0.5368	.8
.3	9.4482	9.9822	9.4660	0.5340	.7
.4	9.4508	9.9820	9.4688	0.5312	.6
.5	9.4533	9.9817	9.4716	0.5284	.5
.6	9.4559	9.9815	9.4744	0.5256	.4
.7	9.4584	9.9813	9.4771	0.5229	.3
.8	9.4609	9.9811	9.4799	0.5201	.2
.9	9.4634	9.9808	9.4826	0.5174	.1
17.0	9.4659	9.9806	9.4853	0.5147	73.0
.1	9.4684	9.9804	9.4880	0.5120	.9
.2	9.4709	9.9801	9.4907	0.5093	.8
.3	9.4733	9.9799	9.4934	0.5066	.7
.4	9.4757	9.9797	9.4961	0.5039	.6
.5	9.4781	9.9794	9.4987	0.5013	.5
.6	9.4805	9.9792	9.5014	0.4986	.4
.7	9.4829	9.9789	9.5040	0.4960	.3
.8	9.4853	9.9787	9.5066	0.4934	.2
.9	9.4876	9.9785	9.5092	0.4908	.1
18.0	9.4900	9.9782	9.5118	0.4882	72.0
	L cos	L sin	L cot	L tan	deg

deg	L sin	L cos	L tan	L cot	deg
18.0	9.4900	9.9782	9.5118	0.4882	72.0
.1	9.4923	9.9780	9.5143	0.4857	.9
.2	9.4946	9.9777	9.5169	0.4831	.8
.3	9.4969	9.9775	9.5195	0.4805	.7
.4	9.4992	9.9772	9.5220	0.4780	.6
.5	9.5015	9.9770	9.5245	0.4755	.5
.6	9.5037	9.9767	9.5270	0.4730	.4
.7	9.5060	9.9764	9.5295	0.4705	.3
.8	9.5082	9.9762	9.5320	0.4680	.2
.9	9.5104	9.9759	9.5345	0.4655	.1
19.0	9.5126	9.9757	9.5370	0.4630	71.0
.1	9.5148	9.9754	9.5394	0.4606	.9
.2	9.5170	9.9751	9.5419	0.4581	.8
.3	9.5192	9.9749	9.5443	0.4557	.7
.4	9.5213	9.9746	9.5467	0.4533	.6
.5	9.5235	9.9743	9.5491	0.4509	.5
.6	9.5256	9.9741	9.5516	0.4484	.4
.7	9.5278	9.9738	9.5539	0.4461	.3
.8	9.5299	9.9735	9.5563	0.4437	.2
.9	9.5320	9.9733	9.5587	0.4413	.1
20.0	9.5341	9.9730	9.5611	0.4389	70.0
.1	9.5361	9.9727	9.5634	0.4366	.9
.2	9.5382	9.9724	9.5658	0.4342	.8
.3	9.5402	9.9722	9.5681	0.4319	.7
.4	9.5423	9.9719	9.5704	0.4296	.6
.5	9.5443	9.9716	9.5727	0.4273	.5
.6	9.5463	9.9713	9.5750	0.4250	.4
.7	9.5484	9.9710	9.5773	0.4227	.3
.8	9.5504	9.9707	9.5796	0.4204	.2
.9	9.5523	9.9704	9.5819	0.4181	.1
21.0	9.5543	9.9702	9.5842	0.4158	69.0
.1	9.5563	9.9699	9.5864	0.4136	.9
.2	9.5583	9.9696	9.5887	0.4113	.8
.3	9.5602	9.9693	9.5909	0.4091	.7
.4	9.5621	9.9690	9.5932	0.4068	.6
.5	9.5641	9.9687	9.5954	0.4046	.5
.6	9.5660	9.9684	9.5976	0.4024	.4
.7	9.5679	9.9681	9.5998	0.4002	.3
.8	9.5698	9.9678	9.6020	0.3980	.2
.9	9.5717	9.9675	9.6042	0.3958	.1
22.0	9.5736	9.9672	9.6064	0.3936	68.0
.1	9.5754	9.9669	9.6086	0.3914	.9
.2	9.5773	9.9666	9.6108	0.3892	.8
.3	9.5792	9.9662	9.6129	0.3871	.7
.4	9.5810	9.9659	9.6151	0.3849	.6
.5	9.5828	9.9656	9.6172	0.3828	.5
.6	9.5847	9.9653	9.6194	0.3806	.4
.7	9.5865	9.9650	9.6215	0.3785	.3
.8	9.5883	9.9647	9.6236	0.3764	.2
.9	9.5901	9.9643	9.6257	0.3743	.1
23.0	9.5919	9.9640	9.6279	0.3721	67.0
.1	9.5937	9.9637	9.6300	0.3700	.9
.2	9.5954	9.9634	9.6321	0.3679	.8
.3	9.5972	9.9631	9.6341	0.3659	.7
.4	9.5990	9.9627	9.6362	0.3638	6
.5	9.6007	9.9624	9.6383	0.3617	.5
.6	9.6024	9.9621	9.6404	0.3596	.4
.7	9.6042	9.9617	9.6424	0.3576	.3
.8	9.6059	9.9614	9.6445	0.3555	.2
.9	9.6076	9.9611	9.6465	0.3535	.1
24.0	9.6093	9.9607	9.6486	0.3514	66.0
	L cos	L sin	L cot	L tan	deg

LOGARITHMS OF TRIGONOMETRIC FUNCTIONS FOR DECIMAL FRACTIONS OF A DEGREE (continued)

deg	L sin	L cos	L tan	L cot	
24.0	9.6093	9.9607	9.6486	0.3514	**66.0**
.1	9.6110	9.9604	9.6506	0.3494	.9
.2	9.6127	9.9601	9.6527	0.3473	.8
.3	9.6144	9.9597	9.6547	0.3453	.7
.4	9.6161	9.9594	9.6567	0.3433	.6
.5	9.6177	9.9590	9.6587	0.3413	.5
.6	9.6194	9.9587	9.6607	0.3393	.4
.7	9.6210	9.9583	9.6627	0.3373	.3
.8	9.6227	9.9580	9.6647	0.3353	.2
.9	9.6243	9.9576	9.6667	0.3333	.1
25.0	9.6259	9.9573	9.6687	0.3313	**65.0**
.1	9.6276	9.9569	9.6706	0.3294	.9
.2	9.6292	9.9566	9.6726	0.3274	.8
.3	9.6308	9.9562	9.6746	0.3254	.7
.4	9.6324	9.9558	9.6765	0.3235	.6
.5	9.6340	9.9555	9.6785	0.3215	.5
.6	9.6356	9.9551	9.6804	0.3196	.4
.7	9.6371	9.9548	9.6824	0.3176	.3
.8	9.6387	9.9544	9.6843	0.3157	.2
.9	9.6403	9.9540	9.6863	0.3137	.1
26.0	9.6418	9.9537	9.6882	0.3118	**64.0**
.1	9.6434	9.9533	9.6901	0.3099	.9
.2	9.6449	9.9529	9.6920	0.3080	.8
.3	9.6465	9.9525	9.6939	0.3061	.7
.4	9.6480	9.9522	9.6958	0.3042	.6
.5	9.6495	9.9518	9.6977	0.3023	.5
.6	9.6510	9.9514	9.6996	0.3004	.4
.7	9.6526	9.9510	9.7015	0.2985	.3
.8	9.6541	9.9506	9.7034	0.2966	.2
.9	9.6556	9.9503	9.7053	0.2947	.1
27.0	9.6570	9.9499	9.7072	0.2928	**63.0**
.1	9.6585	9.9495	9.7090	0.2910	.9
.2	9.6600	9.9491	9.7109	0.2891	.8
.3	9.6615	9.9487	9.7128	0.2872	.7
.4	9.6629	9.9483	9.7146	0.2854	.6
.5	9.6644	9.9479	9.7165	0.2835	.5
.6	9.6659	9.9475	9.7183	0.2817	.4
.7	9.6673	9.9471	9.7202	0.2798	.3
.8	9.6687	9.9467	9.7220	0.2780	.2
.9	9.6702	9.9463	9.7238	0.2762	.1
28.0	9.6716	9.9459	9.7257	0.2743	**62.0**
.1	9.6730	9.9455	9.7275	0.2725	.9
.2	9.6744	9.9451	9.7293	0.2707	.8
.3	9.6759	9.9447	9.7311	0.2689	.7
.4	9.6773	9.9443	9.7330	0.2670	.6
.5	9.6787	9.9439	9.7348	0.2652	.5
.6	9.6801	9.9435	9.7366	0.2634	.4
.7	9.6814	9.9431	9.7384	0.2616	.3
.8	9.6828	9.9427	9.7402	0.2598	.2
.9	9.6842	9.9422	9.7420	0.2580	.1
29.0	9.6856	9.9418	9.7438	0.2562	**61.0**
.1	9.6869	9.9414	9.7455	0.2545	.9
.2	9.6883	9.9410	9.7473	0.2527	.8
.3	9.6896	9.9406	9.7491	0.2509	.7
.4	9.6910	9.9401	9.7509	0.2491	.6
.5	9.6923	9.9397	9.7526	0.2474	.5
.6	9.6937	9.9393	9.7544	0.2456	.4
.7	9.6950	9.9388	9.7562	0.2438	.3
.8	9.6963	9.9384	9.7579	0.2421	.2
.9	9.6977	9.9380	9.7597	0.2403	.1
30.0	9.6990	9.9375	9.7614	0.2386	**60.0**
	L cos	**L sin**	**L cot**	**L tan**	**deg**

deg	L sin	L cos	L tan	L cot	
30.0	9.6990	9.9375	9.7614	0.2386	**60.0**
.1	9.7003	9.9371	9.7632	0.2368	.9
.2	9.7016	9.9367	9.7649	0.2351	.8
.3	9.7029	9.9362	9.7667	0.2333	.7
.4	9.7042	9.9358	9.7684	0.2316	.6
.5	9.7055	9.9353	9.7701	0.2299	.5
.6	9.7068	9.9349	9.7719	0.2281	.4
.7	9.7080	9.9344	9.7736	0.2264	.3
.8	9.7093	9.9340	9.7753	0.2247	.2
.9	9.7106	9.9335	9.7771	0.2229	.1
31.0	9.7118	9.9331	9.7788	0.2212	**59.0**
.1	9.7131	9.9326	9.7805	0.2195	.9
.2	9.7144	9.9322	9.7822	0.2178	.8
.3	9.7156	9.9317	9.7839	0.2161	.7
.4	9.7168	9.9312	9.7856	0.2144	.6
.5	9.7181	9.9308	9.7873	0.2127	.5
.6	9.7193	9.9303	9.7890	0.2110	.4
.7	9.7205	9.9298	9.7907	0.2093	.3
.8	9.7218	9.9294	9.7924	0.2076	.2
.9	9.7230	9.9289	9.7941	0.2059	.1
32.0	9.7242	9.9284	9.7958	0.2042	**58.0**
.1	9.7254	9.9279	9.7975	0.2025	.9
.2	9.7266	9.9275	9.7992	0.2008	.8
.3	9.7278	9.9270	9.8008	0.1992	.7
.4	9.7290	9.9265	9.8025	0.1975	.6
.5	9.7302	9.9260	9.8042	0.1958	.5
.6	9.7314	9.9255	9.8059	0.1941	.4
.7	9.7326	9.9251	9.8075	0.1925	.3
.8	9.7338	9.9246	9.8092	0.1908	.2
.9	9.7349	9.9241	9.8109	0.1891	.1
33.0	9.7361	9.9236	9.8125	0.1875	**57.0**
.1	9.7373	9.9231	9.8142	0.1858	.9
.2	9.7384	9.9226	9.8158	0.1842	.8
.3	9.7396	9.9221	9.8175	0.1825	.7
.4	9.7407	9.9216	9.8191	0.1809	.6
.5	9.7419	9.9211	9.8208	0.1792	.5
.6	9.7430	9.9206	9.8224	0.1776	.4
.7	9.7442	9.9201	9.8241	0.1759	.3
.8	9.7453	9.9196	9.8257	0.1743	.2
.9	9.7464	9.9191	9.8274	0.1726	.1
34.0	9.7476	9.9186	9.8290	0.1710	**56.0**
.1	9.7487	9.9181	9.8306	0.1694	.9
.2	9.7498	9.9175	9.8323	0.1677	.8
.3	9.7509	9.9170	9.8339	0.1661	.7
.4	9.7520	9.9165	9.8355	0.1645	.6
.5	9.7531	9.9160	9.8371	0.1629	.5
.6	9.7542	9.9155	9.8388	0.1612	.4
.7	9.7553	9.9149	9.8404	0.1596	.3
.8	9.7564	9.9144	9.8420	0.1580	.2
.9	9.7575	9.9139	9.8436	0.1564	.1
35.0	9.7586	9.9134	9.8452	0.1548	**55.0**
.1	9.7597	9.9128	9.8468	0.1532	.9
.2	9.7607	9.9123	9.8484	0.1516	.8
.3	9.7618	9.9118	9.8501	0.1499	.7
.4	9.7629	9.9112	9.8517	0.1483	.6
.5	9.7640	9.9107	9.8533	0.1467	.5
.6	9.7650	9.9101	9.8549	0.1451	.4
.7	9.7661	9.9096	9.8565	0.1435	.3
.8	9.7671	9.9091	9.8581	0.1419	.2
.9	9.7682	9.9085	9.8597	0.1403	.1
36.0	9.7692	9.9080	9.8613	0.1387	**54.0**
	L cos	**L sin**	**L cot**	**L tan**	**deg**

LOGARITHMS OF TRIGONOMETRIC FUNCTIONS FOR DECIMAL FRACTIONS OF A DEGREE (continued)

deg	L sin	L cos	L tan	L cot		deg	L sin	L cos	L tan	L cot	
36.0	9.7692	9.9080	9.8613	0.1387	54.0	40.5	9.8125	9.8810	9.9315	0.0685	49.5
.1	9.7703	9.9074	9.8629	0.1371	.9	.6	9.8134	9.8804	9.9330	0.0670	.4
.2	9.7713	9.9069	9.8644	0.1356	.8	.7	9.8143	9.8797	9.9346	0.0654	.3
.3	9.7723	9.9063	9.8660	0.1340	.7	.8	9.8152	9.8791	9.9361	0.0639	.2
.4	9.7734	9.9057	9.8676	0.1324	.6	.9	9.8161	9.8784	9.9376	0.0624	.1
.5	9.7744	9.9052	9.8692	0.1308	.5	41.0	9.8169	9.8778	9.9392	0.0608	49.0
.6	9.7754	9.9046	9.8708	0.1292	.4	.1	9.8178	9.8771	9.9407	0.0593	.9
.7	9.7764	9.9041	9.8724	0.1276	.3	.2	9.8187	9.8765	9.9422	0.0578	.8
.8	9.7774	9.9035	9.8740	0.1260	.2	.3	9.8195	9.8758	9.9438	0.0562	.7
.9	9.7785	9.9029	9.8755	0.1245	.1	.4	9.8204	9.8751	9.9453	0.0547	.6
37.0	9.7795	9.9023	9.8771	0.1229	53.0	.5	9.8213	9.8745	9.9468	0.0532	.5
.1	9.7805	9.9018	9.8787	0.1213	.9	.6	9.8221	9.8738	9.9483	0.0517	.4
.2	9.7815	9.9012	9.8803	0.1197	.8	.7	9.8230	9.8731	9.9499	0.0501	.3
.3	9.7825	9.9006	9.8818	0.1182	.7	.8	9.8238	9.8724	9.9514	0.0486	.2
.4	9.7835	9.9000	9.8834	0.1166	.6	.9	9.8247	9.8718	9.9529	0.0471	.1
.5	9.7844	9.8995	9.8850	0.1150	.5	42.0	9.8255	9.8711	9.9544	0.0456	48.0
.6	9.7854	9.8989	9.8865	0.1135	.4	.1	9.8264	9.8704	9.9560	0.0440	.9
.7	9.7864	9.8983	9.8881	0.1119	.3	.2	9.8272	9.8697	9.9575	0.0425	.8
.8	9.7874	9.8977	9.8897	0.1103	.2	.3	9.8280	9.8690	9.9590	0.0410	.7
.9	9.7884	9.8971	9.8912	0.1088	.1	.4	9.8289	9.8683	9.9605	0.0395	.6
38.0	9.7893	9.8965	9.8928	0.1072	52.0	.5	9.8297	9.8676	9.9621	0.0379	.5
.1	9.7903	9.8959	9.8944	0.1056	.9	.6	9.8305	9.8669	9.9636	0.0364	.4
.2	9.7913	9.8953	9.8959	0.1041	.8	.7	9.8313	9.8662	9.9651	0.0349	.3
.3	9.7922	9.8947	9.8975	0.1025	.7	.8	9.8322	9.8655	9.9666	0.0334	.2
.4	9.7932	9.8941	9.8990	0.1010	.6	.9	9.8330	9.8648	9.9681	0.0319	.1
.5	9.7941	9.8935	9.9006	0.0994	.5	43.0	9.8338	9.8641	9.9697	0.0303	47.0
.6	9.7951	9.8929	9.9022	0.0978	.4	.1	9.8346	9.8634	9.9712	0.0288	.9
.7	9.7960	9.8923	9.9037	0.0963	.3	.2	9.8354	9.8627	9.9727	0.0273	.8
.8	9.7970	9.8917	9.9053	0.0947	.2	.3	9.8362	9.8620	9.9742	0.0258	.7
.9	9.7979	9.8911	9.9068	0.0932	.1	.4	9.8370	9.8613	9.9757	0.0243	.6
39.0	9.7989	9.8905	9.9084	0.0916	51.0	.5	9.8378	9.8606	9.9772	0.0228	.5
.1	9.7998	9.8899	9.9099	0.0901	.9	.6	9.8386	9.8598	9.9788	0.0212	.4
.2	9.8007	9.8893	9.9115	0.0885	.8	.7	9.8394	9.8591	9.9803	0.0197	.3
.3	9.8017	9.8887	9.9130	0.0870	.7	.8	9.8402	9.8584	9.9818	0.0182	.2
.4	9.8026	9.8880	9.9146	0.0854	.6	.9	9.8410	9.8577	9.9833	0.0167	.1
.5	9.8035	9.8874	9.9161	0.0839	.5	44.0	9.8418	9.8569	9.9848	0.0152	46.0
.6	9.8044	9.8868	9.9176	0.0824	.4	.1	9.8426	9.8562	9.9864	0.0136	.9
.7	9.8053	9.8862	9.9192	0.0808	.3	.2	9.8433	9.8555	9.9879	0.0121	.8
.8	9.8063	9.8855	9.9207	0.0793	.2	.3	9.8441	9.8547	9.9894	0.0106	.7
.9	9.8072	9.8849	9.9223	0.0777	.1	.4	9.8449	9.8540	9.9909	0.0091	.6
40.0	9.8081	9.8843	9.9238	0.0762	50.0	.5	9.8457	9.8532	9.9924	0.0076	.5
.1	9.8090	9.8836	9.9254	0.0746	.9	.6	9.8464	9.8525	9.9939	0.0061	.4
.2	9.8099	9.8830	9.9269	0.0731	.8	.7	9.8472	9.8517	9.9955	0.0045	.3
.3	9.8108	9.8823	9.9284	0.0716	.7	.8	9.8480	9.8510	9.9970	0.0030	.2
.4	9.8117	9.8817	9.9300	0.0700	.6	.9	9.8487	9.8502	9.9985	0.0015	.1
40.5	9.8125	9.8810	9.9315	0.0685	49.5	45.0	9.8495	9.8495	0.0000	0.0000	45.0
	L cos	L sin	L cot	L tan	deg		L cos	L sin	L cot	L tan	deg

NATURAL LOGARITHMS

	0	1	2	3	4	5	6	7	8	9	1	2	3	4	5	6	7	8	9
															mean differences				
1.0	0.0000	0100	0198	0296	0392	0488	0583	0677	0770	0862	10	19	29	38	48	57	67	76	86
1.1	0.0953	1044	1133	1222	1310	1398	1484	1570	1655	1740	9	17	26	35	44	52	61	70	78
1.2	0.1823	1906	1989	2070	2151	2231	2311	2390	2469	2546	8	16	24	32	40	48	56	64	72
1.3	0.2624	2700	2776	2852	2927	3001	3075	3148	3221	3293	7	15	22	30	37	44	52	59	67
1.4	0.3365	3436	3507	3577	3646	3716	3784	3853	3920	3988	7	14	21	28	35	41	48	55	62
1.5	0.4055	4121	4187	4253	4318	4383	4447	4511	4574	4637	6	13	19	26	32	39	45	52	58
1.6	0.4700	4762	4824	4886	4947	5008	5068	5128	5188	5247	6	12	18	24	30	36	42	48	55
1.7	0.5306	5365	5423	5481	5539	5596	5653	5710	5766	5822	6	11	17	23	29	34	40	46	51
1.8	0.5878	5933	5988	6043	6098	6152	6206	6259	6313	6366	5	11	16	22	27	32	38	43	49
1.9	0.6419	6471	6523	6575	6627	6678	6729	6780	6831	6881	5	10	15	20	26	31	36	41	46
2.0	0.6931	6981	7031	7080	7129	7178	7227	7275	7324	7372	5	10	15	20	24	29	34	39	44
2.1	0.7419	7467	7514	7561	7608	7655	7701	7747	7793	7839	5	9	14	19	23	28	33	37	42
2.2	0.7885	7930	7975	8020	8065	8109	8154	8198	8242	8286	4	9	13	18	22	27	31	36	40
2.3	0.8329	8372	8416	8459	8502	8544	8587	8629	8671	8713	4	9	13	17	21	26	30	34	38
2.4	0.8755	8796	8838	8879	8920	8961	9002	9042	9083	9123	4	8	12	16	20	24	29	33	37
2.5	0.9163	9203	9243	9282	9322	9361	9400	9439	9478	9517	4	8	12	16	20	24	27	31	35
2.6	0.9555	9594	9632	9670	9708	9746	9783	9821	9858	9895	4	8	11	15	19	23	26	30	34
2.7	0.9933	9969	1.0006	0043	0080	0116	0152	0188	0225	0260	4	7	11	15	18	22	25	29	33
2.8	1.0296	0332	0367	0403	0438	0473	0508	0543	0578	0613	4	7	11	14	18	21	25	28	32
2.9	1.0647	0682	0716	0750	0784	0818	0852	0886	0919	0953	3	7	10	14	17	20	24	27	31
3.0	1.0986	1019	1053	1086	1119	1151	1184	1217	1249	1282	3	7	10	13	16	20	23	26	30
3.1	1.1314	1346	1378	1410	1442	1474	1506	1537	1569	1600	3	6	10	13	16	19	22	25	29
3.2	1.1632	1663	1694	1725	1756	1787	1817	1848	1878	1909	3	6	9	12	15	18	22	25	28
3.3	1.1939	1969	2000	2030	2060	2090	2119	2149	2179	2208	3	6	9	12	15	18	21	24	27
3.4	1.2238	2267	2296	2326	2355	2384	2413	2442	2470	2499	3	6	9	12	15	17	20	23	26
3.5	1.2528	2556	2585	2613	2641	2669	2698	2726	2754	2782	3	6	8	11	14	17	20	23	25
3.6	1.2809	2837	2865	2892	2920	2947	2975	3002	3029	3056	3	5	8	11	14	16	19	22	25
3.7	1.3083	3110	3137	3164	3191	3218	3244	3271	3297	3324	3	5	8	11	13	16	19	21	24
3.8	1.3350	3376	3403	3429	3455	3481	3507	3533	3558	3584	3	5	8	10	13	16	18	21	23
3.9	1.3610	3635	3661	3686	3712	3737	3762	3788	3813	3838	3	5	8	10	13	15	18	20	23
4.0	1.3863	3888	3913	3938	3962	3987	4012	4036	4061	4085	2	5	7	10	12	15	17	20	22
4.1	1.4110	4134	4159	4183	4207	4231	4255	4279	4303	4327	2	5	7	10	12	14	17	19	22
4.2	1.4351	4375	4398	4422	4446	4469	4493	4516	4540	4563	2	5	7	9	12	14	16	19	21
4.3	1.4586	4609	4633	4656	4679	4702	4725	4748	4770	4793	2	5	7	9	12	14	16	18	21
4.4	1.4816	4839	4861	4884	4907	4929	4951	4974	4996	5019	2	5	7	9	11	14	16	18	20
4.5	1.5041	5063	5085	5107	5129	5151	5173	5195	5217	5239	2	4	7	9	11	13	15	18	20
4.6	1.5261	5282	5304	5326	5347	5369	5390	5412	5433	5454	2	4	6	9	11	13	15	17	19
4.7	1.5476	5497	5518	5539	5560	5581	5602	5623	5644	5665	2	4	6	8	11	13	15	17	19
4.8	1.5686	5707	5728	5748	5769	5790	5810	5831	5851	5872	2	4	6	8	10	12	14	16	19
4.9	1.5892	5913	5933	5953	5974	5994	6014	6034	6054	6074	2	4	6	8	10	12	14	16	18
5.0	1.6094	6114	6134	6154	6174	6194	6214	6233	6253	6273	2	4	6	8	10	12	14	16	18
5.1	1.6292	6312	6332	6351	6371	6390	6409	6429	6448	6467	2	4	6	8	10	12	14	16	18
5.2	1.6487	6506	6525	6544	6563	6582	6601	6620	6639	6658	2	4	6	8	10	11	13	15	17
5.3	1.6677	6696	6715	6734	6752	6771	6790	6808	6827	6845	2	4	6	7	9	11	13	15	17
5.4	1.6864	6882	6901	6919	6938	6956	6974	6993	7011	7029	2	4	5	7	9	11	13	15	17

NATURAL LOGARITHMS OF 10^{+n}

n	1	2	3	4	5	6	7	8	9
$\log_e 10^n$	2.3026	4.6052	6.9078	9.2103	11.5129	13.8155	16.1181	18.4207	20.7233

NATURAL LOGARITHMS (continued)

	0	1	2	3	4	5	6	7	8	9	1	2	3	4	5	6	7	8	9
														mean differences					
5.5	1.7047	7066	7084	7102	7120	7138	7156	7174	7192	7210	2	4	5	7	9	11	13	14	16
5.6	1.7228	7246	7263	7281	7299	7317	7334	7352	7370	7387	2	4	5	7	9	11	12	14	16
5.7	1.7405	7422	7440	7457	7475	7492	7509	7527	7544	7561	2	3	5	7	9	10	12	14	16
5.8	1.7579	7596	7613	7630	7647	7664	7681	7699	7716	7733	2	3	5	7	9	10	12	14	15
5.9	1.7750	7766	7783	7800	7817	7834	7851	7867	7884	7901	2	3	5	7	8	10	12	13	15
6.0	1.7918	7934	7951	7967	7984	8001	8017	8034	8050	8066	2	3	5	7	8	10	12	13	15
6.1	1.8083	8099	8116	8132	8148	8165	8181	8197	8213	8229	2	3	5	6	8	10	11	13	15
6.2	1.8245	8262	8278	8294	8310	8326	8342	8358	8374	8390	2	3	5	6	8	10	11	13	14
6.3	1.8405	8421	8437	8453	8469	8485	8500	8516	8532	8547	2	3	5	6	8	9	11	13	14
6.4	1.8563	8579	8594	8610	8625	8641	8656	8672	8687	8703	2	3	5	6	8	9	11	12	14
6.5	1.8718	8733	8749	8764	8779	8795	8810	8825	8840	8856	2	3	5	6	8	9	11	12	14
6.6	1.8871	8886	8901	8916	8931	8946	8961	8976	8991	9006	2	3	5	6	8	9	11	12	14
6.7	1.9021	9036	9051	9066	9081	9095	9110	9125	9140	9155	1	3	4	6	7	9	10	12	13
6.8	1.9169	9184	9199	9213	9228	9242	9257	9272	9286	9301	1	3	4	6	7	9	10	12	13
6.9	1.9315	9330	9344	9359	9373	9387	9402	9416	9430	9445	1	3	4	6	7	9	10	12	13
7.0	1.9459	9473	9488	9502	9516	9530	9544	9559	9573	9587	1	3	4	6	7	9	10	11	13
7.1	1.9601	9615	9629	9643	9657	9671	9685	9699	9713	9727	1	3	4	6	7	8	10	11	13
7.2	1.9741	9755	9769	9782	9796	9810	9824	9838	9851	9865	1	3	4	6	7	8	10	11	12
7.3	1.9879	9892	9906	9920	9933	9947	9961	9974	9988	2.0001	1	3	4	5	7	8	10	11	12
7.4	2.0015	0028	0042	0055	0069	0082	0096	0109	0122	0136	1	3	4	5	7	8	9	11	12
7.5	2.0149	0162	0176	0189	0202	0215	0229	0242	0255	0268	1	3	4	5	7	8	9	11	12
7.6	2.0281	0295	0308	0321	0334	0347	0360	0373	0386	0399	1	3	4	5	7	8	9	10	12
7.7	2.0412	0425	0438	0451	0464	0477	0490	0503	0516	0528	1	3	4	5	6	8	9	10	12
7.8	2.0541	0554	0567	0580	0592	0605	0618	0631	0643	0656	1	3	4	5	6	8	9	10	11
7.9	2.0669	0681	0694	0707	0719	0732	0744	0757	0769	0782	1	3	4	5	6	8	9	10	11
8.0	2.0794	0807	0819	0832	0844	0857	0869	0882	0894	0906	1	3	4	5	6	7	9	10	11
8.1	2.0919	0931	0943	0956	0968	0980	0992	1005	1017	1029	1	2	4	5	6	7	9	10	11
8.2	2.1041	1054	1066	1078	1090	1102	1114	1126	1138	1150	1	2	4	5	6	7	9	10	11
8.3	2.1163	1175	1187	1199	1211	1223	1235	1247	1258	1270	1	2	4	5	6	7	8	10	11
8.4	2.1282	1294	1306	1318	1330	1342	1353	1365	1377	1389	1	2	4	5	6	7	8	9	11
8.5	2.1401	1412	1424	1436	1448	1459	1471	1483	1494	1506	1	2	4	5	6	7	8	9	11
8.6	2.1518	1529	1541	1552	1564	1576	1587	1599	1610	1622	1	2	3	5	6	7	8	9	10
8.7	2.1633	1645	1656	1668	1679	1691	1702	1713	1725	1736	1	2	3	5	6	7	8	9	10
8.8	2.1748	1759	1770	1782	1793	1804	1815	1827	1838	1849	1	2	3	5	6	7	8	9	10
8.9	2.1861	1872	1883	1894	1905	1917	1928	1939	1950	1961	1	2	3	4	6	7	8	9	10
9.0	2.1972	1983	1994	2006	2017	2028	2039	2050	2061	2072	1	2	3	4	6	7	8	9	10
9.1	2.2083	2094	2105	2116	2127	2138	2148	2159	2170	2181	1	2	3	4	5	7	8	9	10
9.2	2.2192	2203	2214	2225	2235	2246	2257	2268	2279	2289	1	2	3	4	5	6	8	9	10
9.3	2.2300	2311	2322	2332	2343	2354	2364	2375	2386	2396	1	2	3	4	5	6	7	9	10
9.4	2.2407	2418	2428	2439	2450	2460	2471	2481	2492	2502	1	2	3	4	5	6	7	8	10
9.5	2.2513	2523	2534	2544	2555	2565	2576	2586	2597	2607	1	2	3	4	5	6	7	8	9
9.6	2.2618	2628	2638	2649	2659	2670	2680	2690	2701	2711	1	2	3	4	5	6	7	8	9
9.7	2.2721	2732	2742	2752	2762	2773	2783	2793	2803	2814	1	2	3	4	5	6	7	8	9
9.8	2.2824	2834	2844	2854	2865	2875	2885	2895	2905	2915	1	2	3	4	5	6	7	8	9
9.9	2.2925	2935	2946	2956	2966	2976	2986	2996	3006	3016	1	2	3	4	5	6	7	8	9
10.0	2.3026																		

NATURAL LOGARITHMS OF 10^{-n} (Note: $\overline{3}.6974 = -3 + 0.6974$)

n	1	2	3	4	5	6	7	8	9
$\log_e 10^{-n}$	$\overline{3}.6974$	$\overline{5}.3948$	$\overline{7}.0922$	$\overline{10}.7897$	$\overline{12}.4871$	$\overline{14}.1845$	$\overline{17}.8819$	$\overline{19}.5793$	$\overline{21}.2767$

RANDOM DIGITS

```
49 57 22 77 94 89 42 96 11 57 04 96 75 84 11 29 01 95 80 35 14 30 27 48 39
88 69 41 36 18 12 20 30 63 45 16 25 83 66 85 97 34 88 34 05 00 39 72 10 21
84 31 94 68 65 44 66 71 61 72 36 19 56 01 31 71 89 13 99 34 18 99 49 00 30
11 84 98 82 70 77 43 94 92 03 93 57 83 52 59 40 30 56 85 73 46 93 41 00 14
54 96 82 91 12 69 89 61 19 78 70 46 38 97 69 80 18 92 45 84 28 62 31 21 47

10 95 42 10 32 76 12 51 77 19 86 88 32 80 34 28 29 02 07 18 41 53 81 17 35
22 78 19 58 21 21 40 14 17 61 23 49 25 56 14 09 61 49 33 51 33 07 20 16 26
86 03 78 91 92 31 35 30 33 89 80 67 50 72 37 44 18 92 83 71 64 74 67 19 80
80 03 20 18 61 67 13 44 81 15 53 99 25 50 10 78 22 45 37 66 08 71 11 02 79
72 75 68 43 05 19 91 20 15 52 39 18 86 02 01 94 82 45 35 74 92 61 28 83 72

79 24 98 22 98 73 96 53 47 80 85 52 14 53 92 96 41 73 55 80 84 47 89 40 58
43 59 13 24 59 60 70 84 80 33 37 66 28 38 91 11 45 46 10 07 97 87 17 90 32
93 31 52 41 21 20 34 79 70 49 38 18 35 47 33 97 01 75 20 75 06 82 27 44 21
63 36 56 42 66 50 69 47 27 61 69 43 11 03 99 33 44 00 33 28 68 17 04 45 45
61 07 84 64 36 94 98 81 04 84 44 61 43 37 33 27 51 82 14 33 14 71 10 87 61

70 84 60 68 44 26 17 44 10 21 64 14 81 26 18 72 87 10 28 37 27 75 10 87 12
68 08 96 22 65 11 97 27 33 46 25 10 30 33 93 82 80 23 71 21 36 32 18 97 35
67 19 80 48 97 14 91 93 36 85 46 78 80 74 60 06 35 03 12 58 41 17 97 25 69
58 03 79 55 24 38 81 07 83 44 79 35 46 38 47 51 99 40 56 90 00 63 93 17 80

93 68 30 74 88 32 41 04 22 58 94 53 11 38 92 36 43 98 38 15 05 10 27 18 13
32 41 21 81 80 74 03 82 66 59 08 52 30 80 51 69 43 41 65 52 96 56 54 12 30
84 97 31 59 66 49 21 56 17 56 62 72 73 87 01 43 48 85 29 28 82 69 18 43 57
63 07 84 46 74 84 42 81 00 61 96 69 92 39 65 36 30 75 44 35 63 28 41 21 95
70 79 55 43 79 35 46 38 47 51 00 33 69 45 98 26 94 03 68 58 70 29 73 41 35
```

Reprinted by permission from "The Compleat Strategyst," by J. D. Williams. Copyright, 1954. McGraw-Hill Book Company, Inc.

HYPERBOLIC SINES $[\sinh x = \frac{1}{2}(e^x - e^{-x})]$

x	0	1	2	3	4	5	6	7	8	9	avg diff
0.0	0.0000	0.0100	0.0200	0.0300	0.0400	0.0500	0.0600	0.0701	0.0801	0.0901	100
.1	0.1002	0.1102	0.1203	0.1304	0.1405	0.1506	0.1607	0.1708	0.1810	0.1911	101
.2	0.2013	0.2115	0.2218	0.2320	0.2423	0.2526	0.2629	0.2733	0.2837	0.2941	103
.3	0.3045	0.3150	0.3255	0.3360	0.3466	0.3572	0.3678	0.3785	0.3892	0.4000	106
.4	0.4108	0.4216	0.4325	0.4434	0.4543	0.4653	0.4764	0.4875	0.4986	0.5098	110
0.5	0.5211	0.5324	0.5438	0.5552	0.5666	0.5782	0.5897	0.6014	0.6131	0.6248	116
.6	0.6367	0.6485	0.6605	0.6725	0.6846	0.6967	0.7090	0.7213	0.7336	0.7461	122
.7	0.7586	0.7712	0.7838	0.7966	0.8094	0.8223	0.8353	0.8484	0.8615	0.8748	130
.8	0.8881	0.9015	0.9150	0.9286	0.9423	0.9561	0.9700	0.9840	0.9981	1.012	138
.9	1.027	1.041	1.055	1.070	1.085	1.099	1.114	1.129	1.145	1.160	15
1.0	1.175	1.191	1.206	1.222	1.238	1.254	1.270	1.286	1.303	1.319	16
.1	1.336	1.352	1.369	1.386	1.403	1.421	1.438	1.456	1.474	1.491	17
.2	1.509	1.528	1.546	1.564	1.583	1.602	1.621	1.640	1.659	1.679	19
.3	1.698	1.718	1.738	1.758	1.779	1.799	1.820	1.841	1.862	1.883	21
.4	1.904	1.926	1.948	1.970	1.992	2.014	2.037	2.060	2.083	2.106	22
1.5	2.129	2.153	2.177	2.201	2.225	2.250	2.274	2.299	2.324	2.350	25
.6	2.376	2.401	2.428	2.454	2.481	2.507	2.535	2.562	2.590	2.617	27
.7	2.646	2.674	2.703	2.732	2.761	2.790	2.820	2.850	2.881	2.911	30
.8	2.942	2.973	3.005	3.037	3.069	3.101	3.134	3.167	3.200	3.234	33
.9	3.268	3.303	3.337	3.372	3.408	3.443	3.479	3.516	3.552	3.589	36
2.0	3.627	3.665	3.703	3.741	3.780	3.820	3.859	3.899	3.940	3.981	39
.1	4.022	4.064	4.106	4.148	4.191	4.234	4.278	4.322	4.367	4.412	44
.2	4.457	4.503	4.549	4.596	4.643	4.691	4.739	4.788	4.837	4.887	48
.3	4.937	4.988	5.039	5.090	5.142	5.195	5.248	5.302	5.356	5.411	53
.4	5.466	5.522	5.578	5.635	5.693	5.751	5.810	5.869	5.929	5.989	58
2.5	6.050	6.112	6.174	6.237	6.300	6.365	6.429	6.495	6.561	6.627	64
.6	6.695	6.763	6.831	6.901	6.971	7.042	7.113	7.185	7.258	7.332	71
.7	7.406	7.481	7.557	7.634	7.711	7.789	7.868	7.948	8.028	8.110	79
.8	8.192	8.275	8.359	8.443	8.529	8.615	8.702	8.790	8.879	8.969	87
.9	9.060	9.151	9.244	9.337	9.431	9.527	9.623	9.720	9.819	9.918	96
3.0	10.02	10.12	10.22	10.32	10.43	10.53	10.64	10.75	10.86	10.97	11
.1	11.08	11.19	11.30	11.42	11.53	11.65	11.76	11.88	12.00	12.12	12
.2	12.25	12.37	12.49	12.62	12.75	12.88	13.01	13.14	13.27	13.40	13
.3	13.54	13.67	13.81	13.95	14.09	14.23	14.38	14.52	14.67	14.82	14
.4	14.97	15.12	15.27	15.42	15.58	15.73	15.89	16.05	16.21	16.38	16
3.5	16.54	16.71	16.88	17.05	17.22	17.39	17.57	17.74	17.92	18.10	17
.6	18.29	18.47	18.66	18.84	19.03	19.22	19.42	19.61	19.81	20.01	19
.7	20.21	20.41	20.62	20.83	21.04	21.25	21.46	21.68	21.90	22.12	21
.8	22.34	22.56	22.79	23.02	23.25	23.49	23.72	23.96	24.20	24.45	24
.9	24.69	24.94	25.19	25.44	25.70	25.96	26.22	26.48	26.75	27.02	26
4.0	27.29	27.56	27.84	28.12	28.40	28.69	28.98	29.27	29.56	29.86	29
.1	30.16	30.47	30.77	31.08	31.39	31.71	32.03	32.35	32.68	33.00	32
.2	33.34	33.67	34.01	34.35	34.70	35.05	35.40	35.75	36.11	36.48	35
.3	36.84	37.21	37.59	37.97	38.35	38.73	39.12	39.52	39.91	40.31	39
.4	40.72	41.13	41.54	41.96	42.38	42.81	43.24	43.67	44.11	44.56	43
4.5	45.00	45.46	45.91	46.37	46.84	47.31	47.79	48.27	48.75	49.24	47
.6	49.74	50.24	50.74	51.25	51.77	52.29	52.81	53.34	53.88	54.42	52
.7	54.97	55.52	56.08	56.64	57.21	57.79	58.37	58.96	59.55	60.15	58
.8	60.75	61.36	61.98	62.60	63.23	63.87	64.51	65.16	65.81	66.47	64
.9	67.14	67.82	68.50	69.19	69.88	70.58	71.29	72.01	72.73	73.46	71
5.0	74.20										

If $x > 5$, $\sinh x = \frac{1}{2}(e^x)$ and $\log_{10} \sinh x = (0.4343)x + 0.6990 - 1$, correct to four significant figures.

HYPERBOLIC COSINES [cosh $x = \frac{1}{2}(e^x + e^{-x})$]

X	0	1	2	3	4	5	6	7	8	9	avg diff
0.0	1.000	1.000	1.000	1.000	1.001	1.001	1.002	1.002	1.003	1.004	1
.1	1.005	1.006	1.007	1.008	1.010	1.011	1.013	1.014	1.016	1.018	2
.2	1.020	1.022	1.024	1.027	1.029	1.031	1.034	1.037	1.039	1.042	3
.3	1.045	1.048	1.052	1.055	1.058	1.062	1.066	1.069	1.073	1.077	4
.4	1.081	1.085	1.090	1.094	1.098	1.103	1.108	1.112	1.117	1.122	5
0.5	1.128	1.133	1.138	1.144	1.149	1.155	1.161	1.167	1.173	1.179	6
.6	1.185	1.192	1.198	1.205	1.212	1.219	1.226	1.233	1.240	1.248	7
.7	1.255	1.263	1.271	1.278	1.287	1.295	1.303	1.311	1.320	1.329	8
.8	1.337	1.346	1.355	1.365	1.374	1.384	1.393	1.403	1.413	1.423	10
.9	1.433	1.443	1.454	1.465	1.475	1.486	1.497	1.509	1.520	1.531	11
1.0	1.543	1.555	1.567	1.579	1.591	1.604	1.616	1.629	1.642	1.655	13
.1	1.669	1.682	1.696	1.709	1.723	1.737	1.752	1.766	1.781	1.796	14
.2	1.811	1.826	1.841	1.857	1.872	1.888	1.905	1.921	1.937	1.954	16
.3	1.971	1.988	2.005	2.023	2.040	2.058	2.076	2.095	2.113	2.132	18
.4	2.151	2.170	2.189	2.209	2.229	2.249	2.269	2.290	2.310	2.331	20
1.5	2.352	2.374	2.395	2.417	2.439	2.462	2.484	2.507	2.530	2.554	23
.6	2.577	2.601	2.625	2.650	2.675	2.700	2.725	2.750	2.776	2.802	25
.7	2.828	2.855	2.882	2.909	2.936	2.964	2.992	3.021	3.049	3.078	28
.8	3.107	3.137	3.167	3.197	3.228	3.259	3.290	3.321	3.353	3.385	31
.9	3.418	3.451	3.484	3.517	3.551	3.585	3.620	3.655	3.690	3.726	34
2.0	3.762	3.799	3.835	3.873	3.910	3.948	3.987	4.026	4.065	4.104	38
.1	4.144	4.185	4.226	4.267	4.309	4.351	4.393	4.436	4.480	4.524	42
.2	4.568	4.613	4.658	4.704	4.750	4.797	4.844	4.891	4.939	4.988	47
.3	5.037	5.087	5.137	5.188	5.239	5.290	5.343	5.395	5.449	5.503	52
.4	5.557	5.612	5.667	5.723	5.780	5.837	5.895	5.954	6.013	6.072	58
2.5	6.132	6.193	6.255	6.317	6.379	6.443	6.507	6.571	6.636	6.702	64
.6	6.769	6.836	6.904	6.973	7.042	7.112	7.183	7.255	7.327	7.400	70
.7	7.473	7.548	7.623	7.699	7.776	7.853	7.932	8.011	8.091	8.171	78
.8	8.253	8.335	8.418	8.502	8.587	8.673	8.759	8.847	8.935	9.024	86
.9	9.115	9.206	9.298	9.391	9.484	9.579	9.675	9.772	9.869	9.968	95
3.0	10.07	10.17	10.27	10.37	10.48	10.58	10.69	10.79	10.90	11.01	11
.1	11.12	11.23	11.35	11.46	11.57	11.69	11.81	11.92	12.04	12.16	12
.2	12.29	12.41	12.53	12.66	12.79	12.91	13.04	13.17	13.31	13.44	13
.3	13.57	13.71	13.85	13.99	14.13	14.27	14.41	14.56	14.70	14.85	14
.4	15.00	15.15	15.30	15.45	15.61	15.77	15.92	16.08	16.25	16.41	16
3.5	16.57	16.74	16.91	17.08	17.25	17.42	17.60	17.77	17.95	18.13	17
.6	18.31	18.50	18.68	18.87	19.06	19.25	19.44	19.64	19.84	20.03	19
.7	20.24	20.44	20.64	20.85	21.06	21.27	21.49	21.70	21.92	22.14	21
.8	22.36	22.59	22.81	23.04	23.27	23.51	23.74	23.98	24.22	24.47	23
.9	24.71	24.96	25.21	25.46	25.72	25.98	26.24	26.50	26.77	27.04	26
4.0	27.31	27.58	27.86	28.14	28.42	28.71	29.00	29.29	29.58	29.88	29
.1	30.18	30.48	30.79	31.10	31.41	31.72	32.04	32.37	32.69	33.02	32
.2	33.35	33.69	34.02	34.37	34.71	35.06	35.41	35.77	36.13	36.49	35
.3	36.86	37.23	37.60	37.98	38.36	38.75	39.13	39.53	39.93	40.33	39
.4	40.73	41.14	41.55	41.97	42.39	42.82	43.25	43.68	44.12	44.57	43
4.5	45.01	45.47	45.92	46.38	46.85	47.32	47.80	48.28	48.76	49.25	**47**
.6	49.75	50.25	50.75	51.26	51.78	52.30	52.82	53.35	53.89	54.43	52
.7	54.98	55.53	56.09	56.65	57.22	57.80	58.38	58.96	59.56	60.15	58
.8	60.76	61.37	61.99	62.61	63.24	63.87	64.52	65.16	65.82	66.48	64
.9	67.15	67.82	68.50	69.19	69.89	70.59	71.30	72.02	72.74	73.47	71
5.0	74.21										

If $x > 5$, cosh $x = \frac{1}{2}(e^x)$, and $\log_{10}$ cosh $x = (0.4343)x + 0.6990 - 1$, correct to four significant figures.

HYPERBOLIC TANGENTS $[\tanh x = (e^x - e^{-x})/(e^x + e^{-x}) = \sinh x/\cosh x]$

x	0	1	2	3	4	5	6	7	8	9	avg diff
0.0	.0000	.0100	.0200	.0300	.0400	.0500	.0599	.0699	.0798	.0898	100
.1	.0997	.1096	.1194	.1293	.1391	.1489	.1587	.1684	.1781	.1878	98
.2	.1974	.2070	.2165	.2260	.2355	.2449	.2543	.2636	.2729	.2821	94
.3	.2913	.3004	.3095	.3185	.3275	.3364	.3452	.3540	.3627	.3714	89
.4	.3800	.3885	.3969	.4053	.4136	.4219	.4301	.4382	.4462	.4542	82
0.5	.4621	.4700	.4777	.4854	.4930	.5005	.5080	.5154	.5227	.5299	75
.6	.5370	.5441	.5511	.5581	.5649	.5717	.5784	.5850	.5915	.5980	67
.7	.6044	.6107	.6169	.6231	.6291	.6352	.6411	.6469	.6527	.6584	60
.8	.6640	.6696	.6751	.6805	.6858	.6911	.6963	.7014	.7064	.7114	52
.9	.7163	.7211	.7259	.7306	.7352	.7398	.7443	.7487	.7531	.7574	45
1.0	.7616	.7658	.7699	.7739	.7779	.7818	.7857	.7895	.7932	.7969	39
.1	.8005	.8041	.8076	.8110	.8144	.8178	.8210	.8243	.8275	.8306	33
.2	.8337	.8367	.8397	.8426	.8455	.8483	.8511	.8538	.8565	.8591	28
.3	.8617	.8643	.8668	.8693	.8717	.8741	.8764	.8787	.8810	.8832	24
.4	.8854	.8875	.8896	.8917	.8937	.8957	.8977	.8996	.9015	.9033	20
1.5	.9052	.9069	.9087	.9104	.9121	.9138	.9154	.9170	.9186	.9202	17
.6	.9217	.9232	.9246	.9261	.9275	.9289	.9302	.9316	.9329	.9342	14
.7	.9354	.9367	.9379	.9391	.9402	.9414	.9425	.9436	.9447	.9458	11
.8	.9468	.9478	.9488	.9498	.9508	.9518	.9527	.9536	.9545	.9554	9
.9	.9562	.9571	.9579	.9587	.9595	.9603	.9611	.9619	.9626	.9633	8
2.0	.9640	.9647	.9654	.9661	.9668	.9674	.9680	.9687	.9693	.9699	6
.1	.9705	.9710	.9716	.9722	.9727	.9732	.9738	.9743	.9748	.9753	5
.2	.9757	.9762	.9767	.9771	.9776	.9780	.9785	.9789	.9793	.9797	4
.3	.9801	.9805	.9809	.9812	.9816	.9820	.9823	.9827	.9830	.9834	4
.4	.9837	.9840	.9843	.9846	.9849	.9852	.9855	.9858	.9861	.9863	3
2.5	.9866	.9869	.9871	.9874	.9876	.9879	.9881	.9884	.9886	.9888	2
.6	.9890	.9892	.9895	.9897	.9899	.9901	.9903	.9905	.9906	.9908	2
.7	.9910	.9912	.9914	.9915	.9917	.9919	.9920	.9922	.9923	.9925	2
.8	.9926	.9928	.9929	.9931	.9932	.9933	.9935	.9936	.9937	.9938	1
.9	.9940	.9941	.9942	.9943	.9944	.9945	.9946	.9947	.9949	.9950	1
3.0	.9951	.9959	.9967	.9973	.9978	.9982	.9985	.9988	.9990	.9992	4
4.0	.9993	.9995	.9996	.9996	.9997	.9998	.9998	.9998	.9999	.9999	1
5.0	.9999										

If $x > 5$, $\tanh x = 1.0000$ to four decimal places.

MULTIPLES OF 0.4343 $(0.43429448 = \log_{10} e)$

x	0	1	2	3	4	5	6	7	8	9
0.0	0.0000	0.0434	0.0869	0.1303	0.1737	0.2171	0.2606	0.3040	0.3474	0.3909
1.0	0.4343	0.4777	0.5212	0.5646	0.6080	0.6514	0.6949	0.7383	0.7817	0.8252
2.0	0.8686	0.9120	0.9554	0.9989	1.0423	1.0857	1.1292	1.1726	1.2160	1.2595
3.0	1.3029	1.3463	1.3897	1.4332	1.4766	1.5200	1.5635	1.6069	1.6503	1.6937
4.0	1.7372	1.7806	1.8240	1.8675	1.9109	1.9543	1.9978	2.0412	2.0846	2.1280
5.0	2.1715	2.2149	2.2583	2.3018	2.3452	2.3886	2.4320	2.4755	2.5189	2.5623
6.0	2.6058	2.6492	2.6926	2.7361	2.7795	2.8229	2.8663	2.9098	2.9532	2.9966
7.0	3.0401	3.0835	3.1269	3.1703	3.2138	3.2572	3.3006	3.3441	3.3875	3.4309
8.0	3.4744	3.5178	3.5612	3.6046	3.6481	3.6915	3.7349	3.7784	3.8218	3.8652
9.0	3.9087	3.9521	3.9955	4.0389	4.0824	4.1258	4.1692	4.2127	4.2561	4.2995

MULTIPLES OF 2.3026 $(2.3025851 = 1/0.4343 = \log_e 10)$

x	0	1	2	3	4	5	6	7	8	9
0.0	0.0000	0.2303	0.4605	0.6908	0.9210	1.1513	1.3816	1.6118	1.8421	2.0723
1.0	2.3026	2.5328	2.7631	2.9934	3.2236	3.4539	3.6841	3.9144	4.1447	4.3749
2.0	4.6052	4.8354	5.0657	5.2959	5.5262	5.7565	5.9867	6.2170	6.4472	6.6775
3.0	6.9078	7.1380	7.3683	7.5985	7.8288	8.0590	8.2893	8.5196	8.7498	8.9801
4.0	9.2103	9.4406	9.6709	9.9011	10.131	10.362	10.592	10.822	11.052	11.283
5.0	11.513	11.743	11.973	12.204	12.434	12.664	12.894	13.125	13.355	13.585
6.0	13.816	14.046	14.276	14.506	14.737	14.967	15.197	15.427	15.658	15.888
7.0	16.118	16.348	16.579	16.809	17.039	17.269	17.500	17.730	17.960	18.190
8.0	18.421	18.651	18.881	19.111	19.342	19.572	19.802	20.032	20.263	20.493
9.0	20.723	20.954	21.184	21.414	21.644	21.875	22.105	22.335	22.565	22.796

EXPONENTIALS (e^n AND e^{-n})

n	e^n	diff	n	e^n	diff	n	e^n (*)	n	e^{-n}	diff	n	e^{-n}	n	e^{-n} (*)
0.00	1.000	10	0.50	1.649	16	1.0	2.718	0.00	1.000	−10	0.50	.607	1.0	.368
.01	1.010	10	.51	1.665	17	.1	3.004	.01	0.990	−10	.51	.600	.1	.333
.02	1.020	10	.52	1.682	17	.2	3.320	.02	.980	−10	.52	.595	.2	.301
.03	1.030	11	.53	1.699	17	.3	3.669	.03	.970	−9	.53	.589	.3	.273
.04	1.041	10	.54	1.716	17	.4	4.055	.04	.961	−10	.54	.583	.4	.247
0.05	1.051	11	0.55	1.733	18	1.5	4.482	0.05	.951	−9	0.55	.577	1.5	.223
.06	1.062	11	.56	1.751	17	.6	4.953	.06	.942	−10	.56	.571	.6	.202
.07	1.073	10	.57	1.768	18	.7	5.474	.07	.932	−9	.57	.566	.7	.183
.08	1.083	11	.58	1.786	18	.8	6.050	.08	.923	−9	.58	.560	.8	.165
.09	1.094	11	.59	1.804	18	.9	6.686	.09	.914	−9	.59	.554	.9	.150
0.10	1.105	11	0.60	1.822	18	2.0	7.389	0.10	.905	−9	0.60	.549	2.0	.135
.11	1.116	11	.61	1.840	19	.1	8.166	.11	.896	−9	.61	.543	.1	.122
.12	1.127	12	.62	1.859	19	.2	9.025	.12	.887	−9	.62	.538	.2	.111
.13	1.139	11	.63	1.878	18	.3	9.974	.13	.878	−9	.63	.533	.3	.100
.14	1.150	12	.64	1.896	20	.4	11.02	.14	.869	−8	.64	.527	.4	.0907
0.15	1.162	12	0.65	1.916	19	2.5	12.18	0.15	.861	−9	0.65	.522	2.5	.0821
.16	1.174	11	.66	1.935	19	.6	13.46	.16	.852	−8	.66	.517	.6	.0743
.17	1.185	12	.67	1.954	20	.7	14.88	.17	.844	−9	.67	.512	.7	.0672
.18	1.197	12	.68	1.974	20	.8	16.44	.18	.835	−8	.68	.507	.8	.0608
.19	1.209	12	.69	1.994	20	.9	18.17	.19	.827	−8	.69	.502	.9	.0550
0.20	1.221	13	0.70	2.014	20	3.0	20.09	0.20	.819	−8	0.70	.497	3.0	.0498
.21	1.234	12	.71	2.034	20	.1	22.20	.21	.811	−8	.71	.492	.1	.0450
.22	1.246	13	.72	2.054	21	.2	24.53	.22	.803	−8	.72	.487	.2	.0408
.23	1.259	12	.73	2.075	21	.3	27.11	.23	.795	−8	.73	.482	.3	.0369
.24	1.271	13	.74	2.096	21	.4	29.96	.24	.787	−8	.74	.477	.4	.0334
0.25	1.284	13	0.75	2.117	21	3.5	33.12	0.25	.779	−8	0.75	.472	3.5	.0302
.26	1.297	13	.76	2.138	22	.6	36.60	.26	.771	−8	.76	.468	.6	.0273
.27	1.310	13	.77	2.160	21	.7	40.45	.27	.763	−7	.77	.463	.7	.0247
.28	1.323	13	.78	2.181	22	.8	44.70	.28	.756	−8	.78	.458	.8	.0224
.29	1.336	14	.79	2.203	23	.9	49.40	.29	.748	−7	.79	.454	.9	.0202
0.30	1.350	13	0.80	2.226	22	4.0	54.60	0.30	.741	−8	0.80	.449	4.0	.0183
.31	1.363	14	.81	2.248	22	.1	60.34	.31	.733	−7	.81	.445	.1	.0166
.32	1.377	14	.82	2.270	23	.2	66.69	.32	.726	−7	.82	.440	.2	.0150
.33	1.391	14	.83	2.293	23	.3	73.70	.33	.719	−7	.83	.436	.3	.0136
.34	1.405	14	.84	2.316	24	.4	81.45	.34	.712	−7	.84	.432	.4	.0123
0.35	1.419	14	0.85	2.340	23	4.5	90.02	0.35	.705	−7	0.85	.427	4.5	.0111
.36	1.433	15	.86	2.363	24			.36	.698	−7	.86	.423		
.37	1.448	15	.87	2.387	24	5.0	148.4	.37	.691	−7	.87	.419	5.0	.00674
.38	1.462	15	.88	2.411	24	6.0	403.4	.38	.684	−7	.88	.415	6.0	.00248
.39	1.477	15	.89•	2.435	25	7.0	1097.	.39	.677	−7	.89	.411	7.0	.000912
0.40	1.492	15	0.90	2.460	24	8.0	2981.	0.40	.670	−6	0.90	.407	8.0	.000335
.41	1.507	15	.91	2.484	25	9.0	8103.	.41	.664	−7	.91	.403	9.0	.000123
.42	1.522	15	.92	2.509	26	10.0	22026.	.42	.657	−6	.92	.399	10.0	.000045
.43	1.537	16	.93	2.535	25			.43	.651	−7	.93	.395		
.44	1.553	15	.94	2.560	26	π/2	4.810	.44	.644	−6	.94	.391	π/2	.208
						2π/2	23.14						2π/2	.0432
0.45	1.568	16	0.95	2.586	26	3π/2	111.3	0.45	.638	−7	0.95	.387	3π/2	.00898
.46	1.584	16	.96	2.612	26	4π/2	535.5	.46	.631	−6	.96	.383	4π/2	.00187
.47	1.600	16	.97	2.638	26	5π/2	2576.	.47	.625	−6	.97	.379	5π/2	.000388
.48	1.616	16	.98	2.664	27	6π/2	12392.	.48	.619	−6	.98	.375	6π/2	.000081
.49	1.632	17	.99	2.691	27	7π/2	59610.	.49	.613	−6	.99	.372	7π/2	.000017
						8π/2	286751.						8π/2	.000003
0.50	1.649		1.00	2.718				0.50	0.607		1.00	.368		

* Note: Do not interpolate in this column.

NORMAL OR GAUSSIAN DISTRIBUTION

x	$p(x) = (2\pi)^{-1/2}\exp(-x^2/2)$	$P(x) = \int_{-\infty}^{x} p(t)\,dt$	x	$p(x) = (2\pi)^{-1/2}\exp(-x^2/2)$	$P(x) = \int_{-\infty}^{x} p(t)\,dt$
0.00	0.39894	0.50000	2.55	0.01545	$0.9^2 4614$
0.05	0.39844	0.51994	2.60	0.01358	$0.9^2 5339$
0.10	0.39695	0.53983	2.65	0.01191	$0.9^2 5975$
0.15	0.39448	0.55962	2.70	0.01042	$0.9^2 6533$
0.20	0.39104	0.57926	2.75	$0.0^2 9094$	$0.9^2 7020$
0.25	0.38667	0.59871			
			2.80	$0.0^2 7915$	$0.9^2 7445$
0.30	0.38139	0.61791	2.85	$0.0^2 6873$	$0.9^2 7814$
0.35	0.37524	0.63683	2.90	$0.0^2 5953$	$0.9^2 8134$
0.40	0.36827	0.65542	2.95	$0.0^2 5143$	$0.9^2 8411$
0.45	0.36053	0.67364	3.00	$0.0^2 4432$	$0.9^2 8650$
0.50	0.35207	0.69146			
			3.05	$0.0^2 3810$	$0.9^2 8856$
0.55	0.34294	0.70884	3.10	$0.0^2 3267$	$0.9^3 0324$
0.60	0.33322	0.72575	3.15	$0.0^2 2794$	$0.9^3 1836$
0.65	0.32297	0.74215	3.20	$0.0^2 2384$	$0.9^3 3129$
0.70	0.31225	0.75804	3.25	$0.0^2 2029$	$0.9^3 4230$
0.75	0.30114	0.77337			
			3.30	$0.0^2 1723$	$0.9^3 5166$
0.80	0.28969	0.78814	3.35	$0.0^2 1459$	$0.9^3 5959$
0.85	0.27798	0.80234	3.40	$0.0^2 1232$	$0.9^3 6631$
0.90	0.26609	0.81594	3.45	$0.0^2 1038$	$0.9^3 7197$
0.95	0.25406	0.82894	3.50	$0.0^3 8727$	$0.9^3 7674$
1.00	0.24197	0.84134			
			3.55	$0.0^3 7317$	$0.9^3 8074$
1.05	0.22988	0.85314	3.60	$0.0^3 6119$	$0.9^3 8409$
1.10	0.21785	0.86433	3.65	$0.0^3 5105$	$0.9^3 8689$
1.15	0.20594	0.87493	3.70	$0.0^3 4248$	$0.9^3 8922$
1.20	0.19419	0.88493	3.75	$0.0^3 3526$	$0.9^4 1158$
1.25	0.18265	0.89435			
			3.80	$0.0^3 2919$	$0.9^4 2765$
1.30	0.17137	0.90320	3.85	$0.0^3 2411$	$0.9^4 4094$
1.35	0.16038	0.91149	3.90	$0.0^3 1987$	$0.9^4 5190$
1.40	0.14973	0.91924	3.95	$0.0^3 1633$	$0.9^4 6092$
1.45	0.13943	0.92647	4.00	$0.0^3 1338$	$0.9^4 6833$
1.50	0.12952	0.93319			
			4.05	$0.0^3 1094$	$0.9^4 7439$
1.55	0.12001	0.93943	4.10	$0.0^4 8926$	$0.9^4 7934$
1.60	0.11092	0.94520	4.15	$0.0^4 7263$	$0.9^4 8338$
1.65	0.10226	0.95053	4.20	$0.0^4 5894$	$0.9^4 8665$
1.70	0.09405	0.95543	4.25	$0.0^4 4772$	$0.9^4 8931$
1.75	0.08628	0.95994			
			4.30	$0.0^4 3854$	$0.9^5 1460$
1.80	0.07895	0.96407	4.35	$0.0^4 3104$	$0.9^5 3193$
1.85	0.07206	0.96784	4.40	$0.0^4 2494$	$0.9^5 4587$
1.90	0.06562	0.97128	4.45	$0.0^4 1999$	$0.9^5 5706$
1.95	0.05959	0.97441	4.50	$0.0^4 1598$	$0.9^5 6602$
2.00	0.05399	0.97725			
			4.55	$0.0^4 1275$	$0.9^5 7318$
2.05	0.04879	0.97982	4.60	$0.0^4 1014$	$0.9^5 7888$
2.10	0.04398	0.98214	4.65	$0.0^5 8047$	$0.9^5 8340$
2.15	0.03955	0.98422	4.70	$0.0^5 6370$	$0.9^5 8699$
2.20	0.03547	0.98610	4.75	$0.0^5 5030$	$0.9^5 8983$
2.25	0.03174	0.98778			
			4.80	$0.0^5 3961$	$0.9^6 2067$
2.30	0.02833	0.98928	4.85	$0.0^5 3112$	$0.9^6 3827$
2.35	0.02522	$0.9^2 0613$	4.90	$0.0^5 2439$	$0.9^6 5208$
2.40	0.02239	$0.9^2 1802$	4.95	$0.0^5 1907$	$0.9^6 6289$
2.45	0.01984	$0.9^2 2857$	5.00	$0.0^5 1487$	$0.9^6 7133$
2.50	0.01753	$0.9^2 3790$			

Note: $0.0^2 9094 = 0.009094$ $0.9^3 0324 = 0.9990324$

$$P(-x) = 1 - P(x) \qquad \int_{-x}^{x} p(t)\,dt = 2P(x) - 1$$

BESSEL FUNCTIONS

TABLE I—$J_n(z)$

n	0	0.1	0.2	0.3	0.4	0.5	0.6	0.7	0.8	0.9
0	1.0000	0.9975	0.9900	0.9776	0.9604	0.9385	0.9120	0.8812	0.8463	0.8075
1	0.7652	0.7196	0.6711	0.6201	0.5669	0.5118	0.4554	0.3980	0.3400	0.2818
2	0.2239	0.1666	0.1104	0.0555	0.0025	-0.0484	-0.0968	-0.1424	-0.1850	-0.2243
3	-0.2601	-0.2921	-0.3202	-0.3443	-0.3643	-0.3801	-0.3918	-0.3992	-0.4026	-0.4018
4	-0.3971	-0.3887	-0.3766	-0.3610	-0.3423	-0.3205	-0.2961	-0.2693	-0.2404	-0.2097
5	-0.1776	-0.1443	-0.1103	-0.0758	-0.0412	-0.0068	+0.0270	0.0599	0.0917	0.1220
6	0.1506	0.1773	0.2017	0.2238	0.2433	0.2601	0.2740	0.2851	0.2931	0.2981
7	0.3001	0.2991	0.2951	0.2882	0.2786	0.2663	0.2516	0.2346	0.2154	0.1944
8	0.1717	0.1475	0.1222	0.0960	0.0692	0.0419	0.0146	-0.0125	-0.0392	-0.0653
9	-0.0903	-0.1142	-0.1367	-0.1577	-0.1768	-0.1939	-0.2090	-0.2218	-0.2323	-0.2403
10	-0.2459	-0.2490	-0.2496	-0.2477	-0.2434	-0.2366	-0.2276	-0.2164	-0.2032	-0.1881
11	-0.1712	-0.1528	-0.1330	-0.1121	-0.0902	-0.0677	-0.0446	-0.0213	+0.0020	0.0250
12	0.0477	0.0697	0.0908	0.1108	0.1296	0.1469	0.1626	0.1766	0.1887	0.1988
13	0.2069	0.2129	0.2167	0.2183	0.2177	0.2150	0.2101	0.2032	0.1943	0.1836
14	0.1711	0.1570	0.1414	0.1245	0.1065	0.0875	0.0679	0.0476	0.0271	0.0064
15	-0.0142	-0.0346	-0.0544	-0.0736	-0.0919	-0.1092	-0.1253	-0.1401	-0.1533	-0.1650

BESSEL FUNCTIONS (continued)

TABLE II—$J_1(z)$

z	0	0.1	0.2	0.3	0.4	0.5	0.6	0.7	0.8	0.9
0	0.0000	0.0499	0.0995	0.1483	0.1960	0.2423	0.2867	0.3290	0.3688	0.4059
1	0.4401	0.4709	0.4983	0.5220	0.5419	0.5579	0.5699	0.5778	0.5815	0.5812
2	0.5767	0.5683	0.5560	0.5399	0.5202	0.4971	0.4708	0.4416	0.4097	0.3754
3	0.3391	0.3009	0.2613	0.2207	0.1792	0.1374	0.0955	0.0538	0.0128	−0.0272
4	−0.0660	−0.1033	−0.1386	−0.1719	−0.2028	−0.2311	−0.2566	−0.2791	−0.2985	−0.3147
5	−0.3276	−0.3371	−0.3432	−0.3460	−0.3453	−0.3414	−0.3343	−0.3241	−0.3110	−0.2951
6	−0.2767	−0.2559	−0.2329	−0.2081	−0.1816	−0.1538	−0.1250	−0.0953	−0.0652	−0.0349
7	−0.0047	+0.0252	0.0543	0.0826	0.1096	0.1352	0.1592	0.1813	0.2014	0.2192
8	0.2346	0.2476	0.2580	0.2657	0.2708	0.2731	0.2728	0.2697	0.2641	0.2559
9	0.2453	0.2324	0.2174	0.2004	0.1816	0.1613	0.1395	0.1166	0.0928	0.0684
10	0.0435	0.0184	−0.0066	−0.0313	−0.0555	−0.0789	−0.1012	−0.1224	−0.1422	−0.1603
11	−0.1768	−0.1913	−0.2039	−0.2143	−0.2225	−0.2284	−0.2320	−0.2333	−0.2323	−0.2290
12	−0.2234	−0.2157	−0.2060	−0.1943	−0.1807	−0.1655	−0.1487	−0.1307	−0.1114	−0.0912
13	−0.0703	−0.0489	−0.0271	−0.0052	+0.0166	0.0380	0.0590	0.0791	0.0984	0.1165
14	0.1334	0.1488	0.1626	0.1747	0.1850	0.1934	0.1999	0.2043	0.2066	0.2069
15	0.2051	0.2013	0.1955	0.1879	0.1784	0.1672	0.1544	0.1402	0.1247	0.1080

BESSEL FUNCTIONS (continued)

TABLE III—$J_2(z)$

n	0	0.1	0.2	0.3	0.4	0.5	0.6	0.7	0.8	0.9
0	0.0000	0.0012	0.0050	0.0112	0.0197	0.0306	0.0437	0.0588	0.0758	0.0946
1	0.1149	0.1366	0.1593	0.1830	0.2074	0.2321	0.2570	0.2817	0.3061	0.3299
2	0.3528	0.3746	0.3951	0.4139	0.4310	0.4461	0.4590	0.4696	0.4777	0.4832
3	0.4861	0.4862	0.4835	0.4780	0.4697	0.4586	0.4448	0.4283	0.4093	0.3879
4	0.3641	0.3383	0.3105	0.2811	0.2501	0.2178	0.1846	0.1506	0.1161	0.0813

TABLE IV—$J_3(z)$

n	0	0.1	0.2	0.3	0.4	0.5	0.6	0.7	0.8	0.9
0	0.0000	0.0000	0.0002	0.0006	0.0013	0.0026	0.0044	0.0069	0.0102	0.0144
1	0.0196	0.0257	0.0329	0.0411	0.0505	0.0610	0.0725	0.0851	0.0988	0.1134
2	0.1289	0.1453	0.1623	0.1800	0.1981	0.2166	0.2353	0.2540	0.2727	0.2911
3	0.3091	0.3264	0.3431	0.3588	0.3734	0.3868	0.3988	0.4092	0.4180	0.4250
4	0.4302	0.4333	0.4344	0.4333	0.4301	0.4247	0.4171	0.4072	0.3952	0.3811

TABLE V—$J_4(z)$

n	0	0.1	0.2	0.3	0.4	0.5	0.6	0.7	0.8	0.9
0	0.0000	0.0000	0.0000	0.0000	0.0001	0.0002	0.0003	0.0006	0.0010	0.0016
1	0.0025	0.0036	0.0050	0.0068	0.0091	0.0118	0.0150	0.0188	0.0232	0.0283
2	0.0340	0.0405	0.0476	0.0556	0.0643	0.0738	0.0840	0.0950	0.1067	0.1190
3	0.1320	0.1456	0.1597	0.1743	0.1891	0.2044	0.2198	0.2353	0.2507	0.2661
4	0.2811	0.2958	0.3100	0.3236	0.3365	0.3484	0.3594	0.3693	0.3780	0.3853

CHAPTER 48

MISCELLANEOUS DATA

PRESSURE–ALTITUDE GRAPH

Design of electrical equipment for aircraft is somewhat complicated by the requirement of additional insulation for high voltages as a result of the decrease in atmospheric pressure. The extent of this effect may be determined from Figs. 1 and 2 and the associated table. (1 inch mercury = 25.4 mm mercury = 0.4912 pound/inch².)

SPARK-GAP BREAKDOWN VOLTAGES

Figure 2 is for a voltage that is continuous or at a frequency low enough to permit complete deionization between cycles, between needle points, or clean smooth spherical surfaces (electrodes ungrounded) in dust-free dry air. Temperature is 25 degrees Celsius and pressure is 760 millimeters (29.9 inches) of mercury. Peak kilovolts shown in the figure should be multiplied by the factors given

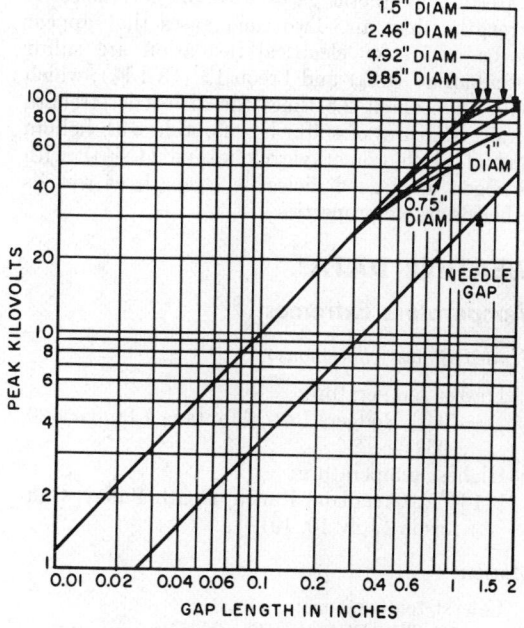

Fig. 2—Spark-gap breakdown voltages.

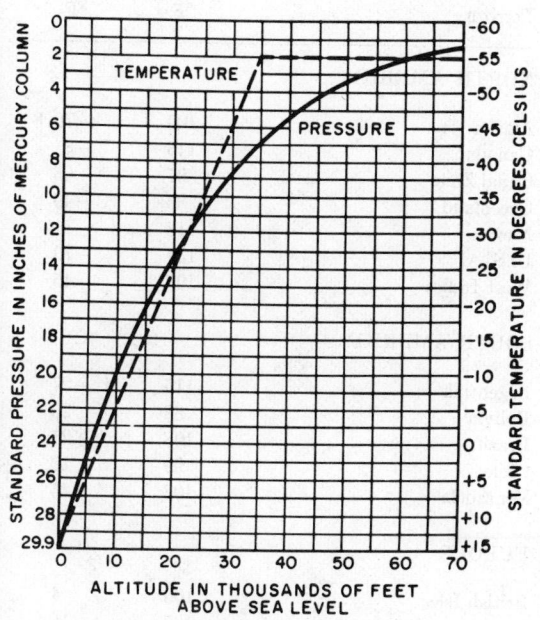

Fig. 1—Pressure as a function of altitude.

Table of Multiplying Factors

Pressure		Temperature in Degrees Celsius					
in. Hg	mm Hg	−40	−20	0	20	40	60
5	127	0.26	0.24	0.23	0.21	0.20	0.19
10	254	0.47	0.44	0.42	0.39	0.37	0.34
15	381	0.68	0.64	0.60	0.56	0.53	0.50
20	508	0.87	0.82	0.77	0.72	0.68	0.64
25	635	1.07	0.99	0.93	0.87	0.82	0.77
30	762	1.25	1.17	1.10	1.03	0.97	0.91
35	889	1.43	1.34	1.26	1.19	1.12	1.05
40	1016	1.61	1.51	1.42	1.33	1.25	1.17
45	1143	1.79	1.68	1.58	1.49	1.40	1.31
50	1270	1.96	1.84	1.73	1.63	1.53	1.44
55	1397	2.13	2.01	1.89	1.78	1.67	1.57
60	1524	2.30	2.17	2.04	1.92	1.80	1.69

in the associated table for atmospheric conditions other than the above.

An approximate rule for uniform fields at all frequencies up to at least 300 megahertz is that the breakdown gradient of air is 30 peak kilovolts/centimeter or 75 peak kilovolts/inch at sea level (760 millimeters of mercury) and normal temperature (25 degrees Celsius). The breakdown voltage is approximately proportional to pressure and inversely proportional to absolute (Kelvin) temperature.

Certain synthetic gases have higher dielectric strengths than air. Two such gases that appear to be useful for electrical insulation are sulfur hexafluoride (SF_6) and Freon 12 (CCl_2F_2), which both have about 2.5 times the dielectric strength of air. Mixtures of sulfur hexafluoride with helium and of perfluoromethylcyclohexane (C_7F_{14}) with nitrogen have good dielectric strength as well as other desirable properties.

WEATHER DATA*

Temperature Extremes

United States (contiguous):

Lowest temperature:
−70°F, Rodgers Pass, Montana (January 20, 1954).
Highest temperature:
134°F, Greenland Ranch, Death Valley, California (July 10, 1913).

Alaska:

Lowest temperature:
−79.8°F, Prospect Creek Camp (January 23, 1971).
Highest temperature:
100°F, Fort Yukon (June 27, 1915).

World:

Lowest temperature:
−90°F, Oimekon, Siberia (February, 1933).
Highest temperature:
136°F, Azizia, Libya, North Africa (September 13, 1922).
Lowest mean temperature (annual):
−14°F, Framheim, Antarctica.
Highest mean temperature (annual)
86°F, Massawa, Eritrea, Africa.

* Compiled in part from "Climate and Man," Yearbook of Agriculture, US Dept. of Agriculture. Obtainable from Superintendent of Documents, Government Printing Office, Washington, D.C. 20402. For additional details refer to "World Weather Records, 1941–1950", US Dept of Commerce Weather Bureau, 1959. Available from the Superintendent of Documents, Washington, D.C. 20402. Also, "Tables of Temperature, Relative Humidity, and Precipitation for the World," Air Ministry Meteorological Office MO 617C, available from HMS Stationery Office, London (1958).

Precipitation Extremes

United States:

Wettest state:
Louisiana—average annual rainfall 57.34 inches.
Dryest state:
Nevada—average annual rainfall 8.60 inches.
Maximum recorded:
Camp Leroy, California (January 22–23, 1943)—26.12 inches in 24 hours.
Minimums recorded:
Bagdad, California (1909–1913)—3.93 inches in 5 years.
Greenland Ranch, California—1.76 inches annual average.

World:

Maximums recorded:
Cherrapunji, India (July, 1861)—366 inches in 1 month. (Average annual rainfall of Cherrapunji is 450 inches).
Baguio, Luzon, Philippines, July 14–15, 1911—46 inches in 24 hours.
Minimums recorded:
Wadi Halfa (Sudan) and Aswan (Egypt) are in the "rainless" area; average annual rainfall is too small to be measured.

World Temperatures

Territory	Maximum °F	Minimum °F
NORTH AMERICA		
Alaska	100	−79.8
Canada	113	−81
Canal Zone	97	63
Greenland	86	−46
Mexico	118	11
U. S. A.	134	−70
West Indies	102	45
SOUTH AMERICA		
Argentina	115	−27
Bolivia	82	25
Brazil	108	21
Chile	99	19
Venezuela	102	45
EUROPE		
British Isles	100	4
France	107	−14
Germany	100	−16
Iceland	71	−6

Territory	Maximum °F	Minimum °F	Territory	Maximum °F	Minimum °F
Italy	114	4	AFRICA		
Norway	95	−26			
Spain	124	10	Algeria	133	1
Sweden	92	−49	Angola	91	33
Turkey	100	17	Egypt	124	31
U. S. S. R. (Russia)	110	−61	Ethiopia	111	32
			Libya	136	35
ASIA			Morocco	119	5
			Rhodesia	112	18
Saudi Arabia	123	35	Somalia	93	61
China	111	−10	Sudan	126	28
East Indies	101	60	Tunisia	122	28
India	120	−19	Union of South Africa	111	21
Iraq	125	19	Zaire	97	34
Japan	101	−7			
Malaysia	97	66	AUSTRALASIA		
Philippine Islands	101	58			
Thailand	106	52	Australia	127	19
Tibet	85	−20	Hawaii	91	51
Turkey	111	−22	New Zealand	94	23
U. S. S. R. (Russia)	109	−90	Samoan Islands	96	61
Vietnam	113	33	Solomon Islands	97	70

Wind-Velocity and Temperature Extremes in North America

Maximum Corrected Wind Velocity (Fastest Single Mile):

Station	Wind (miles/hour)	Temperature, Degrees Fahrenheit	
		Maximum	Minimum
UNITED STATES, 1871–1955			
Albany, New York	71	104	−26
Amarillo, Texas	84	108	−16
Buffalo, New York	91	99	−21
Charleston, South Carolina	76	104	7
Chicago, Illinois	87	105	−23
Bismarck, North Dakota	72	114	−45
Hatteras, North Carolina	110	97	8
Miami, Florida	132	95	27
Minneapolis, Minnesota	92	108	−34
Mobile, Alabama	87	104	−11
Mt. Washington, New Hampshire	188*	71	−46
Nantucket, Massachusetts	91	95	−6
New York, New York	99	102	−14
North Platte, Nebraska	72	112	−35
Pensacola, Florida	114	103	7
Washington, D.C.	62	106	−15
San Juan, Puerto Rico	149†	94	62
CANADA, 1955			
Banff, Alberta	52‡	97	−60
Kamloops, British Columbia	34‡	107	−37
Sable Island, Nova Scotia	64‡	86	−12
Toronto, Ontario	48‡	105	−46

*Gusts were recorded at 231 miles/hour (corrected).
†Estimated.
‡For a period of 5 minutes.

CELSIUS TABLE OF RELATIVE HUMIDITY OR PERCENT OF SATURATION

Dry Bulb

Difference Between Readings of Wet and Dry Bulbs in Degrees Celsius

°C	0.5	1.0	1.5	2.0	2.5	3.0	3.5	4.0	4.5	5	6	7	8	9	10	11	12	13	14	15	16	18	20	22	24	26	28	30	32	34	36	38	40
4	93	85	77	70	63	56	48	41	34	28	15																						
8	94	87	81	74	68	62	56	50	45	39	28	17																					
12	94	89	84	78	73	68	63	58	53	48	38	30	21	12	4																		
16	95	90	85	81	76	71	67	62	58	54	45	37	29	21	14	7																	
20	96	91	87	82	78	74	70	66	62	58	51	44	36	30	23	17	11																
22	96	92	87	83	79	75	72	68	64	60	53	46	40	34	27	21	16	11															
24	96	92	88	85	81	77	74	70	66	63	56	49	43	37	31	26	21	14	10														
26	96	92	88	85	81	77	74	70	67	64	57	51	45	39	34	28	23	18	13														
28	96	92	89	85	82	78	75	72	68	65	59	53	47	42	37	31	26	21	17	13	12												
30	96	93	89	86	82	79	76	73	70	67	61	55	50	44	39	35	30	24	20	16	15	11											
32	96	93	89	86	83	80	77	74	71	68	62	56	51	46	41	36	32	27	23	19	18	15	11										
34	97	93	90	87	84	81	77	74	71	69	63	58	53	48	43	38	34	30	26	22	21	18	15	10									
36	97	93	90	87	84	81	78	75	72	70	64	59	54	50	45	41	36	32	28	24	23	21	16	13	10								
38	97	94	90	87	85	82	79	76	73	71	65	60	56	51	46	42	38	34	30	26	25	23	19	16	13	10							
40	97	94	91	88	85	82	79	76	74	71	66	61	57	52	48	44	40	36	32	29	25	21	17	13	10								
44	97	94	91	88	86	83	80	77	75	73	68	63	59	54	50	47	43	39	36	32	29	23	17	12									
48	97	92	89	86	84	81	78	76	74	69	65	61	56	53	49	45	42	39	35	33	27	21	16	12	11								
52	97	94	92	89	87	84	82	79	77	75	70	66	62	58	55	51	48	44	41	38	35	30	25	20	16	15	11						
56	97	95	92	90	87	85	83	80	78	76	72	68	64	60	57	53	50	46	43	40	38	32	27	23	19	18	14	11					
60	98	95	93	90	88	86	84	81	79	77	73	69	65	62	58	55	52	48	46	43	40	35	30	26	21	18	14	11					
70	98	96	93	91	89	87	85	83	81	79	75	71	68	65	61	58	55	52	50	47	44	40	35	31	27	23	20	17	14	11			
80	98	96	94	92	90	88	86	84	83	81	77	74	71	67	64	61	58	56	53	50	48	43	39	35	31	28	24	20	18	16	14		
90	98	97	95	93	91	89	87	85	84	82	79	76	73	69	67	64	61	58	56	53	51	47	42	39	35	32	28	23	20	18	16	14	
100	99	97	95	93	92	90	88	86	85	83	80	77	74	71	68	66	63	60	58	55	53	49	45	42	38	35	32	29	26	24	22	19	17

Example: Assume dry-bulb reading (thermometer exposed directly to atmosphere) is 20°C and wet-bulb reading is 17°C, or a difference of 3°C. The relative humidity at 20°C is then 74%.

MATERIALS AND FINISHES FOR TROPICAL AND MARINE USE

Corrosion

Ordinary finishing of equipment fails to meet satisfactorily conditions encountered in tropical and marine use. Under these conditions corrosive influences are greatly aggravated by prevailing higher relative humidities, and temperature cycling causes alternate condensation on, and evaporation of moisture from, finished surfaces. Useful equipment life under adverse atmospheric influences depends largely on proper choice of base materials and finishes applied. Especially important in tropical and marine applications is avoidance of electrical contact between dissimilar metals.

Dissimilar metals widely separated in the galvanic series should not be bolted, riveted, etc., without separation by insulating material at the facing surfaces. The only exception occurs when both surfaces have been coated with the same protective metal, e.g., electroplating, hot dipping, galvanizing, etc.

Aluminum, steel, zinc, and cadmium should never be used bare. Electrical contact surfaces should be given copper–nickel–chromium or copper–nickel finish, and, in addition, they should be silver plated. Adjustable-capacitor plates should be silver plated.

An additional 0.000015" to 0.000020" electroplating of hard bright gold over the silver greatly improves resistance to tarnish and oxidation and to attack by most chemicals, lowers electrical resistance, and provides long-term solderability.

Fungus and Decay

The value of fungicidal coatings or treatments is controversial. If equipment is to operate under tropical conditions, greater success can be achieved by the use of materials that do not provide a nutrient medium for fungus and insects. The following types or kinds of materials are examples of nonnutrient mediums that are generally considered acceptable.

Metals
Glass
Ceramics (steatite, glass-bonded mica)
Mica
Polyamide
Cellulose acetate
Rubber (natural or synthetic)
Plastic materials using glass, mica, or asbestos as a filler
Polyvinylchloride
Polytetrafluoroethylene
Monochlortrifluoroethylene

The following types or kinds of materials should not be used, except where such materials are fabricated into completed parts and their use is acceptable to the customer.

Linen
Cellulose nitrate
Regenerated cellulose
Wood
Jute
Leather
Cork
Paper and cardboard
Organic fiberboard
Hair or wool felts
Plastic materials using cotton, linen, or wood flour as a filler

Wood should not be used as an electrical insulator, and its use for other purposes should be restricted to those parts for which a superior substitute is not known. When used, it should be pressure-treated and impregnated to resist moisture, insects, and decay with a waterborne preservative (as specified in Federal Specification TT-W-571), and should also be treated with a suitable fire-retardant chemical.

Finish Application Table

(By Z. Fox. Reprinted by permission from *Product Engineering*, vol. 19, p. 161; January 1948.)

Material	Finish	Remarks
Aluminum alloy	Anodizing	An electrochemical-oxidation surface treatment, for improving corrosion resistance; not an electroplating process. For riveted or welded assemblies specify chromic acid anodizing. Do not anodize parts with nonaluminum inserts. Colors vary: Yellow–green, gray, or black.
	"Alrok"	Chemical-dip oxide treatment. Cheap. Inferior in abrasion and corrosion resistance to the anodizing process, but applicable to assemblies of aluminum and nonaluminum materials.

Material	Finish	Remarks
Copper and zinc alloys	Bright acid dip	Immersion of parts in acid solution. Clear lacquer applied to prevent tarnish.
Brass, bronze, zinc die-casting alloys	Brass, chrome, nickel, tin	As discussed under steel.
Magnesium alloy	Dichromate treatment	Corrosion-preventive dichromate dip. Yellow color.
Stainless steel	Passivating treatment	Nitric-acid immunizing dip.
Steel	Cadmium	Electroplate, dull white color, good corrosion resistance, easily scratched, good thread antiseize. Poor wear and galling resistance.
	Chromium	Electroplate, excellent corrosion resistance and lustrous appearance. Relatively expensive. Specify hard chrome plate for exceptionally hard abrasion-resistive surface. Has low coefficient of friction. Used to some extent on nonferrous metals particularly when die-cast. Chrome-plated objects usually receive a base electroplate of copper, then nickel, followed by chromium. Used for buildup of parts that are undersized. Do not use on parts with deep recesses.
	Blueing	Immersion of cleaned and polished steel into heated saltpeter or carbonaceous material. Part then rubbed with linseed oil. Cheap. Poor corrosion resistance.
	Silver plate	Electroplate, frosted appearance; buff to brighten. Tarnishes readily. Good bearing lining. For electrical contacts, reflectors.
	Zinc plate	Dip in molten zinc (galvanizing) or electroplate of low-carbon or low-alloy steels. Low cost. Generally inferior to cadmium plate. Poor appearance. Poor wear resistance: Electroplate has better adherence to base metal than hot-dip coating. For improving corrosion resistance, zinc-plated parts are given special inhibiting treatments.
	Nickel plate	Electroplate, dull white. Does not protect steel from galvanic corrosion. If plating is broken, corrosion of base metal will be hastened. Finishes in dull white, polished, or black. Do not use on parts with deep recesses.
	Black-oxide dip	Nonmetallic chemical black oxidizing treatment for steel, cast iron, and wrought iron. Inferior to electroplate. No buildup. Suitable for parts with close dimensional requirements as gears, worms, and guides. Poor abrasion resistance.

Material	Finish	Remarks
Steel	Phosphate treatment	Nonmetallic chemical treatment for steel and iron products. Suitable for protection of internal surfaces of hollow parts. Small amount of surface buildup. Inferior to metallic electroplate. Poor abrasion resistance. Good paint base.
	Tin plate	Hot dip or electroplate. Excellent corrosion resistance, but if broken will not protect steel from galvanic corrosion. Also used for copper, brass, and bronze parts that must be soldered after plating. Tin-plated parts can be severely worked and deformed without rupture of plating.
	Brass plate	Electroplate of copper and zinc. Applied to brass and steel parts where uniform appearance is desired. Applied to steel parts when bonding to rubber is desired.
	Copper plate	Electroplate applied before nickel or chrome plates. Also for parts to be brazed or protected against carburization. Tarnishes readily.

PRINCIPAL LOW-VOLTAGE POWER SUPPLIES IN THE WORLD

Territory (Frequency) Voltage

North America:

Alaska (60) 120/240
Bermuda (60) 115/230; some 120/208
Belize (60) 110/220
Canada (60) 120/240; some 115/230
Costa Rica (60) 110/220
El Salvador (60) 110/220
Guatemala (60) 110/240; some 220, 120/208
Honduras (60) 110/220
Mexico (50, 60) 127/220 and other voltages
    Mexico City (50) 125/216
Nicaragua (60) 120
Panama (60) 110/220; some 120/240, 115/230
United States (60) 120/240 and 120/208

West Indies:

Antigua (60) 230/400
Bahamas (60) 115/200; some 115/220
Barbados (50) 120/208; some 110/200
Cuba (60) 115/230; some 120/208
Dominican Republic (60) 115/230
Guadeloupe (50) 127/220
Jamaica (50, some 60) 110/220
Martinique (50) 127/220
Puerto Rico (60) 120/240
Trinidad (60) 115/230
Virgin Islands (60) 120/240

South America:

Argentina (50) 220/380; also 220/440 dc
Bolivia (50, also 60) 220 and other voltages

Territory (Frequency) Voltage

Brazil (50, 60) 110, 220; also other voltages and dc
    Rio de Janeiro (50) 125/216
Chile (50) 220/380; some 220 dc
Colombia (60) 110/220; also 120/240 and others
Ecuador (60) 120/208; also 110/220 and others
French Guiana (50) 127/220
Guyana (50, 60) 110/220
Paraguay (50) 220/440; some 220/440 dc
Peru (60) 220; some 110
Surinam (50, 60) 127/220; some 115/230
Uruguay (50) 220
Venezuela (60, some 50) 120/208, 120/240

Europe:

Austria (50) 220/380; Vienna also has 220/440 dc
Azores (50) 220/380
Belgium (50) 220/380 and many others; some dc
Canary Islands (50) 127/220
Denmark (50) 220/380; also 220/440 dc
Finland (50) 220/380
France (50) 120/240, 220/380, and many others
Germany (Federal Republic) (50) 220/380; also
    others, some dc
Gibraltar (50) 240/415
Greece (50) 220/380; also others, some dc
Iceland (50) 220; some 220/380
Ireland (50) 220/380; some 220/440 dc
Italy (50) 127/220, 220/380 and others
Luxembourg (50) 110/190, 220/380
Madeira (50) 220/380; also 220/440 dc
Malta (50) 240/415

Territory (Frequency) Voltage

Monaco (50) 127/220, 220/380
Netherlands (50) 220/380; also 127/220
Norway (50) 230
Portugal (50) 220/380; some 110/190
Spain (50) 127/220; also 220/380, some dc
Sweden (50) 127/220, 220/380; some dc
Switzerland (50) 220/380
Turkey (50) 220/380; some 110/190
United Kingdom (50) 240/415 and others, some dc
Yugoslavia (50) 220/380

Asia:

Afghanistan (50) 220/380
Burma (50) 230
Cambodia (50) 120/208; some 220/380
Cyprus (50) 240
Hong Kong (50) 200/346
India (50) 230/400 and others, some dc
Indonesia (50) 127/220
Iran (50) 220/380
Iraq (50) 220/380
Israel (50) 230/400
Japan (50, 60) 100/200
Jordan (50) 220/380
Korea (60) 100/200
Kuwait (50) 240/415
Laos (50) 127/220; some 220/380
Lebanon (50) 110/190; some 220/380
Malaysia (50) 230/400; some 240/415
Nepal (50) 110/220
Okinawa (60) 120/240
Pakistan (50) 230/400 and others, some dc
Philippines (60) 110, 220, and others
Saudi Arabia (50, 60) 120/208; also 220/380, 230/400
Singapore (50) 230/400
Sri Lanka (50) 230/400
Syria (50) 115/200; some 220/380
Taiwan (60) 100/200
Thailand (50) 220/380; also 110/190
Vietnam (50) 220/380 future standard
Yemen Arab Republic (50) 220
Yemen, Peoples Democratic Republic (50) 230/400

Africa:

Algeria (50) 127/220, 220/380
Angola (50) 220/380
Dahomey (50) 220/380
Egypt (50) 110, 220 and others; some dc
Ethiopia (50) 220/380; some 127/220
Guinea (50) 220/380; some 127/220
Kenya (50) 240/415
Liberia (60) 120/240
Libya (50) 125/220; some 230/400
Malagasy Republic (50) 220/380; some 127/220
Mauritius (50) 230/400
Morocco (50) 115/200; also 230/400 and others

Territory (Frequency) Voltage

Mozambique (50) 220/380
Niger (50) 220/380
Nigeria (50) 230/400
Rhodesia (50) 220/380; also 230/400
Senegal (50) 127/220
Sierra Leone (50) 230/400
Somalia (50) 220/440; also 110, 230
South Africa (50) 220/380; also others, some dc
Sudan (50) 240/415
Tanganyika (50) 230/400
Tunisia (50) 220/380; also others
Uganda (50) 240/415
Upper Volta (50) 220/380
Zaire (50) 220/380

Oceania:

Australia (50) 240/415; also others and dc
Fiji Islands (50) 240/415
Hawaii (60) 120/240
New Caledonia (50) 220/440
New Zealand (50) 230/400

Notes:

1. Abstracted from "Electric Power Abroad," issued 1963 by the Bureau of International Commerce of the US Department of Commerce. This pamphlet is obtainable from the Superintendent of Documents, US Government Printing Office, Washington, D.C. 20402.

2. The listings show electric (residential) power supplied in each country; as indicated, in very many cases other types of supply also exist to a greater or lesser extent. Therefore, for specific characteristics of the power supply of particular cities, reference should be made to "Electric Power Abroad." This pamphlet also gives additional details such as number of phases, number of wires to the residence, frequency stability, grounding regulations, and some data on types of commercial service.

3. In the United States in urban areas, the usual supply is 60-hertz 3-phase 120/208 volts; in less densely populated areas it is usually 120/240 volts, single phase, to each customer. Any other supplies, including dc, are rare and are becoming more so. Additional information for the US is given in the current edition of "Directory of Electric Utilities," published by McGraw-Hill Book Company, New York, N.Y.

4. All voltages in the table are ac except where specifically stated as dc. The latter are infrequent and in most cases are being replaced by ac. The lower voltages shown for ac, wye or delta ac, or for dc distribution lines, are used mostly for lighting and small appliances; the higher voltages are used for larger appliances.

TABLE 1—NEC CONDUCTOR APPLICATIONS AND INSULATIONS.

Trade Name	Type Letter	Maximum Operating Temperature	Application	Insulation	Outer Covering
Heat-resistant rubber	RH	75°C 167°F	Dry locations	Heat-resistant rubber	Moisture-resistant flame-retardant nonmetallic covering
Heat-resistant rubber	RHH	90°C 194°F	Dry locations	Heat-resistant rubber	Moisture-resistant flame-retardant nonmetallic covering
Moisture- and heat-resistant rubber	RHW	75°C 167°F	Dry and wet locations	Moisture- and heat-resistant rubber	Moisture-resistant flame-retardant nonmetallic covering
Heat-resistant latex rubber	RUH	75°C 167°F	Dry locations	90% unmilled grainless rubber	Moisture-resistant flame-retardant nonmetallic covering
Moisture-resistant latex rubber	RUW	60°C 140°F	Dry and wet locations	90% unmilled grainless rubber	Moisture-resistant flame-retardant nonmetallic covering
Thermoplastic	T	60°C 140°F	Dry locations	Flame-retardant thermoplastic compound	None
Moisture-resistant thermoplastic	TW	60°C 140°F	Dry and wet locations	Flame-retardant moisture-resistant thermoplastic	None
Moisture- and heat-resistant thermoplastic	THW	75°C 167°F	Dry and wet locations	Flame-retardant moisture- and heat-resistant thermoplastic	None
Heat-resistant thermoplastic	THHN	90°C 194°F	Dry locations	Flame-retardant heat-resistant thermoplastic	Nylon jacket
Moisture- and heat-resistant thermoplastic	THWN	75°C 107°F	Dry and wet locations	Flame-retardant, moisture- and heat-resistant thermoplastic	Nylon jacket
Moisture- and heat-resistant cross-linked synthetic polymer	XHHW	90°C 194°F	Dry locations	Flame-retardant cross-linked synthetic polymer	None
Extruded polytetrafluoroethylene	TFE	250°C 482°F	Dry locations	Extruded polytetrafluoroethylene	None
Silicone asbestos	SA	90°C 194°F	Dry locations	Silicone rubber	Asbestos
Flourinated Ethylene Polypylene	FEP	90°C 194°F	Dry locations	Fluorinated Ethylene Propylene	None
Varnished Cambric	V	85°C 185°F	Dry locations	Varnished Cambric	Nonmetallic Covering or lead sheath

POWER SUPPLY WIRING

Electric power supply (mains) wiring is usually controlled for public safety by local or state government boards, based in the USA primarily on the National Electric Code (NEC)* and the National Electric Safety Codes.* Brief extracts from some NEC requirements are given here for convenient reference.

Many products such as wire and cable, fuses, outlet boxes, appliances, etc., are governed within the USA by Underwriters Laboratories (UL) Standards, which specify the terminology used

* American National Standards Institute, Inc., ANSI Standard C1, prepared by the National Fire Protection Association.

* ANSI Standard C2.

for the various classes of an item as well as the safety requirements which must be met by UL approved items. Note that the overall performance of assemblies such as appliances, motors, radio, or television equipment is not covered by NEC or UL standards, which are primarily for personnel safety.

The following tables are provided.

Tables 1 and 2: NEC standard types of insulated wires and cables.

Tables 3 and 4: Allowable currents for several common flexible cords and cables.

Table 4: Derating factors to be applied for ambient temperatures above 30°C (86°F) and for more than 3 conductors in a cable or conduit.

Table 5: Motor starting currents, which determine the overcurrent protection requirements during the starting period.

Table 6: Motor full-load operating currents for usual conditions and speeds.

Guide to Use of Tables

Determine the total equipment load by adding the loads of the various individual items, estimating motor currents according to Table 6 if specific operating-current information is not otherwise available. Any load substantially bigger than this should be interrupted by an overload protective device; normally the next-larger standard fuse or circuit breaker is considered satisfactory.

TABLE 2—NEC FLEXIBLE-CORD DATA.

Trade Name*	Type Letter†	Size Range (AWG)	No. of Conductors	Insulation	Outer Covering
All-rubber parallel cord	SP-3	18–12	2 or 3	Rubber	Rubber
All-plastic parallel cord	SPT-3	18–10	2 or 3	Thermoplastic	Thermoplastic
Lamp cord	C	18–10	2 or more	Rubber	None
Twisted portable cord	PD	18–10	2 or more	Rubber	Cotton or rayon
Vacuum-cleaner cord	SV	18	2 or 3	Rubber	Rubber
Vacuum-cleaner cord	SVT	18–17	2 or 3	Thermoplastic	Thermoplastic
Junior hard-service cord	SJ	18–14	2–4	Rubber	Rubber
Junior hard-service cord	SJO	18–14	2–4	Rubber	Oil-resistant compound
Junior hard-service cord	SJT	18–14	2–4	Rubber or thermoplastic	Thermoplastic
Junior hard-service cord	SJTO	18–14	2–4	Rubber or thermoplastic	Oil-resistant thermo-plastic
Hard-service cord	S	18–2	2 or more	Rubber	Rubber
Hard-service cord	SO	18–2	2 or more	Rubber	Oil-resistant compound
Hard-service cord	ST	18–2	2 or more	Rubber or thermoplastic	Thermoplastic
Hard-service cord	STO	18–2	2 or more	Rubber or thermoplastic	Oil-resistant thermo-plastic
Rubber-jacketed heat-resistant cord	AFSJ	18–16	2 or 3	Impregnated asbestos	Rubber
Rubber-jacketed heat-resistant cord	AFS	18–14	2 or 3	Impregnated asbestos	Rubber
Heater cord	HPD	18–12	2–4	Rubber or thermoplastic with asbestos or all neoprene	Cotton or rayon
Rubber-jacketed heater cord	HSJ	18–16	2–4	Rubber or thermoplastic and asbestos or all neoprene	Cotton and rubber
Jacketed heater cord	HSJO	18–16	2–4	Rubber with asbestos or all neoprene	Cotton and oil-resistant compound
Jacketed heater cord	HS	14–12	2–4	Rubber with asbestos or all neoprene	Cotton and rubber or neoprene
Jacketed heater cord	HSO	14–12	2–4	Rubber with asbestos or all neoprene	Cotton and oil-resistant compound
Parallel heater cord	HPN	18–12	2 or 3	Thermosetting	Thermosetting

*All types shown are recommended for use in damp locations.

†"S" series cords may also be used in pendant applications.

TABLE 3—NEC CURRENT-CARRYING CAPACITY IN AMPERES OF FLEXIBLE CORDS.

Size AWG	Rubber TP, TS Thermoplastic TPT, TSP	Rubber C, PD, E, EO, EN, S, SO, SRD, SS, SSO, SV, SVO, SP Thermoplastic ET, ETT, ETLB, ETP, ST, STO, SRDT, SVT, SVTO, SPT		AFS, AFSJ, HPD, HSJ, HSJO, HS, HSO, HPN	Cotton* CFPD. Asbestos* AFC, AFPD
27†	0.5	—	—	—	—
18		7‡	10§	10	6
16		10‡	13§	15	8
14		15‡	18§	20	17
12		20‡	25§	30	23
10		25‡	30§	35	28
8		35‡	40§	—	—
6		45‡	55§	—	—
4		60‡	70§	—	—
2		80‡	95§	—	—

* Generally used in fixtures exposed to high temperatures, derated accordingly.

† Tinsel.

‡ Three conductor and other multiconductor cords connected so only 3 conductors are current carrying.

§ Two conductor and other multiconductor cords connected so only 2 conductors are current carrying.

Notes:

1. For not more than 3 current-carrying conductors in a cord. If 4 to 6 conductors are used, allowable capacity of each conductor shall be reduced to 80% of values for not more than 3 current-carrying conductors.

2. A conductor used for equipment grounding and a neutral conductor which carries only the unbalanced current from other conductors shall not be considered as current-carrying conductors.

3. Based on room temperature of 30°C (86°F).

Determine the total starting load by using the locked rotor currents computed from Table 5 and the steady-state currents for resistive devices. Make an additional allowance for any large quantities of tungsten lamps, starting transients, and high-inertia loads which will increase the duration of the starting period. The circuit overload protection must be designed to carry this load for the entire starting period. Time-lag fuses or time-delay circuit breakers are usually desirable.

Using the starting currents, determine the voltage drops in the supply circuit; thus be sure that the motor or other device terminal voltage will be adequate at start. Increase the size of the supply conductor or reduce the source impedance if necessary. From the starting and running currents determine the required size of supply conductors.

WIRING OF ELECTRONIC EQUIPMENT AND CHASSIS

There are few official standards for the internal wiring of electronic equipment and chassis. Nevertheless, the following points should be considered.

(**A**) Probable maximum continuous ambient temperature where the wiring is located.

(**B**) Allowable temperature rise of conductor surface under full-load conditions (determines minimum wire size).

(**C**) Maximum voltage to ground or to surrounding metal parts (determines required insulation thickness).

(**D**) Possibility of corona on high-voltage leads; some insulating materials deteriorate rapidly under corona conditions.

(**E**) Need for shield braid on some conductors to reduce noise pickup. Shields must be insulated if positive single-point grounding is to be attained.

(**F**) Skin effect on conductors carrying high radio-frequency currents.

(**G**) Vibration, shock, or relative motion of conductors during normal use of the equipment. Stranded or flexible conductors and adequate clamping or other tie-down conductors and cables may be essential.

(**H**) Wiring should be shielded from the direct heat radiation of high-temperature parts such as electron tubes and power resistors.

(**I**) Wire identification may be required for convenience in manufacture, installation, or servicing.

As a matter of expediency, most electronic equipment employs the smallest conveniently handled wire size (usually 20 to 24 AWG) for most wiring,

TABLE 4—NEC ALLOWABLE CURRENT-CARRYING CAPACITIES IN AMPERES OF CONDUCTORS.

	Copper-Conductor Insulation				Aluminum or Copper-Clad Aluminum Conductor Insulation		
Size AWG	RUW (14-2) T, TW, UF	RH, RHW RUH, (14-2) THW, THWN, XHHW	TA, TBS, SA, FEP, FEPB, RHH, THHN, XHHW*	TFE†	RUW (12-2) T, TW, UF	RH, RHW, RUH (12-2), THW, THWN, XHHW	TA, TBS, SA, RHH THHN, XHHW*
14	15	15	25‡	40	—	—	—
12	20	20	30‡	55	15	15	25§
10	30	30	40‡	75	25	25	30§
8	40	45	50	95	30	40	40
6	55	65	70	120	40	50	55
4	70	85	90	145	55	65	70
3	80	100	105	170	65	75	80
2	95	115	120	195	75	90	95
1	110	130	140	220	85	100	110
0	125	150	155	250	100	120	125
00	145	175	185	280	115	135	145
000	165	200	210	315	130	155	165
0000	195	230	235	370	155	180	185

Correction Factors for Higher Room Temperatures							
C°	°F						
40	104	0.82	0.88	0.91	—	0.82	0.88	0.91
45	113	0.71	0.82	0.87	—	0.71	0.82	0.87
50	122	0.58	0.75	0.82	—	0.58	0.75	0.82
55	131	0.41	0.67	0.76	—	0.41	0.67	0.76
60	140	—	0.58	0.71	.95	—	0.58	0.71

* Dry locations only.

† Nickel or nickel-coated copper only.

‡ For types FEP, FEPB, RHH, THHN, and XHHW, sizes 14, 12, 10, shall be the same as designated for RH, RHW, etc.

§ For types RHH, THHN, and XHHW, sizes 12 and 10 shall be the same as designated for RH, RHW, etc.

Notes:

1. Not more than 3 conductors in raceway or cable.
2. Based on room temperature of 30°C (86°F). See correction factors for higher temperatures.
3. Derating factors—more than 3 conductors in raceway or cable:

Number of conductors	4–6	7–24	25–42	>42
% of current capacity	80	70	60	50

with the larger conductors being installed only for circuits carrying currents greater than that permitted for the "general-use" wire size. With the trend toward compact solid-state integrated-circuit equipment, smaller wire sizes are being used. However, the reduction in wiring bundle size may be small unless the conductor insulation thickness (determined by voltage considerations) can be reduced.

Table 7 gives recommended current ratings for

TABLE 5—NEC MOTOR STARTING-CURRENT DATA.*

Code Letter	Kilovolt-Amperes per Horsepower with Locked Rotor
A	0–3.14
B	3.15–3.54
C	3.55–3.99
D	4.00–4.49
E	4.50–4.99
F	5.00–5.59
G	5.60–6.29
H	6.30–7.09
J	7.10–7.99
K	8.00–8.99
L	9.00–9.99
M	10.00–11.19
N	11.20–12.49
P	12.50–13.99
R	14.00–15.99
S	16.00–17.99
T	18.00–19.99
U	20.00–22.39
V	22.40–up

*Locked rotor currents of motors are useful in determining branch-circuit overcurrent protection requirements and voltage drop at start. These values are indicated by code letters on motor nameplate. Note NEMA standard.

copper and some aluminum based on a 45°C (40°C for wires smaller than 22 AWG) conductor temperature rise due to load current.

Table 7 may be used for the following temperature conditions.

Maximum Allowable Conductor Temperature °C	Maximum Ambient Temperature Around Wire °C	Typical Conductor and Insulation
105	60	Bare or tinned copper or aluminum; polyvinyl-chloride insulation
200	155	Silver-coated copper; FEP or PTFE with FEP jacket insulation*
260	215	Nickel-coated copper; PTFE insulation*

*FEP = Fluorinated ethylene propylene.
PTFE = Polytetrafluoroethylene.

A 60°C ambient temperature around the wiring (20°C internal temperature rise from 40°C (104°F) ambient around the equipment) is typical of some electronic equipment. If higher ambient temperatures, high power, or compact designs with electron tubes or magnetic-core components (except for very low power) are a factor, the temperature in the wiring space should be specifically determined.

TABLE 6—NEC MOTOR FULL-LOAD RUNNING CURRENTS IN AMPERES (USUAL CONDITIONS AND SPEEDS).

Horsepower	Single-Phase AC		3-Phase AC†			DC	
	115 V	230 V*	115 V	230 V*	460 V	120 V	240 V
$\frac{1}{6}$	4.4	2.2	—	—	—	—	—
$\frac{1}{4}$	5.8	2.9	—	—	—	—	—
$\frac{1}{3}$	7.2	3.6	—	—	—	3.1	1.6
$\frac{1}{2}$	9.8	4.9	4	2	1	4.1	2.0
$\frac{3}{4}$	13.8	6.9	5.6	2.8	1.4	5.4	2.7
1	16	8	7.2	3.6	1.8	7.6	3.8
$1\frac{1}{2}$	20	10	10.4	5.2	2.6	9.5	4.7
2	24	12	13.6	6.8	3.4	13.2	6.6
3	34	17	—	9.6	4.8	17	8.5
5	56	28	—	15.2	7.6	25	12.2
$7\frac{1}{2}$	80	40	—	22	11	40	20
10	100	50	—	28	14	58	29
						76	38

*For 208 V, multiply by 1.1; for 200 V, multiply by 1.15.
†Induction type, squirrel cage and wound rotor.

TABLE 7—RECOMMENDED CURRENT RATINGS (CONTINUOUS DUTY) FOR ELECTRONIC EQUIPMENT AND CHASSIS WIRING.

| Wire Size | | Copper Conductor (100°C) Nominal Resistance (Ohms/1000 ft) | Maximum Current in Amperes | | | |
| | | | Copper Wire | | Aluminum Wire | |
AWG	Circular Mils		Wiring in Free Air	Wiring Confined	Wiring in Free Air	Wiring Confined
32	63.2	188.0	0.53	0.32		
30	100.5	116.0	0.86	0.52		
28	159.8	72.0	1.4	0.83		
26	254.1	45.2	2.2	1.3		
24	404.0	28.4	3.5	2.1		
22	642.4	22.0	7.0	5.0		
20	1022	13.7	11.0	7.5		
18	1624	6.50	16	10		
16	2583	5.15	22	13		
14	4107	3.20	32	17		
12	6530	2.02	41	23		
10	10 380	1.31	55	33		
8	16 510	0.734	73	46	60	36
6	26 250	0.459	101	60	83	50
4	41 740	0.290	135	80	108	66
2	66 370	0.185	181	100	152	82
1	83 690	0.151	211	125	174	105
0	105 500	0.117	245	150	202	123
00	133 100	0.092	283	175	235	145
000	167 800	0.074	328	200	266	162
0000	211 600	0.059	380	225	303	190

"Wiring Confined" ratings are based on 15 or more wires in a bundle, with the sum of all the actual load currents of the bundled wires not exceeding 20% of the permitted "Wiring Confined" sum total carrying capacity of the bundled wires. These ratings approximate 60% of the free-air ratings (with some variations due to rounding). They should be used for wire in harnesses, cable, conduit, and general chassis conditions. Bundles of fewer than 15 wires may have the allowable sum of the load currents increased as the bundle approaches the single-wire condition.

RESISTANCE CHANGE WITH TEMPERATURE

The resistance of most conductor materials changes with temperature. Table 7 shows the copper wire resistance at 100°C. Correction factors must be applied to determine the resistance at other temperatures, or for other materials. Thus, from Table 7 determine the copper wire resistance at 100°C (multiply by the conversion factor m of Table 11 for other materials). Use the equation

$$R_t = R_r m[1 + K(t - 100)]$$

where R_t = resistance at desired temperature t, R_r = resistance at 100°C for copper (Table 7), m = material factor for 100°C resistance value (Table 8), K = correction factor (Table 8), and t = desired temperature (°C).

TABLE 8—TEMPERATURE AND MATERIALS CORRECTION FACTORS.

Conductor Material	Material Factor m	Correction Factor K
Soft copper	1.00	0.0039
Hard copper	1.03	0.0038
Copper-clad steel:		
30% conductivity	3.47	0.0044
40% conductivity	2.56	0.0041
Aluminum	1.64	0.0039
Nickel	5.28	0.0047
Nickel-clad copper:		
10%	1.07	0.0038
30%	1.35	0.0036
Silver	0.94	0.0038

WIRE IDENTIFICATION

In a complex wiring assembly, or if both ends of a wire cannot be seen from one station, a means of identifying each lead simplifies manufacture, installation, and servicing. Common identification methods are:

(**A**) Tag each end of a lead with an assigned designation (an alternative method is to print the designation at frequent intervals along the wire insulation).

(**B**) Color code the wires. The wire insulation may be a solid color, color stripes may be spiraled around the wire, or the name of the color (or its numerical code equivalent) may be stamped at frequent intervals along the wire.

Color Coding

The commonly used colors and their numerical codes are:

0	Black	5	Green
1	Brown	6	Blue
2	Red	7	Violet (purple)
3	Orange	8	Gray (slate)
4	Yellow	9	White

While spiral stripes can be applied on top of any basic insulation color, under less favorable viewing conditions it is difficult to distinguish some colors from the basic insulation color. Identification may be slow and subject to error. One or two (sometimes three) colored stripes on a white basic insulation is the preferred combination. To minimize identification errors, the first stripe is made wider than the second (or third), and some rules require that the second stripe be of higher numerical code than the first stripe. If the required variety of wire color codes is not great, the preceding guides should be followed.

Table 9 gives a standard color code used to distinguish by function the various leads in electronic circuits.

In manufacturing practice it is preferred that, at any harness breakout point, all wires of the same color code be connected to the same terminal at that location. When this rule and the wire color coding of Table 9 are both applicable, additional tracers may be used to supplement the primary coding of Table 9.

DIAMETER OF CIRCLE ENCLOSING A GIVEN NUMBER OF SMALLER CIRCLES*

Four of many possible compact arrangements of circles within a circle are shown in Fig. 3. To

*J. Dutka, "How Many Wires Can Be Packed into a Circular Conduit," *Machinery's Handbook*, Industrial Press, Inc., New York City; October 1956.

TABLE 9—COLORS FOR WIRE IDENTIFICATION BY FUNCTION.

Function	Color	Identification No.
Grounds, grounded elements	Black	0
Heaters or filaments	Brown	1
Power supply B+	Red	2
Screen grids	Orange	3
Cathodes and transistor emitters*†	Yellow	4
Control grids and transistor bases†	Green	5
Anodes (plates) and transistor collectors*†	Blue	6
Power supply, negative (−)	Violet (purple)	7
Ac power lines	Gray (slate)	8

*Applies to diodes, semiconductor elements, photoelectric cells, mercury-arc rectifiers, and other elements with operation similar to vacuum tubes and transistors.

†Applies to all types of gas tubes with operation similar to vacuum tubes.

determine the diameter of the smallest enclosing circle for a particular number of enclosed circles all of the same size, three factors that influence the size of the enclosing circle should be considered.

Arrangement of Center or Core Circles

The four most common arrangements of center or core circles are shown in cross section in Fig. 3. It may seem that Fig. 3A would require the smallest enclosing circle for a given number of enclosed circles, but this is not always the case since the most compact arrangement will depend in part on the number of circles to be enclosed.

Diameter of Enclosing Circle When Outer Layer of Circles is Complete

Successive, complete "layers" of circles may be placed around each of the central cores of 1, 2, 3, or 4 circles. The number of circles contained in

A   B   C   D

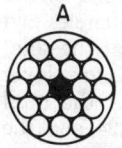 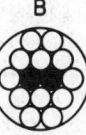  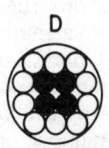

Fig. 3—Arrangements of circles within a circle. *Reprinted with permission from Machinery's Handbook, 17th Edition, Industrial Press, Inc., New York City.*

TABLE 10A—NUMBER OF CIRCLES CONTAINED IN COMPLETE LAYERS OF CIRCLES AND DIAMETER OF ENCLOSING CIRCLE.

				Number of Circles in Center Pattern				
	1	2	3	4	1	2	3	4
No. Complete Layers Over Core, n	Arrangement of Circles in Center Pattern (see Fig. 3)							
	"A"	"B"	"C"	"D"	"A"	"B"	"C"	"D"
	Number of Circles, N, Enclosed				Diameter, D, of Enclosing Circle*			
0	1	2	3	4	d	$2d$	$2.155d$	$2.414d$
1	7	10	12	14	$3d$	$4d$	$4.055d$	$4.386d$
2	19	24	27	30	$5d$	$6d$	$6.033d$	$6.379d$
3	37	44	48	52	$7d$	$8d$	$8.024d$	$8.375d$
4	61	70	75	80	$9d$	$10d$	$10.018d$	$10.373d$
5	91	102	108	114	$11d$	$12d$	$12.015d$	$12.372d$
n	†	†	†	†	†	†	†	†

*Diameter D is given in terms of d, the diameter of the enclosed circles.

†For n complete layers over core, the number of enclosed circles N for "A" center pattern is $3n^2+3n+1$; for "B," $3n^2+5n+2$; for "C," $3n^2+6n+3$; for "D," $3n^2+7n+4$; while the diameter D of the enclosing circle for "A" center pattern is $(2n+1)d$; for "B," $(2n+2)d$; for "C," $[1+2(n^2+n+\frac{1}{3})^{1/2}]d$; and for "D," $[1+(4n^2+5.644n+2)^{1/2}]d$.

Reprinted with permission from Machinery's Handbook, 17th Edition, Industrial Press, Inc., New York City.

arrangements of complete "layers" around a central core of circles, as well as in the diameter of the enclosing circle, may be obtained from Table 10A. Thus, for example, Fig. 3A has a total of 19 circles arranged in two complete "layers" around a central core consisting of one circle; this agrees with the data shown in the left half of Table 10A for $n=2$.

To determine the diameter of the enclosing circle, the data in the right half of Table 10A are used. Thus, for $n=2$ and an "A" pattern, the diameter D is 5 times the diameter d of the enclosed circles.

Diameter of Enclosing Circle When Outer Layer of Circles is not Complete

In most cases it is possible to reduce the size of the enclosing circle from that required if the outer layer were complete. Thus, for example, Fig. 3B shows that the central core consisting of 2 circles is surrounded by 1 complete layer of 8 circles and 1 partial outer layer of 4 circles so that the total number of circles enclosed is 14. If the outer layer was complete, then (from Table 10A) the total number of enclosed circles would be 24 and the diameter of the enclosing circle would be $6d$; however, since the outer layer is composed of only 4 circles out of a possible 14 for a complete second

layer, a smaller diameter of enclosing circle may be used. Table 10B shows that for a total of 14 enclosed circles arranged in a "B" pattern with the outer layer of circles incomplete, the diameter for the enclosing circle is $4.606d$.

Table 10B can be used to determine the smallest enclosing circle for a given number of circles to be enclosed by direct comparison of the "A," "B," and "C" columns. For data outside the range of Table 10B, use the equations in Dr. Dutka's article.

Approximate Equation When Number of Enclosed Circles is Large

When a large number of circles is to be enclosed, the arrangement of the center circles has little effect on the diameter of the enclosing circle. For numbers of circles greater than 10 000, the diameter of the enclosing circle may be calculated within 2 percent from the equation

$$D=d[1+(N/0.907)^{1/2}]$$

where $D=$ diameter of enclosing circle, $d=$ diameter of enclosed circles, and $N=$ number of enclosed circles.

TABLE 10B—FACTORS FOR DETERMINING DIAMETER, D, OF SMALLEST ENCLOSING CIRCLE FOR VARIOUS NUMBERS, N, OF ENCLOSED CIRCLES.*

No. N	Center Circle Pattern "A"	"B"	"C"	No. N	Center Circle Pattern "A"	"B"	"C"	No. N	Center Circle Pattern "A"	"B"	"C"
	Diameter Factor K				Diameter Factor K				Diameter Factor K		
2	3	2	—	34	7	7.083	7.110	66	9.718	9.544	9.326
3	3	2.732	2.155	35	7	7.245	7.110	67	9.718	9.544	9.326
4	3	2.732	3.309	36	7	7.245	7.110	68	9.718	9.544	9.326
5	3	3.646	3.309	37	7	7.245	7.429	69	9.718	9.660	9.326
6	3	3.646	3.309	38	7.928	7.245	7.429	70	9.718	9.660	10.018
7	3	3.646	4.055	39	7.928	7.557	7.429	71	9.718	9.888	10.018
8	4.464	3.646	4.055	40	7.928	7.557	7.429	72	9.718	9.888	10.018
9	4.464	4	4.055	41	7.928	7.557	7.429	73	9.718	9.888	10.018
10	4.464	4	4.055	42	7.928	7.557	7.429	74	10.165	9.888	10.018
11	4.464	4.606	4.055	43	7.928	8	8.024	75	10.165	10	10.018
12	4.464	4.606	4.055	44	8.211	8	8.024	76	10.165	10	10.238
13	4.464	4.606	5.163	45	8.211	8	8.024	77	10.165	10.539	10.238
14	5	4.606	5.163	46	8.211	8	8.024	78	10.165	10.539	10.238
15	5	5.359	5.163	47	8.211	8	8.024	79	10.165	10.539	10.452
16	5	5.359	5.163	48	8.211	8	8.024	80	10.165	10.539	10.452
17	5	5.359	5.163	49	8.211	8.550	8.572	81	10.165	10.539	10.452
18	5	5.359	5.163	50	8.211	8.550	8.572	82	10.165	10.539	10.452
19	5	5.583	5.619	51	8.211	8.550	8.572	83	10.165	10.539	10.452
20	6.292	5.583	5.619	52	8.211	8.550	8.572	84	10.165	10.539	10.452
21	6.292	5.583	5.619	53	8.211	8.810	8.572	85	10.165	10.644	10.866
22	6.292	5.583	6.033	54	8.211	8.810	8.572	86	11	10.644	10.866
23	6.292	6	6.033	55	8.211	8.810	9.083	87	11	10.644	10.866
24	6.292	6	6.033	56	9	8.810	9.083	88	11	10.644	10.866
25	6.292	6.196	6.033	57	9	8.937	9.083	89	11	10.849	10.866
26	6.292	6.196	6.033	58	9	8.937	9.083	90	11	10.849	10.866
27	6.292	6.568	6.033	59	9	8.937	9.083	91	11	10.849	11.214
28	6.292	6.568	6.773	60	9	8.937	9.083	92	11.392	10.849	11.214
29	6.292	6.568	6.773	61	9	9.185	9.083	93	11.392	11.149	11.214
30	6.292	6.568	6.773	62	9.718	9.185	9.083	94	11.392	11.149	11.214
31	6.292	7.083	7.110	63	9.718	9.185	9.083	95	11.392	11.149	11.214
32	7	7.083	7.110	64	9.718	9.185	9.326	96	11.392	11.149	11.214
33	7	7.083	7.110	65	9.718	9.544	9.326	97	11.392	11.440	11.214

*The diameter D of the enclosing circle is equal to the diameter factor, K, multiplied by d, the diameter of the enclosed circles, or $D = K \times d$. For example, if the number of circles to be enclosed, N, is 12, and the center circle arrangement is "C," then for $d = 1\frac{1}{2}$ inches, $D = 4.055 \times 1\frac{1}{2} = 6.083$ inches. *Reprinted with permission from Machinery's Handbook, 17th Edition, Industrial Press, Inc., New York City.*

TORQUE AND HORSEPOWER

Torque varies directly with power and inversely with rotating speed of the shaft, or

$$T = KP/N$$

where T = torque in inch-pounds, P = horsepower, N = revolutions/minute, and K = 63 000 (constant).

TRANSMISSION-LINE SAG CALCULATIONS*

For transmission-line work, with towers on the same or slightly different levels, the cables are assumed to take the form of a parabola instead of

*Reprinted by permission from "Transmission Towers," American Bridge Company, Pittsburgh, Pa.; 1923: p. 70.

their actual form of a catenary. The error is negligible and the computations are much simplified. In calculating sags, the changes in cables due to variations in load and temperature must be considered.

Supports at Same Level (Fig. 4)

The equations used in calculating sags are

$$H = WL^2/8S$$

$$S = WL^2/8H = [(L_c-L)3L/8]^{1/2}$$

$$L_c = L + 8S^2/3L$$

where L = length of span in feet, L_c = length of cable in feet, S = sag of cable at center of span in feet, H = tension in cable at center of span in pounds = horizontal component of the tension at any point, and W = weight of cable in pounds per lineal foot.

If cables are subject to wind and ice loads, W = the algebraic sum of the loads. That is, for ice on cables, W = weight of cables plus weight of ice; for wind on bare or ice-covered cables, W = the square root of the sum of the squares of the vertical and horizontal loads.

For any intermediate point at a distance x from the center of the span, the sag is

$$S_x = S(1-4x^2/L^2).$$

Supports at Different Levels (Fig. 5)

$$S = S_0 = WL_0^2 \cos a/8T$$

$$= WL^2/8T \cos a$$

$$S_1 = WL_1^2/8H$$

$$S_2 = WL_2^2/8H$$

$$L_1/2 = L/2 - (hH \cos a/WL)$$

$$L_2/2 = L/2 + (hH \cos a/WL)$$

$$L_c = L + \tfrac{4}{3}[(S_1^2/L_1) + (S_2^2/L_2)]$$

where W = weight of cable in pounds per lineal foot between supports or in direction of L_0, and T = tension in cable direction parallel with line between supports.

The change l in length of cable L_c for varying temperature is found by multiplying the number

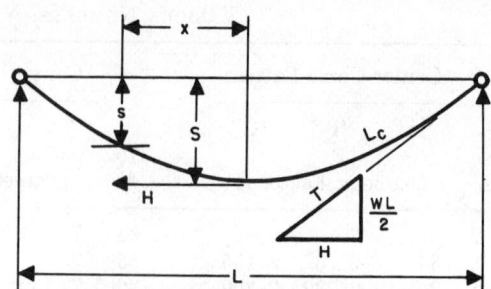

Fig. 4—Supports at same elevations.

of degrees n by the length of the cable in feet times the coefficient of linear expansion per foot per degree fahrenheit c. This is*

$$l = L_c \times n \times c.$$

A short approximate method for determining sags under varying temperatures and loadings that is close enough for all ordinary line work is as follows:

(A) Determine sag of cable with maximum stress under maximum load of lowest temperature occurring at the time of maximum load, and find length of cable with this sag.

(B) Find length of cable at the temperature for which the sag is required.

(C) Assume a certain reduced tension in the cable at the temperature and under the loading combination for which the sag is required; then find the decrease in length of the cable due to the decrease of the stress from its maximum.

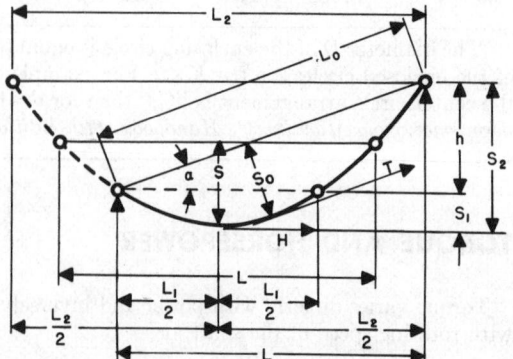

Fig. 5—Supports at different elevations.

*Temperature coefficient of linear expansion is given in Chapter 4, Table 40.

(**D**) Combine the algebraic sum of (**B**) and (**C**) with (**A**) to get the length of the cable under the desired conditions; from this length the sag and tension can be determined.

(**E**) If this tension agrees with that assumed in (**C**), the sag in (**D**) is correct. If it does not agree, another assumption of tension in (**C**) must be made and the process repeated until (**C**) and (**D**) agree.

STRUCTURAL STANDARDS FOR STEEL RADIO TOWERS*

Material

(**A**) Structural steel shall conform to American Society for Testing Materials "Standard Specifications for Steel for Bridges and Buildings," Serial Designation A-7, as amended to date.

(**B**) Steel pipe shall conform to American Society for Testing Materials standard specifications either for electric-resistance welded steel pipe, Grade A or Grade B, Serial Designation A-135, or for welded and seamless steel pipe, Grade A or Grade B, Serial Designation A-53, each as amended to date.

Loading

(**A**) 20-pound design: Structures up to 600 feet in height (unless they are to be located within city limits) shall be designed for a horizontal wind pressure of 20 pounds/foot² on flat surfaces and 13.3 pounds/foot² on cylindrical surfaces.

(**B**) 30-pound design: Structures more than 600 feet in height and those of any height to be located within city limits shall be designed for a horizontal wind pressure of 30 pounds/foot² on flat surfaces and 20 pounds/foot² on cylindrical surfaces.

(**C**) Other designs: Certain structures may be designed to resist loads greater than those described in (**A**) and (**B**). Figure 1 of American Standard A58.1–1955 shows sections of the United States where greater wind pressures may occur. In all such cases, the pressure on cylindrical surfaces shall be computed as being $\frac{2}{3}$ of that specified for flat surfaces.

*Abstracted from "American Standard Minimum Design Loads in Buildings and Other Structures, A58.1–1955," American National Standards Institute, Inc., 1430 Broadway, New York, New York 10018. Also from Electronics Industries Association Standard TR–116; October 1949. Sections on manufacture and workmanship, finish, and plans and marking of the standard are not reproduced here. The section on "Wind Velocities and Pressures" is not part of the standard.

(**D**) For open-face (latticed) structures of square cross section, the wind pressure normal to one face shall be applied to 2.20 times the normal projected area of all members in one face, or 2.40 times the normal projected area of one face for wind applied to one corner. For open-face (latticed) structures of triangular cross section, the wind pressure normal to one face shall be applied to 2.00 times the normal projected area of all members in one face, or 1.50 times the normal projected area for wind parallel to one face. For closed-face (solid) structures, the wind pressure shall be applied to 1.00 times the normal projected area for square or rectangular shape, 0.80 for hexagonal or octagonal shape, and 0.60 for round or elliptical shape.

(**E**) Provisions shall be made for all supplementary loadings caused by the attachment of guys, antennas, transmission and power lines, ladders, etc. The pressure shall be as described for the respective designs and shall be applied to the projected area of the construction.

(**F**) The total load specified above shall be applied to the structure in the directions that will cause the maximum stress in the various members.

(**G**) The dead weight of the structure, and all materials attached thereto, shall be included.

Unit Stresses

(**A**) All parts of the structure shall be so designed that the unit stresses resulting from the specified loads shall not exceed the following values in pounds/inch²:

Axial tension on net section = 20 000 pounds/inch².

Axial compression on gross section:

For members with values of L/R not greater than 120,

$$= 17\,000 - 0.485 L^2/R^2 \text{ pounds/inch}^2.$$

For members with values of L/R greater than 120,

$$= \frac{18\,000}{1 + L^2/18\,000 R^2} \text{ pounds/inch}^2$$

where L = unbraced length of the member, and R = corresponding radius of gyration, both in inches.

Maximum L/R for main leg members = 140
Maximum L/R for other compression members with calculated stress = 200
Maximum L/R for members with no calculated stress = 250
Bending on extreme fibres = 20 000 pounds/inch²
Single shear on bolts = 13 500 pounds/inch²
Double shear on bolts = 27 000 pounds/inch²

Bearing on bolts (single shear) = 30 000 pounds/inch²

Bearing on bolts (double shear) = 30 000 pounds/inch²

Tension on bolts and other threaded parts, on nominal area at root of thread = 16 000 pounds/inch².

Members subject to both axial and bending stresses shall be so designed that the calculated unit axial stress divided by the allowable unit axial stress, plus the calculated unit bending stress, divided by the allowable unit bending stress, shall not exceed unity.

(**B**) Minimum thickness of material for structural members:

Painted structural angles and plates = $\frac{3}{16}$ inch
Hot-dip galvanized structural angles and plates = $\frac{1}{8}$ inch
Other structural members to mill minimum for standard shapes.

(**C**) Where materials of higher quality than specified under "Material" above are used, the above unit stresses may be modified. The modified unit stresses must provide the same factor of safety based on the yield point of the materials.

Foundations

(**A**) Standard foundations shall be designed for a soil pressure not to exceed 4000 pounds/foot² under the specified loading. In uplift, the foundations shall be designed to resist 100 percent more than the specified loading, assuming that the base of the pier will engage the frustum of an inverted pyramid of earth whose sides form an angle of 30 degrees with the vertical. Earth shall be considered to weigh 100 pounds/foot³ and concrete 140 pounds/foot³.

(**B**) Foundation plans shall ordinarily show standard foundations as defined in (**A**). Where the actual soil conditions are not normal, requiring some modification in the standard design, and complete soil information is provided to the manufacturer by the purchaser, the foundation plan shall show the required design.

(**C**) Under conditions requiring special engineering such as pile construction, roof installations, etc., the manufacturer shall provide the necessary information so that proper foundations can be designed by the purchaser's engineer or architect.

(**D**) In the design of guy anchors subject to submersion, the upward pressure of the water should be taken into account.

Wind Velocities and Pressures

Actual Velocity V_a* (miles/hour)	Indicated Velocity V_i (miles/hour)		Pressure P (pounds/foot²) Projected Areas†	
	3-cup Anemometer	4-cup Anemometer	Cylindrical Surfaces ($P=0.0025V_a^2$)	Flat Surfaces ($P=0.0042V_a^2$)
10	9	10	0.25	0.42
20	20	23	1.0	1.7
30	31	36	2.3	3.8
40	42	50	4.0	6.7
50	54	64	6.3	10.5
60	65	77	9.0	15.1
70	76	91	12.3	20.6
80	88	105	16.0	26.8
90	99	119	20.3	34.0
100	110	132	25.0	42.0
110	121	146	30.3	50.8
120	133	160	36.0	60.5
130	144	173	42.3	71.0
140	155	187	49.0	82.3
150	167	201	56.3	94.5

*Although wind velocities are measured with cup anemometers, all data published by the US Weather Bureau since January 1932 includes instrumental corrections and are actual velocities. Prior to 1932 indicated velocities were published.

In calculating pressures on structures, the "fastest-single-mile velocities" published by the Weather Bureau should be multiplied by a gust factor of 1.3 to obtain the maximum instantaneous actual velocities. See p. 48-3 for fastest-single-mile records at various places in the United States and Canada.

† The American Bridge Company equations given here are based on a ratio of 25/42 for pressures on cylindrical and flat surfaces, respectively, while the Electronics Industries Association specifies a ratio of 2/3. The actual ratio varies in a complex manner with Reynolds number, shape, and size of the exposed object.

VIBRATION AND SHOCK ISOLATION

Symbols

$b=$ damping factor

$d=$ static deflection in inches

$E=$ relative transmissibility

$\quad=$ (force transmitted by isolators)/(force transmitted by rigid mountings)

$F=$ force in pounds

$F_0=$ peak force in pounds

$f=$ frequency in hertz

$f_0=$ resonant frequency of system in hertz

$G=$ acceleration of gravity ≈ 386 inches per second^2

$g=$ peak acceleration in dimensionless gravitational units $\quad=\ddot{X}_0/G$

$j=(-1)^{1/2}$, vector operator

$k=$ stiffness constant; force required to compress or extend isolators unit distance in pounds per inch

$r=$ coefficient of viscous damping in pounds per inch per second

$t=$ time in seconds

$W=$ weight in pounds

$x=$ displacement from equilibrium position in inches

$X_0=$ peak displacement in inches

$\dot{x}=$ velocity in inches per second $\quad=dx/dt$

$\dot{X}_0=$ peak velocity in inches per second

$\ddot{x}=$ acceleration in inches per second^2 $\quad=d^2x/dt^2$

$\ddot{X}_0=$ peak acceleration in inches per second^2

$\phi=$ phase angle in radians

$\omega=$ angular velocity in radians per second $\quad=2\pi f.$

Equations

The following relations apply to simple harmonic motion in systems with one degree of freedom. Although actual vibration is usually more complex, the equations provide useful approximations for practical purposes.

$$F = W(\ddot{x}/G) \tag{1}$$

$$F_0 = Wg \tag{2}$$

$$x = X_0 \sin(\omega t + \phi) \tag{3}$$

$$X_0 = 9.77g/f^2 \tag{4}$$

$$\dot{X}_0 = \omega X_0 = 6.28 f X_0 = 61.4g/f \tag{5}$$

$$\ddot{X}_0 = \omega^2 X_0 = 39.5 f^2 X_0 = 386g \tag{6}$$

$$E = \left| \frac{r - j(k/\omega)}{r + j[(\omega W/G) - k/\omega]} \right| \tag{7}$$

$$f_0 = 3.13(k/W)^{1/2} \tag{8}$$

$$b = 9.77r/(kW)^{1/2}. \tag{9}$$

For critical damping, $b=1$.

Neglecting dissipation $(b=0)$, or at $f/f_0 = (2)^{1/2}$ for any degree of damping

$$E = \left| \frac{1}{(f/f_0)^2 - 1} \right|. \tag{10}$$

When damping is neglected

$$k = W/d \tag{11}$$

$$f_0 = 3.13/d^{1/2} \tag{12}$$

$$E = 9.77/(df^2 - 9.77). \tag{13}$$

Acceleration

The intensity of vibratory forces is often defined in terms of g values. From (2), it is apparent, for example, that a peak acceleration of $10g$ on a body will result in a reactionary force by the body equal to 10 times its weight.

If an object is mounted on vibration isolators, the accelerations of the vehicle are transmitted to the object (or vice versa) in an amplitude and phase that depend on the elastic flexing of the isolators in the direction in which the accelerations (dynamic forces) are applied.

Magnitudes

The relations between X_0, $\dot{X}_0$, $\ddot{X}_0$, and f are shown in Fig. 6. Any two of these parameters applied to the graph locates the other two. For example, suppose $f=10$ hertz and peak displacement $X_0=1$ inch. From Fig. 6, peak velocity $\dot{X}_0 = 63$ inches per second and peak acceleration $\ddot{X}_0 = 10g$.

Natural Frequency

Neglecting damping, the natural frequency f_0 of vibration of an isolated system in the vertical direction can be calculated from (12) from the static deflection of the mounts. For example, suppose an object at rest causes a 0.25-inch deflection of its supporting springs. Then

$$f_0 = 3.13/(0.25)^{1/2} = 6.3 \text{ hertz}.$$

Resonance

In Fig. 7, E is plotted against f/f_0 for various damping factors. Note that resonance occurs when $f_0 \approx f$ and that the vibratory forces are then increased by the isolators. To reduce vibration, f_0 must be less than $0.7f$ and it should be as small as $0.3f$ for good isolation.

It is not possible to secure good isolation at all vibrational frequencies in vehicles and similar environments where several different and varying exciting frequencies are present and where the isolators may have to withstand shock as well as vibration. In such cases, f_0 is often selected as about 1.5 to 2 times the predominant f. Vibration in typical vehicles is shown in Table 11.

Although all supporting structures have compliance and may reduce the effects of vibration and shock, the apparent stiffness of many "rigid" mountings is merely a matter of degree, and in conjunction with the supported mass, they can also give rise to resonance effects, thus magnifying the amplitude of certain vibrations.

Damping

Damping is desirable to reduce vibration amplitude when the exciting frequency is in the vicinity of f_0. This occurs occasionally in most installations. Any isolator that absorbs energy provides damping.

It is seldom practical to introduce damping as an independent variable in the design of vibration isolators for relatively small objects. The usual practice is to rely on the inherent damping characteristics of the rubber or other elastic material employed in the mounting. Damping achieved in this way seldom exceeds 5 percent of the amount needed to produce a critically damped system. In vibration isolators for large objects, such as variable-speed engines, the system often can be designed to produce nearly critical damping by employing fluid dashpots or similar devices.

Practical Application

Vibration can be accurately precalculated only for the simplest systems. In other cases the actual vibration should be measured on experimental assemblies using electrical vibration pickups. Complex vibration is often described by a plot of the g values against frequency. These plots usually show several frequencies at which the largest acceler-

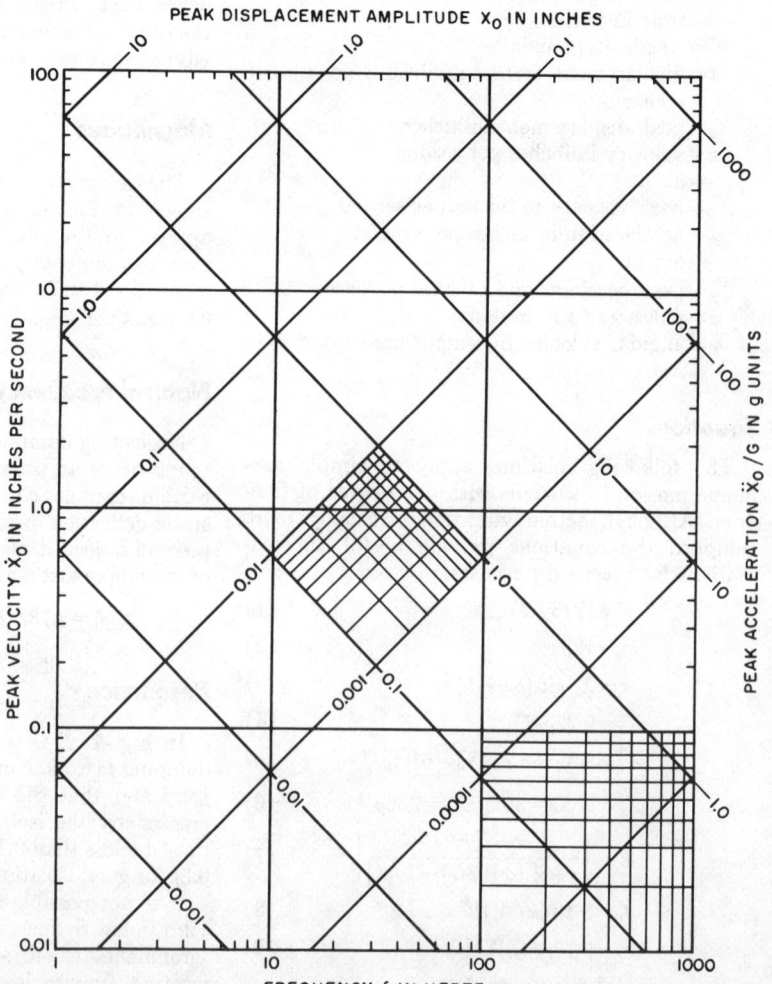

Fig. 6—Relation of frequency and peak values of velocity, displacement, and acceleration.

ations are present. The patterns will vary from place to place in a complicated structure and will also depend on the direction in which the acceleration is measured.

After measuring and plotting vibration in this way, attention can be devoted to reduction of the predominant components using the equations and principles given above as guides in selecting the size, stiffness, damping characteristics, and location of isolators.

Shock

In many practical situations, vibration and shock occur simultaneously. The design of isolators for vibration should anticipate the effects of shock and vice versa.

When heavy shock is applied to a system using vibration isolators, there is usually a definite deflection at which the isolators snub or at which their stiffness suddenly becomes much greater. These actions may amplify the shock forces. To reduce this effect, it is generally desirable to use isolators that have smoothly increasing stiffness with increasing deflection.

Shock protection is improved by isolators that permit large deflections in all directions before the protected equipment is snubbed or strikes neighboring apparatus. The amplitude of vibration resulting from shock can be reduced by employing isolators that absorb energy and thus damp oscillatory movement.

Probabilities of damage to the apparatus itself

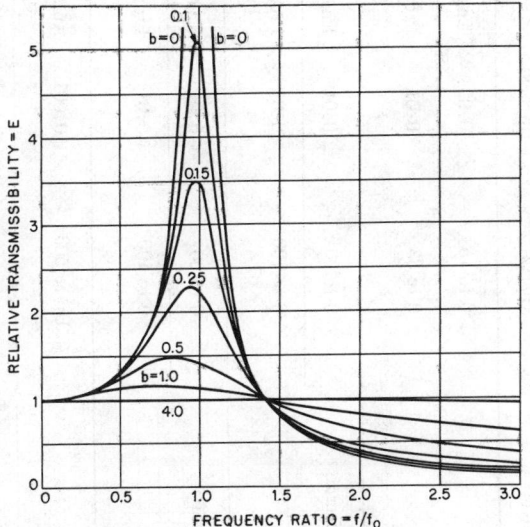

Fig. 7—Relative transmissibility E as a function of the frequency ratio f/f_0 for various amounts of damping b. *By permission from "Vibration Analysis," by N. O. Myklestad. ©1944, McGraw-Hill Book Company, Inc.*

from impact shock can be minimized by:

(**A**) Making the weight of equipment components as small as possible and the strength of structural members as great as possible.

(**B**) Distributing rather than concentrating the weights of equipment components and avoiding rigid connections between components.

(**C**) Employing structural members that have high ratios of stiffness to weight, such as tubes, *I* beams, etc.

(**D**) Avoiding, so far as is practical, stress concentrations at joints, supports, discontinuities, etc.

(**E**) Using materials such as steel that yield rather than rupture under high stress.

US GRAPHIC SYMBOLS

USA Standard Graphic Symbols for Electrical and Electronics Diagrams cover both the communication and power fields. Symbols of primary interest to communications workers are in Fig. 8. They have been abstracted from ANSI Standard Y32.2-1970.* Symbols which also agree with Recommendation No. 117 of the International Electrotechnical Commission are indicated by IEC.

Diagram Types

Block diagrams consist of simple rectangles and circles with names or other designations within or adjacent to them to show the general arrangement of apparatus to perform desired functions. The direction of power or signal flow is often indicated by arrows near the connecting lines or arrowheads on the lines.

Schematic diagrams show all major components and their interconnections. Single-line diagrams, as indicated by that name, use single lines to interconnect components even though two or more conductors are actually required. It is a shorthand form of schematic diagram. It is always used for waveguide diagrams.

Wiring diagrams are complete in that all conductors are shown and all terminal identifications are included. The contact numbers on electron-tube sockets, colors of transformer leads, rotors of adjustable capacitors, and other terminal markings, are shown so that a workman having no knowledge of the operation of the equipment can wire it properly.

Orientation

Graphic symbols are not considered to be coarse pictures of specific pieces of equipment but are

* Obtainable from the American National Standards Institute, Inc., 1430 Broadway, New York, N.Y. 10018.

TABLE 11—VIBRATION IN TYPICAL VEHICLES.

Vehicle	Range of Frequencies (hertz)	Approximate Peak Amplitude (inches)	Nature of Excitation	Usual Choice of Isolator Resonant Frequency
Ships	0 to 15	0.02	Engine vibration in diesel or reciprocating steam drive	6 hertz for vibration isolation in commercial vessels. 27 to 30 hertz for shock isolation on naval vessels. These latter mounts amplify most vibrations to some extent.
	0 to 33	0.01	Propeller-blade frequency = (propeller rpm) × (number of blades)/60	
Piston-engine aircraft	0 to 60	0.01	Engine vibrations	Above 20 hertz. Amplitude of vibrations varies with location in aircraft. Landing shock can be neglected.
	0 to 100	0.01	Propeller vibrations. Aerodynamic vibrations due to buffeting	
Turboprop aircraft	0 to 60	0.01	Engine vibrations = (engine rpm)/60 Also aerodynamic vibrations due to buffeting and turbulence	9 hertz
	0 to 100	0.01	Propeller vibrations	
Jet aircraft	Up to 500	0.001	Audible noise frequencies due to jet wake and combustion turbulence; very little engine vibration	9 hertz

	Frequency (hertz)	Amplitude (inches)	Description	Remarks
Passenger automobiles	1	6	Suspension resonance	25 hertz will usually avoid resonance with wheel hop and suspension resonant frequencies.
	8 to 12	0.02	Unsprung weight resonance (wheel hop)	
	20+	0.002	Irregular transient vibrations due to resonances of structural members with road roughnesses	
Automobile trucks	4	5	Suspension resonance	Above 20 hertz and should not correspond with any structural resonance. It is not advisable to attempt to isolate suspension and unsprung weight resonances.
	20	0.05	Unsprung weight resonance	
	80+	0.005	Structural resonances	
Military tanks	1 to 3	2	Suspension resonance	Similar to automobile truck
	Depends on speed	—	Track-laying frequency $\approx 17.6(\text{speed in mph})/(\text{tread spacing in inches})$	
	100+	0.001	Structural resonances	
Railroad trains	Broad and erratic		Similar to automobiles with additional excitations from rail joints and from side slop in rail trucks and draft gear	20 hertz has been successful in railroad applications. Shock with velocity changes up to 100 inches/second in direction of train occurs when coupling cars or starting freight trains.

true symbols representing the functions of parts in the circuit. Consequently, they may be rotated to any orientation with respect to each other without changing their meanings. Ground, chassis, and antenna symbols, for instance, may "point" in any direction that is convenient for drafting purposes.

Graphic symbols may be correlated with parts lists, descriptions, or instructions by means of reference designations (MIL-STD-16).

Detached Elements

Switches and relays often have many sets of contacts, and these sets may be separated and placed in the parts of the drawing to which they apply. Each separated element should be clearly labeled as part of the basic switch or relay.

Terminals

The terminal symbol need not be used unless it is needed. Thus, it may be omitted from relay and switch symbols. In particular, the terminal symbol often shown at the end of the movable element of a relay or switch should not be considered as the fulcrum or bearing but only as a terminal.

Associated or Future Equipment

Associated equipment, such as for measurement purposes, or additions that may be made later, are identified as such by using broken lines for both symbols and connections.

BRITISH GRAPHIC SYMBOLS

Commonly used British block-diagram graphic symbols are in Fig. 9. They have been abstracted

from British Standard 530: 1948 and Supplement No. 5(1962). The issuing organization is the British Standards Institution, British Standards House, 2 Park Street, London, W. 1, England.

RADIO-SIGNAL REPORTING CODES*

The Comité Consultatif International Radio (CCIR) recommends that the SINPO and SINPFEMO codes be used instead of the older Q, FRAME, RAFISBENQO, and RISAFMONE codes.

A signal report consists of the code word SINPO or SINPFEMO followed by a 5- or 8-figure group respectively rating the 5 or 8 characteristics of the signal code.

The letter X is used instead of a numeral for characteristics not rated.

Although the code word SINPFEMO is intended for telephony, either code word may be used for telegraphy or telephony.

Table 12 defines abbreviations used.

Tables 13 through 16, respectively, give the overall ratings for telegraphy and telephony, plus the SINPO and SINPFEMO codes.

TABLE 12—MEANINGS OF ABBREVIATIONS

Abbreviation or Signal	Definition
WA	Word after... (used after a question mark to request a repetition)
WB	Word before... (used after a question mark to request a repetition)
WD	Word(s) or group(s)
XQ	Prefix used to indicate an operating communication in the fixed service
XXX	When sent 3 times, constitutes the urgency signal
YES	Yes (affirmative)

TABLE 13—OVERALL RATING FOR TELEGRAPHY.

Symbol		Mechanized Operation	Morse Operation
5	Excellent	4-channel time-division multiplex	High-speed Morse
4	Good	2-channel time-division multiplex	100 words/minute Morse
3	Fair	Marginal. Single start–stop printer	50 words/minute Morse
2	Poor	Blocks, XQ's, and call signs readable	Blocks, XQ's, and call signs readable
1	Unusable	Unreadable	Unreadable

*From ITU Radio Regulations, Geneva; 1959.

TABLE 14—OVERALL RATING FOR TELEPHONY.

	Symbol	Operating Condition	Quality
5	Excellent	Signal quality unaffected	Commercial
4	Good	Signal quality slightly affected	
3	Fair	Signal quality seriously affected. Channel usable by operators or by experienced subscribers	Marginally commercial
2	Poor	Channel just usable by operators	Not commercial
1	Unusable	Channel unusable by operators	

TABLE 15—SINPO SIGNAL-REPORTING CODE.

	S	I	N	P	O
		Degrading Effect of			Overall
Rating Scale	Signal Strength	Interference (QRM)	Noise (QRN)	Propagation Disturbance	Readability (QRK)
5	Excellent	Nil	Nil	Nil	Excellent
4	Good	Slight	Slight	Slight	Good
3	Fair	Moderate	Moderate	Moderate	Fair
2	Poor	Severe	Severe	Severe	Poor
1	Barely audible	Extreme	Extreme	Extreme	Unusable

TABLE 16—SINPFEMO SIGNAL-REPORTING CODE.

	S	I	N	P	F	E	M	O
Rating Scale		Degrading Effect of			Frequency of	Modulation		
	Signal Strength	Interference (QRM)	Noise (QRN)	Propagation Disturbance	Fading	Quality	Depth	Overall Rating
5	Excellent	Nil	Nil	Nil	Nil	Excellent	Maximum	Excellent
4	Good	Slight	Slight	Slight	Slow	Good	Good	Good
3	Fair	Moderate	Moderate	Moderate	Moderate	Fair	Fair	Fair
2	Poor	Severe	Severe	Severe	Fast	Poor	Poor or nil	Poor
1	Barely audible	Extreme	Extreme	Extreme	Very fast	Very poor	Continuously overmodulated	Unusable

WORLD TIME CHART

Aleutian Islands, Tutuila, Samoa	Anchorage, Fairbanks, Hawaiian Islands, Tahiti	Los Angeles, San Francisco, Seattle, Juneau	Chicago, Central America (except Panama), Mexico, Winnipeg	New York, Montreal, Miami, Havana, Panama, Bogota, Lima, Quito	Bermuda, Puerto Rico, Caracas, La Paz, Asuncion	Buenos Aires,* Rio de Janeiro, Santos, Sao Paulo, Montevideo	Iceland	Lisbon, Dublin, Algiers, Dakar, Ascension Island	Greenwich Civil Time (GCT) or Universal Time (UT)	London,* Paris,* Madrid,* Brussels, Rome, Berlin, Vienna, Oslo, Stockholm, Copenhagen, Amsterdam, Tunis, Warsaw	Athens, Israel, Ankara, Cairo, Capetown	Moscow,* Ethiopia, Iraq, Malagasy Republic	Bombay, Sri Lanka, New Delhi	Bangkok, Chungking, Chengtu, Kunming	Hong Kong, Manila, Shanghai, Saigon, Taipeh, Celebes	Japan, Adelaide, Korea, Manchuria	Sydney, Melbourne, Brisbane, Guam, New Guinea, Khabarovsk	Solomon Islands, New Caledonia	Wellington,* Auckland*
1:00	2:00	4:00	6:00	7:00	8:00	9:00	11:00	Midnite	0000	1:00	2:00	3:00	5:30	7:00	8:00	9:00	10:00	11:00	11:30
2:00	3:00	5:00	7:00	8:00	9:00	10:00	Midnite	1:00	0100	2:00	3:00	4:00	6:30	8:00	9:00	10:00	11:00	Noon	12:30
3:00	4:00	6:00	8:00	9:00	10:00	11:00	1:00	2:00	0200	3:00	4:00	5:00	7:30	9:00	10:00	11:00	Noon	1:00	1:30
4:00	5:00	7:00	9:00	10:00	11:00	Midnite	2:00	3:00	0300	4:00	5:00	6:00	8:30	10:00	11:00	Noon	1:00	2:00	2:30
5:00	6:00	8:00	10:00	11:00	Midnite	1:00	3:00	4:00	0400	5:00	6:00	7:00	9:30	11:00	Noon	1:00	2:00	3:00	3:30
6:00	7:00	9:00	11:00	Midnite	1:00	2:00	4:00	5:00	0500	6:00	7:00	8:00	10:30	Noon	1:00	2:00	3:00	4:00	4:30
7:00	8:00	10:00	Midnite	1:00	2:00	3:00	5:00	6:00	0600	7:00	8:00	9:00	11:30	1:00	2:00	3:00	4:00	5:00	5:30
8:00	9:00	11:00	1:00	2:00	3:00	4:00	6:00	7:00	0700	8:00	9:00	10:00	12:30	2:00	3:00	4:00	5:00	6:00	6:30
9:00	10:00	Midnite	2:00	3:00	4:00	5:00	7:00	8:00	0800	9:00	10:00	11:00	1:30	3:00	4:00	5:00	6:00	7:00	7:30
10:00	11:00	1:00	3:00	4:00	5:00	6:00	8:00	9:00	0900	10:00	11:00	Noon	2:30	4:00	5:00	6:00	7:00	8:00	8:30
11:00	Midnite	2:00	4:00	5:00	6:00	7:00	9:00	10:00	1000	11:00	Noon	1:00	3:30	5:00	6:00	7:00	8:00	9:00	9:30
Midnite	1:00	3:00	5:00	6:00	7:00	8:00	10:00	11:00	1100	Noon	1:00	2:00	4:30	6:00	7:00	8:00	9:00	10:00	10:30
1:00	2:00	4:00	6:00	7:00	8:00	9:00	11:00	Noon	1200	1:00	2:00	3:00	5:30	7:00	8:00	9:00	10:00	11:00	11:30
2:00	3:00	5:00	7:00	8:00	9:00	10:00	Noon	1:00	1300	2:00	3:00	4:00	6:30	8:00	9:00	10:00	11:00	Midnite	12:30
3:00	4:00	6:00	8:00	9:00	10:00	11:00	1:00	2:00	1400	3:00	4:00	5:00	7:30	9:00	10:00	11:00	Midnite	1:00	1:30
4:00	5:00	7:00	9:00	10:00	11:00	Noon	2:00	3:00	1500	4:00	5:00	6:00	8:30	10:00	11:00	Midnite	1:00	2:00	2:30
5:00	6:00	8:00	10:00	11:00	Noon	1:00	3:00	4:00	1600	5:00	6:00	7:00	9:30	11:00	Midnite	1:00	2:00	3:00	3:30
6:00	7:00	9:00	11:00	Noon	1:00	2:00	4:00	5:00	1700	6:00	7:00	8:00	10:30	Midnite	1:00	2:00	3:00	4:00	4:30
7:00	8:00	10:00	Noon	1:00	2:00	3:00	5:00	6:00	1800	7:00	8:00	9:00	11:30	1:00	2:00	3:00	4:00	5:00	5:30
8:00	9:00	11:00	1:00	2:00	3:00	4:00	6:00	7:00	1900	8:00	9:00	10:00	12:30	2:00	3:00	4:00	5:00	6:00	6:30
9:00	10:00	Noon	2:00	3:00	4:00	5:00	7:00	8:00	2000	9:00	10:00	11:00	1:30	3:00	4:00	5:00	6:00	7:00	7:30
10:00	11:00	1:00	3:00	4:00	5:00	6:00	8:00	9:00	2100	10:00	11:00	Midnite	2:30	4:00	5:00	6:00	7:00	8:00	8:30
11:00	Noon	2:00	4:00	5:00	6:00	7:00	9:00	10:00	2200	11:00	Midnite	1:00	3:30	5:00	6:00	7:00	8:00	9:00	9:30
Noon	1:00	3:00	5:00	6:00	7:00	8:00	10:00	11:00	2300	Midnite	1:00	2:00	4:30	6:00	7:00	8:00	9:00	10:00	10:30
1:00	2:00	4:00	6:00	7:00	8:00	9:00	11:00	Midnite	2400	1:00	2:00	3:00	5:30	7:00	8:00	9:00	10:00	11:00	11:30

Notes: (1) Light-face figures designate AM, bold figures PM. (2) Time is that used at places indicated. In general, this is standard time but for places marked with asterisks it is permanent daylight saving time. Temporary daylight saving time is commonplace but not indicated above. (3) When passing the heavy line going down or to the right, add 1 day. When passing the heavy line going up or to the left, subtract 1 day.

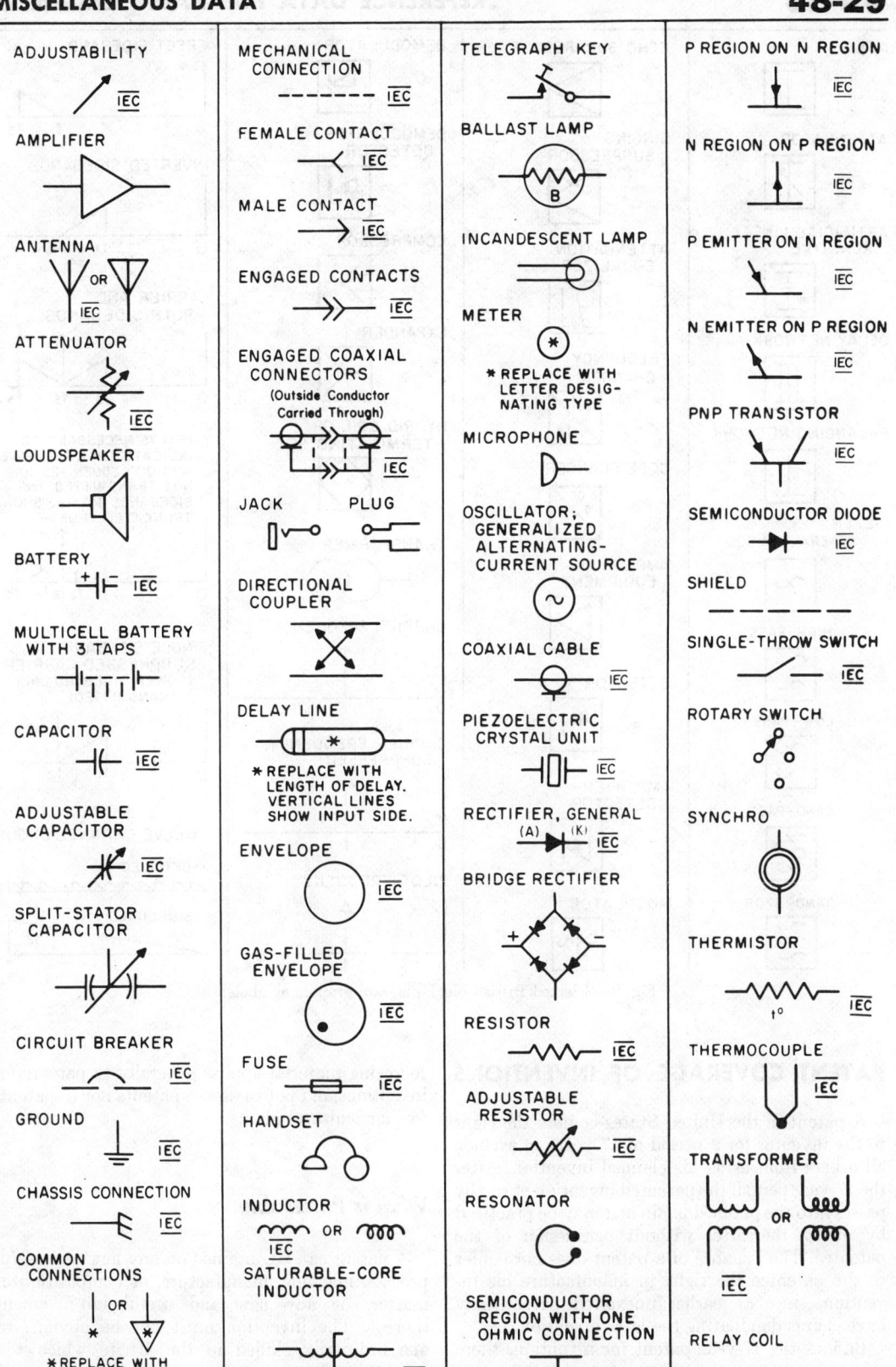

Fig. 8—Selected graphic symbols from ANSI Standard.

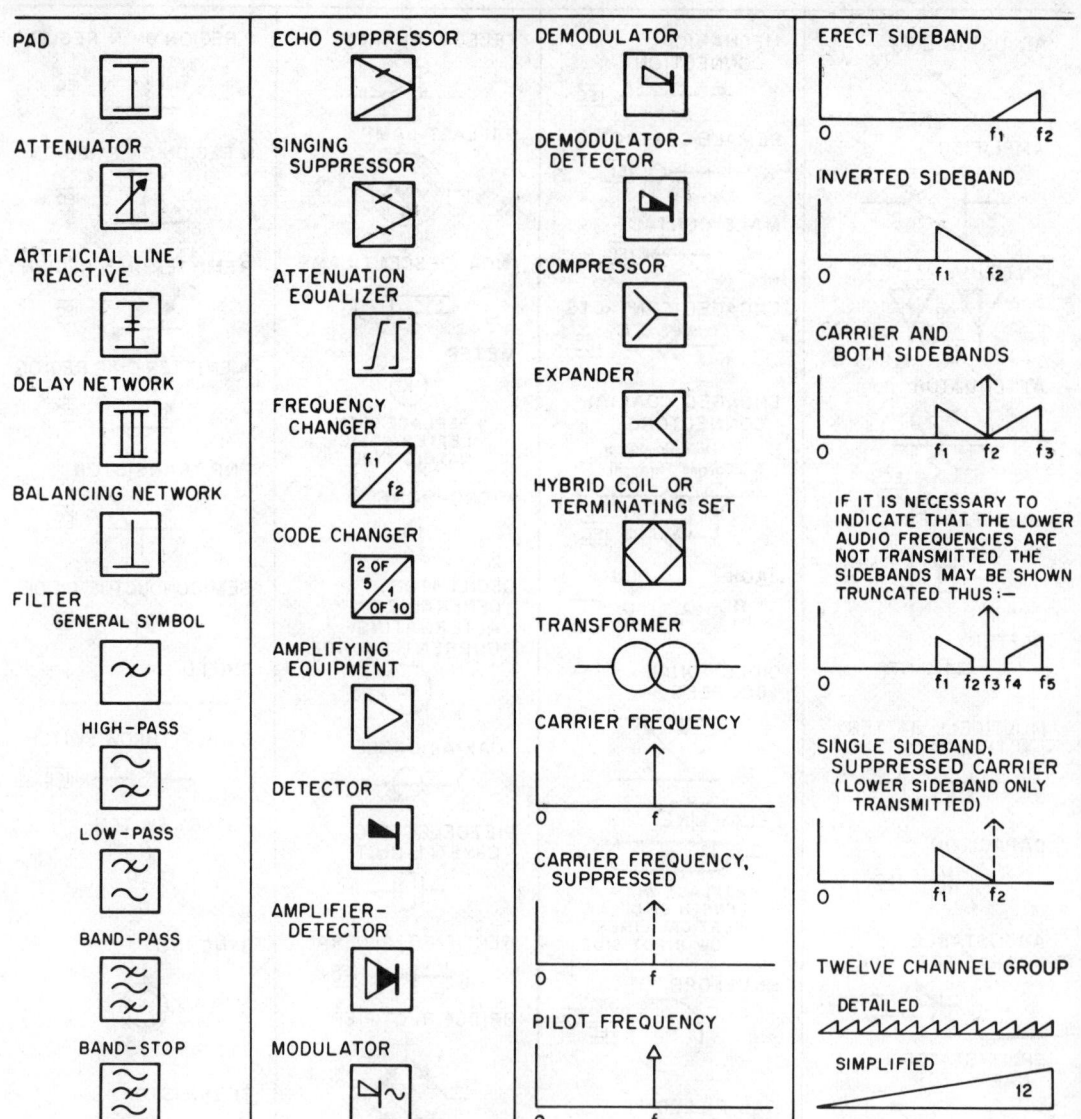

Fig. 9—Selected British block-diagram graphic symbols.

PATENT COVERAGE OF INVENTIONS

A patent in the United States confers the right to the inventor for a period of 17 years to exclude all others from using his claimed invention. After the 17-year period, the patented invention normally passes into the public domain and may be practiced by others thereafter without permission of the patentee. The issuance of a patent does not confer to the patentee the right to manufacture his invention, since an earlier unexpired patent may have claims dominating the later invention.

Besides the 17-year patent for invention, there are design patents for shorter periods that cover the outward artistic configuration of an article of manufacture and patents for new plants. The following material applies generally to patents for inventions and not to design patents nor to patents for horticultural plants.

What is Patentable

A patent can be obtained on any new and useful process, machine, manufacture, or composition of matter, or any new and useful improvement thereof. The invention must not be obvious to one ordinarily skilled in the art to which the invention relates.

In his patent application, the inventor must make the disclosure of his invention sufficiently clear and complete to enable one skilled in the art to build and practice the invention.

Recognizing Inventions

If the improvement or other development is new to the originator and appears either basic or commercially feasible, he should submit a disclosure to his patent attorney for advice. This should include disclosures of new products in the mechanical, chemical, and electrical fields; of new combinations of new and/or old elements that produce a new result, or an old result but with fewer elements; and, in fact, any new improvement in these fields that appears to present a commercial advantage in either cost, durability, or operation. The question of whether the disclosure is a sufficient advancement to be the basis of patent claims depends on a novelty investigation and appraisal by a patent attorney.

Who May be an Inventor

The inventor is the person who originates the idea and causes his mental picture of an embodiment to be reduced to physical form such as a written description or drawings or model. He may draw on the skill of others to complete this physical form of his invention so long as ideas, hints, and suggestions of others are in the regular course of their work as skilled technicians.

A contribution by another beyond ordinary mechanical skill makes the contributor a co-inventor. Employers or supervisors who do not contribute more than ordinary skill should not be identified as co-inventors. On the other hand, a supervisor may convey an idea to another employee and direct its development into a patentable invention and do none of the physical work, yet the supervisor is the true inventor. However, when two or more persons by cross-suggestion conceive and reduce an invention to a physical form, they thereby become joint inventors. If there is real doubt as to whether an invention is sole or joint, the doubt should be resolved in favor of joint.

Making Patentable Inventions

The usual steps of making an invention are:

(**A**) A desired result or problem is first recognized.

(**B**) A concept of an embodiment capable of producing the desired result is visualized. This mental concept should then be followed with a written record of the physical form visualized (drawings and descriptions).

(**C**) Reduction to practice. This may be "constructive" by filing a patent application, or "actual" by building a full-size working embodiment.

Obtaining a Patent

For one to obtain a patent in the United States, the invention must have been made before it was known or used by others in this country, or before it was patented or described by others in any printed publication in this or any foreign country.

An application for patent must be filed within one year from the first date of public use or offer of sale of the invention in this country or any publication in this or any foreign country disclosing the invention, or before the issuance of a foreign patent based on an application filed by the same inventor more than one year before his filing the application for US patent.

Assignment of Inventions

The patent rights to an invention can be assigned and transferred, and this may be done either before or after a patent application is filed or a patent is obtained.

Effect of Publication—Foreign Patents

No public disclosure of an invention should be made before an application for patent is filed on it. The reason for this is that in certain foreign countries, e.g., France, Holland, and Brazil, the law provides that the publication or public use of the invention anywhere in the world before the date of filing of an application for patent makes the idea available to the public and thereby deprives the inventor of any right to a patent in those countries. However, in the United States, one year is allowed following the date of the first publication, or first public use or sale of the invention during which the application for patent may be filed. Since inventors or assignees are often interested in obtaining foreign patents as well as United States patents, the inventor should make certain as a general policy that no publication or public use is made of his invention before a patent application is filed.

The benefit of the United States filing date applies to the obtaining of patents in most important foreign countries, provided the foreign application is filed within one year of the date of filing of the United States application.

Interferences

Occasionally two or more applications are filed by different inventors claiming substantially the same patentable invention. Thus, while a patent application is pending, an interference may be declared by the Patent Office with respect to the application or patent of another inventor. This proceeding is to determine who is rightfully the first inventor, and proof of dates, diligence, and reduction to practice must be established by recorded evidence, such as sketches, description, test data, models, and witnesses.

Engineer's Notebook

The keeping of formal notebook records by engineers facilitates patent applications and prosecution of any subsequent interference cases. The permanently bound type of notebook is preferred, and the engineer should make his original entries therein. Adherence to the following procedures will make the notebook more useful as evidence in legal proceedings.

(**A**) Make entries chronologically. Use ink.

(**B**) Do not leave blank spaces. Draw a line diagonally across unused space on a page. Use both sides of each sheet. Do not skip or remove any notebook pages.

(**C**) Do not erase. Draw a single line through any entries to be cancelled, and initial and date changes made.

(**D**) Make entries directly in notebook. If separate charts, graphs, etc., are a necessary part of an entry, they should be properly signed, witnessed, and dated as well as being referenced on the applicable pages of the notebook. These separate sheets should be securely fastened in the notebook.

(**E**) Make each entry clear and complete.

(**F**) Sign and date each entry on the day it is made.

(**G**) Any entry believed to be sufficiently novel to become the subject of a patent application should be signed and dated by witnesses who understand the subject matter. Sketches, graphs, test data, or other materials related to the invention should be similarly witnessed.

SUMMARY OF MILITARY NOMENCLATURE SYSTEM*

In the Joint Electronics Type Designation System (JETDS), formerly called the "AN" system, nomenclature for electronic equipment consists of a name, followed by a type number.

A type designation assignment for equipment such as a definitive system, subsystem, center, central, set, etc., shall consist of at least an AN, a slant bar, a three-letter equipment designation (Table 17), a dash, and a number. *Example*:

*Adapted from MIL-STD-196C, "Joint Electronics Type Designation System," 22 April 1971 and Notice 1, 20 April 1972. Available from the Superintendent of Documents, Washington, D.C. 20402.

AN/VRC-12 would be a radio communication set installed in a vehicle designed for functions other than carrying electronic equipment.

All groups, including commercial off-the-shelf equipment are identified by a two-letter indicator from Table 18. Applicable equipment indicator letters (Table 17) follow the slant bar to indicate the potential of the group for multiple or peculiar application. *Example*: OE-162/ARC indicates an antenna for aircraft radio-communication equipment. Equipment indicators with a specific model number (e.g., OK-450/TRC-26) are used following the slant bar when the group is peculiar to specific equipment (e.g., AN/TRC-26) with no known potential for other use.

The type designation for units having one end use consists of an indicator (Table 19), a dash, a number, a slant bar, and the equipment it is a part of or used with. *Example*: the receiver portion of the AN/VRC-12 is identified as R-40/VRC-12. If the unit has multiple usage, only those indicators which are common or appropriate are included after the slant bar. *Examples*: A power supply, part of or used with the AN/VRC-12 and AN/VRC-19 would be identified as PP-50/VRC. A power supply, "part of" the AN/VRC-12 and "used with" the AN/VRR-40 would be identified as PP-60/VR.

The system indicator (AN) does not mean that the Army, Navy, and Air Force use the equipment, but simply that the type number was assigned in the JETDS system.

Nomenclature Policy

JETDS nomenclature will be assigned to:

(**A**) Complete systems, subsystems, centers, controls, sets, groups, kits, and units of military design, either definitive or variable in configuration.

(**B**) Groups of articles of either commercial or military design that are grouped for a military purpose.

(**C**) Electronic articles of military design that are part of or used with an item not identified in the JETDS.

(**D**) Commercial articles requiring military identification for use by U.S. Government.

(**E**) Electronic materials of military design which are not part of or used with a set.

JETDS nomenclature will not be assigned to:

(**A**) Articles cataloged commercially except in accordance with paragraph (**D**) above.

(**B**) Minor components of military design for which other adequate means of identification are available.

(**C**) Small parts such as capacitors and resistors.

(**D**) Articles having other adequate identification in joint military specifications.

TABLE 17—SET OR EQUIPMENT INDICATOR LETTERS.

1st Letter (Type of Installation)	2nd Letter (Type of Equipment)	3rd Letter (Purpose)
A Piloted aircraft	A Invisible light, heat radiation	A Auxiliary assemblies (not complete operating sets used with or part of two or more sets or sets series) (inactivated, do not use)
B Underwater mobile, submarine	B Pigeon (do not use)	B Bombing
C Air transportable (inactivated, do not use)	C Carrier	C Communications (receiving and transmitting)
D Pilotless carrier	D Radiac	D Direction finder, reconnaissance, and/or surveillance
	E Nupac (inactivated, do not use)	E Ejection and/or release
F Fixed ground	F Photographic*	
G General ground use (includes two or more ground-type installations)	G Telegraph or teletype	G Fire control or searchlight directing
		H Recording and/or reproducing (graphic meteorological and sound)
	I Interphone and public address	
	J Electromechanical or inertial wire covered	
K Amphibious	K Telemetering	K Computing
	L Countermeasures	L Searchlight control (inactivated, use "G")
M Ground, mobile (installed as operating unit in a vehicle which has no function other than transporting the equipment)	M Meteorological	M Maintenance and/or test assemblies (including tools)
	N Sound in air	N Navigational aids (including altimeters, beacons, compasses, racons, depth sounding, approach, and landing)
P Pack or portable (animal or man)	P Radar	P Reproducing (inactivated, use "H")
	Q Sonar and underwater sound	Q Special, or combination of purposes
	R Radio	R Receiving, passive detecting
S Water surface craft	S Special types, magnetic, etc., or combinations of types	S Detecting and/or range and bearing, search
T Ground, transportable	T Telephone (wire)	T Transmitting
U General utility (includes two or more general installation classes, airborne, shipboard, and ground)		
V Ground, vehicular (installed in vehicle designed for functions other than carrying electronic equipment, etc., such as tanks)	V Visual and visible light	
W Water surface and underwater combination	W Armament (peculiar to armament, not otherwise covered)	W Automatic flight or remote control
	X Facismile or television	X Identification and recognition
	Y Data processing	Y Surveillance (search, detect, and multiple target tracking) and control (both fire control and air control)
Z Piloted and pilotless airborne vehicle combination		

*Not for US use except for assigning suffix letters to previously nomenclatured items.

TABLE 18—GROUP INDICATORS.

Indicator	Family Name	Indicator	Family Name
OA	Miscellaneous groups	ON	Interconnecting groups
OB	Multiplexer and/or demultiplexer groups	OP	Power Supply groups
OD	Indicator groups	OQ	Test Set groups
OE	Antenna groups	OR	Receiver groups
OF	Adapter groups	OT	Transmitter groups
OG	Amplifier groups	OU	Converter groups
OH	Simulator groups	OV	Generator groups
OJ	Consoles and console groups	OW	Terminal groups
OK	Control groups	OX	Coder, decoder, interrogator, transponder groups
OL	Data analysis and data processing groups	OY	Radar Set groups
OM	Modulator and/or demodulator groups	OZ	Radio Set groups

Nomenclature assignments will remain unchanged regardless of later changes in installation and/or application.

Modification Letters

Component modification suffix letters will be assigned for each modification of a component when detail parts and subassemblies used therein are no longer interchangeable, but the component itself is interchangeable physically, electrically, and mechanically.

Set modification letters will be assigned for each modification not affecting interchangeability of the sets or equipment as a whole, except that in some special cases they will be assigned to indicate functional interchangeability and not necessarily complete electrical and mechanical interchangeability. Modification letters will only be assigned if the frequency coverage of the unmodified equipment is maintained.

The suffix letters X, Y, and Z will be used only to designate a set or equipment modified by changing the power input voltage, phase, or frequency. X will indicate the first change, Y the second, Z the third, XX the fourth, etc., and these letters will be in addition to other modification letters applicable.

Developmental Indicators

Experimental Sets: To identify a set or equipment of an experimental nature with the development organization concerned, the following indicators are used within the parentheses:

XA	Aeronautical Systems Division, Wright-Patterson Air Force Base, Ohio
XB	Naval Research Laboratory, Washington, DC.
XC	US Army Signal Engineering Laboratories, The Hexagon, Fort Monmouth, NJ (inactivated, use XE).
XD	Electronic Systems Division, Laurence G. Hanscom Field, Bedford, Massachusetts.
XE	US Army Electronics Laboratories, Fort Monmouth, NJ.
XF	Frankford Arsenal, Philadelphia, Pa.
XG	USN Electronics Laboratory, San Diego, California.
*XH	Aerial Reconnaissance Laboratory, Wright-Patterson Air Force Base, Ohio.
XI	Air Force Armament Laboratory, Eglin Air Force Base, Florida.
XJ	Naval Air Development Center, Johnsville, Pa.
*XK	Flight Control Laboratory, Wright-Patterson Air Force Base, Ohio.
XL	US Army Signal Electronics Research Unit, Mountain View, Cal.
XM	US Army Signal Engineering Laboratories, The Hexagon, Fort Monmouth, NJ (inactivated, use XE).
XN	Department of the Navy, Washington, DC.
XO	US Army Missile Command, Redstone Arsenal, Alabama.
XP	Canadian Department of National Defence, Ottawa, Ontario, Canada.
*XQ	Aeronautical Accessories Laboratory, Wright-Patterson Air Force Base, Ohio.
XR	National Security Agency, Fort George G. Meade, Maryland.
*XS	Electronic Components Laboratory, Wright-Patterson Air Force Base, Ohio.
XT	US Army Security Agency, Arlington Hall Station, Arlington, Va.
XU	USN Underwater Sound Laboratory, Fort Trumbull, New London, Conn.
XV	Air Force Weapons Laboratory, Kirtland Air Force Base, New Mexico.

TABLE 19—UNIT INDICATORS.

Indicator	Family Name	Indicator	Family Name
AB	Supports, antenna	OC*	Oceanographic devices
AM	Amplifiers	OS	Oscilloscopes, test
AS	Antennas, complex and simple	PD*	Prime drivers
AT*	Antennas, simple	PF*	Fittings, pole
BA	Batteries, primary type	PG*	Pigeon articles
BB	Batteries, secondary type	PH*	Photographic articles
BZ	Alarm units	PL	Plug-in units
C	Controls	PP	Power supplies
CA*	Commutator assemblies, sonar	PT	Mapping and plotting units
CB*	Capacitor banks	PU	Power equipments
CG	Cable assemblies, RF	R	Receivers
CK*	Crystal kits	RC*	Reels
CM	Comparators	RD	Recorder–reproducers
CN	Compensators	RE	Relay assemblies
CP	Computers	RF*	Radio-frequency components
CR*	Crystals	RG*	Cables, RF bulk
CU	Couplers	RL	Reeling machines
CV	Converters (electronic)	RO	Recorders
CW	Radomes	RP	Reproducers
CX	Cable assemblies, nonRF	RR	Reflectors
CY	Cases and cabinets	RT	Receiver and transmitter
D	Dispensers	S	Shelters
DA	Loads, dummy	SA	Switching units
DT	Detecting heads	SB	Switchboards
DY*	Dynamotors	SG	Generators, signal
E*	Hoists	SM	Simulators
F	Filter units	SN	Synchronizers
FN*	Furniture	ST*	Straps
FR	Frequency-measuring devices	SU	Optical devices
G	Generators, power	T	Transmitters
GO*	Goniometers	TA	Telephone apparatus
GP*	Ground rods	TB	Towed bodies
H	Head, hand, and chest sets	TC*	Towed cables
HC*	Crystal holders	TD	Timing devices
HD	Environmental apparatus (heating, cooling, etc.)	TF	Transformers
		TG	Positioning devices
ID	Indicators, noncathode-ray tube	TH	Telegraph apparatus
IL*	Insulators	TK*	Tool kits
IM	Intensity-measuring devices	TL*	Tools
IP	Indicators, cathode-ray tube	TN	Tuning units
J	Interface units	TR	Transducers
KY	Keying devices	TS	Test units
LC*	Tools, line-construction	TT	Teletypewriter and facsimile apparatus
LS	Loudspeakers		
M	Microphones	TV*	Testers, tube
MA*	Magazines	TW	Tape units
MD	Modulators, demodulators, discriminators	U*	Connectors, audio and power
		UG*	Connectors, RF
ME	Meters	V	Vehicles
MF*	Magnets or magnetic-field generators	VS*	Signaling equipment, visual
MK	Miscellaneous kits	WD*	Cables, two-conductor
ML	Meteorological devices	WF*	Cables, four-conductor
MT	Mountings	WM*	Cables, multiple-conductor
MU	Memory units	WS*	Cables, single-conductor
MX	Miscellaneous	WT*	Cables, three-conductor
O	Oscillators	ZM	Impedance-measuring devices

*Not for US use except for assigning suffix letters to previously nomenclatured items.

XW	Rome Air Development Center, Rome, New York.
*XY	Weapons Guidance Laboratory, Wright-Patterson Air Force Base, Ohio.
XZ	USN Bureau of Naval Weapons Activities.
XAA	Air Force Ballistic Systems Division, Norton Air Force Base, Cal.
XAE	US Army Electronics Research and Development Activity, Fort Huachuca, Arizona.
XAN	Naval Avionics Facility, Indianapolis, Ind.
XBB	US Army Electronics Command, Proc and Prod Div. Fort Monmouth, NJ.
XCA	US Naval Ammunition Depot, Crane, Ind.
XCC	Air Force Missile Test Center, Patrick Air Force Base, Florida.
XCL	Naval Weapons Center, China Lake, Calif.
XCR	Naval Weapons Center, Corona Laboratory, Corona, Calif.
XDD	US Army Signal Air Defense Engineering Agency, Fort George G. Meade, Md.
XDV	US Naval Weapons Laboratory, Dahlgren, Va.
XGS	Grand Support Equipment Division, Naval Air Engineering Center, Philadelphia, Pa.
XIH	US Naval Ordinance Station Indianhead, Md.
XLW	US Army Limited War Laboratory, Aberdeen Proving Ground, Md.
XMG	Naval Missile Center, Point Mugu, Calif.
XPM	US Army, Project Michigan, Ypsilanti, Michigan.
XSC	US Army Satellite Communications Agency, Fort Monmouth, NJ.
XWH	US Naval Ammunition Depot Earle, Naval Weapons Handling Laboratory, Colts Neck, NJ.
XWO	Naval Ordinance Laboratory, White Oak, Silver Spring, Md.

*Not for Air Force use except for assigning additional developmental designations to previously type-designated items. Use XA for all new equipments.

Example: Radio Set AN/ARC-3( ) might be assigned for a new airborne radio communication set under development. The cognizant development organization might then assign AN/ARC-3-(XA-1), AN/ARC-3(XA-2), etc., type numbers to the various sets developed for test. When the set was considered satisfactory for use, the experimental indicator would be dropped and procurement nomenclature AN/ARC-3 would be officially assigned thereto.

Training Sets: A set or equipment designed for training purposes will be assigned type numbers as follows:

(**A**) A set to train for a specific basic set will be assigned the basic-set type number followed by a dash, the letter T, and a number. *Example*: Radio Training Set AN/ARC-6A-T1 would be the first training set for Radio Set AN/ARC-6A.

(**B**) A set to train for general types of sets will be assigned the usual set indicator letters followed by a dash, the letter T, and a number. *Example*: Radio Training Set AN/ARC-T1 would be the first training set for general airborne radio communication sets.

Parenthesis Indicators: A series of a basic item, i.e. all production and/or nonproduction versions, may be identified by a type designation with an empty parenthesis. *Examples*: AN/APS-25( ) or R-275( )/APS-25. Such an assignment is all inclusive and does not refer to any specific version within the series.

Systems, subsystems, centers, controls, sets, groups, or units with variable parts lists are assigned type designations in the same manner except a parenthetical V (V) is added to the type designation.

Units designed to accept plug-ins which change the function, frequency, or technical characteristics of the type-designated unit will designate with a (P) preceeding the slant bar. The plug-in is not considered a part of the unit.

Examples of JETDS Type Numbers

AN/SRC-3( )	General reference set nomenclature for water surface craft radio communication set number 3.
AN/SRC-3	Original procurement set nomenclature applied against AN/SRC-3( ).
AN/SRC-3A	Modification set nomenclature applied against AN/SRC-3.
AN/APQ-13-T1( )	General reference training set nomenclature for the AN/APQ-13 set.

AN/APQ–13–T1	Original procurement training set nomenclature applied against AN/APQ–13–T1( ).
AN/APQ–13–T1A	Modification training set nomenclature applied against AN/APQ–13–T1.
AN/UPT–T3( )	General reference training set nomenclature for general utility radar transmitting training set number 3.
AN/UPT–T3	Original procurement training set nomenclature applied against AN/UPT–T3( ).
AN/UPT–T3A	Modification training set nomenclature applied against AN/UPT–T3.
T–51( )/ARQ–8	General reference component nomenclature for transmitter number 51, part of or used with airborne radio special set number 8.
T–51/ARQ–8	Original procurement component nomenclature applied against T–51( )/ARQ–8.
T–51A/ARQ–8	Modification component nomenclature applied against T–51/ARQ–8.
RD–31( )/U	General reference component nomenclature for recorder-reproducer number 31 for general utility use, not part of a specific set.
RD–31/U	Original procurement component nomenclature applied against RD–31( )/U.
RD–31A/U	Modification component nomenclature applied against RD–31/U.

INDEX

1